CARBON FIBERS *and their* COMPOSITES

CARBON FIBERS *and their* COMPOSITES

PETER MORGAN

Taylor & Francis
Taylor & Francis Group

Boca Raton   London   New York   Singapore

A CRC title, part of the Taylor & Francis imprint, a member of the
Taylor & Francis Group, the academic division of T&F Informa plc.

Published in 2005 by
CRC Press
Taylor & Francis Group
6000 Broken Sound Parkway NW, Suite 300
Boca Raton, FL 33487-2742

© 2005 by Taylor & Francis Group, LLC
CRC Press is an imprint of Taylor & Francis Group

No claim to original U.S. Government works
Printed in the United States of America on acid-free paper
10 9 8 7 6 5 4 3 2 1

International Standard Book Number-10: 0-8247-0983-7 (Hardcover)
International Standard Book Number-13: 978-0-8247-0983-9 (Hardcover)

This book contains information obtained from authentic and highly regarded sources. Reprinted material is quoted with permission, and sources are indicated. A wide variety of references are listed. Reasonable efforts have been made to publish reliable data and information, but the author and the publisher cannot assume responsibility for the validity of all materials or for the consequences of their use.

No part of this book may be reprinted, reproduced, transmitted, or utilized in any form by any electronic, mechanical, or other means, now known or hereafter invented, including photocopying, microfilming, and recording, or in any information storage or retrieval system, without written permission from the publishers.

For permission to photocopy or use material electronically from this work, please access www.copyright.com (http://www.copyright.com/) or contact the Copyright Clearance Center, Inc. (CCC) 222 Rosewood Drive, Danvers, MA 01923, 978-750-8400. CCC is a not-for-profit organization that provides licenses and registration for a variety of users. For organizations that have been granted a photocopy license by the CCC, a separate system of payment has been arranged.

Trademark Notice: Product or corporate names may be trademarks or registered trademarks, and are used only for identification and explanation without intent to infringe.

Library of Congress Cataloging-in-Publication Data

Catalog record is available from the Library of Congress

Taylor & Francis Group
is the Academic Division of T&F Informa plc.

Visit the Taylor & Francis Web site at
http://www.taylorandfrancis.com

and the CRC Press Web site at
http://www.crcpress.com

Introduction

There is now a wealth of books covering the subject of carbon fibers, some of which are listed in the bibliography, and one may ask, why another book. The aim of this book is to give a practical approach to the subject, with an accent on manufacture and encourage the reader to investigate the references for further information on more theoretical aspects.

The author looks back over the last thirty years and attempts to provide an insight into the problems and attempted solutions of some aspects of the production and testing of carbon fibres and their related products.

Since PAN homopolymer is not used for the production of carbon fibres, the subject is not discussed. Similarly, stabilization in nitrogen is sparingly mentioned.

The Internet has been useful in obtaining background information but, unfortunately, since items are unlikely to be permanent, they cannot be used as a reference. When searching, it must be remembered that there are the English and American ways of spelling (e.g. fibre and fiber), also the use of the word graphite in addition to carbon. An excellent source was an IBM site for US patents, now operated as http://www.delphion.com/.

1. RECOMMENDED TERMINOLOGY FOR THE DESCRIPTION OF CARBON AS A SOLID

1.1. Introduction

In the early years of carbon fibers, there was an unfortunate tendency to specify carbon fibres made at about 2800K as graphite fibers which, in the majority of cases, is simply not true and should be avoided. The term graphite fibres is justified only if three-dimensional crystalline order is confirmed (e.g. by X-ray diffraction measurements). It is considered, at this stage, relevant to define the terminology for the various forms of solid carbon and associated products which are mentioned in this book and requisite information has been used from The International Union of Pure and Applied Chemistry (IUPAC) recommendations, (©1995 IUPAC).

1.2. Description of terms

1. *ACTIVATED CARBON* is a porous CARBON MATERIAL, a CHAR which has been subjected to reaction with gases, sometimes with the addition of chemicals (e.g. $ZnCl_2$), before, during, or after CARBONIZATION.
2. *AMORPHOUS CARBON* is a CARBON MATERIAL without long range crystalline order.
3. *BROOKS AND TAYLOR STRUCTURE IN THE CARBONACEOUS MESOPHASE* refers to the structure of the anisotropic spheres which precipitate from isotropic PITCH during pyrolysis. The structure of the spheres consists of a lamellar arrangement of aromatic molecules in parallel layers, which are perpendicular to the polar axis of the sphere and which are perpendicular to the mesophase-isotropic phase interphase.
4. *BULK MESOPHASE* is a continuous anisotropic phase formed by the coalescence of mesophase spheres. BULK MESOPHASE retains fluidity and is deformable in the temperature range up to about 770K, and transforms into GREEN COKE by further loss of hydrogen or other low molecular weight species.
5. *CARBON* is the sixth element of the Periodic Table of Elements (electronic ground state $1s^2 2s^2 2p^2$).

Note: For description of the various types of CARBON AS A SOLID, the term CARBON should be used only in combination with an additional noun or a clarifying adjective.

6. *CARBON BLACK* is an industrially manufactured COLLOIDAL CARBON material, in the form of spheres as well as of their fused aggregates, with sizes below 1000 nm.
   Note: For historical reasons, CARBON BLACK is popularly, but incorrectly, regarded as a form of SOOT.

7. *CARBON-CARBON COMPOSITE* is a carbon fiber reinforced carbon matrix material. The carbon matrix phase is typically formed by solid, liquid or gaseous pyrolysis of an organic precursor material. The matrix is either a GRAPHITIZABLE CARBON or NON-GRAPHITIZABLE CARBON, and the carbonaceous reinforcement is fibrous. The composite may also contain other components in particulate or fibrous forms.

8. *CARBON FIBERS* are fibers (filaments, tows, yarns, rovings) consisting of at least 92% (mass fraction) CARBON, usually in the NON-GRAPHITIC state.

9. *CARBON FIBERS HT* are CARBON FIBERS with values of Young's Modulus in the range 150–275 or 300 GPa. The term HT, referring to high tensile strength, was previously applied because fibers of this type display the highest tensile strengths.
   Note: The disposition of boundaries between the fiber types is somewhat arbitrary. The tensile strength of CARBON FIBERS is, however, flaw controlled and therefore, the measured values increase strongly as the diameter of the filaments is decreased.

10. *CARBON FIBERS TYPE HM (HIGH MODULUS)* are CARBON FIBRES with a value of Young's modulus (tensile modulus) larger than 300 GPa. Here again, tensile strength is influenced by flaws in the fiber.

11. *CARBON FIBERS TYPE IM (INTERMEDIATE MODULUS)* are related to CARBON FIBERS TYPE HT because of the comparable values of tensile strength, but are characterized by greater stiffness.
    Note: The tensile modulus (Young's modulus) varies ca. 275–350 GPa, but the disposition of the boundaries is somewhat arbitrary. The relatively high tensile strength required is achievable by a significant reduction of the single filament diameter down to about 5 μm. Such small filament diameters are typical of CARBON FIBRES TYPE IM.

12. *CARBON FIBERS TYPE LM (LOW MODULUS)* are CARBON FIBRES with isotropic structure, tensile modulus values as low as 100 GPa and low strength values.

13. *CARBON FIBERS TYPE UHM (ULTRA HIGH MODULUS)* designates a class of CARBON FIBERS having very high values of Young's modulus, greater than 600 GPa.
    Note: Such high values of Young's modulus can be achieved most readily in MESOPHASE PITCH BASED CARBON FIBERS (MPP based carbon fibers).

14. *CARBONACEOUS MESOPHASE* is a liquid crystalline state of PITCH which shows the optical birefringence of disc-like (discotic) nematic crystals. It can be formed as an intermediate phase during thermolysis (pyrolysis) of an isotropic molten PITCH, or by precipitation from PITCH fractions prepared by selective extraction. With continuous heat treatment, the CARBONACEOUS MESOPHASE coalesces to a state of BULK MESOPHASE before solidification to GREEN COKE, with further loss of hydrogen or other low molecular weight compounds.

15. *CARBONIZATION* is a process by which solid residues with increasing content of elemental carbon are formed from organic material, usually by pyrolysis in an inert atmosphere.
    As with all pyrolytic reactions, CARBONIZATION is a complex process in which many reactions take place concurrently, such as dehydrogenation, condensation, hydrogen transfer and isomerization. The final pyrolysis temperature applied controls the degree of CARBONIZATION and the residual content of foreign elements (e.g. at 1200K, the carbon content of the residue exceeds a mass fraction 90 wt%, whereas at 1600K, more than 99 wt.% carbon is found).

16. *CHAR* is a solid decomposition product of a natural or synthetic organic material.

17. *CHARCOAL* is a traditional term for a CHAR obtained from wood, peat, coal or other related natural organic materials.

18. *COAL TAR PITCH* is a residue produced by distillation or heat treatment of coal tar. It is solid at room temperature, consists of a complex mixture of numerous, predominantly aromatic

hydrocarbons and heterocyclics, and exhibits a broad softening range, instead of a defined melting temperature.

19. *COKE* is a solid, high in elemental carbon content and structurally in the NON-GRAPHITIC state. It is produced by pyrolysis of organic material which has passed, at least in part, through a liquid or liquid-crystalline state during the CARBONIZATION process. COKE can contain mineral matter.

20. *COLLOIDAL CARBON* is a PARTICULATE CARBON with particle sizes below ca. 1000 nm in at least one direction.

21. *DIAMOND* is an allotropic form of the element carbon, with cubic structure which is thermodynamically stable at pressures above 6 GPa at room temperature and metastable at atmospheric pressure. At low pressures and temperatures above 1900K in an inert atmosphere, DIAMOND converts rapidly to GRAPHITE. The chemical bonding between the carbon atoms is covalent, with sp^3 hybridization.

22. *EXFOLIATED GRAPHITE* is the product of very rapid heating of relatively large particle diameter (flakes) graphite intercalation compounds, such as graphite hydrogen sulphate. The vaporizing intercalated substances force the graphite layers apart. The EXFOLIATED GRAPHITE assumes an accordian-like shape, with an apparent volume often hundreds of times that of the original graphite flakes.

23. *FIBROUS ACTIVATED CARBON* is an ACTIVATED CARBON in the form of fibers, filaments, yarns or rovings and fabrics or felts. Such fibers differ from CARBON FIBERS, used for reinforcement purposes in composites, in their high surface area, high porosity and low mechanical strength.

24. *FILAMENTOUS CARBON* is a carbonaceous deposit from gaseous carbon compounds, consisting of filaments grown by the catalytic action of metal particles.
    Note: In general, such deposits are obtained at pressures of < 100 kPa in the temperature region 66–1300K on metals such as cobalt, iron and nickel. The filaments may be produced in different conformations, such as helical, twisted and straight.

25. *FULLERENES* were discovered in 1985 [2] and can be considered as another major allotrope of carbon and differ from GRAPHITE and DIAMOND in that they form discrete molecular forms. FULLERENES are symmetrically closed convex GRAPHITE shells, consisting of twelve pentagons and various numbers of hexagons. Since they are arranged in the form of a geodesic spheroid, they were named buckminster fullerenes, after Buckminster Fuller, the architect associated with the geodesic dome, shortened to FULLERENES and known colloquially as bucky-balls.
    The FULLERENE structures are possible in many combinations but, as yet, the most important structures truly established are C_{60}, C_{70}, C_{76}, C_{78} and C_{84}. The first FULLERENE to be discovered was C_{60}, comprising of twenty hexagons and twelve pentagons, with each carbon atom shared by one pentagon and two hexagons, assuming the well known spherical shape of a football.

26. *FURNACE BLACK* is a type of CARBON that is produced industrially in a furnace by incomplete combustion.

27. *GAS PHASE GROWN CARBON FIBERS* are CARBON FIBERS grown in an atmosphere of hydrocarbons with the aid of fine particulate solid catalysts such as iron or other transition metals and consisting of GRAPHITIZABLE CARBON.
    Note: GAS PHASE GROWN CARBON FIBERS transform during GRAPHITIZATION HEAT TREATMENT into GRAPHITE FIBERS. The term vapor grown carbon fibers is also acceptable but, CVD fibers is not acceptable, as it also describes fibers grown by a chemical vapor deposition (CVD) process on substrate fibers.

28. *GLASS-LIKE CARBON* is an AGRANULAR, NON-GRAPHITIZABLE CARBON, with isotropy of its structural and physical properties and with a very low permeability for liquids and gases.
    Note: The terms Glassy Carbon and Vitreous Carbon are trademarks and should not be used and, moreover, as implied, there is no similarity in the structures of silicate glasses, other than the pseudo-glassy appearance of the surface.

29. *GRAPHENE* is a single carbon layer of the graphite structure.

30. *GRAPHITE* is an allotropic form of the element carbon, consisting of layers of hexagonally arranged carbon atoms in a planar condensed ring system (GRAPHENE LAYERS). The layers are stacked parallel to each other in a three-dimensional structure. The chemical bonds within the layers are covalent with sp^2 hybridization. The weak bonds between the layers are metallic with a strength comparable to Van der Waals bonding.
31. *GRAPHITE FIBERS* are CARBON FIBERS consisting mostly of SYNTHETIC GRAPHITE, for which three-dimensional crystalline order is confirmed by X-ray diffraction.
    Note: GRAPHITE FIBERS can be obtained by GRAPHITIZATION HEAT TREATMENT of CARBON FIBERS, if these consist mostly of GRAPHITIZABLE CARBON.
32. *GRAPHITE MATERIAL* is a material consisting essentially of GRAPHITIC CARBON.
33. *GRAPHITE WHISKERS* consist of thin, approximately cylindrical filaments in which GRAPHENE LAYERS are arranged in a scroll-like manner. There is, at least in part, a regular stacking of the layers as in the GRAPHITE lattice. Along the cylinder axis, the physical properties of GRAPHITE WHISKERS approach those of GRAPHITE.
    Note: If there is no three-dimensional stacking order as in GRAPHITE, due to misalignment of the layers caused by their bending, the term CARBON WHISKERS should be used. GRAPHITE WHISKERS and CARBON WHISKERS should be distinguished from more disordered FILAMENTIOUS CARBON.
34. *GRAPHITIC CARBONS* are all varieties of substances consisting of the element carbon in the allotropic form of GRAPHITE, irrespective of the presence of structural defects.
35. *GRAPHITIZABLE CARBON* is a NON-GRAPHITIC CARBON which, upon GRAPHITIZATION HEAT TREATMENT, converts into GRAPHITIC CARBON.
36. *GRAPHITIZATION* is a solid state transformation of thermodynamically unstable NON-GRAPHITIC CARBON into GRAPHITE by means of heat treatment.
    Note: The use of the term GRAPHITIZATION to indicate a process of thermal treatment of CARBON MATERIALS at >2500K, regardless of any crystallinity, is incorrect.
37. *GRAPHITIZATION HEAT TREATMENT* is a process of heat treatment of a NON-GRAPHITIC CARBON, industrially performed at temperatures in the range 2500–3300K, to achieve transformation into GRAPHITIC CARBON.
    Note: The common use of the term GRAPHITIZATION for the heat treatment process only, regardless of the resultant crystallinity, is incorrect and should be avoided.
38. *GRAPHITIZED CARBON* is a GRAPHITIC CARBON with more or less perfect three-dimensional hexagonal crystalline order, prepared from NON-GRAPHITIC CARBON by GRAPHITIZATION HEAT TREATMENT.
    Note: NON-GRAPHITIZABLE CARBONS do not transform into GRAPHITIC CARBON on heat treatment at temperatures above 2500K and therefore, are not GRAPHITIZED CARBONS.
39. *GREEN COKE (RAW COKE)* is the primary solid CARBONIZATION product from high boiling hydrocarbon fractions obtained at temperatures below 900K. It contains a fraction of matter that can be released as volatiles during subsequent heat treatment at temperatures up to approximately 1600K. This mass fraction, the so-called volatile matter, in the case of GREEN COKE is 4–15%, but it also depends on the heating rate.
40. *HEXAGONAL GRAPHITE* is the thermodynamically stable form of GRAPHITE, with an ABAB stacking sequence of the GRAPHENE LAYERS.
    Note: The use of the term GRAPHITE instead of the more exact term HEXAGONAL GRAPHITE may be tolerated in view of the minor importance of RHOMBOHEDRAL GRAPHITE, the other allotropic form.
41. *ISOTROPIC CARBON* is a monolithic CARBON MATERIAL without preferred crystallographic orientation of the microstructure.
42. *ISOTROPIC PITCH BASED CARBON FIBERS* are CARBON FIBERS obtained by CARBONIZATION of isotropic pitch fibers after these have been stabilized (i.e. made non-fusible).
43. *LAMP BLACK* is a special type of CARBON BLACK produced by incomplete combustion of a fuel rich in aromatics that is burned in flat pans. LAMP BLACK is characterized by a relatively broad particle size distribution.

44. *MESOPHASE PITCH* is a PITCH with a complex mixture of numerous, essentially aromatic hydrocarbons containing anisotropic liquid crystalline particles (CARBONACEOUS MESOPHASE), detectable by optical microscopy and capable of coalescence into the BULK MESOPHASE.
45. *MESOPHASE PITCH BASED CARBON FIBERS (MPP BASED CARBON FIBERS)* are CARBON FIBERS obtained from MESOGENIC PITCH after it has been transformed into MESOPHASE PITCH (MPP) at least during the process of spinning, after the spun MESOPHASE PITCH fibers have been made non-fusible (stabilized) and carbonized.
46. *NATURAL GRAPHITE* is a mineral found in nature. It consists of GRAPHITIC CARBON, regardless of its crystalline perfection.
47. *NON-GRAPHITIC CARBONS* are all varieties of solids consisting mainly of the element carbon with two-dimensional long range order of the carbon atoms in planar hexagonal networks, but without any measurable crystallographic order in the third direction (c-direction) apart from more or less parallel stacking.
48. *NON-GRAPHITIZABLE CARBON* is a NON-GRAPHITIC CARBON which cannot be transformed into GRAPHITIC CARBON solely by high temperature treatment up to 3300K under atmospheric or lower pressure.
49. *PAN BASED CARBON FIBERS* are CARBON FIBERS obtained from polyacrylonitrile (PAN) precursor fibers by STABILIZATION TREATMENT, CARBONIZATION and final heat treatment.
50. *PETROLEUM PITCH* is a residue from heat treatment and distillation of petroleum fractions. It is a solid at room temperature, consists of a complex mixture of numerous, predominantly aromatic and alkyl substituted aromatic hydrocarbons, and exhibits a broad softening range instead of a defined melting temperature.
51. *PITCH* is a residue from the pyrolysis of organic material or tar distillation, which is solid at room temperature, consisting of a complex mixture of numerous, essentially aromatic hydrocarbons and heterocyclic compounds. It exhibits a broad softening range instead of a defined melting temperature. When cooled from the melt, pitches solidify without crystallization.
52. *PITCH-BASED CARBON FIBERS* are CARBON FIBERS obtained from PITCH precursor fibers after STABILIZATION TREATMENT, CARBONIZATION, and final heat treatment.
53. *POLYCRYSTALLINE GRAPHITE* is a GRAPHITE MATERIAL with coherent crystallographic domains of limited size regardless of the perfection and preferred orientation (texture) of their crystalline structure.
54. *POLYGRANULAR CARBON* is a CARBON MATERIAL composed of grains, which can be clearly distinguished by means of optical microscopy.
55. *POLYGRANULAR GRAPHITE* is a GRAPHITE MATERIAL composed of grains, which can be clearly distinguished by means of optical microscopy.
56. *PYROLYTIC CARBON* is a CARBON MATERIAL deposited from gaseous hydrocarbon compounds on suitable underlying substrates (CARBON MATERIALS, metals, ceramics) at temperatures ranging 1000–2500K (chemical vapor deposition).
    Note: Pyrocarbon is a trademark and should not be used as a term.
57. *PYROLYTIC GRAPHITE* is a GRAPHITE MATERIAL with a high degree of preferred crystallographic orientation of the c-axis perpendicular to the surface of the substrate, obtained by GRAPHITIZATION HEAT TREATMENT of PYROLYTIC CARBON or by chemical vapor deposition at temperatures above 2500K.
    Note: Pyrographite is a trademark and should not be used as a term.
58. *RAYON-BASED CARBON FIBERS* are CARBON FIBERS made from rayon (cellulose) precursor fibers.
    Note: RAYON-BASED CARBON FIBERS have a more isotropic structure than similarly heat treated polyacrylonitrile (PAN) or MESOPHASE PITCH (MPP) BASED CARBON FIBERS. Their Young's modulus values are, therefore, drastically lower (<100 GPa). RAYON BASED CARBON FIBERS can be transformed into anisotropic CARBON FIBERS with high strength

and Young's modulus values by hot stretching treatment at temperatures of approximately 2800K.
59. *RHOMBOHEDRAL GRAPHITE* is a thermodynamically unstable allotropic form of GRAPHITE with an ABCABC stacking sequence of the layers.
60. *SOOT* is a randomly formed PARTICULATE CARBON material and may be coarse, fine, and/or colloidal in proportions, depending on its origin. SOOT consists of variable quantities of carbonaceous and inorganic solids, together with absorbed and occluded tars and resins.
61. *SYNTHETIC GRAPHITE* is a material consisting of GRAPHITIC CARBON which has been obtained by graphitizing of NON-GRAPHITIC CARBON by chemical vapor deposition (CVD) from hydrocarbons at temperatures above 2500K, by decomposition of thermally unstable carbides, or by crystallising from metal melts supersaturated with carbon.
62. *THERMAL BLACK* is a special type of CARBON BLACK produced by pyrolysis of gaseous hydrocarbons in a preheated chamber in the absence of air.

2. A BIBLIOGRAPHY OF BOOKS ON CARBON FIBERS AND THEIR COMPOSITES

i) Smith EA ed, *Carbon Fibres, Design Engineering Series*, Morgan Grampian, West Wickham, 1970.
ii) Plastics Institute, Carbon fibres; their composites and applications, *Proceedings of the International Conference organized by the Plastics Institute*, London, Feb 2–4 1971.
iii) Gill RM, *Carbon Fibres in Composite Materials*, Iliffe Books, London, 1972.
iv) Jenkins GM, Kawamura K, *Polymeric Carbons-carbon fibre, glass and char*, Cambridge University Press, London, 1976.
v) Institution of Mechanical Engineers, Designing with fibre reinforced materials, *Conference sponsored by the Materials Technology Section of the Applied Mechanics Group of the Institution of Mechanical Engineers*, Mechanical Engineering publications, London, 27–28 September, 1977.
vi) Sittig M ed., *Carbon and Graphite Fibers: Manufacture and Applications*, Vol. 162, Noyes Publications, Park Ridge, 1982.
vii) Donnet JB, Bansal RC, *Carbon Fibers*, Marcel Dekker, New York, 1984.
viii) Watt W, Perov BV eds., *Handbook of Composites*, Vol.1, Strong Fibres, Elsevier, Amsterdam, 1985.
ix) Fitzer E ed., *Carbon Fibres and their Composites*, Springer Verlag, Berlin, 1986.
x) Plastics and Rubber Institute ed., *Carbon Fibres: Technology, Uses and Prospects*, Noyes Publications, 1986.
xi) Delmonte J, *Technology of Carbon and Graphite Fiber Composites*, Krieger Publishing Company, 1987.
xii) Dresselhaus MS, Dresselhaus G, Sugihara K, Spain IL, Goldberg HA, *Graphite Fibers and Filaments, Series in Material Science*, Vol. 5, Springer-Verlag, New York, 1988.
xiii) Donnet JB, Bansal RC, *Carbon Fibers*, International Fiber Science and Technology, Vol. 10, 2nd ed., Marcel Dekker, New York, 1990.
xiv) Ermolenko IN, Lyubliner LP, Gulko NV, Titovets EP, *Chemically Modified Carbon Fibers and their Applications*, Vch Pub, 1990.
xv) Figueiredo JL, Bernardo CA, Baker RTK, Hüttinger KJ ed., *Carbon Fibers Filaments and Composites* (NATO Series E Applied Sciences, Vol. 177), Kluwer Academic Publishers, Dordrecht, 1990.
xvi) Buckley JD, Edie DD ed., *Carbon-Carbon Materials and Composites*, Noyes Publications, 1992.
xvii) Cogswell FN, *Thermoplastic Aromatic Polymer Composites: A study of the Structure, Processing and Properties of Carbon Fibre Reinforced Polyetheretherketone and R*, Combined Book Service Ltd., 1992.
xviii) Savage GG, *Carbon-Carbon Composites*, Chapman and Hall, London, 1992.

xix) Pierson HO, *Handbook of Carbon, Graphite, Diamond and Fullerenes: Properties, Processing and Applications*, Noyes Publications, 1993.
xx) Thomas CR ed., *Essentials of Carbon-Carbon Composites*, The Royal Society of Chemistry, Books Britain, 1993.
xxi) Chung DDL, *Carbon Fiber Composites*, Butterworth Heinemann, Newton, 1994.
xxii) Lovell DR (completed by Starr T), *Carbon and high performance fibres, Directory and Handbook*, Edition 6, Chapman and Hall, London, 1994. Now published by Kluwer Academic Publishers.
xxiii) Peebles LH, *Carbon fibers: Formation, Structure and properties*, CRC Press, Florida, 1995.
xxiv) Lewin M, *Handbook of Fiber Chemistry*, 2nd ed. (International Fiber Science and Technology, Vol. 15), Marcel Dekker, 1998.
xxv) Donnet JB, Wang TK, Peng JC, Rebouillat S, *Carbon Fibers*, Marcel Dekker, New York, 1998.
xxvi) Fitzer E, Manocha LM, *Carbon Reinforcements and Carbon/Carbon Composites*, Springer Verlag, Berlin, 1998.
xxvii) Kelly V. Carbon Fiber. Manufacture and Applications, Elsevier, Amsterdam, 2005.

The Author

Peter Ernest Morgan was born in 1931 and joined Courtaulds Ltd., Coventry from school in 1948. He became an Associate Member of The Royal Institute of Chemistry and worked on materials for chemical plants.

Morgan undertook evaluation of carbon fibers produced in Courtauld's research laboratories. He established procedures to coat carbon fibers with metals using electroless plating techniques and developed a surface treatment process to improve the bond of carbon fibers to epoxy resins.

In 1968, Morgan joined a team at Courtaulds to manufacture carbon fibers and subsequently set up and managed a control laboratory to monitor production. He later became involved with production aspects relating to carbon fiber prepreg and pultrusion. Morgan took on the additional role of chief quality inspector in 1976 and gained approval for the production processes to Defence Standard 05-24, the forerunner of BS 5750. In 1980, he was appointed to Courtauld's Carbon Fibres division board as director with technical responsibilities.

Morgan joined RK Textiles Composite Fibres Ltd. in 1981 as Technical Director and became involved with the design, building, erection and transfer of technology for the sale of carbon fiber plants in Israel, South Korea and India, as well as plants for in-house use.

From 1989 to 1993, Morgan was General Manager/Technical Director of the RK Carbon Fibres Ltd.'s production facility at Muir of Ord in Scotland. In 1993, as Technical Director/Quality Manager, he obtained approval with BSI for BS 5750: Part 2, which later became ISO 9002.

Morgan has written numerous articles on carbon fibers and is the author of ten patents.

Acknowledgments

Grateful thanks are due to my ex-colleagues at Courtaulds—David Carlton, James Darling, John Fagge, Geoff Gould, Tom Heath, John Ludlow and Ed Trewin; together with ex-colleagues at RK Carbon Fibres—Freda Cameron, Stuart Devine, Richard Glanville, Lance Hill, Hayley Robb and Philip Rose.

John Johnson and Neil Turner, formerly of Rolls Royce, and Neil Hancox, previously with AERE Harwell, have been most supportive.

Many universities in the UK have kindly supplied information and thanks are particularly due to Frank Matthews at Imperial College, Tim Mays at Bath, David Johnson at Leeds, and Frank Jones at Sheffield.

Numerous references have been supplied by The British Library and acknowledgment is made to the many companies who have kindly supplied their trade literature and the publications *ACM Monthly*, *Composites Market Reports*, *Nottingham University Composites Club Composite News* and *Reinforced Plastics Journal*.

Finally, I would like to express my gratitude to my wife Doris, for suffering a man with a mission and my daughter Deborah, for occasional proofreading and helpful advice.

Peter E Morgan

Contents

Chapter 1   Structure of the Carbon Atom 1
1.1.   Introduction to the Element Carbon, its Isotopes and Allotropes 1
1.2.   Structure of Carbon 2
    1.2.1.   Structure of the Atom 2
    1.2.2.   Atomic Spectra and Quantum Theory 2
    1.2.3.   Directional Characteristics of Atomic Orbitals 6
    1.2.4.   Hybridization of Atomic Orbitals 7
    1.2.5.   Covalence and Molecular Orbitals 9
References 13

Chapter 2   The Forms of Carbon 15
2.1.   The Allotropes of Carbon 15
2.2.   The Carbon Phase Diagram 16
2.3.   Diamond 17
    2.3.1.   Occurrence, Production and Uses of Diamond 17
        2.3.1.1.   Natural diamonds 17
        2.3.1.2.   High pressure synthetic diamonds 17
        2.3.1.3.   Polycrystalline diamond (PCD) 18
        2.3.1.4.   Chemical Vapor Deposition (CVD) diamond 18
        2.3.1.5.   Diamond-like carbon (DLC) 18
    2.3.2.   Classification of Diamonds 19
    2.3.3.   Identification of Diamond 19
    2.3.4.   The Crystal Structure of Diamond 19
    2.3.5.   The Properties of Diamond 20
        2.3.5.1.   Density 20
        2.3.5.2.   Mechanical properties 22
            2.3.5.2.1.   Hardness 22
            2.3.5.2.2.   Friction 22
            2.3.5.2.3.   Elastic properties 22
            2.3.5.2.4.   Strength 23
        2.3.5.3.   Thermal properties 23
        2.3.5.4.   Optical properties 23
        2.3.5.5.   Electrical properties 23
        2.3.5.6.   Graphitization 23
        2.3.5.7.   Chemical resistance 23
2.4.   Graphite 24
    2.4.1.   Introduction 24
    2.4.2.   Occurrence, Production and Uses of Graphite 24
        2.4.2.1.   Natural graphite 24
        2.4.2.2.   Kish graphite 24
        2.4.2.3.   Synthetic graphite 25
    2.4.3.   Structure of Graphite 27
    2.4.4.   The Properties of Graphite 31
        2.4.4.1.   Density 31
        2.4.4.2.   Mechanical properties 32
            2.4.4.2.1.   Elastic properties 33

		2.4.4.3.	Thermal properties	35
		2.4.4.4.	Electrical properties	35
		2.4.4.5.	Chemical resistance	36
2.5.	Pyrolytic Carbon and Pyrolytic Graphite			38
2.6.	Glass-like Carbon			41
2.7.	Carbon Fibers			42
2.8.	Graphite Whiskers			42
2.9.	Vapor-Grown Carbon Fibers (VGCF) and Catalytic Chemical Vapor-Deposited (CCVD) Filaments			43
2.10.	Other Forms of Carbon			44
	2.10.1.	Carbon Black		44
	2.10.2.	Charcoal		45
	2.10.3.	Coal		45
	2.10.4.	Coke		46
	2.10.5.	Soot		46
2.11.	New Forms of Carbon			46
	2.11.1.	Fullerenes		46
		2.11.1.1.	Discovery and production of fullerenes	46
		2.11.1.2.	Properties and uses of fullerenes	53
	2.11.2.	Carbon Nanotubes		56
		2.11.2.1.	Discovery and production of carbon nanotubes	56
	2.11.3.	Hyperfullerenes		59
2.12.	Summary of Allotropic Forms of Carbon			60
References				60

Chapter 3 History and Early Development of Carbon Fibers 65

3.1.	The Early Inventors			65
3.2.	Work in the USA			66
	3.2.1.	Black 'Orlon'		66
	3.2.2.	Some Early US Carbon Fibers		67
	3.2.3.	More Reent US Carbon Fibers		71
3.3.	Work in Japan			71
	3.3.1.	Early Work in Japan with PAN Precursor		71
	3.3.2.	Work in Japan with Pitch Precursors		72
3.4.	Work in the UK with PAN Precursors			72
	3.4.1.	Work at RAE, Farnborough		72
		3.4.1.1.	The RAE work with carbon fiber and cross-licencing of their patent	72
		3.4.1.2.	Surface treatment	78
		3.4.1.3.	Testing and properties of single filaments and composites	78
		3.4.1.4.	Composite fabrication	79
		3.4.1.5.	Friction and wear	79
	3.4.2.	Work at the Atomic Energy Research Establishment, Harwell		79
		3.4.2.1.	Fiber production	79
		3.4.2.2.	Surface treatment	84
		3.4.2.3.	Testing and properties of single filaments and composites	84
		3.4.2.4.	Carbon fiber reinforced ceramics, glass and cement	86

		3.4.2.5.	Carbon fiber reinforced metal composites.	87
		3.4.2.6.	Composite fabrication and design	89
	3.4.3.	Work at Rolls Royce, Derby		89
		3.4.3.1.	Fiber production	89
		3.4.3.2.	Factors affecting tensile strength of carbon fibers	91
		3.4.3.3.	Resin formulation and composite fabrication	92
		3.4.3.4.	Carbon fiber reinforced metal composites	97
	3.4.4.	Work at Morganite Modmor, London		97
	3.4.5.	Work at Courtaulds, Coventry		98
		3.4.5.1.	Carbon fiber production	98
		3.4.5.2.	Early work with X-ray diffraction to establish structure	100
		3.4.5.3.	Precursor technology	101
		3.4.5.4.	Oxidation stage	103
		3.4.5.5.	Surface treatment	111
		3.4.5.6.	Testing and properties of virgin carbon fiber and composites	112
		3.4.5.7.	Production procedures using carbon fiber	112
		3.4.5.8.	Use and design of carbon fiber in composite materials	113
3.5.	Early UK Prepreggers			114
	3.5.1.	Ciba (ARL) Ltd., Duxford		114
	3.5.2.	Courtaulds Ltd., Coventry		114
	3.5.3.	Fothergill and Harvey Ltd. (F&H), Littleborough		115
	3.5.4.	Rotorway Components Ltd., Clevedon		115
References				115

Chapter 4	Precursors for Carbon Fiber Manufacture			121
4.1.	Introduction			121
4.2.	PAN Precursors			121
	4.2.1.	History		122
		4.2.1.1.	Commercially available PAN fiber	122
	4.2.2.	Requirements for a PAN Precursor		123
	4.2.3.	Homopolymer PAN		125
	4.2.4.	Comonomers		125
	4.2.5.	Methods of Polymerization		130
		4.2.5.1.	Solution polymerization	130
		4.2.5.2.	Aqueous dispersion polymerization	134
	4.2.6.	Methods of Spinning		136
		4.2.6.1.	Wet spinning	136
		4.2.6.2.	Dry spinning	136
		4.2.6.3.	Air gap spinning	136
		4.2.6.4.	Melt spinning	139
	4.2.7.	Processing Stages		141
	4.2.8.	Modification of Spun Fiber		145
		4.2.8.1.	Stretching	145
		4.2.8.2.	Chemical treatment	145
	4.2.9.	Structure of PAN Fibers		146
4.3.	Cellulosic Precursors			148
	4.3.1.	Historical Introduction		148
	4.3.2.	Viscose Rayon Process		150

		4.3.2.1.	Introduction	150
		4.3.2.2.	Steeping stage	150
		4.3.2.3.	Shredding and ageing stages	151
		4.3.2.4.	Xanthation stage	152
		4.3.2.5.	Mixing and ripening stages	152
		4.3.2.6.	Spinning stage	152
		4.3.2.7.	Final treatment stage	153
	4.3.3.	Structure of Rayon Fibers		154
4.4.	Pitch Precursors			156
	4.4.1.	Introduction		156
		4.4.1.1.	Petroleum pitch	157
		4.4.1.2.	Coal tar pitch	158
	4.4.2.	Characterization of the Pitch		158
	4.4.3.	Isotropic Pitches		160
	4.4.4.	Preparation of Mesophase Pitches		161
		4.4.4.1.	Introduction	161
		4.4.4.2.	Production of mesophase by pyrolysis	162
		4.4.4.3.	Production of mesophase by solvent extraction	164
		4.4.4.4.	Production of mesophase by hydrogenation	164
		4.4.4.5.	Production of mesophase by catalytic modification	165
	4.4.5.	Melt Spinning Mesophase Precursor Fibers		166
	4.4.6.	Structure of Pitch Precursor		171
4.5.	Other Precursors			171
References				175

Chapter 5 Carbon Fiber Production using a PAN Precursor ... 185
5.1. Introduction ... 185
5.2. Carbon Fiber Manufacturers ... 185
5.3. World Supply of PAN based Carbon Fiber ... 186
5.4. Manufacturing Costs of PAN based Carbon Fiber ... 187
5.5. Choice of Precursor ... 191
5.6. Desirable Attributes of a PAN based Precursor Polymer and its Subsequent Production ... 192
5.7. Types of PAN based Carbon Fiber ... 194
5.8. A Carbon Fiber Production Line ... 194
    5.8.1. Precursor Station ... 194
    5.8.2. Oxidation ... 195
    5.8.3. Oxidation Plant ... 196
    5.8.4. Removal of Effluent Gases Evolved in the Oxidation Process ... 200
    5.8.5. Oxidized PAN Fiber ... 200
    5.8.6. Low Temperature Carbonization ... 200
    5.8.7. High Temperature Carbonization ... 200
    5.8.8. High Modulus Fiber Production ... 202
    5.8.9. Shrinkage during the Carbon Fiber Process ... 203
    5.8.10. Surface Treatment ... 203
    5.8.11. Sizing ... 203
    5.8.12. Collection ... 203
5.9. Fine Structure and Texture of PAN based Carbon Fibers ... 203

5.10. Aspects of Stabilization 215
    5.10.1. Structure of PAN Fibers Thermally Stabilized at 350°C....................... 218
5.11. Aspects of Carbonization... 221
    5.11.1. Methods of Increasing Fiber Modulus and Effect on Strength............. 225
        5.11.1.1. Hot stretching 225
        5.11.1.2. Effects of neutron irradiation 228
        5.11.1.3. Annealing in the presence of boron 229
    5.11.2. Carbon Fiber Yield 230
5.12. Relation of Carbon Fiber Tensile Properties to Process Conditions 230
5.13. Developments... 232
    5.13.1. Improvements in Carbon Fiber Properties 232
    5.13.2. Alternative Polymer Formulations.. 232
    5.13.3. A Family of Controlled Resistance Carbon Fibers 233
5.14. A Review of the Stabilization of PAN Precursors 234
    5.14.1. Stabilization Schemes of PAN and Associated Observations................ 235
5.15. Mechanisms for the Carbonization Stages of PAN Carbon Fibers 254
References 259

Chapter 6 Carbon Fiber Production using a Cellulosic based Precursor.................... 269
6.1. Introduction... 269
6.2. Current Production.. 272
    6.2.1. Choice of a Suitable Precursor.. 272
    6.2.2. Pyrolysis... 274
    6.2.3. Carbonization 279
    6.2.4. Hot Stretching during Processing of Carbon Fiber 279
    6.2.5. Sizing 280
6.3. Mechanisms for the Pyrolysis and Carbonization Stages of
Cellulosic based Precursors... 280
References 292

Chapter 7 Carbon Fiber Production using a Pitch based Precursor........................... 295
7.1. Introduction... 295
7.2. Choice of Melt Spun Precursor 295
7.3. The Manufacturing Process 296
    7.3.1. Stabilization (thermosetting) of Spun Fiber... 296
    7.3.2. Carbonization 301
    7.3.3. Graphitization.. 303
    7.3.4. Surface Treatment of Pitch based Carbon Fibers 304
7.4. The Structural Ordering and Morphology of Mesophase Pitch Fibers 305
    7.4.1. Mechanisms Associated with the Preparation of Pitch Precursors........... 309
    7.4.2. Mechanisms Associated with the Stabilization
of Pitch Fiber Precursors.. 320
    7.4.3. Mechanisms Associated with the Carbonization of Pitch Fibers............. 321
References 322

Chapter 8 Production of Vapor Grown Carbon Fibers (VGCF) 325
8.1. Introduction... 325

8.2. Preparation of VGCF 325
8.3. Growth Process 334
8.4. Mode of Tensile Failure 339
8.5. Mechanical Properties 339
References 343

Chapter 9 Surface Treatment and Sizing of Carbon Fibers 347
9.1. Introduction 347
9.2. Oxidative Processes 347
    9.2.1. Gas Phase Oxidation 348
    9.2.2. Liquid Phase Oxidation 350
    9.2.3. Anodic Oxidation 352
9.3. Plasma 355
9.4. Non-oxidative Surface Treatment—Whiskerization 356
9.5. Effect of Surface Treatment on Fiber Properties 357
    9.5.1. Introduction 357
    9.5.2. The Effects of Surface Treatment 358
    9.5.3. Summary 362
9.6. Coupling Agents 363
9.7. Sizing Carbon Fiber 363
    9.7.1. Deposition from Solution of a Polymer onto the Fiber Surface 363
    9.7.2. Deposition of a Polymer onto the Fiber Surface by
            Electrodeposition 367
    9.7.3. Deposition of a Polymer onto the Fiber
            Surface by Electropolymerization 369
References 370

Chapter 10 Guidelines for the Design of Equipment for Carbon Fiber Plant 377
10.1. Introduction 377
10.2. Precursor Handling 377
10.3. Drive Systems 379
10.4. Ovens for Oxidation 380
10.5. Removal of Effluent Gases Evolved in the Oxidation Process 383
10.6. Application of an Antistatic Finish 384
10.7. Plaiter Table 384
10.8. LT Carbonization Furnace 384
    10.8.1. LT Furnace Gas Seals 386
    10.8.2. LT Furnace Insulation 387
    10.8.3. Element Materials for LT Furnaces 388
10.9. LT Furnace Exhaust Removal 392
10.10. HT Carbonization Furnace 395
    10.10.1. HT Furnace Gas Seals 396
    10.10.2. HT Furnace Insulation 396
    10.10.3. Element Materials for HT Furnaces 397
10.11. Typical Calculations for the Design of an HT Furnace 398
10.12. Sodium Removal 400
10.13. HM Heat Treatment Furnace 401
    10.13.1. HM Furnace Gas Seals 401

	10.13.2.	HM Furnace Insulation	402
	10.13.3.	HM Furnace Element Design	402
10.14.	Surface Treatment		403
10.15.	Sizing		404
10.16.	Drying		405
10.17.	Online Collection		409
10.18.	Offline Winding		411
10.19.	Packaging		415
10.20.	Exhaust Systems		415
10.21.	Dust Extraction		418
10.22.	Application of Closed Circuit Television (CCTV)		420
References			420

Chapter 11 Operation of Carbon Fiber Plant and Safety Aspects 421

11.1.	Introduction		421
11.2.	Serendipity		421
11.3.	Maintenance		423
11.4.	Protecting Electrical Equipment		423
11.5.	Air Flow Measurement		424
	11.5.1.	Measurement of Pressure	424
	11.5.2.	Determination of Velocity	424
	11.5.3.	Determination of Volume Flow	429
11.6.	Collimation and Spreading of Oxidized and Carbonized Fiber		433
	11.6.1.	Lateral Movement	433
	11.6.2.	Lateral Expansion or Contraction	434
11.7.	Splicing Small Tows		435
11.8.	Drive Systems and Rotating Rollers		436
11.9.	Precursor Creel		438
11.10.	Oxidation Plant		439
11.11.	Pyrolysis Plant		440
11.12.	Low Temperature Carbonization Furnace		440
11.13.	High Temperature Carbonization Furnace		441
	11.13.1.	Calibration of Pyrometer	441
11.14.	High Modulus Furnace		442
11.15.	Surface Treatment		442
11.16.	Sizing		443
11.17.	Winding		443
11.18.	Dealing with Emissions		444
11.19.	Treatment of Cyanide Effluent		444
11.20.	Protecting the Environment		446
11.21.	Safety Committee		448
11.22.	COSH-H Requirements		448
11.23.	Toxicology of Carbon Fibers		449
	11.23.1.	Definitions of Exposure Limits	449
	11.23.2.	Data for UK Exposure Limits for Gaseous Emissions	449
	11.23.3.	Possible Hazards with Carbon and Graphite Fibers	449
11.24.	The Risks of Carbon Fiber Composites in a Fire		450
References			451

Chapter 12 Techniques for Determining the Structure of Carbon Fibers 453
12.1. Introduction 453
12.2. Optical Microscope 453
12.3. Scanning Electron Microscope (SEM) 456
12.4. Transmission Electron Microscope (TEM) 460
12.5. X-ray Diffraction 464
    12.5.1. Convention for Axes in Graphite and Carbon Fibers and Dimensional Notation 464
    12.5.2. Wide Angle X-ray Diffraction 466
    12.5.3. Single Crystal X-ray Diffraction 470
    12.5.4. X-ray Powder Diffraction 470
    12.5.5. Low Angle X-ray Diffraction 473
12.6. Auger Electron Spectroscopy (AES) 473
12.7. X-ray Photoelectron Spectroscopy (XPS or ESCA) 475
12.8. Ultraviolet Photoemission Spectroscopy (UPS) 477
12.9. Infrared Spectroscopy 479
    12.9.1. Introduction 479
    12.9.2. Fourier Transform Infrared Spectroscopy (FTIR) 481
    12.9.3. Fourier Transform Infrared/Attenuated Total Reflectance Spectroscopy (FTIR/ATR) 483
12.10. Electron Energy Loss Spectroscopy (EELS) 483
12.11. Raman Spectroscopy 485
    12.11.1. Surface Enhanced Raman Scattering (SERS) 485
12.12. Secondary Ion Mass Spectrometry (SIMS) 485
    12.12.1. Static SIMS 486
    12.12.2. Dynamic SIMS 489
    12.12.3. Imaging or Microscope SIMS 489
12.13. Scanning Tunnelling Microscopy (STM) 490
12.14. Atomic Force Microscopy (AFM) or Scanning Force Microscopy (SFM) 493
References 494

Chapter 13 Polymer Matrices for Carbon Fiber Composites 501
13.1. Selected Thermoset Resins 501
    13.1.1. Introduction 501
    13.1.2. Phenolic Resins 502
    13.1.3. Polyester Resins 503
    13.1.4. Epoxy Vinyl Ester Resins 507
    13.1.5. Epoxide Resins 508
        13.1.5.1. Bisphenol resins 508
        13.1.5.2. Novalac resins 509
        13.1.5.3. Trifunctional resins 511
        13.1.5.4. Tetrafunctional resins 511
        13.1.5.5. Cycloaliphatic resins 512
        13.1.5.6. New developments 512
        13.1.5.7. Epoxy diluents 513
        13.1.5.8. Characterization of epoxy resins 513
        13.1.5.9. Curing epoxide resins 513
        13.1.5.10. Calculating stoichiometric ratios for epoxy resins and curing agents 519

13.1.6. Cyanate Resins ... 520
13.1.7. Polyimide Resins ... 521
    13.1.7.1. Condensation type polyimides ... 523
    13.1.7.2. Addition type polyimides ... 525
        13.1.7.2.1. The earliest bismaleimides ... 525
        13.1.7.2.2. Bismaleimides ... 527
        13.1.7.2.3. Acetylene (ethynyl) terminated polyimides ... 529
13.1.8. Special Resin Systems ... 530
13.1.9. Introducing Toughness to Thermoset Resin Systems ... 530
    13.1.9.1. Introduction ... 530
    13.1.9.2. Toughening versus flexibilizing ... 531
    13.1.9.3. Types of elastomeric modifiers ... 531
    13.1.9.4. Duplex materials ... 532
    13.1.9.5. Thermoplastic modifiers ... 532
    13.1.9.6. Effect of carbon fiber reinforcement ... 533
13.2. Selected Thermoplastic Resins ... 533
    13.2.1. Introduction ... 533
    13.2.2. Morphology Property Relationships in Semi-crystalline Thermoplastics ... 535
    13.2.3. Polyamide (PA) Resins ... 538
    13.2.4. Polycarbonate (PC) Resin ... 540
    13.2.5. Polyetheretherketone (PEEK) Resin ... 540
    13.2.6. Polyetherimide (PEI) Resin ... 542
    13.2.7. Polyethersulfone (PES) Resin ... 542
    13.2.8. Polyphenylene Sulfide (PPS) Resin ... 543
13.3. Improving the Bond with Carbon Fiber/Thermoplastics ... 543
References ... 544

Chapter 14 Carbon Fiber Carbon Matrix Composites ... 551
14.1. Introduction ... 551
14.2. Selection of Materials for Carbon-Carbon Processing ... 552
    14.2.1. Types of Reinforcement ... 552
        14.2.1.1. Oxidized PAN fiber (opf) ... 552
        14.2.1.2. PAN based carbon fibers ... 552
        14.2.1.3. Pitch based carbon fibers (pbcf) ... 554
        14.2.1.4. Cellulose based carbon fibers ... 555
    14.2.2. Type of Matrix ... 555
        14.2.2.1. Thermosetting resin ... 556
            1. Furan resin ... 556
            2. Phenolic resins ... 557
            3. Polyimide resins ... 557
        14.2.2.2. Thermoplastic matrix precursors ... 558
            1. Pitch ... 558
            2. Other thermoplastic matrices ... 559
14.3. Methods of Processing Carbon-Carbon Matrix Materials ... 560
    14.3.1. Introduction ... 560
    14.3.2. Use of Gas Phase Impregnation and Densification ... 560
        14.3.2.1. Introduction ... 560

		14.3.2.2.	CVI processes	565
			1. Isothermal CVI process	565
			2. Thermal gradient CVI process (TG-CVI)	566
			3. Pressure gradient process	566
			4. Pulse CVD process	566
			5. Possible new routes	566
	14.3.3.	Processing with Thermosetting Resin Matrices		567
		14.3.3.1.	Low pressure impregnation (LPI)	567
		14.3.3.2.	Pressure impregnation and carbonization (PIC)	568
		14.3.3.3.	Hot isostatic pressure impregnation carbonization (HIPIC)	568
14.4.	Some Thoughts on Carbon-Carbon Processing			569
	14.4.1.	Chemical Vapor Deposition		569
	14.4.2.	Liquid Infiltration		572
14.5.	Provision for Providing Oxidation Protection			573
	14.5.1.	Introduction		573
	14.5.2.	The Use of Inhibitors to Provide Oxidation Protection		574
		1. Boron		574
		2. Phosphorus		575
	14.5.3.	The Use of a Barrier Coating		575
		1. Noble metals		575
		2. Silicon coatings		575
	14.5.4.	Other Coating Systems		578
References				578

Chapter 15 Carbon Fiber Reinforced Ceramic Matrices 583

15.1.	Introduction		583
15.2.	Cement, Concrete and Gypsum Matrices		583
	15.2.1. Cement		583
	15.2.2. Concrete		584
	15.2.3. Concrete Additives		584
		15.2.3.1. Silica fume	584
		15.2.3.2. Dispersant	584
		15.2.3.3. Water reducing agent	585
		15.2.3.4. Accelerator	585
	15.2.4. Work Undertaken with Mortar and Concrete		585
	15.2.5. Theory		591
	15.2.6. Fabrication Processes for cfrc		591
15.3.	Glass Matrices		592
	15.3.1. The Glass Matrix		592
	15.3.2. Methods of Preparation of Carbon Fiber Reinforced Glasses		594
		15.3.2.1. Mode of reinforcement	594
		15.3.2.2. Slurry with hot pressing	594
		15.3.2.3. Hot filament winding under tension with hot pressing above the annealing temperature	597
		15.3.2.4. Melt infiltration	597
		15.3.2.5. Sol gel	598
	15.3.3. Work Undertaken with Carbon Fiber Filled Glass Matrices		599
	15.3.4. Coating Carbon Fiber to Improve the Bond to a Glass		601

15.4.	Ceramic Matrices		602
	15.4.1.	Processing Ceramic Matrix Composites	602
	15.4.2.	Types of Ceramic Matrices	602
		15.4.2.1. Oxide matrix materials	603
		1. Alumina (Al_2O_3)	603
		2. Mullite ($3Al_2O_3.2SiO_2$)	603
		3. Zirconia (ZrO_2)	603
		15.4.2.2. Non-oxide matrix materials	603
		1. Silicon carbide (SiC)	603
		2. Titanium carbide (TiC)	604
		3. Boron carbide (B_4C)	604
		4. Titanium boride (TiB_2)	604
		5. Boron nitride (BN)	604
		6. Aluminium nitride (AlN)	604
		7. Silicon nitride (Si_3N_4)	604
	15.4.3.	Fiber Reinforcement	605
	15.4.4.	Processing Techniques	605
		15.4.4.1. Slurry infiltration	605
		15.4.4.2. Slip casting	605
		15.4.4.3. Filament winding	605
		15.4.4.4. Chemical synthesis	606
		1. Sol gel	606
		2. Polymer precursor	607
		15.4.4.5. Melt infiltration	609
		15.4.4.6. In situ chemical reactions	611
		1. CVI (or CVD)	611
		2. Slurry pulse/CVI	612
		3. Hot Isotactic Pressing (HIPing)	613
		4. Reaction bonding	614
		15.4.4.7. Consolidation and densification	615
		1. Sintering	615
		2. Pressureless sintering	615
		3. Hot pressing	615
	15.5.5.	Protective Coatings	615
	15.6.6.	Fracture Mechanics	617
References			617

Chapter 16	Carbon Fibers in Metal Matrices	629
16.1.	Introduction	629
16.2.	Metal Matrix Composites	629
16.3.	Carbon Fiber for Reinforcement of Metal Matrices	629
16.4.	Coating Processes to Improve Wettability	631
	16.4.1. CVD Process	631
	16.4.2. Liquid Metal Transfer Agent (LMTA) Technique	632
	16.4.3. Cementation	632
	16.4.4. Electroless Plating	632
	16.4.5. Electroplating	633
	16.4.6. Solution Coating	633
	16.4.7. Flux	634

	16.4.8.	INCO Ni Coated Carbon Fiber	634
	16.4.9.	Other Coating Processes	635
16.5.	Metal Matrices	635
	16.5.1.	Aluminium	635
	16.5.2.	Magnesium	639
	16.5.3.	Copper	639
	16.5.4.	Nickel	640
	16.5.5.	Lead	640
	16.5.6.	Tin	640
16.6.	Techniques for Fabricating Carbon Fiber Reinforced Metal Matrix Composites	641
	16.6.1.	Factors Influencing Processing of Metal Matrix Composites	641
		16.6.1.1. Capillary effects	641
		16.6.1.2. Fluid flow into the preform	641
		16.6.1.3. Fiber matrix interactions	641
		16.6.1.4. The solidification process	642
	16.6.2.	Processing Methods for Fabricating Metal Matrix Composites	642
		16.6.2.1. Solid state processing methods	643
		1. Powder metallurgy	643
		2. Diffusion bonding	643
		16.6.2.2. Liquid state processing	644
		1. Melt stirring	644
		2. Compocasting or rheocasting	644
		3. Slurry casting	644
		4. Gravity or vacuum casting	644
		5. Pressure casting	644
		6. Squeeze casting	644
		7. Fiber tow (liquid) infiltration	645
		8. Lanxide process	646
		9. Liquid phase hot pressing, liquid phase diffusion bonding or liquid phase sintering	646
		16.6.2.3. Deposition processes	647
		1. Ion plating	647
		2. Plasma spraying	647
	16.6.3.	Fundamental Considerations	647
		16.6.3.1. Capillarity	647
		16.6.3.2. Fluid flow into the preform	648
		16.6.3.3. Fiber matrix interactions	648
		16.6.3.4. The solidification process and matrix microstructure	648
References			649

Chapter 17 Testing of PAN Precursor, Virgin Carbon Fibers, Carbon Fiber Composites and Related Products ... 657

17.1.	Introduction	657
17.2.	Testing of PAN Precursor	657
	17.2.1. Filament Diameter Distribution in PAN Tow	657
	17.2.2. Measurement of Precursor d'tex using the Vibroskop (ASTM D1577)	660

	17.2.3.	Determination of Fiber Moisture Content and Fiber Moisture Regain	660
	17.2.4.	Determination of Residual Solvent (NaSCN) in Courtelle Precursor	661
	17.2.5.	Determination of Sodium Content in the Precursor	661
		1. Atomic absorption spectrophotometer	661
		2. Ion chromatograph	661
	17.2.6.	Determination of the Soft Finish Content in Courtelle Precursor	662
	17.2.7.	Silver Sulphide Staining Test for Checking Structure of a PAN Precursor	662
	17.2.8.	An Experimental Rig for Determination of Precursor Burn-up Temperature	662
17.3.	Testing of Oxidized PAN Fiber (OPF) and Virgin Carbon Fiber		662
	17.3.1.	Mass per Unit Length	662
	17.3.2.	Determination of Density	663
	17.3.3.	Determination of Diameter	666
		1. Mounting a single filament	666
		2. Determining filament diameter using a Watson image shearing eyepiece	667
		3. Determination of filament diameter using a He/Ne laser	668
		4. Calibration of a Stereoscan with a traceable reference standard	669
		5. Preparation of a mini composite (impregnated tow)	670
	17.3.4.	Tensile Testing of Filament	670
		17.3.4.1. Determination of compliance of the tensile test machine system	670
		17.3.4.2. Measurement of filament tensile modulus	671
		17.3.4.3. Measurement of filament tensile strength	674
	17.3.5.	Determination of Oxidized PAN Fiber Finish Content	674
	17.3.6.	Determination of Carbon Fiber Size Content	676
	17.3.7.	Conductivity of a Water Extract	677
	17.3.8.	Skin Core	677
	17.3.9.	Measurement of Electrical Properties	678
17.4.	Carbon Fiber Tow Testing		678
	17.4.1.	Dry Tow Test	678
	17.4.2.	Testing of the Impregnated Tow	679
17.5.	Testing of Carbon Fiber Yarn and Fabric		682
	17.5.1.	Determination of Twist	683
	17.5.2.	Determination of Ends and Picks	683
17.6.	Testing of Matrix		684
	17.6.1.	Fineness of Grind	684
	17.6.2.	Selection of a Suitable Grade of Paper for Resin Coating	684
	17.6.3.	Determination of Gel Time	685
		1. Using the Kofler hotbench	685
		2. Determination of gel time at ambient temperature	686
	17.6.4.	Determination of the Viscosity of a Resin Mix	686
	17.6.5.	Determination of the Epoxy Molar Mass (EMM) of Epoxy Resins	687

		1.	Cetyl trimethylammonium bromide-perchloric acid titration method	687

- 1. Cetyl trimethylammonium bromide-perchloric acid titration method ... 687
- 2. Determination of EMM by potentiometric titration ... 688

17.7. Testing of Carbon Fiber Prepreg ... 688
    17.7.1. Mass per unit Area ... 688
    17.7.2. Volatiles Content ... 688
    17.7.3. Fiber Content ... 688
    17.7.4. Resin Gel Time ... 689

17.8. Testing of Carbon Fiber Composite ... 689
    17.8.1. Introduction ... 689
    17.8.2. Preparation of Composite Specimen from Wet Resins ... 690
    17.8.3. Preparation of Composite Specimen from Prepreg Systems ... 692
    17.8.4. Determination of Carbon Fiber Content ... 693
    17.8.5. Measurement of Tensile Modulus ... 693
    17.8.6. Measurement Tensile Strength ... 695
    17.8.7. Measurement of Strain using Resistance Strain Gages ... 697
    17.8.8. Measurment of Shear Strength ... 699
        17.8.8.1. Interlaminar shear strength ... 699
        17.8.8.2. In-plane shear tests ... 700
            1. The torsion test ... 700
            2. Two-rail or three-rail shear test ... 701
            3. The double V-notch shear (Iosipescu test) ... 702
            4. Tension coupon test ... 702
            5. The $10°$ off-axis test ... 704
    17.8.9. Measurement of Flexural Strength and Modulus ... 706
    17.8.10. Measurement of Uniaxial Compressive Strength and Modulus ... 708
    17.8.11. Testing of Fatigue ... 710
    17.8.12. Measurement of Creep ... 712
    17.8.13. Testing of Impact Behavior ... 714
    17.8.14. Measurement of Interlaminar Fracture Toughness ... 714

17.9. Testing of Carbon Fiber Filled Thermoplastics ... 714
    17.9.1. Measurement of Moisture Content ... 714
    17.9.2. Molding ... 715
    17.9.3. Determination of Melt Flow Index (MFI) ... 717
    17.9.4. Impact Testing of Thermoplastics ... 718

17.10. Instrumental Analysis ... 718
    17.10.1. Optical Microscope ... 718
    17.10.2. Laboratory Furnace ... 720
    17.10.3. Thermal Analysis ... 721
        17.10.3.1. Differential scanning calorimeter (DSC) ... 721
            1. Classical DTA ... 721
            2. Boersma DTA ... 722
            3. DSC ... 722
        17.10.3.2. Thermogravimetric analysis (TGA) ... 725
        17.10.3.3. Dynamic mechanical analysis (DMA) ... 726
        17.10.3.4. Thermomechanical analysis (TMA) ... 729
    17.10.4. Chromatography ... 729
    17.10.5. Infrared Analysis (IR) ... 732
    17.10.6. Elemental Analysis ... 735

17.11.	Non-destructive Testing (NDT)	735
	17.11.1. Ultrasonic Testing	736
	17.11.2. Radiography	738
	17.11.3. Acoustic Emission	738
17.12.	Supplement 1	738
	17.12.1. Sinclair's Loop Test for Filament Testing	738
	1. Tension testing	739
	2. Compression testing	739
References		739

Chapter 18 Statistics and Statistical Process Control (SPC) — 747

18.1.	Frequency Distribution	747
18.2.	Location of Data	748
18.3.	Measures of Dispersion	750
18.4.	Standard Error	751
18.5.	Sample Correlation Coefficient	751
18.6.	Linear Regression	752
18.7.	Normal Distribution	753
18.8.	Weibull Distribution	756
18.9.	Variation	756
18.10.	Control Chart Method	758
18.11.	Statistical Process Control Charts	758
	18.11.1. Average and Range ($\bar{x}$ and R) Chart	760
	18.11.2. Mean and Standard Deviation ($\bar{x}$ and σ) Chart	764
	18.11.3. Median Control Chart	766
	18.11.4. Rules for Detecting Out-of-control Conditions on Control Charts	766
	18.11.5. Cumulative Sum Chart (Cusum)	769
18.12.	Capability Index	770
18.13.	Failure Mode Effect Analysis (FMEA)	771
References		771

Chapter 19 Quality Control — 773

19.1.	Inhouse Testing	773
19.2.	Quality Management and Quality Assurance Standards	773
19.3.	The ISO 9000 Family of Standards and Quality Systems	774
	Para 4.1 Management Responsibility	774
	Para 4.2 Quality System	774
	Para 4.3 Contract Review	774
	Para 4.4 Design Control	774
	Para 4.5 Document Control and Data	774
	Para 4.6 Purchasing	774
	Para 4.7 Control of Customer Supplied Product	774
	Para 4.8 Product Identification and Traceability	774
	Para 4.9 Process Control	774
	Para 4.10 Inspection and Testing	774
	Para 4.11 Control of Inspection, Measuring and Test Equipment	775
	Para 4.12 Inspection and Test Status	775

	Para 4.13	Control of Non-Conforming Product	775
	Para 4.14	Corrective and Preventive Action	775
	Para 4.15	Handling, Storage, Packaging, Preservation and Delivery	775
	Para 4.16	Control of Quality Records	775
	Para 4.17	Internal Quality Audits	775
	Para 4.18	Training	775
	Para 4.19	Servicing	775
	Para 4.20	Statistical Techniques	775

19.4. Quality Gurus ... 775
    19.4.1. The Early Americans ... 776
        19.4.1.1. W Edwards Deeming ... 776
        19.4.1.2. Joseph M Juran ... 778
        19.4.1.3. Armand V Fiegenbaum ... 779
    19.4.2. The Japanese Gurus ... 779
        19.4.2.1. Dr Kaoru Ishikawa ... 779
        19.4.2.2. Dr Genichi Taguchi ... 780
        19.4.2.3. Shigeo Shindo ... 781
    19.4.3. The New Western Group of Gurus ... 782
        19.4.3.1. Philip B Crosby ... 782
        19.4.3.2. Tom Peters ... 783
        19.4.3.3. Claus Møller ... 784

19.5. Quality Circles ... 785
19.6. Total Quality Management ... 786
19.7. Quality Costing ... 788
References ... 789

Chapter 20 Properties of Carbon Fibers ... 791
20.1. The Role of Carbon Fibers ... 791
20.2. Types of Carbon Fibers Available in the World Market ... 792
20.3. Tensile Properties ... 800
20.4. Factors Effecting Composite Strength ... 808
20.5. The Importance of Critical Aspect Ratio ... 810
20.6. Elastic Constants ... 811
20.7. Flexural Properties ... 814
20.8. Effect of Surface Treatment and Sizing on Composite Properties ... 815
20.9. Compression Properties ... 817
20.10. Thermal Properties ... 823
20.11. Thermal Expansion of Carbon Fibers ... 829
20.12. Thermal Conductivity of Carbon Fibers ... 831
20.13. Creep Properties ... 831
20.14. Impact Strength and Fracture Toughness ... 833
20.15. Fatigue Properties ... 834
20.16. Electrical Properties ... 834
20.17. Chemical Resistance ... 836
    20.17.1. Intercalation ... 837
20.18. Friction and Wear ... 837
20.19. Hybrid Composites ... 838
20.20. Some Selected Properties of Composites ... 839
    20.20.1. Thermoplastic Polymer Matrices ... 839

	20.20.2.	Cement Matrices	839

- 20.20.2. Cement Matrices 839
- 20.20.3. Glass and Ceramic Matrices 841
- 20.20.4. Carbon–Carbon 844
- 20.21. Metal Matrices 845
- References 849

Chapter 21 Manufacturing Techniques for Carbon Fiber Reinforced Composites in Thermoset and Thermoplastic Matrices 861

- 21.1. Carbon Fiber Reinforcement and Architecture 861
  - 21.1.1. Virgin Carbon Fiber 861
  - 21.1.2. Non-woven Discontinuous Reinforcement (Staple Fiber) 863
    - 21.1.2.1. Adhesive bonded reinforcements 863
      1. Chopped strand mat (csm) 863
      2. Carbon fiber tissue 864
      3. Carbon fiber paper reinforcement 864
    - 21.1.2.2. Needled mat 864
    - 21.1.2.3. Milled fiber 864
    - 21.1.2.4. Chopped carbon fiber 865
  - 21.1.3. Unidirectional Fabrics 865
    - 21.1.3.1. Non-woven UD fabrics 865
    - 21.1.3.2. Woven UD fabrics 866
      1. Warp UD fabric 866
      2. Weft UD fabric 866
  - 21.1.4. Woven Fabrics (2-D Planar or Biaxial Reinforcement) 866
    1. Plain or square weave 868
    2. Basket (Hopsack) weave 868
    3. Leno weave 869
    4. Mock Leno weave 870
    5. Twill weave 871
    6. Satin weave 871
    7. High modulus (non-crimp) weave 872
  - 21.1.5. Woven Spread Tow 872
  - 21.1.6. Knitted Fabrics 872
    - 21.1.6.1. Weft knitting 874
      1. Plain knitting 874
    - 21.1.6.2. Warp knitting 876
      1. Plain tricot 877
      2. Raschel 877
  - 21.1.7. Inlaid Fabrics 877
  - 21.1.8. Braiding 877
    - 21.1.8.1. Forms of braiding 879
      1. Flat braids 879
      2. Sleevings 879
      3. Wide braided fabric 879
      4. Overbraids 880
    - 21.1.8.2. Braid architecture 880
      1. Biaxial 2-D braid 880
      2. Triaxial 3-D braid 881
  - 21.1.9. 3-D Reinforcements 882

		21.1.9.1.	Multiaxial non-crimp reinforcements	882
			1. Producing a stitched fabric by the simultaneous stitch process	883
			2. Producing a stitched fabric by the weave and stitch process	885
			3. Double bias fabrics...	885
			4. Triaxial weave	885
			5. Quadraxial...	889
		21.1.9.2.	Woven 3-D fabrics	889
		21.1.9.3.	Proprietary 3-D weaving processes	889
		21.1.9.4.	Knitted 3-D fabrics	890
		21.1.9.5.	Braided 3-D multiaxial...	890
		21.1.9.6.	n-D orthogonal blocks	891
		21.1.9.7.	Aztex Inc Z-Fiber^{TM}	893
21.2.	Core Materials			893
21.3.	Manufacturing Processes for Carbon Fibers in Thermoset Matrices..................			894
	21.3.1.	Contact Molding Wet Lay-up		894
		21.3.1.1.	Hand lay-up (contact molding)	895
		21.3.1.2.	Spray lay-up..	895
	21.3.2.	Hot Press Matched Metal Molding		896
		21.3.2.1.	Thermoset dough molding compound (DMC).........	896
		21.3.2.2.	Thermoset bulk molding compound (BMC)	896
		21.3.2.3.	Thermoset sheet molding compound (SMC)	896
	21.3.3.	Resin Transfer Molding (RTM)		897
		21.3.3.1.	Dow AdvRTM^{TM}	898
		21.3.3.2.	Vacuum assisted resin transfer molding (VARTM)	900
		21.3.3.3.	Vacuum infusion processing (VIP)	901
		21.3.3.4.	Seemann Composite Resin Infusion Molding Process (SCRIMP^{TM})	901
		21.3.3.5.	Resin infusion under flexible tooling (RIFT)	901
		21.3.3.6.	Vacuum infusion molding process (VIMP)	901
		21.3.3.7.	SP Resin Infusion Technology (SPRINT^{TM})	901
		21.3.3.8.	Resin film infusion (RFI)	902
	21.3.4.	Sequential Multiport Resin Injection System (SMRIM)....................		904
	21.3.5.	Reaction Injection Molding (RIM)		904
	21.3.6.	Centrifugal Molding		904
	21.3.7.	Preparation of Fiber Preforms..		904
	21.3.8.	Flow and Cure Monitoring of Resin Infusion Processes		904
	21.3.9.	Filament Winding		905
		1. Hoop winding...		906
		2. Helical winding		906
		3. Polar winding		906
		4. Multiaxial winding		907
		5. Variants of multiaxial winding..		907
	21.3.10.	Pultrusion...		909
		1. Reinforcement handling		910
		2. Resin impregnation		911
		3. Pre-die forming..		911
		4. Heated die to shape and cure the resin................................		911
		5. Pulling unit to provide traction...		912

		6.	Cut off saw	912
		7.	Post cure oven	912
	21.3.11.	Prepreg Molding		913
		21.3.11.1.	Prepreg manufacture	913
		21.3.11.2.	Manufacture of composites from prepreg	916
			1. Ply cutting and stacking prepreg	916
			2. Compression molding of prepreg	916
			3. Vacuum bag molding	916
			4. Press-clave molding	917
			5. Autoclave molding	918
			6. Quickstep^{TM} Molding	920
			7. Tube rolling	921
			8. Automatic tape lay-up	921
	21.3.12.	Fiber Placement Systems		921
	21.3.13.	Mold Release		922
		1.	Polyvinyl alcohol (PVA)	923
		2.	Waxes	923
		3.	Internal mold release agents	923
		4.	Silicones	923
		5.	Fluorocarbons	923
		6.	New products	923
21.4.	Carbon Fibers in Thermoplastic Matrices			923
	21.4.1.	The Importance of Critical Aspect Ratio		923
	21.4.2.	Preparation of Thermoplastic Molding Compounds		924
		21.4.2.1.	Sizing carbon fiber with compatible thermoplastic polymer size	924
		21.4.2.2.	Manufacture of thermoplastic molding compound	924
			1. Short fiber process	924
			2. Long fiber process	924
	21.4.3.	Injection Molding		925
	21.4.4.	Film Stacking Process		927
	21.4.5.	Thermoplastic Prepreg		927
		1.	Molding carbon fiber/PEI laminate	928
		2.	Platen pressing of carbon fiber/PEEK laminate	928
	21.4.6.	Thermoplastic Filament Winding		928
	21.4.7.	Thermoplastic Pultrusion		929
	21.4.8.	Continuous Fiber Reinforced Plastic Materials		929
21.5.	Hybrid Composites			929
References				930

Chapter 22 Design 935
22.1. Design Considerations 935
22.2. Micromechanics 935
22.3. Selection of Materials 940
22.4. Elastic Behavior of Multidirectional Laminates 940
22.5. Choice of Composite Manufacturing Method 943
22.6. Bonding and Joining 943
22.7. Fabrication 944

22.8.	Testing and Inspection	944
22.9.	Smart Devices	944
22.10.	Design Cases	945
	22.10.1. Expanding Core Technique	945
	22.10.2. A Yacht Mast	946
References		946
Supplementary Bibliography		947

Chapter 23	The Uses of Carbon Fibers	951
23.1.	Uses of Oxidized PAN Fiber (OPF)	951
	23.1.1. Flameproof Applications	951
	23.1.1.1. Aviation and aerospace	953
	23.1.1.2. Industrial workwear	954
	23.1.1.3. Defence and law enforcement	954
	23.1.1.4. Transportation and furnishings	955
	23.1.1.5. Cable insulation	955
	23.1.2. Friction Materials	955
	23.1.3. Gland Packings	955
	23.1.4. Precursor for PAN based Carbon Fiber and Activated Carbon Fibers	955
23.2.	Uses of Virgin Carbon Fiber	955
	23.2.1. Activated Carbon Fibers (ACF)	955
	23.2.2. Molecular Sieves	958
	23.2.3. Catalysts	958
	23.2.4. Biomedical Applications	958
23.3.	Electrical Applications	960
	23.3.1. Electrical Conduction	960
	23.3.2. Tailored Resistance Carbon Fiber	960
	23.3.3. Cathodic Protection	960
	23.3.4. Elimination of Static	960
	23.3.5. Electrodes	961
	23.3.6. Batteries	962
	23.3.6.1. Lithium Ion Batteries	962
	23.3.7. Fuel Cells	964
	23.3.7.1. Alkaline Fuel Cell (AFC)	965
	23.3.7.2. Proton Exchange Membrane Fuel Cell (PEMFC)	966
	23.3.7.3. Phosphoric Acid Fuel Cell (PAFC)	967
	23.3.7.4. Molten Carbonate Fuel Cell (MCFC)	968
	23.3.7.5. Solid Oxide Fuel Cell (SOFC)	969
	23.3.7.6. Carbon fiber in fuel cells	969
23.4.	Thermal Insulation	970
23.5.	Packing Materials and Gaskets	973
23.6.	Carbon Fibers in Thermoset Matrices	973
	23.6.1. Aerospace	973
	23.6.1.1. Defence aircraft	973
	23.6.1.2. Civil aircraft	973
	23.6.1.3. Helicopters	977
	23.6.1.4. Aero engines	977
	23.6.1.5. Propeller blades	977

	23.6.1.6. Antenna, lightening conductors	979
	23.6.1.7. Gliders and sailplanes	982
	23.6.1.8. Unmanned Aerial Vehicles (UAVs)	982
	23.6.1.9. Stealth aerial vehicles	982
23.6.2.	Space	982
23.6.3.	Rocket Motor Cases	983
23.6.4.	Flywheels	983
23.6.5.	Marine Applications	987
	23.6.5.1. Yachts	987
	23.6.5.2. Submarines	989
	23.6.5.3. Air cushion vehicle	989
23.6.6.	Oil Exploration	989
23.6.7.	Automobile and Racing Car Applications	991
	23.6.7.1. Chassis, body and interior	991
	23.6.7.2. Brakes and clutches	992
	23.6.7.3. Suspension systems	992
	23.6.7.4. Push rods	993
	23.6.7.5. Air bags	993
23.6.8.	Heavy Goods Vehicles and Buses	993
	23.6.8.1. Drive shafts	993
	23.6.8.2. Buses	994
23.6.9.	CNG Storage Cylinders	994
23.6.10.	Motor Bikes	994
23.6.11.	Railways	995
23.6.12.	Engineering and Textile Applications	995
	23.6.12.1. Structural work	995
	23.6.12.2. Robot arms	995
	23.6.12.3. Rollers	995
23.6.13.	Turbine Blades	995
	23.6.13.1. Wind turbine blades	995
	23.6.13.2. Tidal turbine blades	997
23.6.14.	Textile Applications	998
23.6.15.	Chemical and Nuclear Applications	998
23.6.16.	Medical and Prosthetic Applications	998
	23.6.16.1. Hospital equipment	1000
23.6.17.	Dental	1000
23.6.18.	Sports and Leisure Goods	1001
	23.6.18.1. Bicycles, tandem	1001
	23.6.18.2. Bows and arrows	1002
	23.6.18.3. Rifles	1002
	23.6.18.4. Skis and ski sticks	1002
	23.6.18.5. Snowboards	1002
	23.6.18.6. Baseball bats	1002
	23.6.18.7. Cricket bats	1003
	23.6.18.8. Hockey sticks	1003
	23.6.18.9. Golf shafts and heads	1003
	23.6.18.10. Tennis, racquetball, badminton and squash racquets	1004
	23.6.18.11. Snooker and pool cues	1004
	23.6.18.12. Fishing rods and reels	1005

		23.6.18.13.	Hang glider	1005

                    23.6.18.13.   Hang glider.. 1005
                    23.6.18.14.   Canoe paddles 1005
                    23.6.18.15.   Wind surfing.. 1005
         23.6.19.   Musical Instruments and Hi-Fi 1005
                    23.6.19.1.   Loudspeaker cones 1006
                    23.6.19.2.   Carbon fiber cable 1006
                    23.6.19.3.   Satellite reflectors 1006
                    23.6.19.4.   Stringed instruments 1006
                    23.6.19.5.   Bows for cello and violin 1007
         23.6.20.   Other End Uses in Thermoset Matrices 1007
                    23.6.20.1.   Model aeroplanes 1007
                    23.6.20.2.   Knives, fountain pens, watches 1007
                    23.6.20.3.   Precision instruments 1008
                    23.6.20.4.   Tripods 1008
                    23.6.20.5.   Optical instruments 1008
                                  23.6.20.5.1.   Telescopes 1008
                                  23.6.20.5.2.   Binoculars 1009
         23.6.21.   Furniture 1009
         23.6.22.   Carbon Fiber and Wood 1009
23.7.   Carbon Fibers in Thermoplastic Matrices 1009
         23.7.1.   Thermoplastic Molding Compounds 1010
23.8.   Carbon Fibers for Carbon-Carbon Applications 1010
         23.8.1.   Carbon-Carbon Braking Systems 1011
         23.8.2.   Carbon-Carbon Clutches and Limited Slip Differentials 1018
         23.8.3.   Carbon-Carbon in Space 1020
         23.8.4.   Carbon-Carbon for Aircraft 1021
         23.8.5.   Rocket Motor Nozzles and Expansion Tubes 1021
         23.8.6.   Carbon-Carbon in Engines 1022
         23.8.7.   Carbon-Carbon for Biomedical End Uses 1022
         23.8.8.   Carbon-Carbon in Industry 1022
         23.8.9.   Carbon-Carbon as a Dielectric Heat Sink 1023
23.9.   Carbon Fibers in Cement and Concrete 1023
         23.9.1.   Carbon Fibers in Cement and Concrete 1024
         23.9.2.   Carbon Fiber Cement as a Replacement for Asbestos Cement 1024
         23.9.3.   Strengthening of Reinforced Concrete Chimneys, Columns,
                    Beams and Retrofits 1024
         23.9.4.   New Structures with cfrp 1030
23.10.   Carbon Fibers in Glass Matrices 1031
23.11.   Carbon Fibers in Ceramic Matrices 1031
23.12.   Carbon Fibers in Metal Matrices 1031
         23.12.1.   Electromagnetic Interference (EMI) and Heat Dissipation 1031
23.13.   Other End Uses for Carbon Fibers 1032
References 1032

Chapter 24   Looking to the Future 1043
24.1.   The Future 1043
24.2.   The Production Process 1043
         24.2.1.   Precursor Developments 1043
         24.2.2.   Plant Developments 1044

24.3. Carbon Fiber	1044
24.4. Composite Manufacturing Techniques	1045
24.5. Quality Management Standards	1045
24.6. Recycling	1046
24.7. Innovative Developments	1046
24.8. Conclusion	1047
References	1047

Appendix

Appendix 1 Glossory	1049
Appendix 2 The Elements	1061
Appendix 3 The Greek Alphabet	1063
Appendix 4 Some Definitions and Handy Conversion Factors	1065
Appendix 5 ISO Standard Prefixes for SI Units	1067
Appendix 6 Interconversion of Common English and SI Units	1069
Appendix 7 Textile Terminology	1073
Appendix 8 Temperature Estimation from Color	1075
Appendix 9 Humidities over Saturated Salt Solutions	1077
Appendix 10 Wet and Dry Bulb Humidity Table	1079
Appendix 11 Detection of Cyanide [1]	1081
Appendix 12 British Standards on Quality	1083
Appendix 13 Abbreviations used in Spectroscopy and Microscopy	1087
Appendix 14 Typical Properties of Unreinforced Plastic Polymers	1089
Appendix 15 Acronyms for Thermoplastic Polymers	1117
Appendix 16 Companies Involved with Carbon Fibers and their Composites Throughout the World	1119

Index	1133

CHAPTER 1

Structure of the Carbon Atom

1.1 INTRODUCTION TO THE ELEMENT CARBON, ITS ISOTOPES AND ALLOTROPES

In the periodic table, which is an arrangement of chemical elements exhibiting certain regular periodic occurrences in their behavior, the element carbon occupies the sixth position (Appendix 2) and has a molar mass of 12.011 g mol^{-1}. Since 1961, all relative atomic masses have been based on a scale where the atomic mass of ^{12}C (pronounced carbon-12) is exactly 12.00000.

Carbon has seven isotopes and essentially consists of 98.89% of isotope ^{12}C and 1.108% of isotope ^{13}C, which are stable. The other five isotopes ^{10}C, ^{11}C, ^{14}C, ^{15}C and ^{16}C are radioactive, decaying by the emission of a β particle, which can be either negative (an electron) or positive (a positron). Only one isotope ^{14}C has a long half life, which is the time required for the original amount of radioactivity to be reduced by half. An isotope is one of a set of the same species of atoms that have the same atomic number (i.e., the same number of protons and electrons), but have different mass numbers (i.e., different number of neutrons). Where more than one isotope exists, the molar mass is the average of that mixture of isotopes, hence the value for carbon is 12.011 g mol^{-1}. The isotope ^{14}C is formed by cosmic ray bombardment of ^{14}N in the upper atmosphere and is also a by-product of the atomic energy program. ^{14}C has a half life of 5568 years and can be used in tracer techniques, as a research tool to label compounds in organic chemistry, to follow the course of a reaction and for radio-carbon dating. The atmosphere contains 0.05% carbon dioxide that contains a small proportion of ^{14}C, which is incorporated into living tissues and upon the death of that living tissue, no more ^{14}C is assimilated, but it does decay steadily, making it possible to date that item.

Carbon has a unique property of being able to form bonds between its atoms creating stable compounds such as chains, branched chains and rings, termed catenation. Moreover, there are greater than an incredible six million of such known compounds. The compounds of carbon and hydrogen form the basis of organic chemistry.

Carbon has a number of distinct molecular or crystalline forms termed allotropes (or polymorphs), which include graphite, diamond and the more recently discovered fullerenes. Although the allotropes are all built from carbon atoms, they vary in structure to give widely different properties. For example, diamond is the hardest known material, whereas graphite is very soft.

1.2 STRUCTURE OF CARBON

1.2.1 Structure of the atom

It is a fact that all matter is made of atoms and we shall see that matter is a myriad of moving particles [1]. Based on Rutherford's concept, an atom consists essentially of two parts, a relatively heavy positively charged nucleus with a diameter of approximately 10^{-13} cm and surrounded by a diffuse arrangement of electrons, some 10^{-8} cm in diameter, which are extremely distant from the atom's inner core. The actual volume of the nucleus and the electrons within the atom is only about 10^{-12}–10^{-15} of the effective atomic volume. To help envisage this scale, if an atom was magnified to say, the size of a large room, then the nucleus would be a small speck just discernible to the naked eye but, nearly all the mass of the atom would be in that small nucleus [1]. This extranuclear arrangement of electrons determines its chemical behavior and establishes the optical spectra. The nucleus comprises a number of positively charged protons equal to its atomic number (Z) and a number of neutral neutrons (N) equal to the mass number (A) minus the atomic number (Z). The mass number is the nearest whole number to the actual mass and equal to the sum of neutrons and protons, $A = Z + N$.

The physical properties of the three basic particles are given in Table 1.1. However, modern theory has postulated, and in many cases has proved, the existence of about thirty subatomic particles e.g., baryons, mesons, leptons, photons and gravitrons [2]. In addition, all particles have anti-particles, except when the particle is its own anti-particle.

The neutron, postulated by E. Rutherford in 1920 and discovered by J. Chadwick in 1932, can be regarded as a combination of a proton and an electron, but the transition is really more complex, involving mesons and neutrinos (a neutral lepton):

$$\text{Neutron} \xrightleftharpoons[e^-(\text{electron})]{e^+(\text{positron})} \text{Proton}$$

where the positron is the anti-particle of the electron.

1.2.2 Atomic spectra and quantum theory

Energy which can travel through vacuum, such as light and X-rays, is termed electromagnetic radiation and because it has wavelike properties, it will have a wavelength (λ, the distance between two nodes), and a frequency (ν, the number of vibrations per second). If c is the velocity of that radiation, where c is equal to the velocity of light (2.998×10^8 ms^{-1}), then $c = \nu\lambda$. Each type of radiation has a characteristic wavelength and frequency (Figure 1.1) [3] coming in discrete packets, termed quanta that are in the form of

Table 1.1 Physical properties of basic subatomic particles

Name	Mass No.	Mass (kg)	Relative charge	Charge (coulomb)
proton	1	1.673×10^{-27}	+1	$+1.6 \times 10^{-19}$
neutron	1	1.675×10^{-27}	0	0
electron	0	0.911×10^{-31}	−1	-1.6×10^{-19}

STRUCTURE OF THE CARBON ATOM

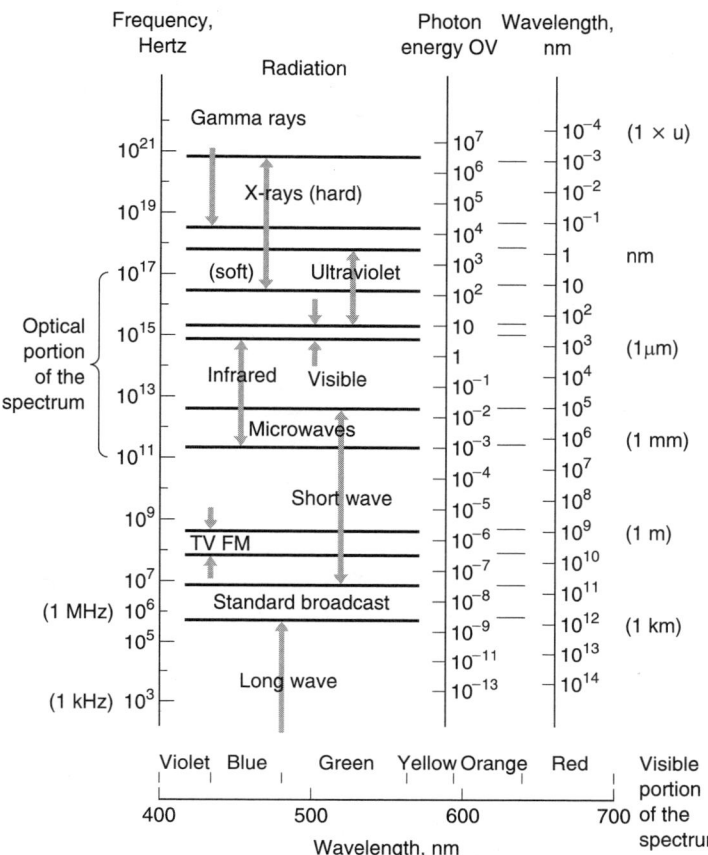

Figure 1.1 The electromagnetic spectrum. *Source:* Reprinted with permission from Pierson HO, *Handbook of Carbon, Graphite, Diamond and Fullerenes*, Noyes Publications, Park Ridge, NJ, 1993. Copyright 1993, William Andrew Publishing.

visible light, termed photons, which are never at rest. Each quantum has associated energy (E) given by $E = h\nu$, where h is Planck's constant (6.626×10^{-34} Js).

A glow discharge occurs when a gas is subjected to a low pressure and high potential, generating a stream of rays that is emitted from the cathode. These rays travel in straight lines, normal to the cathode, possessing sufficient momentum to rotate a small paddle wheel and, also, the rays can be deflected in electric and magnetic fields. J.J. Thompson (1898) showed that the rays were composed of particles with a negative charge, which were later termed electrons.

If cathode rays impinge on matter in a highly evacuated discharge tube, then X-rays are produced, which have no charge, can be diffracted similar to light, but have a much shorter wavelength and are also electromagnetic in character.

G.P. Thompson (1927) showed that electrons, analogous to light and X-rays, can be diffracted using a very thin metallic film and hence, must be associated with wave properties which, interestingly, were shown to be particles by his father some thirty years earlier.

E. Schrödinger (1926), following the earlier work of L. deBroglie (1924), advanced a fundamental equation of wave mechanics. The Schrödinger equation of wave mechanics was developed for waves oscillating in three dimensions with co-ordinates x, y and z:

$$\left(\partial^2/\partial x^2 + \partial^2/\partial y^2 + \partial^2/\partial z^2\right)\psi = -8\pi^2 m/h^2 (E - V)\psi$$

where ψ = the amplitude function, wave function or eigen function
m = mass
E = total energy equal to the sum of kinetic and potential energies
h = Planck's constant
V = potential energy

There is a value of ψ for every energy level and an associated set of quantum numbers. Solutions of this equation produce wave functions (ψ) which are mathematical equations representing the time and space variations in the amplitude of the wave system. This wave characteristic does not imply that there are actual waves associated with the electron but rather, the wave properties using this mathematical relationship determine the probability of finding an electron at a given position within the structure of the atom. Each wave function represents a discrete energy state for an electron within an atom.

At about the same time, Werner Heisenberg introduced another approach to quantum theory, which was subsequently amended in conjunction with Max Born and Pascual Jordan and matrix mechanics was developed.

An electron can assume a dual role of wave and particle functions (wave-particle duality) and W. Heisenberg (1927) introduced his uncertainty principle which dictates that one cannot know both where a particle is and what it is doing, since there is a limit of accuracy with which position and momentum can both be measured, because measuring one alters the other in an unpredictable way. Niels Bohr argued that although wave and particle representations are mutually exclusive, they are not contradictory and instead, are complementary. For most purposes in chemistry, the electron can be regarded as a particle.

The Schrödinger equation, however, does not take into account relativity, which is important when particles move at about the speed of light. So, P.A.M. Dirac (1928) advanced his theory of relativistic quantum theory, introducing time as the fourth dimension.

Quantum theory could now be expressed in two totally different ways: wave mechanics or matrix mechanics and it was not until the early 1930s that John von Neumann supplied a mathematical connection between wave and matrix mechanics. Not surprisingly, the study of quantum mechanics is fraught with the unknown and an eminent American theoretical physicist Richard Feynman said 'I think I can safely say that nobody understands quantum mechanics' [4].

Incandescent solids emit continuous spectra but, gases or vapors, under the same conditions, exhibit spectra which take the form of distinct lines or bands with definite position or wavelength. The line spectra result from atoms and are termed atomic spectra. The bands resulting from molecules can generally be resolved into a series of closely spaced lines termed band spectra. The lines can be accounted for by transitions between a limited number of electronic states of an atom and these transitions give a picture of the possible electron orbits within that atom.

The energy of a body consists of a definite whole number of quanta and that energy can only be taken up or given out in such units, for example a photon, which is a quantum of light. The state of an atom is determined by the possible quantum states of the system and is basically described by four quantum numbers: n, l, m_l and m_s. The principal quantum number (n), distinguishes the energy level and crudely represents the mean distance of the electron from the nucleus, where n can have any integer value from 1–∞. An infinite value corresponds to the complete removal of that electron from the atom and the production of a positive ion, where the ionization potential is the energy required to accomplish such a transition. An atom is normally found in the state of lowest energy, termed the ground state and can 'jump' to another state by absorbing energy. The value of n denotes the electron shell. When $n = 1$, the electrons are in the K shell and when $n = 2$, 3, 4 or 5, the electrons are in

STRUCTURE OF THE CARBON ATOM

the L, M, N or O shell respectively. The maximum number of electrons in a given shell is given by $2n^2$ (where n is the principal quantum number for that shell). Hence, the maximum number of electrons in the K, L, M, N and O shells are 2, 8, 18, 32 and 50 respectively.

The energy due to orbital motion about the nucleus is called the subsidiary or azimuthal quantum number (l) and may have any integer value from 0 to $n-1$. When $l=0$, the electrons in orbit are described as s electrons and when $l=1$, 2, 3 or 4, they are termed p, d, f or g electrons. The designations originate from terms used in early spectra work, namely sharp, principal, diffuse and fundamental.

The wave function associated with the orbital motion of an electron is called an orbital. Electron orbits are not circular but can be considered as periodic motion under the influence of a central atomic nucleus leading to an elliptical orbit. The circle is, however, a special case of an ellipse where the major and minor axes are equal. An electron rotating in an elliptical orbit produces magnetic flux and has an electromagnetic moment. The movement of an electron in an elliptical orbit is not uniform and is greatest when closest to the nucleus. So, according to the theory of relativity, the effective mass of the electron varies throughout its orbit and the path does not return on itself, thereby producing a precessional movement of the electron orbit.

Certain spectral lines are split up in a strong magnetic field (Zeeman effect). The vector quantity describing the orbital angular momentum undergoes a precessional movement and describes a cone about an axis in the direction of the magnetic field (c.f. a gyroscope in a gravitational field) and to explain such phenomena, it was necessary to postulate an additional magnetic quantum number (m_l) which can have any integral value from -1 to $+1$ and has, therefore, $2l+1$ possible orientations. For example, for a p electron $l=1$ and $m_l=-1$, 0 and $+1$.

Due to the spin of the electron, each electron has an added angular momentum and in an applied magnetic field, the vector representing this spin momentum can also orient itself so that its component m_s (spin quantum number) is either in the direction of the field (parallel) or opposed to it (anti-parallel) and has the value $\pm\frac{1}{2}$. For each combination of (n, l, m), two electrons differing only in spin are possible or, putting it another way, two electrons differing in spin may exist for each value of m. Pauli's exclusion principle states that no two electrons in the same atom can have all four quantum numbers the same.

From spectral studies, electrons in an atom arrange themselves into a series of groups or shells (Table 1.2).

Generally speaking, the configurations of these shells are:

K shell $\quad 1s^2$

L shell $\quad 2s^2 \quad 2p^6$

M shell $\quad 3s^2 \quad 3p^6 \quad 3d^{10}$

N shell $\quad 4s^2 \quad 4p^6 \quad 4d^{10} \quad 4f^{14}$

Table 1.2 Allocation of quantum numbers in the first two shells

Shell	n	Orbit	l (0 to $n-1$)	m_l (-1 to $+1$ incl. 0)	s ($\pm\frac{1}{2}$)	Number of electrons in shell	Total number of electrons in orbit
K	1	s	0	0	$\pm\frac{1}{2}$	2	2
L	2	s	0	0	$\pm\frac{1}{2}$	2	
		p	1	$+1$	$\pm\frac{1}{2}$		8
				0	$\pm\frac{1}{2}$	6	
				-1	$\pm\frac{1}{2}$		

The number of electrons in an atom is equal to the atomic number, which for carbon is 6 and the electronic ground state for carbon has the configuration $1s^2$, $2s^2$, $2p^2$. Of the six electrons in the neutral atom, four are available for the formation of chemical bonds in the outer L shell. When electrons enter a level of fixed n and l values, according to Hund's maximum multiplicity rule, the available orbitals are occupied singly until each orbit is so occupied before electron pairing occurs.

The probability of locating an electron at a given site can be calculated or, setting aside the notion of the electron as a particle, we can calculate the probability of distribution or charge density throughout space. Each energy level has a given wave function together with associated quantum numbers and the charge density is calculated from $(\psi)^2$. It is possible from the solution of the Schrödinger equation to have two values for ψ namely, $+\psi$ and $-\psi$ and these are associated with the spin orbitals. So, each orbital can contain a maximum of two electrons which will occupy the same region in space since $(-\psi)^2$ is equal to $(+\psi)^2$. When two electrons are in the same orbital, they are said to have paired spins, whilst a single electron has an unpaired spin. Position can only be described statistically in terms of orbitals (the functions determining spatial distribution) and the shapes of the associated electronic distribution probabilities.

There are regions in space that are reserved for electrons, but not occupied, termed vacant and can be occupied when an electron is excited and raised from a lower energy level.

1.2.3 Directional characteristics of atomic orbitals

Orbitals characterizing an atom differ from each other in terms of r, the distance of the electron from the nucleus. They are also described by their angular distribution in space relative to the nucleus, which is characteristic of a particular orbital type, regardless of the magnitude of the principal quantum number. The electrons can be considered as a charge cloud with varying density and determined by the wave function.

The single $1s$ electron of hydrogen has its most probable location within a sphere whose radius is equal to the radius of orbit of the electron with the nucleus at the center and has no directional characteristics (Figure 1.2).

With p, d and f orbitals, the situation is more complex and they are found in sets of 3, 5 and 7. It is impossible to determine the direction of any one orbital in a given set. However, the axes along which a given set lies are at definite angles to each other in space, with the nucleus at their intersection. Hence p orbitals with the nucleus at the center are at mutually right angles and lie along three cartesian co-ordinates p_x, p_y, p_z, emphasizing the directional character. An orbital, such as a p type, is dumb-bell shaped and has two parts, each situated on either side of a node, which is a region where the probability of finding an electron would be very remote (Figure 1.3). A mathematical function is said to have a node when it changes sign. The two nodes (also termed lobes) are labeled positive and negative representing

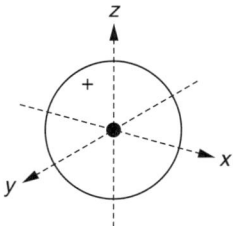

Figure 1.2 Probable location of 1s electron in hydrogen.

STRUCTURE OF THE CARBON ATOM

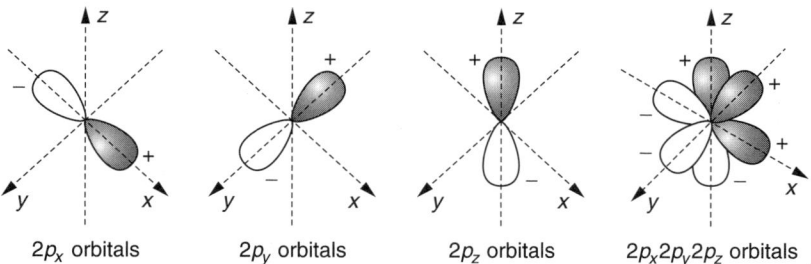

Figure 1.3 Probable location of electrons in 2p orbitals.

the signs of the wave function as determined by the wave equation and it must be emphasized that they do not indicate the sign of the electric charge since, by definition, an electron must possess a negative charge. A simple picture of an electron orbital is to consider the orbital as an electron cloud with the densest region representing the vicinity where the possibility of finding the electron is greatest. d orbitals may be considered to be directed along the slanting edges of a pentagonal pyramid with the nucleus at the apex but, the orbitals are not mutually equivalent. No modular picture is available for f orbitals.

1.2.4 Hybridization of atomic orbitals

The electrons in the outermost orbit are available for bonding to other atoms and are termed valence electrons. L. Pauling and J.C. Slater (1931) developed a method of directed valence bonds which surmised that two atoms are joined with a covalent bond, with one electron being supplied by each atom to form the bond and where the electrons would have a high probability of being found on either atom. The direction of this hybrid bond would correspond to the direction in which the orbital wave functions of the two electrons concerned overlapped as much as possible and the greater this overlap of the wave functions, stronger the bond formed, although total overlap is not possible due to repulsion of the nuclei of the two atoms. With two unpaired spins, carbon would be expected to have a valency of two and although this does exist, it is, however, well established that carbon in the solid state has a valency of four and, to achieve this condition, one s electron is excited into a p state and covalent bonds are formed to give the configuration $1s^2 2s^1 2p_x^1 p_y^1 p_z^1$. The s and p states are so combined to form four new wave functions directed towards the apexes of a regular tetrahedron at a mean angle of 109° 28' to one another (Figure 1.4) and, since each hybrid comprises one s orbital and three p orbitals, they are called sp^3 orbitals. These are stronger than any other combination of s and p orbitals and form the basis of diamond and aliphatic compounds. The hybrid orbital has partial s and p orbital character.

It is possible that hybridization can occur with one s and two p electrons, with the third p electron not taking part and the three wave functions then have their maxima lying in one plane at 120° to one another (Figure 1.5) and are denoted as sp^2 orbitals, which form the basis of graphitic structures and aromatic compounds. If the two p orbitals involved in the hybridization are in the p_x and p_y planes then the three hybridized orbitals will lie in the xy plane and the third uninvolved p_z orbital will be oriented normal to this plane, available to form a π bond. Hybrid orbitals do not represent the structure of a carbon atom, but are formed in the process when carbon bonds with other atoms. Typical arrangements are listed in Table 1.3.

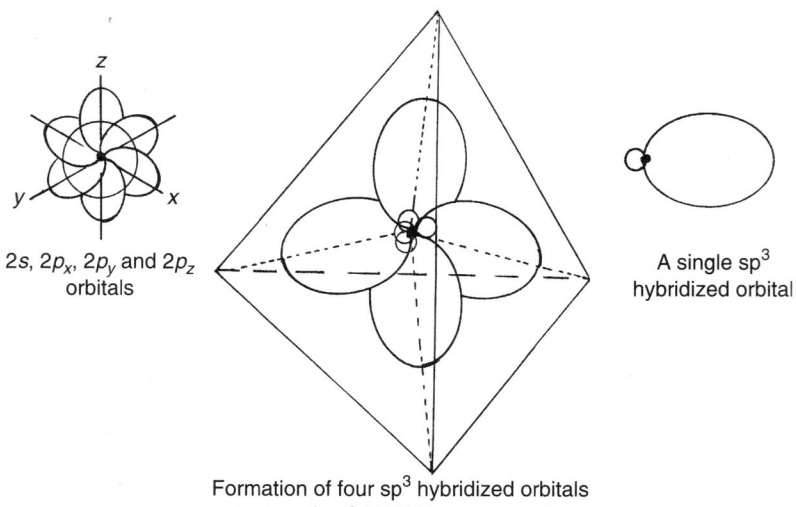

Figure 1.4 Formation of sp^3 hybridized orbitals.

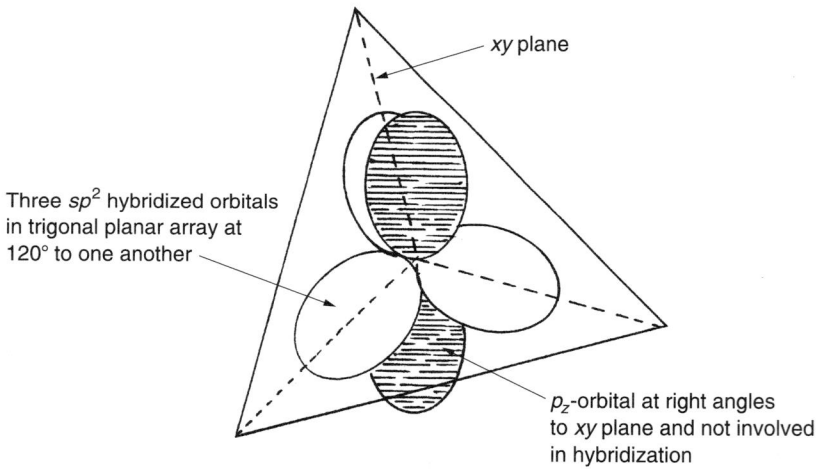

Figure 1.5 Formation of sp^2 hybridized orbitals.

Table 1.3 Arrangements of hybridized orbitals

Hybrid	Number of orbitals	Shape	Angle
sp	2	linear	180°
sp^2	2	planar	120°
sp^3	4	from center of a regular tetrahedron to its apexes	109°28'

STRUCTURE OF THE CARBON ATOM

1.2.5 Covalence and molecular orbitals

The Schrödinger wave equation can only be used for a system with one electron, such as the hydrogen atom and cannot solve for larger atoms and molecules. Two approximations are used; the valence bond and the molecular orbital methods.

The model of two atoms joined with a covalent bond is not used for interpreting molecular orbitals but, instead, consideration is given to the grouping of the atomic nuclei to give the minimum potential energy. As with atoms, it is presumed that there are definite electronic levels in the molecule, similar to those in the atom and N atomic orbitals will give N molecular orbitals.

Since normal quantization of the electrons cannot occur when two atoms approach so close as to produce chemical bonding, in this situation, the principal and azimuthal quantum numbers have little meaning. Corresponding to the classification s, p and d for values of the principal quantum number where $l=0$, 1 or 2, for values of $\lambda=0$, ± 1 or ± 2, where λ is the component of the orbital angular momentum along the line joining the atom centers, we have σ, π and δ electrons. The orbitals are filled up in order of increasing energy as given by z, y, x, etc. and the building-up principle (Aufbau) is used to fill the orbitals exactly the same way as for atoms. Each orbital can accommodate up to two paired electrons and should a choice of orbitals be available, and if they should have almost identical energies, then the orbitals are occupied separately and the electrons have parallel spins.

An s electron in an atom with $l=0$ will have zero orbital angular momentum along the line joining the two atoms. Hence $\lambda=0$ and the s electron becomes a σ electron. On the other hand, a p electron with $l=1$ can have values for λ of -1, 0 or $+1$ and can become a σ or π electron in the molecular orbital. Now for each value of λ, there are two possible values for s (the spin quantum number). Hence, two electrons can be inserted in each σ orbital and four in each π orbital (Table 1.4).

The formation of a molecule such as carbon monoxide from carbon and oxygen can be represented by

$$C[1s^2 2s^2 2p^2] + O[1s^2 2s^2 2p^4] \rightarrow CO[KK(z\sigma)^2(y\sigma)^2(x\sigma)^2(w\pi)^4]$$

In simplistic terms, molecular orbitals can be constructed by the linear combination of atomic orbitals. Constructive interference, termed bonding, enhances the electron density in the internuclear region with an increase in bond strength. In bonding, the energy of the molecule is lowered when occupied by electrons. Destructive interference, termed antibonding, where the energy of the molecule is higher than the two separate atoms, produces a nodal plane between the two nuclei, where the overlapping orbitals have opposite phases.

Table 1.4 Building up electron groups in a molecule

Complete electron shells in atom	Quantum number l	Orbital angular momentum along line joining atom centers λ	Complete electron groups in molecule
s^2	0	0	σ^2
p^6	1	0, ± 1	σ^2, π^4
d^{10}	2	0, ± 1, ± 2	σ^2, π^4, δ^4

To form a σ orbital, it is necessary for atomic orbitals that have cylindrical symmetry to overlap about the axis connecting the two atomic nuclei, which by convention is termed the z axis. It should be noted that a σ orbital is the equivalent molecular version of an s atomic orbital possessing cylindrical symmetry and when viewed along the internuclear axis, looks like an s orbital. Hence, σ orbitals can, for example, be formed from two atoms by the overlap of $2s$ and $2p_z$ orbitals (Figure 1.6, Example 2).

Just as with atoms, the orbitals start to fill up from the level of minimum potential energy but due to lack of accommodation in the molecule, the electron has to occupy a higher principal quantum than the individual atom and the electron is said to be promoted. This promotion requires a relatively large amount of energy and renders the molecule unstable. The respective electron is said to be antibonding (identified with an asterisk e.g., σ*), relating to a repulsive interaction due to the parallel spins of the electrons of the atoms creating the nodal plane. Electrons not requiring energy are termed bonding, with an enhanced probability of being found in the internuclear region and an ability to react strongly with both nuclei, resulting in orbital overlap. The two s electrons in the lowest state are said to be nonbonding. Two or more quantum states having the same energy are said to be degenerate. The $2s$ orbitals overlap to give bonding and antibonding σ orbitals, as do the two $2p_z$ orbitals (Figure 1.6, Examples 1 and 3a, b). The remaining $2p_x$ and $2p_y$ orbitals, which have a nodal plane through the x-axis, overlap to give bonding and antibonding π orbitals (Figure 1.6, Examples 6 and 7).

For simple diatomic molecules of the first ten elements in the periodic table, the sequence of filling the molecular orbitals is: $\sigma 1s$, $\sigma^* 1s$, $\sigma 2s$, $\sigma^* 2s$, $\sigma 2p$, $\pi_y 2p = \pi_z 2p$, $\pi^*_y 2p = \pi^*_z 2p$ and $\sigma^* 2p$, where the relative energies of the σ and π orbitals depend on the molecule and position in the periodic table. So, carbon with one $2s$ and three $2p$ orbitals involved in bonding can hybridize and mix in three ways:

  i. four sp^3 hybrid orbitals
 ii. three sp^2 hybrid orbitals and one p orbital
iii. two sp hybrid orbitals and two p orbitals.

Single, double and triple carbon bonds can be formed by these different types of hybridization. The physical properties of carbon-carbon bonds are given in Table 1.5. Multiple bonds are shorter, due to the extra attraction caused by more electrons being shared, hence pulling the atoms closer until the repulsion of the electron clouds becomes predominant.

Ethene has one double bond ($H_2C=CH_2$) and each carbon atom is connected via three sp^2 orbitals: two spin pair with $1s$ hydrogen electrons and the third is shared with the other carbon atom, giving a total of five σ bonds in one plane. Each carbon atom has, at right angles to this plane, two further p orbitals, each with one electron and these orbitals take no part in the σ hybridization. The p orbitals can overlap sideways and form a π orbital in the form of two dogbone or sausage shaped sections, one above the σ orbital plane and the other below (Figure 1.7) [5], where the plane represents a node for the π orbital. Maximum overlap of the p orbitals prevents rotation about the double bond.

Ethyne has a triple bond (HC≡CH) and each carbon atom is connected by way of sp hybridization. The four atoms lie in a straight line and in this instance, each carbon atom will have two electrons with unpaired spins remaining, each situated in a p orbital, which spin pair to form two π orbitals (Figure 1.8).

Summarizing, if atomic orbitals overlap, then molecular orbitals are formed, where the electron cloud surrounds two or more atoms rather than one, as with atomic orbitals. Nuclei vibrate but with very low amplitude (about 0.01 nm) and can be considered to occupy their

STRUCTURE OF THE CARBON ATOM

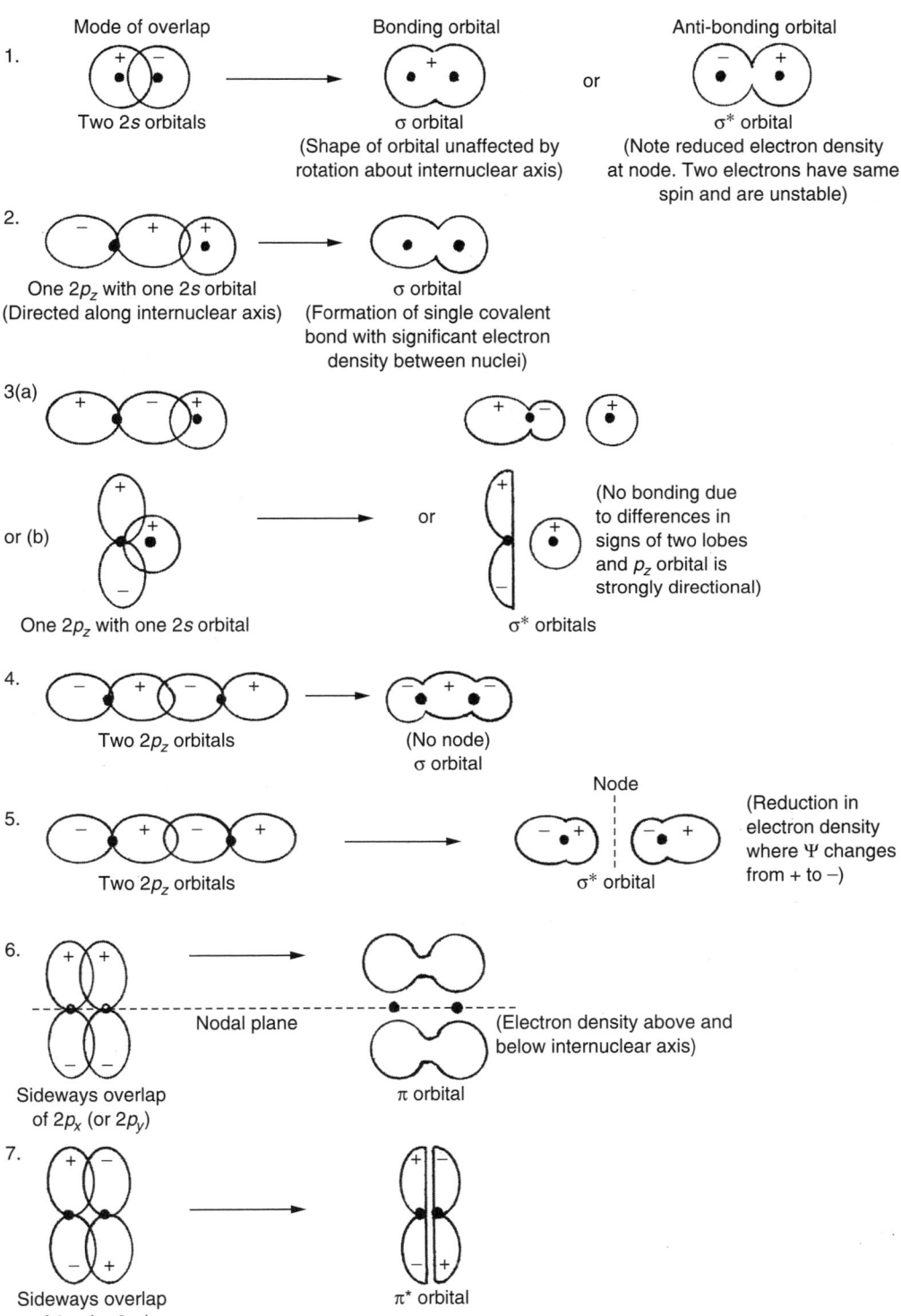

Figure 1.6 Schematic planar representation of σ and π orbitals.

Table 1.5 Physical properties of carbon-carbon bonds

Type of bond	Molecular orbitals involved	Bond length nm	Approximate bond energy kJ mol^{-1}			
			σ	First π	Second π	Total
C—C	σ	0.154	347			347
C=C	σ + π	0.134	347	265		612
C≡C	σ + two π	0.120	347	265	226	838

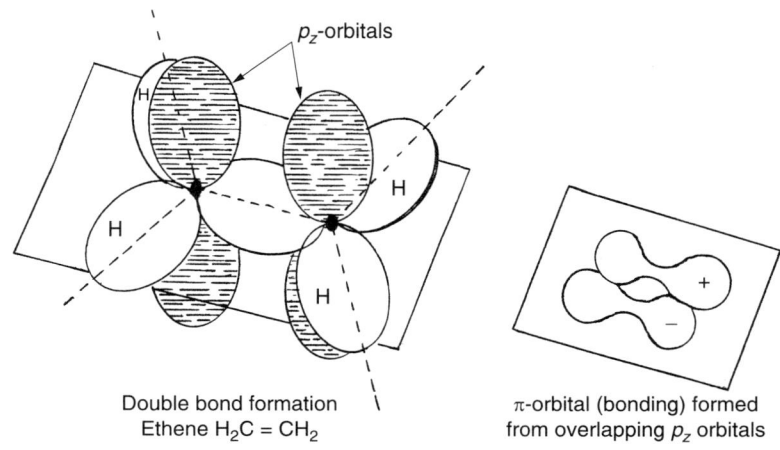

Figure 1.7 Formation of double bond in ethane. *Source:* Reprinted with permission from Savage G, *Carbon-Carbon Composites*, Chapman and Hall, London, 7, 1993. Copyright 1993, Springer.

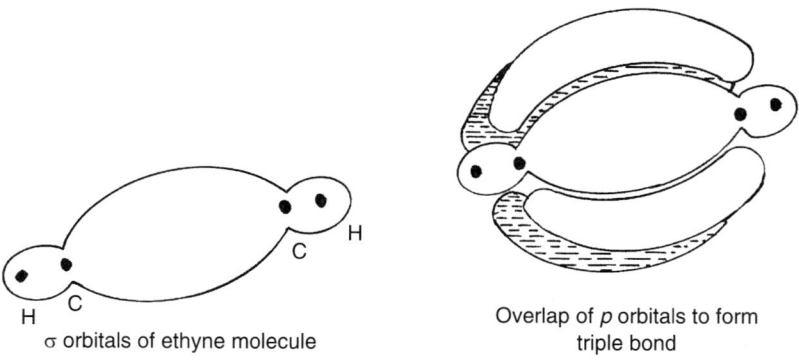

Figure 1.8 Formation of triple bond in ethyne.

mean position. Bond lengths, however, are an order of magnitude larger (0.1–0.3 nm) and the shape of the bond is the shape of the charge cloud associated with that bond.

Practically, the position of the nucleus can be determined by measuring the vibration-rotation spectra and neutron diffraction, whereas the electron cloud can be determined by X-ray scattering (for crystals), electron diffraction (for the gaseous phase), together with electron spin resonance and nuclear magnetic resonance.

REFERENCES

1. Feynman RP, *Six Easy Pieces*, Penguin Books, London, 1995.
2. Polkinghorne JC, *The Particle Play*, WH Freeman and Company, Oxford, 1981.
3. Pierson HO, *Handbook of Carbon, Graphite, Diamond and Fullerenes*, Noyes Publications, Park Ridge, NJ, 263, 1993.
4. Feynman RP, *The Character of Physical Law*, MIT Press, Cambridge, MA, 1967.
5. Savage G, *Carbon-Carbon Composites*, Chapman and Hall, London, 7, 1993.

CHAPTER 2

The Forms of Carbon

2.1 THE ALLOTROPES OF CARBON

Diamond and graphite are the two most common allotropes of carbon comprising infinite-network solids. In diamond, the carbon atoms are sp^3 hybridized and joined by four strong covalent bonds each 0.154 nm long, and joined to four other carbon atoms pointing towards the corners of a regular tetrahedron. Graphite is a layered structure with three 0.1415 nm-long strong bonds within the layer that are sp^2 hybridized with one electron able to take on a dual role, forming co-planar and interplanar bonding. These distributions are termed π and σ respectively. The layers are themselves weakly bonded by what can be described as van der Waals forces, which enable the layers to slide over one another. In 1985, the fullerenes, a new allotropic form of carbon, were discovered [1], comprising both sp^2 and sp^3 hybridization and are true carbon molecules e.g., C_{60}.

The relationship between the allotropes of carbon is shown in Figure 2.1.

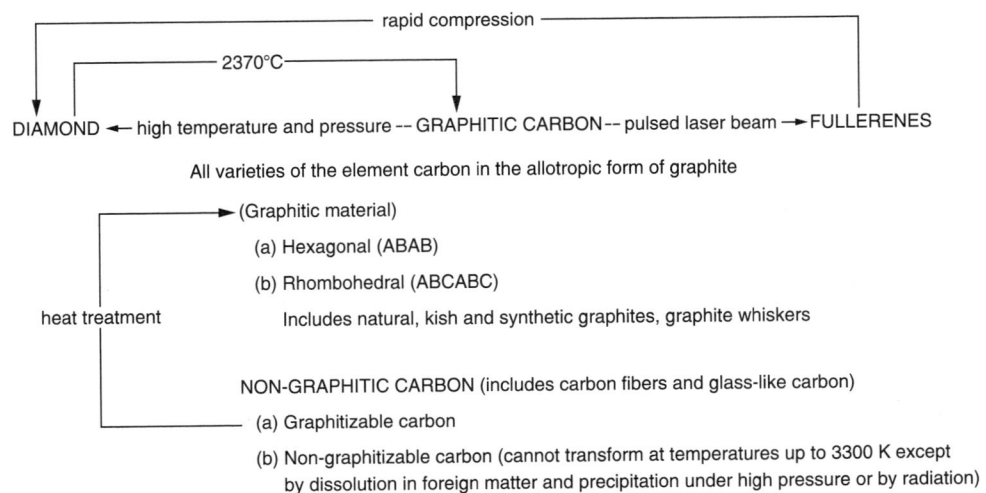

Figure 2.1 The allotropes of carbon.

2.2 THE CARBON PHASE DIAGRAM

The phase diagram for carbon (Figure 2.2) is not clearly defined due to the possible doubtful measurements at the high pressures and temperatures involved. The triple point for graphite (Figure 2.3), where solid, liquid and gas are in equilibrium, is about 4180 K at 10.13 MPa. Carbon vaporizes about 4500 K and 1 kb (100 MPa) or, at 200 K, a pressure of 1000 kb (100 GPa) would be required. Although diamonds can be made at 10 kb (1 GPa) and 1000 K, conversion would be extremely slow and production can be effectively speeded up with catalysts such as nickel, at 100 kb (10 GPa) and 2000 K. Strictly SI does not use multiples and submultiples so $N/m^2 \times 10^6$ preferred to MN/m^2. But this is unwieldy and I have used multiples etc and Pa instead of Nm^{-2} to come into line with composite nomenclature.

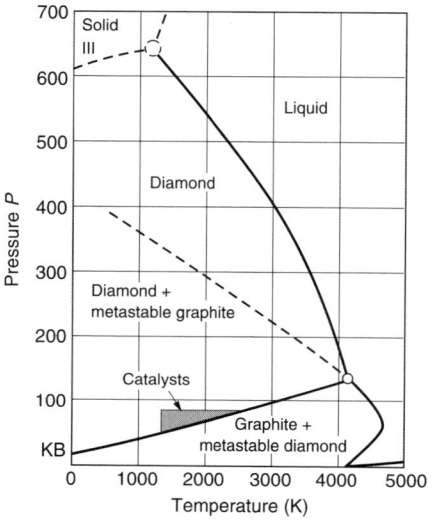

Figure 2.2 Phase diagram of carbon according to Bundy. *Source:* Reprinted from Bundy FP, *J Chem Phys*, 38: 618, 1963.

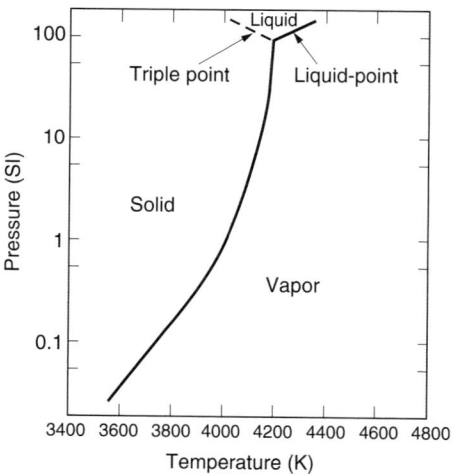

Figure 2.3 Vapor pressure and triple point of graphite. *Source:* Reprinted with permission from Palmer HB, Shelef M, In: Walker RL Jr., ed. *Chemistry and Physics of Carbon*, Marcel Dekker, New York, 4: 1968. Copyright 1968, CRC Press Boca Raton, Florida.

THE FORMS OF CARBON

2.3 DIAMOND

A number of useful books on the properties of diamond have been published [4–8].

2.3.1 Occurrence, production and uses of diamond

2.3.1.1 Natural diamonds

Natural diamonds were reported in India about 700BC, but the technique of cutting and polishing to emphasize their beauty was not established until the late 14th century. The major source of natural diamond is the mineral kimberlite, which is found in igneous rocks and mined in Africa, Australia and Russia. Typically, only some 32 carats (6.4 g) of diamond are produced from 100 tons of ore [4], the diamonds are identified by an X-ray beam, which causes the diamonds to luminesce, facilitating their separation and collection. A very small part of this output is selected for gemstones, which will lose a further 50% of the weight when they are cut and the unselected material is then graded for industrial purposes. Figure 2.4 shows a rough diamond. Although financially some 93% of turnover relates to the gemstone business, it represents only 1% by weight of the diamond output.

In the industrial sector, single crystals present an uninterrupted diamond cutting edge to give a surface finish to the workpiece better than 0.025 µm (1µinch), but new developments with synthetically produced diamonds are rapidly displacing the natural product.

2.3.1.2 High pressure synthetic diamonds

Obviously, with the high cost of natural diamonds, there has always been a desire to manufacture diamonds synthetically. Examination of the phase diagram (Figure 2.2) shows that very high temperatures and pressures are required to synthesize diamond from graphite and, fortunately, this can be avoided by using a metal catalyst that itself acts as a solvent for the diamond and the less soluble diamond phase crystallizes out. Initially, transition metals

Figure 2.4 Rough diamond. *Source:* Courtesy of CSO Valuations AG (De Beers).

such as iron, cobalt and nickel were used but, these have now been displaced by alloys such as Invar (Fe-Ni) or Co-Fe. Pressure can be applied using a hydraulic press and single crystal diamonds up to about 8 mm diameter can be made commercially by this technique. Alternatively, pressure can be generated by the application of a shock wave by detonating TNT, although this procedure yields smaller diamonds with a particle size of about 60 μm. However, making larger diamonds is an expensive process, but synthetic diamonds are consistent and tool blanks can be sawn to give platelets up to 2 mm thick and 8 mm long.

2.3.1.3 *Polycrystalline diamond (PCD)*

Single crystal diamonds are bonded with Co or Ni under high pressure and high temperature to produce PCD, which has superior toughness to single crystal diamond, but due to the presence of cleavage planes, tends to be brittle and may fail under impact. PCD does, however, have a temperature limitation of 700°C due to the presence of the Co or Ni, which can promote reverse synthesis, i.e., conversion of diamond back to graphite. PCD diamond has become a high performance replacement for tungsten carbide tooling.

2.3.1.4 *Chemical Vapor Deposition (CVD) diamond*

Coatings of diamond can be produced on a substrate by the reaction of carbon based gases (such as methane) with atomic hydrogen, in the presence of traces of oxygen, to give polycrystalline diamond, which grows in a columnar structure and is fully dense. CVD diamond has greater thermal stability than PCD diamond, but is relatively brittle and the lack of electrical conductivity precludes the use of spark erosion. There are two formats: a thin film (<30 μm) deposited on, say, a twist drill and thick films (upto 1 mm) deposited on a substrate. Substrates used are refractory materials like Si, W, Ta, Mo, WC, SiC and tool steels. The substrate can be subsequently removed, if desired, to leave a free-standing thick film of polycrystalline diamond. High temperatures in excess of 2000°C are required to produce atomic hydrogen and various techniques are used to attain such temperatures such as resistive heating of a metal filament, an oxy-acetylene torch, or some form of plasma, which can either be an arc or a glow discharge type.

Hollow diamond fibers have been produced by using the CVD process to coat diamond onto a wire, preferably with a relatively low thermal expansion coefficient and having carbide-forming properties [9]. Wires of W, Ti, Ta or copper can be used or fibers of silica and silicon carbide, which are initially abraded with diamond grit to provide nucleation sites. The metals can be etched away in a suitable reagent such as nitric acid for Cu and hydrogen peroxide for W. The diamond deposit has a tensile modulus of about 880 GPa.

Diamond has also been deposited onto carbon fiber [10,11]. Shah and Waite deposited 1–30 μm films of diamond on PAN and pitch carbon fibers with diameter as small as 5 μm by microwave H-plasma assisted chemical vapor deposition. An ultrasonic pretreatment of the fibers in a bath containing methanol and diamond paste was required for successful coating of the fibers.

2.3.1.5 *Diamond-like carbon (DLC)*

DLC is a product which is neither diamond nor graphite. It has properties similar to CVD diamond, but does not require a high temperature substrate. DLC is produced at low pressure by physical vapor deposition, such as ion beam sputtering using argon and a solid

THE FORMS OF CARBON

Table 2.1 Classification of diamonds

Type	Impurities	Occurrence	Special properties
Ia	About 0.1% N_2	Most natural diamonds	N_2 segregated into small platelets
Ib	Contains N_2	Almost all synthetic diamonds	N_2 in dispersed form
IIa	Free of N_2	Very rare in nature	Enhanced thermal and optical properies
IIb	Very pure or add B to synthetic	Extremely rare in nature	Generally blue in color Imparts semiconductivity

Source: Adapted from *Properties of Diamond*, a publication of De Beers Industrial Diamond Division, 1989.

carbon target as the cathode [12]. DLC is, at present, only available in thin coatings, with the added advantage of a very smooth surface but, unfortunately, it does have a maximum usage temperature of about 250°C.

2.3.2 Classification of diamonds

Diamond is composed of the element carbon and only nitrogen and boron with atomic radii smaller than carbon are known, with certainty, to be incorporated in the diamond lattice. Almost every diamond is different, but the following general classification based principally on optical properties (Table 2.1) can be applied.

2.3.3 Identification of diamond

Diamond and graphite cannot be positively identified using electron diffraction techniques alone, but Raman spectroscopy will, however, clearly differentiate the sp^2 bonding of graphite and the sp^3 bonding of diamond.

2.3.4 The crystal structure of diamond

Diamond is the perfect example of an atomic crystal or giant molecule, where there is complete electron pair covalent bonding, which links all atoms in all directions in space. Diamond can occur in several crystal forms and these are classified using the crystallographic notation for the simple planes of a cubic crystal (Figure 2.5). Diamond has three major crystal forms: cubic (100 plane), dodecahedral (110 plane) and the octahedral form (111 plane), which are shown in Figure 2.6. Both cubic and octahedral forms occur in high pressure synthetic diamond and CVD diamond.

In Section 1.2.4, it was shown how the four sp^3 hybridized orbitals of carbon could form a regular tetrahedron (see Figure 1.4) with a bond length of 0.154 nm. Each tetrahedron can then combine with four other tetrahedra to form the strongly bonded covalent structure of diamond (Figure 2.7), which can have either cubic or hexagonal symmetry.

The cubic structure (Figure 2.8) is the most common form and has an ABCABC stacking sequence, where each third layer is identical (Figure 2.9). The unit cell has a total of eight carbon atoms comprising eight at the corners (each shared with eight adjacent cells), six at each face (each shared with an adjacent cell) and four within the cell (positioned on the four diagonals between the corners of the unit cell).

The hexagonal form of diamond is similar to the cubic form, except that it has an ABAB stacking sequence (Figure 2.10) and occurs naturally in the mineral lonsdaleite.

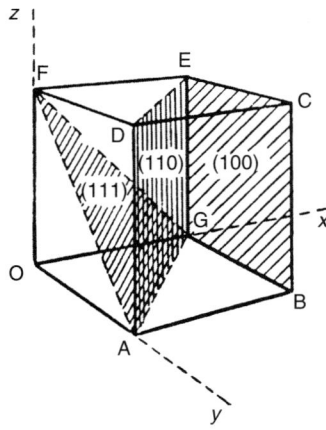

Figure 2.5 Indices of the simple planes in a cubic crystal.

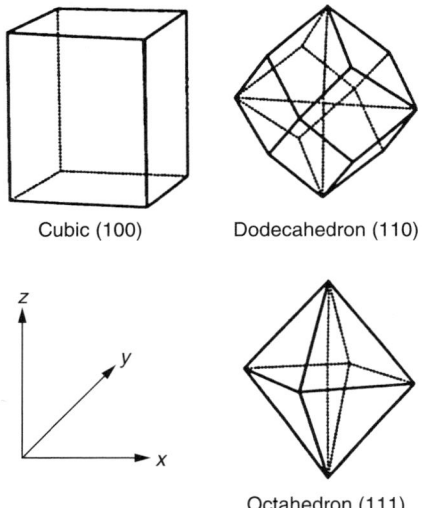

Figure 2.6 The major crystal forms of diamond. *Source:* Reprinted with permission from Pierson HO, *Handbook of Carbon, Graphite, Diamond and Fullerenes*, Noyes Publications, Park Ridge NJ, p. 251, 1993. Copyright 1993, William Andrew Publishing.

Diamond does have impurities which impair the optical performance. These impurities can be either a form of inclusion, or a lattice impurity such as nitrogen or boron, which are adjacent to carbon in the periodic table, have similar atomic radii and therefore, easily fit into the diamond structure. Nitrogen tends to form a nitrogen pair where the nitrogen atoms are adjacent within a unit cell or, alternatively, the nitrogen could form platelets.

2.3.5 The properties of diamond [7]

2.3.5.1 Density

Diamond is an isotropic material with a density of $3.515\,g\,cm^{-3}$ and is more compact than graphite, which has a density of $2.26\,g\,cm^{-3}$. Diamond is unique, having the highest atom

THE FORMS OF CARBON 21

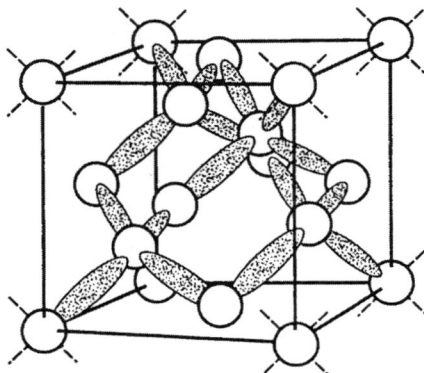

Figure 2.7 Three dimensional representation of sp^3 covalent bonding in diamond. The shaded areas have high electron probabilities when covalent bonding occurs. *Source:* Reprinted with permission from Pierson HO, *Handbook of Carbon, Graphite, Diamond and Fullerenes*, Noyes Publications, Park Ridge NJ, p. 31. 1993, Copyright 1993, William Andrew Publishing.

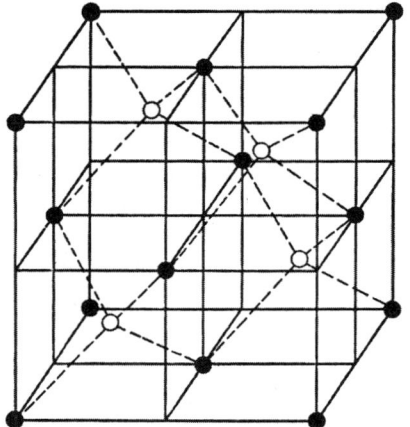

Figure 2.8 Cubic structure of diamond.

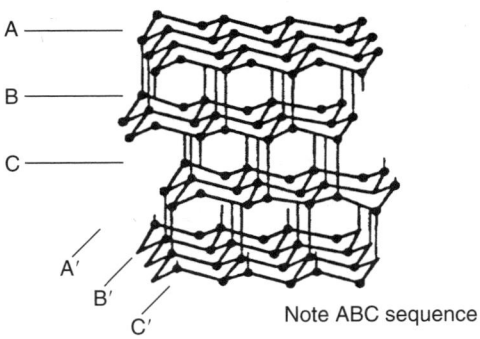

Figure 2.9 ABCABC stacking sequence for cubic form of diamond. *Source:* Reprinted from Cullity BD, *Elements of X-ray Diffraction*, Addison Wesley, Reading MA, 1956.

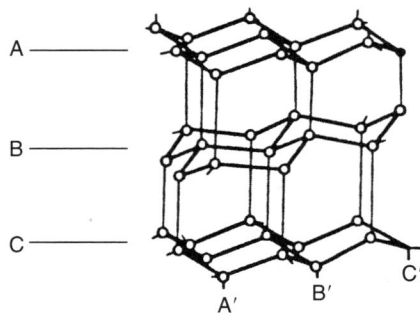

Figure 2.10 ABAB stacking sequence for hexagonal diamond. *Source:* Reprinted from Eggers DF Jr, Halsey GD Jr, *Physical Chemistry*, John Wiley, New York, 1964.

density for any material (0.293 g atom cm^{-3}), which accounts for its hardness, stiffness and resistance to compression.

2.3.5.2 Mechanical properties

2.3.5.2.1 Hardness

Diamond is the hardest material known, with a value of 10 on the Mohs scale, which is a scratch hardness test or, on the Knoop scale, which is an indentation test dependent on the load, indenter shape and the crystal face, giving a value of 5,700–10,400 kg mm^{-2}. The hardness is attributed to the strength of bonding of the atoms in conjunction with the uniformity. There is, however, a possibility that there are 'new' materials, such as carbon nitride (C_3N_4) and compressed C_{60} that may eventually be shown to be harder than diamond.

2.3.5.2.2 Friction

The coefficient of friction (μ) in air is low and about 0.05–0.1. The lubricity is similar to PTFE, but it is highly dependent on the crystal geometry and load. In vacuum, however, μ is 1.0.

2.3.5.2.3 Elastic properties [7]

Diamond has the highest values of elastic moduli of any material and the typical properties are:

C_{11} (longitudinal)	10.76×10^2 GPa
C_{12} (transverse)	1.25×10^2 GPa
C_{44} (shear)	5.77×10^2 GPa
K (bulk)	4.42×10^2 GPa
γ_{21} (Poisson's ratio)	0.104
E_{11} (Young's modulus)	10.50×10^2 GPa

Note: The condition for isotropy is $2C_{44}/C_{11} - C_{12} = 1$, so substituting values for the constants, this ratio becomes 1.21, which means that the Young's modulus does not vary much with orientation.

2.3.5.2.4 Strength

Since diamond behaves as a brittle material, it will readily split, or cleave, normally on the 111 plane, surprisingly, at very high velocities of several thousand meters per second. This property of cleaving is used with great dexterity by skilled diamond workers to transform the crude diamond into a scintillating gemstone. Values for tensile strength are varied and are in the range 10–19 GPa. Shear strengths are very difficult to measure, as is the compressive strength and values for compressive strength are marginally below the tensile strength.

2.3.5.3 Thermal properties

Diamond has the highest thermal conductivity, the value depending on the classification of the diamond. It is highest for a Type IIa diamond (up to $1.5 \times 10^4 \, W\,m^{-1}\,K^{-1}$). The linear thermal expansion is low and is of the order of 0.8×10^{-6} at room temperature.

2.3.5.4 Optical properties

Diamond is the best optical material available and this lack of absorbance accounts for its unique brightness. However, no diamond is perfect and lattice defects combined with impurities, such as nitrogen and boron, give visible coloration and can also cause luminescence. The high cost of diamond has almost completely restricted its use in optical applications.

2.3.5.5 Electrical properties

Diamond is an excellent insulator with a resistivity of $> 10^{14}$ ohm m in the dark but, it does exhibit photoconductivity in UV light. A Type IIb diamond, however, acts as a semiconductor with a resistivity of 0.1–100 ohm m.

2.3.5.6 Graphitization

If diamond is heated in air, a black coating is formed at about 625°C, but this is not true graphitization, which would involve the transition of diamond into graphite without the aid of external agents. In an inert atmosphere, graphitization starts at about 1500°C and a 0.1 carat octahedron is converted completely into graphite at about 2125°C in less than 3 minutes. Interestingly, the process of graphitization involves the removal of one carbon atom at a time from the diamond surface, involving the breakage of three C—C bonds for an octahedral diamond surface and two bonds for a dodecahedral face.

2.3.5.7 Chemical resistance

Diamond is an inert material and is only attacked by strong oxidizing agents such as molten sodium nitrate at about 500°C. Diamond will burn in a jet of oxygen at 720°C to give CO_2. At room temperature, oxygen is absorbed on a diamond surface, which contributes to the low coefficient of friction. Hydrogen is absorbed at a higher temperature of 400°C. At temperatures of about 1000°C, the carbide forming metals W, Ta, Ti and Zr, together with the element B, will react to form the respective carbides. Molten Fe, Co, Mn, Ni, Cr and the Pt group metals act as true solvents for diamond.

2.4 GRAPHITE

There are several most informative textbooks written on graphite [4,13,16] and the journal *Carbon* covers this one element.

2.4.1 Introduction

In the past, much confusion has arisen by the misuse of the term graphite and the Introduction lists the recommended terminology, which should be used for the description of carbon as a solid. Graphite is defined as an allotropic form of the element carbon consisting of layers of hexagonally arranged carbon atoms in a planar condensed ring system (graphene layers). The layers are packed parallel to each other in a three dimensional structure.

2.4.2 Occurrence, production and uses of graphite

2.4.2.1 Natural graphite

Graphite is found in various parts of the world including Brazil, Canada, Czechoslovakia, China, Korea, Madagascar, Mexico, Sri Lanka and the United States and occurs in three forms—flake, crystalline and amorphous—with quite divergent compositions and properties (Table 2.2).The products are milled to a controlled particle size and find application in various forms of lubricants, electrical brushes, batteries and carbon additives, and can be baked with clay for lead pencils.

2.4.2.2 Kish graphite

Kish graphite crystallizes out as flakes in the manufacturing process of cast iron, is a relatively pure form of graphite and is almost identical to natural graphite.

Table 2.2 Properties and characteristics of natural graphite

Property	Type of graphite		
	Flake	Crystalline	Amorphous
Composition			
Carbon (%)	90	96	81
Sulfur (%)	0.1	0.7	0.1
Density (g cm^{-3})	2.29	2.26	2.31
Degree of graphitization (%)	99.9	100	28
002 d-spacing (nm)	0.3355	0.3354	0.3361
Resistivity (Ω cm)	0.031	0.029	0.091
Morphology	Plate	Plate Needle	Granular

Source: Reprinted from Kavanagh A, *Carbon*, 26 (1) 23–32, 1988.

Table 2.3 Typical precursors for carbonization

Precusor product	Empirical formula	MW	MP°C	BP°C	$\rho\,g\,cm^{-3}$	Carbon content %	Typical yield %	Nature of
Aromatic Hydrocarbons								
Coal-tar pitches						75	50	
Coke Petroleum fractions						88	55	Coke
Naphthalene	$C_{10}H_8$	129	81	218	1.14	94		Coke
Anthracene	$C_{14}H_{10}$	178	217	354	1.25	94		Coke
Acenaphthene	$C_{12}H_{10}$	154	96	278	1.02	94		Coke
Phenanthrene	$C_{14}H_{10}$	178	100	340	1.03	94		Char
Biphenyl	$C_{12}H_{10}$	154	71	255	1.18	94		Char
Polymers								
Polyvinyl chloride	$(CH_2CHCl)_n$					41	42	Coke
Polyimide (Kapton)	$(C_{22}H_{10}O_5N_2)_n$					69	60	Coke
Polyvinylidene chloride	$(CH_2CCl_2)_n$					35	25	Char
Polyfurfuryl alcohol	$(C_5O_2H_6)_n$					54	53	Char
Phenolics	$(C_{15}O_2H_{20})_n$					78	60	Char
Polyacrylonitrile (PAN)	$(CH_2CHCN)_n$					68	48	Char
Cellulose	$(C_{12}O_{10}H_{18})_n$					45	20	
Char								

Note: Coke is a graphitizable carbon whereas char is non-graphitizable.
Source: Reprinted with permission from Pierson HO, *Handbook of Carbon, Graphite, Diamond and Fullerenes*, Noyes Publications, Park Ridge NJ, 1993. Copyright 1993, William Andrew Publishing.

2.4.2.3 Synthetic graphite

The manufacture of synthetic graphite basically involves the transition of an organic precursor through a carbonization and graphitization process to yield a char, which may be graphitized with difficulty, or a coke that is a graphitizable carbon (Table 2.3).

Molded graphite is a form of synthetic graphite, manufactured by carbonizing and graphitizing a mixture of a carbon filler with an organic binder and is an exceedingly proprietary process with the working details varying from company to company. It is normally based on the Acheson process and a typical industrial process is shown in Figure 2.11. A carbon filler such as petroleum coke, which is the almost solid residue left in the process of petroleum refining, is processed further by a series of distillation and cracking stages to yield a heavy residue, which is heated for several days to 400–500°C to produce a raw coke. The selection of this raw coke precursor is very important to obtain a good product and needle shaped particles, for example, will enhance the thermal shock resistance. The addition of a 'puffing inhibitor' such as iron facilitates the removal of sulfur and is acceptable in a product destined for electrodes. The coke is further purified by heating to about 1200°C to remove residual hydrogen and to develop the crystal structure. The product is then ground to a particle size dependent on the end use and for a fine graphite, a 'flour' with a particle size <8.4 μm is produced and is subsequently mixed with a carefully selected grade of coal tar pitch, which has a low softening point and high carbon content and acts as a binder. Mixing is carried out at 165–170°C to lower the viscosity of the coal tar pitch to aid penetration. The mixture is then normally cooled to 125°C and extruded through a die to impart a preferred orientation. Alternatively, the mixture can be compression molded in a tungsten carbide die, or to obtain a highly uniform product, albeit an expensive route, it can be isostatically pressed. Sufficient porosity must be retained to permit evolution of volatiles and after shrinkage, the product will have a final porosity of about 25%, which is reduced further by successive pressure

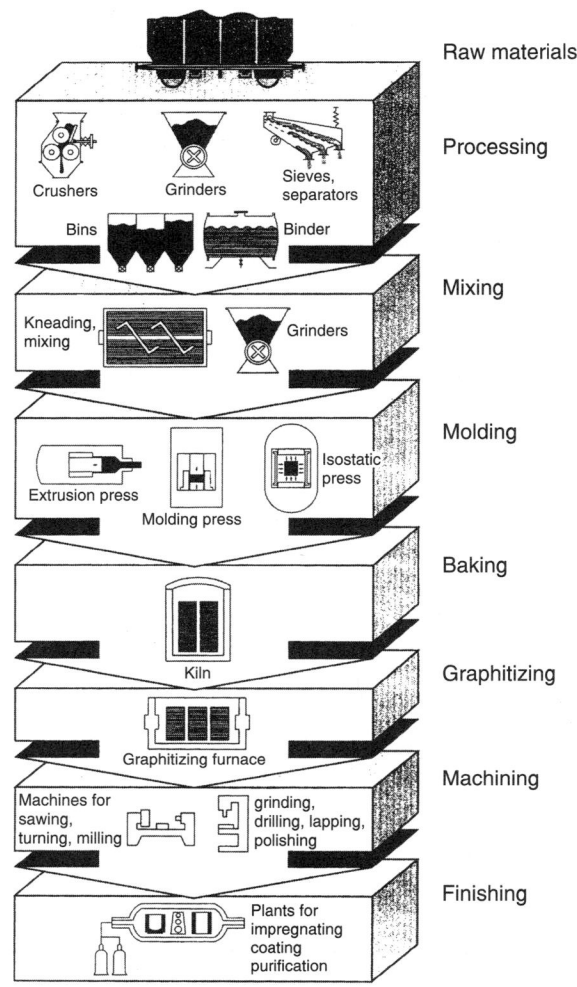

Figure 2.11 Typical manufacturing process for molded graphite. *Source:* Reprinted from SGL Carbon Group technical literature, *Carbon and Graphite Manufacture and Properties*, 1995.

impregnations of a binder using a somewhat lower viscosity version of a coal tar pitch to aid penetration into the pores, or alternatively, to use a phenolic resin. Graphitization is then carried out in a resistance, or an induction heated furnace at 2600–3000°C. Most of the remaining hydrogen is evolved by 1500°C and subsequent heating develops the crystal structure. Cooling is a slow process and takes about two weeks, yielding a product with many faults in the layer planes and some cross linkages. The ash content is about 0.04–0.8% and to obtain a purer product, it is heat treated at 2500°C in a halogen atmosphere (e.g., Cl_2 and F) to remove the impurities as volatile halides, reducing the ash content to about 20 ppm.

To consider what structural changes occur in the formation of graphite, a simplistic overview can be taken by following the course of the graphitization of a pure hydrocarbon in the absence of an oxidizing atmosphere [16]. On heating to 400°C, the hydrocarbon initially darkens, possibly turning black, a process referred to as charring or coking. Condensed ring structures increase in size and progressive dehydrogenation occurs up to

1600°C forming a large network of hexagon rings containing macro-aromatic molecules stacked one above another, but with no crystallographic order perpendicular to these networks. From 1600 to 2800°C, a graphitizable carbon undergoes progressive graphitization with an increasing three-dimensional order. A non-graphitizing carbon with an interlayer spacing of 0.37 nm is, however, not converted fully to graphite even after heating for many hours at 2000°C. The molecular structure of the precursor material affects the type of carbon produced (Table 2.3). Graphitizing-carbons, on the other hand, with an initial interlayer spacing of 0.344 nm become quite well aligned at 2200°C and at lower temperatures, give a mixture of a graphitic structure with an interlayer spacing of 0.355 nm and a turbostratic structure with a spacing of 0.344 nm, normally specifying the mean layer spacing in these instances.

2.4.3 Structure of graphite

Graphite, although it may be pure, can have poorly defined physical properties due to a close association with other forms of carbon, such as char, lampblack and soot. Perfectly crystalline graphite has a brilliant silvery surface and planar morphology but it is dark grey in the polycrystalline form. An ungraphitized carbon does not mark paper, whereas graphite will.

Graphite consists of layers of hexagonally arranged carbon atoms in a planar condensed ring system with each carbon 0.142 nm from its three nearest neighbors. The layers, termed graphene layers, are stacked parallel to each other in a three-dimensional structure. The chemical bonds within the layers, between each carbon atom, are covalent with sp^2 hybridization; each hexagonal array forming six σ bonds and the remaining p orbitals, which are at right angles to the layer planes, take no part in the σ hybridization. The p orbitals of two neighboring carbon atoms overlap sideways and form π orbitals, with three above and three below the σ orbital plane (Figure 2.12) [19]. The π orbitals in each hexagonal array overlap and encircle all six carbon atoms taking up the form of a doughnut and the six electrons become delocalized throughout the π orbital, which lowers the energy and helps to stabilize the molecule. The planes are 0.335 nm apart, denoting that they are held only with weak forces, which are comparable with van der Waals forces. It should be noted that at the edges of the graphene planes there will be 'loose' carbon atoms, or dangling bonds, resembling cut chicken-wire netting.

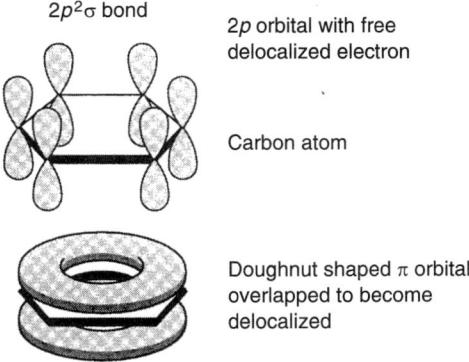

Figure 2.12 An sp^2 hybridized structure of a single hexagonal molecular framework of graphite. *Source:* Reprinted with permission from Baggott JE, *Perfect Symmetry*, Oxford University Press, Oxford, p. 89, 1994. Copyright 1994, Oxford University Press.

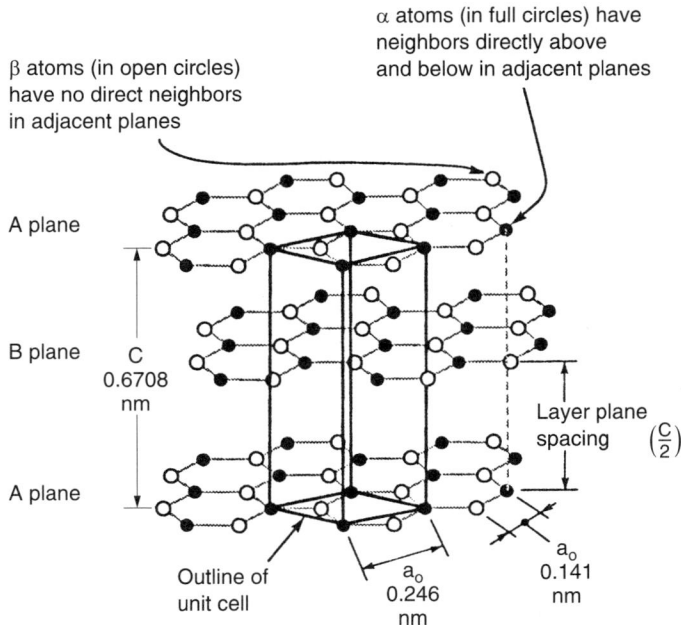

Figure 2.13 Hexagonal unit cell structure of graphite. *Source:* Reprinted with permission from Pierson HO, *Handbook of Carbon, Graphite, Diamond and Fullerenes*, Noyes Publications, Park Ridge NJ, p. 49, 1993. Copyright 1993, William Andrew Publishing.

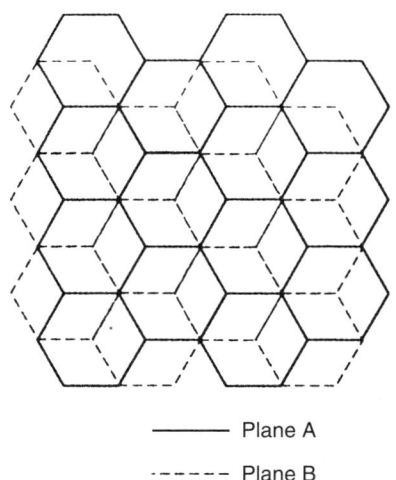

Figure 2.14 Schematic of hexagonal graphite crystal; view is perpendicular to the basal plane. *Source:* Reprinted with permission from Pierson HO, *Handbook of Carbon, Graphite, Diamond and Fullerenes*, Noyes Publications, Park Ridge NJ, p. 45, 1993. Copyright 1993, William Andrew Publishing.

The stacking of the layer planes occurs in two quite similar crystal forms—hexagonal and rhombohedral. Hexagonal is the most common form (Figure 2.13) with an ABABAB packing sequence (Figure 2.14) and the four atoms of the unit cell are at positions: (0, 0, 0), (0, 0, 1/2), (2/3, 1/3, 0) and (1/3, 2/3, 1/2). The rhombohedral form (Figure 2.16) with an ABCABC

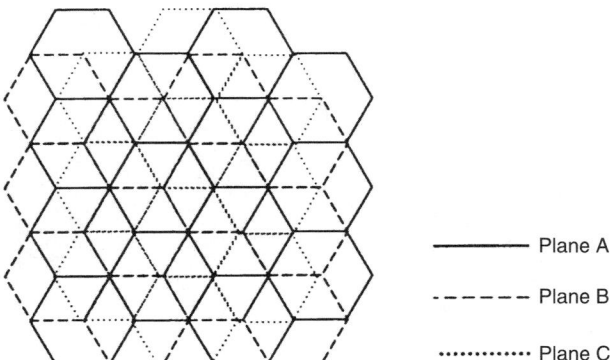

Figure 2.15 Schematic of rhombohedral graphite crystal; view is perpendicular to the basal plane. *Source:* Reprinted with permission from Pierson HO, *Handbook of Carbon, Graphite, Diamond and Fullerenes*, Noyes Publications, Park Ridge NJ, p. 46, 1993. Copyright 1993, William Andrew Publishing.

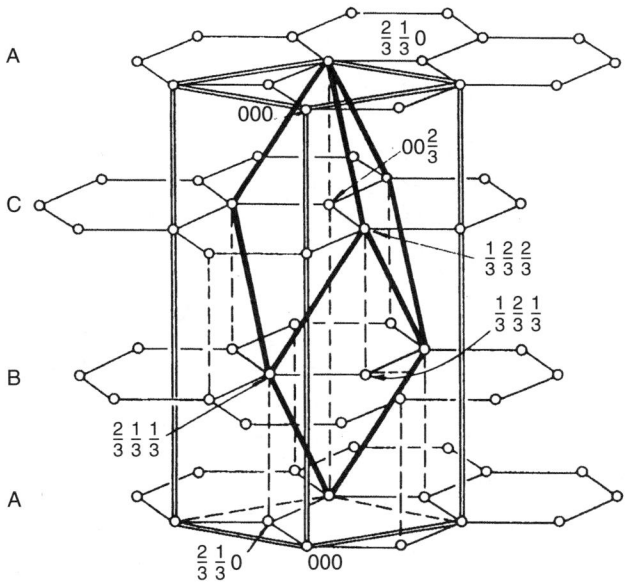

Figure 2.16 Rhombohedral unit cell structure of graphite. *Source:* Reprinted from Reynolds WN, *The Physical Properties of Graphite*, Elsevier, 1968.

packing sequence (Figure 2.15) and the six atoms of the unit cell are at positions (0, 0, 0), (2/3, 1/3, 0), (0, 0, 2/3), (2/3, 1/3, 1/3), (1/3, 2/3, 1/3) and (1/3, 2/3, 2/3). Rhombohedral graphite is always found in association with the hexagonal form and on heating to about 2500°C, the rhombohedral packing is transformed to the hexagonal form. A network of substantially perfect hexagons with parallel layer planes, but with no ordered stacking sequence, is termed turbostratic (Figure 2.17).

Ideal graphite does not exist and the ideal crystal forms invariably contain defects, such as vacancies due to a missing atom, stacking faults and disclination as depicted in Figure 2.18. Other defects include screw and edge dislocations (Figure 2.19). Edge defects find some

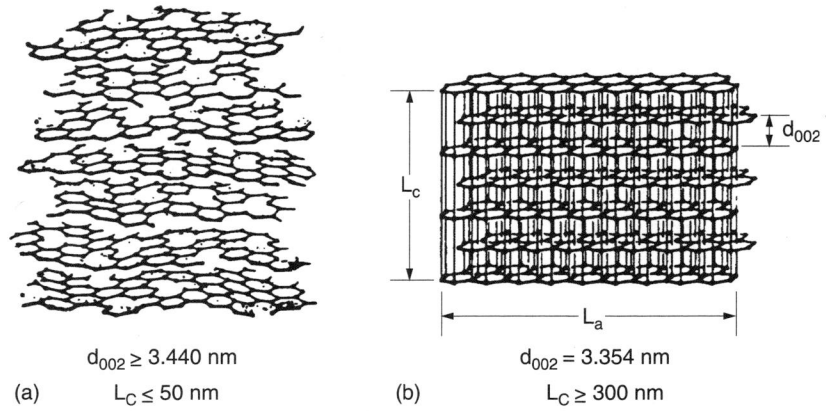

Figure 2.17 Comparison of (a) carbon turbostratic structure with (b) 3-D graphite lattice. *Source:* Reprinted with permission from Bokros JC, Chemistry and Physics of Carbon—A Series of Advances, 5: Walker PL Jr, *Deposition, Structure and Properties of Pyrolytic Carbon*, Marcel Dekker, 1969. Copyright 1969, CRC Press, Boca Raton, Florida.

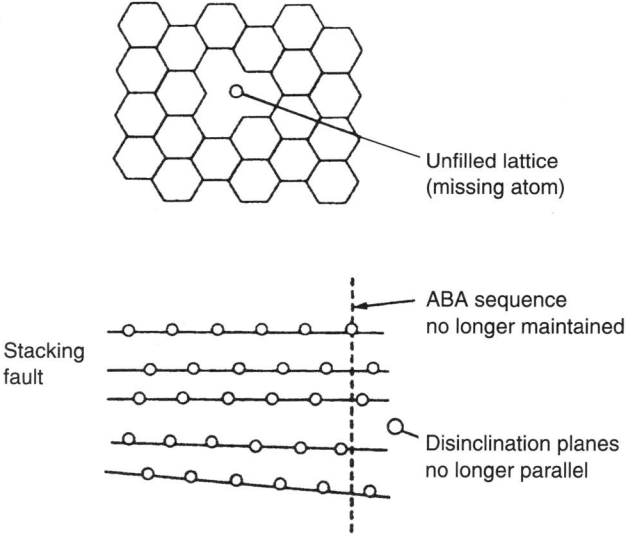

Figure 2.18 Schematic of crystallite imperfections in graphite showing unfilled lattice, stacking fault and disinclination. *Source:* Reprinted with permission from Pierson HO, *Handbook of Carbon, Graphite, Diamond and Fullerenes*, Noyes Publications, Park Ridge NJ, p. 49, 1993. Copyright 1993, William Andrew Publishing.

way of satisfying the electronic valencies [16] and, for example, foreign atoms such as —H, —OH, —O and —O— can be bonded onto these edge defects. However, the network does buckle and bulge to accommodate such groups. Grinding graphite into a powder breaks up some of the bonds and increases the number of defects. Incorporating boron into the structure yields carbons with lowered electrical resistivity and a positive temperature coefficient of conductance. Presumably, boron provides electron acceptor sites. For valency reasons, some of the joins in the net are broken and this could explain why boron aids the graphitization of carbon [21].

THE FORMS OF CARBON

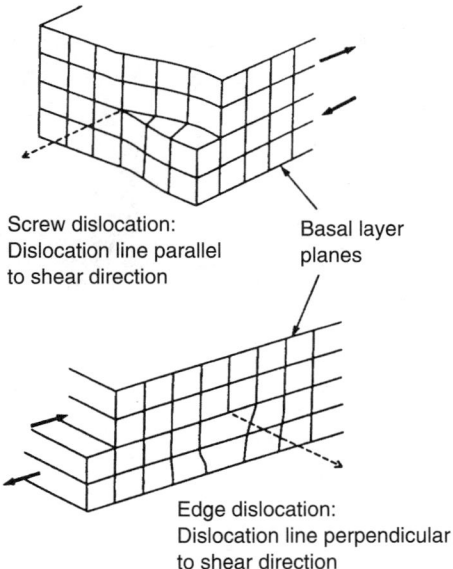

Figure 2.19 Shear dislocations in a graphite crystal. *Source:* Reprinted with permission from Pierson HO, *Handbook of Carbon, Graphite, Diamond and Fullerenes*, Noyes Publications, Park Ridge NJ, p. 49, 1993. Copyright 1993, William Andrew Publishing.

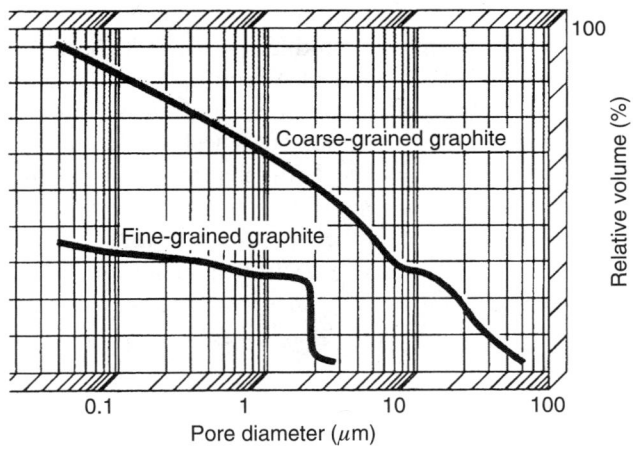

Figure 2.20 Pore volume distribution of pore size in graphite. *Source:* Reprinted from SGL Carbon Group technical literature, *Carbon and Graphite Manufacture and Properties*, 1995.

2.4.4 The properties of graphite

2.4.4.1 Density

Crystallographically perfect graphite has a density of $2.265\,\text{g}\,\text{cm}^{-3}$. Density can give a measure of crystal perfection and in practice, synthetic graphite rarely exhibits a density above $2.0\,\text{g}\,\text{cm}^{-3}$ with some versions as low as $1.6\,\text{g}\,\text{cm}^{-3}$. The density of synthetic graphite is directly related to the pore size (Figure 2.20) and decreasing the particle size, or grain, reduces the relative porosity. The relative grain size of coarse and fine grained graphite is clearly

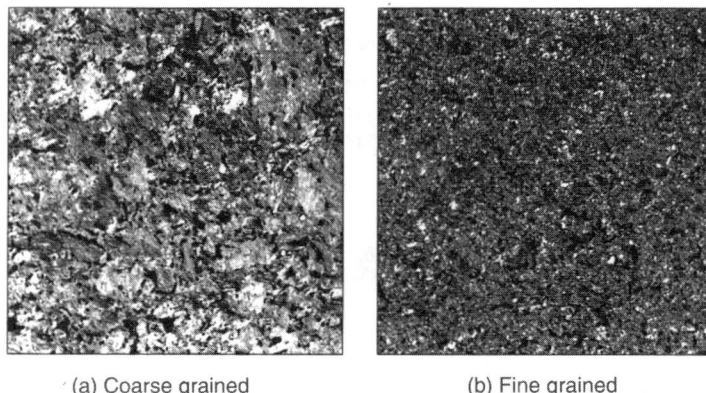

(a) Coarse grained (b) Fine grained

Figure 2.21 Micrographs of graphite sections. *Source:* Reprinted from SGL Carbon Group technical literature, *Carbon and Graphite Manufacture and Properties*, 1995.

shown in Figure 2.21. Density can be increased during the manufacturing process by using a vibration compaction technique and impregnation with resins.

2.4.4.2 Mechanical properties

The mechanical properties are directional and vary with the grain and across the grain. Isostatic pressing gives the most uniform product with the highest density and improved strength, but it is an expensive route. The extruded product aligns the grain in the direction of extrusion, yielding a product with anisotropic properties, but is relatively cheap and is used for the manufacture of parts with a constant cross-section, such as electrodes. Graphite produced by compression molding has the grain perpendicular to the molded direction, having an intermediate cost and is used for making parts with complex shape such as dies or bearings. Sections can, on demand, be molded up to 1.8 m diameter.

Molded graphite is glass-like in behavior and at ambient temperature, fails in a brittle manner. However, at higher temperatures, it is in effect annealed and, unlike other materials, the strength increases with rising temperature. Figure 2.22 shows how the flexural strength at

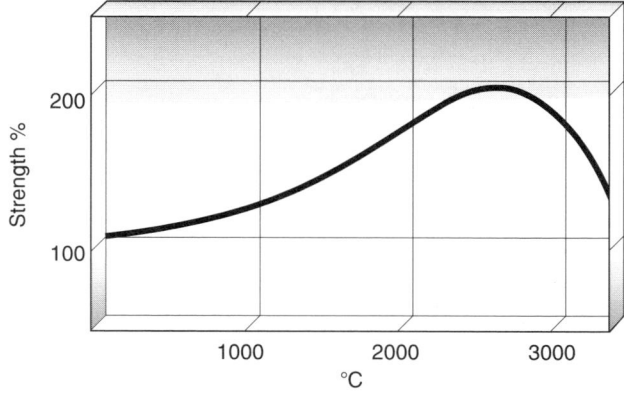

Figure 2.22 Relative change in flexural strength of graphite Type MSY as a function of temperature. *Source:* Reprinted from SGL Carbon Group technical literature, *Carbon and Graphite Manufacture and Properties*, 1995.

THE FORMS OF CARBON

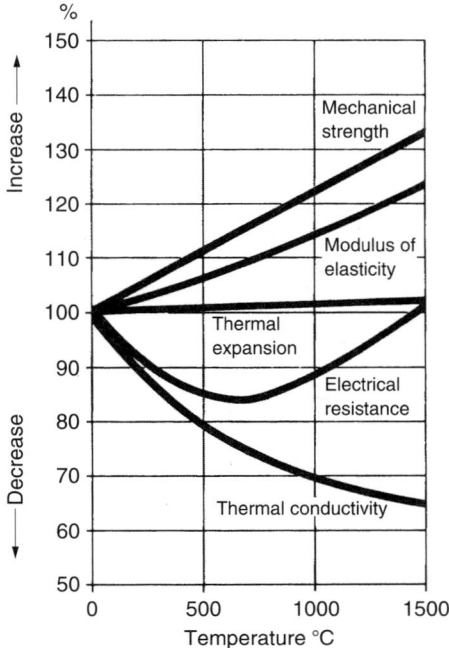

Figure 2.23 Variation of the physical characteristics of electrographite with temperature. *Source:* Reprinted from Conradty C, technical literature, Graphite Electrodes.

room temperature is almost doubled when the temperature is increased to 2500°C, while Figure 2.23 shows the effect of temperature on the other properties.

Generally, the smaller the diameter of the molded section, higher the mechanical

C_{11} (longitudinal)	1060 ± 20 GPa
C_{12} (transverse)	180 ± 20 GPa
C_{13}	15 ± 5 GPa
C_{33}	36.5 ± 1.0 GPa
C_{44} (shear)	0.18–0.35 Gpa*
C_{66}	440 ± 20 GPa
$1/S_{11}$	1020 ± 30 GPa

*The spread in the value for C_{44} is due to the presence of dislocations and the value is ≥ 4.0 GPa for dislocation-free graphite [25].

properties. Consequently, it is important to select the correct diameter for an electrode, as otherwise the superior properties of the outermost section will be machined away. Typical manufacturer's properties of graphite are given in Table 2.4 showing the effect of the method of manufacture, grain size, direction of grain and the size of the section.

2.4.4.2.1 Elastic properties

Elastic constants do pose significant practical problems in their measurement and typical properties [24] measured using compression annealed pyrolytic graphite are:
Note: The condition for isotropy is $\frac{2C_{44}}{C_{11}-C_{12}} = 1$, so substituting the values for these constants, this ratio becomes 0.0004, which means that the Young's modulus varies appreciably with orientation and the material is extremely anisotropic.

Table 2.4 Properties of molded / extruded graphite

Type of product	Manufacturer	Grade	Rod diameter mm	Maximum grain size mm	Density g cm^{-3}	Porosity %	Direction of the grain	Flexural strength MPa	Tensile strength MPa	Young's modulus Gpa	Compressive strength MPa	Specific electrical resistance Ω μm	Thermal conductivity W m^{-1} K^{-1}	Linear coef. of thermal expansion 10^{-6} K^{-1}	Ash content %
Pressed in a mold	Conradty	B497XN	20–85	13	1.75	13	[a]	28	–	–	60	16–22	85	3.5 to 4	0.1
	Ucar	AJL	Bolck	0.15	1.16	9	With	30	30	11.4	68	11	120	2.3	0.16
							Across	25	25	8.5	70	14	100	3.4	–
Extruded	Conradty	CCF/XN	3–9	–	1.76–1.80	15	[b]	40	–	–	60	–	120–130	2.2–2.6	<0.1
			10–25	–	1.75–1.78	15	[b]	20–25	–	–	30–40	–	110–130	1.8–2.5	<0.1
			30–105	–	1.72–1.76	14–16	[b]	20–25	–	–	30–40	–	100–120	1.8–2.5	<0.15
			110–280	–	1.70–1.74	16–18	[b]	18–20	–	–	27–32	–	100–20	1.8–2.5	<0.15
	SGL	MSY	<60	1	1.83	11	With	30	16	15	55	8	160	2	0.15
							Across	17	9	8	50	13	100	3.5	–
			60–130	2	1.78	13	With	24	12	12	45	7	190	2.5	0.15
							Across	14	7	7	40	11	120	3.5	–
		MKS	–	–	1.73	16	[b]	18	–	10	45	10	–	–	–
	UCAR	AGSR	18–72	0.4	1.53–1.58	23–28	With	15–20	10–14	10	22–30	7–10	130–180	1–2	<0.3
							Across	–	–	6	22–30	–	–	–	–
			75–130	0.8	1.60–1.70	23–28	With	12–18	9–13	10	25–33	7–10	130–180	1–2	–
							Across	–	–	6	25–33	–	–	–	–
			150–330	1.6	1.60–1.67	23–28	With	7–14	7–12	10	16–23	7–10	130–180	1–2.5	–
							Across	6–12	4–9	6	16–23	–	–	2.5–4	–
			350–400	3	1.58–1.64	26–32	With	5–8	3–5	10	9–15	7–10	130–180	1–2	–
							Across	4–7	2–5	6	7–15	10–14	90–130	1.5–3	–
			450–600	6	1.53–1.65	26–32	With	4–6	3–5	10	7–13	6.5–10	130–200	0.5–1.5	–
							Across	3–5	2–5	6	7–13	9–14	90–140	1.0–2.5	–
			750–900	6	1.53–1.65	23–28	With	–	–	–	–	6.5–10	130–200	–	–
							Across	–	–	–	–	–	–	–	–
		CS	16–45	0.4	1.60–1.75	15–20	With	15–20	–	14	45–55	7.5–9	140–170	1.5–3	–
							Across	–	–	8	–	9–12	110–140	2.7–4.5	–
			50–450	0.8	1.70–1.80	14–19	With	15–30	8–20	14	35–65	6.5–8.5	150–200	1.5–3	–
							Across	10–20	5–15	8	30–40	8.5–12	110–150	2.7–4.5	–
		ATJ		0.15	1.76	9–13	With	30	30	11.4	68	11	120	2.3	–
							Across	25	25	8.5	70	14	100	3.4	–
Isostatic	Conradty	B650XN	As required	–	1.82	14	–	60	–	–	105	13	80	4.5	<0.1

Notes:
[a] For molded products the grain is perpendicular to the molded direction.
[b] For extruded products the grain is in the extruded direction.

Source: Reprinted from manufacturers' brochures.

THE FORMS OF CARBON

Graphite exhibits good cleavage parallel to the basal planes and because of low surface friction, it is used as a lubricant and in lubricants. The frictional properties are affected by adsorbed impurities on the surface (cf diamond).

2.4.4.3 Thermal properties

Graphite is an excellent refractory material with a melting point of about 4473°C, but it must be under a pressure of some 100 atm, otherwise it just simply sublimes. Since graphite oxidizes slowly in air at about 400°C, it should be protected in an inert atmosphere to operate successfully at elevated temperatures and, it is most important that the inert atmosphere at temperatures in the region of 2500°C is quiescent, since otherwise, the surface will be badly eroded as the gas flow sweeping across the surface continually removes the surface graphite by sublimation.

Specific heat increases with temperature (Figure 2.24) and whilst graphite is considered a good conductor in the basal direction, it is highly anisotropic and is almost an insulator in the direction normal to the basal plane. The thermal conductivity decreases with temperature. The thermal expansion of an extruded rod in a radial direction can be up to three times the value in an axial direction (Figure 2.25) and due care must be taken to accommodate this anisotropy in practical applications.

2.4.4.4 Electrical properties

Graphite is a good conductor of electricity as well as heat and again, is extremely anisotropic, behaving like a metal in the direction of the basal plane and virtually a non-conductor normal to it. In practice, however, molded graphite is quite different in its behavior due to the effect of the different fabrication routes, which can affect the direction of grain, the actual grain size, the porosity and the finished size. The electrical resistivity is an important parameter for graphite when used for the construction of heating elements. Figure 2.26 shows the effect of temperature on a typical electrode material, whilst Figure 2.27 shows the effect of the type of graphite and direction of the grain on the electrical resistivity.

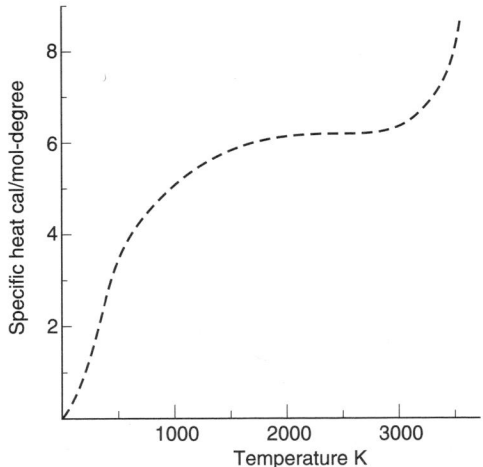

Figure 2.24 Schematic plot of the specific heat of graphite. *Source:* Reprinted with permission from Ubbelohde AR, Lewis FA, *Graphite and its Crystal Compounds*, Oxford, 1960. Copyright 1960, Oxford University Press.

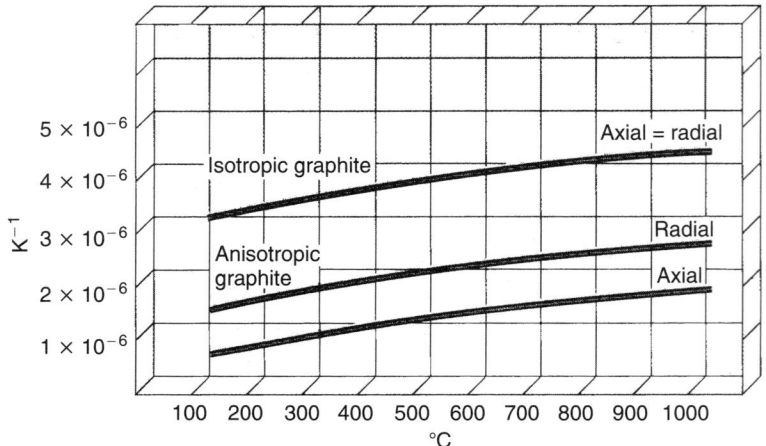

Figure 2.25 Thermal expansion of anisotropic and isotropic graphite. *Source:* Reprinted from SGL Carbon Group technical literature, *Carbon and Graphite Manufacture and Properties*, 1995.

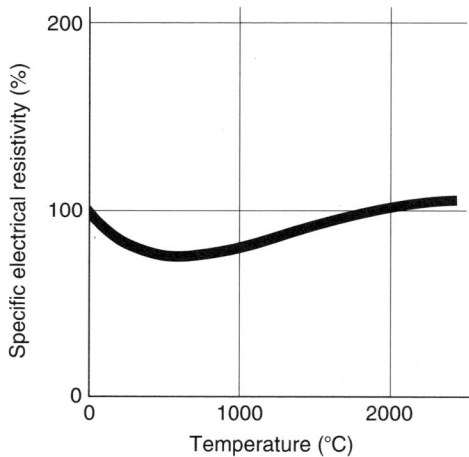

Figure 2.26 Relative change in the specific electrical resistivity of graphite Type MSY as a function of temperature. *Source:* Reprinted from SGL Carbon Group technical literature, *Carbon and Graphite Manufacture and Properties*, 1995.

2.4.4.5 Chemical resistance

Graphite has very good chemical resistance and is used for the construction of special chemical plants such as heat exchangers.

Graphite will form carbides with B, Si, V, Fe, Mo, W and Ta. Oxidizing gases such as oxygen (>150°C), air (>250°C) and steam (>300°C) will attack graphite and such media can be used to manufacture activated carbon. A more graphitic carbon with fewer active sites will be less active.

Graphite reacts to form intercalation compounds when the compound enters between the graphite layer planes, forcing them apart, although heating tends to remove the entire intercalated product, when the graphite reverts to its normal form. Although the layer planes are pushed apart, the distance between the carbon atoms within a sheet remains unaltered. Liquid potassium can form C_8K (Figure 2.28) and $C_{16}K$ with the interlayer spacing increased from 0.335 to 0.54 nm.

THE FORMS OF CARBON   37

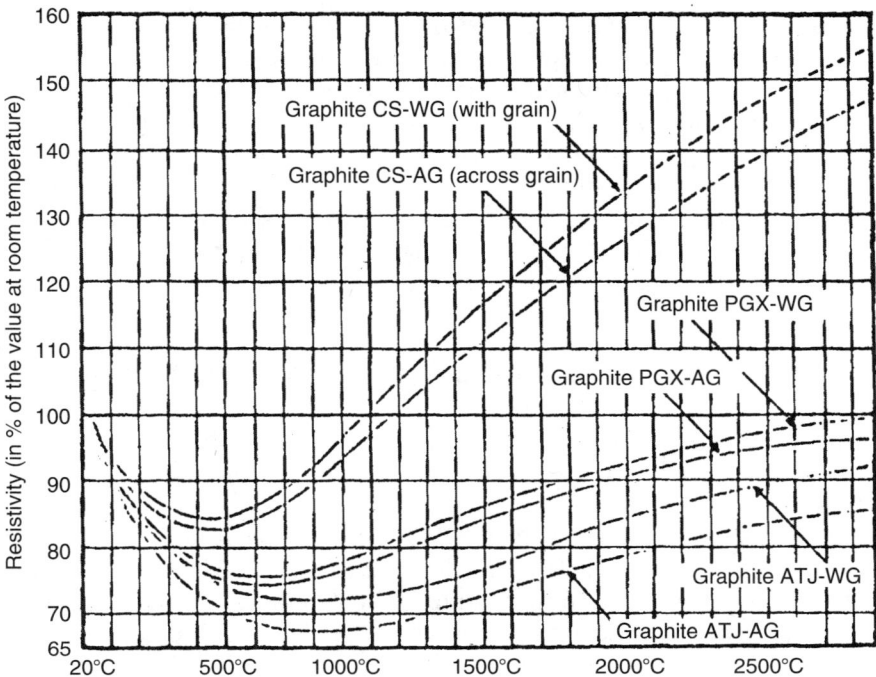

Figure 2.27 Variation of the resistivity at high temperature of various grades of Union Carbide graphite. CS–extruded, PGX–molded, ATJ–fine grain. *Source:* Reprinted from Union Carbide technical literature on graphite.

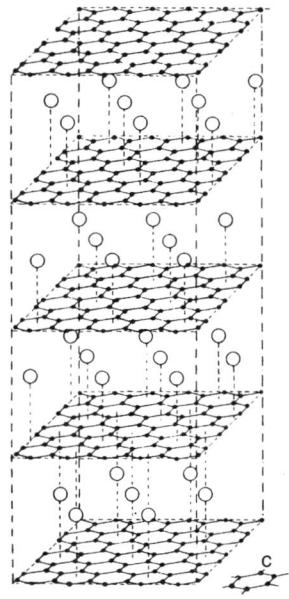

Figure 2.28 Potassium graphite C_8K. *Source:* Reprinted with permission from Ubbelohde AR, Lewis FA, *Graphite and its Crystal Compounds*, Oxford, 1960. Copyright 1960, Oxford University Press.

Sulfuric acid with nitric acid to is a strong oxidizing agent and forms a purple-blue bisulfate, $C_{24}^{+}(HSO_4)^{-}(H_2SO_4)_2$, swelling the graphite (intumescing) with an interlayer spacing of 0.798 nm. This process can be used to purify graphite and to make graphite foil and colloidal graphite. More than sixty halides are known to intercalate graphite [26], such as $FeCl_3$ at 180°C, which gives an interlayer spacing of 0.94 nm.

Heating with a strong oxidizing mixture of potassium chlorate/nitric acid/ sulfuric acid, graphite forms what can be termed graphite oxide, when oxygen and hydrogen are taken up to give a green-brown product $C_8O_2(OH)_2$ (where most of the hydrogen is in the form of water) with an interlayer spacing of 0.6–1.1 nm.

2.5 PYROLYTIC CARBON AND PYROLYTIC GRAPHITE [4]

In the early days of the carbon fiber electric lamp filament, it was the practice to coat carbonized rayon filaments with pyrolytic graphite [27] to improve the mechanical properties of the lamp filament.

Using the chemical vapor deposition (CVD) process, a gas such as methane can be cracked onto, say a carbon substrate and when the temperature is about 1100°C, it forms pyrolytic carbon, which has an isotropic nature. At 1000–1700°C, the carbon deposited has a different form with an intermediate structure, whilst at 1700–2300°C, pyrolytic graphite is formed. The coating can be deposited on the substrate and, if sufficiently thick, can be removed as a free-standing object (cf CVD diamond).

In the 1950s, W. Watt of carbon fiber fame at the Royal Aircraft Establishment (RAE), with A.R.G. Brown, carried out work in coating graphite rods with pyrolytic graphite using methane gas at 2100°C. They obtained a coating with a density of $2.17\,g\,cm^{-3}$ (Table 2.5), which was confirmed by X-ray analysis to have a highly preferred orientation. The density of the material produced at 1700°C, however, was $1.12\,g\,cm^{-3}$ and only increased to $1.52\,g\,cm^{-3}$ when heated at 2800°C, probably due to the initial deposition of soot-like particles, which acted as nuclei for the depositing graphite. Later, with W. Johnson, also of carbon fiber fame, the trio developed this work to study the thermal conductivity of the pyrolytic graphite deposited at 2100°C and further heat treated to 2600°C to obtain uniformity and they found an interesting anisotropy, obtaining for the longitudinal (basal) plane, a value some three hundred times greater than the value for the radial (c-axis) [29]. This work on pyrolytic graphite was to form a valuable background for Watt and Johnson's subsequent work at RAE on PAN-based carbon fibers.

The structure and properties of pyrolytic deposits are influenced by a number of factors, of which temperature is the most important. The size of the crystallites and their degree of preferred orientation increases with temperature, whilst the interlayer spacing decreases [30,31]. Typical values for the density, crystallite diameter and crystallite height are given in

Table 2.5 Density of pyrolytic graphite

Deposition temperature (°C)	Initial density (g cm^{-3})	Density after heating for 2h at 2800°C (g cm^{-3})
2100	2.21	2.24
1900	1.70	1.73
1800	1.14	1.73
1700	1.12	1.52
1600	1.35	2.14

Source: Reprinted from Brown ARG, Hall AR, Watt W, Nature, London 172:1145, 1953.

Table 2.6 Properties of pyrolytic carbons formed from cyclopentadiene initially deposited at 930°C and then heated at 2700°C

Density (g cm^{-3})		L_a Crystallite diameter (10^{-8} cm)		L_c Crystallite height (10^{-8} cm)	
930°C	2700°C	930°C	2700°C	930°C	2700°C
1.926	2.261	39.0	>100	18.6	>200

Source: Reprinted from Cullis CF, Factors affecting the structure and properties of pyrolytic carbons, In: *Petroleum Derived Carbons, ACS Symposium Series*, No.21, 228–236, 1975.

Table 2.6 for pyrolytic carbon formed from cyclopentadiene at 930°C and graphitized at 2700°C. The density of 2.261 g cm^{-3} closely approaches the value for ideal graphite (2.269 g cm^{-3}).

Pyrolytic graphite can be prepared through a CVD process by thermally decomposing a hydrocarbon gas such as methane, ethylene or acetylene at about 1100°C by the induction heating of a substrate undertaken at either a low pressure, when the product tends to be isotropic or, at pressures upto 1 atm (101 kPa), when it is necessary to dilute with a non-reactive gas such as argon or hydrogen to control the deposition process [33].

In a simple form, the reaction can be shown as

$$CH_4 \rightarrow C + 2H_2$$

However, it is really more complex and proceeds via the formation of benzene, various polyaromatic hydrocarbons and is finally deposited as carbon [34]. Other CVD deposition techniques use a fluidized bed [35] and plasma [36]. A variation of the CVD process used for the production of carbon-carbon employs a chemical vapor infiltration (CVI) technique, where the reactive medium diffuses into a porous substrate, such as a 3-D fiber construction, but any by-products formed must be allowed to diffuse outwards, rendering the process extremely slow.

Basically, pyrolytic graphite has a turbostratic structure, which is an aggregate of graphite crystallites that can exist in certain conditions as an almost parallel array, forming a near perfect graphite crystal. By careful control, the structure of the deposit can be columnar, laminar or isotropic. The cleanliness and geometry of the substrate plays an important role in the mode of deposition and Figure 2.29 shows the effect of a surface defect on the structure of the deposit. A columnar structure has the crystallites aligned with the basal planes parallel

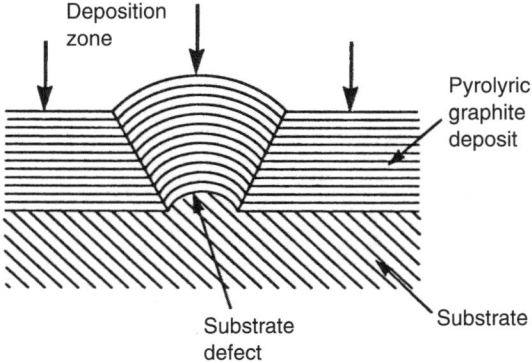

Figure 2.29 Effect of substrate defect on deposited structure of pyrolytic graphite. *Source:* Reprinted with permission from Campbell J, Sherwood EM, *High Temperature Materials and Technology*, John Wiley, New York, 1967. Copyright 1967, The Electrochemical Society, Inc.

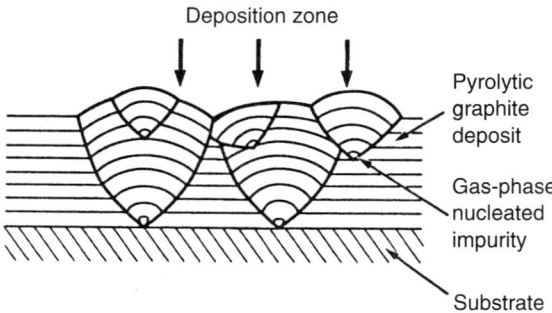

Figure 2.30 Effect of gas phase nucleated impurities on deposited structure of pyrolytic graphite. *Source:* Reprinted with permission from Campbell J, Sherwood EM, *High Temperature Materials and Technology*, John Wiley, New York, 1967. Copyright 1967, The Electrochemical Society, Inc.

to the substrate and as deposition proceeds, the columnar structures widen, becoming cone-shaped with increasing grain size, which is not a desirable factor. This cone formation can be controlled by permitting soot particles to form on the surface, providing new growth sites (Figure 2.30). A laminar structure comprises a number of parallel layers, which can be concentric if deposited on either a fiber, or a particle. Columnar and laminar structures are optically active to polarized light and can be readily graphitized at 2500°C to give a high degree of crystallite alignment and, if annealed further at 2700°C under pressure, will produce highly oriented pyrolytic graphite (HOPG), which is an almost ideal graphite crystal.

The isotropic structure consists of fine grains with no preferred orientation and, since it has virtually no graphitic structure, is termed an isotropic carbon, which does not readily graphitize and shows no optical activity.

The properties of pyrolytic carbon and graphite are given in Table 2.7. The pyrolytic carbon is much harder, stronger, less permeable to gases and can be polished. Consequently,

Table 2.7 Properties of pyrolytic carbon and graphite

Property	Columnar and laminar oriented pyrolytic graphite	Isotropic pyrolytic carbon
Density (g cm^{-3})	2.10–2.24	2.1
Vickers (DPH) hardness (kg mm^{-2})		240–370
Flexural strength (MPa)		350
c direction (across grain)	80–170	Not Applicable
Tensile strength (MPa)		
ab direction (with grain)	110	
Young's modulus (GPa)	28–31	28
Strain to failure (%)		1.2
Thermal conductivity (W m^{-1}K^{-1})		
c direction (across grain)	1–3	
ab direction (with grain)	190–390	
Thermal expansion 0–100°C (10^{-6} m^{-1}K^{-1})		
c direction (across grain)	15–25	
ab direction (with grain)	−1 to 1	
Electrical resistivity (μΩ m^{-1})		
c direction (across grain)	1000–3000	
ab direction (with grain)	4–5	
Gas permeability factor K (cm^2 s^{-1})	10^{-2}–10	10^{-6}–10^{-15}

Source: Adapted from various manufacturers' technical brochures.

it can be used for applications which take advantage of these superior properties, such as rocket nozzles, containers for fissile materials, biomedical devices, abrasion resistant coatings for fiber optic cables and carbon-carbon applications. The graphite version is used for high temperature resistant elements, special equipment used in the semiconductor business and coating fibers to provide a diffusion resistant barrier for metal and ceramic matrices.

2.6 GLASS-LIKE CARBON

It should be noted that Glassy Carbon and Vitreous Carbon have been introduced as trade marks and are versions of glass-like carbon. Glass-like carbon is a form of carbon with a highly disordered non-crystalline structure with a mode of fracture akin to glass. It is made by the carbonization of an aromatic polymer in the solid state, with a 3-D crosslinked structure, generally termed a plastic, resulting in the formation of a char and not a coke. Preferred plastics include polyphenylenes, polyimides, aromatic epoxy formulations, phenolic and furan resins that have a high carbon yield (Table 2.3). The precursor material is initially shaped into an enlarged version of the finished object to allow for processing shrinkage and possibly some after-machining. The heat-up rate depends on the precursor material and artifact thickness, which is normally limited to 4 mm to allow sufficient time for the slow diffusion of the outgassing by-products. Shrinkage and loss of mass occurs mainly at 200–600°C, the artifact shrinking linearly by about 25% and subsequently increasing by about 5% during the final heat treatment process [38]. During the pyrolysis stage, the polymer chains remain intact and unlike coke, do not pass through a mesophase stage.

The structure [39] of glass-like carbon is very disordered, resembling that of a polymer (Figure 2.31), which explains the low density, low thermal and electrical conductivity and why it is isotropic. The crystallites are small, with $L_c < 3$ nm. Many voids, about 1–3 nm in diameter, exist between the graphite layers and act like a molecular sieve and can absorb about 1% of small molecules such as methanol, but have very low permeability to helium, suggesting that helium is too large a molecule to be absorbed into the structure. Glass-like carbon cannot form intercalation compounds and is only attacked by oxygen (above 550°C), some hot melts and powerful oxidizing acids. As a result of this extreme resistance to

Figure 2.31 Structural model of glassy carbon. *Source:* Reprinted from Dübgen R, *Glassy carbon—a material for use in analytical chemistry*, Publication of Sigri, D8901 Meitingen, Germany, 1985.

Table 2.8 Physical properties of 'Sigradur', a glass-like carbon

Properties	Glass-like carbon	
	Made at 1100°C	Made at 2200°C
Bulk density (g cm^{-3})	1.54	1.42
Max. service temperature (°C)	1100	3000
Open porosity (%)	0	0
Permeability coefficient (cm^2 s^{-1})	10^{-11}	10^{-9}
Vickers Hardness (HV$_1$)	340	230
Flexural strength (MPa)	210	260
Compressive strength (MPa)	580	480
Young's modulus (GPa)	35	35
Coefficient of thermal expansion (20–200°C) (K^{-1})	3.5 × 10^{-6}	2.6 × 10^{-6}
Thermal conductivity (30°C) (W K^{-1} m^{-1})	4.6	6.3

Source: Reprinted from Dübgen R, *Glassy carbon—a material for use in analytical chemistry*, Publication of Sigri, D8901 Meitingen, Germany, 1985, from Sigri Technical literature, company is now HTW GmbH, Thierhaupten, Germany.

chemical reagents, glass-like carbon is used for the manufacture of special laboratory analytical equipment such as crucibles and beakers. Other products include a glass-like carbon foam and glass-like carbon microspheres [4].

The physical properties of glass-like carbon are given in Table 2.8.

2.7 CARBON FIBERS

Carbon fibers can be prepared from polymeric precursor materials such as polyacrylonitrile (PAN), cellulose, pitch and polyvinylchloride, which are discussed in detail later. PAN-based carbon fibers predominate and have good strength and modulus properties, whereas carbon fiber can be made with a higher modulus, albeit a lower strength, using a pitch-based precursor.

2.8 GRAPHITE WHISKERS

Many workers handling hydrocarbon gases in equipment at temperatures of 350°–2500°C have observed graphite filaments forming upto 200 µm in diameter and about 50 mm long, ranging from solid filaments to helices and hollow tubes. Graphite whiskers were grown by R. Bacon [40] at the Union Carbide Research Laboratories using a DC arc (75V, 75A) struck between a positive carbon rod electrode, which could be driven downwards (12 mm min^{-1}) as it was consumed, towards a carbon block cathode in an atmosphere of argon at 9.2 kPa and 4173°C. The graphite vaporized from the tip of the anode and roughly 80% condensed on the cathode below, forming a solid block (boule), growing like a stalagmite, yielding whiskers up to 5 µm in diameter and some 30 mm long, which could be extracted by breaking the boule open, yielding needles perpendicular to the concave growing surface. The conditions for the formation of whiskers are close to the triple point of carbon of, at about 1 kPa and 4773°C [41].

The whiskers showed a high degree of crystal perfection, akin to a pure graphite single crystal. Most whiskers had a structure comprising a rolled-up sheet of graphite layer planes forming a scroll along the length of the whisker (Figure 2.32). Bacon proved this structure by passing a large current through the whisker to blow it apart, revealing that the lateral

Figure 2.32 Graphite whisker formed from rolled-up graphite sheet. *Source:* Reprinted from Bacon R, *J Appl Phys*, 31: 283–290, 1960.

dimension of the resultant sheet was many times the circumference of the original whisker. The whiskers were very flexible and strong and could be bent to and fro many times without breaking. Strengths as high as 20 GPa were recorded with a modulus of at least 700 GPa and possibly as high as 1000 GPa. A significant problem in testing was the ability to grip the whisker satisfactorily in the jaws of the testing machine, the low shear strength of graphite contributing to the whisker pulling out of the grip and leaving an outer sheath behind. These high values have established possible targets to be achieved by carbon fibers made from polymers.

2.9 VAPOR-GROWN CARBON FIBERS (VGCF) AND CATALYTIC CHEMICAL VAPOR-DEPOSITED (CCVD) FILAMENTS

Early technology for producing vapor-grown fibers dates back over a century [42], when filaments were grown from a methane/hydrogen mixture pyrolyzed in an iron retort for possible use as a material for electric lamp filaments. It was not until the advent of the electron microscope that a detailed study could be made of this group of fibers and this development is well-reviewed by Baker and Harris [43] and Tibbetts [44] and Gadelle [45] reviews the growth of vapor deposited carbon fibers.

A batch process developed by Tibbetts [46] passed a hydrocarbon/hydrogen mixture over a substrate held at about 1000°C on which iron-containing particles, acting as a catalyst, had been deposited. Figure 2.33 shows in a simple way how filaments are formed by a catalytic particle and growth mechanisms are detailed by Dresselhaus *et al* [47] Properties of VGCF are given in Table 2.9.

A continuous process introduced by Koyama and Endo [50], modified by Hatano and others [51], incorporates ultrafine metallic particles (5–25 nm diameter) in the gas stream via a benzene feedstock or, produces them *in-situ* by decomposition of an organometallic compound such as ferrocene. The continuous process is certainly more prolific, but unfortunately, produces smaller diameter fibers, which at present, cannot match the properties of fiber deposited on a substrate by the batch process.

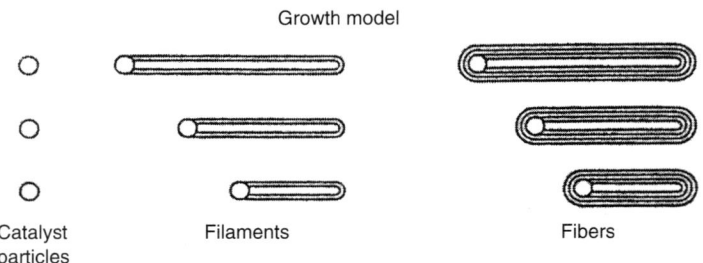

Figure 2.33 Schematic showing growth of vapor-grown carbon fiber in the presence of a catalyst particle. *Source:* Reprinted with permission from Tibbetts GG, Vapor-grown Carbon Fibers, In: Figueiredo JL, Bernardo CA, Baker RTK, Hüttinger KJ, eds. *Carbon Fibers Filaments and Composites*, Kluwer, Dordrecht, 74, 1990. Copyright 1990, Springer.

Table 2.9 Typical properties of VGCF carbon fibers

Property	As-grown value	Heat treated at 2650°C
Density (g cm^{-3})	1.8	2.0
Tensile strength (GPa)	2.9	2.6–2.9
Tensile modulus (GPa)	240	700–760
Thermal conductivity (W cm^{-1}K^{-1})	0.2	30
Electrical resistivity (Ω cm)	10^{-3}	6×10^{-5}

Source: Reprinted with permission from Tibbetts GG, Vapor-grown carbon fibers: status and prospects, *Carbon*, 27: No.3, 745–747, 1989. Copyright 1989, Elsevier. Tibbetts GG, Beetz CP Jr, Mechanical properties of vapor-grown carbon fibers, *J Phys D: Appl Phys*, 20: 292, 1987.

2.10 OTHER FORMS OF CARBON

2.10.1 Carbon Black

Carbon black is a commercially produced form of colloidal carbon, in the form of spheres with fused aggregates below 1 μm. There are several five types:

1. Channel black—produced by many small flames impinging on a cool surface with the spheroidal particles oxidized in air at high temperature, yielding a very fine form of carbon black (about 10 nm).
2. Thermal black—produced by the pyrolysis of gaseous hydrocarbons, including natural gas, in the absence of air, producing larger particles about 100–500 nm.
3. Lampblack—produced by burning an aromatic-rich oil in flat pans in a limited supply of air and subsequently calcined to remove excess oil and aromatic breakdown products, producing a particle size of 100–200 nm with composition and color dependent on the amount of air used in the combustion process.
4. Acetylene black—produced by the thermal decomposition of acetylene at 800°C with a particle size of 3–130 nm.

Carbon black prepared by the combustion of hydrocarbons under oxygen deficient conditions consists of a folded version of the graphite network, formed as the carbon atoms condense similar to the icospiral formation of a soot particle (Figure 2.34). The properties of typical carbon blacks are given in Table 2.10 and they are used as fillers for rubber and base for printing ink.

THE FORMS OF CARBON

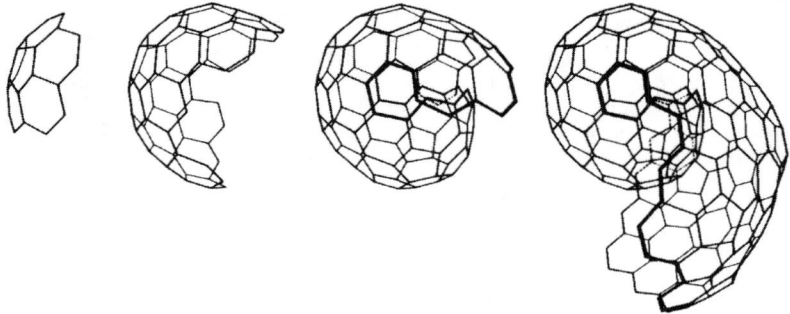

Figure 2.34 The icospiral nucleation mechanism for the formation of soot. *Source:* Reprinted from Kroto H, *Science*, 242: 1139, 1988.

Table 2.10 Typical properties of carbon blacks

Property		Carbon black	Lampblack (dried at 105°C)	Acetylene black
% Composition	H	0.5–1.0	1.20	
	N	0.02–0.09	Traces	
	O	2.5–7.0	7.41	
	S	0.01–0.03	0.66	
CO_2		0.1–1.5		
	CO	0.2–4.0		
	C	Balance	90.67	99
Polymerization products (%)				<1
Ash (%)				0.03–0.04
Moisture (%)				0.05–0.06
Density (g cm^{-3})				2.05
Apparent density (g cm^{-3})				0.02
Surface area (m^2 g^{-1})		25–150		65
Particle size (nm)		10–500		3–130
Oil absorption (cm^3 g^{-1})		0.5–1.5		
Undertone				Blue

Source: Reprinted with permission from Pierson HO, *Handbook of Carbon, Graphite, Diamond and Fullerenes*, Noyes Publications, Park Ridge NJ, 1993. Copyright 1993, William Andrew Publishing.

2.10.2 Charcoal

The porous char of carbon is obtained by the destructive distillation of an organic material such as wood and coconut shells. The product can be activated by heating in steam or CO_2 to increase the surface area and significantly increase the ability to adsorb liquids and gases.

2.10.3 Coal

Coal is produced by an extremely slow dehydrogenation of organic materials by a geological process, initially raising the carbon content by conversion to peat and finally transforming into coal by a geochemical stage, through the action of heat (773°C) and high pressure.

Table 2.11 Typical properties of coke

Property	Premium coke calcined at 1620 K	Regular coke	Metallurgical coke
Real density (g cm^{-3})	2.12–2.14	2.07–2.09	1.95–2.02
Coefficient of thermal expansion (293–773 K) (K^{-1})	1.1×10^{-6}	2.0×10^{-6}	$>3.0 \times 10^{-6}$
Ash (%w/w)	0.05	0.4	8–12
Sulfur (%w/w)	0.6	1.0–1.5	0.6–5.0

Source: Reprinted from Fitzer E, Köchling KH, Boehm HP, Marsh H, *Pure Appl Chem*, 67: (3), 503, 1995, ©1995 (IUPAC).

2.10.4 Coke

Coke is a solid high in carbon, produced by the pyrolysis of an organic material, which has passed, at least in part, through a liquid or liquid-crystalline state and the resulting non-graphitic carbon is graphitizable.

Premium coke made from petroleum tars/residues by a delayed coking process or from refined coal tar pitches yields a product which graphitizes extremely well. Regular coke is a petroleum coke with good graphitizability, whilst metallurgical coke is produced by the carbonization of coal up to 1673°C giving a macroporous product with high strength and a large lump size. Typical properties of cokes are given in Table 2.11.

2.10.5 Soot

Soot is generally formed as an unwanted byproduct of incomplete combustion, or pyrolysis, producing aggregates of spheres, which can be coarse, fine and/or colloidal and can be a mix of solids, tars and resins. When the soot is initially deposited, the dangling bonds around the edge of the small section of graphite confer a measure of instability, which can be corrected by the incorporation of pentagons, but the structure does not close and the structure curves over itself as the particle builds up to a typical soot particle (Figure 2.34).

2.11 NEW FORMS OF CARBON

2.11.1 Fullerenes

2.11.1.1 *Discovery and production of fullerenes*

The story of the buckminster fullerene, C_{60}, is interesting because it reflects the avid enthusiasm of the workers in this field, the feeling of excitement and the race to publish first, petty jealousies that arise between groups, even antagonism, the justness of some workers to fully acknowledge others' achievements, the sheer professionalism of some groups and how luck can play such a significant part, a true reflection of the living world.

Up until 1985, it was considered that there were just two allotropes of carbon and the discovery of a new form of carbon opened up unknown fields in the study of carbon. Hence, it was not long before a publication [54] had been devoted entirely to the specific study of this branch of carbon science. A detailed account of the discovery of this new form of carbon is given by Jim Baggott [19] and textbooks on the fullerenes are now available [55–63].

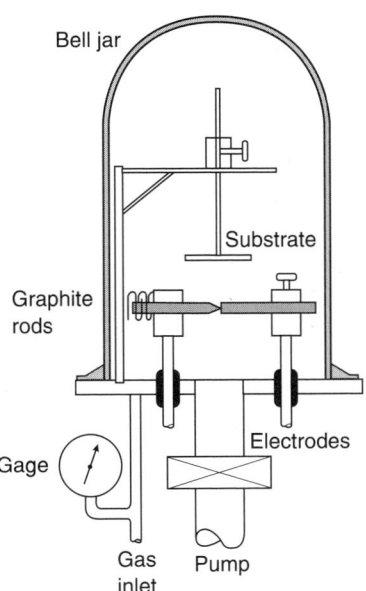

Figure 2.35 The Heidelberg evaporator. *Source:* Reprinted with permission from Baggott JE, *Perfect Symmetry*, Oxford University Press, Oxford, p. 26, 1994. Copyright 1994, Oxford University Press.

At the Max Planck Institute at Heidelberg in 1982, Wolfgang Krätschmer worked in collaboration with Donald Huffman from the University of Arizona, Tucson on a project involved with interstellar dust. They used equipment similar to that used to deposit a thin film of carbon on samples prior to study in an electron microscope (Figure 2.35). An arc was struck between a pair of graphite electrodes, such that the carbon became white hot and evolved fragments of graphite, including atoms of carbon. The pressure was controlled with a series of vacuum pumps, whilst argon or helium could be admitted to cool the carbon vapor. The condensed soot particles were collected on a quartz plate, for subsequent analysis in a spectrophotometer. Their results failed to give any correlation between the relatively strong UV 217 nm band which occurred in interstellar space, but they did find two bumps in the curve, which became known, affectionately, as 'camel humps' and at that time, defied any explanation.

Harry Kroto, from Sussex University, was paying a reciprocal visit to Bob Curl at Rice University, Houston in 1984 and was shown the elaborate and costly equipment (Figure 2.36) used by Curl and his colleague, Rick Smalley, at Rice to investigate Ge and Si semi-conductor clusters. The equipment comprised an Nd:YAG laser, which produced pulses of green light about ten times a second that passed through a quartz window into a steel chamber and were focussed onto a rotating carbon disk. The energy was very intense reaching about 10^7 watts and delivered in a pulse, lasting only about five billionths of a second, blasting atoms from the surface of the target. The atoms began to form clusters and were swept from the chamber at high velocity in the shape of a cone. A skimmer was used to collect the gas from the center of the cone, where it was ionized by an excimer laser, timed to fire just after the Nd:YAG laser, knocking electrons from the clusters and leaving a positive charge. The ions were then deflected down a 1.5 m long tube, with the lightest clusters arriving first at the ion detector of a time-of-flight mass spectrometer.

Kroto was keen to study interstellar molecules and realized that this equipment could be readily adapted to study carbon clusters and obtained tacit approval to undertake the requisite experiments, if and when the equipment became available.

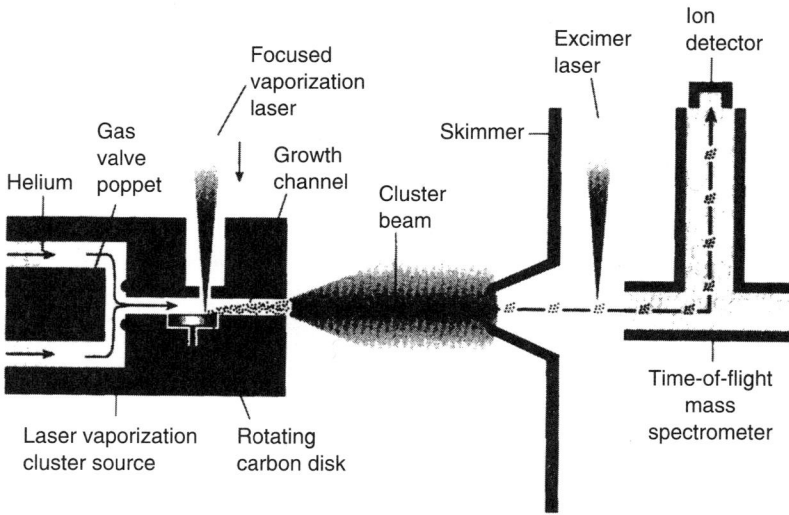

Figure 2.36 Schematic arrangement of pulsed laser equipment used to produce carbon clusters. *Source:* Reprinted from Curl RF, Smalley RE, *Scientific American*, 265: 54–63, October 1991.

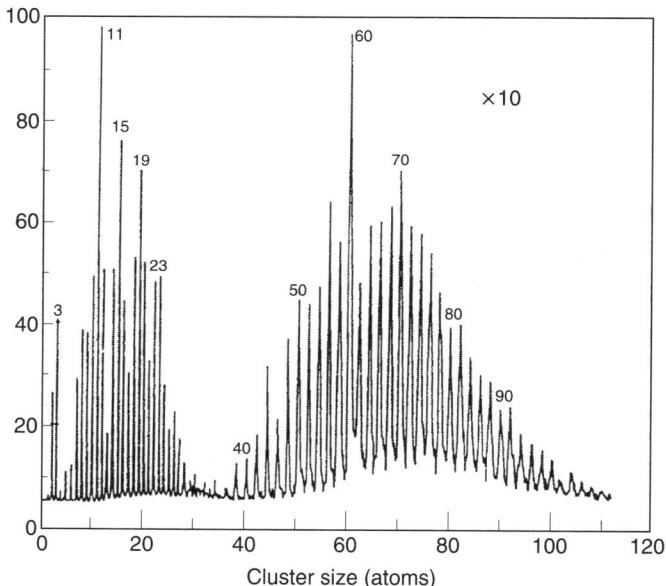

Figure 2.37 Carbon cluster distribution showing prominence of C_{60}. Note that ×10 magnification is used for ions larger than C_{30}^+. *Source:* Reprinted from Rohlfing EA, Cox DM, Kaldor A, Production and characterization of supersonic carbon cluster beams, *J Chem Phys*, 81: 3322, 1984.

Rohlfing, Cox and Kaldor [64] at Exxon using similar equipment, built for them in 1982 by the Rice technicians, obtained a cluster distribution for graphite (Figure 2.37).

Kroto was able to use the Rice equipment in August 1985. Then, after some two weeks working in conjunction with two of Smalley's students, Sean O'Brien and Jim Heath, using graphite in the reactor, they succeeded in tuning the equipment by adjusting the timing of the vaporization laser and extending the length of the cluster zone to produce peaks on the mass spectrum curve (Figure 2.38) corresponding to C_{60} and C_{70}, which was termed the flagpole spectrum. The C_{60} peak was so pronounced that there was obviously some good reason why

THE FORMS OF CARBON 49

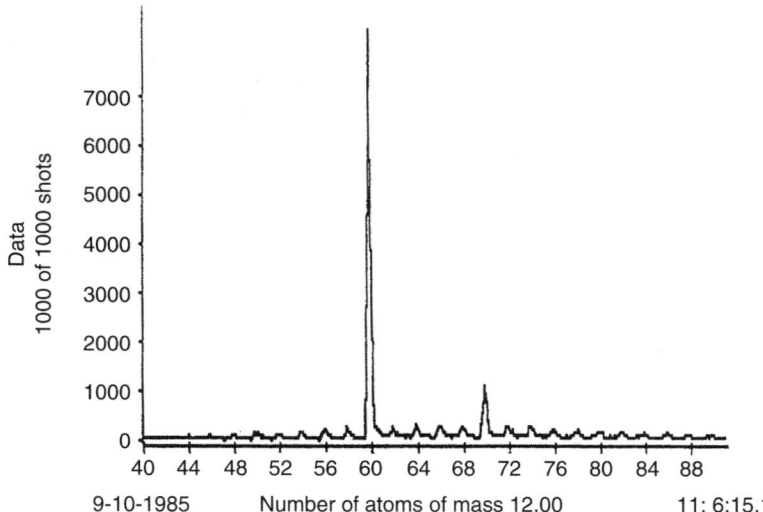

Figure 2.38 The 'flagpole' mass spectrum obtained with conditions optimised for cluster formation. Note the peaks for C_{60} and C_{70}. *Source:* Reprinted with permission from Baggott JE, *Perfect Symmetry*, Oxford University Press, Oxford, p. 64, 1994. Copyright 1994, Oxford University Press.

the clusters preferred to grow forming C_{60}. So what was the structure? Could it be a geodesic sphere? Smalley consulted, but to no avail, the works of Buckminster Fuller [65], the American designer of geodesic domes. Meanwhile, Kroto was convinced that sometime ago he had built his children a cardboard model stardome that had 60 vertices, which he believed incorporated some pentagonal faces. They were unable to come up with a solution whereby a sphere could be constructed from hexagons until Smalley, remembering Kroto's recollection of pentagons, was able to construct a sphere comprising 12 pentagons and 20 hexagons, only to be reminded the following day by colleagues in another department that this was, indeed, the shape of a soccer ball (football). Figure 2.39 depicts a molecular model of C_{60}. They unanimously chose the name buckminsterfullerene, the class later to be termed fullerenes and becoming known, affectionately, as buckyballs. The work was published in 1985 [66] and

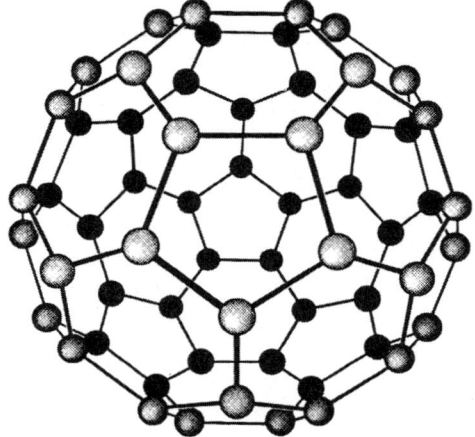

Figure 2.39 A molecular model of soccer ball C_{60}. *Source:* Reprinted with permisison from Baggott JE, *Perfect Symmetry*, Oxford University Press, Oxford, p. 71, 1994. Copyright 1994, Oxford University Press.

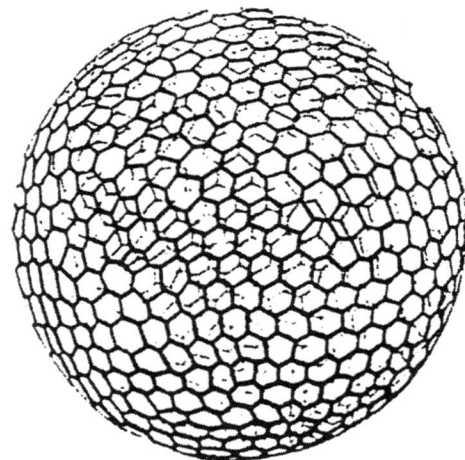

Figure 2.40 Delicate skeleton of single celled micro-organism radiolarian, a natural example of a polyhedral structure composed predominantly of hexagons but with some pentagons and heptagons. *Source:* Reprinted with permission from D'Arcy Thompson W, *Growth and Form*, Cambridge University Press, Cambridge, 708 and 738, 1942. Copyright 1942, Cambridge University Press.

later Curl, Kroto and Smalley received the 1996 Nobel Prize for Chemistry. What is incredible is that DEH Jones, writing in 1966 under the pseudonym 'Daedalus' [67], had actually conceived a giant hollow carbon model and later, in his book [68], referred to the radiolaria, a microscopical sea creature, that frequently had a silica skeleton made up of hexagonal meshes and was depicted in a book by W D'Arcy Thompson (Figure 2.40). Both Jones and Thompson were aware of Euler's Rule that a hexagonal grid, no matter how big, can be closed by the inclusion of exactly 12 pentagons, whilst Euler's Law states that for any polyhedron, the number of corners plus the number of faces minus the number of edges equals 2, thus preventing any polyhedron of ever being made up entirely of hexagons.

In 1970, E Osawa [70], whilst watching his young son playing with a soccer ball clearly observed the bowl- shaped structure of corannulene in its make-up and realized that the ball's design was based on a truncated icosahedron, in which a carbon atom at each of it's 60 vertices could be visualized. He expanded this concept in a book [71] and a translation of the pertinent portions appears in [72]. Russian workers DA Bochvar and EG Galpem had independently named C_{60} as carbo-*s*-icosahedrene [73] and a review of Bochvar and his colleague's work was published in English [74]. These imaginary molecules were not pursued and could be termed premature discoveries.

It was not until 1988 when Krätschmer, spurred on by Huffman's insistence, was able to divert a student, Bernd Wagner, to reinvestigate the camel humps. In the course of this work, Wagner carried out an experiment, reportedly "just for fun", by increasing the pressure about five-fold and produced very prominent camel humps in the UV spectrum, with four strong lines in the IR spectrum, which had been predicted theoretically for C_{60} by several groups of workers. Now, if this soot did contain macro quantities of C_{60}, they had found a way of producing it in a simple apparatus and relatively cheaply. Later, Krätschmer obtained the services of a graduate Konstantinos Fostiropoulos, who after a complete rebuild of the equipment, repeated the experiment using graphite electrodes made from ^{13}C, which should produce an IR spectrum and where the position of the lines would alter slightly since the heavier ^{13}C changes the frequency of vibration. Results consistent with the values expected of a ^{13}C isotope of carbon were obtained [75].

Krätschmer and Huffman called the soot 'Fullerite', but they were always aware that to have a new allotrope of carbon it was necessary to isolate the actual crystal form. By subliming the soot and extracting the sublimate with benzene, Fostiropoulos obtained orange-brown crystals, which could be in the form of hexagonal rods, platelets, or star-shaped flakes. Lowell Lamb, working for Huffman at Tucson, confirmed the UV spectra. They published their results [76], but the definitive X-ray diffraction of a single crystal had still not been achieved, or alternatively, a single line ^{13}C NMR spectrum would have been sufficient to show that C_{60} did, indeed, have a soccer ball structure.

The NMR confirmation came from Kroto's group at Sussex, who were able to separate the mixture of C_{60} and C_{70} by chromatography, obtaining a one-line NMR spectrum for the C_{60} fraction (Figure 2.41) and, after a further chromatographic separation, a five line spectrum for the C_{70} fraction (Figure 2.42). Robert Wilson at IBM obtained a scanning tunneling microscope image (Figure 2.43) of a thin film of fullerite on a gold substrate, clearly showing rows of C_{60} spheres interspersed with somewhat larger slightly elongated molecules of C_{70}. Since molecules of solid C_{60} rotate some 20 billion times a second, the IBM group found [79] that cooling the C_{60} to 77 K reduced the rate of rotation, so that the rotation proceeded in a ratchet fashion, the molecules clicking from one position to the next and they called the transition a 'rotator phase'. This rapid rotation of the C_{60} molecule at room temperature would provide no fixed position for the individual carbon atoms. This gave a

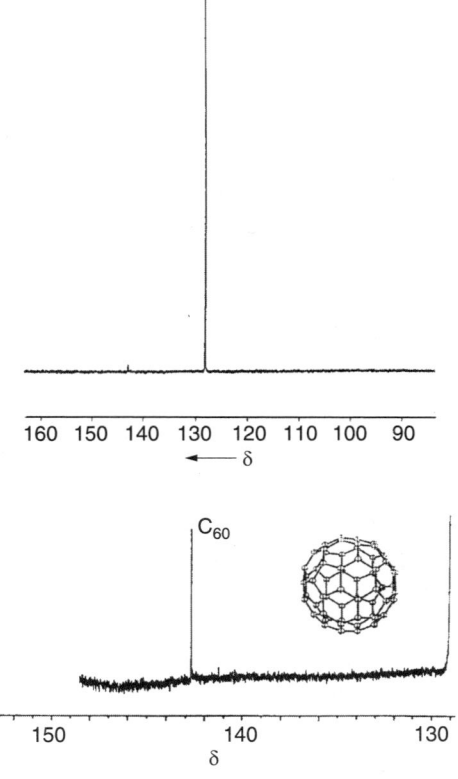

Figure 2.41 The ^{13}C NMR one line proof spectrum. *Source:* Reprinted with permission from Taylor R, Hare JP, Ala'a K Abdul-sada, Kroto HW, *Chem Commn*, 1423, 1990. Copyright 1990, The Royal Society of Chemistry. (a) Upper curve: ^{13}C NMR spectrum, note that the one line proof is the insignificant blip at 142 ppm whereas the significant line at 128 ppm is due to ^{13}C carbon nuclei in natural abundance in the benzene solvent (b) Lower curve: One line is seen more clearly at greater magnification.

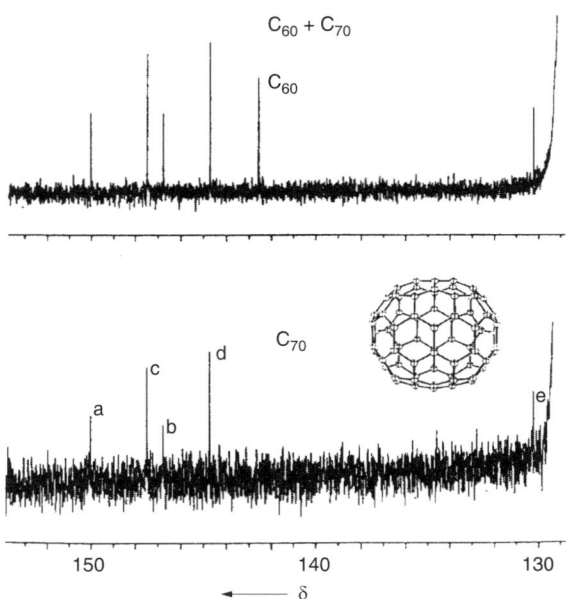

Figure 2.42 The ^{13}C NMR spectrum of C_{70}. *Source:* Reprinted with permission from Taylor R, Hare JP, Ala'a K Abdul-sada, Kroto HW, *Chem Commn*, 1423, 1990. Copyright 1990, The Royal Society of Chemistry. (a) Upper curve: Initial fraction from benzene solution. (b) Lower curve: After further purification to remove last trace of C_{60}.

Figure 2.43 Scanning Tunneling Microscope (STM) image of a thin layer of fullerite on a gold surface showing rows of C_{60} spheres interspersed with a few elongated molecules presumed to be C_{70} present as a contaminant in the fullerite. *Source:* Reprinted from Wilson RJ, Meijer G, Bethune DS, Johnson RD, Chambliss DD, deVries MS, Hunziker HE, Wendt HR, *Nature*, 348: 621, 1990.

ready explanation why it had not been possible to obtain a definitive X-ray diffraction pattern.

Subsequently, various workers have found ways of increasing the 1% yield of fullerite obtained at Heidelberg and it was not long before the University of California at Los Angeles had increased the yield to 14% and laser vaporization of graphite in an oven heated to 1200°C gave yields as high as 40% [80]. More recent developments claim a 94% yield of solvent dissolved fullerenes [81] produced from a carbon arc. An alternative approach using a benzene-oxygen flame, at 9 kPa and diluted with helium, is also reported to give high yields [82]. Relatively cheap reactor equipment is now on the market for organizations wishing to have their own production facilities, or alternatively, fullerite can be purchased direct from laboratory suppliers like Koch Light.

2.11.1.2 Properties and uses of fullerenes

The fullerenes are a family of geometrically closed polyhedral networks that have the general composition C_{20+2n}, comprising 12 pentagons and $(\frac{1}{2}n - 10)$ hexagons [58]. The first member of the family would be C_{20} with just 12 pentagons, but this is a thermodynamically unstable molecule and the next member, C_{22}, cannot be constructed. This is a rapidly developing field and so far C_{60}, C_{70}, C_{76}, C_{78} and C_{84} have been positively identified. Kroto developed the isolated pentagon rule, which stated that each pentagon should be surrounded by a ring of hexagons, thereby isolating each pentagon, imparting relative stability and, as a consequence, each carbon atom is connected to each of its neighboring carbon atoms by two single and one double bond. It has been shown that C_{62}, C_{64}, C_{66} and C_{68} would have abutting pentagons and, therefore, will not comply with the isolated pentagon rule. The incorporation of the pentagon introduces positive curvature to the structure until the network eventually closes (Figure 2.44). This curvature in a conjugated organic molecule causes a change in the hybridization of the σ bonds and this rehybridization, according to Haddon and Raghavachari [83], causes the π bond to be no longer of pure p orbital character and the σ bond hybridization of the fullerenes falls between diamond (sp^3) and graphite (sp^2) depending on the number of carbon atoms (Figure 2.45). The s character in the π orbital becomes progressively less pronounced until it finally reaches zero, when it becomes the ideal graphite crystal. Also, as n grows, the number of possible isomers grows very rapidly.

It is likely that the molecule is formed by the fusion of small aggregates of carbon particles, initially forming linear structures, which close into rings (Figure 2.44). C_{60} aggregates, when absolutely pure, have been shown by X-ray diffraction to have a face-centerd cubic structure, whereas C_{70} has a hexagonal structure.

Whilst benzene is truly aromatic in character, C_{60} is a molecule of ambiguous character [84] and can perhaps be described as a super alkyne, with thirty double bonds wrapped into a ball. C_{70} is, however, aromatic.

Hawkins' group at the University of California at Berkeley [85] produced $C_{60}(OsO_4)$ (4-tert-butylpyridine) (Figure 2.46) a so-called 'ball and stick' structure, the 'stick' preventing the C_{60} molecule from rotating and enabling a definitive X-ray diffraction pattern to be obtained, thereby facilitating the measurement of the carbon- carbon bond lengths within the C_{60} molecule. The C=C was measured as 0.1386 nm and C—C as 0.1434 nm.

C_{60} and C_{70} have been found to occur in nature, but are known to degrade in the presence of light and air (presence of ozone). C_{60} is the softest form of carbon but when compressed to below 70% of its original volume, it is reputed to be harder than diamond. It can be transformed into diamond by rapid compression (1.52 MPa in less than one second) to yield polycrystalline diamond in a carbon matrix. C_{60} has remarkable impact strength, being able to withstand a collision when impacted at 24,135 km h^{-1} onto a stainless steel plate, and just bouncing back.

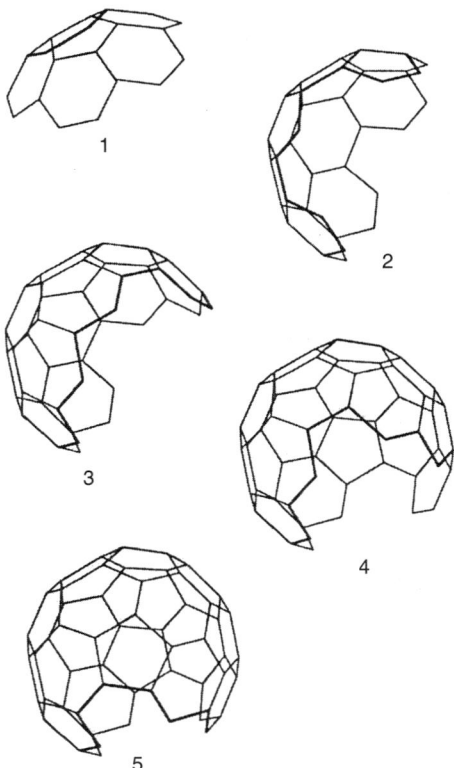

Figure 2.44 Hypothetical growth sequence of a C_{60} molecule. *Source:* Reprinted from Curl RF, Smalley RE, *Scientific American*, 265: 54–63, October 1991.

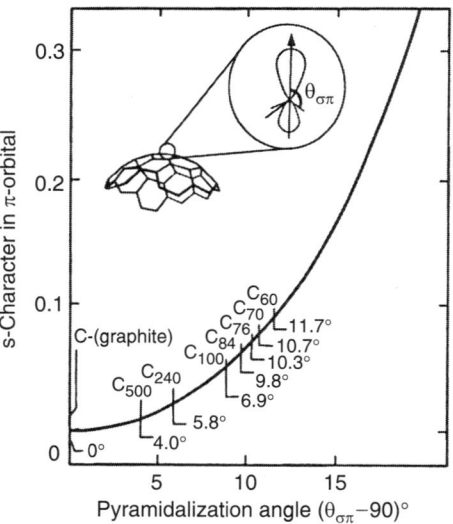

Figure 2.45 Hybridization of fullerene molecules as a function of pyrimidization angle $(\theta_{\sigma\pi} - 90)°$. As the number of carbon atoms increases, the angle is reduced until a planar graphite sheet is formed. *Source:* Reprinted from Haddon RC, *Chem Res*, 25: 127–133, 1992.

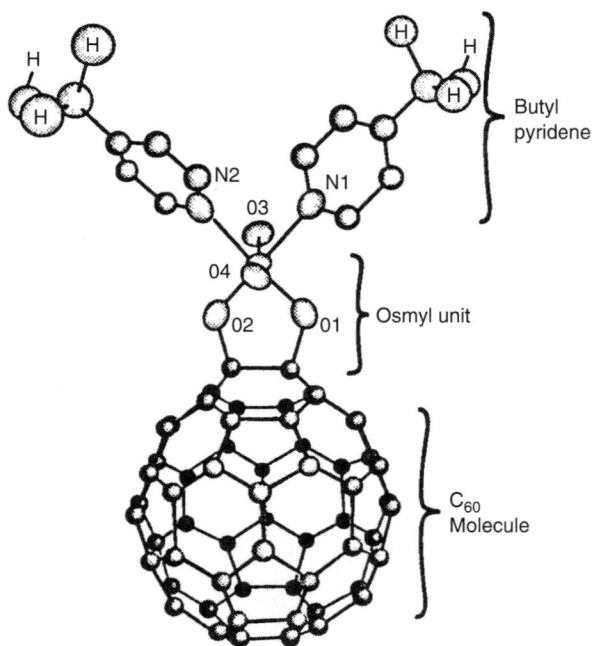

Figure 2.46 'Ball and Stick' structure of $C_{60}(OsO_4)$(4-*tert*-butylpyridine) deduced from X-ray diffraction pattern. The stick structure prevents the C_{60} ball from rotating, enabling detailed information to be obtained from the diffraction pattern. *Source:* Reprinted from Hawkins JM, Axel LTA, Loren S, Hollander FJ, *Science*, 252: 312, 1991.

As discussed, C_{60} has a football structure (Figure 2.39), with a single line in the NMR spectrum, since each carbon atom has an identical bonding environment. It is a mustard colored solid changing through brown to black as the film thickness is increased. Solid C_{60} exists in a needle form with a series of overlapping plates, which can be squares, triangles or trapezia, soluble in aromatic hydrocarbons giving a purple-magenta solution.

C_{60} burns when heated in air to give CO and CO_2. It has a high affinity for electrons due to its thirty double bonds, which can react to give exohedral compounds like $C_{60}H_{18}$, $C_{60}H_{36}$, $(CH_3)_nC_{60}$ ($n = 1$ to 34), where the added groups are outside the C_{60} shell. Halogenated compounds that can be formed, which include $C_{60}F_{40}$, $C_{60}Br_6$, $C_{60}Br_8$ and $C_{60}Br_{24}$. Organometallic complexes with Pd, Pt, Ru and Ir can be formed, such as $C_{60}\{[(C_2H_5)_3P]_2Pt\}_6$. C_{60} is an insulator, but potassium fulleride, K_3C_{60} is a true superconductor at 5 K, but add too much potassium and it becomes an insulator again.

The higher fullerenes beyond C_{60} are described by Diederich and Whetten [86]. C_{70} is shaped like a rugby ball (Figure 2.47), with a five-line structure in the NMR spectrum and is a reddish brown solid, changing to greyish-black with increasing film thickness and giving a deep wine-red solution. C_{76} has a chiral structure with left and right hand forms, one being the mirror image of the other. The solid is a bright yellow-green color, exhibiting the same color in solution. C_{78} has two forms, one is chestnut brown in solution and the other golden yellow, whilst C_{84} is olive-green and has at least two forms (Figure 2.48).

Fullerenes can also form compounds with atoms encapsulated within the cage structure, termed endohedral compounds and designated $M@C_n$, where the @ symbol signifies that the M atom is encapsulated within the C_n cage, e.g., $U@C_{28}$, $Y@C_{60}$, $Y_2@C_{82}$, $La_2@C_{80}$, $La@C_{82}$. Fullerene aggregates are bonded by van der Waals forces (c.f. graphite) and will permit the entrapment of alkali metal ions such as Li, Na, K, Rb and Cs [87].

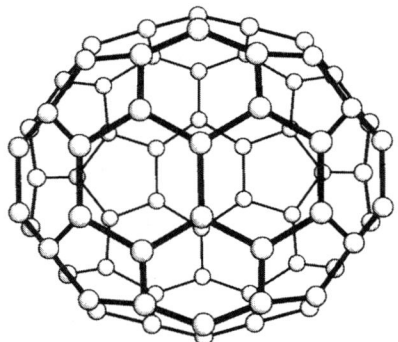

Figure 2.47 Closed cage structure for C_{70}. *Source:* Reprinted with permission from Zhang QL, O'Brien SC, Heath JR, Liu Y, Curl RF, Kroto HW, Smalley RE, *J Phys Chem*, 90: 525, 1986. Copyright 1986, American Chemical Society.

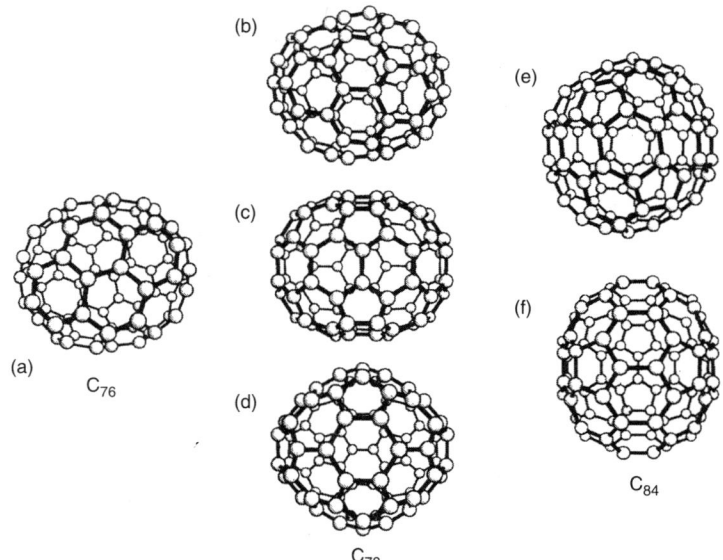

Figure 2.48 The structure of the higher fullerenes. *Source:* Reprinted with permission from Fowler PW, Manolopoulos DE, An Atlas of Fullerenes, *International Series of Monographs on Chemistry*, Clarendon Press, Oxford, Vol 30, 1995. Copyright 1995, Oxford University Press.

Although it is relatively early days, the suggested uses for fullerenes have not been commercially justified and are disappointing, probably due to the chemical reactivity of the fullerene derivatives.

2.11.2 Carbon nanotubes

2.11.2.1 *Discovery and production of carbon nanotubes*

In 1980, prior to the discovery of C_{60}, Sumio Iijima of NEC's Research Laboratory in Japan was examining electron microscope photographs of a thin film of carbon deposited on a substrate from the DC arc discharge evaporation of graphite which is used to prepare samples prior to electron microscopy and he observed, amidst a mass of jumbled lines, the occasional ordered arrangement of lines in the form of a series of concentric lines resembling the section through an onion. Innermost was a central sphere, about 0.8–1.0 nm diameter, with a gap of about 0.34 nm between each of, typically, ten concentric layers [88]. After the discovery of C_{60}, Iijima was convinced that his pictures had revealed a series of fullerenes, one inside the other with a molecule of C_{60} at the center.

Iijima then undertook a series of experiments to investigate the formation of these onion-like structures and found conditions where he could produce fine needles on certain regions of the negative electrode, ranging 4–30 nm diameter and up to 1 μm in length (in context, a human hair is about 70 μm in diameter). A careful study of these needles using electron microscopy revealed that they comprised a series of 2–50 tubes-within-a-tube, with a mean distance of 0.34 nm between the tubes. A more detailed study revealed that the tubes comprised hexagons arranged in a helical spiral pattern corresponding to a ring of some 30 carbon hexagons (Figure 2.49) [89]. This structure was not dissimilar to Bacon's whiskers (Figure 2.32), but the whiskers were in a scroll form, like a rolled-up newspaper, and were formed at 9.32 MPa in argon, whereas Iijima had used only 100 torr (13.4 kPa) of argon. The tips of the needles were usually closed with pentagon-containing hemispheres. Ebbesen and Ajayan significantly increased the yield of nanotubes by using graphite rods of unequal diameter and a higher pressure of 66.7 MPa with helium. A deposit formed on the thicker electrode and was packed with nanotubes, comprising about 25% of the carbon lost, with diameters of 2–20 nm and lengths up to several micrometer [90].

In 1993, Iijima and Ichibashi [91] prepared single wall nanotubes (SWNT) by creating a carbon arc discharge in 1.33 kPa methane, 40 torr argon and iron filings. The iron is melted, vaporized and finally condensed as iron carbide above the cathode. The nanotubes formed as bundles, from which individual SWNTs could be isolated. Bethune *et al* [92] used a different method to produce single wall nanotubes with daimeter of 1.2 nm. Methods

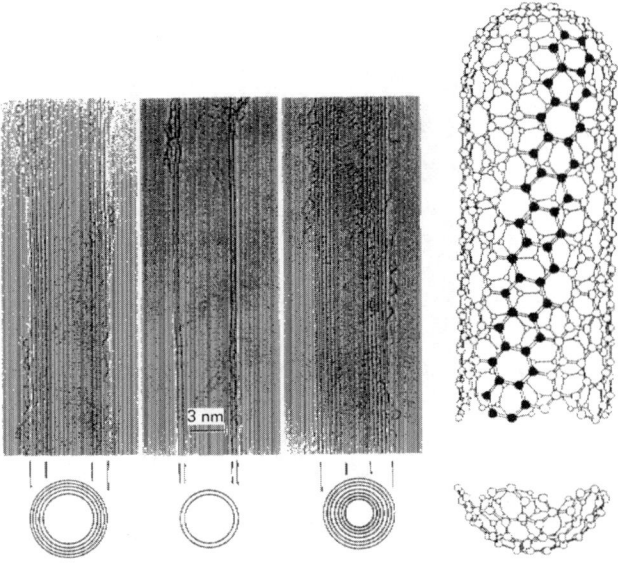

Figure 2.49 Carbon nanotubes. The schematic structure shows hemispherical end caps, each containing six pentagons and showing the helical arrangement of hexagons along the nanotube. *Source:* Reprinted from Iijima S, *Nature*, 354: 56–58, 1991.

have since been described for producing larger scale quantities of single wall nanotubes [89,90,93].

Harris [63] describes the mechanical properties of nanotubes, which have an extraordinary resilience [94,95], with high strength and modulus [96–101] and can be formed into strong ropes [102]. Initially, the mechanical properties were theoretically predicted, suggesting remarkable rigidities [96] and later, modulus values of 1500–5000 GPa were allocated [97], higher than 1060 GPa which is the accepted value for a graphene sheet. Lu [98] calculated 970 GPa, whilst Rubio and colleagues [99] obtained 1240 GPa. The first quantitative measurements were undertaken by Treacy, Ebbesen and Tomanek [97] using Transmission Electron Microscopy (TEM) with values averaging at 1800 GPa. Scanning Probe Microscopy (SPM) permitted the evaluation of single nanotubes, imaged by Atomic Force Microscopy (AFM). Wong, Sheehan and Lieber [101] obtained a value of about 1280 GPa for the Youngs Modulus of a single nanotube. Compressive strength values were estimated to be 100–150 GPa [100], which is, incredibly, at least one hundred times more than any other fiber.

Nanotubes exhibit electrical conductivity and may be used as nanowires [103] and nanoscale transistors in the future. A micro cathode ray tube with an extremely flat display has been created using a set of carbon nanotubes, which probe an area of about 1 mm^2 and a depth of about 0.2 mm.

The internal diameter of a nanotube is 3–6 nm and much attention has been devoted to filling the carbon nanotube with metals or metal oxides [104]. Treating the nanotubes with concentrated nitric acid opens up the ends and metallic elements can then be introduced, including Cr, Ni, Gd and Pb (Figure 2.50). In many cases, the filling reacts with the carbon to form a solid carbide nanorod [106]. Nanotubes can be conducting or insulating, depending on their structure. A switch can be formed if two carbon nanotubes are connected at an angle by using a pentagon ring [107]. The study of nanotubes [108–111] is an exciting and fast growing technology and it is suggested that nanotubes might become the ultimate carbon fibers.

Figure 2.50 Nanotubes filled with molten lead, which upon cooling, crystallizes the lead to a completely new form, which can be described as a molecular scale wire. *Source:* Reprinted from Ajayan PM, Iijima S, *Nature*, 361: 333, 1993.

2.11.3 Hyperfullerenes

In 1992, Daniel Ugarte at the Federal Polytechnic Institute in Switzerland completed some amazing work whilst studying typical carbon soot and exposing it in his electron microscope to a dose of electrons 10–20 times higher than normal. The different forms of graphite particles within the soot were gradually transformed into a series of concentric carbon shells. The larger particles were about 50 nm in diameter and corresponded to some 70 fullerene shells. Ugarte also observed particles with diameters up to several micrometer. This suggested that planar graphite may not be the most stable form of graphite and, under the right conditions, graphite would curl up, incorporating pentagons, hence eliminating the dangling bonds at the extremities of the chicken-wire. This hypothesis was so revolutionary that initially, the submitted paper was rejected by the referees. However, Kroto, who considered this work most important, actually endorsed the paper, which was then published [112]. These particles of fullerenes-within-fullerenes as shown in Figure 2.51 became known as hyperfullerenes, bucky onions or Russian dolls. Ugarte was able to identify C_{60} inside C_{240}

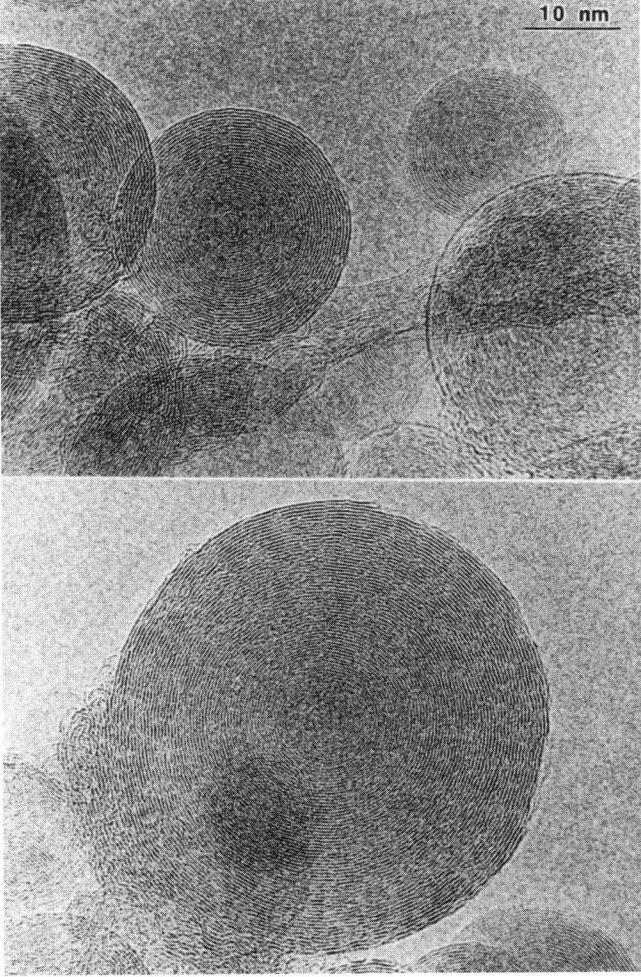

Figure 2.51 Hyperfullerenes consisting of nested fullerenes, one inside the other. *Source:* Reprinted from Ugarte D, *Nature*, 359: 707, 1992.

inside C_{540} inside C_{960}, confirmed by computer simulation and in reality it is highly likely that the spheres are rotating rapidly. The packing of the spheres interacts through van der Waals type forces [113].

2.12 SUMMARY OF ALLOTROPIC FORMS OF CARBON

The allotropes of carbon comprise diamond, which is three dimensional; graphite with its graphene planes that are two dimensional; C_{60} nanotubes, which can be considered as a unit cell of hexagons crystallized in one direction can be termed uni-directional; and finally, spheroidal fullerenes, which are zero dimensional, when considered as a carbon sheet wrapped around a point.

REFERENCES

1. Curl RF, Smalley RE, Fullerenes, *Scientific American*, 265: 54–63, October 1991.
2. Bundy FP, Direct conversion of graphite to diamond in static pressure apparatus, *J Chem Phys*, 38: 618, 1963.
3. Palmer HB, Shelef M, In: Walker RL Jr., ed. *Chemistry and Physics of Carbon*, Marcel Dekker, New York, 4: 1968.
4. Pierson HO, *Handbook of Carbon, Graphite, Diamond and Fullerenes*, Noyes Publications, Park Ridge NJ, 279, 1993.
5. Field JE ed., *The Properties of Diamonds*, Academic Press, London, 1979.
6. Davies G, *Diamond*, Adam Hilger (The publishing arm of The Institute of Physics), Bristol, 1984.
7. *Properties of Diamond*, a publication of De Beers Industrial Diamond Division, 1989.
8. Wilks J, Wilks EM, *Properties and Applications of Diamond*, Butterworth Heinemann, Oxford, 1991.
9. May PW, Rego CA, Thomas RM, Ashfold MNR, Rosser KN, Partridge PG, Everitt NM, Preparation of solid and hollow diamond fibres and the potential for diamond fibre metal matrix composites, *J Mater Sci Lett*, 13: 247–249, 1994.
10. Ting J, Lake ML, Diamond-coated carbon-fiber, *J Mater Res*, 9: No. 3, 636–642, 1994.
11. Shah SI, Waite MM, Diamond deposition on carbon-fibers, *Journal of Vacuum Science & Technology A—Vacuum Surfaces and Films*, 13: No.3, Part 2, 1624–1627, 1995.
12. Wasa K, Hayakawa S, *Handbook of Sputter Deposition Technology*, Noyes Publications, Park Ridge, NJ, 1992.
13. Reynolds WN, *The Physical Properties of Graphite*, Elsevier, 1968.
14. Cullity BD, *Elements of X-ray Diffraction*, Addison Wesley, Reading MA, 1956.
15. Eggers DF Jr, Halsey GD Jr, *Physical Chemistry*, John Wiley, New York, 1964.
16. Ubbelohde AR, Lewis FA, *Graphite and its Crystal Compounds*, Oxford, 1960.
17. Kavanagh A, Schögl R, The morphology of some natural and synthetic graphite, *Carbon*, 26 (1) 23–32, 1988.
18. SGL Carbon Group technical literature, *Carbon and Graphite Manufacture and Properties*, 1995.
19. Baggott JE, *Perfect Symmetry*, Oxford University Press, Oxford, 89, 1994.
20. Bokros JC, Chemistry and Physics of Carbon—A Series of Advances, 5: Walker PL Jr, *Deposition, Structure and Properties of Pyrolytic Carbon*, Marcel Dekker, 1969.
21. Albert AP, Parisot J, *Proceedings of the 3rd Carbon Conference*, Pergammon Press, New York, 1959, p.467.
22. Conradty C, Technical literature, Graphite Electrodes.
23. Union Carbide technical literature on graphite.
24. Blackslee GL, Proctor DG, Seldin EJ, Spence GB, Weng T, Elastic constants of compression-annealed pyrolytic graphite, *J Appl Phys*, 41:8, 3389, 1970.

25. Seldin EJ, Nezbeda CW, *J Appl Phys*, 41:8, 3373–3382, 1970.
26. Ebert LB, Intercalation compounds of graphite, *Annual Rev of Materials Sciences*, 6: 181–211, 1976.
27. Sawyer WE, Man A, US Pat 229,335; June 29, 1880.
28. Brown ARG, Hall AR, Watt W, The density of deposited carbon, *Nature*, London 172:1145, 1953.
29. Brown ARG, Watt W, Powell RW, Tye RP, The thermal and electrical conductivities of deposited carbon, *Br J Appl Phys*, 7: 73, 1956.
30. Brown ARG, Watt W, 1st Ind. Carbon and Graphite Conf. Soc Chem Ind, London, 86, 1958.
31. Blackman LC, Saunders G, Ubbelohde AR, The structure and mechanism of pyrolytic graphite, *Proc Roy Soc*, A264:19, 1961.
32. Cullis CF, Factors affecting the structure and properties of pyrolytic carbons, In: *Petroleum Derived Carbons*, ACS Symposium Series, No.21, 228–236, 1975.
33. Pierson HO, Lieberman ML, The chemical vapour deposition of carbon on carbon fibres, *Carbon*, 13: 159–166, 1975.
34. Lucas P, Marchand A, Pyrolytic carbon deposition from methane: an analytical approach to the chemical process, *Carbon*, 28(1): 207–219, 1990.
35. Pierson HO, *Handbook of Chemical Vapor Deposition*, Noyes Publications, Park Ridge, NJ, 1992.
36. Inspektor A, Carmi U, Raveh A, Khait Y, Avni R, *J Vac Sci Technol*, A4(3): 375–378, 1986.
37. Campbell J, Sherwood EM, *High Temperature Materials and Technology*, John Wiley, New York, 1967.
38. Dübgen R, *Glassy carbon—a material for use in analytical chemistry*, Publication of Sigri, D8901 Meitingen, Germany, 1985.
39. Jenkins GM, Kawamura K, Structure of glassy carbon, *Nature*, 231: 175, 1971.
40. Bacon R, Growth, structure and properties of graphite whiskers, *J Appl Phys*, 31: 283–290, 1960.
41. Bundy FP, Strong HM, Wentdorf RH Jr, In: Walker PL Jr, Thrower PA, eds. *Chemistry and Physics of Carbon*, Marcel Dekker, New York, 213.
42. Hughes TV, Chambers CR, *Manufacture of carbon filaments*, US Patent 405,480, 1889.
43. Baker RTK, Harris PS, The formation of filamentous carbon, In: Walker PL, Thrower PA, eds. *Chemistry and Physics of Carbon*, Marcel Dekker, New York, 14: 83, 1978.
44. Tibbetts GG, Vapor-grown Carbon Fibers, In: Figueiredo JL, Bernardo CA, Baker RTK, Hüttinger KJ, eds. *Carbon Fibers Filaments and Composites*, Kluwer, Dordrecht, 73–94, 1990.
45. Gadelle P, The growth of vapour deposited carbon fibres, In: Figueiredo JL, Bernardo CA, Baker RTK, Hüttinger KJ, eds. *Carbon Fibers Filaments and Composites*, Kluwer, Dordrecht, 95–117, 1990.
46. Tibbetts GG, Lengths of carbon fibres grown from iron catalyst particles in natural gas, *J Cryst Growth*, 73: 431, 1985.
47. Dresselhaus MS, Dresselhaus G, Sugihara K, Spain IL, Goldberg HA, *Graphite Fibers and Filaments*, Springer-Verlag, Berlin, 20–24, 1988.
48. Tibbetts GG, Vapor-grown carbon fibers: status and prospects, *Carbon*, 27: No.3, 745–747, 1989.
49. Tibbetts GG, Beetz CP Jr, Mechanical properties of vapor-grown carbon fibers, *J Phys D: Appl Phys*, 20: 292, 1987.
50. Koyama T, Endo MT, Method for manufacturing carbon fibres by a vapour phase process, Japanese Patent 198258,996, 1985.
51. Hatano M, Ohsaki T, Arakawa K, Graphite whiskers by new process and their composites, *Advancing Technology In Materials and Processes*, National SAMPE Symposium 30: 1467, 1985.
52. Kroto H, Space, stars, C_{60} and soot, *Science*, 242: 1139, 1988.
53. Fitzer E, Köchling KH, Boehm HP, Marsh H, Recommended terminology for the description of carbon as a solid, *Pure Appl Chem*, 67: (3), 503, 1995.
54. *Fullerene Science and Technology*, Braun T ed., Marcel Dekker.
55. Billups WE, Ciufolini MA, *Buckminster Fullerenes*, Vch Publications, 1993.
56. Koruga D, Hameroff S, Withers J, Loutfy R, *Fullerene C60: History, Physics, Nanobiology, Nanotechnology*, Elsevier Science Publications, Amsterdam, 1994.
57. Hirsch A, The Chemistry of the Fullerenes, *Thieme Medical Pub*, 1994.

58. Fowler PW, Manolopoulos DE, An Atlas of Fullerenes, *International Series of Monographs on Chemistry*, Clarendon Press, Oxford, Vol 30, 1995.
59. Ebbeson TW ed., *Carbon Nanotubes, Preparation and Properties*, CRC Press, 1996.
60. Endo M ed., Carbon Nanotubes, *Carbon*, Pergamon Press, Vol.33, 1996.
61. Dresselhaus MS, Dresselhaus G, Eklund P, *Science of Fullerenes and Carbon Nanotubes*, Academic Press, San Diego, 1996.
62. Kroto HW, Walton DRM eds., *The Fullerenes: New Horizons for the Chemistry, Physics and Astrophysics of Carbon*, Cambridge University Press, Cambridge, 1997.
63. Harris PJF, *Carbon Nanotubes and Related Structures: New Materials for the 21st Century*, Cambridge University Press, 1999.
64. Rohlfing EA, Cox DM, Kaldor A, Production and characterization of supersonic carbon cluster beams, *J Chem Phys*, 81: 3322, 1984.
65. Marks R, Buckminster Fuller R, *The Dymaxion World of Buckminster Fuller*, Anchor Press/Doubleday, New York, 1973.
66. Kroto HW, Heath JR, O'Brien SC, Curl RF, Smalley RE, Buckminsterfullerene, *Nature*, 318: 162, 1985.
67. Jones DEH, Pseudonym Daedalus, writing in the column Ariadne, Ariadne, *New Scientist*, 35: 245, 3 November, 1966.
68. Jones DEH, Hollow Molecules, In: *The Inventions of Daedalus*, WH Freeman & Co., Oxford, 118–119, 1982.
69. D'Arcy Thompson W, *Growth and Form*, Cambridge University Press, Cambridge, 708 and 738, 1942.
70. Osawa E, Superaromaticity, *Kagaku (Chemistry)*, 25: 854–863, 1970. (In Japanese).
71. Oshida Z, Osawa E, Aromaticity, *Kypto: Kagaku Dojin*, 1971. (In Japanese).
72. Osawa E, The evolution of the football structure for the C_{60} molecule: a retrospective, In: Kroto HW, Walton DRM eds., *The Fullerenes*, Cambridge University Press, Cambridge, 2–4, 1993.
73. Bochvar DA, Galpern EG, Hypothetical systems: carbo-dodecahedron, *s*-icosahedron, and carbo-*s*-icosahedron, *Dokl Acad Nauk SSSR*, 209: 610–612, 1973.
74. Stankevich IV, Nikerov MV, Bochvar DA, The structural chemistry of crystalline carbon: geometry, stability and electronic spectrum, *Russian Chemical Reviews*, 53: 640, 1984.
75. Krätschmer W, Fostiropoulos K, Huffman DR, The infrared and ultraviolet absorption spectra of laboratory produced carbon dust: evidence for the presence of the C_{60} molecule, *Chem Phys Lett*, 170: 167–170, 1990.
76. Krätschmer W, Lamb Lowell D, Fostiropoulos K, Huffman DR, Solid C_{60}: a new form of carbon, *Nature*, 347: 354, 1990.
77. Taylor R, Hare JP, Ala'a K Abdul-sada, Kroto HW, Isolation, separation and characterization of the fullerenes C_{60} and C_{70}: the third form of carbon, *Chem Commn*, 1423, 1990.
78. Wilson RJ, Meijer G, Bethune DS, Johnson RD, Chambliss DD, deVries MS, Hunziker HE, Wendt HR, Imaging C_{60} clusters on a surface using a scanning tunneling microscope, *Nature*, 348: 621, 1990.
79. Yannoni CS, Johnson RD, Meijer G, Bethune DS, Salem JR, ^{13}C NMR study of the C_{60} cluster in the solid state: molecular motion and carbon shift anisotropy, *J Phys Chem*, 95: 9, 1991.
80. Haufler RE, Chai Y, Chibanti LPF, Conceicao J, Jin C, Wang LS, Marayuma S, Smalley RE, Carbon arc generation of C_{60}, *Mater Res Symp Proc*, 206: 627, 1991.
81. Parker DH *et al*, *Carbon* 30(3): 1167–1182, 1992.
82. Howard JB, Lafleur AL, Makarovsty Y, Mitra S, Pope CJ, Yadav TK, *Carbon* 30(8): 1183–1201, 1992.
83. Haddon RC, Raghavachari K, Electronic structure of the fullerenes: carbon allotropes of intermediate hybridization, *Buckminsterfullerenes*, VCH, 1992.
84. Haddon RC, *Chem Res*, 25: 127–133, 1992.
85. Hawkins JM, Axel LTA, Loren S, Hollander FJ, Crystal structure of osmylated C_{60}. Confirmation of soccer ball framework, *Science*, 252: 312, 1991.
86. Diederich F, Whetten RL, Beyond C_{60}: The higher fullerenes, *Acc Chem Res*, 25:119–216, 1992.
87. Fischer JE, Heiny PA, Smith AB, *Acc Chem Res*, 25: 112–118, 1992.

88. Iijima S, The 60-carbon cluster has been revealed, *J of Phys Chem*, 91: 3466, 1987.
89. Iijima S, Helical microtubules of graphitic carbon, *Nature*, 354: 56–58, 1991.
90. Ebbesen TW, Ajayan PM, Large scale synthesis of carbon nanotubes, *Nature*, 358: 220, 1992.
91. Iijima S, Ichihashi T, Single shell nanotubes of 1 nm diameter, *Nature*, 363: 603–605, 1993.
92. Bethune DS, Kiang CH, deVries MS, Gorman G, Savoy R, Vazquez J, Beyers R, Cobalt-catalyzed growth of carbon nanotubes with single-atomic-layer shells, *Nature*, 363: 605, 1993.
93. Dagani R, Carbon nanotubes: recipes found for simplest variety, *Chem Eng News*, June 21, 1993.
94. Depres JF, Daguerre E, Lafdi K, Flexibility of graphene layers in carbon nanotubes, *Carbon*, 33: 925, 1995.
95. Iijima S, Brabec C, Maiti A, Bernholc J, Structural flexibility of carbon nanotubes, *J Chem Phys*, 104: 2089, 1996.
96. Overney G, Zhong Z, Tomanek D, Structural rigidity and low frequency vibrational modes of long carbon tubules, *Z Phys D*, 27: 93, 1993.
97. Treacy MMJ, Ebbesen TW, Tomanek D, Exceptionally high Young's modulus observed for individual carbon nanotubes, *Nature*, 381: 678, 1996.
98. Lu JP, Elastic properties of carbon nanotubes and nanoropes, *Phys Rev Lett*, 79: 1297, 1997.
99. Hernandez E, Goze C, Bernier P, Rubio A, Elastic properties of C and $B_xC_yN_z$ composite nanotubes, *Phys Rev Lett*, 80: 4502, 1998.
100. Lourie O, Cox DM, Wagner HD, Buckling and collapse of embedded carbon nanotubes, *Phys Rev Lett*, 81: 1638, 1998.
101. Wong EW, Sheehan PE, Lieber CM, Nanobeam mechanics: elasticity, strength, and toughness of nanorods and nanotubes, *Science*, 277:1971, 1997.
102. Thess A, Lee R, Nikolaev P, Dai H, Petit P, Robert J, Xu C, Hee Lee Y, Gon Kim S, Rinzler AG, Colbert DT, Scuseria G, Tomanek D, Fischer JE, Smalley RE, Crystalline ropes of metallic carbon nanotubes, *Science*, 273: 483–487, 1996.
103. Tans SJ, Devoret MH, Dai H, Thess A, Smalley RE, Geerligs LJ, Dekker C, Individual single-wall carbon nanotubes as quantum wires, *Nature*, 386: 474, 1997.
104. Green MLH *et al*, Simple chemical method of opening and filling carbon nanotubes, *Nature* 372: 159, 1994.
105. Ajayan PM, Iijima S, Capillary induced filling of carbon nanotubes, *Nature*, 361: 333, 1993.
106. Dai H *et al*, Synthesis and characterization of carbide nanorods, *Nature*, 375: 769, 1995.
107. Dresselhaus M, Carbon connections promise nanoelectronics, *Physics World* 9: (5), 18, 1996.
108. Ebbesen TW, Carbon nanotubes, *Annu Rev Mater Sci*, 24: 235–64, 1994.
109. Ebbesen TW, Carbon nanotubes, *Physics Today*, 26–32, June 1996.
110. Ebbesen TW ed., *Carbon Nanotubes: Preparation and Properties*, CRC Press, 1996.
111. Endo M ed., Carbon Nanotubes, *Carbon*, Vol. 13, 1996.
112. Ugarte D, Curling and closure of graphitic networks under electron beam irradiation, *Nature*, 359: 707, 1992.
113. Baum R, Fullerenes broaden scientist's views of molecular structure, *Chemical & Engineering News*, 29–34, January 4, 1993.
114. Zhang QL, O'Brien SC, Heath JR, Liu Y, Curl RF, Kroto HW, Smalley RE, Reactivity of large carbon clusters: spheroidal carbon shells and their possible relevance to the formation and morphology of soot, *J Phys Chem*, 90: 525, 1986.

CHAPTER 3

History and Early Development of Carbon Fibers

3.1 THE EARLY INVENTORS

About thirty years ago, carbon fiber was heralded as the new wonder material. However, carbon fiber was certainly not new, although it has since proved to be an extremely useful reinforcement material. Thomas Alva Edison, who made his first sound recordings in 1877, turned his attentions to the incandescent electric lamp and in 1880, patented the use of carbon fiber as filament material for his electric lamp [1]. To achieve this aim, in a period of some 15 months, Edison and his associates in the laboratory (Figure 3.1) at Menle Park, New Jersey, USA built a new type of generator, found a suitable element material and incorporated it in a glass globe operating under a high vacuum [2–4].

To find a satisfactory element material, Edison was reputed to have tried more than 1600 kinds of materials ('paper and cloth, thread, fish line, fiber, celluloid, boxwood, coconut shells, spruce, hickory, hay, maple shavings, rosewood, soft dry wood rotted by fungal attack (punk), cork, flax, bamboo and incredibly the hair out of a red-headed Scotsman's beard'). Until eventually, Edison hit on the idea of carbonizing a loop of ordinary cotton thread, which glowed in a vacuum for more than half of that eventful day of 21 October, 1879. This filament was later replaced by carbonized 'Bristol cardboard' that burned for 170 hours. The incandescent electric lamp had arrived.

The carbon filaments were, at a later date, developed by Edison by dissolving cellulosic materials like natural cellulose or cotton in a solvent, such as zinc chloride, to give a dope which could be extruded through a die into a bath containing a liquid (spin bath), which would regenerate the cellulose in the form of a thread or filament. The filament material was cut into lengths and carbonized in the absence of air in a heated gas furnace. The material was inherently weak and subsequent improvements were made by cracking a layer of pyrolytic carbon onto the surface. This was carried out by placing the filaments in the vapor of a hydrocarbon, such as benzene, and passing an electric current to cause the hydrocarbon to decompose and crack onto the surface of the filaments with resultant improved properties.

Edison also took out a UK patent dated 10 November 1879 [5], which preceded, by some 12 months, a UK provisional patent taken out by J.W. Swann [6] for lamp filaments made of carbon. These patents of Edison and Swann are the first recorded UK patents dealing with carbon fiber. In 1889, Hughes and Chambers [7] patented a process to produce carbon filaments from a mixture of methane and hydrogen pyrolyzed in an iron crucible yielding hair-like carbon filaments, but the process was uneconomic. A further improvement

Figure 3.1 Edison with his team of 'scientific men' at the Menlo Park R&D facility in New Jersey taken in 1876. Edison is seated, wearing cap, and sixth from the left. Note the pipe organ at the back, the focal point for after-hours singing and beer drinking. Those were the days! *Source*: Courtesy of Tom Koba, The Edison Birthplace Museum, Milan, Ohio.

was made in 1909 by Whitney [8], who took a product comprising an impure carbon core surrounded by an outer layer of pyrolytic carbon and heated it in a carbon tube electric furnace to temperatures from 2300°C–3700°C reputedly, converting the outer layer of pyrolytic carbon into a more graphitic form with improved electrical properties, but the fiber remained weak and very brittle. To overcome this brittleness, the filaments were shaped into the intended element construction before the final graphitizing procedure. The use of carbon filaments for electric lamps was relatively short-lived, being replaced *circa* 1910 by the more robust metallic wires such as tungsten.

3.2 WORK IN THE USA

3.2.1 Black 'Orlon'

In 1950, R.C.Houtz [9] heated 'Orlon' (a proprietary polyacrylonitrile (PAN) fiber made by DuPont) for 16–20 hours in air at 200°C. The fiber underwent a series of color changes, passing from white through yellow, brown and finally, black. This fiber was shown to be non-flammable and when plunged into the flame of a Bunsen burner did not burn. Although the fiber glowed and lost about 30% in weight, it did not melt or deform. This black 'Orlon' was the first example of an oxidized PAN fiber (OPF) and it is difficult to believe that almost 10 years were to elapse before the commercial significance of opf was followed up [10–12]. In 1950, it was not understood that the fibers, when exposed in the Bunsen burner flame, had been pyrolyzed to carbon fiber. In the original oxidation process, Houtz postulated a dehydrogenation reaction with the formation of heterocyclic fused rings:

Table 3.1 The physical properties of Black Orlon

Treatment	Dry Tenacity g/denier	Wet Tenacity g/denier	Dry Elongation (%)	Wet Elongation (%)
Control	3.42	3.35	15.7	17.00
16 h at 200°C in air	1.94	1.77	9.3	9.2
16 h at 200°C in N_2	2.87	2.78	12.9	13.7

Source: Reprinted with permission from Houtz RC, J Text Res, 20:786–801, 1950. Copyright 1950, Textile Research Institute.

The theoretical hydrogen content of the Orlon starting material was 5.7% reducing to 2.0% in this condensed ring structure, compared with an actual hydrogen content of 2.3%. The properties in Table 3.1 show the surprisingly good physical properties of black Orlon, suggesting that there was no appreciable polymer chain degradation. The properties remained good after heating for a further 60 hours at 200°C. The black Orlon was insoluble in a number of good PAN solvents. When a load of 0.5 g denier^{-1} was applied, the temperature at which the yarn broke had increased from 172 to 250°C. The fibers had greater moisture absorption and absorbed mineral acids, suggesting the presence of basic groups. When the Orlon was heated in the dry state, it did not undergo a random chain scission reaction as happened when treated with an alkaline solution.

3.2.2 Some early US carbon fibers

The carbon filaments used by Edison had poor mechanical properties but, during the 1950s in the USA, there was an added impetus to make stronger grades of carbon fiber from rayon and other precursors, brought about by the advent of the space program and backed by strong support from US government agencies.

The National Carbon Company, a division of Union Carbide had, by 1959, introduced a carbon cloth made by carbonizing rayon cloth, to be followed by felts and battings in 1960 and in the next year, by development quantities of yarn, as described by Cranch [13]. There were three basic forms of yarn, loosely described as carbonaceous, carbon and graphite (Table 3.2). A commercial cloth (grade WCB) was available with a calculated tensile strength of 60 MPa, whilst unwoven filaments were measured as 345–690 MPa, and a typical yarn (WB-0030) comprised of some 1600 filaments /ply and 10 plies/yarn, with a nominal filament diameter of 4.6 μm.

Later, the National Carbon Company (Table 3.3) had available the following:

1. Grade VYB cloth, which was carbonized up to 1000°C forming a product constituting about 90% carbon, together with oxygen, hydrogen and an ash residue.

Table 3.2 Properties of early National Carbon Company's carbon fiber

Fiber	Volatile content to 2700°C %	Specific Gravity $g\,cm^{-3}$	Linear shrinkage to 2700°C %	Tensile strength MPa	Temperature for 1% weight loss in 8 h °C
Carbonaceous	50	1.4	18	117	100
Carbon	15	1.4	6	186	350
Graphite	0	1.4	0	255	500

Source: Adapted from Cranch GE, Unique properties of flexible carbon fibers, Proceedings of 5th Conference on Carbon, Vol. 1, Pergamon Press, Oxford, 589–594, 1962.

Table 3.3 Properties of some early Union Carbide and Hitco yarns

Yarn	VYB 70-½	VYB 70-1/0	VYB 105-1/5	WYB 125-1/5	CY-1064-2
Supplier	U.C.C.	U.C.C.	U.C.C.	U.C.C.	H. I. Thompson Fiberglass Co.
No. of plies	2	1			2
Filaments per ply	720	720			720
No. of filaments	1440	720	2400	2400	1440
Twist per ply (tpm)	79				98
Twist (tpm)	83	79			244
Weight loss* (%)		10.6			1.3
Analysis: C (%)	77.1	76.5	89.7	98.6	96.4
H (%)	1.7	1.9	0.6	0.3	0.3
N (%)	0.2	0.3			0.4
P (%)	3.0	2.7			0.2
Ash (oxide) (%)	1.3	3.2			1.0
Na (%)		0.6			0.3
Filament Strength (Gpa)	0.53				0.59
Filament Modulus (Gpa)	27		41–55	48–69	31.7
Diameter (μm)	11.4				8.4

*Dried 2 h at 105°C in vacuum.
Source: Reprinted with permission from Bacon R, Pallozzi AA, Slosarik SE, Cabon filament reinforced composites, Proceedings Society Plastics Industry, 21st Technical and Manufacturing Conference, Sect. 8E, 1966, Ezekiel HM, Spain RG, Preparation of graphite fibers from polymeric fibers, J Polymer Sci, 249–265, 1967. Copyright 1967, John Wiley & Sons Ltd.

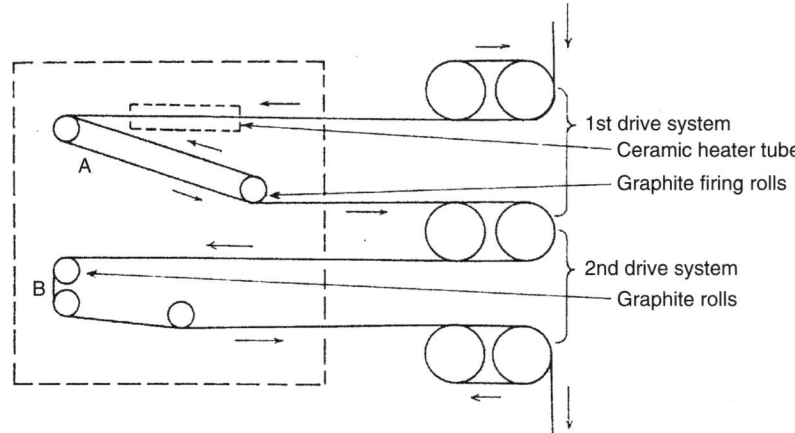

Figure 3.2 Schematic diagram of Hitco's resistive heating equipment for manufacturing carbon fiber. *Source:* Reprinted from Gibson DW, Langlois GB, *Polymer Preprints 9*, No. 2, 1376, 1968.

2. Grade WYB cloth, which was carbonized at temperatures in excess of 2200°C and was a relatively pure carbon (99%) with a so-called graphitic structure, but was essentially a purer form of carbon and was non-graphitizing.

The H.I. Thompson Fiberglass Company (later called Hitco) was also active [15] and D.W. Gibson and G.B. Langlois published their method for producing carbon yarn (Figure 3.2). To start the process, the yarn was sufficiently heated within a ceramic heater tube in an inert atmosphere of nitrogen, until it became electrically conductive. It was then double-wrapped around a pair of graphite electrode rollers through which an electric current was passed to heat the fiber. Once the fiber had been rendered electrically conductive, there was no further need to preheat the fiber, as this was achieved by direct radiation emitting from the lower pass of electrically conductive fiber. The fiber could be stretched by varying the speed ratio of the first drive system. Final heat treatment was achieved by passing over a second pair of driven graphite electrodes, where it could be stretched again.

These products were followed with a range of carbon and graphite felts, used for high temperature insulation materials. The carbon product with a lower thermal and electrical conductivity was useful for thermal insulation in furnaces operating in an inert atmosphere or vacuum [17,18] and certain ablative applications requiring the lower heat conductivity [19]. The relatively pure carbon product, with good lubricity and chemical resistance, found applications for seals and gaskets [20–22] and was also used for controlling the carburizing conditions for certain refractory metals such as V, Nb, and Ta [23].

W.T. Soltes [24] described a technique for the thermal conversion of cellulosic material such as cotton, rayon, hemp or flax in a two stage process in the absence of oxygen to give a textile carbon, which was electrically conductive. This process was subsequently taken up commercially by Union Carbide.

The exact conditions depended on the nature and thickness of the precursor material, but essentially, the first production stage comprised a three zone furnace operating at 180, 350 and 550°C, at a line speed of 0.5 m h^{-1}, with zone lengths adjusted to give residence times of 2, 4 and 3 minutes respectively. After this first carbonization stage, there was an optional water wash aided by gentle agitation to remove water soluble impurities. The second carbonization stage also comprised of a three zone furnace with a dwell time of 30 s at 900°C, 30 s at 1400°C and 15 s in the final unheated cooling zone to prevent the hot carbon bursting into flames when exposed to the atmosphere. Appreciable shrinkage

Table 3.4 Properties of monofilaments

Precursor Material	Diameter μm	Tensile Strength GPa	Specific Resistance μohm cm
1. Invention:			
1100/720 tire cord	5–7	0.74–0.90	2000–5400
1100/480 yarn	6.5–8.5		2100–2500
900/50 yarn	20–25	0.33–0.37	1900–4500
2. Edison	25–158		
3. Whitney	20–160		<160
4. Commercial lamp filaments: (F.J. & J. Planchon, Paris)			
(a)	146	0.19–0.23	4060
(b)	205	0.13–0.15	4400
(c)	500		

Source: Reprinted from Ford CE, Mitchell CV, U.S. Pat., 3,107,152, 1963.

occurred, but the product was flexible and a good conductor of electricity, with an electrical resistivity of 0.827 ohm cm at room temperature.

C.E. Ford and C.V. Mitchell of Union Carbide [25] patented an improved process, whereby the cellulosic material was carbonized under a controlled heating process comprising: $10°C\ h^{-1}$ to $100°C$, $50°C\ h^{-1}$ to $400°C$ and $100°C\ h^{-1}$ to about $900°C$, followed by heating to $3000°C$, using a protective atmosphere of nitrogen or other inert gases when heating through the range $900°C$–$3000°C$ until substantial graphitization had occurred. This treatment could be used for rayon monofilaments, cellulosic yarns or a pre-woven rayon textile material. The product was highly flexible compared with the commercial lamp filaments produced by the Edison or Whitney type processes and, in part, was attributed to the smaller cross-sections (5–25 μm diameter) compared to the much larger diameter (20–500 μm) of the electric lamp filaments. The product also enjoyed the benefits of a significant increase in tensile strength (Table 3.4). There was a distinct difference in the X-ray diffraction patterns, with the newer fibers exhibiting a higher degree of preferred orientation and being more graphitic in nature with consequent lower specific resistance.

In 1964, Union Carbide introduced Thornel 25 in the market (ultimate tensile strength (UTS) 1.25 GPa and Young's modulus (YM) 170 GPa) at a price of more than £1000 per kg. They carbonized a highly oriented viscose rayon to give a fiber with a low modulus of 70 GPa and this fiber was subsequently stretched at a temperature of 2500°C to give a fiber with a modulus of 170 GPa. Afterwards, fibers such as Thornel 50, Thornel 75 and Thornel 100 were introduced with improved moduli (Table 3.5). However, hot stretching was

Table 3.5 Properties of Union Carbide's carbon fiber made from a rayon precursor

Grade	Density g cm^{-3}	Tensile strength GPa	Tensile modulus GPa
Thornel 25	–	1.25	172
Thornel 50	1.63	1.97	345
Thornel 75	1.86	2.59	517
Thornel 100	1.99	3.95	690

Source: Reprinted with permission from Prosen SP, Carbon resin composites, *Fibre Sci Technol*, 3, 81, 1970, Copyright 1970, Elsevier. Epremian E, "Thornel", a new graphite reinforcement, *Seminar on Polyblends and Composites at Polytechnic Institute of Brooklyn*, June 6 and 7, 1969.

prohibitively expensive. The process was discontinued in 1978. Hitco meanwhile produced Hitron HMG 50, an equivalent grade of Thornel 50.

3.2.3 More recent US carbon fibers

In 1972, Hitco produced HG1900, which was made from a Bayer PAN precursor, processed in a unidirectional woven fabric form because it could only be stretched in the warp direction (UTS 1.72 GPa and YM 552 GPa).

In 1970, L.S. Singer of UCC developed a technique to convert pitch to a mesophase stage [28], enabling the product to be spun and after much further work [29,30], enabled UCC, in 1976, to launch a high performance pitch based carbon fiber (HPCF) namely Type P (TS 1.7 GPa, YM 235 GPa), eventually extending the range to P100 (TS 2.2 GPa, YM 724 GPa) and P120 (TS 2.2 GPa, YM 827 GPa).

In 1978, UCC obtained a cross license from Toray and marketed in the USA, the Toray range of PAN based carbon fibers and ultimately, in 1984, produced their own PAN based fiber using the Toray precursor and technology.

3.3 WORK IN JAPAN

3.3.1 Early work in Japan with PAN precursor

A team lead by Shindo of the Industrial Research Institute in Osaka was the first to make carbon fibers from PAN fibers, filing for patents in 1959 [31] and 1962 [32]. The patent disclosed that when a PAN fiber (3 denier Exlan bright fiber made from a copolymer with a 90% acrylonitrile content) was preferably preheated in air at 170–200°C before carbonization, carbon fibers were obtained with a strength of 0.1 GPa. The process did not, however, cover the application of tension during the pre-oxidation stage, but it was established that pre-oxidation, combined with slow heating, prevented chain scission and gave good carbon fiber with an increased yield.

The work was subsequently published in 1961 [33,34], showing that relatively high strength carbon fibers resulted from the pyrolysis of a PAN yarn. Chemical analyzes of the gases evolved from a pre-oxidized sample showed that the total amount of NH_3 and HCN evolved up to 1000°C was about half the total weight loss of the preoxidized sample heated to 1000°C, suggesting that other gaseous products must be evolved and, for example, when the evolved gaseous products were condensed out, an amber liquid was collected, which eventually crystallized out, but was not identified. Since each unit of PAN comprises three carbon atoms, three hydrogen atoms and one nitrogen atom, it would have been expected that if all of the nitrogen was evolved as ammonia then there would have been a direct relationship between the nitrogen and hydrogen evolution, but this was not found to be so and, as shown by analysis, all the nitrogen and hydrogen is not eliminated as ammonia.

Using pre-oxidized fiber it was established that up to 300°C, the strength went down and then increased to 0.49–0.98 GPa at 1000°C, decreasing again to 0.20–0.39 GPa as the temperature reached 3000°C. Above 1000°C, strengths were found to increase with decreasing cross-sectional area, with the greater rate below 5.6 μm diameter, attributed to a smaller core size of weaker fiber and less defects. The elongation dropped from 1% to 0.3% and the Young's modulus increased from 108 GPa to 147 GPa at 3000°C, with a peak value of about 170 GPa at 2000°C, for which no explanation was offered.

There is no doubt that this study sets out many of the ground rules for the present day carbon fiber production process from a PAN precursor but, unfortunately, Shindo and his team failed to exploit the commercial significance of this early work and only patented the process they had used.

3.3.2 Work in Japan with pitch precursors

This work is covered in greater detail in Chapter 4. Initial work by Otani at the Gunma University [35–37] established a method of obtaining a general purpose pitch based carbon fiber (GPCF). The first product was made from a pitch obtained by the pyrolysis of polyvinylchloride at 400°C under nitrogen, which could be easily melt spun at 200–370°C and then converted to an infusible fiber by an oxidative treatment with ozone for 3 h at 60–70°C, finally heating in air by increasing the temperature at a rate of 1.5°C/min up to 260°C, then holding for 1 h, followed by heating in nitrogen up to 1000°C at 5°C/min, holding for 20 min and then cooling. The fiber is similar in structure to a glassy carbon and is isotropic with a consequent low modulus, hot stretching at 2000–2800°C does, however, increase the modulus (440 GPa at 2500°C), but the fibers are not graphitic and have a similar microstructure to rayon and PAN based high modulus carbon fibers. Otani [37] found that an identical pitch based carbon fiber (PBCF) product could be obtained from a blown asphalt by dry distillation at 380°C for 1 h, followed by heat treatment at 270–340°C under vacuum, but an acceptable product could not be prepared from a coal-tar pitch. In 1973, Kureha Chemical Company developed a commercial PBCF process [38–40].

The introduction of HPCF by Union Carbide in the USA initiated intensive research and development to improve the processability of mesophase pitch (MP). Riggs and Diefendorf in the US worked on a neo-mesophase pitch based on a solvent extraction technique. Yamada and co-workers discovered the pre-mesophase pitch [41,42] using hydrogenation followed by a rapid heat treatment, while workers at the Kyushu Industrial Research Institute in Japan hydrogenated an anisotropic pitch (preferably a coal tar pitch), which after heat treatment produced a dormant mesophase pitch [43,44], a process known as the Kyukoshi method and able to produce a type of carbon fiber intermediate between a GP and HP fiber. Mochida and co-workers used a Lewis acid, such as $AlCl_3$, for the catalytic polymerization of an isotropic pitch, but found that an excessive amount of catalyst was required to achieve mesophase formation.

3.4 WORK IN THE UK WITH PAN PRECURSORS

3.4.1 Work at RAE, Farnborough

An excellent account of the work of William Watt, the co-inventor of PAN based carbon fiber, is given in a Biographical Memoir [45] and W. Watt, W. Johnson and R. Moreton describe some of the RAE work [46].

3.4.1.1 *The RAE work with carbon fiber and cross-licencing of their patent*

Independent of the Japanese work, W. Watt, W. Johnson and L.N. Phillips (Figure 3.3) of the Royal Aircraft Establishment at Farnborough (RAE) started work in 1963 on the production of carbon fiber from a PAN precursor. Courtaulds were invited to submit a

Figure 3.3 The co-inventors of carbon fiber at RAE Farnborough, 1968. From left to right: Leslie Phillips, Bill Watts and Bill Johnson. *Source:* Photograph courtesy of Defence Research Agency, Farnborough, England (now Qinetiq).

range of organic fibers, which might be suitable for conversion to carbon fibers. One of these fibers was particularly promising, namely Courtelle, a PAN fiber that had a high melting point and underwent little chain scission during heat treatment. Initially, the work was undertaken with a crimped 4½ denier Courtelle, a commercial PAN fiber containing TiO_2 as a delustrant. Courtaulds then provided some uncrimped 3 denier material without the delustrant and since this was a textile grade of fiber, Courtaulds later developed a special acrylic fiber (SAF), specifically for conversion to carbon fiber. The work at RAE was successful and culminated in Watt, Phillips and Johnson applying for a patent in 1964 [47–49] covering a technique for improving the fiber orientation by restraining the natural shrinkage, or even applying some stretch to the fiber in the oxidation stages of the carbon fiber process, without any subsequent recourse to hot stretching.

The PAN precursor was wrapped around a metal frame, secured and oxidized in the temperature range of 200–250°C, when it was found that if too much fiber had been wound onto the frame, or the oxidation temperature was too high, then a run-away exotherm occurred burning the fibers. Heating slowly to 1000°C at 15°C/h gave a product with a modulus of 150 GPa and this value increased to 380 GPa when heat treated to 2500°C. Shindo's [34] values for strengths and moduli plotted against the RAE values reported by Moreton, Watt and Johnson [50] are shown in Figure 3.4. The strength curve was often quoted thereafter in the literature peaking at about 1500°C but, was later shown to be dependent on the precursor filament diameter and purity. The modulus, however, is an intrinsic property and is dependent on the structure.

Watt and Johnson undertook further work [51] to study the oxidation stage and measured the rate of oxygen uptake at 220°C of 3 denier Courtelle using a Coleman oxygen analyzer (Figure 3.5). PAN fibers heated *in vacuo* turned to a coppery color and did not turn black. When polished, transverse cross sections of partly oxidized fibers were examined using reflected light microscopy, they found a central pale yellow core with an outermost brown annulus, with the pale yellow core diameter, decreasing with time as the thickness of the outer zone increased, suggesting a diffusion controlled oxidation

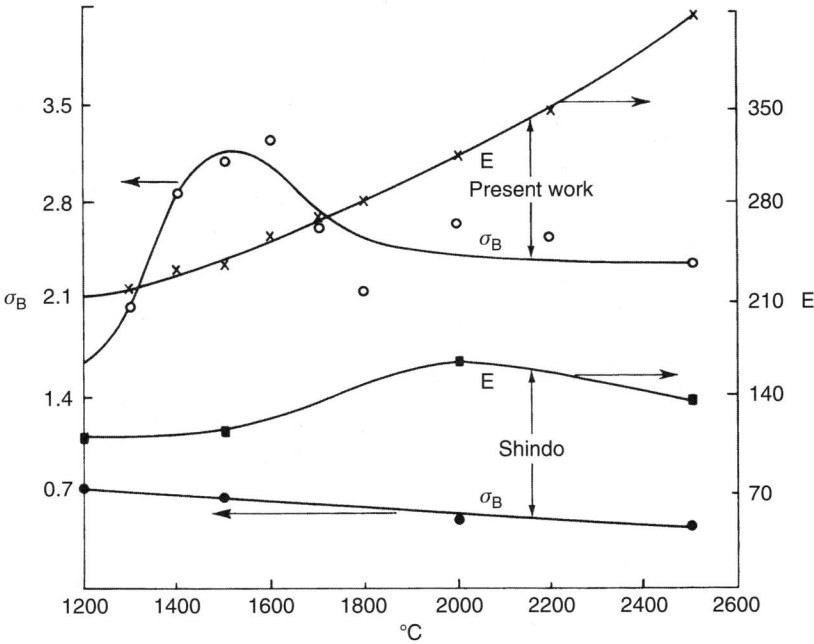

Figure 3.4 Comparison of tensile strength and Young's modulus of carbon fibers Vs Heat treatment temperature as produced by RAE. *Source:* Reprinted from Moreton R, Watt W, Johnson W, *Nature*, 213, 690, 1967, Shindo A, Osaka Kogyo Gijitsu Shikenjo Hokoku, No. 317, 1961.

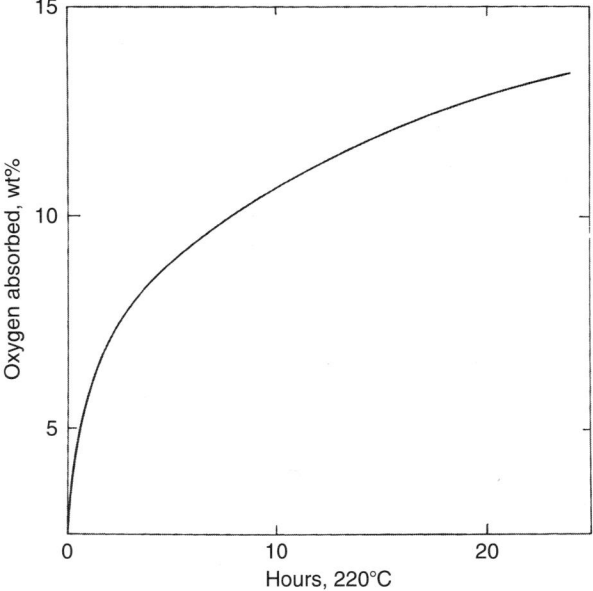

Figure 3.5 Oxygen absorbed by 3 denier Courtelle PAN fiber at 220°C. *Source:* Reprinted from Watt W, Perov BV eds., *Handbook of composites Vol 1 Strong Fibres*, Amsterdam, 344, 1985.

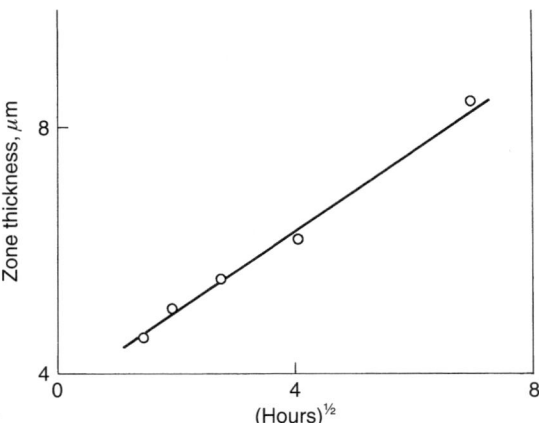

Figure 3.6 Oxidized zone thickness as a fuction of (time)$^{1/2}$ for 3 denier Courtelle PAN fiber in air at 220°C. *Source*: Reprinted from Watt W, Perov BV eds., *Handbook of composites Vol 1 Strong Fibres*, Amsterdam, 345, 1985.

mechanism, which was confirmed by a plot of zone thickness against the square root of time (Figure 3.6).

PAN fibers are not crystalline and when a stretched fiber is heated above its stretching temperature, the fiber shrinks because of entropic length shrinkage, with the fiber wanting to revert to its original disordered state. The RAE work restrained this shrinkage, by securing the fiber on a frame during the oxidation stage. The PAN fibers were stabilized in the oxidation process and there was no necessity to restrict shrinkage in the subsequent carbonization stages. The RAE workers examined the effect of fiber length changes during the stabilization stage and the subsequent effect on the resulting YM. Initially, with 3 denier, Courtelle [51] oxidized at 220°C, with length increases of 0–40% and heat treated without tension to 1000°C and 2500°C, the fibers shrank 13% in length and some 45% in diameter. The YM varied from 155–190 GPa for fiber treated at 1000°C and from 350–420 GPa for fiber treated at 2500°C. Later, the work was repeated with 1.5 denier Courtelle (Figures 3.7, 3.8), when the smaller diameter PAN considerably reduced the oxidation time from about 25 to 5 h [52].

Courtelle and Dralon T (a homopolymer) were pyrolyzed up to 1000°C [53] and the gas evolution curves for HCN and NH_3 are given in Figures 3.9 and 3.10. Nitrogen evolution started at 700°C and was still increasing at 1000°C.

So, summarizing the findings of the 1968 RAE patent, the following important conditions were highlighted [46]:

1. The oxidation temperature must be controlled below the temperature at which a thermal runaway occurs.
2. The degree of oxidation must be sufficient to penetrate to the center of the PAN fiber.
3. Length shrinkage must be restricted during oxidation, or even some stretch applied to the fibers during oxidation.
4. After the oxidation stage, there no tension was needed in any subsequent carbonization stages.

From this batch procedure, RAE developed a small scale laboratory continuous carbon fiber process to demonstrate feasibility. To provide larger quantities for industrial evaluation, a contract was placed, in 1965, with the Atomic Energy Research Establishment, Harwell, who had existing furnaces that could be readily adapted for this application and

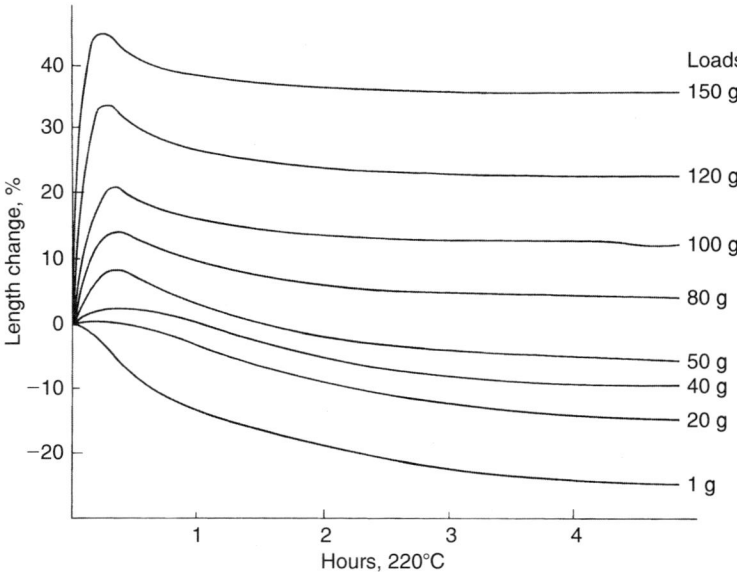

Figure 3.7 Length changes of 700 filament tows of 1.5 denier Courtelle PAN fiber in air at 220°C with different tensile loads. *Source*: Reprinted from Watt W, Johnson W, Conference on High Temperature Resistant Fibres, Amer Chem Soc, Interscience, Atlantic City, 1968.

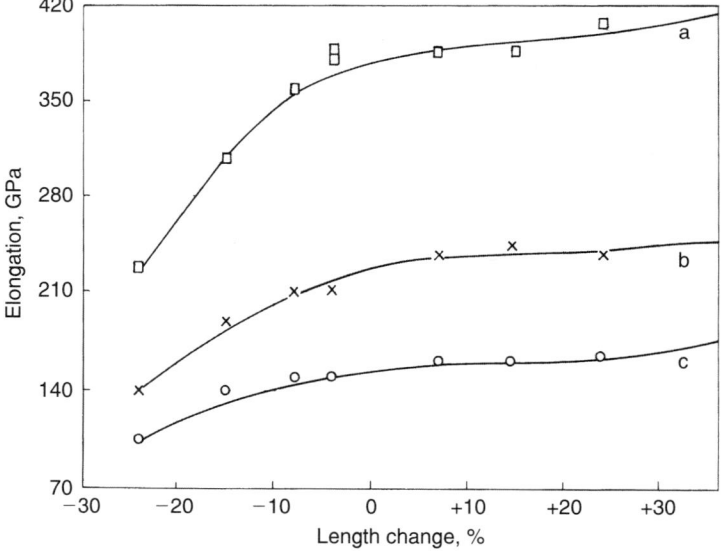

Figure 3.8 Young's modulus of carbon fibers *Vs* Length changes for 700 filament tows of 1.5 denier Courtelle PAN fiber in air at 220°C. (a) Carbonized to 1000°C (b) Heat treated to 1500°C (c) Heat treated to 2500°C. *Source:* Reprinted from Watt W, Johnson W, Conf. High Temperature Resistant Fibres, American Chemical Society, Interscience, Atlantic City, 1968.

initially, Harwell produced a staple fiber in approximately 4.5 kg batches, some 35 cm long [54]. Later, in 1966, the RAE patent was licensed to three British companies to develop carbon fibers commercially: Morganite Research and Development, who were experienced in high temperature technology; Courtaulds Ltd., who had made the precursor; and Rolls

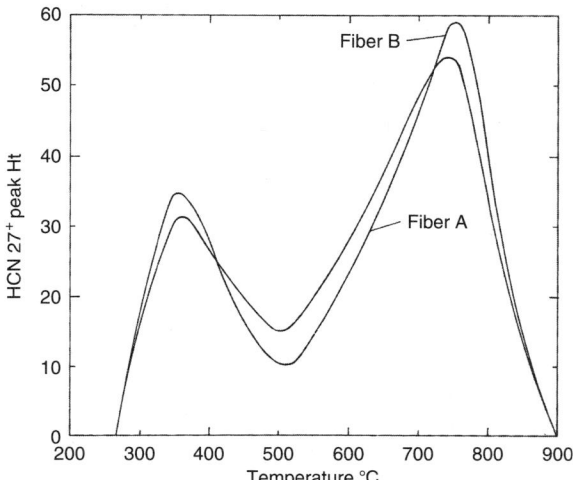

Figure 3.9 HCN evolution measured in mass spec. units (×10 attenuation) during continuous pyrolysis of stabilized fiber per mg of sample. Fiber A—Courtelle (1½denier) Fiber B—Dralon T. *Source:* Reprinted with permission from Watt W, Johnson DJ, Parker E, Pyrolysis and structure development in the conversion of PAN fibres to carbon fibres, *Proc 2nd International Carbon Fibre Conference Plastics institute*, London, Paper No.1, 3–11, 1974. Copyright 1974, Maney Publishing (who administers the copyright on behalf of IOM Communications Ltd, a wholly owned subsidiary of the Institute of Materials, Minerals & Mining).

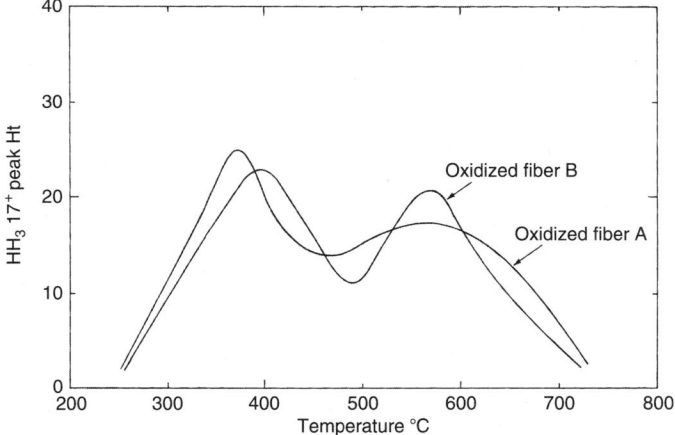

Figure 3.10 NH$_3$ evolution measured in mass spec. units (×10 attenuation) during pyrolysis of stabilized fiber per mg of sample. Fiber A—Courtelle Fiber B—Dralon T. *Source:* Reprinted with permission from Watt W, Johnson DJ, Parker E, Pyrolysis and structure development in the conversion of PAN fibres to carbon fibres, *Proc 2nd International Carbon Fibre Conference Plastics institute*, London, Paper No.1, 3–11, 1974. Copyright 1974, Maney Publishing (who administers the copyright on behalf of IOM Communications Ltd, a wholly owned subsidiary of the Institute of Materials, Minerals & Mining).

Royce, who had already developed their own distillation process, but subsequently changed to the RAE process. Since that time, each company has developed its own process, which may or may not be similar to the original RAE process.

In late 1966, Morganite were asked by RAE to make one ton of carbon fiber, which they supplied by the end of 1967. It was kept in a locked room known as the Black Fort Knox, since at that time, it was worth about £65,000. The fiber was given free to companies, in conjunction with a grant, to evaluate and develop end uses for carbon fiber.

Production of carbon fibers in the USA to the RAE patent was started in 1971 by two companies: Hercules Inc. (who had an arrangement with Courtaulds Ltd.) and Morganite Modmor Inc. (a joint company formed by the Whittaker Corp. and Morgan Crucible).

3.4.1.2 Surface treatment

Wadsworth and Watt patented [55] a method of surface treating carbon fibers by oxidizing in air at 350–850°C, obtaining a weight loss of 0.05–6.0% or, less conveniently, using an oxygen rich, pure oxygen, nitrous oxide or nitrogen dioxide atmosphere. An example quoted for a Type I fiber after heating in air for 1½ h at 550°C had a weight loss of 0.9% and the interlaminar shear strength (ILSS) of an epoxy composite had increased from about 19 to 69 MPa. Work on electrolytic surface treatment was undertaken by Harvey [56], who agreed with the findings of Paul [57] that the level of surface treatment of the fiber was a function only of the number of ions discharged at the fiber surface during its passage through the bath.

Harvey [58] found that the surface area of a surface treated fiber was not significantly different from that of untreated fibers. However, surface treated fibers did absorb less metanil yellow and more methylene blue than untreated fibers of the same type, but this did not necessarily reflect an increase in the ILSS of the composite.

Some work was undertaken [59] to determine the effect of the presence of about 0.2% of sodium found in the Courtauld's PAN fiber, probably present as —COONa groups with some adventitious Na from the NaSCN spin bath solvent. A significant reduction (about 75%) in Na levels was achieved by extracting the PAN fiber with boiling dilute hydrochloric acid, resulting in a slight improvement in composite strength, but with no improvement in YM. In the carbon fiber form, this Na content increased to about 0.5% for a high strength fiber and there was some concern that this may detract the resistance of the composite to hot/wet conditions. Judd and Wright [60] reported that there was some contamination of Na on the surface of a composite, which could be removed by a water wash, but subsequent heating to about 150°C resulted in further contamination. Extraction of the PAN with a 13% NH_4Cl solution was effective in removing about 90% of the Na.

3.4.1.3 Testing and properties of single filaments and composites

Moreton prepared PAN fiber (6% MA) and found that as stretch was increased in the PAN production process, the strength and the Young's modulus of the resulting carbon fiber also increased [61], which is not surprising since the filament diameters were significantly reduced. Preparing a PAN precursor under clean room conditions [46] showed the advantages of limiting adventitious impurities and, when converted at 1400°C, the resultant carbon fiber showed less pronounced gage length effects, whilst the effect was not so pronounced with the sample heat treated at 2500°C.

Many reports were issued on the properties of carbon fiber reinforced plastic (cfrp) and the effect of the environment and Judd investigated the corrosion of carbon fiber in various chemical environments [62]. Curtis issued an extremely useful report on test methods for the measurement of the engineering properties of fiber reinforced plastics under the aegis of Composite Research Advisory Group (CRAG) [63].

3.4.1.4 *Composite fabrication*

R Child [64] developed a resin cure cycle simulator to measure the gel time under autoclave conditions and found that the fiber volume fraction of a laminate made in an autoclave is determined almost entirely by the inherent packing of the fiber and/or prepreg and little can be done during the autoclave cycle to change the natural fiber volume.

A simple, portable and cheap apparatus was developed [64] for vacuum-molding reinforced plastic components and a prototype tape laying machine was also developed [64].

LN Phillips described the film stacking process [64] to produce thermoplastic prepreg by hot pressing thermoplastic coated carbon fiber interleaved with sheets of thin film of the same polymer such as polysuphone and polyethersufhone.

3.4.1.5 *Friction and wear*

Giltrow and Lancaster [65], using chopped carbon fibers in a metal matrix composite, found that they exhibited a lower wear rate when sliding against a tool steel, than the matrix material (Co, Ni, Cu, Ag, Pb) and there was no significant improvement in friction. High temperature performance was unsatisfactory and mechanical properties were degraded when compared with the matrix materials. To a large extent, these problems were overcome by using continuous fiber. Using carbon fiber reinforced plastics, Lancaster and Giltrow [66–73] found that wear under lubricated conditions was influenced by the type of carbon fiber and matrix material, the counterface and its degree of surface roughness, the working temperature and the formation of transferred film.

High strength carbon fibers (Type II) are much more abrasive than the high modulus (Type I) variety [74]. Carbon-carbon composites have high compressive strengths and stiffnesses and friction and wear properties can be modified by the introduction of additives [75].

3.4.2 Work at the Atomic Energy Research Establishment, Harwell

Some of the early work at Harwell, before 1969, was described by Logsdail [76]. The description that follows of the work undertaken by Harwell is not exhaustive but it does serve to illustrate the wide spectrum of the projects undertaken. Some of the topics are enlarged upon in later chapters.

3.4.2.1 *Fiber production*

In 1965, Harwell, at the request of RAE, initiated the production of 35 cm long staple carbon fiber in batches of about 4.5 kg, specifically for use at RAE. Once the commercial UK companies came on stream, Harwell turned their attention to making continuous fiber, at first some 300 m long [77]. The initial production was by a semi-batch process, where the precursor was wound onto metal frames for oxidation and then off-wound and coiled into a vertical pot furnace for subsequent carbonization. A problem associated with this procedure was that the lengths of fiber had in-built kinks, which were points of incipient weakness, where the precursor had come into contact with the metal frame during oxidation.

Table 3.6 Details of large tow textile grade precursors

Brand name	Manufacturer	No. of filaments k	d'tex per filament	Wt % Copolymers	Fiber shape
Acribel	Fabelta (Belgium)			MA	Bean
Acrilan	Monsanto (UK)	160	3.3	7% VA	Bean
Courtelle	Courtaulds (UK)	160	3.3	6% MA, 1% ITA	Round
Creslan	American Cyanamid (US)			MA, MMA up to 1994→ VA	Round
Dolan	Hoechst (Germany)			5% MA	Dog-bone
Crylor	Societé Crylor (France)	90	3.3	6% MMA	Bean
Dralon N	Bayer (Germany)	160		5% MA	Dog-bone
DralonT	Bayer (Germany)	160	3.3	None (homopolymer)	Dog-bone
Euroacril	ANIC (Italy)	200	3.3	5% MA, 5% VA	Off-round
Leacril	Montefibre (Italy)			VA	Off-round
Nitron	Ex. Russia (cf Courtelle)			6% MA, 1% ITA	Round
Orlon	Du Pont (Netherlands)	157	3.3	6% MA	Dog-bone
Tacryl	Superfosfat (Sweden)			8% MA,TAHT	Round
Velicren	SNIA (Italy)	160	3.3	MA	Bean
Vonnel	Mitsubishi (Japan)	160	3.3	VA	Bean
Zefran	Dow Badische (USA)			MA	Round

Key:
- ITA   Itaconic acid
- MA   Methyl acrylate
- MMA   Methyl methacylate
- MVP   Methyl vinyl pyridine
- TAHT   Triacrylohydrothiazine
- VA   Vinyl acetate

Source: Sanders RE, *Chem & Proc Eng*, 100–109, Sep 1968.

Production was aimed at making continuous carbon fiber from crimped textile tow acrylic precursors and these are listed in Table 3.6. Initially, work had been undertaken on the oxidation reaction kinetics and it was established, not surprisingly, that various commercial precursors demanded different processing conditions [79]. The oxidation oven comprised an electrically heated modular design with intermediate drive rollers to permit stretching in the early stages of oxidation and preventing contraction in the later stages. Hot air was circulated with a fan and exhaust gases were continuously removed and replaced by a constant fresh air supply to prevent the build of an explosive mixture. At a reaction temperature of 220°C in air, the main gases evolved in relatively large quantities were HCN, H_2O and CO_2, with CO and NH_3 in smaller amounts. If a thermal runaway, however, did take place, then chain scission occurred and there was a large release of HCN accompanied by the formation of high molecular weight hydrocarbons [80]. The oxidized fiber was collected on spools and interleaved with paper. It was that found raising the temperature slowly was essential, with the larger tow sizes taking due care to avoid folds or thick bunches of fiber, as otherwise a burn-up would occur. It was found that carbon fibers with acceptable mechanical properties could be best produced from oxidized fibers with an oxygen content of 10–12.3%. The fibers after oxidation could be either, batch processed as 350 mm long staple or, processed continuously by carbonizing at 1000°C, followed by heat treatment to 2600°C and then given the standard Harwell wet-hypochlorite surface treatment [81]. Since fibers like Orlon and Acrilan did not have a round cross-section, the carbon fiber produced was tested in a composite form with a 60% V_f and the results normalized to a 100% fiber basis to give typical flexural strengths of 1.76 GPa at 1000°C and 1.45 GPa at 2600°C, with corresponding flexural moduli of 157 GPa and 287 GPa. With better control of the oxidation stage, improved values were obtained at 1100°C: 2.01 GPa and 182 GPa for Courtelle with 2.27 GPa and 171 GPa for Orlon. The effect of heat treatment temperature on the flexural strength and flexural modulus for a range of commercial PAN fibers is shown in Figures 3.11 and 3.12. The flexural strengths peaked at temperatures from 1300–2100°C depending on the precursor, but the flexural moduli increased linearly with

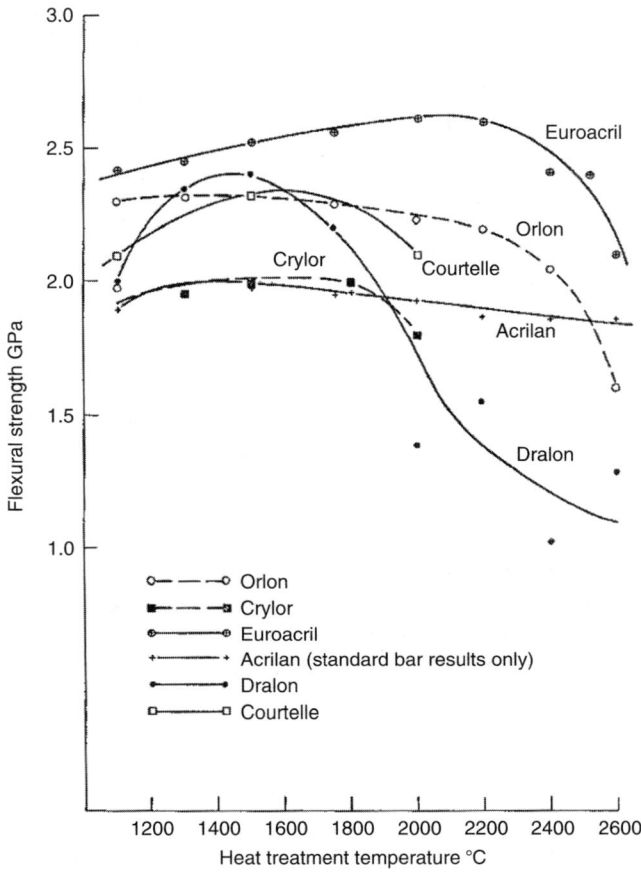

Figure 3.11 Variation of carbon fiber composite flexural strength with heat treatment temperature. *Source:* Reprinted from with permission Hornby J, Kearsey HA, Sharps JW, The preparation of carbon fibres from large-tow textile acrylic fibre: 5. Continuous heat-treatment at temperatures to 2600°C, *AERE–R 8867*, United Kingdom Atomic Energy Authority, Harwell, Jan 1978, Hornby J, Kearsey HA, The preparation of carbon fibres from large-tow textile acrylic fibre: 4. Comparison of different precursors, *AERE–R 8865*, United Kingdom Atomic Energy Authority, Harwell, Dec 1977. Copyright 1977, AEA Technology plc.

temperature and were promoted by increases in tension. The importance of sizing to maintain tow integrity was established. Composite test procedures were outlined by ID Aitken *et al* [83] using the resin system 828/MNA/BDMA utilizing standard bars 330 mm long with 10–12 g fiber which, were later replaced by mini test bars of the same length but thinner (1 mm) and narrower (5 mm) and only containing some 2.5–3.0 g fiber, which were found to give more consistent data. However, the results obtained from the two bars did differ and the rather dubious, but practical, method of applying a correction factor was used to convert from the standard test bar results to mini-bar values (the factor was ×0.9 for flexural modulus). Span:depth ratios were 16:1 for the flexural strength test and 40:1 for flexural modulus. The best carbon fiber properties obtained for the different precursors are given in Table 3.7.

The gas evolution processes that occurred during the formation of carbon fibers up to 1000°C were studied by Bromley [85] and in the region 400–500°C, the main products evolved were NH_3, CO_2, H_2O and CH_4, together with high molecular weight hydrocarbons, but little HCN. In the 750–850°C region, HCN was again evolved.

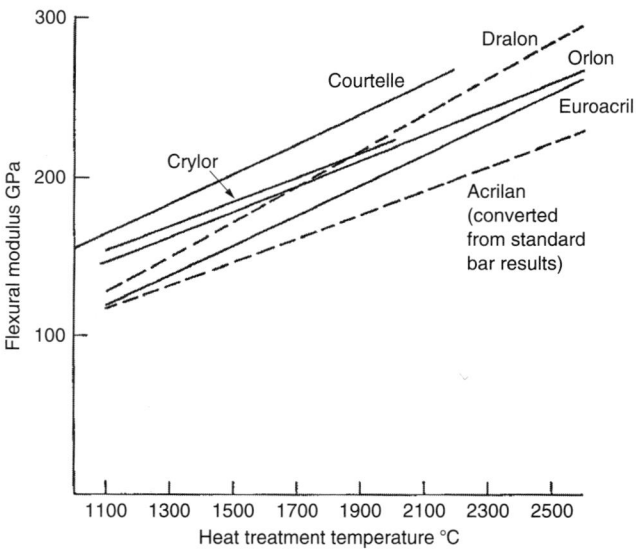

Figure 3.12 Variation of carbon fiber composite flexural modulus with heat treatment temperature. *Source:* Reprinted with permission from Hornby J, Kearsey HA, Sharps JW, The preparation of carbon fibres from large-tow textile acrylic fibre: 5. Continuous heat-treatment at temperatures to 2600°C, *AERE–R 8867*, United Kingdom Atomic Energy Authority, Harwell, Jan 1978. Copright 1978, AEA Technology plc.

Table 3.7 Best carbon fiber properties obtained from textile grade precursors

Precursor	Flexural strength Gpa			Flexural modulus Gpa			Strain to failure for 1100°C material %
	1100°C	1500°C	1700°C	1100°C	1500°C	1700°C	
Acrilan	1.86	–	2.14	124	–	162	1.5
Courtelle	2.46	–	–	162	–	–	1.2–1.5
Crylor	2.14	1.99	1.98	157	187	203	1.4
Dralon	2.44	2.45	1.66	150	177	198	1.4–1.6
Euroacril	2.76	2.70	2.76	145	166	181	1.6–1.9
Orlon	2.50	2.42	2.44	145	173	189	1.5–1.7
Vonnel	2.14	2.15	–	140	170	–	1.5

Source: Reprinted from Hornby J, Kearsey HA, The preparation of carbon fibres from large-tow textile acrylic fibre: 4. Comparison of different precursors, *AERE–R 8865*, United Kingdom Atomic Energy Authority, Harwell, Dec 1977.

In the continuous process, a single oxidized 160 k flat PAN tow band about 100 mm wide, which had been wound on a drum and interleaved with paper, was pulled through tensioning rollers and then through an induction heated graphite tube by an outlet drive to give for Orlon, an optimum carbonization processing time of about 5 min at 1100°C. Inert gas seals at each end of the furnace were purged with argon to prevent ingress of air and the flow served to move the exhaust products from the furnace. The high temperature furnace was a Spembly resistive heated graphite tube, with either graphite felt or carbon black insulation and a water cooled body and ends. Gas seals were provided to prevent ingress of air and argon was used for the inert gas unless arcing occurred, then helium was used.

At temperatures of about 1700°C, Courtelle was prone to produce fiber with internal defects which were attributed to the vaporization of some component within the filament and similar to the defects found in SAF when heated to above 1500°C, as observed by Sharp and Burnay [86].

Harwell developed a laboratory scale pilot plant which was ideal for investigating Courtaulds SAF and established some useful parameters. A 10 k SAF was held under tension continuously in a nine-stage oxidation oven controlled at $220 \pm 2°C$ and the fiber was pulled through by wrapping around a large diameter servo-controlled driven drum. Since oxidized fiber readily absorbs moisture, which Harwell claimed [87] was detrimental in that it reduces the strength of the carbon fiber, the oxidized fiber was dried by passing through an oven at 140°C with a counter current flow of dry N_2 prior to going onward through a heated silica tube carbonizing furnace with a linear temperature gradient of 260–700°C. This furnace was fitted with gas seals to prevent ingress of air, whilst a body flow of inert gas (N_2 or Ar) swept the decomposition products to the hot end of the furnace, where they were removed through a heated outlet tube by an extract system, incorporating a tar trap, thus preventing the deposition of any breakdown products onto the fiber. The fiber then passed through a Heat Treatment (HT) furnace heated to 1500°C, comprising an alumina tube heated with silicon carbide heating elements and fitted with N_2 gas seals at either end. The decomposition products were removed with a countercurrent stream of N_2. The fiber was collected in layers on a driven drum in a helical wind pattern by means of a separately driven traversing mechanism. The carbonizing furnace exhaust system was also used for the decomposition products exiting the HT furnace. It was possible to produce fiber with a filament strength of 3.7 GPa [88] and a YM of 270 GPa. Furthermore, fiber could be produced with a strength of 3.0–3.2 GPa at a much reduced oxidation time of some 20 min, although no details were available. Filament testing was carried out with 25 filaments using a 49 mm gage length in accordance with procedures given by Goggin [89]. Effective fiber strengths can be considerably higher than these values by utilizing a strand test developed by Harwell [90,91]. One surprising fact to emerge was the different behavior of N_2 and Ar in the carbonizing furnace with carry over of carbonizing furnace products occurring with Ar but not with N_2, which could be attributed to the 30% higher viscosity and/or the 50% higher density of Ar.

An analysis of 8000 individual filaments showed that 18% had a filament strength greater than 3.5 GPa, indicating that there was a potential for a fiber with greater strength if certain unknown parameters could be controlled. After a detailed study [92], initial experiments suggested that the gas atmosphere in the HT furnace could be controlling the fiber strength. A ninefold increase of fiber speed in the HT furnace improved the strength from 2.3 to 2.9 GPa, which was attributed to the ninefold increase in the decomposition products that would have occurred. The decomposition products were analyzed using gas chromatography and comprised of H_2, CO, N_2, CH_4, CO_2, NH_3 and HCN. The effect of adding each of these gases (except N_2 and CO_2) was determined. Although CH_4 improved the strength to 3.1 GPa, it formed a sooty deposit, while on the other hand, HCN produced a marked improvement, obtaining strength values up to 3.9 GPa and HCN concentrations greater than 0.7% were found to achieve the maximum effect. It was suggested that surface defects were healed by a process involving the reaction:

$$HCN + H_2 \rightarrow NH_3 + C$$

It was believed that this could be achieved commercially by a pumped recirculation of some of the exhaust gases and/or reducing the furnace purge of N_2. However, contamination with sodium, which was evolved in the HT furnace when using Courtauld's SAF precursor, could make this impracticable. Work was also undertaken at Harwell to produce carbon fibers from cellulosic and pitch precursors.

3.4.2.2 Surface treatment

As continuous carbon fiber came onstream, a continuous surface treatment process was required, which could be used online and Harwell developed an air oxidation process. To obtain an adequate level of surface treatment, the oxidation was regulated to give a weight loss of carbon of 1–2% at the operating temperature of 475–720°C, which required a furnace residence time of about 10 min. Since air oxidation is rapid at these elevated temperatures and difficult to control, the reaction could be poisoned by the addition of 0.5% chlorine to the air [93] with a residence time of 75 sec at 950°C, associated with a weight loss of 3%. The process could be readily monitored by measuring the amount of CO and CO_2 evolved and controlling the emission to equate with the required weight loss.

A continuous electrolytic surface treatment process with the carbon fiber as anode and a copper cathode using a mixed electrolyte of 5% sodium chloride and 0.25% sodium hydroxide was patented [94,95]. The object was to generate free chlorine, which then reacted with the sodium hydroxide to give sodium hypochlorite. This was followed by a batch process [81] for surface treating carbon fiber with a dilute solution of sodium hypochlorite containing preferably 1.0–3.0% available chlorine, using acetic acid to adjust the pH to 4.0–6.0. Treatment was carried out at 40–60°C for times varying 4–16 h depending on the fiber type to be treated followed by washing in water, rinsing with distilled water and drying. It was believed that the reactive species was an oxidizing species derived from the breakdown of un-ionized HOCl. This process was initially used by Morganite for surface treating staple carbon fiber and Courtaulds used their own modified version.

3.4.2.3 Testing and properties of single filaments and composites

Harwell produced many papers on quality assessment techniques covering the testing and properties of carbon fibers and their composites—filament testing was comprehensively undertaken by Goggin [89] and further work examined the effect of gage length. By testing at two different gage lengths, the results could be extrapolated to the critical length of 0.3 mm, which had been determined experimentally (Figure 3.13). Composite testing was covered by Aitken, Rhodes and Spencer [81], whilst Hughes, Morley and Jackson [90,91] developed a very efficient aligned tow test. A comparison of torsion and short beam methods for measuring shear were compared by Reynolds and Hancox [97] with details of the torsion apparatus described by Hancox [98]. The torsion test used solid rods made from Araldite MY750/MNA/BDMA and, because of the absence of tensile and compressive forces, a more accurate value of the shear strength was obtained. The torsional shear modulus and shear strength values are given in Figures 3.14 and 3.15. The torsional shear strength values were dependent on the V_f of fiber, but did not obey the Law of Mixtures. Properties were given by Wells and Colclough [99,100], whilst engineering constants of carbon fiber composites were determined by Reynolds [101] and Goggin [102] (Table 3.8). The thermal and electrical properties were covered by Knibbs [103].

Reynolds, at the Non-destructive Testing Center at Harwell [101], looked at a wide variety of test procedures for carbon fibers and their composites, including measuring modulus by determining the transit time for a short ultrasonic pulse along a single filament. Others included using the Stereoscan to determine fiber volume fractions and determining the density of an image on a photographic plate to obtain fiber volume fractions.

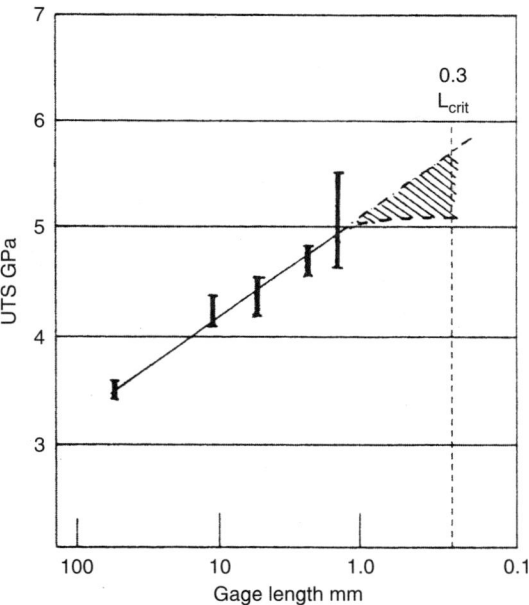

Figure 3.13 The effect of gage length on strength of single carbon fibers. Fibers were tested at two gage lengths and the results extrapolated to the critical length of 0.3 mm. *Source:* Reprinted from Hughes JDH, *J Phys D Appl Phys*, 20, 276–285, 1987.

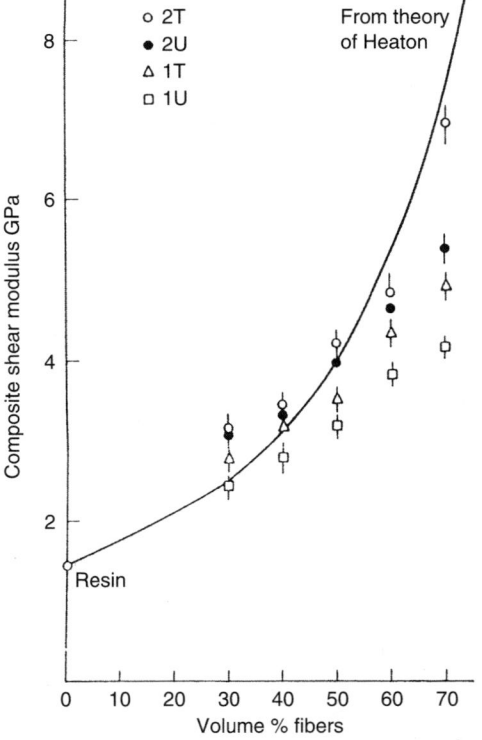

Figure 3.14 The effect of V_f of fibers on the composite shear modulus (The drawn curve is according to the micromechanical theory of Heaton). *Source:* Reprinted with permission from Hancox NL, *J Mater Sci*, 7, 1030–1036, 1972. Copyright 1972, Springer.

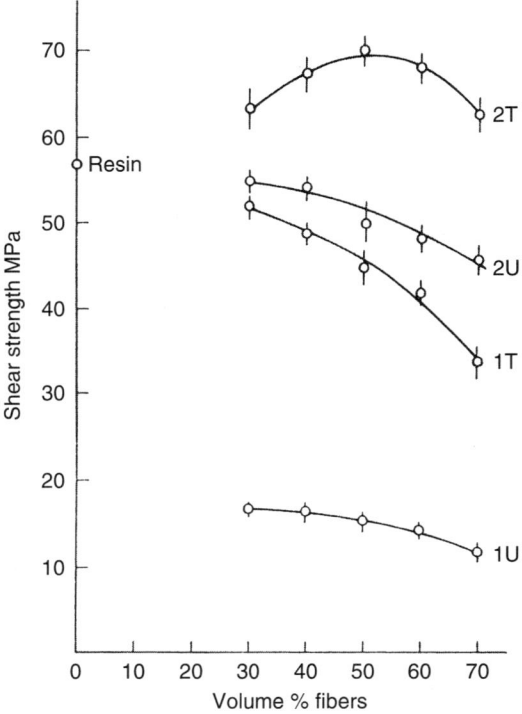

Figure 3.15 The effect of V_f of fibers on the composite shear strength. 1U and 2U are untreated Types I and II fibers, whilst 1T and 2T are treated Types I and II fibers. *Source:* Reprinted with permission from Hancox NL, *J Mater Sci*, 7, 1030–1036, 1972. Copyright 1972, Springer.

Table 3.8 Measured and assumed elastic constants of carbon fiber

		Type of Fiber	Type I	Type II
Engineering constants	E_{33}^*	Longitudinal YM	379 Gpa	228 GPa
	E_{11}	Transverse YM	27.6 Gpa	27.6 GPa
	G_{44}^*	Longitudinal shear modulus	26.2 Gpa	26.2 GPa
	μ_{13}	Longitudinal Poisson's ratio	0.5	0.5
	μ_{13}	Transverse Poisson's ratio	0.28	0.28
Compliance moduli (TN m^{-2})	S_{33}^*		2.6	4.7
	S_{11}		36.3	36
	S_{44}^*		38.3	38
	S_{12}		−10.2	−10.2
	S_{13}		−1.3	−2.4

Source: Reprinted with permission from Goggin PR, *J Mater Sci*, 8, 233, 1973. Copyright 1973, Springer.
*Measured values.

3.4.2.4 *Carbon fiber reinforced ceramics, glass and cement*

Carbon fibers were successfully incorporated in ceramic and glass matrices by Sambell, Bowen, Phillips, and Briggs [104–106]. Briggs [107] has provided a good review of carbon fiber reinforced cement.

3.4.2.5 Carbon fiber reinforced metal composites.

A section within Harwell investigated carbon fiber/metal composites [108,109] and looked at:

1. Coating carbon fibers with metal and producing a composite by hot compaction

The metal was deposited by electro and electroless plating, vacuum deposition and deposition by the decomposition of organometallic compounds.

Electroplating was the preferred method and copper was successfully plated at 3 m/h onto a 10 k tow using a formulated cyanide bath (CuCN, NaCN, Na_2CO_3) and, although giving a porous and slightly uneven product, gave a satisfactory 30% V_f composite after hot pressing, but as the fiber volume fraction increased the strength deteriorated (Figure 3.16). Nickel was also successfully plated onto a Type II untreated carbon fiber using either a Watt's bath ($NiSO_4$, $NiCl_2$, H_3BO_3) or a sulfamate bath ($Ni\{NH_2SO_3\}_2$, $NiCl_2$, H_3BO_3), which produced a brittle plate but made a good composite after compaction.

Copper and nickel were successfully coated using electroless plating techniques to give excellent penetration of the tows with a uniform deposit, but the method proved to be too expensive.

Barrel plating could be used for applying Sn, Pb and Ni. Although the deposits were uniform, the plate tended to be powdery with only moderate adhesion and when hot pressed, contained above average values of oxide.

Vacuum deposition was not successful due to shadowing, with resultant poor penetration of the tow bundle.

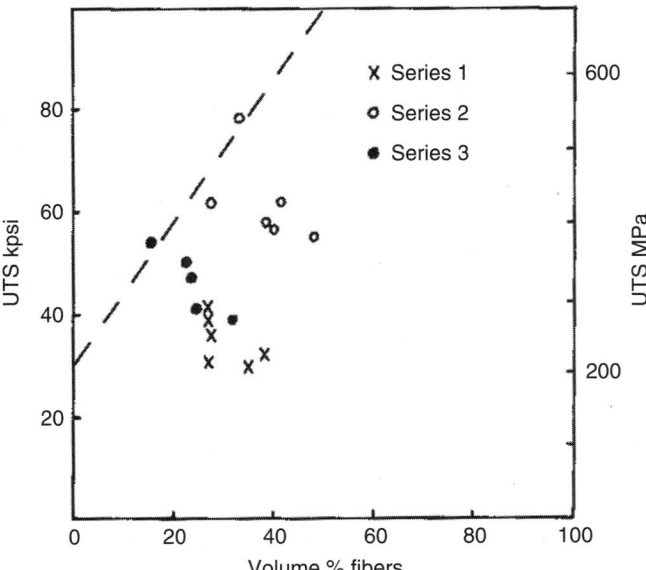

Figure 3.16 Tensile strengths of reinforced Type II electroplated and hot-pressed copper. Broken line as predicted by Law of Mixtures. *Source:* Reprinted with permission from Howlett BW, Minty DC, Old CF, The fabrication and properties of carbon fibre/metal composites, Paper No. 14, International Conference on Carbon Fibres and Applications, The Plastics Institute, London, 1971. Copyright 1971, Maney Publishing (who administers the copyright on behalf of IOM Communication Ltd, a wholly owned subsidiary of the Institute of Materials, Minerals & Mininng).

Good coatings were obtained by the thermal decomposition of tri-isobutyl aluminum but, when hot pressed, some debonding occurred, showing an improvement when isostatic cold pressing was used.

Two techniques were used for hot pressing; one where the coated fiber was placed in a graphite or tool steel die, heated in an electric furnace, or an RF heater, and consolidated by compression in a hydraulic press; and the other involved isostatic pressing, ensuring an even distribution of pressure over the work piece, by packing the coated fiber in a sealed silicone bag, immersing in oil in a pressure chamber, heating if required and finally applying pressure. This technique worked well with less than 1% break-up of fiber and gave good consolidation with only slight porosity. As a rule to minimize fiber break-up, it was found beneficial to use as high a temperature and as low a pressure as possible.

2. Hot working a mixture of metal powders and chopped carbon fiber

This technique was evaluated for 10 μm nominal diameter aluminum powder. Attempting to dry mix with the chopped fiber proved difficult. Eventually a binder (ICI Cranco, polybutylmethacrylate and dibutyl phthalate in a ketone) was used that had sufficient tack, once the solvent had evaporated, to hold the aluminum powder to the dry fiber. The mix was then packed into aluminum cans and the binder removed by baking *in vacuo*. The cans were evacuated, sealed by electron beam welding and extruded at 600–625°C giving extrusions containing 8–25% V_f. Some of the samples were annealed for 1.5 h at 500°C and, although results were below those predicted by the law of mixtures, they were of the same order (Figure 3.17).

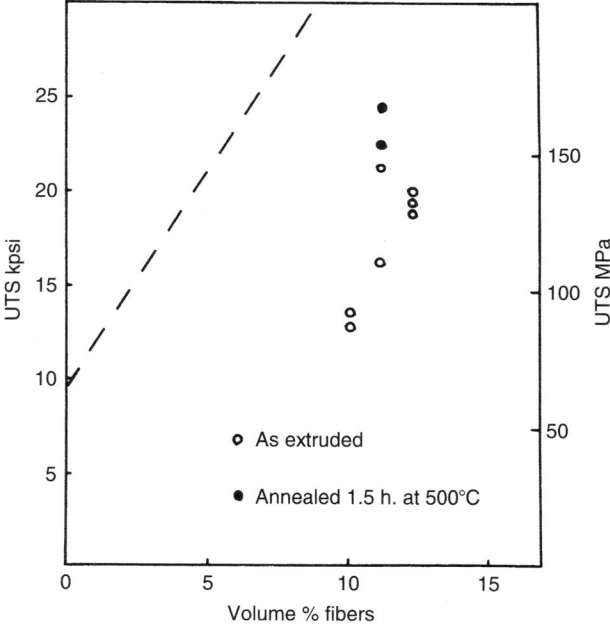

Figure 3.17 Tensile strength plotted against %V_f carbon fiber for aluminum/chopped fiber extrusion. Broken line as predicted by Law of Mixtures. *Source:* Reprinted with permission from Howlett BW, Minty DC, Old CF, The fabrication and properties of carbon fibre/metal composites, Paper No. 14, International Conference on Carbon Fibres and Applications, The Plastics Institute, London, 1971. Copyright 1971, Maney Publishing (who administers the copyright on behalf of IOM Communication Ltd, a wholly owned subsidiary of the Institute of Materials, Minerals & Mininng).

3. Liquid metal infiltration techniques

When using liquid metal infiltration techniques, it was found preferable to initially apply a thin plating coat of copper to the carbon fiber in order to give improved infiltration. The coated fiber was gathered into a bundle and packed into an open silica tube, which was then immersed into the molten matrix alloy, the process being carried out in vacuum with an 8–12% volume fraction of carbon fiber being achieved. A modification of the matrix to produce a wetting but not strongly reacting system is described by Nicholas and Mortimer [109].

Old and Nicholas [110] studied a range of Pb/Sn alloys reinforced with carbon fiber for possible use as bearing materials and published a review [111] on the prospects of metal matrix composites.

3.4.2.6 Composite fabrication and design

Composite fabrication techniques were examined and in particular, the work of TW Thorpe on filament winding should be mentioned, basically for use with the classified centrifuge project for separation of uranium isotopes. Designing with cfrp has been covered by Pearce [112] and more specifically, designing with carbon fiber reinforced metals [113].

3.4.3 Work at Rolls Royce, Derby

3.4.3.1 Fiber production

In the 1950s, Rolls Royce were very active in looking at possible reinforcements which could be used in high temperature composites such as silica reinforced aluminum alloys. In parallel with this work, a group were looking at trying to make fibers from a wide range of covalently bonded materials with the aim of producing very high fiber moduli, such as carbon from a range of polymers, alumina whiskers and fibers and other exotic fiber oxides. The workers realized the advantages that would be gained by using such materials in applications where high specific stiffness was required and in being able to successfully compete with titanium. When, in 1961, Shindo published his work in Japan, this acted as a spur to place more emphasis on carbon fiber work and, initially, Rolls Royce developed their own in-house 1 ton/yr batch distillation process using 9 denier Courtelle, which was stretched down to 3 denier by heating in boiling water. Standage and Prescott [114] heated Courtelle PAN precursor at 1°C/min to 1000°C in an atmosphere of nitrogen and then to 2700°C at 30°C/min producing a fiber with a typical modulus of 221 GPa but, if the heating rates were increased to 5 and 40°C/min respectively, then the modulus plummeted to 52 GPa. Although the initial pyrolysis stage was carried out in an atmosphere of nitrogen, apparently there was a strong possibility that some air may have leaked in, or have been occluded on the fiber. A range of some thirty precursors was examined and some of these are listed in Table 3.6. The pyrolysis up to 400°C is described by Turner and Johnson [115] heating 1.5 denier Courtelle in argon at 1 and 6°C/min. DTA confirmed an exotherm occurring at 190–285°C reaching a maximum at 262°C, with a small amount of NH_3 peaking at 280°C and mainly HCN peaking at 345°C. Further work was carried out taking the temperature to 1000°C, but was not reported. The pyrolzed fiber had stabilized by 500°C and only a further weight loss of 10% occurred when heating was continued up to 1000°C [WN Turner, personal communication, 1999] with the evolution of HCN, CH$_4$, H$_2$ and N$_2$, producing a fiber with an elemental analysis of 86% C, 0.4% H and 8.2% N, with a total weight loss of 50.2%.

Initially, the in-house plant made high modulus fiber, using a bank of specially adapted Spembly furnaces processing the fiber up to 2500°C and it was not until later that the Rolls Royce engineers accepted that they could use carbonized fiber with a modulus of about 200 GPa.

In the period 1966–67, RAE was approached and Rolls Royce was granted an NRDC license to produce carbon fibers by the RAE process. A Rolls Royce patent [116] describes a process with an initial oxidation stage up to 265°C followed by carbonization in an inert atmosphere. I Whitney, MR Rowland and SG Jones developed a continuous in-line carbon fiber manufacture and prepreg unit. The process basically consisted of beaming a warp, comprising an array of collimated 10k PAN fiber tows, stitched across before oxidation to improve the final sheet integrity. Up to eight of these tapes could be processed together through oxidation, followed by carbonization and subsequent coating with a resin system to yield a prepreg tape called Hyfil, for use within the Rolls Royce organization. Various patents were issued covering the manufacture of continuous carbon fiber sheet material [117–119]. Although these patents covered processing up to 1600°C, the final fiber modulus required was dictated by the end use in the aircraft engine and to produce carbon fiber with a modulus of some 200 GPa required a production temperature of about 1200°C and it was not considered prudent to surface treat the fiber, when using a relatively brittle epoxy/novalac matrix. However, Rolls Royce did undertake a wide program of work on surface treatment, establishing types and levels for a range of resin systems, such as Union Carbide's ERLA4617, which was the toughest resin available with Rolls Royce.

In a study of the oxidative stabilization of Courtelle PAN fiber [120], it was suggested that the dimensional stability of the pre-oxidized fiber was due to hydrogen bonding and the structure could be represented as:

Dihydropyridine   4-pyridine   Fully aromatic structure

It shows that there are three different ring structures present: dihydropyridine (x), 4-pyridine (y) and fully aromatic (z); the relative amounts depending on time, temperature and atmosphere of the oxidation conditions.

Although the work undertaken in the RAE patent found that after oxidative stabilization there was no need for tension to be applied in subsequent carbonization and heat treatment stages, the workers at Rolls Royce investigated the effect of stress graphitization [121,122] and were able to apply a stretch of up to 30%, compared with some 200%, which could be applied to a cellulose based carbon fiber. The tensile strength is almost independent of YM for an unstretched fiber. However, for a stretched fiber over the modulus range 275–620 GPa, there was a linear dependence and a 100 GPa increase in modulus corresponded to an increase of 0.50 GPa in strength. Stretching resulted in an increase in the preferred orientation and was accompanied by an increase in the crystal size in the direction of the c-axis (L_c). Table 3.9 gives the change in L_c for changes in stretch ratios and temperature. The

HISTORY AND EARLY DEVELOPMENT OF CARBON FIBERS

Table 3.9 Increase in crystal size at different stretch ratios and temperatures

% Stretch under applied stress	L_c nm				
	1990°C	2140°C	2150°C	2390°C	2970°C
0	5.0	6.3	7.1	12.5	17.0
18				15.0	
20	7.0		11.5		
23					25.0
30		9.0			

Source: Reprinted from Johnson JW, Marjoram JR, Rose PG, *Nature*, 221, 357–358, 25 Jan 1969.

moduli for the stress graphitized PAN was comparable for moduli reported for stress graphitized cellulose and the value of 25 nm for the sample at 2970°C was the highest seen at the time for a PAN based carbon fiber.

PG Rose, whilst with Rolls Royce, took his PhD at the University of Aston in Birmingham and his thesis [123] contains a wealth of information on the physics and chemistry of carbon fibers from PAN precursors and some of this data is highlighted in Chapters 4 and 6.

3.4.3.2 Factors affecting tensile strength of carbon fibers

Rolls Royce attached considerable importance to the presence of defects, found in the precursor and carbon fiber products, and their effect on the tensile strength of the carbon fiber. Johnson [124,125] has shown that carbon fiber can have both internal flaws (Figures 3.18(a) and 3.18(b)) and external flaws (Figure 3.19), revealed by examining the fractured surfaces of filaments when broken in a glycerol-water mixture, which preserved the broken pieces for subsequent examination when using with a scanning reflection electron microscope. Optical examination of the PAN precursor showed that it contained organic and inorganic particulate matter, which formed diconic cavities when the precursor was stretched

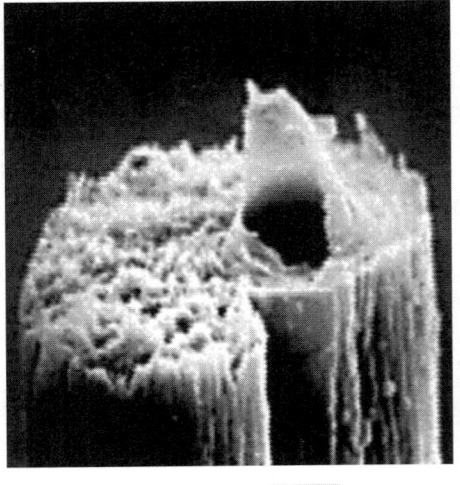

Figure 3.18 (a) Internal flaw responsible for fracture in a high temperature carbon fiber (b) Internal flaw responsible for fracture in a carbonized carbon fiber. *Source:* Photomicrograph kindly supplied by Johnson JW.

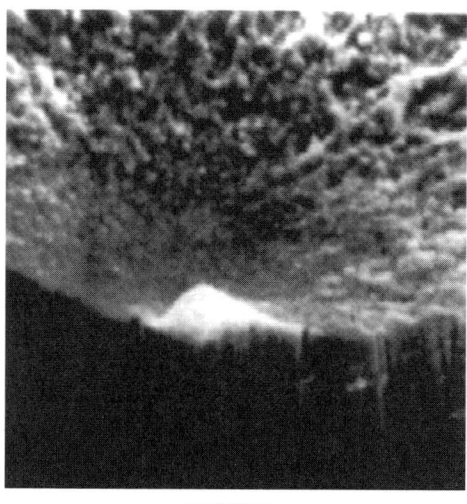

1 μm

Figure 3.19 External flaw responsible for fracture in carbon fiber. *Source:* Photomicrograph kindly supplied by JW Johnson.

in the acrylic fiber production process. A number of surface etching techniques were used to remove surface flaws and a 10 min etch in moist air at 450°C gave optimum strength improvement, as shown by the strength distribution before and after this treatment (Figure 3.20). Although this etching treatment gave some strength improvement, it could not eradicate volume flaws. Further work [126] showed that three types of flaws might operate (Figures 3.21 and 3.22):

1. shows strength distribution due to surface flaws
2. the effect of internal flaws which may predominate or, the two strength ranges for the effect of internal and external flaws that might overlap to give a bimodal distribution and
3. after surface etching, where a new species of surface flaws may have been generated by the etching processor, with an amended structure of the original flaw forming after etching, but characterized by an increased distribution of the fiber strength.

Thorne [127] used optical microscopy to examine internal flaws in acrylic precursor and determined the distribution along the length of the fiber. The flaws comprised of particulate inclusions and voids (Figure 3.23). The internal flaw distribution was an approximate Poisson type, but the mean flaw concentration was variable from 0.03–2.3 flaws/mm. Incorrect processing conditions, such as too fast heating rates, can promote the incidence of flaws. In later years, the precursor manufacturers have significantly improved the quality of their products with a much reduced occurrence of internal defects, whilst the carbon fiber manufacturers have learned how to heal surface defects by cracking carbon into the defect.

3.4.3.3 Resin formulation and composite fabrication

In the early 1960s, Rolls Royce evaluated glass fiber reinforced epoxy composite components in the RB162 vertical take-off gas turbine engine, where the operating temperatures were not too high. To ensure consistency, glass prepreg warp sheets (XP219 and XP253) were obtained from 3M, stamped out, stacked in the requisite order and compression molded [128]. Rolls Royce used Samco Strong hydraulic cutters fitted with

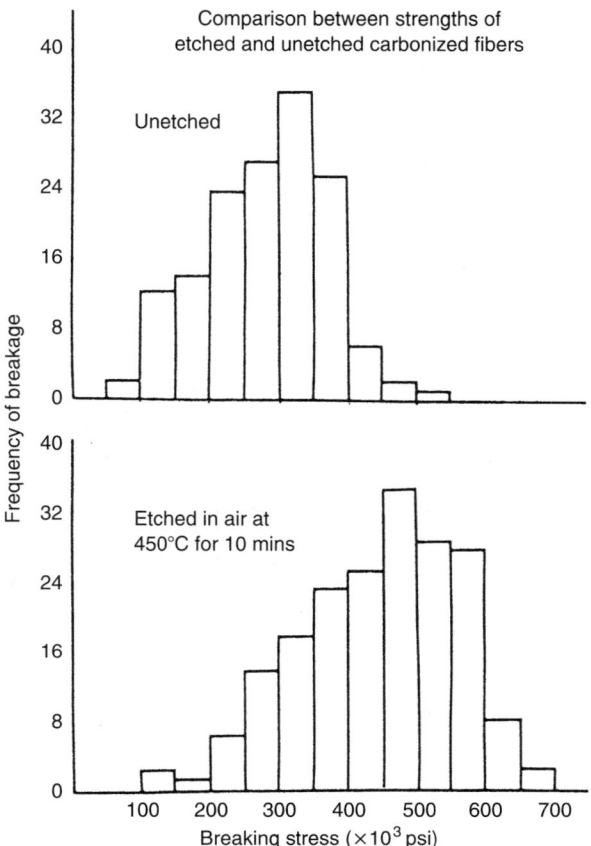

Figure 3.20 Strengths of carbonized fibers after etching in air at 450°C. *Source:* Reprinted with permission from Johnson JW, *Applied Polymer Symposia*, No. 9, 229–243, 1969. Rolls Royce.

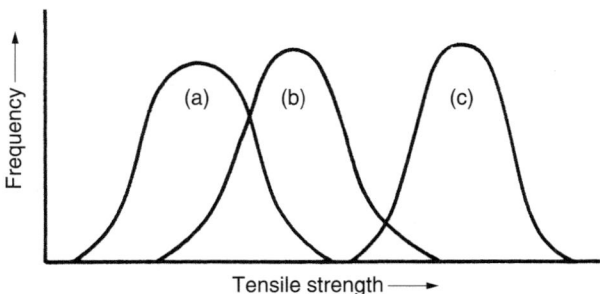

Figure 3.21 Effects of various types of flaw on carbon fiber strength distributions (a) Original surface flaws; (b) Internal flaws; (c) Surface flaws after etching. *Source:* Reprinted with permission from Johnson JW, Thorne DJ, *Carbon*, 7, 659–661, 1969. Copyright 1969, Elsevier.

highly accurate dies to cut the prepreg and specially designed swivelling head power presses built by Power Moulding Ltd. for molding the blades. The design of the press allowed the mold to open so that the top platen opened upwards permitting easy cleaning between cycles. It was necessary for the resin to have a high T_g and initially, an epoxy novalac resin (LY558) cured with a BF_3MEA catalyst gave suitable properties with sufficient tack. It was essential that the composite should be virtually free from voids and this was achieved, in part, by

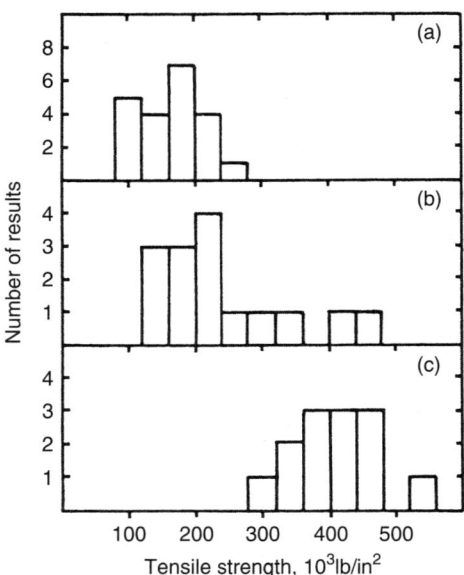

Figure 3.22 Strength distribution of carbon fibers made from Acrilan precursor (a) Unetched; (b) Etched: internal flaw failures; (c) Etched: Surface flaw failures. *Source:* Reprinted with permission from Johnson JW, Thorne DJ, *Carbon*, 7, 659–661, 1969. Copyright 1969, Elsevier.

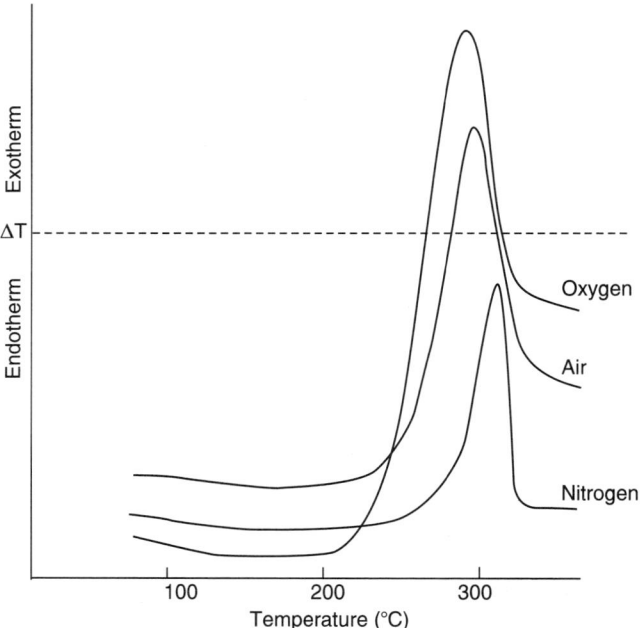

Figure 3.23 Types of internal flaws in acrylic polymer fibers. *Source:* Reprinted with permission from Thorne DJ, *J Appl Poly Sci*, 14, 103–113, 1970. Copyright 1970, John Wiley & Sons Ltd.

bodying up the resin by a partial cure to increase the resin viscosity. For a duty above 200°C the epoxy resin was replaced with a bismaleimide system (Kerimid 601), which unfortunately had poor tack, so the sheets were hot-tacked together.

A bold scheme was embarked upon to establish whether it was feasible to use carbon fiber, which hardly existed at this time, into the Tristar engine. Trial sets of carbon fiber blades were initially made for the Conway LP stage 1 compressor, fitted with electroplated nickel leading edges to prevent erosion, and the blades underwent successful trials when fitted to VC10 aircraft. The next step was to introduce carbon fiber to the larger fan blades of the RB211 engine. This step was not successful due to a number of technical problems, which included complicated geometry, the need for protection against high speed impact erosion, root integrity problems and a serious bird impact problem, so the blades were replaced with heavier forged titanium blades. However, this setback promoted further important developments in composite technology at Rolls Royce.

Problems associated with the use of prepreg for composite fabrication included a high bulk factor and difficulties in obtaining an optimum resin pre-cure to ensure correct consolidation. One approach was to use pressure assisted resin injection, where a fiber preform contained in a mold was evacuated and a fairly mobile resin pumped in under pressure [64]. To help maintain the accurate alignment of the fiber in the perform, it was held in place by a resin binder made of about 4% polysulfone applied as a solution in methylene chloride, drying to remove solvent and then followed by a short treatment at 320°C to fuse the polysulfone onto the fiber. Next, a laminate of primed sheets was prepared, which was about 2.5 times the bulk volume of the finished composite (i.e., when compressed at 0.65 V_f). Two techniques were used to consolidate a preform.

One method, actually used for the experimental RB211 fan blade, was to consolidate the stacked sheets by heating above the T_g (185°C) of the polysulfone, which softened the binder and permitted the preform to be consolidated by the application of a low pressure (0.2 MPa), giving a rigid structure containing fiber at 0.65 V_f, with 5% by volume of polysulfone and about 30% porosity comprising an array of interconnected capillaries. The consolidated preform with accurate fiber alignment could be readily handled, permitted ready evacuation and allowed resin to permeate easily throughout the preform during the resin injection phase. Another advantage was that 3D reinforcement could be introduced during the preform process, whilst the resin binder was above its T_g.

The other method, which was successfully used for the production of nose cones, was to add a measured amount of methylene chloride solvent onto the laminate in the die, applying minimal pressure and when consolidated, warming the die to evaporate the solvent, which boiled at 40°C, again producing a similar preform. The individual laminates could then be peeled apart and relocated to build a de-bulked cone-shaped preform.

The preform was then placed in a mold, a vacuum applied allowing at least 4 h to elapse for all solvent and moisture to be removed and to attain a final vacuum of <0.5 mm Hg. To ensure satisfactory impregnation of the perform, the resin system should have a viscosity of about 0.1 Pa s at the temperature of injection (120–150°C) and a workable gel time at that temperature of about 0.5–4 h. By using a hydraulic ram, pressures of 3–7 MPa could be applied, assisted by a pressurized resin pot (Figure 3.24), which permitted the gradual introduction of resin by applied air pressure at 0.56 MPa until excess resin was carried over from the mold and became visible in the transparent resin pot. The resin was then gelled and cured with applied pressure, before finally releasing the pressure and removing the component for post cure. A wide variety of resins meet these requirements (e.g., a 100/36%w of DGEBA/DDS) and their equipment produced consistent molded parts, free from microporosity, with a uniform fiber dispersion.

In autoclave molding, the resin system should be sufficiently mobile to allow complete penetration of the fiber reinforcement and yet be sufficiently viscous to prevent extraneous resin leaking away during the molding process. So, following discussions between Rolls Royce and Ciba-Geigy, a resin system BSL 314 was developed comprising of a blend of polysulfone in an epoxy resin with a latent catalyst system, (100/35/10 mix of DGEBA/P1700 polysulfone/BF_3MEA) [129]. Ciba-Geigy later replaced BSL314 by a much improved

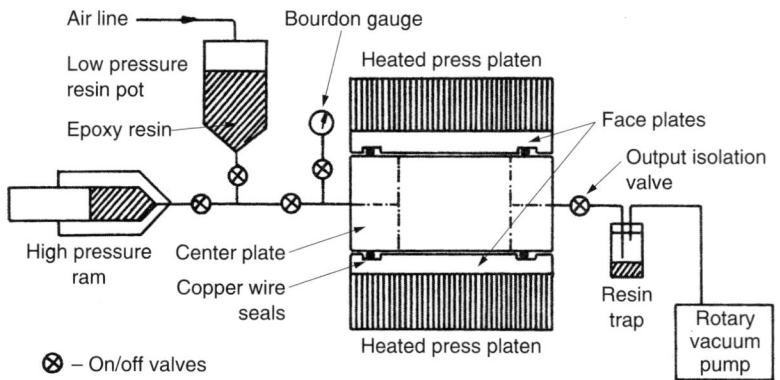

Figure 3.24 Diagram of resin injection equipment. *Source:* Reprinted with permission from Symposium: Fabrication Techniques for Advanced Reinforced Composites, University of Salford, Apr 1980. Copyright 1980, Elsevier.

BSL914 system comprising of a blend of components with high T_g (a tetrafunctional epoxy with polyethersulfone and a dicyandiamide latent catalyst). The polyethersulfone controlled the flow of resin throughout the curing cycle (Figure 3.25) enabling applied pressure to control the size of any voids. The system produces a most interesting two phase cured matrix [130]. Another useful attribute of this system is that limited post-forming can be undertaken after molding, permitting some double curvature.

A method of manufacturing thermoplastic composite materials by a modified film stacking process was patented [131] involving sandwiching sheets of unidirectional carbon

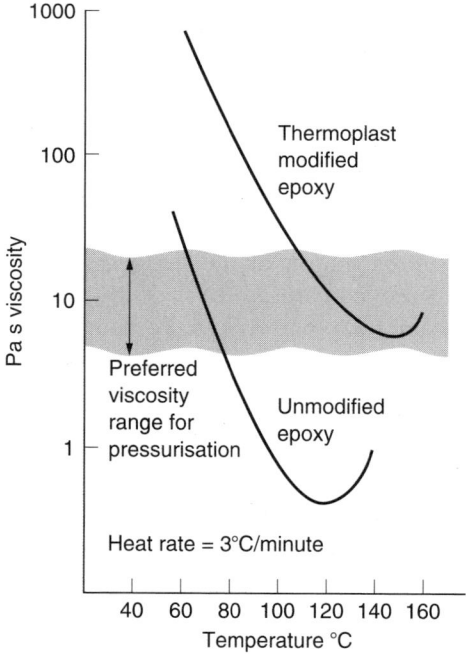

Figure 3.25 Viscosity/temperature curves for unmodified and thermoplastic modified epoxy resins. *Source:* Reprinted from Turner WN, Johnson JW, Hannah CG, Composite materials, Meetham GW ed., *The Development of Gas Turbine Materials.* Applied Science Publishers Ltd., Barking, 121–146,

fiber held in place by cross-stitching with a thermoplastic filament such as polyetheretherketone (PEEK), between sheets of PEEK film and hot pressed at 14 MPa and a temperature of 400°C.

3.4.3.4 Carbon fiber reinforced metal composites

Jackson and Marjoram [132] found that carbon fiber made at 1000°C and 2700°C and then electroplated with nickel, followed by heat treating at 1100°C for 24 h in a vacuum underwent a drastic strength reduction and they found that both types of fiber had recrystallized. The effective crystallite size of the fiber made at 2700°C had increased from some 6 to 50 nm with the angular half width (Z) changing from 18.3 to 6.9°. This work was extended [133] to include electroplated cobalt coatings, which were found to behave in a similar manner to nickel. The carbon fibers underwent structural recrystallization by carbon atoms detaching from the carbon fibers, dissolving in the nickel or cobalt, followed by rapid diffusion through the metal and reprecipitation onto the fibers at a new site and growing in a more fully graphitic form.

3.4.4 Work at Morganite Modmor, London

Morganite quickly filed patents [134–136], later Zbrzezniak filed an intriguing patent [137] using a technique to process carbon fiber continuously by placing the fiber in a cage on rollers within a furnace. Joiner and Findlay gave details of a method using a preferred electrolyte of sodium hypochlorite for continuously surface treating carbon fiber by passing the fiber over a number of grooved graphite anodes with a U-shaped graphite cathode positioned below the fiber [138].

Badami, Joiner and Jones [139] examined the fibrillar fine structure of carbon fiber using X-ray diffraction, whilst these authors with Brydges [140] examined structure and elastic properties. The fiber examined was heat treated to 2600°C, having a filament diameter of 6–8 μm, a modulus of 414 GPa and a strength of 1.72 GPa. Examination showed that the fiber consisted of about 5 nm crystallites with the basal planes highly oriented (±10°) along the fiber axis and a calculated d-spacing of 0.339 nm. The crystallites formed chains along the fiber axis with some over 1 μm long.

Blakelock and Blasdale [141] detailed Morganite Test Methods whilst Blakelock and Lovell [142] and Badami, Joiner and Jones [143] gave physical properties of the Morganite fibers. Use of a Narmco resin system was given [144]. The properties of surface treated Morganite fibers are given in Table 3.10, initially producing $10k$ staple fiber 1 m long with up to 3 tpm of twist, followed later by a mid-length product 30–350 m long, free from permanent twist and then, lengths greater than 350 m long were introduced.

One of the first books on carbon fibers published in 1972 [145] was written by RM Gill, the Technical and Production Director of Morganite Modmor. One very interesting

Table 3.10 Properties of Morganite Modmor carbon fibers

Fiber type	Strength GPa	Modulus Gpa	Short beam shear strength MPa
Type I	1.77–2.26	382–412	55
Type II	2.45–3.14	245–275	75

Source: Reprinted from Modmor's technical literature.

development at Morganite was that they perfected, in conjunction with the Scientific Instrument Research Association of Great Britain (SIRA), with probably initial information supplied by Rolls Royce, a laser diffraction technique for measuring the filament diameters of carbon fibers, which was certainly a much improved method compared with using a microscope fitted with a Watson image-shearing eyepiece.

3.4.5 Work at Courtaulds, Coventry

3.4.5.1 Carbon fiber production

All early work was undertaken with a batch process, starting with a 25 cm staple length and progressing to a 40 cm length. By the end of 1967, a commercial production plant making 1.22 m long staple came onstream. An attribute of the RAE process was that a family of fibers could be made by varying the process conditions and Courtaulds introduced a third carbon fiber member to the RAE family, which they called Type A and made at about 1100°C, which was cheaper to make and gave acceptable bond strengths without surface treatment with brittle resin systems, such as an epoxy novalac (LY558/BF_3MEA). The properties of the Courtaulds' range of carbon fibers are given in Table 3.11.

The batch process comprised positioning the precursor fiber onto a pivoted rotating metal frame, continuously advancing the position of application along the frame to produce two uniform sheets of precursor, one on either side of the frame. A number of these frames were placed in a thermostatically controlled electrically heated oxidation oven, which had provision for extraction of the gases evolved during the process with the continuous admission of fresh air to prevent the build-up of an explosive mixture. The ovens were fitted with explosion relief panels, as run-away exotherms were quite frequent in those early days. The oxidation with the 10 k 1.65 d'tex Courtelle precursor was some 3 h at 225°C and after the oven had cooled, the frames were removed, the oxidized fiber cut from the frames and the cut lengths placed in graphite boats fitted with removable lids, constructed from a graphite tube with the ends blanked off. The boats were then pushed into a furnace by a loading ram which, when withdrawn, permitted the inlet end of the furnace to be sealed and a similar transfer port with a cooling section at the outlet end, and allowed the boats to be recovered without the ingress of air.

Early work for producing high modulus fiber used a resistive heated furnace with a British Acheson grade CS graphite tube element, with power fed to each end through a

Table 3.11 Typical properties of Courtaulds carbon fibers

Courtauld's grade 1967	Courtauld's grade post 1969	RAE equivalent grade	Density g cm^{-3}	Filament diameter μm	Filament tensile strength GPa	Filament tensile modulus GPa	Typical ILSS of composite MPa*
Type A	Grafil A-S	None	1.73	9.0	2.1	190	85
	Grafil A-U						60
Type C	Grafil HT-S	Type II	1.74	8.6	2.2	250	80
	Grafil HT-U						40
Type B	Grafil HM-S	Type I	1.90	8.3	2.0	360	60
	Grafil HM-U						25

*Typical interlaminar shear strengths for 60% V_f carbon fiber composites in Epikote 828/MNA/BDMA
Source: Reprinted from Courtaulds' technical literature.

graphite current feed cone accurately machined to match a water cooled collar, which was clamped around the cone and a bus-bar attached to each collar. Measuring temperatures of the order of 2500°C was difficult and the most satisfactory operating procedure was maintaining a constant level of input power throughout a run (Figure 3.26) rather than measure the temperature (Figure 3.27). Internal and external graphite guard tubes were used and the insulation comprised either a graphite felt or lampblack. Nitrogen was used as the inert gas. Development work with this type of furnace enabled the design to be established for the next generation of furnaces.

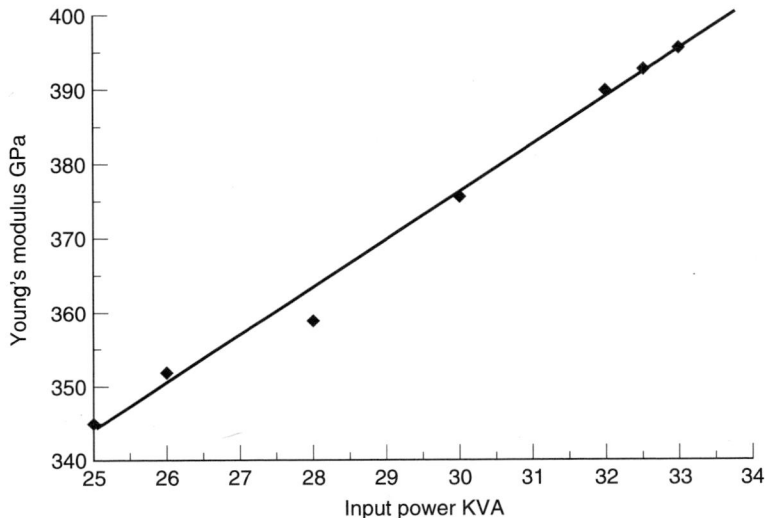

Figure 3.26 Variation of Young's modulus of carbon fiber with input power. *Source:* Kind permission of Acordis UK Ltd (formerly Courtaulds PLC).

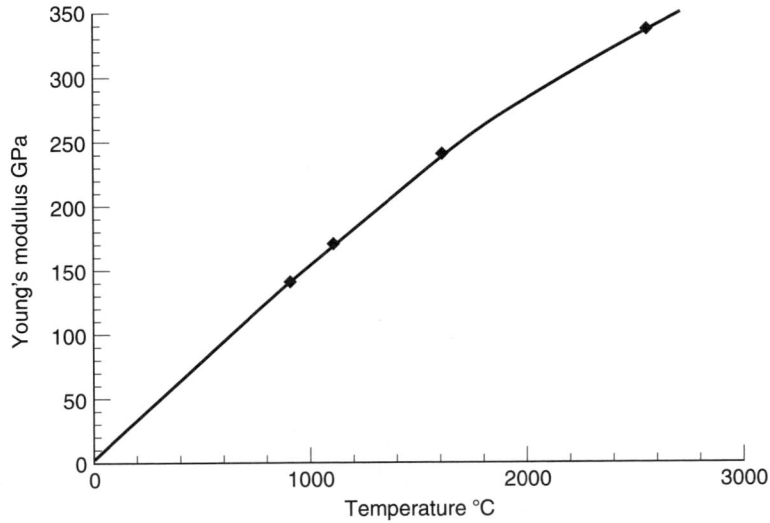

Figure 3.27 Variation of Young's modulus of carbon fiber with measured temperature. *Source:* Kind permission of Acordis UK Ltd (formerly Courtaulds PLC).

When the carbon fiber product was removed from the boat, it was bound with some products of carbonization and formed a stiff tow, but when such a tow was passed between the fingers the tow became limp. This boat process ensured that the fiber was heat treated in proximity to the products of carbonization (also called 'cooking in its own juices'), which certainly did have beneficial properties, presumably healing surface flaws. When continuous fiber first went onstream, the strength of the continuous carbon fiber product was about 0.5 GPa below the strength of the staple product and was attributed to a change in the furnace atmosphere.

There was an urgent requirement to produce a continuous carbon fiber and at first, a process was developed to produce a continuous high modulus material, which was then currently in demand and a continuous HT fiber followed by early 1969. The first stage of a 25 tpa plant came on stream towards the end of 1969, making Grafil A, made at 1100°C and by the end of 1970, the line was in full production for all types of fiber.

3.4.5.2 Early work with X-ray diffraction to establish structure

In 1965, MS Blackie [146] looked at the structure of carbon fibers and found that the individual filaments were derived from a large number of sub-units, called fibrils, originating from the acylic precursor and were approximately 25 nm each in diameter, but of unknown length. The fibrils contained graphitic crystallites and each layer plane of the crystallite was a hexagonal array of carbon atoms, with the layers oriented no more than 10° out of line with the fibril and, hence, the filament axis. The degree of layer plane orientation was determined by X-ray diffraction as produced by a bundle of carbon fiber filaments, measuring the intensity along the meridinal arc. Crystallite sizes were calculated from X-ray diffraction patterns produced by powdered carbon fibers. The width of any arc is inversely proportional to the size of the crystallite structure that produces diffraction. It was found that the L_c dimension tended towards a value of 10 nm and this corresponds to a crystallite in which approximately 30 layer planes are stacked one above another. The L_a dimension tends to a value of 25 nm and this is a measure of the crystallite basal planes. Figure 3.28 shows the increase in crystallite dimensions, L_c and L_a, where treatment temperature and the diameter of the crystallites appears to be limited by the size of the fibrils, each approximately 25 nm in diameter.

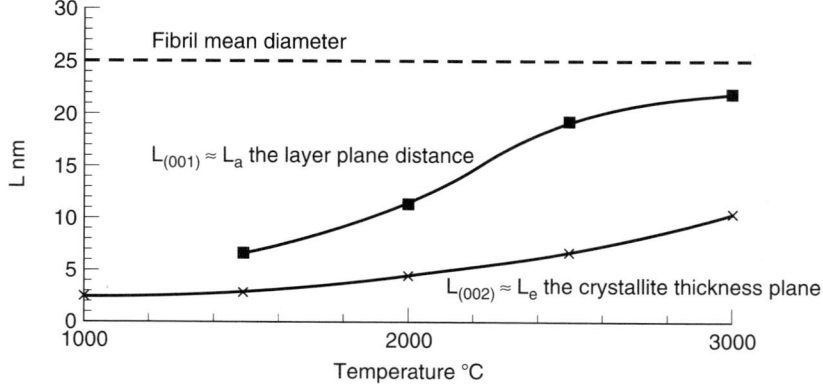

Figure 3.28 The growth of crystallite dimensions with temperature. *Source:* Reprinted from Huxley DV, Carbon fibres, Lecture given to Coventry Textile Society, 20 Nov 1969.

3.4.5.3 Precursor technology

Obviously work on this topic is commercially sensitive and only a general description can be given. Courtelle acrylic fiber is made at Courtaulds (now Acordis), Grimsby and is made from a terpolymer [78], acrylonitrile/methyl acrylate/itaconic acid (AN/MA/ITA) by spinning in a sodium thiocyanate solvent. The ratio of MA/ITA was such that the textile polymer could be readily hot stretched and yet, if required, would have sufficient hot/wet strength to permit wet dying and furthermore, was not over-softened by the household ironing process. The ITA was added to provide a dyesite for basic dyestuffs. It was indeed most fortuitous that the polymer did contain ITA, as this constituent was instrumental in controlling the exotherm in the oxidation stage and promoting cyclization of the nitrile groups. Since the PAN polymer was used for a 320 k textile product, it was contaminated with TiO_2 used as a delustrant, so the first steps were to control any particulate matter by adequate filtration and, ultimately, by the complete isolation of the precursor lines from any textile product. The next stage was to introduce, specifically for the manufacture of carbon fiber, a Special Acrylic Fiber (SAF), which was made at the Coventry works using the same dope but spun on lines with additional filtration and individual dope spinning pumps, enabling precise d'tex control for smaller tows. Initially, the Courtaulds' SAF precursor was 10 k 1.65 d'tex, later 10 k 1.35 d'tex was added to assist making thinner prepreg and in 1978, 1.215 d'tex in 12 k, 6 k and 3 k filaments became available, again facilitating thinner prepregs to be made and coming into line with the Japanese manufacturers.

Figure 3.29 shows thermograms obtained with SAF in oxygen, air and nitrogen. The endotherm near 275°C is more readily visible on the oxygen curve, but it is difficult to be sure that it does occur under inert conditions, which could then have been definitely associated with an oxygen-linked reaction. The onset of the exotherm and its peak move to a lower temperature when oxygen is present and the overall heat evolution is increased by more than a factor of two when oxygen is present.

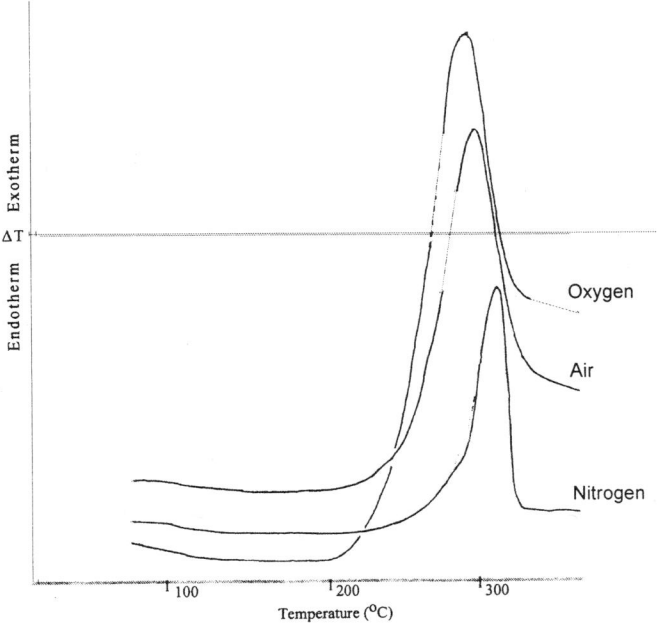

Figure 3.29 Typical thermograms of SAF in oxygen, air and nitrogen.

As explained the early work with SAF remains commercially sensitive and only a generalized account of this work can be given.

With a fixed ITA content raising the MA content increases the temperature of the exotherm peak, whilst ΔH falls. The thermograms in air when MA is introduction into the system has the initial effect of permitting the oxidation reaction to proceed at a lower temperatures, albeit with smaller overall heat evolution, but at greater additions the reaction is retarded and the heat evolution becomes even less.

The effect of increasing the ITA content whilst keeping the MA content constant noticeably reduces the peak temperatures and value of ΔH as the ITA content increases.

Figure 3.34 shows the effect of oxidizing SAF in air at 220, 230 and 240°C and Figure 3.35 shows the change in density as determined by a density gradient column containing carbon

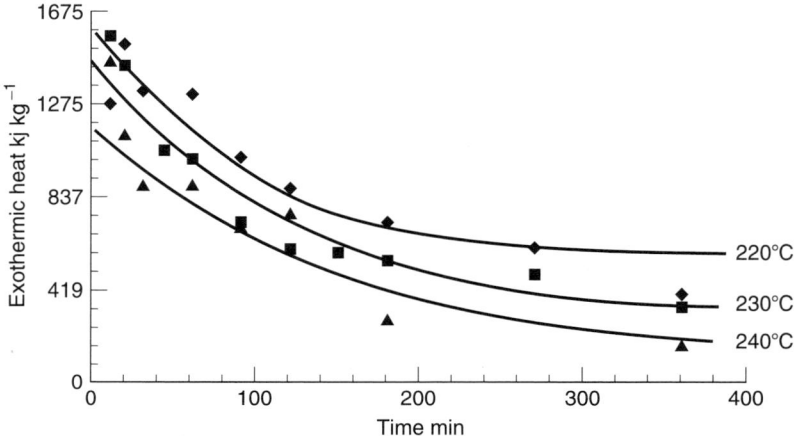

Figure 3.34 Relation between duration and temperature of oxidation and the residual exothermic heat of SAF in air. *Source:* Kind permission of Acordis UK Ltd (formerly Courtaulds PLC).

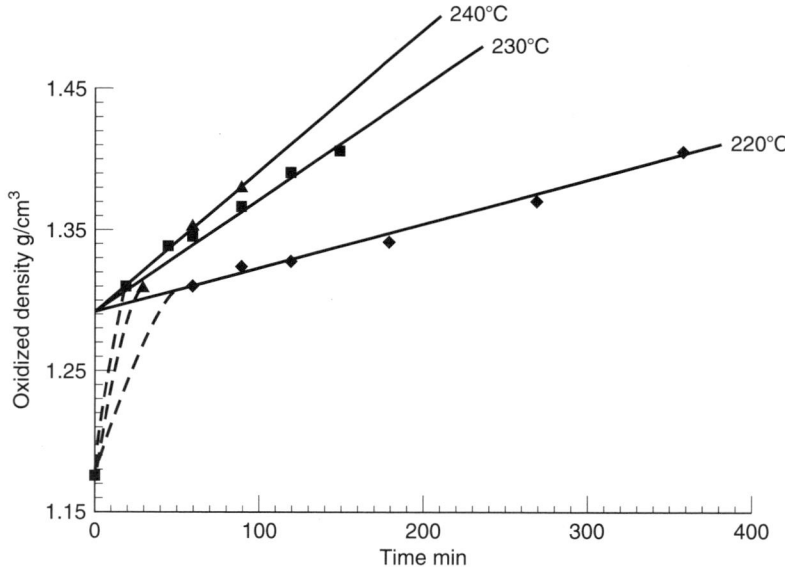

Figure 3.35 Relation between treatment time and temperature and the density of SAF. *Source:* Kind permission of Acordis UK Ltd (formerly Courtaulds PLC).

tetrachloride and *n*-heptane. It can be seen that on heating, the density undergoes an initial relatively rapid rise, followed by a slower increase, which appears to be linear relationship with time in the range of densities studied. The shape of the density time plots would be compatible with an oxidation mechanism in which there is a fairly rapid initial homogeneous reaction involving ring closure between certain nitrile groups in favorable configurations. This will involve some compaction of the structure, hence the increase in density, but must also introduce discontinuities or packing irregularities through which oxygen can diffuse, creating secondary oxidation processes accompanied by the subsequent slower rise in density.

It is of interest to note that extrapolations of the density/time plots intersect at zero time at a density of about 1.29 g cm^{-3}. This could represent the density of the material produced when all the nitrile groups, which are capable of ring closure under moderate heating conditions, have so reacted and, if independent of the oxidation temperature, would extrapolate as shown, at zero time. Density measurements provide a rapid and simple way of following the progress of the oxidation reaction.

Subsequent research work showed that the ITA content was optimum. Since dying occurs principally on the salt form of ITA, about half the carboxylic acid groups are present as the sodium salt to ensure satisfactory dying, producing a polymer containing about 2500 ppm of Na and, despite popular belief, with very little adventitious Na pick-up from the NaSCN solvent. When Courtaulds closed down their carbon fiber business, they stopped production of the SAF, but Acordis remain strong in their range of special textile tow products, which are still made at their Grimsby works.

3.4.5.4 Oxidation stage

When SAF was oxidized, it underwent a loss of orientation shown by the azimuthal orientation increasing from about 25° to about 50°. To examine this effect, 10 k 1.67 d'tex SAF was subjected to tensions of 4 g (effectively unrestrained), tows restrained at constant length and tows with an applied tension of 1.8 kg mass. Samples were taken every 30 min during the 4 h oxidation period at 220°C and were studied by X-ray examination and the filament tensile strength and elongation at break determined (Figures 3.36–3.39). During the first half hour of oxidation, the SAF passed through a plastic state (Figure 3.36) and

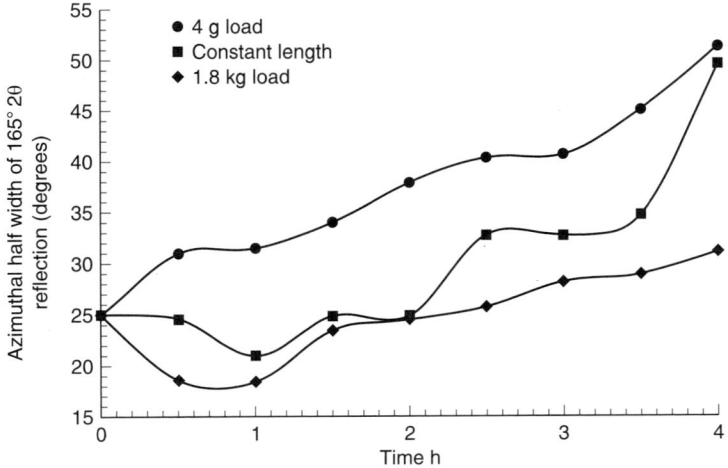

Figure 3.36 Azimuthal half width of 16.5° 2θ reflection (degrees) of SAF oxidized in air at 220°C for different loading conditions. *Source:* Kind permission of Acordis UK Ltd (formerly Courtaulds PLC).

exhibited a tendency to contract, as shown by the samples restrained at constant length and those which were oxidized under 1.8 kg load. The unrestrained fiber during this period decreased in orientation, presumably due to the contraction of the SAF. Thereafter, the shape of the three curves remains essentially the same. The changes of slope may be related to various stages in the oxidation chemistry and that a certain degree of disorder may be necessary for the oxidation to continue at a practical rate and, furthermore, tows subjected to high tension discolored less quickly. The crystallite size perpendicular to the fiber axis (Figure 3.37) increased in the first half hour of oxidation, particularly with applied tension, and then gradually decreased. The filament tensile strength (Figure 3.38) decreased with

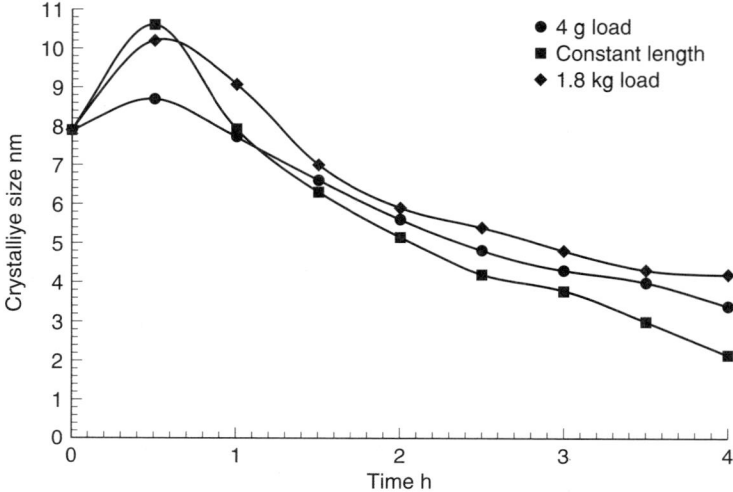

Figure 3.37 Crystallite size perpendicular to fiber axis of SAF oxidized in air at 220°C for different loading conditions. *Source:* Kind permission of Acordis UK Ltd (formerly Courtaulds PLC).

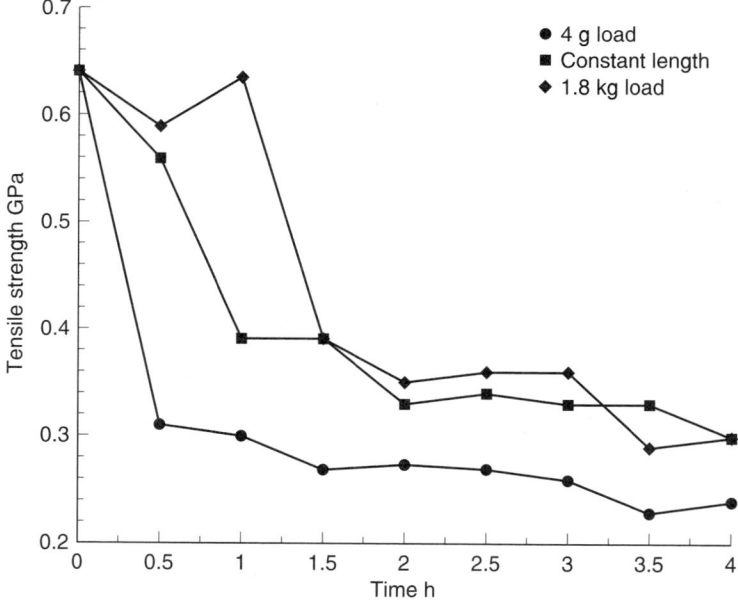

Figure 3.38 Tensile strength of SAF oxidized in air at 220°C for different applied loadings. *Source:* Kind permission of Acordis UK Ltd (formerly Courtaulds PLC).

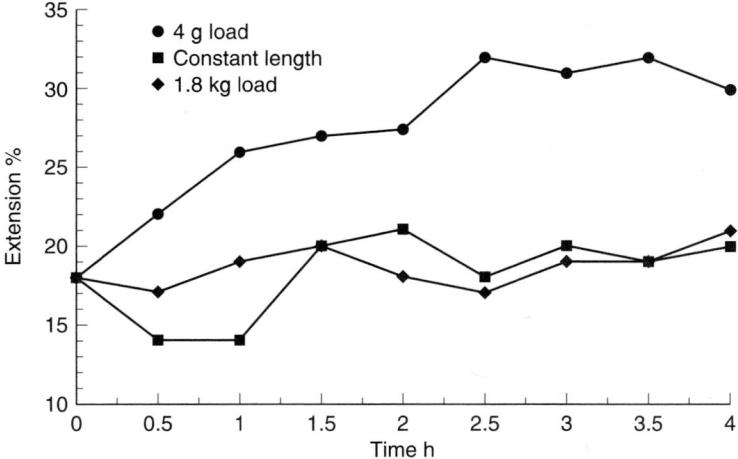

Figure 3.39 The % extension at break of SAF oxidized in air at 220°C for different applied loads. *Source:* Kind permission of Acordis Ltd (formerly Courtaulds PLC).

increasing oxidation tending to level out after 2 h. The extension at break (Figure 3.39) was about 50% higher for the fiber oxidized without restraint during the oxidation.

To better understand the effect of applying tension during oxidation, samples of 10 k 1.67 d'tex SAF were oxidized at 220°C in air for periods up to 4 h, with a nominal weight of 25 g applied to prevent the tow being blown about in the oxidation oven, together with samples which were constrained to constant length. Samples were examined with SEM, X-ray diffraction, DSC and elemental analyzes. Subsequently, oxidized samples were carbonized at 1100°C and the resulting carbon fibers examined by X-ray analysis together with the determination of the tensile properties. Some of this carbon fiber was further carbonized at 1600°C and the tensile properties determined.

SEM analysis revealed there was no substantial difference between the surfaces of the fibers. The relaxed fiber was very micro-cockled. X-ray diffraction data can be represented as a ratio of peak heights (Figure 3.40), which is indicative of the degree of oxidation of the fiber

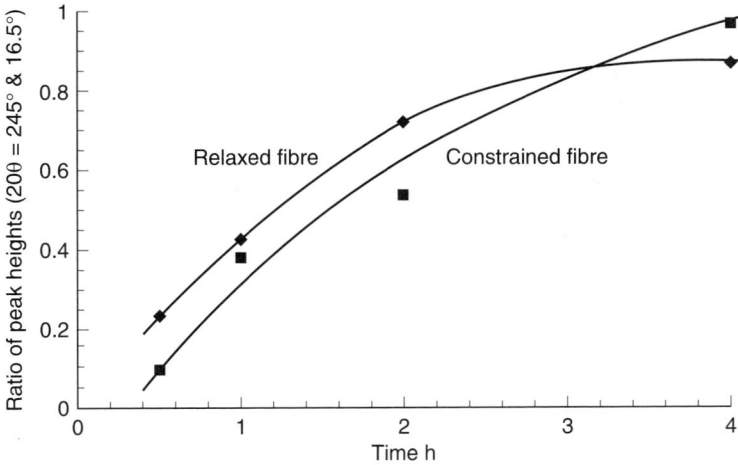

Figure 3.40 Ratio of peak heights for peaks at $2\theta = 24.5°$ and $2\theta = 16.5°$ for SAF oxidized in air at 220°C for relaxed and constrained fiber. *Source:* Kind permission of Acordis Ltd (formerly Courtaulds PLC).

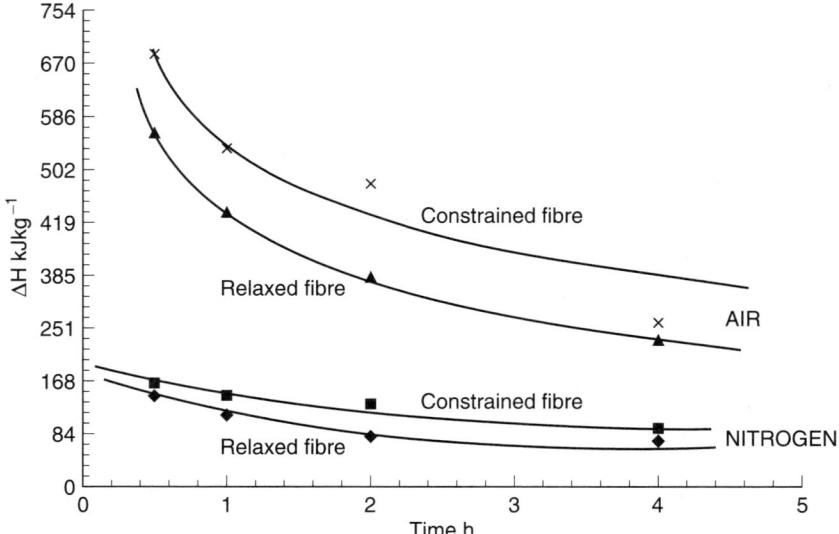

Figure 3.41 DSC study of SAF oxidized in air and nitrogen at 220°C for relaxed and constrained fiber.
Source: Kind permission of Acordis Ltd (formerly Courtaulds PLC).

and together with DSC (Figure 3.41) studies, suggested that for a given time of oxidation, the constrained fiber was less oxidized than the relaxed fiber. However, this was not supported by the oxygen contents, as there was no differentiation, with both increasing at the same rate with time of oxidation (Figure 3.42). This could have been accounted for by the oxygen contents being determined by difference. As would be expected, the carbon, hydrogen and nitrogen contents (Figures 3.43–3.45) decreased with the time of oxidation and there was little difference between the constrained and relaxed fibers.

When converted to carbon fiber at 1100°C, the crystallite size (Figure 3.46) appeared to be independent of the oxidation conditions and the degree of orientation of the constrained fiber was higher than that of the relaxed fiber. The tensile strength and modulus of the

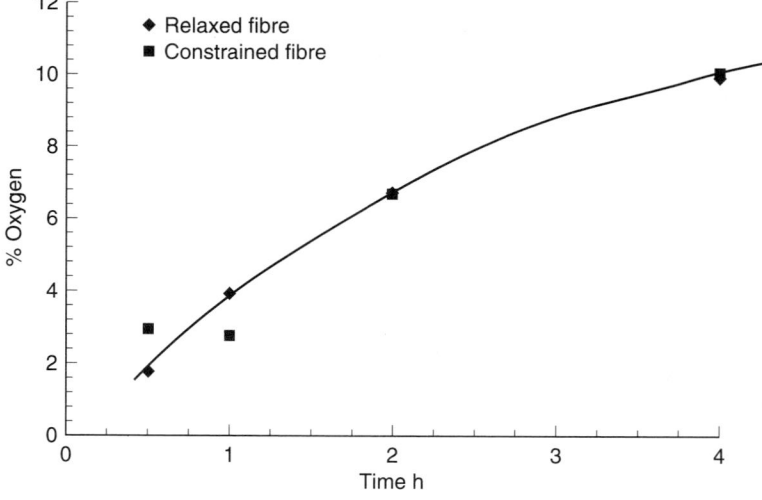

Figure 3.42 Oxygen content (by difference) of SAF oxidized in air at 220°C for relaxed and constrained fiber.
Source: Kind permission of Acordis Ltd (formerly Courtaulds PLC).

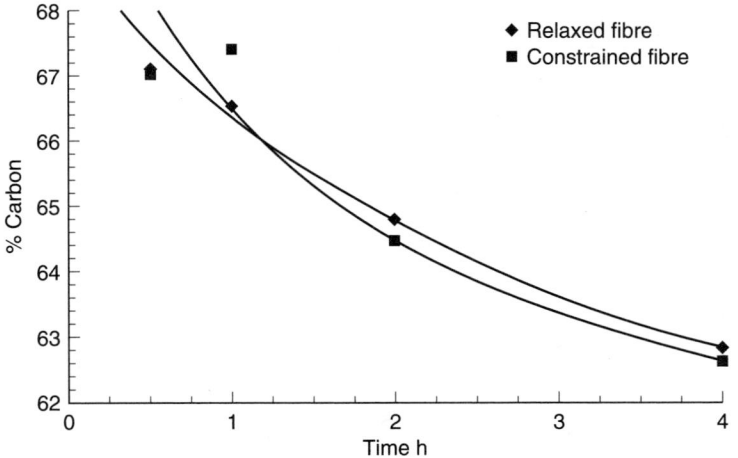

Figure 3.43 Carbon content of SAF oxidized in air at 220°C for relaxed and constrained fiber. *Source:* Kind permission of Acordis Ltd (formerly Courtaulds PLC).

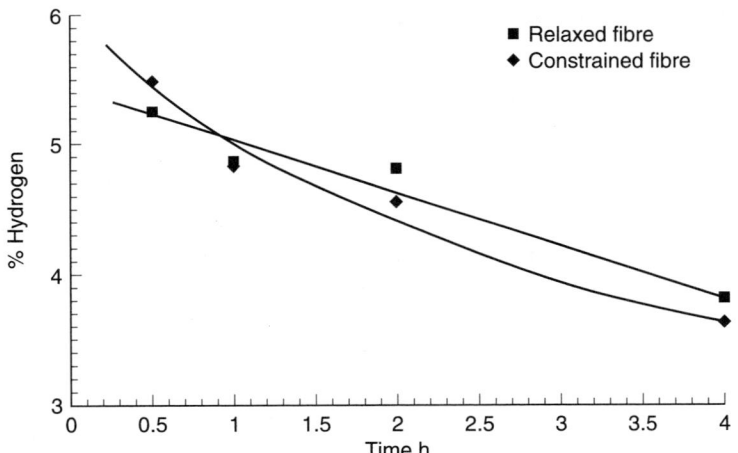

Figure 3.44 Hydrogen content of SAF oxidized in air at 220°C for relaxed and constrained fiber. *Source:* Kind permission of Acordis Ltd (formerly Courtaulds PLC).

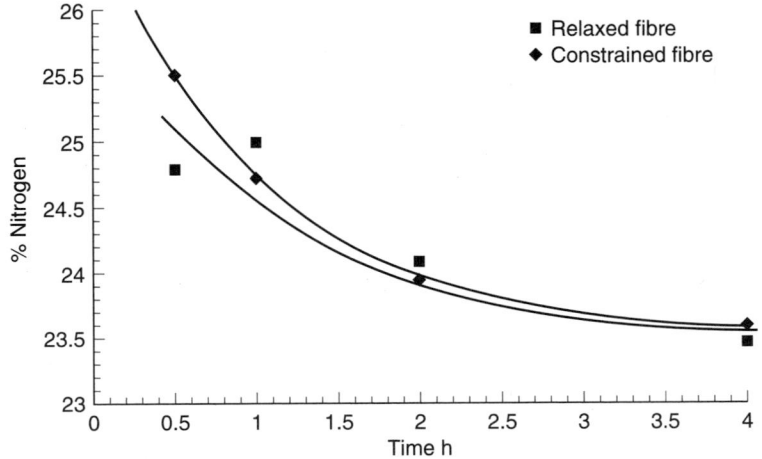

Figure 3.45 Nitrogen content of SAF oxidized in air at 220°C for relaxed and constrained fiber. *Source:* Kind permission of Acordis Ltd (formerly Courtaulds PLC).

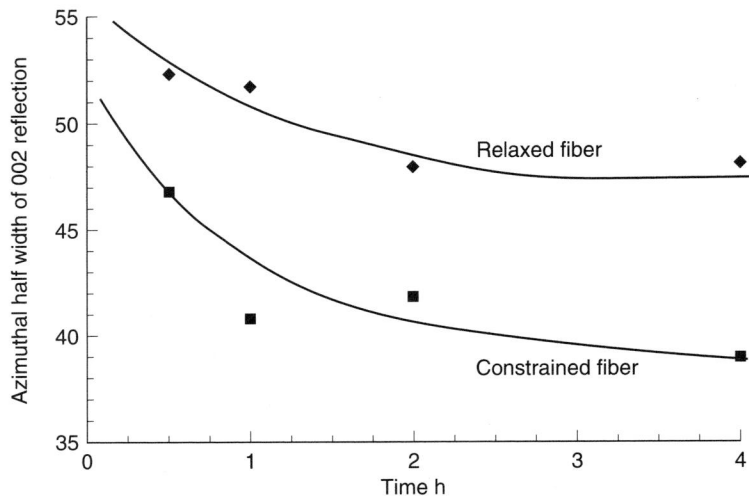

Figure 3.46 Azimuthal half width of 002 reflection for SAF oxidized in air at 220°C for relaxed and constrained fiber. *Source:* Kind permission of Acordis Ltd (formerly Courtaulds PLC).

relaxed fiber were considerably lower than that of the constrained fibers made at 1100 and 1600°C (Figures 3.47, 3.48). Further carbonizing the 1100°C fiber to 1600°C showed no significant improvement in strength, with an expected increase in modulus.

The rate controlling step in the production of carbon fiber from an acrylic precursor is the oxidation stage and G Gould and his research team looked at ways of speeding up this reaction. Various techniques could be used to catalyze the cyclization of PAN, but because the SAF already contained a catalyst comonomer (itaconic acid), the effects were much smaller than those reported in the literature for other acrylic fibers. One of the most promising was treatment with a Lewis acid, $SnCl_4$, which when applied as a solution in diphenyl ether, reduced the residual exotherm of SAF to less than 50 cal g^{-1} in only 6 min, which would normally have taken some 3 h of air oxidation at 220°C to have produced the

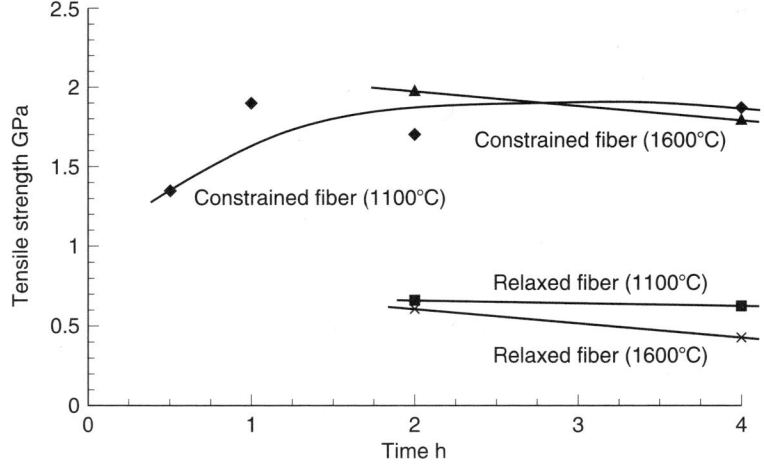

Figure 3.47 Tensile strength of fiber carbonized at 1100°C and 1600°C from Courtauld's SAF oxidized in air at 220°C for relaxed and constrained fiber. *Source:* Kind permission of Acordis Ltd (formerly Courtaulds PLC).

HISTORY AND EARLY DEVELOPMENT OF CARBON FIBERS

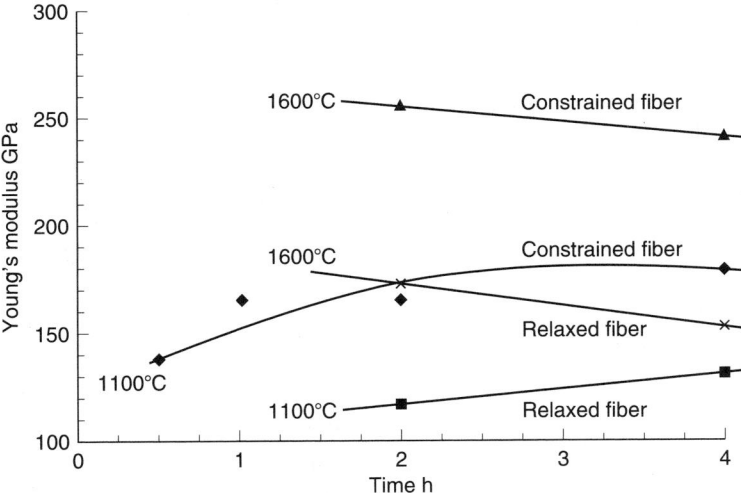

Figure 3.48 Filament Young's modulus of fiber carbonized at 1100°C and 1600°C from Courtauld's SAF oxidized in air at 220°C for relaxed and constrained fiber. *Source:* Kind permission of Acordis Ltd (formerly Courtaulds PLC).

same effect (Figures 3.49, 3.50). Unfortunately, it was not practicable, as $SnCl_4$ readily hydrolyzed and was extremely sensitive to minute quantities of metal impurities.

Most success was obtained using a multi-stage oxidation process. To implement this work, microsections of the oxidized fiber were taken and examined for skin/core structure and the work revealed some interesting facts concerning the oxidation rate:

1. The 1.65 d'tex SAF showed diffuse oxidation zones.
2. Higher ITA levels produced clear and well-defined boundaries.
3. Low d'tex SAF (0.66 d'tex/fil) and SAF treated with sulfonic acids in the finish bath showed no zone formation, gradually darkening in color over the whole cross-section (cf Orlon).

RAE had claimed [147] that zones were formed when cyclization was rapid, so with SAF, oxygen diffusion was rate-determining, whilst with Orlon, the reverse was true, with gradual

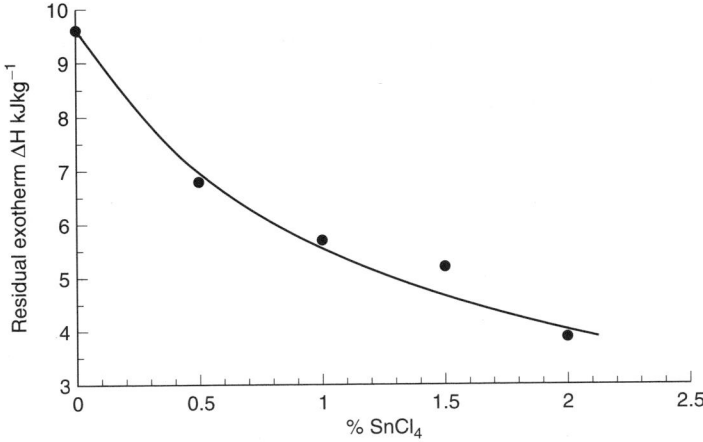

Figure 3.49 $SnCl_4$ concentration Vs ΔH (in N_2) for treatments of the fiber at 230°C for 15 min. *Source:* Kind permission of Acordis Ltd (formerly Courtaulds PLC).

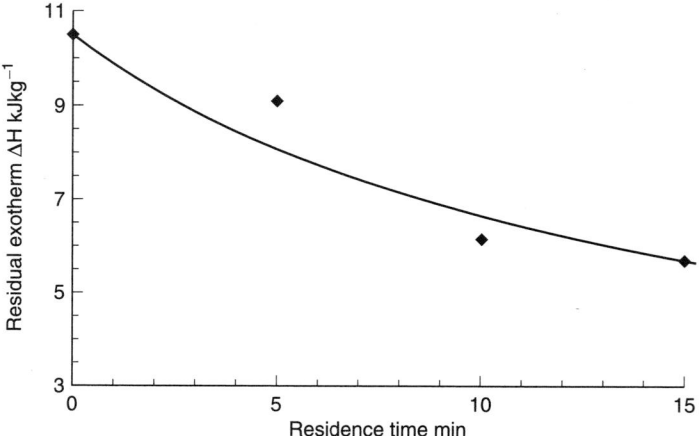

Figure 3.50 Treatment time Vs residual exotherm (in N_2) for treatments of the fiber at 230°C in 1% $SnCl_4$ solution. *Source:* Kind permission of Acordis Ltd (formerly Courtaulds PLC).

darkening of the entire cross-section, without zone formation. Whilst this was basically true, it was an over-simplification and did not explain why the oxidation zone with SAF occurred almost instantaneously to a depth of roughly half the filament radius and did not explain why smaller diameter SAF showed no zone formation.

In joint discussions with J Johnson (Rolls Royce), it was believed that the following reactions occurred:

1. Cyclization of polymer
2. Oxidation of cyclized polymer
3. Oxidation of polymer
4. Cyclization of oxidized polymer

Furthermore, it was believed that prior oxidation of polymer reduced the cyclization rate and prior cyclization of polymer reduced the oxidation rate (rate of oxygen diffusion).

Reference to a series of microsections (Figure 3.51) illustrates the course of the oxidation of 1.65 d'tex SAF in air for up to 180 min at 221°C:

1. In less than 9 min, the outer zone represents rapid oxygen diffusion and oxidation of the polymer.
2. Following a short induction period, the core material cyclizes, i.e., cyclization of the polymer.
3. In the time period 9–60 to 70 min, the outer zone slowly darkens in color without noticeable progression to the center, corresponding to relatively slow cyclization of pre-oxidized polymer.
4. In the time period 70–180 min, the zone boundary moves in towards the center of the filament following a square root of time relationship, which corresponds to a diffusion-controlled oxidation of the cycled polymer. The original outer zone darkens further to black.

So the effect of following a uniform treatment process to complete the oxidation stage resulted in a product with an inhomogeneous structure: the skin is oxidized then cyclized, whilst the core is cyclized and then oxidized. The diffuse boundary can be considered as a third zone, where the rates of the two reactions are fairly compatible. The final structures are not the same and will result in carbon fiber with graphite crystallites exhibiting different orientations relative to the fiber axis and different stacking distances.

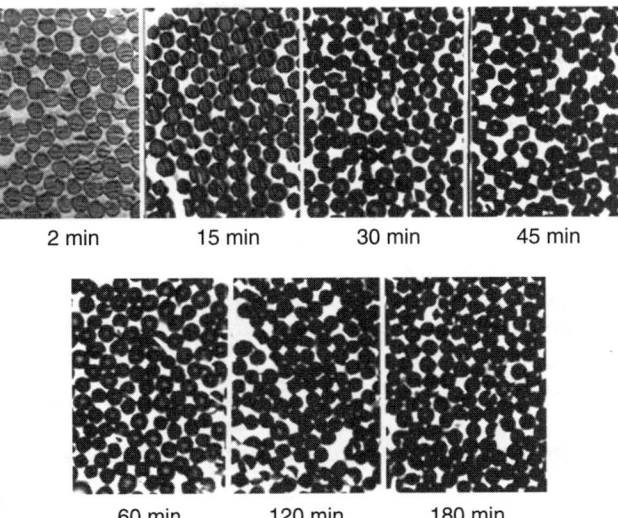

Figure 3.51 Cross-sections of 1.67 d'tex Courtauld's SAF, oxidized in air for periods up to 3 h showing skin/core effect. *Source:* Kind permission of Acordis Ltd (formerly Courtaulds PLC).

The above theory does explain why small diameter SAF did not show oxidation zones. During the short induction period, oxygen has time to penetrate to the center of the filament and oxidize the polymer, followed by the uniform gradual darkening of the filament over the entire cross-section, representing relatively slow cyclization of pre-oxidized fiber. With this smaller diameter fiber, the oxidized structure and presumably the carbonized structure are homogeneous. Whilst faster rates were expected with 0.66 d'tex fiber, because of the smaller diffusion distance, the increase of oxidized fiber density with time is virtually the same as with the 1.65 d'tex material, presumably because prior complete oxidation slowed down the subsequent cyclization reaction. This is similar to Orlon, which had sulfonic acid dye sites that did not initiate the cyclization of the nitrile groups, these oxidized first and required a longer time to stabilize (up to 6 h at 220°C).

3.4.5.5 Surface treatment

Initially, gaseous oxidation was used to surface treat the carbon fiber and air, carbon dioxide and steam were all used. However, air was the simplest operation but was difficult to control and moreover, degraded the tensile and impact strengths, so alternative techniques were investigated. Since only staple fiber was produced in the initial stages of carbon fiber production, a batch process would suffice. A wet treatment using sodium hypochlorite was developed adjusting the pH to about 8.5 by the addition of acetic acid to generate the active species of hypochlorous acid [148]. A boat load of fiber was wrapped in glass cloth and placed in a perforated titanium basket, several baskets were then placed in a polyvinyl chloride (PVC) tank containing the pH adjusted hypochlorite liquor and rocked to and fro by a cam mechanism to ensure uniform agitation. An efficient exhaust system was necessary since chlorine was evolved in the process. The treatment was undertaken at ambient temperature for 12 h with HT fiber and 3 days for HM fiber. After treatment, the fiber was washed in warm water and dried. Although this process was significantly better than air oxidation, the impact strength was halved after only 10 min exposure to the hypochlorite treatment liquor.

When continuous fiber came onstream, an alternative treatment process was required and an electrolytic process offered the best solution, since it could either process fiber collected in tubs, or could be adapted to function as a true online process. Various electrolytes were examined [149,150]:

1. Sulfuric acid: did not lend itself to an operative having to place his hands in the bath.
2. Sodium hydroxide: had a similar problem and was exacerbated by the nuisance of air-borne droplets of entrained sodium hydroxide evolved with the emission of hydrogen gas from the cathodes.
3. Sodium chloride: formed sodium hydroxide with its attendant problems and liberated chlorine.
4. Ammonia or an ammonium salt: problems with free ammonia, but did offer the best choice [57,151].

3.4.5.6 Testing and properties of virgin carbon fiber and composites

In the initial stages, only a few ASTM and no SACMA test procedures were available. So Courtaulds published their Grafil Test Methods [152] (written by JC Darling, PE Morgan and EM Trewin) and the procedures were regularly updated. The test method for interlaminar shear strength used a 5:1 span:depth ratio, which gave lower values than the ASTM 4:1 ratio used in the USA. Initially, all carbon fiber was checked using the filament test, which was later replaced by a tow test developed by Hercules and subsequently modified by Courtaulds to improve the end fixings.

3.4.5.7 Production procedures using carbon fiber

Courtaulds developed many products [153] based on carbon fibers: prepreg as warp sheet and tape, random fiber mats, preferred orientation mats, chopped fiber, thermoset and thermoplastic molding compounds and polytetrafluroethylene (PTFE) impregnated tow. Oxidized PAN fiber was available as continuous tow, tops, yarns, woven and knitted fabrics.

1. Prepreg: Courtaulds part in the prepreg business is described in section 3.6.
2. Pultrusion: Initially, a pultrusion line was run using a caterpillar traction unit and this was replaced by a hand-over-hand unit, which was much kinder to smaller sections. Several epoxy systems were successfully developed [154].
3. Thermoplastic molding compounds: There was a market requirement for carbon fiber filled molding compounds and short length material was desized using a modified proprietary vapor degreasing tank containing 1,1,1-trichloroethane. The desized fiber was then blended with 10% unsized product, sized with a compatible thermoplastic size, such as a polyamide, polycarbonate or polysulfone and cut into 6 mm chips using either a rotary or a guillotine chopper for subsequent incorporation in a thermoplastic matrix.
4. Cement: Incorporating carbon fiber in a cement matrix was difficult due to the size of the cement particles, which tended to be filtered out by the fiber reinforcement, so a cement, Swiftcrete, with a very fine particle size (about 5 µm) was used. The fiber was spread as thinly as possible and sized with a water based compatible size such as sodium carboxymethylcellulose [155,156]. Composites with up to 10% V_f were obtained [157].
5. Plating: Initially, the end-use emphasis was placed on incorporating carbon fiber in metals, so work was undertaken for plating carbon fiber and very successful coatings were achieved using electroless plating techniques [158] (Figures 3.52–3.54), but subsequent incorporation in a metal matrix proved troublesome and the work was dropped in favor of cfrp.

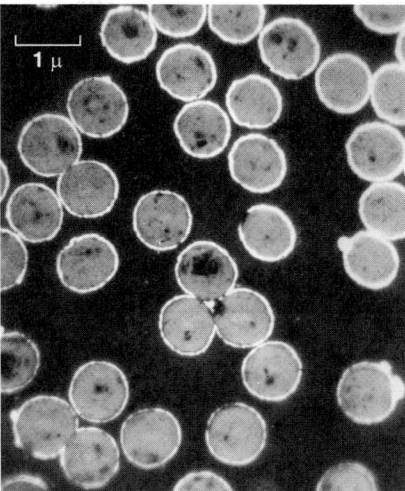

Figure 3.52 Electroless Ni plated carbon fiber. *Source:* Kind permission of Acordis Ltd (formerly Courtaulds PLC).

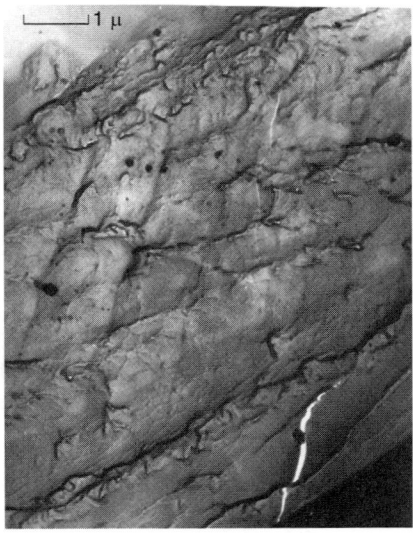

Figure 3.53 Copper plated carbon fiber ×32,000. *Source:* Kind permission of Acordis Ltd (formerly Courtaulds PLC).

An interesting patent was filed [159], where a polymer such as PTFE could be applied by electro-deposition using a process used within the Courtaulds Group to electro-paint the inside of petrol tanks.

3.4.5.8 *Use and design of carbon fiber in composite materials*

R Lucas published papers on utilization and design aspects of carbon fiber [160–162], whilst EM Trewin and RF Turner have written papers on properties and applications of cfrp with particular reference to thermoplastic matrices [163,164].

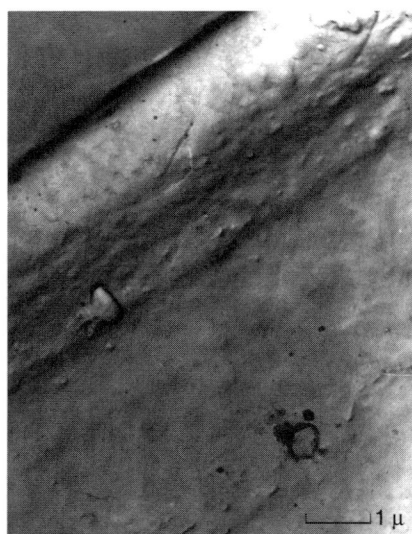

Figure 3.54 Nickel plated carbon fiber ×32,000. *Source:* Kind permission of Acordis Ltd (formerly Courtaulds PLC).

3.5 EARLY UK PREPREGGERS

3.5.1 Ciba (ARL) Ltd., Duxford

Ciba were able to use their considerable expertise in the adhesives field to great effect in the manufacture of epoxy prepreg and concentrated on solventless epoxy systems, although some components were added in a solvent to obtain good dispersion within the resin mix. Ciba's early collaborative work with Rolls Royce resulted in the development and introduction of BSL 914 (a tetrafunctional epoxy, PES, dicy cured system), which gained universal acceptance as a 175°C cure system, with the unique advantage of being able to apply consolidation pressure in the curing cycle without resin bleed in the initial stages of cure. All resin systems were made in-house.

3.5.2 Courtaulds Ltd., Coventry

In some ways, a fiber producer competing in the prepreg market could be construed as unsound commercial practice, but there was no doubt that this gave Courtaulds a good insight into the use of carbon fiber for this application and a better understanding of customers' problems when making cfrp. Added to this was the Hercules' technology, introduced as a result of the Courtaulds/Hercules exchange agreement and the chance to offer 3501–5, 3501–5A and 3501–6 (Ciba 720 tetrafunctional resin based) systems.

At first, the prepreg was made from staple carbon fiber by the hand application of individual tows to a sheet of polyester film, which had been coated with the requisite resin system [165] and attached to a rotatable drum operated with a foot switch. Since the staple fiber had inherent twist, this was removed by the operator as the fiber was applied to the drum, ensuring good spreading. When continuous fiber became available, the fiber was positioned by a traversing mechanism and resin applied to the tow, which was then wound onto a large rotating drum covered with a sheet of release paper [166]. Since the lay-up of the impregnated tow was helical, it was helically cut from the drum to produce a parallel sheet of unidirectional (UD) prepreg. Later, this rather cumbersome equipment was replaced by a purpose-built prepreg line [167] using a multiplicity of parallel tows impregnated by passing over a porous sintered metal applicator, through which a solvent solution of resin was accurately metered. The impregnated tows were supported on release paper and passed through a counter-current of hot air in a drying tunnel to remove all solvent. The solvent system was eventually replaced by hot melt formulations, except for phenolic prepreg, which remained as a solvent system. A number of techniques for spreading carbon fiber were investigated such as an air knife, sonic vibration (cf loudspeaker cone) and lateral stretching of a rubber belt. A number of prepreg plants and/or technology was sold in France, Holland and Japan.

3.5.3 Fothergill and Harvey Ltd. (F&H), Littleborough

They developed a number of in-house resin systems of which Code 69 was the most important, a DDS/BF_3MEA cured 720 resin based system similar to Hercules 3501–5. In the early days, F&H developed considerable expertise making warp sheets by hand lay-up from staple length carbon fiber.

3.5.4 Rotorway components Ltd., Clevedon

This company tended to undertake mostly government contract work and offered resin systems currently in vogue such as an epoxy Novalac (Ciba $LY558/BF_3MEA$), a cycloaliphatic epoxy (Union Carbide ERLA4617/MPD), Shell Epikote DX-209 (a 828/DDM precondensate, BF_3MEA) and Shell Epikote DX-210 (a 828/DDS precondensate, BF_3MEA).

REFERENCES

1. Edison TA, U.S. Pat. 223, 898, 1879.
2. Josephson M, Edison and the incandescent lamp filament, *Scientific American*, 16 Nov, 1959.
3. Josephson M, The invention of the electric light, *Scientific American*, Nov 1959, 98–114.
4. Josephson M, *Edison,* Mc Graw Hill, 1959.
5. Edison TA, Brit. Pat. 4576, 1879.
6. Swann JW, Brit. Pat. 4933, 1880.
7. Hughes TV, Chambers CR, US Pat. 405,480, June 18, 1889.
8. Whitney WR, U.S. Pat. 916, 905, 1909.
9. Houtz RC, Orlon acrylic fibre: chemistry and properties, *J Text Res*, 20:786–801, 1950.
10. Topchiyev AV, Geyderikh MA, Davydov BE, Korgin VA, Krentsel BA, Kustarwvich IM, Polak LL, *Chem & Ind*, 184, Feb 20, 1960.
11. Vosburgh WG, The heat treatment of Orlon acrylic fiber to render it fireproof, *J Text Res*, 30: 882–896, 1960.
12. Anon., *Chem Eng* 68: No.1, 33, Jan 9, 1961.
13. Cranch GE, Unique properties of flexible carbon fibers, *Proceedings of 5th Conference on Carbon*, Vol. 1, Pergamon Press, Oxford, 589–594, 1962.
14. Bacon R, Pallozzi AA, Slosarik SE, Cabon filament reinforced composites, *Proceedings Society Plastics Industry*, 21st Technical and Manufacturing Conference, Sect.8E, 1966.
15. Ezekiel HM, Spain RG, Preparation of graphite fibers from polymeric fibers, *J Polymer Sci*, 249–265, 1967.

16. Gibson DW, Langlois GB, Method for producing high modulus carbon yarn, *Polymer Preprints 9*, No. 2, 1376, 1968.
17. Le Carbone (Great Britain) Ltd., High temperature insulation: fibrous carbon and graphite, *Mach Des Eng*, 2, 63, 1964.
18. T-M Vacuum Products, Heat retainer for 2400°F furnace is water cooled, *Am Mach/Metalwkg Mfg*, 110, No. 15, 164, 1966.
19. Union Carbide Corp., Carbon cloth, *Chem Engng Albany*, 72, No. 23, 122, 1965.
20. Garlock Inc., Graphite filament packing reduces fluid leakage, *Power*, 110, No.2, 98, 1966.
21. The Marlo Co., New packing beats both TFE and graphite-fiber types, *Chem Engng Albany* 75, No. 22, 72, 1968.
22. Raybestos-Manhattan Inc., Graphite filament packings, *Mach Des*, 40, No. 12, 254, 1968.
23. HI Thompson Fiber Glass Co., Graphite cloth aids carburizing, *Steel*, 155, No. 14, 100, 1964.
24. Soltes WJ, U.S. Pat., 3,011,981, 1961.
25. Ford CE, Mitchell CV, U.S. Pat., 3,107,152, 1963.
26. Prosen SP, Carbon resin composites, *Fibre Sci Technol*, 3, 81, 1970.
27. Epremian E, "Thornel", a new graphite reinforcement, *Seminar on Polyblends and Composites at Polytechnic Institute of Brooklyn*, June 6 and 7, 1969.
28. Singer LS, High modulus carbon fiber from pitch, Belg. Pat., 797,543, 1973.
29. Singer LS, High modulus high strength fibers produced from mesophase pitch, U.S. Pat., 4,005,183, 1977.
30. Singer LS, The mesophase and high modulus carbon fibers from pitch, *Carbon*, 16, 408, 1978.
31. Shindo A, Jap. Pat., 28287, 1959.
32. Shindo A, Jap. Pat., 29270, 1962.
33. Shindo A, Osaka Kogyo Gijitsu Shikenjo Kiha, 110:119, 1961.
34. Shindo A, Osaka Kogyo Gijitsu Shikenjo Hokoku, No. 317, 1961.
35. Otani S, On the carbon fiber from the molten pyrolysis products, *Carbon*, 3, 31–38, 1965.
36. Otani S, *Carbon*, 3, 213, 1965.
37. Otani S, Yamada K, Koitabashi T, Yokoyama A, On the raw materials of MP carbon fibre, *Carbon*, 4, 425–432, 1966.
38. Kureha Chemical Ind Co., Jap. Pat., 68–4450, 1965.
39. Kureha Chemical Ind Co., Jap. Pat., 41–15728, 1966.
40. Kureha Chemical Ind Co., Jap. Pat., 69–2511, 1969.
41. Yamada Y, Honda H, Inouse T, Jap. Pat., 58–18421, 1983.
42. Yamada Y, Imamura T, Inouse T, Honda H, Jap. Pat., 58–196292, 1983.
43. Otani S, Jap. Pat., 57–100186, 1992.
44. Tomio A *et al*, Jap. Pat. 59–12286, 1984.
45. Mair WN, Mansfield EH, William Watt 1912–1976, *Biographical Memoirs of Fellows of the Royal Society*, 33, 643–667, 1987.
46. Watt W, Perov BV eds., *Handbook of composites Vol 1 Strong Fibres*, Amsterdam, 1985.
47. Johnson J, Phillips LN, Watt W, Brit. Pat., 1,110,791, Apr 1965.
48. Johnson J, Watt W, Phillips LN, Moreton R, Brit. Pat., 1,166,251, Oct 1966.
49. NRDC, Brit. Pat., 1,166,619, 1969.
50. Moreton R, Watt W, Johnson W, Carbon fibres of high strength and high breakingstrain, *Nature*, 213, 690, 1967.
51. Watt W, Johnson W, Carbon fibres from 3 denier polyacrylonitrile textile fibres, *Paper presented to 3rd Conference on Industrial Carbons and Graphite*, London, 1970.
52. Watt W, Johnson W, Conf. High Temperature Resistant Fibres, American Chemical Society, Interscience, Atlantic City, 1968.
53. Watt W, Johnson DJ, Parker E, Pyrolysis and structure development in the conversion of PAN fibres to carbon fibres, *Proc 2nd International Carbon Fibre Conference Plastics institute*, London, Paper No.1, 3–11, 1974.
54. Peters DM, Breakthrough and early development, Smith EA ed., *Carbon Fibres, Design Engineering Series*, Morgan Grampian, West Wickham, 3–7, 1970.
55. Wadsworth NJ, Watt W, NRDC, Brit. Pat., 1,180, 441, Feb 1970.

56. Harvey J, A simple apparatus for the electrolytic treatment of carbon fibre surfaces, *RAE Technical Report 86077*, Dec 1986.
57. Paul JT Jr, Hercules, Brit. Pat., 1,433,712, 1976.
58. Harvey J, The characteristics of carbon fibre surfaces by dye adsorption, *RAE Technical Memo MAT 231*, Sept 1975.
59. Moreton R, The removal of sodium from polyacrylonitrile precursor fibre and its effect on the mechanical properties of RAE carbon fibre, *RAE Tech Memo MAT 78*, Feb 1970.
60. Judd NCW, Wright WW, Sodium in carbon fibres and their possible effects on composite properties, *RAE Technical Report 75069*, May 1975.
61. Moreton R, Spinning of polyacrylonitrile precursor fibres with reference to the properties of carbon fibres, *3rd Conference, Industrial Carbons and Graphite*, SCI, London, 1970.
62. Judd NCW, Plastics and Polymers, *Conf Suppl*, 5, 258, 1971.
63. Curtis PT, CRAG test methods for the measurement of the engineering properties of fibre reinforced plastics, *RAE Technical Report TR 88012*, Feb 1988.
64. Symposium: *Fabrication Techniques for Advanced Reinforced Composites*, University of Salford, Apr 1980.
65. Giltrow JP, Lancaster JK, *Wear*, 12, 91, 1968.
66. Lancaster JK, *Proc Inst Mech Eng*, 182, 33, 1967–68.
67. Giltrow JP, Lancaster JK, *Proc Inst Mech Eng*, 182, 3N, 147, 1967–68.
68. Lancaster JK, *J Appl Phys*, 1, 549, 1968.
69. Giltrow JP, Lancaster JK, *Wear*, 16, 359, 1970.
70. Lancaster JK, *Wear*, 20, 335, 1972.
71. Lancaster JK, *Wear* 20, 315, 1972.
72. Giltrow JP, *Composites 55*, Mar 1973.
73. Giltrow JP, *ASLE Trans*, 16, 83, 1973.
74. Evans DC, Lancaster JK, Scott D ed., *Material Science Series, Vol 13, Wear of Polymers*, Academic Press, New York, 1979.
75. Lancaster JK, *Proceedings of the 3rd Leeds-Lyon Symposium on Wear of Non-metallic Materials*, Mechanical Engineering Publishers, London, 187–195, 1978.
76. Logsdail DH, Aspects of carbon fibre development at AERE, Harwell, Preston J ed., *High temperature Resistant Fibers from Organic Polymers*, Interscience, New York, 245, 1969.
77. Anon., Production of long lengths of carbon fibres at Harwell, *Rubber and Plastics Age*, 797, Sep 1968.
78. Sanders RE, Acrylic fibres: Process Survey, *Chem & Proc Eng*, 100–109, Sep 1968.
79. Kearsey HA, Chidley BE, Hornby J, Carbon fibre precursor materials: Reaction kinetics of some acrylics in the oxidative degradation stage, *AERE report R6222*, 1969.
80. Kearsey HA, Chidley BE, Hornby J, The oxidation behaviour of acrylic fibres as the first step in carbon fibre manufacture, ACS meeting on Chemistry of fibres and matrices of high performance composites, 28 Mar–2 Apr 1971.
81. Aitken ID, Rhodes G, Spencer RAP, Brit. Pat., 1,353,596, 1971.
82. Hornby J, Kearsey HA, Sharps JW, The preparation of carbon fibres from large-tow textile acrylic fibre: 5. Continuous heat-treatment at temperatures to 2600°C, *AERE–R 8867*, United Kingdom Atomic Energy Authority, Harwell, Jan 1978.
83. Aitken JD *et al*, Conference on Carbon Fibres, Brighton, 1971.
84. Hornby J, Kearsey HA, The preparation of carbon fibres from large-tow textile acrylic fibre: 4. Comparison of different precursors, *AERE–R 8865*, United Kingdom Atomic Energy Authority, Harwell, Dec 1977.
85. Bromley J, Gas evolution processes during the formation of carbon fibres, International Conference on Carbon Fibres, their Composites and Applications, The Plastics Institute, London, 1971.
86. Sharp JV, Burnay SG, High voltage electron microscopy of internal defects in carbon fibres, *Plastics and Polymers Conference Supplement No. 6*, 68, 1971.
87. Bromley J, Improvements in or relating to carbon fibres, Brit. Pat., 1,340,069, 1971.
88. Hughes JDH, Morley H, An experimental rig for continuous production of high strength carbon fibre from polyacrylonitile, *AERE–M3037*, Feb 1981.

89. Goggin PR, Single filament testing of various fibres, *AERE–R7792*.
90. Hughes JDH, Jackson EE, The potential strength of Type II carbon fibre composites—A practical assessment. *AERE–R8090*, Sep 1975.
91. Hughes JDH, Morley H, Jackson EE, Carbon fibre composites which approach theoretical strength, *AERE–R8727*.
92. Hughes JDH, Morley H, The production of high strength carbon fibre from polyacrylonitrile, *AERE–R9328*, Jul 1979.
93. Sach RS, Bromley J, Brit. Pat., 1,255, 005, 1971.
94. Wells H, Colclough WJ, Ger. Pat., Application 1,817,581; 1969.
95. Wells H, Colclough WJ, Brit. Pat., 1,257,022, 1971.
96. Hughes JDH, The evaluation of current carbon fibres, *J Phys D Appl Phys*, 20, 276–285, 1987.
97. Reynolds WN, Hancox NL, *J Phys D*, 4, 1747, 1971.
98. Hancox NL, The use of a torsion machine to measure the shear strength and modulus of unidirectional carbon fibre reinforced plastic composites, *J Mater Sci*, 7, 1030–1036, 1972.
99. Wells H, Colclough WJ, Goggin PR, Some mechanical properties of carbon fibre composites, *AERE–R6149*, 31, HMSO, Jul 1969.
100. Wells H, Colclough WJ, Goggin PR, Section 2C, *Proc 24th Annual Tech Conf of the SPI Reinforced Plastics/Composites Division*, Washington, 1969.
101. Reynolds WN, Structure and mechanical properties of carbon fibres, *Paper 7.2, 3rd Conf Industrial Carbons and Graphite*, SCI, London, 427–430, 1970.
102. Goggin PR, The elastic contents of carbon fibre composites, *J Mater Sci*, 8, 233, 1973.
103. Knibbs RH, Baker DJ, Rhodes G, Section 8F, *Proc of 26th Annual Technical Conference of the SPI Reinforced Plastics/Composites Division*, Washington, 1971.
104. Sambell RAJ, Bowen DH, Phillips DC, Carbon fibre composites with ceramic and glass matrices Part 1 Discontinuous fibres, *J Mater Sci*, 7, 663–675, 1972.
105. Sambell RAJ, Briggs A, Phillips DC, Bowen DH, Carbon fibre composites with ceramic and glass matrices Part 2 Continuous fibres, *J Mater Sci* 7, 676–681, 1972.
106. Phillips DC, Sambell RAJ, Bowen DH, The mechanical properties of carbon fibre reinforced pyrex, *J Mater Sci*, 7, 1454–1464, 1972.
107. Briggs A, Carbon fibre-reinforced cement, *J Mater Sci*, 12, 384–404, 1977.
108. Howlett BW, Minty DC, Old CF, The fabrication and properties of carbon fibre/metal composites, Paper No. 14, International Conference on Carbon Fibres and Applications, The Plastics Institute, London, 1971.
109. Nicholas MG, Mortimer DA, Paper No. 19, International Conference on Carbon Fibres and Applications, The Plastics Institute, London, 1971.
110. Old CF, Nicholas M, Paper No. 14, 2nd International Carbon Fibres Conference, London, 1974.
111. Old CF, Nicholas M, Paper No. 13, 2nd International Carbon Fibres Conference, London, 1974.
112. Pearce DG, Designing with carbon fibre reinforced plastics, *Design Eng*, Feb 1969.
113. Pearce DG, Designing in carbon fibre reinforced metals, *Composites 1*, No. 1, Sep 1969.
114. Standage AE, Prescott R, *Nature*, 211, 169, 1966.
115. Turner WN, Johnson FC, The pyrolysis of acrylic fibre in inert atmosphere I. Reactions up to 400°C, *J Appl Polymer Sci*, 13, 2073–2084, 1969.
116. Rolls Royce Ltd., Fr. Pat. 1,580,443, Sep 1969.
117. Rolls Royce Ltd., Ger. Pat., 1,805,901, Nov 1969.
118. Rolls Royce Ltd., Fr. Pat., 2,006,543, Feb 1970.
119. Rolls Royce Ltd., Fr. Pat., 2,008,173, Mar 1970.
120. Johnson JW, Potter W, Rose PG, Scott G, *Br Polym J*, 4, 527–540, 1972.
121. Rolls Royce Ltd., Br. Pat., 1,174,868, Dec 1969.
122. Johnson JW, Marjoram JR, Rose PG, Stress graphitization of polyacrylonitrile based carbon fibre, *Nature*, 221, 357–358, 25 Jan 1969.
123. Rose PG, A study of the physics and chemistry of the preparation of carbon fibres from acrylic precursors, *Thesis*, University of Aston in Birmingham, Oct 1971.
124. Johnson JW, *Amer Chem Soc Polymer Preprints*, Atlantic City Meeting of Div of Polym Chem, 9(2), 1316, 1968.

125. Johnson JW, Factors affecting the tensile strength of carbon-graphite fibres, *Applied Polymer Symposia*, No. 9, 229–243, 1969.
126. Johnson JW, Thorne DJ, Effect of internal polymer flaws on strength of carbon fibres prepared from an acrylic precursor, *Carbon*, 7, 659–661, 1969.
127. Thorne DJ, Distribution of internal flaws in acrylic fibres, *J Appl Poly Sci*, 14, 103–113, 1970.
128. Turner WN, Johnson JW, Hannah CG, Composite materials, Meetham GW ed., *The Development of Gas Turbine Materials*. Applied Science Publishers Ltd., Barking, 121–146, 1981.
129. McCroft AD, Fibre reinforced articles, Brit. Pat. 1,306,231, Feb 1973.
130. Johnson JW, *Phil Trans Roy Soc*, London, A294, 409, 487, 1980.
131. Lind DJ, Richards J, Rolls Royce Ltd., U.S. Pat., 4,445,951, May 1984.
132. Jackson PW, Marjoram JR, Recrystallization of nickel coated carbon fibres, *Nature*, 218, 83–84, 6 Apr 1968.
133. Jackson PW, Marjoram JR, Compatibility studies of carbon fibres with nickel and cobalt, *J Mater Sci*, 5, 9–23, 1970.
134. Morganite Research & Development Ltd. (Morganite), West German P Appl 1,926,318, UK, 24 May 1968.
135. Morganite, Brit. Pat., 1,271,502, 3 Jul 1968.
136. Morganite, Brit. Pat., 1,271,503, 16 Jul 1968.
137. Zbrzezniak J and Morganite Research & Development Ltd., Fr Pat Applic 2012224, 30 Apr 1970.
138. Morganite, Brit. Pat., 1,326,736, 8 Oct 1969.
139. Badami DV, Joiner JC, Jones GA, Microstructure of high strength, high modulus carbon fibres, *Nature*, 215, 386, 22 Jul 1967.
140. Brydges WT, Badami DV, Joiner JC, Jones GA, The structure and elastic properties of carbon fibres, Preston J ed., *High Temperature Resistant Fibres from Organic Polymers*, Interscience, New York, 255, 1969.
141. Blakelock HD, Blasdale KCA, *Mechanical Property Measurements on Modmor Carbon Fibre-Resin Composites*, Morganite Research & Development Ltd., London, 1968.
142. Blakelock HD, Lovell DR, High modulus reinforcing carbon, *Paper 6B, Proc 24th Ann Tech Conf SPI reinforced plastics/Composites Division*, Society of the Plastics Industry, New York, 1969.
143. Badami DV, Joiner JC, Jones GA, Microstructure, diffraction and physical properties of carbon fibres, *J Phys*, D3, No. 4, 526, 1970.
144. Morganite Research & Development Ltd. and Whittaker Corp., Narmco Materials Division, Modmor Carbon Fibre, *New Scientist*, 43, No. 660, 240, 1969.
145. Gill RM, Carbon Fibres in Composite Materials, Iliffe Books, London, 1972.
146. Huxley DV, Carbon fibres, Lecture given to Coventry Textile Society, 20 Nov 1969.
147. Watt W, Johnson W, *Nature*, 257, No. 5523, 210, Sep 1975.
148. Morgan PE, Courtaulds Ltd., Treatment of carbon fibres, Brit. Pat., 1,238,308, 1971.
149. Chapman DR, Paterson WC, Electrolytically treating filamentary carbon, Brit. Pat., 1,297,946; 1972.
150. White PA, Fibre treatment, Brit. Pat., 1,371,621; 1974.
151. Paul JT Jr, Process for electrolytic treatment of graphite fibres, U.S. Pat., 3,832,297; 1974.
152. Courtaulds Ltd., *Grafil Test Methods*, published by Courtaulds Ltd. Carbon Fibres Division, Coventry.
153. Courtaulds Ltd., *Grafil Data Sheets*, published by Courtaulds Ltd. Carbon Fibres Division, Coventry.
154. Morgan PE, Trewin EM, Watson IP, Some aspects of the manufacture and use of carbon fibre pultrusion, Symposium on Fabrication Techniques for Advanced Reinforced Plastics, Salford, 69–90, Apr 1980.
155. Willats DJ, Morgan PE, Reinforced cement articles, Brit. Pat., 1,425,031; 1976.
156. Morgan PE, Carbon filament tapes, Brit. Pat., 1,425,032; 1976.
157. Waller JA, *Civil Eng Public Works Rev*, 357, Apr 1972.
158. Evans LS, Morgan PE, Coating carbon fibre with metal, Brit. Pat., 1,215,002; 1970.
159. Morgan PE, Lemmon TB, Improvements in coating carbon filaments, Brit. Pat., 1,255,925; 1971.
160. Lucas R, Carbon fibres the versatile performer, *CME Journal*, Apr 1973.

161. Lucas R, Application of carbon fibre composites in general engineering, *Paper 40, Proc 2nd International Conference on Carbon Fibres*, Plastics Institute, London, 1974.
162. Lucas R, *Application of carbon fibres to modern high speed loom sley developments*, 2, 69, Mar 1975.
163. Trewin EM, Turner RF, Carbon fibres—properties and manufacture, *Engineering*, Apr 1980.
164. Trewin EM, Turner RF, Carbon fibres—use in composite materials, *Engineering*, May 1980.
165. Morgan PE, Roberts RW, Composite material, Brit. Pat., 1,254,278; 1969.
166. Willats DJ, Morgan PE, Carbon filament tape, Brit. Pat., 1,340,504; 1971.
167. Morgan PE, Factors in providing quality carbon fibre prepreg materials, *Paper 23, Proc 2nd International Conference on Carbon Fibres*, Plastics Institute, London, 1974.

CHAPTER **4**

Precursors for Carbon Fiber Manufacture

4.1 INTRODUCTION

Since the early work of Edison, many types of precursors have been used to produce carbon fibers, of which polyacrylonitrile (PAN) has proved to be the most popular.

The ideal requirements for a precursor are that it should be easily converted to carbon fiber, give a high carbon yield and allow to be processed economically. The attraction of PAN is that the polymer has a continuous carbon backbone and the nitrile groups are ideally placed for cyclization reaction to occur, producing a ladder polymer, believed to be the first stage towards the carbon structure of the final fiber.

The carbon content of acrylonitrile ($CH_2=CHCN$) is 67.9% and it is not surprising PAN precursors have a carbon yield of some 50–55%, coupled with the ability to produce high modulus fibers.

An acrylic fiber is defined as having acrylonitrile (AN) monomer content greater than 85%. Fibers with AN content less than 85% are termed modacrylics and are not suitable for use as carbon fiber precursors.

A cellulosic precursor $(C_6H_{10}O_5)_n$ has a carbon content of 44.4% but, unfortunately, in practice, the reaction is more complicated than just simple dehydration and the carbon yield is only of the order of 25–30%.

Pitch based carbon fibers, however, do have a higher yield of 85% with a high resultant modulus but, due to their more graphitic nature, they will have poorer compression and transverse properties as compared to PAN based carbon fibers.

Other forms of precursor such as vinylidene chloride and phenolic resins have been investigated and have not been found to be commercially viable.

4.2 PAN PRECURSORS

Two very useful books covering the subject of PAN fibers are by Masson ed. [1] and Frushour and Knorr [2].

4.2.1 History

The development of a PAN fiber was held back because PAN would not melt without decomposition and, consequently, at that time, could not be used in a melt spinning process. It was not until 1942 that a suitable solvent was found, when GH Latham at DuPont discovered dimethyl formamide (DMF) as a possible solvent. Later that year, RC Houtz had evaluated a whole range of solvents and in 1946, patented a series of solvent that might be used [3]. Work on a prior discovery by Bayer was delayed by World War II and in 1950, IG Farben had established NH_4SCN and $ZnCl_2$ as possible solvents and Bayer started their production in 1954.

An organic solvent, soluble in water, could be used for wet spinning into an aqueous bath or dry spinning into a hot environment that removed the solvent by evaporation. DuPont selected the dry spinning route using DMF as the solvent and in 1950, introduced Orlon. The early product was difficult to dye and, by 1952, DuPont had introduced methyl acrylate as a comonomer. DuPont eventually ceased production of acrylic fibers in 1991.

Monsanto started their work with PAN in 1942 and, in 1949, formed a joint venture company with the American Viscose Company calling the new company Chemstrand, which introduced Acrilan using dimethylacetate (DMAc) as the solvent, but the product was fraught with initial problems of fibrillation. The fibrils when abraded on apparel gave areas, which showed with increased whiteness due to the enhanced reflectance. The problem was eventually overcome in 1955 by introducing an annealing stage using low pressure (LP) steam.

Courtaulds initially made a PAN fiber (RL Polymer) by a batch process, to be replaced by Courtelle, made by a continuous polymerization process. The author can remember the day when WG Schmidt found a suitable solvent for the RL acrylic polymer and went running round the laboratory holding a test tube aloft containing a solution of PAN in NaSCN. The process was developed and following two pilot plants at Coventry, a continuous production plant was built at Grimsby in 1959, based on a terpolymer, acrylonitrile/methyl acrylate/itaconic acid (AN/MA/ITA) system using NaSCN as a solvent [4]. An important parallel development was a plant to recover and purify the spent NaSCN, which was fed back into the system.

Courtaulds introduced, specifically for the manufacture of carbon fiber, a Special Acrylic Fiber (SAF), which was made at the Coventry works using the same dope, but spun on production lines with additional filtration and individual dope spinning pumps, enabling precise d'tex control for smaller tows. Courtaulds ceased production of their Special Acrylic precursor (SAF) in 1991.

It is generally recognized that the price of propylene dictates the price of PAN, since acrylonitrile is made from propylene and ammonia. The chemicals for the production of PAN are approximately 45% of the cost of making PAN.

4.2.1.1 Commercially available PAN fiber

The development of acrylic fibers in the 1950s has undergone initial expansion, with the major producers setting up new plants and selling their technology throughout the world followed by changes of ownership and initial producers like DuPont finally bowing out. The present production has been rationalized and Table 4.1 lists the location around the globe of plants producing acrylic fiber.

Table 4.1 World distribution of acrylic fiber plants

Region	Number of Plants	Average Capacity tpa
W. Europe	11	69,000
USA	2	111,000
Japan	6	76,000
S. Korea	2	75,000
Taiwan		279,000
China	15	31,000
Other Countries	26	N/A

Source: Reprinted from Fiber Organon, 70, No.11, November 1999.

Table 4.2 Acrylic precursor manufacturing processes

Manufacturer	Trade name	Polymerization system	Solvent	Typical % polymer
Accordis (Courtaulds)	Courtelle	Continuous solution	NaCNS	10–15
Asahi	Cashmilon	Continuous aqueous dispersion	H_2O/HNO_3	8–12
Hexcel (Hercules/Sumitomo)	Exlan	Continuous aqueous dispersion	$H_2O/NaCNS$	10–15
Mitsubishi	Finel	Continuous aqueous dispersion	$H_2O/DMAc$	22–27
Toho	Beslon	Continuous solution	$ZnCl_2$	8–12
Toray (and Amoco)	Toraylon	Batch solution	DMSO	20–25

4.2.2 Requirements for a PAN precursor

Acrylic precursors for the carbon fiber industry originated from companies that were established commercial scale producers of textile grade acrylic fibers. Hence, the manufacturers that could most readily adapt their existing technology to create a precursor grade material have been most successful (Table 4.2). However, some aspects such as dyeability and a tendency to yellow are not important parameters for a carbon fiber precursor but, because that particular polymer formulation was initially used for other textile end uses, the polymer composition could not be changed. As carbon fibers have developed, the market requirement for suitable precursors has increased and new polymers have been developed specifically for the manufacture of carbon fibers.

To have a better understanding of the requirements for a PAN precursor, it is necessary to consider what properties the PAN polymer should have and what the resultant properties of the spun fiber should be. Some of the more important parameters required to produce an acceptable PAN precursor are:

1. A polymer with an acrylonitrile content less than 85%, or a homopolymer, will not make a satisfactory PAN precursor. The choice of comonomer(s) does have a most important role in the preoxidation stage of the carbon fiber manufacturing process.
2. The molecular weight and molecular weight distribution of the polymer have to be carefully controlled. The number average molecular weight (M_n), and the weight average molecular weight (M_w), together with the polydispersity index (M_w/M_n), control these values, which are

given by:

$$M_n = \frac{n_1 M_1 + n_2 M_2 + n_3 M_3 + \ldots\ldots + n_x M_x}{n_1 + n_2 + n_3 \ldots\ldots n_x}$$

$$\bar{M}_w = \frac{n_1 M_1^2 + n_2 M_2^2 + n_3 M_3^2 + \ldots\ldots + n_x M_x^2}{n_1 M_1 + n_2 M_2 + n_3 M_3 + \ldots\ldots + n_x M_x}$$

where n_x is the number of molecules of weight M_x.

As a general rule [6], the number average molecular weight is 40,000–70,000 g/mole, which is about 1000 repeat units and can be determined by osmotic pressure measurements [7,8], lowering of the vapor pressure, elevation of the boiling point, depression of the freezing point and end group titration or, determination of the sulfur content (if any) in the end groups.

The weight average molecular weight is generally of the order of 90,000–140,000 g/mole and is determined by light scattering methods [9], size exclusion chromatography (which embraces gel permeation chromatography) [10–13] and melt or solution viscosity measurements [14–16]. In practice, the intrinsic viscosity, where the viscosity of a polymer solution is compared with that of the pure solvent, is used as a control method throughout the acrylic fiber industry for measuring molecular weight [17].

The polydispersity index is of the order of 1.5–3.0 and is best determined directly, using size exclusion chromatography.

The number average molecular weight is inversely related to the PAN dye site level, which can be attributed to the presence of polymer dye sites, such as sulfonate, situated at the chain ends. Hence M_n is very sensitive to the fraction of low molecular weight polymer. The weight average molecular weight, on the other hand, relates more to the rheological properties of the PAN polymer and to the actual mechanical properties of the fiber. In practice, it is normal to measure the viscosity at only one concentration and that concentration level will be inversely related to the viscosity average molecular weight. Hence a low molecular weight will be associated with a higher dope concentration, with attendant lower solvent recovery costs. A relative molecular weight of about 100,000, together with an even distribution, results in a precursor fiber with good mechanical properties. Too high a relative molecular weight will give a dope with a high intrinsic viscosity that would be difficult to filter and ultimately, spin. Filtration is critical to keep particulate matter to an absolute minimum since any form of inclusion in the polymer would be a possible site for an intrinsic defect in the resultant carbon fiber.

Tsai and Lin [18] have investigated the effect of molecular weight on the cross-section and properties of PAN precursor and the resulting carbon fiber.

3. The polymer dope concentration in the spinning solvent controls the effectiveness, efficiency and economics of the spinning process, whilst the temperature controls the spinning and can also control the final shape of the spun filament. The highest concentration ensures less solvent to be recovered per kg of fiber produced.
4. The construction and cleanliness of the spinneret has a bearing on the final cosmetics of the precursor fiber.
5. Careful control of the spinning conditions, such as concentration and temperature of all process baths and regulation of the applied stretch are paramount to produce a consistent product with the required mechanical and cosmetic properties.
6. The spun filament should have a relatively fine count (e.g., 1.22 d'tex), which enables the fiber to be heated through to the center at a fairly rapid rate and conversely, permits more readily the dissipation of the heat evolved in the strongly exothermic initial oxidation stage of the carbon fiber manufacturing process. A precursor of about 0.8 d'tex would permit a more uniform structure of the oxidized fiber, but with reduced production. The control of heat flux, reaction rate and temperature of initiation can also be influenced by the choice of comonomer(s) and the actual ratio of these comonomer(s) to the acrylonitrile content.
7. The choice of a suitable finish facilitates the subsequent handling and processing of the precursor fiber and, preferably, the finish should break down during carbonization into gaseous

components. Additional finishes can be used to give extra protection throughout the oxidation and low temperature carbonization stages.

A detailed review of acrylic precursors for carbon fibers is given by Gupta et al [19], some precursor examples are discussed by Rajalingam and Radhakrishnan [20], whilst Bajaj and Roopanwal present an overview of the thermal stabilization of acrylic precursors for the production of carbon fibers [21].

4.2.3 Homopolymer PAN

The homopolymer PAN is not an easy product to process into carbon fiber, since the initial oxidation stage of the carbon fiber process is a difficult reaction to control due to the sudden and rapid evolution of heat, coupled with a relatively high initiation temperature. This rapid surge of heat can cause chain scission with resultant poor carbon fiber properties. As far as is known, homopolymer PAN has never been exploited as a precursor for carbon fiber manufacture. The exothermic reaction can, however, be adequately controlled by suitable comonomers such as itaconic acid.

4.2.4 Comonomers

Comonomers do have a significant effect on the stabilization process, enhancing the segmental mobility of the polymer chains [22,23] resulting in better orientation and mechanical properties of the precursor and resulting carbon fibers. Comonomers can also reduce the temperature of initiation of cyclization [24,25].

Tsai and Lin [26] discuss the effect of comonomer composition on the properties of the PAN precursor and the resulting carbon fiber.

Preferred neutral comonomers for acrylonitrile would be vinyl esters such as vinyl acetate (VAc), MA and methyl methacrylate (MMA), although VAc would not be a suitable candidate for a carbon fiber precursor. These comonomers could be termed plasticizers as they break up the structure, making the polymer more readily soluble in the spinning solvent, improving the quality of spinning, modifying the fiber morphology and where appropriate, improving the rate of diffusion of a dye into the fiber. The neutral comonomers are not intended to take part in the polymerization reaction and, as it happens, AN/MA are an almost ideal pair, with both monomers exhibiting similar polarity, resonance and steric hindrance and in practice, MA does show only a slight tendency to react with acrylonitrile, thereby ensuring a homogeneous polymer. Ogawa studied the effect of MA as a comonomer [27]. Only a minimum quantity of comonomer plasticizer, to allow practicable stretching, should be used.

A list of possible comonomers is given in Table 4.3 [28–73] and it should be noted that it is common practice for companies to use more than one comonomer, e.g., MA and ITA, where, with AN, the three monomers have similar reactivities and so the polymer composition will have more or less the same composition as the monomers in the feed.

Carboxylic acids are very effective comonomers, since the presence of the carboxylic acid affects the ease of oxidation, exothermicity and carbon yield of the precursor and of those studied by Guyot et al [74] and Nagai [75], probably ITA and methacrylic acid have been found to be most effective in reducing the exothermicity. Others studied included acrylic acid, acrylamide and sodium acrylate and the effectiveness of these different comonomers in reducing the initiation temperature of the copolymers, can be expressed in the order itaconic acid > methacrylic acid > acrylic acid > acrylamide [62]. The superiority of ITA

Table 4.3 Possible comonomers for PAN precursors

Class	Comonomer	Structure	References
Acids	Acrylic acid	$CH_2=CHCOOH$	28–31
	Itaconic acid	$CH_2=C(COOH)CH_2COOH$	32–37
	Methacrylic acid	$CH_2=C(CH_3)COOH$	38–50
Vinyl esters	Methyl acrylate (methacrylate)	$CH_2=CHCOOCH_3$	4,51–57
	Ethyl acrylate	$CH_2=CHCOOC_2H_5$	58
	Butyl acrylate	$CH_2=CHCOO(CH_2)_3CH_3$	
	Methyl methacrylate	$CH_2=C(CH_3)COOCH_3$	51,58–60
	Ethyl methacrylate	$CH_2=C(CH_3)COOC_2H_5$	59
	Propyl methacrylate	$CH_2=C(CH_3)COO(CH_2)_2CH_3$	59
	Butyl methacrylate	$CH_2=C(CH_3)COO(CH_2)_3CH_3$	59
	β-hydroxyethyl methacrylate	$CH_2=C(CH_3)COOCH_2CH_2OH$	51
	Dimethylaminoethyl methacrylate	$CH_2=C(CH_3)COOCH_2CH_2N(CH_3)_2$	51
	2-Ethylhexylacrylate	$CH_2=CHCOOCH_2(C_2H_5)(CH_2)_3CH_3$	61
	Isopropenyl acetate	$CH_3COOC(CH_3)=CH_2$	51
	Vinyl acetate	$CH_3COOCH=CH_2$	51,59
	Vinyl propionate	$C_2H_5COOCH=CH_2$	59
Vinyl amides	Acrylamide	$CH_2CHCONH_2$	51,62,63
	Diacetone acylamide	$CH_2=CHCONHC(CH_3)_2CH_2COCH_3$	64
	N-methylolacrylamide	$CH_2CHCONHCH_2OH$	51
Vinyl halides	Allyl chloride	$CH_2=CHCH_2Cl$	51
	Vinyl bromide	$CH_2=CHBr$	65
	Vinyl chloride	$CH_2=CHCl$	66
	Vinylidene chloride (1,1-dichloroethylene)	$CH_2=CCl_2$	66–68
Ammonium salts of vinyl compounds	Quaternary ammonium salt of aminoethyl-2-methyl propenoate	$CH_2=CH(CH_3)COOC_2H_4NH_2$	69,70
Sodium salts of sulfonic acids	Sodium vinyl sulfonate	$CH_2=CHSOONa$	59
	Sodium p-toluene sulfonate (ptsa)		
	Sodium p-styrene sulfonate (SSS)	$CH_2\!=\!CH\!-\!C_6H_4\!-\!SO_3Na$	51,71,72

	Name	Structure	Ref.
	Sodium methallyl sulfonate (SMS)	$CH_2=C(CH_3)CH_2SO_3Na$	51,72
	Sodium 2-acrylamido-2-methyl propane sulfonate (SAMPS)	$CH_2=CHCONH(CH_2)_3SO_3Na$	66,67
Other	Methacrylonitrile	$CH_2=C(CH_3)CN$	51
	2(1-hydroxyalkyl) acrylonitrile	$H_2C=C-X$ $\quad\ \ |$ $\ \ R-CHOH$ (where R = —CH_3 or —C_2H_5 and X = —CN or —COOH)	73
	Allyl alcohol	$CH_2=CHCH_2OH$	51
	Methallyl alcohol	$CH_2=C(CH_3)CH_2OH$	51
	1-vinyl-2-pyrrolidone	(pyrrolidone with N–CH=CH_2)	51
	4-vinylpyridine	(pyridine with CH=CH_2)	51
	2-methylene glutaronitrile	$NCCH_2CH_2C(=CH_2)CN$	51

can be attributed to the presence of two carboxylic acid groups, increasing the possibility of interacting with a nitrile group since, if one carboxylic acid group moves away from an adjacent nitrile group by dipole-dipole repulsion, then it is highly likely that the other carboxylic acid group could move into the vicinity of a nitrile group and could then take part in the cyclization process.

Too much acid comonomer would ensure a safe oxidation reaction, but the reaction could become too slow to be commercially viable. If, however, too little is added, this would establish a higher onset temperature and the oxidation reaction would be likely to be too fast and uncontrollable.

Acid comonomers can initiate the cyclization reaction of the nitrile groups:

Acrylic, methacrylic and itaconic acids also act as plasticizers, so it should not be necessary to add any additional plasticizer comonomer such as MA. However, for textile tow precursor, it may be necessary to introduce either more of the acid comonomer, or add some plasticizer monomer to facilitate the introduction of crimp, which aids packaging of the textile tow.

A differential scanning calorimeter (DSC) plot (Figure 4.1) of homopolymer PAN and copolymer PAN shows how the homopolymer PAN has a very intense and narrow exotherm peak, whilst the presence of ITA broadens this peak, with the onset of oxidation

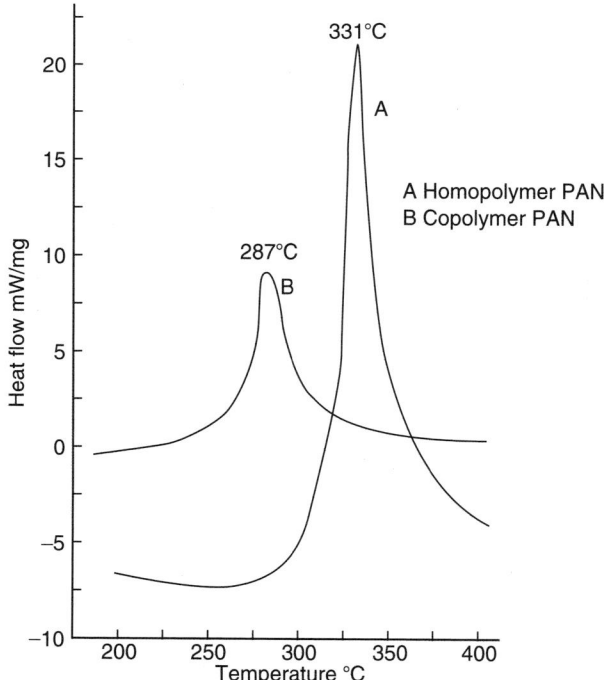

Figure 4.1 Differential Scanning Calorimeter plots of homopolymer and copolymer PAN.

occurring at a lower temperature, clearly showing the advantages, in this instance, of an AN/MA/ITA terpolymer.

Different levels of ITA, in an AN/ITA copolymer have been investigated by Mueller [76], and an increase in the level of ITA was found to result in carbon fibers with better mechanical properties and reduced the optimum oxidation time; ITA also reduces the intensity of the exotherm [76]. Increasing the ITA content, has a substantial effect on the co-polymerization which, unfortunately, when the copolymer is converted to carbon fiber, reduces the carbon yield [74,76] due to some loss of ITA in the carbonization process. Tsai and Lin [77] showed that ITA reduced the initiation and peak temperatures of the exotherm.

Amide comonomers have a less pronounced effect than acid comonomers. Acrylate comonomers do not initiate the cyclization reaction [78,79], whilst MMA results in a lower carbon yield than MA. The bulky side groups of an ester comonomer lower the crystallinity and crystal size but improve the segmental mobility of the polymer chains [22,23]. The size of the ester molecule also affects the structural parameters of the precursor and the resultant carbon fibers have a lower average orientation and mechanical properties [21].

Moreton [68] used vinylidene chloride (VDC) in an attempt to produce a precursor, which would principally oxidize fast and, hopefully, achieve carbonization without oxidation, but unfortunately, the carbon fiber properties obtained were poor. It was hoped that the presence of the VDC monomer units would provide sites for potential cross-linking of the ladder polymers, thereby hastening the oxidation process.

The incorporation of vinyl phosphonates (for example, *bis*-chloroethylvinylphosphonate and *bis*-chlorobutylvinylphosphonate), along with VDC, is claimed in an Asahi patent [67] to reduce the oxidation time.

In conventional PAN textile fibers, dye sites comprising sulfonate groups, together with sulfate groups from the breakdown of the initiators, can attach to the polymer chain ends left over by the free radical initiators. At least 40 such dye sites are required per molecule and should the groups that had attached on the polymer residual chain ends not be sufficient to meet the dyeing requirements, then copolymerizable sulfonate containing comonomers could be incorporated into the polymer chain to provide the additional dye sites and would also increase the hydrophilicity [80]. However, a need for such dye sites should not arise for the manufacture of a precursor fiber unless the sulfonic acid fulfils another role. Sulfonic acids would not be expected to have any beneficial effect, due to the absence of any nucleophilicity necessary to aid the cyclization process. However, ionic monomers, such as the sodium, ammonium or amine salts of sulfonic acids e.g. sodium *p*-styrene sulfonate (SSS), sodium methallyl sulfonate (SMS), sodium *p*-sulfophenyl methallyl ether (SPME) or, sodium 2-acrylamido-2-methylpropane sulfonate (SAMPS) and ethylene sulfonic acid can be added to provide these dye sites along with the sulfonate end groups [1]. The patent literature does cite the use of a number of sulfonates, e.g., Toray used sodium allyl sulfonate, while DuPont used sodium styrene sulfonate, and both Asahi and Bayer [81] have opted for sodium methallyl sulfonate.

Toray introduced a new range of comonomers, based on 2(1-hydroxy alkyl)acrylonitrile, in particular the ethyl and methyl derivatives, with the general formula

$$H_2C=\underset{R-CHOH}{C-X}$$

where R is H, or an alkyl, cycloalkyl, aralkyl or aryl group containing less than 12 carbon atoms and X is —CN, —COOH, a substituted or unsubstituted phenyl or naphthyl group, or a carboxylate ester group of formula —COOR$_1$, where R$_1$ is an alkyl, cycloalkyl, aryl or cycloaryl group containing less than 12 carbon atoms [73]. Toray claims that such comonomers can be oxidized very rapidly under special conditions, e.g., 20 minutes in a nitrous oxide atmosphere activated with UV light, followed by 40 minutes at 140°C associated with an increased carbon yield of 78%.

Courtaulds did introduce a fast oxidizing precursor (80 k, 1.85 d'tex) based on the addition of p-toluenesulfonic acid (PTSA) which had a high burn up temperature (284°C), enabling oxidation to be carried out at a higher temperature (e.g. 45 min at 260°C to give an oxidized density of 1.40 g cm^{-3}) and a lower sodium content of 250 ppm.

4.2.5 Methods of polymerization

Acrylonitrile and its comonomers employing free radical initiation can be polymerized by one of several methods namely:

1. Solution polymerization, when the dope is prepared directly during the process of polymerization
2. Bulk polymerization, which is an auto-catalytic process and rarely used commercially and is therefore not considered;
3. Emulsion polymerization generally restricted to modacrylics and not relevant for a carbon fiber precursor;
4. Aqueous dispersion (slurry) polymerization, which is the most common method used for textile fibers.

Table 4.4 lists the factors influencing the production of a PAN precursor for the manufacture of carbon fibers.

4.2.5.1 Solution polymerization

The main advantage of solution polymerization is that the polymer, as it is formed, goes into solution and it is necessary to remove any unreacted volatile monomers only after polymerization, by distillation. Keeping careful control of any unreacted non-volatile monomers, it is then possible to use the solution of the polymer in the selected solvent, termed a dope, for spinning. The choice of solvent can be either a highly polar organic solvent in order to solvate the highly polar nitrile group, or an inorganic aqueous salt solution (Table 4.5) and the selection affects the level of chain transfer that will occur during polymerization and hence, restricts the magnitude of the relative molecular weight that can be achieved.

The solvents which tend to be used in practice, are those with low chain transfer coefficients and for this reason, DMAC is not used but can, however, be used as a solvent to form a dope for the separated and dried polymer after slurry polymerization. The most common solvents are DMSO, ZnCl$_2$ and NaSCN, although some manufacturers do use DMF, despite its high chain transfer coefficient and it is the preferred solvent for dry-spun fiber due to its relatively lower boiling point (153°C). DMAc, unfortunately, has a 125°C lower ignition point than DMF and consequently, the spinning tube cannot be heated to such a high level, reducing the output by some 30%. The inorganic solvents give higher relative molecular weights due to their lower chain transfer coefficients, but propylene and ethylene carbonates do produce polymers with very high relative molecular weights [82]. DMSO can be used in batch solution processing and can be recovered by distillation, discarding any high boiling components accumulating in the still bottoms. ZnCl$_2$ can be recovered by

Table 4.4 Factors influencing the production of a PAN precursor for the manufacture of carbon fibers

Stage	Process	Advantages	Disadvantages
Method of polymerization	Aqueous dispersion	(i) Impurities left in mother liquor (ii) High percentage conversion (iii) High molecular weight	(i) Re-dissolving process necessary (ii) High gel content (iii) High gas content
	Solution	(i) No drying polymer or re-dissolving required (ii) Reduced gel problems (iii) Low gas content	(i) Impurities retained in the solvent (ii) Must control the solvent impurity level
Polymer process	Batch	(i) Reproducible product (ii) Accurate batch preparation	(i) Wide molecular weight spread
	Continuous	(i) Satisfactory for large scale production	(i) Difficult to control on a small scale (ii) Continuous formation of gels (iii) Variable product quality
Method of spinning	Dry jet (Air gap)	(i) Enhances orientation prior to coagulation (ii) Very good for small tows of fine d'tex (iii) Good mechanical properties	(i) Cannot be used for spinning large tows
	Wet	(i) Satisfactory for spinning large tows with good productivity	(i) Inorganic solvents are corrosive
	Dry	(i) Can be spun into fine d'tex fiber (ii) Can produce profiled and hollow fibers from concentrated spinning solutions	(i) Must use DMF as solvent (ii) Gelling can occur with high concentration solutions (iii) Fibers have dog-bone shape
	Melt	(i) More economic and solvent free	(i) Has core-sheath structure (ii) Presently, not technically viable
	Gel	(i) Can produce high tenacity fibers of high molecular weight	(i) Uses ultra-high molecular weight polymer (ii) More research required

Table 4.5 Typical solvents used for dissolving PAN

Solvent	Formula	Solvent chain Transfer polymer coefficient $C_5(10^4)$	Typical % concentration at 50°C
γ-butyrolactone	(structure)	0.70	–
N,N-dimethylacetamide (DMAC)	$CH_3-CO-N(CH_3)_2$	5.00	22–27
N,N-dimethylformamide (DMF)	$H-CO-N(CH_3)_2$	2.75	28–32
dimethylsulfoxide (DMSO)	$(CH_3)_2S=O$	0.40	20–25
ethylene carbonate (EC)	(structure)	0.43	15–18
aq. sodium thiocyanate (45–55%)	NaSCN	low	10–15
aq. zinc chloride (50%)	$ZnCl_2$	0.006	8–12
nitric acid (65–75%)	HNO_3	–	8–12

Source: Reprinted with permission from Frushhour BG, Knorr RS, Acrylic fibers, Lewin M and Pearce EM eds., *Handbook of Fiber Chemistry*, Marcel Dekker, New York, 869–1070, 1998. Copyright 1998, CRC Press, Boca Raton, Florida.

crystallization and, moreover, it is claimed by Saito and Ogawa [83] that one decided advantage of using aqueous $ZnCl_2$ is that it facilitates the subsequent oxidation of PAN. NaSCN is very corrosive and readily picks up cations such as iron, which can affect the rate of polymerization. The NaSCN solvent is recovered by a multistage evaporation process and, since this builds up impurity levels, it has to be continuously purified by a separate regeneration process in order to maintain control over the impurity level. DMF does confer a yellow coloration to the polymer, which should not, however, be a problem for a carbon fiber precursor, but DMF does have the disadvantages of a high chain transfer constant and residual high boiling components. Organic solvents do place a less severe onus on filtration than either $ZnCl_2$ or NaSCN.

The kinetics of a homogeneous solution polymerization follow the classical kinetics of an addition polymerization reaction, with a new bond being formed between molecules that have not already joined and the reaction proceeding without the elimination of any by-product molecules. Once the reaction has been initiated, the reaction can follow a number of separate stages, which can occur concurrently:

1. Initiation stage

    a. Thermal
    The application of heat produces a free radical, which requires a specific energy of activation and involves a collision factor, which also implies a steric factor depending on the type of monomer, e.g., 2,2′-azo-di-isobutyronitrile (AZDN)

    $$CH_3-\underset{\underset{CN}{|}}{\overset{\overset{CH_3}{|}}{C}}-N=N-\underset{\underset{CN}{|}}{\overset{\overset{CH_3}{|}}{C}}-CH_3 \longrightarrow 2\ CH_3-\underset{\underset{CN}{|}}{\overset{\overset{CH_3}{|}}{C}}{}^* + N_2$$

    An increase in concentration of AZDN will increase the conversion. An increase in temperature will increase the production of free radicals, thereby increasing the probability of terminating a chain, which would then become shorter.

2. Photochemical
   The exposure to light of a given wavelength (such as UV) produces free radicals, e.g.,

$$CH_2\!=\!\underset{CN}{CH} + h\nu \rightarrow CH_2\!=\!{}^*CH\ ({}^*CH\!=\!CH + H^*) \xrightarrow{+CH_2=\underset{CN}{CH}} CH_3\underset{CN}{CH}CH_2\!-\!\underset{CN}{\overset{H}{\underset{|}{C^*}}} \text{ etc}$$

   a. Free radical
      e.g. benzoyl peroxide-Cm

$$O\!-\!O\!-\!O\!-\!CO\!-\! \rightarrow 2^* + 2CO_2$$

   or the free radical can be generated by a redox system, which is generally used with aqueous media, e.g.,

$$Fe^{++} + H_2O_2 \rightarrow Fe^{+++} + HO^- + HO^*$$

   A redox system cannot be used when the solvent is NaSCN since the Fe will react.

   b. Propagation stage
      The free radical species formed in the initiation stage have high reactivity and can add onto the monomer units, leaving a reactive species at the end, the process liberating large amounts of energy.

$$CH_3\!-\!\underset{CN}{\overset{CH_3}{\underset{|}{C^*}}} + CH_2\!:\!\underset{CN}{CH} \rightarrow CH_3\!-\!\underset{CN}{\overset{CH_3}{\underset{|}{C}}}\!-\!CH_2\,\underset{CN}{CH^*}$$

$$CH_3\!-\!\underset{CN}{\overset{CH_3}{\underset{|}{C}}}\!-\!CH_2\,\underset{CN}{CH^*} + CH_2\!:\!\underset{CN}{CH} \rightarrow CH_3\!-\!\underset{CN}{\overset{CH_3}{\underset{|}{C}}}\!-\!CH_2\,\underset{CN}{CH}\!-\!CH_2\,\underset{CN}{CH^*} \text{ etc}$$

   Growth requires much lower energy than the initiation stage and occurs more rapidly, leading to the development of long chains with a life of about 10^{-2} seconds.

   c. Termination stage:
      Termination can occur between a growing chain and

   - Another active chain

$$R(M)_n M^* + R(M)_m M^* \rightarrow RM(M)_{m+n} MR$$

   - A catalyst radical
   - Chain transfer by collision with

      i. An inactivated monomer

$$R(M)_n{}^* + MH \rightarrow R(M)_n H + M^*$$

      reacts with more monomer.

ii. An inactivated polymer

$$R(M)_{n^*} + RM(M)_m R \rightarrow R(M)_n + RM^*(M)_m R$$
$$\text{(growing)} \qquad\qquad\qquad \text{(dead)} \quad \text{(forms branches)}$$

iii. A solvent molecule

$$R(M)_{n^*} + H\text{—}CO\text{—}N(CH_3)_2 \rightarrow R(M)_n H + {}^*CO\text{–}N(CH_3)_2$$

An increase in chain transfer agent will decrease the molecular weight, but will not affect the conversion.

d. By collision with an inhibitor or retarder, e.g., a metal ion

$$Fe^{3+} \rightarrow Fe^{2+}$$

reacts with the free radical forming a stable substance incapable of starting a new chain, hence decreasing the conversion and molecular weight.
e. By reacting with an impurity, such as an amine.
f. By the initial destruction of two original free radicals, so that a polymer is never formed.
g. By radical disproportionation
Two radicals may react by disproportionation to give other radicals and polymer growth then occurs by addition of the monomer to give branched structures:

$$-CH_2CHX- \;+\; -CH_2CHX- \quad \begin{array}{l} -CH_2CH_2X \;+\; CHCH\text{–}X \\ \text{or} \\ CH_3CHX \;+\; -CH_2\text{–}C\text{–}X \end{array}$$

h. Chain transfer stage
Any process by which one polymer chain is terminated (e.g., by a solvent, monomer or additive) and another free radical generated, is known as chain transfer because the new radical can propagate a new chain. Chain transfer cannot start a polymer chain and should not influence the rate of polymerization, but can give shorter chains with a consequent reduction in the relative molecular weight.

The kinetics for copolymer and terpolymer systems are obviously more complicated and involve reactivity ratios expressing the tendency for monomer blocking (reaction with the same monomer), or alternate reaction with the different monomers [84].

A polymerization system can comprise of the AN monomer, copolymer(s), initiator as the source of free radicals, chain transfer agent, retarding agent, color improving agent, chelate agent and agents to adjust pH. Operating a continuous process will entail introducing the recovered volatiles (e.g., a mix of monomers, chain transfer agent and water) and using the recovered solvent can impose further restrictions. When one of the monomers is non-volatile, this means that the unreacted monomer remains in the dope and will probably find its way into the spin bath. Hence, the control of a polymerization reactor is quite complicated and obviously the aim is to run the reaction absolutely steadily with constant monitoring.

4.2.5.2 Aqueous dispersion polymerization

Most acrylic fibers are produced by this method, utilizing a batch or continuous polymerization process, where de-ionized water is the continuous phase and the initiators and dispersants are water soluble (in suspension polymerization they are water insoluble). A redox system will generate free radicals effectively in a water based system at temperatures of about 75°C and an acid pH comprising of:

1. An oxidizer, e.g., ammonium, potassium or sodium persulfate.
2. A reducing agent or activator, e.g., sodium bisulfite, sodium metabisulfite or sulfur dioxide.
3. A catalyst such as 1 ppm on the monomer mix of Fe^{2+} ion as $FeSO_4$.

The reaction is undertaken in a carefully controlled acid environment (about pH 3), when the bisulfite ion predominates and the Fe^{+++} ion formed by oxidation will be soluble. If sulfur dioxide is used, it also helps to maintain the pH and can be used with the metabisulfite. However if SO_2 is not used, then another acid such as sulfuric acid must be used to maintain an acid pH, using sodium bicarbonate to make corrections should it become too acid. Free radicals are formed by one of two reactions:

1. The oxidation of ferrous ion by persulfate ($S_2O_8^{2-}$) ion:

$$Fe^{2+} + S_2O_8^{2-} \rightarrow Fe^{3+} + SO_4^{2-} + SO_4^{1-*}$$

2. The reduction of ferric ion by bisulfite (HSO_3^-) ion:

$$Fe^{3+} + HSO_3^{1-} \rightarrow Fe^{2+} + HSO_3^*$$

The sulfate and sulfonate free radicals react with the monomer and initiate rapid chain growth and the propagation step in aqueous dispersion polymerization is an order of magnitude faster than solution polymerization.

The bisulfite ion takes on a dual role, acting as a reducing agent and as a chain transfer agent.

$$HSO_4^{1-} + M_n^* \rightarrow SO_3^{1-*} + M_nH$$

Now, if the chain reaction is terminated by radical combination, then all polymer chains would have dye sites, formed at each end from sulfate and sulfonate initiator radicals in an equal ratio assuming equal reactivities. Whereas, should a bisulfite ion terminate a chain with a hydrogen atom and start another with a sulfonate radical, then the total dye sites are increased, albeit by chains with one dye site only, and the molecular weight is reduced. To minimize the spread in molecular weight, the ratio of bisulfite to potassium persulfate is maintained at a high level of about 15, which effectively gives a lower conversion, limiting any branching reactions and reducing color formation.

Sodium bisulfite will react with acrylonitrile to form the water soluble sodium salt of β-cyanoethylsulfonic acid (β-sulfoproprionitrile), which is removed in the washing stage.

$$CN-CH=CH_2 + NaHSO_3 \rightarrow CN-CH_2CH_2SO_3Na$$

A solution of the sodium salt of ethylenediaminetetraacetic acid (EDTA), $(NaOOCCH_2)_2NCH_2N(CH_2COONa)_2$, is added, forming a chelate complex with the iron, and acts as a chain stopper.

As the polymer molecules are formed they aggregate, influenced by the degree of agitation and particles about 20–40 μm are removed by a two stage vacuum filtration process, thoroughly washed with de-ionized water and dried to about 0.5% moisture.

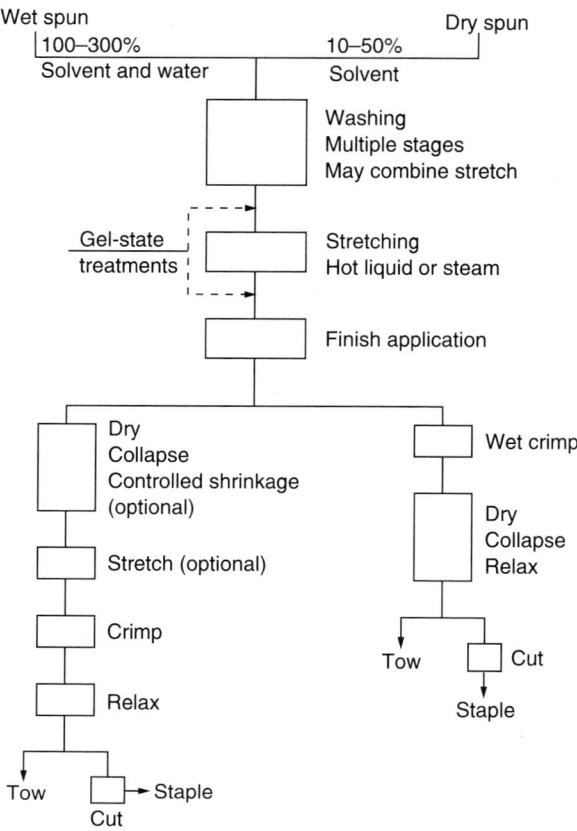

Figure 4.2 Two processing of acrylic fibers. *Source*: Reprinted with permission from Hobson PH, McPeters AL, *Kirk-Othmer: Encyclopedia of Science and Technology*, 1, Interscience, New York, 1978. Copyright 1978, John Wiley & Sons Ltd.

4.2.6 Methods of spinning [85–87]

Wet spinning is used for most commercial PAN based carbon fiber precursor processes and is gradually being replaced by air gap (dry jet wet) spinning. Melt spinning of a plasticized PAN has been used, but has yet to become an acceptable commercial process. Figure 4.2 shows a typical layout of a plant for processing acrylic fiber tow.

Tsai and Wu [89] have determined the effect of the cross-section evenness of PAN precursor on the carbon fiber properties.

4.2.6.1 Wet spinning [90]

1. Dope preparation
   A solution of PAN copolymer is dissolved in a suitable solvent, such as DMF, DMSO, NaSCN or $ZnCl_2$ (Table 4.5) to give a dope. The concentration of the dope (about 15–20%) depends on the molecular weight of the polymer and the solubility in the chosen concentration of solvent. Although a high concentration favors more economic recovery processes, if it is unduly high, the dope will be too viscous to pass through the fine holes of the spinneret (jet) at the required rate and will also be more likely to form gels, which are difficult to filter out and can block the spinneret. When the dope has been made, it is allowed to stand, de-aerated by heating and subjecting to a vacuum and then rigorously filtered in several stages.

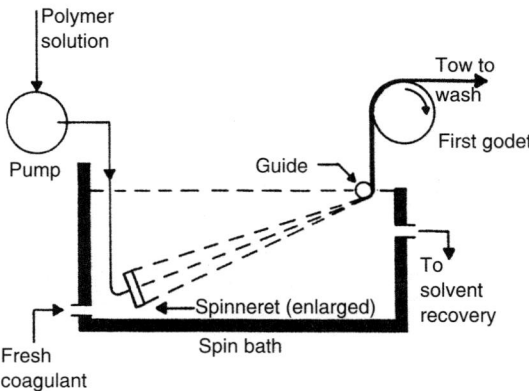

Figure 4.3 Schematic of wet spun coagulation bath. *Source*: Reprinted with permission from Capone GJ, Wet-Spinning Technology, Masson JC ed., *Acrylic Fiber Technology and Applications*, Marcel Dekker, New York, p. 73, 1995. Copyright 1995, CRC Press, Boca Raton, Florida.

2. Coagulation stage

   The dope is pumped with a gear pump, at the required temperature (25–120°C), through a spinneret with a multiplicity of holes, a total of 50–500,000 depending on the application and usually arranged in segments to improve the distribution of the spin bath (0–50°C) across the jet face. The jet is made from a material (e.g., 70 Au, 29.5 Pt, 0.5Rh) that will withstand the chemical environment and the holes (0.05–0.25 mm diameter) with a conical inlet and cylindrical capillary outlet formed by laser drilling. Cleanliness of the jet is most important, since blocked holes will introduce poor cosmetics.

   The jet is immersed in a spinbath, normally of dilute aqueous polymer solvent, to precipitate the PAN polymer. If the bath is too dilute the recovery costs will be high and at the other extreme, if too concentrated, the polymer will not precipitate. The line speed in coagulation is about 3–16 m/min and tension is gradually applied to the newly formed tow with stretching continuing throughout the process to orient the molecular chains along the fiber axis. Figure 4.3 shows a typical wet spun coagulation bath. It is normal to use several jets in the one spinbath.

   The PAN polymer coagulates in the spin bath, initially forming a skin, then the spin bath diffuses through into the center of the filament giving a porous gel network, formed by separation of a polymer solution into a polymer-rich phase and a solvent-rich phase. The structure is a gel network of interconnected polymer fibrils separated by voids of size about 0.3 μm or greater [91]. The rate of diffusion depends on rate, concentration and temperature. The coagulation rate is controlled by the composition and temperature of the spinbath and must be carefully controlled to limit formation of internal pores, surface defects and skin-core [92,93]. Stretching is limited by the rate of coagulation and the speed of fiber drawing, the fibers stretching more readily in the gel state [92,94–96] because the occluded solvent limits the cohesive forces between the molecular chains, thereby allowing the chains to unfold and form an oriented network, with a reasonable distribution of pores and void size [97].

   Irrespective of the solvent or the coagulation system used in wet spinning, a fibrillar structure is formed, generated in an unoriented form during coagulation. The porosity of the structure is controlled, initially by the coagulation conditions and is subsequently reduced by stretching and drying.

   The cross-sectional shape of the filament relates to the volumetric transfer rate of the spin bath liquid (non-solvent) into the fiber versus the outward transfer of the solvent [98,99]. The fiber skin being the first portion to undergo coagulation, limits the volume that the fiber can occupy and, if less volume of solvent diffuses out, the fiber becomes non-circular and progresses towards a kidney bean shape. Each manufacturer uses a different spinning process and the filament cross-section is a useful fingerprint to identify the manufacturing source.

   Figure 4.4 shows typical wet-spun precursor technology using continuous solution polymerization and wet spinning.

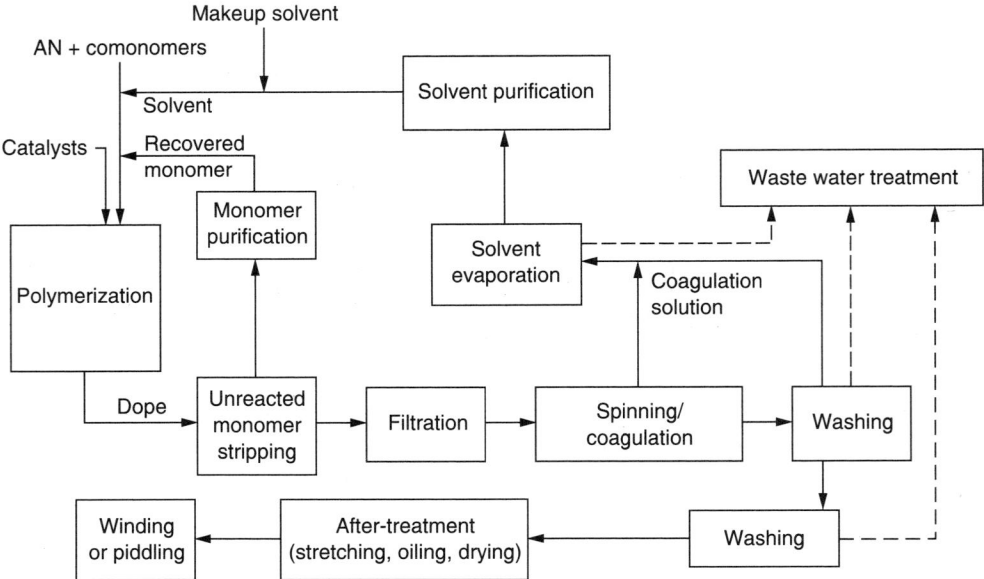

Figure 4.4 Wet spun precursor technology: continuous solution polymerization/wet spinning. Source: reprinted from East G, McIntyre J, Patel G, The dry-jet wet-spinning of an acrylic fiber yarn, *J Text Inst*, 75, 196–200, 1984.

Moreton and Watt [100] showed that PAN fibers spun under clean room conditions gave significantly improved properties of a carbon fiber by reduction of surface flaws when prepared at 2500°C.

4.2.6.2 Dry spinning

In dry spinning, the polymer is dissolved in a suitable solvent such as DMF, and then spun into a tube or cell, where the solvent is evaporated at a temperature above the boiling point of the solvent. The solvent should be economic, non-toxic, readily dissolve the polymer without reaction, have a low boiling point and acceptable heat of vaporization, not generate a static charge and have a low risk of explosion [101]. Dry spinning operates at much faster speeds (1000 m/min) than wet spinning, but the number of filaments in the tow is limited. A spinning tube for the dry spinning process is shown in Figure 4.5.

Dry spinning generates a fiber that initially appears different from typical wet-spun fibers as there is no opportunity for the spin bath to diffuse into the fiber. However, when the unoriented dry-spun fiber is stretched, an oriented fibrillar structure develops, indistinguishable from a stretched wet-spun fiber. Hence, all acrylic fibers, whether dry- or wet-spun are fundamentally similar.

4.2.6.3 Air gap spinning [102,103]

Air gap, or dry jet wet, spinning is a variant of wet spinning and is particularly suitable for spinning precursor material. The jet is positioned close, less than 10 mm, above the spinbath and the filaments extruded vertically into the spinbath liquor. This process permits the dope and spinbath to be at different temperatures and avoids the high stress caused by the dope coagulating at the jet face in wet spinning. Hence, a higher than normal

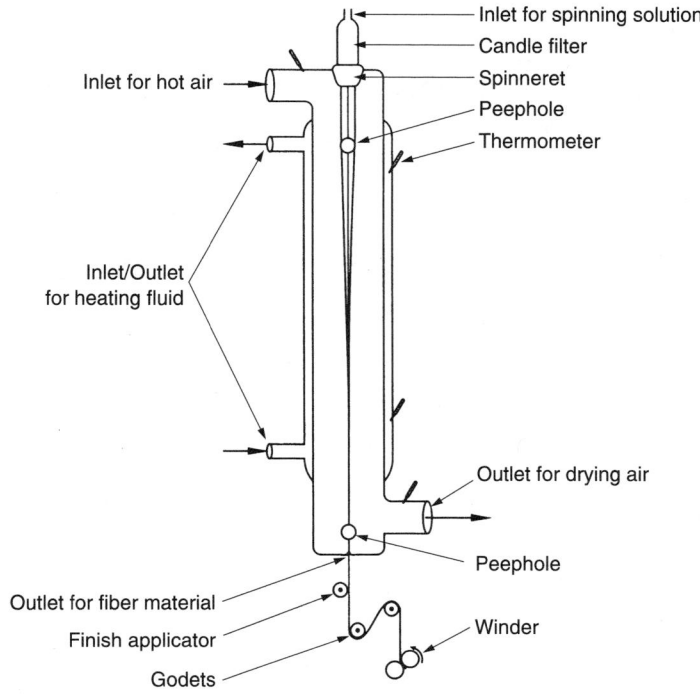

Figure 4.5 Spinning tube for dry spinning process—length 6 to 10 m. *Source*: Reprinted with permission from von Falkai B, Dry spinning technology, Masson JC ed., *Acrylic Fiber Technology and Applications*, Marcel Dekker, New York, 105–165, 1995. Copyright 1995, CRC Press, Boca Raton, Florida.

solids content and faster line speeds can be employed, but the process is limited by the number of holes in the jet and cannot be used for large tows, but is ideal for a $12k$ tow. Orientation is enhanced prior to coagulation and since the spun filament gels before entering the spinbath, the structure is similar to a dry-spun fiber.

4.2.6.4 Melt spinning

PAN starts to cyclize and decompose at a temperature well below its melting point, considered to be about 350°C, which rules out using a conventional melt spinning process and it was discovered that a fiber forming acrylonitrile polymer and water could form a homogeneous single phase fusion melt, which could be extruded into a steam pressurized solidification zone. The phase diagram, shown in Figure 4.6, indicates the amount of water which would be required to produce an extrudable melt, as demarcated by the region DBC, although in practice, a further 7% per polymer unit is added to ensure that all the —CN groups have become uncoupled. As the water content is increased, the polymer melting point decreases until a point is reached when a pure water phase is formed corresponding to the maximum melting point depression and further water added just passes into the aqueous phase [104]. The melting point of PAN is depressed by some 135°C to about 185°C and to obtain these working temperatures and successfully extrude the PAN hydrate, a pressure of 30–70 bar is required [105]. If a comonomer is used, however, the melting point is depressed further, enabling lower temperatures to be used. A statistically distributed polar comonomer, such as a methacrylic ester blocks the nitrile groups so that less water is required to free the —CN dipoles [101].

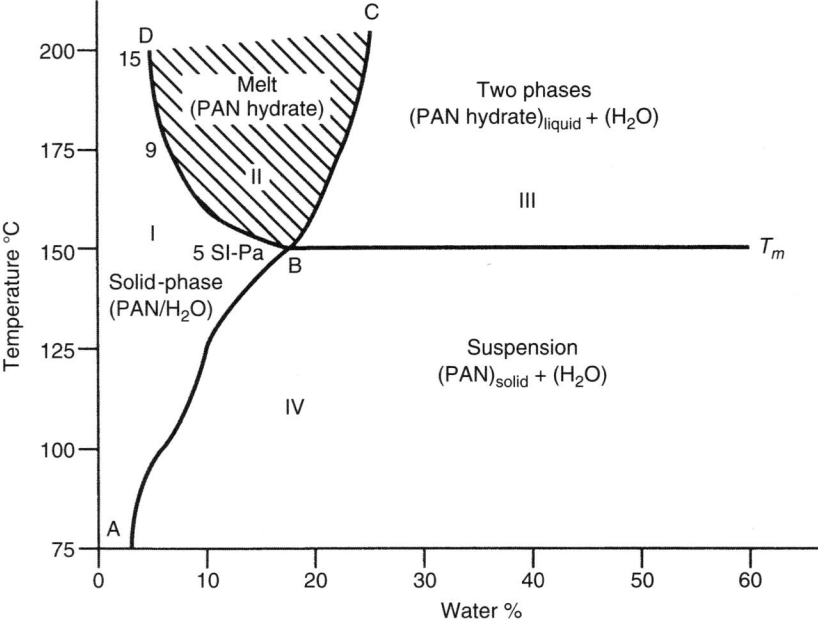

Figure 4.6 Phase diagram polyacrylonitrile/water. *Source*: Reprinted with permission from von Falkai B, Dry spinning technology, Masson JC ed., *Acrylic Fiber Technology and Applications*, Marcel Dekker, New York, 105–165, 1995. Copyright 1995, CRC Press, Boca Raton, Florida.

Although Sumitomo had first introduced a PAN hydrate melt spun fiber in 1948, it was not until about 1969 that American Cyanamid had developed their melt spinning process [106–108], which was later sold to BASF, who continued further developments [109–113].

The original Cyanamid process used a single phase fusion melt of an acrylonitrile copolymer (molecular weight 30,000–60,000) and water and extruded, at about 50 bar, through a spinneret (hole size 60–160 μm) directly into a steam pressurized solidification zone, such that the subsequent release of water from the extrudate avoided any deformation of the extruded filaments and their condition remained such that they could be stretched up to some 25 times. The fiber was cooled with water and passed to the atmosphere by way of a pressure lock.

The use of a copolymer with a lower molecular weight (6,000–16,000) will lower the melting point of the hydrate and facilitate extrusion.

BASF work [109] revealed the use of a polymer containing a minimum of 85% AN, 5–25% acetonitrile and 1–8% of a C_1 to C_4 monohydroxy alkanol, using 12–28% water and an extrusion temperature of 140–190°C. A nitroalkane/water mix used to extrude the polymer was introduced later [110].

Daumit and Ko [114] described the BASF melt spun precursor (MSP) process and outlined the economic advantages (e.g., no need to handle or recover solvents) over a conventional wet spun process. Figure 4.7 shows a schematic layout of this process, which can produce a precursor with a range of cross-sectional shapes such as circular, rectangular, tetralobal and crenulated. Figure 4.8 shows the melt spinning of a PAN hydrate, whilst Table 4.6 gives typical strengths of the MSP based carbon fibers. Melt spun fibers contain more internal and external defects than either wet or air-gap spinning.

DuPont [115] utilized up to 9% of a copolymer with thioether end units derived from a water insoluble mercaptan.

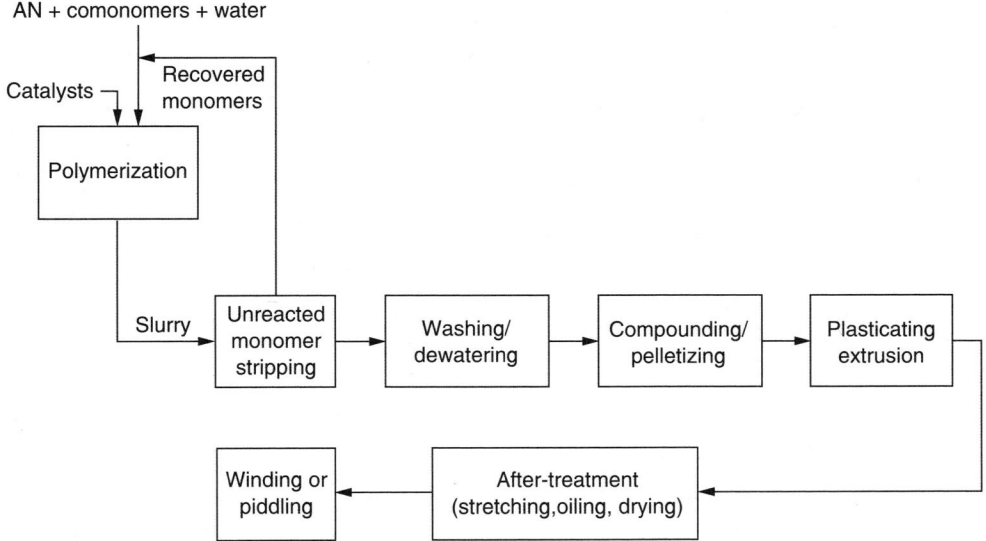

Figure 4.7 Melt spun technology: Aqueous suspension polymerization/melt spinning. *Source*: Reprinted from East G, McIntyre J, Patel G, The dry-jet wet-spinning of an acrylic fiber yarn, *J Text Inst*, 75, 196–200, 1984.

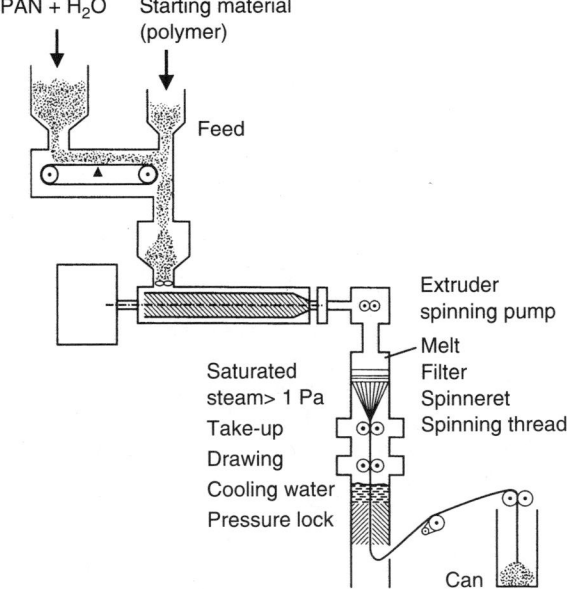

Figure 4.8 Melt spinning of PAN hydrate. *Source*: Reprinted from Porosoff H, Melt-spinning acrylonitrile polymer fibers, American Cyanamid Company, U.S. Pat., 4,163,770, 7 Aug 1979.

4.2.7 Processing stages

Once the fiber has been spun, the remaining processing stages are basically similar, irrespective of the initial spinning route, although handling procedures naturally vary depending whether the process is continuous or not.

Table 4.6 Strengths of melt spun PAN precursor based carbon fibers

Fiber Shape	Strength GPa	Modulus GPa	Elongation %
Circular	3.5–4.0	234–248	1.4–1.7
Ribbon (rectangular)	3.3	234	1.4
Cross (tetralobal)	2.6	234	1.1
Hexalobal (crenulated)	3.1	234	1.3

Source: Reprinted from Daumit GP, Ko YS, A unique approach to carbon fiber precursor development, High Tech—The Way into the Nineties, ed. Brunsch K, Gölden HD, Herkert CM, *Proc of the 7th Int SAMPE Conf*, Munich, Elsevier Science Publishers BV, Amsterdam, June 10–12, 201–213, 1986.

1. Washing

   It is necessary to remove all the solvent from the fiber, usually carried out by counter current washing with hot water in conjunction with fiber stretching.

2. Stretching process

   Stretching or drawing aligns the chains of molecules, imparting strength but reducing the elongation and d'tex. Initially, the temperature is maintained above the wet T_g (about 65°C) and as the solvent content is gradually removed in the washing process, the temperature is increased, permitting application of more stretch. Stretching is achieved by passing the tows over sets of rollers with adjustable speed, permitting a total stretch of up to about ×12 to be applied. Figure 4.9 depicts the action of stretching on the fibrillar network and both the amorphous and crystalline regions become oriented in the direction of the fiber axis and the void regions with entrapped water are elongated (Figure 4.10). The molecules within the chain slide past one another and align parallel, enabling crosslinks to form by dipole-dipole interaction, forming a two dimensional rod structure [116].

3. Finish

   A finish is usually applied as an aqueous emulsion to act as a lubricant and antistat and penetrates into the interior of the fiber if applied before the collapsed stage. Typical finishes are sorbitan esters of long chain fatty acids, polyoxyethylene derivatives and silicones.

4. Drying, collapsing and relaxing

   Drying is undertaken to remove water from the surface of and within the fiber and the actual procedure varies with each commercial process. Figure 4.11 depicts a typical drum drier. As the PAN fiber is dried, it collapses, which is a radial contraction of the fibers, occurring as water is removed from the internal structure of the fiber as the internal pores within the fiber close up. It is believed that the collapse of acrylic fibers follows a two stage process [117]: the first consists of the retraction of stretched molecules to a helical conformation, dependent on the T_g (about 70°C in water at 100°C), the water causing the collapse to occur at a lower temperature due to the increased mobility of the chains. This is followed by a second stage, where junction networks of rods, or fibrils, are formed and the collapsed structure becomes stabilized against reswelling (Figure 4.9). The diameter decreases sharply at about 90°C and the collapse and shrinkage are both essentially complete at about 130°C. A decrease in porosity accompanies collapse and as the network is oriented by drawing, the pores become elongated in the direction of the stretch.

   The relaxation process significantly alters the stress/strain properties of the fiber and can be incorporated in the drying as a continuous stage, or can be carried out by a batch process where the fiber is subjected to a hot wet environment, normally under a slight pressure, in an autoclave, to elevate the boiling point. As the temperature approaches 150°C, water plasticizes the molecules and the internal cracks and fissures fuse together. Silver sulfide staining (Chapter 17) can be used to examine for cracks and fissures.

   Exlan hold a patent [118] for drying PAN in a water swollen state at low temperature (below 70°C) with no subsequent application of finish to give a moisture level, preferably below 2%, producing a precursor fiber with very fine voids and yielding a carbon fiber with superior properties and cosmetics.

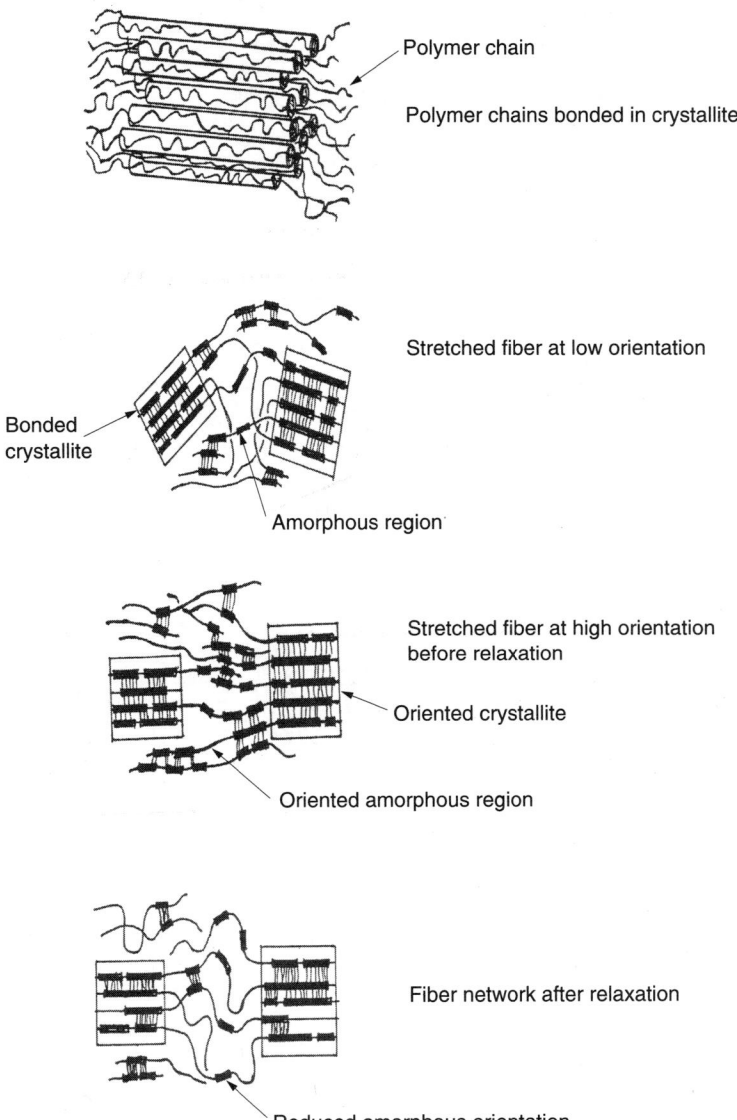

Figure 4.9 Acrylic fiber structural model showing effect of stretching and relaxation. *Source*: Reprinted with permission from Capone GJ, Wet-Spinning Technology, Masson JC ed., *Acrylic Fiber Technology and Applications*, Marcel Dekker, New York, p. 96 and 99, 1995. Copyright 1995, CRC Press, Boca Raton, Florida.

5. Collection

Large tows, above 40 k, are crimped, invariably using a stuffer box crimper (Figures 4.12 and 4.13). Crimping provides interfilament cohesion across the tow and facilitates subsequent handling. The tow is plaited into a cardboard carton, using a controlled longitudinal placement with a progressive transverse movement, to facilitate a stacking sequence, permitting the easy removal of the tow from the package, or alternatively, the product can be baled. Smaller tows, like 3 k and 12 k, are wound directly onto a spool package using precision winders. At this stage, a wetting agent can be applied to help control possible static. Great care must be taken throughout all production stages to maintain good cosmetics, essential for a carbon fiber precursor.

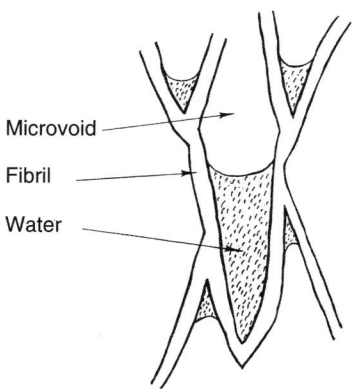

Figure 4.10 Acrylic fiber structure: showing that when fiber is stretched, both the amorphous and crystalline regions are elongated with water trapped within the voids. *Source*: Reprinted with permission from Capone GJ, Wet-Spinning Technology, Masson JC ed., *Acrylic Fiber Technology and Applications*, Marcel Dekker, New York, p. 98, 1995. Copyright 1995, CRC Press, Boca Raton, Florida.

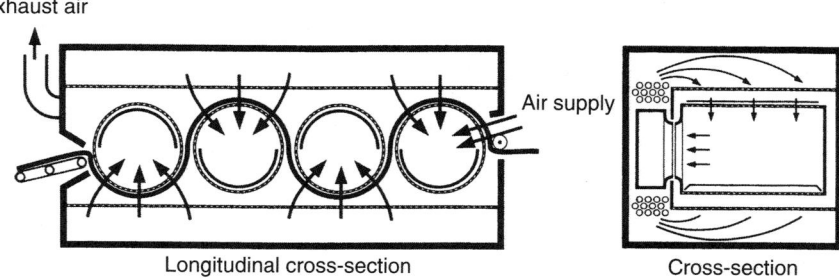

Figure 4.11 Screen drum drier. *Source*: Reprinted with permission from von Falkai B, Dry Spinning Technology, Masson JC ed., *Acrylic Fiber Technology and Applications*, Marcel Dekker, New York, 142, 1995. Copyright 1995, CRC Press, Boca Raton, Florida.

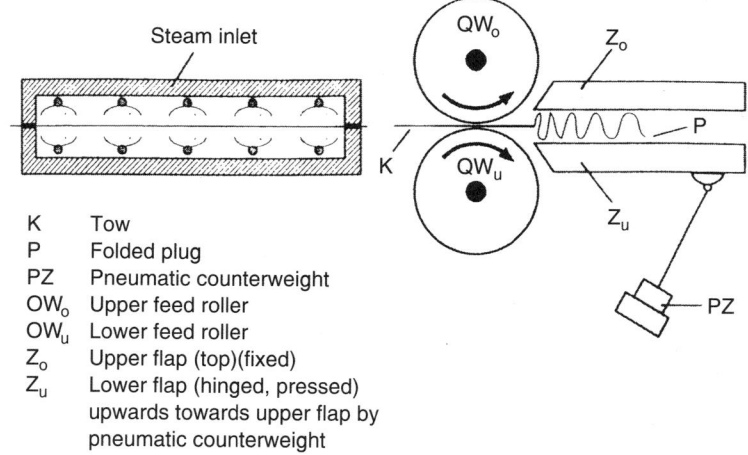

K Tow
P Folded plug
PZ Pneumatic counterweight
OW_o Upper feed roller
OW_u Lower feed roller
Z_o Upper flap (top)(fixed)
Z_u Lower flap (hinged, pressed) upwards towards upper flap by pneumatic counterweight

Figure 4.12 Stuffer box principle. *Source*: Reprinted with permission from von Falkai B, Dry Spinning Technology, Masson JC ed., *Acrylic Fiber Technology and Applications*, Marcel Dekker, New York, 143, 1995. Copyright 1995, CRC Press, Boca Raton, Florida.

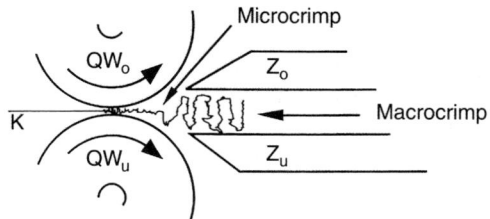

Figure 4.13 Crimp formation (microcrimp) and folding (macrocrimp). *Source*: Reprinted with permission from von Falkai B, Dry Spinning Technology, Masson JC ed., *Acrylic Fiber Technology and Applications*, Marcel Dekker, New York, 143, 1995. Copyright 1995, CRC Press, Boca Raton, Florida.

4.2.8 Modification of spun fiber

4.2.8.1 Stretching

Reducing the filament diameter of a PAN precursor by stretching [119] improves the orientation of the molecular chains, reduces the bulk volume allowing a quicker fiber heat-up rate with better control of a possible exotherm and limiting the formation of skin-core. In conjunction with these beneficial effects, the properties of the resultant carbon fiber are improved, which could be attributed to fewer defects present per unit length, but productivity will be reduced. However, if stretching is overdone, then broken filaments will result, destroying the cosmetics and should the diameter be reduced below a critical value, then the carbon fiber fly would then become respirable and dangerous to health. Bahl has reported that stretching in N_2 is preferable to air [120].

Courtauld's SAF had a very high degree of orientation compared with the Textile product and was attributed to a high stretch ratio resulting in a fiber with a tenacity of about 60 cN/tex and a low extensibility of some 18%. This was reflected in the average crystallite size of SAF being 7.4 nm and about 6.0 nm for the textile product with corresponding azimuthal half width values of 28 and 48° respectively.

Although PAN fiber becomes thermoplastic at about 180°C, stretching is best achieved by following the accepted textile practice of using hot-wet conditions such as hot water or steam [121–126]. Further reduction of filament diameter can be achieved by applying additional stretching stages at higher temperatures by using ethylene glycol at 140°C and glycerol at 180°C [123–131].

4.2.8.2 Chemical treatment

Finishes are applied to the PAN fiber to improve handling and include silicones (modified polysiloxanes) [132] and trimethylol propane-ethylene oxide adduct [133–135]. These finishes are burned off in the latter stages of stabilization, or in the initial stages of the low temperature carbonization furnace and the breakdown products should be volatile to permit removal. At one time, it was common practice to use adventitious sizes applied prior to the stabilization stage to protect the cosmetics of the oxidized fiber during oxidation. These sizes should preferably break down into gaseous components at about 200°C and typical sizes are the ammonium salt of polystyrene maleic anhydride copolymer, ethyl acrylate, ethyl acrylate/methyl methacrylate and polyacrylic acid.

Bajaj and Roopanwal [136] have identified a number of chemical treatments that can be used to modify PAN precursor fibers (Table 4.7). Work undertaken at the NPL in New Delhi has shown that treatment with $KMnO_4$, acetic acid and a combination of both is beneficial in improving tensile strength of carbon fiber from 2.0 to 3.9 GPa when carbonized at 1000°C. Although evidence is provided that all these treatments affect the rate of stabilization, it is not always a commercially viable solution, e.g., $SnCl_4$ (Chapter 3, Section 3.4.5.4).

4.2.9 Structure of PAN fibers

The structure of PAN is generally shown in the planar zig-zag form:

[chemical structure of PAN backbone showing repeating CH2-CH(CN) units in zig-zag form]

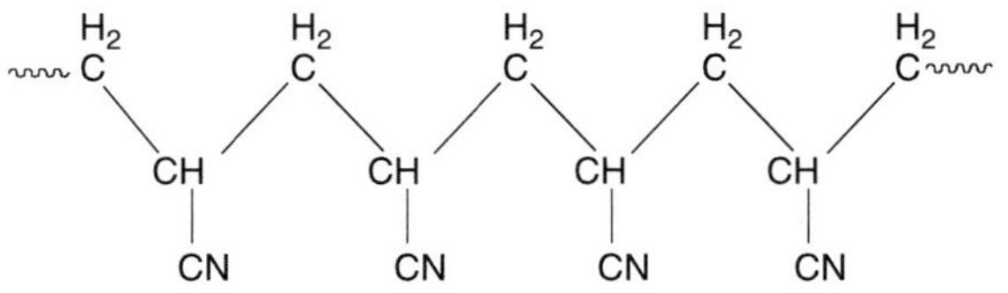

When the substituent (—CN) groups all lie arranged in a random manner about the main carbon chain, then the polymer is said to be atactic and the PAN would be amorphous. However, when the groups all lie on the same side then the polymer is said to be isotactic and would be crystalline and when the groups are spaced alternately on either side of the polymer chain, the polymer is termed syndiotactic. Acrylic fibers have little or no stereoregularity and are not crystalline, but they do have a high degree of order and can be said to be pseudo-crystalline, although no distinct crystalline and amorphous phases can be seen and it would perhaps be better defined as amorphous with a high degree of lateral bonding or as a two dimensional liquid crystalline structure with many defects [155].

The —CN group has a large dipole moment and can be attractive or repulsive. In the planar zig-zag form, the nitrile groups are parallel and maximum repulsion is exerted.

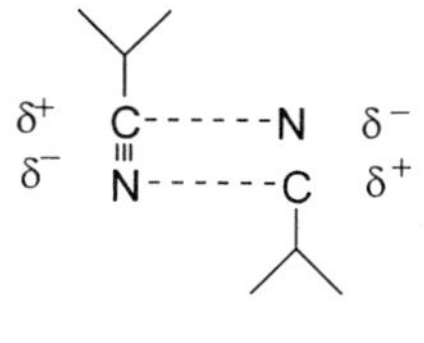

[diagram showing dipole interactions between two C≡N groups with δ^+ and δ^- charges]

Henrici-Olivé and Olivé [156] postulated that the molecule assumed a helical arrangement (Figure 4.14), with the bulk of the molecule contained within a cylinder about 0.6 nm in diameter, with the adjacent nitrile groups positioned as far apart as possible and pointing out from the carbon backbone helix; if some of these chains were to group together and lie alongside each other like sticks of chalk in a box, then some of the nitrile groups would be anti-parallel, exerting a strong attractive force. Hence, repulsive forces would be minimized and attractive forces maximized. However, the net interaction would be repulsive, resulting in an irregular structure with a highly polar atactic chain. Birefringence studies, though, do not support this theory and later X-ray work by Ganster et al [157] suggest that adjacent nitrile groups attract one another and the structure is somewhere between the planar zig-zag and helical forms.

For PAN to be formed into a fiber, it must possess some degree of crystallinity, despite the addition of a comonomer decreasing the crystallinity. Hence Fischer and Fakirov [158] and Hinrichsen [159] have suggested that it is highly likely that a limited two-phase morphology exists for PAN fibers, with lamellar crystalline units being separated by less ordered material of chain-folded lamellae. The unordered domains have been suggested to

Table 4.7 Chemical treatments for PAN precursor fibers prior to stabilization

Class of Treatment	Treatment	References
Inorganic acids	H_2SO_4, HNO_3	137
	H_3PO_4	138
	H_3BO_3	139–141
Organic acids	CH_3COOH	137
Lewis acid compounds	$SnCl_4$, $SnBr_4$, $PbCl_4$, $TiCl_4$, $ZnCl_2$, BF_3	142–143
Bases	Hydroxylamine	137
	Aminophenol quinones	137,144
	Primary amines/Quaternary ammonium salts	137,144
	Aminosiloxanes	137,144
	Hydrazine salts	138
	Urea	137
	Guanidine carbonate	142
Oxidizing agents	Persulfate	139,140
	H_2O_2	139,140
	H_2O_2/Fe^{+++}	145–150
	$K_2Cr_2O_7$	144
	$KMnO_4$	145,151
	$CoCl_2$	147
	$CuCl$	152,153
	$SnCl_2/HCOOH$	137
Organometal compounds	Dibutyltin dimethoxide	154
Transition metal oxides	MnO_4^{1-}	137,142

Source: Reprinted with permission from Ermolenko IN, Lyubliner IP, Gulko NV, Chemically modified Carbon Fibres and their Applications, (translated by EP Titovets) VCH, Germany, Chapter 5, 153, 1990. Copyright 1990, John Wiley & Sons Ltd.

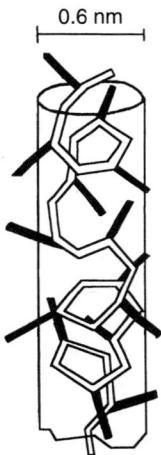

0.6 nm

Figure 4.14 Model showing helical conformation of the PAN molecule. *Source*: Reprinted from Henrici-Olivé G, Olivé S, Molecular interactions and macroscopic properties of polyacrylonitrile and model substances, *Adv Polym Sci*, 32, 123, 1980.

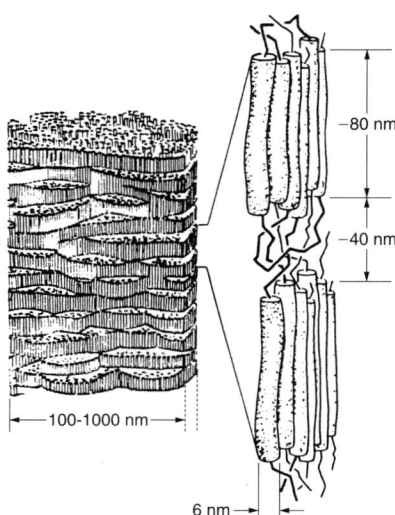

Figure 4.15 Schematic diagram of molecular structure of highly oriented acrylic fibers. *Source*: Reprinted with permission from Warner SB, Uhlmann D, Peebles L, *J Mater Sci*, 10, 758, 1975. Copyright 1975, Springer.

consist of loops, folds, entanglements, chain ends, defects, comonomer sequences, etc. Warner *et al* [160] studied the oxidation of PAN fibers during the first stage of manufacturing carbon fibers and used Bohn *et alia*'s model [161], based on a laterally bonded concept for the partially ordered phase and believed the structure (Figure 4.15) to have a lamellar texture oriented perpendicular to the fiber axis.

Although many attempts have been made to produce a more crystalline polymer with improved fiber tensile properties, success has been limited and, invariably, the techniques used would not be commercially viable. Similarly, attempts by Allen, Ward and Bashir [162] to produce a high modulus PAN fiber (55 GPa) by gel spinning, were not successful and it was concluded that an ultrahigh modulus PAN fiber cannot be made by any technique, due to the intrinsic chain properties, where the strong intramolecular nitrile repulsions cause the PAN to adopt a semi-extended rod-like conformation, thereby not allowing the chain to unravel from its semi-extended form.

4.3 CELLULOSIC PRECURSORS

4.3.1 Historical introduction

The history of cellulosic precursors is interesting because of its association with the early forms of carbon fiber used for electric lamp filaments.

It was Dr. Robert Hooke [163,164], of Hooke's Law fame, who first recorded, in 1664, a possible technique for producing a man-made fiber that could be a substitute for natural silk but, unfortunately, he became too involved with his other inventions and architectural design, helping to rebuild the city of London after the Great Fire, to be able to undertake the requisite trials.

In 1846, Schoenbein invented guncotton, a cellulose tetranitrate which, when dissolved in a 1:7 mixture of ethanol and diethylether, produced collodion, a thick viscous solution. The first patent to produce rayon in England was taken out by George Audemars [165] in 1855, who used collodion made from mulberry bark, and by dipping a needle into the viscous solution and drawing it out he was able to produce filaments and, which hardly surprisingly, was not a commercially viable venture.

At about the same time as Edison was undertaking his work in the USA, Sir Joseph Swan in England had invented an electric lamp and in 1883, produced a precursor material for his carbon fiber lamp filaments by extruding a solution of nitrocellulose dissolved in acetic acid through a small hole into a bath of ethanol and collecting the drawn coagulated fiber, which he later denitrated with a solution of ammonium sulfide, to regenerate the initial cellulose.

Count Hilaire de Bernigaud de Chardonnet, who eventually became known as the father of the rayon industry, had been working on problem diseases associated with the silk worm and, as a consequence, learned the techniques used for spinning natural silk. Years later in 1878, he accidentally dropped a bottle of collodion and the next day whilst trying to clean up the gelatinous product, he observed that it was capable of being drawn into a thread and quickly realized the potential of his discovery, which he patented in 1884 [166] and by 1891, was successfully producing nitrocellulose yarn on a commercial basis. Chardonnet was credited as being the first person to spin a solution of cellulose through a multi-hole jet and produce a cellulose yarn, which was successfully marketed.

In 1890, the French chemist L.H. Despaissis [167] patented the cuprammonium process, which utilized a solution of cellulose in Schweitzer's reagent (ammoniacal copper hydroxide). This process was adopted jointly in 1891 by a German chemist, Dr. Max Fremery and an Austrian engineer, Johan Urban, to make carbon fiber electric lamp filaments. Due to the success of the Chardonnet product as a textile fiber, Fremery and Urban, together with Dr. Emile Bronnert, turned their attention in 1895 to making an artificial silk, which they called Silkimit, taking out a patent in 1897 [168]. By 1898, they were in production and founded the firm Vereinigte Glanzstoff Fabriken (VGF), at Oberbruch and the new product quickly displaced the dangerous and more expensive nitrocellulose, with its high incidents of explosions and fires resulting from the use of an incompletely denitrated product.

Although much safer than the Chardonnet method, the cuprammonium process was still a complex and expensive way to produce a commodity that had to compete with natural fibers. Two English chemists, Charles F. Cross and Edward J. Bevan are associated with the introduction of the viscose process and, together with C. Beadle, patented a process in 1892 [169], for the manufacture of viscose. Their work was assisted by their previous background of cellulose chemistry associated with the paper industry, which had just at that time introduced chemically treated wood pulp as a replacement for cotton/linen rags, previously used as the main raw material. The 1892 patent covered a method of treating wood pulp with caustic soda and other chemicals to produce a golden yellow substance, which they called viscose. A sample of viscose was sent, in 1893, to an amateur physicist C.H. Stearn (a previous collaborator with Joseph Swan), now working with Charles F. Topham on carbon fiber filaments for electric lamps and the successful application of viscose for making filaments culminated in patents taken out in 1898 [170,171]. Encouraged by this success, Stearn encouraged Topham to attempt to make a textile fiber, which necessitated much development work involving viscose aging, filtration, the manufacture of multi-hole spinnerets from platinum and the ingenious means of continuously collecting and twisting the yarn into the acceptable form of a cake or package, by utilizing centrifugal force to throw the yarn onto the inner wall of a rapidly spinning box, called a Topham spinning box [171].

In 1899, Cross and Stearn formed a laboratory-cum-pilot plant at Kew, London to make carbon fiber electric lamp filaments and artificial silk fibers. The national patent rights were gradually sold off and, in 1904, Courtaulds purchased the British rights, setting up a factory in Coventry, which was in production by 1905, becoming so successful that by 1909, they were able to buy out the holders of the American rights and set up a wholly owned subsidiary in the USA, which was to prove a most valuable financial asset. There were many problems to be overcome and it was not until 1911 that real consistency was obtained in the quality of the yarn output. In that year, VGF had bought up the German holders of the viscose rights and abandoned the original cuprammonium process.

In 1924, the name rayon was adopted but, it did include other cellulosic products such as cuprammonium and acetate fibers and the definition was subsequently amended in 1951 to man-made textile fibers and filaments composed of regenerated cellulose.

4.3.2 Viscose rayon process [172,173]

4.3.2.1 Introduction

As continuous fiber is required for use as a precursor for carbon fiber, the additional details for making staple fiber are not considered. The precise technical information for such processes is commercially sensitive and will vary from one firm to another. It is noteworthy that recent technology uses the solubility of cellulose in a 4-methylmorpholine-N-oxide/water mix (VINCEL, LYOCELL in the USA). An outline of a typical viscose process is shown in Figure 4.16.

Rayon is made from cellulose, a naturally occurring polymer, consisting of D-glucose monomer units joined by 1–4 glucosidic bonds, which form an ether linkage by the elimination of water.

The anhydroglucose units arrange themselves in the chair conformation form, each unit turning through 180° about the ether linkage giving a relatively stiff polymer, each cellulose molecule comprising some 800–1000 of these anhydroglucose units.

In each cellulose unit, where n is the degree of substitution, there are $(3n+2)$ alcohol groups located at carbons 2, 3 and 6 of each anhydroglucose unit, with one —OH at each end of the cellulose chain.

4.3.2.2 Steeping stage

The normal source of cellulose is timber, e.g., spruce or eucalyptus, which after debarking, is cut into small chips, treated with calcium bisulfite as a bleach and pressure cooked with steam to remove lignin. The product is washed with water and the cellulose floating on the surface is collected, compressed and given a further bleach and finally pressed into sheets, or boards, of cellulose containing some 30% solids. These sheets, about 95% pure

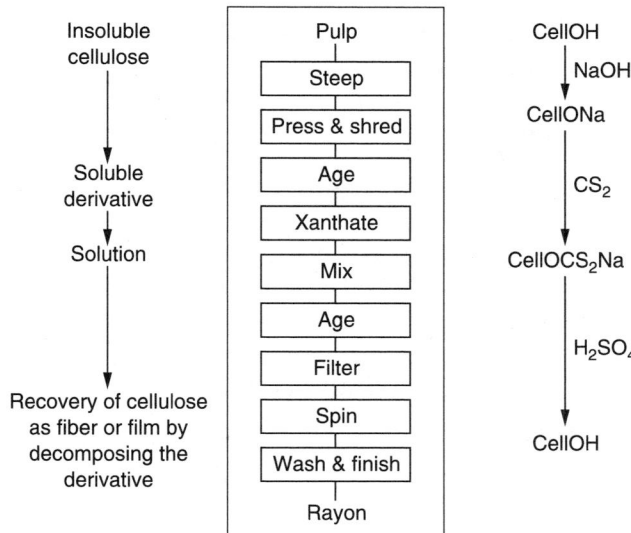

Figure 4.16 Outline of a typical viscose process. *Source*: Reprinted with permission from Dyer J, Daul GC, Rayon Fibers, Lewin M and Pearce EM eds., *Handbook of Fiber Chemistry*, Marcel Dekker, New York, 731, 1975. Copyright 1975, CRC Press, Boca Raton, Florida.

cellulose, are pre-conditioned to control the moisture content, stacked vertically in a hydraulic press and soaked in a solution of about 18% NaOH at ambient temperature. The cellulose swells and is converted to alkali-cellulose (sodium cellulosate) removing some 8% of soluble hemi-celluloses, whilst the flow of NaOH is carefully controlled so that the hemicellulose content in the steeping liquor does not exceed 0.5%.

$$(C_6H_{10}O_5)_{n\,m} + n\,NaOH \rightarrow C_6H_9ONa + n\,H_2O$$
$$\text{cellulose} \qquad\qquad\qquad \text{sodium cellulosate}$$

For every two anhydroglucoside units, only one alcoholate group is formed, mainly at position 2.

Nine allotropes of alkali-cellulose have been identified. After steeping, the hydraulic ram is closed to give a precise alkali/cellulose ratio, removing the excess spent steeping liquor, which can be recovered by dialysis. The alkali-cellulose containing a controlled amount of residual spent liquor is transferred to a shredder.

4.3.2.3 Shredding and ageing stages

The moist board is shredded in a water cooled heavy duty shredding machine and the crumbs transferred to portable storage containers with lids. The containers are moved to a temperature controlled room for the process of ageing to take place. The product reacts with atmospheric oxygen over a period of 1–2 days and, by a chain reaction process, oxidative depolymerization occurs, when the glucose residues fall from about 800 units in the original cellulose molecule to some 350 in the aged product. Higher concentrations of sodium hydroxide and catalysts like Fe, Co and Mn accelerate the depolymerization process while lower temperatures give greater pulp reactivity.

4.3.2.4 Xanthation stage

The aged crumbs, which contain about 30% cellulose, are then transferred to a jacketed, cooled, hexagonal shaped xanthation vessel. A vacuum is applied and about 10% CS_2 added with great care, as it is an extremely flammable substance. The vessel is slowly rotated on its axis, churning the two components for about 3 h to form a deep orange, thick, viscous mass of sodium cellulose xanthate.

$$(C_6H_{10}O_5)_n + nNaOH \longrightarrow (C_6H_9O_4ONa)_n + nH_2O$$
$$\text{cellulose} \qquad\qquad\qquad \text{alkali-cellulose}$$

The color is due to the presence of trithiocarbonate (Na_2CS_3), sulfides and polysulfides.

$$2CS_2 + 6NaOH \rightarrow Na_2CS_3 + Na_2CO_3 + Na_2S + 3H_2O$$

The excess carbon disulfide and other odorous compounds are removed after xanthation by the application of a vacuum, since in practice only about 70% of the requisite CS_2 is required.

4.3.2.5 Mixing and ripening stages

The batches of sodium cellulose xanthate are dissolved in about 5% NaOH solution, transferred to a mixing vessel, blended and rigorously filtered. This gives a solution of viscose, which is initially very thick but, on standing for about 4 days at 15°C, ripens, becoming thinner due to the partial decomposition of the sodium cellulose xanthate, regenerating some cellulose, which is held in an emulsion form giving a solution, which probably comprises molecular and micelle aggregates. The exact degree of ripening is critical to facilitate correct spinning. The ripe viscose, containing about 6.5% of cellulose, is de-aerated by the application of a vacuum, with particular care being taken to avoid the formation of gels.

4.3.2.6 Spinning stage

After the set ripening period, the viscose is pressurized with air through a final filtering stage and metered with a gear pump, by way of a pivoted rounder end (swan neck) to which a multi-holed precious metal (e.g., Au/Pt) spinneret is attached to the open end. The viscose is extruded horizontally into a lead lined viscose spin bath, through which spin bath liquor is circulated at about 50°C. The size of the jet hole is about 0.05–0.1 mm diameter and the holes are drilled using laser equipment. A typical spin bath contains:

H_2SO_4	10%
Na_2SO_4	23%
$ZnSO_4$	0.8%
H_2O	balance

The viscose coagulates in the spin bath, initially forming a skin, then as the spin bath penetrates through into the center of the filament, all the viscose is converted to cellulose, the NaOH is neutralized by the acid spin bath and CS_2 is liberated. Sodium trithiocarbonate, present in the viscose, decomposes to H_2S and CS_2.

$$2\ SC\!\!\begin{array}{l}\diagup SNa \\ \diagdown (OC_6H_9O_4)_n\end{array} + H_2SO_4 \longrightarrow 2\,(C_6H_{10}O_5)_n + 2\,CS_2 + Na_2SO_4$$

$$Na_2CS_3 + H_2SO_4 \rightarrow CS_2 + H_2S + Na_2SO_4$$

The Na_2SO_4 in the spin bath, as well as that formed by neutralization of NaOH, acts as a dehydrating agent, inducing water from the gelled fiber. The $ZnSO_4$ acts as a coagulant, affecting the mode of cellulose regeneration, controlling the diffusion across the membrane of the gelled surface of the filament and having a small contribution as a dehydrating agent. The $ZnSO_4$ influences the strength and produces a filament with a serrated cross section. Other agents like alkyl amines and polyglycols can be added to the bath as modifiers, which generally improve the strength.

This combination of hot H_2SO_4 under reducing conditions, produces an extremely corrosive environment, hence necessitating the use of precious metals for constructing the spinnerets and lead lining for the material of the spin bath.

Whilst in the spin bath, the fiber is passed around godet rollers, with the second godet running faster to impart stretch, which increases the strength, but with the limitation that the extension at break would be lowered. A typical running speed would be about 75 m/min. Depending on the type of spinning machine, in a batch process, the fiber is collected as a cake in a Topham box and subsequently passed through a series of treatment processes to remove any impurities or, alternatively, in a continuous process, accumulating the fiber by passing over a series of incremental rollers and treating *in situ*.

4.3.2.7 Final treatment stage

At this stage, the fiber is contaminated with acid and free sulfur and undergoes a series of operations to effect final purification. Acid is removed by washing with water and sulfur is removed by treatment with about 0.3% NaOH. Sometimes, at this stage, it is bleached with alkaline sodium hypochlorite (0.1% available chlorine) to give a whiter fiber, followed by treatment with sour, which is about 0.2% sulfuric or hydrochloric acid and then, washed in water and finally dried.

A finish (e.g., mineral oil), is applied and the fiber wound onto a bobbin, or if the finish is water based (e.g., sodium oleate/ oleic acid), it is applied prior to the drying stage.

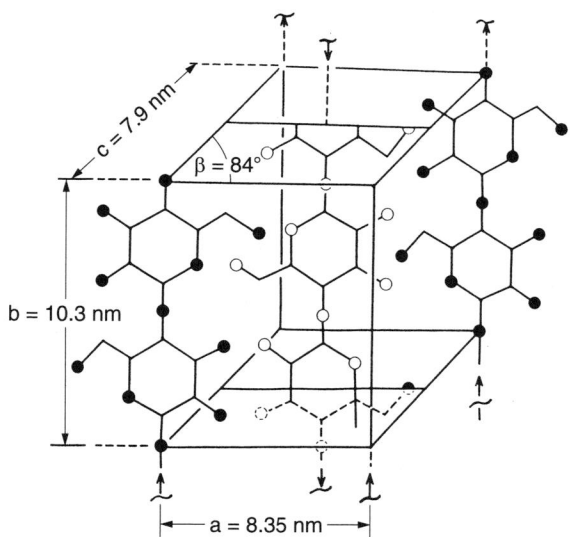

Figure 4.17 Structure of cellulose unit cell (Cellulose I). *Source*: Reprinted with permission from Dyer J, Daul GC, Rayon Fibers, Lewin M and Pearce EM eds., *Handbook of Fiber Chemistry*, Marcel Dekker, New York, 775, 1998. Copyright 1998, CRC Press, Boca Raton, Florida.

4.3.3 Structure of rayon fibers

The basic composition of the final rayon, is the same as the original cellulose (Cellulose I) starting material, with the exception that for rayon, the number of anhydroglucose units is reduced and the regenerated cellulose is in a slightly different allotropic form (cellulose hydrate or Cellulose II). The structure of the original allotropic form (Cellulose I) is shown in Figure 4.17 and the dimensions using X-ray diffraction are given for Cellulose I, II, II and IV in Table 4.8.

The difference in allotropic form can be explained by the rapid crystallization of the cellulose in the regeneration process, whereas in nature, the process is very slow.

It is generally believed that the fine fiber structure comprises fringed micelles and fringed fibrils (Figure 4.18). The presence of zinc in the bath contributes to a skin-core structure (Figure 4.19) and proportions of skin and core can be varied. Skin contains small crystallites and is stronger than the core, which contains fewer, but larger crystallites [175].

Table 4.8 Allotropes of cellulose

Allotrope modification	Unit cell dimensions Axis (A) nm			Angle (β)	Density g cm^{-3}
	a	b	c		
Cellulose I (native cellulose)	8.35	10.3	7.9	80°	1.545–1.562
Cellulose II (general form of rayon)	8.1	10.3	9.1	62°	1.515–1.523
Cellulose III	7.74	10.3	9.9	58°	1.515–1.523
Cellulose IV (high performance rayon)	8.11	10.3	7.9	90°	1.545–1.562

Source: Reprinted with permission from Sisson WA, *Text Res J*, 30(3), 153, 1960. Copyright 1960, Textile Research Institute.

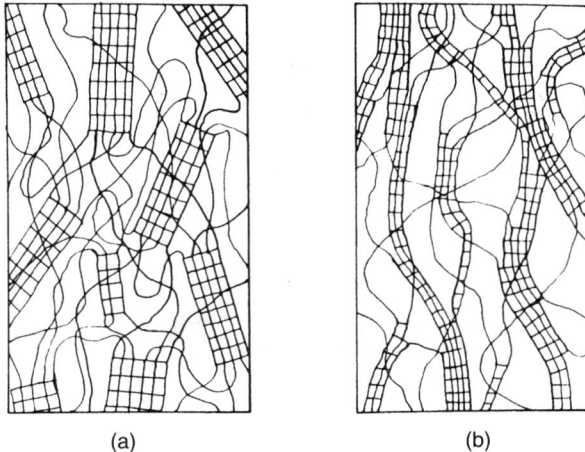

Figure 4.18 Cellulose structure (a) Fringed micelle (b) Fringed fibril. *Source*: Reprinted with permission from Dyer J, Daul GC, Rayon Fibers, Lewin M and Pearce EM eds., *Handbook of Fiber Chemistry*, Marcel Dekker, New York, 777, 1998. Copyright 1998, CRC Press, Boca Raton, Florida.

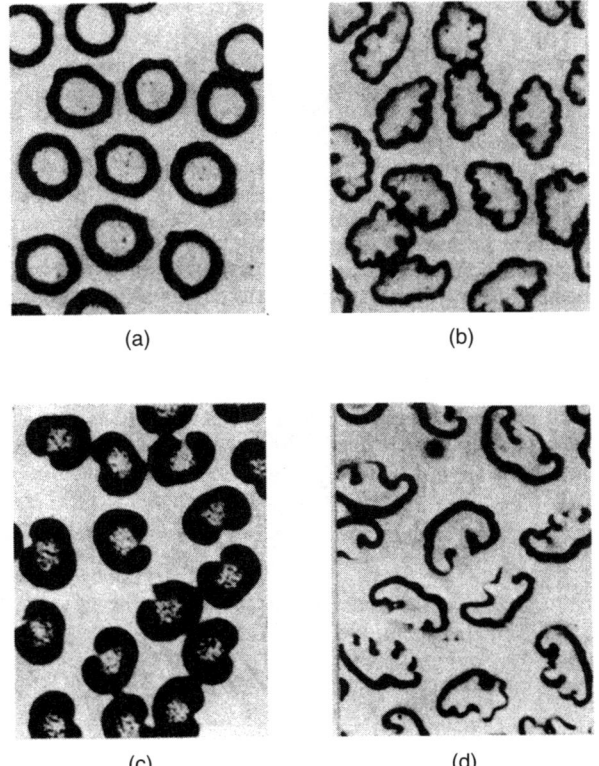

Figure 4.19 Stained rayon cross sections showing skin and core (a) high wet modulus; (b) regular; (c) tire; (d) crimped. *Source*: Reprinted with permission from Dyer J, Daul GC, Rayon Fibers, Lewin M and Pearce EM eds., *Handbook of Fiber Chemistry*, Marcel Dekker, New York, 778, 1998. Copyright 1998, CRC Press, Boca Raton, Florida.

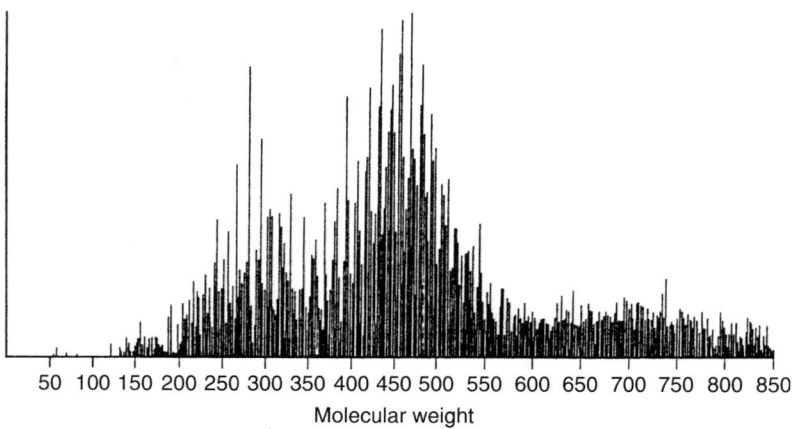

Figure 4.20 Mass spectrum (field desorption mass spectrometry) of petroleum pitch. *Source*: Reprinted from Lewis IC, Chemistry and development of mesophase in pitch, *Journal de Chimie Physique*, 81, 751–758, 1984.

4.4 PITCH PRECURSORS

4.4.1 Introduction

Pitch is a general name for the tarry substance, which is solid at room temperature and can be obtained from one of several sources:

1. Petroleum refining, normally called bitumen, or asphalt in the U.S.A.
2. Destructive distillation of coal
3. Natural asphalt, e.g. from Trinidad
4. Pyrolysis of PVC
5. Pyrolysis of ring compounds, such as naphthalene and anthracene

Pitch is a complex mixture of many hundreds of aromatic hydrocarbons, comprising structures with some three- to eight-membered rings, with alkyl side groups, normally methyl, with an average molecular weight of 300–400 [176], the composition depending on the source and method of pre-treatment. Figure 4.20 is the mass spectrum of a petroleum pitch showing the wide range of molecular species that is present. A similar wide range of species is also present in a pitch prepared from simple aromatic hydrocarbons [177]. The problems arising from the processing of such complicated products are reinforced by the vast amount of research undertaken, particularly in Japan, and amply borne out by the large number of technical papers and patents published, especially on spinning and subsequent stabilization.

In general, Riggs *et al* [178] considered a pitch to be composed of four main classes of chemical compounds:

1. Saturates—low molecular weight aliphatic compounds
2. Naphthene aromatics—low molecular weight aromatics and saturated ring structures
3. Polar aromatics—higher molecular weight and more heterocyclic in nature
4. Asphaltenes—the highest molecular weight fraction in pitch with the highest aromaticity and thermally most stable

PRECURSORS FOR CARBON FIBER MANUFACTURE

Table 4.9 Relative compositions of several oils and pitches

Compound	Asphaltene %	Polar aromatic %	Naphthene aromatic %	Saturate %	Softening point °C
Carbon black oil R	2.5	10.6	69.0	17.9	
EXXON (DAU) bottoms (refinery sludge)	14.5	41.1	18.1	26.3	29
Ashland 240 petroleum pitch	64.4	8.6	25.4	1.6	119
Ashland 260 petroleum pitch	82.7	5.9	11.4	0.0	177

Source: Reprinted with permission from Brooks JD, Taylor GH, The formation of some graphitizing carbon, Walker PL Jr ed., *Chemistry and Physics of Carbon*, Vol 4., Marcel Dekker, New York 168, 243–268. Copyright 1995, CRC Press, Boca Raton, Florida.

Table 4.9 gives the composition of various oils and pitches. The materials richest in asphaltene are the most suitable for conversion to carbon fibers. Manufacturing processes for pitch based carbon fibers are shown in Figure 4.21.

4.4.1.1 *Petroleum pitch*

Petroleum pitch can be obtained from a wide variety of sources such as from the bottoms of catalytic crackers, steam cracking of naphtha and gas oils, and residues from various distillation and refinery processes. The crude product is given an initial refining process, which can include one, or a combination of, several treatments:

1. Prolonged heat treatment to advance the molecular weight of the components
2. Air blowing at about 250°C [180–184]

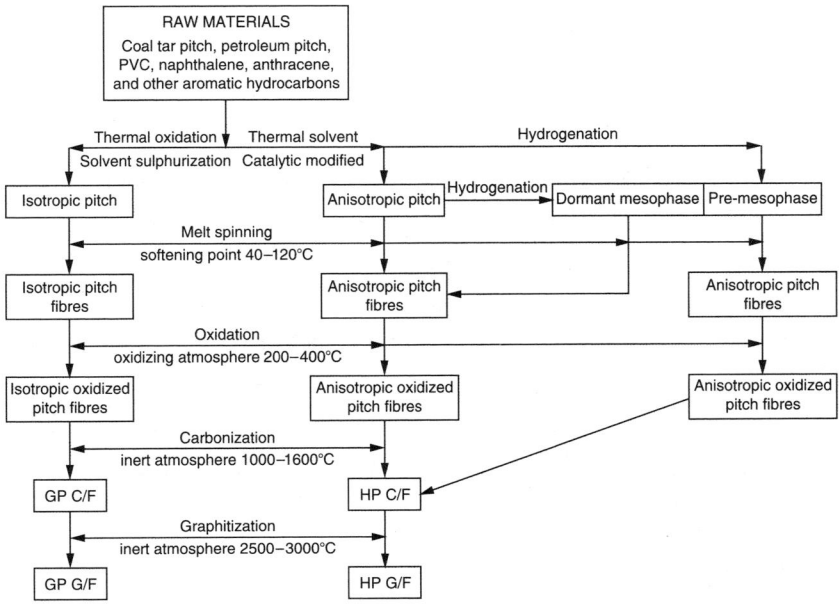

Figure 4.21 Schematic process for the manufacture of pitch-based carbon fibers. *Source*: Reprinted with permission from Brooks JD, Taylor GH, The formation of some graphitizing carbon, Walker PL Jr ed., *Chemistry and Physics of Carbon*, Vol 4, Marcel Dekker, New York, 168, 243–268. Copyright 1995, CRC Press, Boca Raton, Florida.

Table 4.10 The properties of raw and treated pitches

Property	Ashland 240 Petroleum pitch	Ashland Aerocarb 60 Treated petroleum pitch	Ashland Aerocarb 75 Treated petroleum pitch
Softening point °C	119	151	234
T_g °C	80	121	195
Specific gravity g cm^{-3}	1.23	1.24	1.25
Mass % C	92	92	93
Mass % H	6	5	5
Mass % S	2	2	2
Mass % N	0	0	0
Mass % O	0	0	0
% Aromatic H (NMR)	53.5 [BB]		
% Toluene insoluble	6	9	26
Mesophase volume %	<1	<1	<1
% Quinoline insoluble	0.4	<1	<1

Source: Reprinted with permission from Sumner MB, Thermal properties of heavy isotropic petroleum pitches. *Carbon '88, Proceedings of the International Conference on Carbon*, University of Newcastle upon Tyne, 52–54, Sep 18–23, 1988, Copyright 1988, Institute of Physics Publishing. Reprinted with permission from Dickenson EM, Average structures of petroleum pitch fractions by ^{1}H/^{13}C NMR spectroscopy, *Fuel*, 64, 704–706, 1985. Copyright 1985, Elsevier.

3. Steam stripping and the application of a vacuum [185,186] to remove low boiling components
4. Distillation

Longer processing times and higher temperatures yield a product with greater aromaticity, although a petroleum pitch is usually less aromatic than a coal tar pitch. The degree of aromaticity can be roughly quantified by the value for the C/H ratio.

Petroleum bitumens have been investigated as precursors for carbon fibers [187,188]. The properties of typical petroleum pitches are given in Table 4.10.

4.4.1.2 Coal tar pitch

The destructive distillation of coal (e.g., a bituminous coal) to produce coke gives a byproduct of a brown/black oily substance termed tar. Distillation of this tar yields a number of fractions and the fraction produced above 350°C is called coal tar pitch, which can subsequently be air-blown to effect further refinement. The coal tar pitch comprises some two thirds aromatic compounds [191], whilst the remaining compounds are essentially heterocyclic, many of the rings having side groups such as methyl. There are also small amounts of nitrogen, sulfur and oxygen.

The composition of the final pitch depends on the initial source of the tar and the nature of the subsequent treatment processes and typical properties are given in Table 4.11. Coal tar pitches have been investigated as precursors for carbon fibers [192].

4.4.2 Characterization of the pitch

A pitch, as supplied by a manufacturer, is normally characterized by the softening point and the penetration value, hardly sufficient to describe the composition. The petroleum and coal tar industries, unfortunately, have somewhat different methods of further

Table 4.11 Chemical composition of pitch materials.

Property		Petroleum asphalt Showa Sekiyu Co.	Coal-tar pitch Tokyo Gas Co.	PVC pitch Kureha Chemical
Chemical composition (%)	C	85.5	92.3	95.0~92.6
	H	9.69	4.50	5.58~6.78
	S	0.91	0.20	0.00
	N	2.84	1.12	0.00
	O (diff.)	1.06	1.88	0.00~0.62
Atomic ratio C/H		1.36	0.585	0.704~0.878
Mean molecular weight		790	276	555~925
Melting point (°C)		≈83	≈65	165~180

Source: Reprinted with permission from Corbett LW, Reaction variables in the air blowing of asphalt, *Ind and Eng Chem Des Dev*, 14(2), 1975. Copyright 1975, American Chemical Society.

characterization, that are based on the solubilities of the pitch in various solvents, e.g., in the petroleum industry we have:

1. Carboids—insoluble in CS_2.
2. Carbenes—soluble in CS_2 but insoluble in CCl_4.
3. Asphaltenes—insoluble in n-pentane but soluble in CS_2, CCl_4 and C_6H_6.
4. Pre-asphaltenes—soluble in CS_2, CCl_4, C_6H_6 and pyridine.

More generally, Halleux and de Greef [193], Smith [194], McNeil [195], and Diefendorf and Riggs [196], have described solvent fractionation methods of separating the pitch using various boiling solvents. The solvent fractionation processes must be fraught with problems, since the solvents (Table 4.12) vary from highly flammable, to harmful and carcinogenic.

Table 4.12 Boiling points of solvents used to extract pitch

Solvent	Formula	Boiling Point °C
n-pentane	$CH_3(CH_2)_3CH_3$	36
carbon disulfide	CS_2	46
n-hexane	$CH_3(CH_2)_4CH_3$	68
carbon tetrachloride	CCl_4	76
benzene	(benzene ring)	80
toluene	(toluene structure)—CH_3	111
pyridine	(pyridine ring, N)	115
quinoline	(quinoline structure, N)	238

A typical process would be:

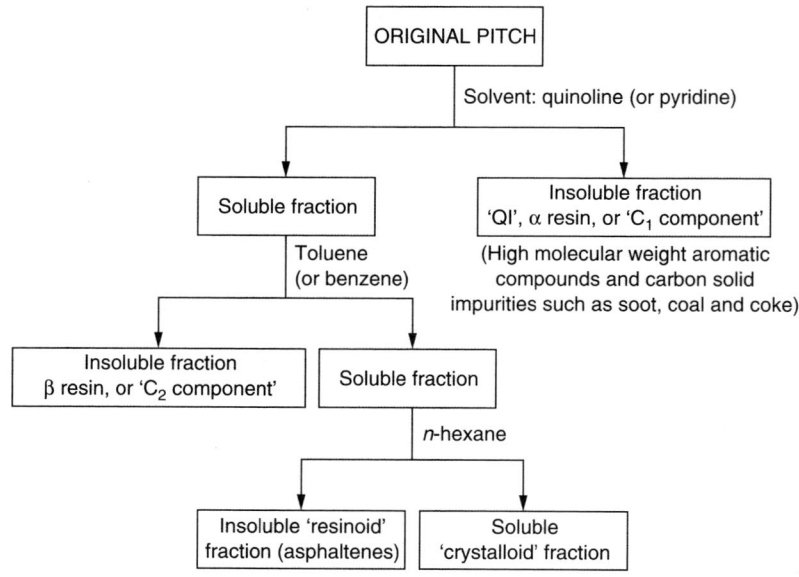

Further characterization can be achieved using the following techniques:

1. Distillation [197]
2. Elemental composition [198]
3. Average molecular weight determination, e.g., by vapor phase osmometry [199]
4. IR and UV spectroscopy [200,201]
5. Column chromatography [202]
6. Gas chromatography (GC), for volatile and low molecular weight components [203,204]
7. High pressure liquid chromatography (HPLC), for larger molecular compounds [205,206]
8. Gel permeation chromatography (GPC), for measuring the molecular weight distribution of larger molecules [207–210]
9. Differential thermal analysis (DTA), for delineating melting, boiling and polymerization reactions [211]. T_g determined by differential scanning calorimetry (DSC) has been used [212] but for high T_g, thermomechanical analysis (TMA) (i.e., penetrometry) is preferred [213,214]
10. Nuclear magnetic resonance (NMR), to establish distribution of hydrogen and carbon [215,216] and NMR with $^1H/^{13}C$ [217] to provide average structure.
11. Mass spectrometry, for measuring molecular weight distribution [218,219]
12. X-ray techniques [220,221]

4.4.3 Isotropic pitches

Isotropic pitches are used to make a general purpose (GP) grade of pitch carbon fiber, which is not graphitic and has poorer properties than the high performance grade (HP), which requires a special treatment process to convert the pitch to a mesophase grade, i.e., an optically anisotropic and graphitic material.

The isotropic pitch has to be treated to generate a product suitable for melt spinning, with a low volatility and filtered to remove solid particles. A good starting material would be Ashland 240, which has low quinoline content. The volatile components can be effectively removed with a wiped film evaporator [222], where a thin film of the molten pitch in the

PRECURSORS FOR CARBON FIBER MANUFACTURE

evaporator is continuously wiped over the heating surface, exposing a fresh surface, thus permitting the efficient removal of some of the volatile components without overheating and, whilst the pitch is molten, any solid impurities are removed by filtration. This refining process raises the softening point and avoids the formation of mesophase, hence increasing the aromaticity of the pitch and making it suitable for melt spinning. Alternatively, prepared pitches like Ashland's Aerocarb 60 and 70 (Table 4.10) can be used.

The preparations of isotropic pitches have been undertaken [182–184,226,227] to make general purpose carbon fibers.

4.4.4 Preparation of mesophase pitches

4.4.4.1 Introduction

Both isotropic and mesophase pitch carbon fibers are made from the same feedstock.

A mesophase pitch suitable for producing high performance carbon fiber must have the following properties [228]:

1. Low ash and metallic ion contents
2. Not contain insolubles, which must be removed by filtration, otherwise would interfere with the spinning and also lower mechanical properties.
3. Must not undergo polymerization during spinning as this would increase the melt viscosity, which in turn would require higher process temperatures and could generate gases and leave bubbles in the spun fiber.
4. The mesophase portion must be able to undergo orientation during the spinning process.
5. The softening point (e.g. 230–280°C) and T_g should be high enough to permit rapid stabilization.
6. The spun fiber must retain sufficient reactivity to undergo the stabilization reaction to prevent fusing of the filaments during subsequent high temperature processing.
7. Have a high carbon yield.

Typical preparation methods of precursor pitch for high performance carbon fibers are presented in Figure 4.22.

An isotropic pitch, when pyrolyzed at about 425°C was shown by Brooks and Taylor [230,231] to produce a liquid/crystal type structure containing domains of highly oriented molecules termed mesophase. Initially, the mesophase forms as a dispersion of ultra-fine

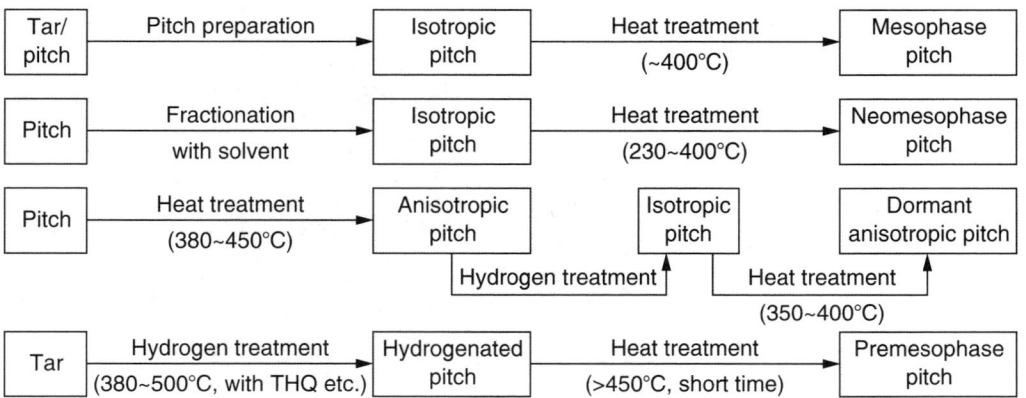

Figure 4.22 Typical preparation methods of precursor pitch for high performance carbon fibers. *Source*: Reprinted from Okuda K, *Trans Mat Res Jpn*, 1, 119–139, 1990.

spheres containing strings of molecular aggregates around a central polar axis and near the edge of the spheres, the strings curve to meet the interface perpendicularly. When the diameter of the spheres has exceeded about 0.5 μm, they exhibit optical anisotropy, when viewed with polarized light. In the initial stages, Zimmer and White [232] envisaged the molecules as mostly planar, or discotic, some 25 μm in diameter, with hydrogen atoms positioned along the edge sites and an average molecular weight of about 2000 (Figure 4.23 shows mainly uncoalesced pitch with the largest particle about 35 μm in diameter). The spheres continue to grow and, as the concentration of the larger molecules increases, more mesophase separates out and the spheres coalesce (Figure 4.24 shows streaky mesophase appearing as a fingerprint pattern) and above 50% mesophase, a phase inversion takes place, when the mesophase becomes the continuous phase, with the isotropic phase dispersed within it [234]. Figure 4.25 shows a fine-structured mosaic pattern, usually only observable under magnifications of 400 times or more. Eventually, the product becomes 100% mesophase and any further treatment would only produce coke.

4.4.4.2 Production of mesophase by pyrolysis

Isotropic pitches can be converted by heat treatment at 350–450°C, to an emulsion of isotropic and mesophase pitches, containing domains of highly oriented polynuclear aromatic hydrocarbons [235–238]. At higher temperatures, the pitch would be rapidly converted to coke. Singer [239] details how Ashland 240 pitch can be transformed from an isotropic pitch to an optically anisotropic fluid phase, by heating in an inert atmosphere for some 40 hours at 400–410°C, producing a yield of about 50% mesophase, which being slightly heavier than the isotropic component, settles at the bottom of the reaction vessel. Unfortunately, the product has an extremely wide molecular weight distribution and is difficult to spin. However, Lewis, McHenry and Singer [240] found that the application of mechanical agitation gave a mesophase fraction with lower molecular weight, forming an emulsion with the isotropic phase, which could be spun more easily. However, emulsions do cause intermittent flow problems during melt extrusion and are also shear sensitive. McHenry [237,241] and Chwastiak [242] overcame this problem by the simultaneous application of a

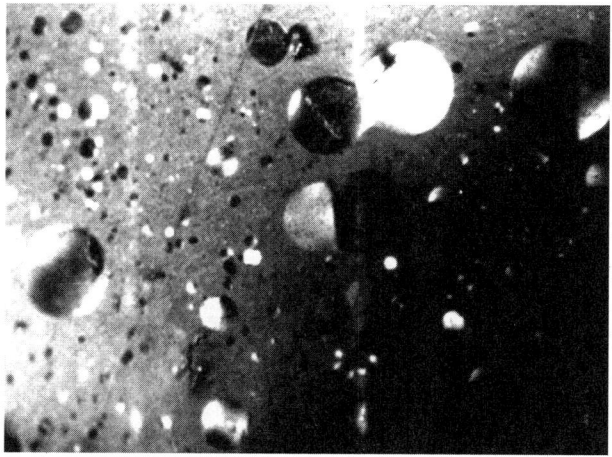

Figure 4.23 Mesophase spheres in a partially converted chemical pitch (polarized light ×400). *Source*: Reprinted with permission from Hornby J, Kearsey HA, Carbon fibres from pitch precursors 2. The preparation of mesophase pitches suitable for spinning into fibres, *AERE Harwell report AERE-M 3029*, Oct 1979. Copyright 1979, AEA Technology plc.

PRECURSORS FOR CARBON FIBER MANUFACTURE

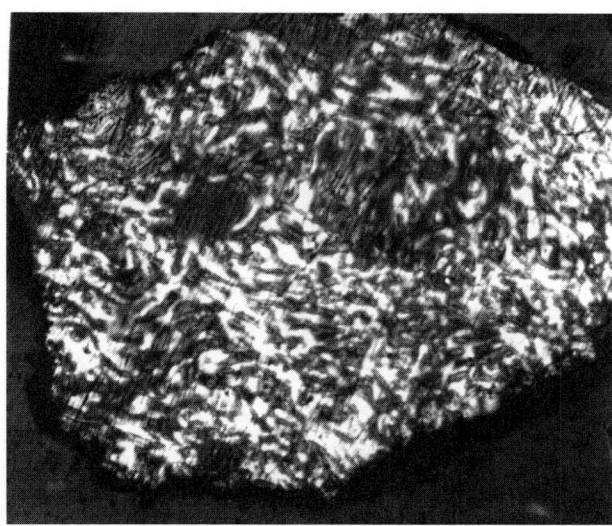

Figure 4.24 Streaky mesophase formed by coalescing spheres (polarized light ×100). *Source*: Reprinted with permission from Hornby J, Kearsey HA, Carbon fibres from pitch precursors 2. The preparation of mesophase pitches suitable for spinning into fibres, *AERE Harwell report AERE-M 3029*, Oct 1979. Copyright 1979, AEA Technology plc.

sparge of an inert gas, which continuously removed the highly volatile components, effectively producing a single phase product, which was 100% mesophase, with over 90% having a molecular weight less than 1500.

A later modification of this process, used in Japan, yielded a less viscous product achieved by cooking at a slightly lower temperature and allowing the heavier mesophase pitch (about 70–80% mesophase) to settle out by standing and then collecting by centrifuging.

Union Carbide [243] claims that a pitch suitable for spinning should preferably contain 40–90% mesophase, with domain sizes greater than 200 µm, which will then produce a fiber with a highly oriented structure. Domain sizes below 100 µm produce stringy clumps, unsuitable for spinning. When the droplets coalesce to form the bulk mesophase, some

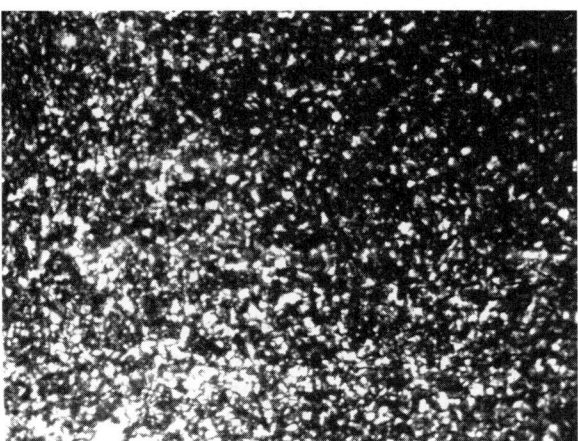

Figure 4.25 Fine structured mesophase pitch (polarized light ×400). *Source*: Reprinted with permission from Hornby J, Kearsey HA, Carbon fibres from pitch precursors 2. The preparation of mesophase pitches suitable for spinning into fibres, *AERE Harwell report AERE-M 3029*, Oct 1979. Copyright 1979, AEA Technology plc.

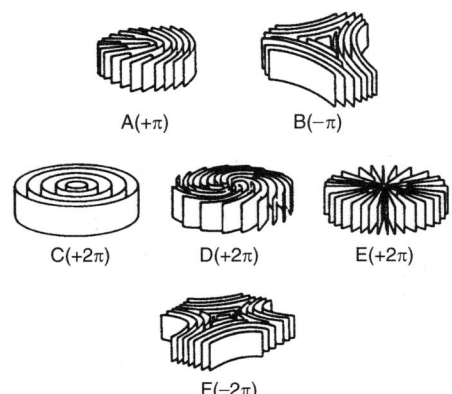

Figure 4.26 Sketches of various wedge dislocations. *Source*: Reprinted with permission from Zimmer JE, White JL, Disclination structures in the carbonaceous mesophase, *Adv in Liq Crysts*, 5, Academic Press, New York, 157, 1982. Copyright 1982, Elsevier.

misalignment occurs, termed disclinations, that can be detected with reflected polarized light in the optical microscope and with the electron microscope [244]. These disclinations are not possible with the spherical systems in the isotropic phase. Schematic representations of disclinations are shown in Figure 4.26 and are assigned with π notations, such that interactions can result in an annihilation (e.g., $+2\pi - 2\pi = 0$) or a combination resulting in the formation of a new disclination (e.g., $+2\pi - \pi = +\pi$).

The effect of mesophase pitches on the tensile modulus of the resultant pitch based carbon fiber has been studied [245] and carbon fiber prepared from an isotropic pitch containing mesophase spheres has also been studied [246].

4.4.4.3 Production of mesophase by solvent extraction

Diefendorf and Riggs [247] utilized solvent fractionation to separate the resinoid fraction prior to heat treatment into a form termed neomesophase, differing from the earlier form, since it did not produce Brooks and Taylor spheres when heated to 240–400°C, instead forming in a matter of minutes, an anisotropic material, which was essentially 100% mesophase. Diefendorf and Riggs believed the original isotropic pitch was a true solution [247,248] but Lafdi and co-workers [249–251] believed the original pitch to be colloidal with small stacks of carbon layers having a diameter <1 nm, suspended in an isotropic medium. On heat treatment, these layers grow and combine into micelles, eventually coalescing into an interpenetrating network of ordered and isotropic phases.

4.4.4.4 Production of mesophase by hydrogenation

A method developed by the Kyushu Industrial Research Institute(known as Kyukoshi in Japan), utilized hydrogenation of a coal tar pitch, which was first de-ashed, followed by distillation as required and then hydrogenated in the presence of a hydrogen donor solvent such as tetrahydroquinoline:

PRECURSORS FOR CARBON FIBER MANUFACTURE

The hydrogenated pitch is heated to remove low volatile constituents, which are returned to the hydrogenation process. The pitch is given a short heat treatment at a relatively high temperature, yielding a pitch which is optically isotropic at the spinning temperature and orients very rapidly during the heat treatment process after spinning to produce a HP carbon fiber.

A variant of this process (the dormant method) uses a pitch with several percent of mesophase initially present in the feedstock, which is hydrogenated and heat treated below 380°C to give a pitch containing isotropic and mesophase pitches, that yield a carbon fiber with properties intermediate between GP and HP carbon fibers. The mesophase fraction, after heating to high temperature, exhibits good spinnability constituting a pitch with large molecules and lower softening point, achieved without degradation.

4.4.4.5 Production of mesophase by catalytic modification

An alternative version of the hydrogenation process is to employ a catalyst. A number of catalytic modification methods are depicted in Figure 4.27. The spinnability of the mesophase pitch can be adjusted by controlling the molecular structure by a co-carbonization reaction [258,259]. Any catalyst used must be removed by filtration.

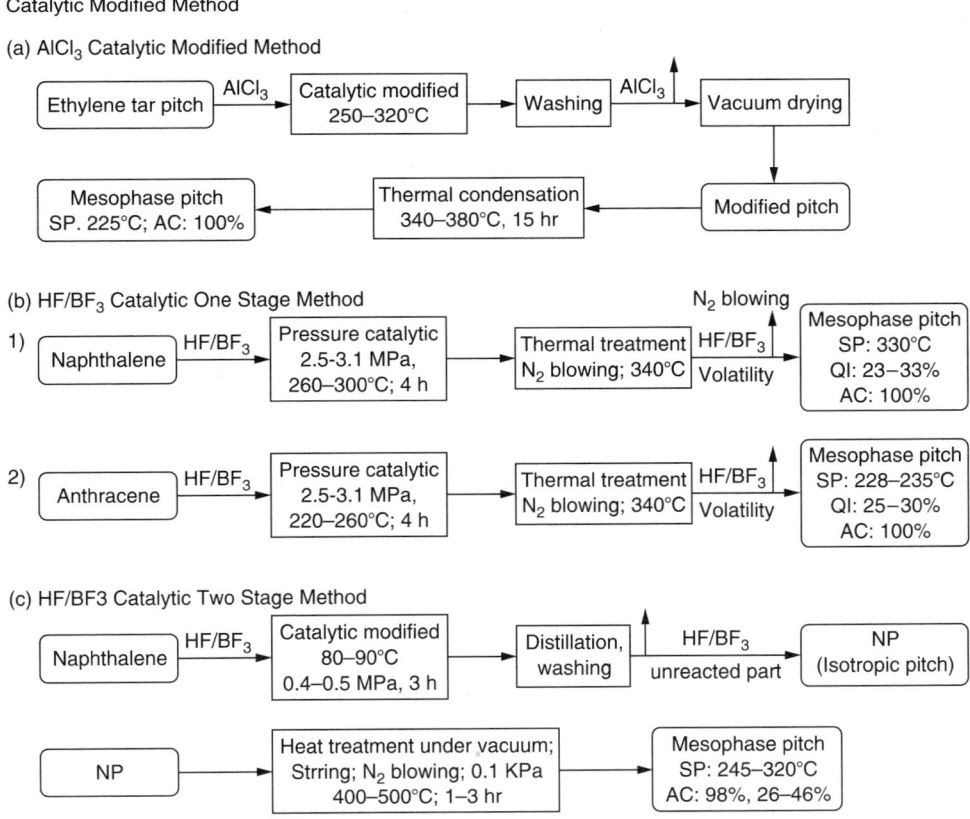

Figure 4.27 Schematic process of catalytic modification methods for preparing mesophase pitch. *Source*: Adapted from Otani S, Jap. Pat., 57–100186, 1982., Lewis IC, Carbon, 16, 425, 1978, Lewis IC, *Fuel*, 66(11), 1527, 1987, Mochida I et al, Carbon, 23(2), 175, 1985, Mochida I, Sone Y, et al, Carbon, 28(2,3), 311,1990, Wang CY, Zheng JM, et al, 22nd Bien Conf on Carbon, Ext Abst and Pro Univ of California, San Diego, 254–255, 16–21 Jul 1995.

4.4.5 Melt spinning mesophase precursor fibers

When producing carbon fibers from pitch, a critical processing parameter is the viscosity of the pitch, which is extremely dependent on the spinning temperature. If pitch was a truly Newtonian fluid, the viscosity would be independent of shear rate, attaining its value almost instantaneously. It would be expected that the ratio of the hot filament diameter to the orifice diameter (die swell ratio) would be less than 1.1 and the molten fluid would not climb the stirring rod (the so called Weissenberg effect) [228].

Nazem [259] showed that a 100% isotropic pitch reached a steady viscosity in about 50 s, whereas a mesophase pitch prepared from it (97% anisotropic) failed to achieve a steady state even after 400 s. However, intermediate pitches prepared with about 75% anisotropy did attain a reasonably constant viscosity at high shear rates and are termed pseudo-Newtonian, which are suitable for spinning. Brooks and Taylor mesophase and neomesophase have also been shown to behave in a similar manner. Figures 4.28 and 4.29 show the dependence of viscosity on temperature for a number of pitches, emphasizing the difference between isotropic and mesophase pitches.

In general, Lewis and Lewis [262] showed that these pseudo-Newtonian pitches follow the Williams, Landel, Ferry (LDF) equation [263]:

$$\log \eta = \log \eta_g - \frac{C_1(T - T_g)}{C_2 + T - T_g} \quad \text{over the temperature range } T_g \text{ to } T_g + 150 \text{ K}$$

where η_g = the viscosity at glass transition temperature (T_g)

C_1 and C_2 are constants showing significant dependence on the composition of the pitch. Although the LDF equation gives a better fit, Otani and Oya [264] used the Andrade equation ($\eta = Ae^{B/T}$) to plot η Vs $1/T$ for two different pitches and when two straight lines were arbitrarily fitted to the data, they intersected at a point, termed the transition temperature (T_s), which was 349°C (Figure 4.30). Fibers spun near the T_s showed radial texture with no wedge, whereas fibers spun at lower temperatures formed a wedge, whilst fibers spun at higher temperatures exhibited either random or onion-like textures. Hamada

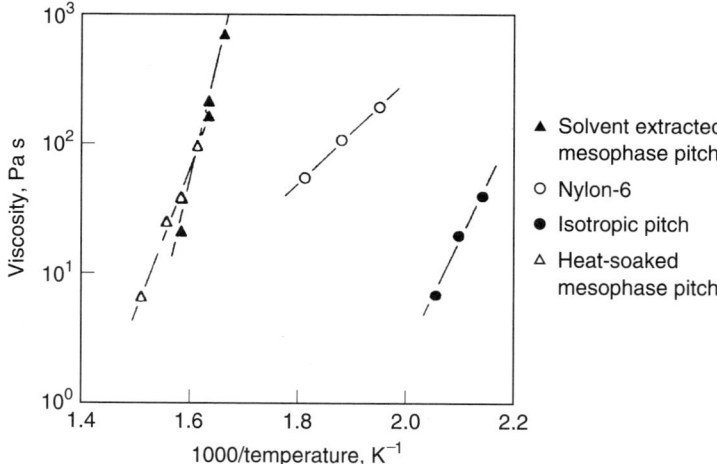

Figure 4.28 The dependence of the viscosity on temperature for various pitches. *Source*: Reprinted with permission from Edie DD, Dunhan MG. Meltspinning pitch based carbon fibers *Carbon* 27, 647, 1989. Copyright 1989, Elsevier.

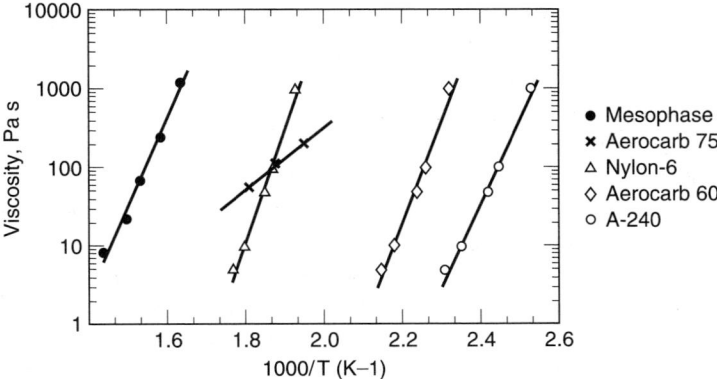

Figure 4.29 The dependency of viscosity on temperature for several isotropic pitches, a mesophase pitch and a typical thermoplastic polymer: Ashland 240 (isotropic petroleum pitch); Aerocarb 60 (isotropic pitch distilled from A240); Aerocarb 75 (isotropic pitch distilled from A240); *Source*: Sumner MB, Thermal properties of heavy isotropic petroleum pitches. *Carbon '88, Proceedings of the International Conference on Carbon*, University of Newcastle upon Tyne, 52–54, Sep 18–23, 1988, Mesophase (produced by pyrolysis of A240); Nylon 6 (a typical melt spur synthetic polymer). *Source*: Reprinted from Whitehouse S, Rand B, Rheology of mesophase pitch from A240. *Carbon '88, Proceedings of the International Conference on Carbon*, University of Newcastle upon Tyne, 175–176, Sep 18–23, 1988.

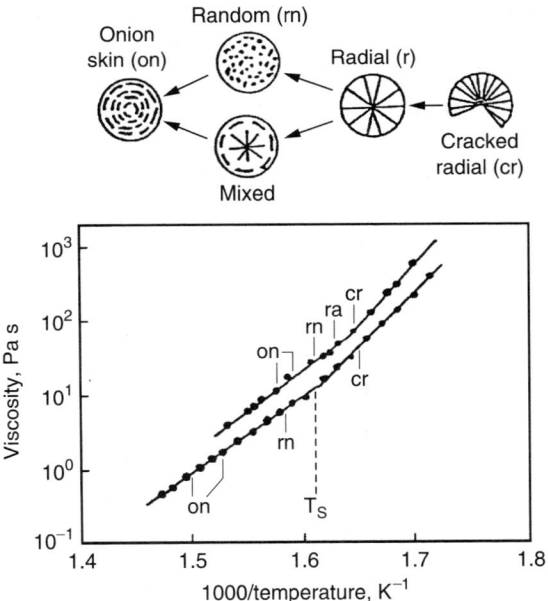

Figure 4.30 Variation of mesophase fiber texture with melt spinning temperature for two different pitches. *Source:* Reprinted with permission from Otani S, Oya A, Progress of pitch based carbon fiber in Japan. Bacha JD, Newman JW, White JL, eds., ACS Symp Ser No. 303, American Chemical Society, Washington, DC, 322, 1986. Copyright 1986, American Chemical Society.

[265] showed that the structural type did not affect the tensile strength or modulus and these parameters may be controlled more by the fine molecular structure. Bourrat and co-workers [266] showed that disclinations persisted during fiber formation, although on a smaller scale and the density of disclinations could well determine the final fiber properties.

Even with good control of the temperature, the tensile stress in the pitch filament will be about 20% of the ultimate strength, whereas this figure is only about 1% of the tensile strength when spinning nylon fiber [267].

A nylon filament reaches its final diameter about 500 mm from the jet face, whereas with pitch, this distance is only 5 mm. In fact, when only 20 mm from the jet, the pitch has cooled to more than 100°C below the T_g and the fiber is very brittle. The rapid neck-down is associated with a high maximum velocity gradient of about $2500\,s^{-1}$ compared with about $500\,s^{-1}$ for nylon. This high velocity gradient creates high tensile stress, close to actual fracture of the filament, on the pitch during spinning. Now a higher softening point will require a higher spinning temperature, which is associated with the increased rate of cooling and higher tensions, explaining why it is more difficult to spin a mesophase pitch. Elevating the temperature of the quench air will slow down the cooling and reduce stress. Increasing the spin temperature will reduce the stress level but too high a temperature can cause thermal breakdown of the pitch, so too high a temperature cannot be used. Yoon et al [268] have shown that increasing the spinning temperature does increase the molecular orientation, so a careful balance has to be chosen.

In practice, spinnerets have a multiplicity of holes, which must be spaced about 1.1 mm apart to avoid interfilament fusing.

Typical commercial melt spinning equipment is shown in Figure 4.31 [269]. Solid chips of pitch are fed into a screw extruder fitted with off-gassing ports and evenly heated to give a uniform feed of molten pitch, which is passed by the latter stages of the extruder to the die head and into a metering pump. This pump helps to minimize any pressure fluctuations created by the rotating screw. The melt is then filtered and forced through a multi-hole spinneret attached to the bottom of the die head. Some initial orientation occurs as the molten pitch passes through the jet capillaries and on emerging from the capillaries, it is cooled by quench air. Whilst still hot, the solid fiber is drawn before wind-up to give a highly oriented precursor fiber requiring no further drawing in any subsequent carbonizing process. Fiber handling in the wind-up process is difficult due to the low strength of the as-spun

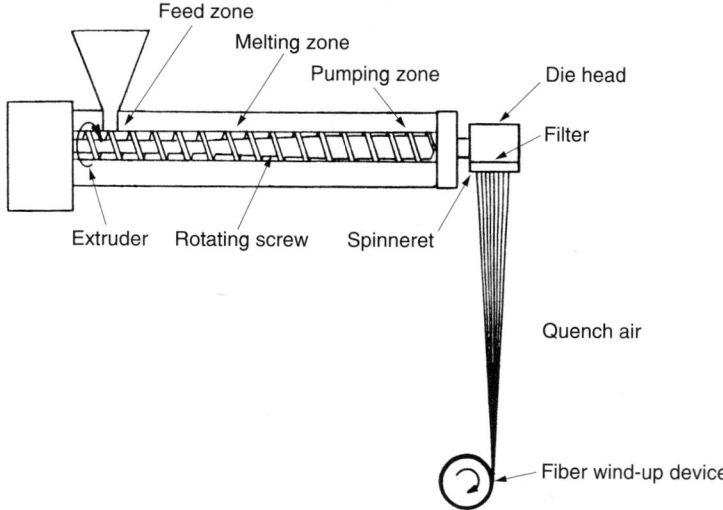

Figure 4.31 Schematic of process for melt spinning mesophase precursor fibers. *Source*: Reprinted with permission from Edie DD, Diefendorf RJ, Carbon fibre manufacturing, Buckley JD, Edie DD eds., *Carbon-Carbon Materials and Composites*, Noyes Publications, Park Ridge, NJ, 1993. Copyright 1993, William Andrew Publishing.

Table 4.13 Average mechanical properties of as-spun and carbonized mesophase fibers

Processing stage	Tensile strength GPa	Tensile modulus Gpa	Elongation %
As spun	0.04	4.7	0.85
Carbonized	2.06	216	0.95

Source: Reprinted with permission from Mochida I *et al*, *Carbon*, 23(2), 175, 1985. Copyright 1985, Elsevier.

mesophase fiber (Table 4.13). Processing speeds up to 1000 m/min are quoted [270,271] and the fiber is drawn down to 8–14 μm diameter. Since there is a mass loss in subsequent carbonization, to obtain a carbon fiber with a final diameter of 10 μm would require an as-spun diameter of about 12 μm. As with PAN fiber, a smaller diameter pitch fiber will produce a carbon fiber with higher tensile strength. Nippon Carbon have introduced a 6 μm diameter pitch based carbon fiber (PBCF) with a strength of about 4 GPa.

Otani *et al* [272] have described how the cross sectional shape of the spinneset hole can control the microstructure and cross sectional shape of the resulting fiber.

Factors such as the kind of pitch, shape of the spinneret, spinning temperature and pressure, all affect the structure of the resultant carbon fiber:

1. A high molecular weight mesophase pitch with no side groups will not spin, since it decomposes before it is sufficiently fluid to flow. Hence, commercially, a mixture of high molecular weight with a few side groups is used. A mesophase pitch is more difficult to spin than an isotropic pitch.
2. Spinneret construction does affect the quality of the PBCF and the spinneret should be made from a material that is readily wetted by pitch and the shape (Figure 4.32) controls the type of flow; a turbulent flow provided by stirring being preferred to a laminar flow (Figure 4.33). Cha has described a typical spinneret structure [274].
3. Increased pressure helps to suppress off-gassing at the spinning stage.
4. Figure 4.30 shows the effect of temperature on the structure and a pitch fiber, with an onion skin structure, is preferred to a radial type structure. Possible cross sectional microstructures of mesophase carbon fibers are given in Figure 4.34 and modification of the flow profile during extrusion can produce a less flow-sensitive product and higher tensile strength. The lines within each section depict carbon layers, which are at least preferentially parallel to the fiber axis.

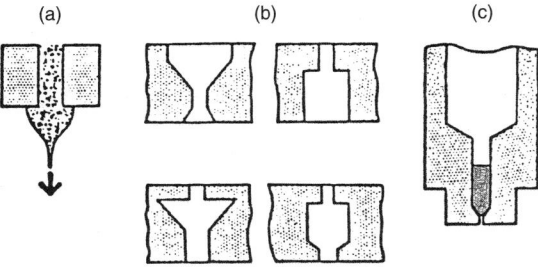

Figure 4.32 Schematic structures of spinnerets (a) Laminar flow disturbed by drop of molten pitch formed on spinneret surface. Jet material with good wettability gives larger drop; (b) Shape of spinneret hole with narrow and wider parts designed to convert from laminar to turbulent flow; (c) Laminar flow disturbed by a plug of stainless steel particles or mesh. *Source:* Reprinted with permission from Otani S, Oya A, In: Kawata K, Umekawa S, Kobayashi A eds. Composites '86 Recent Advances in Japan and the United States. Proc Japan-US CCM III Jpn Soc Compos Mater, Tokyo, 1–23, 1986. Copyright 1986, Japan Society of Composite Materials.

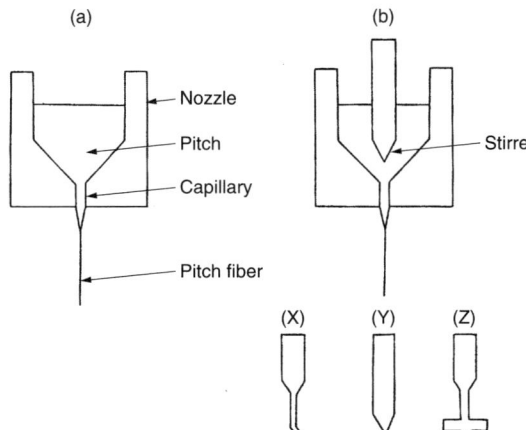

Figure 4.33 Extrusion of pitch fibers through a capillary (a) Without a stirrer; (b) With a stirrer. The microstructure of the fiber is very sensitive to whether a stirrer is used and to the shape (X, Y, Z) of the stirrer. *Source*: Reprinted with permission from Hamada T, Nishida T, Sajiki Y, Matsumoto M, Endo M, Structures and physical properties of carbon fibers from coal tar mesophase pitch, *J Mater Res*, 2, 850, 1987, Copyright 1987, Materials Research Society. Reprinted with permission from Hamada T, Nishida S, Sajiki Y, Furayama M, Tomioka T, *Extended Abstracts of the 18th Biennial Conference on Carbon*, Worcester, MA (American Carbon Society, University Park, PA), 225, 1987. Copyright 1987, The American Carbon Society.

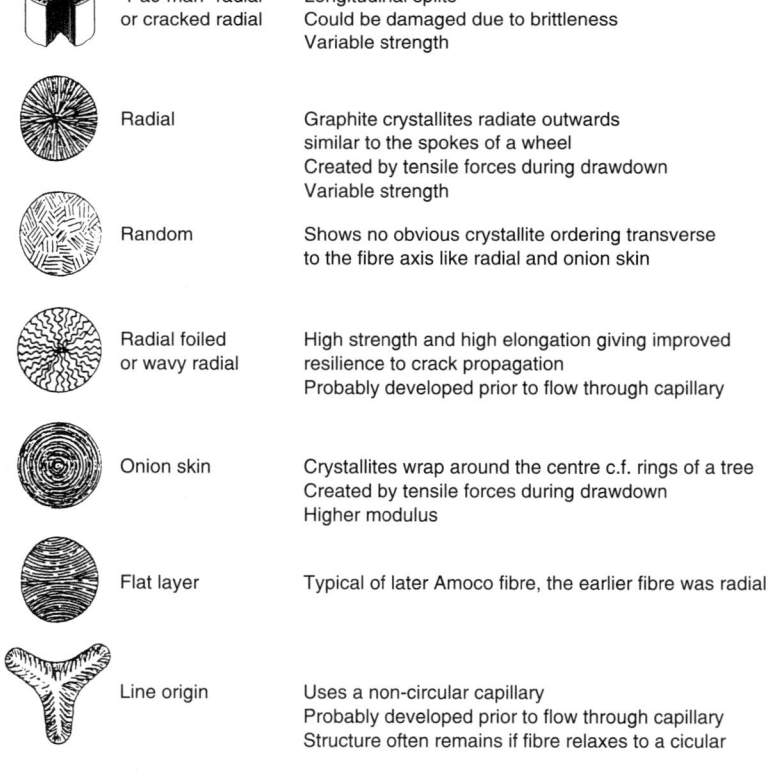

Figure 4.34 Possible cross-sectional microstructures of mesophase carbon fibers. *Source*: Reprinted with permission from Savage G, *Carbon-Carbon Composites*, Chapman and Hall, London, p. 63, 1992. Copyright 1992, Springer.

Due to the fact that the viscosity of the mesophase is extremely temperature dependent, the jet temperature must be accurately controlled, as a change of ±3.5°C at the jet face will produce a ±15% variation in the diameter [270,271].

There is a narrow window of working parameters when processing pitch and they must be very carefully adhered to.

4.4.6 Structure of pitch precursor

Typical structures of compounds found in a coal tar pitch are shown in Table 4.14 and these products will, in the process of making carbon fiber, be converted by pyrolysis to carbon and ultimately, to a three dimensional graphite. The effect of spinning conditions on the structures of pitch based carbon fiber has been investigated [277]. Before carbonization takes place, non-aromatic structures are converted to aromatics, which are the key building blocks for carbon.

Figure 4.35 shows a schematic diagram representative of the type of molecular arrangement one would expect to find within the mesophase.

A typical polynuclear aromatic hydrocarbon in mesophase is shown below:

Molecular weight = 1178
C/H = 1.50
$H_{aromatic}/H_{aliphatic}$ = 1.30
$C_{aromatic}/C_{aliphatic}$ = 6.15

4.5 OTHER PRECURSORS

1. Acrylic resins
   Melt processable acrylic such as BP Amoco Lima Chemicals Barex, an AN-MA copolymer, which can be extruded and Amlon, with a high AN content have been melt spun [279].
2. Aromatic hydrocarbons
   Carbon fibers can be prepared from aromatic hydrocarbons [280] and carbon can be deposited onto a polymer fiber by the pyrolysis of benzene to give a CCVD fiber [281].
3. Coal
   Carbon fibers can be produced from a non-polymeric hydrocarbon extract from coal and the addition of up to 20% of an amine-amide-alcohol or a carboxylic acid with no aromatic groups and at least eight carbon atoms, enables the extrusion of fibers [282]. The viscosity

Table 4.14 Some polynuclear aromatic components in coal tar pitch

Component	Linear structure	Molecular weight	Melting point °C	Boiling point °C	% in coal tar pitch
acenaphthene		$C_{12}H_{10}$	154	91	277	0.3
fluorene		$C_{13}H_{10}$	166	114	296	0.1
phenanthrene		$C_{14}H_{10}$	178	97	340	1.2 includes anthracene
anthracene		$C_{14}H_{10}$	178	216	354	1.2 includes phenanthrene
2-methyl anthracene		$C_{15}H_{12}$	192	207	sublimes	0.3
fluoranthene		$C_{16}H_{10}$	202	109	251	1.5
pyrene		$C_{16}H_{10}$	202	150	>360	1.3
chrysene		$C_{18}H_{12}$	228	254	448	2.7
benz[a]anthracene		$C_{18}H_{12}$	228	158	438	0.7

(*Continued*)

PRECURSORS FOR CARBON FIBER MANUFACTURE

Table 4.14 Continued

Component	Linear structure	Molecular weight	Melting point °C	Boiling point °C	% in coal tar pitch
benzo[a]pyrene	$C_{20}H_{12}$	252	177	–	3.6
dibenzphenanthrene	$C_{22}H_{14}$	278	–	–	–
coronene	$C_{24}H_{12}$	300	442	525	–
dibenzpyrone	$C_{24}H_{14}$	302	–	560 reaction temperature	1.5

Source: Reprinted with permission from Lewis IC, Singer LS, *Preprints of Fuel Div, Am Chem Soc*, 13, 86, 1969. Copyright 1969, American Chemical Society.

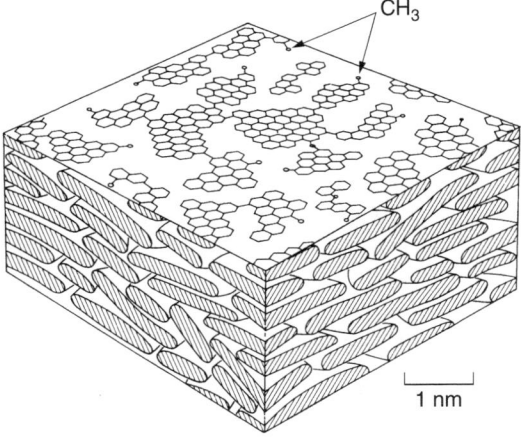

Figure 4.35 Schematic model of the carbonaceous mesophase, a discotic nematic liquid crystal. *Source:* Reprinted with permission from Zimmer JE, White JL, Disclination structures in the carbonaceous mesophase, *Adv in Liq Crysts*, 5, Academic Press, New York, 157, 1982. Copyright 1982, Elsevier.

characteristics can be changed to improve the spinnability by the addition of polypropylene (PP) or polymethylmethacrylate (PMMA) [283]. The spun fibers can then be stabilized by treatment with aqueous bromine solution and further stabilized by oxidation in an oxygen atmosphere [284].

4. Kerosene

The production of carbon fibers from kerosene has been investigated [285].

5. Lignin

Lignin has been investigated [286–289] and adding polyvinylalcohol (PVOH) to an alkaline solution of lignin produces a suitable precursor [290]. Exploding lignin has been used for the preparation of a carbon fiber precursor [291]. Carbon fibers have been prepared by acetic acid pulping of lignin [292,293].

6. Phenol formaldehyde (KYNOL)

The formaldehyde curing of a melt spun novalac resin yields an infusible fiber, which when heat treated in an inert atmosphere, gives a 58% yield of a carbon fiber at 700°C, with a carbon content of 94.5% and a strength of 0.69 GPa. Heat treatment to 1800°C increases the carbon content to 99.96% but decreases the strength to 0.47 GPa [294,295]. The product was not commercially successful.

7. Phenolic resins

A glassy carbon fiber can be produced from a melt spun phenol-hexamine thermoset resin [296,297] or a cured novalac [296,298], without the requirement of a stabilization stage.

8. Poly(bisbenzimidazobenzophenanthroline)

The polymer yarn is stabilized in an oxidizing atmosphere at 200–500°C, followed by rapid heating in an inert atmosphere to 1800–3200°C, heating from 500°C to at least 1800°C in no more than 5 min. [299].

9. Polyacetylenes

Krutchen [300] and a General Electric patent [301] have cited polyacetylenes.

10. Polyamides

Most polyamides, such as Nylon 6 (mp 200°C), tend to melt before achieving complete thermal stability. Nylon 6,6 (mp 260°C) is better [302], but best results are obtained with an aromatic polyamide fiber, NOMEX (mp > 400°C) which can be oxidized at 300°C and can be carbonized to give separable individual fibers [303].

11. Polyaromatics

Fitzer has described polyaromatics with fiber texture as the modern high technology carbon fibers [304].

12. Polybenzimidazole (PBI)

A polybenzimidazole fiber is treated with an acid such as H_2SO_4 to form a polybenzimidazonium salt, which is then oxidized at 400–550°C and can be carbonized in an inert atmosphere at 650–875°C to form a carbonaceous fibrous material with at least 60% w/w carbon. The product can then be oxidized at 450–600°C to give an activated carbonaceous fibrous material [305].

13. Polybutadiene (PB)

Initially fibers are produced by the melt spinning of a polymer containing at least 88% of a butadiene monomer. Cyclization can be achieved by treatment with a solution of a Lewis acid, followed by treatment in a sulfur bath, at 220–270°C, which achieves dehydrogenation and aromatizes the PB. The carbon yield is 70–85%, although yields as high as 95% have been achieved [305].

14. Polydivinylbenzene

Pyrolysis causes chain condensation [306].

15. Polyesters

Courtaulds hold a patent [307] citing polyesters as precursor material.

16. Polyethylene (PE)

PE can be prepared for carbonization by treating with chlorosulfonic acid at 60–90°C, or 98% H_2SO_4, at 100–180°C. Initially, the temperature should be maintained below the softening point of PE and gradually increased as the temperature resistance of the fiber increases [308]. The

structure and development of properties during the conversion of PE precursors to carbon fibers have been investigated [309].
17. Polyimides
   Polyimides have been used as possible precursor materials [310,311].
18. Polymethyl vinyl ketone

   a. Polyoxadiazoles
      The polymer yarn is stabilized in an oxidizing atmosphere at 200–500°C, followed by rapid heating in an inert atmosphere to 1800–3200°C, heating from 500°C to at least 1800°C in no more than 5 min [312].
   b. Poly (*p*-phenylene benzobisthiazole) (PBZT)
      PBZT has been converted to carbon fiber and the morphological aspects investigated [313].
   c. Polythiadiazoles
      The polymer yarn is stabilized in an oxidizing atmosphere at 200–500°C followed by rapid heating in an inert atmosphere to 1800–3200°C, heating from 500°C to at least 1800°C in no more than 5 min [298].

19. Poly(vinylacetylene)
    Poly(vinylacetylene) has been initially investigated as a carbon fiber precursor [314].
20. Polyvinyl alcohol (PVOH)
    Like polyamides, PVOH melts in the early stages of heat treatment [315].
21. Polyvinylchloride (PVC)
    Otani established a method of preparation of an MP carbon fiber by the pyrolysis of a PVC pitch obtained by heating PVC at 400°C in nitrogen [316].
22. Polyvinylidene chloride (PVDC)
    Boucher used PVDC [317,318] and Dow [319] and BASF [320] hold patents.
23. Polyvinylidene fluoride (PVDF)
    PVDF was treated by a dehydrofluorination process to prepare carbon fibers [321].

REFERENCES

1. Masson JC ed., *Acrylic Fiber Technology and Applications*, Marcel Dekker, New York, p. 44, 1995.
2. Frushhour BG, Knorr RS, Acrylic fibers, Lewin M and Pearce EM eds., *Handbook of Fiber Chemistry*, Marcel Dekker, New York, 869–1070, 1998.
3. Houtz RC, Orlan acrylic fibre: Chemistry and Properties, *J Text Res*, 20, 786–801, 1950.
4. Mackenzie HD, Reeder F, Courtaulds Ltd., Improvements in and relating to polyacrylonitrile solutions, Brit.Pat. 944,217, 1963.
5. *Fiber Organon*, 70, No.11, November 1999.
6. Wade B, Knorr R, Polymerization, Masson JC ed., *Acrylic Fiber Technology and Applications*, Marcel Dekker, New York, 1995.
7. Rudin A, *The Elements of Polymer Science and Engineering*, Academic Press, New York, Chapter 3, 1982.
8. Stockmayer WH, Casassa EF, *J Chem Phys*, 20, 1560, 1952.
9. Zimm BH, *J Chem Phys*, 16, 1093, 1948.
10. Moore JC, *J Polym Sci*, A2, 835, 1964.
11. Grubisic Z, Rempp P, Benoit H, *J Polym Sci*, B5, 753, 1967.
12. Weiss AR, Cohn-Ginsberg E, *J Polym Sci*, B7, 379, 1969.
13. Yau WW, Kirkland JJ, Bly DD, *Modern Size-Exclusion Liquid Chromatography*, Wiley, New York, 1989.
14. Huggins ML, *J Am Chem Soc*, 64, 2716, 1942.
15. Kraemer EO, *Ind Eng Chem*, 30, 1200, 1938.
16. Kamide K, Saito M, Viscometric determination of molecular weights, Cooper AR ed., *Determination of Molecular Weight*, Wiley, New York, Chapter 8, 1979.

17. Billmeyer FW Jr, Measurement of molecular weight and size, *Textbook of Polymer Science*, 3rd ed, Wiley, New York, Ch 8, 1984.
18. Tsai JS, Lin CH, The effect of molecular-weight on the cross-section and properties of polyacrylonitrile precursor and resulting carbon-fiber, *Journal of Applied Polymer Science*, 42, No.11, 3045–3050, 1991.
19. Gupta AK, Paliwal DK, Bajaj P, Acrylic precursors for carbon fibers, *J Macromol Sci Rev Macromol Chem Phys*, C31(1), 1–89, 1991.
20. Rajalingham P, Radhakrishnan G, Polyacrylonitrile precursor for carbon-fibers, *J Macromol Sci Rev Macromol Chem Phys*, C31, No.2–3, 301–310, 1991.
21. Bajaj P, Roopanwal AK, Thermal stabilization of acrylic precursors for the production of carbon fibers: An overview, *J Macromol Sci Rev Macromol Chem Phys*, C37(1), 97–147, 1997.
22. Olive GH, Olive S, The chemistry of carbon fibre formation from polyacrylonitrile, *Adv Polym Sci*, 51, 1, 1983.
23. Minagawa M, Okamoto M, Ishizuka O, *J Polym Dci Polym Chy Ed*, 16, 3031, 1978.
24. Bahl OP, Manocha LM, *Angew Makromol Chem*, 48, 145, 1975.
25. Fitzer E, Muller DJ, *Carbon*, 13, 63, 1975.
26. Tsai JS, Lin CH, Effect of comonomer composition on the properties of polyacrylonitrile precursor and resulting carbon-fiber, *Journal of Applied Polymer Science*, 43, No.4, 679–685, 1991.
27. Ogawa H, Studies on the improvement of productivity of high-performance polyacrylonitrile-based carbon-fibers. 1. Effects of comonomer methyl acrylate composition on production of polyacrylonitrile–copolymer-based carbon-fibers, *Nippon Kogaku Kaishi*, No.5, 464–470, 1994.
28. Mitsubishi Rayon Co Ltd., *Japan Kokai Tokkyo Koho*, 62,231,027, 1987.
29. Mitsubishi Rayon Co Ltd., *Japan Kokai Tokkyo Koho*, 62,231,026, 1987.
30. Japan Exlan Co Ltd., U.S. Pat., 4,009,248; Feb 22, 1977.
31. Toray Industries Inc., Jap. Pat., 58,214,534, 1983.
32. Toray Industries Inc., *Japan Kokai Tokkyo Koho*, 58,214,521, 1983.
33. Toray Industries Inc., *Japan Kokai Tokkyo Koho*, 58,214,527, 1983.
34. Toray Industries Inc., *Japan Kokai Tokkyo Koho*, 58,214,535, 1983.
35. Fitzer E, Muller DJ, *Makromol Chem*, 144, 117, 1971.
36. Toray Industries Inc., Jap. Pat., 03,64,514, 1991.
37. Asahi Chemicals Ltd., Jap. Pat., 05,05,244, 1993.
38. Toray Industries Inc., *Japan Kokai Tokkyo Koho*, 59,168,128, 1984.
39. Mitsubishi Rayon Co Ltd., *Japan Kokai Tokkyo Koho*, 60,151,317, 1985.
40. Asahi Chemical Industry, *Japan Kokai Tokkyo Koho*, 61,119,712, 1986.
41. Mitsubishi Rayon Co., Jap. Pat., 63,275,714, 1989.
42. Mitsubishi Rayon Co., Jap. Pat., 63,275,715, 1989.
43. Mitsubishi Rayon Co., Jap. Pat., 63,275,716, 1989.
44. Miraubiahi KK, Jap. Pat., 04,91,230, 1992.
45. Toray Industries Inc., EP 223,199, 1987.
46. Toray Industries., Jap. Pat., 04,240,221, 1992.
47. Toray Industries Inc., Jap. Pat., 02,14,012, 1990.
48. Toray Industries Inc., Jap. Pat., 04,333,620, 1993.
49. Ger D D. 288, 407, 1991.
50. Mitsubishi Rayon Co., Jap. Pat., 05,132,813, 1993.
51. Toray Industries, Jap. Pat., 58,214,526, 1983.
52. Koichi I, Hideo S, *Japan Kokai Tokkyo Koho*, Jap. Pat., 61,289,132, 1982.
53. Mitsubishi Rayon Co., Jap. Pat., 04,240,220, 1992.
54. Mitsubishi Rayon Co., U.S. Pat., 4,695,415, 1987.
55. Mitsubishi Rayon Co., Jap. Pat., 05,140,821, 1993.
56. Mitsubishi Rayon Co., Jap. Pat., 02,139,425, 1990.
57. Japan Exlan Co Ltd., *Japan Kokai Tokkyo Koho*, 58,191,704, 1984.
58. Mitsubishi Rayon Co Ltd., *Japan Kokai Tokkyo Koho*, 59,125,912, 1984.
59. Mitsubishi Rayon Co Ltd., U.S. Pat., 5,051,216, Sep 24, 1991.
60. Mitsubishi Rayon Co Ltd., *Japan Kokai Tokkyo Koho*, 59,125,913, 1984.

61. Tsai Jin-Shy, Lin Chung-Hua, *J Appl Polym Sci*, 43, 679, 1991.
62. Grassie N, McGuchan R, Pyrolysis of polyacrylonitrile and related polymers, *Europ Poly J*, 8, 257, 1972.
63. Mitsubishi Toasty Chemical Inc., *Japan Kokai Tokkyo Koho*, 62,15,329, 1987.
64. Mitsubishi Rayon Co., Jap. Pat., 05,132,813, 1993.
65. Henrici-Olivé G, Olivé S, *Adv Polym Sci*, 51, 36, 1983.
66. Celanese Corp., Brit. Pat., 1,264,026, 1969.
67. Asahi Kasei Kogyo KK, Brit. Pat., 1,435,447, 15 Feb 1974.
68. Moreton R, McLonghlin, Hewins P, *Royal Aircraft Establishment Report No. 76124*.
69. Platonova NV, Klimenko IB, Grachev VI, Kiselev GA, *Vysokomol Soedin*, Ser A, 30, (5), 1056, 1988.
70. Kiselev GA, Robinovich IS, Lysenko AA, Makarevich OI, USSR Pat., 1,065,509, 1984.
71. Le Carbone Lorraine, Brit. Pat. 1,280,850, 1969.
72. Anders RJ, Sweeny W, DuPont, U.S. Pat., 2,837, 500, June 3 1958.
73. Toray Industries Inc., Brit. Pat., 1,254,166.
74. Guyot A, Bert M, Hamoudi A, McNeill I, Grassie N, *Europ Polym J*, 14, 101, 1978.
75. Nagai S, *Bull Chem S Japan*, 36, 1459–1463, 1963.
76. Mueller T, Itaconic acid in carbon fiber precursor, *Intl Fibre J*, 46–50, 3 May 1988.
77. Tsai JS, Lin CH , *J Mater Sci Lett*, 9, 869, 1990.
78. Grassie N, McGuchan M, *Europ Polym J*, 8, 257, 865, 1972.
79. Tsai JS, Lin CH, *J Appl Polym Sci*, 43, 679, 1991.
80. Anders RJ, Sweeny W, DuPont, U.S. Pat., 2,837,500, 3 June 1958.
81. Horst E, Werner P, Eberhard P, Schmidt K, Ger. Pat., 245,885, 1987.
82. Minagawa M, Iwamatsu T, *J Polym Sci Polym Chem*, ed.18, 481, 1980.
83. Saito K, Ogawa H, U.S. Pat., 4,397,831, 1983.
84. Valvassori A, Sartori G, *J Polym Sci*, 5, 28, 1967.
85. Carl W, Acrylic fibres and their familiar modifications, *Chemiefasern/Textilindustrie*, 30/82, 518520, 1980.
86. Prasad G, Wet spinning of acrylic fiber and effects of spinning variables on fiber formation, *Synthetic Fibers*, 616, Jan and Mar 1985.
87. Prasad G, Vaidya A, Acrylic fiber production: developments in wet and dry spinning processes, *Textile Mag*, 4056, Apr 1986.
88. Hobson PH, McPeters AL, *Kirk-Othmer: Encyclopedia of Science and Technology*, 1, Interscience, New York, 1978.
89. Tsai JS, Wu CJ, Effect of cross-section evenness for polyacrylonitrile precursor on properties of carbon-fiber, *J Mater Sci Let*, 12, No.6, 411–413, 1993.
90. Capone GJ, Wet-Spinning Technology, Masson JC ed., *Acrylic Fiber Technology and Applications*, Marcel Dekker, New York, 1995.
91. Grobe V, Mann G, Structure formation of polyacrylonitrile solutions into aqueous spinning baths, *Faserforsch Textiltech*, 19, 49–55, 1968.
92. Ogbolue S, Structure/properties relationships in textile fibres, *Textile Progr, Textile Inst*, 20, 1990.
93. Beder N, Kabanova D, Dvoeglazova I, A method of investigating the structure formation process of unoriented polyacrylonitrile specimens, *Khimicheskie Volokna*, No.2, 31–32, 1986.
94. Wagner W, Comparison of further development of wet and dry spinning processes for acrylic fibre, *Second International Conference on Man-made Fibers*, Beijing, Nov 1987.
95. Eberhard P, Anneliese P, Hartig S, A process for the manufacture of void free and constant cross sectional shaped polyacrylonitrile fibers with high packing densities, East German Pat., No. 78624, 20 Dec 1970.
96. Paul D, A study of spinnability in the wet spinning of acrylic fibers, *J Appl Polym Sci*, 12, 2273–2298, 1968.
97. Padhye MR, Karandikar AV, *J Appl Polym Sci*, 33, 1675, 1987.
98. Grobe V, Meyer K, *Faserforschung Textiltechnik*, 20, 467, 1969.
99. Craig JP, Knudsen JP, Holland VF, Characterization of acrylic fiber structure, *Textile Research Journal*, 32, No. 6, 435–448, 1962.

100. Moreton R, Watt W, The spinning of polyacrylonitrile fibres in clean room conditions for the production of carbon fibres, *Carbon*, 12, 543–554, 1974.
101. von Falkai B, Dry spinning technology, Masson JC ed., *Acrylic Fiber Technology and Applications*, Marcel Dekker, New York, 105–165, 1995.
102. East G, McIntyre J, Patel G, The dry-jet wet-spinning of an acrylic fiber yarn, *J Text Inst*, 75, 196–200, 1984.
103. Baojun Q, Ding P, Zhenqiou W, The mechanisms and characteristics of dry-jet wet-spinning of acrylic fibers, *Adv Polym Technol*, 6, 509–529, 1986.
104. Frushour BG, *Polym Bull.*, 7, 1, 1982.
105. American Cyanamid, U.S. Pat., 4,296,059, 20 Jul 1979.
106. Porosoff H, Melt-spinning acrylonitrile polymer fibers, American Cyanamid Company, U.S. Pat., 4,163,770, 7 Aug 1979.
107. Pfeiffer RE, Roberts RW, Melt-spinning acrylonitrile polymer fiber using spinnerette of high orifice density, American Cyanamid Company, U.S. Pat., 4,220,616, 2 Sept 1980.
108. Pfeiffer RE, Peacher V, Process for melt-spinning acrylonitrile polymer fiber, American Cyanamid Company, U.S. PAT. 4,220,617, 2 Sept 1980.
109. Daumit GP, Ko YS, Slater CR, Venner JG, Young CC, Formation of melt-spun acrylic fibers which are particularly suited for thermal conversion to high strength carbon fibers, BASF, U.S. Pat., 4,921,656; 1 May 1990.
110. Daumit GP, Ko YS, Slater CR, Venner JG, Young CC, Zwick MM, Formation of melt-spun acrylic fibers which are well suited for thermal conversion to high strength carbon fibers, BASF, U.S. Pat., 4,933,128, 12 June 1990.
111. Daumit GP, Ko YS, Slater CR, Venner JG, Young CC, Melt-spun acrylic fibers which are particularly suited for thermal conversion to high strength carbon fibers, BASF, U.S. Pat., 4,981,751, 1 Jan 1991.
112. Daumit GP, Ko YS, Slater CR, Venner JG, Young CC, Zwick MM, Formation of melt-spun acrylic fibers which are well suited for thermal conversion to high strength carbon fibers, BASF, U.S. Pat., 4,981,752, 1 Jan 1991.
113. Daumit GP, Ko YS, Slater CR, Venner JG, Young CC, Melt-spun acrylic fibers possessing a highly uniform internal structure which are particularly suited for thermal conversion to high strength carbon fibers, BASF, U.S. Pat., 5,168,004, 1 Dec 1992.
114. Daumit GP, Ko YS, A unique approach to carbon fiber precursor development, High Tech—The Way into the Nineties, ed. Brunsch K, Gölden HD, Herkert CM, *Proc of the 7th Int SAMPE Conf*, Munich, Elsevier Science Publishers BV, Amsterdam, June 10–12, 201–213, 1986.
115. Cline ET, Cramer FB, DuPont de Nemours EI, Process for melt spinning acrylonitrile polymer hydrates, U.S. Pat., 4,238,442, 9 Dec 1980.
116. Henrici-Olivé G, Olivé S, *Adv Polym Sci*, Springer-Verlag, Berlin, 32, 127, 1983.
117. Dumbleton JH, Bell JP, The collapse process in acrylic fibers, *J Appl Polym Sci*, 14, 2402–2406, 1970.
118. Japan Exlan Co. Limited, Process for producing Carbon Fibers, U.S. Pat., 3,993,719, Nov 1976.
119. Mathur RB, Bahl OP, Matta KV, Nagpal KC, *Carbon*, 26, No.3, 295, 1988.
120. Bahl OP, Mathur RB, Dhami TL, *Mater Sci Eng*, 73, 105, 1985.
121. Mitsubishi Rayon Co., Jap. Pat., 63,35,820, 1988.
122. Mikolajczyk T, Kamecka J, *Polimery*, 36, No.10, 384, 1991.
123. Jap. Pat., 62,299,509, 1988.
124. German DD Pat., 279,275, 1990.
125. Mitsubishi Rayon Co., Jap. Pat., 63,275,713, 1988.
126. Mitsubishi Rayon Co., Jap. Pat., 63,249,715, 1987.
127. Mitsubishi Rayon Co., European Pat., 255,109, 1987.
128. Mitsubishi Rayon Co., Jap. Pat., 63,66,317, 1988.
129. Toray Industries, Jap. Pat., 05,263,313, 1993.
130. Mitsubishi Rayon Co., Jap. Pat., 05,272,005, 1993.
131. Mitsubishi Rayon Co., Jap. Pat., 05,239,712, 1993.
132. Fushie T, Fukui Y, Kobayashi T, *Japan Kokai Tokkyo Koho*, Jap. Pat., 60,181,322, 1985.

133. Tamai H, Adachi Y, Shiromoto S, *Japan Kokai Tokkyo Koho*, Jap. Pat., 62,184,121, 1987.
134. Oda T, Kaneko T, Hattori C, Imai Y, *Japan Kokai Tokkyo Koho*, Jap. Pat., 61,167,024, 1985.
135. Mitsubishi Rayon Co., *Japan Kokai Tokkyo Koho*, Jap. Pat., 60,99,011, 1985.
136. Reference deleted.
137. Ermolenko IN, Lyubliner IP, Gulko NV, Chemically modified Carbon Fibres and their Applications, (translated by EP Titovets) VCH, Germany, Chapter 5, 153, 1990.
138. Mladenov I, *J Polym Sci Polym Chem*, 21, 1223, 1983.
139. Kanebo Ltd., Jap. Pat., 04,119,127, 1992.
140. Brake P, Schuvmus H, Verhoest J, *Inorg Fibers and Composites*, Pergammon Press, New York, 1984.
141. Toray Industries, Jap. Pat., 04,214,414, 1992.
142. Kenje S, Ogawa H, *Nippon Kagaku Kaishi*, 7, 745, 1992.
143. Wang H, *Int SAMPE Tech Conf*, 19, 729, 1987.
144. Mathur RB, Bahl OP, Mittal J, *Carbon*, 30(4), 657, 1992.
145. Ko TH, Lin CH, *J Mater Sci Lett*, 7, 628, 1988.
146. Li J, Pan D, Wu Z, *J China Textile Univ*, Eng ed, 7(4), 13, 1990.
147. Ko TH, Huang LC, *J Mater Sci*, 27, 2429–2436, 1994.
148. Mathur RB, Mittal J, Bahl OP, Sandle NK, Characteristics of $KMnO_4$ modified PAN fibres, its influence on the resulting carbon fibre properties, *Carbon*, 32, (1), 71–77, 1994.
149. Mathur RB, Bahl OP, Mittal J, Advances in the development of high performance carbon fibres from PAN precursor, *Composites Sci Tech*, 51, 223 230, 1994.
150. Ko TH, Chiranairadul P, Ting HY, Lin CH, *J Appl Polym Sci*, 37, 541, 1989.
151. Mathur RB, Mittal J, Bahl OP, *J Appl Polym Sci*, 49, 469, 1993.
152. Mathur RB, Bahl OP, Kundra KD, *J Mater Sci Lett*, 5, 757, 1986.
153. Li J, Pan D, Pan W, Wu Z, *Zhong Fang Daxue Xuebae* (Chinese), 16, 147, 1990.
154. Shieldlin A, Marom G, Zillikha A, *Polymer*, 26(3), 447–451, 1985.
155. Frushour BG, Acrylic polymer characterization in the solid state and solution, Masson JC ed., *Acrylic Fiber Technology and applications*, Marcel Dekker, New York, 198, 1995.
156. Henrici-Olivé G, Olivé S, Molecular interactions and macroscopic properties of polyacrylonitrile and model substances, *Adv Polym Sci*, 32, 123, 1980.
157. Ganster J, Fink HP, Zenke I, Chain confirmation of polyacrylonitrile: a comparison of model scattering and radial distribution functions with experimental wide-angle X-ray scattering results, *Polymer*, 32, 1566, 1991.
158. Fischer EW, Fakirov S, *J Mater Sci*, 11, 899, 1979.
159. Hinrichsen G, *J Polym Sci*, 38, 303, 1972.
160. Warner SB, Uhlmann D, Peebles L, *J Mater Sci*, 10, 758, 1975.
161. Bohn CR, Schaefgen JR, Statton WO, *J Polym Sci*, 55, 531, 1961.
162. Allen RA, Ward IM, Bashir Z, *Polymer*, 35, 2063, 1994.
163. Hooke R, *Micrographia*, Royal Society, London, 1664.
164. Espinasse M, *Robert Hooke*, William Heinemann, London, 1956.
165. Audemars GA, Brit. Pat., 283, 1855.
166. Chardonnet HB, Fr. Pat., 165,349, 1884.
167. Despaissis LH, Fr. Pat., 203,741, 1890.
168. Urban J, Bronnert H, Bronnert E, Fremery M, Ger. Pats., 109,996, 111,313, 119,230, 121,249 and 121,430.
169. Cross CF, Bevan EJ, Beadle C, Brit. Pat., 8700, 1892.
170. Topham CF, Brit. Pat., 12,157, 1902.
171. Topham CF, Brit. Pat., 12,158 1902.
172. Peters RH, Textile Chemistry, Vol 1, *The Chemistry of Fibres*, Elsevier, London, 404–416, 1963.
173. Moncrieff RW, *Man-made Fibres*, Wiley, London, 152–190, 1970.
174. Dyer J, Daul GC, Rayon Fibers, Lewin M and Pearce EM eds., *Handbook of Fiber Chemistry*, Marcel Dekker, New York, 725–801, 1998.
175. Sisson WA, *Text Res J*, 30(3), 153, 1960.
176. Matsumoto T, Mesophase pitch and its carbon fibers, *Pure Appl Chem*, 57, 1553, 1985.

177. Lewis IC, Chemistry and development of mesophase in pitch, *Journal de Chimie Physique*, 81, 751–758, 1984.
178. Riggs DM, Shuford RJ, Lewis RW, Graphite fibers and composites, Lubin G ed., *Handbook of Composites*, Van Nostrand Reinhold Co., New York, 1982.
179. Brooks JD, Taylor GH, The formation of some graphitizing carbon, Walker PL Jr ed., *Chemistry and Physics of Carbon*, Vol 4., Marcel Dekker, New York 168, 243–268.
180. Goppel JM, Knotnerus J, Fundamentals of bitumen blowing, *Fourth World Petroleum Congress Sect 3G, Paper 2*, 1955.
181. Corbett LW, Reaction variables in the air blowing of asphalt, *Ind and Eng Chem Des Dev*, 14(2), 1975.
182. Maeda T, Zeng SM, Tokumitsu K, Mondori J, Mochida I, Preparation of isotropic pitch precursors for general purpose carbon fibers (GPCF) by air blowing. 1. Preparation of spinnable isotropic pitch precursor from coal tar by air blowing, *Carbon*, 31(3), 407–412, 1993.
183. Zeng SM, Maeda T, Tokumitsu K, Mondori J, Mochida I, Preparation of isotropic pitch precursors for general purpose carbon fibers (GPCF) by air blowing. 2. Air blowing of coal tar, hydrogenated coal tar and petroleum pitches, *Carbon*, 31(3), 413–419, 1993.
184. Zeng SM, Maeda T, Mondori J, Tokumitsu K, Mochida I, Preparation of isotropic pitch precursors for general purpose carbon fibers (GPCF) by air blowing. 3. Air blowing of isotropic naphthalene and hydrogenated coal tar pitches with addition of 1,8-dinitronaphthalene, *Carbon*, 31(3), 421–426, 1993.
185. Otani S, Yamada K, Koitabashi T, Yokoyama A, On the raw materials of MP carbon fiber, *Carbon*, 4, 425, 1966.
186. Otani S, Yokoyama A, Characteristic chemical constitution of pitch materials suitable for the MP carbon fiber, *Bull Chem Soc Japan*, 42, 1417, 1969.
187. Kureha Co., U.S. Pat., 3,629,379, 1971.
188. Ungureanu C, Onciu M, Timpu D, Research in carbon fibre's field. 3. Precursors and carbon fibres got from petroleum bitumens, *Materiale Plastice*, 33(1), 57–65, 1996.
189. Sumner MB, Thermal properties of heavy isotropic petroleum pitches. *Carbon '88, Proceedings of the International Conference on Carbon*, University of Newcastle upon Tyne, 52–54, Sep 18–23, 1988.
190. Dickenson EM, Average structures of petroleum pitch fractions by $^1H/^{13}C$ NMR spectroscopy, *Fuel*, 64, 704–706, 1985.
191. Smith FA, Eckle TF, Osterholm RJ, Stichel RM, Manufacture of coal tar and pitches, Hoiberg AJ ed., *Bituminous Materials, Vol. III*, RE Krieger, New York, 57, 1966.
192. Monge JA, la Amoros DC, Solano AL, Oya A, Sakamoto A, Hoshi K, Preparation of general purpose carbon fibers from coal tar pitches with low softening point, *Carbon*, 35(8), 1079–1087, 1997.
193. Halleux A, de Greef H, *Fuel*, 42, 185, 1963.
194. Smith JW, *Fuel*, 45, 233, 1966.
195. McNeil D, The physical properties and chemical structure of coal tar pitch, Holberg WJ ed., *Bituminous Materials, Vol III*, RE Krieger, New York, 139, 1966.
196. Diefendorf RJ, Riggs DM, Forming optically anisotropic pitches, U.S. Pat., 4,208,267, 1980.
197. Franck HG, HH Lowery ed., *Chemistry and Coal Utilization, Supplementary Vol*, Wiley, New York, 592, 1963.
198. van Krevelen DW, *Fuel*, 29, 269, 1950.
199. Phillips G, Wood LJ, *J App Chem* (Lond), 5, 326, 1955.
200. Brooks JD, Steven JR, *Fuel*, 43, 87, 1964.
201. Friedel RA, Kendall DN ed., *Applied Infrared Spectroscopy*, Wiley, New York, 312–343, 1966.
202. Vahrman M, Bangham DH ed., *Progress in Coal Science*, Interscience, New York, 60, 1950.
203. Greinke RA, Lewis IC, *Anal Chem*, 47, 2151, 1975.
204. Schutz RV, Jorgenson JW, Maskarinec MP, Novotny M, *Fuel*, 58, 783, 1979.
205. Bartle KD, Collin G, Stadelhofer JW, Zander M, *J Chem Tech Biotechnol*, 29, 531, 1979.
206. Blumer GP, Kleffner HW, Lucke W, Zander M, *Fuel*, 59, 600, 1980.
207. Edstrom T, Petro BA, *J Poly Sci*, Part C, 21, 171, 1968.

208. Tillmanns H, Ulsamer W, Pietzka G, *Carbon 76 Intl Carbon Conf Preprints 2nd*, 557, 1976.
209. Lewis IC, Petro BA, *J Poly Sci Polym Chem*, ed 14, 1975, 1976.
210. Greinke RA, O'Connor LH, *Anal Chem*, 52, 1877, 1980.
211. Lewis IC, Edstrom T, *J Org Chem*, 28, 2050, 1963.
212. Stadelhofer JW, *Carbon*, 17, 301, 1979.
213. Rand B, Shepherd PM, *Fuel*, 59, 814, 1980.
214. Barr JB, Lewis IC, *Thermochimica Acta*, 52, 297, 1982.
215. Friedel RA ed., *Spectrometry of Fuels*, Plenum Press New York, 1970.
216. Bartle KD, Jones DW, *Fuel*, 48, 21, 1969.
217. Dickinson EM, *Fuel*, 59, 290, 1980.
218. Sharkey AG, Schulz JL, Friedel RA, *Carbon*, 4, 365, 1966.
219. Evans S, Marsh H, *Carbon*, 9, 733 and 747, 1971.
220. Ruland W, *Carbon*, 2, 365, 1965.
221. Simon C, Estrade H, Tchoubar D, Conard J, Carbon, 15, 211, 1977, *Proc 5th London Carbon Conf*, Vol 1, London, 294, 1978.
222. Sawran WR, Turrill FH, Newman JW, Ward C, Process for the manufacture of carbon fibers and feedstock therefor, U.S. Pat., 4,671,864, 1985.
223. Maeda T, Zeng SM, Tokumitsu K, Mondori J, Mochida I, Preparation of isotropic pitch precursors for general-purpose carbon-fibers (GPCF) by air blowing. 1. Preparation of spinnable isotropic pitch precursor from coal-tar by air blowing, *Carbon*, 31(3), 407–412, 1993.
224. Zeng SM, Maeda T, Tokumitsu K, Mondori J, Mochida I, Preparation of isotropic pitch precursors for general-purpose carbon-fibers (GPCF) by air blowing. 2. Air blowing of coal-tar, hydrogenated coal-tar, and petroleum pitches, *Carbon*, 31(3), 413–419, 1993.
225. Zeng SM, Maeda T, Mondori J, Tokumitsu K, Mochida I, Preparation of isotropic pitch precursors for general-purpose carbon-fibers (GPCF) by air blowing. 3. Air blowing of isotropic naphthalene and hydrogenated coal-tar pitches with addition of 1,8-dinitronaphthalene, Carbon, 31: No.3, 421–426, 1993.
226. Alcaniz MJ, Cazorla AD, Linares AS, Oya A, Sakomoto A, Hoshi K, Preparation of general purpose carbon fibers from coal tar pitches with low softening point, *Carbon*, 35(8), 1079–1087, 1997.
227. Korai Y, Ishida S, Watanabe F, Yoon SH, Wang YG, Mochida I, Kato I, Nakamura T, Sakai Y, Komatsu M, Preparation of carbon fiber from isotropic pitch containing mesophase spheres, *Carbon*, 35(12), 1733–1737, 1997.
228. Peebles LH, *Carbon Fibers: Formation, Structure, and Properties*, CRC Press Inc., Boca Raton, p. 29, 1995.
229. Okuda K, *Trans Mat Res Jpn*, 1, 119–139, 1990.
230. Brooks JD, Taylor GH, The formation of graphitizing carbon from the liquid phase, *Carbon*, 3, 185, 1965.
231. Reference deleted.
232. Zimmer JE, White JL, Disclination structures in the carbonaceous mesophase, *Adv in Liq Crysts*, Vol 5, Academic Press, New York, 157, 1982.
233. Hornby J, Kearsey HA, Carbon fibres from pitch precursors 2. The preparation of mesophase pitches suitable for spinning into fibres, *AERE Harwell report AERE-M 3029*, Oct 1979.
234. Collett GW, Rand B, *Fuel*, 57, 162, 1978.
235. Didchenko R, Barr JB, Chwastiak S, Lewis IC, Lewis RT, Singer LS, High modulus carbon fibers from mesophase pitches, *Extended Abstr 12th Biennial Conf on Carbon*, 329, 1975.
236. Lewis IC, Process for producing mesophase pitch, U.S. Pat., 4,032,430, 1977.
237. McHenry ER, Process for producing mesophase pitch, U.S. Pat., 4,026,788, 1977.
238. Lewis IC, McHenry ER, Singer LS, Process for producing mesophase pitch, U.S. Pat., 4,017,327, 1977.
239. Singer LS, High modulus high strength fibers produced from mesophase pitch, U.S. Pat., 4,005,183, 1977.
240. Lewis IC, McHenry ER, Singer LS, Process for producing carbon fibers from mesophase pitch, U.S. Pat., 3,976,729, 1976.

241. McHenry ER, Process for producing mesophase pitch. US Pat 4,026,788; 1977.
242. Chwastiak S, Lewis IC, Solubility of mesophase pitch, *Carbon*, 16, 156–157, 1978.
243. Union Carbide, Brit. Pat., 1,416,614.
244. Zimmer JE, White JL, Disclination structures in the carbonaceous mesophase, *Adv in Liq Crysts*, 5, Academic Press, New York, 157, 1982.
245. Murakami K, Toshima H, Yamamoto M, Effect of mesophase pitches on tensile modulus of pitch-based carbon fibers, 53(3), 73–78, 1997.
246. Korai Y, Ishida S, Watanabe F, Yoon SH, Wang YG, Mochida I, Kato I, Nakamura T, Sakai Y, Komatsu M, *Carbon*, 35(12), 1733–1737, 1997.
247. Diefendorf RJ, Riggs DM, Forming optically anisotropic pitches, U.S. Pat., 4,208,267, 1980.
248. Riggs DM, Carbon fiber from solvent extracted pitch, *Preprints Div Petrol Chem Am Chem Soc*, 29, 400, 1984.
249. Lafdi K, Bonnamy S, Oberlin A, Mechanism of anisotropy occurrence in a pitch precursor of carbon fibers, Part I, Pitches A and B, *Carbon*, 29, 831, 1991.
250. Lafdi K, Bonnamy S, Oberlin A, Mechanism of anisotropy occurrence in a pitch precursor of carbon fibers, Part II, Pitch C, *Carbon*, 29, 849, 1991.
251. Lafdi K, Bonnamy S, Oberlin A, Mechanism of anisotropy occurrence in a pitch precursor of carbon fibers, Part III, Hot stage microscopy of pitch B and C, *Carbon*, 29, 857, 1991.
252. Otani S, Jap. Pat., 57–100186, 1982.
253. Lewis IC, *Carbon*, 16, 425, 1978.
254. Lewis IC, *Fuel*, 66(11), 1527, 1987.
255. Mochida I et al, *Carbon*, 23(2), 175, 1985.
256. Mochida I, Sone Y, et al, *Carbon*, 28(2,3), 311,1990.
257. Wang CY, Zheng JM, et al, *22nd Bien Conf on Carbon*, Ext Abst and Pro Univ of California, San Diego, 254–255, 16–21 Jul 1995.
258. Yoon S, Oka H, et al, et al, *22nd Bien Conf on Carbon*, Ext Abst and Pro Univ of California, San Diego, 46–47, 16–21 Jul 1995.
259. Nazem FF, Flow of molten mesophase pitch, *Carbon*, 20, 345, 1982.
260. Sumner MB, Thermal properties of heavy isotropic petroleum pitches. *Carbon '88, Proceedings of the International Conference on Carbon*, University of Newcastle upon Tyne, 52–54, Sep 18–23, 1988.
261. Whitehouse S, Rand B, Rheology of mesophase pitch from A240, *Carbon '88, Proceedings of the International Conference on Carbon*, University of Newcastle upon Tyne, 175–176 Sep 18–23, 1988.
262. Lewis IC, Lewis IT, Rheological characterization of pitches, *Extended Abstr 20th Bien Conf on Carbon*, 166, 1991.
263. Williams ML, Landel RF, Ferry JD, *J Am Chem Soc*, 77, 3701, 1955.
264. Otani S, Oya A, Progress of pitch-based carbon fiber in Japan, Bacha JD, Newman JW, White JL eds., *ACS Symp Ser No 303*, American Chemical Society, Washington DC, 322, 1986.
265. Hamada T, Nishida T, Sajiki Y, Matsumoto M, Endo M, Structures and physical properties of carbon fibers from coal tar mesophase pitch, *J Mater Res*, 2, 850, 1987.
266. Bourrat X, Roche EJ, Lavin JG, Lattice imaging of disclinations in carbon fibers, *Carbon*, 28, 236, 1990.
267. Edie DD, The effect of processing on the structure and properties of carbon fibers, *Carbon*, 36(4), 345–362, 1998.
268. Yoon SH, Korai Y, Mochida I, Kato I, The flow properties of mesophase pitches and derived from methylnaphthalene and naphthalene in the temperature range of their spinning, *Carbon*, 32, 273–280, 1994.
269. Edie DD, Diefendorf RJ, Carbon fibre manufacturing, Buckley JD, Edie DD eds., *Carbon-Carbon Materials and Composites*, Noyes Publications, Park Ridge, 1993.
270. Edie DD, Pitch and mesophase fibers, Figueiredo J, Bernado CA, Baker RTK, Hüttenger KJ eds., *Carbon Fibers, Filaments and Composites*, Kluwer Academic Publishers, Dordrecht, 647–655, 1990.
271. Edie DD, Dunham MG, Melt spinning pitch based carbon fibers, *Carbon*, 27(5), 647–655, 1989.

272. Otani S, Oya A, Kawata K, Umekawa S, Kobayashi A eds., *Composites '86: Recent Advances in Japan and the United States, Proc Japan-US, CCM-III, Jpn Soc Compos Mater*, Tokyo, 1–10, 1986.
273. Hamada T, Nishida S, Sajiki Y, Furayama M, Tomioka T, *Extended Abstracts of the 18th Biennial Conference on Carbon*, Worcester, MA (American Carbon Society, University Park, PA), 225.
274. Cha Q et al, *Carbon*, 30(5), 739, 1992.
275. Savage G, *Carbon-Carbon Composites*, Chapman and Hall, London, 1992.
276. Lewis IC, Singer LS, *Preprints of Fuel Div, Am Chem Soc*, 13, 86, 1969.
277. Matsumoto M, Iwashita T, Arai Y, Effect of spinning conditions on structure of pitch based carbon fiber, *Carbon*, 31(5), 715–720, 1993.
278. Davidson JA, Jung HT, Hudson SD, Percec S, *Polymer*, 41, 3357, 2000.
279. Mochida I, Yoon SH, Korai Y, Kanno K, Sakai Y, Komatsu M, Carbon-fibers from aromatic-hydrocarbons, *Chemtech*, 25(2), 29–37, 1995.
280. Matsumara K, Takahashi A, Tsukamoto J, *Synth Met*, 11, 9, 1985.
281. Coal Industry (Patents) Brit. Pat., 1,356,566, 8 Sep 1971.
282. Coal Industry (Patents) Brit. Pat., 1,356,567, 8 Sep 1971.
283. Coal Industry (Patents) Brit. Pat., 1,356,569, 22 Dec 1971.
284. Sharon M, Kumar M, Kichambare PD, Ando Y, Zhao XL, Carbon fibers from kerosene, *Diamond Films and Technology*, 8(3), 143–152, 1998.
285. Bayer Brit. Pat., 1,359,764 (West Germany), 16 Apr 1971.
286. Fukuoka Y, *Japan Chem Quart*, 5, 63, 1969.
287. Ger. Pat., 1,952,388, 1971.
288. Sudo K, Shimizu K, A new carbon-fiber from lignin, *J Appl Polym Sci*, 44(1), 127–134, 1992.
289. Mikawa S, *Chem Econ Engg Rev*, 2(8), 43, 1970.
290. Sudo K, Shimizu K, Nakashima N, Yokoyama A, A new modification method of exploded lignin for the preparation of a carbon-fiber precursor, *J Appl Polym Sci*, 48(8), 1485–1491, 1993.
291. Uraki Y, Kubo S, Nigo N, Sano Y, Sasaya T, Preparation of carbon-fibers from organosolv lignin obtained by aqueous acetic-acid pulping, *Holzforschung*, 49(4), 343–350, 1995.
292. Kubo S, Uraki Y, Sano Y, Preparation of carbon fibers from softwood lignin by atmospheric acetic acid pulping, *Carbon*, 36(7–8), 1119–1124, 1998.
293. American Kynol Inc., *Production of Carbon Fiber from Kynol^{TM} Novaloid Precursor Fiber*, New York, Jul 1997.
294. Economy J, Lin RY, *J Mater Sci*, 6, 1151, 1971.
295. Jenkins GM, Kawamura K, *J Mater Sci*, 5, 262, 1970.
296. Coal Industry (Patents), Brit. Pat., 1,228,910, 6 Sep 1968.
297. Kawamura K, *Conf on Ind Carbons and Graphite, Soc Chem Ind*, London, 98, 1970.
298. Ezekiel HM, U.S. Airforce, Dayton, U.S. Pat., 3,635,675, 18 Jan 1972.
299. Krutchen CM, *Paper 2, Papers of 11th Biennial Conference on Carbon*, Gatlenburg, 1973.
300. General Electric Co., Brit. Pat., 1,366,123 (USA), 2 Nov 1970.
301. Rose PG, A study of the physics and chemistry of the preparation of carbon fibres from acrylic precursors, *Thesis*, University of Aston in Birmingham, Oct 1971.
302. Ezekiel HN, *Applied Polymer Symposia*, 9, 315, 1969.
303. Fitzer E, Polyaromatics with fiber texture- the modern high-technology carbon-fibers, *Acta Polymerica*, 41(7), 381–389, 1990.
304. Celanese Corporation, *Production of activated carbon fibers from acid contacted polybenzimidazole fibrous material*, U.S. Pat., 4,460,708, 17 July 1984.
305. Ube Industries, Brit. Pat., 1,451,550.
306. Winslow FH, Baker WO, Pape NR, Matreyek W, *J Polymer Sci*, 16, 101, 1955.
307. Courtaulds Ltd., U.S. Pat., 3,533,741, 1970.
308. Sumitomo Chemical Corp., Brit. Pat., 1,458,571.
309. Zhang D, Sun Q, Structure and properties development during the conversion of polyethylene precursors to carbon fibers, *J Appl Polym Sci*, 62(2), 367–373, 1996.

310. Ezekiel HM, Spain RG, *J Polym Sci*, C(19), 249, 1967.
311. *Amer Chem Soc Div Polymer Chem Polymer Preprints*, 4, 375, 1973.
312. HM Ezekiel, U.S. Airforce, Dayton. U.S. Pat., 3,635,675, 18 Jan, 1972.
313. Jiang H, Desai P, Kumar S, Abhiraman AS, Carbon-fibers from poly(para-phenylene benzobisthiazole) (PBZT) fibers—conversion and morphological aspects, *Carbon*, 29(4–5), 635–644, 1991.
314. Mavinkurve A, Visser S, Pennings AJ, An initial evaluation of poly(vinylacetylene) as a carbon-fiber precursor, *Carbon*, 33(6), 757–761, 1995.
315. Shindo A, Nakanishi Y, Soma I, *Appl Polym Symp*, 9, 305, 1969.
316. Otani S, *Carbon*, 3, 31, 1965.
317. Boucher AE, Cooper RN, Everett DH, *Carbon*, 8, 597, 1970.
318. Adams LB, Boucher EA, Cooper RN, Everett DH, *Papers of the third Conference on Industrial Carbons and Graphite*, London, 478, 1970.
319. Dow Chemical Co., U.S. Pat., 3,840,649, 29 Oct 1971.
320. BASF, Brit. Pat., 1,372,292 (West Germany), 2 Feb 1971.
321. Yamashita J, Shioya M, Nakatani M, Preparation of carbon fibers and films from poly(vinylidene fluoride) using chemical dehydrofluorination, *Carbon*, 36(7–8), 1240–1243, 1998.

CHAPTER 5

Carbon Fiber Production using a PAN Precursor

5.1 INTRODUCTION

A PAN based carbon fiber is produced by the oxidative stabilization of a PAN precursor, normally followed by a two-stage carbonization process, with an added heat treatment stage to manufacture a high modulus fiber. The aim is to remove, in gaseous form, all elements other than carbon, although high strength fibers do contain up to about 5% nitrogen, depending on the manufacturer.

Shindo, of the Industrial Research Institute in Osaka, was the first to make carbon fibers from polyacrylonitrile (PAN) fibers [1], publishing his work in 1961. The process did not, however, cover the application of tension during the pre-oxidation stage, thus limiting the attainable modulus. W. Watt, W. Johnson and L.N. Phillips of the Royal Aircraft Establishment at Farnborough (RAE), started work in 1963 on the production of carbon fiber from a PAN precursor, which successfully culminated in the application for a patent in 1964 [2], covering a technique for improving the fiber orientation by restraining the natural shrinkage, or even applying some stretch to the fiber in the oxidation stages of the carbon fiber process, without any subsequent recourse to hot stretching later.

The use of the term graphite for the higher modulus PAN based carbon fibers is a total misnomer, since, in reality, this form of carbon fiber exists as a structurally disordered carbon (turbostratic), with the disorder improving gradually at treatment temperatures above 2000°C. PAN is a non-graphitizing precursor, yet regrettably, the terms graphite and graphitizing have been widely used with reference to this class of carbon fiber.

A bibliography of books on carbon fibers and their composites is given in the introduction and a number of reviews [3–22], which include information on PAN based carbon fibers have appeared in the literature.

5.2 CARBON FIBER MANUFACTURERS

The manufacturers of carbon fiber in the world can be best presented in diagram form (Figure 5.1), revealing the links between the various companies. Ownership is constantly changing due to numerous acquisitions and mergers.

Although this data excludes production from China, Z. Shen *et al* (Beijing University of Chemical Technology) have quoted a total production capacity of 150 tpa in 1995 in China

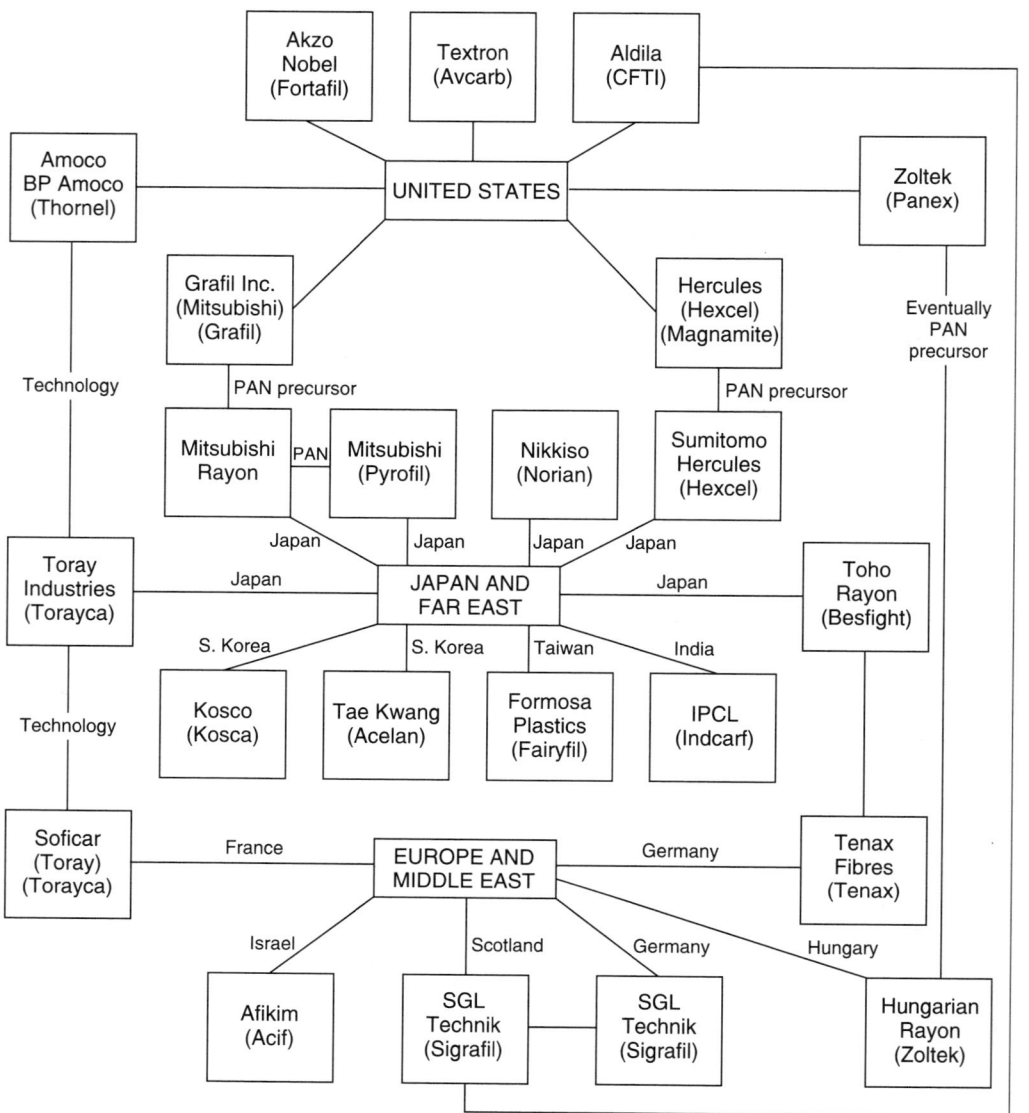

Figure 5.1 World tie-ups among PAN based carbon fiber manufacturers (Russian Federation and China omitted).

for PAN based carbon fibers, with a capability of about 735 tpa of special PAN precursor, of which 200 MT was under construction at the time of the report.

5.3 WORLD SUPPLY OF PAN BASED CARBON FIBER

Table 5.1 gives the output of the major carbon fiber producers for small and large tow for the period 1996–99. The demand for PAN based carbon fiber since its introduction in 1961 is shown in Figure 5.2. Tables 5.2–5.4 give estimates of production of PAN based carbon fiber worldwide, divided between the major geographic regions, consumption in the major market segments and production destined for market sub-segments, whilst Toray have shown in Figure 5.3 the usage of carbon fiber during the period 1985–2005 divided into three

Table 5.1 Carbon fiber production capacity (tpa)

Fiber type	Manufacturer		1996	1997	1998	1999
Small carbon fiber Tow	Amoco		1,200	1,800	1,800	1,800
	FPC		250	750	750	1,750
	Hexcel (Hercules)		1,700	2,000	2,000	2,000
	Mitsubishi	(Japan)	1,000	1,200	2,700	2,700
		(USA)	700	700	700	700
	Toho	(Japan)	2,900	3,300	3,500	3,700
		(Germany)	800	800	1,800	1,900
	Toray	(France)	800	800	800	800
		(Japan)	2,900	2,900	4,700	4,700
		(USA)	0	0	0	1,800
	Sub total		12,250	14,250	18,750	21,850
	Growth rate (%)		22	16	32	17
Large carbon fiber Tow	Aldila		0	600	1,000	1,000
	Fortafil		1,140	2,100	3,500	3,500
	SGL		200	780	950	1,950
	Toray		0	0	300	300
	Zoltek		450	900	1,350	1,800
	Sub total		1,790	4,380	7,100	8,550
	Growth rate (%)		21	145	62	20
	Grand Total of PAN C/F		14,040	18,630	25,850	30,400
	Growth rate (%)		22	33	39	18

Source: Compiled from various sources.

application categories, namely industrial, sports/leisure and aircraft/space. The market share of all composite (including carbon fiber reinforced) shipments in the USA, split into the various manufacturing sectors is shown in Figure 5.4.

5.4 MANUFACTURING COSTS OF PAN BASED CARBON FIBER

The DOE Office of Transportation Technologies' Partnership for a New Generation of Vehicles (PNGV) initiative through Oak Ridge National Laboratory awarded Hexcel Carbon Fibers a development program to define Low Cost Carbon Fiber (LCCF) production technologies. Two papers have been issued, describing the preliminary stages of this program [25,26].

The specifications defined for this work were:

Tensile Strength >2.8 GPa
Tensile Modulus >172 GPa
Strain to Failure $>1\%$
Quantity $>455,000$ kg per year (10^6 lb per year)
Market Price $= \$6.60$–11 per kg (\$3–5 per lb)

The mechanical properties of this specification are almost exactly the properties of an early type of fiber made by Courtaulds, designated Type A, and very similar to a fiber type made by Rolls Royce. These fibers were very sensitive to surface treatment and were used in the untreated form with brittle resin systems like epoxy novalacs.

Based on a manufacturing cost model proposed by Cohn and Das [27] and assuming the use of a 50k large tow PAN precursor to make 10^6 kg per year of carbon fiber on a single line

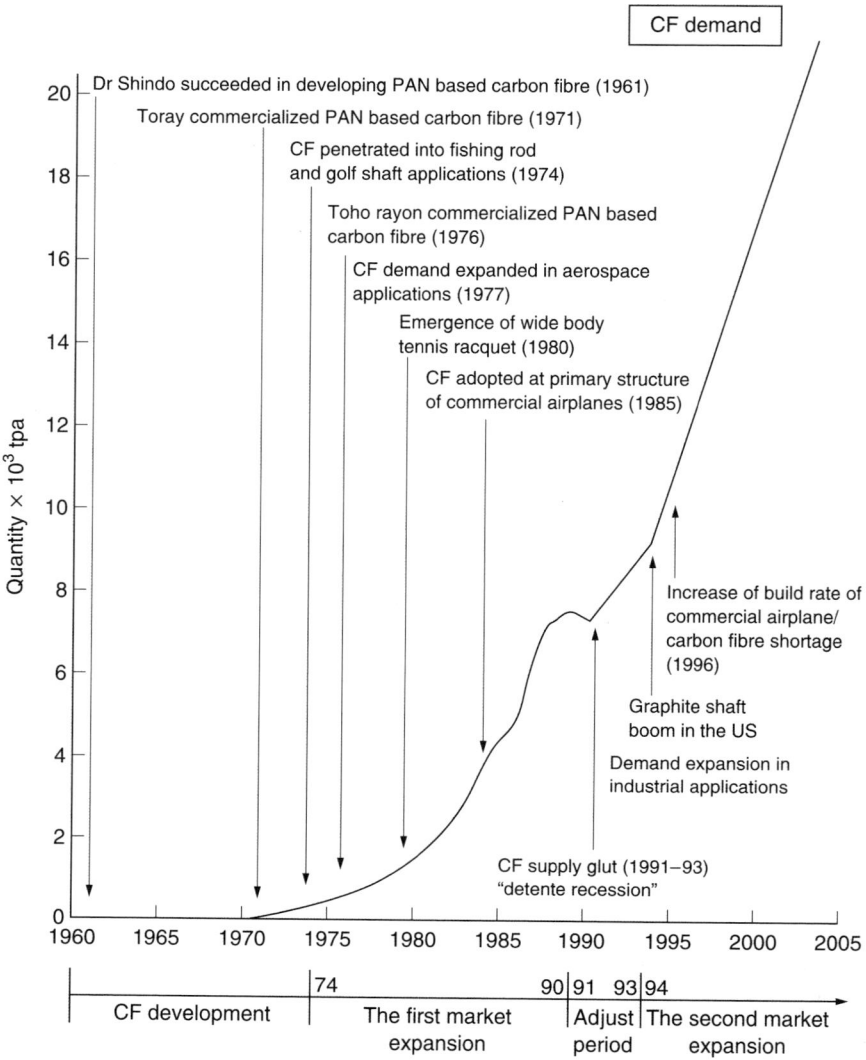

Figure 5.2 Demand for carbon fiber. *Source:* Reprinted with kind permission from Benjamin W, *Composite Market Reports*, Issue 336–3, Nov 1999.

Table 5.2 Estimated World wide production of PAN-based carbon fiber by geographic region (tpa)

Geographic Region	Year 2000 Small tow	Year 2000 Large tow	Year 2005 Small tow	Year 2005 Large tow
North America	4000	2000	5500	4500
Europe	3000	1000	4800	1500
Japan and rest of Asia	4000	1000	5700	2000
Total	11000	4000	16000	8000

Source: Chuck Segal, presented at Intertech Global Outlook for Carbon Fiber Conference, Bordeaux, 16–18 Oct 2001. Copyright 2001, Intertech.

Table 5.3 Estimated worldwide consumption of PAN-based carbon fiber by major market segment (tpa)

Market Segment	Year 2000 Small tow	Year 2000 Large tow	Year 2005 Small tow	Year 2005 Large tow
Aerospace	3400	0	5500	0
Recreation	3500	1000	4000	1000
All Other	4100	3000	6500	7000
Total	11000	4000	16000	8000

Source: Chuck Segal, presented at Intertech Global Outlook for Carbon Fiber Conference, Bordeaux, 16–18 Oct 2001. Copyright 2001, Intertech.

Table 5.4 Estimated worldwide production of PAN-based carbon fiber by potential major market segments (tpa)

Market Segment	Year 2000 Small tow	Year 2000 Large tow	Year 2005 Small tow	Year 2005 Large tow
Electrical	500	2200	500	4500
Infrastructure	1500	200–300	2500	500
Automotive	1000	200	500	1000
Oil & Gas	1000	200–300	2000	500
Miscellaneous	500	200	1000	500
Total	4100	3000	6500	7500

Source: Chuck Segal presented at Intertech Global Outlook for Carbon Fiber Conference, Bordeaux, 16–18 Oct 2001. Copyright 2001, Intertech.

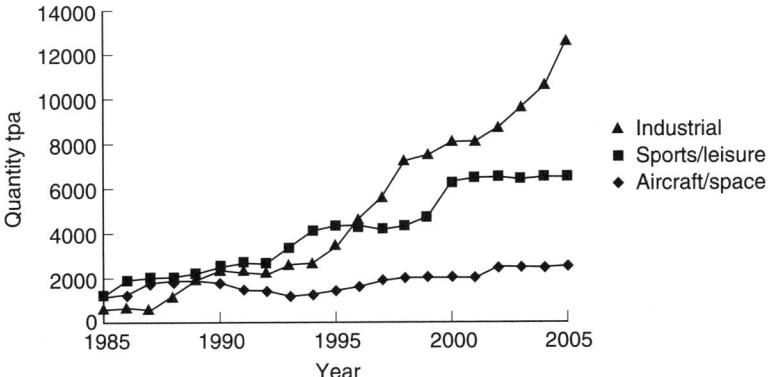

Figure 5.3 Projected worldwide production of carbon fibers. *Source:* Reprinted with permission from Katsumato M, *Keynote Address at Carbon Fibers '98*, San Antonio, Texas. Copyright 1998, Intertech.

handling 212 tows at 180 m/h, the total carbon fiber cost was estimated to be $15.4 per kg ($6.98 per lb). Figure 5.5 shows the breakdown of cost on a weight basis, with no added profit.

The initial conclusions reached in this program were [26]:

1. The man-made fiber industry provides models for high-rate manufacturing processes.
2. Parts of the man-made fiber industry are not profitable and selling prices are often at artificially low levels.
3. Sustainable costs of man-made fiber are unlikely to fall below $1 per kg from any petroleum-based polymer route. Fluctuation in oil price must be continually accepted.

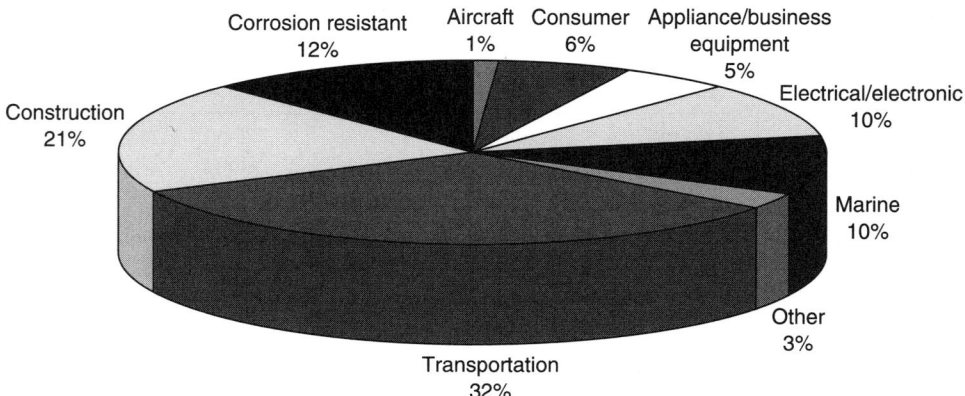

Figure 5.4 US composites shipments for 1998 by market share. *Source:* SPI/Composites Institute.

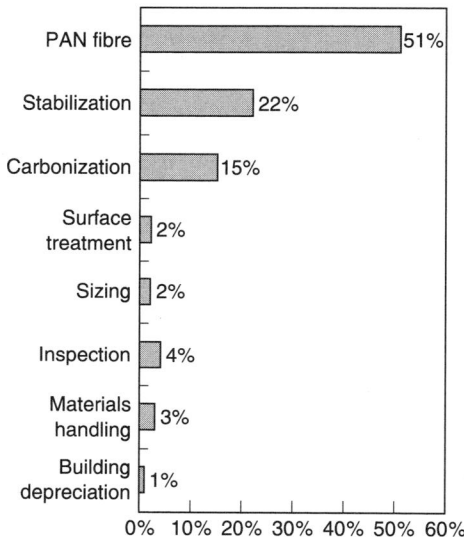

Figure 5.5 Breakdown of costs with no added profit for producing PAN based carbon fiber. *Source:* Reprinted with permission from Leon y Leon CA, O'Brien RA, Dasarathy H, McHugh JJ, Schimpf WC, *Midwest Advanced Materials and Processes Conference*, Dearborn, Sep 12–14, 2000. Copyright, The Society for the Advancement of Material and Process Engineering (SAMPE).

4. It will be difficult to improve upon acrylic textile fiber as a potential filament-based precursor at $1.6 per kg. CF at $11 per kg is likely to be accessible through the use of this material, with appropriate modification, using current best-practice low-cost CF conversion equipment.
5. The scope remaining in lowering the cost by a change of polymer alone is not likely to be more than a further $1–1.50 per kg, assuming the continuous supply of precursor material with guaranteed properties. There is no identified large-scale low cost polymer that is capable of meeting the LCCF property targets at present.
6. If acrylic textile fiber continues to be available and can be successfully converted to LCCF, then thermoformable acrylic polymers will not be favored, unless the sale price of the resin is less than $1 per kg. This may encourage capital investment for new precursor forms.
7. Use of an acrylic polymer in a non-standard roll based product form could lower costs in CF conversion by a larger degree, say below $11 per kg, but end-user acceptance in high-rate forming processes will be critical.

CARBON FIBER PRODUCTION USING A PAN PRECURSOR

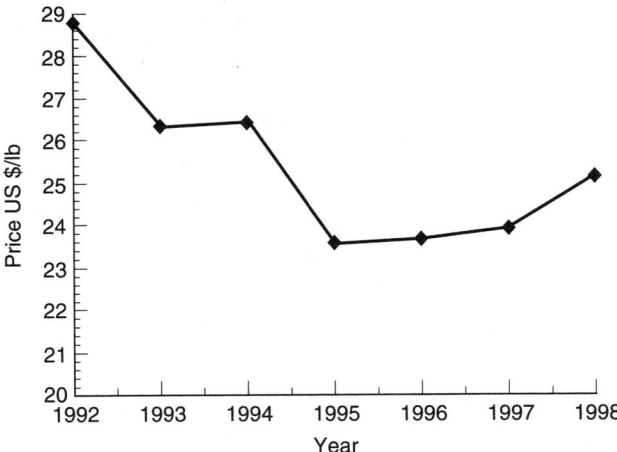

Figure 5.6 Price of carbon fiber over the period 1992–98. *Source:* Based on SACMA information.

8. If a thermoplastic acrylic precursor process is developed, the main technical issues will be overcoming inherent thermoplasticity in stabilization. Chemical treatments or radiation may be helpful here.
9. Non-standard mat or sheet product forms will be less likely to achieve the LCCF mechanical property targets currently sought, as compared to tow-based material, and much optimization will be needed.
10. Use of a non-acrylic polymer in such processes will require a very large development effort, over maybe 5–10 years. There is currently no obvious candidate process today.
11. The prices of carbon fiber, according to SACMA information, for the years 1992–1998 are given in Figure 5.6 and show no tendency to drop below about $23 per kg ($11 per lb).

5.5 CHOICE OF PRECURSOR

Not many textile acrylic fiber production facilities provide a product suitable for use as precursor to make a satisfactory carbon fiber. The major producers of PAN precursor carbon fiber are listed in Table 5.5.

Table 5.5 World producers of PAN carbon fiber precursor

Region	Producer	Product	Availability
Europe	Acordis, UK	Courtelle	Large tow
	Magyar Viscosa Reszventarsursag (Hungarian Rayon Inc.)		Development stage
USA	Hexcel	Exlan (Sumitomo technology)	Small tow
	BP Amoco	Torayca (Toray technology)	Small tow
Japan	Asahi	Cashmillon	Small tow
	Mitsubishi	Pyrofil	Small tow
	Toho Beslon	Besfight	Small tow, developing large tow
	Toray	Toraylon	Small tow
Far East	Formosa Plastics	Fairyfil	
	Tae Kwang, S. Korea	Acelan (Nikkiso technology)	

Many carbon fiber producers have in-house proprietary production capability of PAN precursor material and this includes Hexcel, Mitsubishi, Toho Beslon, Toray and Zoltek. Some companies, such as Toho Beslon, do have more than one type of PAN precursor and this raises the question of whether approval for aerospace should be sought for products, which is quite an expensive and lengthy procedure.

Other carbon fiber producers can obtain their PAN precursor from Acordis UK Ltd., who supply a textile tow precursor, or from Tenax Fibers, who also have a textile tow precursor available to other carbon fiber producers. The polymer formulation and method of processing of each company's precursor material is very specific and will control the final carbon fiber properties.

5.6 DESIRABLE ATTRIBUTES OF A PAN BASED PRECURSOR POLYMER AND ITS SUBSEQUENT PRODUCTION

1. A copolymer must have acrylonitrile content greater than 85%, preferably 90–95%. To the author's knowledge, no commercial exploitation has been made of homopolymer PAN as a precursor for carbon fiber manufacture.
2. The number average molecular weight should be 40,000–70,000 g mole^{-1} preferably, which is about 1000 repeat units.
3. The weight average molecular weight is preferably about 90,000–140,000 and a value of 100,000, with an even distribution, would give a precursor fiber with good mechanical properties. If the molecular weight is too high, then the dope will have a high intrinsic viscosity and would be difficult to filter and, ultimately, spin. Tsai and co workers [28–30] have investigated the effect of molecular weight and uniformity of the cross-section of PAN fiber on carbon fiber properties.
4. A polydispersity index (M_w/M_n) of 1.5–3.0 is preferred.
5. The polymer dope concentration in the spinning solvent is normally about 15% w/w, which controls the effectiveness, efficiency and economics of the spinning process, whilst the temperature controls the spinning and can also control the final cross-sectional shape of the spun filament. The highest dope concentration ensures that lesser solvent has to be recovered per kg of fiber produced, but if too high, the viscosity would make filtration and spinning difficult.
6. Ideally the solvent system should be organic to avoid dissolved metal ion impurities and/or metal ions such as Na and Zn, originating from the salt used as the inorganic solvent e.g. NaSCN or $ZnCl_2$. An organic solvent system is, therefore, preferred and the dope is easier to filter.
7. The choice of comonomers (e.g. acrylic or itaconic acids) can influence the relative ease of processing (Chapter 4). Preferred neutral comonomers for AN would be MA and MMA. A comonomer breaks up the structure and could be termed a 'plasticiser' rendering the polymer more readily soluble in the spinning solvent and improving the quality of spinning.

Only the minimum quantity of a comonomer plasticiser to allow practicable stretching should be used, since in the oxidation and carbonization stages of processing carbon fiber, the presence of a plasticiser is highly undesirable, allowing filament fusion to occur during the oxidation stage, with resultant low carbon fiber strengths, thus reducing the capability of stretching in oxidation and carbonization. The consequence of this latter factor would be the need to use higher furnace temperatures to obtain the requisite modulus, resulting in a change in crystal size with reduced fiber performance, as opposed to increasing the fiber orientation and modulus by the preferred method of effective stretching.

Tsai and Lin [31] have examined the effect of comonomer composition on the properties of PAN precursor and PAN based carbon fibers, whilst Ogawa [32] studied the improvement of MA as comonomer.

8. It is common practice to use more than one comonomer e.g. MA and ITA.
9. Carboxylic acids such as ITA and methacrylic acid (MAA) are very effective comonomers, since they affect the ease of oxidation, exothermicity and carbon yield of the precursor.
10. The optimum level of ITA in an AN/ITA copolymer is chosen to give a carbon fiber with best mechanical properties, a reduced oxidation time and an exotherm intensity conducive with an acceptable carbon yield.
11. AA, MAA and ITA also act as plasticisers and ideally, it should not be necessary to add any additional plasticiser comonomer such as MA but, for textile tow precursor, it is probably necessary to increase the acid comonomer content, or add an additional plasticiser comonomer to facilitate the introduction of crimp, which aids in the packaging of the large textile tow.
12. Freedom from adventitious impurities in the dope by careful selection of raw materials, efficient dope filtration (e.g., the use of membrane filters) and clean room conditions will yield a PAN precursor that will have less faults and hence, produce a carbon fiber with better mechanical properties.
13. A Na free, or low Na, precursor will be easier to process. It is possible to lower the sodium content by washing with an acid, such as HCOOH, prior to oxidation and any remaining traces of acid would breakdown into gaseous components during oxidation.
14. The oxidation process can be speeded up if the precursor radius is reduced to the dimensions of the initial/rapid diffusion boundary, which would probably mean using a precursor of the order of 0.8–0.9 d'tex, giving a carbon fiber with a reduced yield of about 5 μm diameter. Alternatively, a higher d'tex precursor could be used and stretched to meet these conditions albeit with more broken filaments. Bahl *et al* [33] have found that better properties in precursor and carbon fiber are obtained by stretching in N_2 rather than in air (Table 5.6). Bhat *et al* [34] undertook drawing at temperatures higher than 200°C at the onset of stabilization and found that intermediate draw ratios were desirable, whereas excessive drawing produced surface defects, subsequently lowering the strength of the carbon fiber.

A typical stretch applied in the PAN production process is about ×15 at 100°C. A ×2–3 stretch in coagulation helps to align the molecular chains along the fiber axis, raising the YM from about 4 GPA to about 15 GPa.

Smaller diameter precursor filaments will give improved carbon fiber strength properties (Figure 5.7), but these will be at the expense of lower yields. The results follow Weibull's assertion that, by reducing the test volume, there will be a diminution of crack inducing notches of critical length, therefore increasing the strength.

15. The choice of and amount of finish on the precursor can influence the carbon fiber mechanical properties. A further finish can be pre-dried over, or in place of, the normal organic type finish or, alternatively, can be applied prior to the oxidation stage on the carbon fiber plant. Silicones, in conjunction with a good wetting agent, are the best choice, giving protection throughout oxidation and are removed in the LT furnace.
16. It is essential to have strict control of the precursor manufacturing process, with effective quality control to give a consistent uniform product to specification.
17. Good precursor cosmetics will be reflected in the carbon fiber.

Table 5.6 Effect of stretching in air and N_2 on the properties of the drawn PAN and the resultant carbon fibers

Thermal Drawing	Precursor		Carbon Fiber	
	TS GPa	YM GPa	TS GPa	YM GPa
In Air at 230°C	0.51	8.62	1.38	144.8
In N_2 at 220°C	0.57	9.93	2.75	206.9

Source: Reprinted with permission from Bahl OP, Mathur RB, Dhami TL, *Mater Sci Eng*, 73(109), 7937, 1988. Copyright 1988, Elsevier.

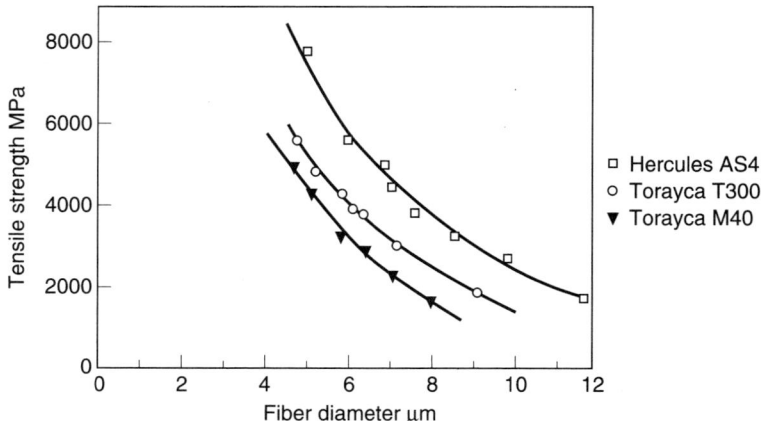

Figure 5.7 Tensile strength of carbon fiber monofilaments as function of fiber diameter. *Source:* Reprinted with permission from Fitzer E, PAN-based carbon fibers-present state and trend of the technology from the viewpoint of possibilities and limits to influence and control the fiber properties by the process parameters, *Carbon*, 27(5), 621–645, 1989. Copyright 1999, Elsevier.

18. These precursor requirements have to be balanced against cost and, the production of larger tows will always be cheaper.

5.7 TYPES OF PAN BASED CARBON FIBER

There are three basic categories of carbon fiber:

1. Large tow: cheaper to manufacture and can be conveniently chopped to a staple form
2. General purpose grades: these have a less stringent product qualification
3. Aerospace grades: these are premium grade products

Each category of carbon fiber is available in several production types allocated by their tensile strength and modulus:

a. General purpose
b. High strength
c. Intermediate modulus
d. High modulus

5.8 A CARBON FIBER PRODUCTION LINE

A carbon fiber production line requires a long workshop. The design of such a plant is described in Chapter 10 and the schematic layout of a typical PAN based carbon fiber plant is shown in Figure 10.1.

5.8.1 Precursor station

To achieve longer production runs, precursor packages will get larger and once a 'one man lift' operation is exceeded, suitable lifting gear must be provided. Creels and improved box handling facilities designed to suit the larger packages, probably with the introduction of robotics for continuous splicing, hence reducing downtime, is required. The golden rule with precursor handling is to have the least number of contact points in order to minimize filament damage.

It is possible that in the future, precursor fiber will be produced on textile beams, but these would be heavy and difficult to manipulate and would necessitate the return of the empty beams to the precursor supplier.

5.8.2 Oxidation

The acrylic precursor is stabilized by controlled low temperature heating (200–300°C) in air to convert the precursor to a form that can be further heat treated without the occurrence of melting or fusion of the fibers. In order to achieve this end, a slow heating rate must be used to avoid run-away exotherms occurring during the stabilization process, exacerbated by the PAN precursor which is a poor conductor of heat.

Stabilization can be achieved isothermally by heating at a constant temperature, but this is time consuming. A more practical method is a stepwise increase in temperature. A third method is a one-step stabilization, with temperature increasing along a tubular furnace [36].

In the oxidation stage, the PAN fiber will increase in density from 1.18 g cm^{-3} to about 1.36–1.38 g cm^{-3} for the oxidized PAN fiber (opf). The actual density will depend on whether the final product is to be used as opf or is required for onward processing to carbon fiber. The final density of an opf product will depend on the opf product specification, whereas for carbon fiber, the density must be at least 1.36 g cm^{-3}, as otherwise, the fiber will tend to pull apart and break on entering the LT furnace. The upper opf density limit for the production of carbon fiber varies with the manufacturer and some manufacturers will use a value as high as 1.40 g cm^{-3}, but others claim that this would produce inferior carbon fiber. The residual exothermicity of SAF heated in air at 230°C (Figure 5.8) after a 3 h treatment, has some 35% exothermic heat remaining in the oxidized fiber.

Warner *et al* [37] kept Courtelle heated in air at 230°C at constant length and measured the tension (Figure 5.9). First, a rapid initial tension developed due to the entropic recovery of the drawn and quenched PAN fiber, which reached a maximum as the T_g is approached (about 140–150°C), followed by a relaxation of the stress down to about the initial stress,

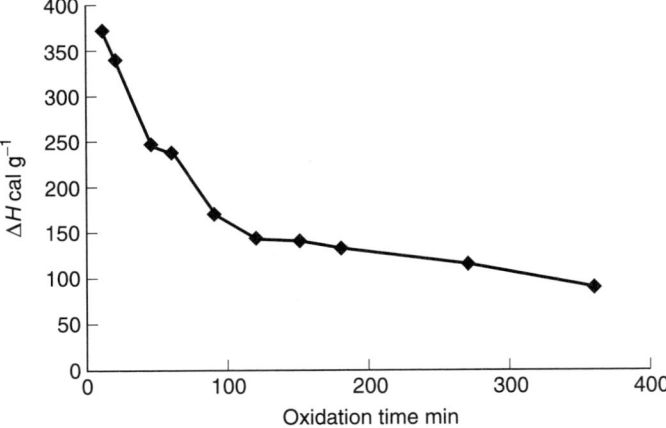

Figure 5.8 Relation between duration of oxidative treatment and residual exothermicity as measured in air with Courtauld's SAF. *Source:* Adapted with kind permission of Acordis UK Ltd (formerly Courtaulds PLC).

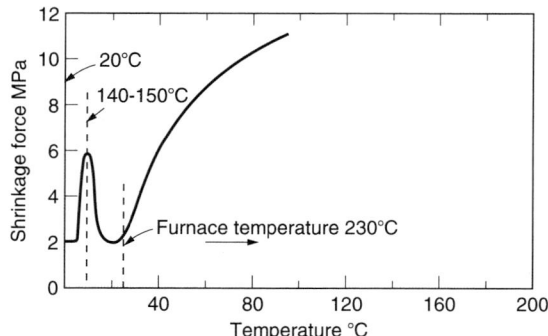

Figure 5.9 Isothermal stress development in air at 230°C of Courtelle, a terpolymer PAN fiber. *Source:* Reprinted with permission from Warner SB, Peebles LH, Jr., Uhlmann DR, *J Mater Sci*, 14, 565, 1979. Copyright 1979, Springer.

succeeded by a slow build-up of stress. This emphasizes the importance of tension control in the oxidation ovens to prevent adjacent passes touching as the fiber expands, or possibly break if too much tension is applied.

Since stabilization is the longest stage in the manufacture of carbon fiber from a PAN precursor, it is not surprising that the study of this subject has aroused considerable effort as evidenced by the literature [38–52].

Ungureanu [53] has chemically modified AN based polymers by thermal treatment in inert and oxidative media.

5.8.3 Oxidation plant

Oxidation is achieved using an oven (Figure 10.5) and passing the fiber through a series of air heated zones, which gradually increase in temperature. The hot air (220–270°C) heats the fiber and provides O_2 for the reaction, besides removing exhaust components and exothermic reaction heat from the fiber. Significant increases in line speeds are achieved by using more than one oven in series to provide additional zones.

In large tow, allowances are made for removal of any crimp in the precursor by taking up any slack in the tow band. In practice and recommended by Jain *et al* [54], oxidation ovens have a series of drives to accommodate variations in shrinkage. Toho Beslon [55] used a system of at least ten 200 mm diameter rollers whose speeds were independently variable.

Since the reaction is strongly exothermic, a uniform temperature distribution must be provided, and noxious gases must be removed in order to prevent the build-up of such gases, which may form an explosive mixture with air. The Threshhold Limit Values (TLV's) for the explosive gases present are HCN (6–41%), CO (12.5–74.2%), NH_3 (15.5–27%) and these gases are continuously displaced throughout the process run using a controlled inlet of fresh air to maintain a safe working atmosphere.

Harwell established that 100 g of Courtelle yielded the products given in Table 5.7. This information enables the concentration of the gases in the oven to be established so that the requisite quantity of air required for dilution can be supplied and an explosive atmosphere can be avoided.

The precursor residence time for a given oxidized fiber density is fixed by the number of passes in the oven(s) using a series of pass back rollers, which can be situated either within the ovens, or outside them. If the passes are vertical, the air flow can be directed along the fiber band, which is preferable, but the pass back rollers then have to be situated within the oven,

Table 5.7 Products formed from 100 g Courtelle

Product	Weight g	Weight g mol	Volume at STP liters	Vol at 230°C liters
HCN	2.16	0.080	1.79	3.30
H_2O	19.6	1.088	24.37	44.89
CO_2	7.5	0.170	3.81	7.02
CO	1.0	0.036	0.81	1.49
NH_3	0.19	0.011	0.25	0.46

Source: Reprinted with permission from Compiled from Bromley J, Gas evolution processes during the formation of carbon fibers, *Int Conf on Carbon Fibers, their Composites and Applications*, The Plastics Institute: London, 1971, Bromley J, Jackson EE, Robinson PS, The carbonization stage of carbon fiber manufacture Part 1: Gas evolution, *United Kingdom Atomic Energy Authority Report*, AERE R6297 Harwell, 1970. Copyright 1970, AEA Technology plc.

as otherwise top outlet slots would permit the ready escape of contaminated hot air by the chimney effect. Unfortunately, internal rollers become hot and can cause interfilament adhesion in the initial stages of oxidation, producing low strength carbon fiber, so cooling the critical rollers becomes necessary.

With horizontal passes, the air flow is directed across the fiber band producing a more harmful effect on the fiber cosmetics. The 320k tows release more exothermic heat, which has to be dissipated by an increased air flow, which in turn could be quite unkind to the smaller tows, but by using variable speed fans, the optimum air flow can be set for each fiber type.

Hughes and Morley [58] were able using a nine stage oxidation unit to reduce the oxidation time to 20 min using Courtauld's 10k SAF. Heine [59], using DSC, showed that a hyperbolic oven temperature profile (Figure 5.10) could be used and could also be simulated in steps, with low temperatures used in the early stages of stabilization and then raised rapidly as the reactants are depleted to obtain a high rate of reaction.

The mass per unit length of fiber exiting an oven is controlled by the ratio of the inlet and exit drive speeds.

Strict control of the exothermic reaction is achieved by maintaining a uniform temperature distribution within the oven of at least ±2°C and avoiding overheating by preventing the fiber from bunching.

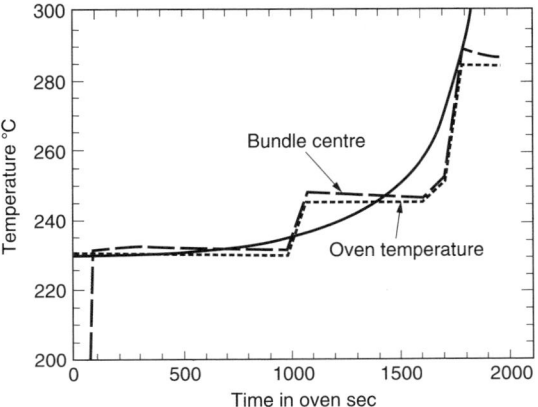

Figure 5.10 Bundle center temperature prediction for oven temperature profile with hyperbolic stages. Note: The smooth curve is the hyperbola approximated by the given temperature ramps indicated by the dashed line. *Source:* Reprinted from Heine M, *PhD Dissertation*, University of Karlsruhe, Germany, 1988.

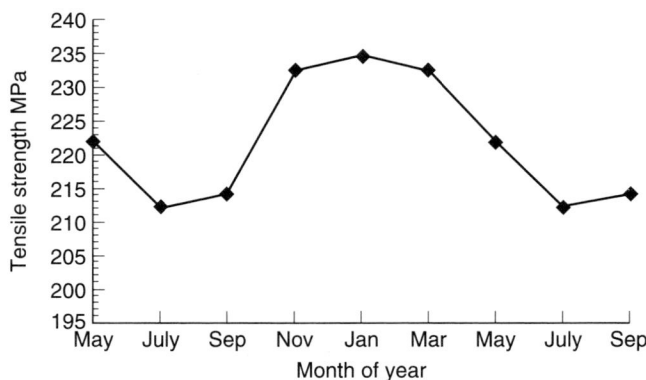

Figure 5.11 Cyclic variation of the tensile strength of oxidized PAN fiber.

Since oxidation ovens require a significant quantity of air, normally drawn from within the building, it is important that a similar quantity of fresh air be drawn into the building from outside to equalize the pressures. This operation is considerably affected by the time of the year when, for example, in summer, more doors and windows would be open to provide additional ventilation. These conditions, if not controlled, can affect the degree of pre-oxidation of PAN fiber (Figure 5.11).

The changes in filament strength, extension and oxidized fiber density that occur throughout a typical oxidation cycle are shown in Figure 5.12.

An alternative approach to oxidation was patented by Toray [60], where the fiber was multi-wrapped, (the number of turns depending on fiber d'tex and line speed), around three pairs of rollers about 50–1000 mm in diameter, with each pair situated one above the other and positioned in a separate cover box compartment to contain the evolved gases. Each pair was progressively staggered laterally and separately driven, to enable the contact time of one pass to be less than 1 s, with one roller slightly canted to ensure self advancing of the fiber around the rollers. The roller surface was heated to 200–400°C and the surrounding air to about 200°C, to give a preferred water uptake of 5–10% by the oxidized fiber. To enable the line to run at the same speed as the precursor line, these heated rollers were controlled at 285, 290 and 305°C and a total contact time of 9.6 min.

Mitsubishi [Nakatani M, Kubayshi T, Imai Y, Yamamoto N, Sasaki S. Process for producing carbon fiber USP 4780,301, 1988] describe a multi oven oxidation treatment for PAN such that fiber density ρn after each oven stage may be maintained as defined by the equation:

$$(\rho_0 - 0.1) + (\rho_k - \rho_0)\frac{\sum_{n=1}^{n} t_n}{\sum_{n=1}^{k} t_n} \leqq \rho_n \leqq (\rho_n + 0.1) + (\rho_k - \rho_0)\frac{\sum_{n=1}^{n} t_n}{\sum_{n=1}^{k} t_n}$$

where
ρ_n = density (g cm^{-3}) of the fiber after the nth treatment stage
ρ_0 = density (g cm^{-3}) of the acrylic precursor fiber
ρ_k = density (g cm^{-3}) of the fiber after completion of pre-oxidation and has a value in the range 1.34–1.40 g cm^{-3}
t_n = the period of the nth stage of pre-oxidation
k = the number of pretreatment stages

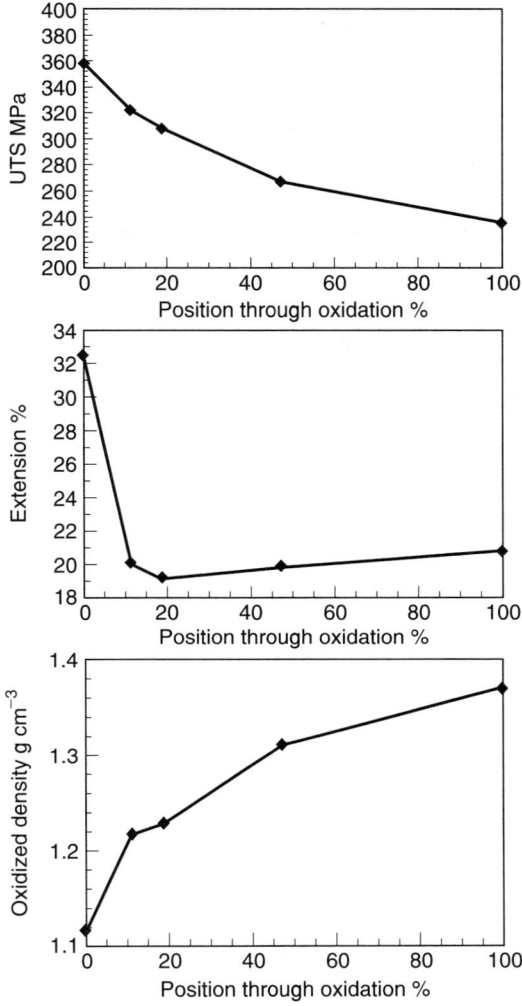

Figure 5.12 Changes in filament strength, extension and oxidized fiber density throughout a typical oxidation cycle.

Toray assessed the efficacy of oxidation in the prevention of biconical hole formation within the carbon fiber during carbonization by mounting the fiber in an epoxy resin, curing, cutting a cross-section with a diamond saw, polishing the end section to 2000 grit, etching the surface with an oxygen plasma for 30 min, coating with a Pd/Pt alloy and examining it through a set of ten SEM photographs, each field displaying about 60–80 filaments. The statistical probability of latent holes (hollow holes with radial cores) detected is calculated. A statistical probability of less than 2% is preferred, although <1% is considered better, while <0.2% is considered ideal, where the fiber has a substantially homogeneous structure with no defects.

In oxidation, there is about a 6–8% weight take up of O_2 [61,62]. Toho, however, suggests that a suitably stabilized fiber will contain 8–12% O_2 [55].

Harwell established that with a Courtelle precursor, approximately 2.5 mol of O_2 entered the fiber for every 1 mol of HCN liberated.

Gases evolved are predominantly NH_3 and HCN, but also include H_2O and CO_2. It should be noted that NH_3 actually retards the reaction, hence the proposed use of acidic components like HCl to react with the NH_3 and speed up the oxidation reaction. However, if temperatures drop below the dew point, severe corrosion of the ovens will ensue in the presence of HCl.

Kiminta [63] investigated the rapid stabilization of acrylic precursors for carbon fibers using NH_3.

5.8.4 Removal of effluent gases evolved in the oxidation process

Gases such as HCN, H_2O, CO_2, CO, NH_3, nitriles and miscellaneous tars and finish are evolved during the oxidation of PAN and it has been the usual practice to pass these hot gases at about 300°C over a heated platinum group metal deposited directly onto a high surface area material coated on a porous ceramic monolith block. The earlier system was rated at a minimum of 95% efficiency, but recent stringent legislation has enforced the use of newly developed proprietary regenerative thermal oxidizers, which are considerably more expensive, but do give a much improved performance.

5.8.5 Oxidized PAN fiber

Oxidized PAN fiber of density 1.37–1.40 $g\,cm^{-3}$ is treated with a water based proprietary antistat finish, dried and collected after plaiting, into boxes positioned on a plaiter table with longitudinal and transverse movements to plait the opf neatly into the boxes without entanglement.

5.8.6 Low temperature carbonization

Some workers have found that the presence of moisture in the opf can reduce the strength of the carbon fiber produced and dry the opf prior to entry into an LT furnace.

The LT furnace can best be described as a tar removal furnace and normally comprises a multizone electrically heated slot furnace (Figure 10.9), purged with N_2 to prevent ingress of air and providing sufficient N_2 flow to remove evolved tars and gases. The temperature in the furnace is gradually increased in the zones to a final temperature of about 950°C, a temperature above which the tars are decomposed leading to the deposition of a sooty product on the fiber, which causes the filaments to stick together and the carbon fiber properties to plummet. The gases evolved during carbonization at temperatures up to 1000°C are listed in Table 5.8, with a brief interpretation of the reactions that occur. Bromley and co-workers [56,57] at Harwell determined the gases evolved during carbonization from 200–1000°C (Figure 5.13) (H_2O, CO_2, NH_3, HCN, H_2, high molecular weight compounds, CO and CH_4). Gas evolution is considerably enhanced if a thermal run-away occurs.

Since the maximum temperature will be 1000°C, it is possible to use a high nickelalloy for the fabrication of the furnace muffle, but the alloy must be carefully chosen to provide adequate strength at operating temperatures and possess adequate resistance to internal and external environments so that it does not corrode.

5.8.7 High temperature carbonization

Basically, the high temperature furnace elevates the temperature in a uniform manner to increase the fiber modulus and a smaller d'tex fiber will give a higher modulus for a given

CARBON FIBER PRODUCTION USING A PAN PRECURSOR

Table 5.8 Carbonization products of oxidized PAN fiber

Temperature °C	Observation	Interpretation
220	HCN evolved and O_2 chemically bonded	Ladder polymer formation and oxidation of polymer
260	Little change. No modulus increase	No chain scission
300	Large CO_2 and H_2O evolution, also CO, HCN and some nitriles. No modulus increase	CO_2 from –COOH groups in oxidized polymer. No cross-linking
400	CO_2, H_2O, CO, HCN and NH_3 evolved. Small evolution of C3 hydrocarbons and nitriles. Modulus increase	Cross-linking by intramolecular H_2O elimination
500	Increased H_2 evolution. Some NH_3 and HCN evolved. Modulus increase	Cross-linking by dehydrogenation
600	Reduced H_2 evolution. HCN and trace N_2	Cross-linking by dehydrogenation
700	N_2, HCN and H_2 evolution. Modulus increase	Cross-linking by dehydrogenation and evolution of N_2
800	Large increase in N_2. H_2 and HCN still evolved. Modulus increase	Cross-linking by evolution of N_2
900	Maximum evolution of N_2, some H_2, traces HCN. Modulus increase.	Cross-linking by N_2 elimination
1000	N_2 evolution decreases to about the same as at 800°C. Trace H_2 evolved. Modulus increase.	Cross-linking by N_2 elimination

Notes: Fiber yield 53.6%, 5.0% nitrogen content.

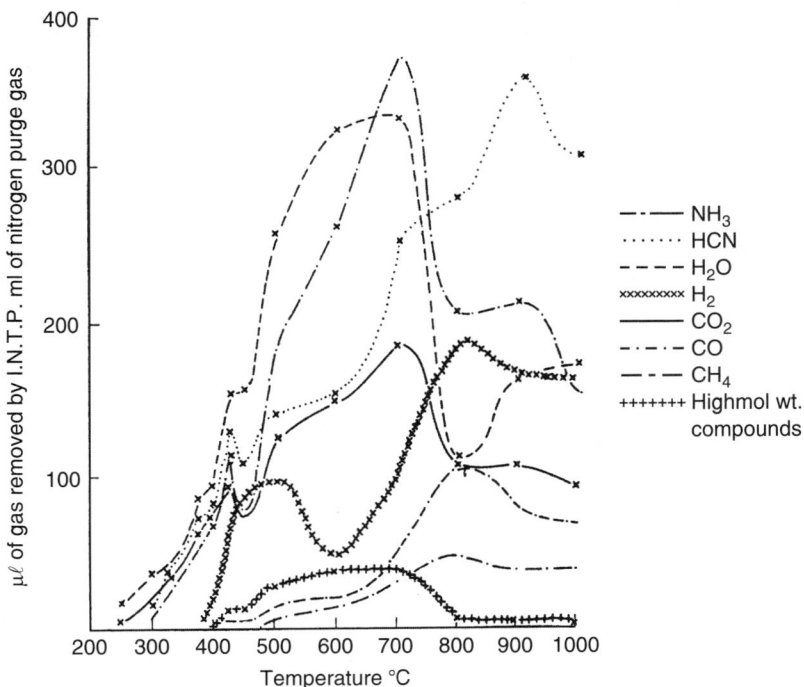

Figure 5.13 Gases evolved during the carbonization of PAN based carbon fiber from 200–1000°C. *Source:* Reprinted with permission from Bromley J, Gas evolution processes during the formation of carbon fibres, *Int Conf on Carbon Fibres, their Composites and Applications*, The Plastics Institute: London, 1971. Bromley J, Jackson EE, Robinson PS, *United Kingdom Atomic Energy Authority Report*, AERE R6297 Harwell, 1970. Copyright 1970, AEA Technology plc.

residence time, but lesser yield. The product formed during carbonization is a good conductor and imposes no limitation on the heating rate by heat transfer [11]. Heating rates above 20°C per minute and temperatures above 1500°C will impair the strength of the resulting carbon fibers.

The presence of Na in the precursor does, however, pose problems in the HT furnace, either by forming elemental Na, or combining with the HCN released in the carbonization process to form NaCN. Consequently, a precursor with little or no Na is preferred.

To prevent ingress of air, a flow of inert gas is introduced at either end of the furnace to produce a gas seal. A body flow of inert gas should be applied at the inlet end to remove any liberated Na via an outlet branch pipe, but the flow should be controlled to maintain an optimum concentration of HCN within the muffle, since the HCN has been found to have the beneficial effect of healing surface flaws on the fiber by cracking to carbon. Sufficient body flow at the outlet end is applied to ensure that waste products exit via the outlet branch pipe.

Generally, N_2 is the preferred gas but Ar can be used, although it is some eight times more expensive, but it does give a carbon fiber with an improved strength, presumably due to the higher density and viscosity of argon.

5.8.8 High modulus fiber production

The high modulus furnace operates at about 2500°C and employs a graphite muffle operating in a carefully controlled inert atmosphere. The conditions must be quiescent, since even low volume gas flow over the hot graphite element is sufficient to continually remove a molecular layer of graphite and cause severe erosion leading to premature failure. Figure 5.14 shows the evaporation rate of graphite in the temperature range 2200–2900°C and it has been seen that a rise of 100°C will give a three-fold increase in evaporation rate.

The added heat treatment provides an improvement in the orientation of the graphite crystallites, giving carbon fibers with a high YM.

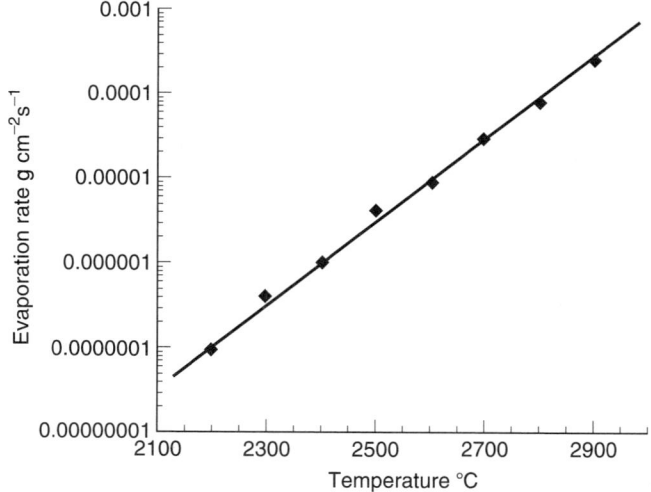

Figure 5.14 Evaporation rate for graphite in the temperature range 2200–2900°C. *Source:* Reprinted from Marmer *et al.* with kind permission of Acordis UK Ltd (formerly Courtaulds PLC).

5.8.9 Shrinkage during the carbon fiber process

Line shrinkage in the carbonization stage is, nominally of the order 5–8%, and determined thus:

$$\% \text{ Shrinkage} = \frac{\text{Drive speed (Exit oxidation)} - \text{Drive speed (collection)}}{\text{Drive speed (Exit oxidation)}}$$

A low shrinkage value signifies a high fiber tension. If the tension is too high, then the fiber will break, and if it is too low, then the increased fiber catenary will permit the fiber to drag along the furnace floors, degrading the fiber cosmetics.

Bahl and Mathur [64] established that for a $6k$ 1.5 denier Toho Beslon PAN (with 3% MA) oxidized for 100 min at 240°C and subsequently heat treated to 1000°C, the best carbon fiber properties were attained with a shrinkage of about 7%. A Toho Beslon [55] patent, however, suggests a shrinkage of 40–70%, whilst a Japan Exlan [65] patent quotes 0–50%.

5.8.10 Surface treatment

It is convenient to use an electrolytic surface treatment process, which permits a good measure of control to be exercised. A suitable water soluble electrolyte is chosen, giving a solution that is readily conductive. The carbon fiber is made the anode and passed close and parallel to graphite cathodes without touching. Faraday's Law applies and 96,500 coulomb will liberate 1 gram equivalent of O_2. The time factor introduces the line speed and it is normal to quote treatment level as the current passing through a unit length of fiber during treatment, usually expressed as $C\,m^{-1}$.

The current requirements for the surface treatment of HM fiber are relatively high and there is an appreciable heating effect, which necessitates cooling of the electrolyte in order to control the rate of treatment. Also, with some electrolytes, cooling is essential to avoid thermal decomposition.

5.8.11 Sizing

Initially solvent based sizes were used which gave excellent inter-filament penetration, but were discontinued on health and cost grounds. Invariably, the present sizes are water based emulsions, preferably using the same chemical class as the ultimate polymer matrix.

5.8.12 Collection

Small tows are best collected using online winders. Winding machines are available that will doff automatically. Large tows are generally plaited into cardboard boxes with the boxes positioned on a plaiter table, smaller tows ($50k$) can be collected in tubs.

5.9 FINE STRUCTURE AND TEXTURE OF PAN BASED CARBON FIBERS

Carbon fibers consist mainly of polyaromatic carbons and effectively exist as structurally disordered (turbostratic) carbon. The crystal structure of a true graphite with a distance of

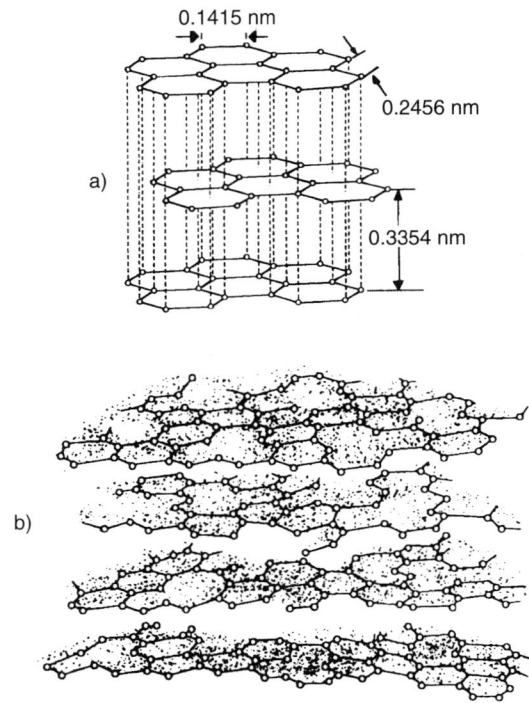

Figure 5.15 a. Crystal structure of graphite crystal. b. Structure of turbostratic carbon. *Source:* Reprinted from Hoffman WP, Hurley WC, Liu PM, Owens TW, The surface topology of non-shear treated pitch and PAN carbon fibers as viewed by STM, *J Mater Res*, 6, 1685, 1991.

0.3354 nm between the layer planes and the characteristic form of a turbostratic carbon with a distance of 0.34 nm between the layer planes are shown in Figure 5.15 a and b respectively. The interlayer distance in turbostratic carbon is always greater than graphite due to the presence of sp^3 bonds. Typical commercially available PAN-based carbon fibers have a layer distance about 0.355 nm. The mean interlayer spacing of PAN-based and MPP-based carbon fibers have been compared by Fitzer [15] with a well graphitizing coke when heat treated at different temperatures (Figure 5.16).

Wide angle X-ray diffraction provides the values of the characterization parameters of PAN-based carbon fibers, and are shown in Table 5.9 [67].

A schematic representation of the development of the layer plane structure by Bennett [68], based on TEM studies, using phase contrast techniques, of longitudinal sections of carbon fibers with varying degrees of heat treatment, is shown in Figure 5.17.

Johnson and Tyson [69], from wide-angle X-ray diffraction evidence, envisaged a structure similar to that depicted for the PAN molecule (Figure 4.15), with ordered zones interspersed with disordered regions. The structure of a carbon fiber ribbon was believed to be a columnar arrangement of misoriented turbostratic graphite crystallites (Figure 5.18). The idealized tetragonal crystallites are stacked above one another, with slight misorientation between the crystals in the direction of the fiber axis, trapping sharp needle-like voids, where the boundaries between the stacks represent the disordered regions. In the core region, extensive folding could occur and Johnson [70] suggested that misoriented crystallites interlink with other oriented and misoriented crystallites, as depicted in Figure 5.19.

Johnson *et al* [71] advanced a block model (Figure 5.20) to demonstrate the interlinked crystallinity of HM carbon fiber showing tilt, twist boundaries, porosity and overlapping boundaries.

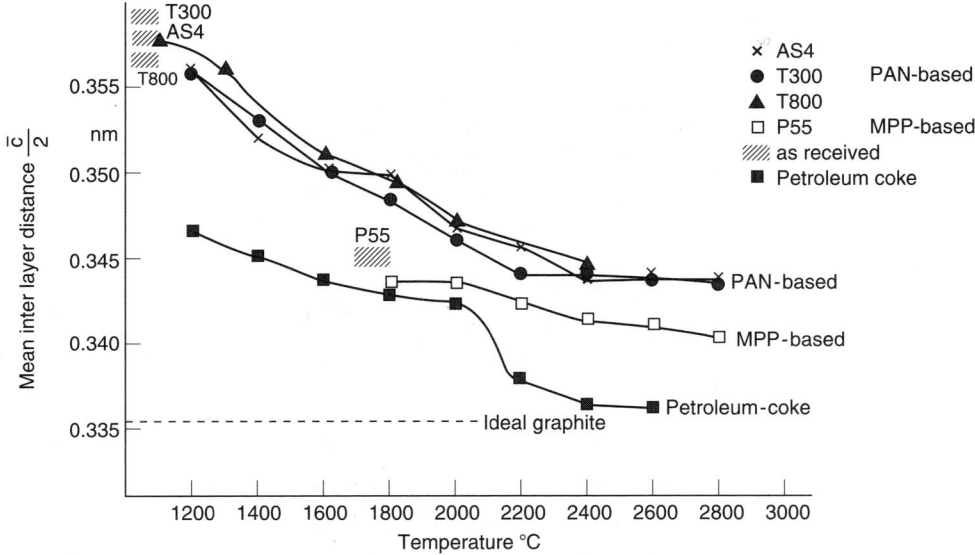

Figure 5.16 The mean interlayer spacing of PAN based and MPP based carbon fibers in comparison with a well graphitizing coke. *Source:* Reprinted with permission from Fitzer E, PAN-based carbon fibers-present state and trend of the technology from the viewpoint of possibilities and limits to influence and control the fiber properties by the process parameters, *Carbon*, 27(5), 621–645, 1989. Copyright 1989, Elsevier.

Table 5.9 Characterization parameters of typical PAN based carbon fibers

Fiber type	Interlayer spacing $c/2$ nm	Stacking size L_c nm	Apparent crystallite length $L_{a\|}$ nm	Apparent crystallite width $L_{a\perp}$ nm	Orientation Z deg
HM	0.340	5.9	9.8	8.0	20.1
IM	0.345	1.9	4.8	7.2	29.8
HS (Toray T1000)	0.348	1.7	2.9	5.2	31.5

Source: Reprinted with permission from Johnson DJ, *Handbook of Polymer-Fibre Composites*, Jones ER ed., Longman Scientific and Technical, Harlow, 24–29, 1994. Copyright 1994, Pearson Education Ltd.

Barnet and Norr [72], using a three dimensional model (Figure 5.21) have shown that the lamellae of the graphite layers in a HM fiber are arranged in an ordered manner in the sheath region, but are more disordered towards the center.

A model was advanced by Bennett [68] (Figure 5.22) and Bennett and Johnson [73] put forward another three-dimensional model (Figure 5.23) depicting the chaotic combination of basic structural units into microdomains containing pores.

Knibbs [74], using polarized light microscopy studied the effect of oxidation of the precursor upon the orientation in the fiber cross-section and found that for limited oxidation, there was little orientation in the graphitized fiber and the center was completely isotropic. He found that with increasing degrees of oxidation, the outer zone formed an oriented layer with the *c*-axes radially oriented corresponding to the lamellar phase. At this stage, the center remains isotropic. With further oxidation, the *c*-axes of the outer zone remained radially oriented, while the inner zone became ordered and its *c*-axes took on a circumferential orientation. After complete oxidation of the fiber, the graphitized material is completely ordered, with all the crystallites adopting the same orientation for the *c*-axes, which are all

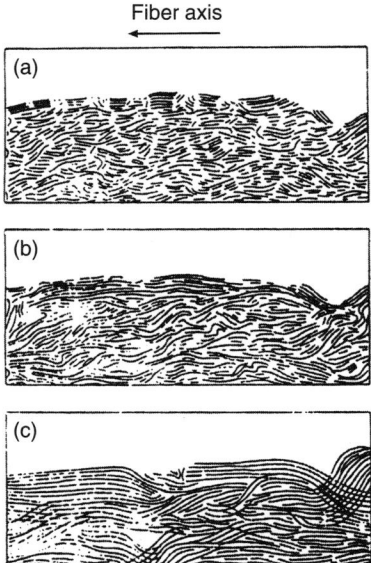

Figure 5.17 Schematic representation of the development of a layer-plane structure from TEM studies. (a) 1000°C; (b) 1500°C; (c) 2500°C. *Source:* Reprinted from Bennett SC, *Strength structure relationships in carbon fibres*, PhD Thesis, University of Leeds, 1976.

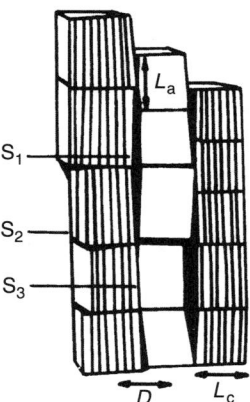

Figure 5.18 Schematic of an idealized diagram for the structure of carbon fiber summarized from X-ray diffraction evidence. S_1, void; S_2, subgrain twist boundary; S_3, intercrystalline boundary. L_c and L_a are thickness and diameters of carbon layer stacks and D the distance between them. *Source:* Reprinted from Johnson DJ, Tyson CN, The fine structure of graphitized fibres, *Brit J Appl Phys (J Phys D)*, 2(2), 787–795, 1969.

radially disposed. (Figure 5.24). Once the lamellae have formed, the carbon between the lamellar zones is subjected to a radial tensile force, since the fiber has to shrink.

This layer plane orientation was supported by Rose's findings [75], although Johnson [76] later interpreted these findings as being the result of strain birefringence caused by residual stresses.

The basic structural unit (BSU) of carbon fibers has been described by Fourdeaux *et al* [77] as a two-dimensional ribbon shaped graphitic layer (Figure 5.25) with no correlation between the directions of the layer borders and direction of the *x*-axis fibers. The ribbons are about 6 nm wide with a length of several hundred nm. Some of these ribbons run parallel to

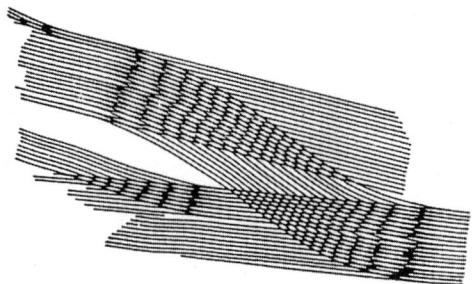

Figure 5.19 Interlinked structure and resulting void in an idealized structure of carbon fiber. *Source:* Reprinted from Johnson DJ, Structure property relationships in carbon fibres, *J Phys D Appl Phys*, 20(3), 285–291, 1987.

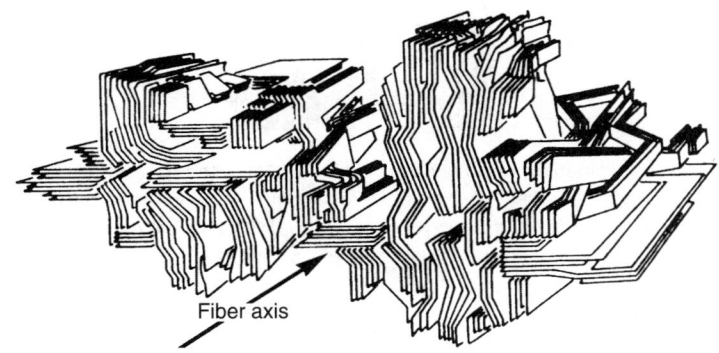

Figure 5.20 Proposed three dimensional structural model of HM carbon fiber depicting interlinked crystallinity. *Source:* Reprinted with permission from Johnson DJ, Crawford D, Oates C, The fine structure of a range of PAN-based carbon fibres, *Extended Abstracts 10th Biennial Carbon Conference*, Bethlehem, PA, 29, 1971. Copyright 1971, American Chemical Society.

form a microfibril with a preferred orientation parallel to the fiber axis. These microfibrils are wrinkled with imperfect packing trapping voids at the boundaries of the microfibrils (Figures 5.26 and 5.27). The voids are about 20–30 nm long and thin (about 1–2 nm across between the microfibrils) and follow the direction of the straighter sections of the ribbons. Fourdeaux and co-workers [77] state that there is a correlation between the size and the perfection of the stacking of the layers, and the orientation of the layer normal to the fiber axis, as shown in Figure 5.27.

Diefendorf and Tokarski [78] subscribe to the view of a wrinkled ribbon of carbon fiber with modulus about 275 GPa, where the ribbons are about 13 layers thick, 4 nm wide and at least a few microns long. For fibers with a modulus of about 750 GPa, the ribbons are about 30 layers thick and 9 nm wide, with almost zero amplitude and essentially parallel to the fiber axis. Figure 5.28 gives a three dimensional view of the ribbon structure with a model shown in Figure 5.29.

Guigon *et al* [79] have defined the BSU shown in Figure 5.30 for a PAN-based high modulus fibers and state that it is very different from the small BSU (<1 nm) found for high strength carbon fibers on the following points:

1. At the molecular level, the BSU is a stack of N carbon layers nearly isometric in shape (with diameter L_a (20–70 nm) and thickness L_c), but bent and rolled around the fiber axis AA′,

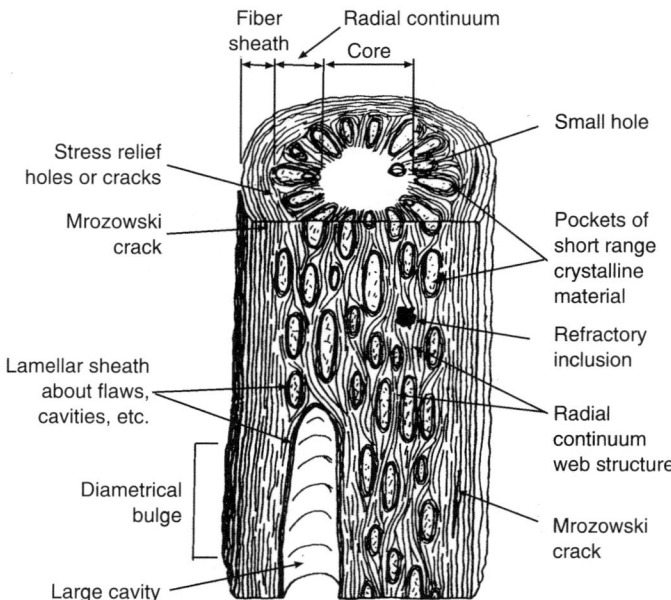

Figure 5.21 A probable three dimensional structural model for a PAN based HM carbon fiber. *Source:* Reprinted with permission from Barnett FR, Norr MK, *Proceedings of the International Conference on Carbon Fibres, their Composites and Applications*, London (Plastics Institute), 32, 1974. Copyright 1974, Maney Publishing (who administers the copyright on behalf of IOM Communications Ltd., a wholly owned subsidiary of the Institute of Materials, Minerals & Mining).

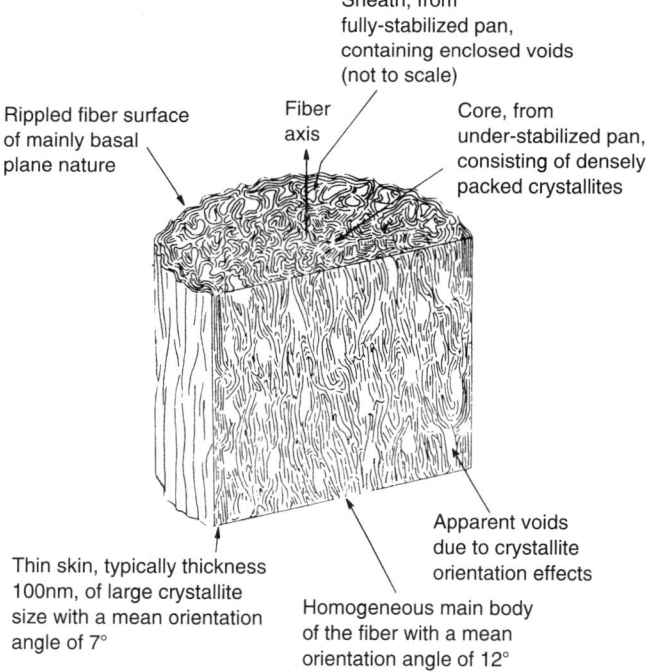

Figure 5.22 Schematic representation of a proposed three dimensional structure of a PAN based HM carbon fiber. *Source:* Reprinted from Bennett SC, *Strength structure relationships in carbon fibres*, PhD Thesis, University of Leeds, 1976.

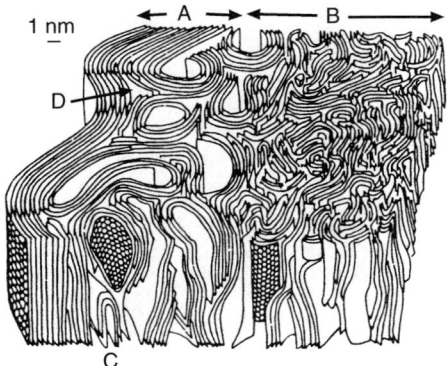

Figure 5.23 A schematic microstructure of PAN based carbon fiber depicting combination of basic structural units into microdomains. A, Skin region; B, Core region; C, A hairpin defect; D, A wedge disclination. *Source:* Reprinted with permission from Bennett SC, Johnson DJ, Strength structure relationships in PAN-based carbon fibres, *5th London International Carbon and Graphite Conference*, Soc Chem Ind, Lond, 377, 1978. Copyright 1978, The Society of Chemical Industry.

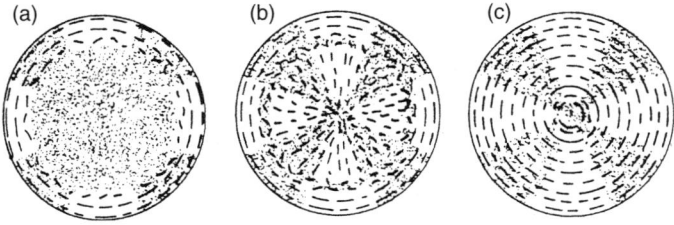

Figure 5.24 Schematic representation of carbon fiber structures obtained from Courtelle precursor. (a) Isotropic center—with an outside skin of oriented crystalline material. (b) Double cross—with the outside showing a different orientation to that of the center. (c) Single cross—where the complete fiber shows one type of preferred orientation. *Source:* Reprinted from Knibbs RH, The use of polarized light microscopy in examining the structure of carbon fibres, *J Microscopy*, 94(3), 273–281, 1971.

Figure 5.25 Sketch of a typical graphene plane in carbon fiber showing vacancy cluster defects. *Source:* Reprinted with permission from Fourdeaux A, Perret R, Ruland W, General structural features of carbon fibres, *Proceedings of the International Conference on Carbon Fibres, their Composites and Applications*, London, Plastics and Polymer Conf Supplement, 57–67, 1971. Copyright 1971 Maney Publishing (who administers the copyright on behalf of IOM Communications Ltd., a wholly owned subsidiary of the Institute of Materials, Minerals & Mining).

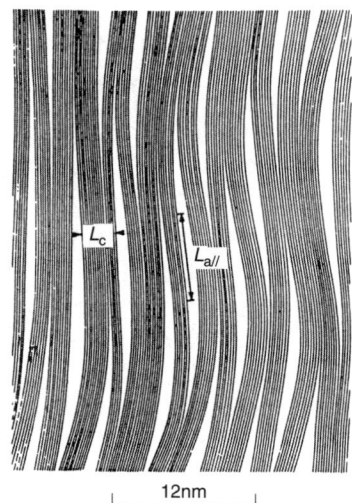

Figure 5.26 Schematic representation of the cross section of a PAN based carbon fiber along the axis direction showing in-plane (L_a) and c-axis (L_c) structural coherence lengths with voids occurring at the boundaries of the microfibrils. *Source:* Reprinted with permission from Fourdeaux A, Perret R, Ruland W, General structural features of carbon fibres, *Proceedings of the International Conference on Carbon Fibres, their Composites and Applications*, London, Plastics and Polymer Conf Supplement, 57–67, 1971. Copyright 1971 Maney Publishing (who administers the copyright on behalf of IOM Communications Ltd., a wholly owned subsidiary of the Institute of Materials, Minerals & Mining).

Figure 5.27 Schematic representation of the correlation between size of the layer stacks (L_c) and orientation of the stacks with respect to the fiber axis. $L_c\ (\phi = 0)$ and $L_c\ (\phi \neq 0)$. *Source:* Reprinted with permission from Fourdeaux A, Perret R, Ruland W, General structural features of carbon fibres, *Proceedings of the International Conference on Carbon Fibres, their Composites and Applications*, London, Plastics and Polymer Conf Supplement, 57–67, 1971. Copyright 1971 Maney Publishing (who administers the copyright on behalf of IOM Communications Ltd., a wholly owned subsidiary of the Institute of Materials, Minerals & Mining).

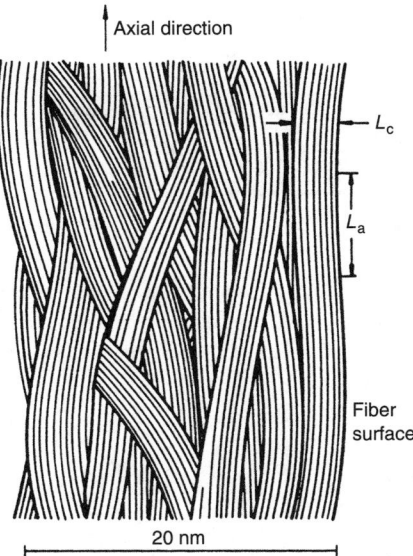

Figure 5.28 Schematic diagram of ribbon structure model for carbon fibers. *Source:* Reprinted with permission from Diefendorf RJ, Tokarsky E, High performance carbon fibers, *Polym Eng Sci*, 15(3), 150–159, 1975. Copyright 1975, Society of Plastics Engineers.

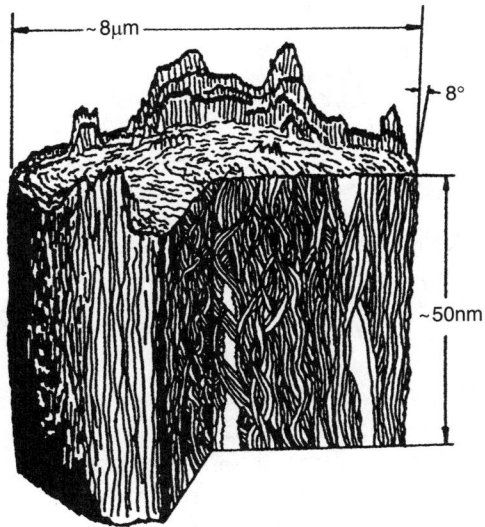

Figure 5.29 Schematic structural model of PAN based Fortafil 5-Y carbon fiber. *Source:* Reprinted with permission from Diefendorf RJ, Tokarsky E, High performance carbon fibers, *Polym Eng Sci*, 15(3), 150–159, 1975. Copyright 1975, Society of Plastics Engineers.

characterized by the transverse radius of curvature r_t (2–13 nm) and the extent of the fold, where a small r_t is associated with an accentuated fold.
2. Microdomains are the next level, with a microtexture up to several hundred nm thick and elongated up to several μm along the fiber axis, limited laterally by pores and dislocations.
3. The highest level of characterization defines the fiber texture. It reflects the changing statistical level of the molecules at a long range.

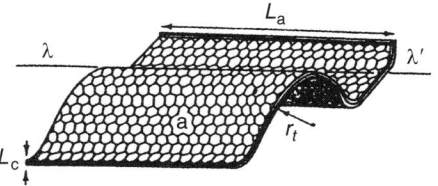

Figure 5.30 The basic structural unit of PAN based HM carbon fiber. *Source:* Reprinted with permission from Guigon M, Oberlin A, Desarmot G, Microtexture and structure of some high modulus PAN-based carbon fibers, *Fiber Sci Technol*, 20, 177–198, 1984. Copyright 1984, Elsevier.

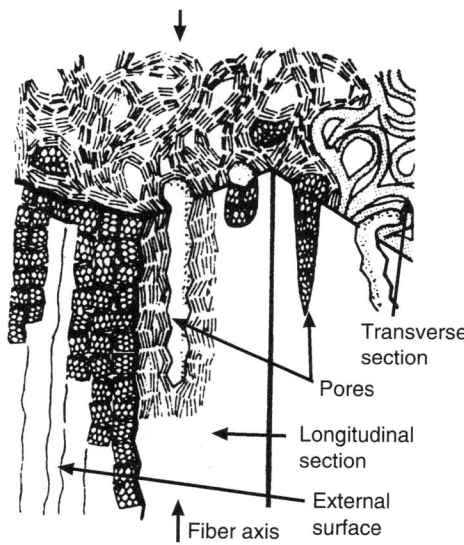

Figure 5.31 A schematic structure of a high strength PAN based carbon fiber. *Source:* Reprinted with permission from Guigon M, Oberlin A, Desarmot G, Microtexture and structure of some high modulus PAN-based carbon fibers, *Fiber Sci Technol*, 20, 177–198, 1984. Copyright 1984, Elsevier.

Microfibrils are also present, with dimensions between a BSU and the texture of fractured carbon fibers observed in micrographs. It is probable that the fine structure within the micofibril controls the compressive strength failure mechanisms.

Guigon *et al* [79] have provided artists' impressions of high strength (Figure 5.31) and high modulus (Figure 5.32) PAN based carbon fibers. The ribbon strips of the carbon crystallite are aligned along the fiber axis, akin to entangled crumpled sheets of newspaper, resulting in irregular pores that are elongated parallel to the fiber axis. The model of HM fiber shows folded and crumpled layers as well as an apparent skin and core effect. Considering the surface of the model (section between double arrows), there are many layers parallel to the fiber surface since r_t is large. As r_t gets larger, the skin becomes more visible since large pore walls tend to be parallel to the surface. Unfortunately, the authors could not provide a true representation since the interlayer spacing is very small (<0.4 nm), whilst the fiber radius is very large (≈4 μm). Hence, the progressive decrease of r_t from the surface to the core cannot be truly represented on the same diagram. This difference in r_t is probably due to release of stresses near the surface, whereas stresses in the core tend to accumulate.

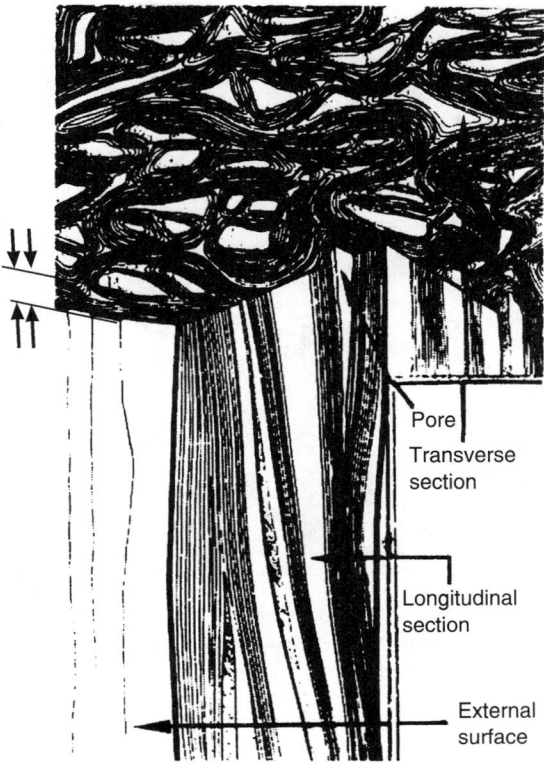

Figure 5.32 Schematic structure of a HM PAN based carbon fiber. *Source:* Reprinted with permission from Guigon M, Oberlin A, Desarmot G, Microtexture and structure of some high modulus PAN-based carbon fibers, *Fiber Sci Technol*, 20, 177–198, 1984. Copyright 1984, Elsevier.

The wrinkled layers of small BSU (<1 nm), associated edge to edge in the high strength fibers, become de-wrinkled in the high modulus fibers.

In high strength fibers, the BSU boundaries are held together by the interlinked structure shown in Figure 5.19.

Figure 5.33 shows a TEM of a longitudinal thin section of PAN based HM carbon fiber taken near the fiber surface. The transverse section (Figure 5.34), which is more difficult to

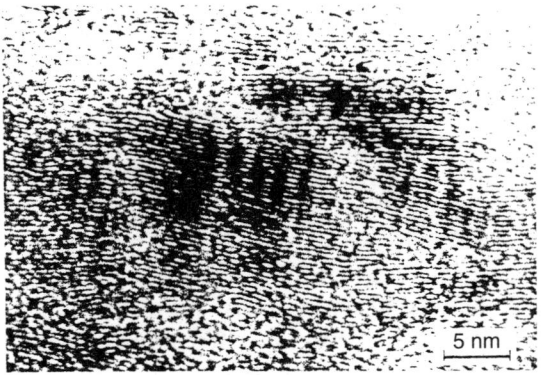

Figure 5.33 Longitudinal section of PAN based HM carbon fiber obtained by TEM. *Source:* Reprinted from Morita K, Murata Y, Ishitani A, Murayama K, Nakajima A, Characterization of commercially available PAN (polyacrylonitrile)-based carbon fibers, *Pure Appl Chem*, 58(3), 455–468, 1986.

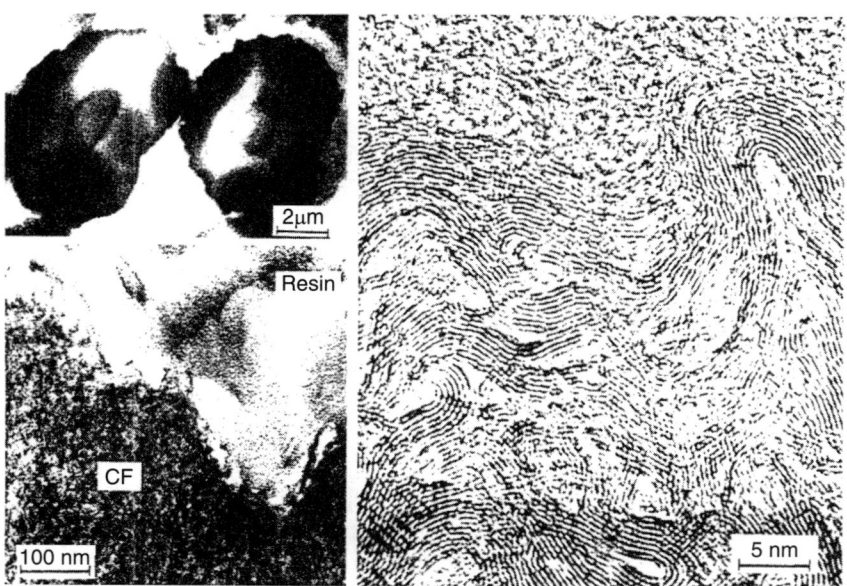

Figure 5.34 Transverse section of PAN based HM carbon fiber obtained by TEM. *Source:* Reprinted from Morita K, Murata Y, Ishitani A, Murayama K, Nakajima A, Characterization of commercially available PAN (polyacrylonitrile)-based carbon fibers, *Pure Appl Chem*, 58(3), 455–468, 1986.

obtain, was prepared by ion milling the resin embedded sample. A fine structure of a boundary and the orientation of small graphite crystallites is also shown.

Although turbostratic graphite predominates, various workers have recorded exceptions and reported evidence of three-dimensional graphite. Shindo [1], in his original work, found evidence of three-dimensional ordering in turbostratic crystallites of fibers carbonized at 2500°C and 3000°C. However, Watt and Johnson found no three-dimensional graphite in their work. Using dark field electron microscopy, Fourdeaux *et al* [77,81] found, in highly oriented carbon fibers heat treated at 2900°C, faint but visible domains with AB ordering about 10–40 nm long (limited to the straight part of the microfibrils) occurring at irregular intervals in the microfibrils and involving some 3–7 layers. Johnson and Tyson [69] found turbostratic graphite in close association with more perfect graphite. Wicks and Coyle [82] used electron diffraction to examine the taper of carbon fiber etched in a flame to a pencil point in order to study the layer plane orientation. Assuming that the etching of the fiber was even, that the surface after etching represented the structure before etching and that the flame treatment did not cause crystal growth or layer plane ordering, it can be stated that the layer plane orientation decreased and the measured crystallite size L_c increased towards the center of the fiber. The surface layers of fiber heated to 2500°C showed three-dimensional graphite and platelets were also found in the body of the fibers. The authors concluded that there was a sheath of circumferentially oriented layer planes enclosing a core of either radial, or random orientation. Hence there was a duplex structure which might contain three-dimensional graphite. The development of these core sheath structures is thought to be due to the extent of stabilization [83] and depends on the treatment temperature [84]. Kowbel *et al* [85] found that the presence or absence of three-dimensional graphite appears to be dependent on whether the fibers are graphitized in the free state or in the presence of a carbon-carbon composite.

Later work by Bennett and Johnson [86] showed that some HM fibers can have skin-core heterogeneity up to about 0.5 μm thick and this is not to be confused with the much thicker

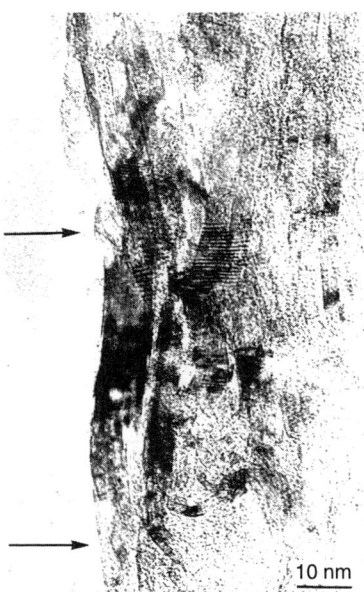

Figure 5.35 Lattice fringe image from a longitudinal section of a HM PAN based carbon fiber depicting a skin region. The fiber surface is indicated by the arrows. *Source:* Reprinted with permission from Johnson DJ, Structural studies of PAN-based carbon fibers, Thrower PA ed., *Chemistry and Physics of Carbon*, Vol 20, Marcel Dekker, New York, 1–58, 1987. Copyright 1987, CRC Press, Boca Raton, Florida.

skin core due to fibers that have not been fully stabilized. Dark field and lattice fringe images and electron diffraction patterns showed (Figure 5.35) that the skin had larger and better oriented crystallites than the core.

Johnson [8] summarized that the crystalline units in high modulus PAN based carbon fibers are elongated, threadlike and fibrillar. The structure was formed from interlocking highly curved sheets of layer planes grouped into microfibrils which were highly oriented to the fiber longitudinal axis. These sheets, however, have no preferred orientation in the cross-section due to their high curvature, but when heated to 2500°C have a thin surface skin of highly ordered graphite.

Fibrils load at different rates and the stress/strain curve increases in slope with increasing strain e.g. YM at 1.3% strain can be over 10% greater than value at 0.3% strain.

5.10 ASPECTS OF STABILIZATION

There are several methods used in practice to measure the degree of oxidation to establish the optimum conditions for stabilization:

1. Measure O_2 content
   Unfortunately this value, determined by elemental analysis, is often established by difference, although Watt and Johnson [88] measured the uptake of O_2 with time of a 3 denier Courtelle at 220°C directly using a Coleman O_2 analyzer (see Figure 3.5). Later, Watt and Johnson [89] determined the O_2 uptake of 1.5 denier Courtelle at 230°C (Figure 5.36). Examination of cross-sections showed a skin core effect after 2 and 4 h of treatment.
   Toho Beslon [55] consider a value of 5–15% $^w/_w$, with 8–12% $^w/_w$ as preferable. They note that a homopolymer PAN contains 0% O_2, whilst an acrylonitrile copolymer is theoretically less

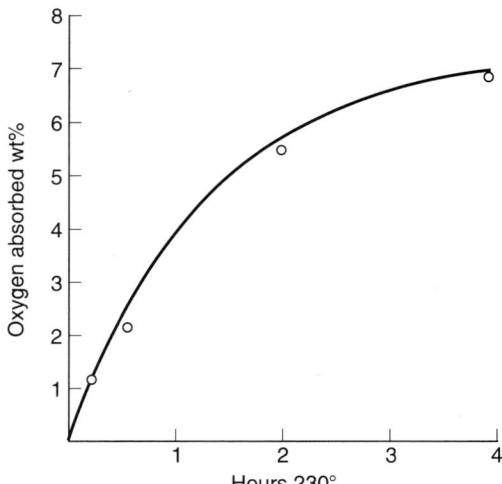

Figure 5.36 Oxygen uptake vs time of 1.5 denier Courtelle in air at 230°C. *Source:* Adapted from Watt W, Johnson W, *Nature*, 257, 210, 1975.

than 3%. Maximum saturation with O_2 gives a value of at least 20%, but values above 15% resulted in carbon fibers with reduced properties and below 5% reduced the yield of carbonized fiber.

2. Measure the oxidized fiber density

    In a production process, this method is much preferred by the author. Takaku *et al* [90] stabilized a AN/MA precursor at 240, 255 and 270°C and observed an increase in oxidized density, which reached a constant value of 1.58 $g\,cm^{-3}$ at the longer stabilization times, obtaining optimum carbon fiber strength with an oxidized density of about 1.375 $g\,cm^{-3}$ (Figure 5.37). Bajaj and Roopanwal [91] attributed the increase in density to closer packing of molecular chains due to cyclization of the nitrile groups.

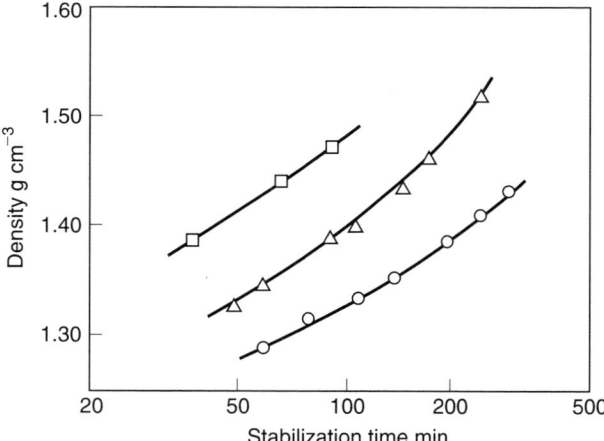

Figure 5.37 Density as a function of stabilization time for acrylic precursor fiber containing AN/MA at stabilization temperatures of ○, 240°C; △, 255°C; □, 270°C. *Source:* Reprinted from Bajaj P, Roopanwal AK, *Polym Sci*, 1, 368, 1994.

3. Measure water uptake of oxidized fiber (equilibrium moisture content)
   Toray [60] measured the water absorbability by placing about 2 g of oxidized fiber in a desiccator for 16 h over a saturated aqueous solution of ammonium sulfate at 25°C to give a controlled humidity of 81% at 25°C. The water absorption is calculated as follows:

$$(W - W_0) \times 100/W_0$$

   where W is the weight after absorption and W_0 the weight after drying for 2 h at 120°C. A value of 5–10% is preferred; <3.5% will make poor carbon fiber and >15% will lower the carbonization yield.

4. Measure chemical shrinkage
   Shrinkage during stabilization is has a physical contribution, termed entropy shrinkage and a chemical contribution, termed the reaction shrinkage. Fitzer et al [11] have used the chemical shrinkage to follow the progress of the chemical stabilization reaction of a 6% MA/ 1% ITA PAN fiber using a thermomechanical analyzer at a heating rate of $3°C\,min^{-1}$. Figure 5.38 shows the *in situ* shrinkage measured during linear heating, together with the derivative of the curve showing two maxima. This graph yields the total shrinkage, showing the physical and chemical components clearly distinguished by a minimum in the first derivative curve. The entropy shrinkage is mostly completed below 200°C and the start and finish of the chemical shrinkage can be identified. The polymer composition, which will influence the result and the amount of entropy shrinkage, depends on the pre-stretching received during PAN manufacture. An increased heating rate will increase the amount of chemical shrinkage, but does not alter the entropy value and the whole temperature range of the chemical reaction is shifted to a higher temperature, an effect known in thermocalorimetry as the Kissinger effect [93]. This work established that for the copolymer studied, the optimum heating rate up to the starting temperature T_i was $5°C\,min^{-1}$. Once the reaction started, the heating rate had to be reduced to $1°C\,min^{-1}$ to avoid overheating the fiber and causing an exothermic reaction. It must be remembered that partly oxidized PAN is a poor conductor of heat and the heat transfer effect allows heat to build up in the filament bundle. DSC measurements were used to establish the maximum temperature that could be used to avoid over-oxidation, which in this case was 270°C.

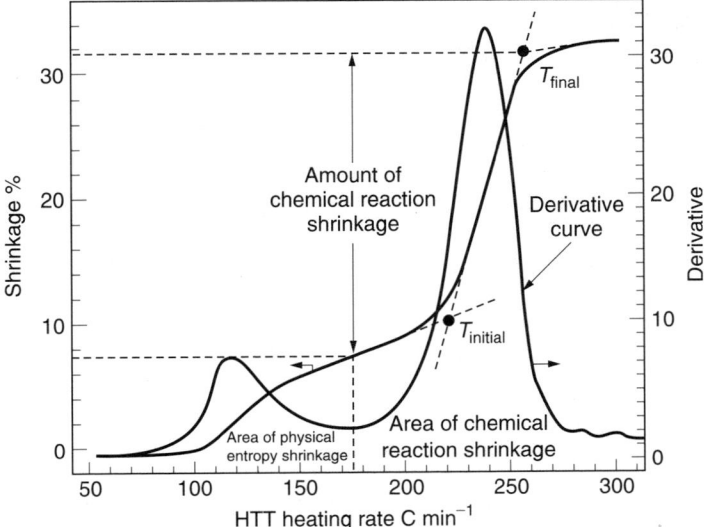

Figure 5.38 *In situ* shrinkage during linear heating of a 6% MA/1% ITA PAN fiber in air using a Mettler thermomechanical analyzer. *Source:* Reprinted with permission from Fitzer E, Heine M, *89th ACS National Meeting*, Miami Beach, USA, 1985. Copyright 1985, American Chemical Society.

5. Achieve of 24% shrinkage to provide 50% cyclization

    In the initial stages of stabilization, a fiber held at constant length will develop tension, contracting about 10% as the T_g of the fiber is approached, reaching a maximum at about 150°C, due to the entropic relaxation of those regions of the drawn PAN structure which possess no lateral crystalline order [54].

    The tension levels out and then starts to increase again, the secondary shrinkage attributed to the melting of the ordered domains as the reaction progresses into the laterally ordered phases. Fitzer and co-workers [94] attributed this secondary shrinkage to the onset of chemical reaction and recommended that for optimum stabilization up to this point, the rate of heating should be considerably reduced. Optimum oxidation of PAN should be carried out to achieve 50% cyclization, which Watt [95] associated with the best tensile properties. Bahl and Manocha [96] further determined that 50% cyclization corresponded to 24% shrinkage and reported an empirical relationship for finding the approximate optimum pre-oxidation time at any temperature, which for copolymer PAN oxidized in air was:

    $$\log t(h) = \frac{5900}{T(K)} - 10.6 + \log\left(\frac{denier}{4.34}\right)$$

6. Employ the aromatization index

    For example, Tsai [97] uses the aromatization index (AI) as a measure of the degree of oxidation.

    $$AI = \frac{(H_V - H_0) \times 100}{H_V}\%$$

    where H_V = exothermic heat of virgin PAN fiber
    H_0 = exothermic heat of oxidized PAN fiber
    and an AI value of 58% was preferred.

7. Establish a model for stabilization

    Dunham and Edie [98] established a mathematical model of the stabilization process for 12–60k PAN fiber and checked theory with experiments using 3k and 12k 1.22 d'tex Courtauld SAF PAN fiber (6% MA, 1% ITA) by embedding a thermocouple in the fiber bundle. The governing equations for the model are based on the rates of chemical reactions, mass balances on reacting species, radial mass transfer and radial heat transfer within the bundle. They showed that the fiber bundle can be as much as 15°C above the stabilization oven temperature and the model predicted the measured temperatures quite well, except for, as would be expected, run-away reaction conditions. Samples stabilized below 230°C did not exhibit a skin core effect, but above 245°C, exhibited distinct skin core differences which were observed by reflected light microscopy. Hence diffusion appears to limit the stabilization rate above 245°C but not below 230°C. Bundles larger than 12k tended to burn when stabilized much above 230°C. The model would not hold for temperatures above 245°C.

    Work undertaken at the NPL, New Delhi [99] showed that during the oxidation of PAN, a second small exotherm occurred with a maximum centered at about 350°C (Figure 5.39). PAN fiber oxidized at 350°C produced a more ordered structure in the opf, which when carbonized gave a carbon fiber with improved mechanical properties (Table 5.10) and moreover, could be rapidly carbonized.

5.10.1 Structure of PAN fibers thermally stabilized at 350°C [9]

When examined by wide angle X-ray the diffractograms showed stabilized fibers to have an additional peak at $2\theta = 13°$ due to a structural entity with 0.68 nm repeat units and the proposed structure is as shown.

CARBON FIBER PRODUCTION USING A PAN PRECURSOR

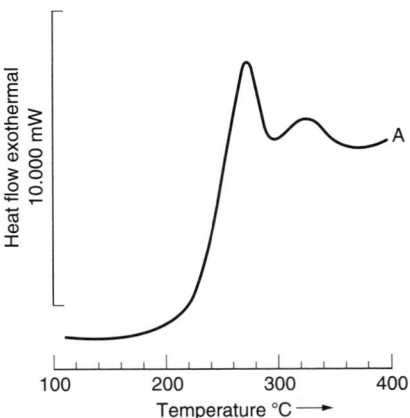

Figure 5.39 DSC scan of PAN precursor showing second small exotherm at about 350°C. *Source:* Reprinted from Mathur RB, Bahl OP, Mittal J, Nagpal KC, *Carbon*, 29(7), 1059, 1991.

Table 5.10 Mechanical properties of carbon fibers prepared from Courtelle PAN stabilized at temperatures 250°C–400°C and carbonized at 1000°C in N_2

	Carbonized at 1000°C	
Oxidation treatment in air	Tensile strength GPa	YM GPa
a. Heated to 250°C at 1°C min^{-1}	1.88	220
b. a + 1 min at 300°C	2.18	214
c. a + 1 min at 325°C	2.2	220
d. a + 1 min at 350°C	2.24	227
e. a + 1 min at 375°C	2.4	234
f. a + 1 min at 400°C	2.4	240

Source: Reprinted with permission from Mathur RB, Bahl OP, Mittal J, Nagpal KC, *Carbon*, 29(7), 1059, 1991. Copyright 1991, Elsevier.

The —C≡N content of the polymer decreases during the exothermic reaction due to ladder polymer formation and Kubasova *et al* [100](Figure 5.40) showed the results for a cast PAN film, which displayed a decrease in nitrile absorbance with time at various temperatures.

Another study shown in Figure 5.41 was undertaken by Johnson *et al* [101], who measured the disappearance of the nitrile stretching band at 2240 cm^{-1} in the infra red spectrum of Courtelle fiber heated up to temperature at about 1°C min^{-1}.

Noh and Yu [102] have stated that the disappearance of the nitrile groups is not complete, but reaches a final value of about 20% residual nitrile groups.

Since O_2 has to diffuse through and at one stage of the oxidation process, the diffusion rate $\propto \sqrt{\text{time}}$, the filament diameter is limited to <10 μm in order to keep the oxidation time within commercial limits.

During oxidation, there is a higher concentration of O_2 in the outer regions of the filament section. During carbonization, the diameter shrinks to half its original value, putting the outer regions into compression, with the inner core in tension (and if half of this fiber is etched away, the fiber curves inward showing that there are tensile stresses in the core).

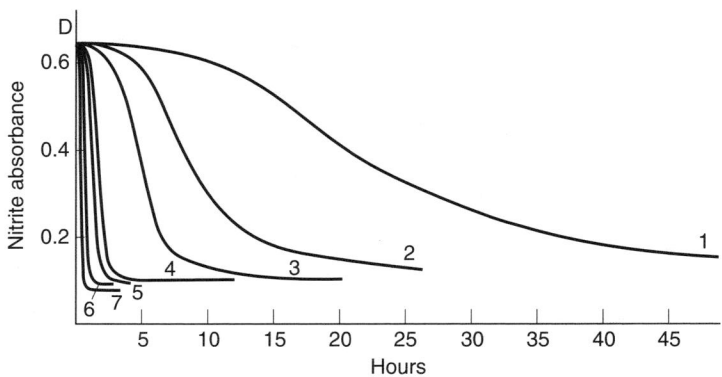

Figure 5.40 Decay of —CN IR absorbancies during stabilization of PAN. 1. 205°C; 2. 215°C; 3. 225°C; 4. 235°C; 5. 245°C; 6. 270°C. *Source:* Reprinted from Kubasova NA, Kusakov MM, Shishkina MV, *Vysokomelek yarnyi Soldininiya*, 3, 193, 1969.

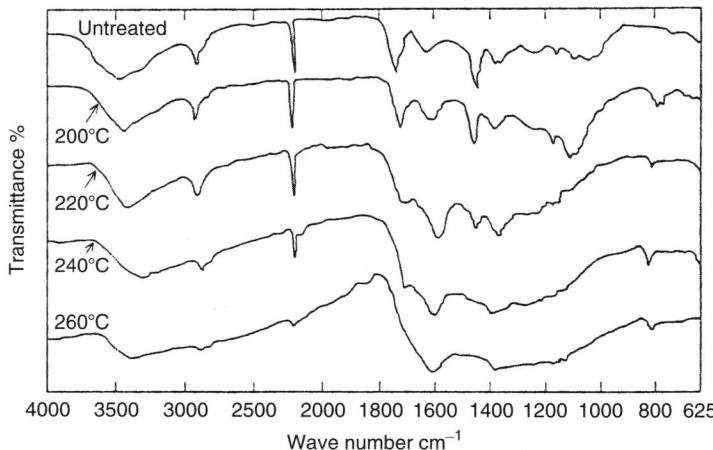

Figure 5.41 IR spectrum of Courtelle treated in air at 1°C min^{-1} showing gradual disappearance of —CN stretching band at 2240 cm^{-1}. *Source:* Reprinted with permission from Johnson JW, Potter W, Rose PG, G Scott G, Stabilization of polyacrylonitrile by oxidative transformation, *Brit Polym J*, 4, 527–540, 1972. Copyright 1972, The Society of Chemical Industry.

With clean dope and clean room conditions, most failures will be on the surface and smaller diameter fiber will have increased strength and lesser flaws [103].

Moreton and Watt [104] describe the spinning of PAN fibers in clean room conditions to avoid contamination with particle impurities (normally 1–3 μm diameter). The benefits of using PAN made under clean room conditions can be clearly seen when testing varying gage lengths of carbon fiber filaments that have been heat treated to 2500°C (Table 5.11). The strength of the clean room fiber was not affected by the gage length, on the other hand, the gage length of control fibers showed a marked effect. These findings were backed by a count of the particle impurities in the PAN fiber, when 0.2 particles per cm were found in clean room fiber and 7.0 particles per cm in the control fiber.

Dhami and co-workers [105] oxidized 6k SAF at 250°C with increasing dwell time (samples A to D) and characterized the fibers by elemental analysis and surface groups by ESCA (Table 5.12). They found that the amount of O_2 on the surface of the opf fibers in the form of surface groups such as C—O and C═O is totally different from that present in the bulk.

CARBON FIBER PRODUCTION USING A PAN PRECURSOR

Table 5.11 The effect of gage length on the tensile strength of carbon fibers heat treated to 2500°C.

Gage length mm	Clean room fibers			Control fibers		
	Average diameter μm	Average strength GPa	Coefficient of variation %	Average diameter μm	Average strength GPa	Coefficient of variation %
10	7.9	2.74	13	6.3	2.19	25
25	7.5	2.76	18	6.3	1.79	32
50	7.5	2.75	27	6.3	1.51	39

Source: Reprinted with permission from Moreton R, Watt W, *Carbon*,12, 543–554, 1974. Copyright 1974, Elsevier.

Table 5.12 Elemental and ESCA analyzes of Oxidized SAF samples

% Method of analysis	O Elemental	O ESCA	N Elemental	N ESCA	C Elemental	C ESCA	H Elemental	Si ESCA	C—O ESCA	C=O ESCA
PAN	4	7.8	23.6	9.5	66.2	82.5	5.8	0.1	27.9	4.3
A	6.9–8.1	11.5	23.0–23.1	15.2	64–65	72.2	4.9–5.0	1.1	33.7	8.1
B	9.8–10.8	13.9	22.2–22.3	13.5	62.6–63.3	71.0	4.4–4.5	1.6	26.5	10.9
C	19.1–20.8	12.5	20.3–20.5	14.0	56.0–56.2	56.0–56.2	3.0–3.1	0.3	26.8	7.7
D	21.0–21.4	13.9	20.6–20.7	15.1	55.5–55.2	55.5–55.2	2.7–2.8	1.4	24.4	8.4

Source: Reprinted from Dhami TL, Mathur RB, Dwivedi H, Bahl OP, Monthioux M, *Pyrolysis behavior of Panex based polymer composites*, Extended Abstracts (Vol II) 23rd Biennial Conference on Carbon July 18–23, 474–475, 1997.

5.11 ASPECTS OF CARBONIZATION

Takaku *et al* [90] established that the density of carbonized carbon fiber decreases with increase in density of the oxidized fiber (Figure 5.42) and the tensile strength of the carbonized fiber suggested that the preferred oxidized fiber density was 1.375 g cm^{-3} (Figure 5.43).

Toho Beslon [55] prefer carbonization to be undertaken for 30 s–30 min at 700–950°C in a N$_2$ atmosphere with a tension of 1 mg denier^{-1} and 40–70% total shrinkage (actual shrinkage

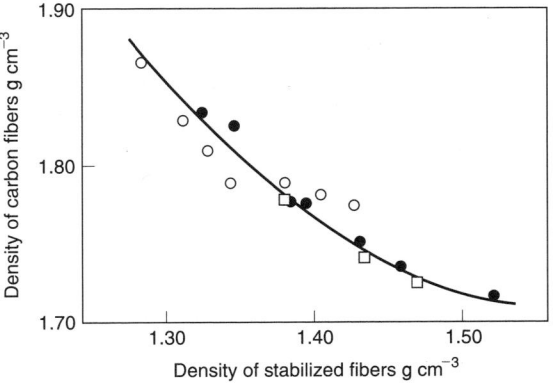

Figure 5.42 Density of carbon fibers as a function of the density of stabilized fibers obtained from an AN/MA acrylic precursor fiber. *Source:* Reprinted with permission from Takaku A, Hashimoto T, Miyoshi T, *J Appl Polym Sci*, 30, 1565; 1985. Copyright 1985, John Wiley & Sons Ltd.

in carbonization is 7–8%) to give a preferred carbon content of 85%, depending on the initial precursor.

The effects of heat treatment during the manufacture of carbon fiber [106–108] have been investigated.

Tsai [109] has compared the performance of one- and two-stage carbonization furnaces, whilst Ko *et al* [110] and Tsai [111] have investigated two-stage carbonization.

Watt [112] states that N_2 evolution occurs in the range 600–1300°C, although this will be influenced by the line speed. Serin *et al* [113] have reported that nitrogen content in carbon fibers induces poor tensile strength in the fibers. Toray T300 carbon fiber was profiled using Electron Energy Loss Spectroscopy (EELS) for the nitrogen content across a longitudinal section of the fiber, including its axis (Figure 5.44), giving N/C concentration varying from 2% on the inside to 7% on the outside of the section. The nitrogen is evolved as molecular N_2 [114]. In the center of the section, the space is compressed and the nitrogen atoms tend to be close and face each other, hence the energy necessary to effect formation of N_2 molecules is lower, promoting enhanced release of N_2 at the center.

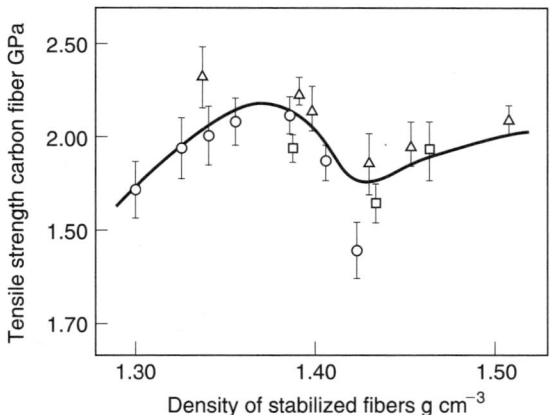

Figure 5.43 The tensile strength of carbon fibers as a function of the density of stabilized fibers from an AN/MA precursor at stabilization temperatures of ○ 240°C; △ 255°C; □ 270°C. *Source:* Reprinted with permission from Takaku A, Hashimoto T, Miyoshi T, *J Appl Polym Sci*, 30, 1565, 1985. Copyright 1985, John Wiley & Sons Ltd.

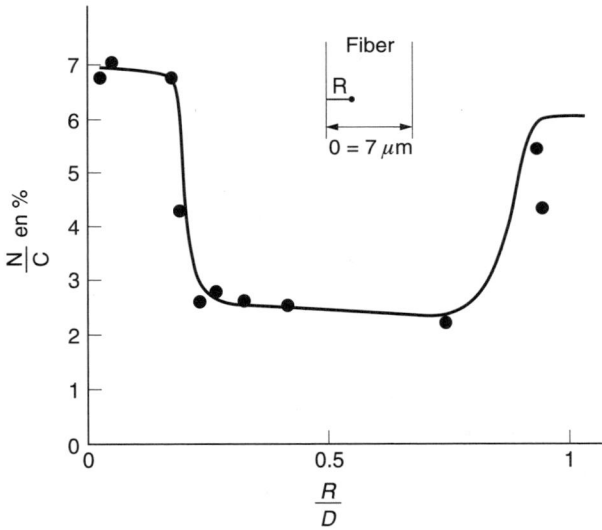

Figure 5.44 Profile of the distribution of nitrogen in a cross-section of Toray T-300 carbon fiber as determined by Electron Energy Loss Spectroscopy (EELS). The position of the area of the surface is measured by the distance R/D, where D is the apparent width of the section in the microscopy image and R is the distance between the edge of the fiber and the point of measure. *Source:* Reprinted with permission from Serin V, Fourmeaux R, Kihn Y, Sevely J, Guigon M, Nitrogen distribution in high tensile strength carbon fibres, *Carbon*, 28(4), 573–578, 1990. Copyright 1990, Elsevier.

Guigon heated T300 fiber at 21°C per min in Ar up to 1100, 1400 and 1700°C with dwell times of 10 min at the temperature and determined the H/C, N/C and N/H ratios (Table 5.13), when molecular N_2 was released, but H_2 was not released until the temperature was above 1400°C. Toray T800 fiber, with a smaller diameter (about 5 μm instead of 7 μm), showed a lower nitrogen content initially, but maintained a constant N/H ratio of 1, which is consistent with good mechanical properties.

Table 5.13 N/H ratios in different PAN based Toray carbon fibers

Fiber	Treatment °C	H/C	N/C	N/H
T300	As received	0.061	0.061	1
	1100	0.062	0.033	<1
	1400	0.061	0.010	<1
	1700	0.045		
T800	As received	0.034	0.035	1

Note: The fibers were heated in Ar from room temperature up to the stipulated temperature at a rate of 21°C min^{-1} and held for 10 min at the final temperature.
Source: Reprinted from Guigon M, Thesis, Université de Technologie de Compiégne, France.

The effect of N_2 on the structure and properties of PAN based carbon fibers has been studied by Tsai [116].

Cullis and Yates [117] have shown that graphite reacts with N_2 between 1400–2000°C noting that the reaction with one grade of graphite peaked at 1600°C whilst another grade the reaction actually increased with temperature; probably due to differences in surface area. A suggested mechanism for the reaction of carbon with N_2 is:

There are forward and reverse reactions [118],

$$2C + N_2 \underset{k'}{\overset{k}{\rightleftarrows}} C_2N_2$$

and as the temperature increases, the mobility of the carbon atoms or other fragments on the surface give more perfect crystallites owing to the movement of the atoms to preferred positions [119]. The deposits have no three dimensional order and are non-graphitizable.

5.11.1 Methods of increasing fiber modulus and effect on strength

5.11.1.1 Hot stretching

Stretching during graphitization produces fibers which differ significantly from unstretched fibers in their mechanical properties.

Hot stretching of cellulosic fibers was undertaken by Union Carbide and reported by Bacon and Schalamon [120] who found that 300% stretching at 2700°C increased the modulus from 70 GPa to 770 GPa. However, the strength did not increase *pro rata* but did increase from 0.7 to 3.5 GPa, with the extension dropping from 1.0 to 0.5%.

It is not surprising that similar work was undertaken with PAN fiber and Johnson, Marjaram and Rose [121] at Rolls Royce, were able to stretch PAN fiber 30% to give a modulus of 620 GPa and, although the fiber did increase in strength, the extension decreased to 0.5%. Johnson [122] at RAE, obtained similar results using a 29% stretch at 2800°C, which increased the modulus from 403 GPa to 656 GPa with an extension of 0.5%.

The increase in preferred orientation on stretching [121] is accompanied by an increase in crystal size in the direction of the *c*-axis, measured by the parameter *L*, was obtained by measuring the line broadening of the graphite (002) diffraction arc. The results for different stretch ratios and temperatures are given in Table 5.14.

More recent work at Swansea University by Isaac and co-workers [123–127], undertaken with Courtauld's 3k SAF heated at 27.3°C min^{-1} up to 3000°C, with a 5 min dwell time, found that the preferred orientation is a function of both the applied stress and the temperature at which stretching is carried out [123] (Figure 5.45), whilst L_c, the apparent crystallite size, is dependent on the heat treatment temperature and is only negligibly affected by the stretching stress (Figure 5.46). The preferred orientation over the temperature range studied was related to the final fiber diameter (Figure 5.47). The properties measured on single filaments after stretching at various temperatures and stresses are given in Table 5.15.

The YM was directly proportional to the induced tension, with some evidence that smaller diameter fibers showed greater improvements [126]. Tensile strength was particularly sensitive to the fiber diameter and extrapolation to 5 μm diameter suggested a tentative tensile strength of 8 GPa might be achieved.

Best fit lines are given for the relationships between YM and the stretched fiber diameter for three fiber types with as received diameters of 10, 6.5 and 5.1 μm are shown in Figure 5.48,

Table 5.14 Increase in crystal size for different stretch ratios and temperatures

Stretch under applied stress %	Temperature °C	L_c nm
0	1990	5.0
20	1990	7.0
0	2150	7.1
20	2150	11.5
0	2390	12.5
18	2390	15.0
0	2140	6.3
30	1140	9.0
0	2970	17.0
23	2970	25.0

Source: Reprinted from Johnson JW, Marjaram JR, Rose PG, Stress graphitization of polyacrylonitrile based carbon fibre, *Nature*, 221, 357–358, 1969.

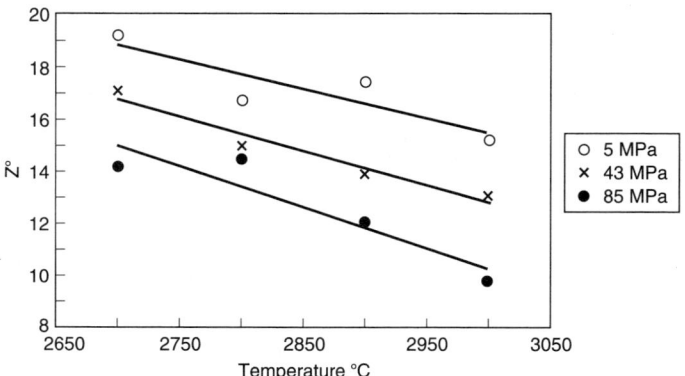

Figure 5.45 Preferred orientation parameter ($Z°$) against heat treatment temperatures for three levels of stress. *Source:* Reprinted with permission from Ozbek S, Isaac DH, Carbon fiber processing: Effects of hot stretching on mechanical properties, *Mater Manuf Process*, 9(2), 179–197, 1994. Copyright 1994, CRC Press, Boca Raton, Florida.

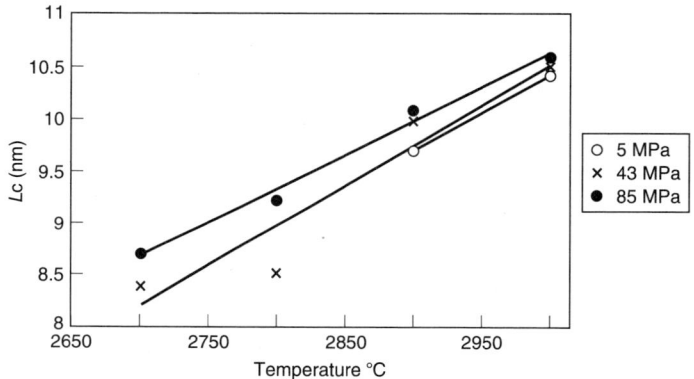

Figure 5.46 Apparent crystallite size L_c against heat treatment temperatures for three levels of stress. *Source:* Reprinted with permission from Ozbek S, Isaac DH, Carbon fiber processing: Effects of hot stretching on mechanical properties, *Mater Manuf Process*, 9(2), 179–197, 1994. Copyright 1994, CRC Press, Boca Raton, Florida.

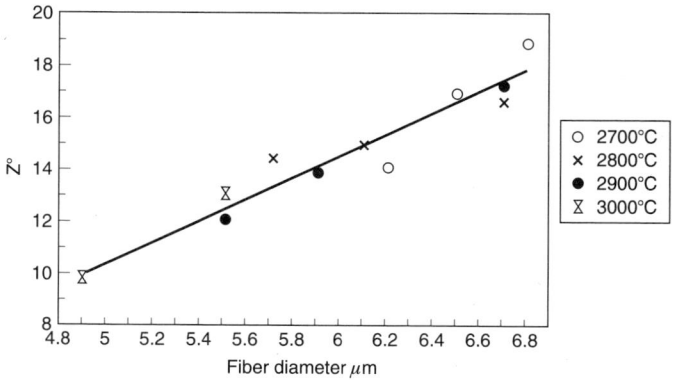

Figure 5.47 Preferred orientation parameter ($Z°$) against fiber diameters after stretching at 2700–3000°C. *Source:* Reprinted with permission from Ozbek S, Isaac DH, Carbon fiber processing: Effects of hot stretching on mechanical properties, *Mater Manuf Process*, 9(2), 179–197, 1994. Copyright 1994, CRC Press, Boca Raton, Florida.

Table 5.15 Properties measured on single carbon fiber filaments following stretching at various temperatures and stresses

Stretching Temperature °C	Stretching Stress Mpa	Total Extension mm	Fiber Diameter μm	Young's Modulus GPa	Preferred Orientation Z°	Crystallite Size L_c nm	Density g cm^{-3}
1300 (3k SAF)	–	–	nominal 7	~180	38	–	1.739
2700	5	2.6	6.8±0.2	335±3	19.1±1.0	–	~1.856
2700	43	6.1	6.5±0.2	389±17	17.1±0.8	8.4±1.1	~1.859
2700	85	11.5	6.2±0.2	468±25	14.2±0.3	8.7±0.5	~1.869
2800	5	2.8	6.7±0.1	359±9	16.7±1.2	–	~1.860
2800	43	6.8	6.1±0.3	428±13	15.0±0.7	8.5±1.0	~1.866
2800	85	14	5.7±0.1	531±5	14.5±2.1	9.2±1.2	~1.878
2900	5	3	6.7±0.1	375±12	17.4±2.4	9.7±0.2	~1.873
2900	43	8.8	5.9±0.2	478±6	13.9±0.4	10.0±0.9	~1.880
2900	85	15	5.5±0.1	583±12	12.1±1.4	10.1±0.4	~1.892
3000	5	3.3	6.2±0.3	401±7	15.2±0.6	10.4±1.0	~1.888
3000	43	10	5.5±0.1	553±30	13.1±3.8	10.5±0.5	~1.908
3000	85	18.5	4.9±0.1	691±18	9.8±0.9	10.6±0.3	~1.930

Source: Reprinted with permission from Amended from Isaac DH, Ozbek S, Francis JG, *Mater Manuf Process*, 9(2), 179–197, 1994. Copyright 1994, CRC Press, Boca Raton, Florida.

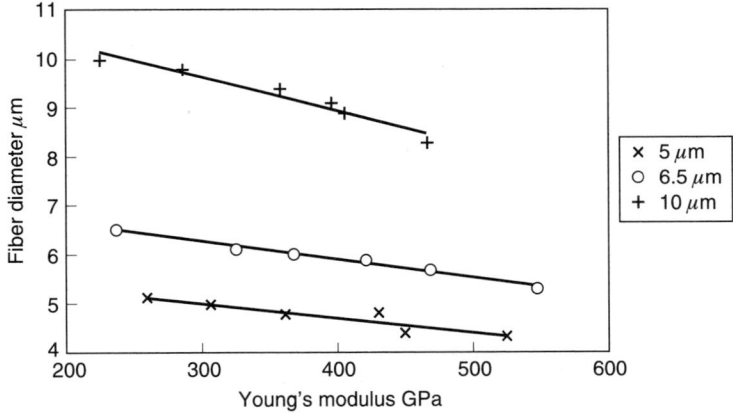

Figure 5.48 The relationship between YM and the filament diameter following hot stretching of the three fiber types. Points for the as received fibers (diameters 10, 6.5 and 5.1 μm) are included and they lie on the best fit straight lines. *Source:* Reprinted with permission from Isaac DH, Ozbek S, Manufacture of carbon fibers from precursors of various diameters, *Mater Manuf Proces*, 9(5), 975–998, 1994. Copyright 1994, CRC Press, Boca Raton, Florida.

clearly showing the improvement of modulus with decreasing diameter [126]. Similarly, the relationship with strength is shown in Figure 5.49 again showing improvement with decreasing diameter.

Isaac and Ozbek [125] found that stretching carbon fibers at high temperatures significantly increased the density (Figure 5.50). The density is closely related (Figure 5.51) to the YM, which is of course related to the temperature.

The high temperature creep was shown to be logarithmic and the major plastic deformation for a 30 min dwell-time occurred during the first 5 min; typically half in 3 min and two thirds in about 5 min [127]. When low extensions are recorded, there is a loss in strength, but at higher strains the strength improved. With an applied load of 0.5 g/filament

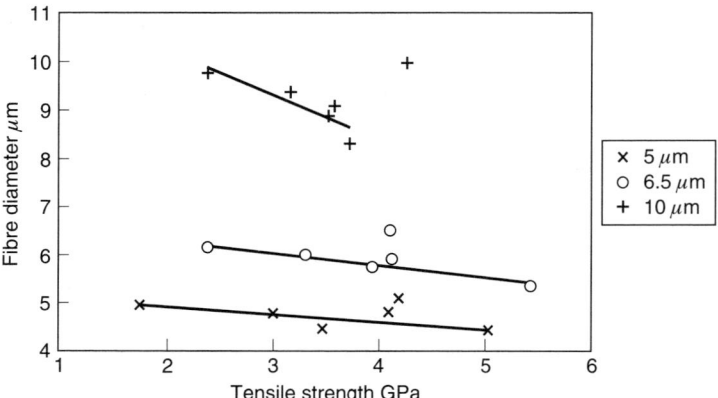

Figure 5.49 The relationship between tensile strength and the filament diameter following hot stretching of the three fiber types. Points for the as received fibers are included to show how heat treatment without significant change in fiber diameter reduces the strength. *Source:* Reprinted with permission from Isaac DH, Ozbek S, Manufacture of carbon fibers from precursors of various diameters, *Mater Manuf Proces*, 9(5), 975–998, 1994. Copyright 1994, CRC Press, Boca Raton, Florida.

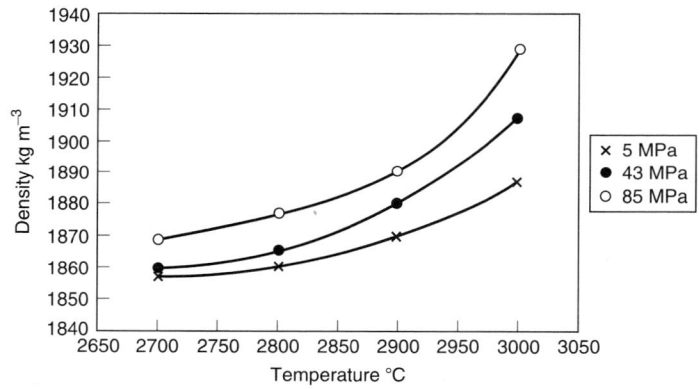

Figure 5.50 Fiber density as a function of processing temperature at different stress levels. *Source:* Reprinted with permission from Isaac DH, Ozbek S, *Density changes in carbon fibers, induced by hot stretching, Carbon '98 22nd Biennial Conference on Carbon, American Carbon Society*, San Diego, CA, 28–29, 1995. Copyright 1995, American Chemical Society.

at 2800°C and dwell time of 30 min, the YM increased from 180 to 700 GPa and the original strength of 3.9 GPa was restored.

The effect of tension during carbonization has been studied by Ogawa [128] and Tsai [129].

5.11.1.2 Effects of neutron irradiation

Allen and co-workers [130] irradiated high strength and high modulus fiber at a flux of 10^{12} n cm^{-2} sec^{-1} (Ni) to a dose of 2.2×10^{17} n cm^{-2} and although the resistivity hardly changed, they observed a 10% increase in strength and modulus. Irradiation causes dislocation pinning without significantly altering the crystal structure and would be expected to increase the strength.

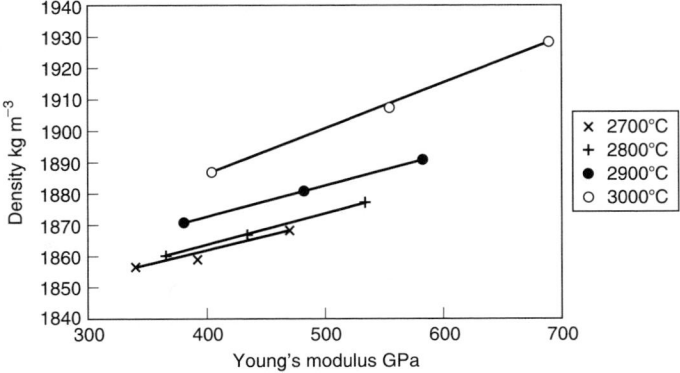

Figure 5.51 Fiber density as a function of YM at different temperatures. *Source:* Reprinted with permission from Isaac DH, Ozbek S, *Density changes in carbon fibers, induced by hot stretching, Carbon '98 22nd Biennial Conference on Carbon, American Carbon Society*, San Diego, CA, 28–29, 1995. Copyright 1995, American Chemical Society.

5.11.1.3 Annealing in the presence of boron

Boron has a smaller atomic radius than carbon and easily fits into the carbon lattice and can, for example, be found in the diamond lattice. The boron lowers the electrical resistivity and presumably acts as an electron acceptor site and, for valency reasons, some of the joins in the net are broken.

Allen *et al* [130,131] showed that boron vapor increased the modulus and conductivity (Table 5.16) by increasing the crystallinity and helping to prevent shear in the crystallites. It is believed that the boron atoms hinder dislocation in the graphite lattice by a type of hardening process akin to the solid solution hardening effect in metallurgy.

The effect of different temperatures on the modulus is shown in Figure 5.52, with the modulus increasing with annealing temperature in the presence of boron. The similarity of the results for the two fiber types is not surprising since they were derived from the same precursor type and heated to the same temperature.

Ezekiel [132] has also shown that boron increases the rate of graphitization and results in an increase in fiber properties. FMI [133] showed a significant increase in both strength and modulus, with HM fiber under tension in an arc plasma, first using BCl_3 and later triethylborane (($C_2H_5)_3B$), which does not form harmful intermediates during pyrolysis. This improvement in properties was believed to be due to the simultaneous healing of surface flaws during graphitization.

Table 5.16 Effect of Boron on tensile modulus and tensile strength

Fiber Type	Boron (3 h at 2750°C) %	YM GPa	TS GPa	Extension %	Resistivity μm cm
Type II (high strength)	0	428	1.81	0.42	700
	1	538	1.79	0.33	250
Type I (high modulus)	0	407	1.88	0.46	770
	1	545	2.17	0.40	250

Source: Reprinted from Allen S, Cooper GA, Mayer RM, *Paper presented at IP and PS Conference on Fibres and Composites*, Brighton, Jun 1969. Reprinted with permission from Allen S, Cooper GA, Johnson DJ, Mayer RM, Carbon fibres of high modulus, *Proc 3rd Conf Industrial Carbons and Graphite*, Soc Chem Ind, London, 456–461, 1970. Copyright 1976, The Society of Chemical Industry.

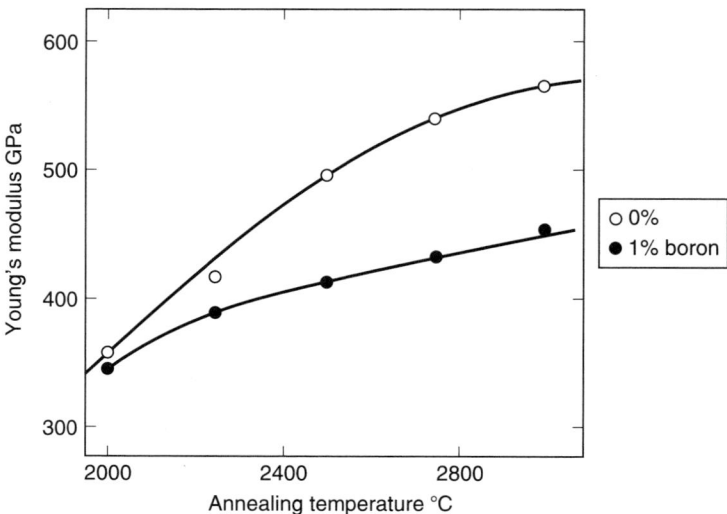

Figure 5.52 The effect of annealing temperature on the modulus in the presence of boron. *Source:* Reprinted from Allen S, Cooper GA, Mayer RM, *Paper presented at IP and PS Conference on Fibres and Composites*, Brighton, Jun 1969.

Later work by FMI workers, Brewster, Nelson and Patton [134], showed that the boron deposits were confined to a surface layer on the fibers about 15 nm thick, strengthening the fibers by healing surface defects such as holes and cracks.

5.11.2 Carbon fiber yield

Although PAN loses constituents in oxidation, there is also an uptake of O_2, providing an overall loss in weight of about 3% during oxidation. Although tar removal might be expected to contribute most of the weight loss in the carbonization process, in reality, the loss up to about 700°C based on the dry oxidized fiber, is about 25%, whilst a further 20% is lost in the furnace up to 1350°C. Yields depend on the composition of the precursor and a typical carbon fiber yield for Courtelle would be of the order of 50–55%.

When making carbon fiber from an SAF precursor, tar removal can be achieved with about a 1.5 min residence time and subsequent carbonization can be attained in less than 1 min, with little effect on strength and only a marginal reduction of modulus. To compensate for this loss in modulus, it would be necessary to increase the carbonization temperature. The effect of residence time on modulus for SAF carbonized at 1350°C and 1610°C is shown in Figure 5.53.

A well stabilized fiber will have a better yield than a fiber with inadequate stabilization [135].

5.12 RELATION OF CARBON FIBER TENSILE PROPERTIES TO PROCESS CONDITIONS

The theoretical modulus of crystalline graphite is about 1060 GPa and hot stretching a PAN based carbon fiber will produce a fiber with a modulus up to about 700 GPa but will not surpass values recorded for pitch based fibers (965 GPa). The theoretical strength of

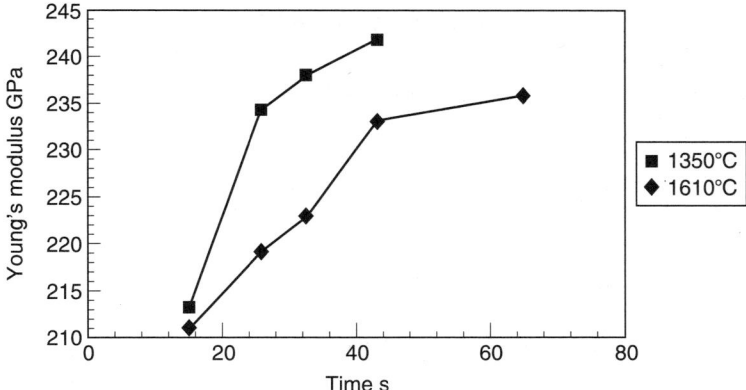

Figure 5.53 Effect of residence time on YM for SAF carbonized at 1350 and 1610°C.

crystalline graphite is approximately 0.1–0.2 of the modulus (i.e. about 100 GPa) but in fact, it is an order of magnitude lower due to the presence of defects, so a value of 20 GPa could be expected. This value, however, is not achieved in practice and at present, a value of about 7.5 GPa has been obtained.

The modulus of PAN based carbon fiber depends on the precursor, the carbon fiber production speed, filament d'tex and carbonization temperature. A smaller d'tex precursor will permit a lower temperature or a faster line speed to be used for a given modulus. The effect of filament decitex at different heat treatment temperatures is shown in Figure 5.54.

Huang and Young [136] have determined the effect of fiber microstructure upon the modulus of PAN and pitch based carbon fibers.

Oberlin et al [137] assume the increase in strength as the lateral cohesion increases is associated with tight bonding between defective BSU sheets (Figure 5.32).

The various properties of PAN based carbon fibers have been investigated [138,139].

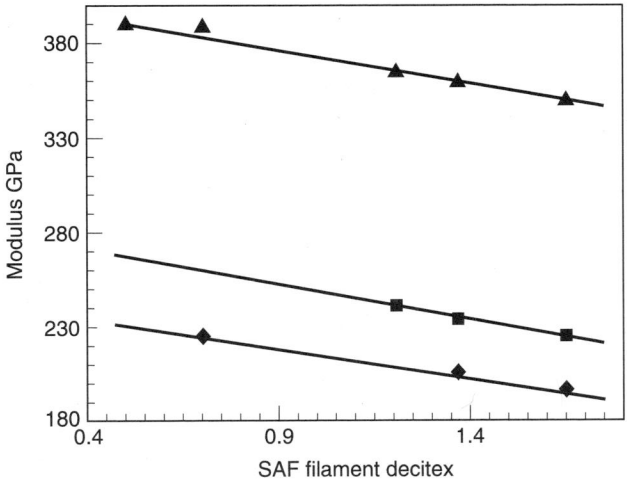

Figure 5.54 The effect of SAF filament d'tex on carbon fiber modulus at various heat treatment temperatures. ◆ 1100°C; ■ 1350°C; ▲ 2500°C.

5.13 DEVELOPMENTS

5.13.1 Improvements in carbon fiber properties

Lin [140] has detailed a number of factors which could help improve the properties of carbon fibers and these are listed in Table 5.17.

5.13.2 Alternative polymer formulations

Various attempts have been made to improve the rate of oxidation of PAN precursors:

1. Workers at RAE used a polymer $(AN)_{25}$, VDC, ITA, which was soluble in NaSCN and it was shown that the vinylidene chloride (CH_2=CCl_2, VDC) performed better than Courtauld's SAF. This finding was surprising since the VDC was added at the expense of the –CN groups, which were considered to be essential for cyanide condensation, cyclization and formation of the six-membered ring (graphite backbone) in oxidation. Clearly cross-linking from one chain to another is equally important.
2. An active cyclization catalyst will have the function of reducing the exothermicity of the reaction, enabling higher temperatures to be used, thus increasing the rate of O_2 diffusion.
3. Grassie suggested the use of a low concentration of a material that would form free in the fiber. Peracetic acid or an azo compound will decompose slowly at 180–200°C and eliminate the slow rate determining initiating step such as:

$$RH(polymer) + O_2 \xrightarrow{200°C} R + HO_2^* \text{ (Intitiation)}$$

$$\left.\begin{array}{l} R^* + O_2 \xrightarrow{rapid} RO_2^* \\ RO_2^* + RH \longrightarrow ROOH + R^* \end{array}\right\} \text{(Propagation)}$$

4. A 3% application of p-toluenesulfonic acid increased the burn-up temperature from 250° to 265°C, enabling the residence time to be reduced.

Table 5.17 Critical factors in the improvement of carbon fiber properties

	Structure Factors		Major Processing Parameters
Basic Structure	Crystal orientation Crystallite content Homogeneity		Precursor • Polymer composition • Appropriate fiber tension • High density and compactness • Prevent filament coalescence
Structure Defects	Interior defects Surface flaws	Voids Impurities Heterogeneity Chemical change Cracks Adhesion	Stabilization Process • Homogeneity (skin core texture) • Extent of applied tension • Stabilization rate Carbonization Process • Rate of temperature rise • Atmosphere of processing • Dust free environment

Source: Reprinted with permission from Lin SS, Recent developments of carbon fiber in Japan, *SAMPE J*, 28(4), 9–19, July/Aug 1992. Copyright 1992, The Society for the Advacement of Material and Process Engineering (SAMPE).

5.13.3 A family of controlled resistance carbon fibers

A whole family of carbon fibers can be made from a PAN precursor by altering the final carbonization temperature, which is illustrated the data obtained by Fitzer and Frohs [141] using Courtauld's 6k SAF, which was stabilized in air at 1°C min^{-1} up to 270°C, achieving a density of 1.4 g cm^{-3}. Carbonization was undertaken at 20°C min^{-1} in Ar. They found that as the carbonization temperature increased, the nitrogen content (Figure 5.55), resistivity (Figure 5.56) and the yield decreased, whilst the modulus (Figure 5.57), strength up to about 1500°C (Figure 5.58) and density (Figure 5.59) increased. The oscillating behavior is attributed to structural healing processes which are interrupted by further N$_2$ outbreaks.

These fibers can be classed by their nitrogen content, with the nitrogen content decreasing as the carbonization temperature was increased (Figure 5.60).

Interestingly, the resistance of fibers made up to about 1000°C will increase on standing in air at room temperature, a phenomenon thought to be due to absorption of O$_2$. This

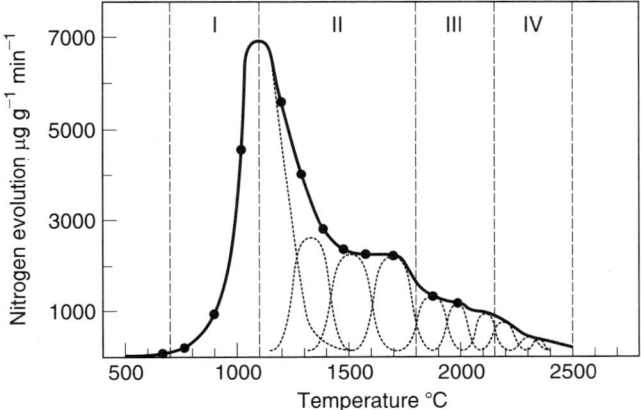

Figure 5.55 N$_2$ evolution as a function of heat treatment temperature, measured by online GC analysis. *Source:* Reprinted with permission from Fitzer E, Frohs W, The influence of carbonization and post heat treatment conditions on the properties of PAN-based carbon fibres, *Presented at Carbon 88*, Newcastle upon Tyne, 298–300, 1988. Copyright 1988, The Insitute of Physics Publishing.

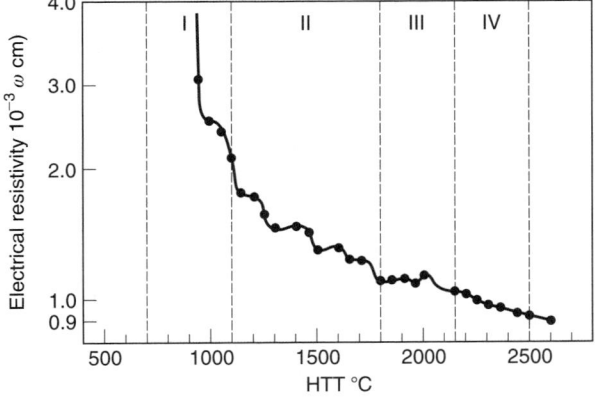

Figure 5.56 Electrical resistivity as a function of heat treatment temperature, measured with four-point method and 20 mm gage length. *Source:* Reprinted with permission from Fitzer E, Frohs W, The influence of carbonization and post heat treatment conditions on the properties of PAN-based carbon fibres, *Presented at Carbon 88*, Newcastle upon Tyne, 298–300, 1988. Copyright 1988, The Insitute of Physics Publishing.

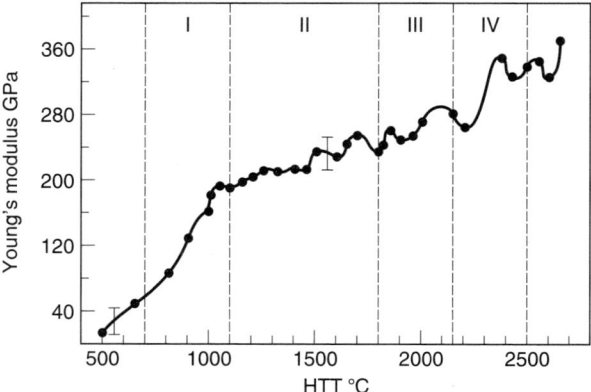

Figure 5.57 YM as a function of heat treatment temperature, measured by single filament test with 30 mm gage length at a crosshead speed of 1 mm min^{-1}. *Source:* Reprinted from Fitzer E, Frohs W, The influence of carbonization and post heat treatment conditions on the properties of PAN-based carbon fibres, *Presented at Carbon 88*, Newcastle upon Tyne, 298–300, 1988.

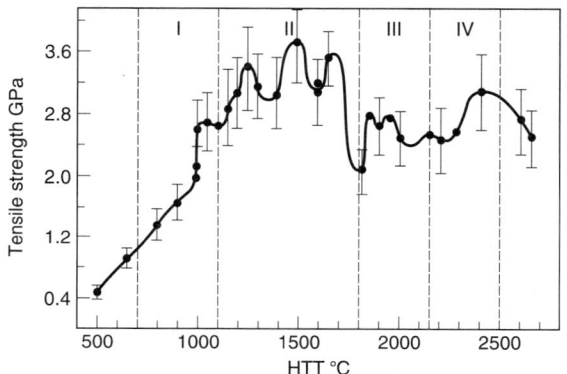

Figure 5.58 Fiber tensile strength as a function of heat treatment , measured by single filament test with 30 mm gage length at a crosshead speed of 1 mm min^{-1}. *Source:* Reprinted with permission from Fitzer E, Frohs W, The influence of carbonization and post heat treatment conditions on the properties of PAN-based carbon fibres, *Presented at Carbon 88*, Newcastle upon Tyne, 298–300, 1988. Copyright 1988, The Insitute of Physics Publishing.

absorption of O_2 is accompanied by evolution of heat, which must be dissipated; otherwise the product will eventually catch fire.

BASF work showed that the bound oxygen content is increased substantially by heating the fiber in air at 240–360°C. Heating the fiber in N_2 limits the O_2 pick up.

5.14 A REVIEW OF THE STABILIZATION OF PAN PRECURSORS

To satisfactorily produce carbon fibers from a PAN precursor, it is necessary to apply an oxidative stabilization pretreatment, normally by a controlled heating process in air for about 2 h at 250°C. Since the reaction is strongly exothermic, a gradual application of heat must be applied, as otherwise the reaction becomes uncontrolled and chain scission will occur, with the fibers finally burning. In the stabilization process, the fibers turn from white to yellow, to golden yellow, to brown and finally black, eventually producing a fiber which is non-flammable.

CARBON FIBER PRODUCTION USING A PAN PRECURSOR

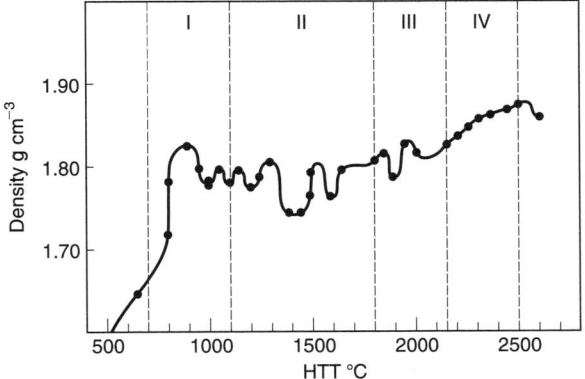

Figure 5.59 Fiber density as a function of heat treatment temperature. *Source:* Reprinted with permission from Fitzer E, Frohs W, The influence of carbonization and post heat treatment conditions on the properties of PAN-based carbon fibres, *Presented at Carbon 88*, Newcastle upon Tyne, 298–300, 1988. Copyright 1988, The Insitute of Physics Publishing.

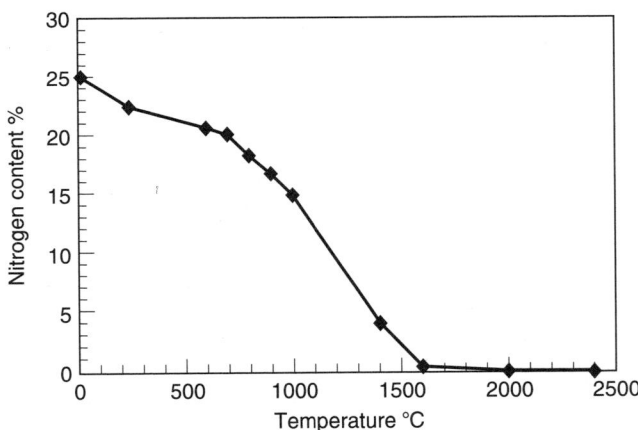

Figure 5.60 The nitrogen content of carbonized PAN fibers up to 2500°C. *Source:* Collated from information supplied by RK Carbon Fibres and Courtaulds.

Although the original work was done nearly fifty years ago and many papers have subsequently been published on the topic of stabilization, the evidence, surprisingly, still remains largely circumstantial and current thoughts suggest that that the mechanism involves a cyclization process with the formation of six-membered rings. A resumé of the various proposals for the possible form that the mechanism of the reaction can take is given, mainly in a historical sequence, together with practical information on the stabilization process. If required, more detail can be found from the references. A paper by Bashir [142] provides an excellent review of the stabilization process.

5.14.1 Stabilization schemes of PAN and associated observations

In 1950, Houtz [143] first observed the color change when a PAN fiber produced by DuPont (Orlon) was heated in air for 16 h at 200°C, the product becoming insoluble in typical PAN solvents, such as dimethylsulfone and not undergoing combustion in a naked

Bunsen burner flame. He postulated a condensed aromatic ring structure (Scheme I) involving a dehydrogenation reaction forming heterocyclic fused rings:-

From the IR spectrum of a PAN powder, Burlant and Parsons (1956) [144] found at that 200 and 260°C, the band in the spectrum assigned to the nitrile group gradually disappeared, coinciding with the appearance of a band attributed to a conjugated —C=N— grouping and they assigned a partially cyclized aromatic structure (Scheme II), a view substantiated by La Combe (1957) [145].

From other IR work, Burlant and Parsons concluded that aromatization of the ring structure occurred at temperatures above 300°C.

Grassie and co-workers (1958) [146] proposed that the hydrogenated naphthyridine structure was formed by the nitrile polymerization reaction alone.

This scheme for a so-called ladder polymer became widely accepted and is normally quoted in textbooks. The conjugated —C=N—C=N— structure occurring during stabilization behaves as a chromophore, which would explain the appearance of color.

In 1958, Schurz and co-workers [147] introduced the concept of an azomethine crosslink (Scheme III), whereby the methine proton is abstracted by the nitrile nitrogen of an adjacent chain and claimed that such a structure was required to explain the insolubility of the stabilized PAN.

Schurz also stated that the UV spectrum was different from that expected of a small molecule heterocyclic compound, which would be similar to substituted pyridines and naphthyridines, resembling six-membered rings arising from cyclization.

Schurz, however, did emphasize that both cyclization and azomethine crosslinking schemes were, at that time, unproven.

Kennedy and Fontana (1959) [148], working with a PAN powder heated in vacuum in the range 230–265°C used IR spectroscopy to establish a change from nitrile groups to a conjugated —C=N—C=N— structure. When heated in air, there was a shortfall in the C, H and N content, which was not explained and the presence of an exotherm at 265°C was attributed the formation of a ladder polymer, triggered by a free radical initiation reaction. They were also the first group to report on the exotherm which occurs.

Grassie and Hay [149] suggested that the intimate packing of chains in AN allowed the reaction to move from molecule to molecule to produce propagation cross-links (Scheme IV), a process inhibited by alkali due to the separation of the molecules, giving deeply colored homogeneous solutions, but this color was found to be additional to the conjugated structure:

Grassie and Hay (1962) [150] heated powdered PAN in a vacuum at 175°C finding that the higher the amount of an acrylic acid copolymer present the faster the —C=N— groups disappeared and, in order to explain this phenomenon, they advanced an ionic mechanism for the initiation and growth of the ladder polymer (Scheme V):

On the basis of IR studies, Conley and Bieron (1963) [151], firmly believed that both the cyclization and azomethine crosslinking mechanisms were incorrect and claimed to have isolated C=C groups on the chain backbone (Scheme VI), which was attributed to hydrogen elimination and the formation of a 1,1,2-trisubstituted olefinic linkage:

Other authors such as Berlin *et al* (1964) [152] and Fester (1965) [154,161] have discussed the formation of the chromophore, also with the elimination of hydrogen, but with the formation of a polyene structure, (Scheme VII) (a conjugated double bond structure with a pendant nitrile group).

Takata and Hiroi (1964) [153], using a model approach, concluded that cyclization was a fair presumption, whilst a later study of model substances by Brandrup and Peebles (1968) [164] suggested that the azomethine structure was unlikely to be the source of the chromophore.

Bell and Mulchandani (1965) [155] studied changes in tensile strength of Courtelle heated in air and N_2 and found that the fiber did not become unduly brittle. They deduced that the degree of crosslinking was not high, since this would be associated with brittleness.

Monohan (1966) [156], when heating PAN at 280–450°C, found the major volatile products to be cyanogen, HCN, AN, acetonitrile, vinylacetonitrile, benzene, toluene, pyridine, 3- and 4-methylpyridine and 1,3,5-triazine. Also 3,6-dimethyl-1,8-naphthyridine was isolated and identified with IR and UV. Ammonia was only found when traces of water were present and the following reaction was demonstrated at room temperature:

$$H_2O + HCN \rightarrow CO + NH_3$$

The only structure (Scheme VIII) of the residue that supported all the obtained microelemental and spectrophotometric data was:

The product 3,6-dimethyl-1,8-naphthyridine was isolated in the reaction crude:

and so supported the naphthyridine formation degradation theory.

Thompson (1966) [157] used DSC to study the thermal behavior of PAN at 250–325°C and observed the exotherm effect in air and N_2, which was contrary to the observations of

Kennedy and Fontana [148], who found the effect only in a vacuum. The exotherm became broader and less intense with decreasing molecular weight of low molecular weight PAN.

Peebles and Brandrup (1966) [158] found that the species responsible for the color developed on oxidation of PAN was due to a random copolymer of polyimine and polynitrone bonds (Scheme IX) and not a system of conjugated olefinic bonds, as previously believed:

Noh and Yu (1966) [159] used IR to study the decay of the nitrile peak of PAN powder at 170, 200 and 230°C, the decay increasing with temperature and the remaining —C≡N groups dropping to some 20% in about 6 h at 230°C, suggesting that these groups were unable to take part in ladder growth. They associated this phenomenon with relatively short ladder polymer runs. Watt [160], however, stated that if the initiation was random, it could proceed to the left or right of the ladder polymer and a gap could occur between the two initiating points propagating outwards and away from each other, or conversely, propagating in opposite directions but towards each other and would, therefore, be unable to join.

Hay (1968) [162] used DSC to study the thermal reactions of PAN. Volatiles were not detected until the temperature reached above 230°C, when least three distinct thermal processes competing with one another were found—coloration by cyanide group polymerization, ammonia evolution from the coloring structures, and chain scission and decomposition.

Reich (1968) [163] suggested that the exotherm is caused by nitrile group polymerization (Scheme X):

Peebles and co-workers (1968) [164] concluded that abnormal structures are present in PAN and the initiation of the coloration reaction takes place at such defects (Scheme XI). The defects originate from a side reaction during the free radical polymerization of PAN.

All free radical prepared polymers are believed to incorporate enamine structures or ketonitrile if the polymerization conditions favored hydrolysis. The rate of yellowing of PAN in air at low temperatures is proportional to the concentration of the above defects.

Brandrup and Peebles (1968) [165] studyied model compounds which contained certain aspects of the structure of PAN and found that the initial attack of O_2 on the polymeric chain was at the methylene hydrogens. The hydroperoxide was formed, which subsequently broke down to yield a β-ketonitrile, which initiated the color forming polymerization of nitrile groups. The conjugated structure then reacted with atmospheric O_2 to partially form a polynitrone (Scheme XII), the final chromophore.

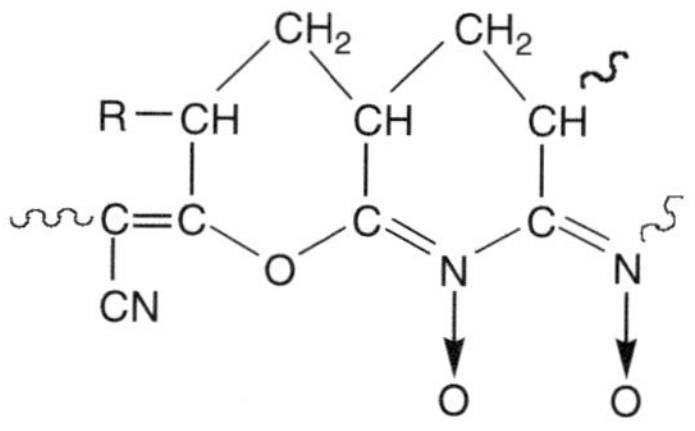

Kubasova et al (1969) [166] showed the decrease of nitrile groups in various cast PAN films was due to the formation of ladder polymer.

Watt (1970) [167] visualized the formation of ladder polymer (Scheme XIII) by analogy with the formation of keto groups during oxidation of hydrocarbon polymers:

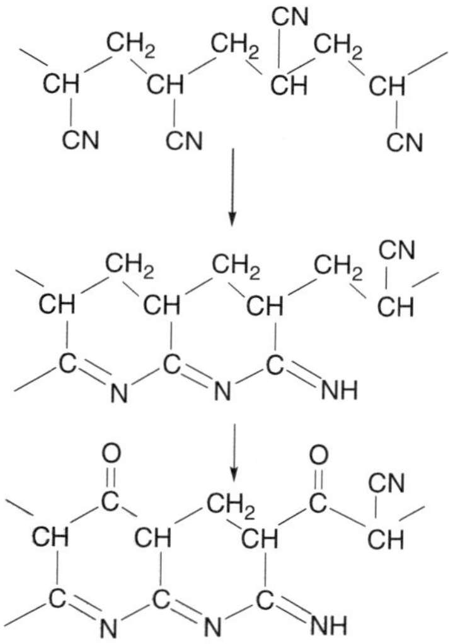

Ulbricht and Makschin (1970) [168] showed that the rate of coloration of some redox initiated polymers depended on the concentration of end groups.

In 1971, Fiedler et al [169] measured TGA, DTA, shrinkage and coloration of PAN in air and N_2. Shrinkage began at 200°C, becoming rapid at 250°C, which in this instance corresponded to the highest temperature without the occurrence of a sudden decomposition. The exotherm was greater in air and was attributed to oxidative stabilization being superimposed on the heat of cyclization.

In 1971, Rose [170] submitted a Thesis on The Preparation of Carbon Fibers and with Watt and Green (1971) [171], provided useful resumés of the pyrolysis of PAN.

Standage and Matkowsky (1971) [172], through IR studies with Dralon T based on the equation

$$(C_3H_3N)_n + \tfrac{1}{2}nO_2 \rightarrow (C_3HNO)_n + 2nH$$

proposed Scheme XIV for an oxidized ladder with O_2 incorporated in the cyclized structure as epoxide or ketonic groups (Scheme XV):

Bailey and Clark (1971) [173], using IR with oxidized Courtelle, detected the following bonds: C=C, C=N, C—N, O—H, N—H, CH_2, C≡N and C=O. The intensities of the —CH_2, —C≡N and —C=O bonds were found to be reduced and they proposed the model structure (Scheme XVI) for partially oxidized fibers:

Fitzer and Müller (1971) [174] proposed that shrinkage occurred as a result of cyclization when PAN was heated at 180–250°C. Intramolecular addition between isotactic neighboring C≡N groups leads to ring formation, whilst intermolecular addition between opposite C≡N groups of neighboring chain molecules leads to acyclic binding (Scheme XVII). These reactions go on side by side, bringing about the cyclization of PAN to an unstable tetrahydronaphthyridine ring system, producing a partial ladder structure on dehydrogenation.

I = intramolecular reaction
II = intermolecular reaction

Johnson et al (1972) [175] found that NaCN rapidly discolored PAN dissolved in DMF (Scheme XVIII). They suggested that in the case of a reaction catalyzed by CN^- ion, initially there was a nucleophilic attack by the CN⁻ on the cyano group, followed by removal of the acidic proton, producing a resonance stabilized carbanion:

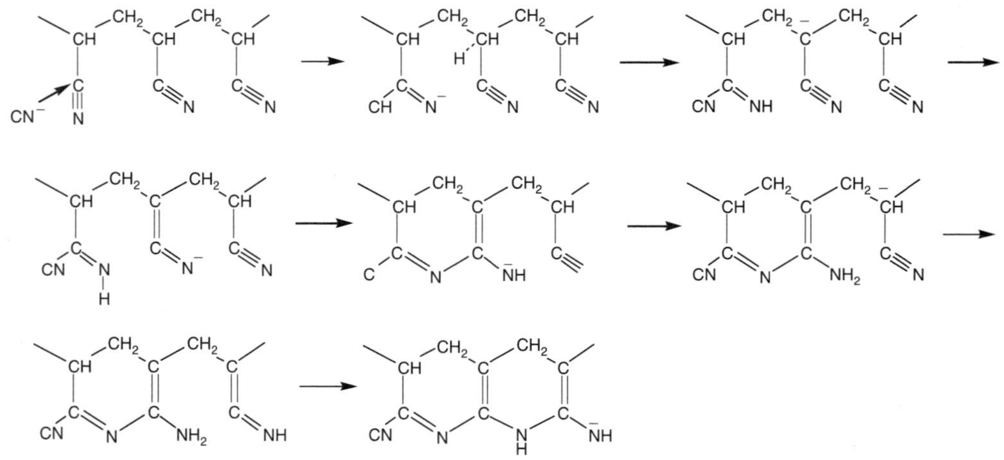

The dihydropyridine structure will readily oxidize to hydroperoxide, which can then break down into ketonic and aromatic structures (Scheme XIX):

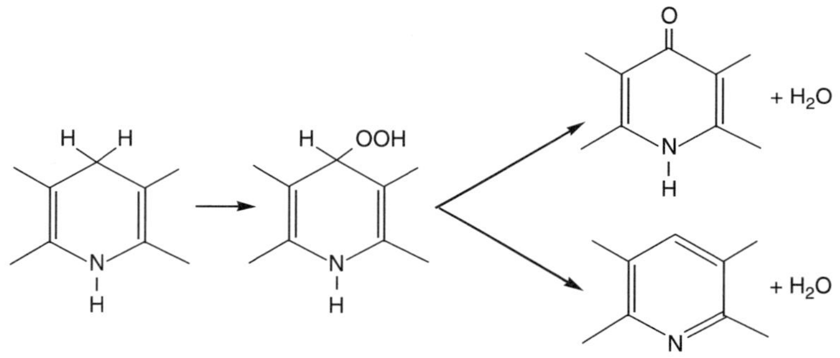

Johnson et al [175] presented an overall structure (Scheme XX) for the pre-oxidized fiber as:

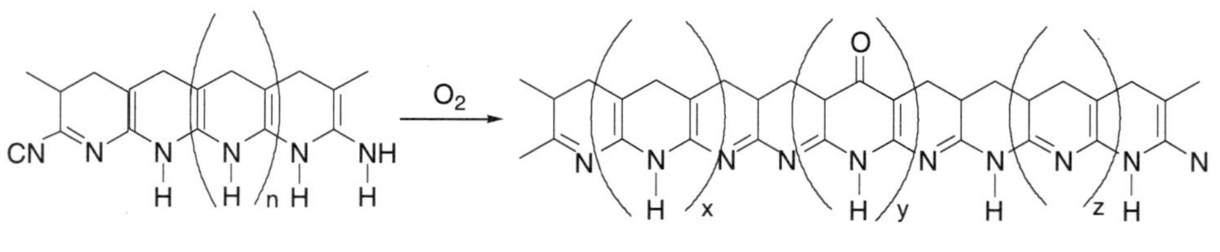

Dihydropyridine   4-pyridine   Fully aromatic structure

Grassie and McGuchan (1970–72) [176–184], using TGA, DTA, TVA (thermal volatilization analysis) and IR, carried out a series of pyrolysis investigations on powder forms of PAN and substituted PAN polymers in order to determine the effect of introducing various copolymers.

O_2 was found to inhibit the thermal polymerization of CN groups, altering the characteristics of the exotherm, whilst in air, the exotherm occurred at higher temperatures [176]. Sharp intense exotherms were found irrespective of the method of polymerization [177]. From IR evidence, CN absorption is substantially reduced during the reaction and some 80% of the CN groups participate [159]. Grassie and McGuchan proposed that the zip length of the reaction is fairly short and followed Scheme XXI:

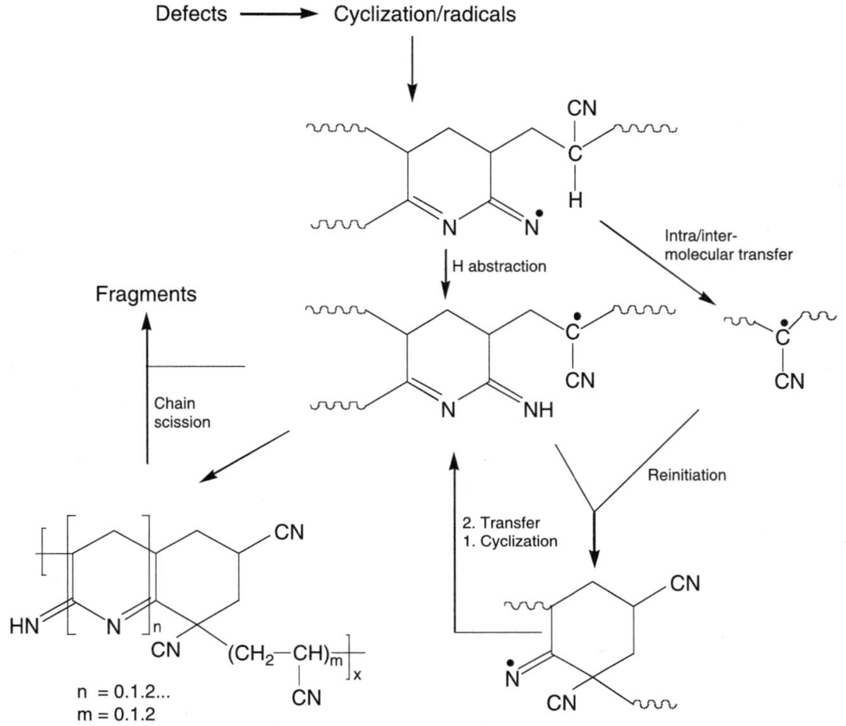

Inorganic salts such as KCN, NaSCN and Na_2S can act as ionic initiators [179], but in the absence of additives, the reaction is free radical in nature.

The oligomerized structure was subsequently slightly revised [180] and is shown in Scheme XXII:

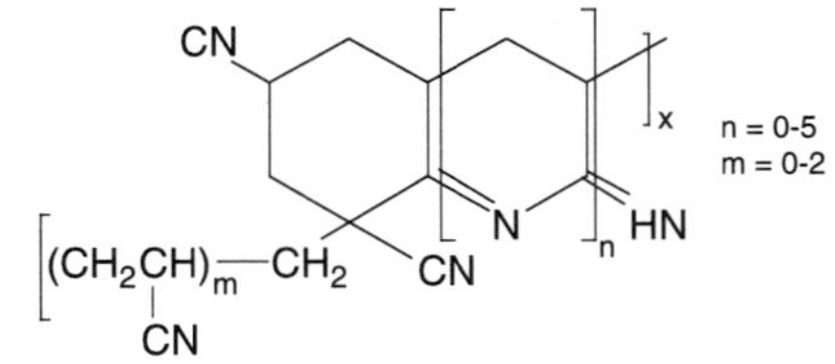

To obtain such a structure, the initiation and termination of the cyclization process must occur at frequent intervals along the chain and it was concluded [180] that termination took place by hydrogen transfer (Scheme XXIII):

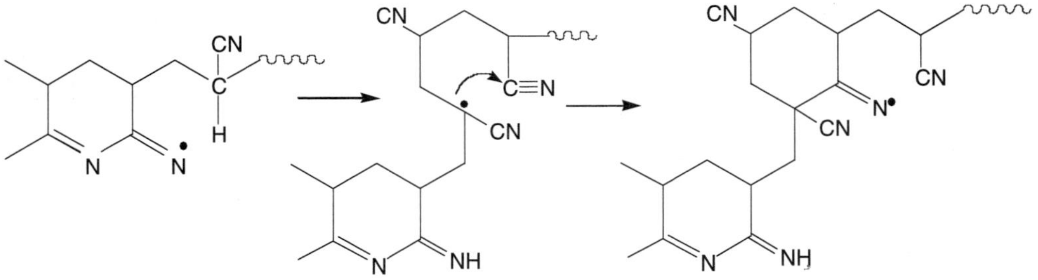

One very interesting aspect of Grassie and McGuchan's work [181] was that certain co-monomers such as ITA acted as initiators for the cyclization (Scheme XXIV):

These co-monomers helped in moderating the stabilization exotherm with rated effectiveness in the order ITA > MAA > AA > acrylamide.

Acrylate and methacrylate co-monomers participate in the nitrile reaction [182] but styrene type monomers act as a blocking agent. Chlorinated co-monomers degrade the AN units at lower temperatures [184] by a dehydrochlorination mechanism (Scheme XXV):

Grassie and McGuchan did not find any evidence that chlorinated comonomers participated in the cyclization reaction [182].

Fitzer and Müller (1973) [185] showed, using DTA and IR, that with homopolymer (Dralon T), O_2 (as air) acts as an inhibitor in the exothermal reaction around 300°C, whereas with the copolymer (Dralon), it had practically no effect.

Clarke and Bailey (1973) [186,187] have proposed a structure (Scheme XXVI) based on elemental and IR analysis for Courtelle fibers heated to 225°C in air for 5 h:

Hydrogen bonding possibilities have been omitted for clarity. The order in which the different species occur cannot be specified, but they are in the correct elemental ratios. The carboxylic acid co-monomer initiates cyclization. The O_2 in the oxidized polymer is present as —OH (probably in β positions) and —C=O groups. The structure can be considered as stiff planar cyclized rings connected with relatively mobile linear segments. The cyclized rings are in different planes and during subsequent pyrolysis in the carbon fiber process, tend to twist and form a more planar structure *en route* to a graphite structure, probably due to a condensation reaction between the hydroxyl groups.

Watt and Johnson (1975) [188] reconciled the various structures proposed for the formation of ladder polymer based on oxidation and tautomeric changes (Scheme XXVII) where the length (n) of the conjugated sequences is about 4–5 monomer units. Chain scission has not been accounted for, but does occur, as shown by the evolution of CO_2, probably from the breakdown of —COOH groups.

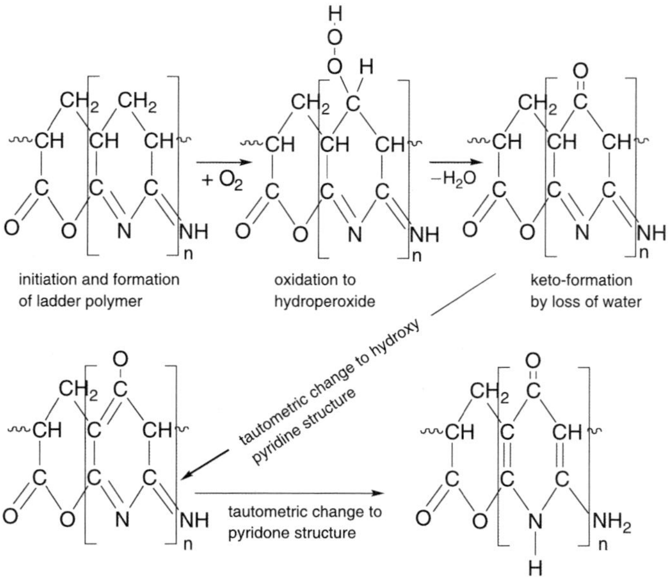

Examination of a thin section of oxidized PAN by an electron probe microanalyzer for O_2 has revealed that the outer brown core was oxidized and relatively rich in O_2 as compared to the inner yellow core [189].

Fitzer and Müller (1975) [190] showed that during stabilization, O_2 can initiate the formation of active centers for cyclization, but can also retard the reaction by increasing the activation energy.

Rascovic and Marinkovic (1978) [191] treated PAN fiber in SO_2 for 3 h at 230°C and then for 1 h at 290°C. The fiber reacted slowly at low temperatures, the rate increasing towards 230°C and becoming rapid at 290°C, to give a cyclized and aromatized product with an increased diameter. Based on DTA, IR and elemental analysis, they proposed the following structure (Scheme XXVIII):

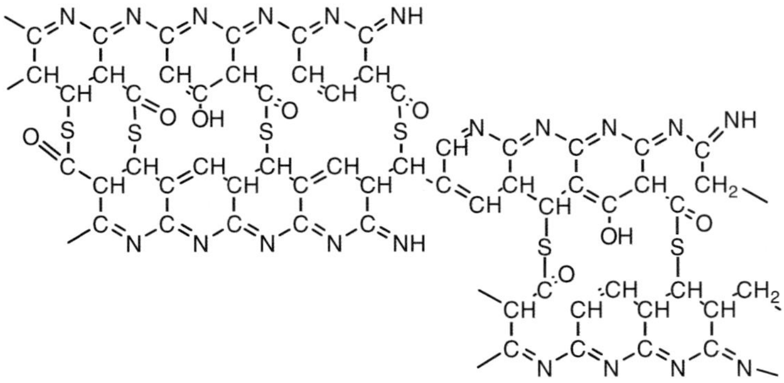

Coleman and Petcavich (1978) [192] used FTIR to propose a route (Scheme XIX) via the initial formation of the Grassie ladder, followed by a tautomeric change to the polycyclic dihydropyridine structure. Oxidation followed, producing carbonyl groups, a prerequisite for the subsequent condensation reaction between adjacent ladder formations to give a final structure proposed by Potter and Scott and more fully reported by

Johnson et al [175]:

Warner et al (1979) [193–196] studied the oxidation of the following 1–4 denier fibers: Acrilan (7% VA), Orlon (6% MA), Dralon T (homopolymer) and Courtelle (6% MA, 1% ITA) using O_2 analysis, microscopic examination of thin cross-sections, SEM and X-rays. The mechanism of oxidative stabilization was found to be basically the same for all acrylic fibers, but was dependent on temperature, fiber diameter and fiber chemistry. The authors have divided the stabilization process into two stages—the first stage they termed prefatory and the second sequent. Prefatory reactions involve initiation and polymerization of the —CN groups to form a reddish brown chromophore (assuming a sequent reaction does not occur simultaneously). The reddish brown material is etched by sulfuric acid and burns. Oxidative reactions can involve the production of hydroperoxides, carboxylates and other active groups, all of which act as initiators for —CN polymerization, forming a polyamine or polyimine chain. Sequent reactions follow when the product of a prefatory reaction is exposed to O_2, quickly turning black. It is not etched by 50% $^w/_w$ sulfuric acid and does not burn. When heated at 220°C, Courtelle had a diffusion controlled rate, but the other fibers showed no zones or cores. If the time for Courtelle was below 8 min, there was no evidence of a two-zone structure. The diffusion of O_2 to reactive sites is limited by previously oxidized material and it is suggested that the slow movement of the oxidized zone towards the center occurs as the O_2 diffuses through the reacted zone, from which time the reaction proceeds in compliance with the classic diffusion-limited ($t^{1/2}$) kinetics. When the prefatory and sequent reactions occur sequentially, the overall stabilization is limited by the rate of the prefatory reaction, but is also limited by the formation of a skin that acts as an O_2 barrier.

The authors [194] examined the stabilization process with fiber of constant length (as in batch processing) and also with fiber fed through the oven at a constant rate (as in continuous processing). In the batch process, during heat up, the fiber tension rises once the T_g has been exceeded and is related to entropic recovery of a previously drawn and quenched fiber. Subsequently, the tension rises again due to stabilization kinetics. These shrinkage reactions can occur in the absence of an oxidizing gas, but speed up in the presence of O_2. When the fiber is continuously processed, the velocity of the fiber varies with the tension and depends on the precursor prehistory: first the fiber shrinks and the diameter increases, then it draws above the T_g and finally, the fiber shrinks again in response to stresses developed due to chemical reactions.

The authors [195] used X-ray diffraction and TEM studies to suggest that PAN fibers have two orders of fine structure within a typical acrylic fibril. Using molecular models, they deduced that the major order was a series of rods parallel to the fibril and the fiber axis, with a diameter about 0.6 nm. In previous chapter (Figure 4.15) The lamellar structure arising from the rods is ordered into a liquid crystal-type array perpendicular to the fibrilar axis, with a repeat spacing of about 12 nm, the rods about 8.0 nm long interspersed with less ordered material (loops, folds, entanglements, chain ends, defects etc.) about 4.0 nm between the rods. The polymer within each rod is twisted into a helical shape, with the —CN groups oriented at various angles to the axis of the rod. It is believed that during oxidative stabilization, selective degradation of this acrylic structure occurs.

Stabilized acrylic fibers are extremely hygroscopic [196], gaining about 8% in moisture and elongating about 15% and can be considered as plasticized material. PAN is strongly plasticized by water.

Olivé and Olivé (1980) [197] have discussed the molecular interactions of PAN and present a rigid irregular helical conformation (Figure 4.14).

Manocha and Bahl (1980) [198] state that during the stabilization of a copolymer in O_2, dehydrogenation and polymerization occur simultaneously, whereas in air, dehydrogenation precedes polymerization. Bahl, Mathur and Kundra (1980) [199] used a sulfur dioxide atmosphere for stabilization.

Olivé and Olivé (1981) [200] presented evidence that intermolecular oligomerization leading to cross-linked structures takes place and at higher temperatures, could be more important than intramolecular reaction leading to ladder polymer.

The oligomerization of —CN groups becomes important above 200°C and since the main chain carbon bonds are the weakest, it is fairly certain that main chain scission occurs and the newly formed radicals initiate the oligomerization. There would appear to be no benefit from an isotactic polymer since the —CN groups would have to be forced into position to react. Adjacent —CN groups in the same chain are held apart by mutual repulsion in the helix-like structure. However, adjacent —CN groups from different polymer chains have strong dipole-dipole interaction bringing the groups into an ideal position for reaction (Scheme XXX) favoring intermolecular reaction:

Chen et al (1981) [201] studied the effect of decolorizing with NaOCl (5.25% available chlorine) on partly or fully stabilized acrylic fibers, heated for various times at 220–250°C. With lightly stabilized fibers, decolorization takes place in two steps—an incubation period with slight decolorization followed by a further incubation period and then gradual decolorization, whereas, with the fully stabilized fiber, the two processes occur simultaneously.

Olivé and Olivé (1983) [202] present a review of the chemistry of formation of carbon fiber from PAN.

Jain and Abhiraman (1983) [203] used thermal analysis and WAXD to study the effect of annealing acrylic precursor fibers for 2 min at 230°C and 4 min–4 h at 270°C. Annealing in the absence of constraint caused significant shrinkage, an increase in the orientation of the ordered phase, but a decrease in the overall orientation. The overall orientation increases if the fiber is constrained and is retained in subsequent annealing without restraint.

Sivy, Gordon and Coleman (1983) [205], studied the IR spectra of the cyclization and oxidation of an acrylamide copolymer in air at 200°C and showed that there were two competing reactions—the initial step being an intramolecular cyclization and crossover, producing the step-ladder polymer initiated by the acrylamide group (Scheme XXXI):

This is followed by a degradation route [151], where a 1,1,2-trisubstituted olefin structure is formed by reaction with O_2 (Scheme XXXII):

The unsaturation introduced permits the elimination of HCN, made possible by the activation of the allylic hydrogens and the cyano groups. A diene structure can then react with an olefinic group of a neighboring chain via a Diels-Alder addition (Scheme XXXIII), which would be an excellent route for the formation of a graphitic structure:

Carbon fibers made from PAN stretched in a hot solution of CuCl were found to have improved mechanical properties by Mathur *et al* (1984) [206].

Takaku and Shimazu (1984) [207] studied the volume contraction of PAN fiber during stabilization at 252°C, reaching a value of 1.57 g cm^{-3} in some 16 h, which is consistent with Scheme XXX.

Fochler *et al* (1985) [208] used ^{13}C NMR and IR to study stabilization and believed that all bands within the IR spectrum could be accounted for by cyclization to a partly aromatic structure with the random distribution of one double bond throughout the ring by tautomerization (Scheme XXXIV):

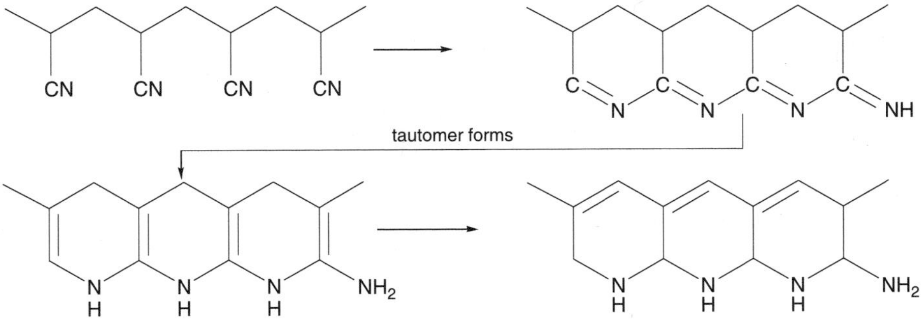

Fitzer, Frohs and Heine (1986) [209] used TMA and DSC to study the optimum heating rate for a 6% MA/1% ITA PAN copolymer. Shrinkage measurements indicated the start and end of the stabilization reaction. The optimum heating rate was 5°C min^{-1} reducing to 1°C min^{-1} when the reaction started, in order to avoid overheating by the exothermic reaction and stopping at 270°C to avoid over-oxidation.

Morita *et al* (1986) [210], using XPS, proposed Scheme XXXV for an air stabilized fiber, although it is, in reality, a more complex three-dimensional structure:

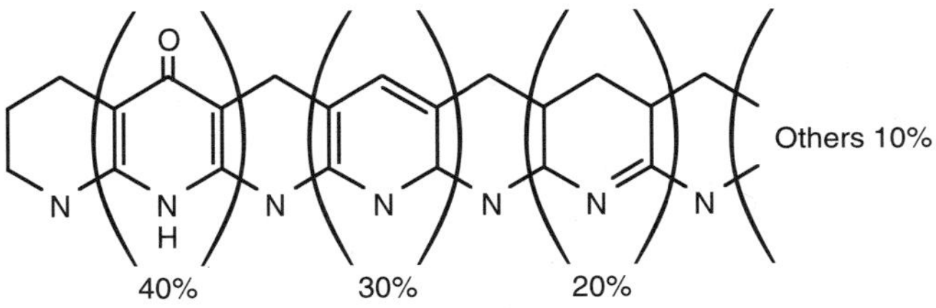

Jain and Abhiraman (1987) [211] reviewed the physical and morphological aspects of the conversion of AN based precursor fibers to carbon fibers, whilst Jain *et al* (1987) [212] studied precursor morphology and thermo-oxidative stabilization as well as sonic modulus, shrinkage, density and mechanical properties of Orlon 43 and a fiber spun from a copolymer of AN and ITA. The diffusion controlled incorporation of O_2 was consistent with the increase in density. Although the rate of stabilization increased with tension, this was probably associated with a reduction in diameter, hence increasing the overall rate of diffusion controlled reactions. The precursor containing ITA had a higher rate of stabilization. They agreed with Warner's [195] model of a PAN precursor consisting of repeat units of both oriented, laterally ordered as well as oriented but laterally disordered domains, with a significant portion of the chain segments in the latter phase forming a bridge between the ordered domains.

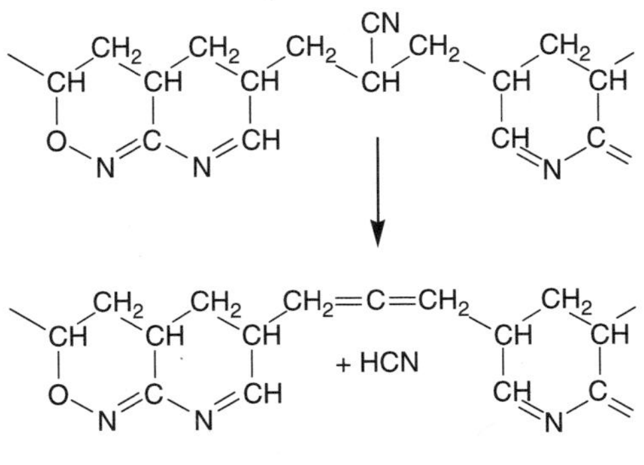

The overall orientation can be increased by stretching at temperatures above those seen in processing the PAN precursor.

Bhat *et al* (1989) [213] showed that the presence of O_2 was essential for reaching the stabilized state and NH_3 acts as an accelerator for the stabilization reactions.

Usami *et al* (1990) [214] studied the thermal degradation of PAN fibers (Nikkiso AN/MA/ITA) using gas chromatography, ^{13}C NMR and FTIR, confirming that the main stabilization reaction was the formation of a ladder polymer and that the contribution of dehydrogenation and/or dehydrocyanide reactions was to form conjugated polyene structures. Scheme XXXVI shows the competitive thermal reactions:

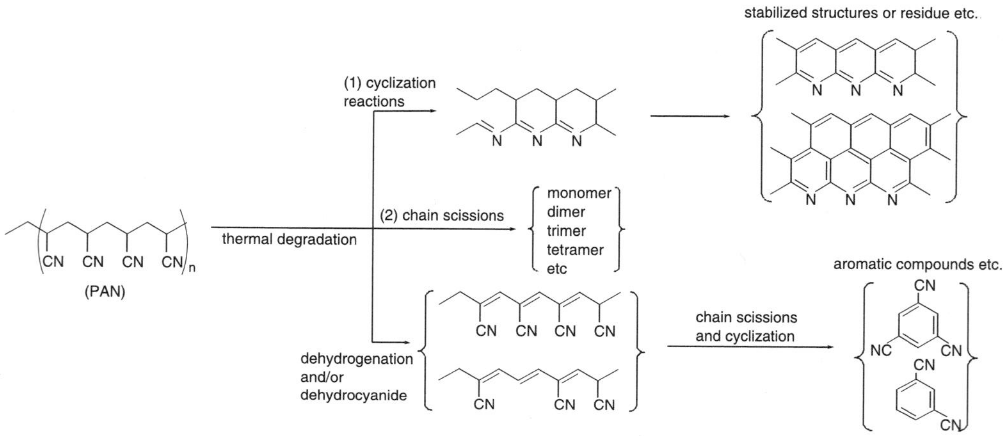

They proposed the model (Scheme XXXVII) for air stabilized fiber, which shows O_2 in the non-aromatic portions. However, this was not found in the spectral analysis of the evolved volatiles.

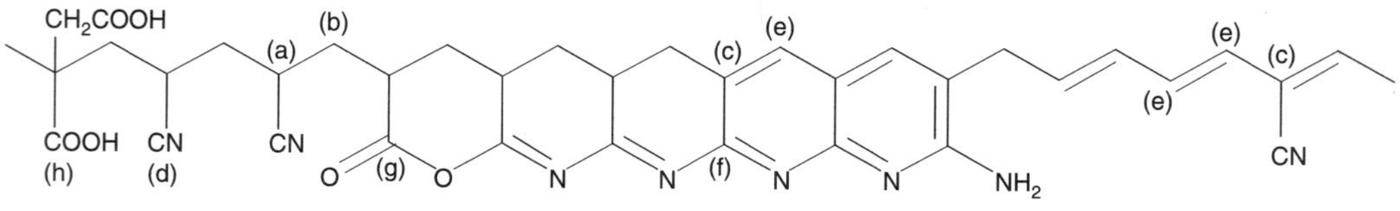

Resonance (ppm)	Functional group (ppm)
(a) 30	CH
(b) 30	CH2
(c) 115	C=C
(d) 122	C≡C
(e) 139	C=C
(f) 153	C=N
(g) 177	C=O
(h) 185	C=O

Bhat *et al* (1990) [215] continued their earlier work [213], showing that the presence of NH_3 could reduce the time of stabilization to almost half that required in air, but the presence of O₂ is essential to complete stabilization. Since NH₃ is one of the byproducts of the cyclization reactions, it would be expected that removing NH₃ would shift the equilibrium favorably and increase the rate of cyclization. According to Hay [162] NH₃ is formed either by the aromatization of the propagating color unit:

or by the chance interaction of two propagating species:

Peebles et al (1990) [216] heated a homopolymer PAN under vacuum at 120–220°C in inert conditions and found using DSC that the polymer evolved NH₃ and HCN at 130°C, the NH₃ acting as an accelerator but not the HCN.

Table 5.18 lists the various means of initiating —CN polymerization given by Peebles et al [204]. The presence of these groups can be either due to adventitious reactive groups or if they have been deliberately introduced as an additive. Several of these mechanisms can occur simultaneously in the stabilization reaction.

Pejanovic and Pavlovic (1990) [217] consider the diffusion of O₂ through fully stabilized fiber to be about 300 times slower than unoxidized fiber.

Deurbergue and Oberlin (1991) [218] categorized various stabilized precursors prior to carbonization by an aromaticity index (AI):

$$\text{AI} = \frac{I_A}{I_A + I_P}$$

Table 5.18 Sources of nitrile initiation

1. Impurities such as residual polymerization products, polymerization inhibitors, catalyst fragments, solvent molecules, etc.
2. The molecular chain end groups [15].
3. Random initiation by the hydrogen atoms alpha to a —CN group [8,37] but disputed by [60].
4. Transformation of a —CN group to an azomethine by hydrogen transfer, which can then initiate polymerization [5,20].
5. The hydrolysis product from the keteneimine group randomly formed during polymerization.
6. The ketonitrile group formed by hydrolysis of the imine group, formed by radical polymerization through a —CN group. The hydrogen alpha to the —CN and —CO groups is a weak acid [22].
7. Hydrolysis of —CN groups to —NH₂ or —COOH due to a low pH of polymerization.
8. Initiation by —NH₃ generated during pyrolysis.

Source: Reprinted with permission from Peebles LH Jr., Peyser P, Snow AW, Peters WC, On the exotherm of polyacrylonitrile: pyrolysis of the homopolymer under inert conditions, *Carbon*, 28, 707, 1990. Copyright 1990, Elsevier.

where I_A is the intensity of the 0.35 nm ladder polymer X-ray diffraction reflection and I_P the intensity of the 0.52 nm PAN reflection.

Dunham and Edie (1992) [219] used bundles of 12,000–60,000 filaments to establish a model of stabilization for Courtauld's SAF PAN based precursor by grouping the many reactions into three major groups of reactions—cyclization, dehydrogenation and oxidation. They found that bundle temperatures could be as much as 15°C above the oven temperature and predicted that incremental oven temperature profiles should be useful in reducing the time taken for stabilization.

Mathur et al (1992) [220] discussed the thermal degradation of PAN fibers. They recommend a heating rate of $5°C\,min^{-1}$ up to 180°C and $2°C\,min^{-1}$ thereafter. The maximum heat treatment temperature should generally be lower than the peak exotherm temperature and a step-wise increase in the temperature is preferred as well as commercially more viable. Although various oxidizing media can be used, air is preferred. During the initial period of stabilization, the sheath or skin becomes stabilized and acts as a barrier for further diffusion of O_2, producing a two-zone morphology. Reducing the diameter of the fiber will assist in stabilization. Over stabilization will lower the strength of oxidized fibers. Since PAN fiber is a poor conductor of heat, heat transfer problems will limit the heating rate. Stretching of the PAN fiber during manufacture causes the decoiling of the molecular structure, which reverses after subsequent heat treatment above the T_g, leading to physical shrinkage of the molecular chains. Chemical shinkage occurs as a result of the cyclization of the —CN groups leading to imperfect ladder formation, which is corrected by applying an optimum tension to ensure that the chains of two different molecules are close enough to permit reaction. The authors favor density to judge the extent of stabilization and X-ray diffraction can provide a fairly good idea. The sharp exotherm of PAN starts at about 200°C and reaches a peak at about 265°C, but careful analysis of a DSC curve has shown the presence of another small, but relatively broad peak, at about 325°C. Normally, stabilization is carried out up to a maximum of 270°C, but as a new approach to stabilization, it is suggested that the thermal stabilization be carried out at a higher temperature, when the following two reactions occur:

Reaction 1

1. Formation of bonds
   Two C—H bonds—in the formation of two molecules of HCN
   Two C—C bonds—cross-linking between atoms 1-2 and 3-4

2. Cleavage of bonds
   Two C—H bonds (X—X)
   Two C—C bonds (Y—Y)

Reaction 2

$$\text{[structure with } H_2, OH, OH \text{ groups]} + \tfrac{1}{2} O_2 \longrightarrow \text{[aromatized structure with } H_2, OH, OH\text{]} + H_2O$$

Reaction 1 causes intramolecular cross-linking of adjacent molecular chains, whilst Reaction 2 produces further aromatization of the fiber structure. The oxidized density increases from a normal 1.35 g cm^{-3} to 1.5 g cm^{-3}. Although it can be argued that this density would be achieved in the early stages of carbonization, which would be in an inert atmosphere, it does appear that the presence of O_2 in the extended oxidation stage does produce carbon fiber with superior mechanical properties.

Grove and Abhiraman (1992) [221] developed a mathematical model for the thermo-oxidative stabilization of acrylic fibers, giving results consistent with the trends observed in practice. However, the model only allows the use of one initiating co-monomer, but it does consider the following set of reactions:

1. reaction with a —CN group of species created through reaction of O_2 with the polymer
2. reaction of an initiating co-monomer with a —CN group
3. self activation among the —CN groups
4. reaction of O_2 with the backbone of the polymer
5. reaction of active —CN groups with unreacted —CN groups (nitrile polymerization)
6. termination of active species through mutual reaction among them

Ko *et al* (1992) [222] treated the precursor with aqueous $KMnO_4$ prior to oxidation and assumed that cyclization was initiated by the permanganate ion.

Tsai (1993) [223] used the AI to classify the degree of oxidation and found that a fiber with a low AI would give a carbon fiber with better mechanical properties when treated by a two stage oxidation process. An oxidized fiber with AI value greater than 50% will itself have poor properties and will, therefore produce carbon fiber with poor properties.

Tsai and Wu (1993) [224] showed that the exothermic heat generated in the continuous oxidation of a PAN precursor could build up on the surface of an internal roller and cause the fiber to overheat, leading to fusion of filaments resulting in carbon fiber with poor mechanical properties.

Beltz and Gustafson (1996) [226] studied the rate of cyclization of homopolymer PAN and an ITA copolymer PAN. Homopolymer PAN cyclized at 210–250°C in N_2, 10.5% and 21% O_2, there was an induction period which shortened as the temperature increased. The copolymer exhibited similar behavior in N_2, with the induction period decreasing with temperature and ITA content. However in an O_2 environment, there was no induction period at any concentration of ITA and the presence of O_2 did exert a retarding effect on the rate of cyclization, which to some extent, was counteracted by an increase in the ITA content.

Although cyclization is the most probable principal reaction associated with the exotherm, it is difficult to visualize since models are presented as two-dimensional, whereas in reality, the models will be a complex three-dimensional structure. Peebles [225] has attempted to portray a schematic three-dimensional intermolecular nitrile polymerization between two adjacent chains:

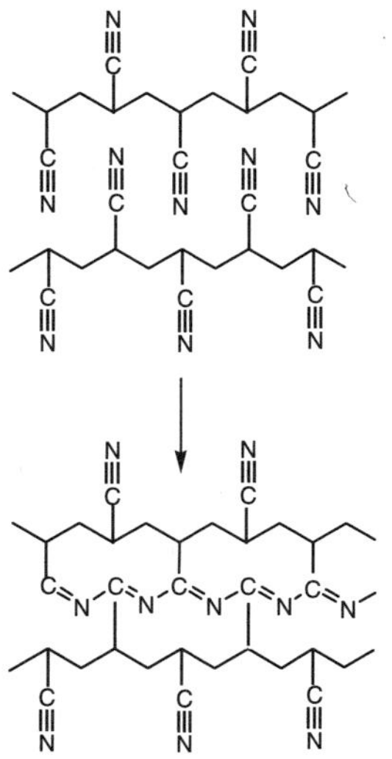

However, the structure will almost certainly involve more than two chains and would be more complex.

Stereoregularity of the acrylic precursor has been shown by Peebles and Snow to be of no importance to the stabilization reaction [225] and one explanation given by Peebles [225] is a rapid imine-enamine tautomerization of the final cyclized unit causing racemization or scrambling of any chiral centers:

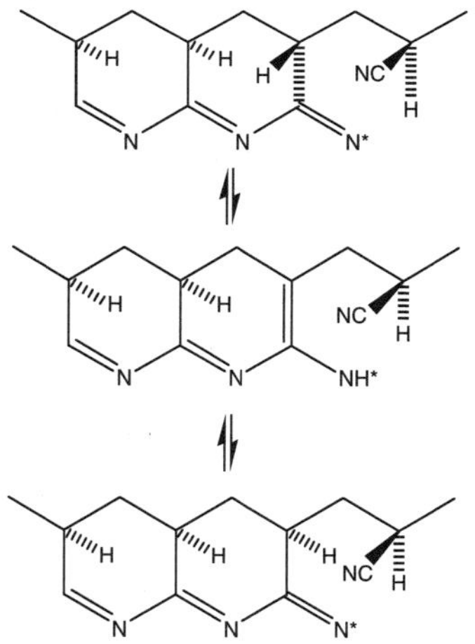

We have certainly broadened our knowledge in the last fifty years but, as yet, there is no definitive answer and it is highly probable that there could be a partial contribution from a number of these proposals.

5.15 MECHANISMS FOR THE CARBONIZATION STAGES OF PAN CARBON FIBERS

Very little is known about the stages of carbonization, with H2O, NH3 and HCN being evolved. Bailey and Clarke (1970) [227] proposed an idealized concept of reactions occurring during carbonization leading to hexagonal structures:

	C %	N %	H %	O %
HOMOPOLYMER PAN	68	26	6	-
STABILIZED PAN	65	22	5	8
CARBONIZED PAN (LHT)	>92	<7	<0.3	<7
CARBONIZED PAN (HHT)	100	-	-	-

The carbonized fiber known as LHT (low heat treated) can have a residual nitrogen content of up to 7% and even when treated at about 1400°C, can still be about 4%.

Watt (1970) [228] undertook the pyrolysis of stabilized (220°C) and unstabilized homopolymer PAN fibers *in vacuo* and analyzed the evolved gases. The HCN evolution suggested two different stages in pyrolysis, the first at about 450°C from the unladdered polymer and a further region of 450–900°C from the laddered structure and resulting in cross-linking.

Watt and Green (1971) [229] added further to this work and found that the unoxidized fiber evolved considerably more NH_3 in the range 300–500°C, which was attributed to a ring closure reaction between adjacent portions of ladder polymer and suggested that the oxidized fiber favored ladder polymer formation and consequently less ring closure. The oxidized fiber evolved CO_2 and H_2O in the early stages of pyrolysis, the presence of CO_2 indicating —COOH groups in the oxidized polymer. From 600°C upwards, N_2 was evolved peaking at 900°C suggesting an intermolecular reaction between adjacent ladder polymer sections:

But Watt believed that this reaction could not account for the evolution of all the N_2 as it would leave a twin ladder ribbon unless, fortuitously, there were favorable steric influences. He was fairly certain that the nitrogen was incorporated in the structure as a substitutional atom such as:

CARBON FIBER PRODUCTION USING A PAN PRECURSOR

At 400°C, C3 hydrocarbons (propylene, propadiene and propane) were evolved, possibly from C3 links in the ring structure:

Watt [160] suggested that the evolution of H_2O at 300–400°C could be by a H_2O elimination reaction:

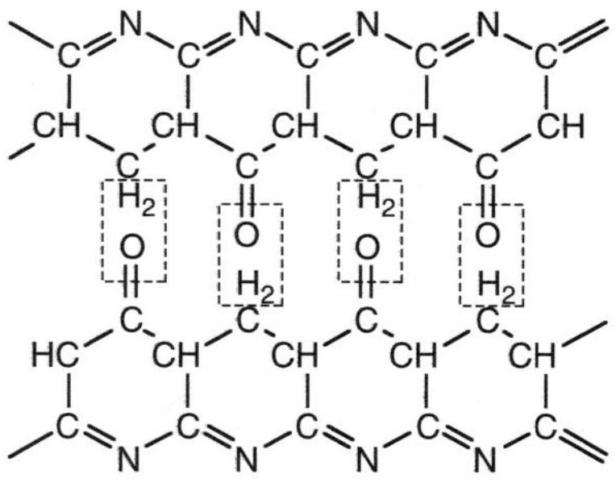

Watts proposed the following schemes to explain the two regimes of gaseous evolution:

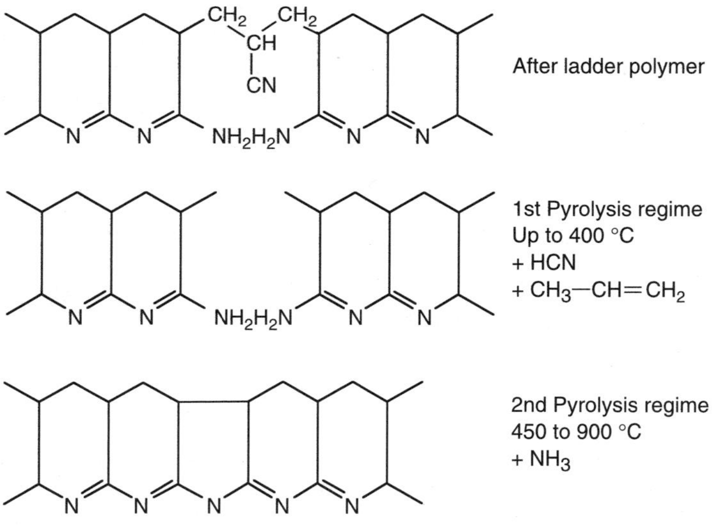

After ladder polymer

1st Pyrolysis regime
Up to 400 °C
+ HCN
+ CH_3—CH=CH_2

2nd Pyrolysis regime
450 to 900 °C
+ NH_3

Watt also postulated that the HCN evolved could be formed by joining ladder ends together or increasing the width of the product by sideways condensation:

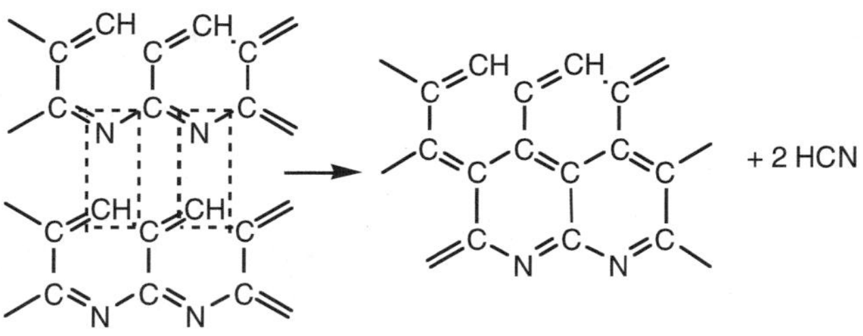

+ 2 HCN

Bromley, Jackson and Robinson at AERE, Harwell (1970) [230] carried out work to establish the nature of gaseous evolution when oxidized Courtelle was heated at 200–1000°C in both batch and continuous processes and N_2 was used to elute the gases from the fiber bundle. The authors reported four distinct regions of gas evolution up to 1000°C.

Region	Description	Approx. Range °C	Peak °C
I	H_2O evolution with little else occurring	50–250	160
II	Associated with an exothermic decomposition reaction	250–350	325
III	H_2O, CO_2, NH_3 evolution with little or no HCN	350–550	400
IV	HCN, H_2, CO and NH_3 evolution with little or no water and CO_2	550–1000	780

In the continuous process, a $10k$ oxidized Courtelle tow was pulled at 7.2 m h^{-1} under tension, first through a furnace at 200°C used as a drier, flushed with a counter-current of N_2 and then, through a second furnace with temperature varying in the range 250–1000°C, with a fixed concurrent N_2 flow. When steady conditions had been reached in the second furnace, a gas sample was taken and results reported as volumes of each gas removed in 10 ml (NTP) of N_2 purge gas.

Stage	Description	Approx. Range °C
I	H_2O evolution with little else occurring.	250
II	Associated with evolution of H_2O, NH_3, HCN and CO_2, increasing to a peak at 425°C.	300–460
III	NH_3 and H_2O evolution increase markedly, but CO_2 and HCN barely increase. There is a rise in the amount of high molecular weight hydrocarbons in the range 400–600°C.	460–600
IV	H_2 shows a marked rise, probably due to cracking of high molecular weight hydrocarbons, accompanied by a rise in HCN and CO production. At 600–700°C, a marked decrease in the NH_3, H_2O and CO_2 evolution seen (may be due to conversion to C, H_2, N_2, HCN and CO.	600–800

It is interesting to note that the release of C, H_2 and N_2 is more or less a gradual increase from 300 to 1000°C, with no drop-off in rate. The C/N ratio for total gas evolved remains close to unity over the entire temperature range, which could be taken to imply that the basic reaction is one of decomposition of —CN groups, or oxidized cyclized forms of this group, leaving the main —CH_2— relatively intact.

The CO_2 is partly derived from the MMA and ITA present in the Courtelle. Carbon dioxide can also arise from hydrolysis of —CN groups by water released elsewhere in the process:

$$-\overset{|}{\underset{|}{C}}-CN \xrightarrow{H_2O} -\overset{|}{\underset{|}{C}}-CO-NH_2 \xrightarrow{H_2O} -\overset{|}{\underset{|}{C}}-H + CO_2 + NH_3$$

Above 600°C, the production of CO_2 ceases and CO formation increases, which could be due to:

$$HCN + H_2O \longrightarrow NH_3 + CO$$

$$C + CO_2 \longrightarrow 2CO$$

$$C + H_2O \longrightarrow CO + H_2$$

$$\begin{array}{c} -\overset{|}{\underset{|}{C}}-H \\ -\overset{|}{\underset{|}{C}}-CO-NH_2 \end{array} \longrightarrow \begin{array}{c} -\overset{|}{C} \\ -\overset{|}{\underset{|}{C}} \end{array} + CO + NH_3$$

The higher molecular weight hydrocarbons (up to C6) are probably produced from side chains and may be copolymer materials attached to the side of the main chain.

REFERENCES

1. Shindo A, *Report 317*, Government Industrial Research Institute, Osaka, Dec 1961.
2. Johnson J, Phillips LN, Watt W, Brit. Pat. 1,110,791, Apr 1965.
3. Goodhew PJ, Clarke AJ, Bailey JE, A review of the fabrication and properties of carbon fibres, *Mater Sci Eng*, 17, 3–30, 1975.
4. Morgan PE, Carbon fibres, Hughes AJ et al, The production of man-made fibres, *Text Pro*, 8(1), 69–77, 1976.
5. Olivé GH, Olivé S, The chemistry of carbon fiber formation from polyacrylonitrile, *Adv Polym Sci*, 51, 1–60, 1983.
6. Guigon M, Oberlin A, Desarmot G, Microtexture and structure of some high modulus PAN-based carbon fibers, *Fiber Sci Technol*, 20, 177–198, 1984.
7. Watt W, Chemistry and physics of the conversion of polyacrylonitrile fibres into high modulus carbon fibres, Watt W and Perov BV eds., Vol 1, *Strong Fibres*, Elsevier, Amsterdam, 327–388, 1985.
8. Johnson W, The structure of PAN-based carbon fibres and relationship to physical properties, Watt W and Perov BV eds., Vol 1, *Strong Fibres*, Elsevier, Amsterdam, 389–444, 1985.
9. Moreton R, The tensile strength of PAN-based carbon fibres, Watt W and Perov BV eds., Vol 1, *Strong Fibres*, Elsevier, Amsterdam, 445–474, 1985.
10. Thorne DJ, Manufacture of carbon fibre from PAN, Watt W and Perov BV eds., Vol 1, *Strong Fibres*, Elsevier, Amsterdam, 4.745–576, 1985.
11. Fitzer E, Frohs W, Heine M, Optimization of stabilization and carbonization treatment of PAN fibers and structural characterization of the resulting carbon fibers, *Carbon*, 24(4), 387–395, 1986.
12. Jain MK, Abhiraman AS, Conversion of acrylonitrile-based precursor fibres to carbon fibres. Part 1 A review of the physical and morphological aspects, *J Mater Sci*, 22, 278–300, 1987.

13. Jain MK, Balasubramanian M, Desai P, Abhiraman AS, Conversion of acrylonitrile-based precursor fibres to carbon fibres. Part 2 Precursor morphology and thermo-oxidative stabilization, *J Mater Sci*, 22, 301–312, 1987.
14. Balasubramanian M, Jain MK, Bhattacharya SK, Abhiraman AS, Conversion of acrylonitrile-based precursor fibres to carbon fibres. Part 3 Thermo-oxidative stabilization and continuous, low temperature carbonization, *J Mater Sci*, 22, 3864, 1987.
15. Fitzer E, PAN-based carbon fibers-present state and trend of the technology from the viewpoint of possibilities and limits to influence and control the fiber properties by the process parameters, *Carbon*, 27(5), 621–645, 1989.
16. Damodaran S, Desai P, Abhiraman AS, Chemical and physical aspects of the formation of carbon-fibers from PAN-based precursors, *J Text Inst*, 81(4), 384–420, 1990.
17. Gupta AK, Paliwal DK, Bajaj P, Acrylic precursors for carbon fibers, *J Macromol Sci Rev Macromol Chem Phys*, C31(2–3), 301–310, 1991.
18. Bashir Z, A critical review of the stabilization of polyacrylonitrile, *Carbon*, 29, 1081, 1991.
19. Rajalingam P, Radhakrishnan G, Polyacrylonitrile precursor for carbon-fibers, *J Macromol Sci Rev in Macromol Chem Phys*, C31(2–3), 301–310, 1991.
20. Peebles LH Jr., Carbon fibres: Structure and mechanical properties, *Int Mat Rev*, 39(2), 75–92, 1994.
21. Bajaj P, Roopanwal RK, Thermal stabilization of acrylic precursors for the production of carbon fibres: An overview, *JMS Rev Macromol Chem Phys*. C37(1), 97–147, 1997.
22. Mittal J, Mathur RB, Bahl OP, *Carbon*, 35, 1713–1722, 1997. (A review paper)
23. Benjamin W, *Composite Market Reports*, Issue 336–3, Nov 1999.
24. Katsumato M, *Keynote Address at Carbon Fibers '98*, San Antonio, Texas.
25. Leon y Leon CA, O'Brien RA, Dasarathy H, McHugh JJ, Schimpf WC, Development of low cost carbon fiber (LCCF) for automotive and non-aerospace applications. Part I: Review of existing and emerging technologies, *Midwest Advanced Materials and Processes Conference*, Dearborn, Sep 12–14, 2000.
26. Smith S, Low cost carbon fiber (LCCF) for automotive and non-aerospace applications. Part II: Precursor processing technologies, *Midwest Advanced Materials and Processes Conference*, Dearborn, Sep 12–14, 2000.
27. Cohn SM, Das S, *A cost assessment of conventional PAN carbon fiber production technology*, Energy Division, Oak Ridge National Laboratory, Tennessee, Unpublished Draft, 1998.
28. Tsai JS, Lin CH, The effect of molecular-weight on the cross-section and properties of polyacrylonitrile precursor and resulting carbon-fiber, *J Appl Polym Sci*, 42(11), 3045–3050, 1991.
29. Tsai JS, Wu CJ, Effect of cross-section evenness for polyacrylonitrile precursor on the properties of carbon-fiber, *J Mater Sci Lett*, 12(6), 411–413, 1993.
30. Tsai JS, Effect of cross-section of polyacrylonitrile fiber on carbon fiber properties, *SAMPE Journal*, 30(6), 51–55, 1994.
31. Tsai JS, Lin CH, Effect of comonomer composition on the properties of polyacrylonitrile precursor and resulting carbon-fiber, *J Appl Polym Sci*, 43(4), 679–685, 1991.
32. Ogawa H, Studies on the improvement of productivity of high-performance polyacrylonitrile-based carbon-fibers ,1. Effects of comonomer methyl acrylate composition on production of polyacrylonitrile copolymer-based carbon-fibers, *Nippon Kagaku Kaishi*, 5, 464–470, 1994.
33. Bahl OP, Mathur RB, Dhami TL, *Mater Sci Eng*, 73(109), 7937, 1988.
34. Bhat G, Daga V, Abhiraman AS, *Proceedings of the Nineteenth Biennial Conference on Carbon*, 258, 1989.
35. Fitzer E, PAN-based carbon fibers-present state and trend of the technology from the viewpoint of possibilities and limits to influence and control the fiber properties by the process parameters, *Carbon*, 27(5), 621–645, 1989.
36. Fitzer E, Gkogkidis A, Jacobsen G, *Ext Abstr International Symp on Carbon*, Carbon Soc Japan, Toyohashi, 284, 1982.
37. Warner SB, Peebles LH, Jr., Uhlmann DR, *J Mater Sci*, 14, 565, 1979.
38. Bhat GS, Cook FL, Abhiraman AS, Peebles LH, New aspects in the stabilization of acrylic fibers for carbon fibers, *Carbon*, 28(2–3), 377–385, 1990.

39. Walenta E, Fink HP, Investigation of the thermal oxidative stabilization of PAN filaments in carbon-fiber processing, *Acta Polymerica*, 41(12), 598–600, 1990.
40. Ko TH, Effects of stabilization on the properties of PAN-based carbon-fibers during air oxidation, *SAMPE Quarterly-Society for the Advancement of Material and Process Engineering*, 22(2), 13–18, 1991.
41. Ko TH, Influence of continuous stabilization on the physical properties and microstructure of pan-based carbon-fibers, *J Appl Polym Sci*, 42(7), 1949–1957, 1991.
42. Deurbergue A, Oberlin A, Stabilization and carbonization of PAN-based carbon-fibers as related to mechanical properties, *Carbon*, 29(4–5), 621–628, 1991.
43. Tsai JS, Effect of drawing ratio during spinning and oxidation on the properties of polyacrylonitrile precursor and resulting carbon-fiber, *J Mater Sci Lett*, 11(3), 140–142, 1992.
44. Tsai JS, Effect of oxidation on the properties of carbon-fiber, *SAMPE Quarterly-Society for the Advancement of Material and Process Engineering*, 24(3), 21–24, 1993.
45. Tsai JS, The relationship between oxidized degree and carbonization temperature for carbon-fiber, *SAMPE Journal*, 29(5), 15–19, 1993.
46. Ogawa H, Effects of comonomer and 2-step oxidation on production of polyacrylonitrile-based carbon-fibers, *Nippon Kagaku Kaishi*, 6, 560–564, 1994.
47. Ogawa H, Effects of heat-treatment temperature, additives and surface treatment conditions on thermal oxidative stability of polyacrylonitrile-based carbon-fibers, *Nippon Kagaku Kaishi*, 10, 933–938, 1994.
48. Ko TH, Day TC, The effect of pre-carbonization on the properties of pan-based carbon-fibers, *Polymer Composites*, 15(6), 401–407, 1994.
49. Tsai JS, Tension effects on the properties of oxidized polyacrylonitrile and carbon-fibers during continuous oxidation, *Polym Eng Sci*, 35(16), 1313–1316, 1995.
50. Ko TH, Li CH, The influence of pre-carbonization on the properties of PAN-based carbon-fibers developed by 2-stage continuous carbonization and air oxidation, *Polymer Composites*, 16(3), 224–232, 1995.
51. Gupta A, Harrison IR, New aspects in the oxidative stabilization of PAN-based carbon fibers, *Carbon*, 34(11), 1427–1445, 1996.
52. Tsai JS, Comparison of batch and continuous oxidation processes for producing carbon fibre based on PAN fibre, *J Mater Sci Lett*, 16(5), 361–362, 1997.
53. Ungureanu C, Carbon fibres researches ,4. Acrylonitriles-based polymers chemical modification by thermal treatment in inert and oxidative media, *Materiale Plastice*, 34(1), 55–62, 1997.
54. Jain MK, Balasubramanian M, Desai P, Abhiraman AS, Conversion of acrylonitrile-based precursor fibres to carbon fibres, Part 2 Precursor morphology and thermooxidative stabilization, *J Mater Sci*, 22, 301–312, 1987.
55. Saito K, Ogawa H, Process for producing carbon fibers, Toho Beslon Co, U.S. Pat., 4069297, Jan 1978.
56. Bromley J, Gas evolution processes during the formation of carbon fibres, *Int Conf on Carbon Fibres, their Composites and Applications*, The Plastics Institute: London, 1971.
57. Bromley J, Jackson EE, Robinson PS, The carbonization stage of carbon fibre manufacture Part 1: Gas evolution, *United Kingdom Atomic Energy Authority Report*, AERE R6297 Harwell, 1970.
58. Hughes JDH, Morley H, An experimental rig for continuous production of high strength carbon fibre from polyacrylonitrile, *AERE Harwell Report*, M 3036, Feb 1981.
59. Heine M, *PhD Dissertation*, University of Karlsruhe, Germany, 1988.
60. Kinoshita Y, Toray Industries Inc, High tensile strength, high Young's modulus carbon fiber having excellent internal structure homogeneity and process for producing same, U.S. Pat., 4065549, Aug 1976.
61. Morita K *et al*, *Ext Abstr International Symp on Carbon*, Carbon Soc Japan, Toyohashi, 321, 1982.
62. Fitzer E, Heine M, G Jacobsen, *Ext Abstr International Symp on Carbon*, Carbon Soc Japan, Toyohashi, 1982, p. 288.
63. Kiminta DM, Rapid stabilization of acrylic precursors for carbon-fibers using ammonia, *Int J Polym Mater*, 23(1–2), 57–65, 1993.

64. Bahl OP, Mathur RB, Effect of load on the mechanical properties of carbon fibres from PAN precursor, *Fibre Sci Technol*, 12, 31–39, 1979.
65. Matsumara Y, Kishimoto S, Ozaki M, Japan Exlan Co, Process for producing carbon fibers, U.S. Pat., 4009991, Mar 1997.
66. Hoffman WP, Hurley WC, Liu PM, Owens TW, The surface topology of non-shear treated pitch and PAN carbon fibers as viewed by STM, *J Mater Res*, 6, 1685, 1991.
67. Johnson DJ, Carbon fibres from PAN-structure, Jones ER ed., *Handbook of Polymer-Fibre Composites*, Longman Scientific and Technical, Harlow, 24–29, 1994.
68. Bennett SC, *Strength structure relationships in carbon fibres*, PhD Thesis, University of Leeds, 1976.
69. Johnson DJ, Tyson CN, The fine structure of graphitized fibres, *Brit J Appl Phys (J Phys D)*, 2(2), 787–795, 1969.
70. Johnson DJ, Structure property relationships in carbon fibres, *J Phys D Appl Phys*, 20(3), 285–291, 1987.
71. Johnson DJ, Crawford D, Oates C, The fine structure of a range of PAN-based carbon fibres, *Extended Abstracts 10th Biennial Carbon Conference*, Bethlehem, PA, 29, 1971.
72. Barnett FR, Norr MK, *Proceedings of the International Conference on Carbon Fibres, their Composites and Applications*, London, (Plastics Institute), 32, 1974.
73. Bennett SC, Johnson DJ, Strength structure relationships in PAN-based carbon fibres, *5th London International Carbon and Graphite Conference*, Society of Chemical Industry, London, 377, 1978.
74. Knibbs RH, The use of polarized light microscopy in examining the structure of carbon fibres, *J Microscopy*, 94(3), 273–281, 1971.
75. Rose PG, *A study of the physics and chemistry of the preparation of carbon fibres from acrylic precursors*, Thesis, University of Aston in Birmingham, Oct 1971.
76. Johnson W, *Nature*, 279, 5709, 1979.
77. Fourdeaux A, Perret R, Ruland W, General structural features of carbon fibres, *Proceedings of the International Conference on Carbon Fibres, their Composites and Applications*, London, Plastics and Polymer Conf Supplement, 57–67, 1971.
78. Diefendorf RJ, Tokarsky E, High performance carbon fibers, *Polym Eng Sci*, 15(3), 150–159, 1975.
79. Guigon M, Oberlin A, Desarmot G, Microtexture and structure of some high modulus PAN-based carbon fibers, *Fiber Sci Technol*, 20, 177–198, 1984.
80. Morita K, Murata Y, Ishitani A, Murayama K, Nakajima A, Characterization of commercially available PAN (polyacrylonitrile)-based carbon fibers, *Pure Appl Chem*, 58(3), 455–468, 1986.
81. Fourdeaux A, Perret R, Ruland W, *J Appl Cryst*, 1, 252, 1968.
82. Wicks BJ, Coyle RA, *J Mater Sci*, 11, 376, 1976.
83. Watt W, Johnson W, *Proc 3rd Conf on Industrial Carbons and Graphite*, Soc Chem Ind, London, 291, 1970.
84. Crawford D, Johnson DJ, *J Microsc*, 94, 51, 1971.
85. Kowbel W, Hippo E, Murdie N, Influence of graphitization environment of PAN based carbon fibers on microstructure, *Carbon*, 27(2), 219–226, 1989.
86. Bennett SC, Johnson DJ, *Carbon*, 17, 25, 1979.
87. Johnson DJ, Structural studies of PAN-based carbon fibers, Thrower PA ed., *Chemistry and Physics of Carbon*, Vol 20, Marcel Dekker, New York, 1–58, 1987.
88. Watt W, Johnson W, Carbon fibres from 3 denier polyacrylonitrile textile fibres, *Proc 3rd Conf on Industrial Carbons and Graphite*, Soc Chem Ind, London, 417, 1970.
89. Watt W, Johnson W, *Nature*, 257, 210, 1975.
90. Takaku A, Hashimoto T, Miyoshi T, *J Appl Polym Sci*, 30, 1565, 1985.
91. Bajaj P, Roopanwal AK, *Polym Sci*, 1, 368, 1994.
92. Fitzer E, Heine M, *89th ACS National Meeting*, Miami Beach, USA, 1985.
93. Kissinger HE, *Anal Chem*, 29, 1702, 1957.
94. Fitzer E, Frohs W, Heine M, Optimization of stabilization and carbonization treatment of PAN fibers and structural characterization of the resulting carbon fibers, *Carbon* 24(4), 387–395, 1986.
95. Watt W, *Carbon* 10, 121, 1972.

96. Bahl OP, Manocha LM, Methods of determining optimum preoxidation time of PAN fibre, *Fibre Sci Technol*, 9, 78–80, 1976.
97. Tsai JS, The relationship between oxidized degree and carbonization temperature for carbon fiber, *SAMPE J*, 29(5), Sep–Oct, 1993.
98. Dunham MG, Edie DD, Model of stabilization for PAN-based carbon fiber precursor bundles, *Carbon*, 30(3), 435–450, 1992.
99. Mathur RB, Bahl OP, Mittal J, Nagpal KC, *Carbon*, 29(7), 1059, 1991.
100. Kubasova NA, Kusakov MM, Shishkina MV, *Vysokomelek yarnyi Soldininiya*, 3, 193, 1969.
101. Johnson JW, Potter W, Rose PG, G Scott G, Stabilization of polyacrylonitrile by oxidative transformation, *Brit Polym J*, 4, 527–540, 1972.
102. Noh J, Yu H, *Polym Lett*, 4, 721, 1966
103. Dorey G, Carbon fibres and their applications, *J Phys D Appl Phys*, 20, 245–256, 1987.
104. Moreton R, Watt W, The spinning of polyacrylonitrile fibres in clean room conditions for the production of carbon fibres, *Carbon*, 12, 543–554, 1974.
105. Dhami TL, Mathur RB, Dwivedi H, Bahl OP, Monthioux M, *Pyrolysis behaviour of Panex based polymer composites*, The Pennsylvania University Conference, American Carbon Society, 474–475, 1997.
106. DK Brown, Phillips WM, Effects of heat-treatment on carbon-fibers, *SAMPE J*, 26(5), 9–14, 1990.
107. Podkopaev SA, Tyumentsev VA, Yagafarov SS, Saunina SI, Influence of heat-treatment conditions on carbon-fiber microstructure, *Russian J Appl Chem*, 67(3), Pt 1, 385–387, 1994.
108. Mittal J, Konno H, Inagaki M, Bahl OP, Carbonization study of stabilized PAN fibres, 3rd *Biennial Conference on Carbon*, Penn State Univ, USA, Session 6D, Jul 1997.
109. Tsai JS, Comparison of one-stage and 2-stage carbonized furnaces forcarbon-fiber, *J Mater Sci Lett*, 12(14), 1114–1116, 1993.
110. Ko TH, Day TC, Peng JA, Lin MF, The characterization of PAN-based carbon-fibers developed by 2-stage continuous carbonization, *Carbon*, 31(5), 765–771, 1993.
111. Tsai JS. Relationship between two-stage carbonization speeds for polyacrylonitrile based carbon fibre, *J Mater Sci Letters*, 15(10), 835–836, 1996.
112. Watt W, *Carbon*, 10, 121, 1972.
113. Serin V, Fourmeaux R, Kihn Y, Sevely J, Guigon M, Nitrogen distribution in high tensile strength carbon fibres, *Carbon*, 28(4), 573–578, 1990.
114. Fitzer E, Muller DJ, *Carbon*, 13, 63, 1975.
115. Guigon M, *Thesis*, Université de Technologie de Compiégne, France.
116. Tsai JS, Effect of nitrogen atmosphere on the structure and properties of a PAN-based carbon-fiber, *Textile Res J*, 64(12), 772–774, 1994.
117. Cullis CF, Yates JG, Reaction of carbon with nitrogen, *Trans Faraday Soc*, 493(60), Part 1, 141–148, Jan 1964.
118. Cullis CF, Yates JG, Thermal reactions of nitrogen, *J Chem Soc*, 549, 2833–2835, Aug 1964.
119. Cullis CF, Yates JG, The structure and mechanism of formation of pyrolytic carbon from cyanogen, *Acta Crystallographica*, 17(11), 1433–1436, Nov 1964.
120. Bacon R, Schalamon W, High temperature resistant fibers from organic polymers, *Appl Poly Symp*, Appl Sci Pub, New York, 285, 1968.
121. Johnson JW, Marjaram JR, Rose PG, Stress graphitization of polyacrylonitrile based carbon fibre, *Nature*, 221, 357–358, 1969.
122. Johnson W, Hot stretching of carbon fibres made from polyacrylonitrile fibres, *Proc 3rd Int Conf on Industrial Carbons and Graphite*, Soc Chem Ind, London, 447–452, 1970.
123. Isaac DH, Ozbek S, Francis JG, Processing of carbon fibers: Texture enhancement induced by hot stretching, *Mater Manuf Process*, 9(2), 179–197, 1994.
124. Ozbek S, Isaac DH, Carbon fiber processing: Effects of hot stretching on mechanical properties, *Mater Manuf Process*, 9(2), 199–219, 1994.
125. Isaac DH, Ozbek S, *Density changes in carbon fibers, induced by hot stretching, Carbon '98 22nd Biennial Conference on Carbon*, American Carbon Society, San Diego, CA, 28–29, 1995.

126. Isaac DH, Ozbek S, Manufacture of carbon fibers from precursors of various diameters, *Mater Manuf Proces*, 9(5), 975–998, 1994.
127. Ozbek S, Isaac DH, Carbon fibres, Effect of processing parameters on mechanical properties, *Processing and Manufacturing of Composite Materials ASME*, PED-Vol49/MD-Vol27, 1991.
128. Ogawa H, Effects of characteristics of oxidized fibers and tension during carbonization on tensile-strength of polyacrylonitrile based carbon-fibers, *Nippon Kagaku Kaishi*, 10, 921–926, 1994.
129. Tsai JS, Tension of carbonization for carbon-fiber, *Polym Eng Sci*, 34(19), 1480–1484, 1994.
130. Allen S, Cooper GA, Mayer RM, *Paper presented at IP and PS Conference on Fibres and Composites*, Brighton, Jun 1969.
131. Allen S, Cooper GA, Johnson DJ, Mayer RM, Carbon fibres of high modulus, *Proc 3rd Conf Industrial Carbons and Graphite*, Soc Chem Ind, London, 456–461, 1970.
132. HM Ezekiel, Effects of boron catalysis during graphitization of selected polymeric fibers, *11th Biennial Conf on Carbon, Extended Abstracts*, Gatlinburg, TN, NTIS Conf–730601, 267–268, 4–8 Jun 1973.
133. Pepper R, Nelson D, Jarmon D, Hotham J, Improved graphite fiber, *Final Technical Report by Fiber Materials Inc for the US Army Mobility Equipment research and Development command Ft Belvoir*, Virginia contract No. DAAK70-77-C-0155, 12 Sep 1978.
134. Brewster EP, Nelson DC, Patton R, Ultra high strength graphite fibers, *28th National SAMPE Symposium and Exhibition*, Anaheim, Apr 1983.
135. Manocha LM, Bahl OP, Jain GC, *Angew Makromol Chem*, 67, 11, 1978.
136. Huang Y, Young RJ, Effect of fiber microstructure upon the modulus of PAN-and pitch-based carbon-fibers, *Carbon*, 33(2), 97–107, 1995.
137. Oberlin A, Molleyre F, Bastik M, *13th Biennial Conference on Carbon*, Irvine, 371, 1977.
138. Mathur RB, Bahl OP, Mittal J, Advances in the development of high performance carbon fibres from PAN precursor, *Composites Sci Technol*, 51, 223–230, 1994.
139. Ogawa H, Effects of precursor process oil and heating rate during carbonization on defects and tensile properties of resultant PAN-based carbon-fibers, *Nippon Kagaku Kaishi*, 10, 927–932, 1994.
140. Lin SS, Recent developments of carbon fiber in Japan, *SAMPE J*, 28(4), 9–19. Jul/Aug 1992.
141. Fitzer E, Frohs W, The influence of carbonization and post heat treatment conditions on the properties of PAN-based carbon fibres, *Presented at Carbon 88*, Newcastle upon Tyne, 298–300, 1988.
142. Bashir Z, A critical review of the stabilisation of polyacrylonitrile, *Carbon*, 29(8), 1081–1090, 1991.
143. Houtz RC, Orlan acrylic fibre: chemistry and properties, *J Text Res*, 20,786–801, 1950.
144. Burlant WJ, Parsons JL, Pyrolysis of acrylonitrile, *J Polym Sci*, 22, 249–256, 1956.
145. LaCombe EM, *J Polym Sci*, 24, 152, 1957.
146. Grassie N, Hay JN, McNeill IC, *J Polym Sci*, 31, 205, 1958.
147. Schurz J, Discoloration effects in acrylonitrile polymers, *J Polym Sci*, 28, 438, 1958.
148. Kennedy JP, Fontana CM, The thermal degradation of polyacrylonitrile, *J Polym Sci*, 39, 501–506, 1959.
149. Grassie N, Hay JN, Thermal discolouration of nitrile polymers, *SCI Monograph*, 13, 184–199, 1961.
150. Grassie N, Hay JN, Thermal colouration and insolubilization in polymers, *J Polym Sci*, 56, 189–202, 1962.
151. Conley RT, Bieron JF, *J App Polym Sci*, 7, 1757, 1963.
152. Berlin AA, Dubindkaya AM, Moshkovski YS, *Vysokomol Soedin*, 6, 1938, 1964; *Polym Sci USSR*, 6, 2145, 1966.
153. Takata T, Hiroi I, *J Polym Sci Part A*, 2, 1567, 1964.
154. Fester W, *Textil Rundschau*, 20(5), 1, 1965.
155. Bell JW, Mulchandani RK, Observations on the thermal degradation of Courtelle acrylic fibre II- Changes in physical properties, *J Soc Dyers & Colourists*, 55–59, Feb 1965.
156. Monahan AR, Thermal degradation of polyacrylonitrile in the temperature range 280-450°C, *J Polym Sci*, Part A-1, 4, 2391–2399, 1966.

157. Thompson EV, The thermal behaviour of acrylonitrile polymers, I. On the decomposition of polyacrylonitrile between 250 and 325°C, *Polym Lett*, 4, 361–366, 1966.
158. Peebles LH Jr., Brandrup J, A chemical means of distinguishing between conjugate and conjugated bonds, *Die Makromolekulare Chemie*, 98(2199), 189–203, 1966.
159. Noh J, Yu H, *Polym Lett*, 4, 721, 1966.
160. Watt W, Chemistry and Physics of the Conversion of Polyacrylonitrile Fibres into High Modulus Carbon Fibres, Watt W, Perov BV eds., *Handbook of Composites Vol 1, Strong Fibres*, Elsevier Science, Amsterdam, 327–387, 1985.
161. Fester W, *Untersuchungen über die Verfärbung von Polyacrylnitril, IV, Polymeranaloge Umsetzungen an durch Erhitzung verfärbtem Polyacrylnitril, J Polym Sci Part C*, Polymer Symposia, 16, 755–763, 1967.
162. Hay JN, Thermal reactions of polyacrylonitrile, *J Polym Sci*, 6, 2127–2135, 1968.
163. Reich L, *Macromolek Rev*, 3, 49, 1968.
164. Peebles LH, Brandrup J, Friedlander HN, Kirby JR, *Macromolecules*, 1, 53, 1968.
165. J Brandrup, Peebles LH Jr., On the chromophore of polyacrylonitrile. IV. Thermal oxidation of polyacrylonitrile and other nitrile containing compounds, *Macromolecules*, 1(1), 64–72, Jan–Feb 1968.
166. Kubasova N, Kusakov N, Shiskina M, *Vsokomolekyarni Soldininiya*, 3, 193, 1969.
167. Watt W, *Proc Int Conf Carbon Soc Chem Ind*, London, 1970.
168. Ulbricht J, Makschin W, *Eur Polym J* 6, 1277, 1970.
169. Fiedler A, Fitzer E, Müller D, *161st Amer Chem Soc Meeting*, Los Angeles, 370, 1971.
170. Rose PG, The Preparation of Carbon Fibres- A study of the physics of and chemistry of the preparation of carbon fibres from acrylic precursors, *PhD Thesis*, University Little Aston, Birmingham, England, 1971.
171. Watt W, Green J, The pyrolysis of polyacrylonitrile, *International conference on Carbon Fibres, their Composites and Applications*, The Plastics Institute, London, 1971.
172. AE Standage, Matkowsky RD, *Eur Polym J*, 7, 775, 1971.
173. Bailey JE, Clarke AJ, *Nature*, 234, 529, 1971.
174. Fitzer E, Müller DJ, *Makromol Chemie*, 144, 117, 1971.
175. Johnson JW, Potter W, Rose PG, Scott G, Stabilization of polyacrylonitrile by oxidative transformation, *Brit Polym J*, 4, 527–540, 1972.
176. Grassie N, McGuchan R Pyrolysis of polyacrylonitrile and related polymers Part I Thermal analysis of polyacrylonitrile, *Eur Polym J*, 6, 1277–1291, 1970.
177. Grassie N, McGuchan R, Pyrolysis of polyacrylonitrile and related polymers Part II The effect of sample preparation on the thermal behaviour of polyacrylonitrile, *Eur Polym J*, 7, 1091–1104, 1971.
178. Grassie N, McGuchan R, Pyrolysis of polyacrylonitrile and related polymers Part III Thermal analysis of preheated polymers, *Eur Polym J*, 7, 1357–1371, 1971.
179. Grassie N, McGuchan R, Pyrolysis of polyacrylonitrile and related polymers Part IV Thermal analysis of polyacrylonitrile in the presence of additives, *Eur Polym J,* 7,1503–1514, 1971.
180. Grassie N, McGuchan R, Pyrolysis of polyacrylonitrile and related polymers Part V Thermal analysis of α-substituted acrylonitile polymers, *Eur Polym J*, 8,243–255, 1972.
181. Grassie N, McGuchan R, Pyrolysis of polyacrylonitrile and related polymers Part VI, *Eur Polym J,* 8, 257, 1972.
182. Grassie N, McGuchan R, Pyrolysis of polyacrylonitrile and related polymers Part VII- Copolymers of acrylonitrile with acrylate, methacrylate and styrene type monomers, *Eur Polym J*, 8, 865–878, 1972.
183. Grassie N, McGuchan R, Pyrolysis of polyacrylonitrile and related polymers Part VIII- Copolymers of acrylonitrile with vinyl acetate, vinyl formate, acrolein and methyl vinyl ketone, *Eur Polym J,* 8, 113–124, 1973.
184. Grassie N, McGuchan R, Pyrolysis of polyacrylonitrile and related polymers Part IX - Copolymers of acrylonitrile with vinyl chloride, vinylidene chloride and α-chloroacrylonitrile, *Eur Polym J*, 8, 507–517, 1973.

185. Fitzer E, Müller DJ, Evidence of separate cyclization and oxidation during heat treatment of PAN by DTA measurements, *ACS Polymer Preprints*, 14, 396–400, 1973.
186. Clarke AJ, Bailey JE, Oxidation of acrylic fibres for carbon fibre formation, *Nature*, 243, 146–150, 1973.
187. Goodhew PJ, Clarke AJ, Bailey JE, A review of the fabrication and properties of carbon fibres, *Mater Sci Eng*, 17, 3–30, 1975.
188. Watt W, Johnson W, Mechanism of the oxidation of polyacrylonitrile fibres, *Nature*, 257, 210–212, 1975.
189. Love G, Cox M, Scott VD, *Mat Res Bull*, 10(8), 1975.
190. Fitzer E, Müller DJ, *Carbon*, 13, 63, 1975.
191. Raskovic V, Marinkovic S, Processes in sulphur dioxide treatment of PAN fibers, *Carbon*, 16, 351–357, 1978.
192. Coleman M, Petcavich R, *J Polym Sci Phys*, 16, 821, 1978.
193. Warner SB, Peebles LH Jr., Uhlmann DR, Oxidative stabilization of acrylic fibers, Part 1 Oxygen uptake and general model, *J Mater Sci*, 14, 556–564, 1979.
194. Warner SB, Peebles LH Jr., Uhlmann DR, Oxidative stabilization of acrylic fibers, Part 2 Stabilization dynamics, *J Mater Sci*, 14, 565–572, 1979.
195. Warner SB, Peebles LH Jr., Uhlmann DR, Oxidative stabilization of acrylic fibers Part 3 Morphology of polyacrylonitrile, *J Mater Sci*, 14, 1893–1900, 1979.
196. Warner SB, Peebles LH Jr., Uhlmann DR, Oxidative stabilization of acrylic fibers, Part 4 Moisture sensitivity, *J Mater Sci*, 14, Letters, 2765–2765, 1979.
197. Olivé GH, Olivé S, Molecular interactions and macroscopic properties of polyacrylonitrile and model substances, *Adv Polym Sci*, 32, 123–152, 1980.
198. LM Manocha, Bahl OP, Role of oxygen during thermal stabilization of PAN fibers, *Fiber Sci Technol*, 13, 199, 1980.
199. Bahl OP, Mathur RB, Kundra KD, Treatment of PAN fibers with SO_2 and development of carbon fibers therefrom, *Fiber Sci Technol*, 13, 155, 1980.
200. Olivé GH, Olivé S, Inter- *versus* intramolecular oligomerization of nitrile groups in polyacrylonitrile, *Polym Bull*, 5, 457–461, 1981.
201. Chen SS, Herms J, Peebles LH Jr., Uhlmann DR, Oxidative stabilizationof acrylic fibres, Part 5 The decolouration reaction, *J Mater Sci*, 16,1490–1510, 1981.
202. Olivé GH, Olivé S, The chemistry of carbon fiber formation from polyacrylonitrile, *Adv Polym Sci*, 51, 1–60, 1983.
203. Jain MK, Abhiraman AS, Oxidative stabilization of oriented acrylic fibres-morphological rearrangements, *J Mater Sci*, 18, 179–188, 1983.
204. Peebles LH Jr., Peyser P, Snow AW, Peters WC, On the exotherm of polyacrylonitrile: pyrolysis of the homopolymer under inert conditions, *Carbon*, 28, 707, 1990.
205. Sivy GT, Gordon B III, Coleman MM, *Carbon*, 21, 573, 1983.
206. Mathur RB, Gupta G, Bahl OP, Dhami TL, Infrared spectral studies of preoxidized PAN fibers incorporated with cuprous chloride additive, *Fiber Sci Technol*, 20, 277, 1984.
207. Takaku A, Shimazu J, Volume contraction and its significance in structural 1319, formation during the thermal stabilization of acrylic fibers, *J Appl Polym Sci*, 29, 1984.
208. Fochier HS, Mooney JR, Ball LE, Boyer RD, Grasselli JG *Spectrochemica Acta*, 41A, 271, 1985.
209. Fitzer E, Frohs W, Heine M, Optimization of stabilization and carbonization treatment of PAN fibres and structural characterization of the resulting carbon fibres, *Pure Appl Chem*, 58, 455, 1986.
210. Morita K, Murata Y, Ishitani A, Murayama K, Ono T, Nakajima A, Characterization of commercially available PAN-based carbon fibers, *Pure Appl Chem*, 58, 455, 1986.
211. Jain MK, Abhiraman AS, Conversion of acrylonitrile-based precursor fibres to carbon fibres, Part 1 A review of the physical and morphological aspects, *J Mater Sci*, 22, 278–300, 1987.
212. Jain MK, Balasubramanian M, Desai P, Abhiraman AS, Conversion of acrylonitile-based precursor fibres to carbon fibres, Part 2 Precursor morphology and thermooxidative stabilization, *J Mater Sci*, 22, 301–312, 1987.

213. Bhat GS, Cook FL, Peebles LH Jr., Abhiraman AS, New aspects in the stabilization of PAN-based precursors for carbon fibers, *18th Biennial Conference on Carbon*, Penn State University, Session 3A, 25–30 Jun 1989.
214. Usami T, Itoh T, Ohtani H, Tsuge S, Structural study of polyacrylonitrile fibers during oxidative thermal degradation by pyrolysis- gas chromatography, solid-state ^{13}C nuclear magnetic resonance and Fourier transform infrared spectroscopy, *Macromolecules,* 23, 2460–2465, 1990.
215. Bhat GS, Cook FL, Abhiraman AS, Peebles LH Jr., New aspects in the stabilization of acrylic fibers for carbon fibers, *Carbon*, 28(2,3), 377–385, 1990.
216. Peebles LH Jr., Peyser P, Snow AW, Peters WC, On the exotherm of polyacrylonitrile, pyrolysis of the homopolymer under inert conditions, *Carbon*, 28(5), 707–715, 1990.
217. Pejanovic S, Pavlovic V, Processing of carbon fibers 1, Oxidative stabilization, *Extended Abstr Int Carbon Conf, Carbone 90*, Paris, 152, 16–20 Jul 1990.
218. Deurbergue A, Oberlin A, Stabilization and carbonization of PAN-based carbon fibers as related to mechanical properties, *Carbon*, 29(4,5), 621–628, 1991.
219. Dunham MG, Edie DD, Model of stabilization for PAN-based carbon fiberprecursor bundles, *Carbon*, 30(3), 435–450, 1992.
220. Mathur RB, Bahl OP, Sivaram P, Thermal degradation of polyacrylonitrile fibres, *Current Science*, 62(10), 662–669, 1992.
221. Grove DA III, Abhiraman AS, A mathematical model of solid-state thermo-oxidative stabilization of acrylic fibers, *Carbon*, 30(3), 451–457, 1992.
222. Ko T-H, Liau S-C, Lin M-F, Preparation of graphite fibers from a modified PAN precursor, *J Mater Sci*, 27, 6071, 1992.
223. Tsai J-S, Effect of oxidation on the properties of carbon fiber, *SAMPE Quarterly*, 21–24, Apr 1993.
224. Tsai J-S, Wu CJ, Exothermic heat effects on polyacrylonitrile and carbon fibers during continuous oxidation, *SAMPE J*, 29(3), 23–26, 1993.
225. Peebles LH, *Carbon Fibers Formation, Structure and Properties*, CRC Press Inc., Boca Raton, p. 15–16, 1994.
226. Beltz LA, Gustafson RR, Cyclization kinetics of polyacrylonitrile, *Carbon*, 34(5), 561–566, 1996.
227. Bailey JE, Clarke AJ, Carbon fibres, *Chemistry in Britain*, 6, 484–489, 1970.
228. Watt W, The pyrolysis of PAN fibres, *3rd Carbon Conf*, Soc Chem Ind, London, 1970.
229. Watt W, Green J, The pyrolysis of polyacrylonitrile, *Intl Conf on Carbon Fibres*, The Plastics Institute, London, 1971.
230. Bromley J, Jackson EE, Robinson PS, The carbonization stage of carbon fibremanufacture Part. Gas Evolution, *Harwell Technical Report*, AERE-R 6297, 1970.

CHAPTER 6

Carbon Fiber Production using a Cellulosic based Precursor

6.1 INTRODUCTION

Several books and reviews have been published which detail the conversion of viscose rayon to carbon fibers [1–7]. Chapter 3 has described how carbon fiber first came onto the scene, way back in 1880, with the introduction of Thomas Edison's electric lamp filaments made from cellulosic precursors. Almost 80 years later, in 1959, the National Carbon Company (a division of Union Carbide) introduced a carbon cloth from a rayon precursor, to be followed by a carbon yarn in 1961. These products are described by Cranch [8]. The best grade on offer was WYB cloth, which was processed at 2200°C and although called graphite, was a form of carbon that was non-graphitizing.

Ford and Mitchell [9] described a process that could be used for rayon monofilaments, cellulosic yarns or a pre-woven rayon textile material. The work detailed a controlled heating process comprising of heating at 10°C per hour to100°C, 50°C per hour to 400°C and 100°C per hour to about 900°C, followed by heating to 3000°C until substantial graphitization had occurred. A protective atmosphere of nitrogen or other inert gases was used when heating through the range 900–3000°C. The product had the benefits of improved tensile strength over the earlier fibers and their X-ray patterns were more graphitic in nature.

Union Carbide claimed that the water content (10–20%) of a cellulosic precursor encouraged the formation of tarry deposits on the individual carbon filaments, causing them to stick together in the carbonization process, giving a weak and brittle product. Besides, only enough inert atmosphere was required to prevent ingress of air into the product and in conventional synthetic graphite production furnaces, this was provided by the use of sacrificial carbonaceous packing material. However, in carbon fiber production, the inert atmosphere was provided by a stream of inert gas, which unfortunately allowed the deposition of soot particles on the fiber which was attributed to the cracking of methane evolved in the degradation of the cellulose. Cross *et al* filed a patent [10] which described the initial drying of the precursor material (e.g. cellulosic cloth for 20 h at 125°C in a gas fired oven), which was then opened and the product covered with a protective layer of coke, quickly closed and the carbonization cycle continued up to 700°C. The carbonization process was carried out by increasing the temperature at a rate of 5°C per hour up to 400°C and then 60°C per hour to 700°C, using a nitrogen purge to remove evolved gases and tars. The oven was cooled and the product was transferred to an electric furnace, where the product was graphitized up to 2700°C in an atmosphere of nitrogen to prevent air ingress.

Tang and Bacon [11] proposed the following simplified stages for the conversion of cellulose to carbon:

Stage I. Physical desorption of about 12% absorbed water (25–150°C) with a small degree of change in lateral order.

Stage II. Dehydration from the —H and —OH fragments present in the cellulose unit (150–240°C). IR shows that —C=O and —C—C— are involved and hence dehydration is essentially intramolecular.

Stage III. Thermal cleavage of the glycosidic linkage and scission of other C=O and some C—C bonds via a free radical reaction (240–400°) which leads to the formation of large amounts of tar, H_2O, CO and CO_2.

Stage IV. Aromatization (400°C and above) where each cellulose unit breaks down into a residue containing four C atoms which then polymerize through condensation reactions involving the removal of —H above 400°C into a C-polymer with an ultimate graphite-like structure.

The year 1965 saw the introduction of the Thornel range of carbon fibers by Union Carbide, starting with Thornel 25, a fiber with a UTS of 1.25 GPa and YM of 170 GPa (25×10^6 lbf/in^2), achieving this modulus by hot stretching at 2500°C, which increased the modulus from 70 to 170 GPa, though at a cost. Later, Thornel 50, 75 and 100 were introduced. Hot stretching was prohibitively expensive and Union Carbide, which later became BP Amoco, discontinued the process in 1978, and replaced it with carbon fibers made from either PAN or pitch precursors.

Strong [12] described the small-scale heat-treatment of rayon precursors for stress graphitization. Since the stress graphitization was to be accomplished at about 2800°C, any non-uniformity in the yarn could lead to a non-uniform stretch or breakage during the stretching process. Therefore, it was essential that the fiber be well supported and transported by rollers. The rayon textile finish was removed by extraction with boiling water and as far as could be ascertained, no flame resistant finish was applied. The pyrolysis process was undertaken with either, (i) continuous treatment for about 7 min in oxygen at 260–280°C or, (ii) a batch treatment for about 20 h in air at 225°C with a 40% weight loss, followed by 7 min in O_2 to give a total weight loss of 50–55%. To start the development of the final carbon structure and improve the strength sufficiently to enable stress-carbonization to be accomplished, a treatment of 1.5 min in N_2 at 350°C was applied. The material was carbonized at 900–2000°C and graphitized at 2800–2900°C. X-ray diffraction showed that the starting material Cellulose II (Table 4.7) was converted to the Cellulose IV structure before degradation commenced. It was observed that during pyrolysis in air, shrinkage occurred at about 25 and 45% weight loss, associated with a tendency to kink if the yarn was not under tension. So, the pyrolysis stage was deliberately limited to postpone the second weight loss stage to the carbonization stage, during which sufficient tension could be applied to prevent kinking.

Ross [13] describes the course of carbonization of viscose rayon and Figure 6.1 shows the weight loss of a typical viscose rayon continuous filament yarn showing a major change in slope in the 230–260°C range, whilst Figure 6.2 shows the variation of carbon content with treatment temperature. The sonic modulus, determined with a Pulse Propogation Meter (PPM) at a frequency of about 10,000 cps, is given in Figure 6.3. The PPM measures the time for a pulse to transmit between two transducers set apart at a known distance, when the gage length is not an additional factor. It should be noted that there is a sharp fall in modulus at about 260°C, which corresponds to the temperature at which maximum weight loss rate is observed. At 260°C, the maximum disruption of the cellulose

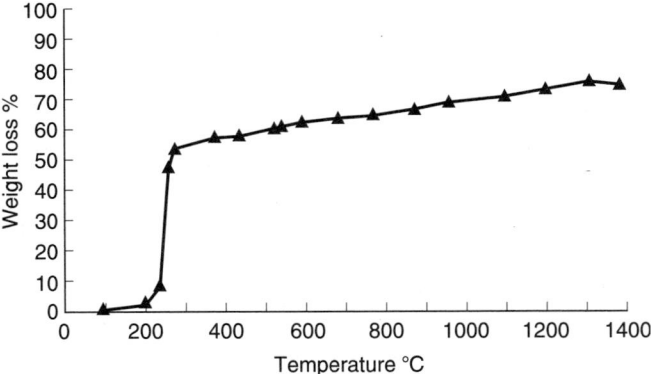

Figure 6.1 Percentage weight loss of a rayon precursor as a function of temperature. *Source*: Reprinted with permission from Ross SE, Observations concerning the carbonization of viscose rayon yarn, *Text Res J*, 38, 906–913, 1968. Copyright 1968, Textile Research Institute.

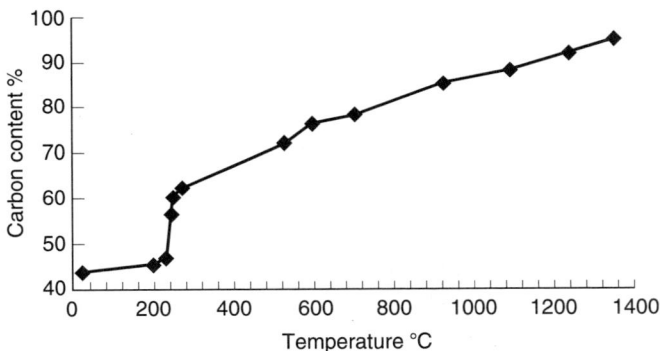

Figure 6.2 Variation of the carbon content of a rayon precursor with treatment temperature. *Source*: Reprinted with permission from Ross SE, Observations concerning the carbonization of viscose rayon yarn, *Text Res J*, 38, 906–913, 1968. Copyright 1968, Textile Research Institute.

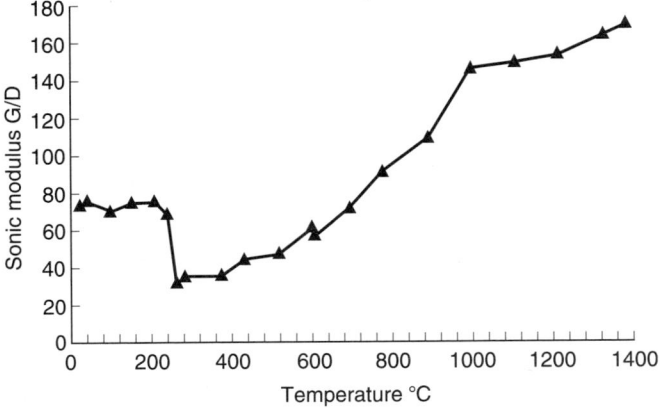

Figure 6.3 Effect of treatment temperature on the sonic modulus of a rayon precursor. *Source*: Reprinted with permission from Ross SE, Observations concerning the carbonization of viscose rayon yarn, *Text Res J*, 38, 906–913, 1968. Copyright 1968, Textile Research Institute.

molecule has occurred, with loss of CO and CO_2 and the formation of four-carbon residues [11,13].

6.2 CURRENT PRODUCTION

There remained a niche market for carbon fibers made from a cellulosic precursor and production was maintained by Polycarbon Inc. in the USA (acquired by SGL Technic in 1980) and RK Carbon Fibers (who also became part of SGL in 1997, when production was transferred to Polycarbon Inc. in the USA). Demand and availability of a suitable precursor has added to production problems.

Little information is available on the production process and a schematic layout of the preparation of carbon fiber from a rayon precursor is shown in Figure 6.4 and each stage will be considered separately.

6.2.1 Choice of a suitable precursor

The basic composition of a rayon precursor, is the same as the original cellulose starting material (Cellulose I), with the exception that the number of anhydroglucose units in the rayon has been reduced and the regenerated cellulose is in a slightly different allotropic form, as cellulose hydrate (Cellulose II). The structure of the original allotropic form (Cellulose I) is shown in Figure 4.17 and the dimensions using X-ray diffraction are given in Table 4.7 for Cellulose types I, II, III and IV. The reason for the difference in allotropic form can be explained by the rapid crystallization of the cellulose in the regeneration process, whereas in nature, the process occurs quite slowly.

To make a continuous carbon fiber, a continuous viscose rayon fiber is required, which immediately imposes a limitation, since the majority of current viscose production is chopped to form a staple product. A further limitation is that specific end users superimpose their own purchasing specifications, like NASA and the atomic energy industry where, for example, the presence of certain trace elements would have to be rigidly controlled. Availability of suitable

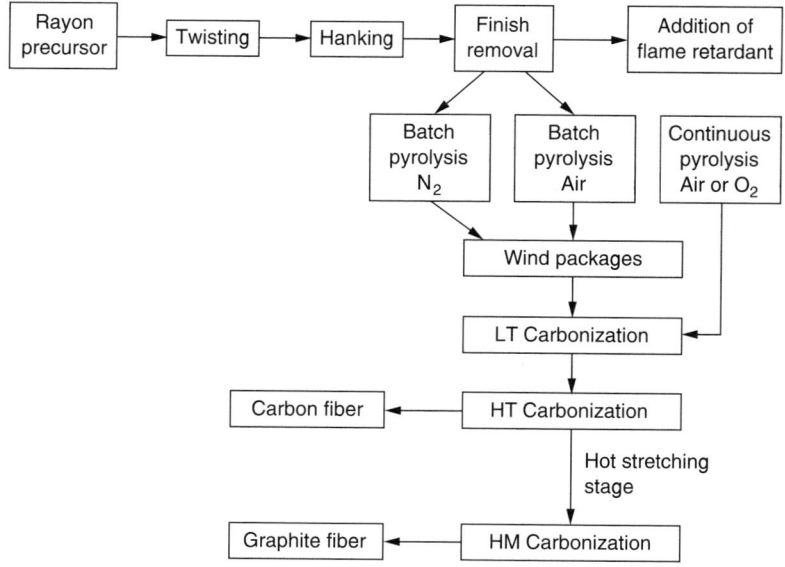

Figure 6.4 Schematic preparation of carbon fiber from a rayon precursor.

precursors have dwindled over the years and manufacturers like Avtex, Celanese Corp. (Fortisan), Industrial Rayon Corp. (Villwyte) and North American Rayon Corp. (NARC) have phased out such products and now, there is probably only one remaining supplier. Generally, suitable yarns had a denier of 1650 with 720 filaments and about 80 tpm of normally Z twist. The SEM photographs in Figures 6.5A–C show cross-sections of typical rayon precursors and the longitudinal striations remain apparent in the final carbon fiber [13].

Figure 6.5 A, B and C: SEM photographs of viscose rayon precursors. *Source*: Courtesy of Polycarbon Inc., Charlotte, North Carolina, USA.

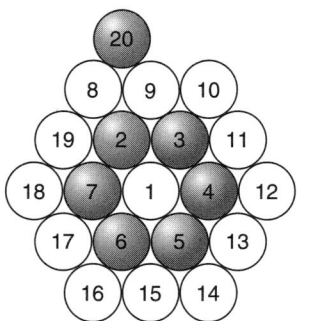

Figure 6.6 Diagram of hexagonal packing arrangement of a 20 ply yarn.

The most common end uses for these carbon fibers are ablative shields and high temperature packing materials, where current practice for the latter is to use the material in the form of 5, 10 or 20 ply yarns, with either S or Z twist. These values are dictated by the customary working practice used in the packing industry and interestingly, this does not follow the practice employed in the ropes industry, where 7 and 19 plies are used. An extra ply, as in 20 plies, would not be locked into the hexagonal structure and would be free to wander (Figure 6.6).

In subsequent pyrolysis, in order to enable the precursor to be uniformly heated and breakdown products to be efficiently removed, the precursor must be presented in the pyrolysis oven as a single tow or wound loosely into a hank form.

Typical rayon precursors will have a textile finish applied by the manufacturer to act as a lubricant to aid spinning and subsequent handling such as weaving. This finish will, at some stage before pyrolysis, have to be removed to avoid the formation of a breakdown product, which would bond individual filaments together and catastrophically lower the strength.

Mineral oil is one such finish which is applied and can be selectively removed by hot solvent followed by drying, to remove all traces of residual solvent. If a water based finish has been applied, then hot water can be used for the extraction process.

6.2.2 Pyrolysis

Pyrolysis should not be considered as a chemical reaction since it is several reactions taking part at the same time.

A TGA analysis of the chosen precursor (Figure 6.7) is helpful to assess a suitable pyrolysis cycle as shown in Figure 6.8. Tang and Bacon [11] determined the weight loss and shrinkage of Fortisan 36 rayon fiber, when heated at 40°C per hour in an argon atmosphere (Figure 6.9), which showed major pyrolytic degradation starting at about 240°C and ending at about 320°C. The weight loss in this instance was over 90% and can be reduced to about 70% with the use of alternative precursors and slower heating rates.

At elevated temperatures, a cellulosic precursor will break down into highly volatile gases, a tarry distillate and a carbon char. Initially, the product releases absorbed water up to about 120°C, followed by a dehydration process up to about 300°C, to form dehydrocellulose. This is accompanied by depolymerization, starting at about 250°C, forming predominantly 1,6-anhydro-β-D-glucopyranose (levoglucosan), the process running concurrently with the latter stages of dehydration [15]:

![Cellulose dehydration reaction scheme showing structures transforming at 120-250 °C dehydration and then 250-400 °C chain splitting dehydration to form levoglucosan]

These two reactions are competitive, but since the withdrawal of the —CH$_2$OH groups starts at 120°C, the early removal of these groups will prevent their subsequent reaction to form the undesirable levoglucosan at 250°C. Hence, holding the temperature in the initial stages below 250°C is an effective way of improving the carbon yield [15]. However, the two reactions cannot be separated and there will always be a temperature region where both reactions will

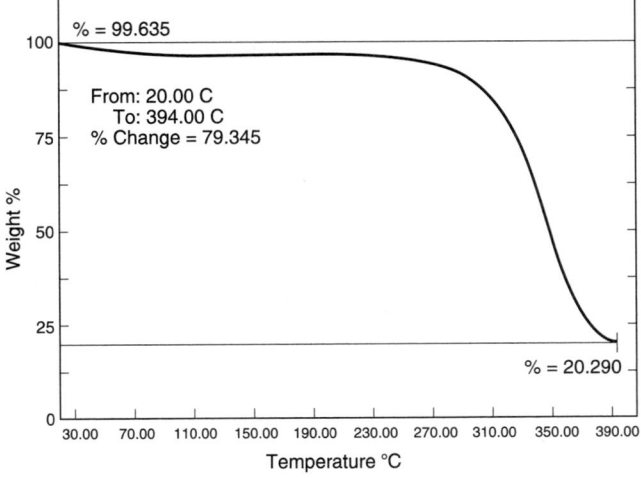

Figure 6.7 Typical TGA curve in N$_2$ for a viscose rayon precursor washed free of applied finish.

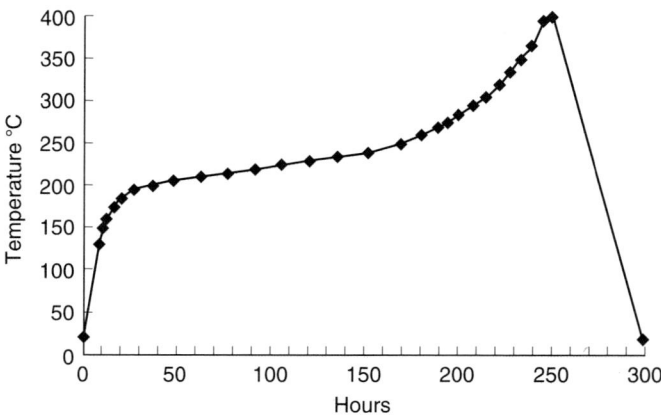

Figure 6.8 Simulated pyrolysis cycle for a viscose rayon precursor heated in N_2.

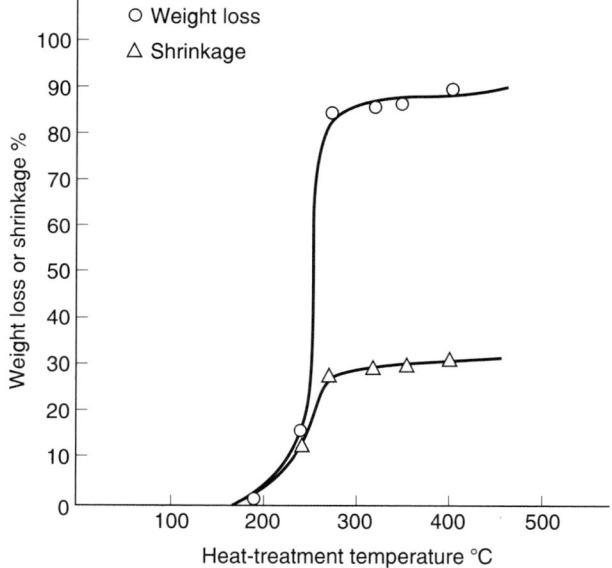

Figure 6.9 Weight loss and shrinkage *vs* temperature of Fortisan 36 fiber when heated at 40°C/h in air. *Source*: Reprinted with permission from Tang MM, Bacon R, Carbonization of cellulose fibers: 1. Low temperature pyrolysis, *Carbon*, 2, 211–220, 1964. Copyright 1964, Elsevier.

occur concurrently. The primary hydroxyl group is essential for the formation of levoglucosan and Parkes *et al* [16] suggested that the levoglucosan, in the presence of oxygen, or air, would decompose to give flammable products. So, any method that would promote the dehydration process relative to the depolymerization process would decrease the flammability of cellulose and be extremely beneficial for the manufacture of carbon fiber.

Schuyten *et al* [17] reported that in the presence of flame retardants, the cellulose did decompose at a lower temperature, at a faster rate and produced a greater amount of char.

Flame retardancy has been widely studied [18–22] for flameproofing cotton and much of that work is directly applicable for viscose rayon. Whilst applications for wearing apparel would be concerned with the effect of any subsequent washing process, this would not be relevant to the pyrolysis of viscose rayon. What is important is the lowering of the

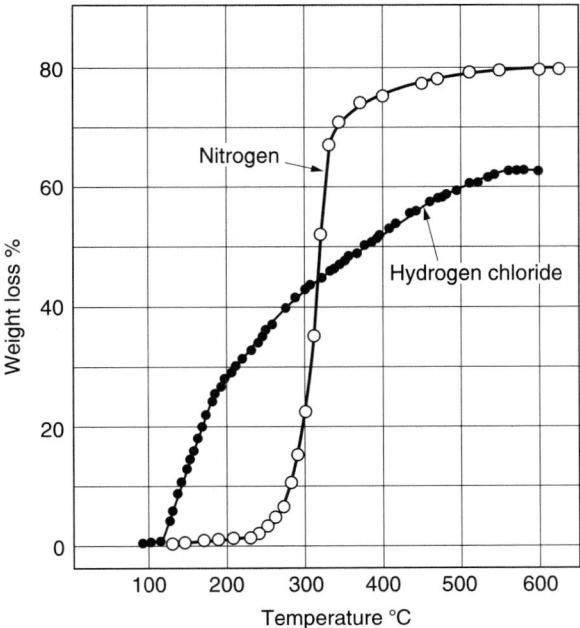

Figure 6.10 Effect of hydrogen chloride vapor on weight loss during pyrolysis of continuous filament high tenacity rayon fiber from Teijin Co. *Source*: Adapted from Shindo A, Nakanishi Y, Soma I, Carbon Fibers from Cellulose Fibers, *Appl Polym Symposia*, No 9, 271–284, 1969.

temperature when water is eliminated and reducing the total loss in mass. The most common finishes are borax ($Na_2B_4.10H_2O$), boric acid (H_3BO_3), urea, diammonium hydrogen phosphate [$(NH_4)_2HPO_4$] or sodium phosphate ($Na_3PO_4.12H_2O$) [23]. These finishes are applied as a solution in water and prior to drying, the excess liquor can be removed by squeeze rollers, which will also promote penetration to the center of the precursor tow. The choice of finish will be dictated by the ability of the final specification to permit the presence of added elements, such as sodium, that would remain in the carbon fiber.

Compounds incorporating nitrogen and phosphorus are well known as effective flame retardants, such as tetrakis (hydroxymethyl) phosphonium chloride (THPC). Phosphorus, together with either halogens or nitrogen, or perhaps a combination of all three, will confer flame retardancy on cellulosic fibers. Phosphorus prevents afterglow, whilst halogens are effective in modifying the volatile products of flaming [24]. Ammonium bromide will decompose with heat and becomes active in the vapor phase. Aqueous ammonium chloride can be used with dilute H_3PO_4 in denatured alcohol. Shindo *et al* [25] found that hydrogen chloride was an effective catalyst (Figure 6.10), obtaining a significant carbon yield, but this will be of doubtful practical significance due to its highly corrosive nature when temperatures of the oven are below the dew point.

Other possible flame retardants are aqueous solutions of nitrogenous salts of strong acids (NH_4Cl), urea, Cs, Na, and K salts (Na_2CO_3), Fe^{++} and Fe^{+++}, alkyl, aryl and halogenated alkyl or aryl phosphates, phosphates, phosphazenes, phosphonates and polyphosphonates [26], but the use of some of these may be restricted due to ecological problems.

Typical commercial flame retardants include:

1. Fyrol 76 (Skauffer Chemical Co.) a precondensate of vinyl phosphonate and *N*-methylolacrylamide with a potassium persulfate catalyst and after application and drying at 120°C, it is cured at 180°C.

2. Proban 210 (Albright & Wilson Ltd.) a derivative of THPC and urea.

$$\begin{array}{c}
-\overset{\overset{\displaystyle O}{\|}}{\underset{\underset{\displaystyle CH_2}{|}}{P}}-CH_2-NH-CO-NH-CH_2-\overset{\overset{\displaystyle O}{\|}}{\underset{\underset{\displaystyle CH_2}{|}}{P}}- \\
\underset{\underset{\displaystyle NH}{|}}{} \qquad\qquad\qquad\qquad\qquad \underset{\underset{\displaystyle NH}{|}}{} \\
\underset{\underset{\displaystyle CH_3}{|}}{} \qquad\qquad\qquad\qquad\qquad \underset{\underset{\displaystyle CH_2}{|}}{} \\
-\overset{\overset{\displaystyle |}{P}}{\underset{\underset{\displaystyle O}{\|}}{}}-CH_2-NH-CO-NH-CH_2-\overset{\overset{\displaystyle |}{P}}{\underset{\underset{\displaystyle O}{\|}}{}}-
\end{array}$$

3. Pyrovatex CP (Ciba Geigy), the *N*-methylol derivative of dimethylphosphonopropionamide reacted with a melamine/formaldehyde condensate used with an acid catalyst such as phosphoric acid:

$$\begin{array}{c} CH_3-O \\ \diagdown \\ P-CH_2-CH_2-CO-NH-CH_2-OH \\ \diagup \\ CH_3-O \end{array}$$

These are after-treatment processes and most of the flame retardant will remain on the surface of the fiber, whereas FR rayon has the flameproof agent distributed uniformly throughout the fiber and might have been an alternative approach, albeit more expensive. Unfortunately, the 'Tris' cancer scare with [*tris*-2,3-dibromopropyl phosphate] and subsequent restrictive US environmental regulations have been responsible for FR rayon falling out of favour.

There are two forms of flame retardants: one where the flame retardant acts as a catalyst and promotes removal of the —OH groups and the second, which actually reacts with the —OH groups [3]. Only a small quantity of catalyst is required, whilst the reactant type requires larger concentrations. When an —OH group is removed, there is a reduction or removal of intermolecular hydrogen bonding, which increases the possibility of the fibers melting, so that a flame retardant that removes —OH bonds and creates covalently bonded crosslinks is desirable.

Lewis acids such as HCl, $ZnCl_2$ and H_3PO_4 [27,28], or $(NH_4)_2HPO_4$ which will decompose to a Lewis acid when heated [23], catalyze the dehydration of cellulose with the possibility of forming covalent cross-links.

Flame retardants which react with the —OH group include atmospheric oxygen. Schwenker [29] has reported that up to 50% of the methylol groups will react to form carboxylic groups. Other reagents which react with the —OH group include chlorosilanes, e.g.,

$$\text{Cell}-\text{OH} + \text{Cl}-\underset{\underset{\displaystyle CH_3}{|}}{\overset{\overset{\displaystyle CH_3}{|}}{Si}}-CH_3 \longrightarrow \text{Cell}-\text{O}-\underset{\underset{\displaystyle CH_3}{|}}{\overset{\overset{\displaystyle CH_3}{|}}{Si}}-CH_3 + \text{HCl}$$

that has reacted to liberate HCl, which then acts as a Lewis acid.

A bifunctional flame retardant such as *bis*-(chlorodiphenyl phosphine) dechlorane would have additional efficiency due to crosslinking [30]. Mixtures of flame retardants can act synergistically, reportedly via the formation of a bifunctional compound which is capable of crosslinking [23,28,31].

It must be appreciated that the route chosen for pyrolysis will be quite dependent on the final application of the carbon fiber and, if admissible, the use of a suitable flame retardant would be preferred, giving a much quicker and more economic route with an increased yield of carbon.

However, in general, this route cannot be adopted and a very slow pyrolysis has to be undertaken in the absence of oxygen, using an inert gas to sweep out the evolved tars and gases, taking particular care to advance the temperature in discrete stages to maintain control over the process. Material processed in the absence of oxygen, when exposed to air, will absorb oxygen with the liberation of heat and if this heat is not correctly dissipated upon removal from the pyrolysis oven, the temperature can build up and a fire may ensue. Poor control of this process will result in very low yields and a product with low strength. Careful control can, however, give yields of about 32%, albeit at a cost, with the process taking about 10 days.

The products of pyrolysis can be burned in a flare or can be fed to an incinerator.

6.2.3 Carbonization

If the pyrolyzed fiber is on hanks, it must now be rewound onto spools for subsequent loading onto a creel prior to carbonization.

It is not feasible to run a continuous process carbonization stage in tandem with a prolonged pyrolysis stage. Therefore, it is practical to run the pyrolysis stage as a batch process and the furnace stages continuously. There are good reasons for separating carbonization into low and high temperature stages, since in the low temperature stage there is some off-gassing, with the requirement to remove gases by an inert gas purge [32], whereas at higher temperatures, the furnace life would be severely curtailed by a high flow of inert gas across the hot element.

6.2.4 Hot stretching during processing of carbon fiber

As mentioned before, the early fiber made from a cellulosic precursor had a low modulus and manufacturers had to resort to hot stretching to orient the fine structure of the fiber and obtain a significant improvement of modulus, albeit at a cost. This approach is not specific to carbon fiber and stretching is widely used to improve properties of many manmade fibers e.g. viscose rayon.

Stretching cannot be undertaken in the pyrolysis stage due to the low strength of the fiber.

Bacon and Smith [33] showed that applying an elongation of 3.6% at high temperature resulted in a 19% increase in modulus. Ezekiel and Spain [34] described methods whereby the carbon fiber could be hot stretched at 2900°C using a resistance heated graphite tube furnace or heating the fiber by passing an electric current through it. Stretching was achieved in the graphitization stage by applying differential speeds to the supply and take-up rollers in conjunction with suspended weights to control the tension.

A novel resistive heating method was described by Gibson and Langlois [35], described in Chapter 3, for the production of Hitco yarn, using a resistance heated furnace to render the fiber electrically conductive and then using resistive heating for the graphitization stage, coupled with differential speed control to facilitate hot stretching.

The modulus of the fibers is dependent on the orientation of the individual graphite planes and the tensile strength is dependent on gage length, but does increase as the modulus increases. High modulus carbon fibers are practically 100% carbon and Bacon and Schalamon [36] report that the structure is of small ribbon-like graphite layers, preferentially

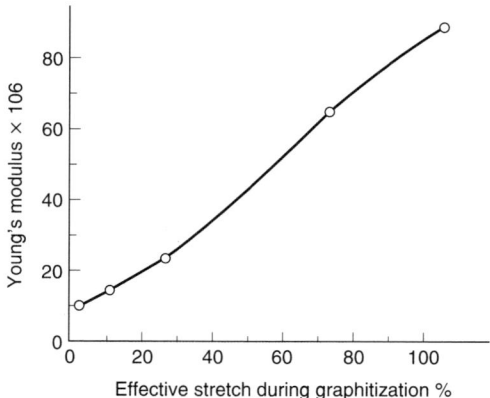

Figure 6.11 Effect on the fiber modulus of stretching rayon fiber precursor up to 2800°C. *Source*: Reprinted with permission from Bacon R, Carbon fibers from rayon precursors, Walker PL, Thrower PA eds., *Chemistry and Physics of Carbon*, Marcel Dekker, New York, 1–102, 1974. Copyright 1974, CRC Press, Boca Raton, Florida.

oriented parallel with the filament axis. About 15–40% of the filament volume consists of preferentially oriented micropores bounded by the graphite layers. The authors were able to increase the modulus from 69 to 690 GPa and the strength from 0.69 to 3.45 GPa by hot stretching, the density increasing from 1.3 to 1.9 g cm^{-3}.

A temperature of about 2800°C is necessary to achieve the requisite extension in a short time using a relatively high load (Figure 6.11) and although lower temperatures can be used, the applied load will be limited in the cooler parts of the furnace and the change in structure will be too slow [37].

Spry [39] claims that treating a rayon yarn previously carbonized at 1200°C followed by treatment at 2800°C with stretching at this latter temperature, produced a six-fold increase in modulus and a three-and-a-half-fold increase in strength.

6.2.5 Sizing

For many applications where the product is to be used as a packing material, it is necessary to apply a size to aid in braiding the product and ensure that the packing material beds down well. These sizes are specific to a given application and can vary from PTFE, polyvinyl pyrrolidone

$$(-CHCH_2-)_n$$

to a flake graphite in a formulated grease.

6.3 MECHANISMS FOR THE PYROLYSIS AND CARBONIZATION STAGES OF CELLULOSIC BASED PRECURSORS

The pyrolysis of cellulosic materials results in the formation of highly volatile materials, a tarry distillate and a carbonaceous char. Early work (1909–10) by Klason [40,41] investigated the thermal degradation of various celluloses, whilst later work concentrated on the treatment of cellulose to render it flameproof [42].

Tamaru [43,44], in 1948, supported by Parks et al [45] in 1955, proposed that the pyrolytic decomposition of cellulose resulted in the breaking of the 1,4-glycoside linkage followed by intramolecular arrangement to form levoglucosan.

Agster [46] stressed the importance of the temperature of degradation and using viscose rayon, proposed the following reaction stages:

1. Primary reaction of hydrolysis below 140°C
2. Air oxidation at 140–160°C
3. Cracking above 160°C

Higgins [47], using IR spectroscopy, found C=C bonds appearing above 280°C and becoming prevalent at 400°C, denoting the onset of an aromatic structure. Major [48], using cotton linters in oxygen at 170°C, found that oxidative degradation occurred primarily in the amorphous regions of the cellulose, but could make no deductions when the reaction was carried out in N_2 under essentially anhydrous conditions. The importance of using IR for examining the breakdown of cellulose was stressed by Hofman [49].

During the early investigations to find a suitable cellulosic precursor for making carbon fiber, many natural fibers were considered, but viscose rayon soon became the preferred material [50]. A major problem in processing such a precursor was the extremely low yields, which could be as little as 10% when using air oxidation as the initial stage, a situation that could be improved by treatment with a flameproofing agent, possibly in conjunction with a catalyst. Alternatively, the reaction could be undertaken in an inert atmosphere, which is then termed pyrolysis and running the process slowly increased yields from 15% to about 30%, but because of such extended treatment times, it was less economic.

Considering these alternative approaches, it is not surprising that they could each have different effects on the mechanism of thermal decomposition.

Konkin [50] divides the chemical processes undergone in the transition of cellulose to carbon fiber into four groups: heterolytic depolymerization, dehydration, homolytic depolymerization and a more thorough thermal decomposition. Since the initial pyrolytic reactions are heterolytic, the course of pyrolysis and the evolved products can be substantially changed by the addition of acidic or alkaline catalysts [51].

Scission of the 1,4-glycosidic linkages in the molecular structure of cellulose readily occurs in the presence of moisture and Shevchenko et al [52] showed that 5% moisture can reduce the degree of polymerization of cellulose by hydrolytic decomposition from about 400 to 120 in 4 h at 240°C. In fact, the variability of the initial moisture content in cellulose can introduce variability in subsequent thermal degradation measurements.

Cellulose contains many hydroxyl groups:

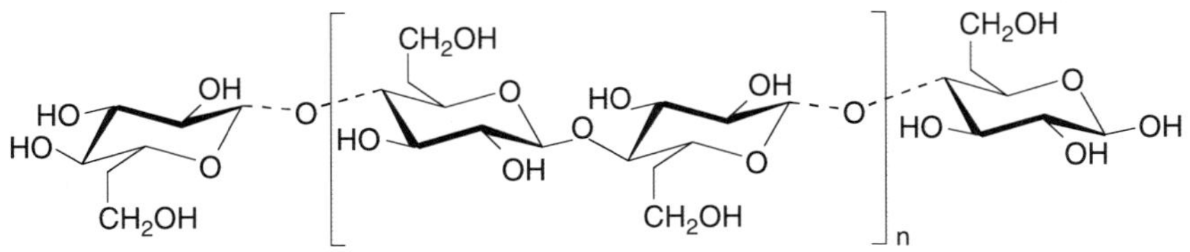

If the pyrolysis reaction is a straight dehydration process then the following reaction (Scheme I) would apply:

$$(C_6H_{10}O_5)_n \rightarrow 6n\ C + 5n\ H_2O$$

Scheme I

However, the theoretical yield of carbon would be 44.2%, a value that is found to be much too high in practice and, consequently, the pyrolysis must be accompanied by other reactions.

Several routes have been suggested for the mode of the elimination of —OH groups:

1. Scheme II — an intramolecular reaction, supported by Tang and Bacon [53]:

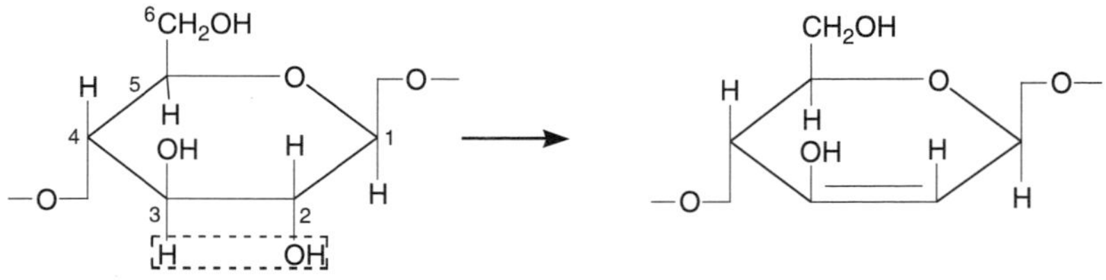

Scheme II

2. Scheme III — a nucleophilic substitution reaction:

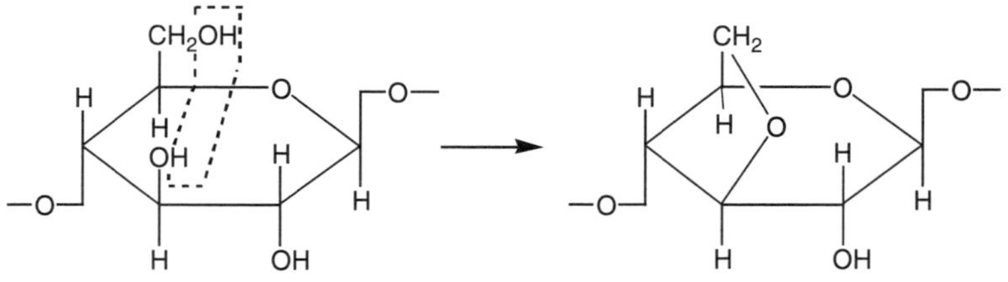

Scheme III

3. Scheme IV — an intermolecular reaction, favoured by Kilzer and Broido [54]:

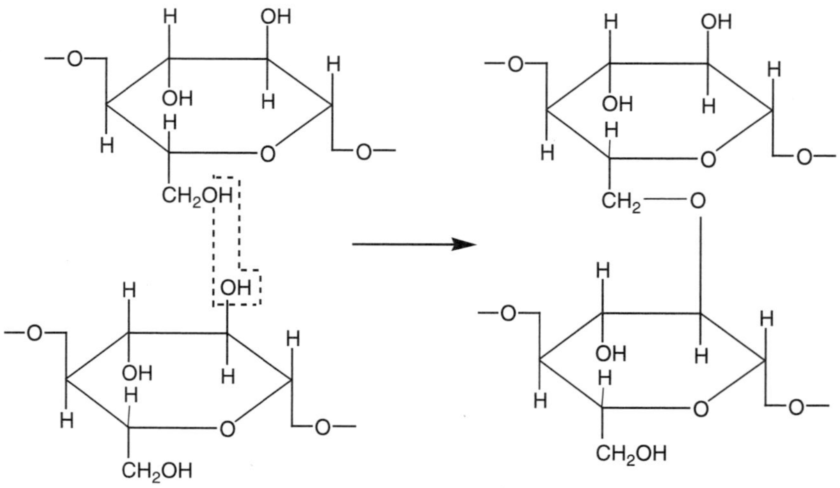

Scheme IV

Shindo et al [55,56], using Teijin high tenacity rayon fiber in the presence of hydrogen chloride, showed that elimination of water commenced at 120°C, whereas in the presence of nitrogen, dehydration was about 100°C higher (Figure 13.3). Fairbridge et al [57], showed that water was the only volatile product of thermal degradation up to 250°C. Tang and Bacon [53], using Fortisan-36 regenerated from cellulose acetate, obtained major pyrolytic degradation above 240°C when heated in argon, becoming complete at about

320°C and proposed the following over-simplistic mechanism accounting only for the major reactions:

Stage 1. Physical desorption of water (25–150°C)
Stage 2. Dehydration from the cellulose unit (150–240°C)
Stage 3. Thermal cleavage of the glycosidic linkage and scission of other C=O bonds and some C—C bonds via a free radical reaction (240–400°C)
Stage 4. Aromatization (400°C and above)

Whereas, in the presence of HCl, Shindo et al [56], proposed:

Stage 1. Physical desorption of water (25–110°C)
Stage 2. Dehydration from the cellulose unit in the amorphous regions (110–160°C) and in crystalline regions (160–185°C)
Stage 3. Elimination of one carbon atom per ring, probably involving opening of furan rings (185–290°C)
Stage 4. Aromatization begins, probably accompanied with further elimination of water and HCHO, although the latter was not detected (290–450°C)
Stage 5. Beginning of carbon structure formation (460°C upwards)

The yield of carbon fiber is reduced by the formation of levoglucosan (1,6-anhydro-β-D-glucopyranose) and Madorky [58] suggested that when a glycoside bond is split, it forms a levoglucosan unit at one end of the molecule and a unit containing four —OH groups at the other end:

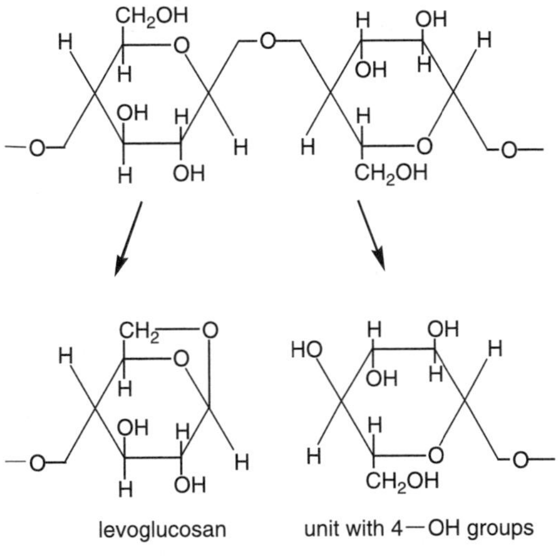

levoglucosan   unit with 4—OH groups

These two reactions are competitive, but since the withdrawal of the —CH_2OH groups starts at 120°C and continues up to about 300°C, the early removal of —CH_2OH groups prevents their subsequent reaction to form levoglucosan at 250°C and above via a dehydration reaction. Hence, holding the temperature in the initial stages below 250°C is an effective way of improving the carbon yield [54]. However, the two reactions cannot be separated and there will always be a temperature region where both reactions occur concurrently.

Levoglucosan, when pyrolyzed at 600°C, gave a range of products listed in Table 6.1 and in the absence of acidic or alkaline catalysts, both fission and dehydration products are formed, but the presence of a catalyst will promote one type of product at the expense of

Table 6.1 Pyrolysis products of levoglucosan

Pyrolysis product	Formula	Yield (%)		
		No catalyst	With $ZnCl_2$	With NaOH
Acetaldehyde	HCHO	1.1	0.3	7.3
Furan	(furan ring)	1.0	1.3	1.6
Acrolein	$H_2C{=}CHCHO$	1.7	<0.1	2.6
Methanol	CH_3OH	0.3	0.4	0.7
2,3-Butanedione	$CH_3COCOCH_3$	0.5	0.8	1.6
2-Butenal	$CH_3C{=}CHCHO$	0.7	0.2	2.2
1-hydroxy-2-propanone	CH_3COCH_2OH	0.8	<0.1	1.1
Glyoxal	HCOCHO	1.4	<0.1	4.9
Acetic acid	CH_3COOH	1.7	0.7	1.5
2-Furaldehyde	(furan-CHO)	0.9	3.0	0.4
5-methyl-2-furaldehyde	(CH_3-furan-CHO)	0.1	0.3	–
Carbon dioxide	CO_2	2.9	6.8	5.7
Water	H_2O	0.7	20.1	14.1
Char	–	3.9	29.0	16.0
Balance (tar)	–	74.3	36.8	40.3

Source: Reprinted with permission from Shafizadeh F, Industrial pyrolysis of cellulosic materials, *Appl Polym Symp*, No. 28, 153–174, 1975. Copyright 1975, Elsevier.

the other [59]. Parks *et al* [60] believed that the flammability of cellulose was largely controlled by the breakdown of levoglucosan to give flammable products and if this was prevented [61–63] the flammability of cellulose would be decreased.

Many workers [40,41,61–65] have identified some of the various volatile pyrolysis products and tars. Klason *et al* [41], in 1910, had already identified materials given in Table 6.2, whilst Robert *et al* [65], using chromatographic techniques identified some 37 compounds, accounting for about 70% of the decomposition products. Some of these products are given in Table 6.3.

The composition of such breakdown products is highly dependent on the precursor, the heating conditions and whether a catalyst is used or not. The term catalyst in this sense is a substance which has an influence on the actual cellulose pyrolysis process. Tamaru [66] investigated the effect of inorganic salts on the pyrolysis and combustion of cellulose. Garn and Denison [67] investigated flame retardants for cotton, which had a direct spin-off for carbon fiber manufacture. Typical treatment products/catalysts that have been investigated

Table 6.2 Volatile products found by Klason

Substance	Formula	Substance	Formula
Water	H_2O	Ethylene	$CH_2{=}CH_2$
Carbon monoxide	CO	Acetone	CH_3COCH_3
Carbon dioxide	CO_2	Acetic acid	CH_3COOH
Methane	CH_4	–	–

Source: Reprinted from Klason P, Heidenstam G, Norlin E, *Z Angew Chem*, 22, 1205, 1909.

Table 6.3 Volatile products found by Robert

Substance	Formula	Substance	Formula
Carbon monoxide	CO	Furfural	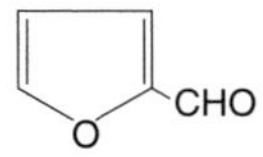 (furan)-CHO
Carbon dioxide	CO_2	Acetone	CH_3COCH_3
Formaldehyde	HCHO	Methylethylketone	$CH_3COC_2H_5$
Methyl alcohol	CH_3OH	Formic acid	HCOOH
Acrolein	$CH_2{=}CHCHO$	Water	H_2O
Proprionaldehyde	C_2H_5CHO	Levoglucosan	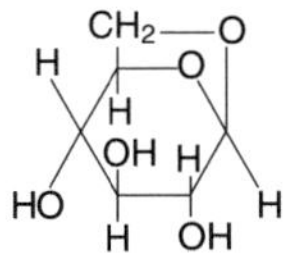 (levoglucosan structure)
n-butyric acid	$CH_3(CH_2)_2COOH$	Lactic acid	$CH_3CH(OH)COOH$
Glyoxal	HCOCHO	5-hydroxymethyl-furfural	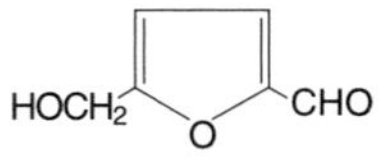 $HOCH_2$-(furan)-CHO

Source: Reprinted with permission from Robert F, Schwenker JR, Louis R, Beck JR, *J Polym Sci*, C(2), 331, 1963. Copyright 1963, John Wiley & Sons Ltd.

include NH_4Cl, borax/diammonium phosphate, borax/boric acid, phosphorus salts, metal chlorides and salts of Fe^{2+} and Fe^{3+}. All catalysts behave similarly and lower the temperature of onset of degradation and decrease the degree of depolymerization of the cellulose.

Capon and Maggs [68] used DTA and TG techniques to investigate the effects of metal chlorides on the thermal decomposition of viscose rayon and found that $ZnCl_2$ and $CdCl_2$ impregnations lowered the rapid decomposition temperature and increased the char yield, appearing to promote the initial dehydration reaction at the expense of the depolymerization process [69].

Ross [70] discussed the weight loss of rayon yarns in an oxidizing atmosphere up to 370°C and the factors influencing the reason why the loss of hydroxyl groups should occur at a much faster rate than that indicated by the weight loss after exposure up to 370°C of one hydroxyl group per mole of water, the figure being nearer to 2.5 hydroxyl groups in practice. Several explanations have been proposed by a number of workers.

Back [71] suggested an auto-crosslinking reaction (Scheme V) when carbonylic groups and radicals are formed initially and after oxidation with periodate, crosslink with a neighboring chain to form hemiacetal:

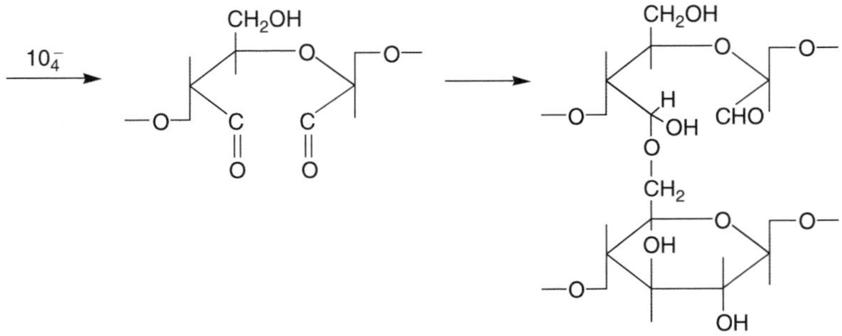

Scheme V

This suggests that this reaction is catalyzed by the acid freed from the flame retardant. This explanation accounts for loss of hydroxyl groups without attendant weight loss, followed by dehydration catalyzed by the presence of the flame retardant, which itself reduces the decomposition temperature, thereby reducing the formation of levoglucosan.

Another approach by Kilzer and Broido [54] suggested the formation of dehydrocellulose by a dehydration reaction, or at higher temperature, depolymerization to give levoglucosan.

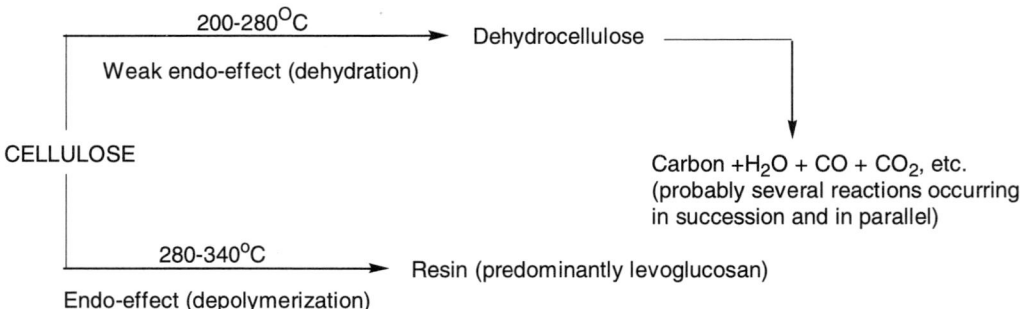

For a possible route via a furan intermediate for the breakdown of the dehydrocellulose see Scheme VI.

Scheme VI

Glucose exists in chair and bed conformations and Byrne et al [72] believe that the conformation of the elementary cellulose unit controls the route in which the reaction will proceed, where if an initial chair conformation exists, then the breakdown proceeds (Scheme VIa) via furfural derivatives, but the bed conformation will form levoglucosan. Unfortunately, the original cellulose conformation cannot be determined.

Scheme VIa

Shindo [56] supports the furfural route (see Scheme VII).
The high carbon yield and low weight of tarry products formed suggest that the furan ring compounds cross-link with one another, or react with a neighboring species, before they escape from the fiber. The mechanism does not explain the formation of saturated ketone groups at 120°C, which were detected by IR. Shindo proposed the elimination of H_2O from the cellulose rings prior to the occurrence of chain scission:

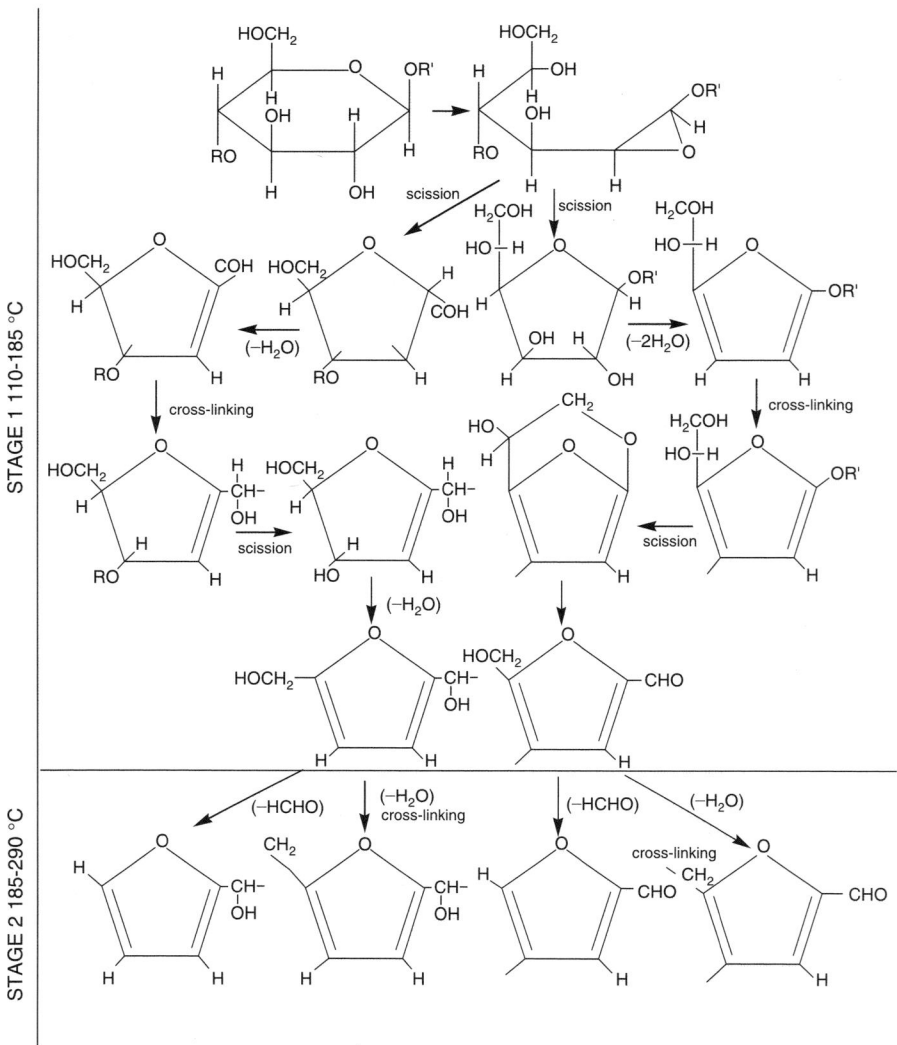

Scheme VII

Tang and Bacon proposed Scheme VIII based on an intramolecular reaction.
The structure of the cellulose has a marked influence on the subsequent decomposition [73–77] and less crystalline materials decompose more readily and in terms of thermal decomposition can be rated viscose cord rayon > viscose continuous filament > viscose rayon fiber > Fortisan fiber > Cotton > hydrocellulose. However, hydrocellulose does not follow this rule.

From X-ray diffraction of films cast from regenerated cellulose (rayon) fibers, based on an assumed four-carbon residue, Bacon and Tang [78] proposed two alternative geometric schemes: Scheme IX, representing longitudinal polymerization, where the four-carbon residues aligned themselves parallel to the b-axis and joined to form a chain along the original cellulose chain direction, with adjacent chains joining to form graphite layers.

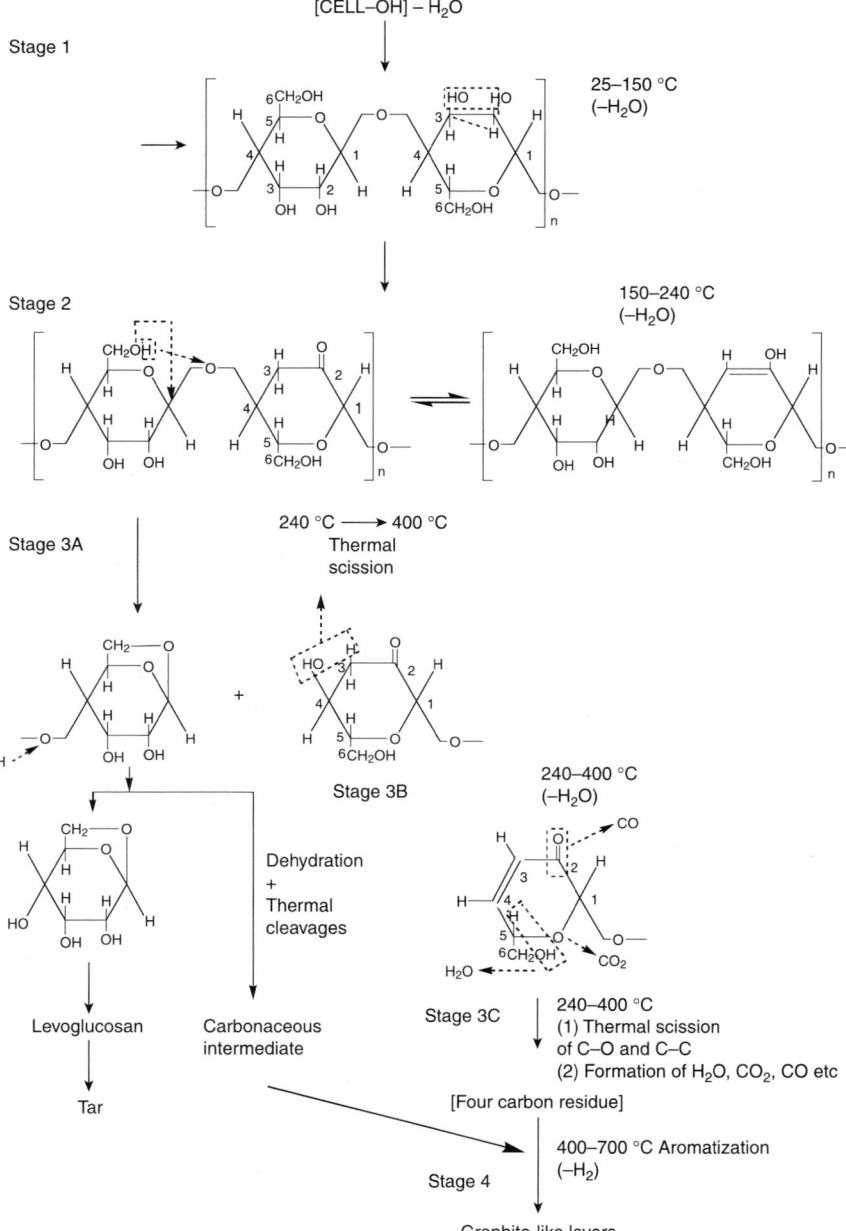

Scheme VIII

Whilst Scheme X is represented by transverse polymerization, where each residue joins with a neighbor, to form a carbon chain in the transverse direction, with adjacent chains joining to form graphite layers. Preference was given to the longitudinal polymerization.

However, Konkin [50] stated that the proposal had not found experimental confirmation and objected to the term polymerization, since it was, more correctly, a series of reactions involving polycondensation, recombination and polyrecombination.

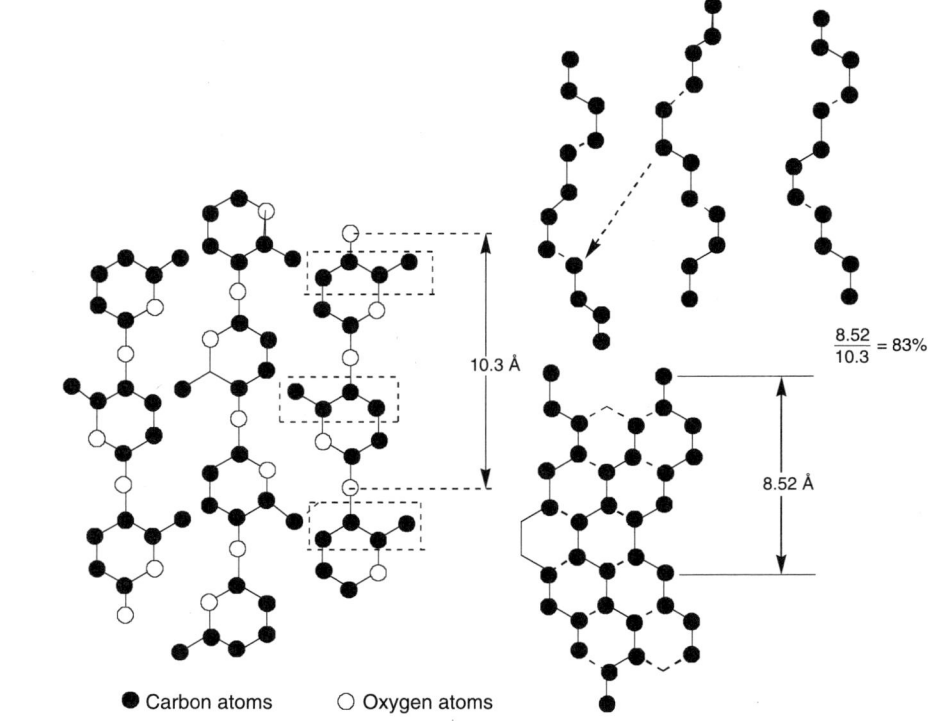

Scheme IX

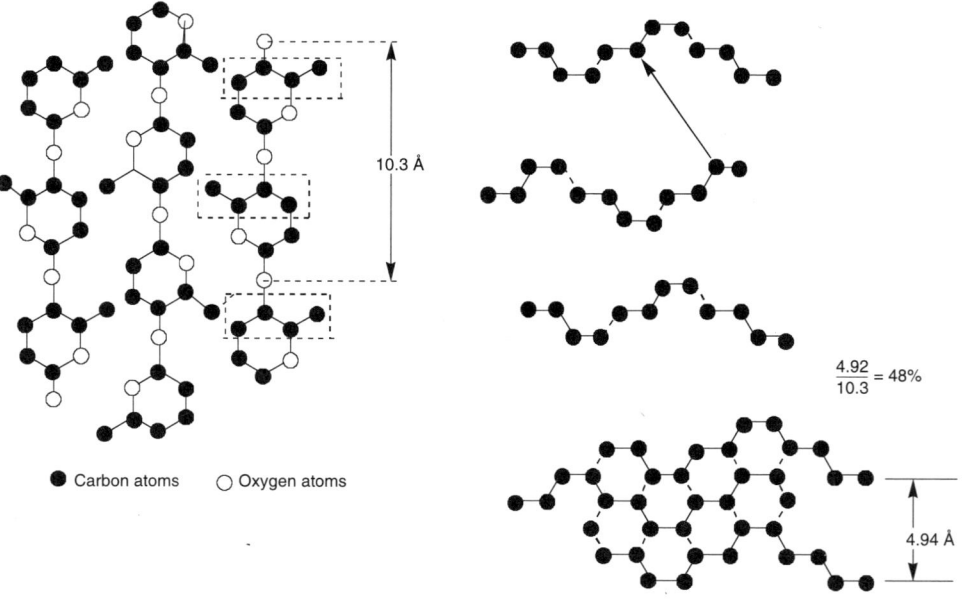

Scheme X

Davidson [79] proposed Scheme XI, which was a rather speculative attempt to suggest transition of a cellulose structure into a graphite structure where, as the treatment temperature increased, the chains became closer, interacting to form polycyclic rings.

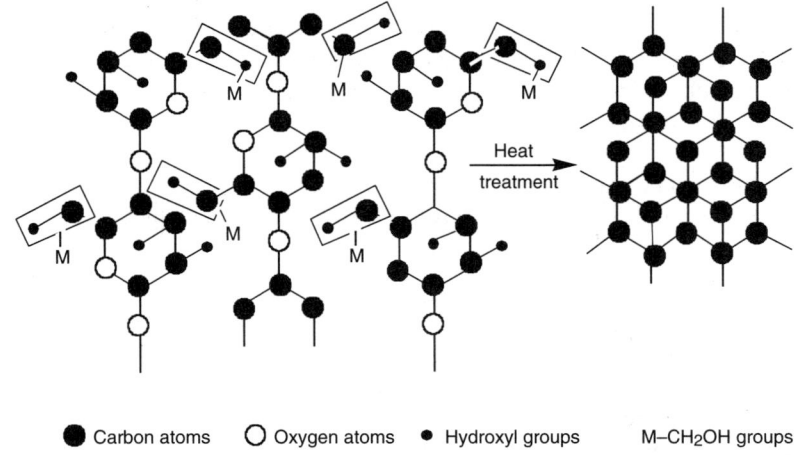

Scheme XI

Davidson proposed this structure

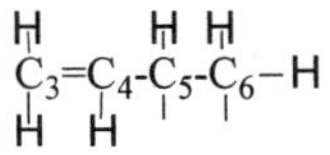

$$\underset{H\ \ H}{\overset{H}{C_3}}=\underset{H\ H}{\overset{H\ H}{C_4\text{-}C_5\text{-}C_6}}\text{-}H$$

for a four-carbon residue.

Davidson and Losty [80], using wheat straw cellulose and heating *in vacuo*, determined that water was evolved at 155°C, peaking at 245°C, whilst CO_2 was evolved at 160°C, peaking at 250°C and CO was first evolved at 200°C, peaking at 250 and 360°C. Tar was noticeable in the range 200–250°C and most of the weight loss occurred in the temperature range 150–450°C, being most rapid below 300°C. Using X-ray diffraction and IR, the authors found that although water was removed, the cellulose molecule remained structurally intact up to about 40% weight loss. An interesting observation was that the pure carbon residue, obtained after carbonization at 1500°C, arose directly from the cellulose molecule and not from cracking of any transitory gaseous products.

Losty and Blakelock [81,82] (Scheme XII), again working with wheat straw cellulose, determined that 40–60% of the weight loss in the primary decomposition process is the elimination of oxygen from the pyranose rings and glycoside links of the original cellulose, forming aromatic platelets about 1.5 nm in diameter, surrounded by a layer of hydrogen. About 60–80% weight loss due to the elimination of H_2 contributes to increasing the free

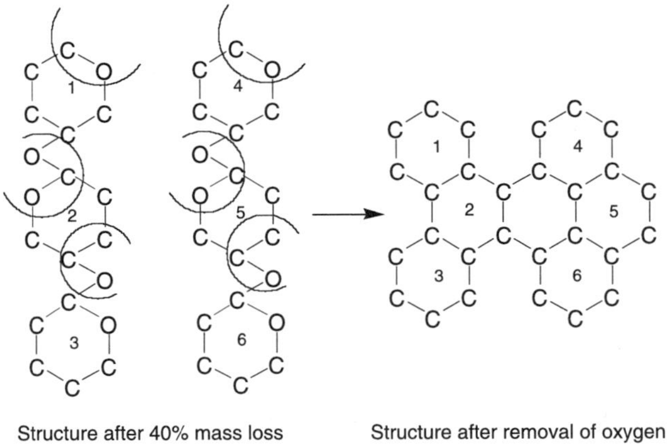

Structure after 40% mass loss Structure after removal of oxygen

Scheme XII

radical concentration until there is insufficient H_2 to isolate the aromatic platelets, when the structure rapidly coalesces and the free radical concentration decreases.

Losty and Blakelock, using electron spin resonance measurements on cellulose heated to 750°C, found:

1. At 350–550°C, there was an increase in the number of electrons available for conduction, but conduction was limited by a low probability of electrons crossing boundaries of aromatic units
2. At 550–700°C, the number of free spins decreases, but occurrence of C—C bonds of aromatic units joining together increases the probability of higher conduction, as found in practice. This observation is consistent with the view that transport of heat is limited by phonon scattering at the boundaries between the aromatic units until they begin to coalesce.

It is interesting to note that SEM observations revealed that the longitudinal striations present on the surface of the original cellulose remained on the surface of the carbon fiber, inspite of the variety of chemical transformations which had occurred.

REFERENCES

1. Gill RM, *Carbon fibres in Composite Materials*, The Plastics Institute, Iliffe Books, London, 20–39, 1972.
2. Bacon R, Carbon fibers from rayon precursors, Walker PL, Thrower PA eds., *Chemistry and Physics of Carbon*, Marcel Dekker, New York, 1–102, 1974.
3. Goodhew PJ, Clarke AJ, Bailey JE, A review of the fabrication and properties of carbon fibres, *Mater Sci and Eng*, 17, 3–30, 1975.
4. Riggs DM, Shuford RJ, Lewis RW, Graphite fibers and composites, Lubin G ed., *Handbook of Composites*, Van Nostrand Reinhold, New York, 196, 1982.
5. Konkin AA, Production of cellulose based carbon fibrous materials, Watt W, Perov BV eds., *Strong Fibres*, Elsevier Science Publishers, Amsterdam, 275–325, 1985.
6. Peebles LH, Carbon Fibers Formation, Structure and Properties, CRC Press Inc., Boca Raton, 1994.
7. Bahl OP, Shen Z, Lavin JG, Ross RA, Manufacture of carbon fibers, Donnet JB, Wang TK, Peng JCM, Rebouillat S eds., Carbon Fibers 3rd Edition, Marcel Dekker, New York, 31–40, 1998.
8. Cranch GE, Unique properties of flexible carbon fibres, *Proc 5th Conf Carbon Vol 1*, Pergamon, New York, 589, 1962.
9. Ford CE, Mitchell CV, Union Carbide Corp., U.S. Pat., 3,107,152, 1963.
10. Cross CB, Ecker DR, Stein OL, Union Carbide Corp., Artificial graphite process, U.S. Pat., 3116975, 1964.
11. Tang MM, Bacon R, Carbonization of cellulose fibers: 1. Low temperature pyrolysis, *Carbon*, 2, 211–220, 1964.
12. Strong SL, Small-scale heat treatment of rayon precursors for stress graphitization, *J Mater Sci*, 9, 993–1003, 1974.
13. Ross SE, Observations concerning the carbonization of viscose rayon yarn, *Text Res J*, 38, 906–913, 1968.
14. Losty HHW, Blakelock HD, The structure and properties of partially pyrolysed cellulose carbon, *Proc 2nd Conf Ind Carbon and Graphite*, 29–35, London, Apr 1965.
15. Kilzer FJ, *A Broido Pyrodynamics*, 2, 151–163, 1965.
16. Parks WG, Antoni M, Petrarca AE, Petrochelli AR, The catalytic degradation and oxidation of cellulose, *Text Res J*, 25, 789, 1955.
17. Schuyten HA, Weaver JW, Reid JD, *Ind Eng Chem*, 47, 1433, 1955.
18. Little RW, *Flameproofing of Textile Fabrics*, Reinhold Publishing, New York, 1947.

19. Lewin M, Sello SB, *Flame Retardant Polymeric Materials*, Vol 1, Lewin M, Atlas SM, Pearce EM eds., Plenum Press, New York, 19–136, 1975.
20. Lewin M, Basch A, *Flame Retardant Polymeric Materials*, Vol 2, Lewin M, Atlas SM, Pearce EM eds., Plenum Press, New York, 1–42, 1978.
21. Lewin M, *Handbook of Fiber Science and Technology, Vol 2, Chemical Processing of Fibers and Fabrics, Part B, Functional Finishes*, Lewin M, Sello SB eds., 1–141, 1984.
22. Calamari TA Jr., Harper RJ Jr., Flame retardants for textiles, Kroschivitz JI, Howe-Green M eds., *Kirk Othmer Encyclopedia of Chemical Technology*, Vol 10, 4th edition, 998, 1993.
23. Mack CH, *Text Res J*, 37, 1063, 1967.
24. Clayton G, Viscose Rayon Fibres, *Text Progress*, 8(1), 9, 1976.
25. Shindo A, Nakanishi Y, Soma I, Carbon Fibers from Cellulose Fibers, Preston J ed., *High temperature Resistant Fibers from Organic polymers*, Interscience, New York, 271, 1969.
26. Dyer J, Daul GC, Rayon Fibers, Lewin M, Pearce EM eds., *Handbook of Fiber Chemistry*, Marcel Dekker, New York, 790, 1998.
27. Shindo A, Nakanishi Y, Soma I, Carbon Fibers from Cellulose Fibers, *Appl Polym Symposia*, No 9, 271–284, 1969.
28. Shindo A, Nakanishi Y, Soma I, *Polymer Preprints*, 9, 1333, 1968.
29. Schwenker RF, Pacsu E, Chemically modifying cellulose for flame resistance, *Ind Eng Chem*, 50(1), 91–96, 1958.
30. Duffy JV, *J Appl Polym Sci*, 15, 715, 1971.
31. Hendrix JE, *J Appl Polym Sci*, 1, 257, 1972.
32. Russell WE, Hogg GR Jr., Thomas EC, Great Lakes Carbon Corp., Heating assembly for heat-treating or graphitizing continuously moving materials and process of heat-treating and/or graphitizing flexible fibrous materials, U.S. Pat., 3,367640, Feb 6 1968.
33. Bacon R, Smith WH, Tensile behaviour of carbonized rayon filaments at elevated temperatures, *Proc 2nd Conf Industrial carbon and Graphite*, SCI, London, 203, 1965.
34. Ezekiel HM, Spain RM, Preparation of graphite fibers from polymeric fibers, *J Polym Sci*, C(19), 211, 1967.
35. Gibson DW, Langlois GB, Method for producing high modulus carbonnyarn, *Polymer Preprints*, 9(2), 1376, 1968.
36. Bacon R, Schalamon R, Physical properties of high modulus graphite fibers made from a rayon precursor, *Applied Polymer Symposia*, No 9, 285–292, 1969.
37. Diefendorf RJ, Tokarsky E, High performance carbon fibers, *Polym Eng Sci*, 15, 150, 1975.
38. Spry WJ, Union Carbide Corp., Process for producing fibrous graphite. U.S. Pat., 3,454,362, Jul 1969.
39. Spry WJ, Union Carbide Corp., Graphite yarn, U.S. Pat., 3,503,708, Mar 1970.
40. Klason P, Heidenstam G, Norlin E, *Z Angew Chem*, 22, 1205, 1909.
41. Klason P, Heidenstam G, Norlin E, *Z Angew Chem*, 23, 1252, 1910.
42. Coppick S, Flameproofing Textile Products, *ACS Monograph 104*, Little RW ed., Reinhold, New York, 1947.
43. Tamaru K, *J Chem Soc Japan, Pure Chem Section*, 69(1–3), 20, 1948.
44. Tamaru K, *J Chem Soc Japan, Pure Chem Section* 69(1–3), 21, 1948.
45. Parks WG, Antoni M, Petrarca AE, Petrochelli AR, The catalytic degradation and oxidation of cellulose, *Text Res J*, 25, 789, 1955.
46. Agster A, *Melliand Textilber*, 37(11), 1338–1344, 1956.
47. Higgins HG, *J Polym Sci*, 28, 645, 1958.
48. Major WD, The degradation of cellulose in oxygen and nitrogen at high temperatures, *TAPPI*, 41(9), 530–537, 1958.
49. Hofman W, Ostrowski T, Urbanski T, Witanowski M, Infra red absorption spectra of products of carbonisation of cellulose and lignin, *Chem Ind*, No 45, 95, 1960.
50. Konkin AA, Production of Cellolose Based Carbon Fibrous Materials, Watt W, Perov BV eds., *Strong Fibres*, Elsevier Science Publishers, Amsterdam, 275–325, 1985.
51. Shafizadeh F, Industrial pyrolysis of cellulosic materials, *Appl Polym Symp*, No. 28, 153–174, 1975.

52. Shevchenko AS, Nepochatykh VI, Bandura NA, Vlasyuk AT, Volkova TG, *Khim Volakna*, No 5, 46, 1978.
53. Tang MM, Bacon R, Carbonization of cellulose fibers—I, Low temperature pyrolysis, *Carbon*, 2, 211–220, 1964.
54. Kilzer FJ, Broido A, *Pyrodynamics*, 2, 151–163, 1965.
55. Shindo A, Nakanishi Y, Soma I, *Polym Prepr*, 9(2), 1333, 1968.
56. Shindo A, Nakanishi Y, Soma I, Carbon Fibers from Cellulose Fibers, *Appl Polym Symposia*, No. 9, 271–284, 1069.
57. Fairbridge C, Ross RA, Sood SP, *J Appl Polym Sci*, 22, 497, 1978.
58. Madorsky SL, *Termicheskoje Razloshenie Organischeskikh Polimerov*, MIR Moscow, 328, 1967.
59. Shafizadeh F, Philpot CW, Ostojic N, *Carbohydr Res*, 16, 279, 1971.
60. Parks WG, Esteve RM Jr., Gollis MH, Guercia R, Petrarca A, Mechanism of Pyrolytic Decomposition of Cellulose, *127th Meeting ACS*, Cincinnati, Apr 1955.
61. Schwenker RF, Pacsu E, Chemically modifying cellulose for flame resistance, *Ind Eng Chem*, 50(1), 91–96, 1958.
62. Madorsky SL, Hart VE, Straus S, Pyrolysis of cellulose in a vacuum, *J Res Nat Bur Stds*, 56(6), 343, 1956.
63. Madorsky SL, Hart VE, Straus S, Thermal degradation of cellulosic materials, *J Res Nat Bur Stds*, 60(4), 343, 1958.
64. Laible RC, *Am Dyestuff Rep*, 47(6), 173–178, 1958.
65. Robert F, Schwenker JR, Louis R, Beck JR, *J Polym Sci*, C(2), 331, 1963.
66. Tamaru K, Pyrolysis and combustion of cellulose in the presence of inorganic salts, *Bull Chem Soc Japan*, 24(4), 164, 1951.
67. Garn PD, Denison CL, *Text Res J*, 47, 485, 1977.
68. Capon A, Maggs FAP, The effect of metal chlorides on the thermal decomposition of viscose rayon, *1st European Symposium on Thermal Analysis*, 176–179, 1975.
69. Weinstien M, Broido A, *Combust Sci Technol*, 1, 287, 1970.
70. Ross SE, Observations concerning the carbonization of viscose rayon yarn, *Text Res J*, 38, 906–913, 1968.
71. Back EL, *Pulp Paper Mag Can Tech*, Sec 1–7, Apr 1967.
72. Byrne GA, Gardiner D, Holmes FH, *J Appl Chem*, 16, 81, 1966.
73. Philip B, Baudisch J, Gaudig A, *Faserforsch Textitechn*, 18, 9, 1976.
74. Philip B, Baudisch J, Gaudig A, *Faserforsch Textitechn*, 18: 461, 1976.
75. Ramiah MV, *J Appl Poly Sci*, 14, 1323, 1970.
76. Basch A, Lewin M, *J Poly Sci Polym Chem*, 11, 3071, 1973.
77. Broido A, Javier-Son AC, Quano AC, Barral EM, *J Appl Polym Sci*, 17, 3627, 1973.
78. Tang MM, Bacon R, Carbonization of cellulose fibers—II. Physical property study, *Carbon*, 2, 221–225, 1964.
79. Davidson, U.S. Pat., 3,104,159.
80. Davidson HW, Losty HHW, The initial pyrolyses of celluloses, *GEC J*, 30, 22–28, 1963.
81. Losty HHW, Blakelock HD, *Proc 2nd Conf Ind Carbon and Graphite*, London, 20–28, Apr 1965.
82. Losty HHW, Blakelock HD, The structure and properties of partially pyrolysed cellulose carbon, *Proc 2nd Conf Ind Carbon and Graphite*, London, 29–35, Apr 1965.

CHAPTER 7

Carbon Fiber Production using a Pitch based Precursor

7.1 INTRODUCTION

The types of pitch based melt spun precursor fibers that may be produced have already been discussed in Chapter 4 and that choice will obviously dictate the type of carbon fiber that can be made. A range of available pitch based carbon fibers is listed in Chapter 20. Table 7.1 lists the properties of Mitsubishi Chemical Corporation's Dialead and Table 7.2 lists the typical properties of the composite materials. An SEM micrograph of Dialead K13C2U at different magnifications is shown in Figure 7.1.

7.2 CHOICE OF MELT SPUN PRECURSOR

An isotropic pitch precursor is used to make a GP grade carbon fiber, whilst a mesophase pitch will produce a HP carbon fiber that can, if necessary, also be made from the same grade of isotropic pitch feedstock. Mesophase pitches can be divided into classes (Chapter 4) on the basis of their method of manufacture—preparation by means of pyrolysis, solvent extraction, hydrogenation or catalyst modification. A variant of the hydrogenation process, termed the Dormant Process, will produce a carbon fiber which is intermediate between GP and HP, with high elongation [1].

Table 7.1 Properties of Mitsubishi Chemical Corporation's Dialea, a 10 μm diameter coal tar pitch based carbon fiber

Property	Grade of Dialead					Single crystal graphite
	K1352U	K1392U	K13B2U	K13C2U	K13D2U	
Tensile strength GPa	3.6	3.7	3.8	3.8	3.7	-
Tensile modulus GPa	620	760	830	900	935	1000
Ultimate elongation %	0.58	0.49	0.46	0.42	0.40	-
Density $g\,cm^{-3}$	2.12	2.15	2.16	2.20	2.21	2.265
Electrical resistivity $\mu ohm\,m$	6.6	5.0	4.1	1.9		0.4
Thermal conductivity ($W\,m^{-1}\,K^{-1}$)	140	210	260	620	800	2000

Source: Reprinted from Mitsubishi technical literature.

Table 7.2 Properties of unidirectional laminates with 60% v/v Mitsubishi Chemical Corporation's Dialead coal tar pitch based carbon fiber in Fiberite 934 resin system

Property	Grade of Dialead				
	KS352U	K1352U	K1392U	K1382U	K13C2U
Longitudinal:					
Tensile strength GPa	2.0	2.0	2.1	2.2	2.2
Tensile modulus GPa	330	380	460	490	560
Compressive:					
Strength Gpa	0.48	0.45	0.40	0.38	0.38
Modulus Gpa	250	310	420	450	560
Shear strength MPa	75	75	70	60	50
Transverse:					
Strength Mpa	40	40	35	30	30
Modulus Gpa	6.3	6.2	6.0	5.5	5.4

Source: Reprinted from Mitsubishi technical literature.

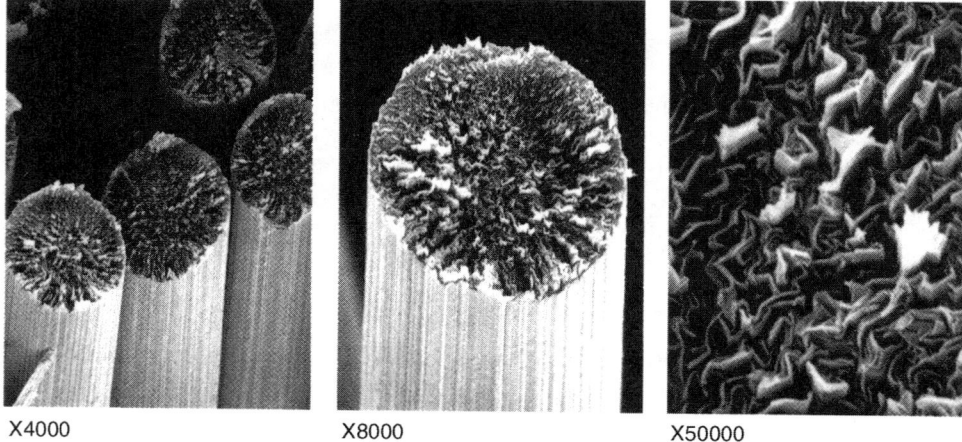

X4000   X8000   X50000

Figure 7.1 SEM of Mitsubishi Chemical Corp. Dialead K13C2U coal tar pitch based carbon fiber.
Source: Reprinted from Mitsubishi Chemical Corp. technical literature.

7.3 THE MANUFACTURING PROCESS

Carbon fibers prepared from a pitch precursor have been discussed by Rand [2], Edie [3] and Bahl et al [4]. A good account of their structure and texture has been given by Oberlin et al [5].

7.3.1 Stabilization (thermosetting) of spun fiber

The pitch precursor fibers, as spun, are very weak and almost without exception, thermoplastic in nature, making it imperative that they are chemically treated to render them suitable for subsequent carbonization. This is best accomplished by some form of oxidation treatment in the gas phase using air, O_2 or an O_2/N_2 mixture, ozone, NO [6], Cl_2 [7], SO_2 or SO_3, although it is possible to use alternative treatments in the liquid phase with HNO_3,

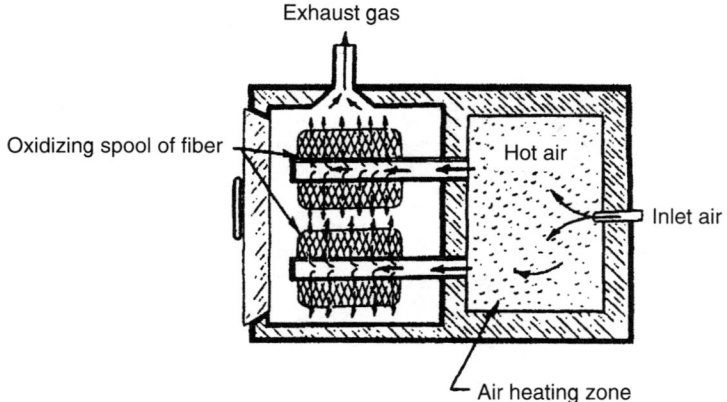

Figure 7.2 On the spool oxidation of mesophase fibers. *Source:* Reprinted from Barnett I, Apparatus for thermal modification of yarns, U.S. Pat., 2,913,802, 1959.

H_2SO_4, H_2O_2 [8] or $KMnO_4$. The simplest, cheapest and most convenient process is air oxidation, which is an exothermic reaction. Air enriched with ozone does permit lower temperatures (60–70°C) to be used initially and could promote crosslinking similar to the ozonization of phenanthrene in solution [9].

There are basically two methods of air oxidation that can be employed, both of which are batch processes:

1. After the fiber has been spun it is wound directly onto a heat resistant spool (Figure 7.2) using a specially designed winder which treats the fiber with special care to prevent damage. The spools are then placed in an oxidation furnace and the fiber oxidized *in situ* on the spools. The spools must be carefully designed so that the oxidizing atmosphere reaches the center of the package to ensure uniform oxidation and the flow rates must be adequate to prevent any build-up of heat from the resulting exothermic reaction.
2. Collect the spun fiber by piddling into a suitable container to facilitate subsequent removal, with the container preferably on a plating table and with the fragile fiber being drawn from the container, spread on a conveyor belt and carried through the oxidation furnace. Large lengths can be processed by using a number of containers strung together by passing the fiber from one container to the next in a continuous manner and processing accordingly. The thickness of the fiber on the belt must be limited to prevent build-up of exothermic heat.

The control of this oxidation process is critical, since under-oxidized fiber will remain partly thermoplastic and when processing in subsequent carbonization at higher temperatures, will permit filaments to fuse together (Figure 7.3), producing a carbon fiber with poor tensile strength. Over-oxidation will produce a brittle product, namely a carbon fiber with very poor tensile strength and will reduce the graphitizability of the pitch [12,13], which is believed to occur due to the formation of quinones

that produce CO when pyrolyzed. Stevens and Diefendorf [14] reckoned that a typical mesophase fiber required an increase of some 6% in mass for adequate stabilization and achieved this in 40 min at 260°C with a mesophase fiber made from a solvent extracted

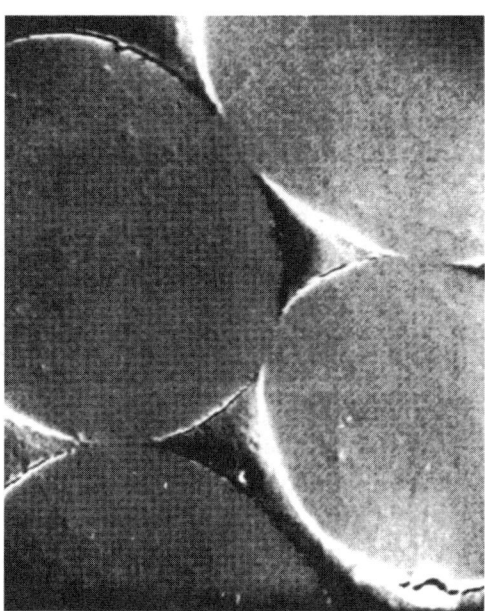

Figure 7.3 Interfilament sticking or fusing. *Source:* Reprinted with permission from Singer LS, Carbon fibres from mesophase pitch, *Fuel*, 60(9), 839–847, 1981. Copyright 1981, Elsevier.

fraction of Ashland 240 pitch, whilst an additional 1 h oxidation treatment only added a further 2% mass.

Lewis, McHenry and Singer [15] state that the time required to effect thermosetting of the fibers will vary with factors such as the chosen oxidizing atmosphere, the temperature employed, the diameter of the fibers, the type of precursor pitch, its mesophase content and the molecular weight distribution. The temperature used to effect thermosetting must not exceed the temperature at which the fibers soften or distort. The maximum temperature is dependent on the pitch precursor, its mesophase content and the molecular weight distribution. The higher the mesophase content and molecular weight distribution of the pitch, the higher will be its softening temperature, and the higher the temperature at which to effect thermosetting, making it possible to effect thermosetting in less time. Fibers with a lower mesophase content and/or lower average molecular weight will require longer times at somewhat lower temperatures to render them infusible. It is considered that a minimum temperature of 250°C is required, whilst temperatures in excess of 400°C can cause melting or excessive loss of carbon by burn-off and treatment should preferably be in the range 275–350°C. At such treatment temperatures, thermosetting can generally be achieved in 5–60 min, but 60 min should not be exceeded, since it is undesirable to render the fibers totally infusible. An isotropic pitch will generally require about 3 h of treatment in a 20% oxygen atmosphere. As the oxidation proceeds, the temperature can be increased to speed up the reaction and typical final temperatures would be around 325–340°C for an isotropic fiber and 300–310°C for an anisotropic fiber.

An isotropic pitch has a lower degree of condensation, higher hydrogen content and a lower softening point, all of which contribute to requiring a longer time to form an oxygen bridge structure via oxidation, dehydrogenation, cross-linking and cyclization.

Singer [16] showed that an oxidized mesophase fiber, after carbonizing at 1600°C, showed a fine domain texture retaining the as spun texture, whereas a fiber carbonized from an un-oxidized fiber showed larger domains, indicating some relaxation of the structure.

Riggs [17] observed that a fiber with larger diameter filaments, spun from a mesophase pitch tended to elongate, whilst smaller diameter filaments shrank, presumably because they were more highly oriented. The elastic modulus of as spun fibers increased with the draw ratio, but decreased with increasing thermosetting temperature and is attributed to a relaxation effect.

A divergence of opinion exists about the benefit of stress applied during the oxidation process. Some workers [18–20] subscribe to the view that since pitch has a large planar molecular structure, any stress applied during oxidation will not be beneficial, whilst others [21–23] believe there is a benefit, since the application of stress will promote the microcrystallinity of the fibers through preferred orientation, thus increasing the fiber strength and modulus. The level of stress has to be carefully chosen, since too great, and the fiber will break. It could be envisaged that fiber wound on a temperature resistant spool would be under some tension in the oxidation process, but it is difficult to conceive that fiber laid on a belt would be under tension.

DSC and T_g studies [24,25] have shown that pitch fibers gain in weight during oxidation and that the reaction was exothermic. Elemental analyzes confirm that the O/C ratio increases, whilst the H/C ratio decreases and the degree of oxidation increases as the temperature is increased. FTIR studies [24–26] show that the peak intensities of $-CH_3$ and $-CH_2ArH$ decreased.

Oxygen diffuses through the sheath to react with the α-hydrogens and IR studies indicate the formation of $-OH$ and $-COOH$ groups. At higher oxidation temperatures, a decrease in the hydroxyl formation is associated with an increase in anhydride formation. After oxidation, Otani [9] found the sheath to contain 30% oxygen and the core, 20%. Mochida et al showed, using electron probe X-ray microanalysis [27], that if insufficient time was taken for stabilization, then a gradient of oxygen through the carbon filament occurred (Figure 7.4). After 15 min, the skin thickness of oxidized fiber was 5 μm, increasing to 9 μm after 30 min and full penetration after 90 min, leaving no core [27,28]. However, filaments with diameter less than 10 μm, in the presence of an adequate oxygen concentration, exhibited no skin core effect whatsoever, which was attributed to the rapid diffusion of oxygen [29].

It may be concluded that stabilization in air causes dehydrogenation, cross-linking and cyclization to take place with the emission of CO, CO_2, H_2O and low molecular weight hydrocarbons, leaving more stable oxygen-bridge structures, which may contain

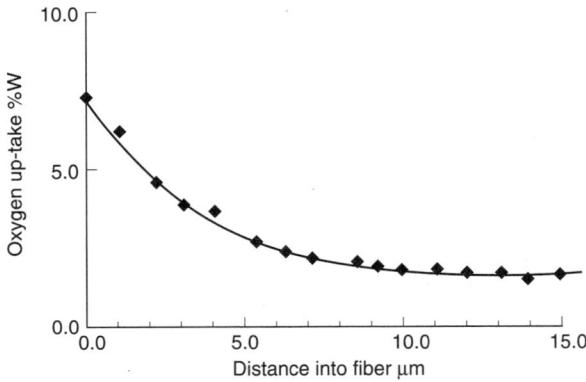

Figure 7.4 Oxygen diffusion along the diameter from the surface to the center of a 30 μm diameter mesophase pitch fiber after stabilization at 300°C for 15 min. *Source:* Reprinted from Mochida I, Toshima H, Korai Y, Hino T, Oxygen distribution in the mesophase pitch fiber after oxidative stabilization, *J Mater Sci*, 24, 389–394, 1989.

acid anhydride, carbonyl group, aromatic ether and phenolic hydroxyl group as depicted [30,31]:

R = alkyl group
A = aromatic group

Lavin [32] showed that –CH$_3$ and hydro groups reacted with carbonyl groups and accelerated the oxidation reaction. The evolution of H$_2$O decreased to zero at a given temperature, increasing again as the temperature was raised and again reducing back to zero, suggesting that there are a specific number of active sites for a particular temperature. Evolution of CO and CO$_2$ was associated with the surface decomposition of oxygenated solid carbon.

Matsumoto and Mochida [33], using ^{13}C NMR with a hydrogenated coal tar pitch mesophase carbon fiber, showed that the initial attack by oxygen was on –CH$_3$ and –CH$_2$– groups, with the gradual formation of carboxyls, esters and aryl carbonyls. Fairly stable cross-links were formed via phenols, ethers and esters. The workers also found that slower heating rates (0.5°C/min instead of 2.0°C/min) produced better mechanical properties and that the final choice would be controlled by the desired target properties and economics.

To avoid filaments sticking together in the subsequent carbonization step, a two-step stabilization, a variant of the oxidation process, can be undertaken where the fiber is deliberately under-oxidized, followed by a solvent extraction with benzene or tetrahydrofuran to remove soluble fractions present in the surface layer of the fiber [34,35].

Other techniques can be used to prevent fibers sticking during stabilization and are generally based on a form of a carbon/graphite lubricant, such as colloidal graphite, or carbon black in aqueous or silicone oil suspension and applied to the fiber prior to stabilization. Although it is reported that suspensions of silica and calcium carbonate

Table 7.3 Average mechanical properties of mesophase fibers before and after stabilization

Mesophase fiber	Tensile strength Gpa	Tensile modulus GPa	Elongation %
As-spun	0.04	4.7	0.85
Stabilized	2.06	216	0.95

Source: Reprinted with permission from Edie DD, Pitch and mesophase fibers, Figueiredo JL, Bernardo CA, Baker RTK, Hüttinger KJ eds., *Carbon Fibers Filaments and Composites*, Kluwer Academic Publishers, Dordrecht, 43–65, 1990. Copyright 1990, Springer.

can also be used, it is unlikely since they would be sources for possible sites of internal flaws which may be formed during carbonization.

The average mechanical properties of mesophase fibers before and after stabilization are given in Table 7.3.

7.3.2 Carbonization

The greatest weight loss occurs in the early stages of carbonization. Therefore, it is advantageous to apply an initial low temperature carbonization stage to avoid disruption of the fiber structure. Normally, about 30 s to 5 min is sufficient time, while suitable treatment would be about 0.5 min at 700°C followed by 0.5 min at 900°C in an inert N_2 atmosphere. A typical carbonizing furnace using a graphite hairpin element and capable of achieving 2000°C is shown in Figure 7.5.

Carbonization is required to remove hetero-atoms like H, N, O and S in the form of H_2O, CO_2, CO, N_2, SO_2, CH_4, H_2 and tars. Above 1000°C, the principal gas evolved is H_2 [8]. Carbonization can be achieved in a string of separate furnaces with individual temperature settings, or one or more furnaces with zoned temperature control. After the initial carbonization stage, the temperature is increased to about 2000°C. Bright and Singer [39] found that at temperatures up to 1000°C, some degradation of the structure occurred with attendant reduction of modulus, but as the temperature increased there was a marked increase in preferred orientation as the hetero-atoms were continuously released, forming a turbostatic graphite-like structure. The authors were unable to establish any relationship with the type of transverse microstructure obtained with the treatment temperature. Singer [38] reported that at 1500°C, mesophase fibers could develop tensile moduli of the order of one third of the theoretical value for graphite. Whilst Riggs *et al* [40] showed that the strength and the modulus (Figures 7.6 and 7.7) of carbon fiber made from a mesophase pitch increased with temperature.

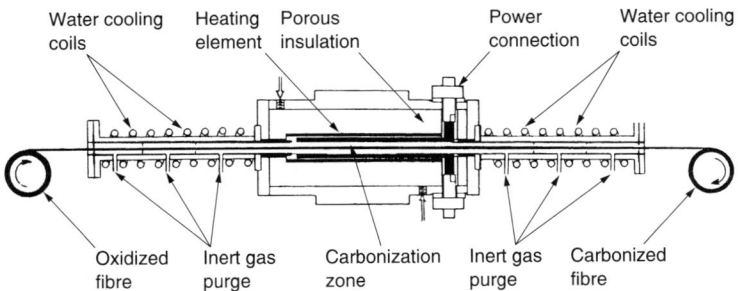

Figure 7.5 Diagram of a hairpin element furnace used to carbonize mesophase pitch fibers. *Source:* Reprinted with permission from Buckley JD, Edie DD Eds., *Carbon Materials and Composites*, Noyes Publications, Park Ridge, NJ, 32, 1993. Copyright 1993, William Andrew Publishing.

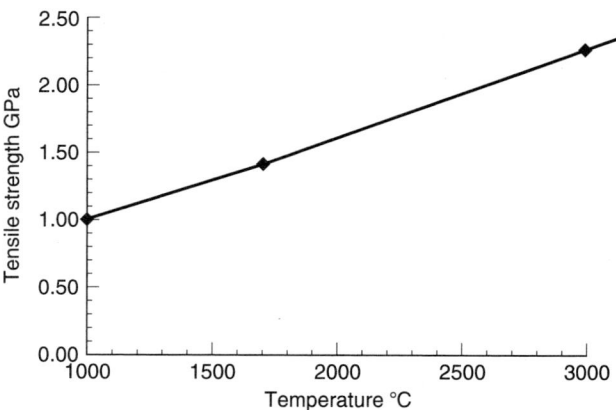

Figure 7.6 Effect of final heat treatment on tensile strength of mesophase fibers. *Source:* Reprinted with permission from Bright AA, Singer LS, *Carbon* 17, 59, 1979. Copyright 1979, Elsevier.

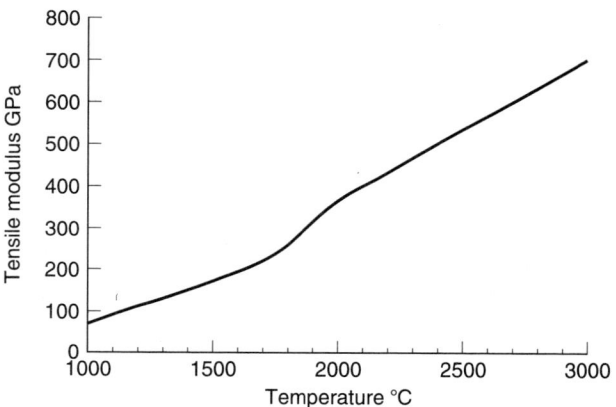

Figure 7.7 Effect of final heat treatment on tensile modulus of mesophase fibers. *Source:* Reprinted with permission from Bright AA, Singer LS, *Carbon* 17, 59, 1979. Copyright 1979, Elsevier.

This has been confirmed by other workers [41,42], whereas with an isotropic pitch based fiber, the strength and modulus change only slightly (Figures 7.8 and 7.9). The graphs clearly show the superior modulus of the pitch based fibers contrasted with the higher strength values obtained with PAN based fibers.

As with PAN based carbon fibers, strengths of mesophase pitch carbon fibers have been shown by Barr *et al* [43] to be limited by flaws. Chwastiak *et al* [44] showed the influence of gage length, with strength almost doubling as gage length was reduced from 20 to 3 mm. Bacon [45] listed the possible reasons for the origin of flaws:

1. Interfilament fusing (Figure 7.3), which could occur during spinning, insufficient oxidation, or thermal processing.
2. Internal voids, caused by gas bubbles in the as spun fiber, termed bloating, or due to the volatilization of foreign matter during the carbonization process. (In practice, gas bubbles should be released by venting the spinneret).
3. Inclusion of foreign particles, due to inefficient cleaning of the precursor pitch.
4. Surface defects, due to mechanical damage or surface contamination.

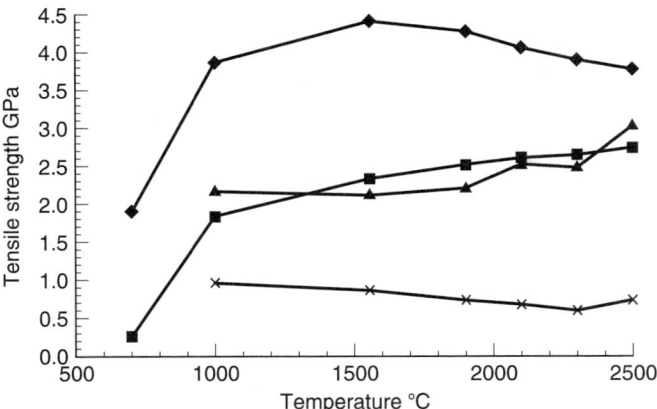

Figure 7.8 Effect of final heat treatment temperature on tensile strength of PAN, mesophase and isotropic pitch fibers. ◆ PAN ▲ MP1 [41] ■ MP2 ∗ IP [42]. *Source:* Reprinted from Matsumoto T, Mesophase pitch and its carbon fibers, *Pure Appl Chem*, 57(11), 1533, 1985, Shen Z, Guo H *et al*, Carbon and Carbonaceous Composite Materials Structure-Property Relationship, *Abstr and Proc*, Malenovice, Czech Republic, 31, Oct 10–13, 1995.

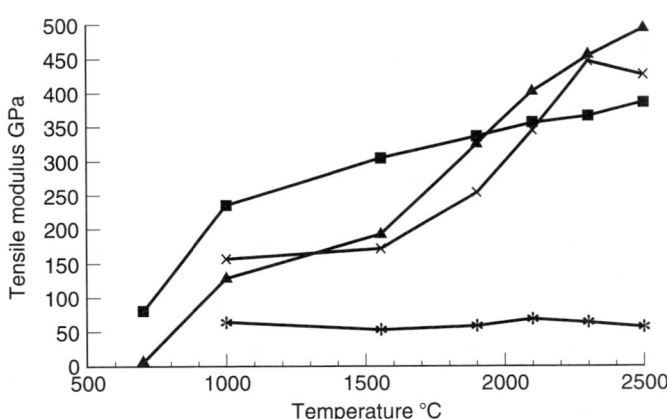

Figure 7.9 Effect of final heat treatment temperature on tensile modulus of PAN, mesophase and isotropic pitch fibers. ■ PAN × MP1 [41] ▲ MP2 ∗ IP [42]. *Source:* Reprinted with permission from Matsumoto T, Mesophase pitch and its carbon fibers, *Pure Appl Chem*, 57(11), 1533, 1985. Copyright 1985, Blackwell Publishers. Shen Z, Guo H *et al*, Carbon and Carbonaceous Composite Materials Structure-Property Relationship, *Abstr and Proc*, Malenovice, Czech Republic, 31, Oct 10–13, 1995.

7.3.3 Graphitization

A mesophase pitch based fiber can be further heat treated in a similar type of furnace, using highly controlled inert atmosphere, at temperatures in the range 2500–3300°C, preferably 2800–3000°C, producing fibers with a high degree of orientation, where the carbon crystallites are parallel to the fiber axis. These fibers are truly graphitic and have a structure characteristic of polycrystalline graphite with a three-dimensional order. A residence time of about 10 s–5 min may be employed. At these high temperatures, furnace lives will be relatively short due to the evaporation of graphite from the surface of the graphite electrodes, which will be aggravated by an excessive flow of inert gas across

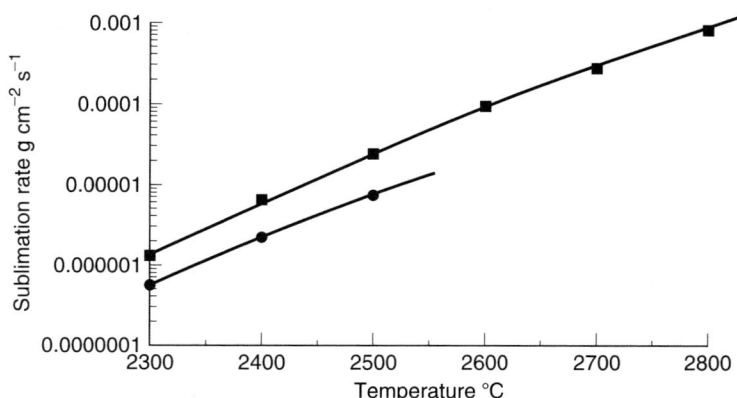

Figure 7.10 Sublimation rate of graphite. ■ Calculated; ● Experimental. *Source:* Adapted from Larsen HL, Graphite elements for high temperature resistor furnaces, *Union Carbide Reprint Part 2, Industrial Heating*, 11, 1962–63.

the hot element surface causing severe element and guard tube erosion. Therefore, these furnaces must be operated under quiescent conditions. The sublimation rate for graphite in the temperature range 2300–2800°C is shown in Figure 7.10.

A furnace with a hairpin element design cannot use argon since the gas ionizes across the element slot at about 2000°C and so, N_2 is generally the preferred inert gas.

7.3.4 Surface treatment of pitch based carbon fibers

If the pitch fibers are to be introduced into a resin matrix, then it would be expected that some form of oxidative treatment would be applied to improve resin wetting and bonding. Companies like Mitsubishi Chemical Corporation do indicate in their literature that shear strength is controllable and typical interlaminar shear strengths are given in Table 7.2. Surface treating a fiber which is graphitic in nature is difficult and great care would have to be taken to ensure that the fiber is not embrittled by the treatment process. When using pitch based carbon fibers in carbon-carbon composites, both the fiber and matrix are brittle materials and a weak interface would be preferred, which, with the application of an applied load, would deflect cracks in the matrix, thus producing debonding at the fiber-matrix interface [47]. The varied forms of oxidative treatment that can be used with carbon fibers are discussed in Chapter 9 but, unfortunately, little information is reported in the literature for pitch-based fibers. Lin and Yip [48] investigated the effect of different treatment conditions on three types of carbon fiber including HM CF, a pitch based fiber from the Toa Nenryo (Tonen) Company, obtaining the highest transverse tensile strength of 19.3 MPa for a 14 h treatment in near boiling 15% H_2O_2 using Epon 828/V–40 resin system. The equivalent values obtained for Hercules IM6 and Toho IM600 were 20.1 and 25.1 MPa, but these values are only about half the value of figures reported by Mitsubishi for their range of Dialead coal tar pitch carbon fibers (Table 7.2). The work showed that the extent of oxidation on the carbon fiber surface was more dependent on the chemical reagents than the surface structures and precursors and, interestingly, the degree of adhesion was not promoted entirely by the amount of oxygen, or carbon-oxygen functional groups, on the fiber surface. Increased treatment did promote surface roughness, enabling more available surface for mechanical and physical interlocking.

7.4 THE STRUCTURAL ORDERING AND MORPHOLOGY OF MESOPHASE PITCH FIBERS

Many factors control the final morphology of a mesophase pitch [1]. The viscosity is extremely dependent on the temperature and controls the microstructure. Stirring the pitch is important and control of flow can govern orientation, the laminar flow promoting radial structure, whilst turbulent flow can be achieved by using spinnerets with variable section. The process of spinning aligns the liquid crystal plates that evolve into planar arrays of carbon atoms [49]. Some stretching is achieved after spinning, as the fiber is drawn in the plastic state before cooling, which confers some axial orientation. The microstructure developed during spinning will invariably be retained after carbonization, unless stress relief occurs during the thermosetting stage. Barnes *et al* [50] have investigated the as-spun orientation as an indication of the graphitized properties of mesophase based carbon fiber. The initial spinning conditions are proprietary and manufacturers have introduced their particular modifications. Typical diagrams of micro cross-sections of mesophase carbon fibers have been shown in Figure 4.34 and Figures 7.11 and 7.12 show typical transverse microstructures.

Fibers with a radial transverse microstructure can exhibit a crack, which Volk [54] reports will tend to spiral around the fiber axis. The crack is able to transform into a true radial structure along a single filament and vice versa (Figure 7.13). White *et al* [55] believe that this is due to the annihilation of disclinations, as depicted in Figure 7.14. Yoon *et al* [56,57] have investigated crack formation in mesophase pitch based carbon fibers.

In 1984, DuPont acquired the Exxon pitch based carbon fiber business and it was highly probable that the DuPont fibers were based on the Exxon patents [58] using a neophase pitch, which had been extracted with solvent, as precursor material. Initial studies revealed an oriented core microstructure [59]. Later work showed E130 to have an oriented core [60], or radial [61], but all workers agreed that there were many disclinations, which is consistent with the low graphitizability of the fiber. Kogure, Sines and Lavin [61] are insistent that

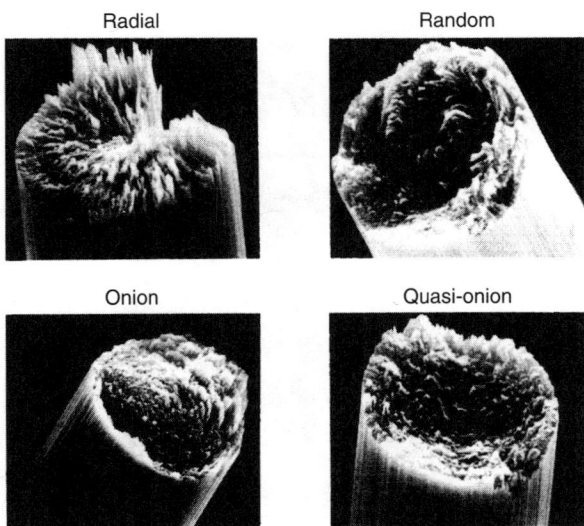

Figure 7.11 Typical transverse microstructures of pitch fibers showing radial, random, onion-skin, and quasi-onion-skin structures. *Source:* Reprinted from Hamada T, Nishida SY, Furuyama M, Tomiaka T, *Extended Abstracts of 18th Biennial Conf on Carbon*, Amer Carbon Soc, Worcester, MA, 225, 1987, Hamada T, Nishida SY, Matsumoka M, Endo M, *J Mater Res*, 2, 850, 1987.

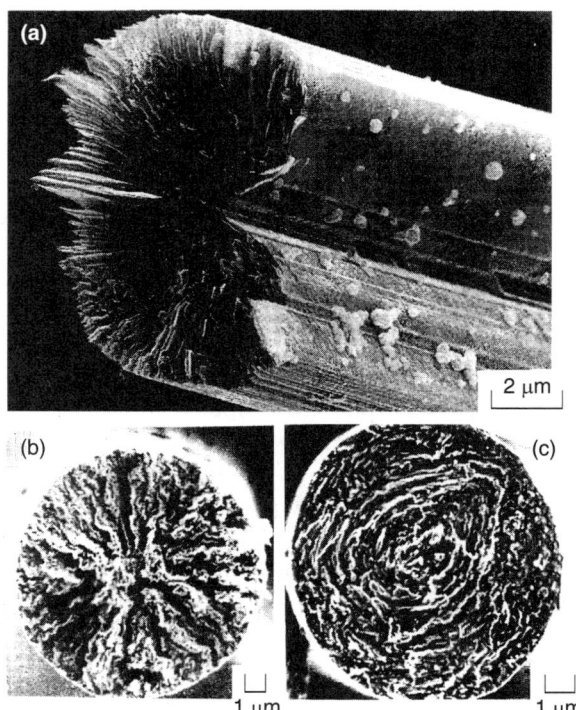

Figure 7.12 SEM images of pitch based carbon fibers. (a) radial fiber with wedge (b) radial fiber (c) concentric fiber. *Source:* Reprinted from Inagaki M, Iwashita N, Hishiyama Y, Kaburagi Y, Yoshida A, Oberlin A, Lafdi K, Bonnamy S, Yamada Y, *Tanso* 147, 57, 1991.

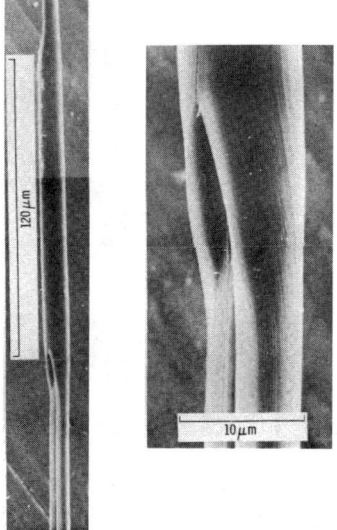

Figure 7.13 Structural transition from random to radial transverse texture in a mesophase pitch carbon fiber after heat treatment up to 2800°C. *Source:* Reprinted from Edie DD, Research into pitch based carbon fibers. In: Genisio M ed., *Recent Research into Carbon–Carbon Composites*, Southern Illinois University, 16–41, 1987.

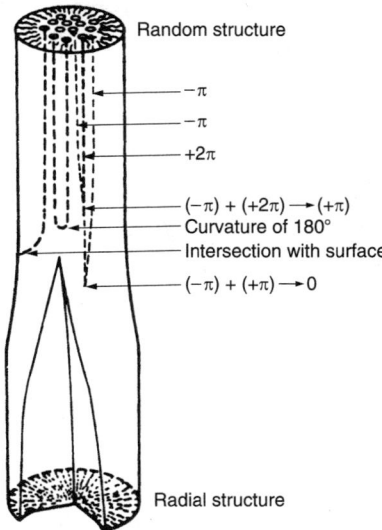

Figure 7.14 Diagram showing possible disclination structures in a transition region of a filament of variable structure. *Source:* Reprinted from Edie DD, Research into pitch based carbon fibers. In: Genisio M ed., *Recent Research into Carbon–Carbon Composites*, Southern Illinois University, 16–41, 1987.

DuPont makes the zigzag radial structure to accommodate residual stresses. So the micrometric zigzags of E130 favor tensile strength but the nanotexture is unfavorable to graphitization. Conoco Inc. took over the business from DuPont and were to initially build a random oriented carbon fiber mat plant based on a mesophase petroleum pitch, but the project was discontinued.

Amoco took over the business from Union Carbide and references to Union Carbide patents suggest that they probably use a gas sparged pitch as the precursor for Thornel fibers. Initially, the fibers were shown by Bright and Singer [39] to have a radial or random micro cross-section. Later, Guigon and Oberlin [62] showed P55S and P75 to have an oriented core with alternate bands of microporous carbon and graphite, attributed to a two phase precursor pitch. The pitch was shown by Lafdi et al [63–66] to contain an isotropic pitch. The P100 grade was shown [62] to be radial with a wedge and comprising flat stiff lamellae and was totally homogeneous. P120 and P130X also have a lamellar construction [67] with P130X showing a selected area diffraction (SAD) of a pure graphite crystal.

Cracked radial and radial structures are reported to have variable strength, whilst flat layer, introduced by Amoco, is considered to be an improvement.

Kashima Oil Company have, by control of the cross-sectional microstructure, introduced Carbonic HM50, HM60 and HM80 with a wavy radial microstructure, which Endo [68] claims does have a higher strength and elongation, with improved resilience to crack propagation due to the arrest of any cracks at the folds. Endo [67] reports that the folds do, however, reduce the modulus of elasticity, electrical conductivity and thermal conductivity compared with the Thornel fibers. A similar strength improvement was found with line-origin by Edie *et al* [68], which was associated by Edie and Fain [69] with flow through non-circular capillaries. A change in capillary cross-section was shown by Matsumoto [70] to result in a fiber which favored an onion-skin structure with a higher modulus. Although a random texture has no obvious transverse order, the graphite planes do remain oriented parallel to the fiber axis.

Summarizing the above data reveals that micro and nano textures favor electrical and thermal conductivities with perfect planar sheets of lamellae and almost infinite graphite

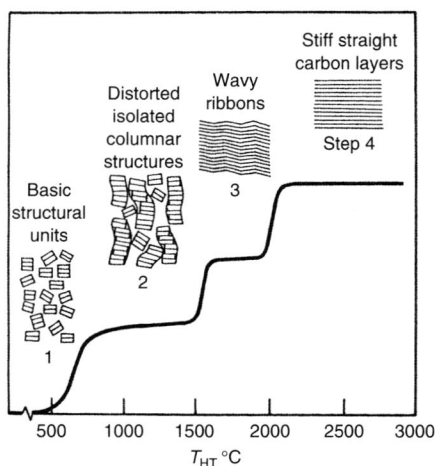

Figure 7.15 Various steps in the graphitization process as a function of heat treatment temperature T_{HT}
Source: Reprinted with permission from Goma J, Oberlin A, Thin solid films, 65, 221, 1980. Copyright 1980, Elsevier.

grains, but are unfavorable to tensile strength due to the planes gliding easily over one another as well as cleavage.

Bennett and Johnson [71], using high resolution TEM, showed that there was a relatively thin (100–150 nm) skin region in fully stabilized fibers, with the crystallite layer planes interlinked in a highly complex manner enclosing an intricate void system.

The cause and prevention of a skin core structure was investigated by Lu *et al* [72], whilst Mochida *et al* [73] determined the structure and properties of a mesophase carbon fiber with skin core carbonized under strain. Guigon and Oberlin [74,75], using TEM, have identified steps in the graphitization process shown in Figure 7.15. For heat treatment temperatures up to 800°C, the structure is basically two parallel layers of piled up distorted columns of ≈1 nm long units in the region where hetero-atoms are released. From 800–1500°C, the lengths of the columns increases with less misorientation and L_c increasing as they become more aligned. From 1500–1900°C, the columnar structures are replaced by wavy ribbons or wrinkled layers, hooking adjacent columns together. By 1700°C, the wavy ribbons start to disappear with the formation of an arrangement of turbostratic layers, with L_a and L_c increasing rapidly. From 1900–2100°C, the layers begin to unwrinkle and by 2100°C, there is a rapid increase in L_a as in-plane structural defects are eliminated as the tilt and twist boundaries disappear with the interlayer spacing at 0.342 nm, accompanied by a dramatic change in 3-D interplanar characteristics such as electronic properties. Above 2100°C, the three dimensional crystal growth occurs with stiff, straight carbon layers, folded parallel to the fiber axis.

The degree of disorder in carbon fiber can be quantified by using TEM and measuring L_a, L_c, and r_t, where L_a and L_c are the in-plane and c-axis coherence lengths of the fibers and r_t is the transverse radius of cuvature for the zigzags in the lamellar ribbon structures. Dresselhaus [37] reports that when correlating structure with mechanical properties, the following relationships apply—the tensile strength will depend linearly on the average ratio (rt/La), whilst the bulk modulus will decrease with increasing $(1/La)$ and the resistivity increases as $(1/La)$. Murakami *et al* [77] investigated the effect of mesophase pitches on the tensile modulus.

Edie [3] believes that since mesophase is lyotropic in nature (where the concentration determines the phase present), modifying the flow profile during extrusion might create a less flaw-sensitive microstructure with consequent increase in tensile strength. An added bonus may be an improvement in the compressive strength of the mesophase fibers, which remains a severe weakness and is only about 20% of the tensile strength, which moreover, was shown by Kumar [78] to become less efficient as the modulus increased and is generally attributed to failure by microbuckling.

The structural characterization of a milled mesophase pitch based carbon fiber was determined by Endo et al [79]. Hollow mesophase pitch based carbon fibers were prepared and their structure determined by Wang et al. [80].

7.4.1 Mechanisms associated with the preparation of pitch precursors

Typical structures of products found in a coal tar pitch have already been listed in Table 4.13 and these compounds will, in the process of making carbon fiber, be converted by pyrolysis to carbon and, ultimately, to a three dimensional graphite. Before carbonization takes place, however, non-aromatic structures are converted to aromatics, which through polymerization are the key building blocks for carbon. Some of this conversion will occur in the preparation of pitch prior to spinning and in order to give a general understanding of the carbonization process, a number of reaction studies with model compounds are described, which are based on articles by Lewis and Singer [81,82].

Table 7.4 lists the thermal reactivity data for a number of polyacenes and it can be seen that the first three members all boil below 400°C, but do not react at their boiling point and atmospheric pressure, whilst tetracene and pentacene, although stable at 400°C, do react at the higher temperatures shown. The reactivity can also be related to the ionization potential, which is a measure of the ease of removal of an electron from the molecule. The lower the ionization potential the more readily a molecule will react and Table 7.5 lists the thermal reactivity of a number of polynuclear aromatic hydrocarbons.

The degree of graphitic character can be established using X-ray techniques and Simon et al [84,85] showed that in the carbonization of anthracene, the average structure of the resultant products for the mesophase stage was represented by condensed anthracene dimers (bisanthene), whilst at 600°C, the average structure for the coke was

Table 7.4 Thermal reactivity data for polyacenes

Structure	bp / Reaction Temperature °C	Residue at 700°C %
benzene	80	0
naphthalene	218	0
anthracene	354	0
tetracene	480	14
pentacene	426	43

Source: Reprinted with permission from Lewis IC, Chemistry of carbonization, *Carbon*, 20(6), 519–529, 1982, Edstrom T, Lewis IC, *Carbon*, 7, 85, 1969. Copyright 1969, Elsevier.

Table 7.5 Thermal reactivity of polynuclear aromatic hydrocarbons

Aromatic hydrocarbon	Reaction temperature °C	Ionization potential eV
(coronene)	637	7.24
(dibenzo[g,p]chrysene)	560	7.07
(picene)	535	6.86
(tetracene)	480	6.64
(pentacene)	411	6.23

Source: Reprinted with permission from Lewis IC, Chemistry of carbonization, *Carbon*, 20(6), 519–529, 1982. Copyright 1982, Elsevier.

non-condensed bisanthene trimers, which formed a large completely condensed aromatic structure at 800°C.

A number of structures, as determined by field desorption mass spectroscopy (FDMS), for polymeric products obtained by the pyrolysis of anthracene have been proposed by Lewis [86], where molecular weights up to 1400 have been identified:

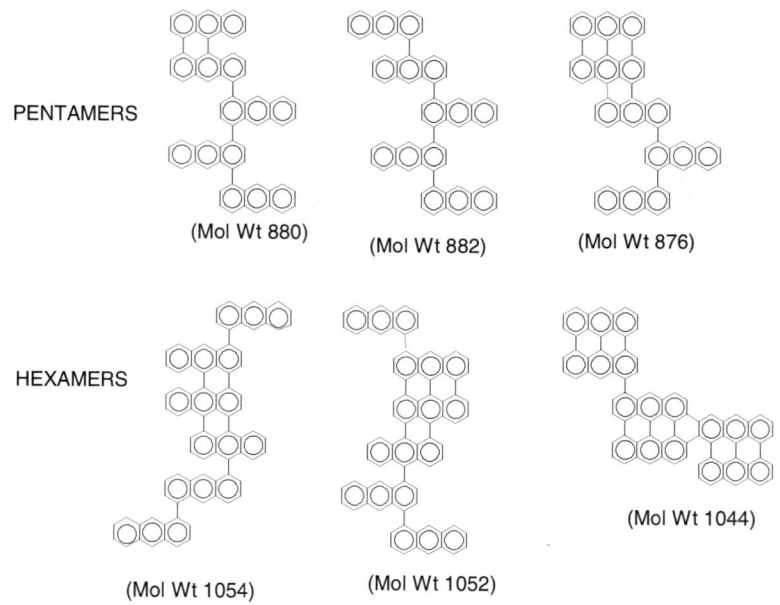

Electron spin resonance (ESR) can be used to study the carbonization process and Lewis *et al* [87] reacted acenaphthalene in an inert solvent (*m*-quinquephenyl), where the formation of a free radical was identified and was attributed to a thermal rearrangement reaction forming an odd alternate perinaphthyl radical. At a later stage in the reaction, a more complex unidentified radical was formed. The formation of stable free radicals such as the naphthanthryl free radical

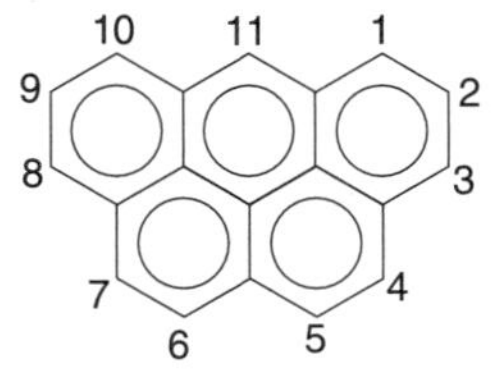

are believed to be associated with aromatic ring systems with an odd number of carbon atoms and the H-atoms. These radical structures are believed to play an important role in pyrolysis.

This is illustrated by Lewis and Singer [87] with naphthalene (I), which forms by rapid polymerization of a non-condensed polymer (III) with three non-condensed naphthalene units with a total of 30 carbon atoms:

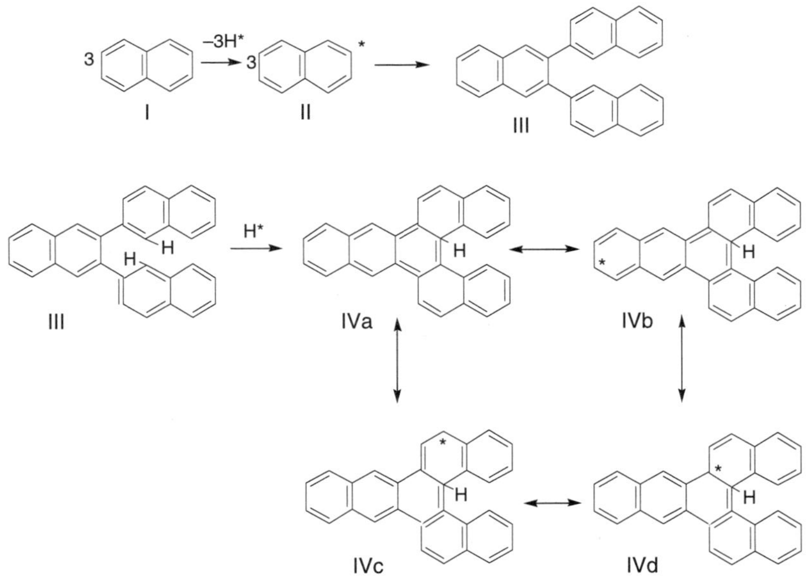

The loss of a single H-atom from this polymer forms a free radical (IV), also with 30 carbon atoms and one sp^3 tetrahedral carbon. The total aromatic π system contains 29 carbon atoms, whilst the unpaired electron is shared by resonance delocalization. Although the loss of a further H-atom would create a fully condensed molecule, it would add little to the resonance stabilization.

Table 7.6 lists the degree of graphitic character of 3000°C carbons, produced from polynuclear aromatic hydrocarbons [83] and are listed in order of decreasing graphitizability. Acenaphthylene, containing one five-membered ring is highly graphitizable with a very low

Table 7.6 The X-ray characterization of 3000°C carbons from polynuclear aromatic hydrocarbons

Component	Linear structure	Molecular weight	mp °C	002 Inter-layer spacing (nm)
9,10-dimethylanthracene	$C_{16}H_{14}$	206	183	0.3354
4,5-benzopyrene	$C_{20}H_{12}$	252	178	0.3350
Acenaphthylene	$C_{12}H_8$	152	91	0.3358
Tetracene	$C_{18}H_{12}$	228	>300	0.3358
Dibenzphenanthrene	$C_{22}H_{14}$	278	–	0.3358
Coronene	$C_{24}H_{12}$	300	442	0.3370
8,9-biphenanthrene	$C_{26}H_{16}$	328	–	0.3373
fluoranthene	$C_{16}H_{10}$	202	110	0.3371

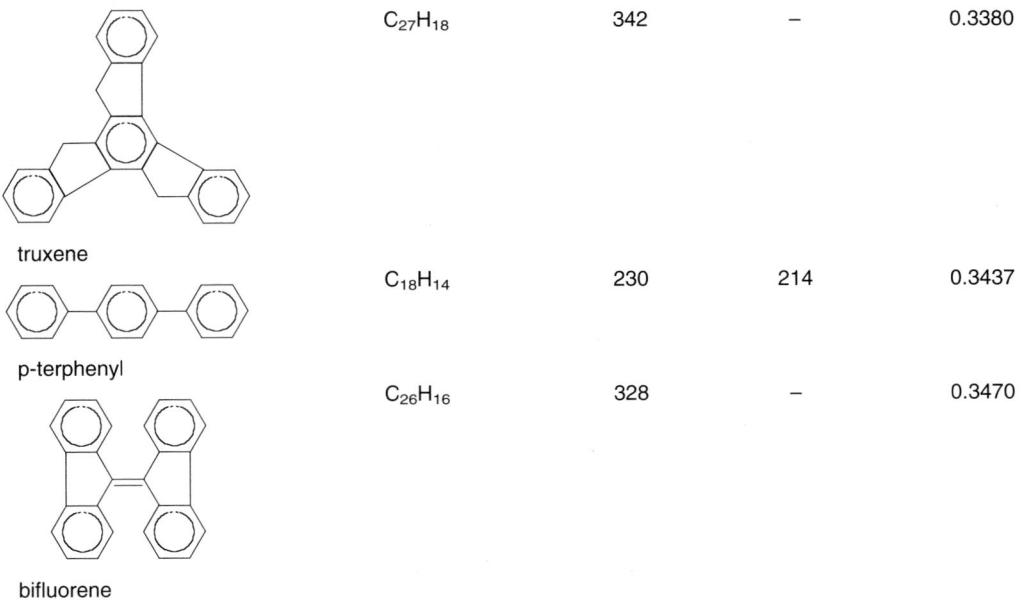

Source: Lang KF, Buffleb H, The X-ray characterization of 3000°C carbons from polynuclear aromatic hydrocarbons. Chem Ber, 94, 1075, 1961.

interlayer spacing, but truxene, with 3 five-membered rings and fluoranthene, with one are less highly graphitizable and form less ordered carbons, whilst terphenyl is non-graphitizing.

Oxygen substituted in an aromatic ring generally gives a less ordered carbon (Table 7.7), although some oxygen containing compounds do graphitize well.

From the preceding paragraphs, it is evident that carbonization is an aromatic growth and a polymerization process which can be envisaged as a general scheme for the pyrolysis of pitch [81] is:

Table 7.7 X-ray characterization of 3000°C carbons from oxygen substituted aromatics

Structure	002-Interlayer spacing (nm)	Structure	002-Interlayer spacing (nm)
	0.3354		0.341
	0.3356		0.341
	0.3373		0.342
	0.3393		0.343

Source: Reprinted with permission from Lewis IC, Chemistry of carbonization, *Carbon*, 20(6), 519–529, 1982. Copyright 1982, Elsevier.

To achieve pyrolysis, selective chemical treatment has to be accomplished and Lewis [81] lists the major reactions which can be used in the pyrolysis of aromatic hydrocarbons:

1. C–H, C–C bond cleavage to form reactive free radicals
2. Molecular rearrangement
3. Thermal polymerization
4. Aromatic condensation
5. Elimination of side chains, H_2

A number of reactions do occur in parallel, but the basic reaction is probably the formation of free radicals by cleavage of C–H or C–C bonds. Bond cleavage of the aromatic molecule can be achieved in two ways. A σ-radical can be produced by breaking an aromatic C–H bond, requiring a bond dissociation energy as high as 420 kJ mol^{-1} [88], but the σ-radical intermediate is very unstable and the free electron is mobilized. The second is an aromatic π-radical, which is considerably more stable, formed by the methyl C–H bond and requiring some 325 kJ mol^{-1}. The unpaired electron is resonance stabilized and simple π-radicals, such as benzyl, have been detected by ESR [89]. The addition polymerization of anthracene has been suggested by Livingstone *et al* [90], in which the π-radical could be generated by a small amount of unstable anthryl σ-radical. The reacting intermediate, once formed, can undergo direct polymerization as in the formation of naphthalene polymer from naphthalene:

These initial reactions involve the loss of hydrogen accomplished by internal hydrogen transfer:

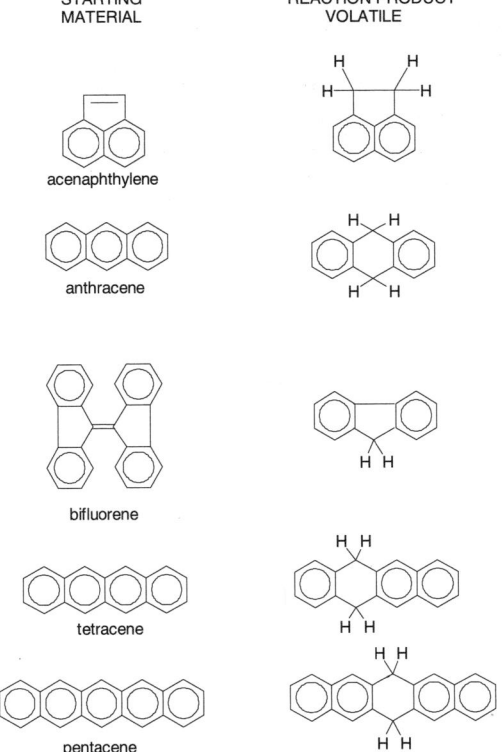

where the pyrolysis of the polynuclear hydrocarbons give hydrogenated products, with the hydrogen added at the most reactive position in the molecule [81].

Thermal rearrangement is another important step in the early stages of carbonization and at this stage, it is difficult to relate the final course of graphitization with the initial starting structure. Lewis [81] illustrates this with the transformations:

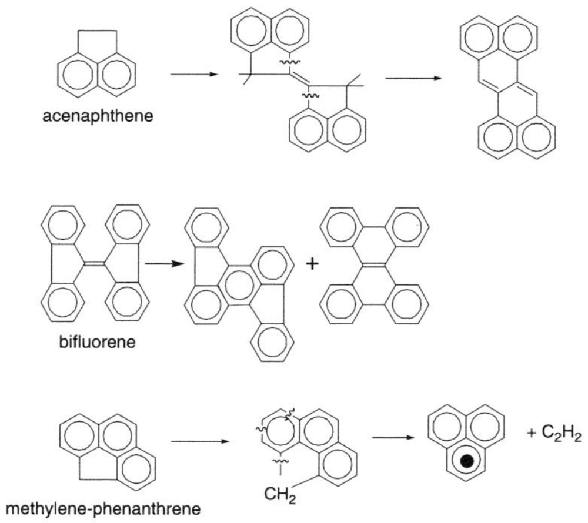

where acenaphthene and biflourene with unstable five-membered rings are transformed to stable six-membered rings without the loss of carbon atoms. However, methylene phenanthrene involves the loss of carbon atoms.

One of the factors making carbonization so complex is the presence of so many possible polymerization sites in an aromatic molecule and the simple dimerization of anthracene [86] may form 11 different reaction products and the number of possible isomeric structures increases rapidly as the reaction progresses.

The site in an aromatic ring at which polymerization should predominantly occur can normally be predicted, but steric effects can override reactivity factors and Lewis [81] shows how the polymerization process for aromatic hydrocarbons can occur in two stages, giving either non-condensed or condensed polymers:

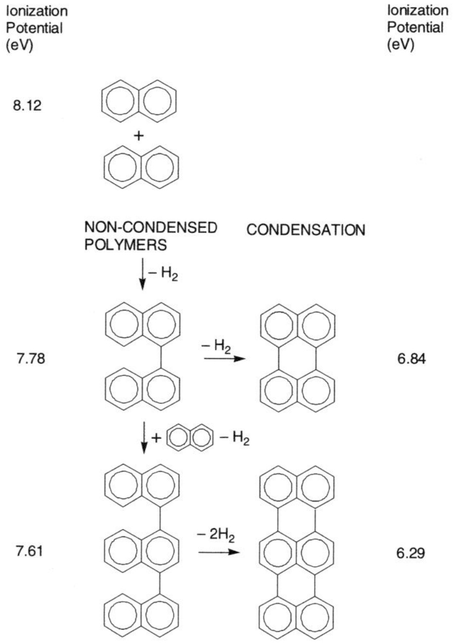

For naphthalene, the loss of two H-atoms gives polymers in which the units are linked by single bonds, whilst the loss of two further H-atoms produces a fully condensed polymer. Although the non-condensed and condensed polymers have virtually the same molecular weights, they vary considerably in structure and properties. The non-condensed polymers are non planar and their reactivities and ionization potentials show little change with increasing size. The condensed polymers, however, are fully planar and show marked changes in reactivity and ionization potential with increasing size. The relative role of these two processes in the pyrolysis of pitch is critical.

The initial reaction products of the thermal pyrolysis of anthracene have been identified [88,91]:

A number of non-condensed dimers are formed and only some of these have the right steric conformation to undergo an addition dehydrogenation reaction to give fully condensed molecules with only six-membered rings involving the most reactive position 9 for this to be achieved.

Similar data exists for the pyrolysis of naphthalene [88,92]:

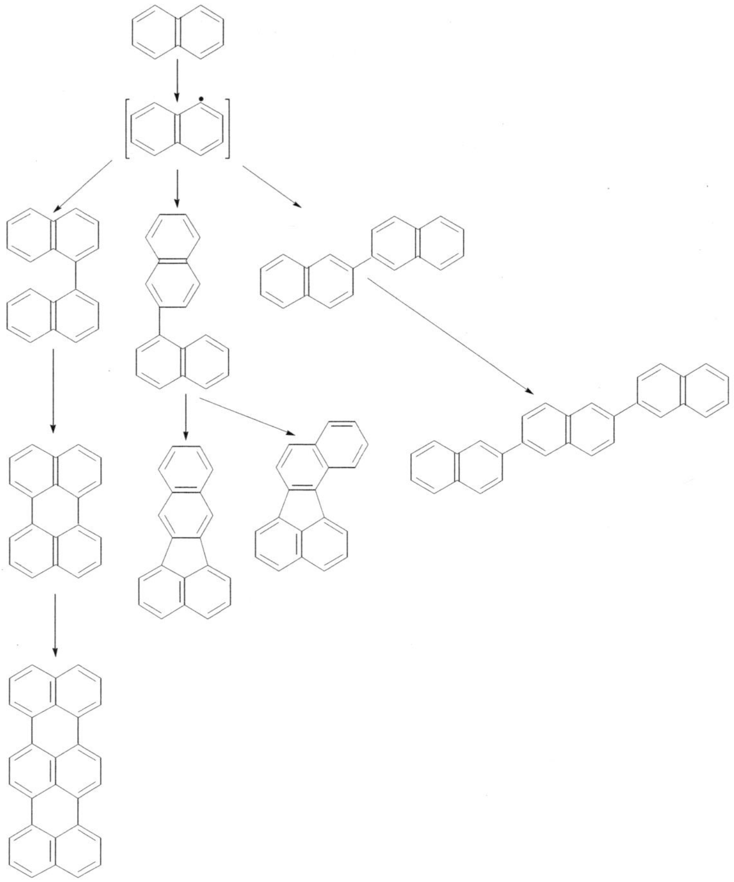

With naphthalene, the most reactive 1-position leads to the formation of fully condensed polymers, containing only 6-membered rings, whilst the less reactive 2-position can inhibit condensation.

Lewis [86] found that naphthalene (I) and anthracene (II) derived pitches are complex mixtures of 2–10 units of the starting material and are only partly condensed, represented by the average structures:

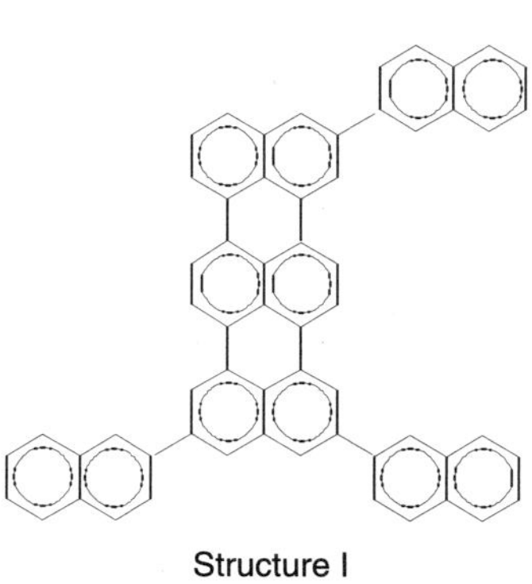

Structure I

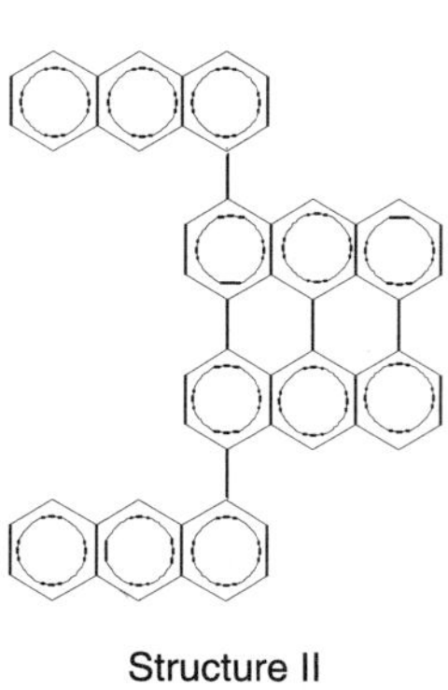

Structure II

The final carbonization process will be represented by a two dimensional planar graphite network and the polymerization of acenaphthylene, which is ideally shaped with the essential reactivity to cross-link in two directions [93,94], gives zethrene a graphite structure without vacancies:

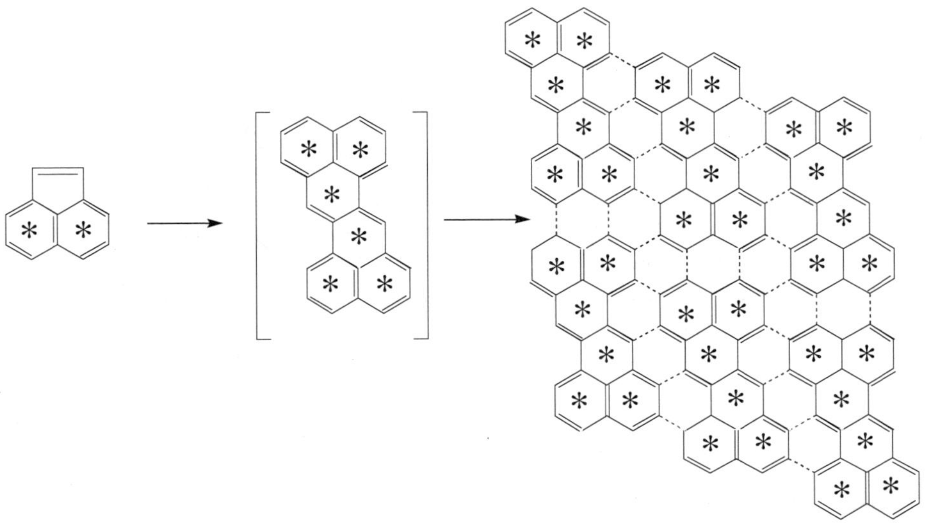

On the otherhand, tetrabenzonaphthalene ($C_{26}H_{16}$) has a non-planar structure and forms vacancies when polymerized [81]:

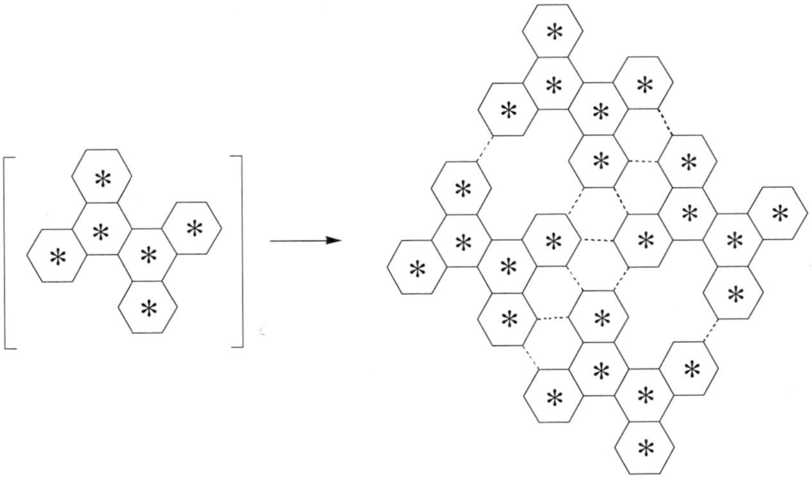

By the use of suitable additives, the carbonization route can be altered considerably and has been aptly reviewed by Fitzer *et al* [95]. Not surprisingly, oxygen has been the most studied and although severe oxidation can significantly reduce the graphitizability of pitch [96,97] and will also inhibit mesophase development [98], mild oxidation can induce dehydrogenative polymerization without impairing subsequent graphitic structure [98].

Singer and Lewis [99,100] identified aryloxy radicals as intermediates when reacting oxygen with aromatic hydrocarbons:

These eventually forming quinones, which pyrolyze with the loss of CO to give non-planar radicals.

Sulphur has also been used extensively [101–105] as an additive and the yield of carbon from acenaphthylene can be increased. The use of too much sulphur resulted in a non-graphitizing carbon, but for a lesser amount, the reverse was true [101]. Greinke and Lewis [104] showed that sulphur incorporated in a strained ring system could be readily eliminated in the early stages of carbonization, but when incorporated in aromatic hydrocarbons, the sulphur containing polymer was stable up to high temperatures (1500°C) [99]. Rand [106] considered sulphur to be undesirable in pitch precursors used for making carbon fiber.

Mochida et al. [107–109] studied the carbonization reaction of aromatic hydrocarbons and heterocyclic compounds using $AlCl_3$ as an additive.

$$AlCl_3 + H_2O \rightleftharpoons H^+AlCl_3(OH)^-$$

The conversion of low molecular weight aromatic hydrocarbons to carbon can be carried out using alkali metal additives [110,111]:

7.4.2 Mechanisms associated with the stabilization of pitch fiber precursors

After the pitch has been prepared and spun, it is usual to undertake a process of stabilization, normally by oxidation in air, which forms cross links between the mesophase particles rendering the fibers infusible. A possible reaction scheme is [30,31]:

R = alkyl group
A = aromatic group

7.4.3 Mechanisms associated with the carbonization of pitch fibers

The process of carbonization aims to eliminate heteroatoms, like H, N and O, enlarging the layer of fused hydrocarbon aromatic rings and forming a turbostratic graphite-like structure, which according to Morida and Kagaka follows a path as shown:

REFERENCES

1. Otani S, Oya I, Composites '86: Recent Advances in Japan and the United States, Proc Japan-US, CCM-III, Kawata K, Umekawa S, Kobayashi A eds., *Japan Soc Composite Mater*, Tokyo, 1–10, 1986.
2. Rand B, Carbon fibres from mesophase pitch, Watt W, Perov BV eds., *Strong Fibres*, Elsevier Science Publishers, Amsterdam, 495–575, 1985.
3. Edie DD, Pitch and mesophase fibers, Figueiredo JL, Bernardo CA, Baker RTK, Hüttinger KJ eds., *Carbon Fibers Filaments and Composites*, Kluwer Academic Publishers, Dordrecht, 43–72, 1990.
4. Bahl OP, Shen Z, Lavin JG, Ross RA, Manufacture of carbon fibers, Donnet JB, Wang TK, Rebouillat S, Peng JCM eds., *Carbon Fibers*, Marcel Dekker, New York, 41–63, 1998.
5. Oberlin A, Bonnamy S, Lafdi K, Structure and texture of carbon fibers, Donnet JB, Wang TK, Rebouillat S, Peng JCM eds., *Carbon Fibers*, Marcel Dekker, New York, 85–159, 1998.
6. German Pat., 2,064,282, 1971.
7. Japan Pat., 69 02,510.
8. German Pat., 2,038,949, 1971.
9. Otani S, *Carbon*, 5, 219, 1967.
10. Barnett I, Apparatus for thermal modification of yarns, U.S. Pat., 2,913,802, 1959.
11. Singer LS, Carbon fibres from mesophase pitch, *Fuel*, 60(9), 839–847, 1981.
12. Kipling JJ, Sherwood JN, Shooter PV, Thompson NR, *Carbon*, 1, 315, 1964.
13. Otani S, *Carbon*, 3, 31, 1965.
14. Stevens WC, Diefendorf RJ, Thermosetting of mesophase pitches: experimental, *Carbon '86 Proceedings of the International Conference on Carbon*, Baden-Baden, 37–39, 1986.
15. Lewis IC, McHenry ER, Singer LS, Process for producing carbon fibers from mesophase pitch, U.S. Pat., 3976729, 1976.
16. Singer LS, Union Carbide, High modulus, high strength carbon fibers produced from mesophase pitch, U.S. Pat., 4,005,183, 1977.
17. Riggs DM, The characterization and kinetic mechanism of mesophase formation in high molecular weight carbonaceous materials, *PhD Thesis*, Rensselaer Polytechnic Inst, Troy, NY, 1979.
18. Okuda K, *Nikkakyo Geppo*, 12(1), 17, 1980.
19. Otani S, *Jidosha Gijutsu*, 34(8), 861, 1980.
20. Sakaguchi Y, *Kagaku Kogaku*, 46: No.3, 145, 1982.
21. Kenichi M, Yoshiro K *et al*, Brit. Pat., 1426502, 1973.
22. Hawthorne HM *et al*, Nature 227: 946, 1970.
23. Nippon Steel Company, Jap. Pat., 58-144123, 1983.
24. Shen Z, Guo H *et al*, 21st *Biennial Conf on Carbon*, 352, 1993.
25. Chi W, Shen Z, *Carbon '96 European Carbon Conference*, 395, 1996.
26. Otani S, Kimura K, *Carbon Fiber*, Kindai Henshu Ltd., 140, 1972.
27. Mochida I, Toshima H, Korai Y, Hino T, Oxygen distribution in the mesophase pitch fiber after oxidative stabilization, *J Mater Sci*, 24, 389–394, 1989.
28. Matsumoto T, Mochida I, *Carbon*, 31(1), 143–147, 1993.
29. Mochida I, Zeng SM, Korai Y, Toshima H, The introduction of a skin core structure in mesophase pitch fibers by oxidative stabilization, *Carbon*, 28, 193, 1990.
30. Shen Z, Qin R *et al*, 36th *Inter SAMPE Symposium and Exhibition*, 36, 1109–1117, 1991.
31. Shen Z, Qin R *et al*, Carbon and Carbonaceous Composite Materials Structure-Property Relationship, *Abstr and Proc*, Malenovice, Czech Republic, 31, Oct 10–13 1995.
32. Lavin JG, Chemical reactions in the stabilization of mesophase pitch-based fibers, *Carbon*, 30, 351–357, 1992.
33. Matsumoto T, Mochida I, A structural study on oxidative stabilization of mesophase pitch fibers derived from coal tar, *Carbon*, 30, 1041, 1992.
34. Mochida I, Zeng SM, Korai Y, Hino T, Toshima H, *Carbon*, 29(1), 23–29, 1990.
35. Zeng SM, Korai Y, I Mochida, Hino T, Toshima H, *Bull Chem Soc Japan*, 63(7), 2083–2088, 1990.
36. Buckley JD, Edie DD Eds., Carbon Materials and Composites, Noyes Publications, Park Ridge, NJ, 32, 1993.

37. Dresselhaus MS, Dresselhaus G, Sugihara K, Spain IL, Goldberg HA, *Graphite Fibers and Filaments*, Springer-Verlag, Berlin, p. 75, 1988.
38. Singer LS, Carbon fibers from mesophase pitch, *Fuel*, 60(9), 839–847, 1981.
39. Bright AA, Singer LS, *Carbon*, 17, 59, 1979.
40. Riggs DM, Shuford RJ, Lewis RW, Graphite fibers and composites, George Lubin ed., *Handbook of Composites*, Van Nostrand Reinhold Co., New York, 1982.
41. Matsumoto T, Mesophase pitch and its carbon fibers, *Pure Appl Chem*, 57(11), 1533, 1985.
42. Shen Z, Guo H et al, Carbon and Carbonaceous Composite Materials Structure-Property Relationship, *Abstr and Proc*, Malenovice, Czech Republic, 31, Oct 10–13, 1995.
43. Barr JB, Chwastiak S, Didchenko R, Lewis IC, Singer LS, High modulus carbon fibers from pitch precursor, *Appl Polym Symp*, Wiley, New York, 29, 161, 1976.
44. Chwastiak S, Barr JB, Didchenko R, *Carbon*, 17, 49, 1979.
45. Bacon R, *Phil Trans Roy Soc*, London, A(294), 437, 1979.
46. Larsen HL, Graphite elements for high temperature resistor furnaces, *Union Carbide Reprint Part 2, Industrial Heating*, 11, 1962–63.
47. Savage G, *Carbon-Carbon Composites*, Chapman and Hall, London, 65, 1993.
48. Lin SS, Yip PW, Surface adhesion of carbon fibers after chemical treatments, *19th Biennial Conference on Carbon*, Penn State University, Session 2A, 244–245, Jun 25–30 1989.
49. Edie DD, Research into pitch-based carbon fibers, Genisio M ed., *Recent Research into Carbon-Carbon Composites*, Southern Illinois Univ, 16–41, 1987.
50. Barnes AB, Dauche FM, Gallego NC, Fain CC, Thies MC, As-spun orientation as an indication of graphitized properties of mesophase based carbon fiber, *Carbon*, 36(7–8), 855–860, 1998.
51. Hamada T, Nishida SY, Furuyama M, Tomiaka T, *Extended Abstracts of 18th Biennial Conf on Carbon*, Amer Carbon Soc, Worcester, MA, 225, 1987.
52. Hamada T, Nishida SY, Matsumoka M, Endo M, *J Mater Res*, 2, 850, 1987.
53. Inagaki M, Iwashita N, Hishiyama Y, Kaburagi Y, Yoshida A, Oberlin A, Lafdi K, Bonnamy S, Yamada Y, *Tanso* 147, 57, 1991.
54. Volk HF, High modulus pitch-based carbon fibers, *Union Carbide Publication adapted from Presentation at Symposium on Carbon Fiber Reinforced Plastics*, Hamburg, 1977.
55. White JL, Ng CB, Buehler M, Watts EJ, Microstructure of mesophase carbon fibres, *Extended Abstracts of 15th Biennial Conf on Carbon*, Amer Carbon Soc, Philadelphia, 310, 1981.
56. Yoon SH, Takano N, Korai Y, Mochida I, Crack formation in mesophase pitch based carbon fibres. 1. Some influential factors for crack formation, *J Mater Sci*, 32(10), 2753–2758, 1997.
57. Yoon SH, Korai Y, Mochida I, Crack formation in mesophase pitch based carbon fibres. 2. Detailed structure of pitch based carbon fibres with some types of open cracks, *J Mater Sci*, 32(10), 2759–2769, 1997.
58. Exxon Research and Engineering, U.S. Pat., 4,005,183, 1980/1981.
59. Roche EJ, Lavin JG, Parrish RG, *Carbon*, 26, 911, 1988.
60. Pennock GM, Taylor GH, Fitzgerald JD, *Carbon*, 31, 591,1993.
61. Kogure K, Sines G, Lavin JG, *Carbon*, 32, 1469, 1994.
62. Guigon M, Oberlin A, *Composites Science and Technology*, 25, 231, 1986.
63. Lafdi K, Bonnamy S, Oberlin A, *Carbon*, 31, 29, 1993.
64. Lafdi K, Bonnamy S, Oberlin A, *Carbon*, 30, 533, 1992.
65. Lafdi K, Bonnamy S, Oberlin A, *Carbon*, 30, 551, 1992.
66. *K Lafdi, S Bonnamy, A Oberlin. Carbon*, 30, 569, 1992.
67. Endo M, Structures of mesophase pitch-based carbon fibres, *J Mater Sci*, 23, 598–605, 1988.
68. Edie DD, Fox NK, Barnett BC, Fain CC, Melt spun non-circular carbon fibers, *Carbon*, 24(4), 477–482, 1986.
69. Edie DD, Fain CC, Melt spun non-circular fibers from mesophase pitch, *Carbon '86, Proceedings of the International Conference on Carbon*, Baden-Baden, FRG, 629–631, 1986.
70. Matsumoto T, Mesophase pitch and its carbon fibers, *Pure Appl Chem*, 57(11), 1553–1562, 1985.
71. Bennett SC, Johnson DJ, *Carbon*, 17, 25, 1979.
72. Lu YG, Wu D, Zha QF, Liu L, Yang CL, Skin-core structure in mesophase pitch based carbon fibers: Cause and prevention, *Carbon*, 36(12), 1719–1724, 1998.

73. Mochida I, Zeng SM, Korai Y, Hino T, Toshima H, Structure and properties of mesophase pitch carbon fiber with a skin core structure carbonized under strain, *J Mater Sci*, 27(7), 1960–1968, 1992.
74. Guigon M, Oberlin A, *Fibre Sci Technol*, 20, 55, 1984.
75. Guigon M, Oberlin A, *Fibre Sci Technol*, 20, 177, 1984.
76. Goma J, Oberlin A, Thin solid films, 65, 221, 1980.
77. Murakami K, Toshima H, Yamamoto M, Effect of mesophase pitches on tensile modulus of pitch based carbon fibers, *Sen-I Gakkaishi*, 53(3), 73–78, 1997.
78. Kumar S, Compressive strength of high performance fibers, *SAMPE Quarterly*, 20(2), 3–8, 1989.
79. Endo M, Kim C, Koraki T, Kasai T, Matthews MJ, Brown SDM, Dresselhaus MS, Tomaki T, Nishimura Y, Structural characterization of milled mesophase pitch based carbon fibers, *Carbon*, 36(11), 1633–1641, 1998.
80. Wang CY, Li MW, Wu YL, Guo CT, Preparation and microstructure of hollow mesophase pitch based carbon fibers, *Carbon*, 36(12), 1749–1754, 1998.
81. Lewis IC, Chemistry of carbonization, *Carbon*, 20(6), 519–529, 1982.
82. Lewis IC, Singer LS, *Preprints of fuel division*, Am Chem Soc, 13, 86, 1969.
83. Edstrom T, Lewis IC, *Carbon*, 7, 85, 1969.
84. Simon C, Estrade H, Tchoubar D, Conard J, *Carbon*, 15, 211, 1977.
85. Simon C, Estrade H, Tchoubar D, Conard J, *Proc 5th Carbon Conf*, London, 1, 294, 1978.
86. Lewis IC, *Carbon*, 18, 191, 1980.
87. Lewis IC, Singer LS, Walker PL Jr., Thrower PL eds., *Chemistry and Physics of Carbon*, Marcel Dekker, New York, 17, 1, 1981.
88. Badger GM, *Prog Phys Organ Chem*, 3, 1, 1965.
89. Carrington A, Smith ICP, *Mol Phys*, 9, 137, 1965.
90. Livingston R, Zeldes H, Conradi MS, *J Am Chem Soc*, 101, 4312, 1979.
91. Lang KF, Buffleb H, *Chem Ber*, 94, 1075, 1961.
92. Lang KF, Buffleb H, Kalowy J, *Chem Ber*, 90, 2888, 1957.
93. Ruland W, *Carbon*, 2, 365, 1965.
94. Singer LS, Lewis IC, *Carbon*, 2, 115, 1964.
95. Fitzer E, Muller K, Schaffer W, Walker PL ed., *Chemistry and Physics of Carbon*, Marcel Dekker, New York, 7, 237, 1971.
96. Kipling JJ, Sherwood JN, Shooter PV, Thompson NR, *Carbon*, 1, 315, 1964.
97. Otani S, *Carbon*, 3, 31, 1965.
98. Barr JB, Lewis IC, *Carbon*, 16, 439, 1978.
99. Lewis IC, Singer LS, *J Phys Chem*, 85, 354, 1981.
100. Lewis IC, Singer LS, Summary of papers presented at the American Carbon Society, 9th Biennial Carbon Conf on Carbon, Boston College, 120, 16–20 Jun 1969.
101. Christu N, Fitzer E, Kalka J, Schafer W, *J Chem Phys Physicochim Biol*, 50, 1959.
102. Blayden HE, Patrick JW, *Fuel*, 49, 257, 1970.
103. Kipling JJ, Shooter PV, Young RN, *Carbon*, 4, 333, 1966.
104. Greinke RA, Lewis IC, *Carbon*, 17, 471, 1979.
105. Lewis IC, Greinke RA, *J Polym Sci Chem*, 20, 1119, 1982.
106. Rand B, Carbon Fibres fron Mesophase Pitch, Watt W, Perov BV eds., *Strong Fibres*, Elsevier Science Publishers, Amsterdam, 516, 1985.
107. Mochida I, Otsuka K, Maeda K, Takeshita K, *Carbon*, 15, 239, 1977.
108. Mochida I, Inoue S, Maeda K, Takeshita K, *Carbon*, 15, 9, 1977.
109. Mochida I, Sakata K, Maeda K, Fujitsu H, Takeshita K, *Fuel Processing Technol*, 3, 207, 1980.
110. Beguin F, Setton R, *Carbon*, 10, 539, 1972.
111. Mochida I, Nakamura E, Maeda K, Takeshita K, *Carbon*, 14, 123, 1976.

CHAPTER 8

Production of Vapor Grown Carbon Fibers (VGCF)

8.1 INTRODUCTION

The term vapor grown carbon fiber (VGCF) is an International Union of Pure and Applied Chemistry (IUPAC) recommendation and Tibbetts [1] believes that this term has won general acceptance for the class of material where a carbonaceous gas, in the presence of a small metal particle acting as a catalyst, forms a carbon filament. However, Dresselhaus and co-authors [2] use the term CCVD filament in their book, which stands for catalytic chemical vapor deposition and is certainly more descriptive of their mode of preparation, but is, unfortunately, not the generally accepted term.

Gadelle [3] suggested identifying three classes of VGCF:

1. Conventional filaments—0.01 μm in diameter and 1 μm in length
2. Very long precursor filaments which are the precursors for VGCF—about 0.01 μm in diameter and ≥ 1 mm in length
3. VGCF resulting from thickening of the precursor filaments—≥ 1 μm in diameter and ≥ 1 mm in length

These classes are depicted schematically in Figure 8.1. Obviously, filaments of just ~0.01 μm diameter will present a serious health risk [5] and the author believes that only thickened fibers would be of any practical use and then, only under carefully controlled conditions. Tibbetts [1] does, however, claim that filaments tend to knit together and become quite difficult to disperse, hence presenting less of a health risk.

8.2 PREPARATION OF VGCF

Valuable contributions on VGCF are given in the book on Carbon Fibers Filaments and Composites [6] and Varshavskii [7] has reviewed the preparation of carbon fibers from the gas phase. The production of vapor grown carbon fibers is also described in the references [8–13]. The first commercial VGCF was produced by Nikkoso in 1991 and called Grasker.

Tibbetts [14] developed a batch process for producing VGCF using the apparatus depicted in Figure 8.2 depositing the filaments on a series of nested semi-cylindrical mullite

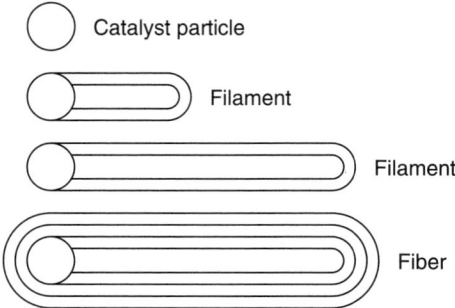

Figure 8.1 Formation of a carbon filament from a catalytic particle and a carbon fiber from a carbon filament. *Source:* Reprinted with permission from Tibbetts GG, Vapor-grown carbon fibers: status and prospects, *Carbon*, 27(5), 745–747, 1989. Copyright 1989, Elsevier.

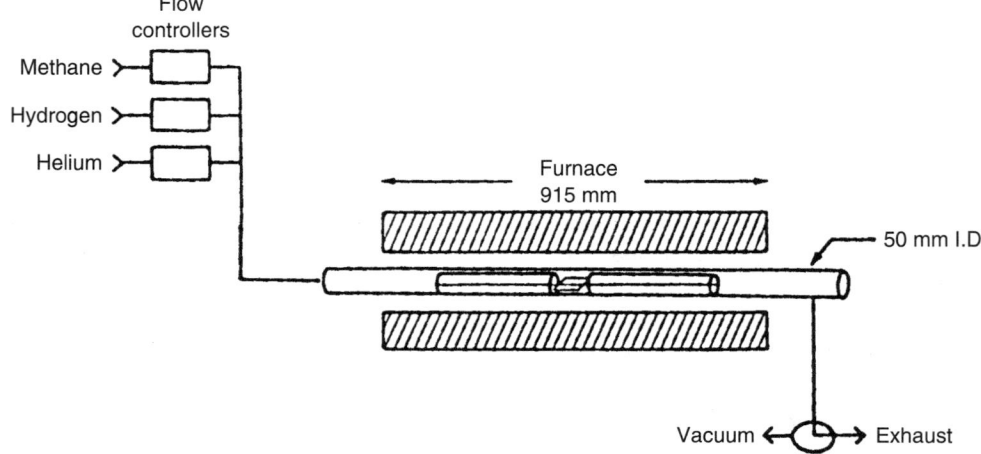

Figure 8.2 An apparatus for growing VGCF at atmospheric pressure. *Source:* Reprinted from Tibbetts GG, Rodda EJ, High temperature limit for the growth of carbon filaments on catalytic iron particles, *Mater Res Soc Symp Proc*, 111, 49, 1988.

tubes of analysis:

Al_2O_3	58.6%
SiO_2	36.8%
Fe_2O_3	0.9%
TiO_2	0.9%
CaO/MgO	0.8%
Na_2O/K_2O	1.32%
Balance	0.68%

These tubes were contained within a mullite muffle, passing a hydrocarbon/hydrogen mixture over the mullite substrate, maintained at about 1000°C, on which iron-containing particles, acting as a catalyst, had been deposited. It was found beneficial to obtain a build-up in the equipment of reactive hydrocarbons (C_2H_x), formed in the process, to promote a faster rate of formation, since methane itself tends to be un-reactive and has poor solubility in the iron particles. The filaments could be made several centimeters in length, with diameters of 7–10 μm and when removed from the substrate, had a variety of forms. At temperatures

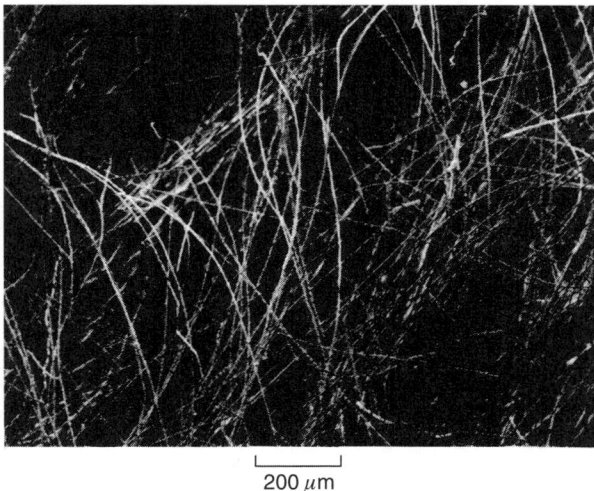

Figure 8.3 Fibers after removal from substrate. *Source:* Reprinted with permission from Tibbetts GG, Vapor-grown carbon fibers, *Carbon Fibers Filaments and Composites*, Figueiredo JL, Bernardo CA, Baker RTK, Hüttinger KJ eds., Kluwer, Dordrecht, 77, 1990. Copyright 1990, Springer.

above 1000°C, they were relatively straight (Figure 8.3), whereas at lower temperatures, they tended to vary and could be twisted, or helical and often had a worm-like, or a vermicular shape. Helium, as an inert gas, helps to prevent convection and adjusting the gas stream carbon content to 15% v/v CH_4 in H_2, with a gas residence time of about 20 s will promote fiber lengthening at about 1000°C and filaments will lengthen to several centimeters in about 10 min. Whilst higher concentrations (30% CH_4 in H_2) and temperatures above 1050°C will promote the formation of carbon on the surface, thus increasing the thickness of the fiber (Figure 8.4a), thinner fibers tend to produce a crenulated surface (Figure 8.4b), due to differential contraction of the layers on cooling. With a gas residence time of about 30 s, fibers can be thickened to 7–10 µm in about 2 h. The surface is free of iron and can be readily graphitized.

Koyama and Endo [17] introduced a continuous process, which was modified by Hatano *et al* [18], where ultrafine catalytic particles (5–25 nm diameter) were incorporated in the feedstock and fed into a reaction chamber (Figure 8.5a) or produced directly in the reactor by decomposition of an organometallic at 1100°C, which decomposes into a suspension of ultrafine catalyst particles (5–25 nm diameter) that are transported by the flow of hydrocarbons and hydrogen gas into the furnace area (Figure 8.5b). The catalyst/hydrocarbon feed ratio determines the diameter and the yield is increased as the fiber diameter is reduced. Once the catalyst particle becomes covered with carbon, oxygen or any form of impurity, the fiber lengthening reaction will stop and fiber thickening takes place. Hence, in order to ensure long filaments, all constituents must be of a high purity. Although in the continuous process, fiber diameters are uniform, they are not as long as fibers produced by the batch process and collect in a sponge-like mass, the filaments being hollow with a diameter of about 0.1–1.5 µm and length about 1 mm. Figure 8.6 shows how a metal catalyst may be dispersed:

a. Fine particles of iron (≤10 nm diameter) in an alcoholic suspension are sprayed on to a substrate and dried by evaporation.
b. A solution of an inorganic salt is sprayed onto a metal substrate and thermally decomposed at ~1000°C. Table 8.1 lists a number of typical salts.

c. An organometallic compound such as the acetylacetonates of Fe, Co and Mn, e.g., Cobalt(III) acetylacetonate ([CH$_3$COCH=C(O—)CH$_3$]$_3$Co)

or ferrocene, Fe(C$_5$H$_5$)$_2$

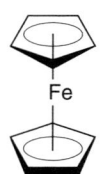

dissolved in a volatile solvent such as benzene is sprayed onto a hot substrate.

The distribution of fiber length for various precursors is shown in Figure 8.7 and the average diameter of fibers with time of exposure at 1150°C is given in Figure 8.8.

Choosing an organometallic will reduce the size of the catalyst particles and increase the growth rate. A growth rate curve can be divided into three regions—an initial accelerating period of growth, a period of constant growth and a slowing down region. Fiber nucleation and elongation starts at about 1000°C and the thickening process starts at a higher temperature when the hydrocarbons are decomposed (Figure 8.9). Tibbetts reckons that fiber lengthening occurs at 1054°C, while Benissad suggests the range to be 1050–1100°C. For a given temperature, Figure 8.10 [6] shows that the rate of filament growth has an inverse square root relationship with the particle size and moreover, as the diameter of the particle increases, so does the melting point (from 1020°C at 6 nm diameter to 1121°C at 25 nm) [21].

The most common catalyst is iron, which can be treated with sulphur, thiophene or hydrogen sulphide, thus lowering the melting point and enabling the catalyst to penetrate the pores of the carbon and producing further sites for growth [23]. Oxides like WO$_3$, Ta$_2$O$_5$, MoO$_3$ and SiO$_2$ will reduce the growth rate and produce hollow filaments, suggesting that lesser number of carbon species will diffuse through the catalyst particle to form the inner core of the filament [24]. Other catalysts are Fe/Ni, Ni, Co, Mn, Cu and Pd. The influence of the catalyst is discussed [25–29].

Some additives like Al$_2$O$_3$ and TiO$_2$ initially provide a barrier on the catalyst particle, but as the temperature is increased, they spall and provide no protective influence. SiO$_2$ reduces the solubility of the carbon and rate of diffusion through the metal catalyst particle [30]. Catalysts can be activated by pretreatment, such as the treatment of iron with steam, where activation is believed to be due to the formation of FeO, which increased the yield tenfold. Gadelle [21] states that a normal Fe concentration is about 5×10^{-6} g cm^{-2}. Organometallic compounds improve the growth rate and a 70% yield can be achieved with a growth rate of 40 μm sec^{-1} using 20% w/w cobalt acetylacetonate and 80% w/w ferrocene.

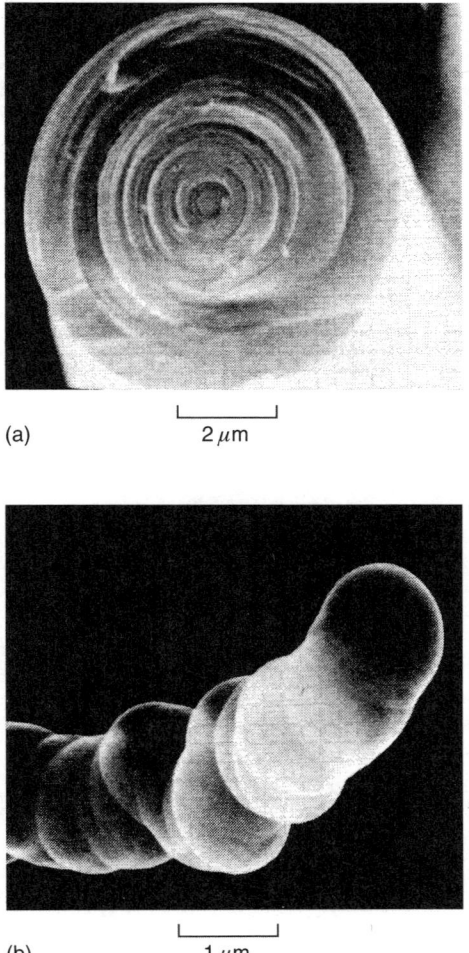

Figure 8.4 (a) A cross-sectional view of a vapor grown fiber which has been exposed to a maximum temperature of 1130°C. (b) SEM of a thinner fiber showing crenulations due to hoop stress. *Source:* Reprinted with permission from Tibbetts GG, Vapor-grown carbon fibers, *Carbon Fibers Filaments and Composites*, Figueiredo JL, Bernardo CA, Baker RTK, Hüttinger KJ eds., Kluwer, Dordrecht, 79, 1990. Copyright 1990, Springer.

The carbonaceous gases used include C_2H_2, C_2H_4, CH_4, natural gas and benzene. Methane is rather unreactive, does not dissolve readily in the Fe particles and is converted to a more reactive species (C_2H_x) with hydrogen. The addition of CO increases the yield and is optimum when the $CO/(CO + H_2)$ volume fraction is 93–95% (Figure 8.11). This is believed to be due to the CO reducing any iron oxides, although in the presence of H_2, CO can react:

$$H_2 + CO \rightarrow H_2O + C$$

The water produced deactivates the iron catalyst and CO_2 is added to the (CO/H_2) mixture in order to enhance the yield and is optimum at 45% with a (77% CO, 19% CO_2 and 4% H_2) mixture (Figure 8.12), where the CO_2 is believed to produce either fine catalyst particles or prevent the aggregation of catalyst particles [31].

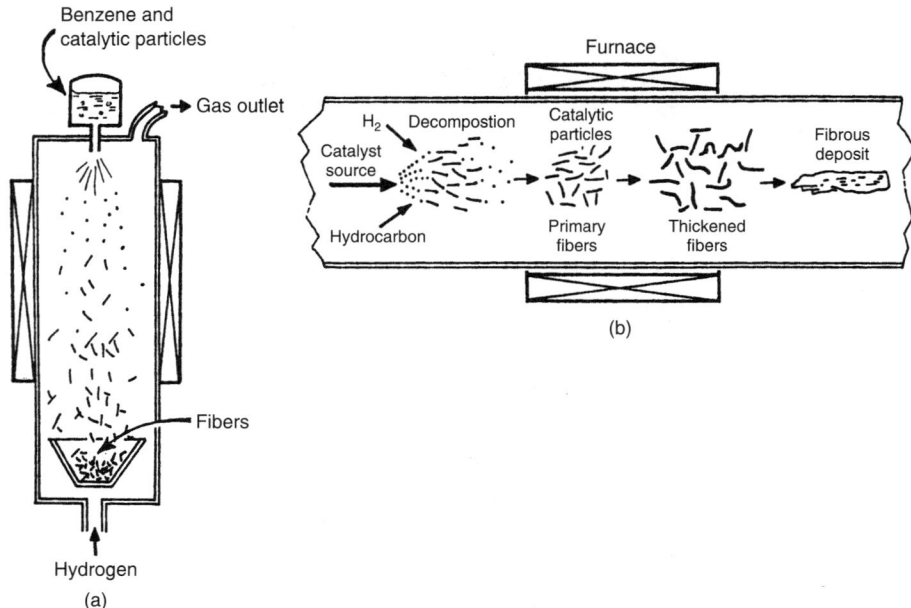

Figure 8.5 Floating catalytic particle methods for three-dimensional growth of fibers in a reaction chamber. (a) Direct introduction of catalytic particles using benzene feedstock. (b) Catalyst introduction by source such as ferrocene. The ferrocene decomposes into a suspension of ultrafine catalyst particles which are transported by the flow of hydrocarbons and hydrogen gas into the furnace area. *Source:* Reprinted with permission from Endo M, Shikata M, *Ohyo Butsuri*, (In Japanese), 54, 507, 1985. Copyright 1985, Ohyo Butsuri.

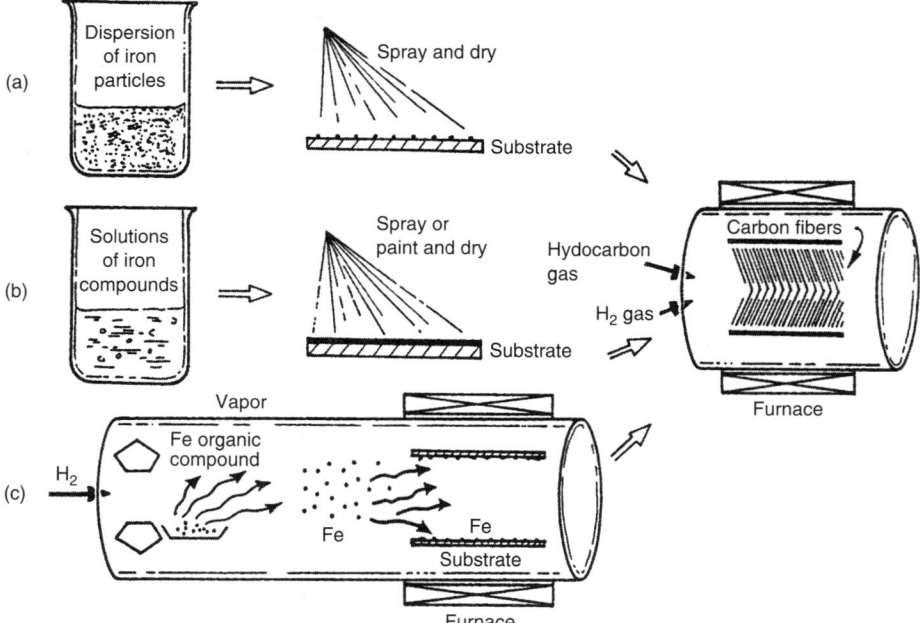

Figure 8.6 Dispersion of iron catalyst particles by three methods: (a) Spray and dry a suspension of iron particles. (b) Thermal decomposition of an inorganic iron compound on heated substrate. (c) Thermal decomposition of organometallic compound on heated substrate. *Source:* Reprinted with permission from Endo M, Shikata M, *Ohyo Butsuri*, (In Japanese), 54, 507, 1985. Copyright 1985, Ohyo Butsuri.

PRODUCTION OF VAPOR GROWN CARBON FIBERS

Table 8.1 Fibers and catalyst particles obtained from various precursors

Precursor	Mean fiber diameter μm	Mean length mm	Mean diameter of the catalyst particles after germination nm
$Fe(NO_3)_3 \cdot 9H_2O$	3	1.11	11.2
$Fe(C_2H_5)_2$ (ferrocene)	5.1	0.55	35.7
$Fe_2(SO_4)_3$	3.2	0.78	25
$Fe(NO_3)_3 \cdot 9H_2O$ + KOH	3.9	1.38	9

Source: Reprinted with permission from Gadelle P, The growth of vapor deposited carbon fibres, *Carbon Fibers Filaments and Composites*, Figueiredo JL, Bernardo CA, Baker RTK, Hüttinger KJ eds., Kluwer, Dordrecht, 110, 1990. Copyright 1990, Springer.

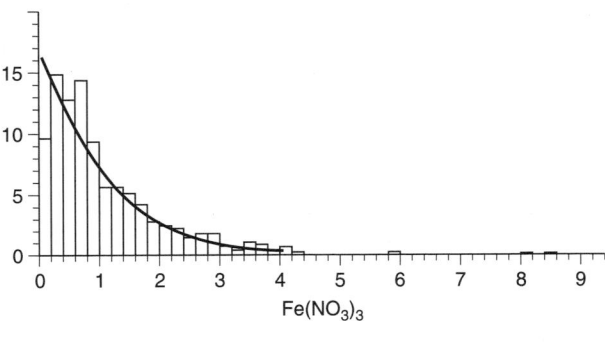

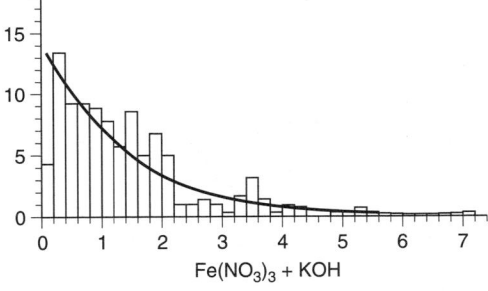

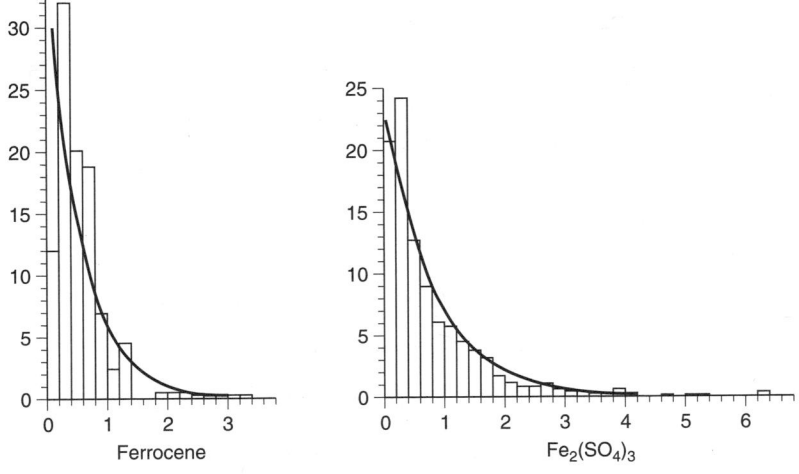

Figure 8.7 Fiber length distributions obtained from various precursors. Units: X-axis in mm. Y-axis in %. *Source:* Reprinted from Benissad F, Gadelle F, Coulon P, Bonnetain L, Influence of catalyst precursor on the lengths of vapor grown carbon fibers, *Proc Int Conf Carbon*, Newcastle upon Tyne, 307–309, 1988.

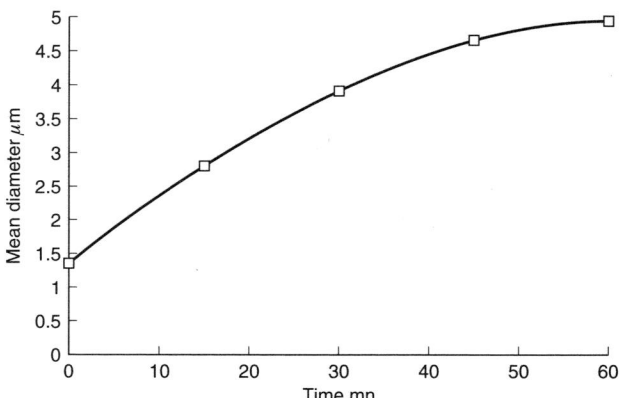

Figure 8.8 Average diameter of fibers *vs* time of exposure at 1150°C. *Source:* Reprinted with permission from Gadelle P, The growth of vapor deposited carbon fibres, *Carbon Fibers Filaments and Composites*, Figueiredo JL, Bernardo CA, Baker RTK, Hüttinger KJ eds., Kluwer, Dordrecht, 113, 1990. Copyright 1990, Springer.

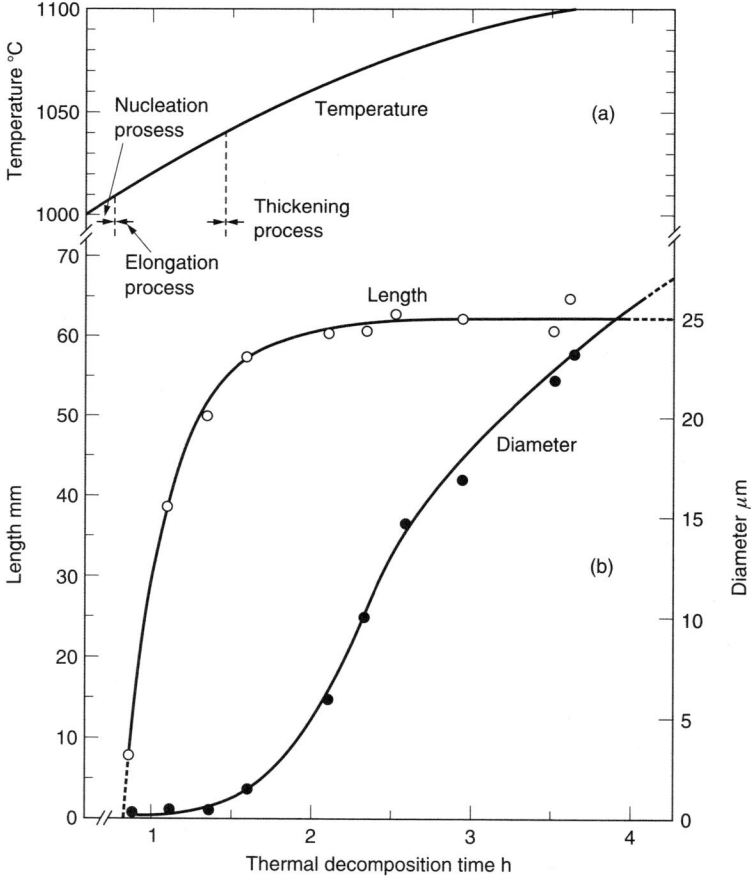

Figure 8.9 Growth curves of carbon fibers from benzene by a CCVD process. (a) Substrate temperature *vs* growth time showing the temperature regimes for the nucleation, elongation and thickening processes. (b) Fiber length and thickness *vs* growth time corresponding to the temperature time schedule shown in (a). *Source:* Reprinted from Endo M, Koyama T, *Kotai Butsuri* (In Japanese), 12, 1, 1977.

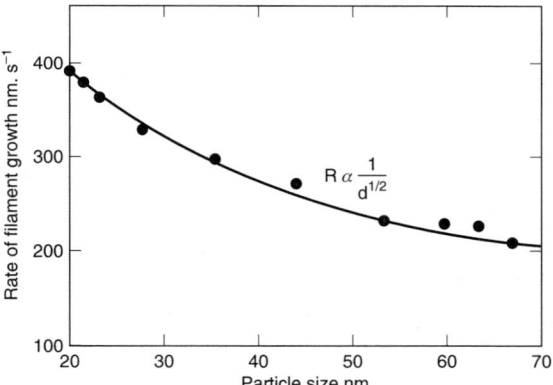

Figure 8.10 Variation of filament elongation with catalyst particle diameter. *Source:* Reprinted with permission from Figueiredo JL, Bernardo CA, Baker RTK, Hüttinger KJ eds., *Carbon Fibers Filaments and Composites*, Baker RTK, *Electron Microscopy Studies of the Catalytic Growth of Carbon Filaments*, Kluwer, Dordrecht, 414, 1990. Copyright 1990, Springer.

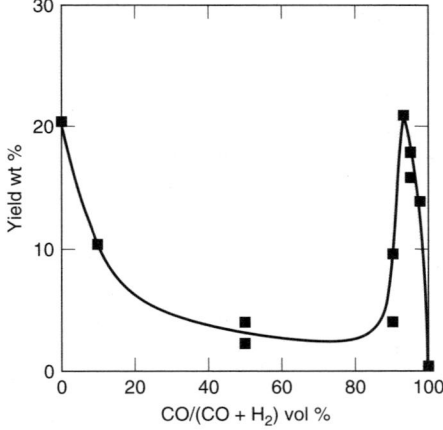

Figure 8.11 Effect of CO content in the CO/H_2 mixture on the carbon yield. *Source:* Reprinted with permission from Ishioka M, Okada T, Matsubara K, Endo M, Formation of vapor-grown carbon-fibers in $CO-CO_2-H_2$ mixtures, *Carbon*, 30(6), 859–863, 1992. Copyright 1992, Elsevier.

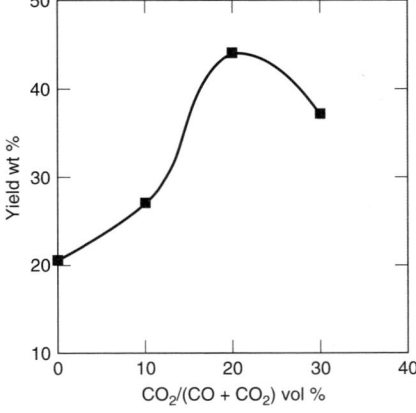

Figure 8.12 Effect of CO_2 content in the $CO/CO_2/H_2$ mixture on the carbon filament yield. *Source:* Reprinted with permission from Ishioka M, Okada T, Matsubara K, Endo M, Formation of vapor-grown carbon-fibers in $CO-CO_2-H_2$ mixtures, *Carbon*, 30(6), 859–863, 1992. Copyright 1992, Elsevier.

When CO is used as the carbon source, it is believed that carbides are involved with filament initiation and possibly, growth.

8.3 GROWTH PROCESS

Many versions of possible growth mechanisms have been proposed. The fiber consists of basal planes nesting together as a series of concentric cylinders so that a transverse cross-section resembles the growth rings of a tree (Figure 8.13) rather than a scroll type of formation as obtained with carbon whiskers.

Gadelle has shown a simplistic view of the stages of fiber growth (Figure 8.14) [21]. Baker et al. [30,33] showed that only part of the catalyst particle was exposed to the hydrocarbon and once that tip became covered, growth was arrested and the mechanism became poisoned (Figure 8.15). It is possible for further growth to occur by replacing the hydrocarbon with H_2 or O_2 and heating for a short time at 700°C (Figure 8.16), permitting existing filaments to continue growing or activating new sites. Oxidation studies of the Ni/C_2H_2 system using controlled atmosphere electron microscopy (CAEM) showed that the central core oxidized first at 600°C, (similar to the behavior of an amorphous or non-graphitic carbon), allowing the catalyst particle in some instances to fall into the tube. At 725°C, the filament skin started to oxidize, which is the temperature at which uncatalyzed gasification of graphite occurs. Hence, Baker and co-workers [34] deduced that the filaments had a duplex structure, with a disordered core surrounded by an outer graphitic skin. These workers suggested that the mechanism involved decomposition of C_2H_2 on the exposed face of the metal particle by an exothermic reaction, forming carbon and releasing H_2. The newly formed carbon dissolves in the particle to form a eutectic of about 4.3% carbon and then diffuses through the particle to be re-deposited at the cooler end via an endothermic reaction, permitting filament growth to occur. Diffusion of carbon is the rate determining step, justified by the excellent agreement between the measured activation energies for filament growth and those for diffusion of carbon through the same metal (Table 8.2).

Vermicular filaments produced at temperatures ≤900°C, depending on the hydrocarbon, have a low density and an easily oxidized core, but are enclosed in a sheath of more graphitic material and show greater resistance to oxidation [33–35]. Filaments made at higher temperatures are more graphitic, with hollow cores and the mode of growth was depicted by

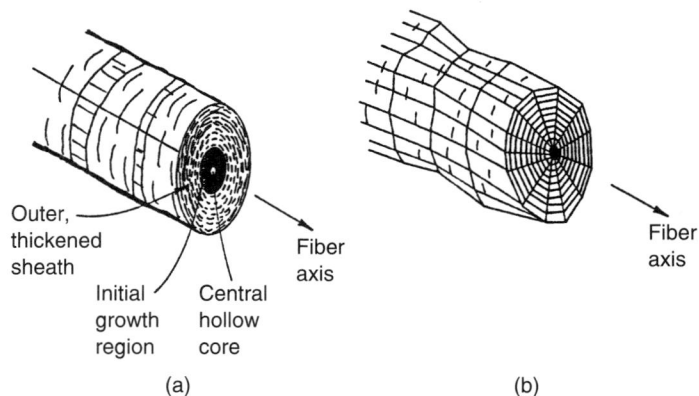

Figure 8.13 Sketch illustrating structure of CCVD filament. (a) As deposited at 1100°C. (b) After heat treatment to 3000°C. *Source:* Reprinted from Dresselhaus MS, Dresselhaus G, Sugihara K, Spain IL, Goldberg HA, *Graphite Fibers and Filaments*, Springer-Verlag, Berlin, 11, 1988.

PRODUCTION OF VAPOR GROWN CARBON FIBERS 335

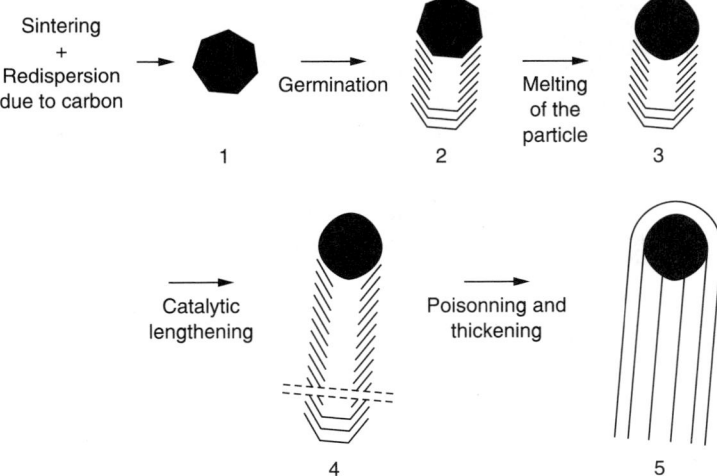

Figure 8.14 Mechanism of fiber growth. (1) Solid catalyst particle. (2) Short filament having grown on a solid particle. (3) Short filament on the liquid particle. (4) Rapid lengthening. (5) Fiber. *Source:* Reprinted with permission from Gadelle P, The growth of vapor-deposited carbon fibres, *Carbon Fibers Filaments and Composites*, Figueiredo JL, Bernardo CA, Baker RTK, Hüttinger KJ eds., Kluwer, Dordrecht, 115, 1990. Copyright 1990, Springer.

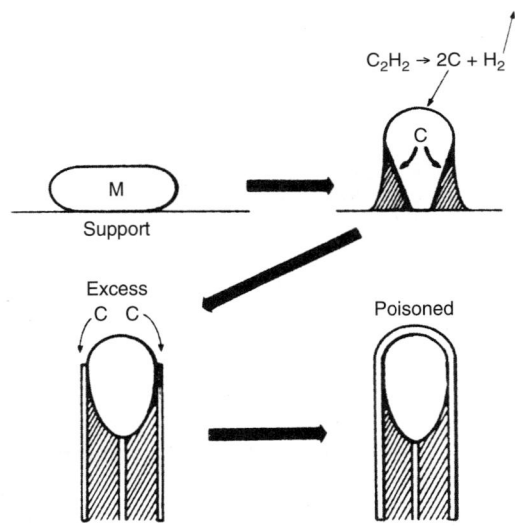

Figure 8.15 The most common mechanism of carbon filament formation from the pyrolysis of acetylene (C_2H_2) on a metal particle (M) where (C) denotes carbon. *Source:* Reprinted with permission from Baker RTK, Electron microscopy studies of the catalytic growth of carbon filaments, *Carbon Fibers Filaments and Composites*. Copyright 1990, Springer. Figueiredo JL, Bernardo CA, Baker RTK, Hüttinger KJ eds., Kluwer, Dordrecht, 419, 1990, Baker RTK, Barber MA, Harris PS, Feates FS, Waite RJ, *J Catal*, 80, 86, 1972.

Oberlin *et al* [36] in Figure 8.17. The fibers derived from benzene are formed at 1100°–1300°C, which is below the melting point of Fe (1534°C) but can be above the melting point of the 4.3% C eutectic (1147°C). The exothermic reaction can add a further 100°C to the reaction temperature, so it is possible, though not in every case, that the metal particle is in the molten state.

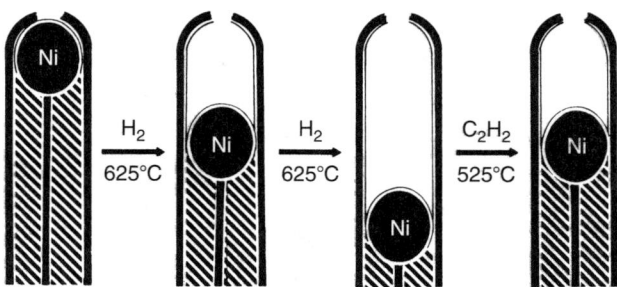

Figure 8.16 Reversible gasification/growth of carbon filaments from nickel particles. *Source:* Reprinted with permission from Baker RTK, Electron microscopy studies of the catalytic growth of carbon filaments, *Carbon Fibers Filaments and Composites*, Figueiredo JL, Bernardo CA, Baker RTK, Hüttinger KJ eds., Kluwer, Dordrecht, 414, 1990. Copyright 1990, Springer.

Table 8.2 Measured activation energies for filament growth with those for carbon diffusion in the corresponding metal catalysts.

Catalyst	Activation Energy kcal mol^{-1}	
	Catalyzed filament growth	Diffusion of Carbon
Ni	34.7	33.0–34.8
α-Fe	16.1	10.5–16.5
γ- Fe	33.9	33.3–37.4
Co	33.0–33.3	34.7
V	27.6	27.8
Mo	38.8	41.0
Cr	27.1	26.5

Source: Reprinted with permission from Baker RTK, Electron microscopy studies of the catalytic growth of carbon filaments, *Carbon Fibers Filaments and Composites*, Figueiredo JL, Bernardo CA, Baker RTK, Hüttinger KJ eds., Kluwer, Dordrecht, 415, 1990. Copyright 1990, Springer.

TEM has shown that the diameter of the hollow tube within the ultra-thin fibers is marginally below that of the diameter of the catalyst particle (Figure 8.18). The dense particle at the tip was identified as Fe_3C and was probably formed during the cooling stage at the end of the process.

Baker and Waite [38] have described a mode of growth termed extruded filament growth (Figure 8.19), where an iron catalyst is deposited on a Pt support and the carbon which is formed by the breakdown of C_2H_2 rapidly diffuses through to form a filament by upward growth.

The introduction of additives to the catalyst particle does influence the mode of growth as shown in Figure 8.20. Sacco [39] states that filaments are generally tubular in shape, but can be in a ribbon or braided form as depicted in Figure 8.21 [48]. The braided filaments are formed from two coils growing in the same direction from different faces of a diamond shaped Ni catalyst particle.

Jaeger and Behrsing [40] used SEM and TEM to investigate thin carbon filaments grown from hydrocarbon (natural gas, CH_4 and benzene) /high purity H_2 mixtures at 700–1100°C on Fe particles (prepared from $Fe(NO_3)_3$) on alumina substrates. Their results were most interesting and at temperatures >1000°C, the filaments are hollow tubes with walls formed from hexagonal carbon layers, which for catalyst particles <50 nm, formed closed, continuous co-axial cylinders parallel to the filament axis. These filaments, which were

PRODUCTION OF VAPOR GROWN CARBON FIBERS

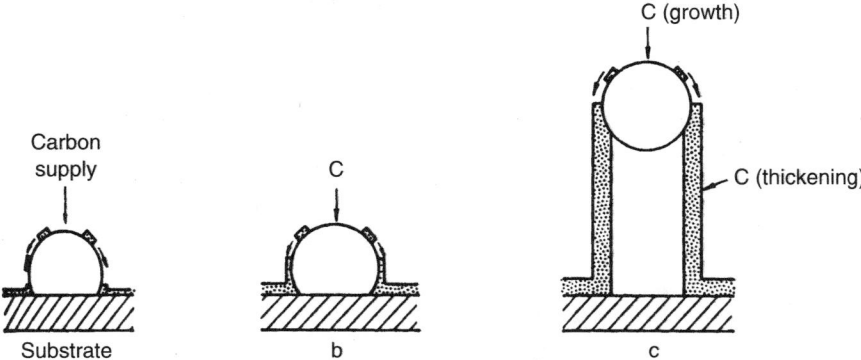

Figure 8.17 Schematic growth mechanism of carbon fibers produced from benzene using a catalyst. *Source:* Reprinted with permission from Oberlin A, Endo M, Koyama T, *Carbon*, 14, 133, 1976. Copyright 1976, Elsevier.

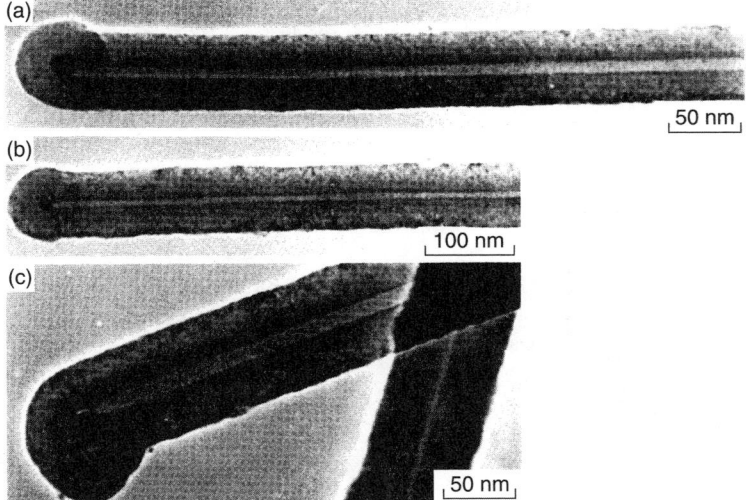

Figure 8.18 TEM photographs of fibers showing tip and hollow tube. *Source:* Reprinted with permission from Endo M, Katoh A, Sugiura T, Shiraishi M, High resolution electron microscopy of vapour grown carbon fibres obtained by ultra-fine fluid catalyst, *Extended Abstracts of the 18th Biennial Conference on Carbon*, Worcester, MA, (American Carbon Society), 151–152, 1987. Copyright 1987, The American Carbon Society.

radially symmetric approximations of single crystals of graphite, had good strength and electrical conductivity. With catalyst particle size >50 nm, however, the layers were inclined at up to $\pm 30°$ to the filament axis, causing the filaments to be very weak.

Filaments produced at temperatures $<1000°C$ were either segmented, consisting of hexagonal carbon layers, conforming to the bottom shape of the iron particle, or were flat ribbons formed by stacking hexagonal carbon layers normal to the filament axis.

The proposed growth from an Fe foil surface is depicted in Figure 8.22. The CO, in the presence of H_2, dissociates depositing the carbon at surface dislocations on the foil and diffuses inwards forming carbides when saturated. The carbides have a large volume and disrupt the foil surface, forming nodules which act as growing sites for the carbon filaments.

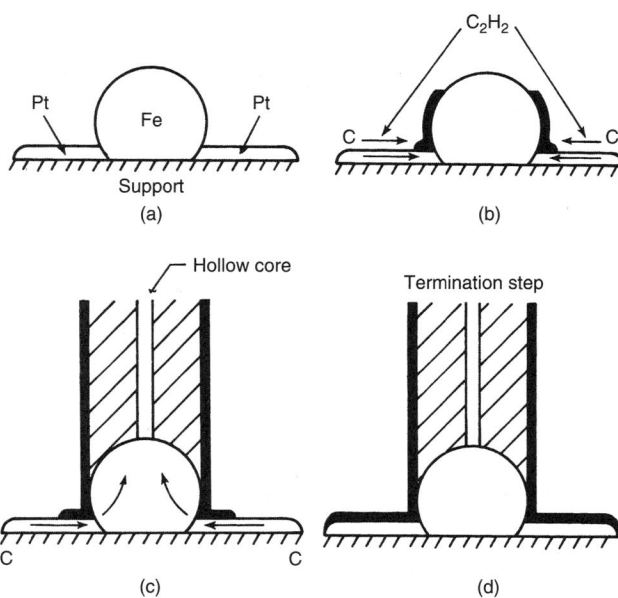

Figure 8.19 Mechanism of carbon filament formation in the Pt/Fe system by extruded filament growth. *Source:* Reprinted with permission from Baker RTK, Waite RJ, Formation of carbonaceous deposits from the platinum–iron catalyzed decomposition of acetylene, *J Catal*, 37, 101–105, 1975. Copyright 1975, Elsevier.

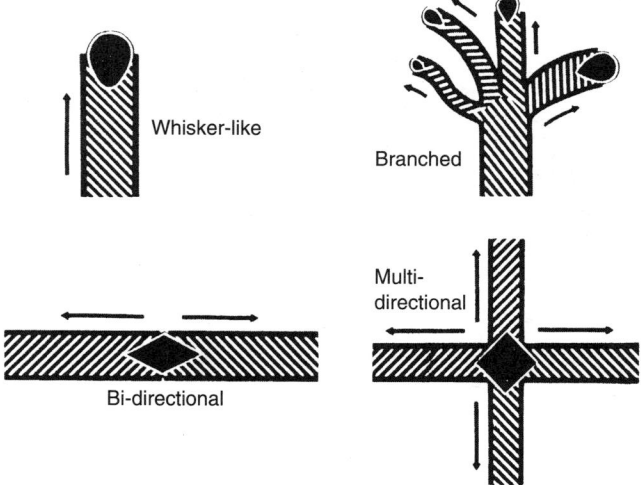

Figure 8.20 Schematic representation of different types of growth observed in carbon filaments showing influence of metal additives to the catalyst particle on the filament growth characteristics. *Source:* Reprinted with permission from Baker RTK, Electron microscopy studies of the catalytic growth of carbon filaments, *Carbon Fibers Filaments and Composites*, Figueiredo JL, Bernardo CA, Baker RTK, Hüttinger KJ eds., Kluwer, Dordrecht, 418, 1990. Copyright 1990, Springer.

The shape of the catalyst particle does have an influence on the mode of growth and Ni, for example does have a favoured crystallographic face, which is Ni(111) for graphite precipitation and the orientation of the catalyst particle is important.

The effect of various surface treatments, including air oxidation [41–44] and the morphology [45–47] are discussed.

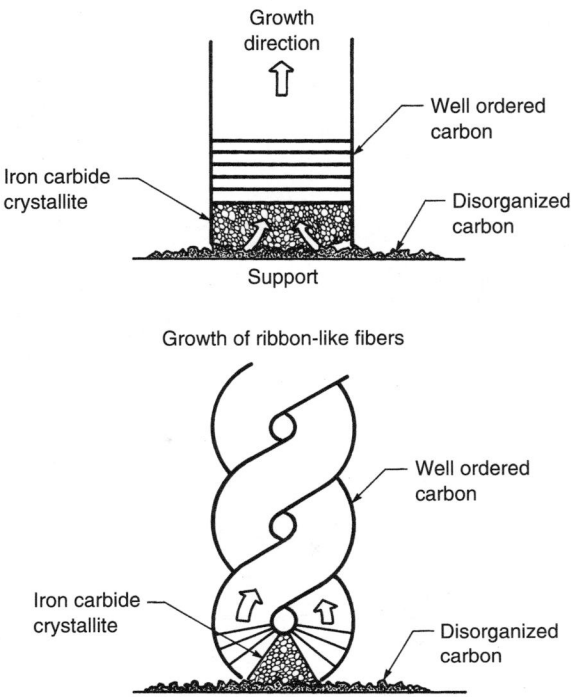

Figure 8.21 Schematic of ribbon and braided carbon filament morphologies. *Source:* Reprinted with permission from Boehm HP, Carbon from carbon monoxide disproportionation on nickel and iron catalysts: Morphological studies and possible growth mechanisms, *Carbon*, 11, 583–590, 1973. Copyright 1973, Elsevier.

8.4 MODE OF TENSILE FAILURE

Filaments do not fail in a catastrophic manner due to the nested basal plane structure and under-go a pull-out or sword-in-sheath failure. Under a tensile load, a surface initiated crack is arrested after progressing through several layers into the fiber but, may well crack rapidly around the circumference. The stress concentration within the fiber increases and the process is repeated at an internal flaw elsewhere along the length until eventually, the diameter left cannot sustain the applied load and a pull-out failure occurs (Figure 8.23). The brittle failures of a PAN and a pitch based carbon fiber are compared in Figure 8.24 with a pull-out failure of a VGCF treated at 2200°C [51].

Madronero *et al* [52] give a diffusion model for the sword-in-sheath failure mode of VGCF.

8.5 MECHANICAL PROPERTIES

In the stress-strain curve of a VGCF, there is a toe-in corresponding to an apparent modulus increase of 28%, termed strain stiffening, which occurs due to the improvement of the orientation of the fiber's graphitic planes as the load is applied. A mean value for E [51] of fibers under 10 μm diameter is 237±49 GPa. Ruland [53] pictured graphite fibers

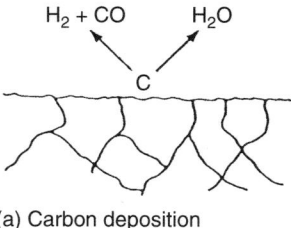
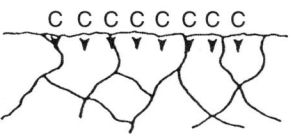

(a) Carbon deposition

(b) Carbon diffusion

(c) Carbide precipitation
1) "Pearlite" like structure
2) Strain causes surface breakup

(d) Surface break-up

(e) Filaments develop from surface nodules

Figure 8.22 Proposed initiation mechanism for development of carbon filaments on an α-iron foil. *Source:* Reprinted with permission from Sacco A Jr., Thacker P, Chang TN, Chiang ATS, The initiation and growth of filamentous carbon from α-iron in H_2, CH_4, H_2O, CO_2 and CO gas mixtures, *J Catal*, 85, 224–236, 1984. Copyright 1984, Elsevier.

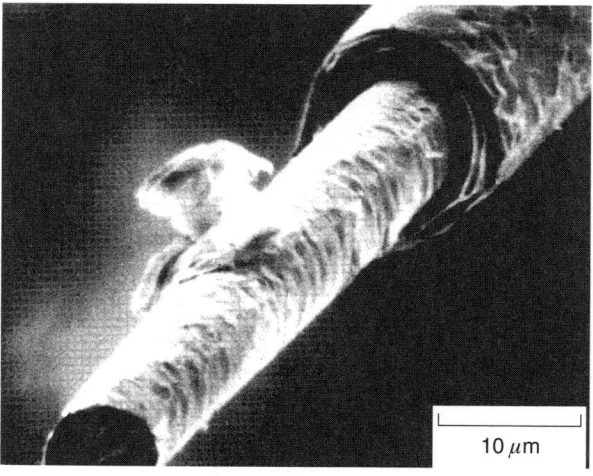

Figure 8.23 SEM of a fiber which has suffered a sword-in-sheath failure. *Source:* Reprinted from Tibbetts GG, Beetz CP Jr., General Motors Research Report No. GMR-5550, 1986 (un-published), Dresselhaus MS, Dresselhaus G, Sugihara K, Spain IL, Goldberg HA, *Graphite Fibers and Filaments*, Springer-Verlag, Berlin, 151, 1988.

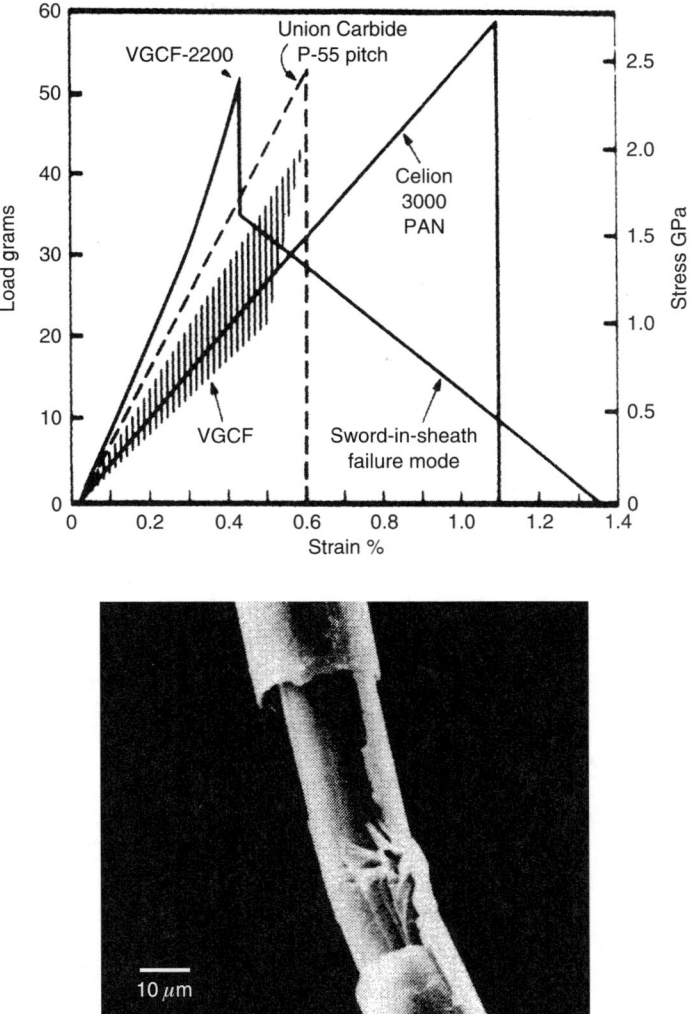

Figure 8.24 Fracture behavior of VGCF compared with that of PAN and pitch based carbon fibers. Also shown is data for VGCF heat treated to 2200°C. The fiber's annular structure prevents brittle failure, yielding instead the sword-in-sheath non-catastrophic failure. *Source:* Reprinted from Tibbetts GG, Beetz CP Jr., Mechanical properties of vapor grown carbon fibers, *J Phys D: Appl Phys*, 20, 292, 1987.

comprising wrinkled graphitic planes nesting together at an angle ϕ from the fiber's axis. The relationship

$$\frac{1}{E} = S_{11}\frac{\cos^2\phi}{\cos\phi} + \frac{\sin^2\phi}{\cos\phi}$$

applies where k is the unwrinkling compliance of the basal planes. A one-parameter fit determines the value of k as 0.0429 GPa^{-1} and a plot of E against fiber diameter (Figure 8.25) shows that E increases as the fiber diameter is reduced, showing an excellent fit for the Ruland theory. E should, however, be independent of the geometry and could be due to the fact that the individual fibers have different structures or simply that the thinner fibers are more graphitized.

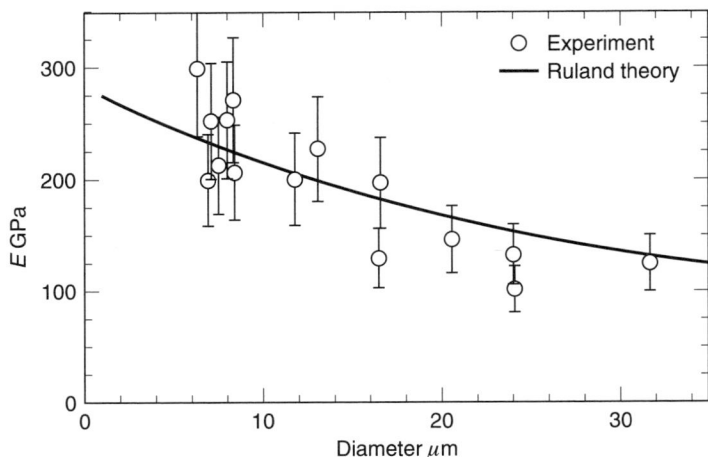

Figure 8.25 Young's modulus of as-grown fibers as a function of the fiber diameter, which have seen a maximum of 1130°C. All of the fibers were grown simultaneously. *Source:* Reprinted from Tibbetts GG, Beetz CP Jr., Mechanical properties of vapor grown carbon fibers, *J Phys D: Appl Phys*, 20, 292, 1987.

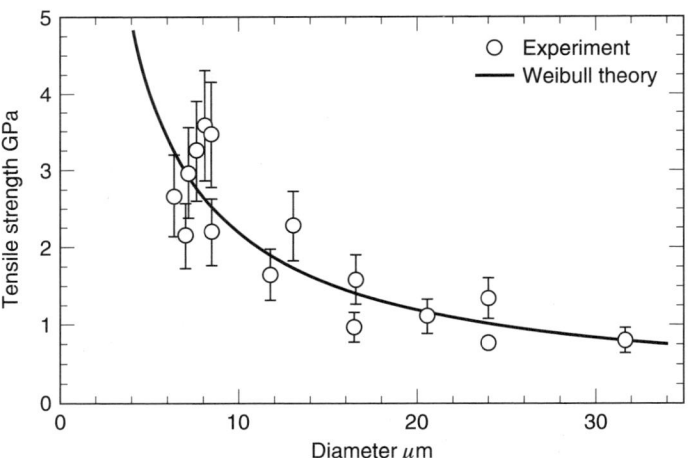

Figure 8.26 Tensile strength as a function of diameter. The optimal Weibull fit for $m = 2.304$ is compared to experimental values. *Source:* Reprinted from Tibbetts GG, Beetz CP Jr., Mechanical properties of vapor grown carbon fibers, *J Phys D: Appl Phys*, 20, 292, 1987.

Weibull [54] introduced a model that used the weakest link principle, where the probability of survival of a fiber of length nL is the same as the nth power of the survival of a fiber of length L. The model is based on the fact if there is a fatal flaw within the tested length, then one can readily find the probability of finding a fatal flaw within the tested volume:

$$F = 1 - \exp\left(-Lr^2\left(\frac{\sigma}{\sigma_o}\right)^m\right)$$

where F is the cumulative probability of failure of a fiber of length L and radius r, σ is the stress on the fiber and σ_o the scale parameter.

Integration of this equation gives the mean strength σ:

$$\sigma = \sigma_o L^{\frac{-1}{m}} r^{\frac{-2}{m}} \Gamma\left(1 + \frac{1}{m}\right)$$

where Γ is the gamma function. Figure 8.26 gives a good fit between tensile strength and diameter when $m = 2.304$. The tensile strength of fibers near 7.5 µm diameter have an average value of 2.92 ± 0.42 GPa.

The physical and mechanical properties of VGCF [55–57] and their use in composites [58,59] are given.

REFERENCES

1. Tibbetts GG, a discussion contribution, GG Tibbetts, Carbon Fibers, Figueiredo JL, Bernardo CA, Baker RTK, Hüttinger KJ eds., *Carbon Fibers Filaments and Composites*, Kluwer, Dordrecht, 560, 1990.
2. Dresselhaus MS, Dresselhaus G, Sugihara K, Spain IL, Goldberg HA, *Graphite Fibers and Filaments*, Springer-Verlag, Berlin, 10, 1988.
3. Gadelle P, a discussion contribution, Tibbetts GG, Carbon Fibers, Figueiredo JL, Bernardo CA, Baker RTK, Hüttinger KJ eds., *Carbon Fibers Filaments and Composites*, Kluwer, Dordrecht, 560, 1990.
4. Tibbetts GG, Vapor-grown carbon fibers: status and prospects, *Carbon*, 27(5), 745–747, 1989.
5. Crocker P, a discussion contribution, Tibbetts GG, Carbon Fibers, Figueiredo JL, Bernardo CA, Baker RTK, Hüttinger KJ eds., *Carbon Fibers Filaments and Composites*, Kluwer, Dordrecht, 561, 1990.
6. Figueiredo JL, Bernardo CA, Baker RTK, Hüttinger KJ eds., *Carbon Fibers Filaments and Composites*, Baker RTK, *Electron Microscopy Studies of the Catalytic Growth of Carbon Filaments*, Kluwer, Dordrecht, 414, 1990.
7. Varshavskii VY, Carbon fibres from the gas phase (review), *Fibre Chemistry*, 28(2), 65–72, 1996.
8. Ishioka M, Okada T, Matsubara K, Formation of vapor-grown carbon-fibers in $CO-CO_2-H_2$ mixtures. 1. Influence of carrier gas composition, *Carbon*, 30(6), 859–863, 1992.
9. Masuda T, Mukai SR, Hashimoto K, The production of long vapor grown carbon-fibers at high growth-rates, *Carbon*, 30(1), 124–126, 1992.
10. Masuda T, Mukai SR, Hashimoto K, The liquid pulse injection technique—a new method to obtain long vapor-grown carbon-fibers at high growth-rates, *Carbon*, 31(5), 783–787, 1993.
11. Wallenberger FT, Nordine PC, Strong, pure and uniform carbon-fibers obtained directly from the vapor-phase, *Science*, 260(5104), 66–68, 1993.
12. Jayasankar M, Chand R, Gupta SK, Kunzru D, Vapor-grown carbon-fibers from benzene pyrolysis, *Carbon*, 33(3), 253–258, 1995.
13. Endo M, Takeuchi K, Kobori K, Takahashi K, Kroto HW, Sarkar A, Pyrolytic carbon nanotubes from vapor-grown carbon fibers, *Carbon*, 33(7), 873–881, 1995.
14. Tibbetts GG, Lengths of carbon fibres grown from iron catalyst particles in natural gas, *J Cryst Growth*, 73, 431, 1985.
15. Tibbetts GG, Rodda EJ, High temperature limit for the growth of carbon filaments on catalytic iron particles, *Mater Res Soc Symp Proc*, 111, 49, 1988.
16. Tibbetts GG, Vapor-grown carbon fibers, Figueiredo JL, Bernardo CA, Baker RTK, Hüttinger KJ eds., *Carbon Fibers Filaments and Composites*, Kluwer, Dordrecht, 77, 79, 1990.
17. Koyama T, Endo MT, Method for manufacturing carbon fibres by a vapour phase process, Japanese Patent 1982-58, 966, 1985.

18. Hatano M, Ohsaki T, Arakawa K, Graphite whiskers by new process and their composites, *Advancing Technology In Materials and Processes, National SAMPE Symposium*, 30, 1467, 1985.
19. Endo M, Shikata M, *Ohyo Butsuri*, (In Japanese), 54, 507, 1985.
20. Benissad F, Gadelle F, Coulon P, Bonnetain L, Influence of catalyst precursor on the lengths of vapor grown carbon fibers, *Proc Int Conf Carbon*, Newcastle upon Tyne, 307–309, 1988.
21. Gadelle P, The growth of vapour deposited carbon fibres, Figueiredo JL, Bernardo CA, Baker RTK, Hüttinger KJ eds., *Carbon Fibers Filaments and Composites*, Kluwer, Dordrecht, 95–110, 113, 115, 1990.
22. Endo M, Koyama T, *Kotai Butsuri* (In Japanese), 12, 1, 1977.
23. Kato T, Haruta K, Kusakabe K, Morooka S, *Carbon*, 30(7), 989–994, 1992.
24. Baker RTK, Chludzinski JJ, *J Catal*, 18, 164, 1980.
25. Ishioka M, Okada T, Matsubara K, Formation of vapor-grown carbon-fibers in $CO-CO_2-H_2$ mixtures. 2. Influence of catalyst, *Carbon*, 30(6), 865–868, 1992.
26. Serp P, Madronero A, Figueiredo JL, Production of vapor-grown carbon-fibers: influence of the catalyst precursor and operating conditions, *Fuel*, 78(7), 837–844, 1999.
27. Cui S, Li YD, Zhang L, Vapor-grown carbon fiber by Ni catalyzed pyrolysis of methane, *Chinese Science Bulletin*, 42(5), 439–440, 1997.
28. Li YD, Chen JL, Chang L, Catalytic growth of carbon fibers from methane on a nickel-alumina composite catalyst prepared from Feitknecht compound precursor, *Applied Catalysis A General*, 163(1–2), 45–57, 1997.
29. Tibbetts GG, Balogh MP, Increase in yield of carbon fibers grown above the iron/carbon eutectic, *Carbon*, 37(2), 241–247, 1999.
30. Baker RTK, Electron microscopy studies of the catalytic growth of carbon filaments, Figueiredo JL, Bernardo CA, Baker RTK, Hüttinger KJ eds., *Carbon Fibers Filaments and Composites*, Kluwer, Dordrecht, 405–415, 418, 1990.
31. Ishioka M, Okada T, Matsubara K, Endo M, *Carbon*, 30(6), 859–863, 1992.
32. Dresselhaus MS, Dresselhaus G, Sugihara K, Spain IL, Goldberg HA, *Graphite Fibers and Filaments*, Springer-Verlag, Berlin, 11, 1988.
33. Baker RTK, Barber MA, Harris PS, Feates FS, Waite RJ, *J Catal*, 26, 51, 1972.
34. Baker RTK, Barber MA, Harris PS, Feates FS, Waite RJ, *J Catal*, 30, 86, 1973.
35. Baker RTK, Harris PS, *J Phys*, E5, 793, 1973.
36. Oberlin A, Endo M, Koyama T, *Carbon*, 14, 133, 1976.
37. Endo M, Katoh A, Sugiura T, Shiraishi M, High resolution electron microscopy of vapour grown carbon fibres obtained by ultra-fine fluid catalyst, *Extended Abstracts of the 18th Biennial Conference on Carbon*, Worcester, MA (American Carbon Society), 151–152, 1987.
38. Baker RTK, Waite RJ, Formation of carbonaceous deposits from the platinum–iron catalyzed decomposition of acetylene, *J Catal*, 37, 101–105, 1975.
39. Sacco A Jr., Carbon deposition and filament initiation and growth mechanisms on iron particles and foils, Figueiredo JL, Bernardo CA, Baker RTK, Hüttinger KJ eds., *Carbon Fibers Filaments and Composites*, Kluwer, Dordrecht, 459–505, 1990.
40. Jaeger H, Behrsing T, The dual nature of vapour-grown carbon fibres, *Composites Science and Technology*, 51, 231–242, 1994.
41. Serp P, Figueiredo JL, An investigation of vapor-grown carbon fiber behaviour towards air oxidation, *Carbon*, 35(5), 675–683, 1997.
42. Serp P, Figueiredo JL, Bertrand P, Issi JP, Surface treatments of vapor-grown carbon fibers produced on a substrate, *Carbon*, 36(12), 1791–1799, 1998.
43. Darmstadt H, Summchen L, Ting JM, Roland U, Kaliaguine S, Roy C, Effects of surface treatment on the bulk chemistry and structures of vapor grown carbon fibers, *Carbon*, 35(10–11), 1581–1585, 1997.
44. Darmstadt H, Roy C, Kaliaguine S, Ting JM, Alig RL, Surface spectroscopic analysis of vapour grown carbon fibres prepared under various conditions, *Carbon*, 36, 1183–1190, 1998.
45. Serp P, Figueiredo JL, A microstructural investigation of vapor-grown carbon fibers, *Carbon*, 34(11), 1452–1454, 1996.

46. van Hattum FWJ, Serp P, Figueiredo JL, Bernardo CA, The effect of morphology on the properties of vapor-grown carbon fibers, *Carbon*, 35(6), 860–863, 1997.
47. van Hattum FWJ, BenitoRomero JM, Serp P, Madronero A, Bernardo CA, Morphological, mechanical and interfacial analysis of vapor-grown carbon fibres, *Carbon*, 35(8), 1175–1183, 1997.
48. Boehm HP, Carbon from carbon monoxide disproportionation on nickel and iron catalysts: Morphological studies and possible growth mechanisms, *Carbon*, 11, 583–590, 1973.
49. Sacco A Jr., Thacker P, Chang TN, Chiang ATS, The initiation and growth of filamentous carbon from α-iron in H_2, CH_4, H_2O, CO_2 and CO gas mixtures, *J Catal*, 85, 224–236, 1984.
50. Tibbetts GG, Beetz CP Jr., General Motors Research Report No. GMR-5550, 1986 (unpublished), Dresselhaus MS, Dresselhaus G, Sugihara K, Spain IL, Goldberg HA, *Graphite Fibers and Filaments*, Springer-Verlag, Berlin, 151, 1988.
51. Tibbetts GG, Beetz CP Jr., Mechanical properties of vapor grown carbon fibers, *J Phys D: Appl Phys*, 20, 292, 1987.
52. Madronero A, Verdu M, Froyen L, Dominguez M, A diffusion model for sword in sheath failure mode in vapor grown carbon fibers, *Advanced Performance Materials*, 4(3), 305–315, 1997.
53. Ruland W, The relationship between preferred orientation and Young's modulus of carbon fibres, *Appl Polym Symp*, 9, 293, 1969.
54. Weibull W, A statistical distribution function of wide applicability, *J Appl Mech*, 18, 293, 1951.
55. Tibbetts GG, Doll GL, Gorkiewicz DW, Moleski JJ, Perry TA, Physical properties of vapor-grown carbon-fibers, *Carbon*, 31(7), 1039–1047, 1993.
56. Jacobsen RL, Tritt TM, Guth JR, Ehrlich AC, Gillespie DJ, Mechanical properties of vapor-grown carbon-fiber, *Carbon*, 33(9), 1217–1221, 1995.
57. Madronero A, Ariza E, Martins M, Mesa G, Froyen L, Improving vapour grown carbon fibres strength by addition of acetylene to the precursor atmosphere, *European Journal of Solid State and Inorganic Chemistry*, 34(5), 445–456, 1997.
58. Ting JM, Lake ML, Vapor-grown carbon-fiber-reinforced carbon composites, *Carbon*, 33(5), 663–667, 1995.
59. Carneiro OS, Covas JA, Bernardo CA, Caldeira G, van Hattum FWJ, Ting JM, Alig RL, Lake ML, Production and assessment of polycarbonate composites reinforced with vapour-grown carbon fibres, *Composites Science and Technology*, 58(3–4), 401–407, 1998.

CHAPTER 9

Surface Treatment and Sizing of Carbon Fibers

9.1 INTRODUCTION

Carbon fibers are produced with high strength and modulus and can be incorporated in a range of matrix materials. When a load is applied to the carbon fiber composite, the stress is transferred from one carbon filament to another via the matrix material. If a weak fiber-resin bond is present, then it will result in poor mechanical properties, such as low interlaminar shear strength (ILSS), which is attributed to a lack of bonding between the resin matrix and fiber filaments. Generally, this problem can be overcome by some form of surface treatment of the fiber, but if the bond is too strong, then the composite may become brittle and weak. However, too little treatment and the composite will remain weak, so it is most important to establish an optimum level of surface treatment for a given fiber and matrix system. High modulus fibers will require maximum treatment in order to give acceptable properties.

Treatments can be applied by a batch or continuous process and obviously, fiber production will favour the latter. Many methods of oxidative surface treatment have been used including gaseous, solution, electrochemical, plasma and catalytic.

Various forms of non-oxidative surface treatments can be used, including the deposition of an active form of carbon, the deposition of pyrolytic carbon and by grafting a polymer onto the fiber surface.

Factors which must be taken into consideration when choosing a system include the length of time that the surface treatment will take, how practical it is to operate, the cost of the treatment, whether the weight loss is significant and, if the process involves a wet system, then the incorporation of a drying stage.

9.2 OXIDATIVE PROCESSES

Initially, carbon fiber was only available in a staple form and batch surface treatment processes had to be used, but with the advent of continuous carbon fiber, these were replaced by continuous processes.

9.2.1 Gas phase oxidation

Air, oxygen diluted with an inert gas, nitrous oxide, nitrous dioxide, ozone, steam or carbon dioxide may be used as the gas phase. Herrick et al [1] treated rayon-based carbon fiber for 16 h in air at 500°C with little improvement in ILSS, but raising the temperature to 600°C gave a 45% improvement in ILSS, although it was accompanied by a serious weight loss. Bobka and Lowell [2] found etching with pure oxygen was non-uniform due to the uneven penetration of the oxygen into the carbon fiber tow bundle and as the process was controlled by the oxygen diffusion rate at the treatment temperature (400–600°C), it resulted in enhanced treatment on the outside of the bundle with little treatment within the bundle. More adequate control was claimed by Wadsworth and Watt [3] with a 0.05–6.0% weight loss, obtaining a good key for the matrix resin establishing pits (30–40 nm deep), which coalesced producing up to 100 nm grooves running parallel to the fiber axis. Sach and Bromley [4] were able to slow the reaction down by introducing an inhibitor such as SO_2, a halogen, or a halogenated hydrocarbon (e.g. CCl_4). Chlorine was preferred as an inhibitor and about 0.5% $^v/_v$ Cl_2 controlled the reaction, reducing the activation energy from 50 to 20 $kcal\,mol^{-1}$. This enabled a higher temperature of 950°C to be used with only a 3% weight loss and the reaction was conveniently controlled by measuring the CO and CO_2 evolved. Novak [5] observed that pitting reduced the fiber strength. Scola and Basche [6] used a treatment time of up to 60 s in N_2 containing 0.1–1.8% O_2 at 1000–1500°C without significant degradation of mechanical properties of the fiber. Air or O_2 diluted with an inert gas was circulated at various flow rates by Clark et al [7], with the treatment being carried out at about 800°C, but the process was very difficult to control due to the high activation energy and a 5–10% weight loss was common. A Type II fiber underwent oxidation at a rate some 600 times greater than a Type I fiber, but the activation energies for both fibers was 171 $kJ\,mol^{-1}$, suggesting that the Type II fiber had a less perfect structure and more active sites. The ILSS for a given weight loss is shown in Figure 9.1 for Type I fiber and TEM confirmed that the pit and groove formation was related to edge planes and imperfections arising from point defects present on the fiber surface or on the exposed basal planes. Pit formation corresponded to a rapid increase in surface area, gradually reducing as the grooves were formed. An increase in ILSS was attributed to this increase in surface area and in part, also due to the formation of surface groups. The pits were adjudged by

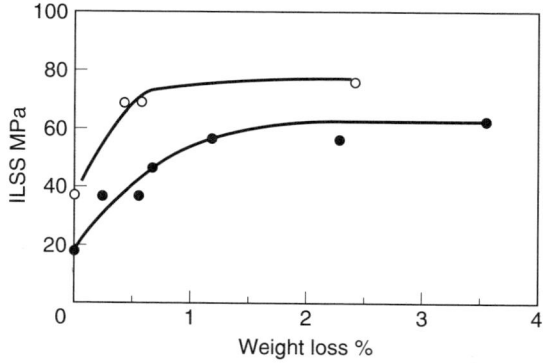

Figure 9.1 ILSS of Type I carbon fiber composite as a function of fiber weight loss. *Source:* Reprinted with permission from Clark D, Wadsworth NJ, Watt W, The surface treatment of carbon fibres for increasing the interlaminar shear strength of cfrp, *Carbon Fibers—Their Place in Modern Technology*, The Plastics Institute, London, 44–51, 1974. Copyright 1974, Maney Publishing (who administers the copyright on behalf of IOM Communications Ltd, a wholly owned subsidiary of the Institute of Materials, Minerals & Mining).

McKee and Mimeault [8] to be produced only in the initial stages of oxidation, with the surface becoming smoother and much thinner as concentric layers of carbon were removed throughout the oxidation process. The presence of Pb or Cu salts deposited on the fiber surface prior to oxidation initiated a catalytic etching, which needed careful control in order to avoid rapid degradation.

An increase in specific surface area and ILSS was obtained by Druin et al [9] by using an inert atmosphere containing a small amount of O_2.

Molleyre and Bastick [10] determined density and specific area to investigate the surface topography during oxidation and found that only the outer graphitic layers were attacked, resulting in an enhancement of the grooves parallel to the fiber axis and producing a higher surface rugosity (ratio of the specific surface area to the geometric area), suggesting that the increase in ILSS is principally due to the increase in interfacial area. The same authors [11] found similar results when treating carbon fibers with CO_2 at 850–925°C, detecting a marked effect of the presence of metallic impurities, such as 640 ppm Fe, resulting in the replacement of the longitudinal striations by a random formation of holes that were attributed to catalytic oxidation.

Untreated, unsized Toray T300 fiber was oxidized in air at 420°C by Vukov [12] with a 0.9% weight loss after 7 h of treatment, obtaining an increase in surface acid groups, which was measured by titration with 1M NaOH (Figure 9.2) and the measured surface area increased from 0.5 to 29 $m^2 g^{-1}$. When heated for 1 h in Ar at 950°C, there was a considerable reduction in the acid group values, whereas the surface area basically remained the same and this was attributed to the reduction in surface oxides with the surface area staying unchanged. Epoxy composites fabricated from fiber, treated for 2 h gave ILSS values increasing from 55 MPa to 70 MPa, but showing no significant increase on post treatment at 950°C. Hence, Vukov postulated that the acid surface groups were an important factor in the fiber/resin interface, forming chemical bonds with the epoxy resin and enhancing the fiber wettability and added that the increase in surface area also assisted in adhesion.

Steam and CO_2 were used by Ryu et al [13] and Alcaniz-Monge et al [14]. The latter used pitch based carbon fibers treated at 1433°C using an equal volume mixture of steam as well as N_2 and CO_2. The effect on the fiber diameter is shown in Figure 9.3 with a significant change using steam but showing only a slight initial loss with CO_2, whilst the effect on the tensile strength (Figure 9.4) shows a continual loss of strength with burn-off in CO_2, whereas

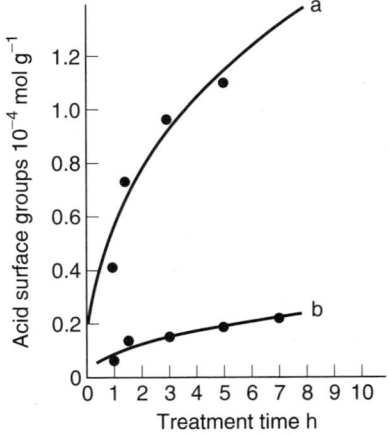

Figure 9.2 Concentration of acid surface groups vs treatment time: a) treated fibers, b) post treated fibers. *Source:* Reprinted with permission from Vukov AJ, *J Serb Chem Soc*, 55, 333, 1990. Copyright 1990, Serbian Chemical Society.

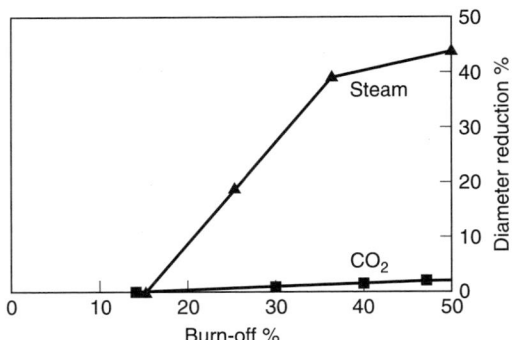

Figure 9.3 Variation of fiber diameter with burn-off. *Source:* Reprinted with permission from Alcaniz-Monge J, Cazorla-Amoros D, Linares-Solano A, Yoshida S, Oya A, *Carbon*, 32, 1277, 1994. Copyright 1994, Elsevier.

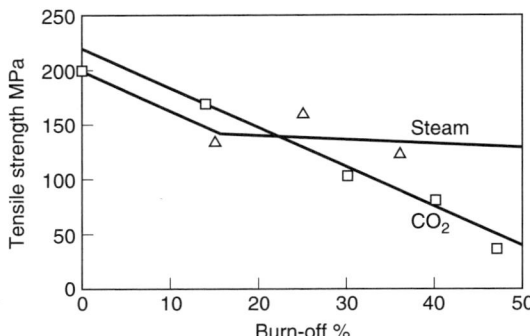

Figure 9.4 Variation of tensile strength with burn-off. *Source:* Reprinted with permission from Alcaniz-Monge J, Cazorla-Amoros D, Linares-Solano A, Yoshida S, Oya A, *Carbon*, 32, 1277, 1994. Copyright 1994, Elsevier.

steam gave an initial loss of up to about 15% and then remained unchanged. The authors believe that steam results in fiber burn-off producing microporosity and a reduction in fiber diameter, whilst CO_2 produces microporosity without a reduction in fiber diameter. However, with CO_2, internal pitting takes place with a significant reducing effect on fiber strength.

9.2.2 Liquid phase oxidation

Wright [15] and others have listed a number of liquid phase oxidizing agents which have been used to surface treat carbon fibers such as HNO_3 [17–22], NaOCl [7,17], $KMnO_4$ [17,23], $NaClO_3$ [17], $Na_2Cr_2O_7$ [17] and $NaIO_4$ [24].

Goan [17], using Fortafil 5Y (YM = 345 GPa) fiber found that the most effective treatment was refluxing in 10% $NaClO_3$ / 25% H_2SO_4 and a composite from Epon 1028/ BF_3 gave a maximum ILSS of 100 MPa (14,500 lbf.in^{-2}) after about 5 min of treatment, with the surface area increasing by about 30%. With increased treatment time, the ILSS showed a progressive fall, whereas the flexural strength increased. Goan [25], using various Fortafil fibers and ERLA 4617/DDM showed different behavior, emphasizing the importance of the choice of resin system (Figure 9.5) and clearly showing the effect of fiber modulus. The impact energy reduced dramatically with treatment time, one fiber type falling from 2.58J to 0.95J in 1 min, underlining the importance of correct balance

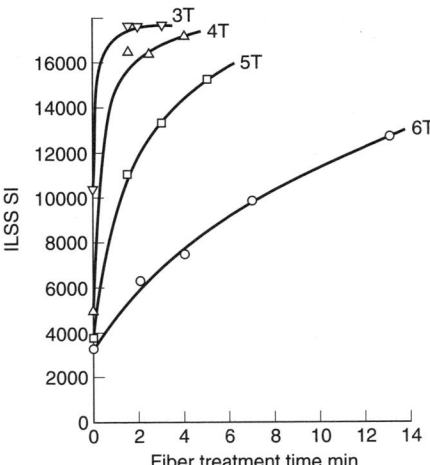

Figure 9.5 Effect of time of surface treatment on properties of various Fortafil composites, Fortafil 3T (YM = 207 GPa), Fortafil 4T (YM = 262 GPa), Fortafil 5T (YM = 331 GPa), Fortafil 6T (YM = 400 GPa). *Source:* Reprinted with permission from Goan JC, Martin TW, Prescott R, The influence of interfacial bonding on the properties of carbon fiber composites, *28th Ann Tech Conf Reinf Plast/Comp Inst*, 28(21–B), 1–4, 1973. Copyright 1973, The Society of Plastic Engineers.

between ILSS and other mechanical properties for a given resin system. However, refluxing in such an aggressive solution can hardly be a practical commercial proposition. Similarly, refluxing with HNO_3 solutions though not practical, can nevertheless, be a useful technique to establish the effect of treatment on the fiber surface. Boiling 68% HNO_3 attacked HT fiber much more vigorously than an HM fiber, which formed only one-tenth of the acidic groups [18]. HM fiber can be readily oxidized by $KMnO_4$ in conc. H_2SO_4 (Hummer's reagent), forming lamellar graphitic oxide with a significant increase in the number of acidic groups formed, but having little effect on the surface area [23]. With a short treatment time, the effect is restricted to the outermost layers, but heating to 200°C causes the lamellar oxide to exfoliate.

A system used commercially was treatment with NaOCl (bleach liquor) solutions, where the active species was HOCl, controlled by adjusting the pH of the solution [26–30].

When Cl_2 is dissolved in water it is rapidly hydrolyzed:

$$Cl_2 + H_2O \rightarrow H^+ + Cl^- + HOCl \text{ (100\% at pH 4–5)}$$
$$\updownarrow$$
$$H^+ + OCL^- \text{ (appreciable above pH 7 and 100\% at pH 10)}$$

Bleach liquor decomposes by one of two methods:

1. $3OCl^- \rightarrow ClO_3^- + 2Cl^-$ (maximum at about pH 7)
2. $2OCl^- \rightarrow O_2 + 2Cl^-$ (catalytic decomposition, practically unaffected by pH)

Adjustment of pH can be achieved by the addition of acetic or phosphoric acid, but too much and free Cl_2 will be evolved, whilst too little fails to liberate sufficient free HOCl (Figure 9.6).

Workers at Harwell [26,31] preferred to use a diluted bleach liquor containing 1–3% available Cl_2, whilst operating at a pH of 4.0–6.0 and a temperature of 40–55°C. Due to the corrosive nature of the solutions and the evolution of toxic Cl_2, titanium was a preferred

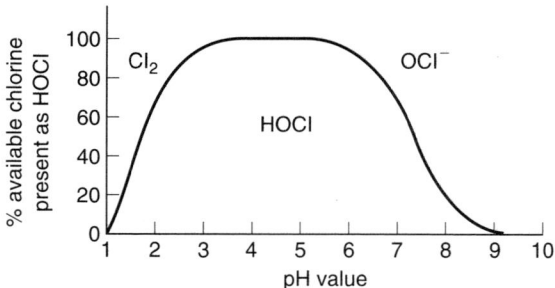

Figure 9.6 Effect of pH on the % available chlorine as HOCl.

Table 9.1 Effect of treatment time on the ILSS of HT staple fiber (60% C/F in 828/MNA/BDMA)

Treatment time h	ILSS (MPa)	Type of shear failure
0	26	Single shear
8	75	Single shear
12	72	Single shear
18	77	Single shear
24	76	Single shear
48	77	Multi-shear, single shear and tension
72	75	Tension

material of construction and it was important to ensure that the carbon fiber surface was continuously replenished with the treatment liquor by using controlled agitation or some form of ultrasonics. Treatment times depended on the fiber type (e.g., 12 h for Type HT and 3 days for Type HM). Treatment must not be prolonged, as otherwise the fiber will be embrittled, manifested as tension failures in the ILSS specimen (Table 9.1).

After treatment, the fiber must be thoroughly washed with warm water to remove deleterious Na ions, followed by rinsing with de-ionized water and finally drying.

9.2.3 Anodic oxidation

When carbon fibers were introduced in a continuous form, anodic oxidation became the favoured route for surface treatment, using the conductive property of carbon fiber to act as an anode in a suitable electrolyte bath. A potential is applied to the fiber, sufficient to liberate O_2 on the surface. The typical equipment used in this process is described in Chapter 10, Section 10.14.

In acid solutions, O_2 is produced at the anode surface by the decomposition of water [32]:

$$2H_2O \rightleftharpoons 4H^+ + 4e^- + O_2$$

For mildly alkaline solutions, a similar process has been represented [33]:

$$H_2O \rightarrow HO + H^+ + e$$
$$2HO \rightarrow H_2O + O$$
$$2O \rightarrow O_2$$

In alkaline solutions, the O_2 is viewed as being produced by the discharge of OH^- ions [32]:

At the anode, $\quad 4OH^- \rightarrow 2H_2O + O_2 + 4e^-$

and at the cathode, $\quad 4e^- + 4H_2O \rightarrow 2H_2 + 4OH^-$

Typical electrolytes include nitric acid (HNO_3) [34], sulfuric acid (H_2SO_4) [35,36], sodium chloride (NaCl) [31,36], potassium nitrate (KNO_3) [37], sodium hypochlorite (NaOCl) [36], sodium hydroxide (NaOH) [34,38–40], ammonium hydroxide (NH_4OH) [41,42], ammonium carbonate ($NH_4(CO_3)_2$) [42], ammonium sulfate (($NH_4)_2SO_4$) [43,44], ammonium bicarbonate (NH_4HCO_3) [42,45], ammonium carbamate ($NH_4NH_2CO_2$) [42], ammonium benzoate ($NH_4C_7H_5O_2$) [42], ammonium dithionate (($NH_4)_2S_2O_6$) [42], ammonium hydrosulfide (NH_4HS) [42], ammonium sulfite (($NH_4)_2SO_2 \cdot H_2O$) [42], ammonium thiosulfate (($NH_4)_2S_2O_3$) [42], ammonium tartrate (($NH_4)_2C_4H_4O_6$) [42] and diammonium hydrogen phosphate (($NH_4)_2HPO_4$) [46].

The concentration of the electrolyte in water must be sufficient to enable the electrolytic process to take place and further electrolyte must be added during the run in order to allow for depletion in the bath due to fiber take-up. A suitable cathode material is a graphite block which is best covered with a fabric bag to prevent the carbon fiber (anode) shorting out. Faraday's laws will apply and the current should be adjusted to provide the correct level of treatment. Type A fiber is readily attacked and a surface dross, akin to colloidal graphite, is formed on the surface of the bath, whereas in order to provide sufficient treatment for a Type HM fiber, relatively heavy currents are required, accompanied by a substantial heating effect, which in turn necessitates cooling.

The choice of electrolyte is very dependent on practical issues. It is necessary for plant operatives to handle fiber in the surface treatment bath and this, therefore, rules out strong acids and alkalis, whilst NaCl will generate Cl_2, which is toxic. Hydrogen liberated at the cathode can form an explosive mixture with air and the fine bubbles can entrap an electrolyte, such as NaOH, providing unpleasant working conditions. Also, NaOH remaining on the fiber in the washing process can act as a water softener and deposit Ca^{2+} ions from the wash water onto the fiber. Ions such as Na^+, if left on the fiber after washing, can cause composites to blister, especially in conditions of high humidity (e.g. boiling water). Hence, it would be most beneficial if any electrolyte remaining on the fiber after washing is removed by volatilization during the drying process. This accents the use of certain ammonium compounds, although the SO_4^{2-}-ion would not be effective due to a higher decomposition temperature.

HM fiber in particular, which necessitates the use of higher treatment currents, can exhibit three types of polarization which can slow the process down:

1. The rate of discharge of OH^- at the anode surface may be greater than the rate of migration of OH^- ions from the bulk of the electrolyte, resulting in a depleted concentration at the interface, which gives rise to concentration polarization.
2. O_2 liberated at the anode surface may be absorbed by it and may inhibit further electrolysis giving rise to absorption polarization.
3. Surface oxidation products may also inhibit further electrolysis resulting in chemical polarization.

These polarization effects can be limited by:

1. Circulation of the electrolyte.
2. Heating the bath to 50°C, this increases the ionic mobility and reduces the surface tension at the fiber surface.

3. Using a multistage treatment process, this helps in providing a fresh surface for treatment.
4. Addition of a surfactant will reduce the surface tension and improve the initial wetting out of the fiber.
5. Use of an AC supply when the AC cycles help to provide a fresh surface. It should be noted that most rectifiers are not true DC and have a slight ripple effect.

Table 9.2 shows the effect of single treatment DC *vs* single treatment AC.

The level of surface treatment is a function only of the number of ions discharged at the fiber surface in its passage through the treatment bath [45]. Harvey [45], using 10% NH_4HCO_3, calculated that at a depth of, say, 100 mm, the potential may fall to about half the value at the surface of the electrolyte, so assuming the current of ions is proportional to the potential, the surface treatment at a depth of 100 mm will only be half as effective as near the surface, with virtually no treatment whatsoever at a depth of 150 mm, supported by visual evidence when observing the rate of O_2 evolution as bubbles released from the fiber surface. This can be circumvented by using a multistage treatment process.

The level of surface treatment must be carefully chosen to satisfy a given fiber type (Figure 9.7) which also shows the influence of the matrix.

Table 9.2 Single treatment DC versus single treatment AC

Treatment	Time min	Current A	Coulombs	% Teepol surfactant	ILSS of 828/MNA/BDMA MPa
DC	1	4	240	–	36
DC	1	4	240	0.1	45
AC	1	4	240	–	56
AC	1	4	240	0.1	68
DC	2	4	480	–	61
DC	2	4	480	–	69
DC	2	4	480	0.1	72
AC	2	4	480	–	72
AC	2	4	480	0.1	68
DC	4	4	960	–	74
DC	4	4	960	–	76
DC	4	4	960	0.1	78
AC	4	4	960	–	66
AC	4	4	960	–	76

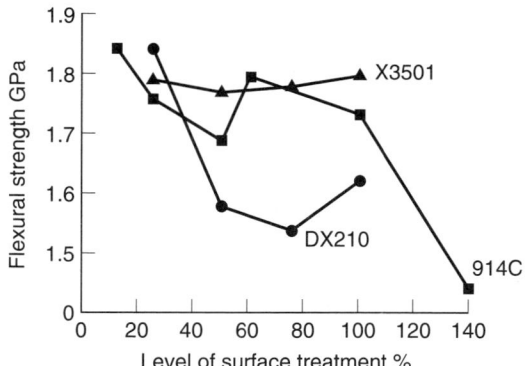

Figure 9.7 Effect of surface treatment on the flexural strength of composites made from Courtaulds HTS fiber and different resin systems.

No consensus has been reached on the roles of physical absorption and chemical bonding when investigating the surface chemistry of carbon fibers and made more difficult by the buried interface. Jones [47] claims that the electrolytic surface treatment process produces a surface on which the known concentration of chemical functionalities cannot be accommodated on the surface of a smooth cylinder. Absorption studies [48] support the fact that erosion could occur and active species can be deposited in the vicinity of intercrystallite voids. Types A and HT fibers have more basal planes that emerge directly to the surface than is the case with HM fiber and hence are more readily surface treated. Hence, it was suggested [49,50] that HM fiber would require an active epoxy group of smaller dimensions that could be accommodated within the micropore.

9.3 PLASMA

Plasma, also called the fourth state of matter, is a partially or fully ionized gas containing electrons, ions and neutral atoms or molecules, where the atoms have so much kinetic energy that the valence electrons are freed by atomic-level collisions [51]. Peebles [52] describes a plasma gas as containing a few parts per million of ions, 2–20% free radicals and a large amount of extremely energetic vacuum-ultraviolet light.

Plasma surface treatment is a dry reaction process and depending on the process conditions, can have the following effects, which may occur concurrently on a carbon fiber surface:

1. Cleans the outside, creating a hydrophilic surface for enhanced bonding, where, if for example O_2 is used as the plasma gas, then the atomic O_2 formed combines with organic contaminants on the surface to form H_2O and CO_2, which are then removed by evacuation during plasma processing leaving a clean surface.
2. Removes the surface layer by a micro-etching process.
3. Penetrates the top few molecular layers (100 about 10 nm) and modifies the surface, creating a new surface chemistry, enabling improved interfacial adhesion in composites.
4. Undertakes CASING (Crosslinks via Activated Species of Inert Gases), where two or more parallel polymer chains are cross-linked using an inert gas, such as O_2 free Ar or He, and in the absence of free radical scavengers such as O_2, byproducts will enable a free radical to form a bond onto an adjacent chain and initiate crosslinking.

Typical gases used to create a plasma include:

1. Air—which is more reactive than N_2 or NH_3; the groups —COOH and —OH have been detected on PAN based fibers, but only —OH on pitch based fibers.
2. O_2—this forms O_2^*, which attacks the edge and basal planes accompanied by an increase in the active and total surface areas. The groups —OH, =O, quinone, —C=O, —COOH, —COOX, —CO_3 and lactone have been detected.
3. NH_3—can introduce aliphatic and aromatic amines (—CNH_2) and some imines (—C=NH), but will only attack edge planes.
4. N_2—has an effect similar to NH_3.
5. Ar—will enhance the active sites but not the total surface area.

Table 9.3 lists some of the plasma that have been investigated.

Model research operating conditions for a plasma reactor could comprise a 13.5 MHz RF generator for excitation with matching network and a power of 100 W, operating at a pressure of 2–50 Pa, with a gas flow rate of 8–40 $cm^3 min^{-1}$ and contact time of 20 s–20 min

Table 9.3 Selected systems used for plasma polymerization

Plasma	Reference
O_2	[53]
	[54]
	[55]
	[56]
	[57]
N_2	[58]
Air	[59]
NH_3	[60,61]
NH_3/N_2	[62]
Acetylene/O_2	[63]
Ethylene/NH_3	[64]
	[65]
Styrene/AN	[66]
Organic gas	[67]
	[68]
	[69]
Allylamine	[70]

[51,53,71]. It is possible to use a microwave-generated plasma [72], a technique in which the grooves present in an untreated M40 fiber are almost completely removed after about 12 s treatment in air [73]. However, it has been suggested that different results may be obtained, depending on whether RF or microwave generation procedures are adopted.

Ultra-high vacuum plasma treatment [74] gives a 143% increase in interfacial shear strength over an untreated Hercules AU-4 fiber compared with a 66% increase for the commercially treated fiber [75].

The plasma treated surfaces may become de-activated when exposed to air due to the recombination of radicals and have to be protected or incorporated immediately into a resin matrix [60].

Chang [51] found that O_2 and NH_3 plasmas improved the interfacial adhesion between carbon fiber (AU-4) and a bismaleimide (BMI) resin (Matrimide5292). Oxygen produced a greater etching effect, but NH_3 was preferred, producing lesser reduction in fiber strength (2% compared with 21%). With O_2 plasma, the improved level of bonding with BMI can be attributed to improved mechanical keying, improved surface wetting and chemical bonding, whilst NH_3 plasma enhanced the surface energy and provided chemical bonding between the amine groups and BMI.

Weisweiler and Schlitter [71] describe the treatment of Hercules AS-4 and Grafil XAS fibers with acetylene/air plasmas. Figure 9.8 shows the dependence of ILSS on the deposition time of the plasma and Figure 9.9 depicts the increased ILSS values obtained with the higher air/acetylene mixtures.

Plasma treatment of carbon fiber has attracted much recent research and a number of papers have been mentioned in the references [76–93].

9.4 NON-OXIDATIVE SURFACE TREATMENT—WHISKERIZATION

Whiskerization involves the growth of minute single crystals, such as SiC, Si_3N_4 and TiO_2 at right angles to the fiber surface in order to promote bonding between the fiber and resin in a composite. Early work with SiC was undertaken by Prosen and Simon [94], working in

SURFACE TREATMENT AND SIZING OF CARBON FIBERS

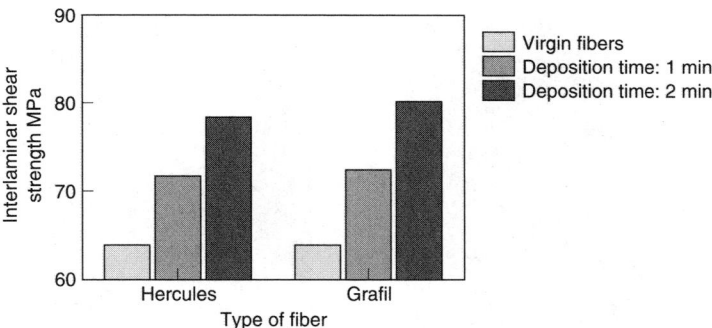

Figure 9.8 Dependence of ILSS of epoxy resin composites on plasma polymer deposition time with 5 sccm acetylene/15 sccm air. *Source:* Reprinted with permission from Weisweiler W, Schlitter K, Surface modification of carbon fibres by plasma polymerization, Figueiredo JL, Bernardo CA, Baker RTK, Hüttinger KJ eds., *Carbon Fibers Filaments and Composites*, Kluwer Academic, Dordrecht, 272, 1989. Copyright 1989, Springer.

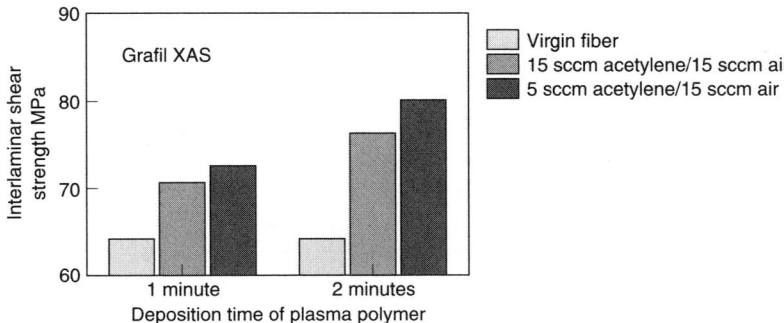

Figure 9.9 Dependence of ILSS of epoxy resin composites containing 60% $^v/_v$ fibers on plasma polymer deposition time. *Source:* Reprinted with permission from Weisweiler W, Schlitter K, Surface modification of carbon fibres by plasma polymerization, Figueiredo JL, Bernardo CA, Baker RTK, Hüttinger KJ eds., *Carbon Fibers Filaments and Composites*, Kluwer Academic, Dordrecht, 273, 1989. Copyright 1989, Springer.

cooperation with Milewski, Shaver and Withers [95]. Prosen reported an increase in the shear strength of graphite fiber composites from 21 MPa to 76 MPa and above, and there is also an improvement in mechanical properties attributed to the increase in surface area. The technique was described by Shaver [96]. The process of growing the whiskers is expensive, which are difficult to wet out with the resin.

Emig *et al* [97] have applied a SiC coating from methyltrichlorosilane onto a pitch based fiber using a molar fraction of 0.5 H_2 to give a nucleated deposit (Figure 9.10) showing an increase in the surface area, but the applied thickness must be over 150 nm to prevent serious oxidation above 750°C.

9.5 EFFECT OF SURFACE TREATMENT ON FIBER PROPERTIES

9.5.1 Introduction

Various workers have made suggestions on the possible effects of surface treatment which include removal of a surface layer, formation of a surface group, acid/base interaction,

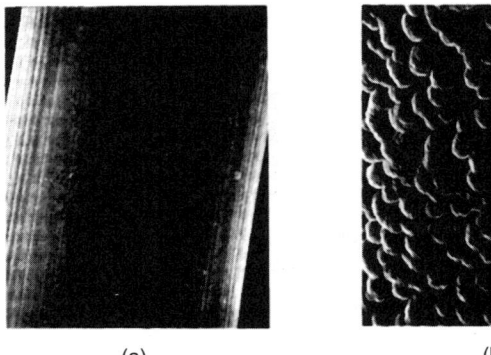

Figure 9.10 SEM photographs of mesophase pitchbased carbon fiber coated with SiC from methyltrichlorosilane: a) No carrier gas, b) 0.5 mol fraction of H_2. *Source:* Reprinted with permission from Emig G, Popovska N, Schoch G, *Thin Solid Films*, 241, 361, 1993. Copyright 1993, Elsevier.

provision of surface roughness aiding mechanical interlocking, covalent bonding, hydrogen bonding, coulomb interaction and van der Waals forces.

9.5.2 The effects of surface treatment

Workers do have divergent opinions about the reasons for improved bonding after surface treatment and a selection is described along with an attempted summary.

Herrick [98] treated Thornel 25 with HNO_3 and found that the ILSS of the resultant composite, the specific surface area and the number of surface oxygen complexes increased. Treatment with H_2 removed the surface oxygen groups accompanied by a distinct reduction in ILSS, suggesting that the surface groups improved the fiber/resin bond.

This work was questioned by Scola and Brooks [99], who, whilst agreeing that the surface area and the amount of surface groups were increased, argued that H_2 reduction of oxidized Hitco HMG 50 did not alter the ILSS, the process having presumably removed the surface groups.

Harvey [100] studied the surface area by gas adsorption (BET) and the adsorption of aqueous solutions of the dyes, methylene blue and metanil yellow, on Courtaulds types A, HT and HM carbon fibers.

Table 9.4 Adsorption of methylene blue and metanil yellow on carbon fibers

Fiber type	Kr BET surface area $\mu mol\, g^{-1}$	Adsorption methylene blue m_{sat} $\mu mol\, g^{-1}$	Adsorption metanil yellow m_{sat} $\mu mol\, g^{-1}$
A-U	0.38	0.22	0.48
A-S	0.45	1.0	0.06
HT-U	0.38	0.31	0.42
HT-S	0.43	0.79	0.38
HM-U	0.42	0.39	0.53
HM-S	0.41	0.39	0.44

Source: Reprinted with permission from Harvey J, The characterization of carbon fibre surfaces by dye adsorption, *RAE Tech Memo Mat*, 231, Sep 1975. Copyright 1975, QinetiQ Ltd.

Harvey found that the surface area, as determined by BET, of untreated fiber was not significantly different from that of the same fiber when surface treated. However, surface treated fibers always adsorbed more methylene blue than metanil yellow for the same fiber type, suggesting that the decrease in adsorption of metanil yellow may be associated with factors which improve the fiber resin bond (Table 9.4).

The surface complexes on graphitic surfaces form only on the edge surfaces and Rand and Robinson [101] found that the external area of HM fiber is increased by HNO_3 oxidation, but there is no difference between HMU and HMS as supplied by Courtaulds. However, both treatments increase the surface active areas by 3–30-fold and also increase the fraction of surface in the form of edge or active sites. A paper by the same authors [102] supports the mechanism of bonding to surface complexes for PAN based carbon fibers, surface treated by Courtaulds.

Fitzer and co-workers [103] showed that the concentration of reactive fiber surface groups on HM fiber is about one magnitude less than for HT fiber, whilst wetting measurements and nitrogen determinations showed that the adhesion between fiber and matrix is at least 50% chemical in nature. The BET surface area of oxidized HT fiber was about forty times that of HM fiber, suggesting that physical adhesion or mechanical interlocking was not a contributory factor.

Drzal et al [104] examined Hercules untreated and proprietary treated A, HT and HM fibers using 828/MPD matrix and suggested a two-part mechanism. Initially, a weak defect laden surface layer is removed, enabling higher shear loads to be supported and then surface oxygen groups are added, which can interact with the polar epoxy matrix.

Rich and Drzal [105] compared the interfacial parameters of Hercules AS6 and IM6 high strain fibers with AS1 and AS4 intermediate strain fibers and found no difference in the atomic surface concentration. Their morphology was smooth, except for AS4 which had a corrugated typography and a high interfacial shear strength, presumably associated with the corrugated surface. The birefringent stress patterns at the interface were identical, suggesting that both the high strain and intermediate strain fibers fail by similar mechanisms.

Zielke, Hüttinger and Hoffman [106–109] examined Hercules AS-4, Toray T800 and M40 surface treated and unsized fibers and compared them with Tenax HT untreated/unsized; treated/unsized; untreated/unsized and oxidized with O_3. Studies were undertaken with SEM, scanning tunneling microscopy (STM), contact angle measurement (CAM), X-ray photoelectron spectroscopy (XPS) and temperature programed desorption (TPD). These workers showed that a real image of surface structure and surface chemistry could only be obtained by removing surface contaminants with boiling water and careful drying of the fiber. FTIR identified a number of molecular components of the degradation products (Table 9.5). Extraction with water increased the surface concentration of –COOH groups, probably

Table 9.5 Molecular components of the degradation products identified with FTIR

Functionality	Assignment region cm^{-1}	Ozone treated	Industrially treated
—OH	3389–3549	Shoulder	Strong
Ethers/—OH	1085–1190	Very strong	Weak
—COOH	3200–3251	Medium	Shoulder
Aliphatics	2849–2859	Weak	Weak
	2920–2925	Weak	Weak
	2956–2961	Weak	Weak
Lactone/—COOH	1723–1724	Very strong	Hidden
	1220–1265	Very strong	Weak
Quinone	1642–1647	Hidden	Strong
Aromatics (C=C)	1585–1600	Weak	Strong
Carboxylate	1413–1423	Medium	Shoulder
Nitrate	1382–1388	Not present	Medium

Source: Reprinted with permission from Zielke U, Hüttinger KJ, Hoffmann WP, Surface oxidized carbon fibers: I. Surface structure and chemistry, *Carbon*, 34(8), 983–998, 1996. Copyright 1996, Elsevier.

resulting from the hydrolysis of anhydrides which were initially formed by a severe drying treatment.

It is suggested that functional groups may also form in the pores and would be controlled by pore diffusion [107]. The work of adhesion after oxidation depends on the type of fiber and is determined by the crystalline perfection of the fiber surface as seen by STM [108].

Preparation of composites from polyetherimide (PEI) and polyethersulfone showed that the ILSS values are controlled by the work of adhesion as measured by the contact angle. Results by XPS of polymer monolayers gives strong evidence for chemical bond formation in the polymer-carbon interface, but bonds are formed at processing temperatures above 300°C, when —COOH groups are decomposed, whereas the —C=O groups are directly involved with bonding epoxies.

Alexander and Jones [110] surface treated Hercules Type A-U fiber electrolytically in an NH_4HCO_3 electrolyte and examined the treated fiber with XPS and TOF SIMS and found four nitrogen containing functionalities:

1. Aza nitrogen equivalent to that found in thermally cyclized PAN, the concentration increased with the level of surface treatment, possibly due to erosion, revealing the PAN precursor resulting from incomplete carbonization.
2. Amide, which is the main nitrogen containing functionality and postulated to form due to the reaction of NH_4^+ with acid groups on the carbon fiber surface.
3. Imide, possibly formed from the reaction of adjacent amide and acid groups.
4. Protonated amine.

Interestingly, Mahy *et al* [44] found that electrolytic oxidation with $(NH_4)_2SO_4$ did not result in the introduction of nitrogen of equivalent concentrations, which may be specific to the NH_4HCO_3 electrolyte or the specific current/voltage regime used.

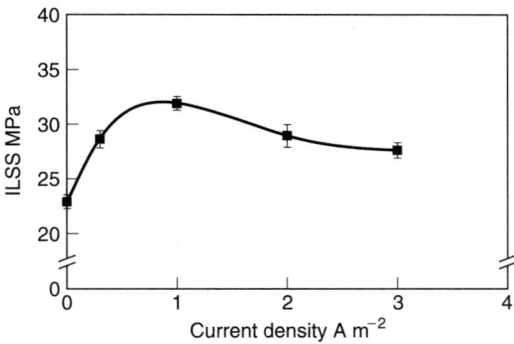

Figure 9.11 ILSS as a function of current density. *Source:* Reprinted with permission from Lee JR, Kim MH, Park SJ, Surface modification of carbon fibres by anodic oxidation and its effect on adhesion, *Key Eng Mater*, 183–187, 1105–1110, 2000. Copyright 2000, Trans Tech Publications.

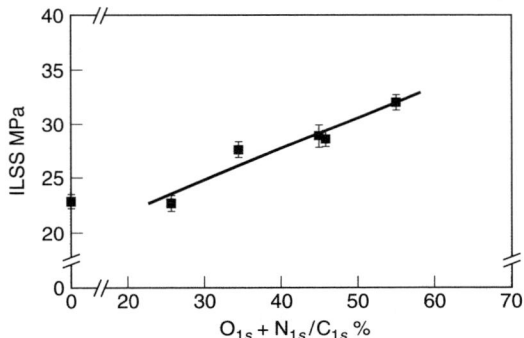

Figure 9.12 ILSS as a function of $O_{1s} + N_{1s}/C_{1s}$ atomic ratio. *Source:* Reprinted with permission from Lee JR, Kim MH, Park SJ, Surface modification of carbon fibres by anodic oxidation and its effect on adhesion, *Key Eng Mater*, 183–187, 1105–1110, 2000. Copyright 2000, Trans Tech Publications.

Lee et al [41] anodically oxidized a PAN carbon fiber (Taekwang, Korea, TZ-307) in 10% NH_4OH and found that the preferred current density was 5 A m^{-2} to give optimum ILSS (Figure 9.11) in a resole phenolic resin (Kangnam CB-8057) and the surface functional groups expressed as an atomic ratio $(O_{1s} + N_{1s}) / C_{1s}$ correlated with the ILSS (Figure 9.12). They believed that the surface functional groups improved the surface energy of the fibers as well as the degree of adhesion by establishing secondary van der Waals forces at the fiber/resin interface [111].

Sherwood, at Kansas State University, has coated electrochemically surface oxidized carbon fibers with a 1% solution of DuPont K3B experimental polyimide in 1-methyl-2-pyrrolidone and using XPS spectra, has gained support for chemical bonding between the surface treated fiber and the resin.

Lin and Yip [112] used three types of carbon fiber: Tonen HM CF, Hercules IM6 and Toho IM600, which were treated with boiling conc. HNO_3, boiling 15% H_2O_2, followed by reduction of the HNO_3 treated fiber with H_2 at 1000°C. They reported that the extent of oxidation was more dependent on the chemical reagent than the fiber type. The degree of adhesion was not substantially promoted by the amount of oxygen or carbon-oxygen functional groups on the surface. They considered the degree of surface roughness, enhanced by the chemical treatment, provided more adhesive surface for mechanical and physical interlocking.

The functional groups determined by Hopfgarten [113] on the fiber surface and at 50 nm below the surface are listed in their relative order of abundance.

At the surface

$$-\overset{|}{\underset{|}{C}}-O-> -\overset{|}{\underset{|}{C}}-OH = -\overset{|}{\underset{|}{C}}-S -> -\overset{O}{\overset{\|}{C}}-OH = -\overset{O}{\overset{\|}{C}}- > -\overset{|}{\underset{|}{C}}-Cl$$

At 500Å depth

$$-\overset{|}{\underset{|}{C}}-O-> -\overset{|}{\underset{|}{C}}-S- = -\overset{O}{\overset{\|}{C}}- > -O-\overset{O}{\overset{\|}{C}}-$$

Hopfgarten found that the main functional groups were —C=O, —COOH and —OH.

An investigation undertaken by Sheffield University using wide angle XPS analysis to examine the surface groups on a range of carbon fibers revealed the following functionalities: C—C/C—H, C—O—C, C(O)$_3$, CNO, and C—NR$_2$ and a discussion of probable assignments are discussed by Alexander and Jones [114].

Table 9.6 gives the atomic percentages of the surface concentrations of elements determined on a range of carbon fibers studied using wide angle XPS spectrum.

The Si is associated in the chemical environment with silica or silicate and Hearn and Briggs [115], using static TOF SIMS, observed the presence of similar SiO$_2$ deposits on an unspecified Hercules carbon fiber and attributed this to silicone residues on the precursor fiber which oxidize during carbonization.

No obvious difference in surface topography and roughness was found between the fiber types using SEM.

Higher surface levels of nitrogen can be associated with the reaction of free NH$_3$ during oxidation via amide formation [116], probably associated with nitrogen picked up from the surface treatment electrolyte bath. The differing nitrogen concentrations may be an effect of using a higher current density.

9.5.3 Summary

Wright [15] has summarized the possible effects of surface treatment:

1. There is little change in the surface area of the carbon fiber, irrespective of the fiber type.
2. A credible reason is the removal of a weak surface layer.

Table 9.6 The atomic percentages of the surface concentrations of elements determined on a range of carbon fibers studied using wide angle XPS spectrum

Atom	RK 30	SGL C30	Grafil A	Grafil A-S	Tenax HTA Untreated	Tenax HTA Treated
C–1s	85.55	87.11	80.6	82.9	94	88
O–1s	8.69	9.52	13.0	13.3	3	9
N–1s	4.90	1.44	6.3	3.9	2	2
Na–1s	0.18	0.10	–	–	0.1	0.3
Si–2p	0.69	1.19	–	–	0.1	0.1
S–2p	–	0.37	–	–	0.1	0.1
Cl–2p	–	0.27	–	–	0.1	0.3
Ca–2p	–	–	–	–	0.1	0.5

Source: Work undertaken by Jones FR, Haddow D, Korgul P, University of Sheffield.

3. Chemical modification of the surface occurs and the following groups have been detected —OH, =O, quinone, =C=O, —COOH, —COOX, —CO$_3$ and lactone.
4. There is an increase in the polar surface free energy but no significant improvement in fiber wetting out due to the presence of these surface groups.
5. Workers are undecided whether chemical bonding occurs between these groups and the bulk resin.

9.6 COUPLING AGENTS

Silane coupling agents are supplied for glass fiber and depend on a –Si–O– oxane bond coupling, but are not successful when applied to carbon fiber. Kenrich introduced a range of titanate liquid coupling agents, such as LICA38, a neo-alkoxy dioctyl phosphato titanate,

$$RO-Ti-\left(-O-\underset{\underset{OH}{|}}{\overset{\overset{O}{\|}}{P}}-O-\overset{\overset{O}{\|}}{P}(OC_8H_{17})_2\right)_3$$

that will withstand the demands of high temperature and work with carbon fiber (Chapter 13, Section 13.3) and there are a range of more expensive organozirconates.

9.7 SIZING CARBON FIBER

The application of a coating is normally termed a size or finish and can be achieved by:

1. Deposition from solution of a polymer
2. Deposition of a polymer onto the fiber surface by electrodeposition
3. Deposition of a polymer onto the fiber surface by electropolymerization
4. Plasma polymerization

Sizing is reputed to improve inter-filamentary adhesion, aid in wetting out the fiber in resin matrices and act as a lubricant to prevent fiber damage during subsequent textile processing such as weaving.

9.7.1 Deposition from solution of a polymer onto the fiber surface

The most common form of sizing is the deposition from solution of a polymer onto the fiber surface. The choice of size depends on the resin matrix and some thermoplastics like PEEK will require the ability to withstand high processing temperatures, so epoxy resins are generally used for epoxy matrices while a polyimide would be required for PEEK. To obtain the right handle, applied sizes must not be tacky or brittle and can be achieved by selecting a resin with a suitable epoxy molar mass to avoid the tacky state, or the epoxy could be partly cured. Some workers believe that a flexible interlayer of size is preferable but others prefer a more brittle system with modulus between that of the fiber and matrix. A number of papers published by Drzal [117–120], have depicted the vicinity of fiber and matrix as a graded three-dimensional region, loosely termed an interphase (Figure 9.13). The proposed mechanism for

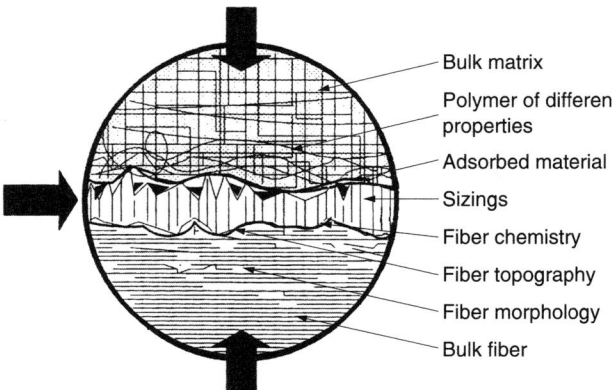

Figure 9.13 Schematic of the interphase region between fiber and matrix. *Source:* Reprinted with permission from Herrera-Franco PJ, Drzal LT, Comparison of methods for the measurement of fibre/matrix adhesion in composites, *Composites*, 23, 2, 1992. Copyright 1992, Elsevier.

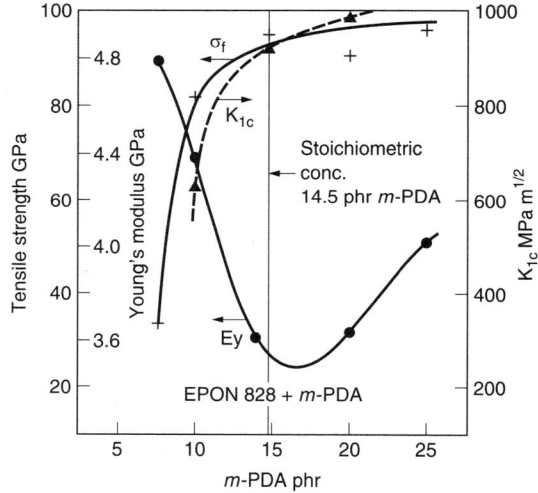

Figure 9.14 The Youngs modulus, tensile strength and interfacial shear strength of Epon 828/*m*-PDA composite by varying the concentration of *m*-PDA. *Source:* Reprinted with permission from Drzal LT, Rich MJ, Koenig ME, Lloyd PF, Adhesion of graphite fibres to epoxy matrices. II. The effect of fibre finish, *J Adhesion*, 16, 133–152, 1983. Copyright 1983, Taylor & Francis Ltd.

this interphase is that since sizes generally contain no hardener, some will diffuse from the matrix into the size layer and hence will contain less than the stoichiometric quantity of curing agent. Drzal et al [122] investigated the possible effects of this curing agent deficiency by measuring the Young's modulus, tensile strength and fracture toughness of a series of composites made from 828/*m*-PDA resin system and varying the *m*-PDA content (Figure 9.14). Their findings were that with less than the stoichiometric quantity (14.5 phr) of *m*-PDA, the system had higher modulus, lower strength and lower fracture toughness. Hence, the coating does not protect the fiber from handling damage and does not improve wetting of the fiber by the matrix resin. So this interphase is a more brittle material and promotes better stress transfer, resulting in a higher interfacial shear strength, but because of the lower fracture toughness, the failure mode changes from interfacial to matrix.

Jones [47] has discussed the structure of the interphase region.

Initially, organic solvents such as 1,1,1-trichloroethane were used for epoxy resins but safety legislation has enforced a change to water based systems. Water based systems can contain an emulsifier and a flexibilizer, which may influence reactions within the interphase. When applied at room temperature, the waterborne resins do not form a continuous film after evaporation of the water, but coalesce to a smooth film when heated above 85°C (Figures 9.15a and 9.15b).

Jones et al [123] found that solvent deposition leads to a strongly bound deposit.

A typical early size was Shell 834, a semi-solid DGEBA resin, which had a vastly better performance when applied in 1,1,1-trichloroethane than any replacement water based system, presumably due to a more efficient wetting out of the fiber bundle (Figure 9.15c).

A series of possible waterborne sizes based on resins with a range of WPE values are given in Table 9.7.

The sizes used for thermoplastic polymers are generally the matrix resin or a lower molecular weight version of the matrix polymer [124,125]. Jones et al [123] have investigated epoxy sizes for a PES matrix and found that when using a brominated DGEBA size, the PES matrix penetrated the size and formed an interphase region, whereas with an epoxy matrix, a strongly bound DGEBA was formed, which appeared to create a weak interface between the sizing resin and matrix. A concept of the interphase structure is shown in Figure 9.16.

Cizmecioglu [125] found that a polycarbonate matrix, when dissolved in a volatile solvent and coated on carbon fiber, after removal of the solvent from at least 65% of the surface, gave a film thickness of 5–100 nm, which facilitated wetting out by the matrix resin during the preparation of the composite. The flexural modulus was improved by 50% and the flexural strength by 37%. However, water based polyester-urethane polymers are now being increasingly used, eliminating the use of organic solvents.

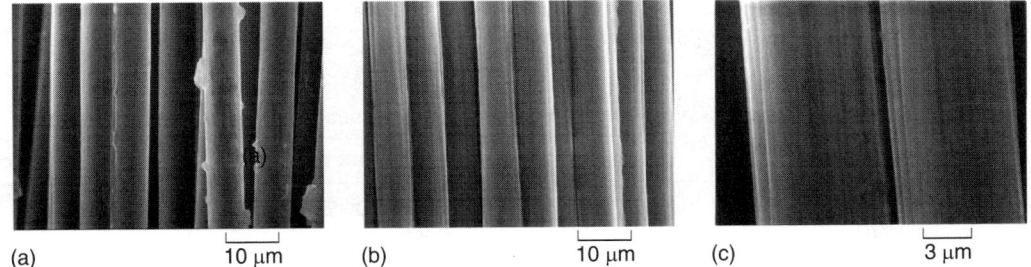

Figure 9.15 A water based size: a) After evaporation of water, b) After treatment above 80°C, c) An organic solvent applied size (samples prepared by Sheffield University).

Table 9.7 Shell waterborne resins suitable for sizing carbon fiber

Product	Description	WPE	% Solids	Viscosity at 25°C cP
Epi-Rez 3510	Liquid Bisphenol A resin (Epon 828 type)	200	60	3000
Epi-Rez 3515	Liquid Bisphenol A resin	250	63	10,000
Epi-Rez 3522	Solid Bisphenol A resin	680	60	10,000
Epi-Rez 3540	Solid Bisphenol A resin (Epon 1007 type)	1800	55	10,000
DPW-3545	Solid Bisphenol A resin (Epon 1007 type)	2000	55	10,000
Epi-Rez 5003	Epoxidized Bisphenol A/Novalac resin (Epon SU-3 type)	200	58	10,000
Epi-Rez 5520	Urethane modified epoxy resin	540	60	10,000
DPW-7010	Bisphenol A based polyester resin	–	60	10,000

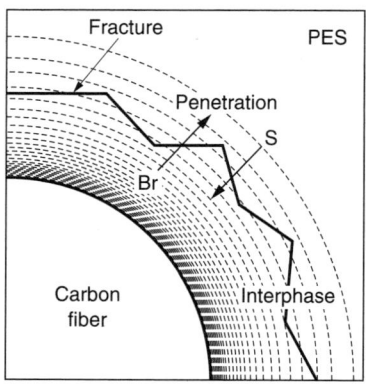
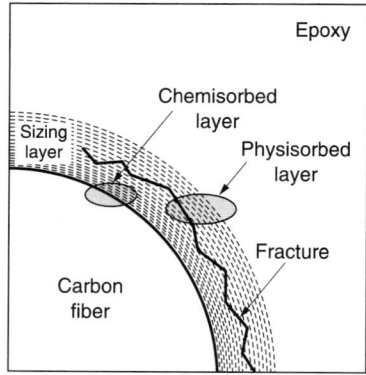

Figure 9.16 The concept of the interphase structure. *Source:* Reprinted with permission from Yumitori S, Wang D, Jones FR, The role of sizing resins in carbon fibre reinforced polyethersulfone (PES), *Composites*, 25, 698–705, 1994. Copyright 1994, Elsevier.

Table 9.8 Some sizes investigated for carbon fibers for deposition from solution

Sizing polymer	Matrix resin	Reference
Polyhydroxyether	ERL2256 (DGEBA)	[24]
Polyphenyleneoxide	ERL2256 (DGEBA)	[24]
Styrene/maleic anhydride (SMA)	LY556	[127,128]
Isoprene/SMA	LY556	[127,128]
Polysulfone	DGEBA/MNA/BDMA	[129]
1,2-polybutadiene	DGEBA/MNA/BDMA	[129]
Silicone (Sylgard 186)	DGEBA/MNA/BDMA	[129]
Cyanamid FM-1000 (nylon epoxy)	Fiberite 976 (TGDDM/DDS)	[130]
Cyanamid MsAF163-2 (elastomer modified epoxy)	Fiberite 976 (TGDDM/DDS)	[130]
Hycar 1300×8 (CTBA) / DGEBA 0164	DER332 (DGEBA)/DDM	[126]
Dow DER 330 (DGEBA)	828/MPD	[122]
Dow DER 332 (DGEBA)	Fiberite 976 (TGDDM/DDS)	[132]

A number of systems investigated as possible carbon fiber sizes are given in Table 9.8. Dauksys [24] attempted to apply a size to form a microductile region adjacent to the interface, which could act as a stress relief medium, a crack inhibitor/arrestor and also attempt to increase the effective transfer length of the fractured fibers. Dauksys, at best, applied a 0.9 $^w/_w$ solution of polyhydroxyether in Cellusolve to Thornel 50 carbon fiber (with the PVA size removed) and from composites prepared from Union Carbide ERL2256 resin, obtaining an 81% increase in ILSS and a 14% increase in flexural strength. However, a 2.3% solution of polyphenylene oxide in benzene obtained no improvement in flexural strength although achieving a 93% increase in ILSS.

Polyphenyleneoxide

Polyhydroxyether

9.7.2 Deposition of a polymer onto the fiber surface by electrodeposition

A preformed polymer with an ionized group attached is said to be electrodeposited when, under an applied voltage, the polymer is attracted to the oppositely charged carbon fiber in an electrolytic cell. A uniform non-conducting layer of polymer with a specified thickness can be applied to the carbon fiber at a controlled deposition rate. Subramanian et al [133–135] used batch and continuous methods and typical operating conditions used for the electrodeposition on Fortafil CG3 (YM = 170 GPa) were 10 V applied for 1 min using a 2.5% solution of the conducting polymer. Some polymers that have been deposited onto a carbon fiber surface by electrodeposition are given in Table 9.9 and Table 9.10 lists systems investigated by Subramanium [138]. The properties of composites obtained by Subramanian [133] using Fortafil CG3 fiber electrodeposited with a range of co-polymers are given in Table 9.11. Interestingly, some properties were better than those obtained with the commercially treated fiber and best results were with the maleic anhydride-styrene copolymer. The commercially treated fiber increased in ILSS but had reduced impact strength, a feature common with composites made with surface treated fibers. Subramanian et al [135] using Fortafil 5 fiber and maleic anhydride/butadiene found an improvement in impact strength, but no improvement in ILSS and is reputed to be due to the occurrence of a crosslinking reaction, resulting in poor interpenetration and producing a weak interphase-matrix bond. However, coating with ethylene/acrylic acid and maleic anhydride/methylvinylether copolymers [134] did permit penetration and showed improved shear and impact strengths.

Table 9.9 Selected systems used for the application of polymers by electrodeposition

Polymer	Reference
Titanium di(octylpyrophoshate)oxyacetate TDPA	[136]
Butadiene-maleic acid co-polymer	[135]
Poly(ethylenecoacrylic acid) and Poly(methylvinylether-co-maleic anhydride)	[134]
Pyrrole	[137]

Table 9.10 Systems investigated by Subramanian *et al* for electropolymerization

Polymer with polymerization time of 2.5 s	Solvent/ electrolyte	Carbon fiber polarity	ILSS MPa[a]
Diacetone acrylamide	H_2SO_4	Cathode	74
Acrylic acid	H_2SO_4	Cathode	73
Methyl methacrylate	$NaNO_3$/DMF	Cathode	65
Styrene	$NaNO_3$/DMF	Cathode	66
Caprolactam	$NaNO_3$/DMF	Anode	58
Styrene/acrylonitrile	$ZnCl_2/CH_2Cl_2$	Cathode	53
Vinyl-terminated butadiene-acrylonitrile copolymer	$NaNO_3$/DMF	Cathode	42
Epon 828/phthalic anhydride	LiCl/DMF	Anode	55

[a]Mean ILSS values with fiber volumes of 45–79% $^v/_v$ Hercules A-U fiber for composites made from 828/MPD. Hercules' ILSS values for A-U and A-S fibers were 69 and 94 MPa respectively.
Source: Reprinted with permission from Subramanian RV, Jakubowski JJ, Williams FD, Interfacial aspects of polymer coating by electropolymerisation, *J Adhesion*, 9, 185–195, 1978. Copyright 1978, Taylor & Francis Ltd.

Table 9.11 The effect of electrodeposited polymer on the properties of 50% $^v/_v$ Fortafil CG-3/Epon828-MPD composites

Electrodeposited polymer	ILSS MPa	Impact Strength kJ m^{-2}	Flexural strength MPa
None (untreated fiber)	34	63	78
None (surface treated fiber)	52	43	96
Maleic anhydride/styrene (1:1)	68	57	110
Maleic anhydride/styrene (1:2)	59	72	110
Maleic anhydride/styrene (1:3)	62	56	100
Maleic anhydride/hexene (1:1)	61	42	100
Maleic anhydride/octadecene (1:1)	52	44	91
Maleic anhydride/methylvinylether MW 500,000	48	86	90
Maleic anhydride/methylvinylether MW 750,000	59	130	95
Maleic anhydride/methylvinylether MW 1,250,000	54	140	86

Source: Reprinted with permission from Subramanian RV, Sundaram V and Patel AK, Electrodeposition of polymers on graphite fibres: Effects on composite properties, *33rd Ann Tech Conf Reinf Plast Comp Div*, Section 20F, 1978. Copyright 1978, Trans Tech Publications.

Earlier, original work by Gynn *et al* [139] electrodeposited an interphase resin with functional groups, previously reacted with an amine to form an amic acid derivative which was rendered water soluble by ionization with a base such as triethylamine then electrodepositing the product and finally thermally closing the ring to form an imide:

Unfortunately, work undertaken with this novel system and Hercules AS-1 fiber showed no improvements.

9.7.3 Deposition of a polymer onto the fiber surface by electropolymerization

Electropolymerization enables the polymerization of monomers in an electrolytic cell, where the carbon fiber can be made the anode or the cathode. The solvent used to dissolve the monomer must act as an electrolyte and be sufficiently conducting to permit a uniform non-conducting layer of polymer to be applied onto the carbon fiber at a controlled deposition rate and specified thickness.

Harris [79] states that the objectives of electropolymerization are to introduce an interlayer which will:

1. absorb energy from the tips of growing cracks, perhaps by blunting them
2. lessen the stress concentrations in the vicinity of the fiber caused by expansion/contraction stresses, or by the application of a load to a composite with fiber and matrix of different elastic properties
3. protect fiber from damage.

Table 9.12 lists a number of systems used for electropolymerization.

Harris and co-workers [79] studied seven systems and found their best results were with either o-diaminobenzene/LiClO$_4$ or p-diaminobenzene/tetraethylammonium tetrafluoroborate ($C_8H_{20}B_4N$) and they could improve the interfacial shear strength of Enka HTA-3000 by 46% over the value for the untreated fiber.

Bell and co-workers [141–144], after an extraordinarily long reaction time of some 2 h, which would not be commercially practical, produced an even 2 μm thick coating of AN/MA copolymer onto Hercules AS-4 fiber and were able to control the composition by adjusting the ratio of monomers, obtaining up to a 40% increase in impact strength, but the ILSS values decreased, possibly due to poor adhesion between the interlayer and matrix. A top coat of vinyl monomer (acrylic acid was selected), which could form chemical bonds between the layers, was applied as the last stage of the process and Figures 9.17 and 9.18 show the effect of this top coat on the impact strength and ILSS, obtaining optimum properties with a coating thickness of about 0.1 μm, though any improvements in properties were marginal.

Table 9.12 Some selected systems used for electropolymerization

Monomer or copolymer	Fiber type	Reference
Acrylic monomers	Hercules AS-4	Chang [131]
Ethylene diamine	Courtauld's XA-U	Barbier [140]
Poly(ethylene-co-acrylic acid)/ Poly(methyl vinyl ether-co-maleic anhydride)	Gt Lakes Fortafil 5	Crasto [134]
Acrylic copolymers	Hercules AS-4	Wimolkiatisak [141]
o-diaminobenzene	Enka HTA-3000	Harris [79]
p-diaminobenzene	Enka HTA-3000	Harris [79]

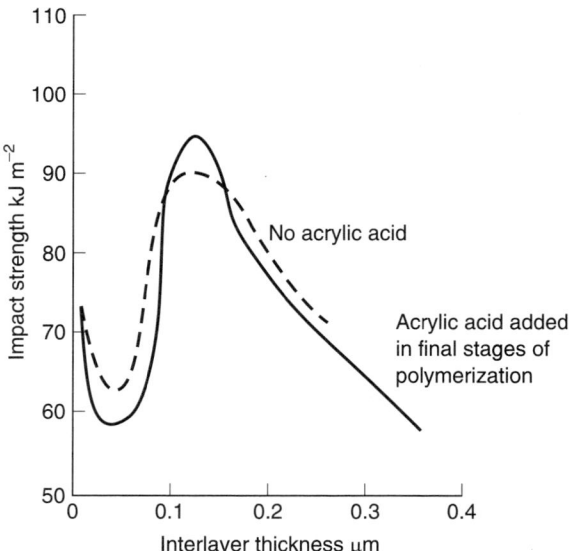

Figure 9.17 Impact strength as a function of interlayer thickness. *Source:* Reprinted with permission from Bell JP, Chang J, Rhee HW, Joseph R, Application of ductile polymeric coatings onto graphite fibres, *Polym Composites*, 8, 46–52, 1987. Copyright 1987, The Society of Plastic Engineers.

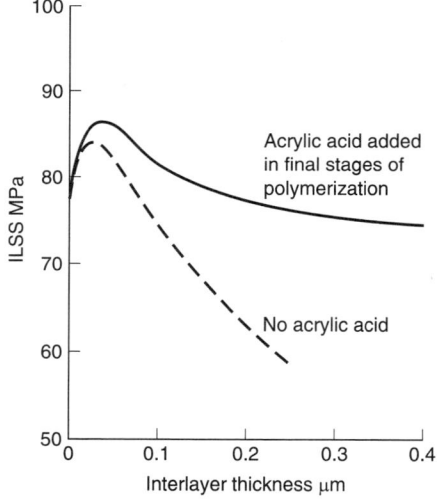

Figure 9.18 ILSS as a function of interlayer thickness. *Source:* Reprinted with permission from Bell JP, Chang J, Rhee HW, Joseph R, Application of ductile polymeric coatings onto graphite fibres, *Polym Composites*, 8, 46–52, 1987. Copyright 1987, The Society of Plastic Engineers.

REFERENCES

1. Herrick JW, Gruber PE, Mansur FT, Surface Treatments for Fibrous Carbon Reinforcements, *AFML-TR-66-178 Part I*, Air Force Materials Laboratory, Jul 1966.
2. Bobka RJ, Lowell LP, Integrated research on carbon composite materials, *AFML-TR-66-310 Part I*, Air Force Materials Laboratory, 1966.
3. Wadsworth NJ, Watt W, Treatment of carbon fibres and composite materials including such fibres, Brit. Pat., 1,180,441, Feb 1968.

4. Sach RS, Bromley J, Improvements in or relating to the treatment of fibrous carbon, Brit. Pat., 1,255,005, Jan 1968.
5. Novak RC, In Composite Materials: Testing and Design, *ASTM Special Technical Publication*, 460, 540, 1969.
6. Schola DA, Bascha M, United Aircraft Corporation, Treatment of carbon fibers, U.S. Pat., 3,720,536, Jun 1970.
7. Clark D, Wadsworth NJ, Watt W, The surface treatment of carbon fibres for increasing the interlaminar shear strength of cfrp, *Carbon Fibers—Their Place in Modern Technology*, The Plastics Institute, London, 44—51, 1974.
8. Mckee DW, Mimeault VJ, Surface properties of carbon fibers, Walker PL, Thrower PA eds., *Chemistry and Physics of Carbon*, Marcel Dekker, New York, 8, 151–241, 1973.
9. Druin ML, Ferment GR, Rad VNP, Enhancement of the surface characteristics of carbon fibers, U.S. Pat., 3,754,957, Aug 1973.
10. Molleyre F, Bastick M, *Traitement de fibres de carbone pas oxydation en phase gazeuse*, Proc Conf Carbon, Deutsche Keram Gesell, Baden-Baden, 500, 1976.
11. Molleyre F, Bastick M, *High Temp-High Pressures*, 9, 237, 1977.
12. Vukov AJ, *J Serb Chem Soc*, 55, 333, 1990.
13. Ryu SR, Jin H, Gondy D, Pusset N, Ehrburger P, Activation of carbon fibers by steam and carbon dioxide, *Carbon*, 31, 841–842, 1993.
14. Alcaniz-Monge J, Cazorla-Amoros D, Linares-Solano A, Yoshida S, Oya A, *Carbon*, 32, 1277, 1994.
15. Wright WW, The carbon fibre/epoxy resin interphase a review—Part I, *Comp Polym* (now called *Polymers and Polym Comp*), 3(4), 231–257, 1990. Note references for this article are given in Part II, [16].
16. Wright WW, The carbon fibre/epoxy resin interphase a review—Part II, *Comp Polym* (now called *Polymers and Polym Comp*), 3(5), 360–391, 1990.
17. Goan JC, Joo LA, Sharpe GE, Surface treatment for graphite fibres, *27th Ann Tech Conf Reinf Plast/Comp Inst*, 27(21–E), 1–6, 1972.
18. Fitzer E, Geigl KH, Hüttner W, Weiss R, Chemical interactions between the carbon fibre surface and epoxy resins, *Carbon*, 18, 389–393, 1980.
19. Manocha LM, Role of fibre surface-matrix combination in carbon fibre reinforced epoxy composites, *J Mat Sci*, 17, 3039–3044, 1982.
20. Wu ZH, Pittman CU, Gardner SD, Nitric acid oxidation of carbon fibres and the effects of subsequent treatment in refluxing aqueous NaOH, *Carbon*, 33(5), 597–605, 1995.
21. Pittman CU, He GR, Wu B, Gardner SD, Chemical modification of carbon fibre surfaces by nitric acid oxidation followed by reaction with tetraethylenepentamine. *Carbon*, 35(3), 317–331, 1997.
22. Perakslis ED, Gardner SD, Pittman CU, Surface composition of carbon fibres subjected to oxidation in nitric acid followed by oxygen plasma, *J Adhesion Sci Technol*, 11(4), 531–551, 1997.
23. Donnet JB, Ehrburger P, Carbon fibre in polymer reinforcement, *Carbon*, 15, 143–152, 1977.
24. Dauksys RJ, Graphite fiber treatments which affect fiber surface morphology and epoxy bonding characteristics, *J Adhesion*, 5, 211–244, 1973.
25. Goan JC, Martin TW, Prescott R, The influence of interfacial bonding on the properties of carbon fiber composites, *28th Ann Tech Conf Reinf Plast/Comp Inst*, 28(21–B), 1–4, 1973.
26. Aitkin ID, Rhodes G, Spencer AP, Improvements in or relating to carbon fibre treatment, Brit. Pat., 1,353,596, 23 Sep 1970.
27. Morgan PE, Treatment of carbon fibres, Brit. Pat., 1,238,308, 14 Aug 1967.
28. Barr JB, Treatment of carbon fibres to improve shear strength in composites, U.S. Pat., 3,791,840, 21 Oct 1970.
29. Druin ML, Diedwardo AH, Parker JA, U.S. Pat., 3,859,187, 7 Jan 1975.
30. Druin ML, Dix R, U.S. Pat., 3,894,884, Jul 1975.
31. Wells H, Colclough WJ, Improvements in or relating to the treatment of fibrous carbon, Brit. Pat., 1,257,022, 3 Jan 1968.
32. Hoare JP, *The electrochemistry of oxygen*, Interscience, 13, 1968.

33. Wessling RA, Settineri WJ, Wagner EH, Studies in cathodic electrodeposition, *American Chemical Society Division of Organic Coatings and Plastics Chemistry*, 31(1), Mar–Apr 1971.
34. Ehrburger P, Herque JJ, Donnet JB, Electrochemical treatment of carbon and graphite fibres, *Proc 4th Int Conf Carbon and Graphite Conf*, Society of the Chemical Industry, London, 201, 1976.
35. Weinberg NL, Reddy TB, Electrochemical oxidation of the surface of graphite fibres, *J Appl Electrochem*, 3, 73, 1973.
36. Joiner JC, Findlay V, Continuous surface treatment of carbon fibre, Brit. Pat., 1,326,736, 8 Oct 1969.
37. Waseem SF, Gardner SD, He GR, Jiang WB, Pittman U, Adhesion and surface analysis of carbon fibres electrochemically oxidized in aqueous potassium nitrate, *J Mater Sci*, 33(12), 3151–3162, 1998.
38. Chapman DR, Paterson WC, Electrolytically treating filamentary carbon, Brit. Pat., 1,297,946, 19 Mar 1969.
39. Ray JD, Steingiser S, Cass RA, Treatment of carbon or graphite fibres and yarns for use in fibre reinforced composites, U.S. Pat., 3,671,411, 3 Mar 1970.
40. White PA, Fibre treatment, Brit. Pat., 1,371,621, 28 Jan 1971.
41. Lee JR, Kim MH, Park SJ, Surface modification of carbon fibres by anodic oxidation and its effect on adhesion, *Key Eng Mater*, 183–187, 1105–1110, 2000.
42. Paul JT Jr., Process for electrolytic treatment of graphite fibres, U.S. Pat., 3,832,297, 9 Mar 1973.
43. King TR, Adams DF, Buttry DA, Anodic oxidation of pitch precursor carbon fibres in ammonium sulfate solutions—Batch screening treatment results, *Comp Sci Technol*, 44(4), 351–359, 1992.
44. Mahy J, Jenneskens LW, Grabandt O, Venema A, Houwelingen GDB, The relation between carbon fiber surface treatment and the fiber surface microstructure, *Surf Interface Anal*, 21, 1–13, 1994.
45. Harvey J, A simple apparatus for the electrolytic treatment of carbon fibre surfaces, *RAE Technical Report*, 86077, Dec 1986.
46. Fitzer E, Popovska N, Rensch HP, Anodic oxidation of carbon fibres in diammonium hydrogen phosphate solution, *J Adhesion*, 36(2–3), 139–149, 1991.
47. Jones FR, Interphase formation and control in fibre composite materials, *Key Eng Mater*, 116, 117, 41–60, 1996.
48. Denison P, Jones FR, Watts JF, *Surf Interface Anal*, 12, 455, 1988.
49. Denison P, Jones FR, Dorey G, Jones LF, Watts JF, McEnaney B, Mays TJ eds., *Carbon '88*, Institute of Physics, London, 1988.
50. Denison P, Jones FR, Watts JF, Ishida H ed., *Interfaces in Polymer, Ceramic, Metal Matrix Composites*, Elsevier, 77–85, 1988.
51. Chang TC, Plasma surface treatment in composites manufacturing, *J Ind Technol*, 15(1), 7, Nov 1998–Jan 1999.
52. Peebles LH Jr., *Carbon Fibers. Formation, Structure and Properties*, CRC Press, Boca Raton, 128, 1994.
53. Morra M, Occhiello E, Garbassi F, Surface studies on untreated and plasma treated carbon fibres, *Composite Sci Technol*, 42, 361, 1991.
54. Jin BS, Lee KH, Choe CR, Properties of carbon fibres modified by oxygen plasma, *Polym Int*, 34(2), 181–185, 1994.
55. Pittman CU, Jiang W, He GR, Gardner SD, Oxygen plasma and isobutylene plasma treatments of carbon fibre: Determination of surface functionality and effects on composite properties, *Carbon*, 36(1–2), 25–37, 1998.
56. Oyama HT, Wightman JP, Surface characterization of PVF sized and oxygen plasma treated carbon fibres, *Surface and Interface Analysis*, 26(1), 39–55, 1998.
57. Bismarck A, Kumru ME, Springer J, Influence of oxygen plasma treatment of PAN based carbon fibres on their electrokinetic and wetting properties, *J Colloid Interface Sci*, 210(1), 60–72, 1999.
58. Farrow GJ, Jones CJ, Adhesion, 45, 29, 1994.
59. Farrow GJ, Atkinson KE, Fluck N, Jones C, *Surf Interface Anal*, 23, 313, 1995.
60. Jones C, Sammann E, *Proc 4th Tech Conf Am Soc Comp*, Blacksburg, 387–396, 1989.

61. Wiertz VB, Bertrand P, Identification of the N-containing functionalities introduced at the surface of ammonia plasma treated carbon fibres by combined TOF SIMS and XPS, *Proc Int Conf on Polymer-Solid Interfaces*, ICPSI-2, Mamur, Aug 12–16, 1996.
62. Jones C, *Comp Sci Technol*, 42, 275–298, 1991.
63. Feih S, Schwartz P, Modification of the carbon fibre matrix interface using gas plasma treatment with acetylene and oxygen, *J Adhesion Sci Technol*, 12(5), 523–539, 1998.
64. Waldman DA, Zou YL, Netravali AN, *J Adhesion Sci Technol*, 9, 1475, 1995.
65. Zou YL, Netravali AN, *J Adhesion Sci Technol*, 9, 1505, 1995.
66. Dagli G, Sung NH, *Polymer Composites*, 10, 109–116, 1989.
67. Commercon P, Wightman JP, Effect of organic gas plasmas on the adhesion of matrix resins to carbon fibers, *J Adhesion*, 47(4), 257–268, 1994.
68. Dilsiz N, Erinc NK, Bayramli E, Akovali G, *Carbon*, 33, 853, 1995.
69. Ebert E, Weisweiler W, NATO ASI Ser, Ser E 230, 287, 1993.
70. Smiley RJ, Delgass WN, *Mater Res Soc Symp Proc*, 305, 129, 1993.
71. Weisweiler W, Schlitter K, Surface modification of carbon fibres by plasma polymerization, Figueiredo JL, Bernardo CA, Baker RTK, Hüttinger KJ eds., *Carbon Fibers Filaments and Composites*, Kluwer Academic, Dordrecht, 272, 273, 1989.
72. Donnet JB, Brendle M, Dhami TL, Bahl OP, *Carbon*, 24, 757, 1986.
73. Donnet JB, Dhami TL, Dong S, Brendle M, *Journal of Physics D: Applied Physics*, 20, 269, 1987.
74. Edie ED, Cano RJ, Ross RA, *20th Biennial Conference on Carbon*, Santa Barbara, Jun 1991.
75. Peng JCM, Rebouillat S, Surface Treatment of Carbon Fibers, Donnet JB, Wang TK, Peng JC, Rebouillat S eds., *Carbon Fibers*, Marcel Dekker, New York, 187, 1998.
76. Su J, Tao X, Wei Y, Zhang Z, Liu L, The continuous cold-plasma treatment of the graphite fibre surface and the mechanism of the modification of the interfacial adhesion, Ishida H ed., *Interfaces in polymer ceramic and metal matrix composites*, Elsevier Science Publishing Co. Inc., New York, 269–277, 1988.
77. Jang BZ, Das H, Hwang LR, Chang TC, Plasma treatments of fiber surfaces for improved composite performance, Ishida H ed., *Interfaces in polymer ceramic and metal matrix composites*, Elsevier Science Publishing Co. Inc., New York, 319–333, 1988.
78. Donnet JB, Dong S, Guilpain G, Brendle M, Carbon fibres: Electrochemical and plasma surface treatment and its assessment, Ishida H ed., *Interfaces in polymer ceramic and metal matrix composites*, Elsevier Science Publishing Co. Inc., New York, 269–277, 1988.
79. Harris B, Braddell OG, Lefebvre C, Verbist J, The surface treatment of carbon fibres by electropolymerization and plasma polymerization, *Plastics Rubber and Composites Processing and Applications*, 18(4), 221–240, 1992.
80. Jones C, Sammann E, The effect of low power plasmas on carbon fibre surfaces, *Carbon*, 28(4), 509–514, 1990.
81. Jones C, Sammann E, The effect of low power plasmas on carbon fibre surfaces. A comparison between low and high modulus PAN based fibres with pitch based fibres, *Carbon*, 28(4), 514–519, 1990.
82. Bascom WD, Chen WJ, Effect of plasma treatment on the adhesion of carbon fibres to thermoplastic polymers, *J Adhesion*, 34(1–4), 99–119, 1991.
83. Yuan LY, Shyu SS, Lai JY, Plasma surface treatments on carbon fibres. 2. Mechanicalproperty and interfacial shear strength, *J Appl Polym Sci*, 42(9), 2525–2534, 1991.
84. Morra M, Occhiello E, Garbassi F, Nicolais L, Surface studies on untreated and plasma treated carbon fibres, *Composites Sci Technol*, 42(4), 361–372, 1991.
85. Yuan LY, Chen CS, Shyu SS, Lai JY, Plasma surface treatments on carbon fibres. 1. Morphology and surface analysis of plasma etched fibres, *Composites Sci Technol*, 45(1), 1–7, 1992.
86. Commercon P, Wightman JP, Surface characterization of plasma treated carbon fibres and adhesion to a thermoplastic polymer, *J Adhesion*, 38(1–2), 55–78, 1992.
87. Allred RE, Schimpf WC, CO_2 plasma modification of high modulus carbon fibres and their adhesion to epoxy resins, *J Adhesive Sci Technol*, 8(4), 383–394, 1994.
88. Dilsiz N, Ebert E, Weisweiler W, Akovali G, Effect of plasma polymerization on carbon fibres used for fibre epoxy composites, *J Colloid Interface Sci*, 170(1), 241–248, 1995.

89. Vaidyanathan NP, Kabadi VN, Vaidyanathan R, Sadler RL, Surface treatment of carbon fibres using low temperature plasma, *J Adhesion*, 48(1–4), 1–24, 1995.
90. Cowbel W, Shan CH, Mechanical behaviour of carbon-carbon composites made with cold plasma treated carbon fibres, *Composites*, 26(11), 791–797, 1995.
91. Chand N, Schulz E, Hinrichsen G, Adhesion improvement of carbon fibres by plasma treatment and evaluation by pull-out, *J Materials Sci Lett*, 15(15), 1374–1375, 1995.
92. Bogoeva G, Mader E, Haussler L, Dekanski A, Characteristics of the surface and interphase of plasma treated HM carbon fibres, *Composites Part A—Appl Sci Manuf*, 28(5), 445–452, 1997.
93. Kettle AP, Beck AJ, O'Toole L, Jones FR, Short RD, Plasma polymerization for molecular engineering of carbon fibre surfaces for optimized composites, *Composites Sci Technol*, 57(8), 1023–1032, 1997.
94. Prosen SP, Simon RA, Composites for Hydro and Aerospace, *Plast Polym*, 36(123), 241, 1968.
95. Milewski JV, Shaver RG, Withers JC, Whiskerized graphite filaments for composites, *Mater Eng*, 67(5), 62, 1968.
96. Shaver RG, Silicon carbide whiskerized carbon fibre, *Pre-print 24E, Materials Conf*, American Inst Chem Engineers, New York, 1968.
97. Emig G, Popovska N, Schoch G, *Thin Solid Films*, 241, 361, 1993.
98. Herrick JW, Resin-fibre interactions in graphite fibre-epoxy composites. *23rd Ann Tech Conf Reinf Plast Comp Div*, Section16-A, 1968.
99. Scola DA, Brooks CS, *25th Soc Plastics Ind Meet*, Washington DC, Feb 1970.
100. Harvey J, The characterization of carbon fibre surfaces by dye adsorption, *RAE Tech Memo Mat*, 231, Sep 1975.
101. Rand B, Robinson R, Surface characteristics of carbon fibres from PAN, *Carbon*, 15, 257–263, 1988.
102. Rand B, Robinson R, A preliminary investigation of PAN based carbon fibre surfaces by flow microcalorimetry, *Carbon*, 15, 311–315, 1977.
103. Fitzer E, Geigl KH, Huttner W, Weiss R, Chemical interactions between the carbon fibre surface and epoxy resins, *Carbon*, 18, 389–393, 1980.
104. Drzal LT, Rich MJ, Lloyd PF, Adhesion of graphite fibres to epoxy matrices. 1. The role of fibre surface treatment, *J Adhesion*, 16, 1–30, 1982.
105. Rich MJ, Drzal LT, Interfacial properties of some high strain carbon fibers in an epoxy matrix, *J. Reinforced Plastics and Composites*, 4, 145–154, March 1988.
106. Zielke U, Hüttinger KJ, Hoffmann WP, Surface oxidized carbon fibers: I. Surface structure and chemistry, *Carbon*, 34(8), 983–998, 1996.
107. Zielke U, Hüttinger KJ, Hoffmann WP, Surface oxidized carbon fibers: II. Chemical modification, *Carbon*, 34(8), 999–1005, 1996.
108. Zielke U, Hüttinger KJ, Hoffmann WP, Surface oxidized carbon fibers: III. Characterization of carbon fiber surfaces by the work of adhesion/pH diagram, *Carbon*, 34(8), 1007–1013, 1996.
109. Zielke U, Hüttinger KJ, Hoffmann WP, Surface oxidized carbon fibers: IV. Interaction with high temperature thermoplastics, *Carbon*, 34(8), 1015–1026, 1996.
110. Alexander MR, Jones FR, Effect of electrolytic oxidation on the surface chemistry of Type A carbon fibres: 3. Chemical state, source and location of surface nitrogen, *Carbon*, 34(9), 1093–1102, 1996.
111. Park SJ, Interfacial Forces and Fields: Theory and Applications, Hsu JP ed., Marcel Dekker, New York, 385, 1999.
112. Lin SS, Yip FW, Surface adhesion of carbon fibres after chemical treatments, *19th Biennial Conference on Carbon*, Session 2A, Penn State University, 244–245, 1989.
113. Hopfgarten F, Surface study of carbon fibres with ESCA and Auger electron spectroscopy, *Fibre Sci Technol*, 11, 67–79, 1978.
114. Alexander M, Jones FR, The effect of electrolytic oxidation upon the surface chemistry of Type A carbon fibres. Part 1—X-ray photoelectron spectroscopy, *Carbon*, 32, 785–794, 1994.
115. Hearn MJ, Briggs D, *Surf Interface Anal*, 17, 421–429, 1991.
116. Alexander M, Jones FR, *Carbon*, 34, 1093–1102, 1996.

117. Drzal LT, *Vacuum*, 41(7–9), 1615–1619, 1990.
118. Drzal LT, *Mater Sci Eng*, A126, 289–293, 1990.
119. Drzal LT, Interfaces Compos, Pantano CG, Chen EJH eds., *Mater Res Soc Symp Proc*, 170, 275–283, 1990.
120. Drzal LT, Treatise on Adhesion and adhesives, Patrick RL, Marcel Dekker, New York, 6, 187–211, 1988.
121. Herrera-Franco PJ, Drzal LT, Comparison of methods for the measurement of fibre/matrix adhesion in composites, *Composites*, 23, 2, 1992.
122. Drzal LT, Rich MJ, Koenig ME, Lloyd PF, Adhesion of graphite fibres to epoxy matrices. II. The effect of fibre finish, *J Adhesion*, 16, 133–152, 1983.
123. Yumitori S, Wang D, Jones FR, The role of sizing resins in carbon fibre reinforced polyethersulfone (PES), *Composites*, 25, 698–705, 1994.
124. Chiang W-Y, C-Yhuang, The effect of polyurethane treated carbon fibre on the properties of polyacetal/carbon fibre composites, *Composite Polymers*, 4(4), 251–256, 1991.
125. Cizmecioglu M, Composites with improved fibre resin interfacial adhesion, U.S. Pat., 4,842,933, 27 Jun 1989.
126. Gerard JF, Characterisation and role of an elastomeric interphase on carbon fibres reinforcing an epoxy matrix, *Polym Eng Sci*, 28, 568–577, 1988.
127. Brie M, Gressus C, Grafting of polymers on carbon fibres, *Fibre Sci Technol*, 6, 47–54, 1973.
128. Brie M, Jouquet G, Riess G, Bourdeaux M, Surface treatment of carbon fibres with alternating and block copolymers, *Int Conf in carbon fibres, their place in modern technology*, Paper 8, Feb 1974.
129. Hancox NL, Wells H, The effects of fibre surface coatings on the mechanical properties of cfrp, *Fibre Sci Technol*, 10, 9–22, 1977.
130. Schwartz HS, Hartness JT, Effect of fibre coatings on interlaminar fracture toughness of composites, Johnston NJ ed., ASTM STP 937, Philadelphia, ASTM, 150–178, 1987.
131. Chang J, *PhD Dissertation*, University of Connecticut, 1986.
132. Peters PWM, Springer GS, Effects of cure and sizing on fibre-matrix bond strength, *J Composite Mater*, 21, 157–171, 1987.
133. Subramanian RV, Sundaram V and Patel AK, Electrodeposition of polymers on graphite fibres: Effects on composite properties, *33rd Ann Tech Conf Reinf Plast Comp Div*, Section 20F, 1978.
134. Crasto AS, Owa S-H, Subramanian RV, The influence of the interphase on composite properties: Poly(ethylene-co-acrylic acid) and Poly(methylvinylether-co-maleic anhydride) electrodeposited on graphite fibres, *Polym Composites*, 9, 78–92, 1988.
135. Crasto AS, Owa S-H, Subramanian RV, Interphase modification in carbon-fibre composites via electrodeposition, Ishida H, Koenig JL eds., *Composite Interfaces*, Elsevier Science Publishing Co. Inc., New York, 133–142, 1986.
136. Own SH, Subramanian RV, Saunders SC, *J Mater Sci*, 21, 3912, 1986.
137. Chiu HT, Lin JS, Electrochemical deposition of polypyrrole on carbon fibers for improved adhesion to the epoxy resin matrix, *J Mater Sci*, 27, 319, 1992.
138. Subramanian RV, Jakubowski JJ, Williams FD, Interfacial aspects of polymer coating by electropolymerisation, *J Adhesion*, 9, 185–195, 1978.
139. Gynn G, King RN, Chappell SF, Deviney ML, Improved graphite fibre adhesion, *The Airforce Wright Avionics Laboratory-TR-81-4096*, Sep 1981.
140. Barbier B, Villatte M, Sanchez M, Désarmot G, *Comptes-Rendus des Sixiémes Journées Nationales Sur Les Composites*, JNC6, AMAC, Editions Pluralis, Paris, 115–130, 1988.
141. Wimolkiatisak AS, Bell JP, *Polymer Composites*, 10, 162–172, 1989.
142. Chang J, Bell JP, Joseph R, Effects of a controlled modulus interlayer upon the properties of graphite/epoxy composite, *SAMPE Quarterly*, 18(3), 39–45, 1987.
143. Bell JP, Chang J, Rhee HW, Joseph R, Application of ductile polymeric coatings onto graphite fibres, *Polym Composites*, 8, 46–52, 1987.
144. Chang J, Bell JP, Shkolnik S, Electrocopolymerisation of acrylonitrile and methyl acrylate onto graphite fibres, *J Appl Polym Sci*, 34, 2105–2124, 1987.

Supplementary references:

These references were obtained after the test had been written but are included for further interest.

145. Xie YM, Sherwood PMA, Experimental investigation of the effects of temperature on the electrochemical treatment of pitch based fibres by potentiostatic and galvanostatic methods, 45(7), 1158–1165, 1991.
146. King JA, Buttry DA, Adams DF, Development and evaluation of surface treatments to enhance the fibre matrix adhesion in PAN based carbon fibre liquid crystal polymer composites 2. Electrochemical treatments, *Polym Composites*, 14(4), 301–307, 1993.
147. Roman I, Novac C, Nita P, Anodic oxidation of the carbon fibres obtained from PAN precursors, *Materiale Plastice*, 32(3–4), 252–257, 1995.
148. Yumitori S, Nakanishi Y, Effect of anodic oxidation of coal tar pitch based carbon fibre on adhesion in epoxy matrix. 2. Comparitive study of three alkaline solutions, *Composites Part A—Appl Sci Manuf*, 27(11), 1059–1066, 1996.
149. Yumitori S, Nakanishi Y, Effect of anodic oxidation of coal tar pitch based carbon fibre on adhesion in epoxy matrix. 1. Comparison between H_2SO_4 and NaOH solutions. *Composites Part A—Appl Sci Manuf*, 27(11), 1051–1058, 1996.
150. Park SJ, Donnet JB, Anodic surface treatment on carbon fibres: Determination of acid base interaction parameter between two identical solid surfaces in a composite system, *J Colloid Interface Sci*, 206(1), 29–32, 1998.
151. Shkolnik S, Barash C, Electrocoating of carbon fibres with polymers 3. Electropolymerization of polyfunctional monomers, *Polymer*, 34(14), 2921–2928, 1993.
152. Singh BP, Nayak N, Kandpal LD, *In situ* electropolymerization of azo-imide resins (chain extended bismaleimides) on carbon fibres, *J Polym Mater*, 13(2), 163–167, 1996.
153. Chen Y, Iroh JO, Electrodeposition of BTBD-ODA-PDA polyamic acid coatings on carbon fibres from non-aqueous solutions, *Polym Eng Sci*, 39(4), 699–707, 1999.

CHAPTER 10

Guidelines for the Design of Equipment for Carbon Fiber Plant

10.1 INTRODUCTION

Carbon fiber production processes are proprietary. In some instances, the plant is designed and constructed in-house, alternatively it is purchased from a specific manufacturer with, or without, input from the purchaser. Such input automatically confers confidentiality, so it is not possible to give exact design details, but a compromise has been taken here and guidelines are given for a PAN production process, the pros and cons of alternative techniques being discussed below.

An important point to remember is that carbon fiber is electrically conductive and great care must be taken to exclude carbon fiber from all electrical equipment and this aspect is covered in the chapter on operation of plant and safety, as are other safety matters such as guarding the nip rollers and dealing with plant emissions.

Design must take into account future servicing and provision being made for plant operatives to safely reach the fiber tow band to deal with production problems such as wraps. In some instances, it may be necessary to employ closed circuit television (CCTV) to facilitate control of inaccessible parts of the production plant.

There are two approaches to the production of small tow carbon fiber:

1. Run a few tows very fast
2. Run multiple tows slowly in order to permit removal of exothermic heat in the oxidation stage.

In the first instance, the equipment is essentially based on processing through a series of heated tubes and this aspect has not been covered whereas, the second requires the utilization of a substantial process plant. A typical plant layout is shown in Figure 10.1 and each item of plant will now be considered together with the appropriate services.

10.2 PRECURSOR HANDLING

Provision must be made for an area with ready access, permitting clean and dry storage of the precursor.

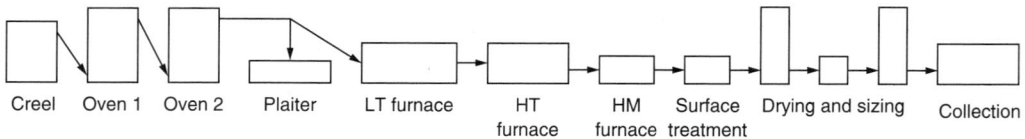

Figure 10.1 A typical carbon fiber plant layout.

1. Handling boxes of large tow

A typical box of 320k PAN precursor will weigh about 180 kg and a fork lift truck, preferably fitted with side lifting clamps, will be required to bring these boxes into the creel area. Local movement of the boxes and final positioning can, however, be readily achieved using a low flat trolley. There is a possibility that in the future, carton weights may be doubled, in which case final positioning might become a problem. Large tow precursor is crimped to give the tow lateral integrity in order to permit removal from the carton. The first task, therefore, is to apply some tension to the tow to help remove part of this crimp and attain a uniform band of fiber. This can be achieved by using a motor driven tow tensioning system equipped with a load cell indicator or, more simply, can be achieved by allowing the precursor fiber to travel upwards some 3–4 m to a series of guide bars. These bars can be straight, or convex, to spread the tow, or concave to reduce the tow width. Angling the bars permits tow movement to position a number of tows alongside each other and produce a uniform band of precursor prior to entry into the oven (Chapter 11). This free run of precursor, up and down, allows tows to be repaired or joined, permitting continuous production. Provision would also have to be made for a motor driven tensioning system in order to have some free run.

2. Creel for small tows

A typical package of 12k PAN will weigh about 20 kg and can be considered to be a one-person lift. However, larger packages up to about 80 kg have been introduced and these require a lifting device to position the package onto the creel spindle. The build of the creel (Figure 10.2) depends on available space (e.g., a restriction on height), but this is usually dictated by the height to which a package can be safely lifted. Generally, creels are placed as near the center line as possible. Creels offset from the center line can introduce a false twist to an individual tow as it unwinds from its package and up to 30 turns of false twist have been

Figure 10.2 Typical creel used for PAN precursor. *Source:* Courtesy of Shelton M, Shelton Machines Ltd., Croft, Leicester, England

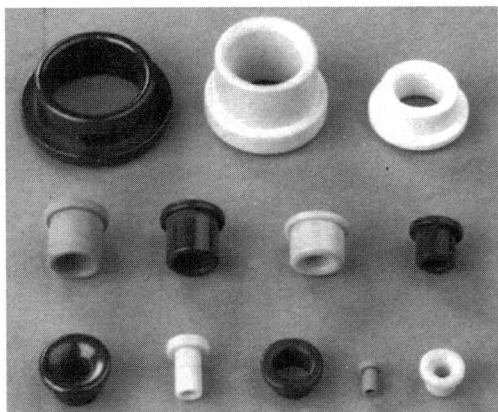

Figure 10.3 Ceramic eyelets. *Source:* Courtesy of Sharkie and Huntbach, Congleton, Cheshire, England

measured with offset creels. This false twist does not travel forward, since it is trapped by the first roller that the tow comes into contact with prior to entry into the oxidation oven. Packages positioned on the center line, with free running spindles, will unwind by the tow's own weight and the tow will sag downwards touching the floor. Hence, to confer some tension, all packages are restrained with a simple adjustable braking device such as a weighted strap drum brake.

Chromium plated pigtails can be used to guide a small tow, which permits easy threading, but they can rust. Alternatively, ceramic eyelets (Figure 10.3) can be used, which have to be individually threaded. These eyelets must be smooth and are best when diamond barrel polished. A good test of their suitability is to thread a short length of PAN precursor through the eyelet and see-saw it in order to quickly establish the likely effect on the fiber cosmetics. It is always sound practice to keep the number of contact points to a minimum. Joining small tows can be carried out using a proprietary air splicer.

If a creel is used to hold precursor fiber that is to be twisted (e.g. rayon), then running speeds are significantly faster and it is best if all ends are fitted with tow tension devices, which will also help detect breakages and indicate when all the fiber has been unwound from the package.

10.3 DRIVE SYSTEMS

Figure 10.4 shows a number of drive systems, which can be used. A rubber covered nip roller (75–80 Shore A hardness) is pushed into contact with the inlet steel roller. The rollers should preferably be chrome plated carbon steel and polished to 0.2 μm finish. A better arrangement is the use of ceramic coated steel rollers, which although more expensive, resists damage from the knives used to cut wraps from the rollers.

Rollers should be about 200–350 mm in diameter in order to minimize bending and the nip roll should be actuated either hydraulically, or by double acting air cylinders, equipped with emergency trip wires and stop buttons, reverting to the nip-open position when operated. When the nip roller comes into contact with the steel roller, it should make intimate contact along its entire length to ensure that the fiber band is gripped uniformly. The simple two roll system has only line contact and is not suitable, whereas a five roll drive is more efficient, but is more expensive and occupies more working space.

A DC drive system with a tachometer feedback is preferred for fiber drives, although AC drives with an inverter have acceptable accuracy and are preferred for controlling fans with

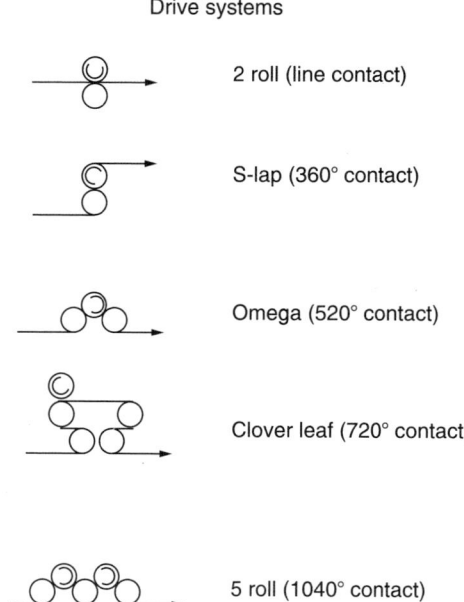

Figure 10.4 Schematic arrangement of drive systems.

variable speed. It should be remembered that if relatively slow line speeds are to be used, perhaps for threading up a line, then it will be necessary to provide secondary air cooling for the motor.

10.4 OVENS FOR OXIDATION

The oxidation stage is required to stabilize the PAN fiber and increases the density of the PAN precursor from $1.18\,\mathrm{g\,cm^{-3}}$ to about 1.36–$1.38\,\mathrm{g\,cm^{-3}}$ for the oxidized PAN fiber (opf). To achieve this, the oxidation process is carried out using an oven (Figure 10.5), passing the fiber through a series of air heated zones, which gradually increase in temperature. The hot air (220–270°C) is necessary to heat the fiber, provide O_2 for reaction, remove exhaust components and reaction heat from the fiber. It is now common practice to use more than one oven to provide additional zones, thus permitting significant increases in line speeds.

Figure 10.5 Typical oxidation ovens. *Source:* Courtesy of RK Carbon Fibres Ltd.

Allowances must be made for the removal of crimp, if any, in the precursor, adjusting for slack in the tow band, coping with a strongly exothermic reaction, providing a uniform temperature distribution, removing noxious gases and preventing the build-up of such gases, which may form an explosive mixture with air. The ovens should be fitted with explosion relief panels of the disintegrating type, venting to atmosphere in order to cope with such an eventuality and designed to lift at 50 mm W.G. [1–4].

The precursor fiber is fed through each oxidation oven in a number of passes to provide the requisite residence time and is guided by a series of pass back rollers, which can be situated either within the ovens, or without. The passes can be vertical or horizontal. Vertical passes readily permit air flow to be directed along the fiber band. This is preferable, but requires the pass back rollers to be situated within the oven, since external pass back rollers would permit the ready escape of contaminated hot air through the outlet slots at the top by the chimney effect. With internal rollers, it is necessary to use a specially adapted bore viewer fitted with quartz iodine illumination and fiber optics that will work at oven temperatures, in order to detect wraps on the internal rollers. The hot rollers can cause interfilament adhesion in the initial stages of oxidation, producing low strength carbon fiber. It is possible to alleviate this problem by passing cooling air through those rollers where interfilament adhesion is likely to be troublesome.

Horizontal passes invariably have air flow across the fiber band, which has a more harmful effect on the fiber cosmetics. It is possible to divide the oven into horizontal chambers, with several passes in each chamber, and directing the air flow longitudinally above and below the fiber bands, using special collection chambers fitted with small fans to gather the air and pass back to the heater bank for recirculation. If the plant is required to process large (e.g. 320k) and, on other occasions, small tows (e.g. 12k), then this could be a problem. The reason is that larger tows release more exothermic heat, which has to be dissipated by an increased air flow and this would confer poo cosmetics for the smaller tows and a solution to this dilemma would be to use variable speed fans for controlling the optimum air flow for each type of fiber. The volume of air flow falls in proportion to the reduction in speed, the pressure in proportion to the square of the speed and the power consumed in proportion to the cube of the speed. It is usual to run oxidation ovens under a slight negative pressure in order to prevent losing heat and the egress of poisonous exhaust gases.

In multi-pass oxidation ovens fitted with external pass-back rollers (Figure 10.6), there are numerous seals at the entry and exit points of the fiber. Although the ovens are operated preferentially under a negative pressure to prevent noxious fumes entering the processing area, there is still a tendency for hot air to leave the upper seals aided by convection. Many forms of seal have been tried, such as limiting the gap with adjustable stainless steel shutters with rounded edges, preferably with the fiber band not making contact, or allowing the fiber to make contact and using a flexible heat resistant membrane such as woven fiberglass. An efficient type of seal is the Perfect Float^{TM}, a proprietary air seal by Litzler, incorporating built-in fans to maintain a balance between the inside and the outside of the oven.

To facilitate cleaning, ovens must have ready access, such as a walkway, which would permit entry, but provided with adjustable lift-up flaps situated on the same levels as the dividing partitions of any horizontal sections. Air-flow should be turbulent to prevent local hot-spots.

In general, rollers that are 1 m wide should be about 150 mm in diameter and if 2 m wide, 350 mm in diameter, in order to ensure sufficient rigidity to restrict bending, if any, to a permissible level. Excessive bow would allow the fiber band to move towards the center line, causing fiber-bunching and a build-up of heat in the thicker tow band, which could eventually result in a fire. Rollers can be fitted with wiper blades to remove stray filaments. Metal doctor blades, though efficient, tend to be noisy and therefore, felt is preferred.

Figure 10.6 Passback rollers for oxidation oven. *Source:* Courtesy of RK Carbon Fibres Ltd.

Free-running pass-back rollers permit the fiber to stretch or shrink without restriction. If any of the rollers are to be driven, they must be manufactured to a specific tolerance on the diameter, since different diameter rollers would give variable line speed.

To allow for the removal of crimp and adjust for any variations in length throughout the oxidation process, it is usual to have several drives, which can be adjusted to provide the requisite tension in different sections of the oven(s). The mass per unit length of fiber exiting an oven is controlled primarily by the ratio of the inlet and exit drive speeds. The exothermic reaction can be controlled by avoiding overheating, preventing the fiber from bunching and maintaining a uniform temperature distribution of at least $\pm 2°C$ by adjusting the air flow to about 1 m/sec using a variable speed fan, but should be adequate to ensure removal of exothermic heat. The exhaust gases are constantly recirculated, normally with a centrifugal type air circulation fan and part of the air stream is continuously removed for subsequent treatment, replaced with clean fresh air to prevent a possible build-up of gases from reaching the explosive limit. It is a wise precaution to mark the fan casing with the direction of rotation of the fan. The air can be heated using a gas heater bank or Inconel sheathed electric heating elements. To ensure that the elements dissipate their heat uniformly, they must be finned and over-temperature devices fitted within the heater banks which would enable the heater power to be cut off.

The uniformity of the air flow within the oven can be checked at room temperature using a hot wire anemometer, although frictional resistance on the recirculating air will add about $20°C$ to the ambient temperature, making it a rather uncomfortable commissioning procedure. Since oxidation is extremely temperature dependent, it is a wise precaution to use duplex thermocouples with the second arm acting as a check thermocouple, or as a possible replacement, in the event of a failure. Type K, Chromel *vs* Alumel (Ni–Cr/Ni–Al) thermocouples are preferred.

The interior of the oven can be subjected to fairly corrosive conditions and should be fabricated from either zinc coated steel, or stainless steel which is a better option, to avoid expensive shut-downs due to corrosion damage in the future, especially where temperatures have dropped below the dew point. Exhaust vapors are kept hot to reduce condensation of tars and aqueous liquors. The efficiency of the oven insulation and checking for heat losses by a conductive path can be undertaken by surveying the oven surface with an infrared heat spy camera.

GUIDELINES FOR THE DESIGN OF EQUIPMENT FOR CARBON FIBER PLANT

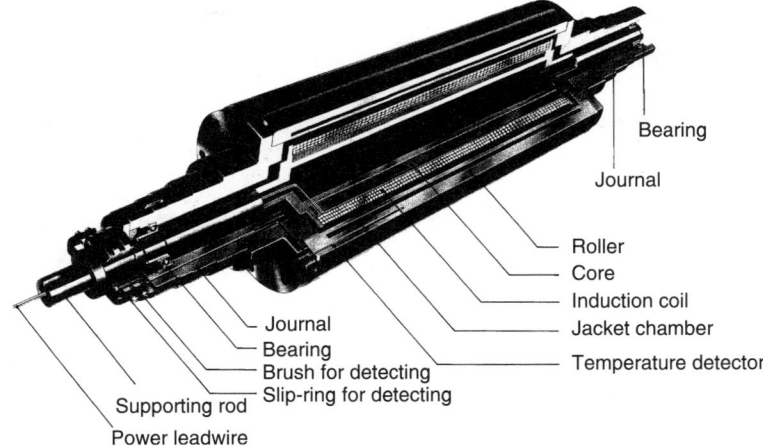

Figure 10.7 Induction heated roller with jacket. *Source:* Reprinted from Technical literature of Tokuden Co. Ltd., Kyoto, Japan

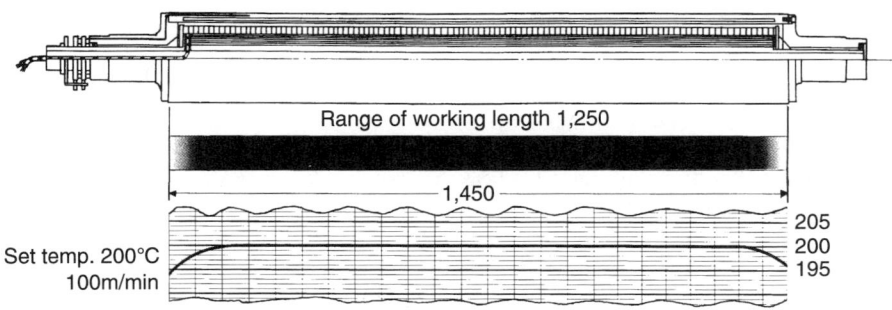

Figure 10.8 Schematic arrangement of jacket roller showing the accuracy of temperature distribution. *Source:* Reprinted from Technical literature of Tokuden Co. Ltd., Kyoto, Japan

Ovens must be provided with a system to deal with a fire, should it occur. Carbon dioxide or steam are probably the most efficient media. It should be remembered that CO_2 is asphyxiating and a flow of CO_2 gas can produce a static discharge. If steam is used, then a steam trap must be employed to remove condensed steam, which would otherwise form a lock of condensed steam and prevent the flow of steam when required in the event of a fire.

Instead of using ovens to oxidize the fiber, a totally different approach would be to pass the fiber around a series of uniformly heated rollers. Figure 10.7 shows a proprietary induction heated roller with a jacket containing a liquid/vapor thermal transfer medium, which will maintain a uniform temperature of ±2°C at 230°C along the length of the roller (Figure 10.8).

10.5 REMOVAL OF EFFLUENT GASES EVOLVED IN THE OXIDATION PROCESS

By-products such as HCN, H_2O, CO_2, CO, NH_3 and miscellaneous tars and finish are evolved during the oxidation of PAN and it is normal practice to pass the hot gases at about 300°C over a heated platinum group metal deposited directly onto a high surface area material, which has been coated on a porous ceramic monolith block. The system is rated at a minimum of 95% efficiency, but more recent proprietary regenerative thermal oxidizers also recover the waste heat and are 99% efficient.

10.6 APPLICATION OF AN ANTISTATIC FINISH

If the product is to be opf, the fiber is usually treated with a proprietary aqueous antistat finish, best applied by a padded roller system and the product subsequently dried, either over heated platens, or heated rollers. It is advisable to fit the application bath with a low level alarm.

10.7 PLAITER TABLE

To collect large opf tow, it is usual to plait each tow into a box fitted with a polyethylene liner, which is positioned on a table that is driven by dual motor actuated mechanisms—one supplying longitudinal movement and the other transverse movement. The stroke of the longitudinal movement matches the length of the box and at the end of the stroke, the motion reverses, whilst the transverse movement spaces the tow apart and reverses when the transverse movement is equal to the width of the box. Alternatively, the plaiter can be driven hydraulically.

The plaiter table is best mounted in the workshop floor, so that the table just clears the deck, making it relatively easy to position and remove cartons. For safety reasons, the area of table movement should be clearly defined on the floor. Plaiter tables mounted above the floor can, for example, be readily damaged by fork lift trucks. It is also useful to mount a set of platform scales in the floor, permitting easy manipulation of heavy cartons.

10.8 LT CARBONIZATION FURNACE

This furnace is required to process the oxidized fiber continuously up to about 1000°C in an inert atmosphere and it is normal to use an electrically heated slot furnace. A typical furnace (Figure 10.9) comprises a high nickel alloy muffle with five or six zones, producing a temperature gradient from ambient to about 1000°C. To prevent oxidation of the fiber by ingress of air, the furnace is provided with inlet and outlet gas seals using N_2 gas as the inert atmosphere. In the process, the fiber loses about 35% weight as gases and tars are formed. The flow of N_2 gas serves as a carrier to remove these products from the furnace via an outlet branch. As a general rule, to provide some cooling for the outlet end of the muffle, the unheated length at the exit of the furnace body is about twice the length of the unheated inlet. Since the fiber is hot when it exits the muffle, probably in excess of 200°C, it is usual to fit the furnace with a water cooled outlet section to prevent the fiber from burning when it enters the plant atmosphere.

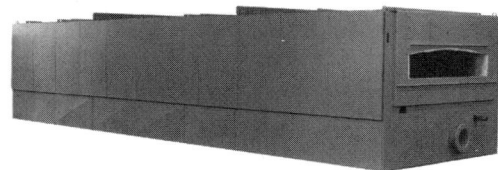

Figure 10.9 Typical LT carbonization furnace (about 8m long). *Source:* Courtesy of Harper International, Lancaster, NY, USA.

Table 10.1 List of high nickel alloys suitable for the construction of LT furnace muffles

Alloy maker	Designation	Elements (%)									
		Ni	Cr	Fe	Si	C	Mn	Al	Cu	P	S	Other
Allegheny ludlum	RA 330 alloy	35	19	43	1.25	0.05						
Inco alloys	INCO alloy 330	34.0–37.0	17.0–20.0	Bal.	0.75–1.50	0.08 max.	2.0 max.			0.030 max.	0.030 max.	
Inco alloys	INCO alloy DS	34.5–41.0	17.0–19.0	Bal.	1.9–2.6	0.10 max.	0.8–1.5		0.5 max.		0.03 max.	0.20 max. Ti
Allegheny ludlum	RA 600 alloy	76	15.5	8	0.2	0.08	0.3					
Inco alloys	INCONEL alloy 600	72.0 min.	14.0–17.0	6.0–10.0	0.5 max.	0.15 max.	1.0 max.		0.5 max.		0.015 max.	
Allegheny ludlum	RA 601 alloy	61.5	22.5	14	0.2	0.05	0.3	1.4				
Inco alloys	INCONEL alloy 601	58.0–63.0	21.0–25.0	Bal.	0.50 max.	0.10 max.	1.0 max.	1.0–1.7	1.0 max.		0.015 max.	
Krupp VDM	NICROFER 6025	Bal.	24.0–26.0	8.0–11.0	0.5 max.	0.15–0.25	0.1 max.	1.8–2.4	0.1 max.			0.1–0.2 Ti, 0.01–0.10 Zr, 0.05–0.12 Y

Figure 10.10 Outlet end of a typical high nickel alloy LT furnace muffle. *Source:* Courtesy of Welding & Brazing Developments (Cannock) Ltd., supplier of specialist and heat resistant fabrications.

The inlet end is relatively cool, which could permit the condensation of tars onto the fiber, resulting in inferior carbon fiber, so it is a useful precaution to fit a tar shield in the roof section of the inlet end to prevent any condensed tar from dripping onto the fiber.

A wide range of high nickel alloys are available for fabrication of the furnace muffle and are listed in Table 10.1. Furnace muffles must combine good thermal conductivity with high strength at operating temperature and resistance to internal and external environments without corrosion. The high nickel alloys develop tightly adherent oxide films that protect the surface from corrosion. A slight amount of internal oxidation occurs, providing a higher chromium content in the surface oxide. The oxidation resistance is advanced further if aluminum is present in the alloy, whilst the nickel and chromium impart resistance to corrosive media and high temperature environments.

The outlet end of a typical muffle is shown in Figure 10.10 and its strength is built in by using longitudinal and lateral corrugations. A typical muffle section is D-shaped, which incorporates a well to accumulate debris and will also allow for the catenary of the fiber when the fiber sags under its own weight in the furnace. The expansion coefficient of a high nickel alloy is about 17 $\mu m \, m^{-1} \, {}^\circ C^{-1}$ and at working temperature, will account for some 80 mm of movement, for which allowance has to be made. The best approach is to fix the muffle at the outlet branch and allow the muffle to expand towards the cool and hot ends, which can be achieved by resting the muffle inside the casing on high nickel alloy, or SiC plate slides, or rollers, and allowing for movement of the inlet and outlet seals.

10.8.1 LT furnace gas seals

Many types of gas seals have been tried, varying from pools of mercury to plastic sausages inflated with nitrogen. However, the usual form of seal is a distribution system for the inert gas, where all inlet piping is of the same diameter and length to ensure uniform flow. Care should be taken to avoid setting up a venturi, which can induce the ingress of air, resulting in an adverse effect on the carbon fiber properties.

The design of seals can be based on gas curtain technology and is influenced by:

1. The dimensions of the slot
2. The angle of flow of the inert gas
3. The velocity of the inert gas

GUIDELINES FOR THE DESIGN OF EQUIPMENT FOR CARBON FIBER PLANT

4. The pressure drop across the seal
5. The temperature and densities of the gases inside and outside the seal

Generally, minimum optimum gas requirements can be achieved with the flow projected at an angle of about 25° to the vertical and directed towards the hot end. Temperature affects the densities of the gases and lower temperatures, associated with higher densities, will help in providing the minimum gas requirements. This fact does rather negate the practice of warming the inert gas supply prior to entry into the gas seal; although an increase in temperature with a resultant increase in the volume, would, to some extent, offset the increased flow requirement. The pressure drop would be small, created by the temperature/density difference on either side of the seal, which is normally air (nominally 78% v/v N_2) on the outside and N_2 on the inside. The presence of turbulence should be minimized and this can be aided by reducing the pressure of the N_2 before it enters the gas seal.

10.8.2 LT furnace insulation

The furnace insulation can be furnace brick or slabs of insulation (e.g. FMI's Ceramaform) of selected thickness, which can be cut to shape with a wood saw and interlocked using lap joints to prevent loss of heat from the furnace interior (Figure 10.11). At the time of fitting, due allowance must be made for the shrinkage factor, which in this case is about 4%. Carbon powder insulation can be used, but is not recommended.

As a result of a flow of thermal energy when heated, a temperature difference develops between the inner and outer surfaces of the refractory wall. Part of this energy is stored within the wall and is known as heat capacity, while the remainder flows through the structure and is lost to the outside air by radiation and convection.

The property which determines the rate of heat flow through a material is known as the thermal conductivity k and for steady heat flow conditions:

$$Q = \frac{kA(t_1 - t_2)}{x}$$

where Q is the quantity of heat flowing between two parallel planes of area A, at a distance x, having a temperature difference (t_1-t_2), k is the thermal conductivity at the mean

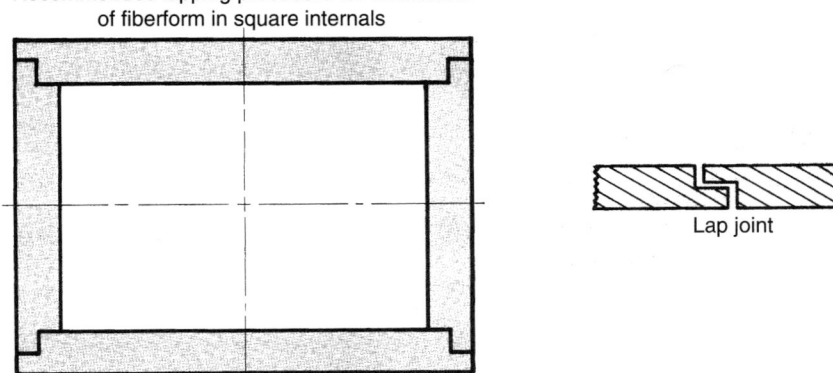

Figure 10.11 Construction of furnace insulation from rigid board insulation material. *Source:* Reprinted from Technical literature of Fiber Material Inc., Biddeford, Maine, USA.

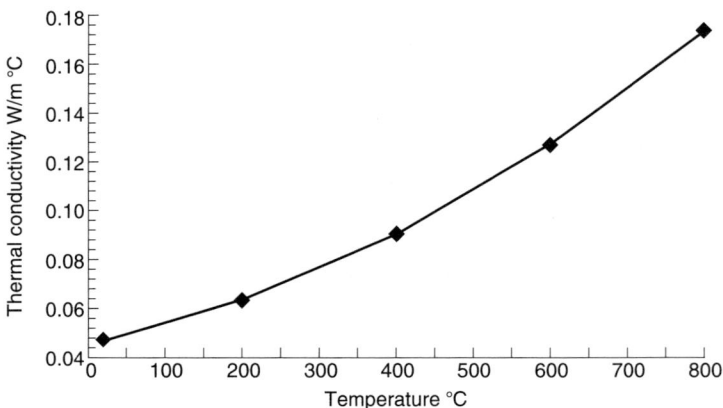

Figure 10.12 Thermal conductivity of FMI's Ceramaform rigid board insulation. *Source:* Reprinted from Technical literature of Fiber Material Inc., Biddeford, Maine, USA.

temperature of t_1 and t_2 [5]. The thermal conductivity is measured in $W\,m^{-2}\,K^{-1}$ or $kcal\,m^{-2}\,h^{-1}/°C$.

Knowing the thermal conductivity (Figure 10.12) of the insulation material at the mean temperature, the quantity of heat passing through a refractory wall is

$$Q = \frac{kA(t_1 - t_2)}{x}$$

and for unit area,

$$Q = \frac{(t_1 - t_2)}{x/k}$$

This equation applies to a simple wall constructed from a single grade of material. For a composite wall, where the same quantity of heat passesthrough each layer of insulation, the equation becomes:

$$Q = \frac{t_1 - t_n}{[(x_1/k_1) + (x_2/k_2) + \cdots + (x_n/k_n)]}$$

where $x_1, x_2, \ldots, x_n$ and $k_1, k_2, \ldots, k_n$ refer to the thickness and thermal conductivity of each layer respectively.

Since the temperature drop across each layer is equal to the product of the heat flow and the actual thermal resistances of the respective layers are

$$t_1 - t_2 = Q\frac{x_1}{k_1} \text{ and } t_2 - t_3 = Q\frac{x_2}{k_2}$$

From these temperature drops, the interface temperature can be calculated. Insulation companies now supply their customers with computer programs for deriving the most suitable thickness of the insulation.

10.8.3 Element materials for LT furnaces

An LT furnace is heated to about 1000°C and generally, the choice of element material is between a wire resistance heating element or SiC (Table 10.2). Alloys like Kanthal A

Table 10.2 Operating temperatures of element materials

Material	Manufacturer's type	Atmosphere	Maximum element operating temperature °C
Fe bal/Cr22/Al5.8	Kanthal A-1	Air	1400
		Nitrogen	–
Fe bal/Cr22/Al5.8	Kanthal APM	Air	1425
		Nitrogen	–
0.02 C/Fe bal/Cr 20.0/Al 5.0/Yt 0.1	Resistalloy Fecralloy	Air	1325
		Nitrogen	1200[a]
0.01 C/Fe bal/Cr 20.0/Al 5.0/Yt 0.4	Resistalloy Fecralloy A	Air	1325
		Nitrogen	1200[a]
Silicon carbide SiC	Kanthal Hot Rod	Air	1625
		Nitrogen	1350[b]
	Kanthal Globar LL	Air	1540
		Nitrogen	1350[b]
	Kanthal Silit	Air	1625
		Nitrogen	1350[b]
Molybdenum disilicide MoSi$_2$	Kanthal Super 1700	Air	1700[c]
		Nitrogen	1600[c]
	Kanthal Super 1800	Air	1800[c]
		Nitrogen	1700[c]
	Kanthal Super 1900	Air	1900[c]
		Nitrogen	1700[c]

Notes
[a]Before use, pre-oxidize for 1–4 h at 1100°C.
[b]At about 1370°C, SiC reacts with N_2 to form Si_3N_4, the reaction becoming rapid at 1400–1500°C. Thereafter, the reaction decreases with increasing temperature and SiC can be used at >1900°C.
[c]MoSi$_2$ becomes ductile at 1200°C. Elements are generally U-shaped and suspended vertically.

(Fe/Cr/Al), resistant to 1350°C, can be used but are rather brittle and are best used with 18Cr/10Ni/3Mo tails, which are welded onto the ends of the Kanthal coils. AERE, Harwell developed Fecralloy^{TM}, a Fe/20Cr/5Al alloy with an addition of 0.1–0.4% Yt, which improved the stability. To prevent sagging, the element wire is coiled and then supported on a ceramic tube of requisite diameter. A furnace manufacturer has reported that the problems most commonly experienced by them with furnaces operating at about 1000°C were loose element connections, leading to sparking and subsequent premature failure. This can be overcome quite easily by using a torque wrench to tighten all connections.

Temperatures can be measured with Chromel vs Alumel thermocouples in Inconel sheaths, which can be earthed to avoid stray pick-up from the heating elements.

In order to heat a furnace to 1000°C, assuming a load of, say, 6 kW operating from 230 V supply, then:

$$V = IR \text{ and watt} = \text{volt} \times \text{ampere}$$

Hence, $R = \dfrac{V^2}{W}$

$$\therefore R = \frac{230^2}{6000} = 8.82 \text{ ohm}$$

This is then corrected using Table 10.3 to obtain the resistance of the cold wire (i.e. as purchased).

$$R_c = \frac{8.82}{1.080} = 8.17 \text{ ohm}$$

Table 10.3 Correction factor for converting resistance of Fecralloy wire from room temperature to working temperature

Temperature °C	Factor	Temperature °C	Factor	Temperature °C	Factor
20	1.000	500	1.032	1000	1.080
100	1.007	600	1.054	1100	1.080
200	1.010	700	1.070	1200	1.081
300	1.016	800	1.077	1300	1.081
400	1.022	900	1.079		

Note: multiply room temperature value by the factor.
Source: Reprinted from technical literature of Resistalloy.

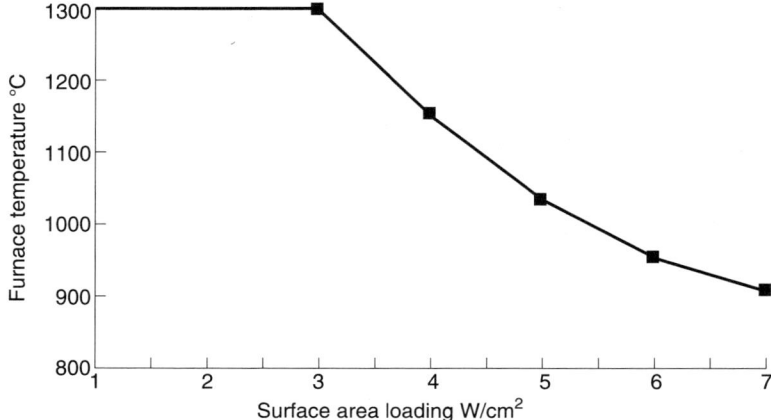

Figure 10.13 Surface loading of Fecralloy wire. *Source:* Reprinted from technical literature of Resistalloy.

Assume the surface loading is $1.5\,\text{W}\,\text{cm}^{-2}$ (Figure 10.13). Calculating the $\text{cm}^2\,\text{ohm}^{-1}$ quantity by dividing the power required by the selected surface area loading and multiplying by the cold resistance,

$$\frac{W}{S \times R_c} = \frac{6000}{1.5 \times 8.17} = 490\,\text{cm}^2\,\text{ohm}^{-1}$$

From the manufacturer's tables (Table 10.4), the nearest suitable standard size wire is 10 swg × 3.00 mm diameter. The resistance at 20°C is 0.190 ohm m^{-1}, so the length required would be

$$\frac{8.17}{0.190} = 43.0\,\text{m}$$

Extra length should be added for connecting to the supply. The wire can be coiled on a mandrel in a lathe and the coiled section can be slipped over the ceramic tube support and stretched to the requisite length.

Kanthal have introduced Kanthal APM (Fe/Cr/5.8Al (An Fe/Cr alloy with 5.8 Al and an addition of Yt) Yt), which is superior to Kanthal A and less brittle.

Since an impervious alloy muffle is used for LT furnace construction, an element material that performs well in an atmosphere of air can be selected. Silicon carbide is one such ideal element material having a working temperature in air of up to 1625°C, which is well above the required operating temperature. Moreover, SiC remains rigid at all operating

GUIDELINES FOR THE DESIGN OF EQUIPMENT FOR CARBON FIBER PLANT

Table 10.4 Extract from resistance data on Fecralloy wire

SWG	Diameter mm	Surface area/length cm² m⁻¹	Length/unit weight m kg⁻¹	Resistance at 20°C ohms m⁻¹	Surface area/ohm cm² Ω⁻¹
8	4.06	128	10.7	0.106	1209
	4.0	126	11.0	0.109	1147
9	3.66	115	13.2	0.130	885
	3.5	110	14.4	0.140	785
10	3.25	102	16.7	0.165	621
	3.0	94.2	19.6	0.193	487
11	2.95	92.7	20.3	0.201	460
	2.91	91.4	20.8	0.208	439
12	2.64	82.9	25.3	0.250	332
	2.59	81.4	26.3	0.260	313
	2.5	78.5	28.2	0.279	281

Source: Reprinted from technical literature of Resistalloy.

temperatures, thus eliminating the need for supports. Kanthal manufacture three classes of SiC for industrial furnaces:

1. Hot Rod—made from high purity α-SiC grains fused as a rod or thick walled tube at >2500°C, with a maximum hot length of 2 m. The cold ends are impregnated with high conductivity Si and aluminized to give a low resistance contact surface for the electrical connections.
2. Globar LL—made by a process similar to Hot Rod, but in three sections with the low resistance cold ends jointed, bonded and then aluminized and available with a hot zone length up to 2.5 m.
3. Silit ED—made in one piece with the outer extremities of cold ends aluminized and available with a hot zone length up to 2.5 m.

Silicon carbide is a semiconductor material with a resistivity much higher than conventional metallic resistance materials. Resistance measurements taken at room temperature will give no indication of the resistance at higher temperatures (Figure 10.14) and values must be determined. These values are measured by the manufacturer at a constant temperature (e.g. 1050°C for Hot Rod and 1070°C for Globar LL and Silit ED elements). Elements to be connected in series should be selected within a resistance range of ±5%, whereas elements connected in parallel can be ±10%.

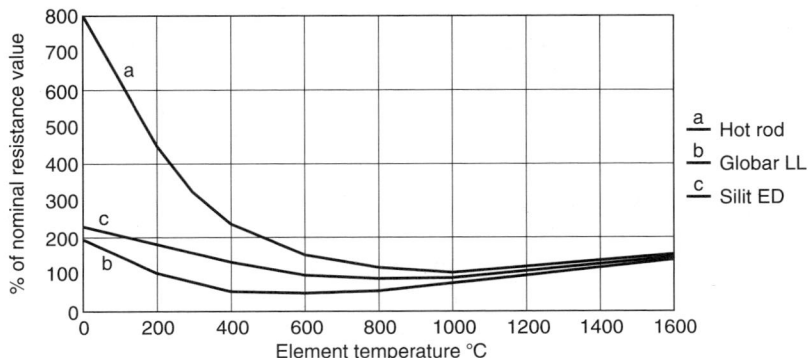

Figure 10.14 Resistance *vs* Temperature of Kanthal SiC elements. *Source:* Reprinted from technical literature of Kanthal.

Table 10.5 Recommended maximum temperatures of use for selected fired refractories

Type of fired refractory	Composition Al_2O_3 SiO_2 Fe_2O_3 TiO_2 CaO Na_2O Bal. MgO K_2O							Maximum service temperature °C
Impervious aluminous porcelain (IAP)	55.3	40.0	0.9	0.9	1.4	1.5	-	1550
Mullite	58.6	36.8	0.9	0.9	0.8	1.32	0.68	1700
Purox recrystallized alumina	99.7	0.05	0.04	0.02	0.03	0.15	0.01	1950

Heating elements made of SiC gradually increase in resistance during their working life and have to be compensated for by using a variable voltage power supply. In the event of an element failure, either the whole group of elements in that zone should be replaced or, the failed element should be replaced by a used element of the same resistance. Optimum loadings for these elements should be 3–8 W cm^{-2} of the hot zone surface area. Elements are mounted in impervious aluminous porcelain (IAP) sleeves in the furnace sidewall insulation (Table 10.5), so that they are isolated from the furnace casing and free to move both axially and radially. Electrical connections can be made with flexible aluminum braid. Temperatures can be recorded with Pt/Pt + 13% Rh thermocouples in ceramic thermowells and control can be achieved with thyristor power packs.

10.9 LT FURNACE EXHAUST REMOVAL

Many methods have been used for disposal of LT furnace exhaust products, varying from scrubbing with water to remove the tars to absorbing the exhaust in a glycol that can subsequently be incinerated. Lately, it has become common practice to directly incinerate the effluent products of carbonization (HCN, H_2O, NH_3, CO, H_2, CH_4, higher hydrocarbons and tars) in a gas- or oil-fired incinerator. More control can be obtained with gas, since a gas-fired proportionaling burner has a wider operating ratio and can more readily cope with supplying maximum fuel at start-up. When the operating temperature has been reached and production has started, the burner will reduce the gas supply since the gases and tars evolved during the process then become the major source of fuel. Larger amounts of fuel become available as fiber throughput increases.

An incinerator (Figure 10.15) is normally a steel body lined with refractory cement, which requires an initial cure followed by a gradual warm-up schedule, in the event of a plant shutdown. The effluent gases are fed tangentially into the incinerator and impinge on the tip of the burner flame. The burner requires primary air supplied by an integral fan unit, whilst secondary air for combustion is admitted and induced into the chamber by the upward draught of the incinerator chimney. The flow of secondary air is adjusted to give a temperature of about 750°C, corresponding to the formation of the least amount of HCN and adequate combustion of other gases. The shell temperature is manipulated to about 180°C, which is above the dew point of any corrosive products of combustion that could condense out. The suction caused by the upward draught in the chimney also draws the effluent gases from the furnace into the incinerator and this flow has to be very carefully controlled to prevent ingress of air into the LT furnace. The exhaust line should be provided with rodding points to keep the pipe free from carbon fiber debris, which may seriously impair the extraction performance.

Generally, such a system has an efficiency of about 90–95% and newer systems are being introduced that are more environment friendly and at least 99% efficient.

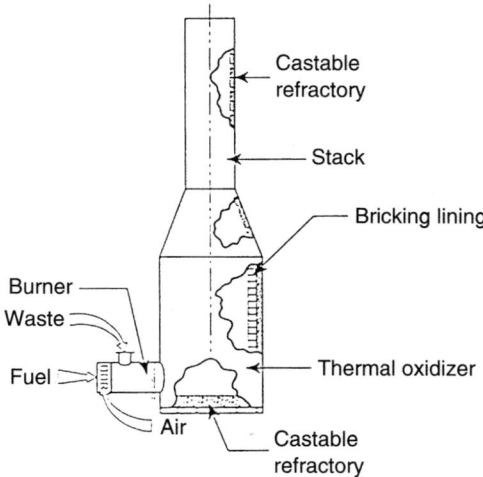

Figure 10.15 A basic thermal oxidation unit. *Source:* Adapted from technical literature of John Zink.

The theory of the combustion process [8] is quite complicated because of the number of possible fuels that can influence the reaction, as well as the presence of moisture. As an example, provided the temperature is adequate, the carbon oxidation chemistries can follow either singly, or simultaneously, the following reactions, all of which occur within a few microseconds:

$$C + O_2 \rightarrow CO_2 \qquad 2C + O_2 \rightarrow 2CO$$
$$C + CO_2 \rightarrow 2CO \qquad CO + \tfrac{1}{2}O_2 \rightarrow CO_2$$
$$C + H_2O \rightarrow CO + H_2 \qquad C + 2H_2O \rightarrow CO_2 + 2H_2$$
$$CO + H_2O \rightarrow CO_2 + H_2$$

When fuels are burned, the N_2 and O_2 in the air are molecularly dissociated at the high flame temperatures and react to form NO_x, which is environmentally unacceptable, being a significant contributor in the generation of smog. Legislation restricts the permitted amount of NO_x in the atmosphere. Lowering the flame temperature does reduce the amount of NO_x formed.

The presence of organic bound nitrogen in the fuel drastically increases the presence of NO_x. If the burning of the bonded nitrogen-bearing products is carried out with sub-stoichiometric oxygen (and ammonia can be added to the hot gases after combustion), the NO_x content is significantly reduced.

Figure 10.16 is a block diagram depicting a two-stage combustion process to dispose of waste products, which, in a single stage combustion process, would have produced a flue gas containing excessive amounts of NO_x. Such a process comprises:

1. Reduction furnace—in which a high temperature reducing (less than stoichiometric air) environment converts the fuel into H_2, H_2O, CO_2 and CO, and converts the NO_x present into N_2.
2. Quench section—which cools the water gas (CO and H_2) to about 760°C by directly contacting it with a cool recycle gas (cooling can also be achieved by injecting water or steam).
3. Thermal oxidizer—which converts the H_2 to H_2O and CO to CO_2.
4. Heat recovery boiler—which produces steam when cooling the flue gas to 177°C.
5. Unlined vent stack

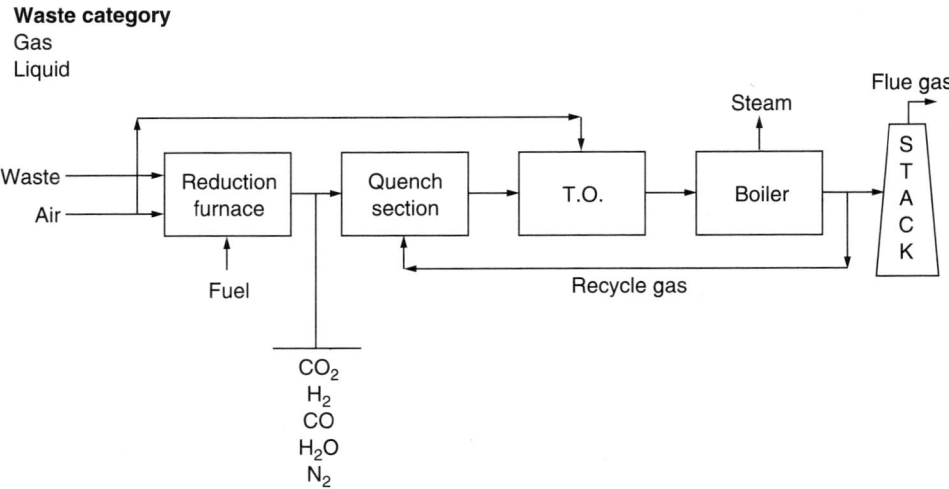

Figure 10.16 Block diagram depicting a two-stage combustion process to dispose of waste products avoiding formation of excessive amounts of NO_x. *Source:* Reprinted from technical literature of John Zink.

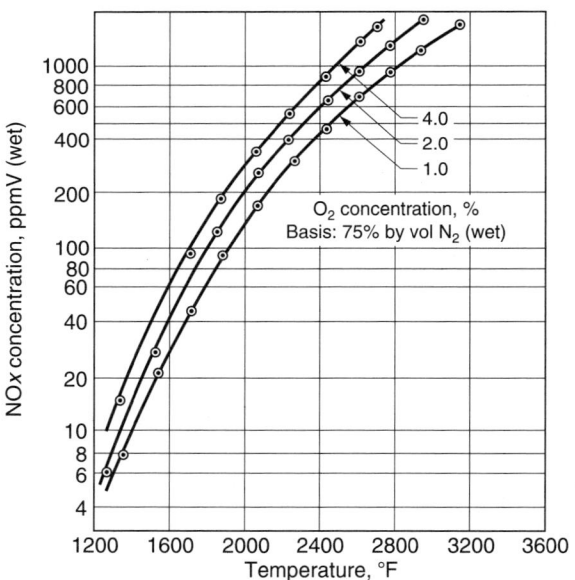

Figure 10.17 Plot of NO_x concentration *vs* temperature with varying amounts of oxygen concentration. *Source:* Reprinted from technical literature of John Zink.

The cooling step between the reducing and re-oxidation stages lowers the thermal oxidizer temperature. Figure 10.17 shows the concentration of NO_x versus temperature for different levels of excess oxygen and it is obviously desirable to operate at the lowest practical temperature. However, there is another consideration, namely, to oxidize the H_2 and CO present to meet acceptable legislative levels and therefore, the optimum temperature selected will be a compromise. Figure 10.18 is a schematic diagram of a typical NO_xIDIZER system.

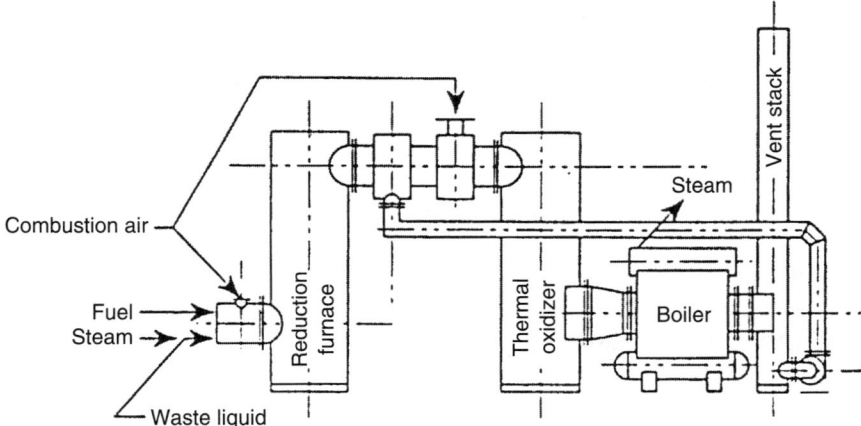

Figure 10.18 Schematic layout of a John Zink Super Lo NO_x boiler system. *Source:* Reprinted from technical literature of John Zink.

Figure 10.19 Typical HT carbonization furnace. *Source:* Courtesy of Harper International, Lancaster, NY, USA.

10.10 HT CARBONIZATION FURNACE

The HT furnace (Figure 10.19) is required to process fiber continuously at temperatures up to about 1500°C in an inert atmosphere. This temperature is too high for the furnace muffle to be constructed from a high nickel alloy, so graphite is used. A rectangular section muffle of the appropriate length with top and bottom electrode chambers can be assembled from blocks of graphite with lap joints (Figure 10.11). To isolate the muffle from the electrode chambers, the muffle joints are glued together using graphite cement. The element chamber can, alternatively, be constructed from blocks of carbon insulation. At such working temperatures, the muffle and element chambers have to be protected, inside and out, with an inert gas such as N_2, which rules out the possibility of using SiC elements, so graphite elements are used. Allowance is made for expansion by supporting the muffle on graphite plate slides or rollers. Either $Pt+13\%Rh/Pt$ thermocouples, or an optical pyrometer (Appendix 12 for calibration details), can be used to measure the temperature and control can be either manual or automatic. If $Pt+13\%Rh/Pt$ thermocouples are used, then they should be protected with ceramic thermowells, which themselves have to be protected with a graphite sleeve against possible attack by Na.

10.10.1 HT furnace gas seals

The gas seals can be of the same design as that used for the LT furnace. However, the body flow of inert gas is considerably reduced, since the LT furnace has to deal with about 30% evolved products, whereas the HT furnace has only to cope with about 15%. Since there is no evolution of tars, a tar shield is not required. However, if the precursor fiber contains Na, then depending on the operating temperature, this can be released in the HT furnace.

10.10.2 HT furnace insulation

The insulation adjacent to the hot muffle will have to be either carbon or graphite and as the temperature is dropped, it can be replaced by a ceramic form, although many subscribe to the view that it is best to keep all insulation black in order to avoid any possible interaction with the water and SiO_2 that may be present in the ceramic insulation, or the Na evolved from the product. Three forms of carbon/graphite insulation can be used, namely, board, felt and powder. The use of powder is not recommended, since there are many handling problems, including possible contamination during use with NaCN.

FMI's Fiberform and Calcarb's product Calcarb are carbon bonded carbon fiber insulation materials (the FMI product is available carbonized up to 1800°C) with a thermal conductivity of $0.15\ W\ m^{-1}\ K^{-1}$, which after stabilization (usually 200° above the maximum working temperature), can be used up to 2750°C. It is essential that this form of insulation is not used above 200°C in an oxidizing atmosphere. Once this product catches fire, it is extremely difficult to extinguish, so the furnace must be protected throughout at all times with an inert gas such as N_2. Fiberform and Calcarb are available as boards, which can be cut with a wood saw and made self supporting by using interlocking and lap-joints techniques (Figure 10.11). At 1500°C, the longitudinal expansion of such insulation in a typical furnace would be of the order of 75 mm.

Carbon and graphite felts are available from a number of manufacturers and are very convenient to use, but relatively expensive. The graphite felt is made by the graphitization of the carbon felt and Figure 10.20 shows the variation of thermal conductivity with temperature.

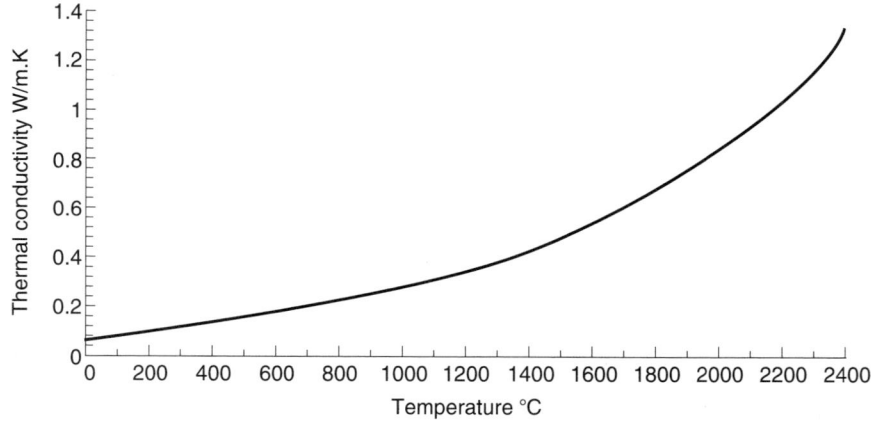

Figure 10.20 Thermal conductivity of Sigratherm felts in nitrogen or argon. a) ≤1400°C (Sigritherm GFA 5) b) >1400°C (Sigritherm GFA 10). *Source:* Reprinted from technical literature of SGL.

One furnace manufacturer lays great emphasis on virtually complete removal of moisture from the insulation before use by flushing with a body flow of N_2 and gradually increasing the temperature, whilst monitoring the dew point of the exiting gases and increasing the temperature when a stipulated dew point has been reached.

It is most essential that air is not allowed to enter this furnace when it is hot since severe erosion of the carbon/graphite will occur. All air in the furnace should be displaced with inert gas before start-up.

10.10.3 Element materials for HT furnaces

Depending on the design, typically 3–6 zones are used. A grade of graphite suitable for operation at the maximum temperature required is selected, which must be available in the requisite length to bridge the width of the furnace. A choice of rod or tube is then made, which has a cross-sectional area that will give the necessary required target resistance for a heating zone, when a number of such elements are connected in series, or parallel, or a combination of both, to give the final power output. Elements can be joined together in series in a U-shape to have extra length in order to provide the requisite resistance, or else, two Us can be connected together to give an M-shape, resulting in connections on the same side (Figure 10.21) of the furnace. Since connections get hot, it is usual to use water cooling on the ends of the elements. A typical water cooled assembly could be made from a hollow copper feed provided with water inlet and outlet pipes and screwed into the end of a graphite power feeder of larger cross-section than the element, which itself is screwed onto the element. The copper cooling section and water supply pipes must be electrically isolated from the furnace body and the cooling sections must be connected to the busbar feed. A 2 m long element will expand by about 20 mm and this expansion is catered for by the use of flexible connectors, which can comprise a bunch of thin copper strips that will accommodate the electrode current *in toto*. The copper strips can be dip coated at the ends with tin to give an integral unit that retains flexibility. When properly made and tightened, a current density of 10 $A\,cm^{-2}$ can be safely used with graphite elements in copper holders. The contact area with the copper should be about twice the cross-sectional area of the electrode. Up to 150 $W\,cm^{-2}$ can be used

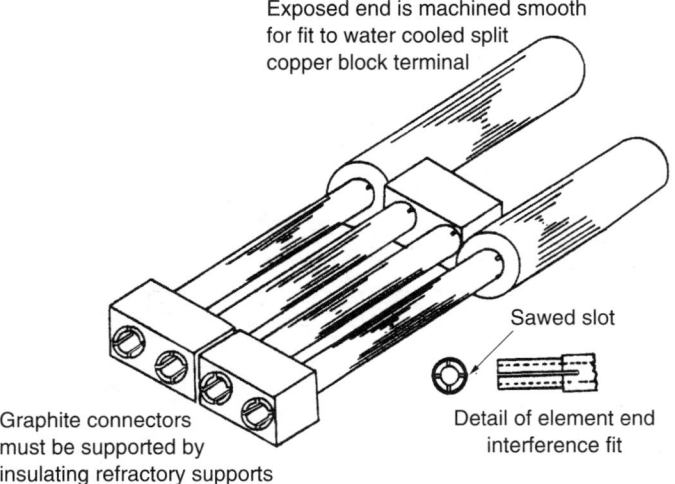

Figure 10.21 Method of joining a number of graphite elements to obtain increased resistance.

with electrodes with <50 mm diameter and 120 W cm^{-2} for diameters ≥50 mm. As a general guide, the power usage comprises about a 25% loss through the power feeders, 10% loss through the insulation, 20% loss with the N_2 and surprisingly, a 45% loss via the fiber.

Kanthal have a range of MoSi2 based Super alloys, which are resistant up to 1700°C in nitrogen, but unfortunately they undergo plastic deformation at high temperature and their use is restricted to U-shaped elements for free vertical suspension. The elements do not age and retain their electrical characteristics; hence it is possible to mix the old and new elements. The resistivity increases sharply with temperature, making it necessary to use voltage control in order to limit the current drawn at low temperatures, but the output must be rated high enough to permit the furnace to heat up from cold.

10.11 TYPICAL CALCULATIONS FOR THE DESIGN OF AN HT FURNACE

This design is a simplified approach for a three zone furnace with a stipulated output of 110 kg h^{-1} of carbon fiber, operating in the range 1000–1500°C, each zone having length and breadth 1.5 m each, using graphite construction throughout (Figure 10.22). Calculations in Table 10.6 are based on the following formulae:

Area of circle $= \pi d^2/4$ where $d =$ diameter

Area of surface of a cylinder $= \pi \times d \times l$ where $l =$ length

Thermal conductivity $=$ W m^{-1} K^{-1} where W = watts, m = meter, K = temperature in kelvin.

The quantity of heat flowing between two parallel planes of area A at a distance x with temperature difference $t_1 - t_2$ is given by:

$$Q = \frac{kA(t_1 - t_2)}{x}$$

Resistivity (specific resistance) $= \Omega$ m $=$ A $\Omega/l\Omega =$ ohm, A $=$ csa

Ohm's law states that $V = IR$, where $V =$ volts, $I =$ current in A, $R =$ resistance in ohm

Watt (J/s) $= V \times I = R \times I^2$ J $=$ joules

Energy (W m^{-2}K^{-4}) $= k \times \varepsilon \times (T_2^4 - T_1^4)$

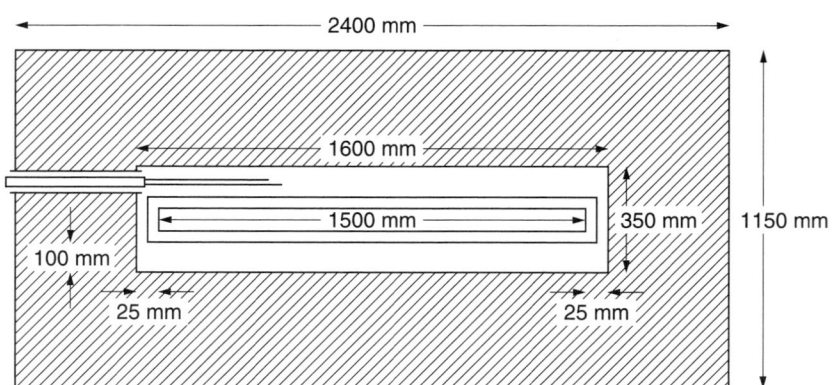

Figure 10.22 Section of model HT furnace.

Table 10.6 Typical calculations for the design of an HT furnace

Zone temperature	Units °C	Zone 1 1000	Zone 2 1200	Zone 3 1400	Total
	Furnace dimensions				
Muffle width	mm	1500	1500	1500	
Muffle depth	mm	100	100	100	
Muffle depth	mm	100			
Zone length	mm	1500	1500	1500	
Muffle thickness	mm	25	25	25	
Zone muffle area	m^2	4.8	4.8	4.8	
Muffle/insulation gap—Top/Bottom	mm	100	100	100	
Muffle/insulation gap—Side	mm	25	25	25	
	Insulation				
Thickness	mm	400	400	400	
Perimeter	m	3.9	3.9	3.9	
Effective area	m^2	5.85	5.85	5.85	
Outer temperature	°C	60	60	60	
Assumed mid temperature	°C	470	570	670	
Thermal conductivity of insulation	$W\,m^{-1}K^{-1}$	0.14	0.14	0.14	
Heat flux	kW	1.92	2.33	2.74	7.00
	Physical constants				
Emissivity ratio for graphite		0.8	0.8	0.8	
Stefan-Boltzmann constant	$W\,m^{-2}K^{-4}$	5.67E-08	5.67E-08	5.67E-08	
Thermal conductivity graphite	$W\,m^{-1}K^{-1}$	45	40	36	
Density of N_2 at STP	$kg\,m^{-3}$	1.25	1.25	1.25	
Specific heat of N_2	$kJ\,kg^{-1}K^{-1}$	1.04	1.04	1.04	
Specific heat graphite	$kJ\,kg^{-1}K^{-1}$	1.51	1.61	1.69	
Specific heat carbon fiber	$kJ\,kg^{-1}K^{-1}$	1.6	1.6	1.6	
	Process conditions				
N_2 flow through muffle	$m^3\,min^{-1}$	0.95	0.95	0.95	
Ambient temperature	°C	20	20	20	
N_2 inlet temperature	°C	5	5	5	
Maximum muffle temperature	°C				1400
Mass throughput of carbon fiber	$kg\,h^{-1}$				110.3
Power to heat carbon fiber to zone temperature	kW	45.7	9.8	10.4	65.9
Mass flow of N_2	$kg\,sec^{-1}$				0.01979
Heat required to warm N_2 to zone temperature	kW	20.5	4.1	4.1	28.7
Total process heat required	kW	66.2	13.9	14.5	94.6
Muffle inner surface temperature	°C	1035	1205	1403	
Muffle outer surface temperature	°C	1043	1207	1405	
	Power feeder (conducted loss)				
Length	mm	400	400	400	
Diameter	mm	40	40	40	
Feeder csa	m^2	1.26E-03	1.26E-03	1.26E-03	
Feeder temperature	°C	70	70	70	
Thermal conductivity	$W\,m^{-1}K^{-1}$	120	120	120	
Conductive loss per feeder	kW	0.35	0.43	0.5	
Loss per zone (8 elements per zone i.e. 16 feeders)	kW	5.6	6.9	8	20.5

(Continued)

Table 10.6 Continued

Zone temperature	Units °C	Zone 1 1000	Zone 2 1200	Zone 3 1400	Total
Element					
Element diameter	mm	20	20	20	
Element length	mm	1550	1550	1550	
Element csa	m²	3.14E-04	3.14E-04	3.14E-04	
Element resistivity	Ω m	1.00E-05	1.00E-05	1.00E-05	
Element resistance	mΩ	49.3	49.3	49.3	
Required power	kW	71.8	20.8	22.5	115.1
Number of elements		8	8	8	
Power per element	kW	8.98	2.60	2.81	
Element surface area	m²	0.0974	0.0974	0.0974	
Element surface temperature	°C	1048	1208	1406	
Element current	A	427	230	239	
Zone current	A	3416	1840	1912	7168
Zone voltage	V	21.0	11.3	11.8	
Power feeder (resistive loss)					
Resistivity	Ω m	1.00E-05	1.00E-05	1.00E-05	
Resistance	mΩ	3.17	3.17	3.17	
Resistive loss per feeder	kW	0.58	0.17	0.18	
Loss per zone (8 elements per zone i.e. 16 feeders)	kW	9.3	2.7	2.9	14.9
Total furnace power	kW				137.0

Source: Original program courtesy of Dr Heath TV.

where k = Stefan Boltzmann constant = 5.67×10^{-8} W m^{-2}K^{-4}
ε = emissivity ratio for graphite (= 0.8)
T_1 = furnace temperature, K
T_2 = element temperature, K

10.12 SODIUM REMOVAL

If Na is present in the precursor, it usually reacts with HCN in the furnace to form NaCN, so it is most beneficial to use a precursor with a low Na content or better, with none. If Na is present, it will be released from the fiber and at working temperature, will vaporize and form NaCN, Na_2CO_3, or could possibly remain as elemental Na. Table 10.7 lists the physical properties of these byproducts.

The body flow of inert gas is used to remove these products from the furnace, avoiding condensation onto the fiber or furnace body. When condensed as a fine powder, the larger

Table 10.7 Properties of solid waste products in the HT furnace

Substance	mp °C	bp °C
Na	98	880
NaCN	564	1496
Na_2CO_3	851	Decomposes

particles can be removed in a cyclone (Figure 10.38), which also has a cooling effect and any products escaping from the cyclone can be absorbed in a water scrubber, which will also acts as a liquid seal pot. This aspect of the process needs very careful control. Conventional blanket filters can be used to remove dust, but changing filters loaded with toxic dust is an unpleasant task.

10.13 HM HEAT TREATMENT FURNACE

When required, some quantity of fiber can be converted to HM type fiber by passing through a furnace heated up to about 2600°C. This requires special care as furnaces at these elevated temperatures have a limited life and must be protected with an inert gas. One type of furnace that can be used is based on a design using a graphite hairpin heating element, which is protected from the furnace insulation with a graphite guard tube (Figure 10.23). The guard tube protects the insulation from any emissions and an inert atmosphere, maintained by purging with an inert gas, protects the fiber, element and insulation. The guard tube also prevents any insulation from shorting across the slot of the element.

10.13.1 HM furnace gas seals

The gas seals can be of the same design as those used for the HT furnace. However, the body flow of inert gas is considerably reduced and must be precisely controlled to obtain quiescent conditions, since a higher flow than required will continually remove the surface layer of carbon vapor from the element and cause severe erosion, considerably reducing the life of the element.

The inert gas could be:

1. Ar—which is expensive (about eight times the price of N_2) and, unfortunately, with a hairpin type of element, ionizes across the element slot at temperatures about 2000°C.
2. He—which is also expensive, does not ionize, but has a high thermal conductivity. Helium can be added to argon to prevent ionization.
3. N_2—which does not show any evidence of reacting with either the fiber or the furnace element.
4. H_2—which can be used but is not recommended on grounds of safety.

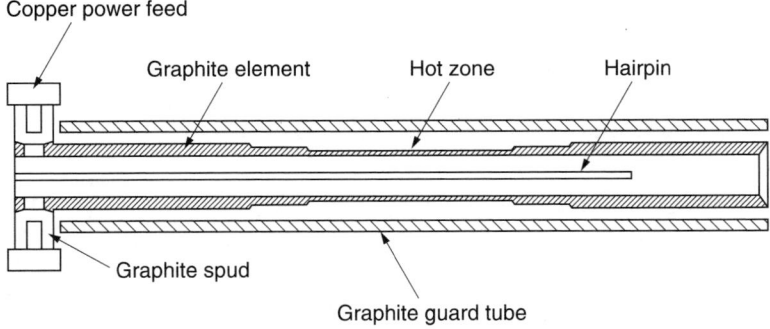

Figure 10.23 Design of HM furnace hairpin element assembly.

10.13.2 HM furnace insulation

Carbon powder can be used for insulation and one way of ensuring a uniform filling of the furnace body is to apply the powder as a slurry in ethanol. Once filled, the alcohol is then removed by gradually applying heat, exhausting the fumes using a slight vacuum and condensing any vapors with a solvent trap. It is advisable to use a carbon powder with a low sulphur content (e.g. made from natural gas) in this method, since a carbon powder made from an oil stock can have high sulphur content, which can be liberated in the furnace. Although it will condense out in the cooler parts of the furnace body, in hotter regions, it can react with, for example, the Ni present in stainless steel to form a nickel-nickel-sulphur eutectic alloy which melts at a temperature of about 750°C, well below the normal maximum temperature of use for a stainless steel.

The preferred form of insulation in this instance would be a carbon or graphite felt (Figure 10.20) conveniently wound onto the muffle guard tube and temporarily secured in place with cellotape, which will burn away when the furnace is heated up.

10.13.3 HM furnace element design

An HM element typically has a tubular design (Figure 10.23) and can be made from a fine grained graphite (e.g. SGL's HLM 85) with a hairpin slot (12.5 mm wide) running about 85% of the total length (1825 mm). The fiber runs through the element (100 mm I/D) and the outside of the element has variable diameter, with the thinnest section (some 500 mm long) with the highest resistance at the center, producing the hot zone. The final resistance of the element is critical and the following procedure should be adopted:

1. Select a solid billet of graphite of the requisite resistance, determined with a four-wire milli-ohmmeter.
2. The cross-section of the billet is profiled by machining, checking the resistance at all stages to ensure that the target resistance (about 5–6 milliohm) is obtained.
3. With the element positioned in the furnace body, non-conductive chocks of the appropriate thickness are inserted into the slot of the hairpin for support, and the graphite power feeds (spuds) are screwed into position, with the total resistance of the element assembly being monitored at each stage. It is usual to lap these spuds into the requisite conical socket of the element prior to assembly, minimizing the contact resistance, thus ensuring that all carbon dust and debris is removed before finally seating. A strap wrench can be used to ensure adequate tightness.
4. The water cooled copper power feeds are then screwed into place and copper shims of the appropriate thickness are used to obtain the correct rotational position for the assembly to the busbar system, which could utilize the copper pipes used for the cooling water. Temperatures up to about 2650°C can be used, but erosion will limit the life of the element. At 2625°C, the total resistance is about 10 milliohms, which rises fairly quickly to about 11 milliohms towards the end of the element's life. Monitoring the transformer voltage and kilowatt readings can be used to calculate the total resistance from from the equation $R = V^2/\mathrm{kW}$.

In one of the designs of HM furnace, the contact resistance of the current feed cone to the element was a controlling factor and up to 30% of the total power loss could be dissipated at these contacts. In operation, a furnace with a 50 mm radial thickness of insulation required a load of 37 kVA to maintain the furnace at 2300°C, about 44% of the heat was lost radially to the outer cooling jacket, about 48% through the end clamps and 8% in the cooled end transfer ports.

Temperature measurement is done by focusing an optical pyrometer onto the element slot in the hot zone via a sighting tube positioned on the furnace body and fitted with a temperature resistant glass window, kept clear of soot with a minimal flow of N_2 (too much will cause severe element erosion).

10.14 SURFACE TREATMENT

The literature lists many ways of surface treating carbon fiber to improve the bond with resin matrices. Obviously, a continuous carbon fiber production process requires a continuous surface treatment process, which can be operated online. Probably, the simplest solution is an electrolytic process, where the carbon fiber is made the anode in a treatment cell using an aqueous electrolyte. The patent literature lists numerous electrolytes, which may be used and that issue is not discussed here. The electrolyte must be a good conductor, relatively cheap, readily washed off the fiber and present no handling problems.

The process must be conducted at a potential level sufficient to generate oxygen at the surface of the carbon fiber (anode).

The treatment current for a specific fiber type can be specified in coulombs per meter ($C\,m^{-1}$). A coulomb is the flow of one ampere current for one second, hence the total coulomb (C) flowing in 1 h would be $3600 \times I\,(C\,h^{-1})$. If m is the total length in meters of all fiber processed in 1 h with n tows running through the surface treatment bath at a speed of s ($m\,h^{-1}$) then,

$$m = sn$$

$$\text{Here} \quad \frac{C}{m} = \frac{3600 \times I}{ns} \quad \text{or} \quad I = \frac{C}{m} \times \frac{ns}{3600}\,A$$

The selected current depends on the number of tows and line speed and if any of these parameters are changed, it will be necessary to adjust the surface treatment current. The value C/m is specific to a particular d'tex type tow and the value of C/m should be adjusted to compensate for any change in the surface area of the tow.

It is surprising that many users control their treatment process at constant voltage, whereas it should be done at constant current. The equipment to perform this duty is freely available for use with the plant ane is used for cathodic protection. The constant voltage equipment is normally rectified AC, not true DC, and there is a ripple effect detectable with a cathode ray oscilloscope, but the ripple is smoothed out to a large extent by the surface treatment cells acting as a large capacitor.

A typical unit cell layout is shown in Figure 10.24. The fiber is passed under a controlled tension over a conductive roller, normally carbon, which is made the anode and is electrically isolated from the main framework. The current is fed to each end of the anode roller via a commutator type device with spring loaded brushes and the feed cables are adjusted to be the same length, thereby assuring an equal current distribution. O-rings fitted around and near each end of the anode rollers act as drip arrestors and prevent the electrolyte from creeping along the roller to the brushes. The cathodes are carbon blocks positioned parallel with the fiber band and can be enveloped within a nylon bag to prevent shorting by stray carbon fiber filaments between the anode and cathode. Whilst O_2 is liberated at the anode, H_2 is liberated at the cathode and due care must be taken in view of its flammable and explosive nature. As the fiber dips into the anode compartment, the potential is quickly diminished and could be reduced by half at a depth of some 100 mm, hence in practice, a number of anode cells are used. Experience with electroplating has shown that the electrolyte must be well circulated

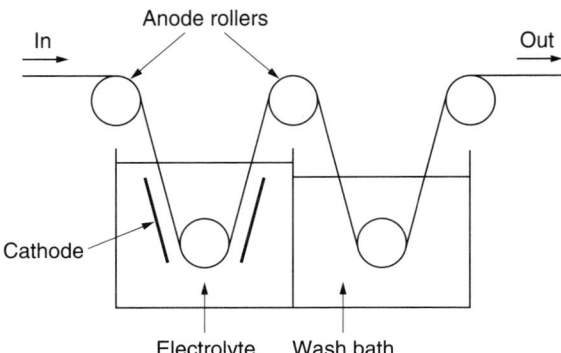

Figure 10.24 A typical unit cell for electrolytic surface treatment of carbon fiber.

and the surface area of the cathode chosen to give uniform current density. The ends of the cathodes are painted with a stopping off lacquer to restrict and balance out the current density at the edges of the tow band. For the same reason, the tow band must be uniformly spread across the width without anygaps, since an uneven current density will ensue resulting in non-uniform surface treatment. The recirculating electrolyte is continuously filtered to remove carbon fiber debris. It is normal practice to surround the working area with a bund to confine any spillage.

The required treatment current measured as $C\,m^{-1}$ depends on the fiber type. An HM fiber needs about twenty times the treatment level used for a high strength fiber. When treating HM fiber, the current levels employed can generate sufficient heat to necesstate cooling of the electrolyte using a carbon block heat exchanger. With increasing currents, polarization can occur, caused by a layer of O_2 on the fiber surface, which stifles further treatment. This can be offset by heating the electrolyte, which reduces the surface tension at the fiber surface and increases the ionic mobility, although the scope is limited, since ammonium salts, for example, would be decomposed above about 50°C.

After surface treatment, it is necessary to wash the fiber free of electrolyte and one advantage of NH_4HCO_3 is that any salt remaining on the fiber after washing will volatilize in the drying stage, which is not the case with $(NH_4)_2SO_4$. It is normal to rinse the washed fiber with demineralized water, especially in areas with hard water supply.If ammonium salts are used as electrolyte, care must be taken to avoid contact with any copper bearing materials, which will corrode.

10.15 SIZING

As mentioned in section 10.16, it is necessary to dry the fiber after surface treatment to provide accurate resin pick-up in the subsequent sizing process. Cost has dictated the use of water based sizes and Figure 10.25 shows a typical sizing rig. It is essential to maintain adequate circulation to ensure uniform size distribution across the width of the fiber band, which can be achieved by allowing the circulating size liquor to flow over a weir extending over the width of the size bath to be returned after filtering to a holding tank fitted with a low level alarm.

It is important to realize that passing the wet fiber over a roller or guide prior to drying will act as a metering device and remove some of the entrained size liquor, depending on the tension applied to the fiber band. This, coupled with the concentration of the size and the openness, or spread, of the carbon fiber tows will control the resin pick-up.

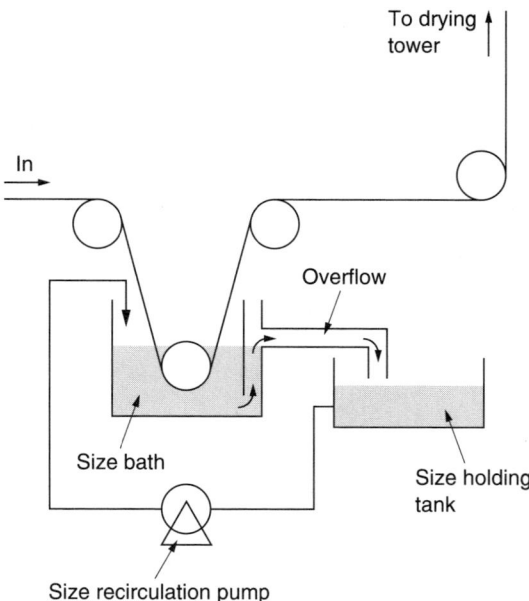

Figure 10.25 Diagram of a typical size bath.

10.16 DRYING

Moisture content can be expressed on a wet basis (kg moisture per kg of fiber plus moisture, or on a dry basis (kg moisture per kg dry weight of fiber). The dry basis will be used throughout this book.

In a carbon fiber process, there are three stages where drying may be required:

1. If oxidized PAN is made, it is necessary to dry it after application of a water-based finish to give a product with controlled moisture content.
2. If a wet stage surface treatment process is used, it is necessary to dry the fiber after surface treatment to provide accurate resin pick-up in the subsequent sizing process.
3. The fiber must be dried after the application of a resin size, which is invariably either in suspension, or in solution, in a water base, to give a product with final moisture content below 0.1%.

Maximum quantity of water is removed by application of mechanical pressure passing the fiber under tension over rollers/guides or between rollers, but this will still leave upto its own weight moisture on the fiber.

Three types of drying methods are used:

1. Passing the fiber over a heated platen
2. Passing through a drying tunnel /tower
3. Passing over a stainless steel heated rotary drum

The platen can be heated either electrically or by steam. The tower can be heated externally by electric heaters, or preferably, a counter current of hot air passed through the tower (Figure 10.26). The drum drier (Figure 10.27) can be heated with a heat transfer medium such as steam, hot oil, or via a rotary seal.

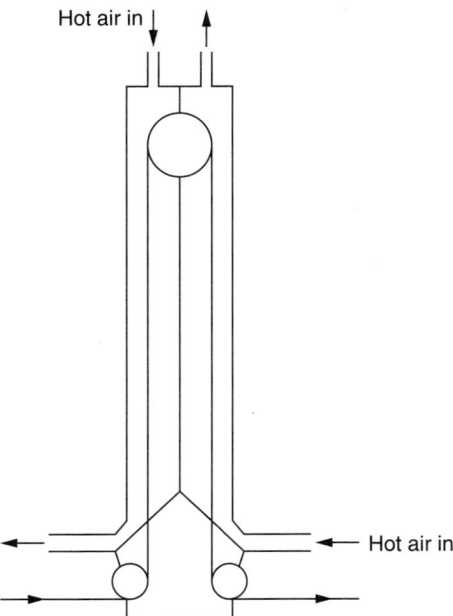

Figure 10.26 Typical drying tower.

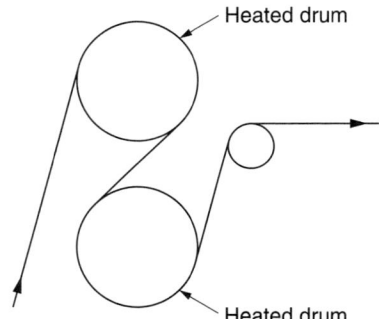

Figure 10.27 Drying carbon fiber with heated drums.

Some problems will be experienced with resin size sticking to the top roller in the tower, or on the surface of the drum, and it is usual to treat these surfaces with a PTFE anti-stick coating. This coating should be replaced at regular intervals by stoving away the old coating above the PTFE breakdown temperature, lightly grit blasting, then spraying on the new coating and finally, baking at the requisite temperature.

The air immediately in contact with the surface of the wet fiber is quickly saturated and must be continually replaced to ensure continuous evaporation. In evaporative drying, four main phases can be identified [6]:

1. The rising rate or warming-up period—during which the heat supplied raises the temperature of the wet fiber to the designed maximum, with a corresponding rise in evaporation rate.
2. The constant rate period—when the temperature is steady, the carbon fiber surface remains wet, and the evaporation rate is constant at the designed maximum, known as the equilibrium moisture content. If the fiber contains more moisture than the equilibrium value, it will dry until its value reaches equilibrium.

3. The first falling rate period—corresponding to the critical moisture content for the fiber, associated with the occurrence of dry patches on the fiber surface and a falling evaporation rate.
4. The second falling rate period—which is governed by the rate of evaporation, migration and diffusion of moisture within the carbon fiber tows. The surface appears dry and air humidity no longer influences the rate.

The drying rate curve (Figure 10.28) is divided into a constant rate period as shown by the portion AB, and the two falling rate periods, BC and CD [9]. Although the amount of moisture removed in the falling rate period is small, the time required for this period is often quite long and hence, will influence the dimensions of the dryer.

For ordinary heat transfer processes, the specific heat at constant pressure (C_p) is defined as the heat in joules required to raise the temperature of one kilogram of air by one degree celsius. At 100°C, is 1.01 kJ kg^{-1}°C^{-1}. For water vapor, it is 2.01 kJ kg^{-1}°C^{-1}, hence C_p is $(1.01 + m)$ for air containing m kg of moisture per kg of mixture. However, this correction is normally neglected, since the heats of evaporation, or condensation, become the predominant factors.

Drying by natural convection is too slow, so forced convection is used:

$$H = h_c \times A \times \Delta t$$

where H = heat transfer rate in joules/sec (watts)
A = area of hot surface in m^2
Δt = temperature difference between the hot surface and the air stream (°C)
h_c = convection heat transfer coefficient in W m^{-2}°C^{-1}

In forced convection, h_c is dependent on the air velocity and the pattern of forced air flow over the surface.

Turbulent flow must be established to obtain a high rate of heat transfer and with turbulent flow parallel to smooth surfaces, h_c is found to be proportional to (mass flow)$^{0.8}$. The convective heat transfer rate h_c is directly linked with the evaporation rate. In the constant rate period of evaporation, the temperature of the wet surface is equal to the wet bulb temperature of the air and the temperature difference available for heat transfer is the difference between this and the dry bulb temperature, that is the wet bulb depression (Δt_w), provided the air is well mixed by turbulence.

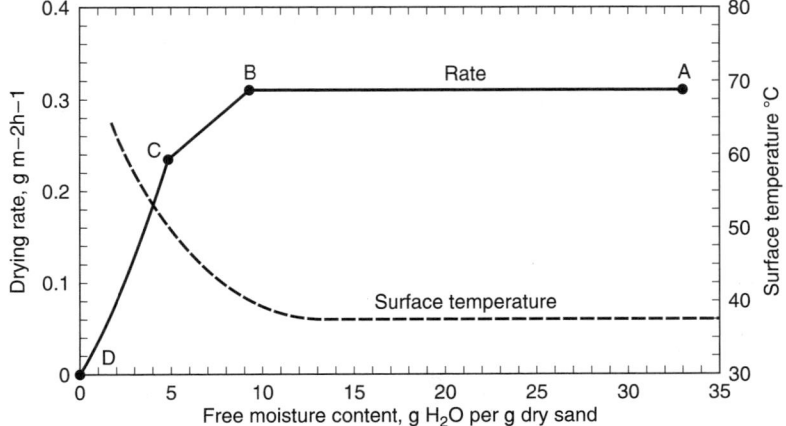

Figure 10.28 Typical drying rate curve. *Source:* Reprinted with permission from Ceaglske NH, Hougen OA, *Trans Am Inst Chem Eng*, 33, 283–312, 1937. Copyright 1937, The American Institute of Chemical Engineers.

Hence heat supplied to A (m^2) of wet surface is:

$$H(\text{kW}) = \frac{h_c}{1000}(\text{kW m}^{-2}\,°\text{C}^{-1}) \times A\,(\text{m}^2) \times \Delta t_w\,(°\text{C})$$

Now the heat required to evaporate M (kg s^{-1}) of vapor with latent heat L (kJ kg^{-1}) is:

$$H(\text{kW}) = M\,(\text{kg s}^{-1}) \times L\,(\text{kJ kg}^{-1})$$

Therefore evaporation rate is

$$M/A = \frac{h_c \times \Delta t_w}{1000 L}\ \text{kg s}^{-1}\text{m}^{-2}$$

In actual cases, heat transfer to the wetted surface by radiation and conduction always occurs and the actual surface temperature will be higher than the wet bulb temperature by about 2°C.

It is difficult to establish true answers without experimental data and, invariably, it is best to undertake drying trials. The possible benefits of air recirculation should not be overlooked.

Figures 10.29 and 10.30 show the effect of drying 20k carbon fiber at 60°C and 105°C, the former being wet with water and the latter sized with an epoxy, where the advantage

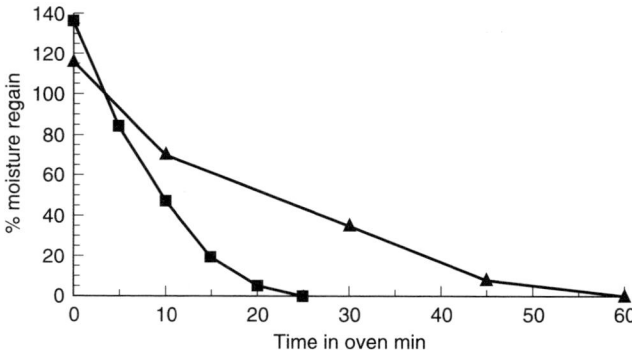

Figure 10.29 Drying 20k carbon fiber wet with water. v Dried at 105°C σ Dried at 60°C.

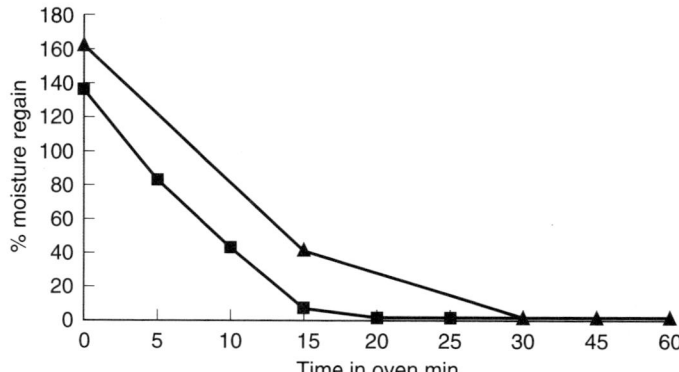

Figure 10.30 Drying 20k carbon fiber sized with a water based epoxy to give 2% size on the fiber. v Dried at 105°C σ Dried at 60°C.

of using the higher temperature can be seen, with no negative effect in the rate when drying the epoxy size.

10.17 ONLINE COLLECTION

Although the initial setup for online winders is expensive (each head costs about £1000), there is considerable savings in labor since it precludes handling of the fiber twice. It is possible to carry out the inspection online.

A typical bank of winders used for the online collection of 3–12k carbon fiber is shown in Figure 10.31 and are fitted with dancing arms to control the tension and speed and to ensures minimum damage to the fiber during the winding process. Large diameter guide rollers, having a pulley design, are used to guide the fiber (Figure 10.32). Another source of damage to the product is the fiber guide, which directs the fiber onto the package. One manufacturer has come up with an ingenious design comprising of a pair of cone guides (Figure 10.33), which are kind to the fiber. To wind heavier tow, the dancing arm can be replaced with a torque motor system.

Online winders are fitted with length counters, which can give an audible warning when the package is to be doffed or, alternatively, a spool package changer can be built in (Figure 10.34). Care must be taken to protect the equipment from carbon fiber fly, which is electrically conductive and winders must be maintained regularly and cleaned daily to remove debris. The mode of action and setting up of the wind pattern is discussed in in the section on offline winding.

The fiber can be collected online without winders in containers, with the process being aided by passing the fiber tow through an air mover, which helps to guide the tow into its respective container. The air movers work on a venturi principle and arranging intermittent air supply will help conserve air. It is essential to ensure that the air supply to the air movers is dry and free from oil, as otherwise the fiber will become contaminated.

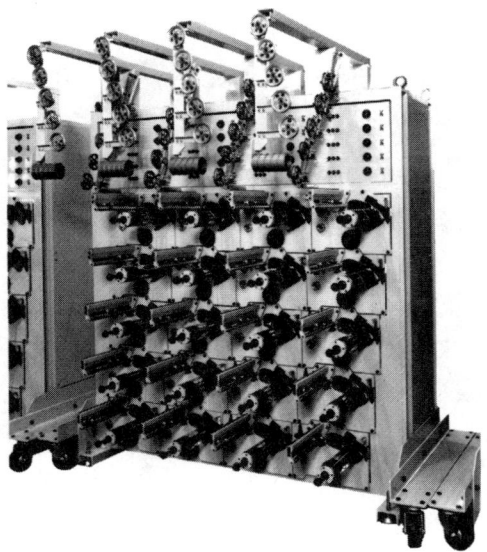

Figure 10.31 Bank of 20 Kamitsu online take-up winders fitted with dancing arms suitable for 3k–12k carbon fiber. *Source:* Courtesy of Izumi International Inc., CA, USA.

Figure 10.32 Close-up of Kamitsu take-up winder dancer roller and guide roller arrangement. *Source:* Courtesy of Izumi International Inc., CA, USA.

Figure 10.33 Kamitsu conical roller traverse guide for self threading and fiber quality. *Source:* Courtesy of Izumi International Inc., CA, USA.

Figure 10.34 Twelve spindle online Kamitsu take-up winder with automatic spool change. *Source:* Courtesy of Izumi International Inc., CA, USA.

10.18 OFFLINE WINDING

A typical rewinder used for carbon fiber is shown in Figure 10.35 and fiber can be rewound at speeds of about 2400 m h^{-1} from another spool or from a container used to collect the fiber online. Rewinding is accomplished with minimum damage to the carbon fiber by using large diameter feed rollers, dancing arms and conical roller traverse guides (Figure 10.36). Carbon fiber is wound without twist, at a tension of about 500 g, onto parallel cardboard spools which normally have an ID of 76 mm.

A precision winder operates at a constant ratio between the speed of the spindle, which holds the package being wound, and the speed of the traversing mechanism, which is a cam driven thread guide. The constant ratio between the speeds of these two components is maintained by the cam, which drives the spindle. However, the yarn speed will not be constant. This is because, as the diameter of the package increases, the yarn speed too increases, but the rpm of the spindle remains constant. The thread guide is subject to wear and can damage the carbon fiber, however as mentioned before, one manufacturer has overcome this problem by using a pair of cones.

The number of winds is the number of revolutions the spindle, or package, makes while the thread guide is making one stroke, e.g. if the spindle makes three revolutions while the thread guide is moving from one end of the package to the other, the package is said to contain three winds.

Since it takes two strokes of the thread guide for each cycle of the cam, the spindle will, therefore, make twice the number of winds for each cam cycle. The smaller the number of winds, greater is the angle from the vertical and sharper is the knuckle at the end of the stroke. Generally speaking, fewer winds will produce straighter ends of the package, offset by having to run at a slower spindle speed. Packages with more winds deliver better when off-wound and produce less scuffing of the fiber surface.

As a package increases in diameter, the tension will build up and deform the package, hence a device known as a differential tension and pressure attachment (dancing arm) is used

Figure 10.35 Kamitsu rewinder for carbon fiber. *Source:* Courtesy of Izumi International Inc., CA, USA.

Figure 10.36 Kamitsu carbon fiber re-winder with large diameter feed rollers and dancing arm. *Source:* Courtesy of Izumi International Inc., CA, USA.

or, alternatively, for larger tows, a torque motor system can be used. As the yarn speed increases due to the increase in package diameter, both tension and pressure must be decreased.

It is important to remember that the first layer of yarn must be wound under sufficient tension and pressure to give a firm base upon which to build the package. Too little pressure or too much tension yields a package with a concave surface, i.e. low in the center and high at the ends.

Gain is applied to avoid successive wraps of fiber winding on top of the preceding wrap. The mechanical means of achieving this is termed the gainer mechanism. Normally, the gain is slightly in advance of the previous wind. The amount of gain depends on the diameter of the tow being wound. In order to achieve an adjustable gainer mechanism, the belt-gain mechanism is used, depending on a combination of wind pulley and belt. This mechanism is used to give a close wound diamond wind.

An open wind is obtained when the gain gives no recognizable pattern on the surface of the package. Although an open wind can be achieved with the belt gain mechanism, it is preferable to use the gear gain since there is no need for further adjustment after the correct open wind has been obtained. With this mechanism, the spindle drives the cam through a train of gears, including change gears for obtaining the desired number of winds.

The calculations for a Leesona 959 machine are as follows:

θ = Helix angle of yarn on package
a = Traverse length (inches)
D = Package diameter (inches)
W = Winds = $W_R/2$
W_R = Wind ratio
N = Number of knuckles to repeat (one end)
R = Number of package revolutions to repeat
d = Axial spacing of yarn on package (inches)

GUIDELINES FOR THE DESIGN OF EQUIPMENT FOR CARBON FIBER PLANT

G = Gain (Deviation of wind ratio (W_R) from whole number)
C, D, E, F = Number of teeth on wind pulleys
T_C = Number of cam turns per cycle

Approximate for small values of G:

$$\cos\theta = \frac{d}{G\pi D} \tag{1}$$

$$\cos\theta = \frac{a}{[a^2 + (W^*\pi D)^2]^{-2}} \tag{2}$$

In these two formulae, W is a whole number and not $\frac{W_R}{2}$.
From (1) and (2),

$$d = \frac{G}{[(W^{*2}/a^2) + (1/\pi^2 D^2)]^{-2}} \tag{3}$$

$$W_R = 2W^* \pm G \quad \text{or} \quad \frac{Wa}{2} = W + \frac{G}{2} \tag{4}$$

Accurate case

$$d = \frac{G}{\left[\left[(W^* \pm \frac{G}{2})/a\right]^2 + (1/\pi^2 D^2)\right]^{-2}} \tag{5}$$

From (4) and (5),

$$d = \frac{G}{[(W_R/2a)^2 + (1/\pi^2 D^2)]^{-2}}$$

This may be written as:

$$G = \frac{d}{N}\left[\frac{W^2}{a^2} + \frac{1}{\pi^2 D^2}\right]^{-2}$$

N is equal to 1 when the wind is equal to a whole number such as 3, 4, 5 or when the wind is equal to a half-number such as $3\frac{1}{2}$, $4\frac{1}{2}$, $5\frac{1}{2}$ etc.
N is equal to 2 for a quarter wind such as $3\frac{1}{4}$, $4\frac{3}{4}$, $5\frac{1}{4}$, $5\frac{3}{4}$ etc.
N is equal to 3 for a one-third wind such as $3\frac{1}{3}$, $3\frac{2}{3}$, $4\frac{1}{3}$, $4\frac{2}{3}$ etc.
N is equal to 4 for a one-eighth wind such as $3\frac{1}{8}$, $3\frac{3}{8}$, $3\frac{5}{8}$, $3\frac{7}{8}$, etc.

As an example, given below are the calculations to find the wind ratio necessary to give a $\frac{1}{8}''$ spacing, with 3 winds, on 10'' traverse and $4\frac{1}{4}''$ OD tube.
For this case, $N=1$, because the wind is equal to a whole number. Therefore,

$$G = \frac{0.125}{10^2}\left[\frac{3^2}{10^2} + \frac{1}{(\pi)^2(4.250)^2}\right]^{-2} = 0.0386509$$

Hence 0.0386509 is the Gain required to give a $\frac{1}{8}''$ spacing.

Now,
Wind ratio $= 2W - G = 6.0000000 - 0.0386509 = 5.9613491$
It is now necessary to find what combination of timing pulleys will give this ratio.

$$\text{Wind ratio} = \frac{\text{Cam pulley}}{\text{Inner idler compound}} \begin{bmatrix} F \\ E \end{bmatrix} \times \frac{\text{Outer idler compound}}{\text{Lay shaft}} \begin{bmatrix} D \\ C \end{bmatrix}$$

$$\times \frac{17}{5} \times 2 \text{ (no. of cam turns)} = 5.9613491$$

For the 981 machine, there is a fixed ratio of $\frac{34}{30} = \frac{17}{15}$.

It is a long and tedious process to calculate the timing pulleys required and this has been worked out on computer and catalogued according to traverse length. For this particular case, looking up the catalogue for a 10″ traverse and a wind ratio of 5.9613491 will give the combination of timing pulleys required.

Similarly, calculations for a Kamitsu EKTW-C winding machine are:

1. Wind Ratio

$$W = 8 \times 2 \times \frac{B}{A} \times \frac{D}{C}$$

For example, for 12k carbon fiber

$$W_1 = 8 \times 2 \times \frac{148}{45} \times \frac{21}{43} = 11.882437$$

2. Pitch (Figure 10.37)

   a. $l_0 = \frac{2L_0}{w}$, where L_0 is the traverse length (for EKTW-C it is 10 in + 1 mm)
   b. $l_1 = -l_0 \times (w - w')$, where $w' =$ integer portion of $w + 1$ (whole number)
      = pitch for every traverse
   c. $l_2 = l_0 - n \times l_1$ ($n = 2$ in Figure 10.37)
   d. $l_3 = l_1 - l_2$
      = pitch for every $(n + 1)$ times, which is considered as pitch

3. Pitch for $w = 11.7333$ (Figure 10.38)

   a. $l_0 = \frac{2 \times 255}{11.7333} = 43.47$
   b. $l_1 = -43.47 \times (11.7333 - 12) = 11.59$
   c. $l_2 = 43.47 - (3 \times 11.59) = 8.69$
   d. $l_3 = 11.59 - 8.69 = 2.90$ i.e. Pitch is 2.90

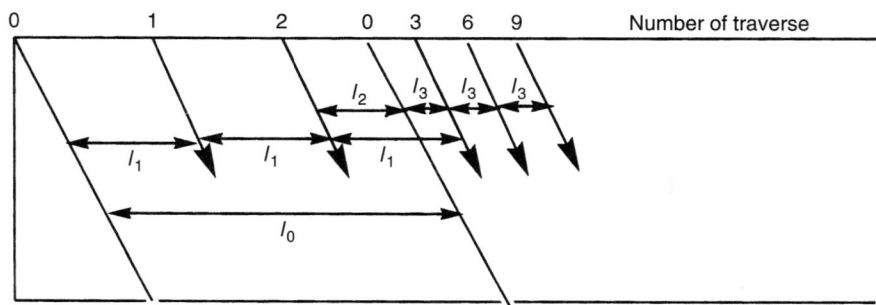

Figure 10.37 Pitch calculation for Kamitsu winder EKTW-C. *Source:* Courtesy of Izumi International Inc., CA, USA.

GUIDELINES FOR THE DESIGN OF EQUIPMENT FOR CARBON FIBER PLANT

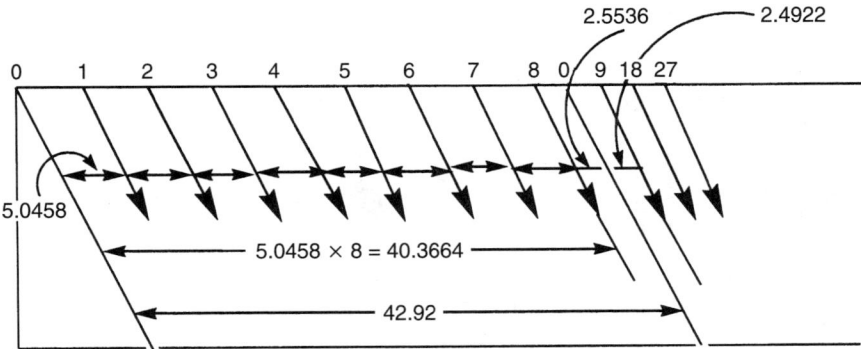

Figure 10.38 Pitch calculation for Kamitsu winder EKTW-C when wind ratio is 11.7333. *Source:* Courtesy of Izumi International Inc., CA, USA.

A program tension control system is described by Hirai [10] and is applicable to carbon fiber.

10.19 PACKAGING

Prior to packaging, the fiber is shrink wrapped and weighed. It is convenient for these latter stages to be handled by a computer, which will organize the printing of labels for the fiber spools, apply bar codes, print labels for the inner and outer packages and produce all the requisite paper work.

10.20 EXHAUST SYSTEMS [6, 7]

If air is to be extracted from a building, it must be remembered that an equivalent quantity of air must be supplied into that building to avoid negative pressure. This is best provided by arranging for a direct line to bring clean fresh air from outside the building, away from any discharge points, and then split to supply air at the point of intake of each individual item of the plant requiring air, taking care to avoid cold draughts playing on operatives.

Figure 10.39 shows the general arrangement of a centrifugal fan with three types of blade, whilst Figure 10.40 depicts an axial flow fan fitted with aerofoil blades with adjustable pitch.

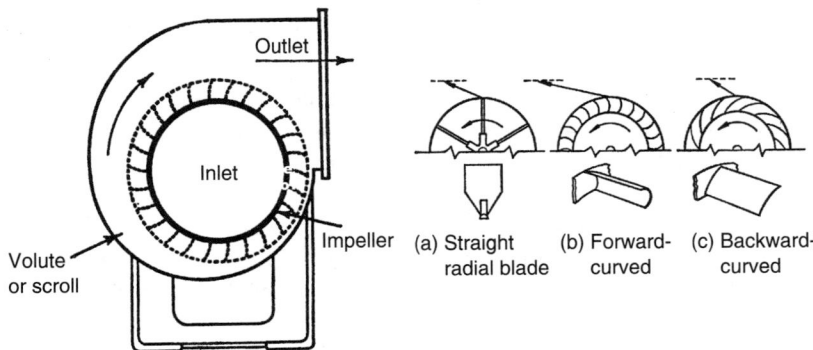

Figure 10.39 General arrangement of a centrifugal fan showing different types of blades. *Source:* Reprinted from Osborne WC, Turner CG, *Woods Practical Guide to Fan Engineering*, Woods of Colchester (a GE Company) Sheffield 1973, 2nd ed., 1974. By kind permission of Fläkt Woods Ltd 2004.

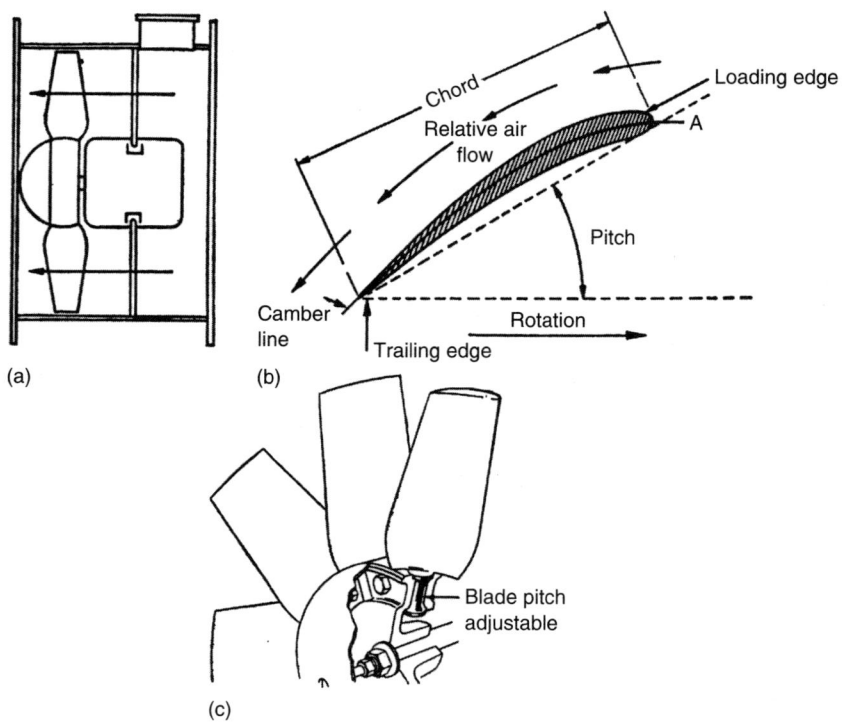

Figure 10.40 Axial flow fans. a) Single stage axial flow fan. b) Typical aerofoil cross section. c) Adjustable pitch axial flow impellor *Source:* Reprinted from Osborne WC, Turner CG, *Woods Practical Guide to Fan Engineering*, Woods of Colchester (a GE Company) Sheffield 1973, 2nd ed., 1974. By kind permission of Fläkt Woods Ltd 2004.

When extracting hot gases (e.g. from above the inlet and outlet ends of oxidation ovens), bifurcated fans should be used to give some protection to the fan motor. Impellors must not be made of aluminum alloys, since these are readily corroded. Other fan systems can be incorporated directly in the duct using axial fans, which for better control, can be fitted with adjustable pitch blades.

In some instances, ductwork will conduct noxious fumes directly to the treatment plant and in others, the extracted air can be discharged to atmosphere. In all cases, any appropriate legislation must be followed. For instance, the point of discharge will be stipulated (e.g. 3 m above the highest point on the roof, preferably at an efflux velocity of about 10m/s) and so on.

Extraction hoods should enclose the source of fumes as completely as possible without impeding production operations. A nominal inlet area and an average air velocity are identified and the product of the two gives the volume to be extracted. A typical canopy is shown in Figure 10.41. The distribution of velocity close to a canopy is shown in Figure 10.42, the figures being air velocities in per cent of the average hood velocity. The surface treatment tank can be designed as shown in Figure 10.43, providing a lateral exhaust system for dealing with fumes, where a jet of clean air is directed across the surface of the tank and exhausted from the opposite side.

If organic solvents, with vapor heavier than air (e.g. 1,1,1-trichloroethane) are used for the application of an epoxy size, then bottom extraction should be used.

As a general rule, swept bends should be used on all trunking and the diameter of the ducts chosen to keep the velocity sufficiently low, so that noise will not be a problem. All

GUIDELINES FOR THE DESIGN OF EQUIPMENT FOR CARBON FIBER PLANT

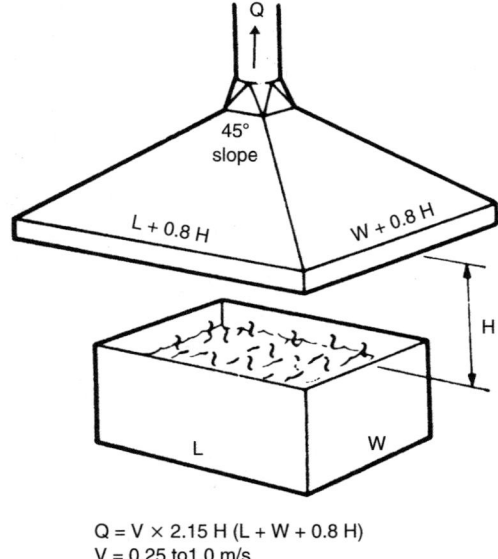

$Q = V \times 2.15 H (L + W + 0.8 H)$
$V = 0.25$ to 1.0 m/s

Figure 10.41 Hood for extraction above an open end of a furnace or above an open bath. *Source:* Reprinted from Daly BB, *Woods Practical Guide to Fan Engineering*, Woods of Colchester (a GE Company) Sheffield 1973, 3rd ed., 72, 1992. By kind permission of Fläkt Woods Ltd 2004.

ductwork should be regularly cleaned to remove carbon fiber debris and provision must be made for easy dismantling.

To size a fan for a given duty, it is necessary to determine the total pressure drop in the system. *Woods Practical Guide to Fan Engineering* [6] details how the pressure drop of each individual element in the system can be determined namely:

Losses at entry to the system
Losses due to friction in duct lengths
Losses at changes of duct area or shape
Losses at bends and changes of direction
Losses at division of flow into branches

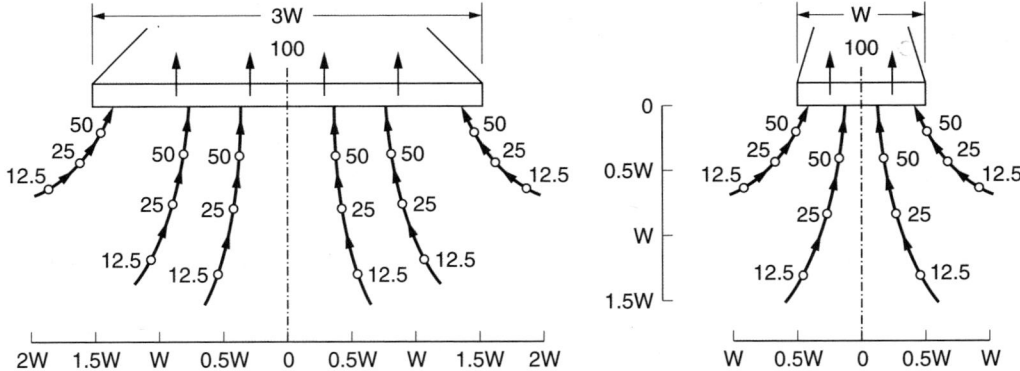

Figure 10.42 Velocities at the inlet to a rectangular hood expressed in percent of the average hood velocity. *Source:* Reprinted from Osborne WC, Turner CG, *Woods Practical Guide to Fan Engineering*, Woods of Colchester (a GE Company) Sheffield 1973, 2nd ed., 39, 1974. By kind permission of Fläkt Woods Ltd 2004.

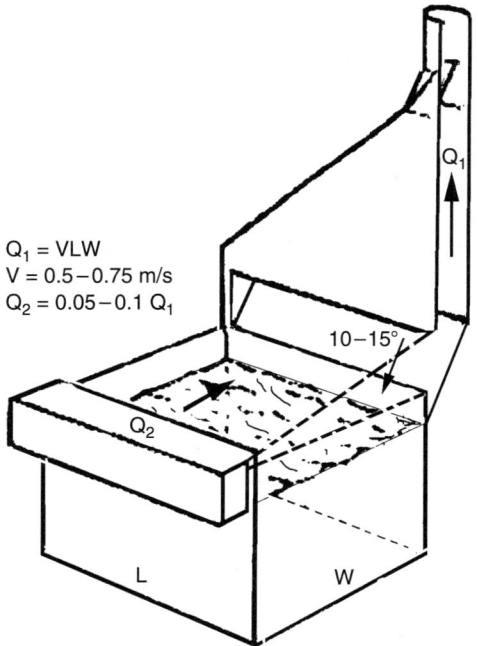

Figure 10.43 Lateral exhaust system for an open surface treatment tank. *Source:* Reprinted from Daly BB, Woods Practical Guide to Fan Engineering, Woods of Colchester (a GE Company) Sheffield 1973, 3rd ed., 72, 1992. By kind permission of Fläkt Woods Ltd 2004.

Losses caused by obstructions
Losses at discharge from the system to atmosphere
Change in atmospheric pressure from inlet to outlet

Fan manufacturers publish pressure/volume characteristics which give a indication of the volume of air moved against a resistance to flow (i.e. the fan develops pressure, which can be used to overcome system resistance). The relationship between pressure drop and flow within a system is illustrated by the square law curve superimposed onto the fan performance graph. Figure 10.44 shows a Fan A, which, at 100 m^3 h^{-1}, developed a pressure of 150 Pa. If only 50% of the volume of air was required through the same system, then Fan B, giving 50 m^3 h^{-1} needs only 37.5 Pa. However, if 50% more air was required, then a larger and faster fan would be required to deliver 150 m^3 h^{-1}, developing 337 Pa, such as Fan C.

10.21 DUST EXTRACTION [7]

Since carbon fiber conducts electricity and presents a hazard for all electrical equipment, it is wise to incorporate a dust extraction process, which should extend throughout the plant. It can be based on a powerful blower, which applies suction to 50 mm uptake pipes situated at strategic positions around the plant, which feed via swept bends into a 75 mm header. At the bottom of each uptake pipe is a flap valve, which permits fiber debris to be hand fed by lifting the flap or, alternatively, connected with a vacuum hose, which can be fitted with a range of cleaning attachments. The header pipe releases some dust into an interceptor cyclone (Figure 10.45) and the remaining dust can be trapped in large filter bags. The system should operate with only one flap valve in use at any one time to ensure optimum performance.

GUIDELINES FOR THE DESIGN OF EQUIPMENT FOR CARBON FIBER PLANT 419

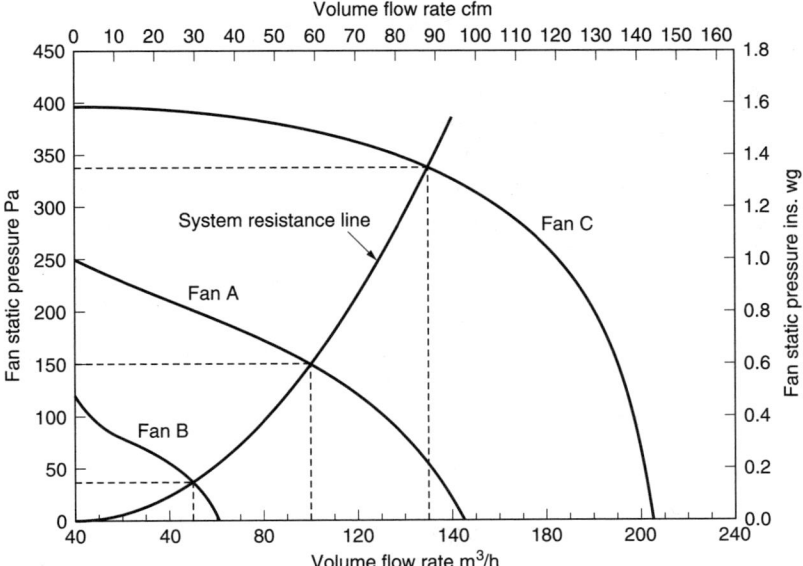

Figure 10.44 Pressure/volume characteristics of three types of fan. *Source:* Reprinted from Mexico Industrial Fans Ltd, Stalybridge, Trade literature.

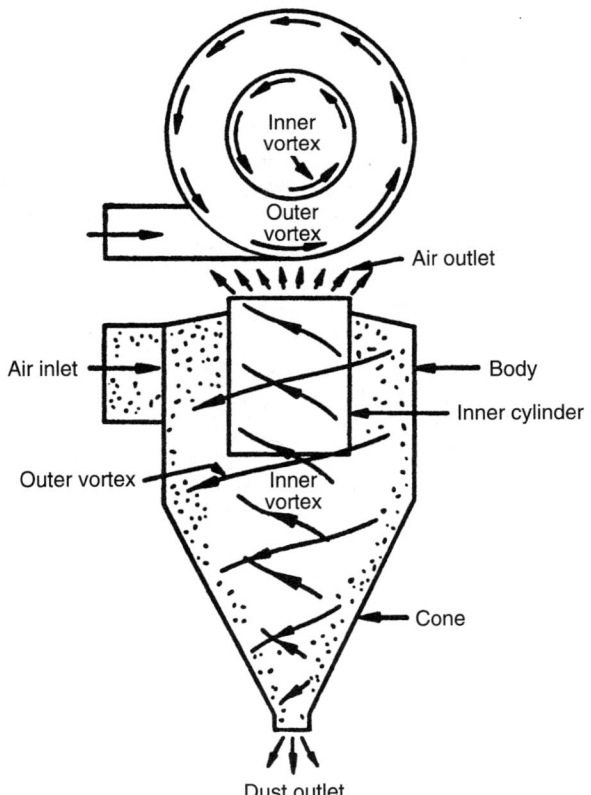

Figure 10.45 Cyclone dust extraction unit. *Source:* Reprinted from Osborne WC, Turner CG, *Woods Practical Guide to Fan Engineering*, Woods of Colchester (a GE Company) Sheffield 1973, 2nd ed., 45, 1974. By kind permission of Fläkt Woods Ltd 2004.

10.22 APPLICATION OF CLOSED CIRCUIT TELEVISION (CCTV)

There are two uses for CCTV, one to constantly monitor inaccessible areas of the plant by using a number of monitors situated around the plant and two, to examine previously recorded tapes while investigating a problem that occurred earlier. Cameras have to be mounted in dustproof camera housings which will prevent ingress of carbon fiber. The connections must use good quality coaxial cable and not exceed the stipulated cable length. Each camera can have several viewing stations, with its own monitor and the master monitor can be switched to any of the camera positions. A multiplexer unit ensures that each position can be observed and simultaneously recorded onto a multitrack timelapse video recorder (VCR). The timelapse VCR condenses the total time of recording from say 24 h to fit into a 3 h tape. Any position that has been recorded can be selected and played back but will have a flicker frequency due to the condensed recording, but the definition will be adequate and the picture perfectly discernible.

REFERENCES

1. Health and Safety Executive, *Evaporating and other ovens*, HS(G) 16, 1981, Health and safety executive, HMSO publication.
2. Cubbage PA, Simmonds WA, An Investigation of Explosion Reliefs for Industrial Drying Ovens. Part I. Top reliefs in box ovens, *Gas Council Research Communications*, GC23 1955, *Trans Inst Gas Eng*, 1955–56.
3. Cubbage PA, Simmonds WA, An Investigation of Explosion Reliefs for Industrial Drying Ovens. Part II. Back reliefs in box ovens: Reliefs in conveyor ovens, *Gas Council Research Communications*, GC43.
4. Brill U, Agarval DC, Alloy 602A—A new alloy for the furnace industry, *Proceedings of the 2nd International Conference on Heat-resistant Materials*, Gatlinburg, 11–14, Sep 1995.
5. Hepworth Refractories Limited, *Calculation of heat transfer through refractory structures by GR Stein Refractories* using ASTM calculation procedure.
6. Daly BB, *Woods Practical Guide to Fan Engineering*, Woods of Colchester (a GE Company) Sheffield 1973, 3rd ed., 72, 1992.
7. Osborne WC, Turner CG, *Woods Practical Guide to Fan Engineering*, Woods of Colchester (a GE Company) Sheffield 1973, 2nd ed., 39, 1974.
8. Reed RD, *Furnace Operations*, Gulf Publishing Company, Houston, 176, 1981.
9. Ceaglske NH, Hougen OA, *Trans Am Inst Chem Eng*, 33, 283–312, 1937.
10. Marshall WR, *Heating, Piping Air Conditioning*, 15(11), 567–572, 1943.
11. Hirai T, Program control system for winding tension force to obtain a uniform residual stress distribution in material, *Proceedings of the 7th International Conference on Composite Materials, 1: Manufacture*, Guangzhou, 101–106, 22–24 Nov 1989.

CHAPTER 11

Operation of Carbon Fiber Plant and Safety Aspects

11.1 INTRODUCTION

The manufacture of carbon fiber from different precursors involves a number of operations that are quite similar. The purpose of this chapter is to provide general guidance on the operation of such plants and their safety aspects.

11.2 SERENDIPITY

One may perhaps wonder why serendipity should be considered under operation of carbon fiber plant but there is a good reason. The word serendipity was coined by Horace Walpole in 1754 and is defined in the dictionary as "the faculty of discovering pleasing or valuable things by chance." However, Louis Pasteur said "In the fields of observation, chance favours only the prepared mind," later to be paraphrased by an eminent American physicist, Joseph Henry, as "The seeds of great discovery are constantly floating around us, but they only take root in minds well prepared to receive them" [1].

The process of discovery can be divided into finding and the recognition of the finding, the latter probably being the more important and difficult phase [2].

There is no doubt that curiosity and perception are important parameters and Lenox [3] has described how these characteristics can be encouraged in the student:

1. Making and recording observations in a notebook.
2. Being flexible in thinking and interpreting results, including the unexpected and not dismissing them as incorrect.
3. Undertaking careful and intensive study of the field under investigation.

Some well known examples of serendipity are listed below:

In 1928, Sir Alexander Fleming discovered penicillin, but it was not until 1939 that Florey and Chain perfected the process of concentration and purification to render its subsequent production commercially viable. The three workers eventually shared the Nobel Prize in Physiology or Medicine in 1945.

In 1938, Plunkett of Du Pont, discovered polytetrafluoroethylene (ptfe), or Teflon as it is commonly known.

In the early 1950s, de Mestral discovered the hook and loop fastener, well known as Velcro [1].

The importance of serendipity cannot be understated and the critical factor is the observation of an effect and, subsequently, establishing a probable reason for it. Some examples experienced by the author relevant to carbon fibers are quoted and, although relatively simple, later interpretation proved to be most helpful.

1. In a certain type of oxidation oven fitted with internal rollers, there were two sets of doors, one set inboard on an inner skin and the other set, external. Once, one of the external doors had not been fully closed, leaving a gap along one edge. Later, it was observed that the carbon fiber produced from the oxidized product adjacent to this partly open door was significantly higher in strength than the fiber sampled elsewhere across the oven. This was eventually traced to the partly open door, which allowed the ingress of cold air. This cooled the internal rollers in the immediate vicinity of the door, thereby preventing individual filaments from fusion welding together, thus significantly improving the strength of the carbon fiber. The effect of allowing ingress of air through all the external doors is depicted diagrammatically in Figure 11.1. Initially, air cooled internal rollers were introduced at the input end of the oxidation oven, where the filaments were more susceptible to fuse together. Later, oxidation ovens designed with external rollers were introduced.
2. A run-away exotherm occurred in one of the oxidation ovens and there was a rush of evolved gases, which dislodged the explosion relief blanket situated at the top of the oven. This accidental repositioning of the blanket was not noticed. However, it happened to coincide with an increase in the carbon fiber strength. Subsequently, a new permanent position was established for the blanket, to take advantage of this bonus in carbon fiber strength.
3. On one occasion, a technical engineer from Rothschild Messinstrumente was demonstrating a rather deluxe portable tension meter for the online measurement of tow tension. Being more accustomed to textile units where line speeds are much higher, he happened to remark, quite naively, that the line speed need not have been slowed down. To facilitate the measurement of tension in individual tows exiting an oxidation oven, a number of tows were diverted over a series of mini parallel bars (known in-house as the five barred gate) and which could later be reintroduced into the main tow band. Afterwards, it was observed that the carbon fiber made from tows which had been diverted was stronger than the fiber in the main tow band. This was eventually attributed to the pressure exerted on the individual filaments as they passed over the bars, forcing the filaments apart and separating any that had become partially welded together

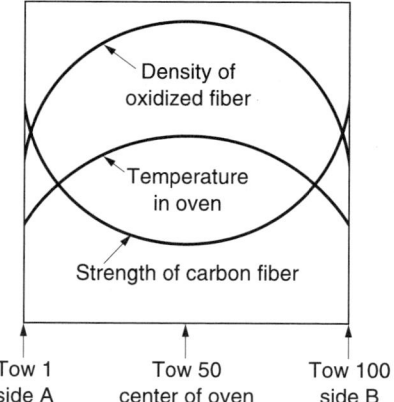

Figure 11.1 Diagramatic representation of the effect of ingress of cold air via external doors of oxidation oven.

during the oxidation stage. This discovery significantly influenced the future design of oxidation ovens.
4. In a prepreg process, there was a demand for a carbon fiber/phenolic prepreg, where the resin was applied as a solution in a C_2H_5OH/H_2O mix. The solvent laden prepreg was supported on a silicone release paper and passed through a drying oven to remove the applied solvent. This prepreg was always troubled with bubbles, which became apparent after the drying process and all attempts to prevent such bubbling were unsuccessful. During a particular run, however, the supervisor found that there was a length of prepreg that was bubble free and eventually traced it to a section where two thicknesses of bottom release paper, resulting from an overlap join which was introduced when a new roll of release paper was attached to the trailing end of the previous roll before it ran out. It was noticed that the short length of prepeg which was supported on this type of double layer release paper was bubble free. This double layer had managed to precisely control the heating rate, had eliminated bubbling and provided a way, albeit more expensive, of producing satisfactory phenolic prepreg.

It is now common practice to use computer control for carbon fiber plant with many attendant advantages, but there is no replacement for the regular visual examination of a working plant and recording pertinent observations.

11.3 MAINTENANCE

All types of maintenance should follow a carefully planned schedule and records must be kept. Some tasks could be weekly, such as topping up the oil level in gear boxes, whereas oil changes could be every six months.

The use of a stethoscope will help to give an early warning of defective bearings and inefficient steam traps. The judicious installation of manometers can indicate when the efficiency of extraction systems or filters has fallen off and help in advancing the cleaning of the respective duct systems, or replacement of filter packs.

If variable speed fans are used, the revolutions should be checked with a tachometer. Safety devices must be regularly inspected (e.g. all chain and drive system guards must be correctly in place). Work outstanding as a result of safety visits should be assessed in order of importance and categories allocated for undertaking such work.

11.4 PROTECTING ELECTRICAL EQUIPMENT

Oxidized PAN fiber is electrically non-conductive and plants dedicated to opf production are not at risk due to electrical shorting. Plants producing and laboratories testing carbon fiber must, however, be protected against stray carbon fiber finding its way into any electrical equipment. Experience has shown that the best approach is to seal all electrical cabinets and pressurize with clean filtered air, preferably detecting when the filter becomes no longer effective by fitting a manometer across the filter. Care must be taken against carbon fiber finding its way through open apertures. For example, one such incident caused serious disruption of the process when carbon fiber had found its way through a cable duct into the main electrical control room and shorted out a set of switchgear. Subsequently, all cables were sealed in their respective ducts prior to exit from the production plant.

Doors of electrical cabinets should be cleaned free of carbon fiber debris using a vacuum cleaner prior to opening or, as a better option, using a factory installed dust extraction system with portable cleaning tool attachments. It must be remembered that if a portable vacuum cleaner is used, although it sucks in at one end it will blow out the other. This air exiting the cleaner must be directed outside, otherwise air currents can cause chaos by disturbing the surrounding carbon fiber dust, which would then become airborne and present a potential threat to any exposed electrical system.

Control rooms housing electrical equipment should be pressurized with clean filtered air and entrances should preferably be fitted with two sets of interlocked doors to seal the factory environment, with only one operable door at a time. All electrical cabinets and consoles should be pressurized with fan units fitted with filters and due care should be taken to ensure that there is sufficient air available to remove excess heat from the cabinets.

Obviously, personnel permitted in such areas must be restricted to a minimum and great care should be taken to avoid carbon fiber adhering to work clothes in such areas. When undertaking electrical repairs, special clean disposable clothing should be worn and when handling switchgear, extra care must be taken. The author has seen a person being thrown across the room housing the electrical switchgear, when these precautions were not followed. Fortunately, the person did not suffer any burns, but was badly shaken. When a flash-over does occur, it is possible that the carbon can be deposited onto an insulating surface, causing tracking and necessitating expensive replacement.

11.5 AIR FLOW MEASUREMENT

11.5.1 Measurement of pressure

The temperature and pressure can be measured in an air stream. The pressure can be the barometric pressure adjacent to the test point and the gage pressure or the pressure difference between the reference atmosphere and the still air within a small hole positioned at the required point in the air stream. If the air is flowing parallel to the surface in which the hole is made, it will be the static pressure of the air stream. These pressures can be measured as follows:

1. A micromanometer—for pressures less than 100 Pa, uses a magnifying device to record the position of a float, with a stem bearing a scale situated in a suction chamber above a large liquid reservoir.It can measure up to 0.2 Pa with an accuracy of ±0.25%, but is prone to the effects of vibration. The modern counterpart is an electronic manometer; having a digital read out with facilities for measuring air pressure, velocity and flow or an analogue display and can directly read air velocities when used with a static Pitot tube.
2. An inclined tube manometer—for pressures in the range 100–1000 Pa, is a very useful general purpose instrument and is ideal for the continuous display of, for example, the condition of a filter (Figure 11.2). The manometer must be carefully leveled and zeroed, ensuring that it is filled with the same fluid that was used for calibration. The fluid can be a refined petroleum oil, dyed red for clarity, which has the advantage of low density (0.784 $g\,cm^{-3}$). Thus, it provides an expanded scale for greater accuracy combined with excellent wetting and draining characteristics and gives a clear and easily readable meniscus. The multiplying factor is the reciprocal of the sine of the angle to the horizontal.
3. A vertical or U-tube manometer—used for pressures in the range 1000–15,000 Pa.
4. A mercury manometer or barometer—used for pressures in the range 15,000–110,000 Pa.

11.5.2 Determination of velocity

The Pitot static tube is the standard method of determining the air velocity within a duct. A variety of designs are available (Figures 11.3, 11.4) and the modified ellipsoidal nose

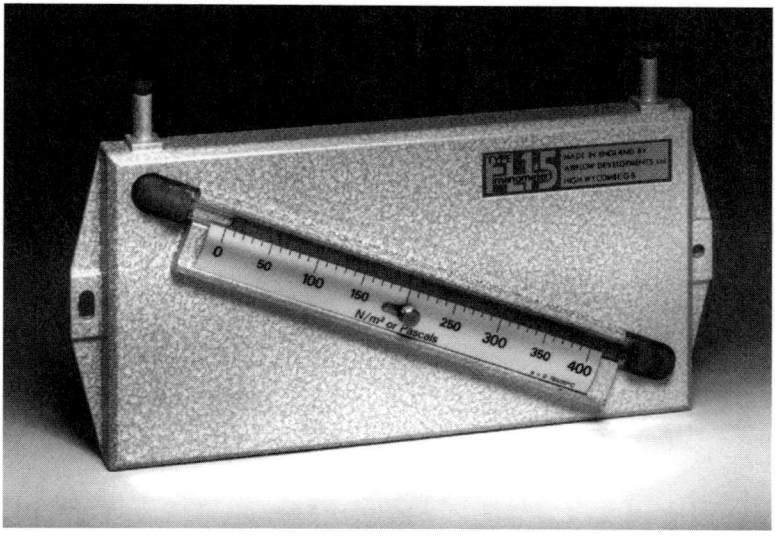

Figure 11.2 Inclined manometer used as a filter loss gage. *Source:* Courtesy of Airflow Developments Ltd.

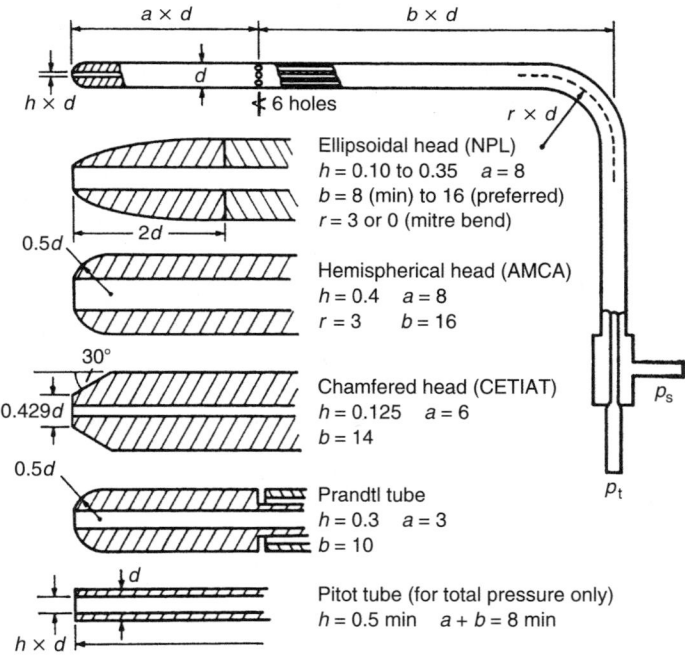

Figure 11.3 Pitot static and other tubes for velocity measurement. *Source:* Reprinted from Daly BB, Woods Practical Guide to Fan Engineering, *Woods of Colchester*, 3rd ed., 230, 1992. By kind permission of Fläkt Woods Ltd 2004.

Pitot static tube, developed by the National Physical Laboratory, is recommended in BS 1042 Part 2.1 for measuring the flow of air in ducts by the traverse method (Figures 11.5a and b). The Pitot tube consists of two concentric tubes with the end turned through 90°, so that after insertion through the duct wall, the tip can face the air stream. The nose senses the total pressure and the ring of holes in the side senses the static pressure, with the inlets terminating in tails that are connected to the opposite sides of a sensitive manometer (Figure 11.6). The

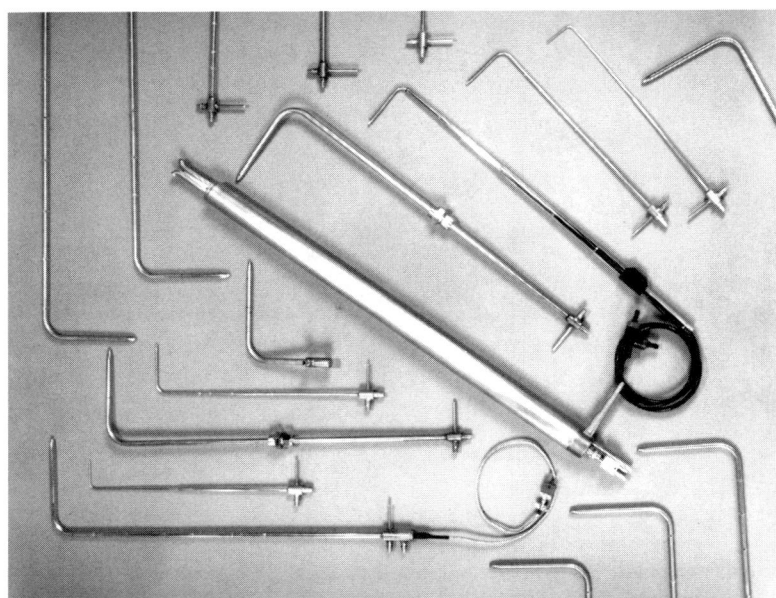

Figure 11.4 Selection of Pitot static tubes. *Source:* Courtesy of Airflow Developments Ltd.

tube must be carefully aligned with the direction of airflow in the duct, preferably within an angle of 5° (a direction pointer helps to achieve correct alignment), whilst spring clip markers simplify the positioning of the tube across the duct when undertaking the survey. The tube diameter must not exceed 15 mm and must not exceed $1/25$ of the duct diameter, or $1/20$ of the smaller side of a rectangular duct (usually, these conditions rule out ducts which have a diameter smaller than 200 mm).

When using a Pitot tube, a pressure is generated in the tube facing the air stream. Air traveling at speed towards the open end of the Pitot is stopped and has to move at right angles into the main air stream. This stoppage of the air converts kinetic energy stored in the moving air into a force that pushes the column of air in the Pitot, thus producing pressure. The holes on the side of the Pitot give the static or reference pressure in the duct (Figure 11.6).

The kinetic energy of a mass of m kg of air moving with velocity v m s^{-1} is $1/2\ mv^2$ J. Hence the kinetic energy of air, which has a mass of ρ kg m^{-3} is $1/2\ \rho v^2$ J m^{-3} when it is moving with a velocity v. This is also its velocity pressure in Pa and

$$\text{Pa} = \frac{\text{kg} \times \text{m}}{\text{s}^2} \times \frac{1}{\text{m}^2} = \frac{\text{kg}}{\text{m}^3}\left[\frac{\text{m}}{\text{s}}\right]^2$$

The convention for nomenclature is:
p_0 = barometric pressure
p_a = absolute pressure = $p_s + p_0$
p_s = static pressure = $p_a - p_0$
p_v = velocity pressure = $1/2\ \rho v^2$
v = velocity
ρ = density
p_t = total pressure = $p_s + p_v = p_s + 1/2\ \rho v^2$
T = temperature in K = $t°\text{C} + 273$, where t = airstream temperature

OPERATION OF CARBON FIBER PLANT AND SAFETY ASPECTS

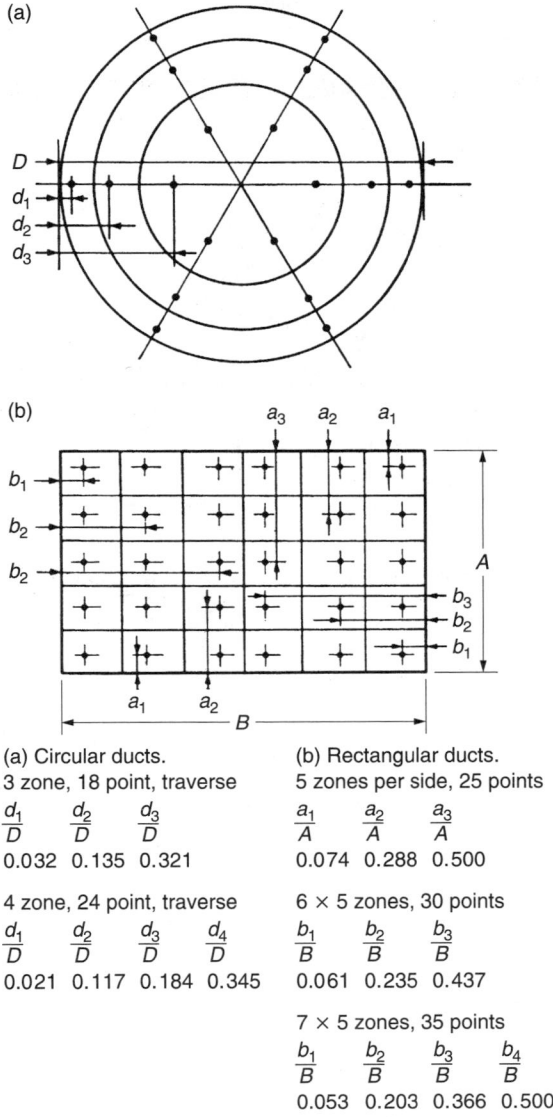

Figure 11.5 Location of velocity measurements in ducts. a. Log linear rule for positions of traverse points for a Pitot survey on three different diameters of a circular duct b. Log Tchebycheff Rule for position of measuring points and traverse lines for Pitot survey of a rectangular duct. *Note:* Minimum ratio of duct diameter to Pitot static tube diameter is 32 BS 1042 Part 2.1. *Source:* Reprinted from Daly BB, Woods Practical Guide to Fan Engineering, *Woods of Colchester*, 3rd ed., 231, 1992. By kind permission of Fläkt Woods Ltd 2004.

Pressures are measured in Pa, velocity in $m\,sec^{-1}$ and density in $kg\,m^{-3}$. The conversion for pressure units is

$$1\,Pa = 1\,N\,m^{-2} = 1\,Nm\,m^{-3} = 1\,J\,m^{-3}$$

The standard equation for calculating velocity from velocity pressure is derived from

$$p_v = \tfrac{1}{2}\rho v^2$$

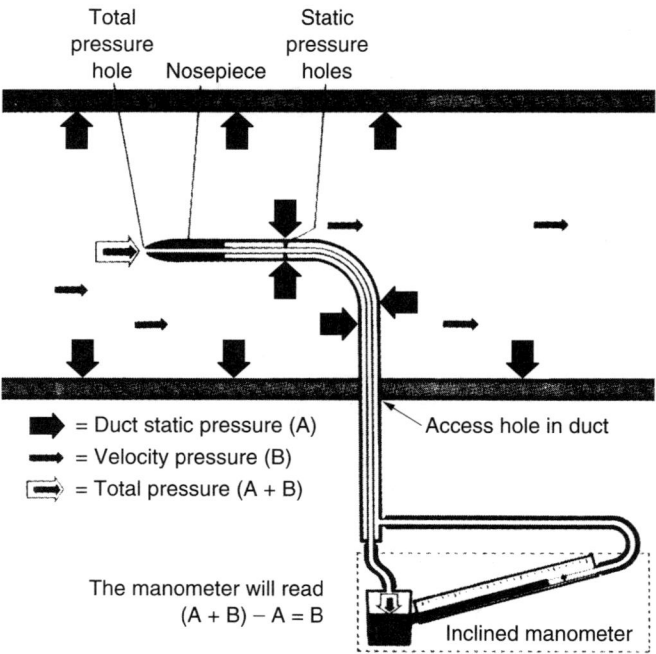

Figure 11.6 Principles of operation of the Pitot static tube. *Source:* Courtesy of Airflow Developments technical literature.

Then $v^2 = \dfrac{2p_v}{\rho}$

Hence $v = \sqrt{\dfrac{2p_v}{\rho}}$

and if the density of air is taken as $1.2\ \text{kg m}^{-3}$ then

$$v = 1.291\sqrt{p_v}$$

For non-standard air conditions this equation becomes:

$$v = 1.291\left[\dfrac{1013.25}{p_0} \times \dfrac{T}{293} \times \dfrac{100{,}000}{100{,}000 + p_s} \times p_v\right]^{-2}$$

Figure 11.7 shows the effect of temperature on the calibration of a static Pitot tube using a pressure transmitter.

A convenient section of duct having a straight parallel-sided section is selected, of not less than six duct diameters or widths upstream and four diameters or widths downstream. Holes slightly larger than the diameter of the Pitot static tube are drilled around the duct in accordance with a recognized standard such as ISO 3966 or BS 1042 Part 2.1 (Figure 11.5a and b). Readings are taken at each of these prescribed points and the square roots of each of the individual readings are averaged.

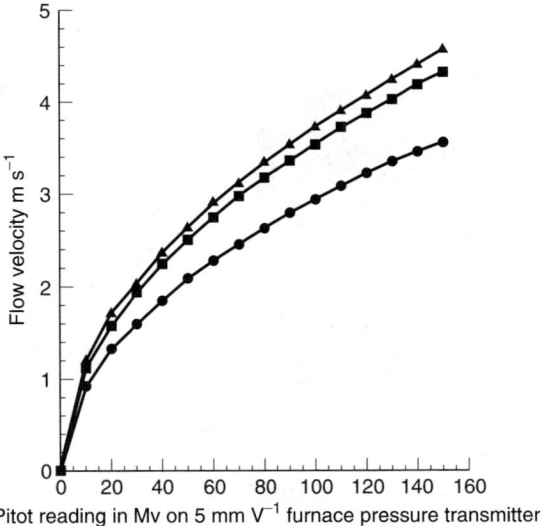

Figure 11.7 Theoretical calibration of static Pitot tube reading on a Furness Controls instrument (5mm H_2O/1 volt pressure transmitter). ◆ 20°C ■ 150°C ▲ 200°C

11.5.3 Determination of volume flow

1. Pitot tube

Having obtained the true velocity from a Pitot survey, multiplying that value by the free area of the duct will give the volume flow.

2. Vane anemometer

Air velocity and volume flow entering area of duct or opening can be determined with the latest type of hand held vane anemometers (Figure 11.8) with an accuracy of ±1% with a resolution of 0.01 m s^{-1} depending on the range. This type of equipment is very useful for checking fresh air inlets in oxidation ovens and a hood can be constructed to enable the instrument to measure the total airflow from the entire cross-sectional area of the air inlet. A well-designed instrument can have yaw and pitch angles of up to ±10% for flow velocity incidence, with a resultant error of <2%. The reading is of true air speed and the impact of the air on the vanes induces them to turn until the blade and mass velocity are in synchronization, allowing the electronics to convert the rotation into a velocity reading, with a direct linear relationship virtually independent of temperature and pressure.

3. Thermal anemometer

This type of instrument (Figure 11.9) has a very small sensing head and is ideal for restricted access measurement and is capable of reading velocities 0–30 m s^{-1} with a resolution down to 0.01 m s^{-1}. Unfortunately, the sensing head is rather prone to damage and depends on temperature, density and composition of the flow media and has an approximate fourth power relationship to velocity, which necessitates considerable computing power to provide an accurate reading across a wide working range.

Figure 11.8 Air velocity and volume flow vane type anemometer. *Source:* Courtesy of Airflow Developments Ltd.

Figure 11.9 A thermal anemometer. *Source:* Courtesy of Airflow Developments Ltd.

4. *Flow measurement with a nozzle or orifice*

This includes orifice plates, ISA nozzles, conical inlets and venturi tubes, all of which depend on the measurement of the pressure drop across a restriction through which the whole of the flow is passed. Figure 11.10 shows a sketch of a typical orifice meter which consists of a concentric square edged circular hole in a thin plate, which is clamped between the flanges of the pipe. The minimum section of the stream tube occurs downstream from the orifice with the formation of a *vena contracta* at section 2. A typical connection would be one diameter

OPERATION OF CARBON FIBER PLANT AND SAFETY ASPECTS

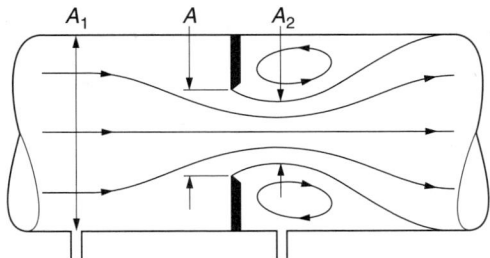

Figure 11.10 Flow characteristics of an orifice plate with a square-edged circular hole. A = cross section of orifice, A_1 = cross section of pipe, A_2 = cross section at *vena contracta*. *Source:* Reprinted with permission from Vernard JK, *Elementary Fluid Mechanics*, Wiley, 423, 1961. Copyright 1961, John Wiley & Sons Ltd.

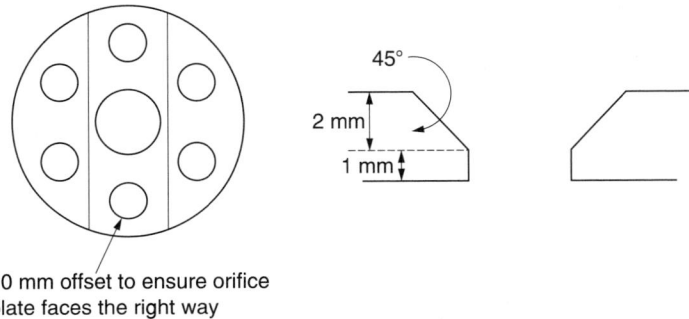

Figure 11.11 Typical construction of an orifice plate with facility for easy removal for cleaning purposes.

upstream and 0.5 diameter downstream. An orifice plate is useful to measure exhaust flow from an HT furnace, which is contaminated with Na. It would be advantageous to regularly clean the orifice plate (Figure 11.11). The orifice plate is strictly an instrument and must be carefully machined from stainless steel with a true block edge. The plate can be removed by slackening the bolts at the flange and sliding the plate out. One bolt hole is offset to ensure that the plate faces the correct way.

The flow rate can be calculated in terms of volume flow,

$$Q_v = \alpha \varepsilon \frac{\pi}{4} d^2 \left[\frac{2\Delta p}{\rho_1} \right]^{-2} \mathrm{m^3 s^{-1}}$$

or in terms of mass flow.

$$Q_m = \alpha \varepsilon \frac{\pi}{4} d^2 [2\Delta p . \rho_1]^{-2} \mathrm{kg\ s^{-1}}$$

where d = orifice throat diameter in m
D = duct diameter at the upstream pressure tapping, m
Δp = upstream static pressure minus the downsteam or throat static pressure, Pa
ρ_1 = density at the upstream tapping, $\mathrm{kg\,m^{-3}}$
Q_v = measured at the upstream density ρ_1, $\mathrm{m^3\,s^{-1}}$
Q_m = constant right through the airway system, $\mathrm{kg\,m^{-3}}$
α = flow coefficient as given in the appropriate standard
ε = expansibility factor as given in the appropriate standard

For testing with atmospheric air, ε may be taken as 1 provided $\Delta p < 1000$ Pa. Also, α can be replaced by CE where
C = coefficient of discharge

$$E = \text{velocity of approach factor} = \frac{D^2}{[D^4 - d^4]^{-2}}$$

Both α and C are dependent on:

$$\beta = \frac{d}{D}, \text{ the diameter ratio}$$

and Re_D or Re_d

Re_D, the duct Reynold's number (based on the average velocity and air conditions in the approach duct D) or

Re_d, the throat Reynold's number (based on the average velocity and air conditions in the orifice opening d).

The Reynold's number need only be known quite roughly, since the changes it produces in α or C are small. When testing with atmospheric air, the approximate formulae given in Table 11.1 can be employed, utilizing either the expected volume flow, Q_v, or the observed pressure difference, Δp. Preliminary values of α and ε can be obtained from Figures 11.12 and 11.13. Once Re has been established, precise values for α and ε can be obtained from the requisite standard.

Table 11.1 Estimation of reynold's number

Quantity	Re_D	Re_d
Exact value	$\dfrac{Dv_D\rho_D}{\mu_D}$	$\dfrac{Dv_d\rho_d}{\mu_d}$
Approximately, given Q_v	$85,000\dfrac{Q_v}{D}$	$85,000\dfrac{Q_v}{d}$
Approximately, given Δp	$86,000\alpha\varepsilon\dfrac{d^2\sqrt{\Delta p}}{D}$	$86,000\alpha\varepsilon d\sqrt{\Delta p}$

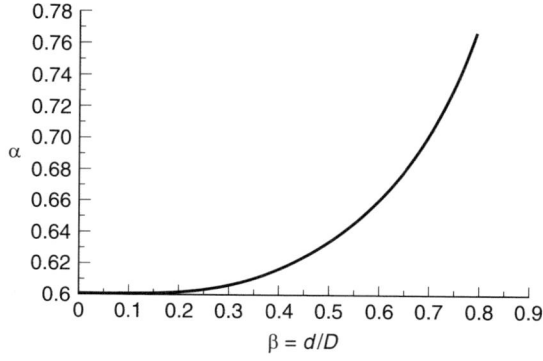

Figure 11.12 Approximate orifice coefficient α for air. *Source:* Reprinted from Daly BB, Woods Practical Guide to Fan Engineering, *Woods of Colchester*, 3rd ed., 246, 1992. By kind permission of Fläkt Woods Ltd 2004.

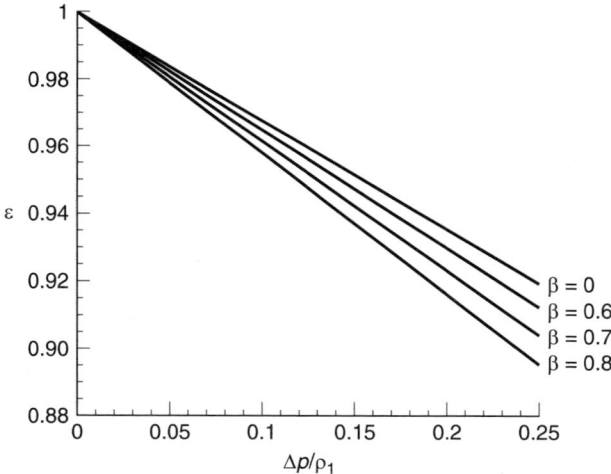

Figure 11.13 Approximate orifice coefficient ε for air. *Source:* Reprinted from Daly BB, Woods Practical Guide to Fan Engineering, *Woods of Colchester*, 3rd ed., 246, 1992. By kind permission of Fläkt Woods Ltd 2004.

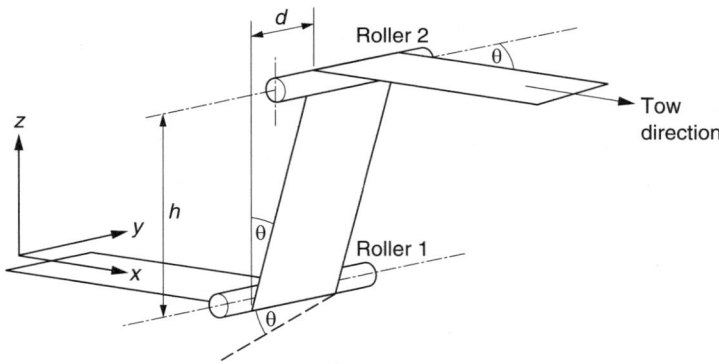

Figure 11.14 Lateral movement of tow band on rollers by canting ($d = h \tan \theta$).

11.6 COLLIMATION AND SPREADING OF OXIDIZED AND CARBONIZED FIBER

In some stage of carbon fiber manufacture, it will be necessary to reposition the tow band, either laterally and/or expand, or condense the width of the tow band.

11.6.1 Lateral movement

The lateral position can be changed by canting a roller (Figure 11.14). The axis of roller R1 is in the horizontal plane, but is inclined at an angle θ to the tow band normal. In general terms, it is not necessary for R1 to be in the plane of the tow band, but it is a convenient simplification in this application. The tow band approaches the roller in the x direction in the (y, z) plane. When passed around, R1 the tow band moves in the vertical (y, z) plane, but because of the inclination of the roller, it moves in a direction at an angle θ to the z axis. If the tow is then passed around a second roller R2, which is parallel to R1, the tow will resume its travel in the x direction, but will move sideways from its original path by a distance d, where $d = h \tan \theta$, h being the vertical separation of the rollers R1 and R2.

11.6.2 Lateral expansion or contraction

Adjustment of the width of a tow band is achieved with a variable bow roller (e.g. a Mount Hope or Wittler roller). When a bowed roller rotates, the flexible covering of the roller expands and contracts as it passes from the inside to the outside of the roller. During the expansion of the covering, a point on the roller surface, viewed perpendicular to the plane of the roller bow, will follow a radial line drawn through the center of curvature of roller bow. A tow free to move along the roller surface will tend to follow such a radial line. Thus, if the tow width a at roller X is known (Figure 11.15 e) and it is required to spread the sheet to a width b using an adjustable bowed roller Y at a distance c from roller X, it

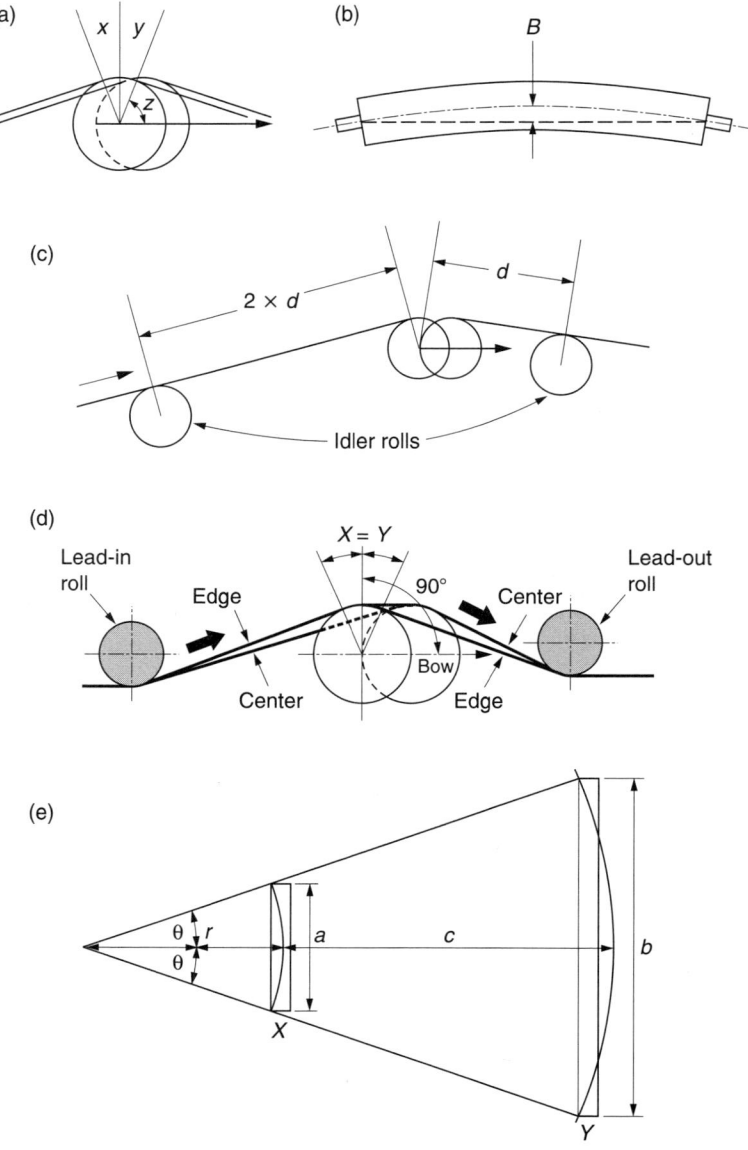

Figure 11.15 Schematic diagrams of bowed rollers ($a/r = \tan \theta$). *Source:* Adapted from Mount Hope and Wittler technical literature.

is possible to compute the radius of curvature of the bowed roller such that tows spreading to the required pattern will follow radial lines. Given this condition, the tows will move naturally on the bowed roller to the desired width. The required radius of curvature R is given by:

$$R = c\left[1 + \frac{a}{b-a}\right] \quad \text{and} \quad \theta = \tan^{-1}\left[\frac{b}{2(r+c)}\right]$$

The roll is installed so that it makes an arc of contact with the web with the apex of the bow, as shown in Figure 11.14A, making the maximum area of contact by arranging the meeting and leaving angles, $x°$ and $y°$, to be equal. This is achieved by bisecting the total arc of contact and drawing a line at 90° to the bisecting line. The axle of the roll should lie in this plane, with the apex of bow, indicated by the arrow (Figure 11.15b), pointing in the direction of travel.

To laterally expand the web, the roller is set with the apex of the bow in the same direction as that of fiber movement. The fiber web will meet the roll on the concave side and leave from the convex side. The distance between the first roller and the expanding roller is made twice as great as the distance (draw) to the second roller (see Figure 11.15c) and this latter distance is kept as short as possible in order to hold the expanded width. The normal mounting position is shown in Figure 11.15d, where angle $x° = y°$ and angle $z° = 90°$. The center and edges of the web travel equal paths.

A uniform spreading action is achieved only when the tow bands entering and leaving the roller are symmetrically distributed relative to the plane of the roller bow.

To laterally contract the web, the apex of the bow is set in a direction opposite to direction of fiber movement. The web will meet on the convex side and leave from the concave side.

11.7 SPLICING SMALL TOWS

Small tows can be satisfactorily joined using a proprietary air intermingler (Figures 11.16 and 11.17). A plentiful supply of oil free air at constant pressure (about 0.4–0.5 MPa) is required, typically 31 $dm^3 s^{-1}$, using relatively large supply pipes (e.g. 25 mm diameter) with as little resistance to flow as possible and, preferably, with a low (5 dm^3) capacity reservoir situated near the splicer. If cutting knives are provided, then knives tipped with WC will have a much longer service life than standard steel ones. The two ends of yarns to be joined are placed in the splicing chamber from opposite sides. When the splicer is activated, a hinged pad clamps the two ends and the waste ends are trimmed off. An air blast is then introduced at very high speed into the chamber. The air is very turbulent and radically disrupts the arrangement of the fibers in the splicing chamber. Those fibers which lie across the opening of the air feed hole are separated by the direct blast, whilst fibers elsewhere are subjected to a chaotic pattern of vortices downstream to the entry point, which produces twisting and intermingling.

The splice has a characteristic form, with a central section essentially unchanged and densely clustered, highly twisted and intermingled on either side the fibers. Each cluster has a small tail where the extreme ends of the spliced yarns have not been fully bound into the structure. When a load is applied to the spliced yarn, the fibers in the clusters slip slightly, until the whole structure stabilizes by the load being taken by the inter-fiber frictional forces. The splice strength is typically 90–95% that of the parent yarn.

Figure 11.16 Pentwyn splicer. *Source:* Courtesy of Pentwyn Splicers Ltd.

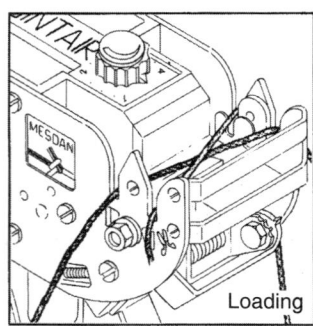 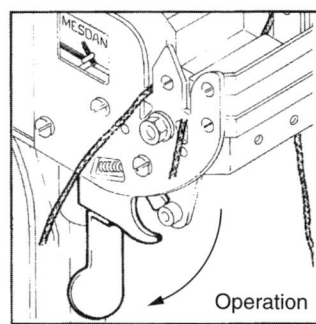 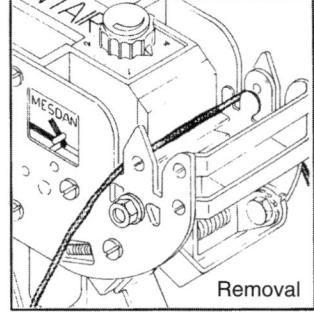

Figure 11.17 Mode of operation of a Mesdan Jointair splicer. *Source:* Adapted from AB Carter Inc. technical literature.

11.8 DRIVE SYSTEMS AND ROTATING ROLLERS

The importance of guard systems for nip rollers has already been emphasized. The sets of nip rollers must close exactly parallel and can be initially checked by trapping a thin piece of paper (e.g. a paper used for rolling cigarettes) to ensure that it is firmly held in the closed nips across the entire width. Rollers not closing properly will permit slippage.

Line speed is an important parameter and can be easily checked with a device dubbed by the author's colleagues the Morganometer (Figure 11.18). A target is placed on the moving fiber band and the movement timed with a stopwatch between two sets of pegs, which can be

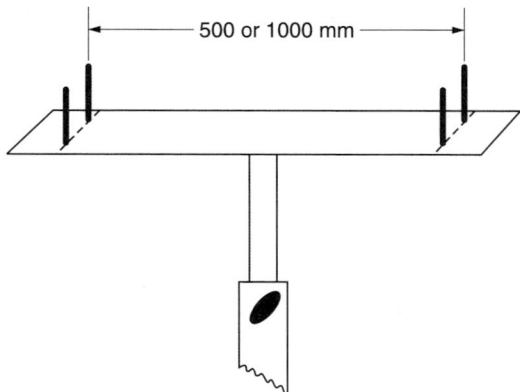

Figure 11.18 Morganometer line speed device.

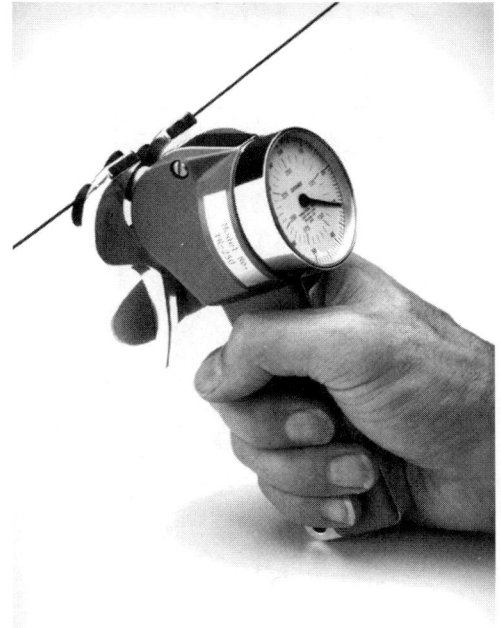

Figure 11.19 Mechanical hand-held trigger tension meter. *Source:* Courtesy of Tensitron Inc., Boulder, CA.

500 or 1000 mm apart, depending on the free run available between equipment. The height is adjusted so that the middle of the pegs is the same height as the fiber band and the eye is sighted in turn and in line with each pair of pegs to avoid parallax as the tip of the target passes by. Alternatively, assuming no fiber slippage, the line speed can be determined by timing, in seconds, one revolution of a drive roller when:

$$\text{Line speed}(\text{m h}^{-1}) = \frac{\text{Circumference of roller(m)} \times 3600}{\text{s per revolution}}$$

All rollers must be rotating and the tow fiber tension can be checked using a hand held fiber tension meter (Figures 11.19–11.21). Alternatively, using an appropriate head and suitable guides, the tension of a single tow can be measured continuously (Figures 11.22

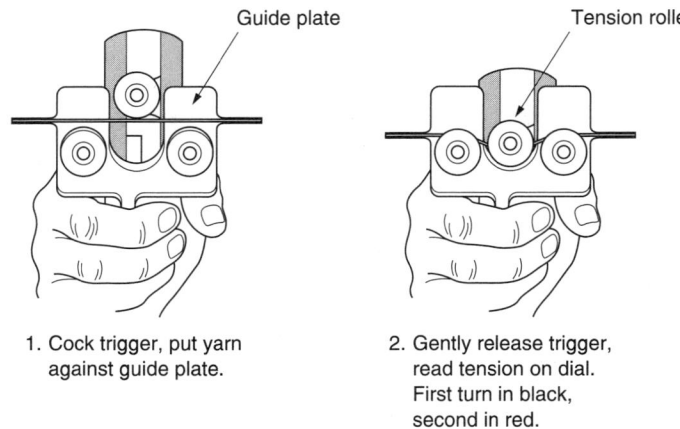

Figure 11.20 Mode of operation of hand held trigger tension meter. *Source:* Courtesy of Tensitron Inc., Boulder, CA.

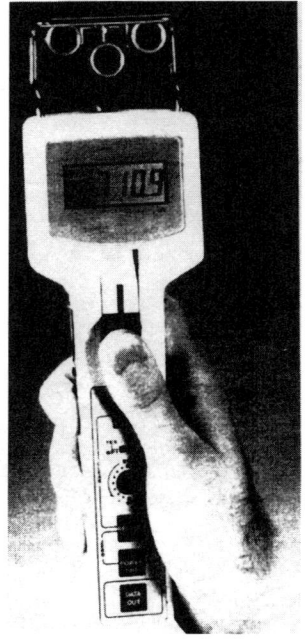

Figure 11.21 Hand held electronic yarn tension meter. *Source:* Courtesy of James H Heal & Co. Ltd., Halifax, England.

and 11.23). With experience, a good indication of fiber tension can be obtained by gently patting the fiber band with the palm of the hand.

11.9 PRECURSOR CREEL

The batch numbers of all precursor spools used on the line should be recorded and a list made of any defects found. The correct degree of braking should be applied to all spools so that they rotate freely and give a uniform band of precursor entering the oxidation oven.

OPERATION OF CARBON FIBER PLANT AND SAFETY ASPECTS 439

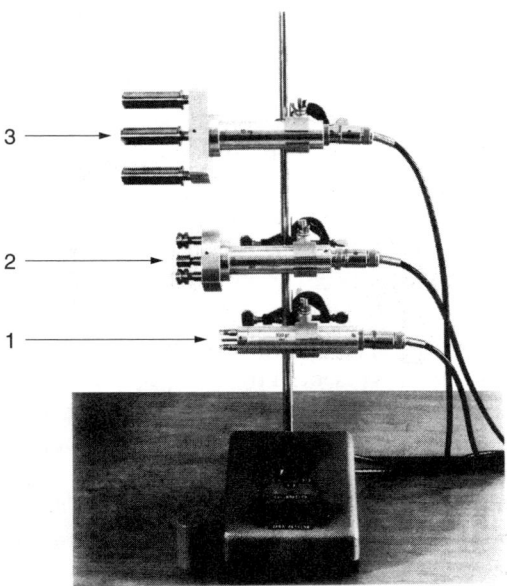

Figure 11.22 Range of tension measuring heads in the range 0–200,000 cN. *Source:* Courtesy of Rothschild, Zurich, Switzerland.

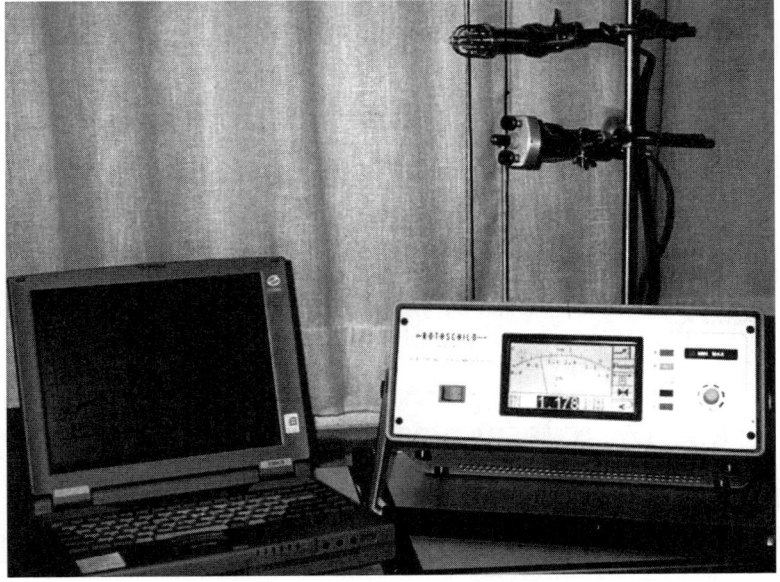

Figure 11.23 Electronic tension meter for continuous measurement of yarn tension. *Source:* Courtesy of Rothschild, Zurich, Switzerland.

11.10 OXIDATION PLANT

It should be established that all fan motors are turning in the correct direction at the correct speed, as excessive air flow with small tow production will give inferior cosmetics. All air inlets and cooling grilles on electric motors must be free of fiber debris.

All pass back rollers must rotate freely and tow tension in the first passes must be adjusted to ensure that the fiber band does not come into contact the floor of the oven. The fiber band must be of uniform thickness, with no gaps and must have a uniform tension. The inlet and outlet fiber speeds in oxidation control the mpul of the opf. The uniformity of colour across the first and second passes of the oxidized PAN tow band will be a good indication of uniform temperature distribution which must be within $\pm 2°C$. The control of the level of oxidation can be ascertained by measuring the final oxidized density.

If temperatures are too high, or the fiber band too thick, an exotherm, and consequently, a fiber burn will occur, where the fiber could glow and smoke, or smolder. The burn causes a collapse of the tow, resulting in a sudden slackening of that section of the fiber band. When the fiber burns, there will be a trailing end (back end of fiber going in) and a leading end (front of the fiber going in) and it is essential to promptly pull back the leading end and splice it in. The trailing end will blow across the oven and should be pulled out and spliced on to an adjacent tow. Care must be taken to ensure that a burn does not spread to the precursor fiber. Procedures for dealing with the more serious stage of a fire in an oven must be provided and CO_2 or steam is probably the most efficient medium to deal with a fire. It should be remembered that CO_2 is asphyxiating and a flow of CO_2 gas can produce a static discharge. If steam is used, then a steam trap must be employed to remove condensed steam, which would otherwise form a lock of condensed steam and prevent its flow when required. Any other type of extinguisher used to deal with a precursor fire must be evaluated prior to use since some agents react with the products of combustion to give a dreadful odor, making the workplace uninhabitable.

Broken filaments traveling around a roller can cause a wrap and unless dealt with quickly, will build up and when cut away will produce two ends, which must be spliced in. All wiper blades should be checked to ensure that they are making correct contact with the rollers.

11.11 PYROLYSIS PLANT

The design of a pyrolysis oven is special and depends on complete sealing, with no ingress of air and with the waste products continuously being removed with a purge of N_2. This process forms a product, which when exposed to air, can absorb O_2 followed by evolution of heat, which must be dissipated, failing which a fire may occur.

11.12 LOW TEMPERATURE CARBONIZATION FURNACE

An ideal way of checking for ingress of air to a furnace is to use a smoke pencil fitted to a rubber bulb, which when depressed, forces air through the smoke tube emitting a fine stream of smoke. Checking must be thorough right across the width, both above and below the fiber band. The author has seen fiber adequately protected below the fiber band, but with air drawn in by a venturi effect above the fiber band, causing the fiber to burn with a blue flame.

In the LT furnace, tars are evolved and unless efficiently removed, can deposit tar on the fiber and form tar needles. A tar shield fitted above the fiber band at the inlet end of the furnace can be beneficial. Correct balancing of the furnace can be achieved by adjusting the body flow of N_2 at the furnace inlet end to a level where smoke is not forced out from the outlet end and, yet, when looking down into the furnace muffle, the atmosphere is not entirely clear, with some wisps of swirling smoke visible.

Tows can break in the LT furnace due to the following reasons.

1. Fiber is underoxidized (oxidized density must be kept above 1.36 g cm^{-3})
2. Ingress of air (requires checking with a smoke pencil e.g. Draeger)
3. Deposition of tar (efficient extraction must be maintained, ensuring that the incinerator is routinely rodded out)
4. Tension is too high (tow tension needs checking with a meter, as also shrinkage).
   The shrinkage of the fiber in the furnace train will be too low if the tension is too high. The and shrinkage is given by

$$\left[\frac{D_{EO} - D_{ST}}{D_{EO}}\right] \times 100\%$$

   where D_{EO} = speed of fiber exiting drive immediately after oxidation, m h^{-1}
   D_{ST} = speed of fiber exiting drive immediately prior to surface treatment, m h^{-1}

   Shrinkage should be of the order of 5–8%. If it is too high, the strength of carbon fiber will fall and because of the increased catenary, the fiber could then drag on the furnace bottoms, impairing fiber cosmetics. A low shrinkage means a higher tension and if too high, will break the fiber.
5. Fiber burns (density of oxidized fiber too low, ingress of air, or insufficient cooling at furnace outlet). Broken tows must be pulled back out of the furnace and spliced on to adjacent tows.

11.13 HIGH TEMPERATURE CARBONIZATION FURNACE

A smoke pencil should be used to check for ingress of air. The furnace can be correctly balanced by adjusting the body flow of N_2 at the furnace inlet end to such a level that, when looking down into the furnace muffle, the atmosphere is not entirely clear but, at the same time, smoke is not forced out from the outlet end. It is essential to maintain a slight cloudy atmosphere, as this permits the correct concentration of HCN to build up within the furnace, significantly improving the carbon fiber strength.

If Na is present in the precursor fiber, it will tend to form NaCN and it is important to wear the correct protective clothing and face mask when dealing with such residues. To conserve water supplies, it is common to use a recirculation water cooling system with a water cooling tower. This water must be treated and routinely checked for legionella bacteria to prevent any outbreak of Legionnaires' disease.

The fiber will burn as it exits the furnace and enters the plant atmosphere unless sufficient cooling is applied.

Modulus is controlled by the temperature of this furnace and can be measured with a thermocouple or pyrometer.

11.13.1 Calibration of pyrometer

Obviously, temperature is an important parameter and pyrometers must be calibrated on a regular basis. Figure 11.24 shows a typical layout for the calibration of a pyrometer. The calibration lamp (GEC) [30], must be calibrated to a National standard e.g. NAMAS, and should be mounted upright in front of the pyrometer, which should be positioned at a distance, d, to enable it to be focused on the lamp filament. It is convenient to mount the pyrometer on a laboratory jack to adjust the height relative to the lamp filament.

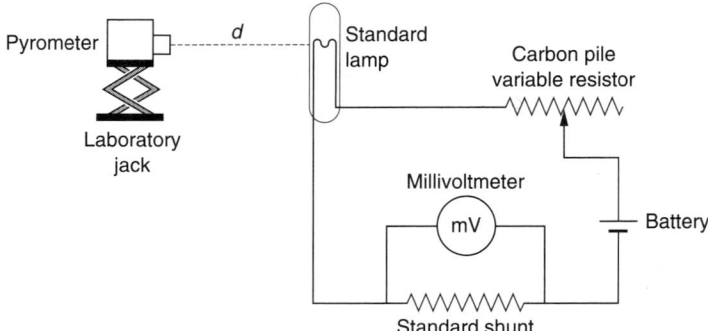

Figure 11.24 Schematic layout of pyrometer calibration facility.

Table 11.2 Typical calibration results for a cylindrical envelope vacuum filament lamp (effective wavelength = 0.66 μm)

Radiance temperature °C	Current A	Uncertainty ±K
800	4.405	4
900	4.910	5
1000	5.565	5
1100	6.355	5
1200	7.260	5
1300	8.275	6
1400	9.375	6
1500	10.535	6

The lamp should be run on DC current supplied by a heavy duty car battery making certain that correct polarity is assigned to the lamp connections. By adjusting the current with the carbon pile variable resistor, the temperature of the lamp can be altered and the vacuum lamp will be within 1°C of final temperature in less than 2 min, and within 0.1°C in about 5 min, over the range 1063–1770°C. The temperature can be ascertained by reading the mV across a standard shunt and using Ohm's law to obtain the current, using the temperature given in the calibration certificate (Table 11.2). It is simplest to adjust the current and obtain a given temperature, which can then be compared with the reading obtained from the pyrometer.

11.14 HIGH MODULUS FURNACE

The high modulus furnace is checked in a manner similar to the HT furnace, with one major exception. The furnace atmosphere should be virtually quiescent, as too high a flow of N_2 will induce severe erosion of the graphite element and seriously curtail its working life. The temperature controls the modulus and, unfortunately, a high modulus demands high temperatures, leading to a significant shortening of element life.

11.15 SURFACE TREATMENT

The purity of the electrolyte must be considered and an additive such as $MgCO_3$, added as a non-caking compound, can have a most deleterious effect, depositing on to the fiber during surface treatment.

With an aqueous surface treatment system, the bath electrolyte level must be checked and the electrolyte concentration be determined by measuring the specific gravity. The temperature, voltage and amperage should be regularly monitored. The surface of the bath should be kept clear of fiber debris, filters cleaned regularly and electrolyte circulation checked. The efficiency of the water wash section can be monitored by determining the conductivity of a water extract of the treated and washed fiber.

Insufficient surface treatment gives a low composite ILSS, whereas if the level is too high, the fiber becomes brittle. The mode of failure of the ILSS specimen can be a useful guide to the degree of treatment.

The fiber cosmetics will be badly affected if a roller at the bottom of the bath does not turn freely. The bearings must be regularly inspected.

To establish that the correct level of current is being applied, the following procedure can be adopted: A probe is constructed that will measure the voltage of, say, a 10 cm length of fiber by recording the voltage when holding the probe connected to a voltmeter against a single fiber tow. This procedure is repeated with selected tows across the fiber band and for each treatment cell. A table of the results can be constructed as follows:

Tow No.	1	20	40	60	80	100	120	140	160	180
					volts/10 cm					
Cell 1	a1	b1	c1	d1	e1	f1	g1	h1	i1	j1
Cell 2	a2	b2	c2	d2	e2	f2	g2	h2	i2	j2
Cell 3	a3	b3	c3	d3	e3	f3	g3	h3	i3	j3
Total volts for three cells	a = a1 + a2 + a3	b	c	d	e	f	g	h	i	j

$$\text{Mean total voltage/10 cm tow} = \frac{a+b+c+d+e+f+g+h+i+j}{\text{number of tow positions}} = V \text{ volts/10 cm}$$

To measure the current, it is now necessary to measure the resistance of a one meter length of identical carbon fiber tow, say $R\ \Omega\,\text{m}^{-1}$, hence resistance of a 10 cm length is $R/10\ \Omega$.

Then, since $V = IR$, the total current I for n tows is

$$I = \frac{10 V_n}{R} \text{A}$$

and this value should agree closely with the recorded ammeter reading.

11.16 SIZING

After an aqueous surface treatment process, it is essential that the fiber is dried after the final surface treatment wash to ensure the correct pick up of size. After sizing, any roller, nip or guide bar that the wet sized fiber passes over will control the level of pick up of the size liquor. It is advantageous to have a level control fitted in the bath to maintain the correct level.

11.17 WINDING

Winding equipment is generally in continuous use and it is essential that it is regularly maintained. Machines should be kept free of fiber debris and a soft bristle paint brush is

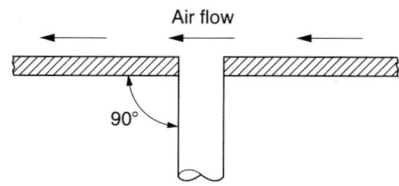

Figure 11.25 Side tube for static pressure measurement.

Table 11.3 Human physiological response to various concentrations of HCN in air

Response	Concentration mg l^{-1}	Concentration ppm
Immediately fatal	0.3	270
Fatal after 10 min	10.2	181
Fatal after 30 min	0.15	135
Fatal after ½–1 h or later, or dangerous to life	0.12–0.15	110–135
Tolerated ½–1 h without, immediate or late effects	0.05–0.06	45–54
Slight symptoms after several hours	0.02–0.04	18–36

Source: reprinted from Patty FA, *Industrial Hygiene and Toxicology*, 2nd ed., Interscience, New York, 199, 1942.

useful. The extraction system must be efficient and all electrical equipment protected from carbon fiber fly. The wind pattern is changed for larger tow sizes such as $40k$.

11.18 DEALING WITH EMISSIONS

All extraction hoods must be correctly designed and the efficiency of the extraction system regularly checked using vertical manometers permanently positioned at strategic points in the extraction ducting. When the static pressure is measured, there must be no velocity against the entry of the tube used for these measurements, which is termed a side tube (Figure 11.25) and the opening must be flush with the inside surface. Routine cleaning of all extraction ducting should ensure an acceptable standard, but if the efficiency is impaired, the relevant ducting must be cleaned. To safeguard the workforce, a planned level of checking concentration levels of HCN, NH_3 and NO_x should be undertaken at defined test points within the plant.

Each shift should have at least one trained First Aider. No food or drink should be taken in the plant. Special care must be taken if NaCN is formed in the HT furnace and all necessary protective gear must be available and worn when dealing with cyanide residues [14,15]. If there is a risk of cyanide poisoning, an antidote should be available to be administered by a qualified practitioner and always replaced before the shelf life has elapsed.

The data in Table 11.3 suggest that concentrations of HCN above 90 ppm may be incompatible with life when exposure periods approach or exceed 30 min [6].

Self-contained breathing apparatus can only be used by trained personnel but others can use air line breathing apparatus or the canister respirator.

11.19 TREATMENT OF CYANIDE EFFLUENT

A recognized treatment for cyanide effluents, for the purpose of detoxification before discharge to a public sewer, is by reaction with NaOCl (bleach liquor), which converts toxic cyanide to a non-toxic cyanate. In order to ensure a smooth and rapid conversion, it is necessary to have a pH value of not less than 11 in the effluent under treatment. Since NaOCl liquor is strongly alkaline, normally no extra alkali will be needed.

$$NaCN + NaOCl + H_2O \rightarrow CNCl + 2NaOH$$

At pH 10 and above, no further hypochlorite is reacted and the cyanogen chloride hydrolyses to the cyanate:

$$CNCl + 2\,NaOH \rightarrow NaCNO + NaCl + H_2O$$

This reaction is not instantaneous $\dfrac{d[CNCl]}{dt} = K \times [CNCl] \times [OH]$

where K varies from 80 $l\,gmol^{-1}\,min^{-1}$ at 0°C to 530 $l\,gmol^{-1}\,min^{-1}$ at 25°C [7,8].

Cyanogen chloride is volatile and almost as toxic as cyanide, so the reaction should be carried out in a closed vessel and the temperature optimized at 20°C. The second reaction involving the hydrolysis of the CNCl is rather slow and is the rate determining step.

The vessel should be well agitated during the treatment and after the reaction, any excess NaOCl is treated with $Na_2S_2O_3$ after determining the amount by titration.

A sample of effluent is checked with phenolphthalein to ensure that the liquor is alkaline; if not, then excess NaOH must be added. The effluent should be treated in batches to ensure so sufficient availability of hypochlorite and the requisite period of contact. About 10 l of NaOCl (14% available chlorine), diluted with 2–3 volumes of water is added per kg of NaCN, which will provide a 20% excess that would require about 0.6 kg of sodium thiosulphate crystals ($Na_2S_2O_3.5H_2O$) for dechlorination, equivalent in its simplest form to:

$$Na_2S_2O_3 + NaOCl \rightarrow Na_2S_2O_4 + NaCl$$

After adding the bleach liquor, it is mixed for a few minutes, allowed to stand for 15 min and a sample taken. The sample is acidified with glacial CH_3COOH, 5–10 cm^3 of 10% KI solution added, followed by a starch indicator. A blue color indicates the liberation of I_2 and the presence of excess bleach liquor. If there is no color change, further bleach liquor must be added and the test repeated. After a treatment time of 12 h, the quantity of excess bleach can be calculated by titrating an aliqout of the treated effluent liqour with a solution of $Na_2S_2O_3$ to be used for the dechlorination by first acidifying with CH_3COOH and adding KI solution and starch indicator.

$$2NaOCl + 4KI + 4\,CH_3COOH \rightarrow 2NaCl + 4\,CH_3COOK + 2I_2 + 2H_2O$$

$$I_2 + Na_2S_2O_3 \rightarrow Na_2S_4O_6 + 2NaI$$

Hence $2NaOCl \equiv 2I_2 \equiv 2Na_2S_2O_3$

i.e., $Na_2S_2O_3 : NaOCl$ is 1 : 1

11.20 PROTECTING THE ENVIRONMENT

A planned level of checking HCN, NH_3, NO_x and particulate matter should be undertaken at prescribed efflux points throughout the plant [9].

1. Particulate matter

If particulate matter is determined in accordance with BS 3405, then the volume of gas sampled must be sufficient to exceed the initial weight of the filter material by 0.3%. A minimum of three test runs must be undertaken, each of 1 h minimum duration each test, comprising a minimum sample volume of 1 m^3. All surfaces leading to the filter medium must be washed three times with CH_3COCH_3 immediately after each test run. The mass collected from these surface washings added to the mass collected on the filter must be used to report the particulate emission. The filter medium must be capable of retaining 99.95% of particles $\geq 0.3 \mu m$.

2. Oxides of nitrogen [10]

Nitrogen in the air can be oxidized to NO and NO_2 by exposure, say, to a firing zone in a burner. In a combustion process, typically 97% of the emission is NO. Normally NO and NO_2 are measured together by converting the NO to NO_2 using ozone generated from an ozonator.

$$NO + O_3 \rightarrow NO_2^* + O_2$$

$$NO_2^* \rightarrow NO_2 + h\nu$$

The reaction forms a characteristic luminescence due to an electronically excited NO_2 molecule decaying to a lower energy state. The intensity of the luminescence is proportional to the concentration of NO. The light emission is detected by a photomultiplier tube, which generates a proportional electronic signal.

To measure the NO_x (NO + NO_2) concentration, the NO_2 must be converted to NO with a molybdenum converter at 325°C, whereas higher temperature converters would also convert any NH_3 to NO. The converted NO is reacted with O_3 to give the NO_x reading.

$$2\,NO_2 \rightarrow 2NO + O_2$$

$$3\,O_2 + Mo \rightarrow 2MoO_3$$

$$3\,NO_2 + Mo \rightarrow 3\,NO + MoO_3$$

The Mo is re-activated with H_2:

$$MoO_3 + 3\,H_2 \rightarrow Mo + 3\,H_2O$$

All samples are taken through a heated sample line and all monitors are calibrated daily using cylinders of gas with two certified concentration ranges (e.g. 200 and 400 ppm) of NO_x. If N_2O has to be determined, a different type of analyzer would be required, that uses an infra red source, narrow bandpass filters and a solid state IR detector.

3. HCN

Daily checks are made on the plant with handheld instruments utilizing a thermodynamic cell, calibrated by the manufacturer at frequent intervals or by using Draeger test tubes. Emission levels are measured with a sample train bubbling a known volume of the dried gas stream through 0.01M NaOH, dividing the quantity into two bubblers connected in series for improved efficiency. The total contents are made up to a known volume and aliquots sampled using a pipette controller to safely handle noxious solutions. The samples are titrated with 0.01 M $AgNO_3$ using 5-(4-dimethylamine benzylidene)rhodanine as an indicator. The cyanide solution changes color from straw to pink. Solutions of KCN made alkaline with NaOH are used as control samples.

$$AgNO_3 + 2KCN + \text{(dimethylaminobenzylidene rhodanine)} \rightleftharpoons K[Ag(CN)_2] + KNO_3$$

Hence $AgNO_3 \equiv 2KCN$ and 1 cm³ of 0.01 M $AgNO_3 \equiv 0.0005204$ g CN^-

As a general guide, the safety level for exposure to HCN for the public (i.e. at the boundary fence of the factory) can be taken as $1/40$ of the threshold limit value (TLV).

$$\text{Safety level of HCN exposure} = 1/40 \times 10 = 0.25 \text{ ppm HCN}$$

A safe emission value would be 10^3 times this value, employing a good safety margin i.e. $0.25 \times 10^3 = 250$ ppm. The provisos for this would be that the emissions from the chimney must have a minimum efflux velocity for dry gases of 15 m s^{-1} and for wet gases 8 m s^{-1}, with a chimney height that is about 2½ times the height of the factory.

A formula used by the H.M. Industrial Pollution Inspectorate is:

$$H^2 = \frac{9M}{20P}$$

where H = Effective height of chimney, m
M = Mass emission, kg day^{-1}
P = Ground level concentration, mg m^{-3}, i.e. maximum concentration in the atmosphere

4. Ammonia

Emission levels are measured by bubbling a sample train of a known volume of the dried gas stream through 0.01M H_2SO_4, dividing the quantity into two bubblers connected in series for improved efficiency. The total contents are made up to a known volume, neutralized with concentrated NH_3 free NaOH and analyzed using a spectrophotometric technique [11] by measuring the absorbance of the blue colored compound (thought to be related to indophenol blue), but appears green against the yellow color of the reagent blank. Ammonia reacts with OCl^- ions generated *in situ* by the alkaline hydrolysis of sodium dichloroisocyanurate, and with sodium salicylate at about pH 12.6 in the presence of sodium nitroprusside, to form a colored compound (thought to be related to phenol blue). The compound is blue, but appears green against the yellow color of the reagent blank. The absorbance of the compound is measured spectrophotometrically at an absorbance of

655 nm and related to the NH_3 concentration in the sample by means of a calibration curve established with solutions of different concentrations of NH_4Cl and expressing the results as N. Sodium citrate is used to mask any possibly interfering cations.

Salicylic acid

Sodium dichloroisocyanurate

Sodium nitroprusside

Sodium dichloroisocyanurate

Indophenol blue

5. Noise

In areas where noise is a problem, the sound level should be established using a sound level meter and if above the stipulated minimum operatives must be provided with ear protection.

11.21 SAFETY COMMITTEE

Safety visits should be made by a cross-section of the workforce and often, the person on the shop floor will be more aware of problems unknown to management. Once a safety visit has been made and a report issued, the management must correct any deficiencies—remember, it is preferable not to wait for an accident to happen as prevention is better than cure.

11.22 COSH-H REQUIREMENTS

Control of Substances Hazardous to Health Regulations 1988 (COSH-H) [12] is a duty required of both employers and employees and is best approached as a stepwise process [13]:

1. Identify the hazardous substances (can be toxic, harmful, corrosive or irritant and can include micro-organisms and dusts)
2. Identify properties and hazards (utilize manufacturer's Hazard Data Sheets) [14]
3. Identify the works activities (in which substances hazardous to health are used or generated)
4. Assess the risks using an assessment process:
   a. Job activity
   b. Hazard to health

c. Existing precautions
d. Present or future risk
e. Action needed

5. Inform employees
6. Reduce risk by simple change [15,16]
7. Control the residual risks
8. Check the effectiveness of control measures
9. Health surveillance [17,18]
10. Records
11. Audit the procedures which have been put in place.

11.23 TOXICOLOGY OF CARBON FIBERS AND EMISSIONS [19]

A safe premise is to consider all dust as nuisance and take due precautions. The type of dust is important and the crucial variety is respirable dust, generally considered to be about 3.5–5.0 μm in size and capable of reaching the alveoli of the lung without being trapped in nasal hair, or trapped *enroute* on the walls of the trachea or bronchia. Particles 7–10 μm in size will not reach the lung and are not considered respirable.

11.23.1 Definitions of exposure limits

1. Permissible Exposure Limit (PEL) is a legally binding airborne exposure limit issued by the Occupational Safety and Health Administration (OSHA)
2. Threshold Limit Value (TLV) is issued by the American Conference of Governmental Industrial Hygienists (ACGIH). TLVs are recommended exposure limits and should be used as guidelines for good practice.
3. A Time Weighted Average (TWA) is for a normal 8 h workday and 40 h work week to which workers can be subjected day after day without adverse effect.

The PEL and TLV-TWA for synthetic graphite is $10\,\mathrm{mg\,m^{-3}}$ of total dust, $5\,\mathrm{mg\,m^{-3}}$ for synthetic graphite respirable dust and $2.5\,\mathrm{mg\,m^{-3}}$ for natural graphite respirable dust. Zustra [20] has specified an air standard of 5 carbon fibers/cc, whilst the US Navy has a limit of 3 carbon fibers/cc.

11.23.2 Data for UK exposure limits for gaseous emissions

In the UK Her Majesty's Stationery Office (HMSO) issue a series of publications relevant to health and safety [9, 11, 12, 14–18]. The occupational exposure limits are taken from the Health and Safety Executive Guidance Note EH 40 [14] and it should be noted that this publication is revised annually.

11.23.3 Possible hazards with carbon and graphite fibers

Several papers have been published on the possible health hazards from carbon fibers [21–28] including *in vitro* (outside the living body and in an artificial environment) and *in vivo* (in the living body) studies and the conclusion reached in 1989 [29] was that PAN based

Table 11.4 Occupational exposure limits

Substance	Long term mel 8 h		Short term mel 10 min	
	ppm	mg m^{-3}	ppm	mg m^{-3}
HCN	–	–	10	10
NaCN	–	5	–	–
NH$_3$	25	17	35	24
CO	50	55	300	330
CO$_2$	5000	1000	1500	27,000
Carbon fiber dust	–	–	–	–

Note: mel = maximum exposure limit
These limits are only intended as a guide and current figures should be consulted in the relevant document publication EH40, which is revised annually.
Source: Reprinted from Health and Safety Executive's Guidance Note EH 40.

carbon fibers did not appear to be a significant inhalation hazard. However, since then, the market has seen the introduction of 5 μm diameter carbon fibers, which may be more of a hazard. Under conditions of the test procedure [23,24], pitch based carbon fibers were considered to be carcinogenic, possibly exacerbated by the use of a benzene extract and Amoco have suggested an insignificant risk of tumor development from chronic skin contact with a pitch based carbon fiber.

Jones *et al* [28] conducted a survey of some 88 workers at Courtauld's, Coventry carbon fiber production plant and chest radiographic evaluations of the workers revealed no dust-related disease.

The most common problem experienced when handling carbon or graphite fibers is an irritation and development of a skin rash. The irritation is due to the attrition of the carbon fibers, which is particularly a problem when chopping carbon fiber. The irritation can be a transient response and in a majority of cases, clears up after a couple of weeks. The more sensitive areas of the skin tend to be most affected, such as between the fingers and backs of the hands and at wrists and neck, where it can be aggravated by the rubbing of tight clothing. The use of suitable protective clothing, which can breathe, coupled with good ventilation and provision of localized extraction points where dust generation can occur, limits the problem.

Certain resins can produce quite severe skin irritation, hence sized fiber can present an additional hazard, especially if the sized fiber fragments pierce the skin. Unfortunately some workers, albeit rare, can become sensitized (an allergic reaction) to this condition and would then have to be given alternative employment not involving sized fiber.

A report by NIOSH [43], investigating a Composites Department at a New Hampshire Ball Bearings factory in Laconia, revealed that the dimensions of airborne carbon fibers were approximately 45 μm in length and 6 μm in diameter. Bulk dust samples comprised long and short carbon fibers with a uniform diameter of approximately 9 μm, with relatively smooth sides and while some fibers had pointed ends, the ends were not particularly sharp or jagged. A few operatives reported a minor transient skin itch, which was attributed to carbon fibers on exposed skin.

11.24 THE RISKS OF CARBON FIBER COMPOSITES IN A FIRE

The hazards have been discussed [31–37] and Bowles [38] has suggested a fire test method for cfrp. The risks of fire are greatest in a confined space such as on board an aircraft, where carbon fiber can be released from a burning carbon composite, presenting electrical hazards [39,40] and there can be hazards from smoke and toxic products of gas combustion [41,42].

REFERENCES

1. Roberts RM, *Serendipity Accidental Discoveries in Science*, John Wiley, New York, 1989.
2. Berson JA, Discoveries missed, discoveries made: creativity, influence and fame in chemistry, *Tetrahedron*, 48, 3–17, 1992.
3. Lenox RS, Educating for the Serendipitous Discovery, *J Chem Educ*, 62, 282, 1985.
4. Daly BB, Woods Practical Guide to Fan Engineering, *Woods of Colchester*, 3rd ed., 230, 231, 246, 1992.
5. Vernard JK, *Elementary Fluid Mechanics*, Wiley, 423, 1961.
6. Patty FA, *Industrial Hygiene and Toxicology*, 2nd ed., Interscience, New York, 199, 1942.
7. Private communication with RG Chappell, Chlor-Chemicals Business, ICI Chemicals & Polymers Ltd., Runcorn.
8. *Chemistry and Industry*, 1232–1236, 1 Oct 1955.
9. *Health and Safety Executive's Guidance Note EH42*, Monitoring Strategies for Toxic Substances (HMSO).
10. Thermo Environmental Instruments Inc. Franklin MA, technical data sheets.
11. Ammonia in Waters, *Methods for the Examination of Waters and Associated Materials*, HMSO publication, 1981.
12. *The Control of Substances Hazardous to Health Regulations*, SI 1988 No. 1657, HMSO with latest amendments, 1988.
13. Pybus RM, *Croner's Guide to COSH-H*, Croner Publications Ltd., Kingston upon Thames, 1988.
14. *Health and Safety Executive's Guidance Note EH40*, Occupational Exposure Limits (HMSO, revised annually).
15. *The Personal Protective Equipment at Work Regulations*, SI 1992 No. 2966, HMSO, 1992.
16. *L25 Personal Protective Equipment at Work: Guidance on Regulations*, HMSO.
17. *Health Surveillance under COSH-H*, HMSO.
18. *Surveillance of People Exposed to Health Risks at Work*, HMSO.
19. SACMA, Safe Handling of Advanced Composite Materials, *Suppliers of Advanced Composite Materials*, Arlington, VA, 35–37, Jul 1991.
20. Zastra M, Evaluation of the potential health hazards associated with the machining of carbon fibre composites, *Master's Thesis*, University of North Carolina, Chapel Hill, NC, 1987.
21. Waritz RS, Carbon Fibers—General Toxicology, *Presented at the Toxicology Forum*, Aspen, Jul 1987.
22. Dahlquist BH, Evaluation of Health Aspects of Carbon Fibers, *Master's Thesis*, Dep Environ Sci Eng, School of Public Health, Univ of North Carolina, Chapel Hill, NC, 1984.
23. Smith LH, Carbon Fibres Health effects. *Preliminary Draft*, Oak Ridge National Laboratory for the US Environmental Protection Agency, 1986.
24. Vu VT, Health Hazard Assessment of Non-asbestos Fibers, *Final Draft*, Office of Toxic Substances, US Environmental Protection Agency, Washington DC, 1988.
25. Holt PF, Horne M, Dust from carbon fibre, *Environ Res*, 17, 199, 1978.
26. Owen PE, Glaister JR, Ballantyne B, Clary JJ, Subchronic inhalation toxicology of carbon fibers, *J Occup Med*, 28, 373, 1986.
27. Martin TR, Meyer SW, Luchtel DR, An evaluation of the toxicity of carbon fiber composites for lung cells *in vitro* and *in vivo*, *Environ Res*, 49, 246, 1989.
28. Jones HD, Jones TR, Lyle WH, Carbon Fibre: Results of a survey of process workers and their environment in a factory producing continuous filament, *Annals of Occupational Hygiene*, 26, 861, 1982.
29. Thompson SA, Toxicology of Carbon Fibers, *Appl Ind Hyg*, 12, 29–33, 1989.

30. Quinn TJ, Barber CR, A lamp as a source of near blackbody radiation for precise pyrometry up to 2700°C, *Metrologia*, Jan 1967.
31. Bell VL, *Source of released carbon fibres*, NASA Langley Research Centre, CP2074, 1974.
32. Carbon fibre study, *NASA Technical Memorandum 78718*, May 1978.
33. Carbon fibre hazard concerns NASA, *Aviation Week & Space Technology*, 47–50, Mar 5 1979.
34. Cornwell AC, Initial tests on the effect of fire of carbon fibre laminates, *RAE Technical Memorandum Mat*, 354, 1980.
35. Bell VL, Potential of release of fibres from burning carbon composites, *NASA Technical Memorandum 80214*, 1980.
36. Koutides DA, Review of thermal properties of graphite composite materials, *NASA Technical Memorandum 100049*, 1987.
37. Carbon and aramid fibre composites, *Technical Notes*, TN A.007.10139, Airbus Industrie, Oct 1991.
38. Bowles KJ, Fire test method for graphite fibre reinforced plastics, *NASA Technical Memorandum 81436*, 1980.
39. Moreton R, Electrical hazards due to the release of carbon fibres into the atmosphere, *RAE MRCC/79/44*, Materials department, 1979.
40. Springer GS, Environmental Effects on Composite Materials, *Electrical hazards posed by graphite fibres*, Chapter 12, Spinger GS ed., CRC Jan 1981.
41. Boeing—Smoke and toxic gas combustion products of composites in aircraft cabins, *Proceedings of the SAMPE 36th International Symposium*, 1991.
42. Kanakia MD *et al*, Fire test methodology for aerospace materials, Thermal and smoke toxicological assessments of graphite and bismaleimide and graphite/epoxy systems. *Project no. 03-5565-001*, Prepared for NASA (contract no. BAS2-10140), Southwest Research Institute, 1980.
43. *General Industrial Health Hazard evaluation*, HETA 95-0207-2592, New Hampshire Ball Bearing, Astro Division, New Hampshire. Aug 1996, NIOSH Publication.

CHAPTER **12**

Techniques for Determining the Structure of Carbon Fibers

12.1 INTRODUCTION

Surface science has now been extended to include well over fifty examination techniques, spurred on by the rapidly developing field of semiconductor devices and assisted by the possibility of attaining ultra high vacuums (UHV). Appendix 13 lists the acronyms for these systems, but the field is too extensive to include all these techniques. Therefore, a few which are most applicable to the study of carbon fibers, have been selected for discussion here. The intention is to briefly describe these techniques, with examples and findings pertinent to carbon fibers, highlighting their advantages and limitations, whilst the reader is referred to other publications for more detailed descriptions.

A feature of all these methods is that the equipment must be sensitive enough to detect the relevant signal above the existing noise level. Nix at Queen Mary, University of London has given an example for surface analysis by considering an area of 1 cm^2, which would have some 10^{15} atoms in the surface layer and to detect impurity atoms at the 1% level, the technique must be sensitive to about 10^{13} atoms. However, for a spectroscopic technique examining a 1 cm^3 of liquid sample containing about 10^{22} molecules, the detection of 10^{13} molecules in this sample would require a sensitivity of 1 ppb, which is indeed very sensitive.

The regimes of surface analysis are shown in Figure 12.1. Surface science offers many techniques, each with advantages and disadvantages, but available databases and user experience must be taken into account.

Morita *et al* [2] have described the characterization of commercially available PAN based carbon fibers using SEM, TEM, EELS, XPS, Raman and FTIR techniques and there are many publications on surface analysis [3,4].

12.2 OPTICAL MICROSCOPE

An optical microscope will provide magnifications of up to a ×1000, albeit with a limited depth of field and the setting up procedure for a microscope required for laboratory work is described in Chapter 17, Section 17.10.1. Typical metallurgical techniques are used for mounting specimens such as hot pressing in a phenolic or alkyd resin or potting in a cold

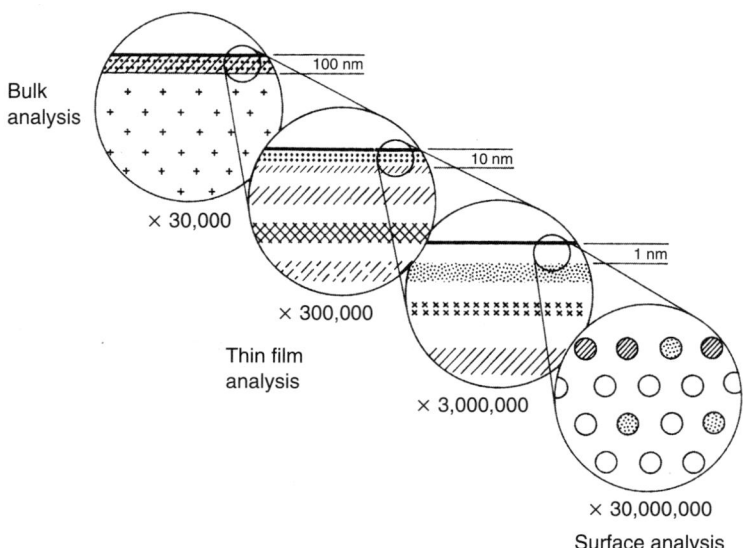

Figure 12.1 The regimes of surface analysis, thin film analysis and bulk analysis. *Source:* Reprinted with permission from Briggs D, Seah SP, Practical Surface Analysis, *Auger and X-ray Photoelectron Spectroscopy*, 2nd ed., Vol.1, John Wiley & Sons, Chichester, 4, 1996. Copyright 1996, John Wiley & Sons Ltd.

setting polyester or epoxy resin and completing the sample preparation by carefully polishing with a diamond paste.

The degree of oxidation of an oxidized PAN fiber sample can be readily determined by preparing a suitable section with a microtome for examination under the microscope and this is detailed in Chapter 17, Section 17.3.8. Typical oxidized PAN microstructures are shown in Figure 3.51.

Graphite is a uniaxial crystal with its optical axis in the c direction. Polarized light that is incident perpendicular to the layer planes of a polished specimen will not be affected and will appear isotropic. If the structure of the carbon fiber is to be deduced, then the Nicols must be exactly crossed, otherwise the intensity maxima and minima are affected [5]. Woodrow [6] showed under crossed Nicols that the maximum intensity occurred when the plane of polarization was parallel to either of the axes and the minimum intensity which occurred on rotation of the sample, was zero. The magnification is normally restricted to ×50 in order to minimize flare due to scattered light. It is possible to use the sensitive tint technique, which uses a transparent birefringent crystal with a selected thickness such that there is a phase change of exactly 1λ in the middle of the visible spectrum. When this retarder plate is placed at 45° between the crossed polarizer, analyzer interference colors of the graphitic lamellae at the surface are produced [5,7]. The color depends on the orientation of the lamellar planes, and yellow and blue indicate that an edge is exposed in the polished surface, whilst purple is indicative of a basal plane. In conjunction with an image analyzing computer, the individual components can be evaluated and an optical texture index (OTI) [8], which is useful in carbon-carbon work. Oberlin *et al* [9] describe the use of a Newton chart to determine the phase shift associated with optical anisotropy.

Two outstanding images of carbon-carbon have been taken under cross polarized light with a first order red compensator (Figure 12.2) and without (Figure 12.3).

Knibbs [5] studied carbon fibers made from Courtelle, Acrilan and Orlon, which were mounted, polished and then, examined using polarized light with crossed Nicols and also using the sensitive tint technique. The Courtelle was processed in three different ways (a, b

TECHNIQUES FOR DETERMINING THE STRUCTURE OF CARBON FIBERS 455

Figure 12.2 Carbon–Carbon under cross polarized light with first order red compensator (50×). *Source:* Photograph courtesy of Anthony Garcia, BF Goodrich, Carbon Products

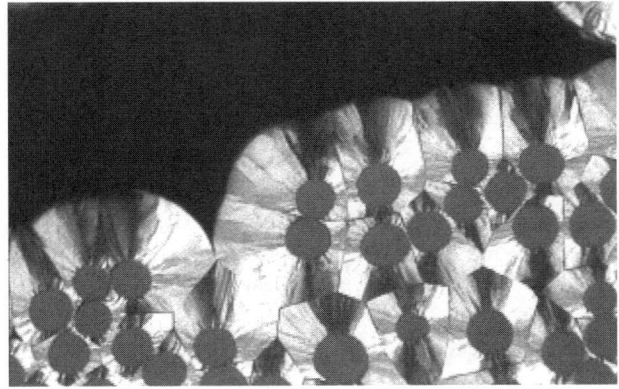

Figure 12.3 Carbon–Carbon under cross polarized light (52×). *Source:* Photograph courtesy of Anthony Garcia, BF Goodrich, Carbon Products

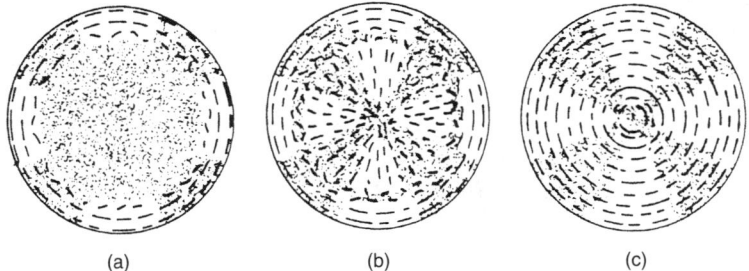

(a) (b) (c)

Figure 12.4 Schematic representation of carbon fiber structures obtained from Courtelle precursor. (a) Isotropic center, with an outside skin of oriented crystalline material. (b) Double cross, with the outside showing a different orientation to that of the center. (c) Single cross, where the complete fiber shows one type of preferred orientation. *Source:* Reprinted with permission from Knibbs RH, The use of polarized light microscopy in examining the structure of carbon fibres, *J Microscopy*, 94(3), 273–281, Dec 1971. Copyright 1971, Blackwell Publishers.

and c) but the final processing stage of all precursors was heat treating to above 2600°C. The schematic appearance of the Courtelle samples is shown in Figure 12.4. The sample (a) had a 60% lower modulus than the other samples and had developed a two zone structure, with the outside layer having some preferred orientation, but an isotropic and non-crystalline central

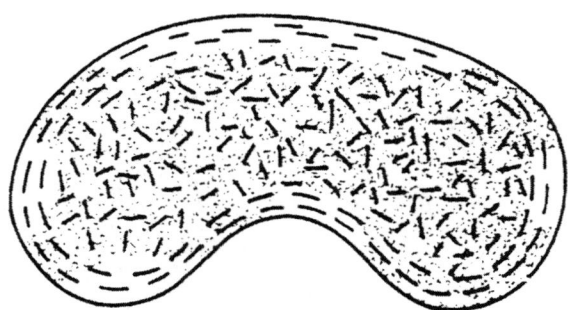

Figure 12.5 Schematic representation of carbon fiber obtained from Acrilan (type C structure). *Source:* Reprinted with permission from Knibbs RH, The use of polarized light microscopy in examining the structure of carbon fibres, *J Microscopy*, 94(3), 273–281, Dec 1971. Copyright 1971, Blackwell Publishers.

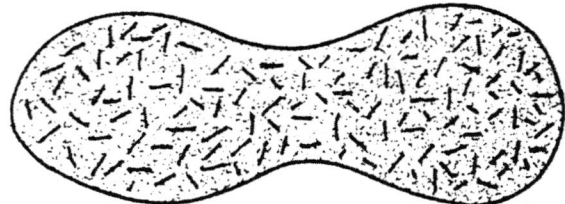

Figure 12.6 Schematic representation of carbon fiber obtained from Orlon (type C structure). *Source:* Reprinted with permission from Knibbs RH, The use of polarized light microscopy in examining the structure of carbon fibres, *J Microscopy*, 94(3), 273–281, Dec 1971. Copyright 1971, Blackwell Publishers.

region; sample (b) had a two zone structure and showed a double cross under crossed Nicols; sample (c) had one zone with radial orientation of the c axes.

Acrilan (Figure 12.5) had a thin skin of crystallites with the c axes aligned radially and Orlon (Figure 12.6) had the whole cross-section optically inactive with complete extinction under crossed Nicols.

Interestingly, Johnson [10] investigated carbon fiber in a glassy carbon matrix and suggested that the strong optical anisotropy in the glassy carbon and also in the carbon fibers themselves was due to strain birefringence.

Nyo *et al* [11] examined PAN based Hitco carbon fibers using polarized light microscopy and found that high modulus fibers have a dual structure, with an onion skin like layer on the surface and a radial in the core (Figure 12.7).

12.3 SCANNING ELECTRON MICROSCOPE (SEM)

Scanning electron microscopes [12] have now become very popular to observe the microstructures on surfaces. With the appropriate equipment, elements contained in extremely small specimen areas can be analyzed.

The SEM, with a typical magnification of ×300,000 and a resolution of 3 nm in the high vacuum (HV) mode and 5 nm in the low vacuum (LV) mode, has a much greater depth of field than either the optical microscope or the transmission electron microscope (TEM).

A beam of electrons (Figure 12.8) is supplied from an electron gun mounted on top of the column actuated by applying a voltage to a filament, usually W in a hairpin gun, or LaB_6 in a field emission gun, causing some electrons to accelerate down the column. The beam is

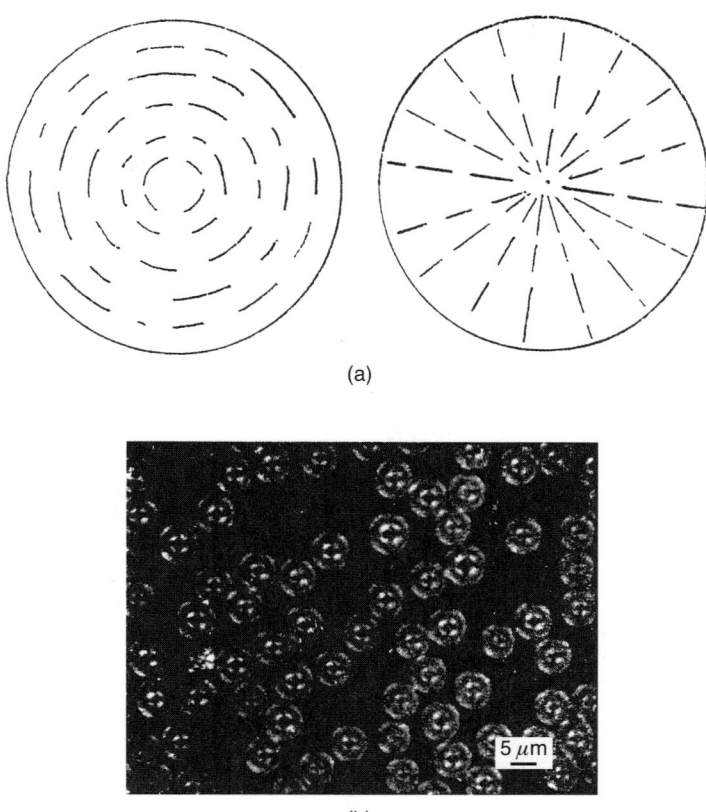

Figure 12.7 Structure of carbon fiber. (a) A schematic illustration of tree trunk or onion skin structure (left) and radial structure (right). (b) A typical optical micrograph of carbon fiber cross sections under polarized light in crossed nicols condition showing maltese cross patterns. Polarizer and analyzers are parallel to picture edges. *Source:* Reprinted from Nyo H, Heckler AJ, Hoernschemeyer DL, Characterizing the structures of PAN based carbon fibers, *24th Nat Symposium*, San Francisco, 179, 51–60, May 8–10.

condensed by passing through a condenser lens and focussed to a fine point onto the sample by the objective lens. To enhance the quality of the image and elicit information, the probe currentis optimized by a zoom condenser lens (Figure 12.9), which eliminates any shift in the observation position or focus when the probe current is changed. The scan coils, actuated by a variable voltage from the scan generator, move the beam to and fro across the specimen and synchronize the beam with a cathode ray oscilloscope, producing the same pattern on a screen. The electron beam hits the sample and produces photons and electrons (Figure 12.10). The secondary electrons generate topographical information, the primary backscattered electrons give information on atomic number and topographical information and X-rays, which can be analyzed, provide details about thickness.

The chamber is always maintained under a vacuum (Figure 12.11) and the degree of vacuum controls the resolution. Latest technology uses UHV, generating a clean low vacuum (10–140 Pa). The electrons colliding with the residual gas molecules ionize and become attracted towards the charged specimen surface, where they are absorbed. This method allows non-conductive specimens to be observed without charge up. In this mode, the detection of secondary electrons becomes impossible, so backscattered electrons are used, reverting to the secondary electrons in the HV mode.

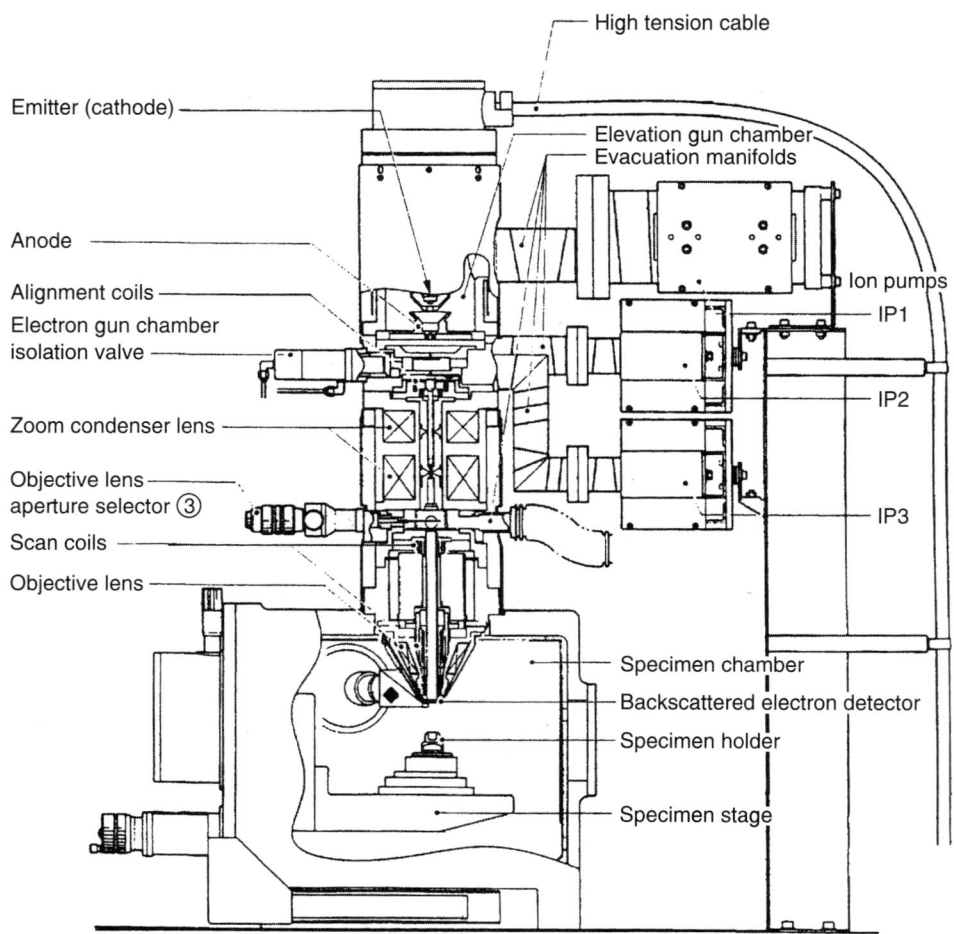

Figure 12.8 Cross-section of Jeol Type FEG scanning electron microscope. *Source:* Courtesy of Jeol Ltd.

In the HV mode it is necessary to prepare the specimen by removing all water, firmly mounting the sample and making non-metallic samples electrically conductive by coating under vacuum with metals like gold.

SEM is a valuable tool for carbon fiber fractography, with no sample preparation required, but polished sections exhibit no structural information and it is necessary to etch the surface. For this purpose, Manders [13] employed electrolytic etching.

It is possible to break carbon fiber filaments in glycerol and recover the broken ends for investigating at the vicinity of fracture with SEM and using X-ray analysis to examine for the presence of chemical impurities, such as TiO_2, which are perhaps responsible for early fracture.

A study of the hydrogenation process in vapor grown carbon fibers using modified SEM fractography was undertaken [14].

Vezie and Adams [15] examined PAN and pitch based carbon fibers with high resolution SEM. The onion skin layering of the carbon matrix surrounding the reinforcing carbon fibers in a carbon-carbon composite is clearly shown in the SEM photograph (Figure 12.12).

TECHNIQUES FOR DETERMINING THE STRUCTURE OF CARBON FIBERS 459

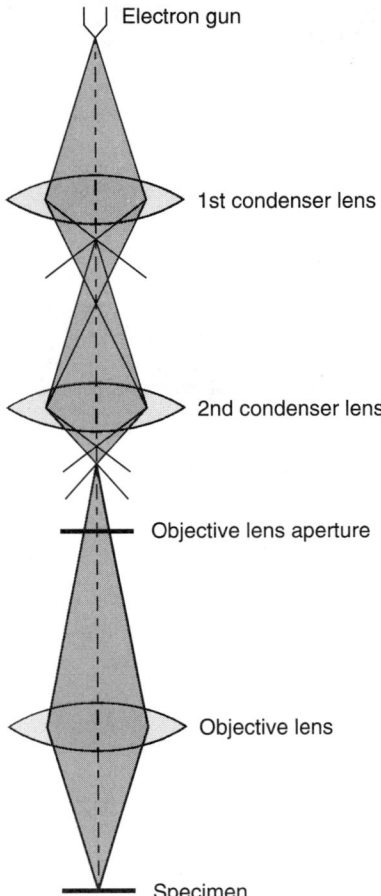

Figure 12.9 Zoom condenser lens. *Source:* Reprinted from technical literature, Jeol Ltd.

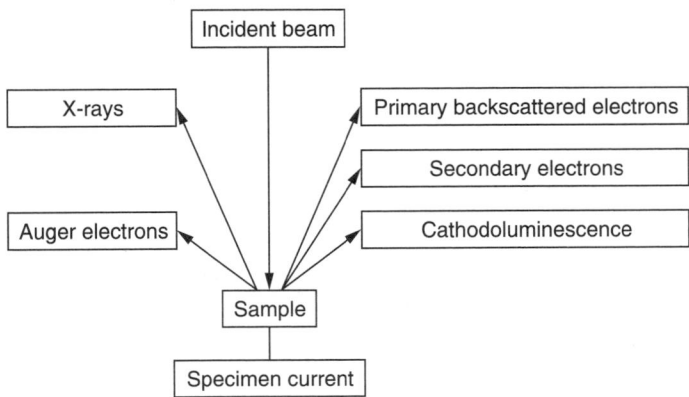

Figure 12.10 Effect of an electron beam striking the sample in SEM.

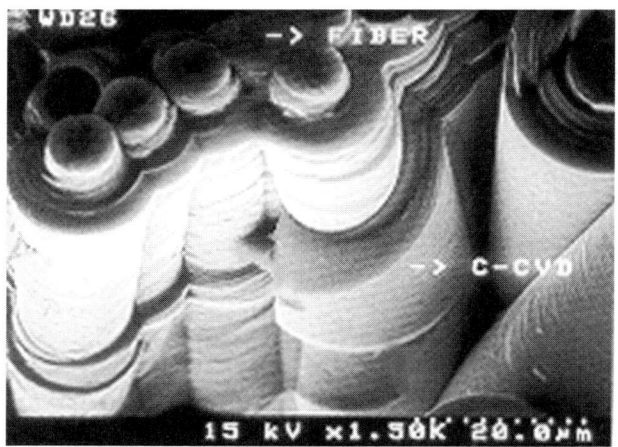

Figure 12.11(a) SEM photograph of carbon matrix showing the onionskin layering of the carbon matrix surrounding the reinforcing carbon fibers. *Source:* Courtesy of SGL Carbon Group.

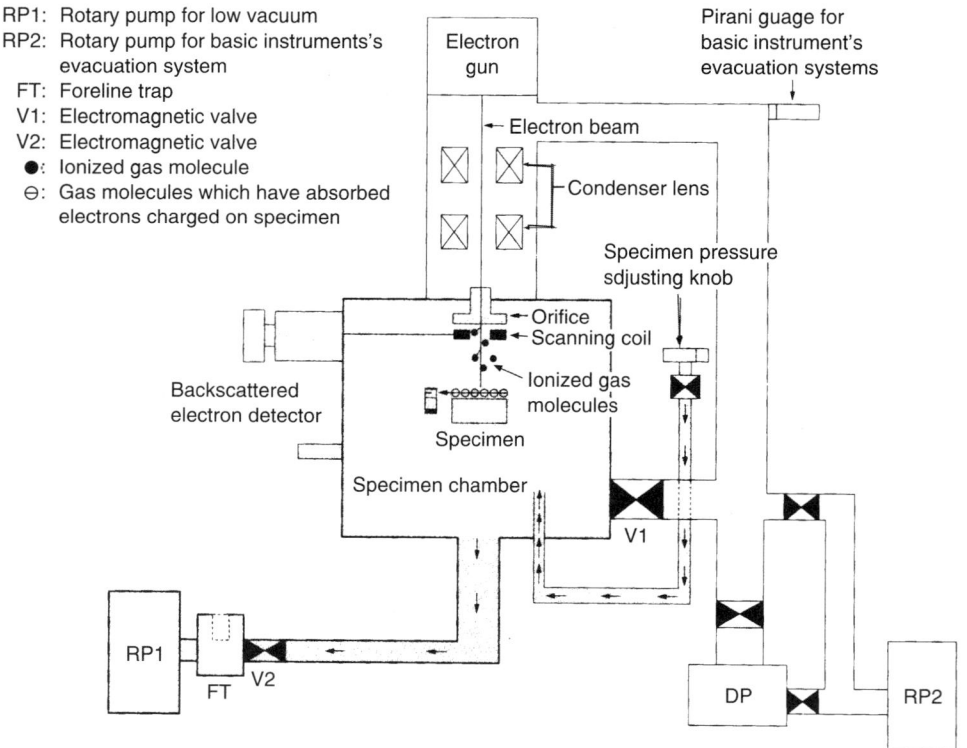

Figure 12.11(b) Vacuum evacuation system of LV SEM. *Source:* Reprinted from technical literature, Jeol Ltd.

12.4 TRANSMISSION ELECTRON MICROSCOPE (TEM)

A transmission electron microscope [16–18] is basically a transmission light microscope using electrons instead of light, but with a resolution some thousand times greater than a transmission light instrument enabling objects as small as a few angstroms to be studied.

At the top of the microscope (Figure 12.13), an electron gun projects a stream of monochromatic electrons through an accelerator, which are deflected into a pair of

TECHNIQUES FOR DETERMINING THE STRUCTURE OF CARBON FIBERS

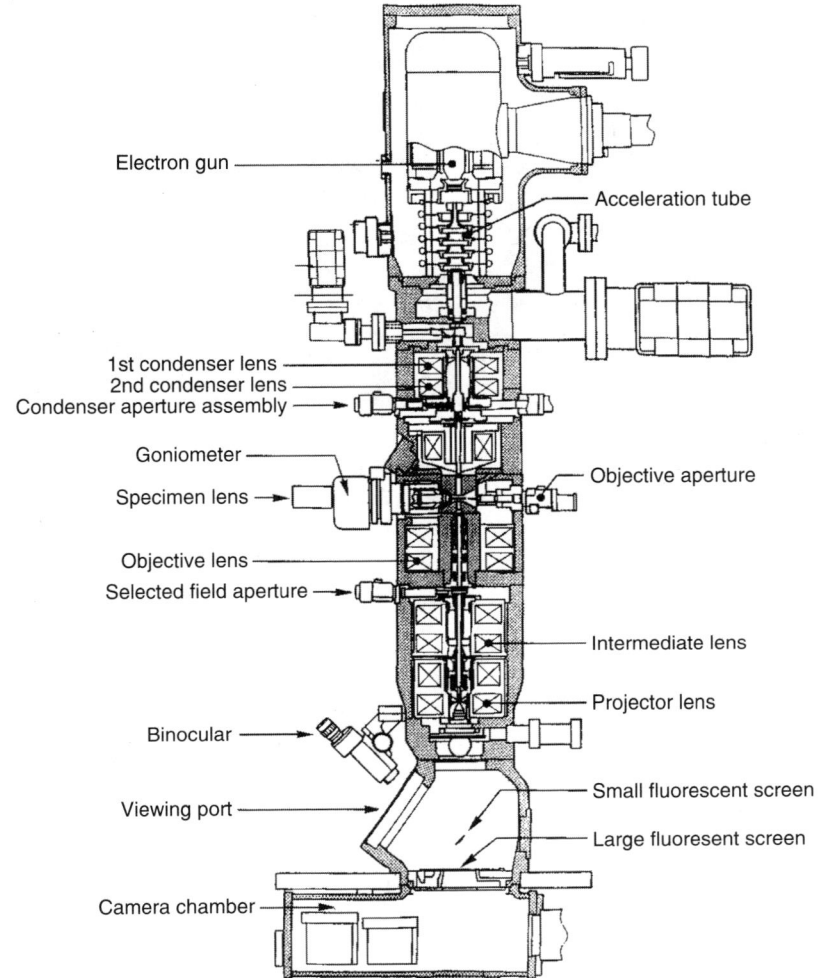

Figure 12.12 Cross-section of a transmission electron microscope. *Source:* Courtesy of Jeol Ltd.

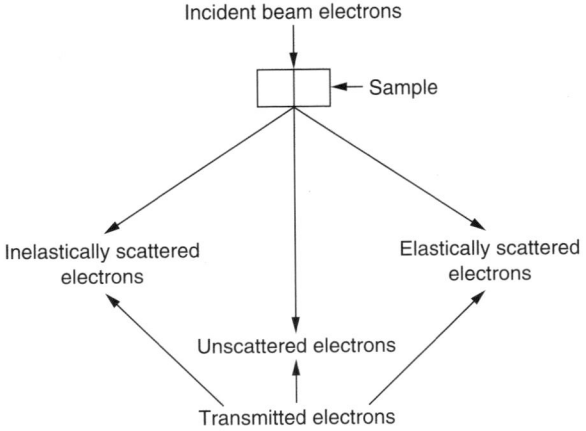

Figure 12.13 Effect of an electron beam passing through a sample in TEM.

Figure 12.14 A 002 lattice fringe image of a transverse section of Courtaulds HMS. *Source:* Reprinted with permission from Guigon M, Oberlin A, Desarmot G, Microtexture and structure of some high modulus PAN based carbon fibres, *Fibre Sci and Technol*, 20, 177–198, 1984. Copyright 1984, Elsevier.

condenser lenses, the first lens controlling the size of the spot and the second determining the intensity or brightness, having the ability to change the size of the spot from widespread to pin point. A condenser aperture eliminates electrons from the axis and the beam then strikes the specimen and is transmitted, passing through the objective lens where it is focussed into an image. Apertures can be used to restrict the beam width. The image passes down the column through the projector lens, where it is magnified and the unscattered electrons then strike a phosphor image screen to give a shadow image (Figure 12.14). The transmission of unscattered electrons is inversely proportional to the specimen thickness. The thicker parts of the image, represented by fewer electrons, appears darker, whilst the lighter parts represent the thinner areas, which permit the transmittance of more electrons.

The instrument can be coupled with energy dispersive X-ray microanalysis, online image processing and particle size measurement. X-ray analysis can detect all elements in the periodic table from Na through U.

Selected samples must be thin (below 100 nm) and specimens are prepared either by grinding or cutting a thin section with a microtome [19,20]. Goodhew [21] has described techniques for the preparation of carbon fibers for SEM and Oberlin *et al* [22] and Pennock and Ogara [23] have also discussed sample preparation. Microtoming can give problems and these have been discussed by Tidjani [24] and Ehrburger [25]. When producing samples from mesophase pitch, it was found that atom-milling produced lesser damage than microtomy [26].

Guigon *et al* [27] studied the microtexture and structure of some nine commercial high modulus PAN based carbon fibers using TEM and found that they are made of basic structural units, which are almost isometric in shape, but folded and entangled parallel to the fiber axis. The less ordered Courtaulds HMS fiber (YM 361 GPa) was more severely folded than the Celion GY70 (YM 430 GPa), Serofim AG25 and AG58 (YM 420 GPa), which were also partially graphitized in some areas. The 002 lattice fringe image of a transverse section of Courtaulds HMS is shown in Figure 12.15 showing numerous random *moiré* fringes due to the superimposition of many pore walls (as indicated by the single arrow). Where the layers exactly fulfil the 002 Bragg condition, there are intense Bragg fringes (as indicated by the double arrow).

A TEM study of HM PAN based carbon fibers was undertaken by Deurbergue and Oberlin [28]. Figure 12.16 shows typical work of Johnson and his colleagues at

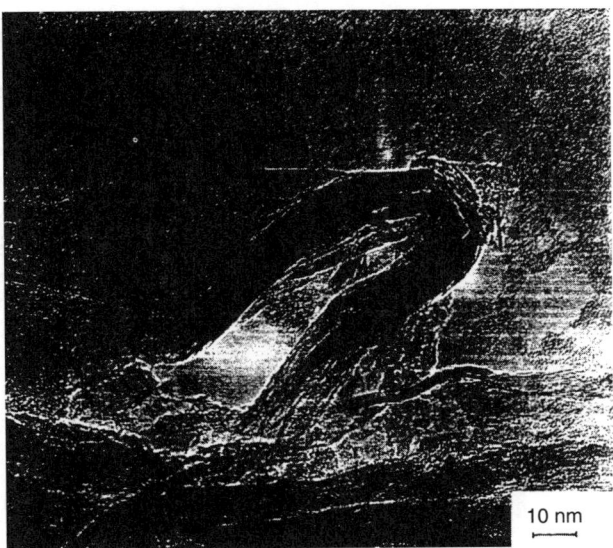

Figure 12.15 Lattice fringe high resolution image from a longitudinal section revealing a hairpin type fold forming a step at the surface of 3 denier fiber oxidized 20 h at 220°C and heat treated to 2500°C. *Source:* Reprinted with permission from Bennett SC, Johnson DJ, Electron microscope studies of structural heterogeneity in PAN based carbon fibres, *Carbon*, 17, 25–39, 1979. Copyrighyt 1979, Elsevier.

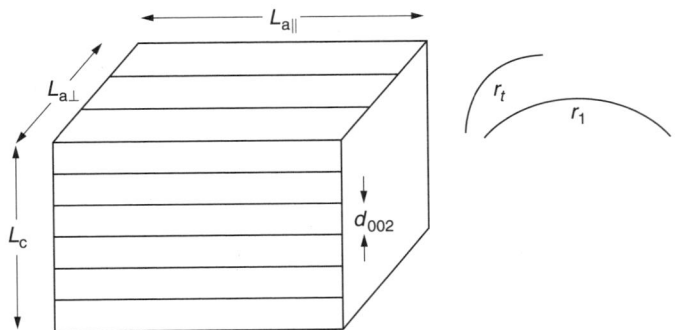

Figure 12.16 Notation used in X-ray diffraction.

Leeds University [29]. They have pioneered the use of TEM studies of carbon fiber and found that for carbon fiber, the electron wavelength is equivalent to the stacking distances between the graphite plates in the crystallites, hence they can be seen. Some of their findings are:

1. A and HT fibers have no sheath and have a homogeneous structure.
2. HM fiber has a very thin (0.1 μm) skin.
3. The orientation of crystallites is random, but with a preferred orientation along the fiber axis. The orientation of Type A and HT fibers are about 45° and 40° respectively.
4. Increasing the carbonization temperature increases the size of the crystallites and their orientation with the fiber axis.
5. The orientation for HM fiber is 25° in the core and 12–15° in the skin, with circumferential orientation. Since the graphite plates are in a plane parallel to the fiber surface, they present a barrier to bonding. This is the reason for the low ILSS for untreated fiber. The edges of the crystallites contain carbon atoms of unsatisfied valency, which provide the sites for bonding to resin.

6. The modulus of HM fiber is high because of the size and orientation of the crystallites in the core. In the skin crystallites are very large (perhaps 0.5 µm long with a stacking height of 20 nm; 20–30 plates) and although the average preferred orientation is 12–15°, certain crystallites are arbitrarily oriented. These cut deeply into the fiber so that the plates are presented approximately parallel to the fiber axis and in this direction, the crystallite has little strength. It is because the crystallites are so large that the odd misoriented one causes a massive stress concentration and leads to early failure by crack propagation and low UTS for HM fiber. This is not the whole story since any fiber treated in excess of 1000°C shows massive internal voids (golf balls on this scale) and surface cracks caused by the volatilization of impurities in the PAN. Clean room spinning does improve the strength, but it is still short of the theoretical value. Hence, the difference in strength at 2500°C between the theoretical and the clean fibers is due to crystallite misorientation.

Klinklin and Guigon [30] studied the interface in carbon fiber reinforced composites by SEM.

12.5 X-RAY DIFFRACTION [31,36]

12.5.1 Convention for axes in graphite and carbon fibers and dimensional notation

The standard designation of lattice directions for graphite is with the c axis normal to the basal planes. The a and b axes lie in the basal plane and include the 120° angle to form a hexagonal unit cell.

For carbon fiber, the X_3 coordinate direction is along the fiber axis, whilst X_1 and X_2 are in a plane normal to the fiber axis and, due to the symmetry about the fiber axis, are equivalent.

The convention for crystallite axes is to assign X_3 along the c axis of the graphite crystal, with the X_2 and X_3 axes lying in the plane delineated by the basal planes.

In X-ray diffraction, the following notation is used (Figure 12.17),

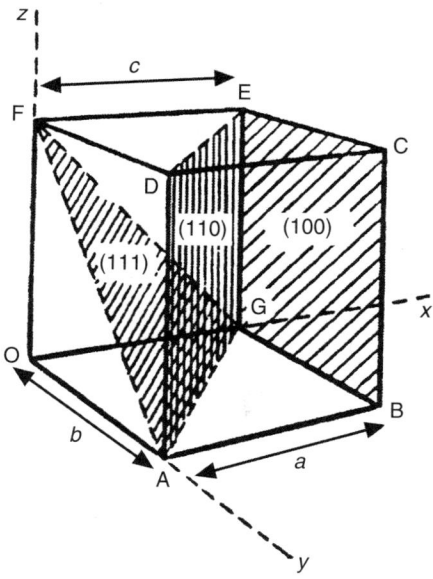

Figure 12.17 The simple cubic crystal.

where d_{002} = interlayer spacing (increases with heat treatment temperature, HTT)
L_c = crystallite size (or more accurately coherent length) perpendicular to carbon layers (increases with HTT, higher for pitch than PAN)
L_a = crystallite size (or more accurately coherent length) parallel to carbon layers (increases with HTT)

Texture preferred orientation of the carbon layers:

$L_{a\|} = L_a$ parallel to fiber axis
$L_{a\perp} = L_a$ perpendicular to fiber axis
r_l = longitudinal radius of curvature
r_t = transverse radius of curvature
f = degree of orientation
v_c = volume fraction of crystallites
v_f = shape and orientation of microvoids
v_a = volume fraction of unorganized or non-crystalline carbon
v_p = volume fraction of microvoids

The orientation of an atomic plane in a crystal lattice is the reciprocal of the fractional intercepts (i.e. how far along the unit cell) which the plane makes with the crystallographic axes. Figure 12.18 depicts a simple cubic crystal, where the distance OG is taken as unity and the Miller Indices are calculated as shown in Table 12.1. The Zig-zag (101) and Armchair (112) faces of the graphite crystal are depicted as shown:

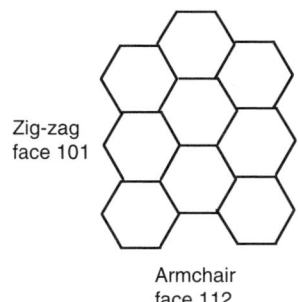

Zig-zag face 101

Armchair face 112

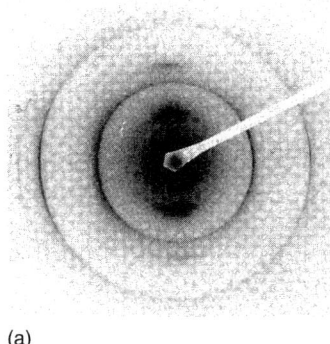
(a)

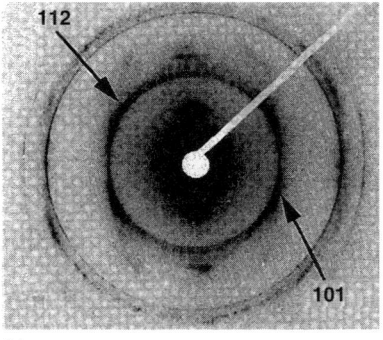
(b)

Figure 12.18 SAD patterns. (a) A turbostratic fiber. (b) A partially graphitized fiber. *Source:* Reprinted with permission from Oberlin A, Bonnamy S, Lafdi K, Structure and Texture of Carbon Fibers, Donnet JB, Wang TK, Peng JCM, Rebouillat S eds. *Carbon Fibers*, Marcel Dekker, New York, 89, 1998. Copyright 1998, CRC Press, Boca Raton, Florida.

Table 12.1 Some Miller Indices of a simple cubic crystal (see Fig. 12.17)

Plane Side	ECBG			AGED			AGF		
	a	*b*	*c*	*a*	*b*	*c*	*a*	*b*	*c*
Intercept length	1	∞	∞	1	1	∞	1	1	1
Reciprocal	1/1	1/∞	1/∞	1/1	1/1	1/∞	1/1	1/1	1/1
As a fraction	1	0	0	1	1	0	1	1	1
Miller indice		(100)			(011)			(111)	

12.5.2 Wide angle X-ray diffraction

When a sample is irradiated, X-rays are diffracted in a way that is characteristic of the compounds present in the sample under analysis. The atomic lattice acts as a three dimensional diffraction grating, causing the X-ray beams to be refracted at specific angles related to interatomic spacings. By measuring the angles and intensities at which diffraction peaks occur, the type and amount of the constituents can be determined.

X-rays interact with the electron cloud which surrounds the nuclei, whereas neutrons react with the nuclei themselves. Neutron diffraction shows up the positions of hydrogen nuclei more clearly than X-rays.

Wide angle X-ray diffraction is the simplest and quickest technique for characterizing structure and has been used to determine the degree of preferred orientation of graphite layer planes parallel to the fiber axis, the average stack height (L_c) and the average layer length (L_a).

X-rays are produced by applying a high voltage (15–60 kV) to a filament (such as a W cathode) in a vacuum, accelerating the electrons into a metal target (like Cu). This produces two types of X-radiation—white radiation and characteristic radiation. It is the latter that is used for X-ray crystallography and is characteristic of the metal target (e.g. the wavelength of Cu $K_{\alpha 1}$ is 0.1540598 nm).

A series of spots, called a diffraction pattern, is created when a parallel beam of X-rays is reflected off repeating planes of atoms (Figure 12.19) and changing the angle of the X-ray beam to the crystal face will change the diffraction pattern. If the specimen is mounted such that the incident rays are parallel to an important axis of the crystal, then the interpretation of the photograph is simplified.

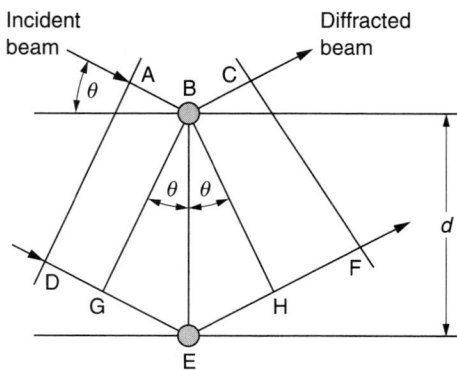

Figure 12.19 Derivation of Bragg's Law.

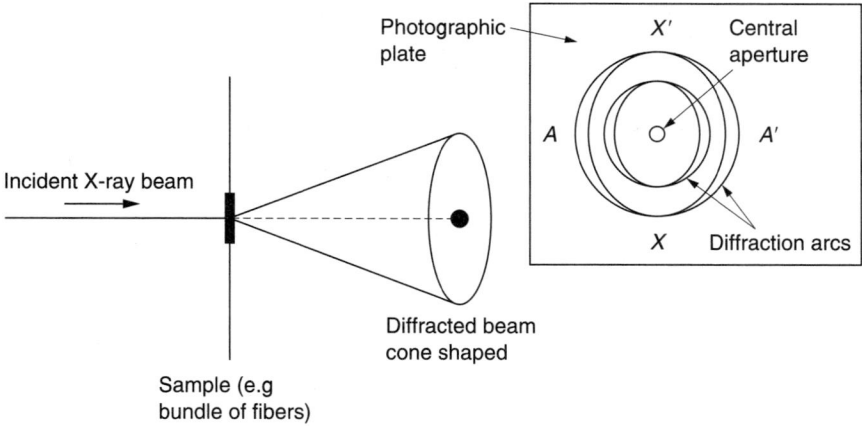

Figure 12.20 X-ray diffraction pattern.

If two waves come together in phase, the maxima and minima of both waves are coincident and the waves reinforce, increasing the intensity of the resultant wave, termed constructive interference (coherent scattering). If the waves are out of phase, then the minimum of one wave combines with the maximum of the other and the wave is destroyed, termed destructive interference (incoherent scattering).

Each unit cell diffracts an incident beam and constructive interference can only occur at points that are an integral multiple of the wavelength. When necessary, the phases of the beams can be made to coincide applying Bragg's Law applies, so that the incident angle equals the reflecting angle (Figure 12.20). The top beam strikes the atom on the surface at A, whereas the second beam continues to the next layer where it is scattered by an atom at E and travels the extra distance, GE + EH. In order to satisfy Bragg's Law, this extra distance will be an integral (n) multiple of the wavelength (λ) for the phases of the two beams to be the same. Hence,

$$\text{Extra distance (GE + GH)} = 2d \sin \theta \text{ or } n\lambda = 2d \sin \theta$$

Particle size is not to be confused with crystallite (or grain size) and a crystallite is the smallest diffracting field in a substance, whereas a particle can comprise of many crystallites.

X-ray scattering occurs in both crystalline and non-crystalline substances. Polymers have some amorphous material, which produces a halo of intensity that forms a low intensity hump when integrated, whereas a crystalline material produces well defined spots or rings, which integrate to give sharp, high intensity peaks. Acrylic precursor fibers and carbonized PAN fibers behave essentially as polycrystalline solids with varying degrees of preferred orientation.

The X-ray diffraction pattern consists of a number of concentric arcs about the axis formed by the X-ray beam (Figure 12.21).

If a film is placed in the path of the refracted beam, then a series of arcs is obtained, but more than those depicted, although the number of arcs is limited by the size of the photographic film and its distance from the specimen. Only two arcs are required for measurement and a larger arc will give greater accuracy. The acrylic precursor fibers exhibit two arcs, one of high intensity (100) and the other of less intensity (200) based on the Miller indices (Table 12.1).The crystallite dimensions of carbonized PAN fibers are determined from the widths of the strongest (002) and the weaker (100) diffraction arcs. The intensity profile along AA' will give the crystal size, whilst the intensity profile around the arc XX' will give

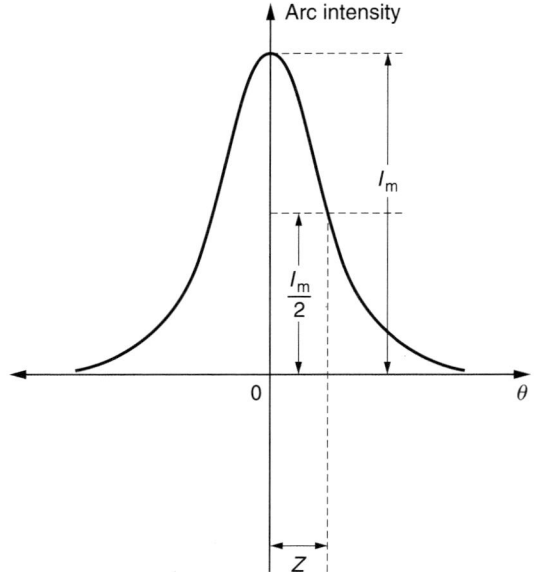

Figure 12.21 The intensity distribution measured around the X-ray diffraction arc.

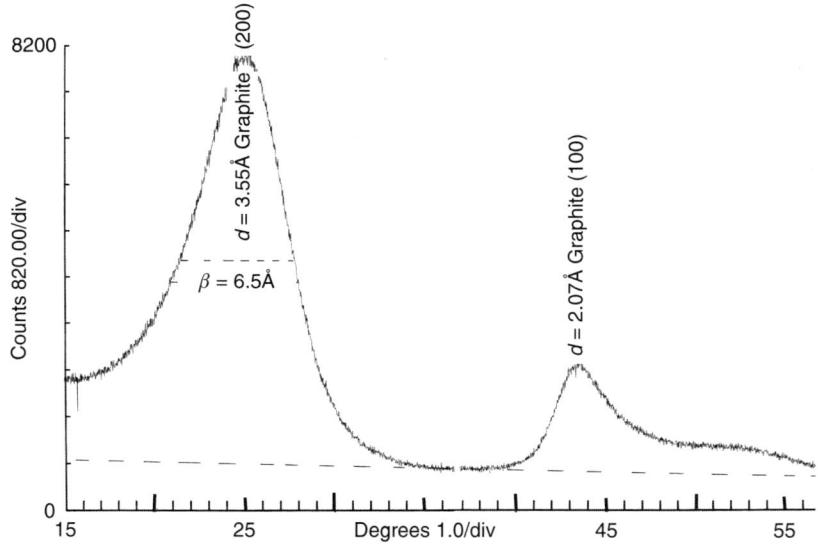

Figure 12.22 X-ray diffraction curve for graphitized carbon made from a cellulosic precursor.

the orientation. With the diffractometer, AA′ is obtained using an equitorial scan and XX′ is obtained by setting the detector to the angle for the peak reading and rotating the sample. The preferred orientation is related to the orientation angle measured as half the angular width of the intensity measured around the diffraction arc at half wave height (Figure 12.22). Bennett et al [33] discuss the evaluation of crystallite size in the electron diffraction and high resolution studies of carbon fibers.

A typical X-ray diffraction chart is shown in Figure 12.23 for carbon fiber made from a cellulosic precursor using a 40 kV source with Cu K_α radiation (0.1540598 nm) and a computer controlled stepping motor travelling at 0.02° per step every 2 s.

TECHNIQUES FOR DETERMINING THE STRUCTURE OF CARBON FIBERS

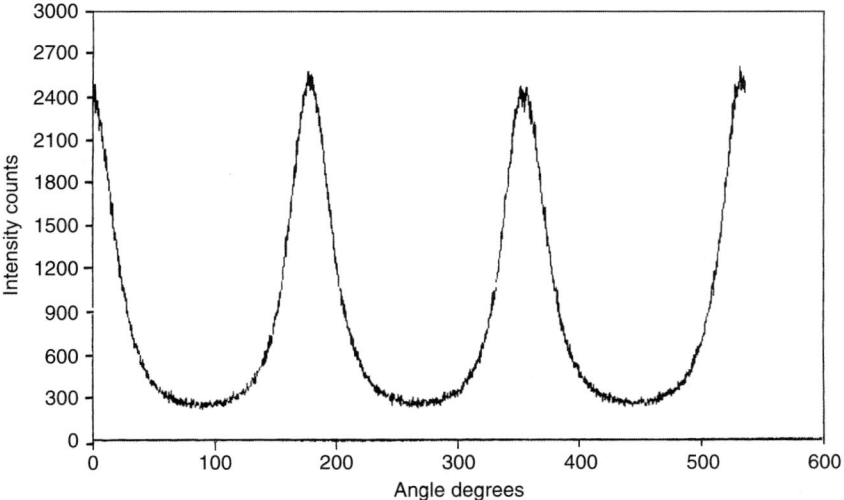

Figure 12.23 Azimuthal scan for SP15 PAN precursor fiber carried out at a position corresponding to the major peak determined in the equitorial scan range 0–500° in steps of 0.335° at 5 s per step using Cu $K_{\alpha 1}$ radiation. *Source:* Graph prepared by Courtaulds Ltd.

Using Bragg's Law,

$$n\lambda = 2d \sin \theta$$

where λ is the X-ray wavelength, d is the interplanar spacing and θ is half the angle of scattering. By using values from Figure 12.23, $\theta = 12.53$, hence 2θ is 25.06°.

The crystallite dimension L_c is calculated from the degree of broadening of the (002) reflection using the Scherrer line broadening equation namely:

$$D_{002} = K\lambda/\beta \cos \theta$$

where D_{002} = crystallite size of the crystallites in the (002) direction
K = Scherrer geometric or shape factor (normally taken to be 1)
β = total line broadening (instrument line broadening must be subtracted), the full width at half maximum height (FWHM) through the (002) reflection
θ = angle of the peak position (Bragg angle), rad

Hence
$$L_c = \frac{0.1540598}{6.5 \times \sin 12.53 \times (\pi/180)} = 1.341 \text{ nm}$$

Some values for L_c and θ obtained for carbon fibers are given in Table 12.2. It should be noted that TEM studies will give higher values for L_c.

X-ray diffraction can be used to monitor the production of PAN (Figure 12.24) and oxidized PAN fiber (Figure 12.25). Mean figures sampled over a four year period for the XRD d-spacings, crystal size and azimuthal half widths for TTP PAN precursor and Panox oxidized PAN fiber are given in Table 12.3.

An HM PAN based carbon fiber does not exhibit any form of three dimensional order, although isolated cases have been reported [10,27], whereas a mesophase pitch based carbon fiber shows a strong (101) arc typical of a more graphitic structure.

Table 12.2 Some values for L_c and θ for carbon fibers

Fiber type	Fiber modulus GPa	L_c nm	$\theta°$	Reference
HM (2560°C)	380	5.7	20	Courtaulds
HM (2650°C)	400	6.3	19	
HM (2700°C)	389 ± 17	–	17 ± 1	Isaac [34]
HM (2700°C)	468 ± 25	8.7 ± 0.5	14.3 ± 0.3	
Type I (≈2550°C)	340	5.3	20	Johnson [35]
Type II (≈1600°C)	240	1.7	41	
Type A (≈1100°C)	230	1.2	41	
P-55 (Mesophase)	Ψ380	12.5	16	Johnson [35]
P-120 (Mesophase)	Ψ825	25.5	8	

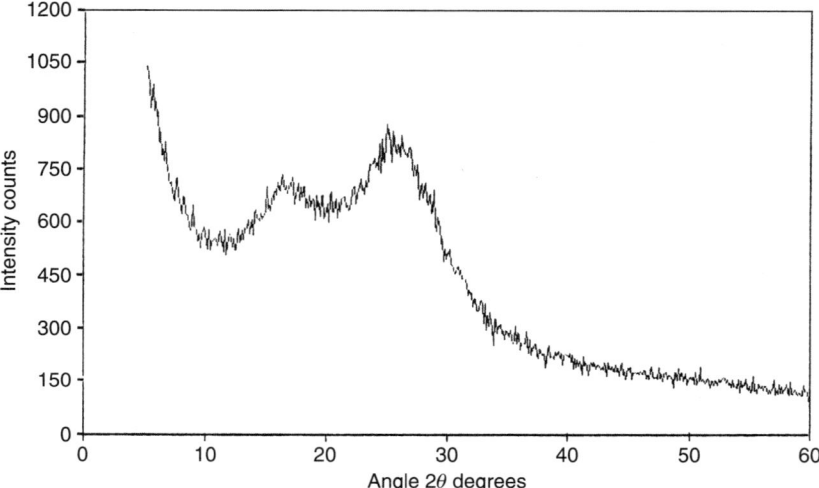

Figure 12.24 Equitorial scan of RK Carbon Fibers' PANOX for 5–60° 2θ with a step size of 0.1° 2θ at 5 s per step using Cu $K_{\alpha 1}$ radiation. *Source:* Graph prepared by Courtaulds Ltd.

Takaku and Shioya [37] have used X-ray measurements to determine the structure of both PAN and pitch based carbon fibers.

12.5.3 Single crystal X-ray diffraction

This method requires a single crystal and is time consuming, but will reveal all information about the structure.

A precision camera uses a coupling between the motions of the crystal and the screened film to give an undistorted pattern, enabling the determination of the unit cell dimensions, which are then set in the computer controlled four-circle diffractometer (Figure 12.26), where the crystal is positioned in the goniometer head and the diffraction intensity is measured with a photomultiplier.

12.5.4 X-ray powder diffraction

Since it is difficult to obtain a large, well formed crystal, grinding the sample to a fine powder is a preferred method of sample preparation and when the beam strikes the powder

TECHNIQUES FOR DETERMINING THE STRUCTURE OF CARBON FIBERS

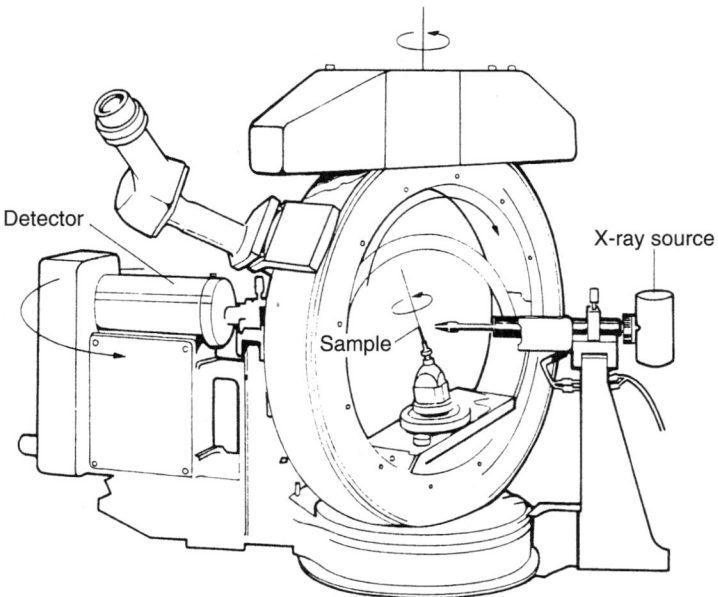

Figure 12.25 The settings of the orientations of the components is controlled by computers each (hkl) reflection is monitored in turn, and this intensities are recorded. *Source*: Reprinted from Atkins PW, Physical Chemistry, Oxford University Press, 627, 1990.

Table 12.3 Mean figures sampled over a four year period for the XRD d-spacings, crystal size and azimuthal half widths of Courtanld's TTP PAN precursor and Panox oxidized PAN fiber

Product	d-spacings nm			Crystal size, 17° and 25° nm			Azimuthal half width°
TTP	0.521	0.302	0.340	5.8	–	–	39
Panox	0.518	0.347	–	–	1.9	1.3	45

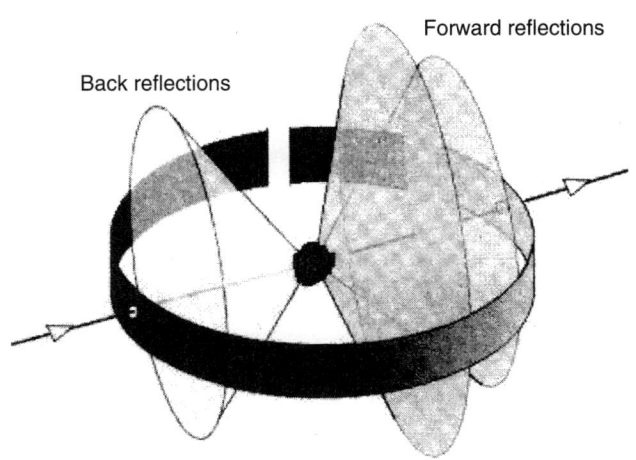

Figure 12.26 X-ray diffraction lines of powdered crystal recorded on film at intersection of each cone. *Source:* Courtesy of Department of Engineering, The University of Liverpool.

all possible crystal representations are offered. The Debye-Scherrer method is used for determining the degree of crystallinity, since all diffraction arcs can be obtained and indexed. It is also used for the determination of mean crystallite dimensions from the widths of some of the diffraction arcs. The ground sample is placed in a heap on a plate or in a glass capillary.

The diffracted beams form continuous cones and a circle of film records the diffraction pattern (Figure 12.27), such that each cone intersects the film and gives a diffraction line shaped as an arc (Figure 12.28).

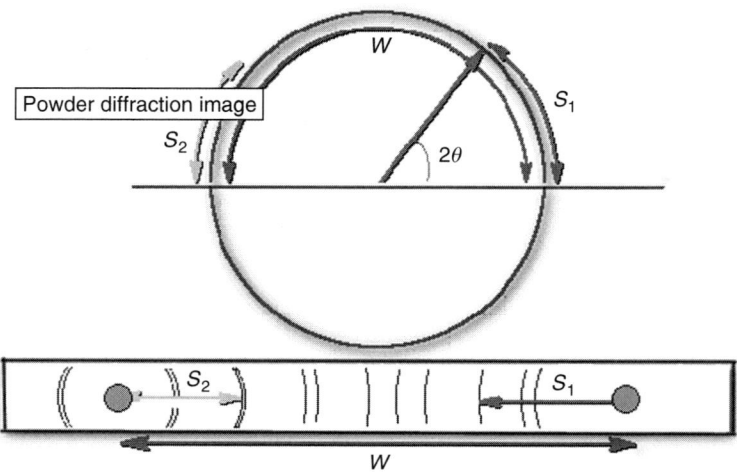

Figure 12.27 X-ray diffraction pattern of powdered crystal showing how each cone intersects the film giving diffraction lines. *Source:* Courtesy of Department of Engineering, The University of Liverpool.

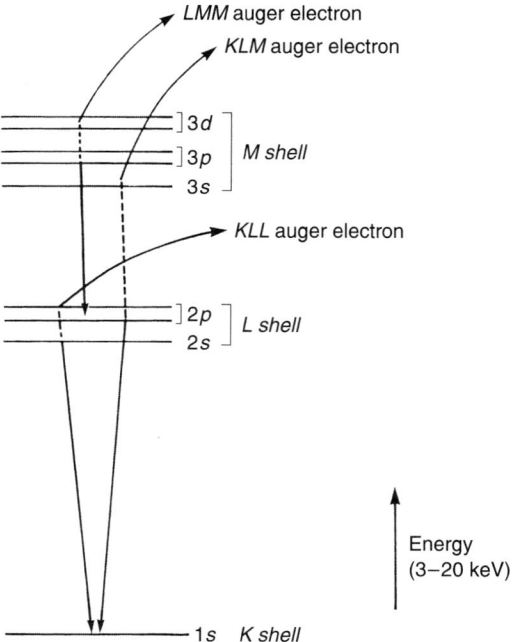

Figure 12.28 The Auger process. *Source:* Reprinted with permission from Savage G, *Carbon-Carbon Composites*, Chapman and Hall, London, 25, 1993. Copyright 1993, Springer.

When the film is laid flat, the distance from the center of the hole to the diffraction line S_1 is measured. When the angle 2θ is greater than 90°, this corresponds to the back reflections and S_2 can be measured. The distance S_1 relates to a diffraction angle of 2θ, which is always the angle between the transmitted and diffracted beams. The distance between the centers of the holes (W) corresponds to a diffraction angle of $\theta = \pi$. So can find θ from:

$$\theta = \frac{\pi S_1}{2W} \quad \text{or} \quad \theta = \frac{\pi}{2}\left(1 - \frac{S_2}{W}\right) \quad \text{and} \quad n\lambda = 2d\sin\theta$$

For a cubic crystal system with interplanar spacing of d,

$$d_{hkl} = \frac{\alpha}{\sqrt{h^2 + k^2 + l^2}}$$

where α is the lattice parameter.

Hence

$$\sin^2\theta = \frac{\lambda^2}{4\alpha^2}(h^2 + k^2 + l^2)$$

Knowing θ, $\sin^2\theta$ can be obtained and multiplying by a selected constant, values for $(h^2 + k^2 + l^2)$ can be obtained. By adjusting to an integer value, the values for h, k, and l can be found in the Miller Indices and hence, α can be calculated.

12.5.5 Low angle X-ray diffraction

The intense low angle scatter from carbon fibers is due to their porous nature and Johnson and Tyson [39] measured the size parameters for the pores and crystallites and found that in HM fiber, sharp edged voids less than 1 nm in width separated the crystallites in the lateral direction and twist or tilt boundaries were held to separate the crystallites in a lateral direction. This work was continued in order to measure the physical properties of carbon fibers [40].

Small angle X-ray scattering in carbon fibers was undertaken by Gupta and co-workers [41]. Two useful publications for more information are [42,43].

12.6 AUGER ELECTRON SPECTROSCOPY (AES)

The principle radiations in surface analysis techniques are given in Table 12.4. Auger electron spectroscopy can be used for high spatial resolution surface analysis [45]. A primary electron beam (3–20 keV and 0.1–1.0 mm diameter) with the energy about five times in excess of the binding energy of an electron in the K shell when incident upon a conducting surface will cause ionization, resulting in the release of a core electron causing a core hole and the release of a photoelectron. The atom relaxes and an electron with a lower binding energy then drops from an outer level into the core hole. The excess energy can be released as a characteristic X-ray photon at that same energy level (X-ray fluorescence), or predominantly given to another electron in the same level, or shallower, causing a second electron to be emitted with retained KE (Auger emission). This leaves the atom double ionized with two vacancies (Figure 12.29). The energy (0–2 keV) related to the Auger electron is characteristic

Table 12.4 Principle radiations in surface analysis techniques

Radiation detected	Input Radiation		
	Electrons	Ions	Photons / X-rays
Electrons	AES {3} EELS HREELS {1}		UPS {3} XPS {3}
Ions		SIMS {10}	

Note: The number of sampling depth monolayers of methods are given in { }.
Source: Reprinted with permission from Briggs D, Seah MP, Practical Surface Analysis, 2nd ed., Vol 1, *Auger and X-ray Photoelectron Spectroscopy*, 2nd ed., Vol. 1, John Wiley & Sons, Chichester, 4, 1990. Copyright 1990, John Wiley & Sons Ltd.

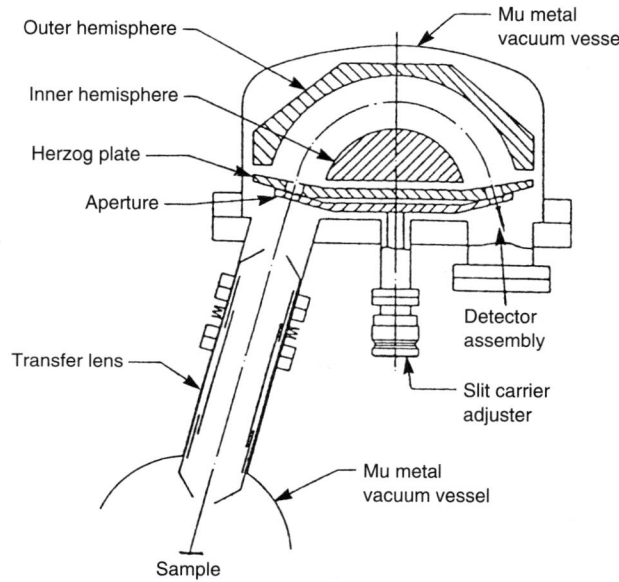

Figure 12.29 Diagram of a Concentric Hemispherical Analyzer (CHA) for XPS with a standard input lens system, which transfers an image of the analysed area on the sample onto the entrance slit to the analyzer, with slight magnification to permit removing the sample from close proximity to the entrance slit of the analyzer permitting greater working space around the sample. *Source:* Reprinted with permission from Coxon P, Krizek J, Humpherson M, Wardell IRM, *Relat Phenomena*, 52, 821, 1990. Copyright 1990, Elsevier.

of the emitting element and can be used to identify it as a function of its KE by using the relationship of the number of electrons *vs* KE, displayed in the derivative mode [47,48].

AES can be used to analyze the top 2–20 atomic layers and in conjunction with ion beam sputtering, can be used for depth profiling. The process must be undertaken in UHV 1.3×10^{-8} Nm^{-2} and surfaces must be sputter cleaned by directing a beam of ions such as Ar at 0.5–5 keV onto the surface. This cleaning process can also erode the sample and expose surface structure. AES can analyze areas as small as 100 nm and up to 2.5 cm^2 and is most sensitive when analyzing elements with low atomic number. However, it cannot detect H_2 or He. Resolution is down to 250 nm (cf SEM at 100 nm).

Hopfgarten [44] has measured the atomic composition of Types I, II, and III carbon fibers and found that for Types I and II, the oxygen content increased till down to a depth of 30 nm. Type III fiber had more surface oxygen, believed to be due to the larger surface area and increased surface roughness.

In Scanning Auger Microscopy (SAM), a focussed electron beam is scanned across a surface to produce Auger electrons and the results are presented as a map on X and Y axes showing the surface distribution of a particular element. Other workers [49–52] have used SAM to determine the distribution of oxygen on the surface of carbon fibers and the structure of the interface in carbon fiber composites [53].

X-ray excited Auger Electron Spectroscopy (XAES) is limited by the flux density of the X-ray source, but conveniently accompanies the photoelectron emission spectrum produced in an X-ray Photoelectron Spectrometer. XAES was used by Desimoni and co-workers [54,55].

12.7 X-RAY PHOTOELECTRON SPECTROSCOPY (XPS OR ESCA)

X-ray photoelectron spectroscopy (XPS), or given its other name Electron Spectroscopy for Chemical Analysis (ESCA), uses X-rays to excite photoelectrons. The emitted electron signal is plotted as a spectrum of binding energies. The photon is absorbed by an atom, molecule or solid leading to ionization and the emission of a core electron. Analysis will reveal the composition from a depth of 2–20 atomic layers and the electronic state of the surface region of the sample. XPS has the ability to identify different chemical states resulting from compound formation, which are revealed by the photoelectron peak positions and shapes.

Typical X-ray sources used are: Mg K_α 1253.6 eV; Al K_α 1486.6 eV and Ti K_α 2040 eV. XPS requires a UHV system (1.3×10^{-8} Nm^{-2}) and is not sensitive to H or He. The energy of the electrons leaving the sample is determined with a Concentric Hemispherical Analyzer (CHA) [56] as shown in Figure 12.30 when attached to an XPS [57].

XPS was used by Brewis et al [58] to determine the levels and nature of oxidation in carbon fiber. ESCA revealed the presence on the carbon fiber surface of =C=O, =C—OH, Na, —SO4, =C=C=, unoxidized nitrogen and silicone type Si. A direct relationship was found to exist between carbon fiber stability and the sodium present as Na_2SO_4 (Figure 12.31). Figure 12.32 shows selected ESCA spectra [59].

The quantification of the fiber surface concentration and distribution of acid groups and chemisorptive sites as a function of oxidative treatment level were determined by Denison et al [60] using a Ba labeling and XPS technique.

XPS was used by Ashwar and Ishwar Bhardwaj [61] to characterize PAN precursor, oxidized for gradually increasing times. A typical XPS survey spectrum is shown in Figure 12.33.

XPS is extremely versatile and has been used for surface analysis [62], including acidity [63] and determination of surface nitrogen [64] and for detecting differences in the surface chemistry and bulk structure of pitch and PAN based fibers [65]. Darmstadt et al [66] used XPS for determining oxygen containing functional groups and found that VGCF had a lower content than that of pitch and PAN based carbon fibers, but was considerably increased after oxidation treatment, becoming closer to that of other carbon fibers. The study of chemical oxidation of carbon fibers [67] and the electrochemical treatment of pitch based fibers [68] can be read in detail in the given references. The fiber matrix chemical interactions in cfrp were undertaken [69]. Nakayama et al [70] used XPS to study the carbon fiber matrix interface and investigated fiber matrix chemical interactions in carbon fiber reinforced composites.

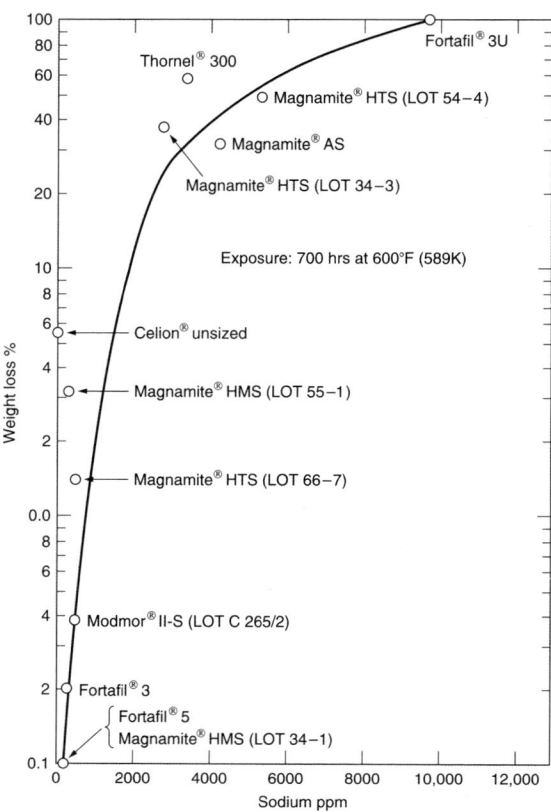

Figure 12.30 Weight loss of virgin carbon fibers in air at 315°C in relation to Na content determined by ESCA. *Source:* Reprinted with permission from Gibbs HH, Wendt RC, Wilson FC, Carbon fibre structure and stability studies, *Polym Eng Sci*, 19(5), 342–349, 1979. Copyright 1979, The Society of Plastic Engineers.

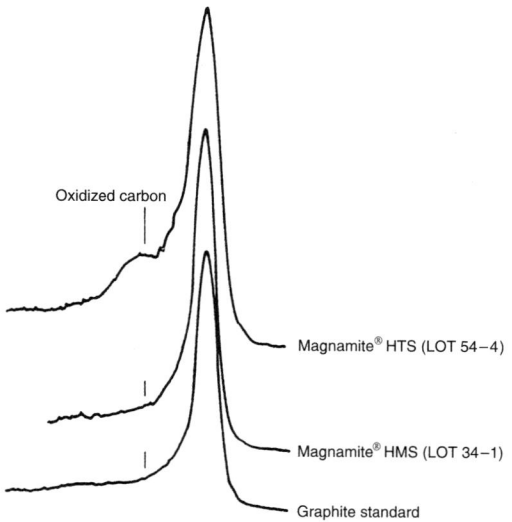

Figure 12.31 Selected ESCA carbon spectra. *Source:* Reprinted with permission from Gibbs HH, Wendt RC, Wilson FC, Carbon fibre structure and stability studies, *Polym Eng Sci*, 19(5), 342–349, 1979. Copyright 1979, The Society of Plastic Engineers.

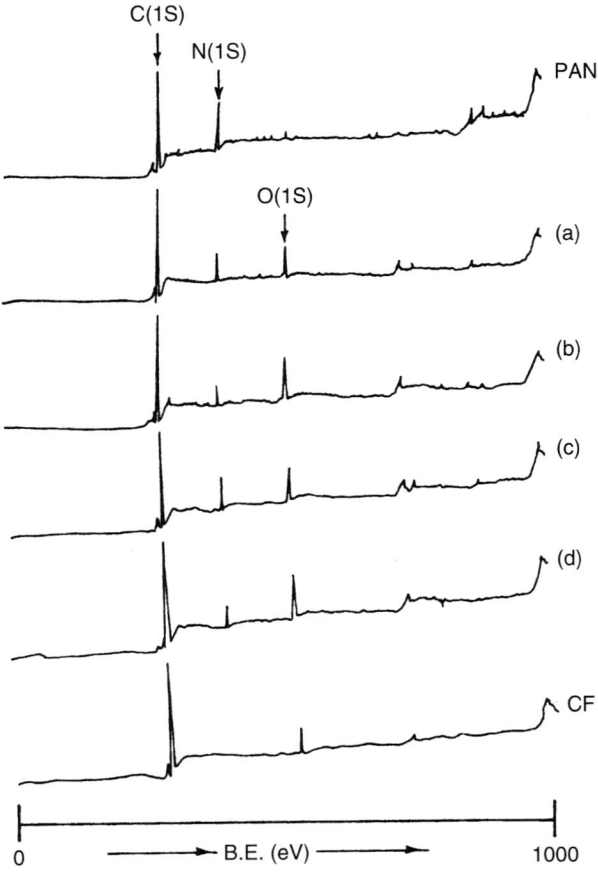

Figure 12.32 XPS survey spectra of PAN carbon fiber precursors oxidized to (a) 35, (b) 40, (c) 45 and (d) 50 min and finished carbon fiber with an epoxy size. *Source:* Reprinted with permission from Bhardwaj A, Bhardwaj IS, ESCA characterization of polyacrylonitrile based carbon fibre precursors during its stabilization process, *J Appl Polym Sci*, 51(12), 2015–2020, 1994. Copyright 1994, John Wiley & Sons Ltd.

12.8 ULTRAVIOLET PHOTOEMISSION SPECTROSCOPY (UPS)

UPS uses a less powerful source of exciting radiation than XPS, such as a HeI discharge lamp emitting radiation of 21.22 eV, or HeII (40.8 eV) for the examination of the valence levels, or shallow core levels, with which the photon will interact, causing ionization and removal of one of these electrons. The low photon energy source is insufficient to excite deep core electrons.

Raymundo-Piñero *et al* [71] have prepared N-containing activated carbon fibers by reacting a melamine resin (containing 45% nitrogen) and a soft petroleum pitch, spinning, stabilizing, carbonizing and activating the resultant carbon fiber with CO_2. The stages of the process were followed with XPS, UPS, EELS and HRTEM.

The UPS spectra obtained are shown in Figures 12.34 and 12.35 using the HeI and HeII energy sources respectively. It was shown that the N atoms promoted condensation and cyclization reactions between molecules, favouring the formation of larger graphene layers.

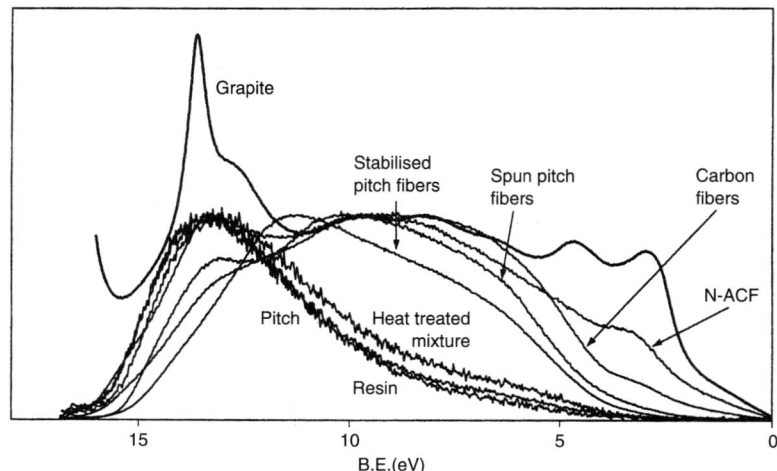

Figure 12.33 UPS HeI valence band spectra for a highly oriented pyrolytic graphite (HOPG), the precursor of the N-ACF and for the materials obtained during the preparation of the N-ACF. (i) Melamine resin (45% N)—valence bond structure corresponds to aromatic structures. (ii) Soft petroleum pitch—absence of valence bond structure and low degree of condensation. (iii) Heat treated mixture—aromatic molecules with a low degree of condensation. (iv) Spun fiber—increased number of aromatic rings. (v) Stabilized pitch fiber—not creating aromatic molecules with a high degree of condensation. (vi) Carbon fiber—important changes in valence bond structure when compared to pitch or stabilized fibers, now comparable to graphite, showing contribution of Π bonds. (vii) Activated carbon fiber—high similarity to graphite indicates gasification reaction which removes the most reactive parts of the carbonaceous material. (viii) Highly oriented pyrolytic graphite (HOPG)—structurally close to the ideal graphite crystal. *Source:* Reprinted with permission from Raymundo-Piñero E, Cazorla-Amoros D, Linares-Solano A, Find J, Wild U, Schlögl R, Structural characterization of N-containing activated carbon fibers prepared from a low softening point petroleum pitch and a melamine resin, *Carbon*, 40, 597–608, 2002. Copyright 2002, Elsevier.

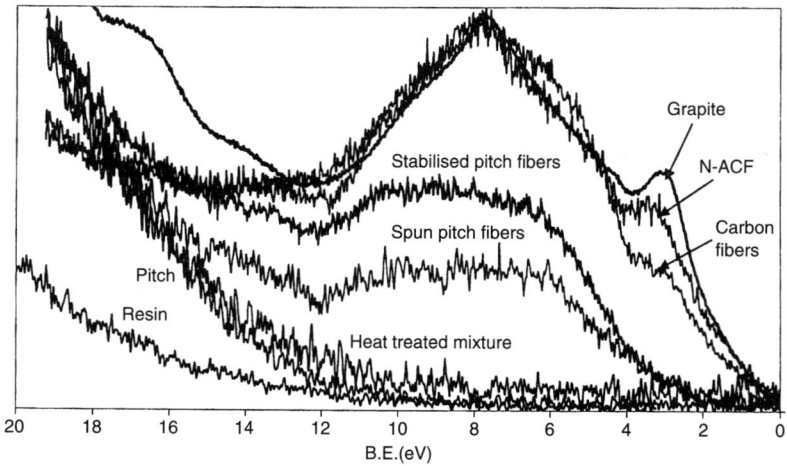

Figure 12.34 UPS HeII valence band spectra for a highly oriented pyrolytic graphite (HOPG), the precursor of the N-ACF and for the materials obtained during the preparation of the N-ACF. *Source:* Reprinted with permission from Raymundo-Piñero E, Cazorla-Amoros D, Linares-Solano A, Find J, Wild U, Schlögl R, Structural characterization of N-containing activated carbon fibers prepared from a low softening point petroleum pitch and a melamine resin, *Carbon*, 40, 597–608, 2002. Copyright 2002, Elsevier.

TECHNIQUES FOR DETERMINING THE STRUCTURE OF CARBON FIBERS

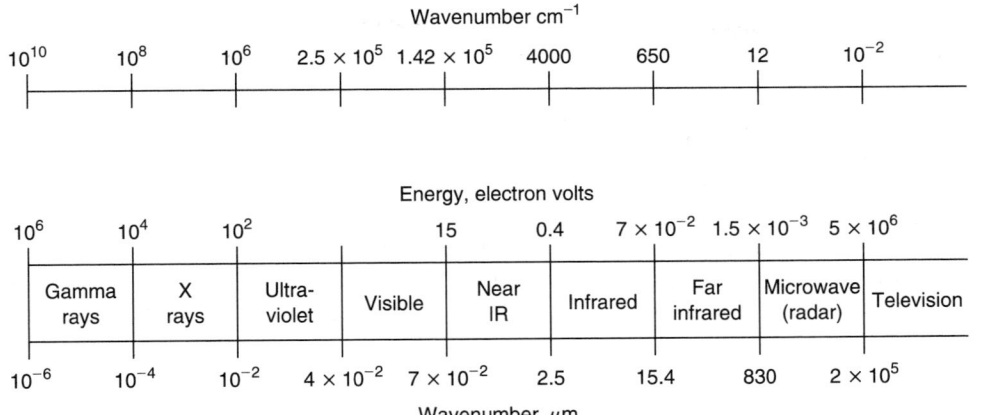

Figure 12.35 The electromagnetic spectrum. *Source:* Reprinted from technical literature of Perkin Elmer Corp.

12.8 INFRARED SPECTROSCOPY

12.8.1 Introduction

If a beam of electromagnetic radiation is passed through a substance, it can either be absorbed or transmitted. If the absorption is accompanied by a transition from one vibrational energy level to another, then the radiation is from the infrared portion of the electromagnetic spectrum. The infrared region extends from about 0.75 μm to almost 1 mm (Figure 12.36), but the segment most often used is 2.5–25 μm, termed the fundamental region.

There are different forms of vibration (e.g. for a molecule B—A—B):

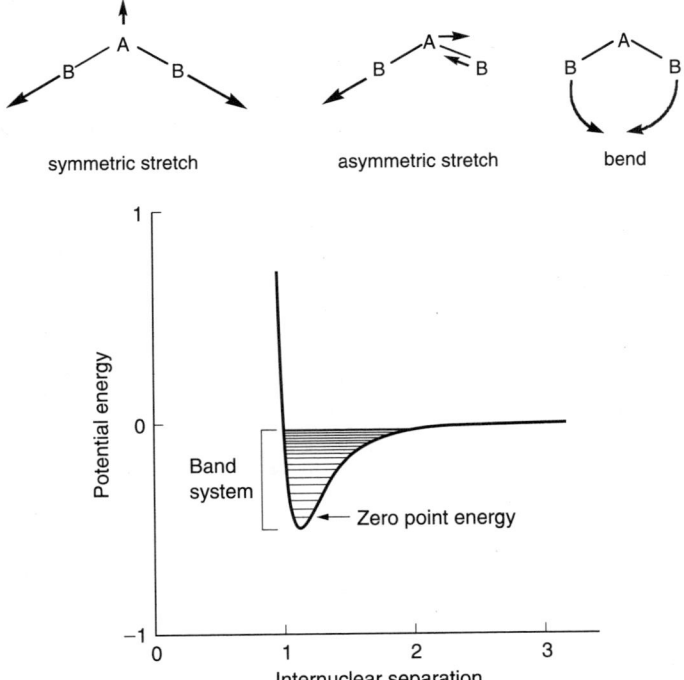

Figure 12.36 Vibration levels of a molecule.

Stretching vibrations have a higher frequency than bending vibrations of the same group. Lower the mass of the atom, the higher the frequency of vibration. The stiffer the bond, higher the frequency of vibration and for groups of atoms with the same mass, a triple bonded group has a higher vibrational frequency than a double bonded group, which in turn, is higher than a single bonded group.

Depending on the energy level, the molecule may absorb a quantum of radiation and pass to a higher level, gaining vibrational energy and thereby increasing the amplitude of the vibration. This vibrational energy equals the frequency of the light causing that transition.

Such vibrational transitions occur at about 10^{14} sec^{-1}, which would be effected by light of wavelength c/λ where c is the velocity of light, so effective light would have a wavelength of

$$\frac{3 \times 10^{10} \text{ cm sec}^{-1}}{10^{14} \text{ sec}^{-1}} = 3 \times 10^{-4} \text{ cm} = 3 \text{ μm}$$

This value occurs in the infrared region of the spectrum and therefore the discipline of dealing with vibrational spectroscopy is called IR spectroscopy, displayed by plotting energy of the IR radiation against the percent of light transmitted. A feature of this branch of spectroscopy is to replace wavelength with frequency, by multiplying the speed of light by the reciprocal of the wavelength. Since c is a constant, this has been simplified to the reciprocal of the wavelength. Hence 3 μm would be

$$\frac{1}{3 \times 10^{-4} \text{ cm}} = 3333 \text{ cm}^{-1}$$

and this is termed wave number. However, it is not a true frequency and to obtain the frequency, it is necessary to multiply by the speed of light.

The IR region generally covers the region 4000–650 cm^{-1}. The typical vibration levels of a molecule are shown in Figure 12.37. It is seen that the lines are not equispaced and converge to a continuum because the shape of the graph is not a true parabola. The different types of vibration possible in a molecule are:

The act of vibration is not sufficient by itself for the molecule to present an IR spectrum and must be associated with a change in the dipole moment of the molecule (the product of

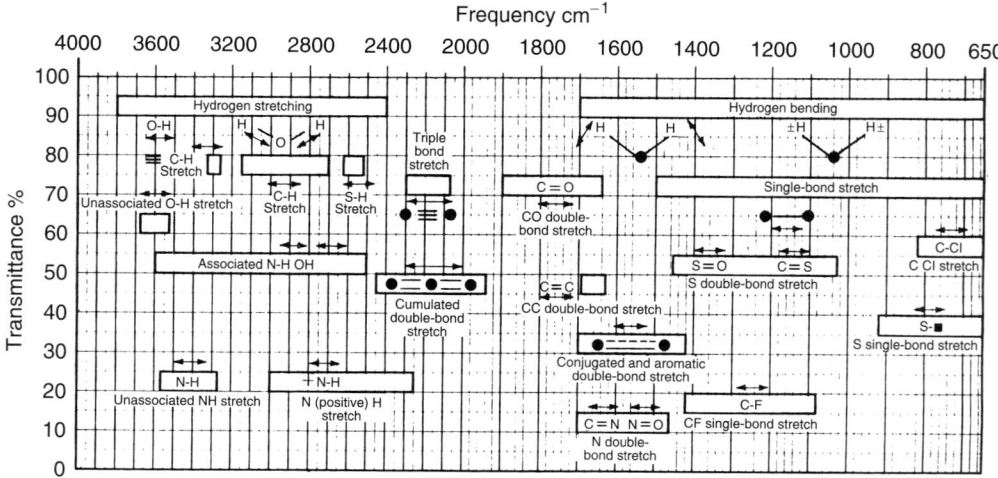

Figure 12.37 Absorption in different regions of the infrared spectrum. *Source:* Reprinted from technical literature of Perkin Elmer Corp.

the distance and one of the two equal point charges of opposite sign separated by that close distance).

The interpretation of IR spectra is helped by the vast library of published spectra, but relies implicitly on the skill of the spectroscopist.

Sample preparation depends on the nature of the substance investigated and liquids should preferably be in a neat 100% form. A drop of the liquid is placed on a highly polished salt plate such as NaCl, KBr or AgCl, a further plate placed on top and the two plates pressed together to ensure a liquid layer <0.1 mm thick. Spacers 0.025 mm thick can be used. To prepare a solution (normally about 0.2 M), the substance is dissolved in a solvent acceptable for IR such as CCl_4 (4000–1230 cm^{-1}), CS_2 (1390–400 cm^{-1}) $CHCl_3$ (but never water) and placed between plates, or in a cell when it is customary to use a reference beam with the solvent in a cell of the same construction, subtracting information from sample results.

When dealing with solids, there are two techniques:

1. The Mull method
   The specimen is finely ground (1–2 µm) in an agate mortar and intimately mixed with a weakly absorbing non-volatile liquid to form a mull and the paste spread between plates. Possible mulling agents are Nujol (a mineral oil), Fluorolube S (perfluorohydrocarbon), hexachlorobutadiene or perchlorokarosene.
   Water must never be used in any of these operations and plates are normally dried in a desiccator.
2. The KBr pellet or disc method
   The sample is finely ground with pure dry KBr, placed in a die, a vacuum applied (130–260 Nm^{-2} of Hg) to remove entrained air and finally pressed in a hydraulic press to form a transparent pellet or disc. The pellet is placed in a pellet holder, positioned in the spectrometer and the spectrum recorded.

A useful technique for determining the —COO/—CN ratio of a PAN precursor is to cast a film of the polymer, about 0.03–0.07 mm thick and compare with a reference sample.

Jones gives summary charts of principal group frequencies in the IR spectra of organic compounds [72], Chia and Ricketts describe the basic techniques and experiments in IR and FTIR spectroscopy [73] and Figure 12.38 shows the absorption in different regions of the IR spectrum.

Yang and Simms [75] used IR to study the raw materials, stabilization and carbonization of carbon fibers made from petroleum pitch and extended this work [76] to examine the distribution of the oxidation products between the surface and the bulk.

Zhu et al [77] studied the cyclization of PAN during the stabilization stage of carbon fiber.

12.8.2 Fourier Transform Infrared Spectroscopy (FTIR) [78,79]

A light source generates energy in the region of interest, ideally truly monochromatic light, which is passed through a beam splitter, directing the light in two directions at mutually right angles, with one beam directed to a stationary mirror and back to the beam splitter whilst the other beam passes to a moving mirror permitting the total path length of this path to be varied with respect to the other beam. The two paths meet at the beam splitter, where they recombine. But, the difference in path lengths gives constructive and destructive interference, creating a interferogram (Figure 12.39). The recombined beam passes through the sample, which absorbs all the wavelengths that are characteristic of the sample's spectrum and subtracts specific wavelengths from the interferogram, allowing the detector to report variations of energy with time for all wavelengths simultaneously. A laser beam is superimposed to provide an instrument reference. The instrument is interfaced with a

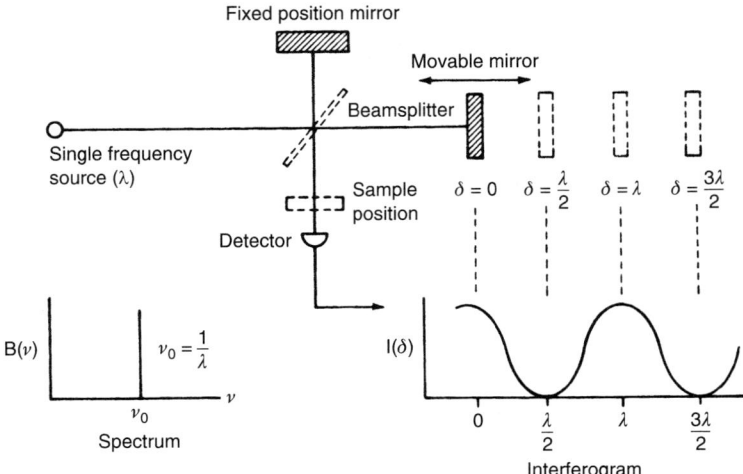

Figure 12.38 Block diagram showing the major components of an FTIR spectrometer. Also shows the spectrum of an infinitely narrow line source and how the interferogram is generated as the moveable mirror is translated. Maxima in the interferogram occur when the retardation is equal to an integral multiple of the wavelength of the source. Minima occur when the retardation is an odd multiple of half wavelengths. *Source*: Reprinted from technical literature of PerkinElmer Corp.

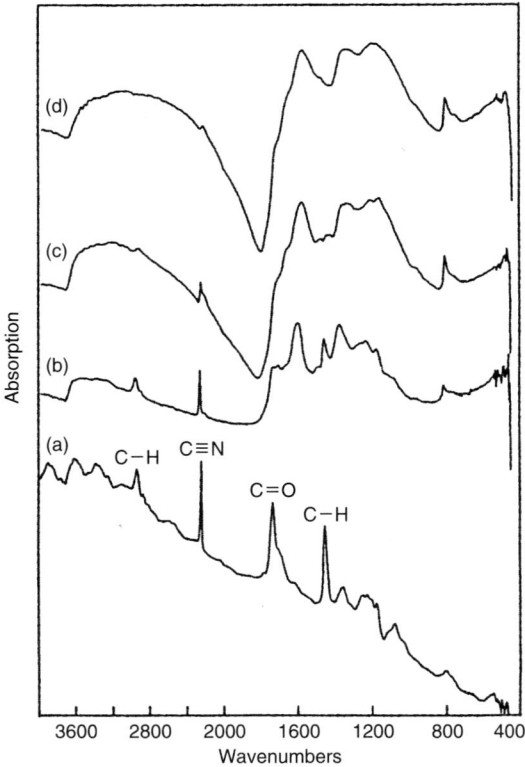

Figure 12.39 Series of transmission FTIR spectra for (a) Nikkoso precursor AN/MA/ITA, (b) Oxidized in air for 20 min 215°C/15 min 235°C, (c) Oxidized in air for 20 min 215°C/180 min 235°C, (d) Oxidized in air for 20 min 215°C/1800 min 235°C. *Source*: Reprinted from Usami T, Itoh T, Ohtani H, Tsuge S, Structural study of polyacrylonitrile fibers during oxidative thermal degradation by Pyrolysis—Gas Chromatography, Solid State ^{13}C Nuclear Magnetic Resonance and Fourier Transform Infrared Spectroscopy, *Macromolecules*, 23, 2460–2465, 1990.

TECHNIQUES FOR DETERMINING THE STRUCTURE OF CARBON FIBERS

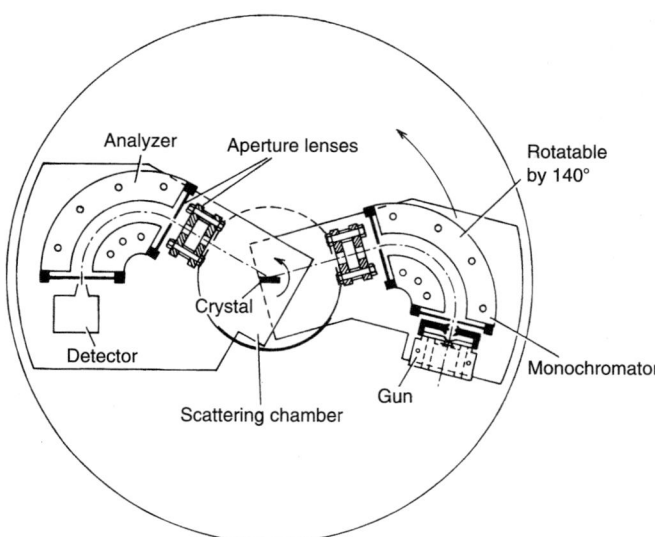

Figure 12.40 Schematic diagram of a HREELS spectrometer. *Source:* Reprinted with permission from Froitheim H, Ibach H, Lehwald S, *Rev Sci Instrum*, 46, 1325, 1975. Copyright 1975, The American Institute of Physics.

computer which converts the energy *vs* time spectrum to an intensity *vs* frequency spectrum using a mathematical function termed a Fourier transform.

So FTIR permits all source energy to reach the specimen, improving the S/N ratio. The longer path length provided by the moving mirror allows higher resolution. The FTIR instrument is, by design, only a single beam instrument [73].

The characterization of carbon fiber surfaces has been undertaken using FTIR [80–84]. The FTIR spectra of a PAN precursor and three differently stabilized PAN fibers obtained by Usami *et al* are shown in Figure 12.40 [81].

12.8.3 Fourier Transform Infrared/Attenuated Total Reflectance Spectroscopy (FTIR/ATR)

FTIR/ATR permits the examination of the surface on a true monolayer scale. Waveguides bevel-cut from silicon wafers transmit the IR radiation through the reactor via BaF viewports. This allows the vibrational modes of surface molecules and atoms to be studied.

12.9 ELECTRON ENERGY LOSS SPECTROSCOPY (EELS)

A beam of electrons of known KE and incident to a specimen is scattered inelastically and loses part of the energy. By analyzing the energy with a spectroscope attached under a TEM or SEM will reveal information on the elemental composition and bonding state. The technique is particularly useful for the transition elements and elements of low atomic number such as Be, B, C, N and O.

A carbon atom has an absorption peak around 284 eV in Electron Energy Loss Spectroscopy (EELS). At a high resolution of about 30 meV (Figure 12.41), it is possible to study the vibrations of the molecules on the surface (HREELS) .

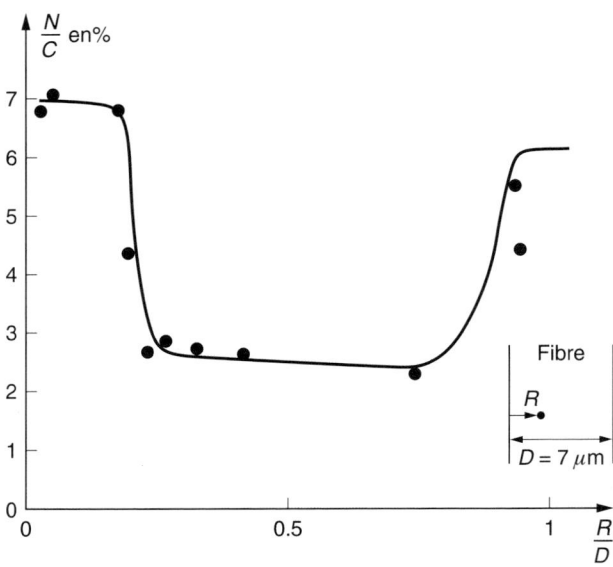

Figure 12.41 Profile of distribution of nitrogen in cross-section of Torayca T-300 carbon fiber, determined by electron energy loss spectrometry (EELS). R/D is the ratio of the distance into the fiber for measurement to the fiber diameter. *Source:* Reprinted with permission from Serin V, Fourmeaux R, Kihn Y, Sevely J, Guigon M, Nitrogen distribution in high tensile strength carbon fibers, *Carbon*, 28, 573, 1990. Copyright 1980, Elsevier.

Details of EELS are provided in [86–88]. Serin *et al* [89], using EELS studied the nitrogen distribution across the diameter of Toray T300 carbon fiber with an accuracy of 10% and found that the nitrogen content in the core is low, about 2%, but was 6–7% in the periphery (Figure 12.42). The higher nitrogen contents were associated with lower tensile strength.

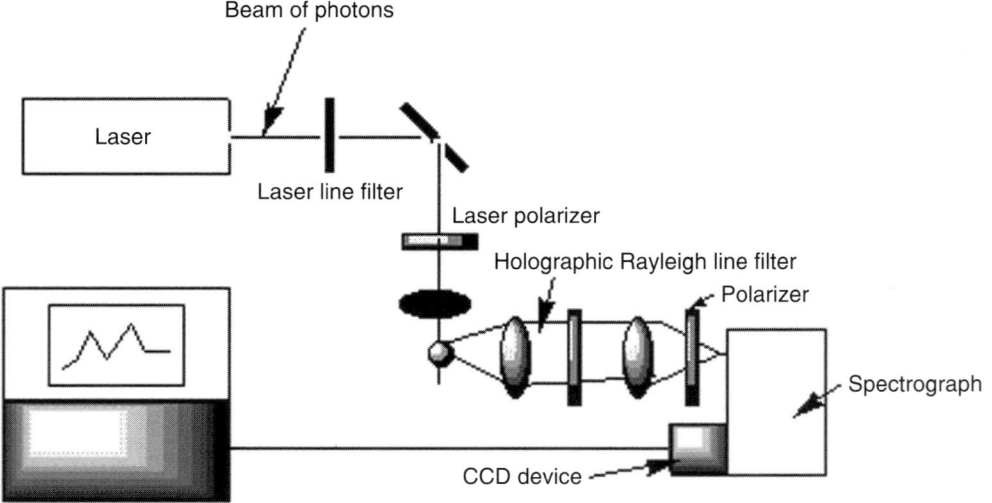

Figure 12.42 Typical RAMAN spectroscopy set up.

12.10 RAMAN SPECTROSCOPY

A photon is scattered by the molecular system either (a) elastically and termed Rayleigh scattering, where the emitted photon is the same wavelength as the absorbing photon or (b) inelastically and termed the Raman effect comprising only about 1 in 10^7 of the incident photons, where the energies of the incident and scattered photons are different. Additional spectral lines appear and the energy is either imparted as rotational and vibrational energies to the scattering molecules, or lost. The spectrum will have the prominent line (Rayleigh) corresponding to the original wavelength of the incident radiation plus additional lines each side of it corresponding to the shorter or longer wavelengths of the altered portion of light. If the energy of the scattered radiation is less than the incident radiation, then this is termed the Stokes line and if the energy is more, anti-Stokes line. Both lines are equally displaced from the Rayleigh line since one vibrational quantum of energy is either gained or lost. The Stokes line is more intense and is normally measured as characteristic of the transmitting substance.

Since the Raman scattering is not very efficient, a high power monochromatic excitation source is required as provided by a laser beam: Ar^+ 488.0 and 514.5 nm; HeNe 633 nm are used, although Nd:YAg 1.064 nm is more intense but FTIR must be used and this overcomes problems with fluorescence.

Raman spectroscopy require little sample preparation and a typical Raman spectroscopy set-up is shown in Figure 12.43.

Although IR and Raman are complementary techniques, Raman spectroscopy can resolve 1–2 μm and can be focussed to a beam about 2.5 μm in diameter, whereas IR cannot be focussed and resolved beyond 10–20 μm.

Fitzer and Rozploch [90] used Raman spectroscopy to undertake lateral measurements of a series of commercial carbon fibers (Figure 12.44). The characteristics of the carbon fibers used in this study are given in Table 12.5. The characterization of graphite fiber surfaces has been determined with Raman spectroscopy [91,92].

12.10.1 Surface enhanced Raman scattering (SERS)

Surface enhanced Raman scattering (SERS) is achieved by depositing islands of silver onto a substrate and obtaining an enhancement of up to as much as 10^3. An optical microscope can be used to focus the laser beam and feed information to a spectrograph for chemical analysis, called the Molecular Optical Laser Examiner (MOLE) (Figure 12.45) and the use of MOLE to examine thin films of graphite graphitized up to 2700°C is shown in Figure 12.46, with only one peak at 158 nm, which is attributed to graphite. Ko [95] also examined graphitization with Raman spectroscopy.

Raman spectroscopy has been used to examine carbon fibers [96], the effect of processing variables [97] and the effect of various treatments [98]. Chase has described Fourier Transform Raman Spectroscopy [99].

12.11 SECONDARY ION MASS SPECTROMETRY (SIMS)

SIMS is the most sensitive of all the commonly employed surface analytical techniques and there are several variations, all requiring UHV. All these technicques are capable of examining specimens up to 50 mm^2.

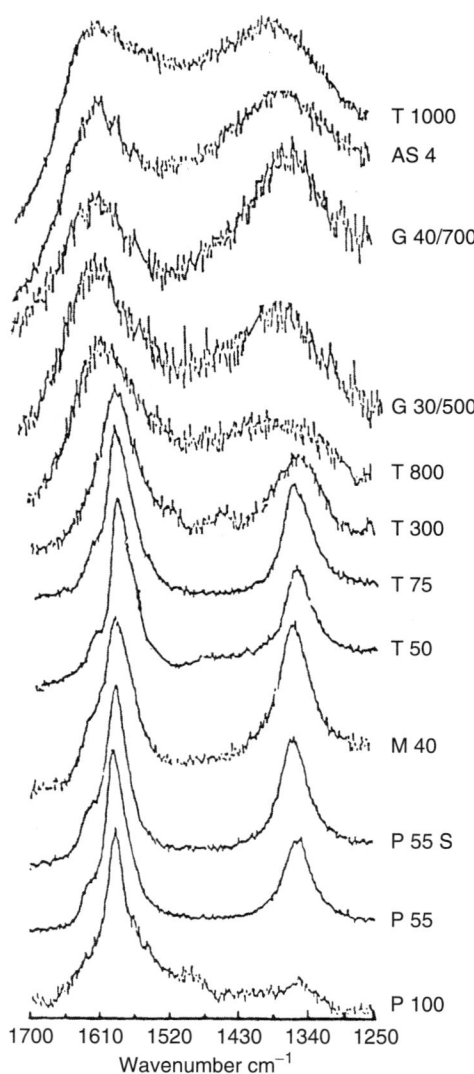

Figure 12.43 Raman spectra of lateral measured carbon fibers. See Table 12.5 for information on fiber types.
Source: Reprinted with permission from Fitzer E, Rozploch F, Laser Raman spectroscopy for the determination of the C—C bonding length in carbon, *Carbon*, 24(5), 594–595, 1988. Copyright 1988, Elsevier.

12.11.1 Static SIMS

Static SIMS is used for sub-monolayer elemental analysis. This technique uses a pulsed beam of Cs or Ga from a liquid metal ion gun. The dose of radiation is limited to minimize any sample damage or chemical modification. The technique has the capability of true monolayer detection (one monolayer is about 10^{15} atoms cm^{-2}). Bombardment of the surface with high energy ions leads to the ejection, or sputtering, of neutral, positively and negatively charged species from the surface, possibly including atoms, clusters of atoms and molecular fragments, but only charged particles are required for the SIMS technique.

TECHNIQUES FOR DETERMINING THE STRUCTURE OF CARBON FIBERS

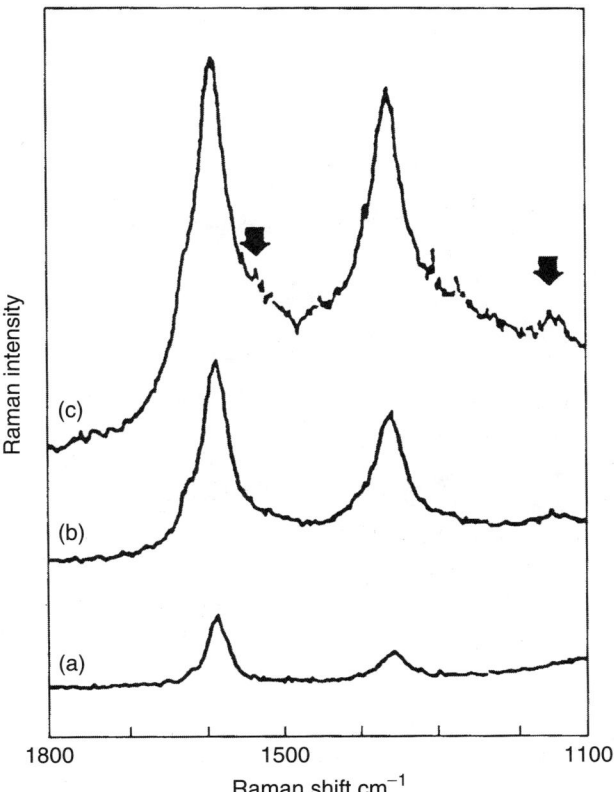

Figure 12.44 (a) Raman spectra of a graphite fiber. (b) Same fiber covered with Ag island films 5.1 nm thickness. (c) of 10.6 nm thickness. *Source:* Reprinted from Ishitani A, Ishida H, Katagiri G, Tomita S, New techniques for the characterization of surfaces and interfaces of carbon fibers, Ishida H, Koenig JL eds., North-Holland, New York, 195, 1986.

Table 12.5 Properties of fibers depicted in Figure 12.43

Fiber Grade	Precursor Type	Strength GPa	Modulus GPa	Heat Treatment Temperature °C
T 1000	PAN	6.37	294	–
AS 4	PAN	4.28	228	1300–1400
G 40/700	PAN	4.97	300	1300–1400
G 30/500	PAN	3.79	234	1300–1400
T 800	PAN	5.49	294	–
T 300	PAN	3.53	230	1300–1400
T 75	Rayon	2.59	517	Stretch graphitized
T 50	Rayon	1.97	345	Stretch graphitized
M 40	PAN	2.74	392	2700
P 55S	Pitch	1.90	379	–
P 55	Pitch	1.90	379	1800
P 100	Pitch	2.41	758	2500

Source: Reprinted with permission from Fitzer E, Rozploch F, *Carbon*, 24(5), 594–595, 1988 Fitzer E, Rozploch F, Laser Raman spectroscopy for the determination of the C—C bonding length in carbon, *Carbon*, 24(5), 594–595, 1988. Copyright 1988, Elsevier.

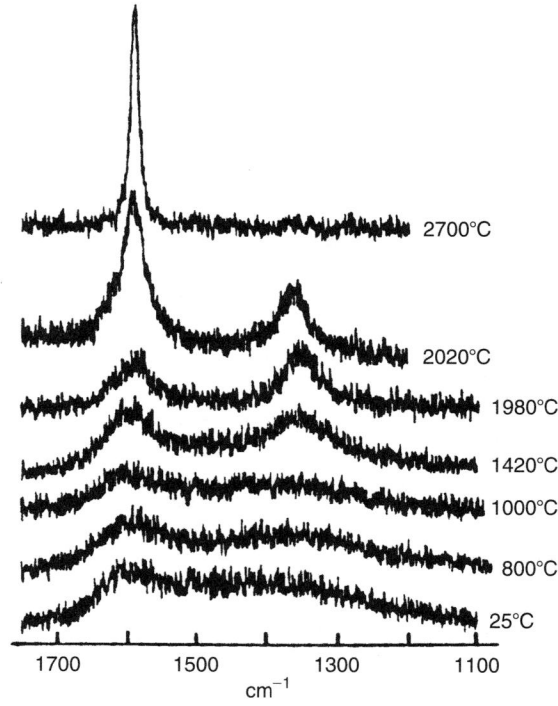

Figure 12.45 Graphitization of thin films followed by MOLE Raman spectroscopy. *Source:* Reprinted from Rouzaud JN, Oberlin A, Beny-Bassez C, *Thin Solid Films*, 105, 75, 1983.

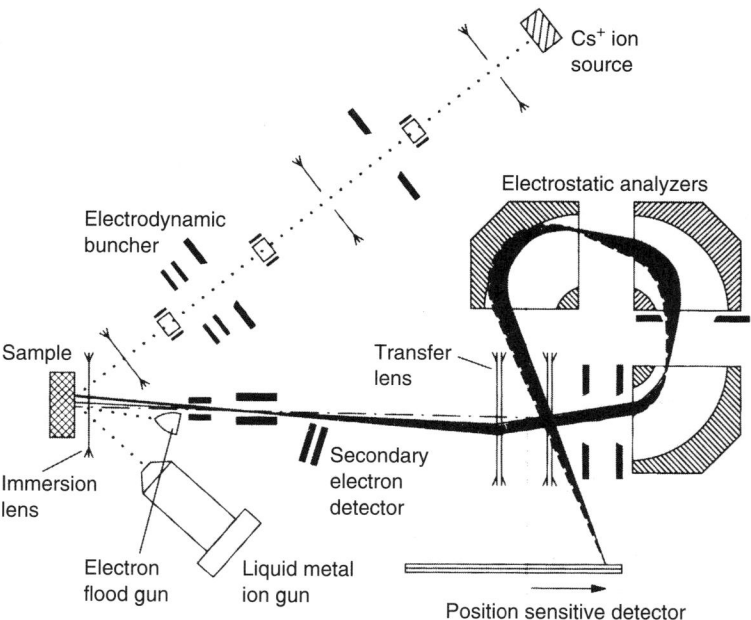

Figure 12.46 Schematic layout of ToF-SIMS equipment. *Source:* Reprinted from Wayne DH, Triple focussing allows real time surface analysis, *Research and Development*, Cahness Publishing Company, 74–77, Aug 1990.

TECHNIQUES FOR DETERMINING THE STRUCTURE OF CARBON FIBERS

12.11.2 Dynamic SIMS

This technique uses a non-pulsed or DC primary ion beam of Cs or Ga ions. Alternating between a DC and a pulsed beam permits the discrete depth profile data to be obtained. In this mode, the technique would be considered destructive in order to obtain the depth concentration profiles.

12.11.3 Imaging or microscope SIMS

This variation of SIMS uses a high current with a large diameter primary ion beam for spatially resolved elemental analysis and is capable of achieving a resolution of 1 µm.

Using Time-of-Flight (ToF) analyzers will significantly improve the resolution and in this instance, it is the molecular ions which reflect the surface composition more closely.

The basic principle of ToF mass spectrometry is to obtain a time separation of ions with the same energy, but different mass [100].

Time of Flight Secondary Ion Mass Spectrometry (ToF-SIMS) is a surface analysis technique used to analyze mass and image constituents that are present on the surface of materials. The equipment (Figure 12.47) uses a pulsed primary ion beam to desorb and ionize species from the sample surface. The resulting secondary ions are accelerated into a mass spectrometer and analyzed by measuring the ToF from the sample surface to the detector. The location and distribution of the species on the surface can be identified and an image shown at the detector. The composition is determined from the mass spectrum. Many different primary sources can be used for ionization:

1. A Cs ion beam covering a large area.
2. A focussed Nd YAg laser operating in the ablation or desorption mode.
3. A micro-focussed liquid metal ion gun used for raster generated imaging.
4. Post ionization used with a, b or c.

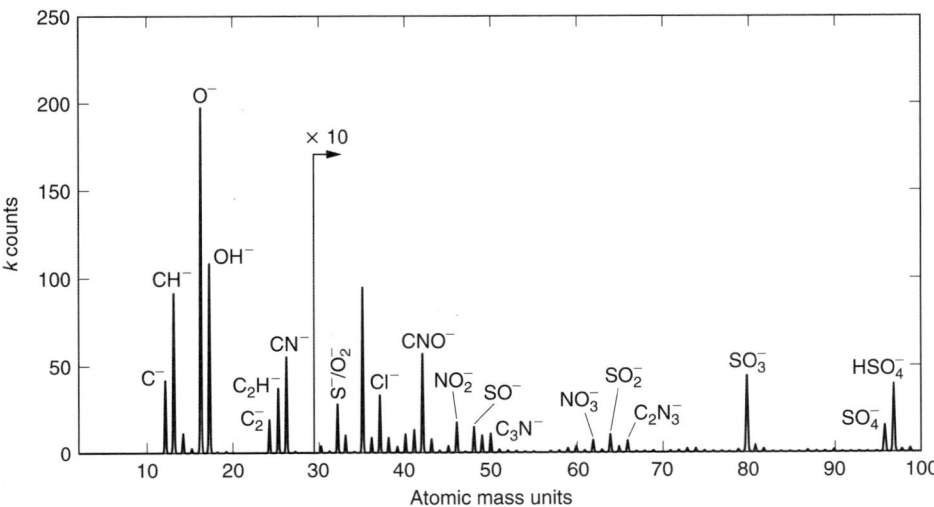

Figure 12.47 Negative ToF-SIMS spectrum from an Hercules AU fiber oxidized electrolytically in 10% NH_4HCO_3 at a level of 250 $C\,m^{-2}$. *Source:* Reprinted with permission from Alexander MR, Jones FR, The chemical environment of nitrogen in the surface of carbon fibres, *Surface and Interface Analysis*, 22, 230–235, 1994. Copyright 1994, John Wiley & Sons Ltd.

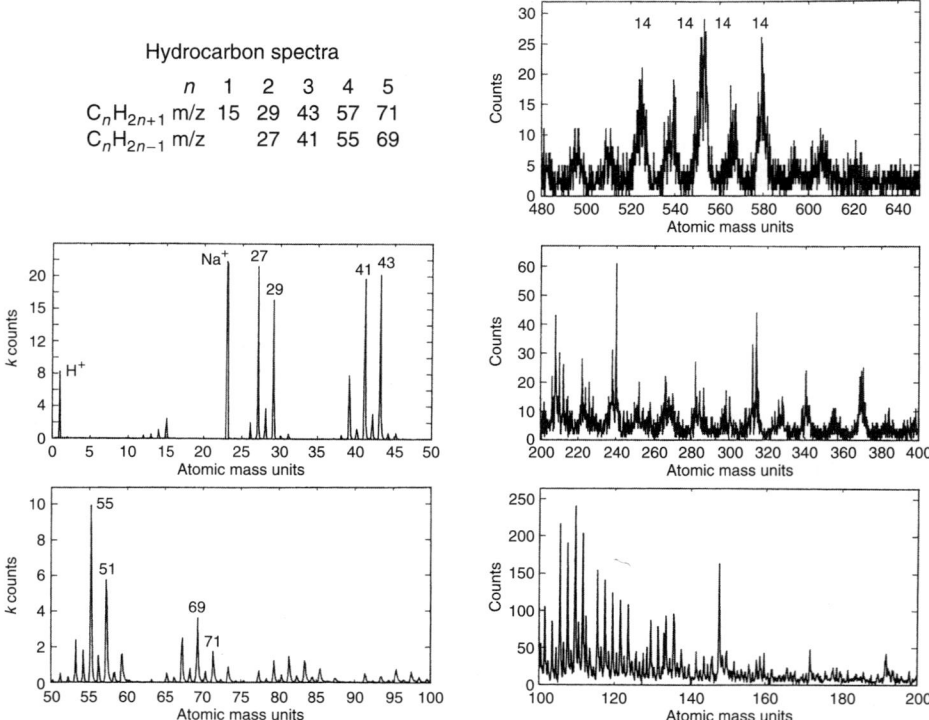

Figure 12.48 Positive ToF-SIMS spectra for RK30 epoxy sized carbon fiber. *Source:* Work undertaken by The University of Sheffield.

Bertrand and co-workers [100] have described the use of ToF-SIMS for examining the interface region of a Hercules AS-4 carbon fiber-polystyrene matrix composite. Triple focussing ToF-SIMS is described by Wayne [101]. Betrand *et al* [102] have used ToF-SIMS to elicit chemical information on the interfacial area of carbon fiber. Weng *et al* [103] used ToF-SIMS in conjunction with XPS to show that the size on Hercules AS-4 contains at least four different compounds—polydimethylsiloxane, dialkyl phthalates, glyceryl monostearate and phenolic antioxidants.

Jones and co-workers have successfully used SIMS in a scanning mode to characterize carbon fiber composite fracture surfaces, obtaining a lateral resolution of 0.2 μm [104,105]. Other literature on the examination of carbon fiber surfaces with ToF-SIMS is also available [106,107].

Alexander and Jones [108] examined the surface of Hercules AU fiber, which had been electrolytically surface treated in 10% NH_4HCO_3, for nitrogen and a typical negative ToF-SIMS is shown in Figure 12.48. Work undertaken by Jones at Sheffield University investigating the surface constituents of RK30 sized carbon fiber is shown in Figures 12.49 and 12.50.

12.12 SCANNING TUNNELLING MICROSCOPY (STM)

The prime use of STM is for the precise three-dimensional measurement of surface topography and is based on the fabrication of an atomically sharp tip with a radius less than 1 nm that is scanned over the surface under investigation (Figure 12.51).

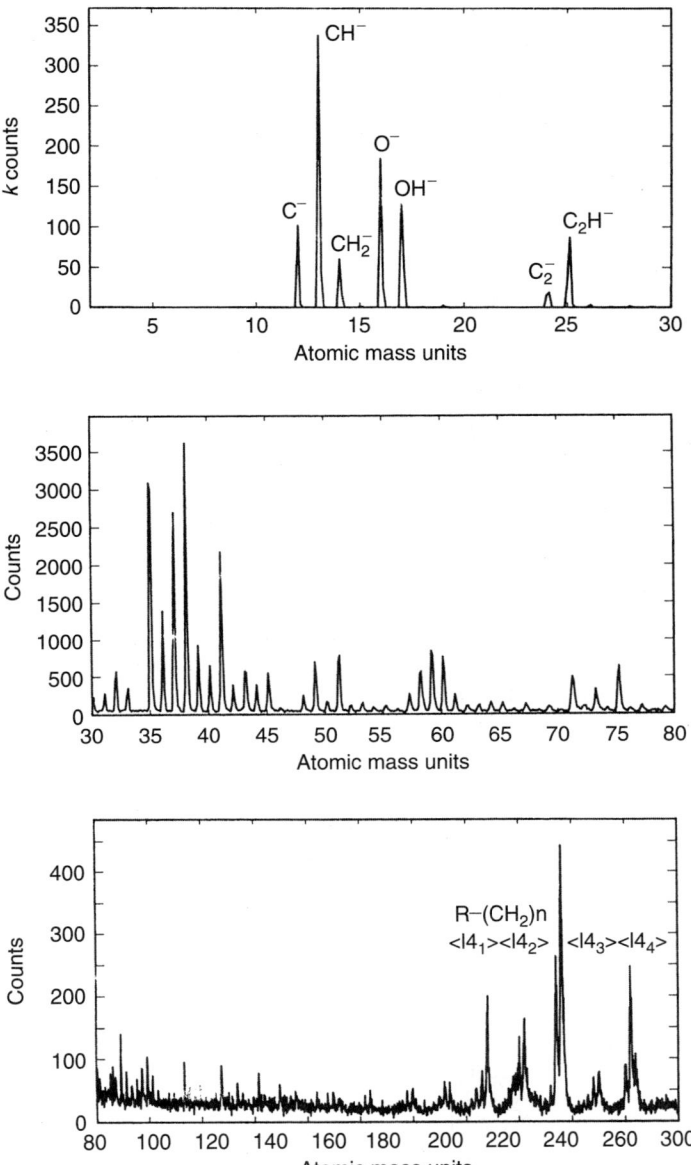

Figure 12.49 Negative ToF-SIMS spectra for RK30 epoxy sized carbon fiber. *Source:* Work undertaken by The University of Sheffield.

It has been established that etched tips may have a small atom-sized protrusion at the apex and it is possible, with infinite care, to prefabricate a pyramid tip by piling three layers of seven, three and finally one atom on top using field ion microscopy techniques. Zhadan [111] has used mechanically sharpened Pt/Ir wires and electrochemically etched W wires. When examining 7 μm diameter carbon fiber filaments proved a difficult practical problem because of the delicate nature of the heads, the problem was overcome by gluing the filament across a steel washer, which was held magnetically onto the microscope stage. The sample is scanned at constant force and contours of constant tunnelling, or the variation of the tunnelling current with position at a fixed sample height, are recorded.

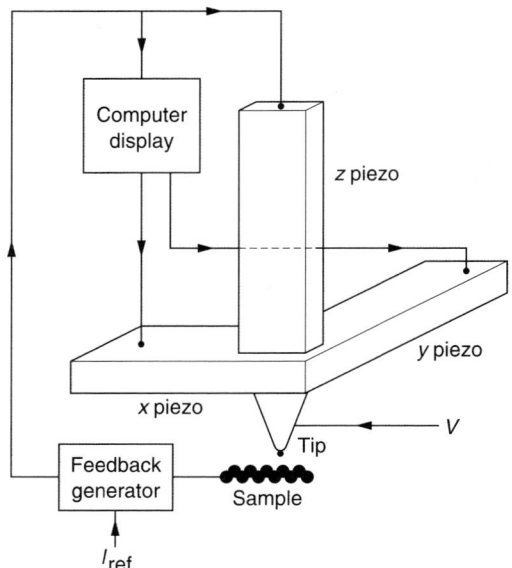

Figure 12.50 Schematic diagram of STM. The z piezo adjusts the tip surface spacing; x and y piezos provide lateral scan. *Source:* Reprinted with permission from Woodruff DP, Delchar TA, *Modern Techniques of Surface Science*, Cambridge University Press, Cambridge, 102, 1994. Copyright 1994, Cambridge University Press.

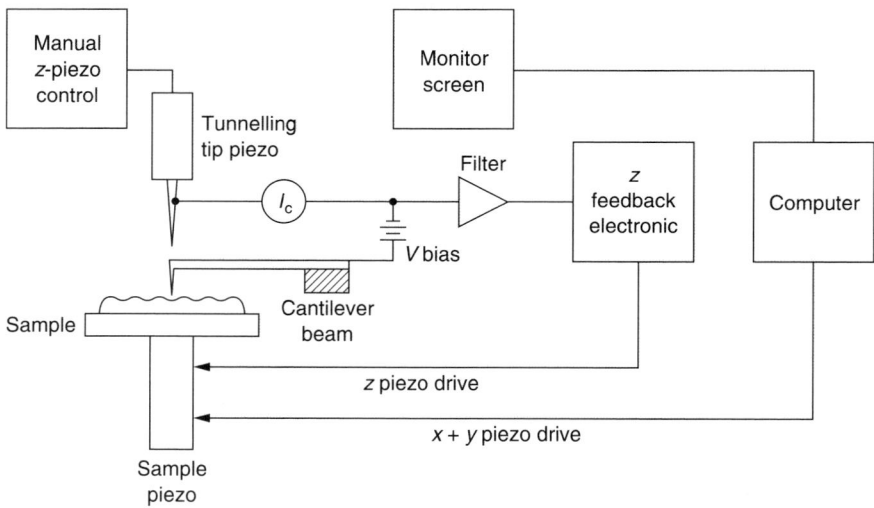

Figure 12.51 Schematic diagram of the AFM showing the auxiliary STM used for detecting movement of the atomic force cantilever. *Source:* Reprinted with permission from Woodruff DP, Delchar TA, *Modern Techniques of Surface Science*, Cambridge University Press, Cambridge, 460, 1994. Copyright 1994, Cambridge University Press.

Measurement of interatomic distances of 0.1 nm–100 μm can be achieved with STM.

It is not necessary for STM to be operated in a vacuum and work can be undertaken at atmospheric pressure and under liquids. At atmospheric pressure, the drag from the meniscus effect of absorbed moisture can be a problem; therefore, some workers prefer to immerse the sample in water.

Donnet *et al* [112–116] have used STM to characterize the surface of carbon fibers, as well as examine carbon fibers made from different precursors, PAN based high strength carbon fibers, and surface treated and activated carbon fibers.

Other workers [117–119] have examined PAN based carbon fibers with STM. Effler *et al* [120] used STM in conjunction with X-ray analyzes to examine mesophase pitch based carbon fibers. Hoffman and co-workers [121] describe the advantages of STM for studying surface treated carbon fibers.

12.13 ATOMIC FORCE MICROSCOPY (AFM) OR SCANNING FORCE MICROSCOPY (SFM)

Scanning Force Microscopy (SFM) is a complimentary scanning technique to STM, but is not limited to conducting samples and therefore, can be used for carbon fibers before interaction with a matrix. It can also be used for characterizing the fiber surface morphology after delamination of the composite. The shape of the tip is very important and they are generally made using micro-lithographic techniques and fall into three categories based on shape and tip radius. A general type is a pyramid form, about 3 µm high with a tip radius of about 30 nm, a longer and thinner tip fabricated from a carbon compound using an electron beam technique for fabrication and a commercial tip fabricated by the latest microlithographic methods with radii down to about 10 nm. Generally, the tip is purchased already attached to the cantilever. Cantilevers are micro-fabricated from Si_3N_4 and scan over the surface applying a constant force of 10^{-8}–10^{-9} N. Instead of the auxiliary STM used to detect the cantilever movement shown in Figure 12.52, later designs have now been perfected, such as laser beams which are deflected off the cantilever beam. It is necessary for the cantilever to have a low mass and is typically gold plated to reflect the laser beam and are only 100 µm in length. Tube piezoceramics are now used to position the tip or the sample with high resolution. Surface topography can be recorded by using a feedback system to measure the up and down movement of the tip or to measure the cantilever deflection.

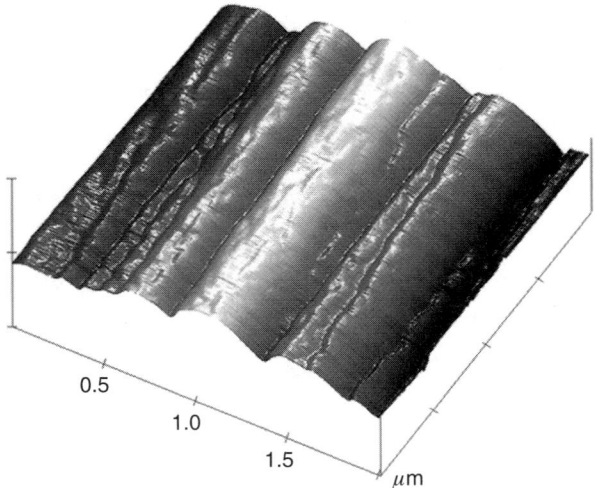

Figure 12.52 AFM image of carbon fiber. *Source:* Reprinted from Zhdan PA, Bors M, Castle JE, *In situ* scanning force microscopy (SFM) study of the electrochemical activations of carbon fibres, *Composite Sci Technol*, 58(3–4), 559–570, 1998.

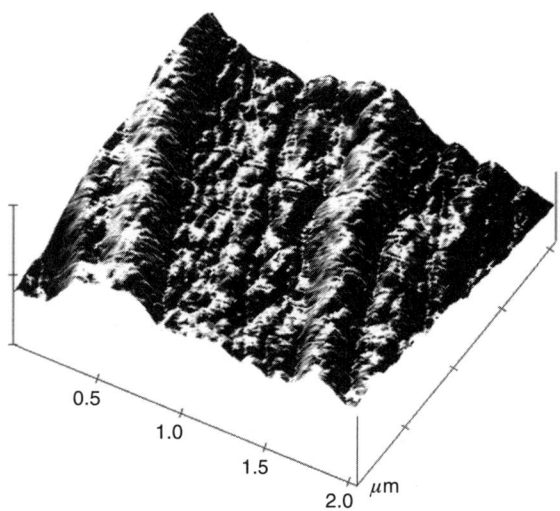

Figure 12.53 *In situ* AFM image of a carb on fiber after electrochemical treatment in NH₄HCO₃ for 10 min.
Source: Reprinted from Zhdan PA, Bors M, Castle JE, *Composite Sci Technol*, 58(3–4), 559–570, 1998.

Resolution for SFM is 0.01–0.1 nm and magnification can be 5×10^2–10^8, compared with magnification of 10–10^5 for SEM.

The translation system must be rigid and avoid resonant frequencies. Stainless steel and Viton rubber stacks can be used to filter out characteristic frequencies. Hong-Qiang Li, of Guelph University has described a unique system to isolate vibration by attaching a large mass to bungy cords, which are firmly anchored to the building and effectively act as a band pass filter, providing safe operation range of 1–100 Hz, enabling atomic resolution to be achieved. AFM images of carbon fiber are shown in Figures 12.52 and 12.53.

Brown *et al* [118] used AFM to study PAN based carbon fiber in air, Hoffman [119] studied surface treated carbon fiber and Effler *et al* [120] examined mesophase pitch based carbon fiber. A number of workers [121–123] have used AFM to study carbon fibers. Zhdan and co-workers [124–126] have described the use of AFM at Surrey University to study carbon fibers and Figures 12.52 and 12.53 show AFM images of carbon fiber.

REFERENCES

1. Briggs D, Seah SP, Practical Surface Analysis, *Auger and X-ray Photoelectron Spectroscopy*, 2nd ed., Vol.1, John Wiley & Sons, Chichester, 4, 1996.
2. Morita K, Murata Y, Ishitani A, Murayama K, Ono T, Nakajima A, Characterization of commercially available PAN (polyacrylonitrile) based carbon fibres, *Pure Appl Chem*, 58, 456–468, 1986.
3. Vickerman JC ed., *Surface Analysis: The Principle Techniques*, John Wiley & Sons, Somerset, NJ, 1997.
4. Riviere JC, Myhra S eds., *Handbook of Surface and Interface Analysis: Methods of Problem Solving*, Marcel Dekker, New York, 1998.
5. Knibbs RH, The use of polarized light microscopy in examining the structure of carbon fibres, *J Microscopy*, 94(3), 273–281, Dec 1971.
6. Woodrow J, Mott BW, Haines HR, Analysis of polarized light reflected from absorbent materials at normal incidence, *Proc Phys Soc B*, 65, 603, 1952.
7. Johnson W, PAN based carbon fibres, Watt W, Perov BV eds., *Strong Fibres*, North Holland, Amsterdam, 400, 1985.

8. Savage G, *Carbon-Carbon Composites*, Chapman and Hall, London, 13–16, 1993.
9. Oberlin A, Bonnamy S, Lafdi K, Structure and Texture of Carbon Fibers, Donnet JB, Wang TK, Peng JCM, Rebouillat S, *Carbon Fibers*, Marcel Dekker, New York, 90–93, 1998.
10. Johnson W, *Nature*, 279(142), 5709, 1979.
11. Nyo H, Heckler AJ, Hoernschemeyer DL, Characterizing the structures of PAN based carbon fibers, 24th *Nat Symposium*, San Francisco, 179, 51–60, May 8–10, 1979.
12. Goldstein J, Yakowitz H, Newbury D, Lifshin E, Colby J, Coleman J, *Practical Scanning Electron Microscopy*, Plenum Press, New York, 1975.
13. Manders PW, Carbon fibre structure by electrolytically etching, *Nature*, 271(142), 5641, 1978.
14. Madronero A, Verdu M, Issi JP, Romero JMB, Barba C, Study by a modified scanning electron microscope fractography of hydrogenation process in vapour grown carbon fibres, *J Mater Sci*, 33(8), 2079–2085, 1998.
15. Vezie DL, Adams WW, High resolution scanning electron microscopy of PAN based and pitch based carbon fiber, *J Mater Sci Lett*, 9(8), 883–887, 1990.
16. Williams DB, *Practical Analytical Electron Microscopy*, Phillips Electronic Instruments Publishing, New Jersey, 1984.
17. Goodhew PJ, *Practical Methods in Electron Microscopy*, Vol II, Elsevier, Oxford, 1985.
18. Loretto MH, *Electron Beam Analysis of Materials*, Chapman and Hall, London, 1984.
19. Bennet SC, Johnson DJ, *Carbon*, 17, 25, 1975.
20. Kowbel W, Hippo E, Murdie N, *Carbon*, 27, 219, 1989.
21. Goodhew PJ, Preparation of carbon fibres for transmission electron microscopy, *J Phys*, E4, 392–394, 1971.
22. Oberlin A, Bonnamy S, Lafdi K, Structure and Texture of Carbon Fibers, Donnet JB, Wang TK, Peng JCM, Rebouillat S, *Carbon Fibers*, Marcel Dekker, New York, 95–96, 1998.
23. Pennock GM, Ogara E, Preparation of carbon fiber sections for light and transmission electron microscopy, *Mater Sci Lett*, 9(7), 847–849, 1990.
24. Tidjani M, *Carbon*, 24, 447, 1986.
25. Ehrburger P, Lahaye J, *Carbon*, 19, 1, 1981.
26. Kowbel W, Don J, *Proc 18th Biennial Conf on Carbon*, Worcester, MA, 286, 19–24 Jul 1987.
27. Guigon M, Oberlin A, Desarmot G, Microtexture and structure of some high modulus PAN based carbon fibres, *Fibre Sci and Technol*, 20, 177–198, 1984.
28. Deurbergue A, Oberlin A, TEM study of some recent high modulus PAN based carbon fibers, *Carbon*, 30(7), 981–987, 1992.
29. Bennett SC, Johnson DJ, Electron microscope studies of structural heterogeneity in PAN based carbon fibres, *Carbon*, 17, 25–39, 1979.
30. Klinklin E, Guigon M, Characterization of the interface in carbon fiber reinforced composites by transmission electron microscopy, *Colloids and Surfaces A-Physicochemical and Engineering Aspects*, 74(92–93), 243–250, 1993.
31. Alexander LE, *X-ray Diffraction Methods in Polymer Science*, Interscience, New York, 1969.
32. Oberlin A, Bonnamy S, Lafdi K, Structure and Texture of Carbon Fibers, Donnet JB, Wang TK, Peng JCM, Rebouillat S eds. *Carbon Fibers*, Marcel Dekker, New York, 89, 1998.
33. Bennett SV, Johnson DJ, Montague PE, Electron diffraction and high resolution studies of carbon fibres, *Proc 4th London Intern Conf Carbon and Graphite*, Soc of Chem Ind, London, 503–507, 1976.
34. Francis JG, Ozbek S, Isaac DH, Microstructural changes in carbon fibres during high temperature processing.?
35. Johnson DJ, Structural studies of PAN-based carbon fibres, Thrower PA ed., *Chemistry and Physics of Carbon*, Vol 20, Marcel Dekker, New York, 1–58, 1987.
36. Johnson DJ, Structure and properties of carbon fibres, Figueiredo JL, Bernardo CA, Baker RTK, Hüttinger KJ eds., *Carbon Fibers, Filaments and Composites*, Kluwer Academic Publishers, Dordrecht, 1989.
37. Takaku A, Shioya M, X-ray measurements and the structure of PAN and pitch based carbon fibers, *J Mater Sci*, 25(11), 4873–4879, 1990.
38. Atkins PW, Physical Chemistry, Oxford University Press, 627, 1990.

39. Johnson DJ, Tyson CN, *J Phys D: Appl Phys*, 2, 787–795, 1969.
40. Johnson DJ, Tyson CN, Low-angle X-ray diffraction and physical properties of carbon fibres, *J Phys D: Appl Phys*, 3, 526–534, 1970.
41. Gupta A, Harrison IR, Lahijani J, Small angle X-ray scattering in carbon fibers, *J Appl Crystallography*, 27(4), 627–636, 1994.
42. Ladd MFC, Palmer RA, *Structure determination by X-ray crystallography*, Plenum Press, New York, 1985.
43. Cullity BD, *Elements of X-ray Diffraction*, 2nd ed., Addison-Wesley, Jun 1978.
44. Hopfgarten F, *Fibre Sci Technol*, 11, 67, 1978.
45. Briggs D, Seah MP, Practical Surface Analysis, 2nd ed., Vol 1, *Auger and X-ray Photoelectron Spectroscopy*, John Wiley, Chichester, 1990.
46. Savage G, *Carbon-Carbon Composites*, Chapman and Hall, London, 25, 1993.
47. Carlson TA, *Photoselection and Auger Spectroscopy*, Plenum Press, New York, 1975.
48. Spruger RW, Haas TW, Grant JT, Quanitative Surface analysis of Materials, *ASTM STP634*, American Society for Testing and Materials, Philadelphia, 1978.
49. Yip PW, Lin SS, Interfaces in Composites, Pantana CG, Chen EJH eds., *Materials Research Society Symposium Proceedings*, Vol 170, Materials Research Society, Pittsburgh, 339, 1990.
50. Lin SS, *J Vac Sci Technol*, 8, 2412, 1990.
51. Vaidyanathan NP, Kabadi VN, Vaidyanathan R, Sadler R, *Adhesion*, 48, 1, 1995.
52. Lin SS, Oxygen concentrations on surfaces of carbon fibers, *Unclassified US Army Materials Technology Laboratory Report*, MTL-TR-89-99, 1989.
53. Cazeneuve C, Castle JE, Watts JF, *J Mater Sci*, 25(4), 1902–1908, 1990.
54. Desimoni E, Cataldi TRI, Ceifidor UB, On the characterization of carbon fiber surfaces by X-ray excited Auger Electron Spectroscopy, *Annali di Chimica*, 82(5–6), 207–218, 1992.
55. Desimoni E, Salvi AM, Casella IG, Damiano D, *Surface Interface Anal*, 20, 909, 1993.
56. Seah MP, Electron and ion energy analysis, Walls JM ed., *Methods of Surface Analysis*, Chapter 3, Cambridge University Press, Cambridge, 1989.
57. Coxon P, Krizek J, Humpherson M, Wardell IRM, *Relat Phenomena*, 52, 821, 1990.
58. Brewis DM, Comyn J, Fowler JR, Briggs D, Gibson VA, Surface treatments of carbon fibres studied by X-ray photoelectron spectroscopy, *Fibre Sci Technol*, 12, 41–52, 1979.
59. Gibbs HH, Wendt RC, Wilson FC, Carbon fibre structure and stability studies, *Polym Eng Sci*, 19(5), 342–349, 1979.
60. Denison P, Jones FR, Watts JF, XPS analysis of barium labelled carbon fibres and interfacial effects in fibre-epoxy composites, *Surface and Interface Analysis*, 9, 431–435, 1986.
61. Bhardwaj A, Bhardwaj IS, ESCA characterization of polyacrylonitrile based carbon fibre precursors during its stabilization process, *J Appl Polym Sci*, 51(12), 2015–2020, 1994.
62. Wang PH, Hong KL, Zhu QR, Surface analysis of polyacrylonitrile based activated carbon fibers by X-ray photoelectron spectroscopy, *J Appl Polym Sci*, 62(12), 1987–1991, 1996.
63. Baillie CA, Watts JF, Castle JE, Determination of the acidity of carbon fiber surfaces by means of X-ray photoelectron spectroscopy adsorption isotherms, *J Mater Chem*, 2(9), 939–944, 1992.
64. Bradley RH, Ling X, Sutherland I, Beamson G, XPS determination of surface nitrogen species on PAN carbon fibers, *Carbon*, 32(1), 185–186, 1994.
65. Xie YM, Sherwood PMA, X-ray photoelectron spectroscopic studies of carbon fiber surfaces. 11. Differences in the surface chemistry and bulk structure of different carbon fibers based on polyacrylonitrile and pitch and comparison with various graphite samples, *Chem Mater*, 2(3), 293–299, 1990.
66. Darmstadt H, Roy C, Kaliaguine S, Ting J-M, Alig RL, Surface spectroscopic analysis of vapour grown carbon fibres prepared under various conditions, *Carbon*, 36, 1183–1190, 1998.
67. Desimoni E, Salvi AM, Casella IG, Damiano D, Controlled chemical oxidation of carbon fibers—An XPS-XAES-SEM study, *Surface Interface Anal*, 20(11), 909–918, 1993.
68. Xie YM, Sherwood PMA, X-ray photoelectron spectroscopic studies of carbon fibers. 14. Electrochemical treatment of pitch based carbon fibers and the surface and bulk structure changes monitored by XPS, XRD, and SEM, *Applied Spectroscopy*, 44(10), 1621–1628, 1990.

69. Weitzsacker CL, Xie M, Drzal LT, Using XPS to investigate fiber matrix chemical interactions in carbon fiber reinforced composites, *Surface and Interface Analysis*, 25(2), 53–63, 1997.
70. Nakayama Y, Soeda F, Ishitani A, XPS study of the carbon fiber matrix interface, *Carbon*, 28(1), 21–26, 1990.
71. Raymundo-Piñero E, Cazorla-Amoros D, Linares-Solano A, Find J, Wild U, Schlögl R, Structural characterization of N-containing activated carbon fibers prepared from a low softening point petroleum pitch and a melamine resin, *Carbon*, 40, 597–608, 2002.
72. Jones RN, Infrared spectra of organic compounds: Summary charts of principal group frequencies, *National Research Council (NRC)*, Bulletin No.6, Ottawa, Reprinted by Perkin-Elmer, 1959.
73. Chia L, Ricketts S, *Basic Techniques and Experiments in Infrared and FT-IR Spectroscopy*, Perkin Elmer, Mar 1988.
74. Hannah RW, Swinehart JS, *Experiments in techniques of Infrared spectroscopy*, Perkin Elmer Corporation, 1967.
75. Yang CO, Simms JR, Infrared spectroscopy studies of the petroleum pitch carbon fibre. 1. The raw materials, the stabilization and carbonization processes, *Carbon*, 31(3), 451–459, 1993.
76. Simms JR, Yang CO, Infrared spectroscopy studies of the petroleum pitch carbon fibre. 2. The distribution of the oxidation products between the surface and the bulk, *Carbon*, 32(4), 621–626, 1994.
77. Zhu Y, Wilding MA, Mukhopadhyay SK, Estimation, using infrared spectroscopy, of the cyclization of polyacrylonitrile during the stabilization stage of carbon fibre production, *J Mater Sci*, 31(14), 3831–3837, 1996.
78. Griffiths PR, de Haseth JA, *Fourier Transform Infrared Spectrometry*, John Wiley, New York, 1986.
79. Perkins WD, *J Chem Ed*, 63, A5, 1986.
80. Ohwaki T, Ishida H, Optimization of the surface characterization of carbon fiber by FT-IR internal reflection spectroscopy, *Applied Spectroscopy*, 49(3), 1997.
81. Usami T, Itoh T, Ohtani H, Tsuge S, Structural study of polyacrylonitrile fibers during oxidative thermal degradation by Pyrolysis—Gas Chromatography, Solid State ^{13}C Nuclear Magnetic Resonance and Fourier Transform Infrared Spectroscopy, *Macromolecules*, 23, 2460–2465, 1990.
82. Yang CO, Simms JR, Characterization of petroleum pitch precursor carbon fibers using Fourier transform infrared spectroscop,. *Abstracts of papers of The American Chemical Society*, 204(1), 53-Fuel, 1992.
83. Ohwaki T, Ishida H, Optimization of the surface characterization of carbon fiber by FT-IR internal reflection spectroscopy, *Applied Spectroscopy*, 49(3), 341–348, 1995.
84. Ohwaki T, Ishida H, Comparison between FT-IR and XPS characterization of carbon fiber surfaces, *J Adhes*, 52(1–4), 167–186, 1995.
85. Froitheim H, Ibach H, Lehwald S, *Rev Sci Instrum*, 46, 1325, 1975.
86. Egerton RF, *Electron Energy Loss Spectroscopy in the Electron Microscope*, Kluwer Academic Publishing, 1996.
87. Ibach H, Mills DL, *Electron Energy Loss Spectroscopy and Surface Vibrations*, Academic Press, 1982.
88. Ibach H, Electron Energy Loss Spectrometers—The Technology of High Performance, Hawkes PW, *Springer Series in Optical Sciences*, 63, 1991.
89. Serin V, Fourmeaux R, Kihn Y, Sevely J, Guigon M, Nitrogen distribution in high tensile strength carbon fibers, *Carbon*, 28, 573, 1990.
90. Fitzer E, Rozploch F, Laser Raman spectroscopy for the determination of the C—C bonding length in carbon, *Carbon*, 24(5), 594–595, 1988.
91. Tuinstra F, Koenig JL, Characterization of graphite fiber surfaces with Raman spectroscopy, *J Compos Mat*, 4, 492–499, 1970.
92. Afanasyeva NI, Jawhari T, Klimenko IV, Zhuravieva TS, Micro-Raman spectroscopic measurements on carbon fibers, *Vibrational Spectroscopy*, 11(1), 79–83, 1996.
93. Ishitani A, Ishida H, Katagiri G, Tomita S, New techniques for the characterization of surfaces and interfaces of carbon fibers, Ishida H, Koenig JL eds., North-Holland, New York, 195, 1986.
94. Rouzaud JN, Oberlin A, Beny-Bassez C, *Thin Solid Films*, 105, 75, 1983.

95. Ko TH, Raman spectrum of modified PAN based carbon fibers during graphitization, 59(4), 577–580, 1996.
96. Meyer N, Marx G, Brzezinka KW, Raman spectroscopy of carbon fibers, *Fresenius J Anal Chem*, 349(1–3), 167–168, 1994.
97. Melanitis N, Tetlow PL, Galiotis C, Characterization of PAN based carbon fibres with laser Raman spectroscopy. 1. Effect of processing variables on Raman band profiles, *J Mater Sci*, 31(4), 851–860, 1996.
98. Cuesta A, Dhamelincourt P, Laureyns J, Martinez Alonso A, Tascon JMD, Effect of various treatments on carbon fiber surfaces studied by Raman microprobe spectrometry, 52(3), 356–360, 1998.
99. Chase B, Fourier Transform Raman Spectroscopy, *Anal Chem*, 59, 881A, 1987.
100. Bertrand P, Weng L-T, Time-of-Flight Secondary Ion Mass Spectrometry (ToF-SIMS), *Mikrochim Acta* [Suppl], 13, 167–182, 1996.
101. Wayne DH, Triple focussing allows real time surface analysis, *Research and Development*, Cahness Publishing Company, 74–77, Aug 1990.
102. Poleunis C, Fallais I, Carlier V, Sclavons M, Bertrand P, Legras R, ToF-SIMS imaging of carbon fiber polymer interface in composite materials, *Proc 10th International Conf on Secondary Ion Mass Spectrometry SIMS X*, Wiley, Münster, Oct 1–6, 1995.
103. Weng LT, Poleunis C, Bertrand P, Carlier V, Sclavons M, Franquinet P, Legras R, Sizing removal and functionalization of the carbon fiber surface studied by combined ToF SIMS and XPS, *J Adhesion Sci Technol*, 9, 859–871, 1995.
104. Denison P, Jones FR, Brown A, Humphrey P, Harvey J, Scanning secondary ion mass spectroscopic studies of the micromechanics and chemical structure in the region of the interface in carbon fibre-epoxy composites, *J Mater Sci*, 23, 2153–2156, 1988.
105. Denison P, Jones FR, Brown A, Humphrey P, Paul AJ, Scanning SIMS spectography of cfrp, Ishida H ed., *Interfaces in Polymer, Ceramic and Metal Matrix Composites*, 239–248, 1988.
106. Hearn MJ, Briggs D, ToF-SIMS studies of carbon fiber surfaces and carbon fiber composite fracture surfaces, *Surface & Interface Analysis*, 17(7), 421 et seq, 1991.
107. Hamerton I, Hay JN, Howlin BJ, Jones JR, Lu SY, Webb GA, Bader MG, Brown AM, Watts JF, ToF-SIMS and XPS studies of carbon fiber surface during electrolytic oxidation in O^{17}/O^{18} enriched aqueous electrolytes, *Chem Mater*, 9(9), 1972–1977, 1997.
108. Alexander MR, Jones FR, The chemical environment of nitrogenin the surface of carbon fibres, *Surface and Interface Analysis*, 22, 230–235, 1994.
109. Woodruff DP, Delchar TA, *Modern Techniques of Surface Science*, Cambridge University Press, Cambridge, 102, 1994.
110. Woodruff DP, Delchar TA, *Modern Techniques of Surface Science*, Cambridge University Press, Cambridge, 460, 1994.
111. Zhadan PA, Grey D, Castle JE, Surface structure of the PAN based carbon fibres studied by the Scanning Probe Microscopy (SPM), *Surface Interface Analysis*, 22, 290–295, 1994.
112. Maganov SN, Cantow HJ, Donnet JB, Characterization of surfaces of carbon fibers by scanning tunneling microscopy, *Polymer Bulletin*, 23(6), 555–562, 1990.
113. Donnet JB, Qin RY, Study of carbon fiber surfaces by scanning tunneling microscopy. 1. Carbon fibers from different precursors and after various heat treatment temperatures, *Carbon*, 30(5), 787–796, 1992.
114. Donnet JB, Qin RY, Study of carbon fiber surfaces by scanning tunneling microscopy. 1. PAN based high strength carbon fibers, *Carbon*, 31(1), 7–12, 1993.
115. Qin RY, Donnet JB, Study of carbon fiber surfaces by scanning tunneling microscopy. 3. Carbon fibers after surface treatments, *Carbon*, 32(2), 323–328, 1994.
116. Donnet JB, Qin RY, Park SJ, Ryu SK, Rhee BS, Scanning tunneling microscopy study of activated carbon fibers, *J Mater Sci*, 28(11), 2950–2954, 1993.
117. Shi DX, Liu N, Yang HQ, Gao JN, Jiang YS, Pang SJ, Wu XB, Ji Z, Scanning tunnelling microscope study of polyacrylonitrile based carbon fibers, *J Mater Res*, 12(10), 2543–2547, 1997.
118. Brown NMD, You HX, A scanning tunnelling microscope study of PAN based carbon fiber in air, *Surface Sci*, 237(1–3), 273–279, 1990.

119. Hoffman WP, Hurley WC, Owens TW, Phan HT, Advantage of the scanning tunnelling microscope in documenting changes in carbon fiber surface morphology brought about by various surface treatments, *J Mater Sci*, 26(17), 4545–4553, 1991.
120. Effler LJ, Fellers JF, Annis BK, Scanning tunnelling microscopy and X-ray analyses on mesophase pitch based carbon fibers, *Carbon*, 30(4), 631–638, 1992.
121. Hoffman WP, Scanning probe microscopy of carbon fibre surfaces, *Carbon*, 30(3), 315–331, 1992.
122. Maganov SN, Gorenberg AY, Cantow HJ, Atomic force microscopy on polymers and polymer related compounds. 5. Carbon fibers, *Polymer Bulletin*, 28(5), 577–584, 1992.
123. Sturm H, Schulz E, Atomic force microscopy with simultaneous ac conductivity contrast for the analysis of carbon fibre surfaces, *Composites Part A—Appl Sci Manuf*, 27(9), 677–684, 1996.
124. Zhdan PA, Grey D, Castle JE, Surface structure of the PAN based carbon fibers studied by the scanning probe microscopy (SPM), 22(1–12), 290–295, 1994.
125. Zhdan PA, Bors M, Castle JE, *In situ* scanning force microscopy (SFM) study of the electrochemical activations of carbon fibres, *Composite Sci Technol*, 58(3–4), 559–570, 1998.
126. Zhdan PA, Nanoscale surface characterization of conducting and non-conducting materials with STM and contact SFM: some problems and solutions, *Surface Interface Anal*, 33, 879–893, 2002.

CHAPTER 13

Polymer Matrices for Carbon Fiber Composites

13.1 SELECTED THERMOSET RESINS

13.1.1 Introduction

Probably about 90% of all thermoset resins used are polyester resins because they are relatively cheap to make and about three quarters of this production are used with some form of reinforcement. Only the resins with better mechanical and high temperature performance are, however, used with the relatively expensive carbon fibers (Table 13.1).

The next most important class of resins used are the epoxide resins, developed over 50 years ago. They are more expensive than polyester resins but have superior mechanical properties and good resistance to alkaline conditions. Epoxies are by far the most widely used polymer matrix for carbon fibers and currently constitute over 90% of the matrix resin material used in advanced composites [1].

The vinyl ester resins have properties that are intermediate between polyesters and epoxides, are easier to process than epoxies, coupled with a better chemical resistance than polyester resins.

Phenolic resins are cured by a condensation reaction using hot press molding and are relatively cheap, having good high temperature performance (230°C), excellent fire and smoke resistance and good resistance to acids.

Bismaleimides and polyimides have good thermal stability, extending the working temperature range of epoxies from about 200°C to 280°C. In the UK, polyimide is not widely used due to the high cost and difficulty in processing. However, bismaleimides are now becoming an important class of thermoset resins.

Johnson [2] enumerates some of the requirements of the resin matrix for a potential application. Firstly, the performance requirements are:

1. High temperature capability (high T_g).
2. Resistance to moisture and other fluids (e.g. aviation fuel).
3. Satisfactory mechanical properties in conjunction with the two points mentioned above (e.g. hot-wet resistance) and any special requirements such as impact strength.
4. Smoke, fire and toxicity considerations.

Table 13.1 Properties of cured resins

Property	Orthophthalic polyester	Isophthalic polyester	Bisphenol-A polyester	Vinylester	Epoxy
Tensile modulus GPa	3.2	3.5	3.2	3.3	2–4
Tensile strength MPa	65–75	70–85	60–75	70–80	80–150
Elongation %	2–4	3–5	2.5–4.0	5–6	1–8

Source: Reprinted from resin supplier's technical literature.

There are also the component manufacturing requirements which are:

1. Shelf life, say 3 months at −18°C, which would probably mean a latent catalyst system with epoxides.
2. Sufficient tack of prepreg for individual plies to adhere to one another, though too much tack is undesirable.
3. Low volatile content to minimize voids. Volatiles can originate from products of a condensation reaction, or from occluded solvent perhaps introduced to improve tack.
4. Adequate but controlled flow under pressure.
5. Easy and economic pressure–temperature cycle with postcuring out of the mold.
6. Freedom from health hazards, particularly relevant with di-amine curing agents.

13.1.2 Phenolic resins [3,4]

Phenolic resins are produced by the reaction of phenols with HCHO. Two types of resin are commercially available:

1. A resol resin formed by reacting a molar excess of HCHO in the presence of an alkaline catalyst followed by further condensation to form ringed systems, stopping the reaction when the desired molecular weight distribution has been achieved. Control of viscosity is achieved by regulating the degree of condensation. Reactivity is believed to be most influenced by the level of methylol phenol groups (—CH_2OH—). The more reactive the resin is to acid catalysis, shorter will be the cure time. If a resol resin is preheated (120–180°C) to reduce the viscosity, the resin will concurrently advance its state of cure or, alternatively a cure can be achieved by using a strong acid catalyst at ambient temperature. A typical acid catalyst is 4–12.5% of *p*-toluene sulfonic acid.

If NaOH is used to provide the alkaline environment, the resol is water soluble, whereas if NH_3 is used, the resol is soluble in an alcohol, such as ethyl or isopropyl alcohol, generally mixed with about 10% water [3]. The alcohol soluble resins can be used for prepreg with a shelf-life of 1 month at room temperature extending, to about 1 year when stored at −20°C. When the prepreg has been made, the solvent has to be removed from the resin without curing the resin, or advancing it to a stage where the melt viscosity is too high to permit adequate resin flow during the final processing. Cure is effected at 130–140°C for about 1 h.

Some phenolic resins are brittle and a number of prepreg formulations are available which have been blended with thermoplastics (e.g., methylated nylon) to confer improved toughness. Tack can be introduced by incorporating polyvinyl butyral (Pilioform BL24), which will also increase the toughness. The structure of an idealized resol is shown:

2. The second type of phenolics are the novalac resins formed by using a molar excess of phenol and an acid catalyst. The simplest form of a novalac resin is Bisphenol F:

Novalac resins cannot condense without the addition of a hardening agent, such as hexamine (hexamethylenetetramine), which will form HCHO:

When a suitable catalyst is employed to suppress the alcohol–epoxy reaction, the novalac resin can react with any epoxy groups through the existing phenolic —OH, forming epoxy-novalacs, e.g., Ciba Araldite EPN1138SP (formerly LY558) and Dow DEN438. The high molecular weight novalac resins are solid materials and are added to epoxies to improve high temperature strength. Lower molecular weight versions are highly viscous at room temperature and can be blended with a diglycidyl ether of bisphenol A (DGEBA).

Phenolic resins cure by a condensation reaction, evolving H_2O and care must be taken to release this water (actually as steam) during the cure cycle, or use high pressures to compress the voids, as otherwise, blistering will occur.

It is interesting to note that if a short term exposure to a hot environment is required, (e.g. a missile part with operating conditions such as 15 min exposure at 400°C, or 1 min at 540°C), a carbon fiber phenolic composite would be the preferred choice, since phenolic resins would have higher thermal/oxidative stabilities than, say PMR-15 or other polyimides and would be cheaper and easier to process.

13.1.3 Polyester resins [3–7]

Unsaturated polyester resins are the product of an esterification reaction between dibasic organic acids (or anhydrides) and a dihydric alcohol (glycol) with the formation of H_2O. One of these groups, generally the acid, contributes the unsaturation to enable crosslinking to take place in the final cure. The most commonly used unsaturated acid is maleic acid, generally used in the anhydride form, since it reacts faster than the acid and releases one less molecule of H_2O during raw material processing:

$$\begin{array}{ccc} HC-COOH & HC-COOH & HC-C=O \\ | & | & | \\ HC-COOH & HOOCC & HC-C=O \end{array}$$

Maleic acid   Fumaric acid   Maleic anhydride
(*cis*-isomer)   (*trans*-isomer)

During esterification, maleic acid readily isomerizes, when reacting with ethylene glycol, into the *trans*-isomer (fumaric acid), with a conversion of about 95%. Hence in this instance, there would be no advantage using fumaric acid in place of maleic anhydride, unless 100% of the *trans*-configuration is required for some specific purpose.

The product is dissolved in a monomer containing vinyl unsaturation such as styrene, with its pendant vinyl group.

$$C_6H_5-CH=CH_2$$

The monomer must be non-aromatic, otherwise crosslinking will not occur. The polyester and reactive diluent crosslink with heat and/or free radical initiation into a solid non-melting network.

Saturated acids can also be used, which can contain an aromatic nucleus and have no pendant double bonds that would react with a peroxide catalyst. Phthalic acid exists in three isomeric forms and one of the isomers, orthophthalic acid, can form an anhydride and is widely used in this form:

Orthophthalic acid   Orthophthalic anhydride   Isophthalic acid   Terephthalic acid
                      (Phthalic anhydride)

Polyester resins derived from orthophthalic anhydride have good all-round properties, whereas isophthalic acid gives a tougher resin with some improvement in chemical resistance. Resins derived from terephthalic acid have properties similar to their isophthalic counterparts, but with a slightly greater heat distortion temperature (HDT).

Ethylene glycol ($HO-CH_2CH_2-OH$) gives a product that tends to crystallize out. Therefore, propylene glycol (propane-1,2-diol) ($HO-CH(CH_3)-CH_2-OH$) is used for most polyester resins. Improved chemical resistance can be achieved using bisphenol A, which may be pre-reacted with ethylene oxide or propylene oxide, or a hydrogenated version of bisphenol A, but all these products require heat curing to obtain maximum chemical resistance.

Bisphenol A
[2,2-bis(4-hydroxyphenyl)propane]
reacted with propylene oxide

Hydrogenated Bisphenol A
[2,2-bis(4-hydroxycyclohexyl)propane]

Flame retardancy can be achieved by using a chlorinated or brominated dibasic acid (or anhydride), such as HET anhydride (hexachloro-endo-methylene-tetrahydrophthalic anhydride), also known as chlorendic anhydride, which contains some 55% of chlorine.

Styrene monomer is extremely volatile and is affected by currents of air in the workshop, producing areas on the work piece that are starved of styrene, as well as contaminating the atmosphere. The level of styrene fumes in the workshop atmosphere has to be very strictly controlled and the workshop should be free from draughts. Vinyl toluene and diallyl phthalate are less volatile, but are more expensive.

Vinyl toluene

Diallyl phthalate

Diallyl phthalate is used to make polyester prepreg, but is difficult to cure at room temperature and hence is restricted to heat cure systems.

Gel coats are used to confer added protection and improve the surface finish. About 0.5% paraffin wax can be added, which will migrate to the surface during the initial stages of cure, forming a protective layer, preventing air inhibition—a condition caused by O_2 inhibiting the resin cure, leaving a film of tacky uncured resin on the surface, which forms unsightly white blotches, when it comes into contact with water. Such protective wax films can also be used to help limit styrene emissions during the cure cycle.

Incorporating some 10–20% of a thermoplastic polymer such as polystyrene (PS), or polymethyl methacrylate (PMMA), will produce low shrinkage resins requiring heat curing.

To prevent self polymerization of an uncatalyzed polyester resin during storage, an inhibitor is added, which donates hydrogen atoms to absorb any free radicals that might be formed. Typical inhibitors would be 100–300 ppm of hydroquinone, or *t*-butyl catechol:

If a further 100 ppm is added, the inhibitor will then increase the gel time and extend the pot life.

Curing polyester resins involves the transformation of the polyester from a liquid to a solid state, taking place on the addition of a catalyst, usually organic peroxide (Table 13.2). To speed up the process, the catalyst has to be activated, either by heating or, by the addition of an accelerator. The crosslinking reaction is exothermic and, since polyester resins are

Table 13.2 List of common peroxide catalysts for curing polyester systems

Type	Structure	Activity	% Active oxygen	Molecular weight
Methylethylketone peroxide	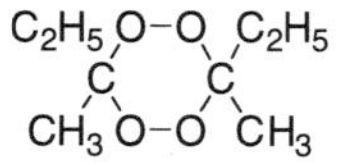 (structure shown)	Low, medium and high	8.5–9.9	176
Benzoyl peroxide	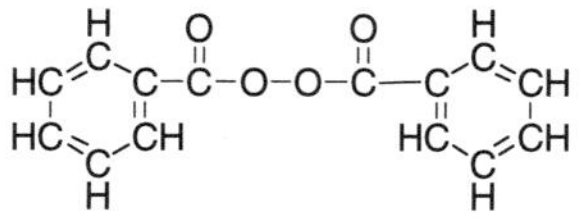 (structure shown)	Medium	3.3	242
Acetyl acetone peroxide	$CH_3-\overset{O}{\underset{\|\|}{C}}-CH_2-\overset{O}{\underset{\|\|}{C}}-CH_3$	Fast	4.1	132
Cyclohexanone peroxide	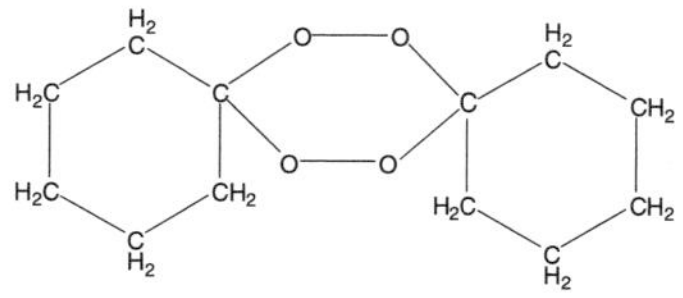 (structure shown)	High	5.1	228
Cumene hydroperoxide	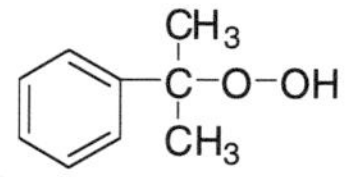 (structure shown)	Special for vinyl ester resins	4.7	152

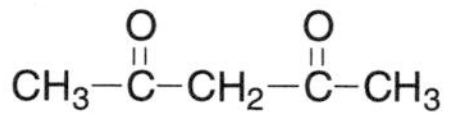

Source: Reprinted from supplier's technical literature.

100% reactive, no byproducts are evolved during the curing process. Precautions must be taken, however, to avoid monomer loss by evaporation and draughts:

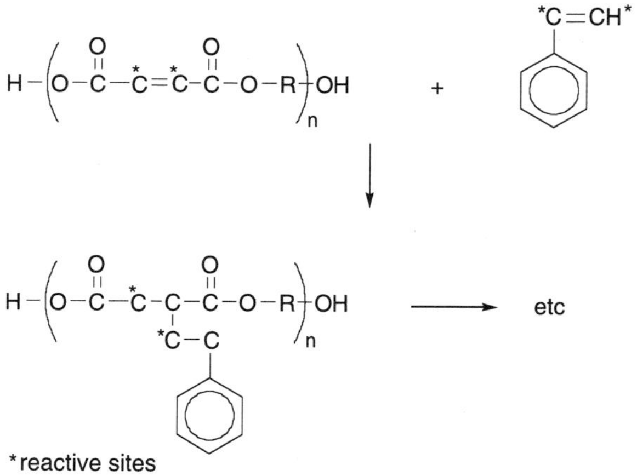

$$H-\left(O-\overset{O}{\underset{\|\|}{C}}-\overset{*}{C}=\overset{*}{C}-\overset{O}{\underset{\|\|}{C}}-O-R\right)_n OH \quad + \quad \overset{*C=CH^*}{\underset{\bigcirc}{}}$$

$$\downarrow$$

$$H-\left(O-\overset{O}{\underset{\|\|}{C}}-\overset{*}{C}-C-\overset{O}{\underset{\|\|}{C}}-O-R\right)_n OH \quad \longrightarrow \quad etc$$
$$\underset{\bigcirc}{\overset{*C-C}{}}$$

*reactive sites

An organic Co compound is generally used as an accelerator and cobalt octoate (10% Co content), or cobalt naphthenate (6% Co content) are the most commonly marketed products. Polyester resins can also be supplied pre-accelerated. The accelerator is always added first and used in conjunction with a peroxide catalyst for room temperature curing. Organic peroxides are hazardous materials and must be handled strictly in accordance with the recommended safety precautions.

The most widely used peroxide catalyst is methyl ethyl ketone peroxide (MEKP), which is really a mix of several peroxides and is available for various activity levels.

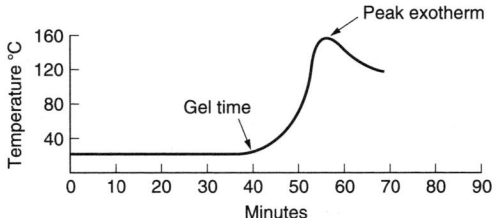

Figure 13.1 Typical gel time curve for a polyester resin. *Source:* Reprinted from Scott Bader Company Limited Publication, *Crystic Polyester Handbook*, Wellingborough, 1994.

Acetyl acetone peroxide provides a faster cure but the reaction is strongly exothermic and is not recommended for thick laminates.

For an elevated temperature cure, benzoyl peroxide (BP) is used with a tertiary amine accelerator, such as dimethylamine (DMA), promoting a more rapid cure once the reaction has started. Post curing will be required.

Cyclohexanone peroxide (CHP) is also a mix of peroxides, with a lower peak exotherm than MEKP and is useful for curing thicker sections without stress or crack formation.

Cumene hydroperoxide (CUHP) was especially developed for vinyl ester resins. It can be used as a blend with BP, when at 70°C, the BP starts the cure, with the hydroperoxide taking over when the temperature reaches 100°C. This catalyst is particularly suitable for thick parts.

The time period for which a polyester resin system remains in liquid state after the addition of the catalyst and accelerator, is termed the gel time (Figure 13.1) and is controlled by the relative amounts of catalyst and accelerator. It should be noted that a certain minimum of catalyst must always be added to neutralize the inhibitor and to achieve a complete cure. Polymerization starts when the retardant has been used up by the peroxide. Ambient temperature significantly affects the gel time and a rise of some 5–8°C can halve the gel time. Gel time should be controlled only by variation of the accelerator content and not by varying the catalyst content.

Post curing at room temperature could take several weeks and this can be reduced by stabilizing at room temperature for two days and then applying an elevated temperature cure for, say 3 h at 80°C. The degree of cure is best checked with a Barcol Hardness Impressor and the cured resin should not be softened by acetone.

13.1.4 Epoxy vinyl ester resins [3–5,8]

Dow Plastics manufactures a range of Derakane epoxy vinyl ester resins and the simplest is made by reacting low molecular weight epoxides with either AA or MAA to produce a chain structure:

$$CH_2=CH-\overset{O}{\underset{\|}{C}}-\left[O-CH_2-\underset{OH}{\underset{|}{CH}}-CH_2-O-\underset{}{\bigcirc}-\underset{CH_3}{\overset{CH_3}{\underset{|}{C}}}-\underset{}{\bigcirc}-O\right]_n-CH_2-\underset{OH}{\underset{|}{CH}}-CH_2-O-\overset{O}{\underset{\|}{C}}-CH=CH_2$$

The polymers have similar characteristics to epoxide resins, but because they have terminal $CH_2=C-$ groups, the products can be crosslinked with vinyl monomers like styrene that are activated with peroxide curing systems such as:

1. MEKP/cobalt octoate/dimethylamine
2. BP/diethylaniline
3. CUHP/cobalt octoate/DMA.

A conventional bisphenol-A polyester resin has ester links and double bonds throughout the molecule, whereas the vinyl ester resin has only two of each per molecule situated at either end of the chain. This reduced number of ester links accounts for the improved chemical resistance, whilst the reduced number of double bonds produces a more resilient structure.

Derakane 411 is the Dow product made from bisphenol-A and the presence of the epoxy group confers additional resistance to alkalies. It can be pre-reacted with epoxy–novalacs forming (Derakane 470) with a higher temperature performance and improved resistance to acids.

A brominated bisphenol-A (Derakane510) confers the maximum degree of fire retardancy, whilst a rubber modified version (Derakane8084) imparts greater toughness and higher elongation.

13.1.5 Epoxide resins [3–5 and 9–13]

Some general articles on the properties of carbon fiber reinforced epoxies have been published in the references given [14–16].

13.1.5.1 Bisphenol resins

Although, strictly speaking, the correct term is epoxide the term epoxy is now in common use.

Epoxy resins contain two or more 1,2-epoxide (oxirane) groups

usually in the form of terminal glycidyl groups

Epoxy resins are generally made from epichlorhydrin (ECH) and diphenylolpropane (DPP), also known as bisphenol-A (BPA):

POLYMER MATRICES FOR CARBON FIBER COMPOSITES

Table 13.3 Typical properties of selected epoxy resins based on bisphenol-A and epichlorhydrin

Shell grade	Form	Viscosity 25°C	Density g cm^{-3}	Epoxy molar mass	Mean molecular weight	Epoxy group content m mol kg^{-1}
828	Liquid	12–14 Pa s	1.16	182–194	380	5260–5420
834	Semi-solid	2.1–2.3^a mPa s	1.18	225–280	470	3800–4250
1001	Solid	6.3–7.9^a mPa s	1.19	450–500	900	2000–2220
1010	Solid	200–300^a mPa s	1.20	4000–6000	–	210–330

^{a}Viscosity of a 40% m/m solution in MEK.
Source: Reprinted from Shell technical literature.

A polymerization reaction occurs and HCl is eliminated giving a series of epoxies where n varies in the range 0–12.

The simplest resin, when $n = 0$, is DGEBA. As n increases, the molecular weight increases from about 340 to 4000, varying from viscous liquids to high melting point solids (Table 13.3).

Instead of measuring the molecular weight, it is normal to measure the epoxide equivalent weight or Epoxy Molar Mass (EMM), which is the equivalent weight in grams of resin containing one epoxide group. Alternatively, it can be expressed as epoxide content, which is the number of equivalents per kilogram of resin.

Commercial resins are not entirely di-epoxides and do contain some free —OH groups due to side reactions.

There are many suppliers of epoxy resin and they market equivalent grades One such example is the standard DGEBA resin, supplied by Ciba Specialty Chemicals Araldite MY750 (6010 in the USA), Dow DER 331 and Shell Epikote 828 (Epon 828 in the USA).

A flame retardant version of bisphenol-A/ECH is available called tetrabromobisphenol-A (TBBA) and the diglycidyl ether is:

13.1.5.2 Novalac resins

A novalac resin is prepared by reacting a phenol with HCHO:

The simplest novalac resin is bisphenol-F, where $n = 0$:

Epoxy novalac resins are prepared by reacting ECH with a novalac resin:

The conventional way of producing epoxy resins with BPA or bisphenol-F produces two epoxy functional systems, one at each end of the molecule. The epoxy novalac, however, yields multifunctional epoxy products (generally 2.5–6.0) (Table 13.4). This type of product, with a high density of reactive groups can give thermosetting systems with extreme toughness, high reactivity and good performance at elevated temperatures, albeit with the introduction of some brittleness due to their highly crosslinked structure. A high temperature cure is necessary to develop the properties of an epoxy novalac to its fullest extent.

Novalac resins can be produced from alkyl phenols, aryl phenols and polyhydroxy phenols to obtain special properties and Table 13.5 lists resins derived from an o-cresolformaldehyde and ECH.

Table 13.4 Properties of Dow Epoxy Novalac resins

Dow grade	Resin form	Functionality	Epoxy molar mass	Viscosity @52°C (mPa s)
DEN 431	Lowest viscosity	2.8	172–179	1100–1700
DEN 438	Highest viscosity	3.6	176–181	34,000–40,000

Source: Reprinted from Dow technical literature.

Table 13.5 Properties of Ciba o-cresolformaldehyde novalac/epichlorhydrin resins

Ciba Grade	Form	Melting point °C	Epoxy molar mass	Mean molecular weight	Functionality	
					Ether type (a)	Epoxide type (b)
ECN 1235	Semi-solid	35	200	540	0.3	1.7(2.7)
ECN 1273	Solid	73	225	1080	1.2	3.8(4.8)
ECN 1280	Solid	80	230	1170	–	4.1
ECN 1299	Solid	99	235	1270	1.6	4.4(5.4)

Source: Reprinted from Ciba technical literature.

R represents chlorhydrins, glycols and/or polymeric ethers. Functionality $a=$ ether type and $b=$ epoxide type, is given in Table 13.5.

Tactix 556, marketed by Ciba Specialty Chemicals is a hydrocarbon epoxy novalac resin having a dicyclopentadiene backbone with ability to retain properties under conditions of moisture and elevated temperature. Tactix 556 can be cured for 3 h at 177°C using 27.6 phr of DDS and then postcured for 2 h at 232°C giving a T_g of 210°C.

13.1.5.3 Trifunctional resins

Tactix 742 is a tris-(hydroxyphenyl) methane based epoxy, which can be cured for 3 h at 177°C using 38 phr DDS and postcured for 2 h at 250°C with a T_g of 299°C to give composites ideal for components near high heat zones such as aircraft engine nacelles.

13.1.5.4 Tetrafunctional resins

Shell Epikote 1031 is a tetrafunctional resin and with appropriate curing agents (e.g., DDS/BF$_3$MEA or MNA/BDMA) produces composites with excellent thermal stability and has been used for filament winding rocket motor cases.

Another very popular tetrafunctional resin, tetraglycidylmethylenedianiline (TGMDA) (e.g., Ciba's MY720), has found widespread use in many prepreg systems used with carbon fiber composites and is available in different grades of purity.

A highly functional resin useful for prepreg formulation is Celanese SU8:

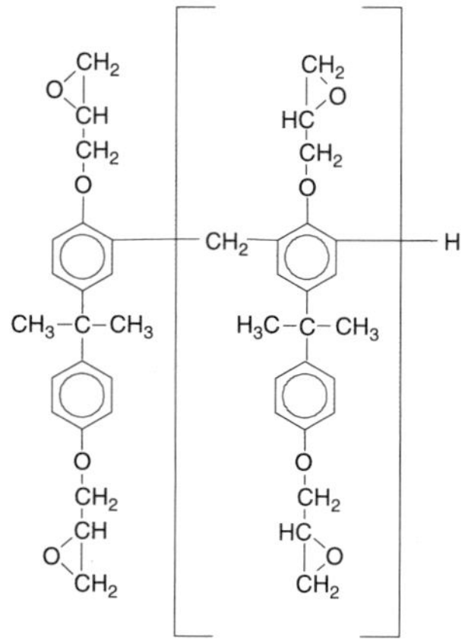

13.1.5.5 Cycloaliphatic resins

Cycloaliphatic epoxies are prepared by converting tetrahydrobenzaldehyde by a Tisotenko condensation to form as an olefin which is epoxidised with a peracid.

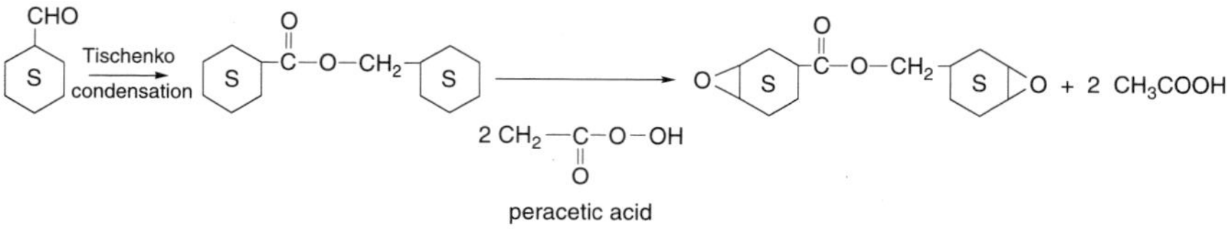

The product, 3,4-cyclohexylmethyl-3,4-epoxycyclohexanecarboxylate, marketed as Ciba CY179, has a very low viscosity and can be used as a diluent, since it is an excellent solvent for other resin families and also is a good solvent for the curing agent 4,4′-diaminodiphenyl-sulfone (DDS). The absence of an aromatic ring is noticeable. Unfortunately, the range of cycloaliphatic resins on the market has been almost terminated due to difficulties in their manufacture.

13.1.5.6 New developments

Sastri and Armistead have described phthalonitrile–carbon fiber composites [17].

An interesting new development uses liquid crystal polymers (LCPs) to produce materials with exceptional physical and mechanical properties. Generally the LCP resin comprises of a polymer chain with structural units (mesogenic groups) which can be incorporated into the polymer backbone, or incorporated as a pendant group, or both. One approach is to use the epoxy resin system [18]:

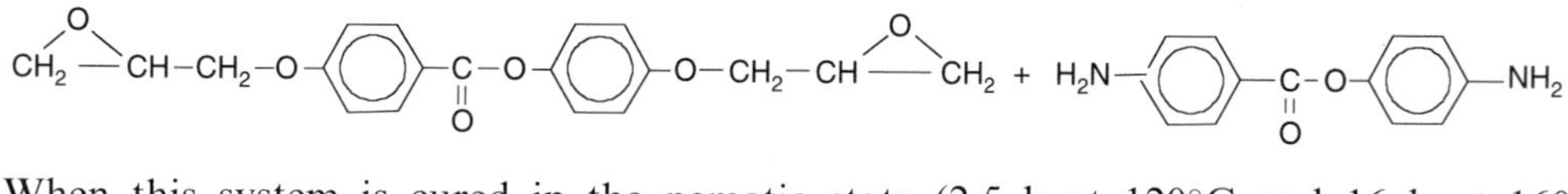

When this system is cured in the nematic state (2.5 h at 120°C and 16 h at 160°C) it provides a cured resin with superior properties to the resin when cured under normal conditions, showing a dramatic improvement of the strain to failure and impact strength.

LCPs have been grafted as a side chain to epoxy matrices to facilitate coupling with carbon fibers [19].

13.1.5.7 Epoxy diluents

The viscosity of a resin system can be reduced by the addition of a diluent, which can be non-reactive (e.g. dibutyl phthalate, benzyl alcohol and furfuryl alcohol) or reactive (a low molecular weight epoxide, triphenyl phosphite or χ-butyrolactone). The χ-butyrolactone can, for example, act as a solvent for a catalyst system used in a solventless system and take part in the reaction. Diluents tend to be rather volatile and could be partially lost in cure systems operating under a vacuum. Flexibilizers can impart toughness but at the expense of mechanical properties, chemical resistance and heat resistance.

Reactive diluents are generally monoglycidyl or diglycidyl ethers. The monoglycidyl ethers have only one epoxide group per molecule and will therefore reduce the functionality and consequently, reduce the crosslink density.

Dow has a series of diluents (e.g., Dow 736 & 732) based on the polyglycols of propylene oxide:

$$CH_2-CH-CH_2-O-[CH(CH_3)-CH_2-O]_n-CH_2-CH-CH_2$$ where $n = 3$ to 9

Two other useful diluents are allyl glycidylether (AGE):

$$CH_2=CH-CH_2-O-CH_2-CH-CH_2$$

and vinyl cyclohexenedioxide (VCHD) marketed as Union Carbide 4206 and Ciba RD-4:

13.1.5.8 Characterization of epoxy resins

Gel permeation chromatograhy (GPC) is used to separate the various components of epoxy resins and these separated components are treated with various chemical reagents and resubmitted for GPC or high performance liquid chromatography (HPLC). Typical reagents used are:

1. Concentrated HCl to convert epoxy groups to chlorhydrin units.
2. Solid NaOH to convert chlorhydrin groups to epoxy groups.
3. Phenol or N-methylaniline to open epoxy rings.
4. KMnO$_4$ to oxidize —CH$_2$— and —OCH$_2$CH(OH)CH$_2$O— bridging units.

13.1.5.9 Curing epoxide resins

Curing of an epoxy resin to give a crosslinked, three dimensional, infusible structure is brought about by using a hardener or curing agent, sometimes with an accelerator to open the epoxide ring [20]. Homopolymerization can also take place through the —OH groups present in the epoxy molecule, reacting with the epoxide groups.

The reaction of epoxides is highly exothermic and best results are obtained from a two stage cure. In the first stage, the small molecules are combined with larger units with a low

Table 13.6 Curing agents for epoxy resins

Name	Molecular weight	Number of active H	Amine H equivalent weight	Calculated phr	Form
DETA	103	5	20.6	10.9	Liquid ($\rho = 0.95$ g cm^{-3})
TETA	146	6	24.4	12.9	Liquid ($\rho = 0.98$ g cm^{-3})
DDM (MDA)	198	4	49.5	26.2	Solid (mp 89°C)
DDS (DADPS)	256	4	57	30	Solid (mp 176°C)
MPD (MPDA)	108	4	27	14.3	Solid (mp 62°C)
HHPA (HPA)	154	–	–	80	Solid (mp 38°C)
MeTHPA	166	–	–	80	Liquid ($\rho = 1.21$ g cm^{-3}) (mp 4°C)
NMA (MNA)	180	–	–	90	Liquid (mp < 12°C)

Source: Reprinted from Dow and Shell technical literature.

degree of crosslinking (termed a B-stage resin), followed by a second stage where the temperature is raised and the clusters are crosslinked to form one big molecule.

There are many curing agents, also termed hardeners, in commercial use (Table 13.6) and when selecting a hardener for a particular end-use, the following points should be considered [2]:

1. Temperature of the cure
2. Compatibility of the hardener with the resin
3. Pot life of the hardener/resin system once mixed
4. Volatility of the hardener
5. Physical and chemical properties of the cured resin
6. Cost

The following examples briefly typify the various categories of hardeners:

1. Aliphatic amine curing agents generate a high exotherm and have a short pot life, but can be cured at room temperature and do not require an intense post cure e.g.

    diethylene triamine (DTA) $NH_2(CH_2)_2NH(CH_2)_2NH_2$
    triethylene tetramine (TETA) $NH_2[(CH_2)_2NH]_2(CH_2)_2NH_2$

    The mode of cure follows this typical reaction sequence:

    $$RNH_2 + CH_2\text{—}CH\text{—}R' \longrightarrow RN\text{—}CH_2\text{—}CHR' \longrightarrow CH_2\text{—}CH\text{—}R' \longrightarrow RN\begin{matrix}CH_2\text{—}CH\text{—}R'\\CH_2\text{—}CO\text{—}R'\end{matrix}$$

    Primary amine   Secondary amine   Completely crosslinked system

    where R = aliphatic, cycloaliphatic, or aromatic.

    The secondary alcohols, although not as active as the epoxide groups, can combine with fatty acids, anhydrides, phenolic and amino resins and epoxide groups in other molecules.

2. Aromatic curing agents have better heat and chemical resistance than the aliphatic amines and, likewise, produce high exotherms. Hence two stage cures are advantageous with a full cure achieved with heat.

a. 4,4′-diaminodiphenylmethane (DDM), also called 4,4′-methylene dianiline (MDA)

$$NH_2-C_6H_4-CH_2-C_6H_4-NH_2$$

One advantage of DDM is that in the presence of phenol or salicylic acid, as an accelerator, it will cure at 0°C. However, for health reasons, there is now a move to replace all systems using this amine with an alternative product.

b. 4,4′-diaminodiphenylsulfone (DDS or DADPS)

$$NH_2-C_6H_4-SO_2-C_6H_4-NH_2$$

DDS can achieve a high HDT, with good retention of strength properties at elevated temperatures. Post cure times can be reduced by using a BF_3MEA catalyst at the expense of a shortened pot life. When used without a catalyst, it is usually used at about 10% excess.

c. m-phenylene diamine (MPD)

$$C_6H_4(NH_2)_2\ \text{(1,3-)}$$

MPD is rather unpleasant to handle, badly staining the skin and decomposing during storage to evolve NH_3. MPD finds limited use, but epoxy composites cured with this hardener do have good chemical resistance.

d. 1-(o-tolyl) biguanide (Shell DX 147)

$$o\text{-}CH_3\text{-}C_6H_4\text{-}NH\text{-}C(=NH)\text{-}NH\text{-}C(=NH)\text{-}NH_2$$

This curing agent can be used at 8–10 phr or 4 phr with 0.3 phr BDMA.

3. Proprietary eutectic mixtures of amines are available that are liquid at room temperature and may be regarded as supercooled liquids. Their major advantage is the relative ease of handling a liquid, permitting mixing without difficulty and extending pot lives because of lower mixing temperature (e.g. Uni Royal's Tonox 60/40, which is 60% crude DDM + 40% MPD).

4. Anhydride curing agents are widely used for curing epoxy resins and are also available as liquids, or low melting point, eutectic anhydride mixtures. Solid anhydrides require heat and considerable mixing to ensure uniform dispersion in the liquid epoxy resin. During cure, large exotherms are not experienced, though a long cure at elevated temperature is involved, which can be reduced by using 1–5 phr of a tertiary amine accelerator such as benzyldimethylamine (BDMA).

$$C_6H_5\text{-}CH_2\text{-}N(CH_3)_2$$

A typical reaction sequence using a tertiary amine as a catalyst is:

$$R_3N + CH_2\overset{O}{-}CH\!\sim \longrightarrow R_3\overset{+}{N}-CH_2-\overset{O^-}{CH}\!\sim \xrightarrow{CH_2\overset{O}{-}CH\sim} R_3\overset{+}{N}-CH_2-CH\!\sim$$
$$\qquad\qquad\qquad\qquad\qquad\qquad\qquad\qquad\qquad\qquad\qquad\qquad\qquad |$$
$$\qquad\qquad\qquad\qquad\qquad\qquad\qquad\qquad\qquad\qquad\qquad\qquad\qquad O-CH_2-CH\!\sim$$
$$\qquad\qquad\qquad\qquad\qquad\qquad\qquad\qquad\qquad\qquad\qquad\qquad\qquad\qquad\qquad\ \ |$$
$$\qquad\qquad\qquad\qquad\qquad\qquad\qquad\qquad\qquad\qquad\qquad\qquad\qquad\qquad\qquad\ \ O^-$$

Another popular accelerator is tris-2,4,6-dimethylaminomethylphenol, marketed by Anchor Chemical as K54 or DMP-30 from Rohm & Haas. It is also used in the form of the hexoate salt, such as Anchor Chemical's K61B, the tris-2-ethylhexoate salt of K54. In the salt form, higher percentages are required, but provide a relatively long pot life. The curing mechanism is believed to be the removal of fatty acid by esterification, with some of the epoxy groups and the tertiary amines thus liberated initiating the polymerization of the epoxy resin.

The curing mechanism of an epoxy resin with an anhydride involves a number of reactions taking place simultaneously. The principal reactions are:

II(I) Opening of the anhydride ring by an —OH group forming a monoester:

I(II) Reaction of the nascent carboxyl group of the monoester formed in (I) with an epoxide group to give a di-ester and an —OH group:

(III) Reaction of epoxide groups with —OH groups to give ether linkages and new —OH groups:

5. Dicyandiamide (dicy) is not soluble in resin systems at low temperatures and is regarded as a latent catalyst, imparting a long shelf life to prepreg systems at ambient temperatures. It exhibits tautomerism

$$NH_2-C(NH_2)=N-C\equiv N \rightleftharpoons NH_2-C(=NH)-NH-C\equiv N$$

Dicy is soluble in 2-methoxyethanol, DMF and an acetone/H_2O mix. In solventless systems, pulverized dicy is used, which is well dispersed into the resin system using either a ball or a 3-roll mill. If the dicy is in suspension in a prepreg system and the particle size is too great, it is possible that the carbon fiber reinforcement will filter the dicy out from the resin matrix, producing dicy starved areas, and result in a partial cure. Dicy can be dissolved prior to use by a technique called spiking, when the dicy/resin mix is held at 120°C for one hour. If lower temperatures are used, say 65°C, the dicy becomes encapsulated within the resin and subsequent heating at 120°C will not allow the dicy to diffuse into the resin, resulting in a product with a T_g some 50°C lower.

Although dicy is considered to be a catalyst, it starts to break down at 130°C to produce curing hardener. The early cure of the dicy/epoxy system involves reaction of the epoxy with all four of the available hydrogens on the dicy and an epoxy/epoxy reaction is catalyzed by the tertiary amine. This primary reaction is exothermic. The final cure reaction, also catalyzed by tertiary amines, is between —OH groups in the partially cured epoxy resin and dicy/cyano groups to form amino groups. It is essential that there are no thick sections in the composite, where an exotherm could produce a temperature rise above 208/210°C, which is the melting point of dicy, and lead to blistering. Degradation starts at 195°C, blisters form at 210°C and serious decomposition takes place at 215°C due to the dicy exploding.

Normally 4–10% (i.e. 4.3–11.1 phr) of dicy is used, but it can be accelerated to speed the cure using an accelerator such as BDMA, or a substituted dicy made by the reaction of an aromatic amine with dicy to improve its solubility in epoxy resins.

6. Substituted ureas can be used for promoting a synergistic effect when curing epoxy resins with dicy. Earlier, 3-(p-chlorophenyl)-1,1-dimethylurea (Monuron made by DuPont) was popular [22,23], but it has now been withdrawn due to health hazards.

$$Cl-\text{C}_6\text{H}_4-NH-\overset{O}{\underset{\|}{C}}-N(CH_3)_2$$

However, 3-(3,4-dichlorophenyl)1,1-dimethylurea (Diuron or DCMU) from Hodogaya is now used:

$$Cl_2\text{C}_6\text{H}_3-NH-\overset{O}{\underset{\|}{C}}-N(CH_3)_2$$

This is used as 5–10 phr and the probable reaction scheme is [24]:
First step:

$$\text{Cl}_2\text{C}_6\text{H}_3-NH-\overset{O}{\underset{\|}{C}}-N(CH_3)_2 + 2\ CH_2-CH-R \longrightarrow (I) + (II)$$

where (I) is the cyclic carbamate and (II) is $(CH_3)_2N-CH_2-CH(OH)-R$.

The anionic polymerization of the oxirane ring which can be catalyzed by (I) and co-catalyzed by (II).

A schematic representation of the effect of changing the ratio of a substituted urea and dicy on the T_g and gel time of an epoxy system is shown in Figure 13.2.

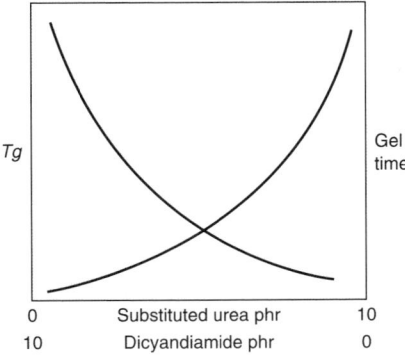

Figure 13.2 Schematic representation of the effect of changing the ratio of a substituted urea and dicy on the T_g and gel time of an epoxy resin.

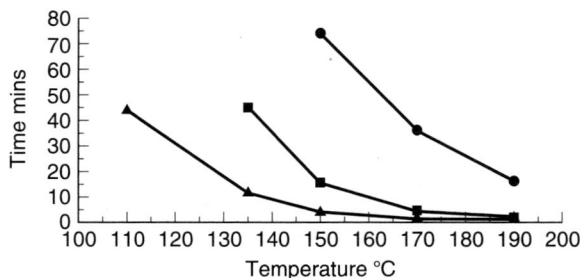

Figure 13.3 Effect of the concentration of BF$_3$MEA on the gel time of 828/DDS resin system at different temperatures. ● 828/36 phr DDS ■ 828/20 phr DDS/1.5 phr BF$_3$MEA (▲ 828/20 phr DDS/2.0 phr BF$_3$MEA.

8. A number of different BF$_3$ complexes are manufactured by Harshaw-Filtrol, including complexes containing monoethylamine, piperidine, benzylamine, aniline and triethanolamine, whilst Anchor has additional complexes based on chloroaniline, isopropylamine and di-N-butylamine. The monoethylamine complex is the most widely used.

$$F-\underset{\underset{F}{|}}{\overset{\overset{F}{|}}{B}}=NH_2CH_2CH_3$$

It is a Lewis acid type catalyst, which initiates the homopolymerization of epoxy resins that predominantly form ether linkages [25]. The complexes are normally used at 3–5 phr and when used as an accelerator in systems involving anhydrides or amines, 1 phr or less. At room temperature, it is stable when mixed into epoxies, resin soluble salt that has essentially no catalytic activity. Due to this latency, most resin systems incorporating BF$_3$MEA can be stored at room temperature for several months without any appreciable change in viscosity. The effect of BF$_3$MEA on the gel time of an 828/DDS resin system at different temperatures is shown in Figure 13.3.

BF$_3$MEA is extremely water soluble and will readily absorb water on exposure to moist air, slowly forming ethyl ammonium fluoborate and hydroxyfluoborates. Hence, it is advisable to store in small containers, which will need to be opened only once or twice. Butanone-2 (MEK) is a preferred solvent since acetone generally contains water.

An interesting BF$_3$ complex made by Allied Chemicals is a BF$_3$–diethylether complex, which is a flammable liquid.

$$F-\underset{\underset{F}{|}}{\overset{\overset{F}{|}}{B}}\rightarrow O\overset{C_2H_5}{\underset{C_2H_5}{}}$$

It provides a convenient means of adding BF$_3$ and controlling its release, as long as diethyl ether is not detrimental to the systems involved.

The latency of the BF$_3$MEA can be improved by replacing with about 2 phr of BCl$_3$octyldimethylamine complex, marketed by Ciba as XU213.

$$\underset{CH_3}{\overset{CH_3}{\diagdown}}N\underset{BCl_3}{\overset{C_8H_{17}}{\diagup}}$$

The activation temperature is 80°C and it will cure at 100°C, but requires a more prolonged cure than BF$_3$MEA (e.g. 2 h at 150°C followed by 2 h at 170°C). Stauffer also market BCl$_3$trimethylamine (BTMA).

9. There are a number of imadazoles that are useful curing agents.

$$\text{1-methyl imidazole (Ciba's DY 070)} \quad \text{Houdry's EMI 24 (2-ethyl-4-methyl imidazole)}$$

EMI 24 is a liquid and can be used at 4–8 phr for moderate to high temperature curing schedules. It is also used as an accelerator for epoxy–anhydride mixtures to provide a long pot life and good elevated temperature properties. It is particularly suitable for filament winding applications.

10. Isophorone diamine (IPD), marketed as Shell DX-139, is a liquid, which will cure satisfactorily at elevated temperature at a dosage of 22 phr, but requires an accelerator for room temperature curing.

13.1.5.10 Calculating stoichiometric ratios for epoxy resins and curing agents [13]

The optimum proportions of resin and curing agent are such that the cured material will have the highest softening temperature and the best combination of mechanical properties.

It is desirable to react the epoxy resin and curing agent at approximately stoichiometric quantities. To achieve this, it is necessary to know the amine hydrogen equivalent weight (AHEW), which is calculated as below:

$$\text{AHEW} = \frac{\text{Molecular weight of amine}}{\text{Number of active hydrogens}}$$

where the active hydrogens are those directly bonded to the nitrogens.

Next, the phr (parts per hundred of resin) of the amine is determined from:

$$\text{phr of amine} = \frac{\text{AHEW} \times 100}{\text{Epoxide equivalent weight of resin}}$$

If there is a blend of resins, then the quantity of curing agent is worked out separately for each constituent part and then added together.

In amine cured systems, the hardener is used at near stoichiometric quantities, but since a tertiary amine has a weak catalytic effect, slightly less than theoretical amounts should be used. In formulations containing anhydrides, less than stoichiometric quantities of curing agent are used, since acid formation is an intermediate stage of the cure that will produce some epoxy homopolymerization. With strong dibasic chlorinated anhydrides, like chlorendic anhydride, the optimum ratio is about 0.50, whilst with the weaker anhydrides, like hexahydrophthalic anhydride, the optimum ratio is in the range 0.70–0.85.

It is normal to determine dosage of catalysts by experiment by assuming one nitrogen atom in the catalyst to be equivalent to 6–8 epoxide groups.

13.1.6 Cyanate resins [21,26]

Cyanate resins have been extensively developed and their high temperature properties are ranked just below the bismaleimide family. Their manufacturing technology was developed by Bayer by the cyanation of phenols to give —Ar(—O—C≡N)$_n$.

Ciba AroCy L-10 is a cyanate ester monomer.

This is a low viscosity liquid, which dissolves many amorphous plastic resins, with high T_gs enabling their ready incorporation into hot melt systems. It is compatible with cyanate ester prepolymers, epoxy resins and bismaleimides.

The triazine ring is formed when the resin cures slowly at <190°C. The reaction can be speeded up by using a trimerization catalyst.

thermoplastic (polycyanurate)

Suitable catalysts are soluble coordination metal carboxylates (naphthenates, octoates) or chelates (acetyl acetonates), predissolved in a liquid alkylated phenol such as nonylphenol, which ensures good dispersion and prevents a localized high concentration of the catalyst producing an uncontrolled exothermic reaction.

Preferred catalysts can be obtained from Shepherd Chemical Co. and Amspec Specialty Chemicals and used with metal concentrations in the range 200–300 ppm of Cu^{++}, Co^{++}, Co^{+++}, Zn^{++} and Mn^{++}, or alternatively, acetyl acetonates can be used in conjunction with 1.5–6 phr nonylphenol. The acetyl acetonates are more latent than the corresponding naphthenates or octoates, and have a typical outlife of up to four weeks at room temperature. Conversion can be achieved at 177°C with a post cure of 1 h at 210°C, followed by 2 h at 250°C.

Cyanate resins can react with epoxies by three mechanisms. There are two types of homopolymerization reaction and one co-reaction involving the formation of oxazoline ring.

a). $3 \text{ R}-\text{O}-\text{C}\equiv\text{N} \xrightarrow{\text{cyclotrimerization}}$ [triazine ring structure]

b). $\text{R}-\text{O}-\text{C}\equiv\text{N} + \text{R}'-\text{CH}-\text{CH}_2 \xrightarrow{\text{co-reaction}}$ oxazoline

c). $n \text{ R}'-\text{CH}-\text{CH}_2 \xrightarrow[\text{polyetherification}]{\text{R}''\text{OH}} H\text{-}(O\text{-}CH\text{-}CH_2)_n\text{-}OR''$

Currently available products are Aro-Cy from Ciba Speciality Chemicals, Primaset PT resins (phenolic triazines via cyanation of phenolic novalacs) from Allied Signal and a cyanate ester based on dicyclopentadiene phenol novalac marketed by Dow. The PT resins have superior thermal performance with thermo-oxidative stability comparable to BMI and PMR resins, coupled with excellent fire resistance. The triazine resin is soluble in ketones, esters, THF and chlorinated solvents.

Mitsubishi Gas Chemical Company Inc. market BT Resin, a blend of a bismaleimide (BMI) and a triazine resin, cured by heating alone or with a zinc octoate, zinc octoate/triethylene diamine, or zinc octoate/N,N-dimethylbenzylamine catalyst system.

Cyanate esters have been used as a toughening network for carbon fiber composites [27] and the mechanism for load transfer discussed [28].

13.1.7 Polyimide resins [1,4,29–32]

Polyimides were introduced to provide outstanding temperature resistance with good thermal ageing properties. A scheme showing the outline history of the development of polyimide resins is given in Figure 13.4. The melting points of polyimides are high and they are difficult to process, as they react quite rapidly at the melt temperature. Moreover, they are only soluble in solvents (Table 13.7), the last traces of which be difficult to remove. They are costly and rather brittle but they do have very good high temperature performance at up to 250°C, because the imide group.

$$-\underset{\underset{O}{\|}}{C}-NH-\underset{\underset{O}{\|}}{C}-$$

This group is very stable and plays no part in the reaction (except if it is the linear thermoplastic type). The reactivity of the remaining types originates from the presence of

Condensation type **Addition type**

Linear polymides (c. 1959) **Maleimides** **Acetylene terminated polymides (ATP)**

DuPont   Pyralin
Monsanto Skybond 700 series (1964 Patent)
                                                    Rhone Poulenc ----

DuPont   NR-150A (adhesive) Kerimid 601
(1972)   NR-150B (matrix) Kerimid 353
                   NR-150B2
Upjohn   2080   Contractor: Contractor:
                                              NASA Lewis ----   US Air Force
                                              Contract with TRW   (1970)
                                              (TRW Patent 1970)   Hughes Aircraft

                                                                                                        HR 600
                                              Contractor: (1975)

                  Ciba-Geigy P13N   German Ministry: Gulf Oil Chemicals
                                P105N   Contact with
                  (Now abandoned)   Technochemie   Thermid 600
                                              Nonbornene   (Patent 1976)
                                              (PMR concept)   Sold by:
                                              PMR-15   Compimide series ----
                                              PMR II   National Starch &
                                              II-30   Technochemie   Chemical
                                              II-50   acquired by
                                              12F-71   Boots (1981)   Thermid IP 600

                                                                         Kerimid resins
                                              Acquired by   acquired from
                                              Shell (c. 1988)   Rhone Poulenc
                                              LARC-160   by Ciba Specialty
                                                                         Chemicals (c. 1990)

                                              Compimide 183   Kerimid B601
                                                          353   701-A (MDA free)
                                                          796   701-B (MDA free)
                                                       15 MRK
                                                       65 FWR Matrimid 5292

Figure 13.4 Development of polyimide systems.

Table 13.7 List of solvents used for polyimide resin systems

Solvent	Chemical name	Structure	Boiling point °C
Diglyme	2-methoxyethylether	$(CH_3OCH_2CH_2)_2O$	162
DMAc	N,N-dimethylacetamide	$CH_3CON(CH_3)_2$	165
EEA	2-ethoxyethylacetate	$HCON(CH_3)_2$	153
NMP	1-methyl-2-pyrolidone	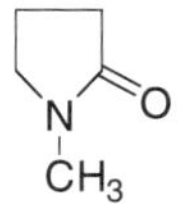 (N-methyl pyrrolidone ring structure)	156
THF	tetrahydrofuran	 (tetrahydrofuran ring structure)	66

Source: Reprinted from Aldrich handbook of fine chemicals.

other unsaturated groups in the molecule. When cured, polyimides do not generate water attracting groups and consequently, will absorb less water than epoxies. This is combined with the relatively high T_gs, which are higher hot wet T_gs as compared to epoxies.

High processing temperatures render conventional epoxy sizes unsuitable and special sizes such as silicones have been used [33–35].

There are two classes of polyimides:

1. Condensation type—such as linear thermoplastic polyimides (NR150)
2. Addition type—such as
   a. Norbornene (nadic) reactive polyimides (P13N, PMR, LARC)
   b. Maleimides (Kerimid, Compimide)
   c. Acetylene terminated polyimides (Thermid)

All these types are used but only norbornene and the maleimides are of any significant commercial interest.

13.1.7.1 *Condensation type polyimides*

In 1965, workers from DuPont first prepared a linear thermoplastic polyimide. When aromatic diamines are reacted with aromatic dianhydrides, aromatic tetracarboxylic acids, or dialkyl esters of aromatic tetracarboxylic acids, in a solvent such as DMF (Table 13.7), they were converted to a polyamic acid, which when heated, cured by cyclization (imide ring closure, or imidization). This ring closure is a condensation process, releasing either water, or an alkyl alcohol, or both [36].

The solvents used are aprotic (i.e. they aid the separation of electric charges but do not supply protons) and they tend to form complexes, which are difficult to breakdown. This makes removal of the last traces of solvent problematic and, furthermore, can also interfere with the imidization process. Removal of the last traces of solvent from the prepreg requires prolonged heating in case of high boiling solvents, which accelerates the imidization process, resulting in an increase in the resin viscosity. So, a combination of all these facts, coupled with the problems associated with the products released during the condensation reaction, resulted in a composite with a high void content (5–10%). This, in turn, resulted in lower than optimum thermal oxidative stability, which severely curtailed any interest in these products.

Although these polyimides are technically thermoplastic, they are practically thermosets because of their high melting points and extreme difficulty in processing, as well as excessively protracted curing schedules.

The Monsanto Skybond resins [37] are based on 3,3′,4,4′-benzophenone tetracarboxylic dianhydride (BTDA) or BTDA reacted with an alkyl alcohol (e.g. ethyl alcohol) and an aromatic diamine (e.g., MDA) or a polyamide acid made by reacting an aromatic diamine with BTDA. The DuPont product, Pyralin, was probably also based on BTDA.

In 1972, DuPont introduced the NR-150A and B series [37,38], which were easier to process and provided composites with void contents <1%. These resins were based on 2,2-bis(3′,4′-dicarboxyphenyl)hexafluoropropane dianhydride (6F), whilst the tetra-acid form (6FTA) is used for formulating polyimide precursor solutions. The success of these resins is attributed to the hexafluoroisipropylidene group, situated between the two anhydride portions of the 6F, not reacting with amines, whilst in BTDA cross-linking does occur [38]. The NR-150 precursor solutions form an amorphous linear polyimide, which is melt-fusible above the T_g and it is, therefore, possible to apply pressure whilst heating above the T_g to compress any volatiles that are evolved during the curing stage, thereby controlling the void content.

13.1.7.2 Addition type polyimides

13.1.7.2.1 The earliest bismaleimides

The first truly addition type polyimide is attributed to Rhone Poulenc, who introduced a bismaleimide, the building block for an addition polyimide, in 1964. Recognizing the potential of the thermoset bismaleimide molding resins produced earlier by Rhone Poulenc, the NASA Lewis Research Center, in the late 1960s, sponsored the development of imide oligomers, which were terminated with nadimide groups. Lubowitz [39,40], working with TRW, developed polyimide systems that were endcapped with endomethylene tetrahydrophthalic anhydride (ETPA) groups. The synthesis was achieved by reacting an ester of benzophenone tetracarboxylic dianhydride (BTDA) with 4,4′-methylene dianiline (MDA) and an ester of nadic anhydride producing a solution in DMF with a molecular weight (M_n) of 1300 g mol^{-1}, which was adjusted by varying the proportions of the reactants.

The product was marketed by Ciba as P13N [41] and had to be processed in a solvent (NMP or THF). The difficulty of removing such solvents resulted in a very stiff prepreg and was the prime reason for abandoning the P13N polyimides and the earliest bismaleimides.

NASA Lewis Research Center further developed the norbornene project, culminating in the PMR concept, where a class of addition type polyimides was formed by the *in situ* Polymerization of Monomer Reactants (PMR) [42].

The monomers used for this first resin comprised the dimethyl ester of 3,3′,4,4′-benzophenonetetracarboxylic acid) (BTDE), methylenediamine (MDA) and monomethyl ester of 5-nobornene-2,3-dicarboxylic acid (NE) dissolved in methyl or ethyl alcohol and reacted in the molar ratio $n:n+1:2$ for BTDE:MDA:NE.

4,4'methylenediamine (MDA)

dimethyl ester of
3.3',4,4'-bezophenonetetracarboxylic acid (BTDE)

monomethyl ester of
5-nobornene-2,3-dicarboxylic acid (NE)

$+ H_2O + CH_3OH$

When $n=2.087$, the formulated molecular weight (FMW) was 1500 and this first generation resin was identified as PMR-15, which is still widely used.

The fiber reinforcement was prepregged with a solution of the monomers in alcohol, followed by removal of the solvent. The prepared prepreg was then heated at 150–200°C involving an *in situ* condensation reaction to form the norbornene-endcapped imide prepolymer. Crosslinking was subsequently achieved by applying pressure and heating at 250–300°C.

Although it was initially believed that the chemistry of the imidization and cross-linking reactions was relatively straightforward, it proved otherwise. Reactions were, in fact, dependent on the actual cure conditions and consequently, the molecular weight of the prepolymer could change considerably.

Free aromatic diamines are used in the PMR-15 concept and some of them (e.g. MDA) are suspected carcinogens. As a safeguard for workers handling these amines, it is a mandatory that the working area be air monitored (e.g. when using any product that contains more than 1% of free MDA). Wherever possible, alternative safe replacements for these harmful amines are now being used.

The thermo-oxidative stability of this first generation material was improved by the introduction of a second generation resin [43]. Initially the BTDE was replaced by the dimethyl ester of 4,4'-(hexafluoroisopropylidene)-bis(phthalic acid) (HFDE), which significantly improved the thermo-oxidative stability, but the initial mechanical properties at 316°C were inferior to PMR-15. This was overcome by replacing the MDA with *p*-phenylenediamine (PPDA). Hence, the monomers used for the second generation resin comprised the dimethyl ester of 4,4'-(hexafluoroisopropylidene)-bis(phthalic acid) (HFDE), *p*-phenylenediamine (PPDA) and monomethyl ester of 5-nobornene-2,3-dicarboxylic acid (NE), in the molar ratio $n:n+1:2$ for HFDE:PPDA:NE. When $n=1.67$, the formulated molecular weight (FMW) was 1267 and the product designated PMR II.

The PMR family have since been joined by other members: PMR-II-30, PMR-II-50, PMR-II-50, PMR-12F-71 and V-cap-12F-71 [29].

The surface groups on commercial carbon fiber surfaces were examined by Electron Spectroscopy for Chemical Analysis (ESCA) [44] and found to contain significant quantities of oxygen present as alcohol (—C—OH), ether (—C—O—C), carbonyl ($\supset$C=O) and carboxylic acid (—C—O—O—) functionalities. Fibers containing the highest oxygen concentration in the form of the above functionalities exhibited the highest shear strength and the least degradation after 500 h at 316°C. This suggested that good adhesion prevented oxygen diffusion at the interface, thereby minimizing degradation.

Xiang and Jones studied the thermal degradation of PMR-15 in the temperature range 382–457°C in both air and N_2 [45] and found that degradation in air occurred at a significantly higher rate. The decomposition of the matrix occurred at about 420°C and was not affected by the surrounding atmosphere. Within the temperature range studied, the expected time limit (t) without major loss in mechanical performance when exposed to air was given by

$$\log t = \frac{-9.46 + 7.31 \times 10^3}{T}$$

where T is the temperature K.

NASA Lewis have developed LARC 160, a melt processable PMR version by replacing the MDA with Jeffamine AP22 (a eutectic blend of MDA type amines) to give a *quasi* melt resin [41,46].

13.1.7.2.2 Bismaleimides

As previously mentioned, Rhone-Poulenc had introduced a series of polyimides based on the reaction of non-stoichiometric quantities of bismaleimides with aromatic diamines [47]. Maleic anhydride and the aromatic diamine form bismaleiamic acid, which undergoes cyclodehydration at 40–50°C in the presence of acetic anhydride and sodium acetate to form the bismaleimide.

Subsequently, a patent [48] was obtained for the monomers, 4,4'-bismaleimidodiphenylmethane with methylene dianiline reacting in the melt, or in solution form. The resultant prepolymer is soluble in aprotic solvents such as NMP or DMF and could therefore be prepregged using solution techniques. This was the basis of Kerimid 601, used as an injection/transfer molding compound or in the solution form as a laminating resin, especially for printed circuit boards. The molding temperature is 170°C and, if postcured at 250°C, the composite will have good high temperature properties up to 200°C, providing the residual solvent level in the prepreg is below 2%.

It is obvious that there will be a large family of bismaleimides, all containing the general bismaleimide building block:

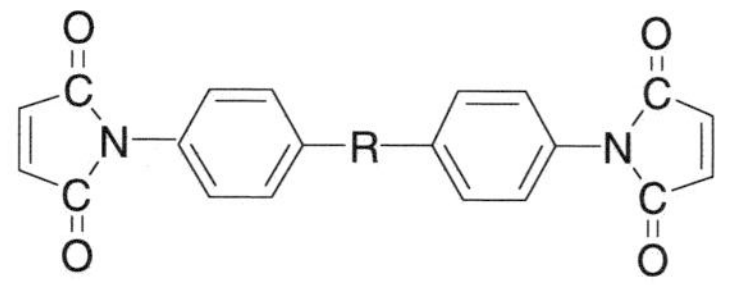

The bismaleimides are highly crosslinked materials, with high modulus, low strength and very low elongations at break all contributing towards their brittleness. A number of routes to enhance their toughness [49] have been pursued including:

   a. Using comonomers such as 2,2'-diallylbisphenol-A (DABA) [50],

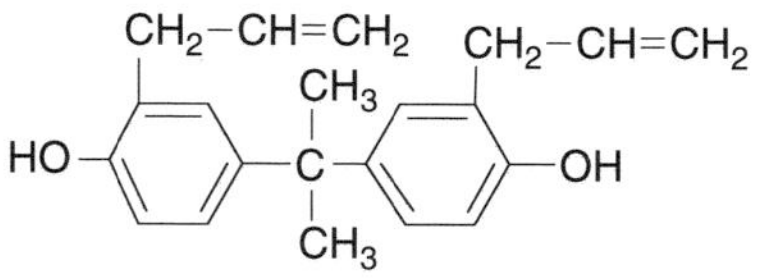

   b. Incorporating a CTBN rubber [51,52] concomitant with a significant loss of high temperature properties and poor long term aging stability at 200°C.
   c. Introducing a thermoplastic resin [53,54] into the matrix such as a thermoplastic polyimide Matrimid 5218 [55,56], polyethersulfone [54–57], polyetherimide [55–56,58–61] and poly (arylene ether) [58–59,61–63]. All these polymers however, cause an attendant loss of processability coupled with poorer solvent resistance.

High temperature stability of BMIs has been studied [64,65] and a technique for the measurement of adhesion between BMI and carbon fiber has been developed [66].

13.1.7.2.3 Acetylene (ethynyl) terminated polyimides

Around 1970, the US Air Force funded a program on acetylene terminated imide oligomers and the first patent was filed by Hughes Aircraft [67] and has been described in the literature [68]. Benzophenone tetracarboxylic anhydride was condensed with an aromatic diamine and 3-ethynylaniline to give either an amide acid soluble in acetone, or a fully imidized structure, soluble in NMP.

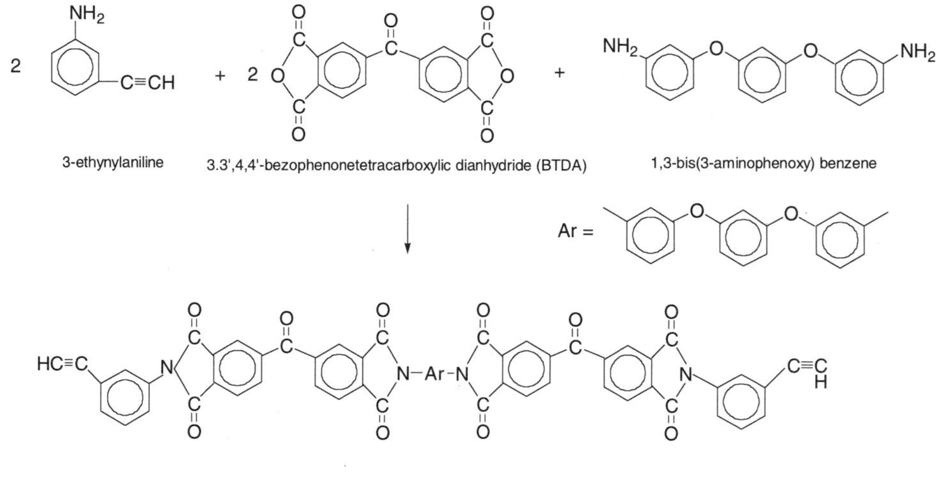

3-ethynylaniline   3.3',4,4'-bezophenonetetracarboxylic dianhydride (BTDA)   1,3-bis(3-aminophenoxy) benzene

Thermid MC-600 a preimidized thermosetting acetylene-capped aromatic polyimide

The first commercial product was introduced in 1975 as Thermid 600 by Gulf Oil Chemicals and is now sold by National Starch and Chemical Corporation. The fully imidized prepolymer has to be processed in NMP, with all its attendant problems. However, the molecular weight can be adjusted to provide a resin which melts at about 200°C and immediately starts to polymerize when molten and so, has a narrow processing window.

[Chemical reaction scheme: 2 × 3-ethynylaniline + 2 × 3,3',4,4'-bezophenonetetracarboxylic dianhydride (BTDA) + 1,3-bis(3-aminophenoxy) benzene → Thermid IP-600 thermosetting acetylene-terminated polyisoimide precursor]

A resin system, Thermid FA-700, based on hexafluoroisopropane-bisphthalic anhydride was introduced and can be easily processed into composites using THF, DMF or DMAc [29].

13.1.8 Special resin systems

Although polyimides are the largest class of high temperature resin systems, other resins have been developed that are used for specialist applications or, alternatively, were never developed commercially, perhaps due to cost considerations.

Ashland Chemical Company (PBOX) are bisoxazoline phenolics and are thermosetting poly(amide ethers), prepared from phenolic resins and bisoxazolines. They can be cured with a catalyst at 175°C and post cured at 225°C.

Bayer introduced the Blendur resins (EPIC) based on the catalyzed reaction of an epoxy resin with an isocyanate.

Dow Chemical introduced benzocyclobutene (BCB), curing without the evolution of volatiles. The reaction proceeds by the four membered benzocyclobutene rings opening to give *o*-quinodimethane structures.

Société Nationale des Poudres et Explosifs (SNPE) have developed the polystyrylpyridine (PSP) resins, which do, however, evolve volatiles during the curing process.

The US Naval Research Laboratory developed the phthalonitrile resins containg four nitrile groups that could be crosslinked to give polymers, which may be phthalocyanines.

13.1.9 Introducing toughness to thermoset resin systems

13.1.9.1 Introduction

Due to the increasing growth of composites in the mid 1980s there was a desire to produce resin systems that were more resistant to impact damage than the grades available at the time.

A design strain limitation of 3000 μcm cm^{-1} had been imposed and this, coupled with the cost of composite manufacture, made various composites non-competitive. This spurred resin manufacturers to introduce a class of tough resins that could meet design strain limits as high as 6000 μcm cm^{-1} [69].

A number of workers [70–74] have reported ways of toughening epoxy matrices, but toughening the matrix does not always translate into enhanced properties of the composite.

13.1.9.2 Toughening versus flexibilizing [75]

The ability of a brittle resin to absorb energy without catastrophic failure can be increased through either flexibilizing or toughening. Flexibilizing is accomplished by compounding the resin with compatible long chain plasticizers, building long chain segments into the resin structure to increase the distance between cross-link sites, or using flexible amine curing agents. Flexibility generally results in a sacrifice of resin strength, stiffness and hardness, especially at elevated temperatures (e.g. hot wet strength).

Toughening, as with flexibilizing, enhances the resin system's ability to resist mechanical and thermal stress. However, unlike flexibilizing, the gross properties of the matrix resins almost unaffected. There is barely any sacrifice of strength, stiffness, hardness, or temperature resistance for the increase in toughness. Toughening is achieved by dispersing a small amount of an elastomer as a discrete phase of microscopic particles embedded in the continuous resin matrix.

The rubbery particles promote the absorption of strain energy by complex interactions, which are thought to involve both craze formation and shear deformations. The overall mechanism is influenced by the size of the microscopic rubbery particles.

Shear deformations are predominant in epoxy resins toughened by small (<0.5 μm) particles, while the crazing mechanism is promoted by larger (1–5 μm) particles. Maximum toughening is generally found under conditions that produce both shear deformations and crazing. Thus, systems containing large and small rubber particles (i.e. a bimodal distribution) highest probability of providing optimum toughness.

The toughening process generally renders the viscosity behavior non-Newtonian (i.e. dependent on the shear rate). Typically, these systems show a decrease in viscosity with an increase in shear rate, or pseudoplasticity. The sensitivity of the viscosity to the strain rate becomes greater as the resin system becomes more advanced.

13.1.9.3 Types of elastomeric modifiers

Rubber toughened epoxy formulations do offer excellent fracture properties [76–81].

However, the presence of the rubber phase can decrease the modulus and thermal stability, while at the same time increasing the water absorption which causes loss of properties at elevated temperatures.

1. Reactive liquid polymers
   Hycar reactive liquid polymers made by BF Goodrich are homopolymers of butadiene, or copolymers of butadiene/acrylonitrile [75]. They may be thought of as long chain dicarboxylic acids with a comonomer acrylonitrile content of 0–27%. They have reactive functional groups (carboxyl, acrylate, vinyl and secondary amine) in both terminal positions of the polymer chain and may also have additional reactive groups pendant on the chain e.g. carboxyl terminated polybutadiene (CTB), carboxyl terminated polybutadiene-acrylonitrile (CTBN), CTBN with pendant, as well as terminal carboxyl groups (CTBNX), amine terminated polybutadiene-acrylonitrile (ATBN), vinyl terminated polybutadiene-acrylonitrile (VTBN),

butadiene-acrylonitrile copolymer terminated with reactive vinyl groups and bearing additional vinyl groups along the chain (VTBNX)—where C = carboxyl, B = butadiene, V = vinyl, N = acrylonitrile, A = amine, X = pendant reactive group, T = terminal reactive groups.

These tougheners can be physically blended into an epoxy resin system or can be capped with an epoxy resin by pre-reaction and then blended with unmodified epoxies and cure agents, usually using triphenyl phosphine as a catalyst system, which promotes the carboxyl-epoxy reaction at temperatures under 120°C [82].

2. Rubber copolymers
   Finaprene, solution polymerized butadiene, butadiene/styrene elastomers, as random and deblock copolymers, are available from Fina Chemicals [83].
3. Polyvinyl butyral derived from polyvinyl alcohol and butyraldehyde can be added to epoxy and phenolic resins to improve impact resistance and peel strength.

13.1.9.4 Duplex materials

Another approach which has been used to introduce added toughness to a composite is to form a duplex material, where a layer, about 0.0125 mm thick, of an extremely high shear strain tough resin system is interposed between the prepreg plies, with the fiber itself embedded in a tough matrix resin [84]. The two resin systems must co-cure, but must remain discrete. The interleaf helps to eliminate stress concentrations and toughness can be almost doubled. Alternatively, a sheet of thin thermoplastic film can be used, or thermoplastic toughener applied to the surface of the prepreg as a fine powder. An aluminum interleaf has been used [85] to increase the fatigue resistance of cfrp and when used in heat sinks, also increases the coefficient of thermal expansion to match the values for typical ceramics used in the manufacture of electronic components.

13.1.9.5 Thermoplastic modifiers

Thermoplastic resins can be added to regulate the viscosity of an epoxy resin and control the flow during cure to prevent resin bleed. Hexcel's 914 system is a blend of tetrafunctional epoxy with polyethersulfone and a dicyandiamide latent catalyst. The polyethersulfone controls the flow of the resin throughout the curing cycle (Figure 3.25), enabling applied pressure to control the size of any voids. The system produces a most interesting two phase cured matrix. Another useful attribute of this system is that limited post-forming can be undertaken after molding, permitting some double curvature to be achieved.

Thermoplastics added to thermoset matrices can also improve the degree of toughness and it is believed that PES and PEEK are used for this application in many of the new generation epoxy systems.

Thermoplastics with a high T_g have been blended with epoxy resins to confer good thermal stability and decrease the tendency to absorb water. PES has been used [86–91] as well as PEI [91–93].

Stenzenberger, in his papers [1,29] for toughening bismaleimide resins, cites the use of thermoplastics such as polyhydantoins [55] (Bayer's Resistofol N, Resistherm PH-10),

Table 13.8 The effect of fiber type and resin system on tensile and compressive strengths and moduli

System	Tensile strength unidirectional (0°) GPa	Tensile modulus unidirectional (0°) GPa	Compressive strength Orientation (0°, ±45°, 90°) GPa	Compressive modulus Orientation (0°, ±45°, 90°) GPa
Toray T300/Narmco 5208	1.52	129	0.52	45
Hercules AS4/Hercules 3501-6	1.98	134	0.66	47
Hercules IM6/Tough resin	2.38	142	0.50	36
HerculesAS4/Thermoplastic	1.90	128	0.36	26

Source: Reprinted with permission from Cam CY, Walker JV, *Toughened Composites*, Johnston NJ ed., *ASTM STP937*, American Society for Testing and Materials, Philadelphia, 9–22, 1987. Copyright 1987, ASTM International.

polyethersuphones [55] (Amoco P1700), polyimides (Ciba-Geigy Matrimid 5218) and polyarylene ethers [58,62].

A reactively terminated poly(ether sulfone) copolymer was added to a triglycidyl-aminophenol epoxy resin system cured with DDS. At low concentrations of added thermoplastic, a single phase microstructure was obtained (up to 8 phr), changing to a particulate microstructure of thermoplastic-rich particles in an epoxy matrix (particles gradually increased in size from 0.2 to 0.4 µm, becoming larger and changing to form elongated ribbons), then to a continuous structure (about 52 phr, with ribbons about 0.5–1.5 µm wide) and then to an inverted form (more than 83 phr, consisting of a second phase of epoxy-rich particles, about 0.1–1.5 µm diameter, in a continuous thermoplastic-rich phase). The single phase microstructure gave no toughening effect. Thereafter, toughness increased with increasing thermoplastic concentration, with no abrupt changes in the toughness with change in the microstructure.

13.1.9.6 Effect of carbon fiber reinforcement

With the introduction of the resins with improved toughness, fiber manufacturers too introduced new grades of carbon fibers with better strength and modulus characteristics, with the fiber failing at a higher strain level. A comparison of the unidirectional strength and stiffness (Table 13.8) [69], showed that a Hercules IM6 composite was 50% stronger than a Toray T300 composite, but its tensile modulus was increased by only 10%. However, when the improved fibers were used in cross ply laminates, these improved properties were not translated across. The diameter of the IM6 fiber was only 5.0 µm, compared with 7.5 µm for T300, thus reducing the cross-sectional area from 0.48 mm^2 to 0.29 mm^2 for AS6 and to 0.27 mm^2 for the IM6 fiber. Although compression tests of unidirectional laminates showed no noticeable change with IM6, when used with the tougher resin systems, such as a thermoplastic matrix, there was a large reduction in compressive strength, also giving lower cross ply compressive strengths than the thermoset systems.

13.2 SELECTED THERMOPLASTIC RESINS

13.2.1 Introduction

Thermoplastic resins are, for convenience, referred to using abbreviations or acronyms listed in Appendix 15. The general properties of a range of thermoplastics, which are all used for injection molding compounds in conjunction with carbon fiber as a reinforcement, are given in Appendix 14. The higher performance polymers such as PES, PEEK, PEI, PPS and PI are also used to make cfrp by the prepreg route. These high performance resins all have a backbone containing elements with a rigid ring structure. The most prevalent unit is provided by a phenylene ring

with extra stiffness provided by biphenyl, a double ring structure

Flexible linkages such as an ether link will permit the chain to rotate,

whilst some groups that are stiff can also have limited mobility such as the bulky sulfone group.

The sulfone group imparts excellent thermo-oxidative stability, thereby elevating the long term use temperature. The phenylene ether segment provides flexibility, giving high toughness, elongation, ductility and ease of melt fabrication. The phenylene sulfone and ether groups confer outstanding hydrolytic stability. A high concentration of sulfone groups (as in PES), however, will attract water, giving high moisture absorption values.

Amorphous thermoplastic polymers have polymer chains in a random coil arrangement without any degree of local order, whereas a semi-crystalline polymer would have some degree of order of the polymer chains. The chains are entangled and because they are not fixed, can slip past one another, whereas a thermoset resin, when subjected to local stress breaks, in a brittle manner. The ability of the thermoplast to dissipate energy by chain slippage confers the property of toughness to the composite.

A thermoplastic matrix [94] does offer certain advantages over a thermoset matrix. No cure is involved, processing is relatively quick, offers an unrestricted shelf life, permits reprocessing and can be thermoformed and welded. The thermoplastic matrix imparts toughness, produces low moisture absorptions and good hot/wet properties, presents less of a health risk, although care must be taken when processing fluoropolymers. Due to high viscosities, thermoplastics do require high processing temperatures, which limit the available processing methods. Unfortunately, thermoplastic prepreg is stiff and non-tacky and cannot be readily draped. The prepreg sheets are tacked together with a heat gun or by using a solvent. The surface treatment of carbon fiber to give optimum properties for different thermoplastic matrices is still in the early days of development. Owing to the high processing temperatures, the epoxy sizes normally applied to carbon fibers for use with conventional epoxy matrices cannot be used and certain specific high temperature polymers, such as a thermoplastic polyimide, are used.

Semi-crystalline thermoplastics (e.g. PEEK) are more efficiently reinforced by carbon fibers than amorphous thermoplasts (e.g. PES) because the fibers act as nucleation sites for the crystallization process and the fiber becomes surrounded by a finely divided microcrystalline structure, which also improves the modulus, especially the flexural modulus [95]. As the crystallinity increases, so does the degree of reinforcement.

Since water acts as a plasticizer, when the polymer absorbs water, the strength and stiffness decrease, but the toughness increases. Although, if the polymer is reinforced with carbon fiber, less water is absorbed since the carbon fiber replaces some of the polymer and will consequently absorb less water. Also, addition of the fiber will increase the dimensional stability with respect to temperature change, since the fiber has a much lower thermal expansion coefficient. On the other hand, a molding that exhibits preferred orientation can create warping, which occurs due to stresses arising from the differences in thermal expansion in the transverse and flow direction.

Fiber reinforcement increases the softening temperature of a polymer by about 10–20°C and is greatest with semi-crystalline polymers. Amorphous polymers do not have a sharp melting point and as the temperature rises, the material gradually softens. The softening is governed by the T_g, which is some 20°C lower than the T_g for the unreinforced polymer. Crystalline polymers have sharp melting points and softening is determined by the melting point (T_m), the melting range, crystallinity and T_g. Higher softening points are favoured by a high T_m, a narrow melting range and high percentage crystallinity, or with fiber reinforcement, a moderate crystallinity. Amorphous resins above their T_g will quickly lose their strength, whereas semi-crystalline materials retain useful levels of strength and stiffness.

Creep becomes more significant above the T_g and the addition of a fiber reinforcement inhibits the molecular mobility and hence improves the creep resistance at higher temperatures. The effect is greater with amorphous polymers, as crystalline polymers will inhibit creep anyway.

A carbon fiber reinforced injection molding compound has less effect on the melt flow than a glass fiber reinforced grade and will produce considerably less abrasion on the injection molding equipment. Reinforcement does, however, confer a higher melt viscosity for a given shear rate, requiring higher processing temperatures and/or higher injection pressures, but shrinkage during processing is reduced.

Other thermoplastic resins can be added to produce a polymer blend in order to achieve a desirable balance of mechanical/thermal properties and cost (e.g. PPO/PS, Noryl from GE). Most blends are two phase systems, where no stress transfer to the dispersed phase is possible. PPO/PS is a special case and forms a single blend T_g between the values of the PPO (210°C) and PS (100°C) depending on the relative proportion of the two polymers (e.g. about 150°C for Noryl).

13.2.2 Morphology property relationships in semi-crystalline thermoplastics

PEEK and PPS are semi-crystalline polymers and the processing conditions can affect the morphology of these thermoplastics, which in turn, can affect the mechanical and chemical resistance properties [96]. Semi-crystalline polymers contain a liquid like amorphous phase and an ordered crystalline phase. Properties of some amorphous and semi-crystalline products are compared in Table 13.9. When solidified from the melt, the crystals form platelets or lamellae, usually nucleating radially from a central point to form birefringent spherulites, which can be seen under a polarizing microscope. In PEEK, the normal spherulite size varies in the range 1–10 µm or even larger, depending on the process history and nucleation density.

Table 13.9 A comparison of amorphous and semi-crystalline thermoplastics

Property	Amorphous thermoplastic	Semi-crystalline thermoplastic
Polymer chain	Random coil arrangement	Degree of local order. Can be annealed to control the level of crystallinity. If too high, polymer is brittle
Action of organic solvents	Dissolved and can be prepregged by the solvent route. Residual solvent can be a problem, reducing T_g and introducing defects. But, can be used to provide a tacky prepreg	Do not dissolve
Prepeg from melt	Difficult	More difficult due to higher processing temperatures.
Melt processing temperature	Melt processing temperature is $T_g + 100°C$ or even less, permitting a higher available T_g. Thermal decomposition occurs at about 420°C	Melt processing temperature is $T_g + 200°C$
Fire resistance	Good	
Smoke resistance	Good	
Non-toxicity	Good	
Toughness	Very good. Toughness increased by water absorption	
	No crystallization to contend with, but undergo free volume annealing, when properties change with time, especially at temperatures just below the T_g	Amorphous regions are less important and ageing is less evident
Change in volume on solidification from the melt	Lower change and less prone to distortion. Composites generate lower levels of stress, but masked if high T_g materials are being used.	
Surface finish	Glossy surface finish	Differential shrinkage between two phases confers a slightly matt finish
Reinforcement with carbon fiber	Less efficiently reinforced	More efficiently reinforced
Creep resistance	Less resistant. Reinforcement improves the creep resistance	More resistant
Fatigue resistance	Less resistant	More resistant
Environmental resistance	Poorer	Outstanding

A type of transcrystalline morphology can also occur during cooling, when columnar crystals preferentially nucleate perpendicular to the reinforcing fibers. This effect is observable under a polarizing optical microscope.

Process conditions have a profound effect on morphology. The less time the material has to crystallize, the lower the crystallinity. A rapid cooling rate, also called quenching, results in the formation of an amorphous resin and can be achieved by transferring the molten material into a cold press with a high thermal mass and quenching below the T_g in less than 10sec. This is normally limited to thin sections that are no more than 1 mm thick. Once solidified, the degree of crystallinity can be increased by annealing at elevated temperatures, which thickens the lamellae as well as allows the formation of newer secondary lamellae, although the spherulite size will essentially remain constant. Pressure during cooling tends to allow spherulites to nucleate at higher temperatures, resulting in smaller spherulites. Also, shear flow during processing can induce transcrystallinity in the absence of any nucleating fibers.

An increased degree of crystallinity and spherulite size normally results in increased tensile strength and yield strength, but lower elongation (note that nylon-6,6 displays the opposite trend). Improved damage tolerance will require relatively small spherulites in the finished part.

Under long term stress conditions, in eithercyclic (fatigue) or static (creep) loadings, intrinsic morphological features can become areas of weakness, e.g., cracks can form in polyethylene at the boundaries of spherulites and within them, suggesting that long term failure is due to separation between lamellae.

Obviously, if spherulite boundaries are areas of weakness, then the occurrence of transcrystallinity must raise doubts. Although transcrystallinity enhances the physical bond between fiber and matrix, in prepregs with a high fiber content, where the fibers are in close proximity, adjacent transcrystalline regions may impinge and the boundary between transcrystalline regions and could act as a long spherulite boundary.

Although to maximize toughness spherulite size and degree of crystallinity should be kept to a low level, solvent and water resistance are maximized through increased crystallinity. Hence, there must be a trade-off between chemical resistance and mechanical properties by manipulating cooling and/or annealing.

These polymers also display the memory effect, such that when heated to slightly above the melting point, the material remembers its premelt chain conformation. This could have implications concerning delamination resistance because, if laminates are processed in the temperature range within which the memory effect is operative, inadequate resin interpenetration through adjacent plies may result.

Normally one would expect the T_g to be the maximum temperature at which the stiffness of these materials can be maintained. Any relaxation below the T_g will result in considerable softening of the material, significantly affecting mechanical properties, (e.g. PC) [97]. If an amorphous material like PET film is held at a temperature just below the T_g even for periods of time as short as 90 min, the material becomes severely embrittled, although the effect is completely reversible if momentarily heated above the T_g. This physical ageing effect is attributed to a gradual loss of segmental mobility due to densification of the amorphous phase with time. Physical ageing occurs with PC and PES and will probably also occur with PEEK. It is noteworthy that since the phenomenon affects only the amorphous regions, maximizing the degree of crystallinity would minimize physical ageing.

A characteristic of amorphous resins is that they have poor resistance to organic solvents (e.g. polysulfone).

The morphology of PEEK has been described by Cogswell [98]. The crystalline structure in semi-crystalline polymers depends on the thermal history such as cooling rate, stress on the system, nucleation sites, as well as molecular structure and mobility of the molecule. Not all the nucleation sites will necessarily be within the resin phase. Heterogeneities can act as nucleation sites and these include fiber dust, gel particles, etc. Hobbs [99] believes that active sites account for most of the nucleation.

HTS type carbon fiber had a turbostratic graphite structure with small (≈ 25 μm) graphite nuclei with a high degree of disorientation, whereas HMS type carbon fiber had much larger (≈ 100 μm) graphite planes that are highly oriented along the fiber axis. These fibers, when incorporated in PP in the hot stage, melted and then slowly cooled until crystallization occurred, showed dramatic differences. The HTS fiber proved to be a poor nucleate for the PP and the spherulites nucleated and grew with about the same frequency as those on the fiber surface. The HMS fiber displayed a high propensity to nucleate PP crystallization and during cooling, brilliant white lines of new nucleated crystals were observed on the fiber surface. This extreme variation in the nucleating ability seems to be a direct result of changes in the size and orientation of the constituent graphite planes. The match between hydrogen

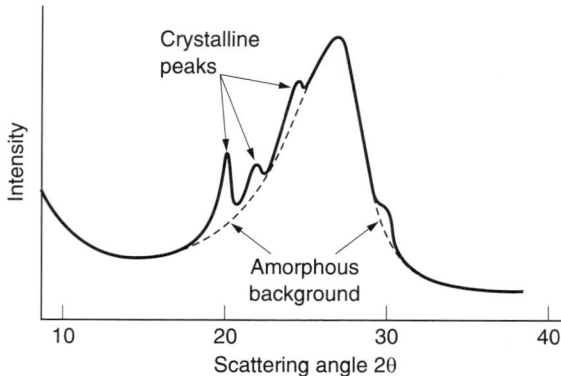

Figure 13.5 WAXS trace of CF/PEEK. *Source:* Reprinted with permission from Cogswell FN, *Thermoplastic Aromatic Composites*, Butterworth-Heinemann, Oxford, 1992. Copyright 1995, Taylor & Francis Ltd.

atoms on the PP and C—C bonds on the substrate, which are low energy sites, enhances nucleation.

Hartness [100], working with XAS and HMS fibers in a PEEK matrix showed similar behavior. The similarities between PEEK and PP are probably greater than the differences in their crystalline structure. Beaumont [101] has shown that with HMS (treated) fiber, there is almost no pull-out, whereas with HMU (untreated) fiber, extensive debonding and pull-out take place. The pull-out lengths can be measured and using an analytical technique outlined by Phillips [102], values can be obtained for the nylon/fiber interfacial bond strength and fracture energies.

The question that establishes the optimum bond between a given fiber and a semi-crystalline polymer, and how that relates to optimum composite properties, is a very important one [103]. It is possible that too strong a bond between fiber and matrix may be just as harmful as too weak a bond.

Oya and Hamada [104] discuss the mechanical properties and failure mechanisms of cfrp. Wide angle X-ray scanning (WAXS) can be used to study crystal structure [105–107] (Figure 13.5). Differential scanning calorimetry (DSC) [108,109] can be used to study the dynamics of crystallization and melting (Figures 13.6 and 13.7) whilst dynamic mechanical analysis (DMA) (Figure 13.8) is a convenient way to display the T_g (143°C) and the T_m (334°C).

13.2.3 Polyamide (PA) resins

The nomenclature for polyamides (nylons) is based on the number of carbon atoms in the amide group, e.g., nylon-6 (PA6) is $[-NH-(CH_2)_5-CO-]_n$. This class of polymer accounts for some 90% of the thermoplastic materials are available with carbon fiber reinforcement and are principally used for molding compounds [110]. Since polyamides readily absorb moisture from the atmosphere, care should be taken before processing to ensure that the polymer has been dried well and in case of excessive absorption, it will be necessary to dry in a vacuum or desiccant oven. When drying in an air circulating oven, the temperature should be limited to 80°C to avoid possible oxidative degradation. PA6 absorbs the maximum amount of water (3.2% at 65% RH and 20°C), whilst PA12 will only absorb 0.4%. Incidentally, PA12 has the lowest density and lowest melting point of the polyamides. The absorption of moisture will result in a lowering of strength and stiffness, but will provide an

POLYMER MATRICES FOR CARBON FIBER COMPOSITES

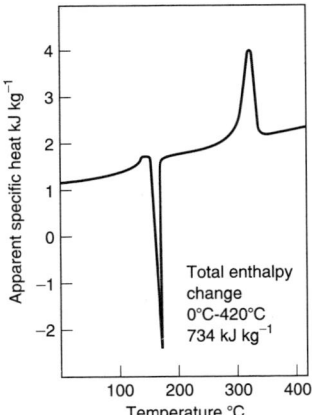

Figure 13.6 DSC heating trace for amorphous PEEK. *Source:* Reprinted with permission from Cogswell FN, *Thermoplastic Aromatic Composites*, Butterworth-Heinemann, Oxford, 1992. Copyright 1995, Taylor & Francis Ltd.

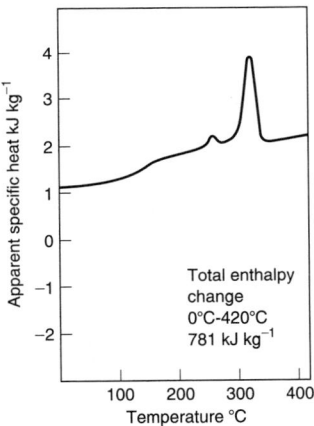

Figure 13.7 DSC heating trace for semi-crystalline PEEK. *Source:* Reprinted with permission from Cogswell FN, *Thermoplastic Aromatic Composites*, Butterworth-Heinemann, Oxford, 1992. Copyright 1995, Taylor & Francis Ltd.

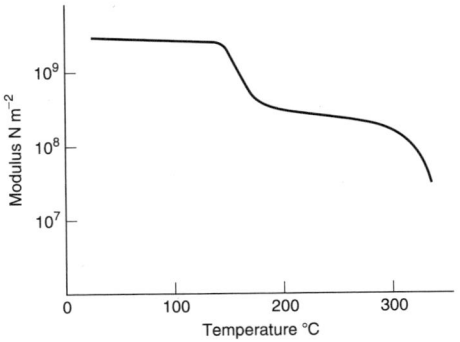

Figure 13.8 Dynamic Mechanical Analysis of PEEK. *Source:* Reprinted with permission from Cogswell FN, *Thermoplastic Aromatic Composites*, Butterworth-Heinemann, Oxford, 1992. Copyright 1995, Taylor & Francis Ltd.

increase in impact strength (toughness). Product dimensions are also affected by moisture absorption, but remain constant once equilibrium conditions have been reached.

Polyamides are readily processed on all types of injection molding machines.

The impact properties of short carbon fiber reinforced PA6.6 have been examined [111] as well as the microstructure of PA6,6 layers in cfrp [112]. The abrasive wear of short carbon-fiber reinforced PA6,6 has been reported [113].

Bessell and Shorthall [114] studied the nucleation and crystallization of nylon 6 reinforced with carbon fiber. The fibers were shown to nucleate a columnar structure at the interface, which was different between HM and HT carbon fiber types and was primarily due to the physical matching of the graphite crystallites. Surface treatment had a pronounced effect, with treated fibers giving small amounts of fiber pull-out with low fracture energies, whereas untreated fiber exhibited extensive pull-out reflected in high fracture energies.

13.2.4 Polycarbonate (PC) resin

$$\left[-O-\underset{\underset{CH_3}{|}}{\overset{\overset{CH_3}{|}}{C}}-O-CO- \right]_n$$

This polymer represents an important matrix for carbon fiber filled injection molding compounds and the molded parts have good tensile strength, impact strength and heat resistance.

PC can be solvent welded using an 8% solution of PC in methylene chloride. It is imperative to thoroughly dry PC prior to use and the permitted residual moisture is restricted to 0.02% (as determined using Karl Fischer reagent) otherwise, hydrolytic degradation of the polymer will occur during processing, producing molecular chain degradation, resulting in impaired mechanical properties as well as surface degradation.

Carbon fiber-PC composites have been characterized [115] and the effect of toughness on compressive strength [116], tensile properties and fracture behavior determined when using PAN based carbon fiber [117].

13.2.5 Polyetheretherketone (PEEK) resin [118]

This polymer is formed by the reaction of hydroquinone, 4,4'-difluorobenzophenone and potassium carbonate in diphenylsulfone at 150–300°C.

HO—⟨⟩—OH + F—⟨⟩—C(=O)—⟨⟩—F + K_2CO_3

↓

$\left[-O-⟨⟩-O-⟨⟩-C(=O)-⟨⟩- \right]_n$ + 2 KF + H_2O + CO_2

Earlier, small amounts of the diphenylsulfone solvent may have remained in the polymer, but over the last decade, product purity has significantly improved. The weight average molecular weight is $M_W = 30,000$ and the number average molecular weight $M_N = 13,000$. The repeat unit is nineteen C atoms, twelve H atoms and three O atoms giving a total molecular weight of 288. Chain termination is by way of an F atom [109]. The chain can rotate at the ether and ketone linkages and takes up a random coil configuration, each molecule entangling with many others, with the chains in the crystalline lamellae packing with ether and ketone groups in any register. Although the formula for PEEK is written in the planar form, the phenylene rings are actually at $\pm 37°$, alternately, along the chain.

The level of crystallinity in PEEK depends on the processing history and the optimum level is 25–40%. Rapid cooling produces an amorphous polymer, which can then be annealed to give the requisite level of crystallinity. Annealing above 300°C should be avoided as this gives a very high level of crystallinity, thereby reducing the toughness.

PEEK has a high melting point (334°C), but provided oxygen is excluded, it will withstand 400°C for more than one hour. Although a maximum service temperature of 120°C would be appropriate for a material with a T_g of 143°C, the maximum working temperature is extended well above the T_g since crystalline resins are self-supporting up to the melting point.

PEEK has outstanding chemical resistance and low water absorption, although it is always recommended to dry for at least 3 h before use. PEEK has become a very important matrix for producing cfrp composites.

Carbon fiber reinforced PEEK has been extensively investigated and the examination of the interface morphology, ranked as most important [119–128]. The ductility at the interface in a carbon fiber/PEEK composite is shown in Figure 13.9. Injection molding compounds of carbon fiber reinforced PAEK were studied [129] and the utilization of high volume fractions of carbon fiber in PEEK has been used to raise the mechanical properties [130]. Lee [131] has reported tensile properties of this composite system. The compressive strengths of cross-plied carbon fiber/PEEK composites are recorded [132]. Help is provided for filament winding thick sections of PEEK cfrp [133].

Failure mechanisms in carbon fiber reinforced PEEK have been studied [134–137]. The importance of a strong fiber/matrix bond has been shown to be a dominant factor in conferring maximum composite toughness [138–140]. Variation of the fiber content changes the role of the matrix phase and the ductilty of the matrix controls the fracture behavior.

Increased crystallinity, induced by annealing, significantly improves the fatigue crack resistance for neat PEEK as well as its short fiber composites, both favouring a high molecular weight matrix [141].

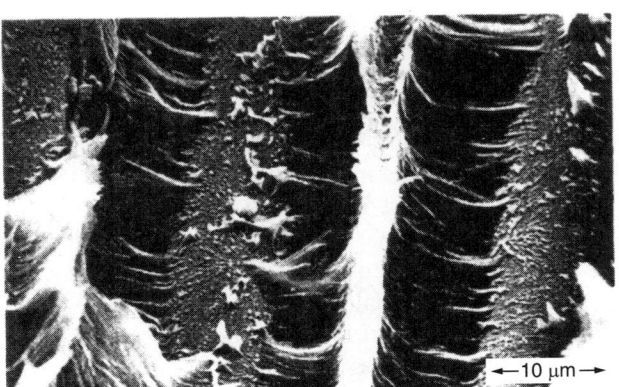

Figure 13.9 Ductility at the interface in carbon a fiber/PEEK composite. *Source:* Reprinted with permission from Cogswell FN, *Thermoplastic Aromatic Composites*, Butterworth-Heinemann, Oxford, 1992. Copyright 1992, Elsevier.

The thermal stability of PEEK/carbon fiber and its effect on consolidation [142], morphological changes [143] and creep [144] have been determined. Gaitonde and Lowson studied the low temperature thermal expansion of carbon fiber/PEEK [145].

Friction and wear of carbon fiber/PEEK have been investigated [146–148]. The thermal joining [149], repair [150] and recycling [151] of carbon fiber/PEEK composites is covered in literature.

13.2.6 Polyetherimide (PEI) resin

PEI is an amorphous thermoplastic resin with a rated continuous temperature of use up to 170°C. It has an limiting oxygen index (LOI) of 47, one of the highest in the commonly used engineering thermoplastics, coupled with a low smoke emission. PEI retains 41% strength retention at 190°C. Like PES it is a tough resin, but it is sensitive to notches and sharp corners.

In common with most thermoplastics, PEI should be thoroughly dried before processing. It is more easily processed than PES or PEEK. It has good chemical resistance, but is soluble in partially halogenated solvents like methylene chloride and trichloroethane.

Solvent induced crystallization [152] and fracture toughness [153] have been studied.

13.2.7 Polyethersulfone (PES) resin

Primarily this product   Low level of polyetherethersulphone also formed

The end groups in a PES polymer chain are chlorine atoms. Other forms of PES are sold which may contain the bisphenol-A moiety, which can compromise thermal stability and chemical resistance.

PES can be used for tens of thousands of hours at temperatures up to 200°C without loss of strength and with negligible changes in dimensions. It has low flammability and starts to soften at 220°C. PES will absorb 0.15% water at 65% RH and 0.3% in boiling water, which can be removed by drying at 150–180°C. It is a tough resin, but is sensitive to notches and sharp corners. PES with a higher molecular weight will have improved chemical resistance, which can be improved further by annealing at 160–170°C.

The polymer can be used to control resin flow when processing epoxy based prepreg and can also be incorporated in resin matrices as a toughener.

The failure mechanisms of carbon fiber reinforced PES have been investigated [154]. Although toughness of the PES composite was lower than comparable results with PEEK, results were independent of the test temperature (−60–100°C), whereas the PEEK composite showed a change to unstable fracture at low temperatures.

13.2.8 Polyphenylene sulfide (PPS) resin

PPS is made by heating *p*-dichlorobenzene with Na_2S in a polar solvent to give poly(thio-1,4-phenylene), simply termed polyphenylene sulfide.

$$Cl-C_6H_4-Cl + Na_2S \longrightarrow [-C_6H_4-S-]_n + 2\,NaCl$$

The PPS polymer can be adjusted to give the required molar mass and has a largely linear structure, with a narrow distribution of chain length of the molecules.

The optimum molding conditions for PPS prepeg have been determined [155]. The prepreg was hot tacked together and placed in a picture frame mold, which was smaller than the final thickness of the laminate, covered with sheets of non-porous release cloth and placed between two grit-blasted stainless steel caul sheets. The complete assembly was placed in a preheated platen press at 316°C. Contact pressure was applied for 4 min followed by a pressure of 0.7–1.0 MPa for an additional 3 min. Initial cooling was achieved by placing in a press at room temperature and allowing to cool to below 38°C (about 1 min). Annealing was carried out at 204°C for 2 h. In common with other semi-crystalline polymers the physical and mechanical properties depend on the morphological structure. Lower levels of crystallinity will produce higher elongation and better toughness. Stiffness, thermal stability and chemical resistance are enhanced by higher levels of crystallinity, at the expense of decreased ductility. Other important parameters are the number and size of the spherulites, the crystalline structure and the crystalline orientation. Orientation can be responsible for anisotropy in the mechanical properties, whilst crystallite type and size may affect the overall mechanical properties. Larger spherulites, for instance, are inherently stiffer but less ductile.

In PPS, a largely amorphous system can be achieved by a quick quench from the melt. Smaller crystallites can then be formed by annealing above the T_g (95°C). Larger crystallites are formed by relatively slow cooling from the molten stage and being less ductile, fail at smaller deformation. Hence, tensile strengths and elongations will be lower. Amorphous material has yields and elongations an order of magnitude more than the annealed material.

In addition to the normal melting point at 275°C, there is a small peak at about 220°C referred to as an annealing peak.

Crystallization and morphology [156–158] have been considered. With materials processed at high temperatures, the type of size used on the carbon fiber is important [159] and the effect of physical aging on the toughness of PES composites [160] has been determined.

13.3 IMPROVING THE BOND WITH CARBON FIBER/THERMOPLASTICS

Initially, a weak interface between matrix and reinforcement was designed to optimize toughness [161]. A crack propagating through a brittle matrix would be deflected from the matrix at the surface of the reinforcing fiber, creating a large amount of free surface and thereby absorbing energy. With thermoplastic matrices, the energy can be dissipated within the matrix. The resin shrinks tightly onto the fiber, but when there is an attempt to move the resin away from the fiber due to transverse flexure or delamination, the advantages of strong adhesion are obvious. There are a number of ways in which strong adhesion can be achieved such as effective wetting of the fiber, chemical bonding, mechanical and crystalline interlocking.

Surface treatment can be tailored to suit a given matrix, but this does mean that the fiber manufacturer would have many surface treatments on the shelf which would not be economic and obviously, some compromise has to be found, although there would be no problem if there is sufficient demand.

Good wetting can be achieved by using a suitable size, but the size must be able to withstand the high processing temperatures used with thermoplastics, which rules out epoxy resins. Some success has been achieved using a dilute solution of the matrix polymer, which cannot be used for semi-crystalline materials like PEEK due to the non-availability of a suitable solvent, so a compromise has to be made sizing with an amorphous polymer. Using a size automatically produces an interphase between the reinforcement and the matrix.

Kenrich introduced a range of titanate liquid coupling agents (LICA), such as LICA12 RO—Ti—(O—P(OC$_8$H$_{17}$)$_2$)$_3$, which will withstand the demands of high temperature thermoplastics and works with carbon fiber reinforcements at dosage levels of about 0.1–0.5% based on the solid content. The titanates are capable of changing the system rheology and careful control of the processing parameters must be maintained to assure adequate internal shear development for proper dispersion. There is also a range of organozirconates, but these are more expensive [162].

REFERENCES

1. Stenzenberger HD, Recent developments of thermosetting polymers for advanced composites, *Composite Structures*, 24, 219–231, 1993.
2. Johnson JW, Resin matrices and their contribution to composite properties, *Phil Trans R Soc Lond*, A294, 487–494, 1980.
3. Dyson RW ed., *Engineering Polymers*, London, Blackie, 1990.
4. Lubin G ed., *Handbook of Composites*, Van Nostrand Reinhold Co., New York.
5. Weatherhead RG, *FRP Technology-Fibre Reinforced Resin Systems*, Applied Science Publishers Ltd., London, 1980.
6. Amoco Chemicals Corporation Bulletin IP-70, *How Ingredients Influence Unsaturated Polyester Properties*.
7. Scott Bader Company Limited Publication, *Crystic Polyester Handbook*, Wellingborough, 1994.
8. Dow Chemical Company, *Technical Literature on Derakane Vinyl Ester Resins*.
9. Lee H, Neville K, *Handbook of Epoxy Resins*, McGraw-Hill Company, New York, 1967.
10. Bruins PF, *Epoxy Resin Technology*, John Wiley, New York, 1968.
11. May CA ed., *Epoxy Resins—Chemistry and Technology*, Marcel Dekker, New York, 1988.
12. Shell Chemicals Publication, *The Long and the Short of Epoxy Resins*, 1992.
13. Dow Chemical Publication, *General Guide—Formulating with Dow Epoxy Resins*.
14. Patel SR, Patel RG, Physicomechanical properties of carbon-fiber reinforced epoxy composites, *Polymer-Plastics Technology and Engineering*, 31(7–8), 705–712, 1992.
15. Patel SR, Patel RG, Carbon-fiber reinforced epoxy composites, *Polym Int*, 30(3), 301–303, 1993.
16. Soni HK, Patel RG, Patel VS, Structural, physical and mechanical-properties of carbon-fiber-reinforced composites of diglycidyl ether of bisphenol-A and bisphenol-C, *Angewandte Makromolekulare Chemie*, 211, 1–8, 1993.
17. Sastri SB, Armistead JP, Keller TM, Phthalonitrile-carbon fiber composites, *Polymer Composites*, 17(6), 816–822, 1996.
18. Meier HM, Braun D, Eigenbach CD, *BMFT Symposium for Material Research, Proceedings*, Hamm, Vol II, 1529, 1988.
19. Lebonheur V, Stupp SI, Coupling carbon-fibers to epoxy matriceswith grafted side-chain liquid-crystal polymers, *Chemistry of Materials*, 6(10), 1880–1883, 1994.
20. Buggy M, Temimhan T, Braddell O, Curing of carbon fibre reinforced epoxy matrix composites, *J Mater Process Technol*, 56(1–4), 292–301, 1996.

21. Jones FR ed., *Handbook of Polymer Fibre Composites*, Polymer Science and Technology Series, Longman Scientific & Technical, Harlow, 1994.
22. Son P, Weber CD, Some aspects of monuron-accelerated dicyandiamide cure of epoxy resins, *J Appl Polym Sci*, 17, 1305–1313, 1973.
23. LaLiberle BR, Bornstein J, Sacher RE, Cure behaviour of an epoxy resin-dicyandiamide system accelerated by monuron, *Ind Eng Chem Prod Res Dev*, 22(2), 261–262, 1983.
24. Iwakura Y, Izawa S, *J Org Chem*, 29, 379, 1964.
25. Wang SP, Garton A, Chemical interactions at the interface between a carbon-fiber and a boron trifluoride-catalyzed epoxy matrix, *J Appl Polym Sci*, 45(10), 1743–1752, 1992.
26. Ciba Specialty Chemicals, *Technical literature on Cyanate Ester resins*.
27. Srinivasan S, Rau AV, Loos AC, McGrath JE, *Abstracts of Papers of the American Chemical Society*, 208(2), 407–pmse, 1994.
28. Armistead JP, Snow AW, Fiber matrix load–transfer in cyanate resin carbon-fiber systems, *Polymer Composites*, 15(6), 385–392, 1994.
29. Stenzenberger HD, Addition polyimides, *Advances in Polymer Science*, Springer Verlag, 117, 65–220, 1994.
30. Shell Resins Technical literature, *Compimide resins for composites*.
31. Ciba Specialty Chemicals, *Product data on Kerimid and Matrimid resins*.
32. National Starch and Chemical, Product data on Thermid resins.
33. Jenkins SD, Emmerson GT, McGrail PT, Robinson RM, *J Adhesion*, 45(1–4), 15–27, 1994.
34. Labronici M, Ishida H, Dynamic mechanical characterization of PMR polyimide/carbon fiber composites modified by fiber coating with silicones, *Composite Interfaces*, 5(3), 257–275, 1998.
35. Labronici M, Ishida H, Effect of the silicone interlayer on mechanical properties of carbon fiber reinforced PMR-15 polyimide composites, *Composite Interfaces*, 5(2), 87–116, 1998.
36. Scroog CE et al, *J Polym Sci*, A3, 1373, 1965.
37. Serafini TT, High temperature resins, G Lubin ed., *Handbook of Composites*, Van Nostrand Reinhold Co., New York.
38. Gibbs HH, *J Appl Polym Symposia*, 35, 207–222, 1979.
39. Lubowitz HR, TRW Systems, U.S. Pat., 3,528,950, 1970.
40. Lubowitz HR, *ACS Org Coat Plast Chem*, 31, 561, 1971.
41. St Clair TL, Jewell RA, *23rd Nat SAMPE Symposium*, 520–527, 1978.
42. Serafini TT, Delvigs P, Lightsey GR, *J Appl Polym Sci*, 16, 905, 1972.
43. Serafini TT, Vannucci RD, Alston WB, *NASA TM-71894 in Proceedings of the 23rd SAMPE National Symposium*, Apr 1976.
44. Schola DA, Vontell JH, Laube BL, Relationship between graphite fiber surface chemistry, PMR-15 composite shear strength and thermo-oxidative stability, *33rd International SAMPE Symposium*, 1506–1518, 7–10 Mar 1988.
45. Xiang ZD, Jones FR, Thermal degradation of an end-capped bismaleimide resin matrix (PMR-15) composite reinforced with PAN-based carbon fibres, *Composites Sci Technol*, 47, 209–215, 1993.
46. St Clair TL, Jewell RA, *8th Nat SAMPE Tech Conf*, 8, 82, 1976.
47. Grundschober F, Sambeth J, Rhone Poulenc, U.S. Pat., 3,380,964, 1968.
48. Bergain M, Combet A, Grosjean P, Rhone Poulenc, Brit. Pat., 1,190,718, 1968.
49. Morgan RJ, Jurek RJ, Yen A, Donnellan T, *Polymer*, 34(4), 835–842, 1993.
50. King J, Chaudhari M, Zahir S, *Nat SAMPE Symp*, 29, 392, 1984.
51. Kinloch AJ, Shaw ST, *ACS PMSE*, 49, 147, 1982.
52. Shaw SJ, Kinloch AJ, *Int J Adhesion Adhesives*, 5, 123, 1985.
53. Raghava RS, *Nat SAMPE Symp*, 28, 267, 1983.
54. Bucknall CB, Partridge I, *Brit Polym J*, 15, 71, 1983.
55. Stenzenberger HD, König D, Herzog M, Römer W, *Int SAMPE Symp*, 33, 1546, 1988.
56. Stenzenberger HD, *Proc 2nd Ann Int Conf on Crosslinked Polymers*, 61, Lucerne, May 30–June 1 1988.
57. Stenzenberger HD, Römer W, Herzog M, König P, *33rd Int SAMPE Symp*, 33, 1546, 1988.
58. Stenzenberger HD, Römer W, Hergenrother PM, B Jensen B, *34th Int SAMPE Symp*, 34, 2054, 1989.

59. Rakutt D, Fitzer E, Stenzenberger HD, *High Performance Polymers*, 2, 133, 1990.
60. Rakutt D, Fitzer E, Stenzenberger HD, *High Performance Polymers*, 3, 59, 1991.
61. Stenzenberger HD, Römer W, Hergenrother PM, Jensen B, Breitigam W, *35th Int SAMPE Symp*, 53, 2175, 1990.
62. Stenzenberger HD, König P, *High Performance Polymers*, 5, 123–127, 1993.
63. Wilkinson SP, Liptak SC, Wood PA, McGrath JE, Ward TC, *36th Int SAMPE Symp*, 35, 482, 1990.
64. Lecoustumer P, Lafdi K, Oberlin A, *Composites Sci Technol*, 52(3), 433–437, 1994.
65. Spratt GR, Akay M, High-temperature stability of bismaleimide carbon-fiber composite materials, *Key Eng Mater*, 99–1, 3–10, 1995.
66. Heisey CL, Wood PA, McGrath JE, Wightman JP, Measurement of adhesion between carbon fibers and bismaleimide resins, *J Adhesion*, 53(1–2), 117–147, 1995.
67. Bilow N, Landis AL, Miller LJ, Hughes Aircraft, U.S. Pat., 3,845,018, 1974.
68. Landis AL, Bilow N, Boshnan RH, Lawrence RE, Aponyi T, *ACS Polym Prep*, 15, 537, 1974.
69. Cam CY, Walker JV, Toughened Composites Selection Criteria, Johnston NJ ed.,*Toughened Composites, ASTM STP937*, American Society for Testing and Materials, Philadelphia, 9–22, 1987.
70. Recker HG, *SAMPE J*, 26(2), 73–78, 1990.
71. Bascom WD, *Polymr Mat Sci Eng*, 62, 676–680, 1990.
72. Recker HG, Altstadt V, Ebele W, Folda T, Gerth D, Heckmann W, Ittemann P, Tesch H, Weber T, *SAMPE J*, 26(2), 73 et seq, 1990.
73. Guigon M, Klinklin E, The interface and interphase in carbon-fiber-reinforced composites, *Composites*, 25(7), 534–539, 1994.
74. Kim YS, Kim SC, Toughening of carbon fiber/thermoset composite by the morphology spectrum concept, *Polymer Composites*, 19(6), 714–723, 1998.
75. *BF Goodrich technical literature on Hycar RLP*.
76. Bascom WD, Cottington RL, Jones RL, Peyser PJ, *J Appl Polym Sci*, 19, 2545, 1975.
77. Kinloch AJ, Shaw SJ, Hunston DL, *Polymer*, 24, 1355, 1983.
78. Pearson RA, Yee AF, *J Mater Sci*, 21, 2475,1986.
79. Kinloch AJ, Hunston DL, *J Mater Sci Lett*, 6, 137, 1987.
80. Huang Y, Kinloch AJ, *J Mater Sci*, 27, 2763, 1992.
81. Low BY, Anderson KL, Vincent M, Gardner SD, Pittman CU, Hackett RM, Toughened carbon fiber/epoxy composites-the relative influence of an elastomer interphase and elastomer dispersed in the matrix, *Composites Eng*, 4437–4457, 1995.
82. Siebert R, Rubber Modified Thermoset Resins, Riew KC, Gillham JK eds., *ACS Advances in Chemistry Series 208*, American Chemical Society, Washington, DC, 179, 1983.
83. *Fina Chemicals technical literature*.
84. Masters JE, Characterization of impact damage development in graphite/epoxy laminates, Johnston NJ ed.,*Toughened Composites, ASTM STP937*, American Society for Testing and Materials, Philadelphia, 1529, 1987.
85. Gunnink JW, Vogelesang LB, Janicki G, Bailey V, Schjelderup H eds., *Proc Int SAMPE Symp and Exhib 35: Advanced Materials: Challenge next decade*, 1708–1721, 1990.
86. Bucknall CB, Partridge JK, *Polymer*, 24, 639, 1983.
87. Bucknall CB, Partridge JK, *Eng Sci*, 26, 54, 1986.
88. Raghava RS, *J Polym Sci Polym Phys*, 26, 65, 1988.
89. Yamanaka K, Inoue T, *Polymer*, 30, 662, 1989.
90. Hedrick JL, Yilgor I, Durek M, Hedrick JC, Wilkes GL, McGrath JE, *Polymer*, 32, 2020, 1991.
91. Murakami A, Saunders D, Oosihi K, Yoshiki T, Saitoo M, Watanabe O, Takezawa M, *J Adhesion*, 39, 227, 1992.
92. Bucknall CB, Gilbert AH, *Polymer*, 30, 213, 1989.
93. Hourston DJ, Lane S, *Polymer*, 33, 1379, 1992.
94. Chung DDL, *Carbon Fiber Composites*, Butterworth-Heinemann, Boston, 1994.
95. Alger MSM, Dyson RW, Dyson RW ed., *Engineering Polymers*, Blackie, London, 1–28, 1990.

96. Lustiger A, Considerations in the utilization of semicrystalline thermoplastic advanced composites, *SAMPE Journal*, 13–16, Sep/Oct, 1984.
97. Sacher E, *J of Macromolecular Sci, Physics Edit*, 9, 163, 1974.
98. Cogswell FN, Microstructure and properties of thermoplastic aromatic polymer composites, *28th National SAMPE Symposium*, 28, 528, 1983.
99. Hobbs SY, *Nature Phys Sci*, 234, 12, 1971.
100. Hartness JT, An evaluation of polyetheretherketone matrix composites fabricated from unidirectional prepreg tape, *SAMPE J*, 26–31, Sep/Oct 1984.
101. Beaumont P, *Lecture notes*, Dept of Engineering, Univ of Cambridge.
102. Phillips DC, *J Mater Sci*, 7, 1175, 1972.
103. Bashtannik TI, Kabak AI, Zinuhov VD, *Mechanics of Composites Materials*, 34(5), 483–488, 1998.
104. Oya N, Hamada H, Mechanical properties and failure mechanisms of carbon fibre reinforced thermoplastic laminates, *Composites Part A-Appl Sci Manuf*, 28(9–10), 823–832, 1997.
105. Blundell DJ, Chalmers JM, MacKenkie MW, Gaskin WF, Crystalline morphology of the matrix of PEE carbon fibre aromatic polymer composites, Part 1: Assessment of crystallinity, *SAMPE Quarterly*, 16(4), 22–30, 1985.
106. Sheu MF, Lin JH, Chung WL, Ong CL, The measurement of crystallinity in advanced thermoplastics, *33rd International SAMPE Symposium*, 1307–1318, 1988.
107. Cebe P, Lowry L, Chung S, Use of scattering methods for characterization of morphology in semi-crystalline thermoplastics, *SPE 47th ANTEC*, 1989.
108. Sichina WJ, Gill PS, *Characterization of PEEK/carbon fibre composites by thermal analysis*, DuPont Instrument Company, Delaware, 1986.
109. Ma CCM, Yur SW, Parameters affecting the crystallization of carbon fibre reinforced Polyetheretherketone, *SPE 47th ANTEC*, 1422–1429, 1989.
110. Petrovan S, Murariu M, Harabagiu L, Avadenei L, Bodron V, Andrei E, Polyamide-based composites for injection- and extrusion-processing, 2. Carbon fibre- and glass balls-reinforced polyamides, *Materiale Plastice*, 32(3–4), 238–243, 1995.
111. Zam Ishak, Berry JP, Impact properties of short carbon-fiber- reinforced nylon 6,6. *Polym Eng Sci*, 33(22), 1483–1488, 1993.
112. Klein N, Marom G, Wachtel E, Microstructure of nylon 6,6 transcrystalline layers in carbon and aramid fibre reinforced composites, *Polymer*, 37(24), 5493–5498, 1996.
113. Tawari US, Bijwe J, Mathur JN, Sharma I, Studies on abrasive wear of carbon- fiber (short) reinforced polyamide composites, *Tribology Int*, 25(1), 53–60, 1992.
114. Bessell T, Shortall JB, The crystallization and interfacial bond strength of nylon 6 at carbon and glass fibre surfaces, *J Mater Sci*, 10, 2035–2043, 1975.
115. Caldeira G, Maia JM, Carneiro OS, Covas JA, Bernardo CA, Production and characterization of innovative carbon fiber polycarbonate composites, *Polymer Composites*, 19(2), 147–151, 1998.
116. Stone PR, Nairn JA, Interfacial toughness and its effect on compression strength in polycarbonate carbon-fiber composites, *Polymer Composites*, 15(3), 197–205, 1994.
117. Zihlif AM, DiLiello V, Martuscelli E, Ragosta G, Tensile properties and fracture behaviour of polycarbonate/PAN based carbon fiber composites, *Int J Polym Mater*, 29(3–4), 211–220, 1995.
118. Cogswell FN, *Thermoplastic Aromatic Composites*, Butterworth-Heinemann, Oxford, 1992.
119. Saiello S, Kenny J, Nicolais L, Interface morphology of carbon-fiber PEEK composites, *J Mater Sci*, 25(8), 3493–3496, 1990.
120. Wang W, Qi ZN, Jeronimidis G, *J Mater Sci*, 26(21), 5915–5920, 1991.
121. Lustiger A, Morphological aspects of the interface in the PEEK-carbon fiber system, *Polymer Composites*, 13(5), 408–412, 1992.
122. Jar PY, Cantwell WJ, Kausch HH, Study of the crystal morphology and the deformation-behaviour of carbon-fiber reinforced PEEK (APC-2), *Composites Sci Technol*, 43(3), 299–306, 1992.
123. Denault J, Vukhanh T, Crystallization and fiber matrix interaction during the molding of PEEK carbon composites, *Polymer Composites*, 13(5), 361–371, 1992.
124. Zhang ZY, Zeng HM, Nucleation and crystal-growth of PEEK on carbon-fiber, *J Appl Polym Sci*, 48(11), 1987–1995, 1993.

125. Saktoun EM, Boudet A, Chabert B, Hilaire B, Bouvart D, Morphology in PEEK carbon-fiber composites observed in transmission electron-microscopy, *Polymer*, 434(12), 2668–2669, 1993.
126. Hachmi BD, VuKhanh T, Crystallization mechanism in PEEK/carbon fiber composites, *J Thermoplastic Mater*, 10(5), 488–501, 1997.
127. Lin SH, Ma CCM, Tai NH, *J Adv Mater*, 28(2), 56–62, 1997.
128. Gao SL, Kim JK, Interphase morphology and fibre pull-out behaviour of carbon fibre PEEK composites, *Key Eng Mater*, 145(1,2), 811–816, 1998.
129. Schmid B, Injection-molding of carbon-fiber reinforced polyaryletherketone, *Kunstoffe-German Plastics*, 82(8), 697–700, 1992.
130. Semadeni M, Zerlik H, Rossini P, Meyer J, Wintermantel E, High fibre volume fraction injection moulding of carbon fibre reinforced polyetheretherketone (PEEK) in order to raise mechanical properties, *Polym Polym Composites*, 6(5), 279–286, 1998.
131. Lee DJ, On studies of tensile properties in injection molded short carbon fiber reinforced PEEK composite, *KSME J*, 10(3), 362–371, 1996.
132. Kominar V, Narkis M, Sigemann A, Vaxman A, Compressive strength of unidirectional and crossply carbon fibre/PEEK composites, *J Mater Sci*, 30(10), 2620–2627, 1995.
133. Colton J, Leach D, Processing parameters for filament winding thick-section PEEK carbon-fiber composites, *Polymer Composites*, 13(6), 427–434, 1992.
134. Dyson IN, Kinloch AJ, Okada A, The interlaminar failure behaviour of carbon-fiber polyetheretherketone composites, *Composites*, 25(3), 189–196, 1994.
135. Hy Yoon, Takahashi K, Mode-I interlaminar fracture-toughness of comingled carbon-fiber PEEK composites, *J Mater Sci*, 28(7), 1849–1855, 1993.
136. Cantwell WJ, Zulkifli R, An investigation into mode-II failure mechanisms in carbon fibre reinforced PEEK, *J Mater Sci Lett*, 16(7), 509–511, 1997.
137. Sivashanker S, Damage growth in carbon fibre PEEK unidirectional composites under compression, *Mater Sci Eng A- Structural Materials Properties Microstructure and Processing*, 249(1–2), 259–276, 1998.
138. Hine PJ, Brew B, Duckett RA, Ward IM, The fracture behaviour of carbon fibre reinforced poly(ether etherketone), *Composites Sci Technol*, 33, 35–71, 1988.
139. Hine PJ, Brew B, Duckett RA, Ward IM, Failure mechanisms in continuous carbon fibre reinforced PEEK composites, *Composites Sci Technol*, 35, 31–51, 1989.
140. Hine PJ, Brew B, Duckett RA, Ward IM, Failure mechanisms in carbon fibre reinforced poly(etheretherketone),II: Material variables, *Composites Sci Technol*, 40, 47–67, 1991.
141. Saib KS, Isaac DH, Evans WJ, Effects of processing variables on fatigue in molded PEEK and its short fiber composites, *Mater Manuf Processes*, 9(5), 829–850, 1994.
142. Phillips R, Glauser T, Manson JAE, Thermal stability of PEEK/carbon fiber in air and its influence on consolidation, *Polymer Composites*, 18(4), 500–508, 1997.
143. Buggy M, Carew A, The effect of thermal aging on carbon-fiber-reinforced polyetheretherketone (PEEK), 2. Morphological changes, *J Mater Sci*, 29(8), 2255–2259, 1994.
144. Katouzian M, Bruller OS, Horoschenkoff A, On the effect of temperature on the creep-behaviour of neat and carbon-fiber-reinforced PEEK and epoxy-resin, *J Composite Mater*, 29(3), 372–387, 1995.
145. Gaitonde JM, Lowson MV, Low-temperature thermal-expansion of PEEK, HTA and some of their composites reinforced with carbon-fibers, *Composites Sci Technol*, 40(1), 69–85, 1991.
146. Ye L, Daghyani HR, Sliding friction and wear of carbon fibre polyetheretherketone comingled yarn composites against steel, *J Mater Sci Lett*, 15(17), 1536–1538, 1996.
147. Hanchi J, Eiss NS, Dry sliding friction and wear of short carbon-fiber-reinforced polyetheretherketone (PEEK) at elevated temperatures, *Wear*, 203, 380–386, 1997.
148. Flock J, Friedrich K, Yuan Q, On the friction and wear behaviour of PAN- and pitch-carbon fiber- reinforced PEEK composites, *Wear*, 229(1), 304–311, 1999.
149. Cantwell WJ, Davies P, Bourban PE, Jar PY, Richard H, Kausch HH, Thermal joining of carbon-fiber reinforced PEEK laminates, *Composite Structures*, 16(4), 305–321, 1990.
150. Cantwell WJ, Davies P, Kausch HH, Repair of impact-damaged carbon-fiber PEEK composites, *SAMPE J*, 27(6), 30–35, 1991.

151. Ramakrishna S, Tan WK, Teoh SH, Lai MO, Recycling of carbon fiber PEEK composites, *Key Eng Mater*, 137, 1–8, 1998.
152. Nelson KM, Seferis JC, Zachmann HG, Solvent-induced crystallization in polyetherimide thermoplastics and their carbon-fiber composites, *J Appl Polym Sci*, 42(5), 1289–1296, 1991.
153. Bullions TA, Mehta RH, Tan B, McGrath JE, Kranbuehl D, Loos A, Mode-I and Mode-II fracture toughness of high-performance 3000g mole^{-1} reactive poly(etherimide)/carbon fiber composites, *Composites Part A-Appl Sci Manuf*, 30(2), 153–162, 1999.
154. Hine PJ, Brew B, Duckett RA, Ward IM, Failure mechanisms in carbon-fiber-reinforced poly(ether sulfone), *Composites Sci Technol*, 43(1), 37–47, 1992.
155. Beever WH, Ryan CL, O'Connor JE, Lou AY, Ryton®-PPS Carbon Fiber Reinforced Composites: The How, When, and Why of Molding, Johnston NJ ed., *Toughened Composites ASTM STP937*, American Society for Testing and Materials, Philadelphia, 319–327, 1987.
156. Lee KH, Park M, Kim YC, Choe CR, Crystallization behaviour of polyphenylene sulfide (PPS) and PPS carbon-fiber composites-effect of cure, *Polym Bull*, 30(4), 469–475, 1993.
157. Konda A, Ohkoshi Y, Takahashi H, Shimizu Y, Crystal-growth of poly(etherether ketone), poly(phenylene sulfide), and poly(ethylene-terephthalate) on the surface of carbon-fibers, *Kobunshi Ronbunshu*, 51(1), 69–72, 1994.
158. Caramaro L, Chabert B, Chauchard J, Vukhanh T, Morphology and mechanical performance of polyphenylenesulfide carbon-fiber composite, *Polym Eng Sci*, 31(17), 1279–1285, 1991.
159. Park M, Lee KH, Choe CR, Jo WH, Effect of sizing materials of carbon fiber on solid-state cure of poly(p-phenylene sulfide), *Polym Adv Technol*, 9(2), 134–137, 1998.
160. Ma CCM, Lee CL, Chang MJ, Tai NH, Effect of physical aging on the toughness of carbon fiber-reinforced poly(etheretherketone) and poly(phenylene sulfide) composites, Part 1, *Polym Composites*, 13(6), 441–447, 1992.
161. Johnston NJ ed., *Toughened Composites ASTM STP937*, American Society for Testing and Materials, Philadelphia, 1987.
162. Sugerman G, Gabayson SM, Chitwood WE, Monte SJ, *Proc 3rd Dev Sci Technol Compos Mater*,, Bunsell AR, Lamicq P, Massiah A eds., European Conference on Composite Materials, ,Elsevier, London, 51–56, 1989.

CHAPTER 14

Carbon Fiber Carbon Matrix Composites

14.1 INTRODUCTION

A carbon-carbon composite is a carbon fiber reinforced carbon matrix material, where the carbon matrix phase is typically formed by the pyrolysis of a solid, liquid or gaseous organic precursor material. The matrix can be either a graphitizable or a non-graphitizable carbon and the carbonaceous reinforcement is in fibrous form. The composite may also contain other components in particulate or fibrous forms.

The real development of carbon-carbon composites started in 1958, with US Air Force sponsored work, which later received a massive boost with the onset of the Space Shuttle Program. Carbon-carbon composites can be tailor-made to give a wide family of products by controlling the choice of fiber type, fiber presentation and matrix.

Many informative publications on carbon-carbon materials and composites are available [1–18], although all production processes remain strictly proprietary.

The precursor fiber type for reinforcing the carbon matrix can be an oxidized PAN fiber (opf), or either a PAN or pitch based carbon fiber. In some instances, for special applications, such as the Shuttle, a cellulose based carbon fiber is used. The reinforcements can be unidirectional; have a random chopped fiber presentation as in a felt format; a woven product from continuous fiber presented in a 2D, 3D, or in a Multi-D format (Section 21.1), or a non-woven carbon fiber. The chosen fiber architecture is most important for a given application and Lei et al [4] describe how, for example, 3-D braiding can be applied to carbon-carbon composites. One of the early forms of near net shape reinforcement used for carbon-carbon aircraft brakes was based on a weft knitted 3-D fabric made by the Pressure Foot® process (Figure 14.1).

The precursor matrix material can be a hydrocarbon gas, a thermoset resin such as a phenolic or furan, or a thermoplastic resin such as a pitch or thermoplastic polymer.

There are two primary methods of manufacturing carbon-carbon. One method is based on chemical vapor infiltration (CVI) also termed chemical vapor deposition (CVD), employing a number of techniques for the infiltration process:

1. Isothermal CVI
2. Thermal Gradient CVI (TG-CVI)
3. Pressure Gradient
4. TG-CVI plus Pulse CVI
5. Other methods

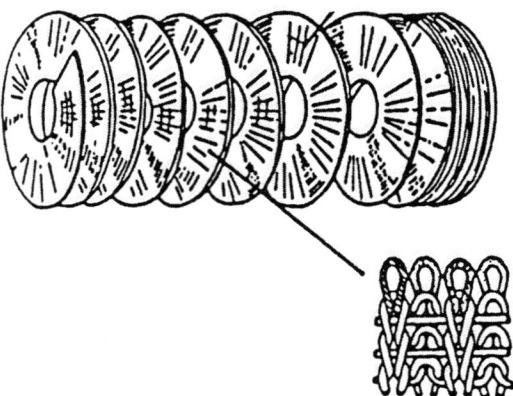

Figure 14.1 Weft knit 3-D fabric made by the Pressure Foot® process producing a near net shaped structure, which when collapsed was used for the construction of carbon-carbon aircraft brakes. *Source:* Reprinted with permission from Williams DJ, New knitting methods offer continuous structures, *Adv Comp Eng Summer*, 12–13, 1978. Copyright 1978, Maney Publishing (who administers the copyright on behalf of IOM Communications Ltd, a wholly owned subsidiary of the Institute of Materials, Minerals & Mining).

The second manufacturing method is based on a liquid phase impregnation process, where the liquid matrix can be either a thermoset or thermoplastic resin, including pitch based resins. There are three basic processes used for impregnation:

1. Low Pressure Impregnation (LPI)
2. Pressure Impregnation and Carbonization (PIC)
3. Hot Isostatic Pressure Impregnation Carbonization (HIPIC)

It is also possible to combine the CVI and LPI processes.

14.2 SELECTION OF MATERIALS FOR CARBON-CARBON PROCESSING

14.2.1 Types of reinforcement

14.2.1.1 Oxidized PAN fiber (opf)

A particular advantage of using opf is that a preform can be more readily processed by conventional textile processes than by using the alternative carbon fiber. The opf route is favoured by BF Goodrich, Dunlop, Honeywell and Messier Bugatti. Chief suppliers of opf are SGL and Zoltek and, whilst Dunlop had their own inhouse facility to manufacture opf at one stage, this has now been discontinued and all supplies are purchased outside.

The blank disks for carbon brakes are produced using modified conventional textile, but are strictly proprietary procedures. Due to the fact that the opf will be converted to a PAN based carbon fiber, the product will be non-graphitic, unsized and untreated.

14.2.1.2 PAN based carbon fibers

Carbon fibers are normally sized for improved handling and if this size is present, it would decompose during the first carbonization stage. Hence it is considered advisable, as a preliminary step, to burn off any size prior to further processing.

Since the matrix and reinforcing fiber in carbon-carbon are both brittle materials, it is preferred to aim for a weak interface in the composite, permitting cracks to travel through the matrix in order to debond at the fiber/matrix interface.

Fitzer and Burger [20] deduced that poor bonding was required, promoting longitudinal fissuring, into which carbon could be deposited to give a better bond and good load transfer.

Boyne et al [21] reported that the flexural strengths of carbon-carbon composites made with Grafil fibers was greatest with HTU and HMU fibers, which showed an unexpected degree of toughness for such brittle materials.

Rand and Robinson [22] showed that although there was little difference in the surface areas of the treated and untreated Grafil fibers, treatment had substantially increased the active surface area and the fraction of the external surface, which is in the form of edge, or active sites. The general understanding was that improvement in ILSS was achieved by bonding to surface complexes.

Thomas and Walker [17] used a series of Courtauld's Grafil fibers, including a range of specially prepared HTS fibers which were surface treated at different levels, which were prepregged with 30 w/o of a phenolic novalac resin and low void composites prepared by curing with hexamine [23]. The phenolic composites were converted to carbon-carbon by pyrolyzing in an inert atmosphere, followed by the deposition of pyrolytic carbon from a hydrocarbon. They found that high strength and high strain treated fibers which were used to produce a tough composite in conventional thermoset matrices, gave brittle low strength carbon-carbon composites.

The surfaces of the fibers were examined by X-ray Photoelectron Spectroscopy (XPS or ESCA) [17] and a strong correlation was found between the surface oxygen of the fiber and the ILSS (measured at a 5:1 span : depth ratio) of the phenolic resin composite. A relationship was found between the longitudinal flexural strength (60:1 span : depth ratio) retained in the carbon-carbon composite and the ILSS of the phenolic composite. Generally, fibers processed above 1500°C and untreated yielded stronger, less brittle carbon-carbon composites. The adhesion of the resin char to the fiber affects the densification and with untreated fibers, the resin char shrinks away from the fiber leaving longitudinal fissures around the fiber, which could subsequently be filled with the pyrocarbon. With treated fiber, the resin char is well bonded and shrinks onto the fiber with no longitudinal gaps, but with regularly spaced transverse cracks.

Fitzer et al [24] found that increasing the bond between carbon fiber and a phenolic matrix decreased the tensile strength and toughness when carbonized at 1000°C. Untreated HM (Type1) fibers exhibit very low adhesion to the resin, giving a low cross-sectional bulk shrinkage, enabling pores to be more readily filled after repeated impregnation steps. HT fibers, on the other hand, have good adhesion. Cross-sectional bulk shrinkage [24] occurred after the first carbonization, which cannot be compensated for by bulk shrinkage in the fiber direction, due to the low compressibility parallel with the fiber axis. A high degree of compression pre-stress is then built up, which reduces strength of the composite.

Surface treated fiber can be detreated by removing the phenolic resin from the prepreg, heating in Ar at 1100°C, re-impregnating with phenolic resin and making a phenolic composite, which possesses a reduced ILSS, leading to improved carbon-carbon properties.

Fitzer [14] also established that PAN based HM or MP pitch fibers should preferably be used without surface treatment as a better translation of fiber properties can be achieved. Fitzer [25] (Figure 14.2) depicted the negative effects of surface treatment and attributed this to the shrinkage behavior of the composite during the first carbonization stage.

Manocha [26,27] reported that graphitized fiber produced a better densified carbon-carbon than ungraphitized carbon fibers. This was attributed to the weaker bond in the HM fibers shrinking away to give voids, that could subsequently be filled with resin, whereas

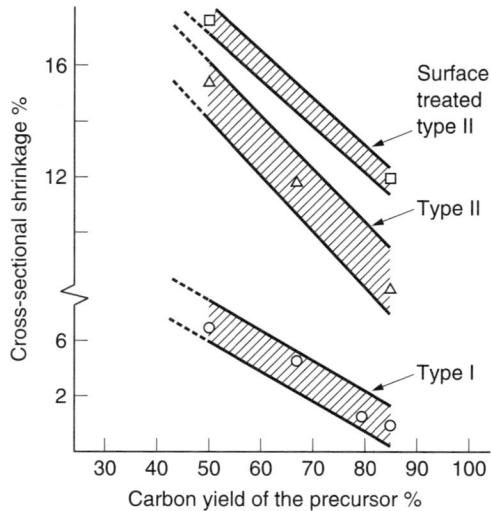

Figure 14.2 Cross-sectional shrinkage as a function of carbon yield in the first carbonization. Unidirectional composites fabricated with PAN based fibers and various matrix precursors. *Source:* Reprinted with permission from Fitzer E, The future of carbon-carbon composites, *Carbon*, 25(2), 163–190, 1987. Copyright 1987, Elsevier.

the polar groups on the less carbonized fiber formed a stronger bond which was less likely to shrink away.

Kowbel and Shan [7], when surface treating carbon fiber to improve the bond to a phenolic precursor, found that after heat treatment to 1000°C, the improved adhesion was reflected in decreased tensile strength and toughness. Hüttinger and Krekel [28] treated Grafil X-AS with polydimethylsiloxane, which when converted to silica above 700°C to give a strong bond at 900°C.

Peebles *et al* [29] reviewed the interaction of interface and matrix and showed that irrespective of the matrix source, the layer planes of the matrix align with the fiber surface [30,31].

Manocha *et al* [32] reported the effect of surface treatment and preferred carbon fiber with a circular cross-section, which was less likely to form stressed areas around the filament periphery [33].

Handling untreated, unsized HT or HM fiber, however, is difficult in practice and is another reason why the opf route is so advantageous.

Graphitization of a carbon-carbon composite has a greater effect on the matrix or fiber and occurs mainly in the matrix as a sheath, some 1–3 μm thick. This thermal expansion stress is termed stress graphitization [34]. Jortner depicts other forms of orientations (Figure 14.3) that have been observed [35].

14.2.1.3 Pitch based carbon fibers (pbcf)

Pitch based carbon fibers are graphitic and will give carbon-carbon composites with higher yields, densities, moduli, thermal conductivities and heat capacities, but will have lower strengths and will be softer than the product made from a PAN fiber. The type of microstructure of the pitch fiber does have an influence on the mode of graphitization of the carbon-carbon composite and a fiber with a parallel sheet-like microstructure pre-stresses the carbon matrix, causing shear and taking up a position in the fiber direction during

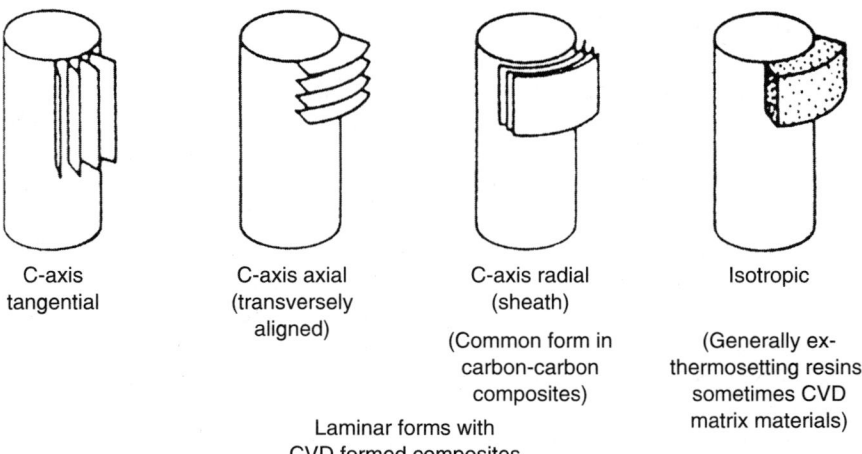

Figure 14.3 Possible matrix orientations of carbon matrix around a fiber. *Source:* Reprinted from Jortner J, Cracking in 3D carbon/carbon composites during processing and effects on performance, *Proc Army Symp Solid Mechanics*, AMMRC MS 76–2, 81, 1976. US Army Publication.

the graphitization process. Stretching the fibers will result in an expansion of the composite, thus increasing the flexural strength [36]. A pitch based fiber with a sheath and core structure does not expand in the fiber direction, and the flexural strength will decrease.

14.2.1.4 Cellulose based carbon fibers

Cellulose based carbon fibers tend to be used for ablative applications, which require lower thermal conductivity and enhanced insulating properties than other carbon-carbon materials. The precursor for this product is rayon, which when carbonized, produces rayon based carbon fibers (RBCF). This type of fiber is usually more porous and weaker, which lowers its effective thermal conductivity. At high temperatures, it oxidizes and/or evaporates without disintegration, producing an evaporative cooling effect.

There is now, to the knowledge of the author, only one supplier of the precursor and Polycarbon is the only one manufacturer of the RBCF. It is expensive to produce, with yields in the region of 30% and it is only existing government contracts which make production feasible.

14.2.2 Type of matrix

The selection of a suitable precursor matrix for a carbon-carbon composite is most important and the following factors should be taken into consideration [14]:

1. The carbon yield must be high, with consequent low weight loss during carbonization.
2. The liquid or molten precursor must have a low viscosity to aid penetration of the matrix.
3. The liquid or molten precursor must readily wet out the carbon fiber reinforcement.
4. Carbonization of the matrix should produce a favourable coke microstructure.

Precursors can be divided into two general categories—thermosetting and thermoplastic resins (including pitches).

14.2.2.1 Thermosetting resin

Kimura [37] selected three kinds of thermosetting resins—furan, diphenyletherformaldehyde and polyimide resins—as matrix precursors to fabricate carbon fiber reinforced carbon composites (C/C composites). After heat treatment at 2000–3000°C, the graphitization process of the matrix was examined by optical microscopy and X-ray diffraction. In the C/C composite derived from a polyimide, the graphite structure was not as well developed as the others. This retarded development is attributed to less internal stress between fibers and matrix as well as to less stretching of the matrix.

When using a thermosetting resin, it is usual to take the carbon fiber reinforcement and prepare a prepreg by impregnating with the chosen resin. Alternatively, a dry preform of opf can be used for conversion to carbon fiber and subsequent initial treatment by the CVI process to provide a preform with some integrity [38]. Polymerization shrinkage will occur, which can be minimized by controlling the rate of temperature rise.

The most promising candidates for a resin matrix are the furfuryl alcohol and phenolic resin systems, which cure by a condensation polymerization releasing water, which will contribute further to porosity and shrinkage.

1. Furan resin

Furfuraldehyde (furan), when reacted with NaOH, forms furfuryl alcohol, which can be polymerized by applying heat in the presence of an acid catalyst (e.g. oxalic acid or maleic anhydride) to initiate the reaction. Once started, the reaction becomes exothermic and can be controlled, or stopped by cooling. The process initially forms difurfuryl alcohol and finally a highly crosslinked resin with a yield, when pyrolyzed at 950°C, of some 50–60% carbon with a density of about 1.55 $g\,cm^{-3}$ and a linear shrinkage of around 20%. In the process, H_2O, CH_4, CO_2, CO and H_2 are evolved [39] and above 450°C, the water evolved in this process actually takes part in the latter stages of the pyrolysis.

Manocha and workers [40] used two types of fibers—an HT fiber (HTA of AKZO company) and an HM fiber (M40 J of Toray Company) to form a furfural alcohol composite.

Fibers without surface treatment had inferior stress-strain properties for samples carbonized at 1000°C, whereas samples graphitized at 3000°C showed better properties

2. Phenolic resins [41]

A phenolic resin developed by Monsanto (Resinox SC-1008), now obtainable from Borden Chemical), gives a comparatively high carbon yield when pyrolyzed. The phenolic resin is typically cured with hexamine in a condensation reaction. The reaction products depend on the initial degree of crosslinking and low molecular weight materials and H_2O are evolved in the region 100–350°C, whilst CO and CH_4 are released above 500°C. Siebold [42] claimed that best properties were obtained by slow pyrolysis up to 1000°C, giving a 54% carbon yield with a density of 1.45 g cm^{-3} and linear shrinkage of 20%.

The pyrolysis chemistry of phenol-formaldehyde polymer [43,44] is:

Interestingly, Kotosonov and co-workers [45] found that a phenol-furfural formaldehyde resin can be graphitized by the application of a pressure (~50 MPa) in the temperature range 400–600°C, but outside these limits, virtually no graphitization can be achieved.

The properties and structure of carbon-carbon composite produced from a phenolic resin have been reported by Ludenbach et al [46].

An HT fiber composite was found to be dimensionally and structurally unstable well below the maximum fiber processing temperature of 1400°C. The fiber shrank (the frozen in process stress relaxes) at temperatures as low as 850°C. The shrinkage of the fiber bundle embedded in phenolic resin during the carbonization process was influenced by matrix shrinkage stresses and pyrolysis products. Above 1000°C, the HTA carbon fiber in carbon-carbon bundles continuously changed its structure. After heat treatment at a temperature of 2800°C, the structure (lattice distance, orientation of the crystallites, crystallite size) was very similar to that of HM fibers.

3. Polyimide resins

Although polyimides [47] give a 60% carbon yield when pyrolyzed, they have not gained wider recognition due to processing problems.

Polyimide (Kapton)

Similarly the H-resins (acetylene terminated oligomers) developed by Hercules had very high carbon yields of up to 95% [48], but were eventually withdrawn due to processing problems when making a prepreg.

Hercules HA 43

14.2.2.2 Thermoplastic matrix precursors

1. Pitch [49–53]

Pitches can be derived from coal tar or petroleum and have been discussed as precursor materials for making pitch based carbon fibers (Chapter 4, Section 4.4). Pitches are oligomers and the composition will depend on the exact source and method of processing. A pitch with a high carbon yield and the ability to flow under high pressure should be selected.

An isotropic pitch gives a carbon yield of about 50%, which increases to about 85% for a mesophase pitch [54] although mesophase pitches are more viscous. Fitzer and co-workers [14,55] found that the yield could be increased by using an additive such as S. High sulphur content was found to favour a non-graphitizing carbon, with the S possibly forming a S containing polymer. A typical pitch would be Ashland's petroleum pitch, A-240. When an isotropic pitch is heated from room temperature, it melts to give a low viscosity material, which devolatilizes as the temperature rises, until a critical temperature is reached when the viscosity increases rapidly [56]. Above 400°C, mesophase spheres about 0.1 μm in diameter form gradually, coalescing to form a continuous phase that favours subsequent graphitization. High pressures cause a reduction in the temperature of the formation of mesophase and 100 MPa is considered optimum [57].

Huttinger and Christ [58] based their studies on liquid phase impregnation of a fiber preform with pitches of different contents of mesophase spherulites. The matrix formed was mainly mesophase because the isotropic mesogenic pitch was pressed out. The composites were stabilized by treatment with air before carbonization of the matrix in order to avoid swelling. After graphitization at 2100°C, flexural strengths up to 650 MPa were measured.

The composition of a pitch is very complex and, together with the processing conditions, controls the nature of the carbon matrix and it is normal to use a blend of pitches to obtain the requisite properties. The carbonization and graphitization of pitch is considered in

CARBON FIBER CARBON MATRIX COMPOSITES

Chapter 7. A preform can be initially impregnated with an epoxy to provide some interfilament adhesion and then impregnated with a pitch [59].

A pitch based resin gives a higher yield than a phenolic resin, is cheaper and when graphitized, has greater density (about 1.9 g cm^{-3}), but does require high pressures for processing.

2. Other thermoplastic matrices

Thermoplastic polymers such as PEEK and PEI have been successfully used [60,61] as alternative polymer matrices, but it is doubtful whether their high cost can ever be justified.

Polyarylacetylene, formed from the polymerization of diethynylbenzene [62,63] is non-graphitizing, but does have a char yield of 88% and can be catalytically graphitized at lower temperatures using carborane ($C_2B_{10}H_{12}$), a boron containing compound.

A new thermosetting resin [64], consisting of condensed aromatic nuclei crosslinked with methylene bridges was prepared from a mixture of pyrene, phenanthrene and 1,4-benzene-dimethanol by heating above 665°C. This resin, named COPNA, adheres well to carbon fibers and the carbon fiber/resin composite (cfrp) prepared by using this resin as a binder exhibits no remarkable changes in mechanical properties after heating at 795°C for 10 h and 845°C for 2 h. The cfrp specimens were converted into carbon-carbon composite by further heating.

Other resins have been investigated and include polybenzimidazole (PBI) [48], which has a 73% carbon yield.

14.3 METHODS OF PROCESSING CARBON-CARBON MATRIX MATERIALS

14.3.1 Introduction

A schematic layout of some methods which can be used to fabricate carbon-carbon composites from different reinforcement materials is shown in Figure 14.4. Material processed at 1100°C by the thermal pyrolysis of a carbon bearing vapor has an isotropic nature and is termed pyrolytic carbon, whereas material processed in the range 1000–1700°C is of a more intermediate nature, becoming increasingly graphitic with rise in temperature, whilst material processed at 1700–2300°C, deposited from a hydrocarbon gas, gives an increased graphitic nature and is loosely termed pyrolytic graphite.

After densification of the preform, the matrix pores become blocked, preventing any more carbon precursor from penetrating the fiber array and it is believed that at this stage that the composite can be cooled quickly, so that the matrix cracks due to a thermal mismatch, allowing the densification process to be continued by filling the cracks so formed. This process can be repeated some 4–6 times.

14.3.2 Use of gas phase impregnation and densification

14.3.2.1 Introduction

Chemical vapor deposition (CVD) involves heating a fiber preform in a gaseous environment to deposit the matrix, present in the gaseous phase, on to the fiber. The term chemical vapor infiltration (CVI) is used to describe CVD densification occurring within the fiber preform as distinct from a simple surface deposition technique. Kohno describes the carbon infiltration of carbon-carbon composites [65].

Although CH_4 requires a temperature above 550°C to initiate carbon deposition, it is widely available and has excellent diffusion properties and is hence, widely used in the CVD process. Diffusion can be increased further by dilution with H_2, He, N_2 or Ar to increase the mean free path (the average distance a molecule moves between two successive collisions) of the CH_4.

Spear [66] outlined a basic model for the different stages that occur during CVD:

1. Forced flow of reactant gases into the reaction vessel
2. Diffusion of reactants through laminar flow boundary layers around the substrate
3. Adsorption of reactants on the surface of the substrate
4. Reaction of adsorbed reactants to give solid products and adsorbed gaseous products
5. Desorption of adsorbed gaseous products

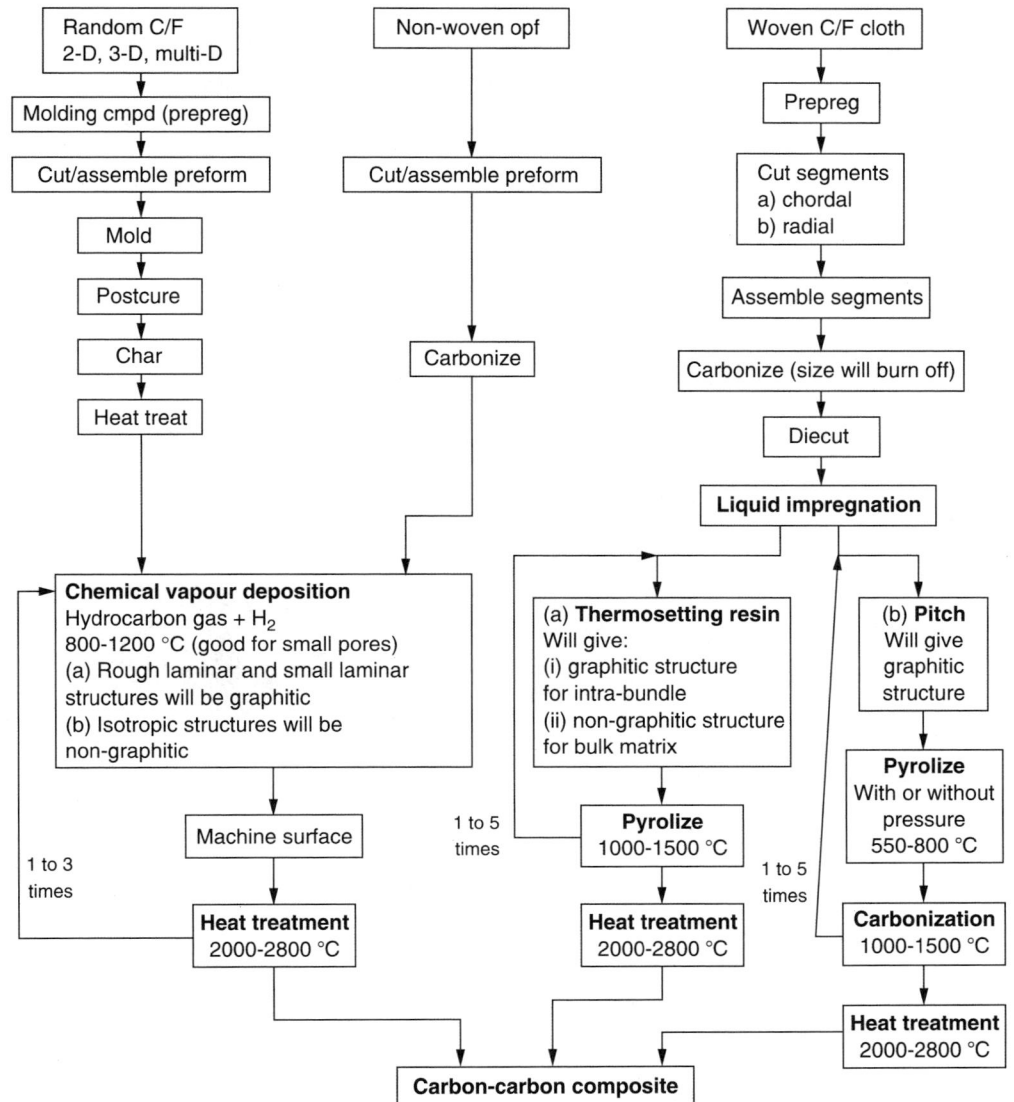

Figure 14.4 Schematic layout of typical methods used to fabricate carbon-carbon composites from different reinforcement materials. *Source:* Adapted from Allied Signals Aerospace technical literature.

6. Diffusion of gaseous products through a boundary layer region
7. Forced flow of gaseous products through a reaction vessel outlet

It is essential to secure the correct balance between diffusion and surface reaction kinetics (Figures 14.5 and 14.6) and a high rate of deposition will tend to block the pores and not fill them. Inevitably, this blocking will occur during the process and it becomes necessary to machine the surface to re-open the pores for subsequent filling. Consequently, to achieve successful densification of the composite, the CVD process must have protracted processing times.

Pierson and Liebermann [68] provided a model showing how the microstructure of the CVI deposit could vary and assumed that CH_4 decomposed to either acetylene (C_2H_2), when

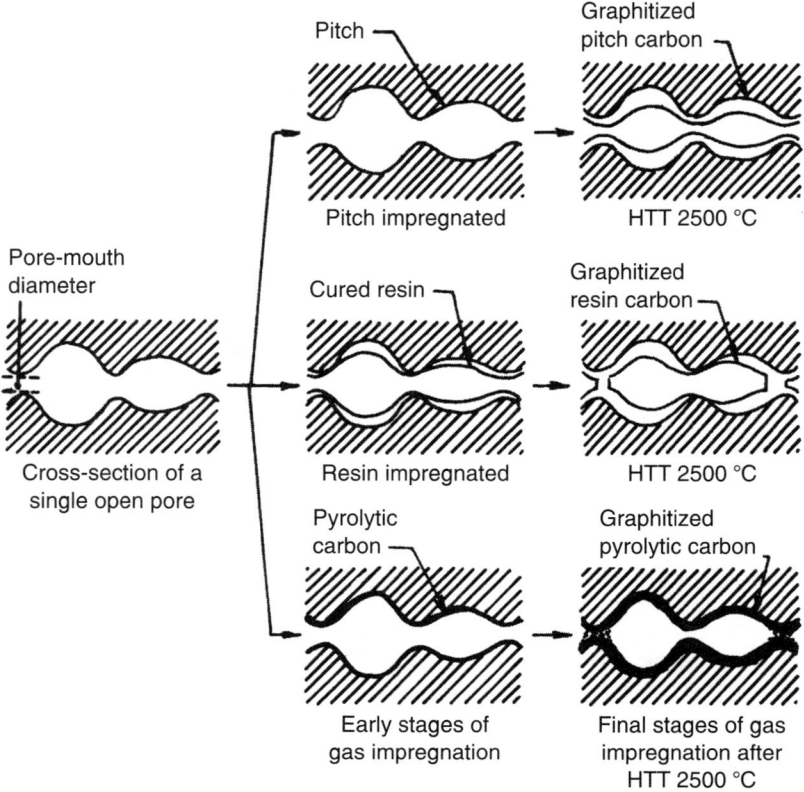

Figure 14.5 Schematic mechanisms of pore filling and pore blocking by liquid impregnation and by chemical vapor deposition. *Source:* Reprinted with permission from Kotlensky WV, *Chem Phys Carbon*, 9, 173, 1973. Copyright 1973, CRC Press, Boca Raton, Florida.

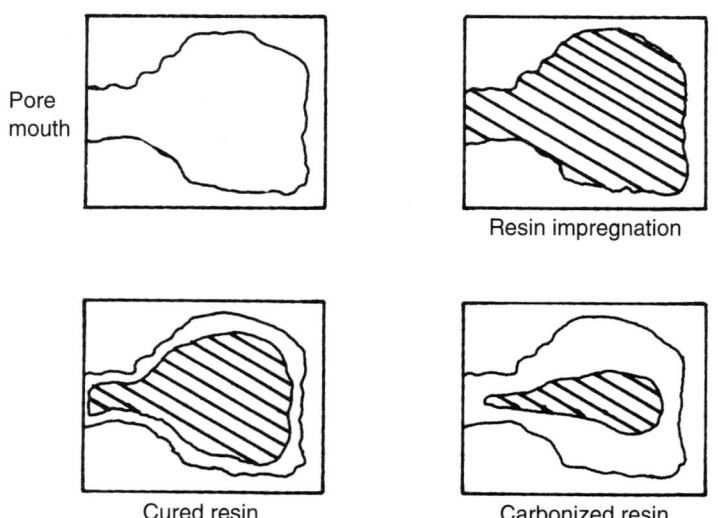

Figure 14.6 Schematic mechanisms of pore filling and blockage during the resin densification of carbon-carbon composites. *Source:* Reprinted with permission from Kotlensky WV, *Chem Phys Carbon*, 9, 173, 1973. Copyright 1973, CRC Press, Boca Raton, Florida.

isotropic carbon is formed, or benzene (C_6H_6), when a smooth laminar type of carbon will be deposited:

1. Smooth laminar deposition—occurring at a low deposition temperature and a high partial pressure of CH_4, with no added H_2, where $[C_2H_2]/[C_6H_6] <5$
2. Rough laminar deposition—occurring at an intermediate deposition temperature and an intermediate partial pressure of CH_4, with some added H_2, where $5<[C_2H_2]/[C_6H_6]<20$.
3. Isotropic deposition—occurring at a high deposition temperature and a low partial pressure of CH_4, with a large amount of added H_2, where $[C_2H_2]/[C_6H_6] >20$.

The typical structure-property relationships of carbon-carbon composites are shown in Table 14.1.

The structural and compositional characterization of carbon-carbon composites has been undertaken by Gunawan and Seraphin [69].

These microstructures are characterized by their optical activity in polarized light as depicted in Figure 14.7 and shown graphically in Figure 14.8. Bokros (Figure 14.9)

Table 14.1 Typical structure-property relationships of carbon-carbon composites

Graphitic/Anisotropic Pitch, CVD	Non-graphitic/ Isotropic PAN, Resin
High thermal conductivity	Low thermal conductivity
High density	Low density
High heat capacity	Low heat capacity
High modulus	Low modulus
Low strength	High strength
Soft	Hard

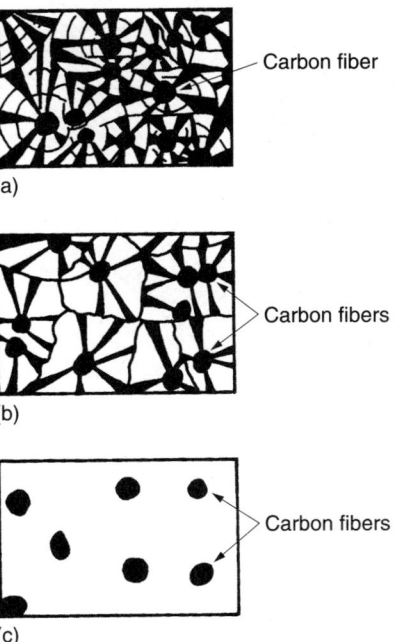

Figure 14.7 Schematic depiction of optical activities of CVD derived carbon microstructure. (a) Rough laminar, (b) Smooth laminar, (c) Isotropic. *Source:* Reprinted with permission from Pierson HO, Liebermann ML, *Carbon*, 13, 159, 1975. Copyright 1975, Elsevier.

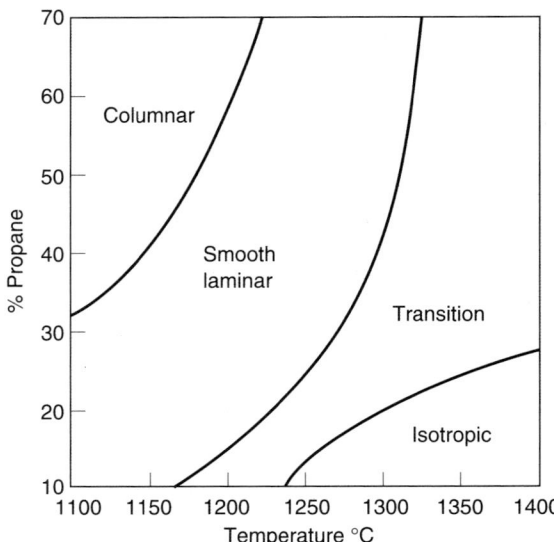

Figure 14.8 Effect of deposition conditions on the microstructure of pyrolytic carbon matrix deposition from propane. *Source:* Reprinted with permission from Oh SM, Lee JY, Structure of pyrolytic carbon matrices in carbon–carbon composites, *Carbon*, 26(6), 763–768, 1988. Copyright 1986, Elsevier.

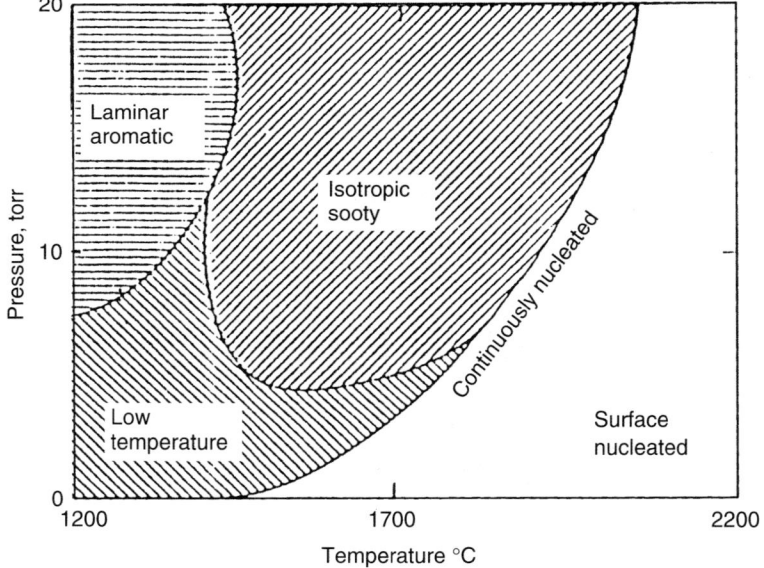

Figure 14.9 Relationship of pyrocarbon microstructure on deposition temperature and gas pressure in CH_4. *Source:* Reprinted with permission from Bokros JC, Walker PL Jr. ed., *Chemistry and Physics of Carbon*, 6, 1, 1969. Copyright 1969, CRC Press, Boca Raton, Florida.

shows the relationship of the pyrocarbon microstructure on the deposition temperature and gas pressure in CH_4. The densities increase with increasing anisotropy in the order $\rho_{\text{rough laminar}} > \rho_{\text{smooth laminar}} > \rho_{\text{isotropic}}$. CVI is best used for the production of thin sections, since the technique deposits carbon within as well as on the surface.

14.3.2.2 CVI processes

There are several types of CVI processes which can be used to process carbon-carbon:

1. Isothermal CVI process

Typically, a carbon preform is placed within an induction furnace susceptor and heated uniformly to about 1100°C, passing the reactant gas over the surface at such a rate as to maintain the surface reaction rate below the diffusion rate. A number of cycles are undertaken, machining away the surface crust prior to the next densification cycle. Dunlop Aviation operate such a system to manufacture carbon-carbon brake material at their Coventry plant (Figure 14.10) and a typical HT furnace used by Hitco is shown in Figure 14.11.

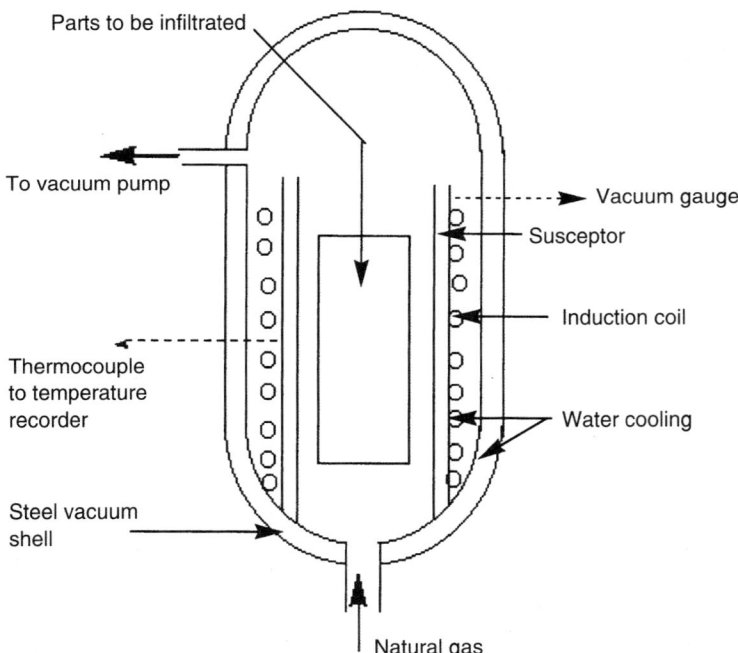

Figure 14.10 Chemical vapor deposition (CVD) furnace. *Source:* Adapted from Dunlop technical literature.

Figure 14.11 Hitco high temperature furnaces used for processing carbon–carbon. *Source:* Reprinted from Hitco technical literature.

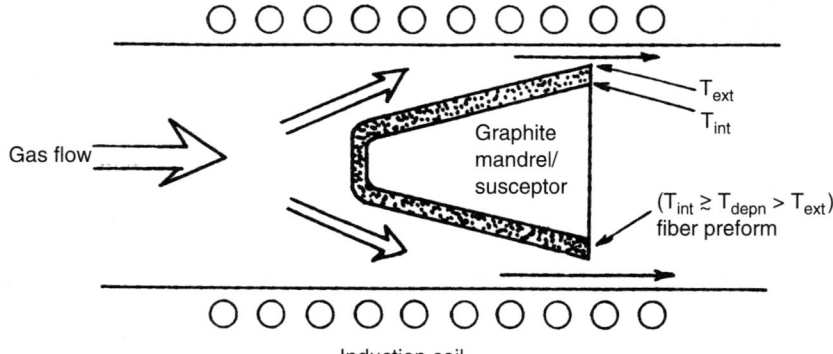

Figure 14.12 The thermal gradient method. *Source:* Reprinted with permission from Savage GG, *Carbon-Carbon Composites*, Chapman and Hall, London, 97, 1992. Copyright 1992, Springer.

2. Thermal gradient CVI process (TG-CVI)

Typically, the carbon preform is positioned in an induction furnace, supported on a mandrel, which acts as a susceptor, and a high gas (CH_4/N_2) flow is passed over the preform surface at such a rate that the surface immediately adjacent to the susceptor is about 500°C hotter than the exterior surface (Figure 14.12). The temperature gradient helps prevent surface build-up. To achieve this result, the fiber preform must have a low thermal conductivity and preferably be in a felt form, in order to control the gas flow and ensure sufficient cooling of the external surface of the preform. The technique has been used for rocket motor cones using a filament wound preform applied directly onto a shaped susceptor. Although the deposition rates are faster than the conventional isothermal method and the process is operated at atmospheric pressure, it is unfortunately, a single item processing procedure.

3. Pressure gradient process

The fiber preform is sealed in a unit and placed in a heated zone. The sealed design ensures that the gas flow passes through the perform, setting up a pressure gradient and depositing carbon with the evolved H_2 being emitted from the outlet end. The deposition rate speeds up with time as the pores become filled and the pressure drops further. The method is not widely used since it is restricted to single item processing and may still require the component to be removed and machined to obtain uniform densification.

4. Pulse CVD process

This technique is a combination of operating the reaction vessel at atmospheric pressure followed by part vacuum to facilitate better penetration of the reactant gases into the pores of the preform. Japanese workers have undertaken work in this area. The commercial difficulties of operating this type of process appear to have been overcome by Daewoo Heavy Industries to give a lower cost and faster densification procedure.

5. Possible new routes

Continuous efforts are being made to improve the manufacturing process. Wright Laboratory's Materials Directorate sponsored work by Aerotherm Corp., Kaiser Aerotech and Lockheed Missile & Space Co. to investigate an alternative approach for products with improved stiffness, but which would be cheaper. The approach used a pitch matrix and provided added axial stiffness to 345 GPa with reduced production time from months to hours.

Carbon Composites International have developed a single step impregnation process using a novel resin and a patented process that is reputed to increase the carbon yield and significantly reduce the cost of carbon-carbon to about $10–50/lb.

In an attempt to make the carbon-carbon production process cheaper, the Georgia Institute of Technology in Atlanta have developed a forced flow/thermal gradient process, which is, reputedly, 30% faster and permits a part of 25 mm thickness to be produced in as little as 8 h time. The C_3H_6, C_3H_8 or CH_4 is forced through the preform under pressure at a temperature of 1200°C. The ensuing temperature gradient in the material ensures a uniform flow of vapor through the preform. This uniform vapor infiltration enables parts of 10 mm thickness to be produced in 8 h. Parts up to 20 mm thick have been produced using this process.

The Across Company, Japan has developed another process wherein a carbon fiber prepreg is coated with graphite precursor powders from coke or pitch. The coating is protected by wrapping with a plastic sleeve during initial weaving or chopping into staple form. The carbon fiber ratio by volume is about 40–60%. The product is then placed in a mold and hot pressed. It is claimed that fewer densification stages are required, but it is not stated what happens to the plastic sleeve, which could remain *in situ*, with the sleeve itself undergoing charring in the subsequent processing. A liquid impregnation increases the densification speed some 50–100 times as compared to the standard process using a gas. A process developed in 1982 by the Division of Military Applications was called Kalamazoo and shelved until 1988, when Textron purchased the license and commenced industrialization of the process. Options were also taken out by SEP, initially using water to validate the process and later, replacing it with a liquid hydrocarbon.

Le Carbone-Lorraine [73] investigated the use of a sol-gel to impregnate the preform with silica. The technique depended on drying the sol to provide a highly microporous silica aerogel around the fiber, enabling the reactant gases to quickly penetrate and deposit carbon at 1000°C and 1 MPa. The silica aerogel was finally removed by subliming the silica at 2500°C.

14.3.3 Processing with thermosetting resin matrices

14.3.3.1 Low pressure impregnation (LPI) [74]

A carbon fiber (normally a mesophase pitch) tow can be impregnated with a thermoset resin system and chopped to a 25–50 mm staple length to give a compression molding compound that can be placed in a mold, heated in a press and cured. LPI is carried out in two discrete stages—carbonization and impregnation. The randomly reinforced composite is carbonized and then vacuum impregnated with additional resin to achieve densification. The impregnation is assisted by the application of about 2 MPa pressure. About four impregnation/carbonization cycles are undertaken to achieve a successful composite. Initially, the density falls and then increases with the ILSS as the number of cycles increases [75].

Alternatively, the carbon fiber reinforcement can be converted to a prepreg using the selected matrix resin system. The advantage of this route is that the architecture of the reinforcement can be selected to give the desired composite properties (e.g. 8-H satin to give improved drape). When the composite is laid up, the direction of reinforcement (e.g. ±45°) can be carefully selected to give the required properties in the composite. The prepreg pack is then cured in a heated press, or an autoclave, prior to subsequent densification. Again, due care must be taken during curing to avoid voids due to evolution of gaseous products.

The composite can be made using the pultrusion process, but this is difficult to operate since the resins cure by a condensation reaction. Filament winding enables the fiber to be placed precisely onto a removeable mandrel and achieve a high fiber content. This technique more suited to handling untreated, unsized fiber and is used to make rocket motor nozzles.

14.3.3.2 Pressure impregnation and carbonization (PIC)

The prepared carbonized composite is initially subjected to a vacuum (400 Pa) and introduced into the liquid resin under a pressure of 1–200 Pa for about 20 h and recarbonized at 1000–2880°C for some 4–12 cycles [76,77].

14.3.3.3 Hot isostatic pressure impregnation carbonization (HIPIC)

Fujioka and co-workers have described the HIPIC process [78]. If a pitch impregnation is undertaken at atmospheric pressure, the carbon content of the resultant composite will only be about 50%, but if a high pressure (100 MPa) is applied, carbon yields are dramatically increased (Figure 14.13), producing a product with a high density (about 1.9 g cm^{-3}) [79]. As the pressure of a coal tar pitch increases, the matrix microstructure changes from a needle-like structure, in which the mesophase is deformed due to bubble percolation, to finally, a coarser more isotropic form [80], with the increased pressure suppressing bubble formation. The application of pressure induces the mesophase to form at lower temperatures, but at pressures above 200 MPa, coalescence of the mesophase does not occur, resulting in inferior mechanical properties [81].

The application of heat to a pitch causes it to soften and flow and hence requires containment during carbonization. The pitch is processed by employing hot isostatic pressure

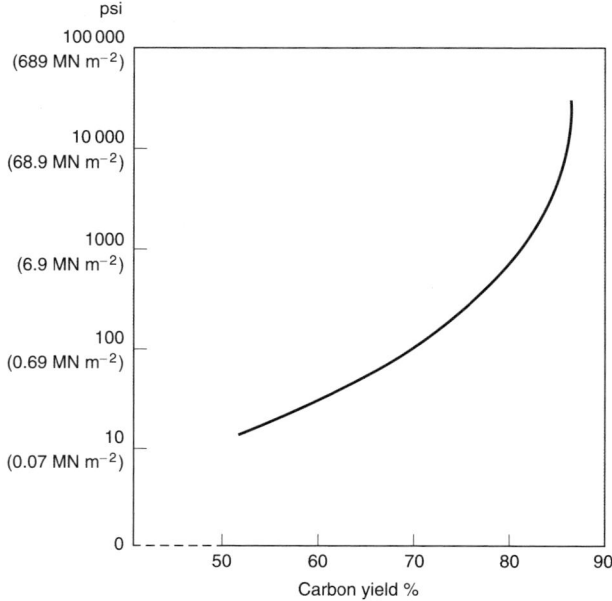

Figure 14.13 Effect of carbonization pressure on carbon yield from petroleum pitch. *Source:* Reprinted with permission from Lachmann WL, Crawford SA, McAllister LE, *Proc Int Conf on Composite Mats*, Met Soc of AIME, New York, 1978. Copyright 1978, The Metallurgical Society of AMIE, now TMS (The Minerals, Metals and Materials Society).

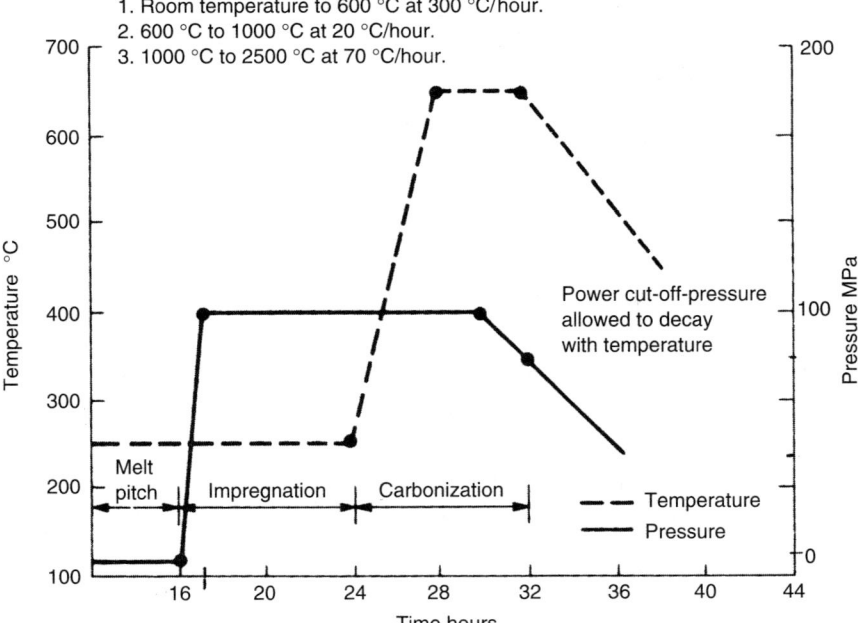

Figure 14.14 Standard HIPIC cycle of pitch at about 100 MPa for impregnation and carbonization. *Source:* Reprinted with permission from Johnson AC, Pinoli PC, Keller RL, Optimization of carbon-carbon processing, *14th Biennial Conference on Carbon*, American Carbon Committee, Penn State University, 238, Jun 1979. Copyright 1979, American Chemical Society.

impregnation carbonization (HIPIC) in a hot gas autoclave following the cycle given in Figure 14.14. Since high pressure can deform the original preform shape, it can be beneficial to introduce an initial forming stage using the resin prepreg technique in conjunction with autoclave molding, followed by the first carbonization stage. Hot pressed composites have flattened pores and 5 Pa pressure can cause vertical cracks [82]. After the HIPIC cycles, the material is removed, surface cleaned and finally graphitized at 2400–2700°C, depending on the original fiber architecture and the properties required, in an Ar atmosphere.

HIPIC can be undertaken with a conventional autoclave but specialist equipment has been developed to contain the liquid pitch (Figure 14.15) using a differential pressure system to ensure that gas movement is from the outside towards the inside [83], avoiding sooty deposits and ensuring control of evolved gases. The pressure ensures that the molten pitch is kept within the pores and also increases the carbon yield. The process is carried out slowly, taking 2–3 days and is a one-off process and expensive to run, but gives a product with good yield and superior mechanical properties.

Another practice is to place the preform in a special metal container or can, which is evacuated and sealed by electron beam welding. The metal container acts like a pressure bag transmitting pressure throughout the workpiece [84].

14.4 SOME THOUGHTS ON CARBON-CARBON PROCESSING

14.4.1 Chemical vapor deposition

In this process, a hydrocarbon gas (CH_4, C_2H_6, C_3H_8, C_6H_6) is thermally degraded onto a hot carbon surface depositing pyrolytic carbon and releasing H_2. Liebermann has reported

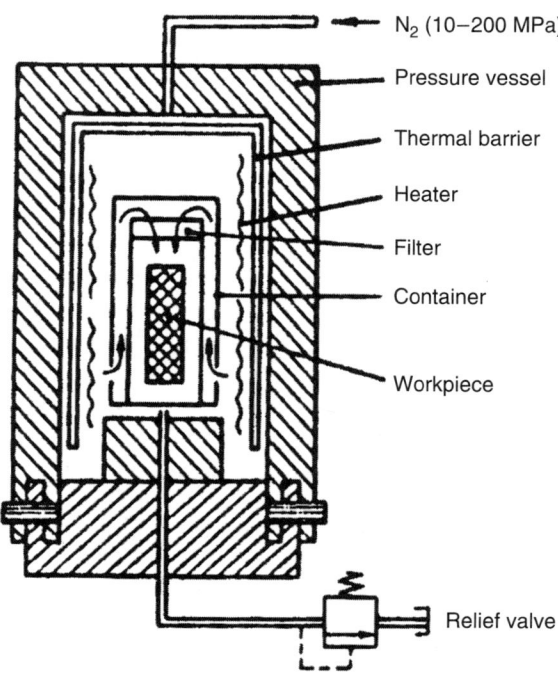

Figure 14.15 Hot Isostatic Pressure Impregnation Carbonization (HIPIC) furnace. *Source:* Reprinted with permission from Hosomura T, Okamoto H, *Mater Sci Eng*, A 143(1–2), 223–229, 1991. Copyright 1991, Elsevier.

the formation of acetylene compounds [85] or aromatic species [86] during the decomposition of CH_4. The mechanisms and morphologies of these depositions have been described by Bokros [87]. The rate of deposition is controlled by the substrate surface area and concentration of the gas, which can be diluted with an inert gas (H_2, He, N_2, Ar) to control the concentration of the cracking gas. As the gas penetrates the pores of the substrate, its concentration decreases, thus establishing a form of concentration gradient which tends to further induce gas to enter the pores, resulting in a deposition gradient within the pore. Savage [88] has shown in Figure 14.16 the balance between diffusion and surface reaction kinetics. There is also a density gradient and to establish higher densities, the composite can be heated to 2700°C, which reopens the pores, with the crystallites rearranging to enable further deposition. Low temperature will favour a heterogeneous deposition, a lengthy and costly process. It will also be necessary to periodically machine away the impervious surface layer that builds up, if the impregnation process is to continue.

The integrated form of the Arrhenius equation is:

$$\ln k = -\frac{E}{RT} + C$$

where C is a constant and a plot of $\ln k$ against $\frac{1}{T}$ will be a straight line. In a typical Arrhenius plot [89] for a gas-solid reaction (Figure 14.17), there are three classical zones or regions:

1. Zone 1 – Chemical zone – at low temperatures, the chemical reaction is slow and the surface reaction rate is uniform on all the pore surfaces. It is the rate controlling and preferred reaction.

CARBON FIBER CARBON MATRIX COMPOSITES

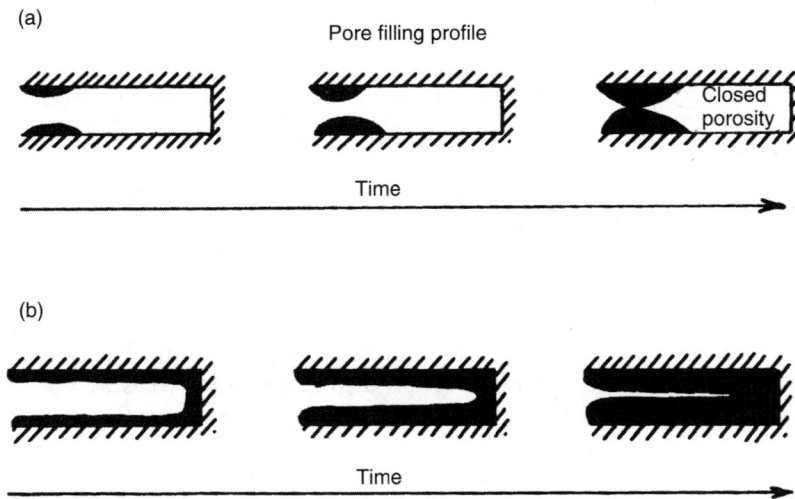

Figure 14.16 The balance of diffusion and surface reaction kinetics: idealized depictions. (a) Surface reaction rate >> diffusion rate (b) Diffusion rate >> surface reaction rate. *Source:* Reprinted with permission from Savage GG, *Carbon-Carbon Composites*, Chapman and Hall, London, 92, 1992. Copyright 1992, Springer.

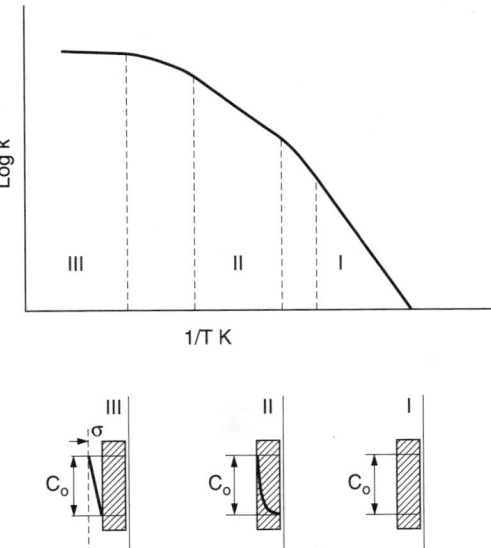

Figure 14.17 A typical Arrhenius plot for a classical reaction between a gas and a porous solid showing the three reaction zones. Zone I – Chemical zone; Zone II – Pore diffusion control; Zone III – Gaseous diffusion control at the exterior of the specimen. *Source:* Reprinted from Heddon K, Wicke E, Some influences on the reactivity of carbon, *Proc of the 3rd Conf on Carbon*, Pergammon Press, New York, 249, 1957.

2. Zone 2 – Pore diffusion control – as the temperature is increased, there is a rapid increase of chemical reaction, which eventually exceeds the rate of supply of the precursor gas into the pores, causing excessive deposition at the outer pore regions and ultimately blocking the pores.
3. Zone 3 – Gaseous diffusion control at exterior of solid – since the precursor gas can no longer penetrate the pores, the gas can only produce a coating on the outside.

14.4.2 Liquid infiltration

A liquid impregnation process essentially provides the means to avoid the closure of narrow pores and Kotlensky [67] has depicted these pore filling mechanisms (Figures 14.5 and 14.6). The liquid impregnation process is normally performed under both vacuum and applied pressure to achieve more efficient pore filling. Molten pitch is a common material selected for liquid impregnation and it is possible to select a fraction of pitch with the most appropriate working viscosity for processing at a given temperature (Figure 14.18). The yield increases with increasing C/H ratio [91]. A pitch can be categorized by measuring its T_g and the minimum viscosity occurs about 100–150°C above the T_g. Normally, as the impregnation cycles proceed, the fraction of pitch can be changed to provide more effective filler for the gaps. Eventually, the surface tension becomes a limiting factor. It is at this stage that the benefits of using a combination of liquid impregnation followed by CVD (or CVI) become apparent, with the latter impregnants penetrating the small pores more readily.

A higher pressure favours an increase in the carbon yield (Figure 14.13), but as the pressure is increased, the deposit becomes more coarse and isotropic, probably due to the evolved gases becoming compressed and being unable to escape. When pitch is carbonized, it evolves gases, a process known as bloating, which can force pitch out of the carbon matrix [92]. Bloating can be overcome by oxidation of the pitch prior to carbonization through treatment with O_2 for up to 100 h at about 220°C, rendering the mesophase fraction infusible and hence permitting carbonization to proceed without disrupting the microstructure [93]. The bulk density of the carbon-carbon composite shows an initial drop in density and then increases with the number of cycles of carbonization impregnation (Figure 14.19), the increase diminishing as the pores become blocked. Graphitization at 2200–3000°C will enable further densification to occur, reaching density levels of 1.84 g cm^{-3} [94]. The flexural strength increases with the number of process cycles and is shown in Table 14.2. The use of mesophase pitch will limit the number of process cycles [95].

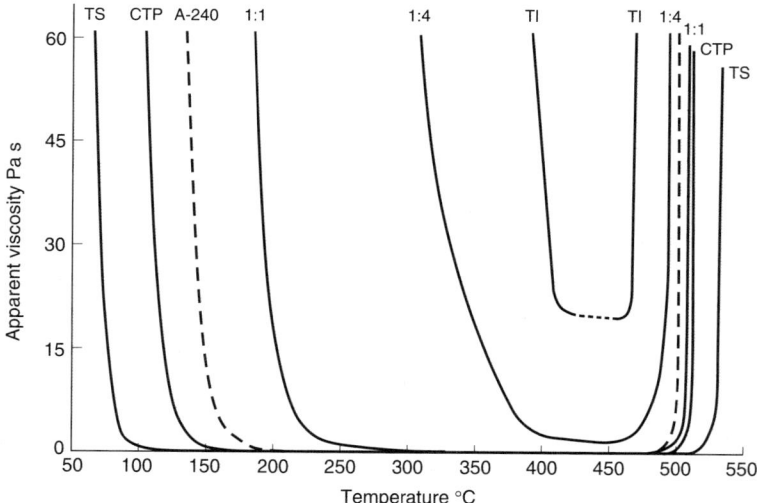

Figure 14.18 Viscosity of various pitch fractions. TS Toluene soluble coal tar pitch; CTP Coal tar pitch; A-240 Ashland A-240; 1:1 Mixture of TS/TI; 1:4 Mixture of TS/TI; TI Toluene insoluble coal tar pitch. *Source:* Reprinted with permission from Bhatia G, Fitzer E, Kompalik D, Mesophase formation in toluene soluble and toluene insoluble coal tar pitch fractions and their mixtures, *International Carbon Conference Extended Abstracts*, Bordeaux, France, 330–331, 1984. Copyright 1984, American Chemical Society.

CARBON FIBER CARBON MATRIX COMPOSITES

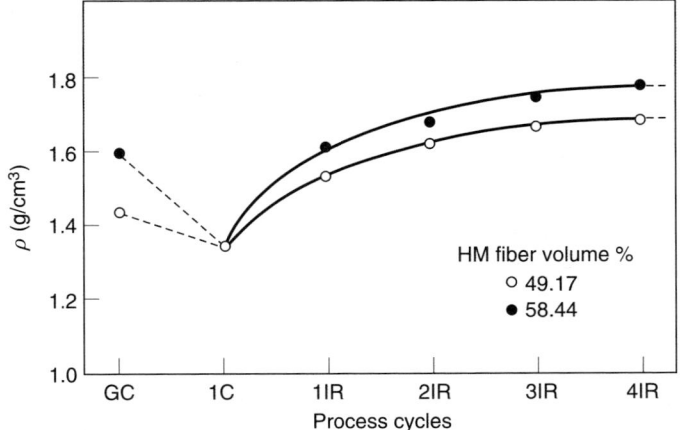

Figure 14.19 The bulk density of carbon-carbon composites after successive process cycles with Torayca M40B and pitch The decrease of density after the first cycle necessitates further processing. GC = green composite; 1C = first carbonization cycle; 1IR, 2IR, 3IR and 4IR = successive impregnation and recarbonization cycles. *Source:* Reprinted with permission from Rhee B, Ryu S, Fitzer E, Fritz W, *High Temp-High Pressures*, 19(6), 677–686, 1987. Copyright 1987, Pion Ltd.

Table 14.2 The flexural strength of carbon-carbon composites as a function of the pitch impregnation recarbonization cycles

Number of cycles	Flexural strength MPa	
	After carbonization at 1000°C	After graphitization at 2600°C
Start	200	180
1	430	240
2	650	400
3	820	530
4	1000	620

Source: Reprinted with permission from Fitzer E, Huttner W, Manocha LM, Influence of process parameters on the mechanical properties of C/C composites with pitch as matrix precursor, *Carbon*, 18, 291–295, 1980. Copyright 1980, Elsevier.

14.5 PROVISION FOR PROVIDING OXIDATION PROTECTION [96–100]

14.5.1 Introduction

Above 370°C, carbon-carbon composites readily oxidize in air,

$$C + O_2 \rightarrow CO_2$$

and if CO_2 can escape freely, this will promote the oxidation process.

The rate of oxidation increases with temperature but decreases with time (Figure 14.20). An increased composite heat treatment temperature (HTT) will reduce the oxidation rate, probably due to a reduced level of impurities, less edge sites and an overall reduction in carbonization stress [102].

Initially, the fiber matrix interface is attacked, resulting in delamination cracking between plies, but as the temperature increases, more severe oxidation attack occurs within the

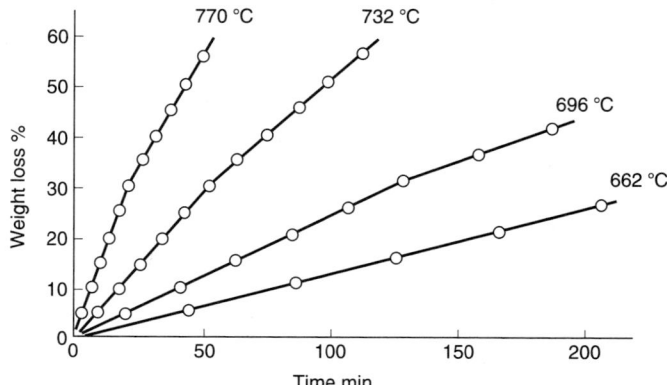

Figure 14.20 Oxidation of carbon-carbon composite (HTT 2800°C) in air (67% HM fiber). *Source:* Reprinted with permission from Yamada E, Kimura S, Shilswa Y, *Trans JSCM*, 6(1), 14, 1980. Copyright 1980, Japan Society for Composite Materials.

fiber bundles, promoting cross bundle cracking [103]. It is, therefore, necessary for many applications at elevated temperatures to apply a protective coating that will provide both oxidation and erosion protection. Basically, there are two modes of protection:

1. Use of inhibitors and sealants to block active sites internally and stifle the rate of oxidation
2. Apply a barrier coating to form a protective outside layer, preventing O_2 moving inwards and carbon moving out.

14.5.2 The use of inhibitors to provide oxidation protection

Gases are evolved from within the carbon-carbon composite during the oxidation process, mainly arising from active sites on the edges of layer planes and vacant sites formed by dislocations in the basal planes [104,105]. These sites can be blocked or poisoned and the composite porosity reduced by using inhibitors based on B, P and halogen compounds which lower the reactivity of C with O_2 [106]. The choice of inhibitor is limited by the temperature of application.

1. Boron

The compound B_2O_3 is widely used for protection up to 1500°C through poisoning sites as well as acting as a glassy sealant [107]. It is a good choice, melting at about 450°C. Although in use it is limited to about 1000°C due to its volatility, as a crack sealant, the working temperature can be extended to 1200°C and to about 1400°C if coated with an outermost layer of SiC. The glass former migrates to the outer surface, sealing cracks and voids *en route*. At 1525°C, the CO liberated as a result of the B_2O_3 reacting with the carbon-carbon surface

$$2B_2O_3 + 7C \rightarrow B_4C + 6CO$$

causes the partial pressure generated to disrupt the glass. Coupled with corrosion of SiC by the borate glass, the maximum temperature of use is limited to 1500°C for short periods. TiO_2 is soluble in B_2O_3, thus increasing the viscosity and helping to prevent volatilization. Boron coatings are susceptible to attack by water at ambient temperatures, forming HBO_2.

CARBON FIBER CARBON MATRIX COMPOSITES

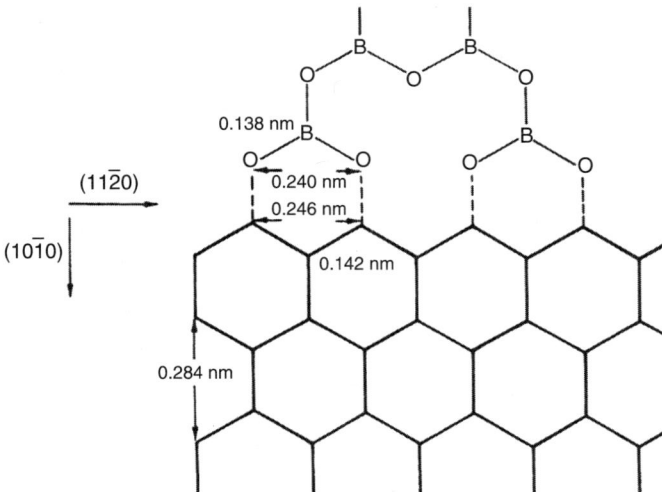

Figure 14.21 Schematic representation of bonding of $(BO_3)_n$ polymer to {10$\bar{1}$0} face of graphite lattice. *Source:* Reprinted with permission from McKee DW, Spiro CL, *Carbon*, 23, 437, 1985. Copyright 1985, Elsevier.

Figure 14.21 shows how boron can be absorbed at vacant edge sites of graphite crystallites and whilst a 3.5% B_2O_3 addition will effect a significant reduction in oxidation, a 10% addition will reduce the oxidation rate by an order of magnitude, but increase the overall weight of the composite.

2. Phosphorus

Organophosphorus compounds, when heated to 400°C, form a phosphate residue [109] that is tenaciously held on the carbon surface, acting as a poison, blocking active sites and reducing the rate of oxidation up to working temperatures of 850–900°C. This temperature can be extended when phosphates are used in conjunction with other inhibitors like B_2O_3, SiO_2 and SiC. Phosphates are not as effective as borates.

14.5.3 The use of a barrier coating

Ideally, a barrier coating should adhere well to the substrate and be free of cracks, with a coefficient of thermal expansion (Figure 14.22) close to the value for the substrate. The coating should have good high temperature resistance (Figure 14.23) and prevent O_2 ingress and C egress. Barrier coatings include noble metals, carbides, borides, nitrides and silicides.

1. Noble metals

Iridium (mp = 2440°C) does not react with C to form carbides (up to 2280°C) and is a superb barrier to O_2 (up to 2100°C) [111] but is very expensive and difficult to apply with poor adhesion to carbon. At high temperatures, it can form volatile Ir_2O_3 and IrO_3, which can cause erosion.

2. Silicon coatings

Pack cementation, or SiC conversion, [112] can be employed, where the materials to be coated are packed in containers holding Si, SiC, Al_2O_3 and MgO powders. These react at

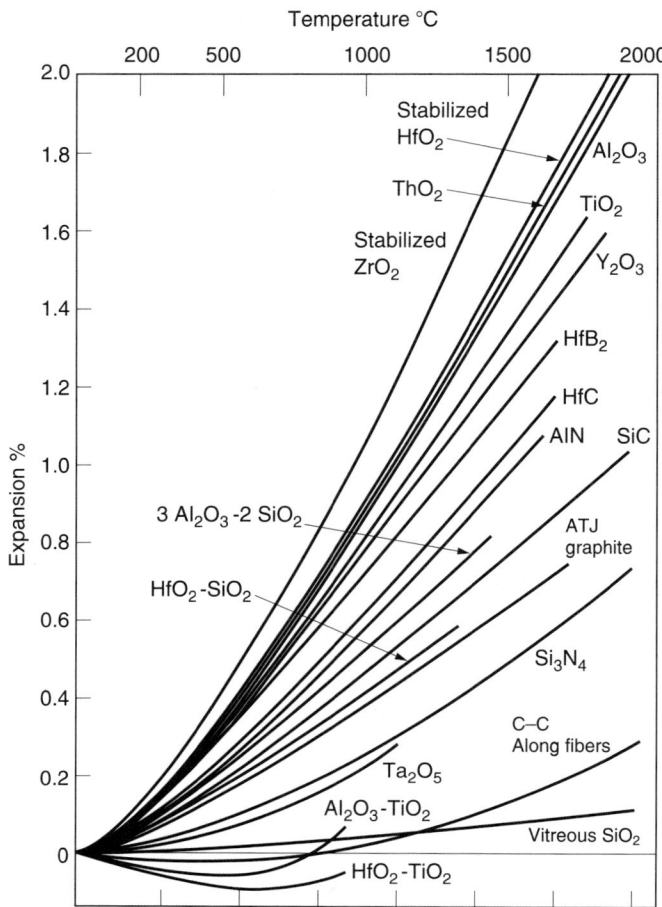

Figure 14.22 Comparison of the thermal expansion characteristics of some refractory materials. *Source:* Reprinted from Lynch JF, Ruderer CC, Duckworth WH, *US Air Force Mat Lab Tech Ref Report*, No. AFML-TR066-52, 1966, Strife JR, Sheehan JE, *Ceramic Bulletin*, 67, 369–374, 1988.

1600°C to form a SiC coating graded from pure carbon on the inside to pure silicon on the outside and achieving a thickness of of about 0.5 mm.

$$Si\ (l\ or\ g) + C \rightarrow SiC$$

$$SiO + 2C \rightarrow SiC + CO\ (g)$$

Molten silicon does, however, have a preference to react with the carbon fibers rather than with the carbon matrix and this can be avoided by applying an initial coat of SiC using the CVD process, giving a more dense coating.

Chemical vapor deposition (CVD) using CH_3SiCl, or CH_3SiCl_3 [113] in the presence of H_2 at 1125°C [114] can be employed to deposit an outer coating of SiC from the gaseous phase, which acts as a primary barrier to O_2 ingress. A B rich inner layer, applied by slurry coating, or the CVD process, provides the means of sealing any cracks in the outer coating.

There is a good thermal expansion match with carbon-carbon composites and silicon ceramics. SiC is limited to use at 1700°C as the layer is disrupted by formation of CO [111].

CARBON FIBER CARBON MATRIX COMPOSITES

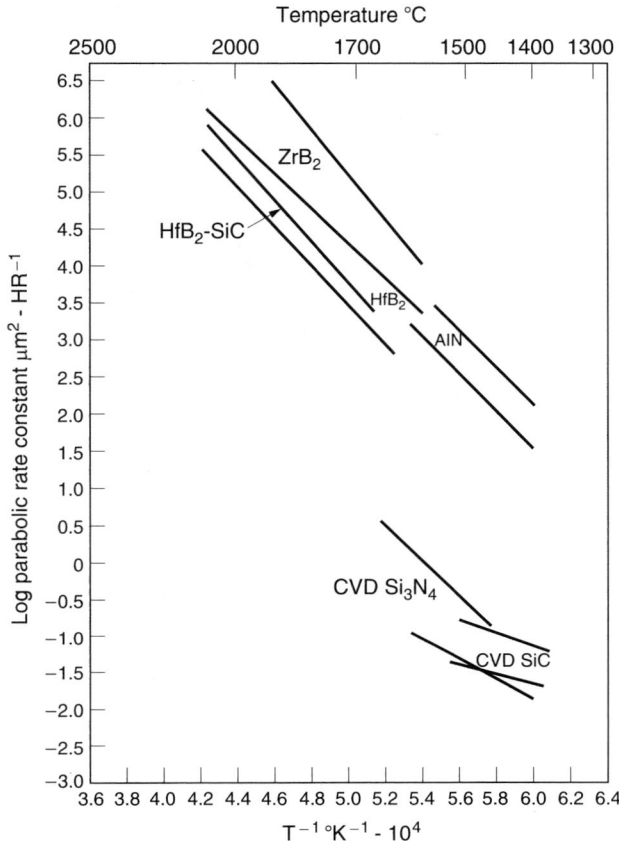

Figure 14.23 Oxidation characteristics of high temperature ceramics. *Source:* Reprinted from Strife JR, Sheehan JE, *Ceramic Bulletin*, 67, 369–374, 1988.

At temperatures below 1150°C, the viscosity of the silica glass is too high and it is unable to flow into any cracks that may develop. However, it is widely used in conjunction with B.

An alternative approach is to use a carbon-silicon matrix via chemical vapor infiltration with carbon and SiC co-deposition from CH_3SiCl_3, C_2H_2, Ar and H_2, by gradually varying the source gases in a semi-continuous process, from C_2H_2 to CH_3SiCl_3. [115]. Although this material seems to be macroscopically homogeneous, microscopic examination shows a gradient variation of the matrix sheaths around each fiber. The oxidation resistance of the material is significantly superior to that of a carbon-carbon material.

A serious problem with a SiC coating is spallation and crack formation due to thermal mismatch stresses. Applying a graded coating—pure carbon on the inside and pure SiC coating on the outside—helps to minimize spallation [116].

General Atomic (GA) Technologies have developed another process termed a preceramic polymer coating (PPC) using a polymeric coating system made by Ethyl Corporation, which has been shown to perform as well as, or better than, the industry standard CVD-SiC system.

Carbon-carbon made with a cellulose based carbon fiber has a coefficient of thermal expansion (CTE) about twice that of a PAN based carbon fiber system. Since the cellulose based carbon-carbon has a CTE fairly close to that of SiC (or Si_3N_4), it is consequently less of a mismatch than a PAN based system. Unfortunately, to take advantage of the improved strength of a PAN based system, it will be necessary to accommodate this serious mismatch of CTEs and will require a sustained program of research to find a solution.

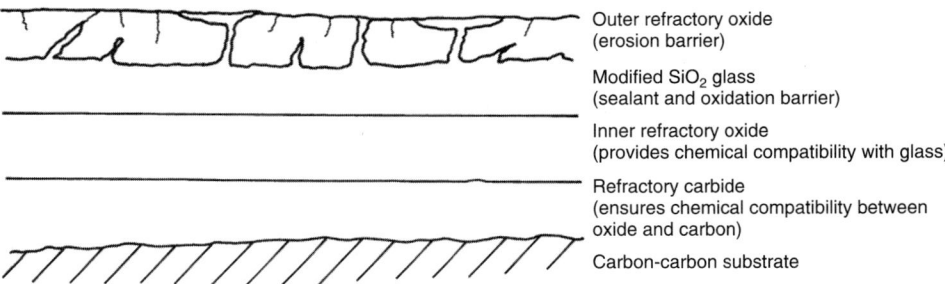

Figure 14.24 A general scheme of a multilayer coating for conferring long term oxidation protection of carbon-carbon at temperatures of 1800°C and above. *Source:* Reprinted from Savage GG, *Carbon-Carbon Composites*, Chapman and Hall, London, 221, 1992.

Silica films can be used as O_2 diffusion barriers and may be applied by impregnation with tetraethylsilicate (TEOS) or by a sol-gel process. Extended use of silica coatings above 1200°C is restricted by recrystallization occurring, with subsequent shrinkage and cracking. SiO_2 has relatively poor wetting characteristics and at high temperatures, SiO_2 must be separated by a hard oxide from the carbon-carbon, as otherwise, it is reduced.

14.5.4 Other coating systems

Oxides in general, are not suitable but

1. ZrO_2 and HfO_2 can be used up to 1700°C
2. $MoSi_2$ and $TiSi_2$ can be used up to 1700°C
3. Si_3N_4 is limited to 1700°C as the layer is disrupted by formation of N_2
4. HfB_2 can be used at temperatures above 1800°C
5. ZrC and ZrB_2 will give limited protection up to 2200°C

A multilayer coating (Figure 14.24) is the best route to impart adequate protection against oxidation and erosion of carbon-carbon composites. The outer refractory oxide coating can be ZrO_2, HfO_2, Y_2O_3 or ThO_2, with a SiO_2 glass inner layer as a sealant and oxidation barrier, an inner refractory oxide layer to provide chemical compatibility with the glass and finally, an inner refractory carbide layer of TaC, TiC, HfC or ZrC to ensure chemical compatibility with the carbon.

REFERENCES

1. Le McAllister, Multidirectionally reinforced carbon/graphite matrix composites, Reinhart TJ ed., *Engineered Materials Handbook*, Vol 1, Metals Park, OH, ASM International, 915, 1987.
2. Diefendorf RJ, Continuous carbon fiber reinforced carbon matrix composites, Reinhart TJ ed., *Engineered Materials Handbook*, Vol 1, Metals Park, OH, ASM International, 911, 1987.
3. Buckley JD, Carbon-carbon an overview, *Ceram Bull*, 67, 364, 1988.
4. Lei C, Ko FK, Pastore C, *Modelling of three dimensional braided carbon-carbon composites*, Dayton, Ohio, 1989.
5. Huettner W, Potential of Carbon-carbon composites in structural materials, Figueiredo, Bernardo CA, Baker RTK, Hüttinger KJ eds., *Carbon Fibers Filaments and Composites*, Kluwer, Dordrecht, 275–300, 1990.

6. Huettner W, Theoretical and practical aspects of liquid phase pyrolysis as basis of the carbon matrix of CFRC, Figueiredo, Bernardo CA, Baker RTK, Hüttinger KJ eds., *Carbon Fibers Filaments and Composites*, Kluwer, Dordrecht, 301–325, 1990.
7. Kowbel W, Shan CH, The mechanism of fiber–matrix interactions in carbon-carbon composites, *Carbon*, 28, 287, 1990.
8. Buckley JD, Edie DD ed., *Carbon-Carbon Materials and Composites*, Noyes Publications, Park Ridge, NJ, 1992.
9. Savage GG, *Carbon-Carbon Composites*, Chapman and Hall, London, 221, 1992.
10. Thomas CR ed., *Essentials of Carbon-Carbon Composites*, The Royal Society of Chemistry, Books Britain, 1993.
11. Chung DDL, *Carbon Fiber Composites*, Butterworth-Heinemann, Boston, 145–176, 1994.
12. Peebles LH, *Carbon Fibers Formation, Structure, and Properties*, CRC Press, Boca Raton, 157–163, 1995.
13. Fitzer E, Manocha LM, Carbon reinforcements and carbon-carbon composites, Baumgartel H, Franck EU, Grünbein W eds., *Physical Chemistry, Materials Science, Chemical Engineering, Condensed Matter and Properties of Materials*, Vol 5, Springer Verlag, New York, 1998.
14. Fitzer E, Manocha LM, *Carbon Reinforcements and Carbon/Carbon Composites*, Berlin, Springer Verlag, 1998.
15. Dhami TL, Bahl OP, Jain PK, Carbon-carbon composites made with oxidized PAN (Panex) fibers, *Carbon*, 33(11), 1517–1524, 1995.
16. Markovic V et al, 15th Biennial Conference on Carbon, Pennsylvania, 272, 1980.
17. Thomas CR, Walker EJ, Effects of PAN carbon fibre surface in carbon-carbon composites, *Proceedings 5th London International Carbon & Graphite Conference*, SCI, London, 520–531, 1978.
18. Christie N, Fitzer E, Kalka J, Schafer WJ, *Chem Phys Physiochem Biol*, 50, 1969.
19. Williams DJ, New knitting methods offer continuous structures, *Adv Comp Eng Summer*, 12–13, 1978.
20. Fitzer E, Burger A, Paper 36, *Proc 1st Int Carbon Fibres Conference*, London, Feb 1971.
21. Boyne L, Cook D, Hill J, Turner K, *Proc 4th Int Carbon and Graphite Conference*, London, 215–230, Sep 1974.
22. Rand B, Robinson R, *Carbon*, 15(4), 257, 1977 and Rand B, Robinson R, *Carbon*, 15(5), 311–315, 1977.
23. Fry M, Thomas CR, Walker EJ, *Reinf Plastics Conf*, Brighton, November, 1976.
24. Fitzer E, Geigl KH, Hüttner W, The influence of carbon fiber surface treatment on the mechanical properties of carbon/carbon composites, *Carbon*, 18, 265, 1980.
25. Fitzer E, The future of carbon-carbon composites, *Carbon*, 25(2), 163–190, 1987.
26. Manocha LM, Bahl OP, Singh YK, *Carbon*, 29(3), 351–360, 1991.
27. Manocha LM, Bahl OP, *Carbon*, 26(1), 13–21, 1988.
28. Hüttinger KJ, Krekel G, Polydimethylsiloxane coated carbon fibers for the production of carbon-fiber reinforced carbon, *Carbon*, 29, 1065, 1991.
29. Peebles LH Jr., Meyer RA, Jortner J, Interfaces in carbon-carbon composites, Ishida H ed., *Interfaces in Polymer, Ceramic and Metal Matrix Composites*, Elsevier Science Publishing Co, New York, 1, 1988.
30. Dubois J, Agache C, White JL, The carbonaceous mesophase formed in the pyrolysis of graphitizable organic compounds, *Metallography* 3, 337, 1970.
31. Zimmer JE, White JL, Disclination structures in the carbonaceous mesophase, *Advances in Liquid Crystals*, Vol 5, Academic Press, New York, 157, 1982.
32. Manocha LM, Yasuda E, Tanabe Y, Kimura S, Effect of carbon fiber surface treatment on mechanical properties of C/C composites, *Carbon*, 26, 333, 1988.
33. Manocha LM, Bahl OP, Singh YK, *Tanso*, 140, 255–260, 1989.
34. Zaldivar RJ, Rellick GS, *Carbon*, 29(8), 1155–1163, 1991.
35. Jortner J, Cracking in 3D carbon/carbon composites during processing and effects on performance, *Proc Army Symp Solid Mechanics*, AMMRC MS 76-2, 81, 1976.
36. Manocha LM, Bahl OP, Singh YK, *Carbon*, 29(3), 351–360, 1991.

37. Kimura S, Graphitization of thermosetting resin derived carbon matrix in C/C composite, *Rep Res Lab Eng Mater*, Tokyo Inst Technol 11, 141, 1986.
38. Markovic V, *Fuel*, 66(11), 1512–1515, 1987.
39. Fitzer E et al, Walker PL Jr. ed., *Chemistry and Physics of Carbon*, 7, 368, 1970.
40. Manocha LM, Yasuda E, Tanabe Y, Kimura S, Effect of carbon fiber surface treatment on mechanical properties of C/C composites, *Carbon*, 26, 333, 1988.
41. Ludenbach G, Peters PWM, Ekenhorst D, Muller BR, Properties and structure of the carbon fibre in carbon/carbon produced on the basis of carbon fibre reinforced phenolic resin, *J Eur Ceramics Soc*, 18(11), 1531–1538, 1998.
42. Siebold RW, Carbonization of phenolic resins, *SAMPE National Symposium*, San Diego, California, 20, 327, 1975.
43. Fitzer E, Schafer W, Yamada S, *Carbon*, 7, 643, 1969.
44. Fitzer E, Schafer W, *Carbon*, 8, 597, 1970.
45. Kotosonov AS, Vinnikov VA, Frolov VI, Ostrenov BG, *Doklady Akad Nauk SSSR*, 185, 1316, 1969 and Effect of mechanical pressure during carbonization of organic polymers on their graphitizability, *Doklady Phys Chem*, 185, 278, 1969.
46. Ludenbach G, Peters PWM, Ekenhorst D, Muller BR, The properties and structure of the carbon fibre in carbon-carbon produced on the basis of carbon fibre reinforced phenolic resin, *J Eur Ceramic Soc*, 18(11), 1531–1538, 1998.
47. Burger A, Fitzer E, Heym M, Terwiesch B, *Carbon*, 13, 149, 1975,
48. Economy J, Jung H, Gogeva T, *Carbon*, 30(1), 81–85, 1992.
49. Christ K, Huttinger KJ, Carbon-fiber-reinforced carbon composites fabricated with mesophase pitch, *Carbon*, 31(5), 731–750, 1993.
50. Zaldivar RJ, Rellick GS, Yang JM, Fiber strength utilization in carbon carbon composites, *J Mater Res*, 8(3), 501–511, 1993.
51. Liedtke V, Huttinger KJ, Mesophase pitches as matrix precursor of carbon fiber reinforced carbon, 1. Mesophase pitch preparation and characterization, *Carbon*, 34(9), 1057–1066, 1996.
52. Liedtke V, Huttinger KJ, Mesophase pitches as matrix precursor of carbon fiber reinforced carbon, 2. Stabilization of mesophase pitch matrix by oxygen treatment, *Carbon*, 34(9), 1067–1079, 1996.
53. Liedtke V, Huttinger KJ, Mesophase pitches as matrix precursor of carbon fiber reinforced carbon, 3. Mechanical properties of composites after carbonization and graphitization treatment, *Carbon*, 34(9), 1081–1086, 1996.
54. Fujiura R, Kojima T, Kanno K, Mochida I, Korai Y, *Carbon*, 31(1), 97–102, 1993.
55. Fitzer E, Huttner W, Manocha LM, Influence of process parameters on the mechanical properties of C/C composites with pitch as matrix precursor, *Carbon*, 18, 291–295, 1980.
56. Fitzer E, The future of carbon-carbon composites, *Carbon*, 25(2), 163–190, 1987.
57. Savage G, *Met Mater*, Inst Met, 4(9), 544–548, 1988.
58. Huttinger KJ, Christ K, Carbon fibre reinforced carbon composites produced with mesophase pitches, Naslain R, Lamon J, Doumeingts D eds., European Association for Composite Materials, American Ceramic Soc, Inc., Ceramic Society of Japan, *High Temperature Ceramic Matrix Composites, Proc. 6th European Conf. on Composite Materials*, Bordeaux, 199–206, Sep 20–24 1993.
59. Chang T, Okura A, *Trans Iron Steel Inst Japan*, 27(3), 229–237, 1987.
60. Savage GG, *Carbon-Carbon Composites*, Chapman and Hall, London, 181–187, 1992.
61. Gray G, Savage GM, *Metals and Materials*, 513, Sep 1989.
62. Zaldivar RJ, Rellick GS, JM Yang, *SAMPE J*, 27(5), 29–36, 1991.
63. Zaldivar RJ, Kobayashi RW, Rellick GS, *Carbon*, 29(8), 1145–1153, 1991.
64. Otani S, Some properties of a condensed polynuclear aromatic resin (COPNA) as a binder for carbon fibre composites, *J Mater Sci*, 21(6), 2027, 1986.
65. Kohno T et al, CVD carbon infiltration on carbon-carbon composites, *1st Japan International SAMPE Symposium and Exhibition*, 1149, Nov 28–Dec 1 1989.
66. Spear KE, *Pure Appl Chem*, 54, 1297, 1982.
67. Kotlensky WV, *Chem Phys Carbon*, 9, 173, 1973.
68. Pierson HO, Liebermann ML, *Carbon*, 13, 159, 1975.

69. Gunawan N, Seraphin S, Structural and compositional characterization of carbon fiber carbon matrix composites, *Microscopy Research and Technique*, 38(5), 544, 1997.
70. Oh SM, Lee JY, Structures of pyrolytic carbon matrices in carbon-carbon composites, *Carbon*, 26(6), 763–768, 1988.
71. Bokros JC, Walker PL Jr. ed., *Chemistry and Physics of Carbon*, 6, 1, 1969.
72. Savage GG, *Carbon-Carbon Composites*, Chapman and Hall, London, 97, 1992.
73. Pajonk GA, Teichner SJ, Aerogels, *Springer Proc Phys 6*, Springer-Verlag, Heidelberg, 163, 1986.
74. Ludenbach G, Peters PWM, Bunk W, Carbon fibre reinforced carbon produced by polymer impregnation and pyrolysis, *Materialwissenschaft und Werkstofftechnik*, 30(4), 185–190, 1999.
75. Rhee B, Ryu S, Fitzer E, Fritz W, *High Temp-High Pressures*, 19(6), 677–686, 1987.
76. Fitzer E, Huttner W, Hartwig G, Met Mat Compos Low Temp, *Proc ICMC Symposium 1978*, 245, 1979.
77. Park HS, Choi WC, Kim KS, *J Adv Mater*, 34, Jul 1995.
78. Fujioka J et al, A new process for manufacturing advanced carbon-carbon composites utilizing hot isostatic pressing, *1st Japan International SAMPE Symposium and Exhibition*, 1160, Nov 28–Dec 1 1989.
79. Lachmann WL, Crawford SA, McAllister LE, *Proc Int Conf on Composite Mats*, Met Soc of AIME, New York, 1978.
80. Burns RL, McAllister LE, *Proc 12th Propulsion Conf*, Palo Alto, 1976.
81. Forest MA, Marsh H, The effects of pressure on the carbonization of pitch and pitch/carbon composites, *J Mater Sci*, 18(4), pp 978–990, Apr 1983.
82. Johnson AC, Pinoli PC, Keller RL, Optimization of carbon-carbon processing, *14th Biennial Conference on Carbon*, American Carbon Committee, Penn State University, 238, Jun 1979.
83. Dietrich H, McAllister LE, *Am Ceram Soc 80th Ann Meeting*, Detroit, 1978.
84. Hosomura T, Okamoto H, *Mater Sci Eng*, A 143(1–2), 223–229, 1991.
85. Liebermann ML, *Proc 3rd Int Conf on CVD*, 3, 95, 1972.
86. Noles GT, Liebermann ML, *J Chromatogr*, 114, 211, 1975.
87. Bokros JC, *Chem Phys Carbon*, 5, 1, 1969.
88. Savage GG, *Carbon-Carbon Composites*, Chapman and Hall, London, 92, 1992.
89. Heddon K, Wicke E, Some influences on the reactivity of carbon, *Proc of the 3rd Conf on Carbon*, Pergammon Press, New York, 249, 1957.
90. Bhatia G, Fitzer E, Kompalik D, Mesophase formation in toluene soluble and toluene insoluble coal tar pitch fractions and their mixtures, *International Carbon Conference Extended Abstracts*, Bordeaux, France, 330–331, 1984.
91. Charit I, Harel H, Fischer S, Marom G, *Thermochim Acta*, 62, 237–248, 1983.
92. Hosty AJ, Rand B, Jones FR, New Materials and Their Applications, *Inst Phys Conf Ser*, Vol 111, Bristol, IOP, 521–530, 1990.
93. Dillon K, Thomas KM, Marsh H, *Carbon*, 31, 1337, 1993.
94. Bahl OP, Manocha LM, Bhatia G, Dhami TL, Aggarwal RK, *J Sci Ind Res*, 50(7), 533–538, 1991.
95. Hosty AJ, Rand B, Jones FR, Inst Phys Conf Ser, Vol 111, New Materials and Their Applications, Bristol, IOP, 521–530, 1990.
96. Figueiredo JL, Bernardo CA, Baker RTK, Huttinger KJ eds., *Carbon Fibers Filaments and Composites*, Kluwer Academic, Dordrecht, 327–336, 1990.
97. Savage GG, *Carbon-Carbon Composites*, Chapman and Hall, London, 193–225, 1992.
98. Buckley JD, Edie DD eds., *Carbon-Carbon Materials and Composites*, Noyes Publications, Park Ridge, 239–266, 1992.
99. Fitzer E, Manocha LM, *Carbon Reinforcements and Carbon/Carbon Composites*, Springer Verlag, Berlin, 281–309, 1998.
100. Chung DDL, *Carbon Fiber Composites*, Butterworth-Heinemann, Boston, 157–164, 1994.
101. Yamada E, Kimura S, Shilswa Y, *Trans JSCM*, 6(1), 14, 1980.
102. Chang HW, Rhee SK, *Carbon*, 16, 17, 1978.
103. P Crocker, B McEnaney, Carbon 29, (7), 881–885, 1991.
104. McKee DW, In Walker PL, Thrower PA eds, *Chemistry and Physics of Carbon*, Vol 1, Marcel Dekker, New York, 1, 1965.

105. Thomas JM, In Walker PL, Thrower PA eds., *Chemistry and Physics of Carbon*, Vol 1, Marcel Dekker, New York, 1981, p. 1.
106. Ehrburger P, In Figueiredo JL, Bernardo CA, Baker RTK, Huttinger KJ eds., *Carbon Fibres Filaments and Composites*, Kluwer Academic, Dordrecht, 327–336, 1990.
107. Ehrburger P, Baranne P, Lahaye J, *Carbon*, 22, 507, 1984.
108. McKee DW, Spiro CL, *Carbon*, 23, 437, 1985.
109. McKee DW, Spiro CL, Lamby EJ, *Carbon*, 22(5), 285, 1984.
110. Lynch JF, Ruderer CC, Duckworth WH, *US Air Force Mat Lab Tech Ref Report*, No. AFML-TR066-52, 1966.
111. Strife JR, Sheehan JE, *Ceramic Bulletin*, 67, 369–374, 1988.
112. Johnson HV, U.S. Pat., 1,948,382.
113. Dickinson RC, In Barkatt A, Verink ED Jr., Smith LR eds., *Mater Res Soc Symp Proc 125 (Materials Stability and Environmental Degradation)*, 3–11, 1988.
114. Buchanan FJ, Little JA, *Surface Coating Technol*, 46(2), 217–226, 1991.
115. Deng J, Wei Y, Liu W, Carbon-fibre-reinforced composites with graded carbon-silicon carbide matrix composition, *J Am Ceramic Soc*, 82(6), 1629–1632, 1999.
116. Yeager RE, Shaw SC, In *Proc Metal and Ceramic Matrix Composite Processing Conf*, Vol 11, US Dept. of Defense Information Analysis Centers, 145–180, 1984.

CHAPTER 15

Carbon Fiber Reinforced Ceramic Matrices

15.1 INTRODUCTION

A ceramic is an inorganic non-metallic solid prepared from powdered materials and can loosely be divided into cement, mortar and concrete, glass and glass ceramics, traditional ceramics, and what may be termed advanced (i.e. high performance, technical, engineering, or fine) ceramics.

Rand and Zeng [1] and Chung [2] give valuable background information on the study of carbon fiber reinforced ceramic matrix composites. Huttinger and Greil [3] review the reinforcement of ceramics by ceramic fibers, Briggs [4] reviews cfrc and Jing Kung [5] reviews the work undertaken on high temperature ceramics and composites at the Shanghai Institute of Ceramics. Cornie [6] discusses processing techniques for fiber reinforced ceramic matrix composites.

15.2 CEMENT, CONCRETE AND GYPSUM MATRICES

15.2.1 Cement

Portland cement, which is not a trade name, is available in several grades, to meet various physical and chemical requirements (see ASTM C–150, which covers eight types of Portland cement and Cement Standards of the World [7]).

A Portland cement is made by calcining a mixture of compounds such as clay, fly ash, iron ore, marl and shale, which contain alumina, Ca compounds, iron oxide and silica at 1500–1650°C. Gypsum is added to control the setting rate and the mixture is ground to a fine powder. A Portland cement can be tricalcium silicate ($3CaO \cdot SiO_2$), dicalcium silicate ($2CaO \cdot SiO_2$), tricalcium aluminate ($3CaO \cdot Al_2O_3$) and tetracalcium aluminoferrite ($4CaO \cdot Al_2O_3Fe_2O_3$) and these compounds, when hydrated, form crystalline and amorphous phases interlocked with a porous gelled network. After mixing, the compounds dissolve and heat of hydration is evolved, followed by a dormant period of several hours and then the cement hardens by a process of hydration. The di- and tri- calcium silicates react with water and form calcium silicate hydrate ($3CaO \cdot 2SiO_2 \cdot 3H_2O$) and calcium hydroxide ($Ca(OH)_2$). The reaction can broadly be represented by:

$$3CaO \cdot SiO_2 + 2CaO \cdot SiO_2 + 5H_2O \rightarrow 3CaO \cdot 2SiO_2 \cdot 3H_2O + 2Ca(OH)_2$$

As the reaction proceeds, cement and water are gradually used up to produce an amorphous gel. To achieve good strength, it is necessary to cure the mass for several days by keeping the surface moist at 10–25°C.

15.2.2 Concrete

Concrete is basically a mix of Portland cement and contains a fine aggregate (coarse sand), a coarse aggregate (gravel or crushed stone) and water. The fine aggregate fills the interstices around the coarse aggregate, making the concrete stronger. A mortar is a concrete without an aggregate and uses fine sand. Optimization of the use of lightweight aggregates in carbon fiber reinforced cement has been examined [8].

Concrete is extremely stiff and good in compression, but quite poor in tension. Steel rebar reinforcement is normally used to make a stronger structure, but because steel is subject to atmospheric corrosion, it has to be shielded with a relatively thick protective layer of concrete, which with time, is removed by weathering and erosion, eventually exposing the steel to attack. Carbon fiber has very good corrosion resistance and would not be affected by any alkalis in the cement. Hence it should be possible to utilize a carbon fiber composite such as carbon fiber/epoxy pultruded rods that would only require a relatively thin coating of concrete. Alternatively, carbon fiber can be incorporated directly into the concrete mix [9].

15.2.3 Concrete additives [2,10–15]

15.2.3.1 Silica fume

Silica fume is made by heating quartz, coal, iron and wood chips to 1800°C and collecting the fume. It is a very fine powder, with a mean diameter of about 0.1 μm (cf cigarette smoke) and a very high specific surface area of about 20,000 $m^2 kg^{-1}$, which is one hundred times finer than cement. When added to a cement paste, it has a microfiller effect, with the particles filling the spaces between the granules of cement. With 15% of added silica fume, there would be 2×10^6 silica fume particles for each grain of cement. The porosity decreases, the compressive strength increases, the bond improves and the number of nucleation sites for hydration products increases. Silica fume also produces a pozzolanic effect, the particles comprising of about 85% amorphous SiO_2, which react with the $Ca(OH)_2$ formed in the hydration reaction to form calcium silicate hydrates:

$$2SiO_2 + 3Ca(OH)_2 \rightarrow 3CaO \cdot 2SiO_2 \cdot 3H_2O$$

This densifies the material and probably accounts for the improved sulphate resistance.

15.2.3.2 Dispersant

Latex can be used as a dispersant and a typical latex would be a styrene-butadiene copolymer e.g. (Dow 460NA) used at 20% w/w of cement. The latex must be used in conjunction with an anti-foam agent such as Dow Corning 2410.

A number of workers have studied the improvements obtained by the addition of latex to mortars [11,16,17] and concrete [18,19].

Latex is expensive and an alternative dispersant is methyl cellulose (Dow Chemical Methocel A15-LV), used at 0.4% w/w of cement in conjunction with 0.13 vol % defoamer (Colloids 1010). The effect of methylcellulose admixture on the mechanical properties of cement has been determined [20].

15.2.3.3 Water reducing agent

A water reducing agent such as the Na salt of condensed naphthalene sulphonic acid (Rohm & Haas Tamol SN), used at 2% w/w of cement, increases the fluidity of the concrete mix for 0.5% fiber reinforcement [10].

The actual water content is critical and should be in the ratio 0.50 water/cement to obtain sufficient fluidity [10]. The wettability of carbon fibers by water has been studied [21].

15.2.3.4 Accelerator

An accelerator reduces the porosity of the concrete [10] and improves the bond of the fiber to the cement. Typical accelerators are triethanolamine, used at 0.06% w/w of cement, or 0.5% potassium aluminum sulphate ($AlK(SO_4)_2$) plus Na_2SO_4 at 0.5% w/w of cement.

15.2.4 Work undertaken with mortar and concrete

Work at Courtaulds [22,23] in the early 1970s attempted to incorporate carbon fiber in a cement slurry, which was difficult due to the size of the cement particles. They tended to be filtered out by the fiber reinforcement, so a cement with a fine particle size (Swiftcrete, an ultra rapid hardening Portland cement with a maximum diameter of about 45 µm) was used and the fiber spread as thinly as possible, using either an air knife, or a water flume and then held in the spread position by sizing with a water based compatible size such as sodium carboxymethylcellulose [22,23]. These larger particles limit the carbon fiber content to about 5% v/v, but in practice, due to a non-uniform distribution, a value of some 12% v/v was attainable.

Waller [24] followed this laboratory work by field trials, hopefully to use the product as decking in multi-storey car parks, but at that time, carbon fiber was considered too expensive for such an application.

Later, workers at Harwell produced samples of carbon fiber reinforced cement (cfrc) by the process of filament winding under tension, giving good fiber alignment and distribution [25].

Work undertaken by Majumdar et al [26,27] at The Building Research Station, Garston showed that the carbon fiber possessed excellent durability in a cement slurry with a modulus of rupture (maximum stress) of 33 MPa and an Izod impact strength of 4 $kJ\,m^{-2}$ for a loading of 1.3% fiber.

Briggs [4] claimed that carbon fiber does not bond to cement, but the surface texture of the fiber is sufficient to provide a form of keying to the gelled cement during the hydration process and suggested that the rough fiber surface was intimately penetrated by the cement gel to give a good frictional bond, making it possible for elastic continuity to be maintained even after cracking.

Later, however, Sugama [28] stated that chemical bonding does exist between carbon fiber and the cement matrix, whilst Yuan et al [29] showed that oxidative surface treatment followed by treatment with KH-550 χ-aminopropyl triethoxysilane (($H_2N(CH_2)_3Si(OC_2H_5)_3$)

increased the flexural strength by 30%, the tensile strength by 48% and the impact strength by 63%. Examination of the IR spectrum revealed the presence of new peaks when treated with silane, indicating a chemical reaction had taken place with the calcium silicon oxide and calcium aluminate in the cement, suggesting KH-550 was an effective coupling agent.

The bond strength in cfrc is improved by using silane-treated carbon fibers [30] and by surface treatment in conjunction with the addition of a polymer to the cement mix [31]. Larson *et al* [32] studied adhesion of cement to carbon fiber in cfrc composites.

The use of silane treated carbon fibers and silane treated silica fume led to increases in tensile strength of cement paste by 56% and modulus and ductility by 39% [30], as compared to the values obtained for cement paste with either as-received carbon fibers or as-received silica fume. Silane treatment of fibers and silica fume contributed about equally to the strengthening. Silane treatment of fibers and silica fume also decreased the air void content. The strengthening and reduction in air void content were less when the fiber treatment involved the use of $K_2Cr_2O_7$ instead of silane, and even less when the treatment involved the use of O_3. The effectiveness of silane is due to its hydrophilic nature.

The tensile strength, modulus and ductility of cfrc paste were increased by O_3 treatment [33] of the fibers prior to their use. Increases were observed irrespective of whether the paste contained methylcellulose, silica fume, or latex. The O_3 treatment involved exposure to the gas (0.3 vol% in air) for 10 min at 160°C. A disadvantage was that ductility was decreased by the O_3 treatment in the case of cement paste with latex.

Linton and co-workers [34] showed that carbon fibers could be dispersed in a cement matrix giving a flexural strength of 20 MPa for a 2% V_f, providing fiber failure was achieved by fracture rather than pull-out, obtained at the expense of a reduction in toughness. Although the load deflection curves are similar, a pitch based fiber gave a lower flexural strength than a PAN based fiber, which was attributed to the initial strength of the virgin fibers (Figure 15.1). Silica fume ensured proper dispersion and bonding, but the presence of sand, induced fiber breakage during mixing (Figure 15.2), producing fibers below the critical length, with a resultant decrease in flexural strength. The minimum carbon fiber content to

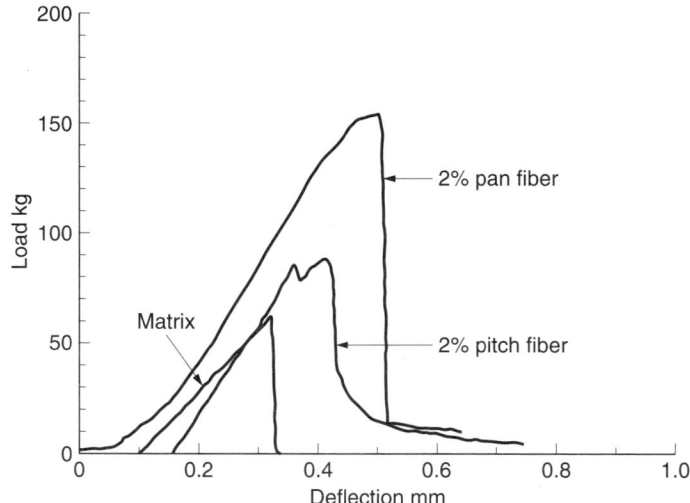

Figure 15.1 Effect of fiber type on composite flexural behavior. Mixing ratios—Silica fume:ordinary Portland cement = 0.195, water:binder = 0.21, sand:binder = 0. *Source:* Reprinted from Linton JR, Burneburg PL, Gartner EM, Bentur A, *Mat Res Soc Symp Proc*, 211, 255–264, 1991.

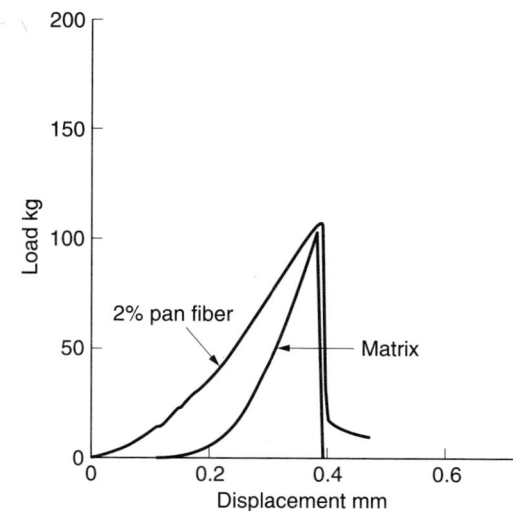

Figure 15.2 Effect of sand on composite flexural behavior. Mixing ratios—Silica fume:ordinary Portland cement = 0.195, water:binder = 0.21, sand:binder = 1. *Source:* Reprinted from Linton JR, Burneburg PL, Gartner EM, Bentur A, *Mat Res Soc Symp Proc*, 211, 255–264, 1991.

provide an increase in flexural strength was 0.1 vol % and the optimum value depended on the type of dispersant used [35]. The beneficial effect of silica fume was affirmed by Katz *et al* [36] with a 370–670% improvement in bond strength as determined by pull-out tests, indicating that better penetration of long and narrow grooves in the carbon fiber played an important role.

Chen and Chung [37] determined the flexural strength and toughness of carbon fiber reinforced mortars and the highest strength was recorded with 2.1% fiber with the addition of latex as a dispersant, irrespective of the presence of sand. It required 4.2% without sand, when using methyl cellulose or methyl cellulose/ silica fume as dispersant. The higher proportion of fiber gave the best flexural toughness.

Sakai *et al* [38] showed that cfrc composite could be used for curtain walls successfully, undergoing wind resistance tests, having far higher fatigue strength than ordinary concrete. However, it was shown that each mixer used to mix the cfrc had its own optimum mixing time to give a uniform dispersion and maximum flexural strength.

Toutanji and workers [39,40] found a significant enhancement of the tensile and flexural strength of cfrc when using Portland cement with 15% silica fume and 1% superplasticizer (by weight of mix) using a water cement ratio of 0.3. The increase in reliability in tension was offset by a decrease in the reliability in bending. Tensile strength values were determined using the cementitious composites axial tensile technique (CCATT) [41], utilizing a Swedish hydraulic tensile tester developed by ASCERA [42]. The CCATT results were compared with predicted theoretical values (Figure 15.3) and were about 4–10% less than predicted values, diverging as the fiber load increased, probably due to insufficient dispersion of fiber bundles into individual monofilaments.

Table 15.1 is a review by Toutanji and co-workers [40] of work undertaken on cfrc composites [43–48] and several workers [49–53] have detailed other work on carbon fiber reinforced mortar. Chung [54] describes work on carbon fiber cement matrix composites.

Furukawa [55] found that effective reinforcement was obtained with 2–4 vol% of chopped carbon fiber, whereas the same effect could be achieved with 0.3–0.5 vol% of continuous carbon fiber.

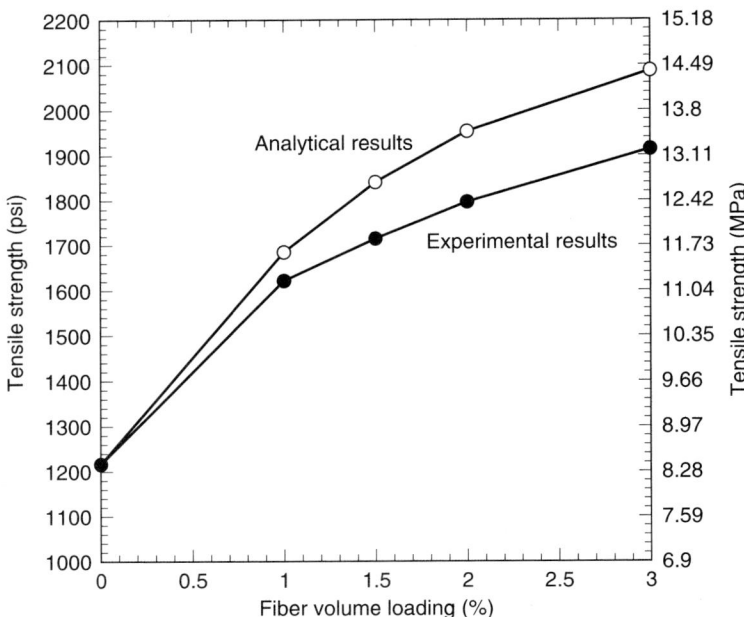

Figure 15.3 Effect of PAN based carbon fibers on the tensile strength of cementitious composites. *Source:* Reprinted with permission from Toutanji HA, El-Korchi T, Katz RN, Leatherman GL, *Cement Concrete Res*, 23, 618–626, 1993. Copyright 1993, Elsevier.

DDL Chung and her research team at the State University of New York at Buffalo have investigated chopped carbon fiber reinforcements (5 mm) at volume fractions up to about 5% [2]. Two mixing approaches were used:

1. Dry mixing—where the fibers are initially mixed with a fine dry aggregate (5 mm), then cement and silica fume added, followed by mixing in a Hobart industrial mixer for about 5 min. The mixing was completed in a concrete mixer to which the coarse gravel aggregate (25 mm) was added [9]. The silica fume helps the dispersion of the carbon fiber and reduces the porosity of the concrete, resulting in an improved bond.
2. It is more practical, however, to employ wet mixing, where the fiber is dispersed in water with a dispersant such as methyl cellulose, which will also require the addition of an anti-foam agent. The addition of about 0.15% of a silica fume on the weight of cement acts as a microfiller, readily occupying the gaps between cement particles. Although latex can be successfully used as a dispersant, it requires addition levels of some 20% (on weight of cement), rendering the formulation very expensive. The optimum fiber length is about 5 mm and longer fibers would be difficult to disperse; a length reduction of about 40% does occur in the Hobart mixer, but there is no reduction in length in the concrete mixer.

The low resistivity of a cfrc enables the concrete to be used as an EMI shield and also function as a smart structure, with a direct relationship between the electrical resistivity and compressive stress [56]. Chung and co-workers have investigated carbon filament cement matrix composites for electromagnetic interference shielding. [57,58].

The addition of 1.7% v/v of a pitch based carbon fiber lowered the electrical resistivity of a gypsum plaster (α-$CaSO_4 \cdot 5H_2O$) and significantly improved the effective EMI shielding [59].

Chung [60] also used a pitch based carbon fiber (mean fiber length 12 mm reducing to 7 mm after mixing in a Hobart mixer) incorporating chemical agents and silica fume with fine

Table 15.1 Summary of the CFRC composite literature review

Source	Fiber type	Fiber %vol	Fiber length mm	W/C ratio	SF/C ratio	Tensile strength % increase	Flexural strength % increase
Ali et al [43]	PAN chopped	3	NA	0.30	NA	70	NA
	PAN continuous	3.7	NA	0.30	NA	500	NA
Akihama et al [44]	PAN chopped	2.1	3	0.47	0.25	210	350
	PAN chopped	4.2	3	0.47	0.25	225	500
	PAN chopped	5.3	3	0.47	0.25	270	510
	Pitch chopped	2.1	3	0.47	0.25	30	125
	Pitch chopped	4.2	3	0.47	0.25	60	210
	Pitch chopped	5.3	3	0.47	0.25	150	270
Ohama et al [45]	Pitch chopped	1	3	0.30	0.40	35	50
	Pitch chopped	3	3	0.30	0.40	180	260
	Pitch chopped	5	3	0.25	0.40	270	340
	Pitch chopped	1	10	0.30	0.40	15	40
	Pitch chopped	10		0.30	0.40	125	150	NA
	Pitch chopped	5	10	0.30	0.40	200	340
Akihama et al [46]	Pitch chopped	1.72	10	0.42	NA	40	NA
	Pitch chopped	3.18	10	0.42	NA	90	NA
	Pitch chopped	3.98	10	0.42	NS	110	NA
	Pitch chopped	2.04	10	0.30	NA	70	NA
	Pitch chopped	4.03	10	0.30	NA	120	NA
Linton et al [34]	PAN chopped	1.65	6	0.37	0.195	NA	165
	PAN chopped	2	6	0.37	0.195	NA	165
	PAN chopped	2	6	0.21	0.195	NA	90
	Pitch chopped	2	6	0.21	0.195	NA	90
Park et al [47]	PAN chopped	1	3	0.30	0.40	120	NA
	PAN chopped	2	3	0.30	0.40	170	NA
	PAN chopped	3	3	0.30	0.40	200	NA
	Pitch chopped	1	3	0.30	0.40	105	NA
	Pitch chopped	2	3	0.30	0.40	150	NA
	Pitch chopped	3	3	0.30	0.40	180	NA
Banthia et al [48]	Pitch chopped	1	6	0.30	NA	NA	250
	Pitch chopped	3	6	0.30	NS	NA	360
	Pitch chopped	5	6	0.30	NA	NA	375

Note: NA = not available   sfp = silica fume in powder form   sfs = silica fume in slurry form   NS = silica fume not specified.
Source: Reprinted with permission from Toutanji HA, El-Korchi T, Katz RN, Cement Concrete Res, 16, 15–21, 1994. Copyright 1994, Elsevier.

and coarse aggregates, which resulted in a concrete with 85% increase in flexural strength, 205% increase of flexural toughness, 22% increase of compressive strength and 83% decrease in electrical resistivity.

Concrete containing short carbon fibers (0.2–0.5 vol %) was found to be an intrinsically smart concrete [61] that can sense elastic and inelastic deformation, as well as fracture. The signal provided is the change in electrical resistance, which is reversible for elastic deformation and irreversible for inelastic deformation and fracture. The pressure exerted by the electrically conducting short fibers is necessary for the concrete to sense elastic or inelastic deformation, but the sensing of fracture does not require fibers. The fibers serve to bridge the cracks and provide a conduction path. The resistance increase is due to conducting fiber pullout in the elastic regime, conducting fiber breakage in the inelastic regime and crack propagation at fracture.

Reactance measurement can be used for self monitoring in carbon fiber reinforced mortar [62], whilst the effect of curing age on the self-monitoring behavior has been established [63].

The cfrc composite can be used as an intrinsically smart material to assess damage during dynamic loading [64,65].

Piezoresistivity [66] was observed in cement matrix composites with 2.6–7.4 vol% unidirectional continuous carbon fibers. The dc electrical resistance in the fiber direction increased upon tensile loading in the same direction, such that the effect was mostly reversible when the stress was below that required for the tensile modulus to decrease. The gage factor was up to 60. The resistance increase was due to the degradation of the interface of the fiber and matrix, which was mostly reversible. Above the stress at which the modulus started to decrease, the resistance abruptly increased with stress/strain, due to fiber breakage. The tensile strength and modulus of the composites were 88% and 84%, respectively, of the calculated values based on the rule of mixtures.

A cement paste formed from 0.5 wt %, 5 mm long carbon fiber reinforced with 15 wt% silica fume was found to be an effective thermistor [67]. Its electrical resistivity decreased reversibly with increasing temperature (1–45°C), with an activation energy of electrical conduction (electron hopping) of 0.4 eV. This is comparable to those of semiconductors (typical thermistor materials) and higher than that of carbon fiber polymer matrix composites. Without carbon fibers, or with latex in place of silica fume, the activation energy was much lower and the resistivity was higher. The voltage range for a linear current-voltage characteristic was wider in the absence of fibers. The current-voltage characteristics of carbon fiber reinforced silica fume cement paste were linear up to 8 V at 20°C.

The Seebeck effect (a difference in temperature at a junction causing a current to flow) in cfrc [68] has been studied.

Addition of carbon fiber to a concrete will increase the thermal conductivity, which can be used to advantage for heated bridge roadways [69].

Carbon fiber reinforced mortar can increase the bonding strength between brick and mortar [70], since the fibers act to both increase mortar strength as well as decrease drying shrinkage. The addition of short carbon fibers in an optimum amount of 0.5% of the cement weight to mortar increased the brick-to-mortar bond strength by 150% under tension and 110% under shear when the gap between the adjoining bricks was fixed, and by 50% under tension and 44% under shear when this gap was allowed to freely decrease due to the weight of the brick above the joint. This effect is attributed to the decrease in drying shrinkage by fiber addition. This decrease was particularly large when curing from 2 to 24 h. The shrinkage at 24 h was decreased by 50% by the fiber addition (0.5% of cement weight). Fibers in excess of the optimum amount reduced the bond strength due to increased porosity in the mortar.

The bonding between old and new concrete can be improved by adding carbon fibers to the new concrete [12], the improved bonding being attributed to the lowering of the drying shrinkage by the addition of the carbon fibers.

15.2.5 Theory

The mechanical properties of a fiber composite depend on many factors [4] including:

1. Fiber properties—dimensions, strength, modulus and stress/strain behavior.
2. Composite make-up—fiber orientation, fiber/matrix interface, type of bonding and degree of consolidation.
3. Matrix properties—strength, modulus and stress/strain behavior.
4. Other factors—changes in properties with time and environment.

Aveston, Mercer and Sillwood [71,72] derived crack theory for cement reinforced with continuous and short fibers and found that the tensile strength of the composite for the aligned fibers was:

$$\sigma_{cu} = \left(1 - \frac{L_c}{2L}\right)\sigma_{fu} V_f$$

where
σ_{cu} = composite ultimate strength
L_c = fiber critical transfer length
L = fiber length
σ_{fu} = fiber ultimate strength
V_f = fiber volume fraction

and $2/\pi$ or 0.5 times this respectively for the two and three dimensional random fibers if $L > L_c$.

Results obtained by these workers for the initial Young's modulus and ultimate tensile strength of cfrc are shown in Figures 15.4 and 15.5.

15.2.6 Fabrication processes for cfrc

A number of processes have been adapted and utilized for fabricating cfrc:

1. The spray-up process employs a continuous tow of carbon fiber, fed to a dual applicator gun where it is both chopped and admixed with cement slurry [10] and finally discharged onto a mold surface.
2. The fiber can be chopped as a separate operation and then admixed with cement slurry before applying to the mold surface [4].
3. Carbon fiber can be impregnated with cement slurry prior to forming sheets, which are placed in a mold and in this form, are particularly useful for curved surfaces [4].
4. The principles of filament winding used for thermosetting resins have been adapted to give a precise approach [73] for fabricating cfrc composites.
5. Pull-pressing uses a modified form of filament winding [74] where a slurry impregnated carbon fiber tow is fed into a mold and consolidated to form sheets or beams.

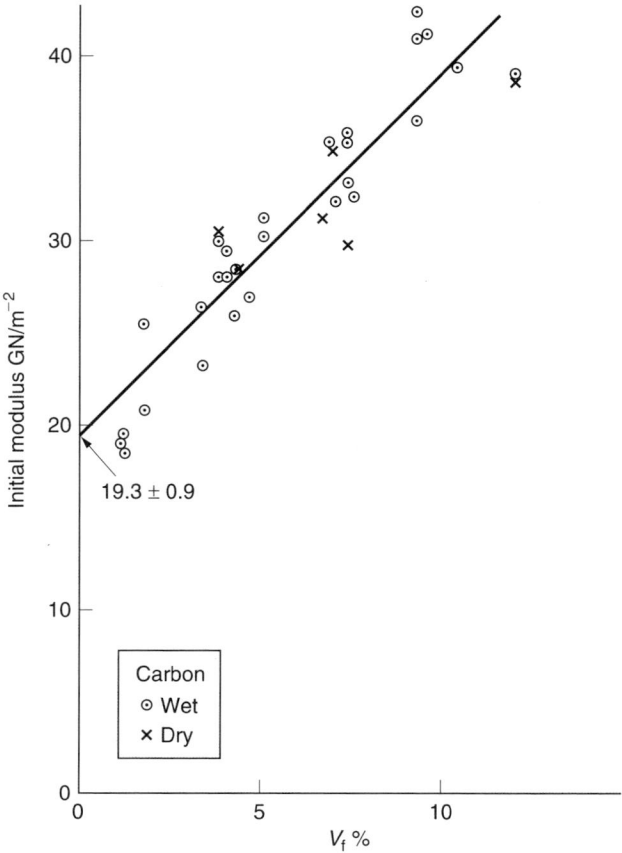

Figure 15.4 Initial Young's modulus of continuous carbon fiber reinforced cement. *Source:* Reprinted from Aveston J, Mercer RA, Sillwood JM, *Conference Proceedings Composites Standards Testing and Design*, NPL 1974, IPC Science & Technology Press, 93–103, 1974.

15.3 GLASS MATRICES [75]

Types of glasses are listed in Table 15.2 and most are based on silica (SiO_2) and if made entirely of SiO_2, is termed a silica glass or vitreous silica. Another class of glass, is termed soda lime and is based on Na_2O and CaO, with the addition of other constituents to obtain specific properties. Borosilicates such as Pyrex, with the addition of Al_2O_3 and B_2O_3 have good chemical resistance and resistance to thermal shock.

15.3.1 The glass matrix

When a liquid glass cools to below its freezing temperature, it does not crystallize but forms a rigid disordered network termed as glass. Solid glass can be considered a highly viscous or rigid liquid and will contain many dislocations. The molecular structure in glass is unable to move and small cracks allow localized stress concentrations to build up, leading to cracks. Glass is not a supercooled liquid and is probably best considered a solid.

Glasses are made by rapidly quenching a melt, but there is no distinct transition glass phase between the melting temperature T_m and the glass transition temperature T_g and in this

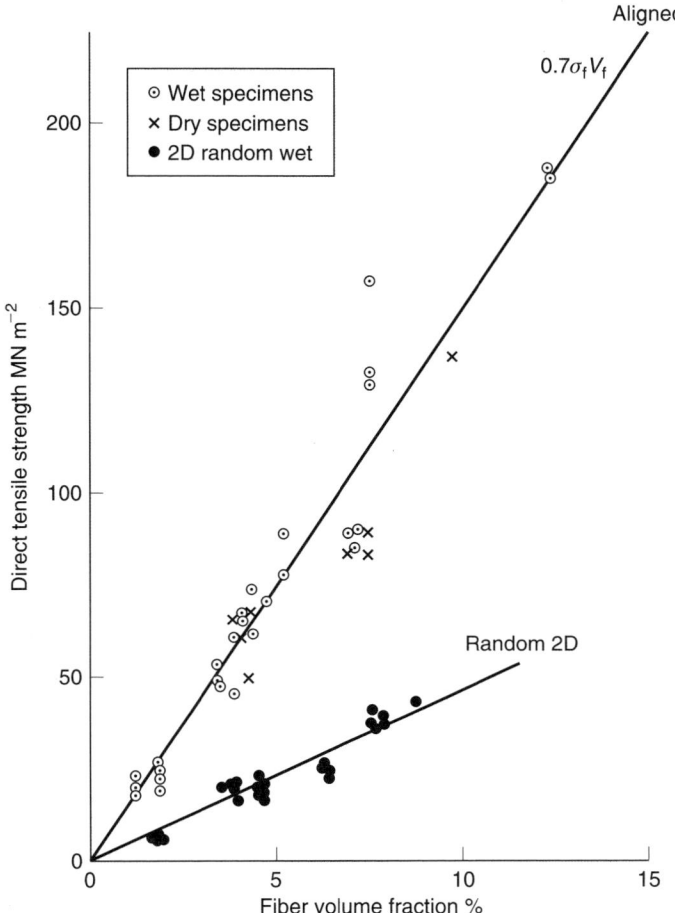

Figure 15.5 Ultimate tensile strength of carbon fiber reinforced cement. *Source:* Reprinted from Aveston J, Mercer RA, Sillwood JM, *Conference Proceedings Composites Standards Testing and Design*, NPL 1974, IPC Science & Technology Press, 93–103, 1974.

range, the amorphous glass does pass through a so called super cooled liquid phase, when the glass can be readily shaped. The T_g changes, depending on how slowly the glass is cooled, probably reaching a minimum value of 270°C.

When carbon fiber is added to a glass matrix, it would be expected to form a composite, with improvements in toughness, strength (when using continuous fiber), increased thermal conductivity and a decreased coefficient of thermal expansion. The glass matrix would not, however, protect the carbon fiber from oxidation in air at about 500°C [76] and the fiber would then have little bonding to the glass matrix.

A recrystallized, glass formed by a process of controlled devitrification (crystallization) is termed a glass-ceramic (or polycrystalline glass).

It is possible to introduce a degree of crystallization into the disordered network of some glasses by, for example, the addition of a nucleating agent (P_2O_5, TiO_2, ZrO_2) and they are then termed glass ceramics e.g. Apoceram [77], Neoceram (Nippon Electric Glass), Pyroceram (Corning), Robax (Schott), Silceram.

The glass ceramic has increased impact strength, hardness and thermal shock resistance as compared to a non-crystalline glass, withstanding a quench into ice/water from red hot temperature and should therefore, be a good candidate matrix for reinforcement with carbon fiber.

The lithium aluminosilicates $Li_2O \cdot Al_2O_3 \cdot SiO_2$ (LAS) are a range of glass ceramics [78] and can be prepared with different crystalline phases possessing specific mechanical properties. The crystallization can be enhanced by the addition of B_2O_3 and delayed by addition of P_2O_5 [79].

Another class of ceramic glasses are termed Y·Mg·Al·Si·O (YMAS), prepared from magnesium aluminum silicate, yttrium silicate, magnesium aluminum oxide and corundum (Al_2O_3).

Another glass ceramic is based on Ca, Al and Si (CAS), $CaO \cdot MgO \cdot Al_2O_3 \cdot SiO_2$ (CMAS) [80].

Oxynitride glass ceramic matrices can be prepared by introducing the nitrogen in the form of Si_3N_4 or AlN, which can be incorporated into an oxide based glass ceramic to give β′-sialon, an oxynitride glass ceramic [81], when crystallized:

$$Al_2O_3 + \alpha\text{-}Si_3N_4 + AlN \rightarrow Si_{6.25}Al_{0.75}O_{0.75}N_{0.75}$$

The oxynitride glass ceramics have improved refractory and mechanical properties as compared to the equivalent oxide materials, due to their increased bond strength. A range of oxynitride glass ceramics of the system M.Si.Al.O.N (e.g. where M can be La, Y or Zr) can be formed, extending the chance of finding compositions which can be crystallized to form single phase glass ceramic products.

Typical glasses and glass ceramics are given in Table 15.2 and Bach and Krause have detailed the Analyzes of the composition and structure of glass and glass ceramics [82].

15.3.2 Methods of preparation of carbon fiber reinforced glasses

15.3.2.1 Mode of reinforcement

A chopped carbon fiber can be used as reinforcement, but in this form, it will be necessary to form a mixture or slurry with the glass and the staple length must be < 3 mm, as otherwise, it will be hard to dissipate. When subjected to pressure in the consolidation process, there will be some fiber flow resulting in a more planar array. Continuous fiber can be adapted to more automated processes such as filament winding [83]. If a water based slurry process is used, it would be expected that a surface treated fiber should wet out more readily by achieving more efficient interfilament penetration of the matrix.

15.3.2.2 Slurry with hot pressing [84,85]

This can be achieved by dispersing chopped fiber in a slurry of a powdered glass matrix, commuted to a size of about 10–15 μm, in a water or propanol vehicle. Wetting agents and organic binders may be added to promote the mixing and consolidation processes. The slurry is vacuum filtered, heated initially to 300°C in an inert atmosphere to remove all volatiles, then to 400–700°C to remove the binder and hot pressed at 700–1000°C in a vacuum, or an Ar atmosphere, thereby avoiding the processing temperature becoming too high, which would promote fiber degradation. Above the working temperature of the glass (Table 15.3), the glass flows and the sufficient pressure must be applied to ensure that it flows into the interstices of the individual filaments [86].

Table 15.2 Glass and glass ceramic matrices

Matrix type	Corning grade	Density g cm⁻³	SiO$_2$	Al$_2$O$_3$	CaO	MgO	B$_2$O$_3$	ZrO$_2$	Na$_2$O	K$_2$O	Li$_2$O	ZnO	TiO$_2$	BaO	Nb$_2$O$_3$	SrO
GLASS MATRICES																
Soda Lime	0071															
Soda Lime	0080															
Soda Lime	0125		72	1	9	4			14							
Soda Lime	9989		65	5.7	5.9	2.9	3.9		14.9	1.2						
Zinc Titanium- cover glass	0211	2.53	64	3			9		7	7		7	3			
Borosilicate- Pyrex	7740	2.53	80.6	2.3			13		4	0.1						
Borosilicate- Pyrex	7789	2.22	81	2			13		3	1						
Borosilicate-High Expansion	7799	2.47	70	6	1	0.5	10		9	1		0.5		2		
Borosilicate	7800															
Aluminosilicate	1723		64	25	+	10								+		
High silica- Vycor	7913	2.18	96.4	0.5			3									
High silica	7930		+				+									
Fused Quartz			>99.5													
GLASS CERAMICS																
Glass Ceramic	9608															
Lithia aluminosilicate types (LAS):																
Glass Ceramic-LAS-I (β- spodumene)			+	+		+		+			+	+		+		
Glass Ceramic-LAS-II (β- spodumene)			+	+		+		+			+	+		+	+	
Glass Ceramic-LAS-III (β-spodumene)			+	+		+		+			+					
Glass Ceramic MAS (Cordierite)			+	+		+								+		
Glass Ceramic BMAS (Ba osmulite)			+	+		+								+	+	
CELSIAN																
Hexacelsian (Hexacelsian)			37.97	13.58										48.45		
Celsian			31.77	37.83			3									27.4

Note: Crystalline form given in brackets. The overall properties will depend on the bonding mechanism. Rapid cooling will favour the non-crystalline form since the time is insufficient to permit the formation of a regular crystalline structure.

Source: Reprinted with permission from Prewo KM, Bacon JF, Glass matrix composites-I Graphite fiber reinforced glass, Noton B ed., *Proc Second Int Conf on Composites*, AIME, 64–74, 1978. Copyright 1978, The Metallurgical Society of AIME, now TMS (The Minerals, Metals and Materials Society).

Table 15.3 Important temperatures for the fabrication of silicate glasses

Type of glass	Annealing temperature[a] °C	Softening temperature[b] °C	Working temperature[c] °C
Soda lime glass	575	685	890
Borosilicate glass	625	820	1155
96% silica glass	1025	1620	2125
Silica glass	1290	1775	2220

[a]Annealing temperature is the temperature at which the glass is cooled in a controlled manner to prevent or remove objectionable stresses.

[b]Softening point is the temperature at which glass, with a viscosity of $10^{7.5}$–10^8 poise will elongate under its own weight [ASTM C-338].

[c]The working temperature is a figure within the working range where the upper end refers to a temperature when the glass is ready for working (viscosity 10^3–10^4 poise) and the lower end refers to a temperature when the glass will just hold its formed shape (viscosity $>10^6$ poise). The viscosity range is generally assumed to be 10^4–$10^{7.6}$ poise.

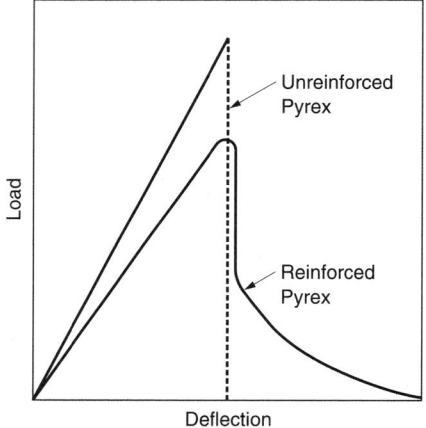

Figure 15.6 Load deflection curve of unreinforced Pyrex and short carbon fiber reinforced Pyrex. *Source:* Reprinted with permission from Sambell RAJ, Bowen DH, Phillips DC, *J Mater Sci*, 7, 663–675, 1972. Copyright 1972, Springer.

The load deflection curve for short carbon fiber reinforced Pyrex is compared with the unreinforced matrix in Figure 15.6 and the contribution of the fiber towards toughening, despite a reduction in flexural strength arising from the poor bond of the fibers with the matrix, is clearly seen.

When using continuous fiber, it is possible to filament wind the fiber through the slurry bath and collect it onto a flat faced mandrel (square or hexagonal), drying the unidirectional impregnated fiber, cutting into sheets, stacking, heating the stack to remove the binder by pyrolysis, placing the stack in a mold and hot pressing at temperatures near or above the softening point. Figure 15.7 shows the effect of temperature on flexural strength of a continuous carbon fiber reinforced borosilicate glass matrix depicting the glass matrix yielding without fracture as it softens at 700°C. Figure 15.8 shows the influence of temperature on the flexural strength of unreinforced and continuous carbon fiber reinforced in a glass ceramic in Ar and how the carbon fiber reinforced glass ceramic retains its strength at higher temperatures.

CARBON FIBER REINFORCED CERAMIC MATRICES

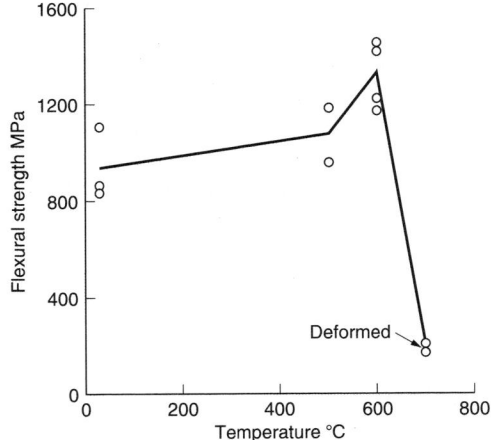

Figure 15.7 Flexural strength of continuous carbon fiber reinforced borosilicate glass in Ar as a function of temperature. *Source:* Reprinted with permission from Prewo KM, Batt JA, *J Mater Sci*, 23, 523, 1988. Copyright 1988, Springer.

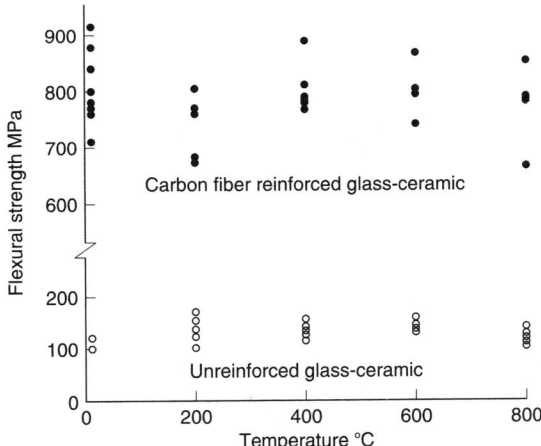

Figure 15.8 Flexural strength of unreinforced and continuous carbon fiber reinforced glass ceramic in Ar at five different temperatures. *Source:* Reprinted with permission from Sambell RAJ, Bowen DH, Phillips DC, *J Mater Sci*, 7, 663–675, 1972. Copyright 1972, Springer.

15.3.2.3 Hot filament winding under tension with hot pressing above the annealing temperature

Impregnated continuous fiber can be wound under tension (about 100 MPa) onto a suitable mandrel without hot pressing when heated to above the glass annealing temperature in an Ar atmosphere.

15.3.2.4 Melt infiltration

This requires heating the coated fiber in Ar until the glass flows into the filament interstices and wets out the fiber. One problem is that at about 1150°C, carbon fiber will react

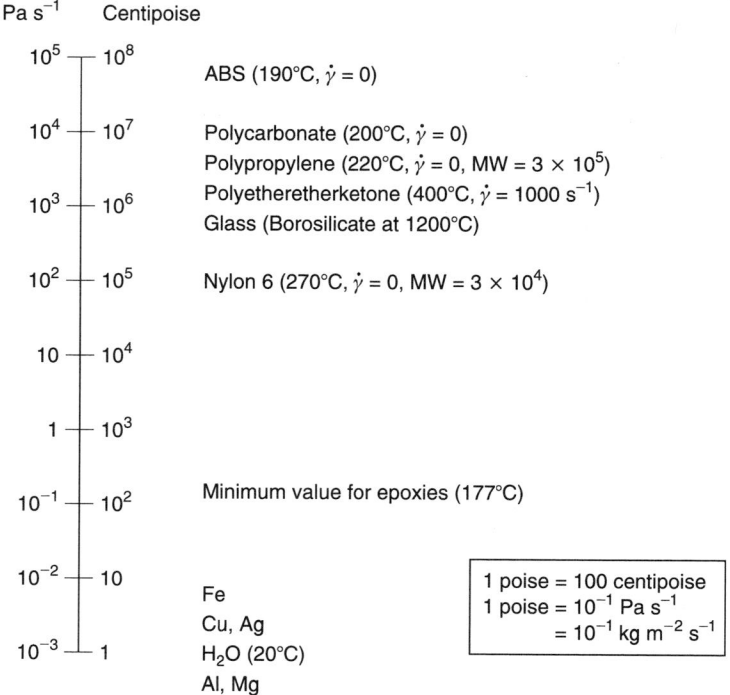

Figure 15.9 Viscosity of selected molten metals, polymers and glasses. *Source:* Reprinted with permission from Cornie JA, Yet-Ming Chiang, Uhlmann DR, Mortensen A, Collins JM, *Ceramic Bulletin*, 65(2), 293–304, 1986. Copyright 1986, The American Ceramics Society.

with SiO_2 [87]:

$$SiO_2 + 3C \rightleftarrows SiC + 2CO$$

To help prevent this reaction, it is possible to coat the fiber with SiC for a silica based matrix. This is a reversible reaction and if the pressure of CO exceeds 9 Pa, then the reaction is forced to the left. Alternatively, if MoO_3 or Nb_2O_5 is added, then CO pressures are generated such that Mo_2C and NbC are preferentially formed at the interface to SiC.

Figure 15.9 shows that molten ceramics are considerably more viscous than molten metals, resulting in low rates of infiltration.

15.3.2.5 Sol gel

Carbon fiber is impregnated with a colloidal solution (sol), allowed to gel, dried at room temperature for 24 h, followed by 24 h at 50°C and heat treated at 300–400°C for 3 h. For a borosilicate matrix, the dried sheets are hot pressed at 900–1200°C in a N_2 atmosphere at 10 MPa [88,89].

In the sol gel process, smaller particles than in the slurry infiltration method are used and can be more easily homogenized into a system, which is nearly a monodisperse. This allows better penetration of the matrix into the fiber interstices, with less fiber damage and, moreover, is achieved at a lower working temperature. Unfortunately, the downside is that excessive shrinkage occurs, necessitating either repeat infiltrations or using a combination of sol gel and slurry methods [1].

15.3.3 Work undertaken with carbon fiber filled glass matrices

Early work with reinforcement of glass matrices with carbon fiber was undertaken by Crivelli-Visconti and Cooper [90] and workers at Harwell—Bowen, Briggs, Phillips and Sambell [91–97]—who demonstrated a significant improvement in toughness, but the composites produced were rather porous, exhibiting a poor bond between the carbon fiber and glass matrix.

The improvement in toughness of a Pyrex matrix reinforced with short carbon fibers is shown in Figure 15.6, where the increase in area under the flexural load-deflection curve is characteristic of a tougher matrix, the strength, however, has decreased. Continuous fibers do tend to give a better strength (Figure 15.7), but the increase is offset by extensive cracking throughout the matrix [98], which is indicative of the poor bond between the carbon fiber and glass matrix. Fitzer [99] has shown (Figure 15.8) that the tensile stress-strain curves of silica glass reinforced with SiC coated and uncoated carbon fibers discloses more when flexural testing was undertaken as tensile testing along the fiber direction did not give results dependent on the fiber-matrix bonding.

This toughness is retained at $-150°C$ [100], but in an environment where heating and cooling occurs, a hysteresis effect can take place, which is more prevalent in a two dimensional reinforced glass than one dimensional and does not take place at all with an unreinforced glass. The hysteresis is attributed to the slippage of the carbon fiber at the glass interface, with the greater effect in a two dimensional reinforced glass attributed to an unmatched resistance in the heating and cooling cycles, whereas the resistance in heating and cooling is the same in a one dimensional composite.

More work done on reinforcing glass with carbon fiber is recorded [101–104].

Carbon fiber oxidizes in air, which limits the strength of a glass composite, since the glass matrix cannot prevent this reaction occurring. Prewo and Batt [76] showed (Figure 15.8) how the flexural strength of a borosilicate glass reinforced with 55% V_f of continuous carbon fiber and heated in an Ar atmosphere exhibited an increase in flexural strength at 500°C, which corresponded to the softening point of the glass, allowing loads to be more readily distributed. At 700°C, however, there is a drastic fall in strength, though without fracture, because the matrix has become very soft. However, an unreinforced glass ceramic retains its strength at higher temperatures (Figure 15.7).

Other work reviewed by Prewo and co-workers [105–108] includes the development of glass ceramic matrices with significantly improved properties.

A twofold increase of bond strengths can be achieved by the addition of MoO_3 or Nb_2O_5 [109], but strength and modulus are reduced. Impurities in the glass matrix can have a dramatic effect on the properties and type of bond formed [110].

The interface between HM carbon fiber and a Pyrex borosilicate glass matrix was Analyzed by Bleay and Scott [111,112] and found to be some 100 nm thick and believed to comprise Na enriched silicon oxycarbide, showing that some reaction had taken place during fabrication. Measurement of the interlaminar shear strength of the composite indicated that this layer was not a source of weakness. Substantial fiber pull-out had occurred, however, exposing clean fiber surfaces and smooth sockets. It was concluded that the interfacial shear process was confined to the outer layer of the fiber. Heat treatment of the composite in air caused preferential oxidation of the fiber, the rate being higher parallel to the fiber axis than perpendicular to it.

Boccaccini [113,114] has reviewed the dispersion reinforcement of glass matrix composite materials. The thermal shock resistance of glass can be improved by the addition of aluminum titanate (Al_2TiO_5) particles [115,116].

An attempt was made to prepare a pseudo three dimensional reinforcement using stretch-broken carbon fibers in a borosilicate glass (Corning 7740 Pyrex) matrix and comparing with a two dimensional reinforcement made from continuous carbon fiber [117]. Six ply, slurry coated, uniaxial and cross ply samples were laid up and hot pressed, preparing the composites with both longitudinal and transverse orientations of the outermost plies held in tension. These composites were fractured in a short beam three-point bending. SEM micrographs showed that the fibers broken due to stretch reduced the degree of delamination and only these fibers prevented brittle failure in the uniaxial transverse (90°) case. Delamination resistance resulted from the interlocking of broken filaments, which were distributed at angles to the fiber axis and also to fibers bridging across the adjacent layers of the two dimensional laminates.

Thermochemical analysis can be used to study interface reactions in carbon fiber reinforced glass matrix composites [118].

Composite materials are not only developed for improvement of mechanical performances, but can also be used for the achievement of a damage monitoring capability. Strengthening and diagnosis of damage are achieved simultaneously using composites consisting of two different materials A and B, where material A, such as carbon fiber, is of high elastic modulus and electrically conductive while material B, such as glass fiber is of high tenacity and electrically insulating. Addition of carbon fibers to a glass fiber reinforced plastic increases the elastic modulus and gives damage monitoring capability by measuring change in electrical resistance, while addition of glass fibers to carbon fiber reinforced plastic avoids sudden fracture. The concept of materials design, integration of structural performance and damage monitoring with simple structures is proposed as Ken-materials. This concept has been extended by Yanagida [119] to the self-diagnosis of a ceramic matrix composite.

The Y.Mg.Al.Si.O (YMAS) matrix has been chosen by a number of workers [120–125] to reinforce with carbon fiber. The structural and textural changes occurring within a glass belonging to the YMAS system have been investigated using TEM, SEM, polarized light optical microscopy, XRD and thermal analysis. Some differences between the carbon (fiber) containing material and the carbon free material were revealed, mainly through local O_2 depletions due to a reduction effect by the carbon. The starting stoichiometry allowed various phases to crystallize during an increasing heat treatment, such as magnesium aluminum silicate as indialite, yttrium silicate, magnesium aluminum oxide as spinel, in addition to corundum.

Bianchi *et al* [123] fabricated and examined unidirectional continuous carbon fiber reinforced (YMAS) glass ceramic matrix composites for dry sliding applications. Different fracture behaviors were recorded using the three-point bend test and fracture surfaces observation. The distance between microcracks, which appeared in the matrix on cooling after hot pressing of the composites as well as changes with the sintering temperature, suggested fiber-matrix reactions.

Zhang and Thompson [126] studied the elimination of cracks in carbon fiber reinforced nitrogen glass composites. A wide range of compositions used as the matrix phase were prepared from powder oxynitride glasses by conventional glass melting techniques and used in the preparation of carbon fiber-reinforced nitrogen glass composites, by means of slurry infiltration, lay up and hot pressing techniques. Microstructural observation showed that some cracks appeared in these materials that were perpendicular to the fiber direction because of the thermal mismatch between fibers and matrix. Various amounts of different forms of ZrO_2 powders were added to this glass matrix to compensate for the thermal mismatch. It was found that with 20 wt% additions of ZrO_2 to a lanthanum sialon glass matrix, the cracks were totally eliminated. Moreover, when up to 40 wt% of ZrO_2 was put in the same glass matrix, the composites possessed significantly improved bending strength and fracture toughness compared with ZrO_2-free samples (the bending strength increasing from 258 to 350 MPa). XRD analysis demonstrated that in all cases, the ZrO_2 was present in the monoclinic form in the final products.

Further work was undertaken by Zhang and Thompson [127] on the reinforcement of nitrogen glass with carbon fiber.

Sung [128] used off-stoichiometric celsian glass ceramic materials with B_2O_3 (27.40 wt% SrO, 37.83 wt% Al_2O_3, 31.77 wt% SiO_2 and 3 wt% B_2O_3) and in conjunction with Park [129], used 48.45 wt% BaO, 13.58 wt% Al_2O_3 and 37.97 wt% SiO_2 as the matrices for graphite fiber reinforced composites. Both the thermal and mechanical properties of the composites were much enhanced compared to those of the unreinforced glass ceramics. These highly improved properties would be induced not only by the properties of the graphite fibers, but also by the good interfacial bonding between the glass ceramic matrix and the fibers.

15.3.4 Coating carbon fiber to improve the bond to a glass

When carbon fiber is incorporated in a typical borosilicate glass matrix, the bond formed between the carbon fiber and glass is poor, producing composites with measured strengths below that predicted by the rule of mixtures. Taylor [130] believed that the poor bond when using HM fiber originated from the highly nonpolar surface formed by the π-electron cloud of the oriented graphite crystals having no tendency to bond with the oxygen in the silicate matrix. The application of a metal film to the carbon fiber should negate the effect of the non-polar surface and provide a metal cation, which should readily combine with the oxygen in the silicate to form a tight oxide bond. The following metals should be effective—Y, Zr, Nb, Mo, Ag, Cd, Ta, W, Zn, Cu, Co, Fe, Mn, Cr, V, Ti, Sc, Al, Mg and Ni.

Taylor took HMS fiber coated by electroplating with 0.5 µm nickel and passed it through a bath containing a slurry of 40 wt% powdered Corning 7740 borosilicate glass dispersed in an aqueous or propanol vehicle. The impregnated fiber was wound onto a square drum and when dried, gave approximately 50% glass coating. The coated fiber was formed into disks, which were stacked in a graphite mold, with Mo foil disks placed at the top and bottom to protect the composite from reacting with the graphite mold. The stacked mold was heated to 300°C in an inert atmosphere to remove all volatiles and consolidation was achieved in a vacuum hot press (1.3×10^{-2} Nm^{-2}, 650°C and 2.54 MPa), where the pressure was applied with a hydraulic ram via a graphite rod. Heating was undertaken at a rate of 5–30°C min^{-1} to achieve the maximum rate of outgassing without allowing the vacuum to reduce beyond 1.0×10^{-1} Nm^{-2}. Finally, a pressure of 10 MPa was applied and the pack heated to 1095°C (giving a viscosity of 10^6 P) then holding for 0.5 h, to ensure that consolidation was achieved at the center of the composite. The composite was cooled by backfilling with Ar gas to speed the cooling process. The flexural strength of the composite increased from 0.53 GPa for the uncoated fiber to 0.75 GPa for the nickel coated fiber.

Other work has been undertaken employing the deposition of Ni on carbon fiber for reinforcing glass matrices [131–135].

A novel processing route uses *in situ* electrophoretic deposition of submicron sized colloidal particles of borosilicate sol onto nickel coated carbon fibers to give a dense fully infiltrated product, which is followed by pressureless sintering. Catastrophic crack growth was prevented by constrained plastic deformation of the interface, fiber debonding and pullout [136].

15.4 CERAMIC MATRICES [137,138]

Ceramics tend to possess ionic and covalent bonding, and SiO_2, for example, can exist in either form—the crystalline form occurs when SiO_2 melt is slowly cooled from above the mp (1723°C), while the amorphous form is obtained if rapid cooling is applied. The type of bonding formed affects the properties (e.g. amorphous ceramics are poorer conductors of heat due to lack of an ordered lattice).

The principal weakness of ceramics is their brittleness and toughening can be achieved by

1. Transformation of the matrix to a more stable phase.
2. Matrix microcracking, which can shield a sub-critical crack.
3. Load transfer from matrix to a reinforcing fiber when the matrix fails.
4. Crack deflection or arrestment due to thermal expansion mismatches. Fiber pull-out occurs to overcome the work when sliding friction occurs.
5. The most effective toughening can be achieved by reinforcing with a continuous fiber. High modulus carbon fibers have the unique property of retaining microstructural stability above 1500°C although they do oxidize, which limits the life, but can be extended by applying a protective coating.

Incorporating carbon fiber in a ceramic matrix will confer the following properties:

1. Improve the toughness.
2. Decrease the coefficient of thermal expansion and increase the thermal conductivity.
3. Increase the electrical resistivity.
4. Reduce the density and hence improve the specific strength and modulus.
5. Reduce the shrinkage that occurs during drying, when a slurry is used to impregnate the fiber.

15.4.1 Processing ceramic matrix composites

A number of reviews have been written on the fabrication of ceramic composites by Signorelli and Carlo [139] in 1985, Schioler and Stiglich [140] in 1986, Cornie et al [6] in 1986, Prewo et al [105] in 1986, Treadway et al [109] in 1989 and Inagaki [141] in 1991.

Unfortunately, many of the techniques used for processing ceramics are proprietary and coupled with some restricted information, limits the available literature on ceramic composites. The methods used for processing are fortunately comparable with those practiced for resin matrix composites.

Basically, processing can be divided into the powder and chemical routes. The powder route includes sintering, slurry impregnation and reaction bonding, whilst the chemical route includes CVI, direct oxidation, sol gel and polymer pyrolysis. Final consolidation and densification must be achieved at high temperature. The powder route produces higher density ceramic matrix composites and the chemical route produces better quality.

15.4.2 Types of ceramic matrices

Table 15.4 lists the most common ceramics used for composite ceramic materials.

A ceramic matrix will provide improved mechanical properties if nanosize (<100 nm) material is added, particles of which are so small that they can be trapped at the grain boundaries or within the grains [142].

Table 15.4 Ceramic matrices used in ceramic composite materials

Name	Formula	Density (g cm^{-3})	Melting point °C	Sintering temperature °C	Maximum temperature of use °C	Covalent character %
Alumina	Al_2O_3	3.97	1977		1750	37
Mullite	$3Al_2O_3.2SiO_2$	2.8	1880	1620	1700	
Silica	SiO_2		1723		1200	49
Magnesia	MgO	3.58	2825	1200		27
Cordierite	$2Mgo.2Al_2O_3.5SiO_2$	2.60			1370	
Silicon carbide	SiC	3.23			1400	89
Boron carbide	B_4C	2.5				
β'- sialon	$Si_{6.25}Al_{0.75}O_{0.75}N_{0.75}$					
Zirconia	ZrO_2	5.89	2709			
Boron nitride	BN	2.27			985 (air)	
Silicon nitride	Si_3N_4	3.44	1900		1500	70
Aluminum nitride	AlN	3.30	Does not melt but dissociates above 2500			
Titanium boride	TiB_2	4.52	≈ 2980	>2000		

15.4.2.1 Oxide matrix materials

1. Alumina (Al_2O_3) [143]

Alumina has a high melting point and is a widely used ceramic, but it does have a high expansion coefficient. When deposited via the sol gel method, excessive shrinkage occurs on drying and firing, which has to be counteracted by the addition of Al_2O_3 [144].

The following reaction with carbon can occur:

$$9C + 2Al_2O_3 \rightarrow Al_4C_3 + 6CO$$

and is best prevented by coating the carbon fiber with SiC.

2. Mullite ($3Al_2O_3.2SiO_2$)

Mullite has a slightly lower melting point than alumina, but the expansion coefficient is low and similar to SiC.

3. Zirconia (ZrO_2) [145–149]

Zirconia can be used with Al_2O_3 to form a series of particulate reinforced composites, with the zirconia particles intergranularly dispersed within the alumina and the composite benefiting by strength retention for many hours at 1000°C. With ZrO_2 rich dispersions, the Al_2O_3 increases strength at room and elevated temperatures and also increases the fracture toughness.

15.4.2.2 Non-oxide matrix materials

1. Silicon carbide (SiC) [150]

Silicon carbide shows great promise as a matrix material, retaining strength up to 1400°C, but unfortunately, it does exhibit brittle behavior up to 2000°C. Carbon fiber has been used for the reinforcement of SiC [151]. The thermal shock resistance is also improved if carbon fiber is used as reinforcement [152]. However, the carbon fiber must be coated to confer a measure of protection against oxidation.

Ansorge [153] produced a carbon fiber reinforced SiC ceramic matrix composite by liquid phase infiltration of Si into a carbon-carbon composite.

Resin transfer molding can be used to produce near net shape ceramic matrix composites, followed by pyrolysis [154].

2. Titanium carbide (TiC)

Titanium carbide can be added to SiC to improve strength and fracture toughness [155].

3. Boron carbide (B_4C) [156]

This material is almost as hard as diamond and has to be fabricated by hot pressing. The B_4C will, however, inhibit the grain growth.

4. Titanium boride (TiB_2)

Titanium boride has a particle size of about 8 μm and can be hot pressed at temperatures >1800°C in a vacuum or 1900°C in Ar, to give almost theoretical density. Sintering aids include C, Fe, Co, Ni, W and WC. For hot isostatic pressing (HIP) at temperatures >2000°C, sintering aids can include Fe, C/Cr, Cr_3C_2. Titanium boride can be reinforced with ZrO_2 to improve the strength and toughness [157].

5. Boron nitride (BN)

Boron nitride may be added to Si_3N_4 to improve shock and electrical resistance [158]. BN is covalently bonded and resistant to sintering without the addition of liquid forming additives, such as Si_3N_4, SiO_2, Y_2O_3 and CaO, which will allow sintering between 1600–1900°C.

6. Aluminum nitride (AlN)

Thermal conductivity is eight times higher than Al_2O_3 and is second only to beryllia. The thermal expansion matches Si and at 200°C, exceeds that of Cu.

A reaction with carbon can occur:

$$3C + 4AlN \rightarrow Al_4C_3 + 2N_2$$

and is best prevented by coating the carbon fiber with SiC.

7. Silicon nitride (Si_3N_4)

It exceeds other ceramics in thermal shock resistance and strength is retained at elevated temperature. If can be sintered, reaction bonded or hot pressed.

Silicon nitride has been reinforced by carbon [159] and Suzuki et al [160] have investigated fiber pullout mechanisms of carbon fiber reinforced Si_3N_4 ceramic composites. Carbon fiber composites are stable in N_2 at high temperatures [161,162]. At present, interest has been generated in SiC-SiC composites. Lundberg et al [163,164] successfully HIPed carbon fiber reinforced Si_3N_4 composites and Guo et al [165] used carbon fiber to reinforce Si_3N_4.

Silicon nitride will react with carbon:

$$3C + Si_3N_4 \rightarrow 3SiC + 2N_2$$

and is best prevented by coating the carbon fiber.

15.4.3 Fiber reinforcement

The temperature resistance of carbon fiber reinforced ceramics is limited by the oxidation of carbon fiber at temperatures above 500°C and it is necessary to provide some form of protective coating. However, the attributes of increased toughness, strength and modulus are worthwhile considerations. Continuous fiber gives improved tensile properties, but may be more difficult to impregnate with the matrix material than discontinuous fiber.

15.4.4 Processing techniques

15.4.4.1 Slurry infiltration

The technique used for slurry infiltration is the same as that used for glass (Section 15.3.2.2) and to improve the densification, it is normal to adopt a multiple slurry impregnation process. The technique is most effective for glass ceramics, although recent attention has concentrated on SiC reinforcement. A limitation when coating continuous fiber is that the impregnated and dried tow has an excess of matrix particles, which persists after densification.

Hillig [166] prepared ceramic composites by infiltration. Grenet et al [167] used long carbon fiber to reinforced Si_3N_4 matrix composites by liquid infiltration of aqueous Si_3N_4 slurry, followed by hot-pressing. They developed a methodology to obtain the maximum volume and uniform infiltration of preforms by optimizing slurry rheology and fiber wetting conditions. They were able to achieve fully infiltrated green forms of 55% theoretical density with about 40 vol% fraction of fibers. The quality of the composites was assessed by microstructural analysis and mechanical characterization.

Slurry processing is less effective with refractory crystalline systems because of the absence of viscous flow. The higher temperature necessary to densify the matrices causes fiber degradation due to oxidation and abrasion of the fiber by the refractory particles [6], although Guo et al [165] achieved lower processing temperatures in a Si_3N_4 matrix and carbon fiber reinforcement by using liquid phase sintering additives.

15.4.4.2 Slip casting

A slip is prepared by ballmilling for several hours and removing any entrained air by evacuation. It is helpful to use a deflocculating agent. Sano, with other workers has reported on slip casting [168–172].

Although slip casting is an economical forming method for fine ceramics, if a gypsum mold is used, this can cause cross contamination. Under a constant casting pressure, the effective pressure will be reduced as the cake thickness increases, producing a green body with a non-uniform density. Sano et al [173] replaced the gypsum mold with a porous resin mold and applied pressure that increased with time to obtain a green body with uniform density using slips of Al_2O_3, Y_2O_3, partially stabilized ZrO_2 and Si_3N_4.

15.4.4.3 Filament winding

Filament winding was used by Matsuo et al [174] to produce dense carbon fiber reinforced hexagonal BN (hBN) composites, using reactive sintering aids such as Si_3N_4, SiO_2 and Y_2O_3, followed by hot pressing at 1800°C in a N_2 flow. The strength of the composites at

1600°C was twice that at room temperature, thought to be due to the large thermal expansion coefficient of hBN in the b axis direction.

The technique is more appropriate for a glass matrix, with its associate viscous flow when processed, since a surface excess of ceramic particles when coating a carbon fiber tow tends to persist in the fiber lay up.

15.4.4.4 Chemical synthesiz

1. Sol gel

When preparing a crystalline ceramic, the purity and particle size of the constituents are most important. The particle size should be <0.1 µm, since a small size will enhance the strength of the ceramic. The sol gel route can be used to prepare a ceramic powder by partial evaporation of a metal alkoxide, creating a stable dispersion of the sol as a gel, which is then dehydrated and calcined to give a fine ceramic powder.

Sol gel processing of ceramics as matrices for fiber reinforced composites offers the flexibility of using polymer composite processing techniques and avoids fiber damage arising from conventional high temperature melt processing infiltration techniques used for glass and ceramic matrix composites. A disadvantage of the sol gel route is high shrinkage and low yield, but the process can be used to apply a thin coating on the fiber as an interfacial layer between fiber and matrix or a thicker coating for subsequent densification into a matrix.

The preparation of Al_2O_3 and $3Al_2O_3.2SiO_2$ sols has been described by Chen et al [175].

Raman et al [176] synthesized SiC incorporated carbon-carbon composites by a sol gel process. The incorporation of oxidation inhibitors or oxidation resistant materials prevents the oxidation of cfrc composites. Thermal studies of such composites with and without SiC showed that oxidation rate was retarded in the presence of SiC. The flexural strength of carbon-carbon composites containing SiC was 45 MPa, the strength of the composite being reduced by crystallization of the SiC and the porous nature of the composite.

Manocha et al [177] developed Si, C and O based matrices through copolymerization of furfuryl alcohol (FFA) and hydrolyzed tetraethylorthosilicate (TEOS), as well as through organic-inorganic hybrid gels from TEOS and 1,4-butanediol (BD). In the FFA-TEOS system, copolymerization with an optimized mole ratio results in a resinous mass. This precursor, when pyrolyzed at 1000°C, results in a Si.O.C type amorphous solid black mass. On heating to 1400–1600°C, crystalline SiC is present. XRD studies showed that the pyrolyzed materials obtained from the BD-TEOS system contained a higher amorphous component even after heat treatment to 1600°C, in contrast to the FFA-TEOS system. Carbon fiber reinforced composites were prepared with these matrices using a prepreg lay up, consolidation and pyrolysis technique. At heating temperatures of around 1400°C, consolidation of the matrix was observed to be maximum, no fiber-matrix reaction was observed and the fibers were not damaged. The composites exhibited non-catastrophic fracture.

Duran et al [178] examined different techniques for reducing porosity by re-infiltration on a carbon fiber reinforced SiC ceramic, using injection of suitable polymers and sol gels.

Wu et al [179] prepared bulk gels corresponding to an oxide composition of $3Al_2O_3.2SiO_2$, from a sol with a high solids yield, using submicron Al_2O_3 powder and colloidal SiO_2 sol as the precursors. When sintered for 2 h at 1300°C, this gave over 99% of theoretical density. Densification occurred by viscous flow of the amorphous SiO_2 matrix containing well dispersed Al_2O_3 particles and after 2 h at 1600°C, almost complete crystallization to $3Al_2O_3.2SiO_2$ had occurred. Uniform unidirectional cfrc composites were produced by a single stage infiltration of fibers using the above sol and hot-pressing either at 1300°, or at 1400°C, to 98 and 97% of the theoretical density respectively. The composites, with a matrix predominantly of SiO_2 glass and dispersed Al_2O_3 particles, exhibited non-brittle failure with mean flexural strengths of 720 and 766 MPa when hot-pressing at 1300 and 1400°C respectively. Extensive fiber pull-out during testing indicated a relatively weak bond between the fibers and the matrix, and TEM revealed no evidence of chemical interactions at the interface.

Hyde [180] at GEC Engineering Research Centre, Stafford describes how a continuous ceramic fiber preform is impregnated with a sol, which then solidifies into a gel. Carbon fiber reinforced silica made in this way has non-zero thermal expansion, but much higher strength than a zero-expansion glass ceramic.

The sol gel route for fiber reinforced ceramic composites was followed by Fitzer and Gadow [181] using SiC coated carbon fibers in sol gel derived SiO_2 and Al_2O_3.

2. Polymer precursor

Some polymer precursors for making ceramics by pyrolysis are listed by Wynn and Rice in Table 15.5 [182].

A polycarbosilane is an organosilicon polymer with a —Si—C— bond and was developed by Yajima [183] for producing SiC fiber (Nippon Carbon's Nicalon) by a melt spinning process as shown below:

Polydimethylsilane $(Si[CH_3]_2)_n$

$$\xrightarrow[\text{Pyrolyze above 1000°C in an inert gas}]{\text{Heat in air to crosslink polymer chain via Si—O—Si bonds}} \text{polycarbosilane}$$

Zhang et al [184] infiltrated a carbon fiber preform with a metal alkoxide followed by gelling to give a stable dispersion (sol) of small particles (<0.1 µm). Partial evaporation, or the addition of a suitable accelerator, gives a three dimensional gelatinous network. The gel is dried to remove water and heat treated to form a ceramic powder.

Tai et al [142] admixed nanosize (30 nm) MTI Corp's SiC powder in toluene with Nichimen Corp. polysilazane (SILACERM NCP-200) using an ultrasonic mixer. It was predicted that the smaller size particles would be trapped at grain boundaries or within the grains. A PAN based carbon fiber cloth (Porcher 3085) was impregnated with this slurry using a multi-impregnation process to densify the composites. The impregnated cloth was dried for 7 h at 115°C and hot pressed up to 300°C under pressure. Laminates were then pyrolyzed under N_2 at 1100°C, 1600°C and 1800°C. Heat treatment at 1800°C was used to further enhance the density of the composites. Above 1600°C, the density was reduced from

Table 15.5 Polymers for making ceramics by pyrolysis

Polymer(s)	Resulting ceramic
Poly(silazanes)	Si_3N_4
Poly(silazanes)	Si.C.N
Polytitanocarbosilane	Si.Ti.C
Poly(carbosilanes)	SiC, Si.C.N
Polysilastyrene	SiC
Carboranesiloxane	SiC-B_4C
Diphenylborosiloxane	Si.O.B
Polyphenylborazole	BN

Source: Reprinted from Wynne KJ, Rice RW, *Ann Rev Mater Sci*, Huggins RA ed., Palo Alto, Annual Reviews, 14, 297–334, 1984.

1.67 to 1.42 g cm^{-3} and higher temperatures promoted the formation of a crystalline structure (β-SiC) and removed volatile constituents. An increase in temperature increased the pore size, whilst the flexural strength was significantly reduced and was attributed to the lower density and higher porosity of the composites, with those containing the nanosize powder showing less strength reduction.

Composites were prepared by He et al [185] using the infiltration of unidirectional carbon fiber with a suspension of α-SiC, polycarbosilane, AlN, Y_2O_3 and xylene. After drying, the tapes were cut, stacked, prepressed and then hot pressed at 25 MPa at 1800°C or 1850°C in an Ar atmosphere. For a fixed AlN content, the flexural and shear strength increased with increasing Y_2O_3 content. The fracture toughness of composites sintered at 1800°C improved with an increase in Y_2O_3 content, but the converse was observed with composites sintered at 1850°C. For a fixed Y_2O_3 content, the mechanical properties increased with increasing AlN content, attributed to strengthening of the grain boundaries and the fiber-matrix interface caused by finer solid solution grains. Both debonding and fiber pull-out were significant in toughening the composites.

PAN based M40JB carbon fiber was used as reinforcement and polycarbosilane as precursor by Zhou et al [186] and continuous carbon fiber reinforced SiC ceramic matrix composites fabricated through polymer precursor pyrolysis methods. Processing conditions such as molding pressure and the number of infiltration/pyrolysis cycles were determined and optimized. The results showed that, as the number of infiltration/pyrolysis cycles increased, the extent of fiber pull-out decreased, the bulk density, fracture toughness and interfacial shear strength increased, but the bending strength initially increased and then decreased. A molding pressure of 15 MPa was found to be suitable.

Hu et al [187] also carried out densification by repeated infiltration, crosslinking and pyrolysis to prepare carbon fiber reinforced composites.

Three types of carbon fibers, PAN based HSCF, pitch based HMCF with a low degree of graphitization and pitch based CF70 with a high degree of graphitization were used by Zheng et al [188] to prepare unidirectional carbon fiber reinforced SiC composites (CF/SiC) by multiple impregnation (5–9 times) with polycarbosilane and subsequent pyrolysis at 1200°C. The experimental results showed that the pyrolytic product of polycarbosilane at 1200°C in N_2 consisted of β-SiC phase in a nearly amorphous state with oxygen as impurity. The impregnation behavior of polycarbosilane for CF70/SiC was different from those of HSCF/SiC and HMCF/SiC. The CF70/SiC showed a flexural strength of 700 MPa, which was higher than those of HSCF/SiC and HMCF/SiC. Moreover, CF70/SiC exhibited non-catastrophic fracture behavior, whereas HSCF/SiC and HMCF/SiC showed brittle fracture behavior. SEM and TEM observation of these composites clarified that HSCF/SiC and HMCF/SiC had strong fiber/matrix interfacial bonding, whereas CF70/SiC had weak interfacial bonding. It was concluded that highly graphitized CF70 was the most suitable reinforcement for the CF/SiC composite.

A novel liquid crosslinkable Si·B·C·N precursor was developed by Kamphowe et al [189] and transformed using the Resin Transfer Molding technique into ultrahigh temperature resistant ceramics by thermolytic conversion in an inert gas atmosphere.

A ceramic matrix from polysiloxane/B mixtures was developed by Suttor et al [190] for carbon fiber reinforced ceramic matrix composites. On pyrolysis, the polymer mixture containing filler is converted to a ceramic matrix, consisting of SiC, BC, B_3N_4, and a silicon oxycarbide (SiOC) glass, without reacting with the carbon fiber. Due to the large volume increase of the B filler on nitridation (142 vol%), no multiple re-infiltration of the structure was necessary to achieve a dense matrix. Tensile strength and interlaminar shear strengths exhibit maxima at a pyrolysis temperature of 1300°C, where extensive fiber pullout is observed.

Hoshii and Kojima [191] prepared three types of diphenylborosilane (PBS), an organometallic compound with Si—O—B bonds, using different Si/B molar ratios and compounded with carbon fiber. Mechanical strength of the composites so formed increased with an increase of Si content in the PBS. When a PBS with a high B content was used, a compact vitreous film of B_2O_3 was formed over the composite, improving the oxidation resistance of the composite.

Three dimensional carbon fiber reinforced SiC and Si_3N_4 composites were fabricated using repeated infiltration of organosilicon slurry under vacuum and pressure by Nakano et al [192]. The open porosity of the infiltrated body was reduced from 40% after the first infiltration to approximately 8% after the seventh cycle. Further porosity reduction to less than 3% was achieved by hot press densification. The maximum values of flexural strength and fracture toughness were 260 MPa and 7.3 MPa m^{-2} for C-Si_3N_4 composites, and 185 MPa and 6 MPa m^{-2} for C-SiC composite.

Sato et al [193] converted a low viscosity perhydropolysilazane as a matrix precursor to Si-N ceramics with a 70–90 wt % yield by pyrolysis at temperatures above 800°C. The carbon fiber preforms of Si-N and SiC-Si-N with unidirectional aligned fiber were prepared by filament winding and the prepreg cured at 100–300°C in a N_2 atmosphere at a pressure of 0.05–0.1 MPa, then fired to 1200°C. To obtain dense composites, the samples were impregnated with polysilazane, followed by curing and firing. The re-impregnation and firing processes were repeated. The densities of the carbon fiber reinforced Si-N and Si-C/Si-N were increased to 90–96 % of the theoretical values. Such superior densification was due to the excellent infiltration and high ceramic yield of the polysilazane. Owing to the high density of the matrix and the adequate shear strength of the fiber/matrix interface, the three-point flexural strength was substantially high, 500–700 MPa. By applying this process, near-net shaped parts were easily fabricated.

Ostertag and Haug [194] also used filament winding to prepare carbon fiber reinforced SiC matrix composites using Si polymers as matrix precursors, which were pyrolyzed to form a SiC matrix.

Seyferth et al [195] have used a liquid silazane precursor to form Si_3N_4.

15.4.4.5 Melt infiltration

Melt infiltration involves using a matrix that is molten at the infiltration temperature and is allowed to wick into the carbon fiber preform and react. The method should provide fully dense and flaw free matrices with little dimensional change offset and cater to all forms of fiber geometry. However, the higher viscosity for ceramic systems (Figure 15.9) compared with metals necessitates the use of higher processing temperatures to ensure adequate infiltration since the higher viscosities will result in lower rates of infiltration. Further, inadequate infiltration may result in fiber damage due to possible chemical reactions. Also, volume change on solidification and thermal expansion mismatch can cause inbuilt tensile stresses. This is more of a problem with brittle ceramics than with ductile matrices.

Another important factor is wetting. With small diameter fibers and increased packing fractions, sufficient capillary pressure must be exerted to ensure satisfactory infiltration. The infiltration velocity is given by the Washburn equation:

$$\frac{dl}{dt} = \frac{(P_a + P_c)d^2}{32\eta l}$$

where $\frac{dl}{dt}$ = infiltration velocity

l = pore length

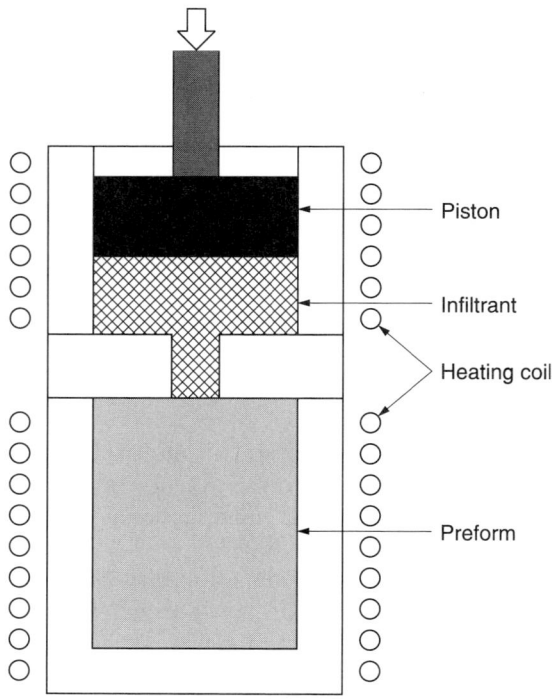

Figure 15.10 Schematic of laboratory scale pressure assisted melt infiltration apparatus for impregnation of fiber preforms. *Source:* Reprinted with permission from Cornie JA, Yet-Ming Chiang, Uhlmann DR, Mortensen A, Collins JM, *Ceramic Bulletin*, 65(2), 293–304, 1986. Copyright 1986, The American Ceramics Society.

P_a = applied pressure
P_c = capillary pressure
d = pore diameter
η = viscosity of fluid

It can be seen that capillary pressure increases approximately inversely in proportion to the fiber spacing, whereas the fluid infiltration rate decreases as the square of this dimension. Wetting may cause an interfacial chemical reaction to occur, or provide a matrix bond which is too strong. Hence it is necessary to provide a controlled degree of wetting by tailoring a specific fiber coating for a given melt infiltration application. Figure 15.10 shows a laboratory scale pressure assisted melt infiltration apparatus to investigate ceramic melt infiltration of a fiber preform.

Singh and Morrison [196] used 3% B in Si, which melts when heated to 1450°C in vacuum and can be infiltrated at 1420°C.

Gadow and Speicher [197] used melt infiltration to form reaction bonded carbon fiber reinforced SiC ceramics by applying an optimized resin and powder mixture in a SMC (Sheet Molding Compound) manufacturing process to produce carbon fiber reinforced preforms. Various designs of resin derived and carbonized preforms were investigated and characterized by pore volume, pore size distribution and melt infiltration experiments. The microstructural morphology of the preforms was modified by mixing ceramic and carbon fillers in the applied resin system, thus varying the shrinkage and pore formation during pyrolysis. Mercury porosimetry was used to study the influence of different fillers on the morphology. The dependence of the filler type and content on the porosity showed that the pore volume fraction and the pore size distribution influenced melt infiltration. This manufacturing technology permitted the combination of short and continuous fibers with the matrix material. The fiber length and content influenced the mechanical and tribological properties as well as the porosity. It was concluded that they influenced the Si infiltration during the reaction bonding process.

15.4.4.6 In situ chemical reactions

1. CVI (or CVD)

CVI is a slow process, taking many weeks [198] and was initially considered for coating fiber. A CVD method will give a uniform coating at temperatures of 1100–1500°C, with the ability to adjust for a specific coating and deposit layers of different compositions. Hence the process is ideally suited for the deposition of coatings, but can be used to deposit a matrix. A SiC matrix can be formed from a mixture of $(CH_3)SiCl_3/H_2$ at 1000°C [199], whilst ZrO_2 can be formed by a CVI route [200]. A schematic diagram of the thermal and pressure gradient method for the CVI of ceramic composites is shown in Figure 15.11. Figure 15.12 shows a schematic for a laser CVD reactor.

It is possible to alter the deposition conditions by adjusting the pressure and temperature gradients to effect a single continuous deposition step. As the reaction proceeds, the pore size is reduced and it is necessary to reduce the infiltration rate, which follows a logarithmic decay [203].

A new technique [204] for studying the progression of densification during the fabrication of composites by CVI has been introduced, which involves the preparation of a carbon-carbon composite and momentarily interrupting the carbon infiltration process (forced flow thermal gradient CVI) at various times to permit the deposition of very thin layers of SiC. Microscopic examination of these layers on a polished cross-section permitted determination

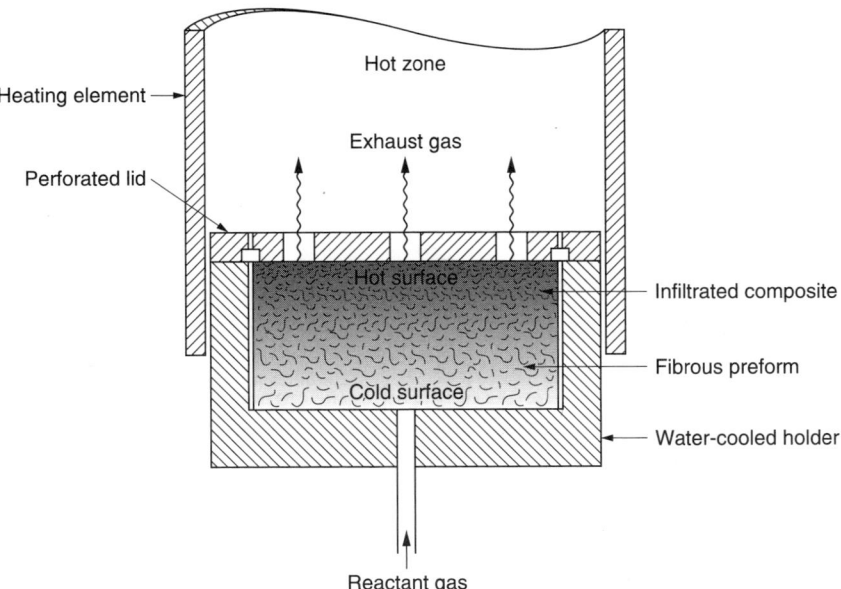

Figure 15.11 Schematic diagram of the thermal and pressure gradient method for CVI of composites. *Source:* Reprinted with permission from Stinton DP, Caputo AJ, Lowden RA, *Am Ceramic Soc Bull*, 65, 347–350, 1986. Copyright 1986, The American Ceramics Society.

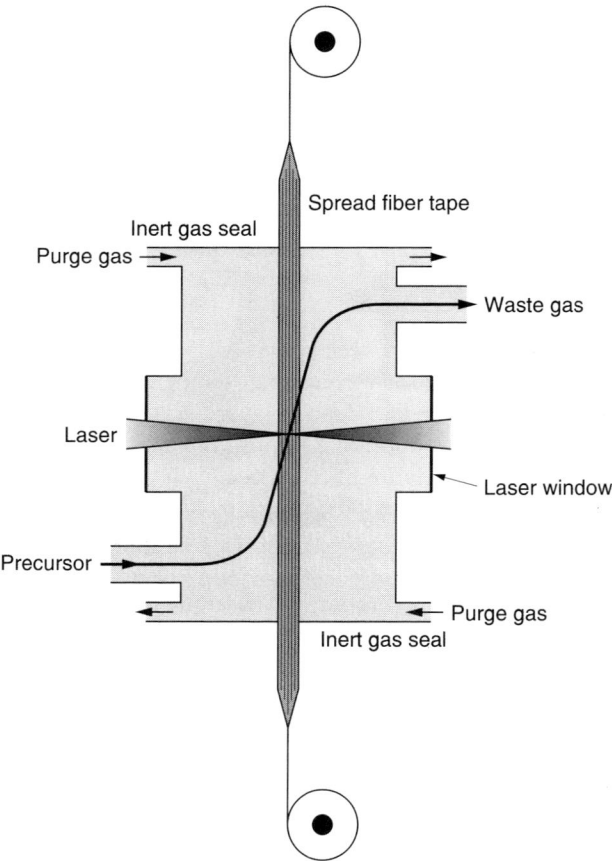

Figure 15.12 Scheme of laser CVD reactor. *Source:* Reprinted with permission from Hopfe V, Weiss R, Meistring R, Brennfleck K, Jäckel R, Schönfeld K, Dresler B, Goller R, CMC, Ceramic Metal Matrix Composites, *Key Eng Mater*, 127–131, 559–566, 1997. Copyright 1997, Trans Tech Publications.

of the extent of infiltration at various locations within the preform as a function of the infiltration time. The technique also distinguishes between open and closed pores and provides information on the existence of temperature gradients within the preform.

Caputo *et al* [205,206] have developed a fast process for the fabrication of ceramic fiber reinforced ceramic composites by CVI.

2. Slurry pulse/CVI

The strength and tensile behavior, at room and high temperatures, as well as the structure of three dimensional carbon fiber/SiC composites, fabricated by the slurry pulse/CVI combined process, were characterized by Suzuki *et al* [207–209]. Carbon fiber preforms, constructed with 4-step braid, 4-step/axial braid, 2-step braid and orthogonal weave, were used as reinforcements of the composites. The composites were fabricated by a process consisting of slurry and dissolved organosilicon polymer infiltrations, followed by the application of pulse CVI.

Carbon fiber reinforced SiC matrix composites suitable for use in hypersonic vehicles and liquid rocket nozzles were developed by the combined process of CVI and precursor impregnation and pyrolysis (PIP) [210]. The physical and mechanical properties of the composites were characterized. The combined process demonstrated a short manufacturing period and the capability of achieving high performance triaxially braided C/SiC composites with high flexural strength (643 MPa) and high fracture toughness (17.91 MPa m^{-2}).

Suzuki et al [208] fabricated three dimensional carbon fiber reinforced SiC matrix composites by a joint process, which consisted of slurry infiltration, organosilicon polymer pyrolysis and subsequent pulse CVI (PCVI) (using the source gas system of $SiCl_4/CH_4/H_2$). The deposition rate in PCVI was over twice as rapid as that using the MTS-H_2 system, though there was little difference in the filled mass at the saturation point between these two. The strength monotonously increased with increase of pulse number, which coincided with the density variation. The structure of the fiber preform had little influence on any final residual porosity. In the case of a composite fabricated using the braid type preform, the flexural strength at 1473K in Ar was about 550 MPa, and the strength at 1773K was over 80% of that at 1473K.

In the fabrication process of three dimensional carbon fiber reinforced SiC matrix composite, Suzuki and Nakano [207] applied PCVI as the final densification process for the specimen, which was made by the joint process of slurry infiltration and organosilicon polymer pyrolysis. The open porosity and bulk density of the specimen changed from 5.3% and 2.63 g·cm^{-3} (relative density of 94%) to 3.5% and 2.67 g·cm^{-3} (relative density of 95%) by the application of PCVI (1173–1223K, total 90,000 pulses). The flexural strength of the specimen increased over 20% (mean value = 153 MPa, maximum value = 174 MPa).

Continuous carbon fiber reinforced SiC composites were prepared by Xu and Zhang [211] using CVI, in which the preforms were fabricated by the three dimensional braid method. For the composites with no interfacial layer, flexural strength and fracture toughness increased with the density of the composites and the maximum values were 520 MPa and 16.5 MPa m^{-2}, respectively. The fracture behavior was dependent on the interfacial bonding between fiber /matrix and fiber bundle/bundle, which was determined by the density of the composites. Heat treatment had a significant influence on the mechanical properties and fracture behavior. The composites with pyrolysis interfacial layers exhibited characteristic fracture and relatively low strength (300 MPa).

3. Hot Isotactic Pressing (HIPing)

Hot isostatic pressing can provide a uniform high density, to almost 100% of theoretical density, providing near net shape components without size limitations. Since the applied pressure is uniform, the density too is uniform. The process can be used directly, or in combination with a preforming step such as injection molding or sintering. The technique has been described in relation to carbon-carbon composites (Chapter 14, Section 14.3.3.3).

Slip infiltrated, HIPed carbon fiber reinforced material, slip infiltrated SiC fiber reinforced nitrided Si_3N_4, and polysilazane solution infiltrated pyrolyzed composites with SiC fibers have been fabricated [159]. All materials exhibited non-brittle fracture.

Graphite fiber composites with aluminum phosphates matrix were prepared by Weng et al [212] and modified with various additives. Composites showed no significant mechanical degradation after heating to 1100°C and the TGA indicated good oxidation resistance up to 900°C in an ambient atmosphere. The high temperature flexural strength of the composite was influenced by fiber treatment. The interface between fiber and matrix, after heating at various temperatures, was affected by thermal mismatch. Unavoidable pores affected mechanical strength at room and elevated temperatures. High temperature properties were improved by infiltrating an organic material, which was then carbonized by HIPing.

Carbon fiber/Si_3N_4 composites were fabricated using slip infiltration of fiber bundles, subsequently stacked in a plaster mold, dried, glass encapsulated and HIPed to form unidirectional composite test bars. Neither chemical reactions between fiber and matrix, nor thermal mismatch cracks in the matrix were observed. The bend fracture behavior of the composites was non-brittle with extensive fiber pullout.

4. Reaction bonding

Unidirectional carbon fiber reinforced SiC/C composites were fabricated by Tani et al [213] using reaction bonding. The prepreg for the composites was prepared by the filament winding method, with a mixed slurry of phenolic resin, silicon powder and ethyl alcohol, then die pressed at 130°C and sintered in Ar. The formation of SiC took place through a reaction between the carbon from the pyrolyzed phenolic resin and the silicon powder above a temperature of about 1320°C and was completed by 1420°C. The bulk density of the composites was about $1.7\,g\,cm^{-3}$. The average flexural strength of the composites with a Si/C ratio of 0.2, 0.4 and 1 was about 310, 250 and 150 MPa respectively.

Plane woven carbon fiber fabric reinforced SiC-C matrix composites were produced at 1450°C by reaction bonding and impregnating with phenolic resin. The relationship between the flexural strength and the open porosity of the composites was dependent upon the heat treatment temperature before impregnation. The strength of composites heat treated at 1000°C, with an open porosity of approximately 15%, was about 300 MPa, whilst the strength of composites heat treated at 1450°C, with an open porosity of approximately 12%, was about 240 MPa. It was concluded that the heat treatment temperatures before the impregnation step might control the interfacial properties between the fiber and the matrix.

Tani et al [214] prepared carbon fiber reinforced SiC composites by reaction bonding.

The composites were fabricated by passing the carbon yarn through a slurry containing silicon powder, phenol resin and ethyl alcohol. SiC was formed by the reaction between the Si and the carbon from the phenol above 1320°C and completed at 1420°C. The carbon fiber was not damaged and fiber pullout was observed on the fracture surface. A bulk density of 1.7 to $1.8\,g\,cm^{-3}$ and a flexural strength of about 130 MPa were achieved. Impregnation with molten Si at 1600°C increased the bulk density, but decreased the flexural strength, due to the reaction of the carbon fiber with the Si.

A carbon fiber preform with SiC and carbon can be infiltrated by molten Si, when the free carbon reacts with excess Si to form solid SiC, providing a multiphase matrix that is hot pressed. During the infiltration process, the Si penetrates the pyrolyzed green body reacting *in situ* with the carbon to form reaction bonded SiC, forming a dense near net shape ceramic body that is ideally suitable for car brake disks. These disks last the lifetime of the car, with the ceramic phase acting as the bearing material and friction determined by the fraction of carbon fiber in the surface of the matrix interface.

Two dimensional woven carbon fiber reinforced SiC-C matrix composites have been produced by reaction bonding [215].

Gibson and Gibson [216] infiltrated a carbon fiber preform with a saturated aqueous solution of tantalum fluoride in sucrose and treated with NH_3 under pressure to convert to the hydrated pentoxide liberating NH_4F. The tantalum pentoxide, when heated, reacted with carbon from the partially carbonized sucrose to give tantalum carbide, with release of CO. Since the yield was low, multiple impregnations were required to give an eventual porosity of 5–10 vol %.

CARBON FIBER REINFORCED CERAMIC MATRICES

15.4.4.7 Consolidation and densification

1. Sintering

Maili et al [217] examined carbon fiber reinforced ceramic composites with a rich carbon interphase by sintering the matrix powder and carbon fibers. A carbon powder sintering method was used by Sakagami et al [218] by infiltrating the carbon fiber with a slurry of powdered matrix, cutting the impregnated tow and pyrolyzing in a mold at 400–700°C to remove the binder and forming the composite by hot pressing.

2. Pressureless sintering

Pressureless sintering can be used to sinter TiB_2 (particle size about 8 μm) followed by hot pressing at >1800°C in a vacuum or 1900°C in Ar. The density of the composite is almost theoretical. Sintering aids that can be used are C, Fe, Co, Ni, W and WC.

3. Hot pressing

Carbon fiber reinforced Si.Al.O.N composite materials were prepared by Dodds et al [219] using slurry infiltration of continuous carbon fiber tows, winding, filter pressing and uniaxial hot pressing using a graphite die in air.

Kikuchi et al [220] determined the reactivities of carbon fibers in hot pressed carbon fiber reinforced SiC ceramic composite.

15.4.5 Protective coatings

Fitzer [221] discusses the application of thin coatings on carbon fibers for high temperature ceramics. The key problem with carbon fibers in composites for high temperatures is damage by heat treatment in oxidizing as well as inert environments. Carbon fibers are the only inorganic fibers that do not lose their strength and stiffness due to recrystallization when heated above 1200°C and preserve their mechanical properties up to temperatures as high as 2000°C, but are limited to non-oxidizing environments. Reaction and diffusion barriers in the form of thin compatible and refractory coatings may solve the oxidation problem in carbon fiber reinforced ceramics, but oxidation resistance in air, however, fails because of the open non-protected cross-sections of the fibers.

A review given by Westwood et al [222] on the oxidation protection of carbon fiber composites identifies the requirement of an effective oxidation protection system for carbon fiber reinforced ceramics and summarizes the work carried out over the last 50 years towards this goal. Carbon fiber reinforced ceramic matrix composites are promising candidates for high temperature structural applications such as gas turbine blades. However, in oxidizing environments at temperatures above 400°C, the carbon fibers are rapidly oxidized. There is, therefore, a need to coat the composite to protect it against oxidation and the most promising coatings are those composed of several ceramic layers that are designed to protect against erosion, spallation and corrosion, in addition to possessing a self-healing capability due to the formation of glassy phases on exposure to O_2.

Vogel et al [223] describes the successful development by Dornier of a cost-effective technique for producing fiber reinforced ceramics and are able to manufacture complex shapes and integrated structures. Carbon fiber reinforced SiC has the required damage tolerant fracture behavior and sufficient reliability with marked weight reduction and increased operating temperatures. The application of these materials is currently limited due to the lack of reliable oxidation protection systems to prevent the carbon fiber from undergoing oxidation at temperatures above 500°C. To overcome these limits, Dornier have developed integrated oxidation protection systems for temperatures up to 1700°C, which can be applied comparatively easily using a slurry or CVD technique. The slurry can be applied by painting, spraying, or dipping, and is, therefore, best suited for complex parts and can be used for repairing used structures. The multilayer oxidation protection systems provide sufficient protection against attack by O_2 in the range 400–1700°C in static air. Protection is afforded for 1400 h at 1400°C and for over 3000 h at 1000°C. Short term thermal shock tests showed no difference at 700, 1200 and 1600°C. The CVD technique is preferred for long term applications and the slurry method for short term applications. The latter offers the best compromise between cost and reliability.

Katzman [224] burnt the size off the carbon fiber and applied an organometallic solution of $Si(OC_2H_5)_4$ agitated ultrasonically, followed by pyrolysis at 1000°C in an atmosphere of NH_3 to give a coating of Si_3N_4.

$$3Si(OC_2H_5)_4 + 4NH_3 \xrightarrow{-1000°C} Si_3N_4 + 12C_2H_5OH$$

or by using $Ti(OC_4H_9)_4$ in an atmoshere of NH_3 and H_2 to give a coating of TiN

$$2Ti(OC_4H_9)_4 + 2NH_3 + H_2 \xrightarrow{-1000°C} 2TiN + 8C_4H_9OH$$

Carbon fiber can be coated with copper phosphate prepared from CuO and H_3PO_4 [225], which after curing for some 4 h forms $Cu(H_2PO_4)_2$ and 24 h gives $CuHPO_4$.

Liu et al [226] prepared C/C-SiC nanomatrix composites by chemical vapor infiltration.

Webster et al [227,228] have shown that yttrium silicate (Y_2SiO_5) possesses the required properties to provide protection of C-SiC components from oxidation for extended periods at 1600°C. Conventional coatings consist of multiple layers of different materials designed to seal cracks by forming glassy phases on exposure to O_2. An attempt was made to develop a coating which would be inherently crack resistant and therefore, not require expensive sealing layers. These requirements can be summarized as low Young's modulus, a low thermal expansion coefficient, good erosion resistance, and low O_2 permeability. They developed protective coatings based on a SiC bonding layer combined with an outer Y_2SiO_5 erosion resistant layer and O_2 barrier. The C-SiC samples were coated using a combination of CVD and slip casting.

Slosarczyk et al [229] used PAN based carbon fibers, both uncoated and coated, with calcium phosphate applied by a sol gel technique. Carbon fiber reinforced hydroxyapatite composites were then prepared by hot pressing hydroxyapatite powder and carbon fibers at 1100°C and 25 MPa for 15 min in an Ar atmosphere. The best strength properties were obtained with the coated fibers and were attributed to the —OH groups on the fiber surface bonding with the calcium phosphate layer.

Carbon fiber reinforced SiC and carbon composites were prepared by Hoshii et al [230], coating the carbon fibers with diphenylborosiloxane (PBS).

Yttrium silicate (Y_2SiO_5) coatings complement SiC coatings for protecting ceramic multilayer composite materials based on carbon fiber reinforced SiC composites (C-SiC).

Dense Y_2SiO_5 100 μm thick coatings were prepared by Aparicio et al [231] by dip coating, using concentrated aqueous slips. Thick, mechanically stable coatings were obtained by sintering in carbon crucibles with a SiC bed in an Ar flow furnace. It was observed that pure Y_2SiO_5 coatings completely separated from the SiC substrates. A high percentage of $Y_2Si_2O_7$

was necessary to fit the thermal expansion coefficients and ensure the stability of the coatings. Oxidation resistance of the coated substrates was investigated by isothermal and stepwise oxidation tests. The coatings provided sufficient protection against oxidation at 1600°C. However, stepwise oxidation tests showed significant weight loss in the temperature range 400–700°C, due to small surface cracks. Efficient oxidation protection in the 400–1600°C range could be obtained using an intermediate coating to seal the cracks and by increasing the Y_2SiO_5-$Y_2Si_2O_7$ coating density by other deposition methods.

An electrodeposition/sintering method was used by Kawai et al [232,233] to fabricate carbon fiber reinforced composites synthesizing carbon-carbon and C-SiC. Two dimensionally woven PAN based carbon fabrics were used as substrates for electrodeposition of the functionally gradient material. By controlling the flexural strength, Young's modulus and thermal expansion coefficient, they were able to choose the condition of electrodeposition and sintering, enabling C-SiC composites to be fabricated with volume fractions of fibers in the range 45–78%. The maximum flexural strength and Young's modulus were 185 MPa and 47.5 GPa with a V_f of 75%, but both properties decreased with a decrease in V_f. Conversely, the thermal expansion coefficient increased with the decrease in V_f, where the value varied in the range $0.2- 2.75 \times 10^{-6}$ K^{-1}.

Electrophoretic infiltration is a novel technique for the fabrication of fiber reinforced composites used by Kooner et al [234]. The technique involves arranging the fibers as one of the electrodes such that deposition of the colloidal ceramic occurs in the fiber preform. This method was investigated for the carbon fiber reinforced Si_3N_4 composite system and produced green composite microstructures with good infiltration uniformity and fiber distribution, with hardly any macro defects.

Electrophoretic deposition (EPD) was also used by Kaya et al [235], followed by pressureless sintering, to produce dense, composites, where the defects were minimized. The process relies on the deposition of a submicron sized colloidal charged particles of borosilicate, onto unidirectionally aligned Ni coated carbon fibers. The metallic Ni interface was determined to be extremely effective in improving the composite mechanical performance, in terms of the non-brittle fracture behavior. Catastrophic crack growth is prevented by mechanisms such as constrained plastic deformation of the interface and fiber debonding and pullout. The proposed processing technique has great potential for fabricating damage tolerant fiber reinforced brittle matrix composites with a ductile interface and minimum defects.

15.4.6 Fracture mechanics

Mechanisms of toughening in ceramic matrix composites are discussed by Rice [236,237] and Marshall [238] deals with failure mechanisms in ceramic matrix composites.

A brief review of microfracture processes and the energy absorption mechanics of fiber reinforced composites is given by Miyajima et al [239]. Fiber pullout is considered to be the most important toughening mechanism. They describe an experimental technique to determine fiber pullout energy, using a 3-point bend specimen. From measurements of fundamental fracture parameters, fracture mechanisms for the fiber pullout processes of carbon fiber reinforced carbon composites are discussed.

REFERENCES

1. Rand B, Zeng RJ, Fibre reinforced ceramic matrix composites, Figueiredo, Bernardo CA, Baker RTK, Hüttinger KJ eds., *Carbon Fibers Filaments and Composites*, Kluwer, Dordrecht, 367–398, 1990.

2. Chung DDL, *Carbon Fiber Composites*, Boston, Butterworth-Heinemann, 177–200, 1994.
3. Huttinger KJ, Greil P, Ceramic composites for applications at extremely high temperatures, *Ceramics Forum Int/Ber DKG*, 69(11/12), 445–460, 1992.
4. Briggs A, Carbon fibre-reinforced cement, *J Mater Sci*, 12, 384–404, 1977.
5. Jing-Kung, Research on high temperature ceramics and composites in the Shanghai Institute of Ceramics, Faenza PV ed., *Science of Ceramics 12*, Proc 12th Int Conf, Saint-Vincent, Italy, 27–30 June, 1983, Ceramurgica, 495, 1984.
6. Cornie JA, Yet-Ming Chiang, Uhlmann DR, Mortensen A, Collins JM, Processing of metal and ceramic matrix composites, *Ceramic Bulletin*, 65(2), 293–304, 1986.
7. *Cement Standards of the World - The European Cement Association*, Brussels, Belgium.
8. Soroushian P, Nagi M, Hsu JW, Optimization of the use of lightweight aggregates in carbon fiber reinforced cement, *ACI Mater J*, 89(3), 267–276, 1992.
9. Chen PW, Chung DDL, Concrete reinforced with up to 0.2-percent of short carbon fibers, *Composites*, 24(1), 33–52, 1993.
10. Qijun Zheng, Chung DDL, Carbon fiber reinforced cement composites improved by using chemical agents, *Cement Concrete Res*, 19, 25–41, 1989.
11. Yang XM, Chung DDL, Latex-modified cement mortar reinforced by short carbon-fibers, *Composites*, 23(6), 453–460, 1992.
12. Pu-Woei Chen, Xuli Fu, Chung DDL, Improving the bonding between old and new concrete by the addition of carbon fibers to the new concrete, *Cement Concrete Res*, 25(3), 491–496, 1995.
13. Fu X, Chung DDL, Carbon-fiber-reinforced mortar as an electrical contact material for cathodic protection, *Cement Concrete Res*, 25(4), 689–694, 1995.
14. Pu-Woei Chen, Xuli Fu, Chung DDL, Microstructural and mechanical effects of latex, methylcellulose and silica fume on carbon fiber reinforced cement, *ACI Mate. J*, 94(2), 147–155, 1997.
15. Xuli Fu, Chung DDL, Combined use of silica fume and methylcellulose as admixtures in concrete for increasing the bond strength between concrete and steel rebar, *Cement Concrete Res*, 28(4), 487–492, 1998.
16. Xuli Fu, Chung DDL, Degree of dispersion of latex particles in cement paste, as assessed by electrical resistivity measurement, *Cement Concrete Res*, 26(7), 985–991, 1996.
17. Shi ZQ, Chung DDL, Improving the abrasion resistance of mortar by adding latex and carbon fibers, *Cement Concrete Res*, 27(8), 1149–1153, 1997.
18. Chen PW, Chung DDL, A comparative study of concretes reinforced with carbon, polyethylene, and steel fibers and their improvement by latex addition, *ACI Mater J*, 93(2), 129–133, 1996.
19. Walters DG, A comparative study of concretes reinforced with carbon, polyethylene, and steel fibers and their improvement by latex addition, Discussion, *ACI Materials Journal*, 94(1), 75, 1997.
20. Xuli Fu, Chung DDL, Effect of methylcellulose admixture on the mechanical properties of cement, *Cement Concrete Res*, 26(4), 535–538, 1996.
21. Lu W, Fu X, Chung DDL, A comparative study of the wettability of steel, carbon, and polyethylene fibers by water, *Cement and Concrete Research*, 28(6), 783–786, 1998.
22. Willats DJ, Morgan PE, Reinforced cement articles, Brit. Pat., 1,425,031, 1976.
23. Morgan PE, Carbon filament tapes, Brit. Pat., 1,425,032, 1976.
24. Waller JA, Civil Eng Public Works Rev, 357, Apr 1972.
25. Briggs A, Bowen DH, Kollek J, Paper No. 17 presented at Int Conf, Carbon fibres, their place in modern technology, The Plastics Institute, London, 1974.
26. Walton PL, Majumdar AJ, Cement-based composites with mixtures of different types of fibres, Composites, 209–216, Sep 1975.
27. Majumdar AJ, Laws V, Fibre cement composites, Research at BRE, Composites, 10(1), 17–27, 1979.
28. Sugama T *et al*, Cement Concrete Res 19, 355, 1989.
29. Yuan CZ, Hua CQ, Feng G, Influence of interface modification in fiber reinforced cement composites on their properties, Mat Res Soc Symp Proc, 211, 209–214, 1991.
30. Xu Y, Chung DDL, Carbon fibre reinforced cement improved by using silane-treated carbon fibres, *Cement Concrete Res*, 29(5), 773–776, 1999.

31. Xuli Fu, Weiming Lu, Chung DDL, Improving the bond strength between carbon fibre and cement by fiber surface treatment and polymer addition to cement mix, *Cement Concrete Res*, 26(7), 1007–1012, 1996.
32. Larson BK, Drzal LT, Sorousian P, Carbon-fiber-cement adhesion in carbon-fiber reinforced cement composites, *Composites*, 21(3), 205–215, 1990.
33. Fu X, Lu W, Chung DDL, Improving the tensile properties of carbon fibre reinforced cement by ozone treatment of the fibre, *Cem Concr Res*, 26(10), 1485–1488, 1996.
34. Linton JR, Burneburg PL, Gartner EM, Bentur A, Carbon fiber reinforced cement and mortar, Mat Res Soc Symp Proc, 211, 255–264, 1991.
35. Chen PW, Chung DDL, Composites, 24(1), 33–52, 1993.
36. Katz A, Li VC, Kazmer A, Bond properties of carbon fibers in cementitious matrix. *J Mater Civil Eng*, 125–128, May 1995.
37. Chen PW, Chung DDL, Ext Abstr Program-Bienn Conf, *Carbon*, 21, 92–93, 1993.
38. Sakai H, Takahashi K, Mitsui Y, Ando T, Awata M, Hoshijima T, Flexural behavior of carbon fiber reinforced cement composite, *American Concrete Institute Report*, SP 142-7, 121–129.
39. Toutanji HA, El-Korchi T, Katz RN, Leatherman GL, Behavior of carbon fiber reinforced cement composites in direct tension, *Cement Concrete Res*, 23, 618–626, 1993.
40. Toutanji HA, El-Korchi T, Katz RN, Strength and reliability of carbon fiber reinforced cement composites, *Cement Concrete Res*, 16, 15–21, 1994.
41. El-Korchi T, Toutanji HA, Katz RN, Leatherman GL, Lucas H, Demers C, Tensile testing of fibre reinforced cementitious composites, *Proc Mater Res Soc*, Pittsburgh, 211, 221, 1991.
42. Hermonssen L, Alderborn J, Burstrom M, Tensile testing of ceramic materials, *High Technology Ceramics*, Elsevier, 1161, 1981.
43. Ali MA, Majumdar AJ, Rayment DL, Carbon fibre reinforcement of cement, *Cement Concrete Res*, 2, 201–212, 1972.
44. Akihama S, Suenaga T, Nakagawa T, Suzuki K, Influence of fibre strength and polymer impregnation on the mechanical properties of carbon fibre reinforced cement composites, *Development in Fibre Reinforced Cement and Concrete*, Proc RILEM Symposium 1988, Sheffield, Paper 2.3, 1988.
45. Ohama Y, Amana M, Endo M, Properties of carbon fiber reinforced cement with silica fume, *Concrete Int Design Construct*, 3, 58–62, 1985.
46. Akihama S, Suenaga T, Nakagawa T, Properties and application of pitchbased carbon fibre reinforced concrete, *Concrete Int Design Construct*, 10, 40–47, 1988.
47. Park SB, Lee BL, Fabrication of carbon fiber reinforced cement composites, *Proc Materials Research Society*, 211, Pittsburgh, 247–255, 1991.
48. Banthia N, Sheng J, Micro-reinforced cementitious materials, *Proc Materials Research Society*, 211, Pittsburgh, 25–32, 1991.
49. Furukawa S, Tsuji Y, Otani S, *Proc 30th Japan Congress Mater Res*, 149–152, 1987.
50. Akihama S, Suenaga T, Banno T, *Int J of Cement Composites Lightweight Concrete*, 8(1), 21–33, 1986.
51. Akihama S, Kobayashi M, Suenaga T, Nakagawa H, Suzuki K, *KICT Report*, 65, Kajima Institute of Construction Technology, Oct 1986.
52. Ohama Y, Sato Y, Endo M, *Proc Asia-Pacific Concrete Technology Conf*, 1986, Institute for International Research, Singapore, 5.1–5.8, 1986.
53. Park SB, Lee PI, Lim YS, *Cement Concrete Res*, 21(4), 589–600, 1991.
54. Chung DDL, Carbon fiber cement-matrix composites, *TANSO*, 190, 300–312, 1999.
55. Furukawa S, Otani S, Kojima A, Miyamoto M, Preparation of cement mortars reinforced with a small amount of carbon fibre paper, *TANSO*, 129, 59, 1987.
56. Chen PW, Chung DDL, Carbon fiber reinforced concrete as a SMART material capable of non-destructive flaw detection, *Smart Mater Struct*, 2, 22–30, 1993.
57. Chiou JM, Zheng Q, Chung DDL, Electromagnetic interference shielding by carbon fiber reinforced cement, *Composites*, 20(4), 379–381, 1989.
58. Xuli Fu, Chung DDL, Submicron carbon filament cement-matrix composites for electromagnetic interference shielding, *Cement Concrete Res*, 26(10), 1467–1472, 1996; 27(2), 314, 1997.

59. Chung DDL, Zheng Q, *Composite Sci Technol*, 36, 1–6, 1989.
60. Chung DDL, *Carbon fibre reinforced concrete*, Final Report, Dept of Mechanical and Aerospace Engineering, State University of New York at Buffalo, (SHRP-ID/UFR-92-605; PB92-186550), 80, 1992.
61. Chen PW, Chung DDL, Carbon-fibre-reinforced concrete as an intrinsically smart concrete for damage assessment during dynamic loading *J Am Ceramics Soc*, 78(3), 816–818, 1995.
62. Xuli Fu, Erming Ma, Chung DDL, Anderson WA, Self monitoring in carbon fiber reinforced mortar by reactance measurement, *Cement Concrete Res*, 27(6), 845–852, 1997.
63. XL Fu, Chung DDL, Effect of curing age on the self-monitoring behavior of carbon fiber reinforced mortar, *Cement and Concrete Research*, 27(9), 1313–1318, 1997.
64. Mundell I, Carbon-fiber smartens up concrete, *New Scientist*, 138(1877), 21, 1993.
65. Chen PW, Chung DDL, Carbon-fiber-reinforced concrete as an intrinsically smart concrete for damage assessment during dynamic loading, *J Am Ceramic Soc*, 78(3), 816–818, 1995.
66. Sihai Wen, Chung DDL, Piezoresistivity in continuous carbon fiber cement-matrix composite, *Cement Concrete Res*, 29(3), 445–449, 1999.
67. Sihai Wen, Chung DDL, Carbon fiber-reinforced cement as a thermistor, *Cement Concrete Res*, 29(6), 961–965, 1999.
68. Sihai Wen, Chung DDL, Seebeck effect in carbon fiber reinforced cement, *Cement Concrete Res*, 29(12), 1989–1993, 1999.
69. Janas VF, In: Carrillo G, Newell ED, Brown WD, Phelan P, eds., Proc Int SAMPE Symp and Exhib 33, *Materials, Pathway to the Future*, 357–368, 1988.
70. Zhu M, Chung DDL, Improving brick-to-mortar bond strength by the addition of carbon fibres to the mortar, *Cement Concrete Res*, 27(12), 1829–1839, 1997.
71. Aveston J, Mercer RA, Sillwood JM, *Conference Proceedings Composites Standards Testing and Design*, NPL 1974, IPC Science & Technology Press, 93–103, 1974.
72. Aveston J, Mercer RA, Sillwood JM, *NPL Report*, S1 No. 90/11/98 (1975).
73. Briggs A, Bowen DH, Kollek J, Paper No.17, *International Conf on Carbon Fibres, Their place in modern technology*, London, Feb 1974.
74. Briggs A *et al*, *Fibres in Civil Engineering*, Shirley Institute Conf, June 1974, Shirley Institute Publication S18.
75. Kingery WD, Bowen HC, Uhlmann DR, *Introduction to Ceramics*, John Wiley & Sons, New York, 1976.
76. Prewo KM, Batt JA, *J Mater Sci*, 23, 523, 1988.
77. Alanyali H, Rawlings RD, Rogers PS, Development of 'Apoceram' glass ceramics, *Brit Ceramic Trans*, 97, 240–245, 1998.
78. Hasselman DPH, *Therm Conduct*, 19, 383–402, 1988.
79. Shyu J-J, Chiang M-T, Sintering and phase transformation in B_2O_3/P_2O_5 doped $Li_2O^{+3>}$-$Al_2O_3^{+3>} \cdot 4SiO_2$ glass ceramics, *J Am Ceramic Soc*, 83(3), Mar 2000.
80. Kim HS, Rawlings RD, Rogers PS, *Br Ceramic Proc*, 42, 59–68, 1989.
81. Brown RM, Edrees HJ, Hendry A, *Br Ceram Proc*, 45, 169–177, 1989.
82. Bach H, Krause D, *Analysis of the Composition and Structure of Glass and Glass Ceramics* (Schott Series on Glass and Glass Ceramics), 1999.
83. Allaire RA, U.S. Pat., 4,976,761, 1990.
84. Sambell RAJ, Bowen DH, Phillips DC, *J Mater Sci*, 7, 663–675, 1972.
85. Rand B, Zeng RJ, In: Figueiredo JL, Bernardo CA, Baker RTK, Huttinger KJ, *Carbon Fibers Filaments and Composites*, Kluwer Academic, Dordrecht, 378, 1990.
86. Tredway WK, Prewo KM, Pantano CG, *Carbon*, 27(5), 717–727, 1989.
87. Benson PM, Spear KE, CG Pantano, Thermochemical analysis of interface reactions in carbon-fiber reinforced glass matrix composites, JA Pask, Evans AG, eds., *Ceramic Microstructures '86 - Role of Interfaces*, Plenum Press, New York, 415, 1987.
88. Gunay V, James PF, Jones FR, Bailey JE, *Br Ceramic Proc*, 45, 229–240, 1989.
89. Gunay V, James PF, Jones FR, Bailey JE, *New Materials and Their Applications 1990*, Inst Phys Conf Ser Vol. 111, IOP, Bristol, 217–226, 1990.
90. Crivelli-Visconti T, Cooper GA, *Nature*, 221, 754–755, 1969.

91. Phillips DC, The Fracture Energy of Carbon Fiber Reinforced Glass, *J Mater Sci*, 7(10), 5–91, 1972.
92. Sambell RAJ, Bowen DH, Phillips DC, Carbon Fibre Composites with Ceramic and Glass Matrices, Part 1-Discontinuous Fibres, *J Mater Sci*, 7(6), 663–675, 1972.
93. Sambell RAJ, Briggs A, Phillips DC, Bowen DH, Carbon Fiber Composites with Ceramic and Glass Matrices, Part 2-Continuous Fibers, *J Mater Sci*, 7(6), 676–81, 1972.
94. Phillips DC, Sambell RAJ, Bowen DH, The mechanical properties of carbon fibre reinforced Pyrex glass, *J Mater Sci*, 7, 1454–1464, 1972.
95. Phillips DC, The fracture energy of carbon fibre reinforced glass, *J Mater Sci*, 7, 1175–1191, 1972.
96. Phillips DC, Interfacial Bonding and the Toughness of Carbon Fiber Reinforced Glass and Glass Ceramics, *J Mater Sci*, 9(11), 1847–1854, 1974.
97. Sambell J, Phillips DC, Bowen DH, Carbon fibres, their composites and applications, *Proc Int Conf*, The Plastics Institute, 105–113, 1974.
98. Habib FA, Cooke RG, Harris B, Br Ceramic Trans J, 89, 115–124, 1990.
99. Fitzer E, High Temp-High Pressures, 18(5), 479–508, 1986.
100. Hasson DF, Fishman SG, *Ceramic Eng Sci Proc*, 11(9–10), 1639–1647, 1990.
101. Prewo KM, Brennan JJ, Layden GK, Fiber reinforced glasses and glass ceramics for high performance applications, *Am Ceramic Soc Bull*, 65, 305, 1986.
102. Prewo KM, Fiber reinforced glasses and glass ceramics, Lewis MH ed., *Glasses and Glass-ceramics*, Chapman and Hall, London, 336–368, 1989.
103. Prewo KM, Evaluation of advanced fibers for the reinforcement of glass and ceramic matrix composites, *1st Japan International SAMPE Symposium and Exhibition*, Nov 28–Dec 1, 875, 1989.
104. Sahebkar M, Schlichting J, Schubert P, Possibility of reinforcing glass by carbon fibers, *Berichte de Deutschen Keramischen Gesellschaft*, 55(5), 265–268, 1978.
105. Prewo KM, Bacon JF, Glass matrix composites-I Graphite fiber reinforced glass, Noton B ed., *Proc Second Int Conf on Composites*, AIME, 64–74, 1978.
106. Prewo KM, Bacon JF, Thompson ER, Graphite fiber reinforced glass, Ahmed I, Noton B eds., *Proc of AIME Conf. Adv Fibers and Composites for elevated temperatures*, 80–93, 1979.
107. Prewo KM, Bacon JF, Dicus DL, Graphite fiber reinforced glass matrix composites, *SAMPE Quarterly*, 10(4), 42, 1979.
108. Prewo KM, Thompson ER, Research on graphite fiber reinforced glass matrix composites, *NASA Contr Rep*, 165711, May 1981.
109. Treadway WK, Prewo KM, Pantano CG, Fiber-matrix interfacial effects in carbon fiber reinforced glass matrix composites, *Carbon*, 27, 717, 1989.
110. Qi D, Pantano CG, Effects of composite processing on the performance of carbon fiber/glass matrix composites, *Proc 16th Annual Conf Composites and Adv Ceramic Materials*, American Ceramic Society, Waterville, 863, 1992.
111. Bleay SM, Scott VD, The role of the interface in the mechanical behavior and oxidation of carbon fiber reinforced Pyrex glass, *Carbon*, 29, 871, 1991.
112. Bleay SM, Scott VD, Microstructure and micromechanics of the interface in carbon fiber reinforced Pyrex glass, *J Mater Sci*, 26, 3544, 1991.
113. Boccaccini AR, Glass matrix composite materials with dispersion reinforcement, a review, Part 1, Historical development and fabrication techniques, Verre, 5(2), 3–14, 1999.
114. Boccaccini AR, Glass matrix composite materials with dispersion reinforcement, a review, Part 2, Properties and applications, Verre, 5(3), 3–11, 1999.
115. Boccaccini AR, Pfeiffer K, Preparation and characterisation of a glass matrix composite containing aluminium titanate particles with improved shock resistance, Glass Sci Technol Glastech Ber, 72, 352–357, 1999.
116. Boccaccini AR, Pfeiffer K, Kern H, Thermal shock resistant Al2TiO5 glass matrix composite, *J Mat Sci Lett*, 18, 1907–1909, 1999.
117. Crosbie GM, Nicholson JM, Deering LA, Pseudo 3-D reinforcement with stretch-broken carbon fibers in a borosilicate glass matrix, JP Singh, Bansal NP eds., Advances in Ceramic-Matrix Composites II, Ceram Trans, American Ceramic Soc Inc, Indianapolis, 46, 211–222, Apr 25–27 1994.

118. Benson PM, Spear KE, Pantano CG, Thermochemical analysis of interface reactions in carbon-fiber reinforced glass matrix composites, JA Pask, Evans AG eds., Ceramic Microstructures '86-Role of Interfaces. Plenum Press, New York, 415, 1987.
119. Yanagida H, Design of self-diagnosis in/with composite materials, Ceramic Eng Sci Proc, 19(4), 567–577, 1998.
120. Bianchi V, Goursat P, Sinkler W, Monthioux M, Menessier E, Carbon fibre reinforced (YMAS) glass-ceramic matrix composites, I. Preparation, structure and fracture strength, *J Eur Ceramic Soc*, 17(12), 1485–1500, 1997.
121. Bianchi V, Goursat P, Menessier E, Carbon-fiber-reinforced YMAS glass-ceramic-matrix composites - IV. Thermal residual stresses and fiber/matrix interfaces, *Composites Sci Technol*, 58(3–4), 409–418, 1998.
122. Sinkler W, Monthioux M, Bianchi V, Goursat P, Menessier E, Carbon fiber-reinforced (YMAS) glass-ceramic matrix composites, II. Structural changes in the matrix with temperature, *J Eur Ceramic Soc*, 19(3), 305–316, 1999.
123. Bianchi V, Fournier P, Platon F, Reynaud P, Carbon fibre-reinforced (YMAS) glass-ceramic matrix composites, Dry friction behavior, *J Eur Ceramic Soc*, 19(50), 581–589, 1999.
124. Bianchi V, Goursat P, Sinkler W, Monthioux M, Menessier E, Carbon fiber reinforced (YMAS) glass-ceramic matrix composites, III. Interfacial aspects, *J Eur Ceramic Soc*, 19(3), 317–327, 1999.
125. Sinkler W, Monthioux M, Bianchi V, Goursat P, Menessier E, Carbon fibre-reinforced (YMAS) glass-ceramic matrix composites, Part 2. Structural changes in the matrix with temperature, *J Eur Ceramic Soc*, 19(3), 305–316, 1999.
126. Zhang E, Thompson DP, Elimination of cracks in carbon fibre reinforced nitrogen glass composites, Bellosi A ed., Basic Science - Trends in Emerging Materials and Applications, Italian Ceramic Society Fourth ECerS Proc 4th European Ceramic Society Conf 4, Riccione, 2–6 October 1995, Faenza Editrice SpA, 193, 1995.
127. Zhang E, Thompson DP, Carbon fibre reinforcement of nitrogen glass, Composites Part A. *Appl Sci Manuf*, 28(6), 581–586, 1997.
128. Sung Y-M, Graphite fibre-reinforced off-stoichiometric $SrO.Al_2O_3.2SiO_2$ glass-ceramic matrix composites, *J Mat Sci Lett*, 18(16), 1315–1317, 1999.
129. Sung Y-M, Park S, Thermal and mechanical properties of graphite fibre-reinforced off-stoichiometric $BaO.Al_2O_3-2SiO_2$ glass-ceramic matrix composites, *J Mat Sci Lett*, 19(4), 315–317, 2000.
130. Taylor MP, Fibre-reinforced composites, U.S. Pat., 4,511,663, Apr 1985.
131. U.S. Pat., 3,607,608.
132. U.S. Pat., 3,681,187.
133. U.S. Pat., 4,256,378.
134. U.S. Pat., 4,263,367.
135. Preno, U.S. Pat., 4,265,968, Apr 1981.
136. Kaya C, Boccaccini AR, Chawla KK, Electrophoretic deposition forming of Ni-coated carbon fibre reinforced borosilicate glass matrix composites, *J Am Ceramic Soc*, 83, 1885, 2000.
137. Kingery WD, Bowen HC, Uhlmann DR, Introduction to Ceramics, John Wiley & Sons, New York, 1976.
138. Reed J, Principles of Ceramic Processing, John Wiley & Sons, New York, 1988.
139. Signorelli R, DiCarlo J, High temperature metal and ceramic composites, *J Metals*, 41–42, Jun 1985.
140. Schioler LJ, Stiglich JJ Jr., Ceramic matrix composites, A literature review, *Am Ceramic Soc Bull*, 65, 289, 1986.
141. Inagaki M, Research and development on carbon/ceramic composites in Japan, *Carbon*, 29, 287, 1991.
142. Tai NH, Lee KP, Hocheng H, Liu CS, Processing and characterisation of carbon fibre reinforced ceramic nanocomposites, Niihara K, Nakano K, Sekino T, Yasuda E eds., High Temperature Ceramic Matrix Composites III. Proc. 3rd Int. Conf. Osaka, 6–9 September 1998, Ceramic Society of Japan, 105–108; Key Engineering Materials, 164–165, 1998.

143. Barta J, Shook WB, Graves GA, Impact stength of alumina composites, *Am Ceramic Soc Bull*, 51(5), 464–470, 1972.
144. Bailey JE, Chen M, James PF, Jones FR, Fibre reinforced alumina ceramic composites by sol-gel processing, ECCM III, 1989.
145. Lange FF, Transformation toughening, Part IV, Fabrication, fracture toughness and strength of Al_2O_3-ZrO_2 composites, *J Mater Sci*, 17, 247–254, 1982.
146. Tien TY, Transformation toughened ceramics - a potential material for light diesel engine applications, AMMRC TR84-26, Army Materials and Mechanics Research Center, MA, Jun 1984.
147. Schioler LJ, Katz RN, Brog T, Lawn BR, Mechanical properties of zirconia toughened alumina, *Ceramic Eng Sci Proc*, 6(7–8), 822–825, 1985.
148. Heuer AH, Hidda LW eds., Advances in Ceramics Vol. 3 Science and technology of zirconia II, Am Ceram Soc, Columbus, 1981.
149. Heuer AH, Claussen N, Ruhle M eds., Advances in Ceramics Vol. 12 Science and technology of zirconia, Am Ceram Soc, Columbus, 1985.
150. Somiya S, Inomata Y, Silicon Carbide Ceramics, Kluwer Academic Publishers, Dordrecht, 1991.
151. Fitzer E, Gadow R, Fiber reinforced silicon carbide, *Am Ceramic Soc Bull*, 65(2), 326–335, 1986.
152. Wallace TC, Cort GE, Damran JJ.
153. Ansorge F, Characterisation of carbon fibre/silicon carbide matrix composites, Naslain R, Lamon J, Doumeingts D eds., *High Temperature Ceramic Matrix Composites*, Proceedings of 6th European Conference on Composite Materials, European Association for Composite Materials, American Ceramic Soc Inc, Ceramic Society of Japan, Bordeaux, 491–498, 20–24 Sep 1993.
154. Sherwood WJ, Whitmarsh CK, Jacobs JM, Interrante LV, Low cost near net shape ceramic matrix composites using resin transfer molding and pyrolysis, *Ceramic Eng Sci Proc*, 20th Annual Conference on Composites, Advanced Ceramics, Materials and Structures-B, Cocoa Beach, 174–183, 1996.
155. Wei GC, Becher PF, Improvements in mechanical properties in SiC by the addition of TiC particles, *J Amer Ceramic Soc*, 67(8), 571–574, 1984.
156. Accountius O, Sisley H, Sheblin S, Bole G, Oxidation resistances of ternary mixtures of the carbides of titanium, silicon and boron, *J Amer Ceramic Soc*, 37(4), 173–177, 1954.
157. Watanabe T, Shoubu K, Mechanical properties of hot pressed TiB_2-ZrO_2 composites, *J Amer Ceramic Soc*, 68(2), C-34–36, 1985.
158. Mazdiyasni KS, Ruh R, High/low modulus Si3N4-BN composite for improved electrical and thermal shock behavior, *J Amer Ceramic Soc*, 64(7), 415–419, 1981.
159. Hirai T, Goto T, Preparation of amorphous Si_3N_4-C plate by chemical vapor deposition, *J Mater Sci*, 16, 17–23, 1981.
160. Suzuki T, Sato M, Sakai M, Fiber pullout processes and mechanisms of a carbon fiber reinforced silicon nitride ceramic composite, *J Mater Res*, 7, 2869, 1992.
161. Fischbach DB, *Carbon Conf*, Baden-Baden, 719–221, 1986.
162. Saruhan B, Ziegler G, *Silic Ind*, 55(1,2), 29–32, 1991.
163. Lundberg R, Pompe R, Carlsson R, Hiped carbon fibre reinforced silicon nitride composites, *Ceram Eng Sci Proc*, 9(7,8), 901–905, 1988.
164. Lundberg R, Pompe R, Carlsson R, Goursat P, Fibre reinforced silicon nitride composites, Naslain R, Harris B eds., Elsevier Applied Science Composites Sci Technol, Elsevier Applied Science; Ceramic Matrix Composites, Components, Preparation, Microstructure & Properties, 37(1/2/3), 165–176, 1990.
165. Guo J-K, Mao Z-Q, Bao C-D, Wang R-H, Yan D-S, Carbon Fiber-Reinforced Silicon Nitride Composite, *J Mater Sci*, 17, 3611–3616, 1982.
166. Hillig WB, Ceramic Composites by Infiltration, *Ceramic Eng Sci Proc*, 5(7–8), 674–683, 1985.
167. Grenet C, Plunkett L, Veyret JB, Bullock E, Carbon fibre-reinforced silicon nitride composites by slurry infiltration, Evans AG, Naslain R eds., High-Temperature Ceramic-Matrix Composites II, Manufacturing and Materials Development, *Ceramic Trans*, American Ceramic Soc Inc., Santa Barbara, 58, 125–130, Aug 21–24 1995.

168. Sano S, Mizuta H, Maeda M, Teranishi H, Oda K, Shibasaki Y, *J Japan Soc Powder and Powder Metall*, 40, 794, 1993.
169. Sano S, Mizuta H, Maeda M, Teranishi H, Oda K, Shibasaki Y, *J Japan Soc Powder and Powder Metall*, 43, 868, 1996.
170. Sano S, Mizuta H, Maeda M, Teranishi H, Oda K, Shibasaki Y, *J Japan Soc Powder and Powder Metall*, 44, 876, 1997.
171. Sano S, Oda K, Ohshima K, Shibasaki Y, In: Advances in the Science and Technology, P Vincenzini ed., *Techna*, 1585–1592, 1995.
172. Sano S, Banno T, Maeda M, Oda K, Shibasaki Y, *J Japan Soc of Powder and Powder Metall*, 104, 984, 1996.
173. Sano S, Hotta Y, Banno T, Oda K, Pressure controlled slipcasting of fine ceramics, *Key Eng Mater*, 161–163, 81–86, 1999.
174. Matsuo Y, Kumagai S, Yasuda K, Mechanical properties of carbon fibre-reinforced hBN matrix composites at elevated temperature, Niihara K, Nakano K, Sekino T, Yasuda E eds., High Temperature Ceramic Matrix Composites III, Proc 3rd Int Conf, Ceramic Society of Japan, Osaka, *Key Eng Mater*, 164–165, 137–140, Sep 6–9 1998.
175. Chen M, Jones FR, James PF, Bailey JE, Inst Phys Conf Ser Vol. 111, *New Mater and their Applications 1990*, IOP, Bristol, 227–237, 1990.
176. Raman V, Bahl OP, Dhawan U, Synthesis of silicon carbide incorporated carbon-carbon composites by sol-gel process, *J Mat Sci Lett*, 14(16), 1150–1152, 1995.
177. Manocha S, Vashistha D, Manocha LM, Sol-gel processing of silicon based matrixes for carbon fibre reinforced ceramic composites, Niihara K, Nakano K, Sekino T, Yasuda E eds., Ceramic Society of Japan, High Temperature Ceramic Matrix Composites III, Proc, 3rd Int Conf, Osaka, *Key Eng Mater*, 164–165, 137–140, Sep 6–9 1998.
178. Duran A, Aparicio M, Rebstock K, Vogel WD, Reinfiltration processes for polymer derived fibre reinforced ceramics, Fuentes M, Martinez-Esnaola JM, Daniel AM eds., Ceramic and Metal Matrix Composites CMMC 96, Trans Tech Publications, Switzerland, *Key Eng Mater*, 127–131, Part 1, 287–294, 1997.
179. Wu J, Chen M, Jones FR, James PF, Characterisation of sol-gel derived alumina-silica matrices for continuous fibre reinforced composites, *J Eur Ceram Soc*, 16(6), 619–626, 1996.
180. Hyde AR, Multidirectional fibre reinforced ceramic matrix composites, *Ceramic Ind Int*, 98(1075), 32–38, 1989.
181. Fitzer E, Gadow R, Fiber reinforced ceramic composites fabricated via the sol/gel route, Conference on Tailoring Multiphase and Composite Ceramics, Penn State Univ, Jul 1985.
182. Wynne KJ, Rice RW, Ann Rev Mater Sci, Huggins RA ed., Palo Alto, Annual Reviews, 14, 297–334, 1984.
183. Yajima S, Silicon carbide fibres, Watt W, Perov BV eds., Strong Fibres, Handbook of Composites, Vol. 1, Kelly A, Rabotnov YN Series eds., Amsterdam, 201–240, 1985.
184. Zhou X, Zhang C, Ma J, Zhou A, The fabrication of CF/SiC ceramic matrix composites by polymer precursor pyrolysis, Nihara K, Nakano K, Sekino T, Yasuda E eds., Ceramic Society of Japan. High Temperature Ceramic Matrix Composites III, Proc 3rd Int. Conf, Osaka, *Key Engineering Materials*, 164–165, 43–47, Sep 6–9 1998.
185. He X-B, Zhang X-M, Zhang CR, Zhou X-G, Zhou A-C, Carbon-fibre-reinforced silicon carbide composites, *J Mat Sci Lett*, 19(5), 417–419, 2000.
186. Zhou X, Zhang C, Ma J, Zhou A, Fabrication of CF/SiC ceramic matrix composites by polymer precursor pyrolysis, Niihara K, Nakano K, Sekino T, Yasuda E eds., High Temperature Ceramic Matrix Composites III, Ceramic Society of Japan, Proc 3rd Int Conf, Osaka, *Key Eng Mater*, 164–165, 43–47, Sep 6–9 1998.
187. Hu H-F, Chen Z-H, Feng C-X, Zhang C-R, Song Y-C, Three-dimensional braided preform reinforced Si-C-N composites prepared by precursor pyrolysis, *Mater Sci Lett*, 17(1), 73–74, 1998.
188. Zheng G, Sano H, Uchiyama Y, Kobayashi K, Suzuki K, Cheng H, Preparation and fracture behavior of carbon fibre/SiC composites by multiple impregnation and pyrolysis of polycarbosilane, *J Ceramic Soc Japan*, 106(12), 1155–1161, 1998.

189. Kamphowe TW, Weinmann M, Bill J, Aldinger F, Preparation of fibre-reinforced Si-B-C-N ceramics by polymer precursor infiltration, *Silic Ind*, 63(11/12), 159–162, 1998.
190. Suttor D, Erny T, Greil P, Goedecke H, Haug T, Fibre-reinforced ceramic-matrix composites with a polysiloxane/boron-derived matrix, *J Am Ceramic Soc*, 80(7), 1831–1840, 1997.
191. Hoshi S, Kojima A, Mechanical properties and oxidation resistivity of carbon fiber/ceramic composites prepared from borosiloxane, *J Mater Res*, 11(10), 2536, 1997.
192. Nakano K, Kamiya A, Nishino Y, Imura T, Chou T-W, Fabrication and characterisation of three-dimensional carbon fibre reinforced silicon carbide and silicon nitride composites, *J Am Ceramic Soc*, 78(10), 2811–2814, 1995.
193. Sato K, Suzuki T, Funayama O, Isoda T, Itoh T, Preparation of carbon fibre reinforced composite by impregnation with perhydropolysilazane followed by pressureless firing, *Ceramic Eng Sci Proc*, 13(9/10), 614–621, 1992.
194. Ostertag R, Haug T, Carbon fibre-reinforced SiC ceramic components by filament winding, Vincenzini P ed., Advanced Structural Inorganic Composites, Proc Satellite Symp 2. Advanced Structural Inorganic Composites, 7th Int Meeting on Modern Ceramics Technologies (7th CIMTEC—World Ceramics Congress) Montecatini Terme, Elsevier Science Publishing Co. Inc., *Mater Sci Monogr*, 68, 469–477, Jun 27–30 1990.
195. Seyferth D, Wiseman GH, Prud'homme C, A Liquid Silazane Precursor to Silicon Nitride, *J Am Ceramic Soc*, 66(1), C–13–C–14, 1983.
196. Singh RN, Morrison WA, U.S. Pat., 4,944,904, 1990.
197. Gadow R, Speicher M, Optimised morphological design for silicon infiltrated microporous carbon performs, *Ceramic Eng Sci Proc*, 21(3), 485–492, 2000.
198. Danforth SC, In: Mostaghaci H ed., Proc Int Symp Adv Processing of Ceramic and Metal Matrix Composites, Halifax, Pergamon, New York, 107–119, 1989.
199. Taylor R, Piddock V, *Mater Sci Forum*, 34–36, 525–530, 1988.
200. Minet J, Langlais F, Quenisset JM, Naslain R, *J Eur Ceram Soc*, 5(6), 341–356, 1989.
201. Stinton DP, Caputo AJ, Lowden RA, Synthesis of fibre reinforced SiC composites by chemical vapor infiltration, *Am Ceramic Soc Bull*, 65, 347–350, 1986.
202. Hopfe V, Weiss R, Meistring R, Brennfleck K, Jäckel R, Schönfeld K, Dresler B, Goller R, Laser based coating of carbon fibres for manufacturing, CMC, Ceramic Metal Matrix Composites, *Key Eng Mater*, 127–131, 559–566, 1997.
203. Naslain R, Langlais F, CVI processing of ceramic-ceramic composite materials, Conf on Tailoring Multiphase and Composite Ceramics, Penn State Univ, Jul 1985.
204. Lackey WJ, Vaidyaraman S, Freeman GB, Agrawal PK, Technique for monitoring densification during chemical vapor infiltration, *J Am Ceramic Soc*, 78(4), 1131–1133,1995.
205. Caputo AJ, Lackey WJ, Fabrication of fiber-reinforced ceramic Composites by chemical vapor infiltration, *Ceramic Eng Sci Proc*, 5(7–8), 654–667, 1984.
206. Caputo AJ, Lackey WJ, Stinton DP, Development of a new, faster process for the fabrication of ceramic fiber-reinforced ceramic composites by chemical vapor infiltration, *Ceramic Eng Sci Proc*, 6(7–8), 694–706, 1985.
207. Suzuki K, Nakano K, Structure and characterisation of SiC matrix composite via slurry and pulse CVI joint process, *Ceramic Eng Sci Proc*, 18(3), 427–433, 1997.
208. Suzuki K, Nakano K, Kume S, Chou TW, Fabrication and characterisation of 3D carbon fibre reinforced SiC matrix composites via slurry and pulse-CVI joint process, *Ceramic Eng Sci Proc*, 19(3), 259–266, 1998.
209. Suzuki K, Nakano K, Ishikawa T, Kanno Y, Chou TW, Characterisation of 3D-carbon fibre reinforced SiC composites, *Ceramic Eng Sci Proc*, 21(3), 493–501, 2000.
210. Zou W, Song M, Wan T, Yan L, Wan K, Triaxially braided C/SiC composites, research and development by the combined process of CVI+PIP, Niihara K, Nakano K, Sekino T, Yasuda E eds., Ceramic Society of Japan High Temperature Ceramic Matrix Composites III, Proc 3rd Int Conf, Osaka, *Key Eng Mater*, 164–165, 213–216, 6–9 Sep 1998.
211. Xu Y, Zhang L, Three-dimensional carbon/silicon carbide composites prepared by chemical vapor infiltration, *J Am Ceramic Soc*, 80(7), 1897–1900, 1997.

212. Weng BJ, Hwung JJ, Chen CI, Hsu SE, Heat resistant composites of graphite fibre reinforced phosphate ceramics, Doyama M, Somiya S, Chang RPH, Kimura S, Kobayashi A, Umekawa S eds., Composites Corrosion/Coating of Advanced Materials, Proc MRS Int Meeting on Advanced Materials 4, Materials Research Society, Tokyo, 135–140, 2–3 Jun 1998.
213. Tani E, Shoubu K, Watanabe T, Carbon fibre-reinforced SiC composites produced by reaction-bonding, *J Ceramic Soc Japan*, 100(4), 596–598, 1992.
214. Tani E, Shobu K, Watanabe T, Carbon fibre-reinforced SiC/C composites produced by reaction-bonding, Naslain R, Lamon J, Doumeingts D eds., High Temperature Ceramic Matrix Composites, European Association for Composite Materials; American Ceramic Soc, Inc.; Ceramic Society of Japan, Proc 6th European Conf on Composite Materials, Bordeaux, 207–213, Sep 20–24 1993.
215. Tani OE, Shobu K, Kishi K, Umebayashi S, Two-dimensional-woven-carbon-fibre-reinforced silicon carbide /carbon matrix composites produced by reaction bonding, *J Am Ceramic Soc*, 82(5), 1355–1357, 1999.
216. Gibson JO, Gibson MG, Production of carbon fiber - tantalum carbide composites, U.S. Pat., 4196230, Apr 1 1980.
217. Maili S *et al*, A study of carbon fibre reinforced ceramic composites with rich carbon interphase, *Aerosp Mater Technol*, 28(3), Jun 1998.
218. Sakagami S *et al*, Carbon-carbon composite by carbon powder sintering method, *1st Japan International SAMPE Symposium and Exhibition*, 1166, Nov 28–Dec 1.
219. Dodds N, Chandler HW, Thompson DP, Influence of processing on the carbon fibre/SiAlON interface, Hoffmann MJ, Becher PF, Petzow G eds., *Silicon Nitride, 93*, Proc Int Conf Silicon Nitride-Based Ceramics, Stuttgart, *Key Eng Mater*, 89–91, 461, Oct 4–6 1993.
220. Kikuchi S, Yasutomi Y, Arakawa H, Reactivities of carbon-fibers in hot-pressed carbon-fiber reinforced SiC ceramic composite, *Nippon Seramikkusu Kyokai Gakujutsu Ronbunshi—Journal of the Ceramic Society of Japan*, 102(5), 456–461, 1994.
221. Fitzer E, Ceramic and coated carbon fibres for high temperature ceramics, Bradley RA, Clark DE, Larsen DC, Stiegler JO eds., ASM International; Oak Ridge National Laboratory, Whisker and Fiber Toughened Ceramics, ASM International; Oak Ridge National Laboratory, Tennessee, Proc Int Conf, 9–52, Jun 7–9 1988.
222. Westwood ME, Webster JD, Day RJ, Hayes FH, Taylor R, Oxidation protection for carbon fibre composites, *J Mat Sci*, 31(6), 1389–1397, 1996.
223. Vogel WD, Hertel D, Rebstock K, Haug T, Integrated oxidation protection of ceramic matrix composites, *Silic Ind*, 60(7/8), 189–192, 1995.
224. Katzman HA, *Mater Manuf Proc*, 5(1), 1–15, 1990.
225. Ermolenko LN, Lyubliner LP, Gulko NV, Chemically Modified Carbon Fibers, Translated by Titovets EP, VCH, Weinheim, 1990.
226. Liu W, Sun S, Li M, Wei Y, Preparation of C/C-SiC nanomatrix composites by chemical vapor infiltration, *J Mater Sci Lett*, 12(12), 886–888, 1993.
227. Webster JD, Westwood ME, Hayes FH, Day RJ, Taylor R, Duran A, Aparicio A, Rebstock K, Vogel WD, Oxidation protection coatings for C/SiC based on yttrium silicate, Baxter J, Cot L, Fordham R, Gabis V, Hellot Y, Lefebvre M, Le Doussal H, Le Sech A, Naslain R, Sevagen A eds., European Ceramic Society Euro Ceramics V, Part 3. Trans Tech Publications, Switzerland, *Key Eng Mater*, 132–136, 1641–1644, 1997.
228. Webster JD, Westwood ME, Hayes FH, Day RJ, Taylor R, Duran A, Aparicio M, Rebstock K, Vogel WD, Oxidation protection coatings for C/SiC based on yttrium silicate, *J Eur Ceramic Soc*, 18(16), 2345–2350, 1998.
229. Slosarczyk A, Klisch M, Blazewicz M, Piekarczyk J, Stobierski L, Rapacz-Kmita A, Hot pressed hydroxyapatite-carbon fibre composites, *J Eur Ceramic Soc*, 20(9), 1397–1402, 2000.
230. Hoshii S, Kojima A, Otani S, Oxidation behavior of CFRC and C/C using diphenylborosiloxane, *J Mater Sci Lett*, 19(2), 169–172, 2000.
231. Aparicio M, Duran A, Yttrium silicate coatings for oxidation protection of carbon-silicon carbide composites, *J Am Ceramic Soc*, 83(6), 1351–1355, 2000.

232. Kawai C, Wakamatsu S, Synthesis of a functionally gradient material based on C/C composites using an electro-deposition method, *J Mater Sci Lett*, 14(7), 467–469, 1995.
233. Kawai C, Wakamatsu S, Fabrication of C/SiC composites by an electrodeposition/sintering method and the control of the properties, *J Mater Sci*, 31(8), 2165–2170, 1996.
234. Kooner S, Campaniello JJ, Pickering S, Bullock E, Fibre reinforced ceramic matrix composite fabrication by electrophoretic infiltration, Evans AG, Naslain R eds., American Ceramic Soc Inc, High-Temperature Ceramic-Matrix Composites II, Manufacturing and Materials Development, Santa Barbara, *Ceramic Trans*, 58, 155–160, Aug 21–24 1995.
235. Kaya C, Boccaccini AR, Chawla KK, Electrophoretic deposition forming of nickel-coated-carbon-fibre-reinforced borosilicate-glass-matrix composites, *J Am Ceramic Soc*, 83(8), 1885–1888, 2000.
236. Rice RW, Mechanisms of toughening in ceramic matrix composites, *Ceramic Eng Sci Proc*, 2(7–8), 661–701, 1981.
237. Rice RW, Ceramic matrix composite toughening mechanisms, An update, *Ceramic Eng Sci Proc*, 5(7–8), 589–607, 1985.
238. Marshall DB, Evans AG, Failure mechanisms in ceramic-fiber/ceramic matrix Composites, *J Am Ceramic Soc*, 68(5), 225–31, 1985.
239. Miyajima T, Sakai M, Fibre pullout and fracture energy of C-fibre/C-matrix composites, Bradt RC, Hasselman DPH, Munz D, Sakai M, Ya Shevchenko V, *Fracture Mechanics of Ceramics 9, Composites, R-Curve Behavior and Fatigue*, Japan Fine Ceramics Center, Proc 5th Int Symp, Nagoya, Jul 15–17 1991, Plenum, 83–95, 1992.

CHAPTER 16

Carbon Fibers in Metal Matrices

16.1 INTRODUCTION

At the time that the author first became involved with carbon fibers, the immediate emphasis on research was to attempt to incorporate carbon fiber in metal matrices (e.g. Figures 3.52–3.54). It was soon realized that this route was beset with many technical problems and attention was switched to polymer matrices, which appeared to be a better way to further the initial sales of carbon fibers. Over the years, much work has now been undertaken with metal matrices but, unfortunately, some of this work is not available for open publication, or is restricted to U.S. citizens only.

16.2 METAL MATRIX COMPOSITES

The books by Chung [1], Peebles Jr. [2] and Matthews [3] give an introduction to the study of metal matrix composites (MMC). The effective reinforcement of metal matrices with carbon fibers will confer the following properties:

1. Increase the strength and modulus at room and elevated temperatures.
2. Reduce the density and hence improve the specific strength and modulus.
3. A high modulus fiber will reduce the coefficient of thermal expansion and increase the thermal conductivity (Table 16.1).
4. Improve the fatigue resistance.
5. Lower the creep.
6. Improve the wear resistance.

It is not surprising that with all these benefits, considerable effort has been expended on achieving reinforcement of metals.

16.3 CARBON FIBER FOR REINFORCEMENT OF METAL MATRICES

Carbon fiber is used in continuous and chopped fiber form and can be made into a preform to obtain near net shape components. The preform can be held together with a binder such as an acrylic polymer or aluminum metaphosphate [4–6]. The more expensive

Table 16.1 Thermal properties of metal matrices and composites

Material	mp °C	Density g cm^{-3}	Axial thermal conductivity W m^{-1}K^{-1}	Axial coefficient of thermal expansion 10^{-6} °C^{-1}
Aluminum	660	2.71	221	23.6
Al/60% P-120 carbon fiber		2.41	419	−0.32
Mg	651	1.74		
Cu	1082	8.91	391	17.6
Cu/60% P-120 carbon fiber		6.23	522	−0.07
Ni	1450	8.86		
Ti	1800	4.50		
Pb	327	11.34		
Sn	232	5.75		

Source: Reprinted with permission from Thaw C, Minet R, Zemany J, Zweben C, *SAMPE J*, 23(6), 40–43, 1987. Copyright 1987, The Society for the Advancement of Material and Process Engineering (SAMPE).

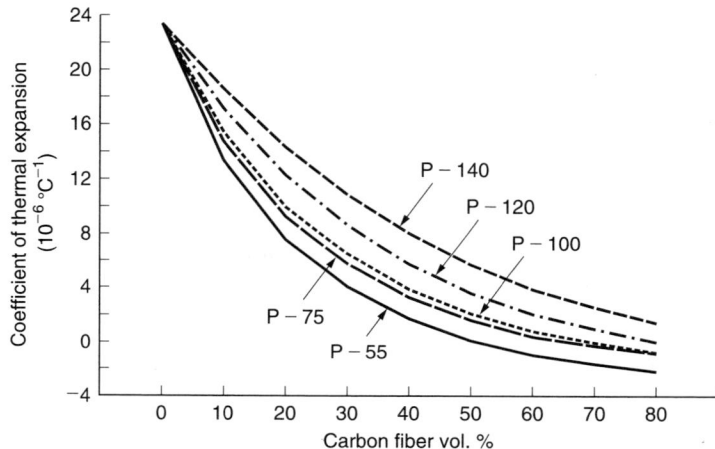

Figure 16.1 Coefficient of thermal expansion versus carbon fiber volume percentage for aluminum reinforced with various kinds of carbon fiber (BP Amoco Thornel P-55, P-75, P-100, P-120 and P-140 in order of increasing modulus) in a cross plied configuration. *Source:* Reprinted with permission from Thaw C, Minet R, Zemany J, Zweben C, *SAMPE J*, 23(6), 40–43, 1987. Copyright 1987, The Society for the Advancement of Material and Process Engineering (SAMPE).

pitch based fibers are generally preferred [7–10] to the PAN based fibers due to the availability of fibers with higher moduli, exhibiting a more graphitic nature with lower thermal expansion (Figure 16.1), higher thermal conductivity (Figure 16.2) and improved corrosion resistance. Ting and Lake [12] utilized the high thermal conductivity of vapor grown carbon fiber to reinforce aluminum composites. Any oxidative surface treatment of the fiber will give an increased amount of carbide in the interphase [13].

Bouix *et al* describe the interface tailoring of carbon fibers [14], whilst Lawcock *et al* [15] considered the effects of the fiber/matrix adhesion on the strength of carbon fiber reinforced metal laminates and the effects of the fiber/matrix adhesion on the impact strength of carbon fiber reinforced metal laminates [16].

If a metal coating is to be applied to carbon fiber, then any gases absorbed on the fiber surface should preferably be removed to prevent delamination occurring during infiltration with the molten matrix and desorption can be achieved by heating the fiber to 1000°C in a vacuum ($<1.3 \times 10^{-4}$ Nm^{-2}) [17].

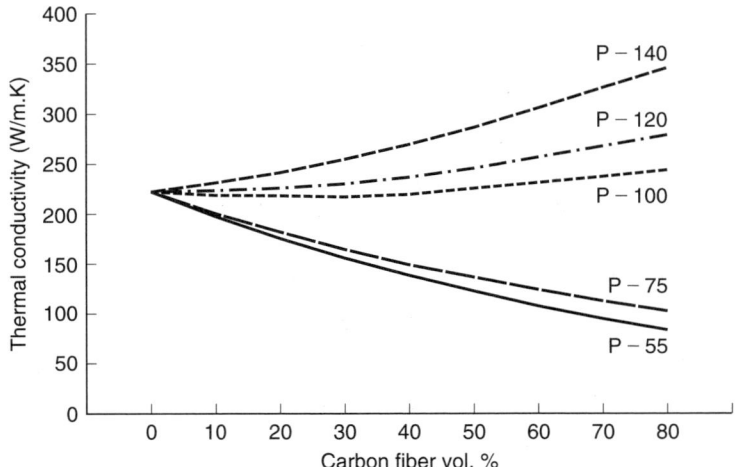

Figure 16.2 Thermal conductivity versus carbon fiber volume percentage for aluminum reinforced with various kinds of carbon fiber (BP Amoco Thornel P-55, P-75, P-100, P-120 and P-140 in order of increasing modulus) in a cross plied configuration. *Source:* Reprinted with permission from Thaw C, Minet R, Zemany J, Zweben C, *SAMPE J*, 23(6), 40–43, 1987. Copyright 1987, The Society for the Advancement of Material and Process Engineering (SAMPE).

16.4 COATING PROCESSES TO IMPROVE WETTABILITY

Carbon reacts readily with several metals near their melting points, especially Al (mp 660°C) and Mg (mp 651°C) to form carbides (e.g. Al_4C_3), which damage the fiber matrix and interface. Hence, research work has been directed towards establishing a coating that would be wetted by molten metals but, on the other hand, would provide a protective coating and prevent the carbon fiber reacting with the molten metals. Coatings of Ni, Cu and Ag are wetted by Al at about 660°C, lowering the mp and viscosity of the matrix and readily dissolving [18]. Delanney et al [19] have reviewed the wetting of solids by molten metals in relation to the preparation of metal matrix composites.

16.4.1 CVD process

1. A coating of TiB_2 can be applied using $TiCl_4/BCl_3$, reducing with Zn vapor at 700°C in an atmosphere of Ar [20–22]. The deposit is not stable in air and must be immersed in molten metal before coming into contact with air. However, it does wet out well in Al, forming $\gamma\text{-}Al_2O_3$ [23,24] between the fiber and matrix, whilst the Mg alloy Al-6061 forms $MgAl_2O_4$ [25]. The oxygen originates from the carbon fiber size, which is deliberately only partly removed prior to the coating procedure [26].
2. A coating of TiC can be adopted for Al and Mg using $TiCl_4$ with CH_4 and a flow of H_2 at 1050°C [27–29].
3. SiC can be deposited using CH_3SiCl_4 with H_2 at 1200°C and Ar as an inert diluent [30], but this high temperature does degrade the fibers. The reaction can be represented as:

$$2CH_3SiCl_4 + H_2 \rightarrow 2SiC + 8HCl$$

However, this is considerably oversimplified and $SiCl_2$ and $-CH_2$ are also formed [31] which can react further:

$$SiCl_2 + H_2 \rightarrow Si + 2HCl \quad \text{and} \quad -CH_2 \rightarrow C + H_2$$

4. BCl_3/H_2 can be used to deposit B_4C [32].
5. TiN can be deposited using $TiCl_4$ with N_2 and a flow of H_2 at 1100°C [29].
6. TiN can also be applied from a solution of $Ti(NMe_2)_4$ in hexane.

16.4.2 Liquid metal transfer agent (LMTA) technique

Carbon fiber is immersed in molten Cu or Sn (liquid metal transfer agent) in which a refractory element (W, Cr, Ti) is dissolved. The molten Cu or Sn must not attack the carbon fiber and the transfer agent is removed by dissolving in liquid Al (e.g. Cu/10%Ti at 1050°C or Sn/1%Ti at 900–1050°C) [33]. This process produces a wire preform, which can be used to fabricate an Al matrix composite.

16.4.3 Cementation

The cementation process can be used to deposit a preferred metal [34]. The carbon fiber is first heat cleaned at 700°C in a vacuum (1.3×10^{-2} to 10^{-3} Nm^{-2}) and activated with 4% CH_3COOH. This is followed by treatment at 95°C with an aqueous salt solution of the metal that is to be deposited (e.g. Cu, Ni or Co) acidified with CH_3COOH, in the presence of a coarse powder (0.5–3 mm) of a more electropositive metal, such as Al, Mg, Zn or Fe. The reaction which takes place can be represented by:

$$3NiSO_4 + 2Mg°\downarrow \rightarrow Mg_2(SO_4)_3 + 3Ni°\downarrow$$

16.4.4 Electroless plating

Electroless plating is carried out without the application of electrical current and the coating is deposited as the result of the controlled reduction of a metal salt in solution, which is catalyzed by the metal being deposited—an autocatalytic process proceeding uniformly with time. Throwing power is exceedingly good, giving a deposit that is quite uniform in thickness, with excellent penetration into the carbon fiber tow bundle. The main metals coated by this technique are Ni and Cu and it must be emphasized that the deposited metal is not pure [35]. Depending on the bath composition, the deposited Ni will contain some 4–13% P when using a hypophosphite reducing agent, whereas using a borohydride ($NaBH_4$) as reducing agent will give a purer Ni with 1.5–5%B.

Nickel can be deposited from $NiCl_2$ using sodium hypophosphite as the reducing agent and controlling the pH with NH_4Cl and sodium citrate acting as buffers to maintain a pH of 9–10 at 85–90°C.

$$NiCl_2 + H_3PO_2 + H_2O \rightarrow Ni°\downarrow + H_3PO_3 + 2HCl$$

Copper can be deposited from $CuSO_4$ by initially activating the fiber with $SnCl_2$, followed by $PdCl_2$ using an ultrasonic bath to obtain good penetration of the fiber bundle, followed by $CuSO_4$ chelated with EDTA (ethylenediaminetetraacetic acid, $\{(HOOCCH_2)_2NCH_2CH_2N(CH_2COOH)_2\}$), buffered with Rochelle salt (potassium sodium tartrate, $\{KOOCCH(OH)CH(OH)COONa\}$) and NaOH, used to maintain the pH at 10–11, with

HCHO as the reducing agent. Care should be taken as HCHO is a known carcinogen [36].

$$SnCl_2 + PdCl_2 \rightarrow SnCl_4 + Pd°\downarrow$$

$$Pd°\downarrow + CuSO_4 \rightarrow Cu°\downarrow + PdSO_4$$

Suzuki et al [37] compared electroless Ni and Cu coated carbon fiber reinforced with Al, fabricated by the centrifugal pressure infiltration method and found the average bending strength of the Cu plated Al composite (fiber V_f 33%) was 491 MPa whilst the Ni coated (fiber V_f 37.6%) was 278 MPa.

Huang and Pai [38] determined the optimum conditions for the electroless plating of Ni on carbon fibers for the EMI shielding of ENCF/ABS composites.

16.4.5 Electroplating

Donovan and Watson-Adams give details of the formation of composite materials by electrodeposition [39] by Ni plating using a $NiSO_4$, $NiCl_2$, H_3BO_3 Watts bath, operated at 37°C, pH 3.5–4.0 and a current density of 1.5 A dm^{-2}. An organic leveling agent was found to introduce a degree of brittleness and addition of Cd was found to introduce a marked leveling effect. Composites were prepared by compacting, using isostatic compression of 28 MPa for 1 h at 400°C. Borruso [40] describes the coating of carbon fibers.

Using carbon fiber made at 1000°C and 2700°C, Jackson and Marjoram electroplated Ni [41,42], and Co [42]. They found that structural recrystallization of the carbon fibers occurred when heated in excess of 1000°C for 24 h, with carbon atoms detaching from the carbon fibers and dissolving in the Ni or Co, then rapidly diffusing through the metal and re-precipitating onto the fibers at a new site and growing there in a more fully graphitic form. More orderly graphitized carbon fibers resist crystallisation better than carbonized fibers, since the stable structure is more resistant to dissolution in the Ni or Co matrix.

A conventional Watts type bath [43] can be used to electroplate Ni using $NiSO_4$, $NiCl_2$ and H_3BO_3—the Watts bath being preferred since it gives deposits with a very low S content, which if present, tends to give rise to temper and temperature brittleness. Abraham and co-workers [44] controlled the pH in the range 3.5–4.0 at 35–39°C and describe studies on Ni coated carbon fibers and their composites. An American Cyanamid patent [45] claims that when electroplating a metal such as Ni, the external voltage must be high enough to dissociate the metal at the core and nucleate the metal through the boundary layer into direct contact with the core. Hence the metal ions are driven through the boundary layer on the surface of the fiber allowing sufficient time to produce a thin firmly adherent coating.

The tensile strengths of electroplated Ni continuous carbon fibers have been determined and examined by Weibull analysis [46].

Copper can be coated on carbon fibers [47] and Zhu and workers have used a three-step electrodeposition process for the fabrication of Cu composites [48].

Cheng and co-workers have coated carbon fiber by chemical silver plating [49].

16.4.6 Solution coating

Carbon fiber is passed through a furnace to burn off any size and then coated by passing through an ultrasonic bath containing a solution of an organosilicon compound in toluene, where the application of ultrasonics aids the penetration into the fiber bundle. The coating on the fiber is then hydrolyzed to an air stable SiO_2 by passing through a chamber into which steam has been injected, finally drying in a furnace with an atmosphere of Ar to vaporize the remaining solvent and H_2O and pyrolyze any unhydrolyzed organosilicon [20–22].

The organometallic can be a range of alkoxides of general formula $M(OR)_x$, where R is a hydrocarbon group and x denotes the oxidation state of the metal atom M, e.g. tetraethylorthosilicate:

$$Si(OC_2H_5)_4 + 2H_2O \rightarrow SiO_2 + 4C_2H_5OH\uparrow$$

$$Si(OC_2H_5)_4 \rightarrow SiO_2 + 2C_2H_5OH\uparrow + 2C_2H_4\uparrow$$

By fixing the temperature and time of immersion in the bath, the uniformity and thickness (70–150 nm) of the amorphous deposit of SiO_2 (plus some carbon) can be controlled by Auger depth profiling [20–22].

When SiO_2 coated carbon fiber (Toray T-300) was infiltrated by molten Mg, the interfacial layer was found to contain MgO and magnesium silicates, $MgSiO_3$ and Mg_2SiO_4. If a high modulus carbon fiber (P-100) was used, then the bond was poor, which was resolved by first burning off any size, then applying a petroleum pitch ultrasonically from a toluene solution and finally evaporating the solvent and pyrolyzing the pitch at 550°C in Ar to give a thin coating of amorphous carbon [20–22].

TiO_2 can be deposited from titanium isopropoxide using the sol gel process [50]. The most effective air stable coating for carbon fiber in an Al matrix is a mixed B-Si oxide applied from organometallic solutions [20–22].

16.4.7 Flux

The function of a flux is to clean the metal surface prior to coming into contact with the molten matrix. A suitable water soluble flux is K_2ZrF_6 [20–22,51–55]:

$$3K_2ZrF_6 + 4Al \rightarrow K_3AlF_6 + 3KAlF_4 + 3Zr$$

$$3K_2ZrF_6 + 4Al \rightarrow 6KF + 4AlF_3 + 3Zr$$

$$Zr + 3Al \rightarrow Al_3Zr$$

$$Zr + O_2 \rightarrow ZrO_2$$

It is postulated that the coating of Al_2O_3 is thin, permitting the K_2ZrF_6 to react with the Al and the K_3AlF_6 product formed in the reaction dissolves any additional Al_2O_3 layer [53]. Although K_2ZrF_6 improves wetting out the carbon fiber, it unfortunately, lowers the tensile strength during the infiltration process [54].

$ZrOCl_2$ can be used to deposit ZrO_2 [56], which is useful for Al matrix composites.

16.4.8 INCO Ni coated carbon fiber

INCO have developed [57] a method of coating carbon fiber with Ni using an adaptation of the carbonyl process, involving the thermal decomposition of nickel carbonyl gas:

$$Ni(CO)_4 \rightleftarrows Ni + 4CO\uparrow$$

Deposits of 99.87% Ni can be coated on unsized carbon fiber to give products coated with 20–55 wt% Ni (Incofiber®).

16.4.9 Other coating processes

1. SiO_2 can be applied by heat cleaning the fiber at 475°C, immersing in an ultrasonically agitated bath of tetraethyl orthosilicate $\{Si(OC_2H_5)\}$ with $SiCl_4$ at 25–30°C. Subsequent treatment with steam converts the silicon compounds to a hydrated form of SiO_2 and H_2O is removed by calcining at 550°C. Silica coated carbon fiber reacts with Mg:

$$2Mg + SiO_2 \rightarrow 2MgO$$

$$2Mg + 2SiO_2 \rightarrow MgSiO_4 + Si$$

$$2Mg + 3SiO_4 \rightarrow 2MgSiO_3 + Si$$

Improved bonding at the interface can be achieved by the deposition of pyrolytic carbon directly onto the carbon fiber [58] and poor adhesion to P-100 fiber and other high modulus fibers can be improved by an intermediate coat of amorphous carbon [59].

2. TiO_2 can be coated on carbon fiber by the sol gel process using 2-propanol $((CH_3)_2CHOH)$ diluted with H_2O and acidified with HCl containing titanium isopropoxide $(Ti[OCH(CH_3)_2]_4)$ combined as a sol. The coating is applied in two stages—the first firing at 400°C in air, followed by 700°C in CO to prevent formation of carbide. This coating, used with an Al matrix, forms an Al_2O_3-Ti layer without the formation of carbides [60]. Alternatively, Ti can be deposited by electroplating using a solution of titanium di(dioctylpyrophosphate) oxyacetate, drying at 80°C and heating at 330°C [61].
3. ZrO_2 can be applied by dipping the fiber in a solution of $ZrOCl_2$, drying at 80°C and heating at 330°C [61].
4. Vacuum deposition has been tried, but due to the problem of shadowing, it is essential to open out the tow as a preliminary step, therefore, it has not proved successful.
5. Coating by thermal decomposition of organometallic compounds has been tried by Harwell, who have used tri-isobutyl aluminum [62].

16.5 METAL MATRICES

Matrices reinforced with carbon fibers include Al, Mg, Cu, Ni, Ti, Pb and Sn alloys. In practice, pure metals are not used and it is common to use alloys possessing improved mechanical properties [63].

The wetting of carbon fibers by molten metals can be improved by the addition of elements such as Mg, Cu, Fe, Pb, In and Tl that form lower melting point alloys [64].

Carbon fiber metal matrix composites are subject to corrosion, as is the unreinforced matrix [65], and must be provided with some form of surface protection.

16.5.1 Aluminum

Aluminum is the most common matrix material and Bushby and co-workers [66] have investigated the manufacture of carbon fiber reinforced Al alloy, whilst the surface reactions of carbon fibers in light weight metal matrix composites have been investigated [67]. Hall and Manrique [68] studied the surface treatment of carbon fibers to be incorporated in Al alloy matrix composites and a combined process of coating and hybridizing has also been employed [69].

Uncoated as well as carbon fibers coated with spin-on-glass and reaction bonded silicon oxycarbide were embedded in Al metal at 1000°C by Balaba et al [70]. While both silica based coatings protected the carbon surface, no wetting was observed, leading to fiber pull-out. However, when the coated fibers were treated with a mixture of Ti and B prior to immersion in the molten Al, complete wetting of the fibers occurred. In the presence of molten Al, the Ti-B coating enabled the exothermic formation of TiB_2 and titanium aluminides, which facilitated the wetting. This reaction is termed ASPIRE (Aluminum Self-Propagating Interfacial Reaction) and in combination with silicon based ceramic coatings, provides a scientific approach to the formation of stable carbon fiber-Al metal matrix composites.

The addition of 1%Pb, or 1%In, or 1%Tl promotes the wetting of Al on carbon fibers [71].

Carbon fiber reinforced Al composites were cast by Chen et al [72], without the application of pressure, using K_2ZrF_6 as the wetting promotion agent. The effect of Si alloying (2, 6 or 12 wt%) on the interfacial reaction and matrix morphology was investigated. Results showed that the interfacial reactions were very active after K_2ZrF_6 treatment due to the diffusion and reaction of Zr in the carbon fiber surface or in the SiC coating. The Si alloying of Al suppressed the interfacial reactions. A perfect interface was achieved in SiC-coated carbon fiber-Al-12 wt% Si composite.

Although carbide formation (Al_4C_3) starts early at about 500°C, it only wets out the carbon at temperatures around 1000°C, when the molten Al is then able to penetrate the oxide film. Wetting out at a lower temperature can, however, be achieved by coating the fiber, or adding alloying elements to lower the melting point of the matrix, which can also reduce carbide formation. The carbides are present as needles projecting through the interphase, with the greater majority forming nuclei at the carbon edge planes. Most of the Al matrix is bonded to surface carbon atoms by relatively low energy interactions. Carbide growth is more dominant with PAN based fibers. Yang and Scott [73] have shown that the formation of carbides is a diffusion controlled mechanism, with the carbon atoms dissociating from the carbon surface into the Al matrix. A typical Al alloy used is Al 6061 (0.04–0.35% Cr; 0.15–0.40% Cu; <0.7% Fe; 0.8–1.2% Mg; <0.15% Mn; 0.40–0.8% Si; <0.15% Tl; <0.25% Zn).

Above 500°C, degradation of the tensile properties occurs because of the rapid formation of Al_4C_3 at the interface [69].

Lacom, Degischer and Schulz [67] have identified three ways of suppressing fiber degradation:

1. Minimize infiltration time and temperature.
2. Replace Al matrix with a Mg based alloy with Al additions to control the amount of interface carbides.
3. Deposit a P compound on the fiber surface to reduce the surface activity.

Torayca M40B-50B and Grafil 12M-AG carbon fibers were desized by heating at 200–420°C in an inert gas and coated with tributylphosphate $\{CH_3(CH_2)_3O\}_3P(O)\}$ and tritolylphophate $\{[CH_3C_6H_4)_3O]_3P(O)\}$. The findings were that an infiltration time of 5 s at 670°C in a pure Al matrix reduced the carbide content to 0.2% (Figure 16.3). The mode of failure and flexural strength depended on the carbide content, with value of 1.5 GPa at 0.2%, close to the value predicted by the Rule of Mixtures (Figure 16.4). By diluting the Al matrix

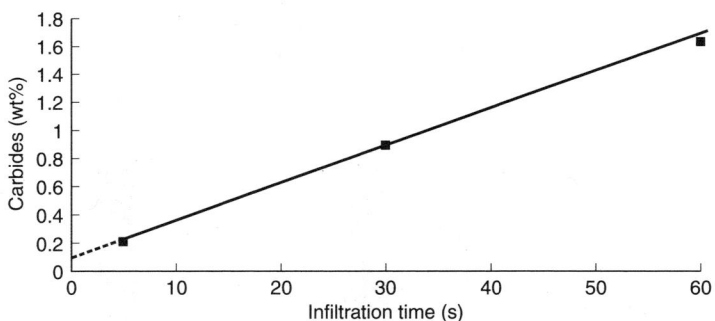

Figure 16.3 Carbide weight fractions in Al/carbon fiber samples as a fraction of infiltration time. *Source:* Reprinted with permission from Lacom W, Degischer HP, Schulz P, Assessment and control of surface reactions of carbon fibres in light weight metal matrix composites, *Key Eng Mater*, 127(l,2), 679–686, 1997. Copyright 1997, Trans Tech Publications.

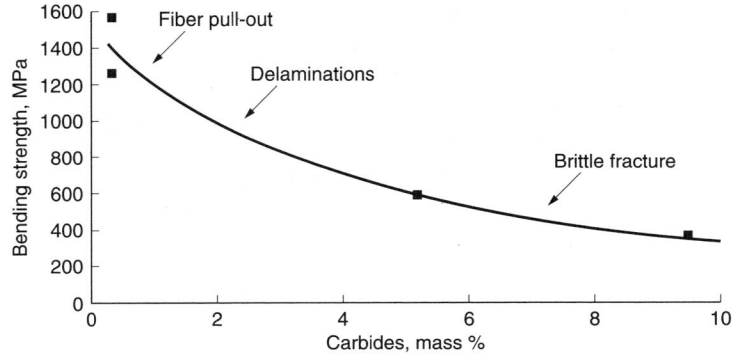

Figure 16.4 Flexural strength of carbon fiber reinforced Al for interface carbides from 0–10% showing different failure mechanisms as determined by fractography. *Source:* Reprinted with permission from Lacom W, Degischer HP, Schulz P, Assessment and control of surface reactions of carbon fibres in light weight metal matrix composites, *Key Eng Mater*, 127(l,2), 679–686, 1997. Copyright 1997, Trans Tech Publications.

with a Mg alloy containing 6–9% Al enabled the achievement of a strength of 1.9 GPa. The P pretreatment increased the temperature of onset of fiber oxidation from 530°C to 700°C, associated with an increase in resistance to the formation of Al_4C_3 (Figure 16.5).

Abraham *et al* [36] have studied Cu coatings on carbon fibers and Zeng has used Al_2O_3 [74] for Al based composites.

To confer a measure of corrosion protection to the Al matrix composite, a surface protective treatment typically used with unreinforced Al can be applied and includes such processes as anodizing with H_2SO_4 to give a 10–30 μm thick oxide coating, which is then sealed using $Na_2Cr_2O_7$.

More effective protection can be achieved by converting the oxide coating to a chromate, phosphate or a complex oxide and even better protection can be achieved by applying a coating of Ni using an electroless plating route, followed by the electrodeposition of an Al-Mn coating [75], whilst limited protection can be achieved with the application of a 10 μm thick polymer coating [76] (Table 16.2).

Kyono and co-workers [77] have studied the compatibility between Torayca carbon fiber and an Al matrix.

The pressure casting technique was used by Cheng and workers [78] to fabricate carbon fiber reinforced Al composites with the hybridization of a small amount of particulates or

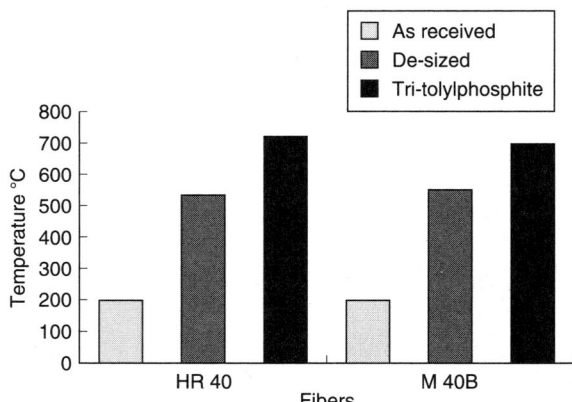

Figure 16.5 Characteristic temperatures for the onset of desizing HR40 and M40B carbon fibers; onset of oxidation of the desized fiber and after impregnation with tri-tolylphoshite. *Source:* Reprinted with permission from Lacom W, Degischer HP, Schulz P, Assessment and control of surface reactions of carbon fibres in light weight metal matrix composites, *Key Eng Mater*, 127(I,2), 679–686, 1997. Copyright 1997, Trans Tech Publications.

Table 16.2 Properties of carbon fiber reinforced 6061 aluminum alloy compared to carbon fiber reinforced epoxy

Property	Graphite/epoxy	P-100/6061	P-55/6061
α_{11} (10^{-5} K^{-1})	−0.080	0.086	0.307
α_{22} (10^{-5}/ K^{-1})	3.67	2.30	2.40
κ_{11} (W m^{-1} K^{-1})	54.0	240.0	98.0
κ_{22} (W m^{-1} K^{-1})	0.7	193.0	98.0
ν_{12}	0.21	0.4	0.27
ν_{21}	0.010	0.031	0.041
E_{11} (GPa)	172	352	213
E_{22} (GPa)	8.07	28	32
E_{12} (GPa)	4.28	14	13

Source: Reprinted with permission from Dacres CM, Reamer SM, Sutula RA, Larrick BF, *Proc Electrochem Soc*, 83(1), 76–95, 1983. Copyright 1983, The Electrochemical Society, Inc.

SiC whiskers. The particulates or whiskers were uniformly distributed among the carbon fibers and preforms prepared from the treated fibers were directly infiltrated by molten Al under applied stress. It was found that the longitudinal tensile strengths of hybrid composites were greatly improved, although their fiber volume fractions were quite low compared to those of conventional composites. With this hybridization method, it is also practical to tailor the fiber volume fraction of composites from 60 to 25 vol%, which is not possible in direct infiltration of fiber preforms by pressure casting. The results obtained lead to the conclusion that particulate or whisker additions do not act directly as reinforcements, but as promoters to improve the infiltration performances of fiber preforms and consequently, increase the strength transfer efficiency of carbon fibers. The addition of particulates or whiskers can also improve other properties of the composites, such as hardness and wear resistance.

Nardone and Strife [79] determined the tensile and compressive behavior of P100 graphite fiber reinforced 6061 Al in the [0/60]s configuration as a function of orientation. The experimentally determined values of the elastic modulus and strength of the composites were compared with those predicted from classical laminate theory. In general, there was good agreement between the predicted and observed values for testing in the longitudinal orientation, but discrepancies existed for testing in the transverse and 45° orientations.

By successive CVD processes, carbon fiber is first coated with a carbon film, which contains a little SiC [80], then with a film of pure SiC and finally with TiB. This sequence of layers ensures that the fiber has a large affinity to Al, while still maintaining its strength.

Short carbon fiber Al matrix composite material can be prepared by extrusion of powder mixtures [81].

Degischer [82] investigated metal matrix composites using foamed Al.

The possibility of increasing the specific properties of recyclable light metals has been described by using ceramic particulates, continuous ceramic or carbon fibers, or by the reduction of weight by foaming the metal. Examples of castings, extrusions and forgings of particulate reinforced (<30 vol%) Al alloys are given and their advantages including stiffness and wear resistance are presented. High volume fractions (>40 vol%) of reinforcements can be produced by gas pressure infiltration of either particulate or fiber preforms. In the case of Al matrix, the specific strength can be increased by a factor of up to 15, and the specific stiffness by a factor of up to 7, whereas for carbon fiber reinforced Mg, the specific strength can be increased even more. The anisotropy of fiber reinforced metal matrix composites is discussed and the possibilities of using cross ply preforms.

The production and properties of reinforced cast metal matrix composites has been described using carbon fiber particles. It is interesting to note that the thermal stresses created by thermal cycling exceed the stress applied during the cycling process due to the mismatch of thermal expansions between the fiber reinforcement and the Al matrix. Hence, the composite creeps during the cooling part of the cycle, which is the reverse to the norm of creep occurring during the application of load [83].

Cheng et al [84–86] have studied the tensile strength and fiber degradation of carbon fiber in Al matrix composites, whilst Friler et al [87] studied strength and toughness.

The fabrication of Cu or Ni coated carbon fiber reinforced Al using a centrifugal pressure infiltration method has been described by a member of NIKKEI Techno-Research [88]. The centrifugal casting technique appears to be an innovative method of fabricating composite materials.

The rate of formation of intermetallic compounds in Al matrix carbon fiber composites has been determined by Okura and Motoki [89].

16.5.2 Magnesium

Magnesium has a density lower than Al, but is more susceptible to corrosion and like Al, requires a protective treatment. Katzman [59] examined suitable treatments for the fabrication of graphite reinforced Mg composites. The tensile properties of coated carbon fiber reinforced Mg composites have been determined by Zhang and co-workers [90]. A study of the compatibility between PAN based carbon fibers and Mg_8Li alloy during the pressure infiltration process was undertaken [91] and Chen and co-workers [92] examined the interface in Mg matrix composites reinforced with carbon fiber. A Ni coated carbon fiber in a Mg matrix will form Mg-Ni compounds and a low melting point eutectic (508°C) [93].

In the Mg alloy AZ-91(9%Al, 1%Zn), a uniform dispersion of $Mg_{17}Al_{12}$ is formed in the Mg rich matrix [26].

16.5.3 Copper

Copper and Cu alloys reinforced with carbon fibers can take advantage of the low electrical resistivity and high thermal conductivity for electrical and electronic applications.

However, the high composite density would be a disadvantage for aerospace. The compatibility of carbon with Cu alloys containing Cr, Ti or V is discussed by Mortimer *et al* [94] and it was found that significant amounts of carbides are formed when Cu is alloyed with Cr, Ti or V to induce wetting of carbon. The lubricity of the carbon fiber does confer good wear resistance, which is most beneficial for electrical brushes [95,96]. Processing routes for carbon fiber reinforced Cu matrix composite are detailed by Le Petitcorps *et al* [97]. Hot pressing Cu coated carbon fiber is the most suitable route for a Cu matrix composite [98].

Hot pressing Cu coated carbon fiber is best accomplished in vacuum or an atmosphere of H_2-N_2 at 700–1000°C using a pressure of 10–25 MPa [99–101].

Despite the high mp of Cu (1082°C), successful reinforcement can be achieved by infiltration [102], although reinforcement is more conveniently attained by hot pressing Cu plated carbon fiber.

Copper can be alloyed with Mo, Cr, V, Fe and Co to improve the wettability [103].

The thermal expansion behavior of unidirectional carbon fiber reinforced Cu matrix composites has been determined by Korb *et al* [104].

16.5.4 Nickel

Unfortunately, Ni catalyzes the crystallization of carbon fibers when heated and the high modulus carbon fibers are more resistant to recrystallization than the high strength types [41,42]. The effect of 1 h of heat treatment in the range 800–1300°C in Ar on the tensile strength, modulus and recrystallization of electroless Ni plated carbon fibers was determined by Warren [105], who showed that the high modulus PAN type carbon fibers were less affected.

Jackson and Marjoram [41] have suggested that the recrystallization of carbon fibers in contact with Ni at about 1000°C depends on the transport of carbon through the Ni. Presland and Walker [106] suggested increased mobility of the carbon at the Ni-carbon interface.

16.5.5 Lead

Carbon fibers have been successfully used to reinforce Pb alloys [107] using liquid metal infiltration and used for the positive electrode of rechargeable lead acid batteries.

16.5.6 Tin

Carbon fiber composites with a Sn-Pb alloy matrix have been studied by Old and Nicholas [108] and Ho [109–112]. Old and Nicholas found that full theoretical strengths could be obtained with matrices in the range Sn 20–40 wt% /Pb reinforced with 9 vol% Type II carbon fiber. Increasing the proportion of Sn decreased the coefficient of friction and improved the wear rate [113].

Ho and Chung [114,115] used unidirectional and continuous carbon fiber Sn matrix composites for the packaging of the high temperature superconductor $YBa_2Cu_3O_7$ delta by diffusion bonding at 170°C and 4 MPa. The Sn acted as the adhesive and increased ductility, normal state electrical conductivity and thermal conductivity. Carbon fibers served to increase the strength and modulus, both in tension along the fiber direction and in compression perpendicular to the fiber layers, though they decreased the strength in compression along the fiber direction. Carbon fibers also served to increase the thermal conductivity and thermal fatigue resistance. At 24 vol% fibers, the tensile strength was approximately equal to the compressive strength perpendicular to the fiber layers. With a further increase of the fiber content, the tensile strength exceeded the compressive strength perpendicular to the fiber layers, reaching 134 MPa at 31 vol% fibers. For fiber contents less than 30 vol%, the compressive ductility perpendicular to the fiber layers exceeded that of the plain superconductor. At 30 vol% fibers, the tensile modulus reached 15 GPa at room temperature and 27 GPa at 77K. The tensile load was essentially sustained by the carbon fibers and the superconducting behavior was maintained after tension, almost to the point of tensile fracture.

16.6 TECHNIQUES FOR FABRICATING CARBON FIBER REINFORCED METAL MATRIX COMPOSITES

A number of useful reviews detailing composite fabrication have been written [43,116–122]. The general methods for forming carbon fiber reinforced metal matrix composites are described [118,123,124].

16.6.1 Factors influencing processing of metal matrix composites

16.6.1.1 Capillary effects

Generally, molten metals do not wet the carbon fiber surface and it is necessary to undertake some form of fiber surface treatment or apply a hydraulic pressure to ensure that the molten metal does penetrate the fiber bundle.

16.6.1.2 Fluid flow into the preform

The viscosity of the molten metal controls the penetration between the fiber interstices and although increasing the temperature will lower the viscosity, it will also increase oxidation of the carbon fiber. The requisite viscosity of molten metals should be about the same as the viscosity of H_2O at 20°C (1 cP) (Figure 15.9). Freeze choking must be avoided during the infiltration process caused when the infiltration velocity drops to a level where the fibers ahead of the infiltration front extract sufficient heat to solidify the metal melt [118].

16.6.1.3 Fiber matrix interactions

Above 500°C, carbon reacts with Al to form large hexagonal plates of Al_4C_3, which degrade the strength [125–128] and moreover, the Al_4C_3 is hygroscopic, affecting corrosion resistance [129]. Alloy additions to the matrix can promote wettability.

The residual strength and impact behavior of the fiber/matrix on carbon fiber reinforced metal laminates has been studied [130,131].

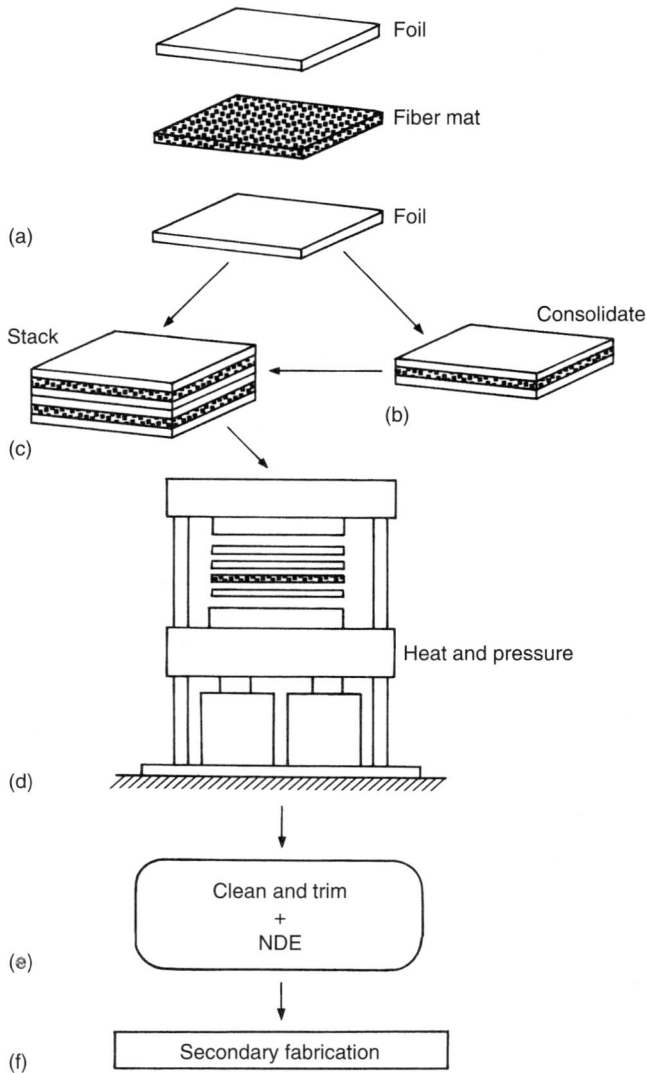

Figure 16.6 Diffusion bonding. (a) starting components: fiber mat and sheets of foil. (b) form ply (sometimes the ply is consolidated). (c) plies are stacked. (d) hot press (e) and (f) finishing. *Source:* Reprinted with permission from Matthews FL, Rawlings RD, *Composite Materials*, CRC Press, Boca Raton, 78–117, 2000. Copyright 2000, CRC Press, Boca Raton, Florida.

16.6.1.4 The solidification process

Porosity in the microstructure is detrimental to the properties of the cast metal matrix and is mainly caused by shrinkage occurring during solidification.

16.6.2 Processing methods for fabricating metal matrix composites

There are three basic routes for manufacturing metal matrix composites, depending on whether the matrix is in the solid, or liquid form as well as on the application of an atomized melt.

CARBON FIBERS IN METAL MATRICES

16.6.2.1 Solid state processing methods

1. Powder metallurgy

Powder metallurgy is particularly suitable when the constituents are chopped fiber and a particulate matrix. This technique has been used for Al, Sn, Pb-Sn, Cu and Ni. The constituents are mixed well in an inert atmosphere, placed in a mold, degassed, pressed and heated.

As a general rule, it is necessary to use as high a temperature as possible, but too much pressure will damage the fiber and it is normal to limit the fiber content to 25%. To obtain uniform pressure, it is preferable to place the materials in a can, which is sealed by electron beam welding, and can then be processed in an isostatic press.

Extrusion and forging are other routes for processing material in particulate form, whilst coextrusion and drawing can be used for coated fiber (wire). In extrusion, a powdered metal and chopped fiber are dispersed, placed in cans and then hot extruded to give a solid bar with aligned fibers.

Yih and Chung [132,133] used coated fillers to make MMC by powder metallurgy.

The low cost route, using powder metallurgy for a Ti matrix composite has been recorded [134].

2. Diffusion bonding

A stack of fiber mat held in place by a polymer binder is interleaved with metal foil and pressed at a low temperature at about 24 MPa to cause bonding in the solid state (Figure 16.7). To avoid oxidation of the metal foil, a vacuum can be used, or the fiber can be metal coated, in which case, metallic foils are not used. The metals Al, Cu, Ni and Ti

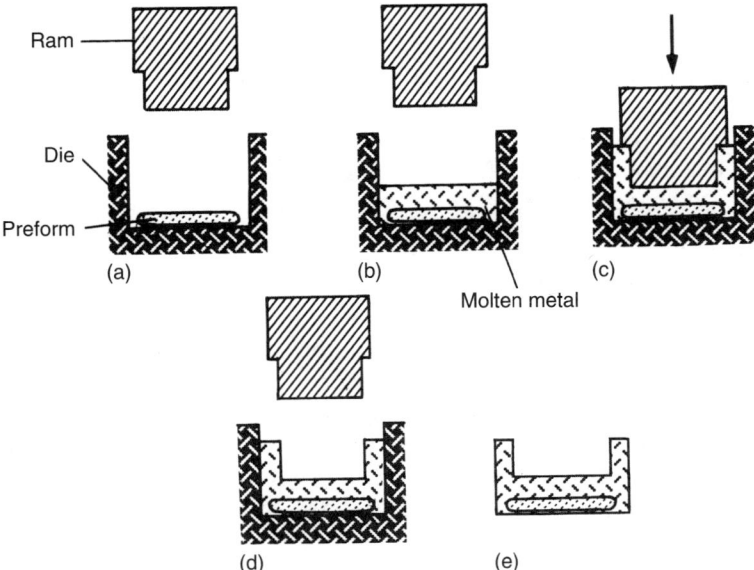

Figure 16.7 Squeeze casting. (a) insert preform into die cavity (b) meter in a precise quantity of alloy (c) close die and apply pressure (d) remove ram (e) extract component. *Source:* Reprinted with permission from Matthews FL, Rawlings RD, *Composite Materials*, CRC Press, Boca Raton, 78–117, 2000. Copyright 2000, CRC Press, Boca Raton, Florida.

can be used to prepare metal matrix composites, but although the process is expensive, it does employ lower temperatures than other forms of liquid state processing.

16.6.2.2 Liquid state processing [136,137]

1. Melt stirring [135]

This is a basic method involving the uniform mixing of chopped fiber into a molten matrix and casting the mix.

2. Compocasting or rheocasting

A modification of melt stirring, termed compocasting (or rheocasting), is used to obtain a suitable slurry by employing powerful agitation of a semisolid (two phase) alloy to yield a non-dendritic phase, which imparts thixotropic properties. The mix is then infiltrated to produce a near net shape component [138–139]. It is desirable that the alloy has a wide solidification range in which it exists as a semisolid. The volume fraction of carbon fiber is restricted to about 20% to obtain effective dispersion.

3. Slurry casting

Pai and workers [140] used a semi solid slurry process for making short carbon fiber dispersed Al alloy matrix composites.

4. Gravity or vacuum casting

Casting by gravity or the use of a vacuum by conventional techniques can be undertaken if the molten alloy wets out the carbon fiber [141–143].

5. Pressure casting

Initially, a vacuum can be applied, followed by application of pressure up to 15 MPa by using a hydraulic ram or a pressurized gas such as Ar [144–146]. This process is used to aid wetting out of large preforms and woven cloth [90,147]. The fiber is limited to 30 vol%, as otherwise, effective penetration of the fiber preform cannot be achieved and the fiber could be displaced and damaged.

6. Squeeze casting

Squeeze casting, depicted in Figure 16.8, is akin to forging by placing the preform and the exact quantity of molten alloy in a matched die, the die is closed and a ram applies a high pressure of 70–100 MPa to penetrate and wet out the preform, allowing the alloy to solidify under pressure [148–157]. Good infiltration is achieved and although high pressures collapse the gas voids and reduce shrinkage porosity, there is a tendency to displace the fiber. This process can only be used for smaller parts due to the limitation of the press size, unless pressure is applied by a gas instead of mechanically by a ram. The preform is placed in a die, the die closed and a vacuum applied to the molten metal, which is then forced by gas into the mold to obtain infiltration and subsequently allowing sufficient dwell time to permit solidification to occur.

The process has been modified by Li *et al* [158] to work under vacuum, incorporating the continuous removal of Al_2O_3 by passing the melt through a Fiberfrax ceramic filter—less oxide will be formed under vacuum.

CARBON FIBERS IN METAL MATRICES

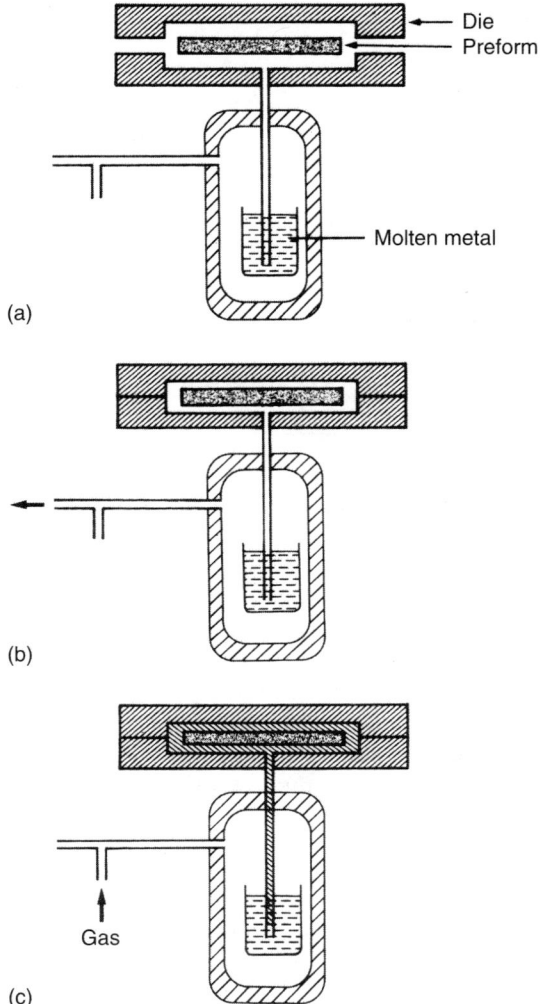

Figure 16.8 Liquid metal infiltration under gas pressure. (a) insert preform and close die (b) evacuate air (c) apply gas pressure and maintain during solidification. *Source:* Reprinted with permission from Matthews FL, Rawlings RD, *Composite Materials*, CRC Press, Boca Raton, 78–117, 2000. Copyright 2000, CRC Press, Boca Raton, Florida.

7. Fiber tow (liquid) infiltration

Figure 16.9 depicts liquid infiltration where the molten metal is evacuated and forced into the preform chamber under pressure, where it infiltrates the perform. The pressure is maintained until solidification has occurred. Yang and Chung [159] and also Chiou and Chung [160] used vacuum infiltration of a liquid metal in an inert gas under pressure.

Infiltration occurs due to capillary attraction within the fiber preform and can be undertaken at atmospheric pressure by passing a continuous tow through a bath of molten metal, but is practical only if the molten alloy wets out the carbon fiber [161–168]. A plated coating used to aid wetting must not be too thin; otherwise it will be dissolved completely before infiltration has been completed. The application of pressure aids infiltration, but if the preform is compressed too much, then infiltration is prevented, limiting the fiber fraction to about 30 vol%. Composites with Pb alloy matrix can be prepared by an infiltration technique

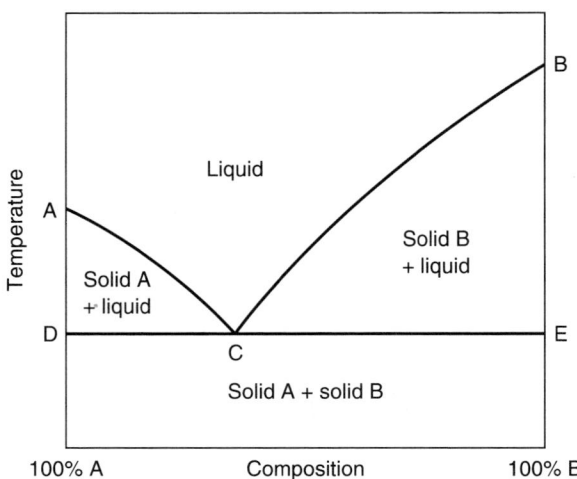

Figure 16.9 Solid-liquid equilibria. Curve ACB—liquidus curve; Curve ADEB—solidus curve. The two curves meet at C the eutectic.

[169]. Ultrasonics has been used to aid the reinforcement of an Al matrix using the infiltration technique [170].

The infiltrated tows are assembled into a preform and consolidated by diffusion bonding or hot molding in the two-phase liquid plus solid region.

A Ti-Cu composite can be achieved by infiltration by heating carbon fiber, Ti wire and Cu ribbon in the region of the liquidus temperature of the Ti-Cu alloy [171].

8. Lanxide process

The Lanxide Corporation (now intellectual property within MSE Inc.) have patented the PRIMEXTM pressureless metal infiltration process and the PRIMEX CASTTM foundry process, which are able to infiltrate ceramic reinforcements with molten metals without application of pressure or vacuum and it is believed that the processes can be used with carbon fiber.

9. Liquid phase hot pressing, liquid phase diffusion bonding or liquid phase sintering

All the above processes [172–179] have also been investigated.

Primex concentrate can be produced by a process involving the pressureless infiltration of Al into a carbon fiber preform. The fibers and metal (powder or foil) are intimately mixed and the temperature raised above the solidus temperature of the metal (Figure 16.6), enabling the use of pressures lower than in the diffusion bonding process, but higher temperatures are associated with fiber degradation. Sakamoto *et al* [180] avoided this problem by using a metal foil of lower solidus than the matrix metal, interleaved alternately between the wire and preform layers, and hot pressing at a temperature between the two solidus temperatures. Rabinovitch *et al* [181] fabricated Al and Mg composites by liquid hot pressing. Using AZ61 (Mg, 6Al, 0.6Zn, 2Mn) and a pressure of 10–20 MPa gave pores, using squeeze casting employing higher pressures gave good infiltration, but with the disadvantage that the fiber could be displaced.

Carbon fiber can also be mixed with metal powders (powder metallurgy), degassed, preferably canned and then heated under pressure (50 MPa) at about 760°C, limiting the fiber content to about 25% by the use of excessive pressure.

CARBON FIBERS IN METAL MATRICES

Guruprasad and workers [182] have described the production and properties of carbon fiber particle reinforced cast metal matrix composites.

16.6.2.3 Deposition processes

1. Ion plating

Ion plating was used by Yoshida and co-workers [183] to make carbon fiber reinforced Al.

A typical ion plating procedure is to spread the carbon fiber with an air knife, feed into a vacuum chamber at about 2 Pa in Ar, apply a potential of 0.5 kV to form an Ar plasma. The Al wire fed is into a crucible where it is evaporated onto the moving fiber. The coated fiber formed by ion plating can be processed by hot pressing/diffusion bonding.

A U.S. Navy patent [184] describes ion plating a mat of graphite fibers with Al and exposing to an organic solvent (methylene chloride, toluene, benzene or heptane), which is absorbed by capillary reaction and the graphite fibers expand and straighten. The excess solvent is removed from the surface and the treated fiber coated with Mg from the vapor at 350–450°C and a composite prepared by densification.

2. Plasma spraying

Metal is plasma sprayed onto continuous carbon fiber, forming discrete droplets of metal, giving rise to an extremely porous composite, which can be consolidated by isostatic pressing.

Wielage and Rahm [185] prepared Al prepregs by using high velocity flame spraying.

SiC may also be applied using a plasma coating technique and by varying the plasma voltage, the modulus of the deposited SiC can be controlled. Therefore, a low modulus deposit can be used to increase the interface strength [186–188].

16.6.3 Fundamental considerations

The infiltration of carbon fiber with molten metals and subsequent solidification depends on [118]:

1. Capillarity
2. Fluid flow into the preform
3. Fiber matrix interactions
4. The solidification process

16.6.3.1 Capillarity

It is difficult to wet carbon fibers with molten Al due to the presence of Al_2O_3 at the liquid vapor interface. Ignoring the effect of such oxides, the pressure difference at the liquid metal front resulting from capillarity effects is given by a version of Kelvin's equation [148–150,189]:

$$p = \frac{4\cos\theta}{r}\sigma_{la}$$

where θ = wetting angle on the fiber in the infiltration atmosphere, often assumed to be 180°
r = curvature at the molten metal front
σ_{la} = surface energy at the liquid vapor interface

Since the liquid metal does not wet the fiber, a hydrostatic pressure is applied to obtain penetration into the fiber bundle.

16.6.3.2 Fluid flow into the preform

An additional amount of pressure must be applied to overcome frictional force. The Washburn equation [190] can be used to model frictional forces:

$$\frac{dl}{dt} = \frac{(P_a + P_c)d^2}{32\eta l}$$

where dl/dt = infiltration velocity
  l = pore length
  P_a = applied pressure
  P_c = capillary pressure (from Kelvin's equation)
  d = pore diameter
  η = viscosity of molten metal

If the applied pressure varies throughout the cycle, then too much pressure will compress the fiber preform and affect the volume fraction.

The viscosity of a molten metal (approximately 1 cP) is important and about the same value as that of water at 20°C (Figure 15.9).

Fiber temperature is another important parameter and temperatures must be above the mp of the metal. Otherwise, a choke will occur and infiltration will be prevented. Fukunaga and Goda [151,152] suggested that a solid layer was first formed around the fibers, hence modifying the fiber diameter and volume fraction.

16.6.3.3 Fiber matrix interactions

Carbon reacts readily with Al near its melting point (660°C) and forms Al_4C_3, which damages the fiber matrix and interface. It is possible to form a SiC coating graded from pure carbon on the inside to pure silicon on the outside, achieving a thickness of about 0.5 mm [191,192]. The Si on the surface promotes bonding to the metal matrix, forming large hexagonal plates progressing into the matrix and partly into the fiber surface [126,127,155,139,167,193–197]. Once this stage has been reached, the fiber strength is degraded [126–128,194,195]. To achieve successful wetting, temperatures must be about 1000°C. Unfortunately, Al_4C_3 is hygroscopic and lowers the corrosion resistance. Various alloy additions can be made such as Si, Al_2O_3 and TiB_2 to improve wettability and reduce fiber degradation [155,139,194,198].

16.6.3.4 The solidification process and matrix microstructure

Shrinkage experienced during solidification is the main culprit for the occurrence of porosity [146,165–167,173,175,199].

West et al [200] state that there are three types of microstructure:

1. A network of dendrites much smaller than the fibers
2. The dendrites nucleate on the fibers and grow into the interfiber spaces (occurring with hypereutectic Al-Si alloys, Al_3Si and Al_3Ni)
3. After growth, a primary phase is formed, found on the fibers and between dendrite branches.

When the MMC is cast, it cools to room temperature and as the thermal expansion is generally greatest in the matrix at this point, tensile stresses can build up in the matrix.

Viala *et al* have determined the effect of Mg on the composition, microstructure and mechanical properties of carbon fiber [201].

REFERENCES

1. Chung DDL, *Carbon Fiber Composites*, Butterworth-Heinemann, Boston, 125–144, 1994.
2. Peebles LH, *Carbon Fibers Formation, Structure, and Properties*, CRC Press, Boca Raton, 164–180, 1995.
3. Matthews FL, Rawlings RD, *Composite Materials*, CRC Press, Boca Raton, 78–117, 2000.
4. Chiou JM, Chung DDL, *J Mater Sci*, 28, 1435–1446, 1993.
5. Chiou JM, Chung DDL, *J Mater Sci*, 28, 1447–1470, 1993.
6. Chiou JM, Chung DDL, *J Mater Sci*, 28, 1471–1487, 1993.
7. Lee RN, Carbon fiber surface properties, Lee SM, ed., *International Encyclopedia of Composites*, Vol 1, VCH Publishers, New York, 241, 1990.
8. Shindo A, Chemical property of carbon fiber surface and interfacial compatibility of composites, Ishida H, Koenig JL, eds., *Composite Interfaces*, North Holland, New York, 93, 1986.
9. Suzuki T, The compatibility of pitch-based carbon fibers with aluminum for the improvement of aluminum-matrix composites, *Composites Sci Technol*, 56(2), 147–153, 1996.
10. Bushby RS, Scott VD, Evaluation of Al-Cu alloy reinforced with pitch-based carbon fibers, *Composites Sci Technol*, 57(l), 119–128, 1997.
11. Thaw C, Minet R, Zemany J, Zweben C, *SAMPE J*, 23(6), 40–43, 1987.
12. JM Ting, Lake ML, Vapor-grown carbon-fiber-reinforced aluminum composites with very high thermal-conductivity, *J Mater Res*, 10(2), 247–250, 1995.
13. Diwanji AP, Hall IW, Fiber and fiber-surface treatment effects in carbon aluminum metal matrix composites, *J Mater Res*, 27, 2093, 1992.
14. Bouix J, Berthet MP, Bosselet F, Favre R, Peronnet M, Viala JC, Vincent C, Vincent H, Interface tailoring in carbon fibers reinforced metal matrix composites, *Journal de Physique*, IV 7(C6), 191–205, 1997.
15. Lawcock GD, Ye L, Mai YW, Sun CT, Effects of fiber/matrix adhesion on carbon fiber reinforced metal laminates - I, Residual strength, *Composites Sci Technol*, 57(12), 1609–1619, 1997.
16. Lawcock GD, Ye L, Mai YW, Sun CT, Effects of fiber/matrix adhesion on carbon-fiber-reinforced metal laminates - II, Impact behaviour, *Composites Science and Technology*, 57(12), 1621–1628, 1997.
17. Bonfield W, The effect of impurity on reinforcement-matrix composites, AG Metcalfe ed., *Composite Materials, Vol 1-Interfaces in Metal Matrix Composites*, Academic Press, New York, 1974.
18. Aggour L, Fitzer E, Heym M, Ignatowitz E, Thin coatings on carbon fibers as diffusion barriers and wetting agents in Al composites, *Thin Solid Films*, 40, 97, 1977.
19. Delanney F, Froyen L, Deruyttere A, Review: the wetting of solids by molten metals and its relation to the preparation of metal matrix composites, *J Mater Sci*, 22, 1, 1987.
20. Katzman H, *Proc Metal and Ceramic Matrix Composite Processing Conf Vol 1*, U.S. Dept. of Defense Information Analysis Centers, 115–140, 1984.
21. Katzman H, Fiber coatings for the fabrication of graphite reinforced magnesium composites, *J Mater Sci*, 22, 144–148, 1987.
22. Katzman H, *Mater Manuf Proc*, 5(1), 1–15, 1990.
23. Kahn IH, The effect of thermal exposure on the mechanical properties of aluminum-graphite composites, *Met Trans*, 7A, 1281, 1976.
24. He C, Zhang G, Wu R, The effects of matrix alloys on the mechanical properties of graphite/aluminum composites, Lin RY, Arsenault RJ, Martins GP, Fishman SG eds., *Interfaces*

in Metal-Ceramics Composites, Warrendale, PA, The Minerals, Metals, & Materials Society, 493, 1990.
25. Lo J, Finello D, Schmerling M, Marcus HL, Interface structure of heat treated aluminum-graphite fiber composites, JE Hack, Amateau MF eds., *Mechanical Behaviour of Metal-Matrix Composites*, The Metallurgical Society of AIME, Warrendale, 77, 1983.
26. Brown LD, Marcus HL, *Proc Metal and Ceramic Matrix Composite Processing Conf Vol 11*, US Dept of Defense Information Analysis Centers, 91–113, 1984.
27. He C, Zhang G, Wu R, The influence of interfacial modification on tensile strength of Gr/Al composites, CG Pantano, Chen EJH eds., *Interfaces in Composites*, Materials Research Society, 170, 257, 1990.
28. He C, Zhang G, Wu R, The effects of matrix alloys on the mechanical properties of graphite/aluminum composites, Lin RY, Arsenault RJ, Martins GP, Fishman SG eds., *Interfaces in Metal-Ceramics Composites*, Warrendale, 493, 1990.
29. Honjo K, Shindo A, Interfacial behaviour of aluminum matrix composites reinforced with ceramic coated carbon fibers, Ishida H, Koenig JL eds., *Composite Interfaces*, North Holland, New York, 101–107, 1986.
30. Honjo K, Shindo A, In: Ishida H, Koenig JL eds., *Proc 1st Compos, Interfaces Int Conf*, North Holland, New York, 101–107, 1986.
31. Sotirchos SV, Papasouliotis GD, Kinetic modelling of the deposition of SiC from methyltrichlorosilane, Besmann TM, Gallois BM, Warren JW eds., *Chemical Vapor Deposition of Refractory Metals and Ceramics II, Materials Research Society Symposium Proceedings*, 250, 35, 1992.
32. Vincent H, Vincent C, Scharff JP, Mourichoux H, J Bouix, *Carbon*, 30(3), 495–505, 1992.
33. Himbeault DD, Varin RA, Piekarski K, In: Monstaghaci H ed., *Proc Int Symp Process Ceram Met Matrix Compos*, Pergamon New York, 312–323, 1989.
34. Kulkarni AG, Pai BC, Balasubramanian N, The cementation technique for coating carbon fibers, *J Mater Sci*, 14, 592, 1979.
35. Goldie W, *Metallic Coating of Plastics*, Vol 1, Electrochemical Publications Ltd, Middlesex, 1968.
36. Abraham S, Pai BC, Satyanarayana KG, Vaidyan VK, Copper coating on carbon fibers and their composites with aluminum matrix, *J Mater Sci*, 27, 3479–3486, 1992.
37. Suzuki T, Umchara H, Hayashi R, Watanabe S, Mechanical properties and metallography of Al matrix composites reinforced by the Cu- or Ni-plating carbon multifilament.
38. Huang CY, Pai JF, Optimum conditions of electrolessNi plating on carbon fibers for EMI shielding effectiveness of ENCF/ABS composites, *Eur Polym J*, 34(2), 261–267, 1998.
39. Donovan PD, Watson-Adams BR, Formation of composite materials by electrodeposition, *Metals Mater*, 3(11), 443, 1969.
40. Borruso M, Coating carbon graphite fibers, *Plating and Surface Finishing*, 79(9), 47, 1992.
41. Jackson PW, Marjoram JR, Recrystallization of nickel-coated carbon fibers, *Nature*, 218, 83, 1968.
42. Jackson PW, Marjoram JR, Compatibility studies of carbon fibers with nickel and cobalt, *J Mater Sci*, 5, 9, 1970.
43. Baker AA, Carbon fiber reinforced metals—a review of current technology, *Mater Sci Eng*, 17, 177, 1975.
44. Abraham S, Pai BC, Satyanarayana KG, Vaidyan VK, Studies on nickel coated carbon-fibers and their composites, *J Mater Sci*, 25(6), 2839–2845, 1990.
45. Morin LG, *Apparatus for the production of continuous yarns or tows comprising high strength metal coated fibers*, U.S. Pat., 4609449, Sep 1986.
46. Soni PR, Rajan TV, Ramakrishnan P, Continuous nickel electroplating of carbon fibers and Weibull analysis of their tensile strengths, *Metals Mater Proc*, 8(2), 187–191, 1996.
47. Bek RY, Zherebilov AF, Copper electrodeposition on carbon fibers, *Russian J Appl Chem*, 68(9), Part 1, 1281–1284, 1995.
48. Zhu Z, Kuang X, Carotenuto G, Nicolais L, Fabrication and properties of carbon fiber-reinforced copper composite by controlled three-step electrodeposition, *J Mater Sci*, 32(4), 1061–1067, 1997.

49. Cheng HM, Zhou BL, Zheng ZG, Wang ZM, Shi CX, Chemical silver plating on carbon-fibers, *Plating and Surface Finishing*, 77(5), 130–132, 1990.
50. Clement JP, Rack HJ, In: *Proc Am Soc Compos Symp High Temp Compos*, Technomic, Lancaster, 11–20, 1989.
51. Schamm S, Rocher JP, Naslain R, In: Bunsell AR, Lamicq P, Massiah A, eds., *Proc 3rd Eur Conf Compos Mater Dev Sci Technol Compos Mater*, Elsevier, London, 157–163, 1989.
52. Patankar SN, Gopinathan V, Ramakrishnan P, *Scripta Metall*, 24, 2197–2202, 1990.
53. Rocher JP, Quenisset JM, Naslain R, Wetting improvement of carbon or silicon carbide by aluminum alloys based on a K_2ZrF_6 surface treatment: Application to composite material casting, *J Mater Sci*, 24, 2697, 1989.
54. Patankar SN, Gopinathan V, Ramakrishnan P, Processing of carbon fiber reinforced aluminum composite using K_2ZrF_6 treated carbon fibers, a degradation study, *J Mater Sci Lett*, 9, 912–913, 1990.
55. Patankar SN, Gopinathan V, Ramakrishnan P, Studies on carbon fiber reinforced aluminum composite processed using pre-treated carbon fibers, *J Mater Sci*, 26, 4196, 1991.
56. Subramanian RV, Nyberg EA, Zirconia and organo-titanate film formation on graphite fiber reinforcement for metal matrix composites, *J Mater Res*, 7(3), 677–688, 1992.
57. Rosenow MWK, Long nickel fibers for EMI shielding, *ETP'99 World Congress Engineering Thermoplastics*, Zurich, Jun 7–9, 1999.
58. Leonhardt G, Kieselstein E, Podlesak H, Than E, Hofmann A, Interface problems in Al matrix composites reinforced with coated carbon fibers, *Mater Sci Eng*, A135, 157–160, 1991.
59. Katzman HA, Fiber coatings for the fabrication of graphite-reinforced magnesium composites, *J Mater Sci*, 22, 144–148, 1987.
60. Clement JP, Rack HJ, Interfacial modification in G/Al metal matrix composites, *Symp on High Temperature Composites*, Proc Am Soc Compos, Technomic Publishing Co, 11, 1989.
61. Subramanian RV, Nyberg EA, Zirconia and organo-titanate film formation on graphite fiber reinforcement for metal matrix composites, *J Mater Res*, 7, 667, 1992.
62. Howlett BW, Minty DC, Old CF, The fabrication and properties of carbon fibre/metal composites, *International Conf on Carbon Fibers, their place in Modern Technology*, The Plastics Institute, London, Paper 14, Feb 1974.
63. Revzin B, Fuks D, Pelleg J, Influence of alloying on the solubility of carbon fibers in Al-based composites, non-empirical approach, *Composites Sci Technol*, 56(l), 3–10, 1996.
64. Interfacial Modifications and Bonding of Fiber Reinforced Metal Composite Material, In: Yosomiya R, Morimoto K, Nakajima A, Ikada Y, Suzuki T eds., *Adhesion and Bonding in Composites*, Marcel Dekker, New York, 235–256, 1990.
65. Friend C, Naish C, O'Brien TM, Sample G, In: J Fueller, ed, *Proc 4th Eur Conf Compos Mater Dev Sci Technol Compos Mater*, Elsevier, London, 307–312, 1990.
66. Bushby RS, Scott VD, Ibbotson AR, Lindsay NJ, Manufacture of carbon fiber reinforced Al alloy, Poursatip A, Street K eds., *ICCM*, Woodhead Publishing, Cambridge, 3–10, 1995.
67. Lacom W, Degischer HP, Schulz P, Assessment and control of surface reactions of carbon fibers in light weight metal matrix composites, *Key Eng Mater*, 127(1,2), 679–686, 1997.
68. Hall IW, Manrique F, Surface-treatment of carbon-fibers for aluminum-alloy matrix composites, *Scripta Metallurgica et Materialia*, 33(12), 2037–2043, 1995.
69. Wang JW, Hong T, Li GY, Li PX, A combined process of coating and hybridizing for the fabrication of carbon fiber reinforced aluminum matrix composites, *Composites, Part A-Appl Sci Manuf*, 28(11), 993–948, 1997.
70. Balaba WM, Weirauch DA, Perrotta AJ, Armstrong GH, Anyalebechi PN, Kauffman S, MacInnes AN, Winner AM, Barron AR, Effect of siloxane spin-on-glass and reaction bonded silicon oxycarbide coatings with a self-propagating interfacial reaction treatment (aspire) in the synthesis of carbon/graphite fiber-reinforced Al metal matrix composites, *J Mater Res*, 8(12), 3192–3201, 1993.
71. Shinoda T, Liu H, Mishima Y, Suzuki T, *Mater Sci Eng*, A146, (1–2), 91–104, 1991.
72. Chen X, Zhen G, Shen Z, TEM study of the interfaces and matrices of SiC-coated carbon fiber/aluminium composites made by the K_2ZrF_6 process, *J Mat Sci*, 31(16), 4297–4302, 1996.

73. Yang M, Scott VD, Carbide formation in a carbon fiber reinforced aluminum composite, *Carbon*, 29, 877, 1991.
74. Zeng Q, Aluminium based composites reinforced with alumina coated carbon fibers, *Mater Sci Technol*, 14(12), 1266–1268, 1998.
75. Aylor DM, Kain RM, *ASTM STP*, 864, 632–647, 1985.
76. Mansfeld F, Jeanjaquet SL, *Corrosion Sci*, 26(9), 727–734, 1986.
77. Kyono T et al, Study on compatibility between carbon fiber ("Torayca") and aluminium matrix, *1st Japan International SAMPE Symposium and Exhibition*, 964, Nov 28–Dec 1, 1989.
78. Cheng HM, Kitahara A, Akiyama S, Kobayashi K, Zhou BL, Fabrication of carbon fiber-reinforced aluminium composites with hybridisation of a small amount of particulates or whiskers of silicon carbide by pressure casting, *J Mat Sci*, 27(13), 3617–3623, 1992.
79. Nardone VC, Strife JR, Mechanical behaviour of [0/60]s p100/6061 Al composites, *J Mater Sci*, 23(1), 194, 1988.
80. Anon, SiC coating of carbon fiber for CFRAl (carbon fiber reinforced aluminium), *Techno Japan Techno Japan*, 20(1), 55, 1987.
81. Simancik F, Jangg G, Degischer HP, Short carbon fiber-aluminum matrix composite material prepared by extrusion of powder mixtures, *Journal de Physique*, IV 3,(C7), Part 3, 1775–1780, 1993.
82. Degischer HP, Innovative light metals, metal matrix composites and foamed aluminium *Mater Des*, 18(4/6), 221–226, 1997.
83. Furness JAG, Clyne TW, *Mater Sci Eng*, A141(2), 199–207, 1991.
84. Cheng HM, Akiyama S, Kitahara A, Kobayashi K, Zhou BL, The tensile strength and fiber degradation of carbon-fiber reinforced aluminum composites, *Scripta Metallurgica et Materialia*, 25(8), 1951–1956, 1991.
85. Cheng HM, Akiyama S, Kitahara A, Kobayashi K, Zhou BL, The tensile strength of carbon-fiber reinforced 6061 aluminum-alloy composites in as-casted and T6-treated states, *Scripta Metallurgica et Materialia*, 26(9), 1475–1480, 1992.
86. Cheng HM, Akiyama S, Kitahara A, Kobayashi K, Zhou BL, Behaviour of carbon-fiber reinforced Al-Si composites after thermal exposure, *Mater Sci Technol*, 8(3), 275–281, 1992.
87. Friler JB, Argon AS, Cornie JA, Strength and toughness of carbon fiber reinforced aluminum matrix composites, *Mater Sci Eng A—Structural Materials Properties Microstructure and Processing*, 162(1–2), 143–152, 1993.
88. Author Affiliation, Tsukuba, Nat, Inst. Materials Chem, Res; Nikkei Techno-Research Co. Ltd, Mechanical properties and metallography of aluminium matrix composites reinforced by the Cu or Ni plating carbon multifilament, *J Mater Res*, 8(10), 2492–2498, 1993.
89. Okura A, Motoki K, Rate of formation of intermetallic compounds in aluminum matrix-carbon fiber composites, *Composite Sci Tech*, 24, 243, 1985.
90. Zhang K, Wang YQ, Zhou BL, Tensile properties of coated carbon fiber reinforced magnesium composites, *Transactions of Non-ferrous Metals Society of China*, 7(3), 86–89, 1997.
91. Kudela S, Gergely V, Jansch E, Hofmann A, Baunack S, Oswald S, Wetzig K, Compatibility between PAN-based carbon-fibers and Mg8Li alloy during the pressure infiltration process, *J Mater Sci*, 29(21), 5576–5582, 1994.
92. Chen Y, Zhang GD, Wu F, Zhu J, Study of the C/Mg interface in magnesium matrix composites reinforced by carbon (graphite) fiber, *Rare Metal Mater Eng*, 26(3), 20–25, 1997.
93. Hall IW, *Metallography*, 20(2), 237–246, 1987.
94. Mortimer DA, Nicholas M, Crispin RM, The compatability of carbon with copper alloys containing chromium, titanium or vanadium, *International Conf on Carbon Fibers, their place in Modern Technology*, The Plastics Institute, London, Paper 15, Feb 1974.
95. Wu Y, Zhang G, In: Wu Y, Gu Z, Wu R eds., *Proc 7th Int Conf on Composite Materials*, Vol 1, Guangzhou, Nov 1989, International Academic Publishers, and Oxford, 463–467, 1989.
96. Kuniya K, Arakawa H, Namekawa T, *Trans Japan Inst Met*, 28(3), 238–246, 1987.
97. LePetitcorps Y, Poueylaud JM, Albingre L, Berdeu B, Lobstein P, Silvain JF, Carbon fiber reinforced copper matrix composites, processing routes and properties, *Key Eng Mater*, 127(1,2), 327–334, 1997.

98. Foster DA, In: Zakrzewski GA, Mazenko D, Peters ST, Dean CD eds., *Proc Int SAMPE Symp and Exhib 34, Tomorrow's Materials, Today*, 1401–1410, 1989.
99. Hutto DA, Lucas JK, Stevens WC, In: *Proc Int SAMPE Symp and Exhib 31, Materials Science Future*, 1145–1153, 1986.
100. Fan DN, Wang YL, Zhang HX, Liu ZN, Li GJ, In: Wu Y, Gu Z, Wu R eds., *Proc 7th Int Conf on Composite Materials*, Vol 1, Guangzhou, Nov 1989, International Academic Publishers, Oxford, 468–474, 1989.
101. Kuniya K, Arakawa H, In: Kawata K, Umekawa S, Kobayashi A eds., *Composites '86, Recent Advances in Japan and the United States, Proc Japan-U.S. CCM-III, Tokyo, Japan Soc Compos Mater*, 465–472, 1986.
102. TW Chow, Kelly A, Okura A, *Composites*, 16(3), 187–206, 1985.
103. Liu H, Shinoda T, Mishima Y, Suzuki T, *ISIJ International*, 29(7), 568–575, 1989.
104. Korb G, Korab J, Groboth G, Thermal expansion behaviour of unidirectional carbon-fiber reinforced copper-matrix composites, *Composites Part A-Appl Sci Manuf*, 29(12), 1563–1567, 1998.
105. Warren R, Anderson CH, Carlsson M, High temperature compatibility of carbon fibers with nickel, *J Mater Sci*, 13, 178, 1978.
106. Presland AEB, Walker PL Jr., *Carbon* 7, 1, 1969.
107. Wang C, Ying M, Yue D, In: Matthews FL, Buskell NCR, Hodgkinson JM, Morton J, *Proc 6th Int Conf Compos Mater 2nd Eur Conf Compos Mater*, Vol 2, Elsevier, London, 2.183–2.188, 1987.
108. Old CF, Nicholas MG, The fabrication and properties of carbon fiber reinforced lead-tin alloys, *International Conf on Carbon Fibers, their place in modern technology*, The Plastics Institute, London, Paper 14, Feb 1974.
109. Ho CT, Carbon-fiber-reinforced tin-lead alloy composites, *J Mater Res*, 9(8), 2144–2147, 1994.
110. Ho CT, Coated carbon-fibers and their composites with tin-lead matrix, *J Mater Sci Lett*, 14(2), 135–138, 1995.
111. Ho CT, Nickel- and Copper-coated carbon fiber reinforced tin-lead alloy composites, *J Mater Sci*, 31(21), 5781–5786, 1996.
112. Ho CT, Wear of carbon fiber-reinforced tin-lead alloy composites, *J Mater Sci Lett*, 16(21), 1767–1770, 1997.
113. Kitamura A, Teraoka T, Sagara R, In: Hayashi T ed., *Proc 4th Int Conf Compos Mater Prog Sci Eng Compos*, Vol 2, 1473–1480, 1982.
114. Ho CT, Chung DDL, Carbon fiber reinforced tin-superconductor composites, *J Mater Res*, 4(6), 1339–1346, 1989.
115. Ho CT, Chung DDL, Carbon fiber reinforced tin-superconductor composites, Bhagat RB, Clauer AH, Kumar P, Ritter AM eds., *Minerals, Metals & Materials Society Metal & Ceramic Matrix Composites, Processing, Modelling & Mechanical Behaviour*, Minerals, Metals & Materials Society Anaheim, 525–533, Feb 19–22, 1990.
116. Kendall EG, Development of metal-matrix composites reinforced with high modulus graphite fibers, Kreider KG ed., *Metal Matrix Composites*, Academic Press, New York, 319, 1974.
117. Chou TC, Kelly A, Okura A, Fiber-reinforced metal matrix composites, *Composites*, 16, 187, 1985.
118. Cornie JA, Yet-Ming Chiang, Uhlmann DR, Mortensen A, Collins JM, Processing of metal and ceramic matrix composites, *Ceramic Bull*, 65(2), 293–304, 1986.
119. Rocher JP, Girot F, Quenisset JM, Pailler R, Naslain R, Procédés d'elaboration des matériaux composites fiberux à matrice d'alliage base aluminium, (Fabrication procedures of fibrous composites with aluminum based alloys), *Memoires Sci Rev Metall*, 69, Feb 1986.
120. Goddard DM, Burke PD, Kizer DE, Bacon R, Harrington WC, Continuous graphite fiber MMCs, Reinhart TJ ed., *Engineered Materials Handbook*, Vol 1, ASM International, Metals Park, 867, 1987.
121. Delannay F, Froyen L, Deruyttere A, Review, the wetting of solids by molten metals and its relation to the preparation of metal matrix composites, *J Mater Sci*, 22, 1, 1987.
122. Mortensen A, Cornie JA, Flemings MC, Solidification processing of metal-matrix composites, *J Met*, 40, Feb 12–19, 1988.

123. Peebles LH, *Carbon Fibers Formation, Structure, and Properties*, CRC Press, Boca Raton, 164–180, 1995.
124. Motoki K, Okura A, Progress in Science and Eng of Composite Materials, Hayashi *et al* eds., *ICCM IV*, Tokyo, 1281, 1982.
125. Li X, Zhang H, Wu R, Proc of the Fifth Int Conf on Composite Materials, Harrigan *et al* eds., *ICCM V*, San Diego, 623, 1985.
126. Kohara Muto, Proc of the Fifth Int Conf on Composite Materials, Harrigan *et al* eds., *ICCM V*, San Diego, 623 and 631, 1985.
127. Blankenburgs G, *J Australian Institute of Metals*, 14(4), 236, 1969.
128. Kimura Y, Mishima Y, Umekawa S, Suzuki T, *J Mater Sci*, 19, 310, 1984.
129. Wu R, Cai W, In: Matthews FL, Buskell NCR, Hodgkinson JM, Morton J eds., *Proc 6th Int Conf Compos Mater 2nd Eur Conf Compos Mater*, Vol 2, Elsevier, London, 2.128–2.137, 1987.
130. Lawcock G, Ye L, Mai Y-W, Sun C-T, Effects of fiber/matrix on carbon fiber reinforced metal laminates, Part 1, Residual strength, *Composites Sci Technol*, 57, 1609–1619, 1997.
131. Lawcock G, Ye L, Mai Y-W, Sun C-T, Effects of fiber/matrix on carbon fiber reinforced metal laminates, Part 2:Impact behaviour. *Composites Sci Technol*, 57, 1621–1628, 1997.
132. Yih P, Chung DDL, Powder metallurgy fabrication of metal matrix composites using coated fillers, *Int J Powder Metall*, 31(4), 335–340, 1995.
133. Yih P, Chung DDL, A comparative study of the coated filler method and the admixtute method of powder metallurgy for making metal matrix composites, *J Mater Sci*, 32, 5321–5333, 1997.
134. Low cost PM route for titanium matrix carbon fiber composites, *Powder Metall*, 39(2), 97–99, 1996.
135. Chen Y, Chung DDL, In situ Al/TiB composite obtained by stir casting, *J Mater Sci*, 31, 399–406, 1996.
136. Mehrabian R, Riek RG, Flemings MC, *Metall Trans*, 5, 1899, 1974.
137. Flemings MC, Mehrabian R, Trans Int Foundry Congress, Moscow 1973, *Am Foundrymen's Soc Trans*, 1, 81, 1973.
138. Levi CG, Abashian GJ, Mehrabian R, Failure Modes in Composites IV, Cornie JA, Crossmann eds., *AIME*, 1977.
139. Russell KC, Comic JA, Ob S-Y, Particulate Wetting and Particle, Solid Interfacial Phenomena in Casting Metal Matrix Composites, *Proceedings of the TMS-AIME Symposium on Interfaces in Metal Matrix Composites*, New Orleans, Mar 1986.
140. Pai BC, Pillai RM, Kelukutty VS, Rao HS, Soman T, Pillai SGK, Sukumaran K, Satyanarayana KG, Ravikumar KK, Gupta AK, Sikand R, Semi-solid slurry process for making short carbon-fiber dispersed aluminum-alloy matrix composites, *J Mater Sci Lett*, 13(17), 1278–1280, 1994.
141. Champion AR, Krueger WH, Hartmann HS, Dhingra AK, *Proc of the 1978 Int Conf on Composite Materials, AIME, ICCM*, 2, 883.
142. Ahmad I, Barranco J, Advanced Fibers and Composites for Elevated Temperatures, Ahmad, Noton eds., *AIME*, 183, 1979.
143. Helminh RC, TS Piwonka, Advanced Fibers and Composites for Elevated Temperatures, Ahmad, Noton, eds, *AIME*, 205, 1979.
144. Abe Y, Kohiri S, Fujiyama K, Ichihi E, Progress in Science and Eng of Composite Materials, Hayashi *et al* eds., *ICCM IV*, Tokyo, 1427, 1982.
145. Banker JG, *SAMPE Quarterly*, 39, Jan 1974.
146. Maire J *et al*, Carbon Fibers, Their Composites and Applications, *Int Carbon Fibers Conf*, London, 107, 1971.
147. Kudela S, Gergely V, Jansch E, Hofmann A, Baunack S, Oswald S, Wetzig K, Compatibility between PAN-based carbon-fibers and Mg_8Li alloy during the pressure infiltration process, *J Mater Sci*, 29(21), 5576–5582, 1994.
148. Fukunaga H, Ohde T, Progress in Science and Eng of Composites, Hayashi *et al* eds., *ICCM IV*, Tokyo, 1443, 1982.
149. Fukunaga H, Kuriyama M, *Bull Japan Soc Mech Eng*, 25(203), 842, 1982.

150. Clyne TW, Bader MG, Proc of the Fifth Int Conf on Composite Materials, Harrigan *et al* eds., *ICCM V*, San Diego, 755, 1985.
151. Fukunaga H, Goda K, *J Jap Inst Metals*, 1, 78, 1985.
152. Fukunaga H, Goda K, *Bull Japan Soc Mech Eng*, 27(228), 1245, 1984.
153. Fukunaga H, Goda K, *Bull Japan Soc Mech Eng*, 28(235), 1, 1985.
154. Ackermann L, Charbonnier J, Desplanches G, Koslowski H, Proc of the Fifth Int Conf on Composite Materials, Harrigan *et al* eds., *ICCM V*, San Diego, 687, 1985.
155. Sawada Y, Bader MG, Proc. of the Fifth Int. Conf. on Composite Materials, Harrigan *et al* eds., *ICCM V*, San Diego, 785–794, 1985.
156. Nakata E, Kagawa Y, Terao H, *Report of the Castings Research Lab*, Waseda University, 34, 27, 1983.
157. Towata S, Yamada S, Ohwari T, *Trans J Inst Metals*, 26(8), 563, 1985.
158. Li Q, Zhang GD, Blucher JT, Cornie JA, Microstructure of the interface and interfiber regions in P-55 reinforced aluminum alloys manufactured by pressure infiltration, Ishida H, ed., *Controlled Interphases in Composite Materials*, Elsevier, New York, 131, 1990.
159. Yang J, Chung DDL, Casting particulate and fibrous metal matrix composites by vacuum infiltration of a liquid metalunder an inert gas pressure, *J Mater Sci*, 24, 3605–3612, 1989.
160. Jeng-Maw Chiou, Chung DDL, Characterization of metal-matrix composites fabricated by vacuum infiltration of a liquid metal under an inert gas pressure, *J Mater Sci*, 26, 2583–2589, 1991.
161. Amateau MF, Progress in the development of graphite-aluminum composites using liquid infiltration technology, *J Composite Mater*, 10, 279, 1976.
162. Mortensen A, Cornie JA, On the infiltration of metal matrix composites, *Metall Trans*, 18A, 1160, 1987.
163. Meyerer W *et al*, *Proc of the 1978 Int Conf on Composite Materials ICCM 2, AIME*, 141, 1978.
164. Harrigan WC, Flowers RH, Failure Modes in Composites IV, Cornie, Crossmann eds., *AIMS*, 319, 1977.
165. Kohyama A *et al*, Proc of the Fifth Int. Conf on Composite Materials, Harrigan *et al* eds., *ICCM V*, San Diego, 609, 1985.
166. Pepper RT, Penty RA, *J Composite Mater*, 8, 29, 1974.
167. Kendall EG, *Composite Materials Vol. 4, Metallic Matrix Composites*, Kreider KG, Academic, 1974.
168. Dacres CM, Reamer SM, Sutula RA, Larrick BF, *Proc Electrochem Soc*, 83(1), 76–95, 1983.
169. Wang C, Ying M, Yue D, In: Matthews FL, Buskell NCR, Hodgkinson JM, Morton J, *Proc 6th Int Conf Compos Mater 2nd Eur Conf Compos Mater*, Vol 2, Elsevier, London, 2.183–2.188, 1987.
170. Cheng HM, Lin ZH, Zhou BL, Zhen ZG, Kobayashi K, Uchiyama Y, Preparation of carbon-fiber-reinforced aluminum via ultrasonic liquid infiltration technique, *Mater Sci Technol*, 9(7), 609–619, 1993.
171. Toloui B, In: Harrigan WC Jr., *Proc 5th Int Conf Compos Mater*, 773–777, 1985.
172. Yajima S *et al*, Composite Materials, Kawata *et al* eds., *Proc US-Japan Conf*, Tokyo, 232, 1981.
173. Tanaka *et al*, Progress in Science and Eng of Composite Materials, Hayashi *et al* eds., *ICCM IV*, Tokyo, 1407, 1982.
174. Morris AWH, Carbon Fibers, Their Composites and Applications, *Int Carbon Fibers Conf*, London, 1971.
175. Harrigan WC, Goddard DM, Proc. of the 1975 Int Conf on Composite Materials, Scala *et al* eds., *ICCM I, AIME*, 849.
176. Toloui B, Proc of the Fifth Int Conf on Composite Materials, Harrigan *et al* eds., *ICCM V*, San Diego, 773, 1985.
177. Kohara S, Muto N, Proc of the Fifth Int Conf on Composite Materials, Harrigan *et al* eds., *ICCM V*, San Diego, 747, 1985.
178. Takahashi S, Proc of the Fifth Int. Conf on Composite Materials, Harrigan *et al* eds., *ICCM V*, San Diego, 1985.
179. Yajima S *et al*, *J Mater Sci Lett*, 15, 2131, 1980.
180. Sakamoto A, Fujiwara C, Tsuzuku T, *Proc Japan Congr Mater Res*, 33, 73–79, 1990.

181. Rabinovitch M, Daux JC, Raviart JL, Mevrel R, Carbon fiber reinforced magnesium and aluminium composite fabricated by liquid hot pressing, *4th European Conf on Composite Materials*, Stuttgart, 405–410, Sep 25–28, 1990.
182. Guruprasad A, Sudhakar A, Seshan S, Production and properties of carbon fiber-particle reinforced cast metal matrix composites, *Metals Mater Proc*, 7(2), 99–106, 1995.
183. Yoshida M, Ikegami S, Ohsaki T, Ohkita T, Studies on ion-plating process for making carbon fiber reinforced aluminum and properties of the composites, *SAMPE Symposium*, San Francisco, 24, 1417, May 1979.
184. Divecha AP, Karmarkar SD, Foltz JV, *Process for producing graphite fiber/aluminium-magnesium matrix composites*, U.S. Pat., 4578287, Mar 1986.
185. Wielage B, Rahm J, Preparation of MMC structures consist of carbon-fiber aluminum prepregs by using the high velocity flame spraying, *Metall*, 48(12), 961–966, 1994.
186. Cornie JA, Argon AS, Gupta V, *MRS Bull*, 16(4), 32–38, 1991.
187. Landis H, *Ph.D. dissertation*, MIT, 1988.
188. Argon AS, Gupta V, Landis KS, Cornie JA, *J Mater Sci*, 24, 1207–1218, 1989.
189. Clyne TW, Bader MG, Cappleman GR, Hubert PA, *J Mater Sci*, 20, 85, 1985.
190. Sutton WH, Whisker Technology, Levitt AP ed., Wiley, 273, 1970.
191. Nutt SR, Wawner FE, *J Mater Sci*, 20, 1953, 1985.
192. Cornie JA, Suplinskas RJ, Debolt H, Surface Enhancement for Metal Matrix Composites, *ONR Contract*, N0014-79-C-0691, Final Report.
193. Schoutens JE, Tempo K, Introduction to Metal Matrix Composites, *MMCCIAC Tutorial Series*, MMC No. 272.
194. Motoki K, Okura A, Progress in Science and Eng of Composite Materials, Hayashi *et al* eds., *ICCM IV*, Tokyo, 1281, 1982.
195. Li X, Zhang H, Wu R, Proc of the Fifth Int Conf on Composite Materials, Harrigan *et al* eds., *ICCM V*, San Diego, 623, 1985.
196. Khan IH, *Met Trans*, 7A, 1281, 1976.
197. Portnoi KI *et al*, *Soviet Powder Met Metal Ceramics*, 20(2), 116, 1981.
198. Towata S, Yamada S, Ohwari T, *Trans J Inst Metals*, 26(8), 563, 1985.
199. Harrigan WC, Flowers RH, Failure Modes in Composites IV, Cornie, Crossmann eds., *AIMS*, 319, 1977.
200. West R, David LD, Djurovich PI, Yu H, Sinclair R, Polysilastyrene, Phenylmethylsilane-dimethylsilane Copolymers as Precursors to Silicon Carbide, *Am Ceramic Soc Bull*, 62(8), 899–903, 1983.
201. Viala JC, Fortier P, Claveyrolas G, Vincent H, Bouix J, Effect of magnesium on the composition, microstructure and mechanical properties of carbon fibers, *J Mater Sci*, 26, 4977, 1991.

CHAPTER **17**

Testing of PAN Precursor, Virgin Carbon Fibers, Carbon Fiber Composites and Related Products

17.1 INTRODUCTION

Personnel involved with testing being asked by production for a retest is a common occurrence. Therefore, a most advisable philosophy is to ensure that the results are true and all the equipment used can be shown to be working correctly. To ensure this commitment, some important parameters must be met prior to testing:

1. Great care must be taken to select the correct sample
2. The samples must be correctly labeled and suitably stored
3. A test procedure should always be used and followed exactly
4. All equipment must be calibrated
5. As an additional check, it is advisable to have reference materials, which can be used as semi-standards
6. Utilize statistics where applicable
7. Draw attention to results not in specification, highlight with a large stamp that says FAILED
8. Ensure that the right people receive copies of the test results
9. Ensure comparison of like with like when comparing test results

Table 17.1 lists the standards used for testing carbon fibers and for precise details, these standards should be used as the reference source. It is proposed to briefly discuss the various techniques used for testing, comparing methods and indicating some of the pitfalls. The emphasis is on the techniques used by carbon fiber manufacturers for routine testing and customer support, and does not cover methods used to obtain design data, which can be obtained by direct reference to the relevant standards, textbooks and references.

17.2 TESTING OF PAN PRECURSOR

17.2.1 Filament diameter distribution in PAN tow

A simple way of checking the variability of the filament diameter in a precursor tow is to use a polished stainless steel plate, which has been drilled with several countersunk holes to accommodate different precursor tow sizes (Figure 17.1). A looped needle threader is pushed

Table 17.1 Testing specifications for carbon fibers

Subject	ASTM Specifications	EN Specifications	ISO Specifications	SACMA Specifications	CRAG Specificatons
FIBER					
1. Mass per unit length		2964	10120	SACMA SRM 13	
2. Density	D1505; D3800		10119	SACMA SRM 15	
3. Diameter		2965	11567		
4. Tensile strength/modulus	D4018		10618; 11566	SACMA SRM 16	
5. Size content			10548	SACMA SRM 14	
6. Thermal oxidation resistance	D4102				
FABRIC					
1. Unit length of yarn			4602		
2. Length	D3773		5025		
3. Width	D3774		5025		
4. Thickness	D1777		4603		
5. Twist of yarn	D1423		1890	SACMA SRM 17	
6. Crimp of yarn	D3883				
7. Count	D1907; D3775				
8. Mass per unit area	D3776		4605		
9. Tensile	D2256; D5034		4606		
10. Burst strength					
MATRIX					
1. Density	D792		1183A		
2. Flexural properties	D790		178		
3. Tensile properties	D638; D882; D3039	527			
4. Compression properties	D695		604		
5. Impact strength	D256; D1822		180/1C		
6. Water absorption	D570		62		
7. Deflection under load temperature	D648		75		
8. Transition temperature	D3418				
9. Flow of thermoplastic	D569				
10. Coefficient of linear thermal expansion	D696				
11. Melt flow index	D1238		1133		
12. Mold shrinkage	D955		2517		
PREPREG					
1. Mass per unit area	D5300	2559			
2. Resin content	C613; D3529			SACMA SRM 23	
3. Volatile matter	D3530	3558			
4. Gel time	D2471; D3532				
5. Resin flow	D3531	2560		SACMA SRM 22	
SPECIMEN PREPARATION					
1. Compression molding					
a) Thermoplastics	D4703		293		
b) Thermosets	D5224		295		
2. Press/autoclave molding	D5687	2565			
3. Injection molding thermoplastics	D3641		294		
TESTING					
1. Fiber content	D3171				Method 1000
2. Density	D1505; D3800	2745			Method 800
3. Void content	D2734				Method 1001
4. Moisture content		3616			
5. Flexural strength & modulus	D790	63, 2562, 2746	178; 14,125*		Method 200
6. Longitudinal tensile strength & modulus (0°)	D3039	2561	527-4; 527-5	SACMA SRM 4	Method 300
7. Transverse tensile strength & modulus (90°)	D3039	2597			Method 301

(Continued)

TESTING OF PAN PRECURSOR, VIRGIN CARBON FIBERS

Table 17.1 Continued

Subject	ASTM Specifications	EN Specifications	ISO Specifications	SACMA Specifications	CRAG Specificatons
8. Longitudinal compression strength & modulus (0°)	C364; D695; D3410	2850, 4585	604; 14,126*	SACMA SRM 1	Method 400
9. In-plane shear strength	D3518; D3846; D4255		6031; 14,130*	SACMA SRM 7	Method 101
10. Interlaminar shear strength	D2344	2563	2580; 4585	SACMA SRM 8	Method 100
11. Impact					
a) Charpy			179		
b) Izod	D256; D4812		180		
12. Creep	D2990		899		
13. Fatigue	D3479				
14. Coefficient of thermal expansion	D696				Method 801
15. Thermal conductivity	C177		8301; 8302		
16. Volume resistivity	D257				

Note: *new standard BS/EN/ISO.

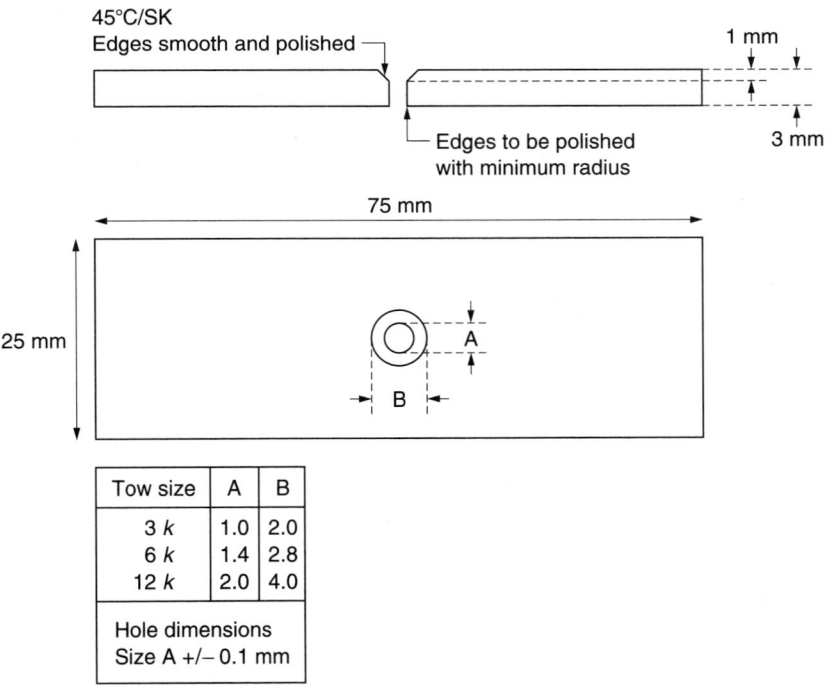

Figure 17.1 Precursor mounting jig.

through the selected hole and a short length of precursor tow passed through the loop of the needle threader and the tow, now doubled up, is pulled through the hole to give a double tow aligned in the hole. The ends are cut off flush on each side of the plate using a sharp new single edge razor blade.

The precursor sandwiched in the plate can then be examined for filament diameter variability using a microscope with transmitted light. Remember, there are two tows positioned alongside each other in the plate.

Table 17.2 Selection of test weights for the Vibroskop

d'tex range	Range to be selected mg	Tension weight mg
0.50–1.25	50	50
0.70–1.75	70	70
1.00–2.50	100	100
2.00–5.00	200	200

17.2.2 Measurement of precursor d'tex using the Vibroskop (ASTM D1577)

The Vibroskop is an electronic instrument, where the fiber to be tested is tensioned with a weight (Vibraclip), which has been selected according to the nominal linear density of the sample (Table 17.2). The weighted fiber is supported over a knife edge and clamped in the Vibroskop, then vibrated with an electrostatic force, which does not affect its natural frequency. An image is shown on a screen and the natural frequency is tuned by adjusting the vibrating length with the traverse knob, regulating the amount of vibration with the amplitude knob. From the conditions at resonance, the machine can give a digital read out of d'tex.

When $d'tex = \dfrac{W \times g \times 10^6}{\lambda^2 f^2}$

where W = weight
λ = wavelength = $2L$
L = length
f = frequency

Once the d'tex has been obtained and the density is known, then assuming that the filaments are round, the mean diameter of the filament would be given by:

$$d = \sqrt{\dfrac{d'tex}{0.007855\rho}}$$

where d = diameter, μm
ρ = density, g cm^{-3}

17.2.3 Determination of fiber moisture content and fiber moisture regain

First of all, it has to be decided which way the moisture is to be expressed, as a content or regain. Use about 10 g of sample in a drying tin fitted with a lid. Dry to constant weight for at least 2 h at 110°C. To avoid breakdown of PAN, do not exceed this temperature. Cool in a desiccator for 30 min before weighing.

$$\text{Moisture content (\%)} = \dfrac{\text{Mass of absorbed water in fiber} \times 100}{\text{Mass of dried fiber}}$$

$$\text{Moisture regain (\%)} = \dfrac{\text{Mass of absorbed water in fiber} \times 100}{\text{Mass of undried fiber}}$$

Oxidized fiber left in an unconditioned atmosphere at room temperature and 60–90% RH, can double its moisture content and at 65% RH, would take up 75% of its final moisture content within 10 min.

If the balance is situated on the plant and vibration is a problem, an extremely effective anti-vibration table can be constructed by placing the balance on a concrete flagstone supported on rubber bungs.

Rapid moisture content can be achieved with a moisture balance, using either an infra red lamp source above the balance pan, or employing an instrument incorporating an internal analytical balance, using a microwave drying system, that can be pulsed to prevent possible degradation.

To determine moisture regain at a given relative humidity level select a salt from Appendix 9 (list of the humidity levels), a saturated solution of which can be used to obtain the required RH. Place the dried fiber (2 h at 110°C) above the saturated salt solution, which has been placed in the bottom of a desiccator, and leave to equilibrate for 72 h. Remove and weigh immediately.

17.2.4 Determination of residual solvent (NaSCN) in Courtelle precursor

Five grams of precursor are refluxed for 1 h in about 200 cm^3 distilled water, cooled, filtered and made up to volume in a 250 cm^3 standard flask. An indicator is prepared by dissolving 500 g of ferric ammonium sulphate crystals in 700 cm^3 distilled water. Now take 200 cm^3 of this concentrated solution, add 400 cm^3 concentrated HNO$_3$ and making up to 1 litre with distilled water. Take a 10 cm^3 aliquot and add 1 cm^3 ferric alum indicator, measure the optical density in a spectrophotometer at 510 µm against a blank of 10 cm^3 distilled water and 1 cm^3 ferric alum indicator. The ppm level of NaSCN can be determined from a calibration graph.

17.2.5 Determination of sodium content in the precursor

1. Atomic absorption spectrophotometer

A successful method employs flame photometry, using an atomic absorption spectrophotometer. Basically, the sample is ashed at 800°C, converted to the chloride with HCl and diluted with ionized water to give a test solution. Calibration solutions containing 1.0, 2.5, 5.0 and 10.0 µg/cm^3 of Na are made up from a commercially available Sodium Stock Solution (containing 1 mg/cm^3 Na). A blank of deionized water/HCl is aspirated into an air/acetylene flame set up with the atomic absorption spectrophotometer, followed by the standard solutions and the sample. A portion of the radiation proportional to the concentration of the sodium is absorbed. The absorption is measured and the concentration can be determined.

2. Ion chromatograph

An alternative method is to use an ion chromatograph, where the sample is ashed for about 10 h at 750°C, the residue dissolved in HNO$_3$, diluted with distilled water to a given volume and an aliquot eluted with 0.02% methane sulphonic acid (CH$_3$SO$_3$H), which forms the Na salt and the conductivity is then checked against known Na standards.

17.2.6 Determination of the Soft Finish content in Courtelle precursor

About 10 g of precursor fiber is weighed and placed in a weighed flask and extracted in a soxhlet with 200 cm^3 redistilled methylene chloride for 3 h (about 9 cycles). After the last cycle, the solvent is allowed to evaporate until about 10–15 cm^3 remains in the flask and the contents of the soxhlet are about to siphon back into the flask. The residual solution in the flask is then evaporated on a water bath in a fume cupboard and finally dried for about 2 h in an oven at 80°C in one continuous operation. After cooling the flask is weighed to determine the quantity of the soft finish.

17.2.7 Silver sulphide staining test for checking structure of a PAN precursor

About 0.1 g of precursor is subjected to 48 h treatment with H_2S in an autoclave at 1.8 MPa pressure and 20°C. The sample is rinsed with water to remove surface H_2S and treated for 20 h with 0.1N $AgNO_3$ at 20°C. The sample is washed with DM water, dried at 40°C and mounted in an ambient cured epoxy system. Cross sections, 0.05–1.0 µm thick, are cut with a microtome and examined with electron microscopy.

17.2.8 An experimental rig for determinination of precursor burn-up temperature

When investigating precursor types, it is advantageous to determine the burn-up temperature, since this information will permit a safe oxidation schedule to be established. An ordinary laboratory oven as shown in Figure 17.2 is modified to permit the ready oxidation of the precursor.

17.3 TESTING OF OXIDIZED PAN FIBER (OPF) AND VIRGIN CARBON FIBER

17.3.1 Mass per unit length

A simple jig can be constructed by fixing a pair of heavy brass hinges onto a board with a slot cut through on the center line, about 75% of the width of an opening half, so

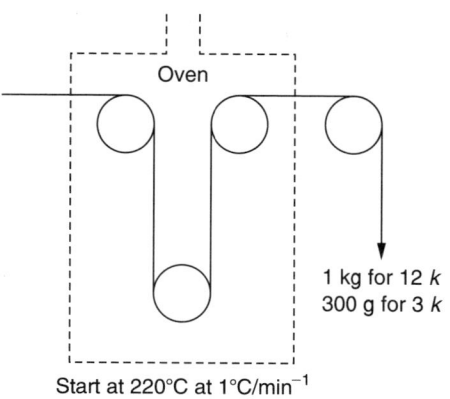

Figure 17.2 Experimental rig for determining precursor burn-up temperature.

17.3.2 Determination of density

It should be noted that oxidized PAN fiber will not burn in air when the density is above 1.37 g cm^{-3}, but care must be taken when undertaking this test.

Initially, densities were measured with a density gradient bottle. Liquids used in this test included distilled water, white spirit, surprisingly liquid paraffin and bromoform. The latter was used for determining the density of HM fiber, but was found to give high results and yellow crystals were observed on the fiber, which, when analyzed were found to be a homologue of bromoform. The supplier could not guarantee that the product would be free of this homologue, so the use of bromoform was discontinued.

The main problem with the density bottle method was eliminating any air entrained on the fiber and many techniques were used to effect removal including centrifuging, the application of vacuum in a desiccator and ultrasonics. Another problem was that technicians tended to carry the filled density bottle by the neck and the warmth of their fingers was sufficient to expand the solvent, with immediate loss through the capillary stopper en-route to the balance room. The solution was to fold a filter paper and wrap around the density bottle neck, gripping the ends of the folded paper.

A sink float method was employed for determining the density of carbon fiber, where about 100 μg of finely chopped fiber was introduced into each of a number of test tubes containing mixed solvents (e.g. CCl$_4$ and 1,2-dibromoethane) to give a range of densities. The tubes were shaken and the tube with the majority of fiber remaining in suspension indicated the fiber density. It was essential that the fiber was chopped finely enough and tweezers were used during transfer into the density tubes to avoid any contact with fingers.

The density gradient column is now widely used and necessitates observing the level to which a test specimen will sink in a column of liquid, the density of which increases uniformly from top to bottom. The column should be about 40–50 mm I/D and normally graduated from 0 at the top to 100 cm at the bottom. Up to three such columns can be positioned in a water bath controlled at $23 \pm 1°C$. A valuable attribute of this method is that the columns, when prepared, can be accurately calibrated with glass beads of known density.

Various liquids are used to cover a spread of working densities (Table 17.3) depending on the density of the test material. A typical density gradient column is shown in Figure 17.3.

Table 17.3 Liquids used to prepare density gradient columns

Light liquid	Heavy liquid	Density range g cm^{-3}
n-Hexane	Carbon tetrachloride	0.66–1.59
Water	ZnCl$_2$ stabilized with HCl	1.00–2.00
Carbon tetrachloride	1,2-dibromoethane (BrCH$_2$CH$_2$Br) (ethylene dibromide)	1.60–2.18
Carbon tetrachloride	Dibromomethane (BrCH$_2$Br) (methylene bromide)	1.60–2.47

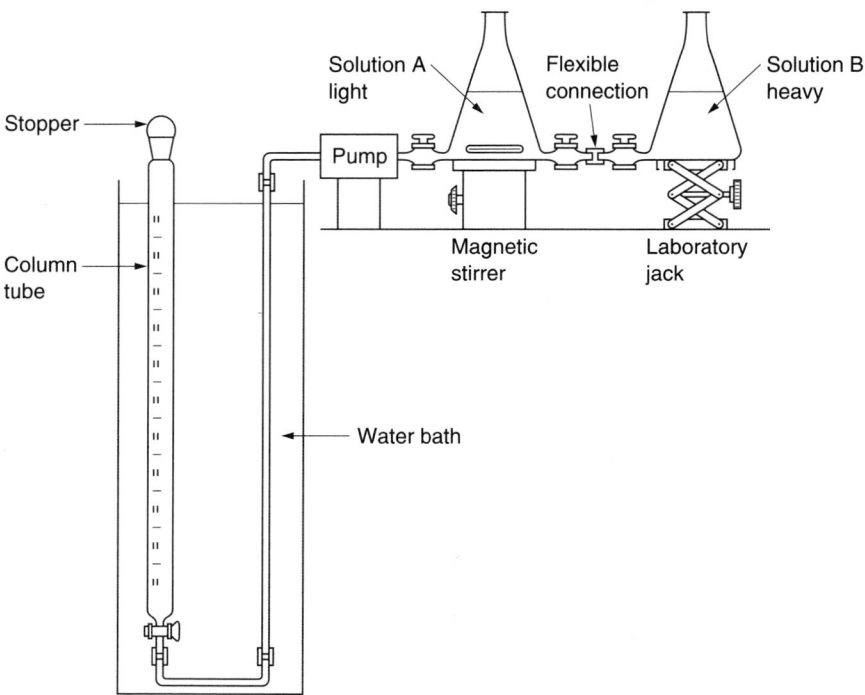

Figure 17.3 Density gradient column.

The principle is for the heavier liquid to enter the flask containing the lighter liquid, where it is mixed with a magnetic stirrer. Connections are made with glass spherical joints and clips or silicone rubber tubing. Initially, one relied on gravity to fill the columns via a capillary to control the flow, but now, this has been replaced by a pump which is used to fill the column at about 1 cm head min^{-1}. Calibration beads of known density contained within the interior of the basket of a shuttlecock are lowered into the column using a nylon line attached to a motor driven hoist, lowering the basket to the bottom at about 1 cm min^{-1} and leaving the basket at the bottom. The position of each bead is determined from the gradations on the column, taking care to avoid parallax errors and the position is plotted against the known density of the glass bead to obtain a calibration curve.

A small sample, about 2–5 mm long, is selected and using rubber gloves (to avoid surface contamination introduced by handling with the bare hands), an identification knot is tied in the sample and the shape of the knot recorded. Before the sample is introduced, it is thoroughly wetted out by total immersion in the liquid with the lowest density for 10 min. After the samples have been inserted in the column using tweezers, a stopper is positioned in the top of the column to prevent solvent evaporation. After a period of 2 h, the position that the sample has reached in the column is read from the calibrations.

As the column becomes choked with spent samples, it can be swept by raising the shuttlecock very slowly (1 cm min^{-1}) using the motor driven hoist, collecting all spent samples together with the calibration beads within the shuttlecock, on its journey upwards. An interval timer, or limit switch, is used to warn when the basket is near the top of the column, as otherwise, the glass calibration beads tend to get scattered. The calibration beads are reintroduced by returning the basket to the bottom at about 1 cm min^{-1} and the column recalibrated and checked using regression analysis to see if the calibration is linear.

Some laboratories use a gas pycnometer to determine the density with Ar or He to intrude into the sample and, not surprisingly, results do differ from methods that use a liquid. The gas pycnometer uses the ideal gas law to determine the volume of a sample and given a known volume of the sample, test chamber and the gas reservoir, together with the change in pressure, the absolute density can be calculated from the volume of the sample and its weight. The method does require a larger sample (0.5–10 g), but the test is non-destructive. Normally 10 iterations are taken to ensure accuracy.

Oxidized PAN fiber has an equilibrium moisture content and Figure 17.4 shows the effect of a moisture regain of 5% and 4% on the measured density, whilst Figure 17.5 shows a correction factor which can be applied to allow for the moisture regain.

The density of PAN based carbon fiber depends on the maximum temperature used for processing, d'tex of precursor and the line speed. Figure 17.6 is a projection of the effect of

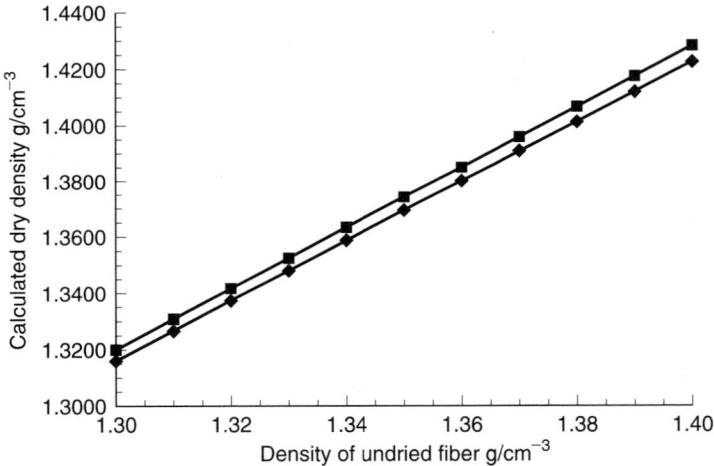

Figure 17.4 The effect of moisture on the measured density of oxidized fiber. ■ moisture regain = 0.05; ◆ moisture regain = 0.04.

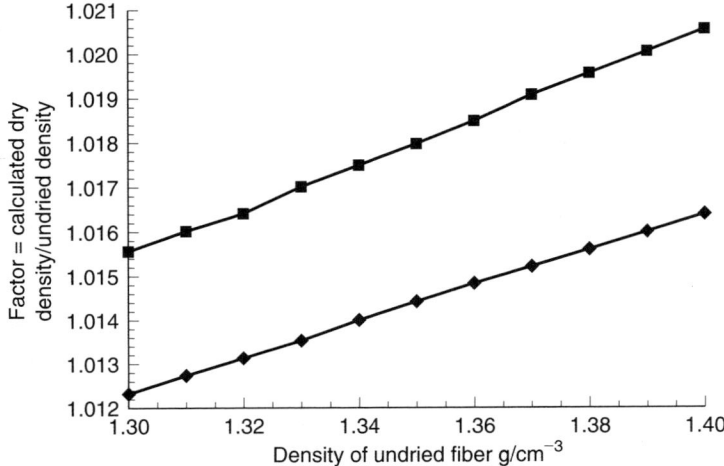

Figure 17.5 *Factor* = (Calculated dry density of opf/Undried fiber density) plotted Vs density of undried opf (g cm^{-3}). ■ moisture regain = 0.05; ◆ moisture regain = 0.04.

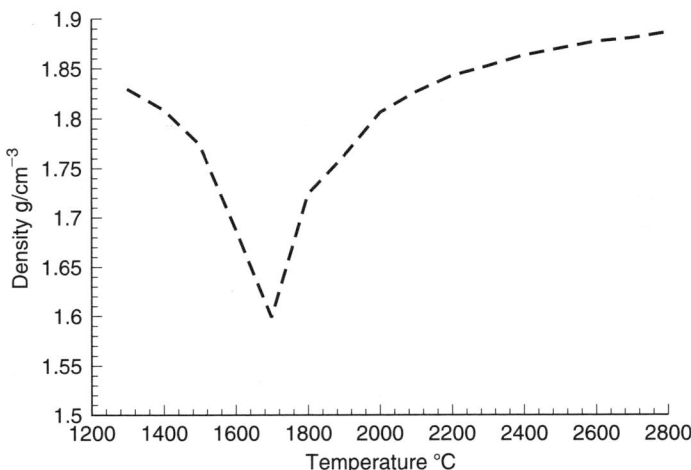

Figure 17.6 Effect of production temperature on density of carbon fiber.

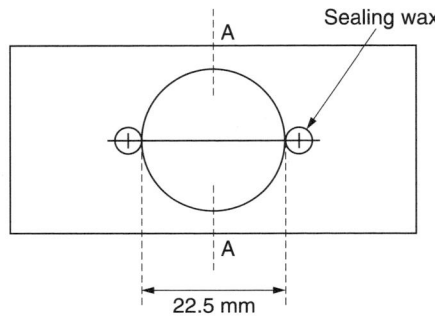

Figure 17.7 Filament test card.

processing temperature on the density of carbon fiber and it will be noted that in the range 1300–2000°C, a given density can be produced at two different temperatures.

17.3.3 Determination of diameter

1. Mounting a single filament

To measure filament tensile properties, it is necessary to mount the filament on a thin card (about 0.3 mm thick) approximately 65×25 mm^2, with a 22.5 mm hole punched out at the center. Gage lengths of 25 and 50 mm are also used by some laboratories. The filament is mounted along the center line across the hole using sealing wax, applied with a soldering iron (Figure 17.7). The blob of wax must be placed exactly on the perimeter of the cut-out to ensure that the correct gage length for the tensile test is achieved. When the filament has been attached at one end care must be taken to avoid rolling the fiber, or introducing twist, when finally positioning the filament on the center line. Schulman [1] describes an ingenious system used at the Air Force Materials Laboratory, where the sealing wax was attached to the grips and melted *in situ* with an integral electric resistance coil.

If a proprietary adhesive is used to mount the filament (e.g. Durafix), beware of new introductions like a new improved version, which in this instance failed to give a satisfactory level of adhesion.

2. Determining filament diameter using a Watson image shearing eyepiece

The image splitting eyepiece, which unfortunately is now no longer manufactured, was made by Vickers Instruments of York and still commonly used, consisted of a special prism assembly, which can be mounted upon a conventional compound microscope. It is essential that the microscope be set up correctly, using transmitted light adjusted for Köhler illumination. The prisms are linked with a micrometer screw, which permits varying their angular relation to each other. When the prism faces are parallel, two images of the object are superimposed and appear as one (Figure 17.8A). When the images of the object are overlapping, the amount of shear is less than the object dimension (Figure 17.8B). If the images of the object are just touching, then the amount of shear is equal to the object dimension (Figure 17.8C). Finally, if the images of the object are apart, then the amount of shear is greater than the object dimension (Figure 17.8D). A green filter such as the Wratten No.60 or a Schott VG9 glass filter can be used. However, a dichroic filter producing one red sheared image and the other green should not be used, since for highest precision, the two images must look and be alike in intensity and contrast. The exact measurement is the amount of shear when going from condition A to condition C. The instrument is calibrated with a Vickers stage micrometer mounted on a glass slide; the scale is 1 mm long with 100 divisions, each division equivalent to 1 µm and each line is 0.0038 mm thick. The measuring accuracy can be as high as 0.1 µm, depending on the Numerical Aperture of the objective in use.

Alternatively, the image splitting eyepiece can be calibrated with W wires of known diameter. The tungsten wires are calibrated with a scanning electron microscope using a standard calibration grating. A problem, however, did occur when it was found that the wires were not truly round and different values could be obtained depending on which way they were mounted. There were problems for technicians using this technique but, with acquired skill, good agreement has been obtained with the laser method (Table 17.4).

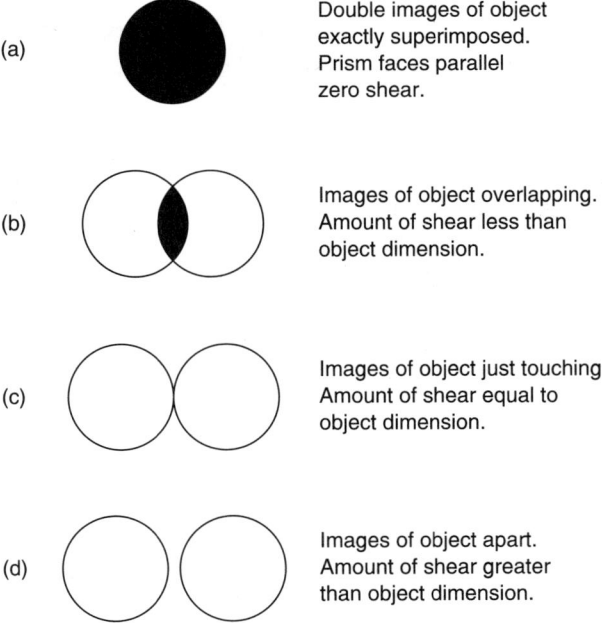

(a) Double images of object exactly superimposed. Prism faces parallel zero shear.

(b) Images of object overlapping. Amount of shear less than object dimension.

(c) Images of object just touching. Amount of shear equal to object dimension.

(d) Images of object apart. Amount of shear greater than object dimension.

Figure 17.8 Images in Watson Shearing Eyepiece.

Table 17.4 Comparison of laser and Vickers image splitting eyepiece methods for determining the filament diameter of oxidized PAN fiber

Sample number	Laser diameter μm	Vickers image splitting eyepiece diameter μm
1	9.99	10.17
2	9.49	8.83
3	11.17	11.28
4	11.63	11.47
5	10.85	12.28
6	11.17	11.28
7	11.17	11.28
8	11.17	10.89
9	11.17	11.47
10	10.85	10.70
11	11.51	11.47
12	11.51	11.66
Mean	10.97	11.07

3. Determination of filament diameter using a He/Ne laser

When a filament is illuminated by a laser, the diffraction pattern will be very similar to a rectangular aperture or slit [2]. The slit approximation theory gives:

$$D_f = \frac{\lambda \times d_s}{d_n}$$

where D_f = diameter of filament
λ = wavelength of the laser
d_s = distance of filament to screen
d_n = distance between the first minima to C/L of the filament

Initially, the laser is checked against the lines on a linewidth mask standard produced by electron beam lithography [3]. The mask comprised 3 lines, each 10 mm long and 7.5, 10.0 and 12.5 ± 0.1 μm wide and 10 mm apart. Since the lines were flat and not spherical, they did not have the problem of specular reflection, which can occur with a W wire.

To undertake a measurement, the line or filament is placed and supported (e.g. a goniometer), so that the beam of the laser is perpendicular to the axis of the line or filament (Figure 17.9). The diffraction pattern is projected onto a screen exactly 30 cm away from the filament. The distance (d_n) from the first minimum point, or node, to the centerline of the filament is determined. To avoid errors when determining this distance, the procedure shown in Figure 17.10 can be adopted.

When time is not of the essence or extreme accuracy is required, the mean diameter of a number of determinations taken at, say, every 10° around the circumference by rotating the filament on a precision rotary table, which will, for example, clearly indicate ovality. The position of the nodes can be sensed using a photodiode array contained within a Line Scan Camera in conjunction with a computer. Chen and Diefendorf have shown that for a section close to circular, five measurements measured every 36° have minimum error [4]. For cross-sections with a re-entrant shape (e.g. dogbone shape for some PAN based fibers and the PACman shape for some pitch based fibers) the laser technique cannot be used [5].

TESTING OF PAN PRECURSOR, VIRGIN CARBON FIBERS

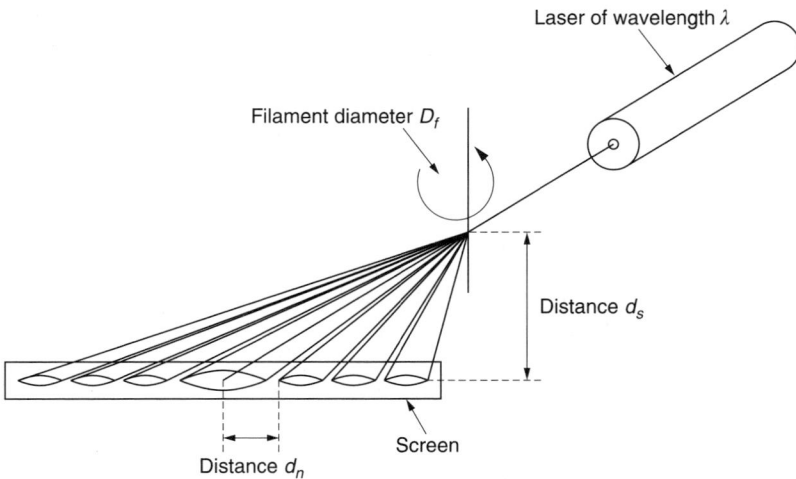

Figure 17.9 Measurement of filament diameter with a laser.

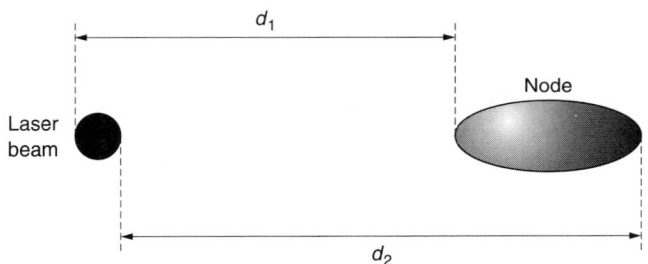

Figure 17.10 Distance to the center of first node from laser beam, $d_n = (d_1 + d_2)/2$.

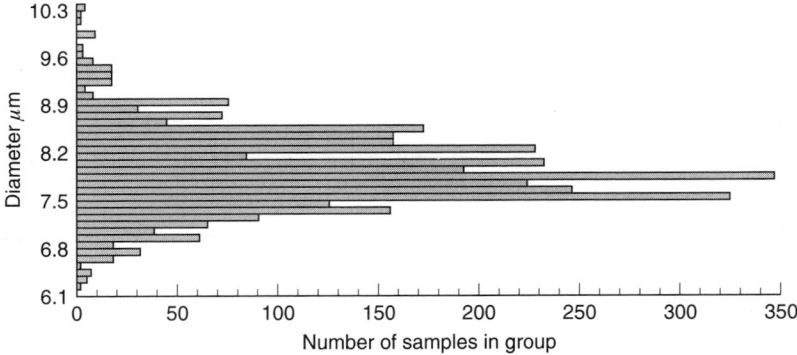

Figure 17.11 Variation of filament diameter of RK10 carbon fiber made from textile tow precursor. Weighted mean diameter = 7.96 μm (3177 samples).

Filament diameter of carbon fiber will bear a direct relationship to the filament diameter of the original precursor and therefore, the textile tow precursor will inevitably have greater variability (Figure 17.11).

4. Calibration of a Stereoscan with a traceable reference standard

This method is generally used for calibrating secondary standards like the diameter of W wires, but can also be used for determining actual filament diameters.

5. Preparation of a mini composite (impregnated tow)

A carbon fiber tow can be impregnated with an epoxy resin and cured. Small lengths of this tow can then be held upright using a special clip (cf a coiled flat spring on edge, similar to that used for mounting metallographic specimens) and mounted in an epoxy resin or acrylic molding compound. If the specimen should topple during the mounting operation the cross-sections will be elliptical and not circular. The section is polished using standard metallographic techniques, starting by wet grinding with SiC papers, gradually reducing the coarseness and finally polishing with diamond paste.

The polished specimen can then be examined with a metallographic microscope and knowing the magnification by using a stage micrometer, the diameter can be determined with an eyepiece fitted with a graticule, or if it is a projection microscope, by placing a rule on the projection screen. Alternatively, from a photograph of known magnification, the area of each filament cross section can be determined with a planimeter. Another technique is to examine the polished mounted specimen using a Leica (Cambridge) Quantimet image analyzing computer, which will determine the mean diameter of the filaments chosen in the field of view. To avoid selecting a preferred field of view, a sampling device can be used, which will acquire a number of arbitrary selections. Another problem is that an edge effect may introduce an error. Using the image analysis method can show the fibers, matrix and voids as light grey, dark grey and black and will record the number of pixels in each level, which can then be expressed as a percentage of the total number of pixels.

17.3.4 Tensile testing of filament

Initially, all carbon fiber tensile testing was by filament testing procedures and was ideal for research work, requiring small quanities of sample material and giving an answer in about 30 min.

A method not used by the author for testing filaments is Sinclair's loop test (Supplement 1 at the end of this Chapter). The procedure has been replaced by some form of tow testing, except when testing oxidized PAN fiber, where filament testing is standard procedure.

Before any testing is undertaken, the tensile test machine must be level, carefully aligned axially and calibrated. When testing filaments, technicians with smooth hands are preferable for this work and girls tend to be better, whilst ex-hairdressers tend to be ideal.

17.3.4.1 Determination of compliance of the tensile test machine system

The system compliance is that portion of the indicated elongation contributed by the load train system and the specimen gripping system. ASTM D 3379 detailed a procedure for the determination of system compliance, but this standard has now been discontinued. It is not possible to use strain gages or extensometers when measuring filament extension, so a true gage length extension has to be obtained from knowledge of the chart speed, crosshead speed and the system compliance. The machine compliance can be obtained by placing a steel bar in a set of grips and applying a load, the same as that used in the filament test (10 cN) and measuring the extension. It can be assumed that the steel bar does not extend at this small load and the extension will be due to the softness of the machine and deflection of the load cell and grips. Unfortunately, the grips used to grip the steel bar are not the same as the grips used for the filament test. The system compliance for machine and grips is best determined experimentally and a series of specimens of different gage lengths are mounted in similar card holders and tensile testing carried out by setting the crosshead and chart recorder speeds to

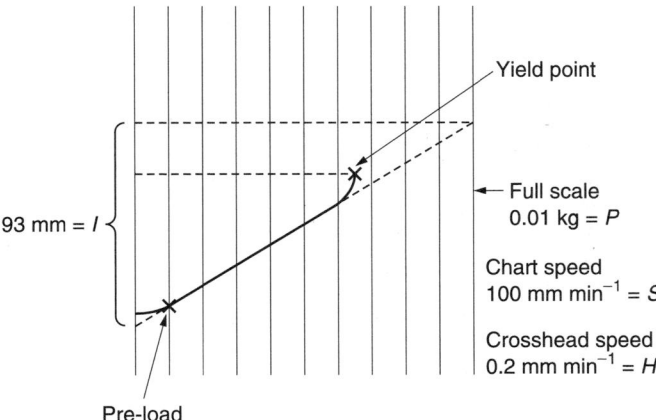

Figure 17.12 Typical filament load-time curve.

initiate failure in about 1 min and the load scale reading to be above 25% of the full scale reading. A typical load-time curve is shown in Figure 17.12 and a line is drawn through the initial straight line part of the curve to intercept the y axis (extension) and up to the full scale load on the x axis. Measure the chart extension and determine the indicated compliance (C_A) for each sample from:

$$C_A = \frac{I \times H}{P \times S}$$

where C_A = indicated compliance, mm N^{-1}
I = total extension extrapolated from load-time curve, mm
H = crosshead speed, mm s^{-1}
P = full scale force, N
S = chart speed, mm s^{-1}

These indicated compliances are now plotted on the y-axis against the gage length on the x-axis. A best fit line is drawn through these points to determine the intercept on the y-axis for zero gage length, or the system compliance C_S (Figure 17.13). The true compliance is given by:

$$C = C_A - C_S$$

where C = true compliance, mm N^{-1}
C_S = system compliance, mm N^{-1}
Note:
Young's Modulus = L/CA
where L = specimen gage length, mm
C = true compliance, mm N^{-1}
A = average filament area, m^2

17.3.4.2 *Measurment of filament tensile modulus*

Figure 17.14 shows the relationship of the distance between atoms and the force between them and an approximation, which is fairly true, is that the straight line portion of the curve is ±1% strain on either side of the neutral or strain free position. It was the work of

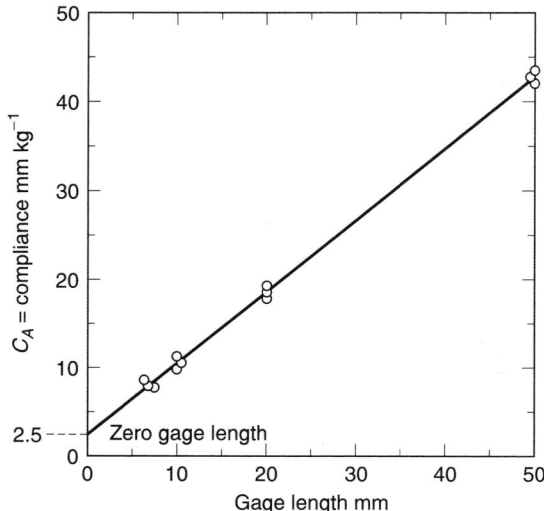

Figure 17.13 Determination of system compliance.

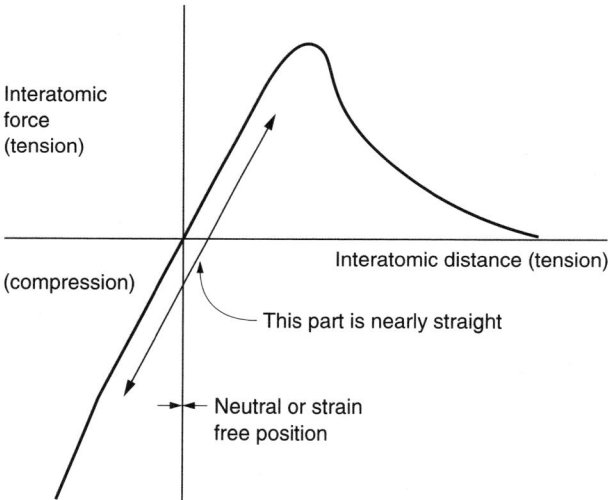

Figure 17.14 Relationship between two atoms and the force between them.

Robert Hooke, who published in 1664 [6,7] in an article entitled Inventions I Intend To Publish, the following anagram *ceiiinosssttuu*. It was not until two years later that he revealed the solution as *Ut tensio sic uis*, that when translated was—As the extension so the force—which can be expressed in the form—stress is proportional to strain and vice versa.

So a fiber like carbon fiber is said to be Hookean and obeys this law, but true to form, there are always exceptions [8–10].

Having mounted the filament on the sample card and the diameter determined, the card is carefully mounted in the lightweight jaws of an Instron test machine. The top jaw has a gimbal mount and should be supported whilst the card is cut through along A...A to free the filament in the test jaws. Using a full scale load of 20 N, a load is applied, at 10 mm min^{-1} for precursor and oxidized PAN fiber and 2 mm min^{-1} for carbon fiber, until the filament breaks. The software package of the Instron test machine calculates the

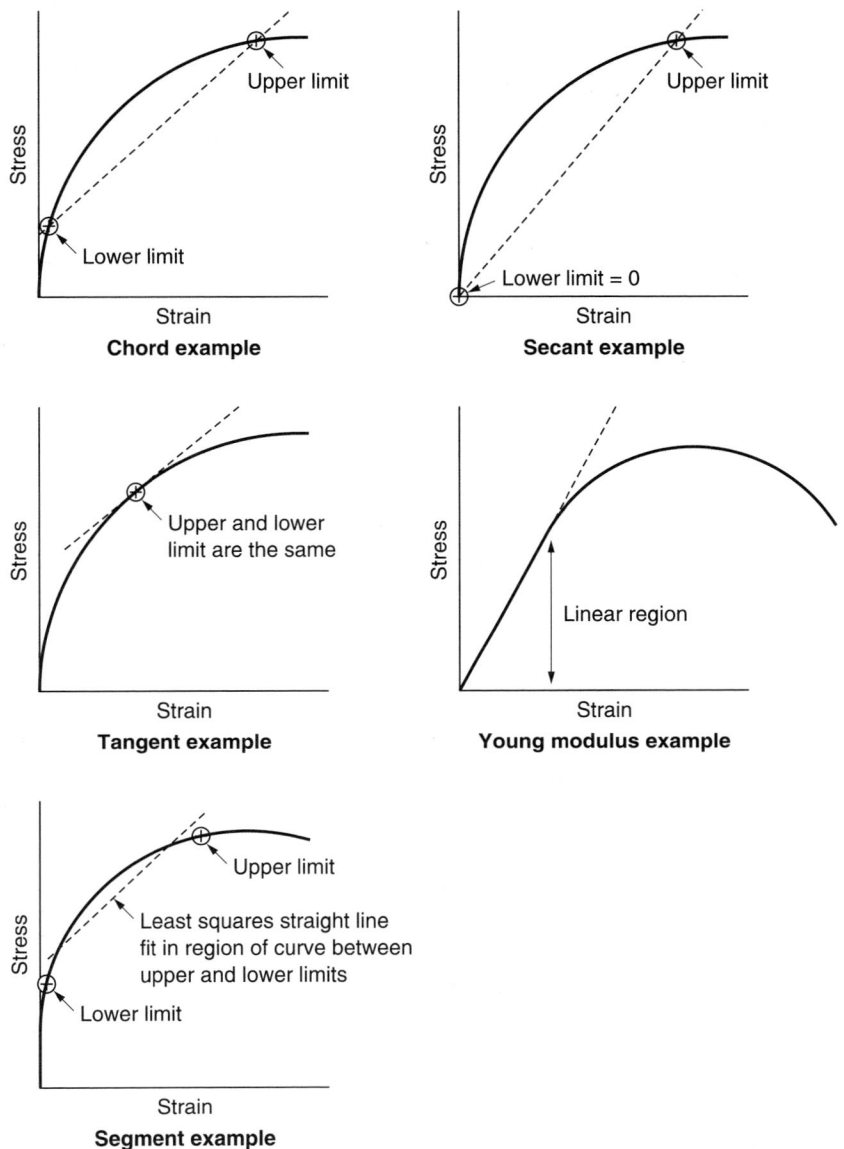

Figure 17.15 Different methods for determining modulus with an Instron test machine.

modulus and tensile strength for a known the filament diameter and is quoted as the mean of 10 determinations.

It is important to establish which method has been used for calculating the modulus and Figure 17.15 depicts examples of Chord, Secant, Tangent, Young's and Segment methods that the Instron software can calculate. Due to the toe in with oxidized PAN fiber, SGL use the Chord Method, while for carbon fiber they use the Secant Method, which in effect, is equivalent to taking the stress and dividing by the strain at a given point. Before any comparisons can be made with different laboratories, it is essential to know which method has been used. Some standards like SACMA work to levels of strain for a given fiber type (Table 17.5).

A generalized stress-strain curve for an acrylic fiber is shown in Figure 17.16.

Table 17.5 SACMA test method parameters for determining composite modulus

Fiber properties at break		Composite modulus measurement		
Elongation %	Microstrain	Elongation %	Microstrain	Maximum preload microstrain
>1.2	$\varepsilon \geq 12000$	0.1–0.6	1000–6000	200
0.6–1.2	$6000 \leq \varepsilon < 12000$	0.1–0.3	1000–3000	100
0.6	$\varepsilon < 6000$	0.05–0.15	500–1500	100

Note: 1% elongation is 0.01 strain, which is 10,000 microstrain.

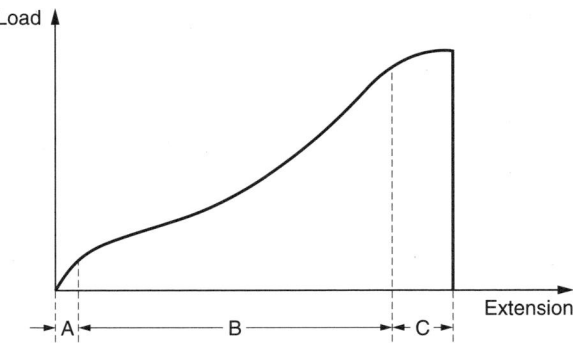

Figure 17.16 A generalized stress-strain curve for an acrylic fiber. Region A – follows Hookean behavior. Region B – thought to be localized chain conformational changes. Region C – only occurs when the fibre is above the T_g and attributed to irreversible flow of chains. *Source:* Reprinted with permission from Rosenbaum S, *J Appl Polym Sci.* 9, 2071, 1965. Copyright 1965, John Wiley & Sons Ltd.

17.3.4.3 Measurment of filament tensile strength

Having determined the filament modulus on the initial part of the stress-strain curve, the load is continued to be applied until the filament breaks. The gage length used affects the strength and a smaller gage length will give higher answers, associated with fewer flaws within the gage length [11], although there may be an element of truth in the suggestion that some of the weaker filaments are broken prematurely due to the increased difficulty of mounting these smaller gage lengths. Moreton assumed that the individual strengths had a normal distribution, although Hitchon and Phillips suggested there may be more than one distribution of flaws [12]. Hughes *et al* [13,14] suggest that the critical gage length is 0.3 mm. Dai and Piggott discuss the strength of carbon fiber as a function of length [15].

Testing filamentary materials is subject to variability and Table 17.6 shows typical scatter obtained when testing Courtauld's TS/11 precursor material for strength, extension, initial modulus and diameter.

17.3.5 Determination of oxidized PAN fiber finish content

Oxidized PAN fiber is treated with a finish to act as a lubricant and this can be extracted with CH_3OH by one of several routes:

1. The slowest route is by Soxhlet extraction following the same procedure as used for the determination of the finish content on Courtelle precursor.

TESTING OF PAN PRECURSOR, VIRGIN CARBON FIBERS

Table 17.6 Filament testing of Courtelle TS/11 precursor

Property		Sample A	Sample B	Sample C	Sample D	Sample E	Sample F	Sample G	Sample H	Sample Mean
UTS (MPa)	Mean	402	408	440	394	405	365	341	417	397
	S.D.	58	59	46	74	61	99	32	102	66
	Min	283	295	378	245	301	194	264	277	280
	Max	385	504	500	475	512	479	384	651	486
Extension (%)	Mean	29.0	22.8	28.0	22.9	26.2	20.9	31.4	27.1	26.0
	S.D.	4.3	5.0	3.6	5.3	5.6	5.8	4.6	3.3	4.7
	Min	22.1	14.7	21.5	13.3	15.8	11.5	25.7	25.0	18.7
	Max	36.9	28.6	34.4	29.3	35.3	27.0	36.1	31.4	32.4
Initial Modulus (MPa)	Mean	1578	2482	2959	2974	4370	2903	3043	2671	2873
	S.D.	407	1010	1679	1426	2316	1483	1413	2062	1475
	Min	1272	1160	1522	1938	2316	1485	1251	745	1461
	Max	5303	4545	6583	5695	7496	5678	5303	4832	5679
Diameter (μm)	Mean	11.7	12.1	12.1	12.3	12.6	11.5	12.8	12.0	12.1
	S.D.	0.7	0.9	0.8	0.6	0.6	1.0	0.3	0.8	0.7
	Min	10.8	10.5	10.5	10.8	11.9	10.0	12.5	10.4	10.9
	Max	13.6	12.8	12.8	12.7	13.6	12.8	13.6	12.7	13.1

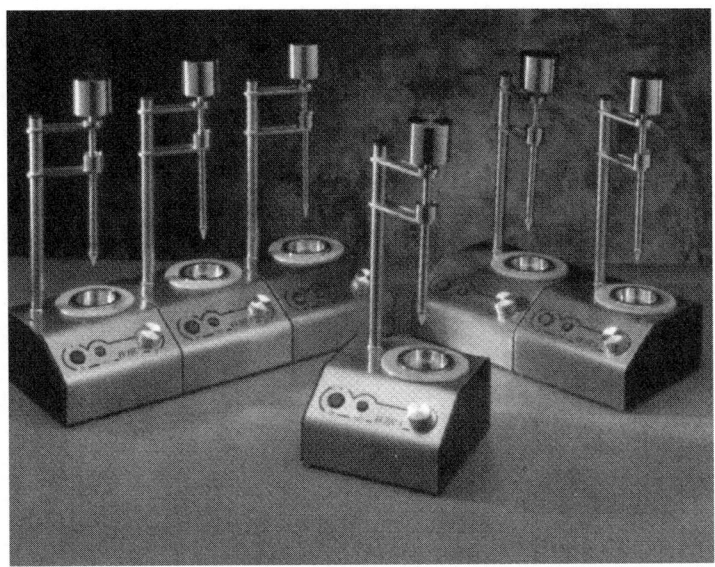

Figure 17.17 WIRA rapid oil extraction equipment.

2. A faster technique is to use a WIRA Rapid Oil Extraction Apparatus (Figure 17.17). A plug of cotton wool is pushed into the extraction tube to retain particulate matter, followed by a 2 g weighed sample of opf, which is tamped down with a weighted plunger to achieve some consolidation. The tube is then filled with CH_3OH and the extract allowed to collect in an Al tray situated on a heated hotplate directly under the extraction tube. To avoid loss of extract by spitting on initial contact with the hot tray, a small amount of solvent is introduced into the hot tray. When all the CH_3OH has passed through the extraction tube, slight pressure is applied to the weighted plunger to express any remaining solvent. The extract is carefully evaporated to remove all solvent taking care not to overheat, since the finish could be decomposed. This process is aided by sitting the Al tray between two rings of Al, which act as a heat sink and ensure

Figure 17.18 Tecator Soxtec fast extraction equipment.

uniform heating. The tray is cooled in a desiccator and then weighed; the result is quoted as the mean of three determinations.

3. There is a fast extraction system available (Tecator Soxtec) (Figure 17.18), which uses hot solvent, heated indirectly by hot circulating oil. It is about five times faster than a conventional cold soxhlet extraction and up to 65% of the solvent can be recovered at the end of the extraction cycle.

17.3.6 Determination of carbon fiber size content

1. Ordinarily, an epoxy size can be determined with butan-1-one (MEK) using a soxhlet extraction for 2 h, dried, cooled in a desiccator and weighed.
2. A quick method of determining size content is by extracting three times in a flask with cold MEK and drying with a UV lamp.
3. Some sizes will necessitate using H_2SO_4/H_2O_2 to effect extraction by chemical breakdown of the size.
4. Glycerol (glycerine) size plus water content can be determined by extraction with water, but since glycerol is hygroscopic, the actual glycerol content is determined by reacting with sodium periodate to produce HCOOH, removing excess periodate by reaction with ethane-1,2-diol and titrating the HCOOH with 0.1M NaOH, using phenolphthalein as indicator [16].

Equation I

$$\underset{\text{glycerol}}{CH_2OH\text{---}CHOH\text{---}CH_2OH} + \underset{\text{sodium periodate}}{2NaIO_4} \rightarrow \underset{\text{formic acid}}{HCOOH} + \underset{\text{formaldehyde}}{2HCHO} + \underset{\text{sodium iodate}}{2NaIO_3} + H_2O$$

Equation II $\underset{\substack{\text{(ethylene glycol)}}}{CH_2OHCH_2OH} \underset{\text{ethene-1,2-diol}}{} + NaIO_4 \rightarrow 2HCHO + NaIO_3 + H_2O$

$1 \text{ cm}^3 \, 0.1M \text{ NaOH} \equiv 0.00921 \text{ g glycerol}$

5. A PTFE size can be determined by burning off for 2 h in a furnace at 700°C using a constant flow of N_2 (20 $dm^3 \, min^{-1}$).

17.3.7 Conductivity of a water extract

This test is useful in detecting the level of ionic impurities on carbon fiber treated with an aqueous surface treatment process. The method involves extracting the carbon fiber for 1 h with distilled water using 20 g distilled water/1 g carbon fiber, transferring to a stoppered flask (to avoid CO_2 pick-up), cooling and measuring the conductivity with a conductivity meter, correcting the reading to 20°C and subtracting the reading for a blank determination. The conductivity meter is calibrated at 20°C with 0.01M KCl, which has a conductivity of 127,800 $\mu S\,m^{-1}$ at 20°C.

$$\text{Conductivity at } 20 = \frac{KC}{1 + 0.02(T - 20)}$$

where K = cell constant at about 0.01M
C = conductance
T = temperature, °C

17.3.8 Skin core

To establish the degree of oxidation in an oxidized PAN fiber, it is advantageous to prepare a section for examination under the microscope. The following technique has been adapted from one used in hospitals to look at sections of bone.

Select a bundle of some 1000 filaments of opf and tie onto a small frame, ensuring that the fiber is taut. Place in a prefabricated tray made from Al foil and immerse with molten Ralwax (a hardened wax from Merck with a congealing point of 55–58°C) applied from an industrial wax heater (Figure 17.19). Place the tray in an oven at 60°C for 30 min to ensure complete wetting out and then cool in a freezer until the wax is hard. Remove from the container and trim with a scalpel to a neat rectangular block that will fit into the jaws of a microtome. Cool the block in a freezer and cut sections 12.5 μm thick with the microtome, forming a continuous ribbon, which is supported with tweezers to effect a clean removal from the blade.

Figure 17.19 Shimalert industrial wax heater.

Sample sections are floated on a pool of 1% gelatin solution (containing a crystal of thymol to prevent mold growth) positioned on a clean microscope slide, transferring the sections with a fine paint brush wetted with the gelatin solution. Shake off any excess gelatin solution and dry for 16 h at 60°C to adhere the sections to the glass slide. At all times, ensure that the face of the slide is not touched to avoid damage. Place two slides back to back in the slot of a Coplin staining jar filled with xylene (flammable) to dissolve away the gelatin. After 10 min, remove from the Coplin jar and shake to remove excess xylene. Place a drop of DPX mountant (a neutral plastic base in xylene, hence compatible with the previous xylene wash stage) on the top of a slide and allow the DPX to run down over the specimen. Avoid using too much mountant. Carefully lower a coverslip over the specimen, taking care to avoid trapping air and finally seal the edges of the coverslip with clear nail varnish. Ensure that the slide is correctly labeled.

Examine the section under a microscope for skin core effects (Figure 3.51).

17.3.9 Measurement of electrical properties

By altering the process conditions, it is possible to manufacture carbon fiber with a wide range of resistivity from below 10 Ω cm to above 10^6 Ω cm. The lower resistivity fibers can be measured with a standard digital multimeter (e.g. a Hewlett Packard HP 34401A), but for the high resistivity fibers, special equipment is required (e.g. the HP 4339B), which will measure in the range 10^3–1.6×10^{16} Ω and uses high voltages up to 1000 V. It is advisable to use a test fixture (HP16339A), which safely holds the specimen and the range is adjusted by selection of an appropriate resistor. Since the resistivity of these fibers can change with time, it is advantageous to use the test fixture to enable rapid measurements. A filament is attached to a standard mounting card in the normal way using sealing wax, but with the blobs of wax are not on the periphery of the hole, but about 40 mm apart, sufficient to enable the filament diameter to be determined. Pieces of thin Al foil are attached to the card with self adhesive tape, making good electrical contact with the filament forming a practical gage length of 30 mm, which is suitable for the test jig. Electrical connection is then made to the Al foil using test leads with alligator clips attached. The resistance is measured with the appropriate meter depending on the resistance. The technique of using a Ag filled conductive epoxy resin to secure the filament and simultaneously fix the gage length takes too long for the epoxy resin to cure and is not suitable.

$$\text{Resistivity} = \frac{\pi d^2 \Omega}{4L} \; \Omega \, \text{cm}$$

where d = diameter, cm
Ω = resistance, ohm
L = gage length, cm

17.4 CARBON FIBER TOW TESTING

17.4.1 Dry tow test [17,18]

To undertake a dry tow test it is necessary to mount a tow in some form of end fixtures, which enable the tow of a given gage length to be gripped in the jaws of a test machine. The machine is best fitted with a UV recorder, which has a fast response time and an integration device, which can compute the area under the stress-strain curve.

Work to fracture (W_f) is the work done under the stress-strain curve:

$$W_f = \frac{\text{Stress} \times \text{Strain}}{2}$$

Since the YM, $E = \dfrac{\text{Stress}, \sigma}{\text{Strain}}$

$\therefore$ strain $= \dfrac{\text{Stress}, \sigma}{E}$

Hence $W_f = \dfrac{\sigma^2}{2E} = \dfrac{\text{Stress} \times \text{Strain}}{2} = \dfrac{1}{2} \times \dfrac{\text{Load}}{\text{c.s.a}} \times \dfrac{\text{Extension}}{\text{Gage length}}$

Now, cross sectional area $= \dfrac{\text{mpul}}{\text{density}}$ and $A = \dfrac{\text{Load} \times \text{Extension}}{2}$

where A = area under the stress-strain curve

Then $\dfrac{\sigma^2}{2E} = \dfrac{A\rho_f}{M_L L_0}$

where M_L = mass per unit length and L_0 = gage length

$$\text{i.e. } \sigma^2 = \frac{2AE\rho_f}{M_L L_0} \qquad (1)$$

Since c.s.a $= \dfrac{M_L}{\rho_f} = \dfrac{N\pi d^2}{4}$

where N = total number of filaments in tow of mean diameter d

$$\text{Hence, } \sigma^2 = \frac{8AE}{\pi d^2 N L_0} \qquad (2)$$

Either by knowing, or assuming, a value for the YM, it is possible to calculate the strength using equations (1) or (2).

In this test, the specimen comprises a multitude of filaments of different strengths and slightly different gage lengths. As the load is applied to the specimen, the weakest filaments with the shortest gage length fail first. Although an individual filament may be weaker, it may have a longer gage length and consequently the load will not be applied till later. In a dry tow test, there is no resin to transfer load and the stress-strain curve gradually reaches a peak and then falls off gradually until the final filament is broken (Figure 17.20).

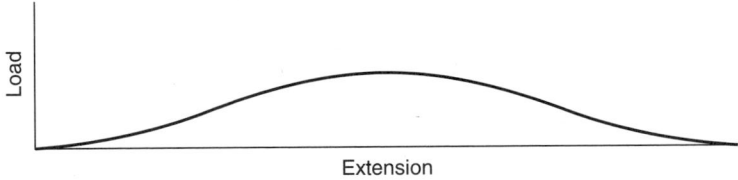

Figure 17.20 Stress-strain curve for dry tow test.

17.4.2 Testing of the impregnated tow

The impregnated tow test can be considered a form of mini composite. It is normal to use an epoxy resin for impregnating the tow and attach some form of end pieces to enable the tow to be gripped in the jaws of a test machine. The resin used for impregnation does contribute to the strength and modulus and corrections can be made (Figures 17.21, 17.22).

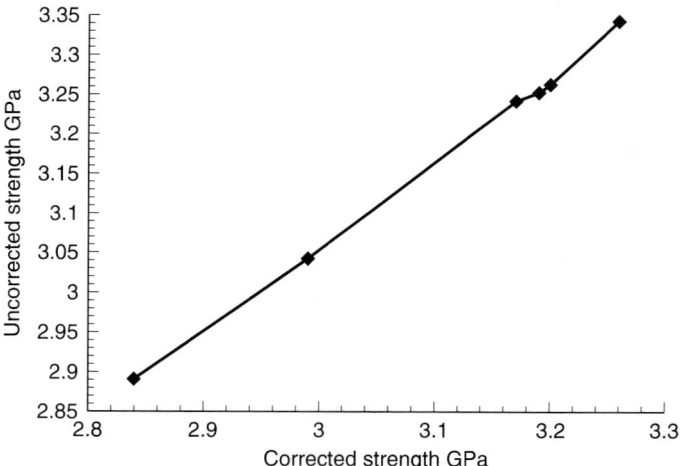

Figure 17.21 Correction for resin content in tensile strength tow test.

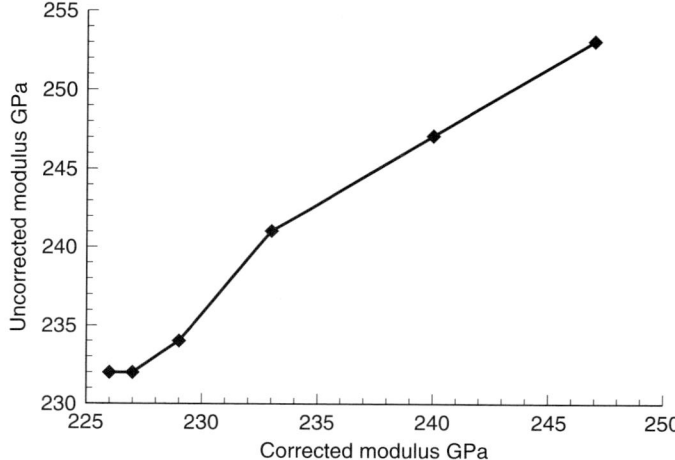

Figure 17.22 Correction for resin content in tensile modulus tow test.

The tow can be impregnated by pulling it through a resin bath diluted with 15–50% of butan-1-one solvent, followed by passage through a graphite die to ensure thorough penetration of the resin solution into the tow bundle as well as removal of the majority of excess resin. The resin must be mixed in glass or stainless steel containers, which can be cleaned after use. The wax used for a waxed cardboard container can inhibit the cure of some resin systems. If an allowance is to be made for the contribution of the resin, the mass per unit length before and after impregnation/removal of solvent/curing is determined to establish the resin content. During the curing process, it is beneficial to apply tension on the tow to align the filaments within it, although when determining modulus, it is not possible to obtain the same degree of alignment as achieved with a single filament. Initially, it was not necessary to use end tabs gripping the impregnated tow with hydraulically actuated grips lined with a soft material, however, as carbon fiber with a higher strength was produced, it became essential to use some form of end tab to transmit the load into the specimen.

The original Hercules end tab was a cast polyamide cylindrical block with a radiused end section to eliminate jaw failure (Figure 17.23). The end tab was gripped between split collets and to ensure that the collets did not slip, a top hat section was added to provide downward

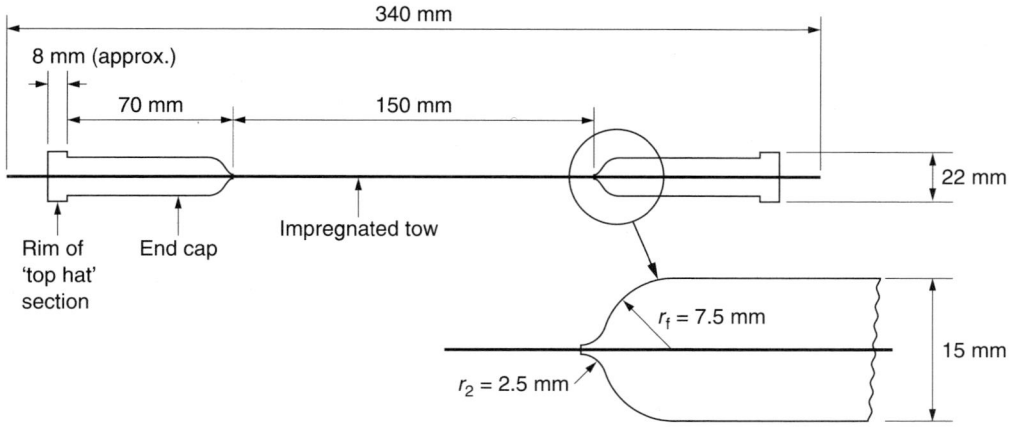

Figure 17.23 Hercules' tow test as modified by Courtaulds.

force to initiate the action of the collets. The method worked well but was labor intensive and it took about 24 h before results were available.

ISO 10618 uses a simplified tab formed by sandwiching the ends of the tow between two pieces of thin card (50×20 mm^2) using an epoxy resin adhesive. RAE developed a method where a 1 mm diameter sock of glass braid (Peribraid) [19], impregnated with an epoxy adhesive, is pulled over the ends of the cured tow specimen and cured in position.

Modulus is determined by the attachment of a light weight 50 mm dual gage extensometer, applying a preload (about 5 kgf) and measuring strain up to some 50 kgf, using a crosshead speed of 5 mm/min. Finally the extensometer is removed and the application of the load continued until the sample breaks.

$$\text{Fiber modulus} = (P_2 - P_1)\frac{I_f \rho_f}{M_f} \times \frac{L_o}{L_2 - L_1} - \frac{E_r \rho_f}{\rho_r}\left[\frac{I_f}{M_f} \times \frac{M_c}{I_c} - 1\right]$$

$$\text{Fiber tensile strength} = \frac{P I_f \rho_f}{M_f} - \varepsilon\frac{E_f \rho_f}{\rho_r}\left[\frac{I_f}{M_f} \times \frac{M_c}{I_c} - 1\right]$$

P = Breaking load
P_1 = Initial load (about 5 kgf)
P_2 = Final load (about 50 kgf)
L_0 = Gage length (50 mm)
L_1 = Length between knife edges of extensometer at load P_1
L_2 = Length between knife edges of extensometer at load P_2
I_f = Length of virgin fiber of mass M_f
I_c = Length of impregnated tow of mass M_c
ρ_f = Density of virgin fiber
ρ_r = Density of cured resin (can assume 1.22 g cm^{-3})
E_r = Tensile modulus of cured resin (can assume 3.10 GPa)
ε = Breaking strain of fiber (can assume 0.015 for high strength fiber and 0.005 for high modulus fiber)

Note:

$$\frac{I_f}{M_f} \times \frac{M_c}{I_c} = \frac{1}{F}$$

where F is the fiber mass fraction of the resinated tow, assume $F = 0.40$.

Table 17.7 Evaluation for a significant outlier

Number in sample	Factor	Number in sample	Factor	Number in sample	Factor	Number in sample	Factor
3	1.15	12	2.41	21	2.73	50	3.13
4	1.48	13	2.46	22	2.76	60	3.20
5	1.71	14	2.51	23	2.78	70	3.26
6	1.89	15	2.55	24	2.80	80	3.31
7	2.02	16	2.59	25	2.82	90	3.35
8	2.13	17	2.62	30	2.91	100	3.38
9	2.21	18	2.65	35	2.98		
10	2.29	19	2.68	40	3.04		
11	2.36	20	2.71	45	3.09		

Note: If $((\bar{x} - \text{outlier})/\sigma)$ is greater than the factor, it is classed as an outlier.

Table 17.8 Effect on the tow test properties of 3k carbon fiber by combining a number of 3k tows to form a multiple tow

Number of tows impregnated	Strength GPa	C of V %	Modulus GPa	C of V %
1 × 1k (by extrapolation)	3.84	–	252	–
2 × 3k	3.70	1.8	244	1.9
3 × 3k	3.54	4.4	236	1.2
4 × 3k	3.30	4.8	227	5.0

It is useful to mount a steel wire (e.g. a mild steel welding rod) in the end caps to use as a secondary standard (the modulus of steel is 210 GPa).

To determine whether a dubious result was a flier, the procedure given in Table 17.7 can be adopted.

Testing 3k tows does present a problem since attaching the extensometer to a single tow is not practical and using more than one tow gives results which are not truly additive (Table 17.8).

17.5 TESTING OF CARBON FIBER YARN AND FABRIC [203]

Yarns can be continuous filament or spun from staple fiber. A measure of the diameter is obtained from the tex (g of 1 km of yarn) although the spun yarn will not be uniform in cross-section. There are two types of twist, designated 'S' and 'Z' as depicted in Figure 17.24. When one measures the twist, it does not take into account the diameter and to allow for this, a twist factor can be introduced where:

$$\text{Tex twist factor} = \text{turns per meter} \times \sqrt{N}, \text{ where } N = \text{tex}$$

Twist increases the effect of cohesion in the yarn and the yarn strength will increase up to a point where any additional twist starts to break the fibers. Twisted yarns can be doubled by plying two yarns, normally of opposite twist, to obtain a further increase in strength. If the yarns are the same twist, the product can be twist lively and when relaxed, any excessive twist will cause the doubled yarn to snarl and curl up on itself.

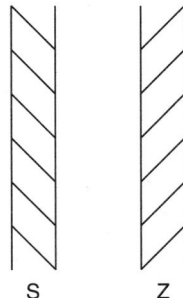

Figure 17.24 Diagrammatic representation of S- and Z- twist.

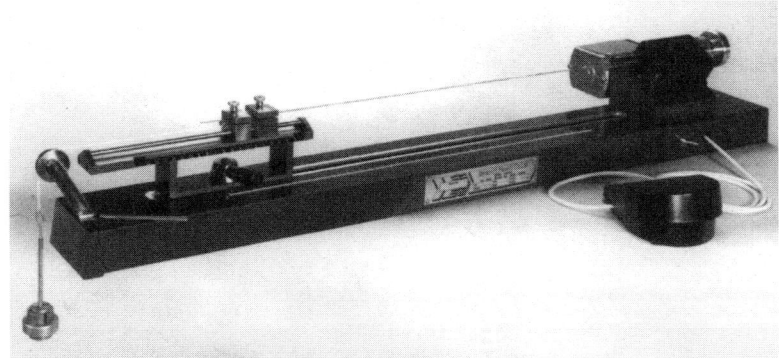

Figure 17.25 Twist tester. *Source:* Courtesy of James H Heal & C. Ltd.

17.5.1 Determination of twist

When sampling, it must be remembered that the sample end may have lost some of its twist and it will be necessary to cut off 1 m before sampling, trapping the yarn between the fingers, but not running it between finger and thumb as this will redistribute the twist. If the yarn is on a package, it must not be sampled end over as this will introduce one turn of twist per package circumference.

There are two types of twist measuring equipment available—one is manual (Figure 17.25) where a fixed length of yarn is held under a slight tension with a mass equivalent to 0.5 tex ±10% and the twist is unraveled by hand turning a wheel. A motor can also be used, until all the twist has been removed while observing the untwisted yarn through a magnifier. The twist level is indicated on a counter.

The other equipment is the WIRA digital twist tester, with an integral microprocessor and is totally automatic.

17.5.2 Determination of ends and picks

A fabric has a number of ends running parallel to the weave and a number of picks, or weft threads, running transverse to the direction of the weave (also known as fills or woofs). To establish the construction of a cloth, it is necessary to determine the number of ends and picks, either by unraveling, or by using a counting glass (Figure 17.26), with the sample resting on an illuminated viewing box.

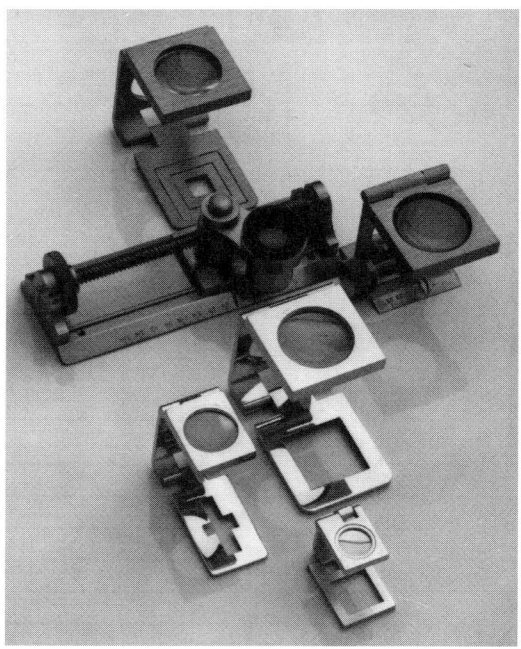

Figure 17.26 Range of piece glasses and pick counters. *Source:* Courtesy of James H Heal & C. Ltd.

17.6 TESTING OF MATRIX

It is necessary to mix resin formulations for a number of applications e.g. sizes, prepreg, pultrusion and filament winding. To check the mixed resin system, a number of tests can be undertaken.

17.6.1 Fineness of grind

If, for example, dicyandiamide is being milled into the product, it is necessary to determine the degree of fineness of the milled dicy using a fineness of grind gage, sometimes called a Hegman gage. It is basically a block of hardened stainless steel with a chamfered channel cut into the block about 0–25 μm deep. A small sample of milled resin is placed in the channel at the deep end and a precision ground hardened doctor blade is drawn at right angles over the resin and towards the shallow end. Holding the gage at grazing incidence and the point where the particles just merge into the film is noted and read on the scale.

17.6.2 Selection of a suitable grade of paper for resin coating

When coating a paper with resin, it is necessary for that coating to be uniform across the entire surface. A convenient way of ascertaining the effectiveness of a grade of paper is to use a Mayer bar, which is a round bar helically wrapped with a wire. The bar is mounted in bearings at either end, with an adjustable device to exert downward pressure, so that when a piece of the test paper trapped below it is pulled uniformly in a horizontal direction, the Mayer bar will rotate and permit the resin, deposited in a pool behind the bar, to be uniformly metered onto the paper surface. The diameter of the wire and the dimensions of the helix control the amount of resin deposited on the paper and a selection of bars will permit the choice of resin film thickness.

17.6.3 Determination of gel time

To assist in the choice of the correct cure conditions, it is advantageous to know the gel time of a resin system.

1. Using the Kofler hotbench

This equipment is a Reichert-Jung hot stage accessory for the Leica-Reichert microscope and can be used without adaption for the determination of gel time. The hot stage (Figure 17.27) is a metal strip constructed from materials of different heat conductivity and heated in such a way that the temperature slope of the hot stage is practically linear. The bench can be calibrated with materials of known melting point and a test substance is selected with a melting point closest to the temperature at which the gel time is to be determined (Table 17.9). To calibrate the hot bench, spread the test substance on the hot surface, allow the test substance to melt and move the pointer over the line of the border between the liquid and solid phase and adjust the index tip of the rider to coincide with the known melting point.

To determine the gel time, place a small amount of resin at the chosen temperature as indicated by the test pointer and measure the time taken for the resin to go stringy.

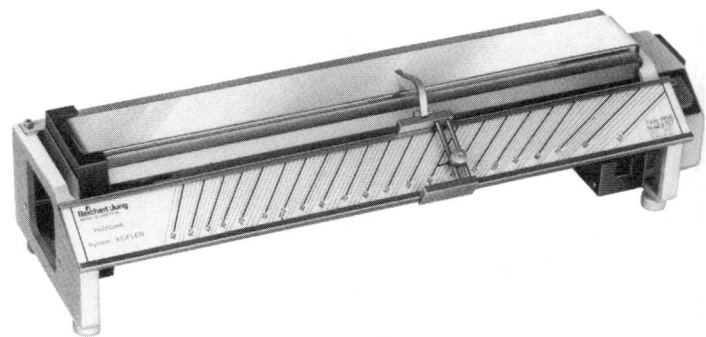

Figure 17.27 Kofler hotbench. *Source:* Courtesy of Reichert-Jung.

Table 17.9 Melting points of calibration substances for the Kofler hotbench

Name	mp °C
Azobenzene	68
Benzil	95
Acetanilide	114.5
Phenacetin	134.5
Benzanilide	165
Succinic acid	189
Dicyandiamide	210
Saccharin	228

2. Determination of gel time at ambient temperature

This gel time equipment made by Shyodu is a low torque synchronous motor to which a specially shaped stirrer is attached. The resin mix is placed in a throw away paper cup and the stirrer placed in position. The timer is started and then stopped when the motor stalls due to gelation occurring as the viscous drag exceeds the motor torque. Care must be taken to ensure that the paper cup is not coated with a wax that will inhibit the cure.

17.6.4 Determination of the viscosity of a resin mix

A most useful form of viscometer is the cone and plate type. A relatively inexpensive version developed by ICI and manufactured by Research Equipment (London) Ltd. (Figure 17.28) can be obtained in several viscosity ranges. Six set temperatures in the range 25–200°C and two different cone angles (0.5° or 2°) are available. Lower the cone onto the plate and allow 10 min to reach working temperature. Adjust the zero, raise the cone and place 3 or 4 drops of standard silicone oil (200 poise) onto the plate, lowering the cone so that it is completely surrounded by the oil and wait 30 sec. Press the button and read the indicated viscosity. The exact position of the head when lowered is most important and sufficient free-play should allow the cone to be lifted off the plate about 0.5 mm. The procedure is repeated with the sample material.

Other more sophisticated instruments are available like the Feranti-Shirley and Haake viscometers. Figure 17.29 shows viscosity data obtained for Hercules resin system 3501-5A using a Feranti-Shirley viscometer.

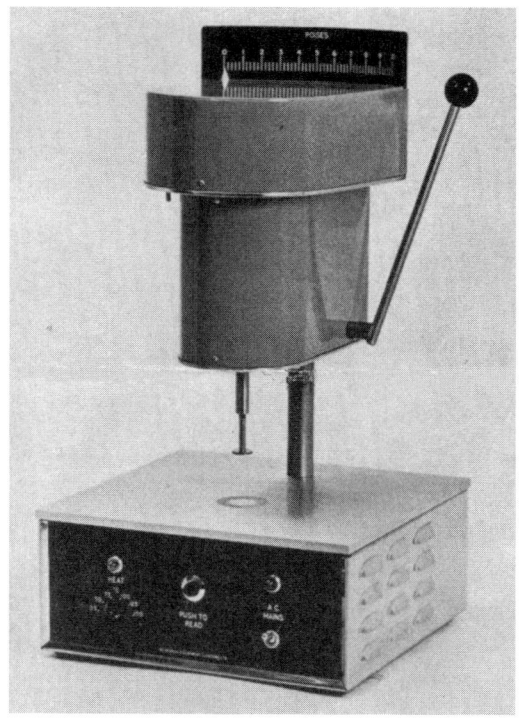

Figure 17.28 Cone and plate viscometer. *Source:* Courtesy of Research Equipment (London) Ltd.

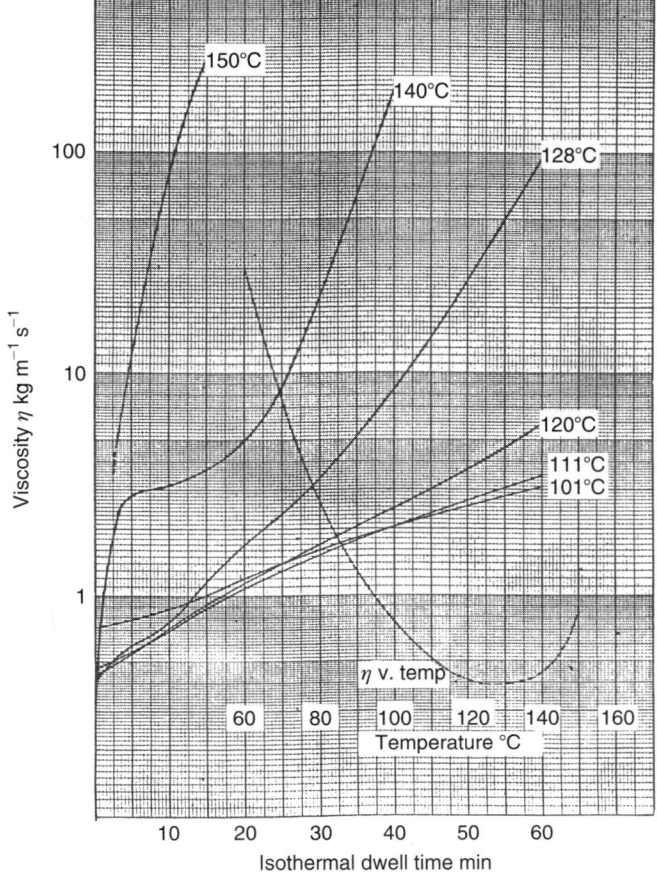

Figure 17.29 Viscosity data for Hercules' resin system 3501-A. Heat-up rate 3°C min^{-1}. *Source:* Determined by Rolls Royce Ltd.

17.6.5 Determination of the epoxy molar mass (EMM) of epoxy resins

The epoxy molar mass is the grams of resin per mole of epoxy group and used to be termed the weight per epoxy equivalent(WEEPY). Individual methods do give different results and the method used should always be quoted.

1. Cetyl trimethylammonium bromide-perchloric acid titration method

The sample is dissolved in a 4:1 mixture of dichloromethane and HCOOH and a stoichiometric excess (2 g) of cetyl trimethylammonium bromide (*N*-hexadecyl-*N*, *N*, *N* trimethylammonium bromide) (CTAB) added and titrated with 0.1N acetous perchloric acid (ASTM E 200) using crystal violet as indicator.

$$\text{EMM (g/mol)} = \frac{1000m}{(V_2 - V_1) \times M}$$

where m = mass of sample, g
V_1 = volume of acetous perchloric acid used for the sample titration, ml
V_2 = volume of acetous perchloric acid used for blank titration, ml
M = molality of acetous perchloric acid at the reagent temperature during the titration.

For each 1°C above this temperature, subtract 0.00011 from the molality as standardized and vice versa.

2. Determination of EMM by potentiometric titration

The sample is dissolved in dichloromethane, an excess of CTAB added, followed by a known excess of 0.1 N perchloric acid in glacial CH_3COOH, which reacts with the quaternary ammonium salt and liberates HBr, which reacts with the epoxy group. The excess acid is back titrated potentiometrically with standard 0.1 N sodium acetate in glacial CH_3COOH. The CH_3COONa is standardized against 0.1 N perchloric acid in glacial acetic acid using Oracet Blue B as indicator. Similarly, 2 g of CTAB is titrated against perchloric acid in glacial CH_3COOH using Oracet Blue B as indicator.

$$\text{EMM (g mol}^{-1}\text{ mol)} = \frac{1000w}{(T_1 \times N_1) - (T_2 \times N_2) - (T_3 \times N_3)}$$

where, T_1 = volume of standard 0.1N perchloric acid solution added to the sample
T_2 = volume of standard 0.1 N CH_3COONa solution required to neutralize the excess perchloric acid solution
T_3 = volume of standard 0.1 N CH_3COONa solution required to neutralize 2 g of CTAB
N_1 = normality of the perchloric acid solution
N_2 = normality of the CH_3COONa solution
w = weight of sample taken, g

17.7 TESTING OF CARBON FIBER PREPREG

17.7.1 Mass per unit area

Often referred to as areal weight, the mass per unit area can be determined by placing a square metal template on the prepreg, cutting round it and weighing it. Knowing the area, the mass per unit area can be calculated. This does not reflect the variation along the length and the width and it is more meaningful to cut rectangular sections along the full length full width and taking the mean areal weight.

17.7.2 Volatiles content

It can be expressed as the percentage loss in weight when a sample of prepreg is freely suspended in an oven at 150°C for 30 min with an epoxy system or 20 min for a phenolic system.

17.7.3 Fiber content

It is determined by extraction of the resin system in a soxhlet apparatus with a suitable solvent such as butanone (note that dicy is not soluble in butanone). If any crosslinking of the resin system has occurred, then it will be necessary to use the H_2SO_4/H_2O_2 digestion system.

17.7.4 Resin gel time

The determination is the same as described in 17.6.3.

17.8 TESTING OF CARBON FIBER COMPOSITE

17.8.1 Introduction

In the composite tests to follow, the reader is referred to text books on classical beam theory should one wish to derive the theory [20–23]. Many excellent publications have been written on the subject of composite materials covering analysis, mechanics, principles and performance [24–49].

Chapter 18 gives a brief introduction to the micromechanics of unidirectional composites and quotes some of the composite properties. Figure 17.30 shows the principal coordinate axes for unidirectional lamina. The unidirectional ply is characterized by:

E_{11} = longitudinal stiffness (elastic modulus)(E_1)
E_{22} = transverse stiffness (E_2)
E_{33} = stiffness (E_3)
G_{12} = in plane shear modulus = shear modulus in 1-2 plane = G_{21}
G_{23} = shear modulus in 2-3 plane = G_{32}
G_{13} = shear modulus in 1-3 plane = G_{31}
ν_{12} = major Poisson's ratio (load in 1-direction, strain in 2-direction)(please note that this is different from ν_{21}, which is the minor Poisson's ratio)
ν_{23} = Poisson's ratio (load in 2-direction, strain in 3-direction) (different from ν_{32})
ν_{13} = Poisson's ratio (load in 1-direction, strain in 3-direction) (different from ν_{31})
σ_{1t} = longitudinal tensile strength
σ_{2t} = transverse tensile strength
σ_{3t} = tensile strength in the 3-direction
σ_{1c} = longitudinal compressive strength
σ_{2c} = transverse compressive strength
σ_{3c} = compressive strength in the 3-direction
τ_{12} = inplane shear failure in 1-2 direction = τ_{21} (interlaminar shear strength)
τ_{23} = inplane shear failure in 2-3 direction = τ_{32}
τ_{13} = inplane shear failure in 1-3 direction = τ_{31}
$\alpha_1, \alpha_2, \alpha_3$ = coefficients of thermal expansion
$\beta_1, \beta_2, \beta_3$ = coefficients of moisture expansion
$\kappa_1, \kappa_2, \kappa_3$ = coefficients of thermal conductivity

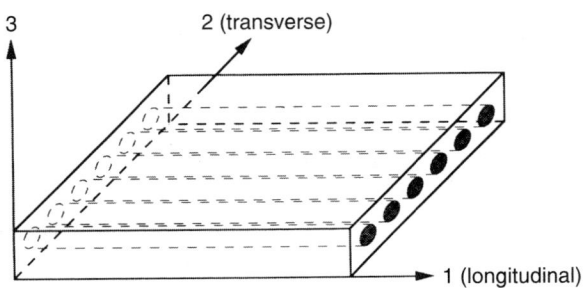

Figure 17.30 Principal coordinate axes for unidirectional lamina.

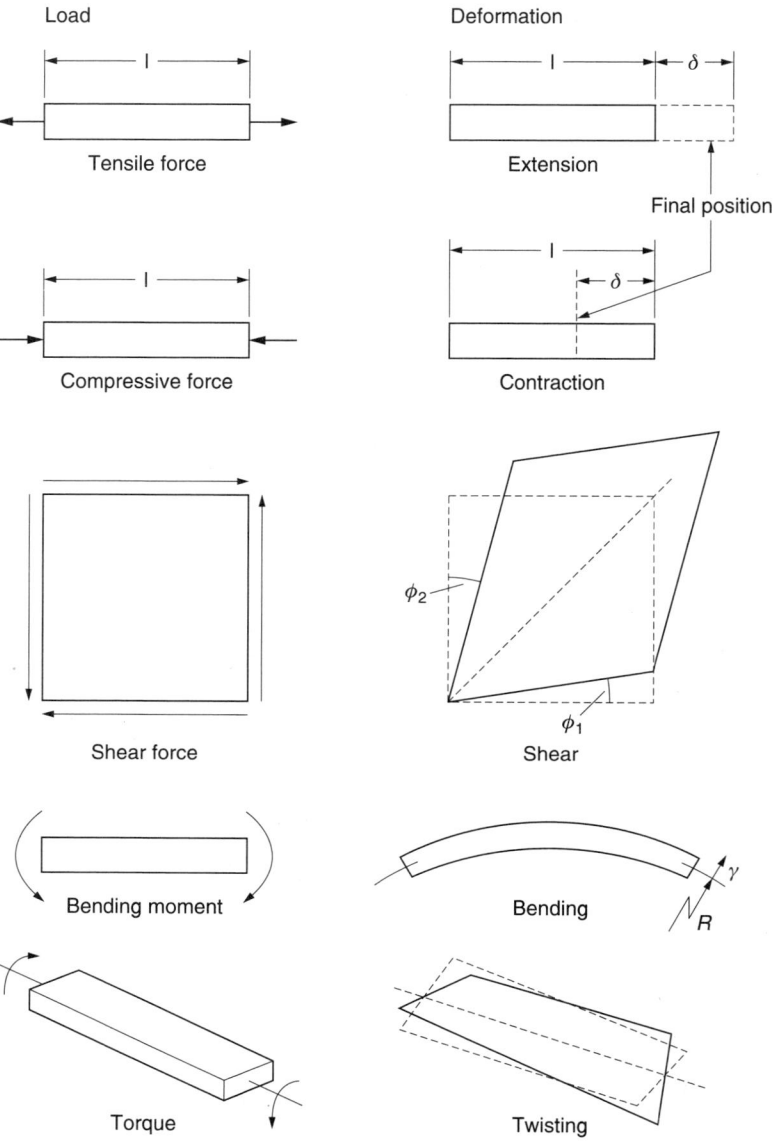

Figure 17.31 The loads with associated deformations of composite materials. *Source:* Reprinted with permission from Matthews EL, Rawlings RD, Chapman and Hall, 208, 1994. Copyright 1994, Springer.

Figure 17.31 [41] shows the different loads which can be applied to composite materials and their associated deformations.

17.8.2 Preparation of composite specimen from wet resins

A matched metal mold must be carefully ground from hardened steel and the finished dimensions carefully checked. The mold, when new, should be coated with silicone grease and baked for 24 h at 200°C, to acquire a patina with an improved release. All assembly set screws and their sockets should be smeared with silicone grease to prevent resin penetration and ensure their effective removal after resin cure. Prior to lay-up, the mold should be evenly coated with a mold release spray, positioned on a hotplate and the bottom of the mold coated with a thin layer of resin applied with a small paintbrush. Tows should be coated with resin heated to 50–60°C to lower the viscosity and achieve thorough penetration and carefully placed in the mold under tension to accomplish the best practical degree of alignment. The most common problems experienced when molding specimens are—failure to have used thoroughly clean molds and set screws that were insufficiently tightened. All molds should be thoroughly inspected for cleanliness before use and it will be found advantageous to use an air operated torque wrench to ensure that the correct and even degree of tightness is used for the set screws. In the early days, when there was little fiber available for test, the length of tow cut for insertion into the mold tended to be too short and once resin had penetrated the free ends of the tow, as the mold was closed, the resin simply extruded outwards together with the fiber. It is therefore important to ensure that the free ends of tow do not become impregnated with resin and there is sufficient dry fiber to achieve some interfilament cohesion to prevent possible extrusion from the mold as it is closed. If G-clamps are used to clamp the mold together during the cure cycle, it is advisable to use the forged version for greater strength and reliability.

Figure 17.32 shows typical packing fraction distributions for circular section fibers and Figure 17.33 shows the spacing between 7 μm diameter filaments for different packing fractions, generally a V_f of 60–65% is used and an accepted practice is to normalize the resin content of the composite to 60% by applying the Law of Mixtures (except for ILSS).

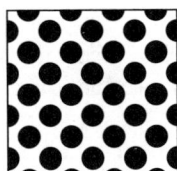 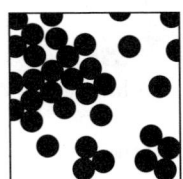 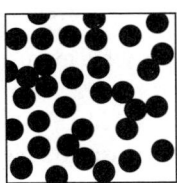

Figure 17.32 Packing fraction distributions for round fibers. (a) Uniform hexagonal packing. (b) Non-uniform clumped distribution. (c) Uniform random packing. *Source:* Reprinted with permission from Cogswell FN, Butterworth-Heinemann, Oxford, 81, 1992. Copyright 1992, Elsevier.

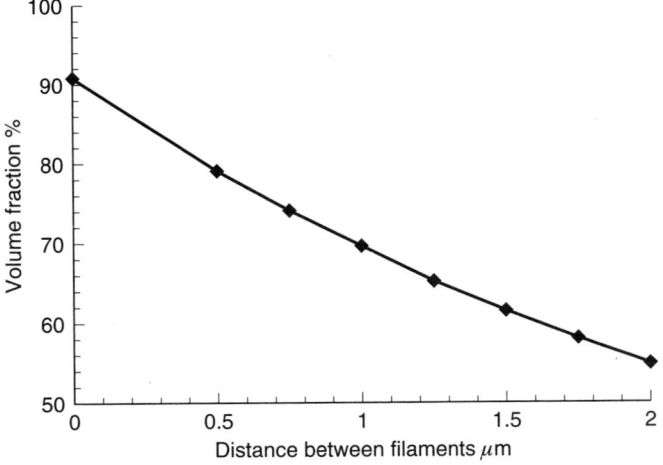

Figure 17.33 Distance between individual 7 μm diameter carbon fiber filaments in a regular hexagonal array at different packing fractions.

The number of tows required when making a 60% V_f composite bar would be:

$$\text{Number of tows} = \frac{0.60 wt\rho}{M_L}$$

where w = width of specimen
t = thickness of specimen
ρ = density of carbon fiber
M_L = mass per unit length of carbon fiber tow

17.8.3 Preparation of composite specimen from prepreg systems

To produce a composite from prepreg, it is necessary to know the areal weight, fiber content of the prepreg and the number of plies required. The number of plies required for a 60% V_f composite will be given by:

$$\text{Number of tows} = \frac{0.60 t\rho}{M_f M_a}$$

where t = thickness of lamina
ρ = density of carbon fiber
M_f = fiber mass fraction in prepreg
M_a = areal weight of prepreg

Molding can be achieved using a matched metal, a press-clave or an autoclave technique. Since the press-clave is a laboratory simulated version of a production autoclave, a brief description will be given (Figure 17.34).

A butyl rubber sheet, or high temperature plastic film, is held taut across two vertically stacked metal frames or rings, which are themselves sealed between the two platens of a press. The requisite number of prepreg plies is formed into a preform, which is sandwiched between sheets of PTFE coated glass cloth, or a non-perforated PTFE film with a perforated PTFE film next to the preform, and the pack placed on a steel plate sprayed with a fluorocarbon release agent and positioned on the bottom heated platen of the press. The preform

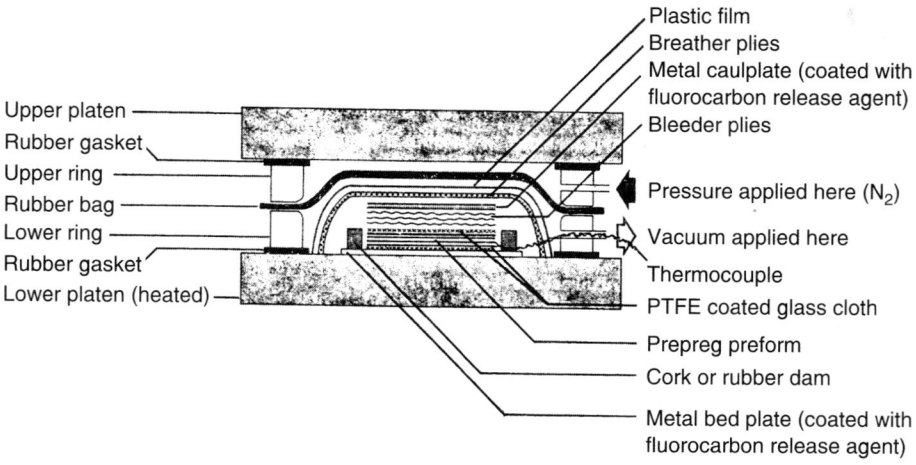

Figure 17.34 Diagrammatic layout of a press-clave.

is surrounded by a dam of rubber/cork composition to contain any extruded resin. The cloth or film is pervious to the resin and can be removed from the cured laminate after cure, leaving a satisfactory finish. A vacuum is applied below the bag to help remove entrained air and any volatile constituents, while pressure is applied above the bag to consolidate the preform. The excess resin liberated is absorbed into a bleeder pack comprising several layers of glass cloth or a non-woven material. To ensure that a uniform pressure is applied to the perform, a metal caul plate coated with a fluorocarbon release agent is used with several layers of glass cloth air breather plies above it to ensure that an even pressure is applied across the caul plate. The rubber bag is protected with plastic film, which is perforated to ensure pressure equalization. The temperature of the bottom platen is adjusted to give the requisite temperature of the preform, which is measured by a thermocouple embedded in the prepreg plies. The thickness of the laminate is controlled by the number and orientation of the prepreg plies, the state of cure of the resin, the pressure applied above the bag and the type and quantity of the bleeder plies.

17.8.4 Determination of carbon fiber content

The mass fraction of cfrp can be reliably determined by chemical digestion of the resin [50,51]. The weighed specimen is placed in a Kjeldahl flask in an efficient fume cupboard utilizing backboard extraction, with the mouth of the flask pointing away from the operator. Concentrated H_2SO_4 is added to the flask, which is then heated to fuming, 50% w/v H_2O_2 is then added dropwise until the solution becomes colorless. The amount added depends on the resin type and an epoxy novalac or phenolic resin will use up to some five times that required for an epoxy resin. After digestion, the carbon fiber is filtered into a weighed sintered glass crucible, washed with distilled water until free from acid, dried and weighed. This process is quite hazardous and a proprietary glass fume extraction/washing system should preferably be attached above the mouths of the Kjeldahl flasks.

17.8.5 Measurement of tensile modulus [52]

There are many versions of tensile specimen and as a routine test procedure, generally, a simple test bar ($150 \times 10 \times 1.0$ mm) is sufficient with end tabs attached, fabricated from prepreg or Al. Preparation of the composite surface by abrasion and subsequent degreasing prior to attaching end pieces is important and end pieces should be held in a jig during the bonding process. If Al is used, then the end pieces should be roughened, or grit blasted and degreased, followed by etching for 10 min at 65°C in:

30 cm^3 distilled H_2O
10 cm^3 concentrated H_2SO_4
1 g $Na_2Cr_2O_7$

and then rinsed thoroughly with deionized water, air dried and then must be used within 4 h of surface preparation.

When the two jaws of the test machine are placed in contact, they must be directly in line to ensure the load will be applied correctly. A 25 mm gage length extensometer is attached and if it is a single channel, it would be very sensitive to exact alignment, whereas a dual gage model is truly averaging (Figure 17.35).

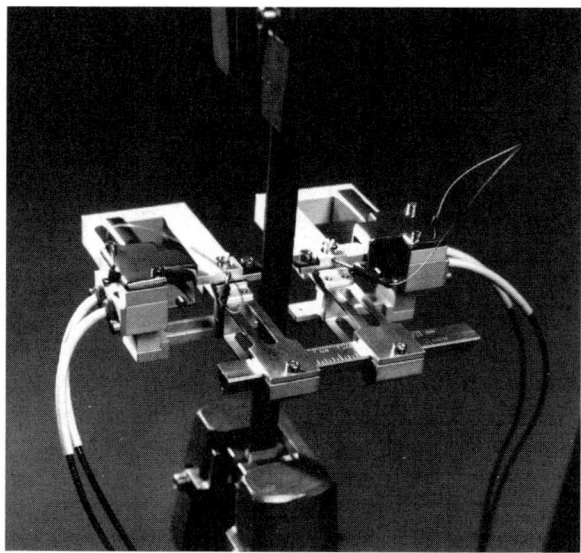

Figure 17.35 Dual gage extensometer. *Source:* Courtesy of Instron Ltd.

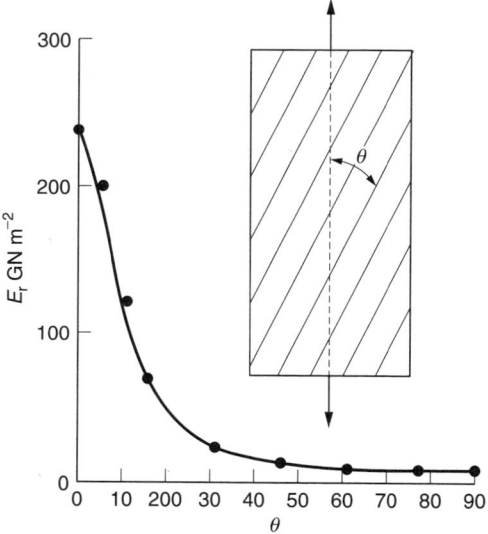

Figure 17.36 The angular dependence of the modulus of a 50% V_f unidirectional cfrp laminate. *Source:* Reprinted with permission from Hull D, Clyne DW, Cambridge University Press, Cambridge, 1996. Copyright 1996, Cambridge University Press.

It is important to check the extensometer knife edges for damage and to ensure that no additional load is applied by the weight of the extensometer cable, which is best supported with a magnetic clip attached to the metal frame of the testing machine. If wedge action grips are used, make certain that the shear pins are intact.

The degree to which the fiber is off axis has an important bearing on the Young's modulus (Figure 17.36) and is derived from:

$$\frac{1}{E_x} = \frac{1}{^1E_c}\cos^4\theta\left(\frac{1}{G_{12}} - \frac{2\upsilon_{12}}{^1E_c}\right)\sin^2\theta\cos^2\theta + \frac{1}{^2E_c}\sin^4\theta$$

TESTING OF PAN PRECURSOR, VIRGIN CARBON FIBERS   695

where $E_x=$ in-plane off-axis modulus and ^1E(longitudinal), ^2E(transverse), G_{12}(shear) and v_{12} ('major' Poisson's ratio) are four independent elastic or engineering constants defining the elastic behavior of the composite. For more details see Hull [35], Jones [36] and Tsai [46]. The angular dependence [53] of the modulus of a unidirectional 50% V_f cfrp lamina is shown in Figure 17.36.

The extensometer can be calibrated with a drum micrometer and it is important that the anvils onto which the extensometer is attached are soft metal, such as mild steel, and not chrome plated, since the extensometer would not be able to bite in sufficiently to prevent slippage.

17.8.6 Measurement tensile strength

To ensure that the tensile specimen will break in the gage length and not at the grips, the specimen is waisted (Figure 17.37) using a high speed (20,000 rpm) router cutter (Figure 17.38). It has been found advantageous to replace the cutter with a small grindstone of the same diameter. The specimen is held in a special template, which the cutter follows

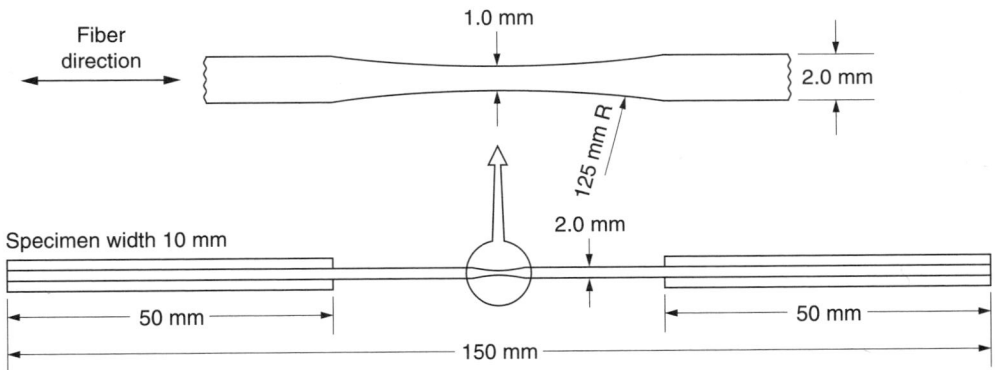

Figure 17.37 Unidirectional tensile composite specimen. *Source:* Courtesy Courtaulds Grafil Test Methods.

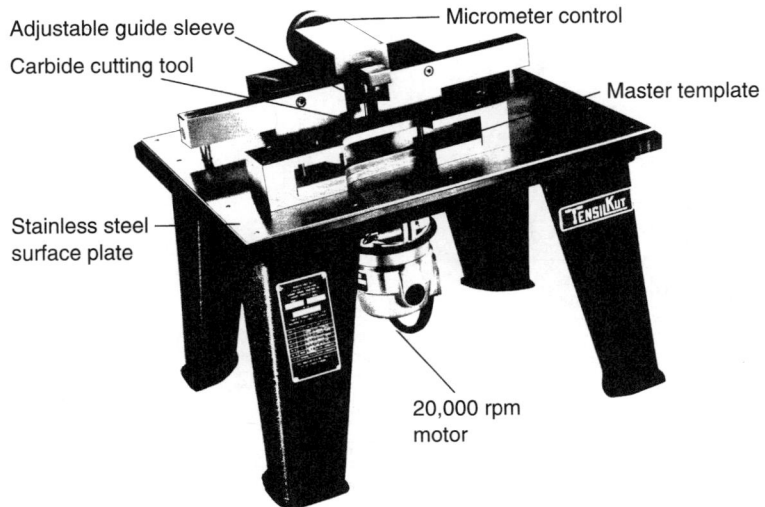

Figure 17.38 High speed router cutter. *Source:* Courtesy of Tensikut Engineering.

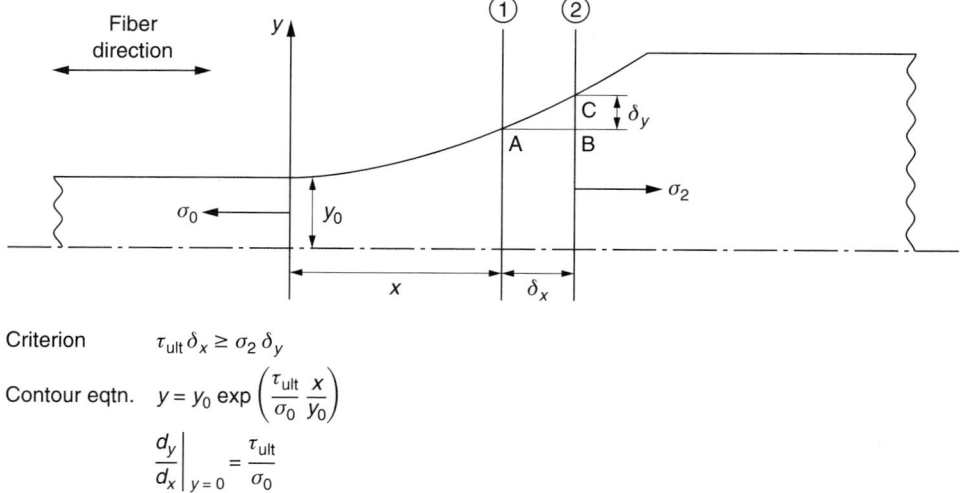

Figure 17.39 Waisting contour of unidirectional cfrp tensile specimen. *Source:* Reprinted from RAE.

exactly as it shapes the tensile specimen. Alternatively, instead of using a router cutter, a 125 mm radius grindstone can be used. This radius was established theoretically by RAE and adjusted practically, so that the same radius could be used for a compression specimen without the latter failing in a buckling mode. The contour equation (Figure 17.39) developed by RAE is:

$$y = y_0 \exp\left(\frac{\tau_{ult}}{\sigma_0} \times \frac{x}{y_0}\right)$$

$$\frac{dy}{dx} = \frac{\tau_{ult}}{\sigma_0 y_0} \exp\left(\frac{\tau_{ult}}{\sigma_0} \times \frac{x}{y_0}\right)$$

$$\frac{d^2 y}{dx^2} = \frac{\tau_{ult}}{\sigma_0} \times \frac{\tau_{ult}}{\sigma_0 y_0} \exp\left(\frac{\tau_{ult}}{\sigma_0} \times \frac{x}{y_0}\right) = \frac{\tau_{ult}^2}{\sigma_0^2 y_0} \exp\left(\frac{\tau_{ult}}{\sigma_0} \times \frac{x}{y_0}\right)$$

where τ_{ult} = shear strength
σ_0 = tensile strength
Now the radius of curvature is given by:

$$\frac{1}{R} = \frac{d^2 y/dx^2}{\pm[1 + (dy/dx)^2]^{3/2}}$$

Hence, $R = \dfrac{\pm\left[1 + \dfrac{\tau_{ult}^2}{\sigma_0^2} \exp\left(\dfrac{2\tau_{ult}}{\sigma_0} \times \dfrac{x}{y_0}\right)\right]^{\frac{3}{2}}}{\dfrac{\tau_{ult}^2}{\sigma_0^2} \exp\left(\dfrac{\tau_{ult}}{\sigma_0} \times \dfrac{x}{y_0}\right)}$

At $y = 0$, $\dfrac{dy}{dx} = \dfrac{\tau_{ult}}{\sigma_0}$
Then,

$$R = \frac{y_0 \left[1 + (\tau_{ult}/\sigma_0)^2\right]^{3/2}}{(\tau_{ult}/\sigma_0)^2}$$

In the above case, $y_0 = 0.5$ mm.

TESTING OF PAN PRECURSOR, VIRGIN CARBON FIBERS

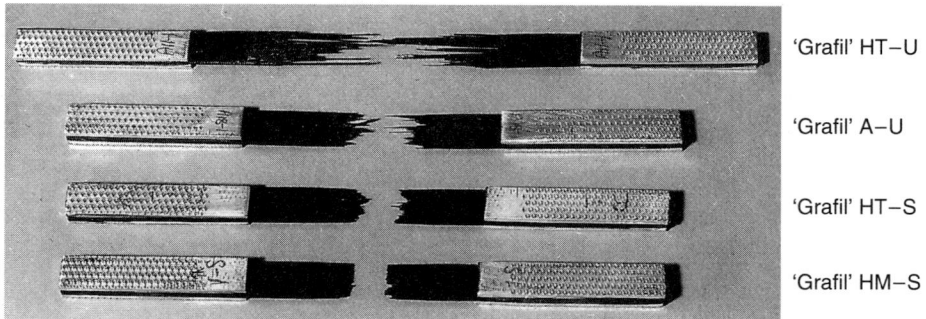

Mode of failure of tensile
composite specimens made from
different types of 'Grafil'

Figure 17.40 Failure modes of cfrp tensile composite specimens made from Grafil HTU, AU, HTS and HMS.

Hence,

$$R = \frac{0.5\left[1 + (\tau_{\text{ult}}/\sigma_0)^2\right]^{3/2}}{(\tau_{\text{ult}}/\sigma_0)^2}$$

Figure 17.40 shows the mode of failure with different fiber types when using a 125 mm radius necked portion and the evidence does suggest that a radius of 175 mm would better accommodate AU and HTU fibers.

When breaking tensile specimens, the operator should be protected against flying debris with a protective screen, which can be a sliding window attached to the frame of the test machine.

17.8.7 Measurement of strain using resistance strain gages [54–56]

Strain gages can be used for the measurement of strain, or displacement, in mechanical testing and can be a wire wound grid or an etched foil pattern, normally having a gage factor of 2 with a resistance of several hundred ohms. High resistances are preferred, since high voltages of the order of 2–4V can be used with corresponding low current, giving improved hysteresis and zero load stability effects. The gages are used in DC bridge circuits (Figure 17.41), or better, in an AC bridge circuit, which has many practical advantages.

Attaching a strain gage to the specimen is an acquired art and the correct surface preparation of the gage and specimen, as recommended by the gage supplier, is paramount. If connections have to be made, it is better to undertake these before the gage is installed, requiring a fine soldering iron and a steady hand, although it is better to use gages with the lead wires already attached.

The strain gage must be accurately aligned and even a 2° misalignment can give a 15% error. If dummy gages are used for temperature compensation, they must also be precisely aligned.

If the gage factor is to be checked on sample gages, then they can be calibrated most conveniently mounted on a flexural specimen (Figure 17.42) against an accurate

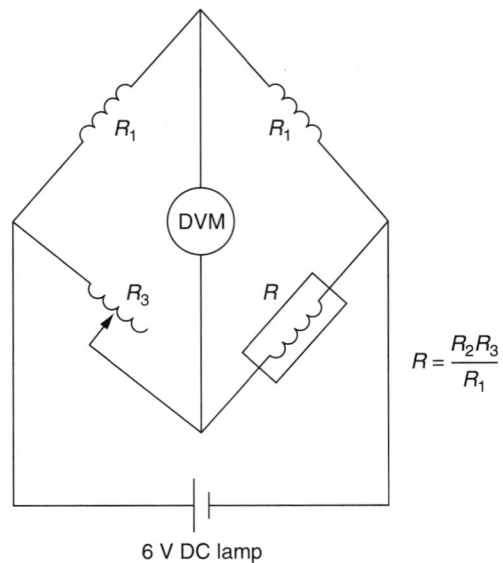

Figure 17.41 Wheatstone bridge DC circuit for a strain gage.

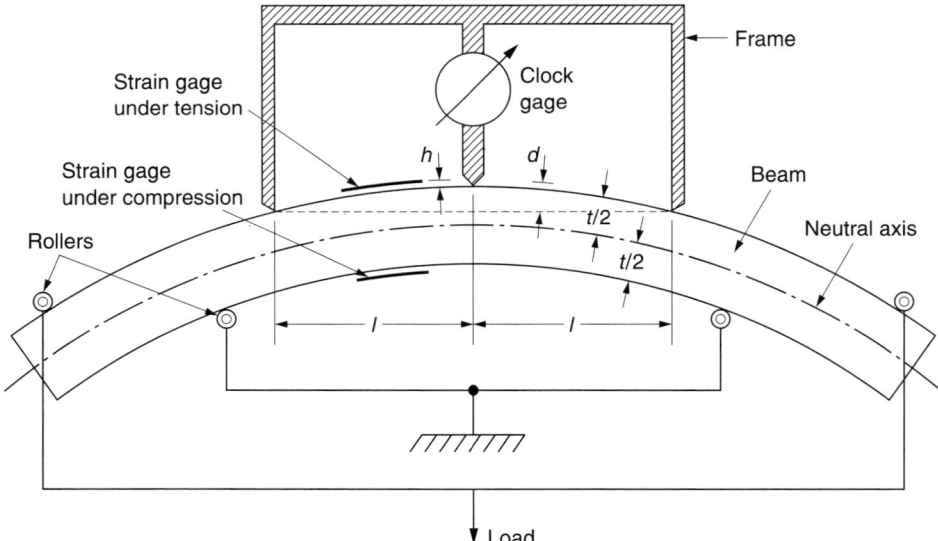

Figure 17.42 Test jig for calibrating strain gages. *Source:* Reprinted from Neubert HKP, *Strain Gauges: Kinds and Uses*, Macmillan, London, 1968.

extensometer. Such a procedure will also check the proficiency of the bonding method.

$$e = \frac{dt}{l^2} \quad e = dt/L^2 \quad \text{and} \quad e = \frac{e't}{(t+2h)}$$

where e = actual surface strain
d = deflection through a distance of $2L$
t = thickness of beam
e' = the apparent strain measured
h = height of strain gage grid above beam surface

17.8.8 Measurment of shear strength

17.8.8.1 Interlaminar shear strength [57–60]

The interlaminar shear strength is the interfacial shear stress or shear strength of the matrix material and is measured with the 3-point bend test(ASTM D 2344), which is ideal for routine testing. Figure 17.43 shows a diagrammatic view of the test and shows how shear, tensile and compressive forces are involved. Shear failure will take place at the midplane in the form of delamination, whilst Figure 17.44 hows how a parabolic shear stress distribution occurs.

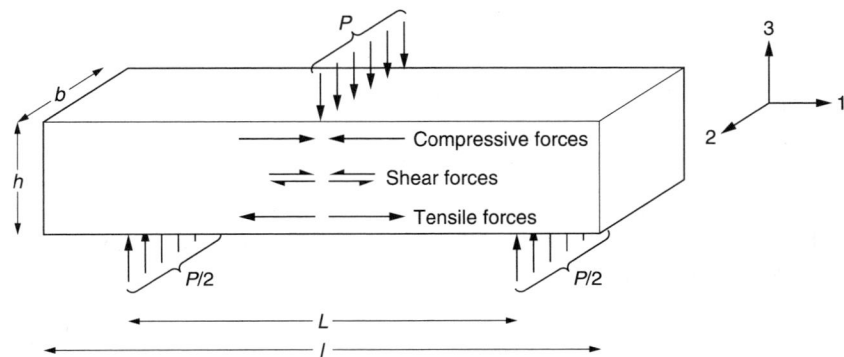

Figure 17.43 Short beam shear test showing test configuration. *Source:* Reprinted from Course on Mechanical Testing of Advanced Fibre Composites, University of London, Imperial College, Sep 1995.

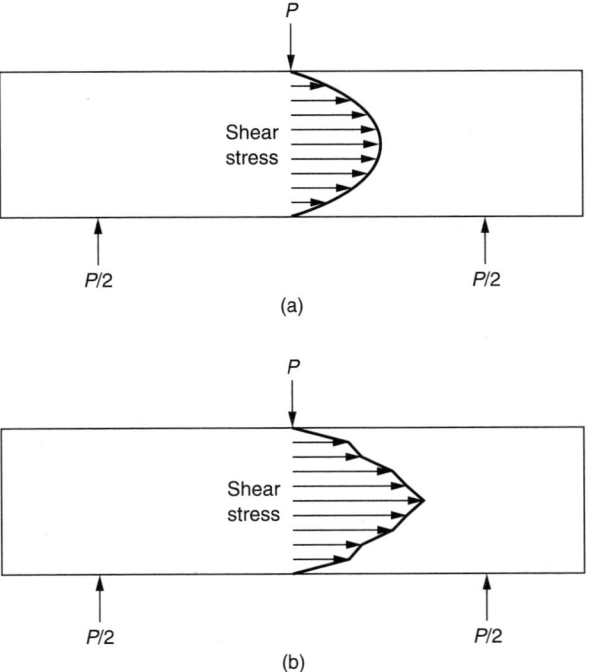

Figure 17.44 Effect of shear stress distribution in (a) homogeneous specimen (b) laminated specimen. *Source:* Reprinted from Course on Mechanical Testing of Advanced Fibre Composites, University of London, Imperial College, Sep 1995.

The shear strength (strictly for a homogeneous specimen) is derived from:

$$\tau \cong \frac{3P}{4bh}$$

where P = force at failure
b = width
h = thickness

If the beam is too long compared to its depth, then a flexural failure (tensile or compressive) can occur at the outer plies of the beam. To ensure that an interlaminar shear failure does occur the span:depth ratio must satisfy:

$$\frac{2L}{h} < \frac{F_1}{\tau}$$

where L = beam span
F_1 = flexural strength of the beam (smaller of F_{1t} or F_{1c})

The span:depth ratio has been chosen to ensure a shear failure. Although the physical validity and accuracy is questionable, useful results can be achieved. However, attention is drawn to the following points:

1. Use a resin system that ensures a shear type failure
2. Ensure that the test jig has been correctly lined up and the loading nose is at right angles to the longitudinal direction of the specimen
3. When comparing ILSS values, be certain to compare like for like, as a 4:1 span/depth ratio will give higher ILSS values than a 5:1
4. Undercure of the resin matrix can also give higher ILSS values
5. Do not apply the Law of Mixtures to correct for fiber content, although the volume fraction of the fibers does affect the ILSS
6. Record the mode of failure. With a valid mode of failure, the calculated shear stress can be termed the Apparent Interlaminar Shear Strength
7. The radii chosen for the loading nose and supports will affect the results and putting rubber under the loading nose avoids surface damage, improving the test reliability [58]

Browning *et al* have described a 4-point shear test [61].

17.8.8.2 In-plane shear tests [62]

The full characterization of a unidirectional composite under an in-plane shear load parallel to the fiber direction will entail the measurement of shear modulus G_{12}, shear strength F_6 and ultimate shear strain γ_6^u. These can be determined by:

1. The torsion test [63,64]

The torsion test is generally considered to be the most accurate shear test and Figure 17.45 shows a thin walled tube subject to axial tension, which gives a state of pure shear and produces reliable measurements for both strength and modulus, but it is costly procedure.

$$(\tau_6)_{max} = \frac{2Tr_0}{\pi(r_0^4 - r_i^4)}$$

$$(\tau_6)_{max} = \psi r_0 = (\varepsilon_x)_{\theta = 45°} - (\varepsilon_x)_{\theta = -45°} = 2(\varepsilon_x)_{\theta = 45°}$$

TESTING OF PAN PRECURSOR, VIRGIN CARBON FIBERS

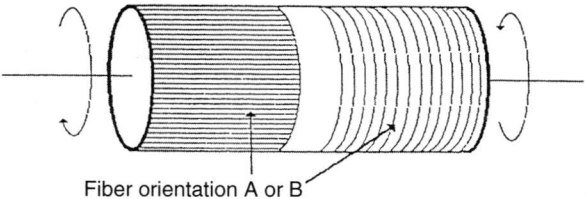

Figure 17.45 Tube subjected to axial torsion. *Source:* Reprinted from Course on Mechanical Testing of Advanced Fibre Composites, University of London, Imperial College, Sep 1995.

where r_i, r_o = inner and outer radii
ψ = angle of twist per unit length
$(\varepsilon_x)\theta = 45°$, $(\varepsilon_x)\theta = -45°$ = surface strains at 45° and −45° with tube axis
As an approximation for a thin walled tube,

$$(\tau_6)_{max} \cong \frac{T}{2\pi \bar{r}^2 h}$$

where the mean radius $\bar{r} = (1/2)(r_0 + r_i)$ and for a solid rod, $r_i = 0$. The shear strain can be measured from the angle of twist or the strains at 45° and/or −45° using strain gages.

2. Two-rail or three-rail shear test [202]

A more practical method is to use the two-rail or three-rail shear test (ASTM D 4255) (Figures 17.46 and 17.47), although Lockwood has questioned the reproducibility [65] of the test. The ASTM is not strictly a standard, but is referred to as a guide.

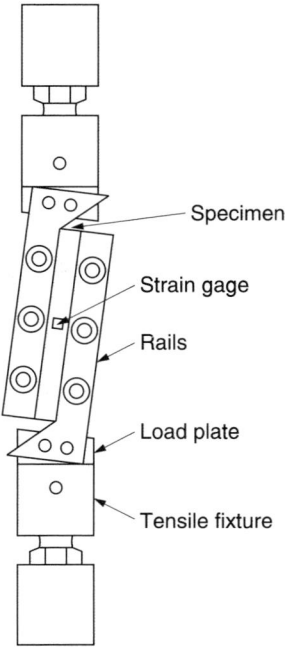

Figure 17.46 Two-rail shear test. *Source:* Reprinted from Course on Mechanical Testing of Advanced Fibre Composites, University of London, Imperial College, Sep 1995.

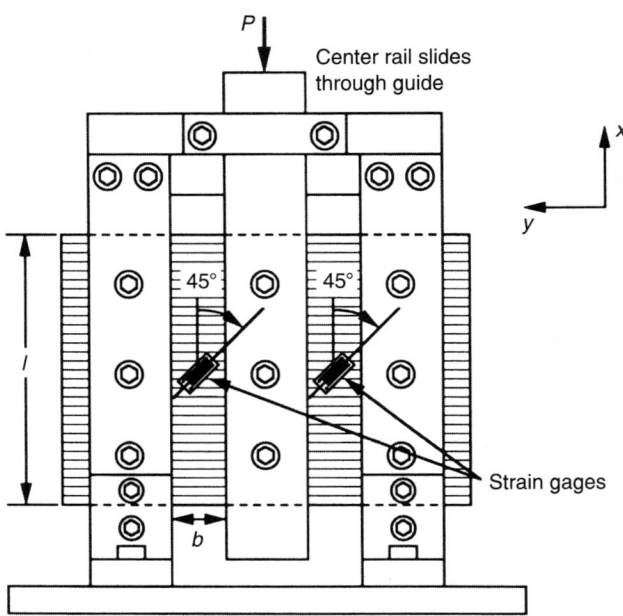

Figure 17.47 ASTM Three rail shear test. (ASTM D4255-83).

The average shear stress applied to the specimen is

$$\tau_6 = \frac{P}{2lh}$$

where P = load
l = specimen length along the rails
h = specimen thickness

The two-rail shear test is only suitable for the determination of shear moduli. The three-rail shear does give a better approximation to pure shear as the compressive load is applied parallel to the clamped edges of the specimen.

3. The double V-notch shear (Iosipescu test) [202]

The Iosipescu test (Figures 17.48 and 17.49) has two opposing V-notches to impose pure shear in the middle of the specimen (ASTM D 5379) and gives results that compare well with the torsion tube method [64,66].

The average shear stress is

$$\tau_6 = \frac{P}{lh}$$

where l = specimen height at notch location
h = specimen thickness

4. Tension coupon test

Figure 17.50 shows the $[\pm 45]_{ns}$ tension coupon test (ASTM D 3518), where the shear strain is measured at $\pm 45°$ to the axis using strain gages where:

$$\gamma_{xy} = \varepsilon_{11} - \varepsilon_{22}$$

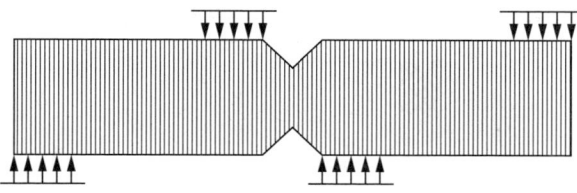
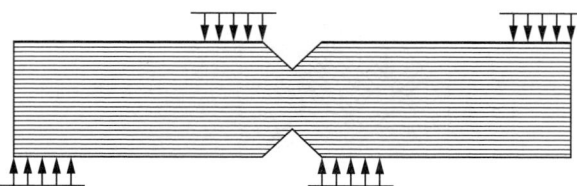

Figure 17.48 Iosipescu shear test configuration. *Source:* Reprinted from Course on Mechanical Testing of Advanced Fibre Composites, University of London, Imperial College, Sep 1995.

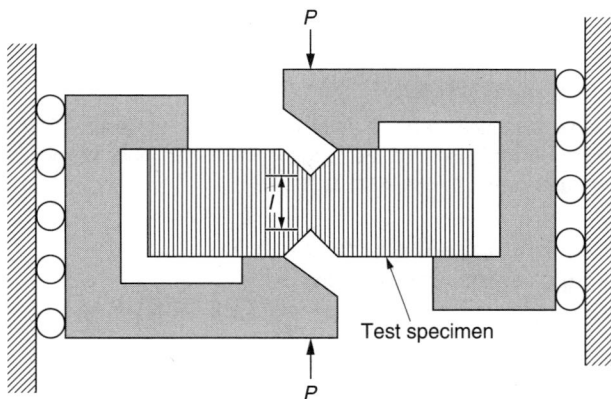

Figure 17.49 Iosipescu shear test jig. *Source:* Reprinted from Walrath DE, Adams DF. The Iosipescu shear test as applied to composite materials, *Exp Mech*, 23, 105–110, 1983.

Output from the strain gages is in volts given by:

$$e = \frac{4V_0}{KV_{in}G}$$

where V_0 = output voltage
K = gage factor
V_{in} = bridge voltage
G = gain
and shear modulus G_{12} is given by:
$G_{12} = \dfrac{P}{2bh(\varepsilon_{11} - \varepsilon_{22})}$ (shear modulus in the 1-2 plane)
where P = tensile force
b = specimen width
h = specimen thickness
ε_{11} = axial strain
ε_{22} = transverse strain

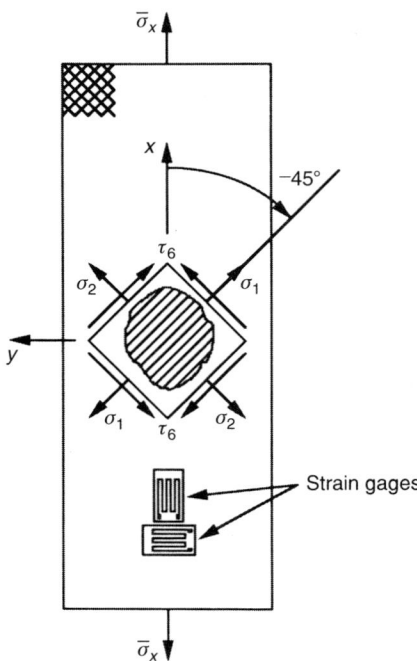

Figure 17.50 [±45]$_{ns}$ angle ply shear test. *Source:* Reprinted from Rosen RW. A simple procedure for experimental determination of the longitudinal shear modulus of unidirectional composites, *J. Composite Materials* 6, 552–554, 1972.

but will require derivation of axial (ε_{11}) and transverse strains (ε_{22}). An equation for modulus can be expressed as:

$$G_{12} = \frac{E_x}{2(1 + v_{xy})}$$

where E and v are the effective Young's modulus and Poisson's ratio respectively of the specimen.

5. The 10° off-axis test

This test [67] is reputed to minimize the effect of longitudinal (σ_1) and transverse (σ_2) tensile stresses on the shear response (Figure 17.51). A two-gage rosette is placed on either side of the specimen, with the two gages of the rosette at ±45° to the fiber direction. The algebraic difference of the strain readings of gages A and B gives the in-plane shear strain:

$$\gamma_6 = \gamma_A - \gamma_B$$

This test is more accurate than the [±45]$_{ns}$ coupon test, but less simple.

When subject to a uniaxial tensile load, three types of stress are experienced—longitudinal, transverse and in-plane shear on the 10° plane. The intralaminar shear strain approaches a maximum value at an angle of about 10° between the fiber direction and applied load, when the shear stress is the major component contributing to fracture. The intralaminar shear strength depends on the shear strengths of polymer matrix and the interfacial bond. Figure 17.52 shows the angular dependence of the fracture strength of 50% V_f.

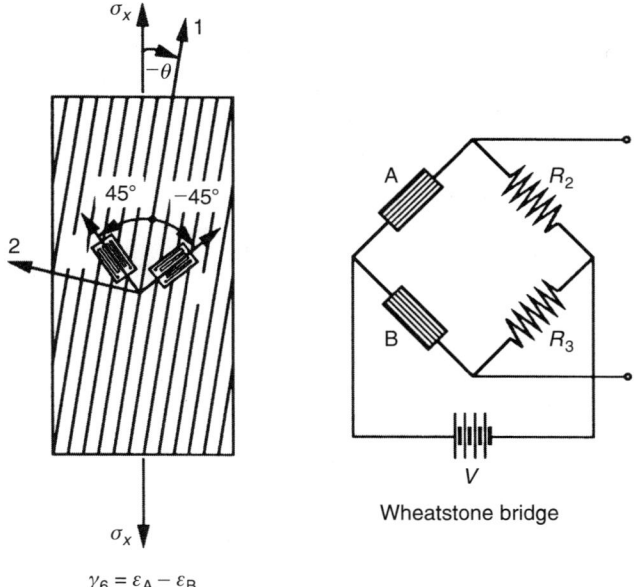

$\gamma_6 = \varepsilon_A - \varepsilon_B$

Figure 17.51 Schematic diagram of 10° off-axis shear test. *Source:* Reprinted from Chamis CC, Sinclair JM, Ten-degree off-axis test for shear properties in fibre composites, *Exp Mech*, 17, 339–346, 1977.

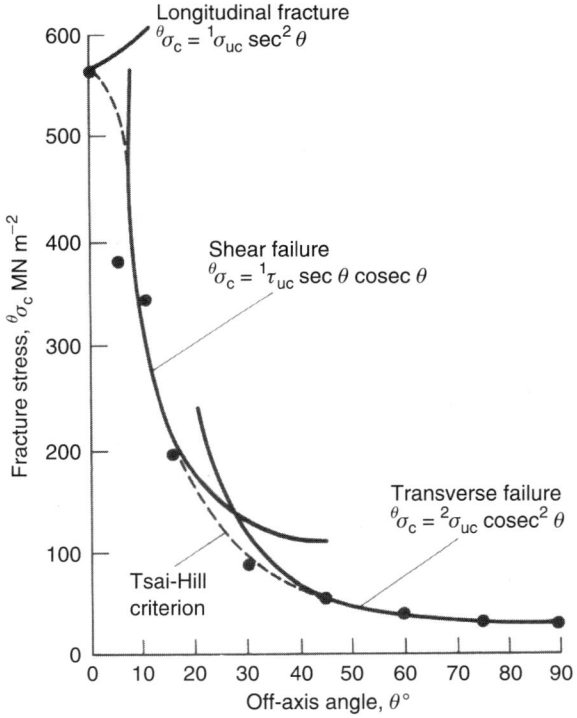

Figure 17.52 The angular dependence of fracture strength of a cfrp unidirectional Type I carbon fiber lamina ($V_f = 0.5$). *Source:* Reprinted from Chamis CC, Sinclair JH, *Exp Mech*, 17, 339–346, 1977 and Tsai-Hill criterion presented by Hull, Hull D, Clyne DW, Cambridge University Press, Cambridge, 1996.

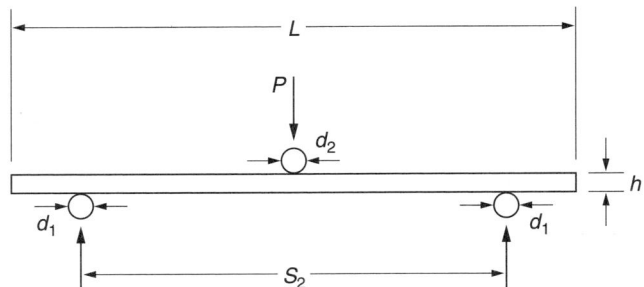

Figure 17.53 A 3-point flexure bending arrangement. *Source:* Reprinted from Course on Mechanical Testing of Advanced Fibre Composites, University of London, Imperial College, Sep 1995.

Type I carbon fiber in epoxy resin system shows three modes of failure—longitudinal fracture, shear failure and transverse failure, and the strength at any given angle is determined by the weakest of these failure modes.

17.8.9 Measurement of flexural strength and modulus [68–71]

Flexural testing is a simple bend test involving no end tabs but, unfortunately, the test conditions do vary widely with the chosen test method, so results must be treated with caution; although the test is useful as a quality control procedure. There are three relevant standards—ASTM D 790M, BSI 2782 Method 1005 and CRAG and all of them cover a 3-point test procedure (Figure 17.53), whilst the ASTM also covers a 4-point procedure (Figure 17.54).

Modulus measurements carried out at higher span/depth ratios are less influenced by the penetration of the loading nose and supports into the specimen. A better technique is to measure the deflection at different span/depth ratios for the same load.

For flexural strength:

$$\text{ASTM (3-point)} \quad s = \left(\frac{3PS_2}{2wh^2}\right)\left(1 + 6\left(\frac{\delta}{S_2}\right)^2 - 4\left(\frac{\delta h}{S_2^2}\right)\right)$$

$$\text{ASTM (4-point)} \quad s = \left(\frac{3PS_2}{2wh^2}\right)\left(1 + 4.7\left(\frac{\delta}{S_2}\right)^2 - 7.04\left(\frac{\delta h}{S_2^2}\right)\right)$$

$$\text{BSI (3-point)} \quad s = \left(\frac{3PS_2}{2wh^2}\right)\left(1 + 4\left(\frac{\delta}{S_2}\right)^2\right)$$

$$\text{CRAG (3-point)} \quad s = \left(\frac{3PS_2}{2wh^2}\right)$$

where s = stress in the outer fibers at mid-span, $N\,m^{-2}$
P = load at break, N
S_1 = load span, m
S_2 = support span, m
w = width of beam, m
h = thickness of beam, m
δ = deflection at center of span, m

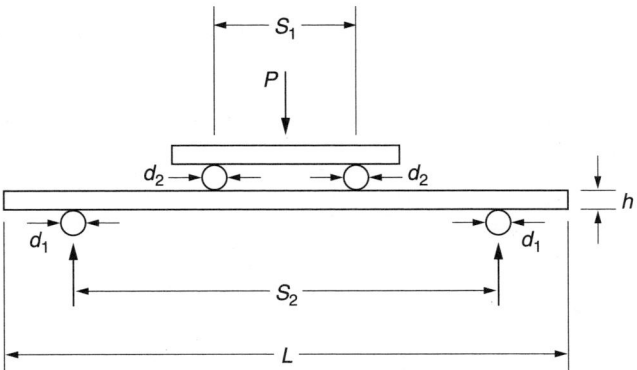

Figure 17.54 A 4-point flexure bending arrangement. *Source:* Reprinted from Course on Mechanical Testing of Advanced Fibre Composites, University of London, Imperial College, Sep 1995.

and for flexural modulus:

$$\text{ASTM, BSI and CRAG (3-point flexure)} \quad E_B = \frac{S_2^3 m}{4wh^3}$$

$$\text{ASTM (4-point flexure)} \quad E_B = \frac{0.21 S_2^3 m}{wh^3}$$

where E_B = tangent modulus of elasticity, $N\,m^{-2}$
m = slope of tangent to initial portion of the load deflection curve, $N\,m^{-1}$

Using the variable span method with 3-point loading:

$$\delta = \frac{PL^3}{4wt^3 E} + C.P$$

where P = load applied
C = softness constant due to machine softness, deformation of sample at load points and shear in the sample

Now the equation can be written in the form

$$y = mx + c$$

where

$$x = L^3$$
$$C = C.P.$$
$$m = \frac{P}{4wt^3 E}$$

Hence m can be obtained by the method of least squares and the value for the modulus can be determined.

Table 17.10 gives the dimensions and test conditions for the ASTM D 790, BSI 2782 Method 1005 and CRAG test standards based on an adopted laminate thickness of 2 mm. The typical values for modulus and strength are given for Ciba Geigy's 913/XAS system [45].

Table 17.10 Dimensions and test conditions for flexure tests based on a laminate thickness of 2 mm

Standard	w mm	h mm	S/h	L mm	d_1 mm	d_2 mm	Test speed mm min^{-1}	Typical value for unidirectional composite	
								E GPa	σ GPa
ASTM 3pt	25	2	16	50	6.4	6.4	0.85	59.0	1.37
3pt	25	2	32	80	6.4	6.4	3.4	103.2	1.44
3pt	25	2	40	100	6.4	6.4	5.3	101.7	1.49
3pt	25	2	60	150	6.4	6.4	12.0	116.7	1.56
4pt,1/3 span pts	25	2	20[a]/60[b]	150	6.4	6.4	13.3	132.4	1.61
4pt,1/4 span pts	25	2	30[a]/60[b]	150	6.4	6.4	12.0	144.9	1.44
BSI 3pt	15	2	16	50	4	10	1.0	73.7	1.46
CRAG 3pt	10	2	40[c]	100	10	25	5.0	114.9	1.72

Note: [a] load roller S_1/h (⅓ span points may also be 5.3, 10.7 and 13.3 mm; ¼ span points may also be 8, 16 and 20 mm;
[b] support roller S_2/h (⅓ span points and ¼ span points may be also 16, 32 and 40 mm;
[c] may also be 25 mm for 90° and 0/90° carbon.
In all these alternatives the total specimen length and testing speed vary accordingly.
Source: Reprinted from Sottos NR, Hodgkinson JM, Matthews FL, *Proc 6th International Conference on Composite Materials*, Imperial College, 1.310–1.320, Jul 20–24, 1987.

A four point bend test will reduce the likelihood of a compression face failure and will provide pure bending between the center loading points.

It should be noted that the strength of fiber placed on the neutral axis does not influence the result. Twist in the fiber will increase both flexural strength and modulus.

17.8.10 Measurement of uniaxial compressive strength and modulus [72–99]

A uniaxial compressive load can be introduced into a specimen by:

1. Direct loading of the specimen end, but this is not suitable for high strength composites due to the specimen ends failing by crushing, which is termed brooming
2. Shear loading of the specimen end
3. The mixed shear and direct loading method

The method of shear loading of the specimen end was used in the original Celanese test method [83], where the tabbed specimen is held in conical grips, held in tapered sleeves accommodated in an outer cylinder (Figure 17.55). A load applied to the sleeve will be transmitted to the specimen by shear through the specimen and grips. This has been adopted in a slightly modified form as ASTM D 3410.

The Illinois Institute of Technology Research Institute (IITRI) developed a modification of the Celanese fixture (Figure 17.56), which replaced the conical grips with tapered grips that fitted into matching pockets in a steel block, aligned by pillar guides fitted with linear bearings.

Adaptions of this version of a compressive test were made by Wyoming University [76]. Purslow and Collings, later modified by Port of RAE [93], used a mixed shear and end loading method, where a specimen is bonded into slots machined in Al end blocks and the value did depend on the type of adhesive used. A similar test modified with an additional

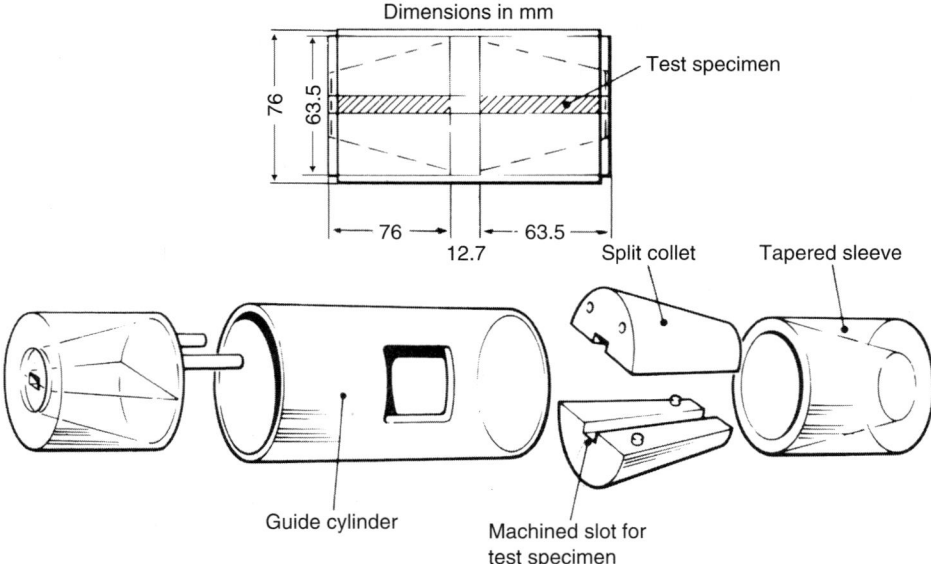

Figure 17.55 The Celanese compression test. *Source:* Reprinted from Courtaulds' Grafil Test Methods.

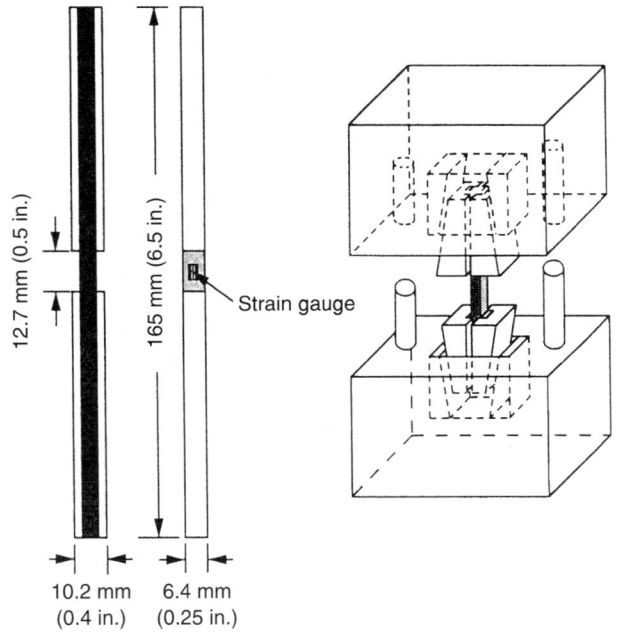

Figure 17.56 Illinois Institute of Technology Research Institute (IITRI) compression test. *Source:* Reprinted from Hofer KE, Rao PN, A new static compression fixture for advanced composite materials, *J. Test Eval* 5, 278–283, 1977.

alignment device was adopted as ASTM D 695, which in its original form was introduced for testing unreinforced plastics [84,85].

The RAE method was modified by Birmingham University [75], where a waisted specimen was clamped at the ends in steel cubes instead of bonding and the method was further improved by Imperial College [84,201] using a four-pillar die set to obtain good alignment.

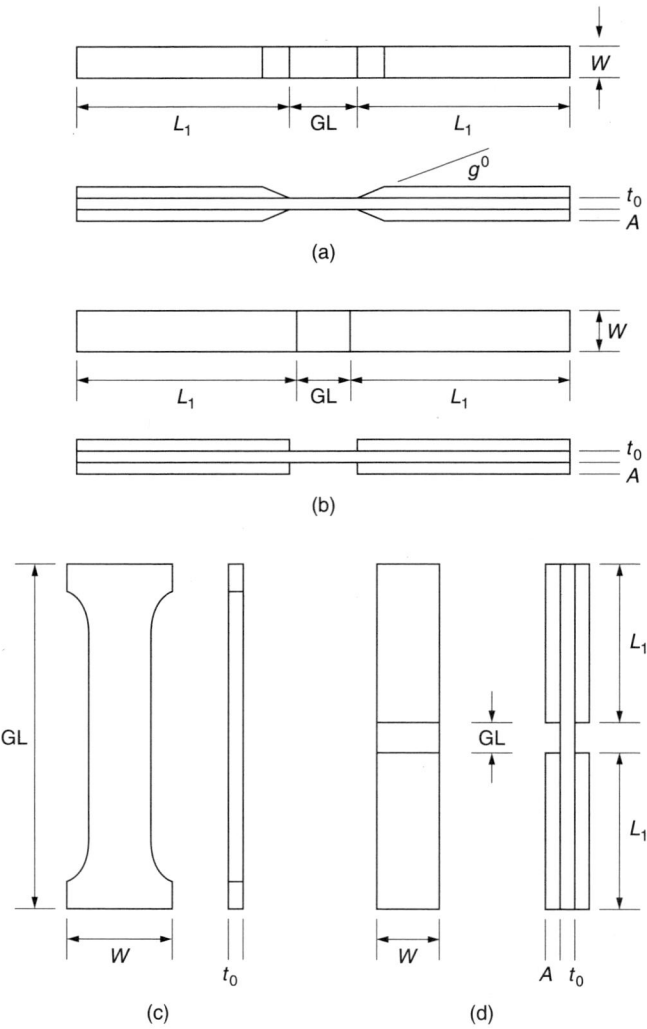

Figure 17.57 Details of compression test specimens. (a) ASTM D3410; (b) CRAG; (c) ASTM D695; (d) ASTM D695 (modified). *Source:* Reprinted from Course on Mechanical Testing of Advanced Fibre Composites, University of London, Imperial College, Sep 1995.

The specimens are shown in Figure 17.57, whilst the dimensions and test conditions are given in Table 17.11 for the three main compression test specimens used for carbon/epoxy composites. The correct alignment of fiber in the specimen, specimen in the jig and the jig in the test machine are paramount.

Compression testing is difficult to undertake because of premature failure due to crushing or buckling and is not used for routine testing.

Kumar and Helminiak [87] have shown a relationship (Figure 17.58) between compressive strength and tensile modulus in carbon fibers.

17.8.11 Testing of fatigue [100–114].

Fatigue is determined under cyclic loading, when a crack is initiated and then grows to a critical size, with failure occuring at stresses below the value obtained for normal static tests.

TESTING OF PAN PRECURSOR, VIRGIN CARBON FIBERS

Table 17.11 Dimensions and test conditions for compression testing of carbon fiber/epoxy

	Celanese test ASTM D-3410	ASTM D 695M		CRAG
		Standard	Modified	
t_0	2	2	2	2
G.L.	12.7	80	5	10
L_1	63.5	–	35	50
W	6.35	19/12	10	10
A	1	–	2	1
End tag material	Steel	–	Carbon/epoxy	Aluminum
End tag profile	9°	–	–	–
Strain measurement	Strain gage	Extensometer	–	Strain gage
Test speed (mm min^{-1})	1.3	1	1	1.26

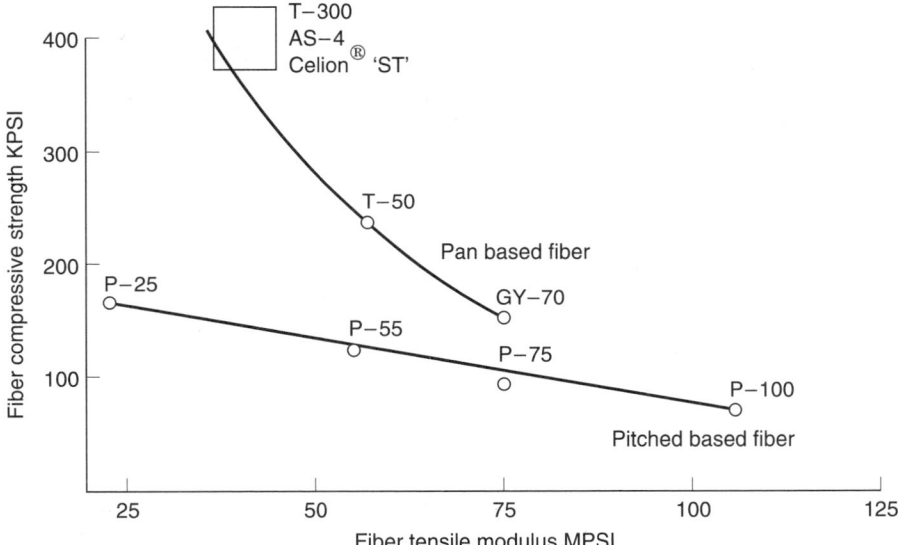

Figure 17.58 Relation between compressive strength and tensile modulus in carbon fibers *Source:* Reprinted from Kumar S, Helminiak TE, *Proceedings of Materials Research Symposium 134*, 363, 1989.

In general, any static test can be adopted for fatigue testing, but fatigue testing is much more demanding. The shape of the test specimen is chosen to fail in the gage length, in a manner akin to the failure of a structural component. Fatigue tests can be undertaken in tensile, flexural, compression and shear modes, generally using a servo-hydraulic test machine. Optical microscopy, ultrasonics, X-ray and IR thermography are used for detecting the onset and study of fatigue failure.

When designing with cfrp, strain levels are kept low (0.3–0.4%) due to possible impact damage and the notch sensitivity of cfrp in compression, so fatigue is not seen as a major problem in aircraft design, except for such components as helicopter rotor blades. A typical tensile fatigue curve is shown in Figure 17.59 for unidirectional cfrp, where peak tensile stress is plotted against the cycles to failure (S-N plot). The effect of fatigue on different carbon fiber types is shown in Figure 17.60. Carbon fibers are not sensitive to fatigue loading, but S-N curves are determined mainly by the strain in the matrix. The weaker fibers fail first due to the presence of flaws introducing a stress concentration in the matrix and at the

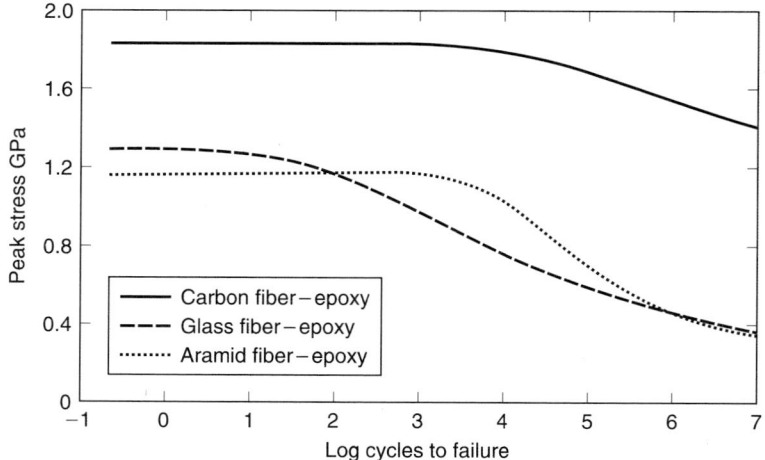

Figure 17.59 S-N diagrams for representative unidirectional DMC composite materials. *Source:* Reprinted with permission from Curtis PT, *RAE Technical Report TR82031*, RAE, Farnborough (now DRA), 1982, Curtis PT, *RAE Technical Report TR86021*, RAE, Farnborough, 1986, Curtis PT, *RAE Technical Report TR87031*, RAE, Farnborough, 1987. Copyright 1987, QinetiQ Ltd.

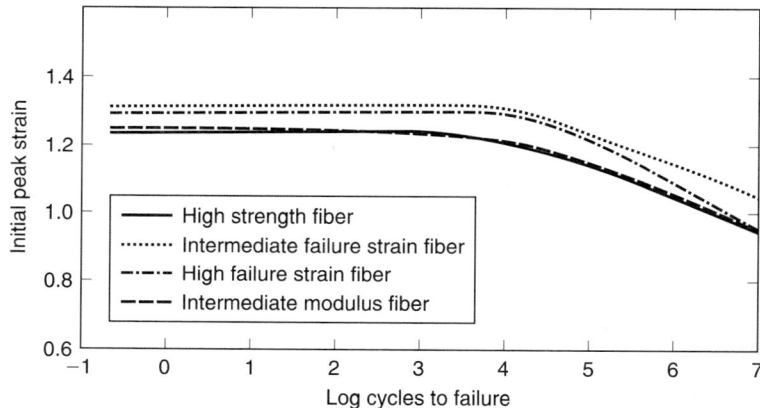

Figure 17.60 S-N fatigue data for unidirectional cfrp showing the limited effect of different fibers with the same epoxy resin. *Source:* Reprinted with permission from Curtis PT, *RAE Technical Report TR82031*, RAE, Farnborough (now DRA), 1982, Curtis PT, *RAE Technical Report TR86021*, RAE, Farnborough, 1986. Copyright 1986, QinetiQ Ltd.

fiber/resin interface, thereby inducing microcracks, which further develop in fatigue, making such fibers ineffective. The tougher resin composites do have increased fatigue sensitivity (Figure 17.61).

17.8.12 Measurement of creep [115,116]

The application of a force to a material over a sustained period of time induces creep and if this force eventually leads to cracking, fracture or rupture, the process is termed stress cracking, static fatigue, or creep rupture and if the environment has speeded up the process, it is termed environmental stress cracking.

The creep of a composite material will depend on the reinforcement and the matrix. Carbon fibers are not much affected by creep, although there will be some loss of strength

TESTING OF PAN PRECURSOR, VIRGIN CARBON FIBERS

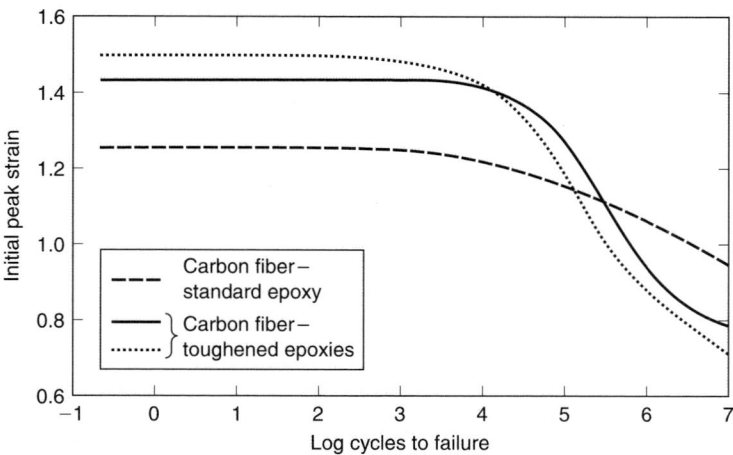

Figure 17.61 S-N data for unidirectional cfrp with the same fibre type in different matrices. *Source:* Reprinted with permission from Curtis PT, *RAE Technical Report TR82031*, RAE, Farnborough (now DRA), 1982, Curtis PT, *RAE Technical Report TR86021*, RAE, Farnborough, 1986, Curtis PT, *RAE Technical Report TR87031*, RAE, Farnborough, 1987. Copyright 1987, QinetiQ Ltd.

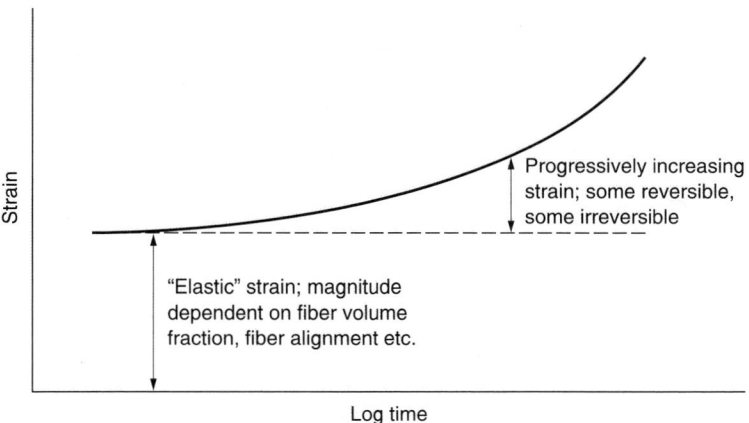

Figure 17.62 The creep response of a composite specimen to an applied stress. *Source:* Reprinted from Course on Mechanical Testing of Advanced Fibre Composites, University of London, Imperial College, Sep 1995.

under sustained loading, but the matrix will exhibit creep, a thermoplastic more so than a thermosetting resin. However, if the fiber is well aligned in a unidirectional composite, it will have good resistance to creep along the fiber axis, but would display creep in torsion and flexure. Plastics do tend to recover when the stress is removed, as long as no damage has occurred. It is normal to plot strain against log time (Figure 17.62) where the initial strain develops rapidly, reflecting the modulus, whereas the time-dependent portion of the curve is associated with a viscoelastic response in the matrix eventually leading to a run-away creep, terminating in creep rupture. An idealized creep equation would be:

$$\varepsilon(t) = \sigma C \int_0^\infty f(\tau)\left(1 - e^{-t/\tau}\right) d\tau$$

where $C =$ a coefficient with the dimensions of elastic compliance
$\tau =$ the retardation time of a single Voigt element

$f(\tau)$ = the distribution function of retardation times
σ = the applied stress

Creep is not well understood and the relevant standards, ASTM D 2990 and BS 4618, only give guidance.

17.8.13 Testing of impact behavior [117–123]

A composite derives its toughness from the ability of the reinforcing fibers to blunt incipient cracks together with the work necessary to either fracture them or pull them out of the matrix. Glass fibers are tougher than carbon, while aramid fibers are tougher than glass and toughened resins can contribute to toughness.

Impact tests are carried out at a fast rate of loading to promote a brittle failure. Izod and Charpy tests are used for isotropic and laminated materials and are not suitable for unidirectional material, or out-of-plane measurements with laminated materials, since notches are not always effective.

Preferred impact test methods are:

1. Out of plane impact testing onto a thin shell structure (e.g. to simulate dropping a tool, or the impact of stones, onto an aircraft structure). There are many variants of this test—the specimen can be simply supported or clamped, the projectile can be dropped or fired ($1–10$ m s^{-1}) and may take the form of a sphere, a cylinder with a hemispherical end or a dart. The energy and/or the test speed can be varied by adjusting the mass or drop height. The test can be instrumented to record the force, energy absorbed and the deflection.
2. Transient tensile and compressive in-plane loads to thin shell structures (e.g. an unexpected service load at a high rate)
3. Sustained in-plane compressive loads to thin and thick walled structures (e.g. crushing under crash conditions)

17.8.14 Measurement of interlaminar fracture toughness [124–139]

Interlaminar shear or delamination occurs in three modes (Figure 17.63) or combinations thereof:

1. Mode I—tensile opening
2. Mode II—in-plane shear
3. Mode III—out-of-plane tearing

The resistance to delamination growth is termed interlaminar fracture toughness and is measured as the strain energy released per unit area of delamination growth (G_I, G_{II} or G_{III}). Pre-cracking of the specimens is undertaken in the selected mode and tests continued, monitoring the position of the crack front with an optical microscope.

17.9 TESTING OF CARBON FIBER FILLED THERMOPLASTICS

17.9.1 Measurement of moisture content

When carbon fiber is compounded with a thermoplastic to produce a carbon fiber reinforced thermoplastic moulding compound, the most important property to control is the

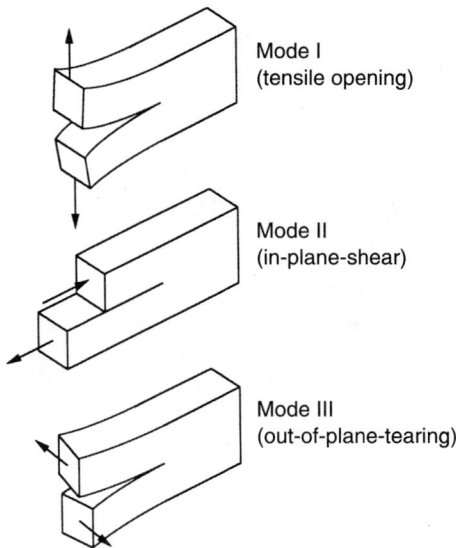

Figure 17.63 Basic delamination modes in composite material.

moisture content of the fiber and of the thermoplastic used. This involves a strict drying schedule. Polycarbonate, for example, has a permissible moisture content of 0.02% by weight, otherwise surface defects and hydrolytic degradation will occur. The drying schedule involves specifying the time of drying at a given temperature and polycarbonate should be dried at a temperature of 120°C.

The moisture content of the polymer granules can be determined by:

1. Titration with Karl Fischer reagent [140]
2. Tomasetti's Volatile Indicator (TVI) test
3. Weighing in combination with drying using an IR, halogen or microwave source.

Only the Karl Fischer method will permit accurate measurements of low moisture contents of about 0.02%.

The TVI test can be used for unreinforced polycarbonate, but will only distinguish between inadequately and adequately dried material. The temperature of the test hotplate is set to the requisite temperature for a given polymer (e.g. $270 \pm 5°C$ for polycarbonate) and a few granules of the test material inserted between two glass slides, which are placed on the hot plate until the material is molten. After a further one minute, the two slides are pressed firmly together until the granules have been flattened to about 10 mm diameter. The slide sandwich is removed from the hotplate, allowed to cool and examined, when the absence of bubbles indicates adequate drying.

17.9.2 Molding

ASTM D 3641 covers the practice of mold design for the injection molding of test specimens from thermoplastic molding compounds and a design option with the mold join on the diagonal is preferred.

For molding, a small injection molding machine (Figure 17.64) is required, which will accept the specimen mold. There must be a facility to dry the molding compound

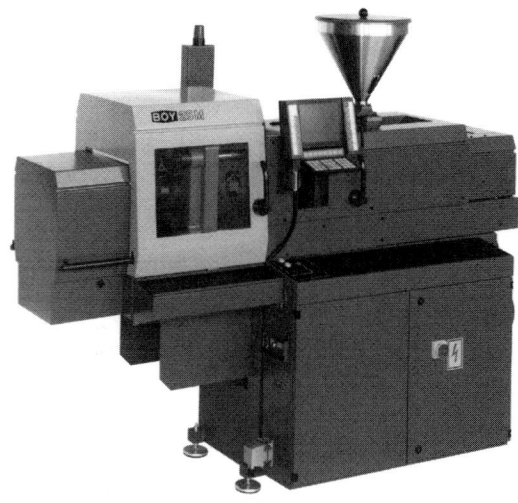

Figure 17.64 Injection molding machine (251 kN clamping force) *Source:* Courtesy of Boy Ltd., Milton Keynes.

Table 17.12 Recommended drying and molding conditions for a range of reinforced thermoplastics

Reinforced polymer	Minimum molding temperature		Barrel temperature °C	Mold temperature °C	Allowable moisture content %	Drying time and temperature (Dew point −18°C)
	Crystalline °C	Amorphous °C				
ABS		215	221–249	49–82	0.06	4h @ 77°C
Acetal	182		193–210	77–107	0.05	2h @ 79°C
Nylon 6	227		232–249	82–93	0.05	4h @ 71°C
Nylon 6.6	260		254–288	82–121	0.05	4h @ 71°C
Nylon 6.12	215		221–243	66–93	0.04	4h @ 71°C
Polycarbonate		255	274–318	93–121	0.02	4h @ 121°C
PBT	238		232–260	82–121	0.02	4h @ 121°C
PET	238		238–260	88–121	0.005	5h @ 121°C
Polyethylene	163		182–216	27–66	0.05	Not required
Polyphenylene sulphide	302		288–321	138–177	0.05	2h @ 121°C
Polypropylene	182		177–238	27–66	0.05	Not required
Polysulphone		304	316–366	143–177	0.05	4h @ 127°C

immediately prior to use (a vacuum oven is preferred). Recommended drying and molding conditions are given in Table 17.12 for a range of carbon fiber reinforced thermoplastic molding compounds.

The pre-dried molding compound is fed into the machine via a feed hopper. If no drier is fitted to the hopper, the material must be used within 30 min of the completion of the drying cycle. A reciprocating screw (with a peripheral screw speed of 0.05–0.2 m s^{-1}) meters a quantity of granules into the barrel, heated by electric band heaters, which progressively increase the temperature of the polymer. Reinforced thermoplastics generally require higher melt temperatures than their unreinforced counterparts. Non-crystalline or amorphous forms do not have a sharp melting point and display a less distinct softening behavior such that indicated barrel temperatures will not correspond to the actual temperature of the melt in the zone of the barrel, which could be 30–40°C lower than that indicated on the respective controller. Conversely, the melt temperature of the nozzle can be 5–15°C higher than the setting for the final barrel zone controller. High back pressures tend to increase the melt temperature due to increased melt shear. Actual melt temperature is best determined by measuring the internal temperature of the exudate from the nozzle. The screw rotates (a screw with a compression ratio less than 2:1 is preferred) and compresses the polymer towards the nozzle, aided by frictional heat generated between the melt and the screw (using a moderate back pressure of 0.3–0.7 $MN\,m^{-2}$ to assist removal of entrapped volatiles and giving better melt uniformity). A small reservoir of molten resin is held at the front of the screw, ready to be injected into the mold and the screw stops rotating.

The mold is secured to two metal blocks called platens, one is stationary the other moves, travelling forwards and backwards with the action of the clamping cylinder operating at high pressure and kept in line with tie bars attached to the lower platen. The two halves of the mold are lined up with dowels, or guide pins, fixed to one half of the mold and passing through bushes in the other half of the mold, keeping them in line at all times. The mold contains a passageway or runner, which connects the cavities to a sprue bush, a tapered hole through which the molten plastic enters the mold. The hollow space in the mold is termed the cavity (termed impression when there is more than one) and the protruding part of the mold is the core. At the start of the molding process, the mold is closed and the screw then rotates to deliver a measured amount of molten plastic, termed a shot. The runner reduces in cross-section as it enters the cavity via the gate. The mold cavity is the part of the mold where the specimen is formed and provision is made for trapped air to escape. To ensure that the mold lines up centrally with the nozzle, the radius of the nozzle should be smaller than the radius of the sprue bush and the hole in the nozzle smaller than the hole in the sprue bush. High pressures of up to 170 $M\,N\,m^{-2}$ (a filled grade will require 10–20% more pressure than an unreinforced polymer) to ensure rapid filling of the cavity, with a follow-up pressure of 5.5–8 $M\,N\,m^{-2}$ to avoid excessive molded-in stresses. The polymer solidifies due to the heat being removed into the cooler walls of the mold. Upon completion of the molding cycle and while still under pressure, the mold is cooled for a predetermined time, usually using water (cf a car engine block), to avoid distortion or built-in stress. The mold is then moved automatically to a position where it opens and ejector pins push all solidified product (specimens, sprue and runner) out of the mold, ready for the next cycle to begin. The sprue is the material that sets in the bush.

The machines can be supplied with microprocessor control units. The molds are heavy and expensive and must be treated with care at all times, observing all safety precautions and always using safety glasses and gloves. After use, the machine should be thoroughly cleaned and purged with polyethylene with a melt flow index of 2.

17.9.3 Determination of Melt Flow Index (MFI)

The MFI serves as a useful measure of the uniformity of the flow rate of a polymer and has been adapted for use with carbon fiber filled thermoplastics. The equipment consists essentially of a thermostatically controlled heated steel cylinder with a WC die at the lower end and a dead weight piston operating within the cylinder.

The equipment, which should be in accordance with ASTM D 1238, is first leveled. The temperature indicating device (previously calibrated as detailed in ASTM D 1238) is set to the requisite temperature (e.g. 300°C for a carbon fiber reinforced polycarbonate) and allowed to stabilize for 15 min. About 5–8 g of predried molding compound is weighed out and over a period of 1 min introduced into the cylinder with frequent tamping to expel air. The anti-wobble piston (mass 100 g) is inserted and after 4 min, the test load (e.g. 5 kg for carbon fiber reinforced polycarbonate) is placed on the top of the piston and the load initially supported with the test load support for 2 min. The support is then removed and five successive 10 sec samples of extrudate collected by cutting through the extrudate with a special knife, guided by special scibe marks on the piston. After cooling, each sample is weighed and the mean weight of the five determinations obtained (say 1.0 g), multiplying by 60 to give the amount of extrudate in 10 min.

The MFI is then expressed as 300°C 5 kg 60 g 10 min. (Ideally, conditions should have been selected to give a flow rate of $0.15-50 \text{ g}(10 \text{ min})^{-1}$).

It is important to keep the equipment clean whilst it is still hot using the special cleaning tools provided and afterwards, the die can be finally cleaned by heating for 2 h in N_2, in a furnace at 600°C.

17.9.4 Impact testing of thermoplastics

Izod and Charpy impact tests can be used for isotropic carbon fiber filled thermoplastic materials as a measure of toughness. The basic principle of these tests is to allow a pendulum of known mass to fall through a known height and strike a specimen at the lowest point of the swing and record the height to which the pendulum continues its swing using an idler pointer on a scale. The Izod test has a cantilevered test specimen, whilst the Charpy employs a supported beam. To cover the whole range of impact energies, it is necessary to have more than one machine, although Ceast have developed models with interchangeable vices and hammers which can cater for Charpy, Izod and Tensile Impact Testing. The Izod test is more generally used and ISO R 180, for example, has established the velocity of the striker as $3.5 \pm 10\% \text{ m s}^{-1}$ with impact energies of 1.0, 2.75, 5.5, 11.0 and 22.0 J. There are four types of specimen, all of which use machined notches formed at an angle of 45°, with a preferred tip radius of 0.25 mm, the notch being cut on the side parallel to the direction of application of the molding pressure. The type of mold, molding machine and type of flow in the mold cavity will influence the impact strength. The value can be expressed by normalizing as energy per unit area behind the notch, which gives a better correlation than normalizing with respect to the notch length.

17.10 INSTRUMENTAL ANALYSIS

17.10.1 Optical microscope

To obtain the best possible image and minimize eye fatigue when using an optical microscope with transmitted light illumination, it is essential that the light path is set up correctly, giving an even illumination and a bright image without glare. To achieve this, the microscope (Figure 17.65) is set up with Köhler illumination requiring that the light source (lamp filament) should be focussed at the aperture diaphragm, the rear focal plane of the objective and at the exit pupil of the eyepiece (i.e. at the iris of the observer's eye). Concurrently, the field diaphragm is focussed at the sample, the eyepiece aperture (reticle plane) and on the retina of the eye.

1. Focus the reticles in the eyepieces by removing the eyepiece, extending the focussing adjustment and looking through the eyepiece, keeping the other eye fully open and trained on a distant object (infinity), then focus the eyepiece to obtain a sharp reticle. Replace the eyepiece. If using a binocular viewing head, also adjust the interpupillary distance to suit ones eyes.

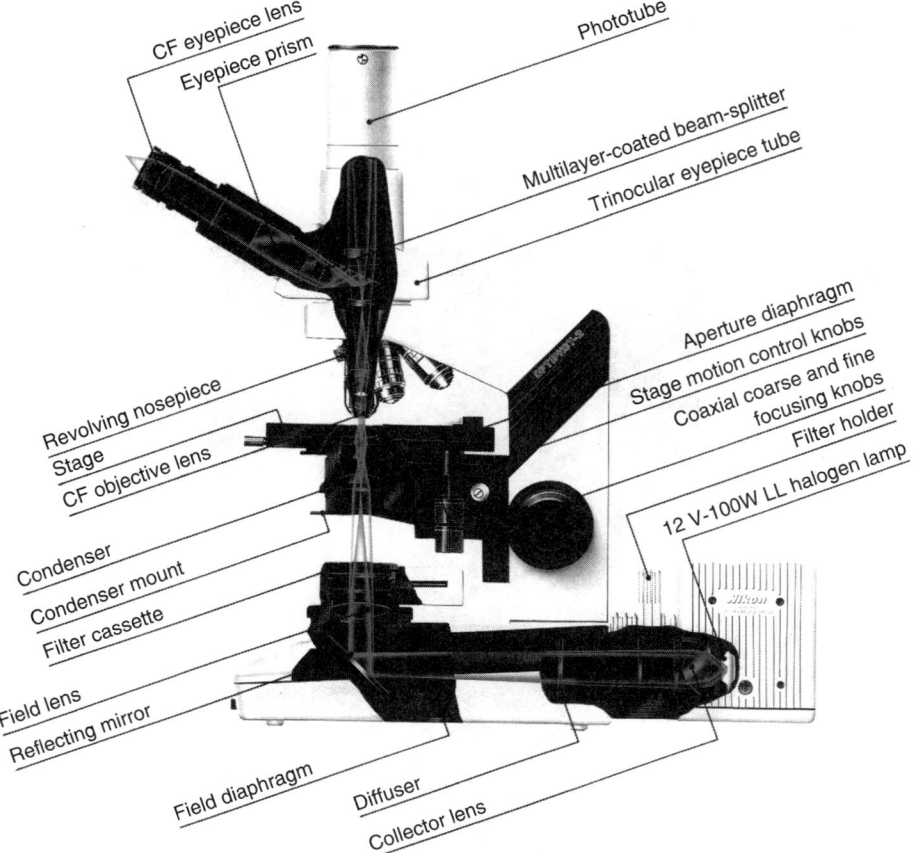

Figure 17.65 Optical microscope. *Source:* Courtesy of Nikon.

2. Turn on the illuminator, place an object on the stage, select the 10× objective by rotating the objective nosepiece, fully open the field diaphragm and check that the sample is illuminated by placing a piece of paper over the sample to see if the light is reaching it.
3. Rack the objective lens so that it is just above the sample without looking through the ocular(s) and then focus on the sample to obtain sharpest detail by looking through the ocular(s) and gradually moving the lens upwards. If the light is too bright, then reduce the setting on the light source rheostat.
4. Start to close the field diaphragm until visible in the field of view. Focus the condensing system until the diaphragm is in sharp focus with no color fringes.
5. Center the image of the field diaphragm using the two knobs on the condenser. Closing down the field diaphragm will assist this procedure.
6. Open up the field diaphragm until its edge is just outside the field of view.
7. Remove the eyepiece and observe the bright disk of light at the back end of the objective. Close down the aperture diaphragm to reduce the size of this disk and center the image of the aperture diaphragm over the rear focal plane of the objective to obtain an optimum balance for contrast and resolution. Adjust diameter of disk to 80–90% wide open.
8. Replace the eyepiece. Adjust the power at the light source to reduce any glare and the microscope should now be correctly illuminated.

It is sometimes beneficial to attach a video camera to the phototube to obtain a scan or succession of images, which can then be recorded onto tape for viewing by a wider audience and for record purposes.

17.10.2 Laboratory furnace

There are a number of determinations which require a suitable laboratory furnace. A furnace will be required for ashing (Figure 17.66) and should be capable of 1000°C, with the facility for admitting fresh air, which can be preheated and directed over the sample without creating a draught and removing any fumes via a chimney (Figure 17.67).

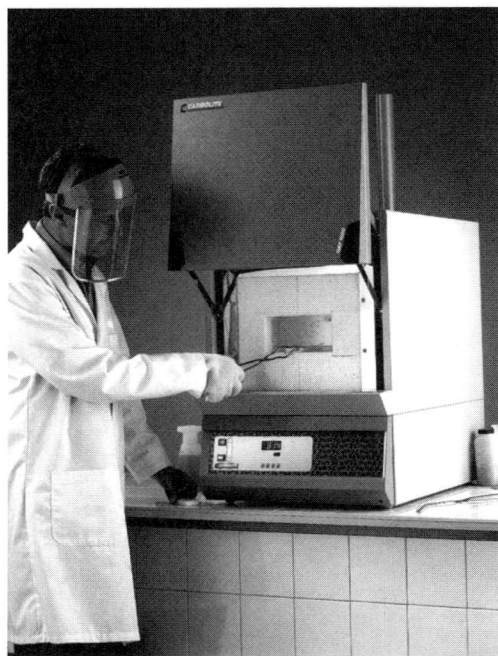

Figure 17.66 Carbolite ashing furnace. *Source:* Courtesy of Carbolite Ltd., Hope Valley, England.

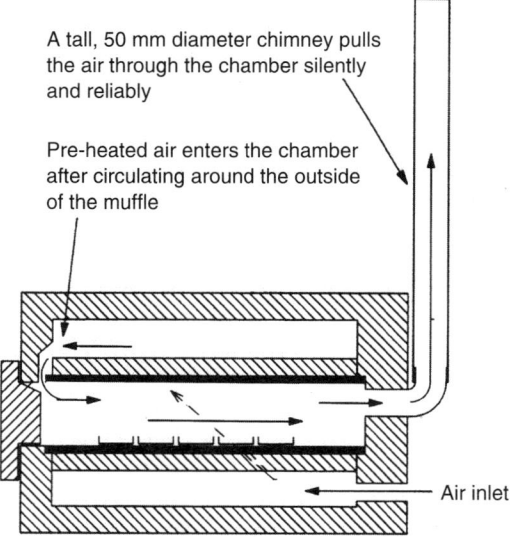

Figure 17.67 Details of air flow in Carbolite ashing furnace. *Source:* Courtesy of Carbolite Ltd., Hope Valley, England.

TESTING OF PAN PRECURSOR, VIRGIN CARBON FIBERS

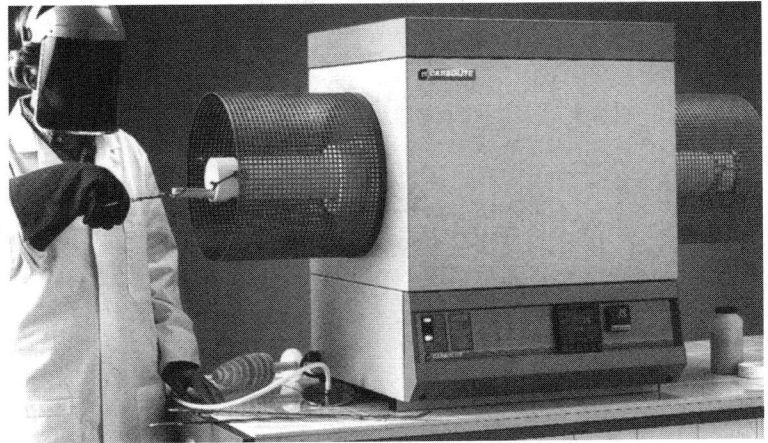

Figure 17.68 Carbolite tubular furnace. *Source:* Courtesy of Carbolite Ltd., Hope Valley, England.

It is useful to have a small furnace, which can be heated to 1500°C and be able to introduce an atmosphere of N_2 into the muffle. A furnace with a mullite muffle and SiC heating elements would be ideal for this application (Figure 17.68). It is important though, that the SiC elements operate in air at these temperatures.

17.10.3 Thermal analysis

Thermal analysis is a technique for characterizing materials by measuring changes in physical or chemical properties resulting from controlled changes in temperature. Systems include DSC, TGA, DMA, TMA and DTA. Companies like Du Pont, Mettler and Perkin Elmer have developed systems, each of which can be used with its own modular control system switching to the thermal analysis system in use, but of course only the system selected can be controlled at any one time.

17.10.3.1 Differential scanning calorimeter (DSC)

DSC belongs to a group of thermal analysis systems used to determine the enthalpy change in the sample under study. Unfortunately, confusion has arisen between the terms DSC and DTA (differential thermal analysis), since DSC is specific to Perkin Elmer instruments [141].

Figure 17.69 depicts the schematic representation of the three principal thermal analysis systems—the classical DTA system with a single heat source and temperature sensors within the sample and the reference; the Boersma DTA with a single heat source having the temperature sensors on the outside of the sample and reference; and the DSC, where the power to individual heaters located in the sample and reference holders is varied continuously in response to sample thermal effects to prevent the development of a differential temperature between the sample and reference channels [142].

1. *Classical DTA*

    If T_S and T_R are sample and reference temperatures respectively then

    $$T_S - T_R = \Delta T$$

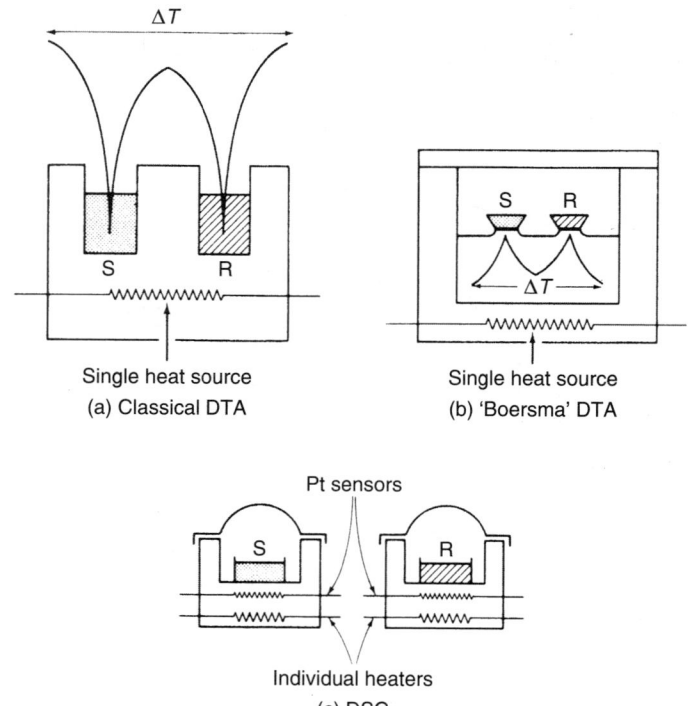

Figure 17.69 Schematic representation of the three principal thermal analysis systems. *Source:* Reprinted from McNaughton JL, Mortimer CT, *Differential scanning calorimetry, IRS; Physical Chemistry Series*, 2, 10, 1975.

is recorded as the ordinate and the abscissa can be T_S, T_R or T_P (program temperature). Classical DTA is practically useless for calorimetric measurements.

2. Boersma DTA

This system can give quite satisfactory calorimetric measurements and a number of successful designs have been based on this system, including two calorimeter cells by DuPont [143].

3. DSC

The differential power supplied is recorded as the ordinate versus the program temperature as the abscissa. The operating principles of the Perkin Elmer system are described [142] and the advantages are indeed very subtle and it is probably true to say that both DSC and DTA instruments can yield valuable information. Figure 17.70 shows a schematic representation of the DSC control loops—one loop controls the average temperature T_P, so that the sample and reference may be increased at a predetermined rate, which is recorded. The second loop ensures that if a temperature difference does occur between the sample and reference, the power imput is adjusted to correct this difference. This is the so-called null-balance principle.

Figure 17.71 shows the thermal resistances in the system, keeping the thermal mass of the sample and reference holders to a minimum, giving validity to the assumption that sample and reference holders are always at the same temperature T_P.

When a thermodynamic change occurs, there is a resultant temperature difference (ΔT) between the two pans, which is proportional to the enthalpy change (ΔH), the heat capacity

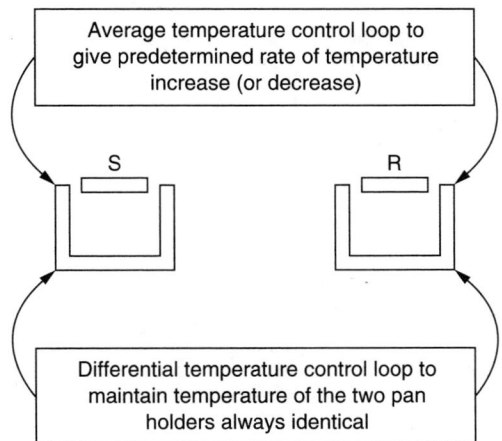

Figure 17.70 Schematic representation of the DSC control loops. *Source:* Reprinted from McNaughton JL, Mortimer CT, *Differential scanning calorimetry, IRS; Physical Chemistry Series*, 2, 10, 1975.

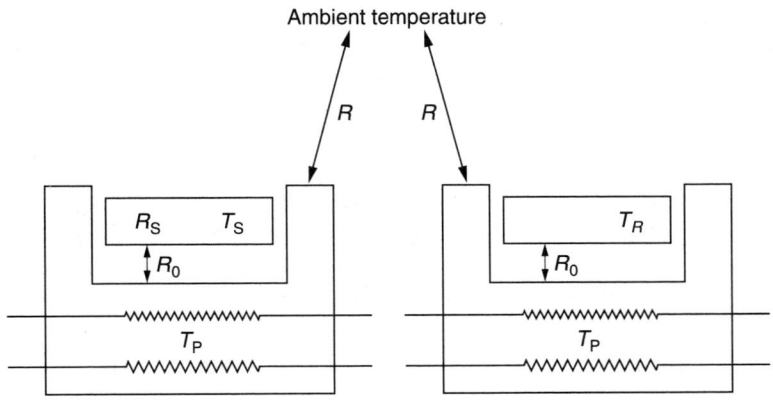

Figure 17.71 Thermal resistances in the DSC system. *Source:* Reprinted from McNaughton JL, Mortimer CT, *Differential scanning calorimetry, IRS; Physical Chemistry Series*, 2, 10, 1975.

(C), and the total thermal resistance (R). The thermal resistance consists of R_0 and R_S, where R_0 is the resistance of the instrument due to the separation of the sample, heaters and platinum resistance thermometers, whilst R_S is the sample resistance and other factors like imperfect thermal contact. In DSC, the differential power input to the two heaters dq/dt is recorded and can be related to true heat flux:

$$\frac{dH}{dt} = -\frac{dq}{dt} + (C_S - C_R)\frac{dT}{dt} - RC_S\frac{d^2q}{dt^2}$$

where dH/dt = rate of absorption of heat per unit time
dq/dt = power input
C_S = heat capacity of sample
C_R = heat capacity of reference
$(C_S - C_R)\,dT/dt$ = displacement of baseline from zero level
$RC_S(d^2q/dt^2)$ = slope of the experimental curve
R = thermal resistance

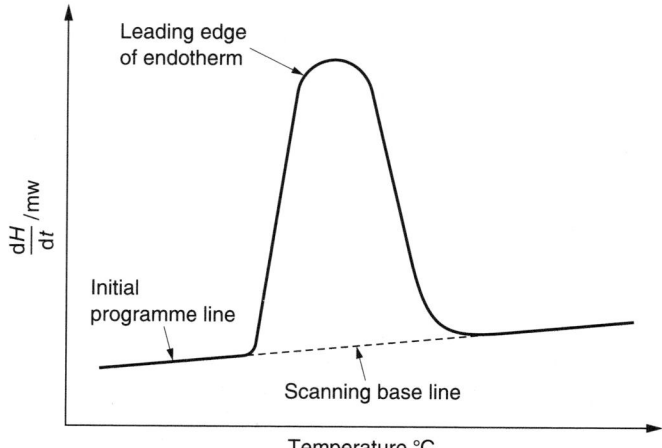

Figure 17.72 Idealized DSC thermogram. *Source*: Reprinted from McNaughton JL, Mortimer CT, *Differential scanning calorimetry, IRS; Physical Chemistry Series*, 2, 10, 1975.

The second term is negligible if sample and reference have comparable heat capacities; whilst the third term, the thermal lag, depends on R, but by reducing the size of the chambers, which will reduce to R_0. Thus using small quantities of material (less than 20 mg) to reduce R_S, will reduce the thermal lag to an acceptable level and all that will be required is $-(dq/dt)$ [144–146].

For optimum resolution, the samples should be thin films or fine granules and are placed in an Al pan, which has a domed lid that is crimped in position. Calibration is achieved with high purity metals, such as In, with accurately known enthalpies of fusion. The heating rate is about $10°C\,min^{-1}$. A thermogram is shown in Figure 17.72 and the base line is drawn in from the point at which the trace begins to depart from the initial program line to the point at which the trace returns to the program line. The area between the interpolated base line and the peak is integrated and used to calculate the calibration constant:

$$k = \frac{\Delta H_{fusion} \times m_C}{A_C} \text{J (unit area)}^{-1}$$

where ΔH_{fusion} = enthalpy of fusion of the calibrant, J g^{-1}
m_C = mass of calibrant, g
A_C = peak area of the calibration thermogram

Once the DSC is calibrated, it can be used for the determination of specific heats, melting points, heats of melting, glass transition temperatures, heats of transition and onset temperatures.

The determination of the melting point of high density polyethylene is shown in Figure 17.73. Figure 17.74 shows the evaluation of the curing process of an epoxy resin, displaying a characteristic exothermic peak as heat is released during the curing process and quantitative calculation of heats of curing and the temperature onset of curing can be made. Figure 17.75 shows the evaluation of the T_g of polystyrene, the detection of a change in the heat capacity of the polymer indicates that the glass transition point has been reached, which takes place over a temperature range and the T_g is taken as the mid-point of the inflectional tangent (Figure 17.82).

Determinations can be made by recording the cooling curve and the shelf life of a prepreg, for example, can be ascertained by this technique. The use of DSC for the characterization of

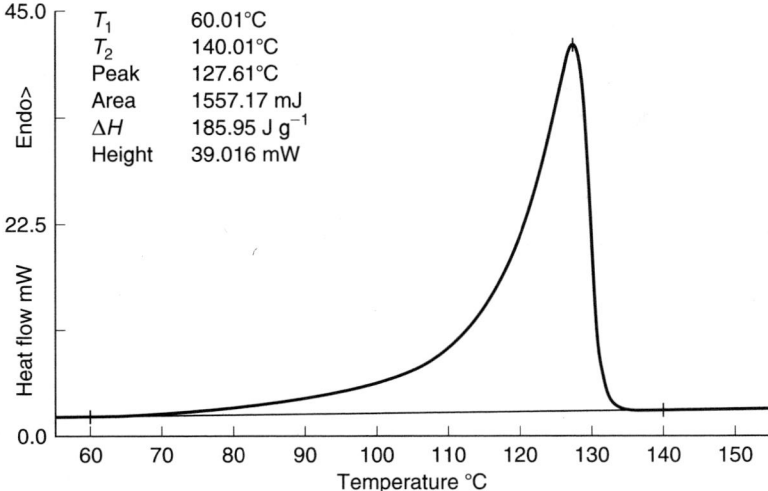

Figure 17.73 Melting point and heat of melting from DSC thermogram. *Source:* Reprinted from Perkin Elmer technical literature.

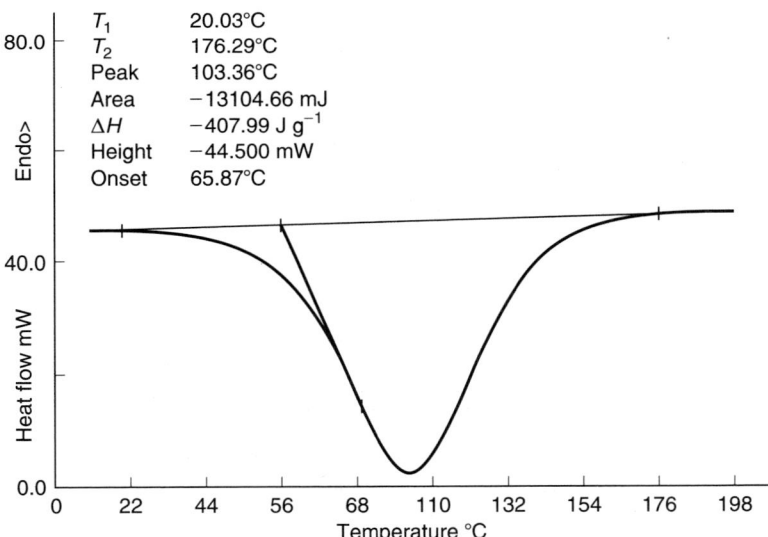

Figure 17.74 Evaluation of the curing process of an epoxy resin system using DSC. *Source:* Reprinted from Perkin Elmer technical literature.

thermosets [147–149], identification of fiber systems [150–151] and use in routine testing and quality control [152] have been discussed.

17.10.3.2 *Thermogravimetric analysis (TGA)*

A thermogravimetric analyzer measures changes in weight in a controlled atmosphere (air or N_2) as a function of temperature and can provide derivative TGA data. DuPont, Perkin Elmer and Stanton-Redcroft all have models differing in layout, but providing similar data. This equipment has proved very useful and Figure 17.76 shows a thermogram of a PAN precursor. A thermogram can also record deterioration, dehydration and evaporation.

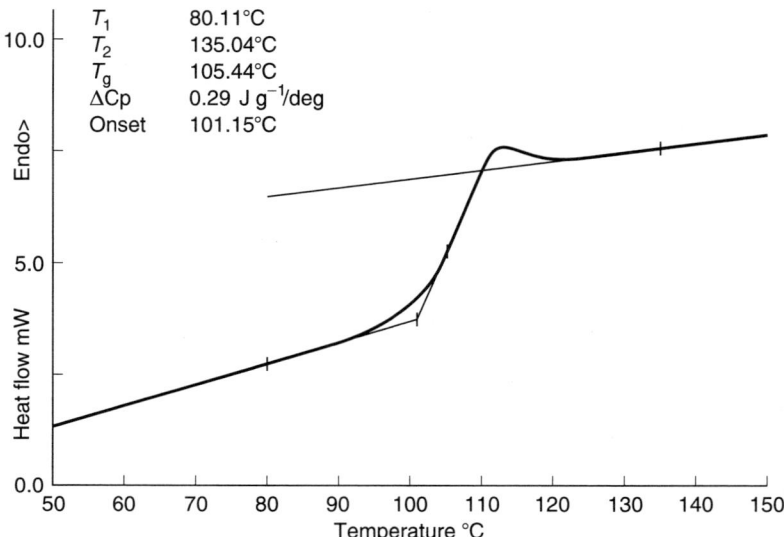

Figure 17.75 Evaluation of the glass transition temperature of polystyrene using DSC. *Source:* Reprinted from Perkin Elmer technical literature.

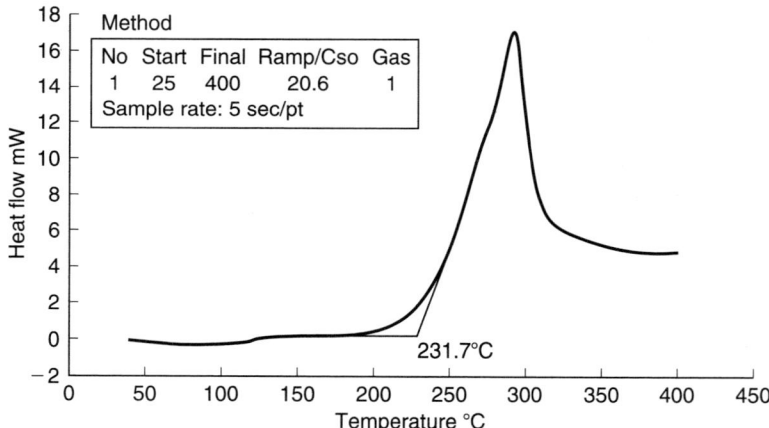

Figure 17.76 Thermogram of PAN precursor using DSC.

17.10.3.3 Dynamic mechanical analysis (DMA) [153–159]

DMA is based on the viscoelastic response of a material subjected to a small oscillatory strain imposed by flexural bending, but can also have the capability of other deformations such as shear. The viscoelasticity of the material is separated into the two components of modulus, E^*—comprising a real part, the elastic modulus (E') and an imaginary part, which is the damping or viscous component (E'').

$$E^* = E' + iE''$$

The sample is rigidly clamped between two arms (Figure 17.77) and becomes part of the resonant system provided by a driver system. The amplitude is measured by a linear voltage

TESTING OF PAN PRECURSOR, VIRGIN CARBON FIBERS

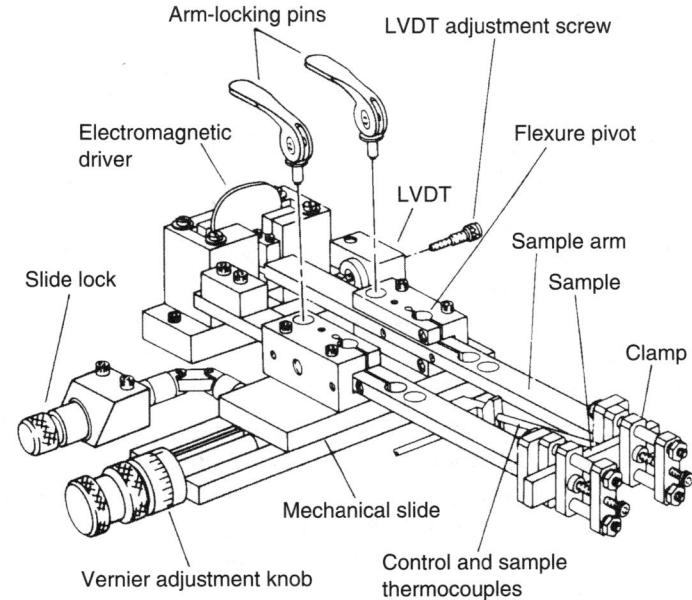

Figure 17.77 Vertical clamp assembly of DuPont 982 Dynamic Mechanical Analysis System. *Source:* Reprinted from DuPont technical literature.

variable differential transformer (LVDT). Any loss of energy due to damping is compensated by the system to provide oscillation at constant amplitude. The frequency of oscillation is directly related to the modulus of the sample and the energy needed to maintain constant amplitude oscillation is a measure of the damping within the sample.

A simplified version of the equations of motion for the DuPont DMA is:

1. Frequency data (for modulus determination)

$$E' = 2(1+\sigma)\left[\frac{8\pi^2 Jf^2 - 2K}{B^2}\right]\frac{(L+\Delta L)}{A}\left[\alpha + \frac{(L+\Delta L)^2}{24(1+\sigma)k^2}\right]\beta(f) \text{ and}$$
$$E' = 2G'(1+\sigma)$$

where σ = Poisson's ratio
A = sample cross-sectional area
k = sample cross-sectional area radius of gyration
$k = T/\sqrt{12}$ for flat samples and $R/2$ for cylindrical samples
B = distance between flexure pivot centers
f = resonant frequency
$\beta(f)$ = instrument compliance factor
α = shear distortion factor
J = moment of inertia
K = pivot sprint constant
L = sample length
ΔL = length compensation
T = sample thickness
E' = tensile modulus

G' = shear modulus
2. Damping data

$$\tan \delta = \frac{[C'V - V_i(f,a)]}{a(f^2 - f_0^2)}$$

where C' = amplitude independent damping constant
a = oscillation amplitude
V = measured damping voltage
$V_i(f,a)$ = instrument damping voltage (a function of frequency and amplitude)
f_0 = free arm resonant frequency
f = measurement frequency

The DMA Data Analyzer includes programs for calibration routines.

Tensile modulus and/or shear modulus values are obtained from the resonant frequency, making corrections for the compliance of the mechanical system—shear distortion for samples clamped in the jaws and end effects due to imperfect clamping.

Tensile loss modulus, shear loss modulus and tan δ values are obtained from the damping signal, again using instrument correction factors.

The Perkin Elmer DMA system is shown schematically in Figure 17.78 and applies the load in 3-point bending to high modulus materials via a linear force motor and can determine measurements up to 1000°C.

Figure 17.79 shows the DMA performance of polyphenylene sulphide undergoing recrystallization, exhibiting a sharp rise in modulus after the T_g (115°C).

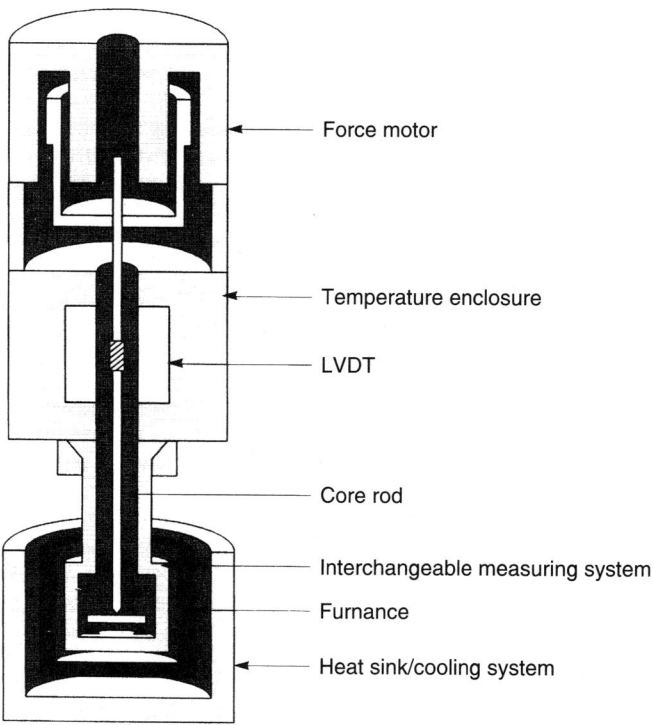

Figure 17.78 Schematic arrangement of Perkin Elmer DMA7 system. *Source:* Reprinted from Perkin Elmer technical literature.

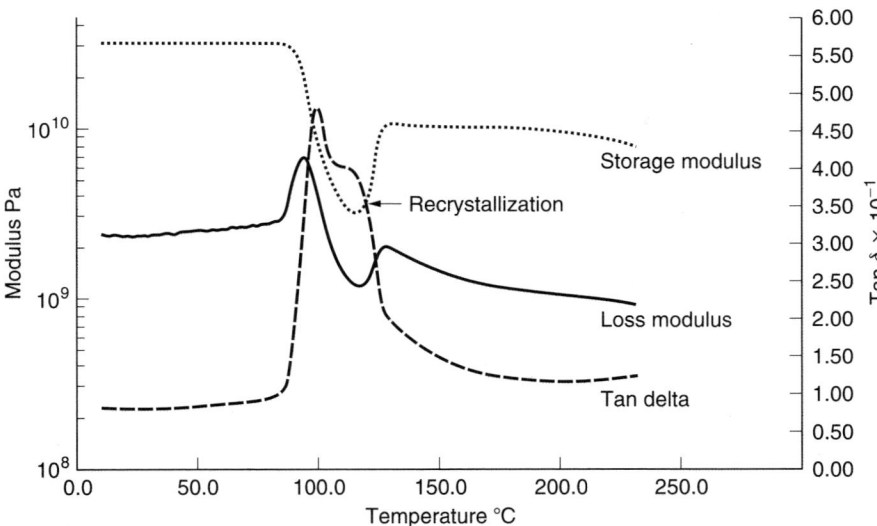

Figure 17.79 DMA performance of polyphenylene sulphide using the Perkin Elmer DMA7 system. *Source:* Reprinted from Perkin Elmer technical literature.

17.10.3.4 Thermomechanical analysis (TMA)

TMA uses a LVDT to measure the linear displacement of the sample held in various probe configurations shown in Figure 17.80, to measure softening by penetration, expansion/contraction, tension and dilatometry. The sample is held in a temperature controlled environment and the sample temperature is measured with a thermocouple in close proximity to the sample. The probe displacement is plotted against sample temperature or time.

Figure 17.81 shows a comparison of TMA with DSC and DMA and Figure 17.82 gives a comparison of TMA with DSC.

17.10.4 Chromatography [160–170]

Gas chromatography (GC) is restricted in application, since only about 15% of materials are volatile and capable of analysis by GC. Gel permeation chromatography (GPC) is more versatile and is a mode of liquid chromatography (LC) in which soluble components of a complex sample can be separated according to molecular size, a characteristic that is closely associated with the molecular weight. However, only soluble fractions of a fully crosslinked thermoset can be evaluated.

GPC separation is achieved by injecting a polymer solution into a flowing stream of solvent (Figure 17.83) and passing through a closely packed bed of carefully selected particles of specific size and porosity for a given application. The molecules with smaller effective size penetrate more pores than molecules with larger effective size and take longer to pass through the column, giving a size separation of the molecules, with the larger molecules exiting the column first. The pore size must be carefully selected, not so large that all molecules enter the pores and not so small that none are able to enter. Columns of different porosity ranges can be used in series to give more effective separation but would increase back pressure and the process is assisted by the use of high pressure pumps. The pump must produce the same flow rate independent of any viscosity difference. The injector must be capable of

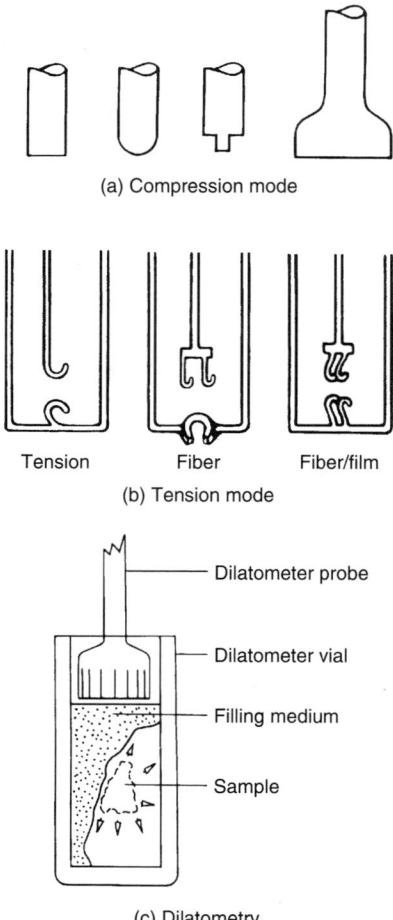

Figure 17.80 Probe configurations to enable DuPont TMA943 system to be used in: (a) compression mode; (b) tension mode; (c) dilatometry. *Source:* Reprinted from DuPont technical literature.

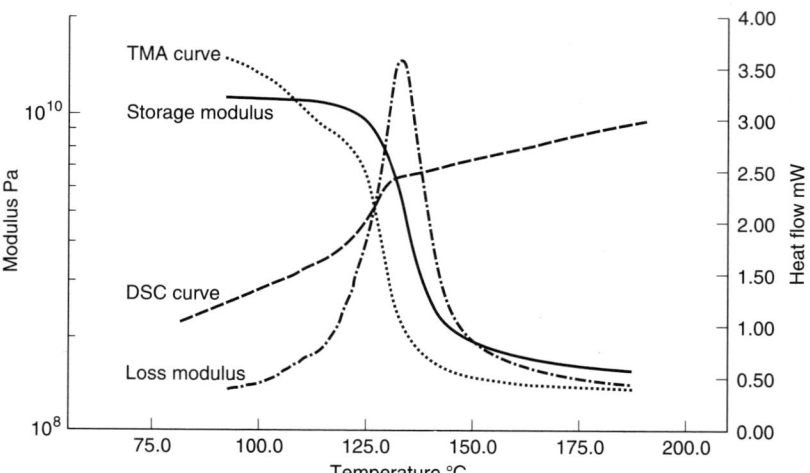

Figure 17.81 Comparison of data from DMA, TMA and DSC of a printed circuit board epoxy resin. *Source:* Reprinted from Perkin Elmer technical literature.

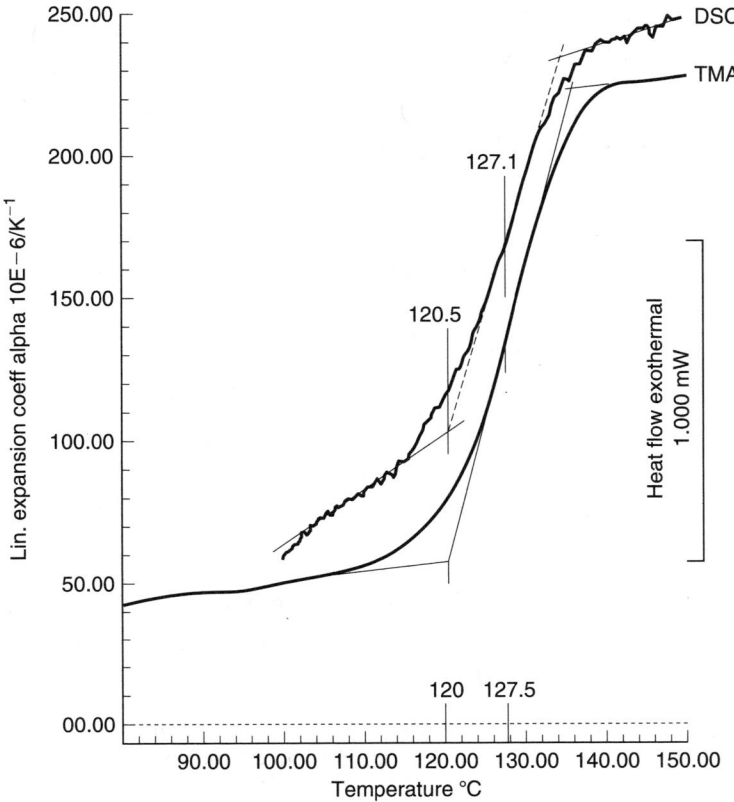

Figure 17.82 The glass transition of a printed circuit board resin (brominated bisphenol resin) determined by DSC and DMA. *Source:* Reprinted from Mettler technical literature.

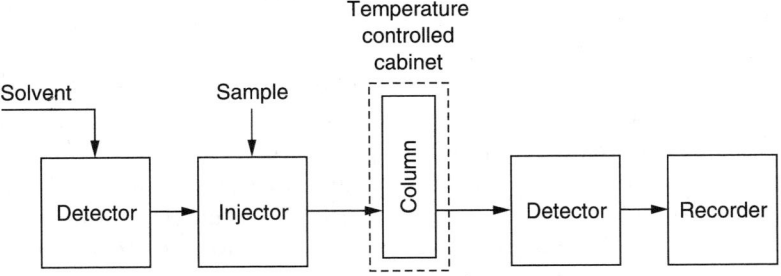

Figure 17.83 Schematic layout of gel permeation chromatography system.

delivering small and large volumes, which could be multiple samples and the introduction of the samples must not disturb the flow of solvent.

A detector is used to monitor the separation process and the response is directly proportional to the concentration. The detector can be based on variable wavelength UV/visible light, fluorescence, polarimetry, electrochemical behavior (for materials that can be readily oxidized), electrical conductivity, refractive index and more recently, the highly successful mass spectrometer.

Figure 17.84 shows a schematic chromatogram where molecules emerge in descending order and peak height is concentration dependent. Region A represents the largest molecules, which is the polymer, followed by the intermediate to low molecular weight components

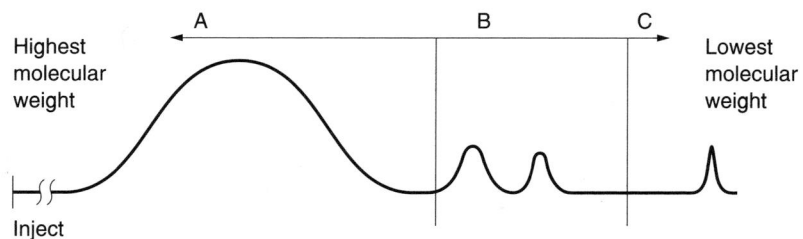

Figure 17.84 Schematic diagram of GPC chromatogram. *Source:* Reprinted from Waters' technical literature.

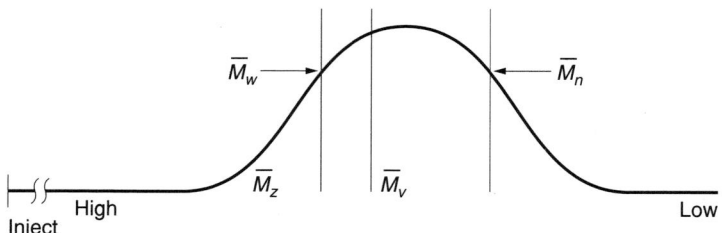

$\overline{M}_z$, $\overline{M}_w$, $\overline{M}_v$, $\overline{M}_n$ can be obtained from GPC data

Figure 17.85 Determination of $\bar{M}_z$, $\bar{M}_w$, $\bar{M}_v$ and $\bar{M}_n$ from GPC data. *Source:* Reprinted from Waters' technical literature.

in region B, such as additives and oligomers and finally in region C, peaks which represent the smallest molecular weight components, like unreacted monomer and contaminants such as moisture.

The distance from the point of injection to the midpoint of the molecular weight distribution curve is a measure of the average molecular weight. If there are peaks in the curve, this suggests a blend of polymers. An initial shoulder in the curve suggests presence of gel, whilst a shoulder at the low end of the distribution suggests addition of some material to the polymer. A skewed distribution indicates a bias towards that end of the molecular weight distribution.

Identification of peaks can be assisted by spiking the sample and repeating the determination. Individual fractions can be collected and subjected to separate analysis, such as UV spectroscopy, for identification of the full structure.

Using known standards, the system can be calibrated to obtain values for the average molecular weight ($\overline{M}_Z$), the weight average molecular weight ($\overline{M}_W$), the viscosity average molecular weight ($\overline{M}_V$) and the number average molecular weight ($\overline{M}_N$) of the polymer and the concentrations of the individual components (Figure 17.85).

17.10.5 Infrared analysis (IR) [171–178]

Infrared is that portion of the electromagnetic spectrum between the visible and microwave regions (Figure 1.1) normally measured in cm^{-1} and the region most used for infrared analysis is 4000–800 cm^{-1}.

The atoms in a molecule vibrate and the bonds between the atoms act like springs. A given molecule has a specific set of vibrational frequencies, which are in the same range as the IR frequencies of electromagnetic vibration.

The IR instrument passes IR radiation through the sample and records which wavelengths have been absorbed and to what extent. The transmittance of the sample is plotted against the frequency (or wavelength) of the radiation to obtain the IR spectrum, which characterizes the sample and its concentration. The instrument is ideal for fingerprinting or identifying substances by matching with a material of proven identity, or may be identified by using one of several libraries of reference materials. The regions in the spectrum for various vibrations are summarized in Figure 17.86.

Liquid samples are contained within a cell, which uses crystal windows such as NaCl, KBr, CaF_2 and BaF_2 and great care must be taken to avoid water, as it can readily etch the cell window.

Fourier Transform Infrared Spectroscopy (FTIR) is based on the Michelson interferometer (Figure 17.87). A monochromatic source is passed through a beam splitter to give

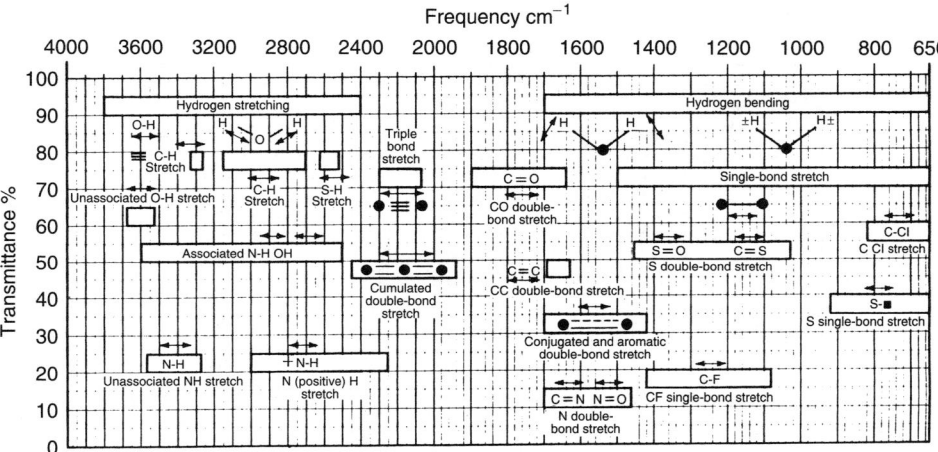

Figure 17.86 Adsorption on different regions of the IR spectrum. *Source:* Reprinted from Perkin Elmer technical literature.

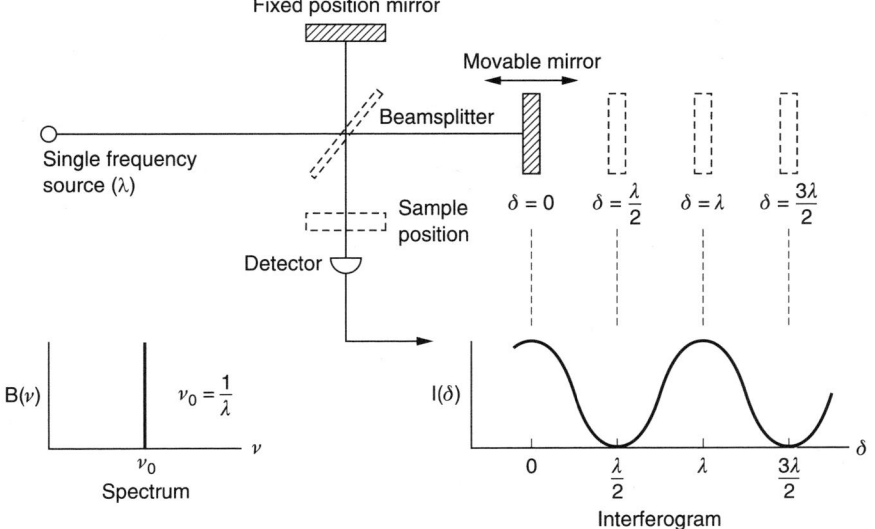

Figure 17.87 Block diagram showing the major components of a FTIR spectrometer. The spectrum of an infinitely narrow line source is depicted and also how the interferogram is generated as the moveable mirror is translated. Maxima in the interferogram occur when the retardation is equal to an integral multiple of the wavelength of the source. Minima occur when the retardation is an odd multiple of half wavelengths. *Source:* Reprinted from Perkin Elmer technical literature.

two equally divided beams. Beam A travels to a fixed mirror, which reflects back to the beam splitter, where a part of the beam is reflected back to the source and the other part passes to a detector. Beam B hits a moveable mirror, which also reflects back to the beamsplitter. Here again, a part of the light is reflected back to the source and the other part to the detector. If the position of the moveable mirror allows beam B to travel exactly the same distance as beam A before striking the detector, then the beams are said to be in phase and reinforce each other. If the optical path lengths are different the beams are out of phase and tend to cancel each other out. The intensity of the radiation at the detector is a cosine function and with the advent of fast computers, it has become possible to apply a Fourier transform to give a conventional spectrum.

Figure 17.88 is an IR fingerprint of oxidized PAN fiber showing the presence of the correct fiber finish.

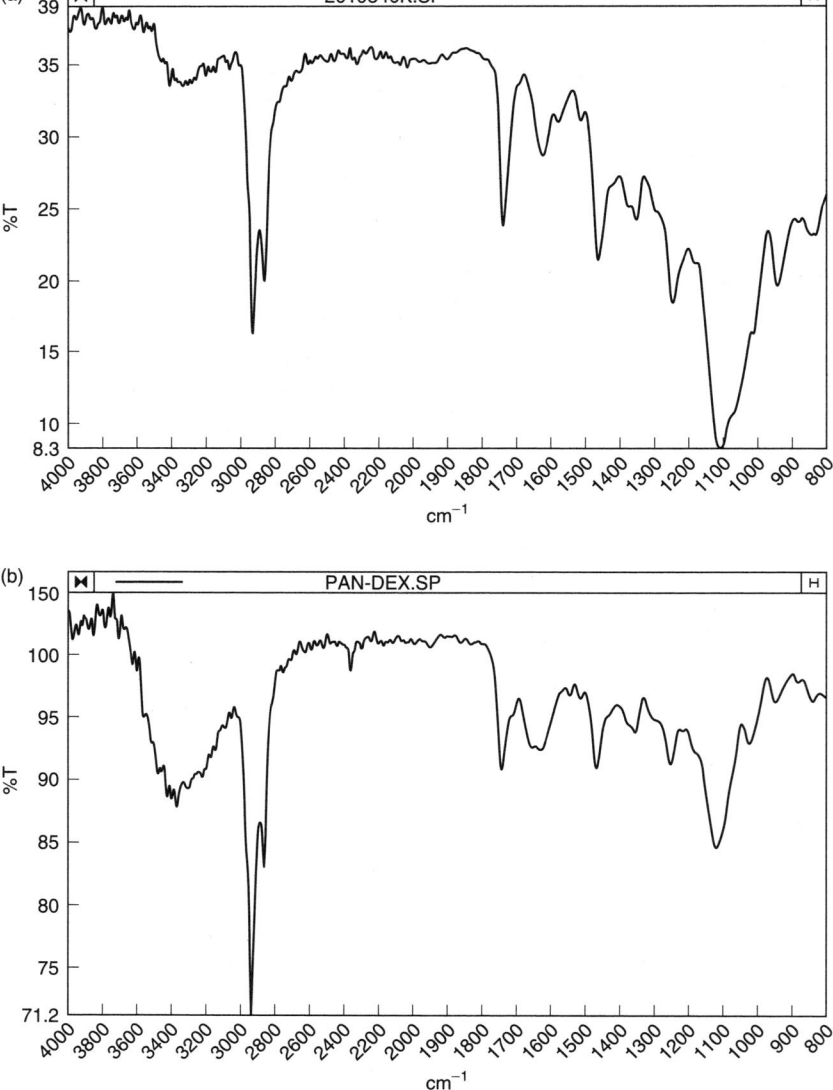

Figure 17.88 Fingerprint of oxidized PAN fiber treated with a proprietary finish. (a) Finish as applied to oxidized PAN fiber. (b) Oxidized PAN fiber with finish.

17.10.6 Elemental analysis

Elemental analyzers are extremely useful and provide a rapid determination (under 5 min) of C, H and N using He or Ar as a carrier gas. The sample is placed in a Sn capsule and first oxidized in a pure O_2 environment in the presence of reagents and a platinized carbon catalyst. The resulting combustion gases (CO_2, H_2O and N_2) are then scrubbed using a heated Cu tube to remove any sulphur products from the oxidation and NaOH to remove acid gases. The carrier gas sweeps the product gases into a chamber where they are mixed and controlled to exact conditions of pressure, temperature and volume (Figure 17.89). The homogenized product gases are then depressurized, passed through a column and separated under steady state conditions using a technique called frontal chromatography, which utilizes the selective retention of the gases to produce a steady state stepwise signal (Figure 17.90). The stepwise series of gases then pass through a thermal conductivity detector system and are measured. Versions are available for the additional determination of sulphur (CHNS) and an oxygen kit can also be provided. The analysis of carbon fiber has been reported [179].

17.11 NON-DESTRUCTIVE TESTING (NDT) [180–186, 190–192]

Generally, NDT is used for composite structures such as pressure vessels, which is outside the scope of this chapter, but there are occasions when it is necessary to assess the effectiveness of a composite lay-up and brief details are given.

The most frequently used NDT techniques for polymer matrix composites are ultrasonics and X-radiography

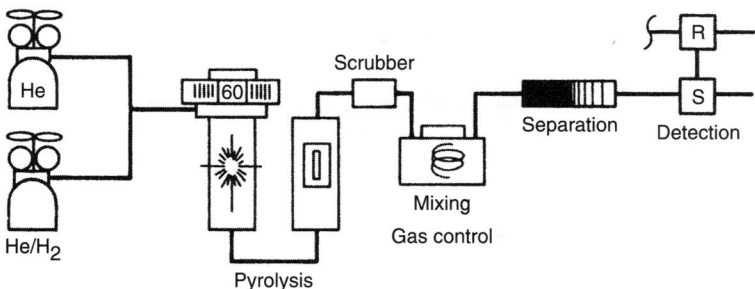

Figure 17.89 Schematic layout of Perkin Elmer CHN elemental analyzer. *Source:* Reprinted from Perkin Elmer technical literature.

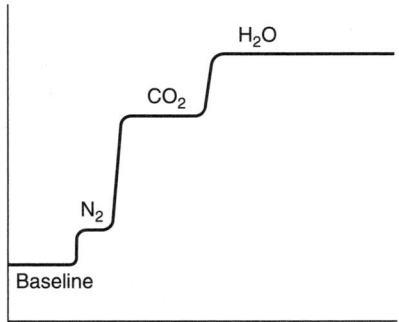

Figure 17.90 Schematic representation of gases issuing stepwise from Perkin Elmer CHN elemental analyzer. *Source:* Reprinted from Perkin Elmer technical literature.

17.11.1 Ultrasonic testing [183–189]

If a composite is subjected to an ultrasonic source, low frequency waves penetrate to a greater depth than a higher frequency source and are more sensitive to defects, but cannot be heard above 20 kHz. A compromise for frequency (1–10 MHz) has to be selected between penetration and sensitivity. A stress wave has to be injected into the composite and the transmitted or reflected beam has to be monitored using piezoelectric transducers, which can send a pulse or receive the vibration as an electric signal. To obtain good results, the transducer must be effectively acoustically coupled. Air is a poor coupling medium, transmitting only 0.03% with a cfrp, whereas water transmits 72.3%. Greases and gels can also be used.

To obtain the most efficient coupling, it is necessary to immerse in a water bath, but this is not always practical and jet-probes have been developed where water is propelled in laminar flow. Alternatively, hand held probes can be used with a grease coupling medium.

To examine a laminate sheet, it is most convenient to measure the attenuation of the ultrasonic beam passing through it and there are three techniques which can be used:

1. Use probes on either side of the laminate
2. Use a single probe positioning a glass plate on the other side to reflect the beam (pulse-echo)
3. More conveniently, use the back surface pulse echo

There are three ways to undertake these ultrasonic scans termed A-, B- and C- scans.

Figure 17.91 shows a typical A-scan using back surface echo, where the amplitude of the signal reflected from the top surface, back surface and defect are displayed on the y-axis of an oscilloscope with time (related to depth) on the x-axis.

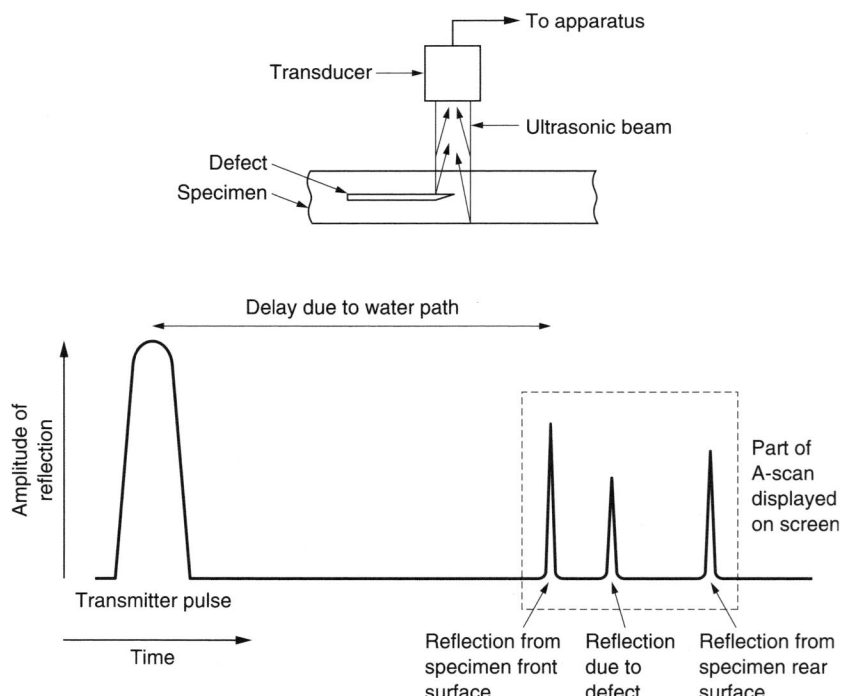

Figure 17.91 A-scan representation of ultrasonic data. *Source:* Reprinted with permission from Matthews EL, Rawlings RD, Chapman and Hall, 421, 1994. Copyright 1994, Springer.

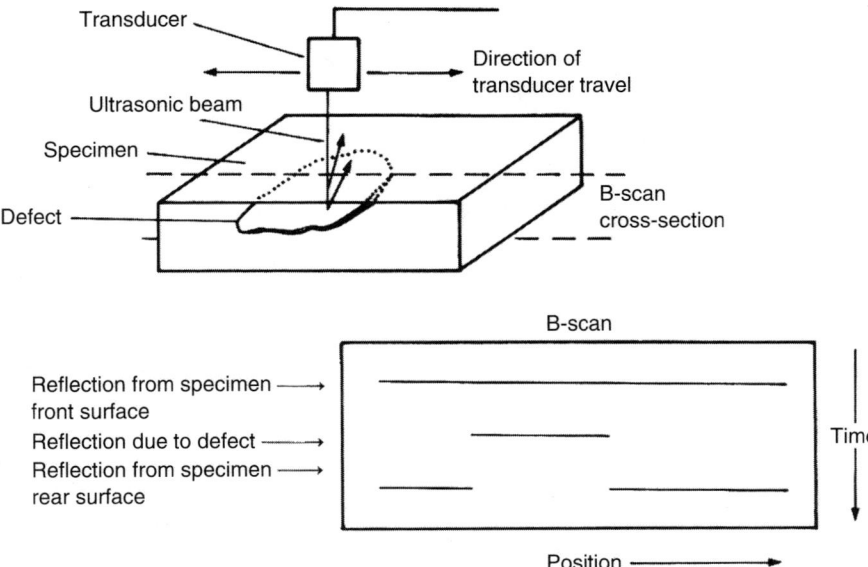

Figure 17.92 B-scan representation of ultrasonic data. *Source:* Reprinted with permission from Matthews EL, Rawlings RD, Chapman and Hall, 421, 1994. Copyright 1994, Springer.

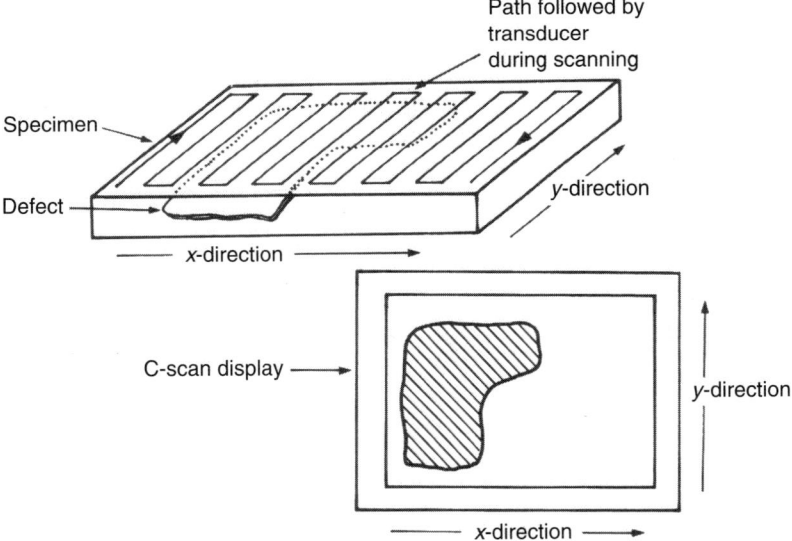

Figure 17.93 C-scan representation of ultrasonic data. *Source:* Reprinted with permission from Matthews EL, Rawlings RD, Chapman and Hall, 422, 1994. Copyright 1994, Springer.

Figure 17.92 shows a B-scan which is basically a series of A-scans taken in a straight line and recorded as a function of position on a storage oscilloscope, giving a picture of the cross-section along the line of scan.

Figure 17.93 shows a C-scan, which is a representation of ultrasonic data where the scanning transducer has been moved to follow a rectilinear raster and the transducer position and amplitude of the signal are recorded on an X-Y plotter. Figure 17.94 depicts the damage incurred in the form of delamination by a cfrp laminate when impacted.

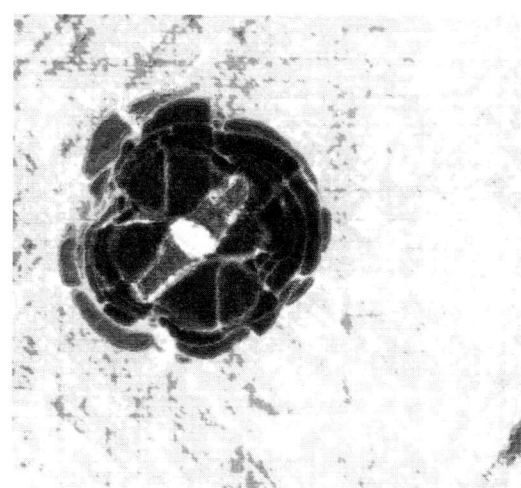

Figure 17.94 Ultrasonic C-scan of cfrp laminate showing an area of impact induced delamination; white indicates negligible damage and black bad damage. *Source:* Reprinted from Cawley P, The sensitivity of the mechanical impedance method of nondestructive testing, *NDT Int*, 30(4), 209–215, Aug 1987.

17.11.2 Radiography

Low energy radiation can penetrate reasonable thicknesses of a polymer matrix composite and 1 MeV X-rays are able to penetrate 500 mm of polymer. The cfrp have low X-ray absorption and consequently, low voltage, low energy X-rays have to be used in conjunction with high contrast recording techniques, which can then detect transverse cracks and inclusions. Unfortunately, carbon fibers and resin matrices have similar mass absorption coefficients and cannot be differentiated. Glass, on the other hand, can be readily detected and can be used in composite construction as a radio-opaque tracer, showing the direction of fiber lay-up.

17.11.3 Acoustic emission [193–195]

Acoustic emission can be used for testing pressure vessels and is not strictly a nondestructive test as it involves listening to the propagation of damage when subjected to an applied stress.

17.12 SUPPLEMENT 1

17.12.1 Sinclair's loop test for filament testing

In order to keep the test volume to a minimum, Sinclair [204] introduced a loop test for glass fiber (Figure 17.95), where the filament is subjected to maximum stress only on the surface in extreme curvature, in tension on the outside and compression on the inside, establishing a very small test volume, in the μm^3 range. The filament is positioned between two glass cover slides and lubricated with glycerine or a light oil. Weights are fastened to the lower end of the filament and the corresponding loop diameters are measured. The test can be used for tension or compression testing.

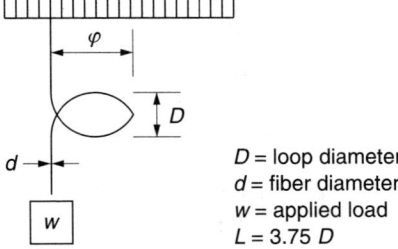

Figure 17.95 Sinclair's loop test for filament testing.

1. Tension testing

$$\sigma = \frac{16wL}{\pi d^3}$$

$$\varepsilon = \frac{4d}{L}$$

$$E = \frac{4wL^2}{\pi d^4}$$

2. Compression testing

$$\sigma = \frac{60wD}{\pi d^3}$$

$$\varepsilon = \frac{16d}{15D}$$

$$E = \frac{\sigma}{\varepsilon}$$

A single fiber recoil method developed by Allen [205] for compression testing involved placing the fiber in tension and cutting it in the middle using an electric spark [206]. This enabled the two fragments to recoil against their corresponding grip ends, producing compression waves when they reached the grip ends and started to return. If the compressive stress is greater than the applied tensile stress, then the filament will fracture at the grip ends.

REFERENCES

1. Schulman S, Methods of single fibre evaluation, *J Polym Sci*, C(19), 211–225, 1967.
2. Perry AJ, Ineichen B, Eliasson B, *Fibre Diameter Measurement by Laser Diffraction*, Chapman & Hall, 1376, 1974.
3. Compugraphics International Ltd., Glenrothes, Scotland.
4. Chen KJ, Diefendorf RJ, In: Hayahashi T, Kawata K, Umekawa S eds., *Progress in Science and Engineering of Composites*, 97–105, 1982.
5. Tzeng SS, Diefendorf RJ, Cross-sectional area measurement of single carbon fibres, *19th Biennial Conference on Carbon*, Session 6A, Penn State University, 298–299, Jun 25–30, 1989.
6. Hooke R, *Micrographia*, Royal Society, London, 1664.
7. Éspinasse M, *Robert Hooke*, William Heinemann, London, 1956.

8. Hawthorne HM, On non-Hookean behaviour of carbon fibres in bending, *J Mater Sci*, 28, 2531–2535, 1993.
9. Huang Y, Young RJ, *J Mater Sci Lett*, 12(2), 92–95, 1993.
10. Shioya M, Hayakawa E, Takaka A, *J Mater Sci*, 31(17), 4521–4532, 1996.
11. Moreton R, The effect of gauge length on the tensile strength of RAE carbon fibres, *Fibre Sci Technol*, 1, 273–284, 1969.
12. Hitchon JW, Phillips DC, *Fibre Sci Technol*, 12, 217, 1979.
13. Hughes JDH, Morley H, Jackson EE, *J Phys D Appl Phys*, 13, 921, 1980.
14. JDH Hughes, The evaluation of current carbon fibres, *J Phys D Appl Phys*, 20, 276–285, 1987.
15. Dai SR, Piggott MR, The strengths of carbon and Kevlar fibres as a function of their lengths, *Composite Sci Technol*, 49(1), 81–87, 1993.
16. Glycerol, *British Pharmacopoeia*, 1, 311–312, 1993.
17. Goggin PR, The use of whole tow tests for quality control of carbon fibres, HCFWP, United Kingdom Atomic Energy Authority, Harwell, 68, 11.
18. Wells H, Colclough WJ, Goggin PR, Some mechanical properties of carbon fibre composites, *AERE -R6149*, United Kingdom Atomic Energy Authority, Harwell, Jul 1969.
19. Suflex Ltd., Risca, Near Newport, Gwent, NP1 6 YD (manufacturer of 'Peribraid').
20. Case J, *Strength of Materials*, Arnold, 1957.
21. Marin J, *Mechanical Behaviour of Engineering Materials*, Prentice Hall, 1962.
22. Morley A, *Strength of Materials*, 1916.
23. Roark RJ, *Formulas for Stress and Strain*, McGraw-Hill, 1965.
24. Adams DF, *Test Methods for Composite Materials*, Technomic, Basel, 1990.
25. Agarwal BD, Btroutman LJ, *Analysis and Performance of Fibre Composites*, 2nd ed., Wiley, New York, 1990.
26. Ashbee KHG, *Fundamental Principles of Fibre Reinforced Composites*, Technomic, Basel, 1989.
27. Broutman LJ, Krock RH eds., *Composite Materials*, 6 vols, Academic, London, 1974.
28. Chawla KK, *Composite Materials, Science and Engineering*, Springer-Verlag, New York, 1987.
29. Chou TW ed., *Structure and Properties of Composites*, VCH, Weinheim, 1993.
30. Daniel IM, Ishai O, *Engineering Mechanics of Composite Materials*, Oxford University Press, Oxford, 1994.
31. Folkes MJ, *Short Fibre Reinforced Thermoplastics*, Research Studies Press, Letchworth, 1982.
32. Halpin JC, *Primer on Composite Materials*, 2nd ed., Technomic, Basel, 1992.
33. Hearle JWS, Grosberg P, Backer S, *Structural Mechanics of Fibers, Yarns and Fabrics*, Wiley, 1969.
34. Hollister GS, Thomas C, *Fibre Reinforced Materials*, Elsevier, London, 1966.
35. Hull D, Clyne DW, *An Introduction to Composite Materials*, Cambridge University Press, Cambridge, 1996.
36. Jones RM, *Mechanics of Composite Materials*, McGraw-Hill, Washington, 1973.
37. Kelly A, MacMillan NH, *Strong Solids*, 3rd ed., Clarendon Press, Oxford, 1986.
38. Kelly A, Rabotnov Yu N, *Handbook of Composites*, 4 vols, Elsevier Science, Amsterdam, 1985.
39. Kelly A, *Concise Encyclopaedia of Composite Materials, Advances in Materials Science and Engineering*, Vol 3, Pergamon, Oxford, 1989.
40. Lubin G, *Handbook of Composites*, van Nostrand Reinhold, New York, 1982.
41. Matthews EL, Rawlings RD, *Composite Materials, Engineering and Science*, Chapman and Hall, 1994.
42. Morley JG, *High-performance Fibre Composites*, Academic Press, London, 1987.
43. Partridge IK ed., *Advanced Composites*, Elsevier, London, 1989.
44. Rubin I ed., *Handbook of Plastic Materials and Technology*, Wiley-Interscience, New York, 1990.
45. Sottos NR, Hodgkinson JM, Matthews FL, A practical comparison of standard test methods using carbon fibre reinforced epoxy, *Proc 6th International Conference on Composite Materials*, Imperial College, 1.310–1.320, Jul 20–24, 1987.
46. Tsai SW, Hahn HT, *Introduction to Composite Materials*, Technomic, Westport, 1980.
47. Vinson JR, Chou TW, *Composite Materials and their Structures*, Applied Science, Barking, 1975.
48. Watt W ed., *New Fibres and their Composites*, Special issue of Proc Royal Soc, A294, 1980.

49. Wendt FW, Leibowitz H, Perrone N eds., *Mechanics of Composite Materials*, Pergamon, Oxford, 1970.
50. Haynes WM, Tolbert TL, Determination of the graphite fibre content of plastic composites, *J Composite Mater*, 3, 709, Oct 1969.
51. Green P, Fiber volume fraction determination of carbon epoxy composites using an acid digestion bomb, *J Mater Sci Lett*, 10(19), 1162–1164, 1991.
52. Krucinska I, Evaluation of the intrinsic mechanical properties of carbon fibers, *Composite Sci Technol*, 41(3), 287–301, 1991.
53. Manders PW, Kowalski IM, The effect of small angular fibre misalignments and tabbing techniques on the tensile strength of carbon fiber composites, *32nd International SAMPE Symposium*, 985–996, Apr 6–9, 1987.
54. Neubert HKP, Resistance strain gauges, *Metron*, 2(4), 123–130, 1970.
55. Neubert HKP, *Strain Gauges: Kinds and Uses*, Macmillan, London, 1968.
56. Perry CC, Lissner HR, *The Strain Gage Primer*, McGraw-Hill, New York, 1962.
57. Sottos NR, Hodgkinson JM, Matthews FL, A practical comparison of standard test methods using carbon fibre reinforced epoxy, Matthews JM et al eds., *Proc ICCM VI and ECCM 21*, Elsevier Applied Science, 310–320, 1987.
58. Cui WC, Wisnom MR, Jones M, Effect of specimen size on interlaminar shear strength of unidirectional carbon fiber-epoxy, *Composites Eng*, 4(3), 299–307, 1994.
59. Harding J, Dong L, Effect of strain rate on the interlaminar shear-strength of carbon-fiber-reinforced laminates, *Composite Science Technology* 51(3), 347–358, 1994.
60. Liu K, Piggott MR, Shear strength of polymers and fibre composites, Part 2: carbon/epoxy pultrusions, *Composites*, 26(12), 841–848, 1995.
61. Browning CE, Abrams FL, Whitney JM, A four point shear test for graphite/epoxy composites, Browning CE ed., *Composite Materials: Quality Assurance and Processing*, ASTM STP 797, American Society for Testing and Materials, Philadelphia, 54–74, 1983.
62. Lee S, Munroe M, In-plane shear properties of graphite/epoxy composites for aerospace applications: evaluation of test methods by the decision analysis method, *Aeronautical Note NAE-AS22*, NRC No.23778, Mechanical Engineering Department, University of Ottawa, Canada, Oct 1984.
63. Sawada Y, Shindo A, Torsional properties of carbon-fibers, *Carbon*, 30(4), 619–629, 1992.
64. Swanson SR, Merrick M, Toombes GR, Comparison of torsion tube and Iosipescu in-plane shear test results for a carbon fibre reinforced epoxy composite, *Composites*, 16, 8220, 1985.
65. Lockwood PA, Results of the ASTM Round Robin on the rail shear test for composites, *Composites Technol Rev*, 3, 83, 1981.
66. Broughton WR, Kumosa M, Hull D, Analysis of the Iosipescu shear test as applied to unidirectional carbon-fiber reinforced composites, *Composite Sci Technol*, 38(4), 299–325, 1990.
67. Chamis CC, Sinclair JH, Ten degree off-axis test for shear properties in fiber composites, *Exp Mech*, 17, 339–346, 1977.
68. Wisnom MR, The relationship between flexural and tensile strength of unidirectional carbon fibre epoxy, *J Composite Mater*, 26(8), 1173–1180, 1992.
69. Buggy M, Carew A, The effect of thermal ageing on carbon-fiber-reinforced polyetheretherketone (PEEK), 1. Static and dynamic flexural properties, *J Mater Sci*, 29(7), 1925–1929, 1994.
70. Wisnom MR, The effect of specimen size on the bending strength of unidirectional carbon fiber-epoxy, *Composite Structures*, 18(1), 47–63, 1991.
71. Wisnom M, The flexural strength of unidirectional carbon fibre-epoxy, *5th European Conference on Composite Materials*, Bordeaux, 165–170, 1992.
72. Adsit NR, Compression testing of graphite/epoxy, Chait R, Papirno R eds., *ASTM STP 734*, American society for Testing Materials, 152–165, 1981.
73. Adsit NR, Compression testing of graphite/epoxy, Chait R, Papirno R eds., *Compression Testing of Homogeneous Materials and Composites*, ASTM STP 808, American Society for Testing and Materials, 175–186, 1983.
74. Clark RK, Lisagor WD, Compression testing of graphite/epoxy composite materials, CC Chamis ed., *Test Methods and Design Allowables for Fibrous Composites*, ASTM STP 734, 34–53, 1981.

75. Barker AJ, Balasundaram V, Compression testing of carbon fibre reinforced plastics exposed to humid environments, *Composites*, 18(3), Jul 1987.
76. Berg JS, Adams DF, An evaluation of composite material compression test methods, *J Comp Technol Res*, 11, 41–46, 1989.
77. Curtis PT, Gates J, Molyneaux CG, An improved engineering test method for the measurement of compressive strength of unidirectional carbon fibre composites, *Technical Report 91031*, DRA, Farnborough, 1991.
78. Carr DJ, Barker AJ, Compressive properties of carbon-fiber reinforced plastics, *Chem Eng Res Design*, 71(A3), 316–318, 1993.
79. Crasto AS, Kumar S, Recoil compression testing of advanced carbon fibers, *35th International SAMPE Symposium*, 318–331, Apr 2–5, 1990.
80. Curtis PT, Gates J, Molyneux CG, An improved engineering test method for the measurement of compressive strength of unidirectional carbon fibre composites, *Composites*, 22(5), 363–368, 1991.
81. Dobb MG, Johnson DJ, Park CR, Compressional behaviour of carbon fibers, *J Mater Sci*, 25(2A), 829–834, 1990.
82. Dobb MG, Guo H, Park CR, Johnson DJ, Structure-compressional property relations in carbon fibers, *Carbon*, 33(11), 1553–1559, 1995.
83. Grimes GC, Experimental study of compression—compression fatigue of graphite/epoxy composites, Chamis CC ed., *ASTM STP 734*, American Society for Testing and Materials, 281–337, 1981.
84. Häberle JG, Matthews FL, Studies on compressive failure in unidirectional cfrp using an improved test method, *Proc ECCM-4 Conf Sept 1980*, Elsevier Applied Science.
85. Hofer KE, Rao PN, A new static compression fixture for advanced composite materials, *J Testing Evaluation*, 5(4), 278–283, 1977.
86. Kumar S, Anderson DP, Crasto AS, Carbon-fiber compressive strength and its dependence on structure and morphology, *J Mater Sci*, 28(2), 423–439, 1993.
87. Kumar S, Helminiak TE, Relation between compressive strength and tensile modulus in carbon fibres, *Proceedings of Materials Research Symposium 134*, 363, 1989.
88. Lamothe RM, Nunes J, Evaluation of fixturing for compression testing of metal matrix and polymer/epoxy composites, Chait R, Papirno R eds., *Compression Testing of Homogeneous Materials and Composites*, ASTM STP 808, American Society for Testing and Materials 241–253, 1983.
89. Melanitis N, Galiotis C, Compressional behaviour of carbon-fibers, 1. Raman spectroscopic study, *J Mater Sci*, 25(12), 5081–5090, 1990.
90. Miwa M, Mori Y, Takeno A, Yokoi T, Watanabe A, Compressive and tensile behaviour of carbon fibres, *J Mater Sci*, 33(8), 2013–2017, 1998.
91. Nakatani M, Shioya M, Yamashita J, Axial compressive fracture of carbon fibres, *Carbon*, 37(4), 601–608, 1999.
92. Ohsawa T, Miwa M, Kawade M, Tsushima E, Axial compressive strength of carbon fiber, *J Appl Polym Sci*, 39(8), 1733–1743, 1990.
93. Port KF, The compressive strength of cfrp, Royal Aircraft Establishment, *Farnborough Tech Report 82083*, 1982.
94. Prandy JM, Hahn HT, Compressive strength of carbon fibers, *SAMPE Quarterly*, Society for the Advancement of Materials and Process Engineering, 22(2), 47–52, 1991.
95. Shinohara AH, Sato T, Saito F, Tomioka T, Arai Y, A novel method for measuring direct compressive proprties of carbon-fibers using a micromechanical compression tester, *J Mater Sci*, 28(24), 6611–6616, 1993.
96. Soutis C, Measurement of the static compressive strength of carbon-fiber epoxy laminates, *Composite Sci Technol*, 42(4), 373–392, 1991.
97. Wisnom MR, The effect of fiber misalignment on the compressive strength of unidirectional carbon-fiber epoxy, *Composites*, 21(5), 403–407, 1990.
98. Woolstencroft DH, Curtis AR, Haresceugh RI, A comparison of test techniques used for the evaluation of the unidirectional compressive strength of carbon fibre reinforced plastics, *Composites*, 12, Oct 1981.

99. Woolstencroft DH, Curtis AR, Haresceugh RI, A comparison of test techniques used for the evaluation of the unidirectional compressive strength of carbon fibre reinforced plastics, *Composites*, 16, 8220, 1985.
100. Buggy M, Dillon G, Flexural fatigue of carbon fiber-reinforced PEEK laminates, *Composites*, 22(3), 191–198, 1991.
101. Bunsell AR, Somer A, The tensile and fatigue behaviour of carbon fibers, *Plastics Rubber and Composites Processing and Applications*, 18(4), 263–267, 1992.
102. Curtis PT, *RAE Technical Report TR82031*, RAE, Farnborough (now DRA), 1982.
103. Curtis PT, An investigation of the mechanical properties of improved carbon fibre materials, *RAE Technical Report TR86021*, RAE, Farnborough (now DRA), 1986.
104. Curtis PT, A review of the fatigue of composite materials, *RAE Technical Report TR87031*, RAE, Farnborough (now DRA), 1987.
105. Gilchrist MD, Kinloch AJ, Matthews FL, Mechanical performance of carbon-fibre and glass-fibre-reinforced epoxy I-beams, III. Fatigue performance, *Composite Sci Technol*, 59(2), 179–200, 1999.
106. Haque A, Raju PK, Monitoring fatigue damage in carbon fiber composites using an acoustic impact technique, *Materials Evaluation*, 56(6), 765–770, 1998.
107. Harris B *et al*, Fatigue behaviour of carbon fibre reinforced plastics, *Composites*, 21(3), 232–242, 1990.
108. Hitchen SA, Ogin SL, Smith PA, Effect of fiber length on fatigue of short carbon-fiber epoxy composite, *Composites*, 26(4), 303–308, 1995.
109. Irving PE, Thiagrarajan C, Fatigue damage characterization in carbon fibre composite materials using an electrical potential technique, *Smart Materials and Structures*, 7(4), 456–466, 1998.
110. Jones CJ *et al*, Environmental fatigue of reinforced plastics, *Composites*, 14, 288–293, 1983.
111. Jones CJ *et al*, Environmental fatigue behaviour of reinforced plastics, *Proc Royal Society of London*, 396, 315–318, 1984.
112. Lifschitz JM, Compressive fatigue and static properties of a UD graphite/epoxy composite, *J Composite Technol Res*, 10(3), 100–106, 1988.
113. Lin CT, Kao PW, Yang FS, Fatigue behaviour of carbon fiber-reinforced aluminium laminates, *Composites*, 22(2), 135–141, 1991.
114. Philips DC, Scott JM, Shear fatigue of unidirectional fibre composites, *Composites*, 8, 233–236, 1977.
115. Nichols ME, Wang SS, Geil PH, Creep and physical ageing in a polyamideimide carbon-fiber composite, *J Macromol Sci Phys*, B29(4), 303–336, 1990.
116. Sturgeon JB, Creep of fibre reinforced thermosetting resins, *Creep of Engineering Materials*, Pomeroy CD ed., Mechanical Engineering Publications Ltd., Chapter 10, 1978.
117. Cantwell WJ, Morton J, The impact resistance of composite materials -a review, *Composites*, 21, 347–362, 1991.
118. Dorey G, Impact damage in composites- development, consequences and prevention, *Proceedings of ICCM VI/ECCM 2*, Vol 3, Mattews FL *et al* ed., Elsevier Applied Science, London, 3.1–3.26, 1987.
119. Dorey G, Bishop SM, Impact performance of carbon fibre/PEEK composites, *Paper No18, International Conference on Impact Testing and Performance of Polymeric Materials*, University of Surrey, Sep 2–3, 1985.
120. Talreja R ed., *Damage Mechanics of Composite Materials*, Technomic, Westport, 1992.
121. Wardle MW, Tokarsky EW, Drop weight impact testing of laminates reinforced with Kevlar, E-glass and graphite, *Composites Technol Rev*, 5(1), 4, 1983.
122. Wardle MW, Zahr GE, Instrumented impact testing of aramid and reinforced composite materials, *Instrumented Impact Testing of Plastics and Composite Materials*, ASTM STP 936, Keesler SL, Adams GL, Driscoll SB, Ireland DR eds., 219–235, 1987.
123. Wyrick DA, Adams DF, Residual strength of carbon/epoxy materials, *Composite Mater*, 749–765, Aug 22, 1988.

124. AA Aliyu, Daniel IM, Effects of strain rate on delamination fracture toughness of graphite/epoxy, *Delamination and Debonding of Materials*, ASTM STP 876, American Society for Testing and Materials, Philadelphia, 336–348, 1985.
125. Arcan L, Arcan M, Daniel IM, SEM fractography of pure and mixed-mode interlaminar fractures in graphite/epoxy composites, *Fractography of Modern Engineering Materials*, Masters J, Au J eds., ASTM STP 948, American Society for Testing and Materials, Philadelphia, 41–67, 1987.
126. Bibo G, Leicy D, Hogg PJ, Kemp M, High temperature damage tolerance of carbon-fiber-reinforced plastics. 1. Impact characteristics, *Composites*, 25(6), 414–424, 1994.
127. Corletto CR, Bradley WL, Mode II delamination fracture toughness of unidirectional graphite/epoxy composites, *Composite Materials: Fatigue and Fracture*, 2nd Vol, ASTM STP 1012, Legace PA ed., American Society for Testing and Materials, Philadelphia, 201–221, 1989.
128. Daniel IM, Strain and failure analysis in graphite/epoxy laminates with cracks, *Exp Mech*, 18, 246–252, 1978.
129. Daniel IM, Shareef I, Aliyu AA, Rate effects on delamination fracture toughness of a toughened graphite/epoxy, *Toughened Composites*, ASTM STP 937, Johnson NJ ed., American Society for Testing and Materials, Philadelphia, 260–274, 1987.
130. Gillespie JW, Carlsson LA, Smiley AJ, Rate dependent Mode I interlaminar crack growth mechanisms in graphite/epoxy and graphite/PEEK, *Composites Sci Technol*, 28, 1–15, 1987.
131. Hitchen SA, Ogin SL, Smith PA, Soutis C, The effect of fiber length on fracture-toughness and notched strength of short carbon-fiber epoxy composites, *Composites*, 25(6), 407–413, 1994.
132. Kishimoto K, Notomi M, Koizumi T, Fracture toughness of short carbon-fiber-reinforced thermoplastic polymide, *Engineering Fracture Mechanics*, 49(6), 943 *et seq*, 1994.
133. Nejhad MNG, Parvizimajidi A, Impact behaviour and damage tolerance of woven carbon-fiber reinforced thermoplastic composites, *Composites*, 21(2), 155–168, 1990.
134. Pavier MJ, Clarke MP, Experimental techniques for the investigation of the effects of impact damage on carbon-fiber composites, *Composite Sci Technol*, 55(2), 157–169, 1995.
135. Ramkumar RL, Whitcomb JD, Characterization of Mode I and Mixed-mode delamination growth in T300/5208 graphite/epoxy, *Delamination and Debonding of Materials*, ASTM STP 876, Johnson WS ed., American Society for Testing and Materials, Philadelphia, 315–335, 1985.
136. Shikhmanter L, Eldror I, Cina B, Fractography of unidirectional cfrp composites, *J Mater Sci*, 24, 167–172, 1989.
137. Stewart M, Feughelman M, The failure mechanism of carbon fibres, *J Mater Sci*, 8, 1119–1122, 1973.
138. Wilkins DJ, Eisenmann JR, Camin RA, Margolis WS, Benson RA, Characterizing delamination growth in graphite/epoxy, *Damage in Composite Materials: Basic Mechanisms, Accumulation, Tolerance, Characterization*, Reifsneider RL ed., ASTM STP 775, American Society for Testing and Materials, Philadelphia, 168–183, 1982.
139. Yeow YT, Morris DH, Brinson HF, The fracture behaviour of graphite/epoxy laminates, *Exp Mech*, 19, 1–8, 1979.
140. Schilt AA, *Moisture Measurement by Karl Fischer Titrimetry*, GFS Chemicals, Powell, 1991.
141. McNaughton JL, Mortimer CT, *Differential scanning calorimetry, IRS; Physical Chemistry Series*, 2, 10, 1975.
142. Gray AP, *Analytical Calorimetry*, Porter RS, Johnson JF eds., Plenum Press, New York, 209–218, 1968.
143. *Dupont Thermal Analysis Bulletin*, No.900–8.
144. Bershstein VA, Egorov VM, *Differential Scanning Calorimetry of Polymers*, Ellis Harwood, 1994.
145. McNaughton JL, Mortimer CT, Differential Scanning Calorimetry, *Thermochemistry and Thermodynamics*, Skinner HA ed., Butterworths, London, 1975.
146. Miller AP, Theory of specific heat of solids, Specific Heat of Solids, Hemisphere, Ho CY ed., Chapter 1, 1988.
147. Barton JM, *Monitoring the cross-linking of epoxide resins by thermoanalytical techniques*, 25–32, 1974.
148. Widmann G, Isothermal curing reactions of an epoxy resin, Thermal analysis, Vol 3, *Proc of the 4th ICTA*, Budapest, 359, 1974.

149. Cassel B, Characterization of thermosets, *28th Pittsburgh Conference*, Cleveland, Mar 1977.
150. Philip WMS, The use of DSC in the identification of synthetic fibers, *Journal Forensic Sciences*, 17(1), 132–140, Jan 1972.
151. Brennan WP, Some applications of differential scanning calorimetry to fibers systems, *Perkin Elmer Thermal Analysis Study*, 6, Jan 1973.
152. Fruh P, Widmann G, Thermal analysis in routine and quality control, *Am Lab*, 14(1), 93, 1982.
153. Hassel RL, Blaine RL, *Proceedings of the SAMPE Symposium*, San Francisco, 1979.
154. Levy PF, Blaine RL, Gill PS, Lear JD, Thermal Analysis- Advances in instrumentation, *Int Lab*, 9, 53–60, 1979.
155. Burroughs P, Leckenby JN, *Analysis of composite materials by thermomechanometry (Dynamic Mechanical Analysis)*, 438–449, 1979.
156. DuPont Technical Literature, *Theory of operation of the DuPont 982 Dynamic Mechanical Analyzer*.
157. Gill PS, Leckenby JN, *Dynamic Mechanical Analysis of advanced composites* (DuPont technical article).
158. Turi E, *Thermal Characterization of Polymeric Compounds*, Academic Press, 1981.
159. Menard KP, *Dynamic Mechanical Analysis: A Practical Introduction*, CRC Press, 1999.
160. Hanai T, *HPLC: A Practical Guide*, Royal Society of Chemistry, Cambridge, 31–47, 1994.
161. Katz E ed., *Handbook of HPLC Chromatographic Science*, Marcel Dekker, 1998.
162. McMaster V, *HPLC: A Practical User's Guide*, Wiley, 1994.
163. Meyer VR, *Practical High Performance Liquid Chromatography*, John Wiley, Chichester, 1994.
164. Neue UD, *HPLC Columns: Theory, Technology and Practice*, John Wiley, 1997.
165. Parriott D ed., *A Practical Guide to HPLC Detection*, Academic Press, Orlando, 1993.
166. Patonay G, *HPLC Detection: Newer Methods*, VCH Publishers, 1993.
167. Pasch H, Trathnigg B, *HPLC of Polmers*, Springer Laboratory, 1998.
168. Sadek PC, *The HPLC Solvent Guide*, Wiley, 2002.
169. Snyder L, Glajch JL, Kirkland J, *Practical HPLC Development*, John Wiley & Sons, New York, 1988.
170. Hagenauer GL, Setton I, Compositional analysis of epoxy resin formulations, *J Liq Chromatogr*, 1(1), 55–73, 1978.
171. Pouchert CJ, *The Aldrich Library of FTIR Spectra*, 3 Vols, Aldrich Chemical, Milwaukee, 1985.
172. Pouchert CJ, *The Aldrich Library of Infrared Spectra*, Aldrich Chemical, Milwaukee, 1981.
173. Roeges NPG, *A Guide to the Complete Interpretation of Infrared Spectra of Organic structures*, John Wiley, Chichester, 1994.
174. Socrates G, *Infrared Characteristic Group frequencies: Tables and Charts*, John Wiley, New York, 1994.
175. Durig JR ed., *Applications of FTIR Spectroscopy Vibrational Spectra and Structure: A Series of Advances*, 18, 1990.
176. Nishikida K, *Selected Applications of Modern FT-IR Techniques*, Academic Press, 1998.
177. Chia L, Ricketts S, *Basic Techniques and Experiments in Infrared and FT-IR Spectroscopy*, Perkin Elmer, Mar 1988.
178. Coleman PB ed., *Practical Sampling Techniques for Infrared Analysis*, CRC Press, Boca Raton, 1993.
179. Culmo F, Krista J, Swanson J, Brennan WP, *Analysis of carbon fiber by the PE 2400 CHN*, Perkin Elmer Publication EAN-14.
180. Boving KG ed., *NDE Handbook, Non-destructive Examination Methods for Condition Monitoring*, Butterworths, 1989.
181. Halmshaw R, *Non-destructive Testing, Metallurgy and Materials Science Series*, Arnold, London, 1989.
182. Mix PE, *Introduction to Non-destructive Testing*, Wiley, 1987.
183. Cawley P, The sensitivity of the mechanical impedance method of nondestructive testing, *NDT Int*, 30(4), 209–215, Aug 1987.
184. Cawley P, Woolfrey AM, Adams RD, *Composites*, 16, 23, 1985.

185. Hill S, Rapid non-destructive testing of carbon fibre reinforced plastics, *Materials World*, 4(8), 450, 1996.
186. Degoeje MP, Wapenaar KED, Non-destructive inspection of carbon fiber reinforced plastics using eddy current methods, *Composites*, 23(3), 147–157, 1992.
187. Birt EA, Damage detection in carbon-fibre composites using ultrasonic Lamb waves, *Insight*, 40(5), 335–339, 1998.
188. Wong BS, Tui CG, Tan KS, Kwan KW, Evaluation of small defects in carbon fiber reinforced composites using ultrasonic pulse echo amplitude, *Insight*, 39(4), 257–263, 1997.
189. Smith RA, Clarke B, Ultrasonic C-scan determination of ply-stacking sequence in carbon fiber composites, *Insight*, 36(10), 741–747, 1994.
190. Birchon D, Non-destructive Testing, *Engineering Design Guides*, Oxford University Press, Oxford, 1975.
191. Bryant LE, McIntire P eds., *Non-destructive Testing Handbook, Radiography and Radiation Testing*, American Society for Non-destructive Testing, 3, 1985.
192. Miller RK, McIntire P eds., *Non-destructive Testing Handbook, Acoustic Emission Testing*, American Society for Non-destructive Testing, 5, 1987.
193. Berthelot JM, Rhazi J, Acoustic emission in carbon fiber composites, *Composite Sci Technol*, 37(4), 411–428, 1990.
194. Bertholot JM, Billand J, *1st Annual Symposium on AE from Reinforced Composites*, Soc Plastics Industry, California, 1983.
195. Fowler TJ, Gray E, Development of an acoustic emission test for FRP equipment, *ASCE Convention and Exposition*, Boston, 1979.
196. Walrath DE, Adams DF. The Iosipescu shear test as applied to composite materials. *Proc. V Intnl Congress on Exp Mech, Soc for Exp. Mechanics*, Bethel, CT, 97–101, 1984.
197. Rosen RW. A simple procedure for experimental determination of the longitudinal modulus of unidirectional composites, *J. Composite Materials* 6, 552–554, 1972.
198. Wang CS, Bai SJ, Rice BP, *Proc of Symp of ACS Div of Polym Mater Sci Eng*, American Chemical Society, Washington, 61, 550, 1989.
199. Rosenbaum S, *J Appl Polym Sci*, 9, 2071, 1965.
200. Cogswell FN, *Thermoplastic Aromatic Composites*, Butterworth-Heinemann, Oxford, 1992.
201. Course on Mechanical Testing of Advanced Fibre Composites, University of London, Imperial College, Sep 1995.
202. Daniel IM, Ishai O, *Engineering Mechanics of Composite Materials*, Oxford University Press, Oxford, 1994.
203. Saville BP, *Physical Testing of Textiles*, CRC Press, Boca Raton, 350, 1998.
204. Sinclair D, *J Appl Phys*, 21, 380, 1950.
205. Allen S, *J Mater Sci*, 22, 853, 1987.

CHAPTER **18**

Statistics and Statistical Process Control (SPC)

18.1 FREQUENCY DISTRIBUTION

The Memory Jogger [2] is a useful pocketbook detailing quality control tools together with management planning tools. Statistics can be regarded as the collection, representation and interpretation of data [1]. In a set of data, it is convenient but not necessary, to arrange the data from smallest to largest (e.g. Table 18.1). It is normal to sample 50–100 data points and to compress this information, the data can be divided into a number of equal class intervals, or groups. The data in Table 18.1 has been grouped into ten classes and the frequency recorded in Table 18.2. If the data has not been sorted, then it is helpful to prepare a tally chart (Table 18.2), by recording a tally mark against the appropriate class for each observation. The tallies are then totaled to obtain the respective frequencies. It is the convention to record the tallies in sets of five with the fifth denoted as a diagonal line through the previous four.

The decision on how big the subgroup should be will depend on cost and obviously, destructive testing of an expensive product would be a limitation.

A simple rule is to take the square root of the number of samples and round to the nearest whole number $k = \sqrt{n}$, which in the above example is $\sqrt{100} = 10$. Alternatively, the sample size can be determined by reference to a table (Table 18.3).

The range for the above example can be calculated as follows:

$$R = X_{max} - X_{min} = 3.79 - 3.20 = 0.59$$

To determine the class width:

$$H = R/k = 0.59/10 = 0.059$$

Rounding this number to the nearest value with the same number of decimal places as the original sample gives $H = 0.06$.

Take the smallest measurement, 3.20 (round down if required), which is the low end of the first class limit; now add the class width: $3.20 + 0.06 = 3.26$.

Therefore, the first class interval would be from 3.20 up to, but not including 3.26 (i.e. 3.20–3.25); the second class interval would be from 3.26 up to, but not including 3.32 (3.26–3.31) and so on.

Table 18.1 A random sample of 100 carbon fiber strengths (GPa) arranged in ascending order

3.20	3.35	3.41	3.45	3.47	3.49	3.52	3.55	3.59	3.64
3.24	3.36	3.41	3.45	3.47	3.50	3.53	3.55	3.59	3.65
3.27	3.37	3.41	3.45	3.48	3.50	3.53	3.55	3.59	3.66
3.29	3.37	3.41	3.45	3.48	3.50	3.53	3.56	3.60	3.66
3.30	3.38	3.41	3.46	3.48	3.50	3.53	3.57	3.60	3.66
3.31	3.39	3.42	3.46	3.49	3.50	3.53	3.57	3.61	3.68
3.32	3.39	3.43	3.46	3.49	3.51	3.54	3.58	3.61	3.69
3.33	3.39	3.44	3.47	3.49	3.51	3.54	3.58	3.61	3.69
3.34	3.40	3.44	3.47	3.49	3.51	3.54	3.58	3.63	3.70
3.34	3.40	3.44	3.47	3.49	3.52	3.54	3.59	3.64	3.79

Table 18.2 Frequency distribution and tally chart of tensile strength values (GPa)

Group	Group range	Number of tallies	Group frequency
1	3.20–3.25	II	2
2	3.26–3.31	IIII	4
3	3.32–3.37	THI III	8
4	3.38–3.43	THI THI III	13
5	3.44–3.49	THI THI THI THI IIII	24
6	3.50–3.55	THI THI THI THI II	22
7	3.56–3.61	THI THI THI	15
8	3.62–3.67	THI II	7
9	3.68–3.73	IIII	4
10	3.74–3.79	I	1

Table 18.3 Determination of sample size

Number of data points Lot size	Typical Number of Classes Sample Size k
<50	5–7
50–99	6–10
100–249	7–12
>250	10–20

The frequency distribution can be expressed as a graph (Figure 18.1) or pictorially as a histogram (Figure 18.2), which is basically a tally chart drawn as a bar chart. If a dot is placed at the mid-point at the top of each rectangle in the histogram and the dots connected, this gives a frequency polygon. It is normal to close the polygon by placing dots on the x-axis, half a class interval, to the left of the lowest class and to the right of the highest class (Figure 18.3).

18.2 LOCATION OF DATA

In order to expand the information from the frequency diagram, the location of the center of the data and its dispersion must be known.

1. Mid-range—which is seldom used, is halfway between the smallest and the largest of the observations. In Table 18.2, it is $(3.20 + 3.79)/2 = 3.50$ GPa.
2. The mode—is little used and is the observation in the sample which occurs most frequently, (e.g. 3.49 GPa in Table 18.2).There can be more than one mode and the system would then be termed multi-modal, usually indicating that the data is from two different sources (e.g. precursor lots).

STATISTICS AND STATISTICAL PROCESS CONTROL (SPC)

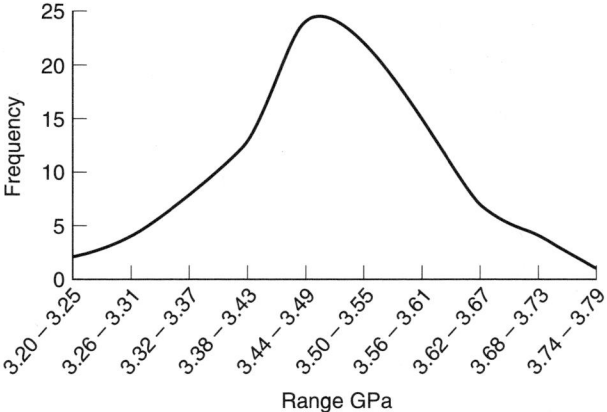

Figure 18.1 Graph depicting frequency distribution.

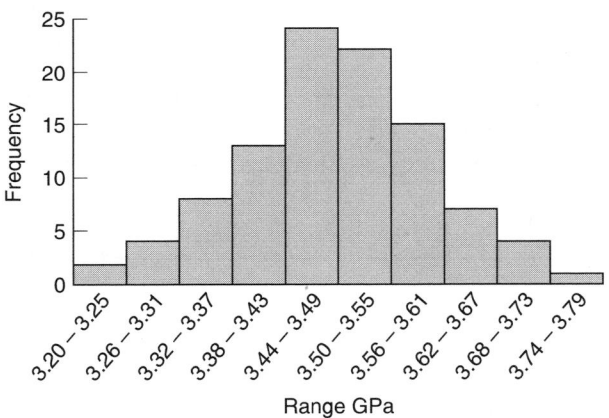

Figure 18.2 Histogram showing frequency distribution.

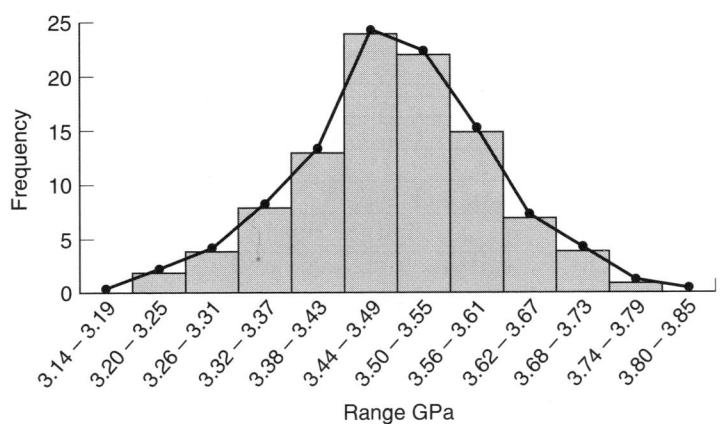

Figure 18.3 Frequency polygon.

3. The median ($\tilde{x}$)—is the middle observation of a number of observations, arranged from smallest to largest when there is an odd number of observations, and half way between the two middle observations when the number is even. Since in the above example there are 100 observations, the median is between the 50th and 51st observations i.e, 3.49 and 3.49, which equals 3.49.
4. Arithmetic mean ($\bar{x}$)—is the sample mean and is the sum of the observations divided by the total number of observations:

$$\bar{x} = \frac{x_1 + x_2 + x_3 + x_4 + \ldots + x_n}{n} = \sum_{i=1}^{n} x_i$$

and is 3.49 in the example, which in this case is the same value as the median. The mean of all means is denoted by $\bar{\bar{x}}$.

18.3 MEASURES OF DISPERSION

The range is the simplest measure of dispersion of data (or measure of variation), but does not consider all the observations and one very small or very large value can have a great effect on the range. Dispersion is best measured using standard deviation (σ) (Greek letter sigma, sometimes lower case s is used). To eliminate negative values, a squared term is used and to obtain a measure of deviation, the average of n such squared deviations of the sample of observations from the sample mean, termed the sample variance (σ^2) is calculated:

$$\sigma^2 = \frac{(x_1 - \bar{x})^2 + (x_2 - \bar{x})^2 + (x_3 - \bar{x})^2 + \ldots + (x_n - \bar{x})^2}{n}$$

The denominator ($n-1$) is, however, preferred to (n), since it gives a more non-biased estimate, helping to compensate for the use of the sample mean rather than the true mean, although for $n > 35$, there is practically no difference. The number ($n-1$) is known as the degrees of freedom of the variance. Hence

$$\sigma^2 = \frac{\sum_{i=1}^{n} (x_i - \bar{x})^2}{n-1}$$

or

$$\sigma = \sqrt{\frac{\sum_{i=1}^{n} (x_i - \bar{x})^2}{n-1}}$$

where i is called the index of summation. Hence, $\sum_{i=1}^{n} (x_i - \bar{x})^2$ means successively replacing i by a sequential number until n is reached in the function $(x_i - \bar{x})^2$. The standard deviation will have the same units as the variable being measured and the value of σ calculated for the 100 results given in Table 18.1 is 0.112 GPa.

If the standard deviation is expressed as a percentage of the mean, then the coefficient of variation (C.V.) is obtained:

$$\text{C.V.} = \frac{100\sigma}{\bar{x}}\% \quad (3.21\% \text{ in above example})$$

18.4 STANDARD ERROR

Supposing a large number of random samples, each containing n individuals, were taken from a population with a normal or near normal distribution then, the standard deviation of this set is termed the standard error of the mean (S.E.) and the following relationship holds:

$$\text{S.E.} = \frac{\text{Standard deviation of the population}}{\sqrt{n}}$$

and

$$\text{Student's } t\text{-distribution} = \frac{\text{Nominal mean} - \text{Standard mean}}{\text{Standard error}}$$

Student's t can be calculated and compared with values in published Statistical Tables [3,4] at the 5% and 1% levels of probability and, if the calculated t exceeds the published value at the 1% level, then a real difference exists and appropriate action must be taken.

18.5 SAMPLE CORRELATION COEFFICIENT

If a random sample of size n is selected and two observations are made (e.g. height and weight) on each member of the sample, giving n pairs of observations, which are denoted by the values $(x_1, y_1), (x_2, y_2), \ldots (x_n, y_n)$, then they can be plotted in a graph by considering the points x and y to be coordinates of a point. This results in a scatter diagram (Figure 18.4). It is difficult to decide whether there is a linear relationship and the sample correlation coefficient (r) is a measure of that linear relationship:

$$r = \frac{\sum(x - \bar{x})(y - \bar{y})}{\sqrt{([\sum(x - \bar{x})^2][(y - \bar{y})]^2)}}$$

The value of r is always between -1 and $+1$, a value of -1 indicates a perfectly linear relationship between x and y, with the value of y decreasing as x increases. A value of $+1$ also indicates a perfectly linear relationship between x and y, but with y increasing as x

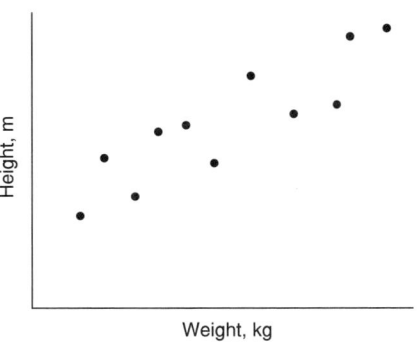

Figure 18.4 A scatter diagram.

increases. If there is no linear relationship, then the value of r will be near zero. A more convenient form of the equation for calculating r is:

$$r = \frac{\sum xy - ((\sum x)(\sum y)/n)}{\sqrt{\left[\sum x^2 - ((\sum x)^2/n)\right]\left[\sum y^2 - ((\sum y)^2/n)\right]}}$$

Hence, in order to find r, the five sums—$\sum x$, $\sum y$, $\sum x^2$, $\sum y^2$ and $\sum xy$—are required.

18.6 LINEAR REGRESSION

It is possible to roughly draw a line through the points in Figure 18.4 by adjusting the line in such a way that there were as many points below the line as above (Figure 18.5). However, it is possible to establish a straight line the Method of Least Squares to establish a straight line, but initially the slope of the line and the value of the y intercept (the value on the y coordinate where the line crosses the y-axis) are required.

Let (x_r, y_r) be a pair of corresponding values in which x_r is correct and y_r is a reading subject to experimental error. If the relationship between x and y is known to be of the form $y = mx + c$, then substituting for x_r, the corresponding value for y can be found as given below:

$$y = mx_r + c \qquad (1)$$

The experimental value is y_r and the error in the experimental value:

$$\text{Error} = y_r - y$$

or

$$= y_r - mx_r - c \qquad (2)$$

The Method of Least Squares adopts the criterion that the sum of the squares of the errors given by equation (2) should be a minimum:

$$\sum (y_r - mx_r - c)^2 = \text{minimum}$$

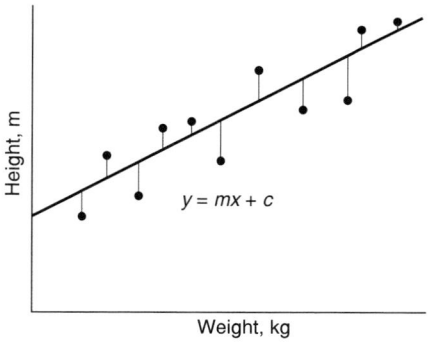

Figure 18.5 Drawing a line through the points in a scatter diagram.

STATISTICS AND STATISTICAL PROCESS CONTROL (SPC)

For functions of one variable, the derivative with respect to the variable equated to zero gives the equation from which the required value of the variable is obtained. With functions of more than one variable, it can be shown that equating the partial derivative with respect to each variable to zero will produce simultaneous equations from which the unknowns may be found. Applying this, the partial derivative with respect to m equated to zero gives:

$$\sum 2(y_r - mx_r - c)(-x_r) = 0$$

or

$$\sum x_r y_r - m \sum (x_r)^2 - c \sum x_r = 0 \qquad (3)$$

Similarly for the variable c:

$$\sum 2(y_r - mx_r - c)(-1) = 0$$

or

$$\sum y_r - m \sum x_r - nc = 0 \qquad (4)$$

where n is the number of pairs of values.

The solution to the simultaneous equations (3) and (4) will give m and c, the unknown values required for the equation of the line $y = mx + c$.

A more convenient form of the equation for calculating m is:

$$m = \frac{\sum x_r y_r - ((\sum x_r)(\sum y_r)/n)}{\sum (x_r)^2 - ((\sum x_r)^2/n)}$$

requiring the five sums—$\sum x_r$, $\sum y_r$, $\sum (x_r)^2$, $(\sum x_r)^2$ and $\sum x_r y_r$.

18.7 NORMAL DISTRIBUTION

A normal distribution, so called because at one time, all other distributions were thought to be abnormal, is often referred to as a Gaussian distribution and is depicted graphically by a frequency/distribution curve, plotted to produce a symmetric bell-shaped graph, which slopes downward on both sides of the maximum, but never quite touches the x-axis (Figure 18.6). In this graph, the mean, mode and median coincide. The measurement of many natural occurrences, such as the height and weight of a population of individuals, or the product made from a machine, follow a natural distribution.

There are other types of non-normal distributions and Figure 18.7 shows a mixed population, which may be due to data coming in from two sources whose properties are different, or if a change in machine setting has been done.

Figure 18.8 shows a skewed distribution, which could be due to a variable on a machine, which when removed, will revert to a normal distribution. Figure 18.9 is typical of a batch, which previously had a wide distribution and has now been sorted. As the sample size is increased, the dispersion becomes less, as shown in Figure 18.10.

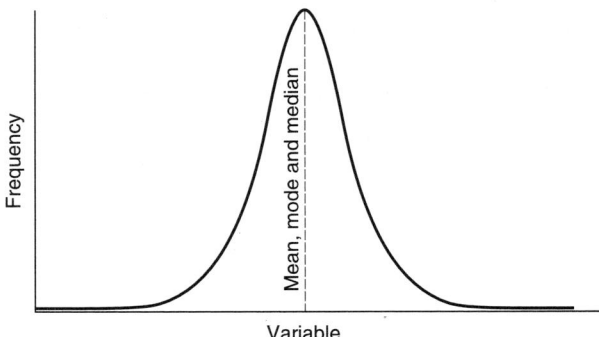

Figure 18.6 A typical Gaussian distribution.

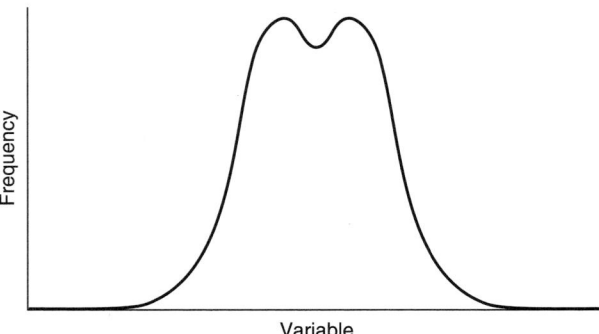

Figure 18.7 Distribution from a mixed population.

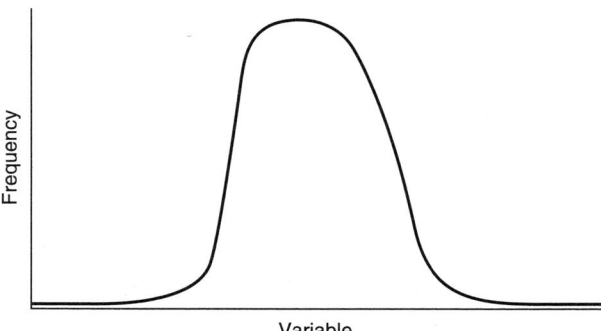

Figure 18.8 A skewed distribution.

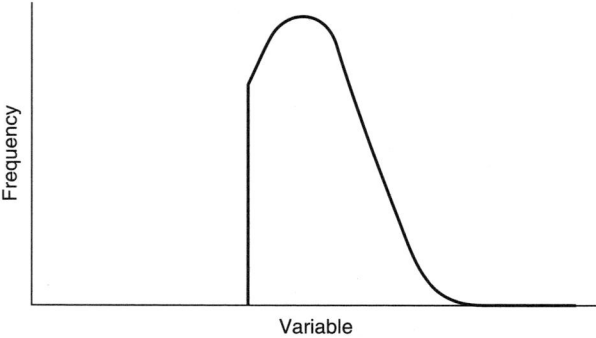

Figure 18.9 A distribution which has been sorted.

STATISTICS AND STATISTICAL PROCESS CONTROL (SPC)

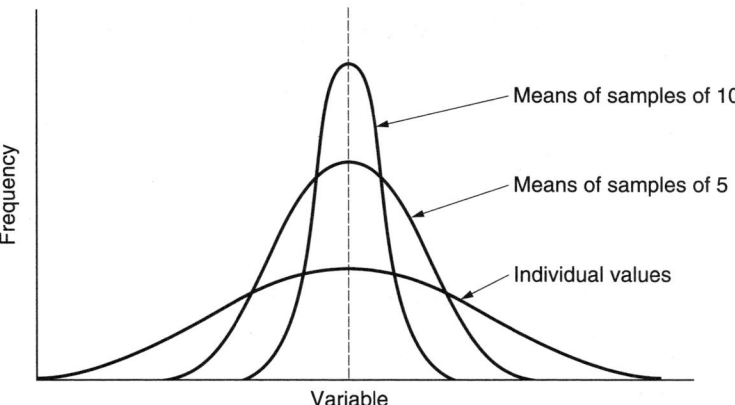

Figure 18.10 Effect of sample size on dispersion.

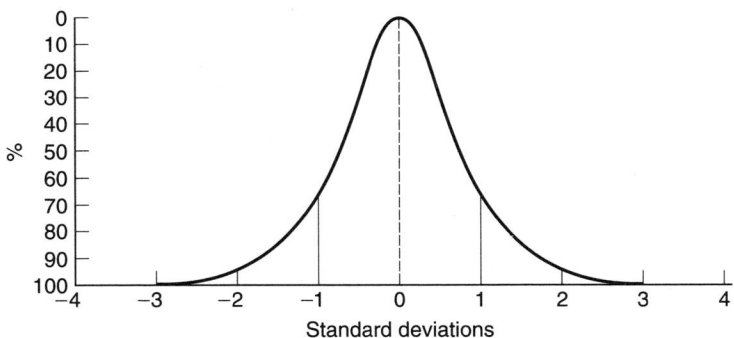

Figure 18.11 Relationship between a normal distribution and standard deviation.

The relationship between a normal distribution and the standard deviation is shown in Figure 18.11 and is an important part of control chart theory and indicates when a process is in statistical control:

68.26% (or about 2/3) fall into the range $\pm 1\sigma$
95.44% (or about 19/20) fall into the range $\pm 2\sigma$
99.73% (all but 1 in 370) fall into the range $\pm 3\sigma$
99.994% (all but 1 in 16,670) fall into the range $\pm 4\sigma$

The output of most production operations will normally be distributed with a variability of $\pm 3\sigma$, or a total of 6σ. A product is normally made with a tolerance of say T, for example $x \pm T$, which is a total tolerance of $2T$. This can be expressed as a capability index, Cp:

$$Cp = 2T/6\sigma$$

and if the value of 6σ is greater than $2T$, there will be some non-conforming products produced. If the value of 6σ is less than $2T$, then the product should be acceptable as long as it is correctly centered, although it must be remembered that some processes are naturally skewed.

18.8 WEIBULL DISTRIBUTION

In 1951, Weibull [5] proposed the use of a mathematical function that, by changing the value of its three parameters, α, β and χ, can cover many shapes and can approximate to a normal distribution under certain conditions. The Weibull distribution is defined by:

$$y = \alpha\beta(X-\chi)^{\beta-1}e^{-\alpha(X-\chi)^\beta}$$

where α = control on the spread of the distribution and can change the curve from a flat to a peaked form β = a shape parameter and can vary in the range 0.5–5. χ (lower case Greek letter gamma) = location parameter and is the smallest possible value of X which is zero for positive random variables.

When $\beta = 1$, the function reduces to an exponential form and when $\beta \approx 3.5$ (with $\alpha = 1$ and $\chi = 0$), it is approximately a normal distribution. It is not necessary to calculate these constants. Using the Weibull probability paper, the fit of the data can be determined as a normal or exponential distribution.

A normal distribution can be represented by:

$$y = \frac{1}{\sigma\sqrt{2\pi}} e^{-\frac{(x-\bar{x})^2}{2\sigma^2}}, \quad \text{where } e = 2.718$$

18.9 VARIATION

When making carbon fiber, the product can deviate due to:

1. Variability within the product, e.g. diameter.
2. Variability between tows produced at the same time, e.g. strength.
3. Variability with time, e.g. humidity, ambient temperature, night shift.

There are essentially four factors which contribute to these variations:

1. Process e.g. voltage variation
2. Raw materials e.g. precursor
3. Operator, depends on many factors such as health, insufficient training etc.
4. Miscellaneous, covering items such as environmental factors (e.g. humidity level affecting moisture pick-up of oxidized PAN fiber), testing equipment (not correctly calibrated).

Almost by definition, no two objects are ever made alike, but as long as the sources of variation are minor, the process will be in control (Figure 18.12). However, should there be a variation of large magnitude (an assignable cause), then the variation will be excessive and out of control, or beyond the expected normal variation (Figure 18.13). To obtain an in-depth understanding of the process, a flow chart should be mapped out to define contributing activities such as operation, inspection, delay or temporary storage, permanent storage and transportation [6].

It is essential to discover what the root cause of a problem is and after a team effort brainstorming session, it is helpful to construct a Pareto Chart, where groups of similar data are arranged in order of magnitude (Figure 18.14). This helps to concentrate on major problems. The so called 80/20 rule suggests that 80% of the problems result from 20% of the causes. A cause and effects analysis is determined and a Fishbone (Ishikawa) diagram

STATISTICS AND STATISTICAL PROCESS CONTROL (SPC)

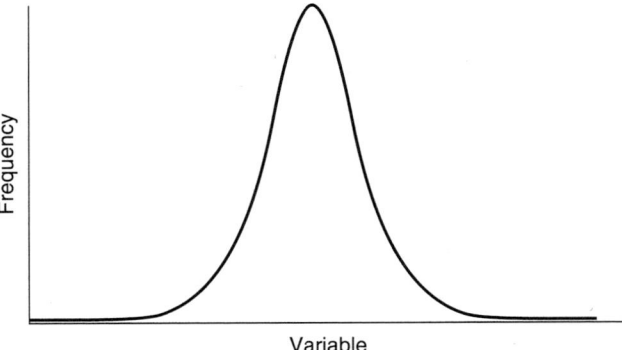

Figure 18.12 A capable process in control.

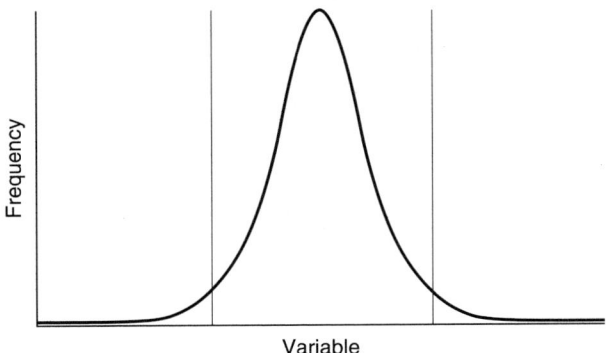

Figure 18.13 An incapable process out of control.

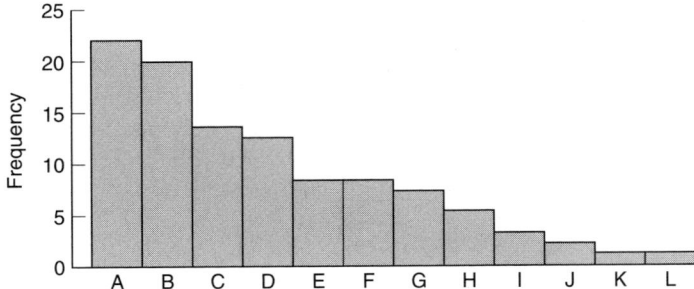

Figure 18.14 A Pareto chart.

(Figure 18.15) is constructed, helping to highlight individual problems. It should then be possible to determine the major cause and provide sufficient information to allow a solution. Typical headings for a Fishbone diagram of a production process are:

$$M = \text{Machines, Methods and Materials}$$
$$P = \text{People, Product, Process, Plant and Programs}$$

The process becomes one of never ending quality improvement and Figure 18.16 shows a typical cycle of improvement.

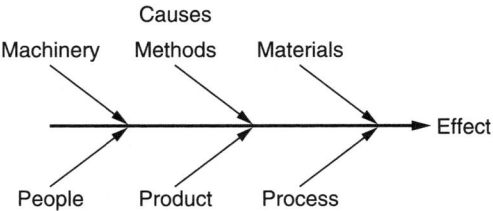

Figure 18.15 A fishbone (Ishikawa) cause and effect diagram.

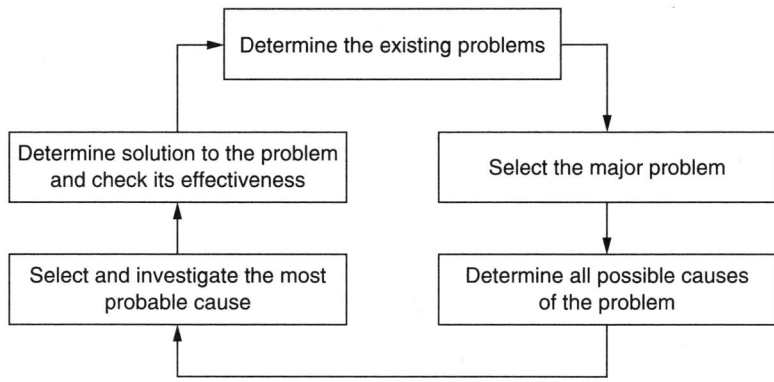

Figure 18.16 Cycle of never-ending quality improvement.

18.10 CONTROL CHART METHOD

The purpose of a control chart is to improve the process and the ultimate aim is to distinguish between a variation that is predictable within limits (a common cause or random variation) and a variation that is not (special cause and is a significant trend). A significant trend could be caused by an inconsistency in raw materials or an irregular process operation. A control chart can, for example, be used for process capability studies, or investigating customer complaints. Once a control chart has achieved its objective, it should be discontinued and efforts should then be directed towards other improvements.

The control chart shown in Figure 18.17 is called an $\bar{x}$ chart and used to control the mean value of samples.

18.11 STATISTICAL PROCESS CONTROL CHARTS

The $\bar{x}$ chart has limitations, since it is based on individual readings and is replaced by a system established on subgroups, where the average value of these subgroups together with their range enables the process spread to be established.

If the process is under statistical control, it will be represented by a normal distribution with a gaussian type curve and from the distribution of the means, the control limits can be calculated.

There are many types of control charts and initially, the choice depends on whether the data is Attribute, which is evaluated in terms of whether it meets a given requirement or not, e.g. pass/fail, go/no go, or whether the data is Variable, where it is assessed in terms of

STATISTICS AND STATISTICAL PROCESS CONTROL (SPC)

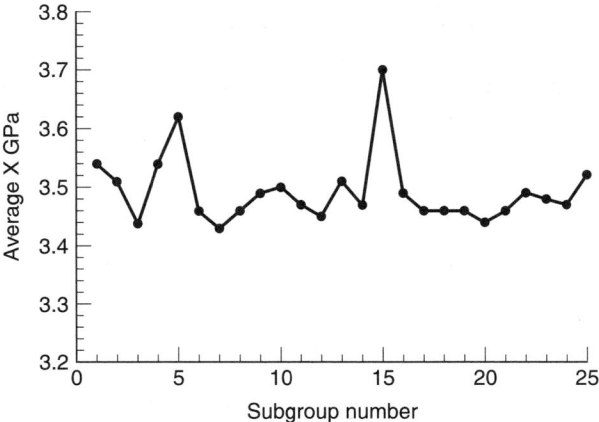

Figure 18.17 A simple $\bar{x}$ chart.

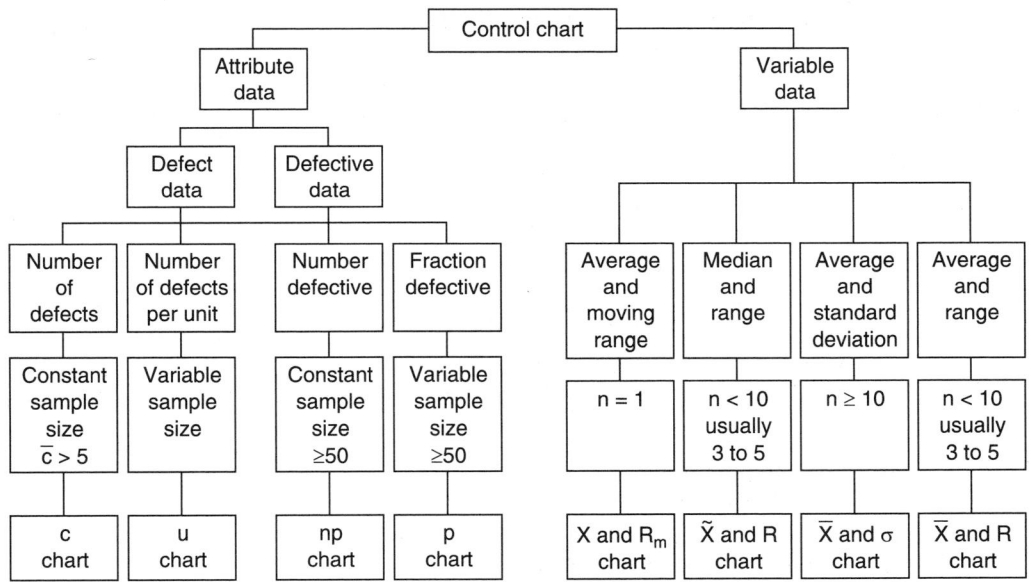

Figure 18.18 Type of control chart.

values on a continuous scale. The control chart can be chosen from Figure 18.18 and since almost all data on carbon fiber plants is classed as variable, this is the aspect considered and if details of charts based on attribute data are required, then the reader is referred to one of many admirable books on statistical control [7–23].

The size of the sample will be influenced by cost and the frequency will depend on the occurrence of patterns. In general, some 20–25 random groups of samples should be taken with unchanged run conditions, varying the time when samples are taken and making notes of any unusual events that might occur. The same ground rules used for the selection of samples can be followed when constructing histograms. These samples will permit the calculation of the requisite statistics to enable the initial control limits to be calculated.

18.11.1 Average and Range ($\bar{x}$ and R) chart

Probably one of the most common charts is the Average and Range ($\bar{x}$ and R) chart and an excellent blank chart is shown in Figure 18.19. Using the figures in Table 18.4 the central values are obtained from the equations:

$$\bar{x} = \frac{\sum_{j=1}^{m} \bar{x}_j}{m} \tag{5}$$

and

$$\bar{R} = \frac{\sum_{j=1}^{m} R_j}{m} \tag{6}$$

where $\bar{x}$ = mean (or average) of the subgroup means
$\bar{x}_j$ = mean of the jth subgroup
m = number of subgroups
$\bar{R}$ = mean (or average) of the subgroup ranges
R_j = range of the jth subgroup

Trial control limits are established at $\pm 3\sigma$ from the central value (Table 18.5) as given by:

$$\text{UCL}_{\bar{x}} = \bar{x} + 3\sigma_{\bar{x}} \tag{7}$$

$$\text{LCL}_{\bar{x}} = \bar{x} - 3\sigma_{\bar{x}} \tag{8}$$

$$\text{UCL}_R = \bar{R} + 3\sigma_R \tag{9}$$

$$\text{LCL}_R = \bar{R} - 3\sigma_R \tag{10}$$

where UCL = upper control limit
LCL = lower control limit
$\sigma_{\bar{x}}$ = standard deviation of the subgroup means ($\bar{x}$s)
σ_R = standard deviation of the range

To simplify these calculations in the $\bar{x}$ chart $3\sigma_{\bar{x}}$ is replaced by $A_2 \bar{R}$, where A_2 is a factor obtained from tables (Table 18.6), i.e.

$$A_2 \bar{R} = 3\sigma_{\bar{x}}$$

This is based on

$$\sigma_{\bar{x}} = \frac{\sigma'}{\sqrt{n}} \quad \text{using} \quad \sigma' = \frac{R}{d_2}$$

where d_2 is a factor for the subgroup size. Therefore

$$3\sigma_{\bar{x}} = \frac{3\sigma'}{\sqrt{n}} = \frac{3}{d_2\sqrt{n}}, \quad \text{hence} \quad A_2 = \frac{3}{d_2\sqrt{n}}$$

STATISTICS AND STATISTICAL PROCESS CONTROL (SPC)

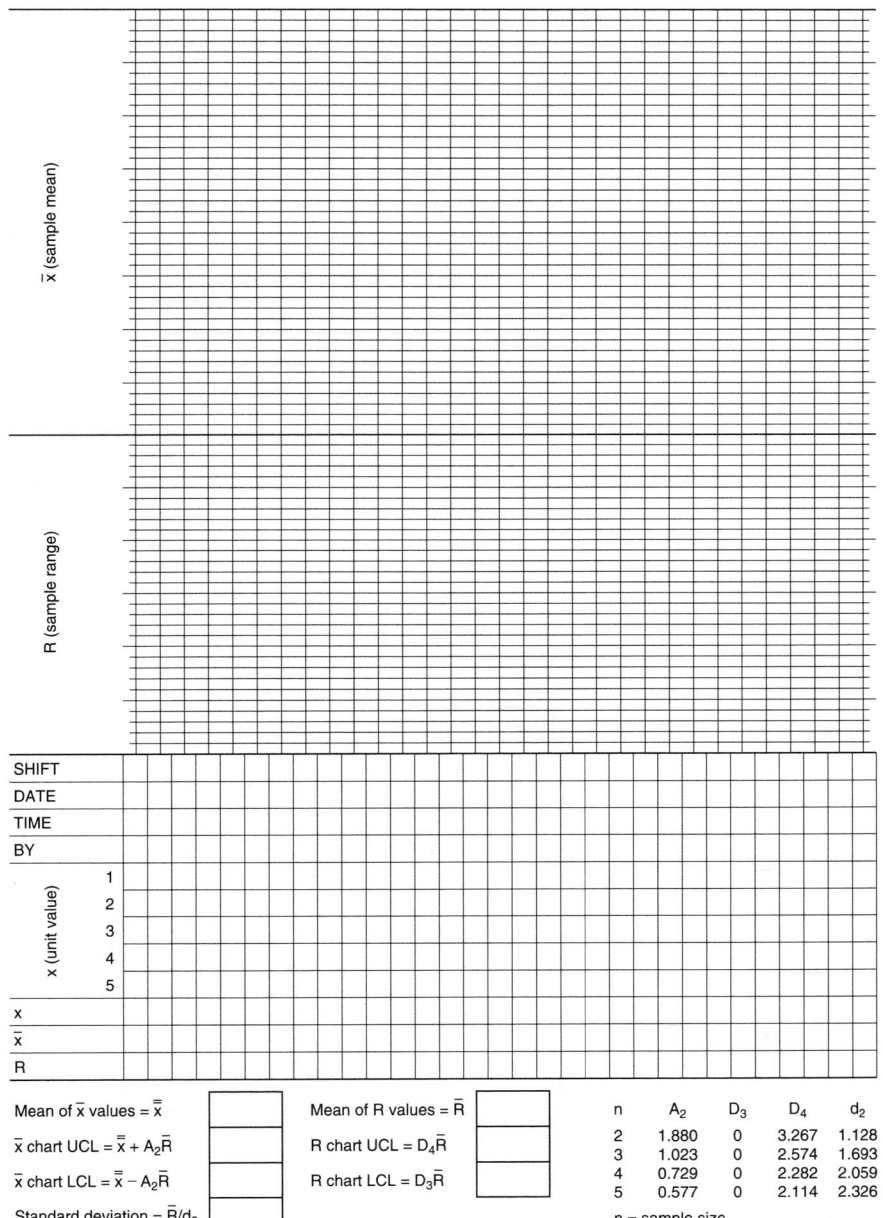

Figure 18.19 Blank form for plotting a Mean and Range Chart. *Source:* Copyright 2004 SMMT Ltd Statistical Process Control – A Guide for Business Improvement.

Table 18.4 Specimen results for control charts

Sub-group	Date	Individual Measurements				Average $\overline{X}$	Range R	Median $\tilde{X}$	Standard Deviation	Comments
		x_1	x_2	x_3	x_4					
1	01-Dec	3.53	3.70	3.32	3.60	3.54	0.38	3.47	0.1609	
2	02-Dec	3.41	3.50	3.63	3.49	3.51	0.22	3.50	0.0911	
3	03-Dec	3.35	3.58	3.31	3.53	3.44	0.27	3.44	0.1325	
4	04-Dec	3.46	3.49	3.55	3.65	3.54	0.19	3.52	0.0838	
5	05-Dec	3.54	3.68	3.69	3.58	3.62	0.15	3.63	0.0741	Power dip
6	06-Dec	3.47	3.41	3.58	3.36	3.46	0.22	3.44	0.0947	
7	07-Dec	3.46	3.50	3.48	3.27	3.43	0.23	3.47	0.1063	
8	08-Dec	3.66	3.39	3.37	3.43	3.46	0.29	3.41	0.1340	
9	09-Dec	3.61	3.20	3.64	3.50	3.49	0.44	3.56	0.2009	
10	10-Dec	3.52	3.44	3.45	3.59	3.50	0.15	3.49	0.0698	
11	11-Dec	3.55	3.66	3.34	3.33	3.47	0.33	3.45	0.1623	
12	12-Dec	3.61	3.30	3.50	3.40	3.45	0.31	3.45	0.1330	
13	13-Dec	3.59	3.51	3.37	3.57	3.51	0.18	3.54	0.0993	
14	14-Dec	3.54	3.41	3.48	3.45	3.47	0.13	3.47	0.0548	
15	15-Dec	3.66	3.69	3.64	3.79	3.70	0.15	3.68	0.0666	New lab technician
16	16-Dec	3.49	3.47	3.59	3.41	3.49	0.18	3.48	0.0748	
17	17-Dec	3.29	3.53	3.47	3.55	3.46	0.26	3.50	0.1183	
18	18-Dec	3.54	3.45	3.52	3.34	3.46	0.20	3.49	0.0903	
19	19-Dec	3.50	3.41	3.48	3.44	3.46	0.09	3.46	0.0403	
20	20-Dec	3.49	3.47	3.57	3.24	3.44	0.33	3.48	0.1417	
21	21-Dec	3.39	3.53	3.44	3.47	3.46	0.14	3.46	0.0585	
22	22-Dec	3.46	3.38	3.51	3.61	3.49	0.23	3.49	0.0963	
23	23-Dec	3.54	3.49	3.50	3.39	3.48	0.15	3.50	0.0638	
24	24-Dec	3.53	3.51	3.40	3.45	3.47	0.13	3.48	0.0591	
25	25-Dec	3.56	3.42	3.49	3.60	3.52	0.18	3.53	0.0793	
Sum						87.32	5.53	87.39	2.4865	
Mean						3.49	0.22	3.5	0.0995	

Table 18.5 Formulae for calculating control limits

Type of Control Chart	Plotted value	Upper Control Limit	Lower Control Limit
Average and Range $\overline{X}$ and R	$\overline{X}$ R	$UCL_{\overline{X}} = \overline{\overline{X}} + A_2\overline{R}$ $UCL_R = D_4\overline{R}$	$LCL_{\overline{X}} = \overline{\overline{X}} - A_2\overline{R}$ $LCL_R = D_3\overline{R}$
Average and Standard Deviation $\overline{X}$ and σ	$\overline{X}$ σ	$UCL_{\overline{X}} = \overline{\overline{X}} + A_3\overline{\sigma}$ $UCL_\sigma = B_4\overline{\sigma}$	$LCL_{\overline{X}} = \overline{\overline{X}} - A_3\overline{\sigma}$ $LCL_\sigma = B_3\overline{\sigma}$
Median and Range $\tilde{X}$ and R	$\tilde{X}$ R	$UCL_{\tilde{X}} = \overline{\tilde{X}} + A_2\overline{R}$ $UCL_R = D_4\overline{R}$	$LCL_{\tilde{X}} = \overline{\tilde{X}} - A_2\overline{R}$ $LCL_R = D_3\overline{R}$
Individuals and a Moving Range X and R_m	X R_m	$UCL_X = \overline{X} + E_2\overline{R}_m$ $UCL_{Rm} = D_4\overline{R}_m$	$LCL_X = \overline{X} - E_2\overline{R}_m$ $LCL_{Rm} = D_3\overline{R}_m$

In the R chart, the range $\overline{R}$ is used to estimate σ_R. The simplified trial control limits now become:

$$\text{UCL}_{\overline{x}} = \overline{X} + A_2\overline{R} \qquad (11)$$

$$\text{LCL}_{\overline{x}} = \overline{X} - A_2\overline{R} \qquad (12)$$

$$\text{UCL}_R = \overline{R} + D_4\overline{R} \qquad (13)$$

$$\text{LCL}_R = \overline{R} - D_3\overline{R} \qquad (14)$$

STATISTICS AND STATISTICAL PROCESS CONTROL (SPC)

Table 18.6 Table of constants for computing the 3σ control limits

n	2	3	4	5	6	7	8	9	10
A_2	1.880	1.023	0.729	0.577	0.483	0.419	0.373	0.337	0.308
$\tilde{A}_2$	(1.88)	1.187	(0.80)	0.691	(0.55)	0.509	(0.43)	0.412	(0.36)
A_3	2.659	1.954	1.628	1.427	1.287	1.182	1.099	1.032	0.975
B_3	0	0	0	0	0.030	0.118	0.185	0.239	0.284
B_4	3.267	2.568	2.266	2.089	1.970	1.882	1.815	1.761	1.716
C_4	0.7979	0.8862	0.9213	0.9400	0.9515	0.9594	0.9650	0.9693	0.9727
d_2	1.128	1.693	2.059	2.326	2.534	2.704	2.847	2.970	3.078
D_3	0	0	0	0	0	0.076	0.136	0.184	0.223
D_4	3.267	2.574	2.282	2.114	2.004	1.924	1.864	1.816	1.777
E_2	2.659	1.772	1.457	1.290	1.184	1.109	1.054	1.010	0.975

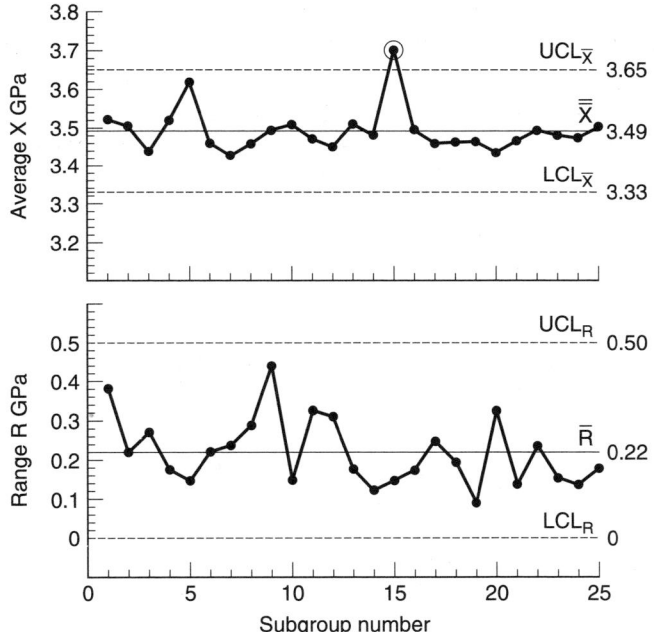

Figure 18.20 Mean and Range Chart ($\bar{X}$ and R).

where A_2, D_3, and D_4 are factors that vary with the subgroup size (n) and are given in Table 18.6.

In the $\bar{x}$ chart, the control limits are symmetrical about the central value. However, with an R chart and a subgroup size of 6 or less, the lower control limit would be negative, which is not permitted. So, putting $D_3 = 0$ places the LCL at zero.

Equations (5) and (6) are used to calculate $\bar{x}$ and $\bar{R}$, then using the appropriate constants from Table 18.6 for the requisite subgroup size in the equations (11)–(14) enables the initial control limits to be calculated and placed on the chart (Figure 18.20).

More than likely, the initial study will reveal that the process is not in control and examination of Figure 18.20 reveals that subgroup 15 is an out-of-control point. Table 18.4 reveals a new lab technician as the probable cause. All points with an assignable cause

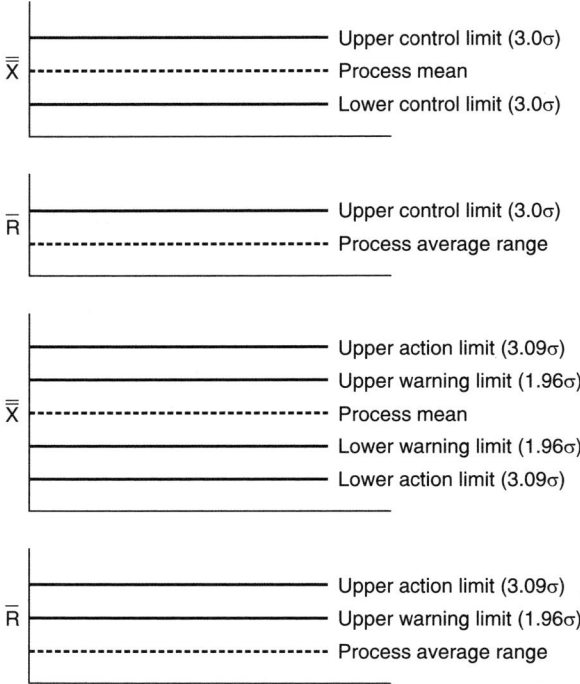

Figure 18.21 Two systems used for Control charts.

should be eliminated and a new $\bar{x}$ is calculated. Since there are no out-of-control points on the range chart, the corresponding value(s) of R will remain.

$$\text{So} \quad \bar{x}_{\text{new}} = \frac{\sum(\bar{x} - \bar{x}_d)^2}{n - n_d}$$

where $\bar{x}_d$ = sum of discarded subgroup means
n_d = number of discarded subgroups

The value of $\bar{x}_{\text{new}}$ can be used to calculate new control limits, which are used to report the results of future subgroups, if or when a desired change has been made in the process.

Traditionally, control charts were set up with control limits set at $\pm 3\sigma$ and warning limits set at $\pm 2\sigma$ corresponding to $\pm 0.135\%$ and $\pm 2.275\%$. If the control limits, however, are set at $\pm 3.09\sigma$ and warning limits at $\pm 1.96\sigma$, this corresponds to $\pm 0.1\%$ and $\pm 2.5\%$. These limits are preferred to the more awkward values previously used and hence, the subscripts 0.001 and 0.025 are used in statistical tables, although in practice there is little real difference (Figure 18.21) whilst Figure 18.22 shows the distribution curve for these revised limits.

It should be noted that specification limits are a customer's requirement and are not related to whether a process is in control or not.

18.11.2 Mean and Standard Deviation ($\bar{x}$ and σ) chart

Although the range chart is most commonly used for assessing variation, standard deviation is more accurate, although for subgroup sizes less than 10, both charts show the

STATISTICS AND STATISTICAL PROCESS CONTROL (SPC)

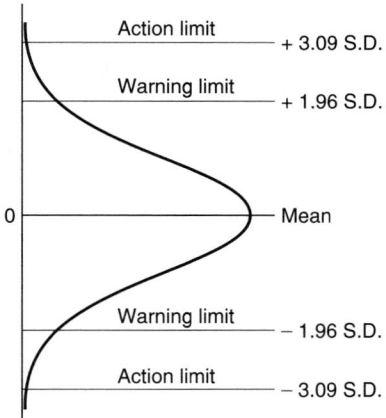

Figure 18.22 Alternative method of showing control limits.

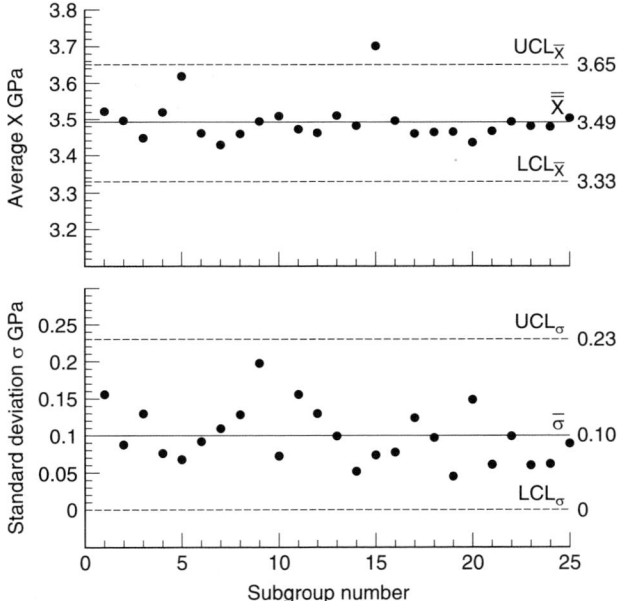

Figure 18.23 A typical Mean and Standard Deviation Chart.

same variation. The formulae used for the calculation of the trial control limits are:

$$\bar{\bar{x}} = \frac{\sum_{j=1}^{m} \bar{x}_j}{m} \quad \text{and} \quad \bar{\sigma} = \frac{\sum_{j=1}^{m} \sigma_j}{m}$$

$$\text{UCL}_{\bar{x}} = \bar{\bar{x}} + A_3 \bar{\sigma}$$

$$\text{LCL}_{\bar{x}} = \bar{\bar{x}} - A_3 \bar{\sigma}$$

$$\text{UCL}_{\sigma} = B_4 \bar{\sigma}$$

$$\text{LCL}_{\sigma} = B_3 \bar{\sigma}$$

A typical Mean and Standard Deviation chart is shown in Figure 18.23.

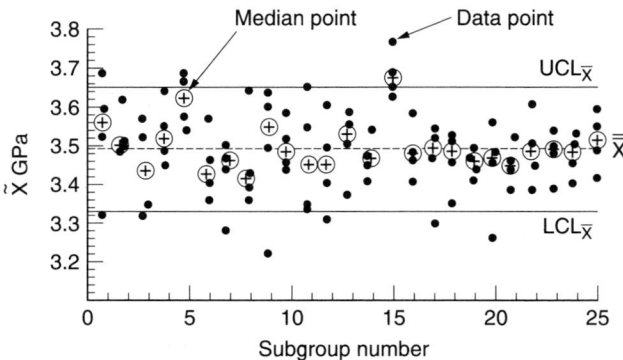

Figure 18.24 Median Control Chart.

18.11.3 Median control chart

Median control charts are plotted in conjunction with Range charts in a similar manner to ($\bar{x}$ and R) charts. The individual measurements from each subgroup are plotted giving a vertical line. For an odd number of values, the middle value is marked, and for even number of values, a mark is placed midway between the two central values. The medians are connected with a solid line. Each subgroup's median ($\tilde{x}$) and range (R) are plotted on the chart. The mean of the subgroup medians ($\bar{\tilde{x}}$) is calculated and drawn on the chart as the central line (Figure 18.24).

The mean of the ranges ($\overline{R}$) are calculated and the control limits found using:

$$\text{UCL}_{\tilde{x}} = \bar{\tilde{x}} + A_2 \overline{R}$$

$$\text{LCL}_{\tilde{x}} = \bar{\tilde{x}} - A_2 \overline{R}$$

$$\text{UCL}_R = D_4 \overline{R}$$

$$\text{LCL}_R = D_3 \overline{R}$$

These control limits are then drawn on the chart.

There is less arithmetic to be done with a median chart, which can be maintained by operators, but it does not solve any extreme values in a subgroup.

18.11.4 Rules for detecting out-of-control conditions on control charts

The following are the types of out-of-control conditions that might occur on control charts:

1. Outside control limits (Figure 18.25).
2. An unnatural pattern (e.g. values that are widely spread or too close to the mean).

To help establish whether there is an unnatural pattern, it is convenient to divide the chart into six equal bands, three between the central value and the UCL and three between the central value and the LCL. In a normal pattern, there will be about 34% in each of the two bands adjacent to the central value, about 13.5% in the middle bands and some 2.5% in the outside bands. If there is a significant divergence from normal, then it would be ranked

STATISTICS AND STATISTICAL PROCESS CONTROL (SPC)

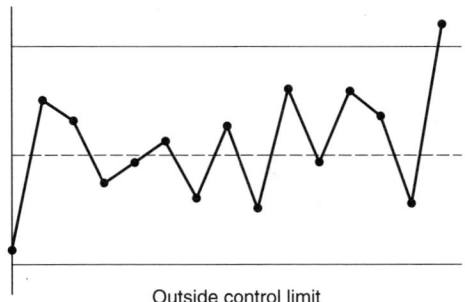

Figure 18.25 Value outside Upper Control Limit.

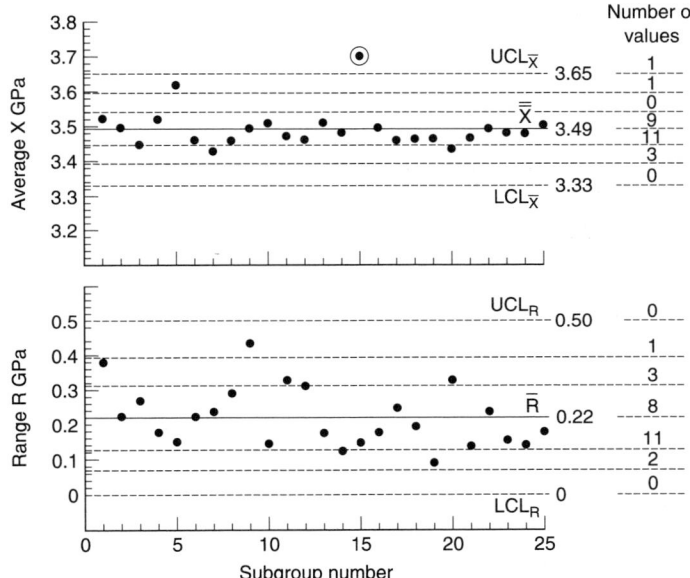

Figure 18.26 Method of detection of an unnatural pattern in a Control Chart.

as out-of-control (e.g. if in a period of ten or more points, more than a third lie outside the central third between the central value and either of the control limits) (Figure 18.26).

If values are too close to the mean, this could be due to a sudden improvement in the process capability, incorrect reading, a faulty measuring instrument, or just a chance fluctuation. If the values are too close to the control limits, this could be due to errors in plotting the values.

3. A run with seven consecutive values above or below the mean (Figure 18.27).

A run is normally indicative that the process mean, or central point of the distribution has moved.

4. A trend, or steady change, with seven ascending or descending values (Figure 18.28).

A trend is normally indicative that something has happened to either shift the process mean, or the central point of the distribution, such as deterioration of equipment due to wear.

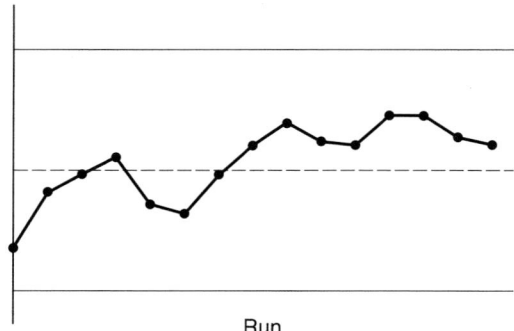

Figure 18.27 A control chart out of control due to a Run.

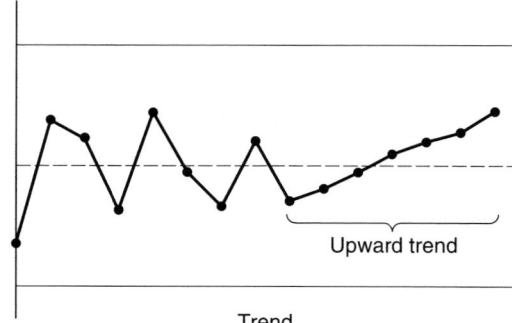

Figure 18.28 A control chart out of control due to a Trend.

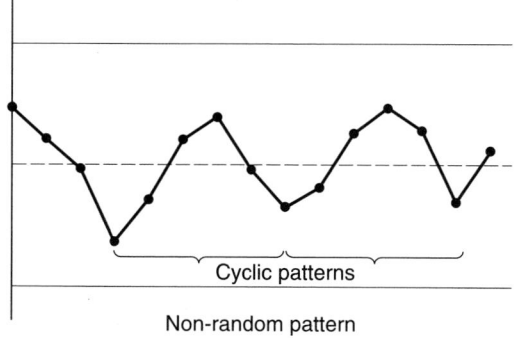

Figure 18.29 A control chart out of control due to a Cyclic Variation.

A trend on the R chart is less common—a downward trend could be due to improvement in worker skill, whereas an opposite trend could be due to a decrease in worker skill due to tiredness etc. An improvement in incoming materials could also be a reason for a trend.

5. A non-random pattern (Figure 18.29).

A cyclic variation can be associated with reasons such as a fluctuating power supply, cold conditions early in the morning and so on.

STATISTICS AND STATISTICAL PROCESS CONTROL (SPC)

All these different patterns should be investigated when they occur and corrected, or if beneficial, checked to see whether a permanent improvement can be established. Inspection error can often be the cause of out-of-control patterns, such as equipment incorrectly calibrated, sampling from more than one universe and calculation errors.

18.11.5 Cumulative Sum chart (Cusum)

A cusum chart is a very useful technique for establishing a gradual change or trend. The technique is readily adapted to a continuous process and permits retrospective corrective action to be taken. A target mean (k) is selected, normally the average, and the difference between the target mean and the value is determined and the cumulative sum c_n is calculated:

$$c_1 = (\bar{x}_1 - k)$$

$$c_2 = (\bar{x}_2 - k) + (\bar{x}_1 - k) = \bar{x}_2 - k + c_1$$

$$c_3 = (\bar{x}_3 - k) + c_2$$

$$c_n = (\bar{x}_n - k) + c_{n-1}$$

Hence, each sum is obtained from its predecessor by the simple operation of adding on the new difference $(\bar{x}_n - k)$ (Table 18.7). The cumulative sum is plotted against the sample

Table 18.7 Specimen results for Cusum chart

Sample Number	Strength (GPa)	Value -k	Cumulative Sum
1	3.47	−0.03	−0.03
2	3.50	0	−0.03
3	3.44	−0.06	−0.06
4	3.52	0.02	−0.04
5	3.63	0.13	0.15
6	3.44	−0.06	0.07
7	3.47	−0.03	−0.09
8	3.41	−0.09	−0.12
9	3.56	0.06	−0.03
10	3.49	−0.01	0.05
11	3.45	−0.05	−0.06
12	3.45	−0.05	−0.1
13	3.54	0.04	−0.01
14	3.47	−0.03	0.01
15	3.68	0.18	0.15
16	3.48	−0.02	0.16
17	3.50	0	−0.02
18	3.49	−0.01	−0.01
19	3.46	−0.04	−0.05
20	3.48	−0.02	−0.06
21	3.46	−0.04	−0.06
22	3.49	−0.01	−0.05
23	3.50	0	−0.01
24	3.48	−0.02	−0.02
25	3.53	0.03	0.01
Mean	3.50		−0.01
Std. Dev.	0.06		0.07

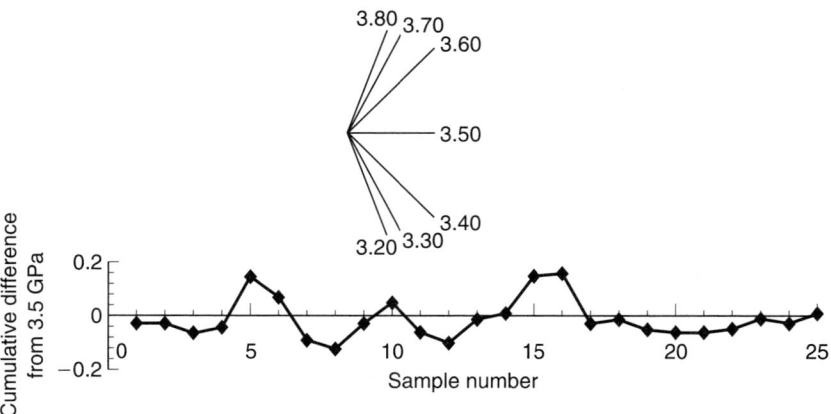

Figure 18.30 A Cusum Chart.

number. A change in slope, either negative or positive, is indicative of a change in process and the greater the magnitude of the slope, the greater the effect. The position of the line on the chart is not relevant and if the line runs off the chart, then the target mean is re-adjusted. It is important to select the correct scales for the construction of the chart, since an exaggerated scale can emphasize the slope, whereas a compressed scale will conceal a change in slope. It is normal to arrange for 1 unit on the x-axis to be equal to 2σ on the y-axis. As a guide, a key is included to give an indication of the magnitude of the slope, which can be expressed in either units or numbers of σ. A typical cusum chart with keys is shown in Figure 18.30.

A cusum chart is very useful to indicate, say, when an HM furnace element is coming towards the end of its life, or there has been a change in the precursor.

18.12 CAPABILITY INDEX

A way of assessing a process to produce to a given total tolerance is to measure its Capability Index (Cp), where $Cp = 2T/6\sigma$. For a process to be capable, Cp must be 1 or more. The Cp value compares the spread of $2T$ with the spread of 6σ, but unfortunately, does not take into account the actual location. However, Cp values are used extensively, but the shortcomings can be overcome by using Cp_k, which compares the spread between the upper tolerance and the process mean ($\bar{\bar{x}}$) with the actual spread from the process mean (3σ), or the spread between the process mean ($\bar{\bar{x}}$) and the lower tolerance with the actual spread of the process from the process mean (3σ).

Hence,

$$Cp_k = \text{minimum value of } \frac{\text{USL} - \bar{\bar{x}}}{3\sigma} \text{ or } \frac{\bar{\bar{x}} - \text{LSL}}{3\sigma}$$

A process with Cp_k less than 1 is not capable, barely capable when equal to 1, acceptable at 1.33 and is considered safe when equal to 2.

If only part of the process is considered, such as an individual machine performing one operation only, machine capability can be considered, but if the machine carries out

Table 18.8 Rankings used for FMEA analysis

Ranking	Severity S	Occurrence O Possible failure rate	Cp_k	Detection D
10	Very high	>1 in 2	<0.33	Absolute certainty of non-detection
9	Very high	1 in 3	≥0.33	Very low
8	High	1 in 8	≥0.51	Low
7	High	1 in 20	≥0.67	Low
6	Moderate	1 in 80	≥0.83	Moderate
5	Moderate	1 in 400	≥0.1.00	Moderate
4	Moderate	1 in 2,000	≥0.1.17	High
3	Low	1 in 15,000	≥0.1.33	High
2	Low	1 in 150,000	≥0.1.50	Very high
1	Minor	<1 in 1,500,000	≥0.1.67	Very high

more than one operation, it must be considered as a process. For assessing the capability of a machine:

$$C_m = \frac{2T}{8\sigma} \quad \text{and} \quad \frac{\text{USL} - \bar{\bar{x}}}{4\sigma} \quad \text{or} \quad \frac{\bar{\bar{x}} - \text{LSL}}{4\sigma}$$

and the minimum C_{mk} is considered to be 1.33.

Some sources have extended the range limit from $\bar{x} \pm 3\sigma$ to $\bar{x} \pm 4\sigma$.

18.13 FAILURE MODE EFFECT ANALYSIS (FMEA)

FMEA is a method of determining problem areas in either a process or a product by developing a list of potential failure modes, ranked according to their effect on the customer, end user or the next in line in the manufacturing process and establishing a priority system for corrective action considerations.

Each stage in a process, or part of a product, is graded on a scale of 1 to 9 in three categories (Table 18.8):

S = severity of failure an assessment of the seriousness of the effect
O = the likelihood of the occurrence of a failure
D = probability of detection of any failure

The Risk Priority Number (RPN) is the product of the Severity (S), Occurrence (O), and Detection (D):

$$\text{RPN} = S \times O \times D$$

and is used to rank the order of concerns in the process in Pareto fashion and will give a value in the range 1–1,000. The team should undertake corrective action to reduce the calculated risk. When implementing control charts, the results of the FMEA should be used to decide which process should be controlled.

REFERENCES

1. Moroney MJ, *Facts from Figures*, Penguin, London, 1978.
2. Brassard M, Ritter D, *The Memory Jogger*^{TM} *II*, Goal/QPC, Methuen, Ma, 1994.

3. Fisher RA, Yates F, *Statistical Tables for Biological, Agricultural and Medical Research*, Oliver & Boyd, Edinburgh, 6, 1963.
4. American Society for Quality Control Statistics Division, *Glossary and Tables for Statistical Quality Control*, American Society for Quality Control, 1996.
5. Weibull W, A statistical distribution function of wide applicability, *J Appl Mech*, 18(3), 293–297, Sep 1951.
6. Duran JM, Gryna FM, *Quality Control Handbook*, McGraw-Hill, 1988.
7. Currie RM, *Work Study*, Pitman, London, 1973.
8. Abbott JC, *Practical Understanding of Capability by Implementing Statistical Process Control*, Robert Houston Smith, 1999.
9. Besterfield D, Control Charts for Variables, *Quality Control*, Prentice-Hall, 1997.
10. Blank R, *The SPC Troubleshooting Guide*, Quality Resources, 1998.
11. Bothe DR, *Measuring Process Capability, Techniques and Calculations for Quality and Manufacturing Engineers*, McGraw-Hill, 1997.
12. Evans JR, *Statistical Process Control for Quality Improvement, A Training Guide to Learning Statistical Process Control*, Prentice-Hall, 1991.
13. Keats JB, Montgomery DC, *Statistical Applications in Process Control*, Marcel Dekker, 1996.
14. McMillen N, *Statistical Process Control and Company-wide Quality Improvement*, IFS Publications, 1991.
15. Oakland JS, *Statistical Process Control*, Butterworth-Heinemann, 1986.
16. Oakland JS, *Statistical Process Control*, DTI, London, 1992.
17. Owen M, *SPC and Continuous Improvement*, IFS Publications, Bedford, 1989.
18. Owen M, *SPC, A Guide to Non-manufacturing Applications*, IFS Publications, Bedford, 1991.
19. Queensberry Charles P, Queensberry CP, *SPC Methods for Quality Improvements*, John Wiley, 1997.
20. *SPC Explained*, SPC Services International Ltd., Witham.
21. Thompson JR, Koronacki J, *Statistical Process Control for Quality Improvement*, Chapman & Hall (Turpin Distribution Services), 1993.
22. Wheeler DJ, *Understanding Statistical Process Control*, SPC Press, 1992.
23. Wise SA, Fair DC, *Innovative Control Charting, Practical SPC Solutions for Today's Manufacturing Environment*.
24. *SMMT Limited Statistical Process Control – A Guide for Business Improvement*, London, 2004.

CHAPTER **19**

Quality Control

19.1 INHOUSE TESTING

To monitor production of carbon fibers, each manufacturer has their own inhouse test procedures. The primary aim of these procedures is to ensure the uniformity of products, with all measured properties within the stipulated manufacturing range. In the developmental stages, only a few ASTM standards but no SACMA test procedures were available. So, Courtaulds developed and published their Grafil Test Methods [1], which were regularly updated. Major carbon fiber users and government agencies, however, had their own buying specifications, which in many instances incorporated their own test procedures. There was consequently a great demand for unified national specifications but, unfortunately, each country tended to write their own specifications e.g. U.S.A. (ASTM), Great Britain (BSI), Japan (JIS) and Germany (DIN), which led to great confusion.

19.2 QUALITY MANAGEMENT AND QUALITY ASSURANCE STANDARDS

In the initial stages of carbon fiber production in the UK, the carbon fiber process was approved under a series of Defense Standards (e.g. Def. Stan. 05-24), which were approved by two government departments—AQD Harefield, dealing with the fiber and Woolwich Arsenal, dealing with resin systems. These early standards formed the basis of BS 5750 introduced in 1979 and BS5750: Part 2, for instance, became the replacement for Def. Stan. 05-24. Later, BS 5750 was incorporated in an international standard BS EN ISO 9000, with BS EN ISO 9002 replacing BS5750: Part 2. Fortunately, the ISO series became adopted worldwide. There are now over 100,000 organizations registered to the ISO 9000 series throughout the world. At present, accreditation can be achieved through one of about thirty five approved companies. There is a whole range of standards that supports an organization's quality program and Appendix 12 lists the current British Standards associated with quality.

QS 9000 was developed by Chrysler, Ford and General Motors and was adopted by a number of truck manufacturers. This standard was based on BS EN ISO 9000 and defined the essential quality requirements for suppliers of production parts, materials and services to the automotive industry.

BS EN ISO 14001 is a specification with guidance for use of an environmental management system, which is becoming widely accepted, but progress has been relatively slow.

19.3 THE ISO 9000 FAMILY OF STANDARDS AND QUALITY SYSTEMS [2]

The basic difference between ISO 9001 and ISO 9002 is that the former includes the supplier's capability to design. Both standards use the same numbering system and a simplified guide for the Quality requirements [personal communication with Alex McClean, Quality Management Consultant, Inverness] would be:

Para 4.1 Management Responsibility

The supplier must be committed to quality, meeting their customer's requirements, and having trained people to do it.

Para 4.2 Quality System

The supplier must have a written system of routines to ensure that they get it right the first time, every time.

Para 4.3 Contract Review

The supplier must be certain that they know what their customers want and that they are capable of doing it.

Para 4.4 Design Control

The supplier must have and maintain documents to ensure control and verification of the design of the product so that specified requirements are met.

Para 4.5 Document Control and Data

The supplier must ensure that everybody has the documents needed to do the job. Nobody must have obsolete documents.

Para 4.6 Purchasing

The supplier must buy only from those people who they know will be able to provide in accordance with the supplier requirements and the vendors must match quality and delivery needs.

Para 4.7 Control of Customer Supplied Product

The supplier must check and take care of material issued by the customer.

Para 4.8 Product Identification and Traceability

The supplier must, whenever needed, to be able to trace any material from start to finish.

Para 4.9 Process Control

All work processes must be planned and controlled through training and instructions to ensure that mistakes are not, and cannot be, made.

Para 4.10 Inspection and Testing

The supplier system is designed to prevent problems. However, to err is human; therefore inspection and testing at key points is made to make sure that the process is running satisfactorily.

QUALITY CONTROL

Para 4.11 Control of Inspection, Measuring and Test Equipment

The required accuracy of all equipment and instruments used to measure or test the products must be ensured and the equipment maintained in that condition.

Para 4.12 Inspection and Test Status

Whenever inspection or testing is undertaken, the test status of the work piece must be identified.

Para 4.13 Control of Non-Conforming Product

If something goes wrong, then the necessary steps must be taken to ensure that the non-conforming or faulty material does not get into the finished product.

Para 4.14 Corrective and Preventive Action

In order to service customer complaints, the supplier must find out where the error occurred, and prevent it happening again. If there are other trends towards non-conformity, these must be investigated to get to the root of the problem and corrective and preventive measures must be taken to ensure that the non-conformity does not recur or occur, respectively.

Para 4.15 Handling, Storage, Packaging, Preservation and Delivery

All material must be treated with the respect it deserves. This should take into account correct handling and handling material safely.

Para 4.16 Control of Quality Records

The supplier must keep records of the quality of their work to show themselves and their customers that they have met the requirements.

Para 4.17 Internal Quality Audits

The supplier must make sure that they remain efficient through regular checking of the quality system, by asking the questions—Are we good enough, could we do better?

Para 4.18 Training

The supplier must define everybody's job and provide the necessary training to ensure that the job is well done.

Para 4.19 Servicing

Servicing deals with the control of after-sales and routine servicing.

Para 4.20 Statistical Techniques

By using recorded data, QC reports and other similar reports, the supplier must statistically check the consistency of their product characteristics and help solve problems.

19.4 QUALITY GURUS [3]

A Guru is an acknowledged leader and Quality Gurus are indeed charismatic people who get their messages across. These messages, however, are not always in agreement and do change with the advent of time, but their rules can be adapted by giving careful consideration to the requirements of one's own company.

Three distinct genre of Gurus can be distinguished since the 1950s:

1. The Early Americans who delivered their messages to the Japanese in the early 1950s.
2. The Japanese school of the late 1950s onwards.
3. The New Western Wave of the 1970–1980s, concentrating on Quality Awareness.

19.4.1 The Early Americans [3]

19.4.1.1 W Edwards Deeming

He can perhaps be considered the No.1 Guru and he applied the concepts of the statistician Walter Shewhart's work in preparation for the 1940 U.S. Census and achieved productivity improvements [4]. In the immediate postwar period, his work was well received by engineers, but tended to be disregarded by the management. After the war, Deeming was sent to Japan as an adviser for their forthcoming Census and his Quality Control techniques, now regarded as part of TQM, were readily appreciated. He had now broadened the work of Shewhart to include non-manufacturing and human variation. Deeming encouraged the Japanese to extend their approach beyond statistics and introduced the Deeming or PDCA (Plan, Do, Check, Action) cycle (Figure 19.1).

It was not until the late 1970s that he made any headway in the West, assisted by his own books [5,6] followed by a range of books by other authors explaining his philosophies [7–11]. The new work had a strong bias towards the part that management had to play, rather than towards statistics. As early as 1950, Deeming was stressing that the consumer was the most important part of the production line and portrayed this with the Joiner Triangle (Figure 19.2), using the word obsession to express his belief in quality and the wish to delight the customer, rather than merely satisfy. The Joiner Triangle showed that quality was achieved with teamwork and a scientific approach.

By 1980, Deeming had introduced his 14 points:

1. Create constancy of purpose
2. Adopt new philosophy
3. Cease dependence on inspection

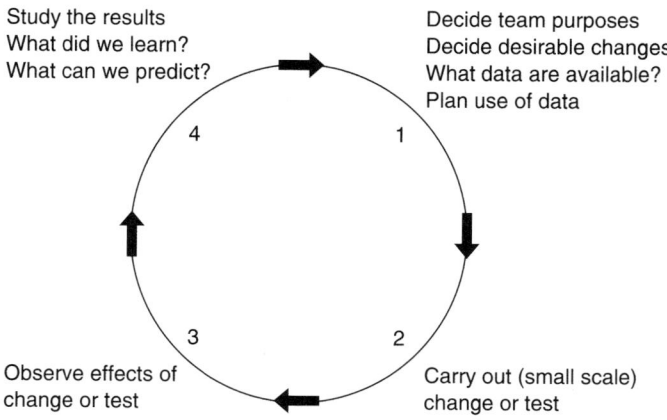

Figure 19.1 Plan, Do, Check, Action (PDCA) Cycle (according to Deeming).

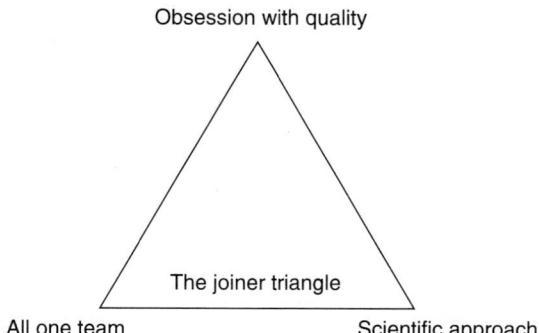

Figure 19.2 The Joiner Triangle (according to Deeming).

4. End awarding business on price
5. Improve constantly and forever the system of production and service
6. Institute training on the job
7. Institute leadership
8. Drive out fear
9. Break down barriers between departments
10. Eliminate slogans, exhortations and numerical targets
11. Eliminate quotas or work standards and management by objectives or numerical goals
12. Remove barriers that rob people of their right to pride of workmanship
13. Institute a vigorous education and self-improvement program
14. Put everyone in the company to work to accomplish the transformation

Deeming also introduced a list of so called Deadly Diseases:

1. A lack of constancy of purpose
2. Emphasis on short term profits
3. Evaluation of performance, merit-rating, or annual review
4. Mobility of management
5. Management by use only of visible figures

Deeming then introduced his seven-point action plan for change:

1. Management struggles over the 14 Points, the Deadly Diseases and obstacles, and plans direction
2. Management takes pride and develops courage for the new direction
3. Management explains to the people in the company why change is necessary
4. Divides each company activity into stages, identifying each stage of the activity as the customer of the previous stage
5. Starts as soon and as quickly as possible to construct an organization to guide continual quality improvement
6. Everyone takes part in a team to improve the input and output of any stage
7. Embarks on construction of organization for quality (uses knowledgeable statisticians)

After some 60 years of experience, Deeming then introduced his four-part System of Profound Knowledge:

1. Appreciation for a system
2. Knowledge of statistical theory
3. Theory of knowledge
4. Knowledge of psychology

19.4.1.2 Joseph M Juran

This is the second Guru and he became known internationally in 1951, when he published his first book, Quality Control Handbook [12], which has since become the standard world reference book on quality management. Other books were published [13,14]. He visited Japan in 1954 and gave a series of lectures and seminars postulating that quality control should be conducted as an integral part of management control.

Juran introduced his Quality Planning Road Map which featured the following:

1. Identify the customers
2. Determine the needs of those customers
3. Translate those needs into understandable language
4. Develop a product that can respond to those needs
5. Optimize the product features so as to meet the needs of the company as well as the customer
6. Develop a process which is able to produce the product
7. Optimize the process
8. Prove that the process can produce the product under operating conditions
9. Transfer the process to Operations

Juran, identified other external and internal customers besides the end customer, necessitating a fitness of use of the interim product for subsequent internal customers illustrated by the Quality Spiral (Figure 19.3). In order to meet the demands of quality awareness, he proposed that action should be 90% substance and 10% exhortation and not the reverse:

1. Establish specific goals to be reached
2. Establish plans for reaching the goals
3. Assign clear responsibility for meeting the goals
4. Base the rewards on the goals achieved

Juran believes there are no short cuts to quality and the majority of quality problems are the fault of poor management, rather than poor workmanship on the shop-floor and therefore, long term training should start at the top.

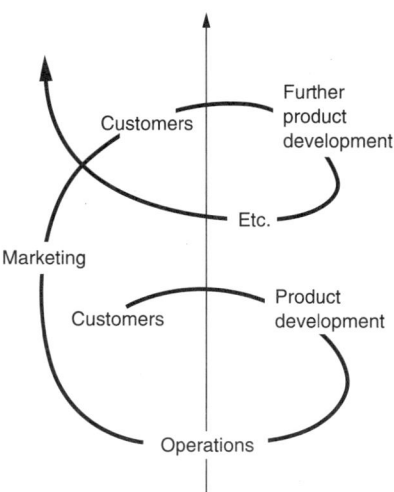

Figure 19.3 The Quality Spiral (according to Juran).

19.4.1.3 Armand V Fiegenbaum

This is the third Guru and is best known as the originator of Total Quality Control and became known to the Japanese in the 1950s through his books [15,16]. He moved towards quality control more as a business method rather than a system based on technical methods, which he considered as just a segment of a comprehensive quality control program. Fiegenbaum stressed that quality means—best for customer use and selling price—and control was a management tool with four steps:

1. Setting quality standards
2. Appraising conformance to these standards
3. Acting when standards are exceeded
4. Planning for improvements in the standards

The control aspects can be classified as:

1. New-design control
2. Incoming material control
3. Product control
4. Special process studies

He believed that statistical methods should be used wherever they would be helpful. Success depended on the involvement of the entire plant organization and he stressed the need for complete support from top management.

In the 1990s, Fiegenbaum formulated ten benchmarks:

1. Quality is a company-wide process
2. Quality is what the customer says it is
3. Quality and cost are a sum, not a difference
4. Quality requires both individual and team zealotry
5. Quality is a way of managing
6. Quality and innovation are mutually dependent
7. Quality is an ethic
8. Quality requires continuous improvement
9. Quality is the most cost-effective, least capital-intensive route to productivity
10. Quality is implemented with a total system connected with customers and suppliers

19.4.2 The Japanese Gurus [3]

19.4.2.1 Dr Kaoru Ishikawa

He is probably best known for his association with the Quality Circle Movement in Japan, which is described later (Section 19.5). Ishikawa published three books, which were all translated into English [17–19]. To prioritize quality improvements, he made good use of Pareto diagrams (Figure 18.14) together with Cause and Effect diagrams, also called Ishikawa or Fishbone diagrams (Figure 19.4), which were constructed via open group communication. He also advocated the use of control charts and scatter diagrams.

Ishikawa, following the visits of Deeming and Juran, was associated with the company-wide Quality Control Movement that had started in Japan in the period 1955–1960, when it was implied that quality does not only mean the quality of the product, but also of after

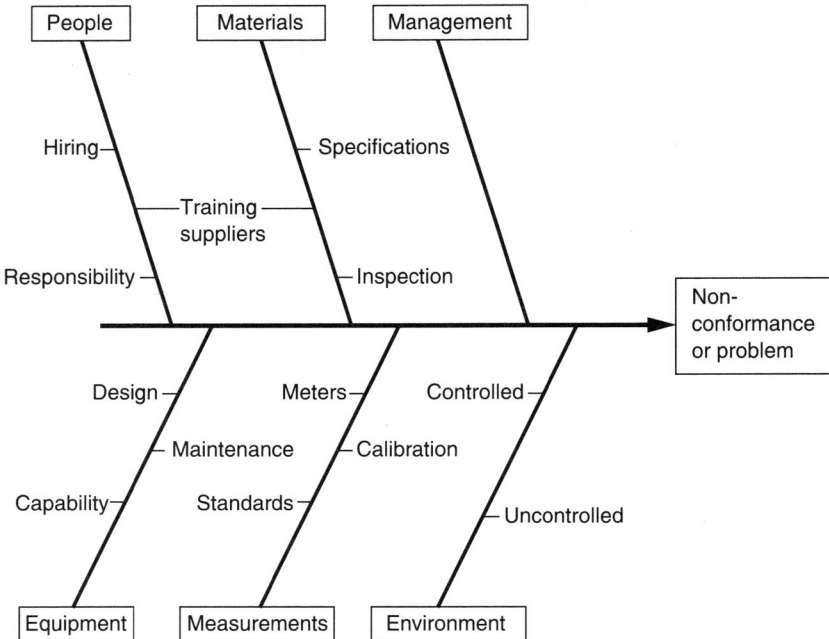

Figure 19.4 Cause and Effect Diagram (according to Ishikawa).

sales service, quality of management, the company itself and the human being and has the following effects:

1. Product quality is improved and becomes uniform, defects are reduced
2. Reliability of goods is improved
3. Cost is reduced
4. Quantity of production is increased, and it becomes possible to make rational production schedules
5. Wasteful work and rework are reduced
6. Technique is established and improved
7. Expenses for inspection and testing are reduced
8. Contracts between vendor and purchaser are rationalized
9. The sales market is enlarged
10. Better relationships are established between departments
11. False data and reports are reduced
12. Discussions are carried out more freely and democratically
13. Meetings are operated more smoothly
14. Repairs and installation of equipment and facilities are done more rationally
15. Human relations are improved

19.4.2.2 Dr Genichi Taguchi

This is the second member of the Japanese group who had an excellent background of statistics and from 1950 onwards, developed his own methods and published several books [20,21], whilst other authors described the Taguchi methodology [22–25]. When Taguchi visited the U.S.A. in 1980, a number of American companies became involved in his methodology.

QUALITY CONTROL

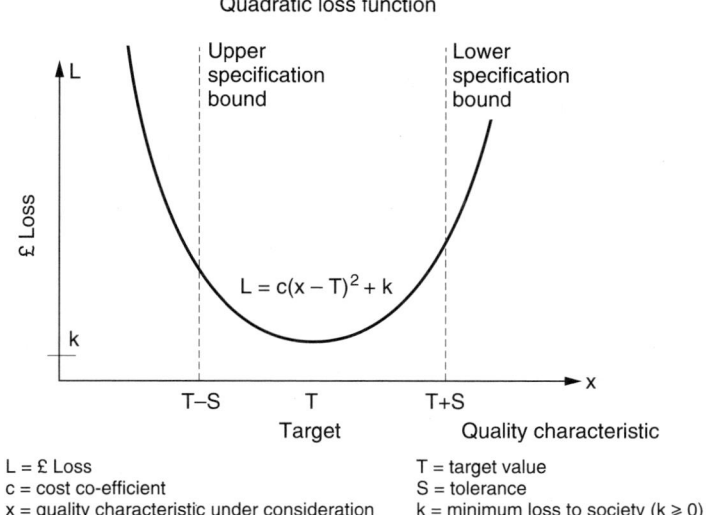

Figure 19.5 Quadratic Loss Function (according to Taguchi).

Basically, Taguchi works in terms of a quality loss function rather than quality and defined quality loss as the loss imparted by the product to society from the time it is shipped. By taking a target value as the best possible value for the quality characteristic under investigation, he used a simple quadratic loss function, with deviations from the target to show that a decrease in loss was associated with an increase in quality (Figure 19.5). It will be noted that a loss will occur even when the product is within the permitted specification, but this loss is minimal when the product is on target.

An important aspect of the Taguchi methodology is that it can be applied offline in design, or online in production. The offline quality control can be broken down into:

1. System design
2. Parameter design
3. Tolerance design

Taguchi methodology is used mainly by engineers and is therefore tailored with a strong engineering bias. Hence in addition to the control variables, noise variables which disrupt production are also considered. In attempting to optimize the product to get quality characteristics on target and to minimize the spread about that target by controlling. The Signal to Noise Ratio, through selecting the control setting that gives minimum noise, there is a direct tie up with SPC.

Following Taguchi's early work [19] with Orthogonal Arrays, he was able to express them in a systematic form to provide a route whereby an engineer could select a minimum number of prototypes that would be necessary for experimentation.

19.4.2.3 Shigeo Shindo

He was a professional management consultant and first applied statistical quality control in 1951 and by 1954, had investigated some 300 companies. He undertook training at Toyota Motor Co., Matsushita Electrical Industrial Co. and at Mitsubishi Heavy Industries and reduced the time for hull assembly of a 65,000 ton super tanker from 4 months to

2 months. The system soon spread to every shipyard in Japan. During 1961–1964 he developed the Poka-Yoke, mistake proofing or Zero Defects concept. He visited Europe in 1973 and the U.S.A the following year. Shindo wrote a number of books, some of which were translated into English [26–28] and *Absolutely Zero Loss* by Copeland W [29] is relevant.

Shindo realized that statistical quality control methods would not reduce defects to zero, but by using Poka-Yoke and source inspection systems, established that defects could be reduced to zero, necessitating stopping the process whenever a defect occurred, defining the cause and preventing that defect from recurring. This implies essentially identifying and correcting errors at source before they have become defects. Hence, statistical sampling was no longer required.

As aids to this system, Shingo developed simple Poka-yoke mechanical, or physical, devices, which prevented incorrect assembly and could also indicate when parts had been incorrectly assembled. These devices, in conjunction with inspection to prevent defects occurring, constituted Zero Quality Control.

Certainly, stopping a carbon fiber line in the event of a defect occurring would be a major issue and it is doubtful if Shindo's methods can be fully adapted to carbon fiber manufacture.

19.4.3 The New Western Group of Gurus [3]

19.4.3.1 Philip B Crosby

He is an American and probably the most charismatic of all the Gurus and is best known for his concepts—Do It Right First Time and Zero Defects. He has written two books [30,31]. Crosby's approach has received much criticism from other workers, but basically, he believes that people can make mistakes and a company should not assume, from the onset, that people will not do so. Crosby believes that one should create a core of quality specialists within the company with a strong emphasis on the top-down approach (Figure 19.6) and implicitly believes that management is entirely responsible for quality.

Crosby has specified the Four Absolutes of Quality Management as:

1. Quality is defined as conformance to requirements, not as goodness or elegance
2. The system for causing quality is prevention, not appraisal
3. The performance standard must be Zero Defects, not a that's-close-enough policy
4. The measurement of quality is the Price of Non-conformance, not indices

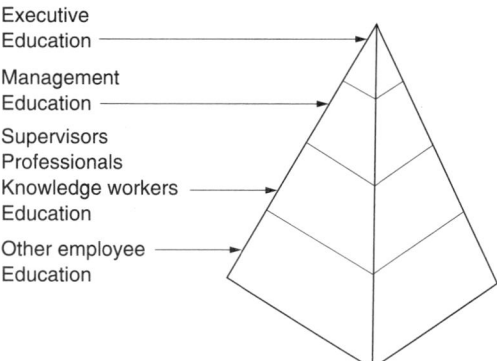

Figure 19.6 Crosby approach to Education.

QUALITY CONTROL

To implement the Quality Improvement Process, he detailed the following fourteen steps to Quality Improvement:

1. Make it clear that management is committed to quality
2. Form quality improvement teams with senior representatives from each department
3. Measure processes to determine where current and potential quality problems lie
4. Evaluate the cost of quality and explain its use as a management tool
5. Raise the quality awareness and personal concern of all employees
6. Take actions to correct problems identified through previous steps
7. Establish process monitoring for the improvement process
8. Train supervisors to actively carry out their part of the quality improvement program
9. Hold a Zero Defects Day to let everyone realize that there has been a change and to reaffirm management commitment
10. Encourage individuals to establish improvement goals for themselves and their groups
11. Encourage employees to communicate to management the obstacles they face in attaining their improvement goals
12. Recognize and appreciate those who participate
13. Establish quality councils to communicate on a regular basis
14. Do it all over again to emphasize that the quality improvement program never ends

Crosby added to this list to broaden the concept into becoming an Eternally Successful Organization:

1. People routinely do things right the first time
2. Change is anticipated and used to advantage
3. Growth is consistent and profitable
4. New products and services appear when needed
5. Everyone is happy to work there

The book *Right First Time* by Frank Price [32] provides good reading.

19.4.3.2 Tom Peters

He is also an American, has published books [33–35] describing his analyzes of a number of successful American companies and has prescribed ways of bringing about a Management Revolution. Peters prefers the term leadership to management and proposes a new role, cheerleader and facilitator, brought about by Managing by Wandering About (Figure 19.7),

Figure 19.7 Managing by Wandering About—The Technology of the Obvious (according to Peters).

enabling the leader to keep in touch with customers, innovation and people. Whilst the leader wanders, three major activities occur:

1. Listening—suggests caring
2. Teaching—values must be transmitted face to face
3. Facilitating—able to give on-the-spot help

By the late 1980s, Peters concentrated on the four familiar areas—customers, innovation, people and leadership and offered 45 prescriptions, all calling for urgent radical reform. He described 12 attributes or traits of a quality revolution:

1. Management obsession with quality
2. Passionate systems
3. Measurement of quality
4. Quality is rewarded
5. Everyone is trained for quality
6. Multi-functional teams
7. Small is beautiful
8. Create endless Hawthorne effects (new goals, new themes and new events as the antidote to the 12–18 month doldrums)
9. Parallel organization structure devoted to quality improvement
10. Everyone is involved
11. When quality goes up, costs go down
12. Quality improvement is a never ending journey

19.4.3.3 Claus Møller

This is final Guru in the New Western Group and is a Danish business economist, who has been able to successfully adapt his concept of Personal Quality to suit Japanese and Russian cultures. Møller has published a book [36] and identifies three vital areas—Productivity, Relations and Quality.

He identifies two standards of Personal Quality—the ideal performance level (IP) and the actual performance level (AP) and presents Twelve Golden Rules to help improve the AP level:

1. Set personal quality goals
2. Establish your own personal quality account
3. Check how satisfied others are with your efforts
4. Regard the next link as a valued customer
5. Avoid errors
6. Perform tasks more effectively
7. Utilize resources well
8. Be committed
9. Learn to finish what you start
10. Control your stress
11. Be ethical, maintain your integrity
12. Demand quality

Møller also identifies 17 hallmarks of a quality company:

1. Focus on quality development
2. Management participation in the quality process
3. Satisfied customers/users
4. Committed employees

5. Long-term quality development
6. Clearly defined quality goals
7. Quality performance rewarded
8. Quality control perceived positively
9. Next person in work process is a valued customer
10. Investments in personnel training and developments
11. Prevention/reduction of mistakes
12. Appropriate decision level
13. Direct route to end users
14. Emphasis on both technical and human quality
15. Company actions directed towards customer needs
16. Ongoing value analysis
17. Company recognition of its role in society

It will be seen that there are contradictions between the different philosophies, which is really not surprising. However, there are many common features and it will be necessary to custom build a quality process for one's company.

19.5 QUALITY CIRCLES [3]

Quality Circles, although surprisingly, had its roots in the U.S.A. in the 1950s, is a topic which evokes widely opposing views. Eminent authorities like Dr Juran have thrown doubt on their effectiveness in the West and warns that there are no short cuts to quality, whereas Dr Ishikawa is best known as a pioneer of the Quality Circle movement in Japan in the early 1960s.

The Japanese do have a paternalistic style of management and select a company, not a job. Another difference that the Japanese have is a high sense of loyalty to the company, which provides a very stable work force. A Japanese company does not recruit specialists, but rather employs generalists, who are trained and it is not until middle management that the generalists become specialists. In the western world, it is virtually the opposite, recruiting specialists and then training them as generalists. These differences probably account for why the Japanese have developed a high degree of sophistication in their treatment of consensus management, taking more time in the decision making process than companies in the West, but much less time in implementing those decisions.

For Quality Circles to succeed, there must be total commitment from upper management, so that whatever the Circle has suggested will be implemented , or a good reason given why it could not, thereby providing encouragement for the Circle members. Circle members receive no direct financial reward for their improvements.

In Japan, a typical Circle comprises 5–10 voluntary workers from the same work area, led by a foreperson, a deputy foreperson or a leader from one of the workers. The aims of the Circle are:

1. Contribute to the improvement and development of the project
2. Have mutual respect of human relations and help build a happy and contented workshop providing job satisfaction
3. Use human potential to the full

The Circle members have been taught and are proficient in statistical quality control covering:

1. Pareto charts
2. Cause and effects diagrams

3. Check sheets
4. Histograms
5. Scatter diagrams
6. Control charts and graphs

Whenever possible, the members implement their own solutions, or put strong pressure on management to introduce them.

In Japan, there are now over 10 million Quality Circle members. However, not all Circles are a success and failure can nearly always be attributed to management, either through lack of interest, or too much intervention. It is crucial to impress on people that Circles do not remove the management's decision making power. Reasons for the high rate of success in Japan can be attributed to their participative style and their high level of employee training. Although at first sight, benefits may appear to be minor, added together, they represent substantial improvements [37].

Circles hold regular meetings in company time, usually meeting for about one hour once a week or fortnight, although in Japan, meetings are often extended after hours.

A circle is usually launched by a member of management bearing a title like Facilitator or Coordinator, whose job is to train the circle leaders, to promote the circle concept on the shop floor and among middle management (with the help of top management and forepersons) and to help co-ordinate the activities of different circles.

In the West, when launching a Quality Circle, careful adaptation to a Western approach must be employed and there is a practice when forming a Circle to enlist the help of consultants, who can also assist in training. Some circles include management specialists and others invite them to attend their meetings for specialist advice.

In the US and Europe, probably the most opposition to Quality Circles has come from middle management, who believe that they are threatened. Suitable training would, however, help to avoid this situation.

19.6 TOTAL QUALITY MANAGEMENT [38]

Total Quality Management (TQM) is a way of managing an organization so that every job, every process, is carried out right, first time and every time and affects everyone. Many books have been written on TQM [39–48] and much advice has been delivered by the Gurus, not always in consensus. TQM involves three major components:

1. A quality assurance system
2. Quality tools and techniques
3. Teamwork

TQM can be aided by employing the guidance and experience of a consultant, but it is user driven and ideas for improvement must come from those with the knowledge and experience of the methods and techniques.

Teamwork plays a most essential part of the implementation of TQM. The person identifying a given problem, and the head of that particular department, should decide when to form a team to investigate the given problem. A team normally comprises up to seven members, formed from people who are affected by the problem, possessing specialized and local knowledge of the topic to be investigated. A team organizer and spokesperson (who can be the same person) are chosen by the team.

All members of the team are equally important and all must have their say and be entirely frank when presenting their contribution. Complete agreement of any action has to be achieved and duly recorded by the Organizer.

QUALITY CONTROL

The object of the team is to work together to improve Quality and the following principles should be observed:

1. Quality is conformance to requirements
2. Every activity can be described by the work process model
3. The measuring standard for quality is money (Cost of Non-conformance)
4. The personal attitude and performance standard is Zero Defects
5. The systematic way to quality is through prevention in all areas of work (do it right, first time, every time)

To achieve quality, probably the most important key is to fully understand the customer's requirements, whether internal or external.

To successfully participate, all employees, including directors and managers, should be trained to:

E   Evaluate—the situation and define their objectives
P   Plan—to fully achieve those objectives
D   Do—implement the plans
C   Check—that the objectives are being achieved
A   Amend—take corrective action if they are not

This is the so called Helix of Never-Ending Improvement (Figure 19.8).

A resumé of what the Gurus believe needs to be done to implement TQM is [adapted from 3,38]:

1. Develop a systematic approach to manage the implementation of TQM
2. Management commitment and employee awareness are essential and Deeming's philosophy, Peters' Top Twelve Traits, Crosby's Zero Defects and Møller's Personal Quality will be useful to encourage the correct attitudes
3. Adopt the philosophy of zero errors/defects to change the culture to Crosby's Right First Time
4. The awareness should be backed up by facts and figures, Costs of Quality can be used to measure progress as advocated by Juran and Crosby
5. Train the people to understand the customer-supplier relationship, the internal customer emphasized by Juran and Crosby and the external customer by Peters and Deeming.
6. Simple tools can be used for problem solving and improvement, as outlined by Ishikawa
7. More technical tools to control industrial design and manufacturing can be employed using Taguchi methods (Tsai has described the optimization of carbon-fiber production using the Taguchi method [49])
8. Management tools should be used to achieve quality, such as Crosby's Zero Defect approach and the concepts of Company Wide Quality and Total Quality associated with Ishikawa and Fiegenbaum

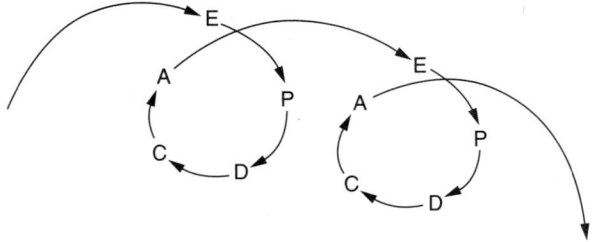

Figure 19.8 The Helix of Never-Ending Improvement.

9. Employ teamwork to improve communication and problem solving and cross-functional teams, as advocated by Peters and Crosby and Quality Circles by Ishikawa
10. Recognize that improvement of the systems has to be managed
11. Constantly educate and retrain

Finally, achievement of TQM within a company [38] can be achieved by:

1. Effective Leadership

   a. Developing clear beliefs and objectives
   b. Developing clear and effective strategies
   c. Identifying critical processes
   d. Reviewing management structure
   e. Encouraging effective employee participation

2. Quality aspects

   a. Satisfying customer's needs
   b. Getting close to customers
   c. Planning to do all jobs right first time
   d. Agreeing on expected performance standards
   e. Implementing company-wide quality improvement
   f. Measuring performance
   g. Measuring quality mismanagement and firefighting
   h. Demanding continuous improvement
   i. Recognizing achievements

19.7 QUALITY COSTING [3,50]

The very term Cost of Quality is misleading and perhaps should really be Cost of Non-quality and is defined by Phil Crosby as the expense of doing things wrong. It is generally recognized that the cost of quality is 15–30% of sales revenue and can be directly attributed to weaknesses in management, arising from interactions between people in their daily work. All companies employ people, so they will incur quality costs. The cost of quality includes rework, scrap, stock write-offs, downtime, changes in production schedules, engineering changes, design changes, warranty claims and management time. Also, the customer can incur costs through poor product and reliability, eventually leading to further losses to the manufacturer as the market share falls.

Setting up a total quality system will result in reduced operating costs and Dr Armand V Feigenbaum believes that quality has become the single most important force leading to organizational success and growth.

Taguchi's loss function can be used to evaluate design decisions on a financial basis to establish whether additional costs in the production process would be worthwhile in the market place.

If quality is defined in the sense of Conformance of Manufacture with the Design Specification, then this is primarily associated with manufacture. If the meaning is broadened to include Conformance to the Defined Requirements of the customer (design and/or marketing functions), then this takes in both manufacture and design and broadens the quality function from control to assurance. Finally, one can encompass marketing, or market research, so that the quality assurance function monitors and influences the company's definition of its customers' requirements.

REFERENCES

1. Courtaulds Ltd., *Grafil Data Sheets*, Courtaulds Ltd. Carbon Fibres Division, Coventry.
2. Gillespie H. *ISO 9000 for the Chemical Process Industry*. McGraw-Hill, 1997.
3. Bendell A, *The Quality Gurus*, DTI, London, 1991.
4. Deeming WE, *Statistical Adjustment of Data*, Constable, Dover, 1964.
5. Deeming WE, *Out of Crisis*, Cambridge University Press, 1988.
6. Deeming WE, *Quality, Productivity, and Competitive Position*, MIT Center for Advanced Engineering Study, Cambridge University, 1982.
7. Gitlow HS, Gitlow SJ, *Deeming Guide to Quality and Competitive Position*, Prentice-Hall Inc., 1987.
8. Mann NR, *The Keys to Excellence: The Story of the Deeming Philosophy*, Prestwick Books, Los Angeles, 1985.
9. Neave HR, *The Deeming Dimension*, SPC Press, Knoxville, 1990.
10. Scherkenbach WW, *The Deeming Route to Quality and Productivity, Road Maps and Road Blocks*, CEE Press Books, Washington DC, 1986.
11. Walton M, *The Deeming Management Method*, Mercury Books, London, 1989.
12. Juran JM, Gryna FM, *Quality Control Handbook*, McGraw-Hill, New York, 1988.
13. Juran JM, *Juran on Planning for Quality*, The Free Press, New York, 1988.
14. Juran JM et al, *Quality Planning and Analysis: from Product Development Through Use*, McGraw-Hill, New York, 1980.
15. Fiegenbaum AV, *Total Quality Control*, McGraw-Hill, New York, 1991.
16. Fiegenbaum AV, *Quality Control: Principles, Practices and Administration*, McGraw-Hill, 1951.
17. Ishikawa K, *Guide to Quality Control*, Asian Productivity Organisation, Tokyo, 1976.
18. Ishikawa K, *What is Total Quality Control? The Japanese Way*, Prentice-Hall, 1988.
19. Ishikawa K, *Introduction to Total Quality Control*, Chapman and Hall, 1991.
20. Taguchi G, *Introduction to Quality Engineering*, Asian Productivity Organisation, Tokyo, 1986.
21. Taguchi G, *Systems of Experimental Design*, Unipub/Kraus International Publications and American Supplier Institute, 1978.
22. Bendell A ed., *Taguchi Methods*, Elsevier Science Publishers Ltd., London, 1989.
23. Bendell A et al ed., *Taguchi Methods: Applications in World Industry*, IFS Publications, Bedford, 1989.
24. Bendell A et al ed., *Taguchi Methodology Within Total Quality*, IFS Publications, Bedford, 1990.
25. Tsai JS, Optimization of carbon fibre production using the Taguchi Method, *J Mater Sci*, 30(8), 2019–2022, 1995.
26. Shingo S, *The Sayings of Shigeo Shingo- Key Strategies for Plant Improvement*, (trans), Productivity Press, Cambridge, 1985.
27. Shingo S, *Non-stock Production*, (trans), Productivity Press, Cambridge, 1988.
28. Shingo S, *Zero quality control: Source Inspection and the Poka-yoke system*, (trans), Productivity Press, Cambridge, 1986.
29. Copeland W, *Absolutely Zero Loss*, Absolutely Zero Loss, 1997.
30. Crosby PB, *Quality is Free*, McGraw-Hill, New York, 1979.
31. Crosby PB, *Quality Without Tears*, McGraw-Hill, New York, 1984.
32. Price F, *Right First time: Using Quality Control for Profit*, Wildwood House, 1986.
33. Peters TJ et al, *In Search of Excellence*, Harper and Row, 1982.
34. Peters TJ, Ashton N, *A Passion for Excellence*, Random House Inc., New York, 1985.
35. Peters TJ, *Thriving on Chaos*, AA Knopf Inc., 1987.
36. Møller C, *Personal Quality*, Time Manager International A/S, Denmark, 1987.
37. Quality Control Circles save Lockheed nearly $3 million in two years, *Quality Magazine*, May 1977.
38. Mortiboys R, Oakland J, *Total Quality Management and Effective Leadership*, DTI, London, 1991.
39. Caplen RH, *A Practical Approach to Quality Control Business Books*, Century Arrow, London, 1972.
40. Creech W, *The Five Pillars of TQM: How to Make Total Quality Management Work for You*, Penguin Group, 1995.
41. Evans JR, Lindsay WM, *Management and Control of Quality*, South Western College Publishing, 1995.

42. Oakland J, *Total Quality Management*, Heinemann, Oxford, 1989.
43. Omachonu V, Ross JE, *Principles of Total Quality*, Kogan Page, 1994.
44. Ott ER, *Process Quality Control-Troubleshooting and Interpretation of Data*, McGraw-Hill, New York, 1975.
45. Pyzdek T, *What Every Engineer Should Know About Quality Control*, Marcel Dekker, 1988.
46. Slater K, Harrison PW, ed., *Physical Testing and Quality Control, Textile Progress*, Textile Institute, 1993.
47. Tapiero CS, *Management and Control of Quality*, Chapman and Hall, 1995.
48. Taylor JR, *Quality Control Systems: Procedures for Planning Quality Control Programmes*, McGraw-Hill, 1988.
49. Tsai JS, Optimization of carbon-fiber production using the Taguchi method, *J Mater Sci*, 30(8), 2019–2022, 1995.
50. B Dale, J Plunkett, Case for Costing Quality, London, DTI, 1989.

CHAPTER 20

Properties of Carbon Fibers

20.1 THE ROLE OF CARBON FIBERS

In order to understand the role that carbon fibers can play, the attributes of carbon fibers must be considered:

1. Available in many grades and forms with wide-ranging properties
2. High modulus, especially pitch based fiber
3. Good strength, especially PAN based fiber
4. Low density, giving good specific properties
5. Good thermal stability in the absence of O_2
6. High thermal conductivity, assisting good fatigue properties
7. Low thermal expansion coefficient
8. Excellent creep resistance
9. Good chemical resistance and does not wick
10. Low electrical resistivity
11. Biocompatibility
12. No significant inhalation problem with filament diameters down to 5 μm

The disadvantages are:

1. Relatively high cost, but prices have been falling and more emphasis is now placed on using large tows
2. Low strain to failure with attendant handling problems
3. Compressive strength is lower than tensile strength and larger diameter fiber does not give improved compression properties
4. Poor impact strength of composites
5. Care required during handling carbon fiber, since it is electrically conducting and can cause havoc with electrical systems
6. Oxidizes in air at temperatures above 450°C
7. Exhibits anisotropy in the axial and transverse directions

20.2 TYPES OF CARBON FIBERS AVAILABLE IN THE WORLD MARKET

The first commercial carbon fibers were based on viscose rayon, a cellulosic precursor, but Polycarbon is now the only current producer of this type of carbon fiber. The properties of rayon-based carbon fibers are listed in Table 20.1. A difficult fiber to produce with a low yield, its main use is in existing space programs.

Oxidized PAN fiber is made by a slightly modified first stage of the PAN based carbon fiber process and the properties are listed in Table 20.2. Oxidized PAN fiber, with a density above 1.37 g cm^{-3}, is non-flammable.

Tables 20.3A and B list the properties of PAN based carbon fibers and as they are not graphitic, this will limit the modulus attainable, but strengths are greater than pitch based fibers. The introduction of the Intermediate Modulus fibers was very useful. Torayca T1000 is the strongest carbon fiber in the world and Torayca M70J has the highest tensile modulus of any PAN based carbon fiber.

Fibers from a pitch precursor are graphitic and for a given process temperature, can attain higher moduli than PAN based fibers, approaching the value for the graphite crystal ($\approx$1000 GPa). Table 20.4 gives the properties for pitch based carbon fibers.

The properties of carbon fiber are quite dependent on the structure, in particular, the crystallite size as defined by the coherent length perpendicular (L_c) and parallel (L_a) to the carbon layers. These values increase as the heat treatment temperature increases and for a given process temperature, L_c is higher for a pitch based fiber (Figure 20.1), increasing with temperature at a steady rate, whereas a PAN based fiber increases sharply above 2300°C.

Young's modulus is an intrinsic property and is governed by the orientation of the graphitic crystallites relative to the fiber axis. The lower this angle, greater is the modulus. L_a is a measure of the crystallite basal planes and increases with temperature (Figure 20.2) and the modulus (Figure 20.3). The orientation for HM fiber is 25° in the core and 12–15° in the skin, with circumferential orientation.

Carbon fiber properties are continually being improved and the values quoted in the tables can serve as a guide.

Early work with SAF PAN based carbon fiber quickly established a relationship between the Young's modulus of the carbon fiber and the production temperature. An early plot of attainable YM for a measured process temperature is given in Figure 20.4. The values obtained for the tensile strength and modulus by Moreton, Watt and Johnson, appended with Shindo's results, are shown in Figure 3.4. Some 15 years later, Matsumoto [2] compared the tensile properties of a mesophase pitch fiber with a PAN based carbon fiber produced at different process temperatures (Figure 20.5). The modulus increases steadily with temperature, but the strength peaks at about 1575°C. As the quality of the PAN precursor has been improved and its diameter reduced, this has enabled carbon fibers to be produced with higher strengths, with a diameter of about 5 μm. The smaller the carbon fiber diameter, greater is the strength (Figure 5.7).

This fact explains the introduction of grades of carbon fiber with improved tensile properties having filament diameters of about 5 μm, whereas other earlier grades have diameters of about 7 μm.

Ozbek and Isaac [3] found that significant increases in modulus (Figure 20.6) could be recorded with PAN based carbon fibers by hot stretching (loads of up to 0.5 g filament^{-1} at temperatures of 2600°C and 2800°C for times up to 30 min), but there was a loss in strength following such heat treatment. The use of higher strains could restore the fiber strength, although in practical terms, the long treatment times to obtain this increased modulus could be a disadvantage. It is probable that this improvement in strength is due to a reduction in the filament diameter during hot stretching, hence there would be a

Table 20.1 Properties of cellulose based carbon fibers

Company	Grade	Carbon Assay minimum %	Ash Content maximum %	Breaking Strength minimum kg	Specific Gravity minimum	Mpul g m^{-1}	Tex	Yield m kg^{-1}	Filaments per ply nominal	Twist tpm	Plies	Moisture Nominal %	Filament Tensile strength GPa	Filament Tensile modulus GPa
Polycarbon Inc.														
Carbon yarns	C-5	95	1.0	9.3	1.40	0.32	310	3125	720	90	5	1.0	0.76	41
	C-10	95	1.0	11	1.42	0.68	680	1472	720	72	10	1.0	0.76	41
	C-20	95	1.0	20	1.44	1.28	1280	782	720	86	20	1.0	0.76	41
Graphite yarns	G-5	99	1.0	7.7	1.35	0.31	310	3263	720	86	5	1.0	0.72	41
	G-10	99	1.0	10	1.35	0.62	620	1615	720	86	10	1.0	0.66	41
	G-20	99	1.0	14	1.40	1.26	1260	794	720	80	20	1.0	0.66	41
RK Carbon Fibers Grayon														
Carbon yarns	CA 5	96	1.0	10	1.4	0.31	310	3225	720	90	5	0.75	0.82	34
	CA10	96	1.0	13.6	1.4	0.63	630	1587	720	80	10	0.75	0.82	34
	CA 20	96	1.0	25	1.4	1.30	1300	769	720	75	20	0.75	0.82	34
Graphite yarns	RG 5	99.5	1.0	10	1.4	0.30	300	3286	720	90	5	0.75	0.82	41
	RG 10	99.5	1.0	13.6	1.4	0.62	620	1612	720	80	10	0.75	0.82	41
	RG 20	99.5	1.0	2	1.4	1.29	1290	775	720	75	20	0.75	0.82	41

Note: The yarns are available with S or Z twist and a range of sizes such as graphite dispersions, PVA and PTFE. When SGL acquired RK Carbon Fibers in 1997, the RK products were integrated by Polycarbon.
Source: Reprinted from manufacturer's technical literature.

Table 20.2 Properties of oxidized PAN fibers

Supplier	Trade Name	Tow Size k	Density g cm^{-3}	Tenacity CN Tex^{-1}	UTS MPa	Elongation %	Filament Diameter μm	Moisture Content %
Afikim	Thermopan	320	1.40	16	245	18	11	9
Akzo Nobel	Fortafil o.p.f.	200	1.40	16	NA	19	NA	6
Korea Steel Chemical	Oxipan	320	1.40	12	NA	19	12	6
IPCL	Indcarf o.p.f.	320	1.40	12	NA	19	12	6
SGL Technic (RK Carbon Fibers)	Panox B	320,s,t,x	1.37–1.39		200	19	11	8
	Panox M		1.38–1.40		200	18	11	8
Textron	Avox							
Toho Rayon	Pyromex	s,y	1.35–1.45	16–23	NA	15–25	13–15	6 to 10
Universal Carbon Fibers	Panotex	y,f	To Customer Requirements					
Zoltek	Pyron 139	40,80, 160,320	1.39	NA	310	32	13	4 to 5

Key: NA not available; f fabric; s staple; t tops; x special tow sizes; y yarn.
Source: Reprinted from manufacturer's technical literature.

Table 20.3A Properties of PAN based carbon fibers

Company	Trade name	Tow size k	TS GPa	YM GPa	Diameter μm	Elongation %	Density g cm^{-3}	Minimum carbon content %
AKZO	Fortafil 502, 503,504,505	40	3.80	231	6	1.64	1.80	
	506,507,508,509	50	3.45	217	7	1.59	1.80	
	510,511,512,513	80	3.80	231	6	1.64	1.80	
	555,556	58	3.80	231	6.2	1.65	1.80	
Afikim	Acif IS	3,6,12, 40,320	2.5	230	6.8	1.3	1.78	93
	HT	3,6,12,40	2.9	230	6.8	1.4	1.78	95
	XHT	3,6,12	3.3	230	6.8	1.55	1.78	95
	HM	3,6,12	2.2	335	6.6	0.75	1.86	99.5
Cytec Carbon Fibers LLC	Thornel T300	1,3,6,12	3.75	231	7.0	1.4	1.76	92
	T300C	3,6,12	3.75	231	7.0	1.4	1.76	92
	T650/35	3,6,12	4.28	255	6.8	1.7	1.77	94
	T650/35C	12	4.28	255	6.8	1.7	1.77	94
Asahi-Kasei	Hi-Carbolon	3,6,12	4.31	230	7.0	1.87	1.78	
Fortafil Fibers Inc.	Fortafil 502, 503,504,505	40	3.80	231	6	1.64	1.80	
	506,507,508,509	50	3.45	217	7	1.59	1.80	
	510,511,512,513	80	3.8	231	6	1.64	1.80	
Grafil Inc. (Mitsubishi Rayon Co.)	Grafil 34-700	12,24	4.50	234	7	1.9	1.80	
	34-700WD	12	4.50	234	7	1.9	1.80	
	34-600, 34-600WD	48	4.15	234	7	1.8	1.80	
	Pyrofil TR40	1	4.70	235	7	2.0	1.80	
	TR30S	3,6	4.41	235	7	1.9	1.79	
	TR50S	12,24	4.90	240	7	2.0	1.82	
	TRH50	12,24	4.90	255	7	1.9	1.80	

(Continued)

Table 20.3A Continued

Company	Trade name	Tow size k	TS GPa	YM GPa	Diameter μm	Elongation %	Density g cm^{-3}	Minimum carbon content %
	MR35E	12	4.41	295	7	1.5	1.75	
	MR40	12	4.41	295	6	1.5	1.76	
	MR50	12	5.40	290	6	1.8	1.80	
	MS40	12	4.61	345	6	1.3	1.77	
	HR40	12	4.41	390	6	1.1	1.82	
	HS40	12	4.41	450	5	1.0	1.85	
Hexcel	AS4	3,6,12	4.28	228		1.87	1.78	
	AS4C	3,6,12	4.35	231	6.9	1.88	1.78	94
	AS4D	12	4.69	245	6.7	1.92	1.79	
	IM4	12	4.80	276	6.7	1.74	1.78	
	IMC	12	5.52	290		1.90	1.80	
	IM6	12	5.59	279	5.2	2.00	1.76	94
	IM7 (5000 Spec)	6	5.18	276	5.2	1.87	1.78	94
	IM7 (6000 Spec)	6						
	IM7 (5000 Spec)	12	5.52	276	5.2	2.01	1.78	94
	IM7 (6000 Spec)	12	5.76	290	5.1	1.99	1.79	94
	IM7 (5000 Spec)	12	5.76	292		2.00	1.80	94
	IM7 (6000 Spec)	12						
	IM7C	12	5.52	290	5.4	1.90	1.80	94
	IM8	12	5.59	304	5.1	1.84	1.79	94
	IM9	6	6.07	290	4.4	2.10		94
	IM9	12	6.14	290	4.4	2.10	1.80	
	IMC	12	5.52	290	5.4	1.90	1.80	94
	PV36/700	12	4.69	248		1.9		
	PV42/800	12	5.52	290	5.4	1.9	1.80	94
	PV42/850	12	5.76	292	4.4	1.97		
	UHM	3,6,12	3.45	441		0.8	1.87	
IPCL	Indcarf 25	3,6,12	Min. 2.50	215–240	6.8	1.05–1.40	1.78	93
	30	3,6,12	Min. 3.00	220–240	6.8	1.25–1.60	1.78	95
Kosco	Kosca GP250	12	2.8	220	6.8	1.3	1.80	93
	HS300	12	3.3	230	6.8	1.4	1.80	95

Source: Reprinted from manufacturer's technical literature.

Table 20.3B Properties of PAN based carbon fibers

Company	Trade name	Tow size k	TS GPa	YM GPa	Diameter μm	Elongation %	Density g cm^{-3}	Minimum carbon content %
SGL Technic (RK Carbon Fibers)	C10	60,160,320, 400,410	Min. 2.0	180–240	8	1.0	1.75	>95
	C25	60,160,320, 400,410	Min. 2.5	215–240	8	1.05–1.40	1.78	>95
	C30	60,160,320, 400,410	Min. 3.0	220–240	7.0	1.25–1.60	1.78	>95
	T18	320	0.9–1.4	40–60	9–11	>1.9	1.68–1.74	
	T16	320	1.1–1.6	60–80	9–11	>1.2	1.76–1.82	

(Continued)

Table 20.3B Continued

Company	Trade name	Tow size k	Filament properties					Minimum carbon content %
			TS GPa	YM GPa	Diameter μm	Elongation %	Density g cm^{-3}	
Soficar	Torayca	See Toray						
Tenax Fibers	Tenax HTA	1,3,6,12,24	3.95	238	7	1.5	1.77	
	HTS	1,3,6,12,24	4.3	238	7	1.5	1.77	
	STS	24	4.00	240	7	1.5	1.79	
	UTS	12,24	4.70	240	7	2.0	1.80	
	IMS 3131	12	4.12	295	6.4	1.4	1.76	
	IMS 5131	12,24	5.60	290	5.0	1.9	1.80	
	HMA	6,12	3.00	358	6.75	0.7	1.77	
	UMS 2526	12,24	4.56	395	4.8	1.1	1.78	
	UMS 3536	12	4.50	435	4.7	1.1	1.81	
Textron	Avcarb HC		2.07	207		1		88–92
	HCB		1.90	262		0.72		99.5
Toho	Besfight HTA	12	3.92	235	7	1.6	1.77	
	ST3	3,6,12	4.41	235	7	1.9	1.77	
	IM400	6,12	4.31	295	6.4	1.5	1.75	
	IM500	12	5.00	300	5.0	1.7	1.76	
	HM30	6,12	4.30	295	6.4	1.5	1.75	
	HM35	3,6,12	2.74	343	6.7	0.8	1.79	
	HM45	6,12	3.10	441	6.4	0.48	1.90	
	UM40	6,12	2.55	392	6.6	0.65	1.83	
Toray*	Torayca T300	1,3,6,12	3.53	230	7	1.5	1.76	93
	T300J	3,6,12	4.21	230	7.0	1.8	1.78	94
	T400H	3,6	4.41	250	7	1.8	1.80	94
	T600S	24	4.31	230		1.9	1.79	
	T700S	12,24	4.90	230	7	2.1	1.80	93
	T700G	12,24	4.90	240		2.1	1.80	
	T800H	6,12	5.49	294	5	1.9	1.81	96
	T1000G	12	6.37	294	5	2.2	1.80	
	M35J	6,12	4.70	343	6	1.4	1.75	99
	M40J	6,12	4.41	377	5	1.2	1.77	99
	M46J	6,12	4.21	436	5	1.0	1.84	99
	M50J	6	4.12	475	5	0.8	1.88	99
	M55J	6	4.02	540	5	0.8	1.91	99
	M60J	3,6	3.82	588	4.7	0.7	1.94	99
	M30S	18	5.49	294	6.5	1.9	1.73	98
	M40	1,3,6,12	2.74	392	6.5	0.7	1.81	99
Zoltek	Panex 33	48,160,320	3.80	228	7.2	1.6	1.81	94

*TS & YM results are based on the tow test.
Source: Reprinted from manufacturer's technical literature.

reduced number of flaws in a smaller unit volume. However, Jones and Duncan [4] noted that a smaller diameter fiber would have a higher proportion of sheath material than a fiber with larger diameter and the more graphitic nature of this outer sheath would suggest higher mechanical properties.

It might be expected that the properties of carbon fiber could approach those of the graphite single crystal, where the YM is of the order of 1000 GPa and

PROPERTIES OF CARBON FIBERS

Table 20.4 Properties of pitch based carbon fibers

Company	Trade name	Tow size k	Filament properties					Minimum carbon content
			TS GPa	YM GPa	Diameter μm	Elongation %	Density g cm^{-3} %	
Amoco	Thornel P25	2,4	1.38	159	11	0.90	1.90	
	P55S	2,4	1.90	379	10	0.50	2.00	97
	P75	2	2.10	517	10	0.40	2.00	99
	P100	2	2.41	758	10	0.32	2.16	99
	P100S	2	2.07	758	10	0.27	2.16	99
	P100HTS	2	3.62	724	10	0.50	2.17	99
	P120	2	2.41	827	10	0.29	2.17	99
	P120S	2	2.24	827	10	0.29	2.17	99
	K1100X	2	3.10	966	10	0.30	2.20	
Ashland Oil Mitsubishi Kasei	Dialead K133	4	2.35	441		0.53	2.08	
	K135	2,4	2.55	539		0.47	2.1	
	K137	4	2.65	637		0.42	2.12	
	K139	2	2.75	735	10	0.37	2.14	
	K223	4	2.84	225	10	1.21	2.00	
	K321	2,4	1.96	176		1.08	1.90	
Nippon Graphite Fiber Corp.	Granoc XN50A[1]	0.5,2	3.83	520	10	0.7	2.14	
	XN70A	0.5,1,2	3.63	720	10	0.5	2.16	
	XN80A	1,2	3.63	785	10	0.5	2.17	
	XN85A	1,2	3.63	830	8.5	0.4	2.17	
	YS50A[2]	4.5	3.83	520	7	0.7	2.14	
	YS70A	4.5	3.63	720	7	0.5	2.16	
	YS90A	3	3.63	880	7	0.4	2.19	
	YS50	3,4.5	3.73	490	7	0.8	2.09	
	YS60	3,4.5	3.53	590	7	0.6	2.12	
	YS70	3,4.5	3.53	690	7	0.5	2.14	
	YS80	3,4.5	3.53	785	7	0.5	2.15	
	YT-50-10S	1	4.05	490	6	0.8		
Osaka Gas	Donacarbo	2	3.00	500	9	0.6	2.10	
Petoca	Carbonic HM50	2	2.75	490	10	0.56	2.16	
	HM60	2	2.94	588	10	0.50	2.17	
	HM70	1,2	2.94	686	10	0.43	2.18	
Tonen Corp.	Forca FT500	3	3.00	500	10	0.6	2.14	
	FT700	3	3.30	700	10	0.5	2.16	

Notes: 1. XN grades are based on petroleum pitch.
2. YS grades are based on coal tar pitch.
Source: Reprinted from manufacturer's technical literature.

the theoretical strength would be expected to about one-tenth of the YM, i.e. 100 GPa (Table 20.5). The strengths are, however, well below this theoretical figure, but the YM of high modulus pitch based fiber types can approach 1000 GPa (e.g. Du Pont when in production of a mesophase pitch based fiber obtained a value of 894 GPa for their E-130 grade). Since a PAN based fiber is not graphitic, its modulus will be lower at a given production temperature as compared to a pitch based fiber, but theoretical strengths are higher.

A graphitizable carbon (pitch based carbon fiber) undergoes progressive graphitization in the range 1600°C–2800°C, with an increasing three-dimensional order, whereas a

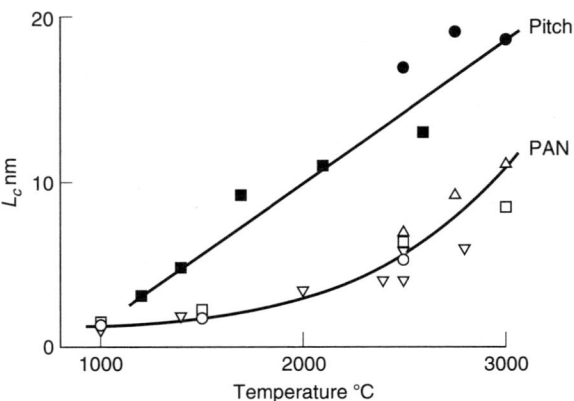

Figure 20.1 Plot of L_C against heat treatment temperature for pitch-based and PAN-based carbon fibers. *Source:* Reprinted with permission from Takaku A, Shioya M, J Mater Sci, 25, 4873, 1990. Copyright 1990, Springer.

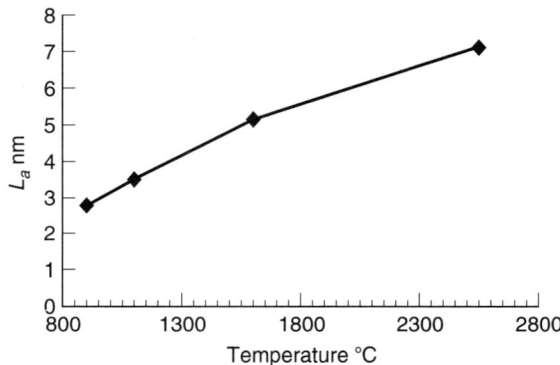

Figure 20.2 Relationship of L_a with temperature. *Source:* By kind permission of Acordis UK Ltd (formerly Courtaulds PLC).

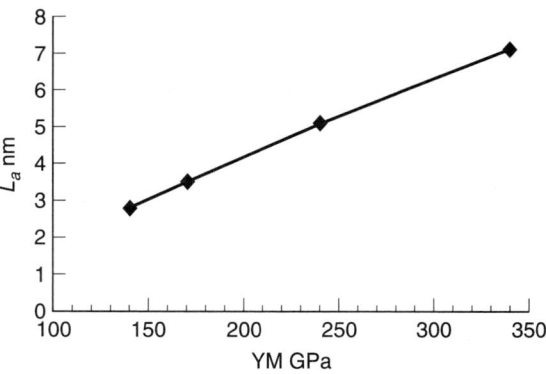

Figure 20.3 Relationship of L_a with Young's modulus. *Source:* By kind permission of Acordis UK Ltd (formerly Courtaulds PLC).

non-graphitizing carbon (PAN based carbon fiber) is never fully converted to graphite, even after heating for many hours at 2000°C. The distance between the layer planes in the crystal structure of a true graphite is 0.3354 nm, but in a turbostratic form of carbon, the distance is always greater than crystal graphite due to the presence of sp^3 bonds.

PROPERTIES OF CARBON FIBERS

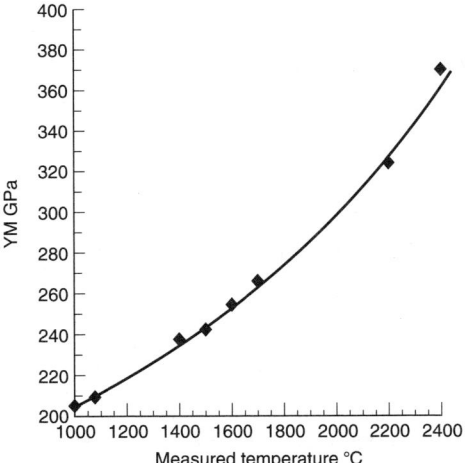

Figure 20.4 Young's modulus of a PAN based carbon fiber at given measured temperatures. *Source:* By kind permission of Acordis UK Ltd (formerly Courtaulds PLC).

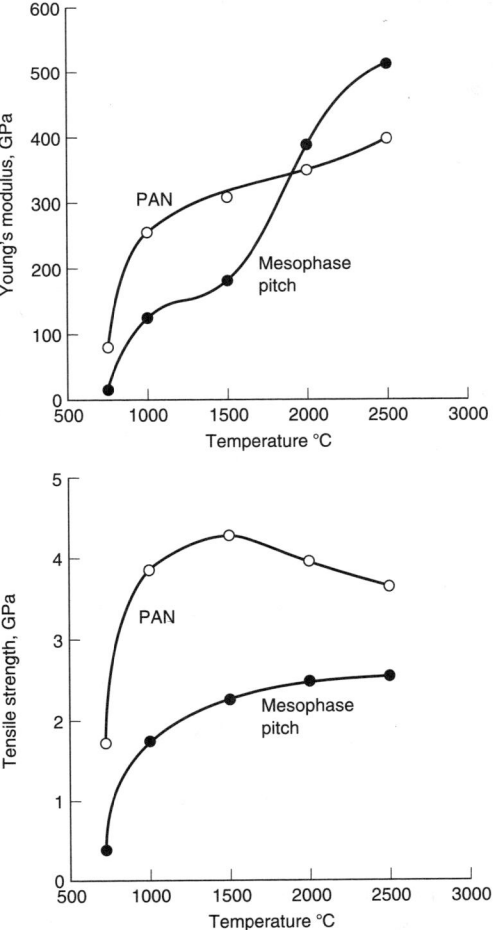

Figure 20.5 Tensile strength and Young's modulus for PAN based and mesophase pitch based carbon fibers *vs* final heat treatment temperature. *Source:* Reprinted from Matsumoto T, *Pure Appl Chem*, 57, 1553, 1985.

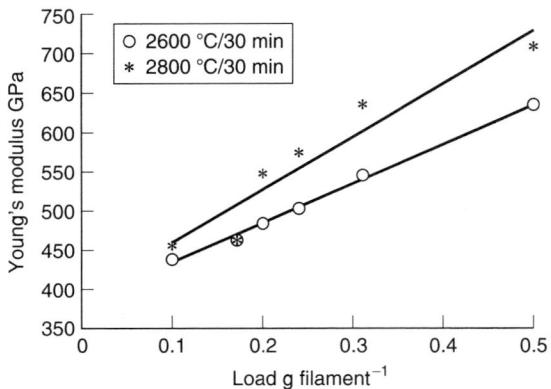

Figure 20.6 The effect of load on Young's modulus of fibers stretched for 30 min at 2600°C and 2800°C. *Source:* Reprinted with permission from Ozbek S, Isaac DH, *Proc Manuf Composite Mater, ASME*, 307–320, 1991. Copyright 1991, ASME.

Table 20.5 Comparison of the values for tensile modulus and strength for a graphite single crystal and PAN and pitch based carbon fibers

	Theoretical value for graphite single crystal	Type HT PAN based carbon fiber	Type IM PAN based carbon fiber	Type HM PAN based carbon fiber	Type UHM Pitch based carbon fiber
Young's modulus (E)	1000 GPa	250 GPa	290 GPa	350 GPa	735 GPa
Tensile strength (σ)					
a) $\sigma_{theoretical} = E/10$	100 GPa	25 GPa	29 GPa	35 GPa	73.5 GPa
b) σ typical value		3.5 GPa (14%)	6.1 GPa (21%)	4.5 GPa (13%)	2.8 GPa (4%)

Source: Reprinted with permission from Fitzer E, Kunkele F, *High Temperatures-High Pressures*, 22(3), 239–266, 1990. Copyright 1990, Pion Ltd.

A comparison of the mechanical properties of some carbon fibers with other reinforcing fibers is given in Table 20.6. By plotting the strength against the modulus, the breaking strain is a useful parameter to illustrate differences in their properties (Figure 20.7). The mechanical properties of PAN and mesophase pitch based carbon fibers are given in Table 20.7.

20.3 TENSILE PROPERTIES

Testing has been covered in Chapter 17 and it is most important to compare the test results of like with like. Cost and time taken will always be a consideration when testing for routine control purposes and for more meaningful results, it may be necessary to undertake expensive extended composite testing.

Initially, filament testing was used, but this demanded considerable operator skill, with ability to handle fine filaments and required many tests to get a reasonable statistical result, which tended to be operator dependent. However, it did produce relatively speedy results, ideal for research purposes. Tow testing subsequently became the preferred test procedure and Table 20.8 shows a comparison of the test procedures for very early Courtaulds fiber.

It is not surprising that the single filament test gives the highest value for modulus, since it is a single filament correctly aligned, whereas with a tow test, it is not possible to align

Table 20.6 A comparison of the mechanical properties of reinforcing fibers

Class GPa	Manufacturer GPa	Type g cm^{-3}	Strength GPa	Modulus GPa	Density	Specific strength	Specific modulus
Carbon fibers	Toray	T300 (SM)	3.53	230	1.76	2.01	131
	Toray	T1000 (HT)	7.06	294	1.82	3.88	162
	Hercules	IM8 (IM)	5.45	303	1.8	3.03	168
	Hercules	UHM (HM)	3.45	441	1.87	1.85	236
	Amoco	P75 (UHM)	2.10	517	2.00	1.05	258
	Mitsubishi	K137 (UHM)	2.65	637	2.12	1.25	300
	Amoco	P120 (UHM)	2.41	827	2.17	1.11	381
	Amoco	K1000X (UHM)	3.10	966	2.20	1.41	439
Boron fibers	Avco	102 μm on a tungsten core	3.52	400	2.57	1.37	156
Aramid fibers	DuPont	KEVLAR 29 (Meta)	3.62	58	1.44	2.51	40
	Akzo nobel	TWARON 80 (Para)	2.80	80	1.44	1.94	56
	DuPont	KEVLAR 49 (Meta)	2.76	120	1.45	1.90	83
	DuPont	KEVLAR Hm (149) (Meta)	2.40	160	1.47	1.63	109
Ceramic fibers							
Alumina	ICI	SAFFIMAX SD	2	300	3.30	0.61	91
Alumina-boria-silica	3M Co.	NEXTEL 440	2	186	3.00	0.67	62
Alumina-silica	Thermal ceramics	KAOWOOL	1.4	120	2.56	0.55	47
Silicon-carbide	Nippon carbon	Nical;on	2.7	193	2.55	1.06	76
Glass & quartz							
E-glass	Vetrotex	E Glass	3.4	73	2.60	1.31	28
R-glass	Owens corning	R Glass	4.4	86	2.55	1.73	34
S-glass	Owens corning	S 2	4.5	86	2.49	1.81	35
Quartz	Quartz Et Silice	99.99% SiO_2	3.6	69	2.20	1.64	31
Metal fibers							
Aluminum		7075-T6 Alloy	0.50	71	2.00	0.25	36
Titanium		6AL-4V Alloy	1.14	110	4.46	0.26	25
Steel		A663 (Hot rolled)	0.40	207	7.86	0.05	26
HSLA steel		High strength, low alloy steel	0.80	207	7.86	0.10	26
Polymer fibers							
Polyamide-imide	Rhone-Poulenc	KERMEL 235	2.5 N tex^{-1}	—	1.34	—	—
PBI	Hoechst	PBI	0.38	5.5	1.43	0.27	4
PE	Allied Signal	SPECTRA 1000	3.09	172	0.97	3.18	177

(Continued)

Table 20.6 Continued

Class GPa	Manufacturer GPa	Type g cm^{-3}	Strength GPa	Modulus GPa	Density	Specific strength	Specific modulus
PEI	Teijin	PEI	2.8 g denier^{-1}	–	1.27	–	–
PEEK	Nitto Boseki	TEXXES	1.5–3.2 g denier^{-1}	–	1.30	–	–
PPS	Shakespeare	MX 505	0.43	21.8	[1.35]	[0.32]	[16]
Whiskers							
Boron carbide	Third Millenium	B4C	–	–	2.52	–	–
Potassium titanate	Otsuka	TISMO	7	280	3.30	2.12	85
Silicon carbide	Tokai	TOKAWHISKER	–	–	3.20	–	–
Silicon nitride	Tateho	97% Purity	1.4	387	3.18	0.44	122
Titanium nitride	Third Millenium	90% Purity	–	–	5.40	–	–
Graphite		Graphite whisker	20	680	2.26	8.85	301

PROPERTIES OF CARBON FIBERS

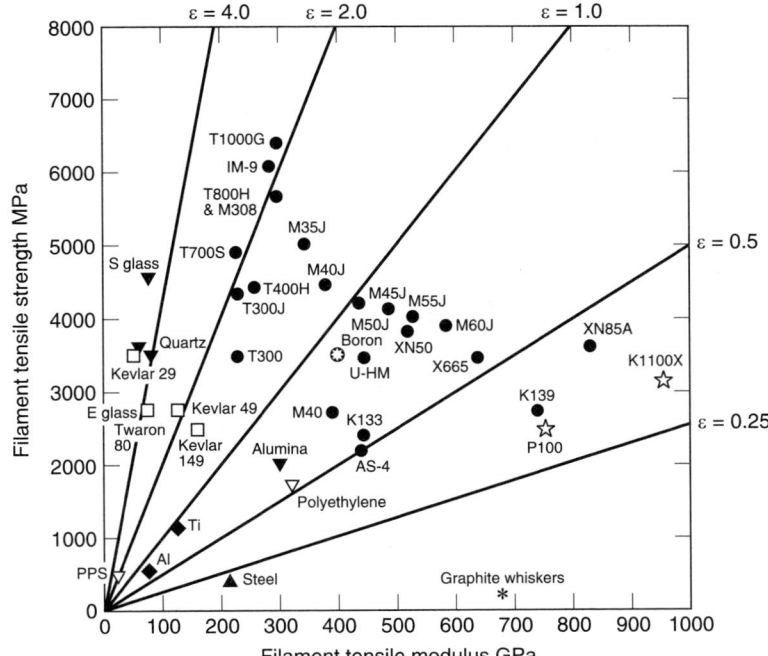

Figure 20.7 Fiber strength as a function of the modulus for carbon, ceramic, metal, polymer fibers and graphite whisker. • Carbon fiber, ☆ Cellulose based carbon fiber, □ Aramid fiber, ▼ Ceramic and glass fibers, ◆ Metal fiber, ∗ Graphite whiskers, ▽ Organic polymer fiber.

all the filaments within the tow and even less so, when producing a composite bar, although filament winding does help.

Kowalski [7] contradicted these results and obtained a lower modulus for single filaments than for strand or composite tests (Table 20.9), but does say that the lower filament moduli may be due to not taking into account the instrument compliance.

It is interesting that when multiple $3k$ tows were tow tested, there was a definite trend for the modulus to fall as the number of $3k$ tows was increased (Table 20.10), again reflecting the degree of filament alignment within the tow(s).

A further comparison of tow and composite test results for early Grafil E/XAS fiber is given in Table 20.11 for different tow sizes and in each case, the strand tensile moduli are higher than the corresponding normalized composite moduli.

Flaws within and on a filament greatly influence the tensile strength and as the filament gage length is decreased, the resulting number of flaws within the gage length is reduced and therefore, the strength increases (Figure 20.8). The effect of gage length on testing PAN based carbon fibers has been reported by Moreton [10] and Barry [11]; for pitch fibers by Bacon and Schalamon [12]; and for cellulose based carbon fibers by Chwastiak et al [13]. Westbury and Drzal [14] considered that the minimum practical gage length was 5 mm, but in a composite, the minimum length (critical length) which can sustain a load is about 3 mm [15].

Westbury and Drzal used the single fiber fragmentation test to measure tensile strength at short gage lengths, embedding the filament in polycarbonate resin using solvent casting to avoid any residual strain. Figure 20.9 shows results obtained with Hercules AS-4 fiber embedded in polycarbonate and epoxy matrices showing good agreement of the standard tensile test with the strain free polycarbonate, but a displaced curve of the same slope due to the filaments being pre-tensioned.

Table 20.7 Properties of unidirectional carbon fiber composites

Units	Property	CourtauldsXA	Hercules IM-7	Hercules AS-4	Hercules AS-4	Hercules AS-4	Toray T300	Toray T300	Toray T300	Toray T300	Toray T800	Toray T800H	Toray M46J	Toray M46J	Toray M60J	Toray M60J
Resin		1	2	3	4	5	6	7	8	8	7	8	7	8	7	8
%	v/o	60	60	60	60	62	62	60	60	60	60	60	60	60	60	60
Mg m^{-3}	ρ	1.57	1.54	1.56	1.56	1.66	1.57	1.54	1.54	1.54	1.57	1.57	1.59	1.59	1.65	1.65
Gpa	E_{lt}	138	175	148	148	148	133	125–140	125	125	150–170	160	245–260	260	310	330
Mpa	σ_{lt}	2000	2924	2137	2137	2137	1510	1760–1800	1760	1760	2840–2900	2840	2200–2500	1960	1960	1760
%	ε_{lt}						1.13	1.3	1.3	1.3	1.6	1.55	0.9	0.7	0.6	0.5
GPa	E_{tt}		10				9	8.8–9.0	7.8	7.8	8.8–9.0	7.8	6.9–7.5	6.9	5.9	5.9
MPa	σ_{tt}		71				34	80	80	80	65–70	80	45	45	30	30
%	ε_{tt}						0.358	1.0	1.0	1.0	0.8–0.9	1.0	0.6		0.6	
GPa	E_{lf}	126		134	128	127		120			150		235	245		320
MPa	σ_{lf}	180		1724	1793	1793		1700			1800		1450			
GPa	E_{lc}	126	172				130	125	125			145	230			
MPa	σ_{lc}	1350	1600				1435	1370–1400	1570	1570	1570–1600	1570	1030–1050	880	880	780
%	ε_{lc}						1.19		1.2							
	v_{lt}	0.25					0.31	0.34	0.34	0.34		0.34				
	v_{tl}	0.02					0.027									
GPa	G_{lt}								4.4	4.4		4.5		3.9		3.9
MPa	τ_{lt}								98	98		98		59		39
MPa	ILSS	100	112	127	127	120		100	110	100	100	100	90	80	70	70

Note: Resin systems used were:
1. Ciba Geigy 913; 120°C cure epoxy
2. Courtaulds toughened epoxy; 175°C cure
3. Hercules epoxy high temperature performance
4. Hercules epoxy high temperature performance; T_g 190°C
5. Hercules epoxy high temperature performance; T_g 215°C.
6. US Polymeric bismaleimide resin
7. Toray epoxy 2500; 121°C cure
8. Toray epoxy 3631; 177°C cure, semi-toughened

Source: Reprinted with permission from Hancox NL, Mayer RM, Design Data for Reinforced Plastics—A Guide for Engineers and Designers, Chapman & Hall, London, 118, 1994. Copyright 1994, Springer.

PROPERTIES OF CARBON FIBERS

Table 20.8 Comparison of filament, tow and composite tests

Test	Property	Grafil HM-S staple	Grafil HT-S staple	Grafil A staple	Grafil A continuous
Single Filament	Tensile strength	2.23 (3.4)	2.61 (4.0)	2.32 (6.9)	2.43 (6.3)
[Mean of 10]	Tensile modulus	395 (4.1)	268 (3.4)	231 (7.0)	243 (5.5)
Tow	Tensile strength	2.09(9.6)	2.59 (8.0)	2.18 (8.2)	2.49 (8.2)
[Mean of 3]	Tensile modulus	382 (4.3)	248 (3.1)	198 (6.0)	190 (3.4)
Composite (normalized)	Tensile strength	1.94 (10.7)	2.38 (6.0)	1.37 (6.7)	2.49 (11.9)
(828/MNA/BDMA)	Tensile modulus	346 (4.7)	225 (6.2)	179 (2.9)	183 (4.0)
[Mean of 3]	Flexural strength	1.48 (6.5)	2.40 (5.5)	1.87 (6.1)	2.45 (5.4)
	Flexural modulus	315 (5.0)	222 (3.8)	185 (2.1)	177 (3.2)

Note: Coefficients of variation quoted in brackets.
Source: Reprinted from Courtaulds technical data.

Table 20.9 Non-linearity by least squares analysis of stress strain data for single filaments, tows and unidirectional composites under the conditions of zero stress at zero strain

Test	Property	Amoco T-40	Toray T-700
Single Filament	Initial tangent modulus E_0 (GPa)	259 (4.2)	235 (5.5)
	Secant modulus E_{sec} (GPa)	271	245
	No. of tests	25	25
Tow	Initial tangent modulus E_0 (GPa)	267 (1.4)	243 (1.3)
	Secant modulus E_{sec} (GPa)	281	256
	No. of tests	13	19
Composite (normalized)	Initial tangent modulus E_0 (GPa)	263 (0.9)	237 (1.0)
(828/MNA/BDMA)	Secant modulus E_{sec} (GPa)	280	253
	No. of tests	32	12

Note: Coefficients of variation quoted in brackets.
E_{sec} is the secant modulus taken between strain of 0.1 and 0.6.
Source: Reprinted with permission from Kowalski IM; Whitcombe JD ed., *ASTM STP972*, ASTM, Philadelphia, 1988. Copyright 1991, ASTM International.

Table 20.10 Testing multiple $3k$ Grafil XA-S tows

	$2 \times 3k$ XA-S fiber tows	$3 \times 3k$ XA-S fiber tows	$4 \times 3k$ XA-S fiber tows
Tow test tensile strength (GPa) ($n=8$)	3.70 (1.8)	3.54 (4.4)	3.63 (4.8)
Tow test tensile modulus ($n=8$)	244 (1.9)	236 (1.2)	227 (5.0)

Source: Reprinted from Courtaulds technical data.

Table 20.11 Early quality control data summary of strand test and composite test for Grafil E/XA-S fiber

Test Property	Tow Test						Composite Test					
	Tensile Strength GPa			Tensile Modulus GPa			Tensile Strength GPa			Tensile Modulus GPa		
Tow size	$10k$	$12k$	$3k$	$10k$	$12k$	$3k$	$10k$	$12k$	$3k$	$10k$	$12k$	$3k$
Decitex	1.35	1.215	1.215	1.35	1.215	1.215	1.35	1.215	1.215	1.35	1.215	1.215
Mean value	3.21	3.33	3.27	233	238	235	3.25	3.43	3.15	215	216	216
CV (%) within batch	9.3	3.1	8.8	2.8	0.5	2.3	5.4	7.9	5.4	1.9	2.5	2.4
Number of fiber batches	77	5	16	77	5	16	22		11	22		11

Source: Reprinted from Courtaulds technical data.

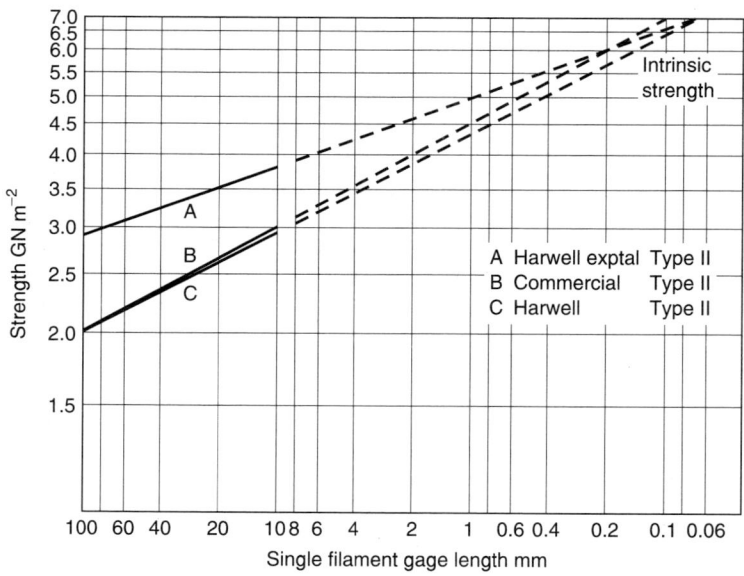

Figure 20.8 Tensile strength Vs gage length of Type II (HT) carbon fibers. *Source:* Reprinted with permission from Goggin PR, A method of measuring the quality of carbon fibres, *AERE R-7790*. Copyright, AEA Technology plc.

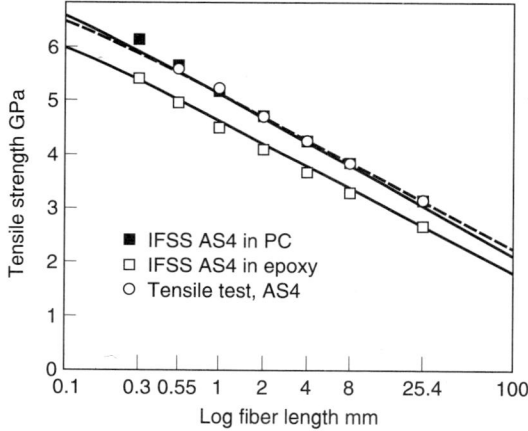

Figure 20.9 Fiber tensile strength Vs long gage length for single filament tests in polycarbonate and epoxy matrices. *Source:* Reprinted with permission from Westbury MC, Drzal LT, *J Composite Technol Res*, 13(1), 22–28, 1991. Copyright 1991, ASTM International.

Strictly, the stress-strain curve is not linear and the modulus does tend to increase as the strain becomes larger, an effect termed strain stiffening. This effect dictates that the strain area selected for the modulus determination must be clearly stated; otherwise the determined modulus of an identical fiber can be different (Figure 20.10).

The variation of tensile strength occurs both along the length of the tow as well as tows sampled across the fiber band of a batch. The Weibull distribution can be used to describe this distribution [16] as:

$$F = 1 - \exp\left\{-\left(\frac{L}{L_0}\right)\left(\frac{\sigma - \sigma_u}{\sigma_0}\right)^m\right\}$$

PROPERTIES OF CARBON FIBERS

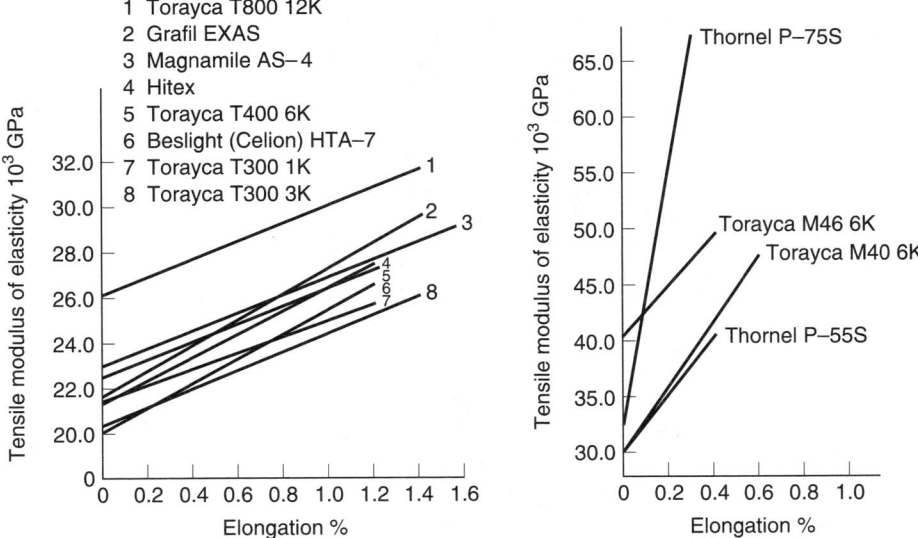

Figure 20.10 Strain stiffening. *Source:* Reprinted from Toray technical literature.

where L = gage length
L_0 = standard gage length
m = Weibull modulus
σ = tensile strength
σ_0 = a scaling parameter
σ_u = an arbitrary parameter normally set to 0
F = cumulative probability of failure

When there are a large number of data points, this equation can be written in the form:

$$\ln\ln\left(\frac{1}{1-F}\right) = m\ln[\sigma - \sigma_u] + m\ln\left[\frac{\left(\frac{L}{L_o}\right)^{\frac{1}{m}}}{\sigma_0}\right]$$

and m and σ_0 can be determined from the straight line plot.

Hitchon and Phillips [17] assumed a bimodal distribution of flaw sizes and if the data was not linear, then Own *et al* [18] suggested considering a bimodal log-normal density distribution which is given by:

$$W = \frac{1}{(2\pi)^{1/2}\sigma_1\sigma}\exp\left(-\frac{[\ln(\sigma) - \mu]^2}{2\sigma_1^2}\right)$$

where σ_1 = log-normal standard distribution
μ = log-normal mean tensile strength

Owen *et al* concluded that the internal flaws controlled the higher strength distribution, whereas the surface flaws controlled the lower strength distribution.

Recently, filament testing was undertaken by Dobb *et al* [19] for tensile and compressive filament testing (Table 20.12).

Table 20.12 Mechanical properties of various carbon fibers (single filament tests)

Fiber Type	Manufacturer	Diameter μm	Tensile strength GPa	Tensile modulus GPa	Tensile strain %	Compressive strength GPa	Compressive strain %
PAN	Grafil HMS	7.6 (0.4)	2.1 (0.5)*	320 (40)	0.63 (0.08)	0.8	0.26
Toho	S-g (C$_2$)	6.8 (0.2)	3.0 (0.7)	220 (20)	1.35 (0.26)	1.5	0.69
(X)		4.7 (0.1)	3.4 (0.8)	370 (45)	0.93 (0.18)	1.4	0.38	
(Y)		4.7 (0.3)	3.1 (0.4)	330 (40)	0.96 (0.11)	1.3	0.40	
(Z)		4.5 (0.1)	3.5 (0.6)	280 (30)	1.25 (0.20)	1.9	0.67	
Toray T1000		5.2 (0.1)	5.7 (1.1)	255 (20)	1.87 (0.34)	2.2	0.87	
MP	Thornel P-25	8.6 (1.0)	1.4$^\otimes$	140	0.9	0.9	0.64
	P-55	10.3 (0.6)	2.1	380	0.5	0.5	0.13
	P-75	9.7 (0.7)	2.0	500	0.5	0.5	0.11
	P-100	9.8 (0.9)	2.2	690	0.4	0.4	0.06
	P-120	10.6 (1.3)	2.2	820	0.3	0.3	0.04

Note: Standard deviations are in parenthesis.
*Figures are calculated by assuming that the Young's moduli for tension and compression are equal.
⊗The tensile properties for MP based fibers are manufacturer's data.
Source: Reprinted from Dobb MG, Guo H, Johnson DJ, Park CR, Structure–compressional property relations in carbon fibres, *Carbon*, 33(11), 1553–1559, 1995.

20.4 FACTORS EFFECTING COMPOSITE STRENGTH

Johnson [20] has suggested that the tensile failure of carbon fibers is due to the presence of defects and the structure suggests that anisotropic effects may be ignored and the size of the basal plane cracks estimated from a simple Griffith relation:

$$c = \frac{\gamma}{\pi} \times \frac{2E}{\sigma^2}$$

where γ = surface free energy
c = size of the basal plane cracks

Cottrell [21] has estimated the value of the surface free energy as 4.2 J m^{-2}. Hence Reynolds [22] has drawn curves of defect size *vs* fiber strength (Figure 20.11) for Type I (HM) and Type II (HT) fibers, indicating the size of the defect which has to be eliminated to obtain a given strength.

Goggin [9] examined the strengths of three samples of Type II carbon fiber produced from the same feedstock and showed that they converged to an intrinsic fiber strength of 6.5–7.0 GPa at 0.1 mm gage length (Figure 20.8).

Coleman [23] showed that the strength of a tow of fibers will be less than the mean fiber strength as determined by the number of low strength fibers, effectively reducing the tow cross-section and is related to the strength dispersion (Figure 20.12),

$$\left(\frac{\text{standard deviation}}{\text{mean filament strength}}\right)$$

which was typically 0.2 at a 5 mm gage length, suggesting a maximum tow strength of some 70% of the mean filament strength at this gage length. By applying correction factors to the figures in Figure 20.8, Hughes and Jackson [24] predicted the intrinsic tow strength at this gage length to be about 6.5 GPa, as shown in Figure 20.13. Hughes *et al* [25] used a

PROPERTIES OF CARBON FIBERS

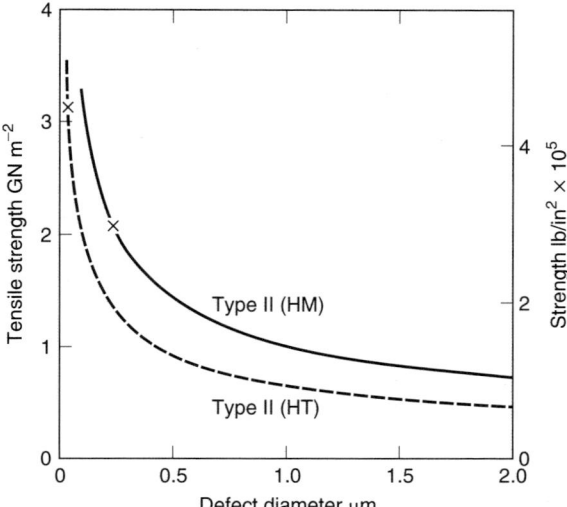

Figure 20.11 Theoretical strength Vs defect size for Type I (HM) and Type II (HT) carbon fibers. *Source:* Reprinted with permission from Reynolds WN, *Proc 3rd Conference Industrial Carbons and Graphites*, SCI, London, 427–430, 1971. Copyright 1971, The Society of Chemical Industry.

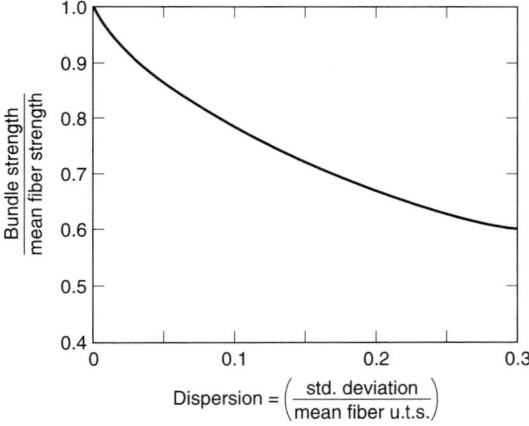

Figure 20.12 Tow strength as a function of strength dispersion. *Source:* Reprinted with permission from Coleman BD, On the strength of classical fibres and bundles, *JMPS*, 7, 60, 1958. Copyright 1958, Society of Manufacturing Engineers.

gage length of 3 mm and obtained an estimated strength up to 5 GPa, whilst Jones and Johnson [26] used a loop method and obtained an estimated strength of 7 GPa and Tsushima [27] used a knotted loop method and reported 6 GPa.

There are three factors which are particularly important in controlling the strength of composites, apart from the basic fiber and matrix properties namely:

1. Critical length (the minimum length of fiber embedded in the resin to avoid pull-out when an axial load is applied, calculated to be about 0.1 mm)
2. The fiber volume fraction
3. The alignment of the fiber

The tensile properties of carbon fibers have been investigated [28–32]. Tibbets *et al* [33] determined the properties of VGCF.

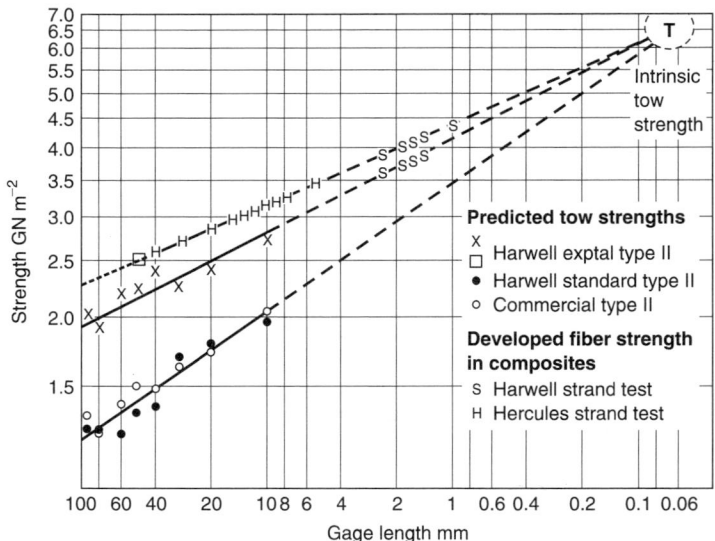

Figure 20.13 Predicted tow strengths and effective gage lengths in composites. *Source:* Reprinted with permission from Hughes JDH, Jackson EE, The potential strength of Type II carbon fibre composites, A practical assessment, *AERE-R8090*, Sep 1975. Copyright 1975, AEA Technology plc.

20.5 THE IMPORTANCE OF CRITICAL ASPECT RATIO

The improvement in mechanical properties of a fiber filled thermoplastic composite is due to the ability of the fiber to withstand a higher proportion of the mechanical load than the matrix that it replaces. A critical aspect ratio (length to diameter ratio) has been identified for adequate load transference [34]. This critical aspect ratio relates to a single fiber embedded in, and completely wet by, the matrix. Fibers shorter than the critical aspect ratio will pull out of the matrix, therefore the full reinforcing effect of the fiber will not be utilized. At the critical aspect ratio, both the fiber and the matrix will fracture along the same failure plane. Further increases in fiber aspect ratio will not lend additional strength to the composite, since the failure will be the same as that experienced at the critical aspect ratio. The critical ratio is given by:

$$\left(\frac{L}{D}\right)_C = \frac{\sigma_f}{2\tau_1}$$

where $(L/D)_C$ represents an ideal value

σ_f = tensile strength of fiber

τ_1 = interfacial shear strength between the fiber and the matrix (strongly influenced by processing conditions and degree of adhesion between the fiber and the matrix)

In actual practice, the fibers must overlap each other, even if all are oriented in the same direction, in order to provide homogeneity within the part. At any given failure plane, all the fibers will then be able to share the mechanical stress equally. In practice, maximum utilization of short fibers occurs when the fibers have an aspect ratio that is ten times the critical aspect ratio [35]. Under these conditions, uniaxially oriented short fibers can have tensile strengths equal to 95% of that of the continuous fiber uniaxially oriented composites. In randomly oriented fiber filled composites, the tensile strength is less than one third that of composites in which fibers are uniaxially oriented.

PROPERTIES OF CARBON FIBERS

In actual practice, the fibers are not always well bonded to the matrix, further reducing the effectiveness of the reinforcement. A strong fiber-matrix bond increases τ_1 and reduces $(L/D)_C$. During the compounding and processing of the fiber filled composites, considerable fiber breakage can occur; hence it is important to minimize fiber breakage in order to maintain maximum utility of the fibers. Fiber fragments below $(L/D)_C$ weaken the structure just as the incorporation of a filler would. Fragments between $(L/D)_C$ and $10(L/D)_C$ contribute fractionally to improving the strength of the matrix.

An equation predicting the tensile strength of short fiber filled composites is:

$$\sigma_c = \sigma_m V_{fm} + V_{ff}\varepsilon_0\varepsilon_1$$

where σ_c = tensile strength of the composite
σ_m = tensile strength of the matrix
σ_f = tensile strength of the fiber
V_{fm} = volume fraction of the matrix
V_{ff} = volume fraction of fibers
ε_0 = efficiency factor accounting for fiber orientation ($1 \geq \varepsilon_0$)

   $\varepsilon_0 = 1$ for uniaxially oriented fibers
   $\varepsilon_0 = 0.33$ for random in-plane oriented fibers
   $\varepsilon_0 = 0.167$ for three dimensional randomly oriented fibers

ε_1 = efficiency factor accounting for partial reinforcement ($0 > \varepsilon_1 > 0.95$)

For many fiber-polymer systems $(L/D)_C$ is in the range 10–50. From this analysis, it is evident that fiber length is important to the development of maximum tensile properties in the composite. It is also apparent that changes in the composite tensile strength are monotonically dependent on fiber concentration.

A typical carbon fiber used for reinforcing thermoplastics is 7 µm diameter and some 6 mm long and gives an L/D of 857. Examination of the carbon fiber after removal of the polymer from the molded composites by digestion with H_2O_2/H_2SO_4 reveals that considerable fiber breakage has occured. The average aspect ratio of the fibers in the molded parts is reduced to approximately 35:1. The critical aspect ratio for carbon fibers in Nylon-6,6 has been established as 12.5:1[35] and compounding and molding conditions strongly influence the degree to which fiber integrity is maintained. Higher viscosity, higher melting polymers exert more viscous force on fibers during compounding and molding than lower viscosity, lower melting polymers.

Bigg [36] concluded that short fibers can be effective reinforcing materials as long as they are wet by the matrix and are longer than 10 times their critical aspect ratio fibers and when below the critical aspect ratio, they behave as fillers. Moreover, there will be no gain exceeding the value of $10(L/D)_C$, except a more conductive electrical network can be produced using longer fibers rather than shorter fibers.

20.6 ELASTIC CONSTANTS

The independent characteristics usually measured are as follows:

E_1 = Longitudinal tensile modulus
E_2 = Transverse tensile modulus
v_{12} = Transverse contraction on uniaxial extension (major Poisson's ratio)
v_{21} = Longitudinal contraction in transverse extension (minor Poisson's ratio)

v_{23} = Thickness contraction on transverse extension
G_{12} = In-plane shear modulus
G_{23} = Through thickness shear modulus

The axial YM is determined by dynamic resonance methods which are at very low strain and the determined values are lower than values measured at higher levels of strain, the reason being the stress strain curve for carbon fiber is not linear.

The representative elastic properties of carbon fiber, E-glass and aramid fibers in unidirectional fiber reinforced epoxy resins are given in Table 20.13.

Elastic constants for carbon fiber as determined by Reynolds are given in Table 20.14, whilst values determined by Goggin are given in Table 20.15.

Wagoner and Bacon [39] determined the properties of a number of composites of surface treated carbon fibers (Table 20.16). Deviations from trends probably reflect that different precursors have different structural features.

The value for the shear modulus g for the shear between the planes oriented normal to the c-axis was calculated by Northolt *et al* [40] from (Table 20.17)

$$\frac{1}{E} = \frac{1}{e_1} + \frac{(\cos^2 \varphi)}{g}$$

The transverse mechanical properties of carbon fibers were measured by Kawabata [41] (Table 20.18).

Table 20.13 Representative elastic properties of unidirectional fiber reinforced epoxy resins

Material	Fiber volume fraction V_f	E_{11} GPa	E_{22} GPa	v_{12}	G_{12} GPa
CFRP (A-S carbon fiber)	0.66	140	8.96	0.30	7.10
CFRP (IM-6 carbon fiber)	0.65	200	11.10	0.32	8.35
GFRP (E-glass fiber)	0.46	35	8.22	0.26	4.10
KFRP (Kevlar-49 fiber)	0.60	76	5.50	0.33	2.35

Table 20.14 Estimated elastic constants for a single fiber of Type I (HM) and for a crystallographically perfectly oriented fiber

Constant		Type I (HM) Fiber	Ideal Fiber
S_{11}	m²TN	35.5	23.2
S_{33}		2.55	0.98
S_{13}		−0.38	−0.2
S_{12}		−7	−17.3
S_{44}		40.4	66.9
$S_{66} = 2(S_{11} - S_{12})$		85	81
C_{11}	GPa	29.3	99
C_{33}		393	1040
C_{13}		5.2	35
C_{12}		5.8	74
C_{44}		24.8	14.9
$C_{66} = \frac{1}{2}(C_{11} - C_{12})$		11.8	12.3

Note: It has been assumed that the ideal fiber contains no free dislocations or porosity. If it is fully graphitized, it may be expected to contain dislocations which increase S_{11}, S_{12}, and S_{44} by a factor ∼3, with corresponding reductions in C_{11}, C_{12} and C_{44}. S_{33}, S_{13}, C_{33} and C_{13} would not be affected in this case.

Source: Reprinted with permission from Reynolds WN, Structure and mechanical properties of carbon fibres, *Proc 3rd Conference Industrial Carbons and Graphites*, SCI, London, 427–430, 1971. Copyright 1971, The Society of Chemical Industry.

Table 20.15 Measured and assumed elastic constants of carbon fibers

	Type I carbon fiber (HM)	Type II carbon fiber (HT)
Engineering constants:		
E_{33} (Longitudinal Young's modulus)*	379 Gpa	228 GPa
E_{11} (Transverse Young's modulus)	27.6 GPa	27.6 GPa
G_{44} (Longitudinal shear modulus)*	26.2 GPa	26.2 GPa
v_{13} (Longitudinal Poisson's ratio)	0.5	0.5
v_{12} (Transverse Poisson's ratio)	0.28	0.28
Compliance moduli $TN^{-1}m^2$		
S_{33}*	2.6	4.7
S_{11}	36.3	36
S_{44}*	38.3	38
S_{12}	−10.2	−10.2
S_{13}	−1.3	−2.4

* Measured values.
Source: Reprinted with permission from Goggin PR, The elastic constants of carbon fibre, J Mater Sci, 8, 233–244, 1973. Copyright 1973, Springer.

Table 20.16 Computed carbon fiber properties of some Amoco carbon fibers based on composite properties

Composite specimen	E_{af} GPa	G_{af} GPa	v_{af}	E_{tf} GPa	G_{tf} GPa	α_{af} $\mu\varepsilon/c$	α_{tf} $\mu\varepsilon/c$
P-55/1962H	305	15.6	0.32	10.8	3.0	−1.37	12.1
P-55/1962L	300	12.9	0.29	10.7	3.2	−1.37	11.6
P-75/1962H	450	13.3	0.24	8.8	2.5	−1.49	12.2
P-75/1962L	449	12.7	0.23	8.8	2.5	−1.44	12.7
P-100/1962H	770	23.6	0.30	7.1	2.1	−1.48	9.4
P-100/1962L	775	20.6	0.22	6.8	2.0	−1.47	9.5
T-50/934H	358	18.1	0.13	10.3	3.3	−1.28	6.6
T-50/934L	354	16.4	0.15	10.2	3.3	−1.17	6.8
T-300X/9B (heat stabilized)	341	19.0	0.13	10.3	3.5	−1.2	7.0
T-300/9B #17	204	22.2	0.26	14.7	5.0	−0.67	8.9
T-300/9B #87	205	23.4	0.27	14.4	4.9	−0.67	8.8
T-650/42/9B	243	23.1	0.29	13.8	5.0	−0.84	7.8

Note: Resin systems used were Fiberite 934, Amoco 9B and ERLX-1962.
Source: Reprinted with permission from Wagoner G, Bacon R, 19th Biennial Conference on Carbon, Penn State University, Session 6A, 296–297, 1989. Copyright 1989, The American Carbon Society.

Table 20.17 Values of crystal YM (e_1) perpendicular to the c-axis, shear modulus and the correlation coefficient (R^2)

Type of fiber	Source	No. of samples	Crystal YM e_i to the c-axis GPa	Shear modulus g GPa	Correlation coefficient R^2
PAN	Northolt	4	820±90	34±7	–
	Calculated	17	611	39	0.90
MP	Northolt	3	870±70	12±2	–
	Calculated	8	776	15	0.88
Rayon	Northolt	45	930±320	14.1±0.8	–
	Calculated	45	900	14.3	0.82

Source: Reprinted with permission from Northolt MG, Veldhuizen LH, Jansen H, Tensile deformation of carbon fibres and the relationship with the modulus for shear between the basal planes, Carbon, 29, 1267, 1991. Copyright 1991, Elsevier.

Table 20.18 Transverse mechanical properties of carbon fibers

Fiber Type		PAN (Toray)				PITCH (UCC)				
Grade		T300	T400	M30	M40	P25	P55	P75	P100	P120
Mean Filament Diameter	μm	6.4	5.6	5.4	6.2	10.7	8.7	9.6	8.8	9.6
Transverse modulus E_T	GPa	6.03	10.08	8.76	7.56	9.95	6.75	4.85	4.07	3.08
Longitudinal modulus E_L	GPa	234.6	309.8	281.1	343.2	126.2	160	167.2	276.9	379.0
E_{LS}	GPa	308.1	383.9	356.0	512.8	190.1	503	528.0	758.0	761.9
Transverse strength σ_{BT}	GPa	2.73	3.34	2.58	0.95	0.64	0.34	0.22	0.13	0.079
σ_{BL}	GPa	3.08	5.19	4.02	3.82	2.45	2.35	4.73	3.24	3.43

Source: Reprinted with permission from Kawabata S, Measurement of the transverse mechanical properties of high performance fibres, *J Text Inst*, 81(4), 432–447, 1990. Copyright 1990, The Textile Institute.

Darby et al [42] investigated the effect of broken or cut plies in a 6k XAS reinforced PES matrix. To calculate the stress distribution of a break in a single carbon fiber under tension, the shear lag method of Cox [43] was used and the ineffective length δ_f of a single broken fiber was obtained from:

$$\frac{\delta_f}{d_f} = \frac{1}{2} \left[\frac{\left(1 - v_f^{\frac{1}{2}}\right)}{v_f^{\frac{1}{2}}} \frac{E_f}{G_m} \right]^{\frac{1}{2}} \cosh^{-1} \left[\frac{1 + (1 - \varphi)^2}{2(1 - \varphi)} \right]$$

where d_f is the fiber diameter G_m is the shear modulus of the matrix, E_f is the Young's modulus of the fiber, V_f is the volume fraction of the fiber, and φ is some fraction of the maximum fiber stress. The value of φ is somewhat arbitrary, but a value of 0.9 is often used and is the value chosen. For the system 6k XAS in a PES matrix, the appropriate constants are:

$d_f = 6.8 \times 10^{-3}$ mm
$G_m = 0.93$ GPa
$E_f = 230$ GPa
$V_f = 0.5$

giving the ineffective length of a single broken fiber as:

$\delta_f = 11.65$ and $d_f = 0.079$ mm

The work showed that when comparing the ineffective length of a discontinuous ply with that of a cut single fiber, for a fixed volume fraction and a fiber of 6×10^{-3} mm diameter, $\delta_f = 0.08$ mm and the value for a cut ply is an order of magnitude larger than the ineffective length of a single fiber. Hence, it can be assumed that continuous plies uniformly carry the load and failure occurs when their tensile strength is exceeded, but the value is lower than the average stress carried by continuous plies, suggesting that there is a stress concentration near the defect.

20.7 FLEXURAL PROPERTIES

Wisnom [44] finds that flexural strengths are higher than either tensile or compressive strengths and there is a significant decrease in strength with specimen size (Figure 20.14). The typical design data for unidirectional carbon fiber epoxy (XA-S/913) is a tensile strength of 2.00 GPa, a compressive strength of 1.07 GPa and a flexural strength of 2.46 GPa. Although it might be expected that the flexural strength would be lower than the tensile or compressive

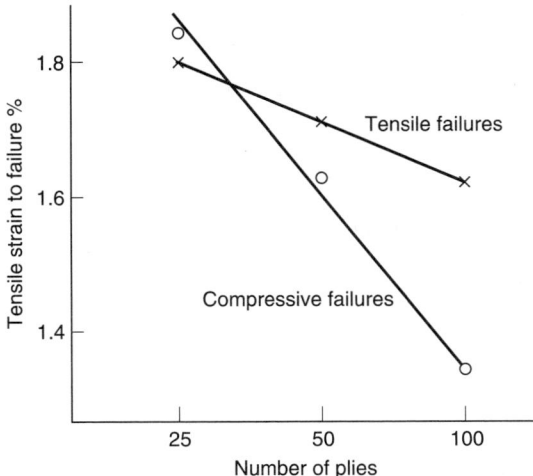

Figure 20.14 Effect of specimen size on flexural strength. *Source:* Reprinted with permission from Wisnom MR, The flexural strength of unidirectional carbon fibre epoxy, *5th European Conference on Composite Materials*, Elsevier, Bordeaux, 165–170, 1992. Copyright 1992, Sage Publications.

strength, it is not and failure usually occurs on the tensile side, despite the supposedly much higher strength in tension than compression. The Weibull theory does predict a higher strength in bending than in tension, but it assumes that failure initiates from a critical defect, whereas many unidirectional composites fail gradually in bending [45].

20.8 EFFECT OF SURFACE TREATMENT AND SIZING ON COMPOSITE PROPERTIES

Carbon fibers are surface treated by the manufacturers using proprietary processes to improve the bond between resin and fiber. For control purposes, it is usual to measure the three point interlaminar shear strength (ILSS) and the choice of resin, volume fraction, span/depth ratio, mode of failure and time of treatment (Table 9.1) are important parameters. As the fiber modulus increases, longer treatment times will be required to obtain the same level of ILSS and overtreatment can result in loss of properties, particularly impact strength (Figure 20.15).

The effect of surface finishes on the properties of carbon fiber composites has been discussed by Wright [47,48] and Peebles [49].

Work by Drzal and co-workers [50] has attempted to determine the interfacial shear strength length of Hercules AU-1 and AU-4 fibers, as well as when surface treated together with the effect of a 100–200 nm coating of DER 330 (DGEBA) epoxy size. A single fiber was encapsulated in an 828/m-PDA epoxy matrix cured for 2 h at 75°C and 2 h at 125°C and subsequently, tension applied. The tensile forces are transferred to the fiber through shear forces at the fiber resin interface and as the maximum strain to failure of the fiber is much less than the matrix, the fiber fractures into small fragments within the matrix. The process is continued until fracture no longer occurrs and the minimum length of the fiber fragments can be obtained (critical shear transfer length), allowing the calculation of the interfacial shear strength.

When dealing with a brittle matrix, the composite failed at the first or second break [51]. The surface treatment process increased the interfacial shear strength by about 60% (Table 20.19) [50]. It was found that the atomic surface concentrations of oxygen did not

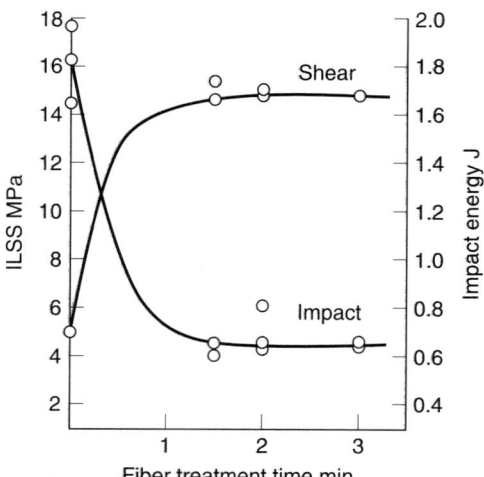

Figure 20.15 Effect of fiber treatment level on shear and impact properties with Fortafil 41 fiber/ERLA 4617-DDM epoxy resin system. *Source:* Reprinted from Goan JC, Martin TW, Prescott R, *28th Ann Tech Conf Reinf Plast/Comp Inst 28*, 21B, 1–4, 1973.

Table 20.19 Interfacial shear strength of Hercules AU-1 and AU-4 PAN carbon fibers in an epoxy matrix

Fiber	Interfacial shear strength MPa
Hercules AU-1	48
Hercules AS-1	74
Hercules AU-4	37
Hercules AS-4	61

Source: Reprinted with permission from Rich MJ, Drzal LT, *J Reinf Plast Composites*, 7(2), 145–154, 1988. Copyright 1988, Sage Publications.

account for the improved shear strength of the AS-1 fiber. The surface oxygen groups only accounted for some 10% of the improvement. They concluded that the finish contained less than the stoichiometric quantity of curing agent, if any, and thereby created a layer with a higher modulus and lower fracture toughness than the bulk resin, hence promoting better stress transfer and consequently, a higher interfacial shear strength. Absorption of water [52] makes the interlayer more compliant than the matrix and acts to reduce the amount of damage occurring at the interface.

A more comprehensive investigation was undertaken by Herrera-Franco and Drzal [53] and the authors suggested that the differences observed in the tests were due to the manner in which each test reacted to the conditions that prevailed at the interface region. The short beam shear is only an estimate of the interlaminar shear, whereas the Iosipescu does produce a state of pure shear (Table 20.20).

Drzal and Madhukar [54] undertook the detailed measurement of the properties of AU-4 (untreated), AS-4 (treated) and AS-4C (treated and sized) carbon fiber/epoxy composites to evaluate the effect of fiber/resin adhesion (Table 20.21).

If the interfacial shear strength is weak, this results in early failure of the composite. Conversely, if the treatment is too strong, then the composite becomes brittle. Increasing the fiber matrix adhesion does enhance the composite compressive strength.

Table 20.20 Summary of interfacial shear strength values

Test	Hercules AU-4 fiber/828/m-PDA MPa	Hercules AS-4 fiber/828/m-PDA MPa	Ratio AS-4/AU-4
Single fiber fragmentation	37.20	68.30	1.8
Microbond	23.44	50.30	2.1
Micro-indentation	55.54	71.53	1.3
$(\pm 45)_{3s}$ tension	37.16	72.19	1.9
Iosipescu shear	54.98	95.28	1.7
Short beam shear	47.52	84.04	1.8

Source: Reprinted with permission from Herrera-Franco PJ, Drzal LT, Comparison of methods for the measurement of fibre/matrix adhesion in composites, *Composites*, 23, 2, 1992. Copyright 1992, Elsevier.

Table 20.21 Property results for carbon fiber/epoxy composites

Test	AU-4 epoxy	AS-4 epoxy	AS-4C epoxy with finish
$[0°]_{12}$ Tensile modulus, E_{11} (GPa)	130±9	138±5	150±9
$[0°]_{12}$ Tensile modulus, σ_1^f (Mpa)	1403±107	1890±143	2044±256
$[0°]_{12}$ Compressive modulus, E_{1C} (GPa)	131±8	126±9	153±8
$[0°]_{12}$ Compressive strength, σ_{1C}^f (MPa)	679±116	911±180	1174±207
$[0°]_{12}$ 3-point flexural modulus, E_{1B} (MPa)	154±6	136±11	147±5
$[0°]_{12}$ 3-point flexural modulus, σ_{1C}^f (GPa)	1662±92	1557±102	1827±52
$[90°]_{12}$ Tensile modulus, E_{22} (GPa)	8.9±6	9.8±0.6	10.3±0.6
$[90°]_{12}$ Tensile strength, σ_2^f (MPa)	18.0±3.9	34.2±6.2	41.2±4.7
$[90°]_{12}$ Flexural modulus, E_{2B} (GPa)	10.2±1.5	9.9±0.5	10.7±0.6
$[90°]_{12}$ Flexural strength, σ_{2B}^f (MPa)	21.4±5.8	50.2±3.4	75.6±14.0
$[\pm 45°]_{3s}$ In-plane shear modulus G_{12} (GPa)	9.1±1.5	6.2±0.5	6.0±0.2
$[\pm 45°]_{3s}$ In-plane shear τ_{12}^f strength (GPa)	37.2±1.8	72.2±12.4	97.5±7.4
Iosipescu in plane shear modulus, G_{12} (GPa)	7.2±0.5	6.4±1.0	7.9±0.4
Iosipescu in plane shear strength, τ_{1C}^f (MPa)	55.0±3.0	95.6±5.1	93.8±3.3
Short beam interlaminar shear strength, τ_{12}^f (MPa)	47.5±5.4	84.0±7.0	93.2±3.8

Source: Reprinted with permission from Drzal LT, Madhukar M, *J Mater Sci*, 28, 569, 1993. Copyright 1993, Springer.

When considering off-axis properties, the fiber matrix adhesion has a greater effect on the transverse flexural strength than the transverse tensile strength.

A number of workers have published results on cfrp using the Iosipescu test [55], 4- point shear test [56], effect of specimen size [57], effect of strain rate [58], in-plane shear test [59], shear strength of pultruded product [60], rail shear test [61] and torsion test [62,63].

20.9 COMPRESSION PROPERTIES

Many workers [64–87] have published results on the compression testing of cfrp. Kumar and Helminiak [88] have reported that the ratios of compressive strength to tensile strength, when evaluated from composite data, can vary in the range 0.2–1.0. It is normal to employ the rule of mixtures and a favourable relationship was reported by Ewins and Ham [89] (Figure 20.16). However, the application of the rule of mixtures for composites does not always give meaningful results and can be overcome by measuring the compressive strength of a single filament embedded in a resin.

Several methods have been used to determine the compression strength of carbon fiber and its composites—The Loop Test [90]; Single Filament in a Beam [91]; Critical Length under Compression [92]; Micro-compression [93,94]; Fiber Recoil [95–97]; Piezo Method [98,99]; Raman Spectroscopy [100–102]; Composite [103,104] and Mini-composite [105].

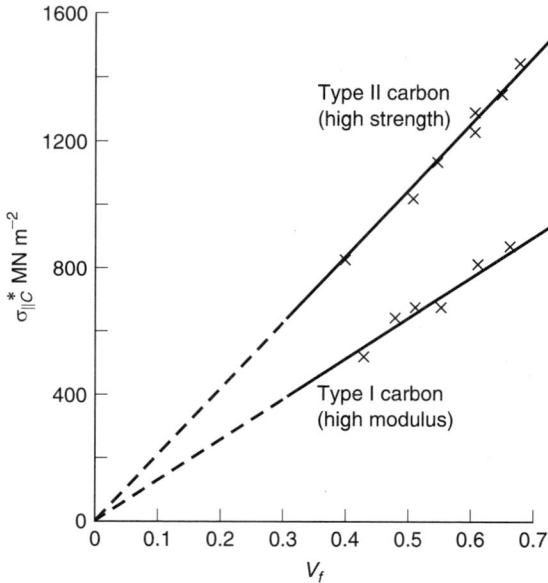

Figure 20.16 Effect of V_f on longitudinal compressive strength of carbon fiber epoxy resin laminate at 20°C. *Source:* Reprinted with permission from Ewins PD, Ham AC, Royal Aircraft Establishment Technical Report 73057, 1973. Copyright 1989, The American Carbon Society.

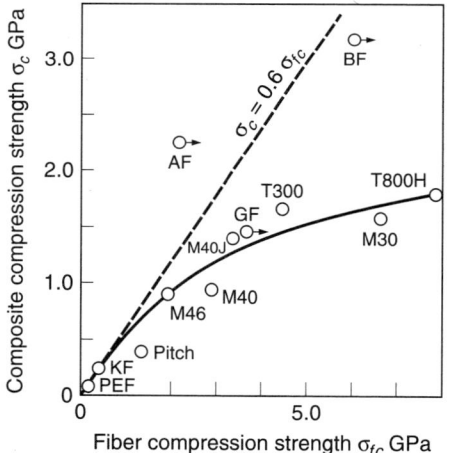

Figure 20.17 Compression strength of single carbon fibers *Vs* composite compression strength. *Source:* Teprinted from Toray technical literature.

It has been suggested that the compressive strength of a composite may be a system property rather than a material property. The single fiber compressive strength tends to increase as the fiber strength when measured by the single fiber loop test. But there is considerable deviation of the composite and single filament compressive strengths (Figure 20.17) and an increase in fiber strength does not assure an increase in composite compressive strength. The composite compressive strength is much lower than the tensile strength, about one half, and the compressive strength of a pitch based carbon fiber is lower than a PAN based carbon fiber.

The compressive stress decreases with increased anisotropy, which can be represented by the ratio of fiber tensile modulus to the torsional modulus (E_f/G_f) (Figure 20.18). Kumar

PROPERTIES OF CARBON FIBERS

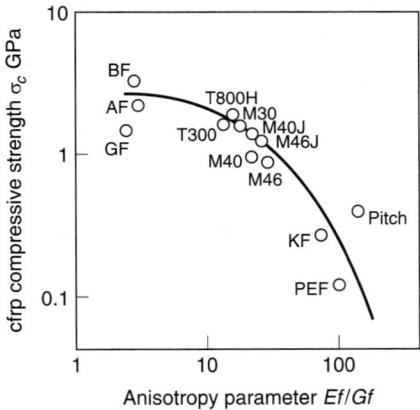

Figure 20.18 Anisotropy parameter Vs composite compression strength. *Source:* Reprinted from Toray technical literature

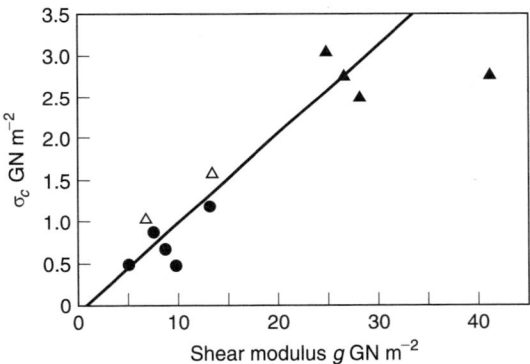

Figure 20.19 Relation between the compressive strength (σ_c) and the shear modulus (g) of carbon fibers. *Source:* Reprinted with permission from Northolt MG, Veldhuizen LH, Jansen H, *Carbon*, 29, 1267–1279, 1991. Copyright 1991, Elsevier.

[106] has shown that the compressive strength decreases with increasing Young's modulus (Figure 17.58), whilst Northolt *et al* [107] showed that the compressive strength increases as the shear modulus (Figure 20.19).

Dobb *et al* [108,109] studied PAN and mesophase pitch based carbon fibers by measuring their compressional properties using the recoil test and have stated that the compression strength of carbon fibers is influenced by a combination of various structural features, rather than by any single feature. The compressive strength increases linearly with increasing amounts of both D_c, the intracrystalline disorder (Figure 20.20) and D_d, the intercrystalline disorder (Figure 20.21).

D_c was calculated from:

$$D_c = \left[1 - \frac{d_{\text{turb}} - d_{\text{obs}}}{d_{\text{turb}} - d_{\text{gra}}}\right] \times 100\%$$

where d_{turb}, d_{gra} and d_{obs} are the interlayer spacings $c/2$ in a turbostratic carbon, perfect graphite and the actual carbon fiber respectively (d_{gra}=0.335 mn and d_{turb}= 0.350nm).

D_d is resolved by rotating the specimen continuously in the X-ray scan over the range 10–65° and traces resolved into peaks (ordered fraction) and background (disordered

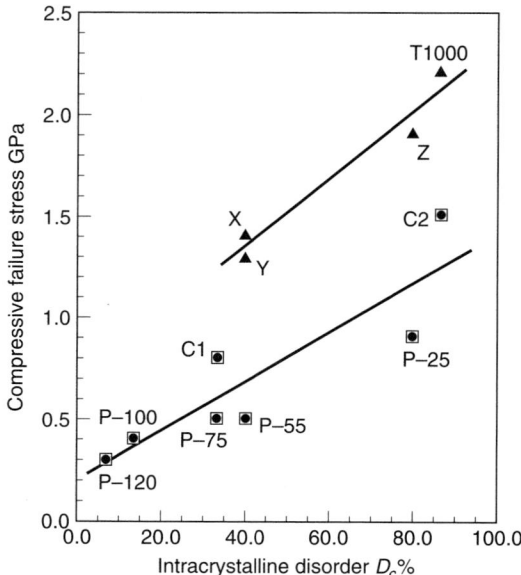

Figure 20.20 Compressive failure stress of carbon fibers Vs intracrystalline disorder (D_c). *Source:* Reprinted with permission from Dobb MG, Guo H, Johnson DJ, Park CR, *Carbon*, 33(11), 1553–1559, 1995. Copyright 1995, Elsevier.

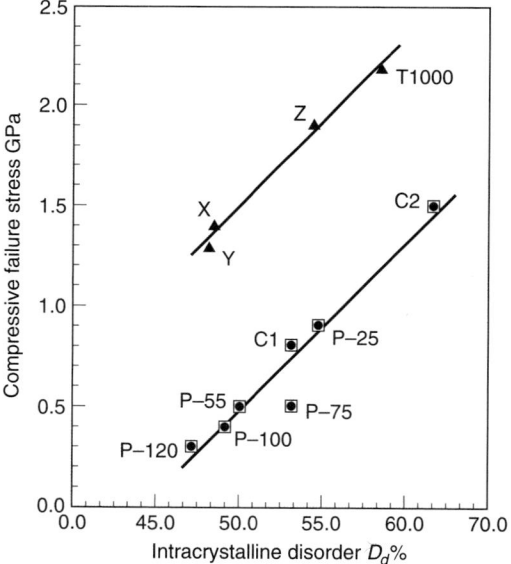

Figure 20.21 Compressive failure stress of carbon fibers Vs intercrystalline disorder (D_d). *Source:* Reprinted with permission from Dobb MG, Guo H, Johnson DJ, Park CR, *Carbon*, 33(11), 1553–1559, 1995. Copyright 1995, Elsevier.

fraction), where D_d is the area under the background, expressed as a fraction of the total area. The values for D_c and D_d are given in Table 20.22 [109].

Figures 20.20 and 20.21 clearly show the dependence of the compressive failure stress on both the intracrystallite as well as the intercrystallite disorder (D_c and D_d). In each case, two distinct fiber groups were identified, one was for the newly developed PAN based fibers with their improved compressive strengths and the other, the more conventional PAN and

PROPERTIES OF CARBON FIBERS

Table 20.22 Structural parameters of various carbon fibers

Fiber Type	Code	D_d %	D_c %	$L_{\alpha\parallel}$ nm	$L_{\alpha\perp}$ nm	L_c nm	Z deg	σ_c GPa
PAN	HM-S	55.8	33.3	4.6	7.4	5.9	20.1	0.8
	C_2	66.0	86.7	3.0	5.3	1.7	36.1	1.5
	X	50.2	40.0	4.8	7.2	4.7	21.2	1.4
	Y	49.9	40.0	4.8	7.0	4.3	22.0	1.3
	Z	57.5	80.0	3.4	5.9	2.6	30.8	1.9
	T1000	62.2	86.7	2.9	5.2	1.7	31.5	2.2
MP	P-25	57.7	80.0	3.2	5.9	2.5	34.3	0.9
	P-55	52.0	40.0	10.0	13.1	11.7	16.5	0.5
	P-75	55.8	33.3	11.4	13.0	14.6	11.1	0.5
	P-100	51.0	13.3	55.3	16.4	23.7	11.0	0.4
	P-120	48.6	6.7	61.6	18.6	25.5	8.3	0.3

D_d intercrystallite disorder; D_c intracrystallite disorder; $L_{\alpha\parallel}$ crystallite length along the α-axis; $L_{\alpha\perp}$ crystallite width perpendicular to the α-axis; L_c crystallite width perpendicular to the c-axis; Z preferred orientation; σ_c compressive strength.
Source: Reprinted with permission from Dobb MG, Guo H, Johnson DJ, Park CR, *Carbon*, 33(11), 1553–1559, 1995. Copyright 1995, Elsevier.

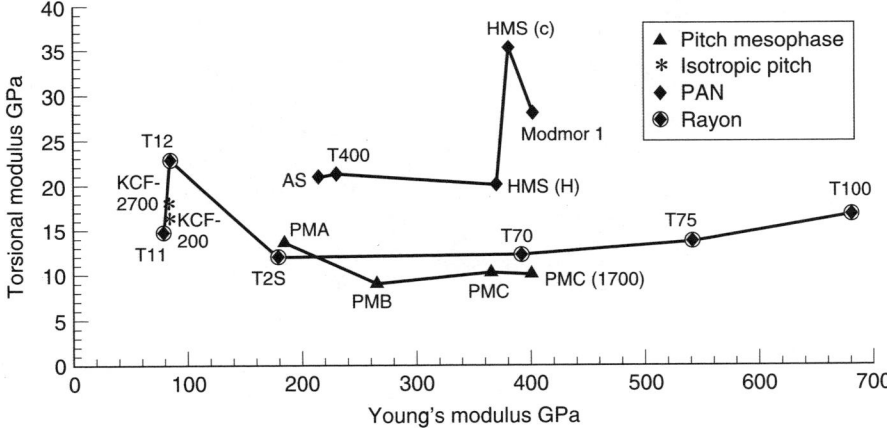

Figure 20.22 Torsional modulus Vs Young's modulus for various carbon fibers. *Source:* Graph plotted from data adapted from Donnet JB, Bansal RC, 2nd ed., Marcel Dekker, New York, 267–366, 1990.

MP based fibers due to the more homogeneous distribution of microstructural features. It is suggested that all crystallite dimensions should be limited to below 5 nm, whilst maintaining good orientation.

The SEM observations revealed that PAN based fibers failed by a buckling mechanism, whereas MP based fibers failed by shear. The T1000 fiber exhibited the highest compressive failure strength and tensile strain to break. Compressive strengths of carbon fibers are related to the ability to absorb energy, as shown in Figures 20.20 and 20.21.

The torsional modulus is shown in Figure 20.22 and appears to be governed by the structure of the microsection increasing in the order—mesophase pitch based carbon fiber < rayon based carbon fiber < isotropic pitch based carbof fiber < PAN based carbon fiber.

Dobb *et al* [108] used the recoil test and examined the broken filaments with SEM and obtained shear bands in high modulus MP fibers (Figure 20.23A) and kink bands in PAN based carbon fiber (Figure 20.23B). Parry and Wronski [111] show a kink band produced by 4-point loading a pultruded carbon fiber epoxy section (Figure 20.24).

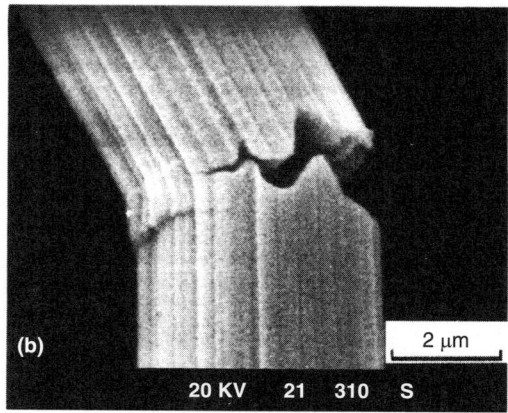

Figure 20.23 A. Shear band in high modulus MP based carbon fiber after moderate deformation. B. Kink band in PAN based carbon fiber after recoil compression under high deformation. *Source:* Reprinted with permission from Dobb MG, Johnson DJ, Park CR, *J Mater Sci*, 25, 829, 1990. Copyright 1990, Springer.

Figure 20.24 Polished section of a carbon fiber/epoxy resin pultruded specimen showing a kink band formed in the compression zone of a 4-point bend test. *Source:* Reprinted with permission from Parry TV, Wronski AS, *J Mater Sci*, 16, 439–450, 1981. Copyright 1981, Springer.

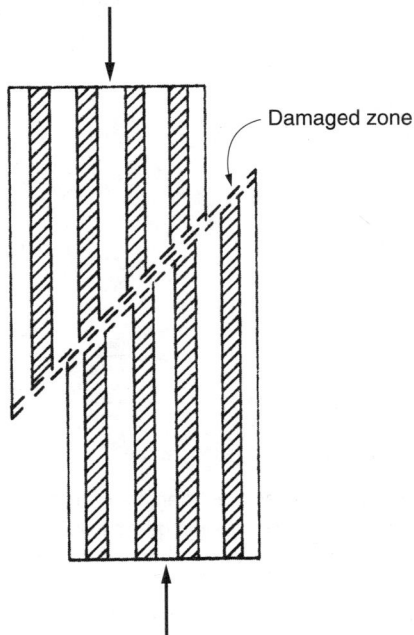

Figure 20.25 Schematic view of shear failure due to longitudinal compressive stresses. *Source:* Reprinted with permission from Hull D, Cambridge University Press, Cambridge, 160, 1981. Copyright 1981, Cambridge University Press.

Deformation occurs by microbuckling with a shear mechanism, forming kink bands on the fracture surface at 45° to the fiber axis. The onset of buckling is temperature dependent and although Tonen MP fails without buckling at 25°C, it does buckle at 140°C. West [112] found that the compressive strength is approximately proportional to the square of the T_g.

PAN type carbon fibers like AS-4, IM6 and IM7 fail by transverse shear, with no microbuckling. Ewins and Ham [89] proposed that the shear stresses in the laminae could cause a shear mode of failure (Figure 20.25). If the shear strength is less than the buckling strength, then shear failure will occur in preference to buckling. However, Ewins and Potter [114] showed that microbuckling does occur as the temperature approaches the T_g (Figure 20.26). The failure starts as a crack on the tensile side of the bent fiber, which propagates to a kink band on the compressive side. The failure transition is dependent on the shear modulus of the matrix and the shear strength of the fiber. Increase in fiber strength and the uptake of water, which decreases resin modulus, will have the same effect. Figure 20.27 shows an SEM micrograph of the fracture surface of a carbon fiber/epoxy laminate after a fiber buckling mode failure due to longitudinal compressive stress showing tension and compressive failure.

It was suggested that larger diameter carbon fiber could exhibit higher compressive strength, but Toray have found no evidence for this (Figure 20.28).

20.10 THERMAL PROPERTIES [115–118]

Early work [119–124] showed that the thermo-oxidative instability of carbon fibers affected the stability of high temperature laminates at 300°C.

Carbon fibers oxidize in air and Figure 20.29 shows the weight loss of Grayon fibers made from a viscose rayon precursor, where the more graphitic version has the best oxidation

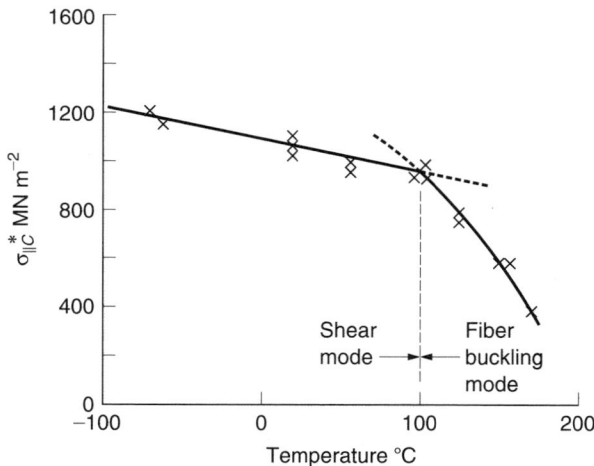

Figure 20.26 Variation of longitudinal compressive strength of carbon fiber epoxy resin laminate with temperature showing transition from shear mode to buckling mode failure. Failure depends on the shear modulus of the matrix and shear strength of the fibers and a similar effect is observed with the uptake of water. *Source:* Reprinted with permission from Ewins PD, Potter RT, *Phil Trans R Soc London*, A294, 507–517, 1980. Copyright 1980, The Royal Society of Chemistry.

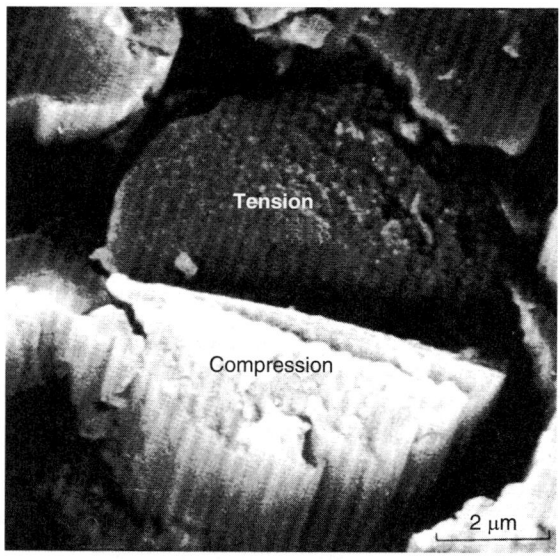

Figure 20.27 SEM micrograph of the fracture surface of a carbon fiber epoxy resin laminate after a fiber buckling mode failure due to a longitudinal compression stress viewed under high magnification showing both tension and compressive fracture in a single fiber. *Source:* Reprinted with permission from Ewins PD, Potter RT, *Phil Trans R Soc London*, A294, 507–517, 1980. Copyright 1980, The Royal Society of Chemistry.

resistance. The thermal stability of oxidized PAN and PAN based carbon fiber is compared with an aramid in Figure 20.30. An oxidized PAN fiber will lose more weight as the temperature increases (Figure 20.31) and the reported loss will include any absorbed moisture in the oxidized PAN (about 7%).

The thermal oxidative behavior in air at 250 and 300°C of several grades of PAN based carbon fibers was studied by Gourdin [125] and the weight losses at 250 and 300°C as a

PROPERTIES OF CARBON FIBERS

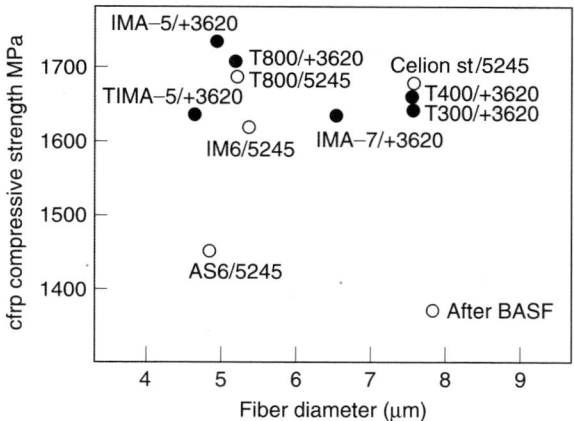

Figure 20.28 Dependency of compression strength on fiber diameter. *Source:* Reprinted from Toray technical literature.

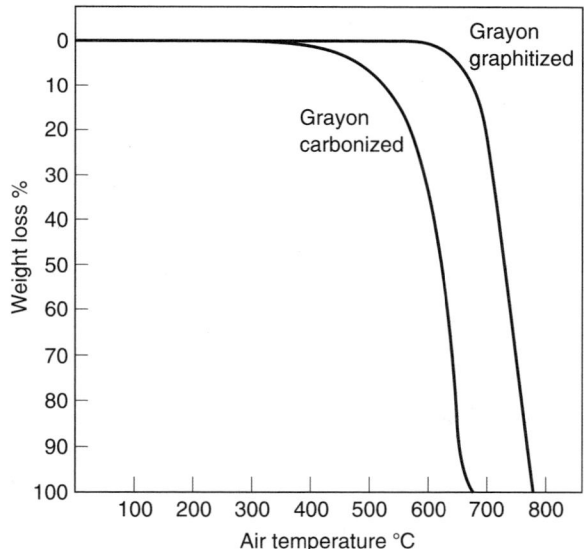

Figure 20.29 Grayon (carbonized rayon) carbon fiber weight loss *Vs* temperature in air. *Source:* Reprinted from RK Carbon Fibre's technical literature.

function of time in air are shown in Figures 20.32 and 20.33. The analyzes of the fibers are given in Table 20.23. Gourdin's results did not support any correlation of thermo-oxidative properties of the fiber with the Na content, since Hercules AS-4 fiber had a high Na content, but good thermo-oxidative resistance, whereas Toray T300B had a low Na content but poor stability. However, fibers produced at the highest production temperatures, such as HTS and HMS, did have the best thermo-oxidative resistance.

Schola, of United Technology Research Center, believes that it is not necessary to use the most thermo-oxidatively stable fiber to achieve a good high temperature system. Although T-40 has a greater thermo-oxidative stability than AS4 or Celion G30-500, the retention of properties is better with these low modulus fibers. The thermo-oxidative stability is not directly related to the presence of impurities such as Na or K. Rather, it is a function of the

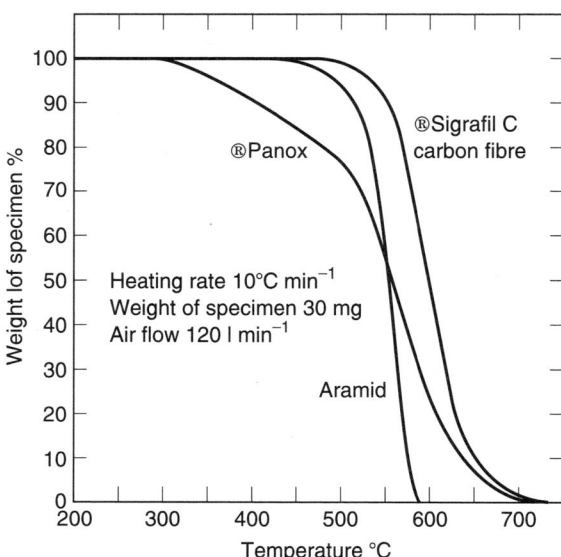

Figure 20.30 Thermal stability of oxidized PAN and PAN based carbon fiber. *Source:* Reprinted from SGL Carbon Group technical literature.

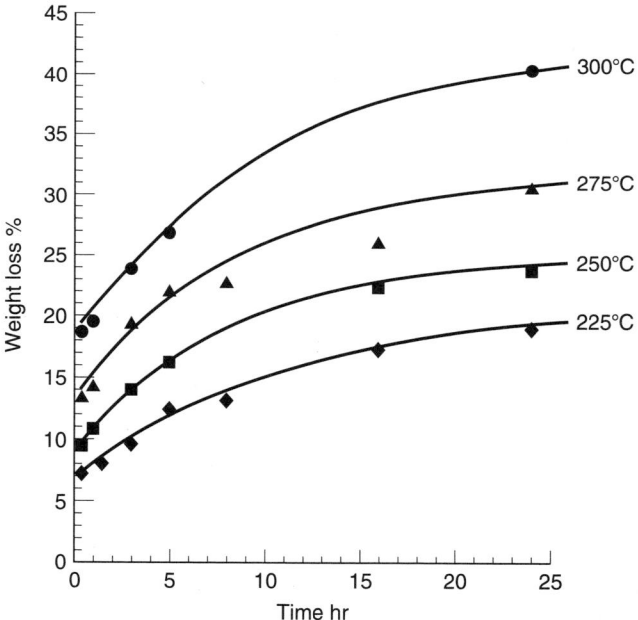

Figure 20.31 Weight loss of Grafil 'O' ($\rho = 1.35\,\text{g cm}^{-3}$) with time and temperature (7% of this weight loss is attributed to moisture loss). *Source:* Reprinted from Courtaulds technical literature.

amount of oxygen at the filament surface and to what extent the surface has been treated. Untreated fiber does not stand up well, but properly treated AS4, G30-500 or an equivalent grade, exhibit good thermo-oxidative stability. Schola believes the resin size protects the fiber; hence a resin size with a high degree of stability, such as PMR-15, would be beneficial.

PROPERTIES OF CARBON FIBERS

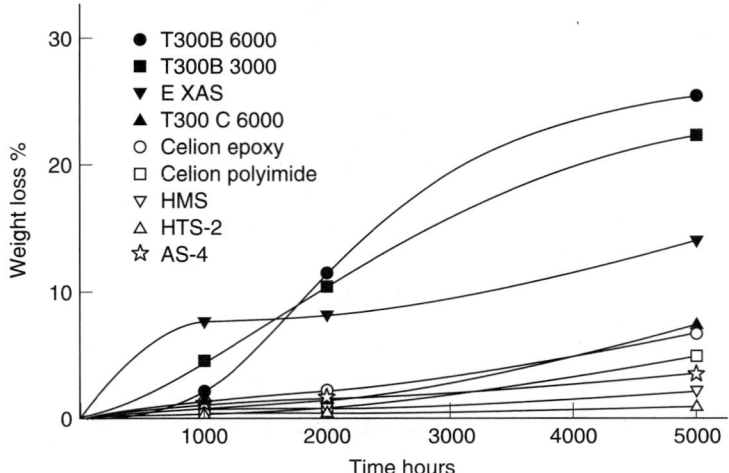

Figure 20.32 Weight loss of carbon fibers after ageing at 250°C. *Source:* Reprinted from Gourdin C, *SAMPE*, Bourdeaux, 49–61, Oct 17–20 1983.

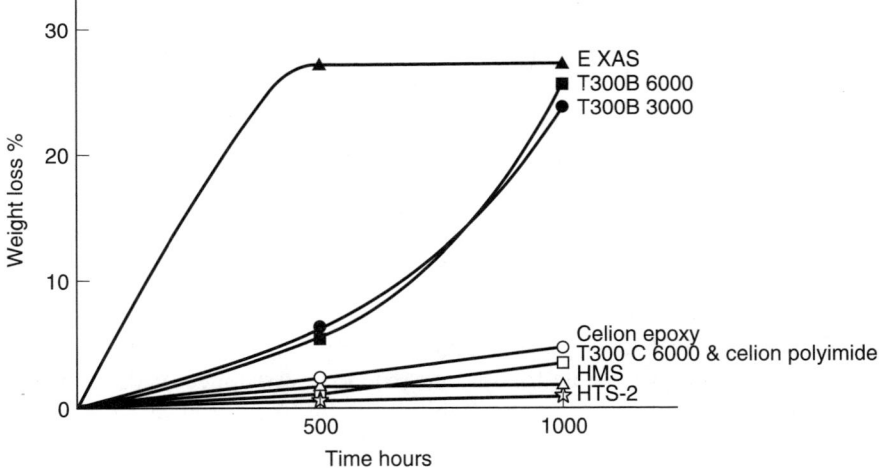

Figure 20.33 Weight loss of carbon fibers after ageing at 300°C. *Source:* Reprinted from Gourdin C, *SAMPE*, Bourdeaux, 49–61, Oct 17–20 1983.

Table 20.23 Chemical analyzes of carbon fibers

Fiber	YM GPa	C %	Elements ppm							Density g cm^{-3}	Size %	Wt loss after 2000 h @ 250°C
			Na	Si	Mg	Ca	Zn	Fe	Al			
Celion 6000	241	93.0	–	144*	180*	180*	90*	–	–	1.77*	1.1*	2.7
Hercules HTS-2	274	99.2	300	200	10	10	<30	60	20	1.68	0	0.2
Hercules AS-4	234	94.7	10^4	200	35	35	<30	5	6	1.81	1.0	1.6
Courtaulds XAS	233	94.9	10^4	200	40	40	<30	30	30	1.71	1.8	7.9
Toray T300B	232	93.4	200	200	10	10	<30	10	10	1.72	1.2	11.0
Toray T300C	225	95.1	<30	80	<10	<10	<30	<5	10	1.74	0.9	1.5

Note: *Taken from Celanese Technical Data Sheet.
Source: Reprinted from Gourdin C, Ageing of carbon fibres of various origins, *SAMPE*, Bourdeaux, 49–61, Oct 17–20, 1983.

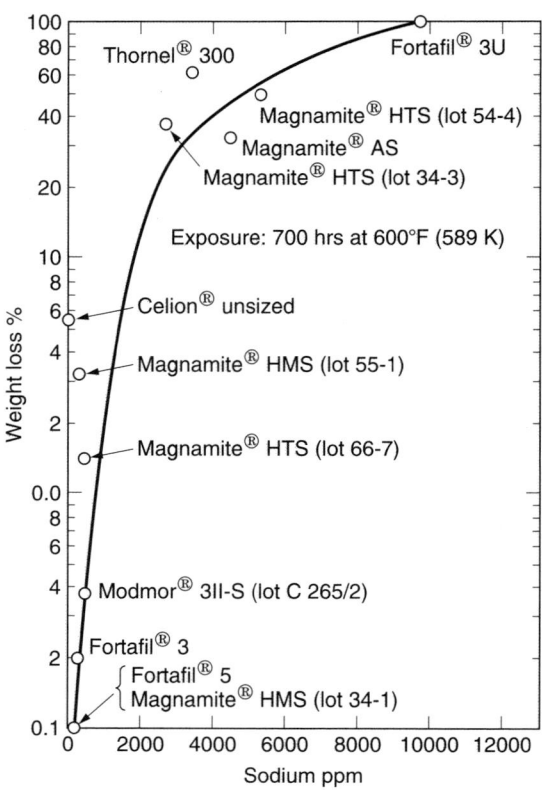

Figure 20.34 Relationship between total sodium content and bare carbon fiber stability. *Source:* Reprinted with permission from Gibbs HH, Wendt RC, Wilson FC, *33rd Ann Tech Conf SPI*, 1978. Copyright 1978, The Society of Plastics Engineers.

Gibbs [126] and McMahon [127] postulated that Na_2SO_4 from the PAN precursor spinning operation was responsible for thermo-oxidative resistance of the carbon fiber; the higher the sodium content, lower is the stability of the fiber (Figure 20.34). The Na_2SO_4 in conjunction with O_2 from the air was postulated to oxidize the carbon to CO_2 and in turn reduce to SO_2 and Na_2O. The author, however, is unaware if the sulphate salt could be the probable cause, as the PAN fiber production process in question uses a production step to remove the sulphate and the Na is more likely to be present as the salt of itaconic acid.

The Toray T300 and Courtaulds XAS both have large weight losses after 1000 h at 300°C, but then XAS appears to be stabilized, whereas the T300 has a gradual, but continuous, weight loss. The XAS, after ageing, had a much larger concentration of Na at the surface than within the fiber. All fibers with high carbon content had good resistance but no definite relationship was found between the thermo-oxidative resistance and the Na content.

The oxidation rate of various carbon fibers heated for 3 h in air at 500°C does show some relationship with the Na content (Figure 20.35) and certainly fibers with low Na showed the best thermo-oxidative resistance and are associated with fibers with the highest carbon content.

Page and Duquette [128] showed that the oxidation resistance of the carbon fibers in a glass matrix was related to the microstructure of the glass reinforcement and the adherence of the glass to the carbon fibers. Coating the fibers with a SiC precursor by CVD was shown to improve the oxidation resistance of the fibers.

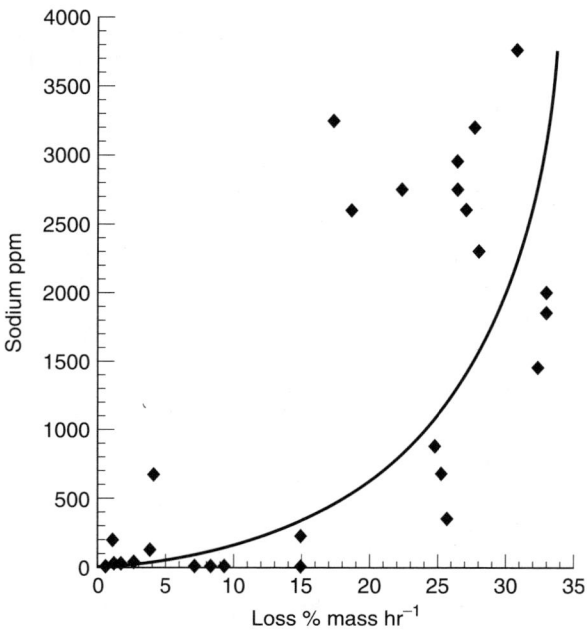

Figure 20.35 Oxidation rate of various carbon fibers for 3 h at 500°C in air Vs Na content (Na content measured by atomic absorption spectroscopy). *Source:* Reprinted from Courtaulds technical literature.

20.11 THERMAL EXPANSION OF CARBON FIBERS

Extensive work has been undertaken determining the thermal expansion of carbon fibers [129–131].

The relationship of the coefficient of thermal expansion with Young's modulus is shown in Figure 20.36 and is believed to be due to the dependence of both properties on crystal structure.

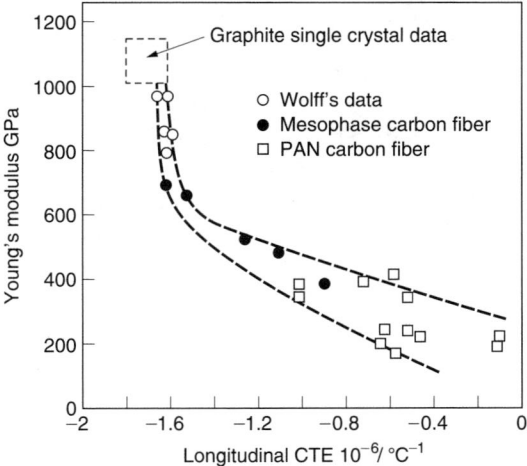

Figure 20.36 Correlation between Young's modulus and the coefficient of thermal expansion for a selection of carbon fibers. *Source:* Reprinted with permission from Wolff EG, Stiffness-thermal expansion relationship in high modulus carbon fibres, *J Composite Mater*, 21, 81, 1987. Copyright 1987, Sage Publications.

Pirgon et al [133] determined the linear thermal expansion in a direction parallel to the principal axis of HTS fiber and ERLA 4617-BDMA matrix at 80–400 K to be $-9.1 \times 10^{-7} \mathrm{K}^{-1}$.

Further work by Rogers et al [134] using the same system found that, at higher temperatures, there were marked changes in the temperature dependence of the dimensional behavior, resulting from the relaxation accompanying the softening of the resin.

Yates et al [135] determined the influence of the fiber volume fraction using HTS fiber in a DLS351/BF$_3$400 matrix at 90–500 K. The linear thermal expansion coefficients of the fiber at room temperature were found consistent with $-10 \times 10^{-7} \mathrm{K}^{-1} < \alpha_{\|}^{f} < -9 \times 10^{-7} \mathrm{K}^{-1}$ in the direction parallel, and $0.5 \times 10^{-5} \mathrm{K}^{-1} < \alpha_{\perp}^{f} < 1.9 \times 10^{-5} \mathrm{K}^{-1}$, in the direction perpendicular to their length.

Yates et al [136] determined the influence of the fiber resin type and found that, with HTS fiber, Fothergill's Code 69 system and DLS 351/BF$_3$400 at 90–500 K, composites based on DLS 351/BF$_3$400 and Code 69 may be employed at temperatures about 90° higher than the ERLA 4617-mPDA system investigated in earlier work. The T_g of Code 69 was found to be 507K. A rapid rise in the linear thermal expansion coefficient of cross-plied laminates indicates danger of causing high residual strains in complex cfrp structures by post curing at too high a temperature.

Work by Yates et al [137] to determine the effects of multidirectional lay-up quoted values for the room temperature linear thermal expansion coefficients for the 0° and 90° directions of the tridirectional laminate (−45°, 0°, +45°) to be $-3.3 \times 10^{-7} \mathrm{K}^{-1}$ and $5.3 \times 10^{-6} \mathrm{K}^{-1}$ respectively.

In work investigating the influence of the matrix curing characteristics, Yates et al [138] used the following systems—Derakane 411-48/MEKP (where the overall contraction exceeded 6%); Epikote 828/MNA/BDMA (with an overall contraction during cure of less than 2%); and Ciba XB2878A/XB2878B (with an overall contraction of the resin during cure approaching 6%).

Dootson et al [139] found that the thermal expansion characteristics of UD cfrp produced by standard production techniques are reproducible from one batch to another and although the linear thermal expansion coefficients of UD test specimens may differ by several orders of magnitude according to fiber type, these differences are probably not significant in the majority of practical applications, because the mean values are all within $-10^{-6} \mathrm{K}^{-1} < \alpha < 10^{-6} \mathrm{K}^{-1}$, provided that they are not kept above room temperature for an extended period.

The influence of the distribution between fiber orientations upon the thermal expansion characteristics was determined by Parker et al [140], who found that a knowledge of the linear thermal expansion coefficients parallel and perpendicular to the fibers of a UD reinforced lamina, together with that of the pure resin and the approximate values of the elastic constants, allowed a sufficiently good prediction of the influence of the fiber distribution on the linear thermal expansion coefficients of (0°, 90°) and (±45°, 90°) laminates for most practical purposes, but (0°, ±45°) laminates are less satisfactory. Another finding was the greater sensitivity of the linear thermal expansion coefficients of composites based on the resin Fiberdux 914C to the specimen thermal history compared with those of structurally similar composites based on ERLA 4617/mPDA, DLS351/BF$_3$400 and Code 69.

The influence of fiber weave in fabric reinforcement was studied by Rogers et al [141], who used Morganite Type II and Grafil E/XAS carbon fibers in plain weave, 2 × 2 twill weave and 5-shaft satin weave and UD fiber with a DLS351/BF$_3$400 matrix. The ratio of fiber tow densities in the principal fiber directions, the crimp in the reinforcing fibers and the laminate stacking sequence, all influenced the magnitudes and temperature dependencies of the linear thermal expansion coefficients. In addition, the response of the dimensions to changes of temperature was studied in detail. The linear thermal expansion coefficients of

PROPERTIES OF CARBON FIBERS

Grafil E/XAS in directions parallel, $\alpha_\parallel^f$, and perpendicular $\alpha_\perp^f$ to the fiber axis were estimated as $\alpha_\parallel^f = -2.6 \times 10^{-7} \, K^{-1}$ and $\alpha_\perp^f = 2.6 \times 10^{-5} \, K^{-1}$.

Under commonly encountered conditions of moisture content, the presence of voids and slight imperfections in fiber lay-up were found by Yates et al [142] to have little practical effect on thermal expansion data.

Ozbek et al [143] found the thermal expansion of HM PAN based carbon fiber treated at 2800°C to be negative below about 400°C and positive at higher temperatures. The variation from $\sim -1 \times 10^{-6} \, K^{-1}$ at 150°C to $\sim +1.5 \times 10^{-6} \, K^{-1}$ at 1000°C is within the expected range. Above 2000°C, there is a divergence between the heating and cooling curves due to time dependent permanent plastic deformation.

20.12 THERMAL CONDUCTIVITY OF CARBON FIBERS

The thermal conductivity measurements of some glass fiber and carbon fiber reinforced plastics have been determined [144].

Pilling et al [145] determined the transverse thermal conductivities of Morganite HMS and HTS carbon fibers in a DX210/BF$_3$400 matrix at 180 K, 225 K and 270 K and were found to be 2.0, 3.1 and 5.7 W m^{-1} K^{-1} respectively.

The thermal conductivity of Curlon is of particular interest for use as battings and Figure 20.37 shows the effect of the batting density on the thermal conductivity.

20.13 CREEP PROPERTIES

Ozbek and co-workers [143] investigated creep behavior using a series of tows stretched at three constant loads and four final temperatures and the results are shown in Figure 20.38. The creep was logarithmic and the inverse of creep rate was plotted as a function of time for varying temperatures (Figure 20.39) and of time for varying loads (Figure 20.40). Creep will affect measurement of α above ~1500°C. Since Arrhenius plots could not be drawn, it was concluded that the rate determining process was an internal accommodation, related to differential thermal expansion, rather than thermal activation.

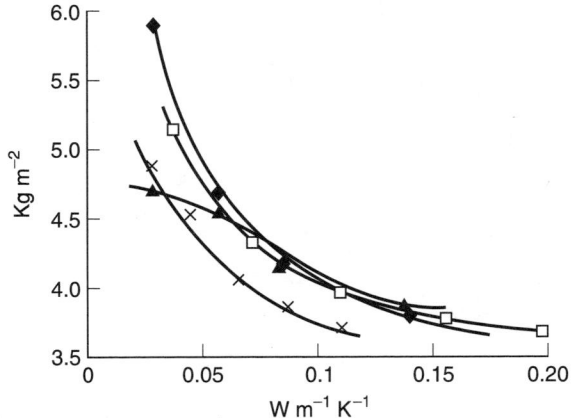

Figure 20.37 Thermal conductivity Vs density of Curlon battings. ◆ Sample A (SP-6), □ Sample B (SP-8), ▲ Sample A + Aluminum foil (SP-6 + foil). Foil weight not included in the density. × Sample C (Micro). *Source:* Reprinted from RK Carbon Fibres technical literature.

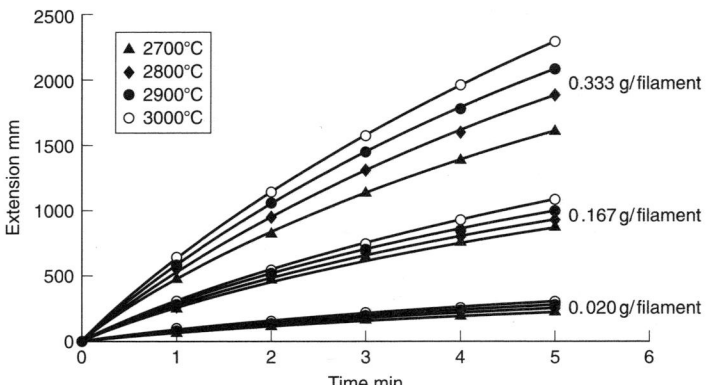

Figure 20.38 Creep extension as a function of time for four temperatures and three loads. *Source:* Reprinted with permission from Ozbek S, Jenkins GM, Isaac DH, Thermal expansion and creep of carbon fibres, *20th Biennial Conf on Carbon*, American Carbon Society, Santa Barbara, 270–271, 1991. Copyright 1991, The American Carbon Society.

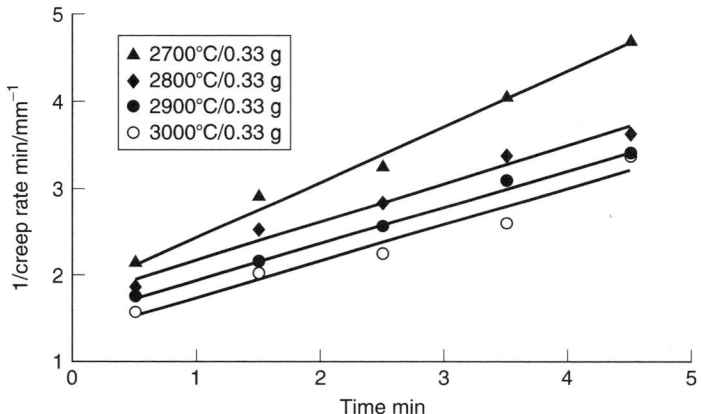

Figure 20.39 Inverse of creep rate as a function of time for varying temperatures. *Source:* Reprinted from Ozbek S, Jenkins GM, Isaac DH, Thermal expansion and creep of carbon fibres, *20th Biennial Conf on Carbon*, American Carbon Society, Santa Barbara, 270–271, 1991.

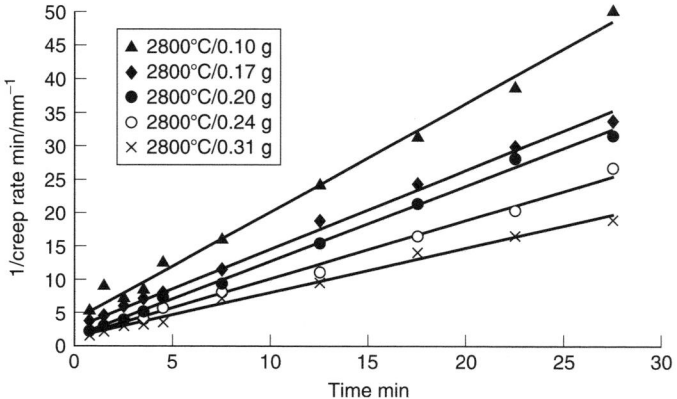

Figure 20.40 Inverse of creep rate as a function of time for varying loads. *Source:* Reprinted from Ozbek S, Jenkins GM, Isaac DH, Thermal expansion and creep of carbon fibres, *20th Biennial Conf on Carbon*, American Carbon Society, Santa Barbara, 270–271, 1991.

20.14 IMPACT STRENGTH AND FRACTURE TOUGHNESS

Work on impact strength and fracture toughness is listed [147–153], as well as fractography [154–168]. The impact strength of carbon fiber composites as determined by the Charpy test are given in Table 20.24 [169]. The impact strengths in the transverse direction are low since there are few, if any, fibers in this direction to provide strength and hence, impact strength.

The effect of the fiber type on impact strength is shown in Figure 20.41, demonstrating the poor performance of carbon fiber and the advantage of using a hybrid construction.

Table 20.24 Charpy impact data for carbon fiber composites

	Impact energy kJ cm^{-3}		
Directional of lay-up	Longitudinal	Transverse	Cross ply
Ciba 914C epoxy (high temp resistant)	20	0.25	5
Ciba 920C (impact resistant)	20	1	4
Kerimid 601 polyimide	71	0.25	50
ICI PES (solvent applied)	76	2.5	50
ICI PES (melt applied)	110	2.5	75
GE PEI	26	2	38
ICI PEEK	73	2	40

Source: Reprinted with permission from Hancox NL, Mayer RM, *Design Data for Reinforced Plastics—A Guide for Engineers and Designers*, Chapman & Hall, London, 118, 1994. Copyright 1994, Springer.

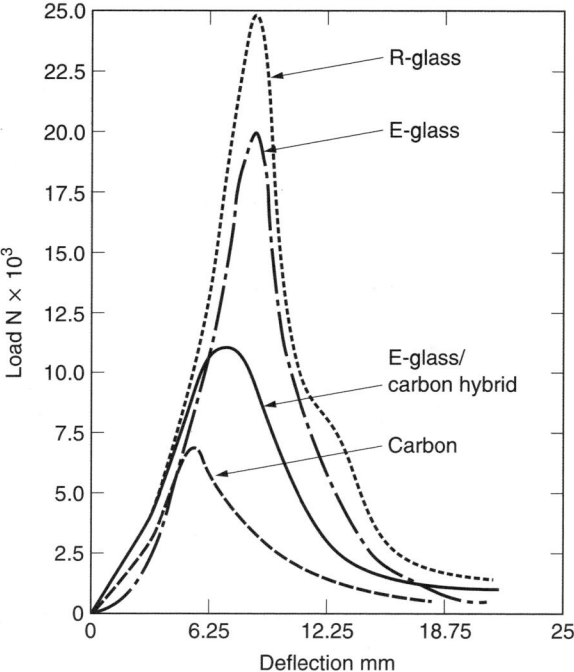

Figure 20.41 The effect of fiber type and hybrid construction on impact. *Source:* Reprinted with permission from Walker RA, *Impact Database*, PERA International, UK. Copyright, PERA International.

Nakano *et al* [171] prepared continuously aligned carbon fiber reinforced SiC, Si_3N_4-Sialon and Si_3N_4 matrix composites by slurry impregnation followed by hot pressing. Highest values of the dynamic elastic-plastic fracture toughness and dynamic stress intensity factors when measured at room temperature using a computer aided Charpy impact testing system for the SiC matrix composites.

The Izod impact test for measuring the fracture energy of fiber reinforced ceramics was used by Kamiya and Wada [172], who prepared uniaxial carbon fiber reinforced Si_3N_4 composite by hot pressing and impact specimens were machined from the hot pressed discs, with their long axes parallel to the fiber direction. Selection of the appropriate notch geometry and loading condition resulted in mode I failure and the fracture energy was measured to be 62 $kJ\,m^{-2}$, about 20 times that of monolithic Si_3N_4.

20.15 FATIGUE PROPERTIES

Properties resulting from fatigue testing are given in references [173–185].

There are a number of parameters that influence fatigue—the type of fiber and matrix; strain to failure and the strengths of fiber and matrix; the laminate configuration and the cycling frequency. Figure 17.59 shows the superior fatigue performance of carbon fiber epoxy laminate over glass and aramid.

Varying the carbon fiber type in the same resin matrix has little effect upon the fatigue behavior (Figure 17.60), but efforts to improve resin toughness have resulted in improved static strength but poorer fatigue properties (Figure 17.61).

20.16 ELECTRICAL PROPERTIES

Wang and Chung [186] have observed apparent negative electrical resistance in interfaces between layers of carbon fibers in composite material in a direction perpendicular to the fiber layers.

The magneto-resistance of carbon fibers made from coal tar mesophase pitch treated at 2500°C becomes lower and L_C (002) becomes smaller as the diameters get smaller [187].

A patent by Matthews and Ko [188] describes how the electrical resistivity of a PAN based carbon fiber heat treated at 650°–1050°C increases with time when aged in air and the magnitude of this increase is proportional to the initial resistance—the higher the initial resistance, greater is the increase. BASF established that the stability can be enhanced by treatment in air for 8–48 h at 260–285°C.

A family of resistive fibers (Rescar) was investigated by RK Carbon Fibers and to measure resistivity, the correct equipment must be used (Chapter 17, Section 17.3.9). Figure 20.42 shows a range of resistivities that can be obtained by varying the process temperature and Figure 20.43 shows the change in resistance when that fiber is exposed to air. The presence of size significantly reduces the rate of ageing.

$$\text{Resistivity} = \frac{\pi d^2 \Omega}{4L} \text{ohm cm}$$

where d = diameter, cm
ω = resistance, ohm
L = gage length, cm

PROPERTIES OF CARBON FIBERS

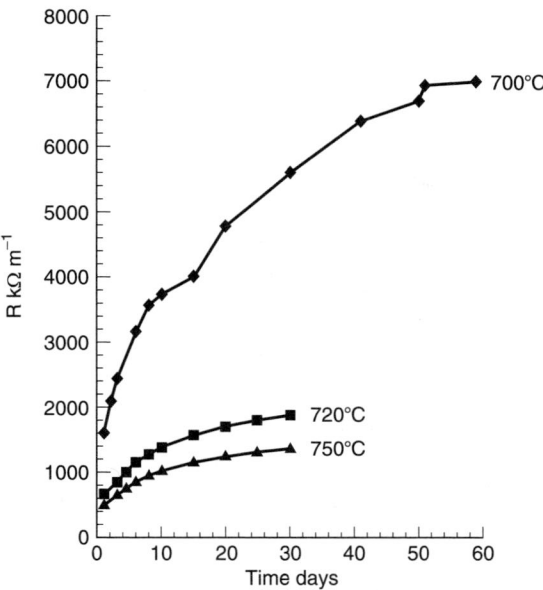

Figure 20.42 Resistance *Vs* time for Rescar fiber made at different temperatures using 46k tow. *Source:* Reprinted from RK Carbon Fibres technical literature.

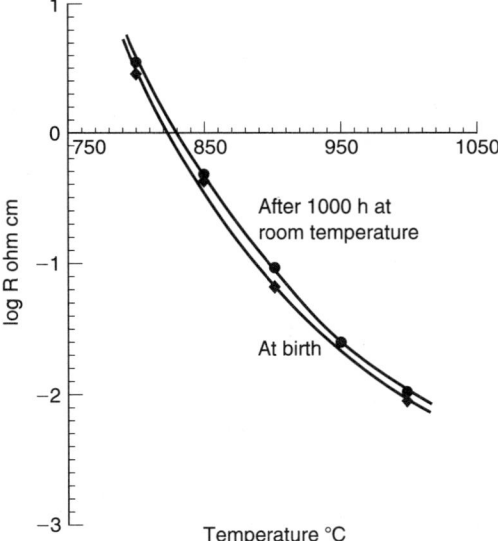

Figure 20.43 Log resistance *Vs* temperature for Rescar fiber when exposed to air. *Source:* Reprinted from RK Carbon Fibres technical literature.

Figure 20.44 shows a log plot of resistance against time. Three equations for the change in resistance can be deduced:

At 700°C $\quad \log R = 0.375 \log t + 3.21$

At 720°C $\quad \log R = 0.35 \log t + 2.77$

At 750°C $\quad \log R = 0.33 \log t + 2.695$

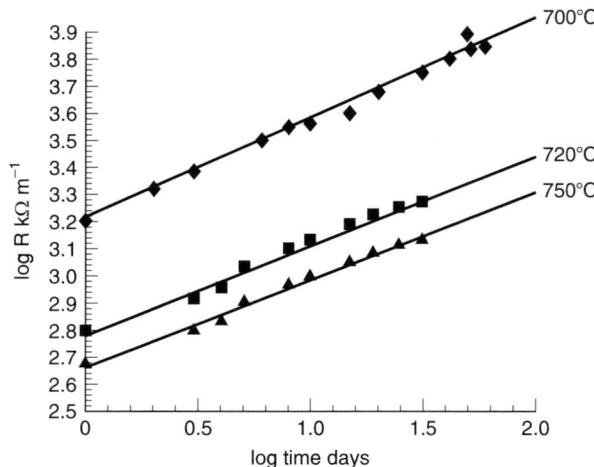

Figure 20.44 Log plot resistance *Vs* time for fiber made at different temperatures. *Source:* Reprinted from RK Carbon Fibres technical literature.

Lerner [189] found that oxidized PAN fibers subjected to heat treatment temperatures of 715–945 K are semiconductors. The room temperature conductivity is dominated by the contributions of impurity states, but these are not related to defects in the polymer. There is a decrease in conductivity on ageing in air caused by a decrease in the electron-phonon scattering time. The conductivity increases at temperatures above 473 K, as the samples aged in air outgas.

The electrical resistance of cfrp was found to be closely related to the mechanical deformation and damage due to a change of cross-sectional area and length [190].

The electrical properties of pitch based semi-conductive carbon fibers have been determined by Okubo *et al* [191] and Gerteisen [192] examined carbon fiber reinforced thermoplastics with controlled surface resistivity. The techniques used for measuring the conductivity in carbon fibers have been described by Maslii and Panasenko [193].

Agari *et al* determined the influence of carbon fiber form and fiber loading on the electrical conductivity of polyethylene composites [194].

20.17 CHEMICAL RESISTANCE

The chemical resistance of oxidized PAN fiber in some common reagents is given in Table 20.25 and the outgassing of oxidized PAN fiber is given in Table 20.26.

The influence of moisture on the mechanical properties and failure behavior of cfrp composites has been studied by Selzer and Friedrich [195] and their behavior in a marine environment by Alias and Brown [196].

Table 20.25 Chemical resistance of oxidized PAN fiber

Concentration	H_2SO_4	HCl	NaOH	NH_4OH	HNO_3	Acetic acid	H_2O_2	Chromic acid
1%	Fair	Good	Fair	Good	Good	Good	Good	Fair
10%	Fair	Good	Poor	Fair	Good	Good	Fair	Poor
Conc.	V. poor	Poor	(20%) Poor	Fair	V. poor	Good	Fair	V. poor

Source: Reprinted from Courtaulds technical data.

Table 20.26 Outgassing of oxidized PAN fiber

Temperature °C	Time min	HCN Degassed µg	HCN/g Grafil 'O' as a %
250	5	770	0.077
	10	1190	0.119
	15	2900	0.290
300	5	4120	0.412
	10	7600	0.760
	15	8000	0.800
350	2	3180	0.318
	5	18600	1.860
	10	20200	2.020
400	1	17200	1.720
	2	21700	2.170

Source: Reprinted from Courtaulds technical data.

The chemical resistance of carbon fiber reinforced PEEK and PPS composites have been determined [197].

Enomoto et al [198] evaluated the corrosion resistance of carbon fibers in hot phosphoric acid.

20.17.1 Intercalation

Carbon fibers can react to form intercalation compounds when the reactant (intercalate) enters between the graphite layer planes, forcing them apart. Although the layer planes are pushed apart, the distance between the carbon atoms within a sheet remains unaltered. Hence the volume of carbon fiber per unit volume is reduced and the resistivity is decreased. Intercalation is generally restricted to carbon fibers with a graphitic structure. Typical intercalates are Br_2, liquid K and $FeCl_3$. Tressaud et al [199] have reported the intercalation of carbon fibers with fluorine.

20.18 FRICTION AND WEAR

The mechanism of friction and wear of carbon fiber reinforced metals has been studied [200–204].

Bianchi et al [205] fabricated unidirectional continuous carbon fiber reinforced (YMAS) glass ceramic matrix composites for dry friction applications. Pitch based fibers (P55) and PAN based fibers (T400H and M40) were used and friction and wear tests carried out using a disc-on-disc tribometer. The tribological behavior of the composites is mainly linked to the fragility of the matrix and to the microcracking induced during preparation of these composites. The fiber graphitization rate does not seem to influence the tribological behavior of the composites. The mechanisms of formation of the third body are associated with the fragmentation of the matrix and fibers. A velocity accommodation mechanism occurs by the shearing of the pressed powder bed in lift zones. This study has shown the potential of carbon fiber reinforced (YMAS) glass ceramic matrix composites in dry friction, notably with regard to the low wear rate.

20.19 HYBRID COMPOSITES

A hybrid is generally considered to be a composite with two or more types of reinforcement, although there can also be a hybrid matrix [206]. In this chapter, the most common form of hybrid which is two types of fiber is considered. The hybrid effect is a term used to describe the apparent synergistic improvement of composite strength to a value more than that predicted by the rule of mixtures. However, negative effects have also been observed. A typical hybrid could be glass/carbon in a common matrix. The hybrid may be formed by alternate plies, interspersed tows or intermingled fibers. Alternatively, a hybrid could comprise a woven cloth with say, one fiber in the weft and the second fiber in the warp. Short and Summerscales [207] have reviewed hybrid composites.

Considering equal fiber fractions of a unidirectional HM fiber/E-glass under axially applied tension, the stiffness of the hybrid E_H confirmed by experiment can be found from:

$$E_H = E_1 V_1 + E_2 V_2$$

The strength of the carbon does appear to be greater in the hybrid than in a 100% carbon fiber composite. This cannot be explained by the law of mixtures (line AD in Figure 20.45) and Bader [208] states that this is due to the difference in strains of the two fibers (0.01% for HM carbon and 0.028% for E-glass). Hence, when the hybrid is subject to tension, the carbon fiber will fail in the region of 0.01% strain, but the glass continues to bear load. It is necessary to consider the relative proportions of each fiber and if the glass predominates, it will continue to sustain the load, but if the composite has a higher carbon loading, then the glass cannot continue to carry the load and the composite fails. In Figure 20.45, the points A and D represent the failure stresses of all-glass and all-carbon plies. The line BD represents the stress at which the carbon plies fail and AE, for the glass. The point C represents the change in failure behavior. At glass contents below C, the carbon will be expected to fail when the applied strain reaches the failure strain of the carbon when there is insufficient glass to carry the load. The line AC denotes the stress in the composite. When there is more glass, then although the carbon will fail, there is sufficient glass to sustain the load until failure of the glass occurs as predicted by CD. It has been observed experimentally that failure stresses occur above the focus ACD, giving rise to a positive hybrid effect. The explanation can be due to carbon fibers having a negative coefficient

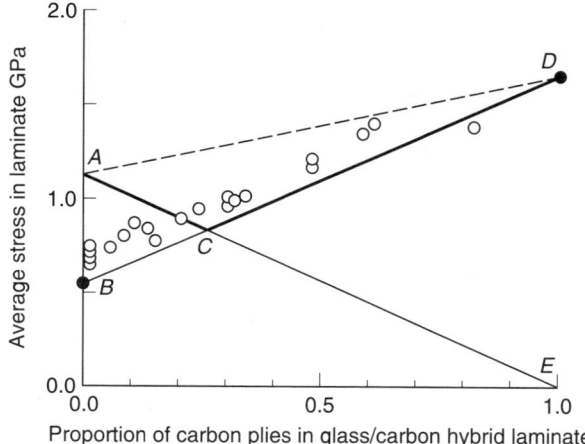

Figure 20.45 The strength of hybrid glass/carbon composites. *Source:* Reprinted with permission from Manders PW, Bader MG, *J Mater Sci*, 16, 2233–2245, 1981. Copyright 1981, Springer.

of thermal expansion, whereas the glass fiber and resin matrix have much larger positive coefficients. Hence when the resin is cured at elevated temperature and cooled, the carbon fiber will be forced into compression and axial tensile stresses will develop in the glass. When tension is applied to the composite, the compression in the carbon fiber must first be negated, producing an apparent increase in the strain to failure. This is not believed to be the complete explanation, as short lengths of fiber have a higher tensile strength than longer lengths due to probability of the occurrence of more flaws and so, a statistical approach [209] provides a more plausible explanation.

The hybrid effect has been studied [210] with carbon fibers and glass [211–217], carbon fibers and aramid [218–220], carbon fibers and PEEK [221] and carbon fibers and polyethylene [222–232].

The sandwich composite is a special form of hybrid and the sandwiched material itself may or may not be a composite. Hybrids can lower the cost, improve properties and enable composite properties to be tailored.

20.20 SOME SELECTED PROPERTIES OF COMPOSITES

Useful publications covering the properties of composite materials are by Chung [233], Peebles [49] and Donnet [110].

20.20.1 Thermoplastic polymer matrices

Buggy and Carew have reported the effect of thermal ageing of carbon fiber/PEEK [234]. The theoretical modulus based on the Law of Mixtures for a 61% high strength carbon fiber (YM = 227 GPa) and 39% PEEK (YM = 3.9 GPa) is:

$$E_{theory} = 0.61 \times 227 + 3.96 \times 3.6 = 140$$

Hence in practice, the laminate had achieved 98% of the theoretical value.

Mei and Chung [235], using electrical resistance measurement studied the T_g and melting behavior of carbon fiber reinforced thermoplastic composites.

The stiffness of high strength carbon fiber/PEEK composites at 23°C based on a 61%v/v (APC-2) is given in Table 20.27 [236].

20.20.2 Cement matrices

Carbon fiber cement composites have been studied [237–245].

Linton et al [246] showed (Figure 20.46) that the flexural strength of a pitch carbon fiber (Ashland fiber modulus 220 GPa and strength 2.76 GPa) composite was considerably lower than a PAN (Zoltek fiber modulus 45 GPa and strength 0.55 GPa) composite but load-deflection curves were similar. The strength of the fibers are fully utilized in these composites. The use of silica fume prevented weakening due to matrix porosity and microcracking and enabled the carbon fibers to be better dispersed. The presence of sand had little effect on the unreinforced matrix but reduced the strength of the composite (Figure 20.47), believed to be due to breakage of the filaments during the mixing stage, producing fibers that are shorter than the critical length.

Aveston et al [247] prepared Grafil A reinforced Swiftcrete cement composites and the results of tension testing (Figure 20.48) showed considerable scatter due to the difficulty in

Table 20.27 The stiffness of carbon fiber/PEEK @ 23°C based on a 61% v/v high strength carbon fiber (APC-2)

Property	Value
E_1 (GPa)	137 (2)
E_2 (GPa)	9.4 (0.5)
E_3 (GPa)	9.1*
ν_{12}	0.33 (0.01)
ν_{13}	0.32 (0.02)
ν_{21}	0.04 (0.01)
ν_{23}	0.40 (0.10)
ν_{31}	0.04*
ν_{32}	0.40 (0.10)
G_{12} (GPa)	5.1 (0.5)
G_{13} (GPa)	4.7*
G_{21} (GPa)	5.1*
G_{23} (GPa)	3.2
G_{31} (GPa)	4.6*
G_{32} (GPa)	3.2*

Note: *indicates estimated value and 95% confidence limits are quoted in brackets.
Source: Reprinted with permission from Zhen Mei, Chung DDL, Glass transition and melting behavior of carbon fibre reinforced thermoplastic composite, studied by electrical resistance measurement, *Polymer Composites*, 21(5), 2000. Copyright 2000, The Society of Plastics Engineers.

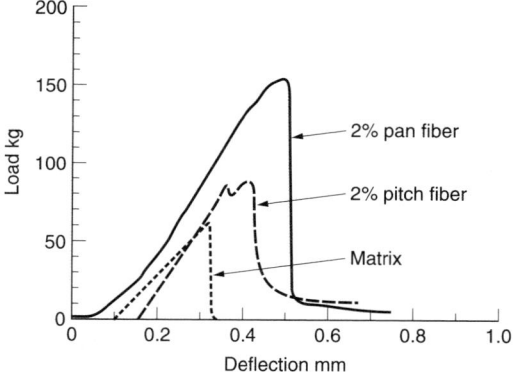

Figure 20.46 Effect of fiber type on composite flexural behavior of a reinforced carbon fiber cement composite.
Source: Reprinted from Linton JR, Berneburg PL, Gartner EM, Bentur A, Carbon fibre reinforced cement and mortar, *Mater Res Symp Proc*, 211, 1991.

making accurate strain measurements on cement composites in direct tension. However, modulus values for the fiber E_f of 192 GPa and the matrix (from the intercept) E_m of 20 GPa were obtained. The strength values are given in Figure 20.49.

Toutanji et al [239] and significantly enhanced the tensile strength properties of cementitious composites reinforced with a low fraction of chopped carbon fiber (YM 235 GPa and UTS 3.80 GPa). The values shown in Figure 20.50 compared favourably with predicted theoretical results based on the law of mixtures.

Kim and Park [248] investigated the effects of three types of carbon fiber shapes (C shape, round and hollow shape) on the tensile and flexural strength development of randomly

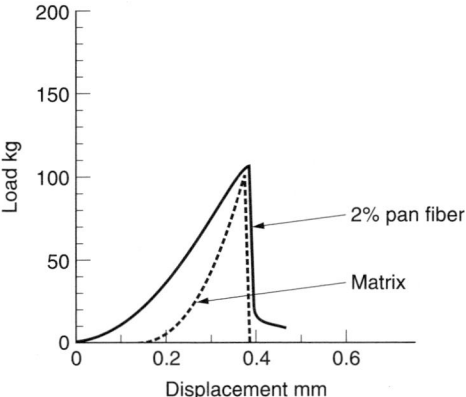

Figure 20.47 Effect of sand on composite flexural behavior of a reinforced carbon fiber cement composite. *Source:* Reprinted from Linton JR, Berneburg PL, Gartner EM, Bentur A, Carbon fibre reinforced cement and mortar, *Mater Res Symp Proc*, 211, 1991.

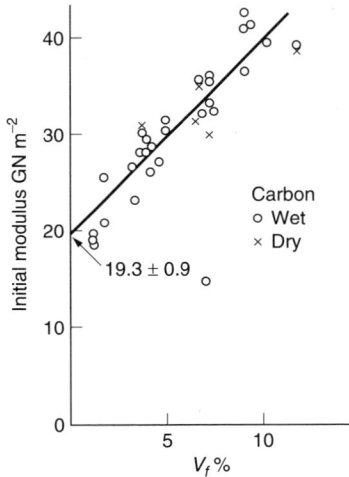

Figure 20.48 Initial Young's modulus of continuous carbon fiber reinforced cement. *Source:* Reprinted from Aveston J, Mercer RA, Sillwood JM, Fibre reinforced cements—Scientific foundations for specifications, *Composites-Standards Testing and Design Conf Proc Composites Standards Testing and Design*, NPL, IPC Science and Technology Press, 174, 93–103, 1974.

oriented carbon fiber reinforced lightweight cement composites (CFRLC). The C-shape CFRLC (C-CFRLC) showed higher tensile and flexural strength development than any other shape. A C-CFRLC fiber volume loading of 3%, in particular, increased the tensile and flexural strength by about 40% compared to round shape CFRLC (R-CFRLC). Hollow-shape CFRLC (H-CFRLC) showed a slightly higher tensile and flexural strength than R-CFRLC, whilst C-CFRLC presented stronger fiber-matrix interfacial adhesion, due to mechanical anchorage into the matrix, than any other fibers. Silica fume significantly influenced the increase in tensile and flexural strength for the CFRLC.

20.20.3 Glass and ceramic matrices

Ziegler *et al* [249] prepared carbon fiber reinforced ceramic matrix composites by infiltration of fiber preforms using the polymer infiltration pyrolysis (PIP) technique.

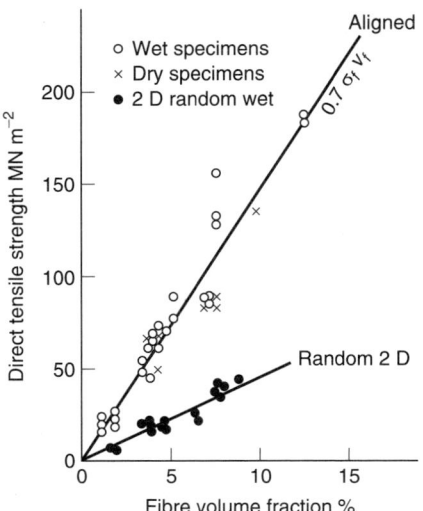

Figure 20.49 Ultimate tensile strength of carbon fiber reinforced cement. *Source:* Reprinted from Aveston J, Mercer RA, Sillwood JM, Fibre reinforced cements—Scientific foundations for specifications, Composite-Standards Testing and Design Conf. *Proc Composites Standards Testing and Design*, NPL, IPC Science and Technology Press, 174, 96, 1974.

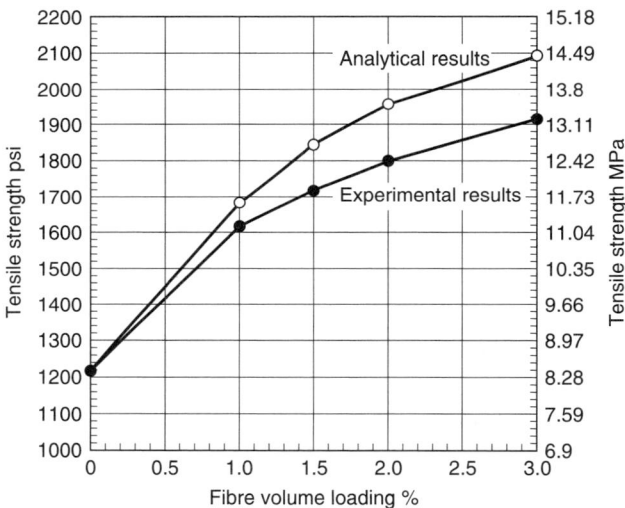

Figure 20.50 Effect of PAN based carbon fibers on the tensile strength of cementitious composite. *Source:* Reprinted with permission from Toutanji HA, El-Korchi T, Katz RN, Leatherman GL, Behaviour of carbon fibre reinforced cement composites in direct tension, *Cement Concrete Research*, 23, 618–626, 1993. Copyright 1993, Elsevier.

They measured the thermal expansion coefficient (CTE) and strength at 1000°C; the C/SiCN composites showed anisotropic expansion behavior. Structural changes in the matrix, when tempered above the processing temperature of 1000°C, were responsible for an increase of the CTE values. Bending strength showed a maximum value after 5 PIP cycles. The strength at 1000°C was slightly higher than at room temperature. The strength values were clearly strongly influenced by the formation of internal stresses due to thermal mismatch of fiber and matrix.

Hegeler and Bruckner [250] have reported the mechanical properties of carbon fiber reinforced glasses and Saewong and Rawlings have determined the erosion of carbon fiber reinforced glass-ceramic composites [251].

Zheng et al [252] determined the properties of carbon fiber reinforced SiC composites, fabricated by multiple impregnation of polycarbosilane and subsequent pyrolysis at 1200°C, were improved by adding B powders (0.1 µm) to the matrix. Three types of carbon fibers, HSCF (PAN based, $E = 240$ GPa), HMCF (pitch based, $E = 277$ GPa) and CF70 (pitch based, $E = 700$ GPa), were used as reinforcements. It was found that the addition of B had only slight effect on the mechanical properties of HSCF/SiC and CF70/SiC, but a significant effect on the properties of HMCF/SiC. The HMCF/SiC, without B addition, showed brittle fracture mode, with a flexural strength of 270 MPa, whereas HMCF/SiC with B addition showed non-brittle fracture mode, with a flexural strength of 690 MPa. XRD analysis of B heat treated at 1200°C in N_2 showed formation of BN, which is regarded to have a weak bonding with fiber or matrix. The additives lowered the direct contact area between the carbon fiber and SiC matrix, hence weakening the interfacial bonding strength of CF/SiC.

Xu et al [253] prepared three dimensional continuous carbon fiber reinforced SiC composites by CVI and the microstructure and mechanical properties were investigated. For composites (C/SiC) with no pyrolytic carbon interfacial layer, the mechanical properties (flexural strength, flexural elastic modulus, shear strength, and fracture toughness) increase with increasing composite density. High density (2.1 $g\,cm^{-3}$) C/SiC composites exhibited high fracture toughness (16.5 $MPa\,m^{-2}$), but brittle fracture behavior because of strong fiber/matrix bonding. Low density composites showed a non-catastrophic failure mode with long fiber bundle pull-out. The composites (C/PyC/SiC) with a pyrolytic carbon interfacial layer exhibited good mechanical properties and typical failure behavior.

Unidirectional carbon fiber reinforced SiC composites were prepared from PAN based HSCF, pitch based HMCF, CF50 and CF70 carbon fibers, through nine or twelve cycles of impregnation of polycarbosilane and subsequent pyrolysis at 1200°C [254]. The polycarbosilane derived matrix was found to be β-SiC with a crystallite size of 1.95 nm. The mechanical properties of the composites were evaluated by 4-point bending tests. CF50/SiC and CF70/SiC exhibited high strength and non-brittle fracture mode, with multiple matrix cracking and extensive fiber pullout, whereas HSCF/SiC and HMCF/SiC exhibited low strength and brittle fracture with almost no fiber pullout. Values of flexural strengths of CF70/SiC and CF50/SiC were 967 MPa and 624 MPa, respectively. The relatively lower strength of CF50/SiC was attributed mainly to shear failure of CF50/SiC during bending tests.

Modal fracture energies of carbon fiber reinforced pitch derived carbon composites heat treated at 1000°C and 1200°C were evaluated [255] and the tensile fracture energies of the composites were 0.92 and 1.4 $kJ\,m^{-2}$ respectively, the shear fracture energies were 0.020 and 0.030 $kJ\,m^{-2}$, respectively.

Tani and Shobu [256] fabricated unidirectional carbon fiber reinforced SiC composites with a polymer pyrolysed and reaction bonded SiC matrix. In spite of a low density of about 1.85 $mg\,m^{-3}$, the flexural strength of the composites was about 530 MPa. Addition of polymer decreased the open porosity and increased the elastic modulus of the composites.

The toughness of boride particle and boride particle-carbon fiber reinforced SiC composites was evaluated [257] by pre-cracked (SEPB) and notched beams (SEVNB) from room temperature to 1773 K. Both methods gave a decreased value of toughness with TiB_2/SiC composite at 1773 K, from 5.3 to 3.7 $MPa\,m^{-2}$. In the case of TiB_2-CF/SiC, the toughness of SEPB increased, from 7.3 to 8.0 $MPa\,m^{-2}$, even though the toughness of

SEVNB decreased at 1723 K from 4.8 to 4.2 MPa m^{-2}. The results suggest that fiber bridging is an effective toughening mechanism at high temperature. A material with toughness independent of temperature was developed by using fiber bridging.

Yoshida et al [258] fabricated unidirectional and two-directional carbon fiber reinforced SiC composites (UD-Cf/SiC and 2D-Cf/SiC, respectively) by hot pressing to investigate the influence of fiber orientation angle (θ) to the direction of applied load on flexural strength and fracture toughness. The UD-Cf/SiC composite with $\theta = 0°$ showed highest strength and highest fracture toughness. They were 950 MPa at room temperature, 1050 MPa at 1450°C and 30 MPa m^{-2}, respectively. The flexural strength of UD-Cf/SiC composites with $\theta = \theta°$ were estimated by using flexural strengths of UD-Cf/SiC composites with $\theta = 0°$ and $\theta = 90°$ and Tai-Hill Theory. Flexural strength and fracture toughness of 2D-Cf/SiC composites with the orientation angle of $+\theta/-\theta/+\theta/-\theta\ldots$ were higher than those of UD-Cf/SiC with the same orientation angle of $+\theta/+\theta/\ldots$.

Klug et al [259] applied alternating bending loads to SiC and carbon fiber reinforced alkaline earth borosilicate glass samples. The mechanical properties were determined by a 3-point bend test after 1000 stress reversals, and compared to untreated samples. It was determined that all the composites could be subjected to an alternating bending load up to their elastic limits without failure or loss of mechanical properties. Alternating loads in excess of this limiting value led to rapid fatigue of SiC fiber reinforced glass, whereas the carbon fiber reinforced material had greater tolerance, but homogeneity of the fiber distribution was important.

SiC composites reinforced with unidirectional carbon fibers were prepared by impregnation with a slurry containing SiC, AlB$_2$, polysilastyrene and toluene, followed by hot pressing [260]. The flexural strength and fracture toughness were measured at room temperature and at high temperature in vacuum. Strength and fracture toughness were 420 MPa and 13 MPa m^{-2} at room temperature, and 600 MPa and 20 MPa m^{-2} at 1400°C to 1600°C. The improved mechanical properties at high temperatures were attributed to the suppression of crack propagation and to increased fiber pull-out due to softening of the fiber-matrix interface.

20.20.4 Carbon–carbon

The properties of carbon–carbon composites are discussed by Savage [261] and Chung [233] and by Hüttner [262]. Carbon–carbon composites are much stronger than graphite components and properties are improved further with a three dimensional construction. Peters et al [263] determined the strength of unidirectional carbon fiber reinforced carbon matrix composites with fiber treated at different oxidation levels, after each step of the production cycle. An impregnated bundle had a strength 1.9–4.3 times the strength of a loose bundle, but after carbonisation, the strength was less than that of a loose bundle. This was attributed to the formation of defects in the fiber due to shrinkage during carbonisation. The defects were larger with a good fiber/matrix bond in the green material. Low strength of the composite was not caused by stress concentration effects.

Flexural strength and modulus of carbon–carbon increase with the density, the strength increasing up to a value of about 500 MPa. The flexural modulus has a value about twice that of resin matrix composites [264].

Sato and co-workers [266] prepared three types (A, B and C) of carbon–carbon composites (Table 20.28) and found that all composites had superior mechanical properties to a control sample (G) of isostatically pressed fine grained graphite. The PAN based C/C composite (B) was superior to the other composites (A and C) despite having a lower fiber

PROPERTIES OF CARBON FIBERS

Table 20.28 Mechanical properties of three carbon/carbon composites compared with graphite

		A	B	C	G
Bulk density (g cm^{-3})		1.68	1.77	1.57	1.76
Young's modulus (GPa)		13.5	26.3	17.0	10.5
Vickers hardness (5 kg load)		135	163	–	172
Bending strength (MPa)		65.7	96.9	–	39.6
Tensile strength (MPa)	RT	35.7	55.4	68	28
	800°C	43.4	65.4	88	30
	1600°C	42.0	50.4	102	37
	2400°C	62.7	83.0	111	44
Fracture toughness (MPa m^{-1})	RT	2.96	3.44	4.0	0.8
	800°C	2.82	3.58	5.5	0.8
	1600°C	4.64	6.75	6.1	1.0
	2400°C	5.30	12.9	7.0	1.9
Thermal diffusivity (mm^{2}s^{-1})		62.4	56.6	–	48.0
Thermal shock resistance (W mm^{-1})		≈148	≈155	≈171	50±6
Thermal shock fracture toughness (Wmm^{-2})		≈779	≈805	≈856	33±3

A Pitch based carbon fiber felt (47% w/w)/carbon composite (HTT 3000°C).
B PAN based carbon fiber felt (34% w/w)/carbon composite (HTT 3000°C).
C Stacked rayon based carbon fiber cloths/carbon composite (HTT 3000°C).
G Isostatically molded fine grained graphite (HTT 3000°C).
Source: Reprinted from Sato S, Kurumada A, Iwaki H, Komatsu Y, *Carbon*, 27(6), 791–801, 1989.

content, C except the rayon based composite (C) did have superior tensile strength. The low thermal shock resistance of the rayon based composite (C) was attributed to interlaminar failure between the plies of woven reinforcement, whereas the felt composites (A and B) had higher values, possibly attributed to the adventitious effect of partial 3-D strength imparted by the felts.

Typical properties of carbon–carbon composites are given in Table 20.29 and figures for the Dunlop products are quoted in Table 20.30.

The mechanical properties are significantly affected by the heat treatment temperature (Figure 20.51) and the level of surface treatment (Figure 20.52). Manocha et al. [267] showed that a carbon-carbon composite when carbonized at 1000°C will exhibit significant differences when graphitized at 2700°C as depicted in Fig. 20.51. There is a 54% in creak in flexural strength, 40% decrease in ILSS and a 932 increase in flexural modulus. These results are strongly indicative of a change in the fiber matrix interaction. The increase in modulus can be probably attributed to the further graphitization of the pitch based fiber assisted by envelopment within the carbon matrix. Hüttner [262] states that the toughness peaks at a level of about 2400°C and values for thermal and electrical conductivities are given in Table 20.31.

The thermal conductivity of vgcf has a particularly high value of about 1900 W m^{-1}K^{-1} at 25°C [265] and when incorporated in a carbon matrix can produce a composite with a thermal conductivity >1000 W m^{-1}K^{-1}.

20.21 METAL MATRICES

Soni *et al* [269] determined the tensile behavior of carbon fiber reinforced Al matrix composites. The thermal expansion behavior of unidirectional carbon fiber reinforced Cu matrix composites was undertaken by Korb *et al* [270].

Table 20.29 Typical properties of carbon–carbon materials

Property	Carbenix 2100	Carbenix 4000	Balanced 2-D T-300	Unbalanced 2-D T-300	3-D T-300 Woven	Pitch Non-woven
Fiber	P-25	Panox	T-300	T-300	T-300	P-100
Fiber type	Pitch	PAN	PAN	PAN	PAN	Pitch
Fiber heat treatment (°C)	1300	1800	2200	2200	2200	2200
Fiber modulus (GPa)	172		352	352	352	690
Reinforcement	chopped fiber	needled felt	balanced 8HS fabric	3:1, 5HS fabric	Anie interlock	needled fabric
Type of lay-up	random-molded	radial/cordal	0/90°	0°		quasi-isotropic
Fiber volume (%)	35	25–28	46–50	55–60	35	30
Type of matrix	char/CVD	CVD	char/filler/CVD	char/CVD	char/CVD	CVI
Tensile strength (MPa) x	41–62	55–69	27.6–31.0	48.3–51.7	9.7	
y				152–172	90	
Modulus (GPa) x	6.9–10.3	6.9–10.3	83–90	145–166	35	
Compressive strength (MPa) x	120–140	120–150	140–150			
Iosopecu shear (MPa) xy			45–55		38	
Double notch shear (MPa)			14–21	12–17		
Interlaminar tensile (MPa)			7–10	7–9	17	
Thermal conductivity x,y W/m°C z			100–120			300–400
			30–50			250–350
Coeff Thermal Expansion x,y 10^{-6} °C^{-1} z	2.0–2.3	0.9	0.9			
	2.2–2.7	4.5	4.5			

PROPERTIES OF CARBON FIBERS

Table 20.30 Physical and mechanical properties of a range of Dunlop carbon–carbon composites

Property	Units	CA7	CA8	CB5	CB7	CC3	CD4
Density	g cm^{-3}	1.75–1.8	1.75–1.8	1.70–1.75	1.70–1.75	1.75–1.80	1.80
Tensile strength axial	MPa	70	55	80	70	120	250
Tensile modulus axial	GPa	30	30	40	35	60	70
Compressive strength axial	MPa	150	100	170	130	190	350
Compressive strength transverse	MPa	110	100	110	100	130	130
Compressive modulus axial	GPa	20	15	20	15	30	60
Compressive modulus transverse	GPa	2.0	1.9	2.0	1.9	3	2
Flexural strength transverse	MPa	120	85	135	120	250	400
Flexural modulus transverse	GPa	28	28	30	29	50	70
Interlaminar shear strength axial	MPa	17	9	17	12	25	10
Thermal conductivity axial	W m^{-1}K^{-1}	20	120	25	150	30	40
Thermal conductivity transverse	W m^{-1}K^{-1}	8	30	8	40	9	9
Thermal expansion axial RT to 500°C	W m^{-1}K^{-1}	0.06	0.05				
1000°C	W m^{-1}K^{-1}	0.2	0.2				
1500°C	W m^{-1}K^{-1}	0.3	0.3				
Thermal expansion transverse RT to 500°C	W m^{-1}K^{-1}	0.7	0.6				
1000°C	W m^{-1}K^{-1}	1.3	1.2				
1500°C	W m^{-1}K^{-1}	1.7	1.6				
Lay up		±22.5° Fabric	±22.5° Fabric	±90° Fabric	±90° Fabric	±90° Fabric	U.D Rod

Source: Reprinted from Dunlop technical literature.

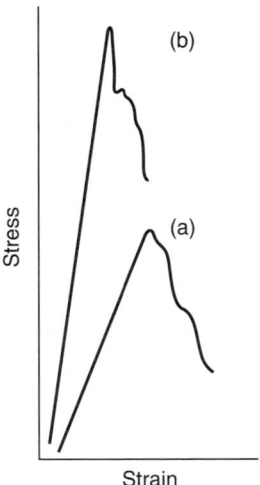

Figure 20.51 Generalized tensile stress–strain curves for dc composites made from pitch based carbon fiber fabric (a) as carbonized (b) graphitized composites. *Source:* Reprinted from Manocha LM, Bahl OP, Singh YK; Jones FR ed., *Proc Int Conf Interfacial Phenomena in Composite Materials '89*, Butterworth, 310–315, 1989.

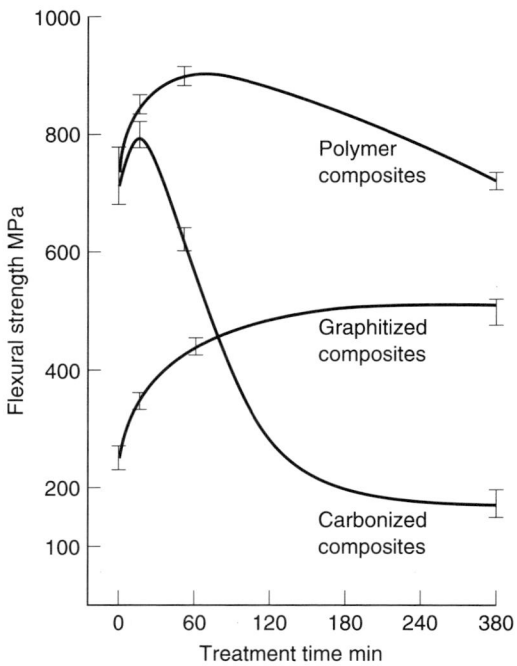

Figure 20.52 The effect of time of surface treatment on the flexural strength of carbonized and graphitized carbon–carbon. *Source:* Reprinted from Manocha LM, Bahl OP, Singh YK; Jones FR ed., *Proc Int Conf Interfacial Phenomena in Composite Materials '89*, Butterworth, 310–315, 1989.

Table 20.31 Thermal conductivity and electrical resistivity parallel and perpendicular to the laminates of two dimensional weave carbon–carbon composites

Heat treatment temperature °C	Thermal conductivity W m^{-1}K^{-1}		Electrical resistivity μω m	
	‖	⊥	‖	⊥
1200	36–43	4–7	33–37	98–114
2800	127–134	39–46	8–12	68–81

Source: Reprinted with permission from Hüttner W; Figueiredo JL, Bernardo CA, Baker RTK, Hüttinger KJ eds., *Carbon Fibers Filaments and Composites*, Kluwer Academic Publishers, Dordrecht, 1990. Copyright 1990, Springer.

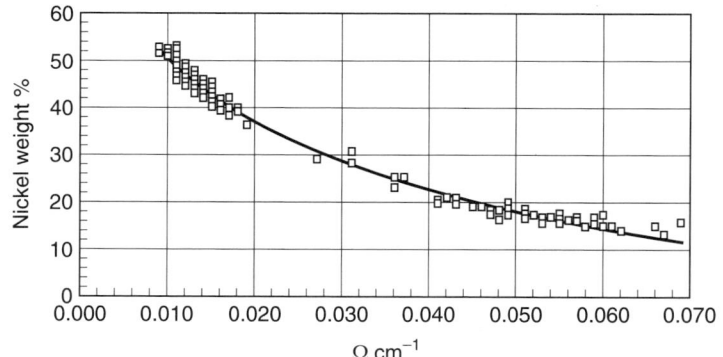

Figure 20.53 Conductivity of nickel coated carbon fiber as a function of wt% Ni. *Source:* Reprinted from Rosenow MWK, *ETP'99 World Congress Engineering Thermoplastics*, Zurich, Jun 7–9, 1999.

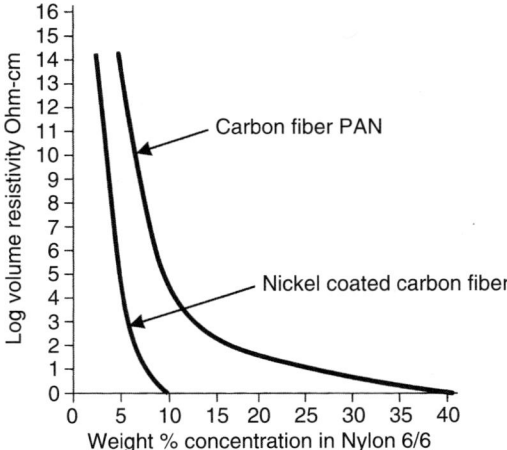

Figure 20.54 Effect of the concentration of PAN based carbon fiber and Ni coated carbon fiber on the volume resistivity of Nylon-6,6. *Source:* Reprinted from RTP technical literature.

Bell and Hansen [271] determined the properties of Ni coated carbon and Kevlar fibers produced by decomposition of nickel carbonyl.

Figure 20.53 shows the conductivity of Ni coated carbon fiber as a function of the weight per cent of Ni.

The effect of the concentration of PAN based carbon fiber and Ni coated carbon fiber on the volume resistivity of Nylon-6,6 is shown in Figure 20.54.

REFERENCES

1. Takaku A, Shioya M, *J Mater Sci*, 25, 4873, 1990.
2. Matsumoto T, *Pure Appl Chem*, 57, 1553, 1985.
3. Ozbek S, Isaac DH, Carbon fibres: Effect of processing parameters on mechanical properties, PED 49:/MD27, *Proc Manuf Composite Mater, ASME*, 307–320, 1991.
4. Jones BF, Duncan RG, *J Mater Sci*, 6, 289, 1971.
5. Fitzer E, Kunkele F, High Temp-High Pressures, 22(3), 239–266, 1990.
6. Hancox NL, Mayer RM, *Design Data for Reinforced Plastics—A Guide for Engineers and Designers*, Chapman & Hall, London, 118, 1994.
7. Kowalski IM, Composite Materials, Testing and Design, Whitcombe JD ed., *ASTM STP972*, ASTM, Philadelphia, 1988.
8. Kowalski IM, Composite Materials, Testing and Design, ASTM STP972, Whitcombe JD ed., ASTM, Philadelphia, 1988.
9. Goggin PR, A method of measuring the quality of carbon fibres, *AERE R-7790*.
10. Moreton R, *Fibre Sci Technol*, 1, 273, 1969.
11. Barry PW, *Fibre Sci Technol*, 11, 245, 1978.
12. Baconr, Schalamon W, *Appl Polym Symp*, 9, 285, 1969.
13. Chwastiak S, Barr J, Didchenko R, *Carbon*, 17, 49, 1979.
14. Westbury MC, Drzal LT, *J Composite Technol Res*, 13(1), 22–28, 1991.
15. Powell PC, *Engineering with Polymers*, Chapman & Hall, London, 176–177, 1988.
16. Beetz CP Jr., *Fibre Sci Technol*. 16, 45–59, 1982.
17. Hitchon JW, Phillips DC, *Fibre Sci Technol*, 12, 217, 1979.
18. Own SH, Subramanian RV, Saunders SC, *J Mater Sci*, 21, 3912, 1986.

19. Dobb MG, Guo H, Johnson DJ, Park CR, Structure–compressional property relations in carbon fibres, *Carbon*, 33(11), 1553–1559, 1995.
20. Johnson JW, *Am Chem Soc Symp on Fibres of Thermally Resistant Organic Polymers*, Atlantic City, 1968.
21. Cottrell AH, *Proc Royal Soc*, A276,1, 1963.
22. Reynolds WN, Structure and mechanical properties of carbon fibres, *Proc 3rd Conference Industrial Carbons and Graphites*, SCI, London, 427–430, 1971.
23. Coleman BD, On the strength of classical fibres and bundles, *JMPS*, 7, 60, 1958.
24. Hughes JDH, Jackson EE, The potential strength of Type II carbon fibre composites, A practical assessment, *AERE-R8090*, Sep 1975.
25. Hughes J, Morley H, Jackson E, *J Phys D, Appl Phys*, 13, 921, 1980.
26. Jones WR, Johnson JW, *Carbon*, 9, 645, 1971.
27. Tsushima E, *Proc 34th Int SAMPE Symp*, 2042, 1989.
28. Tsai JS, Wu CJ, Control of Young's Modulus of carbon fibre, *J Mater Sci Lett*, 13(4), 272–274, 1994.
29. Hamada H, Oya N, Yamashita K, Maekawa ZI, Tensile strength and its scatter of uni-directional carbon fibre reinforced composites, *J Reinf Plastics Composites*, 16(2), 119–130, 1997.
30. Kobets LP, Deev IS, Carbon fibres structure and mechanical properties, *Composite Sci Technol*, 57(12), 1571–1580, 1997.
31. Ibarra L, Macias A, Palma E, Mechanical properties of composite materials consisting of short carbon fibre and thermoplastic elastomers, *Kautschuk Gummi Kunstoffe*, 48(3), 180–184, 1995.
32. Jacobsen RL, Tritt TM, Guth JR, Ehrlich AC, Gillespie DJ, *Carbon*, 33(9), 1217–1221, 1995.
33. Tibbetts GG, Doll GL, Gorkiewicz DW, Moleski JJ, Perry TA, Dasch CJ, Balogh MJ, Physical properties of vapour grown carbon fibres, *Carbon*, 31(7), 1039–1047, 1993.
34. Pegoraro M, Pagani G, Clerici P, Penta A, Interaction between short sub-critical length fibres and polymer matrix, *Fibre Sci Technol*, 10, 263, 1977.
35. Folkes MJ, *Short Fibre Reinforced Thermoplastics*, Research Studies Press, Letchworth, 1982.
36. Bigg DM, Characteristics of short, conductive fibre reinforced injection mouldable composites, *J. Ind Fabrics*, 2(3), 4–14, 1983–1984.
37. See 22.
38. Goggin PR, The elastic constants of carbon fibre, *J Mater Sci*, 8, 233–244, 1973.
39. Wagoner G, Bacon R, Elastic constants and thermal expansion coefficients of various carbon fibres, *19th Biennial Conference on Carbon*, Penn State University, Session 6A, 296–297, 1989.
40. Northolt MG, Veldhuizen LH, Jansen H, Tensile deformation of carbon fibres and the relationship with the modulus for shear between the basal planes, *Carbon*, 29, 1267, 1991.
41. Kawabata S, Measurement of the transverse mechanical properties of high performance fibres, *J Text Inst*, 81(4), 432–447, 1990.
42. Darby MI, Richards JM, Yates B, Effect of discontinuous plies on the tensile strength of CFRP laminates, *J Mater Sci Lett*, 4, 203–206, 1985.
43. Cox HL, *Brit J Appl Phys*, 3, 72, 1952.
44. Wisnom MR, The flexural strength of unidirectional carbon fibre epoxy, *5th European Conference on Composite Materials*, Elsevier, Bordeaux, 165–170, 1992.
45. Wisnom MR, The relationship between tensile and flexural strength of unidirectional composites, *J Composite Mater*, 26(8), 1173–1180, 1992.
46. Goan JC, Martin TW, Prescott R, The influence of interfacial bonding on the properties of carbon fibre composites, *28th Ann Tech Conf Reinf Plast/Comp Inst, 28*, 21B, 1–4, 1973.
47. Wright WW, The carbon fibre/epoxy resin interphase—a Review—Part 1, *Composite Polym (now called Polym and Polym Composites)*, 3(4), 231–257, 1990 (Note references for this article are given in Part II of reference 47).
48. Wright WW, The carbon fibre/epoxy resin interphase a review—Part II, *Composite Polym (now called Polym and Polym Composites)*, 3(5), 360–391, 1990.
49. Peebles LH Jr., *Carbon Fibers, Formation, Structure and Properties*, CRC Press, Boca Raton, 113–120, 1994.

50. Rich MJ, Drzal LT, *J Reinf Plast Composites*, 7(2), 145–154, 1988.
51. Jacques D, Favre JP, Determination of the interfacial shear strength by fibre fragmentation in resin systems with a small rupture strain, *6th International Conf on Composite Mater*, Matthews FL, Buskell NCR, Hodgkinson JM eds., Elsevier Applied Science, New York, 5, 471, 1987.
52. Drzal LT, Rich MJ, Koenig MF, Adhesion of graphite fibres to epoxy matrices, III. The effect of hygrothermal exposure, *J Adhesion*, 18, 49–72, 1985.
53. Herrera-Franco PJ, Drzal LT, Comparison of methods for the measurement of fibre/matrix adhesion in composites, *Composites*, 23, 2, 1992.
54. Drzal LT, Madhukar M, *J Mater Sci*, 28, 569, 1993.
55. Broughton WR, Kumosa M, Hull D, Analysis of the Iosipescu shear test as applied to unidirectional carbon-fibre reinforced composites, *Composite Sci Technol*, 38(4), 299–325, 1990.
56. Browning CE, Abrams FL, Whitney JM, A four point shear test for graphite/epoxy composites, Browning CE ed., *Composite Materials, Quality Assurance and Processing, ASTM STP 797*, American Society for Testing and Materials, Philadelphia, 54–74, 1983.
57. Cui WC, Wisnom MR, Jones M, Effect of specimen size on interlaminar shear strength of unidirectional carbon fibre-epoxy, *Composites Eng*, 4(3), 299–307, 1994.
58. Harding J, Dong L, Effect of strain rate on the interlaminar shear-strength of carbon-fibre-reinforced laminates, *Composite Science and Technology*, 51(3), 347–358, 1994.
59. Lee S, Munroe M, In-plane shear properties of graphite/epoxy composites for aerospace applications, evaluation of test methods by the decision analysis method, *Aeronautical Note NAE-AS22, NRC No.23778*, Mechanical Engineering Department, University of Ottawa, Oct 1984.
60. Liu K, Piggott MR, Shear strength of polymers and fibre composites. 2. Carbon/epoxy pultrusions, *Composites*, 26(12), 841–848, 1995.
61. Lockwood PA, Results of the ASTM Round Robin on the rail shear test for composites, *Composites Technol Rev*, 3, 83, 1981.
62. Sawada Y, Shindo A, Torsional properties of carbon-fibres, *Carbon*, 30(4), 619–629, 1992.
63. Swanson SR, Merrick M, Toombes GR, Comparison of torsion tube and Iosipescu in-plane shear test results for a carbon fibre reinforced epoxy composite, *Composites*, 16, 8220, 1985.
64. Hofer KE, Rao PN, A new static compression fixture for advanced composite materials, *J Testing Eval*, 5(4), 278–283, 1977.
65. Häberle JG, Matthews FL, Studies on compressive failure in unidirectional cfrp using an improved test method, *Proc ECCM-4 Conf*, Elsevier Applied Science Sep 1980.
66. Adsit NR, Compression testing of graphite/epoxy, Chait R, Papirno R eds., *ASTM STP 734*, American society for Testing Materials, 152–165, 1981.
67. Woolstencroft DH, Curtis AR, Haresceugh RI, A comparison of test techniques used for the evaluation of the unidirectional compressive strength of carbon fibre reinforced plastics, *Composites*, 12, 82–83, 1981.
68. Port KF, The compressive strength of cfrp, Royal Aircraft Establishment, *Farnborough Tech Report 82083*, 1982.
69. Adsit NR, Compression testing of graphite/epoxy, Chait R, Papirno R eds., *Compression Testing of Homogeneous Materials and Composites, ASTM STP 808*, American Society for Testing and Materials, 175–186, 1983.
70. Lamothe RM, Nunes J, Evaluation of fixturing for compression testing of metal matrix and polymer/epoxy composites, Chait R, Papirno R eds., *Compression Testing of Homogeneous Materials and Composites, ASTM STP 808*, American Society for Testing and Materials, 241–253, 1983.
71. Woolstencroft DH, Curtis AR, Haresceugh RI, A comparison of test techniques used for the evaluation of the unidirectional compressive strength of carbon fibre reinforced plastics, *Composites*, 16, 8220, 1985.
72. Clark RK, Lisagor WB, Compression testing of graphite/epoxy composite materials, Chamis CC ed., *Test Methods and Design Allowables for Fibrous Composites, ASTM STP 734*, 34.
73. Barker AJ, Balasundaram V, Compression testing of carbon fibre reinforced plastics exposed to humid environments, *Composites*, 18(3), Jul 1987.

74. Berg JS, Adams DF, An evaluation of composite material compression test methods, *J Composite Technol Res*, 11, 41–46, 1989.
75. Dobb MG, Johnson DJ, Park CR, Compressional behaviour of carbon fibres, *J Mater Sci*, 25(2A), 829–834, 1990.
76. Melanitis N, Galiotis C, Compressional behaviour of carbon-fibres, 1. Raman spectroscopic study, *J Mater Sci*, 25(12), 5081–5090, 1990.
77. Ohsawa T, Miwa M, Kawade M, Tsushima E, Axial compressive strength of carbon fibre, *J Appl Polym Sci*, 39(8), 1733–1743, 1990.
78. Wisnom MR, The effect of fibre misalignment on the compressive strength of unidirectional carbon-fibre epoxy, *Composites*, 21(5), 403–407, 1990.
79. Curtis PT, Gates J, Molyneaux CG, An improved engineering test method for the measurement of compressive strength of unidirectional carbon fibre composites, *Technical Report 91031*, DRA, Farnborough, 1991.
80. Prandy JM, Hahn HT, Compressive strength of carbon fibres, *SAMPE Quarterly*, 22(2), 47–52, 1991.
81. Soutis C, Measurement of the static compressive strength of carbon-fiber epoxy laminates, *Composite Sci Technol*, 42(4), 373–392, 1991.
82. Carr DJ, Barker AJ, Compressive properties of carbon-fibre reinforced plastics, *Chem Eng Res Des*, 71(A3), 316–318, 1993.
83. Kumar S, Anderson DP, Crasto AS, Carbon-fibre compressive strength and its dependence on structure and morphology, *J Mater Sci*, 28(2), 423–439, 1993.
84. Shinohara AH, Sato T, Saito F, Tomioka T, Arai Y, A novel method for measuring direct compressive proprties of carbon-fibres using a micromechanical compression tester, *J Mater Sci*, 28(24), 6611–6616, 1993.
85. Dobb MG, Guo H, Johnson DJ, Park CR, Structure-compressional property relations in carbon fibres, *Carbon*, 33(11), 1553–1559, 1995.
86. Miwa M, Mori Y, Takeno A, Yokoi T, Watanabe A, Compressive and tensile behaviour of carbon fibres, *J Mater Sci*, 33(8), 2013–2017, 1998.
87. Nakatani M, Shioya M, Yamashita J, Axial compressive fracture of carbon fibres, *Carbon*, 37(4), 601–608, 1999.
88. Kumar S, Helminiak TE, The Materials Science and Engineering of Rigid-Rod Polymers, Adams WW, Eby RK, McLemore DE eds., *MRS Symp Proc*, 134, 363–374, Pittsburgh, 1989.
89. Ewins PD, Ham AC, The nature of compressive failure in unidirectional carbon fibre reinforced plastics, *Royal Aircraft Establishment Technical Report 73057*, 1973.
90. Furuyama N, Higuchi M, Kubomura K, Jiang H, Kumar S, Compressive properties of single filament carbon fibres, *J Mater Sci*, 28, 1611, 1993.
91. Hawthorne HM, Teghtsoonian E, Axial compression fracture in carbon fibres, *J Mater Sci*, 10, 41, 1975.
92. Ohsawa T, Miwa M, Kawade M, Tsushima E, Axial compressive strength of carbon fibre, *J Appl Polym Sci*, 39, 1733, 1990.
93. Shinohara AH, Sato T, Saito F, Tomioka T, Arai Y, *J Mater Sci*, 28, 6611, 1993.
94. Macturk KS, Eby RK, Adams WW, Characterization of compressive properties of high performance polymer fibres with a new micro-compression apparatus, *Polymer*, 32, 1782, 1991.
95. Allen SR, Tensile recoil measurement of compressive strength for polymeric high performance fibres, *J Mater Sci Eng*, 22, 853, 1987.
96. Wang CS, Bai SJ, Rice BP, Axial compressive strengths of high performance fibres by tensile recoil technique, *Polym Mater Sci Eng*, 61, 550, 1989.
97. Dobb MG, Johnson DJ, Park CR, *J Mater Sci*, 25, 829, 1990.
98. De Teresa SJ, Piezoresistivity and failure of carbon filaments in axial compression, *Carbon*, 29, 397, 1991.
99. Macturk KS, Eby RK, Adams WW, *Polymer*, 32, 1782, 1991.
100. Melanitis N, Galiotis C, Compressional behaviour of carbon fibres, Part 1. A Raman spectroscopic study, *J Mater Sci*, 25, 5081, 1990.
101. Vlattas C, Galiotis C, *Polymer*, 32, 1788, 1991.

102. Everall N, Lumsden J, Fundamental reproducibility of Raman band positions and strain measurements of high modulus carbon fibres—the effect of laser induced heating, *J Mater Sci*, 26, 5269, 1991.
103. Kumar S, Anderson DP, Crasto AS, *J Mater Sci*, 28, 423, 1992.
104. Curtis PT, Gates J, Molyneux CG, An improved engineering test method for measurement of compressive strength of unidirectional carbon fibre composites, *Composites*, 22, 363, 1991.
105. Prandy JM, Hahn HT, *SAMPE Quarterly*, 22, 47–52, 1991.
106. Kumar S, Structure and properties of high performance polymeric and carbon fibres—an overview, *SAMPE Quarterly*, 20, 3, 1989.
107. Northolt MG, Veldhuizen LH, Jansen H, *Carbon*, 29, 1267–1279, 1991.
108. Dobb MG, Johnson DJ, Park CR, Structure compressional property relations in carbon fibres, *J Mater Sci*, 25, 829, 1990.
109. Dobb MG, Guo H, Johnson DJ, Park CR, Structure compressional property relations in carbon fibres, *Carbon*, 33(11), 1553–1559, 1995.
110. Donnet JB, Bansal RC, *International Fiber Science and Technology 10 (Carbon Fibers)*, 2nd ed. Marcel Dekker, New York, 267–366, 1990.
111. Parry TV, Wronski AS, Kinking and tensile compressive and interlaminar shear failure mechanisms in cfrp beams tested in flexure, *J Mater Sci*, 16,439–450, 1081.
112. West AR, *J Mater Sci Lett*, 18, 2026, 1981.
113. Hull D, *An introduction to composite materials*, Cambridge University Press, Cambridge, 160, 1981.
114. Ewins PD, Potter RT, Some observations on the nature of fibre reinforced plastics and the implications for structural design, *Phil Trans R Soc London*, A294, 507–517, 1980.
115. Freeman W, Campbell MD, Thermal expansion characteristics of graphite reinforced composite materials, *ASTM STP 497*, Philadelphia, American Society for Testing and Materials, 121–142, 1972.
116. Mijovic J, Gsell TC, Calorimetric study of poyetheretherketone (PEEK) and its carbon-fibre composite, *SAMPE Quarterly*, 21(2), 42–46, 1990.
117. Schwarz G, Krahn F, Hartwig G, Thermal-expansion of carbon-fibre composites with thermoplastic matrices, *Cryogenics*, 31(4), 244–247, 1991.
118. Tsai CL, Daniel IM, Method for thermomechanical characterization of single fibres, *Composites Sci Technol*, 50, 7–12, 1994.
119. Wang ASD, Pipes RB, Ahmadi A, Thermoelastic expansion of graphite/epoxy unidirectional and angle-ply composites, *Composite Reliability ASTM STP580*, American Society for Testing and Materials, Philadelphia, 574–585, 1975.
120. Gibbs HH, Wendt RC, Wilson FC, *33rd Annual Technical Conference SPI*, 1978.
121. Gibbs HH, Wendt RC, Wilson FC, Carbon fibre structure and stability studies, *Polym Eng Sci*, 19(5), 342–349, Apr 1979.
122. McMahon PE, 23rd *SAMPE*, 150–159, 1978.
123. McMahon PE, *ASTM STP658*, 254–266, 1978.
124. Gourdin C, *SNPE Technical Reports*, 34/81, 155/81 and 54/82 DRET Contract No. 80/195.
125. Gourdin C, Ageing of carbon fibres of various origins, *SAMPE*, Bourdeaux, 49–61, Oct 17–20, 1983.
126. Gibbs HH, Wendt RC, Wilson FC, *33rd Ann Tech Conf SPI*, 1978.
127. McMahon PE, *23rd SAMPE*, 150–159, 1978 and *ASTM STP658*, 254–266, 1978.
128. Page KL, Duquette DJ, Oxidation behaviour of carbon reinforced glass matrix composites, *Ceramic Int*, 23(3), 209–213, 1997.
129. Freeman W, Campbell MD, Thermal expansion characteristics of graphite reinforced composite materials, *ASTM STP497*, American Society for Testing and Materials, Philadelphia, 121–142, 1972.
130. Wang ASD, Pipes RB, Ahmadi A, Thermoelastic expansion of graphite/epoxy unidirectional and angle-ply composites, *Composite Reliability ASTM STP580*, American Society for Testing and Materials, Philadelphia, 574–585, 1975.
131. Schwarz G, Krahn F, Hartwig G, Thermal-expansion of carbon-fibre composites with thermoplastic matrices, *Cryogenics*, 31(4), 244–247, 1991.

132. Wolff EG, Stiffness-thermal expansion relationship in high modulus carbon fibres, *J Composite Mater*, 21, 81, 1987.
133. Pirgon O, Wolstenholm GH, Yates B, Thermal expansion at elevated temperatures IV. Carbon fibre composites, *J Phys D, Appl Phys*, 6, 309–321, 1973.
134. Rogers KF, Phillips LN, Kingston-Lee DM, Yates B, Overy MJ, Sargent JP, McCalla BA, The thermal expansion of carbon fibre reinforced plastics, *J Mater Sci*, 12, 718–733, 1977.
135. Yates B, Overy MJ, Sargent JP, McCalla BA, Kingston-Lee DM, Phillips LN, Rogers KF, The thermal expansion of carbon fibre reinforced plastics, Part 2. The influence of fibre volume fraction, *J Mater Sci*, 13, 433–440, 1978.
136. Yates B, McCalla BA, Sargent JP, Rogers KF, Phillips LN, Kingston-Lee DM, The thermal expansion of carbon fibre reinforced plastics, Part 3. The influence of fibre resin type, *J Mater Sci*, 13, 2217–2225, 1978.
137. Yates B, McCalla BA, Sargent JP, Rogers KF, Kingston-Lee DM, Phillips LN, The thermal expansion of carbon fibre reinforced plastics, Part 4. Ply multidirectional effects, *J Mater Sci*, 13, 2226–2232, 1978.
138. Yates B, McCalla BA, Phillips LN, Kingston-Lee DM, Rogers KF, The thermal expansion of carbon fibre reinforced plastics, Part 5. The influence of matrix curing characteristics, *J Mater Sci*, 14, 1207–1217, 1979.
139. Dootson M, Sargent JP, Wolstenholm GH, Yates B, Time and temperature effects in the thermal expansion characteristics of carbon fibre reinforced plastics, *Composites*, 73–78, Apr 1980.
140. Parker SFH, Chandra M, Yates B, Dootson M, Walters BJ, The influence of distribution between fibre orientations upon the thermal expansion characteristics of carbon fibre reinforced plastics, *Composites*, 281–287, Oct 1981.
141. Rogers KF, Kingston-Lee DM, Phillips LN, Yates B, Chandra M, Parker SFH, The thermal expansion of carbon fibre reinforced plastics, Part 6. The influence of fibre weave in fabric reinforcement, *J Mater Sci*, 16, 2803–2818, 1981.
142. Yates B, Rogers KF, Kingston-Lee DM, Phillips LN, The thermal expansion of carbon fibre reinforced plastics, Part 7. Technological implications, *J Mater Sci*, 17, 1880–1888, 1982.
143. Ozbek S, Jenkins GM, Isaac DH, Thermal expansion and creep of carbon fibres, *20th Biennial Conf on Carbon*, American Carbon Society, Santa Barbara, 270–271, 1991.
144. McIvor SD, Darby MI, Wostenholm GH, Yates B, Banfield L, King R, Thermal conductivity measurements of some glass fibre reinforced and carbon fibre reinforced plastics, *J Mater Sci*, 25(7), 3127–3132, 1990.
145. Pilling MW, Yates B, Black MA, Tattersal P, The thermal conductivity of carbon fibre reinforced plastics, *J Mater Sci*, 14, 1326–1338, 1979.
146. See 145.
147. Dorey G, Bishop SM, Impact performance of carbon fibre/PEEK composites, Paper No. 18, *International Conference on Impact Testing and Performance of Polymeric Materials*, University of Surrey, Sep 2–3 1985.
148. Wardle MW, Tokarsky EW, Drop weight impact testing of laminates reinforced with Kevlar, E-glass and graphite, *Composites Technol Rev*, 5(1), 4, 1983.
149. Dorey G, Impact damage in composites—development, consequences and prevention, *Proceedings of ICCM VI/ECCM 2*, Vol 3, Mattews FL et al ed., Elsevier Applied Science, London, 3.1–3.26, 1987.
150. Wyrick DA, Adams DF, Residual strength of carbon/epoxy materials, *Composite Mater*, 22, 749–765, Aug 1988.
151. Cantwell WJ, Morton J, The impact resistance of composite materials—a review, *Composites* 21, 347–362, 1991.
152. Talreja R ed., *Damage Mechanics of Composite Materials*, Westport, Technomic, 1992.
153. Wardle MW, Zahr GE, Instrumented impact testing of aramid and reinforced composite materials, Keesler SL, Adams GL, Driscoll SB, Ireland DR eds., *Instrumented Impact Testing of Plastics and Composite Materials, ASTM STP 936*, 219–235, 1987.
154. Daniel IM, Strain and failure analysis in graphite/epoxy laminates with cracks, *Exp Mech*, 18, 246–252, 1978.

155. Yeow YT, Morris DH, Brinson HF, The fracture behaviour of graphite/epoxy laminates, *Exp Mech*, 19, 1–8, 1979.
156. Wilkins DJ, Eisenmann JR, Camin RA, Margolis WS, Benson RA, Characterizing delamination growth in graphite/epoxy, Reifsneider RL ed., *Damage in Composite Materials, Basic Mechanisms, Accumulation, Tolerance, Characterization, ASTM STP 775*, American Society for Testing and Materials, Philadelphia, 168–183, 1982.
157. Aliyu AA, Daniel IM, Effects of strain rate on delamination fracture toughness of graphite/epoxy, *Delamination and Debonding of Materials, ASTM STP 876*, American Society for Testing and Materials, Philadelphia, 336–348, 1985.
158. Ramkumar RL, Whitcomb JD, Characterization of Mode I and Mixed-mode delamination growth in T300/5208 graphite/epoxy, Johnson WS ed., *Delamination and Debonding of Materials, ASTM STP 876*, American Society for Testing and Materials, Philadelphia, 315–335, 1985.
159. Arcan L, Arcan M, Daniel IM, SEM fractography of pure and mixed-mode interlaminar fractures in graphite/epoxy composites, Masters J, Au J eds., *Fractography of Modern Engineering Materials, ASTM STP 948*, American Society for Testing and Materials, Philadelphia, 41–67, 1987.
160. Daniel IM, Shareef I, Aliyu AA, Rate effects on delamination fracture toughness of a toughened graphite/epoxy, Johnson NJ ed., *Toughened Composites, ASTM STP 937*, American Society for Testing and Materials, Philadelphia, 260–274, 1987.
161. Gillespie JW, Carlsson LA, Smiley AJ, Rate dependent Mode I interlaminar crack growth mechanisms in graphite/epoxy and graphite/PEEK, *Composites Sci Technol*, 28, 1–15, 1987.
162. Corletto CR, Bradley WL, Mode II delamination fracture toughness of unidirectional graphite/epoxy composites, Legace PA ed., *Composite Materials, Fatigue and Fracture, 2nd Vol, ASTM STP1012*, American Society for Testing and Materials, Philadelphia, 201–221, 1989.
163. Shikhmanter L, Eldror I, Cina B, Fractography of unidirectional cfrp composites, *J Mater Sci*, 24, 167–172, 1989.
164. Nejhad MNG, Parvizimajidi A, Impact behaviour and damage tolerance of woven carbon-fibre reinforced thermoplastic composites, *Composites*, 21(2), 155–168, 1990.
165. Bibo G, Leicy D, Hogg PJ, Kemp M, High temperature damage tolerance of carbon-fibre-reinforced plastics, 1. Impact characteristics, *Composites*, 25(6), 414–424, 1994.
166. Hitchen SA, Ogin SL, Smith PA, Soutis C, The effect of fibre length on fracture-toughness and notched strength of short carbon-fibre epoxy composites, *Composites*, 25(6), 407–413, 1994.
167. Kishimoto K, Notomi M, Koizumi T, Fracture toughness of short carbon-fibre-reinforced thermoplastic polymide, *Engineering Fracture Mechanics*, 49(6), 943 et seq, 1994.
168. Pavier MJ, Clarke MP, Experimental techniques for the investigation of the effects of impact damage on carbon-fibre composites, *Composite Sci Technol*, 55(2), 157–169, 1995.
169. Stori A, Magnus E, Marshall IH ed., Paper 24, *Composite Structures*, Elsevier Applied Science, London, 1983.
170. Walker RA, *Impact Database*, PERA International, UK.
171. Nakano K, Kamiya A, Yamauchi S, Kobayashi T, Fracture toughness of carbon fibre reinforced ceramic composites, Japan Fine Ceramics Center, *Fracture Mechanics of Ceramics 9. Composites, R-Curve Behaviour and Fatigue*, Bradt RC, Hasselman DPH, Munz D, Sakai M, Shevchenko VY eds., *Proc 5th Int Symp*, Nagoya, 15–17 Jul 1991, *Fracture Mechanics of Ceramics*, Plenum, 9, 123–132, 1992.
172. Kamiya N, Wada S, *J Ceramic Soc Japan*, 100(4), 581–584, 1992.
173. Philips DC, Scott JM, Shear fatigue of unidirectional fibre composites, *Composites*, 8, 233–236, 1977.
174. Grimes GC, Experimental study of compression—compression fatigue of graphite/epoxy composites, Chamis CC ed., *ASTM STP734*, American Society for Testing and Materials, 281–337, 1981.
175. Jones CJ et al, Environmental fatigue of reinforced plastics, *Composites*, 14, 288–293, 1983.
176. Jones CJ et al, Environmental fatigue behaviour of reinforced plastics, *Proc Royal Society of London*, 396, 315–318, 1984.
177. Lifschitz JM, Compressive fatigue and static properties of a UD graphite/epoxy composite, *J Composite Technol Res*, 10(3), 100–106, 1988.

178. Harris B et al, Fatigue behaviour of carbon fibre reinforced plastics, *Composites*, 21(3), 232–242, 1992.
179. Buggy M, Dillon G, Flexural fatigue of carbon fibre-reinforced PEEK laminates, *Composites*, 22(3), 191–198, 1991.
180. Lin CT, Kao PW, Yang FS, Fatigue behaviour of carbon fibre-reinforced aluminium laminates, *Composites*, 22(2), 135–141, 1991.
181. Bunsell AR, Somer A, The tensile and fatigue behaviour of carbon fibres, *Plastics Rubber and Composites Processing and Applications*, 18(4), 263–267, 1992.
182. Hitchen SA, Ogin SL, Smith PA, Effect of fibre length on fatigue of short carbon-fibre epoxy composite, *Composites*, 26(4), 303–308, 1995.
183. Haque A, Raju PK, Monitoring fatigue damage in carbon fibre composites using an acoustic impact technique, *Materials Evaluation*, 56(6), 765–770, 1998.
184. Irving PE, Thiagarajan C, Fatigue damage characterization in carbon fibre composite materials using an electrical potential technique, *Smart Materials and Structures*, 7(4), 456–466, 1998.
185. Gilchrist MD, Kinloch AJ, Matthews FL, Mechanical performance of carbon-fibre and glass-fibre-reinforced epoxy I-beams, III. Fatigue performance, *Composite Sci Technol*, 59(2), 179–200, 1999.
186. Wang S, Chung DDL, Apparent negative electrical resistance in carbon fibre composites, *Composites*, 30, 579–590, 1999.
187. Hamada T, Nishida T, Sajiki Y, Furuyama M, Tomioka T, The diameter dependence on magneto-resistance and L_c of pitch-based carbon fibres, *Extended Absracts 18th Biennual Conf Carbon*, 225–226, 1987.
188. Matthews JGV, Ko YS, U.S. Pat., 4,938,941, Jul 3 1990.
189. Lerner NR, Electrical conductivity and electron-spin resonance in oxidatively stabilized polyacrylonitrile subject to elevated temperature, *J Appl Phys*, 52(11), 6757–6762, Nov 1981.
190. Song DY, Park JB, Takeda N, Failure behaviour and electrical property of CFRP and CFGFRP, *Key Eng Mater*, 183–187, 1129–1134, 2000, Published in *Fracture and Strength of Solids*, 2000.
191. Okubo A, Miyayama M, Yanagida H, Electrical properties of pitch based semi-conductive carbon fibres, *J Ceramic Soc Japan*, 103(9), 954–959, 1995.
192. Gerteisen SR, Carbon fibre reinforced thermoplastics with controlled surface resistivity, *Eng Plastics*, 9(2), 126–134, 1996.
193. Maslii AI, Panasenko AV, Measuring the conductivity in carbon fibres, *Russian J Electrochem*, 34(5), 476–478, 1998.
194. Agari A, Ueda A, Nagai S, *J Appl Polym Sci*, 52, 1223–1231, 1994.
195. Selzer R, Friedrich K, Mechanical properties and failure behaviour of carbon fibre reinforced polymer composites under the influence of moisture, *Composites Part A—Appl Sci Manuf*, 28(6), 595–604, 1997.
196. Alias MN, Brown R, Corrosion behaviour of carbon fibre composites in the marine environment, *Corrosion Sci*, 35(1–4), 395–402, 1993.
197. Ma CCM, Lee CL, Tai NH, Chemical resistance of carbon fibre reinforced poly(ether ether ketone) and poly(phenylene sulfide) composites, *Polymer Composites*, 13(6), 435–440, 1992.
198. Enomoto M, Ohashi T, Hata T, Susuki Y, Maeda M, Sato H, Evaluation of corrosion resistance of carbon fibres in hot phosphoric acid, *Denki Kagaku*, 65(4), 320–325, 1997.
199. Tressaud A, Gupta V, Piraux L, Lozano L, Marquestaut E, Flandrois S, Marchand A, Bahl OP, Fluorine intercalated carbon fibres 1. Structural and transport properties, *Carbon*, 32(8), 1485–1492, 1994.
200. Hisakado T, Kumehara H, Suda H, Kusaka S, Mechanism of friction and wear of carbon-fibre reinforced metals, 2. Effects of humidity, *J Japanese Soc Tribologists*, 35(5), 337–342, 1990.
201. Hisakado T, Urano W, Mechanism of friction and wear of carbon-fibre reinforced metals—effect of normal load, *J Japanese Soc Tribologists*, 36(10), 785–791, 1991.
202. Hisakado T, Kurosu T, Suda H, Mechanism of friction and wear of carbon fibre reinforced metals—Effect of environment, *J Japanese Soc Tribologists*, 43(8), 702–709, 1998.

203. Wielage B, Dorner A, Podlesak H, The resistance of carbon fibre aluminum—composites to frictional wear, *Praktische Metallographie-Practical Metallography*, 35(l), 21–30, 1998.
204. Wielage B, Dorner A, Wear behaviour of carbon fibre aluminium-composites during abrasive loading, *Praktische Metallographie-Practical Metallography*, 35(7), 350–358, 1998.
205. Bianchi V, Fournier P, Platon F, Reynaud P, Carbon fibre-reinforced (YMAS) glass–ceramic matrix composites, dry friction behaviour, *J Eur Ceramic Soc*, 19(5), 581–589, 1999.
206. Laity MA, Smith PA, Bader MG, Curtis PT, *Composites*, 23, 387, 1992.
207. Short D, Summerscales J, *Composites*, 10, 215, 1979 and *Composites*, 11, 33, 1980.
208. Manders PW, Bader MG, The strength of hybrid glass/carbon composites, Part 1. Failure strain enhancement and failure mode, *J Mater Sci*, 16, 2233–2245, 1981.
209. Manders PW, Bader MG, The strength of hybrid glass/carbon composites, Part 2, A statistical model, *J Mater Sci*, 16, 2246–2256, 1981.
210. Aveston J, Sillwood JM, Synergistic fibre strengthening in hybrid composites, *J Mater Sci*, 11, 1877–1883, 1976.
211. Stevanovic MM, Stecenko TB, Mechanical-behavior of carbon and glass hybrid fibre reinforced polyester composites, *J Mater Sci*, 27(4), 941–946, 1992.
212. Zhu XY, Li ZX, Jin YX, Laminar fracture-behavior of (carbon glass) hybrid fibre reinforced laminates,1. Laminar fracture process, *J Eng Fracture Mechanics*, 44(4), 545, 1993.
213. Zhu XY, Li ZX, Jin YX, Laminar fracture-behavior of (carbon glass) hybrid fibre reinforced laminates, 2. Laminar fracture criterion, *J Eng Fracture Mechanics*, 44(4), 553, 1993.
214. Miwa M, Horiba N, Effects of fibre length on tensile-strength of carbon glass-fibre hybrid composites, *J Mater Sci*, 29(4), 973–977, 1994.
215. Zhu X, Li Z, Jin Y, Shaw WJD, Creep-behavior of a hybrid fibre (glass/carbon) reinforced composite and its application, *Composite Sci Technol*, 50(4), 431–439, Jan 1994.
216. Tan TTM, Nieu NH, Hybrid carbon-glass fibre vinyl ester resin composites, *Angewandte Makromolekulare Chemie*, 234, 53–58, 1996.
217. Marston C, Gabbitas B, Adams J, The effect of fibre sizing on fibres and bundle strength in hybrid glass carbon fibre composites, *J Mater Sci*, 32(6), 1415–1423, 1997.
218. Harel H, Aronhime J, Schulte K, Friedrich K, Marom G, Rate-dependent fatigue of aramid-fibre carbon-fibre hybrids, *J Mater Sci*, 25(2b), 1313–1317, 1990.
219. Aronhime J, Harel H, Gilbert A, Marom G, The rate-dependence of flexural shear fatigue and uniaxial compression of carbon-fibre and aramid-fibre composites and hybrids, *Composites Sci Technol*, 43(2), 105–116, 1992.
220. Duvis T, Papaspyrides CD, Skourlis T, Polyamide coating on carbon-fibres and potential application in carbon kevlar epoxy hybrid composites, *Composites Sci Technol*, 48(1–4), 127–133, 1993.
221. Jang J, Kim H, Improvement of carbon fibre/peek hybrid fabric composites using plasma treatment, *Polymer Composites*, 18(1), 125–132, 1997.
222. Peijs A, Venderbosch RW, Lemstra PJ, Hybrid composites based on polyethylene and carbon-fibres, 3. Impact resistant structural composites through damage management, *Composites*, 21(6), 522–530, 1990.
223. Peijs A, Catsman P, Govaert LE, Lemstra PJ, Hybrid composites based on polyethylene and carbon-fibres, 2. Influence of composition and adhesion level of polyethylene fibres on mechanical-properties, *Composites*, 21(6), 513–521, 1990.
224. Peijs A, Venderbosch RW, Hybrid composites based on polyethylene and carbon-fibres, 4. Influence of hybrid design on impact strength, *J Mater Sci Lett*, 10(19), 1122–1124, 1991.
225. Ward IM, Comments on hybrid composites based on polyethylene and carbon-fibres, *Composites*, 22(4), 341, 1991.
226. Peijs A, Vanklinken EJ, Hybrid composites based on polyethylene and carbon-fibres, 5. Energy-absorption under quasi-static crash conditions, *J Mater Sci Lett*, 11(8), 520–522, 1992.
227. Peijs A, Dekok JMM, Hybrid composites based on polyethylene and carbon-fibres, 6. Tensile and fatigue behavior, *Composites*, 24(1), 19–32, 1993.
228. Jang JS, Moon SI, Impact and behavior of carbon fibre/ultra-high modulus polyethylene fibre hybrid composites, *Polymer Composites*, 16(4), 325–329, 1995.

229. Banerjee AN, Saha N, Mitra BC, Flexural behavior of unidirectional polyethylene carbon fibres PMMA hybrid composite laminates, *J Appl Polym Sci*, 60(1), 139–142, 1996.
230. Park RC, Jang JS, Impact behavior of carbon fibre polyethylene fibre hybrid composite, the effect of surface treatment of polyethylene fibre, *Polymer Composites*, 19(5), 600–607, 1998.
231. Saha N, Banerjee A, Dynamic mechanical study on unidirectional polyethylene carbon fibres, PMMA hybrid composite laminates, *J Appl Polym Sci*, 67(9), 1631–1637, 1998.
232. Li Y, Xian XJ, Choy CI, Guo MI, Zhang ZG, Compressive and flexural behavior of ultra-high-modulus polyethylene fibre and carbon fibre hybrid composites, *Composites Sci Technol*, 59(1), 13–18, 1999.
233. Chung DDL, *Carbon Fiber Composites*, Butterworth-Heinemann, Newton, 1994.
234. Buggy M, Carew A, The effect of thermal ageing on carbon-fibre-reinforced polyetheretherketone (PEEK), 1. Static and dynamic flexural properties, *J Mater Sci*, 29(7), 1925–1929, 1994.
235. Zhen Mei, Chung DDL, Glass transition and melting behavior of carbon fibre reinforced thermoplastic composite, studied by electrical resistance measurement, *Polymer Composites*, 21(5), 2000.
236. Jones FR ed., *Handbook of Polymer Fibre Composites, Polymer Science and Technology Series*, Longman Scientific & Technical, Harlow, 1994.
237. Park SB, Experimental-study on the engineering properties of carbon fibre reinforced cement composites, *Cement Concrete Research*, 21(4), 589–600, 1991.
238. Yuan CZ, Hua CQ, Feng G, Influence of interface modification in fibre reinforced cement composites on their properties, *Mat Res Soc Symp Proc*, 211, 209–214, 1991.
239. Toutanji HA, El-Korchi T, Katz RN, Leatherman GL, Behaviour of carbon fibre reinforced cement composites in direct tension, *Cement Concrete Research*, 23, 618–626, 1993.
240. Toutanji HA, El-Korchi T, Katz RN, Strength and reliability of carbon fibre reinforced cement composites, *Cement Concrete Research*, 16, 15–21, 1994.
241. Katz A, Li VC, Kazmer A, Bond properties of carbon fibres in cementitious matrix, *J Mater Civil Eng*, 125–128, May 1995.
242. Shui ZH, Li JZ, Huang FP, Yang DP, Study on the electrical properties of carbon fibre cement composite (CFCC), *Journal of Wuhan University of Technology—Materials Science Edition*, 10(4), 37–41, 1995.
243. Sakai H, Takahashi K, Mitsui Y, Ando T, Awata M, Hoshijima T, Flexural behaviour of carbon fibre reinforced cement composite, *American Concrete Institute Report SP 142-7*, 121–129.
244. Takeda K, Mitsui Y, Murakami K, Sakai H, Nakamura M, Flexural behaviour of reinforced concrete beams strengthened with carbon fibre sheets, *Composites Part A—Appl Sci Manuf*, 27(10), 981–987, 1996.
245. Norris T, Saadatmanesh H, Ehsani MR, Shear and flexural strengthening of R/C beams with carbon fibre sheets, *J Structural Eng*, ASCE 123(7), 903–911, 1997.
246. Linton JR, Berneburg PL, Gartner EM, Bentur A, Carbon fibre reinforced cement and mortar, *Mater Res Symp Proc*, 211, 1991.
247. Aveston J, Mercer RA, Sillwood JM, Fibre reinforced cements—Scientific foundations for specifications, *Composites-Standards Testing and Design Conf Proc Composites Standards Testing and Design*, NPL, IPC Science and Technology Press, 174, 93–103, 1974.
248. Kim TJ, Park CK, Flexural and tensile strength developments of various shape carbon fibre-reinforced lightweight cementitious composites, *Cement Concrete Res*, 28(6), 955–960, 1998.
249. Ziegler G, Lucke J, Richter I, Suttor D, C-fibre reinforced composites with polymer-derived matrix, microstructure, thermal properties, strength, Baxter J, Cot L, Fordham R, Gabis V, Hellot Y, Lefebvre M, Le Doussal H, Le Sech A, Naslain R, Sevagen A eds., *European Ceramic Society, Euro Ceramics V*, Part 3. Trans Tech Publications, Switzerland, 1870–1873, 1997 *Key Eng Mater*, 132–136.
250. Hegeler H, Bruckner R, Mechanical-properties of carbon fibre-reinforced glasses, *J Mater Sci*, 27(7), 1901–1907, 1992.
251. Saewong P, Rawlings RD, Erosion of carbon fibre reinforced glass-ceramic composites, *J Mater Sci Lett*, 18, 1915–1919, 1999.
252. Zheng GB, Sano H, Uchiyama Y, Kobayashi K, Effects of boron addition on the mechanical properties of carbon fibre reinforced SiC composites, Suzuki H, Komeya K, Uematsu K eds.,

Japan Society for the Promotion of Science, *Novel Synthesis and Processing of Ceramics, Proc Int Symp*, Kurume, 393–398, Oct 26–29, 1997, *Key Eng Mater*, Vol 159–160.
253. Xu Y, Cheng L, Zhang L, Yan D, Mechanical properties and microstructural characteristics of carbon fibre reinforced silicon carbide matrix composites by chemical vapour infiltration, Niihara K, Nakano K, Sekino T, Yasuda E eds., Ceramic Society of Japan, *High Temperature Ceramic Matrix Composites III, Proc 3rd Int Conf*, Osaka, Sep 6–9 1998, 73–76, *Key Eng Mater*, Vol 164–165.
254. Zheng GB, Sano H, Uchiyama Y, Kobayashi K, Cheng HM, Properties of carbon fibre/sic composites fabricated through impregnation and pyrolysis of polycarbosilane, *J Mater Sci*, 34(4), 827–834, 1999.
255. Yasuda K, Tanaka S, Matsuo Y, New evaluation method of modal fracture energy of fibre-reinforced ceramic matrix composites, *J Mater Sci*, 34(10), 2331–2334, 1999.
256. Tani E, Shobu K, Effect of polymer addition on the strength of carbon fibre reinforced reaction-bonded sic matrix composites, Evans AG, Naslain R eds., American Ceramic Soc, Inc, *High-Temperature Ceramic-Matrix Composites II, Manufacturing and Materials* Development, *Ceramic Trans*, Santa Barbara, 58, 255–260, Aug 21–24 1995.
257. Yamada K, Matsubara M, Suzuki N, Fukuda S, Matsumoto M, High temperature toughening mechanisms of particle and particle-fibre reinforced SiC composites, Evans AG, Naslain R eds., American Ceramic Soc, Inc, *High-Temperature Ceramic-Matrix Composites I, Design, Durability and Performance, Proc 2nd Int Conf, Ceramic Trans*, Santa Barbara, 57, 413–418, Aug 21–24 1995.
258. Yoshida H, Miyata N, Naito K, Ishikawa S, Yamagishi C, Influence of orientation angle of fibre on mechanical properties in uni-directional and two-directional carbon fibre reinforced SiC composites, *J Ceramic Soc Japan*, 102(11), 1016–1021, 1994.
259. Klug T, Bornhoft H, Bruckner R, Alternating bending load behaviour of unidirectionally fibre-reinforced glasses, *Glastech Ber*, 65(8), 207–215, 1992.
260. Nakano K, Kamiya A, Ogawa H, Nishino Y, Fabrication and mechanical properties of carbon fibre reinforced silicon carbide composites, *J Ceramic Soc Japan*, 100(4), 472–475, 1992.
261. Savage G, *Carbon–Carbon Composites*, Chapman and Hall, London, 1993.
262. Hüttner W, Potential of carbon/carbon composites as structural materials, Figueiredo JL, Bernardo CA, Baker RTK, Hüttinger KJ eds., *Carbon Fibers Filaments and Composites*, Kluwer Academic Publishers, Dordrecht, 1990.
263. Peters PWM, Ludenbach G, Pleger R, Weiss R, Influence of matrix and interface on the mechanical properties of unidirectional carbon/carbon composites, *J Eur Ceramic Soc*, 13(6), 561–569, 1994.
264. Rhee B, Ryu S, Fitzer E, Fritz W, High Temperatures—High Pressures, 19(6), 677–686, 1987.
265. Sapp JW Jr., Bowers DA, Dinwiddie RB, Burchell TD, *Ext Abstr Program-Bienn, Conf Carbon*, 30, 644–645, 1991.
266. Sato S, Kurumada A, Iwaki H, Komatsu Y, *Carbon*, 27(6), 791–801, 1989.
267. Manocha LM, Bahl OP, Singh YK, In: Jones FR ed., *Proc Int Conf Interfacial Phenomena in Composite Materials '89*, Butterworth, 310–315, 1989.
268. Lake ML, Hickok JK, Brito KK, Begg LL, *Proc Int SAMPE Symp Exhib 35, Advanced Materials, Challenge Next Decade*, 960–969, 1990.
269. Soni PR, Rajan TV, Ramakrishnan P, Tensile behaviour of carbon fibre reinforced aluminium matrix composites, *Metals Mater Proc*, 7(4), 267–273, 1996.
270. Korb G, Korab J, Groboth G, Thermal expansion behaviour of unidirectional carbon-fibre reinforced copper-matrix composites, *Composites Part A-Appl Sci Manuf*, 29(12), 1563–1567, 1998.
271. Bell JAE, Hansen G, Properties of nickel coated carbon and Kevlar fibres produced by decomposition of nickel carbonyl, *Proceedings of 23rd International SAMPE Technical Conference*, New York, 1183–1193, Oct 22–24, 1991.
272. Rosenow MWK, Long nickel fibres for EMI shielding, *ETP'99 World Congress Engineering Thermoplastics*, Zurich, Jun 7–9, 1999.

CHAPTER 21

Manufacturing Techniques for Carbon Fiber Reinforced Composites in Thermoset and Thermoplastic Matrices

21.1 CARBON FIBER REINFORCEMENT AND ARCHITECTURE

This section covers PAN, pitch and cellulose based carbon fibers, but does not include vapor grown carbon fibers, although such fibers can be spun into staple yarns and converted to chopped fiber. A classification designating the various levels of carbon fiber architecture is outlined in Figure 21.1.

Andersson [1] defines a fabric as an integrated fibrous structure produced by fiber entanglement, or yarn interlacing, interlooping, intertwining, or multiaxial placement. Hearle and Du [2] show how rigid fiber assemblies can be formed by the interaction of textile technology and composite engineering in Figure 21.2.

3-D textile preforms [3–6] are particularly important for shaping a reinforcement prior to inclusion in, say, RTM and carbon–carbon composites.

Since carbon fiber is electrically conductive, it is good practice for processors working with carbon fiber, to have designated work areas set aside, so that all electrics can be adequately protected from carbon fiber fly and an efficient air filtration system built into the workplace to safeguard operatives. Oxidized PAN fiber is not electrically conductive, however, and does not have these problems.

With carbon reinforcements, it is possible to incorporate two or more different yarns (e.g. an aramid and carbon, or glass and carbon), to form a hybrid carbon reinforcement, enabling a designer to incorporate properties into a composite that fall between the performance levels of the chosen fibers. So the introduction of an aramid fiber into a carbon fabric reinforcement would improve the impact resistance, albeit at the expense of other mechanical properties. Special scissors are available to cut aramid fibers.

21.1.1 Virgin carbon fiber

A tow of carbon fibers is a plurality of non-twisted continuous carbon filaments, about 5–10 μm in diameter, with typical tow sizes of $3k$, $6k$ and $12k$, normally surface treated and sized. Larger and less expensive carbon fiber tows, such as $24k$, $40k$, $48k$, $80k$, $160k$, $320k$, $400k$ and $410k$ (Tables 20.3, 20.4) are now becoming more widely available.

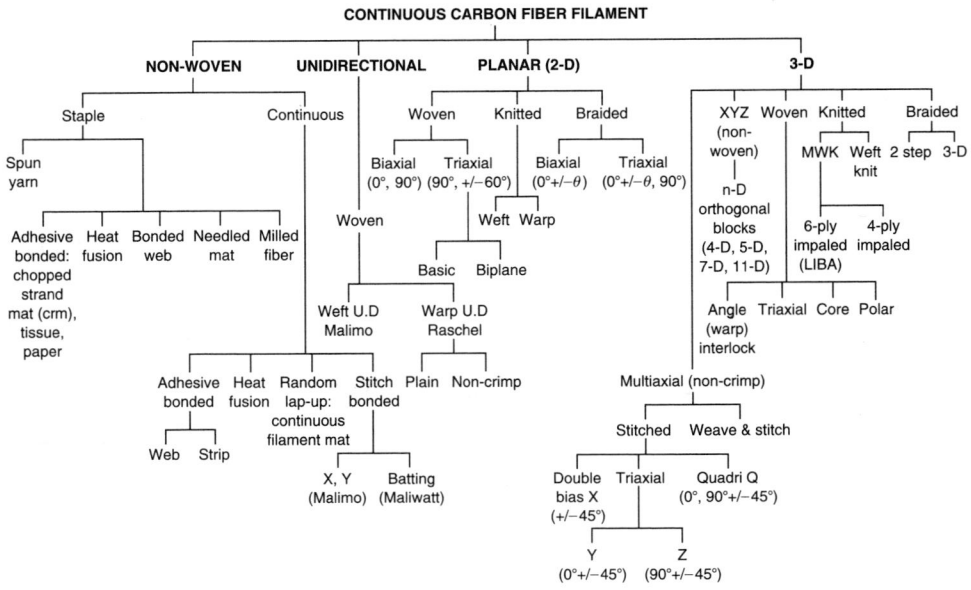

Figure 21.1 Classification of carbon fiber architecture.

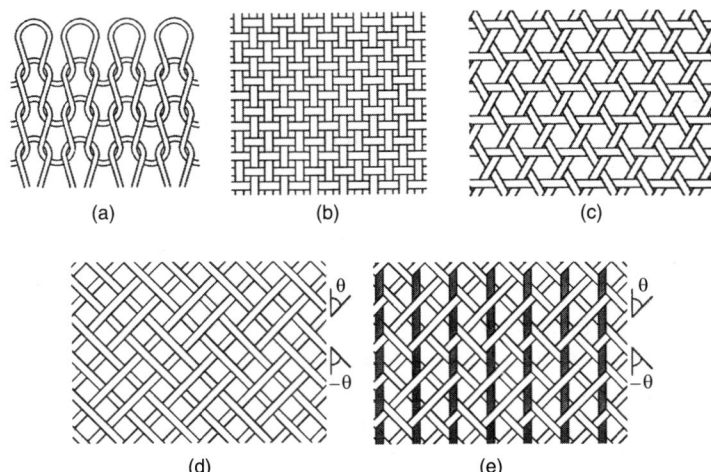

Figure 21.2 Interaction of textile technology and composite engineering to form rigid fiber assemblies. *Source:* Reprinted with permission from Hearle JWS, Du GW, *J Text Inst*, 81(4), 1980. (a) Knitted structure with repeated reversal of yarn directions. (b) Simple plain weave fabric with threads continuous in X and Y directions. (c) Triaxial weave fabric. (d) Braid with yarns at $\pm\theta$ to axis. (e) Triaxial braid. Copyright 1980, The Textile Institute.

Hexcel categorize their grades of carbon fiber according to the modulus, as shown below:

Grade	Modulus psi $\times 10^6$	GPa
Commercial Grade (CG)	30–32	207–221
Standard Modulus (SM)	32–35	221–241
Intermediate Modulus (IM)	38–43	262–246
High (HM) and Ultra High Modulus (UHM)	≥48	≥331

MANUFACTURING TECHNIQUES FOR CARBON FIBER REINFORCED COMPOSITES

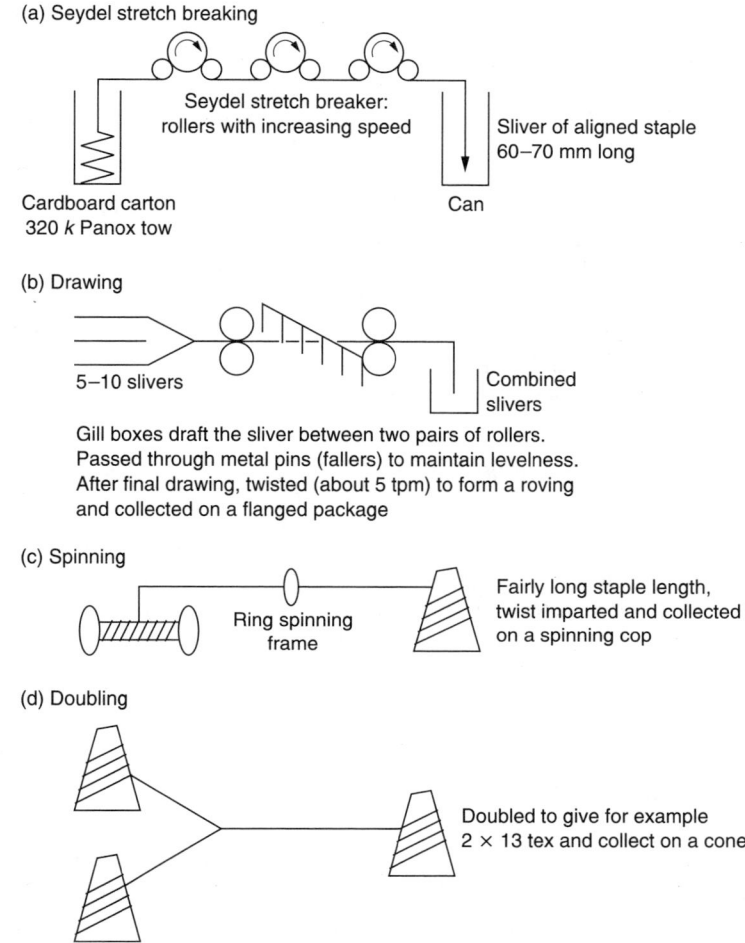

Figure 21.3 Processing route for converting 320k Panox to yarn.

A filament yarn (or a flat yarn) is a continuous tow that has been twisted to aid future processing. The direction of twist in a yarn is designated as Z or S twist—Z twist if the spirals around the axis slope in the same direction as the middle portion of the letter Z and S if the spirals slope in the same direction as the middle portion of the letter S (Twist in Appendix 1). A Z twist is normally applied to a single tow, whereas plied yarns tend to have S twist. A twisted tow can give improved composite mechanical properties, such as flexural strength.

A large tow of opf can be chopped to give a staple product which can then be converted into a finer spun yarn using a modified Worsted spinning route (Figure 21.3).

21.1.2 Non-woven discontinuous reinforcement (staple fiber)

21.1.2.1 Adhesive bonded reinforcements

1. Chopped strand mat (csm)

A carbon fiber csm consists of a random array of chopped fibers held together by a thermoset or thermoplastic binder. A carbon fiber csm is compatible with polyester and epoxy resin systems and can be incorporated into carbon preforms and held in position by sewing. Composites made from a csm will have low volume fractions, of about 20–25% v/v.

2. Carbon fiber tissue

An air knife can be used to spread a carbon fiber tow and form a thin veil of carbon fiber or tissue, which can be used to hold a resin rich gel coat on the surface of a fiberglass laminate construction to give a high surface finish concurrent with added microcrack suppression, as well as provide additional chemical resistance. It will give better performance than a glass tissue due to the absence of wicking.

An electrically conductive veil can be made using metal coated carbon fiber, to provide a wide range of surface conductivity, from electrostatic dissipation levels to electromagnetic and radio frequency shielding levels.

3. Carbon fiber paper reinforcement

A paper based on carbon fiber can be made using a conventional paper making plant, where chopped fiber is dispersed in a liquid carrier (normally water) containing wetting and binding agents. The dispersed fiber is removed from the slurry by vacuum deposition onto a perforated screen, washed and the carbon paper removed from the screen and dried. If a water flume is used to spread the fibers, then it is possible to give a product with more than 80% fiber orientation. Paper is used for specialist applications like loudspeaker cones.

21.1.2.2 Needled mat

A carbon fiber needled mat is dry laid, having no binder, with the chopped fibers being held together by entanglement. The entanglement is done by using a machine that passes a fiber web under a reciprocating beam holding a large number of barbed needles, which penetrate the fiber web and entangle the fibers as the hooks are withdrawn. Needled mats can be incorporated into preforms. This type of process was adapted by SEP to form Novoltex® [7], based on the entanglement of a stack of woven fabric layers by needle punching a non-woven web through a number of layers to bind them together. The use of opf will minimize damage to the fibers. After needling, the preforms are heat treated to convert them into carbon fiber, coated with a pyrocarbon interface, machined to near net shape, followed by a succession of SiC CVI densification stages and machined to produce the final part.

Fukuta [8,9], in Japan, has developed a similar type of process using fluid jets to create fiber entanglement in place of needles.

21.1.2.3 Milled fiber

Initially, an end use was sought for short lengths of carbon fiber, but has now developed into a defined end product. Fiber is sized, cut into a short staple form and milled to give a carbon fiber powder, which is sieved and has become important for incorporating in resin matrices, to confer EMI/RFI shielding, and in automobile air bags, to provide an electrical path for rapid ignition of the explosive charge.

Milled carbon fibers are typically 30–3000 µm long, averaging about 300 µm with a mean L/D ratio of 30.

21.1.2.4 Chopped carbon fiber

Chopped carbon fiber is generally cut 3 or 6 mm long and sized with glycerol for incorporating in processes using a water carrier or a thermoplastic size such as polyurethane for integrating with thermoplastic matrices. The fiber can be chopped with a reciprocating knife, or a rotary blade (lawn mower type). Cutting to a length of 1 mm can be achieved by chilling the fiber in liquid air prior to cutting.

21.1.3 Unidirectional fabrics

Definitions do vary, but a unidirectional fabric is normally taken as a fabric with at least 90% of the fiber weight in one direction, which can be in either the warp or the weft direction. The primary fiber can be held in position by bonding using a meltable thermoplastic, hot melt or thermosetting adhesive, weaving or stitching. Fibers are straight and uncrimped and mechanical properties are only bettered by prepreg, which has no secondary material other than the resin matrix system holding the tows in position.

21.1.3.1 Non-woven UD fabrics

A non-woven carbon fabric is made from fibers held in position by a binder, which can include melt bonding powders and meltable fibers, thermosetting resins and thermoplastic polymers.

1. Heat fusion by applying a fine thermoplastic web (e.g. Polyester) over the entire surface; the bonded web is flat and has no crimp, but may be more difficult to wet out with resins.
2. Strips of thermoplastic web are laid as a fine weft (Figure 21.4); the bonded strip is flat and has zero crimp, but certain resins can soften and loosen the adhesive strip. The resin strip does tend to be tacky and so the prepared fabric is stored interleaved with polyethylene film.
3. Stitch bonding can be used to hold together an assembled warp or weft UD fabric. To confer adequate stability, a mat or tissue is usually placed on the fabric prior to stitching.

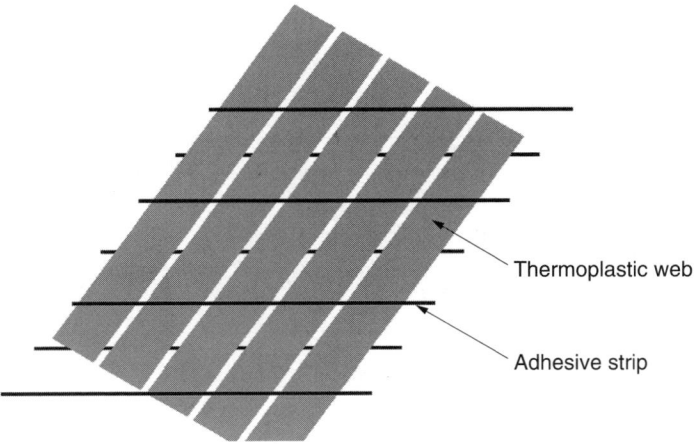

Figure 21.4 Non-woven UD fabric with adhesive bonding strips. *Source:* Reprinted from SP Systems, Newport, IOW, technical data with kind permission.

21.1.3.2 Woven UD fabrics

1. Warp UD fabric

This fabric has the primary fiber in the 0° (warp) direction.

- In a plain weave UD (Figure 21.5), the secondary fiber is a woven weft, which will introduce a certain measure of crimp that can be minimized by using a low weight weft fiber or increasing the spacing between the wefts, both actions being at the expense of stability.
- A non-crimp UD weave fabric (Figure 21.6) utilizes two light weight secondary fibers, a plain weave weft, but with a further secondary fiber running in the warp direction and placed between each of the primary warps. This latter fiber takes up the crimp and gives a fabric with the primary fiber straight and virtually free from crimp. It is not, however, possible to cut the fabric into strips.

2. Weft UD fabric

This fabric has the primary fiber in the 90° (weft) direction.

21.1.4 Woven fabrics (2-D Planar or biaxial reinforcement) [11,12]

Table 21.1 gives a list of some woven carbon fiber manufacturers.

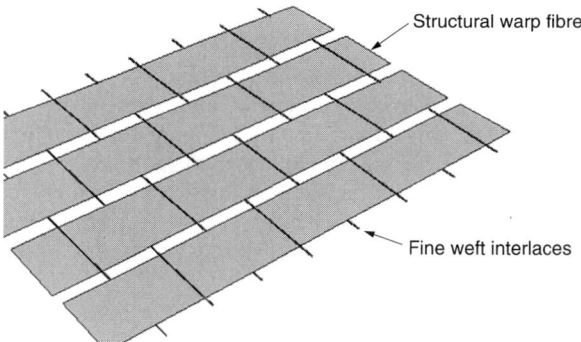

Figure 21.5 Plain weave UD construction. *Source:* Reprinted from SP Systems, Newport, IOW, technical data with kind permission.

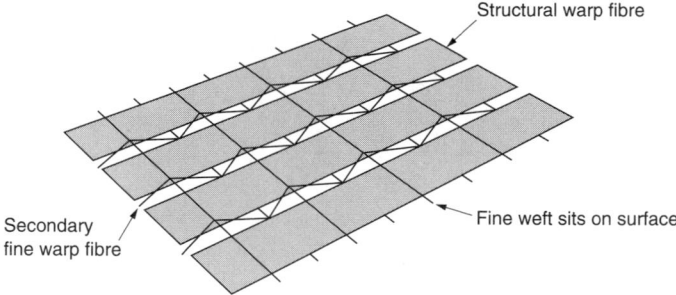

Figure 21.6 Non-crimp weave UD fabric. *Source:* Reprinted from SP Systems, Newport, IOW, technical data with kind permission.

Table 21.1 List of some woven carbon fiber manufacturers

Company	Type of Weave*								
	Plain	Basket	Twill	5-Harness Satin	8-Harness Satin	Leno	Multi-Axial	Tape/Braid	Non-Woven
3-Tex Inc., Cary, NC, USA							X		
A&P Technology, USA								X	
AGL, Cannes -LaBocca, France									
Angeloni, Mestre, Italy									
Atelier Textile du Gier (ATG), Saint-Matin-la -Plaine, France								X	
Bally Ribbon Mills, USA	X			X	X		2D, 3D	X	
Barthels-Feldhoff, Wuppertal, Germany									
BGF Industries Inc., USA	X		X	X	X				
BTI Europe, Andover, UK							X		
BP Amoco Polymers	X				X				
Carr Reinforcements Ltd., Stockport UK	X		X	X					
Chomerat Cie, (Les Fils d'Auguste), France	X								
CP Films (Courtaulds Aerospace) Ltd., UK	X			X	X				
Cramer C & Co., USA	X		X	X	X				
CS Interglas AG, Erbach, Germany	X		X		X				
Devold AMT AS, Langevag, Norway							X		
EMS Chemie, Switzerland									Commingled
Eurocarbon BV, Sittard, Netherlands								X	
Fiber Materials Inc., USA	X	X		X	X				
Fothergill Engineered Fabrics, Littleborough, UK									
FTS SRL, Italy	X								
Gamma Tensor, Spain									
Heinsco, Rochdale, UK									
Hexcel Fabrics, Villeurbanne, France (Brochier)	X		X	X	X				
Hollingsworth & Vose Co., USA									X
Kaiser Aerotech							2D		
FA Kuempers, Rheine, Germany									
Mitsubishi Rayon Co. Ltd., Japan	X				X				
Mutual Industries Inc., USA	X	X		X	X				
North American Textiles Inc., USA	X		X		X				
Plastic Developments Ltd., UK	X		X	X					
Plastic Reinforcement Fabrics Ltd., UK	X								
Porcher Industries, Badinieras, France	X		X	X	X				
RTM Composites, Montferrat, France								X	
Saertex, Saerbeck, Germany									
Schappe Techniques, Charnoz, France									
Seal, legnano, Italy									
Selcom SRL, Fregano (TV), Italy	X						X		
Sigmatex (UK) Ltd., Runcorn, UK	X		X	X					

(Continued)

Table 21.1 Continued

Company	Type of Weave*								
	Plain	Basket	Twill	5-Harness Satin	8-Harness Satin	Leno	Multi-Axial	Tape/Braid	Non-Woven
Siltex, Julbach, Germany									
SGL Technik GmbH, Germany	X		X		X				
SP Systems Ltd., UK	X		X	X					
Swiss Silk Zurich, Switzerland	X		X						
Technical Fiber Products, UK									X
Techniweave Inc., USA	X	X		X	X				
Tencate Advanced Composites BV, Nijverdal, Netherlands	X		X	X	X				
Textile Products Inc., USA	X	X	X	X	X	X			
Textile Technologies Inc., USA	X	X	X	X	X				
Textron Speciality Materials, USA	X			X	X				
Tissa Glasbweberei AG, Switzerland	X								
Toho Rayon Co. Ltd., Japan	X				X				
Toray Industries Inc., Japan	X		X		X				
Universal Carbon Fibers Ltd., UK	X								
Verseiday Industrie Textilen GmbH, Germany	X			X					
Vom Baur Sohn, Germany								X	
Wade Fibers (Pty) Ltd., South Africa									Stitched
Wolffe, Oyonnax, France									
Woven Structures Inc., USA	X			X	X				
Zoltec Corp., USA	X	X			X				

* These are the six most popular weaves. Companies listed may offer other fabric constructions.

A series of woven products can be created by positioning yarns at 0° and 90° to one another and interlacing to form a series of regular geometric patterns. In a woven fabric, the yarns running parallel to the direction of weaving are termed warp, also called ends. The yarns running transverse to the direction of weaving are termed weft, also termed picks, fill or woof. The three basic weaves are plain, twill and satin.

1. Plain or square weave

The plain weave pattern (Figure 21.7) is the simplest and most commonly used woven fabric, with each warp end passing alternately under and over each weft pick. The fabric is symmetrical, has good structural integrity, exhibiting optimum fiber stability, with reasonable porosity. The weave is tight, providing in-plane resistance and consequently, has poor drape and is unsuitable for compound curvatures. The high level of crimp reduces the mechanical properties by about 15%. It is not used for heavy yarns due to excessive crimp.

2. Basket (Hopsack) weave

This is a modification of plain weave in which two or more ends and picks weave as one, moving one yarn for each end or pick (Figure 21.8). It is more dense than plain weave and more flexible, tends to abrade more readily and is not as strong as a plain weave. If two warp yarns pass over and under two weft yarns, this gives a 2×2 basket construction. The first

MANUFACTURING TECHNIQUES FOR CARBON FIBER REINFORCED COMPOSITES 869

Figure 21.7 Plain weave fabric. *Source:* Reprinted from SP Systems, Newport, IOW, technical data with kind permission.

Figure 21.8 Basket weave fabric. *Source:* Reprinted from SP Systems, Newport, IOW, technical data with kind permission.

number identifies the number of weft yarns over which the warp yarns float and the second number indicates the number of weft yarns under which the warp yarn passes. The most common types of weaves are 2×2, 4×4 and 8×8.

3. Leno weave

Warp yarns cross over each other, interlacing with one or more filling yarns thus locking them together (Figure 21.9) and are commonly used for edge locking on wide fabrics.

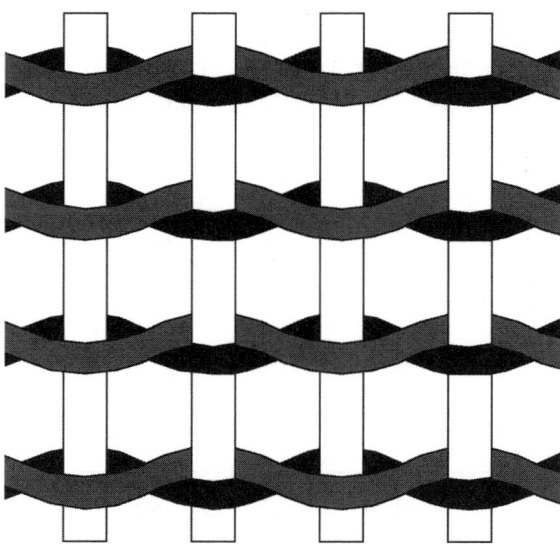

Figure 21.9 Leno weave fabric. *Source:* Reprinted from SP Systems, Newport, IOW, technical data with kind permission.

Figure 21.10 Mock Leno weave fabric. *Source:* Reprinted from SP Systems, Newport, IOW, technical data with kind permission.

A variant is gauze weave, which is an open mesh type with only one filling yarn, whereas leno has more than one.

4. Mock Leno weave

This weave is a version of plain weave in which occasional warp yarns, at regular intervals and usually several yarns apart, interlace every two or more fibers (Figure 21.10). A similar pattern occurs in the weft direction to give a fabric with increased thickness, a rougher surface and additional porosity.

MANUFACTURING TECHNIQUES FOR CARBON FIBER REINFORCED COMPOSITES

5. Twill weave

The number of warp ends and weft picks passing over each other determines the pattern, moving one yarn for each end or pick, high density with good drapeability (Figure 21.11) can be obtained. It gives highest retention of fiber strength and modulus and is recognized by parallel diagonal ridges, usually as 2×2, 3×3, or 3×1 (crowfoot) constructions.

6. Satin weave

Each end or pick passes over N yarns and under one crossing yarn, described as $N+1$ (e.g. 8 shaft passing over seven and under one) (Figure 21.12). The long floats do tend to snag

Figure 21.11 Twill weave fabric. *Source:* Reprinted from SP Systems, Newport, IOW, technical data with kind permission.

Figure 21.12 4-Harness satin weave fabric. *Source:* Reprinted from SP Systems, Newport, IOW, technical data with kind permission.

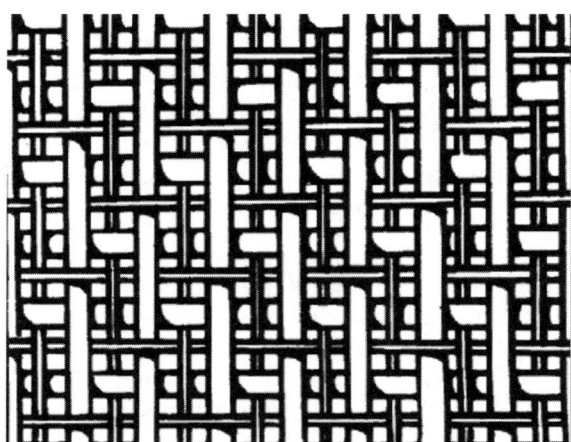

Figure 21.13 High modulus weave fabric. *Source:* Reprinted from Bridon Composites technical literature.

easily. There is a warp and a weft face, both smooth. Crimp is kept to a minimum, giving good translation of strength and modulus. This weave shows the highest bidirectional strength, is very pliable with high density and consequently, high fiber volume fraction and low porosity. It has extremely good drapeability, but may be less stable than twill. The most common satin weaves are 4HS, 5HS and 8HS.

7. High modulus (non-crimp) weave

The warp and weft yarns are positioned without being interlaced (Figure 21.13). A second set of finer warp and weft yarns binds them together, but does not contribute to the mechanical performance of the fabric. This eliminates the crimp and shear factor.

21.1.5 Woven spread tow

Patented processes have been developed to spread 12 or 24k carbon fiber tow to about four or five times its original width to give products which can be woven, giving low areal weights. They show increased surface smoothness with less pronounced cross-over defects when prepregged and molded, due to fewer interlacing points, increased fiber floats and less crimp. Basically, there are two manufacturing techniques. One of the techniques uses an air knife, employing hot air to soften certain proprietary sizes, which allows easier spreading. Careful control of the fiber tension on either side of the air knife gives a fiber catenary permitting adequate spreading, operating up to 10 m min^{-1} (Harmoni Industry Inc.). The other process (Teknomax Corp.) uses a water flume with circulating hot water, which extracts some of the water based fiber size and permits easier spreading, operating up to 10 m min^{-1}. The water is continually replenished to avoid resin build-up and the spread fiber is then dried on an electrically heated drum. The spread fiber from either process can then be woven, normally utilizing a plain weave.

Oxeon AB, in their weaving process, utilize 6 g m^{-2} polyester to stabilize the fabric.

21.1.6 Knitted fabrics [13–15]

A knitted fabric is produced by interlooping flexible yarns of one yarn set. The yarn is formed into loops and the newly formed loops are inter-meshed with previous knitted loops. Each loop in a knitted fabric is called a stitch. The knitted fabrics vary in complexity and are highly conformable and drapeable, available as a flat, tubular or sandwiched construction. The fabrics may be with or without stretch and with a closed or open structure. Fabrics are generally resistant to ravelling at cut edges, thus simplifying cutting and handling operations. The technology is divided into weft (Figure 21.14) and warp (Figure 21.15) knitting. A course is a horizontal row of knitted stitches running widthwise (cf weft in a woven fabric) and a wale is a vertical column of stitches running lengthwise (cf warp in

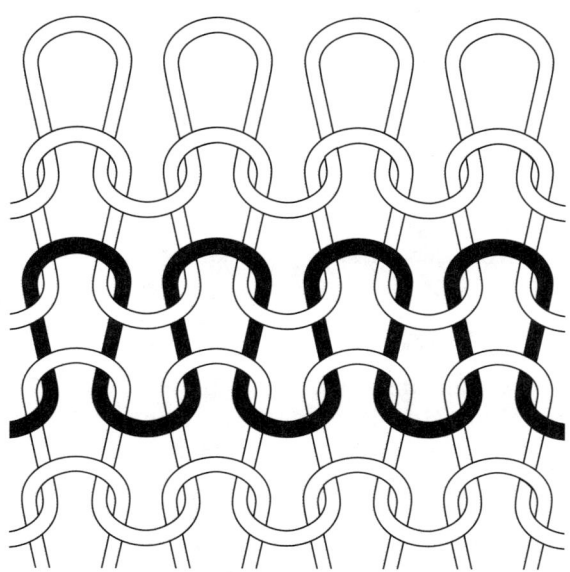

Figure 21.14 Example of weft knitting (Plain). *Source:* Reprinted from Raz S, *Industrial Knitted Fabric Design and International Trends*, IFAI, Boston, Nov 1992.

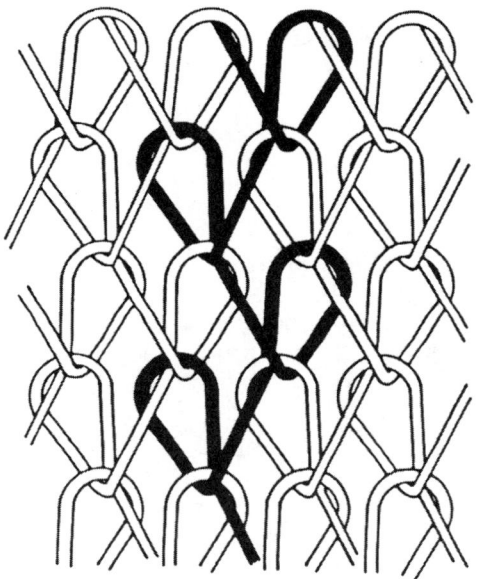

Figure 21.15 Example of warp knitting. *Source:* UMIST.

a woven fabric), with the appearance of a ridge of raised threads. Each wale is associated with its own needle. Gage is used to express the closeness of the needles—the higher the gage number, finer the fabric.

Knitting is an expensive process, but is assuming an important role in the production of 3-D fabrics. The products can be carefully designed to provide the requisite weight and degree of openness. However, the degree of openness does make it impossible to achieve high fiber volume fractions.

21.1.6.1 Weft knitting

As the name implies, yarns run horizontally across the width of the fabric. The knitting needles form the loops in horizontal courses, with one loop built on top of the other, and are generally made from a single yarn. Weft knits can be made flat on a flat bed machine, or tubular on a circular knitting machine. Weft knits are basic loop structures and can be stretched in both directions (Figure 21.16).

There are three basic types of weft knitting—plain, rib and purl.

1. Plain knitting (Figure 21.17)

a. Jersey or single knit—The plain knit is the basis of jersey fabric. If a yarn breaks in a jersey knit, the fabric adjoining stitches will unravel, producing a run. A jersey knit fabric is unbalanced and will curl at the edges. To confer stability, threads can be inserted into the structure and held in place by the loops. With a single knit, the inlay is plated on the reverse of the fabric, whilst in a double knit, the inserted threads are within the structure.

A jersey knit has vertical ribs (wales) on the face, which is the smooth side and horizontal ribs (courses) on the back, which is textured.

b. Rib knitting—A rib knit fabric has alternate wales of plain stitches (giving a raised stitch) and purl stitches (giving a recessed stitch) on both sides of the fabric giving distinct lengthwise rib effects on both sides of the fabric. It is termed a 1 × 1 rib when there is alternately one wale of plain stitches with one wale of purl stitches, on both sides of the fabric. A rib knit fabric lies flat and does not curl at the edges.

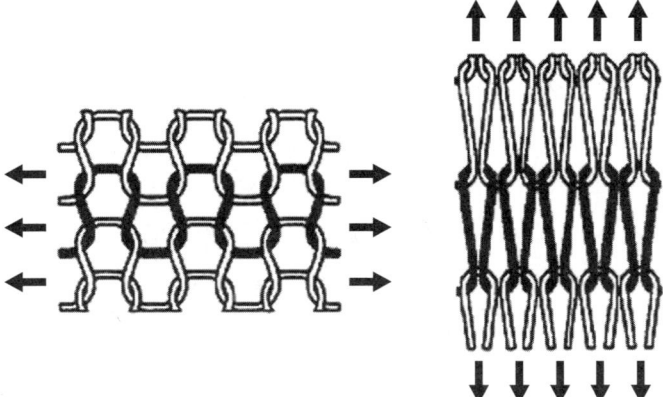

Figure 21.16 Effect of stretch on weft knitting. *Source:* Reprinted from Raz S, *Industrial Knitted Fabric Design and International Trends*, IFAI, Boston, Nov 1992.

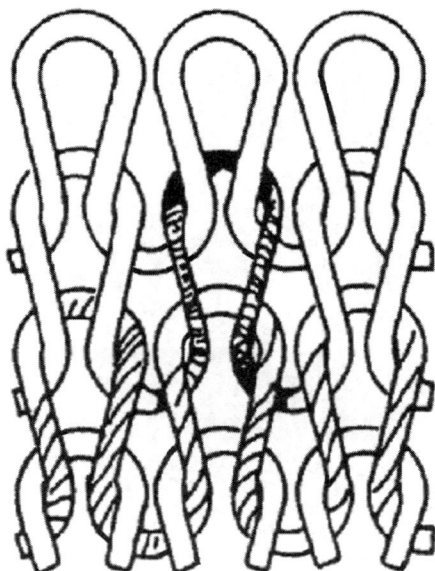

Figure 21.17 Weft knit plain construction. *Source:* Reprinted with permission from Buckley JD, Edie DD eds., Noyes Publications, Park Ridge, 1992. Copyright 1992, William Andrew Publishing.

Figure 21.18 Weft knit purl construction. *Source:* Reprinted with permission from Buckley JD, Edie DD eds., Noyes Publications, Park Ridge, 1992. Copyright 1992, William Andrew Publishing.

c. Purl knitting (Figure 21.18)—A purl knit fabric has alternate courses of plain stitches (recessed) and purl stitches (raised). A purl knit fabric lies flat and does not curl and has the greatest elasticity in the length direction.

The stability of a weft knit can be improved by inserting threads within a double knit construction (Figure 21.19) or by plating threads on the reverse side of a single knit fabric (Figure 21.20). To insert threads in the length of a weft knit does, however, require special machine attachments.

Figure 21.19 Warp inserted tows within a double knit weft construction. *Source:* Reprinted from Raz S, *Industrial Knitted Fabric Design and International Trends*, IFAI, Boston, Nov 1992.

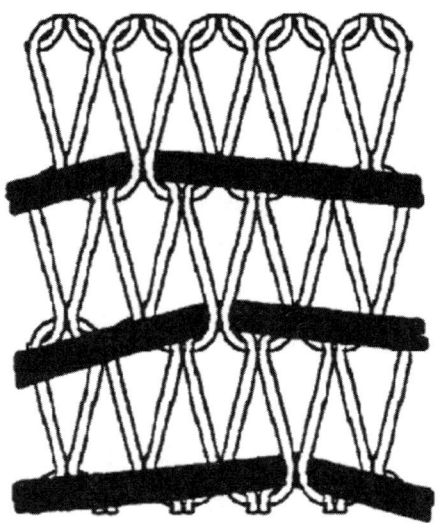

Figure 21.20 Warp inserted tows on reverse side of a single knit weft construction. *Source:* Reprinted from Raz S, *Industrial Knitted Fabric Design and International Trends*, IFAI, Boston, Nov 1992.

21.1.6.2 Warp knitting [15,17,18]

Warp knitting is a very flexible system and involves a set of yarns running parallel (cf the warp on a weaving machine). The yarns form a vertical loop in one course and then move diagonally to the next wale to make a loop in the following course, zig-zagging from side to side along the length. Each stitch in a course is made by a different yarn and each needle is provided with at least one yarn via a yarn guide. Warp knitting will produce a flat fabric with straight edges and once the machine has been set up, will produce at a fast rate.

MANUFACTURING TECHNIQUES FOR CARBON FIBER REINFORCED COMPOSITES

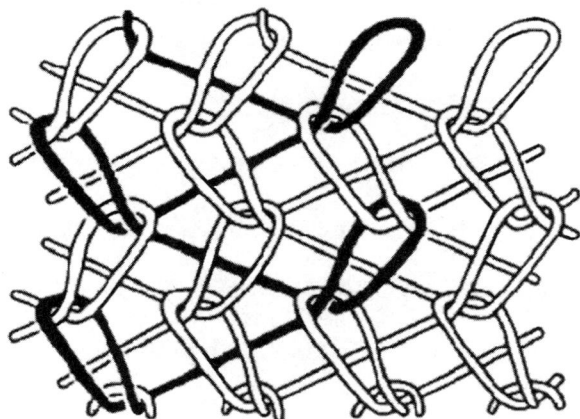

Figure 21.21 Warp knit plain tricot construction. *Source:* Reprinted with permission from Buckley JD, Edie DD eds., Noyes Publications, Park Ridge, 1992. Copyright 1992, William Andrew Publishing.

The two primary types of warp knitting are tricot and Raschel.

1. Plain tricot (Figure 21.21)

It has vertical ribs (wales) on the face and courses (horizontal ribs, cf crow's feet) on the back. A tricot knitting machine has a flat bed, with each warp yarn knitted by one needle and all needles (normally beard type) mounted in one needle bar and acting in unison.

2. Raschel

It involves columns of chain stitching with yarns traversing between the chain stitches. The principal differences with tricot are that the needles are latch needles and there are 4–48 guide bars, permitting greater variation of knits.

Pastore *et al* [19] have discussed the design and analysis of warp knit fabrics for composites.

21.1.7 Inlaid fabrics

The tows in inlaid warp fabrics are not crimped and are held in place by a looped fine thread. Tows can run in the warp direction (unidirectional) or in the weft direction (transverse filament) as in Figure 21.22. The insertion of a weft yarn requires a weft insertion device [20]. Tows can be balanced biaxially using two sets of tows to give a stable fabric with good mechanical properties and Figure 21.23 clearly shows the absence of crimp. A fabric with extremely good stability in all directions is provided by a multiaxial structure [21] with four sets of tows held by looped threads (Figure 21.24). To insert the diagonal yarns, the machine is required to be fitted with a rotating platform.

21.1.8 Braiding [22,23]

In many ways, braiding can be compared to filament winding, but cannot achieve such high volume fractions in a composite. A braid is a system of three or more yarns intertwined such that no two yarns are twisted around one another. Braiding machines can be made with

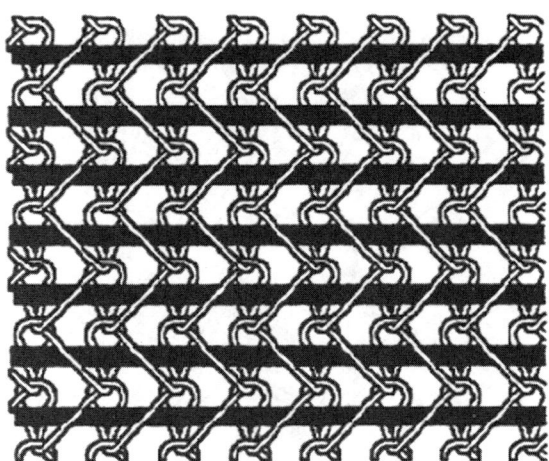

Figure 21.22 Weft inserted yarn in warp knitting. *Source:* Reprinted from Raz S, *Industrial Knitted Fabric Design and International Trends*, IFAI, Boston, Nov 1992.

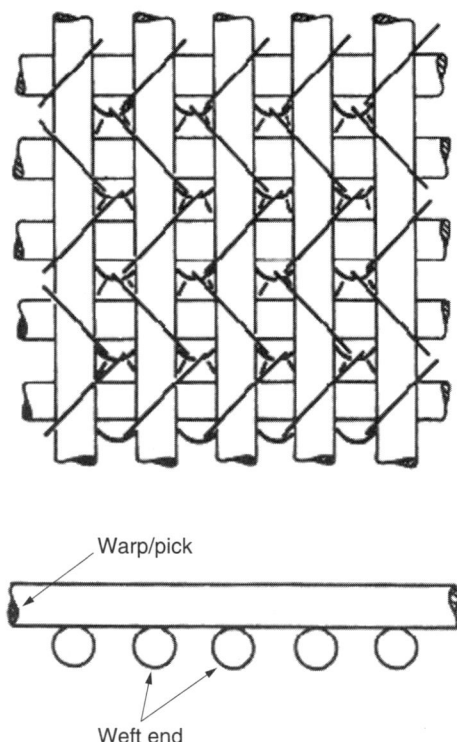

Figure 21.23 Warp/weft inlay fabric showing zero crimp. *Source:* Reprinted from Raz S, *Knitted Fabrics—Guide to Technical Textiles*, Karl Mayer, Obertshausen, 1988, with kind permission.

24–800 carriers (13.7 m diameter). The machine circumference increases in proportion to the number of carriers, whilst the braid area is proportional to the the square of the number of carriers. When a braid is opened out, in practical terms, it is equivalent to a fabric woven on the bias. Braids tend to be made to order and are supplied on reels or piddled into boxes. Hybrid formats are available, incorporating aramid and glass.

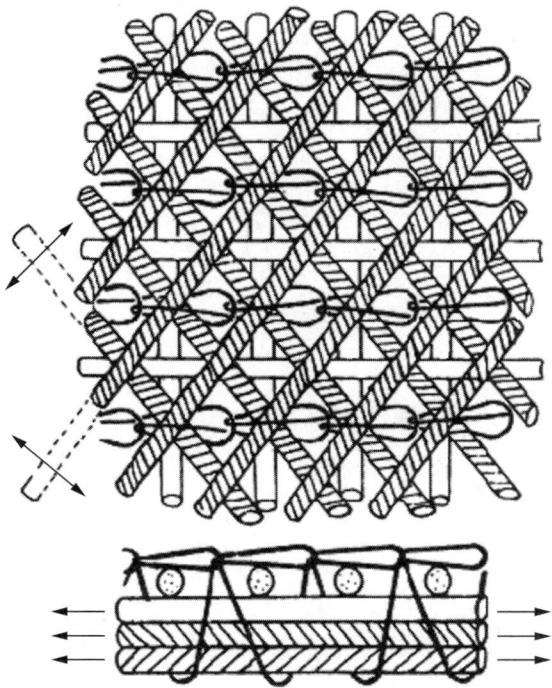

Figure 21.24 Multiaxial inlaid fabric with four sets of tows held by looped threads. *Source:* Reprinted from Raz S, *Knitted Fabrics—Guide to Technical Textiles*, Karl Mayer, Obertshausen, 1988, with kind permission.

Using computer controlled 3-D braiding, near-net shape structures such as I-beams, T-sections, hat-shaped and rocket motor exit cones can be produced, where the braid can conform to almost the exact shape of the sections. The final preforms should require no trimming or finishing before molding. Composites made from braiding have good shear resistance and torsional rigidity with very good resistance to fatigue and impact damage.

21.1.8.1 Forms of braiding

There are four main forms of braiding:

1. Flat braids

These use one set of yarns, where each yarn is interwoven in a zig-zag pattern with every other yarn from the set, to insert the diagonal yarns (Figure 21.25).

2. Sleevings

These use two sets of continuous yarn, one applied clockwise and the other counter-clockwise (Figure 21.25), where each fiber from one set is interwoven with every fiber from the second set in a continuous spiral pattern.

3. Wide braided fabric

This is produced on machines permitting a biaxial lay-up (+45°, −45°) or triaxial such as (0°, +45°, −45°) and (0°, +60°, −60°) in the one layer. Interestingly, a (0°, +60°, −60°) lay-up ensures equal mechanical properties in all directions.

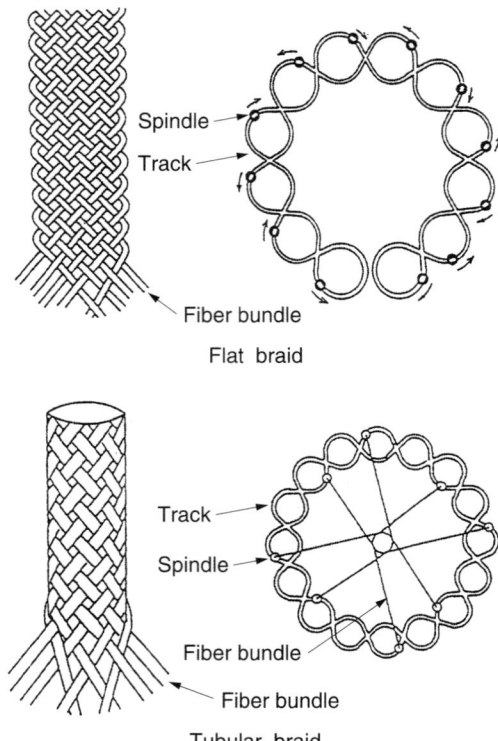

Figure 21.25 Flat and tubular braids and their tracks. *Source:* Reprinted with permission from Fujita A, Hamada H, Maekawa Z, Uozumi T, Okauji T, Ohno E, *Proceedings, TexComp-2*, Leuven, Belgium, May 17–19, 1994. Permission from Katholieke Universiteit Leuven.

4. Overbraids

Braids can be manufactured onto a mandrel made out of a foam, a low melting point metal, or a blow molded thermoplastic. Braiding permits computer controlled lay-up (Figure 21.26) and is used for making consistent complex shapes with accurate control of fiber angles, between 20 and 85°. A true 90° angle can be achieved by overwrapping using additional equipment. Fibers fed in the 0° direction help to control bending loads. The preform can then be impregnated with resin by Resin Transfer Moclding.

21.1.8.2 Braid architecture

There are two main forms of braid architecture:

1. Biaxial 2-D braid

A biaxial braid represents the most common form of braid. The construction of a flat braid is shown in Figure 21.27 where the processing parameters are defined as—θ, the braiding angle, which is half the angle of the interlacing between yarn systems, also called bias angle and fiber angle; pick spacing, which is the distance between interlacing points; d, the diameter of the braid. Tightness of braid is controlled by the frequency of the interlacings and the fiber tex. The process permits the introduction of an axial yarn between the bias yarns without crimping occurring.

Figure 21.26 Eurocarbon's 96 carrier computer controlled over braider. *Source:* Reprinted from Eurocarbon BV, Netherlands, technical literature, with kind permission.

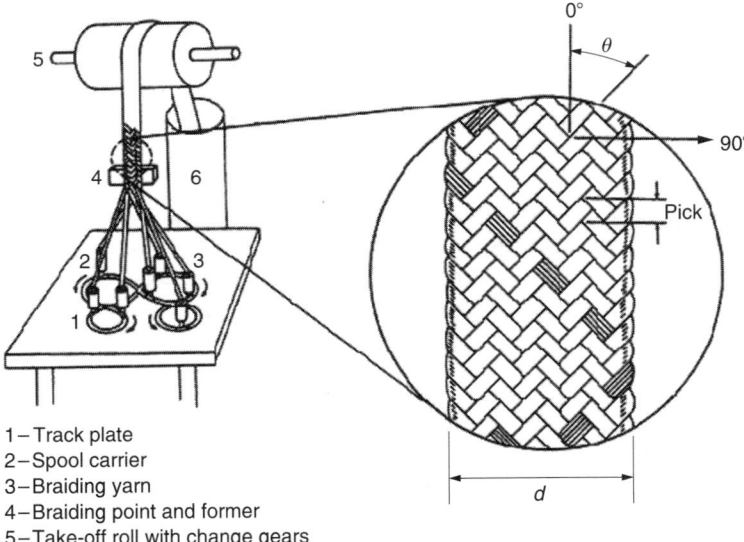

1 – Track plate
2 – Spool carrier
3 – Braiding yarn
4 – Braiding point and former
5 – Take-off roll with change gears
6 – Delivery can

Figure 21.27 A typical flat braiding machine and diagram of braid construction. *Source:* Reprinted with permission from Ko FK, Braiding, ASM International, Metals Park, 519–528, 1987. Copyright 1987, ASM International.

2. Triaxial 3-D braid

Triaxial braid is made with three sets of yarn whose intersections with a $0°$, $+60°$, $-60°$ setup form equilateral triangles, ensuring equal strength in every direction (Figure 21.28). The third yarn in a sleeving locks the diameter and prevents the braiding from expanding or

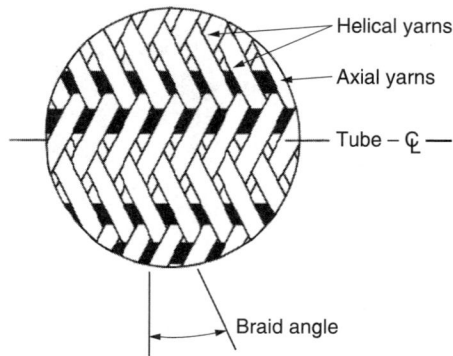

Figure 21.28 Triaxial braid construction. *Source:* Reprinted with permission from Burns RL; Edie ED eds., *Carbon–Carbon Materials and Composites*, Noyes Publications, Park Ridge, 197–222, 1992. Copyright 1992, William Andrew Publishing.

contracting, imparting bending and torsional stiffness to a composite. Masters *et al* [27] have reported on the properties of triaxially braided composites. Fujita *et al* [24] have described the fabrication and mechanical properties of I-beam composites. Due to the significant increase in fiber volume of yarns laid in the axial direction, there will be added crimp in the bias yarns.

21.1.9   3-D reinforcements [28–33]

21.1.9.1   *Multiaxial non-crimp reinforcements [11,34]*

The multiaxial architecture permits double bias fabrics ±45° (X-designation); triaxial with 0°/±45° (warp triaxial, Y-designation) and 90°/±45° (weft triaxial, Z-designation); and quadraxial fabrics with 0°/90°/±45° (Q-designation) (Figure 21.29). It is possible to have ±60° and ±45°, but these constructions necessitate major machine changes. Manufacturers can join layers together with an adhesive, or more probably, stitch with a yarn such as polyester or glass. The polyester stitching may give a poor bond with some resin systems. Stitched multiaxial fabrics have better mechanical properties than woven fabrics, since the fibers are always straight and not crimped.

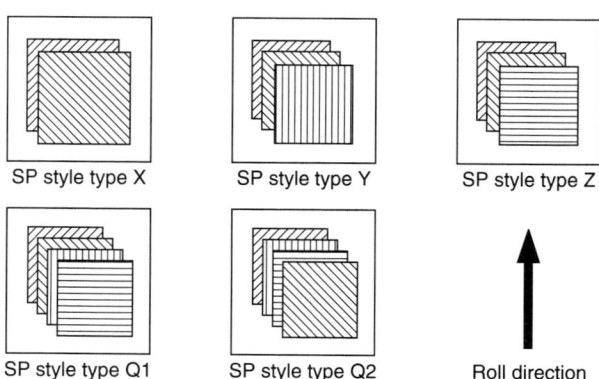

Figure 21.29   Forms of multiaxial construction. *Source:* Reprinted from SP Systems, Newport, IOW, technical data, with kind permission.

MANUFACTURING TECHNIQUES FOR CARBON FIBER REINFORCED COMPOSITES

The most common fabric is (+45°, 0°, −45°), in which the 0° lay-up contains 50% of the fiber, with the other 50% spread equally between the ±45° layers. Angles of ±60° and ±30° do necessitate major changes to the machine setup, which becomes time consuming and expensive. The width of a given machine is fixed (e.g. 1270 mm) and produces good quality fabrics above 1000 g m^{-2}.

The fabric is penetrated by a stitch bar containing up to 700 needles and the method of stitching used is important:

1. The stitch gage is the frequency of knitting across the fabric width and smaller gages provide greater stability, improving the drape, and is suitable for fabrics needing more handling.
2. A smaller stitch length will improve drape, but raises production costs.
3. The stitch style affects the handling of the fabric. A chain stitch (linear) gives good drapeability, whilst a tricot stitch (zig-zag) gives a more stable fabric. A modified tricot stitch can provide optimum drape and stability.

After stitching, the fabric can be slit to specified widths for use in, say, pultrusion.

Multilayered fabrics can be constructed, where up to eight layers can be assembled, with the warp yarns travelling to and fro from the bottom surface of the combined lay-up to the top, as shown in Figure 21.30a, whilst Figure 21.30b has additional in-plane strength due to the warp stuffer yarns, giving a quasi-3-D fabric.

1. Producing a stitched fabric by the simultaneous stitch process [34]

Multiaxial architecture of stitched reinforcements permit up to 5 plies of virtually any angle in the range 0–90°, such as 45, 90, 30, 60 and 22°, where the 0° ply runs in the direction

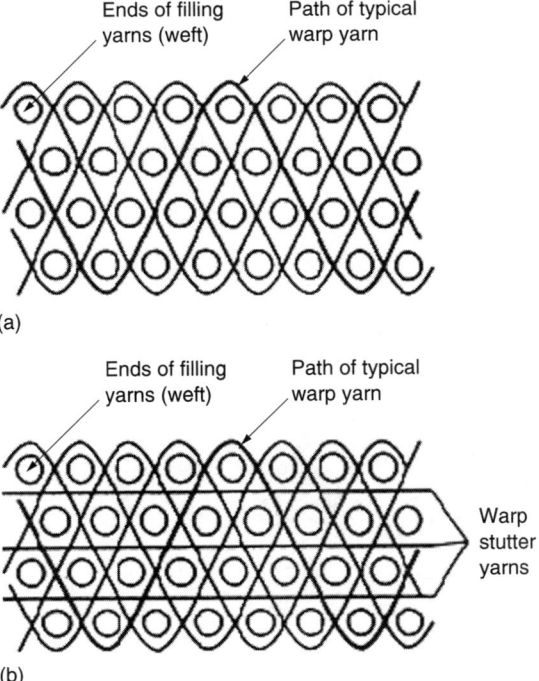

Figure 21.30 Geometry of angle (warp) interlock fabric. *Source:* Reprinted with permission from Savage GG, Chapman and Hall, London, 76, 1992. (a) No added stuffer yarn. (b) With added warp stuffer yarn (quasi 3-D fabric). Copyright 1992, Springer.

of the warp and is fed from a creel, whilst the cross plies are fed by carriages, which move to and fro across the warp direction. The stitching is carried out on special machines based on the knitting process.

The WIMAG warp knitted multiaxial layer (Liba system), which is a multiaxial weft insertion process carried out on Karl Mayer machines (Figure 21.31) and the NVG a stitch bonded variable layer (MALIMO) process undertaken on Malimo machines (Figure 21.32) are examples. The NVG process displays lower mechanical properties due to fiber damage which occurs during the insertion of the 0° filaments via a small tube that is strongly bent during insertion into the knitting elements.

The Liba process is non-impaled (or non-impregnated) and has the stitches inserted around the insert yarns (Figure 21.33). The Malimo process is termed impaled (or impregnated structure) with stitches piercing the insert yarns. The non-impaled knitted structure is useful for high modulus fibers, with good fiber positioning and little build up of stress in the load bearing system. The impaled fabric can give 3-D toughened structures through interlocking by stitch bonding.

The Liba initially used modified lace making machinery, which produced a thick glass fiber fabric and when adapted for carbon fiber, used a low weight thermoplastic thread to

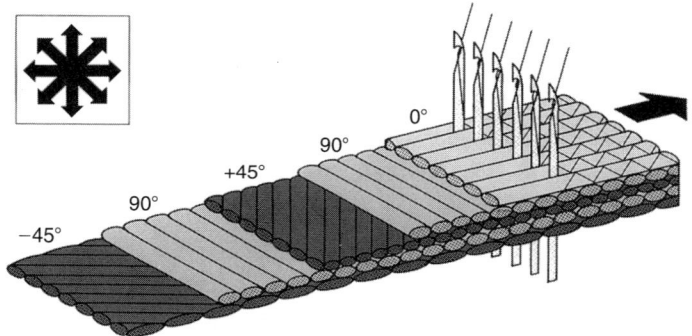

Figure 21.31 Liba version of simultaneous stitch. *Source:* Reprinted from SP Systems, Newport, IOW, technical data, with kind permission.

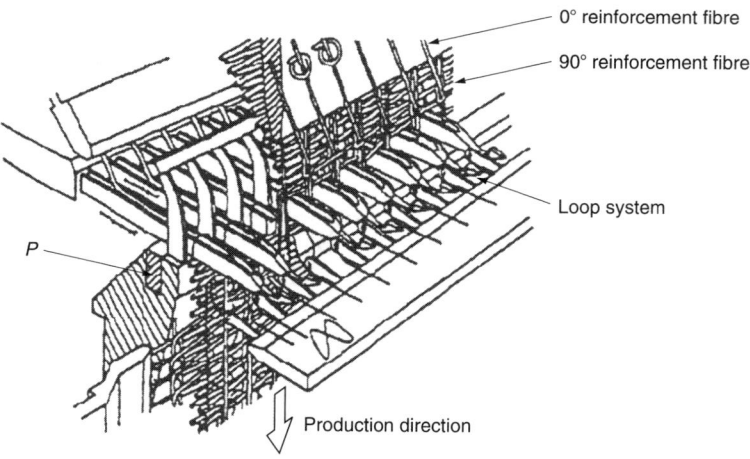

Figure 21.32 Malimo stitch bonding process. *Source:* Reprinted with permission from Hörsting K, Wulhürst B, Franzke G, Offerman P, *SAMPE J*, 29(1), 7–11, Jan–Feb 1993. Copyright 1993, The Society for the Advancement of Material and Process Engineering (SAMPE).

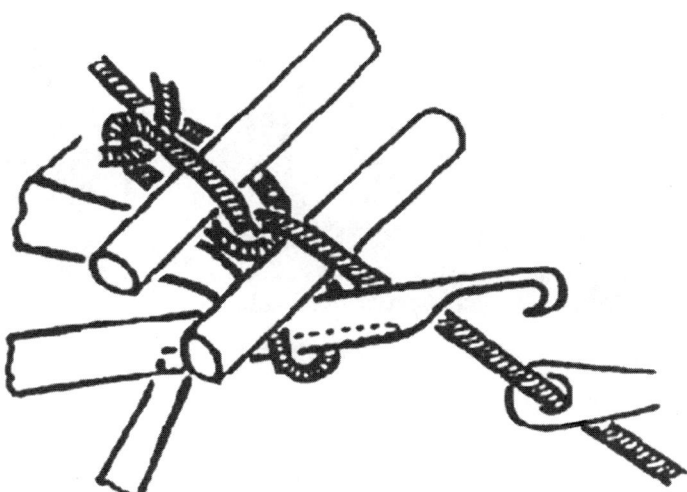

Figure 21.33 The principle of warp knitting with weft inserts (Karl Mayer machine). *Source:* Reprinted from Raz S, *Knitted Fabrics—Guide to Technical Textiles*, Karl Mayer, Obertshausen, 1988, with kind permission.

hold the tows together with a minimum of crimp. The Liba process maintains closely spaced straight tows without gaps and is able to achieve a laminate with 65% V_f using the autoclave process. Mills *et al* [37] at Cranfield University, in conjunction with BTI Europe, have made modifications permitting 12, 24 and 48k carbon fiber to be spread to fine areal weights with no gaps. Such a construction previously necessitated the use of more expensive 3 and 6k carbon fiber. Polished cross-sections of equivalent unidirectional tape and Liba NCF laminates are shown in Figure 21.34.

2. Producing a stitched fabric by the weave and stitch process

There is another method of producing ±45° stitched fabrics by weaving a weft unidirectional fabric and then pulling the fiber on a special machine to introduce a 45° skew (Figure 21.35). The process does introduce a measure of crimp when the fiber is skewed, but heavy tows can be used, keeping costs down. A quadraxial lay-up could consist of (+45°, 0°, 90°, −45°), where the 0° and 90° can be a conventional woven fabric used with the skewed +45° and −45° layers and after assembly, the three layers are then stitched together. The disadvantage is that the woven layer will introduce a measure of crimp.

3. Double bias fabrics

Double bias fabrics (X-designation) with ±45° reinforcements have good conformability, torsional strength and enhanced surface cosmetics and can be used to provide off-axis strength for pultrusion products.

4. Triaxial weave [38]

Triaxial weave originated in the 1970s and feature a 90±60° hexagonal planar arrangement. A more popular construction is a unidirectional + and − layer, with either a weft (90°) or a warp (0°) unidirectional layer. All three layers are normally combined by stitchbonding into a single fabric, which gives better drapeability than an adhesive bonded fabric.

Specialist multiaxial equipment for weaving triaxial fabrics is shown in Figure 21.36.

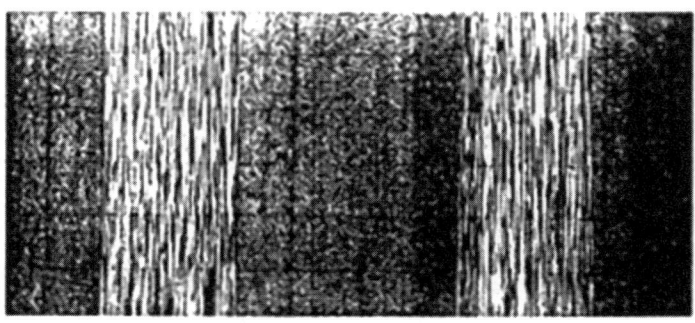

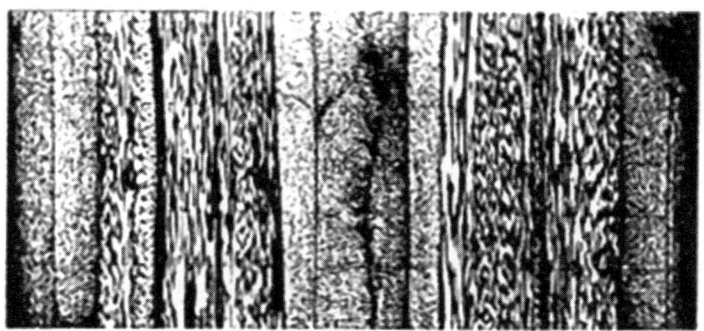

Figure 21.34 Polished cross-sections of equivalent UD tape and Liba NCF lamintes. (a) Unidirectional prepreg composite laminate. (b) Liba NCF composite laminate. *Source:* Reprinted with permission from Mills A, Burley G, Backhouse R, *44th International SAMPE Symposium*, May 23–27, 1999. Copyright 1999, The Society for the Advancement of Material and Process Engineering (SAMPE).

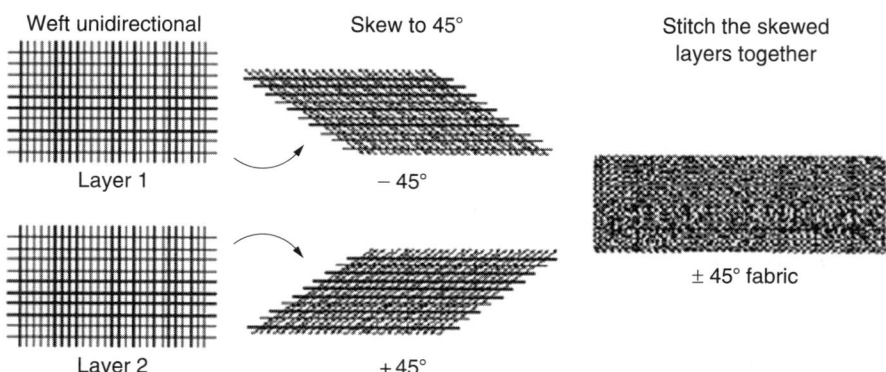

Figure 21.35 Multiaxial fabric prepared by Weave and Stitch method. *Source:* Reprinted from SP Systems, Newport, IOW, technical data, with kind permission.

a. Basic weave—Basic triaxial (Figure 21.37) is similar to plain weave with each end going over and under the next yarn. Basic basket uses doubled ends.

b. Biplain weave—This weave uses tripled ends, with each single end going over one and under the other of one crossing pair, but over both other pairs (Figure 21.38).

MANUFACTURING TECHNIQUES FOR CARBON FIBER REINFORCED COMPOSITES

Figure 21.36 Brunswick Technologies Inc., Andover equipment for weaving triaxial fabrics. *Source:* Courtesy Brunswick Technologies Inc.

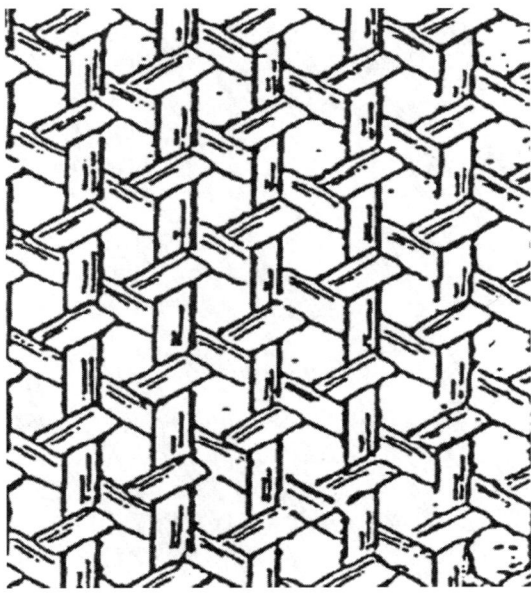

Figure 21.37 Structural geometry of triaxially woven fabric (basic weave). *Source:* Reprinted with permission from Ko FK; Buckley J, Edie DD eds., *Carbon–Carbon Materials and Composites*, Noyes Publications, Park Ridge, 71–104, 1993. Copyright 1993, William Andrew Publishing.

Triaxially woven fabric composites are isotropic in modulus and anisotropic in strength [39]. The tensile properties of the basic triaxial woven composites are superior to biplain and biaxial woven fabric composites. A multilayer triaxially woven fabric is shown in Figure 21.39.

Figure 21.38 Structural geometry of triaxially woven fabric (biplain weave). *Source:* Reprinted with permission from Ko FK; Buckley J, Edie DD eds., *Carbon–Carbon Materials and Composites*, Noyes Publications, Park Ridge, 71–104, 1993. Copyright 1993, William Andrew Publishing.

Figure 21.39 Multilayer triaxially woven fabric. *Source:* Reprinted with permission from Ko FK; Buckley J, Edie DD eds., *Carbon–Carbon Materials and Composites*, Noyes Publications, Park Ridge, 71–104, 1993. Copyright 1993, William Andrew Publishing.

5. Quadraxial

Quadraxial fabrics are quasi-isotropic, providing strength in all four fiber axes.

21.1.9.2 Woven 3-D fabrics [29,40–43]

A problem encountered with composites prepared from 2-D fabrics is that cleavage can subsequently occur between layers, leading to premature failure of the composite. This is a problem that does not occur with 3-D structures due to the lack of laminae.

Hill and co-workers [34] have described the use of woven integrated structures for engineering preforms.

Early work by Hitco used angle (warp) interlock with thick woven fabrics [44]. Soden and Hill [45] have described the fabrication of a diverse range of flat and shaped preforms using conventional weaving, undertaken on an electronically controlled Jacquard power loom with a CAD design package developed at Ulster University [46].

The yarns within a cross-section can be classified depending on their interlinking characteristics:

1. Inplane interlacers always remaining in their layer of origin and following a designated interlacing weave pattern.
2. Inplane stuffers that are straight uncrimped yarns lying in their layer of origin and not partaking in any interlacing.
3. Through the thickness interlinking yarns, which do contribute to the Z-axis proportion by forming connections to other layers that penetrate a proportion of, or total, fabric thickness.

The diverse range of 3-D weave architectures can be categorized as:

1. Integrated structures—tend to be based on plain, twill and satin weaves with through thickness interlinks.
2. Warp binding or orthogonal type architecture—have uncrimped yarn paths bound together by yarns penetrating the total thickness from surface to surface, interspersed with groups of stuffer warps. This type has poor drapeability.
3. Shaped preforms [45]—a preform can be produced by weaving a shaped reinforcement in a flat form and then opening or folding after removal from the loom to give the required shape. Folding a preform can form gaps, which will give resin rich areas in the composite, but these can be eliminated by careful initial design. In multilayer architecture, twill and satin weaves are preferred, since plain weaves exhibit high crimp levels.

21.1.9.3 Proprietary 3-D weaving processes

Magnaweave was an early 3-D structure and described by Ko [47] with details of mechanical composite properties [48].

Autoweave^{TM} [49] was a process developed by Brochier in France and licenced exclusively to Avco/Textron in the USA. The radial reinforcement in Autoweave^{TM} is a screw-like reinforced phenolic resin rod, which is produced as a continuous stock, cut to length and inserted by computer control into an expendable low cost phenolic foam mandrel. The axial and circumferential reinforcements, which can be either dry fiber or a prepreg, are then positioned precisely in the radial corridors to produce a 3-D preform.

A 3-D sandwich structure, sometimes referred to as Parabeam, consists of two bi-directional woven fabrics mechanically connected with vertical woven plies. The fabric

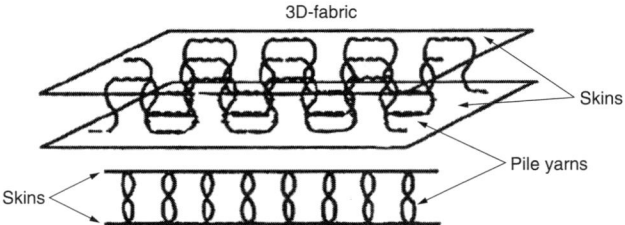

Figure 21.40 Schematic view of a 3-D fabric: the two skins are interconnected by pile yarns in a one step weaving process. *Source:* Reprinted with permission from Ivens J, Vandeurzen PH, van Vuure AW, Verpoest I, Ko FK, Meerding K, *Proceedings, TexComp-2*, Leuven, 540–547, May 17–19, 1994. Permission from Katholieke Universiteit Leuven.

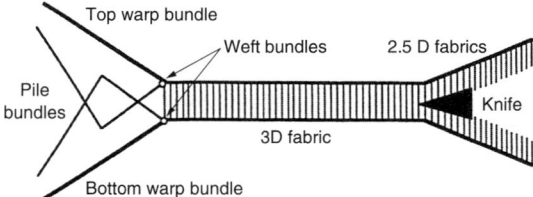

Figure 21.41 Production process for 3-D and 2.5-D fabrics. *Source:* Reprinted with permission from Verpoest I, Ivens J, van Vuure AW, Efstratiou V, *Proceedings, TexComp-2*, Leuven, May 17–19, 1994. Permission from Katholieke Universiteit Leuven.

has a preset space between the two surface decks and this hollow core can be filled with materials such as foam. This structure was developed at the Universities of Leuven and Zaragoza, where two layers of 3-D fabrics were connected by orthogonal threads (Figure 21.40) using velvet weaving technology and since a 3-D fabric was required, the pile threads were intentionally not cut (Figure 21.41). However, the pile threads were erroneously cut, producing two pieces of velvet fabric, which were jokingly dubbed 2.5D and this terminology appears to have been universally adopted. These so called 2.5D fabrics are stated to overcome the problem of delamination in composites [50].

21.1.9.4 Knitted 3-D fabrics [51]

Weft flat knitting machines, in conjunction with computer control, can produce a variety of 3-D structures from one yarn including boxes, cones and spheres. In warp knitting, it is normal to use an adapted Raschel machine fitted with two needle bars and several guide bars. Each needle bar produces a flat fabric, with the two flat fabrics simultaneously connected forming a sandwich, which can be readily impregnated with resin.

21.1.9.5 Braided 3-D multiaxial [52–55]

Braiding is termed 3-D when three or more systems of braiding yarns are involved to form a three-dimensional braid. Jacquard braiding uses a mechanism which facilitates connecting groups of yarns to braid different patterns simultaneously. Atlantic Research Corporation (ARC) have developed a form of 3-D braiding (Through the Thickness Braiding) and can utilize a binder to make a rigid preform. General Electric developed Omniweave and SEP developed Scoudid.

Figure 21.42 shows a schematic view of 3-D tape made by the braiding process.

MANUFACTURING TECHNIQUES FOR CARBON FIBER REINFORCED COMPOSITES

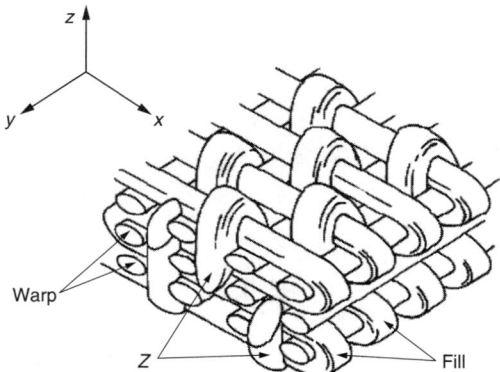

Figure 21.42 Schematic view of 3-D tape made by the braiding process. *Source:* Reprinted with permission from Burns RL, Manufacturing and design of carbon-carbon composites, Buckley JD, Edie ED eds., *Carbon–Carbon Materials and Composites*, Noyes Publications, Park Ridge, 197–222, 1992. Copyright 1992, William Andrew Publishing.

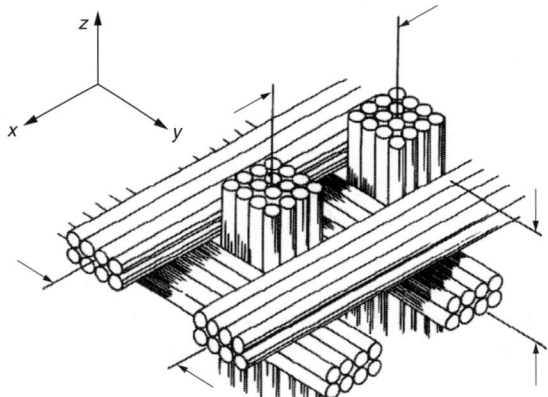

Figure 21.43 Typical 3-D block construction. *Source:* Reprinted with permission from Burns RL, Manufacturing and design of carbon-carbon composites, Buckley JD, Edie ED eds., *Carbon–Carbon Materials and Composites*, Noyes Publications, Park Ridge, 197–222, 1992. Copyright 1992, William Andrew Publishing.

21.1.9.6 n-D orthogonal blocks

A typical 3-D block construction is shown in Figure 21.43. Higher in-plane shear moduli can be achieved with 4-D (Figure 21.44) and 5-D (Figure 21.45) architectures. Using a polar 3-D architecture (Figure 21.46) provides axial tensile and compressive strengths in the Z-direction, whilst the hoop fibers provide hoop tensile strength and the radial fibers transmit radial compressive and torsional shear strength.

Hiroka *et al* [56] have established that orthogonal woven composites have better compressive and flexural fatigue strengths than 2-D composites.

Cahuzac [57] divides the processes used by Aerospatiale for making 3-D preforms into three categories:

1. Processes using a network of rods—metallic wire rods set in thick pierced plates are used to simulate the longitudinal direction of the weave. The first process stage is on a weaving machine, when yarns are inserted between the rods in the Z direction accompanied by

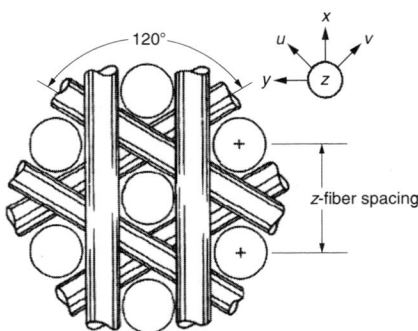

Figure 21.44 Schematic view of 4-D construction. *Source:* Reprinted with permission from Burns RL, Manufacturing and design of carbon-carbon composites, Buckley JD, Edie ED eds., *Carbon-Carbon Materials and Composites*, Noyes Publications, Park Ridge, 197–222, 1992. Copyright 1992, William Andrew Publishing.

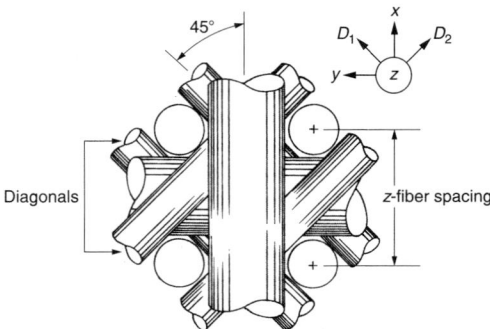

Figure 21.45 Schematic view of 5-D construction. *Source:* Reprinted with permission from Burns RL, Manufacturing and design of carbon-carbon composites, Buckley JD, Edie ED eds., *Carbon-Carbon Materials and Composites*, Noyes Publications, Park Ridge, 197–222, 1992. Copyright 1992, William Andrew Publishing.

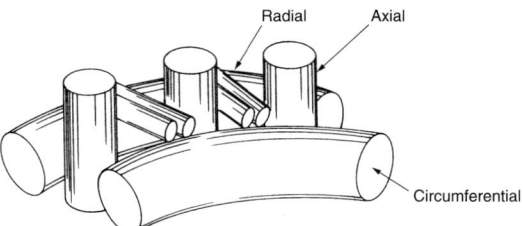

Figure 21.46 Three dimensional cylindrical weave construction. *Source:* Reprinted with permission from Burns RL, Manufacturing and design of carbon-carbon composites, Buckley JD, Edie ED eds., *Carbon-Carbon Materials and Composites*, Noyes Publications, Park Ridge, 197–222, 1992. Copyright 1992, William Andrew Publishing.

compaction/vibration to provide the correct bulk density. The second stage uses a special needle to push the rod out of the preform, but not out of the plate. On its travels, the eye of the needle catches the correct yarn ends from the bottom of the lay-up and pulls them through to give a preform. It is possible to make orthogonal 3-D blocks, polar 3-D cylinders and cones and shapeable 3-D cylinders. A numerical controller can be used to place the yarns between the rods.
2. Fully computerized processes using a mandrel. A foam mandrel with implanted metallic pins is the basis of the Novoltex process.
3. Weaving a thick fabric to produce a 2.5-D fabric.

An orthogonal 3-D block can be made by a weaving machine fitted with two thick pierced stacked blocks holding a series of rods in the Z direction. Two lateral rows of special needles situated under the top block interweave the X and Y yarns, which are compacted by pushing down after each layer. Once the requisite length of the fiber stack is achieved, it is removed from the weaving machine and positioned on a lacing machine, which is set to replace the rods with yarn to give the required preform.

Hearle and Du [2] discuss the interaction of textile technology and composite engineering and Brookstein [58] discusses braided and woven interlocked fiber architecture.

21.1.9.7 Aztex Inc Z-Fiber™

This type of through thickness pinning process invented by Foster-Miller Inc. inserts small diameter composite or metal rods through the thickness of the composite, using a special tool and can be applied to most thermoset matrix resins and dry fiber performs, to give a stiffened and reinforced product resistant to delamination without over-designing with extra thickness. The pins can be arranged at the required pitch and length in a consumable foam sheet, which is placed over the uncured lay-up under the vacuum bag prior to cure. With standard heat and pressure in the autoclave, the foam collapses gradually driving the pins into the structure [59]. The rod diameter (0.25 mm) is small enough not to significantly degrade the laminate strength and there are no knots or loops to disrupt the fiber distribution.

21.2 CORE MATERIALS [10]

If a laminate with a thick section has to be used, it is common practice to achieve weight saving by using a core of lightweight material (e.g. foam) with a density less than the reinforcement and matrix materials and is generally sandwiched between two laminates. The laminates are ideally bonded to the core material with a suitable film adhesive. Surface skins on core materials should be roughened to improve adhesion of the film and all open cell structures must be sealed with resin. Some of the core materials are honeycombs, which can vary in cell shape (e.g. hexagonal or rectangular), cell size and core thickness. The honeycomb sections can be machined to final configuration using 3-, 4- and 5-axis numerically controlled machines. It is important that the core has good moisture resistance and the open edges are sealed to prevent ingress of moisture. A sandwich construction with a core thickness equal to the total skin thickness will be, for only a 3% increase in weight, about 7 times more rigid and 3.5 times as strong. The laminate faces take up tensile and compressive forces, whilst the core transfers the shear forces. The core will also contribute some resistance to impact damage. Table 21.2 gives the properties of some core materials.

1. Closed cell PVC foam—is a chemical hybrid of PVC and polyurethane and there are basically three types:

   a. Crosslinked—are better mechanical properties and elevated temperature performance. PVC crosslinked foams must be resin sealed before use.
   b. Uncrosslinked (linear)—are tougher, more flexible and easier to heat form around curves.
   c. Toughened—addition of plasticiser as toughening agent contributes to some loss of mechanical properties.

2. Polystyrene (PS) foam—are rarely used for high performance composites.

Table 21.2 Properties of core materials

Property	Density kg m^{-3}	Shear modulus	Shear strength	Resistance to smoke and flame LOI	Water absorption %	Form Ability	Maximum Temperature of use °C
THERMOSETS							
Polyisocyanurate foam	30–45	poor	fair	poor (24)	good	good	93–121
Syntactic phenolic foam	42	very good	fair	very good (40)	very good	very good	130
THERMOPLASTICS							
PVC foam	40–270	poor	fair	poor (40)	good	poor	65–80
Polycarbonate honeycomb	48–320						95
PMI (polymethacrylamide) (Rohacell®)	20–300			good	good	good	XT grade can be co-cured up to 190
Graphitic foam (MER Corp)	16–620						
OTHER FORMS							
Balsa foam	112–320	very good	good	good	poor	poor	
Aramid /phenolic coated honeycomb	24–200	poor	good	very good	poor	poor	

3. Polyurethane (PU) foam—has only moderate mechanical properties.
4. Polymethylmethylacrylamide (PMI) foam—are expensive but the foams have good high temperature properties, enabling their use with elevated temperature curing prepregs.
5. Styrene/acrylonitrile copolymer (SAN) foam—is thermoformable, with higher toughness and elongation than PVC foams, hence superior impact properties. They tend to replace linear PVC foams due to superior high temperature performance and improved properties, especially toughness.
6. Polyetherimide/polyethersulphone (PEI) foam—is expensive but has excellent high temperature performance coupled with good fire resistance.
7. Polycarbonate (PC) foam—has good heat resistance, mechanical properties and impact properties.
8. Nomex honeycomb cores (an aramid paper coated with a phenolic resin)—are more than twice as expensive as a foam core, but have good mechanical properties with fire retardant properties.
9. Balsa honeycomb—is cheap but readily soaks up resin, incurring a weight penalty.
10. Al honeycomb—is not too expensive, has good mechanical properties, but can be corroded by sea water.

Akay and Hanna [60] have compared foam core and honeycomb core epoxy sandwich panels.

21.3 MANUFACTURING PROCESSES FOR CARBON FIBERS IN THERMOSET MATRICES [61]

There are now many processes available and normally referred to by an acronym and these have been listed in Table 21.3.

21.3.1 Contact molding wet lay-up

Contact molding is the process of applying reinforcement, preferably pre-cut to size, onto a mold treated with a release agent to produce a molding, which will have only one smooth

Table 21.3 Acronyms associated with plastic molding processes

Acronym	Description
CIM	Compression injection molding
CRTM	Continuous resin transfer molding
GAIN	Gas assisted injection molding
LM	Liquid molding
LTVB	Low temperature bag molding
RFI	Resin film infusion
RIFT	Resin fusion under flexible tooling
RIM	Reaction injection molding
RIRM	Resin injection recirculation molding
RRIM	Reinforced reaction injection molding
RTM	Resin transfer molding
SCRIMP	Seemann composite resin infusion molding process
SMRIM	Sequential multi port resin injection system
SPRINT	SP resin infusion technology
SRIM	Structural reaction injection molding
TERTM	Thermal expansion resin transfer molding
VARI	Vacuum assisted resin injection
VARTM	Vacuum assisted resin transfer molding
VIMP	Vacuum infusion molding process
VIP	Vacuum infusion processing
VRTM	Vacuum resin transfer molding

surface, the one that is in contact with the mold. The application can be by hand lay-up, or spray lay-up; both techniques use the same mold preparation and the first production stage can entail the application of a resin gel coat. Generally, polyester resins or room temperature cure epoxies are used. Sufficient time must elapse before removal from the mold, say 24 h, and where possible, it is advisable to support the structure after removal from the mold and complete the cure at elevated temperature. With glass reinforcement, the final glass content will be about 20% v/v.

21.3.1.1 Hand lay-up (contact molding)

The applied reinforcement, such as mat, is placed on the mold and well wetted out with the chosen resin system, which is normally applied by brush, and consolidation effected by thorough brushing with a mohair roller followed by rolling with a split-washer roller to remove entrapped air. The process is rarely undertaken with carbon fiber reinforcement. Although contact molding will be used for some time in the future, there is a restriction in Scandinavia on styrene emission to be not greater than 10 ppm and this form of stringent environmental legislation will pressurize users to look closely at closed mold processes such as RTM, VARTM, vacuum bagging and cold compression molding.

21.3.1.2 Spray lay-up

A level of automation using a twin gun system can be achieved by applying the reinforcement as a roving, which is chopped in one gun and the cut staple combined with resin that has been admixed in the correct proportions in the other gun. The admixing in the gun can be carried out, with latest equipment, within the gun to curtail emissions, and the chopped reinforcement/mixed resin spluttered onto the mold. Emissions are reduced by using a controlled spray program, which limits the surface area of the wet resin. Entrapped air is removed by rolling with a mohair roller followed by a split washer roller. The process is useful for the application of a glass backing reinforcement to a mold but is rarely used for carbon fiber composite.

21.3.2 Hot press matched metal molding

A machined matched metal mold (Figure 21.47) treated with a release agent is used to contain the resin/reinforcement and pressure applied to the stops to give the requisite laminate thickness by using a hydraulic press (10–50 MPa) to consolidate the reinforcement and remove any entrained air along with excess resin. Subsequently, the resin is cured by the application of heat. This method is used to prepare laboratory composite samples for routine quality control purposes (Section 17.8.2). DMC, BMC and SMC can be processed by this method.

21.3.2.1 Thermoset dough molding compound (DMC)

Generally used with a glass reinforcement, a dough is produced using a high shear mixer by mixing the resin, catalyst and about 10% v/v of approximately 10 mm chopped reinforcement, a release agent (such as zinc stearate) together with filler and pigment. After mixing, the dough is formed into a rope or billet. DMC has a shelf life that is less than that of SMC.

21.3.2.2 Thermoset bulk molding compound (BMC)

Similar to a DMC, but with an improved resin, such as an isophthalic polyester, although the terminology is restricted to the UK, as in the USA, a BMC is the same product as a DMC.

21.3.2.3 Thermoset sheet molding compound (SMC)

To aid processing flat products, it is helpful to use a sheet molding compound that can be made in a similar manner to a DMC, but with a higher concentration of reinforcement (20%

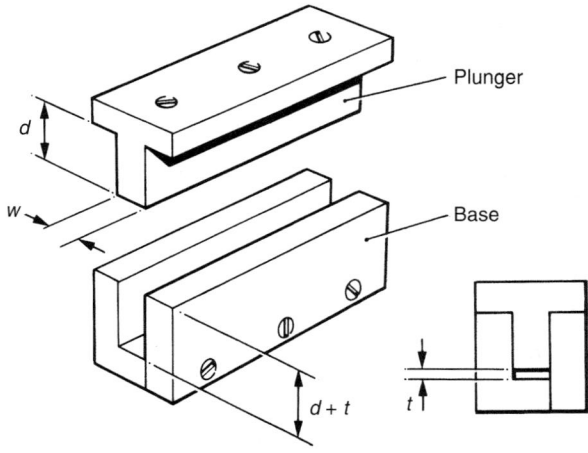

Figure 21.47 Matched metal mold for laboratory test composites.

v/v) and a longer staple length (about 25 mm). The catalyzed resin is deposited on a polyethylene release film and the reinforcement consolidated in it by adding a top release film and passing the sandwich through a series of rollers to give an SMC of the required thickness. The SMC will have a shelf life of about 3–6 months at room temperature. Hexcel have marketed HexMC, a 2000 g m^{-2} sheet containing 57% v/v carbon fiber in an epoxy with an in-built release agent, which can be cured in 5 min.

21.3.3 Resin transfer molding (RTM) [62,63]

Resin transfer molding (RTM), sometimes referred to as a closed mold process, simply entails transferring the resin through a port, or series of ports, under moderate pressure (0.35–0.70 MPa) into a closed and clamped mold in which the reinforcement has already been positioned (Figure 21.48). The process can be used with polyester, polyester/urethane, vinyl ester, epoxy, phenolic, and modified acrylic resins. Cure is effected with heat and a resin with a viscosity of about 200–600 cP will penetrate all the surfaces of the mold cavity when used with a vacuum (Vacuum Assisted Resin Injection (VARI)) to assist resin flow and reduce formation of voids. It is a good method to achieve high volume fractions for complex shapes and has become an accepted operation in the aircraft industry, assuming significant practical importance for a number of reasons:

1. Repeatable high quality composite production with a high quality surface finish
2. Parts can be produced quickly at a competitive cost provided the production run is long enough to offset high tooling costs
3. High performance resin systems have been developed for RTM
4. Environmentally friendly due to the use of a closed mold
5. Ability to use automated preforms and can incorporate braiding

When injecting resin into an RTM mold, two methods may be used to control resin flow— measuring the resin velocity or the resin pressure. Control of resin pressure provides greater repeatability and Plastech TT have achieved this by measuring the pressure and adjusting the speed of the resin pump, preferably using a digital system with PLC and an electro-pneumatic pressure regulator to control pump speed and, hence, the downstream pressure.

Probably the best material for a mold is a vinyl ester resin tooling resin system, applied with a gel-coat and backed up with a suitable resin system, both with a heat distortion

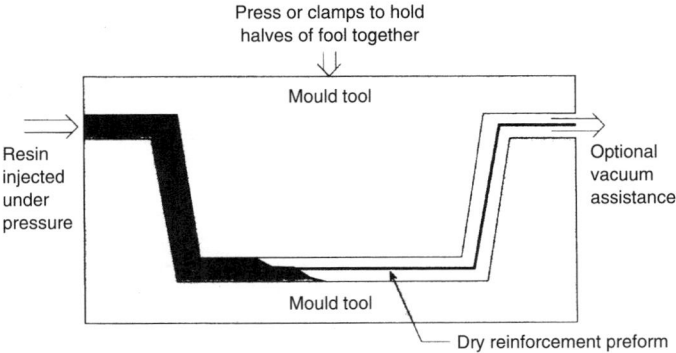

Figure 21.48 Resin transfer molding. *Source:* Reprinted from SP Systems, *Composite Engineering Materials— Product Information Handbook; Guide to Composites*, Newport, IOW, technical data, with kind permission.

temperature of 150°C. A minimum draft angle of 1° is included to permit removal from the mold [65]. Typically, a mold is some 7–10 mm thick, with a high temperature core and a total thickness of about 20 mm, backed up with a steel box section tailored frame. Heating is provided by an embedded electrical heater blanket or implanted pipes for water heating/cooling. The design of the mold must accommodate resin shrinkage in the part, which could render the mold impossible to open upon completion of the cure.

The matched molds will require special sealing for RTM and vacuum molding, as shown in Figure 21.49. The dynamic seal is the preferred construction. Resin is normally introduced at a central position of the mold cavity shape and can be introduced at the top or bottom of the mold, whichever is most convenient.

Since 1967, Dowty Aerospace Propellers have continually developed a very effective production method [66] for the manufacture of propellers using RTM claiming advantages over the prepreg route (Table 21.4).

A dry fiber preform is used with the layers of fibers/fabric holding it together with a powder binder. This initial preform is consolidated under heat and pressure to form a rigid preform, which is then placed in the blade mold and a core of low density polyurethane foamed *in situ*, followed by applying layers of carbon fiber at ±45° using a braiding machine.

21.3.3.1 Dow AdvRTM^{TM}

Dow-United Technologies Composite Products Inc. (Dow-UT) (a partnership of The Dow Chemical Company and United Technologies Corporation) introduced Advanced Resin Transfer Molding (AdvRTM^{TM}), a high specification RTM process and in 1999, the

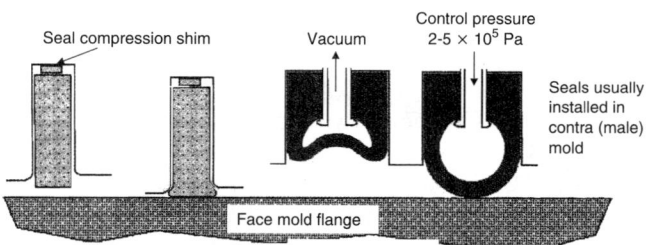

Figure 21.49 Seals for RTM. *Source:* Reprinted with permission from Harper A, *Reinforced Plastics*, 28–34, Feb 2001. Left – passive seal with adjusting shim, Right – dynamic seal with pressure control. Copyright 1999, Elsevier.

Table 21.4 RTM process advantages

RTM	Prepreg
Low material cost	High material cost (approximately 2× cost of RTM)
Minimal material storage problem	Prepreg has short life at room temperature and must be stored in a freezer
Close control on part thickness and fiber content	To achieve accurate thickness control, must use matched metal molds and computer controlled presses
Thick complex parts can be molded in one shot	Difficult to mold complex parts in one shot

Source: Reprinted with permission from Abraham D, McCarthy R, Design of polymer composite structural components for manufacture by resin transfer moulding (RTM), *SAMPE Europe Conference and Exhibition. Proc of 20th International SAMPE Europe Conference of the Society for the Advancement of Materials*, Paris, France, 407–415, 1999. Copyright 1999, The Society for the Advancement of Material and Process Engineering (SAMPE).

company was purchased by GKN Westland Aerospace. The process has become an alternative to prepreg for composite manufacture and over the years, has been refined to become accepted by leading aerospace companies like Boeing for the Raptor F-22 advanced tactical fighter, enabling program molding to construct the carbon fiber aft boom fairings in one piece, effecting a 47% reduction in weight resulting in a 55% reduction in cost. The key for a successful process is the initial preparation of the preform and utilizing computer controlled robotics to provide high quality parts, which are worthy successors to Al and Ti.

Epoxy matrices with improved hot/wet performance and toughness were developed for prepreg applications, but the application remained labor intensive. The outstanding virtue of RTM was the achievement of significant cost savings. Ideally, the viscosity of a resin should be <500 cP to effect bulk flow and ensure efficient fiber penetration, but using conventional prepeg resin systems, raising the temperature to sufficiently lower the viscosity considerably shortened the pot life.

With two-part resin systems, the standard procedure was to weigh and mix resin components in a heated pressure pot, degas by the application of vacuum and transfer under pressure into a mold. A variant of this system is to use two-part meter mix equipment.

A significant advance was made by the introduction of 3M's PR500 one-part resin system, which could be used with melt on demand equipment (such as Graco), enabling the resin container to be used as the reservoir [67]. The advantages of this system are—no mixing required, so no extra vessels to clean and no problems weighing out; environmentally friendly; no degassing required and controlled heat exposure of the resin. Typical melt on demand equipment is shown in Figure 21.50. The use of a resin weir enables a thin film of resin to be heated providing efficient heat transference.

A typical manufacturing procedure for PR500 when using a heated resin pot would be:

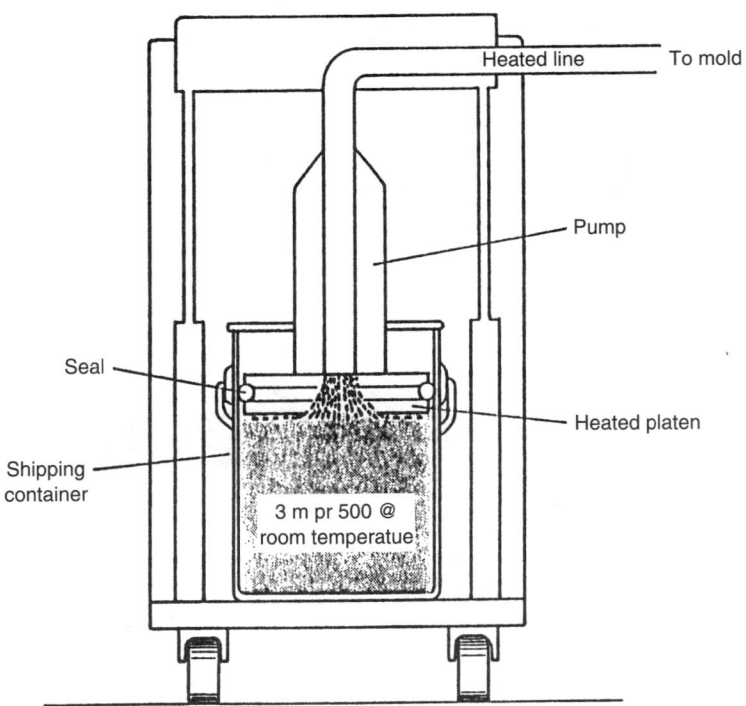

Figure 21.50 Melt on demand unloader. *Source:* Reprinted from Sundsrud GJ, *SME Composites in Manufacturing Conference*, Pasadena, Jan 19, 1993.

1. Maintain pot temperature at 106°C, when the resin viscosity will be about 400 cP, permitting the resin to be readily pumped into the mold
2. Degas resin for 5 min
3. Inject resin into the mold at 160°C, when the resin viscosity will drop below 45 cP
4. Increase mold temperature to 177°C and cure for 2 h. No post cure is required

Another important aspect with RTM is the assembly of the preform and 3M introduced a compatible resin tackifier (PT500), which could be used to tack plies of fabric together prior to being placed in the mold. The normal procedure is to uniformly apply about 5% w/w to the surface of the preform and place it for 1 min in a non-air circulating oven at 100°C. Alternatively, each ply can be tacked in place with a hot iron, with the aid of release separator film to avoid contamination with silicones.

$3M^{TM}PR520$ is a one-part resin with higher toughness than PR500. Hexcel have introduced RTM6, a premixed resin system for advanced resin transfer molding.

21.3.3.2 Vacuum assisted resin transfer molding (VARTM) [68]

Vacuum assisted resin transfer molding is not the same as applying a vacuum with RTM because:

1. Parts can be fabricated on a single sided open mold, which can be male or female
2. The fiber reinforcement is placed in the open mold, enclosed within a bag and impregnated with the resin in conjunction with a vacuum

A typical lay-up is shown in Figure 21.51. Traditional VARTM processing entailed opening and closing the resin ports manually, but this has been replaced by sequential injection automation, with resin flow monitored by film sensors such as SMARTweave.

The major advantages of VARTM processes compared with conventional autoclave processes are lower tooling cost and no size constraints like there is with an autoclave. However, the autoclave could achieve higher volume fractions, but this has been overcome by using stitching and debulking methods to achieve preforms that are near-net shape, with little further compaction required during processing.

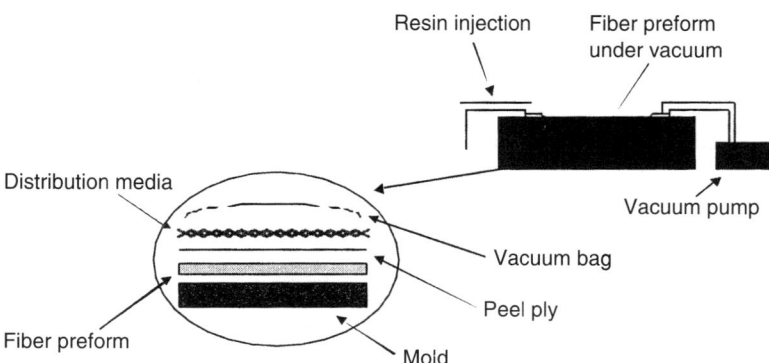

Figure 21.51 Schematic of VARTM process with lay-up detail. *Source:* Reprinted with permission from Heider D, Hofmann C, Gillespie JW Jr., *Proceedings 45th International SAMPE Symposium/Exhibition*, Long Beach, May 21–25, 2000. Copyright 2000, The Society for the Advancement of Matrerial and Process Engineering (SAMPE).

21.3.3.3 Vacuum infusion processing (VIP) [69]

In this process, the mold is filled with reinforcement, covered with a vacuum bag, evacuated by vacuum which draws resin into the mold and through the reinforcement until it is saturated. Vinylester resins are particularly suitable for this application. A technique developed by Intermarine Savannah uses grooved cores which facilitate the rapid and even distribution of resin throughout the laminate. This process is quicker than hand lay-up, enables a higher fiber fraction to be attained and provides a cleaner working environment.

21.3.3.4 Seemann Composite Resin Infusion Molding Process (SCRIMP™)

SCRIMP™ is a patented variant of the VARTM process with patent rights owned by TPI Technology. Unfortunately, the process is expensive to licence ($25,000), so the economics of the process have to be carefully ascertained. The patents describe a flexible cover incorporating a highly permeable disposable flow medium placed over the upper surface of the preform to ensure resin distribution. Flow occurs through the preform in the SCRIMP process. It is performed under a high vacuum, effectively removing air from the lay-up before the resin (50–1000 cP) is infused, resulting in a low void composite. This process can be advantageously used with carbon preforms and cores, obtaining 50–70% fiber weight, depending on the fiber architecture. The process can be used for large structures with laminates up to 6″ (15 cm) thick as well as simple $1/8''$ (3.2 mm) laminates. Large, virtually void free structural shapes can be made that are inherently repeatable with pinhole free surfaces. The closed system traps VOC emissions.

21.3.3.5 Resin infusion under flexible tooling (RIFT)

RIFT or vacuum resin infusion is a hybrid of RTM and vacuum bag molding. In RIFT, one mold face of RTM is replaced by a flexible polymeric bag and resin is introduced into the mold by vacuum only. The process is better known in the USA as SCRIMP, which is a patented process. During resin infusion, the initial dry fiber compaction affects the permeability to resin flow and is time dependent, but when the resin enters the fibers, it appears to act as a lubricant and leads to further compaction.

21.3.3.6 Vacuum infusion molding process (VIMP)

This process eliminates the throwing away of components like bags, breathers and release films and hence can achieve cost savings. It is a closed system and environmental friendly, achieving near net utilization of resin. The resin is introduced at the center, hence reaching the surface last, achieving an excellent bond with the gel coat. Infusion takes some 15 min with a vacuum of about 2.40 kPa.

21.3.3.7 SP Resin Infusion Technology (SPRINT™) [70,71]

SP Resins have introduced SPRINT™ materials, consisting of a layer of reinforcement on either side of a pre-catalyzed resin film and is available, if required, as a pack incorporating a syntactic core (Figure 21.52). SPRINT™ can be used with conventional and low temperature cure epoxies. The materials are laid up in a mold and vacuum bagged, similar to a

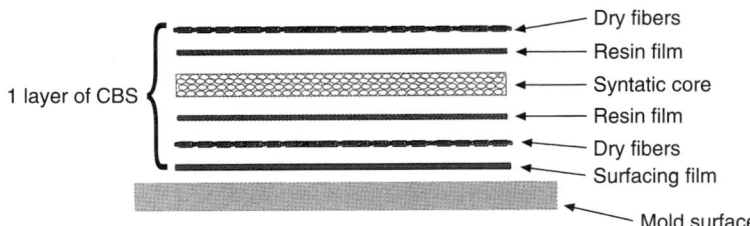

Figure 21.52 SP System's SPRINT™ pack incorporating a syntactic foam. *Source:* Reprinted from Ness D, Jones DT, *FRC2000 Conference*, Newcastle, Sep 13–14, 2000.

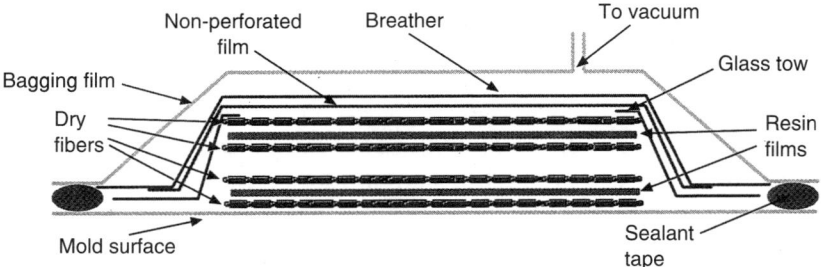

Figure 21.53 SP System's SPRINT™ system. *Source:* Reprinted from Ness D, Jones DT, *FRC2000 Conference*, Newcastle, Sep 13–14, 2000.

conventional prepreg (Figure 21.53), but because the materials are dry, air is more readily removed on application of a vacuum. On raising the temperature, the resin softens and flows into the air-free reinforcement to give a laminate with a very low void content (<0.5%).

SPRINT™ allows heavier weight reinforcement to be used when compared with prepreg technology and conforms readily during processing. Resin flow is more easily controlled than in the resin infusion process and the resin only has to flow through the thickness of the fiber to achieve full impregnation.

21.3.3.8 Resin film infusion (RFI)

This process was pioneered by Seferis [72] at the University of Washington, to impregnate fabrics to a fiber volume fraction equivalent to prepreg by sequential vacuum impregnation, followed by a pressure curing cycle (Figure 21.54). The process was adapted by McDonnell Douglas to impregnate Liba non-crimp fiber stack preforms. In addition, a through thickness aramid stitching was used (Figure 21.55) and the total bulk necessitated a thick film RFI approach, as the resin could only be applied at each face of the assembled fabric preform.

For cases where additional stitching is not required, Mills *et al* [37] at Cranfield University, developed a single sided prepreg approach by applying the resin film at a constant thickness to each layer of fabric. A semi-solid resin film coated on a release paper is placed in contact with the dry non-crimp fabric and rolled to form a single-sided prepreg. These coated fabrics are then stacked, vacuum bagged to remove air from the dry fabrics and then heated to permit the resin to melt and flow into the fabrics and finally cure (Figure 21.56). This approach produced laminates with a low void content, typically 0.25%, and allowed a net resin approach to be used.

MANUFACTURING TECHNIQUES FOR CARBON FIBER REINFORCED COMPOSITES

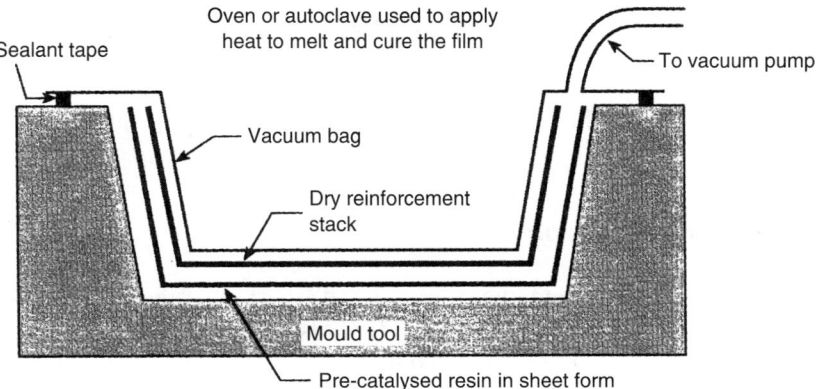

Figure 21.54 Resin film infusion. *Source:* Reprinted from SP Systems, Newport, IOW, technical data, with kind permission.

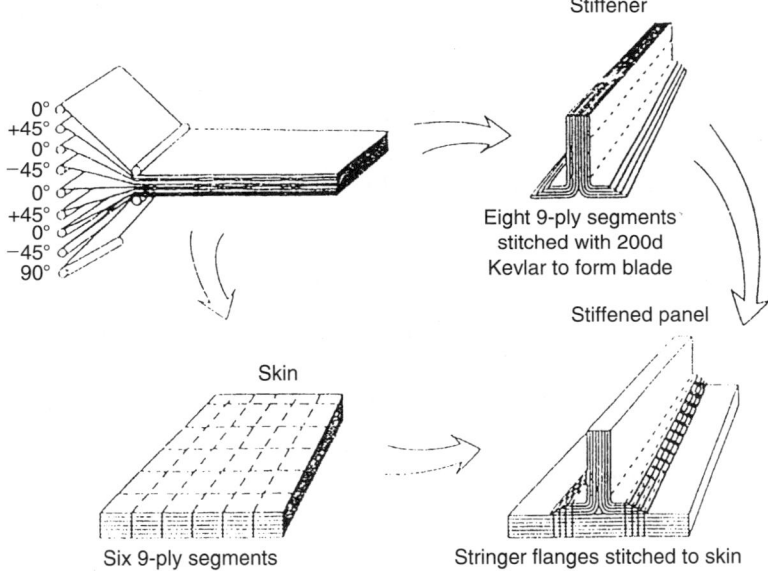

Figure 21.55 NASA- McDonnell Douglas stitched preform manufacturing approach. *Source:* Reprinted from Mills AR, *Aeronaut J Royal Aeronaut Soc*, 539–545, Dec 1996.

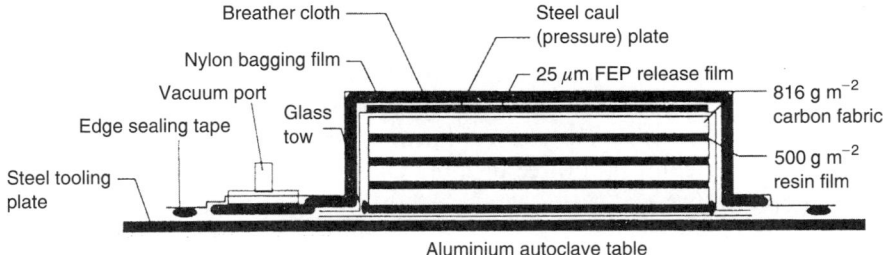

Figure 21.56 Autoclave lay-up for single sided prepreg resin infusion. *Source:* Reprinted with permission from Mills A, Burley G, Backhouse R, *44th International SAMPE Symposium*, May 23–27, 1999. Copyright 1999, The Society for the Advancement of Matrerial and Process Engineering (SAMPE).

21.3.4 Sequential multiport resin injection system (SMRIM)

Design Evolution 4 have developed a sequential multiport resin injection system (SMRIM) that allows controlled resin injection geometry using sequential filling of multiple ports, permitting the resin flow to be changed at any one of the ports.

21.3.5 Reaction injection molding (RIM) [84]

Two or more liquids are mixed in a resin pot and fed at low pressure into a closed mold, where they react. This technique can be used to feed catalyzed resin into a pultrusion die. The process can be adapted to reinforced reaction injection molding (RRIM) by introducing milled or chopped fiber.

21.3.6 Centrifugal molding

This process involves spraying a chopped fiber/resin mix from a gun attached to a boom that reciprocates along the fiber axis, discharging onto the inner surface of a rotating hollow mandrel. The speed of rotation can generate sufficient centrifugal force for the reinforcement to be wetted out by the resin and ensures that the resin does not drain due to gravity. After two passes, a roller passes along the mold to remove entrained air. The equipment can be computer controlled and can comprise more than one arm.

This technique, termed rotational molding (e.g. Engel Process), can be used with thermoplastic materials such as polyethylene, polypropylene, nylon, polycarbonate and polyurethane. The powdered plastic is placed in a mold and heated externally in an oven chamber to about 220–400°C, whilst it is rotated around both vertical and horizontal axes. As the powder melts, it adheres to the inner mold surface, building up an even layer over the surface of the mold. The mold, whilst still rotating, is withdrawn from the heated chamber and moved into a cooling chamber and the component removed after solidification [85].

21.3.7 Preparation of fiber preforms [77]

The preparation of a fiber preform can be greatly assisted by the presence of 2–5% of a thermoplastic binder on the surface, which can be softened by heat, allowing the preform to be set to the requisite shape and compacted to consolidate it [73]. Multiaxial stitched preform reinforcements for RTM fabrication were investigated by Harris *et al* [74]. Stitching technology can confer resistance to delamination, enable the preform to be handled without distortion, allow attachment of peel failure resistant stringers and permit compaction to near net shape [75,76]. The best stitch is a modified type of lock stitch [75], since the knot is formed on the surface of the preform, with the least disruption to the internal fiber structure. Carbon fiber imposes a speed restriction on the stitching process [76], which is expensive since there is a need to stitch components whilst conforming to their final molded curvature [76], necessitating computer control in three or more axes, each with separate needle thread and bobbin thread applicators.

21.3.8 Flow and cure monitoring of resin infusion processes

Flow and cure monitoring is a vital aspect of composites manufacture and being able to recognize the position and extent of the cure is important, as resin flow is responsible for the final mechanical properties of the composite part and good control can reduce mold design and process development times. There are two techniques that should be mentioned, namely Dielectric Cure Monitoring and SMARTweave Sensing.

Dielectric cure monitoring is based on the traditional AC excitation of the liquid resin, where one system uses an electrode sensor embedded in the resin (embedded inter-digitated —IDEX) and the other uses a reusable tool mounted sensor (tool mount sensor— TMS). The dipoles in the resin system orient as a function of cure, enabling an assessment to be made of viscosity changes, loss factor and state of cure.

The monitoring of the cure of epoxy composite materials has been evaluated by several workers [78–83]. A DC based flow and cure monitoring system called SMARTweave has recently been developed jointly by the Army Research Laboratory (ARL) and the University of Deleware-Center for Composite Materials (UD-CCM). Micromet Instruments have obtained an exclusive license to develop and market a commercial version based on this system. The system consists of a sensor grid, an electronics package, and a Windows based LabVIEW software program to control, record and display the sensor program. The sensor is laid up in the RTM mold and used to map the flow of resin during mold filling by rapid measurements of the electrical properties at each junction of the grid. The grid comprises two orthogonal sets of conductive filaments, separated by one or more preform layers. As the mold fills with resin, the gaps between the filament planes are filled with conductive resin, which completes an electrical circuit to provide a map of the part filling process. The subsequent resin gel and onset of cure can also be detected. The data obtained near the end of the cure is not as sensitive as with AC measurements.

SMARTweave [82,83] is very useful for molding thick parts and can be used with RTM, SCRIMP®, vacuum assisted RTM and other resin infusion processes.

21.3.9 Filament winding [86–90]

Filament winding (Figure 21.57) entails wrapping a rotating mandrel with continuous fiber, with the ability to alter the wind geometries and achieve different architectures. The fiber can be prepreg, wound wet or, more commonly nowadays, wound dry and used in

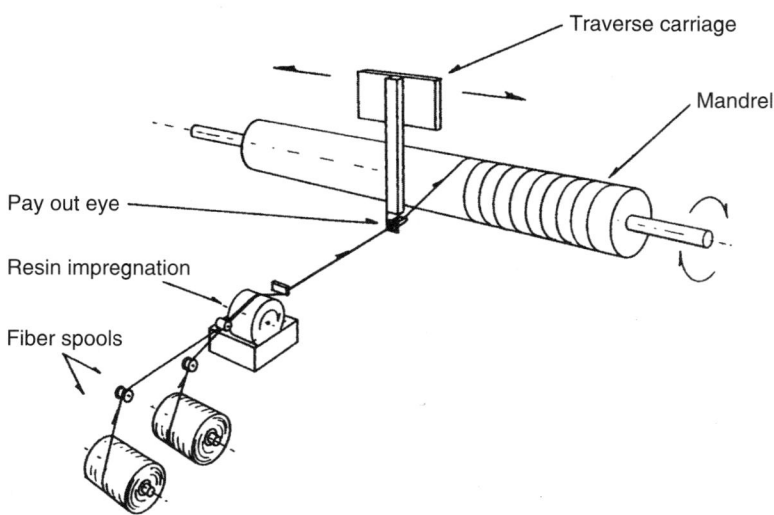

Figure 21.57 Basic filament winding system. *Source:* Reprinted with permission from Shaw-Stewart D, Filament winding-materials and engineering, *Materials and Design* 6, No. 3, 1985. Copyright 1985, Elsevier.

Table 21.5 Filament winding machine manufacturers

Manufacturer	Address
Automation Dynamics	Signal Hill, CA, USA
Bolenz and Schäfer	Bauhofstrasse 2, D-35239, Steffenberg, Germany
Composites Machines Company Inc. (CMC), (now Zoltek Ltd.)	Salt Lake City, Utah, USA
Drostholm Products	Vedbaek, Denmark
Dura-Wound Inc.	Washougal, Washington, USA
En Tec, (now Zoltek Ltd.)	Salt Lake City, Utah, USA
McClean Anderson	Schofield, WI, USA
MFL Composite, (Bought out Goldsworthy)	Saint Etienne, France
Mikrosam MFL Composite	Prilep, Macedonia Saint Etienne, France
Plastex Manustin (MATRA)	France
Precision Engineering	USA
Pultrex Ltd.	Clacton-on Sea, Essex, UK

conjunction with processes like RTM. Filament winding equipment is commercially available from a number of manufactures (Table 21.5).

There are basically three types of filament winding—hoop or circumferential, helical and multi-directional winding.

1. Hoop winding

This is a special case of helical winding and is used only at angles greater than 70°. The simplest type of machine is the lathe type, comprising a rotating mandrel coupled with a reciprocating feed eye which controls the ratio of the rotational speed to the reciprocating speed (Figure 21.58). For every rotation of the mandrel, the fiber will advance a given distance, which is called the winding pitch.

2. Helical winding

In Figure 21.59, the winding pitch is greater than the band width of the fibers, hence the pattern starts open and gradually fills in to cover the mandrel. Helical winding encompasses all angles, approaching 0° (fiber laid axially along the mandrel) and 90° (zero pitch) representing the two extremes. A winding angle of 45° provides equal strength in the longitudinal and hoop directions.

3. Polar winding

This type of winding (Figure 21.60) is used to wind most axial fibers for domed pressure vessels. The arm rotates about a vertical axis and can be set to different angles of tilt. Gearing provides mandrel rotation relative to the arm.

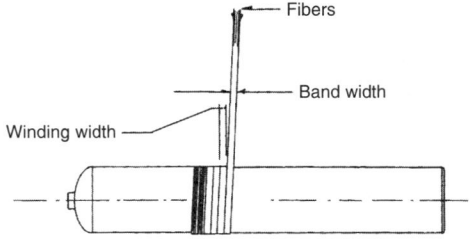

Figure 21.58 Circumferential or hoop winding. *Source:* Reprinted with permission from Shaw-Stewart D, Filament winding-materials and engineering, *Materials and Design* 6, No. 3, 1985. Copyright 1985, Elsevier.

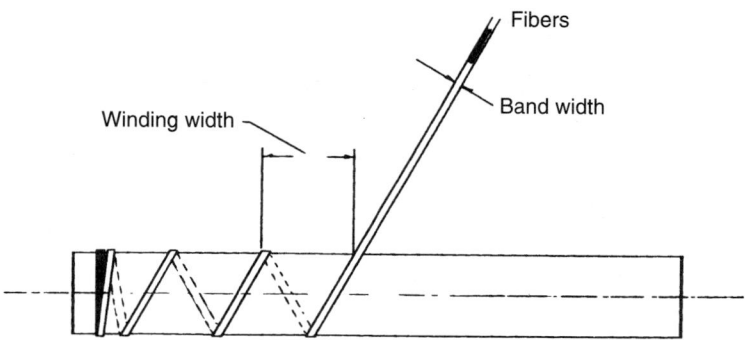

Figure 21.59 Helical winding. *Source:* Reprinted with permission from Shaw-Stewart D, Filament winding-materials and engineering, *Materials and Design* 6, No. 3, 1985. Copyright 1985, Elsevier.

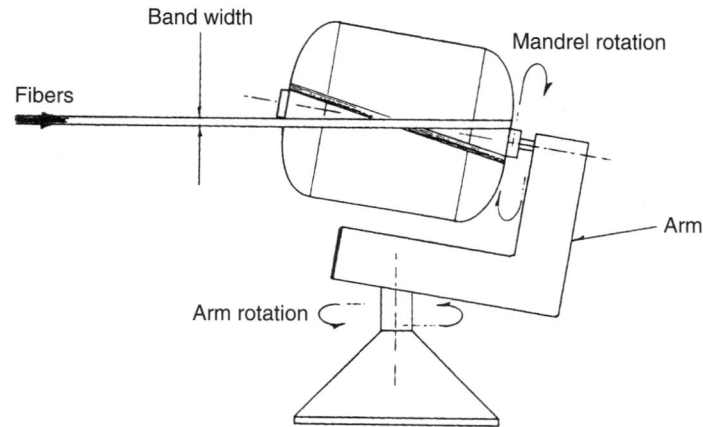

Figure 21.60 Polar winding. *Source:* Reprinted with permission from Shaw-Stewart D, Filament winding-materials and engineering, *Materials and Design* 6, No. 3, 1985. Copyright 1985, Elsevier.

4. Multiaxial winding

Modern technology has developed complicated Multiaxial winding machines, which are numerically controlled and their design has tended to merge with fiber or tape placement machines. A 5-axis filament winding machine is shown in Figure 21.61, where the spindle is located in the headstock, rotating in one direction only. To lay 0° axial fibers along the mandrel, the spindle is indexed in discrete increments becoming a fully controlled axis A. The traverse carriage is large, carrying other axes designated X, whilst a vertical slide Z moves on the X carriage and a horizontal ram moves in and out the Y axis. Mounted on the Y ram is a rotary unit, B axis. Machine slides can be controlled to increments of 0.025 mm and speeds up to 50 m min^{-1} can be achieved.

5. Variants of multiaxial winding

There are many variants of these winding machines, such as the Racetrack model [88] (Figure 21.62), which features a feed eye that completely orbits the mandrel in a path that resembles a racetrack. Reciprocating the feed eye backwards and forwards, the machine has the capabilities of a lathe type winding machine and in addition, longitudinal winding, if the track is slightly tilted, causing the feed eye to completely orbit the mandrel. A helical pattern can be obtained by adjusting the ratio of the orbiting eye to the mandrel. This machine is

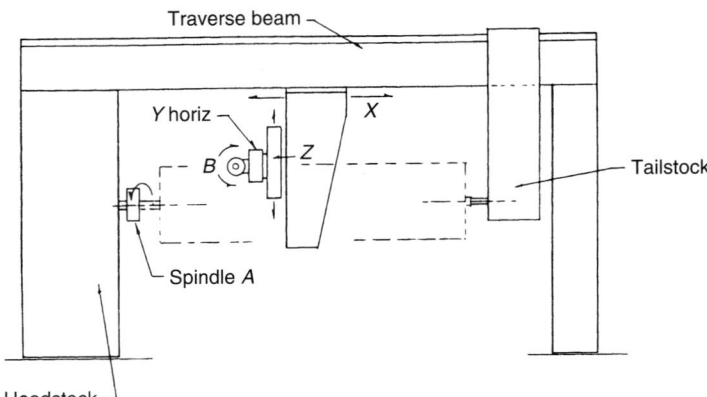

Figure 21.61 Pultrex 5-axis filament winding machine. *Source:* Reprinted with permission from Shaw-Stewart D, Filament winding-materials and engineering, *Materials and Design* 6, No. 3, 1985. Copyright 1985, Elsevier.

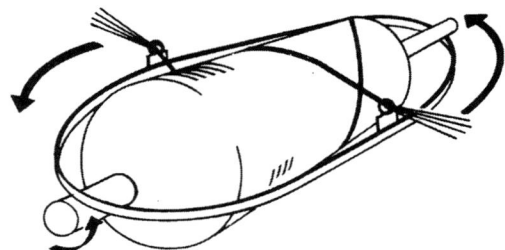

Figure 21.62 Racetrack model filament winding machine. *Source:* Reprinted with permission from Goldsworthy WB, *Conference Filament Winding*, The Plastics Institute, London, Oct 17–18, 1967. Copyright 1967, Maney Publishing (who administers the copyright on behalf of IOM Communications Ltd, a wholly owned subsidiary of the Institute of Materials, Minerals & Mining).

quite versatile, but is sensitive to changes in mandrel diameter and has tension control problems imposed by a moving filament supply.

Another variant is the Whirling arm type winder (Figure 21.63) employing a whirling arm to achieve longitudinal winding and using a separate carriage to apply circumferential and helical wrapping.

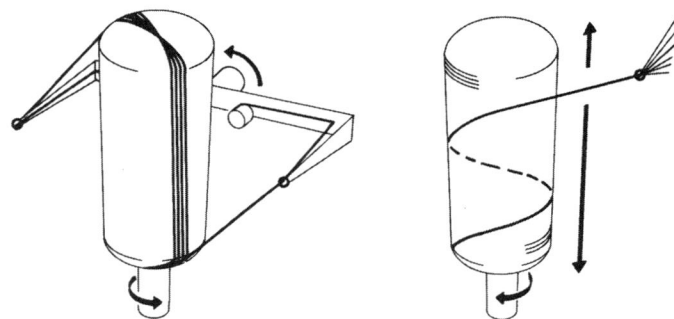

Figure 21.63 Whirling arm type filament winding machine. *Source:* Reprinted with permission from Goldsworthy WB, *Conference Filament Winding*, The Plastics Institute, London, Oct 17–18, 1967. Copyright 1967, Maney Publishing (who administers the copyright on behalf of IOM Communications Ltd, a wholly owned subsidiary of the Institute of Materials, Minerals & Mining).

If prepreg is used in filament winding, it can be applied as a single tow, several tows laid side by side, or in a tape form. Dugger and Hirt [91] have described how to utilize PMR-15 powder coated carbon fiber towpreg. When winding with wet resin, the dry tow(s) are pulled over rollers immersed in a resin bath, the resin can be heated to lower the viscosity to aid good wet out (350–1500 cP). Excess resin can be removed with a doctor blade, or alternatively, the resin can be metered onto a roller and transferred to the fiber. The pot life must be long enough to allow fabrication of the desired component without the complications of a rapidly advancing cure.

The fiber tension is very important as this controls the resin pick-up, normally 35–40% volume fraction. Tensions are about 1N/1000 filaments for wet winding and 3N/1000 filaments for other types of winding. If the tension is too high, the fiber does not spread and is damaged by abrasion in the guide and if too low, produces waviness in the applied fiber. The type of fiber size and size content must be carefully chosen to help achieve good resin wet out. The principal matrix materials are epoxy, polyester and vinylester resins, but thermoplastic prepregs such as PEEK can also be applied.

When the thickness of the composite exceeds a certain value, it may be necessary to apply the cure in two stages.

The final wound structure is removed from the winding machine, and the mandrel/workpiece is placed in an oven for curing. Various mandrels are used and can be made of hard chrome plated and polished ground steel, which may be of a segmented collapsible construction, but must have no concave curvature. If the part has complex geometry, then fusible low melting point alloys, a soluble plaster that can be removed with hot water after curing, or an inflatable bag can be used. In some instances, such as compressed natural gas (CNG) tanks, the mandrel is left in position as a liner.

21.3.10 Pultrusion [92–94]

Pultrusion is a process for producing continuous lengths of reinforced plastic shapes with constant cross-sectional area. Pultrusion equipment is commercially available from a number of manufactures (Table 21.6) and a diagram of a typical pultrusion system is shown in Figure 21.64. Figure 21.65 shows an 8 MT machine with a hand-over-hand gripper system.

Table 21.6 Pultrusion machine manufacturers

Manufacturer	Address
Bedford Reinforced Plastics Inc.	Bedford, PA, USA
Entec Composite Machines Inc. (ECM), (now Zoltek Ltd.)	Salt Lake City, Utah, USA
Creative Pultrusions Inc. (CP)	Alum Bank, PA, USA
Sales channel: Martin Pultrusion Group	Twinsburg, Ohio, USA
Drostholm Products,	Vedbaek, Denmark
Engineering Tech Inc. (EnTec), (now Zoltek Ltd.)	Salt Lake City, UT, USA
MFL Composite	Saint Etienne, France
Minze Fibron	
Morrison Molded Fiberglass Co.,(MMFG), (now called Strongwell)	Bristol, Virginia, USA
Nordic Industrie A/S	Skodje, Norway
PTI	Twinsburg, Ohio, USA
Pultrex Ltd.	Clacton-on Sea, Essex, UK
Pultron Co.	Gisborne, New Zealand
Pultrusion Dynamics, (a Division of Creative Pultrusions Inc.)	Oakwood Village, Ohio, USA
Strongwell Corp.	Bristol, VA, USA
Tigsons Technologies Inc.	Mississauga, Ontario, Canada

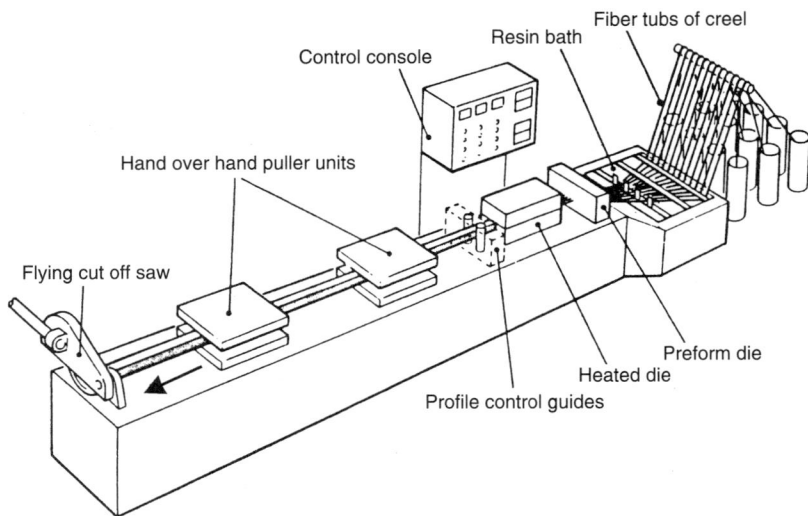

Figure 21.64 Schematic pultrusion unit with hand-over-hand puller units.

Figure 21.65 Pultrex P8000 pultrusion machine—each gripper has two upper clamps (each 11,000 kg). *Source:* Reprinted from Pultrex technical literature.

The basic pultrusion process can be divided into the following operations:

1. Reinforcement handling

A suitable creel positions the requisite number of tows, with minimal damage, prior to entry into the resin bath. If tows are supplied from containers, a christmas tree with ceramic eyelets will be required to direct the tows to the resin bath. In some pultruded sections, smaller tows are used in parts where the profile shape does not permit the use of larger tows. Higher size contents (2–5% w/w) will permit easier handling and minimize fiber damage. Tows can be joined by knotting, but knots must be staggered to ease their passage through the die. Hybrid constructions can be used (e.g. with glass and aramid) and transverse properties can be introduced with an In-feed system (Figure 21.66) using woven cloth, knitted

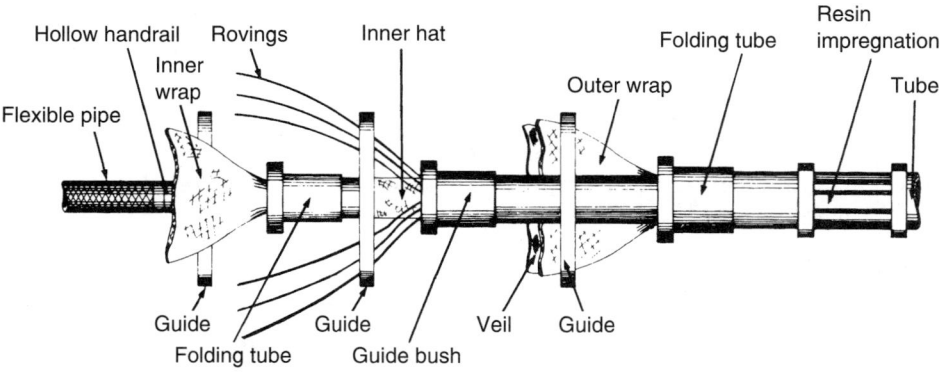

Figure 21.66 Pultrex In feed system. *Source:* Reprinted with permission from Parmenter J, A process which has some pull in industry, *CME*, Jul/Aug 1984. Copyright 1984, Institution of Mechanical Engineers.

fabrics, braid, or mat. The number of tows (n) required to produce a pultruded part of given cross-section ($c.s.a.$) with 65% V_f can be calculated from:

$$n = \frac{0.65 \times c.s.a. \times \rho_f}{m_{\text{pul}}} \text{ tows}$$

2. Resin impregnation

The dimensions of resin baths are restricted to minimize the volume of catalyzed resin and can be heated to control the resin viscosity to promote fiber wetting, although this will reduce the working life of the bath. To facilitate lacing up, the roller assembly in the resin bath is manufactured in two parts—a lower fixed set of rollers submerged in the resin bath and a moveable upper set, under which the fiber is positioned. The assembly is then pressed down to push the fiber into the bath to contact with the lower set of rollers. This system facilitates an easy lace-up procedure and ensures good compaction to expel all air and promote fiber wetting. Alternatively, the fiber can be passed over a drum upon which the correct amount of resin has been metered and adjusted by a doctor blade.

3. Pre-die forming

A preforming die gently shapes the material and removes all but about 10% of the excess resin prior to entry into the pultrusion die.

4. Heated die to shape and cure the resin

The pultrusion die can be made from polished chromium plated tool steels, or when pultruding epoxies, a high chromium content tool steel. The die must be accurately lined up and its length typically 300–1000 mm, which is governed by the size of the section being pulled, the pulling speed and the resin system. Longer dies require greater pulling forces due to the increased frictional drag and a die lubricant, such as zinc stearate, can be added to the resin mix to help reduce frictional resistance, but which may interfere with any subsequent composite bonding process. The die inlet is tapered at 7–10°, with well rounded edges to prevent fiber fracture. The excess resin exudes from the inlet end of the die, causing the entering fiber bundle to swell, eventually attaining equilibrium with the process conditions. Adding this exuded resin to the resin bath will curtail the life of the bath.

Cartridge or plate heaters are preferred for heating the die to a uniform temperature within ±1°C and maintaining a temperature gradient along the die to avoid premature gelation, while taking into account any exotherm. An RF (Radio Frequency wave generator) unit can be used to either heat the fiber entering the die or the die/resin. The die must be preheated prior to commencement of pultrusion.

Shrinkage during polymerization reduces die forces and should always be arranged to be greater than the thermal expansion caused by the temperature rise.

5. Pulling unit to provide traction

The pultruded product is cooled prior to the traction unit, which can be a counter rotating caterpillar unit, or preferably, a hand-over-hand reciprocating clamp type unit, since the caterpillar unit requires the tracks to be fitted with machined rubber pads to accommodate each pultruded profile. The hand-over-hand unit grips above and below and while one unit is pulling, the other unit returns to position, ready to take over the role of pulling. Typical line speeds vary in the range 1.5–100 m h^{-1}, depending on the section(s) being produced. The pulling forces depend on the type of machine which are available upwards to some 30 MT.

6. Cut off saw

Once the pultruded section has left the die and cooled sufficiently, it is clamped and a flying saw moves along with the clamped section to cut off required lengths. Extra long lengths can be accommodated by feeding the pultrusion out through a door, window or hatch at the end of the building.

7. Post cure oven

For optimum properties, all pultruded sections will require post curing and care must be taken to ensure adequate support along the entire pultruded length to prevent deformation occurring in the post cure oven.

Attaching a Pullwinding machine (Figure 21.67) to a pultrusion machine enables a combination of pultrusion and continuous filament winding to be accomplished.

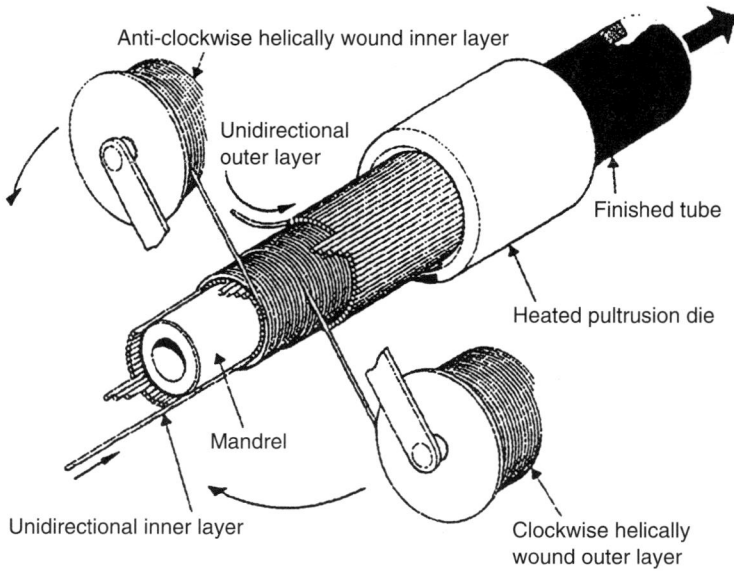

Figure 21.67 Pultrex Pullwinding machine. *Source:* Reprinted from Pultrex technical literature.

Studies with the hybrid pultrusion of glass and carbon at the University of Mississippi have shown that the flexural stiffness of a glass/epoxy composite could be significantly improved by the addition of carbon fiber, symmetrically distributed at the outer surfaces of the composite beams, but perhaps this is not surprising, since the glass occupies the neutral axis.

JAMCO, a Japanese company, has developed a machine which will pultrude curved L and T stiffeners, but it is slow.

21.3.11 Prepreg molding

21.3.11.1 Prepreg manufacture

A prepreg comprises a fiber reinforcement, embedded in an uncured thermoset polymer matrix, or a thermoplastic matrix (Section 21.4.5). The reinforcement can be unidirectional or fabric based.

Table 21.7 lists companies who manufacture carbon fiber prepreg equipment. Prepreg manufacturers do make their own in-house equipment (Figure 21.68) and a typical layout of a prepreg machine is shown in Figure 21.69.

Prepreg can be made using wet resins, molten resin, or resin coated film. The technology of mixing, coating resins and making prepreg is proprietary.

Table 21.7 Prepreg machine manufacturers

Manufacturer	Address
American Tool	Wabash, Indianna, USA
Applied Composites Engineering Ltd.	Pelham Barn, Millcourt, Shenstone, S. Staffs., UK
Automation Dynamics	Signal Hill, CA, USA
California Graphite Machines Inc.	Corona, California, USA
Century Design Inc.	San Diego, California, USA
T.H. Dixon Company Ltd. (coating equipment)	Letchworth, Herts., UK
Hoffman & Schwabe (coating equipment)	Krefeld, Germany
Olbrich	Hanover, Germany
Vits Machinebau GmbH	Langenfeld/Rhld., Germany
Western Advanced Engineering Co. (WAECO)	Anaheim, CA, USA
Zappa (agents in UK are Web Processing (M/c) Ltd. Whaleybridge, Stockport)	Italy

Figure 21.68 SP System's in-house prepreg machine. *Source:* Courtesy SP Systems, IOW.

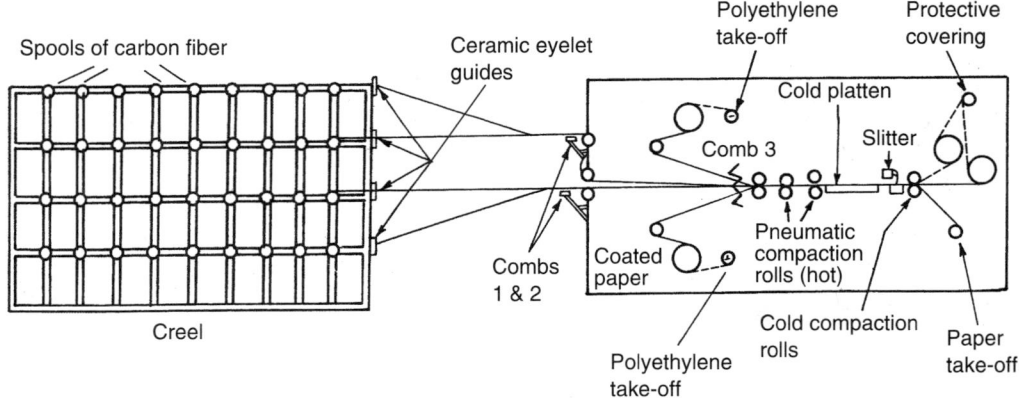

Figure 21.69 Typical prepreg machine layout.

Hardeners such as dicyandiamide need to be ground using equipment like a three-roll mill (Figure 21.70) to ensure proper distribution. Mixing is important and high shear and static mixers are used. Wet resins dissolved in solvents can be coated by many techniques (Figure 21.71) normally onto a silicone coated release paper, supplied with a differential

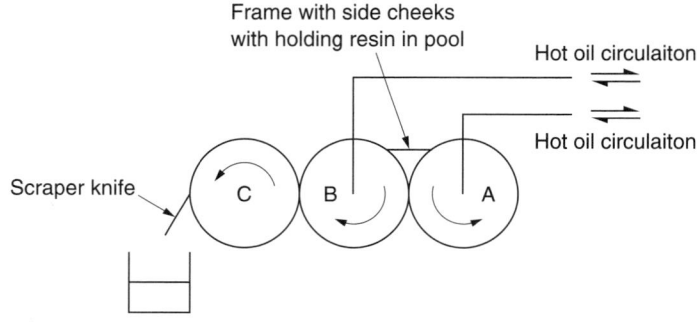

Figure 21.70 A three roll mill.

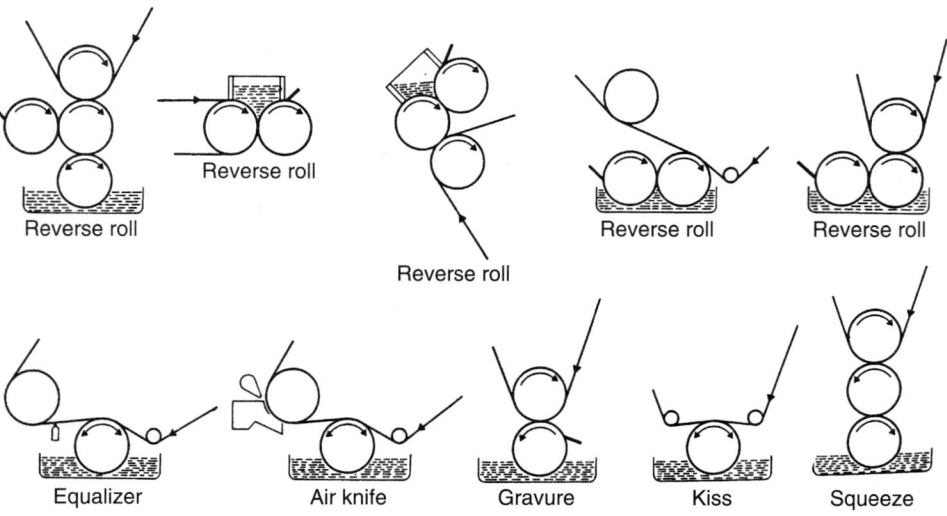

Figure 21.71 Coating techniques used on a Dixon coater. *Source:* Reprinted from Dixon technical literature.

release to enable the resin coated paper to be rolled up. Therefore, when unwound, it will not stick to the reverse side, which has been coated with the higher release. Alternatively, a second release paper/polyethylene film can be used. The release paper must withstand the highest temperature used in the process.

Solventless resin systems are melted and coated onto paper, the resin film thickness being controlled with a knife—a procedure which can be carried out on, or off, line.

The requisite number of correctly tensioned carbon fiber tows are fed from a creel (Figure 21.69) and positioned across the width by passing through a comb set up for correct spacing (using a lazy tongs device), to give the preferred areal weight of prepreg. Smaller tows, such as $3k$, permit the manufacture of thinner prepreg.

The fiber and/or resin can be heated using infrared heaters and Figure 21.72 shows the radiation energy achieved at different temperatures and the wavelength that is required.

There is a move towards using larger tows which are cheaper and these may need expanding, which can be achieved with an air knife (Figure 21.73), where air is fed to the base of the air knife and into a plenum. From this inner chamber, the air passes through a transverse slit 0.115 mm wide. The gap between the base of the air knife and the top of the fiber inlet side is adjustable to obtain the best spreading conditions (approximately 2.5 mm).

Manufacturing methods are outlined in Figure 21.74 and if solvent is used, it is necessary to have a drying tower (Figure 21.75) to remove solvent to a sufficiently low level to obtain an acceptable void content in the composite. The manufacturer adjusts the level of prepreg

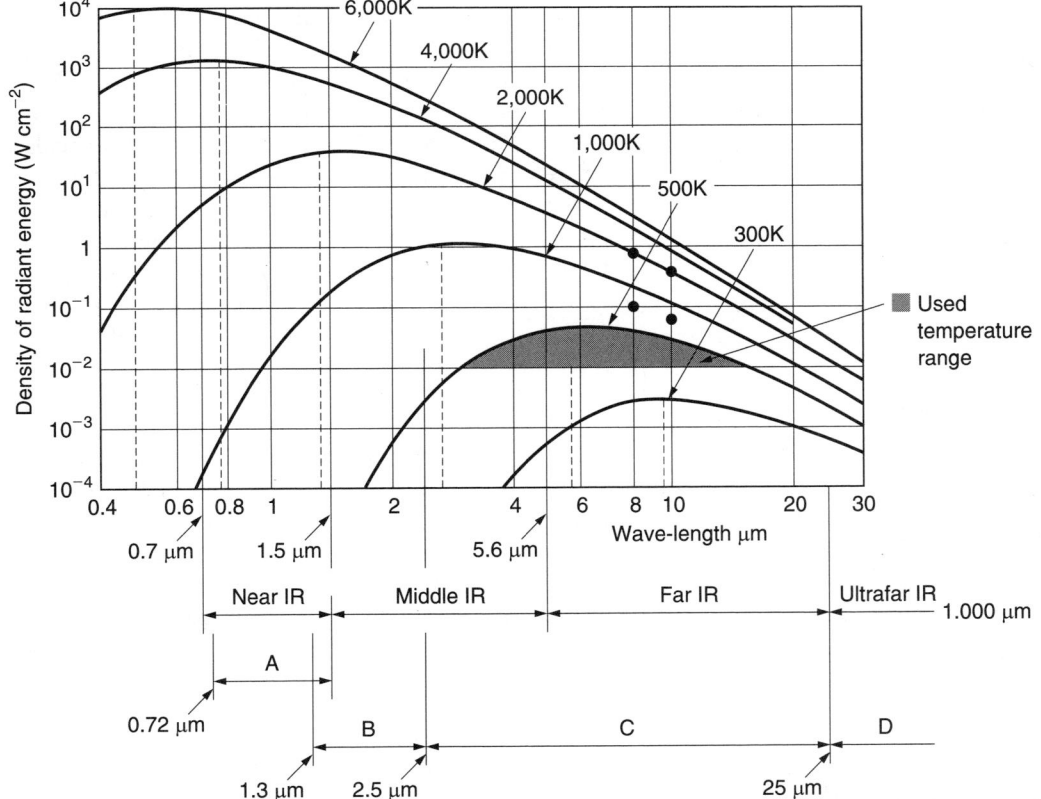

Figure 21.72 Distribution in wavelength within infra-red region at different temperatures. *Source:* Reprinted from Caratsch technical literature.

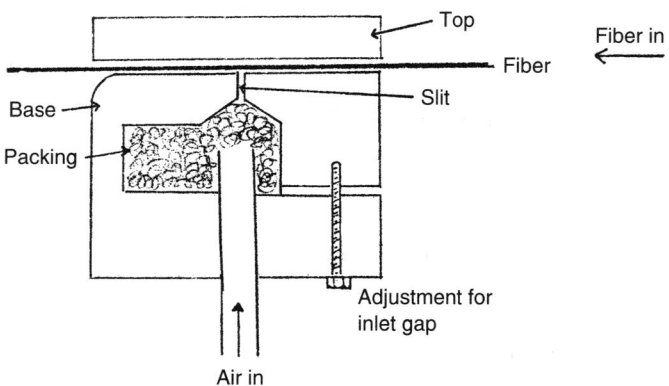

Figure 21.73 Diagram of an air knife.

tack to facilitate this lay-up procedure by controlling the level of procure, or by the use of suitable resins in the prepreg formulation.

21.3.11.2 Manufacture of composites from prepreg

1. Ply cutting and stacking prepreg

Automatic cutting machines are widely used in the aerospace industry to cut prepreg and include equipment based on reciprocating knives or chisels, lasers, ultrasonic and water jet machines. Water jet machines operate at 2000–3500 bar with water forced through a small orifice, travelling at a speed of Mach 2, cutting cfrp up to 5 mm thick at about 1 mm s^{-1}, a performance that cannot be matched by a laser. Computer programs decide the most efficient way to cut prepreg to avoid waste and in order to minimize handling, attention is paid to the position that the cut ply occupies. Once a ply has been cut, bar codes can provide a convenient way to retain identification. Plies of prepreg may be stacked in a specified manner prior to curing. It is normal to build up a stack with each ply laid in the correct position to obtain desired laminate properties by using aids such as Mylar polyester film templates, with the lay-up procedure displayed on a VDU. The major aircraft companies are developing automated collation. Preferably, kits are assembled ready for immediate use to avoid storing at $-20°C$.

2. Compression molding of prepreg

This process uses the same technique as hot press matched metal molding (Section 21.3.2), but Johnson [95] describes a technique where pressing to stops is not used. Preforms were made using Kerimid 601 polyimide resin as an adhesive, precuring at 135°C, placing some 45 prepared components in the mold, molding for 20 min at 175°C and a pressure of 7 MPa. Molding was achieved entirely without the use of stops, obtaining a void free composite by maintaining sufficient hydraulic pressure within the fluid resin until gelation fixed the composite structure.

3. Vacuum bag molding

A recent introduction to prepreg technology is the use of a resin system such as 3M's Scotchply™ SP 381, which can be used in a one side tacky (OST) form. Processing using a vacuum bag enables an easy path for air to exit between the plies prior to resin saturation and achieve a composite with less than 0.4% voids using only a vacuum cure.

MANUFACTURING TECHNIQUES FOR CARBON FIBER REINFORCED COMPOSITES

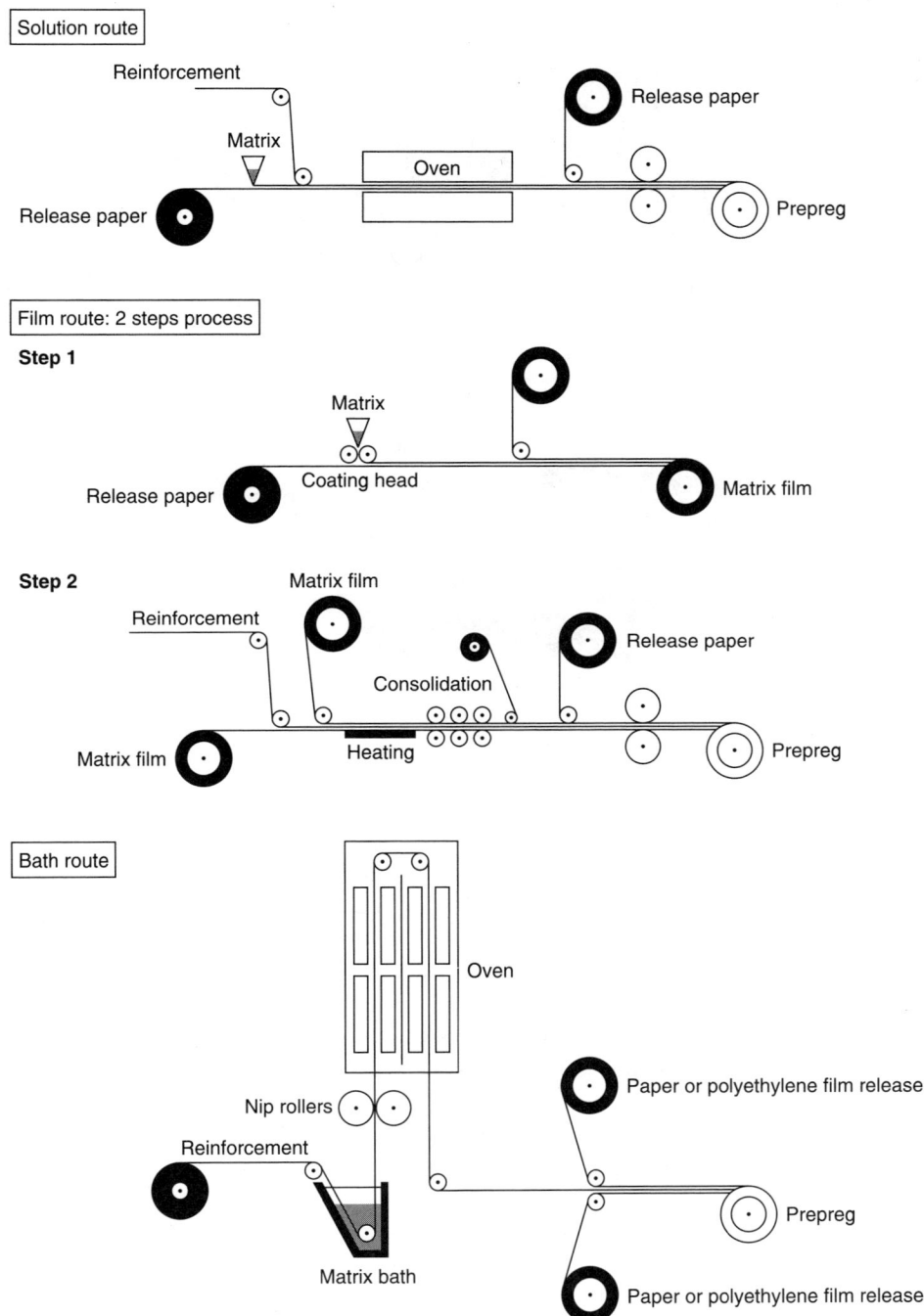

Figure 21.74 Outlines of manufacturing routes for making prepreg. *Source:* Reprinted from Ciba technical literature.

4. Press-clave molding

The press-clave is described in Section 17.8.3 and is a laboratory simulation of an autoclave using press platens as the heating source. The temperature of the bottom platen is adjusted to give the requisite temperature of the preform, which is measured by a

Figure 21.75 Prepreg drying tower. *Source:* Courtesy CP Films Ltd., Runcorn, Cheshire.

thermocouple embedded in the prepreg plies. The thickness of the laminate is controlled by the number and orientation of the prepreg plies, the state of cure of the resin, the pressure applied above the bag and the type and quantity of the bleeder plies.

5. Autoclave molding

An autoclave is a heated pressure vessel which accommodates the bagged workpiece, which can be placed under a vacuum to remove air and volatiles and to which additional pressure can be applied by pressurizing the vessel. Autoclaves are expensive and the size available imposes a limit on the size of the workpiece, but the quality of the component is excellent.

A typical vacuum bag lay-up is shown in Figures 21.76 and 21.77, whilst Figure 21.78 depicts a characteristic autoclave cure cycle.

$$\text{Number of bleed plies} = \left[W_R - \left(\frac{W_F \times \rho_R \times V_R}{\rho_F \times V_F}\right)\right]\frac{N_P}{A}$$

where A = Absorbency of bleed layer, $\mathrm{g\,m^{-2}}$
V_R = % resin volume (= $100 - \%V_F$)
W_R = Resin areal weight in prepreg, $\mathrm{g\,m^{-2}}$
ρ_R = Resin density, $\mathrm{g\,cm^{-3}}$
W_F = Fiber areal weight in prepreg, $\mathrm{g\,m^{-2}}$
ρ_F = Fiber density, $\mathrm{g\,cm^{-3}}$
V_F = % fiber volume (as selected)
N_P = Number of plies of prepreg in stack

MANUFACTURING TECHNIQUES FOR CARBON FIBER REINFORCED COMPOSITES

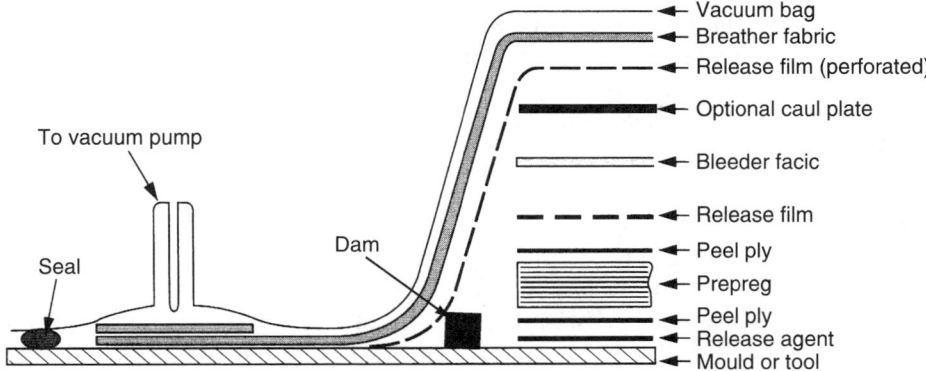

Figure 21.76 Diagram of a typical vacuum bag lay-up.

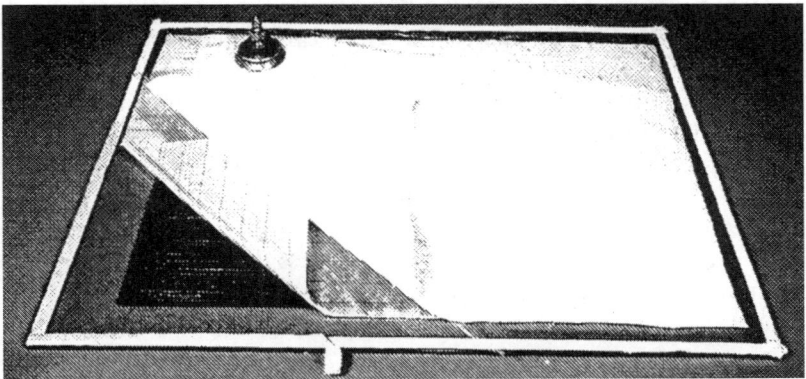

Figure 21.77 A vacuum bag lay-up. *Source:* Reprinted from SP Systems, Newport, IOW, technical data, with kind permission.

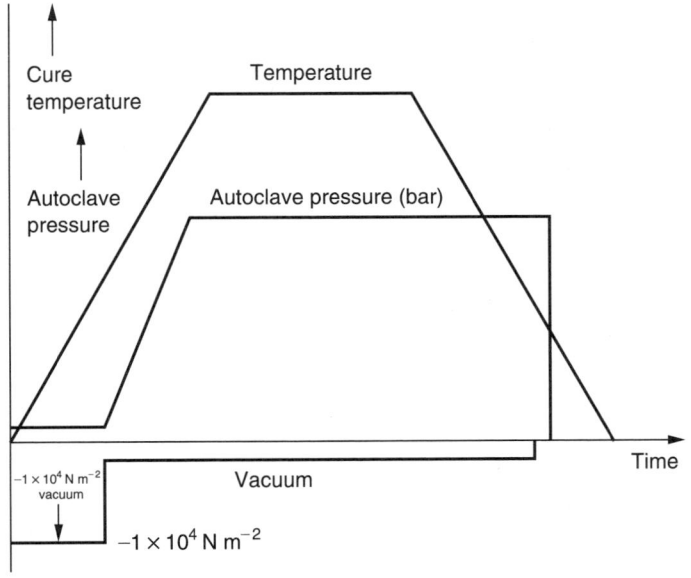

Figure 21.78 Diagram depicting a typical autoclave cure cycle.

An epoxy prepreg will provide a composite with higher mechanical properties than a comparable epoxy hand lay-up resin and with less voids (0–2%) compared with 5–8% voids in a wet laminate. However, the prepreg does require to be stored under refrigeration, requires a well constructed mold and must be cured at 70°C or above.

6. Quickstep^{TM} Molding [96]

A unique system which can be described as fluid filled, balanced pressure, heated, floating mold technology was created by Neil Graham, Quickstep Technologies Pty. Ltd., Applecross, Western Australia, with the scientific back-up of CSIRO, and patented in 1994.

The laminate is trapped between the two halves of a free floating rigid, or semi-rigid, mold, floating in a balanced circulating pressurized closed loop heat transfer medium (Figure 21.79), which can be water (up to 105°C) or oil (up to 200°C). A vacuum (0.098 MPa) is applied to the bagged lay-up and the heat transfer medium is vibrated to remove air from the lay-up, followed by compaction by the application of pressure (11–20 kPa) and finally heated to cure the laminate. The mold can be described as a clam shell, with the bottom half bonded around the edges to a pressure chamber and the top half bonded via a flexible membrane attached to the top of the upper pressure chamber, with the two halves clamped together with simple quick action toggle clamps. Since the system is one of balanced pressure, the laminate, including a core if used, can be compressed without distortion.

Thermoset (epoxy, vinyl ester and phenolic resins) or thermoplastic prepreg can be used, as well as wet resin/dry fiber, to give composites with less than 2% voids. An autoclave is not required and epoxy composites can be produced some 10 times quicker.

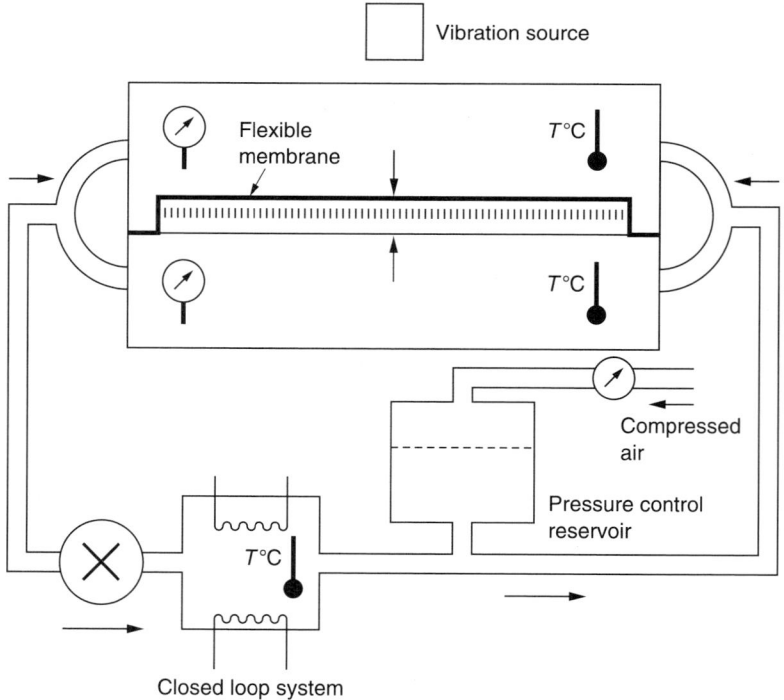

Figure 21.79 Schematic layout of the Quickstep^{TM} process. *Source:* Reprinted from Quickstep technical literature.

7. Tube rolling

This is a manual lay-up method, where prepreg tape is rolled onto a metal mandrel on a hot table. The mandrel can be truncated when required for fishing rod blanks, ski poles and aerials. When wrapping is complete, the lay-up is over-wrapped with shrink wrap tape, which contracts when heated and applies pressure to the assembly. The wrapped mandrels are suspended vertically in an oven for curing to ensure straightness of the component.

8. Automatic tape lay-up

This technology is widely used and excellent machines are made by Cincinnati (formerly Cincinnati Milacron) and Ingersoll. It is possible to convert tape lay-up machines to fiber placement units by the addition of a suitable head for running tows, special software and hardware for heating control. Since the action of these machines is similar to fiber placement, they are discussed below.

21.3.12 Fiber placement systems [97–100]

Fiber placement is basically a hybrid of filament winding and tape laying and is a continuous process for fabricating composite shapes, by laying thermoset or thermoplastic preimpregnated tows (in tow or slit tape form) onto a mandrel or tool having concave or convex surfaces. There is no limit to the fiber angle and tows can be added or removed as and when necessary.

Cincinnati's Viper fiber placement head (Figure 21.80) accurately collimates width (3.2 mm) and position (± 1.3 mm), dispenses, clamps, cuts and restarts up to 32 tows, compacting the prepreg tow directly onto the lay-up surface. During the placement of a course, each tow is dispensed at its own speed, allowing each tow to conform to the surface of the part such that when the head is following a curved path, the outer tows of the fiber band will pull more length than the inner tows. A seven axis capability (Figures 21.81, 21.82)

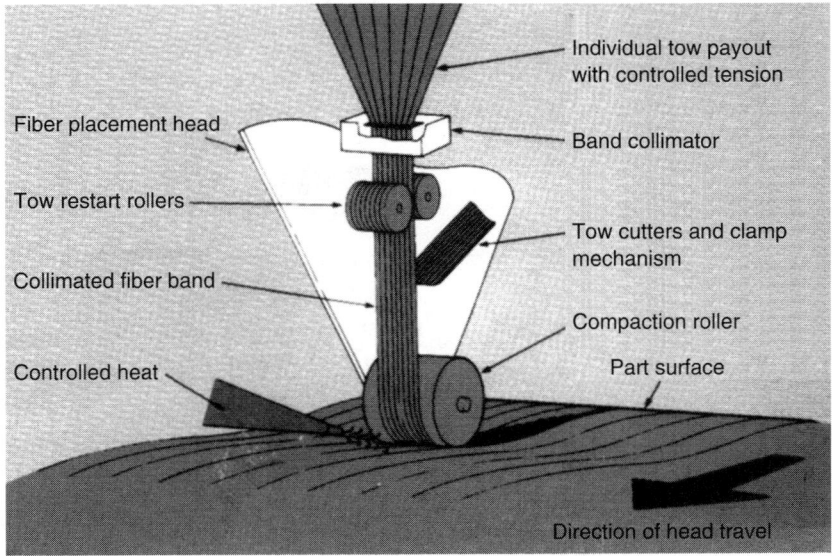

Figure 21.80 Processing head for Viper 3000 7-axis fiber placement system. *Source:* Courtesy Cincinnati Machine.

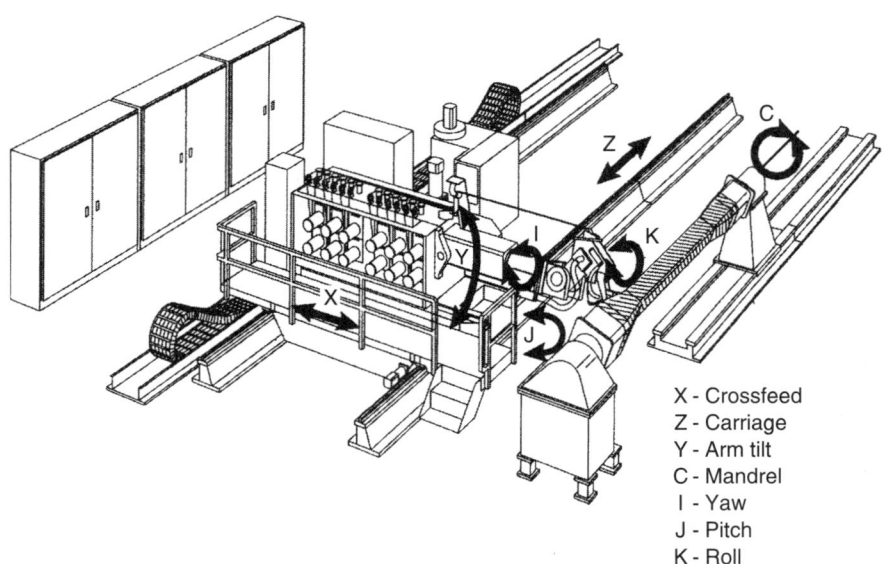

Figure 21.81 Viper FPS-200 CNC 7-axis fiber placement system. *Source:* Courtesy Cincinnati Machine.

Figure 21.82 Viper FPS-3000 fiber placement machine. *Source:* Courtesy Cincinnati Machine.

positions the tows at desired angles at up to 30 m min^{-1} to maximize the strength and minimize weight.

21.3.13 Mold release

There are basically two types of mold release agents—an external release, applied to the mold and an internal release, incorporated in the resin mix.

It is most important to choose an effective mold release and obviously, the mold must have originally been correctly designed to achieve an effective release, free from porosity, polished and most importantly, clean. The mold release must uniformly wet out the mold surface and bond to it, yet have similar polarity to the wet resin and be able to separate from it. The release must have sufficient heat resistance to withstand the processing temperature and only trials will confirm the suitability of a given release. A satin finish release agent can reduce pre-release associated with higher shrink resin systems. A green mold is generally given special treatment to avoid initial sticking problems.

When applicable, a release agent should be fully cured before molding. Recently, there has been a trend to use water based products in a move to go green and change from solvent systems. Mold releases can be divided into the following groups:

1. Polyvinyl alcohol (PVA)

It is a safe product soluble in a water base, but insoluble in styrene and used over waxed molds.

2. Waxes

A typical wax is Carnuba wax, a well established external release and this class of release agent has been upgraded by modern technology, although some formulations may contain silicone.

3. Internal mold release agents

These are incorporated in the resin mix and are very convenient for sheet and dough molding compounds and processes like pultrusion. Typical release agents would be salts of fatty acids such as aluminum or zinc stearate, which would also act as a die lubricant in pultrusion. Paraffin wax can be used with polyester resins, where air inhibition is a problem, where the paraffin wax migrates out from the resin during the curing process and forms a protective film on the resin surface, preventing contact with air. Internal releases, however, can present problems if the molding is to be subsequently painted or used with an adhesive.

4. Silicones

Silicones are used for releasing epoxy resins and can be used for conditioning new metal molds, but are not used with polyesters due to pre-release problems.

5. Fluorocarbons

This class includes ptfe based materials and are easy to use and useful for laboratory molds.

6. New products

Manufacturers are introducing new proprietary products designed for specific end uses, such as resin transfer molding, that are silicone and stearate free.

21.4 CARBON FIBERS IN THERMOPLASTIC MATRICES

21.4.1 The importance of critical aspect ratio [101]

Critical aspect ratios are important and have been discussed in Chapter 20.5.

Table 21.8 Critical aspect ratios for Hercules and Hysol-Grafil carbon fibers

Matrix	Hercules AS-1	Hercules AS-4	Grafil XAS
DGEBA/MPD	42	55	32
Polycarbonate	119	108	54
Polyphenylene oxide		121	55
Polyetherimide	84	93	55
Polysulphone		121	55
PPO/Polystyrene (75/25)		206	61

Source: Reprinted with permission from Pegoraro M, Pagini G, Clerici P, Penta A, Interaction between short subcritical length fibres and polymer matrix, *Fibre Sci Technol*, 10, 263, 1977. Copyright 1977, Elsevier.

Critical aspect ratios for Hercules and Hysol-Grafil carbon fibers have been determined (Table 21.8) [104]. Miwa and Endo [105] have discussed the critical fiber length and tensile strength for epoxy composites.

21.4.2 Preparation of thermoplastic molding compounds [101,103]

21.4.2.1 *Sizing carbon fiber with compatible thermoplastic polymer size*

To aid dispersion of the carbon fiber in the polymer matrix, it is usual to apply a polymer compatible size, normally a lower molecular weight version of the polymer can be used, preferably in a water base (e.g. a polyurethane). If the fiber has an epoxy size, this must be removed by solvent extraction in a solvent degreasing plant. Under some conditions, it is possible to blend 10% epoxy sized fiber. If the thermoplastic polymer is to be used at high temperatures (e.g. PEEK), then the size must be temperature resistant, such as a polyimide.

21.4.2.2 *Manufacture of thermoplastic molding compound*

The properties of the most common thermoplastics that are used for reinforcing with carbon fibers are given in Appendix 14.

1. Short fiber process [106]

Sized chopped fiber (Section 21.1.2.4) can either be premixed with thermoplastic polymer granules or better, fed separately into a twin screw extruder, where the polymer is melted in the first part of the barrel, the chopped fiber pellets are introduced via a side feeder, thoroughly mixed in the latter part of the barrel, the product is then discharged into a single screw cross head extruder and extruded through a 6 mm die into a water bath, where it solidifies to a continuous tow in rod form, passed over an air knife to remove excess water and pelletized. The pellets are then spin dried to remove the majority of entrained water and finally dried in an oven (Figure 21.83). Mixing and compounding of polymers [107–109] and twin screw compounding [110,111] have been described.

2. Long fiber process

An alternative process is to feed continuous fiber into the extruder instead of pellets. This process was originally operated by Turner Brothers Asbestos, who sold the know-how to

MANUFACTURING TECHNIQUES FOR CARBON FIBER REINFORCED COMPOSITES

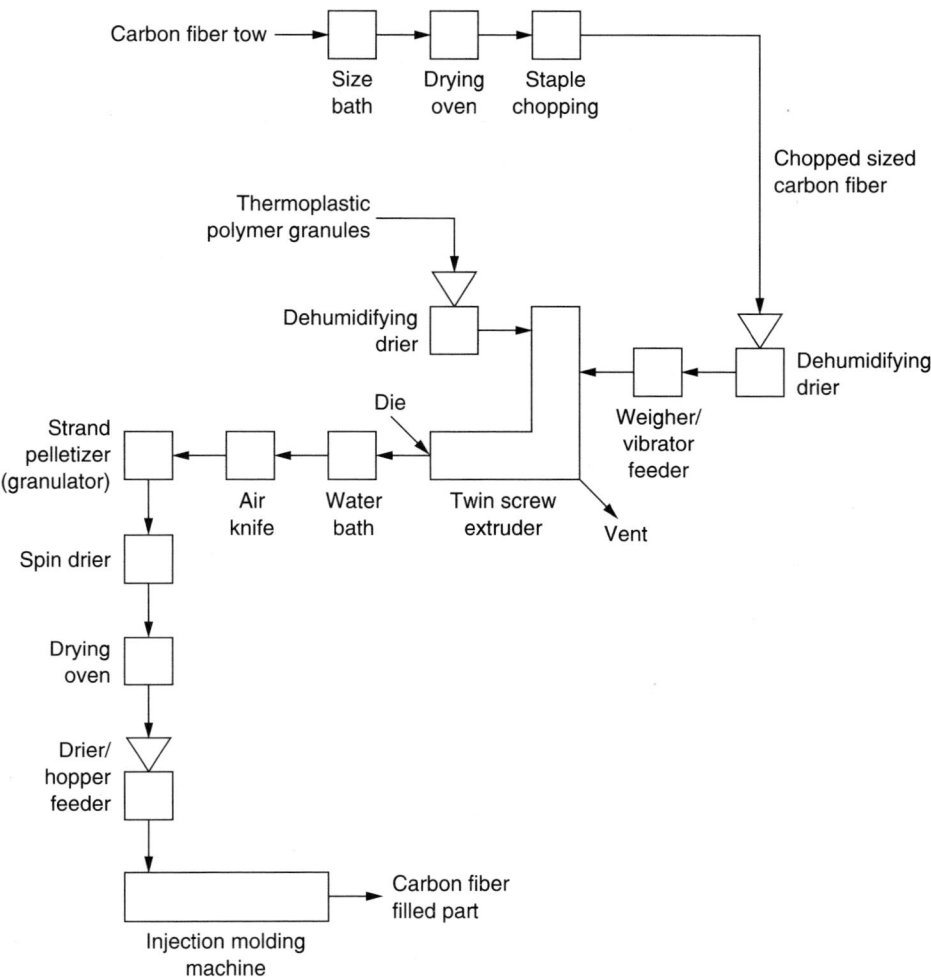

Figure 21.83 Typical route to manufacture a carbon fiber filled thermoplastic molding compound using a twin screw compounder.

Norsk Hydro. The process supposedly does less damage to the reinforcement, although in practice, there would appear to be little advantage.

21.4.3 Injection molding

There are two types of machines that can be used for injection molding—one fitted with a plunger and the other a screw type injection unit, the latter being preferred for carbon fiber applications. Section 17.9.2 explains the functions of the individual parts of an injection molding machine and describes the terminology used for this application.

The process is similar except carbon fiber reinforced thermoplastic polymer pellets are used.

Table 21.9 suggests recommendations for obtaining satisfactory moldings and probably, the most important parameter requiring strict control during injection molding is keeping the dew point of the drying air below $-20°C$, preferably by using a dehumidifier type drier and ensuring that the manufacturer's drying schedule is rigorously followed. It is preferable to

Table 21.9 Recommendations for molding thermoplastic molding compounds

	Situation									Remedial action
Brittleness	Burn Marks	Discoloration	Flashing	Sink marks	Short shots	Splay	Sticking	Surface roughness	Warpage long term	
Yes	Yes	Yes				Yes		Yes		Predry material
				[+]	[+]					Adjust feed
	[−]		[−]	[+]	[+]		[−]	[+]	[−]	Injection pressure
				[+]	[+]			[+]		Injection rate
			[−]	[+]						Injection hold time
							[+]			Cure time
[−]	[−]	[−]	[−]	[+]	[+]			[+]	[+]	Mold temperature
			[−]	[−]	[+]	[−]		[+]	[+]	Stock temperature
			[+]							Clamp pressure
Yes		Yes						Yes		Contamination check
Yes		Yes						Yes		Regrind check
							Yes	Yes		Polish mold
	Yes			Yes	Yes					Vent mold

[+] = Increase.
[−] = Decrease.

gate into the region with the greatest wall thickness and if there is more than one part, then gate symmetrically. If the holding time has been found to be too short, then for different hold times, weigh the parts after removing the sprue and select the holding time for which the weight does not increase. Ensure that the hot runner is thermally isolated from runners and nozzles.

21.4.4 Film stacking process

Film stacking [112–115] comprises sizing the fiber reinforcement with 15–20% of the requisite thermoplastic to form a prepreg sheet with sufficient integrity to hold together, then interleaving this with a thin film (about 0.1 mm thick) of the thermoplastic material and compacting in a mold with heat and pressure to give a void free laminate. The technique is restricted to thermoplastics which are soluble in a suitable solvent and yet have a high softening point and can be used with PES, PSU and PC, but not PEEK.

21.4.5 Thermoplastic prepreg

The carbon fiber has to be coated with a suitable size to aid wetting out and this size can normally be a lower molecular weight version of the thermoplastic polymer applied in a suitable solvent, if possible water based, e.g. low molecular weight nylon, PC and PSU, whilst polyurethane type sizes are also useful. Care must be taken to choose a size that will withstand the processing conditions and for PEEK, it is necessary to use a polyimide size.

A pre-plied laminate is built up in the required format to provide necessary strength and stiffness for the specific application. The predried laminates are consolidated in large presses at high pressure and temperature. Typical thermoplastic resins offered by TenCate Advanced Composites are PEI, useful for most applications and PPS, where resistance to MEK or Skydrol is required. The following forming techniques can be used for thermoplastic prepreg:

1. Thermofolding—a blade heater permits combined pressure and heating to achieve thermofolding
2. Compression molding—the laminate is heated beyond its softening temperature using hot air or infrared and pressed into the required shape. A heated platen press can be used to initially heat the laminate.

Typical molds for this application are shown in Figure 21.84 and best quality is obtained with matched metal chrome plated steel.

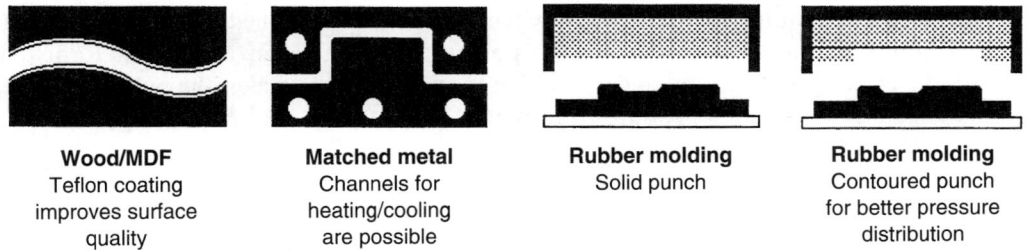

| Wood/MDF | Matched metal | Rubber molding | Rubber molding |
| Teflon coating improves surface quality | Channels for heating/cooling are possible | Solid punch | Contoured punch for better pressure distribution |

Figure 21.84 Molds for preparing thermoplastic laminates. *Source:* Reprinted from Tencate Advanced Composites technical literature.

1. Molding carbon fiber/PEI laminate

This is an amorphous polymer and heating and cooling rates are not critical. Blanks are cut using water-jet and molded to shape by heating on both sides using an infrared oven (395°C for PPS) and placing in a preheated silicone rubber mold (170°C for PPS) with a rubber coated male tool using a re-consolidation pressure of 0.6 MPa. The part is cooled under pressure for 30 s, as a slight increase in temperature occurs after de-molding due to the heat stored in the laminate. Almost net shaped parts can be obtained and NC machining can be used for complex parts. Resistance and induction welding are among the welding methods that can be successfully used to join parts together. Ultrasonic inspection is used to assure a void free product.

2. Platen pressing of carbon fiber/PEEK laminate

A prepreg stack is laid up using a hot soldering iron to tack plies together and the stack placed between release coated high temper Al foils, which can be contained within a picture frame with mirror finish stainless steel plates (at least 1.5 mm thick) on the top and bottom, or preferably in a matched mold. The pressing is in three distinct stages—heating, consolidating and cooling. The surface temperature of the platens must be 370–390°C at a nominal pressure of 10 KPa allowing 1 min per ply, to a maximum of 30 min. Consolidation is then effected by applying 0.1–0.2 MPa for 5 min and finally, the stack is cooled under pressure down to 200°C at about $40°C\,min^{-1}$, achieved by transferring to a press at 180°C where cooling takes about 5 min. With PEEK, this cooling cycle is most important in order to maintain crystallinity.

21.4.6 Thermoplastic filament winding [116,117]

Thermoplastics do not bond well to carbon fiber and the correct size must be selected to withstand the thermoplastic processing conditions, such as a polyimide size for processing PEEK. Since covalent bonding is not applicable for bonding thermoplastic resins to carbon fiber, it is possible to achieve bonding by differential shrinkage occurring due to the difference in fabrication temperature and the temperature of use [118], whereas Hoffman [119] believes that microroughness is important.

When filament winding with thermoplastics, it is difficult to obtain effective impregnation of the fiber reinforcement, as the thermoplastic melts at a temperature some 2–3 orders of magnitude greater than melt processable thermoset matrices, and consequently, the process is much slower. When using the polymer in powder form, a fluidized bed can be used to impregnate a carbon fiber tape, passing the partially impregnated tape under pressure around heated pins to effect improved wetting out and then winding onto a mandrel using heat and pressure to effect final impregnation. The polymer can also be applied using an extruder to effect melting. Alternatively, providing the polymer is available in a filament form (e.g. PEEK), it can be commingled with carbon fiber, rapidly heated to melt the polymer, passed around heated pins to apply pressure to fully wet out the fiber, and finally, wound onto a mandrel. With PEEK, the correct cooling schedule must be followed to ensure crystallinity. This process can be operated more quickly than the melt process, where the polymer is melted in an extruder, but it is more expensive and limited by the availability of the thermoplastic in filament form.

Flemming *et al* [120] have described the production of aligned short carbon fiber reinforced thermoplastic prepreg.

21.4.7 Thermoplastic pultrusion

Dow, using Fulcrum thermoplastic composites technology and Isoplast thermoplastic polyurethanes, have developed a new thermoplastic pultrusion process.

21.4.8 Continuous fiber reinforced plastic materials

Applied Fiber Systems produce Towflex® by passing carbon fiber through a fluidized bed powder coating chamber and melt fusing the adhering thermoplastic particles to give a uniform distribution. In a molding process, the coated fiber wets out and consolidates quickly. The matrix resins include PP, PA6, PPS, PEI and PEEK and can be applied in a range of resin contents. The product is highly flexible and can be used for fabric, braid and laminated panels.

21.5 HYBRID COMPOSITES

A hybrid composite [121,122] can contain more than one type of reinforcement and/or more than one type of matrix, with the objective of improving or lowering the cost of the basic composite [123]. The second reinforcement may be a fiber (continuous or chopped), particles or whiskers. The fiber reinforcement can be in the same laminae and interspersed using any textile process such as weaving, or in different laminae, interspersing plies to obtain the desired mechanical/ physical properties. A sandwich composite is a special case which has an interlayer of a material such as Al foil or a honeycomb. The matrix may be different for each type of reinforcement, or added to infiltrate the reinforced matrix (e.g. a thermoplastic resin such as PSU) to confer controlled viscosity in the matrix, or an elastomer (e.g. CTBN) for increased toughness.

There can be a hybrid effect when a property, such as tensile strength, can be higher than that predicted by the rule of mixtures (Figure 21.85).

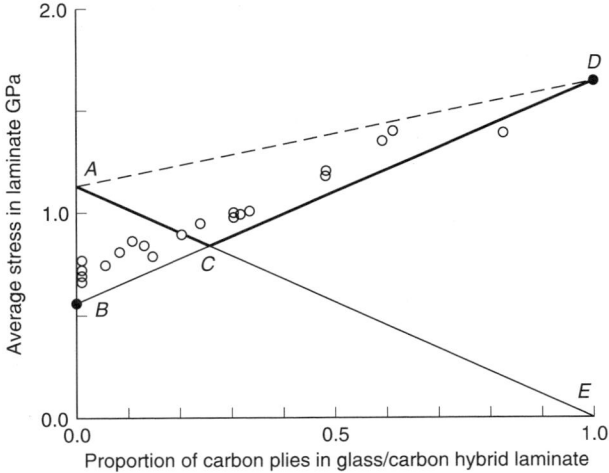

Figure 21.85 Illustration of the hybrid effect. The line of AD is the simple Rule of Mixtures prediction. ACD indicates the predicted strengths based on the failure strains of the two components, and BD is the stress at which the least extendable (carbon) plies are expected to fail. The data points, which lie above this predicted locus, indicate a positive hybrid effect. *Source:* Reprinted with permission from Manders PW, Bader MG, *J Mater Sci*, 16, 2233, 1981; 16, 2246, 1981. Copyright 1981, Springer.

REFERENCES

1. Andersson CH, Yarn handleability and weft inserted warp knit non-crimp fabrics, *Proceedings, TexComp-2*, Leuven, Belgium, May 17–19, 1994.
2. Hearle JWS, Du GW, Forming rigid fibre assemblies, the interaction of textile technology and composite engineering, *J Text Inst*, 81(4), 1990.
3. Ko FK, Advanced textile structural composites, Moran-Lopez JL, Sanchez JM eds., *Advanced Topics in Materials Science Engineering*, Plenum Press, New York, 1993.
4. Ko FK, Textile preforms for carbon-carbon composites, Buckley J, Edie DD eds., *Carbon–Carbon Materials and Composites*, Noyes Publications, Park Ridge, 71–104, 1993.
5. Ko FK, Du GW, Processing of textile preforms, Gutowsk TG ed., *Advanced Composite Manufacturing*, John Wiley, New York, 1997.
6. Ko FK, Du GW, Textile preforming, Peters ST ed., *Handbook of Composites*, Chapman and Hall, London, 1998.
7. Geoghegan PJ, *3rd Textile structural Composites Symposium*, Philadelphia, 1988.
8. Fukuta K, Aoki E, *15th Textile Structural Composites Symposium*, 1986.
9. Fukuta K, Aoki E, Onooka R, Magatsuka Y, *Bull Res Inst Polym Text*, 131, 159, 1982.
10. SP Systems, *Composite Engineering Materials—Product Information Handbook—Reinforcement Materials*, Newport, IOW.
11. Dow NF, Ramnath V, Walter RB, *Analysis of woven fabrics for reinforced composite materials—Technical Final report*, NASA Cr-178275, 1987.
12. Tzeng SS, Lin WC, Mechanical behavior of two-dimensional carbon fiber fabrics reinforced carbon matrix composites, *Key Eng Mater*, 145(1,2), 877–882, 1998.
13. Knitted fabric composites-properties, Jones FR ed., *Handbook of Polymer—Fibre Composites*, Longman Scientific and Technical, Essex, 48, 1994.
14. Knitted reinforcements Jones FR ed., *Handbook of Polymer—Fibre Composites*, Longman Scientific and Technical, Essex, 341, 1994.
15. Raz S, *Industrial Knitted Fabric Design and International Trends*, IFAI, Boston, Nov 1992.
16. Buckley JD, Edie DD eds., *Carbon–Carbon Materials and Composites*, Noyes Publications, Park Ridge, 1992.
17. Kawabata S, Warp-knitted direct oriented structures for pre-shaped composites, *J Text Inst*, 81, 432–447, 1990; Also in Andersson CH, Eng K, Zäh W, Stähl J-E, *Proceedings, TexComp-2*, Leuven, May 17–19, 1994.
18. Ko FK, Kutz J, Multi axial warp knit for advanced composites, *4th Annual ASM/ESD ACCE*, Dearborn, Sep 13–15, 1988.
19. Pastore CM, Whyte DW, Soebreto HB, Ko FK, Design and analysis of multiaxial warp knit fabrics for composites, *J Ind Fabrics*, 5(1), 84–91, Sep 1986.
20. Ko FK, Krauland K, Scardino F, Weft insertion warp knit for industrial applications, Hybrid composites, *J Ind Fabrics*, 1(3), 26, 1983.
21. Raz S, *Knitted Fabrics—Guide to Technical Textiles*, Karl Mayer, Obertshausen, 1988.
22. Ko FK, Braiding, *Engineered Materials Handbook, Vol 1, Composites*, ASM International, Metals Park, 519–528, 1987.
23. Ko FK, Pastore C, Head A, *Atkins and Pearce Handbook of Industrial Braids*, Covington, 1989; Also in, *Proceedings, TexComp-2*, Leuven, May 17–19, 1994.
24. Fujita A, Hamada H, Maekawa Z, Uozumi T, Okauji T, Ohno E, I-beam composites with three-dimensional braiding structure- fabrication and mechanical properties, *Proceedings, TexComp-2*, Leuven, Belgium, May 17–19, 1994.
25. Eurocarbon BV, Netherlands, *Technical literature*.
26. Burns RL, Manufacturing and design of carbon-carbon composites, Buckley JD, Edie ED eds., *Carbon-Carbon Materials and Composites*, Noyes Publications, Park Ridge, 197–222, 1992.
27. Masters JE, Foye RL, Pastore CM, Gowayed YA, Mechanical properties of triaxially braided composites, Experimental and analytical results, *Contract Report 189572*, NASA, Jan 1992.
28. Miravete A ed., *3-D Textile Reinforcements in Composite Materials*, Woodhead Publishing Ltd., 1999.

29. Ko FK, Three dimensional fabrics for structural composites, Chou TW, Ko FK eds., *Textile Structural Composites, Composite Materials Series*, Vol 3, Elsevier Science, Amsterdam, 129–169, 1989.
30. Kelly A ed., Three dimensional fabrics for composites, *Concise Encyclopedia of Composite Materials*, Revised Edition, Pergammon Press/Elsevier, 297–305, 1994.
31. Miravete A ed., 3-D textile reinforcements in composite materials, *3-D Textile Reinforcements in Composite Materials*. (To be published 1999).
32. Pastore C, Ko FK, Near net shape manufacturing of composite engine components by 3-D architecture, *Gas Turbine and Aeroengine Congress and Exhibition*, Toronto, Jun 4–8, 1989.
33. Verpoest I, Ivens J, van Vuure AW, Efstratiou V, Research in textile composites at K.U. Leuven, *Proceedings, TexComp-2*, Leuven, May 17–19, 1994.
34. Hill BJ, McIlhagger R, McLaughlin P, Weaving multilayer fabrics for reinforcement of engineering components, *Composites Manuf*, 4(4), 227–232, 1993; Also in *Proceedings, TexComp-2*, Leuven, May 17–19, 1994.
35. Savage GG, *Carbon-Carbon Composites*, Chapman and Hall, London, 76, 1992.
36. Hörsting K, Wulhürst B, Franzke G, Offerman P, New types of textile faqbrics for fiber composites, *SAMPE J*, 29(1), 7–11, Jan–Feb 1993.
37. Mills A, Burley G, Backhouse R, Innovative materials and manufacturing processes for the cost effective manufacture of composite airframe structures, *44th International SAMPE Symposium*, May 23–27, 1999.
38. Hamada H, Fujita A, Maekawa Z, Ohishibashi H, Mechanical properties of triaxial woven fabric composites, *Proceedings, TexComp-2*, Leuven, May 17–19, 1994.
39. Fujita A, Hamada H, Maekawa Z, Tensile properties of carbon fiber triaxial woven fabric composites, *J Composite Mater*, 27(15), 1428–1442, 1993; Also in *Proceedings, TexComp-2*, Leuven, May 17–19, 1994.
40. Bogdanovich A, Singletary J, Yushanov S, 3-D woven fabric preforms and composites, experimental characterization and predictive analysis, *Proceedings, TexComp-5*, Leuven, Sep 18–20, 2000.
41. Ivens J, Vandeurzen PH, van Vuure AW, Verpoest I, Ko FK, Meerding K, Modeling of the skin properties of 3-D fabric sandwich composites, *Proceedings, TexComp-2*, Leuven, 540–547, May 17–19, 1994.
42. Ko FK, Three dimensional fabrics for composites, *Textile structural Composites*, Elsevier, 1989.
43. Processing of textile preforms, Gutowski TG ed., *Advanced Composites Engineering*, Wiley Interscience, 1997.
44. Hitco, *Thick woven fabrics*, U.S. Pat., 1 296 369, Nov 15, 1972.
45. Soden JA, Hill BJ, Conventional weaving of shaped preforms on conventional looms, *Composites Part A*, 29(7), 757–762, 1998.
46. Soden JA, Hill BJ, McIlhagger R, Miller L, An integrated computer system for the design and analysis of 3D woven engineering composites, *Proceedings of the International Conference on Advanced Composites* (ICAC'98), Egypt, 227–236, Dec 15–18, 1998.
47. Ko FK, Three dimensional fabrics for composites, An introduction to the 'Magnaweave' structure, Hayashi et al, eds., *Progress in Science and Engineering of Composites*, ICCM-V, Fourth International Conference on Composites, 982, 1982.
48. Ko FK, Evaluation of the mechanical properties of 'Magnaweave' composites, *Proceedings, National Specialists' Meeting*, American Helicopter Society, Philadelphia, March 1983.
49. Rolincik PG Jr., Autoweave^{TM}—A unique automated 3D weaving technology, *SAMPE J*, 40–47, Oct 1987. Also in *Proceedings, TexComp-2*, Leuven, May 17–19, 1994.
50. McGoldrick C, Morel J, Wevers M, Verpoest I, 2.5D fabrics for delamination resistant composite structures, *Composites* (France), 31(3), 284–290, 1991.
51. Raz S, Three-dimensional knitted structures for technical uses, *Technical Textiles International*, 16–18, May 1993.
52. Laourine E, Schneider M, Wulfhorst B, Pickett A, Production and analysis of 3-D braided textile preforms for composites, *Proceedings, TexComp-5*, Leuven, Sep 18–20, 2000.

53. Kostar TD, Chou TW, Design and automated fabrication of 3-D braided preforms for advanced structural composites, *Computer Aided Design in Composite Material Technology III*, Elsevier Science, 63–78, 1992.
54. Ko FK, Tensile strength and modulus of a three dimensional braid, Whitney JM ed., *Composite Materials Testing and Design*, 7th Conf, ASTM STP 893, ASTM, Philadelphia, 392–403, 1986.
55. Ko FK, Soebroto HB, Lei C, 3-D net shaped composites by the 2-step braiding process, *33rd Int SAMPE Symp*, 33, 912–932, Mar 7–10, 1988.
56. Hiroka T, Yasuda J, Iwasaki Y, The characteristics of 3-D orthogonal woven fabric reinforced composites, *36th Int SAMPE Symp*, 151–160, 1991.
57. Cahuzac G, Aerospatiale Processes for 3-D preforms, *Proceedings, TexComp-2*, Leuven, May 17–19, 1994.
58. Brookstein DS, Interlocked fibre architecture, braided and woven, *35th Int SAMPE Symp*, Apr 1990.
59. Mills AR Manufacturing technology development for aerospace composite structures, *Aeronaut J Royal Aeronaut Soc*, 539–545, Dec 1996.
60. Akay M, Hanna R, A comparison of honeycomb-core and foam-core carbon fiber epoxy sandwich panels, *Composites*, 21(4), 325–331, 1990.
61. Weatherhead RG, *FRP Technology Fibre Reinforced Resin Systems*, Applied Science Publishers Ltd., London, 1980.
62. Marsh G, Process refinement key to RTM success, *Reinforced Plastics*, 26–30, Feb 1999.
63. Benjamin W, Beckwith S, Resin Transfer Molding (RTM) Technologies, *SAMPE Monograph 3*.
64. SP Systems, *Composite Engineering Materials—Product Information Handbook; Guide to Composites*, Newport, IOW.
65. Harper A, Closed mould processing, *Reinforced Plastics*, 28–34, Feb 2001.
66. Abraham D, McCarthy R, Design of polymer composite structural components for manufacture by resin transfer moulding (RTM), *SAMPE Europe Conference and Exhibition. Proc of 20th International SAMPE Europe Conference of the Society for the Advancement of Materials*, Paris, France, 407–415, 1999.
67. Sundsrud GJ, Advantages of a one-part resin system for processing aerospace parts by resin transfer molding (RTM), *SME Composites in Manufacturing Conference*, Pasadena, Jan 19, 1993.
68. Heider D, Hofmann C, Gillespie JW Jr., Automation and control of large scale composite parts by VARTM processing, *Proceedings 45th International SAMPE Symposium/Exhibition*, Long Beach, May 21–25, 2000.
69. Anderson M, Gebart BR, Landström S, Langström R, Development of guidelines for the vacuum infusion process, *FRC2000 Conference*, Newcastle, Sep 13–14 2000.
70. Ness D, Jones DT, SP resin infusion technology (SPRINT), *FRC2000 Conference*, Newcastle, Sep 13–14, 2000.
71. SP Systems, SPRINT^{TM} SP resin infusion technology, *Advantage Magazine*, 4–5, Jan 2001.
72. Ahn KJ, Seferis JC, *SAMPE Quarterly*, Jan 1990.
73. Jones WR, Johnson JW, A resin injection technique for the fabrication of aero-engine composite components, *Symposium, Fabrication Techniques for Advanced Reinforced Plastics*, Salford, 40–47, Apr 22–23, 1980.
74. Harris H, Schinske N, Krueger R, Swanson B, Multi-axial stitched preform reinforcements for RTM fabrication, *36th Int SAMPE Symp*, Covina, Apr 15–18, 1991.
75. Benson Dexter H, Harris E, Johnston N, Recent progress in NASA Langley textile reinforced composites research programme, *2nd NASA Advanced Composites Technology Conference*, Langley, 1992.
76. Hawley AV, *53rd Annual Conference on Mass Properties Engineering*, SAWE, Long Beach, May 1994.
77. Dexter HB, Development of textile reinforced composites for aircraft structures, *4th International Symp for Textile Composites*, Kyoto, Oct 12–14, 1998.

78. Kim JS, Lee DG, Online cure monitoring and viscosity measurement, *J Mater Proc Technol*, 37(1–4), 405–416, 1993.
79. Maistros GM, Partridge IK, Dielectric monitoring of cure in a commercial carbon fibre composite, *Composite Sci Technol*, 53(4), 355–359, 1995.
80. Maistros GM, Partridge IK, Monitoring autoclave cure in commercial carbon fibre/epoxy composites, *Composites Part B-Engineering*, 29(4), 355–359, 1995.
81. Kim JS, Lee DG, Measurement of the degree of cure of carbon fiber epoxy composite materials, measurement, *J Composite Mater*, 30(13), 1436–1457, 1996.
82. Fink BK, Walsh SM, DeSchepper DC, McCullough RL, Don RC, Waibel BJ, Gillespie JW Jr., Advances in resin transfer molding flow monitoring using SMARTweave sensors, *Proceedings ASME Materials Div*, 69–2, 999–1015, 1995.
83. Vaidya UK, Jadhav NC, Hosur MV, Gillespie JW, Fink BK, Assessment of flow and cure monitoring using direct current and alternating current sensing in vacuum-assisted resin transfer molding, *Smart Materials and Structures*, UK, 9(6), 727–736, 2000.
84. Stanford JL, Powell JR, Wilkinson AN, Structural composites formed by Reaction Injection Moulding, *FRC2000 Conference*, Newcastle, Sep 13–14, 2000.
85. Crawford RJ, Rotational Moulding, *RAPRA Review Report No. 71*, 86, 1993.
86. Peters S, Humphrey D, Foral R, Filament Winding Composite Structures Fabrication, 2nd ed. *SAMPE*.
87. Shaw-Stewart D, Filament winding-materials and engineering, *Materials and Design* 6, No. 3, 1985.
88. Goldsworthy WB, The effect of proper equipment selection on filament wound end product economics, *Conference Filament Winding*, The Plastics Institute, London, Oct 17–18, 1967.
89. Rosato DV, *Filament Winding, Its Development, Manufacture, Applications, and Design*, New York, Interscience Publishers, 1964.
90. Olaofsson K, Manufacturing analysis of wet filament winding, *2nd International Convention for Filament Winding Technology*, Brussels, Oct 17–19, 2001.
91. Dugger JA, Hirt DE, *Polymer Composites*, 17(3), 492–496, 1996.
92. Morgan PE, Trewin EM, Watson IP, Some aspects of the manufacture and use of carbon fibre pultrusion, *Symposium, Fabrication Techniques for Advanced Reinforced Plastics*, University of Salford, UK, 69–90, Apr 22, 23, 1980.
93. Jones BH, Jakway W, Goldsworthy Engineering Inc, MM&T—Pultruded composite structural elements, *US Army Aviations systems Command Final Report*, USAAMRDL-TR-76-5, Dec 1976.
94. Parmenter J, A process which has some pull in industry, *CME*, Jul/Aug 1984.
95. Johnson FC, The compression moulding of composite aero engine components with elevated thermal stability, *Symposium, Fabrication Techniques for Advanced Reinforced Plastics*, University of Salford, 10–18, Apr 22–23, 1980.
96. Graham DF, Space age materials come back to earth, *Reinforced Plastics*, 48–50, May 2001.
97. Turner MR, Rudd CD, Long AC, Middleton V, McGeehin P, Automated fibre lay-down techniques for preform manufacture, *Proceedings of the 4th International Conference on Automated Composites*, Nottingham, 431–438, Sep 6–7, 1995.
98. Turner MR, Rudd CD, Long AC, Middleton V, Automated fibre lay-down techniques for preform manufacture, *Proceedings of the 5th International Conference on Automated Composites*, Glasgow, 245–252, Sep 6–7, 1997.
99. Turner MR, Rudd CD, Long AC, Middleton V, McGeehin P, Net-shape preform manufacture using automated fibre placement, *Advanced Composites Letters*, 4(4), 121–124, 1995.
100. Turner MR, Rudd CD, Long AC, Middleton V, Tow placement for dry fibre preforms, *Composites*, A30, 1105–1121, 1999.
101. Bigg DM, Characteristics of short, conductive fibre reinforced injection mouldable composites, *J Ind Fabrics*, 2(3), Winter 4–14, 1983–1984.
102. Pegoraro M, Pagini G, Clerici P, Penta A, Interaction between short subcritical length fibres and polymer matrix, *Fibre Sci Technol*, 10, 263, 1977.
103. Folkes MJ, *Short Fibre Reinforced Thermoplastics*, Research Studies Press, Letchworth, 1982.

104. Bascom WD, Yon KJ, Jensen RM, Cordner L, The adhesion of carbon fibers to thermoset and thermoplastic polymers, *J Adhesion*, 34, 79, 1991.
105. Miwa M, Endo I, Critical fibre length and tensile strength for carbon fiber epoxy composites, *J Mater Sci*, 29(5), 1174–1178, 1994.
106. De SK, White JR eds., Short Fiber Polymer Composites, Structure Property Relations, Lee SM ed., *International Encyclopedia of Composites*, Vol 5, VCH, New York, 130, 1991.
107. Manas-Zlocower, Tadmor Z eds., *Mixing and Compounding of Polymers, Theory and Practice*, Hanser Publishers, Munich, 1994.
108. Cheremisinoff N, *Polymer Mixing and Extrusion Technology*, Marcel Dekker, New York, 1997.
109. Todd DB ed., *Plastics Compounding—Equipment and Processing*, Hanser Publishers, Munich, 1998.
110. Gras D, *Plastics Technology*, 8, 40, Feb 1972.
111. Mack WA, *Plastics Technology*, 11, 45, Feb 1975.
112. Phillips LN, Fabrication of reinforced thermoplastics by means of the film stacking technique, *Symposium, Fabrication Techniques for Advanced Reinforced Plastics*, Apr 22–23, 1980, University of Salford, IPC Science and Technology Press, Guildford, 101–107, 1980.
113. Hogan PA, The production and uses of film stacked composites for the aerospace industry, *SAMPE Conference*, New York, 1980.
114. Willats DJ, Advances in the use of high performance continuous fibre reinforced thermoplastics, 5th international SAMPE Conference, Montreaux, *SAMPE J*, 6–10, Sep/Oct 1984.
115. Phillips LN, Murphy DJ, Properties of carbon fibre reinforced thermoplastics moulded by the film-stacking method, Royal Aircraft Est, *Technical Report TR-76140*, 1976.
116. Jacob A, Automation sets filament winding on the right path, *Reinforced Plastics*, 48–50, Sep 1998.
117. Schlottermüller M, Thermoplastic filament winding, current research and future potentials, *2nd International Convention for Filament Winding Technology*, Brussels, Oct 17–19, 2001.
118. DiLandro L, Pegorara M, Carbon fiber-thermoplastic matrix adhesion, *J Mater Sci*, 22, 1987, 1980.
119. Hoffman WP, Scanning probe microscopy of carbon fiber surfaces, *Carbon*, 30, 315, 1992.
120. Flemming T, Kress G, Flemming M, A new aligned short carbon fiber reinforced thermoplastic prepreg, *Advanced Composite Materials*, 5(2), 151–159, 1996.
121. Short D, Summerscales J, *Composites*, 10, 215, 1979.
122. Short D, Summerscales J, *Composites*, 11, 33, 1980.
123. Chung DDL, Ch 10 Hybrid composites, *Carbon Fiber Composites*, Butterworth-Heinemann, Boston, 201–209, 1994.
124. Manders PW, Bader MG, *J Mater Sci*, 16, 2233, 1981.
125. Manders PW, Bader MG, *J Mater Sci*, 16, 2246, 1981.

CHAPTER 22

Design

22.1 DESIGN CONSIDERATIONS

The outstanding design properties of carbon fiber reinforced plastics are their high strength to weight and stiffness to weight ratios. Correctly designed cfrp can be stronger and stiffer than a steel part of similar thickness and offer a weight reduction of about 35%. A cfrp has excellent fatigue strength and very good chemical resistance. However, being relatively brittle, the composites have no yield behavior and have poor impact strength. If fasteners are used, then Al cannot be used due to galvanic corrosion and the more expensive Ti has to be used.

22.2 MICROMECHANICS

The calculation of a set of elastic moduli for a composite using knowledge of the elastic moduli of the fiber (Table 22.1) and matrix properties may be considered as the initial phase in the design of a composite [1].

To obtain relations between mathematical and engineering constants in a unidirectional composite, the tension and shear effects for normal stress and shear loading are as depicted in Figure 22.1.

Anisotropic elastic theory is dealt with by Daniel and Ishi [2] and Jones [3], whilst Hull [4] gives an excellent introduction to the physical interpretation of non-isotropic bodies. Laminates are not isotropic (i.e. having the same properties in every direction) and can therefore be considered as orthotropic (having three mutually perpendicular planes of symmetry with different properties in three mutually perpendicular directions). The stresses at a point can be represented by the three normal stresses, σ_{11}, σ_{22} and σ_{33} and three shear stresses, τ_{23}, τ_{31}, and τ_{12}, where the first suffix refers to the direction normal to the plane in which the stress is acting and the second suffix, to the direction in which the stress is acting (Figure 22.2). The corresponding strains have the notation ε_{11}, ε_{22}, ε_{33}, γ_{23}, γ_{31} and γ_{12}. A contracted notation is used when the two suffixes are the same value, so σ_{11} becomes σ_1 and ε_{33} becomes ε_3. The stress components on the faces of the cube are taken as positive and may be considered as the force per unit area exerted by the material outside the cube upon the material inside. Hence, normal tensile stresses are positive and normal; compressive forces are negative (Table 22.2).

Table 22.1 The properties of AS-4 type high strength carbon fiber

Stiffness and strength		
E_1	Axial tensile modulus (GPa)	227
E_2	Transverse tensile modulus (GPa)	15
G_{12}	Axial shear modulus (GPa)	20
G_{23}	Torsional shear modulus (GPa)	5
σ_1	Axial tensile strength (GPa)	3.65
Poisson's ratio		
γ_{12}	Transverse contraction with axial extension	0.25
γ_{23}	Transverse contraction with transverse extension	0.40
γ_{21}	Axial contraction with transverse extension	0.013
Thermal properties		
α_1	Coefficient of axial thermal expansion (°C^{-1})	-1.2×10^{-6}
α_2	Coefficient of transverse thermal expansion (°C^{-1})	12×10^{-6}
K_1	Thermal conductivity along the fiber (W (m°C)$^{-1}$)	~16
K_2	Thermal conductivity across the fiber (W (m°C)$^{-1}$)	~3
C_p	Specific heat (J (kg°C)$^{-1}$) @23°C	0.75
	@143°C	0.99
	@380°C	1.42
ρ	Density (kg m^{-3})	1.780

Source: Reprinted with permission from Wagoner G, Bacon R, Elastic constants and thermal expansion coefficients of various carbon fibers, *19th Biennial Conference on Carbon*, Penn State University, Session 6A, 296–297, 1989. Copyright 1989, American Chemical Society.

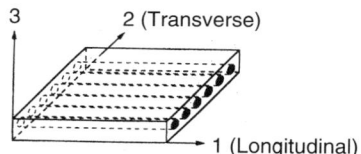

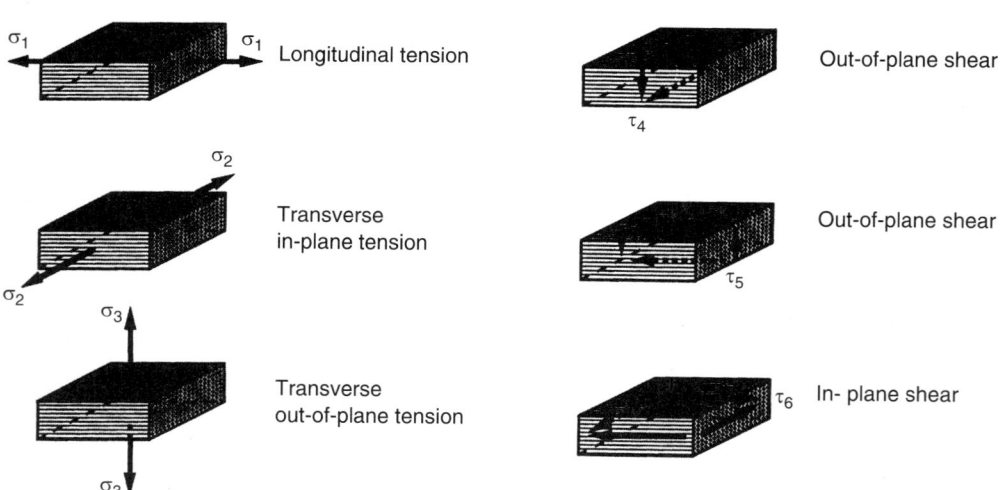

Figure 22.1 Tension and shear effects in a unidirectional composite.

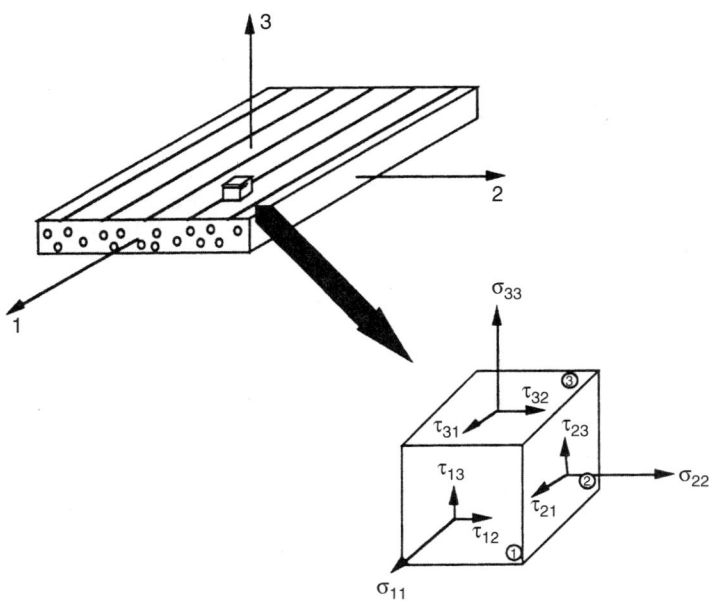

Figure 22.2 Principal directions and stress components for an orthotropic material (G_{12} is the in-plane shear modulus as in plane No. 3 above.) *Source:* Reprinted from University of London, Imperial College of Science, Technology and Medicine, Course entitled *Mechanical Testing of Advanced Fibre Composites*, 1995.

Table 22.2 Representative elastic properties of unidirectional fiber reinforced epoxy resins

Material	Fiber volume fraction V_f	E_{11} GPa	E_{22} GPa	ν_{12}	G_{12} GPa
CFRP (A-S fiber)	0.66	140	8.96	0.30	7.10
CFRP (IM-6 fiber)	0.65	200	11.10	0.32	8.35
GFRP (E-glass fiber)	0.46	35	8.22	0.26	4.10
KFRP (Kevlar-49 fiber)	0.60	76	5.50	0.33	2.35

Nine stress components are used to define the state of a stress at a point: σ_1, σ_2, σ_3, τ_{23}, τ_{31}, τ_{12}, τ_{32}, τ_{13} and τ_{21}. By considering moments about the center of the unit cube, it can be shown that for equilibrium at any point

$$\tau_{23} = \tau_{32} \quad \tau_{31} = \tau_{13} \quad \text{and} \quad \tau_{12} = \tau_{21}$$

Hooke's law gives $\sigma = E\varepsilon$ where E is the Young's modulus.

The normal strain transverse to the applied stress is $-\nu\varepsilon$, where ν is Poisson's ratio.

The shear modulus G is defined by

$$\tau = G\gamma$$

where γ is the engineering shear strain.

Hooke's law, in a generalized form, using the contracted notation can be written as

$$\sigma_i = \sum_{j=1}^{6} C_{ij}\varepsilon_j$$

where $i, j = 1, 2, 3, 4, 5, 6$. The σ_i are the stress components and ε_j are the strain components. C_{ij} is called the stiffness matrix. Since $C_{ij} = C_{ji}$, the first equation relating to stress and strain can be written as:

$$\sigma_1 = C_{11}\varepsilon_1 + C_{12}\varepsilon_2 + C_{13}\varepsilon_3 + C_{14}\gamma_{23} + C_{15}\gamma_{31} + C_{16}\gamma_{12}$$

and the second equation as:

$$\sigma_2 = C_{12}\varepsilon_1 + C_{22}\varepsilon_2 + C_{23}\varepsilon_3 + C_{24}\gamma_{23} + C_{25}\gamma_{31} + C_{26}\gamma_{12}$$

and so on.

To explain anisotropic elasticity theory, a knowledge of matrix algebra is required. It has been dealt with in a simple manner and the matrices have been clearly explained in sufficient detail by Matthews and Rawlings [6]. Matrix algebra is a convenient way of handling relationships between stresses and strains and the six equations can be represented in matrix notation as:

$$\begin{bmatrix} \sigma_1 \\ \sigma_2 \\ \sigma_3 \\ \tau_{23} \\ \tau_{31} \\ \tau_{12} \end{bmatrix} = \begin{bmatrix} C_{11} & C_{12} & C_{13} & C_{14} & C_{15} & C_{16} \\ C_{12} & C_{22} & C_{23} & C_{24} & C_{25} & C_{26} \\ C_{13} & C_{23} & C_{33} & C_{34} & C_{35} & C_{36} \\ C_{14} & C_{24} & C_{34} & C_{44} & C_{45} & C_{46} \\ C_{15} & C_{25} & C_{35} & C_{45} & C_{55} & C_{56} \\ C_{16} & C_{26} & C_{36} & C_{46} & C_{56} & C_{66} \end{bmatrix} \begin{bmatrix} \varepsilon_1 \\ \varepsilon_2 \\ \varepsilon_3 \\ \gamma_{23} \\ \gamma_{31} \\ \gamma_{12} \end{bmatrix}$$

The inverse matrix S_{ij} can be drawn for the compliance, which is the reciprocal of the stiffness.

An orthotropic material is called transversely isotropic when one of its principal planes is a plane of isotropy, i.e. at every point there is a plane on which the mechanical properties are the same in all directions [2]. Unidirectional carbon fibers packed in a hexagonal array with a relatively high volume fraction can be considered transversely isotropic, with the 2–3 plane normal to the fibers as the plane of isotropy (Figure 22.2). For a transversely isotropic material, it should be noted that the subscripts 2 and 3 (for a 2–3 plane of symmetry) in the material constants are interchangeable. Hence

$$C_{12} = C_{13} \quad C_{22} = C_{33} \quad S_{12} = S_{13} \quad \text{and} \quad S_{22} = S_{33}$$

Also subscripts 5 and 6 are interchangeable, thus:

$$C_{55} = C_{66} \quad \text{and} \quad S_{55} = S_{66}$$

$$C_{44} = \frac{C_{22} - C_{23}}{2}$$

This simplifies the matrix to:

$$\begin{bmatrix} \sigma_1 \\ \sigma_2 \\ \sigma_3 \\ \tau_4 \\ \tau_5 \\ \tau_6 \end{bmatrix} = \begin{bmatrix} C_{11} & C_{12} & C_{12} & 0 & 0 & 0 \\ C_{12} & C_{22} & C_{23} & 0 & 0 & 0 \\ C_{12} & C_{23} & C_{22} & 0 & 0 & 0 \\ 0 & 0 & 0 & \frac{C_{22}-C_{23}}{2} & 0 & 0 \\ 0 & 0 & 0 & 0 & C_{55} & 0 \\ 0 & 0 & 0 & 0 & 0 & C_{55} \end{bmatrix} \begin{bmatrix} \varepsilon_1 \\ \varepsilon_2 \\ \varepsilon_3 \\ \gamma_4 \\ \gamma_5 \\ \gamma_6 \end{bmatrix}$$

with the inverse relationships:

$$\begin{bmatrix} \varepsilon_1 \\ \varepsilon_2 \\ \varepsilon_3 \\ \gamma_4 \\ \gamma_5 \\ \gamma_6 \end{bmatrix} = \begin{bmatrix} S_{11} & S_{12} & S_{12} & 0 & 0 & 0 \\ S_{12} & S_{22} & S_{23} & 0 & 0 & 0 \\ S_{12} & S_{23} & S_{22} & 0 & 0 & 0 \\ 0 & 0 & 0 & 2(S_{22}-S_{23}) & 0 & 0 \\ 0 & 0 & 0 & 0 & S_{55} & 0 \\ 0 & 0 & 0 & 0 & 0 & S_{55} \end{bmatrix} \begin{bmatrix} \sigma_1 \\ \sigma_2 \\ \sigma_3 \\ \tau_4 \\ \tau_5 \\ \tau_6 \end{bmatrix}$$

where σ = stress
ε = strain
C = stiffness component
S = compliance component
τ = shear stress
γ = transverse shear strain

Hence, to define the elastic properties of the fiber, five independent components of elastic modulus are required—Axial Young's modulus (E_{11} or E_α); Shear modulus (G_{11} or G_α); Transverse Young's modulus (E_{22} or E_t); Transverse Shear modulus (G_{22} or G_t) and the Axial Poisson ratio (ν_{12} or ν_α).

The relations of mathematical and engineering components can be derived from an orthotropic material element subjected to uniaxial tensile loading in the longitudinal direction (σ_1):

$$\varepsilon_1 = S_{11}\sigma_1$$
$$\varepsilon_2 = S_{12}\sigma_1$$
$$\varepsilon_3 = S_{13}\sigma_1$$
$$\gamma_4 = \gamma_4 = \gamma_6 = 0$$

and from engineering considerations:

$$\varepsilon_1 = \frac{\sigma_1}{E_1} \quad \varepsilon_2 = \frac{\gamma_{12}}{E_1}\sigma_1 \quad \varepsilon_3 = \frac{\gamma_{13}}{E_1}\sigma_1$$

$$\gamma_4 = \gamma_4 = \gamma_6 = 0$$

and

$$S_{44} = \frac{1}{G_{23}} \quad S_{55} = \frac{1}{G_{13}} \quad S_{66} = \frac{1}{G_{12}}$$

$$C_{44} = \frac{1}{S_{44}} \quad C_{55} = \frac{1}{S_{55}} \quad C_{66} = \frac{1}{S_{66}}$$

whilst
$$C_{44} = G_{23} \quad G_{12} = G_{13} \quad C_{66} = G_{12}$$

and in the special case of transversely isotropic material with the 2-3 plane as the plane of symmetry,

$$E_2 = E_3 \quad G_{12} = G_{13} \quad \gamma_{12} = \gamma_{13}$$

22.3 SELECTION OF MATERIALS

A layer (lamina or ply) of fiber reinforced composite is strong along the fiber direction, but considerably weaker in all off-fiber directions. Hence, in practice, it is necessary to use a number of layers oriented in different directions to withstand loadings from multiple angles.

A lay-up notation is used to specify a laminate structure: $[\pm\alpha_a \pm\beta_b \pm\chi_c]_{ds}$, where α, β, χ are the angles of the UD fiber in relation to a reference axis; the integers a, b, c indicate the number of layers, d is the number of times the bracketed construction is repeated and s, when present, indicates that the whole lay-up is symmetrical about a center plane.

Figure 22.3 shows the large in-plane anisotropy for the elastic constants E and G of a Type I carbon fiber in an epoxy matrix tested at various angles to the fiber direction. The interlaminar shear stress is shown as a function of ply angle in Figure 22.4.

22.4 ELASTIC BEHAVIOR OF MULTIDIRECTIONAL LAMINATES

The behavior of a multidirectional laminate will depend on the properties and stacking sequence of the individual layers.

The classical lamination theory will predict the behavior of a thin laminate, provided it is not a hybrid and the following assumptions are met when the laminate is extended or bent:

1. Normals remain straight and do not bend; remain unstretched and keep the same length; remain normal and always make a right angle to the neutral plane.
2. Perfect bonding between the layers is attained.

The material will have a given strength expressed as stress or strain, beyond which it fails. In order to postulate the failure, it is necessary to have a failure criterion with an associate theory to be able to effect a satisfactory design. Such theories include maximum stress, maximum strain, Tsai-Hill (based on deviatoric strain energy theory) and Tsai-Wu (based on interactive polynomial theory). The Tsai-Wu theory is the most commonly used.

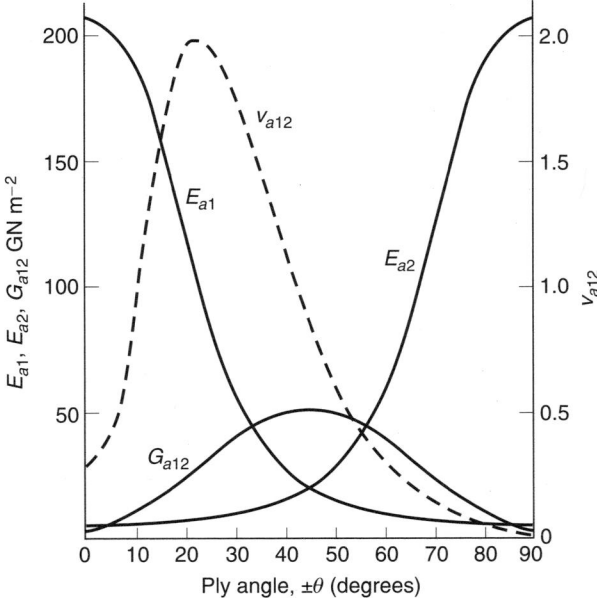

Figure 22.3 Elastic constants of $\pm\theta$ angle ply of Type I high modulus carbon fiber epoxy laminates. At $\theta = 0°$, $E_{a1} = E_{11}$; $E_{a2} = E_{\perp}$; and $G_{a12} = G_{\#}$. The maximum value of G_{a12} occurs at $\theta = \pm 45°$. Source: Reprinted from Sinclair JH, Chamis CC, Proceedings of the 35th SPI/RP Annual Technology Conf, Paper 12A, Society of the Plastics Institute, New York, 1980.

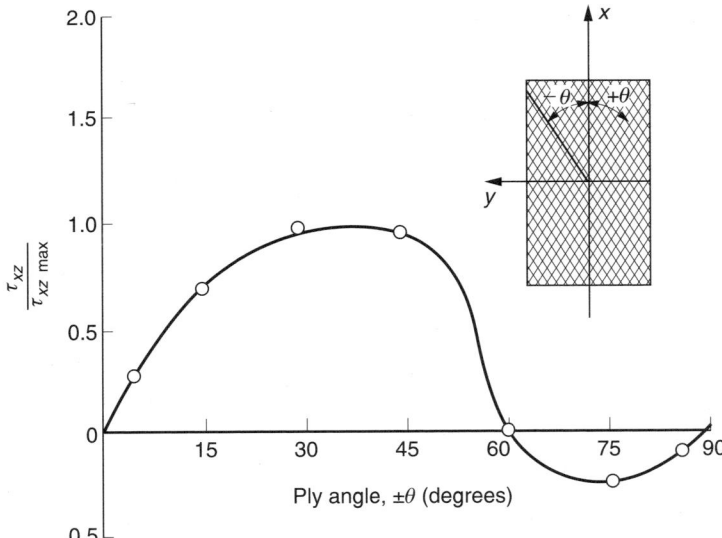

Figure 22.4 Interlaminar shear stress as a function of ply angle for Type I high modulus carbon fiber epoxy resin laminate tested in uniaxial tension in x-direction. The maximum interlaminar shear strength occurs at $\theta = 35°$ and the stresses are zero at 0, 60 and 90°. Source: Reprinted with permission from Pipes RB, Pagano NJ, Interlaminar shear stress in composite laminates under axial tension, J Composite Mater Sci, 13, 2131–2136, 1978. Copyright 1978, Sage Publications.

Since the laminate comprises a number of laminae oriented in different directions with respect to each other, having the same stress-strain relations, the stress-strain equation of the kth layer of the laminate is as given by Hull [4] as:

$$\begin{bmatrix} \sigma_x \\ \sigma_y \\ \tau_{xy} \end{bmatrix}_k = \begin{bmatrix} \overline{Q}_{11} & \overline{Q}_{12} & \overline{Q}_{16} \\ \overline{Q}_{12} & \overline{Q}_{22} & \overline{Q}_{26} \\ \overline{Q}_{16} & \overline{Q}_{26} & \overline{Q}_{66} \end{bmatrix}_k \begin{bmatrix} \varepsilon_x \\ \varepsilon_y \\ \gamma_{xy} \end{bmatrix}_k$$

where $\overline{Q}_{ij}$ is called the transformed reduced stiffness matrix.

The $\overline{Q}_{ij}$ has to be evaluated for each layer, hence for a given strain distribution, the stress in each layer can be evaluated.

Physical observations led to the development of criteria which distinguish between fiber failure (fracture of the fibers) and inter-fiber failure (fracture of the matrix). These fiber failure modes are depicted in Figure 22.5. Different mathematical formulations are used for these modes of failure and it is most important that the designer knows exactly which type of failure is occurring. Examples of such criteria are Simple Puck, Modified Puck and Hashin. An Action Plane Criterion was developed by Puck and in conjunction with a non-linear degradation model and is probably one of the best models currently available. Using this theory, Material SA, Belgium have developed Composite Star design software and materials database. The software incorporates a graph engine which can display the failure envelope together with the constant and variable stress vector.

There are many versions of software programs available for the analysis and design of composite laminates and laminated structural elements. ESAComp is one such version, initiated by the European Space Agency, covering fiber/matrix mechanics, plies, laminates, plates and stiffened panels, beams and columns, bonded joints and mechanical joints. The software can interface with the widely used finite element software packages.

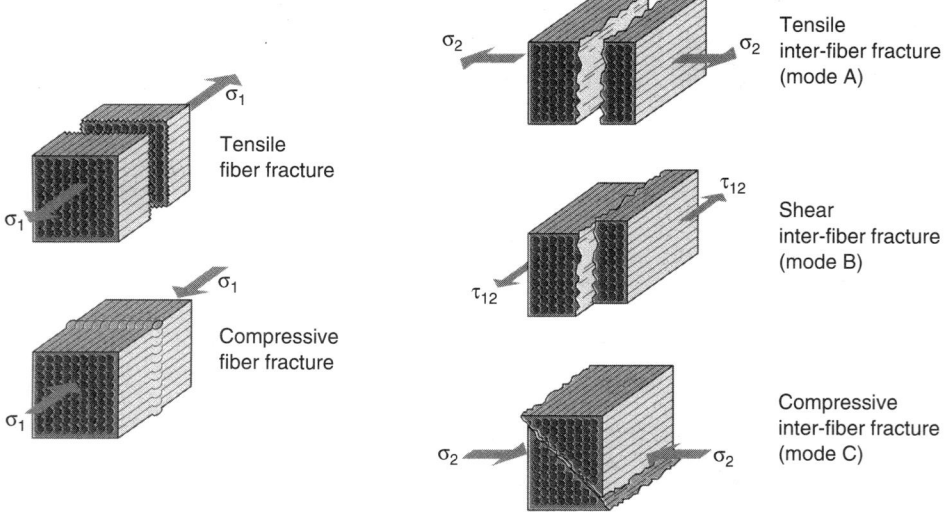

Figure 22.5 Fiber failure and inter-fiber failure modes. *Source:* Reprinted with permission from Laval C, Composites design in the real world, *Reinforced Plastics*, 50–53, Sep 2003. Copyright 2003, Elsevier.

HyPerComp Engineering Inc., for example, use finite element methods to analyze the cylinder geometry of a pressure vessel dome to determine the liner hoop stress at service pressure.

Delft University of Technology (TUD), in conjunction with the Netherlands Organization for Applied Scientific Research (TNO), have developed Kolibri, a composite design package aimed at non-specialist composite designers and students, where only a background knowledge of finite element analysis is required. The package is available from Centre of Lightweight Structures, Delft.

22.5 CHOICE OF COMPOSITE MANUFACTURING METHOD

The choice of composite manufacturing method has been discussed in Chapter 21 and Mazumdar has described the manufacture of composites in detail [11].

Obviously, cost is a controlling factor and to compete against Al, the lowest cost form of carbon fiber must be used in conjunction with an automated high speed lay-up process. These criteria have been exemplified by Mills and co-workers [12], working at Cranfield University, who have achieved these aims by developing a machine that can achieve a prepreg output of about 40 $kg\,h^{-1}$.

22.6 BONDING AND JOINING

There are a number of useful publications which discuss structural joints and adhesives in detail [13–16].

As a rule, adhesive bonding should only be used if it can be undertaken in the controlled conditions of a manufacturing plant and with skilled labor. To ensure a good bond it is necessary to:

1. Maximize the bond area
2. Aim for compression and shear loads in the adhesive joint
3. Attempt to avoid peel and tensile stresses

Surface cleaning is critical and depending on the choice of adhesive, can include solvent cleaning, application of a primer, surface abrasion and peel of the ply embedded in the laminate.

Adhesive bonded joints are not weakened by drill holes, as there is no local stress concentration, and can accommodate tolerances in the component part and act as a sealant with very good fatigue resistance.

Moussiax and Lügering [17] have outlined factors that may help in selecting the type of adhesive:

1. Toughened epoxies—have highest mechanical and best temperature performance with good resilience, but require good surface roughening and are best cured with external heating.
2. Two-component polyurethanes—are easy to apply and provided the environmental conditions are not too critical, they have good toughness and very good durability in European climate and in wet conditions.
3. Cyano-acrylates—have the fastest cycle time using moisture in the air for cure, but are not in common use. They have the ability to bond to most surfaces with little preparation and are used mainly for GRP assemblies.

4. Two-pack methacrylate—have a fast cure and are strong and flexible, but the onset of cure is sudden, reaches peak exotherm in about 3 min and heating will speed up the onset of cure [18].

Bonded joints are limited by the shear strength of the adhesive (about 2800 MPa for epoxies). The highest stress is at the edges and for a single bonded joint, can be 3–10 times the stress in the remaining 90% of the joint. When joining metal to composite, machining grooves in either part to within 5 mm of the edge will limit the effects of edge stress.

22.7 FABRICATION

The fabrication of composites is covered by the *Handbook of Composites*, Vol 4 [19], whilst Chen [20] has discussed drilling cfrp. Kim and Lee [21] have studied ultrasonic vibration cutting of cfrp and Davies *et al* [22] have reported the joining and repair of cfrp.

22.8 TESTING AND INSPECTION

An important aspect of design is inspection and testing and Campbell *et al* [23] have given an overview of the causes of porosity in cfrp and Adams has covered Test Methods [24].

22.9 SMART DEVICES

Optical fiber sensors can be used within, say a carbon fiber mast structure to measure strain, whereas if electronic strain gages were used, they would generally be attached outside a structure and would require an array of exposed connecting wires to the instrumentation. The optical fiber sensor, however, can be built into the structure and can measure the strain within the composite material. The sensor comprises a number of Bragg diffraction gratings, wherein stress causes the spacing between the grating lines to alter, thereby changing the wavelength of light reflected through the optical fiber. This change is detected by an array of photodiode detectors.

The fiber in an optical fiber sensor acts as the microtransducer with a Bragg grating situated within the core of the fiber, operating as a spectral reflector. The grating is photowritten by a UV laser containing zones which have refractive index n varying with a spatial frequency Λ. The zones form a grating, reflecting the characteristic frequency λ_B. Any modification of n or Λ has a proportional incidence on the reflected wavelength. The sensor can be used to measure *in situ* deformations and temperatures.

Insensys has developed a small, lightweight strain measurement system which can be used for aerospace, civil structures, marine, oil and gas industry, wind energy and research applications, enabling up to one hundred sensors to be multiplexed on one optical cable. Sensors can be bonded on the surface or embedded during composite lay-up with a sensor range of ± 6000 microstrain.

The magnitude and distribution of a compressive force between two mating or impacting surfaces can be detected by inserting Super Pressurex (Sensor Products Inc. East Hannover, USA) a tactile force indicating sensor. The sensor is a 0.1–0.2 mm thick PET film coated with a layer of microcapsules, which rupture on the application of pressure—the intensity of color increasing with increasing pressure—and are available in a range of sensitivities to accommodate a wide range of pressures. After a test run, the film can be removed and examined to establish the pressure variation across the contact area. These sensors can be

DESIGN 945

used for composite lay-up (e.g. honeycomb bonding), for checking the parallelism of rollers and platens and for establishing the effectiveness of a gasket seal.

22.10 DESIGN CASES

22.10.1 Expanding core technique

Design does not have to be complicated and can center around simple constructional methods as depicted in Figures 22.6, 22.7 and 22.8 based on an RK Technologies method,

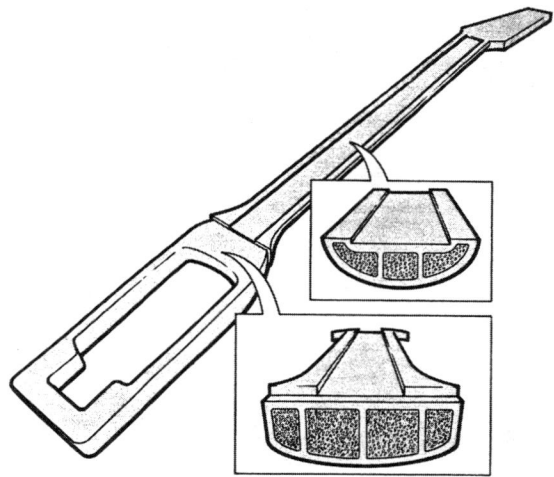

Figure 22.6 Guitar neck fabricated using expanded core technique. *Source:* Reprinted with permission from George J, Technology puts carbon fibre into mass markets, *Eureka*, Franks-Hall, 35–37, Mar 1985. Copyright 1985, Franks Hall.

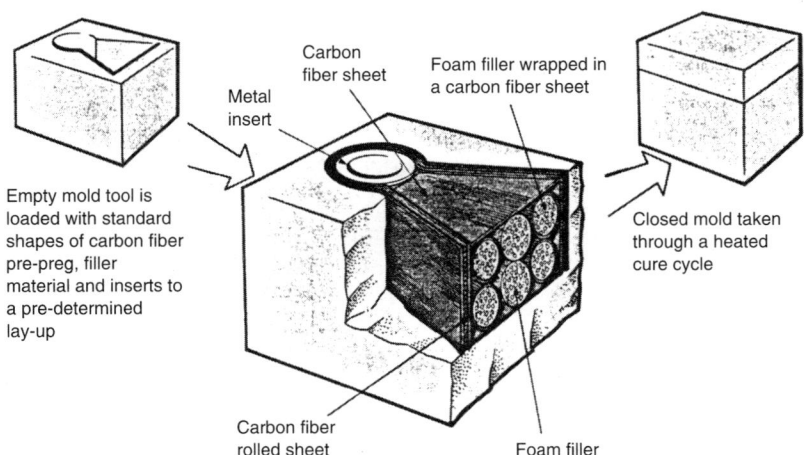

Figure 22.7 Expanded core technique showing detail of carbon fiber prepreg inserts. *Source:* Reprinted with permission from George J, Technology puts carbon fibre into mass markets, *Eureka*, Franks-Hall, 35–37, Mar 1985. Copyright 1985, Franks Hall.

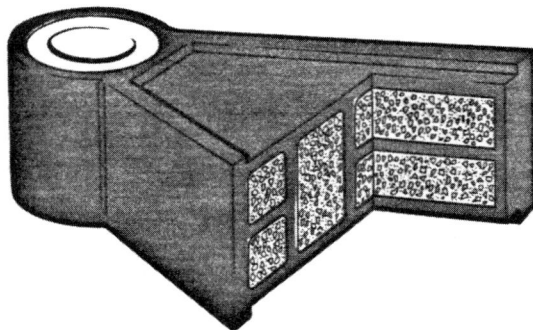

Mold unloaded and finished component deflashed

Figure 22.8 Detail of finished component after deflashing. *Source:* Reprinted with permission from George J, Technology puts carbon fibre into mass markets, *Eureka*, Franks-Hall, 35–37, Mar 1985. Copyright 1985, Franks Hall.

where the filler is an expanding foam based on cork granules combined with a resin and a gassing agent. In the curing process, gas (N_2) is evolved together with water vapor from the cork granules, which force the reinforcement to the extremities of the mold [25].

22.10.2 A Yacht mast

Richard Glanville started a boat building business in South Africa and based his design on the calculations of the inertia of a section of the combined plies (laminate) using a HP handheld calculator. His design was for a Freewing rig with an unstayed, fully rotating wing mast. Its aerodynamic advantage was a very high lift to drag ratio and was efficient and easy to operate, with the mast rotating about a bearing. Glanville now operates Freewing Masts from Clachnaharry Works Lock, Inverness in the UK and all calculations are now undertaken using LUSAS finite element analysis software developed by Finite Element Analysis Ltd., which could, for example, also be used for civil and structural work, including bridges.

The mast, fitted to the Safari catamaran, is 13 m long (from above the boom to the masthead). The chord length (from the leading edge to the trailing edge) at the base is 760 mm and tapers to 240 mm at the masthead.

REFERENCES

1. Dean GD, Turner P, The elastic properties of carbon fibres and their composites, *Composites*, 174–180, Jul 1973.
2. Daniel IM, Ishi O, *Engineering Mechanics of Composite Materials*, Oxford University Press, Oxford, 1994.
3. Jones RM, *Mechanics of Composite Materials*, McGraw-Hill, Washington DC, 1973.
4. Hull D, *An Introduction to Composite Materials*, Cambridge University Press, Cambridge, 1981.
5. University of London, Imperial College of Science, Technology and Medicine, Course entitled *Mechanical Testing of Advanced Fibre Composites*, 1995.
6. Matthews FL, Rawlings RD, *Composite Materials*, CRC Press, Boca Raton, Appendix, 448–455, 2000.
7. Wagoner G, Bacon R, Elastic constants and thermal expansion coefficients of various carbon fibers, *19th Biennial Conference on Carbon*, Penn State University, Session 6A, 296–297, 1989.

8. Sinclair JH, Chamis CC, Prediction of fiber composite mechanical behaviour made simple, *Proceedings of the 35th SPI/RP Annual Technology Conf*, Paper 12A, Society of the Plastics Institute, New York, 1980.
9. Pipes RB, Pagano NJ, Interlaminar shear stress in composite laminates under axial tension, *J Composite Mater Sci*, 13, 2131–2136, 1978.
10. Laval C, Composites design in the real world, *Reinforced Plastics*, 50–53, Sep 2003.
11. Mazumdar SK, *Composites Manufacturing*, CRC Press, Boca Raton, 2000.
12. Mills A, Burley G, Backhouse R, Innovative materials and manufacturing processes for the cost effective manufacture of composite airframe structures, *44th International SAMPE Symposium*, May 23–27, 1999.
13. Lees WA, *Adhesives in engineering design*, The Design Council, London, 1984.
14. Adams RD ed., *Structural Joints in Engineering*, Elsevier, New York, 1983.
15. Kinloch AJ ed., *Structural Adhesives, Developments in Resins and Primers*, Elsevier, London, 1986.
16. Matthews FL ed., *Joining Fibre Reinforced Plastics*, Elsevier, London, 1986.
17. Moussiaux E, Lügering A, Adhesive bonding brings lasting performance, *Reinforced Plastics*, 46–50, Jun 2000.
18. Sauer J, New developments in composite bonding, *Reinforced Plastics*, Feb 16–20, 2001.
19. Kelly A, Mileiko ST eds., *Fabrication of Composites, Handbook of Composites*, Vol 4, Elsevier, Amsterdam, 1983.
20. Chen WC, Some experimental investigations in the drilling of carbon fibre reinforced plastic (cfrp) composite laminates, *Int J Machine Tools Manuf*, 37(8), 1097–1108, 1997.
21. Kim JD, Lee ES, A study of ultrasonic vibration cutting of carbon fibre reinforced plastics, *Int J Adv Manuf Technol*, 12(2), 78–86, 1996.
22. Davies P, Cantwell WJ, Jar PY, Bourban PE, Zysman V, Kausch HH, Joining and repair of a carbon fibre reinforced thermoplastic, *Composites*, 22(6), 425–431, 1991.
23. Campbell FC, Mallow AR, Browning CE, Porosity in carbon fiber composites—An overview of causes, *J Adv Mater*, 26(4), 18–33, 1995.
24. Adams DF, *Test Methods for Composite Materials*, Technomic, Basel, 1990.
25. George J, Technology puts carbon fibre into mass markets, *Eureka*, Franks-Hall, 35–37, Mar 1985.

SUPPLEMENTARY BIBLIOGRAPHY

The following references are listed in chronological order:

1. **Design**

    (i) Tetlow R, Structural engineering, design and applications, Marcus Langley ed, *Carbon Fibres in Engineering*, McGraw-Hill, Maidenhead, 108–159, 1973.
    (ii) Tsai SW, Composites Design 1985, *Think Composites*, Dayton, 1985.
    (iii) West GH, *Engineering Design in Plastics—Data and Applications Guide*, Plastics Research Institute, London, 1986.
    (iv) Richardson T, *Composites, A Design guide*, Industrial Press, New York, 1987.
    (v) Phillips LN (ed.), *Design with Advanced Composite Materials*, The Design Council, London, 1989.
    (vi) Edwards K, *Rethinking the design process, Advanced Composites Engineering*, Sep 1989.
    (vii) Smith RS, *Design of Marine Structures in Composite Materials*, Elsevier, Barking, 1990.
    (viii) Carlsson LA, Gillespie JW, *Delaware Composites Design Encyclopaedia*, Technomic, Basel, 6, 1989–1991.
    (ix) Chou TW, *Microstructural Design of Fibre Composites*, Cambridge University Press, Cambridge, 1992.
    (x) Ashby MF, *Materials Selection in Mechanical Design*, Pergamon, Oxford, 1992.
    (xi) Mayer RM, *Design with Reinforced Plastics*, The Design Council, London, 1993.

(xii) Hancox NL, Mayer RM, *Design Data for Reinforced Plastics*, Chapman and Hall, London, 1994.
(xiii) Quinn JA, *Composites Design Manual*, J Quinn Associates Ltd., Liverpool, 1995.
(xiv) Rosato DV, *Designing with Reinforced Plastics*, Hanser/Gardner, Cincinnati, 1997.
(xv) Owen MJ, Middleton V, Jones IA, *Integrated Design and Manufacture Using Fibre-Reinforced Polymeric Composites*, CRC Press, Boca Raton, 2000.
(xvi) Quinn JA, *Composites Design Manual*, 3rd Edition, James Quinn Associates Ltd., Liverpool, 2002.
(xvii) Quinn JA, *The Design Manual of Engineered Composite Profiles*, Fibreforce, Runcorn.
(xviii) Kelly SA, Zwebor C eds., *Comprehensive Design Materials*, 6 vols, Elsevier Science.
(xix) Akberov SD, Guz AN, *Mechanics of Curved Composites*, Kluwer Academic Publishers, 1999.
(xx) *Composite design, Users guide for short carbon fibre composites*, Zoltek
(xxi) Clarke JL, *Eurocomp Design Code and Handbook* for the structural use of polymer composites—a design code of recommended practice for the design of structures made of advanced composite structures. ASCE Civil Eng. Database, 1997.
(xxii) McAllister LE, Lachman WL, Walter L, Multidirectional carbon-carbon composites, fabrication of composites, Kelly A, St Mileiko eds., *Handbook of Composites*, Elsevier Science Publ Co Inc, 4, 109–175, 1983.
(xxiii) Hyer MWW, *Stress analysis of fiber-reinforced composite materials*, McGraw Hill, 1998.

2. Composite materials

(xxiv) Hollister GSF, Thomas C, *Fibre Reinforced Materials*, Elsevier, London, 1966.
(xxv) Wendt FW, Leibowitz H, Perrone N eds., *Mechanics of Composite Materials*, Pergamon Oxford, 1970.
(xxvi) Broutman LJ, Krock RH, *Composite Materials*, Academic, London, 6, 1974.
(xxvii) Vinson JR, Chou I-W, *Composite Materials and their Structures*, Applied Science, Barking, 1975.
(xxviii) Tsai SW, Hahn HT, *Introduction to Composite Materials*, Technomic, Westport, 1980.
(xxix) Lubin G, *Handbook of Composites*, van Nostrand Reinhold, New York, 1982.
(xxx) Kelly A, Rabotnov YN, *Handbook of Composites*, Vols 1-4, Elsevier Science, Amsterdam, 1985.
(xxxi) Shook G, *Reinforced Plastics for Commercial Composites Source Book*, ASM, Metals Park, Ohio, 1986.
(xxxii) Chawla KK, *Composite Materials-Science and Engineering*, Springer-Verlag, New York, 1987.
(xxxiii) Morley JG, *High–performance Fibre Composites*, Academic Press, London, 1987.
(xxxiv) Partridge IK, *Advanced Composites*, Elsevier, London, 1989.
(xxxv) Lee SM, *Dictionary of Composite Materials Technology*, Technomic, Basel, 1989.
(xxxvi) Ashbee KHG, *Fundamental Principles Of Fibre Reinforced Composites*, Technomic, Basel, 1989.
(xxxvii) Kelly A, *Concise Encyclopaedia of Composite Materials, Advances in Materials Science and Engineering*, Pergamon, Oxford, 3, 1989.
(xxxviii) Middleton DH ed., *Composite Materials in Aircraft Structures*, Harlow, Longman, 1990.
(xxxix) Mallick PK, Newman S eds., *Composites Materials Technology*, Hanser, London, 1990.
(xl) Agarwal BD, Broutman LJF, *Analysis and Performance of Fibre Composites*, 2nd ed, Wiley, New York, 1990.
(xli) Kelly A, *Strong Solids*, 2nd ed, Clarendon Press, Oxford, 1973, 3rd Edn, with MacMillan, NH, 1990.
(xlii) Halpin JC, *Primer on Composite Materials*, 2nd ed., Technomic, Basel, 1992.
(xliii) Talreja R, *Damage Mechanics of Composite Materials*, Elsevier, Barking, 1992.
(xliv) Noakes K, *Successful Composite Techniques*, 2nd ed., Osprey, London, 1992.

(xlv) Chou TW, *Structure and Properties of Composites*, VCH, Weinheim, 1993.
(xlvi) Murphy J, *Reinforced Plastics Handbook*, 2nd ed, Elsevier Advanced Technology, Oxford, 1998.
(xlvii) *Introduction to Composites*, 4th Edition, Composites Institute, Society of the Plastics Industry, New York, NY, 1998.
(xlviii) Pagano NJ, *Mechanics of Composite Materials*, Kluwer Academic Publishers, 1994.
(xlix) Materials. *Handbook* (MIL17), covers polymer matrix composites, metal matrix composites (limited to US citizenship/green card status) and ceramic matrix composites, First three volumes cover polymer matrix composites:

Volume 1   Guidelines for the Characterization of Structural Materials   June 2002
Volume 2   Materials Properties   June 2002
Volume 3   Materials Usage, Design and Analysis   June 2002
Volume 4   Metal Matrix Composites   June 2002
Volume 5   Ceramic Matrix Composites   June 2002
Volume 6   Structural Sandwich Composites   Planned

CHAPTER 23

THE USES OF CARBON FIBERS

Carbon fibers have now been available for some 35 years and in that time, many significant improvements have been made. The replacement of cellulose based precursors with PAN precursors provided carbon fibers with good strength and improved the moduli, followed later by the intermediate modulus PAN based range, whilst the pitch based fibers were available with very high moduli and were truly graphitic. In the 1980s, the US Government made it mandatory for military applications to use PAN precursor made in the USA, which heralded an increase in production of PAN precursor in the US.

As the world political situation changed, the requirement for carbon fibers in military applications has decreased, entailing major defense cuts. The emphasis has shifted to commercial applications, which have grown extensively. Therefore, it is not surprising that carbon fibers are involved in a whole gamut of applications. Developments do occur very rapidly in the composites field and some of these applications may now have been discontinued or replaced, but serve to illustrate the diverse applications of carbon fiber.

23.1 USES OF OXIDIZED PAN FIBER (OPF)

Although this represents the first stage of making carbon fiber from a PAN precursor, the production of opf does not follow exactly the same route and is tailor made for a given end use.

23.1.1 Flameproof applications

The opf, with an oxidized density greater than 1.38g cm^{-3}, is non-flammable and finds many uses as a non-flammable material, which includes a replacement material for asbestos. Initially, a large (e.g. 320k) opf tow is converted into an opf yarn via a modified Worsted spinning route (Figure 21.3), the opf spun yarn can then be woven into a fabric. Typical properties of the opf spun yarns are given in Table 23.1 and are compared with asbestos in Table 23.2.

The measurement of relative flammability can be ascertained by determining the Oxygen Index (ASTM D2863), often referred to as the Limiting Oxygen Index (LOI), when a carefully regulated, known mixture of N_2 and O_2 is passed over the sample, which has been ignited with a 12 mm natural gas, butane or propane flame (Figure 23.1). The LOI can be defined as the highest oxygen value at which the specimen does not burn for more than 3 min, or over a length of 50 mm. In general, LOI values for flammable fibers are <25, whilst flame

Table 23.1 Typical properties of opf (Panox) spun yarns

Yarn count		Tenacity	Breaking strain
w.c.	Nm	cN tex^{-1}	%
1/10	11/1	12.9	15
2/10	11/2	13.5	16
1/20	22/1	13.1	14
2/20	22/2	14.1	14
1/36	40/1	9.4	13
2/36	40/2	10.1	10

Source: Reprinted from RK Carbon Fibres technical literature.

Table 23.2 Typical properties of opf (Panox) spun yarns compared with asbestos

Property	Typical value for Panox oxidized PAN fiber (opf)	Typical value for Asbestos
Diameter (μm)	11	1
Tensile strength (cN tex^{-1})	18	130
Tensile elongation (%)	17	3
Tensile modulus (cN tex^{-1})	100	5000
Density (g cm^{-3})	1.40	2.5
Moisture regain (% at 65% RH and 20°C)	9	2.3 @ 85%RH
Specific heat (J g^{-1} K^{-1})	1.26	1.13
Thermal conductivity (W m^{-1} K^{-1})	0.14	0.21
Electrical resistivity (Ω cm)	10^{11}	10^{15}

Source: Reprinted from RK Carbon Fibres' technical literature.

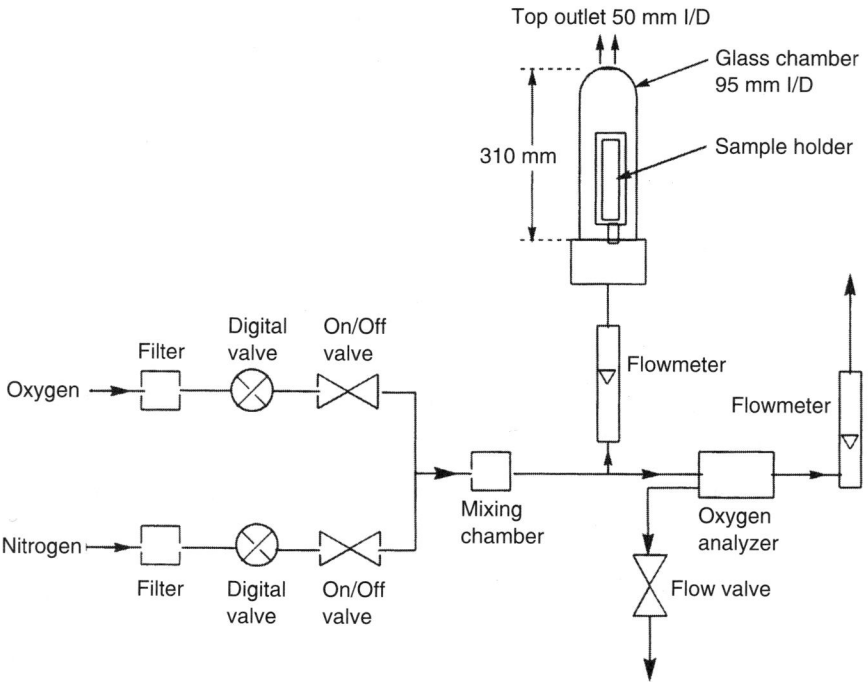

Figure 23.1 Line diagram of LOI apparatus.

Table 23.3 The limiting oxygen indices (LOI) of some materials in fiber form

Fiber		Typical LOI (%)
Oxidized PAN fiber (e.g. Panox)	55	Self extinguishing
Poly(benzimidazole) (PBI)	37	Self extinguishing
Kynol (phenolic)	33	Self extinguishing
Nomex (*meta*-aramid)	28.5	Self extinguishing
Conex	30	Self extinguishing
Kevlar (*para*-aramid)	29	Self extinguishing
Wool	25	Self extinguishing
Polyester	22	Not self extinguishing
Nylon 6.6	21	Not self extinguishing
Urethane foam	17	Not self extinguishing

retardant fibers are 25–28 and flame resistant fibers have LOI values in excess of 28 [1]. Flameproof fibers are completely unaffected by heat. The LOI value can only be used as an indication, since the presence of additives, or the way the sample is supported, can significantly affect the result. The LOI of a number of materials are given in Table 23.3. Although the LOI value for opf is high, when it is admixed with an aramid to improve the abrasion resistance and/or wool to confer a better feel, then this will reduce the overall LOI.

23.1.1.1 Aviation and aerospace [2]

There is no doubt that aircraft are an extremely safe mode of mass transport but, unfortunately, when an accident does occur, it tends to be a major disaster and possibly, the two main sources of fatality are crash impact and fire. The fire may be initiated by an electrical fault, neglected cigarettes, or the ignition of the fuel spilling from the aircraft following the crash. Danger from fire can be a result of the heat of the fire itself, from asphyxiation, or from poisoning by toxic gases released by the combustion process.

The Federal Aviation Authority in the USA commissioned NASA Ames Research Laboratory to examine the problems of aircraft interior flammability and they initially concentrated on limiting the effects of post crash fuel fire. The intense radiated heat ignites curtains, seats and decorative panels and it was reasoned that since the seats contained urethane foam, which will burn and generate asphyxiating gases, any improvement in upgrading the flammability resistance of the seats would provide a significant improvement in safety performance in the event of a post crash fire.

Since it was necessary to retain the inner foam core and the outer decorative fabric, it was decided to use a series of protective layers between the two components that would protect the foam from the effects of the heat.

Companies like Lantor Universal Carbon Fibres [3] have developed special fabrics for aircraft seating without exacting heavy weight penalties (about an extra 1 kg per seat) and a typical system could comprise:

1. Dress cover (outer facing)—100% wool with a flame retardant finish applied to it
2. Charformer/fireblocker layer—a 250–300 $g m^{-2}$ layer of opf has been found ideal for this purpose, acting as a char former, dissipating the heat evolved during combustion by ablation and, as a fire blocker, preventing radiant heat from impinging directly on the underlying structure
3. Air gap layer—acts as an insulating barrier, shielding the core from the heat conducted from the surface, either by the structure of the opf woven cloth or, by using a double layer of opf fabric with a felt backing to give added improvement. The latter shows better results

4. Dissipating layer—a conductive barrier, which can be an Al foil bonded to the felt
5. Flame retardant foam seat—it has been found necessary to use a knitted fabric fitted over the foam to aid placement and subsequent removal of the protective layers above it.

23.1.1.2 Industrial workwear

The opf is able to maintain a barrier against 900°C flame for over 5 min and provides outstanding protection against molten metal splash and welding sparks. The low thermal conductivity provides enhanced protection by reducing the rate of temperature rise through the fabric structure and absorbing and radiating heat. Unlike asbestos, opf does not fibrillate and does not cause a health and safety hazard.

The opf is also used in knitted cuffing, gloves and survival suit fabrics in the offshore industry.

23.1.1.3 Defense and law enforcement

Firefighter's jackets use opf and being fire and petrol resistant, opf can also be used as a protective cover for body armor. The opf can sustain protection for a few minutes and the fire test with Panox protective gear (Figure 23.2) demonstrates this effectiveness.

Figure 23.2 Panotex clothing bathed in fire. Source: Courtesy of Lantor Universal Carbon Fibres Ltd.

23.1.1.4 Transportation and furnishings

The opf based fireblocker fabrics have widespread use as aircraft seating and other forms of seating in cinemas, theatres, airports and conference centers. It is used by Daimler, Iveco and Scania as a firewall and bonnet liner for turbo diesel trucks, which experience a high temperature environment.

23.1.1.5 Cable insulation

The high electrical resistance and ability to withstand heat degradation makes a textile braided opf an ideal material for electrical cable insulation in civil and armed forces applications, such as in a submarine, where fire is a special hazard.

23.1.2 Friction materials

Chopped opf can be incorporated as an asbestos replacement into a phenolic resin to produce friction linings for automotive clutch and brake units.

23.1.3 Gland packings

Companies like AW Chesterton Co. use opf, in the form of 10 or 20 ply spun yarns, which are impregnated with PTFE and/or graphite pastes and when braided, produce a satisfactory packing material, suitable as an asbestos replacement, able to withstand working pressure and sustained working up to 260°C.

23.1.4 Precursor for PAN based carbon fiber and activated carbon fibers

The opf is basically the first stage of the manufacture of PAN based carbon and activated fibers and these are discussed separately.

23.2 USES OF VIRGIN CARBON FIBER

23.2.1 Activated carbon fibers (ACF)

Rebouillat *et al* [4] and Suzuki [5] give good reviews of activated carbon fibers. Traditionally, activated carbon granules are made by the carbonization of a product such as coconut shells, which due to their physical granular form, tend to be difficult to handle and the development of an activated woven cloth by the British Chemical Defence Establishment at Porton Down [6,7] via the controlled heat treatment of a woven rayon cloth offers many advantages. The activated charcoal cloth (ACC) product was made under licence in 1977, by Charcoal Cloth Ltd. One such process used a 1.8 m wide fabric, reducing to about 1.0 m at the end of the process. To aid carbonization, the cloth was treated with a solution of chemicals to confer a measure of flame retardancy. As explained in Chapter 6, there are two forms of flame retardant—one where the flame retardant acts as a catalyst and promotes removal of the —OH groups and the other form, which actually reacts with the —OH groups. Lewis acids such as HCl, $AlCl_3$, $ZnCl_2$ and H_3PO_4, or $(NH_4)_2HPO_4$ (which itself decomposes to a Lewis acid when heated), will catalyse the dehydration of cellulose.

After pre-treatment, the cloth is dried and carbonized up to about 850°C in an atmosphere of N_2, using a heating rate of about 20°C min^{-1} and at this stage, the fabric is extremely brittle and is unable to withstand applied tension or rubbing. At 850–1000°C, steam or CO_2, is introduced to activate the fiber and sweep away the tars. The process of activation helps free the pores from occluded tars to give an apparent pore volume of about 0.5 cm^3 g^{-1}. One problem associated with the use of chemicals is that they can leave a residue, which may be unacceptable for certain specific end uses.

Toho Beslon developed a process to produce activated fiber from PAN using a Fe salt [8,9]. The PAN fiber, usually in the form of a felt or fabric (about 150 g cm^{-2}), is first pre-oxidized under tension in air, or O_2, at 150–300°C, using a lower temperature for a PAN fiber with comonomer content >6%, to prevent individual filaments sticking together. If, however, a Fe compound was initially incorporated in the PAN fiber, then sticking could be averted. A whole range of copolymers can be used and the polymer will contain 1000–2000 repeat units. Higher comonomer content favours easier PAN spinning, making stretching easier and increasing the yield and strength of the activated carbon fiber, but the tendency for filaments to stick will then increase. Either divalent or trivalent Fe compounds can be used, such as $FeCl_2$ or $FeCl_3$. The process can be further improved by undertaking a multistage pre-oxidation treatment. If the final Fe content is above 1%, the fiber is extracted with an organic acid (e.g. acetic, tartaric, or oxalic acid) to reduce the Fe level, preferably to below 0.3%, which prevents subsequent overactivation and loss of strength.

The oxidized fiber can then be treated with a chemical activation reagent such as $ZnCl_2$, H_3PO_4 or HCl at 700–1000°C or, alternatively, gaseous activation can be undertaken in CO_2, NH_3 or steam in the presence of N_2 from 700–1300°C. The product has a high N_2 content (up to 15%) with a specific surface area of some 300–2000 m^2 g^{-1} and a fiber strength of 0.25–0.35 GPa.

Later, Toho introduced up to 0.3% P or B into the precursor [10] to aid processing. Various studies on the preparation of PAN based activated carbon fiber have been undertaken [11,12], including work with a hollow fiber [13]. The study of the activation stage has also been reported [14–17].

Courtaulds originally made a product from viscose rayon called Asgard, believed to be produced from a doped viscose fiber, or by treating the viscose fiber with a solution of diammonium dihydrogen phosphate and urea to give an uptake of 1–2% P on the fiber. This treated material was heated to give a black product, stable at temperatures up to 150°C and able to withstand a 1300°C bunsen burner flame for 4 min. The product could be used alone, blended with aramids, or used as a precursor for the production of activated carbon fabrics.

Activated fibers can also be made from a phenolic resin fiber (Kynol) [18,19] and pitch fiber [5].

Other work [20] used a hybrid of two fabrics, utilizing a PAN or phenolic fiber to supply strength and, yet, can be subsequently activated, whilst the second fiber, normally a cellulosic fiber, can be activated by a different process giving a surprisingly stronger product but, as expected, capable of adsorbing a wider range of molecules.

The activation process confers a high internal surface area to the carbon in a well defined pore structure, which can be determined by measuring the quantity of N_2 which is adsorbed under controlled conditions, known as the Apparent BET N_2 Surface Area [21], measured in m^2 g^{-1}. This internal surface area can be related to the pore volume (cm^3 g^{-1}). The higher these BET values, the greater will be the adsorption capacity.

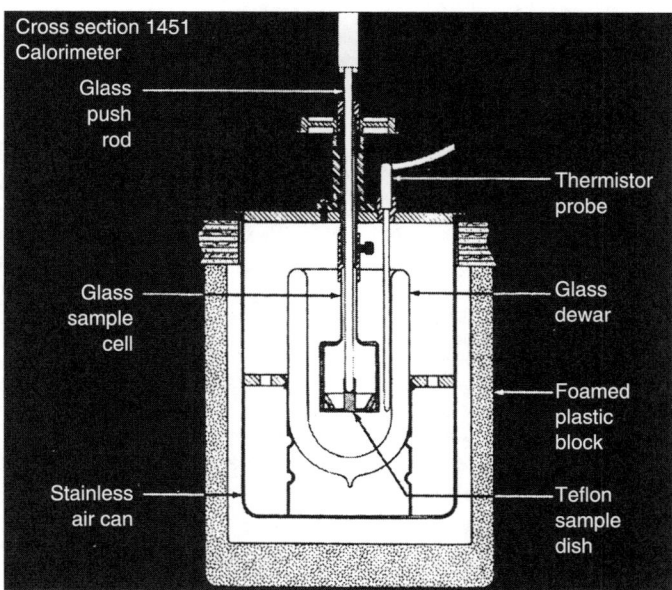

Figure 23.3 Cross section of a Parr Solution Calorimeter. *Source*: Reprinted from Parr Instrument Co., Illinois, USA, technical literature.

As a chemical is adsorbed, its heat content changes and that heat is released as the heat of adsorption, which is known as the heat of wetting. To measure the efficiency of the activation process, it is usual to measure the heat of wetting (HOW) with a Solution Calorimeter (Figure 23.3), which has been calibrated applying the precise and reproducible exothermic reaction of tris(hydroxymethyl)aminomethane (($HOCH_2$)$_3CNH_2$) and $0.1N$ HCl. Typical adsorbates used for assessing the degree of activation are CCl_4, *n*-hexane, benzene, toluene, di-*n*-butyl phthalate and silicone oil (2 cP). Dresselhaus *et al* [22] have described new characterization techniques for activated carbon fibers.

One reason that ACC is so effective is the shape of the pores, which are slit-shaped, long and narrow (90% are <2 nm wide) and able to accommodate quite large molecules, with the ability to permit the escape of pre-adsorbed water and allowing other molecules to be preferentially adsorbed. The time taken to fill the pores determines the effective filter life, whilst the total amount adsorbed determines the saturation vapor adsorption. In most cases, the ACC can be regenerated, without damage to its structure. Typical properties of ACC are given in Table 23.4.

The measurement of HOW alone will not indicate the effectiveness of the ACC. It is the performance measured under dynamic conditions that is the crucial parameter. In general, the ACC [24] can adsorb many organic molecules, especially in the form of vapors. Organic impurities in aqueous solution can be adsorbed as well as phosphates and nitrates. Most importantly, bacteria and viruses are also adsorbed.

Defense applications for ACC include NBC (Nuclear, Biological, Chemical) face masks, respirators, filters in closed air conditioning systems such as tanks, submarines and aircraft, protective clothing and combat gear.

Industrial applications include air conditioning, odor control, solvent recovery, liquid purification such as removal of Cl_2 from water, enzyme and catalyst support with the ability to maintain uniform controllable heat over a large area, recovery of gold and other precious metals from waste liquors (if the ACC is made an anode, the polarity can be reversed to recover the metal), control of premature ripening of products such as bananas (when they ripen during storage, they emit ethylene, which causes fruit in the immediate vicinity to

Table 23.4 Typical properties of Activated Carbon cloth

Property	Typical Value
Construction of cloth	1/1 plain weave
Weight	110 g m^{-2}
Thickness	0.5 mm/layer
Density	0.21 g cm^{-3}
Breaking strength warp	10 Ncm^{-1} width
weft	15 Ncm^{-1} width
Apparent internal surface area	1225 m^{2}g^{-1}
External surface area (microscopic)	1 m^{2}g^{-1}
Pore volume (total)	0.5 cm^{3}g^{-1}
Heat of wetting (di-n-butyl phthalate)	25 Jg^{-1}

Source: Reprinted with permission from Edwards W, *Carbonizable fabrics of activated, carbonized fibers and differently activated or unactivated fibers*, US Pat., 4,714,649, 1987, Freeman JJ, McLeod Al, Nitrogen BET surface area measurement as a fingerprint for the measurement of pore volume in active carbons, *Fuel*, 62, 1090–1091, Sept 1983. Copyright 1983, Elsevier.

prematurely ripen), retardation of tarnishing of metal objects in display cabinets and protection of artifacts from external contamination.

In medical applications, it is used in bandages to remove offensive odors, for the control of odor with permanently artificial openings in the body (ostomy), reduction of adsorption of anesthetics in face masks used by operating room personnel and for blood purification.

23.2.2 Molecular sieves

Oak Ridge National Laboratory (ORNL) has developed a carbon fiber composite molecular sieve designed specifically to absorb CO_2 emitted from coal fired power plants and gas turbines. Petroleum pitch based chopped fiber is bonded with a phenolic resin and activated in steam, O_2, or CO_2 at 850°C, to form a product with a large surface area and pore volume with mesopores of 2–50 nm, capable of absorbing CO_2. There are also macropores (50–100 μm) which allow sufficient fluid flow with low pressure drop. It also has potential to be used for removal of CO_2 from natural gas for fuel cells. [25].

23.2.3 Catalysts

A porous carbon fiber carbon composite with a density >0.2 g cm^{-3} with a significant volume of mesopores (2–50 nm) and macropores (50–100 μm) which allow excellent fluid flow with minimal pressure drop has potential as catalyst support. Fortafil P200 PAN based carbon fiber was slurried in water with a phenolic resin, vacuum molded, dried at 50°C, cured for 3 h at 130°C and carbonized in a flow of N_2 at 650°C. In this process, O_2, CO_2, H_2O and CO were trapped inside the micropores during carbonization and gasified the surface of the PAN fibers, resulting in large surface areas (572 m^2 g^{-1}) and mesopore volumes (1 cm^3 g^{-1}), a possible candidate for a catalyst support [25–27].

23.2.4 Biomedical applications

The use of carbon materials and carbon fiber reinforced composites for medical applications is examined [28,29]. Carbon fibers are first considered, looking at their physical and chemical properties *in vitro* (including the mechanical properties of carbon braids); the properties and histology of two commercial carbon fibers (Torayca M-40-3000; AGH-IWCh) *in vivo* (within a living organism); and the histology of carbon cloth *in vivo*. The use of carbon-carbon composites (amorphous when using phenol-formaldehyde resin as the matrix precursor; crystalline when using pitch as the matrix precursor) in medical applications is then examined.

Any material used in a surgical medical application has to be proved safe and the material to be implanted for an extended period of time must be non-toxic, non-carcinogenic and unaltered by the body environment. Where a mechanical movement is involved, the implant must have good fatigue resistance and be unaffected by corrosion due to body fluids and the implanted material must not suffer a rejection process.

Jenkins, at the University Hospital of Wales, Cardiff worked on implants [30,31] for the heel tendons of sheep and rabbits and knee ligaments for sheep using HM carbon fiber, chosen because of its higher purity and probable better biocompatability. A $10k$ tow was twisted and doubled and subjected to pressure steam sterilization prior to use. The results were quite successful and after two months, it was virtually impossible to tell which limb the sheep had been operated on, as there was no tissue rejection and the filaments acted as a scaffold for collagenous growth. After 3 months, the twisted tow was completely engulfed in new tissue (neotendon), which forced apart the filaments in the tow and assisted the transfer of load from the carbon fiber filaments through to the newly formed collagen tissue. The cells tended to grow in a spiral fashion along the longitudinal axis of the individual filaments. Since the HM fiber was difficult to manipulate, it was replaced with Type A fiber, which was stronger, more readily handled and exhibited no ensuing biocompatability problems.

Veterinary applications were pursued at the University of Bristol on racehorses. When a racehorse ruptured a tendon, the effect was far reaching and usually resulted in the horse being destroyed, so there was every encouragement to establish whether carbon fibers could induce tendon and ligament growth in racehorses. The work of Goodship *et al* [32] was most successful. They used A-S fiber if the implant was required to be mechanically functional and HM-S fiber otherwise. The size of the lesion determined whether $2 \times 10k$, or $4 \times 10k$ twisted tows were used, implanting the twisted tows deep into an incision in the tendon, ensuring that they were completely buried. A single twisted tow was implanted, at the same time, into the other limb to combat compensatory weight transfer from the injured limb. The horse could walk freely after one week, the swelling reducing with time and exercise. The horse could resume training after nine months and was so successful that two thirds of the treated horses were able to race again. The HM-S fiber produced less swelling in the post-operative period and the smaller diameter of the HM-S fiber may have had a beneficial effect. More recent work with racehorses has been reported by Reed and co-workers [33].

A hernia repair with carbon fiber [34], ligament repairs [35–38] and the introduction of carbon fiber pads in the knee [39–41] have also been reported. One problem with the knee pad is that the carbon fiber does tend to get ground away and not all surgeons favour carbon fiber implants, preferring to use items from a tissue bank.

Park and Vasilos [42] fabricated carbon fiber reinforced calcium phosphate composites, made by hot pressing to give a ceramic with significantly improved ductility albeit with a slight decrease in ultimate flexural strength. The failure strains of the composites and the monolithic calcium phosphate ceramics were 0.36 and 0.21%, respectively. It was demonstrated that carbon fiber reinforced calcium phosphate composites would be good biomaterials for bone replacement.

23.3 ELECTRICAL APPLICATIONS

23.3.1 Electrical conduction

In the 1960–1970 era, when PAN carbon fiber was first introduced, obviously creature comfort was paramount and carbon fiber was introduced into toilet seats as a heating element. It was also used in wall panels for room heating. The modern equivalents of these end uses are a Gorix electro-conductive textile, a PAN based woven carbon fiber capable of acting as a flexible heating element, portable heating unit, large area temperature sensor, an electrical switching function, temperature management system, warning and control devices. Gorix is used for self heating diving suits (regulated at 34±0.2°C), controlling the temperature without a thermostat and without hot spots across an entire area. A typical suit comprises a strip on each forearm to keep the hands warm, a further pair sewn into the calf sections to keep the feet warm and a fifth patch sewn into the lower back to warm the kidneys.

Gorix used in car seats enables uniform heat to be applied and controlled with a slider. The idea has been considered for possible use in carpets as a replacement for central heating. Gorix can also be used for clothing, heater beds and blankets, heated spinal injury board, alpine recovery stretcher and thermal boots/gloves.

Carbon fiber can be used for carbon fiber brushes in electrical apparatus [43].

23.3.2 Tailored resistance carbon fiber

It is possible to make carbon fiber with a range of resistivities, so that a fiber can be custom made at a given process temperature to provide the requisite resistivity for a given application. A list of typical resistivities is given in Chapter 20. This type of fiber can be used as a low signature type (Radar Absorbing Materials, RAM) to avoid detection by radar [43a]. Since the fiber is lightweight and does not corrode, it can be advantageously used to leak a charge on a power line. In cities, television ghost images due to multiple reflections can be solved by coating with an absorbent layer. The addition of about 5% to brake pad formulations can significantly reduce the wear and improve the life of the brake pad.

23.3.3 Cathodic protection

The addition of 0.53–1.1%v/v short (5 mm) isotropic pitch carbon fiber in mortar applied to steel reinforced concrete decreased the contact resistivity and volume resistivity of the new mortar, enabling a satisfactory electrical contact material to be made for the cathodic protection of steel reinforced old mortar or concrete [44].

23.3.4 Elimination of static

Carbon fiber has been incorporated in vinyl tiles to dissipate static and is particularly useful for hospital flooring. Also, to assist dissipation of static, carpets have been manufactured containing carbon fiber in the backing material and computer casings made from a molding compound containg carbon fiber.

23.3.5 Electrodes

This application is hardly likely to use much carbon fiber, but a most interesting application is connected with a micro-ion-to-phoretic assembly, commonly manufactured from five or seven multi-barreled assemblies, where the glass tubes are fused by gripping the ends in chucks and pulling the assembly in a heated zone, whilst rotating one end to fuse the individual lengths of glass together. Recently *theta* or quadrant glass blanks have become available and are more convenient. If carbon fiber is incorporated in the tube, this provides a conductive path, giving excellent signal to noise ratio when recording extracellular recording of neuronal firing. J Millar and others [45–49] have described a technique for making these electrodes by wetting the fiber tow with acetone to facilitate pull-through and inserting a connecting wire, effecting good electrical contact with an electroconductive silver glue (Figure 23.4). A tip of a Carbostar-7 electrode is shown in Figure 23.5 and the correct tip length is achieved by spark etching [50,51]. The stiffness of the carbon fiber provides adequate stability and the sharp tip permits easy tissue penetration. The extracellular spikes are typically a few hundred microvolts in amplitude and are generated by action potentials across the membranes of neurons. The great advantage is that the activity of the neurons can be recorded without damage to them.

These carbon fiber electrodes can also be used for voltametric analysis of transmitters *in vivo*, but the electrodes must not be spark-etched as they become electrically noisy, instead they can be plated with silver [52].

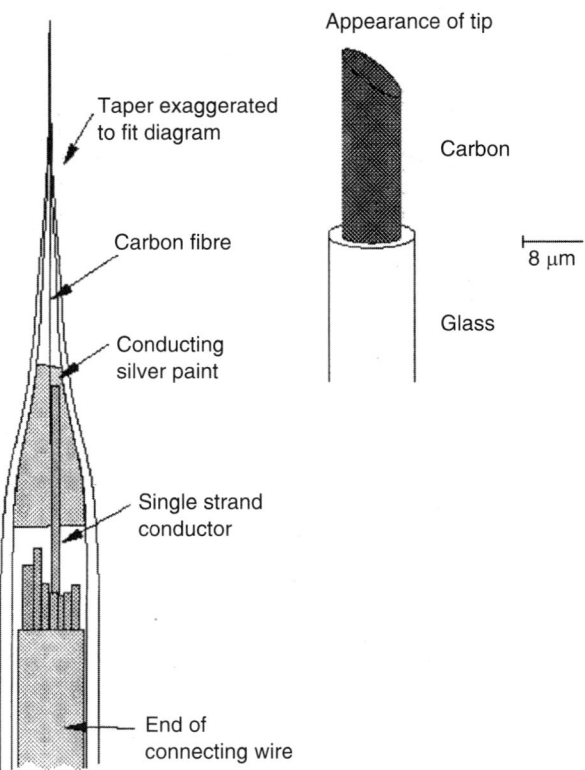

Figure 23.4 Single barrel carbon fiber microelectrode. *Source*: Reprinted from Millar J, Dept. of Physiology, Queen Mary & Westfield College, London, with kind permission.

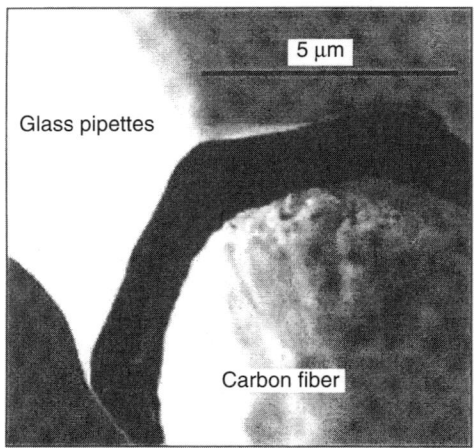
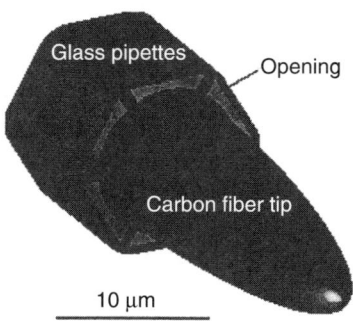

Figure 23.5 Carbostar -7 electrode based on SEM. Carbon fiber is closely surrounded by six fused-together micropipettes forming a seal around the protruding carbon fiber allowing combined micro-iontophoresis and extracellular recording. *Source*: Reprinted from Kation Scientific, Minneapolis, USA, technical literature.

23.3.6 Batteries

A battery is an electrochemical device used to store chemical energy and release it in the form of electricity and comprises two dissimilar metals forming the anode and cathode, which are immersed in an electrolyte that transmits ions (Figure 23.6).

Japan manufacturers dominate the rechargeable battery market with a 75% share of Ni-Cd and 99% of Li ion.

23.3.6.1 Lithium ion batteries

Lithium is a light metal and can deliver a relatively high voltage (3.6V), with a high capacity, holding twice the power when compared with a Ni-Cd or Ni-metal hydride (NiMH) battery on a weight basis and one and a half times on a volume basis. Moreover, it has no memory effect.

A problem initially experienced with rechargeable Li batteries was that during a recharge cycle, the Li metal tended to form dendrites (metal whiskers) that shorted the electrodes and limited the number of charge/recharge cycles. It also had an incipient fire risk [53].

THE USES OF CARBON FIBERS

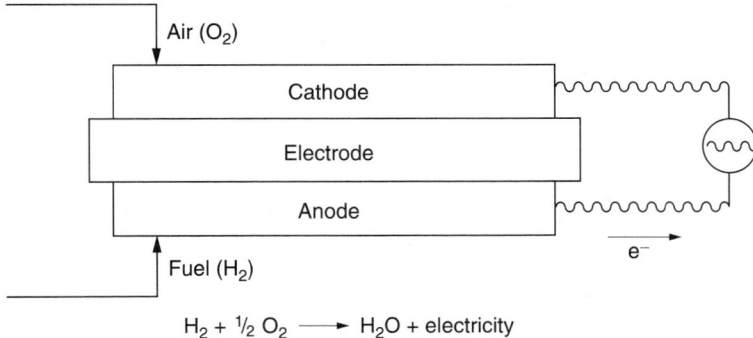

Figure 23.6 Simple fuel cell.

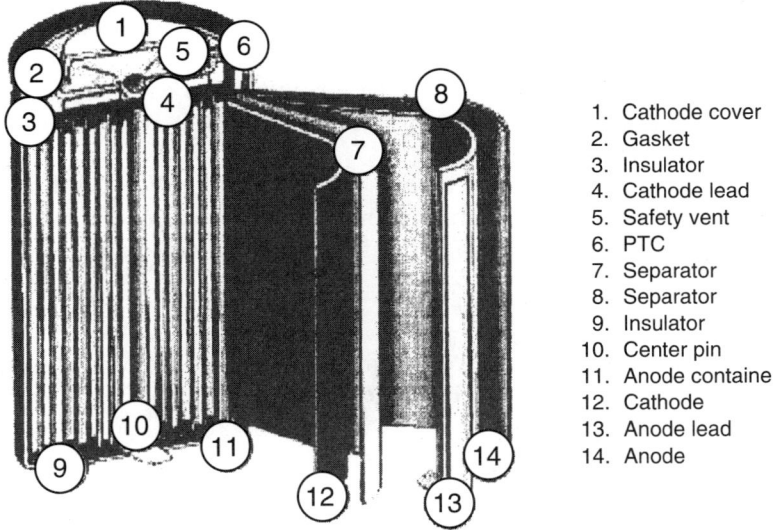

1. Cathode cover
2. Gasket
3. Insulator
4. Cathode lead
5. Safety vent
6. PTC
7. Separator
8. Separator
9. Insulator
10. Center pin
11. Anode container
12. Cathode
13. Anode lead
14. Anode

Figure 23.7 Construction of a Li ion battery. *Source*: Reprinted from Sony technical literature.

The problem was overcome by the development of Li ion cells based on Li intercalation compounds, which can donate or accept Li ions without depositing Li or any other solid.

Initially, Li batteries used a liquid electrolyte, necessitating the use of a robust case for safety. It is now used in the ionized form. Figure 23.7 shows a typical Li ion cell utilizing a Li_2O cathode and a carbon compound anode separated by a microporous membrane, using a non-aqueous electrolyte such as a Li salt dispersed in a mixture of alkyl carbonates. Since the non-aqueous electrolytes can be flammable, Valence Technology has developed the Li ion polymer battery using liquid lithium ion electrochemistry in a matrix of conductive polymers that eliminate free electrolyte within the cell.

Lithium ion cells are manufactured by AEA Technology plc, AGM Batteries Ltd., BYD Battery Co., Electrofuel, Hitachi/Maxell, Lithium Technology, Matsushita Battery, Sanyo, Sony Energytec, Yardney and Yuasa/ACEP. The batteries are used for cell phones, laptop computers and other portable electronic devices.

A chemical reaction affecting cell safety can occur between a Li containing carbon and an electrolyte at high temperature under conditions of mechanical or electrical abuse.

MacNeil and Dahn, at Dalhousi University, believe that carbon fiber is a good candidate for the anode in Li ion cells, with the proviso that an efficient way of packing the fibers needs to be found which should offer optimum geometry for safety and performance.

Amorphous carbon fiber, appropriately heat treated, has a high discharging capacity for anode material in Li ion batteries. Pitch based carbon fiber has been used for anodes in Li rechargeable batteries [54,55].

Florida Atlantic University is investigating carbon fiber materials as potential anodes for reversibly storing Li, with the objective of developing a binderless electrode structure. Also work is on to use carbon fiber as the positive electrode material in which both cations (Li^+) and anions are reversibly inserted in carbonaceous materials.

23.3.7 Fuel cells

Fuel cells have become a most promising new technology for providing energy and a rapid growth in usage is predicted. A fuel cell is an electrochemical energy device that converts hydrogen fuel, in the presence of O_2 from the air, into electrical energy, heat and water. The attributes of fuel cells are zero emissions of particulates, CO_2, CO, NO_x, as well as being virtually noiseless (60 dB at 30 m).

Hydrogen is the primary fuel source for a fuel cell and can be used directly or derived from a hydrocarbon fuel such as natural gas, methanol and hydrocarbon fuels through the process of reforming [56]. The reforming process converts the fuel source to H_2 using steam:

$$CH_4 + H_2O \leftrightarrows CO + 3H_2$$

and can be undertaken external to the cell, or internally, where the products introduced into the cell are electrochemically oxidized. Hydrogen can also be derived from organic matter of plant and animal origin (biomass).

The purity of H_2 supplied to a given fuel cell system will depend on that particular fuel cell technology. Each cell is limited to about 1.23 V, so to achieve the necessary power requirement, a number of cells are stacked and a collection of these fuel cell stacks can be used to produce a power plant. The conversion of the chemical energy of a fuel into electrical energy without combustion is quite efficient and remarkably clean.

In the search for a fuel cell for a new generation of vehicles, there has been a concerted effort for the development of partnerships, with vehicle manufacturers working closely with government departments and research organizations to overcome technical barriers with innovative approaches to produce fuel cells suitable for use in concept vehicles, such as the Daimler Chrysler Jeep commander, Ford P2000 and the GM Precept. Simultaneously, there have been improvements in the clean-up of emissions like CO and S, which are also pertinent for running the fuel cells more efficiently. In an automobile, H_2 can be stored as a metal hydride (which is the simplest and safest), or as a cryogenic liquid (but this will occupy three times the volume of gasoline) and finally, as a compressed gas (which is expensive). Since there is no established distribution system for H_2, fuel cell cars will initially use a catalytic reformer to generate H_2 from a liquid fuel such as methanol.

If a reformate is used, catalytic converters can remove emissions which cause smog, but CO_2 cannot be readily removed from on board the vehicles.

Carbon fibers have featured in several aspects of this work, but due to its proprietary nature, only brief details are available. Certainly, by looking at the recent patent literature, the field of using carbon fiber for fuel cells is very active—The University of California have used treated carbon fibers to improve performance [57]; Hyperion Catalysts have used

THE USES OF CARBON FIBERS

carbon fibrils for Li battery electrodes [58]; Sandia Corp describe a method of preparing carbon materials for use as electrodes in rechargeable batteries [59]; Mitsubishi Chemical detail an electrode material for a non-aqueous solvent secondary battery [60]; Mitsubishi Gas Chemical describe a method of producing isotropic pitch carbon fibers and carbon materials for non-aqueous solvent secondary battery [61]; and Petoca describe the use of mesophase pitch based carbon fiber for use as the negative electrode of a secondary battery [62].

There are basically five types of fuel cells, which are characterized by the type of electrolyte used [63].

23.3.7.1 Alkaline Fuel Cell (AFC)

This type of fuel cell uses 35–50% KOH as the electrolyte at about 70–100°C (Apollo used 85% KOH at 250°C) [64]. Water is produced at the anode twice as fast as it is used up at the cathode. The cathode is the electrically positive terminal and it is towards this positive terminal that the electrons flow.

$$\text{Anode} \quad H_2 + 2(OH)^- \rightarrow 2H_2O + 2e^-$$
$$\text{Cathode} \quad \tfrac{1}{2}O_2 + H_2O + 2e^- \rightarrow 2(OH)^-$$
$$\text{Cell} \quad H_2 + \tfrac{1}{2}O_2 \rightarrow H_2O$$

A schematic cell is shown in Figure 23.8. Since the KOH electrolyte absorbs CO_2, it is necessary to use a pure H_2 source, so a reformate cannot be used. On-board H_2 storage causes no emissions and alkali doped carbon nanotubes have a high storage capacity.

Alkaline cells are made by Alternative Fuel Systems Ltd., Slinfold, UK.

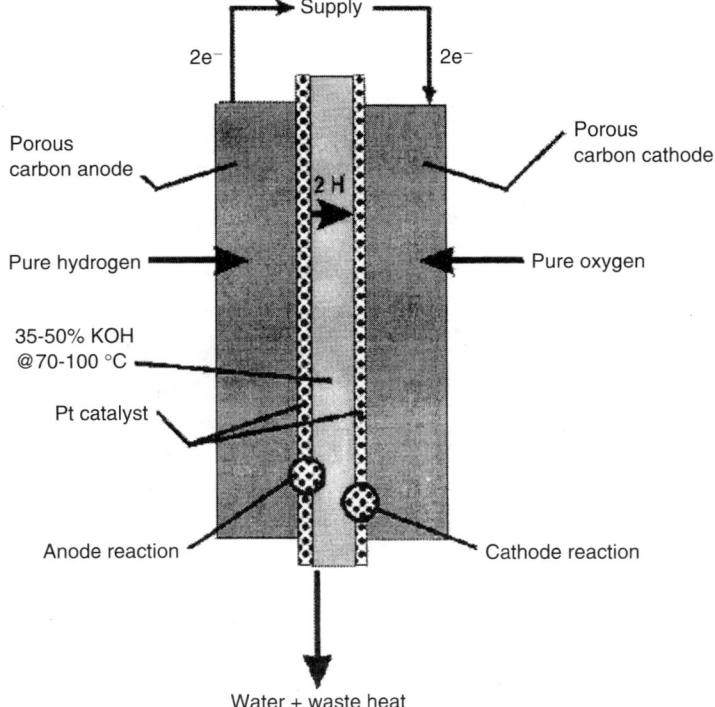

Figure 23.8 Alkaline fuel cell.

23.3.7.2 Proton Exchange Membrane Fuel Cell (PEMFC)

This is also called the Solid Polymer Fuel Cell (SPFC) and the Direct Methanol Fuel Cell (DMFC) is included in this classification. These cells use a solid perfluorinated sulfonated polymer ion exchange membrane (e.g. DuPont Nafion) [65] in the form of a thin plastic film, which serves as the electrolyte in the PEM fuel cell operating at 50–100°C.

$$-[(CF_2 CF_2)_n - CF_2 CF]_x-$$
$$|$$
$$OCF_2 C FOCF_2CF_2 - SO_3H$$
$$|$$
$$CF_3$$

Dow Chemical Co., Asahi Chemical Co. and Chloride Engineers Ltd. make a similar product. The polymer membrane only conducts H^+ when fully hydrated and one solution to this problem has been to use carbon fiber wicks (Figure 23.9). SGL Carbon is making fiber based gas diffusion layers in a roll form. The cell is sensitive to low levels of CO, which can be removed by a Pt/Ru catalyst, but the high cost restricts use. However, a later technique to increase the surface area has reduced the cost by a factor of 75.

The cell must be operated below 100°C to ensure that the by-product H_2O does not evaporate faster than it is produced and the membrane is hydrated. A proton exchange membrane permits the H^+ generated at the anode to pass through to the cathode where they combine with O_2 to form H_2O.

The reactions can be represented by:

Anode $H_2 \rightarrow 2H^+ + 2e^-$ (made possible with a Pt/Ru catalyst)
Cathode $\frac{1}{2}O_2 + 2H^+ + 2e^- \rightarrow H_2O$
Cell $H_2 + \frac{1}{2}O_2 \rightarrow H_2O$

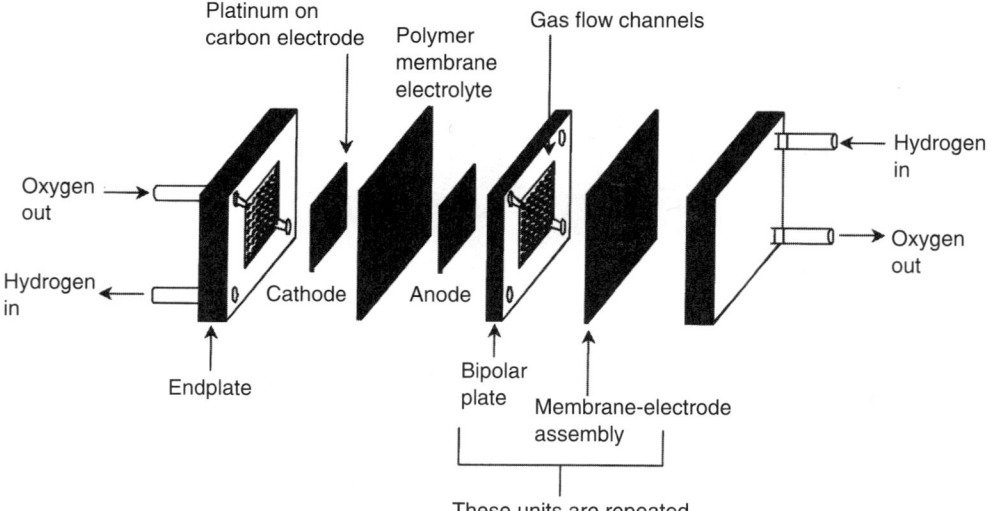

Figure 23.9 PEM fuel cell.

PEM fuel cells are manufactured by Allied Signal Aerospace, Analytic Power, Energy Partners, Fuel Cell Energy Inc., H Power Corp., Ida Tech, International Fuel Cells and Proton Energy Systems Inc. in the USA; Ballard Power Systems, in Canada (the world leader in PEM fuel cells); Daimler Benz and Siemens AG in Germany; Alternative Systems Ltd. in the UK and de Nora S.p.A in Italy.

23.3.7.3 Phosphoric Acid Fuel Cell (PAFC)

This cell uses stabilized phosphoric acid on a SiC matrix as the electrolyte, operating at 160–210°C. Platinum or alloys of Pt are used as the catalyst at both electrodes supported on carbon black. The fuel must be H_2 rich and contain <2% CO. The efficiency is 36–89%.

$$\text{Anode} \quad H_2 \rightarrow 2H^+ + 2e^-$$

$$\text{Cathode} \quad \tfrac{1}{2}O_2 + 2H^+ + 2e^- \rightarrow H_2O$$

$$\text{Cell} \quad H_2 + \tfrac{1}{2}O_2 \rightarrow H_2O$$

A typical phosphoric acid fuel cell is shown in Figure 23.10.

Water is produced at the cathode, whereas in the alkali fuel cell, it is produced at the anode. At present, a 200 kW PAFC costs about $3000 per kW. The leading manufacturer is International Fuel Cells (Toshiba/United Technologies).

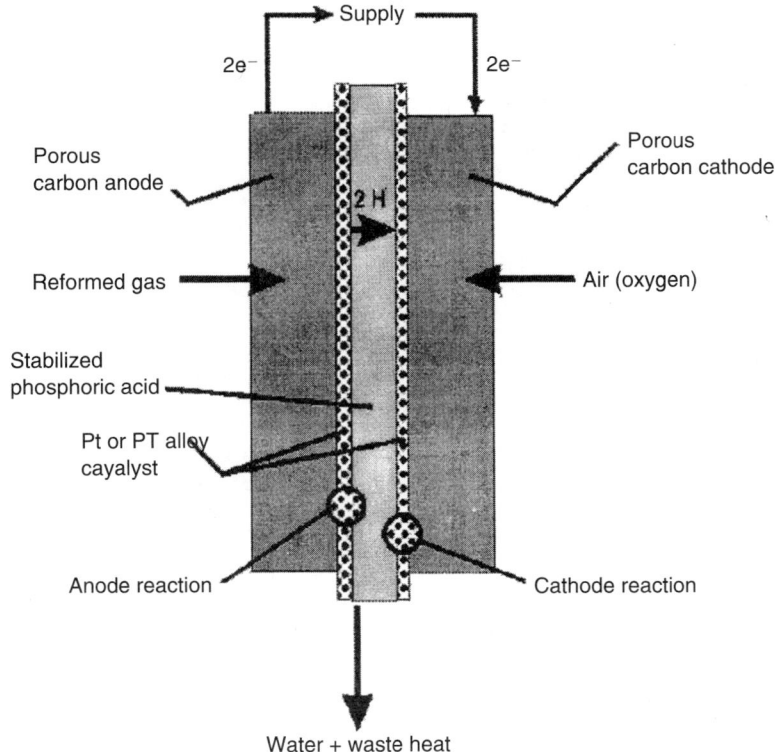

Figure 23.10 Phosphoric acid fuel cell.

23.3.7.4 Molten Carbonate Fuel Cell (MCFC)

This type of cell uses lithium, sodium and/or potassium carbonate electrolyte soaked in a ceramic matrix of $LiAlO_2$ and operates at 650°C, a temperature that is necessary to achieve sufficient conductivity of the electrolyte, but with the dual benefits of achieving a higher efficiency (60% and 85% if the waste heat is used) and not requiring a noble metal catalyst. However, corrosion due to the molten carbonate is a problem.

The high working temperature eliminates the need for a separate reformer and CH_4 can be steam reformed, where the reaction occurs simultaneously with the electrochemical oxidation of the H_2 within the anode compartment. The heat required for steam reforming is supplied by the heat generated in the cell reaction and improved conversion is favoured by operating the process at 5×10^5 Nm^{-2}. A schematic layout of a MCFC is shown in Figure 23.11 showing deployment of the Ni catalyst.

The electrochemical reactions which occur are:

$$\text{Anode} \quad H_2 + CO_3^{--} \rightarrow H_2O + CO2 + 2e^-$$

$$\text{Cathode} \quad \tfrac{1}{2}O_2 + CO_2 + 2e^- \rightarrow CO_3^{--}$$

$$\text{Cell} \quad H_2 + \tfrac{1}{2}O_2 + CO_2 \text{ (atcathode)} \rightarrow H_2O + CO_2 \text{ (atanode)}$$

It is normal practice to recycle the CO_2 generated at the anode to the cathode, where it is consumed.

The commercialization of this type of electrode is being followed by Energy Research Corporation (ERC), International Fuel Cells Corporation and M-C Power Corporation (MCP)in the USA; Brandstofel Nederland (BCN), Deutsche Aerospace AG and Ansaldo (Italy) in Europe; and Hitachi, Ishikawajima-Harima Heavy Industries, Mitsubishi Electric Corporation in Japan.

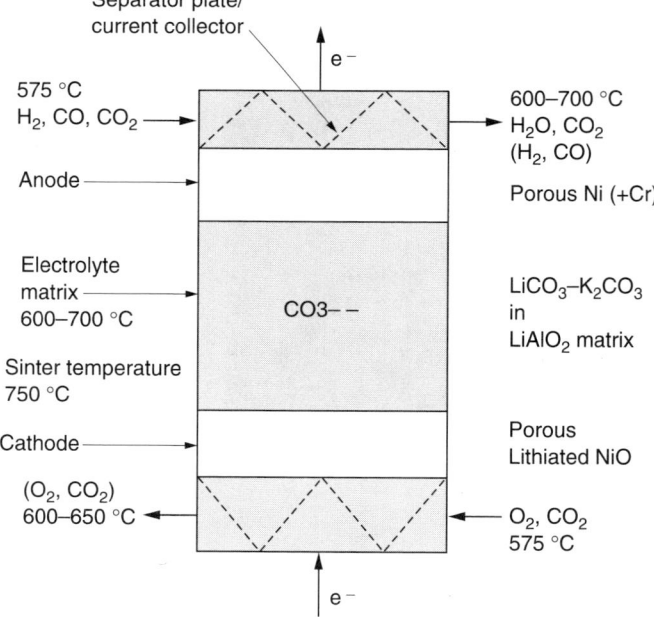

Figure 23.11 Schematic layout of a molten chloride fuel cell (MCFC).

23.3.7.5 Solid Oxide Fuel Cell (SOFC)

This is a cell where the electrolyte is Y_2O_3 stabilized with ZrO_2, operating at 800–1000°C. SOFCs have an efficiency similar to MCFCs. The high operating temperature allows internal reforming, but suitable materials of fabrication are an issue. The CO is generated from the water gas shift reaction that occurs within the cell.

The reactions occurring are:

$$\text{Anode} \quad H_2 + O^{--} \rightarrow H_2O + 2e^-$$

$$\text{Cathode} \quad \tfrac{1}{2}O_2 + 2e^- \rightarrow O^{--}$$

$$\text{Cell} \quad H_2 + \tfrac{1}{2}O_2 \rightarrow H_2O$$

Westinghouse Electric Corporation manufacture solid oxide fuel cells and Figure 23.12 shows a schematic cross-section of such a cell, typically tubular in design. The original design had a porous zirconia support tube which was replaced by using the air electrode (cathode) as the basic support. Pressurizing the cell increases the efficiency to about 70%. Westinghouse has designed a 100 kW unit comprising 1152 tubular cells 150 cm long and 22 mm in diameter. Flat plate designs are under investigation in the USA by Allied Signal Aerospace Co., Ceramatec Inc. and Ztek Inc.

23.3.7.6 Carbon fiber in fuel cells

Fuel cells have a great potential for energy saving and provide a cleaner energy conversion, but at present are too expensive (about $20 per W). For a fuel cell using H_2, the efficiency decreases with increasing temperature. The O_2 (air) supply must be free from compressor oil entrainment.

Each type of fuel cell tends to be specific to a given manufacturer and information on the construction is highly proprietary. Ogawa and Shimizaki [66] studied the effects of carbonization and surface treatment conditions on the performance of redox flow battery using PAN based carbon fibers.

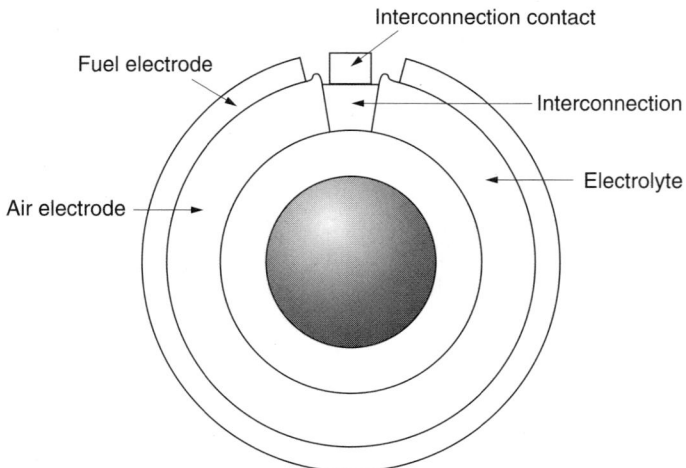

Figure 23.12 Westinghouse solid oxide fuel cell (SOFC).

Carbon fiber can certainly be used for PEM fuel cells. Lydall have introduced Lyflex GDL (gas diffusion layer), a flexible microporous carbon nonwoven, for PEM fuel cells [67]. Graftech have patented a flexible graphite composite material for a PEM fuel cell electrode.

A combination of Pt/C/Teflon has been coated on a carbon cloth and hot pressed to a membrane obtaining 0.45 $g\,cm^{-2}$ Pt [68] and used in PEM cells.

The Proton Exchange Membrane fuel cell manufactured by Ballard Power systems comprises of a stack of some 100–200 carbon plates acting as electrodes and ionizing H_2 and O_2 at 90°C, which powers the fuel cells. The intended application is in trucks and automobiles.

Toray market a porous carbon paper and carbon cloth which can be pretreated with Teflon in a hydrophobic form and can then be used directly for electrode manufacturing. It is able to conduct a current, support a catalyst and allow gases and liquids to pass through.

23.4 THERMAL INSULATION

Companies like Fiber Materials Inc. (Fiberform) and Calcarb in the UK have developed carbon bonded carbon fiber (CBCF) insulation board made by mixing chopped viscose rayon fibers about 2 mm long together with ground reworked CBCF, into an aqueous slurry, adding a phenolic resin and binder. It is then filtered and vacuum-formed into slabs, dried and carbonized at 1000°C. The process gives a 50% yield and is followed by a further heat treatment to a temperature about 200°C above the maximum intended use temperature to minimize outgassing (Figure 23.13). The final product has a carbon content >99.9% and the insulative properties are governed by the porosity. The properties of CBCF are given in Table 23.5.

The material must not be used in the presence of air above 200°C. These products are particularly useful for vacuum furnace insulation [70] and can be used in vacuum or in the presence of an inert gas up to 2750°C.

Carbon felts can be used as an insulation material, made from scrap material which is relatively cheap, where further carbonization continues in use. Under these conditions, the possibility of shrinkage and outgassing products has to be accommodated. Superior, but expensive, carbon and graphite felts can be made by carbonizing long small diameter organic filaments, which are very stable and have low shrinkage.

In aircraft, fire protection can be offered to the cabin by using a layer of insulating batting between the aircraft skin and the cabin interior panels. The fuselage insulation is placed in bags to resist moisture penetration and the bags are fitted between the outer skin and interior trim panels.

Dow Chemical Co. developed a high-performance product based on PAN carbon fiber type materials using non-conducting carbonaceous curly fibers. The technology produced permanently bulked carbonaceous fibers with 65–85% carbon and bulked carbon fibers with over 85% carbon, which could be used as the basis for lightweight (3.2–16 $kg\,m^{-3}$) fire retardant insulation and for fire blocking panels.

Some 15% by weight of the curly fiber in a polyester batting will render the mixed batting self-extinguishing, with a density of 12.8 $kg\,m^{-3}$, although the effect will be dependent on the batting density—a lighter weight batting would require more of the curly fiber. Dow termed the product EDF® (Experimental Dow Fiber) [71–74] and an exclusive licence to this technology was purchased by RK Carbon Fibres in 1991. RK continued development and changed the name to Curlon® (for the bulked version) and Lineon® (for the straight fiber) and processing was subsequently transferred to the Orcon Corp., San Francisco. Typical properties of Curlon and Lineon fibers are given in Table 23.6.

THE USES OF CARBON FIBERS

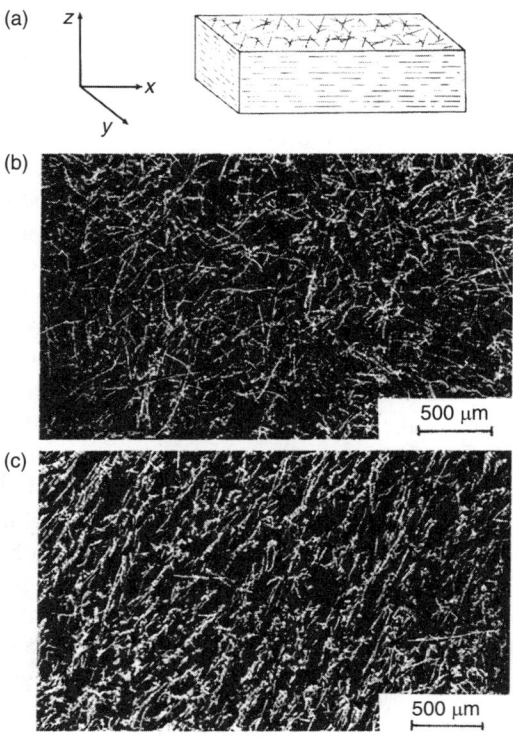

Figure 23.13 Orientation of the fibers in carbon bonded carbon fiber material. (a) Schematic representation with the axis system used to denote planes and directions, (b) Micrograph in a plane perpendicular to the application of vacuum i.e. xy plane, (c) Micrograph in a plane parallel to the application of vacuum e.g. zx with x direction at 55° to the bottom of the micrograph. *Source*: Reprinted from Davies IJ, PhD thesis, London, 1992.

Table 23.5 Properties of CBCF (density 0.17 g cm^{-3})

Property	x/y direction	z direction
Young's modulus (MPa)	105.6	9.57
Compressive strength (MPa)	0.78	0.60
Flexural strength (MPa)	1.03	0.15
Tensile strength (MPa)	0.48	0.08
Coefficient of thermal expansion 0–1000°C (K^{-1})	3.0×10^{-6}	2.8×10^{-6}
Electrical resistivity (Ωm)	1.1×10^{-3}	4.1×10^{-3}

Source: Reprinted from Calcarb Ltd. technical literature and Matthews FL, Rawlings RD, *Composite Materials*, CRC Press, 152, 2002.

Table 23.6 Typical properties Curlon and Lineon fibers

Property	Value
Filament diameter (μm)	9
Tensile strength (GPa)	0.6
Tensile modulus (GPa)	20.0
Breaking strain (%)	3.0
Density (g cm^{-3})	1.5

Source: Reprinted from RK Carbon Fibres technical literature.

The fiber offers fire resistance and high thermal insulation together with low smoke emission, low electrical conductivity and a weight saving. Current applications include aircraft fuselage thermal insulation, aircraft fire blockers, fire protective clothing, personal insulation and fire retardant insulation boards for special lightweight applications. Table 23.7 lists the vertical burn test results for several Curlon® fiber blends.

Work undertaken at Auburn University [75] describes the basic technology as a heat treatment of a crimped opf fiber in an inert atmosphere at a temperature of 600–700°C. The work involved preparation of battings for an improved insulating layer in, for example, military clothing and started with a cloth woven made from $12k$ opf, heat treated at 600°C, deknitted, staple cut to about 75 mm, opening up the staple and blending with polyester, rumbling in a prefeeder followed by carding. The opened fibers were then made into battings using a Rando Webber and bonding was achieved by passing through an oven at 165°C. Batt densities of 1.6–24 $kg\,m^{-3}$ were achieved, controlling the density by the compression applied during the thermal bonding stage. Due to the spring-like crimp, the battings were very resilient, exhibiting hardly any permanent set after repeated loadings.

The performance is due to the non-conducting nature of the fiber, whilst the high temperature emissivity as a black body radiator provides a cooling mechanism in a high temperature flame. Since Curlon is about 9 μm in diameter, it can be classed as a microfiber and, therefore, is a good insulator. The loss on ignition of the basic fiber is about 55%. About 25% of a polyester fiber binder can be incorporated to thermobond the lightweight battings and a water repellent coating is added to enhance the water repellency.

The insulation properties of typical Curlon battings are given in Table 23.8 and are compared with the insulation value of current fiberglass aircraft insulation.

Table 23.7 Results of vertical burn tests with blends of Curlon

Fiber blend	Test result
15% Curlon/85% polyester	Pass
20% Curlon/10% polyester/70% polypropylene	Pass
10% Curlon/10% polyester/80% cotton	Pass
40% carbon fiber/60% polyester	Fail
40% *para*-aramid/60% polyester	Fail
40% *meta*-aramid/60% polyester	Fail
50% oxidized PAN fiber/50% polyester	Fail

Sample conditions:
Felt thickness 2.54 cm; felt density 6.7–9.6 $kg\,m^{-3}$;
Vertical burn test, 90° to FTM 5903 and FAR 25.853b
Source: Reprinted from RK Carbon Fibres technical literature.

Table 23.8 The comparative properties of 2.54 cm thick Curlon and glass insulation felts

Product	Felt density $kg\,m^{-3}$	Fiber diameter μm	K value $W\,m^{-1}K^{-1}$
Curlon	3.20	9	0.043
	3.20	5	0.036
Glass	6.70	2	0.039
	9.60	2	0.035

Source: Reprinted from RK Carbon Fibres technical literature.

Comparing this with the best aircraft glass insulation, which has a K value of 0.039 $Wm^{-1}K^{-1}$ at 6.72 kgm^{-3}, shows that the Curlon has the potential for a weight savings of 33–50% in aircraft insulation and related applications.

It is believed that the fire resistance is enhanced by the ability to act like a flame arrester [76].

When the flame hits the batting, the flammable fiber pulls away from the flame leaving a Curlon gauze in front of the batting. The Curlon gauze does not shrink or pull back, only slowly oxidizing in the flame. It is non-melting and has an insulating property and through its good emissivity, throws some of the heat back as light.

The Curlon and Lineon fibers are non-fibrillating and skin tests have shown no allergic reactions or irritation.

Curlon fiber can also be used to provide a measure of sound insulation in aircraft. The aerospace market is currently being driven by the most recent notice of proposed rule making (NPRM) covering flame propagation and burn through [77]. The FAA has cited the performance level of Curlon batting in small scale testing to simulate post crash fuel fires. The recommended route for aircraft with 20 or more passenger seats is to use 5 cm of fiberglass with 2.5 cm of Curlon batt (which meets the proposed requirement for burn through protection) enclosed in a metallized PVF film bag.

23.5 PACKING MATERIALS AND GASKETS

Carbon fiber reinforced packing materials are favoured by Garlock Inc., since they have higher thermal and oxidative stability than opf and Garlock require a packing material to withstand 260°C. If opf is used at a higher temperature, carbonization takes place with loss in volume and off-gassing, which can cause a blow-out.

Richard Klinger was the first UK manufacturer to introduce an asbestos-free calendered gasket material and developed a product (Klingersil C4500) comprising of PAN based carbon fibers and a nitrile rubber binder, which was resistant to saturated steam at 290°C and possessed excellent oil and chemical resistance, particularly suitable for use in highly alkaline environments. The product had excellent resistance to creep and gasket sealing stresses were similar to compressed asbestos fiber materials. Other manufacturers of carbon fiber reinforced gasket material are Garlock Inc., Reinz and James Walker.

23.6 CARBON FIBERS IN THERMOSET MATRICES

23.6.1 Aerospace

23.6.1.1 Defense aircraft

Typical structural items fabricated for a defense aircraft from cfrp are shown in Figure 23.14. The Eurofighter/Typhoon (Figure 23.15) aircraft uses an IM7/epoxy prepreg.

23.6.1.2 Civil aircraft

Airbus Industrie was the first civil aircraft manufacturer to use carbon prepregs for parts of the primary structure and cfrp was used in the tail fin of the Airbus A300 (Figure 23.16). Typical structural items fabricated from cfrp for a civil aircraft are shown in Figure 23.17.

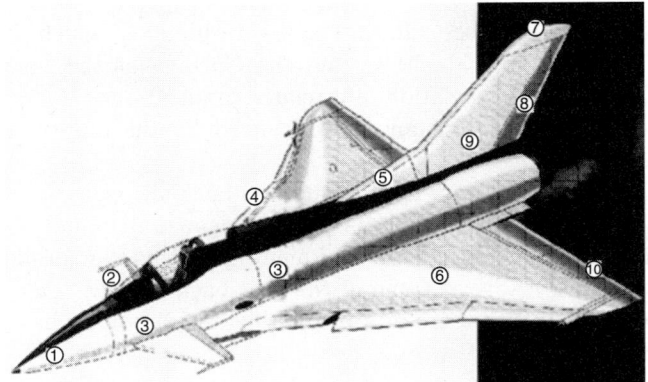

Figure 23.14 Schematic defense fighter showing where cfrp is used. Radar transparent Apple, to military and civil aircraft. 2. Foreplane Canard Wings—Epoxy carbon prepregs, 3. Fuselage Panel Sections—Epoxy carbon prepregs, 4. Leading Edge Devices—Epoxy carbon and glass prepregs, 5. Fin Fairings—Epoxy glass and carbon prepregs, 6. Wing Skins and Ribs—Epoxy carbon and glass prepegs, 8. Rudder—Epoxy carbon prepreg, 9. Fin—Epoxy carbon/glass prepreg, 10. Flying Control Surfaces—Epoxy carbon and glass prepregs. *Source:* Reprinted from Hexcel technical literature.

Figure 23.15 Eurofighter/Typhoon uses IM7/epoxy prepreg. Source: Reprinted from Hexcel technical literature.

Figure 23.16 Airbus Industrie A300 civil aircraft. *Source:* Reprinted from SP Systems technical literature, with kind permission.

THE USES OF CARBON FIBERS

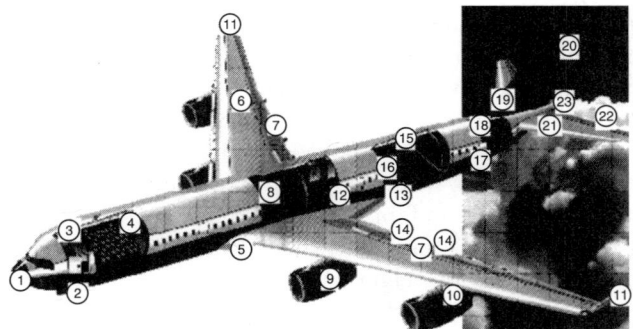

Figure 23.17 Schematic civil aircraft showing the areas where cfrp is used. 2. Landing Gear Doors and Leg Fairings—Glass/carbon prepregs, 5. Wing to Body Fairing—Carbon/glass/aramid prepregs, 6. Wing Assembly (Trailing Edge Shroud Box)—Carbon/glass prepregs, 7. Flying Control Surfaces—Ailerons, Spoilers, Vanes, Flaps—Glass/carbon/aramid prepregs, 9. Engine Nacelles and Thrust Reversers—Carbon/glass prepregs, 10. Pylon Fairings—Carbon/glass prepregs, 11. Winglets—Carbon/glass prepregs, 12. Keel Beam—Carbon prepregs, 14. Flaptrack Pairings—Carbon/glass prepregs, 18. Pressure Bulkhead—Carbon prepregs, 19. Vertical Stabilizer—Carbon/glass/aramid prepregs, 20. Rudder—Carbon/glass prepregs, 21. Horizontal Stabilizer—Carbon/glass prepregs, 22. Elevator—Carbon/glass prepregs, 23. Tail Cone—Carbon/glass prepregs. *Source*: Reprinted from Hexcel technical literature.

Airbus A380 and A400M have used cfrp for the ailerons, flap track fairings, outer flaps, main and center landing gear doors, main landing gear leg fairing door, nose landing gear doors, central torsion box, belly fairing skins, upper deck floor beams, passenger floor panels and struts, pressure bulkhead, apron, horizontal stabilizer outer box, vertical stabilizer and tail cone. Some 1.5 tons are saved from the weight of the structure by using carbon fiber for the central wing box.

Mills [78,79] has given a brief history of composites for airframe manufacturing and the part that carbon fiber played in early development. Probably the most notable were the McDonnell Douglas DC-10 rudder in 1968, the Deutsche Airbus A310 fin box in 1979 and the A320 horizontal stabilizer by Deutsche Airbus/CASA in 1984. These constructions were justified on lifetime ownership cost reduction through lower weight. The DC-10 rudder saved 38% weight and suffered little damage over a 15 year service life.

After the 1980s, applications tended to tail off due to the high costs of composite manufacture, so cost became the driving force for all new developments. The IM fiber types provided structures with higher stiffness and tape laying machines were developed to create double curvature and multi-axial non-crimp fabrics were stitched into preforms for impregnation by RTM. Triaxial braiding, combined with stitching and processed by RTM offers potential for future developments.

To reduce the cost of fabricating exit guide vanes on turbojet engines, Allied Signal has used braided carbon fiber impregnated with epoxy resin using a multicavity RTM tool instead of a compression molded prepreg tape.

Woven carbon fiber reinforced phenolic skins bonded to an aramid honeycomb are made by Hexcel for the flooring in the cabin and flight areas of Airbus aircraft (Aerospatiale has a 37.9% stake in the Airbus consortium).

A one-piece aircraft horizontal stabilizer has been demonstrated by Raytheon Aircraft Company using RTM and was 2.32 m long with a tapered wing structure with internal spars to prevent buckling. The spars were built on mandrels, producing five cells to give the spars, which were covered with dry carbon fiber reinforcement, placed in the mold and then covered with carbon fiber fabric to give a single piece structure. Carbon fiber braid was chosen as the ideal reinforcement for the spars since it could be precisely positioned and kept in place. Some spars had two layers of braid.

Lockheed Martin used a carbon fiber braid preform and RTM processing to produce a 90 kg demonstrator carbon composite tail, reducing the number of parts from 13 to 1 and eliminating over a 1000 fasteners with a 60% reduced production cost.

Scaled Composites designed and fabricated the record breaking Voyager aircraft that flew around the world non-stop and non-refueled with a crew of two. To achieve these aims in a time of nine days, the structural weight had to be as low as possible, and the structure as well as the spars, were graphite tape skins on a Nomex honeycomb core.

In 1998, Scaled Composites, an aerospace research company, unveiled the Proteus aircraft (Figure 23.18), designed for long duration high altitude operations, intended for piloted as well as unmanned aerial vehicle missions.

Raytheon Aircraft have designed the Premier I cabin around a sandwich of inner and outer carbon fiber layers around a layer of honeycomb material. The carbon fiber/epoxy tows were applied by a Cincinnati Machine Viper 7-axis fiber placement system, producing a structure 20% lighter than an Al one (Figure 23.19).

Figure 23.18 Proteus high altitude multi mission aircraft. *Source:* Courtesy of Scaled Composites Inc.

Figure 23.19 Cabin of Proteus aircraft fabricated by Viper 7-axis fiber placement system. *Source:* Courtesy of Cincinnatti Machine.

THE USES OF CARBON FIBERS

Boeing have reported that each projected 7E7 twin aisle passenger jet scheduled for 2007 will use 25 tons of toughened carbon fiber/epoxy laminate and sandwich material.

23.6.1.3 Helicopters

The Bell Boeing tiltrotor V-22 Osprey was the first military aircraft, with the prototype built almost entirely of solid laminate composite materials making its first flight in 1989. The airframe was 25% lighter than one using metal counterparts. The aft-fuselage was built using a fully automated fabrication process with Hercules 'Towpreg' applied along six axes to conform to the structure's complex shape.

Typical structural items fabricated from cfrp for a helicopter are shown in Figure 23.20. The tail rotor drive shaft for the Eurocopter Tigre helicopter (Figure 23.21) is filament wound by Urenco using cfrp.

23.6.1.4 Aero engines

Typical items fabricated from cfrp for an aero engine are shown in Figure 23.22. Parts of the Dornier 328 engine nacelle fabricated from cfrp are shown in Figure 23.23.

Figure 23.24 shows the parts of C-17, MD-11 and M-90 aircraft fabricated with cfrp.

Airbus A380 and A400M have used cfrp for pylon fairings and nacelles cowlings.

23.6.1.5 Propeller blades

Abraham and McCarthy [80] describe how RTM has been used successfully at Dowty for the manufacture of propellers. The composite blade construction is shown in Figure 23.25.

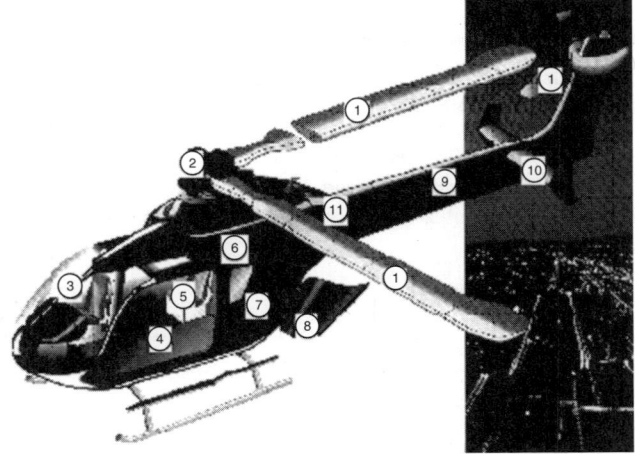

Figure 23.20 Schematic helicopter showing the areas where cfrp is used. 1. Rotor Blades—Prepregs/carbon/glass honeycombs, 2. Rotor Hub—Carbon epoxy prepregs, 3. Glazing Bars—Epoxy carbon/glass prepregs, 6. Engine/body Fairings and Access Panels—Epoxy/BMI glass/carbon/aramid prepregs, 7. Fuselage—Carbon and glass prepregs, 8. Main and Cargo Doors—Epoxy carbon/glass prepreg, 9. Boom and Tail Section—Epoxy carbon/glass prepreg, 10. Horizontal Stabilizers—Epoxy glass/carbon/aramid prepregs, 11. Fuselage Panels—Epoxy carbon/glass prepreg. *Source*: Reprinted from Hexcel technical literature.

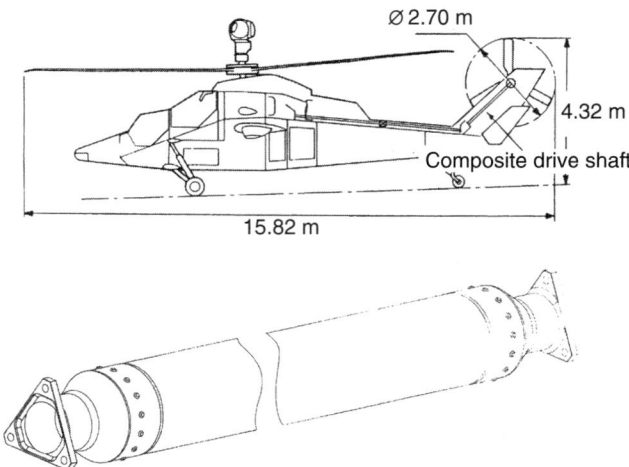

Figure 23.21 Tigre helicopter drive shaft. *Source:* Courtesy of Urenco.

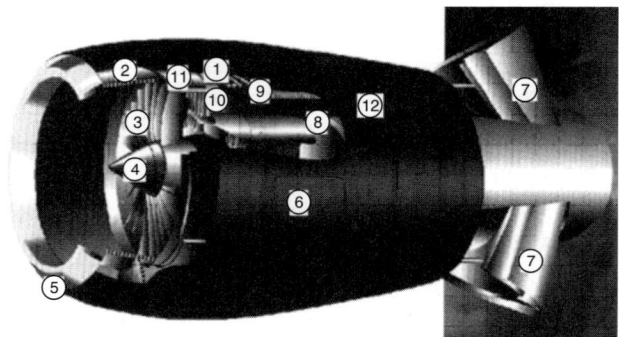

Figure 23.22 Schematic aero engine showing the areas where cfrp is used. 1. Electronic Control Unit Casing—Epoxy carbon prepregs, 2. Acoustic Lining Panels—Carbon/glass prepregs, 3. Fan Blades—Epoxy carbon prepregs or Resin Transfer Molding (RTM) construction, 6. Engine Access Doors—Woven and UD carbon/glass prepregs, 7. Thrust Reverser Buckets—Epoxy woven carbon prepregs or RTM materials, 8. Compressor Fairing—BMI/epoxy carbon prepreg, 9. Bypass Duct—Epoxy carbon prepreg, 10. Guide Vanes—Epoxy carbon RFI/RTM construction, 11. Nacelle Cowling—Carbon/glass prepregs. *Source:* Reprinted from Hexcel technical literature.

The design has an outstanding service safety record with no blade losses over 20 years and 75 million flying hours. Spars made of mainly UD carbon fiber take the main load and resist centrifugal and bending loads on the blade, extending from root to tip and reducing in thickness towards the tip as the loads are reduced. The outer shell has carbon fiber wound at ±45° to the blade axis. The structural foam core permits transference of shear stress between the two composite spar beams. The spars form via a transition into a cylinder and connect to a metal outer sleeve using a system of glass fiber wedges; with the introduction of an inner sleeve totally enclosing the annular wedge (Figure 23.26). Lightening protection is achieved by an Al braid running along both faces, from the tip to the metal outer sleeve. A Hercules C130J transporter aircraft fitted with Dowty propeller blades is shown in Figure 23.27.

THE USES OF CARBON FIBERS

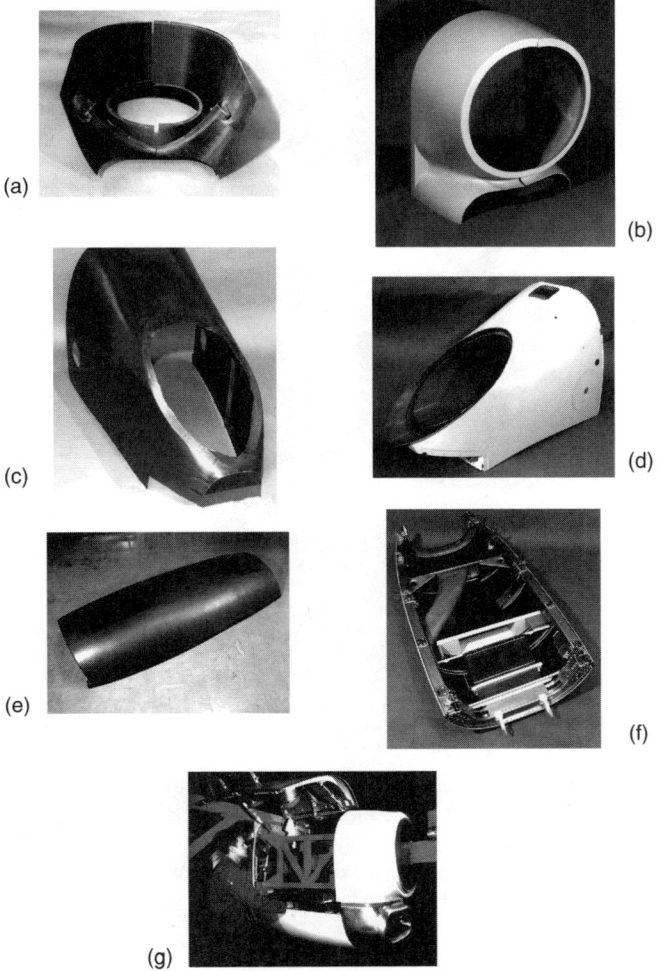

Figure 23.23 The cfrp parts of a Dornier 328 engine nacelle. (a) Reverse side of forward cowl, (b) Nacelle forward cowl, (c) Nacelle aft cowl, (d) Nacelle aft cowl completed, (e) Nacelle lower cowl, (f) Nacelle lower cowl fitted out, (g) Assembling nacelle. *Source*: Courtesy of GKN.

A selection of rotor blades and propellers made by GKN is shown in Figure 23.28 using a carbon fiber/glass hybrid, tailored to give the right strength, stiffness, ballistic balance and failure mode. Composites enable the manufacture of complicated shapes necessary to give lift control characteristics and allow complicated geometry to be made relatively easily and moreover, possess a virtually infinite fatigue life.

Scaled Composites used an all graphite construction for a two blade controllable pitch unit for the Raptor unmanned aerial vehicle, which could be converted to a manned configuration.

23.6.1.6 Antenna, lightening conductors

Antennae can be manufactured by filament winding carbon fiber, utilizing a process used for making fishing rod blanks (e.g. the Shakespeare process, where, as the section is tapered,

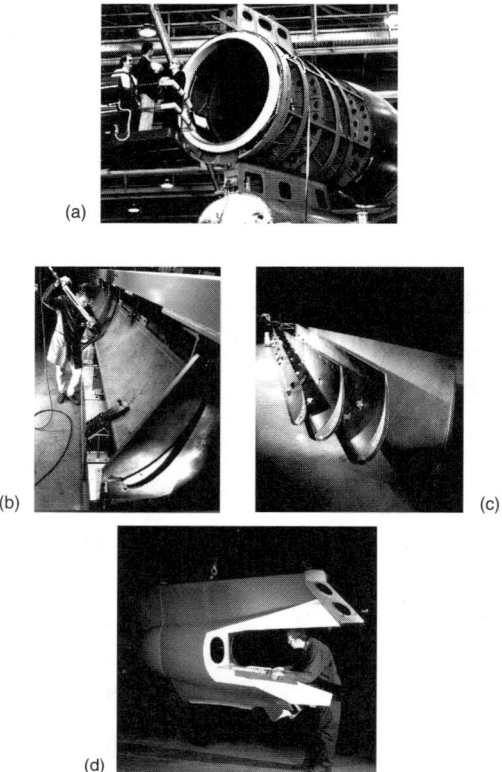

Figure 23.24 The cfrp parts fabricated for C-17, MD-11, and MD-90 aircraft. (a) No.2 tail mounted engine air intake duct on MD-11 aircraft, about 3 m in diameter and 6 m long made of Rohacell foam stiffeners, (b) MD-11 aircraft vanes with carbon fiber skin & ribs and Rohacell foam core, (c) C-17 aircraft vanes with carbon fiber outer skin & ribs and Rohacell foam core, (d) MD-90 tailfin tip with carbon fiber outer skins and metallic internal structure. *Source:* Courtesy of GKN.

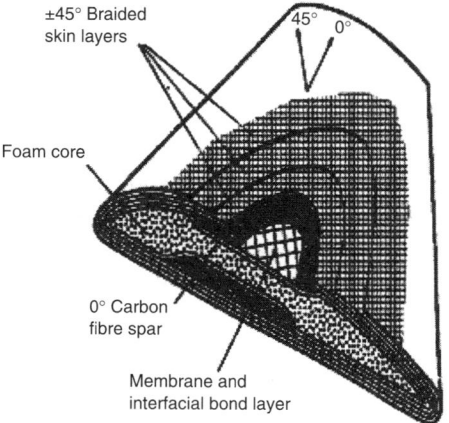

Figure 23.25 Composite propeller blade construction used by Dowty Rotol. *Source:* Reprinted from Abraham D, McCarthy R, Design of polymer composite structural components for manufacture by resin transfer moulding (RTM). SAMPE 20th Jubilee Europe Conference & Exhibition, in Paris, 13–15th April, 407–415, 1999.

THE USES OF CARBON FIBERS 981

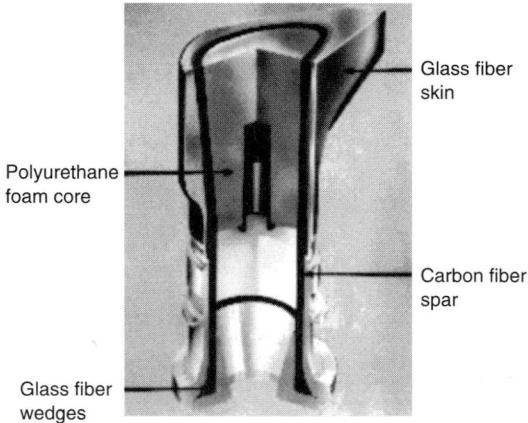

Figure 23.26 Dowty Rotol propeller blade root construction. *Source:* Reprinted from Abraham D, McCarthy R, Design of polymer composite structural components for manufacture by resin transfer molding (RTM). SAMPE Europe Conference & Exhibition, 407–415, 1999.

Figure 23.27 Hercules C-130J aircraft with Dowty Rotol propellers. *Source:* Reprinted from Abraham D, McCarthy R, Design of polymer composite structural components for manufacture by resin transfer molding (RTM). SAMPE Europe Conference & Exhibition, 407–415, 1999.

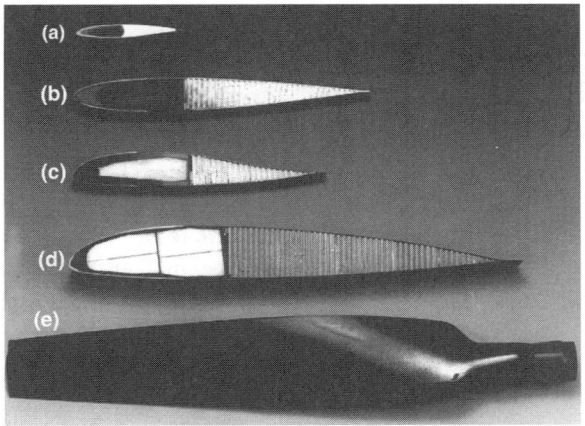

Figure 23.28 Selection of rotor blades and propellers made from carbon fiber/glass hybrid. *Source:* Courtesy of GKN.

the individual tows are systematically cut); the aerials are hung vertically and cured that way to ensure straightness.

23.6.1.7 Gliders and sailplanes

Slingsby has been associated with sailplanes and gliders for many years and was the first company to use grp in production aircraft, as early as 1953, and was probably the first company to use cfrp materials for main structural members. [81].

23.6.1.8 Unmanned Aerial Vehicles (UAVs)

Unmanned aerial vehicles are becoming strategically quite important, operate under a blanket of extreme secrecy and have proved their value in the Kosovo conflict. The Global Hawk, for example, has an Al fuselage frame but the nacelle, radomes, rear empennage, stabilizers and wing/body fairings are all constructed of carbon fiber and fiberglass composites. These UAVs can be built almost an order of magnitude cheaper than a fighter aircraft and can be operated safely from a remote location. It is envisaged that future UAVs could be constructed from a minimum of 90% advanced composites, including carbon fiber.

Lightweight winged satellites powered by the sun achieved low weight by using a carbon epoxy framework of sparsand ribs covered with polyester film. Typical of such flying wings is the Helios, developed by Aero-Vironment and is a remotely piloted aircraft having a wingspan of 75 m, with the majority of its components made from carbon fiber, carbon/epoxy and Kevlar, which has reached a height of 25 km [82].

23.6.1.9 Stealth aerial vehicles

At present, aerial vehicles such as the B-2 Bomber, F-22 Raptor, Tacit Blue and F-117A, employ stealth (low-observable) technologies and it is expected that all new manned and unmanned combat aircraft will do so (e.g. Sikorsky's RAH-66 Comanche helicopter). Stealth aircraft have a low profile, with no right angles which are strong radar reflectors. No surfaces reflect radar directly back and this is ensured by covering with a coating of a radar absorbing material, which can be attained by matching the wave impedance with a resistive material of about 377 Ω, which is where carbon fiber could be of use.

23.6.2 Space

Typical structural items fabricated from cfrp for use in space are shown in Figure 23.29.

Carbon fiber epoxy composites predominate in space applications [83] and items like antenna dishes and support structures on spacecraft exploit the stiffness and dimensional stability of cfrp. An automatically unfurlable/retractable antenna was fabricated using light cfrp ribs to support a gold plated Mo mesh opening out to 5 m diameter.

Orbital Science have built the X-34 rocket plane 17.6 m long, with a wingspan of 8.5 m, as a re-useable test bed and capable of speeds up to Mach 8, featuring an all composite primary and secondary structure. The fuselage components use carbon fiber epoxy skins with an Al honeycomb core.

Tomita has described an 8 m collimation mirror fabricated from cfrp [84] and Di Vita [85] has shown how filament winding is a key technology for space propulsion. Blasi [87] has described the use of tape wrapping for the ARIANES booster.

THE USES OF CARBON FIBERS

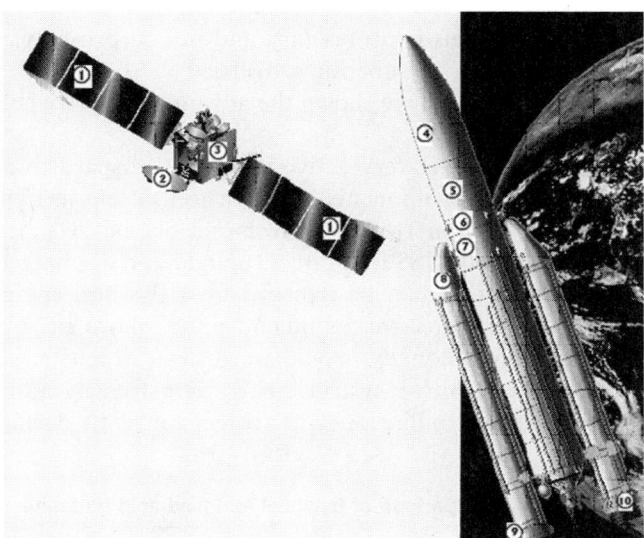

Figure 23.29 Schematic space vehicle showing the areas where cfrp is used. 1. Solar Panels—Epoxy carbon prepreg, 2. Satellite Structures—Carbon prepreg, 4. Fairings—Carbon prepregs 5. External Payload Carrier Assembly (SPELTRA)—Carbon prepregs, 6. EPS Ring—Epoxy/carbon prepreg or RTM, 7. Front Skirt—Carbon prepreg, 8. Yoke—Epoxy carbon filament winding, 10. Heat Shield—Carbon prepreg/high temperature resistant glass fabric. *Source:* Reprinted from Hexcel technical literature.

An attribute of cfrp is its low thermal expansion and combined with structural rigidity, a graphite/epoxy composite was selected for the main structure of the Hubble Telescope [86]. The successor to the Hubble will be the James Webb Space Telescope, with backing struts and mirror support struts made from a carbon hybrid composite to operate at temperatures in the region of $-235°C$.

23.6.3 Rocket motor cases

Starchaser Industries have launched Nova, a 11 m tall rocket fitted with cfrp cowlings [87]. Man Technologies have developed a carbon composite booster case for Ariane 5 by laying up dry non-crimp fabrics on a cylindrical mandrel, using resin injection under vacuum and curing in an oven at ambient pressure.

23.6.4 Flywheels

A flywheel is a device for storing energy in a rotating mass and uses a high strength composite rotor mounted on a shaft, rotating in a vacuum with a motor attached to the shaft, which in the charging mode, turns the rotor up to speed and in the discharging mode, acts as a generator to covert the rotor's kinetic energy to electrical energy. The rotor is mounted on special bearings for stabilization and to minimize energy loss, is enveloped in a container acting as a vacuum vessel and safety entrainment. Carbon fiber/epoxy flywheels are claimed to have considerable potential for storing energy, with a regenerative efficiency of at least 85%. Work on flywheels is proprietary and not surprisingly, each organization claims to have developed flywheels of superior properties.

Regenerative Power and Motion have shown the advantages of flywheel source of energy over lead acid batteries in Table 23.9.

NASA Glenn Research Center is very active with an aerospace flywheel development program and they depict the key components of a flywheel system in Figure 23.30.

The International Space Station (ISS) has rechargeable Ni- H_2 batteries, which are charged from solar arrays whilst the ISS is in direct sunlight and discharged when the station is in shadow. It is hoped that they can be replaced by a flywheel energy storage system, when the flywheels will be able to recover as much as 80% more energy than when using electrochemical batteries and additionally, would weigh less. Magnetic bearings suspended in vacuum are fitted with a control system, with an extremely rapid response time, of the order of a fraction of one millisecond, to correct for shaft deviations and stabilize

Table 23.9 Comparison of flywheel and lead acid batteries

Property	3 kwh Flywheel	50 kwh Flywheel
Height (m)	0.31	0.76
OD (m)	0.31	0.76
ID/OD ratio	0.75	0.75
Weight (kg)	10.4	168
Equivalent number of lead acid batteries (total weight)	5 off (>250 lb) (>113 kg)	>80 (>4000 lb) (>1814 kg)

Source: Reprinted from Regenerative Power and Motion technical literature.

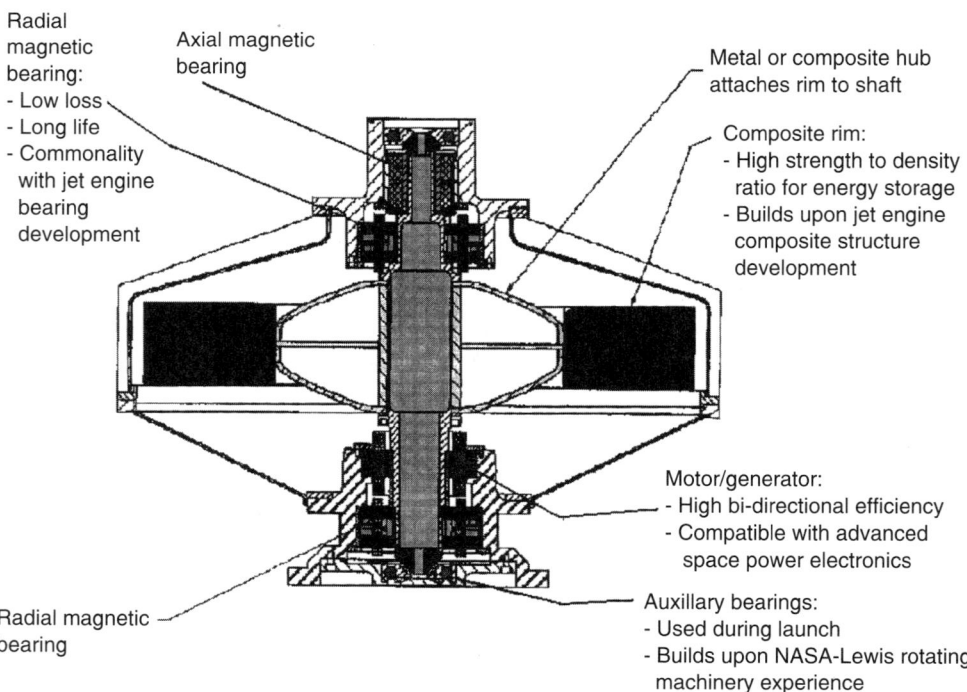

Figure 23.30 Key components of an aerospace flywheel. Source: Reprinted from NASA Glenn Research Center.

the rotating shaft, permitting flywheels to rotate at 60,000 rpm. The University of Texas Center for Mechanics are planning to supply a 10% glass/90% carbon fiber flywheel to US Flywheels that will be able to withstand operating at 60,000 rpm and would use a composite housing to provide protection and contain a sealed vacuum prior to launch into space.

The University of Texas Electromechanical Center, in conjunction with Allied Signal, are fabricating a flywheel assembly weighing some 9980 kg, with a flywheel about 125 cm diameter made primarily of IM carbon fiber/epoxy with some glass. The flywheel is made by filament winding 12 concentric rings, each 76 cm long and bonded together.

Penn State University has used concentric rings of glass and carbon that are either press fit at the interfaces, or separated with a compliant polyurethane interlayer to provide radial stress relief during operation. The outermost rings are fabricated from HS carbon fiber, whilst lower strength carbon fiber is used for the inner rings, with a glass fiber ring innermost. The rings are fabricated by wet filament winding, wrapping in a hoop direction onto a heated mandrel. To overcome the problem of wavy fibers and poor consolidation in the thick rotors, the epoxy resin is continuously gelled and cured during the winding process, carefully controlling the conditions to maintain just about 12.5 mm of uncured resin on the rotor surface [88].

At this stage of development, it is doubtful whether a flywheel incorporated in an electrical highway vehicle can replace lead acid batteries due to problems with vibration and safety, but the performance of a lead acid battery deteriorates in hot wet conditions, requires frequent maintenance and has a life expectancy of only about 4 years. The aim of US Flywheel Systems, however, is to drive a car with a flywheel system. It is reckoned that some 16 units would be required to fulfil this objective and adequate protection would have to be supplied, since, in the event of a wheel disintegrating, it would dissipate its energy into hot fluff and high speed dust. A very sophisticated computer control system is used to filament wind each wheel with a high fiber content of 86% w/w [89].

As the need for uninterruptible power supply (UPS) systems increases, there is a developing market here for flywheel systems using carbon fiber rotors. The faster a flywheel spins, the greater is the energy storage. A 275 kg steel rotor rotating at 8000 rpm would generate 900 W h of energy, whilst a cfrp flywheel weighing only 68 kg can spin at 22,500 rpm and store about 2700 W h of energy. Increasing the speed of rotation to 100,000 rpm would increase the storage capacity by a factor of 10.

Acumentrics produce a series of online uninterruptible power supply designed around a carbon fiber composite flywheel energy storage system contained within a steel housing.

The combined efforts of American Flywheel Systems and Trinity Flywheel Power (AFST) have become established in flywheel energy storage technology and rotors are made from carbon fiber by filament winding and a current production rotor spins at 44,000 rpm with a projected service life of 20 years.

Indigo Energy, in conjunction with Toray Industries, have developed a CF/epoxy flywheel rotating at 40,000 rpm and capable of storing 3000 W h of electricity, reputedly maintenance free for 20 years and for possible use in electronic industries.

The Nippon Oil Company have described a flywheel energy storage system with the rotor filament wound from ultrahigh modulus pitch based carbon fiber/epoxy composite, with a regenerative efficiency of over 85% [90]. In order to rotate the flywheel at high speed, carbon fibers have been used with an increasing modulus in the radial direction from the inner side of the flywheel to the outer side (XN40, 390 GPa; to XN50, 490 GPa; to XN70, 690 GPa). The unit produced 3.09 kWh at 35,600 rpm.

EADS Composites Atlantic Ltd, has filament wound a CF/E flywheel weighing 180 kg.

Flywheel Energy Systems Inc. of Canada has developed a series of high performance flywheels for electromechanical batteries [91]. The flywheels depicted in Figure 23.31 shows a range producing 75–96 Wh kg^{-1}, to accommodate various market needs and by changing the axial length, can provide the ability to store between 300–30,000 Wh of energy. All wheels were wet filament wound using S2glass/(Mitsubishi & GrafilInc.) carbon fiber and epoxy matrix. The flywheels are assembled to a flex rim hub and arbor interface for spin testing in an air turbine (Figure 23.32).

Urenco Power Technologies have developed their high speed composite flywheel technology based on experience gained from in-house high speed centrifuge for the nuclear industry, using filament winding and carbon fiber. Their KESS storage system has a predicted cycling capacity of at least 10 million cycles, which equates to about 20 years of service. The applications of Urenco flywheel energy storage systems include smoothing cyclic loads (e.g. rolling mills); increasing the performance of wind generation farms and smoothing the peak demands on the railways.

Figure 23.31 Series 45 Flywheel depicting Mk 4, Mk 3, Mk 2 and Mk 1 models developed during 1994–1997. Fabricated by wet filament winding, unidirectional, reinforced with S2 glass/Mitsubishi or Grafil carbon fiber with an epoxy matrix. *Source:* Courtesy of Flywheel Energy Systems Inc., Canada.

Figure 23.32 Series 45 Mk 4 flywheel assembled to flex rim hub and arbor interface for spin testing in an air turbine (specific energy 100 Wh kg^{-1} at linear velocity of rim tip speed of 1000 ms^{-1}. *Source:* Courtesy of Flywheel Energy Systems Inc., Canada.

THE USES OF CARBON FIBERS

23.6.5 Marine applications

The power from a ship's engine can be transmitted to the propeller by a shaft and Centra Antriebe Kishey in Germany and Geislinger in Austria supply filament wound cfrp shafts, which, because they are lighter, can be made longer than conventional steel shafts.

The design of the humble Laser dinghy is carefully controlled by stringent class rules, leaving little scope for the use of cfrp except for the construction of the tiller and tiller extension, where the slim profile offered by cfrp is an obvious choice [92].

23.6.5.1 Yachts

It is now common practice for all top boats built for the America's Cup to be fabricated from cfrp with a honeycomb core using dry prepreg methods. Goss Challenges have built a racing catamaran (Team Philips) (Figure 23.33) using carbon fiber reputed to be the largest carbon composite structure ever built in Europe (36.5 m long × 21 m wide, with an unstayed mast 39 m high) [93]. The designer Adrian Thompson of Paragon Mann relied heavily on finite element analysis using SRDC I-DEAS. Initial trials with the vessel revealed a jointing problem, which was resolved, but subsequent problems necessitated abandoning the vessel, perhaps emphasizing the need to undertake this aspect of construction in a very controlled manner.

Ellen MacArthur, who became the fastest woman circumnavigator in Kingfisher, an 18 m raceboat, utilized carbon fiber/Nomex for hull and deck, Kevlar skinned cfrp frame for the blister cabin, mast and deck spreaders molded in HM cfrp and a cfrp boom, emphasizing the advantages of mixed construction.

The 38 m long catamaran PlayStation set a new world record from Miami to New York, with hulls and beams constructed from cfrp/Al honeycomb and a carbon fiber mast, which was some 45 m above the water.

Reputedly the tallest mast (53.33 m) in sailing history was fabricated in two halves by the Consolidated Yacht company for the Zeus super yacht, using carbon/epoxy prepreg, vacuum bagged and cured in a special oven tailormade from foil insulated polyurethane

Figure 23.33 The 33.6 m Team Philips catamaran. *Source:* Courtesy of SP Systems.

foam board and an Al plenum curing up to 82°C. The two halves were initially fastened mechanically along their length and then bonded together curing at 82°C [94].

Ordinarily, tall masts are made in several sections due to limitations of the curing oven, but a mast 59 m long was fabricated from carbon fiber prepreg in one piece for the yacht Hyperion by Royal Huisman (Figure 23.34), utilizing the composites workshop as a giant oven and curing at 80°C. Figure 23.35 shows the mast.

The latest claim is for the 75 m super yacht, Mirabella V, built by VT Shipbuilding and reputed to be the largest composite vessel and largest single masted sailing yacht yet built. The 90 m composite mast was built by VT Halmatic by joining five sections of >25mm thick tube made from UD carbon/epoxy prepreg and bonded together with an epoxy adhesive. The upper deck used carbon/vinyl ester skins over a foam core and the bulwarks and

Figure 23.34 Hyperion 2 yacht. *Source:* Courtesy of SP Systems.

Figure 23.35 Hyperion 2 yacht mast—59 m long fabricated from cfrp. *Source:* Courtesy of SP Systems.

superstructure were all reinforced with UD carbon fiber to carry the stress around the various deck openings.

The D-shaped fenders of Swedish and Danish Lifeboat Rescue Institutions have been made from shaped foam reinforced with layers of carbon fiber and coated with a polyurethane elastomer, providing buoyancy and stability and also deflecting the waves, thereby enhancing visibility.

Hall Spars use heat cured prepreg products and claim that autoclave curing is superior to vacuum bag curing for the manufacture of carbon fiber spars, enabling pressures of up to seven times the effective pressure of vacuum bagging to be used. Since a carbon rig is lighter than an Al mast extrusion, its pitching moment is reduced and it also pumps less in a seaway.

23.6.5.2 Submarines

Two submarines capable of diving to a depth of 6000 m were constructed using a sandwich construction of a special syntactic core and carbon fiber prepreg, to withstand the extreme pressures and were used to enable underwater shots when making the film *Titanic*.

23.6.5.3 Air cushion vehicle

Aker Finnyards Oy used a foam sandwich structure, utilizing an epoxy resin infusion process with carbon fiber reinforcement to make the superstructure of a prototype air cushion vehicle.

23.6.6 Oil exploration

The US Department of the Interior has depicted the types of deepwater development systems basically into two categories:

1. Bottom supported and vertically moored structures (Figure 23.36)
2. Floating production and subsea systems (Figure 23.37)

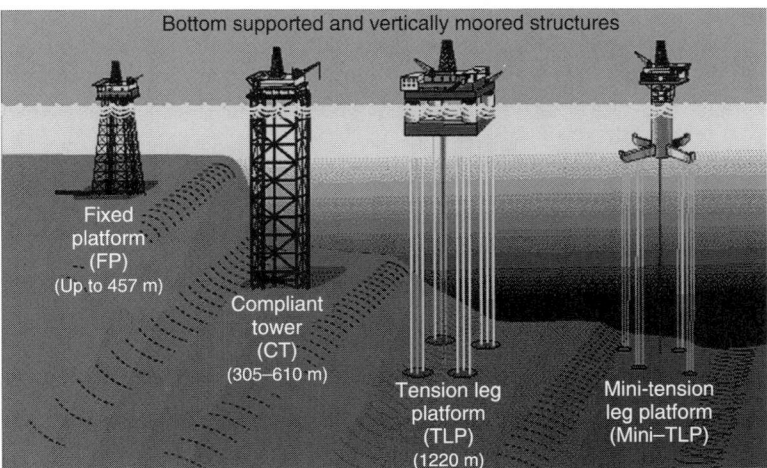

Figure 23.36 Bottom structured and vertically moored structures for oil exploration. *Source:* Reprinted from US DOI.

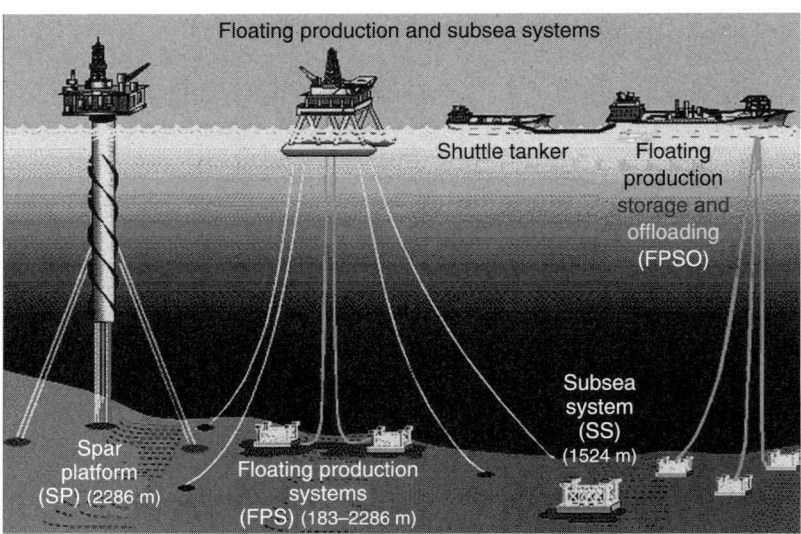

Figure 23.37 Floating production and subsea systems for oil exploration. *Source:* Reprinted from US DOI.

A brief description of each type is given to better understand how cfrp can be used in their construction.

1. Fixed Platform (FP)—a jacket comprising a tall tubular steel vertical section supported by piles driven into the seabed, with the crew accommodated on a platform and operation capability at depths up to 460 m.
2. Compliant Tower (CT)—built to withstand large lateral forces and normally operates at 305–610 m.
3. Tension Leg Platform (TLP)—a floating structure held in place by tensioned tendons securely fixed with templates to the seabed. As long as vertical motion is limited, it can operate up to 1220 m.
4. Mini-tension Leg Platform (Mini-TLP)—a low cost mini-tension leg platform, it can also be used as a satellite for early deepwater production.
5. SPAR Platform (SPAR)—large diameter single vertical cylinder equipped with a deck, three types of riser and a hull moored with 6–20 taut catenary systems fixed to the seabed. It is presently used at 915 m but could be extended to 2285 m.
6. Floating Production System (FPS)—a semi-submersible unit anchored with wire rope and chain. The surface deck receives oil from production risers designed to accommodate platform motion and used at a water depth of 185–2285 m.
7. Subsea System (SS)—sends oil to an FPS or TLP system and can be used in water depths greater than 1525 m.
8. Floating Production, Storage and Offloading System (FPSO)—a large tanker type vessel moored to the sea floor. Periodically, it offloads to a smaller shuttle tanker.

The oil industry is now showing great interest in deepwater production in areas like the Gulf of Guinea, Brazil and the Gulf of Mexico, where the most promising production is at a depth of 1220 m. The deposits are vast and although initial investment is very large, the cost of the recovered oil is cheaper than shallow water operations. Petrobras/RB Falcon has reached a depth of 2777 m in the Brazil location and obviously, as depth increases, the risk factor becomes greater and the oil companies will require smart systems, which can be built into composite structures. As the water depth increases, cfrp risers, tendons and rig elements offer considerable savings over conventional materials. The standard platform installation is not suitable and would be extremely costly to modify. A tension leg platform, using carbon fiber composite cables as tethers, becomes cost competitive in water deeper than 1600 m [95]. The cable developed by Freyssinet, Institut Français du Pétrole and Doris Engineering comprises nineteen 6 mm diameter pultruded carbon fiber/epoxy rods arranged in a hexagonal array and built up in modules.

Automated Dynamics Corporation have used PES and PEEK/carbon fiber thermoplastic prepreg for the manufacture of high temperature pipe and down hole components.

Garseth [96] has described filament winding applications in the offshore industry.

Deepwater Composites, a joint venture between Aker Knaerner and ConocoPhillips, produces Comptether, carbon fiber reinforced 6 mm diameter pultruded rods for use as tethers in part of the mooring system used for tension leg platforms. About 781 rods bundled together into 13 strands of 85 or 31 rods were used. To facilitate spooling, the assembled tendon, strands and profiles are given a helical twist during the assembly process.

Spencer Composites have used filament winding to fabricate drilling risers about 15 m long and 560 mm diameter, by overwinding carbon fiber and fiberglass onto Ti tubing, costing 40% less, weighing 20% less and with an expected fatigue life some three times that of the Ti tubing it is replacing.

Fiberspar have introduced a continuous fiberglass and/or carbon fiber/epoxy pipe over a polyethylene liner, which can be pulled through existing steel pipe to effect a repair in pipelines carrying corrosive fluids.

23.6.7 Automobile and racing car applications

A bonnet fabricated from cfrp can reduce the weight from 18 to 7.25 kg and utilize the stiffness of carbon fiber. Similarly, GM's Corvette used carbon fiber, producing a hood weighing only 9.3 kg and saving 4.8 kg from the standard fiberglass SMC. Adam [97] has detailed the use of cfrp in automotive applications.

23.6.7.1 Chassis, body and interior

In 1980, John Barnard at McLaren International created the world's first carbon fiber chassis Grand Prix winner, the McLaren MP4.

Concept cars such as the Ford Prodigy HEV and the aXcess Australia have used cfrp in their construction and there is no doubt that such innovative designs will help advance the use of composite materials in the automotive industry.

The Daimler Chrysler Concept vehicle, Crossfire, uses a one-piece carbon fiber body on an Al frame and the Daimler Chrysler's Willys2, akin to a jeep, also has a one-piece carbon fiber body on an Al frame, with a removable carbon fiber hardtop.

The Lamborghini Murcielago utilizes cfrp for the body and the Porsche Carrera GT uses cfrp for the body and structural elements.

Aston Martin has used carbon fiber in their Vanquish model for the transmission tunnel, windshield pillars and parts of the sub-frame.

On-Shore has used nickel plated carbon fiber for laminated components, including the steering wheel and grab handle.

DSM have developed a carbon fiber SMC for use in making low weight semistructural and structural automotive components which can weigh as much as 55% less than those made of traditional SMC.

Figure 23.38 Thrust SSC on Black Rock Desert, Nevada. *Source:* Photo courtesy of CB Hill.

Figure 23.39 Thrust SSC in hangar on Black Rock Desert, Nevada. *Source:* Photo courtesy of CB Hill.

Thrust SSC (Super Sonic Car), in 1998, became the first land vehicle to break the sound barrier, achieving 764 mph across the Black Rock desert in Nevada (Figures 23.38, 23.39). Cytec's carbon fiber prepreg was used around the forward position of Thrust's engine nacelles, nose cone, air inlet structures and cockpit canopies.

23.6.7.2 Brakes and clutches

Carbon-carbon is used for the construction of brakes and clutches (Section 23.8.1 and 23.8.2).

23.6.7.3 Suspension systems

The Diamler Chrysler ESX3 represents a new generation of automobile design and incorporates a rear torsional axle system, sprung by elastomeric elements on a splined shaft housed inside a square carbon fiber tube, the shaft connecting to a carbon fiber and Al trailing arm that holds the wheel assembly.

BMW are using carbon fiber in the BMW Z22 model for reinforcing side frames, roof, tailgate and flooring, enabling 20 components to replace 80 pieces with 50% less weight than a steel body.

23.6.7.4 Push rods

The cfrp push rods are 70% lighter than metal push rods, reducing noise and increasing engine efficiency [98].

23.6.7.5 Air bags

Side air bags upon inflation, deploy into position to cushion a vehicle occupant's head in the event of a side impact. BMW use carbon braid, enabling the bag to shrink in length and expand in diameter, but only braid made on an extremely large braider (e.g. A&P megabraider) would satisfy these requirements. The braid tube must permit the bag to condense to a small volume and fit in the door trim and yet permit it to expand to a tubular structure without bursting.

23.6.8 Heavy goods vehicles and buses

Leaf springs fabricated from cfrp weigh only 20% as compared to one made of steel and yet have the same spring rate and load carrying capacity. Leaf springs are essentially beams in bending, with minor torsional loads and localized edge loads, hence most of the fiber can be oriented along the length of the leaf.

A hybrid of glass and carbon bumper has been designed for heavy trucks and buses, giving a corrosion resistant product some 30 kg less in weight, with superior vibration resistance.

23.6.8.1 Drive shafts

Carbon fiber has been used for fabricating drive shafts for many years and initially, cost was a serious drawback, but with lower costs and development of fibers with improved mechanical properties, it is estimated that over 12,000 composite shafts are now in service in the industry today. Carbon fiber produces lightweight shafts with good corrosion resistance and excellent torsional strength and fatigue resistance; reduced need for balance weights; better damping characteristics and torsional compliance, which reduces shock loads on gears and universal joints. The mechanical properties enable single span shafts up to 6.5 m in length and not requiring a center bearing to reduce vibration. The fiber is filament wound in a helix to provide optimum properties.

Strongwell manufactures driveshafts for Spicer, who has replaced an Al tube with an overwrap of glass and carbon fiber construction by using a combination of filament winding and pultrusion. Starting with a filament wound glass tube, the carbon fiber is pultruded over it, with the axial presentation of the carbon fiber providing optimum properties for bending stiffness. This construction is used for Class 8 heavy trucks with a span of 3.3 m and is likely to be extended to nearly twice the length for other heavy trucks. The Nissan 350Z has a carbon fiber drive shaft.

23.6.8.2 Buses

North American Bus Industries are using composite bus structures fabricated in Hungary and assembled in the US, made from multiaxial knitted E-glass, but employing some triaxial woven carbon fiber to reinforce areas around windows and roof, permitting standard bus windows to be used.

23.6.9 CNG storage cylinders

Over 10% of buses operating in the US are now powered by natural gas and carbon fiber reinforced CNG cylinders are the likely candidates. Light duty vehicles are generally fitted with two or three pressure vessels, whilst heavy duty vehicles tend to be fitted with four to six. The cylinder types are classified:

1. Type II—metallic vessel body with composite hoop strap (generally fiberglass)
2. Type III—metallic liner with full composite overwrap (mainly carbon fiber)
3. Type IV—all composite construction (mainly carbon fiber)

All composite construction CNG vessels are made by IMPCO Technologies, Lincoln Composites and Quantum Technologies Worldwide in the US and MCS Cylinder Systems GmbH and Ullit in Europe.

Dynatek Industries manufactures cylinders for CNG powered vehicles using carbon fiber/glass composite, offering great savings for transit buses. They are used for the Ford Focus Fuel-Cell Vehicle (FCV).

Du Vall [99] compares the cost of wet filament winding versus prepreg filament winding for Type II and Type IV CNG cylinders and Funck [100] discusses high pressure vessels for compressed gas.

Lincoln Composites manufacture a range of on board vehicle tanks for CNG. The tanks are made by overwrapping a polymeric liner such as polyethylene with a composite and a hybrid mix of glass and carbon fiber is used to improve impact resistance.

Dynatek Industries, based in Calgary, has found it advantageous to use a selected heat treatment to Al liners to provide higher strength and then overwrapping with high strength carbon fiber.

A conformable tank has been developed by ATK Thiokol Propulsion Co. for on-board storage of H_2 for fuel cell vehicles, comprising of a polymer lining, a carbon fiber inner layer and an impact resistant outer layer.

America Technical Center has used a seamless metallic liner overwrapped with carbon fiber for storing 5000 psi hydrogen in a fuel cell.

23.6.10 Motor bikes

Synergy Motorsports have introduced the first carbon fiber chassis for a motocrosser, with the weight saving enabling a 4-stroke engine to be fitted and the weight of the motorcycle is less than that of a comparable 2-stroke machine. Carbon fiber was used for the frame, swing-arm, cradle, air box and several protective covers.

Britten, a New Zealand motorcycle company, has used a composite subchassis made of carbon fiber/Kevlar Derakane epoxy vinyl ester resin laminate for their Britten V1000 racing machine. The Aprila 250 uses cfrp wheels molded in two halves from a 2×2 HS twill weave fabric. HiPer Technology use carbon fiber reinforced nylon for their motocross vehicles.

23.6.11 Railways

Gears fabricated from carbon fiber reinforced nylon thermoplastic molding compound have been used in gearboxes of railway engines.

The front end of the Japanese Shinkansen high speed train is based on a PMI core and carbon fiber/epoxy laminates.

23.6.12 Engineering and textile applications

Fabricating the rotor bow of a wire bunching (stranding) machine from cfrp lowers the weight and allows higher operating speeds.

Icotec have developed a range of composite floating anchor nuts, which snap into a base plate mounted on a structure. One version is based on a carbon fiber/PEEK construction, providing low weight and is able to match the coefficient of expansion of the surrounding PEEK composite material.

23.6.12.1 Structural work

Carbon fiber reinforced composites have been used for structural upgrade and life extension of cast iron struts [101] and cfrp has been used for strengthening of tunnel supports in the London Underground [102], where a 0.5 ton carbon fiber beam carries the same load as a 9 ton steel beam.

23.6.12.2 Robot arms

Lightweight carbon fiber composite has made its debut on the top entry beam of a robot arm [103].

23.6.12.3 Rollers

High modulus carbon fiber can be used to improve the performance of industrial rollers and limit deflection.

23.6.13 Turbine blades

23.6.13.1 Wind turbine blades

Wind based energy is currently the lowest cost per kWh of all the renewable energy sources, but remains more expensive than hydrocarbon based systems. Europe has 75% of the world's existing wind turbines and in 2003, there were some 1030 turbines operating in the UK, generating 588 MW with the intention of increasing output of offshore wind farms to 6 GW by the year 2010. The British Wind Energy Association calculates the present cost is about 2.5p per kWh, which is approximately the same price as a coalfired plant. It is estimated that there is a total world capacity of some 23,300 MW and an output of 1.75 MW would provide about 1000 homes with electricity, saving more than 2,000 tons of CO_2 emissions.

Typical operating sites for wind turbines are mountainous or coastal areas. Wind speeds vary with geographic location and are typically 6 m s^{-1} (13 mph) in Europe, but could be 11 m s^{-1} (25 mph) in New Zealand, producing virtually twice the annual production. It is recommended that the tower is a minimum of 10 m above and situated twice that distance from any obstructions, and the higher the tower, the higher the wind speed. Wind turbine power plants have become quite reliable and two blade units have given way to a three blade design. A typical modern wind farm is featured in Figure 23.40.

About 90% of wind turbine manufacturers are European, with Denmark, Germany and Spain taking active roles [104]. Blades have to be resistant to destruction in high winds by shedding load [105] and large blades tend to use a pitch stall mechanism. The blade is turned edge-on into the wind until they are driven into an aerodynamic stall, diverting the wind and lessening the load on the rotor. Marsh explains that with small wind turbines this would be too expensive and a solution has been found by tailoring the blade's aero-elasticity using carbon fiber, laid off axis by 15°, ensuring that the leading edge of the blade is stretched in higher winds, tending to twist the blade. Jacobs [104] states that the most common design is a three bladed stall controlled constant speed model, followed by a controlled version pitch. Rotors, rotating at 15–25 rpm, are presently up to about 66 m diameter and mounted on steel towers 25–80 m high. It is predicted that the next generation of machines will feature rotor diameters of up to 110 m. SP Systems believes that once blades become over 40 m long, it will then be necessary to use carbon fiber.

Figure 23.40 A typical modern wind farm featuring NEG Micron machines. *Source:* Courtesy of SP Systems.

Table 23.10 Atlantic Orient Corporation's categorization of wind generator sizes

Type	Rotor Size m	Electricity Produced
Micro	5–1.25	20–300 W
Mini	1.25–2.75	300–850 W
Household	2.75–7	0.85–10 kW
Industrial	7–30	10–100 kW
Utility	30–90	100 kW–4 MW

Source: www.aocwind.net

To avoid coming into contact with the support towers, longer blades have to be cantilevered away and as the blade length increases beyond, say 50 m, it will probably be necessary to use cfrp or a glass/carbon hybrid and the cost will be the deciding factor. An average wind speed is 8 m s^{-1} and any changes are an important factor in controlling the cost. A decrease of about 1 m s^{-1} results in a cost increase of 33%, whilst an increase of 1 m s^{-1} provides a reduction of 25%. The power generated is a function of the cube of the wind speed.

Wind generators come in a range of sizes (Table 23.10) with rotors as small as 0.5 m diameter to as large as 90 m diameter. The smaller sizes produce a DC supply, whilst the larger industrial and utility designs produce an AC supply.

Vestas, the world's largest wind turbine manufacturer, markets a range of turbines from 660 kW–2.0 MW and have a model with 47 m diameter blades, generating a nominal output of 660 kW at 1650 rpm at an optimum wind speed of about 15 m s^{-1}.

In Denmark, Vestas have a joint venture with a 40% share of Gamesa Eolica s.a. in Spain, who manufacture carbon fiber wind turbine blades. Another front runner is Danish manufacturer NEG Micon A/S, who acquired the Danish company Wind World in 1997 and NedWind of the Netherlands in 1998.

Wind for the next millennium is discussed [106] and Gallet and Hamlyn describe a new approach in filament winding for large complex parts [107].

Sound level, both mechanical from gearbox, or generator and aerodynamic from swish, when a blade passes the tower has to be taken into account when situated near dwellings.

The Aeromax Corp. has developed stealth acoustic carbon fiber blades, molding the cfrp to slim aerodynamic profiles. The safety aspect is enhanced by using cfrp, which keeps the rotational inertia low even at high speeds. The possibility of lightening strike will have to be accommodated and the longer blades may have to be made on site due to difficulties in transporting by road. It is interesting that the factory of Aerolaminates on the Isle of Wight, UK is situated next to a navigable river, permitting transportation by sea.

There is, however, a possible market for small wind machines to supply individual homes and outlying farms when cfrp becomes a feasible proposition.

23.6.13.2 *Tidal turbine blades*

Marine current turbines derive their power from the tides, which unlike wind flow, is in two directions only. A major difference is that since water is some 800 times denser than air, the thrust on the blades is so much greater and the blades tend to be shorter and sturdier. The turbines are designed to operate in a current of 4.5 knots (2.3 m s^{-1}) but a 1 MW turbine operating at full power would have to withstand some 900 tons s^{-1} of water passing through its blades, which is indeed onerous [107a].

Composites are an ideal choice for the blades and Marine Current Turbines Ltd. are operating a twin axial flow rotor 11 m diameter trial unit driving a generator via a gearbox with a coefficient of performance of 40–45% and an energy capture some 27% better than expected. To achieve these results, the blades are stiffened using carbon fiber in the main spar, surrounded by a grp envelope made in two halves and bonded together to give the blade the requisite hydrodynamic shape, with the grp envelope supported by carbon fiber reinforced ribs attached to the main spar. Although this is early days, the system does appear to have great promise.

23.6.14 Textile applications

A picking stick is a lever which drives the shuttle back and forth in a fly shuttle weaving loom operating at 100 cycles min^{-1}, accelerating a 0.5 kg shuttle to 13 $m\,s^{-1}$ over a 0.3 m distance. Initially, picking sticks were made of densified laminated wood and had poor fatigue resistance, with a life of some 3–6 months and the heavy stick restricted the maximum loom operating speed. Picking sticks fabricated from pultruded carbon fiber epoxy composite were 66% less in weight, operated with 3 dB less noise and had a life of up to 3 years, permitting a 10% speed increase.

Other textile components fabricated from carbon fiber include heald (heddle) frames (these position the warp to allow a shuttle with weft to pass through), lay bars, flyer arms, needles, sinkers, guide bars and faller bars. Due to the controlled expansion, the components do not change overnight when the shed temperature drops. Lucas [108] describes the application of carbon fibers to modern high speed loom sley developments.

23.6.15 Chemical and nuclear applications

A full scale plant for the enrichment of U using gas centrifuges would require many thousands of rotors and cfrp is ideal to enhance the performance by enabling the longest length rotors and highest practical operational speeds to be used.

23.6.16 Medical and prosthetic applications

Carbon fiber has found many uses in biomedical applications [109–139].

Roland Christensen established Applied Composite Technology, a company that makes artificial feet, marketed as Flex-Foot, that uses carbon fiber epoxy prepreg, which was so effective that a sports event participant with an artificial foot ran the 100 m in 11.3 s in the 1996 Atlanta Paralympic Games. These artificial feet are sold throughout the world and production has exceeded some 20,000 [140].

Braided carbon fiber is used by Ossur, an Icelandic company, to manufacture a custom fitted socket for an artificial limb, enabling the sock to conform to the changing contour of the socket. The braid is pre-impregnated with a water-activated polyurethane resin and sealed in a watertight package. When required, the prosthetist places a silicone sleeve over the limb (below the knee), activates the resin with water, positions the wet braid over the silicone sleeve and the resin sets in about 4 min, achieving a 90% cure in 45 min.

A Blatchford & Sons in the UK, market a wide range of prosthetic devices and in the late 1970s, introduced carbon fiber into their prosthetic devices. The cfrp used in their Endolite equipment possessed superior strength, light weight and was easily formed to difficult shapes.

THE USES OF CARBON FIBERS

They prepared a preform from woven carbon fiber or unidirectional prepreg comprising many overlapping individual patterns, molded carefully in a controlled process to avoid fiber buckling. The finished molding is then CNC machined using an in-house CAD system. The cfrp possesses excellent fatigue resistance, easily surpassing metallic limb systems. Blatchford is the largest UK consumer of CF prepreg outside the aerospace industry, with an output as high as 45,000 components per annum (Figure 23.41–23.43).

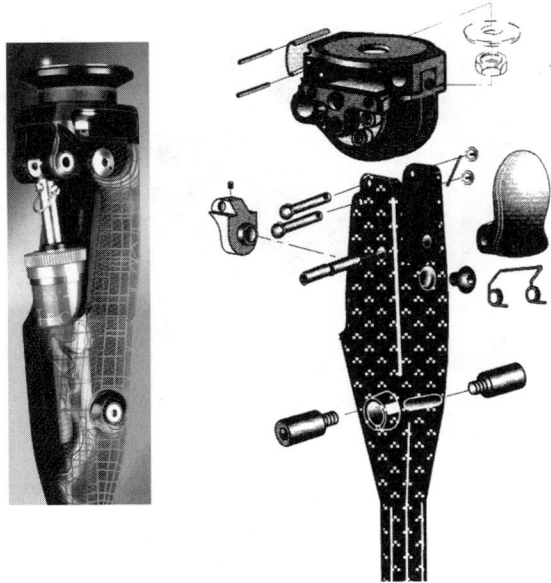

Figure 23.41 Blatchford's 160 Universal shin range. *Source:* Courtesy of Chas A Blatchford & Sons Ltd., Basingstoke, UK.

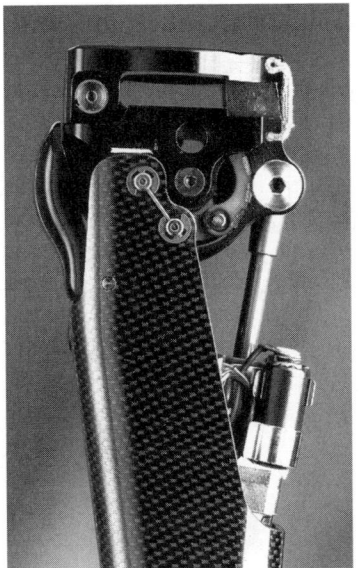

Figure 23.42 Blatchford's Stance flex stabilized knee. *Source:* Courtesy of Chas A Blatchford & Sons Ltd., Basingstoke, UK.

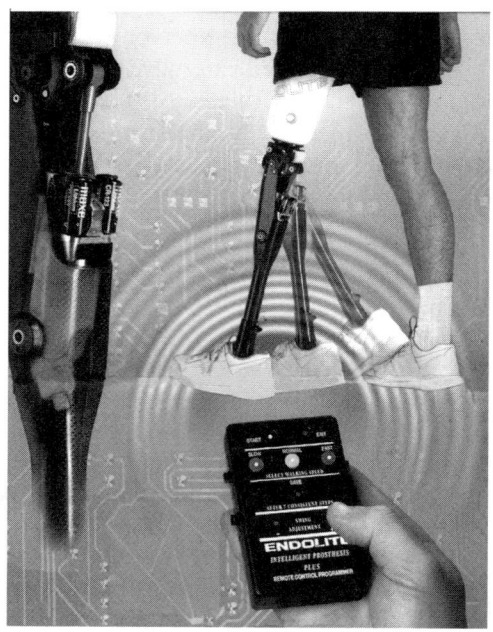

Figure 23.43 Blatchford's The Intelligent Prosthesis Plus. Automatically adjusts the swing of the knee to match the individual amputee's walking speeds. *Source:* Courtesy of Chas A Blatchford & Sons Ltd., Basingstoke, UK.

Huettner [141] discusses the role of carbon fiber composites in state-of-the-art ceramics in surgery, such as reinforced carbon shaft endoprosthesis, and finds that cfrp is suitable for the construction of endoprosthesis shafts having high static and dynamic strength. Claes [142] describes experimental investigations on hip prostheses with carbon fiber reinforced carbon shafts with ceramic heads, finding cfrp more suitable than stainless steel.

23.6.16.1 Hospital equipment

A patient's exposure to X-rays is primarily governed by the structural materials of the equipment and by using cfrp, significant reductions in exposure levels are possible because the absorption level of cfrp is only a seventh that of Al. Hence cfrp is used for X-ray and surgical tables, stretchers and film cassettes [143,144]. Ultrasonic scanner brackets which operate at low temperatures are made from carbon fiber epoxy composite. A light weight stretcher is made by The Sekura Rubber Company using cfrp and Ti.

Busch SA have adapted a Küschall Fusion wheelchair, with a cfrp right angled joint connecting two dissimilar profiled cross-sections, avoiding the occupant making difficult bone embrittling bending operations and also acting as a vibration damper, helping to absorb shocks and vibrations.

23.6.17 Dental

Carbon fiber has been used for many applications in dentistry [145–157].

23.6.18 Sports and leisure goods

An important aspect of many sports goods is the feel of the item, such as when an angler has a bite at the end of the line, or the response of a golf club. Hence, many grades of carbon fiber are available for the designer to use, which makes cfrp an admirable choice as a material of construction.

23.6.18.1 Bicycles, tandem

Kestrel produced the first carbon fiber bike frame in 1986 and the first all-carbon mountain bike frame in 1988. The Talon Sl road bike has a carbon fiber frame weighing only 1.07 kg, using a modular monocoque frame fabricated from three individually molded bladder parts.

A carbon fiber reinforced nylon resin (Towflex®) has been used to make a fabric for the construction of a mountain bike monocoque frame by compression molding in an Al mold, satisfying weight, durability and manufacturing requirements.

A carbon fiber bicycle wheel rim is fabricated by Sun Rims saving 400 g over an Al rim and with 30% less carbon than that achieved by competitors using carbon fiber braid. A core of foam is rolled into a circle to fit the mold rim periphery and a braid sleeve drawn over the core with the ends overlapped. Extra pieces of carbon fiber prepreg are positioned at high stress points, the reinforcement is impregnated with epoxy resin and the lay-up is then compression molded. The pressures are sufficient for the overlap joint to be uniformly incorporated into the rim thickness.

Carbon fiber braid has been used by Centro de Materials Compositos Europa to produce a tubular bicycle frame using RTM.

Trek are renowned for success in the Tour de France with their carbon composite frames and Klein Bikes, owned by Trek, have cfrp chainstays, seat stays and forks [158].

Giant Bicycles claims 17 years of experience in building carbon fiber bikes. The monocoque frames are built with carbon fiber/Kevlar/epoxy resin using a bladder molding technique. The majority of their frames are, however, built with Al, using carbon forks and seat posts.

EPX Bicycles produces bikes with 100% carbon fiber, using a bladder molding technique with 8 layers of CF/E twill weave fabric and up to 20 layers in high stress areas.

The Italian company, Gruppo SPA, market carbon fiber frames, forks, rear chainstays and seat stays, taking advantage of the low weight and vibration damping qualities of cfrp.

Vyatek Sports has introduced Isogrid, using carbon fiber braid to form a grid pattern of ribs which can be molded internally or externally to a part, providing improved resistance to buckling and has been used for handlebars and seatposts. The same company has developed BiFusion tubes, incorporating carbon fiber tubing with co-molded metal endpieces, that can be welded to a metal tube ensuring that during the welding process, the cfrp portion is adequately cooled.

Dave Lloyd supplies carbon fiber frames and bottom brackets, handlebars, head stems and seatpins made by Vision Tech from carbon fiber.

Enders customizes tandem bicycles using a carbon fiber/epoxy frame, utilizing a monocoque construction with primary load bearing in the main beam, producing a laterally stiff frame with vertical compliance to dampen road shock.

Tandem bikes have been made from cfrp [159] and have been specially customized for team riders [160].

23.6.18.2 Bows and arrows

The cfrp reinforced bows provide additional strength and allow the arrow to travel faster and further. In 1971, Hoyt USA first used carbon fiber limbs in a bow. Easton Technical Products market a range of arrows incorporating carbon fiber, using a carbon fiber core with high strength carbon fiber to provide hoop stiffness, with one model bonded onto a precision Al core.

23.6.18.3 Rifles

Christensen Arms, Utah has marketed a rifle with a carbon fiber barrel, utilizing the following benefits of carbon fiber: 3–4 times stronger and 4–5 times stiffer than steel; the barrel casing is about 5 times lighter than steel and the barrel accuracy is not affected by temperature, shooting straight hot or cold, with an additional 25% barrel life due to the improved heat dissipation. Horgesheimer [161] explains how the barrel comprises a steel liner fitted with a high modulus graphite/epoxy casing about 25 mm in diameter, fitted into place by high pressure. In the construction, the steel barrel is stretched inside the liner and, in conjunction with the stiff outer barrel, reduces the barrel vibrations and hence significantly increases the accuracy. Barrel droop caused by heat is reduced. All these benefits contribute to a reputed 25–50% increase in accuracy and the company guarantee 1.25 cm groups at about 100 m.

23.6.18.4 Skis and ski sticks

A carbon/glass Bisphenol-F epoxy reinforcement was used by Glasforms to filament wind an alpine ski pole which resulted in about 15% lighter weight as compared to an Al shaft and an increase in strength of 25%. The tube was made with 3-ply axial (0°) carbon fiber to maximize the stiffness and strength to weight, with the axial plies separated by a ply of circumferential E-glass and another of S-2 glass wound at ±85°.

Atomic Skis uses carbon fiber reinforcement in their skis and have developed a ski with an aluminized top-skin material.

Blizzard Skis employs Kompressor Technology in their skis, with a design incorporating a carbon fiber rod positioned in a channel to produce a stable short ski.

Fischer uses carbon fiber in their higher performance skis but does not use prepreg. Instead, it employs a dry reinforcement utilizing its Air Carbon and Power Vacuum construction.

Head Skis utilizes a Full Carbon Jacket, comprising a wood core torsion box with a cfrp top sheet.

Salewa have introduced a ski–binding, incorporating a carbon fiber frame to produce one of the lightest frame bindings, weighing only 1.2 kg.

23.6.18.5 Snowboards

A snowboard will contain about two thirds composite material and although glassfiber is used as the primary reinforcement, there is potential for carbon fiber.

23.6.18.6 Baseball bats

Carbon fiber is used for baseball bats.

23.6.18.7 Cricket bats

An early application in 1970 for carbon fiber by Gunn and Moore [162] was the use of a number of pieces of narrow pultruded cfrp strips as the springing in the cricket bat handle, replacing the steel, reputedly giving a more powerful drive.

23.6.18.8 Hockey sticks

Although wood is the traditional material for hockey sticks, composites do provide a more durable stick with the ability to efficiently transfer energy and reduce player fatigue. The sticks can be produced as a single piece construction with combined shaft and blade, or a two-piece unit utilizing a separate shaft (cf a golf shaft). Carbon fiber is used in the more expensive sticks. Bending Branches produce a wood/carbon fiber stick.

Exel Oy manufacture carbon/glass replacement blades and are using triaxial braiding to produce a stick which can be tailored with a number of flex points, each with its own stiffness rating.

23.6.18.9 Golf shafts and heads

A carbon shaft weighs about 40% less than a conventional steel shaft, which allows the golf club designer to add weight to the club head and achieve a greater driving distance, yet still save some weight.

The author remembers when carbon fiber was first used for making golf club shafts when it was jokingly stated that it enabled one to hit the ball another 20 yards into the rough, however the popularity of cfrp golf club shafts is expected to continue as players strive to achieve greater distance and accuracy. In 1998, Karl Woodward, using a Jordan Golf thermoplastic shaft (Figure 23.44) fitted with a Ti head enabled him to hit the ball a world record distance of 376.2 yards (344.0 m). The shaft is molded using a gas injection process, resulting in a central core hollowed out by the gas injection and is subsequently wrapped with several layers of carbon fiber prepreg at specific angles to add the required torque and stiffness. When tested, the energy transfer within the shaft produced an increase of 6.4%

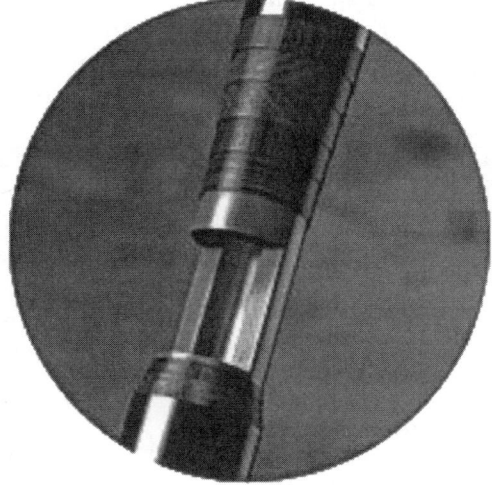

Figure 23.44 Jordan golf thermoplastic shaft. *Source:* Jordan Golf technical literature.

in ball velocity, equating to a further 15 yards (13.7 m) with a 85 mph (137 kmph) swing or 25 yards (23 m) at 115 mph (185 kmph).

The golf club business is very fickle and new materials always attract players wishing to be kitted out with the latest technology. There is no doubt that there has been much research undertaken to improve the golf shaft and head in recent years.

The choice of equipment for golf is obviously a personal issue and some users would claim that a stronger player would be better with steel shafts. However, there is no doubt that from the onset, carbon fiber has made great inroads into golf, perhaps confusing, since they are termed graphite shafts. Interestingly, Yonex introduced a graphite club head in 1990 but the product was withdrawn in 1998.

Phoenix TPC makes a thermoplastic golf shaft from AS-4, IM-6 or IM-7 with PPS, which apparently benefits players with a middle to low handicap. Golf clubs made from thermoplastic shafts are reputed to have less shock, yet matching the thermoset carbon fiber shafts in distance.

Callaway Golf retail a large compression cured carbon fiber composite (C4) head attached to a 53 g graphite shaft [163].

23.6.18.10 Tennis, racquetball, badminton and squash racquets

Dunlop Sports were early leaders in the use of carbon fiber in tennis racquets, using a thermoplastic molding compound [164], but this was replaced by carbon fiber prepreg. Latest racquets have advanced considerably in their design and performance and the so called Muscle Weave racquets are made with twisted braided carbon fiber to give a stronger, more powerful frame and are available in versions for the power hitter with a fast swing speed, the player with a slower swing speed and oversize. CAD/CAM programs are now used to design the products in the UK but the racquets are made in the Far East. A cfrp racquet will maintain its high performance beyond the endurance life of conventional racquets.

Prince, now part of the Benetton Group, has been involved with mass production of tennis racquets in the USA from the early 1970s and has expanded production into Mexico. Prince introduced the oversize racquet in 1976, followed by the longbody version, acquiring Grafalloy and Grafamex in 1983, and expanding into squash racquets in 1987. Versions of racquets have been introduced utilizing carbon fiber and Ti in their construction.

Wilson, in its design of tennis racquet, has introduced models with a Center of Percussion (i.e. a no shock point or, as termed by some manufacturers, the sweet spot) at different positions on the racquet head to accommodate the requirements of players.

Wilson and Head favour a mixed construction of Ti and carbon fiber for their racquetball racquets.

Unfortunately, racquets using carbon fiber in their manufacture have been universally termed graphite, which is technically incorrect and it is interesting to note that Prince has introduced Graphite Extreme^{TM} branding in 1998 to differentiate between the use of a premium grade of graphite (carbon) fiber, which could well be an intermediate modulus version.

Pro Kennex states that it uses a mix of 70% ultrahigh modulus graphite and 30% fiberglass in one of their tennis racquets. Badminton racquets are made from 100% carbon fiber.

23.6.18.11 Snooker and pool cues

The strength and stiffness of carbon fiber adapts well to the construction of snooker and pool cues, although wood is probably the preferred material.

23.6.18.12 Fishing rods and reels

The Orvis Company markets a specialist rod using thermoplastic coated HM carbon fiber to improve the impact resistance and a resin which incorporates thermoplastic and Ti nano-ceramic binders to overcome the inherent brittleness of HM fibers. The rods are lightweight, limiting angler fatigue and have improved casting accuracy.

Loop Tackle markets relatively cheap compression molded chopped carbon fiber fly reels, which are lightweight, with virtually zero thermal expansion and water absorption and excellent corrosion resistance to sea water.

Shikari uses a combination of E-glass and carbon fiber to produce a rod lighter in weight than an all glass rod and has models incorporating standard and HM carbon fiber.

Carbon fiber rods are substantially lighter than fiberglass, or bamboo, and this reduced weight in a typical fly rod increases the response and handfeel of the rod, but more significantly, the damping effect improves the csrp casting distance and accuracy.

23.6.18.13 Hang glider

Center of Gravity Inc. has developed a hang glider harness with a rigid carbon fiber frame in an aerodynamic shape and a smooth finish. The pilot can cause some 30% of the drag on a hang glider and utilizing a tapered tail and solid surface (no profile) in the design contributes a 10% increase in efficiency in straight flight.

23.6.18.14 Canoe paddles

Olympic competition canoe paddles with tapered oval shafts weighing less than 200 g are made by Zaveral Racing Equipment. Two layers of carbon fiber epoxy prepreg are laid at 0 and 90° to the shaft axis and a braided carbon sleeve is then pulled down over the prepreg using a 75 mm diameter ±45° braid, which draws down to the 33 mm diameter shaft, producing a ±18° braid angle with desired bending strength. The paddle is then taped with oriented PP film and oven cured.

For the recreational canoeist, a 425 g blade is made using RTM, which enables the dry braid to be kept in position until it is thoroughly wetted out by the resin.

23.6.18.15 Wind surfing

A carbon/glass hybrid epoxy construction is used by Glasforms to make a stiff, resilient and light weight tube for the masts and battens of wind surfers. The tapered masts and the constant cross-section battens are made by filament winding with the continuous hybrid reinforcement using a multiple construction of axial and biaxial plies.

23.6.19 Musical instruments and Hi-Fi

An early application for carbon fiber was by Decca, to make a carbon fiber dust brush to clean vinyl records, but the author has always considered this a dubious practice due to the possibility that stray carbon filaments could cause damage to electrical circuitry of the Hi-Fi system. Expensive tone arms using cfrp provided increased rigidity with low weight.

23.6.19.1 Loudspeaker cones

Castle Acoustics has developed a 170 mm bass/midrange, which uses a cone fabricated from woven carbon fiber reinforced resin to produce a diaphragm, which is light and stiff, with negligible energy storage potential, resulting in rapid sound transmission and dynamic sound quality, with virtually no coloration.

Sound Domain market 10 (25 cm) and 12 in (30 cm) subwoofers with woven carbon fiber cones and cast Al baskets.

Sennheiser has used a carbon fiber housing, acting as an excellent absorber of energy and cutting out non-musical resonances.

23.6.19.2 Carbon fiber cable

Carbon fiber has been used for Hi-Fi cables.

23.6.19.3 Satellite reflectors

Nippon Granoc high modulus pitch based carbon fiber has been found to be an ideal choice for the manufacture of Ku-Ka band reflectors due to a near zero coefficient of thermal expansion over the operational temperature, and low electrical resistivity and high reflectivity at high RF frequency.

A light weight graphite fabric has been developed specifically for satellite reflectors [165].

23.6.19.4 Stringed instruments

In 1985, RK Carbon Fibres was involved in the utilization of carbon fiber for the production of the necks of guitars using a novel expanding core technique. This entailed wrapping the expanding core with prepreg and carefully stacking the rolled sheets in prescribed positions in a mold, which was then heated. The core, when heated, expanded by about 50%, forcing the material into all parts of the closed mold, a web forming at the point of contact of each wrapped roll. The core material was formulated from cork granules, phenolic micro-balloons and epoxy resins blended together with curing and blowing agents give the requisite viscosity. The moisture in the cork and the N_2 evolved from the blowing agent were sufficient to ensure adequate internal pressure to consolidate the lay-up within the mold, forming a lightweight foam core.

Ovation Guitars were created by Charlie Kaman, a man with a background in aerospace, who developed the roundback guitar based on the structural arch, with the aim of providing rigidity without the need for braces inside the guitar. The shape was designed to reflect the sound towards the soundhole to give better projection and improved volume. Some 20 years ago, Ovation introduced the first carbon fiber top making it half the weight of a spruce top and many times stronger.

Modulus Guitars, in their Genesis range, uses a compression molded carbon fiber element to improve load carrying capacity of the neck of resonant tone woods.

Rainsong claim that 99% of the sound of a guitar comes from the top (soundboard). Damping gives a guitar its tonal character and a carbon fiber/epoxy soundboard has almost constant damping across the acoustic spectrum rather than increasing rapidly at higher frequencies. The guitar also possess the ability to produce louder trebles as compared to a wooden guitar. In an effort to reduce the weight of the soundboard, Rainsong have eliminated traditional soundboard bracing.

Kuau Technology of Hawaii's Rain Song guitars claim that carbon fibers do not absorb high frequencies, whilst aramid fibers dampen unwanted tinniness.

The Zeidler Carrera mandolin uses a stainless steel truss rod with additional carbon fiber reinforcement.

23.6.19.5 Bows for cello and violin

Spiccato Bows and Coda Bows are companies that manufacture carbon fiber bows. The carbon fiber stick does not change in length with temperature, humidity or altitude and one of the models can be adjusted to allow the player to change the flexibility of the stick according to personal taste.

23.6.20 Other end uses in thermoset matrices

23.6.20.1 Model airplanes

It is not surprising to find that carbon fiber has been used for making parts of model airplanes and Northeast Sailplane Products market the quite majestic looking radio controlled Graphite F3J sailplane with a wingspan of 3.1 m (Figure 23.45), carbon fiber is used in the fuselage, wing tips, and trailing edges. Northeast also uses a carbon fiber tube for the FKV Jonny Bee to help achieve light weight.

23.6.20.2 Knives, fountain pens, watches

Probably as a fashion trend, some models of Boker Klotzli Linerlock knives have been fitted with carbon fiber handles, one of them also being gold plated cfrp. Another fashion trend is the Dunhill carbon fiber/palladium fountain pen, whilst Sensa introduced a pen with Ni plated carbon fiber.

The German company Junghens make watches with carbon fiber watch cases.

Figure 23.45 Radio controlled Model Graphite F3J model sailplane employing carbon fiber in the fuselage wing tips and trailing edges. *Source:* Reprinted from Northeast Sailplane Products technical literature.

Figure 23.46 Spectrometer body. A laboratory instrument body which relies on the near zero coefficient of expansion of carbon fiber to give temperature range stability for its operation; replaced a 6 mm thick welded steel plate construction. *Source:* Courtesy of GKN.

23.6.20.3 Precision instruments

An extendable micrometer can be advantageously manufactured from cfrp with the following advantages over its metal counterpart—less weight, no corrosion, less error due to flexing and less affected by body temperature. It is interesting that a set of measuring rods fabricated from cfrp were requested to be used for the scientific dimensional study of the pyramids (pyramidology).

A spectrometer body made from cfrp (Figure 23.46) relies on the near zero coefficient of expansion of carbon fiber to give temperature range stability for its operation and replaces a 6 mm thick welded steelplate construction.

23.6.20.4 Tripods

Gitzo have introduced carbon fiber tripods which have proved most advantageous for nature photographers, being more stable and lighter in weight than similar Al tripods.

23.6.20.5 Optical instruments

23.6.20.5.1 Telescopes

The Large Millimeter Telescope in the State of Puebla, Mexico designed by Simpson, Gumpertz & Heger of Arlington MA, is a 50 m diameter, fully steerable radio telescope, which uses cfrp tube members to minimize deformations due to gravity and thermal loads. It also has low coefficient of thermal expansion and a stiffness to weight ratio that is 40% better than steel or Al.

The main tube, trusses and secondary tube of the amateur portable Uti 8in Newtonian telescope are made from carbon fiber, providing a stable platform for the optics, with virtually no distortion or sagging in the optical path.

Antebi descibes the use of cfrp in the design of radio telescope structures [166].

23.6.20.5.2 Binoculars

Nikon feature several sports binoculars with carbon fiber shells.

23.6.21 Furniture

In Malaysia, an Australian company, Talon, manufactures elegant, durable and weatherproof ultra lightweight tables and chairs from metallic coated carbon fiber incorporated in an epoxy matrix. Composite Design (Australia) also manufactures a range of lightweight furniture from cfrp.

23.6.22 Carbon fiber and wood

Tingley of Wood Science & Technology (WSTI), Corvalis, Oregon patented a technique for reinforcing engineered wood products with a thin sheet of fiber reinforced polymer composite, which could include carbon fiber to contribute stiffness. The glue-laminated wood (glulam) can be used for structural beams in many applications, ranging from I-beams to skateboards and plywood. Many of these applications require stiffness and carbon fiber is the material of choice. The use of FRP provides added strength at a lower cost as compared to wood alone. The use of cfrp reduces the quantity of wood by 20–40%, giving a lighter product that represents some 2% of the total glulam market.

23.7 CARBON FIBERS IN THERMOPLASTIC MATRICES

Carbon fiber reinforced PEI thermoplastic floor panels have been developed for the Gulfstream V corporate jet, replacing a heavier Al design [167]. The floor panels comprised preconsolidated carbon fiber/PEI skins bonded to each side of a Nomex/phenolic honeycomb and processed by thermoforming and ultrasonic welding.

The cfrp has been used in harsh and demanding environments for oil and gas conduits [168]. Rolatube Technology are in early development stages of thermoplastic composite bi-stable reeled (BRC) tubing, preimpregnating UD carbon fiber tows with HD PE or PP by rolling to form a tape and welding along an overlapped seam.

Polycarbonate over-molded with a carbon fiber reinforced PEEK has been used by Entegris to make linear wafer carriers for the semiconductor industry.

EMS-Chemie and Schappe Techniques have produced a flexible carbon- polyamide 12 hybrid yarn by blending PA 12 staple fibers with stretch broken carbon fibers with a staple length of about 80 mm. This hybrid material can be compression molded, giving void free composites.

Ten Cate Advanced Composites produces Cetex, a continuous carbon fiber/PPS composite, widely used for structural aircraft components such as the vertical rudder nose ribs for the A330-200 Airbus.

The Westland Group/GKN have undertaken production of many components from thermoplastic prepreg. Griffiths, Damon and Lawson [169] have described their earlier work and Figures 23.47 and 23.48 depict some of these items.

Vybron Composites manufacture profiles by pultrusion/pulforming using CF/PPS for air compressor blades and guitar neck, CF/nylon with a braided exterior for orthopedic end uses and CF/PEEK for aerospace profiles.

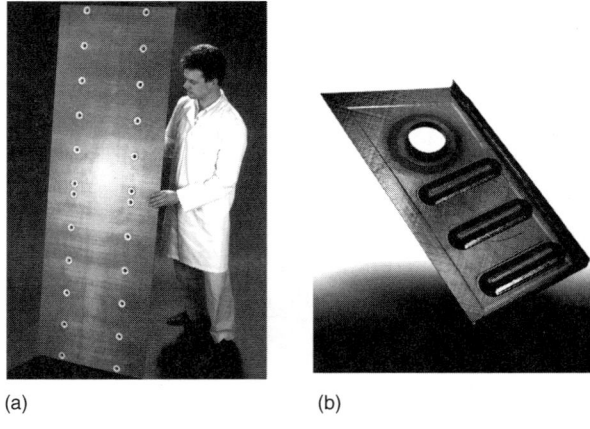

Figure 23.47 A selection of carbon fiber thermoplastic components. (a) Carbon fiber reinforced APC thermoplastic aircraft floorpanel (EH101 Merlin) Makes use of improved toughness and damage tolerance of thermoplastic materials, (b) Bulkhead or wing stiffener made from carbon fiber/PEEK thermoplastic Made in one piece apart from the ring stiffener, (c) Underslung carbon fiber APC2 reinforced thermoplastic tailplane. This item was made over 10 years ago and was the largest piece of flying thermoplastic in the world Fins are fabricated from Ultem 1000 (PEI). *Source: Courtesy of GKN.*

23.7.1 Thermoplastic molding compounds

Carbon fiber is used to reinforce thermoplastic molding compounds, to improve the structural integrity and enhance the wear resistance by increasing the thermal conductivity and creep resistance, thus improving the working pressure velocity of the matrix. It also enables the dissipation of static electricity. The load and velocity capacity (PV) of a bearing material is expressed as a product of the load (P in psi based on the projected bearing area) and the linear velocity (V in ft min^{-1}) of the part. The limiting pressure velocity (LPV) is the maximum PV a material can withstand before it experiences a severe failure, while the working PV is defined as $LPV/2$.

Table 23.11 lists the companies manufacturing carbon fiber filled thermoplastic molding compounds and indicates the types of polymer matrices available.

23.8 CARBON FIBERS FOR CARBON-CARBON APPLICATIONS

The USA is the predominant user of carbon-carbon components, with some 80% of the market split more or less equally between aircraft brakes, re-entry vehicles and rocket nozzles.

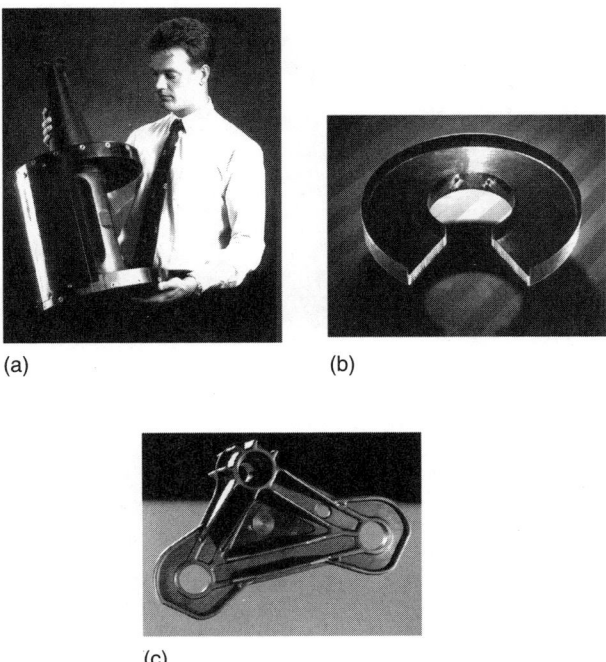

Figure 23.48 Further selection of carbon fiber thermoplastic components. (a) Kinetic Energy vehicle made from carbon fiber/thermoplastic Gave a substantial weight saving using UD construction whereas woven carbon fiber would probably double the weight, (b) Part of a Kinetic Energy vehicle made from carbon fiber/PEEK Fabricated with UD carbon fiber and difficult to make due to low extension of carbon fiber. Easier to make with woven material which would have increased the weight to equal properties obtained with UD, (c) Prototype carbon fiber/IXEF flying control lever Injection molded with chopped fiber. Has the same strength and stiffness as its metal counterpart as well as same weight. Utilizing the geometry of the part to achieve properties. *Source:* Courtesy of GKN.

A reader requiring detailed information on carbon-carbon materials and composites is referred to the following books [170–172] and the information given here is to provide an idea of the extensive range of end uses that have been developed for this product.

Schmidt [173–175] has described unique applications for carbon-carbon composite materials.

23.8.1 Carbon-carbon braking systems

Over 60% of carbon-carbon manufactured is used for braking systems in airplanes, high speed trains, racing cars, motorcycles and tanks. The principal brake manufacturers are Aircraft Braking Systems Corporation (ABSC), BF Goodrich (BFG includes Super-Temp), Dunlop Aviation, Honeywell Aircraft Landing Systems (Allied Signal and Bendix), Messier-Bugatti (includes Carbone Industrie and SEP) and SGL Carbon Group (includes HITCO). Each company has developed its own system of manufacture of carbon-carbon and the information is proprietary.

Aircraft brakes are made up of multiple disks of carbon-carbon, comprising rotors attached to the wheel and stators attached to the brake assembly. The braking action is achieved by hydraulically pressing the disks together. Stimson and Fisher [176] have

Table 23.11 List of carbon fiber filled thermoplastic molding compounds suppliers

Company	Trade name	ABS	ECTFE	ETFE	FEP	PA	PAA	PAI	PBT	PC	PE	PEEK	PEI	PEEKK	PEKK	PES	PET	PFA	PK	POM	PP	PPA	PPE	PPO	PPS	PPSU	PS	PSU	PSOFE*	PU	PVDC	PVDF	THV	TPI	TPX
Akzo Engineering Plastics Inc., USA	Elecrafil	×			×				×	×	×					×	×			×	×			×			×								
Albis, Germany																																			
Atochemie, France	Rilsan					×																													
Bay Mills Ltd., Canada	Baycomp					×				×		×													×										
Delta Environmental Products Inc., USA																																			
EMS Chemie AG, Germany	Grilamid					×																													
Ensige, Germany		×		×	×	×			×	×	×	×	×			×	×			×	×				×		×								
Ferro Eurostar S.A., France					×	×			×							×	×		×	×	×				×				×						
General Electric Plastics, USA	Lexan																																		
Hydro Polymers, UK (Thermofil)	Arpylene					×																													
Junkyong, Korea																																			
Kolon																																			
Lati Spa, Italy						×				×																									

Table 23.11 (cont'd.) List of carbon fiber filled thermoplastic molding compounds suppliers

Company	Trade name	ABS	FE	FEP	PA	PAA	PAI	PBT	PC	EK	PE	PEI	EKK	KK	PES	PET	PFA	PK	POM	PP	PPA	PPE	PPO	PPS	SU	PS	PSU	PSOFE*	PU	PVDC	DF	THV	TPI	TPX
Lehmann & Voss, Germany																																		
LNP Corp., USA (also have plants in Eurpoe and Malaysia)	Thermo-comp	x	x		x				x	x	x	x			x	x			x	x		x		x		x								
Mapril S.A., Portugal																																		
Mitsubishi Rayon Corp., Japan	Pyrofil				x				x	x	x				x							x		x							x			
Mitsui & Corp., Japan																																		
Nippon Steel Chemical Company, Japan																																		
Phillips 66, USA	Ryton																																	
Polychemie, Germany																																		
Polymer Composites Inc., USA	Vari-cut																																	
Raycon Industries, USA																																		
RTP Company, USA	RTP	x	x	x	x				x	x	x	x			x	x			x	x	x	x		x		x			x			x	x	x
Soddy, Taiwan																																		
Tekno Polimer, Turkey																																		
Thermofil Inc., USA	Thermofil	x			x				x		x				x	x			x	x		x		x		x								
Toray Industries Inc., Japan		x			x	x																												
Victrex, UK												x																						

*Note: PTFE is not a thermoplastic polymer.

PSOFE is not a thermoplastic polymer.

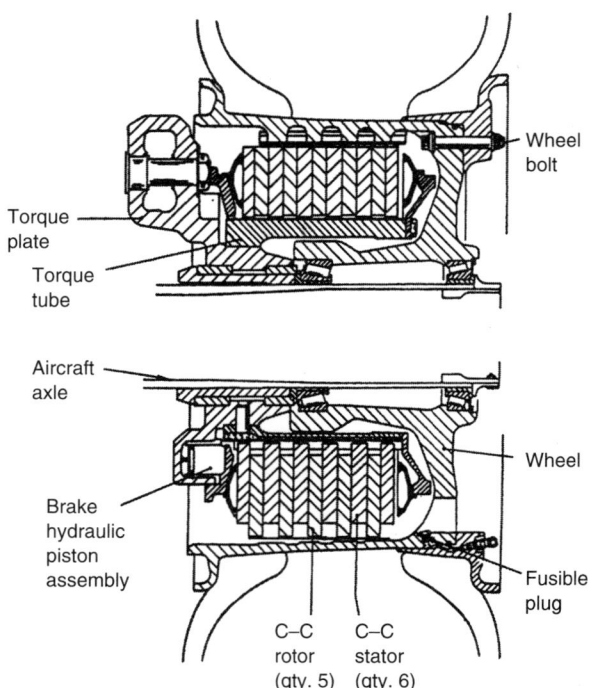

Figure 23.49 A section of a Concorde wheel and brake assembly. *Source:* Reprinted with permission from Stimson IL, Fisher R, Design and engineering of carbon brakes, *Phil Trans R Soc Lond*, A294, 583–590, 1980. Copyright 1980, The Royal Society of Chemistry.

Table 23.12 Typical properties of brake disk materials—carbon-carbon *Vs* steel

Property	Carbon-carbon composite	Steel
Specific heat ($Jg^{-1}K^{-1}$)	1.42	0.59
Tensile strength (MPa)	66	410
Impact resistance (J)	0.7	110
Strain to failure (%)	0.55	33
Thermal conductivity ($Jm^{-1}s^{-1}K^{-1}$)	10–150	59
Coefficient of linear expansion ($10^6\ K^{-1}$)	0–8	14

described the design and engineering of carbon brakes and a section of the Concorde wheel and brake assembly is shown in Figure 23.49. Typical properties of a carbon-carbon composite are compared with steel in Table 23.12 and with high strength graphite in Table 23.13. The carbon-carbon brakes offer a 40% weight saving over steel and their heat capacity is some two and a half times greater. A typical brake temperature of an aircraft on landing could be 500°C, but in an emergency abandoned take-off, the temperature could reach 1300°C, which would be quickly absorbed by the carbon-carbon acting as a heat sink. The thermal loading of a disk is severe and the frictional forces on the brake disk cause it to deflect radially up to some 3.5% of the diameter. In operation, the brake disks contact at local points only, forming an annulus of contact as the disk rotates and wear occurs, releasing carbon-carbon. The band then moves to a new position.

ABSC began testing carbon brakes in the laboratory leading to the first installation, in 1972, of a carbon brake on an F-15A aircraft. ABSC originally purchased their

THE USES OF CARBON FIBERS

Table 23.13 Properties of high strength graphite compared with a carbon-carbon composite

Property	High strength graphite	Carbon-carbon composite
Density (g cm^{-3})	1.82	1.6
Flexural strength (MPa)	55	100
Compressive strength (MPa)	125	300
Thermal conductivity (Jm^{-1}s^{-1}K^{-1}) perpendicular	85	30
parallel		100

carbon-carbon from HITCO, but now produce all their requirements in-house, which is believed to be from a pitch based carbon fiber.

In 1972, BFG, at their Super-Temp Division in Santa Fe Springs, were the first company to develop the carbon-carbon brake and they can now handle carbon-carbon components up to 2.44 m in diameter and 3.66 m in length, providing heat treatment up to 2760°C. Carbon/phenolic preforms are manufactured at the Aerostructures Group at Riverside, whilst carbonization and densification are undertaken at the Carbon Products Division at Pueblo. Since BFG use the CVD process to manufacture their brakes, they are able to refurbish worn disks using the CVD densification process followed by re-machining, thereby doubling the life of the disk.

The SuperTemp process was licensed to Dunlop, who was the first to introduce carbon-carbon brakes into regular airline service, fitting the VC10 in 1973 and the Concorde in 1974. The physical and mechanical properties of a range of Dunlop carbon-carbon composites are given in Table 23.14. A set of carbon-carbon aircraft brake disks is shown in Figure 23.50.

Honeywell Aircraft Landing Systems balances the customer's need and the material property requirements and markets a range of carbon-carbon products.

Table 23.14 Typical properties for carbon-carbon friction material produced for automotive brake rotors, pads and clutch plates

Property		Value x and y direction	z direction
Tensile strength x and y direction		70–100 MPa	5 MPa
Youngs modulus x and y direction		40 Gpa	
Flexural strength		140–179 Mpa	
Flexural modulus		50 Gpa	
Density			1.72–1.8 g cm^{-3}
Coefficient of friction	Under 100°C	0.15–0.35	
	Over 100°C	0.35–0.45	
Emissivity			0.8–0.95
Heat capacity	24°C	0.7 Jg^{-1}K^{-1}	
	511°C	1.6 Jg^{-1}K^{-1}	
	1100°C	2.0 Jg^{-1}K^{-1}	
Thermal conductivity	23°C	0.125 W cm^{-1}K^{-1}	0.046 W cm^{-1}K^{-1}
	300°C	0.188 W cm^{-1}K^{-1}	0.066 W cm^{-1}K^{-1}
	500°C	0.126 W cm^{-1}K^{-1}	0.075 W cm^{-1}K^{-1}
Coefficient of thermal expansion	100°C	0.3×10^{-6}	1.14×10^{-6}
	800°C	5×10^{-6}	10×10^{-6}
Thermal stability			Sublimes above 2500°C

Figure 23.50 Carbon-carbon aircraft brake disks. *Source:* Courtesy of SGL Carbon Group.

SEP first investigated carbon-carbon as a candidate for brakes in 1972 [177] and became involved with aircraft brakes from 1978, whilst Carbone Industrie became involved from 1986. SEP was able to draw heavily on its own experience gained developing carbon-carbon for rocket propulsion, having produced commercial quantities of carbon-carbon from 1974 onwards. Carbon-carbon brakes were tested by Carbone Industrie on Formula One racing cars in 1997, enabling an F1 racing car to decelerate without fading from 350 to 100 km h^{-1} in three seconds, the mechanical characteristics improving as the temperature rose and lasting up to some six times longer than a conventional braking system. The braking performance of carbon-carbon at high speeds is exceptionally good and CI has developed brakes for France's TGV trains, which operate in commercial service at speeds up to 350 km h^{-1}. It is likely, however, that carbon/SiC brake disks may offer improved performance in this application.

The friction, or efficiency factor (μ), of a brake varies during the braking process and initially, with a carbon-carbon brake, μ increases as the operating temperature rises, with fading occurring only at extreme temperatures, when other materials would have long since failed. For instance, each brake in an Airbus jetliner absorbs 70 MJ of energy, stopping the aircraft in 1,300 m with the peak temperature of the carbon-carbon rising to 2,300°C. Typical energies absorbed by each brake for various transport systems are given in Table 23.15. Some carbon-carbon braking systems used in aircraft are listed in Table 23.16.

Table 23.15 Energy absorbed per brake in various systems of transport

Application	Normal braking energy absorbed kJ kg^{-1}	Emergency braking energy absorbed kJ kg^{-1}
Concorde	850	2200
Airbus	400–1000	2400
Mirage 2000	800–1000	1650
Falcon 900	1000	1800
F1 racing car	500	1000
Motorcycle	400	700
Tank	450–800	1000
TGV high speed train	500	1000

Table 23.16 Some carbon-carbon braking systems

ABSC	BFG	Dunlop Aviation	Honeywell	Messier-Bugatti	SGL
F-14	Concorde (1972)	VC10	Boeing 767-300	Mirage F1 (1971)	F-14
F-15A	C-5B	Concorde	Boeing 777	Mirage 2000 (1979)	F-15
Bombardier Challenger 604	Lear Fan 2100	Boeing 757	MD-11	Airbus A310-300 (1985)	F-16
Raytheon Horizon	Grumman X-29A	British Aerospace 146 RJ	Airbus A330	Airbus A300-600 (1985)	FSX
IAI Galaxy	Boeing Vertol 234/414 helicopter	British Aerospace ATP	Airbus A340	Falcon 900 (1991)	B-1B
Sino-Swearingen SJ-30	Embraer AMX	McDonnell Douglas AV8B	F-15	Falcon 2000 (1993)	Airbus A310
Visionaire Vantage	Embraer 135	Aeritalia/Aerospatiale ATR72	F-18A/B	Rafale (1994)	Fokker F-100
F-117	Embraer 140	Gulfstream IV-SP	F-18E/F	Airbus A319 (1996)	Saab SF340
Stealth Fighter	Embraer 145	Gulfstream V	F-22	Airbus A320 (1996)	F117 Stealth Fighter
Swedish JAS-39	de Haviland Dash 8	BAE Regional Jet	Dornier 328	Boeing B767 (1999)	F1 race cars
Taiwanese IDF	Navy's S3	CASA C295		ATR 42	
Japanese F-2	P3-Orion	Fairchild Dornier 328JET		F1 and GT race cars	
	F-16	Eurofighter/Typhoon		Motor cycle and sidecar	
	Boeing 747–400	Maglev train (Japan)			
	Boeing 767–400ER	Thrust SSC car			
	Boeing 777–200	F1 race cars			
	Boeing 777–300				
	Airbus A321				
	Airbus A340				
	Airbus (A340–500)				
	Airbus (A340–600)				
	Airbus (A380)				
	Space Shuttle				

In 1975, General Dynamics awarded Goodyear Aerospace Corporation the carbon brake contract for the F-16 fighter jet aircraft and the carbon-carbon was supplied by HITCO. Goodyear discontinued support for F1 racing cars in 1986 and the following year, Goodyear aerospace became Loral Defense Systems.

HITCO, following the successful development of carbon-carbon brakes for military and commercial aircraft, investigated carbon-carbon as brake materials in high performance racing cars. This development is recorded by Gibson and Taccini [178,179]. The brakes comprised one disk and two pads. Team Brabham became interested in carbon-carbon brakes in the late 1970s and in 1980, obtained encouraging results in the Long Beach F1 Grand Prix race and entered into an agreement with HITCO to develop a carbon brake for F1 racing circuits. Initial trials were retrofits of the cast iron brake rotors, which would always be a compromise. Racing car brakes are used repetitively and an major initial problem was the boiling of the brake fluid due to the heat dissipated by the high surface temperature of the carbon-carbon. This was subsequently accommodated by increasing the air flow to the rotor facings and reducing the heat conduction from the pads by positioning an asbestos/ phenolic insert between the caliper piston and brake pads. Investigations revealed a critical temperature of about 700°C, which if exceeded, would dramatically increase pad wear due to increased surface oxidation. At 120°C, wear rates for rotor and pad were 0.06 and 0.09 mm per 100 km, but if the temperature rose to 220°C, wear became 1.89 and 1.26 mm per 100 km. Hence, the brakes were quite satisfactory on a light/medium braking circuit (e.g. decelerations from 160 (257 kmph) to 80 mph (129 kmph)), but were not adequate on heavy to severe braking circuits, where the wear rate on front wheel pads increased by a factor of 3. Driver's experience found that the carbon-carbon brakes required less effort, had no fade, required less distance for deceleration and showed an improvement in acceleration after severe braking due to a significant weight advantage. An agreement in 1984 with Automotive Products plc saw further improvements to the brakes, such as using a larger diameter rotor (280 mm, now fixed at 278 mm) and increased disk thickness (28 mm), which resulted in less wear due to the higher mass heat sink effect which resulted in lowered surface temperature, combined with improved radial cooling within the disk. AP Racing, Coventry currently supplies two types of carbon-carbon material for their brakes—H Material, which has good wear rate and high strength for very high engine speeds (over 12,000 rpm) such as F1 and S Material, which is cheaper, with an exceptionally low wear rate, but does have a lower strength and is suitable for lower engine speeds (under 10,000 rpm). Figure 23.51 is a typical AP carbon-carbon disk and pads for a GT car. Carbon brakes have been used for Grand Prix and Superbike motorcycles and direct bolt on replacement kits have become available for OE braking equipment.

23.8.2 Carbon-carbon clutches and limited slip differentials

In the early 1980s, carbon-carbon clutches were tried for drag racers, but it was not until 1984, when HITCO supplied carbon-carbon clutch plates to Tilton Engineering Inc. for evaluation, that any real progress was made. Tilton then developed 2- and 3- plate clutches [178,179]. Advantages are weight saving (about 40%); improved wear resistance; resistance to hydraulic fluids; reduction in moment of inertia, centrifugal force and kinetic energy, permitting enhanced engine response and faster gear shifting; and reduced weight at the flywheel end. The clutches are supplied in many configurations and a typical 3-plate clutch is shown in Figure 23.52. As the temperature increases, the clutch grips better, but above 400°C, the wear rate increases due to oxidation. It is believed that as the disk rotates faster, the oxidation debris is thrown from the friction surface, reducing the wear rate,

THE USES OF CARBON FIBERS

Figure 23.51 GT carbon-carbon disk and pads. The disk is 380mm diameter and is the largest disk used by AP. *Source:* Courtesy of AP Racing.

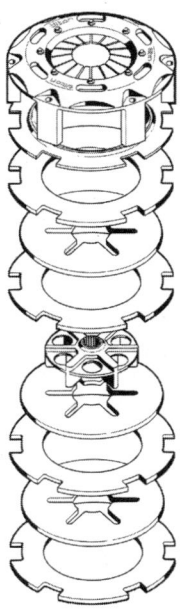

Figure 23.52 Tilton 3plate carbon-carbon clutch. *Source:* Reprinted with permission from Gibson DW, Taccini GJ, *5th Annual SAE Colloquium on Brakes*, Atlantic City, 1–12, Oct 5–8 1987. Copyright 1987, SAE International.

whereas at lower velocities, the particles remain on the friction surface and increase the rate of wear.

AP Racing has developed a series of clutches and currently supplies nine out of eleven F1 racing teams. Clutches are made with up to 9 plates (4 driven and 5 stationary) (Figure 23.53).

HITCO, working in conjunction with the Eaton Corporation, has supplied carbon fiber fabric densified with CVD carbon, which is then bonded to a metal backing plate for use as

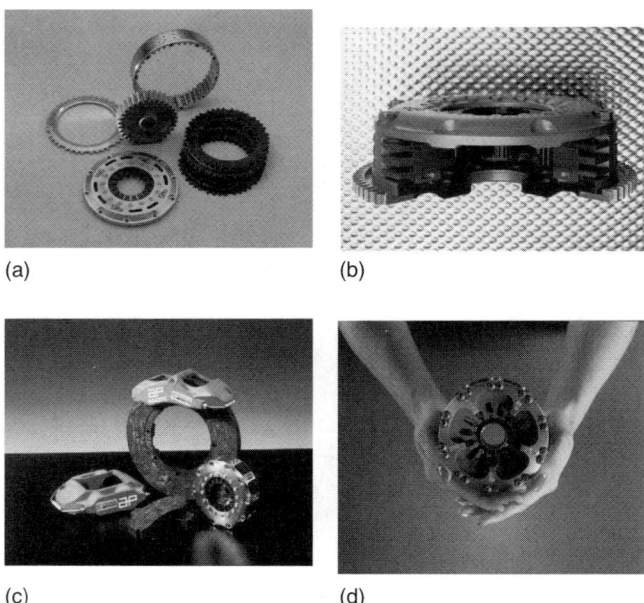

Figure 23.53 A selection of AP carbon-carbon clutches. (a) A ring type carbon-carbon clutch showing carbon stack, adaptor ring and cover that houses the diaphragm spring drive hub and main pressure plate, (b) Cushion flywheel, 140 mm diameter twin plate carbon-carbon clutch; predominantly used in Touring Car formulas, (c) Formula 1 brakes and clutch, 4 and 6 piston brake calipers; 280 mm diameter carbon disk with carbon-carbon pads; 115 mm diameter triple plate push type carbon-carbon clutch, (d) 97 mm F1 clutch, depicts the latest carbon-carbon clutch being used in F1 racing. It has a push type design and is a triple plate clutch. *Source:* Courtesy of AP Racing.

oil immersed limited slip differential. They are used in rear-wheel drive cars and light trucks. Carbon disks are positioned behind each side gear and preloaded with a central spring assembly. Pyrolytic carbon bonded to steel discs are alternated with non-patterned steel disks, creating a clutch pack that restricts the relative wheel rotation and provides more driving force to the wheel with the greatest traction.

23.8.3 Carbon-carbon in space [180]

The shuttle rockets, when taking off, exceed speeds of 27,000 km h^{-1} and the temperature generated on the wing leading edges and nose cone reaches 1,370°C, whilst on re-entry, the temperature is even higher, reaching 1,650°C. Carbon-carbon is an ideal material for these applications, resisting thermal shock from −158°C in space to 1,650°C on re-entry. Another virtue of carbon-carbon is that as the temperature increases, so does the strength. For ablative applications, it is important that the carbon-carbon has a low thermal conductivity and this is best provided by a carbon fiber made from a cellulosic precursor (Rayon Based Carbon Fiber-RBCF). Initially, there was a choice of possible routes:

1. Avtex supplied continuous rayon fiber to Highland Industries for weaving
2. Woven fabric was then supplied to:

    a. Amoco, who carbonized the fabric (e.g. to type WCA) and submitted fabric to Hexcel for prepregging

b. Hitco, who carbonized the fabric (e.g. to type CCA3)
   c. Polycarbon, who carbonized fabric (e.g. to type CSA)

3. The carbonized fabrics were submitted to LTV for fabrication of ablative panels.

Times have changed—Avtex no longer supply continuous rayon; Amoco has stopped production of RBCF and LTV (Vought Missiles) has faded out but has recently re-emerged as Vought Aircraft Industries. However, Polycarbon does remain a supplier of RBCF.

In the 1970s, carbon-carbon leading edges and nose cones were required for the space shuttle and were made using a rayon based precursor. The nose cone, for example, comprised of a two-dimensional lay-up using a phenolic carbon prepreg, which was positioned in a mold, cured, trimmed to size and pyrolyzed. The composite was then impregnated three times with furfuryl alcohol and pyrolyzed after each impregnation. To confer a measure of oxidation resistance, the pack conversion process was used, applying a SiC powder [181] on the inside and outside surfaces and firing at 1,650°C to form a surface layer. Since the SiC is brittle and has a divergent coefficient of expansion, the SiC layer does tend to spall, so it was further protected with repeated applications of tetraethyl-orthosilicate $[Si(OC_2H_5)_4]$ (tetraethoxysilane), using a strong mineral acid as catalyst, which when gelled and heated, formed a fluxed SiO_2 glaze [182], which further reduced the exposed area of carbon.

The carbon-carbon leading edges of the shuttle transfer the aerodynamic forces to the Al substructure and to ensure that the temperature of the Al did not exceed 180°C, a layer of insulation was positioned between the carbon-carbon and Al. The carbon-carbon leading edge can safely withstand the impact from meteors without fracturing.

23.8.4 Carbon-carbon for aircraft

An aircraft capable of reaching Mach II requires a high performance turbojet engine and carbon-carbon with high temperature and load bearing capability has been used successfully for making a turbine wheel capable of operating at 40,000 rpm. To confer extra strength, an extra impregnation/pyrolysis cycle was introduced. The bond between fibers and matrix must not be too strong, which would result in brittleness and at the same time, must not be too weak, which would allow the fibers to debond. In a 2-D lay-up, the principal mode of failure is due to interlaminar shear, failing by delamination between the plies. The strength parallel to the fiber is some 15 times greater than in the transverse direction, hence if failure is expected transversely, a 3-D structure might be preferred.

Carbon-carbon has also been used for ice protection on aircraft.

23.8.5 Rocket motor nozzles and expansion tubes

For rocket motor nozzles and expansion tubes, FMI utilizes a 3-D woven carbon cloth, after examining the weave integrity by using X-rays. The woven structure is pressure impregnated with molten pitch to fill all the interstices, followed by densification and pyrolyzation. Carbon-carbon will endure flame temperatures up to 3,300°C. Figure 23.54 shows a carbon-carbon Peacekeeper missile exit cone.

SEP uses a carbon/SiC composite for liquid rocket engine nozzles which can withstand an operating temperature of 1975°C and 2275°C for a few seconds.

Figure 23.54 Carbon-carbon Peacekeeper missile exit cone. *Source:* Courtesy of SGL Carbon Group.

23.8.6 Carbon-carbon in engines

1. Bearings and seals—carbon-carbon has good wear, friction and thermal characteristics.
2. Valve guides—carbon-carbon has a low coefficient of friction, typically 0.1, over a wide temperature range and it is hoped that current research will negate the use of the engine oil lubricant, which could improve engine power. A near-frictionless grade of carbon-carbon (GlisCarb) has been developed by SGL, offering uniform frictional characteristics in the temperature range 25–800°C, and is associated with extremely good wear resistance.
3. Pistons—conventional Al pistons have a density of 2.7 $g\,cm^{-3}$, but carbon-carbon, with a density of 1.75 $g\,cm^{-3}$, offers a significant weight saving, which can be translated into engine power. In addition, the carbon-carbon offers good strength retention at elevated temperatures, permitting the design of piston to have a better fit, giving a reduction in unburned fuel and providing a greener environment. The low coefficient of thermal expansion suggests that a carbon-carbon piston could be used with a carbon-carbon liner to produce a ringless piston.

23.8.7 Carbon-carbon for biomedical end uses

Carbon-carbon has good compatibility with living tissue as well as chemical inertness and has been used for implants.

23.8.8 Carbon-carbon in industry

Carbon has been used in the chemical industry as a material for construction, offering very good corrosion resistance but, unfortunately, it is mechanically weak and is readily broken (e.g. when cleaning the tubes of a carbon tube heat exchanger)—a situation that can be solved by replacing with carbon-carbon, which has the chemical resistance of graphite combined with the mechanical strength of metals. It can be used for column packings in distillation columns, distillation trays and supports, sparger tubes, feed pipes, mist eliminators, thermowells and pump impellers. The product is resistant to a wide range of chemicals including HF, HCl, HBr, H_2S, H_2SO_4, H_3PO_4 and NaOH.

Carbon-carbon is also used for furnace heating elements, hot gas ducts, hot press dies, molds, items for handling hot glass in the glass making industry and mechanical fasteners including bolts, nuts and studs (e.g. from M6 to M20 made by the Across Co.).

23.8.9 Carbon-carbon as a dielectric heat sink

Applied Sciences Inc. have developed a low cost substrate material called Black●Ice by making a VGCF semi-aligned mat using a fixed catalytic process [183] and converting to a preform with a furfuryl alcohol binder. A CVI, or a molten pitch infiltration technique is then used to densify the preform to produce a carbon matrix composite [184]. The cycle is repeated several times to seal the pores and provide an electrically conductive composite of outstanding conductivity (900 $W m^{-1} K^{-1}$), whereas other brands of carbon-carbon vary as 102–550 $W m^{-1} K^{-1}$. A dielectric coating of diamond [185,186] and SiC or SiO_2 can be applied if required, to give an electrically insulating layer on the carbon-carbon.

23.9 CARBON FIBERS IN CEMENT AND CONCRETE

Chung and co-workers [187–191] have investigated piezoresistivity in carbon fiber composites and how the effect can be used to advantage. Piezoresistivity was observed in cement-matrix composites with 2.6–7.4 vol% unidirectional continuous carbon fibers [191]. The DC electrical resistance in the fiber direction increased upon tensile loading in the same direction, such that the effect was mostly reversible when the stress was below the level of tensile modulus reduction. The increase in resistance was due to fiber-matrix interface degradation, which was mostly reversible. Above the stress at which the modulus started to decrease, the resistance abruptly increased with stress/strain, due to fiber breakage.

Carbon fiber has been found to be an effective thermistor [192–194], such as a cement paste reinforced with chopped carbon fiber (about 5 mm long) with silica fume (15 wt% cement). Its electrical resistivity decreased reversibly with increasing temperature (1–45°C), with activation energy of electrical conduction (electron hopping) of 0.4 eV. This value is comparable to semiconductors (typical thermistor materials) and is higher than that of carbon fiber polymer matrix composites. The current-voltage characteristics of carbon fiber reinforced silica fume cement paste were linear up to 8 V at 20°C.

There are many cases where carbon fiber has been recorded as a smart material [195–203]. Concrete containing short carbon fibers (0.2–0.5 vol%) was found to be an intrinsically smart concrete [201] that could sense elastic and inelastic deformation, as well as fracture. The signal provided is the change in electrical resistance, which is reversible for elastic deformation and irreversible for inelastic deformation and fracture. The presence of electrically conducting short fibers is necessary for the concrete to sense elastic or inelastic deformation, but the sensing of fracture does not require fibers. The fibers serve to bridge the cracks and provide a conduction path. The increase in resistance is due to conducting fiber pullout in the elastic regime, conducting fiber breakage in the inelastic regime and crack propagation at fracture.

Damage in carbon fiber and its polymer matrix and carbon matrix composites has been determined by electrical resistance measurement [204–212].

23.9.1 Carbon fibers in cement and concrete

Carbon fiber reinforced mortar has been used for composite slab construction [213].

Sakai *et al* [214] have used a cfrc for curtain walls, providing adequate wind resistance with higher fatigue strength than ordinary concrete.

23.9.2 Carbon fiber cement as a replacement for asbestos cement

Carbon fiber has made significant advances in replacing asbestos in asbestos cement products in two basic applications—structural covering and flues [215].

In structural covering, the carbon fiber confers a waterproof effect and provides mechanical reinforcement, which is so effective that the amount of reinforcement previously used for asbestos can be reduced by an order of magnitude. To aid subsequent dispersion in an aqueous medium, 5% of glycerol size is applied to a large tow carbon fiber, which is then chopped (6–12 mm long) and admixed with about 5% of other bulking fibers, selected fillers and an antifoam agent. The improvement in mechanical properties has been so good that longer sections of structural covering can be manufactured (Figure 23.55). For flues, carbon fiber is the only material that can be incorporated to provide the necessary resistance to withstand heat up to 450°C with adequate mechanical reinforcement. Carbon fiber, unlike other types of fiber, can provide this temperature resistance, coupled with resistance to the action of alkalis (pH 12–13).

23.9.3 Strengthening of reinforced concrete chimneys, columns, beams and retrofits

A good review of strengthening of reinforced concrete structures is given by Ballinger, Maeda and Hoshijima [216]. Repair and strengthening of concrete structures are discussed [217–220] as is the use of carbon fiber strands and rods or bars [221–223]. Mitsubishi

Figure 23.55 Carbon fiber/cement roofing. *Source:* Courtesy of SGL Carbon Group.

Chemical Corp. produce Leadline, made from Dialead pitch based carbon fibers, which has been used for rods and tendons for pre-stressing concrete.

Reinforced concrete deteriorates with time due to corrosion of the steel reinforcement and environmental effects on the concrete, excessive loading due to earthquakes and wind, coupled with increased loads on, say bridges, due to heavy traffic. These situations necessitate repairs. When steel had been used previously, there were a number of attendant drawbacks such as increased weight of steel—the steel plates had to be welded together—and there was considerable increase in overall thickness due to the protective jacket of concrete.

In Japan, work was expedited by the necessity to repair extreme damage caused by earthquakes, particularly the 1948 Fukai and 1978 Miyagiken Oki. Work concentrated on the use of carbon fiber prepreg and carbon fiber tow. The carbon fiber had a strength of 2.45 GPa and modulus of 236 GPa and was one fifth the weight of steel. However, due regard had to be paid to possible UV degradation and moisture permeability of the resin, changes in ambient temperature, flammability, smoke and toxicity.

In Japan, the original bridge column specifications did not provide sufficient strength against earthquakes and a new specification [224] to avoid brittle failure provided for the shear strength to be higher than the flexural strength, maintaining the lateral resistance when the bridge columns were subjected to a large deformation. This lateral resistance was provided using reinforcement applied to the bottom portion of the column. The seismic design strength was 1.64 GPa, based on an assured strength of 2.45 GPa (i.e. a 1.5 safety factor). Sheets of cfrp were bonded in the vertical direction to provide adequate flexural strength, and the shear strength and ductility were achieved by circumferentially wrapping sheets and/or fiber tows [225].

Kikukawa et al [226] describe the use of carbon fiber textiles for the reinforcement of concrete floor slabs.

Eight 7 m diameter hollow concrete columns, 30–60 m high, supporting the Sakawa River Bridge have been jacketed with a carbon fiber wrap using carbon fiber from Tonen, Mitsubishi and Toray, with concrete-steel splice joints to increase the flexural strength.

Chimneys were reinforced using a similar principle, but employing automated equipment. Initially, attachments such as lightening conductors were removed, then badly damaged areas patched, followed by bonding cfrp sheets in the vertical direction, circumferentially wrapping cfrp sheets or carbon fiber tow and finally repositioning any attachments [227]. At this stage, only one prestressed concrete highway bridge and the floor of an apartment building had been repaired with bonded cfrp materials.

However, work in Switzerland was at a more advanced stage, where strengthening beams by bonding cured cfrp plates to effect repairs [228] was being carried out. In 1991, the multispan prestressed concrete box girder Ibach Bridge in Lucerne County was repaired by bonding 2 mm thick × 150 mm wide cfrp strips to the underside of the damaged beam. The repair was undertaken at night using a lightweight mobile man-lift platform.

In 1996, the Co-op City store in Winterthur, Switzerland was expanded to take two new freight elevators by a specialty contractor Sika AG, who had started to specialize in concrete strengthening work in 1994. The installation of the elevators entailed reinforcing the neighboring concrete in the vicinity of the newly cut floors. This was achieved at the top and bottom of the floors by bonding pultruded epoxy composite plates, which weighed about one fifth of the conventional steel reinforcement—12 mm thick and 100 mm wide for steel against 1.2 mm × 100 mm for the cfrp. Moreover, since the cfrp plates were thinner, they could be threaded under existing pipes and electrical conduits to achieve the requisite crossovers to satisfy the requirements of biaxial reinforcement. The cfrp plates were pultruded by Stesalit AG, Zullwil, Switzerland. The concrete was initially grit blasted and ground and a thin layer of adhesive applied to the concrete and the cfrp plate.

In the USA, initial work in repairing bridge columns used glass and aramid fiber reinforced plastic, and buildings were repaired with bonded steel plates, but it is now believed that these will be replaced by cfrp.

To prevent cracking in prestressed concrete sheet piles, it is necessary to place the reinforcement near the tension surface of the concrete to provide effective crack control. If steel is used, it requires a substantial thickness of concrete cover to provide protection against corrosion, whereas Makizumi *et al* [229] found that carbon fiber net required only one seventh of the cover to provide effective crack control, where the transverse strands of netting play an important role in resisting the applied tensile force. The netting had a mesh size of 20 mm and each strand was $3 \times 18k$ tows of pitch type carbon fiber (10 μm diameter, with strength 1.96 GPa and 176 GPa modulus), impregnated with 40% epoxy resin and fully cured.

Californian earthquakes, such as Whittier 1987, Loma Prieta 1989 and Northridge 1994, demonstrated the vulnerability of older reinforced concrete bridge columns to failure under seismic demands [230].

XXsys Technologies Inc., a San Diego company, specializes in the retrofit of concrete structures using cfrp and has automated the process with a wrapper that applies carbon fiber by wet winding to a column to augment the vertical reinforcement and utilizes portable oven curing technology. The process is used for seismic and corrosion retrofits and a typical corrosion retrofit process is shown in Figures 23.56 and 23.57.

Figure 23.56 XXsys Technologies corrosion retrofit process (part A). (a) Typical corroded bridge structure with cap beam showing extreme corrosion, (b) Corroded structure gritblasted and concrete repaired prior to composite application, (c) Robo-Wrapper can wind column diameters up to 1 m, (d) Robo-Wrapper applying composite material with automatic control, (e) Robo-Wrapper in operation. *Source:* Courtesy of XXsys Technologies Inc., San Diego.

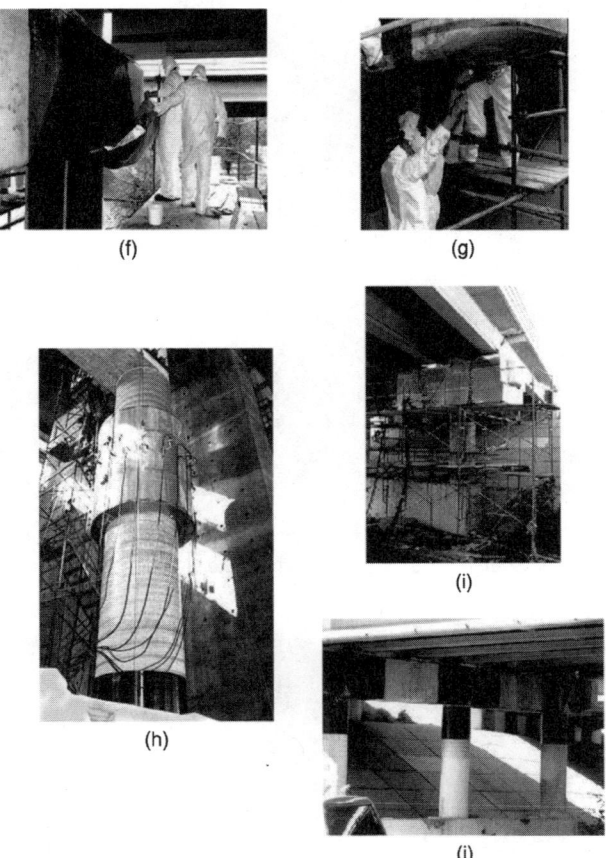

Figure 23.57 XXsys Technologies corrosion retrofit process (part B). (f) Manual application of UD fabric, (g) Adhesive is applied to concrete prior to UD fabric being applied to lower cap beam with adhesive applied between layers, (h) Robo curing system (55 kW m^{-2}), (i) Curing process in operation (2 h), (j) Application complete. Cap beam retrofit and column retrofit. *Source:* Courtesy of XXsys Technologies Inc., San Diego.

The use of bonded cfrp plates is now an accepted cost-effective process in the UK as an alternative to replacement and other traditional methods of strengthening. The cfrp plates can be bonded to concrete, masonry, timber, cast/wrought iron and steel [231–237]. The substrate must be prepared by grit blasting and then vacuumed to remove any dust. The adhesive is applied to both the substrate and the cfrp plate, manually offering the plate up to the substrate, pressed and rolled into position.

The first early steel bridge (built in 1936) to be reinforced with cfrp plates was Slattocks Canal Bridge, Rochdale. Mouchel (a consulting group) (Figure 23.58) used two 4 mm thick ×100 mm wide ×7.5 m long cfrp plates, which were factory bonded prior to bonding on site to the steel beams with Exchem Resiflex adhesive, using temporary support clamps to hold up the plates during cure. Throughout the entire operation, the bridge was carrying traffic.

Hythe Bridge, Oxfordshire (Figure 23.59) is an historic cast iron structure built in 1861 and in 1999, it was upgraded to increase the load carrying capacity from 7.5 tons to 40 tons. The work would have required excessively thick (70 mm) cfrp plates to reinforce the flanges, so Mouchel decided to bond the prestressed cfrp plate, in order to mobilize the locked-in dead load stresses in the cast iron beams, thereby enabling them to carry live loads. Work was undertaken by repair specialists Balvac (part of the Specialist Holdings Division of Balfour

Figure 23.58 The cfrp plates on Slattocks Canal Bridge, Rochdale being temporarily supported with clamps after bonding. *Source:* Courtesy of Dr Sam Luke, Advanced Engineering, Mouchel.

Figure 23.59 Hythe Bridge, Oxford comprising of two 7.5 m spans supported on a central pier strengthened by bonding four prestressed cfrp plates to each of the inner eight beams on both spans. *Source:* Courtesy of Dr Sam Luke, Advanced Engineering, Mouchel.

Beatty Ltd.) and the cfrp plates were prestressed by anchoring one end of each plate and applying a tension of 18 tons using a stressing jack prior to bonding (Figure 23.60). This is the first metal structure in the world where prestressed cfrp plates have been used.

A new development, especially for cast iron bridges, where stiffness is more critical than strength, is to use high modulus cfrp plates and these have been used for a number of projects in the UK such as the Redmile Canal Bridge (Figure 23.61). This used plates 14 mm thick tapering to a thickness no more than 2–3 mm to reduce end peeling stresses, obviating the need for bolting. These plates are manufactured from prepreg and are considerably more expensive than a pultruded plate, but a pultruded section of at least a thickness of 30 mm would have been required and would also have been more expensive to instal.

In the UK, there have been few applications for shear strengthening but during an assessment of an open spandrel bridge in Leeds by Tony Gee & Partners, it became apparent that one of the arch ribs was deficient in shear capacity and was rectified by bonding layers of carbon fiber fabric to the prepared concrete surface, where the fabric assists in arresting diagonal cracks which are limited by the development of bond strength on either side of the crack. Debonding can occur at strains as low as 0.004.

Figure 23.60 Prestressing device mounted on beam. *Source:* Reprinted with permission from Luke S, Composites gain ground in civil engineering, *Reinforced Plastics*, 34–42, Jun 2000. Copyright 2000, Elsevier.

Figure 23.61 Redmile canal bridge strengthened with high modulus cfrp plates. *Source:* Reprinted with permission from Luke S, Composites gain ground in civil engineering, *Reinforced Plastics*, 34–42, Jun 2000. Copyright 2000, Elsevier.

The Pioneer Centre, a Grade II listed structure, was converted to a residential home and it was necessary to introduce a number of staircases to connect the split levels of each unit. The concrete floor was relatively thin (152 mm) and the slabs were strengthened in flexure by bonding 1.2 mm thick pultruded Sika Carbodur laminates onto the bottom and top surfaces of the carefully cleaned concrete substrate prior to cutting the requisite openings (Figure 23.62).

Structural Preservation Systems has an alliance with Master Builders Inc. (MBI) who supply the MBrace Composite Strengthening System using a paper-thin carbon fiber sheet. The system was originally developed by Tonen with Forca carbon fiber [237,237a]. In conjunction with Carl Walker Construction, they applied 32,400 m^2 of MBrace to strengthen 5,200 double tee stems in the parking facility at Pittsburgh Airport. SPS has strengthened several silos by inserting carbon fiber bars into a series of grooves cut into each silo.

The Grayson bridge over the Little Sandy River in east Kentucky has been repaired by applying thin sheets of CF/epoxy to the steel girders with an epoxy adhesive and applying a final protective coat of epoxy, thereby avoiding the expensive replacement of a new bridge.

Figure 23.62 Pioneer Center, Nunhead, London. Reinforced concrete slab trimmed with cfrp reinforcement prior to cutting new opening. Courtesy Tony Gee.

The 8-span concrete Ebey Island bridge near Everett, WA had spalled badly and in 1999, after removing loose concrete and grit blasting exposed steel rebars, carbon fiber/epoxy plies were applied to the webs both longitudinally and transversely to effect a speedy repair.

23.9.4 New structures with cfrp [241]

Herning bridge, an 80 m footbridge in Denmark, will be built as a cable stayed structure with a tower holding the cfrp reinforced concrete deck by means of 40 mm diameter pultruded cfrp cables tested to a capacity of 100 tons. The bridge will be monitored electronically with all cable stays equipped with strain gages, whilst the cfrp reinforcement in the deck will have vibration wire gages. A change in pitch will reveal a change in stress [239]. The estimates for the cost of a new bridge built in steel or cfrp is given in Table 23.17.

A glass/carbon fiber hybrid structure using Hetron resin has been used by Morrison Fiber Glass Co. to manufacture pultruded I-beams for an 18 m long footbridge.

Mitsubishi Chemical manufactures a seismic retrofit material Replark and have formed a joint venture with Fiberite in the USA. Replark is a UD carbon fiber tape with low epoxy resin content with a fiberglass scrim backing to facilitate handling in the field. The application entails the preparation of the concrete, application of an epoxy primer, putty and resin, followed by the requisite number of layers of Replark tape and additional resin, to form a cured composite and finally, the application of a surface coat to improve the appearance. Fiberite use type $12k$ T300 or AS4C carbon fiber and the Replark is offered in areal weights from 175–300 $g\,m^{-2}$.

Table 23.17 Estimates of cost for bridge with steel or cfrp reinforcement

Cost item	Steel $	cfrp at present price $	cfrp at 50% reduction of price $
Construction	662,000	858,000	704,000
Net Present Value replacement of membrane	164,000	0	0
Net Present Value additional maintenance	165,000	214,000	215,000
Total Net Present Value	991,000	1,072,000	919,000

Source: Reprinted from COWI, Denmark.

23.10 CARBON FIBERS IN GLASS MATRICES

Boccaccini and Gevorkian have advocated carbon fiber reinforced glass matrix composites as self lubricating materials for wear applications in a vacuum [242].

23.11 CARBON FIBERS IN CERAMIC MATRICES

Hughes Aircraft [243] patented a graphite fiber reinforced silica matrix composite, comprising of carbon fibers bonded together in a matrix of silica, boron phosphate and β-spodumene modified with a minor amount of an alkaline earth metal oxide. The extremely low (nearly zero) coefficient of thermal expansion coupled with the moderate thermal conductivity and low density of the composite make the composite particularly suitable as a substrate material for high energy laser mirrors.

23.12 CARBON FIBERS IN METAL MATRICES

23.12.1 Electromagnetic interference (EMI) and heat dissipation [244]

The problem of electromagnetic interference and dissipation of heat emitted from electronic products such as computers has become a challenge and EMI is said to exist whenever undesirable voltages or currents are present, adversely influencing the performance of a device. The problem can be alleviated either by making the insulative plastic conductive or by incorporating a metal coating to the surface of the plastic. The plastic may be made conductive by incorporating carbon fiber or a metal coated carbon fiber in a thermoplastic molding compound [245–248].

1. Carbon fiber as the conductive material—high carbon fiber content (30 – 40% w/w) will provide 40 dB of shielding in the 30 MHz–1 GHz range.

    Shielding effectiveness (dB) = $20 \log E_i/E_f$
    where E_i = measured field when material under test is absent
    E_f = measured field when material under test is present [249]

2. Metal coated carbon fiber as the conductive material—metal coated carbon fiber has been used for providing lightening strike protection, EMI/RFI shielding and static dissipation. End uses include Green chaff, Stealth applications, de-icing and power grid counter measures. Composite Materials, LLC has introduced Compmat MCG, a 99.9% pure Ni coated graphite fiber, available in continuous, chopped, mat or tape forms. The chopped form can be supplied with a variety of sizings to optimize dispersion in each type of molding resin.

    a. Plated carbon fiber—electroplated or electroless plated deposits are not pure and are limited to the thickness that can be applied (Table 23.17).
    b. INCO Ni coated carbon fiber—INCO have developed [247] a method of coating carbon fiber with Ni using an adaptation of the carbonyl process by the thermal decomposition of nickel carbonyl gas:

    $$Ni(CO_4) \leftrightarrows Ni + 4CO \uparrow$$

Deposits of 99.87% Ni can be coated on unsized carbon fiber to give products with 20–55% w/w Ni (Incofiber®). For electronic applications, to obtain an optimized target level

Table 23.18 Incoshield long fiber concentrates

Incoshield® Long fiber Ni concentrate	Dry blending polymers
Incoshield® ABS	PVC, ABS, PC, PC/ABS
Incoshield® PA6	PA6, PA66, PA12
Incoshield® PA12	PA6, PA66, PA12
Incoshield® PBT	PBT, PC
Incoshield® PC	PC, PBT
Incoshield® PE	HDPE, ABS, PC
Incoshield® PEI	PEI, PS
Incoshield® PMMA	PVC, PC/ABS, ABS
Incoshield® PPS	PPS

Source: Reprinted from Inco Special Products technical literature.

Table 23.19 EMI shielding effectiveness at a loading of 15% w/w Ni coated carbon fiber independent of thermoplastic polymer

Frequency MHz	Shielding effectiveness dB
30	83
100	91
300	94
1000	104

of 42–48% [248], thermoplastic compatible size can be applied to the Ni coated carbon fiber, which has been chopped to 6 mm length and the chopped product incorporated in a thermoplastic matrix (Table 23.18) to give a concentrate (Incoshield®). This composite contains 60% Ni, which can then be dry blended with a compatible thermoplastic molding compound and finally injection molded [250] to provide EMI shielding for thermoplastic parts. Using this technique, longer carbon fiber reinforcement (6 mm) can be utilized, giving more effective utilization.

Maximum shielding attenuation is achieved with about 15% Ni coated carbon fiber (Table 23.19).

Nickel coated carbon fiber can be used as a conductive plate for fuel cell plates, ice trays and automotive mirror housings; an electronic housing to provide EMI shielding for computers, cellular phones, anti-lock brakes, coaxial cable and telecommunication; as EMI shielding at electronic board level for computers, cellular phones and 900 MHz phones. Bell and Hansen [251] have described the use of Ni coated fibers for aerospace applications.

23.13 OTHER END USES FOR CARBON FIBERS

Chung and Wang have investigated the concept of opto-electronic and electronic devices made from carbon fiber polymer matrix structural composite [252] and as capacitors [253].

REFERENCES

1. Horrocks AR, Tunc M, Price D, The burning behaviour of textiles and its assessment by oxygen-index methods, *Textile Progress*, 18(1–3), 1–205, 1989.
2. Heath TV, Flammability of aircraft interiors, *Presentation by RK Textiles Composites Fibres Ltd.*, Manchester, Aug 19 1981.

3. Saville N, *32nd International SAMPE Symposium*, 1347, 1987.
4. Rebouillat S, Peng JCM, Donnet JB, Ryu SK, Carbon Fibre Applications, Donnet JB, Wang TK, Rebouillat S, Peng JCM eds., *Carbon Fibers*, Marcel Dekker, New York, 526–533, 1998.
5. Suzuki M, Activated carbon fiber-fundamentals and applications, *Carbon*, 32(4), 577–586, 1994.
6. Aily A, Maggs A, GB Pat., 1,301,101, 1969.
7. A Aily, A Maggs, GB Pat., 1,310,011, 1973.
8. Toho Beslon Co., Japanese Pat., 132,193, 1976.
9. Toho Beslon Co., *A process for producing fibrous activated carbon*, UK Pat., 2062599 A, 1981.
10. Toho Beslon Co., *Method for the manufacture of activated carbon fibre*, UK Pat., 2099409 A, 1982.
11. Shimazaki K, Hirai M, Studies on preparation of polyacrylonitrile based activated carbon fiber,1, *Nippon Kagaku Kaishi*, 7, 739–744, 1992.
12. Shimazaki K, Studies on the development of polyacrylonitrile based activated carbon fiber,6. Preparation of polyacrylonitrile based activated carbon fiber (PAN-ACF) having high mesopore volume, *Nippon Kagaku Kaishi*, 7, 807–812, 1993.
13. Yang MC, Yu DG, Influence of precursor structure on the properties of polyacrylonitrile based activated carbon hollow fiber, *J Appl Polym Sci*, 59(11), 1725–1731, 1996.
14. Wang PH, Conversion of polyacrylonitrile fibers to activated carbon fibers, effect of pre-oxidation extent, *J Appl Polym Sci*, 62(10), 1771–1773, 1996.
15. Ko TH, Chiranairadul P, Lu CK, Lin CH, The effects of activation by carbon dioxide on the mechanical properties and structure of PAN-based activated carbon fibers, *Carbon*, 30(4), 647–655, 1992.
16. Wang PH, Yue ZR, Liu J, Conversion of polyacrylonitrile fibers to activated carbon fibers, Effect of activation, *J Appl Polym Sci*, 60(7), 923–929, 1996.
17. CazorlaAmoros D, AlcanizMonge J, LinaresSolano A, Characterization of activated carbon fibers by CO_2 adsorption, *Langmuir*, 12(11), 2820–2824, 1996.
18. Kuraray Chemicals, Japanese Pat., 7,583, 1980.
19. Economy J, Lin RY, *J Mater Sci*, 6, 1151, 1971.
20. Edwards W, *Carbonizable fabrics of activated, carbonized fibers and differently activated or unactivated fibers*, US Pat., 4,714,649 , 1987.
21. Freeman JJ, McLeod AI, Nitrogen BET surface area measurement as a fingerprint for the measurement of pore volume in active carbons, *Fuel*, 62, 1090–1091, Sept 1983.
22. Dresselhaus MS, Fung AWP, Rao AM, Divittorio SL, Kuriyama K, Dresselhaus G, Endo M, New characterization techniques for activated carbon fibers, *Carbon*, 30(7), 1065–1073, 1992.
23. Capon A, Maggs FAP, Robins GA, The mechanical properties of charcoal cloth, *J Phys Appl Phys*, 13, 897–907, 1980.
24. Technical literature from Charcoal Cloth (International) Ltd., Tyne and Wear, UK.
25. Burchell TD, Carbon fiber composite molecular sieves, *Proc 8th Annual Fossil Energy Mater Conf*, Pub Oak Ridge National Laboratory, Oak Ridge, May 10–12, 1994, CONF-9405143, ORNL/FMP-94/1, 63–70, Aug 1994.
26. Klett JW, Burchell TD, Carbon fiber carbon composites for catalyst supports, *Proc 22nd Conf on Carbon*, Pub American Carbon Society, San Diego, July 1995.
27. Kimber GM, Fei YQ, Physical properties of carbon fiber composites for catalytic applications, *Abstracts of Papers of the American Chem Soc*, 211(2), 20–CATL, 1996.
28. Carlton WD, Some bio-medical applications of carbon fibre, *SAMPE meeting*, London, Sep 23–24th 1980.
29. Blazewicz M, Chlopek J, Wajler C, Kus WM, Gorecki A, Carbon composite biomaterials, *Mater Eng*, 7(2/3), 339–351, 1996.
30. Jenkins DHR, et al, *J Bone Joint Surgery*, 1, 59–8, Feb 1977.
31. Johnson Nurse C, Jenkins DHR, *Br J Surg*, 67, 1980.
32. Goodship AE, et al, *Veterinary Record*,106, 217–221, 1980.
33. Reed KP, Vandenberg SS, Rudolph A, Albright JA, Casey HW, Marino AA,Treatment of tendon injuries in thoroughbred racehorses using carbon-fiber implants, *Journal of Equine Veterinary Science*, 14(7), 371–377, 1994.
34. Carbon-Fibers And Hernia Repair, *Lancet*, 336(8721), 976, 1990.

35. Demmer P, Fowler M, Marino AA, Use of carbon-fibers in the reconstruction of knee ligaments, *Clinical Orthopaedics and Related Research*, 271, 225–232, 1991.
36. Aithal VK, Jenkins DHR, Stabilization of a dislocating spastic hip with a carbon-fiber ligament, *Acta Orthopaedica Scandinavica*, 63(6), 679, 1992.
37. Schweitzer G, Carbon-fiber replacement of knee ligaments, *South African Medical Journal*, 84(4), 236, 1994.
38. Becker HP, Rosenbaum D, Zeithammel G, Gnann R, Bauer G, Gerngross H, Claes L, Tenodesis versus carbon fiber repair of ankle ligaments, a clinical comparison, *Clinical Orthopaedics and Related Research*, 325, 194–202, 1996.
39. Muckle DS, Minns RJ, Biological response to woven carbon-fiber pads in the knee - a clinical and experimental-study, *Journal of Bone and Joint Surgery-British Volume*, 72(1), 60–62, 1990.
40. Nicholson P, Mulcahy D, Curtin B, McElwain JP, Role of carbon fibre implants in osteochondral defects of the knee, *Irish Journal of Medical Science*, 167(2), 86–88, 1998.
41. Meister K, Cobb A, Bentley G, Treatment of painful articular cartilage defects of the patella by carbon-fibre implants, *Journal of Bone and Joint Surgery-British Volume*, 80B(6), 965–970, 1998.
42. Park K, Vasilos T, Characteristics of carbon fibre-reinforced calcium phosphate composites fabricated by hot pressing, *J Mater Sci Lett*, 16(12), 985–987, 1997.
43. Bates JJ, Carbon fibre brushes for electrical machines, Marcus Langley ed., *Carbon Fibres in Engineering*, McGraw-Hill, Maidenhead, 194–221, 1973.
43a. Lederer PG, An introduction to Radar absorbent materials (RAM), Royal Signals and Radar Establishment, *Malvern Report No. 85016*, Feb 1986.
44. Fu X, Chung DDL, Carbon fiber reinforced mortar as an electrical contact material for cathodic protection, *Cement Concrete Res*, 25(4), 689–694, 1995.
45. Armstrong-James M, Millar J, Carbon fibre microelectrodes, *J Neuroscience Methods*, 1, 279–287, 1979.
46. Armstrong-James M, Fox K, Millar J, A method for etching the tips of carbon fibre microelectrodes, *J Neuroscience Methods*, 2, 431–432, 1980.
47. Anderson CW, Cushman MR, A simple and rapid method for making carbon fiber microelectrodes, *J Neuroscience Methods*, 4, 435, 1981.
48. Millar J, Simultaneous *in vivo* voltammetric and electrophysiological recording with carbon fiber microelectrodes, Conn PM, ed., *Electrophysiology and microinjection, Methods in Neurosciences*, Academic Press, San Diego, 4, 143–154, 1991.
49. Fu J, Lorden JF, An easily constructed carbon fiber recording and microiontophoresis assembly, *J Neuroscience Methods*, 68, 247–251, 1996.
50. Williams JEG, Millar J, Kruk ZL, A comparison of cut and spark-etched electrodes for fast cyclic voltammetry, *Br J Pharmocol*, 107, 1992.
51. Williams JEG, Millar J, Kruk ZL, Preparation of spark-etched nodes for fast cyclic voltammetry, *Br J Pharmocol*, 107, 1992.
52. Millar J, Williams GV, Ultra-low noise silver-plated carbon fiber microelectrodes, *J Neuroscience Methods*, 25, 59–62, 1988.
53. Page D, Stretching rechargeable battery technology, *Popular Electronics*, 14, 57–58, Jul 1997.
54. Takami N, Satoh A, Hara M, Ohsaki T, Rechargeable Li-ion cells using graphitized mesophase pitch-based carbon fiber anodes, *J Electrochem Soc*, 142(8), 2564–2571, 1995.
55. Suzuki K, Iijima T, Wakihara M, Electrode characteristics of pitch-based carbon fiber as an anode in lithium rechargeable battery, *Electrochimica Acta*, 44(13), 2185–2191, 1999.
56. NC Thomas, The role of hydrogen as a future fuel, *Science Progress*, 72(285), 37–52, 1988.
57. U.S. Pat., 5,874,166, Jan 23rd 1999.
58. U.S. Pat., 5,879,836 Mar 9th 1999.
59. U.S. Pat., 5,882,621 Mar 16th 1999.
60. U.S. Pat., 5,906,900 May 25th 1999.
61. U.S. Pat., 5,994,980 Aug 31st 1999.
62. U.S. Pat., 5,951,959 Sep 14th 1999.
63. Larminie J, Dicks A, *Fuel Systems Explained*, J Wiley & Sons, 2000.
64. Service RF, *Science*, 285, 682, 1999.

65. Connolly DJ, Gresham WF, U.S. Pat., 3282875, 1966.
66. Ogawa H, Shimazaki K, Studies on the improvement of productivity of high-performance polyacrylonitrile based carbon fiber, 8. Effects of carbonization and surface treatment conditions on the performance of redox flow battery using polyacrylonitrile based carbon fibers, *Nippon Kagaku Kaishi*, 12, 1112–1117, 1994.
67. Segit P, Quah M, High speed manufacturing of carbon gas diffusion layers in PEM fuel cells, 2000, *18th Annual Membrane Technology/Separations Planning Conference*, Newton, Dec 4–5, 2000.
68. Gottesfeld S, Zawodzinskiin TA, In: Alkire RC, Gerische H, Kolb DM, Tobias CW eds., *Advances Electrochemical Science and Engineering*, Wiley/VCH New York, 5, 197, 1997.
69. Matthews FL, Rawlings RD, *Composite Materials, Engineering and Science*, Woodhead Publishing Ltd. and CRC Press, 152, 2002.
70. Bielefeldt I, A space age material for vacuum furnace insulation, *Heat Treating Magazine*.
71. Johnson WD, McCullough FP, Light weight, nonwoven materials for fire blocking applications, *Presented at the Aircraft Interior Materials/Fire Performance*, sponsored by The Wichita State University, Apr 4 and 5 1989.
72. Giroux JM, Lichon RJ, Thermal insulation for aircraft fuselage, a new way to save weight, *Presented at the Aircraft Interior Materials/Fire Performance*, sponsored by The Wichita State University, Apr 4 and 5 1989.
73. McCullough FP Jr., Hall DM, *Carbonaceous fibers with spring like deflections and method of manufacture*, U.S. Pat., 4,837,076, Jun 6 1989.
74. Broughton RM, Hall DM, McCullough FP, Development of a new carbonaceous fiber and its uses in non-woven battings, *TAPPI Non-wovens Conference*, Atlanta, 1993.
75. Broughton R Jr., Hall D, Brady P, Shanley L, Slaten BL, The use of a new carbonaceous fiber in thermal insulative battings, *INDA JNR*, 5(4), 38–42, 1994.
76. Smith N, Carbonaceous fibers for fire-retardant insulation, Presented at the TAPPI 1992 Nonwoven Fibers, Properties, Characteristics, and Applications Short Course, *Tappi Journal* 76(4), 176–180, Apr 1993.
77. National Archives and Records Administration, Federal Register Part II, Department of Transportation, Federal Aviation Administration 14 CFR Part 25, *et al*, Sep 20, 2000.
78. Mills AR, Manufacturing technology development for aerospace composite structures, *Aeronaut J Royal Aeronaut Soc*, 539–545, Dec 1996.
79. Mills AR, *Automation of carbon fibre preform manufacture for affordable aerospace applications*, Cranfield University Paper, 1998.
80. Abraham D, McCarthy R, Design of polymer composite structural components for manufacture by resin transfer moulding (RTM), *SAMPE Europe 20th Jubilee Conference and Exhibition*, Paris, 407–415, 13–15 April 1999.
81. Simmons M, Slingsby Sailplanes, Gowood Press, 1996.
82. Marsh G, 'Winged satellites' rely on reinforced plastics, *Reinf Plastics*, 86–90, Mar 2001.
83. Marsh G, Space-a special needs environment, *Reinf Plastics*, 26–30, Jan 2000.
84. Tomita T *et al*, Development of a 8-metre collimation mirror of CFRP, *1st Japan International SAMPE Symposium and Exhibition*, 1710, Nov 28–Dec 1 1989.
85. DiVita G, Filament winding, a key technology for space propulsion, *2nd International Convention for Filament Winding Technology*, Brussels, Oct 17–19 2001.
86. Chapman RD, What kind of material is the Hubble Telescope made of?, *Composite Mater*, May 1996.
87. Blasi R, Tape wrapping of ARIANE 5 booster, *2nd International Convention for Filament Winding Technology*, Brussels, Oct 17–19 2001.
88. Gabrys CW, Bakis CE, Fabrication of thick filament wound carbon epoxy rings using in-situ curing, Manufacture and quality, *Proc Am Soc Composites*, 9th Technical Conf, Technomic, 1090–1097, 1994.
89. Hively W, Reinventing the wheel, *Discover Magazine*, Aug 1996.
90. Ikeda T, *The 73rd JSME Spring Annual Meeting*, 3(96–1), 471–472, 1996.
91. Department of Natural Resources Canada CANMET alternative Energy Division, *Report AED-0397-02*, Rev 1, Apr 1997.

92. Commander MW, Carbon fibre composites for marine applications, *Materials World*, 7(7), 403–405, Jul 1999.
93. Jacob A, Racing catamaran relies on carbon fibre, *Reinforced Plastics*, 36–42, Mar 2000.
94. Stover D, Building the world's tallest carbon fiber mast, *High Performance Composites*, Nov/Dec, 1993.
95. Use of carbon fiber composites for tension leg platform tethers, *Offshore*, 60(5), 156, May 2000.
96. Garseth S, Filament winding applications in the offshore industry, *2nd International Convention for Filament Winding Technology*, Brussels, Oct 17–19 2001.
97. Adam H, Carbon fibre in automotive applications, *Mater Design*, 18(4–6), 349–355, 1997.
98. Gilchrist MD, Curley L, Manufacturing and ultimate mechanical performance of carbon fibre reinforced epoxy composite suspension push rods for a Formula 1 racing car, *Fatigue & Fracture of Engineering Materials & Structures*, 22(1), 25–32, 1999.
99. Du Vall FW, Cost comparisons of wet filament winding versus prepreg filament winding for Type II and Type IV CNG cylinders, *SAMPE J*, Mar/Apr 2001.
100. Funck R, High pressure vessels for compressed gas, *2nd International Convention for Filament Winding Technology*, Brussels, Oct 17–19 2001.
101. Moy SSJ, Barnes F, Moriarty J, Dier AF, Kenchington A, Iverson B, Structural upgrade and life extension of cast iron struts using carbon fibre reinforced composites, *FRC2000 Conference*, Newcastle, Sep 13–14 2000.
102. Moy SSJ, Barnes F, Moriarty J, Dier AF, Kenchington A, Iverson B, Strengthening of tunnel supports using carbon fibre composites, *FRC2000 Conference*, Newcastle, Sep 13–14 2000.
103. Lightweight carbon fiber composite debuts on top-entry beam robot arm, *Modern Plastics*, 74(6), 109, 1997.
104. Jacobs A, Wind energy-the fuel of the future?, *Reinforced Plastics*, 20–24, Feb 2000.
105. Marsh G, Reinforced plastics transform 'small wind'market, *Reinforced Plastics*, 22–26, Jan 2002.
106. 1999 European wind energy conference, Wind for the next millennium. *Proceedings of the European Wind Energy Conference*, Nice, Mar 1–5th 1999, James and James Science Publishers Ltd., London 1250, 1999.
107. Gallet C, Hamlyn A, New approach in filament winding of large complex parts, *2nd International Convention for Filament Winding Technology*, Brussels, Oct 17–19 2001.
107a. Marsh G, Tidal turbines harness the power of the sea, *Reinforced Plastics*, 48(6), 44–47, Jun 2004.
108. Lucas R, Application of carbon fibres to modern high speed loom sley developments, 2, 69, Mar 1975.
109. Ring ND, Benford JM, *Bio Med Eng*, 6, Jan 17–21 1971.
110. Bokros JC, Carbon biomedical devices, *Carbon*, 15, 1977.
111. Hastings GW, *J Phys E*, 13,1980.
112. Bradley JS, Hastings GW, Johnson Nurse C, *Biomaterials*,1 Jan 1980.
113. Bradley JS, Hastings GW, *Bio Eng in North Staffs*, MED Ltd., 9, 3,1980.
114. Nelham RL, Future Priorities in Orthotics and Prostheses Practices, *ISPO UK Scientific Meeting*, University College, London, 27 Mar 1980.
115. Jenkins GW, University of Swansea, *Bio-medical applications of Carbons and Graphites, Physics in Medicine and Biology*, Pub Inst of Physics, Aug 1980.
116. Ducheyne P, Topoleski LDT, Cuckler JM, *Reinforced bone cement, method of production thereof and reinforcing fiber bundles therefor*, U.S. Pat., 4963151, Oct 1990.
117. Ali MS, French TA, Hastings GW, Rae T, Rushton N, Ross ERS, Wynn-Jones CH, Carbon-fiber composite bone plates - development, evaluation and early clinical-experience, *Journal of Bone and Joint Surgery-British Volume*, 72(4), 586–591, 1990.
118. Kwarteng KB, Stark C, Carbon-fiber reinforced peek (APC-2 AS-4) composites for orthopedic inplants, *SAMPE Quarterly*, 22(1), 10–14, 1990.
119. Berard C, Delmas MC, Locqueneux F, Vadot JP, Anticalcaneus carbon-fiber orthosis for children with myelomeningocele, *Revue De Chirurgie Orthopedique et Reparatrice de L Appareil Moteur*, 76(3), 222–225, 1990.

120. Ruka MP, Tungekar MF, Kuit J, Hofstra W, Partial replacement of the esophageal muscle layers by a carbon-fiber prosthesis, *European Surgical Research*, 23(1), 35 et seq, 1991.
121. Minns RJ, Sutton RA, Carbon-fiber pad insertion as a method of achieving soft-tissue augmentation in order to reduce the liability to pressure sore development in the spinal-injury patient, *British Journal of Plastic Surgery*, 44(8), 615–618, 1991.
122. Dubkova VI, Burya AI, Ermolenko IN, Krinitsky AP, A study of biological compatibility of polyamide composites reinforced by element carbon-containing fibers, *Doklady Akademii Nauk Belarusi*, 36(2), 136–139, 1992.
123. Pemberton DJ, Mc Kibbin B, Savage R, Tayton K, Stuart D, Carbon-fiber reinforced plates for problem-fractures, *Journal of Bone and Joint Surgery-British Volume*, 74(l), 88–92, 1992.
124. Pemberton DJ, Evans PD, Grant A, Mckibbin B, Fractures of the Distal Femur in the Elderly Treated with a Carbon-Fiber Supracondylar Plate, Injury-International, *Journal of the Care of the Injured*, 25(5), 317–321, 1994.
125. Brittberg M, Faxen E, Peterson L, Carbon-fiber scaffolds in the treatment of early knee osteoarthritis - a prospective 4-year follow-up of 37 patients, *Clinical Orthopaedics and Related Research*, 307, 155–164, 1994.
126. Klasson BL, Carbon-Fiber and Fiber Lamination in Prosthetics and Orthotics - Some Basic Theory and Practical Advice for the Practitioner, *Prosthetics and Orthotics International*, 19(2), 74–91, 1995.
127. Cooper PR, A carbon fiber reinforced polymer cage for vertebral body replacement, technical note - comment, *Neurosurgery*, 41(5), 1206, 1997.
128. Heim M, Yaacobi E, Azaria M, A pilot study to determine the efficiency of lightweight carbon fibre orthoses in the management of patients suffering from post-poliomyelitis syndrome, *Clinical Rehabilitation*, 11(4), 302–305, 1997.
129. Allcock S, Ali MA, Early failure of a carbon-fiber composite femoral component, *Journal of Arthroplasty*, 12(3), 356–358, 1997.
130. Sonntag VKH, A carbon fiber reinforced polymer cage for vertebral body replacement, technical note - comment, *Neurosurgery*, 41(5), 1206, 1997.
131. Benzel EC, A carbon fiber reinforced polymer cage for vertebral body replacement, technical note - comment, *Neurosurgery*, 41(5), 1206, 1997.
132. Ciappetta P, Boriani S, Fava GP, A carbon fiber reinforced polymer cage for vertebral body replacement, technical note, *Neurosurgery*, 41(5), 1203–1206, 1997.
133. Lewandowskaszumiel M, Komender J, Gorecki A, Kowalski M, Fixation of carbon fibre-reinforced carbon composite implanted into bone, *J Mater Sci—Materials in Medicine*, 8(8), 485–488, 1997.
134. Brooke NSR, Rorke AW, King AT, Gullan RW, Preliminary experience of carbon fibre cage prostheses for treatment of cervical spine disorders, *British Journal of Neurosurgery*, 11(3), 221–227, 1997.
135. Korkala O, Syrjanen KJ, Intrapelvic cyst formation after hip arthroplasty with a carbon fibre-reinforced polyethylene socket, *Archives of Orthopaedic and Trauma Surgery*, 118(1–2), 113–115, 1998.
136. Wang A, Lin R, Stark C, Dumbleton JH, Wear behavior of carbon fiber reinforced PEEK composites in total joint replacements, *Abstracts of Papers of the American Chemical Society*, 216(2), 254–PMSE, 1998.
137. Besnard P, Goutallier D, ACL surgical repair augmented with carbon fibers, long term follow-up on clinical and radiographical outcome, *Revue de Chirurgie Orthopedique et Reparatrice de L Appareil Moteur*, 84(2), 162–171, 1998.
138. Tullberg T, Failure of a carbon fiber implant - A case report, *Spine*, 23(16), 1804–1806, 1998.
139. Wang A, Lin R, Polineni VK, Essner A, Stark C, Dumbleton JH, Carbon fiber reinforced polyether ether ketone composite as a bearing surface for total hip replacement, *Tribology International*, 31(11), 661–667, 1998.
140. Wharton T, Utah Byways, Town's Fancy Footwork is World Famous, *Salt Lake Tribune*, Feb 5 2000.

141. Huettner W, Carbon-fibre-reinforced carbon shaft-endoprosthesis - state of the art Ceramics in Surgery, *Proc. 2nd Int. Symp, on Bioceramics*, Lignano Sabbiadoro, Italy, 16–19 June, 1982, Vincenzini P ed., Amsterdam, Elsevier, 225, 1983.
142. Claes L, Experimental investigations on hip prostheses with carbon-fibre-reinforced carbon shafts and ceramic heads Ceramics in Surgery, *Proc. 2nd Int. Symp, on Bioceramics*, Lignano Sabbiadoro, Italy, 16–19 June, 1982, Vincenzini P ed., Amsterdam, Elsevier, 243, 1983.
143. Dance DR, Lester SA, Carlsson GA, Sandborg M, Persliden J, The use of carbon fibre material in radiographic cassettes, estimation of the dose and contrast advantages, *Br J Radiology*, 70(832), 383–390, 1997.
144. Brennan PC, Hourihan SP, The cost-effectiveness of carbon fibre cassettes in mobile chest radiography, *European Radiology*, 8(2), 301–305, 1998.
145. Manley TR, Bowman AJ, Cook M, *Br Dental J*, Jan 21 1979.
146. Malquarti G, Berruet RG, Bois D, Prosthetic use of carbon fiber-reinforced epoxy-resin for aesthetic crowns and fixed partial dentures, *Journal of Prosthetic Dentistry*, 63(3), 251–257, 1990.
147. Louis JP, Dabadie M, Fibrous carbon implants for the maintenance of bone volume after tooth avulsion - 1st clinical-results, *Biomaterials*, 11(7), 525–528, 1990.
148. Belous NK, Samuskevich VV, Ermolenko IN, Effect of fibrous carbon on setting of phosphate dental cement and its properties, *Russian Journal of Applied Chemistry*, 66(12,1), 2047–2051, 1993.
149. Viguie G, Malquarti G, Vincent B, Bourgeois D, Epoxy/carbon composite resins in dentistry - mechanical properties related to fiber reinforcements, *Journal of Prosthetic Dentistry*, 72(3), 245–249, 1994.
150. Purton D, Payne J, Comparison of carbon-fiber and stainless-steel root-canal posts, *Journal of Dental Research*, 74(3), 750, 1995.
151. Purton DG, Love RM, Rigidity and retention of carbon fibre versus stainless steel root canal posts, *International Endodontic Journal*, 29(4), 262–265, 1996.
152. Fredriksson M, Astback J, Pamenius M, Arvidson K, A retrospective study of 236 patients with teeth restored by carbon fiber reinforced epoxy resin posts, *Journal Of Prosthetic Dentistry*, 80(2), 151–157, 1998.
153. MartinezInsua A, DaSilva L, Rilo B, Santana U, Comparison of the fracture resistances of pulpless teeth restored with a cast post and core or carbon–fiber post with a composite core, *Journal Of Prosthetic Dentistry*, 80(5), 527–532, 1998.
154. Quintas AF, Araujo MAJ, Bottino MA, Snelart N, Carbon fibre posts, effect of surface treatment on core retention, *Journal Of Dental Research*, 77(SIB), 344, 1998.
155. Martinez A, Dasilva JL, Santana U, Rilo B, Mora MJ, Comparison of carbon fibre post and cast dowel-core, *Journal of Dental Research*, 77(5), 1240, 1998.
156. Mannci F, Vicochi A, Ferrari M, Carbon fiber versus cast posts, a two years' recall study, *Journal of Dental Research*, 77(5), 1259, 1998.
157. Mannocci F, Vichi A, Ferrari M, Watson TF, Confocal microscope and SEM evaluation of carbon fibre posts restorations, *Journal of Dental Research*, 77(SIB), 2232, 1998.
158. Nelson R, Bike frame races carbon consumer goods forward, *Reinforced Plastics*, 36–40, Jul/Aug 2003.
159. Michaeli W, Goedel M, Schlegel W, Tandem bicycle frames made from carbon fibre reinforced plastics, *Kunstoffe- German Plastics*, 82(5), 416–419, 1992.
160. Tandem bike customized to team riders, *Reinforced Plastics*, 6, Dec 1998.
161. Hogesheimer J, Graphite rifles, Intron Corporation, July 2000.
162. Gunn and Moore Ltd., Carbon fibre in cricket bats, *Composites*, 1(4), 200, 1970.
163. Buck M, Sporting goods applications for TMC, *Carbon Fiber 2002 Conference*, Raleigh, Oct 21–23 2002.
164. Haines RC, *et al*, *Proc Inst Mech Eng*, 197B, 71, May 1983.
165. Matsumoto T, Saba M, Ishikawa G, Kiuchi N, Watanabe A, Light weight graphite fabric for satellite reflectors. Congress Jina, Nice, France, 12–14 Nov, 2002.

166. Antebi J, The use of cfrp in the design of radio telescope structures, *32nd International SAMPE Technical Conference*, Boston, Nov 5–9 2000.
167. Offringa A, Davies CR, Gulfstream V floors-primary aircraft structure in advanced thermoplastics, *J Adv Mater*, 2–10, Jan 27 1996.
168. Mayer C, Reinforcements for harsh and demanding environments, cutting-edge thermoplastic composite material solutions for oil and gas conduits, *2nd International Convention for Filament Winding Technology*, Brussels, Oct 17–19 2001.
169. Griffiths GR, Damon JW, Lawson TT, Manufacturing techniques for thermoplastic matrix composites, *SAMPE J*, 32–35, Sep/Oct 1984.
170. Buckley JD, Edie DD ed., *Carbon-Carbon Materials and Composites*, Noyes Publications, 1992.
171. Savage GG, *Carbon-Carbon Composites*, Chapman and Hall, London, 1992.
172. Thomas CR ed., *Essentials of Carbon-Carbon Composites*, The Royal Society of Chemistry, Books Britain, 1993.
173. Schmidt DL, Unique applications of carbon-carbon composite materials (Part 1), *SAMPE J*, 35(3), May/Jun 1999.
174. Schmidt DL, Unique applications of carbon-carbon composite materials (Part 2), *SAMPE J*, 35(4), Jul/Aug 1999.
175. Schmidt DL, Unique applications of carbon-carbon composite materials (Part3), *SAMPE J*, 35(5), Sep/Oct 1999.
176. Stimson IL, Fisher R, Design and engineering of carbon brakes, *Phil Trans R Soc Lond*, A294, 583–590, 1980.
177. Cullerier JL, The carbon-carbon story, from rocket propulsion to high-performance brakes, *GEC Alsthom Technical Review*, 8, 23–34, 1992.
178. Gibson DW, Taccini GJ, Carbon/carbon friction materials for commercial brakes and clutches, *5th Annual SAE Colloquium on Brakes*, Atlantic City, 1–12, Oct 5–8 1987.
179. Gibson DW, Taccini GJ, Carbon/carbon friction materials for dry and wet brake and clutch Applications, *40th Annual Earthmoving Industry Conference*, Peora, 1–6, Apr 11–13 1989.
180. Klein AJ, Carbon-carbon composites, Advanced Materials and Processes inc. *Metal Progress*, 186, 64–67, Nov.
181. Dickinson RC, Carbon-carbon composites, Fabrication and properties and selected experiences, Materials stability and environmental degradation, Barkatt A, Verink ED Jr., Smith ER eds., *Materials Research Society Symposium Proc*, 125, 3–13, 1988.
182. Shuford DM, *Enhancement coating and process for carbonaceous substrates*, U.S. Pat., 4,471,023, Sep 11 1984.
183. Lake ML, *Mater Technol*, 1996.
184. Ting JM, Lake ML, *Carbon*, 33(5), 663, 1995.
185. Ting JM, Lake ML, *Diamond/Carbon-carbon composite useful as an integral dielectric heat sink and method for making same*, U.S. Pat., 5,389,400, Feb 14 1995.
186. Ting JM, Lake ML, *Method for making a Diamond/Carbon-carbon composite useful as an integral dielectric heat sink*, U.S. Pat., approved 1996.
187. Shoukai Wang, Chung DDL, Piezoresistivity in continuous carbon fiber polymer matrix composite, *Polymer Composites*, 21(1), 13–19, 2000.
188. Xiaojun Wang, Chung DDL, Short carbon fiber reinforced epoxy coating as a piezoresistive strain sensor for cement mortar, *Sensors and Actuators*, A71(3), 208–212, 1998.
189. Xiaoping Shui, Chung DDL, Piezoresistive carbon filament polymer-matrix composite strain sensor, *Smart Mater Struct*, 5, 243–246, 1996.
190. Xiaojun Wang, Chung DDL, Short carbon fiber reinforced epoxy as a piezoresistive strain sensor, *Smart Mater Struct*, 4, 363–367, 1995.
191. Wen S, Chung DDL, Piezoresistivity in continuous carbon fibre cement-matrix composite, *Cement Concrete Res*, 29(3), 445–449, 1999.
192. Shoukai Wang, Chung DDL, Carbon fiber polymer-matrix composite interfaces as thermocouple junctions, *Composite Interfaces*, 6(6), 519–530, 1999.
193. Wen S, Chung DDL, Carbon fibre-reinforced cement as a thermistor, *Cement Concrete Res*, 29(6), 961–965, 1999.

194. Sihai Wen, Shoukai Wang, Chung DDL, Carbon fiber structural composites as thermistors, *Sensors and Actuators*, A78, 180–188, 1999.
195. Xiaojun Wang, Chung DDL, Continuous carbon fiber epoxy-matrix composite as a sensor of its own strain, *Smart Mater Struct*, 5, 796–800, 1996.
196. Xuli Fu, Erming MA, Chung DDL, Anderson WA, Self-monitoring in carbon fiber reinforced mortar or concrete, *Cement Concrete Res*, 27(6), 845–852, 1997.
197. Xuli Fu, Chung DDL, Vibration damping admixtures for cement, *Cement Concrete Res*, 26(1), 69–75, 1996.
198. Xuli Fu, Chung DDL, Self monitoring of fatigue damage in carbon fiber reinforced cement, *Cement Concrete Res*, 26(1), 15–20, 1996.
199. Pu-Woei Chen, Chung DDL, Carbon fiber reinforced concrete as an intrinsically Smart concrete for damage assessment during dynamic loading, *J Am Ceramic Soc*, 78(3), 816–818, 1995.
200. Pu-Woei Chen, Chung DDL, Concrete as a new strain/stress sensor, *Composites, Part B*, 27B, 11–23, 1996.
201. Chen P-W, Chung DDL, Carbon-fibre-reinforced concrete as an intrinsically smart concrete for damage assessment during dynamic loading, *J Am Ceramic Soc*, 78(3), 816–818, 1995.
202. Chen PW, Chung DDL, Carbon fiber reinforced concrete as an intrinsically smart concrete for damage assessment during static and dynamic loading, *ACI Materials Journal*, 93(4), 341–350, 1996.
203. Xiaojun Wang, Xuli Fu, Chung DDL, Strain sensing using carbon fiber, *J Mater Res*, 14(3), 790–802, 1999.
204. Shoukai Wang, Chung DDL, Apparent negative electrical resistance in carbon fiber composites, *Composites, Part B*, 30(6), 579–590, 1999.
205. Shoukai Wang, Chung DDL, Electrical behavior of carbon fiber polymer-matrix composites in the through-thickness direction, *J Mater Sci*, 35(1), 91–100, 2000.
206. Xiaojun Wang, Shoukai Wang, Chung DDL, Sensing damage in carbon fiber and its polymer-matrix and carbon-matrix composites by electrical resistance measurement, *J Mater Sci*, 34(11), 2703–2714, 1999.
207. Xiaojun Wang, Chung DDL, Real-time monitoring of fatigue damage and dynamic strain in carbon fiber polymer-matrix composite by electrical resistance measurement, *Smart Mater Struct*, 6, 504–508, 1997.
208. Xiaojun Wang, Chung DDL, Sensing delamination in a carbon fiber polymer-matrix composite during fatigue by electrical resistance measurement, *Polymer Composites*, 18(6), 692–700, 1997.
209. Xiaojun Wang, Xuli Fu, Chung DDL, Electromechanical study of carbon fiber composites, *J Mater Res*, 13(11), 3081–3092, 1998.
210. Xuli Fu, Chung DDL, Contact electrical resistivity between cement and carbon fiber, its decrease with increasing bond strength and its increase during fiber pull-out, *Cement Concrete Res*, 25(7), 1391–1396, 1995.
211. Sihai Wen, Chung DDL, Effect of carbon fiber grade on the electrical behavior of carbon fiber reinforced cement, *Carbon*, In press.
212. Pu-Woei Chen, Chung DDL, Carbon fiber reinforced concrete as an electrical contact material for Smart structures, *Smart Mater Struct*, 2, 181–188, 1993.
213. Bayasi MZ, Zeng J, Composite slab construction utilizing carbon fiber reinforced mortar, *ACI Structural Journal*, 94(4), 442–446, 1997.
214. Sakai H, Takahashi K, Mitsui Y, Ando T, Awata M, Hoshijima T, Flexural behaviour of carbon fiber reinforced cement composite, *American Concrete Institute Report*, SP 142–7, 121–129, 1999.
215. High-Performance Fiber-Reinforced Concrete Thin Sheet Products, Originally presented at the ACI Convention in Chicago, Ill., March, 1999, *American Concrete Institute OSP*, 190, 115, 2000.
216. Ballinger C, Maeda T, Hoshijima T, Strengthening of reinforced concrete chimneys, columns and beams with carbon fiber reinforced plastics, *American Concrete Institute Report*, SP 138–15, 233–248.
217. Greenfield TK, Carbon-fiber laminates for repair of concrete structures, *Materials Performance*, 34(3), 36–38, 1995.

218. Meier U, Strengthening of structures using carbon fibre epoxy composites, *Construction and Building Materials*, 9(6), 341–351, 1995.
219. Advanced carbon fiber reinforced polymer composite aids in bridge repair, *ITE Journal-Institute of Transportation Engineers*, 67(6), 19, 1997.
220. High Performance Fiber reinforced Concrete in Infrastructural Repair and Retrofit, *American Concrete Institute OSP185*, 185, 2000.
221. Maissen A, CAM deSmet, Prestressed concrete using carbon fibre reinforced plastic (CFRP) strands, *Materials and Structures*, 31(207), 175–177, 1998.
222. Soudki KA, Green MF, Clapp FD, Transfer length of carbon fiber rods in precast pretensioned concrete beams, *PCI Journal*, 42(5), 78–87, 1997.
223. Abdelrahman AA, Rizkalla SH, Serviceability of concrete beams prestressed by carbon-fiber reinforced-plastic bars, *ACI Structural Journal*, 94(4), 447–457, 1997.
224. Japan Road Association, *Design specifications for highway bridges*, (Part 5 on Seismic Design), 1990.
225. Higashida N, Kobatake Y et al, Retrofit of reinforced bridge columns with carbon fibers, *Proceedings of the 45th annual meeting*, Japan Civil Engineering Association, 1990.
226. Kikukawa K, Mutoh K, Ohya H, Ohyama Y, Tanaka H, Watanabe K, Flexural reinforcement of concrete floor slabs by carbon fiber textiles, *Composite Interfaces*, 5(5), 469–478, 1998.
227. Katsumata H, Yagi K, Applications of retrofit method with carbon fiber for existing reinforced concrete structures, Obayashi Corp. and Mitsubishi Kasei Corp. *Presented at 22nd joint UJNR panel meeting on Repair and Retrofit of Existing Reinforced Concrete Structures*, National Institute for standards and Testing (NIST), Gaithersburg, 1990.
228. Meier U, Kaiser H, Strengthening of structures with cfrp laminates, *Speciality Conference on Advanced Composite Materials in Civil engineering Structures*, Las Vegas, American Society of Civil engineers, New York, Jan 1991.
229. Makizumi T, Sakamoto Y, Okada S, Control of cracking by use of carbon fiber net as reinforcement for concrete, *American Concrete Institute Report*, SP 138–18, 287–295.
230. Marsh G, Seismic retrofit provides opportunities for FRP, *Reinforced Plastics*, 38–43, Mar 1998.
231. Meier U, Rehabilitation and retrofitting of existing structures through external bonding of thin carbon-fiber sheets, *Materials and Structures*, 28(176), 105–106, 1995.
232. Garden HN, Hollaway LC, Thorne AM, A preliminary evaluation of carbon fibre reinforced polymer plates for strengthening reinforced concrete members, *Proceedings of the Institution of Civil Engineers-Structures and Buildings*, 122(2), 127–142, 1997.
233. Garden HN, Hollaway LC, An experimental study of the influence of plate end anchorage of carbon fibre composite plates used to strengthen reinforced concrete beams, *Composite Structures*, 42(2), 175–188, 1998.
234. Luke S, Skwarski A, Strengthening with carbon fibre plates, *Reinforced Plastics*, 48–50, Mar 1998.
235. Mitsui Y, Murakami K, Takeda K, Sakai H, Study on shear reinforcement of reinforced concrete beams externally bonded with carbon fiber sheets, *Composite Interfaces*, 5(4), 285–295, 1998.
236. Kimura K, Kobatake Y, Long-term performance of reinforced concrete beams with carbon fiber sheet, *Composite Interfaces*, 5(4), 297–303, 1998.
237. Materials - parking garage repair completed with carbon fiber, *Civil Engineering*, 69(3), 24, 1999.
237a. McConnell V, Composites make progress in reinforcing concrete, *Reinforced Plastics*, 40–46, Jul/Aug 1999.
238. Luke S, Composites gain ground in civil engineering, *Reinforced Plastics*, 34–42, Jun 2000.
239. Swiatecki S, Building better bridges with cfrp, *Reinforced Plastics*, 44–47, Mar 1998.
240. Moriarty J, Barnes F, The use of carbon fiber composites in the London Underground Ltd, Civil Infrastructure Rehabilitation Program, *SAMPE J*, 34(2), 23–28, 1998.
241. Holloway LC, Leeming MB eds., *Strengthening of reinforced concrete structures*, Woodhead Publishing, Cambridge, 1999.
242. Boccaccini AR, Gevorkian G, Carbon fibre reinforced glass matrix composites; self lubricating materials for wear applications in a vacuum, *Glastech Berichte*, 2001 (In the press).

243. Hughes Aircraft, *Graphite fibre reinforced silica matrix composite*, UK Pat. Appl. GB2208076 A.
244. Chung DDL, Materials for electromagnetic interference shielding, *J Mater Eng Performance* 9(2), 2000.
245. Xiaoping Shui, Chung DDL, Submicron nickel filaments made by electroplating carbon filaments as a new filler material for electromagnetic interference shielding, *J Electron Mater*, 24(2), 107–113, 1995.
246. Rosenow MWK, Bell JAE, Injection moldable nickel coated carbon fibre concentrate for EMI shielding applications, *Proceedings of the SPE 55th Annual Technical Conference*, ANTEC'97, April 27th–May 2nd, 1492–1498, 1997.
247. Rosenow MWK, Long nickel fibers for EMI shielding, *ETP'99 World Congress Engineering Thermoplastics*, Zurich, Jun 7–9 1999.
248. Rosenow MWK, Bell JAE, Injection moldable Nickel Coated Carbon Fibre Concentrate for EMI shielding applications, *Proceedings of the SPE 55th Annual Technical Conference*, ANTEC '97, 1183–1193, Apr 27–May 2 1997.
249. Tesche FM, Ianvoz MV, Karlsson T, *EMC Analysis Method and Computational Models*, John Wiley & Sons, 1997.
250. Rosenow MWK, Injection molding of long carbon fiber reinforced thermoplastic materials, *Proceedings of the 40th International SAMPE Symposium and Exhibition*, 1534–1541, May 8–11 1995.
251. Bell JAE, Hansen G, Nickel coated fibres foraerospace applications, *Proceedings of the 24th International SAMPE Technical Conference*, Toronto, T902-T911, Oct 20–22 1992
252. Chung DDl, Shoukai Wang, Carbon fiber polymer-matrix structural composite as a semiconductor and concept of optoelectronic and electronic devices made from it, *Smart Mater Struct*, 8, 161–166, 1999.
253. Xiangcheng Luo, Chung DDL, Carbon fiber polymer–matrix composites as capacitors, *Composites Sci Technol*, In press.

CHAPTER 24

Looking to the Future

24.1 THE FUTURE

One fact is certain for carbon fiber—the future is black.

In the first chapter, the initial description of individual particles could be replaced by a number of String Theories, where particles are envisaged as tiny vibrating strings with each string tied down to a membrane and every sub-particle is associated with a much heavier particle, termed a Smarticle, possessing super symmetry.

Certainly, the future of carbon fibers may be strongly influenced by nanofibers and Pyrograf Products Inc., a subsidiary of Applied Sciences Inc., have built a plant to produce Pyrograf III-R a vapor grown carbon nanofiber.

24.2 THE PRODUCTION PROCESS

The world market for carbon fibers has always been very fickle and in the past has proved problematic for some companies.

Conoco considered that carbon fibers were overengineered for many applications and built a plant for random oriented carbon fiber mat, which could readily be used by dispersing in matrices like asphalt, carbon, concrete, plastics etc. Although the plant was acceptable, the technology of converting a chopped mat to more useable forms was not economic and the process was shut down.

To serve future markets some companies have obtained long term deals like Zoltek with Vestas Wind Systems AS of Denmark.

24.2.1 Precursor developments

Can new PAN precursors be developed which are totally free from Na, cheaper and have definitive rate determining stages? However, such introductions would involve expensive re-qualification procedures. The reduction of flaws by more effective dope filtration would improve the attainable carbon fiber tensile strength, but there is probably no more latitude for improvement via this route.

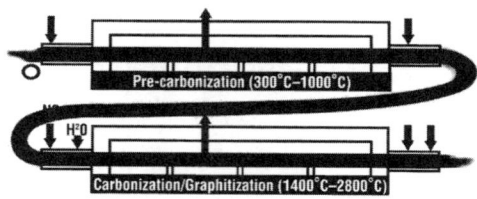

Figure 24.1 Dual fiber slot furnace combining LT and HT operations. *Source:* Courtesy Harper International Corporation.

Reducing filament diameter would reduce the gradient of properties across the filament diameter, but further reductions, below say 4 μm diameter, would almost certainly have associated health risks.

Along more conventional lines, Mitsubishi Gas Chemical Co. are producing a synthesized mesophase pitch AR, which can be used for carbon–carbon and carbon fibers.

24.2.2 Plant developments

Harper have made a combined LT/HT furnace, one furnace sitting atop the other (Figure 24.1), although innovative it is difficult to perceive how this could be easily operated.

The Composite Materials Technology Group at Oak ridge National Laboratory are investigating the microwave heating of PAN precursor in a plasma instead of conventional thermal processing, which could cut the processing costs by 50% and could result in a significant 20% reduction in the total cost of carbon fiber.

24.3 CARBON FIBER

Lukjanova and Lovzova [1] have described how carbon fibers occur in nature. Certainly, more research on surface treatment should help to apportion a specific treatment for a given matrix. The introduction of better sizes would be expected, probably applied by electrodeposition, or electropolymerization, to provide better control during application. The surface treatment of carbon fiber may be tailored for a given resin matrix and similarly, the size can be chosen to perform with a given resin matrix or production process. Hydrosize Technologies Inc. has developed a special size for vinyl esters [2].

Any improvement in the application of a coating, which would facilitate the wetting of metals, would be most welcome.

Improvements in the compressive properties of a PAN based fiber would be beneficial, but may be at odds with the compressive properties which depend on a well disordered structure.

The Defense Evaluation and Research Agency (DERA), (now Qinetiq) UK has developed a continuous hollow PAN based carbon fiber with an outside diameter of approximately 20 µm and a wall thickness of approximately 5 µm [3]. Weight for weight, the new fiber can carry 30% more load than solid fibers without buckling. This fiber should be much more resistant to buckling than a narrower standard fiber. In addition, it is possible to incorporate a core material down the center of the fiber.

Fain and Eddie have developed a process to produce a hollow C-section carbon fiber by extruding a carbonaceous anisotropic liquid precursor through a spinneret with a C-section capillary, followed by carbonizing in an inert atmosphere for 1–5 min at 600–1000°C and 5–10 min at 1550–1600°C.

Sailor Research has electrosynthesized carbon fibers, depositing small fibers on a Cu cathode by the electroreduction of CCl_4 in acetonitrile, using a Pt anode at a current density of 3 mA cm^{-2}. The fibers are 0.1–5 µm in diameter with aspect ratios of 2–100, amorphous and containing about 75–85% carbon with about 5–10% each of N_2, O_2 and Cl_2. The materials are highly porous and can accommodate large ions.

24.4 COMPOSITE MANUFACTURING TECHNIQUES

In the techniques used for manufacturing composites, the effect of atmospheric pollution is an important consideration and one-pack resin systems, RTM processing, etc. will be significant local issues.

Electron Beam (EB) curing has undergone considerable development over the last few years and companies are introducing resin systems particularly suitable for this process. The addition of photo-initiators, coupled with improved accelerators, has helped to reduce the cost of EB equipment. EB curing, as distinct from UV curing, can rapidly penetrate deep (25 mm) the composite component, of the order of seconds in one pass, at low temperatures and in a controlled manner. Lower processing temperatures permit the use of cheaper molds for manufacture of advanced composite parts. Matrix systems can be cured using either a free radical system or a cationic one, although a thermal post cure may be required for the latter. Typical cationic initiators include Union Carbide UV6990 and UVI 6974; Sartomer CD1012, General Electric Silicones OPPI and Sandia National Laboratories DW1.

24.5 QUALITY MANAGEMENT STANDARDS

The year 2000 saw the projected revision of ISO 9000, permitting organizations to go beyond simple compliance with Quality Management System requirements for certification purposes. The former ISO 9000 family comprised over 20 standards which has been reduced to 3:

ISO 9000:2000 (QMS-Fundamentals and vocabulary)
ISO 9001:2000 (QMS-Requirements)
ISO 9004:2000 (QMS-Guidance for performance improvement)

24.6 RECYCLING

An important aspect of the composite world is the issue of recycling [4]. Kemmochi et al [5,6] examined the possibility of closed loop material recycling for fiber reinforced thermoplastic composites and Allred et al [7] have developed a tertiary recycling process.

Adherent Technologies Inc. [8] has developed a process for the reclamation of carbon fibers from carbon/epoxy composites. It has studied the depolymerization of thermoset carbon fiber reinforced epoxy matrix composites using a low temperature (20 min at 325°C) catalytic tertiary recycling reclamation process and has been able to obtain a product with 99.8% carbon and 0.2% residual resin, with only a loss of about 8.6% in fiber tensile strength. The process can be economically viable, provided sufficient scrap feedstock is available. Possible applications for the recovered fiber include thermoplastic and thermoset molding compounds.

In the UK, Nottingham University has undertaken trials utilizing a fluidized bed [9], operating at about 500°C in conjunction with a secondary combustion chamber, which is needed for the breakdown of the resin system.

24.7 INNOVATIVE DEVELOPMENTS

Warren [10] has discussed the use of carbon fiber in future vehicles and Schmidt et al [11] have outlined the unique applications of carbon–carbon composite materials.

In space, solar sails attached to a spacecraft reflect photons of sunlight and harness their momentum; the sail must however be super-lightweight to benefit from the massless particles bouncing off the sail and must be extremely reflective to maximize the propulsive force. More energy is delivered when closer to the sun. The force on the sail will constantly accelerate the spacecraft over time and will eventually exceed the velocity of a rocket. Energy Science Laboratories has recently developed a carbon fiber web with an areal weight of 5 g m^{-2}. The mesh is a network of criss-crossed carbon fiber nanotubes, which will not require a cumbersome deployment structure, since it can be folded and packed, but will jump out into position when released (cf crumpled silk). Additionally, the sail can be propelled by lasers heating the sail, making it glow orange/white (2600°C), while the web still maintains its integrity.

Nanoporous carbon fibers and carbon nanotubes exhibit a high H_2 storage capacity [12]. Carbon nanotubes have also been demonstrated to store Li efficiently. The textile industry is likely to be hugely impacted by nanotechnology [13].

Carbon nanotubes can be either semiconducting or conducting (metallic), depending on the chirality and diameter of the nanotube. These nanotubes can conduct heat as efficiently as diamond and electricity as efficiently as copper.

To improve the mechanical properties of activated carbon fiber, a dual viscose and PAN based carbon filament has been developed [14], with the core made of PAN based fiber and an external layer of viscose based fiber, which can readily be activated with CO_2. The central core comprises of about 50 discontinuous PAN filaments held together by a continuous filament of PVOH and concentrically wrapped by discontinuous viscose filaments using a friction spinning process. The dual filament is carbonized in an inert atmosphere at 900°C and activated by CO_2 at 900°C. Another way of achieving a coating of cellulose would be to apply a solution of cellulose in a mix of water and 4-methylmorpholine-N-oxide.

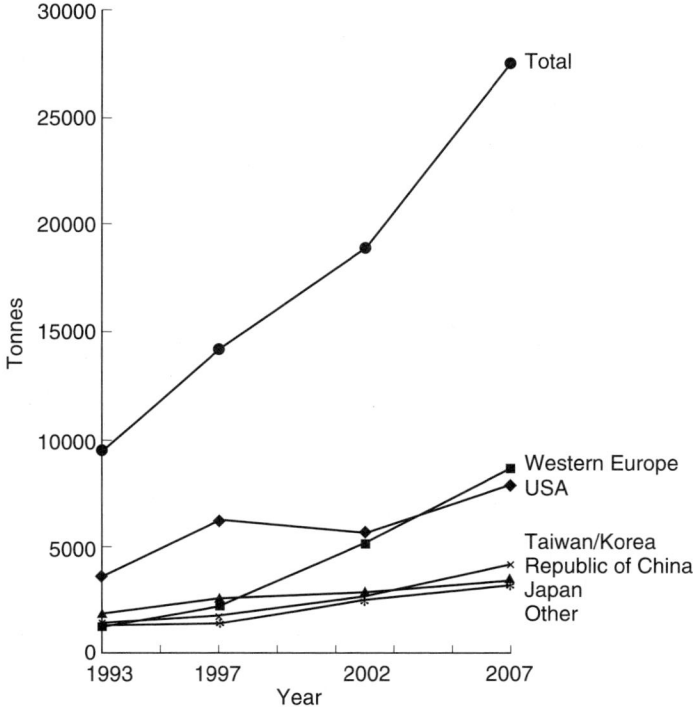

Figure 24.2 World consumption of carbon fiber. *Source:* Adapted from SRI Consulting, source based on contact with industry.

24.8 CONCLUSION

The cost of carbon fiber will always be an important factor and has been an issue that has been continually championed by Zoltek.

The predicted world consumption of carbon fiber is shown in Figure 24.2 and SRI Consulting have projected that the total consumption of carbon fibers will exceed 23,000 tonnes in 2007. Ed Trewin [15] projected that the global market for PAN-based carbon fiber would be 28,600 tonnes/year by 2007 which is in good agreement iwth the SRL forecast. Trewin makes a speculative projection further ahead to 2010 and suggests the global market might reach 35,000 tonnes/year by 2015. However, this would require substantial additional investment in manufacturing capacity to enable production to keep pace with such a level of demand. In the past carbon fiber supply/demand has been notoriously volatile and the market continues to be frustratingly difficult to forecast accurately with frequent swings from over-capacity to supply shortages, affecting price levels and take-up rates in new and emerging applications. As a result there is a general reluctance to forecast with any confidence further than five years ahead.

REFERENCES

1. Lukjanova VT, Lovzova RV, Carbon fibres in nature, *Carbon*, 32(5), 777–783, 1994.
2. Mason KF, Sizing and surface treatment: the keys to carbon fiber's future, *High Performance Composites*, 12(2), 38–43, Mar 2004.
3. Cooper M, Less is more in the race to be strong, *New Scientist*, 23, Apr 13 1996.

4. Marsh G, Facing up to the recycling challenge, *Reinforced Plastics*, 22–26, Jun 2001.
5. Kemmochi K, Takayanagi H, Nagasawa C, Takahashi J, Hayashi R, Manifestation models for closed-loop material recycling in carbon fiber reinforced thermoplastics (CFRTP), *High Technology Composites in Modern Applications*, Paipetis SA, Youtsos AG eds., 203–213, Sep 1995.
6. Kemmochi K, Takayanagi H, Nagasawa C, Takahashi J, Hayashi R, Possibility of closed loop material recycling for fiber reinforced thermoplastic composites, *Advanced Performance Materials*, 2(4), 385–394, Oct 1995.
7. Allred RE, Coons AB, Simonson R, Properties of carbon fibers reclaimed from composite manufacturing scrap by tertiary recycling, *Proc 28th Int SAMPE Tech Conf*, Seattle, Nov 4–7, 1996.
8. Allred RE, Busselle LD, Shoemaker JM, Catalytic process for the reclamation of carbon fibers from carbon/epoxy composites, *SPE Annual Recycling Conference*, 1999.
9. Pickering SJ, Yip H, Kennerley JR, Kelly R, Rudd CD, The recycling of carbon fibre composites using a fluidised bed process, *FRC2000 Conference*, Newcastle, Sep 13–14, 2000.
10. Warren CD, Carbon fiber in future vehicles, *SAMPE J*, Mar/Apr 2001.
11. Schmidt D, Davidson K, Theibert L, Unique applications of carbon-carbon composite materials (Part 2), *SAMPE J*, Mar/Apr 1999.
12. Baker RTK, Synthesis, properties and applications of graphite nanofibers. *R&D status and trends in nanoparticles, nanostructured materials and nanodevices in the United States*, Siegel RW, Hu E, Roco MC eds., International Technology Research Institute, World Technology (WTEC) Division, Loyola College, Balitmore, NTIS #PB98-117914.
13. A TM contributor, *Microscopic revolution*, 10–12, TM Oct 2001.
14. Ehrburger P, Chéret D, Choserot A, Dziedzinl P, Gransard S, Dual viscose and PAN-based carbon filaments: Mechanical and porous properties, 23rd *Bienial Conf on Carbon Pennsylvania State University*, 456–457, July 13–18, 1997.
15. Trewin EM, Specsult Ltd., Balsall Common, Coventry.

APPENDIX 1

Glossary

Accelerator (promoter): When mixed with a thermosetting resin in conjunction with a catalyst, speeds up, or promotes the curing process.

ACOTEG: Advanced Composites Technology Group (Europe), a consortium of B.Ae, MBB and Aerospatiale. The consortium adopts EN series test methods and proposes new ones to AECMA.

Activation: An oxidation process, which increases the surface area of carbon to facilitate the process of adsorption.

Additive: Any substance added to a resin system prior to cure, to induce curing (e.g. catalyst, accelerator) or, to improve properties (e.g. filler, flame retardant).

Adsorption: The taking-up of one substance at the surface of another, can be chemisorption or physisorption.

AECMA: *Association Europèen des Constructeurs de Material Aerospatial.*

AFNOR: *Association Français de Normalization* (France).

Allotrope: The same atoms of a given element connected in a different bonding arrangement to give a different form (e.g. diamond, graphite and buckminsterfullerene are allotropes of carbon).

Amorphous: Substance has no definite repeating pattern within its atomic structure. There could be small regions of order but as a whole, there is significant disorder (e.g. a semi-crystalline polymer).

Anemometer: Instrument for measuring the velocity of air.

Angleply laminate: Any balanced laminate consisting of ±θ plies where θ is an acute angle with respect to the reference direction.

Anisotropic: Having different properties along the three different axes (e.g. unidirectional and multi-directional laminates, filament winding and pultrusion). Opposite of isotropic.

ANSI: American National Standards Institute.

Aramid: Aromatic polyamide, made from a linear polymer containing recurring amide groups (—CO—NH—) joined directly to two aromatic rings.

Areal weight: Weight of fiber per unit area of prepreg.

ASME: American Society of Mechanical Engineers.

Aspect ratio: Ratio of fiber length to diameter.

ASTM: American Society for Testing Materials.

Autoclave molding: Heat and pressure applied to a composite by placing it in an autoclave.

Bag molding: Consolidation of prepreg in a mold by application of fluid pressure through a flexible membrane.

Balanced laminate: Any laminate containing one ply of $-\theta$ orientation with respect to the laminate principal axis for every identical ply with a $+\theta$ orientation. Minimizes distortion on demolding.

Barcol hardness: Hardness value obtained by measuring resistance to penetration of a spring loaded hardened steel point, giving reading of 0–100 on a Barcol Impressor. Used to measure degree of cure of plastic polymer.

Batch: Total amount of material produced during a production run.

Bleeder cloth: Usually made of glass fabric or non-woven, used to absorb a calculated amount of excess resin from prepreg when pressure is applied in a vacuum bag lay-up.

BP: British Patent.

BPF: British Plastics Federation.

Breather Fabric: Permits air and volatiles to be removed when vacuum is applied to vacuum bag assembly.

BMC: Bulk molding compound.

BMI: Bismaleimide, a polyimide resin that cures by an addition reaction.

BS: British Standard.

BSI: British Standards Institution.

B-staging: Partial cure of a thermoset resin.

BVID: Barely visible impact damage.

CAA: The UK Civil Air Authority (cf FAA).

CAD: Computer aided design.

Carbon black: Finely divided carbon made by burning hydrocarbons (e.g. CH_4) under conditions in which combustion is incomplete, contains up to 95% carbon.

Catalyst (hardener): Promotes rate of cure of a thermoset resin.

Catenary: Tendency for parts of a tow to sag lower than others when that tow is held horizontally under tension.

CBCF: Carbon bonded carbon fiber.

CEM: Centrifugal molding or rotational molding.

CEN: *Comité Européen de Normalisation.*

CFA: Composites Fabricators Association.

CFMA: Carbon Fiber Manufacturers Association (Japan).

CFMC: Carbon Fiber Management Council, an industry group representing the manufacturers of carbon fiber formed by the CFA to replace SACMA.

cfrp: Carbon fiber reinforced plastic (polymer).

CG-BPF: Composites Group of the British Plastics Federation.

Chiral structure: Structure having left and right handed forms.

CIM: Compression injection molding.

CIRTM: Co-injection resin transfer molding.

Coefficient of variation (CoV): Value obtained by expressing standard deviation as percentage of mean.

CMC: Ceramic matrix composite.

COM: Contact molding.

Co-mingled yarn: Hybrid yarn made with two types of materials intermingled into a single yarn (e.g. a thermoplastic yarn with carbon fiber).

Compliance: Reciprocal of stiffness.

Composite: Combination of reinforcement and matrix resin.

APPENDIX 1: GLOSSARY

Compression molding: Molding of thermoset plastics, reinforcement fiber and matrix resin placed in mold cavity, mold closed, heat and pressure applied, until material has cured or achieved final form.

Compressive strength: Maximum stress a material can sustain under crush loading. Compressive strength of materials that do not shatter in compression, amount of stress required to deform material to arbitrary amount. Compressive strength calculated by dividing maximum load by original cross-sectional area of specimen.

Condensation reaction: Polymerization reaction in which byproduct(s) (e.g. water) are evolved.

Conductivity: The reciprocal of resistivity.

Confidence limits:

$$\pm L = \frac{\sigma_e t}{\sqrt{n}}$$

where L = confidence limits (usually 95%)
σ_e = estimate of true standard deviation i.e. σ
n = number of specimens
t = Student's t

Contact molding: Application of reinforcement and resin by hand to a mold without pressure.

Count: Measure of fineness of yarn.

Coulomb: Charge transported when current of 1 ampere flows for 1 second.

Covalent bond: Strong bond holding atoms or molecules together by sharing pairs of electrons; one pair, single bond; two pairs, double bond; three pairs, triple bond.

Cover factor: Measure of degree of openness of fabric, measured in both warp and weft directions. At higher values, fabric becomes stiffer, drapes less easily.

CRAG: Composite Research Advisory Group, M.o.D., UK. Proposed a series of test methods recommended by UK Defense establishments and aerospace industry companies.

Crimp: Holding untwisted tow integral by application of small folds or corrugations across tow in regular manner, at elevated temperature, pressure.

Crimp percentage: Wavy path followed by warp and weft when interlaced in a fabric, crimp percentage is measure of waviness,

$$\text{Crimp percentage} = \frac{(\text{Straightened length} - \text{Crimped length}) \times 100}{\text{Crimped length}}$$

Crystalline solid: Having same repeating unit throughout entire structure, can be composed of individual molecules, or ranging network of atoms.

CSA: Canadian Standards Association.

CSM: Chopped strand mat.

Cure: Irreversible change of thermosetting resin by polymerization usually with aid of heat, catalyst, accelerator, with or without pressure.

Cusum: Cumulative sum.

CVD: Chemical vapor deposition, method for growing solids where gaseous precursor containing fragments of desired solid is decomposed and deposited on desired surface.

CVI: Chemical vapor infiltration.

Daylight: Distance in open position between moving and fixed platens of press.

Degrees of freedom: $n - 1$, known as degrees of freedom of variance.

Delamination: Separation of plies in laminate.

Denier: Mass in grams of 9000 meters of filament.

Dielectric: Non-conductor of electricity.

DIN: *Deutsches Institut für Normung eV* (Germany).

Discontinuous reinforcement: Reinforcement fibers may be whiskers, chopped or milled fibers.

DMC: Dough molding compound, contains all components for complete final cure, also called bulk molding compounds.

DMMC: Discontinuous metal matrix composite.

DNV: *Det Norske Veritas.*

DoD: US Department of Defense.

DOE: US Department of Energy.

Doff: Sub-batch.

Doff-out: Removal of fiber from production line at any stage of process.

Drape: Ability of fabric to conform to requisite shape (e.g. double curvature).

DSC: Differential scanning calorimeter.

Dwell time: Time required to equalize tool and component temperatures and to initiate controlled prepreg cure.

E: Young's modulus (modulus of elasticity).

E-beam curing: Electron beam curing for composites.

EC: European Community.

EFTA: European Free Trade Association.

Elastic constants: Material, when subjected to stress in one of three modes produces three elastic constants.
>Longitudinal—Young's modulus
>Shear—Rigidity modulus
>Compression—Bulk modulus
>Emulsion

Emulsion: Colloidal suspension of one liquid in another.

EN: *Comité de Européen Normalisation* (European Standard).

End: Untwisted bundle, or strand, of continous filaments.

Endoscope: Instrument used for examining internal cavities, normally based on fiber optics.

Equilibrium moisture content: Percentage of water in substance which exerts vapor pressure equal to partial vapor pressure of surrounding atmosphere.

$$\text{Extension} = \frac{\text{Elongation} \times 100}{\text{Initial length}} \%$$

Extrusion: Formation of section by forcing material through die, using screw driven ram.

FAA: Federal Aviation Authority, US government agency responsible for all aspects of US civil aviation.

Fatigue strength: Maximum cyclic stress withstood by material given number of cycles before failure occurs, or residual strength after being subjected to fatigue.

Fiber: Filament form from which yarns, fabrics and braids are made by spinning, knitting, weaving, braiding etc.

Fiber content: Amount of fiber in composite, expressed as % volume or % mass.

Fiber fineness: Measure of cross-sectional area of strand, fine fiber can be spun to finer counts than coarse fiber, will be more uniform, stronger.

Fiber fraction: Fiber content expressed as volume or mass fraction.

Fick's Law of Diffusion: Rate of diffusion in given direction, proportional to negative of concentration gradient.

Filament winding (FIW): Pulling dry fiber through resin bath (wet winding) or using prepreg tape or tow (dry winding), winding onto rotating mandrel.

Filler: Inert material added to resin as bulking agent to lower cost or to confer special property (e.g. improve physical properties).

Fineness of grind gage: See Hegman gage.

Finish: Formulated liquid product applied to fiber to confer certain properties (e.g. antistat, textile processing aid or to maintain tow integrity).

Flexural strength: Maximum fiber stress developed in specimen just before cracking or breaking in flexure test.

Foam core: Low density foamed plastic sandwiched between reinforced plastic skins to confer stiffness, individual cells may be closed or interconnected.

Folded yarn: Also known as ply twisting, formed from two or more yarns twisted together to increase strength and/or improve appearance. Spun staple fiber yarns are ply twisted in direction opposite to spinning twist to give resultant composite yarn with virtually zero twist.

FRM: Fiber reinforced metal.

FRP: Fiber reinforced plastic.

G: Shear modulus (Modulus of rigidity).

Gel time: Time taken for resin system to gel at given temperature after addition of curing agents.

Glass: Amorphous or disordered material (cf liquid in structure) with no true melting point, softening with increase in temperature.

Godet: Roller used in the viscose rayon process to guide rayon tow, when using pairs rotating at different speeds, will impart differential stretch to fiber.

GST: See T_g—glass transition temperature.

Hank: Continuous length of yarn wound as circular structure, subsequently collapsed to form flattened loop with two sides touching.

Hardener: Curing agent.

HDT: Heat distortion temperature, temperature at which test bar deforms to specified distance under prescribed load.

Hegman gage: Fineness of grind gage, hardened stainless steel block with one or two channels ground into its top surface with depth graded 0–25, 50 or 100 μm.

Heterogeneous: Comprised of different elements or constituents, consisting of more than one phase, properties vary from point to point.

HIP: Hot isostatic pressing.

HIPIC: Hot isostatic pressure impregnation carbonization.

HM: High modulus fiber. Tensile modulus >350 GPa.

Homogeneous: Of uniform character throughout, properties same at every point.

HS: High strength fiber.

Humidity: Absolute mass of water present in uniform volume of moist air (e.g. $g\,m^{-3}$). Relative humidity is ratio of actual vapor pressure to saturated vapor pressure at same temperature, expressed as %.

Hybrid: Composite containing two or more types of reinforcement, one may reinforce/enhance other.

ILSS: Interlaminar shear strength.

IM: Intermediate modulus fiber. Tensile modulus approximately 300 GPa.

IM: Injection molding.

IMC: Injection molding compound.

Impact strength: Energy to fracture specimen subjected to shock loading as in an impact test, indication of toughness of material.

Inhibitor: Substance which retards a chemical reaction.

Initial modulus: Measures of resistance to extension at low forces.

Initiator: Substance which promotes the curing process.

Intercalation compound: A compound formed by substance (e.g. Br_2) entering between and pushing apart the layer planes of graphite.

ISO: International Organization for Standardization, also prefixes a standard.

Isotope: Atoms of same element having different masses, have different number of neutrons but atomic numbers (i.e. the number of protons) remain same.

Isotropic: Having uniform properties in all directions in plane e.g. SMC, DMC, mat and fabric (not UD) thermosets. Opposite of anisotropic.

Izod test: Notched bar impact test.

JISC: Japanese Industrial Standards Committee.

K: Bulk modulus (modulus of compression).

Kish: Form of solid graphite separated from product floating on surface of molten bath of cast iron, or pig iron, high in carbon.

Lamella: A thin plate.

Laser: Light amplification by stimulated emission of radiation.

Law of Mixtures: See Rule of Mixtures.

Lay up: Content and arrangement of reinforcement in laminate following lay up code, designated stacking sequence.

Liquid crystal: Pure liquid, turbid and like crystal, anisotropic over definite temperature range above freezing point.

LOI: Limiting oxygen index, minimum oxygen content in admixture with nitrogen to sustain combustion of specified material under strictly controlled conditions.

Lot: Processed set of precursor from single production line.

Mandrel: Mold or jig onto which uncured composite system is filament wound or wrapped.

Matrix: Phase in which reinforcement phase is embedded and transmits loads between reinforcing fibers, matrix can be a polymer, metal, ceramic or carbon.

Mean:

$$\text{Arithmetic mean} = \frac{x_1 + x_2 + x_3 + \ldots + x_n}{n}$$

Median: Middle value of series of values arranged in order of magnitude.

Melt spinning: Formation of continuous filaments by extrusion of molten polymer.

Mesomorphous: Existing in state of aggregation midway between true crystalline and completely irregular amorphous state.

Mild steel: Hot rolled carbon steel containing about 0.04% carbon.

Miller Indices: Integers determining orientation of crystal plane in relation to three crystallographic axes.

MMC: Metal matrix composite.

Mode: Value which occurs most often.

Modulus of Elasticity: Rate of change of strain as function of stress. Slope of straight line portion of stress–strain curve, Depending on the type of loading, several types of moduli—compressive,

APPENDIX 1: GLOSSARY

flexural, shear, tensile and torsion. Modulus used alone generally refers to tensile modulus of elasticity.

Modulus of Rigidity: Rate of change of strain as function of stress in specimen subjected to shear or torsion loading.

Moisture content: Mass of water in material expressed as percentage of the total mass.

Moisture regain: Mass of water in material expressed as percentage of oven-dry mass.

Molality: Number of gram moles of solute per 1000 g solvent.

Molar volume:

$$22.41 \times \frac{T}{273} \times \frac{1013}{P}$$

where T = absolute temperature (K)
P = atmospheric pressure (mb)

NASA: National Aeronautiocals and Space Administration (USA).

NBS: National Bureau of Standards (US Department of Commerce, Washington, DC, USA).

NDE: Non-destructive evaluation.

NDT: Non-destructive testing.

Near-net-shape: Original formation of part to shape that is as close to desired final shape as possible, requiring few finishing operations.

Nematic: Mesomorphous substance whose atoms or molecules are oriented in parallel lines.

NIST: National Institute for Standards and Technology.

NPL: National Physical Laboratory (Teddington, Middlesex, UK).

NSAI: National Standards Authority Ireland.

NTP: Normal Temperature and Pressure, now called STP.

Orthotropic: Having three mutually perpendicular planes of elastic symmetry at each point.

Out-life: Period of time prepreg material remains with unchanged properties, in handleable form outside of the specified storage conditions.

Oxygen Index: See LOI.

Package: Fiber wound in prescribed manner to form tight cylindrical structure on outside of cardboard tube.

PAN: Polyacrylonitrile.

phr:

$$\text{Parts per hundred of resin, \% constituent} = \frac{\text{phr}}{100 + \text{phr}}$$

Pick: Single weft thread in fabric, also called fill or woof.

Plasma: Ionized gaseous discharge with no resultant charge.

Plie: Multiplicity of parallel filaments with twist applied.

Ply: One of the layers that make up a stack or laminate.

PMC: Polymer matrix composite.

Poisson's ratio: Ratio of lateral strain to axial strain in axially loaded specimen.

Polymer: Molecule composed of linked repeating units (called monomers).

Post cure: Final processing of laminate at elevated temperature to complete the cure process.

Pot life: Useable life of resin system at working temperature.

ppm: Parts per million.

PPM: Prepreg molding.

Precursor: Product (e.g. PAN fiber) used at start of process, eventually converted to final product.

Prepreg: Fibrous reinforcement held in position with polymeric matrix. May be sheet, tape or, tow. If the matrix is thermoset resin, may be partly cured to induce right level of tack, a condition known as B-stage.

PRM: Press molding.

Promoter: Accelerator.

Pyrolysis: Decomposition of substance by heat.

Pyrometer: Instrument for measuring high temperatures.

Quiescent: Motionless conditions in furnace atmosphere.

RAE: Royal Aircraft Establishment at Farnborough, later called DRA, Defense Research Agency, Farnborough, then DSTL, Defence Evaluation and Research Agency and QinetiQ.

Regression: Model of the relationship between expected value of random variable and values of one or more possible related variables.

Relative density: Originally called Specific Gravity, ratio of mass of given volume of substance to mass of equal volume of water at 4°C.

Relaxation: Continued extension without further application of force.

Release agent: Plastic film or spray used to facilitate the removal of molded composite from mold.

Resistivity: Erroneously called specific resistance.

$$R = \frac{\rho L}{A}$$

where R = resistance in ohms of fiber
L = length in meters length
A = uniform cross section in m^2
ρ = resistivity in Ωm

RFI: Resin film infusion

RIFT: Resin infusion under flexible tooling.

RIM: Reaction injection molding.

RRIM: Reinforced reaction injection molding.

RTM: Resin transfer molding, fiber reinforcement is placed in mold, then catalyzed resin is transferred, procedure eliminates secondary bonding operations.

Rule of Mixtures: Properties P depend on respective contributions of fibers P_f and matrix P_m, fiber volume fraction V_f, geometric effectiveness G_e, rule of mixtures assumes that contributions of components can be summed linearly in proportion to respective fractions,

$P = P_f G_e V_f + P_m (1 - V_f)$. When P_f, G_e, V_f are large, fibers dominate resultant property, converse if matrix dominates and P_f and V_f are low.

Type of lay-up	G_e	Typical V_f
Uni-directional filaments	1.0	Up to 65%
or fabric		Up to 55%
Bi-directional fabric	0.5	30–45%
Short length fibers	0.4	20–40%

Hence for a UD composite, $X_c = X_f V_f + X_m V_m$

Run: Series of consecutive lots all manufactured to same specification.

SACMA: Suppliers of Advanced Composite Materials Association, USA. Ceased to exist after June 2000 but Test Methods and Safe Handling publications available from SAMPE and Composites

Fabricators Association (CFA), the latter has formed a Carbon Fibre Management Council (CFMC), open to all PAN-based carbon fiber manufacturers.

SAMPE: Society for the Advancement of Materials & Process Engineers.

SCRIMP™: Seemann Composite Resin Infusion Molding Process.

Selvedge (Or selvage): Edging of piece of cloth so finished to prevent unravelling.

S.G: Specific gravity, see relative density.

Shear strength: Maximum shear stress that can be sustained by material before rupture, determined in a torsion test, equal to the torsional strength. A much simpler but less acurate method is the interlaminar shear strength test (ILSS).

Shrinkage

$$\text{Shrinkage} = \frac{100(L_0 - L_1)}{L_0}\%$$

where L_0 = length before
L_1 = length after

Size: Formulated resin product applied to fiber to confer certain properties (e.g. improve inter-filamentary adhesion, aid wetting-out in resin matrices), acts as lubricant to prevent fiber damage during subsequent textile processing.

SMC: Sheet molding compound.

Snarling: Tendency of twist lively ply to double back on itself and form series of loops.

Specific property: Ratio of that property to the relative density (specific gravity).

Splice: Joining a tow with either glue join or by the use of air intermingler.

Spontaneous combustion: Ignition of substance without direct application of flame.

SD: Standard deviation = the square root of mean of squares of deviations of values from their mean, has same units as value being measured.

Stacking sequence: Configuration showing ply composition and exact location or sequence of various plies.

SAA: Standards Australia.

Staple: Discontinuous filaments made by chopping continuous material into discrete uniform lengths.

Stefan-Boltzmann Law: Total radiation emitted from black body per unit area per unit time is proportional to fourth power of its absolute temperature.

STP: Standard Temperature and Pressure
0°C and 101.325 $kN\,m^{-2}$ (standard atmosphere equivalent to pressure exerted by column of mercury 760 mm high at 0°C).

Strain (tensile): Elongation divided by initial length.

Stress: Force per unit area.

Student's t:

$$t = \frac{\text{nominal mean} - \text{sample mean}}{\text{standard error}}$$

Stuffer box: Box into which continuous filament is packed closely in intermittent continuous operation resulting in yarn becoming crimped, yarn is heated to set crimp.

Superconductivity: Property of some materials when cooled to very low temperatures to have no resistance. In this state, material resists penetration of magnetic field.

Surface treatment: Process applied to fiber to modify surface (e.g. etching and/or attachment of chemical groups), to improve fiber bond in resin matrices.

swg: Standard wire gage.

Synergism: Co-operative action of two additives such that combined effect is greater than sum when taken independently.

Tenacity:

$$\frac{\text{Breaking force (in N or cN)}}{\text{Linear density (in tex, dtex, or denier)}}, \text{ measured in g/denier or g/tex.}$$

Tex: Mass in grams of 1 km of filament.

Tex twist factor: Direct system, equals turns/meter × $\sqrt{\text{tex count}}$.

T_g: Glass transition temperature, approximate mid-point of the temperature range over which glass transition takes place; a reversible change in amorphous polymers, from a partially crystalline to viscous or rubbery condition, or vice versa; not a phase transition.

TGA: Thermogravimetric analysis.

Thermoplastic: Material that can be repeatedly softened by heating, and can be shaped by molding or extrusion when soft.

Thermoset: Polymer that can be irreversibly hardened (cured) when heated with or without addition of other chemical agents.

Top: Tow converted by cutting or stretch breaking, called tow to top conversion.

Tow: Multiplicity of parallel filaments.

Transition point: Temperature at which substance is converted from one crystalline form to another solid form.

TSC: Thermoplastic sheet compound, can be regarded as thermoplastic prepreg, has no shelf life problems.

Twist: Measure of spiral turns per unit length given to yarn or strand hold constituent filaments together—S twist if spirals slope in same direction as middle portion of letter S, Z twist if spirals slope in same direction as middle portion of letter Z.

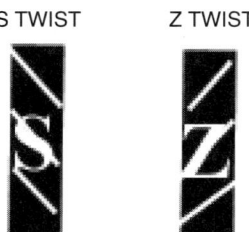

Twist factor:

$$\text{Indirect system, also called twist multiplier} = \frac{\text{tpi}}{\sqrt{\text{cotton count}}}$$

UD: Unidirectional aligned fibers (i.e. all fibers aligned in single direction).

UHM: Ultra high modulus.

UL: Underwriters Laboratories (Inc.).

Vacuum bag: Means of consolidation of laminate by enclosing prepreg in heat resistant bag in autoclave.

VARI: Vacuum assisted resin injection.

APPENDIX 1: GLOSSARY

VARTM: Vacuum assisted resin transfer molding.

Viscose: Cellulose xanthate formed by treating cellulose with sodium hydroxide and dissolving in carbon disulfide.

Void: Pocket of enclosed gas or air trapped within composite.

VPE: vinyl polyester.

Warp: Run parallel to direction of weaving in woven fabric, also called ends.

Weft: Run transverse to direction of weaving in woven fabric, also called picks, fill or woof.

Whisker: Short single crystal fiber with length to diameter ratio greater than 10.

White spirit: Petroleum distillate.

Work factor: Also known as work of rupture; area under stress strain curve divided by breaking stress multiplied by breaking strain, If fiber obeys Hooke's Law, stress/strain curve is straight line, then this factor is 0.5.

XMC: Form of SMC with aligned fiber bundles.

Yarn: Assembly of staple fibers or filaments produced by spinning establishes fiber cohesion by imparting twist to drafted roving.

Yarn count: Number indicating mass per unit length per unit mass of yarn, various systems used which be stated (e.g. cotton, worsted).

Yield point: End of elastic region.

Yield strength: Lowest stress at which material undergoes plastic deformation, below this point, material is elastic.

Young's modulus: Force per unit area divided by extension per original gage length.

APPENDIX 2

The Elements

ELEMENT	SYMBOL	ATOMIC NUMBER	MOLAR MASS g mol^{-1}
Actinium	Ac	89	227.03
Aluminum	Al	13	26.98
Americium	Am	95	241.06
Antimony	Sb	51	121.75
Argon	Ar	18	39.95
Arsenic	As	33	74.92
Astatine	At	85	210
Barium	Ba	56	137.34
Berkelium	Bk	97	249.08
Beryllium	Be	4	9.01
Bismuth	Bi	83	208.98
Boron	B	5	10.81
Bromine	Br	35	79.91
Cadmium	Cd	48	112.40
Calcium	Ca	20	40.08
Californium	Cf	98	251.08
Carbon	C	6	12.01
Cerium	Ce	58	140.12
Cesium	Cs	55	132.91
Chlorine	Cl	17	35.45
Chromium	Cr	24	52.01
Cobalt	Co	27	58.93
Copper	Cu	29	63.54
Curium	Cm	96	247.07
Dysprosium	Dy	66	162.50
Einsteinium	Es	99	254.09
Erbium	Er	68	167.26
Europium	Eu	63	151.96
Fermium	Fm	100	257.10
Fluorine	F	9	19.00
Francium	Fr	87	223
Gadolinium	Gd	64	157.25
Gallium	Ga	31	69.72
Germanium	Ge	32	72.59
Gold	Au	79	196.97
Hafnium	Hf	72	178.49
Helium	He	2	4.00
Holmium	Ho	67	164.93
Hydrogen	H	1	1.008
Indium	In	49	114.82
Iodine	I	53	126.90
Iridium	Ir	77	192.2
Iron	Fe	26	55.85
Krypton	Kr	36	83.80
Lanthanum	La	57	138.91
Lawrencium	Lr	103	262
Lead	Pb	82	207.19
Lithium	Li	3	6.94
Lutetium	Lu	71	174.97
Magnesium	Mg	12	24.31
Manganese	Mn	25	54.94
Mendelevium	Md	101	258.10
Mercury	Hg	80	200.59
Molybdenum	Mo	42	95.94
Neodymium	Nd	60	144.24
Neon	Ne	10	20.18
Neptunium	Np	93	237.05
Nickel	Ni	28	58.71
Niobium	Nb	41	92.91
Nitrogen	N	7	14.01
Nobelium	No	102	259
Osmium	Os	76	190.20
Oxygen	O	8	16.00
Palladium	Pd	46	106.4
Phosphorus	P	15	30.97
Platinum	Pt	78	195.09
Plutonium	Pu	94	239.05
Polonium	Po	84	209
Potassium	K	19	39.10
Praseodymium	Pr	59	140.91
Promethium	Pm	61	146.92
Protactinium	Pa	91	231.04
Radium	Ra	88	226.03
Radon	Rn	86	222
Rhenium	Re	75	186.2
Rhodium	Rh	45	102.91
Rubidium	Rb	37	85.47

(*Continued*)

The Elements (Continued)

ELEMENT	SYMBOL	ATOMIC NUMBER	MOLAR MASS g mol^{-1}	ELEMENT	SYMBOL	ATOMIC NUMBER	MOLAR MASS g mol^{-1}
Ruthenium	Ru	44	101.07	Thallium	Tl	81	204.37
Samarium	Sm	62	150.35	Thorium	Th	90	232.04
Scandium	Sc	21	44.96	Thulium	Tm	69	168.93
Selenium	Se	34	78.96	Tin	Sn	50	118.69
Silicon	Si	14	28.09	Titanium	Ti	22	47.90
Silver	Ag	47	107.87	Tungsten	W	74	183.85
Sodium	Na	11	22.99	Uranium	U	92	238.03
Strontium	Sr	38	87.62	Vanadium	V	23	50.94
Sulfur	S	16	32.06	Xenon	Xe	54	131.30
Tantalum	Ta	73	180.95	Ytterbium	Yb	70	173.04
Technetium	Tc	43	98.91	Yttrium	Y	39	88.91
Tellurium	Te	52	127.60	Zinc	Zn	30	65.37
Terbium	Tb	65	158.92	Zirconium	Zr	40	91.22

APPENDIX 3

The Greek Alphabet

Capital letter	Small letter	Name
A	α	alpha
B	β	beta
Γ	γ	gamma
Δ	δ	delta
E	ε	epsilon
Z	ζ	zeta
H	η	eta
Θ	θ	theta
I	ι	iota
K	κ	kappa
Λ	λ	lambda
M	μ	mu
N	ν	nu
Ξ	ξ	xi
O	o	omicron
Π	π	pi
P	ρ	rho
Σ	σ	sigma
T	τ	tau
Y	υ	upsilon
Φ	φ	phi
X	χ	chi
Ψ	ψ	psi
Ω	ω	omega

APPENDIX 4

Some Definitions and Handy Conversion Factors

Force = Mass × acceleration
Newton = The force acting on a mass of one kilogram which imparts an acceleration of one meter per second square to that mass.

The acceleration due to gravity is $9.81 \, \text{m s}^{-2}$.

Hence the force experienced by an object of mass $1/9.81$ kg (i.e. 101.9 g, approximately the mass of an apple), falling under gravity will be 1 N. Therefore 1 cN = 1.019 gf

Pascal (Pa) = A unit of stress equal to $1 \, \text{N m}^{-2}$

$1 \, \text{GPa} = 1 \, \text{GN m}^{-2} = 10^3 \, \text{MN m}^{-2} = 10^3 \, \text{N mm}^{-2} = 10^3 \, \text{MPa} = 10^9 \, \text{N m}^{-2} = 10^9 \, \text{Pa}$
$\quad\quad\quad = 10^4 \, \text{bar(b)} = 100 \, \text{hb}$
$\quad\quad\quad = 102 \, \text{kgf mm}^{-2} = 1.02 \, \text{kgf cm}^{-2}$

$10^6 \, \text{lbf in}^{-2} = 6.895 \, \text{GPa} = 0.6895 \, \text{hb}$
$\quad\quad\quad\quad = 0.7031 \, \text{kgf mm}^{-2}$

$1 \, \text{kgf mm}^{-2} = 0.807 \, \text{MPa} = 0.9807 \, \text{hb}$
$\quad\quad\quad\quad = 1442 \, \text{lbf in}^{-2}$

APPENDIX 5

ISO Standard Prefixes for SI Units

Multiplication Factor	Prefix	Symbol
One million million (billion) $=10^{12}$	tera	T
One thousand million $=10^{9}$	giga	G
One million $=10^{6}$	mega	M
One thousand $=10^{3}$	kilo	k
One hundred $=10^{2}$	hecto	h
Ten $=10^{1}$	deca	da
Unity $=10^{0}$	-	-
One tenth $=10^{-1}$	deci	d
One hundredth $=10^{-2}$	centi	c
One thousandth $=10^{-3}$	milli	m
One millionth $=10^{-6}$	micro	μ
One thousand millionth $=10^{-9}$	nano	n
One million millionth $=10^{-12}$	pico	p

APPENDIX 6

Interconversion of Common English and SI Units

LENGTH

1 angstrom (Å)	=	10^{-10} m			
1 microinch (μin)	=	0.0254 μm			
1 thou (or mil)	=	25.4 μm	1 μm	=	0.03937 thou in
1 inch (in)	=	25.4 mm	1 mm	=	0.03937 in
	=	2.54 cm	1 cm	=	0.3937 in
			1 m	=	39.37 in
1 foot (ft)	=	0.3048 m	1 m	=	3.2808 ft
1 yard (yd)	=	0.9144 m	1 m	=	1.0936 yd
1 mile	=	1.6093 km	1 km	=	0.6214 mile

AREA

1 square inch (in^2)	=	645.2 mm^2	1 mm^2	=	0.001550 in^2
1 square foot (ft^2)	=	0.09290 m^2	1 m^2	=	10.76 ft^2
	=	929.0 cm^2			
1 square yard (yd^2)	=	0.8361 m^2	1 m^2	=	1.1960 yd^2

VOLUME AND CAPACITY

1 cubic inch (in^3)	=	16.39 cm^3	1 cm^3	=	0.06102 in^3
1 cubic foot (ft^3)	=	0.02832 m^3	1 m^3	=	35.31 ft^3
	=	28.32 dm^3	1 dm^3	=	0.03531 ft^3
1 fluid ounce (fl oz) ($^1/_{20}$ pint)	=	28.41 cm^3			
1 pint (pt)	=	0.5683 dm^3			
1 imperial gallon (gal)	=	4.54609 dm^3			
	=	4.54596 litres	1 l	=	0.2200 gal
1 US gallon	=	3.78541 dm^3			
	=	3.78530 litres	1 l	=	0.2642 US gal

VELOCITY

1 ft min^{-1}	=	0.00508 m s^{-1}			
1 ft s^{-1}	=	0.3048 m s^{-1}	1 m s^{-1}	=	3.281 ft s^{-1}

MASS

1 ounce (oz)	=	28.35 g	1 g	=	0.03527 oz
1 pound (lb)	=	0.4536 kg	1 kg	=	2.205 lb

1 ton (2240 lb)	=	1016.1 kg
	=	1.0161 tonne
1 US ton (2000 lb)	=	907.18 kg

MASS PER UNIT LENGTH

1 oz in^{-1}	=	1.116 kg m^{-1}	1 kg m^{-1}	=	0.8961 oz in^{-1}
1 lb ft^{-1}	=	1.488 kg m^{-1}	1 kg m^{-1}	=	0.6720 lb ft^{-1}
1 lb yd^{-1}	=	0.4961 kg m^{-1}			

LENGTH PER UNIT MASS

1 yd lb^{-1}	=	2.016 m kg^{-1}	1 m kg^{-1}	=	0.4960 yd lb^{-1}

MASS PER UNIT AREA

1 oz ft^{-2}	=	305.2 g m^{-2}			
1 oz yd^{-2}	=	33.91 g m^{-2}			
1 lb ft^{-2}	=	4.882 kg m^{-2}	1 kg m^{-2}	=	0.2048 lb ft^{-2}

VOLUME RATE OF FLOW

1 ft^3 min^{-1}	=	1.699 m^3 h^{-1}	1 m^3 h^{-1}	=	0.5886 ft^3 min^{-1}
1 gal h^{-1}	=	4.546 dm^3 h^{-1}	1 dm^3 h^{-1}	=	0.2200 gal h^{-1}

DENSITY

1 lb in^{-3}	=	27.68 g cm^{-3}	1 g cm^{-3}	=	0.03613 lb in^{-3}
1 lb ft^{-3}	=	16.02 kg m^{-3}	1 g l (kg m^{-3})	=	0.06243 lb ft^{-3}
1 lb gal	=	99.78 kg m^{-3}			

FORCE

1 lbf	=	4.448 N	1 N	=	0.2248 lbf
	=	0.457 kgf	1 kgf	=	9.8067 N
1 tonf	=	9.964 kN			

FORCE PER UNIT LENGTH

1 lbf in^{-1}	=	175.1 N m^{-1}

PRESSURE AND STRESS

1 lbf in^{-2} (psi)	=	6.895 kPa (k N m^{-2})	1 N m^{-2}	=	0.0001450 lbf in^{-2}
	=	0.0703 kgf cm^{-2}	1 atm	=	101.33 k N m^{-2}
1 ton in^{-2} (tsi)	=	15.44 Mpa	1 torr	=	133.32 N m^{-2}
1 ft water	=	2.989 kPa			
1 in mercury	=	3.386 kPa			
	=	33.86 mbar (mb)			
1 bar	=	10^5 Pa (N m^{-2})			

APPENDIX 6: INTERCONVERSION OF COMMON ENGLISH AND SI UNITS

VISCOSITY

1 poise (P) = 0.1 kg m^{-1} s^{-1}
 = 0.1 Pa s

ENERGY AND WORK

1 Btu = 1.055 kJ
 = 252.0 gcal
1 cal = 4.1868 J

POWER

1 hp = 745.7 W (J s^{-1})
1 kWh = 3.6 MJ

THERMAL CONDUCTIVITY

1 Btu ft ft^{-2} h^{-1} °F^{-1} = 1.731 W m^{-1} K^{-1} (J m m^{-2} s^{-1} K^{-1})
1 Btu in ft^{-2} h^{-1} °F^{-1} = 0.1442 W m^{-1} K^{-1}
 = 0.00413 cal cm cm^{-2} s^{-1} K^{-1}
1 Btu in ft^{-2} s^{-1} °F^{-1} = 519.2 W m^{-1} K^{-1}

SPECIFIC HEAT

1 Btu lb^{-1} °F^{-1} = 4.187 J g K^{-1} 1 J g K^{-1} = 0.2388 Btu lb^{-1} °F^{-1}

TEMPERATURE

1 °C = $[(1.8)°C + 32]°F$
T K (Kelvin) = $T°C + 273.15$
T °F = $(°F - 32) \times \frac{5}{9} °C$

ELECTRICAL UNITS

Resistance (Ω)
Conductance = $\frac{1}{\text{Resistance}}$ = $\frac{1}{\Omega}$ Siemens (S) or mho (υ)

Resistivity = Specific resistance = $\Omega m^2 m^{-1}$ = Ωm

 = $\frac{A\Omega}{L}$

 = $\frac{mpul\Omega}{L}$

If $L = 1$ m, of resistance Ω then:

Resistivity = $\frac{mpul\Omega}{\rho}$

APPENDIX 7

Textile Terminology

Known Yarn Count	Constant into which the known yarn count is divided in order to obtain the equivalent yarn count in the other system	
	Tex	Denier
Worsted	885.8	7972
Woolen (Yorkshire skein)	1938	17442
Cotton	590.5	5315
Glass (USA and UK)	4961	44649

Tex = mass in g of 1 km of filament
Denier = mass in g of 9000 m of filament
Worsted = number of 560 yd hanks per lb
Woolen (Yorkshire) = number of 256 yd skeins per lb

APPENDIX 8

Temperature Estimation from Color

Temperature °C	Color
532	Barely discernible red
566	Blood red
635	Dark cherry red
677	Medium cherry red
746	Cherry red
843	Bright cherry red
899	Red-orange
941	Orange
996	Yellow
1079	Light yellow
1204	White

APPENDIX 9

Humidities over Saturated Salt Solutions

Saturated Salt Solution	Temperature		
	20°C	25°C	30°C
	(Relative Humidity (%))		
Potassium sulfate	97	97	96
Potassium nitrate	94	92	91
Potassium chloride	87	85	85
Ammonium sulfate	81	80	80
Sodium chloride	76	75	75
Sodium nitrite	-	65	63
Ammonium nitrate	69	62	59
Sodium dichromate	56	54	52
Magnesium nitrate	56	53	52
Potassium carbonate	44	43	43
Magnesium chloride	34	33	33
Potassium acetate	23	22	22
Lithium chloride	13	12	12
Potassium hydroxide	10	8	7

Source: Adapted from GWC Kaye and Laby Tables of Physical Constants.

APPENDIX 10

Wet and Dry Bulb Humidity Table

Wet bulb depression °C	Dry bulb Temperature °C							
	10°C	12°C	14°C	16°C	18°C	20°C	22°C	24°C
				Relative Humidity(%)				
0.5	94	94	95	95	95	96	96	96
1.0	88	89	90	90	91	91	92	92
1.5	82	83	84	85	86	86	87	88
2.0	76	78	79	81	82	83	83	84
2.5	71	73	74	76	77	78	79	80
3.0	65	68	70	71	73	74	76	77
3.5	60	63	65	67	69	70	72	73
4.0	54	57	60	62	65	60	68	69
4.5	49	53	56	58	60	62	64	66
5.0	44	48	51	54	56	59	61	62
5.5	39	43	47	50	52	55	57	59
6.0	34	38	42	46	49	51	54	56
6.5	29	34	38	41	45	48	50	52
7.0	24	29	33	37	41	44	47	49
7.5	19	24	29	34	37	41	44	46
8.0	14	20	25	30	34	37	40	43
8.5	9	16	24	26	30	34	37	40
9.0	5	11	17	22	27	30	34	37
9.5	–	7	13	18	23	27	31	34
10.0	–	3	9	15	20	24	28	31
10.5	–	–	5	11	16	21	25	28
11.0	–	–	2	8	13	18	22	26

Source: Adapted from GWC Kaye and Laby Tables of Physical Constants.

APPENDIX 11

Detection of Cyanide [1]

From time to time, it may be necessary to detect for cyanide contamination in waste liquors or solids. This spot test is quite useful and is best carried out in conjunction with a blank of distilled water and a known cyanide for comparison.

Test Solutions

1. Solution T—dissolve 1g of chloramine-T in 100 cm³ of distilled water.

Chloramine-T (sodium N-chloro-p-toluene sulphonamide

2. Solution A—prepare a saturated solution by taking 0.5 g of 3-methyl-1-phenyl-5-pyrazalone in 100 cm³ of distilled water. Heat the solution to 60°C with stirring. Cool to room temperature and filter.

3-Methyl-1-phenyl-5-pyrazalone

3. Solution B—dissolve 0.1 g of *bis*-(3-methyl-1-phenyl-5-pyrazalone) in 100 cm³ of pyridine. It will take several minutes for this material to dissolve. Prepare solution daily.

Bis-(3-Methyl-1-phenyl-5-pyrazalone) Supplied by: Eastman Kodak Co.
(3,3'dimethyl-1,1'-di-phenyl[4,4'-bi-2-pyrazoline]-5,5'dione)

Test Procedure

1. Introduce a 2–3 cm^3 sample in solution or suspension into a small test tube.
2. Adjust pH to between 6 and 7 using NaOH or CH$_3$COOH as required.
3. Add 4 drops of Solution T. Mix and allow to stand for 2–3 min.
4. Prepare the pyridine-pyrazalone reagent by adding 5 cm^3 of Solution A to 1 cm^3 of Solution B.
5. Add 5 cm^3 of this mixed reagent to the sample solution.
6. A blue color indicates the presence of cyanide. The color may take up to 30 min to develop, depending on the cyanide concentration.

The cyanide (CN$^-$) is converted to cyanogen chloride (CNCl) at pH < 8, without hydrolyzing to the cyanate, by reaction with the Chloramine-T which acts as a source of hypochlorous acid (HOCl). After the reaction is complete, the CNCl forms a blue dye (pyrazole blue) on the addition of the pyridine-pyrazalone reagent.

Pyrazole-blue

REFERENCE

1. Stevens F, Fischer G, Macarthur D, *Analysis of Metal Finishing Effluents*, Draper, 1968.

APPENDIX 12

British Standards on Quality

	Description of standard
Quality Management and Quality Assurance	
BS 5750: Part 8 (EN 29004-2, ISO 9004-2)	Quality systems. Guide to quality management and quality systems elements for services
BS EN ISO 9000-1	Quality management and quality assurance standards. Guidelines for selection and use
BS EN ISO 9000-2	Quality management and quality assurance standards. Generic guidelines for the application of ISO 9001, ISO 9002, ISO 9003
BS EN ISO 9000-3	Quality management and quality assurance standards. Generic guidelines for the application of ISO 9001 to the development, supply, installation and maintenance of computer software
BS EN ISO 9001	Quality systems. Model for quality assurance in design, development, production, installation and servicing
PD ISO/TS 16949	Quality systems. Automotive suppliers. Particular requirements for the application of ISO 9001
BS EN ISO 9002	Quality systems. Model for quality assurance in production, installation and servicing
PD 9002	Register of firms, products and services approved under the BS 9000 system
BS EN ISO 9003	Quality systems. Model for quality assurance in final inspection and test
BS EN ISO 9004-1	Quality management and quality assurance standards. Guidelines
PD 6584	Model for a supplier quality assurance assessment questionnaire
BS ISO 10005	Quality management and quality system elements. Guidelines for quality plans
BS ISO 10007	Quality management. Guidelines for configuration management
BS ISO 10013	Guidelines for developing quality manuals
PD 6642 (ISO/IEC TR 17010)	General requirements for bodies providing accreditation of inspection bodies
Definitions and Concepts	
BS EN ISO 8402	Quality management and quality assurance. Vocabulary
BS 4778-2	Quality vocabulary. Quality concepts and related definitions
BS 4778-3.1	Quality vocabulary. Availability, reliability and maintainability terms. Guide to concepts and related definitions
BS 4778-3.2	Quality vocabulary. Availability, reliability and maintainability terms. Glossary of international terms
BS 5233	Glossary of terms used in metrology (Incorporating BS 2643)
BS EN ISO 8402	Quality management and quality assurance. Vocabulary
Dependability	
BS 5750-8 (EN 29004-2, ISO 9004-2)	Quality systems. Guide to quality management and quality system elements
BS 5750-14 (EN 60300-1, ISO 9000-4)	Quality systems. Guide to dependability programme management

Economics of Quality
BS 6143-1 — Guide to the economics of quality. Process cost model
BS 6143-2 — Guide to the economics of quality. Prevention, appraisal and failure model
PD ISO/TR 10014 — Guidelines for managing the economics of quality

Product and Service Design
BS 7000-2 — Design management systems. Guide to managing the design of manufactured products
BS 7000-3 — Design management systems. Guide to managing service design
BS 7000-4 — Design management systems. Guide to managing design in construction
BS 7000-10 — Design management systems. Glossary of terms used in design management

Specifications
BS 7373 — Guide to the preparation of specifications

Total Quality Management
BS 7850-1 — Total quality management. Guide to management principles
BS 7850-2 (ISO 9004-4) — Total quality management. Guidelines for quality improvement

Auditing the Quality System
BS EN 30011-1 — Guidelines for auditing quality systems. Auditing
BS EN 30011-2 — Guidelines for auditing quality systems. Qualification criteria for quality systems auditors
BS EN 30011-3 — Guidelines for auditing quality systems. Management of audit programmes

Quality Assurance Requirements for Measuring Equipment
BS EN 30012-1 — Quality assurance requirements for measuring equipment. Metrological confirmation system for measuring equipment

Maintainability and Quality
BS 6548-1 — Maintainability of equipment. Guide to specifying and contracting for maintainability
BS 6548-2 — Maintainability of equipment. Guide to maintainability studies during the design phase
BS 6548-3 — Maintainability of equipment. Guide to maintainability, verification and the collection, analysis and presentation of maintainability data
BS 6548-4 — Maintainability of equipment. Guide to the planning of maintenance and maintenance support
BS 6548-5 — Maintainability of equipment. Guide to diagnostic testing
BS 6548-6 — Maintainability of equipment. Guide to statistical methods in maintainability evaluation

Reliability and Quality
BS 5760-0 — Reliability of systems, equipment and components. Introductory guide to reliability
BS 5760-1 (EN 60300-2) — Reliability of systems, equipment and components. Dependability programme elements and tasks
BS 5760-2 — Reliability of systems, equipment and components. Guide to the assessment of reliability
BS 5760-3 — Reliability of systems, equipment and components. Guide to reliability practices: examples
BS 5760-4 — Reliability of systems, equipment and components. Guide to the specification clauses relating to the achievement and development of reliability in new and existing items
BS 5760-5 — Reliability of systems, equipment and components. Guide to failure modes, effects and criticality analysis (FMEA and FMECA)
BS 5760-6 — Reliability of systems, equipment and components. Guide to programmes for reliability growth
BS 5760-7 — Reliability of systems, equipment and components. Guide to fault free analysis
BS 5760-8 — Reliability of systems, equipment and components. Guide to assessment of reliability of systems containing software
BS 5760-10.1 — Reliability of systems, equipment and components. Guide to reliability testing. General requirements
BS 5760-10.2 — Reliability of systems, equipment and components. Guide to reliability testing. Design of test cycles

APPENDIX 12: BRITISH STANDARDS ON QUALITY

BS 5760-10.3	Reliability of systems, equipment and components. Guide to reliability testing. Compliance test procedures for steady-state availability
BS 5760-10.5	Reliability of systems, equipment and components. Guide to reliability testing. Compliance test plans for success ratio
BS 5760-11	Reliability of systems, equipment and components. Collection of reliability, availability, maintainability and maintenance support data from the field
BS 5760-12	Reliability of systems, equipment and components. Guide to the presentation of reliability, maintainability and availability predictions
BS 5760-13.1	Reliability of systems, equipment and components. Guide to reliability test conditions for consumer simulation for indoor portable equipment
BS 5760-13.2	Reliability of systems, equipment and components. Guide to reliability test conditions for consumer equipment. Conditions providing a high degree of simulation for equipment for stationary use in weather protected locations
BS 5760-13.3	Reliability of systems, equipment and components. Guide to reliability test conditions for consumer equipment. Conditions providing a low degree of simulation for equipment for stationary use in partially weather protected locations
BS 5760-13.4	Reliability of systems, equipment and components. Guide to reliability test conditions for consumer equipment. Conditions providing a low degree of simulation for equipment for portable and non-stationary use
BS 5760-13.5	Reliability of systems, equipment and components. Guide to reliability test conditions for consumer equipment. Ground mobile equipment Low degree of simulation
BS 5760-14	Reliability of systems, equipment and components. Guide to formal design review
BS 5760-20	Reliability of systems, equipment and components. Guide to the specification of dependability requirements
BS EN 61078 (BS 5760-9)	Reliability of systems, equipment and components. Guide to the block diagram technique

Environmental Management Systems

BS EN ISO 14001	Environmental management systems. Specification with guidance for use

Standards and Quality Management: an Integrated Approach

PD 3542	Standards and quality management. An integrated approach

Testing Laboratory

BS 7501 (EN 45001)	General criteria for the operation of testing laboratories
BS 7513 (EN 45013)	General criteria for certification bodies operating certification of personnel

Sampling and Statistical Process Control

BS 6000 (ISO TR 8550)	Guide for the selection of an acceptance sampling system, scheme or plan for inspection of discrete items in lots
BS 6001-1 (ISO 2859-1)	Sampling procedures for inspection by attributes. Introduction to the BS 6001 attribute sampling scheme
BS 6001-2 (ISO 2859-2)	Sampling procedures for inspection by attributes. Specification for sampling plans indexed by limiting quality (LQ) for isolated lot inspection
BS 6001-3 (ISO 2859-3)	Sampling procedures for inspection by attributes. Specification for skip-lot procedures
BS 6001-4 (ISO 8422)	Sampling procedures for inspection by attributes. Specification for sequential sampling plans
BS 6002-1	Sampling procedures for inspection by variables. Specification for single sampling plans indexed by acceptable quality level (AQL). Known standard deviation
BS 6002-4.1	Sampling procedures for inspection by variables. Specification for sequential sampling plans for percent nonconforming. Known standard deviation
BS 7782 (ISO 7870)	Control charts. General guide and introduction
BS 7783 (ISO 7966)	Acceptance control charts

APPENDIX 13

Abbreviations used in Spectroscopy and Microscopy

AEAPS	Auger Electron Appearance Potential Spectroscopy
AES	Auger Electron Spectroscopy
AFM	Atomic Force Microscopy (or SFM)
APS	Appearance Potential Spectroscopy
ARUPS	Angle-Resolved Ultraviolet Photoelectron Spectroscopy
ATR	Attenuated Total Reflection (or MIR)
BIS	Bremmstrahlung Isochromat Spectroscopy
BF	Bright Field
CHA	Concentric Hemispherical Analyzer
CITS	Current Imaging Tunneling Spectroscopy
CMA	Cylindrical Mirror Analyzer
CPD	Contact Potential Difference
DAPS	Disappearance Potential Spectroscopy
DF	Dark Field
DRIFTS	Diffuse Reflectance IR Spectroscopy
EAPFS	Extended Appearance Potential Fine Structure
EDX	Energy Dispersive X-ray Analysis
EELS	Electron Energy Loss Spectroscopy
EIS	Electron Impact Spectroscopy
ESCA	Electron Spectroscopy for Chemical Analysis (or XPS)
ESD	Electron Stimulated Desorption
ESDIAD	Electron Stimulated Desorption Ion Angular Distribution
ESFD	Electron Stimulated Field Desorption
EXAFS	Extended X-ray Absorption Fine Structure
EXELFS	Extended Energy Loss Fine Structure
FEM	Field Emission Microscopy
FIM	Field Ion Microscopy
FTRAIS	Fourier Transform Reflection Absorption Infrared Spectroscopy
HEIS	High Energy Ion Scattering
HREELS	High Resolution Electron Energy Loss Spectroscopy (or VELS)
HREM	High Resolution Electron Microscopy
HRHAS	High Resolution Helium Atom Scattering
IEE	Induced Electron Emission
ILS	Ionization Loss Spectroscopy
IMBS	Inelastic Molecular Beam Scattering
INS	Ion Neutralization Spectroscopy
IPES	Inverse Photoemission Spectroscopy
IRAS	Infrared Reflection-Absorption Spectroscopy
KRIPES	k-Resolved Inverse Photoemission Spectroscopy
LAMMS	Laser Microprobe Mass Spectrometry
LEED	Low Energy Electron Diffraction
LEIS	Low Energy Ion Scattering
MBE	Molecular Beam Epitaxy

MBS	Molecular Beam Spectroscopy
MEED	Medium Energy Electron Diffraction
MEIS	Medium Energy Ion Scattering
MIR	Multiple Internal Reflection Spectroscopy
MOLE	Molecular Optical Laser Examiner
NEXAFS	Near-Edge X-ray Absorption Fine Structure
OM	Optical Microscopy
PES	Photoelectron Spectroscopy
PESIS	Photoelectron Spectroscopy of Inner Shells
PESOS	Photoelectron Spectroscopy of Outer Shells
PhD	Photoelectron Diffraction
PSD	Photon Stimulated Desorption
PSTM	Photon Scanning Tunneling Microscopy
RAIRS	Reflection Absorption Infrared Spectroscopy
RFA	Retarding Field Analyzer
RHEED	Reflection High Energy Electron Diffraction
SAD	Selected Area Electron Diffraction
SAM	Scanning Auger Microscopy
SANS	Small Angle Neutron Scattering
SAXS	Small Angle X-ray Scattering
SEM	Scanning Electron Microscopy
SERS	Surface Enhanced Raman Scattering
SEXAFS	Surface EXAFS (Extended X-ray Absorption Fine Structure)
SFM	Scanning Force Microscopy (or AFM)
SIMS	Secondary Ion Mass Spectroscopy
SPIES	Surface Penning Ionization Electron Spectroscopy
SPM	Scanning Probe Microscopy
STM	Scanning Tunneling Microscopy
STS	Scanning Tunneling Spectroscopy
SXAPS	Soft X-ray APS (Appearance Potential Spectroscopy)
SXRD	Surface X-ray Diffraction
SXW	Standing X-ray Wavefield Absorption (or XSW)
TEAS	Thermal Energy Atom Scattering
TEM	Transmission Electron Microscopy
ToFSIMS	Time of Flight Secondary Ion Mass Spectroscopy
TPD	Temperature Programmed Desorption
UHV	Ultra High Vacuum
UPS	Ultraviolet Photoelectron Spectroscopy
VELS	Vibrational ELS (or HREELS)
XAES	X-ray Excited Auger Electron Spectroscopy
XANES	X-ray Absorption Near-Edge Structure
XPS	X-ray Photoelectron Spectroscopy (or ESCA)
XSW	X-ray Standing Wavefield absorption (or SXW)

Adapted from Woodruff DP, Delchar TA, *Modern Techniques of Surface Science*, Cambridge University Press, pp xvii & xviii, 1994.

APPENDIX 14

Typical Properties of Unreinforced Plastic Polymers

1. ABS

1. Material   ABS
2. Formula

    Acrylonitrile butadiene styrene
    Made by blending acrylonitrile styrene copolymer and butadiene acrylonitrile rubber or inter-polymerizing polybutadiene with styrene and acrylonitrile.

3. Crystal state   Amorphous
4. Cost index   0.67
5. Ease of molding   Good flow
6. Properties

Property	Value
Density	1.04–1.06 g cm^{-3}
Tensile strength at yield	41–62 MPa
Tensile modulus	2.1–2.7 GPa
Elongation at break	13–26%
Flexural strength	66–79 MPa
Flexural modulus	2.14–3.10 GPa
Compressive strength	55–86 MPa
Compressive modulus	–
Impact Notched Izod	20–30 kJ m^{-2}
Unnotched Izod	
Chemical resistance	Good, better than SAN. Degrades in UV. Attacked by strong oxidizing acids. Dissolves in esters, ketones and unsaturated alkyl halides.
Water absorption	0.2%
Thermal	
Melting point	–
T_g	105°C
HDT (1.82 MPa)	75–93°C
Max. continuous temperature of use	65–70°C
Coefficient of linear thermal expansion	0.84–0.96 × 10^{-4} K^{-1}
Thermal conductivity	–

The cost index is based on the price of PA6 (Nylon 6) for 1000 kg.

Oxygen Index	–
Electrical volume resistivity	10^{16} Ωcm
7. Processing	
Mold shrinkage	0.7%
Drying temperature	80°C
Drying time	2–4 h
Mold temperature	70°C
Melt temperature	220–260°C
8. Dimensional stability	Excellent surface finish

9. Trade names
   BASF 'Terluran'
   DSM 'Ronfalin'
   Bayer 'Lustran'
   General Electric Plastics 'Cycolac'
   Bayer 'Novadur'

2. ETFE

1. Material   ETFE   Ethylenetetrafluoroethylene
2. Formula
   Made by copolymerization of ethylene
   and tetrafluoroethylene

$$CH_2{=}CH_2 + CF_2{=}CF_2 \rightarrow [{-}(CH_2)_2(CF_2)_2{-}]_n$$

3. Crystal state	–
4. Cost index	10.2
5. Ease of molding	Needs care
6. Properties	
Density	1.70 g cm^{-3}
Tensile strength at yield	45 MPA tendency to creep at high loads
Tensile modulus	0.8 GPa
Elongation at break	150%
Flexural strength	69 MPa
Flexural modulus	1.3 GPa
Compressive strength	–
Compressive modulus	–
Impact Notched Izod	–
Unnotched Izod	–
Chemical resistance	Chemically inert with excellent weathering resistance
Water absorption	Very low < 0.02%
Thermal	
Melting point	270°C
T_g	–
HDT (1.82 MPa)	74°C
Max. continuous temperature of use	150°C with no load
Coefficient of linear thermal expansion	–
Thermal conductivity	–
Oxygen Index	–
Electrical volume resistivity	–
7. Processing	
Mold shrinkage	1.5–2.0% reduced by a factor of 10 when reinforced with C/F
Drying temperature	–
Drying time	–
Mold temperature	–
Melt temperature	–
8. Dimensional stability	Limited but improved by C/F addition
9. Trade names	Hoechst 'Hostaflon'

3. FEP

1. Material   FEP   Fluorinated ethylene propylene
2. Formula   Made by copolymerization of tetrafluoroethylene
                                (TFE) and hexafluoropropylene (HFP)

APPENDIX 14: TYPICAL PROPERTIES OF UNREINFORCED PLASTIC POLYMERS

$$CF_2=CF_2 + CF_2=CF-CF_3$$

3. Crystal state	Partially crystalline
4. Cost index	10.2
5. Ease of molding	Difficult; can only use a screw injection machine using a high processing temperature and a low injection rate. Use high nickel alloys for m/c parts
6. Properties	
Density	2.12–2.17 g cm^{-3}
Tensile strength at yield	10–12 MPa
Tensile modulus	0.55 GPa
Elongation at break	–
Flexural strength	–
Flexural modulus	–
Compressive strength	–
Compressive modulus	–
Impact Notched Izod	–
Unnotched Izod	–
Chemical resistance	Swells and degrades in benzyl chloride, DMF and fuming nitric acid. Swollen by chlorofluorohydrocarbons. Stress cracks in basic substances with pH above 11.5
Water absorption	Very low
Thermal	
Melting point	250–280°C
T_g	–
HDT (1.82 MPa)	54°C
Max. continuous temperature of use	205°C
Coefficient of linear thermal expansion	1.1×10^{-4} K^{-1}
Thermal conductivity	0.2 W m^{-1}K^{-1}
Oxygen Index	95–96%
Electrical volume resistivity	10^{18} Ωcm
7. Processing	
Mold shrinkage	4%
Drying temperature	–
Drying time	–
Mold temperature	Min. 130°C typical 180–200°C
Melt temperature	Up to 410°C typical 360–370°C
8. Dimensional stability	Precision molding is virtually impossible due to high shrinkage
9. Trade names	Hoechst 'Hostaflon FEP'

4. PA46

1. Material	PA46	Polyamide 46 (Nylon 46)
2. Formula		Made by condensation of 1.4 diaminobutane and adipic acid

$$NH_2(CH_2)_4NH_2 + HOOC-C-(CH_2)_4-COOH \rightarrow [-NH-(CH_2)_4-NH-CO-(CH_2)_4-CO]_n$$

3. Crystal state	High degree of crystallinity (approx. 70% cf 50% for PA66)
4. Cost index	2.45
5. Ease of molding	Cycle times considerably reduced due to very fast crystallization
6. Properties	
Density	1.18 g cm^{-3}
Tensile strength at yield	100 MPa
Tensile modulus	3.30 GPa
Elongation at break	40%
Flexural strength	150 MPa
Flexural modulus	3.00 GPa
Compressive strength	–
Compressive modulus	–
Impact Notched Izod	10 kJ m^{-2}
Unnotched Izod	No break

Chemical resistance	Excellent resistance to oils and greases. Attacked by strong mineral acids and absorbs polar solvents
Water absorption	3.7%
Thermal	
Melting point	295°C
T_g	–
HDT (1.82 MPa)	160°C
Max. continuous temperature of use	163°C for 2500 h
Coefficient of linear thermal expansion	$0.8–1.0 \times 10^{-4} K^{-1}$
Thermal conductivity	$0.22\ W\,m^{-1} K^{-1}$
Oxygen Index	27%
Electrical volume resistivity	$10^{15}\ \Omega cm$
7. Processing	
Mold shrinkage	1.5–2.0%
Drying temperature	–
Drying time	–
Mold temperature	80°C
Melt temperature	310–320°C
8. Dimensional stability	Excellent flow in the melt with molding showing no signs of flashing
9. Trade names	DSM 'Stanyl'

5. PA6

1. Material   PA6   Polyamide 6 (Nylon 6)
2. Formula   Made by self condensation of ε-caprolactam (aminocaproic acid)

$$NH_2(CH_2)_5 COOH \rightarrow [-NH-(CH_2)_5-CO-]_n$$

Note: the number 6 relates to the number of carbon atoms in the amide group

3. Crystal state	Crystalline
4. Cost index	1.00
5. Ease of molding	Can be processed in all commercial injection molding machines with correctly designed injection units
6. Properties	
Density	$1.13–1.14\ g\,cm^{-3}$
Tensile strength at yield	80–85 MPa (45 MPa when wet)
Tensile modulus	3.1 GPa (1.0 GPa when wet)
Elongation at break	20–50%
Flexural strength	1.05 MPa
Flexural modulus	2.80 GPa
Compressive strength	–
Compressive modulus	–
Impact Notched Izod	$6.5–10\ kJ\,m^{-2}$ (No break when wet)
Unnotched Izod	–
Chemical resistance	Good resistance to alcohols, ketones, esters, ethers, hydrocarbons, dilute bases, oils and greases
Water absorption	9–10%
Thermal	
Melting point	218–220°C
T_g	40–60°C
HDT (1.82 MPa)	75–77°C
Max. continuous temperature of use	>180°C
Coefficient of linear thermal expansion	$0.7–1.0 \times 10^{-4}\ K^{-1}$
Thermal conductivity	$0.33\ W\,m^{-1} K^{-1}$
Oxygen Index	–
Electrical volume resistivity	$10^{12} – 10^{15}\ \Omega cm$
7. Processing	
Mold shrinkage	0.65–0.8%

APPENDIX 14: TYPICAL PROPERTIES OF UNREINFORCED PLASTIC POLYMERS

Drying temperature	65°C
Drying time	Up to 6 h
Mold temperature	60–80°C
Melt temperature	230–280°C
8. Dimensional stability	In 1 yr 0.1% change in dimensions by moisture absorption from air

9. Trade names
   Allied Chemical 'Capron6'
   DuPont 'Zytel PA6'
   BASF 'Ultramid B'
   EMS 'Grilon'
   Bayer 'Durethan PA6'
   DSM 'Akulon PA6'
   Nyltech 'Sniamid PA6'
   Hoechst 'Celanese Nylon PA6'

6. PA66

1. Material   PA66   Polyamide 66 (Nylon 66)
2. Formula   Made by the condensation of hexamethylene diamine and adipic acid

$$NH_2(CH_2)_6NH_2 + HOOC-(CH_2)_4-COOH \rightarrow [-NH-CH_2)_6NHCO(CH_2)_4CO-]_n$$

(Copolymers with nylon 6 are also available). This PA has the greatest hardness, rigidity and highest resistance to abrasion and heat deformation

3. Crystal state	Crystalline (approx. 50%)
4. Cost index	1.11
5. Ease of molding	Excellent with good flow
6. Properties	
Density	1.13–1.14 g cm^{-3}
Tensile strength at yield	80–82.7 MPa (60 MPa when wet)
Tensile modulus	3.2 GPa (1.6 GPa when wet)
Elongation at break	40–80% (250% when wet)
Flexural strength	117–118 MPa (54 MPa when wet)
Flexural modulus	2.79–2.90 GPa (1.18 GPa when wet)
Compressive strength	–
Compressive modulus	–
Impact Notched Izod	6 kJ m^{-2}
Unnotched Izod	–
Chemical resistance	Good resistance to alcohols, ketones, esters, ethers, hydrocarbons, dilute bases, oils and greases
Water absorption	8–9%
Thermal	
Melting point	257–260°C
T_g	55°C
HDT (1.82 MPa)	75–77°C
Max. continuous temperature of use	>200°C
Coefficient of linear thermal expansion	$0.8 \times 10^{-4} K^{-1}$
Thermal conductivity	0.33 W m^{-1}K^{-1}
Oxygen Index	24–26%
Electrical volume resistivity	10^{12} Ωcm
7. Processing	
Mold shrinkage	0.25–2.0% depending on direction
Drying temperature	65°C
Drying time	6 h
Mold temperature	60–80°C
Melt temperature	260–305°C
8. Dimensional stability	1 yr 0.1% change in dimensions by moisture absorption from the air

9. Trade names
   Allied Signal 'Capron 6'
   DSM 'Akulon PA66'
   Asahi (Mitsui) 'Leona'
   EMS 'Grilon'
   Bayer 'Durethan PA66'
   Hoechst 'Celanese Nylon PA66'
   BASF 'Ultramid A6'
   Nyltech 'Technyl PA66'

7. PA612

1. Material PA612 Polyamide 612 (Nylon 612)
2. Formula

$$[-NH-(CH_2)_6-NH-CO-(CH_2)_{10}-CO-]_n$$

3. Crystal state	Crystalline
4. Cost index	–
5. Ease of molding	–
6. Properties	
Density	1.06 g cm^{-3}
Tensile strength at yield	61 MPa
Tensile modulus	2.70 GPa
Elongation at break	100%
Flexural strength	–
Flexural modulus	–
Compressive strength	–
Compressive modulus	–
Impact Notched Izod	3.5 kJ m^{-2}
Unnotched Izod	–
Chemical resistance	–
Water absorption	3%
Thermal	
Melting point	218°C
T_g	–
HDT (1.82 MPa)	90°C
Max. continuous temperature of use	–
Coefficient of linear thermal expansion	1.31 × 10^{-4} K^{-1}
Thermal conductivity	–
Oxygen Index	–
Electrical volume resistivity	10^{15} Ωcm
7. Processing	
Mold shrinkage	1.3%
Drying temperature	80°C
Drying time	Dry until <0.2% moisture
Mold temperature	–
Melt temperature	–
8. Dimensional stability	–
9. Trade names	DuPont 'Zytel PA 612'

8. PA11

1. Material PA11 Polyamide 11 (Nylon 11)
2. Formula   Self condensation of ε-aminoundeconoic acid

$$NH_2(CH_2)_{10}COOH \rightarrow [-NH-(CH_2)_{10}-CO-]_n$$

3. Crystal state	Crystalline
4. Cost index	3.31
5. Ease of molding	–
6. Properties	
Density	1.03 g cm^{-3}

Tensile strength at yield	40 MPa
Tensile modulus	–
Elongation at break	300%
Flexural strength	45 MPa
Flexural modulus	1.00 GPa
Compressive strength	–
Compressive modulus	–
Impact Notched Izod	12–29 kJ m^{-2}
Unnotched Izod	No break
Chemical resistance	High resistance to attack by most chemicals
Water absorption	1.9% (lowest of the nylons)
Thermal	
Melting point	183–187°C
T_g	–
HDT (1.82 MPa)	50°C
Max. continuous temperature of use	100°C peaks up to 140/150°C
Coefficient of linear thermal expansion	0.85 × 10^{-4}K^{-1}
Thermal conductivity	–
Oxygen Index	–
Electrical volume resistivity	10^{14} Ωcm
7. Processing	
Mold shrinkage	–
Drying temperature	–
Drying time	–
Mold temperature	–
Melt temperature	–
8. Dimensional stability	1% moisture uptake causes dimensions to vary <0.25%
9. Trade names	Ato-Chemie 'Rilsan B'

9. PA12

1. Material   PA12   Polyamide 12 (Nylon 12)
2. Formula   Made by the polycondensation of laurolactam (cyclododecanone isooxime)

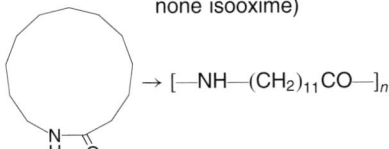 → [—NH—(CH$_2$)$_{11}$CO—]$_n$

3. Crystal state	Semi-crystalline
4. Cost index	3.18
5. Ease of molding	The secondary pressure, sprue and gate dimensions must be large enough to compensate for a loss in volume due to crystallization
6. Properties	
Density	1.01–1.02 g cm^{-3} lightest polyamide available on market
Tensile strength at yield	31–45 MPa
Tensile modulus	1.08 GPa
Elongation at break	300%
Flexural strength	25 MPa
Flexural modulus	0.55 GPa
Compressive strength	–
Compressive modulus	–
Impact Notched Izod	8–12 kJ m^{-2}
Unnotched Izod	No break
Chemical resistance	Strong acids cause hydrolytic breakdown. Dissolves in conc. mineral acids, phenols, methanolic CaCl$_2$ solution and highly halogenated acetic acid
Water absorption	1.4% It is the polyamide with second lowest water absorption

Thermal
  Melting point   174–178°C
  T_g –
  HDT (1.82 MPa)   45–48°C
  Max. continuous temperature of use   90–110°C
  Coefficient of linear thermal expansion $1.2–1.3 \times 10^{-4} K^{-1}$
  Thermal conductivity   –
Oxygen Index   –
Electrical volume resistivity $10^{12} – 10^{14}$ Ωcm

7. Processing
  Mold shrinkage   –
  Drying temperature   70°C, tends to oxidize at temperatures >80°C
  Drying time   8–16 h
  Mold temperature   40–60°C tends to stick at higher temperatures
  Melt temperature   220–280°C
8. Dimensional stability   High
9. Trade names
  Elf-Atochem 'Rilsan A'   EMS 'Grilamid'

10. PA

1. Material   PA   Polyamide Super tough grade
2. Formula   A version of PA 66 insensitive to notching, scratches, does not concentrate stress and does not become brittle at −20°C
3. Crystal state   –
4. Cost index   1.67
5. Ease of molding   Similar to PA 66
6. Properties
  Density   1.08 g cm^{-3}
  Tensile strength at yield   50 MPa
  Tensile modulus   2.00 GPa
  Elongation at break   60%
  Flexural strength   –
  Flexural modulus   –
  Compressive strength   –
  Compressive modulus   –
  Impact   Notched Izod   80 kJ m^{-2}
        Unnotched Izod   Did not break
  Chemical resistance   Good resistance to chemicals, solvents, oils and greases
  Water absorption   6.7%
  Thermal
    Melting point   263°C
    T_g –
    HDT (1.82 MPa)   66°C
    Max. continuous temperature of use   –
    Coefficient of linear thermal expansion $1.6 \times 10^{-4} K^{-1}$
    Thermal conductivity   –
  Oxygen Index   –
  Electrical volume resistivity 10^{14} Ωcm
7. Processing
  Mold shrinkage   1.7%
  Drying temperature   80°C
  Drying time   To a moisture content <0.2%
  Mold temperature   –
  Melt temperature   –
8. Dimensional stability   Good superior finish to PA6
9. Trade names
  BASF 'Ultramid'
  DuPont 'Zytel ST'

APPENDIX 14: TYPICAL PROPERTIES OF UNREINFORCED PLASTIC POLYMERS 1097

11. PA HTN

1. Material   PA HTN   Polyamide HTN (high temperature nylon)
2. Formula   Partly an aromatic polyamide, can be based on PA 6/6T.
                                          Copolymer of ethylenediamine and terephthalic acid

$$NH_2(CH_2)_6NH_2 \; + \; \text{(terephthalic acid: benzene ring with COOH groups at 1,4 positions)}$$

3. Crystal state   –
4. Cost index   –
5. Ease of molding   –
6. Properties
   Density $1.10 \, \text{g cm}^{-3}$
   Tensile strength at yield   76 MPa (72 MPa wet)
   Tensile modulus   –
   Elongation at break   23–40% (40 MPa wet)
   Flexural strength   100 MPa (96 MPa wet)
   Flexural modulus   2.4 GPa (2.4 GPa wet)
   Compressive strength   –
   Compressive modulus   –
   Impact   Notched Izod $11 \, \text{kJ m}^{-2}$
               Unnotched Izod   No break
   Chemical resistance   Very good
   Water absorption   0.4%
   Thermal
     Melting point   320°C
     T_g 125°C
     HDT (1.82 MPa)   130°C
     Max. continuous temperature of use   –
     Coefficient of linear thermal expansion $0.8 \times 10^{-4} \, \text{K}^{-1}$
     Thermal conductivity   –
   Oxygen Index   –
   Electrical volume resistivity $2.3 \times 10^{16} \, \Omega \text{cm}$
7. Processing
   Mold shrinkage   0.8–0.9%
   Drying temperature   110°C
   Drying time   6 h
   Mold temperature   70–150°C
   Melt temperature   298°C
8. Dimensional stability   Excellent high temperature stability
9. Trade names
   Bayer 'Ultramid T'
   Mitsui 'Arlen'
   Du Pont 'Zytel HTN'

12. PA (amorphous)

1. Material   PA(amorphous)   Amorphous polyamide (amorphous nylon)
2. Formula

$$H_2NH_2C\text{-}\bigcirc\text{-}CH_2NH_2 \; + \; HOOC(CH_2)_4COOH \; \rightarrow \; \left[-HNH_2C\text{-}\bigcirc\text{-}CH_2NH\text{-}\overset{O}{\underset{\|}{C}}\text{-}(CH_2)_4\text{-}\overset{O}{\underset{\|}{C}}\text{-} \right]_n$$

3. Crystal state   Amorphous
4. Cost index   –
5. Ease of molding   –
6. Properties
   Density $1.18 \, \text{g cm}^{-3}$
   Tensile strength at yield   100 MPa

Tensile modulus	3.40 GPa
Elongation at break	15%
Flexural strength	–
Flexural modulus	–
Compressive strength	–
Compressive modulus	–
Impact Notched Izod	4 kJ m^{-2}
Unnotched Izod	No break
Chemical resistance	–
Water absorption	7%
Thermal	
Melting point	125°C
T_g	–
HDT (1.82 MPa)	104°C
Max. continuous temperature of use	–
Coefficient of linear thermal expansion	0.6×10^{-4} K^{-1}
Thermal conductivity	–
Oxygen Index	–
Electrical volume resistivity	10^{14} Ωcm
7. Processing	
Mold shrinkage	–
Drying temperature	–
Drying time	–
Mold temperature	–
Melt temperature	–
8. Dimensional stability	–
9. Trade names	EMS 'Grivory'

13. PAA

1. Material   PAA — Polyarylamide
2. Formula — Covers a range of compounds produced from a polyarylamide namely poly[*m*-xylylene adipimide, formed by the polycondensation of *m*-xylylene diamine. (MXDA) and adipic acid and in accordance with PA nomenclature termed PA MXD6

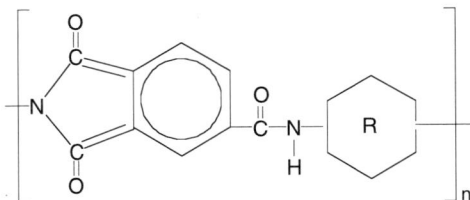

3. Crystal state — Semi-crystalline
4. Cost index — –
5. Ease of molding — Can be processed on both ram and screw injection molding machines

6. Properties
Density	–
Tensile strength at yield	–
Tensile modulus	–
Elongation at break	–
Flexural strength	–
Flexural modulus	–
Compressive strength	–
Compressive modulus	–
Impact Notched Izod	–
Unnotched Izod	–

Chemical resistance	Similar to PA6 and PA66
Water absorption	Water is absorbed more slowly and less than PA6 and PA66
Thermal	
Melting point	–
T_g	90°C
HDT (1.82 MPa)	85–100°C
Max. continuous temperature of use	120–230°C
Coefficient of linear thermal expansion	–
Thermal conductivity	–
Oxygen Index	27.5%
Electrical volume resistivity	–
7. Processing	
Mold shrinkage	–
Drying temperature	80°C
Drying time	About 6 h, moisture content must be <0.4%
Mold temperature	100°C
Melt temperature	235–240°C
8. Dimensional stability	–
9. Trade names	Solvay 'Ixef MXD6'

14. PAI

1. Material	PAI	Polyamideimide
2. Formula		Grade 4203L contains 3% TiO_2 and ½% fluorocarbon.

3. Crystal state	–
4. Cost index	–
5. Ease of molding	Can use conventional reciprocating screw injection with high shear
6. Properties	To develop optimum physical properties post cure: 24 h, 165°C; 24 h, 245°C and 24 h, 260°C
Density	1.42 g cm^{-3}
Tensile strength at yield	192 MPa
Tensile modulus	4.9 GPa
Elongation at break	15%
Flexural strength	244 MPa
Flexural modulus	5.0 GPa
Compressive strength	220 MPa
Compressive modulus	4.0 GPa
Impact Notched Izod	142 Jm^{-1}
Unnotched Izod	1062 Jm^{-1}
Chemical resistance	Very good. May be attacked by saturated steam, strong bases and some high temperature acid systems (e.g. benzene sulfonic acid)
Water absorption	0.28–0.33%
Thermal	
Melting point	–
T_g	275°C
HDT (1.82 MPa)	278°C
Max. continuous temperature of use	220°C
Coefficient of linear thermal expansion	0.31 × 10^{-4} K^{-1}
Thermal conductivity	0.26 W m^{-1}K^{-1}

Oxygen Index	43–45%
Electrical volume resistivity	2×10^{15} Ωcm

7. Processing

Mold shrinkage	0.75%
Drying temperature	120°C
Drying time	8 h
Mold temperature	230°C
Melt temperature	330–360°C

8. Dimensional stability — Very stable. Some swelling in water, drying restores original dimensions.
9. Trade names — Amoco 'Torlon'

15. PBT or PBTP

1. Material   PBT or PBTP — Polybutylene terephthalate
2. Formula — Made by the condensation of 1,4-butanediol and terephthalic acid. One of a family of thermoplasticpolyesters.

$$HO-(CH_2)_4-OH + \rightarrow HOOC-\langle\bigcirc\rangle-COOH \rightarrow \left[-\overset{O}{\underset{\|}{C}}-\langle\bigcirc\rangle-\overset{O}{\underset{\|}{C}}-O-(CH_2)_4-O-\right]_n$$

3. Crystal state	Semi-crystalline
4. Cost index	1.22
5. Ease of molding	Simple using standard injection molding machines. Good flow properties

6. Properties

Density	1.3 g cm^{-3}
Tensile strength at yield	55–60 MPa
Tensile modulus	2.5–2.8 GPa
Elongation at break	>50%
Flexural strength	85–90 MPa
Flexural modulus	2.8 GPa
Compressive strength	–
Compressive modulus	–
Impact Notched Izod	–
Unnotched Izod	–
Chemical resistance	No known solvents at room temperature. Limited resistance to dilute acids. Not resistant to aqueous alkalis. Prone to hydrolysis but will sustain brief contact with hot water. No stress cracking observed.
Water absorption	0.5%

Thermal

Melting point	220°C
T_g	–
HDT (1.82 MPa)	57–70°C
Max. continuous temperature of use	200°C for short cycles
Coefficient of linear thermal expansion	$1.3–1.6 \times 10^{-4}$K^{-1}
Thermal conductivity	–
Oxygen Index	20–24%
Electrical volume resistivity	10^{16} Ωcm

7. Processing

Mold shrinkage	0.7–2.0%
Drying temperature	80–120°C
Drying time	2–4 h
Mold temperature	40–80°C
Melt temperature	250–275°C

APPENDIX 14: TYPICAL PROPERTIES OF UNREINFORCED PLASTIC POLYMERS

8. Dimensional stability   Excellent surface finish, can anneal for 24 h at 120°C
9. Trade names
   BASF 'Ultradur'
   EMS 'Grilpet'
   Bayer 'Pocan'
   General Electric Plastics 'Valox'
   DuPont 'Crastin'
   Ticona 'Celanex' and 'Vandar'

16. PC

1. Material   PC   Polycarbonate
2. Formula

$$\left[-O-\underset{}{\bigcirc}-\underset{CH_3}{\overset{CH_3}{C}}-\underset{}{\bigcirc}-O-\underset{\underset{O}{\parallel}}{C}- \right]_n$$

3. Crystal state   Amorphous
4. Cost index   1.06–1.31
5. Ease of molding   A medium melt grade is recommended
6. Properties
   Density   1.20 $g\,cm^{-3f}$
   Tensile strength at yield   63 MPa
   Tensile modulus   2.25–2.40 GPa
   Elongation at break   120–125%
   Flexural strength   90 MPa
   Flexural modulus   2.30 GPa
   Compressive strength   75 MPa
   Compressive modulus   –
   Impact   Notched Izod   25 $kJ\,m^{-2}$
               Unnotched Izod   No break
   Chemical resistance   Can stress crack
   Water absorption   0.35%
   Thermal
      Melting point   –
      T_g 148°C
      HDT (1.82 MPa)   130°C (141°C after annealing)
      Max. continuous temperature of use   125°C
      Coefficient of linear thermal expansion $0.7 \times 10^{-4} K^{-1}$
      Thermal conductivity   0.2 W/m^2
   Oxygen Index   27%
   Electrical volume resistivity 10^{16} Ωcm
7. Processing
   Mold shrinkage   0.5–0.8%
   Drying temperature   120°C
   Drying time   2–4 h
   Mold temperature   80–90°C
   Melt temperature   280–310°C
8. Dimensional stability   Good over wide temperature range
9. Trade names
   Bayer 'Apec HT' and 'Makrolon'
   Indemitsu Chemical 'Indemitsu'
   DSM 'Xantar'
   Mobay Chemical 'Merlon'
   General Electric Plastics 'Lexan'

17. PEEK

1. Material   PEEK — Polyetheretherketone
2. Formula — poly(oxy-1,4-phenyleneoxy-1,4-phenylenecarbonyl-1,4-phenylene)

3. Crystal state — Semi-crystalline (typically 35% crystalline)
4. Cost index — 21.6
5. Ease of molding — Standard reciprocating screw injection molding machines can be used Process on clean machines, avoid purging

6. Properties
   Density — 40% crystalline 1.32 g cm^{-3}, amorphous 1.26 g cm^{-3}
   Tensile strength at yield — 92 MPa
   Tensile modulus — 3.6 GPa (secant)
   Elongation at break — >60%
   Flexural strength — 170 MPa
   Flexural modulus — 4.1 GPa
   Compressive strength — 119 MPa
   Compressive modulus — 3.6 GPa
   Impact Notched Izod — 80 Jm^{-1}
           Unnotched Izod — No break
   Chemical resistance — Very good. Dissolves in conc. H_2SO_4 Attacked by halogens, halogen acids, phenol, sodium, hot MEK and nitrobenzene

   Water absorption — 0.5%
   Thermal
     Melting point — 334°C
     T_g — 143°C
     HDT (1.82 MPa) — 152°C
     Max. continuous temperature of use — 260°C
     Coefficient of linear thermal expansion — $<T_g$ 0.47, T_g 1.08 × 10^{-4}K^{-1}
     Thermal conductivity — 0.25 W m^{-1}K^{-1}
   Oxygen Index — 24% (0.9 mm thick), 35% (3.2 mm thick)
   Electrical volume resistivity — 4.9 × 10^{16} Ωcm

7. Processing
   Mold shrinkage — 0.7–1.2% (depends on degree of crystallinity)
   Drying temperature — 150°C or 180°C
   Drying time — 3 h or 2 h
   Mold temperature — 160–200°C
   Melt temperature — 360–370°C
8. Dimensional stability — Can anneal for 4 h at 200°C
9. Trade names
   Victrex 'PEEK'
   Film 'STABAR'
   Fiber 'ZYEX'

18. PEI

1. Material   PEI — Polyetherimide
2. Formula

APPENDIX 14: TYPICAL PROPERTIES OF UNREINFORCED PLASTIC POLYMERS 1103

  3. Crystal state   Amorphous
  4. Cost index   5.04
  5. Ease of molding   Outstanding moldability. Can mold sections as thin as 0.25 mm, Easier to process than PES or PEEK

  6. Properties
       Density   1.34 g cm^{-3}
       Tensile strength at yield   103 MPa
       Tensile modulus   2.94 GPa
       Elongation at break   60%
       Flexural strength   142 MPa
       Flexural modulus   3.24 GPa
       Compressive strength   137 MPa
       Compressive modulus   2.84 GPa
       Impact   Notched Izod   50 Jm^{-1} displays notch sensitivity
                   Unnotched Izod   1300 Jm^{-1}
       Chemical resistance   Unlike other amorphous resins has good Chemical resistance. Soluble in aprotic solvents, methylene chloride and trichloroethane

     Water absorption   0.28%
     Thermal
       Melting point   –
       T_g 215°C
       HDT (1.82 MPa)   200°C
       Max. continuous temperature of use   170°C
       Coefficient of linear thermal expansion 0.56×10^{-4} K^{-1}
       Thermal conductivity   0.22 W m^{-1} K^{-1}
     Oxygen Index   47%
     Electrical volume resistivity 6.7×10^{17} Ωcm
  7. Processing
     Mold shrinkage   0.4%
     Drying temperature   150°C
     Drying time   4 h (until <0.05% H$_2$O)
     Mold temperature   65–175°C (95°C is optimum)
     Melt temperature   340–425°C
  8. Dimensional stability   Outstanding
  9. Trade names   General Electric Plastics 'Ultem'

19. PES

  1. Material   PES   Polyethersulfone (polyarylsulfone)
  2. Formula   Primarily   With a low level of polyetherethersulfone. Other forms of PES are sold which may contain Bisphenol A moiety which can compromise thermal stability and chemical resistance.

$$\left[\!\!\!\bigcirc\!\!\!-\!\!\overset{\overset{\displaystyle O}{\|}}{\underset{\underset{\displaystyle O}{\|}}{S}}\!\!-\!\!\bigcirc\!\!\!-O \right]_n$$

$$\left[\!\!\!\bigcirc\!\!\!-\!\!\overset{\overset{\displaystyle O}{\|}}{\underset{\underset{\displaystyle O}{\|}}{S}}\!\!-\!\!\bigcirc\!\!\!-O-\!\!\bigcirc\!\!\!-O \right]_n$$

  3. Crystal state   Amorphous
  4. Cost index   4.84
  5. Ease of molding   Reciprocating screw injection machine recommended with 400°C max., cylinder temperature and injection pressure of 1000–1500 kg cm^{-2}

6. Properties
  Density — 1.37 g cm^{-3}
  Tensile strength at yield — 83 MPa
  Tensile modulus — 2.66 GPa
  Elongation at break — 40–80%
  Flexural strength — 111–129 MPa
  Flexural modulus — 2.55–2.90 GPa
  Compressive strength — 100 MPa
  Compressive modulus — 2.68 GPa
  Impact Notched Izod — 65 Jm^{-1}
       Unnotched Izod — No break
  Chemical resistance — Good except for ketones, ethers and certain chlorinated solvents. Attacked by conc. oxidizing mineral acids. Soluble in highly polar organic solvents e.g. DMSO. Susceptible to environmental stress cracking in some organic solvents
  Water absorption — 0.4%
  Thermal
    Melting point — –
    T_g — 220–225°C
    HDT (1.82 MPa) — 204°C
    Max. continuous temperature of use — 170–180°C
    Coefficient of linear thermal expansion — 0.49×10^{-4} K^{-1}
    Thermal conductivity — 0.16 W m^{-1} K^{-1}
  Oxygen Index — 39.3%
  Electrical volume resistivity — 1.7×10^{15} Ωcm
7. Processing
  Mold shrinkage — 0.6%
  Drying temperature — 140°C or 150°C or 160°C
  Drying time — 6 h   3.5 h   3 h
  Mold temperature — 140–165°C
  Melt temperature — 345–390°C
8. Dimensional stability — Does absorb water, increases 0.15% at 20°C, 65%RH in 4 months
9. Trade names
  Amoco 'Radel A'
  BASF 'Ultrason E'
  Sumitomo 'Sumika Excel'

20. PFA

1. Material PFA — Perfluoroalkoxy
2. Formula — Copolymer of tetrafluoroethylene and perfluorinated co-components
3. Crystal state — Partially crystalline
4. Cost index — 14.5
5. Ease of molding — Normally used for 'lost' mold but conventional molding is becoming increasingly important. Use high nickel alloy materials

6. Properties
  Density — 2.12–2.17 g cm^{-3}
  Tensile strength at yield — 13–16 MPa
  Tensile modulus — 0.27 GPa
  Elongation at break — 250–300%
  Flexural strength — –
  Flexural modulus — 0.59–0.63 GPa
  Compressive strength — –
  Compressive modulus — –
  Impact Notched Izod — –
       Unnotched Izod — –
  Chemical resistance — Attacked by certain halogen complexes containing fluorine, molten sodium and potassium

APPENDIX 14: TYPICAL PROPERTIES OF UNREINFORCED PLASTIC POLYMERS 1105

    Water absorption   >0.03%
    Thermal
      Melting point   305–310°C
      T_g 90°C and −80°C
      HDT (1.82 MPa)   50°C
      Max. continuous temperature of use   260°C
      Coefficient of linear thermal expansion 1.2×10^{-4} K^{-1}
      Thermal conductivity   0.19–22 W m^{-1}K^{-1}
    Oxygen Index   95%
    Electrical volume resistivity 10^{16} Ωcm
7. Processing
    Mold shrinkage   –
    Drying temperature   –
    Drying time   –
    Mold temperature   'Lost' mold 200–210°C;
                                       conventional 250°C
    Melt temperature   Up to 420°C
8. Dimensional stability   –
9. Trade names
    DuPont 'Teflon PFA'
    Hoechst 'Hostaflon PFA'

21. PK
1. Material   PK   Aliphatic Polyketone
2. Formula   Polymerization of ethylene and carbon monoxide comono-
                                       mers to produce a perfectly alternating carbon monoxide/
                                       ethylene copolymer
3. Crystal state   Semi-crystalline
4. Cost index   1.22
5. Ease of molding   Easy to mold and flows well
6. Properties
    Density   1.24 g cm^{-3}
    Tensile strength at yield   60 MPa
    Tensile modulus   1.4 GPa
    Elongation at break   350%
    Flexural strength   55 MPa
    Flexural modulus   1.4 GPa
    Compressive strength   –
    Compressive modulus   –
    Impact   Notched Izod   20 kJ m^{-2}
                Unnotched Izod   –
    Chemical resistance   Excellent chemical resistance to hydrocarbons, solvents, salt
                                       solutions, weak acids and bases
    Water absorption   0.5% at 50% RH
    Thermal
      Melting point   220°C
      T_g 15°C
      HDT (1.82 MPa)   100°C
      Max. continuous temperature of use   Short term up to 180°C
      Coefficient of linear thermal expansion 1.1×10^{-4}K^{-1}
      Thermal conductivity   0.25 W m^{-1}K^{-1}
    Oxygen Index   21%
    Electrical volume resistivity 10^{13} Ωcm
7. Processing
    Mold shrinkage   2%
    Drying temperature   Pre-drying generally not required
    Drying time   –
    Mold temperature   80°C
    Melt temperature   240–250°C

8. Dimensional stability   Good mold definition and surface quality, no annealing required
9. Trade names   Shell 'Carilon'

22. POM
1. Material   POM   Polyoxymethylene (polyacetal)
2. Formula   'Delrin' is an acetal homopolymer whereas, 'Ultraform' is a linear chained random copolymer of trioxane and another monomer made from trioxane the trimer of Formaldehyde

$$(-O-CH_2-O-CH_2-O-CH_2-)_n (-O-CH_2-O-CH_2-CH_2-O-CH_2-)_n$$

      homopolymer   copolymer   trioxane

3. Crystal state   Semi-crystalline
4. Cost index   1.31
5. Ease of molding   All commercial machines with correctly designed injection unit
6. Properties
   Density   1.41 g cm^{-3} (heavier than most plastics)
   Tensile strength at yield 65 MPa, Excellent dynamic fatigue strength
   Tensile modulus   2.30–2.90 GPa, Disappointing creep resistance with C/F
   Elongation at break   6–12%
   Flexural strength   50–75 MPa
   Flexural modulus   2.5–3.0 GPa
   Compressive strength   –
   Compressive modulus   –
   Impact   Notched Izod   5–6 kJ m^{-2} avoid sharp notches
            Unnotched Izod   15–18 kJ m^{-2}
   Chemical resistance   Environmental stress cracking unknown. Poor UV resistance. Rendered unserviceable by strong mineral acids
   Water absorption   0.65–0.8% Can be used for plumbing
   Thermal
     Melting point   164–167°C
     T_g −65°C
     HDT (1.82 MPa)   101–106°C
     Max. continuous temperature of use 100°C (80°C under load)
     Coefficient of linear thermal expansion $1.2 \times 10^{-4} K^{-1}$
     Thermal conductivity   –
   Oxygen Index   –
   Electrical volume resistivity 10^{15} Ωcm
7. Processing
   Mold shrinkage   1.8–2.2%
   Drying temperature   100–110°C
   Drying time   2–3 h
   Mold temperature   60–90°C (use 90–120°C for precision parts)
   Melt temperature   185–230°C
8. Dimensional stability   Stress relieve by annealing at 150°C, accelerates post crystallization
9. Trade names
   BASF 'Ultraform'
   Mitsubishi 'Iupital'

APPENDIX 14: TYPICAL PROPERTIES OF UNREINFORCED PLASTIC POLYMERS

DuPont 'Delrin P'
Nyltech 'Sniata POM'
Ticona 'Hostaform'

23. PP

1. Material   PP   Polypropylene
2. Formula   Homopolymer. About 90–95% isotactic (i.e. all methyl groups lie on the same side of the carbon atom. There are also grades which are block co-polymers and random co-polymers

$$\left[\begin{array}{cc} H & CH_3 \\ | & | \\ -C - & C- \\ | & | \\ H & H \end{array} \right]_n$$

3. Crystal state — Semi-crystalline
4. Cost index — 0.22
5. Ease of molding — Can be processed on all modern injection molding machines with screw plasticization

6. Properties
   - Density — $0.91 \, g\,cm^{-3}$
   - Tensile strength at yield — 34–40 MPa
   - Tensile modulus — 1.45–2.0 GPa
   - Elongation at break — 20–>50%
   - Flexural strength — 49 MPa
   - Flexural modulus — 1.50 GPa
   - Compressive strength — 38–48 MPa
   - Compressive modulus — –
   - Impact   Notched Izod — 1.5–$4.5 \, kJ\,m^{-2}$
   -            Unnotched Izod — –
   - Chemical resistance — Swollen by aliphatic and aromatic hydrocarbons and halogenated hydrocarbons. Attacked by strong oxidizing agents e.g. conc. HNO_3
   - Water absorption — >0.01%
   - Thermal
     - Melting point — 160–165°C
     - T_g — –
     - HDT (1.82 MPa) — 55–65°C
     - Max. continuous temperature of use — –
     - Coefficient of linear thermal expansion — 6.9–$10.5 \times 10^{-5}/mm^{-1}\,°C^{-1}$
     - Thermal conductivity — $0.17 \, W\,m^{-1}\,K^{-1}$
   - Oxygen Index — –
   - Electrical volume resistivity — $0.93 \times 10^{18} \, \Omega\,cm$

7. Processing
   - Mold shrinkage — 1.0–2.5%
   - Drying temperature — 120°C
   - Drying time — 3 h
   - Mold temperature — 20–60°C
   - Melt temperature — 200–280°C

8. Dimensional stability — Generally produces moldings with a glossy surface

9. Trade names
   - Amoco polypropylene Elf 'Appryl'
   - BASF 'Novolen'
   - Hoechst 'Targor'
   - 'Procom'
   - Shell(Montedison) 'Montell'
   - DSM 'Stamylan P'

24. PPO or PPE

1. Material PPO or PPE Polyphenylene oxide (also polyphenylene ether)
2. Formula Made from 2,6-xylenol. Modified with high impact PS (HIPS) as 'Noryl' and 'Noryl GTX' with added PA. Modified with SBS as 'Luranyl'

[chemical structure: benzene ring with two CH_3 groups and O linkage, repeating unit n]

3. Crystal state — Amorphous
4. Cost index — 'Noryl' 1.24 and 'Noryl GTX' 1.61
5. Ease of molding — Reciprocating screw m/c preferred with Ni plated or stainless steel parts
6. Properties
   Density — $1.06 \, \text{g cm}^{-3}$
   Tensile strength at yield — 45–55 MPa
   Tensile modulus — –
   Elongation at break — 30–50%
   Flexural strength — –
   Flexural modulus — 2.0–2.2 GPa
   Compressive strength — –
   Compressive modulus — –
   Impact Notched Izod — $10\text{–}14 \, \text{kJ m}^{-2}$
           Unnotched Izod — –
   Chemical resistance — –
   Water absorption — Very low (lowest of any engineering plastic)
   Thermal
     Melting point — –
     T_g — 210°C
     HDT (1.82 MPa) — 106–119°C
     Max. continuous temperature of use — –
     Coefficient of linear thermal expansion — $0.65 \times 10^{-4} \, \text{K}^{-1}$
     Thermal conductivity — –
   Oxygen Index — 22%
   Electrical volume resistivity — $10^{15} \, \Omega\text{cm}$
7. Processing
   Mold shrinkage — 0.5–0.7% Ability to reproduce intricate detail
   Drying temperature — 104–110°C preferably use closed loop dehumidifying system
   Drying time — 2–4 h. Do not exceed 8 h, otherwise loss of physical properties
   Mold temperature — 66–93°C
   Melt temperature — 282–310°C
8. Dimensional stability — ±0.002 in/in tolerance
9. Trade names
   GE Plastics 'Noryl' (PPO + HIPS) and 'Noryl GTX' ('Noryl' + PA)
   BASF 'Luranyl' (PPO + SBS)

25. PPS

1. Material PPS Polyphenylene sulfide
2. Formula

[chemical structure: benzene ring with two CH_3 groups and S linkage, repeating unit n]

APPENDIX 14: TYPICAL PROPERTIES OF UNREINFORCED PLASTIC POLYMERS 1109

3. Crystal state	Semi-crystalline, depends on thermal history. Can have up to 65% crystalline phase (highest for thermoplastic material)
4. Cost index	5.10
5. Ease of molding	Easily processed on conventional reciprocating screw machines
6. Properties	
Density	1.35 g cm^{-3}
Tensile strength at yield	86 MPa
Tensile modulus	–
Elongation at break	3–6%
Flexural strength	145 MPa
Flexural modulus	4.14 GPa
Compressive strength	–
Compressive modulus	–
Impact Notched Izod	0.7 J
Unnotched Izod	8–16 J
Chemical resistance	Excellent resistance to chemicals. No known solvent below 200°C. Inherently self extinguishing
Water absorption	0.01–0.02%
Thermal	
Melting point	260–285°C
T_g	88°C (amorphous phase)
HDT (1.82 MPa)	104–110°C
Max. continuous temperature of use	200–240°C
Coefficient of linear thermal expansion	0.4–0.42 × 10^{-4}K^{-1}
Thermal conductivity	–
Oxygen Index	48%
Electrical volume resistivity	10^{16} Ωcm
7. Processing	
Mold shrinkage	1.0–1.2%
Drying temperature	149–177°C (do not exceed 260°C, can reduce melt flow values)
Drying time	2–3 h
Mold temperature	–
Melt temperature	304–343°C
8. Dimensional stability	
9. Trade names	
GE Plastics 'Supec'	
Phillips Petroleum 'Ryton'	
Ticona 'Fortron'	
Solvay 'Primef' (only supply glass filled)	

26. PPSu

1. Material	PPSu	Polyphenylsulfone
2. Formula		

$$\left[\begin{array}{c} \text{structural formula} \end{array} \right]_n$$

3. Crystal state	Amorphous
4. Cost index	9.82
5. Ease of molding	Since polymer has a high softening point must use rapid injection speeds to fill the mold before thick section of frozen skin develops
6. Properties	
Density	1.29 g cm^{-3}
Tensile strength at yield	70 MPa
Tensile modulus	2.30 GPa

Elongation at break	60–120%
Flexural strength	91 MPa
Flexural modulus	2.40 GPa
Compressive strength	99 MPa
Compressive modulus	1.73 GPa
Impact Notched Izod	694 Jm^{-1}
Unnotched Izod	No break
Chemical resistance	Very good chemical resistance except for ketones and conc. acids. Exceptional resistance to hydrolysis
Water absorption	0.37%
Thermal	
Melting point	–
T_g	220°C
HDT (1.82 MPa)	207°C
Max. continuous temperature of use	190°C
Coefficient of linear thermal expansion	0.56×10^{-5} K^{-1}
Thermal conductivity	0.35 W m^{-1}K^{-1}
Oxygen Index	44%
Electrical volume resistivity	$>10^{15}$ Ωcm

7. Processing

Mold shrinkage	0.6–0.7%				
Drying temperature	177°C	or	149°C	or	135°C
Drying time	Min. 2.5 h		4 h		3.5 h
Mold temperature	149–163°C				
Melt temperature	366–393°C				

8. Dimensional stability –
9. Trade names Amoco 'Radel R'

27. PS

1. Material   PS   Polystyrene
2. Formula

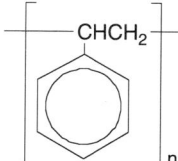

3. Crystal state –
4. Cost index   0.36 and 0.38 for high impact grade
5. Ease of molding –
6. Properties

Density	1.05 g cm^{-3}
Tensile strength at yield	35–69 MPa
Tensile modulus	3.30 GPa
Elongation at break	2%
Flexural strength	67–103 MPa
Flexural modulus	2.76–3.45 GPa
Compressive strength	79–110 MPa
Compressive modulus	–
Impact Notched Izod	–
Unnotched Izod	–
Chemical resistance	–
Water absorption	0.1%
Thermal	
Melting point	–
T_g	85–100°C
HDT (1.82 MPa)	72–80°C

APPENDIX 14: TYPICAL PROPERTIES OF UNREINFORCED PLASTIC POLYMERS

Max. continuous temperature of use	–
Coefficient of linear thermal expansion	$0.8 \times 10^{-4} K^{-1}$
Thermal conductivity	–
Oxygen Index	18%
Electrical volume resistivity	$>10^{15}$ Ωcm

7. Processing

Mold shrinkage	0.45%
Drying temperature	–
Drying time	–
Mold temperature	40°C
Melt temperature	230°C

8. Dimensional stability —
9. Trade names
BASF 'Polystyrol' GP and impact modified grades
Shell 'Carinex'

28. Psu

1. Material   PSu   Polysulfone
2. Formula

$$\left[\underset{O}{\overset{O}{\underset{\|}{\overset{\|}{S}}}} \text{-phenyl-} O \text{-phenyl-} \underset{CH_3}{\overset{CH_3}{C}} \text{-phenyl-} O \right]_n$$

3. Crystal state	Amorphous
4. Cost index	3.76
5. Ease of molding	Reciprocating screw machines preferred

6. Properties

Density	1.24 g cm^{-3}
Tensile strength at yield	70.3 MPa
Tensile modulus	2.48 GPa
Elongation at break	50–100%
Flexural strength	106.2 MPa
Flexural modulus	2.69 GPa
Compressive strength	–
Compressive modulus	–
Impact Notched Izod	69 Jm^{-1}
Unnotched Izod	–
Chemical resistance	Highly resistant to mineral acids and alkalis. Attacked by ketones, chlorinated hydrocarbons and aromatic hydrocarbons. Stress cracking by some organic solvents
Water absorption	0.3% Excellent steam resistance up to 140°C
Thermal	
Melting point	–
T_g	180–220°C
HDT (1.82 MPa)	174°C
Max. continuous temperature of use	150°C
Coefficient of linear thermal expansion	$0.56 \times 10^{-4} K^{-1}$
Thermal conductivity	0.26 W m^{-1}K^{-1}
Oxygen Index	30%
Electrical volume resistivity	5×10^{16} Ωcm

7. Processing

Mold shrinkage	0.7%
Drying temperature	135°C
Drying time	3.5 h
Mold temperature	150–160°C

Melt temperature —
8. Dimensional stability — High but, since amorphous, has poor surface finish
9. Trade names
   Amoco 'Udel'
   BASF 'Ultrason S'

29. PTFE
1. Material   PTFE — Polytetrafluoroethylene
2. Formula — Has a helical structure to accommodate the large F atom

$$[-CF_2-]_n$$

   Note PTFE is not a thermoplastic polymer
3. Crystal state — Semi-crystalline. Has three crystalline phases at atmospheric pressure. When polymerized is of 90–95% crystallinity. When heated above mp, becomes amorphous. Most fabrications are 50–75% crystalline
4. Cost index — 3.26
5. Ease of molding — Cannot be injection molded, must be sintered
6. Properties
   Density — 2.1–2.2 $g\,cm^{-3}$ (2.302 for 100% crystalline and 2.00 for amorphous)
   Tensile strength at yield — 20.6–34.3 MPa (rarely used in tension)
   Tensile modulus — 0.75 GPa
   Elongation at break — 250–400%
   Flexural strength — No break
   Flexural modulus — —
   Compressive strength — 118 MPa
   Compressive modulus — 0.4 GPa
   Impact   Notched Izod — 160 Jm^{-1} Excellent impact properties with well prepared samples
             Unnotched Izod
   Chemical resistance — Excellent, chemically inert and unaffected by all known chemicals except alkali metals, fluorine and fluorine compounds
   Water absorption — —
   Thermal
     Melting point — 327°C (sintered), 332–346°C (unsintered) decomposes above 400°C
     T_g
     HDT (1.82 MPa) — 56°C
     Max. continuous temperature of use — 260°C
     Coefficient of linear thermal expansion — $0.95 \times 10^{-4} K^{-1}$ (not constant)
     Thermal conductivity — 0.25 $W\,m^{-1}K^{-1}$
     Oxygen Index — —
     Electrical volume resistivity — $>10^{18}$ Ωcm
7. Processing
   Mold shrinkage — —
   Drying temperature — —
   Drying time — —
   Mold temperature — Preformed, sintered and cooled. Does not Melt, hence voids may be formed
   Melt temperature — —
8. Dimensional stability — 1–1.8% volume change in range 10–25°C
9. Trade names
   DuPont 'Teflon'
   Hoechst 'Hostaflon TF'
   ICI 'Fluon'

APPENDIX 14: TYPICAL PROPERTIES OF UNREINFORCED PLASTIC POLYMERS

30. PVDF

1. Material   PVDF — Polyvinylidenefluoride
2. Formula — Obtained by the polymerization of vinylidene fluoride by two methods: (a) Suspension polymerization producing S-PVDF (TypeII PVDF)- 'Solef' is of this type and is more crystalline with fewer structural defects (b) Emulsion polymerization producing E-PVDF (Type I PVDF) PVDF can be co-polymerized

$$[-CH_2-CF_2-]_n$$

3. Crystal state — Crystalline (>50%) There are 3 forms α(II), β(I) and γ(III)
4. Cost index — –
5. Ease of molding — Can be readily processed
6. Properties
   - Density — 1.77–1.79 g cm^{-3}
   - Tensile strength at yield — 53–57 MPa
   - Tensile modulus — 2.2–2.8 GPa
   - Elongation at break — 20–60%
   - Flexural strength — 77–94 MPa
   - Flexural modulus — 2.5–2.8 GPa
   - Compressive strength — 75–85 MPa
   - Compressive modulus — 2.15–2.90 GPa
   - Impact   Notched Izod — Tough
   -          Unnotched Izod — –
   - Chemical resistance — Swells in strongly polar solvents (e.g. acetone and ethyl acetate. Soluble in aprotic solvents like DMF and DMAC. Can stress crack
   - Water absorption — <0.04%
   - Thermal
     - Melting point — 172–175°C
     - T_g — –40°C
     - HDT (1.82 MPa) — 105–115°C
     - Max. continuous temperature of use — 150°C (140°C when stressed)
     - Coefficient of linear thermal expansion — 1.2–1.45 × 10^{-4} K^{-1}
     - Thermal conductivity — 0.19 W m^{-1} K^{-1}
   - Oxygen Index — 44%
   - Electrical volume resistivity — 1.4–5 × 10^{14} Ωcm
7. Processing
   - Mold shrinkage — ≈3% (6.4 mm thick) but very consistent
   - Drying temperature — –
   - Drying time — –
   - Mold temperature — 66°C
   - Melt temperature — 220–260°C
8. Dimensional stability — Good, can stress relieve at 150°C followed by slow cooling
9. Trade names
   - Elf Atochem 'Kynar' 460
   - Solvay 'Solef' 1000 series

31. PVDF copolymer

1. Material   PVDF copolymer — Polyvinylidene fluoride copolymer
2. Formula — Vinylidene fluoride copolymerized with other fluorinated monomers to give a polymer with higher fluorine content and increased flexibility
3. Crystal state — Crystalline but less so than the homopolymer
4. Cost index — 5.92
5. Ease of molding — Readily molded
6. Properties

Density | 1.76–1.79 g cm^{-3}
Tensile strength at yield | 15–35 MPa
Tensile modulus | 0.35–1.1 GPa
Elongation at break | 200–450%
Flexural strength | 112 MPa
Flexural modulus | 1.1 GPa
Compressive strength | 43 MPa
Compressive modulus | 1.00 GPa
Impact   Notched Izod | Has increased impact strength
        Unnotched Izod | –
Chemical resistance | Has additional chemical compatibility to high pH solutions
Water absorption | 0.04%
Thermal
  Melting point | 134–169°C
  T_g | –
  HDT (1.82 MPa) | 35–52°C
  Max. continuous temperature of use | –
  Coefficient of linear thermal expansion | 1.4–1.8 × 10^{-4} K^{-1}
  Thermal conductivity | 0.18 W m^{-1} K^{-1}
Oxygen Index | 42%
Electrical volume resistivity | >10^{14} Ωcm

7. Processing
   Mold shrinkage | –
   Drying temperature | –
   Drying time | –
   Mold temperature | 49–93°C
   Melt temperature | 140–145°C
8. Dimensional stability | –
9. Trade names
   Elf Atochem 'Kynarflex' copolymer
   Solvay 'Solef' copolymer 11000 series

32. THV

1. Material   THV | Tetrafluoroethylene, hexafluoropropylene, vinylidene fluoride
2. Formula | A terpolymer (there are three grades)
3. Crystal state | –
4. Cost index | –
5. Ease of molding | –
6. Properties
   Density | 1.95–1.98 g cm^{-3}
   Tensile strength at yield | 23–24 MPa
   Tensile modulus | –
   Elongation at break | 500–600%
   Flexural strength | –
   Flexural modulus | 0.05–0.21 GPa
   Compressive strength | –
   Compressive modulus | –
   Impact   Notched Izod | –
           Unnotched Izod | –
   Chemical resistance | Good resistance to inorganic & organic acids, halogen compounds, hydrocarbons, alcohols and ethers

   Water absorption | –
   Thermal
     Melting point | 115–180°C (depends on grade)
     T_g | 420–440°C
     HDT (1.82 MPa) | –
     Max. continuous temperature of use | –
     Coefficient of linear thermal expansion | –
     Thermal conductivity | –

APPENDIX 14: TYPICAL PROPERTIES OF UNREINFORCED PLASTIC POLYMERS

Oxygen Index	75%
Electrical volume resistivity	–
7. Processing	
Mold shrinkage	–
Drying temperature	–
Drying time	–
Mold temperature	–
Melt temperature	–
8. Dimensional stability	–
9. Trade names	3M/Hoechst 'THV'

33. TPI

1. Material TPI	Thermoplastic polyimide
2. Formula	–
3. Crystal state	Semi-crystalline (maximum crystallinity 45%)
4. Cost index	–
5. Ease of molding	Highly processable
6. Properties	
Density	1.33 g cm^{-3} (amorphous), 1.38 g cm^{-3} (35% crystalline)
Tensile strength at yield	94 MPa
Tensile modulus	–
Elongation at break	90%
Flexural strength	140 MPa
Flexural modulus	3.0 GPa
Compressive strength	122 MPa
Compressive modulus	2.2 GPa
Impact Notched Izod	9 kgcm cm^{-1}
Unnotched Izod	–
Chemical resistance	–
Water absorption	0.34%
Thermal	
Melting point	–
T_g	250°C
HDT (1.82 MPa)	238°C
Max. continuous temperature of use	–
Coefficient of linear thermal expansion	0.55 × 10^{-4} K^{-1}
Thermal conductivity	0.15 kcal/m.h.K
Oxygen Index	47%
Electrical volume resistivity	10^{17} – 10^{18} Ωcm
7. Processing	
Mold shrinkage	0.83%
Drying temperature	180°C or 200°C
Drying time	10 h min. 5 h
Mold temperature	Min. 170°C, preferably 200°C
Melt temperature	390–420°C
8. Dimensional stability	–
9. Trade names	Mitsui 'Aurum' Grade 450

34. TPX

1. Material TPX	Polymethylpentene
2. Formula	Prepared by polymerization of 4-methylpentene-1

$$-[CH_2C_4\,[CH_2C_4(CH_3)_2]\,]_n-$$

3. Crystal state	Crystalline
4. Cost index	1.84
5. Ease of molding	Readily molded on single or twin extruders

6. Properties
   Density — 0.83 g cm^{-3} lowest density of all commercial plastics
   Tensile strength at yield — 18–28 MPa
   Tensile modulus — 0.6–1.6 GPa
   Elongation at break — 25–120%
   Flexural strength — 25–50 MPa
   Flexural modulus — 0.6–1.7 GPa
   Compressive strength — –
   Compressive modulus — –
   Impact   Notched Izod — 3–4 kgcm cm^{-1}
             Unnotched Izod — 10 > 50 kgcm cm^{-1}
   Chemical resistance — Excellent, since no polar functional groups. Swells in trichloroethylene. Highly resistant to water and steam
   Water absorption — 0.01%
   Thermal
     Melting point — 221–239°C
     T_g — 20–30°C
     HDT (1.82 MPa) — 45–55°C
     Max. continuous temperature of use — –
     Coefficient of linear thermal expansion — $1.17 \times 10^{-4} K^{-1}$
     Thermal conductivity — 4.9–5.3×10^{-4} cal.cm/sec.cm^2
   Oxygen Index — –
   Electrical volume resistivity — 10^{16} Ωcm
7. Processing
   Mold shrinkage — 1.5–3.0%
   Drying temperature — Does not require predrying, inherently free of moisture problems
   Drying time — –
   Mold temperature — 20–80°C
   Melt temperature — 270–330°C
8. Dimensional stability — –
9. Trade names — Mitsui 'TPX'

APPENDIX 15

Acronyms for Thermoplastic Polymers

ABS	Acrylonitrile butadiene styrene copolymer
CA	Cellulose acetate
CAP (or CP)	Cellulose acetate proprionate
CTFE	Trifluoro chloro ethylene copolymer
EC	Ethyl cellulose
ECTFE	Ethylene monochloro trifluoroethylene
EPS	Expanded polystyrene
ETFE	Ethylene tetrafluoroethylene copolymer
EVA	Ethylene vinyl acetate
FEP	Fluorinated ethylene-propylene copolymer
HDPE	High density polyethylene
HIPS	High impact polystyrene
LCP	Liquid crystal polymer
LDPE	Low density polyethylene
PA	Polyamide
PAA	Polyaryl amide
PABM	Polyamino bismaleimide
PAEK	Polyaryl ether ketone (Polyketone)
PAI	Polyamide imide
PAS	Polyarylsulfone
PBT (or PBTP)	Polybutylene terephthalate
PC	Polycarbonate
PE	Polyethylene
PEC	Polyester carbonate
PEEK	Polyetherether ketone
PEI	Polyetherimide
PEK	Polyether ketone
PEKEKK	Polyether ketone ether ketone ketone
PEKK	Polyetherketone ketone
PES	Polyethersulfone
PET (or PETP)	Polyethylene terephthalate
PFA	Perfluoro alkoxy polymer
PIB	Polyisobutylene
PK	Aliphatic polyketone
PMMA	Polymethyl methacrylate
POM	Polyoxy methylene (Polyacetal)
POP	Polyphosphonate
PP	Polypropylene
PPA	Polyphthalamide

PPE (or PPO)	Polyphenylene ether
PPO (or PPE)	Polyphenylene oxide
PPS	Polyphenylene sulfide
PPSU	Polyphenyl sulfone
PS	Polystyrene
PSP	Polystyrl pyridine
PSU (or PSF)	Polysulfone
PTFE	Polytetrafluoroethylene (NB not a thermoplastic)
PTMT	Polytetramethylene terephthalate
PU (or PUR)	Polyurethane
PVAC	Polyvinyl acetate
PVC	Polyvinyl chloride
PVDC	Polyvinylidene chloride
PVDF	Polyvinylidene fluoride
PVOH	Polyvinyl alcohol
SAN	Styrene acrylonitrile copolymer
SB	Styrene-butadiene-styrene (block copolymer)
THV	Tetrafluoroethylene-hexafluoropropylene-vinylidene fluoride
TPE	Thermoplastic polyester elastomer
TPI	Thermoplastic polyimide
TPX	Polymethyl pentene
TPU (or TPUR)	Thermoplastic polyurethane

APPENDIX 16

Companies Involved with Carbon Fibers and their Composites Throughout the World

1. A & P Technology, Cincinnati, Ohio, USA—formerly located at Covington, Kentucky, manufacture carbon fiber braid. Have the world's largest braider, an 800 carrier machine.
2. Acordis Group—a joint venture with Hoechst AG, Germany for their viscose and acrylic fiber, production was 72.5% owned by Courtaulds, but the venture with Hoechst ended in March 1998 as a prerequisite to purchase of Courtaulds plc by Akzo Nobel NV, Holland in July, 1998. The acrylic production plant was renamed Acordis Acrylic Fibres. Great Lakes Carbon Corporation was acquired by Akzo from Horsehead Industries and produces a range of Fortafil carbon fibers, now part of the Acordis Group, trading under Fortafil Fibers Inc.
3. Adherent Technologies Inc. Albuquerque, NM, USA—an R&D company which developed tow spreading for up to $50k$ carbon fiber to give 20 gsm at 100 $m h^{-1}$ for resin film or wet impregnation processes.
4. Advanced Composites Group, Heanor, Derbyshire, UK—founded in 1975. Advanced Composites Materials Ltd. is a prominent prepregger in the UK, with a second factory, Advanced Composites Materials Inc., in Tulsa, Oklahoma, USA. It is the third largest prepregger in the world.
5. Aerolor, France—a joint venture company formed in 1975 between Aerospatial and Le Carbone Lorraine to make carbon–carbon composites, mainly by densification of felts or cloth in a liquid or gaseous phase. Main applications are ablative refractory materials for brakes. They have licensed their technology to Hercules.
6. Aerovac Ltd., Sandbeds, Keighley, West Yorkshire, UK—a member of the EUMECO Group, provides consumables for vacuum processing.
7. Afikim Carbon Fibers, Kibbutz Afikim, Israel—acquired technology and plant from RK Textiles Composite Fibres Ltd. in the UK to produce a range of PAN based ACIF carbon fibers.
8. AGL, Cannes-la-Bocca, France—weaver
9. AIK Advanced Composites GmbH, Kassel, Germany—produce a range of Pyropreg prepreg.
10. Aircraft Braking Systems, Akron, Ohio, USA—ABSC, formerly Loral, is a leading brake supplier. Carbon–carbon supplied by Hitco.
11. Airtech International Inc., Huntington Beach, California, USA—manufacture and distribute vacuum bagging materials.
12. Akzo-Fortafil Fibers Inc., Rockwood, Tennessee, USA—Enka acquired Great Lakes Carbon Corporation, which was in turn purchased by Akzo, who subsequently merged with Nobel. They manufacture a range of Fortafil carbon fibers using Courtauld's TTP precursor.
13. Akzo Nobel Faser AG, Germany—formerly Enka, a German company, purchased by the Dutch company Akzo, who then merged with the Swedish company Nobel. Obtained technology from Toho Beslon, who in 1983, obtained a controlling interest and by 1996, owned

about 80% share. The company is now called Tenax Fibers and manufacture a range of Tenax carbon fibers.

14. Aldila Inc., Rancho Bernardo, San Diego, California, USA—a US based company, the world's leading supplier of carbon fiber golf clubs at six separate facilities: there are two in Poway and Rancho Bernardo in San Diego and three others located in Tijuana, Mexico and the most recent, at Zhuhai, China. Manufacture prepreg and have constructed a new facility, Aldila Materials Technology Corporation at Evanston, Wyoming which came on stream in 1998, to manufacture large tow carbon fiber. Announced a joint venture with SGL, who have acquired a 50% interest in Aldila's carbon fiber manufacturing operation.
15. Allied Signal Aerospace—produce carbon–carbon brakes from oxidized PAN using a CVD process and from pitch based carbon fiber using a pitch/phenolic resin impregnation process. During 1998, acquired DuPont Lanxide Composites. At the end of 1999, merged with Honeywell under the Honeywell name.
16. American Cyanamid Co., Wayne, New Jersey, USA—acquired Electro-Metalloid Corp., who had developed carbon fiber metal coating technology. Produce a range of metal coated (e.g. nickel, silver and gold) Cycom carbon fibers. Company now called Cytec Engineering Materials Inc., Wallingford, CT, USA. In 1958, American Cyanamid produced Creslan acrylic fiber, now made by Sterling Fibers in Pace, Florida.
17. American Materials and Technology Corp., Los Angeles, California, USA—AMT is the parent of Culver City Composites and Grafalloy, which makes golf club shafts. Acquired by Cytec Industries in 1998.
18. Amoco Performance Products Inc., Greenville, South Carolina, USA—acquired Union Carbide's carbon fiber production at Greenville, where a range of carbon fibers from pitch, cellulosic and PAN precursors were made under the trade name of Thornel. PAN precursor is also produced there. In the early 1990s, also acquired BASF's carbon fiber production. BASF, in 1985, had acquired the Celion carbon fiber production from Celanese at Rockhill, North Carolina. Meltspun PAN was also produced at Rockhill, but this was discontinued in favour of using precursor from Toho Rayon. Became BP Amoco and was purchased by Cytec Industries Inc.
19. Angeloni, Mestre, Italy—weaver
20. Applied Composite Technology (ACT), Fayette, Utah, USA—manufacture composite products, a leading producer of prosthetic limbs. Also produce rifle and pistol barrels.
21. Applied Fiber Systems, Clearwater, Florida, USA—manufacture Towflex, which includes continuous carbon fiber powder coated with a variety of thermoplastics, such as Nylon-6, PP, PET, PES, PEI, PPS and PEEK, available as a flexible towpreg suitable for hot filament winding. Acquired by Hexcel in April 2001.
22. Asahi Kasei Carbon Fiber Co., Tokyo, Japan—Asahi Chemical Industry Co. and Nippon Carbon Co. formed a joint venture company in 1981 called Asahi-Nippon Carbon Fiber Co. Produce a PAN based carbon fiber Hi-Carbolon. Acquired by Mitsubishi.
23. Ashland Chemical Company, Ashland, Kentucky, USA—produced a range of Carboflex petroleum pitch based general performance carbon fibers in 1992 and sold the plant in 1993 to Anshan East Asia Carbon Fibers Inc., China. Ashland also sell prepared pitch as a precursor.
24. Ashland Speciality Chemical Co., Dublin, Ohio, USA—polyester resins including Hetron vinylester resin.
25. Atelier Textile du Gier (ATG), St Matin la Plaine, France—manufacture carbon fiber tapes and braids.
26. August Kremple Soehne, Vaihingen/Enz, Germany—produce a range of prepreg.
27. AVCO, Lowell, Massachussetts, USA—acquired by Textron in 1984, now called Textron Speciality Materials. Produce a range of in-house Avcarb PAN based carbon fibers which are woven and converted into carbon–carbon composite using thermoset resin and CVD processes.
28. Avtex Fibers Inc., Front Royal, Virginia, USA—make a continuous carbonizable rayon.
29. Badische Anilin Soda Fabrik (BASF), Ludwigshafen, Germany—acquired the Celanese carbon fibers plant at Charlotte, North Carolina, USA in 1985 and subsequently sold to Amoco. BASF purchased the PAN melt spinning technology from Cyanamid and in late 1988, were reported to

be producing non-circular carbon fibers with superior compression and shear strengths via this route. Through BASF Structural Materials Inc., Anaheim, California, operated Narmco, a speciality prepreg manufacturer. BASF's structural materials business was acquired by Cytec in 1992. Cytec was formerly the prepreg arm of Cyanamid.
30. Bally Ribbon Mills, Bally, PA, USA—weave tapes and narrow width fabrics; have developed a 3-D weaving method.
31. Barthels-Feldhof, Wuppertal, Germany—manufacture braided ribbons and tubes.
32. Bendix, South Bend, Indianna, USA—part of Allied Signal and produces carbon–carbon brakes, originally from chopped pitch based carbon fiber coated with a phenolic resin. Mintex and BBA Group (UK) are affiliated companies. At the end of 1999, Allied Signal merged with Honeywell under the Honeywell name and in 2001, was acquired by General Electric.
33. BF Goodrich (BFG), Aerospace Division, Alto Pueblo, Colorado, USA—the Super Temp Division originally developed the carbon–carbon brake system. This process was licensed to Dunlop. BFG are licensed by SEP to make carbon–carbon using laminated carbon fiber felts with cross-ply reinforcement and licensed by Messier-Bugatti to make brakes. Merged with Rohr Inc., Chula, Vista, CA, a producer of aircraft engine components, in 1997. Original plant situated in Santa Fe springs CA, main plant in Pueblo CO, with additional capacity at new facility in Spokane, WA.
34. BGF Industries, Greensboro, NC, USA—weave carbon fiber fabrics.
35. Borden Chemicals, Sully, South Glamorgan, UK—manufacture phenolic resins.
36. BP Chemicals (Hitco) Inc., Gardena, California, USA—British Petroleum (BP), UK acquired Bristol Composite Materials in 1979, but sold to GKN-Westland in 1997. In 1986, acquired Hitco from Owens Corning and sold to SGL Carbon in 1997. BP Amoco's carbon fiber business was sold to Cytec Fiberite in September, 2001.
37. Bristol Composite Materials, Avonmouth, Avon, UK—facility established from the plant acquired by a consortium in 1972 from Rolls Royce who had shut down their carbon fiber manufacturing operation. BCME originally made Hyfil carbon fiber, which was converted online to prepreg. Toray took a 49% interest in 1976 and British Petroleum took over BCM's interest in 1979. The carbon fiber production facility was shut down in 1980. BP acquired Toray's interest at the end of 1980 and continued with prepreg operation. At about that time, BP purchased Hitco and US Polymeric, forming BP Advanced Composites at Avonmouth. In 1988, it absorbed BCM within this organisation, which was renamed BP Chemicals in 1988. Prepreg production was stopped in 1990.
38. British Petroleum, In the late 1960s, introduced Amlon, a thermoformable product that could be extruded into fibers, without cost of associated solvent recovery.
39. Bryte Technologies Inc., San Jose, California, USA—manufacturer of carbon fiber prepreg and now a member of Royal Ten Cate.
40. BTI Europe (Brunswick Technologies), Andover, Hants, UK—has manufacturing facilities in Maine and Texas. Manufacture multiaxial reinforcement. Acquired in 2000 by Certain Teed Corp., a subsidiary of St Gobain, France.
41. Cape Composites Inc., San Diego CA, USA—produce composite prepreg products. Acquired by Zoltek in November 1999 and renamed SP Systems (USA).
42. Carbon–Carbon Advanced Technology Inc. (CCAT), Fort Worth, Texas, USA—originally formed by ex-Vought Missiles employees and now owned by Alco Standard. Produce 2-D carbon–carbon composites from phenolic prepreg.
43. Carbon Composites International (CCI), Tucson, Arizona, USA—CCI is a subsidiary of Materials and Electrochemical Research (MER) Corp. and produces carbon–carbon composites using a proprietary process requiring only several hours to fabricate a component, instead of two months.
44. Carbone Industrie (CI), Villeurbanne, France—formed in 1985 as a joint subsidiary of GEC Alsthom and Société Européenne de Propulsion (SEP) to develop, produce and market carbon–carbon composites. SEP then sold CI an exclusive ten year licence for a gaseous densification process to make carbon–carbon materials. CI produce Sepcarb carbon–carbon and in 1997, were acquired by Messier-Bugatti, a carbon–carbon brake manufacturer.

45. Carr Reinforcements Ltd., Stockport, UK—weave carbon fiber cloth.
46. Celanese Advanced Engineering Composites, Rock Hill, South Carolina, USA—Celanese first became active in the field of carbon fiber technology in 1966. A Celanese subsidiary, Narmco Materials Company, sold prepreg. Produced Celion carbon fibers at Summit, New Jersey and GY70 carbon fiber, with a modulus of 70 million psi, was noteworthy. Celion carbon fiber s have very good oxidation resistance. Also produced Celiox, an oxidized PAN fiber. In 1976, acted as distributor for Toho Beslon Co. and subsequently, produced fiber using Toho's technology. In 1985, the plant was sold to BASF, together with the Narmco prepreg operation and BASF subsequently sold to Amoco.
47. Century Design Inc., San Diego, California, USA—manufactured prepreg machines since 1972, including solution coating, S-wrap machines, uni-directional prepreg and cantilevered machines.
48. Charcoal Cloth (International) Ltd., Tyne and Wear, UK—in 1997, acquired licence from the British Chemical Defense Establishment (CDE), Porton Down to manufacture activated charcoal cloth. Acquired in 1996 by the Calgon Carbon Corporation, USA and is now a division of the European Operating Unit, Chemiviron Carbon Ltd., Brussels, Belgium.
49. Chomarat (Les Fils D'Auguste), Mariac, France—manufacture carbon fiber reinforcements.
50. Ciba Composites, Duxford, UK—Bonded Structures Division of Ciba Geigy manufactured prepreg, noted for BSL914, an epoxy modified with a thermoplastic resin which controlled the resin flow during the curing cycle. Own Brochier in France, a weaving and prepreg facility. Ciba Composites merged with Hexcel in 1996 and are now called the Hexcel Corporation. Ciba resins are handled by Ciba Speciality Chemicals.
51. Ciba Speciality Chemicals, Duxford, Cambridge, UK—manufacture epoxy and high performance resins and are rated in top three in the world. Manufacturing facility of Ciba Speciality Chemicals, Basel, Switzerland.
52. Conoco Carbon Fibers, Houston, Texas, USA—acquired pitch carbon fiber technology from DuPont and are building a plant in Ponca City, Oklahoma to produce a random oriented carbon fiber mat. The company has been renamed Conoco Cevolution. Project was expanded in 2001, completed in 2002, but abandoned in 2003.
53. Coton Freres, France—a prepreg manufacturer.
54. Courtaulds, Coventry, UK—a company closely associated with the early development of PAN based carbon fibers at the Royal Aircraft Establishment (RAE) at Farnborough by providing PAN precursor (Courtelle). Subsequently, developed SAF (Special Acrylic Fiber) precursor specifically for carbon fiber manufacture. One of the original three UK carbon fiber producers operating the RAE process. Marketed Grafil carbon fibers. Formed an arrangement with Hercules in the USA in 1967. Licensed a process from Fiber Materials Inc. (FMI) to make Apollo high strength carbon fiber. Built Courtaulds-Grafil Inc. at Sacramento in the USA. Initially had a tie-up with Dexter Hysol and in 1983, formed Hysol-Grafil. In 1991, stopped production of carbon fibers in the UK and SAF, and sold the American carbon fibers plant to Mitsubishi. Courtaulds continued to make a PAN based Textile Tow Precursor (TTP) at their Grimsby plant in the UK. In April 1994, Courtaulds formed a joint venture with Hoechst AG, Germany for their viscose and acrylic fiber production and was 72.5% owned by Courtaulds, but the venture with Hoechst ended in March 1998 as a prerequisite when Akzo Nobel NV, Holland purchased Courtaulds plc in July, 1998. The acrylic production plant was renamed Acordis Acrylic Fibers.
55. Courtaulds Performance Films, Runcorn UK—Courtaulds bought Fothergill & Harvey in 1973. Company was previously called Fothergill-Symonds, becoming part of Courtaulds Aerospace Ltd. Manufacture prepreg fabric. CP Films are now called Solutia, a spin-off from Monsanto.
56. CMT Developments Co., Japan—a joint venture company originally set up between Toray Industries Inc. and Mitsui Coke Co. for research and development to produce pitch based carbon fiber. Company was dissolved in the latter part of the 1980s.
57. CP Films Ltd., Runcorn, Cheshire, England—was part of Courtaulds, weavers carbon fiber cloth and manufacturers of a range of prepregs.

58. Creative Pultrusions Inc., Alum Bank, Pennsylvania, USA—established in 1973, has two manufacturing facilities located at Alum Bank and Roswell, New Mexico. Acquired Pultrusions Dynamics, Oakwood Village, Ohio in 1997 as its Pultrusion Technology Center.
59. CS Interglas, Erbach, Germany—weavers carbon fiber fabrics.
60. Culver City Composites Corp., Los Angeles, California, USA—a subsidiary of The American Materials & Technologies Corp. (AMT). Culver City acquired Ferro Corp. and SP Systems. Manufacture primarily phenolic prepreg. Acquired by Cytec Industries Inc. in 1998.
61. Custom Composite Materials Inc., Atlanta, Georgia, USA—manufacture Towflex.
62. Cytec Fiberite Ltd., Wrexham, UK—Cyanamid acquired Fothergill & Harvey (F & H) and in 1983, Courtaulds purchased F & H. Cyanamid was then renamed Cyanamid Aerospace and in 1993, became the European arm of Cytec Engineered Materials Inc., Havre de Grace, USA. Manufacture a range of film adhesives and also Cycom and Rigidite prepregs. In 2001, changed name to Cytec Engineered Materials Inc. and acquired the carbon fiber business of BP Amoco.
63. Cytec Industries Inc., West Patterson, New Jersey, USA—in 1993, acquired BASF's structural materials business at Anaheim, California, which included Reliable Manufacturing Co. Manufacture a range of film adhesives and also Cycom and Rigidite prepregs. In 1997, acquired all the assets of Fiberite Inc., becoming the world's second largest prepreg supplier and the largest supplier of structural adhesives, doing business as Cytec Fiberite Inc. In 1998, acquired American Materials & Technologies Corp and in 2001, acquired the carbon fiber business of BP Amoco.
64. Devold AMT AS, Langevag, Norway—produce multiaxial carbon fiber reinforcements.
65. Dow Chemical Company, Midland, Michigan, USA—manufacture epoxy and vinylester resins.
66. Dowty Aerospace Propellers, Gloucester, England—produces composite propeller blades utilizing a fail-safe root retention system.
67. DSM Composite Resins (DSM BASF Structural Resins), Zwolle, Netherlands—manufacture polyester and vinylester resins. Combined its saturated polyester resin business with BASF in 1996.
68. Dunlop Standard Aerospace Group, Coventry, UK—presently owned by a British private equity group, Doughty Hanson & Co., who took over from The British Tyre and Rubber Company in 1998. Dunlop have licensed the Super Temp technology from BF Goodrich to produce carbon–carbon and in 1973, fitted multi-plate carbon–carbon brakes to the VC10 aircraft, and to Concorde in the following year. Dunlop predominantly use the CVD process. Acquired a small plant from Courtaulds (originally from RK Textiles Composites Fibres Ltd.) to manufacture oxidized PAN fiber, but now purchase all their opf requirements. Carbon–carbon brakes have been fitted to the Boeing 757, British Aerospace 146 & ATP, Mc Donnell Douglas AV8B and the Aeritalia/Aerospatial ATR72.
69. DuPont Advanced Composites Corp., USA—in 1984, DuPont purchased Exxon's carbon fiber facility and later sold to Conoco.
70. Elf Atochem, Paris, France—manufacture organic peroxides for curing polyester resins.
71. EMS Chemie, Domat, Switzerland—manufacture co-mingled yarns and prepreg.
72. Enka AG, Obernberg, Germany—initially Enka were agents for Nippon Carbon's carbon fiber products. Enka carbon fibers was purchased by Akzo, who merged with Nobel. Use Toho Rayon technology. In 1983, Toho acquired a controlling interest. Now called Tenax Fibers and in 1996, Toho increased their holding to some 80%. Produce a range of Tenax carbon fibers.
73. Entec Composite Machines Inc., Salt Lake City, Utah, USA—manufacture filament winding and pultrusion machines. Taken over at the end of 1999 by Zoltek Corp. and combined with Composite Machines Co. Have built the largest filament winding machine, capable of winding a part some 55 m in length and 7 m in diameter.
74. Epsilon Composites, Gaillan, France—pultrude carbon fiber tubulars and profiles with 12 pultrusion lines. Have capability of pultruding thermoplastics including PEEK.
75. Eurocarbon BV, Sittard, Netherlands—manufacture carbon fiber woven tapes, fabrics and braided reinforcements.

76. European Aeronautics Defense and Space Company (EADS), Netherlands—formed in 1999 by a merger of Diamler-Chrysler Aerospace and Aerospatiale Matra to form the third largest aerospace company, after Boeing and Lockheed Martin.
77. Fabric development Inc., Quakertown, PA, USA—incorporates Lydall Manning and Textile Products and is the world's largest independent weaver of high performance fabrics.
78. Ferro Corp., Composites Div., Los Angeles, California, USA—manufacture a range of Ferropreg prepreg, acquired by Culver City.
79. FA Kümpers, Rheine, Germany—manufacture true multiaxial paramax fabrics and 3-D braidings.
80. Fiber Innovations Inc., Norwood, Massachusetts, USA—a thermoplastic prepreg manufacturer.
81. Fiberite Inc., Tempe, Arizona, USA—an established prepreg manufacturing facility of Beatrice Chemicals, was taken over by Imperial Chemical Industries (ICI) in the UK and then, by the partnership of DLJ Merchant Banking and Carlisle Enterprises, who had acquired the Hercules carbon fiber facility and Simmaco, France, a thermoset bulk moulding compound manufacturer. Formed a joint company with Mitsubishi Chemical to manufacture seismic retrofit and structural reinforcements, using Mitsubishi's Replark materials, which would be made by Fiberite at Greenville, Texas. Acquired DuPont's Advanced Materials Systems, who manufactured aramid and PEKK thermoplastic sheet materials. However, in 1997, the prepreg business was acquired by Cytec Industries Inc. and the product line for composite materials, together with structural prepreg technology was acquired by Hexcel Corp. In 1997, launched a milled carbon fiber project mainly for use as a static dissipating filler.
82. Fiber Materials Inc., Bideford, Maine, USA—a high technology company working with many aspects of high temperature processing, carbon–carbon composites and 3-D weaving. Distributed Amoco (Thornel) range of carbon fibers in the UK.
83. Fibreforce Composites Ltd., Runcorn, Cheshire, UK—pultrude carbon fiber. Was originally part of the Bridon Group in the UK, but was taken over by Shell, who also acquired Pultrex at Clacton on Sea, the other leading UK pultruder. The company was taken over in 1994 by Lemame, which is part of Pacific Composites, an Australian company.
84. Formosa Plastics, Taiwan—plant built by a US company. Started production of PAN based carbon fiber in 1985.
85. Fortafil Fibers Inc., Knoxville, TN, USA—manufacture carbon fibers. Became part of Akzo NV. Now a member of the Acordis Group. Have plant in Rockwood, TN and a European Sales office in Wuppertal, Germany.
86. Fothergill Engineered Fabrics, Littleborough, England—originally called F & H. Produce woven fabrics and were one of the original UK carbon fiber prepregers, initially making prepreg warp sheets from staple carbon fiber. Acquired by Cyanamid in 1973, becoming Cyanamid Aerospace, and Courtaulds acquired the textile side, with the prepreg part of the business eventually becoming Cytec Engineering Masterials Inc. in 1997.
87. FTS SRL, San Maurizio, Italy—weavers of carbon fiber fabrics.
88. Fuji Oil Co., Japan—originally produced carbon fiber in 1984 from pitch.
89. General Atomic (GA) Technologies, San Diego, California, USA—apply CVD coatings (based on SiC with boron inhibitors) to carbon–carbon. Have licensed the technology to BFG, Chromalloy, Kaiser Aerotech and Rohr Industries.
90. Gepem, Villeurbanne, France—formed in 1973 by Société Européene de Propulsion (SEP) and Compagnie Electro-mécanique (CEM). The latter was taken over by Alsthom Atlantique in 1984. Gepem's initial assignment was to produce carbon reinforcements for the fabrication of SEP's ablative composites and in the early 1980s, took on the additional job of making carbon–carbon brake discs. They became defunct when Carbone Industrie was formed.
91. Goodyear Aerospace Corporation, Akron, Ohio, USA—originally called Goodyear Aircraft Corp, renamed in 1963. Received contract in 1975 to supply carbon–carbon brakes for the F-16 fighter jet. Became Loral Defense Systems Div. in 1987 and in 1996, Lockheed-Martin Tactical Defense Systems Div.

92. Grafil Inc., Sacremento, California, USA—a production facility established by Courtaulds in the USA to produce Grafil carbon fibers. Was originally associated with Dexter Hysol. When Courtaulds shut down their UK operation, the US plant was acquired by Mitsubishi Rayon Corporation (MRC), from whom Grafil Inc. obtain their PAN precursor. Supply Grafil and Pyrofil carbon fibers. In 1998, bought a 20% share in Structil SA, a pultrusion company within the Groupe SNPE. Sell prepreg through Newport Adhesives and Composites.
93. Great Lakes Carbon Corporation, Tennessee, USA—was acquired by Akzo from Horsehead Industries and produce a range of Fortafil carbon fibers. Now part of the Acordis Group and trade under Fortafil Fibers Inc.
94. GKN Westland Aerospace, Cowes, Isle of Wight, UK—GKN purchased Westland Helicopter with production facilities at Yeovil and the Structures Division at Cowes, Isle of Wight. The latter, in 1997, purchased BP Advanced Composites Division, who manufacture aircraft components, at Avonmouth, formerly called Bristol Composites Materials. At the end of 1998, acquired Dow-United Technologies Composite Products Inc. The company operate RTM and advanced RTM processes.
95. Gunei Chemical Industry Co., Takasaki, Japan—produce Kynol, a phenol resin based product which they convert into carbon fiber.
96. Heinsco, Rochdale, UK—manufacture a range of prepregs.
97. Hercules Aerospace Espana SA (HAESA), Paria, Madrid, Spain—established by Hercules in 1989 to make prepreg in Spain.
98. Hercules Aerospace Co., Magna, Utah, USA—originally had an agreement with Courtaulds in 1969, and in 1971 produced Magnamite carbon fibers from PAN by the RAE process. Courtaulds supplied their SAF precursor. Started production of prepreg in 1972. Hercules had an arrangement with Sumitomo, Japan in 1979, for the supply of Exlan, an alternate PAN precursor, forming a joint venture company, Sumika Hercules, subsequently Hercules obtained the sole ownership of the PAN production facility at Decatur, Alabama increasing the output to two lines. Started to produce IM6 fiber in 1982, followed by IM7 and IM8 in 1984 and 1987 respectively. Intended to build a joint venture company with Pechiney-Ugine Kuhlmann (PUK) using Sumika-Hercules PAN precursor in 1983, but the venture was abandoned due to over-capacity of carbon fiber in Europe at that time. Subsequently, PUK joined the Soficar project with Elf Aquitaine and Toray Industries. Have a prepreg facility in Parla, Madrid, Spain. The Hercules' carbon fiber facility was purchased by Fiberite, Winona (owned by Carlyl Enterprises) but in June, 1998, became part of Hexcel.
99. Hexcel Carbon Fibers, Salt Lake City, Utah, USA—previously called Hecules Aerospace Co. Manufacture carbon fibers.
100. Hexcel Composites, Duxford, Cambridge, UK—previously called Ciba Bonded Structures. Manufacture a wide range of epoxy prepregs and are now the largest prepregger in the world. Have adopted Hexply as the trademark for all their prepregs.
101. Hexcel Corporation Inc., Pleasanton, California, USA—a well known prepreg manufacturer, who has now merged with Ciba Composites, UK becoming the largest prepreger in the world, having acquired the Hercules carbon fiber facility at Magna, Utah and the PAN precursor facility at Decatur, Alabama. In 1998, acquired Fiberite's composite materials product line, together with structural prepreg technology. The company has been organized into eight business units, which include resin transfer moulding and pultrusion. Acquired Clark-Schwebel Inc. in 1988.
102. Hexcel Fabrics, Villeurbanne, France—previously called Brochier. Supply a range of woven carbon fiber fabrics.
103. Hitco Technologies, Gardena, California, USA—originally HI Thompson Inc., closely associated with early work on carbon fiber, a leading producer of carbon–carbon. Was part of Armco Inc. and included US Polymeric. Owens Corning purchased Hitco from Armco in 1985, which in turn was subsequently acquired by the BP Corporation in 1986. Produces carbon–carbon brakes using carbon fiber cloth, made with a variety of densification techniques. Have four other factories producing nozzles, aircraft components, insulation and marine sonar bow domes. Was acquired by SGL Technik in 1997.

104. Hollingsworth & Vose Co., East Walpole, USA—manufacture non-woven carbon fiber mat.
105. ICI Fiberite, Winona, Maine, USA—in 1985, ICI America purchased the Fiberite subsidiary of Beatrice Foods Co. Ltd, which has subsequently been acquired by Carlyle Enterprises. Fiberite is a leading US producer of carbon fiber prepregs.
106. Indemitsu Kosan Co., Ltd., Tokyo, Japan—Japanese State oil company, operates a petroleum pitch based carbon fibers plant.
107. Indian Petrochemicals Corporation (IPCL), Baroda, India—a Government of India undertaking. Technology and plant supplied by RK Textiles Composite Fibres. IPCL produce a range of Indcarf carbon fibers.
108. Isola Werke AG, Dueren, Germany—produce a range of Durapreg carbon fiber prepreg.
109. Isovalta AG, Wiener Neudorf, Austria—produce a range of Airpreg carbon fiber prepreg.
110. Kaiser Aerotech, San Leandro, California, USA—produce K-Carb, a 2-D carbon–carbon composite.
111. Kashima Oil Company, Japan—see Petoca.
112. Kowasaki Steel Corporation, Chiba, Japan—produced a general purpose coal tar pitch carbon fiber using the KMFC trade mark, but stopped production in 1994.
113. Kobe Steel Company, Kobe Hyogo, Japan—manufacture carbon–carbon by hot isostatic pressure impregnation carbonization (HIPIC) route.
114. Korea Steel Chemical Company (KOSCO), Pohang, South Korea—a company within the Daewoo Group. Acquired technology and some carbon fiber manufacturing plant from RK Textiles Composite Fibers in 1986. Have now been acquired by Pohang Steel Company (POSCO) and the production of Kosca carbon fibers is now limited.
115. A. Krempel Soehne GmbH & Co., Vaihingen, Germany. Pultrude Wacosit composite sections.
116. Kureha Chemical Ind. Co. Ltd., Fukushima, Japan—produce Kureha carbon fiber from a petroleum pitch precursor.
117. Lantor Universal Carbon Fibres, Cleckheaton, West Yorkshire, England—previously called Universal Carbon Fibres. Supply fire protection and high performance textiles and yarns with trade name Panotex.
118. Le Carbone Lorraine, Gennevilliers, France—manufacture of Rigilor carbon fibers, now ceased. Manufacture Aerolor carbon–carbon.
119. Lewcott Corp., Millbury, Massachusetts, USA—manufacture phenolic prepreg.
120. LNP Corp., Exton, Philadelphia, USA—manufacture a range of carbon fiber filled thermoplastic moulding compounds. Also have production facilities in the Netherlands and Malaysia.
121. LTV Aerospace, USA—produce rayon based carbon–carbon composites for the leading edge of the shuttle orbiter.
122. 3M Company, St Paul, Minnesota, USA—manufactured a range of Scotchply prepreg. Sold out composites business to Cytec Industries in 2001.
123. McClean Anderson, Schofield, Wisconsin, USA—a division of Isami, manufacturer of filament winding machines.
124. Messier-Bugatti, Vélizy, France—a subsidiary of SNECMA. Messier started research into carbon–carbon brakes in 1968 and by 1971, Dassault had used Messier brakes in the Mirage F1. In 1997, merged with Carbone Industrie. Messier is now the world's leading supplier of carbon brakes for aircraft and racing cars (F1 and GT).
125. Mitsubishi Chemical Corp., Tokyo, Japan—produce a range of Dialead coal tar pitch based carbon fibers, with increasing thermal conductivity as the modulus is increased. Possess substantial know-how for making carbon–carbon. Offer the Replark line for retrofit of concrete structures.
126. Mitsubishi Oil Co., Kowasaki, Japan—originally produced carbon fiber from petroleum pitch.
127. Mitsubishi Rayon, Toyohashi, Japan—Mitsubishi produce a PAN precursor in Japan, which they sell onward to Grafil Inc., which is now owned by Mitsubishi. Have also acquired Asahi Nippon's carbon fiber operation and provide an alternate PAN precursor for this activity.

APPENDIX 16: CARBON FIBERS AND THEIR COMPOSITES

128. Monsanto Company, St Louis, Missouri, USA—in 1949, formed Chemstrand with American Viscose to produce Acrilan acrylic fiber; Monsanto acquired American Viscose's equity in 1961 and changed the name to Monsanto in 1964. The chemical businesses became Solutia Inc. in 1997.
129. Morganite Modmor, Battersea, London, UK—was a subsidiary of Morgan Crucible and one of the three original UK producers adopting the RAE process. Formed a joint company in the USA with the Whittaker Corporation associated with the Narmco Corporation. Ceased production of Modmor carbon fibers when the Battersea site was closed down.
130. Morrison Molded Fiberglass (MMFG), Bristol, Virginia and Chatfield, Minnesota, USA—now called Strongwell. Operate pultrusion plants.
131. Narmco, USA—a well known prepreg producer in the USA that was taken over by BASF, then Amoco and now owned by Cytec, which used to be the prepreg arm of Cyanamid.
132. Newport Adhesives & Composites Inc., Irvine, California, USA—associated with Grafil Inc. and manufacture a range of carbon fiber prepreg and also factor parent company Mitsubishi Rayon Corporation's Pyrofil prepreg.
133. Nikkiso Co. Ltd., Tokyo, Japan—a producer of Grasker graphite whiskers. Nikkiso coat graphite whiskers with titanium by CVD and then compact using hot isostatic pressing (HIP). Developed Norian, a PAN based carbon fiber spun in NaSCN using dry/wet spinning and high stretch carbonization for high strength.
134. Nippon Carbon Corporation, Tokyo, Japan—initially had a joint venture with Asahi and produce Carbolon carbon fiber. Taken over by Mitsubishi.
135. Nippon Graphite Fiber Corporation, Tokyo, Japan—the carbon fiber business of Nippon Petrochemicals was forwarded to Nippon Oil Co. in 1993 and in 1995, Nippon Oil and Nippon Steel formed a joint company, Nippon Graphite Fiber Corporation, a pitch based carbon fiber business. This company make the world's smallest diameter (6 μm) pitch based carbon fiber. Produce Granoc yarn; the XN grades are based on petroleum pitch and YS grades are based on coal tar pitch. Produce woven fabric and supposedly have the lightest carbon fiber fabric at 60 gsm made from $1k$ pitch fiber (YT50 yarn).
136. Nitto Boseki Co. Ltd., Tokyo, Japan—produce a pitch based carbon fiber.
137. North American Rayon Corporation (NARC), USA—manufacture a continuous carbonizable rayon.
138. Osaka Gas Co. Ltd., Osaka, Japan—plant was a joint venture with Dainippon Ink and completed in 1984. Produce Donacarbo carbon fiber from pitch.
139. Performance Fabrics, Greensboro, MC, USA—manufacture prepreg.
140. Performance Polymers—a division of Ciba Speciality Chemicals, has purchased the product lines and technology in 1997 for Quatrex and Tactix resins from Dow Chemical Company.
141. Peroxid Chemie-Groupe Laporte, Pullach, Germany—produce catalysts and accelerators for polyester resins.
142. Petoca, c/o Kashima Oil Co. Ltd., Tokyo, Japan—produce Carbonic, a pitch based carbon fiber. Offer a series of Melblon products made from mesophase pitch-based graphite fiber mat possessing superior lithium intercalation, which is milled. A boron doped grade is available.
143. Phoenixx TPC Inc., Dighton, MA, USA—manufacture thermoplastic prepreg tape up to 30.5 cm wide. Acquired Quadrax's golf shaft assets pertaining to thermoplastic golfshafts.
144. Plastic Developments Ltd., Bolton, UK—manufacture carbon fiber U/D tapes.
145. Polycarbon Inc., Charlotte, North Carolina, USA—was a subsidiary of Sigri Carbon Corp., now SGL Technik GmbH. Produces carbon fiber from a rayon based precursor mainly for the packing industry.
146. Porcher Industries, Bourgoin-Jallieu Cedex, France—weave carbon fiber.
147. Primco Ltd., Middleton, Manchester, UK—produce a range of Primco prepregs.
148. Pultrex Ltd., Colchester, Essex, UK—produce pultrusion and filament winding machines. Acquired by Shell in 1994, then acquired by Pacific Composites, Australia. In 2003, equipment manufacturing side was purchased by Colin Leek (Pultrex MD) and Douglas Curtis Machine Tools, who manufactured the machines on behalf of Pultrex. The pultrusion manufacturing remained with Pacific Composites.

149. Pultrusion Technology Inc., USA—purchased by Morrison Molded Fiber Glass (MMFG), Bristol, Virginia, which is part of Shell Polymers and Catalyst Enterprises (SPACE). PTI is the world's largest producer of turnkey pultrusion equipment systems. MMFG is the world's largest producer of engineered composite parts made by the pultrusion process.
150. RAE – The Royal Aircraft Establishment, Farnborough, England, was a Ministry of Defense (MOD) organization that was the home for the invention of carbon fiber in the UK, later it became DERA – the Defense Evaluation and Research Agency. In July 2001, DERA separated into two organisations: (a) all the defense laboratories and capabilities within DERA became DSTL, the Defense, Science and Technology Laboratory and (b) QinetiQ, Britain's largest independent science and technology company.
151. Quadrax Advanced Materials Systems Inc., Portsmouth, Rhode Island, USA—a prepreg manufacturer.
152. Reglass Spa., Minerbio, Italy—a prepreg manufacturer.
153. Richmond Aircraft Products, Norwalk, CA, USA—supplier of vacuum processing materials and Rohacell foam.
154. Rohr Inc., Vista, CA, USA—Rohr work with General Electric making components for jet engines. Formed a joint venture company in 1989 with Hercules, called Refractory Technology Aerospace Components (RTAC), who manufacture carbon–carbon composites for rocket motors, missiles, turbine engines and aerospace. Acquired by B.F. Goodrich in 1997.
155. Rolls Royce, Hucknal, UK—an early manufacturer of carbon fibers for in-house use in jet engines. Carried out excellent work on the technology of producing carbon fibers from PAN precursors. One of the original three UK companies making carbon fiber from a PAN precursor by the RAE process. Produced a carbon fiber, which was converted on-line to Hyfil carbon fiber epoxy prepreg. Plant subsequently sold to Bristol Composite Materials.
156. RK Carbon Fibres (RKCF), Muir of Ord, Ross-shire, Scotland, UK—the company originated as RK Textiles Composite Fibres (RKTCF) Ltd., which commenced R&D work on oxidized PAN and carbon fibers in 1973 at Stockport, initially concentrating on making opf from Courtauld's Textile Tow Precursor (TTP), since the Special Acrylic Fibre (SAF) was not available to RK at the time. The opf was called Panox and the first commercial plant was commissioned at Muir of Ord in 1981, called RK Textiles (Scotland) Ltd. At that time, Coats Patons took a 25% interest in the company, extending this to 75% in 1982. That same year, RKCF started production of carbon fibers in an adjacent factory using RAE technology, tending to concentrate on large tow fibers. Later, Bridon acquired an interest in RKCF. RKTCF started making carbon fiber plant for in-house use and selling plant and carbon fiber technology abroad. In 1985, the Coats Patons and Bridon interests were purchased and RK Holdings Ltd. was formed. Enichem acquired a 50% interest in 1987, later to be bought back by RK, and the company was then acquired by SGL Technik in January 1997, becoming SGL Technic Ltd. (Carbon Fibres). It is the only remaining UK carbon fiber producer and probably, the largest manufacturer of opf in the world. The company has specialized in making large tow carbon fiber and also makes a range of carbon fibers with controlled electrical properties. RK made Grayon carbon fibers from a rayon precursor, but this business has been transferred to Polycarbon in the USA, within the SGL group. RK developed hot melt prepreg equipment, employing a unique S-wrap technique, but have now withdrawn from prepreg manufacture and sold in-house machines to the Advanced Composites Group. Have developed Curlon flame retardant and low weight insulation battings and felts, besides Lineon, a continuous fiber form and had a facility in the USA for the manufacture of felts. However, this facility has now been closed and fiber and know-how supplied to Orcon Corp., San Francisco for onward processing.
157. RTM Composites, Montferrat, France—manufacture carbon braid.
158. RTP Co., Winona, Minnesota, USA—manufacture carbon fiber filled thermoplastic moulding compounds.
159. Saertex Wagener GmbH & Co. KG, Saerbeck, Germany—manufacture non-crimp carbon fabrics.

160. Schappe Techniques Sarl (STS), Charnoz, France—use stretch breaking process to manufacture opf, and carbon fiber yarns and material for carbon–carbon preforms. Market a range of Graphilite precision cut fiber fillers from opf and carbon fibers for composite materials and conductive coatings.
161. Schunk Group, Heuchelheim/Giessen, Germany—involved in the graphite and ceramics business and make carbon–carbon brake shoes and bipolar sheets for fuel cells.
162. Scott Bader, Wollaston, UK—produce polyester resins.
163. Seal Spa., Legnano, Italy—was called Texipreg GmbH, manufacture a range of prepreg materials.
164. Selcom SRL, Fregona (TV), Italy—produce UD and multidirectional carbon fiber fabrics.
165. SGL Carbon Composites, Gardena, California, USA—formerly Hitco Technologies, acquired by SGL Carbon in 1997. Specialises in carbon–carbon composites and is the largest producer of carbon–carbon brakes.
166. SGL Carbon Group, Meitingen, Germany—comprises Sigri Electrographit, which was part of the Hoechst Group and market carbon fiber as Sigrafil and oxidized PAN fiber as Sigrafil O. Manufacture carbon–carbon products. Polycarbon in the USA, was purchased in 1980. In 1997, acquired RK Carbon Fibres Ltd., Hitco Technologies in the U.S.A and PG Lawton, a UK carbon fiber specialist. SGL Technik is a 100% subsidiary of the SGL Carbon Group, which is the world's largest producer of carbon and graphite products comprising sixteen production plants divided into graphite electrodes (42%), speciality graphite products (39%), carbon products (8%) and technical products (11%). About 15% of this production comprises of fibers, fabrics and felts.
167. SGL Technic Ltd., Carbon Fibres, Muir of Ord, Ross-shire, Scotland—up to January 1997 was RK Carbon Fibres Ltd. In July 1999, announced a joint venture with Aldila Inc., where SGL have acquired a 50% interest in Aldila's carbon fiber manufacturing operation.
168. Showa Denko KK, Omachi, Japan—have developed a gaseous phase process to produce carbon fibers.
169. Sigmatex UK Ltd., Runcorn, Cheshire—weave carbon fiber fabrics.
170. Siltex, Julbach, Germany—produce carbon braid.
171. Société Européenne de Propulsion (SEP), Suresnes, France—SEP is a subsidiary of SNECMA and is a leading carbon–carbon producer. In 1985, SEP and GEC Alsthom formed Carbone Industrie, a major carbon–carbon producer with a trade name Sepcarb.
172. Soficar, Abidos, France—a joint venture of Toray (70%) and Elf Acquitaine & Pechiney (Atochem) formed in 1982, producing a range of Torayca carbon fibers. Technology and precursor supplied by Toray. Until the plant came onstream in 1985, marketed Toray carbon fibers in Europe. Initial line produced high strength carbon fibers and was supplemented by a second line coming onstream in 1992, producing high and intermediate modulus carbon fibers. The French and Toray products are equivalent. Paris based Sales Division provides Europe with French and Japanese fibers. In 2004, increased production capacity from 880 tpa to 2600 tpa.
173. Solutia Inc. St Louis, Missouri, USA—formerly the Monsanto Company, manufacture an acrylic fiber.
174. Spencer Composites Corp., Sacramento, California, USA—manufacture filament wound composite structures up to 4.5 m in diameter and 18 m in length.
175. SP Structural Polymer Systems Ltd., Newport, Isle of Wight, UK—SP Systems was initially bought by Montecatini, part of the Montedison Group, who then acquired Ferro. However Montecatini then bowed out and SP Systems became a stand alone company, with Ferro incorporated in Culver City Composites, previously called SP Systems Inc., a separate US company. The US company has now been acquired by American Materials and Technology Co. (AMT). A significant development is the introduction of a RefNet, an information source based on their website and a free CD-ROM acting as an offline version for access to company and product information, including SP System's Composite Materials Handbook. SP were acquired by Zoltek in November 1999, and then sold to a management buy-out in November

2000, to be subsequently acquired by Gurit-Heberlein; became part of Gurit Material Systems in September 2002.
176. Stackpole Carbon Fibers, Lowell, Massachusetts, USA—a manufacturer of oxidized PAN fiber, Pyron and Panex carbon fibers. Acquired by Zoltek in 1987.
177. Sterling Fibers Inc., Pace, Florida, USA—a subsidiary of Sterling Chemicals Inc. The second largest producer of acrylic fiber (Creslan) in the US.
178. Stesalit AG, Zullwil, Switzerland—manufacture Stesapreg prepreg and is a member of the Isola Group, who produce carbon fiber pultrusion.
179. Strongwell, Bristol, VA, USA—previously called Morrison Molded Fiberglass, operate pultrusion plants.
180. Structil (Groupe SNPE), Vert le Petit, France—a subsidiary of SNPE, produces a range of prepreg products by hot melt process. Also has eight pultrusion machines. In 1998, Grafil Inc. purchased a 20% share.
181. Sulzer Composites, Winterthur, Switzerland—manufacture thermoplastic prepregs. Sold to the Gurit-Heberlein Group in 2002 and relocated to Flurlingen.
182. Sumika-Hercules, Saidai, Japan—a joint venture of the Sumitomo Chemical Co., Japan with Hercules Inc., USA, to produce PAN precursor and sell carbon fiber prepreg. Now part of the Hercules-Hexcel group.
183. Syncoglas NV, Zele, Belgium—manufacture carbon fiber reinforcements.
184. Technical Fibre Products (TFP), Kendal, Cumbria, UK—produce carbon fiber mats and veils.
185. Tenax Fibers GmbH & Co. KG, Wuppertal, Germany—Enka Carbon Fibers was purchased by Akzo, who merged with Nobel. Use Toho Rayon technology and all precursor is obtained from Mishima. Toho acquired a controlling interest in 1983, increasing their holding to about 80% in 1996. The company is now called Tenax Fibers and produce a range of Tenax carbon fibers. Previously made prepreg but sold their machine to Primco.
186. Tenax Fibers GmbH & Co KG, Wuppertal, Germany—an affiliate of Toho Rayon Co., Ltd. Japan. Produce a range of PAN based carbon fibers under the trade name of Tenax.
187. Tencate Advanced Composites BV, Nijverdal, Netherlands—produce a range of Cetex prepreg materials. In 1999, acquired Bryte Technologies Inc.
188. Textron Systems, Wilmington, Massachusetts, USA—formerly known as Avco and later as Textron Speciality Materials. Market Avcarb carbon fiber and Avox oxidized PAN fiber. The carbon–carbon and metal matrix composite parts are at Lowell, MA.
189. Thiokol-TCR Composites Corp., Ogden, Utah, USA—manufacture prepreg tow for filament winding, non-woven, UD tapes, braids and fabrics. In 1998, changed name to Cordant Technologies Inc.
190. Ticona UK Ltd., Milton Keynes, UK—a member of the Hoechst Group. Produce a range of thermoplastic polymers.
191. Tissa Glasweberei AG, Oberkulm, Switzerland—weave carbon fiber fabrics.
192. Toa Nenryo Kogyo KK, Kowasaki, Japan—produce a pitch based carbon fiber from Tonen's fluid catalytic cracking operations at Kawasaki.
193. Toho Rayon Co., Mishima, Japan—a large carbon fiber producer in Japan, sold as Besfight produced from their own PAN precursor. Licensed their carbon fiber technology in 1980 to Celanese Corp., USA, and in 1983, to Enka AG at Wuppertal, Germany. Toho acquired a controlling interest in 1983, increasing their holding to about 80% in 1996. The company is now called Tenax Fibers and produce a range of Tenax carbon fibers. Formed a joint venture company Toho Badische Structural Materials Co. Ltd. with BASF AG and also BSM Inc. in the USA. In 1999, Teijin Ltd., Osaka, Japan acquired a majority share of Toho Rayon. Became Toho Tenax Co. Ltd.
194. Tonen Corp., Tokyo, Japan—produce a range of Forca pitch based carbon fibers.
195. Toray Industries Inc., Tokyo, Japan—the world's largest producer of PAN based carbon fibers and the undisputed leader in modern carbon fiber technology. Production of Torayca T300 carbon fiber started in 1978. Cross licensed technology to Union Carbide in the USA in 1978, and carbon fiber production in 1984. Formed a joint company (Soficar) with Elf Acquitaine Pechiney in France and assumed marginal control (70%) in 1989, producing fiber at Abidos.

When Amoco acquired the carbon fiber manufacturing facility of Union Carbide in 1986, Toray transferred the licensing agreement to Amoco. Toray manufacture their own PAN precursor at Ehime. Toray is a leading carbon fiber prepreg manufacturer and has production facilities at Ehime, Japan. Toray Carbon Fibers America Inc. was established in 1997 at Decatur, Alabama and produce T700SC, T800SC, M30SC and M30GC carbon fibers and also distribute the whole range of Toray brand carbon fiber.

196. Union Carbide, Danbury, CT, USA—pioneered the use of cellulosic precursors to make Thornel carbon fibers and entered the market in 1965 but, unfortunately, had to resort to hot stretching to obtain properties to be able to compete with PAN based products. In 1973, started the manufacture of carbon fibers from a pitch base. Entered into a technical licensing agreement with Toray in 1978 to produce carbon fiber in the USA based on Toray's PAN technology. The carbon fiber facility was acquired by Amoco and the carbon fibers are distributed by FMI Composites in the UK.
197. Universal Carbon Fibres Ltd., Cleckheaton, UK—undertake stretch breaking and produce oxidized PAN yarn which can be woven and subsequently carbonized to produce a carbon fabric. Now renamed Lantor Universal Carbon Fibres.
198. US Polymeric, USA, A prepreg manufacturer acquired by British Petroleum and then by Fiberite Ltd.
199. Vermont Composites, Bennington, Vermont, USA—formerly known as Courtaulds Aerospace, manufacture composite components for diagnostic imaging applications.
200. Verseidag-Indutex GmbH, Krefeld, Germany—produce a range of Chemtex prepreg.
201. Vom Baur Sohn, Wuppertal, Germany—produce carbon hemmed tapes.
202. Westinghouse Electric Corp., Bedford, Pennsylvania, USA—produce phenolic prepreg.
203. Victrex, Thornton Cleveleys, Lancashire, UK—produce PEEK polymer.
204. Vom Baur Sohn, Wuppertal, Germany—produce carbon tapes.
205. Wade Fibres (PTY) Ltd., Gardenview, South Africa—produce stitched carbon fibers.
206. Wood Science & Technology (WSTI), Corvalis, Oregon, USA—Zoltec and Dow Chemical Co. have an agreement with WSTI, who will use WSTI patents and manufacture thin sheets of composite laminate for the reinforcement of engineered lumber products using Zoltek's carbon fiber and Dow's Derakane epoxy vinyl ester resins.
207. YLA Inc. Advanced Composite Materials, Benicia, California, USA—company was originally Young Lee and Associates. Produce a range of specialist prepregs. Perstorp AB acquired majority equity in 1994.
208. Zoltek, St. Louis, Missouri, USA—acquired Stackpole in 1987, to make Pyron oxidized PAN fiber and Panex carbon fibers. When Courtaulds, Coventry terminated their carbon fiber business in 1992, Zoltek purchased one of their carbon fiber production lines. In 1995, acquired Magyar Viscosa Reszventarsursag (Hungarian Rayon Inc.), a textile acrylic fiber producer situated near Budapest and licensed to use the DuPont acrylic process. Will also produce carbon fiber from this site in Hungary. Produce carbon fiber at St. Charles, MO and have established a carbon fiber manufacturing facility in Abilene, TX. At the end of 1999, acquired SP Systems, Isle of Wight, a prepreg company; Cape Composites Inc., San Diego, CA, a prepreg company; Composites Machine Company Inc. (CMC) and Engineering Tech Inc. (EnTec), both of Salt Lake City, UT and filament winding and pultrusion machine manufacturers. In May 2000, acquired a significant stake in Hardcore Composites Operations LLC, New Castle, Deleware, a leading manufacturer of engineered composite structures, such as bridges, for the civil infrastructure market. After one year, SP systems were sold back to former management.

Index

A

Acheson process, 25
Acilan, 80
Acrylic fiber plants, world distribution of, 123
Acrylic precursor manufacturing processes, 123
Acrylic resins, as precursor, 171
Acrylonitrile, 101–103
Acrylonitrile/methyl acrylate/itaconic acid, 101–103
Activated carbon cloth, typical properties of, 958
Activated carbon fibers, use of carbon fiber, 955–958
Aero engines, use of carbon fiber, 977
Aerospace, uses of carbon fibers, 953–954, 973–982
Aerospace grades, polyacrylonitrile based carbon fiber, 194
Air bags, use of carbon fiber, 993
Aircraft, carbon-carbon for, 1021
Air cushion vehicle, use of carbon fiber, 989
Air flow measurement, 424–432
    electrolytic process, 403–404
    mercury manometer or barometer, 424
    micromanometer, 424
    safety issues, 424–432
    vertical manometers, 444
    vertical or U-tube manometer, 424
Air intermingler, proprietary, 435
Air oxidation, pitch based precursor, 297–300
Alkaline fuel cell, use of carbon fiber, 965
Allotropes, carbon, 1, 15
Alternative polymer formulations, 232
Aluminum, metal matrices, 603, 635–639
    aluminum self-propagating interfacial reaction, 636
    coated, 636
    fiber degradation, suppression of, 636–637
    pressure casting technique, 637–638
    uncoated, 636
Aluminum nitride, matrices, 604
Aluminum self-propagating interfacial reaction, metal matrices, 636
Amide, 129
Amine cured systems, 519
Amine hydrogen equivalent weight, 519
Ammonia, safety issues, 447–448
Amorphous thermoplasts, 534–535
Anisotropic elastic theory, 935–940
Anodic oxidation, 352–355
Antenna, lightening conductors, use of carbon fiber, 979–982
Antibonding, 9–10
Antistatic finish, application of, equipment design for, 384
Anti-Stoke line, 485
Aqueous dispersion polyacrylonitrile precursor, 130
Aqueous dispersion polymerization, 134–136
Architecture, carbon fiber, 861–863
Aromatic condensation, pitch based precursor, 314
Aromatic hydrocarbons, as precursor, 171
Arrhenius plot, 570–571
Asbestos cement, carbon fiber cement as replacement for, 1024
Asymmetric stretch, 479–480
Atlantic Orient Corporation, wind generator sizes, categorization of, 997
Atom, carbon, structure of, 1–14
Atomic Energy Research Establishment, Harwell, early work with polyacrylonitrile precursors, 79–89
Atomic orbitals
    directional characteristics, 6–7
    hybridization of, 7–8
    overlapping of, 10–12
Atomic spectra, carbon atom, 2–8
Atoms, distance between, force between them, relationship of, 671–672
Audemars, George, 148
Autoclave molding, 95, 918–920
Automatic tape lay-up, 921
Automobile, applications, use of carbon fiber, 991–993
Automotive brake rotors, carbon-carbon friction material produced for, properties for, 1015
Aviation, uses of carbon fibers, 953–954
Axes, convention, dimensional notation, carbon fibers, graphite, 464–466
Axial flow fan, exhaust systems, 415
Azimuthal quantum number, 5
Aztex Inc. Z-fiber, 893

B

Badminton racquets, use of carbon fiber, 1004
Balsa honeycomb, 895
Barrier coating, 575–578
    chemical vapor deposition, 561, 576
    noble metals, 575
    oxides, 578
    preceramic polymer coating, 577
    silicon coating, 575–578
Baseball bats, use of carbon fiber, 1002

Basic structural unit, polyacrylonitrile based carbon fibers, 206–214
Batch process, 43
Batteries, use of carbon fiber, 962–964
Benzocyclobutene, 530
Bevan, Edward J., 149
Biconical hole formation, prevention of, oxidation plant, 199
Bicycles, tandem, use of carbon fiber, 1001
Bifurcated fans, 416
Binoculars, use of carbon fiber, 1009
Biomedical end uses, carbon-carbon for, 1022
Bismaleimides, 525–527
Bisoxazoline phenolics, 530
Black orlon, 66–67
Bleach, treatment of cyanide effluent, 444–445
Blendur resins, 530
Bloating, liquid infiltration, 572
Block model, polyacrylonitrile-based carbon fibers, 204
Boiling points, solvents used to extract pitch, 159
Bonding, 9–12, 943–944
Bond to glass matrices, coating carbon fiber, 601
Boron, annealing in presence of, 229–230
Boron carbide (V_4C), matrices, 604
Boron nitride, matrices, 604
Bragg's Law, 467
Braided 3-D multiaxial, 890
Braiding, 877–882
Brake
    energy absorbed, various systems of transport, 1016
    use of carbon fiber, 992
Brake disk materials-carbon-carbon vs. steel, properties of, 1014
Braking systems, carbon-carbon, 1011–1018
Brittle matrix, dealing with, 815–816
Buckyballs, 49
Bucky onions, 59
Building-up principle, 9
Buses, use of carbon fiber, 994

C

Cable insulation, uses of carbon fibers, 955
Calibration lamp, 441–442
Canoe paddles, use of carbon fiber, 1005
Capability index, 770–771
Carbon
    heat treatment carbonization furnace, 396
    new forms of, 46–60
Carbonaceous yarn, 68
Carbon atom, 2
    allotropes, 1
    atomic orbitals
        directional characteristics, 6–7
        hybridization of, 7–8
        overlapping of, 10–12
    atomic spectra, 2–8
    building-up principle, 9
    carbon-carbon bonds, 12
    catenation, 1
    constructive interference, 9–12
    covalence, 9–12
    destructive interference, 9–10
    electrons, 2
    glow discharge, 3
    Heisenberg uncertainty principle, 4
    hybridized orbitals, arrangements of, 8
    isotopes, 1
    matrix mechanics, 4
    molecular orbitals, 9–12
    nucleus, 2
    Pauli's exclusion principle, 5
    periodic table, defined, 1
    principle quantum number, 4–5
    quantum numbers, allocation of, 5
    quantum theory, 2–8
    Rutherford's concept, 2
    Schrödinger equation of wave mechanics, 3–4, 9
    subatomic particles, physical properties of, 2
    subsidiary number, 5
    valence electrons, 7
    wave mechanics, 4
    Zeeman effect, 5
Carbon black, 44–45
Carbon-carbon applications, carbon fibers for, 1010–1023
Carbon-carbon bonds, 12
Carbon-carbon braking systems, 1017
Carbon-carbon clutches, use of carbon fiber, 1018–1020
Carbon-carbon composite
    high strength graphite, compared, properties of, 1015
    structure-property relationships of, 563
Carbon-carbon matrix materials, processing, 560–569
Carbon-carbon processing, 552–560, 569–573
    carbon fiber reinforcement, 533
    furan resin, 556–557
    materials for, 552–560
    matrix materials, 560–569
    phenolic resins, 557
    pitch, 558–559
    polyimide resins, 557–558
    thermoplastic matrix, 534
    types of matrix, 556–560
Carbon cloth, activated, typical properties of, 958
Carbon fiber, 42
    architecture, 861–863
    carbon matrix composites, 551–582
    in composite materials, 113
    coupling agents, 363
    early development of, 65–120
    elastic constants, 86
    emissions, toxicology of, 449–450
    future developments, 1044–1045
    impregnation of, matrices, 591
    oxidation stage, 103–111
    phenolic matrix, bond between, 553

INDEX

plating, 112
properties, 68, 791–860
role of, 791
sizing, 363–370
structure, determination techniques, 453–500
surface treatment, 111–112
tensile strength, factors affecting, 91–92
testing specifications for, 658–659
types of, 792–800
uses of, 951–1042 (*See also* Uses of carbon fibers)
virgin, testing, 112
Carbon fiber cable, use of carbon fiber, 1006
Carbon fiber composites, polymer matrices for, 501–550
Carbon fiber composites in fire
 lateral expansion or contraction, 434–435
 lateral movement, 433
 pressure measurement, 424
 velocity, determination of, 424–429
 volume flow, determination of, 429–432
Carbon fiber composite testing, 689–714
 composite specimen, from wet resins, 690–692
 law of mixtures, 691
Carbon fiber in fuel cells, 969–970
Carbon fiber manufacture, precursors for, 121–184
 cellulosic precursors, 148–156
 pitch precursors, 156–171
Carbon fiber manufacturers, 185–186
Carbon fiber plant, 377–420
Carbon fiber prepreg testing, 688–689
Carbon fiber production, 98–100, 185–268
 carbonization, 279
 carbonization stages, cellulosic based precursors, 280–282
 cement, 112
 choice of suitable precursor, 272–274
 current production, 272–280
 hot stretching during processing of carbon fiber, 279–280
 line, 194–203
 prepreg, 112
 procedures using, 112–113
 pultrusion, 112
 pyrolysis, 274–279
 sizing, 280
 thermoplastic molding compounds, 112
 using cellulosic based precursor, 269–294
 using pitch based precursor, 295–324 (*See also* Pitch based precursor)
 using polyacrylonitrile precursor, 185–268 (*See also* Polyacrylonitrile precursor)
Carbon fiber properties, improvements in, 232
Carbon fiber reinforced composites
 braiding, 877–882
 centrifugal molding, 904
 contact molding wet lay-up, 894–896
 continuous fiber reinforced plastic materials, 929
 core materials, 893–894
 3-D reinforcements, 882–883
 fiber placement systems, 921–922
 filament winding, 905–909
 film stacking process, 927
 flow, cure monitoring of resin infusion processes, 904–905
 hot press matched metal molding, 896
 hybrid composites, 929
 importance of critical aspect ratio, 923–924
 injection molding, 925–927
 inlaid fabrics, 877
 knitted fabrics, 872–877
 mold release, 922–923
 non-woven discontinuous reinforcement, 863–865
 preparation of fiber preforms, 904
 preparation of thermoplastic molding compounds, 924–925
 prepreg molding, 913–921
 pultrusion, 909–913
 reaction injection molding, 904
 resin transfer molding, 897–903
 sequential multiport resin injection system, 904
 thermoplastic filament winding, 928
 in thermoplastic matrices, 923–929
 thermoplastic prepreg, 927–928
 thermoplastic pultrusion, 929
 in thermoset matrices, manufacturing processes, 894–923
 unidirectional fabrics, 865–866
 woven fabric (2-D planar or biaxial reinforcement), 866–872
 woven spread tow, 872
Carbon fiber reinforced metal composites, 97
Carbon fiber reinforced plastic, 78
Carbon fiber reinforcement, 861–863
Carbon fibers in metal matrices, 629–656
Carbon fibers in thermoplastic matrices, 923–929
Carbon fibers in thermoset matrices, 973–1009
Carbon fiber yield, 230
Carbonization, 221–230, 279
 electron energy loss spectroscopy, 222
 pitch based precursor, 301–303
 pitch fibers, 321
 precursors for, 25
Carbonization furnace, 395–398
Carbonization stages
 carbon fiber production, cellulosic based precursors, 280–282
 mechanisms for, 254–259
 polyacrylonitrile carbon fibers, 254–259
Carbon matrices, 551–582
Carbon matrix composites, carbon fiber, 551–582
 carbon-carbon processing, 569–573
 materials for carbon-carbon processing, 552–560
 matrix materials, carbon-carbon processing, 560–569
 oxidation protection, 573–578
Carbon nanotubes, 56–58, 1046
Carbon phase diagram, 16
Carbon powder, furnace insulation, 402

Carbon yarn, 68
Carboxylic acids, 125–128
Catalysts, use of carbon fiber, 958
Catalytic chemical vapor-deposited filaments, 43–44
Catenation, 1
Cathodic protection, use of carbon fiber, 960
Celanese test, 708
Cello, bows for, use of carbon fiber, 1007
Cellulose, 270
Cellulose based carbon fibers, 556
Cellulosic based precursor, carbon fiber production using, 269–294
Cellulosic material, thermal conversion, technique, 69–70
Cellulosic precursors, 148–156
  ageing stages, 151–152
  allotropes of cellulose, 154
  carbon fiber manufacture, 148–156
  for carbon fiber manufacture, 148–156
  final treatment stage, 153–154
  historical introduction, 148–150
  mixing, ripening stages, 153
  nitrocellulose yarn, 149
  rayon, defined, 150
  shredding, 151–152
  spinning stage, 152–153
  steeping stage, 150–151
  structure of rayon fibers, 154–156
  viscose, 149
  viscose rayon process, 150–154
  xanthation stage, 152
Cement, carbon fiber reinforced, 86, 1023–1030
Cement composite, carbon fiber reinforced, 589
Cement matrices, 839–841
Cement matrix composites, matrices, 587
Centrifugal fan, exhaust systems, 415
Centrifugal molding, 904
Ceramic, heat treatment carbonization furnace, 396
Ceramic matrices, 583–628, 841–844
  carbon fiber reinforced, 583–628
  carbon fibers in, 1031
  in ceramic composite materials, 603
  chemical route, 602
  powder route, 602
  processing ceramic matrix composites, 602
  reinforced, carbon fiber, 583–628 (See also Reinforced ceramic matrices)
Ceramics, carbon fiber reinforced, 86
Channel black, 44
Charcoal, 45
Charpy test, 833
Chassis, body, interior, use of carbon fiber, 991–992
Chemical applications, use of carbon fiber, 998
Chemical resistance, 836–837
Chemical resistance of diamond, 23–24
Chemical resistance of graphite, 36–38
Chemical vapor infiltration processes, 565–567
Chemical vapor deposition, 38, 560
Chemical vapor deposition diamond, 18

Chemical vapor infiltration technique, 39
Chemstrand, acrilan using dimethylacetate (DMAc) solvent, 122
Chopped carbon fiber, metal powders, hot working mixture, 88
Chromatography, 729–732
Ciba (ARL) Ltd., Duxford, prepreggers, early work in U.K., 114
Civil aircraft, use of carbon fiber, 973–977
Classical differential thermal analysis, 721–722
Closed cell PVC foam, 893
Closed circuit television, equipment design for, 420
Clover leaf drive systems, 380
Clutches, use of carbon fiber, 992
CNG storage cylinders, use of carbon fiber, 994
Coal, 45–46, 171–174
Coal tar pitch, 158, 172–173
Coke, 46
Collection, production line, carbon fiber, 203
Column chromatography, petroleum pitch, 160
Comonomers, 125–130, 192–193
Compocasting, metal matrix, 644
Composite manufacturing, 943, 1045
Composite materials, carbon fiber in, 113
Composites
  carbon fiber reinforced, 861–863
  fabrication, 89
  properties of, 839–845
  testing, 78, 84–86
Composite strength, factors effecting, 808–810
Compression molding, 916, 927
Compression properties, 817–823
Compression testing, 711, 739
Compressive strength, uniaxial, modulus, measurement of, 708–710
Concrete
  carbon fibers in, 1023–1030
  matrix, 585–591
Concrete additives, 584–585
Concrete matrices, 583–592
Condensation type polyimides, 523–524
Constructive interference, 9–12
Contact angle measurement, 359–360
Contact molding wet lay-up, 894–896
Continuous fiber reinforced plastic materials, 929
Continuous process, 43
Control chart method, 758
Control charts, specimen results for, 762
Controlled atmosphere electron microscopy, vapor grown carbon fibers, 334
Controlled resistance carbon fibers, 233–234
Control limits, formulae for calculating, 762
Control of manufacturing process, polyacrylonitrile based precursor polymer, 193
Control of Substances Hazardous to Health Regulations 1988, 448–449
Conventional autoclave, use of, 569
COPNA, 559
Copper metal matrices, 639–640

Core materials, 893–894
Coupling agents, 363
Courtaulds Ltd., Coventry, 98–113, 122
Courtelle, 73–75, 80, 82, 89–91, 122, 197, 675
Covalence, 9–12
Creep measurement, 712–714
Creep properties, 831–832
Cricket bats, use of carbon fiber, 1003
Critical aspect ratio, importance of, 810–811
Crosby, Philip B., 782–783
Cross, Charles F., 149
Cross-linking, pitch based precursor, 299–300
Crystal structure of diamond, 19–20
Cubic crystal, Miller indices, 466
Cumulative sum chart, 767–770
Cuprammonium process, 149
Curing agents, 514, 519
Curlon, insulation felts, comparative properties, 972
Curlon fibers, properties, 971
Cusum chart, specimen results for, 769
Cut off saw, 912
Cyanate resins, 520–521
Cyanide effluent, 444–445
  barometer, 424
  bleach liquor, treatment with, 444–445
  mercury manometer, 424
  safety issues, 444–445
Cycloaliphatic resins, 512

D

Dancing arms, online collection, 409
Dark field electron microscopy, polyacrylonitrile based carbon fibers, 214
Data location, 748–750
de Chardonnet, Hilaire, de Bernigaud, 149
Deeming, W. Edwards, 776–777
Defense, uses of carbon fibers, 954
Defense aircraft, use of carbon fiber, 973
Dehydrochlorination mechanism, polyacrylonitrile precursors, 244
Density gradient columns, liquids used to prepare, 663
Density of diamond, 20–22
Density of graphite, 31–32
Design, joining, 943–944
Design cases, 945–946
Design considerations, 935–950
  bonding, 943–944
  carbon fiber plant, 377–420
  for carbon fiber plant, 377–420
  composite manufacturing method, 943
  design cases, 945–946
  fabrication, 944
  inspection, 944
  joining, 943–944
  materials selection, 940
  micromechanics, 935–940
  multidirectional laminates, elastic behavior of, 940–943
  smart devices, 944–945
  testing, 944
Despaissis, L. H., 149
Destructive interference, 9–10
Development of carbon fibers, early, 65–120
Diamond, 17–24
  classification of, 19
  crystal structure, 19–20
  identification of, 19
  production of, 17–19
  properties of, 20–24
  uses of, 17–19
Diamond-like carbon, 18–19
Dielectric cure monitoring, 905
Dielectric heat sink, carbon-carbon as, 1023
Diels-Alder addition, polyacrylonitrile precursors, 248
Differential scanning calorimetry, 538, 722–725
  polyacrylonitrile precursors, 128
  thermal analysis, 721–725
Differential thermal analysis, petroleum pitch, 160
Diffusion bonding, metal matrix, 643–644
Dimethyl formamide (DMF), solvent for, 122
Dip coating, matrices, 616
Dipole-dipole interaction, polyacrylonitrile precursors, 247–247
Dispersion measures, 750
Double V-notch shear, 702
Dow AdvRTM, 898–900
Dow epoxy novalac resins, properties of, 510
Drive shafts, use of carbon fiber, 993
Drive systems, 379–380
Drying, equipment design for, 405–409
Drying conditions, reinforced thermoplastics, 716
Dry mixing, matrices, 588
DTA. *See* Differential thermal analysis
Dual filament, 1046
Dunlop carbon-carbon composites, 847
Dupont, Orlon, 122
Dust extraction, 418–419
Dynamic mechanical analysis, 538, 726–729

E

Early development of carbon fibers
  inventors, 65–66
  Japan, work in, 71–72
  United Kingdom
    atomic energy research establishment, Harwell, 79–89
    Ciba (ARL) Ltd., Duxford, 114
    Courtaulds, Coventry, 98–113
    Courtaulds Ltd., Coventry, 114–115
    early prepreggers, 114–115
    Fothergill and Harvey Ltd. (F&H), Littleborough, 115
    Morganite Modmor, London, 97–98
    polyacrylonitrile precursors, 72–113
    RAE, Farnborough, 72–79

Early development of carbon fibers (*continued*)
  Rolls Royce, Derby, 89–97
    Rotorway components Ltd., Clevedon, 115
  United States, work in, 66–71
Edison, Thomas Alva, 65
Effluent gases
  in oxidation process, removal of, 200
  removal, production line, carbon fiber, 200
  removal of, equipment design for, 383–384
Eighty/twenty rule, 756
Elastic constants, 811–814
Elastic constants of carbon fibers, 813
Elastic properties of diamond, 22
Elastic properties of graphite, 33–35
Electrical applications, uses of, 960–970
Electrical equipment, protecting, 423–424
Electrical properties, 834–836
  of carbon fibers, 834, 836
  of diamond, 23
  of graphite, 35–36
  measurement, 678
  of pitch, 836
  of resistive fibers, 834
Electrodeposition, deposition of polymer onto fiber surface by, 367–369
Electrodes, use of carbon fiber, 961–962
Electromagnetic interference, use of carbon fiber, 1031–1032
Electron beam (EB) curing, future developments, 1045
Electron energy loss spectroscopy, 477
Electrons, carbon atom, 2
Electron spectroscopy for chemical analysis, 527
Electron spin resonance, pitch based precursor, 311
Electrophoretic deposition, matrices, 617
Electrophoretic infiltration, matrices, 617
Electroplating, 87
Electropolymerization, systems used for, 369
Element materials for heat treatment furnaces, 397–398
Emissions, carbon fibers, toxicology, 449–450
Engineering applications, use of carbon fiber, 995
Engines, carbon-carbon in, 1022
Epoxide resins, 508–520
Epoxy diluents, 513
Epoxy molar mass, epoxy resins, 687–688
Epoxy resins, curling agents, 514
Equipment design
  antistatic finish, application of, 384
  carbonization furnace, 384–392
  closed circuit television, 420
  drive systems, 379–380
  drying, 405–409
  dust extraction, 418–419
  effluent gases, removal of, 383–384
  exhaust systems, 415–418
  heat treatment furnace, 395–398, 401–403
  offline winding, 411–415
  online collection, 409–410
  ovens for oxidation, 380–383
  packaging, 415
  plaiter table, 384
  precursor handling, 377–379
  sizing, 404–405
  sodium removal, 400–401
  surface treatment, 403–404
Euler's Rule, 50
Evaporative drying, rising rate period, 406
Exhaust removal, 392–394
Exhaust systems, 415–418
Expansion tubes, use of carbon fiber, 1021
Exposure limits for gaseous emissions, 449
Extraction hoods, 416
Extruded graphite, properties of, 34

F

Failure mode effect analysis, 771
Fatigue testing, 710–712
Fiber modulus, increasing, 225–230
Fiber placement systems, 921–922
Fiber preforms, preparation of, 904
Fiber recoil, 817
Fiber reinforcement, 605
Fiber tow (liquid) infiltration, metal matrix, 645–646
Fiegenbaum, Armand V., 779
Field desorption mass spectroscopy, pitch based precursor, 310
Filament winding, 605–606, 905–909
Filament winding machine manufacturers, 906
Film stacking process, 96–97, 927
Fire, risks of carbon fiber composites in, 450–451
Fischer, Karl, 715
Fishbone diagram, 756–757, 779
Fishing rods, reels, use of carbon fiber, 1005
Five roll drive systems, 379
Flameproof applications, 951–955
Flame retardants, 278
Flexural properties, 814–815
Flexural strength, increase in, 586
Flexure tests, based on laminate thickness, 708
Flow
  cure monitoring of resin infusion processes, 904–905
  pressure drop, relationship between, exhaust, 418
Flow chart, 756
Fluorocarbons, 923
Flywheel
  lead acid battery, comparison of, 984
  use of carbon fiber, 983–986
Forms of carbon
  Acheson process, 25
  carbon black, 44–45
  carbon fibers, 42
  carbon nanotubes, 56–58
  carbon phase diagram, 16
  catalytic chemical vapor-deposited filaments, 43–44
  charcoal, 45
  chemical vapor deposition (CVD) process, 38
  chemical vapor infiltration (CVI) technique, 39

coal, 45–46
coke, 46
diamond, 17–24
Euler's Rule, 50
fullerenes, 46–56
glass-like carbon, 41–42
graphite, 24–38
graphite whiskers, 42–43
hyperfullerenes, 59–60
Knoop scale, 22
Mohs scale, 22
pyrolytic carbon, 38–41
pyrolytic graphite, 38–41
soot, 46
van der Waals forces, 15, 55, 60
vapor-grown carbon fibers, 43–44
Fothergill and Harvey Ltd. (F&H), Littleborough, prepreggers, early work in U.K., 115
Fourier transform infrared/attenuated total reflectance spectroscopy, 483
Fourier transform infrared spectroscopy, based on Michelson interferometer, 733–734
Fourier transform raman spectroscopy, 485
Four-point shear test, 817
Fourteen steps to quality improvement, 783
Fracture toughness, 833–834
Free aromatic diamines, use of, 526
Frequency distribution, 747–748
Friction, 22, 837
Fuel cells, use of carbon fiber, 964–970
Fullerenes, 46–56
    ball, stick structure, 53
    camel humps, 47, 50–51
    discovery of, 46–53
    Euler's Rule, 50
    isolated pentagon rule, 53
    NMR confirmation, 51
    properties, 53–56
    uses of, 53–56
    van der Waals forces, 55
Furan resin, 556–557
Furnace
    carbonization, equipment design for, 384–392
    heat treatment, equipment design for, 401–403
Furnishings, uses of carbon fibers, 955, 1009
Future developments
    carbon fiber, 1044–1045
    carbon nanotubes, 1046
    composite manufacturing techniques, 1045
    dual filament, 1046
    electron beam (EB) curing, 1045
    innovative developments, 1046
    ISO 9000, projected revision of, 1045
    plant developments, 1044
    precursor developments, 1043–1044
    production process, 1043–1044
    quality management standards, 1045–1046
    recycling, 1046
    solar sails, 1046

G

Gas adsorption, 358–359
Gas chromatography, 729
Gaseous emissions, United Kingdom exposure limits for, 449
Gaseous evolution, polyacrylonitrile carbon fibers, carbonization stages, 257–259
Gas evolution process, 81
Gaskets, use of carbon fibers, 973
Gas phase impregnation, densification, 560–567
Gas phase oxidation, 348–350
Gaussian distribution, 753
Gel permeation chromatography, 729
Gel permeation chromatography separation, 729–731
Gel time, 685–686
General purpose grades, polyacrylonitrile based carbon fiber, 194
General purpose pitch based carbon fiber, 72
Gland packings, uses of carbon fibers, 955
Glass, carbon fiber reinforced, 86
Glass-ceramic matrices, 593–594
Glass insulation felts, comparative properties, 972
Glass-like carbon, 41–42
Glass matrices, 592–595
    carbon fibers in, 599–601, 1031
    damage monitoring capability, achievement of, 600
    scanning electron microscope, 600
    thermal shock resistance, improvement of, 599
Gliders use of carbon fiber, 982
Glow discharge, carbon atom, 3
Golf shafts, heads, use of carbon fiber, 1003–1004
GPC. See Gel permeation chromatography
Graphite, 24–38, 397
    heat treatment carbonization furnace, 396
    properties of, 24, 31–38, 40
    structure, 27–31
    term as misnomer, 185
Graphite felt, furnace insulation, 402
Graphite hairpin, 402
Graphite whiskers, 42–43
Graphite yarn, 68
Graphitization
    diamond, 23
    pitch based precursor, 303–304
Grassie ladder, 245–247
Gravity casting, metal matrix, 644
Grind fineness, testing, 684
Growth rate curve, vapor grown carbon fibers, 328
Guncotton, 148
Gypsum matrices, 583–592

H

Hand lay-up (contact molding), 895
Hang glider, use of carbon fiber, 1005
Hardness of diamond, 22
Heat dissipation, use of carbon fiber, 1031–1032

Heated die to shape, cure resin, 911–912
Heat treatment furnace, 83, 395–398
    calculations for design of, 398–400
    element design, equipment design for, 402–403
    element materials, equipment design for, 397–398
    equipment design for, 395–398, 401–403
    gas seals, equipment design for, 396, 401
    graphite, 396
    insulation, equipment design for, 396–397, 402
Heavy goods vehicles, buses, use of carbon fiber, 993–994
Heisenberg uncertainty principle, 4
Helical winding, 906
Helicopters, use of carbon fiber, 977
Helix of never-ending improvement, 787–788
Hemolytic depolymerization, 281
Heterolytic depolymerization, 281
Hi-Fi, use of carbon fiber, 1005–1007
Highly oriented pyrolytic graphite, 40
High modulus fiber production, 202
High modulus furnace, 442
    aqueous surface treatment system, 443
    carbon fiber fractography, 458
    interlaminar shear strength, 443
    safety issues, 442
High pressure liquid chromatography, petroleum pitch, 160
High pressure synthetic diamonds, 17–18
High resolution transmission electron microscope, pitch based precursor, 308
High spatial resolution surface analysis, 473–474
High strength graphite, carbon-carbon composite, compared, properties of, 1015
High temperature carbonization, 200–202
High temperature carbonization furnace, safety issues, 441
High temperature creep, polyacrylonitrile based carbon fibers, 227–228
History of carbon fibers
    early inventors, 65–66
    Japan, work in, 71–72
    United Kingdom
        atomic energy research establishment, Harwell, 79–89
        Ciba (ARL) Ltd., Duxford, 114
        Courtaulds, Coventry, 98–113
        Courtaulds Ltd., Coventry, 114–115
        early prepreggers, 114–115
        Fothergill and Harvey Ltd. (F&H), Littleborough, 115
        Morganite Modmor, London, 97–98
        polyacrylonitrile precursors, 72–113
        RAE, Farnborough, 72–79
        Rolls Royce, Derby, 89–97
        Rotorway components Ltd., Clevedon, 115
    United States, work in, 66–71
        black orlon, 66–67
        early US carbon fibers, 67–71
        recent US carbon fibers, 71

Hitco, properties of early yarn, 68
HM furnace insulation, 402
Hockey sticks, use of carbon fiber, 1003
Homopolymer polyacrylonitrile, 125
Hooke, Robert, 148
Hooke's law, 148, 672, 937
Hoop winding, 906
Horizontal passes, oxidation plant, 197
Hospital equipment, use of carbon fiber, 1000
Hot compaction, producing composite by, 87–88
Hot isostatic pressure impregnation carbonization, 568–569
Hot pressing, 615
Hot press matched metal molding, 896
Hot stretching, 225–228
    polyacrylonitrile based carbon fibers, 225–228
    during processing of carbon fiber, 279–280
Hybrid composites, 838–839, 929
Hybrid effect, defined, 838
Hybridized orbitals, arrangements of, 8
Hydraulic press, compression in, 88
Hyperfullerenes, 59–60
Hysol–Grafil carbon fibers, critical aspect ratios for, 924

I

ILSS. *See* Interlaminar shear strength
Impact strength, 833–834
Incoshield long fiber concentrates, 1032
Industrial workwear, uses of carbon fibers, 954
Industry, carbon–carbon in, 1022–1023
Infrared analysis, 732–734
Infrared spectroscopy, 479–483
Infrared thermography, fatigue testing, 711
Inhibitors, provide oxidation protection, 574–575
Injection molding, 925–927
Inlaid fabrics, 877
Innovative developments, 1046
Inorganic acids, polyacrylonitrile precursor fibers, 147
Inorganic aqueous salt, as solvent, polyacrylonitrile precursors, 130
In-plane shear test, 700–706, 817
In situ chemical reactions, 611–614
    chemical vapor infiltration (or CVD), 611–612
    densification, progression of, 611–612
    hot isotactic pressing (HIPing), 613–614
    reaction bonding, 614
    slurry pulse/chemical vapor infiltration, 612–613
In situ electrophoretic deposition, matrices, 601
Interlaminar fracture toughness, measurement of, 714
Interlaminar shear strength, 350–352, 358, 360–361
Internal mold release agents, 923
Inventors, early, 65–66
Ion plating, 647
Iosipescu test, 817
Ishikawa, Dr. Kaoru, 779–780
Ishikawa diagram, 756–757, 779

INDEX   1141

ISO 9000 family of standards and quality systems, 774–775
Isophorone diamine, curing agent, 519
Isostatic pressing, 88
Isothermal chemical vapor infiltration process, 565
Isotopes, carbon atom, 1
Isotropic, 558
Isotropic deposition, 563
Isotropic pitches, 160–161
Itaconic acid, 101–103
Izod and Charpy impact tests, 718
Izod impact test, 834

J

Japan, early work in, 71–72
Johnson, Bill, 73
Juran, Joseph M., 778

K

Kerosene, polynuclear aromatic component, 174
Keto groups, formation of, polyacrylonitrile precursors, 240
Kimberlite, 17
Kish graphite, 24–25
Knitted 3-D fabrics, 890
Knitted fabrics, 872–877
    jersey or single knit, 874
    plain knitting, 874–876
    plain tricot, 877
    purl knitting, 875–876
    raschel, 877
    rib knitting, 874
    weft knitting, 874–876
    wrap knitting, 876–877
Knives, fountain pens, watches, use of carbon fiber, 1007
Knoop scale, 22
Kofler hotbench, 685
Kyukoshi method, 72

L

Ladder polymer, polyacrylonitrile carbon fibers, carbonization stages, 256–258
Laminates, orthotropic, 935
Lampblack, 44
Lanxide process, metal matrix, 646
Large tow, polyacrylonitrile based carbon fiber, 194
Law enforcement, uses of carbon fibers, 954
Lead acid battery, flywheel, comparison of, 984
Leadership, 788
Lead metal matrices, 640
Leisure goods, use of carbon fiber, 1001–1005
Length counters, online collection, 409
Levoglucosan, pyrolysis production of, 284
Lewis acid compounds, polyacrylonitrile precursor fibers, 147

Lignin, polynuclear aromatic component, 174
Linear regression, 752–753
Linear voltage variable differential transformer, 726–727
Line speed, morganometer, 436–437
Liquid infiltration, 572–573
Liquid metal infiltration techniques, 89
Liquid phase diffusion bonding, metal matrix, 646–647
Liquid phase hot pressing, metal matrix, 646–647
Liquid phase oxidation, 350–352
Liquid phase sintering, metal matrix, 646–647
Liquid state processing, metal matrix, 644–647
Lithium ion batteries, use of carbon fiber, 962–964
Long fiber process, 924–925
Loop test, 817
Losipescu test, 702
Loudspeaker cones, use of carbon fiber, 1006
Low angle x-ray diffraction, 473
Low interlaminar shear strength, 347
Low temperature carbonization, 200
Low temperature carbonization furnace, safety issues, 440–441

M

Magnesium metal matrices, 639
Managing by wandering, 783–784
Manufacturers
    filament winding machine, 906
    prepreg machine, 913
    pultrusion machine, 909
    thermoplastic molding compound, 924–925
Manufacturing costs, polyacrylonitrile based carbon fiber, 187–191
Marine applications, use of carbon fiber, 987–989
Mass spectrometry, petroleum pitch, 160
Materials selection, design considerations, 940
Matrices
    carbon, 551–582
    ceramic, 583–628
    metal, 629–656
    polymer, 501–550 (*See also* Polymer matrices)
    thermoplastic, 861–934
    thermoset, 861–934
Matrix materials, carbon-carbon processing, 560–569
Measures of dispersion, 750
Medical applications, use of carbon fiber, 998–1000
Meiller, Claus, 784–785
Melt flow index, determination of, 717–718
Melt infiltration, matrices, 609–611
Melt spinning mesophase precursor fibers, 166–171
Melt spun polyacrylonitrile precursor based carbon fibers, 142
Melt spun precursor, pitch, 295–296
Melt stirring, metal matrix, 644
Mercury manometer, cyanide effluent, 424
Mesophase, catalytic modification, production, 165
Mesophase pitches
    hydrogenation, 164–165

Mesophase pitches (*continued*)
  mechanical agitation, 162–163
  preparation of, 161–165
  pyrolysis, production by, 162–164
  solvent extraction, 164
Metal catalyst, dispersement of, vapor grown carbon fibers, 327–328
Metal composites, carbon fiber reinforced, 87–89, 97
Metal matrices, 845–849
  aluminum, 636–638
  carbon fibers in, 629–656, 1031–1032
    aluminum, 635–639
    capillarity, 647–648
    capillary effects, 641
    carbon fiber reinforced metal matrix composites, 641–649
    compocasting, 644
    copper, 639–640
    deposition processes, 647
    diffusion bonding, 643–644
    fiber matrix interactions, 641, 648
    fiber tow (liquid) infiltration, 645–646
    fluid flow into the preform, 641, 648
    gravity casting, 644
    ion plating, 647
    lanxide process, 646
    lead, 640
    liquid phase diffusion bonding, 646–647
    liquid phase hot pressing, 646–647
    liquid phase sintering, 646–647
    liquid state processing, 644–647
    magnesium, 639
    matrix microstructure, 648–649
    melt stirring, 644
    metal matrices, 635–641
    metal matrix composites, 629
    nickel, 640
    plasma spraying, 647
    powder metallurgy, 643
    pressure casting, 644
    processing methods, 642–647
    reinforcement of metal matrices, carbon fiber for, 629–631
    rheocasting, 644
    slurry casting, 644
    solidification process, 642, 648–649
    solid state processing methods, 643–644
    squeeze casting, 644
    thermal properties, metal matrices, 630
    tin, 640–641
    vacuum casting, 644
    wettability, coating processes to improve, 631–635
  composites, factors influencing, 641–642
  factors influencing composites, 641–642
  wettability, coating processes to improve, 631–635
Metal powders, chopped carbon fiber, hot working mixture, 88

Method of least squares, 752
Methyl acrylate, 101–103
Micromechanics, 935–940
Miller indices, cubic crystal, 466
Mitsubishi Chemical Corporation, Dialea, 295
Model airplanes, use of carbon fiber, 1007
Mohs scale, 22
Moisture content measurement, 714–715
Molecular Optical Laser Examiner, 485
Molecular orbitals, 9–12
Molecular sieves, use of carbon fiber, 958
Molecular weight, number average, polyacrylonitrile based precursor polymer, 192
Molecular weight determination, pitch, 160
Molten carbonate fuel cell, use of carbon fiber, 968
Molten pitch, 572
Monofilaments, properties of, 70
Monomer reactants, polymerization of, 525–527
Morganite Modmor, London, early work with polyacrylonitrile precursors, 97–98
Morganite Modmor carbon fibers, properties, 97
Morganometer, line speed, 436–437
Mortar, concrete, matrix, 585–591
Motor bikes, use of carbon fiber, 994–995
Molded graphite, properties of, 34
Molding, 715–717
  carbon fiber/polyetherimide laminate, 928
  conditions, reinforced thermoplastics, 716
  thermoplastic molding compounds, recommendations for, 926
Mold release, 922–923
Mullite matrices, 603
Multiaxial winding, 907
Multidirectional laminates, elastic behavior of, 940–943
Multi oven treatment, oxidation plant, 198
Multi-wrapped approach, oxidation plant, 198
Musical instruments, use of carbon fiber, 1005–1007

N

Naphthyridine formation degradation theory, 238
National Carbon Company, properties of early carbon fiber, 68
Natural diamonds, 17
Natural graphite, 24
N-D orthogonal blocks, 891–893
Neutron irradiation, 228–229
Nickel metal matrices, 640
Nitrile initiation, sources of, 251
Noise, safety issues, 448
Nomex honeycomb cores, 895
Non-destructive testing, 735–738
Non-oxidative surface treatment-whiskerization, 356–357
Non-oxide matrix materials, matrices, 603–604
Non-woven discontinuous reinforcement, 863–865
Non-woven UD fabrics, woven UD fabrics, 866
Normal distribution, 753–755

INDEX

Nuclear applications, use of carbon fiber, 998
Nuclear magnetic resonance, petroleum pitch, 160
Nucleus, carbon atom, 2
Null-balance principle, thermal analysis, 722

O

Occupational exposure limits, carbon fibers, emissions, 450
Offline quality control, 781
Offline winding, 411–415
Oil exploration, use of carbon fiber, 989–991
Omega drive systems, 380
One-part resin system, 898–900
Online collection, 409–410
Optical fiber sensor, 944–945
Optical instruments, use of carbon fiber, 1008–1009
Optical microscope, 453–456
Optical microscopy, fatigue testing, 711
Optical properties of diamond, 23
Organic acids, polyacrylonitrile precursor fibers, 147
Organometal compounds, polyacrylonitrile precursor fibers, 147
Orlon, 80, 82
Ourtaulds, Courtelle, 122
Outside control limits, 766
Ovens for oxidation, 380–383
Oxidation plant, 196–200, 439–440
Oxidation processes, production line, carbon fiber, 195–196
Oxidation protection, 573–578
Oxidative processes, 347–355
    gas phase oxidation, 348–350
    interlaminar shear strength, 348, 350–352
    liquid phase oxidation, 350–352
    pure oxygen, 348–350
Oxides of nitrogen, safety issues, 446
Oxidized polyacrylonitrile fiber, 66, 200, 792
    carbonization products of, 201
    chemical resistance of, 836
    density, determination of, 663–666
    mass per unit length, 662–663
    outgassing of, 837
    picks, 683
    production line, carbon fiber, 200
    properties of, 794
    testing, 662–678
    twist, 683
    uses of, 951–955

P

Packaging, equipment design for, 415
Packing materials, 973
PAN. See Polyacrylonitrile
Paper for resin coating, grade of, 684–685
Parabeam, 889–890
Pareto chart, 756, 779
Particulate matter, safety issues, 446

Pauli's exclusion principle, 5
PEEK. See Polyetheretherketone
Penetration value, petroleum pitch, 158
Periodic table, defined, 1
Perkin Elmer system, thermal analysis, 721–722
Permissible Exposure Limit, carbon fibers, emissions, 449
Peroxide catalysts for curing polyester systems, 506
Personal quality, standards of, 784
Peters, Tom, 783–784
Petroleum pitch, 157–158
Phenol formaldehyde, polynuclear aromatic component, 174
Phenolic matrix, carbon fiber, bond between, 553
Phenolic resins, 502–503
    gel coats, 505
    novalac resins, 503
    polynuclear aromatic component, 174
    saturated acids, 504
    vinyl monomers, crosslinking with, 507
Phenylene sulphone group, 534
Phillips, Leslie, 73
Phosphoric acid fuel cell, use of carbon fiber, 967
Phthalonitrile resins, 530
Pitch based carbon fiber, 554–555, 818–823
    properties of, 797
    surface treatment of, 304
Pitch based precursor, carbon fiber production using, 295–324
    air oxidation, 297–300
    aromatic condensation, 314
    carbonization, 301–303, 321
    cross-linking, 299–300
    cross-sectional microstructure, 307
    cyclization, 299–300
    dehydrogenation, 299–300
    electron spin resonance, 311
    field desorption mass spectroscopy, 310
    final carbonization process, 319–320
    gage length, influence of, 302
    graphitic character, degree of, 309–313
    graphitization, 303–304
    high resolution transmission electron microscope, 308
    manufacturing process, 296–304
    melt spun precursor, 295–296
    mesophase pitch fibers, 305–321
    Mitsubishi Chemical Corporation, Dialea, 295
    molecular rearrangement, 314
    oriented core microstructure, 305–307
    pitch based carbon fibers, surface treatment of, 304
    pitch fiber precursors, 320–321
    polyacenes, thermal reactivity data for, 309
    preparation of pitch precursors, 309–320
    radial transverse microstructure, 305–307
    side chains, 314
    stabilization (thermosetting) of spun fiber, 296–301

Pitch based precursor, carbon fiber production using (*continued*)
    stress, benefit of, 299
    thermal polymerization, 314
    thermal reactivity, 309
    time required, factors effecting, 298–299
    x-ray characterization, 312–314
Pitch precursor, 156–171, 320–321
    asphaltenes, composed of, 156
    boiling points, solvents used to extract pitch, 159
    carbon fiber manufacture, 156–171
    carbonized mesophase fibers, mechanical properties, 169
    catalytic modification, 165
    characterization of pitch, 158–160
    coal tar pitch, 158
    collapsing, 142
    early work in Japan, 72
    hydrogenation, 164–165
    isotropic pitches, 160–161
    melt spinning mesophase precursor fibers, 166–171
    mesophase pitches, 161–165
    naphthene, composed of, 156
    oils, 157
    petroleum pitch, 157–158
    pitch materials, chemical composition, 159
    polar aromatics, composed of, 156
    pyrolysis, 162–164
    raw pitch, properties of, 158
    relaxing, 142
    saturates, composed of, 156
    solvent extraction, 164
    structure of, 171
    treated pitch, properties of, 158
    treated pitches, properties of, 158
Plaiter table, 384
Planar zig-zag form, 146
Plant developments, future developments, 1044
Plant safety issues, 421–452
    air flow measurement, 424–432
    carbon fiber composites in fire, 424–429, 433–435
    Control of Substances Hazardous to Health Regulations 1988, 448–449
    drive systems, rotating rollers, 436–438
    dust extraction, 418–419
    electrical equipment, protecting, 423–424
    fire, risks of carbon fiber composites in, 450–451
    furnace gas seals, 401
    furnace insulation, 402
    gaseous emissions, United Kingdom exposure limits for, 449
    high modulus furnace, 442
    high temperature carbonization furnace, 441
    lateral movement, 433
    low temperature carbonization furnace, 440–441
    maintenance, 423
    oxidation plant, 439–440
    precursor creel, 438–439
    risks of carbon fiber composites in fire, pressure measurement, 424
    safety committee, 448
    sizing, 443
    splicing small tows, 435–436
    toxicology of carbon fibers, emissions, 449–450
    treatment of cyanide effluent, 444–445
    winding, 443–444
Plasma, 355–356
Plasma spraying, metal matrix, 647
Plastic molding processes, acronyms associated with, 895
Platen pressing of carbon fiber/polyetheretherketone laminate, 928
Ply cutting, stacking prepreg, 916
Poka-yoke, 782
Polarized light microscopy, polyacrylonitrile based carbon fibers, 205–206
Polar winding, 906
Polyacenes, thermal reactivity data for, 309
Polyacetylenes, 174
Polyacrylonitrile, 42, 121
Polyacrylonitrile based carbon fiber, 203–215, 792–798, 818
Polyacrylonitrile based precursor polymer, number average molecular weight, 192
Polyacrylonitrile carbon fibers
    carbonization stages, 254–259
    epoxy matrix, 816
Polyacrylonitrile fiber
    commercially available, 122–123
    structure of, 146–148
    thermal degradation of, 252–253
Polyacrylonitrile precursor
    alternative polymer formulations, 232
    boron, annealing in presence of, 229–230
    carbon fiber manufacturers, 185–186
    carbon fiber yield, 230
    carbonization, 221–230
    carbonization stages, mechanisms for, 254–259
    choice of precursor, 191–192
    controlled resistance carbon fibers, 233–234
    desirable attributes, 192–194
    developments, 232–234
    early work in Japan, 71–72
    fine structure, 203–215
    improvements, carbon fiber properties, 232
    manufacturing costs, polyacrylonitrile based carbon fiber, 187–191
    production line
        collection, 203
        effluent gas removal, 200
        high modulus fiber production, 202
        high temperature carbonization, 200–202
        low temperature carbonization, 200
        oxidation, 195–196
        oxidation plant, 196–200
        oxidized polyacrylonitrile fiber, 200
        precursor station, 194–195

shrinkage, 203
sizing, 203
surface treatment, 203
requirements for, 123–125
stabilization, 215–221, 234–254
tensile properties, 230–231
texture, 203–215
types, 194
world supply, polyacrylonitrile based carbon fiber, 186–187
Polyacrylonitrile precursor fibers, chemical treatments for, 147
Polyacrylonitrile precursors, 72–113, 121–148
ageing stage, 151–152
air gap spinning, 138–139
allotropes of cellulose, 154
aqueous dispersion polymerization, 134–136
carbon fiber, production procedures using, 112–113
carbon fiber production, 98–100
cement, carbon fiber reinforced, 86
ceramics, carbon fiber reinforced, 86
chain transfer stage, 134
chemical treatment, 145–146
chemical treatments, prior to stabilization, 147
coagulation stage, 137–138
collection, 143
commercially available, 122–123
comonomers, 125–130
composite fabrication, 79
composite materials, carbon fiber in, 113
composites, 78, 84–86, 89
dope preparation, 136
drying, 142
dry spinning, 138
early work with in U.K.
  at atomic energy research establishment, Harwell, 79–89
  at Courtaulds, Coventry, 98–113
  at Morganite Modmor, London, 97–98
  at RAE, Farnborough, 72–79
  at Rolls Royce, Derby, 89–97
fiber production, 79–83, 89–91
final treatment stage, 153–154
finish, 142
free radical, 133
friction, 79
glass, carbon fiber reinforced, 86
history, 122
homopolymer polyacrylonitrile, 125
initiation stage, 132–133
melt spinning, 139–140
metal composites, carbon fiber reinforced, 87–89, 97
mixing, ripening stages, 153
oxidation stage, 103–111
photochemical, 133
polymerization, 130–136
possible comonomers for, 126–127
precursor technology, 101–103

processing stages, 141–145
propagation stage, 133
requirements for, 123–125
resin formulation, composite fabrication, 92–97
shredding stage, 151–152
single filaments, testing, 78, 84–86
solution polymerization, 130–134
spinning methods, 136–140
spinning stage, 152–153
spun fiber modification, 145–146
steeping stage, 150–151
stretching, 142, 145
structure of fibers, 146–148
structure of rayon fibers, 154–156
surface treatment, 78, 84, 111–112
tensile strength of carbon fibers, factors affecting, 91–92
termination stage, 133–134
virgin carbon fiber, testing, properties of, 112
viscose rayon process, 150–154
washing, 142
wear, 79
wet spinning, 136–138
xanthation stage, 152
x-ray diffraction, early work with, 100
Polyacrylonitrile precursor testing, 657–662
atomic absorption spectrophotometer, 661
fiber moisture content, 660–661
filament diameter distribution, 657–660
ion chromatograph, 661
precursor burn-up temperature, 662
precursor d'tex using vibroskop, 660
residual solvent, in Courtelle precursor, 661
silver sulfide staining test, 662
sodium content in precursor, 661
soft finish content in Courtelle precursor, 662
Polyamide (PA) resins, 538–540
Polyamides, 174
Polyaromatics, 174
Polybenzimidazole, 174
Poly(bisbenzimidazobenzophenanthroline), 174
Polybutadiene, 174
Polycarbonate (PC) foam, 895
Polycarbonate (PC) resin, 540
Polycrystalline diamond, 18
Polydispersity index, polyacrylonitrile based precursor polymer, 192
Polydivinylbenzene, 174
Polyester resins, 503–507
Polyesters, 174
Polyetheretherketone, 535–537, 559
Polyetherimide, 559
Polyetherimide/polyethersulphone foam, 895
Polyetherimide resin, 542
Polyethersulphone resin, 542
Polyethylene, 174–175
Polyimide, 175
Polymer, 123–124

Polymer dope concentration
  polyacrylonitrile based precursor polymer, 192
  spinning solvent, 124
Polymer infiltration pyrolysis, 841–842
Polymerization methods, polyacrylonitrile precursors, 130–136
Polymer matrices
  amine cured systems, 519
  amine hydrogen equivalent weight, 519
  amorphous thermoplasts, 534–535
  benzocyclobutene, 530
  bismaleimides, 525–527
  bisoxazoline phenolics, 530
  blendur resins, 530
  carbon fiber reinforcement, effect of, 533
  condensation type polyimides, 523–524
  curing agents, 514, 519
  cyanate resins, 520–521
  cycloaliphatic resins, 512
  differential scanning calorimetry, 538
  duplex materials, 532
  dynamic mechanical analysis, 538
  electron spectroscopy for chemical analysis, 527
  epoxide resins, 508–520
  epoxy diluents, 513
  free aromatic diamines, use of, 526
  monomer reactants, polymerization of, 525–527
  phenolic resins, 502–503
  phenylene sulphone group, 534
  phthalonitrile resins, 530
  polyamide (PA) resins, 538–540
  polycarbonate (PC) resin, 540
  polyester resins, 503–507
  polyetheretherketone, 535–537
  polyetheretherketone (PEEK) resin, 540–542
  polyetherimide (PEI) resin, 542
  polyethersulphone (PES) resin, 542
  polyphenylene sulfide, 535
  polyphenylene sulfide (PPS) resin, 543
  polystyrylpyridine, 530
  polyvinyl butyral, 532
  quenching, 536
  reactive liquid polymers, 531–532
  rubber copolymers, 532
  semi-crystalline thermoplastics, 536
  sulphone group, 534
  thermoplastic matrix, 534
  thermoplastic modifiers, 532–533
  thermoplastic resins, 533–543
  thermoset matrix, 534
  thermoset resins, 501–533
  toughness, 532
  unsaturated polyester resins, 503–504
  vinyl monomers, crosslinking with, 507
Polymer precursor, matrices, 607–609
Polymethylmethylacrylamide (PMI) foam, 895
Polymethyl vinyl ketone, 175
Polynitrone, formation of, 240

Polynuclear aromatic components, in coal tar pitch, 172–173
Polyphenylene sulfide, 535, 543
Polystyrene (PS) foam, 893
Polystyrylpyridine, 530
Polyurethane (PU) foam, 895
Poly(vinylacetylene), 175
Polyvinyl alcohol, 175, 923
Polyvinyl butyral, 532
Polyvinylchloride, 175
Polyvinylidene chloride, 175
Polyvinylidene fluoride, 175
Portland cement, matrices, 583–584
Post cure oven, 912–913
Powder metallurgy, metal matrix, 643
Precision instruments, use of carbon fiber, 1008
Precursor creel, safety issues, 438–439
Precursor developments, future developments, 1043–1044
Precursor for polyacrylonitrile based carbon fiber, activated carbon fibers, 955
Precursor handling, 377–379
Precursor radius, reduction of, polyacrylonitrile based precursor polymer, 193
Precursor station, production line, carbon fiber, 194–195
Pre-die forming, 911
Pre-oxidized fiber, 71
Prepreggers, early work in U.K., 114–115
Prepreg manufacture, 913–916
Prepreg molding, 913–921
Press-clave molding, 917–918
Pressure assisted resin injection, 95
Pressure casting, metal matrix, 644
Pressure drop, flow, relationship between, exhaust systems, 418
Pressureless sintering, 615
Pressure measurement, carbon fiber composites in fire, 424
Principle quantum number, 4–5
Process control, statistical, 747–772. *See also* Statistics
Production line, carbon fiber
  collection, 203
  effluent gas removal, 200
  high modulus fiber production, 202
  high temperature carbonization, 200–202
  low temperature carbonization, 200
  oxidation, 195–196
  oxidation plant, 196–200
  oxidized polyacrylonitrile fiber, 200
  precursor station, 194–195
  shrinkage, 203
  sizing, 203
  surface treatment, 203
Production process, future developments, 1043–1044
Production temperature, Young's modulus of carbon fiber, relationship between, 792
Propeller blades, use of carbon fiber, 977–979
Properties of carbon fibers, 791–860

Prosthetic applications, use of carbon fiber, 998–1000
Proton exchange membrane fuel cell, use of carbon fiber, 966–967
Pulling unit provide traction, 912
Pull-pressing, matrices, 591
Pulse chemical vapor deposition process, 566
Pultruded product, shear strength of, 817
Pultrusion, 909–913
Push rods, use of carbon fiber, 993
Pyrolysis, 274–279
    carbon fiber production, mechanisms for, 280–282
    flame retardancy, 276–279
    polymers for making ceramics by, 607
Pyrolytic carbon, 38–41
Pyrolytic graphite, 38–41
Pyrovatex CP, commercial flame retardants, 278

Q

Quality assurance standards, 773
Quality circles, 785–786
Quality control, 773–790
    Crosby, Philip B., 782–783
    Deeming, W. Edwards, 776–777
    early Americans, 776
    Fiegenbaum, Armand V., 779
    inhouse testing, 773
    inspection, 775
    Ishikawa, Dr. Kaoru, 779–780
    ISO 9000 family of standards and quality systems, 774–775
    Japanese gurus, 779
    Juran, Joseph M., 778
    Meiller, Claus, 784–785
    new western Group of gurus, 782–785
    Peters, Tom, 783–784
    quality circles, 785–786
    quality costing, 788
    quality gurus, 775–785
    quality management, quality assurance standards, 773
    Shindo, Shigeo, 781–782
    Taguchi, Dr. Genichi, 780–781
    test status, 775
    total quality management, 786–788
Quality costing, 788
Quality gurus, 775–785
Quality management standards, future developments, 1045–1046
Quality Planning Road Map, 778
Quality spiral, 778
Quality systems, standards, ISO 9000 family, 774–775
Quantum numbers, allocation of, 5
Quantum theory, 2–8
Quickstep molding, 920

R

Racing car applications, use of carbon fiber, 991–993
Racquetball racquets, use of carbon fiber, 1004
Radiations in surface analysis techniques, 474
RAE, Farnborough, early work with polyacrylonitrile precursors, 72–79
Rail shear test, 817
Railways, use of carbon fiber, 995
Raman effect, 485
Raman spectroscopy, 817
Raw pitch, properties of, 158
Rayleigh scattering, 485
Rayon, defined, 150
Rayon precursor, Union Carbide carbon fiber from, properties of, 70
RC Houtz, 122
Reaction injection molding, 904
Reactive liquid polymers, 531–532
Recirculation water cooling system, 441
Recycling, 1046
Reinforced ceramic matrices, carbon fiber, 583–628
    alumina (Al_2O_3), 603
    aluminum nitride, 604
    Arrhenius plot, 570–571
    bond to glass, coating carbon fiber, 601
    boron carbide (V_4C), 604
    boron nitride, 604
    carbon fiber, impregnation of, 591
    carbon fiber cement matrix composites, 587
    carbon fiber reinforced cement, 585
    carbon fiber reinforced mortar, 590
    cement, 583–584
    ceramic matrices, 602–617
    chopped, 591
    concrete, 584–591
    concrete additives, 584–585
    consolidation, densification, 615
    conventional autoclave, use of, 569
    densification, 608
    dip coating, 616
    dry mixing, 588
    electrophoretic deposition, 617
    electrophoretic infiltration, 617
    fabrication processes, 591–592
    fiber reinforcement, 605
    filament winding, 591, 605–606
    flexural strength, increase in, 586
    fracture mechanics, 617
    glass-ceramic, 593–594
    glass matrices, 592–594
        carbon fiber filled, 599–601
        damage monitoring capability, achievement of, 600
        scanning electron microscope, 600
        thermal shock resistance, improvement of, 599
    high shrinkage, 606
    low yield, 606
    melt infiltration, 609–611
    mortar, 585–591
    mullite ($3Al_2O_3 \cdot 2SiO_2$), 603
    non-oxide matrix materials, 603–604
    polymer precursor, 607–609

Reinforced ceramic matrices, carbon fiber (*continued*)
  Portland, 583–584
  preparation methods, 594–598
  protective coatings, 615–617
  pull-pressing, 591
  silane treated carbon fibers, use of, 586
  silicon carbide (SiC), 603–604
  silicon nitride (Si_3N_4), 604
  in situ chemical reactions, 611–614
  in situ electrophoretic deposition, 601
  slip casting, 605
  slurry infiltration, 605
  sol gel, 606–607
  spray-up process, 591
  titanium boride (TiB_2), 604
  titanium carbide (TiC), 604
  unidirectional carbon fiber, infiltration of, 608
  wet mixing, 588
  wetting, 609–610
  YMAS matrix, 600
  zirconia (ZrO_2), 603
Reinforced plastic carbon fiber, 78
Reinforcement handling, 910
Relation of carbon fiber tensile properties to process conditions, 230–231
Resin coating, grade of paper for, 684–685
Resin film infusion, 902
Resin formulation, composite fabrication, 92–97
Resin impregnation, 911
Resin infusion under flexible tooling, 901
Resin transfer molding, 897–903
Resistance strain gages, measurement of strain using, 697–698
Rheocasting, metal matrix, 644
Risk priority number, 771
Risks of carbon fiber composites in fire, pressure measurement, 424
Robot arms, use of carbon fiber, 995
Rocket motor cases, use of carbon fiber, 983
Rocket motor nozzles, use of carbon fiber, 1021
Rollers
  safety issues, 436–438
  use of carbon fiber, 995
Rolls Royce, Derby, early work with polyacrylonitrile precursors, 89–97
Rotating rollers, safety issues, 436–438
Rotorway components Ltd., Clevedon, preppregers, early work in U.K., 115
Rough laminar deposition, 563
Rubber copolymers, 532
Ruland theory, 341
Rule of mixtures for composites, 817
Russian dolls, 59
Rutherford's concept, carbon atom structure and, 2

S

SACMA test method parameters, for determining composite modulus, 674

SAF. *See* Special Acrylic Fiber
Safety committee, 448
Safety issues
  air flow measurement, 424–432
  carbon fiber composites in fire
    lateral expansion or contraction, 434–435
    lateral movement, 433
    pressure measurement, 424
    velocity, determination of, 424–429
    volume flow, determination of, 429–432
  Control of Substances Hazardous to Health Regulations 1988, 448–449
  drive systems, rotating rollers, 436–438
  dust extraction, 418–419
  electrical equipment, protecting, 423–424
  fire, risks of carbon fiber composites in, 450–451
  furnace gas seals, 401
  furnace insulation, 402
  gaseous emissions, United Kingdom exposure limits for, 449
  high modulus furnace, 442
  high temperature carbonization furnace, 441
  lateral movement, 433
  low temperature carbonization furnace, 440–441
  maintenance, 423
  oxidation plant, 439–440
  precursor creel, 438–439
  risks of carbon fiber composites in fire, pressure measurement, 424
  safety committee, 448
  sizing, 443
  splicing small tows, 435–436
  toxicology of carbon fibers, emissions, 449–450
  treatment of cyanide effluent, 444–445
  winding, 443–444
Sailplanes, use of carbon fiber, 982
Sample correlation coefficient, 751–752
Sample size, determination of, 748
Satellite reflectors, use of carbon fiber, 1006
Scanning auger microscopy, 475
Scanning electron microscope, 359–360, 362, 456–460, 462, 483, 494
Scanning tunneling microscopy, 359–360, 490–493
Schrödinger equation of wave mechanics, 3–4, 9
Secondary ion mass spectrometry, 485
  dynamic SIMS, 489
  imaging or microscope SIMS, 489–490
  static SIMS4, 486–488
Seemann Composite Resin Infusion Molding Process, 901
Semi-crystalline thermoplastics, 536
Sequential multiport resin injection system, 904
Servo-hydraulic test machine, fatigue testing, 711
Seven-point action plan for change, 777
Shear strength measurement, 699–706
Shindo, Shigeo, 781–782
Short fiber process, 924
Shrinkage, 203
Sigradur, glass-like carbon, 42

Silane treated carbon fibers, use of, 586
Silicon carbide
    in equipment design, 390–392
    matrices, 603–604
Silicones, 923
Silicon nitride, matrices, 604
Sinclair's loop test for filament testing, 738
Single crystal x-ray diffraction, 470
Single fiber fragmentation test, 803
Single filament test, 78, 84–86, 800–803
Single phase fusion melt, carbon fiber, 139–140
Sintering, 615
Sizing, 363–370
    carbon fibers, surface treatment, 347–376
    with compatible thermoplastic polymer size, 924
    deposition from solution onto fiber surface, 363–367
    effect on composite properties, 815–817
    electrodeposition, deposition of polymer onto fiber surface by, 367–369
    electropolymerization, 369–370
    equipment design for, 404–405
    production line, carbon fiber, 203
    safety issues, 443
    Shell waterborne resins, 365
    sizes, 366
Skewed distribution, 753
Skin core, 677–678
Skis, ski sticks, use of carbon fiber, 1002
S-lap drive systems, 380
Slip casting, matrices, 605
Slip differentials, use of carbon fiber, 1018–1020
Slurry casting, metal matrix, 644
Slurry infiltration, matrices, 605
Slurry pulse/chemical vapor infiltration, 612–613
Small tow, approaches to production of, 377
Smart devices, 944–945
SMARTweave Sensing, 905
Smooth laminar deposition, 563
Snooker, pool cues, use of carbon fiber, 1004
Sodium removal, 400–401
Softening point, petroleum pitch, 158
Solar sails, 1046
Sol gel, matrices, 606–607
Solid oxide fuel cell, 969
Solids, techniques for, mull method, 481
Solid waste products, heat treatment furnace, 400
Solution polymerization, polyacrylonitrile precursors, 130–134
Solvent system, polyacrylonitrile based precursor polymer, 192
Sorted distribution, 753
Soxhlet extraction, 674
Space
    carbon–carbon in, 1020–1021
    use of carbon fiber, 982–983
Special Acrylic Fiber, 122
Specimen size, effect of, 817
Spinnerette construction

    carbon fiber factors affecting structure of, 169
    cleanliness of, 124
Spinning conditions, control of, 124
Spinning methods, 136–140
Spinning solvent, polymer dope concentration in, 124
Splicing small tows, 435–436
Sports goods, use of carbon fiber, 1001–1005
Spray lay-up, 895–896
Spray-up process, matrices, 591
SP Resin Infusion Technology (SPRINT), 901–902
Spun fiber modification, carbon fiber precursor, 145–146
Spun filament, fine count, 124
Squash racquets, use of carbon fiber, 1004
Squeeze casting, metal matrix, 644
Stabilization of polyacrylonitrile precursors, 234–254
Stainless steel heated rotary drum, over a, 405–406
Standard deviation, normal distribution, relationship between, 755
Standard error, 751
Standards, 773–790. *See also* Quality control
Static, elimination of, use of carbon fiber, 960
Statistical process control charts
    attribute data, 758–759
    average, range chart, 760–764
    mean deviation chart, 764–766
    median control chart, 766
    rules for detecting out-of-control conditions on control charts, 766–769
    standard deviation chart, 764–766
    variable data, 758–759
Statistical quality control, proficient in, 785–786
Statistics
    capability index, 770–771
    control chart method, 758
    cumulative sum chart, 767–770
    eighty/twenty rule, 756
    failure mode effect analysis, 771
    fishbone (Ishikawa) diagram, 756–757
    flow chart, 756
    frequency distribution, 747–748
    Gaussian distribution, 753
    linear regression, 752–753
    location of data, 748–750
    measures of dispersion, 750
    method of least squares, 752
    mixed population, 753
    non-random pattern, 768
    normal distribution, 753–755
    outside control limits, 766
    Pareto chart, 756
    risk priority number, 771
    sample correlation coefficient, 751–752
    skewed distribution, 753
    sorted distribution, 753
    standard deviation, normal distribution, relationship between, 755
    standard error, 751
    unnatural pattern, 766–767

Statistics (contiuned)
   variation, 756–758
   Weibull distribution, 756
Stealth aerial vehicles, use of carbon fiber, 982
Stearn, C. H., 149
Step-ladder polymer, polyacrylonitrile precursors, 248
Stereoscan calibration, with traceable reference standard, 669
Stokes line, 485
Strain rate, effect of, 817
Strength of diamond, 23
Stress graphitization, 90–91, 270
Stretch ratios, increase in crystal size, 91
Stringed instruments, use of carbon fiber, 1006–1007
Structural work, use of carbon fiber, 995
Structure of atom, 2
Structure of carbon, 1–14
Structure of carbon fibers
   anti-Stoke line, 485
   asymmetric stretch, 479–480
   bend, 479
   Bragg's Law, 467
   determination of, 453–500
   electron energy loss spectroscopy, 477
   Fourier transform infrared/attenuated total reflectance spectroscopy, 483
   Fourier transform raman spectroscopy, 485
   high modulus furnace, 442
   high spatial resolution surface analysis, 473–474
   infrared spectroscopy, 479–483
   low angle x-ray diffraction, 473
   Miller indices, cubic crystal, 466
   Molecular Optical Laser Examiner, 485
   optical microscope, 453–456
   radiations in surface analysis techniques, 474
   Raman effect, 485
   Rayleigh scattering, 485
   scanning auger microscopy, 475
   scanning electron microscope, 456–460, 483, 494
   scanning tunneling microscopy, 490–493
   secondary ion mass spectrometry, 485
   single crystal x-ray diffraction, 470
   solids, techniques for, mull method, 481
   Stokes line, 485
   surface enhanced Raman scattering, 485
   symmetric stretch, 479–480
   Time of Flight Secondary Ion Mass Spectrometry, 489–490
   Time-of-Flight (ToF) analyzers, 489
   transmission electron microscope, 456, 460–464, 483
   ultra high vacuums, 453
   vibrational transitions, 480
   wide angle x-ray diffraction, 466–470
   winding, 443–444
   x-ray diffraction, 464–473
   x-ray excited auger electron spectroscopy, 475
   x-ray photoelectron spectroscopy, 477, 490
Structure of graphite, 27–31
Styrene/acrylonitrile copolymer (SAN) foam, 895

Subatomic particles, physical properties of, 2
Submarines, use of carbon fiber, 989
Subsidiary number, 5
Surface enhanced Raman scattering, 485
Surface treatment, 347–376
   effect of fiber properties, 357–363
   equipment design for, 403–404
   non-oxidative surface treatment-whiskerization, 356–357
   oxidative processes, 347–355
   plasma, 355–356
   production line, carbon fiber, 203
   sizing, effect on composite properties, 815–817
Suspension systems, use of carbon fiber, 992–993
Swann, J.W., 65
Swept bends, exhaust, 416–417
Symmetric stretch, 479–480
Synthetic graphite, 25–27
System of profound knowledge, 777

T

Taguchi, Dr. Genichi, 780–781
Tailored resistance carbon fiber, 960
Tautomeric changes, polyacrylonitrile precursors, 244–245
Tautomerization, polyacrylonitrile precursors, 249
Teamwork, 786–787
Telescopes, use of carbon fiber, 1008
TEM. See Transmission electron microscope
Temperature, production, Young's modulus of carbon fiber, relationship between, 792
Temperature programmed desorption, 359–360
Tennis racquets, use of carbon fiber, 1004
Tensile failure, vapor grown carbon fibers, 339–343
Tensile modulus, 693–695
Tensile properties, 800–808
Tensile strength
   carbon fibers, factors affecting, 91–92
   measurement, 695–697
Tensile strength values, distribution, 748
Tensile test machine system, determination of compliance of, 670–671
Textile applications, use of carbon fiber, 995, 998
Textile grade precursors, carbon fiber properties from, 82
Texture, polyacrylonitrile precursor, 203–215
Thermal analysis, 721–729
Thermal anemometer, volume flow determination, 429
Thermal black, 44
Thermal conductivity, 831
Thermal conversion, cellulosic material, technique, 69–70
Thermal expansion, 829–831
Thermal gradient chemical vapor infiltration process, 566
Thermal insulation, uses of carbon fibers, 970–973
Thermal polymerization, pitch based precursor, 314
Thermal properties, 823–829

INDEX

Thermal properties of diamond, 23
Thermal properties of graphite, 35
Thermal reactivity, pitch based precursor, 309
Thermofolding, 927
Thermomechanical analysis, 729
Thermoplastic filament winding, 928
Thermoplastic matrices, 534, 861–934, 1009–1010
    Arrhenius plot, 570–571
    continuous fiber reinforced plastic materials, 929
    film stacking process, 927
    impervious surface layer, 570
    importance of critical aspect ratio, 923–924
    injection molding, 925–927
    polymer, 839
    precursors, 558–560
    preparation of thermoplastic molding compounds, 924–925
    thermoplastic filament winding, 928
    thermoplastic prepreg, 927–928
    thermoplastic pultrusion, 929
Thermoplastic modifiers, 532–533
Thermoplastic molding compounds
    carbon fiber filled, suppliers, 1012–1013
    use of carbon fiber, 1010
Thermoplastic prepreg, 927–928
Thermoplastic pultrusion, 929
Thermoplastic resins, 533–543
Thermoplastics, 714–718
Thermoset bulk molding compound, 896
Thermoset dough molding compound, 896
Thermoset matrices, 534, 861–934
    carbon fibers in, 973–1009
    centrifugal molding, 904
    contact molding wet lay-up, 894–896
    end uses in, 1007–1009
    fiber placement systems, 921–922
    filament winding, 905–909
    flow, cure monitoring of resin infusion processes, 904–905
    hot press matched metal molding, 896
    manufacturing processes, 894–923
    mold release, 922–923
    preparation of fiber preforms, 904
    prepreg molding, 913–921
    pultrusion, 909–913
    reaction injection molding, 904
    resin transfer molding, 897–903
    sequential multiport resin injection system, 904
Thermoset resins, matrix, 501–533
Thermoset sheet molding compound, 896–897
Thermosetting, spun fiber, pitch based precursor, 296–301
Thermosetting resin, 556–558
Thermostatically controlled heated steel cylinder, 717–718
Thornel 25, 70
Three-dimensional graphite, 214
Three-dimensional reinforcements, 882–883
    autoweave, 889

basic weave, 886
biplain weave, 886
double bias fabrics, 885
integrated structures, 889
magnaweave, 889
multiaxial non-crimp reinforcements, 882–889
NVG stitch bonded variable layer (MALIMO) process, 884
proprietary 3-D weaving processes, 889–890
quadraxial, 889
shaped preforms, 889
simultaneous stitch process, producing stitched fabric by, 883–885
triaxial weave, 885–888
warp binding or orthogonal type architecture, 889
weave, stitch process, producing stitched fabric by, 885
WIMAG wrap knitted multiaxial layer (Liba system), 884–885
woven 3-D fabrics, 889
Tidal turbine blades, use of carbon fiber, 997–998
Time-of-flight analyzers, 489
Time of flight secondary ion mass spectrometry, 489–490
Tin metal matrices, 640–641
Titanium boride (TiB_2), matrices, 604
Titanium carbide (TiC), matrices, 604
Topham, Charles F., 149
Topham box, 149, 153
Torsion test, 817
Total pressure drop, determination of, exhaust, 417
Total quality control, 779
Total quality management, 786–788
Toughness, thermoset resin systems, 530–533
Tow fiber tension, safety issues, 437–438
Tow testing, 678–682
    dry tow test, 678–679
    Hercules end tab, 680–681
    impregnated tow, 679–682
Tow textile grade precursors, 80
Toxicology, carbon fibers, emissions, 449–450
Transition metal oxides, polyacrylonitrile precursor fibers, 147
Transmission electron microscope, 456, 460–464, 483
Transportation, uses of carbon fibers, 955
Transverse mechanical properties, carbon fibers, 814
Tripods, use of carbon fiber, 1008
Tube rolling, 921
Turbine blades, use of carbon fiber, 995–998
Turbostratic graphite, polyacrylonitrile based carbon fibers, 214
Turbostratic structure, 39–40
Two-part resin systems, 899
Two roll drive systems, 379
Types of matrix, 556–560
    hot isostatic pressure impregnation carbonization, 568–569
    thermosetting resin, 556–558

U

Ultra high vacuums, 453
Ultrasonic testing, 736–738
Ultraviolet spectroscopy, pitch, 160
Uncertainty principle, of Heisenberg, 4
Unidirectional carbon fiber
  composites, properties of, 804
  infiltration, 608
Unidirectional fabrics, 865–866
Unidirectional fiber reinforced epoxy resins, elastic properties of, 937
Uniform temperature distribution, oxidation plant, 196–197
Union Carbide, properties of early yarn, 68
United Kingdom
  early prepreggers, 114–115
  polyacrylonitrile precursors, 72–113
    at atomic energy research establishment, Harwell, 79–89
    at Courtaulds, Coventry, 98–113
    at Morganite Modmor, London, 97–98
    at RAE, Farnborough, 72–79
    at Rolls Royce, Derby, 89–97
United States, early work in, 66–71
  black orlon, 66–67
  early US carbon fibers, 67–71
  recent US carbon fibers, 71
Unmanned aerial vehicles (UAVs), use of carbon fiber, 982
Unsaturated polyester resins, 503–504
Uses of carbon fibers, 951–1042
  carbon-carbon applications, carbon fibers for, 1010–1023
  cement, carbon fibers in, 1023–1030
  ceramic matrices, carbon fibers in, 1031
  concrete, carbon fibers in, 1023–1030
  electrical applications, 960–970
  electrodes, 961–962
  gaskets, 973
  glass matrices, carbon fibers in, 1031
  metal matrices, carbon fibers in, 1031–1032
  oxidized polyacrylonitrile fiber, 951–955
  packing materials, gaskets, 973
  thermal insulation, 970–973
  thermoplastic matrices, carbon fibers in, 1009–1010
  thermoset matrices, carbon fibers in, 973–1009
  virgin carbon fiber, 955–959

V

Vacuum assisted resin injection, 897–898
Vacuum assisted resin transfer molding, 900
Vacuum bag molding, 916
Vacuum casting, metal matrix, 644
Vacuum infusion molding process, 901
Vacuum infusion processing, 901
Valence electrons, 7
van der Waals forces, 15, 55, 60
Vane anemometer, volume flow determination, 429
Vapor grown carbon fibers, 43–44, 325–346
  batch process, 325–327
  continuous process, 327–328
  controlled atmosphere electron microscopy, 334
  conventional filaments, 325
  growth process, 334–339
  growth rate curve, 328
  metal catalyst, dispersement of, 327–328
  precursor filaments, 325
  preparation, 325–334
  properties of, 44
  Ruland theory, 341
  scanning electron microscope, 336
  tensile failure, 339–343
  transmission electron microscope, 336
  vermicular filaments, 334–336
  weakest link principle, 342–343
Variable bow roller, 434–435
Velocity determination
  orifice, flow measurement with, 430–432
  Pitot static tube, 424–428
Vermicular filaments, vapor grown carbon fibers, 334–336
Vertical burn tests, blends of curlon, 972
Vertical manometers, 444
Vertical passes, oxidation plant, 196–197
Vibrational transitions, 480
Vibroskop, selection of test weights for, 660
Vickers stage micrometer, 667
Vinyl esters, 125
Vinyl monomers, crosslinking with, 507
Violin, bows for, use of carbon fiber, 1007
Virgin carbon fiber, 662–678, 955–959
Viscose rayon, 792
Viscosity, resin mix, determination of, 686–687
Volume flow determination
  nozzle, flow measurement with, 430–432
  thermal anemometer, 429
  vane anemometer, 429

W

Warp UD fabric, weft UD fabric, 866
Water cooling system, recirculation, 441
Water extract, conductivity of, 677
Watson image shearing eyepiece, determining filament diameter using, 667–668
Watts, Bill, 73
Wave mechanics, 4
Waxes, 923
Weakest link principle, vapor grown carbon fibers, 342–343
Weibull distribution, 756

Weight average molecular weight, polyacrylonitrile based precursor polymer, 192
Wettability, coating processes to improve, metal matrices, 631–635
Whiskerization, non-oxidative surface, 356–357
Wide angle x-ray diffraction, 466–470
Wind generator sizes, Atlantic Orient Corporation, categorization of, 997
Winding, safety issues, 443–444
Wind surfing, use of carbon fiber, 1005
Wind turbine blades, use of carbon fiber, 995–997
WIRA Rapid Oil Extraction Apparatus, 675–676
Wire resistance heating element, furnace, 388–390
Wood, carbon fiber and, 1009
World distribution, acrylic fiber plants, 123
World supply, polyacrylonitrile based carbon fiber, 186–187
Woven carbon fiber manufacturers, 867–868
Woven fabric, 866–872
    basket (Hopsack) weave, 868–869
    high modulus (non-crimp) weave, 872
    Leno weave, 869–870
    mock Leno weave, 870
    plain or square weave, 868
    satin weave, 871–872
    twill weave, 871
Woven spread tow, 872

X

Xanthation stage, cellulosic precursors, 152
X-ray
    petroleum pitch, 160
    pitch based precursor, 312–314
X-ray diffraction, 100, 464–473
X-ray excited auger electron spectroscopy, 475
X-ray photoelectron spectroscopy, 249–250, 359–360, 477, 490, 553

Y

Yachts, use of carbon fiber, 946, 987–989
Yarn, 68
YMAS matrix, 600
Young's modulus
    carbon fiber, 792
    coefficient of thermal expansion, relationship between, 829

Z

Zeeman effect, 5
Zero defects concept, 782
Zirconia, matrices, 603